ALGAE ABSTRACTS

A Guide to the Literature

Volume 1
To 1969

Volume 1: To 1969

Volume 2: 1970 – 1972

ALGAE ABSTRACTS
A Guide to the Literature

Volume 1
To 1969

Prepared from material supplied by
Water Resources Scientific Information Center
Office of Water Resources Research
Department of the Interior
Washington, D.C.

IFI/PLENUM · NEW YORK – WASHINGTON – LONDON

Library of Congress Catalog Card Number 73-84004
ISBN 0-306-67181-6

Published 1973 by IFI/Plenum Data Corporation
A Subsidiary of Plenum Publishing Corporation
227 West 17th Street, New York, N. Y. 10011

United Kingdom edition published by Plenum Press, London
A Division of Plenum Publishing Company, Ltd.
Davis House (4th Floor), 8 Scrubs Lane, Harlesden, NW10 6SE, London, England

Printed in the United States of America

INTRODUCTION

Algae Abstracts is the first in a series of bibliographies on water resources and pollution published by IFI/Plenum Data Corporation in cooperation with the Water Resources Scientific Information Center (WRSIC). It is produced wholly from the information base comprising material abstracted and indexed for *Selected Water Resources Abstracts.*

The bibliography is divided into volumes according to the publication dates of the source documents. Volume 1 contains 569 abstracts covering publication dates up to and including 1969; Volume 2 contains 730 abstracts covering the years 1970 to 1972. The material included in this bibliography represents computer selections based on the presence of a form of the word "alga" somewhere in the referenced citation. Substantively, the material typifies WRSIC's "centers of competence" approach to information support of the Office of Water Resources Research (OWRR) of the Department of the Interior. Most of the references in this bibliography are the work of the center of competence on eutrophication at the University of Wisconsin.

The indexes refer to the WRSIC accession number, which follows each abstract. The *Significant Descriptor Index* is made up of a fraction of the total descriptors and identifiers by which each paper has been indexed. It represents weighted terms that best describe the information content; this status is indicated by the asterisks which precede them. The *General Index* includes all the remaining descriptors and identifiers by which each paper in this bibliography has been indexed. Through permutation, each word in a multiple-word descriptor or identifier is made to file in its normal alphabetical order, giving access to each abstract by all conceivable routes. To use the indexes, scan the middle rank for a few keywords describing your subject matter interest, then note the WRSIC accession numbers at the right margin. These numbers locate the full record in the bibliography, which is arranged in ascending accession number order.

Each volume, including a complete *Author Index,* is a self-contained entity, offering ready access to the pertinent literature for the time period it spans.

CONTENTS

ABSTRACTS

Publication dates up to and including 1969

The abstracts are presented in order of increasing accession number.
This number appears as the last line of each entry.

INTERTIDAL COMMUNITIES AS MONITORS OF POLLUTION,

WASHINGTON UNIV. DEPT. OF CIVIL ENGINEERING.

R. T. OGLESBY, AND DAVID JARMISON.

ASCE PROC, J SANIT ENG DIV, VOL 94, NO SA3, PAP 6008, PP 541-550, JUN 1968. 5
FIG, 14 REF, 1 APPEND.

DESCRIPTORS:
*POLLUTANTS, *ESTUARIES, *TIDAL WATERS, ECOSYSTEMS, ESTUARINE
ENVIRONMENT, MARINE ALGAE, SALINITY, TEMPERATURE, MONITORING, AERIAL
PHOTOGRAPHY, INFRARED RADIATION, DYE RELEASES, BEACHES, SANDS, BIOTA,
WASTE WATER(POLLUTION), CURRENTS(WATER).

IDENTIFIERS:
*INTERTIDAL COMMUNITIES, *WASTE DISCHARGES, FLORA, FAUNA, *BIOLOGICAL
COMMUNITIES, INTERTIDAL ALGAE, PUGET SOUND, FLOURESCEIN DYE.

ABSTRACT:
ECOLOGICAL SURVEYS OF INTERTIDAL ZONES ARE PROPOSED TO BE THE MOST
ECONOMICAL WAY TO ASSESS THE BIOLOGIC IMPACT OF MANY WASTES DISCHARGED
INTO ESTUARIES AND COASTAL MARINE ENVIRONMENTS. SPECIES DIVERSITY AND
COMMUNITY STRUCTURE ARE THE MEASURES SELECTED; THESE ARE DETERMINED
THROUGH A CAREFULLY DESIGNED AND STATISTICALLY VALID PROGRAM OF
SAMPLING. AERIAL PHOTOGRAPHY IS USED IN DEVELOPING THE SAMPLING PROGRAM
WITH INFRARED EKTACHROME FILM PROVIDING DATA ON DISTRIBUTION AND, TO
SOME EXTENT, THE NATURE OF INTERTIDAL FLORA. DYE TRACERS IN THE
EFFLUENT ARE IMAGED, ON TIME-SEQUENCE, COLOR AERIAL-PHOTOGRAPHS, AND
PROVIDE INFORMATION ON PATTERNS OF WASTE DISPERSAL. THE METHODOLOGY
PROPOSED IS APPLICABLE ONLY WHERE EITHER SPATIAL OR TEMPORAL CONTROLS
CAN BE ESTABLISHED.

FIELD 05C

ACCESSION NO. W68-00010

REMOVAL OF ALGAL NUTRIENTS FROM RAW WASTEWATER WITH LIME,

METCALF AND EDDY, BOSTON, MASS.

J. C. BUZZELL, AND C. N. SAWYER.

WATER POLL CONTROL FED JOUR, VOL 39, NO 10, R16-R24 OCT 1967.

DESCRIPTORS:
*ACTIVATED SLUDGE, ALGAE, EUTROPHICATION, LIME, *NITROGEN, HYDROGEN ION
CONCENTRATION, *PHOSPHORUS, SEWAGE TREATMENT, *OXIDATION LAGOONS.

IDENTIFIERS:
*NUTRIENTS, PONDS, PRIMARY TREATMENT, SECONDARY TREATMENT,
STABILIZATION PONDS, SEWAGE LAGOONS.

ABSTRACT:
A LABORATORY EVALUATION OF THE APPLICATION OF LIME TO RAW DOMESTIC
WASTEWATER AS A MEANS OF ENHANCING THE EFFECTIVENESS OF PRIMARY
TREATMENT, SPECIFICALLY TO REDUCE PHOSPHORUS LEVELS. LIME TREATMENT
EFFECTED AN 80- TO 90-PERCENT REMOVAL OF TOTAL PHOSPHORUS, WITH GREATER
THAN 97-PERCENT REMOVAL OF SOLUBLE, INORGANIC FORMS. LIME TREATMENT
REMOVED 50 TO 70 PERCENT OF THE BOD, APPROXIMATELY 25 PERCENT OF THE
TOTAL NITROGEN, AND DESTROYED 99.9 PERCENT OF THE COLIFORM BACTERIA.
LIME DOSAGE WAS CONTROLLED BY PH. THE VOLUME OF SLUDGE PRODUCED WAS
APPROXIMATELY 1 PERCENT OF THE VOLUME OF WASTEWATER TREATED. THE LIME
TREATMENT EFFLUENTS HAD NITROGEN: PHOSPHORUS RATIOS OF 25:1 OR GREATER.
PRELIMINARY STUDIES INDICATED THAT STABILIZATION PONDS SHOULD SERVE
WELL IN POLISHING THE LIME-TREATED WASTEWATER, REMOVING 75 TO 80
PERCENT OF THE REMAINING PHOSPHORUS.

FIELD 05D

ACCESSION NO. W68-00012

LAKE EUTROPHICATION PROBLEM AND PROGRESS,

SOUTH DAKOTA STATE UNIV., BROOKINGS.

A. C. FOX.

MIN 152ND MTG, MO BAS INTER-AGENCY COMM, APPEND F, PP F-1/F-6, MAR 1968. 4
REF.

DESCRIPTORS:
*EUTROPHICATION, SEQUENCE, LAKES, CHEMICALS, FISHING, SPORT FISHING,
RECREATION, SILT, NUTRIENTS, AQUATIC POPULATIONS, OLIGOTROPHY, *FISH
POPULATIONS, PLANT POPULATIONS, ROUGH FISH, NITROGEN, PHOSPHORUS,
DIATOMS, *ALGAE, CHLOROPHYTA, CYANOPHYTA, LAND MANAGEMENT, WATER
POLLUTION SOURCES, DREDGING, *POLLUTION ABATEMENT, DIKES, SOIL
CONSERVATION.

IDENTIFIERS:
DESMINS, FLAGELLATES.

ABSTRACT:
THE SUBSTANTIAL INFLOW OF SILT AND NUTRIENTS CAUSES A LAKE TO AGE
RAPIDLY AND TO MOVE FROM OLIGOTROPHY TO EUTROPHY AND SO AFFECT THE
GAME-FISH POPULATION. THE ADDITION OF NITROGEN AND PHOSPHORUS CAUSES
LAKES TO CHANGE FROM ONE ONLY SPARSELY POPULATED, MOSTLY BY DESMINS, TO
ONE IN WHICH DIATOMS REPLACE THE DESMINS, THESE IN TURN REPLACED BY
FLAGELLATES AND GREEN ALGAE, AND FINALLY BY BLUE-GREEN ALGAE,
ALARMINGLY IN TREMENDOUS 'BLOOMS'. THE ZOOLOGICAL SEQUENCE IS FROM THE
GENERALLY OLIGOTROPHI TROUT LAKES, TO THE LAKES POPULATED BY BASS,
PIKE, PERCH, AND PANFISH, TO THE EUTROPHIC POPULATION OF ROUGH FISH.
MOST OF THE CAUSATIVE NUTRIENTS ARE THE RESULT OF AGRICULTURAL
POLLUTION RESULTING FROM POOR LAND PRACTICES AND LIVESTOCK FEEDLOT
OPERATIONS. ONCE THERE IS EUTROPHY, REMEDIES ARE EXPENSIVE, SUCH AS
EFFORTS TO DILUTE THE LAKE BY DREDGING OR DIKING. THE CONTROL OF INFLOW
OF HARMFUL NUTRIENTS AND SILT CAN BE DONE INEXPENSIVELY, AS BY
INITIATING SOIL CONSERVATION PRACTICES AND BY TREATING DOMESTIC AND
INDUSTRIAL SEWAGE. CONTROL BY HARVESTING ALGAE IS NOT WITHIN REACH AS
YET BUT FISH CAN BE HARVESTED AND MARKETED ECONOMICALLY TO REDUCE THE
TOTAL AMOUNT OF NUTRIENTS IN THE WATER. CONTROL BEST STARTS WITH LAKES
IN EARLY EUTROPHY.

FIELD 05C

ACCESSION NO. W68-00172

THE USE OF ALGAE IN REMOVING NUTRIENTS FROM DOMESTIC SEWAGE,

WASHINGTON UNIV., SEATTLE.

R. H. BOGAN.

TRANS 1960 SEM ON ALGAE AND METRO WASTES, ROBT A TAFT SANITARY ENGG CENTER,
CINCINNATI, OHIO, TECH REPT W61-3, PP 140-147, 1961. 8 P.

DESCRIPTORS:
*EUTROPHICATION, *ALGAE, NUTRIENTS, DOMESTIC WASTES, *SEWAGE TREATMENT,
SEWAGE, BIOLOGICAL TREATMENT, SOLID WASTES, MICROBIOLOGY, ACTIVATED
SLUDGE, RATES, CHLORELLA, HYDROGEN ION CONCENTRATION, PLANT TISSUES,
SCENEDESMUS, LIMNOLOGY, WATER PROPERTIES, WATER TEMPERATURE,
ALKALINITY, *PHOTOSYNTHESIS, LIGHT PENETRATION, PHOSPHORUS, LAGOONS.

IDENTIFIERS:
STIGLEOCLONIUS, INSOLUBLE WASTES, CULTURE MEDIA.

ABSTRACT:
THE SOLUBLE ORGANIC PHOSPHORUS IN SEWAGE MAY BE CONVERTED BY BIOLOGICAL
OR CHEMICAL MEANS INTO RECOVERABLE INSOLUBLE MATTER. THE LATTER MEANS
HAS RECEIVED MORE ATTENTION TO DATE. THE RATE OF REMOVAL BY BIOLOGICAL
MEANS IS A FUNCTION OF CELL-TISSUE SYNTHESIS WHICH VARIES MARKEDLY WITH
THE TYPE OF ORGANISM. THE MIXED MICROBIAL CULTURES PROVIDED BY THE
ACTIVATED SLUDGE PROCESS APPEAR TO BE THE MOST RAPID MECHANISM.
CHLORELLA, SCENEDESMUS, AND STIGLEOCLONIUS WERE GROWN IN RAW AND
TREATED SEWAGE FOR THE EXPERIMENTS, AND THE AUTHOR STUDIED THEIR ROLE
IN ADJUSTING PH. TEMPERATURE, CELL-TISSUE CONCENTRATION, COMPOSITION OF
CULTURE MEDIA, AND ALKALINITY INFLUENCE THE RATE OF PHOTOSYNTHETIC PH
ADJUSTMENT. WITH ADEQUATE LIGHT, THE MINIMUM VALUE BEING 100-200 F C,
RAPID EXTRACTION OF PHOSPHORUS BY PHOTOSYNTHETIC BIOLOGICAL ACTIVITY IS
POSSIBLE. A HIGH-RATE PROCESS WAS DEVELOPED IN THE LABORATORY BY WHICH
SOLUBLE PHOSPHORUS REDUCTIONS OF 90 PERCENT OR MORE WERE ACHIEVED IN
CONTACT TIMES OF AS LITTLE AS 6-12 HOURS. TRANSFERRING THE PILOT-PLANT
STUDY TO LAGOONAL WATERS 3-4 FT DEEP CREATED LIGHTING
PROBLEMS--SUNLIGHT COULD NOT PENETRATE DEEPLY ENOUGH THROUGH THE ALGAL
GROWTH TO CAUSE AN ADEQUATE PHOTOSYNTHETIC RESPONSE.

FIELD 05D

ACCESSION NO. W68-00248

INDICES OF GREAT LAKES EUTROPHICATION,

A. M. BEETON.

PUBL GREAT LAKES RES DIV, (IN PRESS), VOL 14, 1966.

DESCRIPTORS:
*EUTROPHICATION, NUTRIENTS, NITROGEN, PHOSPHORUS, PLANKTON, DISSOLVED
OXYGEN, HYPOLIMNION, GREAT LAKES, LAKE ERIE, STATISTICS, GROWTH RATES,
ALGAE, FRESH WATER FISH.

ABSTRACT:
INDICES OF EUTROPHICATION WERE CITED: (1) INCREASES IN NITROGEN AND
PHOSPHORUS; (2) CHANGES IN SPECIES COMPOSITION AND AN INCREASE IN THE
ABUNDANCE OF PLANKTON; (3) DECREASES IN THE DISSOLVED OXYGEN CONTENT OF
BOTTOM WATERS; (4) CHANGES IN THE FISH POPULATION; (5) THE REPLACEMENT
OF BOSMINA COREGONI BY B. LONGIROSTRIS; AND (6) EXTENSIVE GROWTHS OF
CLADOPHORA. OTHER CHANGES SUCH AS INCREASES IN TDS AND MAJOR IONS ARE
REGARDED AS REPRESENTATIVE OF ENVIRONMENTAL CHANGES AND NOT NECESSARILY
INDICES OF EUTROPHICATION. THERE ARE FEW OFFSHORE DATA ON NUTRIENTS
(NITROGEN AND PHOSPHORUS) FROM THE GREAT LAKES OTHER THAN LAKE ERIE,
AND EVEN THESE DATA ARE QUESTIONABLE. CHANGES IN THE RATES OF GROWTH OF
FISH SHOULD BE VIEWED WITH CAUTION WHEN RELATING THEM TO EUTROPHICATION
INASMUCH AS MANY ENVIRONMENTAL VARIABLES MAY BE OF INFLUENCE AS WELL AS
AN INCREASE IN NUTRIENTS.

FIELD 05C

ACCESSION NO. W68-00253

PLANKTONIC ALGAE AS INDICATORS OF LAKE TYPES, WITH SPECIAL REFERENCE TO THE
DESMIDIACEAE,

MINNESOTA UNIV., MINNEAPOLIS.

A. J. BROOK.

LIMNOL OCEANOGR, VOL 10, PP 403-411, 1965. 9 P.

DESCRIPTORS:
*EUTROPHICATION, AQUATIC ALGAE, *LAKES, *CLASSIFICATION, DESMIDS,
OLIGOTROPHIC, *PHYTOPLANKTON, BRITISH ISLES, BIOINDICATORS, *ESTIMATING
EQUATIONS, PERIOD OF GROWTH.

ABSTRACT:
AFTER REVIEWING THE WORK OF NYGARD (1949) ON THE 'QUOTIENT SYSTEM' OF
LAKE CLASSIFICATION, THE AUTHOR DESCRIBES THE USE OF DESMIDS AS
INDICATORS OF LAKE TYPES. NYGARD STATES THAT QUOTIENTS CAN ONLY BE
DETERMINED FROM SAMPLES COLLECTED DURING THE TIME OF GREATEST ALGAL
DEVELOPMENT, FROM MID-MAY THROUGH SEPT. A BIG PROBLEM APPEARS TO BE THE
CORRECT IDENTIFICATION OF ALL OF THE PLANKTON SPECIES. NYGARD'S
'COMPOUND QUOTIENT' (1949 AND 1953) IS: MYXOPHYCEAE + CHLOROCOCCALEA +
CENTRALES + EUGLENINEAS/DESMIDIACEAS. SLIGHT ERRORS IN DETERMINING THE
NUMBER OF DESMIDS WILL ALTER GREATLY THE DIMENSIONLESS NUMBERS DERIVED
FROM THE QUOTIENTS AND THUS WILL BIAS THE INVESTIGATOR AS TO THE
TROPHIC STATUS OF THE LAKES. EUTROPHIC WATERS SOMETIMES CONTAIN AS MANY
PHYTOPLANKTON SPECIES AS DO OLIGOTROPHIC WATERS. IN THE AUTHOR'S
OPINION, THE COMPOSITION OF THE PHYTOPLANKTON IS NOT NECESSARILY
ALTERED BY AN INCREASE IN NUTRIENTS; FURTHER, THAT A BLOOM MAY NOT FORM
IF THE INCREASE IN NUTRIENTS IS A VERY SLOW ONE. A STUDY OF 300 LAKES
IN THE BRITISH ISLES INDICATES THAT DESMIDS ARE MOST COMMON (59
PERCENT) IN OLIGOTROPHIC WATERS, BUT ARE NOT LIMITED TO THEM; THEY
OCCUR, ALSO (24 PERCENT) IN WATERS CLASSED AS EUTROPHIC AND HAVING
QUOTIENTS OF 2.0 OR MORE.

FIELD 05C

ACCESSION NO. W68-00255

A SYNERGISTIC APPROACH TO PHOSPHORUS REMOVAL,

DORR-OLIVER INC, STANFORD, CONNECTICUT.

ROBERT J. SHERWOOD.

J OF CIV ENG, AM SOC CIV ENG, PP 32-35, MAY 1968. 4 P, 3 FIG, 1 TAB, OF REF.

DESCRIPTORS:
*EUTROPHICATION, SEWAGE, *PHOSPHORUS, PHOSPHATES, *SEWAGE TREATMENT, DOMESTIC WASTES, *ALGAE, WATER PROPERTIES, BIOCHEMICAL OXYGEN DEMAND, *BACTERIA, ACTIVATED SLUDGE, IRON COMPOUNDS, LIME, OXYGEN, TERTIARY TREATMENT, COST COMPARISONS.

ABSTRACT:
THE PHOSPHORUS IN SEWAGE HAS BEEN HELD RESPONSIBLE FOR THE EXCESSIVE ALGAL GROWTHS IN LAKES INTO WHICH DOMESTIC SEWAGE IS DISCHARGED, BUT THE EXTENT TO WHICH IT MUST BE ELIMINATED IS NOT KNOWN. THE DESIGN FACTORS IN A PHOSPHATE-REMOVAL PLANT ARE: THE RATIO OF SOLUBLE TO TOTAL PHOSPHATE; WATER ALKALINITY AND HARDNESS; BOD REMOVAL REQUIRED; PHOSPHORUS REMOVAL REQUIRED; AND THE RATIO OF SOLUBLE BOD TO TOTAL BOD. THE BIOCHEMICAL REMOVAL OF PHOSPHORUS FROM WASTES BY THE ACTIVATED SLUDGE METHOD IS EFFECTIVE IN CERTAIN RATIOS OF INFLUENT BOD TO PHOSPHORUS. THE REMOVAL RATE IS HIGH WHEN THE SLUDGE AGE IS 0.5-1 DAY, BUT IS GREATLY REDUCED WHEN THE AGE IS 15-25 DAYS. ALGAE CAN HELP IN SEWAGE STABILIZATION IN THIS CYCLE: ALGAE PRODUCE OXYGEN WHICH COMBINES WITH ORGANIC MATTER; THIS, BY BACTERIAL ACTION RELEASES CARBON DIOXIDE, AMMONIA, PHOSPHATE, AND WATER, WHICH ARE THEN USED BY THE ALGAE TO PRODUCE MORE OXYGEN. TERTIARY TREATMENT BY THE ADDITION OF CHEMICALS IS EFFECTIVE, BUT ITS EXPENSE INCREASES GREATLY AFTER THE INITIAL PHOSPHORUS REMOVAL BY THE FIRST APPLIED DOSES. LIME IS A LESS EXPENSIVE ADDITIVE THAN ALUM OR AN IRON SALT AND IS ALSO EFFECTIVE IN PRIMARY PHOSPHATE COAGULATION.

FIELD 05C

ACCESSION NO. W68-00256

INDUCED EUTROPHICATION - A GROWING WATER RESOURCE PROBLEM,

FWPCA, US DEPT. OF THE INTERIOR, CINCINNATI.

A. F. BARTSCH.

TRANS 1960 SEM ON ALGAE AND METROPOLITAN WASTES, ROBT A TAFT SANITARY ENG CENTER, TECH REPT W61-3, PP 6-9, 1961. 4 P.

DESCRIPTORS:
*EUTROPHICATION, SEWAGE, FISH POPULATIONS, PLANKTON, *CYANOPHYTA, WASHINGTON, DC, *ESTUARIES, HUMAN POPULATION, SEWAGE EFFLUENTS, AQUATIC POPULATIONS, NUISANCE ALGAE, DUCKS(WILD), OYSTERS, CHANNELS, POLLUTION ABATEMENT, HUMAN POPULATION.

ABSTRACT:
A FAMILIAR THREAD OF SIMILARITY AND REPETITION CAN BE NOTED IN EACH CASE OF EUTROPHICATION. SOMETIMES IT INCLUDES ALL, BUT MORE COMMONLY ONLY SEVERAL, OF THE FOLLOWING STEPS: (1) INTRODUCTION OF RAW OR TREATED, SEWAGE; (2) REPLACEMENT OF PRIZED DEEP-WATER TROUT OR WHITEFISH BY LESS DESIRABLE KINDS; (3) INCREASE IN PLANKTON OR FREE-FLOATING PLANT AND ANIMAL LIFE; AND (4) EXPLOSIVE SEASONAL APPEARANCE OF THE CYANOPHYTE, OSCILLATORIA RUBESCENS. THE POTOMAC RIVER BELOW WASHINGTON, DC, IS ANOTHER OF THE SUSCEPTIBLE BODIES OF WATER. HERE IT IS A TIDAL REACH, WHERE THE WATER WASHES BACK AND FORTH, AND IS AS MUCH LIKE A LAKE AS A RIVER. THE IMMEDIATE TRIBUTARY LAND AREA IS OCCUPIED BY A CONCENTRATED SEWERED POPULATION OF 1.7 MILLION PEOPLE. LIMITED INTERMITTENT STUDIES DURING THE LAST FEW YEARS REVEAL THAT ALGAL POPULATIONS ARE MUCH GREATER NOW THAN IN THE PAST. BLOOMS OF CYANOPHYTES ARE COMMON AND MAY BECOME OBJECTIONABLE IN SOME OF THE BAYS AND COVES. LONG ISLAND DUCKLINGS ALSO CONTRIBUTE TO THE ENRICHMENT. THE AREA NORMALLY RAISES FINE OYSTERS BUT THE OYSTER BEDS BECOME POLLUTED UNLESS CHANNELS ARE DREDGED TO ALLOW THE FLUSHING OF THE BAY BY THE TIDES.

FIELD 05C

ACCESSION NO. W68-00461

SCIENTIFIC STUDIES AND CHEMICAL TREATMENT OF THE MADISON LAKES,

CITY OF MADISON, MADISON, WISC.

BERNHARD DOMOGALLA.

A SYMPOSIUM ON HYDROBIOLOGY, UNIV OF WISCONSIN PRESS, MADISON, PP 303-309, 1941. 7 P, 3 FIG, 4 TAB, 8 REF.

DESCRIPTORS:
*EUTRIPHICATION, *LAKES, WISCONSIN, COPPER SULPHATE, NUISANCE ALGAE, *ALGAL CONTROL, WEED CONTROL, *INVESTIGATIONS, ZOO-PLANKTON, SPRAYING, BIOCONTROL, INFECTION, NUTRIENTS, ALGAE, NITROGEN, PHOSPHORUS, BICARBONATES, *AQUATIC POPULATIONS.

IDENTIFIERS:
SAPROLEGNIA.

ABSTRACT:
STUDIES OF CHEMICAL TREATMENTS APPLIED TO THE MADISON LAKES INDICATED: (1) SPRAYING WITH COPPER SULPHATE AND OTHER CHEMICAL TREATMENTS WAS FOUND TO BE THE MOST EFFECTIVE AND ECONOMICAL METHOD OF CONTROLLING GROWTHS OF OBNOXIOUS ALGAE AND WEEDS. (2) STUDIES SHOWED THAT THE CHEMICAL TREATMENTS AS THEY WERE MADE DID NOT SERIOUSLY INJURE THE ZOOPLANKTON. (3) THE CHEMICAL TREATMENTS FOR CONTROL OF ALGAE ALSO CONTROLLED SAPROLEGNIA. (4) CHEMICAL SPRAYING ALSO BROUGHT 'SWIMMER'S ITCH' AND FUNGUS INFECTIONS UNDER CONTROL. (5) STUDIES OVER AN 18-YR PERIOD INDICATE THAT NITROGEN COMPOUNDS, PHOSPHORUS, AND BICARBONATES ARE AMONG THE FACTORS THAT TEND TO PROMOTE THE EXCESSIVE GROWTH OF ALGAE AND LARGE AQUATIC PLANTS. CHEMICAL TREATMENT WITH COPPER SULPHATE REDUCED THE NUMBERS OF ALGAE IN THE MADISON LAKES BUT DID NOT ELIMINATE THEM. THESE ALGAE IDENTIFIED IN LAKE MONONA IN 1940 WERE, IN DESCENDING ORDER OF ABUNDANCE: ANABAENA, MICROCYSTIS, PEDIASTRUM, LYNGBA, STAURASTRUM, FRAGILARIA, HYDRODICTYON, AND PANDORINA.

FIELD 05C

ACCESSION NO. W68-00468

FERTILIZATION AND ALGAE IN LAKE SEBASTICOOK, ME.

FWPCA, DEPT. OF THE INTERIOR, CINCINNATI.

REPT OF THE TECH ADVISORY AND INVESTIGATIONS ACTIVITIES OF THE TECH SERVICES PROGRAM, FED WTR POLL CONTR ADM, ROBT A TAFT ENGG CENTER, CINCINNATI, OHIO, 122 P, 1966.

DESCRIPTORS:
*EUTROPHICATION, MAINE, NUTRIENTS, AQUATIC POPULATIONS, WATER QUALITY, LAKE MORPHOMETRY, DOMESTIC WASTES, PLANT GROWTH SUBSTANCES, LIMNOLOGY, PLANT GROWTH REGULATORS, CYANOPHYTA, ENVIRONMENTAL EFFECTS, CURRENTS(WATER), INDUSTRIAL WASTE, CORES, DIATOMS, ODOR-PRODUCING ALGAE, CHLOROPHYLL, CHEMTROL, COPPER SULFATE, LAKES, NITROGEN.

IDENTIFIERS:
MICROCYSTIS, ANABAENA.

ABSTRACT:
IN 1965, A STUDY WAS MADE OF LAKE SEBASTICOOK, ME, BY THE TECH ADVISORY AND INVESTIGATIONS ACTIVITY AND THE ME WTR IMPR COMM. THE AREA OF THE LAKE IS 1,736 HA, ITS MAX DEPTH IS 17.7 M, AND ITS MEAN DEPTH IS 6 M. SAMPLING WAS DONE IN FEB, MAY, JUL, AND OCT IN THE EFFORT TO DETERMINE THE MAJOR SOURCES OF NUTRIENTS, TO ASSESS THEIR SIGNIFICANCE, AND TO RECOMMEND THEIR CONTROL. DATA INDICATE THAT ABOUT 98% OF THE NUTRIENTS COME FROM SOURCES OTHER THAN LAND DRAINAGE, THE BULK (75%) FROM DOMESTIC AND INDUSTRIAL (INCLUDING WOOLEN MILLS AND A POTATO-PROCESSING PLANT) WASTES. EXTENSIVE BLOOMS OF THE BLUE-GREEN ALGAE, MICROCYSTIS AERUGINOSA AND ANABAENA SPP, PRODUCE OBJECTIONABLE MATS AND ODORS. THE AVERAGE DETENTION TIME IN THE LAKE IS 3.5 YR. PHOSPHORUS IS JUDGED TO BE THE MOST SIGNIFICANT NUTRIENT AND IS FOLLOWED IN IMPORTANCE BY NITROGEN. THE PHOSPHORUS INPUT EXCEEDS ITS OUTPUT. A LIMITED ANALYSIS OF A 48 CM CORE SHOWED AN INCREASE IN DIATOMS IN THE TOP FEW CM. CHLOROPHYLL CONTENT FLUCTUATES WIDELY. STATEMENTS BY RESIDENTS AND OTHERS INDICATE THAT MOST OF THE CHANGES IN THE LAKE OCCURRED WITHIN THE PAST 10-20 YR. CHEMICAL CONTROL BY THE USE OF COPPER SULFATE IS NOT RECOMMENDED BECAUSE IT WOULD BE ONLY TEMPORARY AND TOO EXPENSIVE, UP TO $40,000/YR.

FIELD 05C

ACCESSION NO. W68-00470

CHANGES OF EUTROPHICATION AND BIO-PRODUCTION MEASURED BY THE BIOMASSTITRE OF
  TEST ALGAE (GERMAN),

BUNDESGESUNDHEITSAMT, INSTITUT FUR WASSER-, BODEN- UND LUFTHYGIENE, BAD
  GODESBERG, W. GERMANY.

G. BRINGMANN, AND R. KUHN.

GESUNDHEITS-INGENIEUR, HEFT 2.79, PP 50-54, 1958. 5 P, 10 FIG, 1 TAB.

   DESCRIPTORS:
     *EUTROPHICATION, TEST ANALYSIS, *LABORATORY TESTS, TEST PROCEDURES,
     NURTRIENTS, RIVERS, LAKES, *GROWTH RATES, *WATER TEMPERATURE, SEASONAL,
     TEST PROCEDURES, SCENEDESMUS, *ALGAE.

   IDENTIFIERS:
     BERLIN(WEST), GERMANY.

   ABSTRACT:
     THE ARTICLE DESCRIBES A CONTINUATION OF THE WORK OF BRINGMANN AND KUHN
     IN WHICH A BIO-ASSAY TECHNIQUE WAS USED TO ESTIMATE THE INFLUENCES OF
     ENRICHMENT ON THE STREAMS AND OTHER BODIES OF WATER IN THE BERLIN
     (WEST) WATER NET. BY COMPARING WATER TEMPERATURES AND THE BIOPRODUCTION
     OF THE TEST ALGAE, THE AUTHORS NOTED THAT THE HIGHEST PRODUCTION
     OCCURRED DURING THE WINTER WHEN WATER TEMPERATURES WERE LOW, AND VICE
     VERSA. AS BASED UPON MEAGER DATA--ABOUT 9 SAMPLINGS A YEAR FOR
     TEMPERATURE AND 7 SAMPLINGS A YEAR TO DETERMINE
     BIO-PRODUCTION--TEMPERATURE AND PRODUCTION WERE USUALLY FOUND TO BE OUT
     OF PHASE. VALUES WERE COMPARED FOR THE INTERVAL FROM MAY 1956 TO APRIL
     1957. SCENEDESMUS WAS USED AGAIN AS THE TEST ORGANISM.

   FIELD 05C

   ACCESSION NO. W68-00472

THE ALGAE-TITRE AS A MEASURE OF EUTROPHICATION OF WATER AND MUD (GERMAN),

INSTITUT FUR WASSER-BODEN-UND LUFTHYGIENE, BERLIN-DAHLEM, GERMANY.

G. BRINGMANN, AND R. KUHN.

GESUNDEITS-INGENIEUR, VOL 77, PP 374-381, 1956. 8 P.

   DESCRIPTORS:
     *EUTROPHICATION, ALGAE, GERMANY, VOLUMETRIC ANALYSIS, LIMNOLOGY, GROWTH
     RATES, *LABORATORY TESTS, *SCENEDESMUS, DIATOMACEOUS EARTH, MUD, LAKES,
     CULTIVATION, CHANNEL MORPHOLOGY, SEWAGE INPUT, BIOLOGICAL PROPERTIES,
     *WATER QUALITY, TURBIDITY, PLANT POPULATIONS.

   IDENTIFIERS:
     BERLIN, *ALGAE-TITRE.

   ABSTRACT:
     ALGAE OF THE GENUS SCENEDESMUS WERE INJECTED INTO MUD SAMPLES AND INTO
     WATER SAMPLES FILTERED FREE OF ALGAE. THESE WERE THEN CULTIVATED UNDER
     CONSTANT CONDITIONS FOR UNIFORM PERIODS, AFTER WHICH THE TURBIDITY, A
     FUNCTION OF THE NUMBER OF ORGANISMS, WAS DETERMINED. THE NEPHELOMETRIC
     VALUE IS GENERALLY EQUIVALENT TO MG/1 OF INFUSORIAL SOIL. THE
     EQUIVALENT VALUE OF INFUSORIAL SOIL SERVED AS A MEASURE OF TURBIDITY
     AND IS TERMED 'ALGAE-TITRE.' THE RELATION OF THE ALGAE-TITRE TO THE
     DISCHARGE CHANNEL AND TO THE SEWAGE INPUT WAS DEMONSTRATED. EXPERIMENTS
     SHOWED THAT ALGAE-TITRE MAY SERVE AS A BIOLOGICAL MEASURE OF THE
     EUTROPHICATION OF WATER AND MUD. DATA WERE DERIVED FROM SAMPLES TAKEN
     AT 41 PLACES IN THE BERLIN WATER NET DURING 14 DAYS FROM JULY 1955 TO
     JAN 1956. THE AVERAGE VALUES OF ALGAE-TITRE WERE THEN COMPARED ON
     SEVERAL BASES. THE AUTHORS HAVE USED THE METHOD IN ROUTINE TESTS OF
     WATER QUALITY.

   FIELD 05C

ACCESSION NO. W68-00475

EUTROPHICATION PROBLEMS AND PROGRESS,

PACIFIC NORTHWEST WATER LABORATORY, FWPCA, CORVALLIS, OREGON.

A. F. BARTSCH.

MIN 152ND MTG, MO BAS INTER-AGENCY COMM, BISMARCK, N DAK, APPEND G, PP
G-1/G-7, MAR 1968.

DESCRIPTORS:
*EUTROPHICATION, LAKES, *NUTRIENTS, UNITED STATES, PHOSPHORUS,
NITROGEN, SEWAGE, INDUSTRIAL WASTES, DOMESTIC WASTES, SEWAGE TREATMENT,
HARVESTING OF ALGAE, BOTTOM SEDIMENTS, WEED CONTROL, SOLAR RADIATION,
*WATER POLLUTION CONTROL, ALGICIDES, BIOCONTROL, *WATERSHED MANAGEMENT,
*ENVIRONMENTAL EFFECTS, CHEMTROL, ALGAE.

IDENTIFIERS:
CRATER LAKE, ORE, GROWTH RATES, WATER PROPERTIES.

ABSTRACT:
THERE ARE ALL DEG OF EUTROPHICATION IN THE 100,000 SMALL LAKES OF THE
USA. CITIZENS AND THEIR AGENCIES ARE INCREASINGLY AWARE OF THE THREAT.
THE DENSE ALGAL GROWTH OF ADVANCED EUTROPHICATION IS THE RESULT OF
ADDITION OF CERTAIN NUTRIENTS TO WATER: ONLY .4 OZ OF PHOSPHORUS OR 1
LB OF NITROGEN MIXED WITH AN ACRE FT OF WATER WILL SUPPORT AN
OBJECTIONABLE GROWTH. MAN CONTRIBUTES GREAT QUANTITIES OF NUTRIENTS IN
SEWAGE, INDUSTRIAL WASTES, AND THE FERTILIZERS WASHED FROM FARMLANDS. 4
NUTRIENT-CONTROL MEASURES ARE: (1) MECHANICAL REMOVAL OF PLANTS FROM
THE WATER; (2) CONTROLLING PLANT GROWTH BY THE ADDITION OF CERTAIN
CHEMICALS; (3) BIOLOGICAL CONTROL BY INTRODUCING PLANT-KILLING FUNGI
AND VIRI; AND (4) ECOLOGICAL CONTROL BY: (A) REDUCING THE INPUT OF
PHOSPHORUS AND NITROGEN (SEWAGE DIVERSION, PERHAPS); (B) TREATMENT OF
SEWAGE TO REMOVE MOST OF THE PHOSPHORUS BEFORE DISCHARGE TO A LAKE; (C)
DILUTION OF THE NUTRIENTS ALREADY PRESENT BY FLUSHING OF LAKE (MOSES
LAKE, WASH); AND (D) REMOVING NUTRIENTS BY HARVESTING ALGAE (NOT NOW
FEASIBLE) AND WATERWEEDS (NOW POSSIBLE). ESTIMATES ARE THAT IF
WATERWEED HARVEST CAN BE 3X THE NUTRIENT INPUT, EUTROPHICATION CAN BE
CONTROLLED. BOTTOM-SEDIMENT/NUTRIENT INTERCHANGE, AND THE EFFECT OF
REDUCING INSOLATION ARE BEING STUDIED.

FIELD 05C

ACCESSION NO. W68-00478

ALGAE BLOOMS IN LAKE ZOAR, CONN,

GENERAL DYNAMICS CORP., GROTON, CONN.; WATER RESOURCES COMM., HARTFORD, CONN.

R. J. BENOIT, AND J. J. CURRY.

TRANS 1960 SEM ON ALGAE AND METRO WASTES, ROBT A TAFT SANITARY ENGG CENTER,
CINCINNATI, OHIO, TECH REPT W61-3, PP 18-22, 1961. 5 P.

DESCRIPTORS:
*EUTROPHICATION, *POLLUTION ABATEMENT, DRAINAGE SYSTEMS, FLOW RATES,
CONN STREAM FLOW, ALGAE, AQUATIC ALGAE, CURRENTS(WATER), LIMNOLOGY,
LAKES, *PILOT PLANTS, *NUISANCE ALGAE, ROOTED AQUATIC PLANTS,
*RESERVOIRS, HYDROELECTRIC PLANTS, WEED CONTROL, WATER PROPERTIES,
PHOSPHORUS, CHEMICAL ANALYSIS.

IDENTIFIERS:
HYDRODICTYON, ELODEA, CERATOPHYLLUM, MICROCYSTIS, ANABAENA.

ABSTRACT:
LAKE ZOAR, CONN, WAS IMPOUNDED IN 1919 BEHIND THE STEVENSON DAM ACROSS
THE HOUSATONIC RIVER. THE DRAINAGE AREA ABOVE THAT POINT IS ABOUT 1,545
SQ MI. DURING HIGH-FLOW STAGE (2,500 CFS) THE LAKE WATERS ARE REPLACED
EVERY 9 DAYS; DURING LOW-FLOW PERIODS (AUG AND SEPT) THE REPLACEMENT
TIME IS 40 DAYS. BY 1947 THE INCREASE IN ALGAL GROWTH HAD BEEN SO GREAT
AS TO CREATE A SERIOUS NUISANCE. ROOTED AQUATIC PLANTS WERE FIRST
DESIGNATED AS THE NUISANCE-CAUSING ORGANISMS BUT ALGAL BLOOMS WERE
FOUND TO BE EQUALLY IMPORTANT. HYDRODICTYON OCCURS IN MID-JUL IN GREAT
QUANTITIES IN THE SHALLOW, QUIET WATERS AND IS INTIMATELY ASSOCIATED
WITH SUCH AQUATIC WEEDS AS ELODEA AND CERATOPHYLLUM. CURRENTS MOVE
FLOATING MATS OF WEEDS FROM ONE PLACE TO ANOTHER. IN THE EARLY SUMMER,
LARGE QUANTITIES OF MICROCYSTIS AND ANABAENA APPEAR IN THE UPPERMOST
4-FT ZONE OF LAKE WATER. PHOSPHORUS CONCENTRATIONS MEASURED IN THE
WATERS OF THE LAKE AND SOURCE STREAMS RANGE FROM 12-41 PPB AND AVERAGE
ABOUT 25 PPB. THE POLLUTION-CONTROL AGENCIES CONSTRUCTED A PILOT PLANT
ON THE PLAN SUGGESTED BY LEA, ROHLICH, AND KATZ (1954). IT WAS PARTLY
DESTROYED BY A FLOOD AND, DESPITE APPARENTLY SUCCESSFUL OPERATION, WAS
NOT REBUILT BECAUSE AN INVESTIGATION BY AN INDEPENDENT AGENCY FOUND AN
UNEXPLAINED INCREASE IN PHOSPHORUS CONTENT.

FIELD 05C

ACCESSION NO. W68-00479

NUTRIENTS AND ALGAE IN LAKE SEBASTICOOK, MAINE,

FWPCA, CINCINNATI, OHIO.

K. M. MACKENTHUN, L. E. KEUP, AND R. K. STEWART.

JOUR WATER POLL CONTROL FED, 40, 2 PART 2, R72-R81, FEB 1968. 10 P, 6 FIG, 3 TAB, 9 REF.

DESCRIPTORS:
ALGAE, EUTROPHICATION, LAKES, MICROORGANISMS, NITROGEN, NUTRIENTS, PHOSPHORUS, PLANKTON, PLANTS(ORGANISMS).

IDENTIFIERS:
LAKE SEBASTICOOK, ME, MAINE, PERIODIC VARIATIONS, SEASONAL VARIATIONS, SOURCES.

ABSTRACT:
FOUR SEASONAL FIELD STUDIES WERE MADE ON LAKE SEBASTICOOK, MAINE, AND ITS SURROUNDING AREA, TO DETERMINE THE CAUSE OF NUISANCE ALGAL GROWTHS, IDENTIFY MAJOR NUTRIENTS SOURCES, ASSESS THEIR SIGNIFICANCE, AND RECOMMEND CONTROL MEASURES. IN FEBRUARY TOTAL NITROGEN VALUES WERE 3.3 MG/1 IN THE LAKE'S SURFACE WATERS AND 6.2 MG/1 IN THE PROFUNDAL WATERS. VALUES WERE LOWER DURING OTHER SEASONS. THERE WAS NO REDUCTION IN NITROGEN PASSING THROUGH THE LAKE. THE LAKE RECEIVED 8,000 LB (3,630 KG) OF TOTAL PHOSPHORUS ANNUALLY, DISCHARGED 4,150 LB (1,880 KG), AND RETAINED 48 PERCENT. PHYTOPLANKTON WERE AS HIGH AS 2,260 LB/SURFACE ACRE (2,540 KG/HA) (WET WEIGHT) DURING THE MAXIMUM GROWING PERIOD. AN 80 PERCENT REDUCTION IN CONTROLLABLE INFLOWING PHOSPHORUS WOULD CORRECT THE NUISANCE CONDITIONS.

FIELD 05C

ACCESSION NO. W68-00481

PLANKTON ALGAE AS INDICATORS OF THE SANITARY CONDITION OF A STREAM,

FWPCA, CINCINNATI, OHIO.

F. J. BRINLEY.

BIOLOGICAL STUDIES, OHIO RIVER POLLUTION SURVEY II, SEWAGE WKS. J, VOL 14, PP 152-159, 1942. 8 P.

DESCRIPTORS:
*EUTROPHICATION, PLANKTON, *ALGAE, CHEMICAL PROPERTIES, BACTERIA, ORGANIC POLLUTANTS, *STREAMS, EUGLENA, NUTRIENTS, PLANT POPULATIONS, HISTORY, WATER PURIFICATION, *WATER UTILIZATION, DECOMPOSING ORGANIC MATTER, *STREAM POLLUTION.

IDENTIFIERS:
*DOMESTIC WATER USE.

ABSTRACT:
THE AUTHOR COMPARES THE POPULATION OF CERTAIN SPECIES OF PLANKTONIC ALGAE WITH THE CHEMICAL AND BACTERIOLOGICAL INDICES OF THE DEG OF ORGANIC POLLUTION IN A FLOWING STREAM. THE COMPARISON SHOWS THAT LARGE NUMBERS OF CHRYSOCOCCUS AND CRYPTOMONAS INDICATE THAT THE DECOMPOSITION OF THE ORGANIC MATTER IN THE STREAM HAS BEEN COMPLETED BY NATURAL PROCESSES AND THE STREAM MAY BE CONSIDERED CLEAN BUT NOT THAT IT CAN BE PUT TO DOMESTIC USE WITHOUT FURTHER TREATMENT. LARGE NUMBERS OF EUGLENA, TRACHELOMONAS AND PHACOTUS INDICATE THAT THE WATER UPSTREAM IS HEAVILY POLLUTED, AND THAT BACTERIA HAVE CHANGED THE ORGANIC MATTER TO AVAILABLE PLANT FOOD TO PRODUCE A RICH MEDIUM. THE PLANKTON POPULATION SERVES AS A BETTER INDICATOR OF THE HISTORY OF A STREAM THAN IT DOES OF THE SANITARY CONDITION OF THE STREAM AT ANY SINGLE PLACE. MOST WATERS, EVEN HEAVILY POLLUTED ONES, WILL CONTAIN A FEW OF THESE FORMS: THEREFORE, THEIR PRESENCE DOES NOT NECESSARILY INDICATE PURIFIED WATER NOR DOES THEIR ABSENCE INDICATE A POLLUTED CONDITION.

FIELD 05B

ACCESSION NO. W68-00483

FERTILIZATION OF LAKES--GOOD OR BAD,

MICH DEPT OF CONSERV.

ROBERT C. BALL.

MICH CONSERV, VOL 17, NO 9, PP 7-14, SEPT 1948. 8P, 1 FIG.

DESCRIPTORS:
*EUTROPHICATION, *LAKES, NUTRIENTS, OXYGEN REQUIREMENTS, ICED LAKES,
*FERTILIZATION, WATER FLEAS, LARVAE, GROWTH RATES, *FRESH WATER FISH,
MICHIGAN, TROUT, WINTERKILLING, FISHKILL, NUISANCE ALGAE, ALGAE,
SEASONAL, LIMNOLOGY, LIGHT PENETRATION, OXYGEN, SPORT FISHING, ON-SITE
INVESTIGATIONS.

IDENTIFIERS:
ARTIFICIAL FERTILIZATION, COMMERCIAL FERTILIZERS.

ABSTRACT:
STUDIES OF 4 LAKES IN THE NORTHERN PART OF THE SOUTHERN PENINSULA OF
MICHIGAN INDICATE THAT INORGANIC FERTILIZERS APPLIED TO THE WATERS MAY
BENEFIT THE FISH POPULATIONS IN SOME LAKES, BUT EXERT A HARMFUL EFFECT
IN OTHERS. 2 PAIRS OF LAKES WERE STUDIED, 1 OF EACH PAIR RECEIVED THE
INORGANIC FERTILIZER AND THE OTHER DID NOT. THE EXPERIMENTS HAD BEEN
CARRIED ON FOR 3 YEARS BY 1948. THE ADDITION OF COMMERCIAL FERTILIZER
TO SMALL PONDS INCREASES THE NUMBER OF ORGANISMS UPON WHICH YOUNG FISH
FEED. THE NUTRIENTS IN THE FERTILIZER PROVIDE FOOD FOR ALGAE, THE FOOD
FOR WATER FLEAS AND INSECT LARVAE; THESE ORGANISMS IN TURN, ARE A FOOD
SOURCE FOR BOTH YOUND AND ADULT GAME FISH. THE ADDITION OF FERTILIZER
SEEMS TO INCREASE THE TOTAL NUMBER OF SMALL GAME FISH BUT DOES NOT
RESULT IN GROWTH TO SIZES OF INTEREST TO GAME FISHERMEN. ARTIFICIAL
FERTILIZER INCREASES THE GROWTH OF ALGAE. IT ALSO FAVORS THE GROWTH OF
LESS DESIRABLE SPECIES OF GAME FISH AT THE EXPENSE OF TROUT. THE HEAVY
ALGAL GROWTH OF OVER-FERTILIZED LAKES, TENDS TO DEVELOP A SERIOUS
OXYGEN DEFICIENCY WHEN ICED OVER IN WINTER; EXCLUSION OF AIR AND MUCH
OF THE SUNLIGHT CAUSES THE ALGAE TO DIE AND, IN THEIR DECAY, TO USE UP
THE OXYGEN NEEDED BY FISH; SEVERE WINTERKILL OF GAME FISH RESULTS.
UNTIL MORE IS KNOWN, THE USE OF FERTILIZERS IN SMALL LAKES SHOULD BE
AVOIDED.

FIELD 05C

ACCESSION NO.   W68-00487

ELEVEN YEARS OF CHEMICAL TREATMENT OF THE MADISONLAKES--ITS EFFECT ON FISH AND
FISH FOODS,

CITY OF MADISON AND WIS STATE LAB OF HYGIENE.

BERNHARD DOMOGALLA.

TRANS AMER FISH SOC, VOL 65, PP 115-121, 1935. 7 P, 1 FIG, 3 TAB.

DESCRIPTORS:
*EUTROPHICATION, *LAKES, WISCONSIN, ALGAE, *NUISANCE ALGAE, SPRAYING,
SEASONAL, *ALGAL CONTROL, FISHKILL, ROOTED AQUATIC PLANTS, GLACIAL
LAKES, PLANKTON, EPIDEMICS.

IDENTIFIERS:
SECCHI DISC, VISIBILITY TESTS, CONTROL METHODS.

ABSTRACT:
LAKE MONONA IS THE SECOND OF FOUR GLACIAL LAKES IN THE VALLEY OF THE
YAHARA RIVER NEAR MADISON, WIS. COMPLAINTS OF OBNOXIOUS ODORS AND ALGAL
GROWTHS LED IN 1925 TO STUDIES OF POSSIBLE REMEDIAL MEASURES. IT WAS
FOUND THAT PHOSPHORUS, NITROGEN, AND BICARBONATES WERE THE NUTRIENTS
THAT CAUSED ALGAL OVERGROWTHS. TO CONTROL THESE GROWTHS, COPPER
SULPHATE WAS FIRST APPLIED BY DRAGGING A BAG FILLED WITH IT THROUGH THE
WATER. MUCH MORE EFFECTIVE WAS THE SYSTEMATIC SPRAYING OF THE LAKE WITH
THE SAME CHEMICAL. THE SPRAYING REDUCED THE GROWTH OF CLADOPHORA AND
ULOTHRIX IN MAY; ANABAENA IN JUN; MICROCYSTIS, HYDRODICTYON, AND
APHANIZOMENON IN JUL AND AUG; AND APHANIZOMENON, CLADOPHORA, AND
ULOTHRIX IN SEPT. ROOTED WEEDS WERE CUT FROM UNDER 600 AC OF THE 5.4 SQ
MI OF THE LAKE. WEEDS IN THE DEEPER WATERS WERE NOT AFFECTED BY THE
CHEMICAL SPRAYING. VISIBILITY TESTS SHOWED DEPTHS OF 4.3 M IN LAKE
MONONA AS COMPARED WITH 0.6 AND 2.4 M IN OTHER LAKES IN THE GROUP. AN
EPIZOOTIC AMONG FISHES INFECTED WITH SAPROLEGNIA WAS PARTICULARLY
SEVERE IN 1930; HOWEVER, THE SPRAYING OF LAKE MONONA GREATLY REDUCED
THE FISHKILL THERE.

FIELD 05C

ACCESSION NO.   W68-00488

NUTRIENT BUDGET: RATIONAL ANALYSIS OF EUTROPHICATION IN A CONNECTICUT LAKE,

DEPT OF SOILS CLIMATOL, THE CONN AGR EXP STATION, NEW HAVEN, CONN.

C. R. FRINK.

ENVIRONMENTAL SCI AND TECH, VOL 1, NO 5, PP 425-428, MAY 1967. 4 P.

DESCRIPTORS:
*EUTROPHICATION, *LAKES, CONNECTICUT, WATER POLLUTION, *NUTRIENT
REQUIREMENTS, *BOTTOM SEDIMENTS, ON-SITE DATA COLLECTIONS, AQUATIC
PLANTS, WATERSHED, ALGAE, AQUATIC VEGETATION, NITROGEN, PHOSPHOROUS,
AVAILABLE NUTRIENTS, SEDIMENT CORES.

IDENTIFIERS:
*SEDIMENTS, *NUTRIENT BUDGET, CYCLING NUTRIENTS.

ABSTRACT:
THE SIGNIFICANCE OF NUTRIENT CONTAMINATION OF WATER BY MAN CAN BE
EVALUATED ONLY BY COMPARISON WITH ALL OTHER SOURCES. IN AN EXEMPLARY
EUTROPHIC LAKE IN NORTHWESTERN CONNECTICUT, THE NUTRIENT INPUT FROM A
LARGELY FORESTED WATERSHED WITH NO OVERT SOURCE OF POLLUTION WAS
ADEQUATE TO SUPPORT THE OBSERVED ABUNDANT VEGETATIVE GROWTH. IN
ADDITION, A VAST ACCUMULATION OF NUTRIENTS WAS FOUND IN THE LAKE BOTTOM
SEDIMENTS; THE UPPER CENTIMETER OF SEDIMENT CONTAINS AT LEAST 10 TIMES
THE ESTIMATED ANNUAL INPUT OF NITROGEN AND PHOSPHORUS. MOREOVER, THIS
RESERVOIR OF NUTRIENTS IN THE SEDIMENTS SHOULD BE CAPABLE OF SUPPORTING
PLANT GROWTH FOR SOME TIME EVEN IF ALL NUTRIENTS COULD BE EXCLUDED FROM
THE LAKE. APPARENTLY, THE ABUNDANT WEEDS AND ALGAE IN THIS LAKE ARE THE
RESULT OF NATURAL EUTROPHICATION WHICH MAN WILL BE HARD PRESSED TO
ALTER.

FIELD 05C

ACCESSION NO. W68-00511

EUTROPHICATION OF LAKES BY DOMESTIC DRAINAGE,

A. D. HASLER.

ECOL, VOL 28, PP 383-395, 1947. 13 P, 1 TAB, 2 FIG.

DESCRIPTORS:
*EUTROPHICATION, ALGAE, AQUATIC PLANTS, BIBLIOGRAPHIES, CYANOPHYTA,
FRESH WATER FISH, LAKES, *LAKE ZURICH, *MUNICIPAL WASTES, NUTRIENTS,
NUISANCE ALGAE, OLIGOTROPHY, PHYTOPLANKTON, *REVIEWS, *WATER POLLUTION
EFFECTS.

ABSTRACT:
THE REPORTS OF PURPOSEFUL FERTILIZATION OF LAKES ARE VERY MEAGER, BUT
INADVERTENT EUTROPHICATION HAS BEEN RECORDED FOR MANY LAKES,
PRINCIPALLY THOSE IN EUROPE. A CLASSIC EXAMPLE IS LAKE ZURICH IN WHICH
2 BASINS ARE SEPARATED BY A NARROW PASSAGE. THE LARGER UNTERSEE (141 M
DEEP) WAS ONCE DECIDEDLY OLOGOTROPHIC BUT NOW TRENDS STRONGLY TOWARD
EUTROPHY. THE SHALLOWER OBERSEE (50 M DEEP) SHOWS VIRTUALLY NO EFFECTS
OF EUTROPHY. THE UNTERSEE IS RIMMED BY NUMEROUS COMMUNITIES (COMBINED
POPULATION OF 110,000) (ZURICH CONTRIBUTES NO SEWAGE) BUT THE OBERSEE
LACKS SUCH COMMUNITIES. EVIDENCES OF EUTROPHICATION OF THE UNTERSEE
DATE BACK TO LAST CENTURY AND ARE THE INCREASE OF NUISANCE ALGAE; THE
DISAPPEARANCE OF DESIRABLE FISH AND REPLACEMENT BY COARSE FISH; AND
INCREASES IN THE CHLORIDE ION. THE AUTHOR REVIEWS THE FERTILIZATION
PROCESS IN NUMEROUS LAKES IN EUROPE AND THE UNITED STATES. HE CONCLUDES
THAT FERTILIZATION ALMOST CERTAINLY HASTENS THE EXTINCTION OF A LAKE
AND THUS LIMITS ITS BENEFITS TO MAN TO A RELATIVELY SHORT PERIOD.
(BYRNE-WISC)

FIELD 05C, 02H

ACCESSION NO. W68-00680

BIOLOGICAL REMOVAL OF NUTRIENTS FROM TREATED SEWAGE: LABORATORY EXPERIMENTS,

WISCONSIN UNIV.

G. P. FITZGERALD, AND G. A. ROHLICH.

VERH INTER VER LIMNOL, VOL 15, PP 597-608, 1964. 12 P, 3 FIG, 2 TAB, 26 REF.

DESCRIPTORS:
*EUTROPHICATION, LABORATORY TESTS, BIOLOGICAL TREATMENT,
BIBLIOGRAPHIES, CHLORELLA, *SEWAGE, EFFLUENTS, CARBON DIOXIDE, *ALGAE,
*HYDROGEN ION CONCENTRATION, PHOSPHORUS, NITROGEN, GROWTH RATES,
STABILIZATION PONDS, SEWAGE DISPOSAL.

ABSTRACT:
THE AUTHORS DEMONSTRATED BY LABORATORY TESTS THAT IN A RELATIVELY SHORT
PERIOD OF TIME (13 DAYS) THE NUTRITIONAL VALUE OF TREATED SEWAGES, BOTH
SECONDARY EFFLUENT AND STABILIZATION POND WATERS, WAS MORE CLOSELY
RELATED TO THE AMOUNTS OF AMMONIA NITROGEN THAN TO PHOSPHATE
PHOSPHORUS. THEIR STUDIES SHOWED THAT THE TYPE AND QUANTITY OF ALGAL
GROWTH ATTAINED IN A STABILIZATION POND WILL BE CONTROLLED BY THE
ENVIORNMENT. 1% INOCULATIONS FROM LAKE WATERS WITH VARIED ALGAL
POPULATIONS INTO STABILIZATION POND WATER HAD NO APPARENT EFFECT ON
LATER GROWTH OF THE ALGAE. THE EXPERIMENTS INDICATED THAT THE RATE OF
NUTRIENT UTILIZATION AND GROWTH OF CHLORELLA IN PRIMARY OR SECONDARY
EFFLUENTS APPEARS TO BE INDEPENDENT OF THE ORGANIC MATTER PRESENT.
BECAUSE THE CARBON DIOXIDE PRESENT EFFECTS PH OF THE SEWAGE, THE CARBON
DIOXIDE USED BY ALGAE RESULTS IN HIGHER PH VALUES AND IN THE
CONSIDERABLE LOSS OF SOLUBLE PHOSPHATE PHOSPHORUS IN THE MEDIUM. THE
PRECIPITATED PHOSPHORUS CAN BE REDISSOLVED WHEN THE PH OF THE MEDIUM
DECREASES. (BYRNE-WISC)

FIELD 05D

ACCESSION NO. W68-00855

THE INFLUENCE OF THE MINERAL COMPOSITION OF THE MEDIUM ON THE GROWTH OF
PLANKTONIC ALGAE. PART I. METHODS AND CULTURE MEDIA,

LONDON UNIV., QUEEN MARY COLL., ENGLAND.

S. P. CHU.

J ECOL, VOL 30, NO 2, PP 284-325, AUG 1942. 42 P, 8 FIG, 22 TAB, 43 REF.

DESCRIPTORS:
*EUTROPHICATION, PHYTOPLANKTON, BIBLIOGRAPHIES, *ALGAE, *CHEMICAL
PROPERTIES, NUTRIENTS, LAKES, TESTS, NITROGEN, PHOSPHORUS, CALCIUM,
MAGNESIUM, SODIUM, SILICA, DIATOMS, RESISTANCE, *PLANT GROWTH
REGULATORS, FRESH WATER LAKES, LABORATORY TESTS.

ABSTRACT:
THE AUTHOR REPORTS THAT ALL OF THE ALGAE USED IN A TEST SERIES, WITH
BUT ONE EXCEPTION, GROW EQUALLY WELL WHEN THE MEDIA WERE SUPPLIED WITH
NITROGEN SOURCES FROM NITRATE SALTS OR AMONIUM SALTS, SO LONG AS THE
NITROGEN CONCENTRATION WAS WITHIN THE OPTIMUM RANGE; IN LOWER NITROGEN
CONCENTRATIONS, GROWTH WAS GENERALLY BETTER WHEN THE SOURCE WAS A
NITRATE SALT. THE EXPERIMENT WAS ONE IN WHICH 14 PLANKTONIC ALGAE WERE
MAINTAINED IN CULTURE MEDIA SIMULATING THE CHEMICAL COMPOSITIONS OF
NATURAL FRESH WATERS. ALGAE WERE FOUND TO RESPOND WITH CONSIDERABLE
DIFFERENCE TO CONCENTRATIONS OF CALCIUM, MAGNESIUM, POTASSIUM, SODIUM,
AND SILICA. THE CALCIUM REQUIREMENT WAS OFTEN LOWER IN MEDIA WITH
HIGHER MAGNESIUM CONCENTRATIONS, WHILE EXCESSIVE POTASSIUM INCREASES
TOLERANCES FOR HIGHER CONCENTRATIONS OF CALCIUM AND MAGNESIUM.
APPRECIABLE QUANTITIES OF SILICA WERE FOUND TO BE NECESSARY ONLY FOR
THE GROWTH OF DIATOMS; SOME ALGAE, IN FACT, WERE INHIBITED BY THE
PRESENCE OF DISSOLVED SILICA. THERE WAS A WIDER RANGE IN PHOSPHORUS
CONCENTRATION WHEN NITROGEN WAS DERIVED FROM A NITRATE SOURCE RATHER
THAN FROM AN AMMONIUM SALT. (BYRNE-WISC)

FIELD 05C

ACCESSION NO. W68-00856

RADIOACTIVE PHOSPHORUS AND EXCHANGE OF LAKE NUTRIENTS,

F. R. HAYES, AND C. C. COFFIN.

ENDEAVOUR, VOL 10, PP 78-81, 1951. 4 P, 3 FIG, 8 REF.

DESCRIPTORS:
*EUTROPHICATION, *LAKES, ACID LAKES, BIBLIOGRAPHIES, *PHOSPHORUS
RADIOISOTOPE, ON-SITE DATA COLLECTIONS, FISHES, PLANKTON, ALGAE, FROGS,
MOSSES, HYPOLIMNION, THERMOCLINE, *ION EXCHANGE, WATER CIRCULATION,
NUTRIENT ABSORPTION, *CHEMICAL PROPERTIES, EPILMNION, BOTTOM SEDIMENTS.

ABSTRACT:
THE AUTHOR'S CONCLUDE FROM FIELD EXPERIMENTS THAT THE DISTRIBUTION OF
IONS IN WATER TAKES PLACE FAR MORE RAPIDLY (FROM 10-100 TIMES) THAN
PREVIOUSLY SUPPOSED, AND THAT THE REMOVAL OF MATERIALS FROM THE WATERS
DOES NOT TAKE PLACE SEPARATELY, BUT PERHAPS IS AN EXCHANGE WITH
SUBSTANCES IN CONTACT WITH THE WATER. PHOSPHORUS 32 WAS ADDED FIRST TO
THE EPILIMNION OF A HIGHLY STRATIFIED ACID-BOG LAKE LESS THAN 1 AC IN
AREA AND 22 FT IN DEPTH. THE UPTAKE BY ZOOPLANKTON, SPONGES, SPHAGNUM,
AND ALGAE WAS IMMEDIATE; FISHES, FUNDULUS, AND NOTEMIGONUS SHOWED NO
MEASURABLE UPTAKE FOR SEVERAL DAYS, SUGGESTING THAT THEIR INTAKE WAS BY
FEEDING ON PLANKTON; NO EVIDENCE WAS FOUND THAT THE ADDITIVE REACHED
THE HYPOLIMNION. CERTAIN PLANTS CAN CONCENTRATE PHOSPHORUS UP TO 40,000
TIMES ITS CONCENTRATION IN THE SURROUNDING WATER. IN THE SECOND
EXPERIMENT, THE TRACER WAS ADDED TO THE HYPOLIMNION OF THE SAME LAKE.
THE ISOTOPE MOVED NO MORE THAN 5 MPH THROUGH THE COLD BOTTOM WATER AND
THE MOVEMENT WAS LARGELY LATERAL. AN 8-AC PRIMITIVE LAKE, WITH LITTLE
VEGETATION AND NO SPHAGNUM, WAS THE SITE FOR THE THIRD EXPERIMENT. ITS
WATERS WERE COMPLETELY MIXED FROM TOP TO BOTTOM. THE TRACER WAS ADDED
AT THE SURFACE, AND THE SURFACE AND BOTTOM WATERS, BOTTOM SEDIMENTS,
AND PLANTS QUICKLY REACHED EQUILBRIUM. (BYRNE-WISC)

FIELD 05C

ACCESSION NO. W68-00857

ON THE KINETICS OF PHOSPHORUS EXCHANGE IN LAKES,

DALHOUSIE UNIV., NOVA SCOTIA, CANADA.

F. R. HAYES, J. A. MCCARTER, M. L. CAMERON, AND D. A. LIVINGSTONE.

J ECOL, VOL 40, PP 202-216, 1952. 15 P, 7 FIG, 2 TAB.

DESCRIPTORS:
*EUTROPHICATION, OLIGOTROPHIC, *LAKES, *PHOSPHORUS RADIOISOTOPE,
*TURNOVERS, NUTRIENTS, PHOSPHORUS, INFLUENT SEWAGE, EQUILIBRIUM,
MATHEMATICAL MODELS, ON-SITE DATA COLLECTIONS, BOTTOM SEDIMENTS,
BIBLIOGRAPHIES, ABSORPTION, FISHES, ALGAE.

ABSTRACT:
ANY SINGLE OR PERIODIC APPLICATION OF A FERTILIZER IS NOT LIKELY TO
HAVE ANY MORE THAN A BRIEF EFFECT ON THE WATERS TO WHICH IT WAS
APPLIED, HOWEVER GREAT ITS QUANTITY. THIS CONCLUSION WAS REACHED FROM
STUDIES OF THE DATA COLLECTED AFTER PHOSPHORUS 32 WAS ADDED TO THE
WATER OF A SMALL LAKE NEAR HALIFAX, NOVA SCOTIA. THE TRACER WAS RAPIDLY
LOST BY UPTAKE BY AQUATIC ORGANISMS AND BOTTOM SEDIMENTS (MUD). THE
ADDED PHOSPHORUS INCREASED THE NORMAL WATER CONTENT BY ONLY 0.25%. THE
ACTIVE EXCHANGE OF THE PHOSPHORUS RADIOISOTOPE AND THE PHOSPHORUS IN
THE LAKE ORGANISMS AND BOTTOM SEDIMENTS SOON BECAME A SINGLE
EQUILIBRIUM SYSTEM. THE TURNOVER TIME FOR PHOSPHORUS IN THE WATER WAS
5.4 DAYS, AND THAT FOR THE PHOSPHORUS IN ORGANISMS AND MUD WAS 39 DAYS.
ONLY THE SUPERFICIAL LAYER 1 MM THICK WAS INVOLVED IN THE INTERCHANGE.
EUTROPHICATION CAN BE INDUCED BY THE CONSTANT ADDITION OF SEWAGE TO A
LAKE; THIS KEEPS THE NUTRIENTS CONTINUOUSLY ABOVE THE EQUILIBRIUM LEVEL
FOR ORGANISMS AND MUDS. STOPPAGE OF SEWAGE DISCHARGE WOULD LOWER THE
PHOSPHORUS LEVEL AND THE LAKE WOULD BECOME OLIGOTROPHIC. (BYRNE-WISC)

FIELD 05C

ACCESSION NO. W68-00858

ANTIBIOTIC ASPECTS OF COPPER TREATMENT OF LAKES,

WISCONSIN UNIV., DEPARTMENT OF ZOOLOGY.

ARTHUR D. HASLER.

TRANS WISCONSIN ACAD OF SCI, ARTS AND LETTERS, VOL 39, PP 97-103, 1949. 7 P,
18 REF.

DESCRIPTORS:
*EUTROPHICATION, *LAKES, *COPPER SULFATE, *ALGAL CONTROL, *TOXINS,
ALGAE, RESISTANCE, FISH, AMPHIBIANS, INVERTEBRATES, AQUATIC
POPULATIONS, BIBLIOGRAPHIES, BOTTOM SEDIMENTS, WATERSHED MANAGEMENT,
HARVESTING OF ALGAE, WATER ZONING.

IDENTIFIERS:
PHYSIOLOGICAL EFFECTS.

ABSTRACT:
LAKES, IN A COMPARATIVELY SHORT TIME, MOVE TOWARD A EUTROPHIC STAGE IN
WHICH MASSIVE ALGAL DEVELOPMENTS MAY BE ANTICIPATED. SUCH ALGAL GROWTHS
INHIBIT MANY OF THE RECREATIONAL AND OTHER HUMAN USES MADE OF THE LAKES
AND OF THEIR WATERS. THE SPRAY APPLICATION OF COPPER SULFATE SOLUTION
HAS BEEN FOUND EFFECTIVE IN INHIBITING THE DEVELOPMENT OF ALGAL BLOOMS.
HOWEVER, THE ADDITION OF COPPER THROUGH SPRAYS MAY INTRODUCE A TOXIC
CONDITION. THE RESULTS OF SHORT EXPOSURES OF FISH TO COPPER SALT
TOXICITY DO NOT REFLECT THE IMPACT OF TOXIC EFFECTS OVER LONGER
PERIODS. ALGAE, FISH, INVERTEBRATES, AND AMPHIBIANS ARE ALL SUBJECT TO
INHIBITION OR DESTRUCTION BY THE ABSORPTION OF COPPER SALTS. THE AUTHOR
SUGGESTS THAT STUDIES BE UNDERTAKEN TO SUBSTITUTE OTHER CONTROL
SUBSTANCES FOR A PERMANENT POISON THAT HAS BEEN EMPLOYED TO COMBAT A
TEMPORARY NUISANCE. HARVESTING OF CROPS OF ALGAE AND LARGE AQUATIC
PLANTS MIGHT BE A CONTROL MECHANISM, AS MIGHT ZONING OF LAKES FOR
CERTAIN SPECIFIC USES, AND VARIOUS KINDS OF WATERSHED CONTROLS.
(BYRNE-WISC)

FIELD 05C

ACCESSION NO. W68-00859

LIMNOLOGICAL STUDIES IN CONNECTICUT. 7. A CRITICAL EXAMINATION OF THE SUPPOSED
RELATIONSHIP BETWEEN PHYTOPLANKTON PERIODICITY AND CHEMICAL CHANGES IN LAKE
WATERS,

YALE UNIV., OSBORN ZOOL LAB.

G. EVELYN HUTCHINSON.

ECOL, VOL 25, NO 1, PP 3-26, JAN 1944. 24 P, 6 FIG, 37 REF.

DESCRIPTORS:
*EUTROPHICATION, *LAKES, PHYTOPLANKTON, ALGAE, SEASONAL, CYANOPHYTA,
*INORGANIC COMPOUNDS, PHOSPHORUS, BIBLIOGRAPHIES, NITRATES, NUTRIENTS,
PHOSPHATES, SILICATES, IRON OXIDES, *PLANT POPULATIONS, REVIEWS, LIGHT
INTENSITY, WATER TEMPERATURE, DIATOMS, CHLOROPHYTA.

IDENTIFIERS:
DESMIDS.

ABSTRACT:
AS THE RESULT OF HIS STUDIES OF LINSLEY POND, A SMALL LAKE IN CONN, THE
AUTHOR CONCLUDES THAT THERE IS IN GENERAL, NO CLEAR CUT CORRELATION
BETWEEN THE CHEMICAL CONDITION OF THE WATER AND THE QUALITATIVE
COMPOSITION OF THE PHYTOPLANKTON. THE PHYSIOLOGICAL CONDITION OF A
POPULATION AND ITS RELATION TO POPULATIONS OF OTHER SPECIES IS LIKELY
TO EXPLAIN MANY OF THE APPARENT INCONSISTENCIES BETWEEN THE SEASONS AND
DIFFERENT LAKES. THE SPRINGTIME APPEARANCE OF DYNOBRYON DIVERGENS IN
LINSLEY POND IS TO BE CORRELATED WITH A RISE IN THE RATIO OF NITRATE TO
PHOSPHATE, BUT IS INDEPENDENT OF VARIATIONS IN SOLUBLE SILICATE.
INVARIABLY THE DYNOBRYON POPULATION SEEMS TO INCREASE AFTER THE MAJOR
COMPONENTS OF THE SPRING MAXIMUM HAVE DECLINED. THE POND GIVES NO
EVIDENCE THAT THE APPEARANCE OF ASTERIONELLA FORMOSA DEPENDS ON THE
LEVEL OF INORGANIC NUTRITION. THE BLUE-GREEN ALGAL BLOOMS OF LATE
SUMMER ARISE WHEN THE INORGANIC NUTRIENTS ARE PRACTICALLY EXHAUSTED. IN
SMALL EUTROPHIC LAKES, ALL OF THE INORGANIC NUTRIENTS, EXCEPT
PHOSPHORUS AND COMBINED NITROGEN, ARE NORMALLY PRESENT IN GREAT EXCESS.
AS FOR IRON, ONLY PARTICULATE FERRIC IRON IS LIKELY TO BE IMPORTANT.
(BYRNE-WISC)

FIELD 05C

ACCESSION NO. W68-00860

ECOLOGICAL STUDY OF THE EFFECTS OF STRIP MINING ON THE MICROBIOLOGY OF STREAMS,

KENTUCKY UNIV., LEXINGTON.

HARRY D. NASH, AND RALPH H. WEAVER.

RESEARCH REPORT 18, KENTUCKY WATER RESOURCES INST., LEXINGTON, 1968. 35 P, 19 TAB.

DESCRIPTORS:
STRIP MINES, DRAINAGE, FERROBACILLUS, WATER TEMPERATURE, HYDROGEN ION CONCENTRATION, FUNGI, YEASTS, PYRITE, ALGAE, CYANOPHYTA, DOMINANT ORGANISM, NUTRIENTS, RAINFALL-RUNOFF RELATIONSHIPS, GEOGRAPHICAL REGIONS, KENTUCKY, ACID STREAMS, MICROFLORA.

IDENTIFIERS:
TAXONOMIC GROUPS, MOLECULAR FILTER, CHEMOSYNTHETIC, SAPROPHYTIC, PLATE COUNT, FILAMENTOUS, UNICELLULAR, NATURAL RECOVERY.

ABSTRACT:
THE MICROFLORA OF CANE BRANCH OF BEAVER CREEK IN MCCREARY COUNTY, KENTUCKY, WHICH DRAINS AN AREA THAT WAS STRIP-MINED BETWEEN 1955 AND 1959, WAS STUDIED AND COMPARED WITH THAT OF HELTON BRANCH WHICH DRAINS A COMPARABLE AREA WHERE THERE HAS BEEN NO MINING. DIFFERENCES INCLUDE: THE ESTABLISHMENT OF FERROBACILLUS FERROOXIDANS, FOR WHICH PROCEDURES WERE DEVELOPED FOR DIRECT COLONY ISOLATION FROM THE STREAM; FEWER SAPROPHYTIC BACTERIA; MORE NUMEROUS AND MORE DIVERSIFIED FILAMENTOUS AND UNICELLULAR FUNGI; AND CHARACTERISTIC DIFFERENCES IN ALGAL FLORA. REPRESENTATIVES OF 42 GENERA OF FILAMENTOUS FUNGI WERE IDENTIFIED. OF THESE, 21 WERE ISOLATED ONLY FROM CANE BRANCH. REPRESENTATIVES OF FIVE GENERA OF UNICELLULAR FUNGI WERE FOUND. ONE, RHODOTORULA, WAS FOUND CONSISTENTLY IN CANE BRANCH BUT NEVER IN HELTON BRANCH. FROM 1966 TO 1968, BUMILLERIA APPEARS TO HAVE ESTABLISHED ITSELF AS THE DOMINATE ALGA IN CANE BRANCH AT SOME DISTANCE DOWNSTREAM FROM THE STRIP-MINE DRAINAGE AREA. SEASONAL DIFFERENCES IN THE MICROFLORA APPEAR TO BE RELATIVELY INSIGNIFICANT, EXCEPT FOR THE ALGAE. (AUTHOR)

FIELD 05C

ACCESSION NO. W68-00891

FACTORS AFFECTING ACCELERATED EUTROPHICATION OF FLORIDA LAKES,

FLORIDA UNIV., GAINESVILLE, FLORIDA WATER RESOURCES RESEARCH CENTER.

HUGH D. PUTNAM.

OFFICE OF WATER RESOURCES RESEARCH, PROJECT COMPLETION REPORT A-002-FLA, AUGUST 1968. 123 PP, 36 FIG, 28 TAB, 37 REF.

DESCRIPTORS:
*EUTROPHICATION, ALGAL BLOOMS, WATER BLOOM, AQUATIC ALGAE, AGING(BROLOGICAL), *AQUATIC PRODUCTIVITY, LIMNOLOGY, *NUTRIENTS, BIO MASS, OLIGOTROPHY, *PRIMARY PRODUCTIVITY, TROPIC LEVEL, WATER QUALITY, WASTE ASSIMILATIVE CAPACITY, STANDING CROP, ANNUAL TURNOVER, FISH POPULATIONS.

ABSTRACT:
EUTROPHICATION HAS BEEN INVESTIGATED IN AN EXPERIMENTAL LAKE UNDER STRESS WITH A PREDETERMINED NUTRIENT LOADING RATE (500 MG N AND 42.7 MG P/M 3 YR). OBJECTIVES ARE TO DETERMINE VALID PARAMETERS IN ASSESSING TROPHIC LEVELS IN SHALLOW, SUBTROPICAL LAKES AND TO DEVELOP NEW CRITERIA IF NECESSARY; TO CONSTRUCT A MATHEMATICAL MODEL OF EUTROPHICATION USING SYSTEMS ANALYSES; AND TO IDENTIFY CONTROLLING FACTORS IN EUTROPHICATION. THUS FAR THE EXPERIMENTAL LAKE HAS PHOSPHORUS AND AMMONIA NITROGEN IN EXCESS OF LEVELS IN AN ADJACENT CONTROL LAKE, BUT NO CLEAR CUT BIOLOGICAL DIFFERENCES BETWEEN THE TWO LAKES HAVE BEEN NOTED. PRIMARY PRODUCTION AND PLANKTON CHLOROPHYLL HAVE REMAINED LOW (1 - 190 MGC PER M 3 DAY - 1, 1 - 18 MG CHL A/M 3). BIOASSAYS FOR LIMITING NUTRIENTS REVEAL THAT PHOSPHORUS COMMONLY LIMITS PRODUCTION IN THE EXPERIMENTAL LAKE WHILE MANY NUTRIENTS (PHOSPHORUS, TRACE METALS, VITAMINS, SILICA) HAVE CYCLIC BEHAVIORS IN THE CONTROL LAKE. THE MICROBIOTA OF THE LAKE IS VARIED, BUT A MEASURABLE CROP OF DINOBRYON, SYNURA, PERIDINIUM UMBONATUM AND STENTOR AMETHYSTINUM PERSISTS AT TIMES. TERTIARY TROPHIC LEVELS ARE BEING INVESTIGATED USING THE PLANKTIVOROUS FISH LABIDESTHES SICCULUS. CHANGES IN ZOOPLANKTON BIOMASS AND THE LABIDESTHES POPULATION ARE BEING MONITORED. (AUTHOR)

FIELD 02H, 05C

ACCESSION NO. W68-00912

EUTROPHICATION IS BEGINNING IN LAKE MICHIGAN,

FEDERAL WATER POLLUTION CONTROL ADMIN.

A. F. BARTSCH.

WATER AND WASTE ENG, VOL 5, NO 9, PP 84-87, SEPT 1968. 4 P, 3 FIG, 1 TAB, 3
REF.

DESCRIPTORS:
*EUTROPHICATION, *LAKE MICHIGAN, ALGAE, NUISANCE ALGAE, PLANKTON,
BENTHIC FAUNA, SLUDGE WORMS, BLOODWORMS, WATER POLLUTION.

IDENTIFIERS:
ACCELERATED EUTROPHICATION.

ABSTRACT:
A STUDY OF THE PHYSICAL, CHEMICAL, AND BIOLOGICAL STATUS OF LAKE
MICHIGAN REVEALS SOME EARLY SYMPTOMS OF ACCELERATED EUTROPHICATION. IN
THE OFFSHORE AREAS, TWO CHANGES IN SPECIES COMPOSITION OF ZOOPLANKTON
WERE SEEN. BOSMIA LONGIROSTRIS REPLACED B. COREGONI, AND DIAPTOMUS
OREGONENIS HAS APPEARED. BOTTOM-DWELLING ANIMAL POPULATIONS HAVE
INCREASED MARKEDLY. TOTAL DISSOLVED SOLIDS INCREASED 30 MG PER L IN THE
LAST NINETY YEARS, SULFATE 13 MG PER L, AND CHLORIDE 6 MG PER L.
BETWEEN 1955 AND 1966, OXYGEN CONTENT DECREASED SLIGHTLY. IN THE
INSHORE AREAS, CHANGES ARE DRASTIC. ATTACHED AND FLOATING ALGAE,
INCLUDING LARGE GROWTHS OF CLADOPHORA, ARE FREQUENTLY AT NUISANCE
LEVELS. AT GREEN BAY, PHOSPHATE CONCENTRATION VARIES FROM 0.07-0.6 MG
PER L, AND IN MILWAUKEE HARBOR IT AVERAGES 0.44 MG PER L AND IS
SOMETIMES AS HIGH 1.4 MG PER L. A SHIFT OF BOTTOM POPULATIONS TO DENSE
CONCENTRATIONS OF POLLUTION-TOLERANT SPECIES SUCH AS BLOODWORMS AND
SLUDGE WORMS HAS OCCURRED. LAKE MICHIGAN'S RETENTION OF NITROGEN IS 81%
AND OF PHOSPHATE, 95%, SO THE PROBLEM WILL NOT BE SOLVED QUICKLY BY
ELIMINATING INPUT. IMMEDIATE CONTROL OF NUTRIENT INPUT IS, HOWEVER,
NECESSARY TO PREVENT FURTHER DETERIORATION. (KNAPP-USGS)

FIELD 05C

ACCESSION NO. W68-01244

THE EFFECTS OF STRIP MINING ON THE MICROBIOLOGY OF A STREAM FREE FROM DOMESTIC
POLLUTION,

KENTUCKY UNIVERSITY, DEPARTMENT OF MICROBIOLOGY, LEXINGTON.

RALPH H. WEAVER, AND HARRY D. NASH.

REPRINTS OF PAP 2D SYMP ON COAL MINE DRAINAGE RES, MELLON INST, PP 80-97, MAY
1968. 16 P, 5 TAB, 17 REF.

DESCRIPTORS:
*ACID MINE WATER, *AQUIATIC MICROORGANISMS, *ALGAE, *AQUATIC BACTERIA,
*AQUATIC FUNGI, YEASTS, HYDROGEN ION CONCENTRATION, IRON, SULFATE,
TEMPERATURE, KENTUCKY.

IDENTIFIERS:
ACID WASTE MICROFLORA, MCCREARY COUNTY, KENTUCKY.

ABSTRACT:
THE MICROFLORA OF CANE BRANCH, MCCREARY CO., KENTUCKY, WHICH DRAINS A
STRIP-MINED AREA, AND HELTON BRANCH, WHICH DRAINS A SIMILAR BUT
NON-MINED AREA, WERE STUDIED TO DETERMINE THE BIOLOGICAL EFFECTS OF
ACID MINE DRAINAGE. NUMBERS AND TYPES OF BACTERIA, FUNGI, YEASTS, AND
ALGAE, FROM BOTH SURFACE AND BOTTOM SAMPLES, AS WELL AS TEMPERATURE AND
PH WERE STUDIED TO LEARN THE CHANGES OF ECOLOGY WITH DILUTION AND
RECOVERY. THESE DATA ARE TABULATED BY SEASON AND STREAM. IRON OXIDIZING
BACTERIA APPEAR TO BE INDIGENOUS IN ACID MINE WATER AND DIRECTLY
ASSOCIATED WITH ACID FORMATION. FERROBACILLUS FERROXIDANS WAS FOUND IN
CANE BRANCH IN ALL SEASONS, BUT ONLY IN SUMMER IN HELTON BRANCH. THE
LOWERING OF PH TO 3.0-4.1 FROM 6.3-6.7, WITH AN INCREASE IN SULFATES
AND ALMOST TOTAL ELIMINATION OF BICARBONATE ALKALINITY, RESULTED IN THE
ALTERATION OF MICROFLORA IN CANE BRANCH, PRIMARILY THE ESTABLISHMENT OF
F. FERROXIDANS. THE ROLE OF FUNGI IN STREAM CHEMISTRY IS AS YET
UNKNOWN. CANE BRANCH ALGAL GROWTH AND DIVERSITY OF TYPES INCREASED WITH
DISTANCE FROM THE SURFACE OF POLLUTION WHERE BUMILLERIA, WHICH IS FOUND
ONLY IN STREAMS WITH ACID MINE WASTE, WAS THE ONLY GENUS PRESENT.
(KNAPP-USGS)

FIELD 05C

ACCESSION NO. W69-00096

THE HIDDEN FLORA OF A LAKE,

HEBREW UNIVERSITY OF JERUSALEM.

MENACHEM RAHAT, AND INKA DOR.

HYDROBIOL, VOL 31, FASC 2, PP 186-192, 1968. 8 P, 1 TAB, 27 REF.

DESCRIPTORS:
*LAKES, PLANTS, *ALGAE, *SALINITY, IRRIGATION EFFECTS, DRAINAGE
EFFECTS, *SAMPLING, ALGAE TOXINS, SALINE WATER, EQUILIBRIUM, BIOLOGICAL
COMMUNITIES, *AQUATIC ALGAE, HALOPHYTES.

ABSTRACT:
A SURVEY WAS CONDUCTED TO DETERMINE HIDDEN ALGAL FLORA OF LAKE KINNERET
(SEA OF GALILEE) IN THE JORDAN RIFT VALLEY, ISRAEL, WHICH WOULD
DOMINATE IN A RISE OF SALINITY DUE TO DISTURBANCE OF PRESENT
EQUILIBRIUM FROM DRAINAGE FOR IRRIGATION. THE CHRYSOMONAD PRYMNESIUM
PARVUM WHICH CAUSED TOXIC BLOOMS WAS OF PARTICULAR INTEREST. WATER
SAMPLES WERE SELECTIVELY ENRICHED FOR HALOPHILIC ALGAE WITH AN
INORGANIC SALT MEDIUM. A RADICAL CHANGE WAS OBSERVED. ALGAE RARE OR
UNKNOWN TO THE LAKE WERE PRESENT. REPRESENTATION OF THE ALGAL GROUPS
WAS AS FOLLOWS: 7 SPECIES BELONGED TO CHLOROPHYCEAE, 8 SPECIES TO
DRATOMEAE AND 15 TO CYANOPHYCEAE. PRYMNESIUM PARVUM WAS PRESENT IN 6 OF
THE 27 SAMPLES. THUS IT WAS INDICATED THAT A SUFFICIENT RISE OF THE
LAKES SALINITY WOULD RESULT IN A PREDOMINANCE OF HALOPHILIC ALGAE AND
THE PRESENCE OF PRYMNESIUM PARVUM FORMS A POTENTIAL DANGER OF TOXIC
BLOOMS. (AFFLECK-ARIZ)

FIELD 05C, 02H

ACCESSION NO. W69-00360

AN EARTH-SMELLING COMPOUND ASSOCIATED WITH BLUE-GREEN ALGAE AND ACTINOMYCETES,

CALIFORNIA UNIV., BERKELEY.

LLOYD L. MEDSKER, DAVID JENKINS, AND JEROME F. THOMAS.

ENVIRON SCI TECHNOL, VOL 2, NO 6, PP 461-64, JAN 1968. 4 P, 4 FIG, 14 REF.

DESCRIPTORS:
*WATER ANALYSIS, *ODORS, *ODOR PRODUCING ALGAE, EUTROPHICATION,
ALCOHOLS, ACTINOMYCETES, CYANOPHYTA, SPECTROSCOPY, GAS CHROMATOGRAPHY,
ANALYTICAL TECHNIQUES, WATER POLLUTION SOURCES.

IDENTIFIERS:
PROTON MAGNETIC RESONANCE SPECTROMETRY, OSCILLATORIA TENUIS, RAMMAN
SPECTROSCOPY, MASS SPECTROMETRY, SYMPLOCA MUSCORUM.

ABSTRACT:
A COMPOUND, WITH EARTHY-MUSTY ODOR SIMILAR TO THAT OFTEN CAUSING
PROBLEMS IN FOOD AND WATER SUPPLIES, WAS ISOLATED BY PREPARATIVE GAS
CHROMATOGRAPHY FROM A CULTURE OF ACTINOMYCETES (NO 18) AND CULTURES OF
2 SPECIES OF CYANOPHYTES, SYMPLOCA MUSCORUM AND OSCILLATORIA TENUIS.
ACTINOMYCETES WERE GROWN IN MEDIUM (M SUB 1 B SUB 2); CYANOPHYTES, IN
HUGHES MEDIUM 11 (16 LITERS AT 26-28 DEG C AND 200-300 FT-CANDLES OF
LIGHT). THE FORMER YIELDED CA 200 MICROG/1 AFTER 8-12 DAYS AT ROOM
TEMP; SYMPLOCA, 1 MG/VESSEL AFTER 4-6 WK. CYANOPHYTE CULTURES WERE
SHOWN TO BE FREE OF CONTAMINATING ACTINOMYCETES. MASS SPECTRUM OF THE
COMPOUND SHOWED IT TO BE IDENTICAL WITH THAT OF GEOSMIN (C SUB 12 H SUB
22 O). VARIOUS ANALYTICAL TECHNIQUES (DEGRADATION, BROMINATION,
HYDROGENATION, ANALYTICAL GAS CHROMATOGRAPHY, HIGH RESOLUTION MASS
SPECTROMETRY, INFRARED SPECTROMETRY AND PROTON MAGNETIC RESONANCE
SPECTROMETRY) SHOWED THAT GEOSMIN IS A DIMETHYL SUBSTITUTED, SATURATED,
TERTIARY ALCOHOL WITH 2 RINGS IN WHICH THE HYDROXYL GROUP IS VERY
STERICALLY HINDRED. STRUCTURE OF THE RING SYSTEM IS UNKNOWN, BUT RAMMAN
SPECTROSCOPY INDICATES THAT IT MAY BE SIMILAR TO ALPHA-PINENE. GEOSMIN
IS A COLORLESS, VISCOUS LIQUID, WITH CAMPHORISH ODOR (WHEN
CONCENTRATED) WHICH BECOMES EARTHY UPON DILUTION. A THRESHOLD ODOR OF
0.05 MICROG/1 IS REPORTED. (EICHHORN-WISC)

FIELD 05A, 05C

ACCESSION NO. W69-00387

PLANKTONIC PHOTOSYNTHESIS AND THE ENVIRONMENT OF CALCIUM CARBONATE DEPOSITION
IN LAKES,

MINNESOTA UNIV., MINNEAPOLIS. LIMNOLOGICAL RESEARCH CENTER.

ROBERT O. MEGARD.

INTERIM REP 2, UNIV MINNESOTA, JAN 1968. 47 P, 7 FIG, 6 TAB, 33 REF. OWRR
PROJ A-008-MINN.

DESCRIPTORS:
*PHOTOSYNTHESIS, *CALCIUM CARBONATE, AQUATIC ENVIRONMENT, AQUATIC
PRODUCTIVITY, EUTROPHICATION, *LAKES, LIMNOLOGY, *PLANKTON, ALGAE,
BIBLIOGRAPHIES, CHEMICAL PRECIPITATION, IONS, BIOCHEMICAL OXYGEN
DEMAND, MINNESOTA, WATER CHEMISTRY, BIODEGRADATION.

ABSTRACT:
CALCIUM AND ALKALINITY CONCENTRATIONS INCREASE IN A PRODUCTIVE LAKE
DURING WINTER AND THE WATER AT ALL DEPTHS BECOMES UNDERSATURATED WITH
CALCIUM CARBONATE. SOON AFTER ICE MELT, THE ENTIRE WATER COLUMN BECOMES
OVERSATURATED WITH CALCIUM CARBONATE, AND CALCIUM AND ALKALINITY
CONCENTRATIONS DECREASE DURING SPRING, SUMMER, AND AUTUMN. WHEN THE
LAKE BECOMES THERMALLY STRATIFIED, CARBONATE OVERSATURATION INCREASES
AT DEPTHS IN WHICH PHOTOSYNTHESIS OCCURS, BUT THE WATER AT GREATER
DEPTHS BECOMES UNDERSATURATED. CARBONATE SATURATION APPEARS CONTROLLED
BY THE BALANCE BETWEEN CARBON DIOXIDE ASSIMILATION BY PLANKTONIC
PHOTOSYNTHESIS AND CARBON DIOXIDE RELEASE BY RESPIRATION. SEASONAL
CHANGES OF CALCIUM CONCENTRATIONS ARE BALANCED BY CHEMICALLY EQUIVALENT
ALKALINITY CHANGES, BUT MAGNESIUM CONCENTRATIONS DO NOT FLUCTUATE.
PHOTOSYNTHESIS OCCURS IN BOTH THE EPILIMNION AND THE METALIMNION OF A
DEEPER, LESS PRODUCTIVE LAKE. THESE ARE THE ONLY DEPTHS OVERSATURATED
WITH CALCIUM CARBONATE DURING OPEN-WATER SEASON.

FIELD 02H, 05C

ACCESSION NO. W69-00446

LIMNOLOGY, PRIMARY PRODUCTIVITY, AND CARBONATE SEDIMENTATION OF MINNESOTA LAKES,

MINNESOTA UNIV., MINNEAPOLIS.

ROBERT O. MEGARD.

LIMNOLOGICAL RES CENTER, INTERIM REP 1, JAN 1967. 69 P, 25 FIG, 11 TAB, 7
REF. OWRR PROJECT A-008-MINN.

DESCRIPTORS:
SEDIMENTATION RATES, *EUTROPHICATION, *LAKES, *PRIMARY PRODUCTIVITY,
*PLANKTON, ALGAE, *PHOTOSYNTHESIS, PRODUCTIVITY, WATER QUALITY,
*LIMNOLOGY, MINNESOTA, WATER ANALYSIS, CALCIUM CARBONATE, HYPOLIMNION.

ABSTRACT:
THE REPORT DESCRIBES INVESTIGATIONS OF THE REGIONAL LIMNOLOGY OF MINN.
THE WORK HAS INVOLVED 3 PHASES: (1) A MOBILE LABORATORY WAS DESIGNED
AND OUTFITTED; (2) STANDARD FIELD AND LABORATORY METHODS WERE MODIFIED
TO MAKE THEM MORE COMPATIBLE WITH THE MOBILITY OF THE LABORATORY; AND
(3) A STUDY WAS BEGUN OF THE PRIMARY PRODUCTIVITY AND GENERAL LIMNOLOGY
OF 14 LAKES IN VARIOUS PARTS OF THE STATE. THE LABORATORY PERMITS WATER
SAMPLES TO BE ANALYZED AT LAKESIDE BEFORE NUTRIENT CONCENTRATIONS ARE
ALTERED BY BIOLOGICAL ACTIVITY AND PROVIDES FACILITIES FOR MEASURING
THE GROWTH RATES OF ALGAE AND DETERMINING THE FERTILITY OF LAKE WATER.
THE OXYGEN LIGHT-DARK-BOTTLE METHOD IS BEING USED TO COMPARE ALGAL
PRODUCTIVITY. THE RESPONSE OF THE CARBONATE SATUROMETER IN FRESH WATER
WAS TESTED, AND THE INSTRUMENT WAS USED TO DETERMINE THE DEGREE OF
CALCIUM CARBONATE SATURATION AT DIFFERENT TIMES OF THE YEAR AND AT
VARIOUS DEPTHS IN THE LAKES. THE SEASONAL CHEMICAL CHANGES IN THE LAKES
ARE DESCRIBED AND THE PRODUCTIVITY DATA ASSEMBLED, BUT THERE ARE NOT
YET ENOUGH DATA FOR A DETAILED COMPARISON OF THE LAKES.

FIELD 02H, 05C

ACCESSION NO. W69-00632

SURFACE-WATER CHEMISTRY OF SOME MINNESOTA LAKES, WITH PRELIMINARY NOTES ON DIATOMS,

MINNESOTA UNIV., MINNEAPOLIS.

ROBERT C. BRIGHT.

LIMNOLOGICAL RES CENTER, BELL MUSEUM NATUR HISTORY, UNIV MINN INTERIM REP 3, APR 1968. 51 FIG, 44 REF, APPEND, DISC. OWRR PROJECT B-001-MINN.

DESCRIPTORS:
BIBLIOGRAPHIES, *MINNESOTA, *LAKES, *DIATOMS, *ALGAE, ECOLOGY, SOUTH DAKOTA, FIELD INVESTIGATIONS, WATER POLLUTION, WATER SAMPLING, *WATER ANALYSIS, *CHEMICAL ANALYSIS, PH, SAMPLING, ON-SITE DATA COLLECTIONS, LIMNOLOGY.

ABSTRACT:
A COMPREHENSIVE FIELD STUDY WAS MADE OF DIATOM ECOLOGY IN LAKES OF MINN AND ADJACENT PARTS OF SOUTH DAKOTA. THE PROGRAM CONSISTED OF 4 PHASES: (1) IDENTIFICATION AND DETERMINATION OF THE GEOGRAPHIC DISTRIBUTION AND RELATIVE ABUNDANCE OF DIATOMS IN THE REGION, (2) CORRELATION OF DIATOM DISTRIBUTION AND RELATIVE ABUNDANCE WITH REGIONAL PATTERNS OF SURFACE-WATER CHEMISTRY, (3) CORRELATION OF INFORMATION FROM THE FIRST 2 PHASES WITH THE COMPOSITION OF THE DIATOM FLORA IN SURFACE SEDIMENTS FROM EACH LAKE, AND (4) APPLICATION OF CONCLUSIONS FROM THE FIRST THREE PARTS TO FOSSIL DIATOM FLORAS AND PALEOECOLOGY OF LAKES. MINNESOTA AND ADJACENT SOUTH DAKOTA COMPRISE THE BEST AREA IN NORTH AMERICA FOR SUCH A BROAD STUDY BECAUSE OF THE WIDE VARIETY OF LAKE TYPES AND THEIR WIDE RANGE OF SALINITY. THE ENSUING DISCUSSIONS DEAL MAINLY WITH FALL SURFACE-WATER CHEMISTRY, POSSIBLE ORIGINS OF ITS REGIONAL PATTERN, ITS CORRELATION WITH SIMILAR DATA FROM OTHER AREAS, AND ITS CONTROLLING FACTORS.

FIELD 02H, 05C

ACCESSION NO. W69-00659

THE ECOLOGICAL SIGNIFICANCE OF CELLULOLYTIC BACTERIA IN QUABBIN RESERVOIR,

CLARK UNIV., WORCESTER, MASS.

JOHN T. REYNOLDS.

PROJ COMPLETION REP, JAN 1968. 11 P. OWRR PROJECT A-007-MASS.

DESCRIPTORS:
*ECOLOGY, BACTERIA, ALGAE, PHYTOPLANKTON, *LAKES, *CELLULOSE, DISTRIBUTION PATTERNS, MEMBRANE PROCESSES, MASSACHUSETTS.

ABSTRACT:
ECOLOGICAL SIGNIFICANCE OF CELLULOLYTIC BACTERIA IN A LARGE RESERVOIR IN CENTRAL MASSACHUSETTS WAS STUDIED. BACTERIA WERE ENUMERATED USING MEMBRANE FILTER TECHNIQUES: CELLULOLYTIC FORMS WERE GROWN OUT ON A MODIFICATION OF SANBORN'S CHINA BLUE-CELLULOSE DEXTRIN AGAR, AND HETEROTROPHS ON A MINERAL SALT-MUD EXTRACT AGAR AND ON MPH AGAR. SUSPENDED CELLULOSE AND HEMICELLULOSE WERE CONCENTRATED BY FILTRATION AND ESTIMATED BY SPECTROPHOTOMETRIC DETERMINATIONS AFTER TREATMENT WITH PHENOL AND SULFURIC ACID. ALGAE WERE ENUMERATED AND TYPED USING STANDARD PROCEDURES, AND ESTIMATES OF THE AMOUNTS OF CELLULOSE PRODUCED BY THE PHYTOPLANKTON OBSERVED WERE MADE AFTER DETERMINATIONS OF THE AMOUNTS OF CELLULOSE PRESENT IN KNOWN CONCENTRATIONS OF ALGAE REPRESENTATIVE OF THE TYPES FOUND IN THE RESERVOIR. TOTAL NUMBERS OF BACTERIA VARIED BETWEEN LESS THAN 1 PER ML OF SAMPLE TO A HIGH OF 170 PER ML OF SAMPLE. PRESUMPTIVE CELLULOLYTIC BACTERIA ACCOUNTED FOR FROM LESS THAN 0.5 TO 52% OF THE TOTAL BACTERIAL POPULATIONS MEASURED IN SEVERAL DIFFERENT AREAS BEFORE, DURING AND AFTER AN ALGAL BLOOM. NUMBERS OF PRESUMPTIVE CELLULOLYTIC BACTERIA APPEAR TO HAVE NO CONSISTANT RELATION WITH ESTIMATES OF AMOUNTS OF CELLULOSE. ESTIMATES OF CELLULOSE CONTENT OF THE WATER VARIED BETWEEN 360 TO 2750 MICROGRAMS PER LITER OF SAMPLE.

FIELD 02H, 05C

ACCESSION NO. W69-00976

THE EFFECT OF SELECTED HERBICIDES UPON THE GROWTH OF PLANKTONIC FRESH WATER
ALGAE AND THEIR PERSISTENCE IN SURFACE WATERS,

LOUISIANA POLYTECHNIC INSTITUTE, RUSTON, BOTANY AND BACTERIOLOGY DEPT.

OTTO WASMER, JR.

LOUISIANA WATER RESOUR RES INST, LOUISIANA STATE UNIV, BATON ROUGE, MAY 1967.
19 P, 11 FIG, 5 REF. OWRR PROJECT A-003-LA.

DESCRIPTORS:
*2-4 D, *HERBICIDES, *PHYTOPLANKTON, *PHYTOTOXICITY, ALGAL POISONING,
SCENEDESMUS, CHLORELLA, *EUGLENA, ALGAL CONTROL, BACTERICIDES, GROWTH
RATES.

ABSTRACT:
THE PURPOSES OF THE INVESTIGATION WERE TO DETERMINE EFFECT OF SELECTED
HERBICIDES UPON THE RATE OF GROWTH OF CERTAIN FRESH WATER PLANKTONIC
ALGAE; TO DETERMINE EFFECT OF SELECTED HERBICIDES UPON THE PHYSIOLOGY
OF THESE ALGAE; AND TO DETERMINE MORPHOLOGICAL CHANGES IN ALGAE EXPOSED
TO HERBICIDE PRESENCE. DUE TO LIMITATIONS IMPOSED BY TIME, CERTAIN
RESTRICTIONS IN THE INVESTIGATIONAL PROCEDURES WERE NECESSARY. A SIGNAL
SPECIES OF FRESH WATER ALGAE WAS UTILIZED IN THE COMPLETE
INVESTIGATION, ALTHOUGH 2 OTHER SPECIES WERE EXAMINED IN THE
MORPHOLOGICAL STUDIES. A SINGLE HERBICIDE WAS USED IN ALL PHASES OF THE
STUDY. FINALLY, THE PHYSIOLOGICAL STUDY WAS CONFINED TO THE EFFECT OF
HERBICIDE PRESENCE ON RATE OF RESPIRATION IN ALGAL CELLS.

FIELD 05G

ACCESSION NO. W69-00994

POLLUTION OF WATERWAYS IN THE ELLIOT LAKE AREA, ONTARIO,

WESTERN ONTARIO UNIV. LONDON, ONTARIO, CANADA.

RICHARD P. W. MCCUTCHEON.

ONTARIO GEOGR, NO 2, PP 25-33, 1968. 9 P, 1 FIG, 5 REF.

DESCRIPTORS:
*WATER POLLUTION, *MINE WASTES, *RADIOACTIVE WASTES, *MUNICIPAL WASTES,
AQUATIC ALGAE, URANIUM RADIOISOTOPES, RADIUM RADIOISOTOPES.

IDENTIFIERS:
ELLIOT LAKE, ONTARIO.

ABSTRACT:
RAPID DEVELOPMENT OF URANIUM MINING IN THE ELLIOT LAKE AREA, ONTARIO,
CAUSED WIDESPREAD RADIOLOGICAL AND SEWAGE POLLUTION OF THE SERPENT
RIVER SYSTEM. THE WIDELY USED ACID LEACHING PROCESS OF URANIUM ORE...
TREATMENT PRODUCES LARGE AMOUNTS OF WASTE, BOTH SOLID AND LIQUID. FOR
EACH TON OF ORE, NEARLY 1 TON OF TAILINGS AND AT LEAST 2 TONS AND AS
MUCH AS 5 TONS OF WATER, ACIDS, AND NEUTRALIZERS ARE PRODUCED. NOT ALL
THE URANIUM, AND NONE OF THE RADIUM-226, IS EXTRACTED FROM THE ORE, SO
THE TAILINGS PILES CONTRIBUTE SOME RADIOACTIVITY TO WATER PASSING
THROUGH THEM. SOME LAKES AND STREAMS CONTAIN MORE THAN THE SAFE AMOUNT
OF RADIUM. LEAKS IN THE REFINING SYSTEMS ALSO CONTRIBUTE POLLUTION. ALL
SEWAGE IS TREATED, BUT LAKE AND STREAM POLLUTION BY INCREASED OXYGEN
DEMAND AND INCREASE OF NUTRIENTS IS COMMON AND HAS MADE ONE FORMER
SWIMMING AREA OBNOXIOUS. NO FISH CAN LIVE IN ANGEL AND HORNE LAKES, AND
ALGAE ARE PRESENT IN EXCESSIVE AMOUNTS. (KNAPP-USGS)

FIELD 05G

ACCESSION NO. W69-01165

EFFECTS OF WASTES ON RECEIVING WATER--SOME COMPUTATIONAL AND SCIENTIFIC
PROBLEMS.

PP 49-56 OF ENTRY NO. W69-01269 IN 05G.

DESCRIPTORS:
BIOCHEMICAL OXYGEN DEMAND, ALGAL TOXINS, ORGANIC WASTES, POLLUTANT
IDENTIFICATION, POLLUTANTS, QUALITY CONTROL, WATER POLLUTION, *WATER
POLLUTION SOURCES, SEPARATION TECHNIQUES, *ANALYTICAL TECHNIQUES,
CLASSIFICATION, *MODEL STUDIES, WASTE IDENTIFICATION, BIOLOGICAL
PROPERTIES, PHYSICAL PROPERTIES, WASTE WATER(POLLUTION), WATER QUALITY,
WATER POLLUTION EFFECTS, IMPAIRED WATER QUALITY, WASTES.

ABSTRACT:
COMPUTATIONAL AND SCIENTIFIC PROBLEMS INVOLVED IN DETERMINING THE
EFFECT OF WASTES ON RECEIVING WATER ARE DISCUSSED. THE NEED FOR AN
EFFICIENT AND FLEXIBLE MEANS OF CALCULATION OF POLLUTANTS IS NOTED AND
PROBLEMS INVOLVED IN DEVISING SUCH A MEANS ARE DISCUSSED. TWO BASIC
PROBLEMS ARE PRESENTED. THE FIRST IS THE CONCEPTUAL PROBLEM OF DEVISING
A MODEL WHICH COMPREHENDS RELEVANT RELATIONSHIPS AND IS AMENDABLE TO
RAPID CALCULATION. THE SECOND IS THE DETERMINATION OF THE SPECIFIC
NUMERICAL EFFECT OF A CHANGE IN ONE VARIABLE ON THE OTHERS. IT IS
CONCLUDED THAT ACCURATE KNOWLEDGE OF PHYSICAL AND BIOLOGICAL
RELATIONSHIPS IN WATER BODIES IS AN ESSENTIAL PREREQUISITE TO THE
DESIGN AND OPERATION OF OPTIMUM WASTE-DISPOSAL SYSTEMS. RESEARCH GEARED
TO EXPLORING A MORE DEFINITELY EMPIRICAL APPROACH TO WATER PHENOMENA IS
ENCOURAGED.

FIELD 05C

ACCESSION NO. W69-01273

MAJOR SOURCES OF NUTRIENTS FOR ALGAL GROWTH IN WESTERN LAKE ERIE,

FEDERAL WATER POLLUTION CONTROL ADMIN., GROSSE ILE, MICH.

GEORGE L. HARLOW.

PUBLICATION NUMBER 15, GREAT LAKES RES DIV, UNIV MICH, PP 389-394, 1966. 6 P,
2 FIG, 5 TAB, 9 REF.

DESCRIPTORS:
*EUTROPHICATION, *LAKE ERIE, *NUTRIENTS, ALGAE, WATER POLLUTION
SOURCES, NITROGEN COMPOUNDS, PHOSPHATES, GREAT LAKES, MICHIGAN.

IDENTIFIERS:
LAKE ST CLAIR, ROUGE RIVER, HURON RIVER, RAISIN RIVER, DETROIT RIVER,
MAUMEE RIVER.

ABSTRACT:
DUE TO AN EXCESS OF NUTRIENTS, WESTERN LAKE ERIE HAS DEVELOPED PROLIFIC
BLOOMS OF ALGAE CONTRIBUTING TO AND INDICATING THE ACCELERATED
ENRICHMENT OF THIS VALUABLE NATURAL RESOURCE. NUTRIENT CONCENTRATIONS
AND QUANTITIES INCREASE AS THE WATERS FROM LAKE ST CLAIR PASS BY THE
METROPOLITAN COMPLEX OF DETROIT TO WESTERN LAKE ERIE. THE SOURCES OF
WASTE WHICH CONTRIBUTE TO THESE INCREASES ARE PRESENTED, AS WELL AS THE
RELATIVE QUANTITIES FROM EACH WASTE SOURCE. MUNICIPAL WASTES CONTRIBUTE
89% OF TOTAL PHOSPHATES AND 86% OF TOTAL NITROGEN RECEIVED INTO LAKE
ERIE FROM SOUTHEAST MICHIGAN, WHEREAS THE CONTRIBUTION OF PHOSPHATE
FROM LAND DRAINAGE IS A MINOR FACTOR. NITROGEN RUNOFF FROM LANDS
CONTRIBUTES MORE SIGNIFICANTLY TO NUTRIENT LOADING THAN DOES PHOSPHATE
FROM THE SAME SOURCE. IN LAKE ERIE, CONCENTRATIONS OF NUTRIENT
COMPOUNDS CLOSER TO THE MOUTH OF THE DETROIT RIVER ARE HIGHER THAN AT
STATIONS FURTHER OFFSHORE, EXCEPT IN THE CASE OF TWO OFFSHORE STATIONS
THOUGHT TO BE INFLUENCED BY THE MAUMEE RIVER. RESULTS OF NUTRIENT
DETERMINATIONS FROM ALL STATIONS REPORTED FOR LAKE ERIE SHOW
CONCENTRATIONS GREATER THAN THOSE THOUGHT TO BE CRITICAL FOR TRIGGERING
BLOOMS OF ALGAE.

FIELD 05B

ACCESSION NO. W69-01445

EUTROPHICATION AND WATER POLLUTION,

FEDERAL WATER POLLUTION CONTROL ADMINISTRATION, WASHINGTON.

EDWARD J. MARTIN, AND LEON W. WEINBERGER.

PUBLICATION NUMBER 15, GREAT LAKES RES DIV, UNIV MICH, PP 451-469, 1966. 19 P, 4 FIG, 8 TAB, 45 REF.

DESCRIPTORS:
*EUTROPHICATION, *NUTRIENTS, POLLUTANTS, POLLUTION ABATEMENT, NUISANCE ALGAE, PHOSPHATES, NITRATES, RIVERS, SEWAGE TREATMENT, WATER POLLUTION SOURCES, PRIMARY TREATMENT, SECONDARY TREATMENT, TERTIARY TREATMENT, ELECTRODIALYSIS, ION EXCHANGE, ACTIVATED SLUDGE, DENITRIFICATION.

ABSTRACT:
AUTHORS DEFINE THE MECHANISM OF EUTROPHICATION AND EVALUATE PROBABLE SOURCES AND NATURE OF THE POLLUTIONAL LOADS RESPONSIBLE FOR ACCELERATED EUTROPHICATION. THIS INVOLVES CONSIDERATIONS OF TYPES AND QUANTITIES OF NUTRIENT MATERIAL, ESPECIALLY NITROGEN AND PHOSPHORUS, IN TERMS OF CONTROL AND THE FREQUENCY AND INTENSITY OF ALGAL BLOOMS. ON RIVERS, BETWEEN 1961 AND 1963, 49% OF 130 STATIONS REPORTED ALGAL COUNTS WHICH COULD BE DEFINED AS BLOOMS; 55% OF 345 STATIONS REPORTED VALUES OF PHOSPHATE EQUALING OR EXCEEDING 0.2 MG/1. WASTES RESULTING FROM MUNICIPAL, AGRICULTURAL, AND INDUSTRIAL USES OF WATER CONTRIBUTE SIGNIFICANT AMOUNTS OF NUTRIENTS THAT MAY PROMOTE EUTROPHICATION IN WATERWAYS. ABOUT 235 TO 2,350 TONS OF PHOSPHATE ARE DISCHARGED DAILY TO RECEIVING WATERS FROM MUNICIPAL WASTE SOURCES IN THE US ALONE. MAXIMUM EFFICIENCES OF SEVEN TREATMENT PROCESSES FOR REMOVING NITROGEN AND PHOSPHORUS ARE REVIEWED. 50% OF TOTAL PHOSPHORUS AND 80% OF TOTAL NITROGEN ARE REMOVED BY USE OF AN ACTIVATED-SLUDGE PROCESS, WHILE 99+% OF THE PHOSPHORUS AND NITROGEN ARE REMOVED WITH A TREATMENT SEQUENCE USING ION EXCHANGE. THESE DATA INDICATE THAT TREATMENT CAPABILITIES ARE AVAILABLE TO CONTROL THE DISCHARGE OF NUTRIENT COMPOUNDS FROM POINT SOURCES OF WASTES AND TO TREAT THEM TO ALMOST ANY DESIRED DEGREE.

FIELD 05F

ACCESSION NO. W69-01453

THE BIOTIC RELATIONSHIPS WITHIN WATER BLOOMS,

WISCONSIN UNIV., MADISON. HYDRAULIC AND SANITARY LAB.

GEORGE P. FITZGERALD.

D F JACKSON, ALGAE AND MAN, PLENUM PRESS, NEW YORK, PP 300-306, 1964. 7 P, 41 REF.

DESCRIPTORS:
*EUTROPHICATION, SUCCESSION, WATER QUALITY, ALGAL CONTROL, ALGICIDES, DOMINANT ORGANISMS, GRAZING, WATER POLLUTION EFFECTS, DIATOMS, CHLOROPHYTA, CYANOPHYTA, NUISANCE ALGAE, WISCONSIN, NUTRIENTS, INDICATOR SPECIES, EUGLENA, CHLAMYDOMONAS.

IDENTIFIERS:
*BLOOMS, *BIOTIC FACTORS, SPECIES DIVERSITY, LAKE WAUBESA, LYNGBYA, MICROCYSTIS, APHANIZOMENON, ANABAENA, MELOSIRA, CERATIUM, OSCILLATORIA, EUDORINA.

ABSTRACT:
ALGAL SUCCESSIONS, IN WHICH NEARLY UNISPECIFIC POPULATIONS OF ALGAE ALTERNATE WITH MORE HIGHLY DIVERSIFIED ASSEMBLAGES OF SPECIES, ARE CHARACTERISTIC OF MORE NATURAL AQUATIC ENVIRONMENTS AS WELL OF IMPOUNDMENTS, SUCH AS SEWAGE STABILIZATION PONDS, WITH NO LACK OF THE USUALLY CONSIDERED NUTRIENTS. THE USUAL SEQUENCE OF SPECIES, ESPECIALLY IN EUTROPHIC ENVIRONMENTS, CONSISTS OF A VERNAL BLOOM OF DIATOMS, AN EARLY SUMMER GROWTH OF GREEN ALGAE, AND A LATE SUMMER BLOOM OF CYANOPHYTES (BLUE-GREEN ALGAE). AFTER TREATMENT OF A WATER BLOOM WITH ALGICIDES TO ALLEVIATE NUISANCE CONDITIONS, THE RAPID REPLACEMENT OF DOMINANT ALGAL SPECIES BY OTHERS EMPHASIZES THAT FACTORS WHICH INFLUENCE DOMINANCE MAY BE VERY SUBTLE. HIGH POLLUTIONAL LOADS FROM SEWAGE SOURCES NOT ONLY INCREASE ALGAL GROWTH, BUT MAY ALSO INFLUENCE SPECIES COMPOSITION AND ALGAL SUCCESSION. EVIDENCE SUGGESTS THAT SECONDARY EFFLUENTS MAY ACT TO SUPPRESS SUCCESSION AND REINFORCE A REDUCTION IN SPECIES DIVERSITY LONGER THAN WOULD NORMALLY BE EXPECTED. SINCE NUTRIENT STATUS OF AN AQUATIC ENVIRONMENT DOES NOT ALWAYS EXPLAIN THE COMPOSITION OF A BLOOM, ONE MUST OFTEN CONSIDER THE EFFECTS OF BIOTIC FACTORS SUCH AS GRAZING BY THE FAUNA, ACTIVITIES OF BACTERIA AND FUNGI, AND THE EXCRETION OF BIOLOGICALLY ACTIVE COMPOUNDS BY THE ALGAE THEMSELVES. (WIS)

FIELD 05C

ACCESSION NO. W69-01977

EFFECTS OF PESTICIDES ON FRESHWATER ORGANISMS,

CLEMSON UNIV., CLEMSON. DEPT. OF ENTOMOLOGY AND ZOOLOGY.

JOHN K. REED, LAMAR E. PRIESTER, AND RUDOLPH PRINS.

WATER RESOURCES INSTITUTE, TECHNICAL COMPLETION REPORT MAY 1, 1965 - JUNE 30, 1968. 31 P, 4 TAB. OWRR PROJECT A-001-SC.

DESCRIPTORS:
ALGAE, PROTOZOA, *AQUATIC INSECTS, *AQUATIC DRIFT, FISH FOOD ORGANISMS, FISH PHYSIOLOGY, FISH TAXONOMY, DDT, PHOSPHOTHIOATE PESTICIDES, HERBICIDES, *PESTICIDE RESIDUES, DIATOMS, CRAYFISH, POLLUTANT IDENTIFICATION, *WATER POLLUTION EFFECTS, GAS CHROMATOGRAPHY.

IDENTIFIERS:
SOUTH CAROLINA, LYOPHILIZATION, *PESTICIDE METABOLISM, *PESTICIDE ACCUMULATION.

ABSTRACT:
LISTS OF ARTHROPODS, REPTILES, AMPHIBIANS AND FISH FOUND IN NORTHWESTERN SOUTH CAROLINA STREAMS BY THE AUTHORS ARE GIVEN. THE ECOLOGY OF THESE FORMS WAS AFFECTED BY THE PRESENCE OF ORGANIC PESTICIDE POLLUTION IN CERTAIN STREAMS. THE POLLUTANTS FOUND WERE DDT, DDE, BHC, TOXAPHENE, ALDRIN AND DIELDRIN. THE HIGHEST LEVEL OF PESTICIDE WAS IN THE ORGANISMS, THEN BOTTOM MUD WITH THE LEAST IN WATER. LABORATORY STUDIES SHOWED SELECTED SPECIES OF ALGAE AND PROTOZOA CONCENTRATED DDT AND PARATHION FROM WATER BUT BREAKDOWN PRODUCTS WERE NOT DETECTABLE. TRIFLURALIN WAS BOTH CONCENTRATED FROM WATER AND METABOLIZED BY GOLDFISH. DRIFTING FOOD ORGANISMS OF FISH IN THE FIELD WERE FOUND TO BE INFLUENCED BY THE SEASON OF THE YEAR, WEATHER, AND POLLUTION.

FIELD 05C

ACCESSION NO. W69-02782

PRIMARY PRODUCTIVITY STUDIES IN ONONDAGA LAKE, NEW YORK,

SYRACUSE UNIV., N. Y. DEPT OF CIVIL ENGINEERING.

DANIEL F. JACKSON.

FOR PUBLICATION IN 'XVII VERHANDLUNGEN INTERNATIONALE VEREINIGUNG FUR THEORETISCHE UND ANGEWANDTE LIMNOLOGIE', 1968. 19 P. NY DEPT. OF HEALTH C-19638.

DESCRIPTORS:
*EUTROPHICATION, *WATER POLLUTION SOURCES, *WATER POLLUTION EFFECTS, *PRIMARY PRODUCTIVITY, ALGAE, BIOMASS, CALCIUM COMPOUNDS, CHLORIDES, SALINE LAKES, HEAVY METALS, INDUSTRIAL WASTES, LAKES, PHOTOSYNTHESIS, PHYTOPLANKTON, CISCO, CARP, DIATOMS, CHLOROPHYTA, CHLAMYDOMONAS, CHLORELLA.

IDENTIFIERS:
*ONONDAGA LAKE, ONONDAGA COUNTY, SYRACUSE, WARBURG RESPIROMETER, CLADOPHORA, ENTEROMORPHA, LEPOCINCLIS, CHLOROGONIUM, SCENEDESMUS, CYCLOTELLA, STEPHANODSICUS.

ABSTRACT:
ONONDAGA LAKE, WHOSE SOUTH END IS WITHIN THE CITY LIMITS OF SYRACUSE, NY, IS SMALL, BRACKISH, AND GROSSLY POLLUTED. IT RECEIVES DAILY 200,000,000 LITERS OF PARTIALLY TREATED SEWAGE, 172 METRIC TONS OF CALCIUM CHLORIDE, 32 METRIC TONS OF CALCIUM SULFATE, ONE METRIC TON OF GREASE AND OIL, AND 22 KILOGRAMS OF HEAVY METALS. TO ESTABLISH A BASE LINE FOR DETERMINING EXTENT OF EUTROPHICATION, AUTHOR ESTABLISHED TWO STATIONS: I AT NORTH END OF LAKE, WHICH IS SUBJECT TO INORGANIC POLLUTION; AND V, A SOUTHERN STATION INFLUENCED BY OUTFALL FROM A PRIMARY TREATMENT SEWAGE PLANT. BETWEEN MAY AND OCTOBER 1967, MONTHLY DETERMINATIONS WERE MADE OF ALGAL COUNTS, ASH-FREE DRY WEIGHTS, AND RESPIRATORY AND PHOTOSYNTHETIC RATES (BY WARBURG MANOMETRY). PHYTOPLANKTON WAS DOMINATED BY SPECIES OF CHLOROPHYTES AND DIATOMS; VERY CYANOPHYTES WERE COUNTED. PATTERNS OF DOMINANCE DID NOT DIFFER BETWEEN STATIONS. AT 1% LEVEL OF CONFIDENCE, THERE WERE SIGNIFICANT DIFFERENCES BETWEEN STATIONS. AS COMPARED WITH I, VALUES AT V WERE GREATER IN THE FOLLOWING (SAMPLE NUMBERS IN PARENTHESES): (N=3) NITRATE-NITROGEN, AMMONIUM-NITROGEN, PHOSPHATES; (N=6) ALGAL NUMBERS, ASH-FREE DRY WEIGHTS; (N=98) PHOTOSYNTHETIC RATE, RESPIRATORY RATE. DETAILED LIMNOLOGICAL STUDIES ARE NEEDED BEFORE THE EXPENDITURE OF LARGE SUMS FOR POLLUTION ABATEMENT PROJECTS.

FIELD 05C, 02H

ACCESSION NO. W69-02959

ADSORPTION OF PHOSPHORUS BY LAKE SEDIMENT,

CONNECTICUT AGRICULTURAL EXPERIMENTAL STATION, NEW HAVEN.

ROBERT D. HARTER.

SOIL SCI SOC OF AMER PROC, VOL 32, NO 4, PP 514-518, JULY-AUG 1968. 5 P, 5
  FIG, 1 TAB, 12 REF.

DESCRIPTORS:
  *ADSORPTION, *PHOSPHORUS COMPOUNDS, *LAKES, *SEDIMENTS,
  *EUTROPHICATION, ALGAE, CLAYS, SILICATES.

IDENTIFIERS:
  EXTRACTABLE PHOSPHORUS.

ABSTRACT:
  TO STUDY THE MECHANISM OF SORPTION OF PHOSPHOROUS BY EUTROPHIC LAKE
  SEDIMENTS, BETWEEN 0 AND 2.2 MG P WAS ADDED TO 0.1-G SEDIMENT SAMPLES
  AT A 1:50 SEDIMENT-SOLUTION RATIO. AFTER EQUILIBRIUM HAD BEEN
  ESTABLISHED, P REMAINING IN SOLUTION WAS DETERMINED, AND ADSORBED P WAS
  EXTRACTED WITH 0.5N AMMONIUM FLUORIDE AND 0.1N NAOH. WHEREAS ALL P
  ADSORBED INTO THE NAOH-EXTRACTABLE FRACTION APPEARED TO OCCUR AS AN
  IRON PHOSPHATE, AMMONIUM FLUORIDE APPARENTLY EXTRACTED P BONDED BY TWO
  DIFFERENT MECHANISMS. WHEN LESS THAN ABOUT 0.1 MG P WAS ADDED, AMMONIUM
  FLUORIDE EXTRACTED A TIGHTLY BONDED FORM OF P, PROBABLY OCCURRING AS AN
  ALUMINUM PHOSPHATE. WHEN MORE THAN 0.1 MG P WAS ADDED, ADDITIONAL P IN
  A MORE LOOSELY BONDED FORM WAS ABSORBED INTO THE AMMONIUM
  FLUORIDE-EXTRACTABLE FRACTION. THE LOOSELY BONDED P APPEARED TO BE
  INDEPENDENT OF AL CONTENT OF THE SEDIMENT, AND COULD BE REMOVED BY
  SUCCESSIVE WATER EXTRACTION. THE CAPABILITY OF THE SEDIMENT TO ADSORB
  CONSIDERABLE LOOSELY BONDED P MEANS THAT LARGE INFLUXES OF P INTO THE
  LAKE MAY BE HELD TEMPORARILY AND SUBSEQUENTLY RELEASED TO GROWING
  PLANTS AND ALGAE. (KNAPP-USGS)

FIELD 02H, 05C

ACCESSION NO. W69-03075

NUTRIENT SOURCES FOR ALGAE AND AQUATIC WEEDS,

WISCONSIN UNIV., MADISON. ENGINEERING EXPERIMENT STATION.

GEORGE P. FITZGERALD.

IN: DETECTION OF LIMITING OR SURPLUS NUTRIENTS IN ALGAE, 9 DECEMBER 1968. 21
  P. WP - 297.

DESCRIPTORS:
  *ALGAE, *CYCLING NUTRIENTS, *ENZYMES, *ESSENTIAL NUTRIENTS, *NITROGEN
  CYCLE, ANALYTICAL TECHNIQUES, AQUATIC ALGAE, AQUATIC MICROBIOLOGY,
  AQUATIC PRODUCTIVITY, AQUATIC WEEDS, BALANCE OF NATURE, BIOASSAY,
  CHLOROPHYTA, CYANOPHYTA, ENVIRONMENTAL EFFECTS, EUTROPHICATION,
  LIMNOLOGY, NITROGEN COMPOUNDS, NITROGEN FIXATION, NUTRIENT
  REQUIREMENTS, PHOSPHORUS COMPOUNDS, PHYSIOLOGICAL ECOLOGY,
  PHYTOPLANKTON, POLLUTANT IDENTIFICATION, RAIN, RAINFALL-RUNOFF
  RELATIONSHIPS, RAIN WATER, ROOTED AQUATIC PLANTS, SEWERS, WATER
  POLLUTION, WATER POLLUTION CONTROL, WATER POLLUTION EFFECTS, WATER
  POLLUTION SOURCES.

IDENTIFIERS:
  LAKE MENDOTA(WIS).

ABSTRACT:
  SIMPLE BIOASSAYS FOR LIMITING OR SURPLUS NITROGEN AND PHOSPHORUS
  CONDITIONS IN ALGAE OR AQUATIC WEEDS HAVE BEEN USED TO FOLLOW THE
  NUTRITION OF ALGAE AND AQUATIC WEEDS DURING SUMMER OF 1968. OF
  PARTICULAR INTEREST HAVE BEEN THE SOURCES OF NITROGEN AND PHOSPHORUS
  AVAILABLE TO THESE PLANTS. IT HAS BEEN SHOWN THAT DURING SUMMER RAIN
  COULD BE THE MAJOR SOURCE OF AVAILABLE NITROGEN TO CLADOPHORA SP IN
  LAKE MENDOTA, WIS. INCREASES IN PHOSPHORUS ASSOCIATED WITH RAINFALLS
  WERE DETECTED BUT WERE NOT AS DRAMATIC AS INCREASES IN NITROGEN. IN
  ADDITION, STUDIES HAVE SHOWN THAT MIXED BLOOMS OF PLANKTONIC ALGAE DO
  NOT SHARE THEIR NUTRIENTS WITH OTHER ALGAE EVEN WHEN ONE SPECIES MAY
  HAVE A SURPLUS AND ANOTHER BE NUTRIENT-LIMITED. COMPARISONS OF SURFACE
  AND SUBSURFACE PHYTOPLANKTON HAVE INDICATED THAT AT CERTAIN TIMES
  SURFACE PLANKTON COULD BE NUTRIENT-LIMITED WHILE THE SAME SPECIES FROM
  SUBSURFACE SOURCES HAD ADEQUATE OR SURPLUS NUTRIENTS. FALL OVERTURN AS
  A NUTRIENT SOURCE HAS BEEN USED TO DEMONSTRATE HOW SIMILAR CHANGES CAN
  OCCUR IN THE NUTRITION OF DIFFERENT TYPES OF ALGAE. THE RESULTS OF
  THESE STUDIES ARE USED TO POINT OUT SOME OF THE IMPORTANT FACTORS
  INVOLVED IN FIELD STUDIES OF THE NUTRITION OF AQUATIC PLANTS.

FIELD 05B

ACCESSION NO. W69-03185

FIELD AND LABORATORY EVALUATIONS OF BIOASSAYS FOR NITROGEN AND PHOSPHORUS WITH
ALGAE AND AQUATIC WEEDS,

WISCONSIN UNIV., MADISON. ENGINEERING EXPERIMENTAL STATION.

GEORGE P. FITZGERALD.

IN: DETECTION OF LIMITING OR SURPLUS NUTRIENTS IN ALGAE, 9 DECEMBER 1968. 12
P. WP - 297.

DESCRIPTORS:
*CHLOROPHYTA, *CYCLING NUTRIENTS, *ENZYMES, *ESSENTIAL NUTRIENTS,
*NITROGEN CYCLE, *NUTRIENT REQUIREMENTS, ALGAE, ANALYTICAL TECHNIQUES,
AQUATIC ALGAE, AQUATIC MICROBIOLOGY, AQUATIC PRODUCTIVITY, AQUATIC
WEEDS, BIOASSAY, CYANOPHYTA, ENVIRONMENTAL EFFECTS, EUTROPHICATION,
LIMNOLOGY, NITROGEN COMPOUNDS, NITROGEN-FIXATION, PHOSPHORUS COMPOUNDS,
PHYSIOLOGICAL ECOLOGY, PHYTOPLANKTON, POLLUTANT IDENTIFICATION, RAIN,
RAINFALL-RUNOFF RELATIONSHIPS, RAIN WATER, ROOTED AQUATIC PLANTS, WATER
POLLUTION, WATER POLLUTION CONTROL, WATER POLLUTION EFFECTS, WATER
POLLUTION SOURCES, SEWERS.

ABSTRACT:
THE RATE OF AMMONIA ABSORPTION IN THE DARK AND THE AMOUNT OF
ORTHOPHOSPHATE EXTRACTED BY BOILING WATER TREATMENTS HAVE BEEN USED TO
FOLLOW THE TRANSIENT NATURE OF THE NITROGEN AND PHOSPHORUS NUTRITION OF
ALGAE AND AQUATIC WEEDS. THE MEASUREMENT OF THE ENZYMATIC ACTIVITY OF
ALGAL ALKALINE PHOSPHATASE ALSO USED TO FOLLOW PHOSPHORUS NUTRITION HAS
BEEN SHOWN TO BE INHIBITED BY THE PRESENCE OF 5 MG/L OR MORE OF
PHOSPHATE-PHOSPHORUS, BUT THE ACTUAL ENZYMATIC ACTIVITY OF THE ALGAE IS
NOT INHIBITED BY PHOSPHATE-PHOSPHORUS AND IS REDUCED ONLY BY GROWTH (OR
DILUTION) UNDER CONDITIONS OF SURPLUS AVAILABLE PHOSPHORUS. IT IS
SUGGESTED THAT ONLY TERMINAL PORTIONS OF AQUATIC WEEDS BE USED FOR
NUTRITIONAL BIOASSAYS BECAUSE OF NUTRITIONAL DIFFERENCES BETWEEN YOUNG
AND OLD PORTIONS OF THE SAME PLANT. AVAILABLE NITROGEN COMPOUNDS
RELATED TO RAINFALL HAVE BEEN SHOWN TO INFLUENCE THE NITROGEN
NUTRITIONAL STATUS OF SPIROGYRA AND CLADOPHORA. IN APPLICATIONS OF
BIOASSAYS THE IMPORTANCE OF TESTING EACH SPECIES OF PLANTS SEPARATELY
IS SHOWN BY CONTRASTING RESULTS OBTAINED WITH NITROGEN-FIXING
(PHOSPHORUS-LIMITED) AND NON-FIXING (NITROGEN-LIMITED) BLUE-GREEN ALGAE
FROM THE SAME ENVIRONMENT.

FIELD 05A

ACCESSION NO. W69-03186

SOME FACTORS IN THE COMPETITION OR ANTAGONISM BETWEEN BACTERIA, ALGAE, AND
   AQUATIC WEEDS,

   WISCONSIN UNIV., MADISON. ENGINEERING EXPERIMENT STATION.

   GEORGE P. FITZGERALD.

   IN: DETECTION OF LIMITING OR SURPLUS NUTRIENTS IN ALGAE, 9 DECEMBER 1968. 19
      P. WP-297.

      DESCRIPTORS:
         *ALGAE, *ALGAL TOXINS, *AQUARIA, *AQUATIC WEEDS, *INHIBITION,
         *PHYSIOLOGICAL ECOLOGY, *PHYTOTOXICITY, ALGAL CONTROL, ALGAL POISONING,
         ANALYTICAL TECHNIQUES, AQUATIC ALGAE, AQUATIC BACTERIA, AQUATIC
         MICROBIOLOGY, BALANCE OF NATURE, BIOASSAY, CHLOROPHYTA, CYANOPHYTA,
         DIATOMS, CYCLING NUTRIENTS, ENVIRONMENTAL EFFECTS, ENZYMES, ESSENTIAL
         NUTRIENTS, NITROGEN COMPOUNDS, NUISANCE ALGAE, NUTRIENT REQUIREMENTS,
         ̄̄̄̄̄POUNDS, PHYTOPLANKTON, PLANKTON, POLLUTANT
                                 PLANTS, SEWERS, SEWAGE, SEWAGE
                                 ATER POLLUTION, WATER POLLUTION EFFECTS,

                        IN THE POPULATIONS OF AQUATIC WEEDS AND
                        THAT AQUATIC WEEDS HAVE ANTAGONISTIC
                        N. NUTRITIONAL STUDIES IN THE LABORATORY
                        AQUATIC WEEDS MYRIOPHYLLUM SP,
                        D (LEMNA MINOR L); LIQUID CULTURES OF
                        CKSON VARIETY); AND CULTURES OF THE
                        OPHORA SP AND PITHOPHORA OEDOGONIUM (MONT)
                        LY FREE OF EPIPHYTES OR COMPETING
                        S ARE NITROGEN LIMITED. FIELD OBSERVATIONS
                        MED THAT THE GROWTH OF EPIPHYTES ON THE
                        DITIONS OF SURPLUS AVAILABLE NITROGEN
                        AT THIS ANTAGONISTIC ACTIVITY MAY BE DUE TO
                        WHICH THE AQUATIC WEEDS OR FILAMENTOUS
                        TH OF CONTAMINATING ALGAE BY COMPETITION
                        MPOUNDS AVAILABLE. HOWEVER, THE PRESENCE OF
                        ICH HAVE SELECTIVE TOXICITY TO CERTAIN ALGAE
                        IPLE FACTORS EXIST. AUTHOR DISCUSSES THE
                        CH AS NITROGEN-LIMITED GROWTH, OF
                        AE WITH BACTERIA HAVING SELECTIVE TOXICITIES
                        N ENVIRONMENTAL CONDITIONS.

                        HATES IN ALGAL CULTURES,

                        R CHEMISTRY LAB.

   G. FRED LEE, NICHOLAS   ERI, AND GEORGE P. FITZGERALD.

   INT J AIR WATER POLLUTION, VOL 9, PP 715-722, 1965. 8 P, 1 FIG, 5 TAB, 8 REF.

      DESCRIPTORS:
         *ALGAE, *ANALYTICAL TECHNIQUES, *CYCLING NUTRIENTS, *EUTROPHICATION,
         *PHOSPHATES, *PHYSIOLOGICAL ECOLOGY, AQUATIC ALGAE, AQUATIC
         MICROBIOLOGY, CHLOROPHYTA, CYANOPHYTA, ESSENTIAL NUTRIENTS,
         ENVIRONMENTAL EFFECTS, NITROGEN COMPOUNDS, NUTRIENT REQUIREMENTS,
         PHOSPHORUS COMPOUNDS, PHOSPHORUS, PHYTOPLANKTON.

      ABSTRACT:
         TOTAL PHOSPHATE ANALYSIS PROCEDURES WERE EVALUATED. IT WAS FOUND THAT
         THE STANDARD METHODS (1960) PROCEDURE OF ACID HYDROLYSIS IN 1 ML
         SULFURIC ACID (30%) WAS SUITABLE FOR PYROPHOSPHATES AND
         TRIPOLYPHOSPHATES BUT NOT FOR TOTAL PHOSPHATE ANALYSIS OF ALGAE OR
         ALGAL EXTRACTS. ASHING (600 DEG C, 1 HR), PERCHLORIC ACID DIGESTION
         (SULFURIC ACID PLUS NITRIC ACID PLUS PERCHLORIC ACID), AND SULFURIC
         PLUS NITRIC ACID DIGESTION GAVE SATISFACTORY RESULTS. SUPERNATANTS FROM
         THE GREEN ALGA CHLORELLA PYRENOIDOSA (WIS 2005) DID NOT AFFECT
         ORTHOPHOSPHATE ANALYSES, BUT A CONSISTANT ERROR (LOSS) WAS FOUND WHEN
         THE SUPERNATANT OF THE BLUE-GREEN ALGA MICROCYSTIS AERUGINOSA (WIS
         1036) WAS TESTED. (EICHHORN-WIS)

      FIELD 05A

      ACCESSION NO. W69-03358

THE EFFECT OF ALGAE ON BOD MEASUREMENTS,

WISCONSIN UNIV., MADISON. HYDRAULIC LAB., AND WISCONSIN UNIV, MADISON.
SANITARY LAB.

GEORGE P. GITZGERALD.

J WATER POLLUTION CONTROL FED, VOL 36, NO 12, PP 1524-1542, DEC 1964. 19 P, 7
FIG, 4 TAB, 18 REF. ENGINEERING EXPERIMENT STATION REPRINT NO. 728.

DESCRIPTORS:
*BIOCHEMICAL OXYGEN DEMAND, *CHLORELLA, *PHOTOSYNTHETIC OXYGEN, ALGAE,
ANALYTICAL TECHNIQUES, AQUATIC ALGAE, AQUATIC BACTERIA, AQUATIC
MICROBIOLOGY, BIOASSAY, CHLOROPHYTA, CYANOPHYTA, ENVIRONMENTAL EFFECTS,
ESSENTIAL NUTRIENTS, FREEZE DRYING, FREEZE-THAW TESTS, HARVESTING OF
ALGAE, NUISANCE ALGAE, NUTRIENT REQUIREMENTS, PHYSIOLOGICAL ECOLOGY,
PHYTOPLANKTON, SEWAGE BACTERIA, SEWAGE TREATMENT, WATER POLLUTION,
WATER POLLUTION CONTROL, WATER POLLUTION EFFECTS, WATER POLLUTION
TREATMENT.

ABSTRACT:
THE 5-DAY, 20 DEG C BOD OF THE GREEN ALGA CHLORELLA PYRENOIDOSA (WIS
2005) WAS COMPARED WHEN THE ALGAE HAD BEEN GROWN IN DIFFERENT CULTURE
MEDIA. THE RELATIVE GROWTH AND RELATIONSHIPS BETWEEN GROWTH
MEASUREMENTS ARE PRESENTED. THE ACTIVITY OF CHLORELLA IN BOD TESTS WAS
INDEPENDENT OF THE VOLUME OF SETTLED SEWAGE SEED. ALGAE FROM PURE
CULTURES HAD THE SAME BOD AS ALGAE FROM CULTURES SEEDED WITH SEWAGE
BACTERIA; THE BOD OF CULTURE SUPERNATANTS WAS DECREASED TO ONE-FIFTH BY
BACTERIA ADDED TO THE CULTURES. DEAD ALGAE HAD 4 TIMES THE BOD OF LIVE
ALGAE. OXYGEN PRODUCED BY ALGAE AFTER 2-HR INCUBATION IN LIGHT WAS USED
TO FOLLOW THE PHOTOSYNTHETIC CAPACITY OF ALGAE DURING INCUBATION TESTS
IN THE DARK FOR AS LONG AS 16 DAYS. CHLORELLA COULD STILL PRODUCE
OXYGEN AFTER 16 DAYS IN THE DARK AND FOR AT LEAST 7 DAYS AFTER THE BOD
BOTTLES BECAME ANAEROBIC. THE BOD OF CHLORELLA VARIED BETWEEN 0.09 AND
0.19 MG OXYGEN USED PER MILLIGRAM (DRY WEIGHT) OF SUSPENDED SOLIDS PER
LITER FOR 5 DAYS AT 20 DEG C. THE RESULTS FOR ALGAE FROM DIFFERENT
CULTURE MEDIA AND DIFFERENT SPECIES ARE PRESENTED. (EICHHORN-WISC)

FIELD 05A

ACCESSION NO. W69-03362

DETECTION OF LIMITING OR SURPLUS NITROGEN IN ALGAE AND AQUATIC WEEDS,

WISCONSIN UNIV., MADISON. WATER CHEMISTRY LAB.

GEORGE P. FITZGERALD.

J PHYCOL, VOL 4, NO 2, PP 121-126, 1968. 6 P, 4 FIG, 1 TAB, 13 REF.

DESCRIPTORS:
*ALGAE, *ANALYTICAL TECHNIQUES, *AQUATIC WEEDS, *BIOASSAY, *ESSENTIAL
NUTRIENTS, *NITROGEN COMPOUNDS, AQUATIC ALGAE, AQUATIC MICROBIOLOGY,
CHLAMYDOMONAS, CHLOROPHYTA, CYANOPHYTA, DIATOMS, CYCLING NUTRIENTS,
ENVIRONMENTAL EFFECTS, AMMONIA, AMMONIUM COMPOUNDS, NITROGEN FIXATION,
NUTRIENT REQUIREMENTS, NUTRIENTS, PHYSIOLOGICAL ECOLOGY, PHYTOPLANKTON,
ROOTED AQUATIC PLANTS, BIOINDICATORS.

ABSTRACT:
THE RATE OF AMMONIUM-NITROGEN ABSORPTION BY ALGAE AND AQUATIC WEEDS IN
THE DARK HAS BEEN SHOWN TO BE 4 TO 5 TIMES GREATER FOR PLANTS WHICH ARE
NITROGEN-LIMITED THAN FOR PLANTS WITH SUFFICIENT AVAILABLE NITROGEN.
EIGHT SPECIES OF GREEN ALGAE, 2 BLUE-GREEN ALGAE, 2 DIATOMS, AND 3
AQUATIC WEEDS WERE USED TO DEMONSTRATE THE USEFULNESS OF THE TEST IN
DETERMINING IF AVAILABLE NITROGEN WAS IN SURPLUS OR LIMITED SUPPLY IN A
PARTICULAR ENVIRONMENT. THE TEST WAS SHOWN NOT TO DIFFERENTIATE BETWEEN
BLUE-GREEN ALGAE CAPABLE OF FIXING NITROGEN (4 SPECIES) FROM MEDIA WITH
NITRATE-NITROGEN OR WITHOUT COMBINED NITROGEN. THE FACTORS INFLUENCING
THE RESULTS OF AMMONIUM-NITROGEN ABSORPTION TESTS HAVE BEEN
INVESTIGATED. IN ORDER TO DIFFERENTIATE BETWEEN PLANTS WITH SUFFICIENT
AVAILABLE NITROGEN AND THOSE WHICH ARE NITROGEN-LIMITED, THE RATE OF
AMMONIUM-NITROGEN ABSORPTION (0.1 MG NITROGEN) OVER 1-HR INCUBATION IN
THE DARK BY 10-20 MG OF ALGAE OR AQUATIC WEED TISSUES IS MEASURED. THE
RELATIVELY SIMPLE ANALYSIS FOR AMMONIUM-NITROGEN IN THE SAMPLES MAKES
IT VERY EASY TO FOLLOW THE CHANGING NITROGEN NUTRITION OF PLANTS IN
CULTURES WITH A LIMITED NITROGEN SUPPLY OR IN THE PRESENCE OF POSSIBLE
NITROGEN SOURCES. (EICHHORN-WISC)

FIELD 05C

ACCESSION NO. W69-03364

COPPER IN LAKE MUDS FROM LAKES OF THE MADISON AREA,

WISCONSIN UNIV., MADISON. STATE LAB. OF HYGIENE.

M. STARR NICHOLS, THERESA HENKEL, AND DOROTHY MCNALL.

TRANS WIS ACAD SCI ARTS LETT, VOL 38, PP 333-350, 1946. 18 P, 6 FIG, 9 TAB, 4 REF.

DESCRIPTORS:
*LAKES, *MUD, *COPPER, COPPER SULFATE, ALGAL CONTROL, ALGICIDES, TRACE ELEMENTS, CORES, WISCONSIN, BOTTOM SEDIMENT, BENTHIC FAUNA.

IDENTIFIERS:
*MADISON(WIS), NUISANCE ODORS, LAKE MENDOTA(WIS), LAKE MONONA(WIS), LAKE WINGRA(WIS), LAKE KEGONSA(WIS), LAKE WAUBESA(WIS), PRECIPITATION.

ABSTRACT:
THE AMOUNT OF COPPER SULFATE APPLIED ANNUALLY FROM 1925-1944 TO MADISON, WIS, LAKES IS GIVEN. MUCH OF THE COPPER SULFATE ADDED TO LAKE WATERS OF NOTABLE ALKALINITY WAS PRECIPITATED AS A BASIC COPPER COMPOUND. FROM LAKE MENDOTA, LAKE MONONA, LAKE WINGRA, LAKE WAUBESA AND LAKE KEGONSA THE BOTTOM MUDS AT VARIOUS WATER DEPTHS WERE ANALYZED FOR TOTAL AND SOLUBLE COPPER. CORE SAMPLES OF LAKE MUDS FROM THE VARIOUS LAKES WERE ALSO TAKEN. LAKE MONONA HAS RECEIVED COPPER SULFATE TREATMENT FOR ALGAL CONTROL FOR A MUCH LONGER PERIOD THAN THE OTHER LAKES, AND THIS IS REFLECTED IN THE HIGHER CONCENTRATIONS OF COPPER FOUND IN THE LAYERS OF MUD OF LAKE MONONA. THE GREATEST CONCENTRATION OF COPPER IS FOUND IN THE DEEPER PART OF THE LAKE. THE COPPER CONTENT OF THE LAKE MONONA CORE PENETRATED TO A DEPTH OF 8 FEET. ACTION BY BURROWING FAUNA MAY HAVE CAUSED MIXING OF THE UPPER LAYERS TO THIS DEPTH. IT APPEARS THAT BY FAR THE GREATEST AMOUNT OF COPPER APPLIED REMAINS AS A DEPOSIT IN THE MUD OF THE LAKE. (EICHHORN-WISC)

FIELD 05A

ACCESSION NO. W69-03366

STREAM POLLUTION BY ALGAL NUTRIENTS,

HARVARD UNIV., CAMBRIDGE, MASS. DIV. OF ENGINEERING AND APPLIED PHYSICS.

WERNER STUMM, AND JAMES J. MORGAN.

SANITARY ENGINEERING REPRINT NO. 45. TRANS 12TH ANN CONF SANITARY ENG, UNIV KANSAS, PP 16-26, 1962. 11 P, 3 FIG, 2 TAB, 19 REF.

DESCRIPTORS:
*STREAMS, *WATER POLLUTION EFFECTS, *CYCLING NUTRIENTS, *EUTROPHICATION, ALGAE, AQUATIC ALGAE, AQUATIC BACTERIA, AQUATIC MICROBIOLOGY, BALANCE OF NATURE, WISCONSIN, BIOASSAY, BIOCHEMICAL OXYGEN DEMAND, ENVIRONMENTAL EFFECTS, ENVIRONMENTAL SANITATION, ESSENTIAL NUTRIENTS, HARVESTING OF ALGAE, INHIBITION, LIGHT PENETRATION, NITROGEN COMPOUNDS, PHOSPHORUS COMPOUNDS, NUTRIENT REQUIREMENTS, PHOTOSYNTHETIC OXYGEN, WATER POLLUTION CONTROL, WATER POLLUTION SOURCES, WATER POLLUTION TREATMENT, POLLUTION IDENTIFICATION, SEWAGE, SEWAGE DISPOSAL, SEWAGE EFFLUENTS, SEWAGE LAGOONS.

IDENTIFIERS:
MADISON(WIS).

ABSTRACT:
SINCE ORGANISMS ARE CONSUMED AS WELL AS PRODUCED, SIZE OF STANDING BIOMASS MAY BEAR LITTLE RELATION TO ORGANISM ACTIVITY RATE. AEROBIC BIOLOGICAL SEWAGE TREATMENT MINERALIZES OXIDIZABLE ORGANIC SUBSTANCES BUT ELIMINATES ONLY 20-50% OF NITROGEN AND PHOSPHORUS COMPOUNDS. IN SEWAGE TREATMENT, CARBON BECOMES LIMITING BEFORE NITROGEN AND PHOSPHORUS ARE INCORPORATED INTO SLUDGE. INCREASE OF PHOSPHORUS FROM DETERGENTS MAY REACH POINT WHERE PHOSPHORUS COULD NO LONGER BE CONSIDERED A LIMITING FACTOR. SEWAGE STABILIZATION PONDS REPRESENT MIXTURE OF BACTERIAL DECOMPOSITION AND ALGAL GROWTH, BUT, TO REMOVE NITROGEN AND PHOSPHORUS, ALGAE MUST BE SEPARATED FROM POND EFFLUENT. DECOMPOSITION OF PHYTOPLANKTON WHICH SETTLED TO BOTTOM OF LAKE HELPS INCREASE NUTRIENT CONTENT OF BOTTOM WATERS DURING STAGNATION PERIODS. NOT ALL NUTRIENTS ARE REGENERATED. MADISON, WIS, LAKES RETAIN 30-60% OF NITROGEN RECEIVED. CONTRASTED WITH STABILIZATION PONDS, LAKE STAGNATION TENDS TO SEPARATE AUTOTROPHS (ALGAE) FROM HETEROTROPHS (BACTERIA AND ANIMALS). RELATION BETWEEN FERTILIZATION AND ALGAE IS INDICATED BY ASSUMPTION THAT EVERY ION OF PHOSPHORUS ADDED TO WATER, IF COMPLETELY UTILIZED FOR PHOTOSYNTHETIC PRODUCTION, CAUSES USE OF 16 NITROGEN ATOMS AND 106 ATOMS OF CARBON ALGAL PROTOPLASM. COMPLETE OXIDATION OF ORGANIC MATTER CONTAINING ONE ATOM OF PHOSPHORUS REQUIRES 150 MOLECULES OF OXYGEN. NUTRIENT REMOVAL METHODS ARE REVIEWED. (EICHHORN-WISC)

FIELD 05C

ACCESSION NO. W69-03369

EFFECT OF WATER SAMPLE PRESERVATION METHODS ON THE RELEASE OF PHOSPHORUS FROM ALGAE,

WISCONSIN UNIV., MADISON. WATER CHEMISTRY LAB.

G. P. FITZGERALD, AND S. L. FAUST.

ENGINEERING EXPERIMENT STATION REPRINT NO. 995. LIMNOL OCEANOGR, VOL 12, NO 2, PP 332-334, APRIL 1967. 3 P, 2 TAB, 9 REF.

DESCRIPTORS:
*ANALYTICAL TECHNIQUES, *CYCLING NUTRIENTS, *EUTROPHICATION, *PHOSPHORUS COMPOUNDS, *WATER POLLUTION SOURCES, ALGAE, ALGAL CONTROL, ALGAL POISONING, ALGICIDES, AQUATIC ALGAE, BIOASSAY, CHLOROPHYTA, CYANOPHYTA, DIATOMS, DISINFECTION, ENVIRONMENTAL EFFECTS, ESSENTIAL NUTRIENTS, FREEZE DRYING, FREEZE-THAW TESTS, INHIBITION, NUTRIENT REQUIREMENTS, PHOSPHORUS, PHOSPHATES, PHYSIOLOGICAL ECOLOGY, PHYTOPLANKTON.

ABSTRACT:
RECOMMENDED METHODS FOR PRESERVING WATER SAMPLES BEFORE ANALYSIS WERE USED TO DEMONSTRATE THAT IF ALGAE ARE NOT REMOVED FROM THE SAMPLES BEFORE PRESERVATION, PHOSPHATE WILL BE RELEASED BY TREATED ALGAE. THE AMOUNT OF PHOSPHATE-PHOSPHORUS (IN MILLIGRAMS PER 100 MG ALGAE) RELEASED BY DIFFERENT ALGAL SPECIES VARIED FROM 0.01-0.017 FOR OVERNIGHT REFRIGERATION AT 3-5 DEG C TO 0.38-0.78 AFTER 60-MIN EXTRACTION IN BOILING WATER BATH. FREEZING (-15 DEG C) OVERNIGHT, CHLOROFORM SATURATION, AND AN ORGANIC MERCURY ALGICIDE ALSO CAUSED THE RELEASE OF PHOSPHATE-PHOSPHORUS. IT IS CONCLUDED THAT THE RELEASE OF PHOSPHORUS FROM ALGAE COULD BE SIGNIFICANT UNLESS THE ALGAE ARE REMOVED BEFORE THE SAMPLES ARE PRESERVED FOR STORAGE. (EICHHORN-WISC)

FIELD 05B

ACCESSION NO. W69-03370

THE EFFECT OF DILUTION MEDIA ON THE BOD OF ALGAE,

WISCONSIN UNIV., MADISON. HYDRAULIC LAB; AND WISCONSIN UNIV., MADISON. SANITARY LAB.

MARY M. ALLEN, GEORGE P. FITZGERALD, AND GERARD A. ROHLICH.

ENGINEERING EXPERIMENT STATION REPRINT NO. 697. J WATER POLLUTION CONTROL FED, VOL 36, NO 8, PP 1049-1056, AUG 1964. 8 P, 3 TAB, 11 REF.

DESCRIPTORS:
*BIOCHEMICAL OXYGEN DEMAND, *PHOTOSYNTHETIC OXYGEN, *CHLORELLA, ALGAE, ANALYTICAL TECHNIQUES, AQUATIC ALGAE, AQUATIC BACTERIA, AQUATIC MICROBIOLOGY, BIOASSAY, CHLOROPHYTA, CYANOPHYTA, DIATOMS, ENVIRONMENTAL EFFECTS, ESSENTIAL NUTRIENTS, HARVESTING OF ALGAE, NUISANCE ALGAE, NUTRIENT REQUIREMENTS, PHYSIOLOGICAL ECOLOGY, PHYTOPLANKTON, SEWAGE, SEWAGE BACTERIA, SEWAGE TREATMENT, WATER POLLUTION, WATER POLLUTION CONTROL, WATER POLLUTION EFFECTS, WATER POLLUTION TREATMENT.

ABSTRACT:
THE 5-DAY, 20 DEG C BOD OF CHLORELLA PYRENOIDOSA (WIS 2005) AVERAGED 0.11 MG OF OXYGEN USED PER MILLIGRAM OF ALGAE WHEN TESTED IN BOD DILUTION WATER BUT WAS 0.25 MG OF OXYGEN USED PER MILLIGRAM OF ALGAE WHEN TESTED IN FRESH LAKE WATER. THE PHOTOSYNTHETIC CAPACITY (MILLIGRAMS OF OXYGEN PRODUCED IN 2 HOURS IN LIGHT PER MILLIGRAM OF ALGAE) OF ALGAE INCUBATED 5 DAYS IN THE DARK IN LAKE WATER WAS NEARLY THE SAME AS ALGAE INCUBATED IN DILUTION WATER. THE ADDITION OF MAJOR OR MINOR NUTRIENTS AND ADDED BUFFER CAPACITY TO STANDARD DILUTION WATER DID NOT INCREASE THE ALGAL BOD. THE ORGANISMS PRESENT IN LAKE WATERS HAD LITTLE EFFECT ON ALGAL BOD. WHEN BOD WAS MEASURED IN ALGAL CULTURE MEDIUM, THE SAME RATE AS IN DILUTION WATER WAS OBTAINED. DEIONIZED LAKE WATER MADE UP TO STANDARD DILUTION WATER GAVE SAME BOD OF ALGAE AS IN DILUTION WATER PREPARED WITH DISTILLED WATER. IT IS CONCLUDED THAT SOME HEAT-STABLE FACTOR THAT CANNOT PASS THROUGH ION EXCHANGE COLUMN AFFECTED THE ALGAL BOD OR THAT THE NATURAL RATIOS OF IONS IN LAKE WATERS ARE ESSENTIAL FOR MAXIMUM ACTIVITY OF ALGAE IN BOD. (EICHHORN-WISC)

FIELD 05A

ACCESSION NO. W69-03371

EXTRACTIVE AND ENZYMATIC ANALYSES FOR LIMITING OR SURPLUS PHOSPHORUS IN ALGAE,

WISCONSIN UNIV., MADISON. WATER CHEMISTRY LAB; AND WISCONSIN UNIV., MADISON.
MCARDLE MEMORIAL LAB.

GEORGE P. FITZGERALD, AND THOMAS C. NELSON.

ENGINEERING EXPERIMENT STATION REPRINT NO. 909. J PHYCOL, VOL 2, NO 1, PP
32-37, 1966. 6 P, 1 FIG, 2 TAB, 33 REF.

DESCRIPTORS:
*ANALYTICAL TECHNIQUES, *BIOASSAY, *ENZYMES, *NUTRIENT REQUIREMENTS,
*PHOSPHORUS COMPOUNDS, ALGAE, AQUATIC ALGAE, AQUATIC MICROBIOLOGY,
AQUATIC PRODUCTIVITY, AQUATIC WEEDS, CHLOROPHYTA, CYANOPHYTA, DIATOMS,
CYCLING NUTRIENTS, ENVIRONMENTAL EFFECTS, ESSENTIAL NUTRIENTS,
NUTRIENTS, PHOSPHORUS, PHOSPHATES, PHYSIOLOGICAL ECOLOGY,
PHYTOPLANKTON, ROOTED AQUATIC PLANTS, EUTROPHICATION.

ABSTRACT:
AN EXTRACTIVE PROCEDURE FOR DETECTION OF SURPLUS-STORED PHOSPHORUS
(LUXURY CONSUMPTION) IN ALGAE AND AN ENZYMATIC ANALYSIS FOR CONDITIONS
OF PHOSPHORUS-LIMITED GROWTH IN ALGAE HAVE BEEN EVALUATED. A SIMPLE
60-MIN BOILING WATER EXTRACTION OF ALGAE KNOWN TO CONTAIN SURPLUS
PHOSPHORUS SEPARATES ESSENTIAL PHOSPHORUS COMPOUNDS AND SURPLUS-STORED
PHOSPHORUS COMPOUNDS. SURPLUS PHOSPHORUS COMPOUNDS CAN BE MEASURED IN
THE EXTRACT AS ORTHOPHOSPHATE. EXTRACTS OF ALGAE LIMITED IN THEIR
GROWTH BY THE AMOUNT OF AVAILABLE PHOSPHORUS CONTAIN LITTLE OR NO
ORTHOPHOSPHATE. LIMITATION OF ALGAL GROWTH BY PHOSPHORUS SUPPLY INDUCES
THE ENZYME ALKALINE PHOSPHATASE. THE ACTIVITY OF THIS ENZYME CAN BE
MEASURED AT PH 9 USING PHOSPHORUS-NITROPHENYLPHOSPHATE AS SUBSTRATE.
ALGAE WHICH WERE PHOSPHORUS-LIMITED AND CONTAINED NO EXTRACTABLE
ORTHOPHOSPHATE HAVE AS MUCH AS 25 TIMES MORE ALKALINE PHOSPHATASE
ACTIVITY THAN ALGAE WITH SURPLUS AVAILABLE PHOSPHORUS. COMPARATIVE
RESULTS ARE PRESENTED OF EXTRACT AND ENZYME ANALYSES OF 9 SPECIES OF
ALGAE FROM LABORATORY CULTURES AND ALGAE FROM 3 LAKES BEFORE AND AFTER
INCUBATION IN LIMITING OR SURPLUS PHOSPHORUS IN THE LABORATORY.
CONCENTRATIONS OF MORE THAN 0.08 MG PHOSPHATE-PHOSPHORUS PER 100 MG OF
ALGAE EXTRACTED INDICATE ALGAE CONTAINED SURPLUS PHOSPHORUS.
(EICHHORN-WISC)

FIELD 05A

ACCESSION NO. W69-03373

STRIPPING EFFLUENTS OF NUTRIENTS BY BIOLOGICAL MEANS,

WISCONSIN UNIV., MADISON. SANITARY LAB.

GEORGE P. FITZGERALD.

TRANS OF THE SEMINAR ON ALGAE AND METROPOLITAN WASTES, ROBERT A TAFT SAN ENGR
CTR, CINCINNATI, 1961. PP 136-139, 5 FIG.

DESCRIPTORS:
*NUTRIENT REQUIREMENTS, *EUTROPHICATION, *WATER POLLUTION CONTROL,
*WATER POLLUTION TREATMENT, ALGAE, ALGAL CONTROL, ANALYTICAL
TECHNIQUES, AQUATIC ALGAE, AQUATIC MICROBIOLOGY, BALANCE OF NATURE,
BIOASSAY, BIOCHEMICAL OXYGEN DEMAND, CHLOROPHYTA, CYCLING NUTRIENTS,
ENVIRONMENTAL EFFECTS, ESSENTIAL NUTRIENTS, ENVIRONMENTAL ENGINEERING,
HARVESTING ALGAE, NITROGEN COMPOUNDS, PHOSPHORUS COMPOUNDS,
PHOTOSYNTHETIC OXYGEN, PHYSIOLOGICAL ECOLOGY, PHYTOPLANKTON, SEWAGE,
SEWAGE BACTERIA, SEWAGE DISPOSAL, SEWAGE EFFLUENTS, SEWAGE LAGOONS,
SEWAGE TREATMENT, WATER POLLUTION, WATER POLLUTION SOURCES.

ABSTRACT:
LABORATORY EXPERIMENTS WITH THE GREEN ALGA CHLORELLA PYRENOIDOSA (WIS
2005) HAVE SHOWN THIS ALGA WOULD FIRST ABSORB THE AMMONIUM-NITROGEN,
THEN NITRITE-NITROGEN AND NITRATE-NITROGEN FROM SECONDARY SEWAGE
EFFLUENTS. LESS THAN 0.5 MG/LITER OF ANY OF THE NITROGEN OR PHOSPHORUS
OF THE EFFLUENT REMAINED IN SOLUTION AFTER 17 DAYS OF CULTURE. GROWTH
OF CHLORELLA PARALLELED NITROGEN AND PHOSPHORUS REMOVAL IN PRIMARY AND
SECONDARY EFFLUENT. GROWTH IN SECONDARY EFFLUENT COULD BE STIMULATED BY
THE ADDITION OF CARBON DIOXIDE TO AIR. THE OPERATION OF A 1/2-ACRE
STABILIZATION POND FOR NUTRIENT REMOVAL BY ALGAE IN THE POND INDICATED
THAT A SUCCESSION OF ALGAL SPECIES TOOK PLACE DESPITE A CONTINUOUS
SUPPLY OF NUTRIENTS. THE AVERAGE NITROGEN REMOVAL THROUGHOUT THE YEAR
IS ABOUT 30%, WITH SUMMER REMOVALS REACHING ABOUT 70%. THERE WERE ONLY
33 DAYS DURING 1956 AND 76 DAYS DURING 1957 WHEN NITROGEN REMOVAL
EXCEEDED 50%. PHOSPHORUS REMOVAL COINCIDED WITH PERIODS OF HIGH PH.
DURING EARLY WINTER, POND EFFLUENT CONTAINED HIGHER LEVELS OF
PHOSPHORUS THAN INFLUENT DUE TO DISSOLUTION OF PRECIPITATED PHOSPHORUS.
(EICHHORN-WISC)

FIELD 05F

ACCESSION NO. W69-03374

RELATIVE ABSORPTION OF STRONTIUM AND CALCIUM BY CERTAIN ALGAE,

ARIZONA UNIV., TUCSON. DEPT. OF AGRICULTURAL CHEMISTRY AND SOILS.

WALLACE H. FULLER, AND JAMES E. HARDCASTLE.

REPRINT FROM SOIL SCI SOC AMER PROC, VOL 31, NO 6, PP 772-774, NOV-DEC 1967. 3 P, 2 FIG, 2 TAB.

DESCRIPTORS:
*SOIL ALGAE, *AQUATIC ALGAE, ALGAE, ARID CLIMATES, SEMIARID CLIMATES, *ABSORPTION, *CALCIUM, SOIL CONTAMINATION, *STRONTIUM RADIOSOTOPES, CYTOLOGICAL STUDIES.

IDENTIFIERS:
NUTRIENT MEDIA.

ABSTRACT:
SOIL ALGAE APPEAR IN ABUNDANCE IN ARID AND SEMIARID LANDS WHERE THEY FORM SURFACE CRUSTS OF VARYING DENSITY AND CLING TO STONE SURFACES AND LIME OUTCROPS. THE UPTAKE OF RADIOSTRONTIUM BY 7 FRESH-WATER AND DESERT ALGAE WITH RESPECT TO THEIR FUNCTION IN CONCENTRATING RADIOACTIVE CONTAMINATION PRODUCTS OF URANIUM FISSION WAS EVALUATED. IN ADDITION, THE INTERRELATIONSHIPS BETWEEN THE UPTAKE BY ALGAE OF STRONTIUM AND CALCIUM WERE STUDIED. THE OBJECTIVE WAS TWOFOLD: (1) TO EVALUATE THE CAPACITY OF CERTAIN ARID LAND ALGAE TO DISTINGUISH BETWEEN CALCIUM AND STRONTIUM IN THEIR METABOLISM, AND (2) TO EVALUATE THE EXTENT TO WHICH CALCIUM ENTERS INTO A COMPETITIVE POSITION IN THE ABSORPTION AND CONCENTRATION OF STRONTIUM BY ALGAE CELLS. UPTAKE ALWAYS FAVORED CALCIUM OVER STRONTIUM WHEN THE TWO ELEMENTS WERE PRESENT IN NUTRIENT MEDIA TOGETHER THEREBY MODIFYING THE HAZARD OF STRONTIUM ENTERING THE FOOD CHAIN. (AFFLECK-ARIZ)

FIELD 05B

ACCESSION NO. W69-03491

ENVIRONMENTAL REQUIREMENTS OF BLUE-GREEN ALGAE.

FEDERAL WATER POLLUTION CONTROL ADMINISTRATION, CORVALLIS, ORE.; PACIFIC NORTHWEST WATER LAB, AND WASHINGTON UNIV., SEATTLE.

PROCEEDINGS OF A SYMPOSIUM CORVALLIS ORE., SEPT 23-24, 1966. OCT, 1967, 111 P.

DESCRIPTORS:
*CYANOPHYTA, *EUTROPHICATION, *PLANKTON, ALGAE, *NUTRIENT REQUIREMENTS.

IDENTIFIERS:
*BLUE-GREEN ALGAE, *ALGA BLOOMS, ALGAE PHYSIOLOGY.

ABSTRACT:
THE PROBLEM OF ACCELERATED EUTROPHICATION HAS MANY FACETS, AND SOME ARE BEING EXAMINED THROUGH RESEARCH TO FIND THE KEYS FOR SOLUTION. THIS SYMPOSIUM WAS HELD TO FULFILL THE NEED TO UNDERSTAND BETTER THE ENVIRONMENTAL REQUIREMENTS OF BLUE-GREEN ALGAE. IT CONTAINS EIGHT PAPERS BY SCIENTISTS WHO ARE CONCERNED WITH THE ECOLOGICAL CONTROL OF ALGAE BLOOMS. CURRENT KNOWLEDGE ON THE SUBJECT HAS BEEN EVALUATED AND GAPS IDENTIFIED FOR FUTURE INVESTIGATION. THE PAPERS ARE FOLLOWED BY A GENERAL DISCUSSION AND A SPECIFIC DISCUSSION OF NUTRIENT MEASUREMENT AND NUISANCE CONTROL. (SEE ALSO W69-03513 THRU W69-03518)

FIELD 05C

ACCESSION NO. W69-03512

WHY STUDY BLUE-GREEN ALGAE,

WASHINGTON UNIV., SEATTLE. DEPT. OF ZOOLOGY.

W. T. EDMONSON.

SYMPOSIUM ON ENVIRONMENTAL REQUIREMENTS OF BLUE-GREEN ALGAE. FEDERAL WATER
  POLLUTION CONTROL ADMIN, PACIFIC NORTHWEST WATER LAB, CORVALLIS, ORE, PP
  1-6, OCT 1967. 6 P, 11 REF.

DESCRIPTORS:
  *CYANOPHYTA, *EUTROPHICATION, *PHYSIOLOGICAL ECOLOGY, *WATER POLLUTION
  EFFECTS, ALGAE, ALGAL CONTROL, HARVESTING, AQUATIC ALGAE, AQUATIC
  ANIMALS, AQUATIC MICROBIOLOGY, AQUATIC PRODUCTIVITY, BALANCE OF NATURE,
  CYCLING NUTRIENTS, ENVIRONMENTAL EFFECTS, ENVIRONMENTAL SANITATION,
  ESSENTIAL NUTRIENTS, INHIBITION, LIGHT INTENSITY, LIMNOLOGY, NITROGEN
  COMPOUNDS, NUISANCE ALGAE, NUTRIENT REQUIREMENTS, PHOSPHORUS COMPOUNDS,
  PHOTOSYNTHETIC OXYGEN, PHYTOPLANKTON, PLANKTON, POLLUTANT
  IDENTIFICATION, SEWAGE EFFLUENTS, SEWAGE DISPOSAL, WATER POLLUTION
  CONTROL, WATER POLLUTION SOURCES.

IDENTIFIERS:
LAKE WASHINGTON.

ABSTRACT:
  IN 1950, L WASHINGTON HAD ONLY 1,500,000 CUBIC MICRONS PER MILLILITER
  OF PHYTOPLANKTON, 15% OF WHICH WERE BLUE-GREEN ALGAE. IN 1962, IT HAD
  10,500,000 CUBIC MICRONS PER MILLILITER OF WHICH 95% WERE BLUE-GREEN
  ALGAE, AND THE APPEARANCE OF THE LAKE STIMULATED THE CONSTRUCTION OF AN
  EXPENSIVE SEWAGE DIVERSION PROGRAM. THE BLUE-GREENS ATTRACT ATTENTION
  BECAUSE OF THE PUBLIC NUISANCE THEY CAUSE DUE TO THEIR ABILITY TO FLOAT
  AND TO CONCENTRATE DOWNWIND; THEY HAVE SPECIAL NUTRITIONAL PROPERTIES,
  ARE CONSPICUOUS IN EXTREME KINDS OF HABITATS, AND ARE IMPORTANT IN
  FORMING AND STABILIZING SOIL. THE COLONIAL SPECIES OF BLUE-GREENS ARE
  LESS READILY EATEN BY SOME ZOOPLANKTON, WHICH MAY BE A FACTOR IN THEIR
  SURVIVAL. THE RELATIONSHIP BETWEEN HIGH NUTRIENT WATERS AND PRESENCE OF
  BLUE-GREEN ALGAE IS DISCUSSED, AND IT IS POINTED OUT THAT THE
  BLUE-GREEN, OSCILLATORIA AGARDHII, REMAINED IN L WASHINGTON, WHEN
  NITROGEN AND PHOSPHORUS LEVELS WERE VERY LOW. THE KEY TO DOMINANCE OF
  ALGAL SPECIES MAY BE IN THE PROPORTION OF NUTRIENTS MORE THAN IN
  QUANTITY. MORE INFORMATION IS NEEDED CONCERNING ANTIBIOTIC SUBSTANCES
  PRODUCED BY ALGAE. (FOR MAIN ENTRY SEE W69-03512)

FIELD 05C

ACCESSION NO. W69-03513

PROBLEMS IN THE LABORATORY CULTURE OF PLANKTONIC BLUE-GREEN ALGAE,

MANCHESTER COLLEGE, NORTH MANCHESTER, IND.

WILLIAM R. EBERLY.

SYMPOSIUM ON ENVIRONMENTAL REQUIREMENTS OF BLUE-GREEN ALGAE, FEDERAL WATER
POLLUTION CONTROL ADMIN, PACIFIC NORTHWEST WATER LAB, CORVALLIS, ORE, PP
7-34, OCT 1967. 28 P, 15 FIG, 1 TAB, 16 REF.

DESCRIPTORS:
*CYANOPHYTA, *PHYSIOLOGICAL ECOLOGY, *PHYTOPLANKTON, *CULTURES, ALGAE,
AQUATIC PRODUCTIVITY, CYCLING NUTRIENTS, ENVIRONMENTAL EFFECTS,
ESSENTIAL NUTRIENTS, EUTROPHICATION, INHIBITION, LIGHT INTENSITY, LIGHT
PENETRATION, LIMNOLOGY, NITROGEN COMPOUNDS, NUTRIENT REQUIREMENTS,
PHOTOSYNTHETIC, OXYGEN, PLANKTON, WATER POLLUTION EFFECTS.

IDENTIFIERS:
ALGAL TAXONOMY, ALGAE IDENTIFICATION.

ABSTRACT:
SURFACE BLOOMS ARE USUALLY DOMINATED BY EITHER CHROOCOCCALES
(MICROCYSTIS) OR FILAMENTOUS FORMS (ANABAENA OR APHANIZOMENON), WHEREAS
DEEP-WATER BLOOMS ARE NEARLY ALWAYS OSCILLATORIA. THE AUTHOR IS
CONCERNED WITH THE OSCILLATORIA ASSOCIATED WITH DEEP-WATER OXYGEN
MAXIMA (10-12 DEG C, LIGHT 1% OF SURFACE INTENSITY). GRAPHS ARE
PRESENTED SHOWING THE RATES OF GROWTH (OD) FOR A NUMBER OF OSCILLATORIA
ISOLATES AND THE EFFECTS OF INITIAL INOCULATION LEVEL, PH, SOIL
EXTRACT, TEMPERATURE, AND LIGHT INTENSITY IN ASM AND ASM-1 MEDIA. THE
TAXONOMIC IMPLICATIONS OF THESE FINDINGS ARE DISCUSSED. ALSO CONSIDERED
IS THE PROBLEM OF HOW ALGAE IN NATURE MAINTAIN THEIR POSITION, WHEREAS
IN LABORATORY CULTURES CONSIDERABLE TURBULENCE IS REQUIRED TO KEEP THEM
SUSPENDED. OTHER PROBLEMS REVIEWED ARE THE RELATION BETWEEN BACTERIA
AND ALGAE, THE ALMOST UNIALGAL BLOOMS OCCURRING IN NATURE, AND THE
RELATIONS BETWEEN LIGHT AND TEMPERATURE. AUTHOR POINTS OUT THAT WITH AN
INCREASE OF NUTRIENTS (EUTROPHICATION) LIGHT AND TEMPERATURE OPTIMAL
MIGHT BE RAISED WITH THE RESULT THAT THE MAXIMUM CONCENTRATION OF ALGAE
WOULD MOVE CLOSER TO THE SURFACE. SEASONAL DOWNWARD MOVEMENTS OF A
BLOOM OF OSCILLATORIA DURING THE SUMMER DEPLETION OF NUTRIENTS IS
DESCRIBED. (FOR MAIN ENTRY SEE W69-03512)

FIELD 05C

ACCESSION NO. W69-03514

ASPECTS OF THE NITROGEN NUTRITION OF SOME NATURALLY OCCURRING POPULATIONS OF
BLUE-GREEN ALGAE,

ALASKA UNIV., COLLEGE. INST. OF MARINE SCIENCE.

VERA A. (DUGDALE) BILLAUD.

SYMPOSIUM ON ENVIRONMENTAL REQUIREMENTS OF BLUE-GREEN ALGAE, FEDERAL WATER
POLLUTION CONTROL ADMIN, PACIFIC NORTHWEST WATER LAB, CORVALLIS, ORE. PP
35-53, OCT 1967. 19 P, 5 FIG, 6 TAB, 19 REF.

DESCRIPTORS:
*CYANOPHYTA, *NITROGEN COMPOUNDS, *NUTRIENT REQUIREMENTS,
*PHYSIOLOGICAL ECOLOGY, ALGAE, ANALYTICAL TECHNIQUES, AQUATIC ALGAE,
AQUATIC MICROBIOLOGY, AQUATIC WEEDS, BALANCE OF NATURE, BIOASSAY,
CHLOROPHYTA, DIATOMS, CYCLING NUTRIENTS, ENVIRONMENTAL EFFECTS,
ESSENTIAL NUTRIENTS, EUTROPHICATION, INHIBITION, LIGHT INTENSITY, LIGHT
PENETRATION, LIMNOLOGY, NITROGEN CYCLE, NITROGEN FIXATION, NUTRIENTS,
PHOTOSYNTHETIC OXYGEN, PHYTOPLANKTON, PLANKTON, ROOTED AQUATIC PLANTS,
ALASKA.

IDENTIFIERS:
SMITH LAKE(ALASKA).

ABSTRACT:
IT IS WELL ESTABLISHED THAT NITROGEN-FIXING BLUE-GREEN ALGAE INCREASE
THE NITROGEN CONTENT OF WATERS, AND IT IS ASSUMED THAT THIS NITROGEN IS
AVAILABLE UPON DEGRADATION, TO OTHER ALGAL SPECIES. EXPERIMENTS IN
SMITH LAKE, CENTRAL ALASKA, INDICATE THAT AN ANABAENA POPULATION
DEVELOPS WITHIN A FEW DAYS AFTER ICE BREAKUP. DURING THE SUMMER, A
STEADY, LOWER LEVEL OF APHANIZOMENON GROWTH OCCURS. THE INITIAL
NITROGEN USED BY ANABAENA IS AMMONIUM-NITROGEN; THEN NITROGEN FIXATION
BECOMES IMPORTANT, AT ONE POINT ACCOUNTING FOR HALF THE NITROGEN
ASSIMILATED. THE FIRST ANABAENA FILAMENTS GROWING ON AMMONIUM-NITROGEN
HAVE NO HETEROCYSTS; THEY APPEAR AT THE TIME NITROGEN FIXATION OCCURS.
THE POPULATION OF APHANIZOMENON IS VERY STABLE AND APPEARS TO BE
CONTROLLED BY GRAZING BY DENSE POPULATION OF ZOOPLANKTON. NITROGEN
FIXATION IN SIGNIFICANT RATES ONLY OCCURS DURING THE ANABAENA BLOOM. IN
ANOTHER LAKE, APHANIZOMENON OCCURS IN ANOXIC, DEEPER WATERS. THE
PRESENCE OF COMBINED NITROGEN SOURCES DOES NOT APPEAR TO PREVENT
NITROGEN-FIXING BLOOMS FROM DEVELOPING. (FOR MAIN ENTRY SEE W69-03512)

FIELD 05C

ACCESSION NO. W69-03515

ENVIRONMENTAL REQUIREMENTS OF THERMOPHILIC BLUE-GREEN ALGAE,

OREGON UNIV., EUGENE, DEPT. OF BIOLOGY.

RICHARD W. CASTENHOLZ.

SYMPOSIUM ENVIRONMENTAL REQUIREMENTS OF BLUE-GREEN ALGAE, FEDERAL WATER
POLLUTION CONTROL ADMIN, PACIFIC NORTHWEST WATER LAB, CORVALLIS, ORE. PP
55-79, OCT 1967. 25 P, 4 FIG, 1 TAB, 53 REF.

DESCRIPTORS:
*CYANOPHYTA, *PHYSIOLOGICAL ECOLOGY, *SALINE WATER, *THERMAL WATER,
ALGAE, AQUATIC ALGAE, AQUATIC MICROBIOLOGY, BALANCE OF NATURE,
BIOASSAY, CHLOROPHYTA, DIATOMS, ENVIRONMENTAL EFFECTS, INHIBITION,
LIGHT, LIGHT INTENSITY, LIGHT PENETRATION, LIMNOLOGY, PHOTOSYNTHETIC
OXYGEN, TEMPERATURE, THERMAL POLLUTION, THERMAL SPRINGS, THERMAL
STRESS.

ABSTRACT:
THERMOPHILIC ALGAE ARE DESIGNATED AS THOSE WHOSE OPTIMUM TEMPERATURE IS
ABOVE 45 DEG C. THE GREEN MATS FORMED IN HOT SPRINGS ABOVE 45 DEG C MAY
PERSIST BECAUSE OF THE ABSENCE OF GRAZERS. THERMOPHILIC ALGAE MUST BE
ABLE TO WITHSTAND THE VERY HIGH LIGHT INTENSITIES OF THE SHALLOW SPRING
WATERS. THERE IS A GREAT SIMILARITY IN THE FLORA OF HOT SPRINGS
THROUGHOUT THE WORLD. HOT SPRINGS PROVIDE VERY CONSTANT TEMPERATURE
HABITATS, BUT SOME BLUE-GREEN THERMOPHILES CAN WITHSTAND TEMPERATURE
SHOCKS AS GREAT AS ROOM TEMPERATURE TO 70 DEG C WITHOUT EFFECT. THE
HIGHEST TEMPERATURE FOR CONSISTENT SURVIVAL APPEARS TO BE 74-75 DEG C,
BUT THIS LIMIT MAY DEPEND UPON WATER CHEMISTRY. DETAILS OF OBSERVATIONS
ON TEMPERATURES AS RELATED TO HABITATS ARE PRESENTED AND REVIEWED. THE
CORRELATIONS OF GROWTH AND LIGHT INTENSITIES FOUND IN FIELD AND
LABORATORY STUDIES ARE DISCUSSED. (FOR MAIN ENTRY SEE W69-03512)

FIELD 05C

ACCESSION NO. W69-03516

GROWTH REQUIREMENTS OF BLUE-GREEN ALGAE AS DEDUCED FROM THEIR NATURAL DISTRIBUTION,

HUMBOLDT STATE COLL., ARCATA, CALIF.

WILLIAM C. VINYARD.

SYMPOSIUM ON ENVIRONMENTAL REQUIREMENTS OF BLUE-GREEN ALGAE, FEDERAL WATER POLLUTION CONTROL ADMIN, PACIFIC WATER LAB, CORVALLIS, ORE, PP 81-85, OCT 1968. 5 P, 2 REF.

DESCRIPTORS:
*EUTROPHICATION, *CYANOPHYTA, *PHYSIOLOGICAL ECOLOGY, ALGAE, ANALYTICAL TECHNIQUES, AQUATIC ALGAE, AQUATIC MICROBIOLOGY, BALANCE OF NATURE, CYCLING NUTRIENTS, ENVIRONMENTAL EFFECTS, ESSENTIAL NUTRIENTS, CHLOROPHYTA, DIATOMS, INHIBITION, LIGHT, LIGHT INTENSITY, LIGHT PENETRATION, LIMNOLOGY, NUTRIENT REQUIREMENTS, NUTRIENTS, PHOTOSYNTHETIC OXYGEN, PHYTOPLANKTON, SALINE WATER, ECOLOGICAL DISTRIBUTION, CALIFORNIA.

ABSTRACT:
ORGANIC POLLUTION ALMOST ALWAYS INSURES AN OVERABUNDANCE OF BLUE-GREEN ALGAE. ALSO, EXCESSIVE GROWTH OF CYANOPHYTES OCCURS DURING PERIODS WHEN HIGH TEMPERATURES AND LIGHT INTENSITIES PREVAIL. BECAUSE NOT ALL SPECIES OF CYANOPHYTES HAVE NUISANCE SIGNIFICANCE, THEIR TAXONOMY ASSUMES SOME IMPORTANCE. AUTHOR INDICATES THAT KNOWLEDGE RELATING TO TAXONOMY IS DEFICIENT AND THAT MODERN STUDIES RELATING SYSTEMATICS TO PHYSIOLOGICAL AND OTHER CULTURAL REQUIREMENTS ARE NEEDED. INFORMATION PERTAINING TO THE GEOGRAPHICAL DISTRIBUTION OF THESE ALGAE IS ALSO REQUIRED. SPECIES OF BLUE-GREENS PREDOMINATE IN SUMP PONDS OF NON-BRINE OIL FIELDS CONTAINING CRUDE OILS AND TARS, SOME OCCURRING AT THE BOTTOM UNDER THE HEAVY SCUM OF BLACK TAR, CONDITIONS WHICH PRESENT THE ALGAE WITH LITTLE OR NO LIGHT AND MINIMAL POSSIBILITY OF GASEOUS DIFFUSION INTO OR OUT OF THE MEDIUM. IN FRESHWATER LAGOON, A RELATIVELY COOL COASTAL ENVIRONMENT NEAR ORICK, CALIF, A NEARLY CONTINUOUS BLOOM, INCLUDING MICROCYSTIS, COELOSPHAERIUM AND ANABAENA, OCCURRED FOR SOME FIVE YEARS. (FOR MAIN ENTRY SEE W69-03512)

FIELD 05C

ACCESSION NO. W69-03517

RECENT ADVANCES IN THE PHYSIOLOGY OF BLUE-GREEN ALGAE,

SCRIPPS INSTITUTION OF OCEANOGRAPHY, LA JOLLA, CALIF.

OSMUND HOLM-HANSON.

SYMPOSIUM ON ENVIRONMENTAL REQUIREMENTS OF BLUE-GREEN ALGAE, FEDERAL WATER POLLUTION CONTROL ADMIN, NORTHWEST WATER LAB, CORVALLIS, ORE, PP 87-96, OCT 1967. 10 P.

DESCRIPTORS:
*AQUATIC MICROBIOLOGY, *PHYSIOLOGICAL ECOLOGY, *CYANOPHYTA, ALGAE, ANALYTICAL TECHNIQUES, AQUATIC ALGAE, BIOASSAY, CYCLING NUTRIENTS, ENVIRONMENTAL EFFECTS, ESSENTIAL NUTRIENTS, EUTROPHICATION, FREEZE DRYING, FREEZE-THAW TESTS, FREEZING, LIGHT INTENSITY, LIGHT PENETRATION, LIMNOLOGY, NITROGEN COMPOUNDS, NITROGEN FIXATION, NUTRIENTS, NUTRIENT REQUIREMENTS, PHOTOSYNTHETIC OXYGEN, PHYTOPLANKTON, CHLOROPHYTA.

ABSTRACT:
SOME OF THE SUBJECTS DISCUSSED ARE THE MORPHOLOGY OF BLUE-GREEN ALGAE COMPARED TO RELATED ALGAE OR BACTERIA, THE UNDEFINED MANNER IN WHICH BLUE-GREEN ALGAE MOVE, AND THEIR MICROELEMENT NUTRITION. PRESENTED ARE DETAILS OF WORK ON THE RESISTANCE OF BLUE-GREEN ALGAE TO FREEZING AND FREEZE-DRYING, ESPECIALLY ALGAE ISOLATED IN ANARCTICA. ADDITIONS OF HORSE SERUM, GLYCEROL, MILK POWDER, ETC, MAY HELP THE SURVIVAL OF GREEN ALGAE; BUT THEY DO NOT HELP BLUE-GREENS. THE QUESTION IS RAISED AS TO WHAT FORM OF CARBON THE BLUE-GREENS CAN UTILIZE; AT PH 11 MOST CARBON IS PRESENT AS CARBONATE IONS. RESULTS OF EXPERIMENTS ON THE ORGANIC NUTRITION OF NOSTOC ARE PRESENTED WHICH INDICATE THAT THIS ALGA CAN UTILIZE GLUCOSE IN THE LIGHT BUT LITTLE OR NONE IN THE DARK. (FOR MAIN ENTRY SEE W69-03512)

FIELD 05C

ACCESSION NO. W69-03518

'PREDICTING DIURNAL VARIATIONS IN DISSOLVED OXYGEN CAUSED BY ALGAE IN ESTUARINE WATERS, PART I',

STANFORD UNIV., STANFORD, CALIFORNIA.

RICHARD C. BAIN, JR.

PROCEEDINGS OF THE NATIONAL SYMPOSIUM ON ESTUARINE POLLUTION, AUGUST 23-25, 1967, PP 250-279.

DESCRIPTORS:
*PHYTOPLANKTON, *DISSOLVED OXYGEN, *ALGAE, *DIURNAL DISTRIBUTION, *CALIFORNIA, ESTIMATING EQUATIONS, OXYGENATION, STANDING CROP, MICROENVIRONMENT, AQUATIC LIFE, ENVIRONMENT, MICROORGANISMS, WATER CHEMISTRY, PLANTS, SOUTHWEST U. S., REGIONS, PACIFIC COAST REGIONS, GEOGRAPHICAL REGIONS, PHOTOSYNTHESIS, CHEMICAL REACTIONS, ZOOPLANKTON, ANIMALS, AQUATIC ANIMALS, PHOTOSYNTHETIC OXYGEN, GASES, OXYGEN, OXYGEN DEMANDS, EUTROPHICATION.

ABSTRACT:
EUTROPHIC ENVIRONMENTS ARE OFTEN DOMINATED BY PLANKTONIC ALGAL POPULATIONS (PHYTOPLANKTON) WHICH CAN CAUSE DIURNAL VARIATIONS IN DISSOLVED OXYGEN CONCENTRATIONS THROUGH RESPIRATORY ACTIVITY AND PHOTOSYNTHESIS. PHOTOSYNTHETIC OXYGENATION AND RESPIRATORY DEOXYGENATION RATES OF ESTUARINE PHYTOPLANKTON WERE MEASURED AT VARIOUS STANDING CROP (CHLOROPHYLL) LEVELS. OXYGEN PRODUCTION AND CONSUMPTION RATES FOR ACTIVELY GROWING PHYTOPLANKTON POPULATIONS WERE RELATED TO STANDING CROP AT 20 C AND NONLIMITING LIGHT. VARIATIONS IN ALGAL PHOTOSYNTHETIC PRODUCTION RATE AS RELATED TO LIGHT ADAPTION, AGE OF CELLS, NUTRITION, TEMPERATURE, AND ALGAL TYPE ARE DISCUSSED. LIGHT-PRODUCTION RELATIONSHIPS (BASED ON OCEANOGRAPHIC LITERATURE) WERE USED TO ESTIMATE TOTAL PRODUCTION OF A WELL-MIXED SYSTEM. STREETER-PHELPS EQUATIONS WERE MODIFIED TO INCLUDE PHYTOPLANKTON PRODUCTION AND RESPIRATION RATES IN FORMULATIONS DESIGNED TO PREDICT DISSOLVED OXYGEN CONCENTRATIONS OVER A 24 HOUR PERIOD. AN EXAMPLE IS GIVEN, AND THE RESULTING DISSOLVED OXYGEN PREDICTION IS COMPARED WITH FIELD MEASUREMENTS FROM A TIDAL REACH OF THE SAN JOAQUIN RIVER, CALIFORNIA.

FIELD 05C

ACCESSION NO. W69-03611

SEATTLE'S EFFORTS IN RESTORATION OF BAYS AND ESTUARIES,

GARY W. ISAAC, AND CURTIS P. LEISER.

32ND NORTH AMERICAN WILDLIFE AND NATURAL RESOURCES CONFERENCE, SAN FRANCISCO, CALIFORNIA, TRANSACTIONS, MARCH 13-15, 1967, PP 127-137.

DESCRIPTORS:
*WASHINGTON, *SEWAGE TREATMENT, *WATER POLLUTION CONTROL, PACIFIC NORTHWEST U. S., CIVIL ENGINEERING, WASTE TREATMENT, ENGINEERING, ALGAE, PLANTS, CONTROL, POLLUTION ABATEMENT, ABATEMENT, PACIFIC COAST REGION, REGIONS, GEOGRAPHICAL REGIONS, BAYS, BODIES OF WATER.

ABSTRACT:
CITIZENS IN THE SEATTLE AREA, WHICH IS COMPRISED OF 14 CITIES, INITIATED THE LEGISLATION AND VOTED INTO BEING A METROPOLITAN MUNICIPAL CORPORATION, SPECIFICALLY TO DEAL WITH THE POLLUTION PROBLEM, BUT ABLE ALSO TO EXPAND TO ALLEVIATE OTHER REGIONAL DIFFICULTIES AS THEY WOULD ARISE. AN OPERATING TREATMENT AGENCY AND FOUR ENGINEERING FIRMS COMPOSED A MASTER PLAN WHICH IS DETAILED IN THIS PAPER. THE COST OF THE PLAN WAS ESTIMATED AT $125,000,000. THE POLLUTION CONTROL PROGRAM WILL CONTINUE IN THREE MAJOR PARTS: (1) EMISSION CONTROLS APPLIED TO INDUSTRIAL WASTE DISCHARGES AT THE SOURCE, (2) TREATMENT PLANT CONTROL AND MONITORING BY PERSONNEL ATTACHED PERMANENTLY TO PLANT OPERATIONS, (3) A SEPARATE FORCE TO MONITOR THE RECEIVING WATERS FOR BIOLOGICAL NUTRIENT AND PHYSICAL PARAMETERS.

FIELD 05G, 05D

ACCESSION NO. W69-03683

'NUTRIENT ASSIMILATION IN A VIRGINIA TIDAL SYSTEM',

STANFORD UNIV., CALIF.

MORRIS L. BREHMER.

PROCEEDINGS OF THE NATIONAL SYMPOSIUM ON ESTUARINE POLLUTION, AUGUST 23-25,
   1967, PP 218-237.

   DESCRIPTORS:
      *WATER PROPERTIES, *NUTRIENTS, *VIRGINIA, NITROGEN COMPOUNDS,
      PHOSPHORUS, NUISANCE ALGAE, SALTS, TIDES, SOUTHEAST U. S., REGIONS,
      GEOGRAPHICAL REGIONS, COASTAL PLAINS, ATLANTIC COASTAL PLAIN, INORGANIC
      COMPOUNDS, METALS, PLANKTON, AQUATIC LIFE, ZOOPLANKTON, AQUATIC
      ANIMALS, ANIMALS, APPALACHIAN MOUNTAIN REGION.

   ABSTRACT:
      THIS STUDY WAS CONDUCTED TO ASSESS NUTRIENT ASSIMILATION AND
      PHYTOPLANKTON RESPONSE IN THE TIDAL JAMES RIVER ESTUARY AND IN THE
      NANSEMOND ESTUARY, A TRIBUTARY TO THE JAMES. THE ASSIMILATION OF
      NITROGEN AND PHOSPHORUS WAS MEASURED FOR A PERIOD OF ONE YEAR IN EACH
      OF THE ABOVE TWO RIVER SYSTEMS. IT WAS DETERMINED THAT THE NUTRIENT
      ASSIMILATION CAPACITY OF ESTUARINE WATERS VARIES SEASONALLY, BEING
      GREATEST IN THE WINTER, EVEN THOUGH FRESH WATER DISCHARGE LEVELS MAY
      REMAIN NEARLY CONSTANT. DATE INDICATE THAT WATER CONTAINING DISSOLVED
      SOLIDS OF MARINE ORIGIN MAY BE ABLE TO ASSIMILATE HIGHER NUTRIENT
      LEVELS THAN FRESH WATER WITHOUT PRODUCING AQUATIC NUISANCE CONDITIONS.

   FIELD 05C

   ACCESSION NO. W69-03695

THE EFFECT OF INDUCED TURBULENCE ON THE GROWTH OF ALGAE,

   GEORGIA INST. OF TECH., ATLANTA. SCHOOL OF ENGINEERING AND GEORGIA INST. OF
      TECH., ATLANTA. WATER RESOURCES CENTER.

   LAWRENCE W. OLINGER.

   GEORGIA INST TECH, WRC REP 0468, DEC 1968. 81 P, 20 FIG, 2 TAB, 47 REF, 4
      APPEND. 5T1-WP-62-03 AND 04, FWPCA.

      DESCRIPTORS:
         *ALGAE, *PHOTOSYNTHESIS, *GROWTH RATES, *TURBULENCE, CHLORRELA,
         SCENEDESMUS, KINETICS, ECOSYSTEMS.

      IDENTIFIERS:
         ALGAL GROWTH KINETICS, MONOD GROWTH EQUATION, INCIDENT LIGHT
         UTILIZATION.

      ABSTRACT:
         A LABORATORY STUDY WAS MADE OF THE RELATIONSHIP BETWEEN TURBULENCE AND
         UTILIZATION OF INCIDENT LIGHT BY ALGAL CULTURES. IT WAS FOUND THAT THE
         MAXIMUM GROWTH RATE CONSTANT OF THE MONOD GROWTH EQUATION INCREASED
         WITH INCREASING TURBULENCE AND THAT THE RATE AT WHICH CONSTANT GROWTH
         OCCURRED WAS A FUNCTION OF TURBULENCE. WITH CONSTANT ILLUMINATION OF
         600 FT - CANDLES AND VARIABLE TURBULENCE, THE GROWTH RATE CONSTANT
         VARIED FROM 1.65 TO 8.5. (KNAPP-USGS)

      FIELD 05C

      ACCESSION NO. W69-03730

MICROBIOLOGY OF OCEANS AND ESTUARIES,

MIAMI UNIV., FLA. INST. OF MARINE SCIENCE.

E. J. F. WOOD.

AMSTERDAM, ELSEVIER PUBL CO OCEANOGR SER, 1967. 319 P, 29 FIG, 8 PLATE, 19
  TAB, 521 REF.

    DESCRIPTORS:
     *MICROBIOLOGY, *AQUATIC MICROBIOLOGY, *ESTUARIES, *MARINE
     MICROORGANISMS, MARINE ALGAE, MARINE ANIMALS, MARINE BACTERIA, MARINE
     FUNGI.

    IDENTIFIERS:
     ESTUARINE MICROBIOLOGY.

    ABSTRACT:
     A TEXTBOOK ON MARINE AND ESTUARINE MICROBIOLOGY IS PRIMARILY INTENDED
     FOR UNDERGRADUATE STUDENTS IN MARINE BIOLOGY AND OCEANOGRAPHY AND MAY
     ALSO BE OF USE TO RESEARCH WORKERS IN MARINE MICROBIOLOGY. THE PROBLEMS
     USUALLY ENCOUNTERED IN STUDY, INCLUDING SAMPLING TECHNIQUES, TAXONOMY,
     AND QUANTIFICATION, ARE DISCUSSED THOROUGHLY. ENVIRONMENTAL
     RELATIONSHIPS AND MODIFICATIONS BY ORGANISMS OF THEIR ENVIRONMENTS ARE
     EXPLAINED. TAXONOMY IS DISCUSSED WITH REFERENCES TO THE MOST
     SIGNIFICANT TAXONOMIC WORKS AND WORKERS IN EACH GROUP. NUTRITION AND
     REPRODUCTION ARE CONSIDERED IN DETAIL. ECONOMIC ASPECTS SUCH AS
     FOULING, THE RELATION OF BORERS TO FUNGI, MICROBIAL CORROSION,
     FERMENTATION, ROTTING OF CORDAGE AND STRUCTURES, FISH SPOILAGE,
     SHELLFISH POISONING, AND GEOBIOLOGICAL ASPECTS AND ORE FORMATION ARE
     DISCUSSED. (KNAPP-USGS)

    FIELD 02L, 05C

    ACCESSION NO. W69-03752

WATER POLLUTION IN THE GREAT LAKES BASIN,

    FEDERAL WATER POLLUTION CONTROL ADMINISTRATION, GREAT LAKES REGION, CHICAGO,
    ILL.

    H. W. POSTON.

    LIMNOS, VOL 1, NO 1, PP 6-11, SPRING 1968. 6 P, 6 PHOTO.

    DESCRIPTORS:
     *GREAT LAKES, *WATER POLLUTION, *WATER POLLUTION CONTROL, WATER
     POLLUTION EFFECTS, EUTROPHICATION, FISHKILL, ALGAE, OXYGEN SAG, WATER
     QUALITY, CHLORIDES, PHOSPHORUS, SULFATES, HARDNESS(WATER).

    IDENTIFIERS:
     GREAT LAKES BASIN, BACTERIAL POLLUTION.

    ABSTRACT:
     THE CAUSES OF WATER POLLUTION OF THE GREAT LAKES BASIN, PRESENTLY
     AVAILABLE REMEDIES, AND RECOMMENDATIONS FOR ACTION IN POLLUTION CONTROL
     ARE DISCUSSED. THE LAKES, PARTICULARLY LAKE ERIE, ARE BECOMING
     EUTROPHIC BECAUSE OF WASTE DISCHARGES. TERTIARY TREATMENT WITH
     PHOSPHORUS REMOVAL WILL ALLEVIATE THE PROBLEM AND ALSO REDUCE
     BIOCHEMICAL OXYGEN DEMAND. CHLORIDES, SULFATES, AND HARDNESS ARE
     INCREASING. BACTERIAL POLLUTION IS A SERIOUS PROBLEM IN TRIBUTARIES,
     AND IN LOCAL ZONES IN THE GREAT LAKES NEAR POPULATION CENTERS WHERE
     RECREATIONAL DEMAND IS ALSO HIGHEST. ELIMINATION OF COMBINED SEWERS AS
     WELL AS BETTER SEWAGE TREATMENT WILL HELP. CHEMICAL CONTAMINATION
     CAUSED BY INDUSTRIAL WASTE DISPOSAL IS WIDESPREAD. OXYGEN LEVELS ARE
     LOW IN STREAMS, SMALL LAKES, AND THE CENTRAL BASIN OF LAKE ERIE. AN
     UNUSUAL FORM OF POLLUTION IS THE OVER-POPULATION OF ALEWIVES IN LAKE
     MICHIGAN. PERIODIC FISHKILLS LITTER THE SHORE. ONE OF THE MAIN
     DIFFICULTIES IN SOLVING GREAT LAKES PROBLEMS IS DIVERSITY AND NEED FOR
     COORDINATION OF GOVERNMENT AGENCIES IN THE BASIN. CREATION OF THE WATER
     QUALITY STANDARDS AND THE GREAT LAKES BASIN COMMISSION SHOULD HELP.
     (KNAPP-USGS)

    FIELD 05B, 05C

    ACCESSION NO. W69-03948

NUTRIENT AND POLLUTANT RESPONSE OF ESTUARINE BIOTAS,

STANFORD UNIV., CALIF.

JAMES B. LACKEY.

IN PROCEEDINGS OF THE NATIONAL SYMPOSIUM ON ESTUARINE POLLUTION, AUGUST 23-25, 1967, PP 188-217.

DESCRIPTORS:
*NUTRIENTS, *MARINE MICROORGANISMS, *POLLUTANTS, RECREATION, WATER SPORTS, ENVIRONMENT, AQUATIC ENVIRONMENT, ESTUARINE ENVIRONMENT, WATER QUALITY, SEWAGE BACTERIA, ENVIRONMENTAL EFFECTS, AQUATIC LIFE, AQUATIC MICROORGANISMS, MICROORGANISMS, SESTON, INVERTEBRATES, SHELLFISH, WASTE DISPOSAL, SEWAGE DISPOSAL, WATER CHEMISTRY, ECOLOGY, WATER POLLUTION, WATER POLLUTION CONTROL, CONTROL, ALGAE, PLANTS, COMMERCIAL FISHING, FISHING, INDUSTRIES, COMMERCIAL SHELLFISHING, ANIMALS, AQUATIC ANIMALS.

ABSTRACT:
THE BEHAVIOR OF MICROBIOTA--ALGAE AND PROTOZOA--AS AFFECTED BY COMMERCIAL, RECREATIONAL, AND METROPOLITAN USES OF ESTUARINE WATERS ARE OUTLINED. SOME OF THE FINDINGS PRESENTED AT THE SYMPOSIUM ARE AS FOLLOWS: (1) ESTUARIES AND BAYS ARE OF GREAT ECONOMIC AND RECREATIONAL VALUE, FOR NURSERIES, FISHING GROUNDS, SHELLFISHERIES, CRUSTACEAN PRODUCTION, BOATING, AND BATHING. (2) HUMAN ACTIVITY OFTEN MODIFIES THEIR PREFERRED ECOLOGY BY ADDING NUTRIENTS OR POLLUTANTS. (3) THE SUSPENDED (PLANKTON) MICROSCOPIC PLANTS AND ANIMALS, AND THOSE OF THE INTERFACE (BENTHOS) ARE MOST EASILY STUDIED QUALITATIVELY AND QUANTITATIVELY. (4) SOME POLLUTANTS, SUCH AS SILT, SHARPLY REDUCE THE MICROSCOPIC POPULATIONS. (5) THE ROLE OF THE ENGINEER WITH REGARD TO ESTUARINE STUDIES IS THAT OF SEEING THAT ADEQUATE BIOLOGICAL STUDIES ARE MADE, SO THAT HE IN TURN CAN DESIGN AND CONSTRUCT THE NECESSARY TREATMENT PLANTS AND OUTFALLS WHICH WILL PREVENT OVERENRICHMENT OR POLLUTION.

FIELD 05C, 02L

ACCESSION NO. W69-04276

EXPERIMENTS ON THE UTILIZATION OF NITROGEN IN FRESH WATER,

FRESHWATER BIOLOGICAL ASSOCIATION, AMBLESIDE (ENGLAND).

WINIFRED PENNINGTON.

J ECOL, VOL 30, PP 326-340, 1942. 1 FIG, 8 TAB, 18 REF, 2 APPEND, DISC.

DESCRIPTORS:
*NITROGEN COMPOUNDS, *NITROGEN UTILIZATION, *FRESH WATER, ALGAE, NITRATE AMMONIUM, NITRITE, OXIDATION-REDUCTION POTENTIAL, CYCLING NUTRIENTS, EUTROPHICATION.

ABSTRACT:
THE EXPERIMENTS DESCRIBED WERE CARRIED OUT ON MIXED CULTURES OF ALGAE AND BACTERIA IN A RICH CULTURE SOLUTION CONTAINING NITRATE OR AMMONIUM. NO ATTEMPT WAS MADE TO DIFFERENTIATE BETWEEN THE ROLES OF ALGAE AND BACTERIA IN NITROGEN UTILIZATION. AMMONIUM AND NITRATE WERE BOTH UTILIZED, BUT AMMONIUM APPEARED TO BE USED MORE RAPIDLY. NITRITE WAS PRODUCED FROM BOTH AMMONIUM AND NITRATE, AND THERE WAS EVIDENCE THAT AMMONIUM WAS PRODUCED FROM NITRATE AND NITRATE FROM AMMONIUM. IN SOME EXPERIMENTS THERE WAS A MARKED LOSS OF NITROGEN FROM THE CULTURES, APPARENTLY DUE TO LIBERATION OF NITROGEN. (KONRAD-WISC)

FIELD 05C, 02K, 02H

ACCESSION NO. W69-04521

DISTRIBUTION, ENVIRONMENTAL REQUIREMENTS AND SIGNIFICANCE OF CLADOPHORA IN THE
  GREAT LAKES,

ONTARIO WATER RESOURCES COMMISSION, TORONTO, BIOLOGY BRANCH.

JOHN H. NEIL, AND GLENN E. OWEN.

PUBLICATION NUMBER 11, GREAT LAKES RES DIV, MICHIGAN UNIV, PP 113-121, 1964.
  4 FIG, 2 TAB, 2 REF.

DESCRIPTORS:
  *ALGAE, *ALGAL CONTROL, *ESSENTIAL NUTRIENTS, *EUTROPHICATION,
  *CLADOPHORA, *NUISANCE ALGAE, *SEWAGE, ALGICIDES, AQUATIC ALGAE,
  AQUATIC PRODUCTIVITY, BIOASSAY, CYCLING NUTRIENTS, ENVIRONMENTAL
  EFFECTS, HARVESTING ALGAE, INHIBITION, LIGHT PENETRATION, LIMNOLOGY,
  NITROGEN COMPOUNDS, NUTRIENT REQUIREMENTS, NUTRIENTS, PHOSPHORUS
  COMPOUNDS, PHYSIOLOGICAL ECOLOGY, POLLUTION IDENTIFICATION, SEWAGE
  DISPOSAL, WATER POLLUTION, WATER POLLUTION CONTROL, WATER POLLUTION
  EFFECTS, WATER POLLUTION SOURCES.

ABSTRACT:
  EXCESSIVE GROWTHS OF CLADOPHORA SP ALONG CERTAIN SECTIONS OF THE GREAT
  LAKES SHORELINE CREATE SERIOUS NUISANCE CONDITIONS WHICH AFFECT THE USE
  OF WATER FOR RECREATIONAL, INDUSTRIAL, AND MUNICIPAL PURPOSES.
  INFORMATION ON THE ECOLOGY OF THIS ALGAE WAS COLLECTED AS PART OF A
  STUDY DIRECTED TOWARD THE DEVELOPMENT OF CONTROL MEASURES. THE PRESENCE
  OF CLADOPHORA SP IS DEPENDENT ON A SUITABLE SUBSTRATE FOR ATTACHMENT,
  WATER MOVEMENT, ADEQUATE LIGHT, AND NUTRIENTS IN EXCESS OF THOSE
  NORMALLY AVAILABLE IN THE WATERS OF THE UPPER GREAT LAKES. LAKES
  ONTARIO AND ERIE HAVE SUFFICIENT INHERENT FERTILITY TO SUPPORT MARGINAL
  GROWTHS, BUT, WHERE LOCAL NUTRIENT SOURCES ARE AVAILABLE, PRODUCTION
  INCREASES. APPLICATION OF PHOSPHORUS TO A LOCATION PROVIDING SUITABLE
  PHYSICAL CONDITIONS BUT DEVOID OF CLADOPHORA SP RESULTED IN THE
  ESTABLISHMENT OF A SIZEABLE AREA OF GROWTH. THE RESULTS OF ATTEMPTS AT
  CONTROL ARE ALSO DISCUSSED. (FITZGERALD-WIS)

FIELD 02H, 05C

ACCESSION NO. W69-04798

UNUSUAL PHOSPHORUS SOURCE FOR PLANKTON ALGAE,

TEXAS A AND M UNIV., COLLEGE STATION. DEPT. OF BIOLOGY.

WALTER ABBOTT.

ECOLOGY, VOL 38, NO 1, P 152, 1957.

DESCRIPTORS:
  *PHOSPHORUS COMPOUNDS, *PHYTOPLANKTON, *WATER POLLUTION SOURCES,
  *CYCLING NUTRIENTS, ALGAE, ANALYTICAL TECHNIQUES, CLAY LOAM, CLAYS,
  COLLOIDS, EUTROPHICATION, IMPOUNDMENTS, LAKES, LOAMS, RESERVOIRS,
  TEXAS.

IDENTIFIERS:
  HARRIS COUNTY, TEX, POLYPHOSPHATES, ORGANIC PHOSPHATES.

ABSTRACT:
  ALTHOUGH LAKE HOUSTON, A NEWLY IMPOUNDING LAKE IN EASTERN HARRIS
  COUNTY, TEXAS, SUPPORTS PHYTOPLANKTON POPULATIONS FLUCTUATING
  TWENTY-FOLD, PHOSPHATE-PHOSPHORUS IN CENTRIFUGATE FROM SAMPLES OF WATER
  FROM TEN STATIONS WAS NOT DEMONSTRABLE BY A TECHNIQUE SENSITIVE TO 1
  GAMMA/LITER. IN A SERIES OF 18 SAMPLINGS, WATER CONTAINING COLLOIDAL
  MATTER, AFTER OXIDATION WITH PERCHLORIC ACID, CONTAINED A MEAN OF 85
  GAMMA/LITER (RANGE, 32-285), HIGHEST VALUES OCCURRING IMMEDIATELY AFTER
  HEAVY RAINS. FILTRATE FROM PASSAGE THROUGH SEITZ NO. 14 FILTER,
  SIMILARLY TREATED, SHOWED ONLY BARELY DETECTABLE TRACES OF
  PHOSPHATE-PHOSPHORUS. AUTHOR OFFERS THESE DATA AS EVIDENCE THAT
  PLANKTONIC ALGAE MAY DERIVE NUTRITIVE PHOSPHORUS FROM COMPLEX
  POLYPHOSPHATES OR IN THE SUSPENDED COLLOIDAL CLAY, ORIGINATING FROM
  SOILS OF SURROUNDING WATERSHEDS, WITHOUT BENEFIT OF INTERMEDIATE STAGE
  OF DISSOLVED PHOSPHATE. (EICHHORN-WIS)

FIELD 05B

ACCESSION NO. W69-04800

ALGAL GROWTH AQUEOUS FACTORS OTHER THAN NITROGEN AND PHOSPHORUS,

ROBERT A. TAFT SANITARY ENGINEERING CENTER, CINCINNATI, OHIO.

KENNETH M. MACKENTHUN, AND WILLIAM MARCUS INGRAM.

PUBLICATION NO WP-24, FEDERAL WATER POLLUTION CONTROL ADMIN, DEPT OF
INTERIOR, WASHINGTON, DC, 1966.

DESCRIPTORS:
*ALGAE, *BIBLIOGRAPHIES, *EUTROPHICATION, *PLANT GROWTH, ANALYTICAL
TECHNIQUES, ANTIBIOTICS, BIOASSAY, NORON, CALCIUM, CHLORINE, COBALT,
COPPER, ECOLOGY, ENVIRONMENTAL EFFECTS, GRAZING, IRON, MANGANESE,
MOLYBDENUM, PLANT GROWTH REGULATORS, PREDATION, PRODUCTIVITY, TRACE
ELEMENTS, SODIUM, VIRUSES, VITAMINS, WATER CHEMISTRY, WATER POLLUTION
CONTROL, WATER POLLUTION SOURCES, WATER QUALITY.

IDENTIFIERS:
AUTOINHIBITORS, EXTRACELLULAR PRODUCTS, HORMONES, AUXINS, METABOLITES,
INHIBITION, SILICON, VANADIUM, ZINC, GALLIUM.

ABSTRACT:
ALTHOUGH RESEARCHERS HAVE EMPHASIZED THE ROLE IN EUTROPHICATION OF THE
MAJOR PLANT NUTRIENTS, NITROGEN AND PHOSPHORUS, THERE IS LITTLE DOUBT
THAT OTHER CHEMICAL AND BIOTIC INFLUENCES ON PRODUCTION OF MACROPHYTES
AND PHYTOPHANKTON ARE IMPORTANT, THOUGH POORLY UNDERSTOOD. THIS
SELECTED BIBLIOGRAPHY OF 399 ENTRIES, COMPILED FOR ENGINEERS AND
SCIENTISTS CONFRONTED WITH APPLIED CONTROL INVESTIGATIONS RELATED TO
POLLUTANTS, FOCUSES ATTENTION ON SUCH FACTORS AND THEIR EFFECTS ON
ALGAE. THE REPORT IS ORGANIZED INTO SECTIONS, AS FOLLOWS, WITH NUMBERS
OF CITATIONS INDICATED: GENERAL REFERENCES, 15; ANTIBIOTICS, 31;
AUTOINHIBITORS, 11; ENVIRONMENTAL FACTORS, 25; EXTRACELLULAR PRODUCTS,
31; HORMONES AND AUXINS, 13; INHIBITING METABOLITES, 24; PREDATION AND
GRAZING, 24; GENERAL REFERENCES ON TRACE METALS AND COMPOUNDS, 29;
BORON, 14; CALCIUM, 2; CHLORINE, 5; COBALT, 7; COPPER, 5; GALLIUM, 1;
IRON, 2; MANGANESE, 5; MOLYBDENUM, 12; SILICON, 7; SODIUM, 8; VANADIUM,
3; ZINC, 17; MIXED ELEMENTS, 15; VIRUSES AND PARASITES, 11; VITAMIN
ASSAY METHODS, 22; VITAMIN OCCURRENCE, 22; ORGANISMIC REQUIREMENTS FOR
VITAMINS, 27; SYNTHESES OF VITAMINS, 11. CITATIONS ARE DATED FROM 1903
TO 1966, OF WHICH 69 CITATIONS ARE FROM 1960-1966. (EICHHORN-WIS)

FIELD 05C, 10

ACCESSION NO. W69-04801

MINERAL REQUIREMENTS FOR THE GROWTH OF ANABAENA SPIROIDES IN VITRO,

OREGON STATE UNIV., CORVALLIS. DEPT. OF BOTANY.

SHERRY L. VOLK, AND HARRY K. PHINNEY.

CAN J BOT, VOL 46, PP 619-630, MAY 1968. 4 FIG, 4 TAB, 67 REF.

DESCRIPTORS:
*ALGAE, *ANALYTICAL TECHNIQUES, *BIOASSAY, *ESSENTIAL NUTRIENTS,
*NUTRIENT REQUIREMENTS, *PHYTOPLANKTON, AQUATIC ALGAE, AQUATIC
MICROBIOLOGY, AQUATIC PRODUCTIVITY, CYANOPHYTA, ENVIRONMENTAL EFFECTS,
EUTROPHICATION, LIMNOLOGY, NITROGEN COMPOUNDS, NITROGEN FIXATION,
NUISANCE ALGAE, NUTRIENTS, PHOSPHORUS COMPOUNDS, PHYSIOLOGICAL ECOLOGY,
PLANKTON, OREGON.

IDENTIFIERS:
KLAMATH LAKE.

ABSTRACT:
ANABAENA SPIROIDES KLEBAHN ISOLATED FROM UPPER KLAMATH LAKE, OREGON,
WAS CULTURED IN GERLOFF'S MODIFICATION OF CHU'S NO 10 MEDIUM.
VARIATIONS OF THE CONCENTRATIONS OF NUTRIENTS WERE STUDIED TO DETERMINE
MINIMAL REQUIREMENTS FOR MAXIMUM GROWTH. MAXIMUM GROWTH IN THE
NUTRITION TESTS VARIED BETWEEN 30 AND 80 MG/LITTER DRY WEIGHT. IRON
PROVIDED IN THE ALKALINE-STABLE CHELATE ETHYLENEDIAMINE
DI-ORTHO-HYDROXYPHENYLACETATE WAS REQUIRED IN LOWER CONCENTRATION (0.3
PPM) THAN IN GERLOFF'S MEDIUM (IRON CITRATE, 1.2 PPM). THE MINIMUM
CONCENTRATION OF NUTRIENTS PROVIDING FOR MAXIMUM GROWTH: MAGNESIUM,
0.125 PPM; PHOSPHORUS, 0.52 PPM; CALCIUM, 10 PPM; SULFUR, 3 PPM;
POTASSIUM, 2.5 PPM. NITROGEN WAS FIXED BY THIS ORGANISM. UREA INHIBITED
GROWTH (AT 16 PPM NITROGEN) WHILE GLYCINE MIGHT HAVE STIMULATED GROWTH.
CARBONATE DID NOT AFFECT GROWTH. NO REQUIREMENT OF MINOR ELEMENTS WAS
FOUND FOR THE AMOUNT OF GROWTH ATTAINED. (FITZGERALD-WIS)

FIELD 05C

ACCESSION NO. W69-04802

DEVELOPMENT AND STATUS OF POND FERTILIZATION IN CENTRAL EUROPE,

WISCONSIN UNIV., MADISON. DEPT. OF ZOOLOGY.

JOHN C. NEESS.

TRANS AMER FISH SOC, VOL 76, PP 335-358, 1949. 24 P, 1 FIG, 73 REF.

DESCRIPTORS:
*CYCLING NUTRIENTS, *ENVIRONMENTAL EFFECTS, *ESSENTIAL NUTRIENTS,
*FISH, *FISHERIES, ALGAE, AQUATIC ALGAE, AQUATIC MICROBIOLOGY, AQUATIC
WEEDS, BALANCE OF NATURE, EUTROPHICATION, FISH FARMING, FISH ROOD
ORGANISMS, FISH GENETICS, FISH HANDLING FACILITIES, FISH MANAGEMENT,
LIMNOLOGY, NITROGEN COMPOUNDS, NITROGEN CYCLE, NUTRIENT REQUIREMENTS,
LIME, PHOSPHORUS COMPOUNDS, PHYSIOLOGICAL ECOLOGY, PLANKTON, SEWAGE,
SEWAGE LAGOONS.

IDENTIFIERS:
EUROPE.

ABSTRACT:
A REVIEW OF RECENT EUROPEAN LITERATURE ON POND FERTILIZATION IS OFFERED
AS A FOUNDATION FOR CERTAIN PHASES OF AMERICAN POND-CULTURE TECHNIQUE.
THE FERTILIZING SUBSTANCES, LIME, POTASSIUM, PHOSPHORUS, NITROGEN, AND
CARBON, ARE DISCUSSED INDIVIDUALLY IN THE LIGHT OF THEIR USEFULNESS IN
INCREASING FISH YIELDS AND WITH REGARD TO THE IMPORTANT MECHANISMS
THROUGH WHICH THEY ARE ABLE TO EFFECT THESE INCREASES. EUROPEAN
TECHNIQUES IN THE EMPLOYMENT OF FERTILIZER ARE DESCRIBED WHERE THESE
ARE DIFFERENT FROM CORRESPONDING AMERICAN PROCEDURES. IN ADDITION, AN
ATTEMPT HAS BEEN MADE TO INTEGRATE DIVERGENT ASPECTS OF THE SUBJECT
THROUGH A CONSIDERATION OF CERTAIN FUNDAMENTALS OF TROPHIC METABOLISM
IN PONDS, THE FUNCTION OF POND SOILS AS CONSERVATORS OF NUTRIENTS, AND
MICROBIOLOGICAL PROCESSES RELATING TO THE THEORY OF FERTILIZATION. A
SYNOPSIS OF THE HISTORY OF POND FERTILIZATION IN CENTRAL EUROPE AND A
GENERAL CRITICISM OF RESEARCH METHODS ARE INCLUDED. (FITZGERALD-WIS)

FIELD 05G

ACCESSION NO. W69-04804

NITROGEN AND PHOSPHORUS IN WATER; AN ANNOTATED BIBLIOGRAPHY OF THEIR BIOLOGICAL
EFFECTS.

PUBLIC HEALTH SERVICE, WASHINGTON, D. C.

PUBLIC HEALTH SERVICE PUBLICATION NO 1305, WASHINGTON, D. C., 1965. 111 PP.

DESCRIPTORS:
*BIBLIOGRAPHIES, *EUTROPHICATION, *NITROGEN COMPOUNDS, *NUTRIENTS,
*PHOSPHORUS COMPOUNDS, *WATER POLLUTION EFFECTS, ALGAE, ANIMALS,
AQUATIC PLANTS, BIOMASS, FERTILIZATION, LAKES, LIMNOLOGY, MIDGES,
NITRATES, NITROGEN, NITROGEN CYCLE, NITROGEN FIXATION, PHOSPHATES,
PHOSPHORUS, PRODUCTIVITY, STANDING CROP, STREAMS, WATER CHEMISTRY,
WATER POLLUTION CONTROL, WATER POLLUTION SOURCES, WATER QUALITY.

IDENTIFIERS:
DUCKS, NITRATE POISONING.

ABSTRACT:
THIS SELECTED AND ANNOTATED BIBLIOGRAPHY OF 171 ENTRIES WAS INTENDED BY
ITS AUTHOR FOR ENGINEERS AND SCIENTISTS WHO ARE FACED WITH PREDICTING
LIMNOLOGICAL CHANGES RESULTING FROM PLANT NUTRIENT LOADINGS TO STANDING
BODIES OF WATER, EMPHASIZING CRITICAL CONCENTRATION VALUES FOR ALGAL
DEVELOPMENT AND EFFECTS OF FERTILICATION UPON AQUATIC LIFE. IT IS
INDEXED FOR MAJOR CATEGORIES WITH NUMBERS OF CITATIONS INDICATED AS
FOLLOWS: ALGAE, 86; ANIMALS, 4; CHEMISTRY, 12; EUTROPHICATION, 5;
ARTIFICIAL FERTILIZATION, 32; FERTILIZATION BY DUCKS, 4; LAKES, 18;
MIDGES, 1; NITRATE POISONING, 8; NITROGEN FIXATION BY ALGAE, 7;
NITROGEN AND PHOSPHORUS, 194; ORGANISM STANDING CROPS, 11; HIGHER
AQUATIC PLANTS, 6; PRODUCTION, 12. DATES CITED RANGE FROM 1918 TO 1965
WITH 70 REFERENCES FROM 1960-1965. PUBLICATION INCLUDES A GENERAL
INTRODUCTION, SUBJECT INDEX, AUTHOR INDEX, AND LISTS OF DEFINITIONS AND
EQUIVALENTS AT BOTH BEGINNING AND END OF THE BIBLIOGRAPHY.
(EICHHORN-WIS)

FIELD 05C, 10

ACCESSION NO. W69-04805

OBSERVATIONS ON THE RESPONSE OF SOME BENTHONIC ORGANISM TO POWER STATION
  COOLING,

S. MARKOWSKI.

JOURNAL OF ANIMAL ECOLOGY, VOL 29, NO 2, 1960, PP 349-357.

    DESCRIPTORS:
      *POWERPLANTS, *THERMAL POLLUTION, *TEMPERATURE, *COOLING WATER,
      CHEMICAL PROPERTIES, ECOLOGY, SALINITY, BENTHOS, AQUATIC LIFE, ALGAE,
      PLANTS, WATER PROPERTIES, WATER TEMPERATURE, OUTLETS, HYDROELECTRIC
      PLANTS, ELECTRIC POWERPLANTS, ENGINEERING STRUCTURES, INDUSTRIAL
      PLANTS, STRUCTURES, AFTERBAYS, WATER TYPES, INTAKES.

    ABSTRACT:
      THE AUTHOR REPORTS OBSERVATIONS MADE ON THE SETTLING OF BENTHONIC
      ORGANISMS ON EXPERIMENTAL SLABS PLACED IN THE INTAKE AND IN THE OUTFALL
      WATER OF THE POWER STATION LOCATED AT THE CAVENDISH DOCK. DATA INDICATE
      THAT OUTFALL AREA IS MORE CONDUCIVE TO GROWTH AS THE BENTHONIC FORMS
      APPEAR EARLIER THAN IN THE INTAKE. NO ALGAE GROWTH WAS FOUND IN THE
      INTAKE, BUT THERE WAS VERY PROLIFIC GROWTH IN THE OUTFALL. SPECIFIC
      COMPOSITION OF BENTHONIC INVERTEBRATES WERE SIMILAR IN BOTH INTAKE AND
      OUTFALL SLABS. IN ADDITION, DENSER ANIMAL POPULATIONS WERE FOUND IN THE
      INTAKE THAN IN THE OUTFALL. GENERAL CHARACTERISTICS OF THE ENVIRONMENTS
      AND THE FACTORS INTRODUCED BY THE POWER STATION INTO THE AQUATIC MEDIUM
      ARE DISCUSSED.

    FIELD 05C

    ACCESSION NO. W69-05023

SEASONAL VARIATION IN CONTENT OF NITROGENOUS COMPOUNDS AND PHOSPHATE IN THE
  WATER OF TAKASUKA POND, SAITAMA, JAPAN,

  TOKYO IMPERIAL UNIV. (JAPAN). GEOGRAPHICAL INST.

  S. YOSHIMURA.

ARCHIV FUR HYDROBIOLOGIE, VOL 24, PP 155-176, 1932. 6 FIG, 13 TAB, 72 REF.

    DESCRIPTORS:
      *NITROGEN COMPOUNDS, *PHOSPHORUS COMPOUNDS, *CYCLING NUTRIENTS, WATER
      CHEMISTRY, WATER ANALYSIS, PHOSPHORUS, PONDS, PHYTOPLANKTON,
      EUTROPHICATION, LAKES, LIMNOLOGY, AQUATIC ENVIRONMENT, AQUATIC
      PRODUCTIVITY, HYPOLIMNION, ALGAL CONTROL, SEASONAL, ANNUAL TURNOVER.

    IDENTIFIERS:
      TAKASUKA POND(JAPAN), LAKE MENDOTA(WIS), PHYTOPLANKTON CROPS, TROPHIC,
      TROPHOLYTIC, NITROGEN ANNUAL BUDGET, PHOSPHORUS ANNUAL BUDGET.

    ABSTRACT:
      SMALL, EUTROPHIC TAKASUKA POND, JAPAN, HAS BEEN ANALYZED BOTH
      VERTICALLY AND SEASONALLY (1928 TO 1931) FOR NITRATE, FREE AMMONIA,
      ALBUMINOID NITROGEN AND SOLUBLE PHOSPHATE, AND THE DATA CORRELATED WITH
      PHYTOPLANKTON CYCLES. ORGANIC NITROGEN IS INVERSELY PROPORTIONAL TO
      INORGANIC NITROGEN. INORGANIC NITROGEN AND SOLUBLE PHOSPHATE ARE
      GENERATED AT THE END OF STAGNATION CAUSED BY DECOMPOSITION OF
      WARM-WATER PHYTOPLANKTON AND SOME AQUATIC PLANTS. SEASONAL CHANGE IN
      SOLUBLE PHOSPHATE IN INFLOWING WATERS AFFECTS THE PHYTOPLANKTON
      PRODUCTIVITY. DATA CONCERNING THE HYPOLIMNION IN THE STAGNANT PERIOD
      (APRIL TO END OF OCTOBER) ARE PRESENTED. CALCULATIONS ARE OFFERED FOR
      ANNUAL BUDGETS FOR NITROGEN AND PHOSPHORUS IN THE TROPHIC ZONE,
      TROPHOLYTIC ZONE, AND POND AS A WHOLE. COMPARISONS ARE MADE WITH LAKES
      AND PONDS OF THE WORLD, INCLUDING LAKE MENDOTA, WISCONSIN. THE LIMITING
      FACTOR FOR PHYTOPLANKTON CROP IN TAKASUKA IS SAID TO BE 'PHOSPHATE AT
      TIMES AND AMMONIA IN OTHERS.' (MCCOY-WISCONSIN)

    FIELD 02H, 05C

42

    ACCESSION NO. W69-05142

DEMONSTRATION OF A TOXIN FROM APHANIZOMENON FLOS-AQUAE (L.) RALFS,

NEW HAMPSHIRE UNIV., DURHAM; NATIONAL MARINE WATER QUALITY LAB., KINGSTON, R. I.

PHILIP J. SAWYER, JOHN H. GENTILE, AND JOHN J. SASNER, JR.

CANADIAN JOURNAL OF MICROBIOLOGY, VOL 14, NO 11, PP 1199-1204, 1968. 7 P, 1 TAB, 4 FIG, 23 REF. OWRR PROJECT A-013-NH.

DESCRIPTORS:
CYANAPHYTA, ALGAL CONTROL, *ALGAL TOXINS MEMBRANE, NORTHEAST U.S.

IDENTIFIERS:
AQUATIC ALGAE, WATER POLLUTION, *TOXICITY, BIOLOGICAL MEMBRANES, HUMID AREA.

ABSTRACT:
A POTENT TOXIN WAS EXTRACTED FROM A NATURAL POPULATION OF THE BLUE-GREEN ALGA, APHANIZOMENON FLOS-AQUAE. THE TOXIN IS THERMO- AND ACID-STABLE; ALKALINE LABILE; SOLUBLE IN WATER AND ETHANOL; INSOLUBLE IN ACETONE, ETHER, AND CHLOROFORM; AND READILY DIALYZABLE. PRELIMINARY STUDIES SHOW THAT THE ALGA CONTAINS A TOXIN WHICH, WHEN RELEASED FROM LYSED CELLS, OPERATES AT THE MEMBRANE LEVEL, DESTROYING EXCITABILITY WITHOUT ALTERATION OF THE TRANSMEMBRANE RESTING POTENTIAL.

FIELD 05C

ACCESSION NO.   W69-05306

NITROGEN CYCLE IN SURFACE AND SUBSURFACE WATERS,

WISCONSIN UNIV., MADISON.

S. WITZEL, E. MCCOY, O. J. ATTOE, L. B. POLKOWSKI, AND K. T. CRABTREE.

WATER RESOURCES CENTER, UNIV., OF WISCONSIN, TECHNICAL COMPLETION REPORT, DECEMBER 1968. 65 P, 15 TAB, 12 FIG, 27 REF. OWRR PROJECT B-004-WIS.

DESCRIPTORS:
*DOMESTIC ANIMALS, *WASTES, FERTILIZERS, *ESSENTIAL NUTRIENTS, *NITRIFICATION, CROPS, TOXICITY, SOIL POROSITY, IRRIGATION, GROUND WATER, RUNOFF, FROZEN GROUND, SOIL EROSION, WATER POLLUTION, *DENITRIFICATION, PHOSPHORUS COMPOUNDS, ALGAE, AQUATIC PLANTS.

IDENTIFIERS:
*NITROGEN CYCLE, *FARM WASTE, GROUND WATER, *NITRATES, EUTROPHICATION, *WATER POLLUTION SOURCES, AGRICULTURAL WATERSHEDS.

ABSTRACT:
AUTOTROPHIC AND HETEROTROPHIC NITRIFICATION HAVE BEEN STUDIED WITH 191 SAMPLES OF SHALLOW WATER FROM STREAMS, FARM PONDS AND ADJACENT SOILS. OF 47 CHOSEN FOR REPEATED TESTS, 45 PRODUCED NO SUB 2-N RANGING 5-154 MICROGRAM/ML, AV. 48 MICROGRAM/ML. ONLY 2 PRODUCED NO SUB 3-N AT 33 AND 46 MICROGRAM/ML. OF 167 STOCK CULTURES OF SOIL FUNGI, THE MAIN PRODUCERS WERE IN THE ASPERGILLUS FLAVUS-ORYZAE AND A. WENTII GROUPS (75% YIELDED 65-100 MICROGRAM/ML OF NO SUB 3-N) AND THE PENICILLIUM GENUS (21 OF 24 SPECIES YIELDED 7-19 MICROGRAM/ML OF NO SUB 3-N FROM NO SUB 2-N, NOT FROM ORGANIC N). NITRIFIERS OF NO SUB 2, NO SUB 3 TYPE WERE FOUND IN 5 OTHER GENERA. RESIDUAL NO SUB 3 FOLLOWING CROP MATURITY MIGRATES DOWNWARD FROM 12 INCHES TO AQUIFER DEPTH AT 20 FEET OVER WINTER ON WAUPUN AND PLAINFIELD SOILS, RESPECTIVELY. ONE WISCONSIN COMMUNITY HAD 86 WELLS WITH 34.5% UNSAFE CONTAINING HIGH NO SUB 3 AND ANOTHER HAD 550 WELLS WITH 1/3 UNSAFE. SURFACE WATERS RECEIVED LESS THAN 5.7 LB. N AND 2.53 LB. P PER A. IN FLOOD FLOWS FROM A 1346 A. WATERSHED; 3.62 LB. N AND 1.14 LB. P PER A. FROM 3 FARM AREAS TOTALING 246 A. LANCASTER PLOTS RECEIVING 15 TONS DAIRY COW WASTES PER A. LOST 19.8% MORE N AND 11.3% MORE P WHEN APPLIED ON FROZEN GROUND IN A YEAR OF HIGH WINTER RUNOFF.

FIELD 05B, 05G

ACCESSION NO.   W69-05323

THE EFFECTS OF ALGAE ON WATER QUALITY,

SYRACUSE UNIV., N. Y. DEPT. OF CIVIL ENGINEERING.

DANIEL F. JACKSON.

WATER QUALITY RES SYMP, LST PROC, NEW YORK STATE DEPT HEALTH, FEB 20, 1964.
23 P, 5 FIG, 2 TAB, 53 REF.

DESCRIPTORS:
*ALGAE, *WATER QUALITY, *EUTROPHICATION, COLOR, TASTE, ODOR, TOXINS,
DISSOLVED OXYGEN, LAKES, RIVERS, LIGHT PENETRATION, FLUORESCENCE,
PHOTOSYNTHESIS, BLUE-GREEN ALGAE, PHYTOPLANKTON, OHIO RIVER,
PENNSYLVANIA, WATER SUPPLY.

IDENTIFIERS:
PHOTOSYNTHETIC RATE, PHYTOPLANKTON DYNAMICS, EXTRACELLULAR PRODUCTS,
ALGAL CELLS.

ABSTRACT:
HISTORICAL OCCURRENCES OF COLORED WATER CAUSED BY ALGAL GROWTH ARE
CITED. INVESTIGATIONS (1797) REVEALED THAT VOLVOCALES CAUSED RED WATER,
WITH FIRST RECORDED STUDY ON PHYTOPLANKTONIC DEVELOPMENT (OSCILLATORIA)
CONDUCTED BY DECANDOLLE (1825). IN PYMATUNING RESERVOIR, PENNSYLVANIA,
CYANOPHYTES BLOOM ANNUALLY, RESTRICTING LIGHT PENETRATION TO LESS THAN
46 CM DURING JULY AND AUGUST, RETARDING GROWTH OF ROOTED AQUATIC
PLANTS, AND EXPOSING BOTTOM SOIL TO WAVE-ACTION EROSION AND CONSEQUENT
SUSPENSION OF NUTRIENT-RICH MATERIALS. TABULAR DATA ON ODORS AND TASTES
ARE DISCUSSED. AUTHOR DETERMINED PERCENTAGE OF VIABLE CELLS TO VALIDATE
COMMON ASSUMPTION THAT DEGREE OF ODOR IS ASSOCIATED WITH QUANTITY OF
ALGAL CELLS. AVERAGE VIABLE FRACTION FOR 1,700 MEASUREMENTS ON OHIO
RIVER SAMPLES (1960-61) WAS 59%; MEASUREMENTS WERE BASED ON CHLOROPHYLL
FLUORESCENCE. DEAD ALGAL CELLS, WHICH CREATE ODORS AND TASTE, SHOULD BE
DETERMINED IN PLANKTON ALGAE STUDIES. DISCUSSION OF ALGAL CONTRIBUTION
TO DISSOLVED OXYGEN INDICATES THAT EACH SPECIES HAS ITS OWN
MINIMUM/MAXIMUM PHOTOSYNTHETIC RATE INFLUENCED BY ENVIRONMENT, AND
TOTAL OXYGEN PRODUCTION FOLLOWS A PHYLOGENETIC HIERARCHY. LITERATURE
REVIEW AND EXPERIMENTS SHOW THAT ALGAE CONTRIBUTE SIGNIFICANT AMOUNTS
OF DISSOLVED OXYGEN. AUTHOR'S REVIEW OF ALGAL EXCRETION AND SECRETION
INCLUDES DISCUSSION OF TOXINS. HE DISCUSSES THEORETICAL EQUATION TO
EVALUATE PHYTOPLANKTON DYNAMICS REPORTED BY RILEY AND VON ARX (1949).
(KERRIGAN-WIS)

FIELD 05C

ACCESSION NO. W69-05697

USE OF POTASSIUM PERMANGANATE FOR CONTROL OF PROBLEM ALGAE,

WISCONSIN UNIV., MADISON. HYDRAULIC AND SANITARY LAB.

GEORGE P. FITZGERALD.

J AMER WATER WORKS ASSOC, VOL 58, NO 5, PP 609-614, 1966. 1 FIG, 1 TAB, 20 REF.

DESCRIPTORS:
*NUISANCE ALGAE, *ALGICIDES, *COPPER SULFATE, *WATER QUALITY CONTROL, ALGAL CONTROL, SLIME, TASTE-PRODUCING ALGAE, ODOR-PRODUCING ALGAE, DIATOMS, CYANOPHYTA, CHLOROPHYTA, IRON, MANGANESE, EUTROPHICATION, WATER POLLUTION CONTROL.

IDENTIFIERS:
*POTASSIUM PERMANGANATE, ALGICIDAL TESTS, ALGISTATIC TESTS, SLIMEPRODUCING ALGAE, PHOTOMICROGRAPHS, MICROCYSTIS AERUGINOSA, ANABAENA CIRCINALIS, GLOEOTRICHIA ECHINULATE, OSCILLATORIA RUBESCENS, OSCILLATORIA CHALYBIA, HYDRODICTYON RETICULATUM, DICTYOSPHAERIUM PULCHELLUM, GOMPHONEMA, DINOBRYON.

ABSTRACT:
AUTHOR TREATED FOLLOWING SPECIES OF ALGAE WITH VARIOUS CONCENTRATIONS OF POTASSIUM PERMANGANATE AND COPPER SULFATE TO DETERMINE CONCENTRATIONS REQUIRED TO KILL CELLS AFTER 12 AND 72 HOURS' TREATMENT AND TO PREVENT GROWTH: (BLOOMFORMING CYANOPHYTES) MICROCYSTIS AERUGINOSA, ANABAENA CIRCINALIS, GLOEOTRICHIA ECHINULATA, OSCILLATORIA RUBESCENS; (FILTER-CLOGGING CYANOPHYTE) OSCILLATORIA CHALYBIA; (FILTER-CLOGGING CHLOROPHYTES) HYDRODICTYON RETICULATUM, DICTYOSPHAERIUM PULCHELLUM; (DIATOM) PROBABLY GOMPHONEMA. TOXICITY OF COPPER SULFATE WAS HIGHLY VARIABLE; CYANOPHYTES EXCEPT OSCILLATORIA WERE KILLED AT CONCENTRATIONS OF 0.1 PPM; O RUBESCENS WAS KILLED AFTER 4 HOURS AT 0.4 PPM; GROWTH OF O CHALYBIA AND HYDRODICTYON WAS INHIBITED. BY 0.1-0.2 PPM BUT ALGICIDE REQUIRED 0.6-0.8 PPM FOR 72 HOURS; INHIBITION OF DICTYOSPHAERIUM AND THE DIATOM REQUIRED 0.4 PPM, WHILE ALGICIDE REQUIRED MORE THAN 8 PPM AND 72 HOURS. WITH POTASSIUM PERMANGANATE, ALL SPECIES EXCEPT DICTYOSPHAERIUM WERE KILLED BY 1-5 PPM. AUTHOR SPECULATES THAT VARIATIONS IN EFFECTIVENESS OF COPPER MAY RESULT FROM VARIATIONS IN METABOLIC RATES, METABOLIC PROCESSES, OR RATES OF ABSORPTION AMONG THE SEVERAL SPECIES. HE SUGGESTS THAT POTASSIUM PERMANGANATE TREATMENT OF RAW-WATER RESERVOIRS MAY BE MULTIPLY EFFECTIVE FOR ALGICIDE FOR ELIMINATION OF TASTES AND ODORS RESULTING FROM DECOMPOSED ALGAE AND OTHER SOURCES AND FOR REMOVAL OF IRON AND MANGANESE. (EICHHORN-WIS)

FIELD 05F

ACCESSION NO. W69-05704

LIST OF REFERENCES ON CONTROL OF AQUATIC PLANTS, INCLUDING ALGAE,

CHIPMAN CHEMICAL CO., INC., BOUND BROOK, N. J. RESEARCH DIV.

MARGARET GREENWALD.

(INTERNAL PUBLICATION), CHIPMAN CHEMICAL CO, INC, BOUND BROOK, NJ, JAN 1956. 22 P, 716 REF.

DESCRIPTORS:
*BIBLIOGRAPHIES, *ALGAL CONTROL, *AQUATIC WEED CONTROL, *EUTROPHICATION, AQUATIC PLANTS, HERBICIDES, PESTICIDES, HARVESTING, WATER QUALITY CONTROL, WATER POLLUTION CONTROL, WATER POLLUTION EFFECTS.

IDENTIFIERS:
BIOCIDES.

ABSTRACT:
ALTHOUGH THIS BIBLIOGRAPHY COMPRISES SOME 716 ENTRIES, IN ITS PREFACE THE COMPILER STATES THAT IT IS NEITHER COMPLETE NOR SELECTED. HOWEVER, EMPHASIS WAS PLACED ON RECENT WORKS. 263 OF THE DATED WORKS WERE PUBLISHED BETWEEN 1950 AND 1966. EARLIEST CITATION IS DATED 1896, AND THE LIST IS ALMOST EXCLUSIVELY FROM THE LITERATURE IN ENGLISH. IN MOST CASES, BIOLOGICAL AND ECOLOGICAL STUDIES WERE OMITTED, BUT SOME HANDBOOKS, GENERAL TEXTS, AND A FEW MISCELLANEOUS REFERENCES ARE INCLUDED WHERE THEY HAVE BEEN USEFUL IN PLANT IDENTIFICATION OR THE UNDERSTANDING OF CONDITIONS ASSOCIATED WITH PLANT CONTROL OR GIVE INFORMATION ON PROBLEMS RELATED TO PLANT CONTROL (E.G., TOXICITY OF HERBICIDES TO ANIMAL ORGANISMS). (EICHHORN-WIS)

FIELD 05G, 04A

ACCESSION NO. W69-05705

SUPPLEMENT TO THE LIST OF REFERENCES ON CONTROL OF AQUATIC PLANTS, INCLUDING ALGAE,

CHIPMAN CHEMICAL CO., INC., BOUND BROOK, N. J. RESEARCH DIV.

MARGARET GREENWALD.

(INTERNAL PUBLICATION), CHIPMAN CHEMICAL CO, INC, BOUND BROOK, N. J., JAN 1957. 12 P, 258 REF.

DESCRIPTORS:
*BIBLIOGRAPHIES, *ALGAL CONTROL, *AQUATIC WEED CONTROL, *EUTROPHICATION, AQUATIC PLANTS, HERBICIDES, PESTICIDES, HARVESTING, WATER QUALITY CONTROL, WATER POLLUTION CONTROL, WATER POLLUTION EFFECTS, FUNGI, AQUATIC FUNGI.

IDENTIFIERS:
BIOCIDES.

ABSTRACT:
THIS SUPPLEMENT RECORDS CORRECTIONS TO COMPILER'S ORIGINAL BIBLIOGRAPHY, AND ADDS 258 ENTRIES TO THE ORIGINAL 716. ORIGINAL COVERAGE OF AQUATIC PLANTS AND ALGAE IS EXTENDED TO INCLUDE A FEW REFERENCES TO AQUATIC FUNGI. NEW CITATIONS ARE FOR LITERATURE PUBLISHED BETWEEN 1922 AND 1957, 134 HAVING BEEN PUBLISHED DURING THE PERIOD 1950-1957. (EICHHORN-WIS)

FIELD 05G, 04A

ACCESSION NO. W69-05706

NEARSHORE PHYTOPLANKTON POPULATIONS IN THE GRAND HAVEN, MICHIGAN VICINITY DURING THERMAL BAR CONDITIONS,

MICHIGAN UNIV., ANN ARBOR. GREAT LAKES RESEARCH DIV.

E. F. STOERMER.

PROC. 11TH CONF. GREAT LAKES RES., VOL 11, PP 137-150, APR 1968. 14 P, 13 FIG, 1 TAB, 17 REF.

DESCRIPTORS:
*PHYTOPLANKTON, *LAKE MICHIGAN, *WATER CIRCULATION, NUISANCE ALGAE, THERMAL POLLUTION, DIFFUSION, DIATOMS, CHLOROPHYTA, CHRYSOPHYTA, DINOFLAGELLATES, CYANOPHYTA.

IDENTIFIERS:
*THERMAL BAR.

ABSTRACT:
IN LATE APRIL 1967, THE THERMAL BAR PHENOMENON HAD PROFOUND EFFECTS ON BOTH THE ABUNDANCE AND SPECIES COMPOSITION OF PHYTOPLANKTON IN NEARSHORE LAKE MICHIGAN WATERS IN THE VICINITY OF GRAND HAVEN, MICHIGAN. LANDWARD OF THE THERMAL BAR POPULATIONS AVERAGED BETWEEN 1500 AND 2000 CELLS/ML COMPARED WITH AVERAGE VALUES OF ABOUT 400 CELLS/ML IN WATER ENTERING THE LAKE FROM THE GRAND RIVER AND ABOUT 350 CELLS/ML IN WATER OUTSIDE THE THERMAL BAR. THE HIGHEST POPULATIONS OBSERVED, OCCURRED IN THE INTERFACE BETWEEN INSHORE AND OFFSHORE WATERS. THE RELATIVE ABUNDANCE OF MOST SPECIES OF ALGAE WAS STRIKINGLY DISSIMILAR IN THE DIFFERENT WATER MASSES. FOUR GENERAL TYPES OF SPECIES DISTRIBUTION WERE NOTED: (1) SPECIES WHICH WERE ABUNDANT IN THE HARBOR FLORA, RARE IN THE INSHORE WATERS, AND VERY RARE OR LACKING IN THE OFFSHORE WATERS; (2) SPECIES WHICH WERE ABUNDANT IN THE INSHORE WATERS, RARE IN THE HARBOR FLORA, AND RARE OR LACKING IN THE OFFSHORE WATERS; (3) SPECIES WHICH WERE ABUNDANT IN THE OFFSHORE WATERS, RARE IN THE INSHORE AREA AND LACKING IN THE HARBOR FLORA; (4) SPECIES WHICH WERE MOST ABUNDANT IN THE INTERFACE WATER WITH SECONDARY ABUNDANCE PEAKS IN EITHER OF THE OTHER HABITATS. IT APPEARS THAT THE PRIMARY EFFECT OF THERMAL ADDITIONS TO LAKE MICHIGAN IS THROUGH CONSTRAINT ON CIRCULATION WHICH TRAPS NUTRIENTS NEAR SHORE.

FIELD 02H, 05C

ACCESSION NO. W69-05763

DEATH OF A LAKE,

DAVID C. FLAHERTY.

QUEST, VOL 6, NO 4, PP 6-15, DEC 1968. 10 P, 9 FIG.

DESCRIPTORS:
*LAKES, *EUTROPHICATION, FERTILIZATION, FISH, *LIMNOLOGY, FISHKILL,
NUTRIENTS, *ALGAE, *ALGAL CONTROL, ALGAL POISONING, ALGICIDES, AQUATIC
ALGAE, AQUATIC ENVIRONMENT, AQUATIC PLANTS, AQUATIC POPULATIONS,
CONSERVATION, WATER QUALITY, WATER POLLUTION, WATER RESOURCES.

IDENTIFIERS:
ALGAL NUTRIENTS.

ABSTRACT:
THE NATURAL PROCESS OF LAKE EUTROPHICATION IS BEING ACCELERATED BY THE
ACTIVITIES OF MAN. SEWAGE DISCHARGES, WASTE EFFLUENTS, AND AGRICULTURAL
FERTILIZERS DRAINING INTO A LAKE PRODUCE AN OVERABUNDANCE OF NUTRIENTS
IN THE LAKE. THE ABUNDANCE OF NUTRIENTS GREATLY STIMULATES THE GROWTH
OF ALGAE AND OTHER WATER PLANTS, BUT SETS OFF A DESTRUCTIVE CHAIN
REACTION. AT FIRST, THE ABUNDANT FOOD CHAIN WILL BENEFIT THE FISH, BUT
IN SEVERE CASES, THE ALGAE MAY GROW IN SUCH NUMBERS THAT WHEN THEY DIE,
THEY DEPLETE THE OXYGEN SUPPLY AND RELEASE TOXIC SUBSTANCES. THE RESULT
MAY BE A LOSS OF A SPARKLING FRESH LAKE, SLIME INFESTATION, BAD FLAVOR
AND ODOR, DISCOLORATION OF WATER, STAGNATION, AND MASSIVE FISHKILL.
CLEAN-WATER AND POLLUTED-WATER ALGAE ARE DESCRIBED AND METHODS FOR
ALGAE CONTROL ARE COVERED. LIMNOLOGY STUDIES AT WILLIAMS LAKE AND ROCK
LAKE, WASH, ARE DISCUSSED. MANY BODIES OF WATER APPEAR DESTINED TO DIE
LONG BEFORE THEIR NATURAL TIME, PRODUCING AN ECONOMIC LOSS AMOUNTING TO
MILLIONS OF DOLLARS. THE LOSS IN AESTHETIC VALUES OF LAKES CANNOT BE
MEASURED. (USBR)

FIELD 05C, 02H

ACCESSION NO. W69-05844

AVAILABILITY OF IRON AND MANGANESE IN SOUTHERN WISCONSIN LAKES FOR THE GROWTH
OF MICROCYSTIS AERUGINOSA,

WISCONSIN UNIV., MADISON. DEPT. OF BOTANY.

GERALD C. GERLOFF, AND FOLKE SKOOG.

ECOLOGY, VOL 38, NO 4, PP 551-556, OCT 1957. 7 TAB, 10 REF.

DESCRIPTORS:
*EUTROPHICATION, *IRON, *LAKES, *MANGANESE, *WISCONSIN, ALGAE, CALCIUM,
CHELATION, CYANOPHYTA, NUISANCE ALGAE, NUTRIENT REQUIREMENTS,
NUTRIENTS, TOXICITY, WATER CHEMISTRY, WATER POLLUTION EFFECTS.

IDENTIFIERS:
*NUTRIENT AVAILABILITY, *ALGAL GROWTH, *MICROCYSTIS AERUGINOSA,
ANTAGONISMS, ALGAL NUTRITION, ALGAL PHYSIOLOGY, LAKE MENDOTA(WIS), LAKE
MONONA(WIS), LAKE WAUBESA(WIS), SPAULDING POND(WIS), LAKE DELAVAN(WIS),
PEWAUKEE LAKE(WIS), NORTH LAKE(WIS), LAKE NAGAWICKA(WIS).

ABSTRACT:
CELLULAR CONTENTS WERE COMPARED OF IRON AND MANGANESE IN MICROCYSTIS
AERUGINOSA, COLLECTED FROM BLOOMS IN SEVEN SOUTHERN WISCONSIN LAKES,
WITH CRITICAL LEVELS OF THESE ELEMENTS FOR MAXIMUM GROWTH AS DETERMINED
IN THE LABORATORY. CRITICAL LEVELS WERE APPROXIMATELY 100 PPM IRON AND
4 PPM MANGANESE. RANGE OF ALGAL IRON CONTENT FOR LAKES SAMPLED WAS
164-543 PPM; MANGANESE, 13-232 PPM. FOR 36 SAMPLES FROM EUTROPHIC LAKE
MENDOTA COLLECTED AT SURFACE AND AT 10 AND 0.5 METERS ABOVE BOTTOM AT
12 DATES DURING ALL SEASONS FOR ONE YEAR, IRON CONCENTRATIONS IN WATER
HAD RANGE 0.000-0.083 PPM; MANGANESE, 0.035-0.673 PPM. CELLULAR
CONTENTS OF IRON AND MANGANESE WERE SUFFICIENTLY IN EXCESS OF SUCH
CRITICAL VALUES THAT IT SEEMS UNLIKELY THAT EITHER IS IMPORTANT IN
DEVELOPMENT OF MICROCYSTIS BLOOMS IN LAKES. ANTAGONISTIC RELATIONSHIPS
BETWEEN MANGANESE AND IRON AND BETWEEN CALCIUM AND MANGANESE WERE
DEMONSTRATED AND CONTRIBUTED TO DIFFICULTIES IN DETERMINING LEVELS OF
IRON AND MANGANESE REQUIRED FOR NUTRITION. SUCH ANTAGONISMS WERE
CONSIDERED AS INSUFFICIENTLY SERIOUS TO SIGNIFICANTLY AFFECT ACCURACY
OF THEIR STUDIES AND CONCLUSIONS DERIVED THEREFROM. MANGANESE APPEARS
TO ATTAIN CONCENTRATIONS IN LAKE WATER SUFFICIENTLY HIGH TO BE TOXIC TO
ALGAE WERE IT NOT FOR ANTAGONISM OF CALCIUM, ALSO PRESENT IN RELATIVELY
HIGH CONCENTRATIONS. (EICHHORN-WISC)

FIELD 05C, 02H

ACCESSION NO. W69-05867

CELL CONTENTS OF NITROGEN AND PHOSPHORUS AS A MEASURE OF THEIR AVAILABILITY FOR GROWTH OF MICROCYSTIS AERUGINOSA,

WISCONSIN UNIV., MADISON. DEPT OF BOTANY.

GERALD C. GERLOFF, AND FOLKE SKOOG.

ECOLOGY, VOL 35, NO 3, PP 348-353, JULY 1954. 5 FIG, 7 TAB, 8 REF.

DESCRIPTORS:
*CHEMICAL ANALYSIS, *NITROGEN, *PHOSPHORUS, ALGAE, CYANOPHYTA, NITRATES, NITROGEN COMPOUNDS, NUISANCE ALGAE, NUTRIENT REQUIREMENTS, NUTRIENTS, PHOSPHORUS COMPOUNDS, WATER CHEMISTRY, WATER POLLUTION EFFECTS, WATER QUALITY, WISCONSIN, LAKES, EUTROPHICATION.

IDENTIFIERS:
*CHEMICAL COMPOSITION, *MICROCYSTIS AERUGINOSA, *NUTRIENT AVAILABILITY, ALGAL GROWTH, ALGAL NUTRITION, ALGAL PHYSIOLOGY, KJELDAHL PROCEDURE, LUXURY CONSUMPTION, NUTRIENT REMOVAL, SPAULDING'S POND(WIS), LAKE WAUBESA(WIS), GREEN LAKE(WIS).

ABSTRACT:
QUANTITATIVE RELATIONSHIPS WERE STUDIED BETWEEN TOTAL NITROGEN AND PHOSPHORUS CONTENTS OF MICROCYSTIS AERUGINOSA GROWN IN NUTRIENT SOLUTIONS CONTAINING THESE ELEMENTS IN QUANTITIES VARYING FROM HIGHLY DEFICIENT TO EXCESSIVE. CELLULAR CONTENT OF NITROGEN OR PHOSPHORUS INCREASES WITH EXTERNAL SUPPLY OVER A WIDE RANGE, FOR A MAJOR PORTION OF WHICH THE AMOUNT OF GROWTH REMAINED PRACTICALLY CONSTANT. THUS, ABOVE FAIRLY CONSTANT CRITICAL LEVELS OF NITROGEN AND PHOSPHORUS REQUIRED FOR MAXIMUM YIELD, LARGE CELLULAR EXCESSES ACCUMULATED, REPRESENTING LUXURY CONSUMPTION. ALTHOUGH NOT INFLUENCING GROWTH, THESE EXCESSES REFLECT ABUNDANCE OF EXTERNAL SUPPLIES. NITROGEN AND PHOSPHORUS CONTENTS OF MICROCYSTIS COLLECTED FROM BLOOMS IN THREE WISCONSIN LAKES, WHEN COMPARED WITH CULTURES GROWN IN LABORATORY, SHOWED THAT CELLULAR CONTENTS OF THESE ELEMENTS IN COLLECTED SAMPLES WERE IN RANGE OF LUXURY CONSUMPTION AND SUFFICIENTLY ABOVE CRITICAL LEVELS TO SUGGEST THAT NEITHER PHOSPHORUS NOR NITROGEN LIMITED ALGAL GROWTH IN THESE LAKES. NITROGEN OCCURRED IN RELATIVELY SMALLER EXCESS THAN PHOSPHORUS. CELLULAR EXCESSES WERE SUFFICIENT TO PERMIT CONSIDERABLE GROWTH WHEN ALGAE WERE TRANSFERRED TO MEDIA DEFICIENT IN THESE NUTRIENTS. EFFICACY OF PHOSPHORUS REMOVAL FROM CULTURAL EFFLUENTS MAY BE DOUBTFUL UNLESS LEVELS ARE ALSO MUCH REDUCED, BOTH IN OTHER INFLOWING WATER AND IN 'SINKS' WITHIN LAKES. (EICHHORN-WISC)

FIELD 05C, 02H

ACCESSION NO. W69-05868

USE POLLUTION TO BENEFIT MANKIND,

VITRO/HANFORD ENGINEERING SERVICES, RICHLAND, WASH.

J. LEON POTTER.

OCEAN IND, VOL 4, NO 5, PP 94-97, MAY 1969. 4 P, 1 FIG, 2 TAB, 4 REF.

DESCRIPTORS:
*WASTE TREATMENT, *AQUICULTURE, ALGAE, YEASTS, BACTERIA, FISH, PROTEINS, OIL WASTES, OILY WATER, SULFITE LIQUORS, ULTIMATE DISPOSAL, WATER REUSE, FISH FARMING.

IDENTIFIERS:
*OIL SPILLS, PROTEIN AQUICULTURE.

ABSTRACT:
ORGANIC WASTE, HEATED WATER, SEWAGE NUTRIENTS, OIL WASTES, AND GARBAGE MAY BE USED AS NUTRIENTS FOR GROWING ALGAE, BACTERIA, AND YEASTS TO PRODUCE PROTEINS. WASTE LOW-GRADE PETROLEUM IS USED IN A PLANT AT LAUERA, FRANCE. A POUND OF BACTERIA, FEEDING ON CRUDE SO WORTHLESS IT IS USUALLY BURNED, CAN PRODUCE 10 LBS OF PROTEIN A DAY. WASTE EFFLUENTS RICH IN CARBOHYDRATES, SUGARS AND INORGANIC NUTRIENTS MAY BE COLLECTED AND SEPARATED INTO SOLIDS AND LIQUIDS. THE SOLIDS ARE GOOD FERTILIZERS AND ANIMAL FEED ADDITIVES, WHILE THE SOLUBLE NUTRIENTS MAY BE USED BY TORULA YEAST FOR PROTEIN PRODUCTION. HEATED WATER DISCHARGES MAY BE USED TO INCREASE THE PRODUCTION OF FISH. THE COST OF PROTEIN PRODUCED FROM WASTES IS ABOUT 3 CENTS PER LB, COMPARED WITH FARM-GROWN PROTEIN PRICES OF ABOUT 10 CENTS. ANALYSES OF TORULA YEAST AND ALGAE ARE TABULATED. (KNAPP-USGS)

FIELD 05D, 05E

ACCESSION NO. W69-05891

RIVER WATER QUALITY FOR ARTIFICIAL RECHARGE,

NEBRASKA UNIV., LINCOLN. DEPT. OF CIVIL ENGINEERING.

LEON S. DIRECTO, AND MERLIN E. LINDAHL.

J AMER WATER WORKS ASS, VOL 61, NO 4, PP 175-180, APR 1969. 6 P, 8 FIG, 1 TAB, 8 REF. RES SUPPORTED BY NEBR UNIV. WATER RESOURCES CENTER.

DESCRIPTORS:
*WATER QUALITY, *ARTIFICIAL RECHARGE, *DATA COLLECTIONS, PIT RECHARGE, WATER SUPPLY, MUNICIPAL WATER, SEDIMENT LOAD, CONDUCTIVITY, CALCIUM, MAGNESIUM, ALKALINITY, SULFATES, CHLORIDES, SODIUM, IRON, DISSOLVED SOILDS, PHOSPHATES, NITROGEN, ALGAE, COLIFORMS.

IDENTIFIERS:
PLATTE RIVER, LINCOLN(NEBR).

ABSTRACT:
THE PLATTE RIVER WAS SAMPLED IN 1966-67 AT 3 LOCATIONS NEAR ASHLAND, NEBRASKA TO INVESTIGATE SEASONAL VARIATIONS IN ITS PHYSICAL, CHEMICAL, AND BIOLOGICAL QUALITIES IN A STUDY OF THE FEASIBILITY OF USING THE RIVER WATER FOR RECHARGE OF THE AQUIFER SUPPLYING THE WELLS FOR THE CITY OF LINCOLN. ALL THE SAMPLES WERE TAKEN USING THE EQUAL TRANSIT RATE METHOD OF DEPTH INTEGRATION. STREAMFLOW DATA WERE OBTAINED FROM USGS GAGING STATION RECORDS. THE MEAN SPECIFIC CONDUCTANCE OF 439 MICROMHOS/CM WAS EXCEEDED 60% OF THE TIME. THE MEAN SOLIDS CONCENTRATION WAS 1090 MG/1. THIS IS EXPECTED TO CAUSE CLOGGING PROBLEMS IN USE OF THE WATER FOR RECHARGE UNLESS SOLIDS ARE REMOVED. THE SODIUM ABSORPTION RATIO AVERAGED 1.05 MG/1 AND IRON CONCENTRATION 0.41 MG/1, NEITHER OF WHICH SHOULD CAUSE ANY PROBLEM. PHOSPHATE CONCENTRATION AVERAGED 1.07 MG/1 WITH AN ALGAE COUNT OF 910, WHICH COULD RESULT IN HEAVY ALGAL GROWTH ON SAND RECHARGE BEDS. COLIFORM DENSITY WAS 6,610/100 ML AT ASHLAND. ALL THE COLLECTED DATA, INCLUDING TEMPERATURE, PH, ALKALINITY, AND HARDNESS ARE SHOWN GRAPHICALLY. RIVER WATER IS SIMILAR IN CHEMICAL QUALITY TO THE GROUNDWATER IN THE WELL FIELD, AND SHOULD CAUSE NO COMPATIBILITY PROBLEMS. THE HIGH SOLIDS, ALGAE, AND COLIFORM CONCENTRATIONS MAY CAUSE CLOGGING. (KNAPP-USGS)

FIELD 05B, 04B

ACCESSION NO. W69-05894

LIMNOLOGICAL INVESTIGATIONS OF TEXAS IMPOUNDMENTS FOR WATER QUALITY MANAGEMENT PURPOSES - THE USE OF ALGAL CULTURES TO ASSESS THE EFFECTS OF NUTRIENT ENRICHMENT ON THE HIGHLAND LAKES OF THE COLORADO RIVER, TEXAS,

TEXAS UNIV., AUSTIN.

BERT A. FLOYD, E. GUS FRUH, AND ERNST M. DAVIS.

TEXAS CENTER FOR RESEARCH IN WATER RESOURCES, TECHNICAL PROJECT REPORT NO 2, JANUARY, 1969. 45 P, 3 TAB, 4 FIG, 23 REF. OWRR PROJECT B-020-TEX.

DESCRIPTORS:
HIGHLAND LAKES, COLORADO RIVER, TEXAS, *NUTRIENT ENRICHMENT TESTS, *NITROGEN, *PHOSPHORUS, *IRON, *LIMNOLOGY, *ALGAE.

ABSTRACT:
THIS STUDY WAS AN INVESTIGATION OF THE EFFECTS OF INORGANIC NITROGEN, PHOSPHORUS, AND IRON ON ALGAL GROWTH IN THE HIGHLAND LAKES OF THE COLORADO RIVER, TEXAS. RIVER WATER FROM THE UPPER AND LOWER REACHES OF THE RESERVOIR SYSTEM WAS USED AS MEDIA FOR UNIALGAL CULTURES IN LABORATORY CONTROLLED GROWTH EXPERIMENTS. DIFFERENCES IN GROWTH WERE MEASURED BETWEEN A CONTROL AND SAMPLES ENRICHED WITH THE THREE NUTRITIVE SUBSTANCES. THESE MEASUREMENTS SHOWED INCREASED ALGAL GROWTH WITH THE ADDITION OF NITROGEN AND PHOSPHORUS, BUT NO INCREASE WITH EITHER OF THE THREE ELEMENTS SEPARATELY. ALSO, DIFFERENCES IN GROWTH AND NUTRIENT REQUIREMENTS WERE FOUND BETWEEN GREEN AND BLUE-GREEN SPECIES. FINALLY, AN INCREASE IN THE FERTILITY OF THE WATER AS IT FLOWS THROUGH THE RESERVOIR SYSTEM WAS NOTED.

FIELD 05C

ACCESSION NO. W69-06004

MEASURING ESTUARINE POLLUTION,

FEDERAL WATER POLLUTION CONTROL ADMINISTRATION, WASHINGTON, D. C. ESTUARINE
AND OCEANOGRAPHIC PROGRAMS BRANCH.

T. A. WASTLER.

OCEANOLOGY INT, PP 43-45, MAY-JUNE 1969. 3 P, 3 PHOTO.

DESCRIPTORS:
*ESTUARIES, *ESTUARINE ENVIRONMENT, *INSTRUMENTATION, *POLLUTION
IDENTIFICATION, *POLLUTION ABATEMENT, TIDAL WATERS, MIXING, SEA WATER,
FRESH WATER, SHORES, WATER CIRCULATION, FLOODS, DENSITY DURRENTS,
MEASUREMENTS, SEWAGE EFFLUENTS, TEMPERATURE, OXYGEN, SALINITY, ALGAE,
MECHANICAL EQUIPMENT.

IDENTIFIERS:
*ESTUARINE POLLUTION.

ABSTRACT:
THIS ARTICLE DESCRIBES BRIEFLY THE BASIC ELEMENTS OF ESTUARY POLLUTIONS
AND THE COMMON TECHNIQUES USED FOR THE DETECTION AND MAPPING OF THESE
POLLUTANTS. IN GENERAL THE STUDY SHOWS THAT ACTUAL TECHNIQUES OF DATA
COLLECTION IN THE ESTUARINE ZONE MUST BE DESIGNED TO COPE WITH THE
UNIQUE DEMANDS OF THE ESTUARINE ENVIRONMENT. THE INCREASING
SOPHISTICATION OF INSTRUMENTATION AND DATA ANALYSIS PROCEDURES,
COMBINED WITH THE NATIONAL CONCERN ABOUT ESTUARINE POLLUTION, OFFERS
HOPE, ACCORDING TO THE AUTHOR, THAT A VIABLE NATIONAL BASIC DATA
COLLECTION SYSTEM MAY BE DEVELOPED SOON TO SURVEY THE COASTAL ZONES.
(GABRIEL-USGS)

FIELD 05A

ACCESSION NO. W69-06203

RAIN AS A SOURCE OF VITAMIN (B-12),

WASHINGTON UNIV., ST LOUIS, MO. DEPT. OF BOTANY.

BRUCE C. PARKER.

NATURE, VOL 219, NO 5154, PP 617-618, AUG 10, 1968. 1 TAB, 15 REF.

DESCRIPTORS:
*RAIN, *VITAMINS, ALGAE, BIOASSAY, BIOINDICATORS, CHLAMYDOMONAS,
CARBON, ECOSYSTEMS, EUGLENA, EUTROPHICATION, ORGANIC MATTER,
PHYTOPLANKTON, PRIMARY PRODUCTIVITY, SOIL CHEMISTRY, WATER CHEMISTRY,
WATER POLLUTION EFFECTS, WATER POLLUTION SOURCES.

IDENTIFIERS:
*VITAMIN B-12, ALGAL GROWTH, CYANOCOBALAMIN, EUGLENA GRACILIS Z,
MICROBIOLOGICAL ASSAY, SPECIES COMPOSITION.

ABSTRACT:
FORMS AND CONCENTRATIONS OF VITAMIN B-12 IN NATURAL AQUATIC ECOSYSTEMS
ARE IMPORTANT IN THAT THEY MAY LIMIT PRIMARY PRODUCTIVITY AND INFLUENCE
THE SPECIES COMPOSITION OF PHYTOPLANKTONIC POPULATIONS. AUTHOR REPORTS
STUDIES STEMMING FROM THE OBSERVATION THAT AFTER SPRING RAINS,
PHYTOPLANKTONIC CHLAMYDOMONAS BECAME DOMINANT ORGANISMS IN SMALL
EXPERIMENTAL PONDS CONCOMMITANTLY WITH INCREASES IN THEIR VITAMIN
CONTENT. RAIN, COLLECTED IN STAINLESS STEEL CONTAINERS, WAS FILTERED
THROUGH MILLIPORE FILTERS (PORES, 2 MICRONS), AND ASSAYED FOR VITAMIN
B-12 MICROBIOLOGICALLY UTILIZING EUGLENA GRACILIS, Z STRAIN. IN NINE
COLLECTIONS TAKEN DURING 13 MONTHS, CONCENTRATIONS RANGED FROM
NEGLIGIBLE TO 20.0 PICOGRAMS/MILLILITER. FRAGMENTARY EVIDENCE SUPPORTS
THE HYPOTHESIS THAT VITAMIN ORIGINATES FROM SOIL PARTICLES BORNE INTO
ATMOSPHERE. AUTHOR POINTS OUT THAT A HYPOTHETICAL LAKE CONTAINING
EUGLENA, AND RICH IN ALL NUTRIENTS EXCEPT B-12, WOULD SUPPORT GROWTH OF
A MILLION CELLS/SQUARE CENTIMETER OF LAKE SURFACE AFTER A RAIN OF 1
CENTIMETER CONTAINING 20 PICOGRAMS OF VITAMIN/MILLILITER. (EICHHORN-WIS)

FIELD 05B

ACCESSION NO. W69-06273

HARVESTING UNDERWATER WEEDS,

WISCONSIN UNIV., MADISON. DEPT. OF MECHANICAL ENGINEERING.

DONALD F. LIVERMORE.

WATER WORKS ENG, PP 118-120, 151, FEB 1954. 4 FIG, 4 REF.

DESCRIPTORS:
*AQUATIC WEED CONTROL, *EUTROPHICATION, *HARVESTING, AQUATIC PLANTS,
AQUATIC WEEDS, ECONOMIC FEASIBILITY, LAKES, NUISANCE ALGAE, WATER
POLLUTION CONTROL, NUTRIENTS, WATER POLLUTION EFFECTS, WATER POLLUTION
SOURCES, WISCONSIN.

IDENTIFIERS:
LAKE MENDOTA(WIS), LAKE MONONA(WIS), LAKE KEGONSA(WIS), LAKE
WAUBESA(WIS), MADISON(WIS), UNIVERSITY OF WISCONSIN, YAHARA RIVER(WIS),
NUTRIENT REMOVAL.

ABSTRACT:
ONE APPROACH TO EUTROPHICATION CONTROL IS REPEATED HARVESTING OF A
NUTRIENT-LADEN CROP, ITS REMOVAL FROM THE WATER, AND ITS UTILIZATION,
TRANSFORMATION, OR TRANSFER FROM THE THREATENED DRAINAGE BASIN. AUTHOR
PRESENTS DESIGN FOR A MECHANICAL HARVESTER, DEVELOPED AT UNIVERSITY OF
WISCONSIN, WHICH WILL CUT AND COLLECT UNDERWATER WEEDS. HE DESCRIBES
OPERATION OF PROTOTYPE MODEL WHICH WAS FUNCTIONAL TO DEPTH OF FIVE
FEET, OPERATED AT SPEEDS OF 1-2 MILES/HOUR, AND COULD REMOVE OVER FOUR
TONS OF DRAINED WEEDS/HOUR. COST OF OPERATING MACHINERY OF THIS TYPE
WOULD NECESSARILY BE QUITE HIGH. AUTHOR ESTIMATES, FOR ERA WHEN THIS
PAPER WAS WRITTEN, THAT INITIAL INVESTMENT FOR ONE HARVESTING MACHINE,
THREE 'SHUTTLE' BARGES FOR TRANSPORTATION OF HARVESTED PLANTS, AND
DOCKS AND UNLOADING FACILITIES WOULD BE IN RANGE OF $20,000 AND THAT AN
ANNUAL OPERATING BUDGET OF $10,000-$15,000 WOULD BE REQUIRED TO COVER
FIXED AND OPERATING CHARGES. MECHANICAL HARVESTING OF MACROPHYTES WOULD
NOT ONLY REMOVE PLANT NUTRIENTS BUT WOULD ALSO ELIMINATE SOME
UNDESIRABLE END EFFECTS OF EUTROPHICATION BY CLEARING BOATING AND
SWIMMING AREAS AND BY ELIMINATING SCUM AREAS WHICH DEVELOP FROM
ENTRAPMENT BY EMERGENT PLANTS OF ALGAE AND OTHER FLOATING DEBRIS.
AUTHOR SUGGESTS THAT HARVESTED LACUSTRINE PLANTS MAY HAVE POTENTIAL
COMMERCIAL VALUE. (EICHHORN-WIS)

FIELD 05G, 04A

ACCESSION NO. W69-06276

COBALT AS AN ESSENTIAL ELEMENT FOR BLUE-GREEN ALGAE,

WISCONSIN UNIV., MADISON. DEPT. OF BOTANY.

OSMUND HOLM-HANSEN, GERALD C. GERLOFF, AND FOLKE SKOOG.

PHYSIOLOGIA PLANTARUM, VOL 7, PP 665-675, 1954. 3 FIG, 5 TAB, 8 REF.

DESCRIPTORS:
*COBALT, *CYANOPHYTA, *ESSENTIAL NUTRIENTS, *TRACE ELEMENTS, ALGAE,
EUTROPHICATION, MOLYBDENUM, NITROGEN FIXATION, NUISANCE ALGAE,
NUTRIENTS, VITAMINS, WATER POLLUTION EFFECTS, WATER POLLUTION SOURCES,
VITAMIN B.

IDENTIFIERS:
*CYANOCOBALAMIN, ALGAL GROWTH, ALGAL NUTRITION, ALGAL PHYSIOLOGY,
CALOTHRIX PARIETINA, COCCOCHLORIS PENIOCYSTIS, DIPLOCYSTIS AERUGINOSA,
NOSTOC MUSCORUM, VANADIUM, VITAMIN B-12.

ABSTRACT:
AUTHORS, EXPERIMENTING WITH NOSTOC MUSCORUM, DEMONSTRATED THAT COBALT
IS AN ESSENTIAL NUTRIENT ELEMENT FOR THIS CYANOPHYTE AS WELL AS FOR
THREE ADDITIONAL SPECIES: NITROGEN-FIXING CALOTHRIX PARIETINA; AND THE
NON-FIXERS, COCCOCHLORIS PENIOCYSTIS AND DIPLOCYSTIS AERUGINOSA. IN
NOSTOC, QUANTITATIVE REQUIREMENT IS QUITE LOW, INCREASES IN YIELD AND
NITROGEN CONTENT OBTAINING FROM COBALT CONCENTRATIONS FROM 0.002 TO
0.40 MICROGRAMS/LITER, THE LATTER LEVEL ORDINARILY REQUIRED FOR OPTIMUM
GROWTH. THESE DATA AND OTHER EVIDENCE INDICATE THAT THE FOLLOWING
ESSENTIALITY SPECIFICATIONS HAVE BEEN MET: ELEMENT HAS POSITIVE EFFECT
ON TOTAL GROWTH; IT EXERTS A POSITIVE PHYSIOLOGICAL EFFECT ON ALGAE; IT
IS NOT REPLACEABLE BY ANOTHER ELEMENT; DEFICIENCY IS REVERSIBLE IN
INCIPIENT STAGES; AND RESPONSE CAN BE DEMONSTRATED IN REPRESENTATIVE
NUMBER OF SPECIES. THERE IS EVIDENCE THAT COBALT REQUIREMENT IS
INDEPENDENT OF NITROGEN SOURCE ALTHOUGH IT IS MUCH EASIER TO
DEMONSTRATE COBALT RESPONSE IN THE TWO NITROGEN FIXERS THAN IN THE TWO
NON-FIXERS, A CONDITION WHICH QUITE LIKELY RESULTS FROM LOWER
REQUIREMENT IN THE LATTER. PRELIMINARY EXPERIMENTS SHOW THAT ADDITION
OF AS LITTLE AS 0.075 MICROGRAM OF VITAMIN B-12 PER LITER OF CULTURE
ELIMINATES REQUIREMENT FOR ADDED COBALT AND RESULTS IN OPTIMUM YIELD
REFERABLE TO ADDED COBALT. (EICHHORN-WIS)

FIELD 05A

ACCESSION NO. W69-06277

WASTES, WATER, AND WISHFUL THINKING: THE BATTLE OF LAKE ERIE,

CASE WESTERN RESERVE UNIV., CLEVELAND.

ARNOLD W. REITZE, JR.

CASE W RES L REV, VOL 20, NO 1, PP 5-86, NOV 1968. 82 P, 448 REF.

DESCRIPTORS:
*OHIO, *LAKE ERIE, *WATER POLLUTION CONTROL, *POLLUTION ABATEMENT, POLLUTANTS, WASTES, THERMAL POLLUTION, TOXINS, PESTICIDE RESIDUES, PHOSPHATES, ALGAE, SEWAGE, WATER POLLUTION, WATER LAW, WATER POLLUTION EFFECTS, WATER POLLUTION SOURCES, WATER POLLUTION TREATMENT, FINANCING, GRANTS, FEDERAL GOVERNMENT, STATE GOVERNMENTS, WATER QUALITY STANDARDS.

ABSTRACT:
LAKE ERIE IS SERIOUSLY POLLUTED, AND ITS CONTINUED USE AS A PUBLIC WATER SUPPLY IS IN JEOPARDY. THE POLLUTANTS INCLUDE SEWAGE, PHOSPHATES, ORGANIC CHEMICALS, AND PETROLEUM PRODUCTS. THE FEDERAL GOVERNMENT HAS PASSED POLLUTION STATUTES WHICH DATE BACK TO THE 19TH CENTURY, BUT THE FIRST SERIOUS ATTEMPT TO CLEAN UP LAKE ERIE BEGAN IN 1965, UNDER THE FEDERAL WATER POLLUTION CONTROL ACT. UNDER THIS ACT, THE STATES ARE REQUIRED TO SET WATER QUALITY STANDARDS FOR THEIR NAVIGABLE WATERWAYS. EACH JURISDICTION MUST MAINTAIN THE EXISTING QUALITY OF THE WATER AND MAY NOT ALLOW TREATABLE WASTES TO BE DISCHARGED INTO THE WATER IN AN UNTREATED STATE. OHIO HAS ESTABLISHED POLLUTION STANDARDS, BUT THESE STANDARDS ARE CRITICIZED BECAUSE THEY ASSUME THE WATER DOES NOT HAVE TO BE PURE AND BECAUSE THEY ARE TOO VAGUE AND INCONSISTENT. THE POLLUTION ABATEMENT OF LAKE ERIE REQUIRES THE EFFORTS OF FEDERAL AND STATE ADMINISTRATIVE AGENCIES AND OF THE COURTS. THE ATTEMPT AT POLLUTION ABATEMENT OVER THE LAST THREE YEARS HAS ENDED IN FAILURE. OHIO EXPENDS ONLY $239,000 YEARLY ON ITS RIVER SANITATION PROGRAM; APPROXIMATELY TEN BILLION DOLLARS ARE NEEDED TO CLEAN LAKE ERIE. (HOFFMAN-FLA).

FIELD 05G, 05B, 06E

ACCESSION NO. W69-06305

THE AGING OF LAKES,

WISCONSIN UNIV., MADISON. WATER CHEMISTRY LAB.

G. FRED LEE, AND E. GUS FRUH.

IND WATER ENG, VOL 3, NO 2, PP 26-30, FEB 1966. 2 FIG, 12 REF.

DESCRIPTORS:
*EUTROPHICATION, *LAKES, ALGAE, AQUATIC WEEDS, CYANOPHYTA, FISHKILL, FERTILIZATION, LIMITING FACTORS, NITROGEN COMPOUNDS, NUTRIENTS, ODOR, OLIGOTROPHY, OXYGEN, OXYGEN DEMAND, PHOSPHORUS COMPOUNDS, PLANT GROWTH, PRIMARY PRODUCTIVITY, RIVERS, TASTE, THERMAL STRATIFICATION, WASTE TREATMENT, WATER QUALITY, WATER POLLUTION CONTROL, WATER POLLUTION EFFECTS, WATER POLLUTION SOURCES.

IDENTIFIERS:
*CULTURAL EUTROPHICATION, RECREATIONAL USE, WASTE EFFLUENTS.

ABSTRACT:
A LAKE'S FERTILIZATION IS DETERMINED TO A LARGE EXTENT BY THE RATE OF NUTRIENT INFLUX. THIS SITUATION IS ANALOGOUS TO SOILS; MORE FERTILIZER, MORE CROPS. WHEN LAKES HAVE EXCESSIVE ALGAE AND/OR ROOTED AQUATIC VEGETATION ALONG THE SHORELINE, IT IS A SIGN THAT THIS FERTILIZATION PROCESS IS IN MOTION. FREQUENTLY, THE INFLUENCE OF MAN IN THE LAKE'S WATERSHED RESULTS IN INCREASED RATES OF NUTRIENT INFLUX WHICH BRING ABOUT AN ACCELERATED RATE OF STANDING CROP PRODUCTION IN THE LAKE. THIS RESULTS IN A DETERIORATION OF THE WATER QUALITY. DOMESTIC AND INDUSTRIAL WATER TREATMENT COSTS ARE INCREASED BECAUSE OF ADDED COST TO REMOVE RESULTANT TASTE AND ODORS AND INCREASED NUMBERS OF ORGANICS FROM THE EUTROPHIC WATER. SUCH RECREATIONAL PURSUITS AS BOATING, FISHING, AND SWIMMING CAN ALSO BE SERIOUSLY IMPAIRED. ODOR ARISING FROM DECAYING MATS OF ALGAE CAN ALSO SERIOUSLY DECREASE THE ECONOMIC VALUE OF SHORELINE PROPERTY. THE LARGE PRODUCTIVITY OF EUTROPHIC LAKES CAN LEAD TO THE ABSENCE OF DISSOLVED OXYGEN IN CERTAIN PARTS OF THE HYPOLIMNION, ELIMINATING SUCH COLD-WATER SPORT FISHES AS TROUT AND CISCO. THUS, THE CULTURAL EUTROPHICATION OF NATURAL WATERS HAS BECOME ONE OF THE MAJOR WATER MANAGEMENT PROBLEMS OF TODAY. (LEE-WIS)

FIELD 02H, 05C

ACCESSION NO. W69-06535

SIMULATION OF AN AQUATIC ECOSYSTEM,

WASHINGTON STATE UNIV., PULLMAN. DEPT. OF ZOOLOGY; AND WASHINGTON STATE
   UNIV., PULLMAN. DEPT. OF INFORMATION SCIENCE.

RICHARD A. PARKER.

BIOMETRICS, VOL 24, NO 4, PP 803-822, DEC 1968. 9 FIG, 3 TAB, 18 REF.

DESCRIPTORS:
   *SYSTEMS ANALYSIS, *ECOSYSTEMS, *SIMULATION ANALYSIS, WATER POLLUTION
   EFFECTS, LAKES, PHOSPHORUS, TROPHIC LEVEL, PRODUCTIVITY, MODEL STUDIES,
   SALMON, AQUATIC ENVIRONMENT, PHOSPHATES, ALGAE, DAPHNIA, GROWTH RATES,
   PREDATION.

IDENTIFIERS:
   KOOTENAY LAKE, BRITISH COLUMBIA, CANADA, AQUATIC ECOSYSTEM MODEL,
   SIMULATION MODEL, PHOSPHORUS EFFECT, CLADECERA.

ABSTRACT:
   A DETERMINISTIC MODEL SIMULATING AQUATIC ECOSYSTEM OF KOOTENAY LAKE,
   BRITISH COLUMBIA, WAS DEVELOPED CONSIDERING ONLY CONDITIONS IN UPPER 10
   METERS OF SOUTH ARM, WHICH RECEIVES HIGH PHOSPHATE LOAD FROM TRIBUTARY
   RIVER. SOLAR RADIATION WAS EXPRESSED AS WEEKLY PHOTOPERIOD CHANGES;
   SEASONAL TEMPERATURE CURVE DERIVED FROM MEANS OF MEASUREMENTS AT THREE
   DEPTHS. FOUR DIFFERENTIAL EQUATIONS SIMULATING CHANGE RATES IN
   PHOSPHORUS CONCENTRATION, ALGAE ABUNDANCE, CLADOCERAN DENSITY, AND
   KOKANEE SALMON BIOMASS WERE SIMULTANEOUSLY APPLIED. SEVERAL APPROPRIATE
   CONSTANTS WERE USED. CHANGE IN PHOSPHATE CONCENTRATION WAS ASSUMED TO
   BE LINEAR FUNCTION OF PHOSPHATE INPUT TO SOUTH ARM, OUTPUT FROM ARM,
   AND CHANGE RATES IN ALGAL AND CLADOCERAN POPULATIONS. CHANGES IN
   BIOLOGICAL COMPONENTS WERE DERIVED ESSENTIALLY FROM DIFFERENCE BETWEEN
   GROWTH AND DEATH RATES, WHICH WERE ASSUMED TO BE LINEARLY DEPENDENT ON
   COMBINATION OF FACTORS SUCH AS PHOTOPERIOD, PHOSPHATE LEVEL,
   TEMPERATURE, AND PREDATION. MODEL SIMULATED ALGAL AND CLADOCERAN CYCLES
   REASONABLY WELL BUT NOT PHOSPHATE CYCLE. 80% PHOSPHATE LEVEL REDUCTION
   DELAYED ALGAL PEAK 9 WEEKS AND DECREASED OTHER COMPONENTS MORE. 30%
   ALGAL GROWTH RATE REDUCTION DELAYED ALGAL AND CLADOCERAN PEAKS BY 1-2
   WEEKS AND DECREASED KOKANEE ABUNDANCE. 30% ALGAL DEATH RATE DECREASE
   PRODUCED NO SIGNIFICANT CHANGES. INCREASING CLADOCERAN'S ALGAL
   PREDATORY RATES HARDLY DISPLACED VARIOUS PEAKS BUT DEPRESSED ALL
   POPULATIONS. (TERAGUCHI-WIS)

FIELD 02H, 05C

ACCESSION NO. W69-06536

APPLICATION OF AN ELECTRONIC PARTICLE COUNTER IN ANALYZING NATURAL POPULATIONS OF PHYTOPLANKTON,

CORNELL UNIV., ITHACA, N. Y. DEPT. OF BOTANY.

HUGH F. MULLIGAN, AND JOHN M. KINGSBURY.

LIMNOL OCEANOGR, VOL 13, NO 3, PP 499-506, JULY 1968. 6 FIG, 3 TAB, 20 REF.

DESCRIPTORS:
*PHYTOPLANKTON, *PONDS, *EUTROPHICATION, ALGAE, BIOMASS, CHLOROPHYLL, NEW YORK.

IDENTIFIERS:
*COULTER COUNTER, *POPULATION DYNAMICS, CELLULAR SIZE, CHLORELLA VULGARIS, CORNELL UNIVERSITY, ELECTRONIC PARTICLE COUNTING, EUDORINA ELEGANS, EUGLENA GRACILIS, FOREST CENTRIFUGE, GLENODINIUM PULVISCULUS, HAEMOCYTOMETER COUNTING, ITHACA(NY), GLENODINIUM QUADRIDENS, OOCYSTIS PUSILLA, OOCYSTIS SOLITARIA, PANDORINA MORUM, PEDIASTRUM BORYANUM, PLEODORINA CALIFORNICA, SCENEDESMUS BIJUGA, SCHROEDERIA JUDAYI, SEDGWICK-RAFTER METHOD, SPHAEROCYSTIS SCHROETERI.

ABSTRACT:
UTILIZING COULTER COUNTER WITH ORIFICE OF 100-MICRON DIAMETER, AUTHORS ANALYZED, FROM 10 SMALL MORPHOMETRICALLY IDENTICAL PONDS, PHYTOPLANKTONIC POPULATIONS BY 10 SIZE INTERVALS (MEAN VOLUME, IN MICRONS, BY INTERVAL: 9.1, 23.8, 49.5, 89.3, 191, 396, 715, 1530, 3170, 5720). THEY COMPARED DETERMINATIONS OF BIOMASS BY COULTER COUNTING (B) IN MILLIGRAMS/LITER, WITH CHLOROPHYLL-A (C) IN MILLIGRAMS/CUBIC METER AND DRY WEIGHT (D) IN MILLIGRAMS/LITER. FOR EXAMPLE, IN A SINGLE POND FOR FIVE DATES, RATIO B:C HAD VALUES 0.215, 0.270, 0.94, 0.302, 0.66; AND B:D, 1.77, 1.76, 2.32, 1.82, 1.81. WHILE COUNTER CANNOT DISTINGUISH LIVING FROM DEAD PARTICLES, NOR SPECIES FROM SPECIES, ITS USE IN SIZING AND COUNTING PLANKTONIC POPULATIONS YIELDS VALUES OF BIOMASS WITH TIME WHICH CONSTITUTE HIGHLY RESOLVED AND CHARACTERISTIC PROFILE OF POND DYNAMICS. PEAK BIOMASSES CAN OFTEN BE CORRELATED WITH A DOMINANT PHYTOPLANKTER, OBSERVABLE BY DIRECT MICROSCOPIC EXAMINATION. BIOMASS DETERMINED IN THIS WAY IS MORE PRECISE THAN DETERMINATIONS OF CHLOROPHYLL OR DRY WEIGHTS, AND METHOD IS MORE EFFICIENT THAN ESTIMATIONS UTILIZING HAEMOCYTOMETER. AUTHORS EMPHASIZE THAT METHOD IS PARTICULARLY EFFECTIVE WHEN USED IN CONJUNCTION WITH STANDARD METHODS. (EICHHORN-WIS)

FIELD 07B, 05C

ACCESSION NO. W69-06540

CULTURAL EUTROPHICATION IS REVERSIBLE,

WISCONSIN UNIV., MADISON. LAB. OF LIMNOLOGY.

ARTHUR D. HASLER.

BIOSCIENCE, VOL 19, NO 5, PP 425-431, MAY 1969. 3 FIG, 1 TAB, 7 REF.

DESCRIPTORS:
*EUTROPHICATION, *FERTILIZATION, *LIMNOLOGY, *PHYTOPLANKTON, FISH, GREAT LAKES, LEGISLATION, NITROGEN, NUISANCE ALGAE, PHOSPHORUS, WASHINGTON, WISCONSIN, CALIFORNIA.

IDENTIFIERS:
*CULTURAL EUTROPHICATION, LAKE MONONA(WIS), LAKE TAHOE(CALIF), LAKE WASHINGTON(WASH), ZURICHSEE(SWITZERLAND).

ABSTRACT:
VARIOUS HUMAN ACTIVITIES CONTRIBUTE TO EXCESSIVE ENRICHMENT (CULTURAL EUTROPHICATION) OF WATERS. AMONG THE SYMPTOMS OF CULTURAL EUTROPHICATION ARE NUISANCE BLOOMS OF ALGAE, INCREASED NUTRIENT LEVELS, DEPLETION OF HYPOLIMNETIC OXYGEN, INCREASED TURBIDITY, AND CHANGES IN THE SPECIES-COMPOSITIONS OF PHYTOPLANKTON, INVERTEBRATES, AND FISHES. A CASE HISTORY IS PRESENTED (ZURICHSEE) DOCUMENTING CHANGES IN LIMNOLOGICAL CONDITIONS, AS WELL AS ADDITIONAL RELEVANT DATA FROM NORTH AMERICAN LAKES. AS METHODS OF SOLVING THE PROBLEMS OF CULTURAL EUTROPHICATION, AUTHOR SUGGESTS THAT MORE EXTENSIVE DISSEMINATION OF INFORMATION, INVOLVEMENT OF ALL CITIZENS AND CITIZEN GROUPS IN WORKSHOPS AND SEMINARS, AND ENLIGHTENED LEGISLATION. THE WISCONSIN RESOURCES LAW (1965) IS GIVEN AS AN EXAMPLE OF LEGISLATION AND THE PHILOSOPHY OF ZONING IS DISCUSSED. THREE EXAMPLES ARE GIVEN (LAKES MONONA, TAHOE, WASHINGTON) WHERE SUCCESS IS CURBING CULTURAL EUTROPHICATION HAS BEEN ACHIEVED. METHODS OF CONTROL AND HARVESTING AQUATIC PLANTS AND FISH ARE DISCUSSED; ALSO CHEMICAL CONTROL AND SEWAGE UTILIZATION. TABLE GIVES ESTIMATES OF NITROGEN AND PHOSPHORUS INPUT TO WISCONSIN SURFACE WATERS. SEE ALSO W68-00680. (VOIGTLANDER-WIS)

FIELD 05G, 02H

ACCESSION NO. W69-06858

THE REMOVAL OF DISSOLVED PHOSPHATE FROM LAKE WATERS BY BOTTOM DEPOSITS,

A. V. HOLDEN.

VERH INTERNAT VEREIN LIMNOL, VOL 14, PP 247-251, JULY 1961. 4 FIG, 2 REF.

DESCRIPTORS:
*PHOSPHATE, *LAKES, *CYCLING NUTRIENTS, *SEDIMENT-WATER INFERFACE, MUD-WATER INTERFACE, PHOSPHORUS, ADSORPTION, EUTROPHICATION, ALGAE, MACROAQUATIC PLANTS, FERTILIZATION.

ABSTRACT:
PHOSPHATE FERTILIZER WAS ADDED TO SCOTTISH LOCHS AND PHOSPHATE LOSS FROM THE WATER MEASURED. AT LOW LEVELS OF PHOSPHATE ADDED (45 MICROGRAMS/LITER) THE EXPOTENTIAL EXPRESSION DC/DT = KC DESCRIBED PHOSPHATE DECLINE; AT HIGHER LEVELS, THE EXPRESSION WAS ALTERED TO ACCOUNT FOR PHOSPHATE SATURATION OF THE SEDIMENT. REMOVAL RATES WERE EXPRESSED AS KD VALUES, MICROGRAMS PHOSPHATE/CENTIMETER SQUARED, PER MILLIGRAMS PHOSPHATE/LITER PER DAY. A KD VALUE OF 6.5 FOR LOCH KINARDOCHY AS COMPARED WITH 0.6 FOR CORES OVERLAIN BY WATER (30 TO 40 CM) AND LACKING ALGAE OR MACROPHYTES, SUGGESTED THAT ATTACHED FLORA LARGELY ACCOUNTED FOR PHOSPHATE UPTAKE IN THE LOCH. IN LABORATORY EXPERIMENTS INVOLVING HIGH PHOSPHATE UPTAKE (700 MICROGRAMS/CM SQUARED), PHOSPHATE INCREASE UP TO 200 MILLIGRAM PHOSPHATE PERCENT OCCURRED IN THE SEDIMENT AEROBIC ZONE (UPPER 20 MILLIMETERS); SOME PHOSPHATE PENETRATED TO 80 MILLIMETERS, INTO ANAEROBIC ZONE. ABOUT 6% OF PHOSPHATE RETAINED IN THE AEROBIC ZONE WAS DECINORMAL HYDROCHLORIC ACID EXTRACTABLE. DEVELOPMENT OF ANEROBIC ZONES DECREASED ACID-EXTRACTABLE PHOSPHORUS, ALTHOUGH AUTHOR DOES NOT STATE WHETHER REOXIDATION OCCURRED BEFORE EXTRACTION. PHOSPHORUS GRADIENTS FROM PRODUCTIVE, UNFERTILIZED LOCHS, WERE SIMILAR TO FERTILIZED LOCHS INCUBATED 48 MONTHS. (ARMSTRONG-WIS)

FIELD 02H, 05C

ACCESSION NO. W69-06859

PROVISIONAL ALGAL ASSAY PROCEDURE,

JOINT INDUSTRY-GOVERNMENT TASK FORCE ON EUTROPHICATION, NEW YORK.

G. P. FITZGERALD, JOSEPH SHAPIRO, C. R. GOLDMAN, O. R. ARMSTRONG, AND JACK MYERS.

PUBLISHED BY JOINT INDUSTRY/GOVERNMENT TASK FORCE ON EUTROPHICATION, P O BOX 3011, GRAND CENTRAL STATION, NEW YORK, 62 PP, FEB 1969. 7 FIG, 5 TAB, 26 REF.

DESCRIPTORS:
*ALGAE, *BIOASSAY, NUTRIENTS, EUTROPHICATION, WATER POLLUTION EFFECTS, WATER POLLUTION SOURCES, BIOINDICATORS, NITROGEN FIXATION.

IDENTIFIERS:
*ALGAL GROWTH, NUTRIENT AVAILABILITY, ALGAL GROWTH POTENTIAL, SELENASTRUM CAPRICORNUTUM, ANABAENA FLOS-AQUAE, MICROCYSTIS AERUGINOSA, ALGAL CULTURES, CHEMOSTATS, CARBON-14.

ABSTRACT:
THE PROCEDURE DESCRIBED WAS DEVELOPED BY A TEAM OF INTERNATIONAL EXPERTS UNDER THE SPONSORSHIP OF THE JOINT INDUSTRY/GOVERNMENT TASK FORCE ON EUTROPHICATION AND REPRESENTS THE FIRST STEP TOWARD DEVELOPING RELIABLE AND REPRODUCIBLE TESTS TO DETERMINE THE CAPACITY OF VARIOUS AQUATIC ENVIRONMENTS TO GROW ALGAE. IT IS EMPHASIZED THAT THE PROCEDURES PRESENTED ARE PROVISIONAL AND ARE NOT STANDARDIZED TESTS. THREE PROCEDURES ARE DESCRIBED: A BOTTLE TEST, A CONTINUOUS-FLOW CHEMOSTAT TEST, AND AN IN SITU TEST. DETAILED DESCRIPTIONS OF EQUIPMENT REQUIRED, LABORATORY PROCEDURES, ANALYTICAL TECHNIQUES, AND SAMPLE PREPARATION ARE GIVEN. INDICATOR ORGANISMS SELECTED FOR USE IN THE TESTS ARE SELENASTRUM CAPRICORNUTUM, A GREEN ALGA; ANABAENA FLOS-AQUAE, A BLUE-GREEN ALGA CAPABLE OF FIXING NITROGEN; AND MICROCYSTIS AERUGINOSA, A BLUE-GREEN ALGA NOT CAPABLE OF FIXING NITROGEN. (UTTORMARK-WIS)

FIELD 05A, 05C

ACCESSION NO. W69-06864

ALGAL CULTURE: FROM LABORATORY TO PILOT PLANT.

CARNEGIE INSTITUTION OF WASHINGTON PUBLICATION 600, 357 PP, WASHINGTON, DC,
MARCH 1961. 103 FIG, 59 TAB, 297 REF.

DESCRIPTORS:
*ALGAE, *RESEARCH AND DEVELOPMENT, BIBLIOGRAPHIES, CHLORELLA,
CHLOROPHYTA, DIATOMS, LIGHT, LIPIDS, NUTRIENT REQUIREMENTS, ORGANIC
MATTER, PHOTOSYNTHESIS, TEMPERATURE, TROPICAL REGIONS.

IDENTIFIERS:
*ALGAL CULTURE, *LABORATORY STUDIES, *MASS CULTURE, *PILOT-PLANT
STUDIES, ALGAL GROWTH, ALGAL NUTRITION, ALGAL PHYSIOLOGY, AMERICAN
RESEARCH AND DEVELOPMENT CORP(MASS), ARTHUR D LITTLE INC(MASS),
CARNEGIE DEPT OF PLANT BIOLOGY, CARNEGIE INSTITUTION OF WASHINGTON,
CHEMICAL COMPOSITION(ENGLAND), FOOD SOURCES, GREENHOUSE STUDIES, GROWTH
KINETICS, INDUSTRIAL RAW MATERIALS(ISRAEL), LIGHT-ENERGY CONVERSION,
MINERAL NUTRITION, RESEARCH CORPORATION OF NY, STANFORD RESEARCH
INSTITUTE(CALIF), STEROLS.

ABSTRACT:
RESEARCH SPONSORED BY THE INSTITUTION IS SUMMARIZED, BOTH IN-HOUSE AND
CONTRACTUAL, ON MASS CULTURE OF ALGAE IN ITS BASIC ASPECTS AND IN THE
DEVELOPMENTAL AND ENGINEERING PHASES OF HARVESTING AND UTILIZATION AS
FOOD. THE SCOPE IS INDICATED BY THE SUBJECT-MATTER OF CHAPTERS BY
VARIOUS AUTHORS: CURRENT STATUS OF LARGE-SCALE CULTURE OF ALGAE; NEED
FOR NEW FOOD-SOURCES; BIOLOGY OF ALGAE SUMMARIZED; ALGAL GROWTH
CHARACTERISTICS AS APPLIED TO MASS CULTURE; EFFICIENCY OF LIGHT-ENERGY
CONVERSION IN CHLORELLA; PHOTOSYNTHESIS BY CHLORELLA IN FLASHING LIGHT;
EFFECT OF DIURNALLY INTERMITTENT LIGHT ON CHLORELLA; INORGANIC
NUTRITION OF ALGAE; LABORATORY EXPERIMENTS ON CHLORELLA CULTURE AT
INSTITUTION'S LABORATORY; PRODUCTION OF ORGANIC MATTER BY CHLOROPHYTES
AND DIATOMS; NON-STERILE GREENHOUSE CULTURE; ALGAL ACCUMULATION OF
LIPIDS; BRITISH EXPERIMENTS; TROPICAL CULTIVATION OF ALGAL COMPLEXES;
ISRAELI EXPERIMENTS; CHLORELLA GROWTH KINETICS; PRE-PILOT-PLANT AND
PILOT-PLANT STUDIES IN MASS CULTURE; CHEMICAL COMPOSITION OF ALGAE;
NUTRITIONAL VALUES; ALGAE AS INDUSTRIAL RAW MATERIALS; STEROL CONTENT
OF ALGAE. A SELECTED GENERAL BIBLIOGRAPHY AND LIST OF LITERATURE CITED
IS INCLUDED. (EICHHORN-WIS)

FIELD 05C

ACCESSION NO. W69-06865

MADISON METROPOLITAN SEWERAGE DIST V COMM ON WATER POLLUTION (DISCHARGE OF
SEWAGE INTO LAKE WATER).

260 WIS 229, 50 NW 2D 424-439 (1951).

DESCRIPTORS:
*WISCONSIN, *SEWAGE EFFLUENTS, *WATER POLLUTION CONTROL, *LOCAL
GOVERNMENTS, STATE GOVERNMENTS, BIOCHEMICAL OXYGEN DEMAND,
ODOR-PRODUCING ALGAE, POLLUTION ABATEMENT, SEWAGE TREATMENT, SEWAGE
DISPOSAL, WASTE DISPOSAL, WATER POLLUTION, LAKES, STREAMS, SEWAGE,
REMEDIES, JUDICIAL DECISIONS, LEGAL ASPECTS.

ABSTRACT:
PETITIONER SEWERAGE DISTRICT, A PUBLIC CORPORATION, DISCHARGED THE
EFFLUENT FROM ITS SEWAGE TREATMENT PLANTS INTO SURROUNDING LAKES. THE
INORGANIC CHEMICALS CONTAINED IN THE EFFLUENT CONTRIBUTED SUBSTANTIALLY
TO THE ABNORMAL GROWTH OF A TYPE OF ALGAE WHICH CREATED OBJECTIONABLE
CONDITIONS IN THOSE LAKES. RESPONDENT, A PUBLIC AGENCY ACTING UNDER
POWERS GRANTED BY THE LEGISLATURE ORDERED PETITIONER TO MODIFY AND
IMPROVE ITS SEWAGE DISPOSAL SYSTEM SO AS TO ELIMINATE THE ALGAE
NUISANCE. THE CIRCUIT COURT AFFIRMED THE LOWER COURT'S ORDER AND
PETITIONER APPEALED. THE SUPREME COURT HELD THAT THE RIGHT OF A
MUNICIPALITY TO EMPTY ITS SEWAGE INTO A STREAM OR RIVER WAS MERELY A
LEGISLATIVE LICENSE REVOCABLE WHENEVER PUBLIC HEALTH AND SAFETY
REQUIRED. THE RESPONDENT DID NOT HAVE TO STATE EXPRESSLY THAT APPELLANT
WAS CREATING A NUISANCE WHEN ITS FINDINGS SUBSTANTIALLY DEMONSTRATED
THAT A NUISANCE EXISTED AS A RESULT OF APPELLANT'S OPERATIONS.
(GABRIELSON-FLA)

FIELD 06E, 05B

ACCESSION NO. W69-06909

UPTAKE AND ASSIMILATION OF NITROGEN IN MICROECOLOGICAL SYSTEMS,

GEOLOGICAL SURVEY, MENLO PARK, CALIF.

G. G. EHRLICH, AND K. V. SLACK.

ASTM STP NO. 448, SYMP ON MICROORGANIC MATTER IN WATER, AMER SOC TESTING
MATER 71ST ANNU MEETING, JUNE 23-28, 1968, SAN FRANCISCO, PP 11-23, FEB
1969. 13 P, 5 FIG, 2 TAB, 29 REF.

DESCRIPTORS:
*NITROGEN COMPOUNDS, *NUTRIENTS, *ALGAE, *MICROORGANISMS,
BIODEGRADATION, NITRIFICATION, DENITRIFICATION, ORGANIC COMPOUNDS,
AMMONIA, NITRITES, NITRATES, YEASTS, BACTERIA, AQUATIC BACTERIA, WATER
POLLUTION EFFECTS.

IDENTIFIERS:
NITROGEN ASSIMILATION.

ABSTRACT:
A STUDY OF BIOLOGICAL NITROGEN REMOVAL UTILIZED 2 RECIRCULATING
LABORATORY STREAMS. THE SOLE NITROGEN SOURCE IN 1 STREAM WAS CALCIUM
NITRATE AND IN THE OTHER WAS YEAST EXTRACT. NITRITE AND NITRATE
INCREASED SLIGHTLY. BACTERIA AND COMBINED NITROGEN DECREASED RAPIDLY
AFTER THE INITIAL RISE. IN THE STREAM WITH THE INORGANIC NITROGEN
MEDIUM, ABOUT HALF OF THE NITRATE DISAPPEARED IN 2 WEEKS, AND NITRATE
WAS UNDETECTABLE AFTER THE 3RD WEEK. BACTERIAL COUNTS WERE RELATIVELY
LOW AND CONSTANT. PERIPHYTON INCREASED IN BOTH STREAMS AS NUTRIENTS
DECREASED, REACHING QUASI-STEADY STATES AFTER ABOUT 4 WEEKS. THE RATE
OF BIOMASS INCREASE WAS GREATER IN THE INORGANIC MEDIUM. NITRATE WAS
DIRECTLY ASSIMILATED BY ALGAE. ORGANIC NITROGEN WAS CONVERTED TO
AMMONIA BY PROTEOLYTIC BACTERIA. AMMONIA FROM ORGANIC COMPOUNDS WAS
PARTLY ASSIMILATED BY ALGAE AND PARTLY NITRIFIED BY BACTERIA. NITRATE
OF BACTERIAL ORIGIN WAS ASSIMILATED BY ALGAE. NITRIFICATION APPARENTLY
WAS NOT OF MAJOR IMPORTANCE IN CONVERTING ORGANIC NITROGEN TO ALGAL
BIOMASS. LABORATORY STREAMS SIMULATE NATURAL STREAMS IN DIURNAL CYCLES
OF PH AND DISSOLVED OXYGEN, IN PERIPHYTON GROWTH RATE AND BIOMASS, AND
IN ASSIMILATION OF NUTRIENTS. USE OF MODEL SYSTEMS TO STUDY NATURAL
PROCESSES UNDER CONTROLLED CONDITIONS APPEARS JUSTIFIED. (KNAPP-USGS)

FIELD 05C, 05B

ACCESSION NO. W69-06970

EUTROPHICATION -- CAUSES AND EFFECTS,

MICHIGAN UNIV., ANN ARBOR. DEPT. OF SANITARY AND WATER RESOURCES ENGINEERING.

J. A. BORCHARDT.

J AMER WATER WORKS ASS, VOL 61, NO 6, PP 272-275, JUNE 1969. 4 P, 2 FIG, 1
TAB, 1 PHOTO, 4 REF.

DESCRIPTORS:
*EUTROPHICATION, *WATER POLLUTION SOURCES, *LAKES, NUTRIENTS,
FERTILIZERS, PHOSPHATES, NITRATES, ALGAE, GREAT LAKES.

IDENTIFIERS:
EUTROPHICATION PREVENTION, ECOLOGIC IMBALANCES.

ABSTRACT:
THE CAUSES AND EFFECTS OF ACCELERATED EUTROPHICATION OF LAKES AND
STREAMS ARE OUTLINED. THE IMPORTANT NATURAL FACTORS INVOLVED ARE
GEOLOGY, LAKE BASIN GEOMETRY AND SIZE, TYPE AND SIZE OF WATERSHED, AND
LATITUDE. THE MOST IMPORTANT SINGLE VARIABLE IS MAN'S TREATMENT OF THE
LAKE. ACCELERATED EUTROPHICATION IS A RESULT OF ACCELERATED ADDITION OF
PHOSPHORUS AND NITROGEN NUTRIENTS, NEARLY ALWAYS CAUSED BY SOME FORM OF
HUMAN ACTIVITY. FERTILIZERS, MUNICIPAL WASTES, INDUSTRIAL WASTES, AND
RECREATIONAL WASTES ARE THE MOST COMMON NUTRIENT SOURCES. THE MOST
VISIBLE AND OBNOXIOUS RESULT OF EUTROPHICATION IS ALGAE GROWTH.
CORRECTION OF EUTROPHIC CONDITIONS IS ALMOST IMPOSSIBLE. PREVENTION IS
NECESSARY TO AVOID THE PROBLEM IN THE FUTURE. (KNAPP-USGS)

FIELD 05C, 02H

ACCESSION NO. W69-07084

AN ECOLOGICAL STUDY OF THE ALGAE OF THE RIVER MOOSI, HYDERABAD (INDIA) WITH
SPECIAL REFERENCE TO WATER POLLUTION - 1. PHYSICO-CHEMICAL COMPLEXES,

OSMANIA UNIV., HYDERABAD, (INDIA). DEPT. OF BOTANY.

V. VENKATESWARLU.

HYDROBIOLOGIA, VOL 33, NO 1, PP 117-143, MAR 1969. 26 P, 18 FIG, 2 TAB, 55
REF.

DESCRIPTORS:
*ECOLOGY, *ALGAE, *RIVERS, *WATER POLLUTION, *WATER POLLUTION SOURCES,
IRRIGATION, AQUATIC ALGAE, WATER ANALYSIS, CARBONATES, NITRATES,
CHLORIDES, AMMONIUM COMPOUNDS, OXYGEN, SILICATES, ORGANIC MATTER, AIR
TEMPERATURE, AEROBIC BACTERIA, WATER TEMPERATURE, IRON.

IDENTIFIERS:
*INDIA, MOOSI RIVER ECOLOGICAL STUDY.

ABSTRACT:
THIS INVESTIGATION IS CONCERNED WITH THE ECOLOGY OF THE RIVER MOOSI,
HYDERABAD (INDIA) USING THE DATA EXTENDED OVER A PERIOD OF TWO YEARS
WITH THE COLLECTION OF MONTHLY SAMPLES. IT DEALS MAINLY WITH THE
PHYSICO-CHEMICAL COMPLEXES PRESENT IN THE UNPOLLUTED AND POLLUTED
WATERS OF THE RIVER. THE INTERRELATIONSHIPS AND FLUCTUATIONS EXISTING
BETWEEN THE FOLLOWING FACTORS HAVE BEEN INVESTIGATED: (1) CHLORIDE AND
WATER CURRENT; (2) NITRATE, DISSOLVED OXYGEN, AND FREE AMMONIA; (3)
NITRATE AND ALBUMINOID AMMONIA; (4) FREE AMMONIA AND ALBUMINOID
AMMONIA; (5) DISSOLVED OXYGEN, OXIDIZABLE ORGANIC MATTER, AND WATER
TEMPERATURE; AND (6) TOTAL IRON AND PH. (GABRIEL-USGS)

FIELD 05B, 02I

ACCESSION NO. W69-07096

THE ROLE OF ALGAE IN THE BIOLOGICAL TREATMENT OF WATER (FRENCH),

IRCHA (FRANCE). MICROBIOLOGY SERVICE.

R. CABRIDENC, AND H. LEPAILLEUR.

TERRES AND EAUX, REV INT DE L'HYDRAUL, VOL 22, NO 58, PP 12-18, JAN-MAR 1969.
7 P, 7 FIG, 1 TAB, 25 REF.

DESCRIPTORS:
*ALGAE, *BIOLOGICAL TREATMENT, *WATER TREATMENT, *POLLUTANT
IDENTIFICATION, *POLLUTANT ABATEMENT, EFFLUENTS, STREAMFLOW,
DESALINATION, TERTIARY TREATMENT, LAGOONS, OXYGENATION, ABSORPTION,
ECOLOGY, WATER PURIFICATION, CARBON, BICARBONATES, PHOTOSYNTHESIS,
NUTRIENTS, MICROBIOLOGY.

IDENTIFIERS:
WATER PURIFICATION BY ALGAE.

ABSTRACT:
THE ROLE OF ALGAE, AS A DESALINATION AND PURIFICATION AGENT, WAS
INVESTIGATED ON THE BASIS OF EARLIER PUBLICATIONS AND RECENT
EXPERIMENTS CONDUCTED BY THE AUTHORS. THE STUDY SHOWS THAT OXYGEN
LIBERATED IN THE COURSE OF PHOTOSYNTHESIS LEADS TO THE OXYGENATION OF
AQUATIC MEDIA AND CONSIDERABLE REDUCTION OF PATHOGENIC GERMS. THE STUDY
ALSO SHOWS THAT THERE IS COMPETITION BETWEEN ALGAE AND OTHER
MICROORGANISMS; HOWEVER, IN THE PRESENT STATE OF KNOWLEDGE, THE PROBLEM
OF INTERACTION DUE TO THE PRODUCTION BY ALGAE OF INHIBITING OR
ACTIVATING SUBSTANCES IS STILL NOT CLEARLY UNDERSTOOD. THE DEGREE OF
PURIFICATION OF WATER BY THE LAGOONING PROCESS, IN MANY CASES IS
EQUIVALENT TO, OR GREATER THAN THAT OBTAINED BY OTHER BIOLOGICAL
PROCESSES. (GABRIEL-USGS)

FIELD 05C

ACCESSION NO. W69-07389

MICROBIAL TRANSFORMATIONS OF MINERALS,

RENSSELAER POLYTECHNIC INST., TROY, N. Y. DEPT. OF BIOLOGY.

HENRY L. EHRLICH.

PROC RUDOLFS RES CONF, RUTGERS UNIV, NEW BRUNSWICK, NJ. PRINCIPLES AND
APPLICATIONS IN AQUATIC MICROBIOLOGY, HEUKELEKIAN, H AND DONDERO, NORMAN C
(EDS), JOHN WILEY AND SONS, NEW YORK, PP 43-60, 1964. 5 FIG, 43 REF, DISC.

DESCRIPTORS:
*MICROBIOLOGY, *MINERALOGY, *MICROORGANISMS, *WATER POLLUTION, *WATER
PURIFICATION, *BACTERIA, *ENZYMES, CORROSION, OXIDATION,
REDUCTION(CHEMICAL), HYDROLYSIS, CHELATION, FUNGI, ALGAE, PROTOZOA,
LICHENS, PLANTS, DIATOMS, CORAL, MOLLUSKS, MINE ACIDS, COAL MINES,
ACIDITY, PYRITE, SULFUR COMPOUNDS, MOLYBDENUM, ROTIFERS, ECOSYSTEMS,
PHOTOSYNTHESIS, ENVIRONMENT, AQUEOUS SOLUTIONS, TERRESTRIAL HABITATS,
IRON COMPOUNDS, MANGANESE, TROPHIC LEVEL.

IDENTIFIERS:
*MINERAL TRANSFORMATION, METAZOA, RADIOLARIA, FORAMINIFERA,
THIOBACILLUS-FERROBACILLUS, THIOBACILLUS DENITRIFICANS, W VIRGINIA,
PENNSYLVANIA, MARCASITE, CHALCOCITE, COVELLITE, SPHALERITE, MILLERITE,
ORPIMENT, TETRAHEDRITE, BORNITE, CHALCOPYRITE, CHEMOSYNTHESIS,
CONCRETIONS, ARTHROBACTER, PYROPHOSPHATE, ORTHOPHOSPHATE, DISMUTATION,
TANNINS.

ABSTRACT:
BIOLOGICAL PROCESSES OF SOME MICROORGANISMS, ESPECIALLY BACTERIA, CAUSE
TRANSFORMATION OF MINERAL MATTER. INVOLVED ARE DIRECT ENZYMIC
INTERACTION--OXIDATIONS AND REDUCTIONS, HYDROLYSIS, CHELATE
DESTRUCTION--AND, INDIRECT NONENZYMIC INTERACTION--INSOLUBLE MATTER
CORROSION BY METABOLICALLY PRODUCED ACID, INORGANIC IONS PRECIPITATION
BY OTHER, METABOLICALLY-PRODUCED INORGANIC IONS, ADSORPTION ONTO CELL
SURFACES, AND METAL CHELATE FORMATION. THE BACTERIAL CHEMOSYNTHETIC
AUTOTROPHS USE MINERAL OXIDATION AS A CHEMICAL ENERGY SOURCE, REDUCING
POWER FOR THE SYNTHESIS OF ORGANIC MATTER FROM CARBON DIOXIDE AND
WATER. OXIDATION OF METAL SULFIDES AND OF FE (II) TO FE (III) TAKES
PLACE. BACTERIAL PHOTOSYNTHETIC AUTOTROPHS USE MINERAL OXIDATION MERELY
AS A SOURCE OF REDUCING POWER IN CARBON DIOXIDE ASSIMILATION, DEPENDING
ON RADIATION FOR ENERGY. THE ENZYME, INORGANIC PYROPHOSPHATASE, PRESENT
IN BACTERIA, IS PROBABLY HYDROLYZED TO ORTHOPHOSPHATE. PYROPHOSPHATE
CAN COMPLEX MN (III) AND HYDROLYSIS COULD LEAD TO THE ION
PRECIPITATION. ENZYMIC BREAKDOWN OF THE CHELATING AGENT BY BACTERIA MAY
RESULT IN PRECIPITATING THE METAL IONS. THE NONENZYMIC INTERACTIONS
DEPEND ON END PRODUCTS OF METABOLISM, ACIDS WHICH ACT ON ACID-SOLUBLE
MINERALS. SOME PRECIPITATIONS MAY FORM AMORPHOUS OR CRYSTALLINE
AGGLOMERATES. IN ADSORPTION, IRON AND MANGANESE OXIDES POSSIBLY FORM
INTO ENCRUSTATIONS. IN CHELATION A METABOLICALLY PRODUCED ORGANIC
COMPOUND MAY CO MPLEX ONE OR ANOTHER METAL ION. (JONES-WISC)

FIELD 05C

ACCESSION NO. W69-07428

PROBLEMS OF BIOLOGICAL OCEANOGRAPHY.

AKADEMIYA NAUK URSR. INST. OF BIOLOGY OF SOUTHERN SEAS.

AVAILABLE FROM CLEARINGHOUSE AS AEC-TR-6940 AT $3.00 IN PAPER COPY AND $0.65 IN MICROFICHE. TRANSL VOPR BIOOKEANOGRAFFI MATERIALY II MEZHDUNARODNOGA OKEANOGR KONGR 30 MAY-9 JUNE 1966. NAUKOVA DUMKA, KIEV, 1967. US ATOMIC ENERGY COMM, TRANSLATION SERIES AEC-TR-6940. 94 PP, 13 FIG, 18 TAB, 99 REF.

DESCRIPTORS:
*CONFERENCES, *OCEANOGRAPHY, *TRANSLATIONS, ALGAE, CRUSTACEANS, BENTHOS, BRINE SHRIMP, ECOSYSTEMS, FISH, PHYTOPLANKTON, PLANKTON, RADIOACTIVITY, RADIOACTIVITY EFFECTS, RADIOECOLOGY, SORPTION, WATER POLLUTION EFFECTS, WATER POLLUTION SOURCES.

IDENTIFIERS:
*BIOLOGICAL OCEANOGRAPHY, *TRANSACTIONS, BLACK SEA, CONTACT ZONES, ARTEMIA, BALANUS, DESORPTION, EMBRYOLOGY, ENERGY TRANSFORMATIONS, HETEROTROPHY, ORGANIC POLLUTION, MARINE ORGANISMS, MARINE ASSOCIATIONS, MOSCOW, PETROLEUM POLLUTION, PHOTOSYNTHETIC PIGMENTS, PRODUCTION BIOLOGY, RUSSIA, SECOND INTERNATIONAL OCEANOGRAPHIC CONGRESS, HYPONEUSTON.

ABSTRACT:
TROPICAL COVERAGE OF 16 INDIVIDUAL CONTRIBUTIONS ARE AS FOLLOWS: PLANKTONIC ECOSYSTEMS OF THE BLACK SEA; PRODUCTION RATE OF HETEROTROPHS; TRANSFORMATION OF SUBSTANCES AND ENERGY BY AMPHIBIONTIC CRUSTACEA IN BLACK SEA; ENERGY TRANSFORMATION BY ARTEMIA, SEASONAL DYNAMICS OF ALGAL PHOTOSYNTHETIC PIGMENTS IN BLACK SEA; BENTHOS DEVELOPMENT IN MEDITERRANEAN BASIN; ECOLOGICO-GEOGRAPHIC DEVELOPMENT OF CONTACT ZONES IN SOUTHERN SEAS; INFLUENCE OF PETROLEUM PRODUCTS ON PHYTOPLANKTON OF SEAS; INFLUENCE OF ORGANIC POLLUTION ON DEVELOPMENT OF BALANUS; MARINE RADIOECOLOGY AND OCEANOGRAPHY; RADIOECOLOGICAL SIGNIFICANCE OF HYPONEUSTON; EFFECTS OF RADIOACTIVITY ON EMBRYOLOGY OF FISHES; KINETICS OF ACCUMULATION AND EXCHANGE OF RADIOISOTOPES BY MARINE ALGAE; ARTIFICIAL RADIOACTIVITY OF MARINE ORGANISMS; DISTRIBUTION OF RADIOACTIVITY IN MARINE ORGANISMS AND ASSOCIATIONS; AND SORPTION AND DESORPTION OF RADIONUCLIDES IN BOTTOMS OF SHALLOW SEAS. (EICHHORN-WISC)

FIELD 05C

ACCESSION NO. W69-07440

GROWTH CHARACTERISTICS OF ALGAE IN RELATION TO THE PROBLEMS OF MASS CULTURE,

CARNEGIE INSTITUTION OF WASHINGTON, STANFORD, CALIF. DEPT. OF PLANT BIOLOGY.

JACK MYERS.

EXTRACTED FROM ALGAL CULTURE: FROM LABORATORY TO PILOT PLANT, JOHN S BURLEW, EDITOR. CARNEGIE INSTITUTION OF WASHINGTON PUBLICATION 600, WASHINGTON, DC. 4 FIG, 2 TAB.

DESCRIPTORS:
*GROWTH RATES, *ALGAE, PHOTOSYNTHESIS, EUTROPHICATION, LIGHT, TEMPERATURE, CARBON DIOXIDE, NUTRIENTS, BIOLOGICAL PROPERTIES, WATER POLLUTION EFFECTS.

IDENTIFIERS:
*MASS CULTURE, *ALGAL GROWTH, CHLORELLA PYRENOIDOSA(EMERSON STRAIN), AUTOINHIBITORS.

ABSTRACT:
BIOLOGICAL SIGNIFICANCE OF THE CONDITIONS AFFECTING ALGAL GROWTH IS DISCUSSED WITH SPECIAL REFERENCE TO CHLORELLA PYRENOIDOSA (EMERSON STRAIN). CONSIDERATION IS GIVEN TO THE EFFECT OF CONTROLLED AND MEASURABLE CONDITIONS, AND THESE RESULTS ARE EXTRAPOLATED TO DESCRIBE THE CHARACTERISTICS OF HIGH-DENSITY CULTURES. THE EFFECT OF LIGHT AS A LIMITING FACTOR IS DISCUSSED ALSO. IN PARTICULAR, GROWTH RATES ARE DESCRIBED AS FUNCTIONS OF (1) TEMPERATURE, (2) LIGHT INTENSITY, (3) CONCENTRATION OF CARBON DIOXIDE, (4) COMPONENTS OF THE NUTRIENT MEDIUM, (5) AUTOINHIBITORS, AND (6) CHARACTERISTICS OF THE ORGANISM. A SPECIFIC GROWTH RATE OF 1.96 PER DAY IS CONSIDERED TO BE ABOUT THE MAXIMUM RATE FOR C PYRENOIDOSA AND IT IS CONCLUDED THAT THIS VALUE IS NOT ESTABLISHED BY LIMITATIONS OF PHOTOSYNTHESIS. (UTTORMARK-WISC)

FIELD 05C

ACCESSION NO. W69-07442

PREDICTING DISSOLVED OXYGEN VARIATIONS CAUSED BY ALGAE,

FEDERAL WATER POLLUTION CONTROL ADMINISTRATION, ALAMEDA, CALIF. CENTRAL
PACIFIC RIVER BASINS PROJECT.

RICHARD C. BAIN, JR.

ASCE PROC, J SANIT ENG DIV, VOL 94, NO SA5, PAP 6155, PP 867-881, OCT 1968.
15 P, 8 FIG, 2 TAB, 26 REF, APPEND.

DESCRIPTORS:
*DISSOLVED OXYGEN, *ALGAE, *RESPIRATION, *ESTUARIES, *EUTROPHICATION,
*CALIFORNIA, WATER POLLUTION, REAERATION, NUTRIENTS, OXYGEN SAG,
AERATION, BIOCHEMICAL OXYGEN DEMAND, PHOTOSYNTHETIC OXYGEN.

IDENTIFIERS:
STREETER-PHELPS EQUATION, DISSOLVED OXYGEN VARIATIONS, SAN JOAQUIN
RIVER, POTOMAC ESTUARY, SAN FRANCISCO BAY.

ABSTRACT:
EUTROPHIC ENVIRONMENTS ARE OFTEN DOMINATED BY PLANKTONIC ALGAL
POPULATIONS (PHYTOPLANKTON) WHICH CAUSE DIURNAL VARIATIONS IN DO
CONCENTRATIONS THROUGH RESPIRATORY ACTIVITY AND PHOTOSYNTHESIS.
PHOTOSYNTHETIC OXYGENATION AND RESPIRATORY DEOXYGENATION RATES OF
ESTUARINE PHYTOPLANKTON WERE MEASURED AT VARIOUS STANDING CROP
(CHLOROPHYLL) LEVELS. OXYGEN PRODUCTION AND CONSUMPTION RATES FOR
ACTIVELY GROWING PHYTOPLANKTON POPULATIONS WERE RELATED TO STANDING
CROP AT 20 DEG C AND NON-LIMITING LIGHT. VARIATIONS IN ALGAL
PHOTOSYNTHETIC PRODUCTION RATE, AS RELATED TO LIGHT ADAPTION, AGE OF
CELLS, NUTRITION, TEMPERATURE AND ALGAL TYPE, ARE CONSIDERED. THE
STREETER-PHELPS EQUATION WAS MODIFIED TO INCLUDE PHYTOPLANKTON
PRODUCTION AND RESPIRATION RATES IN FORMULATIONS DESIGNED TO PREDICT DO
CONCENTRATIONS OVER A 24-HR PERIOD. AN EXAMPLE IS GIVEN, AND THE
RESULTING PREDICTIONS ARE COMPARED WITH FIELD MEASUREMENTS FROM A TIDAL
REACH OF THE SAN JOAQUIN RIVER, CALIFORNIA. (KNAPP-USGS)

FIELD 05C

ACCESSION NO. W69-07520

THE GALLOPING GHOST OF EUTROPHY,

NEW HAMPSHIRE WATER SUPPLY AND POLLUTION CONTROL COMMISSION.

TERRENCE P. FROST.

APPALACHIA, VOL XXXVII, NO 1, PP 25-36, JUNE 1968. 1 FIG, 4 PLATES.

DESCRIPTORS:
*EUTROPHICATION, *LAKES, GLACIATION, URBANIZATION, DOMESTIC WASTES,
NUTRIENTS, PHOSPHORUS, ALGAE, ALGICIDES, HARVESTING OF ALGAE,
ENVIRONMENT, CYANOPHYTA.

IDENTIFIERS:
*AESTHETIC DETERIORATION, *ADVANCED WASTE TREATMENT, *NUTRIENT REMOVAL,
AGRICULTURAL DRAINAGE, APPALACHIAN MOUNTAIN CLUB HUTS, TEGERNSEE,
GERMANY, EXPANDING TECHNOLOGY, OVERPOPULATION.

ABSTRACT:
THE INCREASED INFLUENCE OF MAN ON THE ENVIRONMENT HAS CAUSED A CHANGE
IN THE DEFINITION OF EUTROPHICATION FROM A NATURAL AGING PROCESS OF
LAKES DUE TO TOPOGRAPHY, DEGREE OF FERTILITY, SEDIMENT LOADS FROM
MINERAL AND ORGANIC SOLIDS, SUBSURFACE GEOLOGY, AND WEATHERING, TO AN
ACCELERATED AGING DUE TO MAN'S INFLUENCE. INCREASING URBANIZATION ALONG
WITH OUR CHANGING LIFE STYLE, IS SUGGESTED AS THE CAUSE FOR 'GALLOPING
EUTROPHY.' THE EFFECT IS THE PROLIFERATION OF ALGAL SPECIES, HARMLESS
AND DIVERSIFIED IN SMALL NUMBERS; NOXIOUS AND CONFINED TO PROFUSION OF
A FEW HARDY BLUE-GREEN SPECIES IN OVERPOPULATION. THE RESULT IS
AESTHETIC DETERIORATION, FISH KILLS, SEPTIC ODORS, DISCOLORATION OF
OBJECTS IN WATER, ORGANIC DEBRIS BUILDUPS ON BOTTOMS, SCUMS AND MATS.
SUGGESTED METHODS OF CONTROL INCLUDE CHEMICAL TREATMENT WITH ALGICIDES
OR HARVESTING ORGANISMS, BUT THESE ELIMINATE SYMPTOMS WITHOUT REACHING
THE CAUSES. REAL NEED IS NUTRIENT REMOVAL FROM SEWAGE EFFLUENT BY A
VARIETY OF TECHNIQUES, COAGULATION (LIME OR ALUM), REVERSE OSMOSIS, ION
EXCHANGE, ELECTRODIALYSIS, DISTILLATION, HYDROPHONICS, AND IRRIGATION.
MENTION IS ALSO MADE OF METHODS AND IMPROVEMENTS IN WASTE DISPOSAL AT
APPALACHIAN MOUNTAIN CLUB HUTS AND THE TEGERNSEE RECREATION AREA IN
GERMANY. AUTHOR'S CONCLUSION: THE HEART OF THE PROBLEM LIES IN
UNHARNESSED TECHNOLOGY AND PEOPLE POLLUTION. (KETELLE-WISC)

FIELD 05C

ACCESSION NO. W69-07818

SEAWEED EXTRACTS AS FERTILISERS.

PORTSMOUTH COLL. OF TECH. (ENGLAND). SCHOOL OF PHARMACY.

G. BLUNDE, S. B. CHALLEN, AND D. L. WOODS.

J SCI FOOD AGRIC, VOL 19, PP 289-293, JUNE 1968, 1 FIG. 8 TAB, 26 REF.

DESCRIPTORS:
*FERTILIZERS, *CYCLING NUTRIENTS, EUTROPHICATION, ALGAE, PLANT GROWTH
SUBSTANCES, PLANT GROWTH, CARBOHYDRATES.

IDENTIFIERS:
*SEAWEED EXTRACTS, ASCOPHYLLUM NODOSUM, ALGINIC ACID, FUCUS
VESICULOSUS, GROWTH FACTORS, LAMINARIA SACCHARINA, MANNITOL, MUSTARD
GROWTH TEST, ORGANIC GROWTH FACTORS, POLYSACCHARIDES.

ABSTRACT:
IN THE MUSTARD-GROWTH TEST, THE GROWTH-PROMOTING EFFECTS OF AQUEOUS
EXTRACTS OF LAMINARIA SACCHARINA, FUCUS VESICULOSUS AND ASCOPHYLLUM
NODOSUM WERE LARGELY DUE TO THE METAL CATIONS PRESENT, BUT THE EFFECTS
WERE MODIFIED BY ORGANIC SUBSTANCES IN THE EXTRACTS. AMINO ACIDS AND
MANNITOL IN SEAWEED EXTRACTS HAD LITTLE EFFECT ON PLANT GROWTH, AND
COMPOUNDS EXTRACTED WITH ORGANIC SOLVENTS WERE ONLY PARTIALLY
RESPONSIBLE FOR THE MODIFICATION OF THE GROWTH-PROMOTING EFFECT. THE
POLYSACCHARIDE ALGINIC ACID AND ITS SALTS SEEMED TO BE THE MAIN ORGANIC
COMPOUNDS RESPONSIBLE FOR REDUCING THE EFFECTS OF METALS AND MAY HAVE
COMPETED WITH THE PLANTS BY ION EXCHANGE FOR THE METALS IN THE EXTRACT.
(KONRAD-WISC)

FIELD 05C

ACCESSION NO. W69-07826

ALGAE AND MAN.

SYRACUSE UNIV., N.Y. DEPT. OF CIVIL ENGINEERING.

FROM SYMP ON ALGAE AND MAN, NATO ADVANCED STUDY INST, LOUISVILLE, KY, 1962.
DANIEL F. JACKSON, (ED). PLENUM PRESS, NEW YORK, 1964. 90 FIG, 48 TAB, 1000
REF.

DESCRIPTORS:
*EUTROPHICATION, *ALGAE, BIOASSAY, BIOLOGICAL COMMUNITIES, CULTURES,
DIATOMS, ECOLOGY, NUISANCE ALGAE, PHAEOPHYTA, PHOTOSYNTHESIS,
PHYTOPLANKTON, PRIMARY PRODUCTIVITY, SYSTEMATICS, WATER POLLUTION
EFFECTS, WATER POLLUTION SOURCES, WATER SUPPLY.

IDENTIFIERS:
BENTHIC ALGAE, BLOOMS, CLOSED SYSTEMS, CYTOLOGY, ENVIRONMENTAL
CONDITIONS, EXTRACELLULAR PRODUCTS, GAS EXCHANGE, GREEN PLANTS, MASS
CULTURE, MEDICINE, METABOLIC PATTERNS, MICROALGAE, MICRONUTRIENTS,
PHYCOLOGY, SPACE RESEARCH, TOXIC ALGAE.

ABSTRACT:
THIS COMPENDIUM IS BASED ON A SERIES OF LECTURES PRESENTED AT THE NATO
ADVANCED STUDY INSTITUTE, THE FIRST TO BE HELD IN THE UNITED STATES,
WHICH MET AT THE UNIVERSITY OF LOUISVILLE, KENTUCKY, JULY 22-AUGUST 11,
1962. THE SERIES BEARS ON THE ROLE OF ALGAE IN EUTROPHICATION OF
NATURAL WATERS AND ON THEIR POSSIBLE UTILIZATION FOR CLOSED
LIFE-SUPPORT SYSTEMS IN THE SPACE AGE. THE SCOPE OF THE WORK IS
INDICATED BY NAMES OF AUTHORS AND TOPICS OF THEIR CONTRIBUTIONS: G. W.
PRESCOTT, SYSTEMATICS; C VAN DEN HOEK, TAXONOMY; TYGE CHRISTENSEN,
GROSS CLASSIFICATION; MARGARET ROBERTS, CYTOLOGY OF PHAEOPHYTA; G. E.
FOGG, ENVIRONMENTAL CONDITIONS AND ALGAL METABOLIC PATTERNS; CLYDE
EYSTER, MICRONUTRIENT REQUIREMENTS; A. G. WURTZ, PROBLEMS OF CULTURE;
F. E. ROUND, ECOLOGY OF BENTHIC ALGAE; RUTH PATRICK, DIATOM
COMMUNITIES; JAMES B. LACKEY, ECOLOGY OF PLANKTONIC ALGAE; JACOB
VERDUIN, PRINCIPLES OF PRIMARY PRODUCTIVITY; C. MERVIN PALMER, ALGAE IN
WATER SUPPLIES; OTTO SKULBERG, EUTROPHICATION AND BIOASSAY; GEORGE P.
FITZGERALD, BIOTIC RELATIONS IN BLOOMS; PAUL R. GORHAM, TOXIC ALGAE;
MARCEL LEFEVRE, EXTRACELLULAR PRODUCTS; DAVID AND MORTON SCHWIMMER,
ALGAE AND MEDICINE; RICHARD J. BENOIT, MASS CULTURE FOR GAS EXCHANGE;
AND F. EVENS, FUTURE OF PHYCOLOGY. WSELECTED INDIVIDUAL CONTRIBUTIONS
TO THE VOLUME WILL BE ABSTRACTED INDIVIDUALLY. (SEE ALSO W69-01977).
(EICHHORN-WISC)

FIELD 05C

ACCESSION NO. W69-07832

ALGAL PROBLEMS RELATED TO THE EUTROPHICATION OF EUROPEAN WATER SUPPLIES, AND A
BIO-ASSAY METHOD TO ASSESS FERTILIZING INFLUENCES OF POLLUTION ON INLAND
WATERS,

NORSK INSTITUTT FOR VANNFORSKNING, BLINDERN.

OLAV M. SKULBERG.

ALGAE AND MAN, JACKSON, DANIEL F, ED, PLENUM PRESS, N Y, PP 262-299, 1964. 16
   FIG, 9 TAB, 76 REF.

   DESCRIPTORS:
     *ALGAE, *BIOASSAY, *EUTROPHICATION, *FERTILIZATION, *INLAND WATERWAYS,
   *WATER POLLUTION EFFECTS, *WATER SUPPLY, BIOINDICATORS, CALCIUM,
   CHLORIDES, CONDUCTIVITY, CULTURES, HYDROGEN ION CONCENTRATION, IRON,
   LAKES, MAGNESIUM, NUTRIENTS, NUISANCE ALGAE, OLIGOTROPHY, ODOR,
   PHYTOPLANKTON, SULFATES, SEDIMENTATION, SEDIMENTS, TASTE, WATER
   CHEMISTRY, WATER QUALITY.

   IDENTIFIERS:
     *EUROPE, ALGAL GROWTH, ANKISTRODESMUS, BACILLARIOPHYCEAE,
   CHLOROPHYCEAE, CHRYSOPHYCEAE, CARTERIA, COMPARATIVE STUDIES, CRUCIGENIA
   RECTANGULARIS, DICERAS CHODATI, DINOPHYCEAE, FILTER BLOCKING,
   HETERETROPHS, ITALY, LAKE BORREVANNET NORWAY, LAKE LUGANO ITALY, LAKE
   MARIDALSVANNET NORWAY, LAKE ZURICH, SWITZERLAND, NITELV RIVER NORWAY,
   NORWAY, OSCILLATORIA RUBESCENS, OSLO NORWAY, PERMANGANATE VALUES,
   SCHIZOPHYCEAE, SELENASTRUM CAPRICORNUTUM, TABELLARIA FLOCCULOSA.

   ABSTRACT:
     THE AUTHOR LISTS AMONG THE PROBLEMS CAUSED BY EXCESSIVE ALGAL GROWTH,
   FILTER-BLOCKING, IMPARTATION OF TASTE AND ODOR TO WATER SUPPLIES,
   DISCOLORATION OF WATER, PROVISION OF NUTRIENTS FOR UNDESIRABLE
   HETEROTROPHS IN PIPES, AND FORMATION OF SEDIMENTS IN LAKES. TO
   ILLUSTRATE DIFFERENCES IN TROPHIC STATUS AMONG LAKES, HE COMPARES TWO
   LAKES IN NORWAY: LAKE MARIDALSVANNET (3.9 SQUARE KILOMETERS; MAXIMUM
   DEPTH, 45 METERS) WHICH PROVIDES OSLO WITH DRINKING WATER, IS
   OLIGOTROPHIC, AND SHOWS LITTLE ANNUAL VARIATION IN ITS LOW
   CONCENTRATIONS OF DISSOLVED NUTRIENTS. LAKE BORREVANNET (1.8 SQUARE KM;
   MAX DEPTH, 16 METERS) IS EUTROPHIC AND SHOWS IMPORTANT ANNUAL CHANGES
   WITH RESPECT TO CHEMISTRY AND WATER QUALITY. AUTHOR STATES THAT LAKE
   LUGANO, NORTHERN ITALY, HAS EUTROPHIED DUE TO POLLUTION ENRICHMENT,
   CITING CLASSIC EXAMPLE OF LAKE ZURICH AS EVIDENCE THAT LACUSTRINE
   TROPHIC CHANGES ARE BEST RECORDED IN THE SEDIMENTS. UTILIZING
   SELENASTRUM (OR ANKISTRODESMUS) CAPRICORNUTUM, AN ORGANISM REQUIRING
   ONLY A FEW DAYS TO ATTAIN PEAK GROWTH, AUTHOR DEVISED A BIOASSAY TO
   ASSESS EFFECTS AND CONCENTRATIONS OF NUTRIENTS IN NATURAL WATERS. SUCH
   AN APPROACH WOULD APPEAR TO HAVE ITS GREATEST VALUE IN COMPARATIVE
   LIMNOLOGICAL STUDIES. (SEE ALSO W69-07832). (EICHHORN-WISC)

   FIELD 05C

   ACCESSION NO. W69-07833

TRANSFORMATIONS IN INFILTRATION PONDS AND IN THE SOIL LAYERS IMMEDIATELY UNDERNEATH,

RESEARCH INSTITUTE FOR PUBLIC HEALTH ENGINEERING TNO, THE HAGUE (NETHERLANDS). WATER, SOIL, AND AIR DIV.

J. K. BAARS.

PROC RUDOLFS RES CONF, RUTGERS UNIV, NEW BRUNSWICK, NJ. PRINCIPLES AND APPLICATIONS IN AQUATIC MICROBIOLOGY, HEUKELEKIAN, H AND DONDERO, NORMAN C (EDS), JOHN WILEY AND SONS, INC, NEW YORK, PP 344-365, 1964. 8 FIG, 2 TAB, 9 REF, DISC.

DESCRIPTORS:
*INFILTRATION, *PONDS, *SOIL, GROUNDWATER BASINS, WATER SUPPLY, MUD, BENTHOS, HYDROGEN, BACTERIA, AEROBIC BACTERIA, METABOLISM, CANALS, SANDS, CHLORINATION, MICROORGANISMS, CHEMICAL ANALYSIS, TEMPERATURE, NITRATES, PLANTS, AMMONIA, HARDNESS(WATER), ACIDITY, OXYGEN, ALGAE, SELF-PURIFICATION, BACTERIOPHAGE, RESERVOIRS, SPORES, NITRITES, IRON, ORGANIC MATTER, SUBSOIL, VELOCITY, OXIDATION, DENITRIFICATION, SULFATES, REDUCTION(CHEMICAL), FERMENTATION, METHANE, HYDROGENATION, CHEMICAL OXYGEN DEMAND, BIOCHEMICAL OXYGEN DEMAND, POTASSIUM COMPOUNDS, CARBON DIOXIDE, PROTOZOA, SEWAGE, SLUDGE, E COLI.

IDENTIFIERS:
*TRANSFORMATIONS, AMSTERDAM, HAARLEM, LEYDEN, THE HAGUE, MINERALIZATION, CLAY LENSES, RHINE RIVER, AMSTERDAM RHINE, POLDERS, COMPOSITION, PSEUDOMONAS, MYCOBACTERIUM, BACILLUS SUBTILIS, BACILLUS MYCOIDES, BACILLUS MESENTERICUS, DIATOMEA, CLADOPHORA, FLAGELLATA, CILIATES, RANUNCULUS CIRCINATUS, POTAMOGETON PUSILLUS.

ABSTRACT:
WATER FROM STORAGE PONDS, FILTERED THROUGH BOTTOM MUD AND A BODY OF SAND, WAS MEASURED FOR CHLORIDE, FORMS OF NITROGEN, OXYGEN, ORGANIC MATTER, ALGAE, AND BACTERIAL CONTENT. TO STUDY BENTHOS, WATER WAS ANALYZED FROM WELLS CONSTRUCTED AT DISTANCES FROM A POND IN THE DIRECTION OF FLOW. NOT ONLY DISSOLVED ORGANIC MATTER, AS DETERMINED BY THE POTASSIUM PERMANGANATE METHOD, IS MINERALIZED, BUT MUCH MORE OXIDANT (FREE OXYGEN PLUS NITRATE OXYGEN) IS USED, APPARENTLY IN THE BENTHOS. THE HIGH OXIDANT CONSUMPTION INDICATES THAT THE REDUCTION OF ORGANIC MATTER IS INTENSE. ORGANIC SUBSTANCES, DETECTABLE BY TASTE, ARE NOT COMPLETELY ELIMINATED. WHEN THERE IS A CONSTANT SUPPLY OF NEW OXIDANT, DEAD BIOLOGICAL MATERIALS MAY BE MINERALIZED AEROBICALLY. IF THIS SUPPLY IS TERMINATED, ANAEROBIC TRANSFORMATIONS MAY DOMINATE. PROBABLY BACTERIA ARE STRONGLY ADSORBED TO THE SAND GRAINS IN CONCENTRATIONS RESULTING IN INTENSE STRUGGLE FOR LIFE AND DECREASE RAPIDLY WITH INCREASING DISTANCE FROM THE POND. SAMPLES FROM THE BOTTOM OF THE PONDS DOWNWARD SHOWED PRONOUNCED DECREASE IN BACTERIA AT 0.5-METER DEPTH. A SMALL AMOUNT OF SILT IN THE WATER MAY CAUSE CLOGGING OF THE SOIL. THE INTERMITTENT INFILTRATION SYSTEM AFFORDS THESAME PERMEABILITY OF THE SAND EACH YEAR WITH THE CONSEQUENT USE OF AVAILABLE PURIFYING ACTORS. (JONES-WIS)

FIELD 05D

ACCESSION NO. W69-07838

EFFECTS OF TEMPERATURE ON THE SORPTION OF RADIONUCLIDES BY A BLUE-GREEN ALGA,

DU PONT DE NEMOURS (E. I.) AND CO., AIKEN, S. C. SAVANNAH RIVER LAB.

R. S. HARVEY.

AVAILABLE FROM CLEARINGHOUSE AS CONF 670503 AT $3.00 IN PAPER COPY AND $0.65
IN MICROFICHE. SYMPOSIUM ON RADIOECOLOGY, PROC 2ND NATL SYMP, ANN ARBOR,
MICH, PP 266-269, MAY 15-17, 1967. 7 FIG, 2 TAB, 5 REF.

DESCRIPTORS:
  *ALGAE, *RADIOISOTOPES, *SORPTION, *TEMPERATURE, SOUTH CAROLINA,
CESIUM, STRONTIUM RADIOISOTOPES, ZINC RADIOISOTOPES, IRON, MANGANESE,
BENTHIC FLORA, STREAMS, CULTURES, COBALT RADIOISOTOPES.

IDENTIFIERS:
  SAVANNAH RIVER LABORATORY, PLECTONEMA BORYANUM, REACTOR.

ABSTRACT:
  THE EFFECTS OF TEMPERATURE DIFFERENTIAL (15 DEGREES CELSIUS) ON
SORPTION OF CESIUM-137, ZINC-65, IRON-59, COBALT-57, AND MANGANESE-54,
BY THE FILAMENTOUS CYANOPHYTE, PLECTONEMA BORYANUM, ARE REPORTED. THE
ALGA WAS COLLECTED FROM REACTOR EFFLUENT STREAMS. UNIALGAL CULTURES
WERE DEVELOPED. CULTURES WERE GROWN CONCURRENTLY AT FOUR WATER
TEMPERATURES AND SAMPLES WEIGHED AND RADIO-ASSAYED AFTER EXPOSURES OF
INCREASING TIMES. THIS SPECIES GREW BEST BETWEEN 30 C AND 40 C. GROWTH
WAS NOT AFFECTED BY LOW CONCENTRATIONS OF RADIONUCLIDES IN THE MEDIUM.
RADIONUCLIDE CONCENTRATIONS AT THE VARIOUS WATER TEMPERATURES WERE
COMPARED. FOR A GIVEN WATER TEMPERATURE, SORPTION LEVELS DIFFERED FOR
THE RADIONUCLIDES STUDIED BECAUSE OF VARIANCES IN SPECIFIC ACTIVITY AND
BIOLOGICAL DEMAND FOR VARIOUS ELEMENTS. THE ESSENTIAL ELEMENTS OF
MANGANESE-54, XENON-65, COBALT-57, AND IRON-59, WERE SORBED TO HIGHER
LEVELS THAN WERE THE NONESSENTIAL ELEMENTS, STRONTIUM-85 AND
CESIUM-137, PERHAPS DUE IN PART TO THEIR PHYSICAL STATE; ONLY
CESIUM-137 AND STRONTIUM-85 WERE PRESENT, MAINLY IN IONIC FORM. SINCE
SORPTION LEVELS FOR THE VARIOUS RADIONUCLIDES WERE RAISED OR LOWERED BY
FACTORS LESS THAN 2.5 BY THE TEMPERATURE DIFFERENTIAL, THE CONCLUSION
WAS DRAWN THAT NONLETHAL VARIATIONS IN WATER TEMPERATURE HAVE NO MAJOR
INFLUENCE ON THE SORPTION BY P BORYANUM OF THE RADIONUCLIDES TESTED.

FIELD 05C

ACCESSION NO. W69-07845

ACCUMULATION OF RADIUM-226 IN TWO AQUATIC ECOSYSTEMS,

UTAH STATE UNIV., LOGAN. DEPT. OF WILDLIFE RESOURCES.

SUSAN S. MARTIN, WILLIAM T. HELM, AND WILLIAM F. SIGLER.

AVAILABLE FROM CLEARINGHOUSE AS CONF 670503 AT $3.00 IN PAPER COPY AND $0.65
IN MICROFICHE. SYMPOSIUM ON RADIOECOLOGY, PROC 2ND NATL SYMP, ANN ARBOR,
MICH, PP 307-318, MAY 15-17, 1967. 10 FIG, 14 REF, DISC. CONF-670503.

DESCRIPTORS:
 *AQUATIC ENVIRONMENTS, *ECOSYSTEMS, *RADIUM RADIOISOTOPES, *URANIUM
 RADIOISOTOPES, RIVERS, MILLS, COLORADO, WASTES, BIOLOGICAL COMMUNITIES,
 SEDIMENTS, BIOTA, FISH, UTAH, ALGAE, INSECTS, INVERTEBRATES, BACKGROUND
 RADIATION, FLOCCULATION, POLLUTANTS, SAMPLING.

IDENTIFIERS:
 ANIMAS RIVER, DURANGO, SAN MIGUEL RIVER, DOLORES RIVER, URAVAN, COTTUS
 SPP, RHINICHTHYS OSCULUS, SUCKERS, ICTALURUS PUNCTATUS, TRICHOPTERA,
 ODONATA, EPHEMEROPTERA, PLECOPTERA, DIPTERA, COLEOPTERA, HEMIPTERA,
 MOLLUSCA, RAFFINATE, THORIUM, TAILINGS.

ABSTRACT:
 RADIUM-226, IN URANIUM PROCESSING WASTES, SERIOUSLY AFFECTS STREAM
 BIOTA. ITS ACCUMULATION IN BIOTIC COMPONENTS WAS FOLLOWED FOR THREE
 YEARS. THE ANIMAS RIVER, COLORADO, AFTER DIVERSION OF THE MOST TOXIC OF
 A MILL'S EFFLUENTS, SHOWED EXCELLENT RECOVERY BIOLOGICALLY. WATER,
 SEDIMENT, ALGAE, INVERTEBRATES, AND FISH, WERE SAMPLED ABOVE AND BELOW
 MILL SITES. COMPARISON OF BACKGROUND LEVELS AND RADIUM ACCUMULATION
 SHOWED DIFFERENCES IN UPTAKE BY VARIOUS FISH SPECIES. A CONCENTRATION
 FACTOR (THE RATIO OF THE CONCENTRATION OF A PARTICULAR RADIONUCLIDE IN
 THE ORGANISM TO ITS CONCENTRATION IN THE AQUEOUS MEDIUM) SHOWED THE
 MEAN RADIUM-226 CONTENT DECREASING WITH DOWNSTREAM DISTANCE. DURING THE
 LAST TWO YEARS, RADIUM CONTENT OF SEDIMENTS AVERAGED ONLY ABOUT THREE
 TIMES BACKGROUND LEVELS, EARLIER SAMPLES CONTAINED UP TO TWENTY TIMES,
 WHILE STILL EARLIER WORK INDICATED UP TO SEVERAL HUNDRED TIMES THE
 BACKGROUND LEVELS OF RADIUM-226. THE SAN MIGUEL-DOLORES RIVER SYSTEM
 WAS IN POOR BIOLOGICAL CONDITION DUE PRIMARILY TO CHEMICAL POLLUTION
 FROM A URANIUM MILL. BIOTA SAMPLES CONTAINED INCREASING AMOUNTS OF
 RADIUM-226 AS DISTANCE BELOW THE MILL INCREASED, ALTHOUGH RADIUM-226
 CONTENT OF THE WATER AND SEDIMENTS DECREASED WITH DISTANCE. APPARENTLY
 ORGANISMS COULD NOT SURVIVE LONG ENOUGH IN THE UPPER POLLUTION ZONE TO
 ACCUMULATE THE AMOUNTS OF RADIUM-226 WHICH MIGHT OTHERWISE BE FOUND.
 (SEE ALSO VOL. 2, NO. 18, FIELD 5C, W69-07441) (JONES-WISC)

FIELD 05C

ACCESSION NO. W69-07846

EFFECT OF HANFORD REACTOR SHUTDOWN ON COLUMBIA RIVER BIOTA,

BATTELLE-NORTHWEST, RICHLAND, WASH. PACIFIC NORTHWEST LAB.

D. G. WATSON, C. E. CUSHING, C. C. COUTANT, AND W. L. TEMPLETON.

AVAILABLE FROM CLEARINGHOUSE AS CONF 670503 AT $3.00 IN PAPER COPY AND $0.65
IN MICROFICHE. SYMP ON RADIOECOLOGY, PROC 2ND NATIONAL SYMP, MAY 15-17,
1967, ANN ARBOR, MICH, NELSON, DANIEL J AND EVANS, FRANCIS C (EDS). US
ATOMIC ENERGY COMM, DOC CONF 670503, PP 291-299. 8 FIG, 23 REF, DISC.

DESCRIPTORS:
  *BIOTA, *COLUMBIA RIVER, CHROMIUM, PHOSPHORUS RADIOISOTOPES, TROPHIC
  LEVEL, ZINC RADIOISOTOPES, IRON, SEDIMENTS, FISH, RADIOECOLOGY,
  EFFLUENTS, ECOSYSTEMS, PLANKTON, ALGAE, PERIPHYTON, INVERTEBRATES,
  ADSORPTION, PHYTOPLANKTON, SUCKERS, SHINERS, CADDISFLIES, TEMPERATURE,
  FLOW, DIATOMS, MANGANESE.

IDENTIFIERS:
  *HANFORD(WASH), *REACTOR, RADIONUCLIDES, ASSIMILATION, BA-LA,
  HALF-LIFE, PTYCHOCHEILUS OREGONENSIS, ACROCHEILUS ALUTACEUS, ULOTHRIX,
  CLADOPHORA, PROSOPIUM WILLIAMSONI.

ABSTRACT:
  CLOSURE OF THE HANFORD REACTORS FOR AN EXTENDED PERIOD CAUSED RAPID AND
  EXTENSIVE DECLINE IN CONCENTRATION OF RADIONUCLIDES. CHROMIUM-51
  (CR-51) AND PHOSPHORUS-32 (P-32) DECREASED TWO TO THREE ORDERS OF
  MAGNITUDE IN LOWER TROPHIC LEVELS. THE RAPID CHANGE OF P-32 WAS
  PROBABLY DUE TO ITS RELATIVE SHORT PHYSICAL HALF-LIFE AND ITS RAPID
  TURNOVER IN AQUATIC ORGANISMS. THIS WAS CHIEFLY A BIOLOGICAL PROCESS,
  NOT MERELY SURFACE ADSORPTION. ALTHOUGH CR-51 WAS ABUNDANT, ITS LOW
  CONCENTRATION IN HIGHER LEVELS INDICATES ITS MINOR BIOLOGICAL
  IMPORTANCE. ITS RELATIVELY HIGH CONCENTRATIONS IN PLANKTON, ALGAE, AND
  INVERTEBRATES, WAS PROBABLY DUE TO ADSORPTION RATHER THAN ASSIMILATION.
  DECLINES IN CONCENTRATIONS OF ZINC-65 (ZN-65), MANGANESE-54, AND
  IRON-59, WERE MUCH LESS. THEIR LESSER DECLINE IN BIOTA AS OPPOSED TO
  RIVER WATER SUGGESTED THAT THESE NUCLIDES EITHER ARE TURNED OVER AT A
  VERY SLOW RATE IN THE ORGANISMS, OR THAT THEY WERE AVAILABLE FROM PARTS
  OF THE ECOSYSTEM OTHER THAN WATER. IN FISH P-32 WAS LOST RAPIDLY, ZN-65
  SLOWLY. THE SLOWER RATE OF CHANGE OF RADIONUCLIDES BY FISH INDICATED
  DIFFERENCE IN UPTAKE ROUTES. NEAR EQUILIBRIUM CONCENTRATIONS OF
  RADIONUCLIDES IN MOST RIVER ORGANISMS WERE APPROACHED WITHIN TWO OR
  THREE WEEKS AFTER RESUMPTION OF REACTOR OPERATION.
  (JONES-WISC)

FIELD 05C, 05B

ACCESSION NO. W69-07853

THE ROLE OF TUBIFICID WORMS IN THE TRANSFER OF RADIOACTIVE PHOSPHOROUS IN AN
AQUATIC ECOSYSTEM,

WESTERN MICHIGAN UNIV., KALAMAZOO.

BERT K. WHITTEN, AND CLARENCE J. GOODNIGHT.

AVAILABLE FROM CLEARINGHOUSE AS CONF 670503 AT $3.00 IN PAPER COPY AND $0.65
IN MICROFICHE. SYMP ON RADIOECOLOGY, PROC 2ND NATIONAL SYMP, MAY 15-17,
1967, ANN ARBOR, MICH, NELSON, DANIEL J AND EVANS, FRANCIS C (EDS). US
ATOMIC ENERGY COMM, DOC CONF 670503, PP 270-277. 2 FIG, 2 TAB, 20 REF.

DESCRIPTORS:
  *TUBIFICIDS, *ECOSYSTEMS, *PHOSPHORUS RADIOISOTOPES, *TRANSFER, WATER,
  BACTERIA, SEDIMENTS, MINNOWS, DETRITUS, CYCLING NUTRIENTS, PLANKTON,
  ALGAE, INVERTEBRATES, TURBIDITY, WORMS, SUNFISHES.

IDENTIFIERS:
  RADIOAUTOGRAPHS, HALF-LIFE, CONCENTRATION FACTOR, LIMNODRILUS SPP,
  AEOLOSOMA HEMPRICHI, ESCHERICHIA COLI, MYRIOPHYLLUM, LEPOMIS
  MACROCHIRUS, PIMEPHALES NOTATUS.

ABSTRACT:
  ACCUMULATION OF RADIOPHOSPHORUS BY TUBIFICID WORMS (COMMON HABITAT
  BOTTOM SEDIMENTS OF STREAMS AND LAKES) FROM WATER, BACTERIA, AND
  SEDIMENT, WAS STUDIED. TUBIFICID WORMS TOOK UP MORE PHOSPHORUS FROM
  WATER AND BACTERIA THAN FROM SEDIMENT. THESE WORMS WERE ABLE TO
  ACCUMULATE PHOSPHORUS FROM BOTH SOLUBLE AND ORGANIC PARTICULATE
  SOURCES. RADIOAUTOGRAPHS DEMONSTRATED THAT MUCH OF THE ACTIVITY WAS IN
  THE TISSUES OF THESE WORMS AND NOT SIMPLY ADSORBED. NO SIGNIFICANT
  ACCUMULATION FROM RADIOACTIVE ORTHOPHOSPHATE, WHICH WAS ADSORBED ONTO
  STERILE SEDIMENTS, WAS OBSERVED. BLUEGILLS AND BLUNTNOSE MINNOWS FED
  TUBIFICID WORMS LABELLED WITH RADIOPHOSPHORUS ACCUMULATED
  RADIOPHOSPHORUS IN THEIR TISSUES. THEORETICAL CALCULATIONS BASED ON
  THESE FEEDING STUDIES DEMONSTRATED THAT AN EQUILIBRIUM WAS APPROACHED
  AFTER TWO WEEKS IN THE TISSUES OF THESE FISH AND WAS MAINTAINED THROUGH
  THE DURATION OF THE EXPERIMENT. TUBIFICID WORMS, BECAUSE OF THEIR
  ABUNDANCE IN SOME AQUATIC ECOSYSTEMS, MAY HAVE AN IMPORTANT ROLE IN
  TRANSFER OF RADIOACTIVE PHOSPHORUS FROM WATER AND DETRITUS TO OTHER
  COMPONENTS OF THE AQUATIC ECOSYSTEM, SUCH AS FISH. THESE WORMS APPEAR
  TO FUNCTION IN CONJUNCTION WITH BACTERIA IN THE RECYCLING OF
  RADIOPHOSPHORUS FROM DETRITUS AND SEDIMENTS. THIS EXPERIMENTATION IS
  INTENDED TO SUGGEST POSSIBLE RELATIONSHIPS WHICH MAY OCCUR IN THE
  ENVIRONMENT.
  (JONES-WISC)

FIELD 05C

ACCESSION NO. W69-07861

RADIONUCLIDE CYCLING BY PERIPHYTON: AN APPARATUS FOR CONTINUOUS IN SITU
MEASUREMENTS AND INITIAL DATA ON ZINC-65 CYCLING,

BATTELLE-NORTHWEST, RICHLAND, WASH. PACIFIC NORTHWEST LAB.

C. E. CUSHING, AND N. S. PORTER.

AVAILABLE FROM CLEARINGHOUSE AS CONF 670503 AT $3.00 IN PAPER COPY AND $0.65
IN MICROFICHE. SYMP ON RADIOECOLOGY, PROC 2ND NATIONAL SYMP, MAY 15-17,
1967, ANN ARBOR, MICH, NELSON, DANIEL J AND EVANS, FRANCIS C (EDS). US
ATOMIC ENERGY COMM, DOC CONF-670503, PP 285-290. 3 FIG, 1 TAB, 12 REF.

DESCRIPTORS:
*PERIPHYTON, *CYCLING NUTRIENTS, *ZINC RADIOISOTOPES, STREAMFLOW,
SAMPLING, ALGAE, ECOSYSTEMS, ENVIRONMENT, RETENTION, PHOTOPERIODISM,
LIGHT INTENSITY, VELOCITY, TEMPERATURE, CHEMICAL ANALYSIS,
PHYSIOLOGICAL ECOLOGY, OXYGEN, CARBON DIOXIDE, CARBON RADIOISOTOPES,
DETRITUS, COLUMBIA RIVER, ELECTRONICS, SCALING.

IDENTIFIERS:
*RADIONUCLIDE, ULOTHRIX, UPTAKE, HALF-LIFE, DETECTOR.

ABSTRACT:
THIS PART OF THE COLUMBIA RIVER PROGRAM CONCERNS THE ROLE OF THE
PERIPHYTON COMMUNITY IN CYCLING RADIONUCLIDES. DIFFICULTIES IN STUDYING
A LARGE RIVER WITH EXTREMELY UNSTABLE HYDROGRAPHY HAVE LED TO
DEVELOPMENT OF A SYSTEM HAVING ADVANTAGES OF LABORATORY CONTROLLED
EXPERIMENTS WHILE MAINTAINING SOME SEMBLANCE OF NATURAL CONDITIONS. IT
PERMITS THE CONTINUOUS MEASUREMENT OF UPTAKE AND CYCLING OF
RADIONUCLIDES BETWEEN STREAM PERIPHYTON AND A CONTROLLED AQUEOUS
ENVIRONMENT. THE SYSTEM WAS DESIGNED TO AVOID INHERENT DIFFICULTIES IN
AQUARIA STUDIES, THAT IS, THE NECESSITY OF DESTROYING OR SUB-SAMPLING
THE COMMUNITY FOR SEQUENTIAL ANALYSES, THE LACK OF A CONTINUOUS FLOW OF
WATER OVER ALGAE TO SIMULATE STREAM CONDITIONS, AND INABILITY TO
MAINTAIN A FIXED AMBIENT RADIONUCLIDE CONCENTRATION BECAUSE OF
IMMEDIATE UPTAKE AND RECYCLING BY THE ORGANISMS. APPARATUS PERMITS
EVALUATION OF THE EFFECT OF INDIVIDUAL PHYSICAL AND CHEMICAL
ENVIRONMENTAL FACTORS IN EITHER AN OPEN ONE-PASS SYSTEM OR IN A CLOSED
RECIRCULATING SYSTEM. PRELIMINARY RESULTS OF SIX UPTAKE AND TWO
RETENTION EXPERIMENTS USING ZINC-65 IN THE CLOSED SYSTEM ARE PRESENTED.
DESPITE COMMUNITY TYPE, THAT IS MATURE OR 'YOUNG', THE TIME OF
APPROXIMATE EQUILIBRIUM IS AROUND 20 HOURS AND THE ACTIVITY ACCUMULATED
ON A WEIGHT BASIS IS ABOUT 0.021 NANNOCURIES/MILLIGRAM DRY WEIGHT.
(JONES-WISC)

FIELD 05C, 07B

ACCESSION NO. W69-07862

REMOVAL OF NITROGENOUS COMPOUNDS FROM WASTEWATERS,

CARNEGIE-MELLON UNIV., PITTSBURGH, PA.

WILLIAM R. SAMPLES.

WATER REUSE, AMER INST CHEM ENG PROGR SYMP, SER NO 78, VOL 63, PP 223-229, 1967. 7 P, 1 FIG, 35 REF.

DESCRIPTORS:
*WASTE WATER TREATMENT, *WATER REUSE, *NITROGEN, NITROGEN COMPOUNDS, OXIDATION, BIODEGRADATION, NITRIFICATION, DENITRIFICATION, NUTRIENTS, TERTIARY TREATMENT, ALGAE, AMMONIA.

IDENTIFIERS:
NITROGEN REMOVAL.

ABSTRACT:
NORMAL TREATMENT OF OUR WATERBORNE WASTES HAS NOT IN THE PAST BEEN CONCERNED WITH REMOVAL OF NITROGENOUS MATERIALS. IN RECENT YEARS, HOWEVER, WASTE DISCHARGES CONTAINING COMPOUNDS OF NITROGEN HAVE BEEN INCRIMINATED IN THE FERTILIZATION OF STREAMS AND LAKES, CAUSING POSSIBLE HEALTH HAZARDS, AND IN GENERAL, CAUSING A DECREASE IN WATER QUALITY FOR MANY USES. A REVIEW OF MANY INVESTIGATIONS INTO MECHANISMS FOR REMOVAL OF NITROGENOUS MATERIALS IN WASTE EFFLUENTS IS GIVEN. THE MECHANISMS REVIEWED INCLUDE BOTH BIOLOGICAL AND CHEMICAL METHODS. NITROGENOUS MATERIALS MAY BE REMOVED FROM WASTE EFFLUENTS BY ANY ONE OF SEVERAL METHODS. THE MOST APPLICABLE AT THE PRESENT TIME APPEAR TO BE DENITRIFICATION, INCORPORATION BY ALGAE, AND AMMONIA STRIPPING. THE SELECTION OF THE PROPER METHOD FOR NITROGEN REMOVAL IN ANY PARTICULAR CASE WILL DEPEND ON THE LOCAL CIRCUMSTANCES, INCLUDING OTHER TREATMENTS REQUIRED, DEGREE OF NITROGEN REMOVAL DESIRED, CHEMICAL QUALITY OF THE WATER, FURTHER USES FOR THE WATER, AND MANY OTHERS. THE REMOVAL OF NITROGEN FROM EFFLUENTS WILL BE EXPENSIVE BUT WILL BY JUSTIFIABLE IN AN INCREASING NUMBER OF INSTANCES. (KNAPP-USGS)

FIELD 05D

ACCESSION NO. W69-08053

A DESIGN PROCEDURE FOR BIOLOGICAL NITRIFICATION AND DENITRIFICATION,

TEXAS UNIV., AUSTIN.

W. WESLEY ECKENFELDER, JR.

WATER REUSE, AMER INST CHEM ENG PROGR SYMP, SER NO 78, VOL 63, PP 230-234, 1967. 5 P, 6 FIG, 1 TAB, 6 REF, APPEND.

DESCRIPTORS:
*DENITRIFICATION, *BIODEGRADATION, *WATER REUSE, *RECLAIMED WATER, NITROGEN COMPOUNDS, NITROGEN, AMMONIA, NUTRIENTS, BIOCHEMICAL OXYGEN DEMAND, TERTIARY TREATMENT, ALGAE, BACTERIA, ACTIVATED SLUDGE.

IDENTIFIERS:
NITROGEN REMOVAL.

ABSTRACT:
THE REMOVAL OF NITROGEN FROM SEWAGE AND INDUSTRIAL WASTEWATERS IS ASSUMING INCREASING IMPORTANCE AS THE WASTE LOADINGS FROM URBAN AND INDUSTRIALIZED AREAS INCREASE. UNOXIDIZED NITROGEN EXERTS ON OXYGEN DEMAND ON THE RECEIVING WATERS AND OXIDIZED NITROGEN SERVES AS A NUTRIENT SOURCE FOR ALGAL GROWTH. IT IS POSSIBLE TO REMOVE NITROGEN BY BIOLOGICAL MEANS BY FIRST OXIDIZING AMMONIA TO NITRATES FOLLOWED BY DENITRIFICATION IN WHICH MICROORGANISMS REDUCE THE NITRATE TO NITROGEN GAS. STUDIES WHICH HAVE BEEN CONDUCTED ON THIS PROCESS TO DATE IN LABORATORY, PILOT-PLANT, AND, TO A LIMITED EXTENT, PLANT-SCALE INVESTIGATIONS ARE REVIEWED. THE THEORY OF THE PROCESS OPERATION IS DEVELOPED FROM BASIC MICROBIOLOGICAL CONSIDERATIONS. A DESIGN EXAMPLE FOR A SEWAGE TREATMENT PLANT IS PRESENTED TO ILLUSTRATE THE DEVELOPMENT OF THE PROCESS. (KNAPP-USGS)

FIELD 05D

ACCESSION NO. W69-08054

INFLUENCE OF ENVIRONMENTAL FACTORS ON THE CONCENTRATIONS OF ZN-65 BY AN
EXPERIMENTAL COMMUNITY,

BUREAU OF COMMERCIAL FISHERIES, BEAUFORT, N. C. RADIOBIOLOGICAL LAB.

T. DUKE, J. WILLIS, T. PRICE, AND K. FISCHLER.

AVAILABLE FROM CLEARINGHOUSE AS CONF 670503 AT $3.00 IN PAPER COPY AND $0.65
IN MICROFICHE. SYMP ON RADIOECOLOGY, PROC 2ND NATIONAL SYMP, MAY 15-17,
1967, ANN ARBOR, MICH, NELSON, DANIEL J AND EVANS, FRANCIS C (EDS). US
ATOMIC ENERGY COMM, DOC CONF 670503, PP 355-362. 3 FIG, 1 TAB, 18 REF.

DESCRIPTORS:
 *ZINC RADIOISOTOPES, *ENVIRONMENTAL EFFECTS, SALINITY, TEMPERATURE, SEA
WATER, OYSTERS, CLAMS, CRABS, NORTH CAROLINA, SEDIMENTS, RADIOACTIVITY,
MARINE ANIMALS, ESTUARINE ENVIRONMENT, FALLOUT, NUCLEAR EXPLOSIONS,
BIOTA, SPECTROMETERS, BACTERIA, ALGAE, SANDS, HYDROGEN ION
CONCENTRATION.

IDENTIFIERS:
 *CONCENTRATIONS, AEQUIPECTEN IRRADIANS, PIVERS ISLAND(N C), DETECTOR,
BEAUFORT(N C), CRASSOSTREA VIRGINICA, MERCENARIA MERCENARIA, PANOPEUS
HERBSTII.

ABSTRACT:
 CAPACITY OF ORGANISMS TO CONCENTRATE ELEMENTS FROM SEAWATER IS AFFECTED
BY CHANGES IN ENVIRONMENTAL FACTORS THAT ALTER THE PHYSIOLOGICAL
CONDITION OF THE ORGANISMS OR THE PHYSICAL-CHEMICAL PROPERTIES OF
ELEMENTS. THIS KNOWLEDGE IS NEEDED TO EVALUATE POTENTIAL HAZARDS FROM
RADIOACTIVE POLLUTION. ZINC-65 (ZN-65) ENTERS THE BIOCHEMICAL CYCLE OF
THE ESTUARINE ENVIRONMENT IN FALLOUT FROM NUCLEAR WEAPONS AND IN
EFFLUENT FROM NUCLEAR REACTORS, OCCURRING IN MEASURABLE AMOUNTS IN
ESTUARINE BIOTA, SEDIMENTS, AND WATER. A POLYFACTORIAL APPROACH IS
REQUIRED TO STUDY THE CONCENTRATION OF RADIOISOTOPES. DIFFERENCES IN
SALINITY, TEMPERATURE, PH, AND ZINC IN SEAWATER SIGNIFICANTLY AFFECTED
THE CONCENTRATION OF ZN-65 OF A COMMUNITY OF OYSTERS, CLAMS, MUD CRABS,
AND SCALLOPS, AND THEIR SEDIMENT SUBSTRATE. A FACTORIAL ANALYSIS OF
VARIANCE SHOWED NO INTERACTIONS AMONG THE ENVIRONMENTAL FACTORS. HIGH
SALINITY AND ZINC CONCENTRATION SUPPRESSED THE CONCENTRATION OF ZN-65
IN ANIMALS AND SEDIMENT, WHEREAS HIGH TEMPERATURE AND PH HAD THE
OPPOSITE EFFECT. TEMPORAL VARIATION IN CONCENTRATION FACTORS FOR
CONTROL ANIMALS AND SEDIMENT WAS SIGNIFICANT DURING THE EXPERIMENT.
SUCH VARIATION MUST BE CONSIDERED WHEN REPORTING CONCENTRATION FACTORS
FOR RADIONUCLIDES IN MARINE ORGANISMS. DATA FROM THIS EXPERIMENT
REPRESENT A FIRST STEP TOWARD PREDICTION OF THE FATE OF A RADIOISOTOPE
RELEASED INTO AN ESTUARY.
(JONES-WISC)

FIELD 05C

ACCESSION NO. W69-08267

DISTRIBUTION OF RADIONUCLIDES IN THE ENVIRONMENT OF ENIWETOK AND BIKINI ATOLS,
   AUGUST 1964,

WASHINGTON UNIV., SEATTLE. COLL. OF FISHERIES.

A. D. WELANDER.

SYMP ON RADIOECOLOGY, PROC 2ND NATL SYMP, MAY 15-17, 1967, ANN ARBOR, MICH,
   NELSON, DANIEL J AND EVANS, FRANCIS C (EDS). US ATOMIC ENERGY COMM, DOC
   CONF 670503, PP 346-354. 8 TAB, 7 REF, DISC. AVAILABLE FROM CLEARINGHOUSE
   AS CONF 670503 AT $3.00 IN PAPER COPY AND $0.65 IN MICROFICHE.

DESCRIPTORS:
   *RADIOISOTOPES, *DISTRIBUTION, ANIMALS, PLANTS, BOTTOM SEDIMENTS,
   SOILS, NUCLEAR EXPLOSIONS, COBALT RADIOISOTOPES, CESIUM, STRONTIUM
   RADIOISOTOPES, MANGANESE, ALGAE, IRON, BIRDS, RADIOACTIVITY,
   INVERTEBRATES, PLANKTON, FISH, SEA WATER, GAMMA RAYS, RADIOCHEMICAL
   ANALYSIS, TRITIUM, CARBON RADIOISOTOPES, BIOTA, CLAMS, FALLOUT, PACIFIC
   OCEAN, SPECTROMETERS.

IDENTIFIERS:
   *ENIWETOK ATOLL, *BIKINI ATOLL, RUTHENIUM-106, ANTIMONY-125,
   GROUNDWATER, BISMUTH-207, VERTEBRATES, PLUTONIUM-239, RATS,
   ZIRCONIUM-95-NIOBIUM, EUROPIUM-155, SILVER-110M, CERIUM, GUETTARDA,
   IPOMOEA, PISONIA, PANDANUS, MUSCLE, BONE, LEAVES, SHELL, RHODIUM-102.

ABSTRACT:
   RADIONUCLIDE ANALYSES WERE MADE OF MORE THAN 2000 SAMPLES OF ANIMALS,
   PLANTS, WATER, BOTTOM SEDIMENTS, AND SOILS, COLLECTED AT ENIWETOK AND
   BIKINI ATOLLS, SITE OF 59 NUCLEAR TESTS BETWEEN 1946 AND 1958.
   COBALT-60 WAS FOUND IN ALL SAMPLES BUT WAS DOMINANT IN THE MARINE
   ENVIRONMENT. CESIUM-137 (CS-137) AND STRONTIUM-90, HOWEVER,
   PREDOMINATED IN TERRESTRIAL ENVIRONMENTS. ALL SAMPLES CONTAINED TRACES
   OF MANGANESE-54, ALTHOUGH THERE WERE SLIGHTLY LARGER AMOUNTS OF THIS
   RADIONUCLIDE IN LAND PLANTS AND LAND INVERTEBRATES. RUTHENIUM-106 AND
   ANTIMONY-125 WERE DETECTED IN SIGNIFICANT AMOUNTS IN GROUNDWATER AND
   SOIL, WITH TRACES IN ANIMALS AND PLANTS. TRACE AMOUNTS OF BISMUTH-207
   WERE ALSO FOUND IN MOST SAMPLES, AND CERIUM-144 USUALLY IN ALGAE, SOILS
   AND LAND PLANTS. THERE WERE COMPARATIVELY LARGE AMOUNTS OF IRON-55 IN A
   NUMBER OF SAMPLES, ESPECIALLY IN VERTEBRATES. PLUTONIUM-239 WAS FOUND
   IN SOIL AND IN SKINS OF RATS AND BIRDS, IN THE FEW SAMPLES ANALYZED.
   SOILS CONTAINED HIGHEST LEVELS OF RADIOACTIVITY, FOLLOWED BY
   INVERTEBRATES, GROUNDWATER, SHOREBIRDS, PLANTS, RATS, PLANKTON, ALGAE,
   FISH, BOTTOM SEDIMENTS, SEAWATER, AND SEABIRDS. IN LAND SAMPLES, CS-137
   PROVED TO BE BEST MEASURE OF DISTRIBUTION OF RADIOACTIVITY IN
   BIOENVIRONMENTAL SAMPLES. HIGHEST RADIOACTIVITY WAS FOUND IN SAMPLES
   COLLECTED NEAR TEST SITES.
   (JONES-WIS)

FIELD 05C

ACCESSION NO. W69-08269

MEASUREMENTS OF BACKGROUND RADIATION IN AQUATIC HABITATS IN ALASKA,

DARTMOUTH COLL., HANOVER, N. H. DEPT. OF BIOLOGICAL SCIENCES; AND ARMY
  TERRESTRIAL SCIENCES CENTER, HANOVER, N. H.

G. E. LIKENS, AND P. L. JOHNSON.

SYMP ON RADIOECOLOGY, PROC 2ND NATL SYMP, MAY 15-17, 1967, ANN ARBOR, MICH,
NELSON, DANIEL J AND EVANS, FRANCIS C (EDS). US ATOMIC ENERGY COMM, DOC
CONF 670503, PP 319-328. 3 FIG, 2 TAB, 21 REF, DISC. AVAILABLE FROM
CLEARINGHOUSE AS CONF 670503 AT $3.00 IN PAPER COPY AND $0.65 IN MICROFICHE.

DESCRIPTORS:
  *BACKGROUND RADIATION, *AQUATIC HABITATS, *ALASKA, LAKES, RIVERS, HOT
  SPRINGS, TERRESTRIAL HABITATS, DEPTH, AIR, SEDIMENTS, GEOLOGIC
  FORMATIONS, ARCTIC, TUNDRA, SCALING, GRAVELS, MUD, SANDS, SHALES,
  BOTTOM SEDIMENTS, CATTAILS, RADIOACTIVITY, INTERFACES, ORGANIC MATTER,
  RADIOISOTOPES, GAMMA RAYS, URANIUM RADIOISOTOPES, RADIUM RADIOISOTOPES,
  POTASSIUM RADIOISOTOPES, SPECTROMETERS, EROSION, FALLOUT, LITTORAL,
  ALGAE, BIOTA, GENETICS, PLANKTON.

IDENTIFIERS:
  DETECTOR, ALASKA, HUDEUC LAKE, PULLIN LAKE, GRAPHITE LAKE, OLD JOHN
  LAKE, PINGO LAKE, TWELVE MILE LAKE, BRANT LAKE, IMIKPUK LAKE, NORTH
  MEADOW LAKE, GYTTJA, DUFF, OOZE, COBBLES, EQUISETUM, THORIUM-232, YUKON
  RIVER, NEBESNA RIVER, MOOSE CREEK, ARCTIC CIRCLE HOT SPRINGS, RADON,
  CENTRAL CITY(COLO), MUTATION, BETA RAYS, COSMIC RADIATION, STEESE
  HIGHWAY, POINT BARROW.

ABSTRACT:
  HIGH HUMAN WHOLE-BODY COUNTS OF ESKIMO POPULATION IN NORTHERN ALASKA
  INTRODUCED THROUGH FALLOUT PARTICLES FROM NUCLEAR WEAPONS TESTING
  PROMPTED STUDY OF BACKGROUND RADIATION IN VARIOUS ALASKAN LAKES.
  MEASUREMENTS OF IONIZING RADIATION, MADE DURING 1965 IN 9 LAKES, 3
  RIVERS, 1 HOT SPRINGS AREA, AND ADJACENT TERRESTRIAL SUBSTRATES IN
  ALASKA, INDICATED THAT BACKGROUND RADIATION IN THESE HABITATS VARIED
  GREATLY. THE LAKES WERE CHARACTERIZED BY A SMALL AMOUNT OF IONIZING
  RADIATION AT MID-DEPTHS AND INCREASING QUANTITIES NEAR THE AIR AND
  SEDIMENT BOUNDARIES. MOREOVER, RADIOACTIVITY OF SEDIMENTS NEAR SHORE
  WAS 3.2 TIMES GREATER AS THAT NEAR THE CENTER. THIS PATTERN OF
  BACKGROUND RADIATION HAS BEEN RELATIVELY CONSISTENT IN NATURAL LAKES OF
  DIVERSE GEOGRAPHIC AREAS; HENCE, A MODEL FOR BACKGROUND RADIATION IN
  FRESHWATER LAKES IS PROPOSED. VARIATIONS IN THIS MODEL WERE A FUNCTION
  OF LOCAL DIFFERENCES IN GEOLOGIC SUBSTRATES AND INPUT OF ALLOCHTHONOUS
  MATERIALS. RADIOACTIVITY FROM SEDIMENTS IN ARCTIC TUNDRA PONDS WAS DUE
  TO NATURALLY OCCURRING RADIONUCLIDES WITH PROBABLE EFFECTS ON BIOTA.
  AUTHORS SPECULATE THAT ORGANISMS LIVING IN OR NEAR SUBLITTORAL
  SEDIMENTS OF LAKES IN RECENT TIMES MIGHT UNDERGO MORE FREQUENT GENETIC
  CHANGES THAN PLANKTON OR ORGANISMS IN DEEPWATER SEDIMENTS. (JONES-WISC)

FIELD 05C

ACCESSION NO. W69-08272

THE PHOSPHORUS AND ZINC CYCLES AND PRODUCTIVITY OF A SALT MARSH,

GEORGIA UNIV., ATHENS. DEPT. OF ZOOLOGY.

L. R. POMEROY, R. E. JOHANNES, E. P. ODUM, AND B. ROFFMAN.

SYMP ON RADIOECOLOGY, PROC 2ND NATL SYMP, MAY 15-17, 1967, ANN ARBOR, MICH, NELSON, DANIEL J AND EVANS, FRANCIS C (EDS). US ATOMIC ENERGY COMM, DOC CONF 670503, PP 412-419. 3 FIG, 1 TAB, 33 REF.

DESCRIPTORS:
*ZINC RADIOISOTOPES, *PHOSPHORUS RADIOISOTOPES, *SALT MARSHES, *CYCLING NUTRIENTS, GEORGIA, SEDIMENTS, BACTERIA, DETRITUS, ESTUARIES, PLANKTON, ECOSYSTEMS, SINKS, TRACERS, CRABS, SHRIMP, MULLETS, FISH, ALGAE.

IDENTIFIERS:
DUPLIN RIVER(GA), SPARTINA ALTEMIFLORA, ORGANISMS, TURSIOPS TRUNCATUS, BREVOORTIA TYRRANUS, CONCENTRATION PROCESSES.

ABSTRACT:
BY SYNTHESIZING RESULTS OF EXPERIMENTS USING PHOSPHORUS-32 (P-32) AND ZINC-65 (ZN-65) WITH EARLIER WORK IN GEORGIA SALT MARSHES, THE CYCLES OF THE ELEMENTS P AND ZN ARE DESCRIBED QUANTITATIVELY. MARSH GRASS AND SEDIMENTS DOMINATE BOTH CYCLES. THE UPPERMOST METER OF SEDIMENTS CONTAINS ENOUGH P TO SUPPORT SPARTINA PRODUCTION FOR 500 YEARS AND ENOUGH ZN FOR 5000 YEARS. SPARTINA PRODUCTION REMOVES P AND ZN FROM SUBSURFACE (REDUCED) SEDIMENTS AND INTRODUCES THEM INTO THE WATER (VIA BACTERIAL UTILIZATION AND SUBSEQUENT UTILIZATION BY DETRITUS FEEDERS) AT A RATE THAT REPLACES TOTAL WATER P IN A MONTH AND TOTAL WATER ZN IN A YEAR. SIGNIFICANT PART OF P IS EXPORTED FROM THE MARSH IN ORGANISMS AND DETRITUS. THE P AND ZN IN ESTUARINE WATER ARE IN EQUILIBRIUM WITH PLANKTON, BACTERIA, AND SURFACE (OXIDIZED) SEDIMENTS. EQUILIBRIUM STRONGLY FAVORS THE SEDIMENTS. SUBSURFACE-SEDIMENT COMPARTMENT OF P AND ZN IS PROBABLY REPLACED BY CONVERSION OF OXIDIZED SEDIMENTS TO REDUCED SEDIMENTS THROUGH CREEK MEANDERING. OXIDIZED SEDIMENTS ARE IN EQUILIBRIUM WITH WATER RECEIVING INPUTS OF P AND ZN FROM LAND AND SEA. TRANSFER OF P AND ZN FROM DEEP SEDIMENTS TO WATER BY SPARTINA EXPLAINS THE HIGH CONCENTRATION OF THESE ELEMENTS IN THE WATER. (JONES-WIS)

FIELD 05C

ACCESSION NO. W69-08274

A COMPARISON BETWEEN THE UPTAKE OF RADIOACTIVE AND STABLE ZINC BY A MARINE UNICELLULAR ALGA,

COMITATO NAZIONALE PER L'ENERGIA NUCLEARE, LA SPEZIA (ITALY).

M. BERNHARD, AND A. ZATTERA.

SYMP ON RADIOECOLOGY, PROC 2ND NATL SYMP, MAY 15-17, 1967, ANN ARBOR, MICH, NELSON, DANIEL J AND EVANS, FRANCIS C (EDS). US ATOMIC ENERGY COMM, DOC CONF 670503, PP 389-398. 5 FIG, 7 REF, DISC. AVAILABLE FROM CLEARINGHOUSE AS CONF 670503 AT $3.00 IN PAPER COPY AND $0.65 IN MICROFICHE.

DESCRIPTORS:
*ZINC RADIOISOTOPES, *MARINE ALGAE, *RESINS, SEA WATER, PHYSICOCHEMICAL PROPERTIES, TRACERS, DISTRIBUTION, CHELATION, DETERGENTS, RADIOACTIVITY, AMINO ACIDS.

IDENTIFIERS:
*RADIONUCLIDE UPTAKE, *PHAEODACTYLUM TRICORNUTUM, EDTA, LIGURIAN SEA, EUROPEAN MEDITERRANEAN.

ABSTRACT:
ORGANIC SUBSTANCES OCCURRING IN SEAWATER OR INTRODUCED AS CONTAMINANTS, E.G., DETERGENTS, MAY CHELATE ZINC AND INFLUENCE ITS AVAILABILITY TO ALGAE. UPTAKE OF ZINC-65 (ZN-65) AND STABLE ZINC BY A MARINE UNICELLULAR ALGA (PHAEODACTYLUM TRICORNUTUM) AND BY A CHELATING RESIN (CHELEX-100) WAS STUDIED IN BATCH CULTURES DURING A CERTAIN TIME INTERVAL. IN ALL EXPERIMENTS, THE DISTRIBUTION PATTERN OF ZN-65 AND STABLE ZINC BETWEEN PARTICULATE MATTER (ALGAE OR CHELATING RESIN) AND NATURAL AND ARTIFICIAL SEAWATER MEDIUM DIFFERED. WHEN ZN-65 WAS ADDED AS IONIC ZINC, PROPORTIONATELY MUCH MORE RADIOACTIVE ZINC WAS TAKEN UP THAN STABLE ZINC. HOWEVER, WHEN ZN-65 IS ADDED AS ZN-EDTA-COMPLEX TO A BATCH CONTAINING SEAWATER AND CHELEX, STABLE ZINC WAS INITIALLY TAKEN UP AT A HIGHER RATE THAN WAS ZN-65. THESE PRELIMINARY RESULTS DO NOT EXPLAIN THE DIFFERENCE IN DISTRIBUTION OF RADIOACTIVE AND STABLE ZINC, BUT THE DATA SHOW THAT UPTAKE RATE OF THE STABLE ISOTOPE DOES NOT NECESSARILY FOLLOW UPTAKE RATE OF THE RADIOACTIVE ISOTOPE IF THE TWO ARE PRESENT IN DIFFERENT PHYSICAL-CHEMICAL STATES. THIS POSSIBILITY MUST BE TAKEN INTO ACCOUNT IN TRACER STUDIES AND IN PREDICTIONS OF THE DISTRIBUTION OF RADIOISOTOPES IN ENVIRONMENT BASED ON SPECIFIC ISOTOPE CONTENT APPROACH. (JONES-WIS)

FIELD 05C

ACCESSION NO. W69-08275

ALGAE: AMOUNTS OF DNA AND ORGANIC CARBON IN SINGLE CELLS,

CALIFORNIA UNIV., SAN DIEGO, LA JOLLA. INST. OF MARINE RESOURCES.

OSMUND HOLM-HANSEN.

SCIENCE, VOL 163, PP 87-88, JANUARY 1969. 1 FIG, 19 REF.

DESCRIPTORS:
*ALGAE, *CARBON, *CARBON CYCLE, *CYTOLOGICAL STUDIES, CULTURES,
FLUOROMETRY, BIOMASS, OCEANOGRAPHY.

IDENTIFIERS:
*DNA, *ORGANIC CARBON, *CELL SIZE, EUKARYOTES, MONOCHRYSIS LUTHERI,
NAVICULA PELLICULOSA, GONYAULAX POLYEDRA, 3,5-DIAMINOBENZOIC ACID
DIHYDROCHLORIDE, INFRARED GAS ANALYSIS, COULTER COUNTER, EUGLENA
GRACILIS, CACHONINA NIEI, SKELETONEMA COSTATUM, DUNALIELLA TERTIOLECTA,
AMPHIDINIUM CARTERI, SYRACOSPHAERA ELONGATA, THALASSIOSIRA FLUVIATILIS,
DITYLUM BRIGHTWELLII, BIOLOGICAL OCEANOGRAPHY, ECOLOGICAL TECHNIQUES.

ABSTRACT:
UTILIZING INFRARED GAS ANALYSIS FOR CARBON AND FLUOROMETRY FOR
DESOXYRIBONUCLEIC ACID (DNA), AUTHOR ANALYZED THE FOLLOWING VARIOUSLY
SIZED SPECIES OF UNICELLULAR ALGAE: MONOCHRYSIS LUTHERI, NAVICULA
PELLICULOSA, GONYAULAX POLYEDRA, CACHONINA NIEI, SKELETONEMA COSTATUM,
DUNALIELLA TERTIOLECTA, AMPHIDINIUM CARTERI, SYRACOSPHAERA ELONGATA,
THALASSIOSIRA FLUVIATILIS, DITYLUM BRIGHTWELLII. CARBON CONTENT VARIED
FROM 10 TO 6000 PICOGRAMS PER CELL, AND TOTAL CELLULAR DNA IS DIRECTLY
PROPORTIONAL TO CELL SIZE. CONTENT OF DNA IS EQUAL TO APPROXIMATELY
1-3% OF CELLULAR ORGANIC CARBON. DATA AND RELATIONSHIPS DEVELOPED IN
THIS STUDY SUGGEST THAT DNA-DETERMINATIONS ARE OF POTENTIAL VALUE IN
DETERMINING LIVING BIOMASS IN ECOLOGICAL STUDIES. (EICHHORN-WIS)

FIELD 02K, 05C

ACCESSION NO. W69-08278

LAKE TERMINOLOGY: WATER BLOOM,

MICHIGAN STATE UNIV., EAST LANSING.

J. O. VEATCH, AND C. R. HUMPHRYS.

BULL MICH AGRIC COLL EXP STATION, EAST LANSING, P 241, 1964. 1 FIG.

DESCRIPTORS:
*EUTROPHICATION, ALGAE, DUCKS, COLOR, FISHKILL, LAKES, ODOR, TOXICITY,
WATER POLLUTION EFFECTS, WATER QUALITY, CATTLE.

IDENTIFIERS:
*DEFINITIONS, *WATER BLOOM, TOXIC ALGAE, RECREATIONAL USE.

ABSTRACT:
THIS LEXICON OF LAKE TERMINOLOGY DEFINES A WATER BLOOM AS: 'A PROLIFIC
GROWTH OF PLANKTON. A BLOOM OF ALGAE MAY BE SO DENSE THAT IT IMPARTS A
GREENISH, YELLOWISH, OR BROWNISH COLOR TO THE WATER. THE GROWTH MAY BE
SO CONCENTRATED IN SOME PARTS OF A LAKE THAT IT INTERFERES WITH
SWIMMING AND BOATING. THE ALGAE NOT ONLY IMPARTS A DISAGREEABLE ODOR,
BUT IT MAY BE A CAUSE OF FISH MORTALITY, AND SOME SPECIES MAY BE
POISONOUS TO CATTLE AND DUCKS AND A MENACE TO DRINKING WATER SUPPLIES.'
THIS ENTRY INCLUDES AN AERIAL PHOTOGRAPH OF AN ALGAL BLOOM CONCENTRATED
IN BAY OF A LAKE. (EICHHORN-WIS)

FIELD 05C

ACCESSION NO. W69-08279

PRACTICAL ALGAE CONTROL METHODS FOR NEW HAMPSHIRE WATER SUPPLIES,

NEW HAMPSHIRE WATER SUPPLY AND POLLUTION CONTROL COMMISSION, CONCORD.

TERRENCE P. FROST.

J NEW HAMPSHIRE WATER WORKS ASSOC, 7 P, APRIL 1960. 6 REF.

DESCRIPTORS:
*ALGAE CONTROL, *WATER SUPPLY, *NEW HAMPSHIRE, ALGAE, ODOR, TASTE,
LIGHT TEMPERATURE, NUTRIENTS, COPPER SULFATE, STRATIFICATION.

IDENTIFIERS:
*CORRECTIVE TREATMENT, *PREVENTATIVE TREATMENT, ALGAE NUISANCE, ALGAE
BLOOM, ALGAE GROWTH, WATER DENSITY, CONVECTION CURRENTS, ALGAE
TREATMENT COSTS, DINOBRYON, CONCORD(N H), PEANACOOK LAKE(N H).

ABSTRACT:
AUTHOR BRIEFLY REVIEWS CHARACTERISTICS OF ALGAE AND ASSOCIATED PROBLEMS
IN NEW HAMPSHIRE WATER SUPPLIES, OUTLINING PREVENTATIVE MEASURES
TOGETHER WITH EASIEST AND MOST ECONOMICAL CONTROLS USED TO DATE.
SYMPTOMS ARE UNDESIRABLE TASTE AND ODOR IN WATER SUPPLIES, BLOOMS IN
RESERVOIRS, FOULED PIPES, AND CLOGGED FILTERS. OCCURRENCE OF ALGAE IS
INFLUENCED BY LIGHT, TEMPERATURE, CONVECTION CURRENTS, DENSITY
STRATIFICATION (SPRING AND FALL TURNOVERS CIRCULATE BOTTOM MATERIALS
AND ALGAE, OFTEN INITIATING ALGAL BLOOM), AND NUTRIENTS WHICH PROMOTE
ALGAL GROWTH. IDEALLY, EACH WATERWORKS SHOULD PERIODICALLY EXAMINE
SAMPLES MICROSCOPICALLY TO MONITOR ALGAE SPECIES AND APPLY PREVENTATIVE
TREATMENT BEFORE NUISANCE LEVELS ARE REACHED. ALGAL CONTROL IN MOST OF
NEW HAMPSHIRE'S 76 PUBLIC SURFACE WATER SUPPLIES IS PRACTICABLE AND
INEXPENSIVE. PENACOOK LAKE, SUPPLYING WATER FOR CONCORD, HAS A PROBLEM
CAUSED PRIMARILY BY THE SPECIES DINOBRYON; THIS LAKE COULD BE TREATED
WITH COPPER SULFATE TO A DEPTH OF TEN FEET (3,380 ACRE FEET) FOR $340.
MUNICIPALITIES CAN OBTAIN HELP FROM NEW HAMPSHIRE'S WATER POLLUTION
COMMISSION OR SANITARY ENGINEERING DIVISION FOR IDENTIFICATION OF
ORGANISMS AND RECOMMENDED DOSAGES OF COPPER SULFATE. USE OF
CHLORINATION AND COPPER SULFATE SEEM THE BEST AVAILABLE MEANS AND ONLY
ACCEPTABLE CHEMICALS FOR TREATING THEIR WATER SUPPLIES. (KETELLE-WIS)

FIELD 05C, 05F

ACCESSION NO. W69-08282

THE 'WORKING' OF THE MADISON LAKES,

WILLIAM TRELEASE.

WISCONSIN ACAD SCI, ARTS AND LETTERS, VOL 7, PP 121-129, 1889. 1 FIG, 1
PLATE, 60 REF.

DESCRIPTORS:
*EUTROPHICATION, CYANOPHYTA, ODOR, SCUM, FISHING, WISCONSIN,
BIBLIOGRAPHIES, LAKES, NUISANCE ALGAE, BACTERIA, WATER POLLUTION
EFFECTS.

IDENTIFIERS:
*MADISON(WIS), *'WORKING' OF LAKES, *WATER-BLOOM, 'BREAKING' OF LAKES,
CHROOCOCCACEAE, NOSTOCHINEAE, CLATHROCYSTIS AERUGINOSA, COELOSPHAERIUM
KUTZINGIANUM, ANABAENA FLOS AQUAE, ANABAENA MENDOTAE, ANABAENA HASSALI,
LYNGBYA NOLLEI, GLOCOTRICHIA PISUM, NOSTOC VERRUCOSUM, LAKE
MENDOTA(WIS), LAKE MONONA(WIS), LAKE WAUBESA(WIS), ANABAENA CIRCINALIS.

ABSTRACT:
THE MADISON (WISCONSIN) LAKES WERE OBSERVED FOR SIGNS OF EUTROPHICATION
FROM 1882-1887. SYMPTOMS OBSERVED INCLUDED GREENISH-YELLOW SCUM OR FINE
GRANULES SUSPENDED IN WATER, ODORS ASSOCIATED WITH ALGAL DECOMPOSITION,
STAINING OF SHORELINES, AND DECREASED FISH BITING. THE ALGAE GENERALLY
APPEARED IN MID-SUMMER AND PERSISTED FOR ONLY A FEW DAYS BEFORE BEING
BROKEN UP AND DISPERSED BY BREEZES. SUCH PHENOMENA HAVE BEEN REFERRED
TO AS THE 'WORKING' OR 'BREAKING' OF LAKES AND ARE SYNONYMOUS WITH THE
EUROPEAN TERM 'WATER-BLOOM'. THE SYMPTONS, CAUSED BY THE OCCURRENCE OF
MEMBERS OF THE ALGAE GROUP CYANOPHYCEAE, ARE FAVORED IN STAGNANT WATER
BODIES AND USUALLY OCCUR AFTER PROLONGED HOT SPELLS. THE TWO PRINCIPAL
SUBDIVISIONS OF THE GROUP ARE CHROOCOCCACEAE AND NOSTOCHINEAE. OF THE
FORMER, CLATHROCYSTIS AERUGINOSA AND COELOSPHAERIUM KUTZINGIANUM WERE
THE PRINCIPAL PROBLEM SPECIES IN THE MADISON LAKES, AND FROM THE
LATTER, ANABAENA FLOS AQUAE AND ANABAENA CIRCINALIS WERE MOST COMMON.
ANABAENA MENDOTAE, WHICH IS MORE TOLERANT OF COLD, PREDOMINATES LATE IN
THE SEASON. PAPER INCLUDES A LIST OF SIXTY REFERENCES ON WATER-BLOOMS.
ONLY PRINCIPAL PAPERS TREATING RECURRENCE OF BACTERIA IN QUANTITY ARE
INCLUDED, MOST OF THEM HAVE LITTLE BOTANICAL VALUE BUT SERVE AS A
NUCLEUS FOR OTHER REFERENCES AVAILABLE IN 1889. (KETELLE-WIS)

FIELD 02H, 05C

ACCESSION NO. W69-08283

OLEFINS OF HIGH MOLECULAR WEIGHT IN TWO MICROSCOPIC ALGAE,

HOUSTON UNIV., TEX. DEPT. OF CHEMISTRY; HOUSTON UNIV., TEX. DEPT. OF
BIOLOGY; AND HOUSTON UNIV., TEX. DEPT. OF BIOPHYSICAL SCIENCES.

E. GELPI, J. ORO, H. J. SCHNEIDER, AND E. O. BENNETT.

SCIENCE, VOL 161, NO 3842, PP 700-701, AUGUST 1968. 2 FIG, 1 TAB, 23 REF.

DESCRIPTORS:
*ALGAE, CYANOPHYTA, CHRYSOPHYTA, CULTURES, DIAGENESIS, SEDIMENTS, GAS
CHROMATOGRAPHY, LIPIDS, OIL, OIL SHALE.

IDENTIFIERS:
*BOTRYOCOCCUS BRAUNII, *ANACYSTIS MONTANA, *HYDROCARBONS, *OLEFINS,
GREEN RIVER FORMATION, CHLOROPHYLL DERIVATIVES, RECENT SEDIMENTS,
PRECAMBRIAN SEDIMENTS, CHLORELLA PYRENOIDOSA, FUCALES, ALKANES,
ALKENES, ALIPHATIC HYDROCARBONS, PETROLEUM CRUDES, MASS SPECTROMETRY,
N-HEPTADECANE, MONOENES, DIENES, TRIENES.

ABSTRACT:
UTILIZING GAS CHROMATOGRAPHY AND MASS SPECTROMETRY, AUTHORS STUDIED THE
HYDROCARBON COMPOSITION OF TWO SPECIES OF ALGAE: BOTRYOCOCCUS BRAUNII,
A GOLDEN-BROWN (CHRYSOPHYTE), AND ANACYSTIS MONTANA, A BLUE-GREEN
(CYANOPHYTE). HYDROCARBONS OF THE CHRYSOPHYTE ARE ALKENES WITH ONE,
TWO, OR THREE DOUBLE BONDS; AND ODD NUMBERS OF CARBON ATOMS (IN RANGE,
17-33). DIOLEFINS WITH 27, 29 OR 31 CARBON ATOMS PREDOMINATE, THAT OF
29 CARBONS BEING MAJOR COMPONENT. DISTRIBUTION IN THE CYANOPHYTE
COMPRISES PRIMARILY MONOENES WITH CARBON ATOMS IN RANGE, 19-29, THE
MAJOR COMPONENT BEING A MONOENE OF 27 CARBONS. HEPTADECANE, ONLY
PARAFFIN FOUND, CONSTITUTED 11.5% OF HYDROCARBONS FOUND IN ANACYSTIS.
CONSIDERATION OF INFORMATION DEVELOPED IN THIS STUDY, TOGETHER WITH
DATA REPORTED ELSEWHERE IN THE LITERATURE, LED AUTHORS TO THE FOLLOWING
CONCLUSIONS: HYDROCARBONS OF HIGH MOLECULAR WEIGHT ARE FOUND IN
CONTEMPORARY COUNTERPARTS OF ANCIENT MICROSCOPIC ORGANISMS; SUCH
HYDROCARBONS ARE PRESENT IN HIGH CELLULAR CONCENTRATIONS (0.1-0.3% OF
DRY CELL WEIGHT); THEY POSSESS HIGH DEGREE OF UNSATURATION; THEY SHOW
MARKED ODD OVER EVEN PREDOMINANCE; AND THEY SHOW ACTUAL AND POTENTIAL
SIMILARITIES WITH PATTERNS OF ANCIENT AND MODERN SEDIMENTS AND OIL
SHALES. (EICHHORN-WIS)

FIELD 05A

ACCESSION NO. W69-08284

EUTROPHICATION,

J. FOEHRENBACH.

JOURNAL WPCF, VOL 41, NO 6, PP 1029-1036, JUNE 1969. 53 REF.

DESCRIPTORS:
*EUTROPHICATION, *REVIEWS, DOCUMENTATION, BIBLIOGRAPHIES, PHOSPHORUS
COMPOUNDS, NITROGEN COMPOUNDS, BIOINDICATORS, PRODUCTIVITY, WASTE
TREATMENT, WATER POLLUTION EFFECTS, WATER POLLUTION SOURCES, CYCLING
NUTRIENTS, COST COMPARISONS, ECONOMICS, ALGAE, ALGAL CONTROL,
ALGICIDES, SEDIMENTS, DISSOLVED OXYGEN, OXYGEN DEMAND, CYANOPHYTA,
DIATOMS, SAMPLING, LAKE MICHIGAN.

IDENTIFIERS:
CULTURAL EUTROPHICATION, SPECIES DIVERSITY, CHEMICAL PRECIPITATION.

ABSTRACT:
AUTHOR REVIEWS 53 STUDIES CONTRIBUTED TO THE EUTROPHICATION LITERATURE
DURING THE PERIOD 1966-1968. SUBJECT COVERAGE, WITH NUMBER OF
LITERATURE CITATIONS PER SUBJECT IN PARENTHESES, ARE: USE OF FERTILIZED
WATER TO INCREASE BIOLOGICAL PRODUCTION (8); IDENTIFICATION OF
NUTRIENTS (6); OXYGEN KINETICS (3); ALGAL CONTROL (2); SOURCES OF
NITROGEN AND PHOSPHORUS (4); ROLE OF SEDIMENTS (2); RESULTS OF HUMAN
CULTURAL ACTIVITIES (2); EFFECTS OF POLLUTIONAL LOAD (2); CYANOPHYTE
BLOOMS AND DECREASE IN ALGAL SPECIES DIVERSITY (2); EUTROPHICATION
INDICATORS IN LAKE MICHIGAN (2); SAMPLING AND DIATOM-INDICATORS (2);
NUTRIENTS, THEIR CRITICAL LEVELS AND CYCLING (4); ADVANCED WASTE
TREATMENT TECHNIQUES WITH COST ESTIMATES (7); CHEMICAL TREATMENT FOR
NUTRIENT REMOVAL WITH COST ESTIMATES (7). (EICHHORN-WIS)

FIELD 05C

ACCESSION NO. W69-08518

SORPTION OF RADIOACTIVE NUCLIDES BY SARGASSUM FLUITANS AND S NATANS,

TEXAS A AND M UNIV., COLLEGE STATION; AND KANSAS UNIV., LAWRENCE.

JOHN E. SIMEK, J. A. DAVIS, C. E. DAY III, AND ERNEST E. ANGINO.

AVAILABLE FROM CLEARINGHOUSE FOR SCI AND TECH INFORMATION AS CONF-670503.
SYMP ON RADIOECOLOGY, PROC 2ND NATL SYMP, MAY 15-17, 1967, ANN ARBOR, MICH.
NELSON, DANIEL J AND EVANS, FRANCIS C (EDS). US ATOMIC ENERGY COMM, DOC
CONF 670503, P 505-508. 2 FIG, 5 REF. DISC.

DESCRIPTORS:
*RADIOISOTOPES, *ALGAE, *SORPTION, ABSORPTION, GULF OF MEXICO, ATLANTIC
OCEAN, RADIOACTIVITY, GAMMA RAYS, POTASSIUM RADIOISOTOPES, MANGANESE,
CESIUM, RADIUM RADIOISOTOPES, URANIUM RADIOISOTOPES.

IDENTIFIERS:
*SARGASSUM FLUITANS, *SARGASSUM NATANS, SEAWEED,
ZIRCONIUM-95-NIOBIUM-95, PRE-BOMB MATERIAL, RUTHENIUM-106-RHODIUM-106,
RUTHENIUM-103-RHODIUM-103, CERIUM-144-PRASEODYMIUM-144.

ABSTRACT:
SARGASSUM NATANS AND S FLUITANS, FREE-FLOATING MARINE ALGAE, CAN
EFFECTIVELY CONCENTRATE CERTAIN RADIONUCLIDES. IN THE EVALUATION OF
RADIONUCLIDES FOUND ASSOCIATED WITH THESE TWO SPECIES, AN ANSWER WAS
SOUGHT AS TO THE MANNER OF THIS ASSOCIATION. FLOATING SAMPLES WERE
COLLECTED FROM THE GULF OF MEXICO AND FROM NORTH ATLANTIC AREAS. AN
ANALYSIS OF SARGASSUM MATERIAL COLLECTED PRIOR TO 1940 (PRE-BOMB)
PROVIDED A POSITIVE MEANS OF COMPARING THE ACTIVITY ADDED TO THE
SPECTRUM BY NUCLEAR ACTIVITY. TESTS UTILIZING ACID, BASE, AND NEUTRAL
WASHES, WERE DEVISED TO DETERMINE WHETHER ADSORPTION OR ABSORPTION WAS
THE MAJOR NUCLIDE-PLANT ASSOCIATION. WASHINGTON WITH UNCONTAMINATED,
DOUBLY DISTILLED WATER, RESULTED IN REMOVAL OF ONLY SMALL AMOUNTS OF
POTASSIUM-40 (K-40), PROBABLY DUE SIMPLY TO A WASHING OUT OF SEA SALTS.
SEPARATE, SHORT TERM (UP TO 1 HOUR) WASHED OF SAMPLES WITH SLIGHTLY
ACIDIC AND BASIC SOLUTION FAILED TO INDICATE ANY ACTIVITY OTHER THAN
THAT OF K-40. HOWEVER, ACID AND BASE WASHES FOR 24 HOURS PRODUCED
DEFINITE EVIDENCE OF ACTIVITY; OVER THIS PERIOD, BREAKDOWN OF THE PLANT
MATERIAL BECAME OBVIOUS. RESULTS SUGGEST THAT RADIONUCLIDES IN QUESTION
ARE ABSORBED, OR METABOLICALLY FIXED, RATHER THAN ADSORBED TO SARGASSUM
NATANS AND S FLUITANS.
(JONES-WIS)

FIELD 05C

ACCESSION NO. W69-08524

THE NATURE OF THE DISTRIBUTION OF TRACE ELEMENTS IN LONGNOSE ANCHOVY (ANCHOA
  LAMPROTAENIA HILDEBRAND), ATLANTIC THREAD HERRING (OPISTHONEMA OGLINUM
  LASUEUR), AND ALGA (UDOTEA FLABELLUM LAMOUROUX),

PUERTO RICO NUCLEAR CENTER, MAYAGUEZ.

ROBERT Y. TING, AND V. ROMAN DE VEGA.

AVAILABLE FROM CLEARINGHOUSE FOR SCI AND TECH INFORMATION AS CONF-670503.
  SYMP ON RADIOECOLOGY, PROC 2ND NATL SYMP, MAY 15-17, 1967, ANN ARBOR, MICH.
  NELSON, DANIEL J AND EVANS, FRANCIS C (EDS). US ATOMIC ENERGY COMM, DOC
  CONF 670503, P 527-534. 5 FIG, 15 REF, DISC.

DESCRIPTORS:
    *ALGAE, *HERRINGS, COPPER, IRON, COBALT, MANGANESE, ATLANTIC OCEAN,
    PUERTO RICO, SPECTROSCOPY, DISTRIBUTION, CALCIUM, TROPHIC LEVELS, FOOD
    WEBS.

IDENTIFIERS:
    *TRACE ELEMENTS, *ANCHOA LAMPROTAENIA, *OPISTHONEMA OGLIMUN, UDOTEA
    FLABELLUM, ZINC, NICKEL, PUERTO RICO, LA PARGUERA, JOYUDA.

ABSTRACT:
    QUALITATIVE AND QUANTITATIVE INFORMATION ON STABLE ELEMENTS CONTAINED
    IN ORGANISMS AND THEIR ENVIRONMENT IS ESSENTIAL IN PREDICTING
    DISTRIBUTION PATTERNS OF ARTIFICIALLY INTRODUCED RADIOISOTOPES OF THE
    SAME ELEMENTS. LITTLE IS KNOWN OF THE DISTRIBUTION AND VARIABILITY OF
    VARIOUS ELEMENTS AND THEIR RADIOISOTOPES WITHIN SPECIES OF MARINE
    ORGANISMS SAMPLED FROM THE SAME AREA AT A GIVEN TIME. TO STUDY THEIR
    TRANSFER THROUGH FOOD WEBS AND TO EVALUATE CONCENTRATION FACTORS AT
    EACH LEVEL, TRACE AMOUNTS OF THE ELEMENTS COPPER, IRON, AND ZINC, WERE
    DETERMINED IN LONGNOSE ANCHOVY (ANCHOA LAMPROTAENIA HILDEBRAND) AND IN
    ALGA (UDOTEA FLABELLUM LAMOUROUX) FROM LA PARGUERA AND JOYUDA, PUERTO
    RICO. LEVELS OF COBALT, COPPER, IRON, MANGANESE, NICKEL, AND ZINC IN
    ATLANTIC THREAD HERRING (OPISTHONEMA OGLINUM LASUEUR) FROM LA PARGUERA,
    WERE ALSO DETERMINED. ANALYSES WERE ACCOMPLISHED WITH ATOMIC ABSORPTION
    SPECTROSCOPY. CONCENTRATIONS OF THE ELEMENTS IN INDIVIDUALS OF THE SAME
    SPECIES SAMPLED WITHIN THE SAME LOCALITY AT THE SAME TIME DID NOT
    APPROXIMATE NORMAL DISTRIBUTION; HOWEVER, DISTRIBUTIONS OF THE
    LOG-TRANSFORMED CONCENTRATIONS CLOSELY APPROXIMATED A NORMAL CURVE IN
    MOST CASES. WITH EXPANDED NUCLEAR TECHNOLOGY, RADIOISOTOPES IN THE
    ENVIRONMENT MAY APPROACH LEVELS FOR WHICH KNOWLEDGE OF THE DISTRIBUTION
    PATTERNS OF THESE MATERIALS WILL BE OF PRIMARY IMPORTANCE.
    (JONES-WIS)

FIELD 05C

ACCESSION NO. W69-08525

PHOSPHORUS TURNOVER BY CORAL REEF ANIMALS,

GEORGIA UNIV., ATHENS. DEPT. OF ZOOLOGY; AND NORTH CAROLINA UNIV., CHAPEL
HILL. DEPT. OF ENVIRONMENTAL SCIENCES AND ENGINEERING.

L. R. POMEROY, AND E. J. KUENZLER.

AVAILABLE FROM CLEARINGHOUSE FOR SCI AND TECH INFORMATION AS CONF-670503.
SYMP ON RADIOECOLOGY, PROC 2ND NATL SYMP, MAY 15-17, 1967, ANN ARBOR, MICH,
NELSON, DANIEL J AND EVANS, FRANCIS C (EDS). US ATOMIC ENERGY COMM, DOC
CONF 670503, PP 474-482. 1 FIG, 3 TAB, 34 REF, DISC.

DESCRIPTORS:
*PHOSPHORUS, *CORAL, *REEFS, *MARINE ANIMALS, FISH, ECOSYSTEMS, CYCLING
NUTRIENTS, TEMPERATURE, LIGHT QUALITY, CHLOROPHYLL, ZOOPLANKTON, ALGAE,
POLYCHAETA, CRUSTACEA, MOLLUSCA, ECHINODERMATA, PACIFIC OCEAN.

IDENTIFIERS:
ENIWETOK ATOLL, TURNOVER TIME, TRIDACNA CROCEA, WOODS HOLE
OCEANOGRAPHIC INSTITUTION, ACROPORA HYACINTHUS, POCILLOPORA EYDOUXI,
PORITES LOBATA, LEPTASTREA PURPUREA, FUNGIA FUNGITES, ZOOXANTHELLAE,
ACANTHURUS TRIOSTEGUS, CONUS EBRAEUS.

ABSTRACT:
CORAL REEFS ARE PARTICULARLY INTERESTING ECOSYSTEMS IN WHICH TO STUDY
NUTRIENT CYCLING, BECAUSE OF THEIR GREAT DIVERSITY OF LIFE AND THEIR
HIGH RATES OF ORGANIC PRODUCTION IN A NUTRIENT-DEPLETED ENVIRONMENT.
PHOSPHORUS (P) CONTENT AND ELIMINATION RATES WERE MEASURED FOR DOMINANT
REEF ANIMALS AT ENIWETOK ATOLL, AND FLUX OF P THROUGH THE ORGANISMS WAS
EXPRESSED AS TURNOVER TIME. HERBIVOROUS FISHES APPEAR TO RECEIVE IN
THEIR DIET JUST ENOUGH P FOR GROWTH AND REPRODUCTION; THIS IS REFLECTED
IN A SOMEWHAT LOWER EXCRETION RATE AND LONGER TURNOVER TIME THAN WOULD
BE PREDICTED FOR ORGANISMS OF THEIR SIZE, PARTICULARLY THE JUVENILES.
CARNIVOROUS ANIMALS AND DEPOSIT FEEDERS RECEIVE EXCESS P IN THEIR DIET,
AND THEIR TURNOVER TIME DOES NOT DIFFER FROM ECOSYSTEMS WITH A MORE
ABUNDANT P SUPPLY. OF 5 SPECIES OF REEF CORALS EXAMINED, 4 SHOWED A
VERY LONG TURNOVER TIME AND LITTLE FLUX OF P TO THE ENVIRONMENT.
TRIDACNA CROCEA SHOWED A TYPICAL TURNOVER TIME FOR A MOLLUSC OF ITS
SIZE, BUT SEEMED TO LOSE MOST P VIA INCORPORATION IN LIVING
ZOOXANTHELLAE, WHICH ARE SUBSEQUENTLY LOST. THIS SAMPLING OF THE FLUX
OF P IS NECESSARILY INCOMPLETE ALTHOUGH ORGANISMS KNOWN TO BE DOMINANT
HAVE BEEN SELECTED FOR STUDY.
(JONES-WIS)

FIELD 05C

ACCESSION NO. W69-08526

NATURAL AND POLLUTION SOURCES OF IODINE, BROMINE, AND CHLORINE IN THE GREAT
LAKES,

MICHIGAN UNIV., ANN ARBOR. DEPT. OF METEOROLOGY AND OCEANOGRAPHY.

MARY A. TIFFANY, JOHN W. WINCHESTER, AND RONALD H. LOUCKS.

J WATER POLLUT CONTR FEDERATION, VOL 41, NO 7, P 1319-1329, JULY 1969. 11 P,
10 FIG, 5 TAB, 13 REF. CONTRACT NO AT(11-1)-1705AEC).

DESCRIPTORS:
*WATER QUALITY, *GREAT LAKES, *TRACE ELEMENTS, *WATER POLLUTION
EFFECTS, CHLORIDES, HALOGENS, ALGAE, WATER POLLUTION SOURCES, WATER
CHEMISTRY, WATER ANALYSIS.

IDENTIFIERS:
*BROMINE, *IODINE.

ABSTRACT:
THIS STUDY INVOLVES THE DETERMINATION OF TRACE ELEMENTS I, BR, AND CL
IN THE GREAT LAKES, USING NEUTRON ACTIVATION ANALYSIS OF 90 WATER
SAMPLES FROM LAKE SUPERIOR AND ITS TRIBUTARY STREAMS, LAKE MICHIGAN,
SOUTHERN LAKE HURON, LAKE ST. CLAIR, WESTERN LAKE ERIE, AND NORTHERN
LAKE ONTARIO. POSSIBLE POLLUTION BY BROMINE THROUGH AN ATMOSPHERIC
ROUTE IS OF INTEREST BECAUSE OF THE COMBUSTION OF LEADED GASOLINE.
IODINE DEFICIENCY IN LAKE WATER MAY BE RELATED TO THYROID DISORDERS
AMONG MARINE FISH WHICH HAVE BECOME ADAPTED TO FRESH WATER, AND ALGAE
MAY OFFER COMPETITION FOR THE AVAILABLE IODINE. CHLORINE IS A
NOTICEABLE CONTAMINANT EXCEPT IN LAKE SUPERIOR. THE LAKE SUPERIOR
STREAMS APPEAR TO REPRESENT A GOOD AVERAGE OF THE ATMOSPHERIC INPUTS OF
I, BR, AND CL. (KNAPP-USGS)

FIELD 05A, 02K

ACCESSION NO. W69-08562

BIOLOGICAL CONCENTRATION OF PESTICIDES BY ALGAE,

NORTH TEXAS STATE UNIV., DENTON. DEPT. OF BIOLOGY.

B. DWAIN VANCE, AND WAYMON DRUMMOND.

J AMER WATER WORKS ASS, VOL 61, NO 7, P 360-362, JULY 1969. 3 P, 2 TAB, 11 REF. GRANT NO CC00269(PHS).

DESCRIPTORS:
*PESTICIDE RESIDUES, *ALGAE, BIOASSAY, BIOINDICATORS, FISHKILL, HAZARDS, PESTICIDE KINETICS, PESTICIDE TOXICITY, POLLUTANTS, WATER POLLUTION EFFECTS, ALDRIN, DIELDRIN, ENDRIN, DDT.

IDENTIFIERS:
PESTICIDE BIOACCUMULATION.

ABSTRACT:
ALGAE, THE BASE CONSTITUENTS OF THE AQUATIC FOOD CHAIN, CONCENTRATE PESTICIDES AND ARE MUCH MORE RESISTANT TO TOXIC EFFECTS THAN THE HIGHER ORGANISMS THAT EAT THEM. ALGAL CULTURES WERE GROWN IN THE PRESENCE OF DDT, ALDRIN, DIELDRIN, AND ENDRIN, IN CONCENTRATIONS OF 5-20 MICROGRAM PER ML. THE ALGAE CONCENTRATED THE PESTICIDES 100 TIMES OR MORE IN ALL THE TESTS. THE DATA ARE TABULATED. PRACTICALLY NO DEGRADATION OF THE PESTICIDES OCCURRED. (KNAPP-USGS)

FIELD 05C, 05B

ACCESSION NO. W69-08565

CHEMICAL FERTILIZERS ON LAKE WATERS (IN GERMAN),

VON HEINZ AMBUHL.

SUMMARIES IN ENGLISH AND FRENCH. GAS- UND WASSERFACH, VOL 107, NO 14, P 357-363, APR 1966. 6 FIG, 2 TAB, 18 REF.

DESCRIPTORS:
*EUTROPHICATION, *NUTRIENTS, *LAKES, PONDS, FERTILIZERS, ALGAE, NITROGEN, PHOSPHORUS, WATER POLLUTION EFFECTS, WATER POLLUTION SOURCES, WATER POLLUTION CONTROL, OXYGEN BALANCE, SEWAGE, RUNOFF.

IDENTIFIERS:
LAKE LUCERNE, FERTILIZED FIELDS, SUBTERRANEAN DRAINAGE, MECHANICAL AND BIOLOGICAL PURIFICATION, SWITZERLAND.

ABSTRACT:
THE POLLUTION OF LAKES AND PONDS IS IN A LARGE PART DUE TO INORGANIC SALTS PROMOTING ALGAL GROWTH AND SUBSEQUENT PROCESSES OF DECOMPOSITION. AS REVEALED BY STUDIES OF LAKE LUCERNE, A FEW PARTS PER MILLION OF FERTILIZER SALTS, PARTICULARLY PHOSPHATES, ARE SUFFICIENT TO STIMULATE THE GROWTH OF ALGAE AND UPSET THE OXYGEN BALANCE. A LARGE FRACTION OF EUTROPHYING SUBSTANCES IS BROUGHT TO LAKES FROM FERTILIZED FIELDS VIA RUN-OFF AND SUBTERRANEAN DRAINAGE. THE AMOUNT OF PHOSPHORUS FROM THESE SOURCES IS EQUAL TO THAT SUPPLIED BY THE NORMAL DISCHARGES OF SEWAGE; THE AMOUNT OF NITROGEN DELIVERED FROM FIELDS EXCEEDS THAT CONTRIBUTED BY SEWAGE. ON THE AVERAGE, SWISS LAKES RECEIVE 0.3 TO 0.5 KG OF PHOSPHORUS AND ABOUT 45 KG OF NITROGEN PER HECTARE, PER YEAR; THE LATTER FIGURE IS DERIVED FROM A VERY WIDE AMPLITUDE. THIS INFLOW OF NUTRIENTS IS SUFFICIENT TO EUTROPHY LAKES, BUT THE ONLY CORRECTIVE MEASURE AVAILABLE AT THIS TIME IS PURIFICATION OF THE SEWAGE BY MECHANICAL AND BIOLOGICAL METHODS ELIMINATING PLANT NUTRIENTS. (WILDE-WIS)

FIELD 05C

ACCESSION NO. W69-08668

OBSERVATIONS AND EXPERIENCES IN THE CONTROL OF ALGAE,

TERRENCE P. FROST.

J NEW HAMPSHIRE WATER WORKS ASSOC, 10 P, DEC 1963. 1 TAB, 12 REF.

DESCRIPTORS:
*NUISANCE ALGAE, *ALGAL CONTROL, *NEW HAMPSHIRE, *COPPER SULFATE,
SEWAGE EFFLUENTS, SMELTS, FISHKILL, CRUSTACEANS, BOTTOM SEDIMENTS.

IDENTIFIERS:
*NEW HAMPSHIRE WATER POLLUTION COMMISSION, ALGAL BLOOMS, NUTRIENT LOAD,
SYNTHETIC DETERGENTS, FISH MANAGERS, FISH YIELDS, UNDERWATER SHOCK
WAVE, ANABAENA, PHYGON XL(DICHLONE), NEW HAMPSHIRE, LAKE WINNISQUAM,
LONG POND, CONCORD, BOSTON LOT RESERVOIR, LEBANON, WADLEIGH STATE PARK,
LAKE SHATUTAKEE, MASCOMA LAKE, GLEN LAKE.

ABSTRACT:
AUTHOR OFFERS VARIOUS OBSERVATIONS AND EXPERIENCES RELATED TO ALGAL
CONTROL OPERATIONS IN NUTRIENT RICH WATERS BY THE NEW HAMPSHIRE WATER
POLLUTION COMMISSION. SINCE 1961, ABOUT 6,000 ACRES/YEAR IN LAKE
WINNISQUAM AND OTHERS HAVE BEEN TREATED. OVERENRICHMENT OF THE LAKES IS
DUE TO EFFLUENTS FROM MUNICIPAL AND INDUSTRIAL TREATMENT PLANTS NOT
EQUIPPED TO STRIP NUTRIENTS BECAUSE OF INCREASED OPERATIONAL COSTS. IN
CONTROLLING ALGAE WITH COPPER SULFATE, THE COMMISSION HAS ENCOUNTERED
SOME OPPOSITION FROM FISH MANAGERS WARY OF ITS EFFECTS ON FISH
POPULATIONS. TESTS RUN ON ADULT SMELT AND EGGS SHOWED LEVELS APPLIED IN
ALGAL CONTROL WERE NOT LETHAL TO FISH AND MAY IN FACT INCREASE FISH
YIELDS IN SOME LAKES. A SIDE EFFECT OF COPPER SULFATING IS THE
INCREASED CRUSTACEAN POPULATION WHICH IS SUPPRESSED BY LARGE ALGAL
COMMUNITIES. SO FAR IN NEW HAMPSHIRE, SAMPLING OF BOTTOM MUDS HAS SHOWN
NO ACCUMULATIONS OF COPPER EXCEEDING THE LEVELS IN UNTREATED LAKES.
AUTHOR MENTIONS MANY SUCCESSFUL PROJECTS AND ONLY ONE CATASTROPHE, A
MASSIVE FISHKILL IN BOSTON LOT RESERVOIR, RESULTING FROM A
MISCALCULATION OF VOLUME AND DOUBLING DOSE OF COPPER SULFATE. THE
COMMISSION PLANS TO EXPERIMENT WITH COPPER SULFATING IN THE FUTURE FOR
PREVENTIVE CONTROL. (KETELLE-WIS)

FIELD 05G

ACCESSION NO. W69-08674

RELATION OF PHOSPHATES TO EUTROPHICATION,

NORTH CAROLINA UNIV., CHAPEL HILL. DEPT. OF ENVIRONMENTAL SCIENCES AND
ENGINEERING.

CHARLES M. WEISS.

J AMER WATER WORKS ASS, VOL 61, NO 8, P 387-391, AUG 1969. 5 P, 2 FIG, 4 TAB,
22 REF.

DESCRIPTORS:
*EUTROPHICATION, *LAKES, *PHOSPHATES, NUTRIENTS, ALGAE, WATER POLLUTION
EFFECTS, WATER POLLUTION SOURCES, AQUATIC PRODUCTIVITY, LIMNOLOGY,
PLANKTON, PRIMARY PRODUCTIVITY, TROPHIC LEVEL.

IDENTIFIERS:
LAKE TAHOE.

ABSTRACT:
THE RELATIONSHIP OF PHOSPHATES IN WATER TO PRIMARY PRODUCTION IS
EXAMINED, AND THE PHOSPHORUS CYCLE IN NATURAL LAKE BIOLOGICAL SYSTEMS
IS OUTLINED. ALTHOUGH HIGHER THAN NORMAL MACRONUTRIENT LEVELS WILL
GENERALLY PRODUCE AN INCREASE IN LEVELS OF PRODUCTIVITY OR SHIFT OF
TROPHIC STATUS, THE SPECIFIC INTERRELATIONSHIPS OF EACH NUTRIENT FACTOR
ARE EXCEEDINGLY COMPLEX AND REMAIN UNCLEAR. THE BIOLOGICAL STORAGE OF
PHOSPHORUS BY ALGAE IN EXCESS OF IMMEDIATE GROWTH REQUIREMENT, ITS
SECRETION AS SOLUBLE ORGANIC PHOSPHORUS, AND ITS REUSE WOULD APPEAR TO
BE A SIGNIFICANT FACTOR IN SATISFYING NUTRIENT LEVELS, PARTICULARLY FOR
THOSE SPECIES WITH LOW PHOSPHORUS REQUIREMENTS. INCREASES IN TROPHIC
STATUS ARE GENERALLY ASSOCIATED WITH INVASION OF BLUE-GREEN ALGAE,
WHICH HAVE CAPABILITY FOR NITROGEN FIXATION WITH RESULTING ENHANCEMENT
OF NITROGEN LEVELS AND ACCELERATION OF THE RATE OF EUTROPHICATION. THE
RATE OF TROPHIC RESPONSE OF EACH LAKE OR RESERVOIR TO CHANGES IN
MACRONUTRIENT LEVELS INVOLVES NOT ONLY NUTRIENT CHANGE, BUT ALSO THE
TOTAL CHEMICAL AND PHYSICAL CHARACTERISTICS OF THE PARTICULAR BASIN.
THE CHEMICAL NATURE OF THE PARTICULAR PHOSPHORUS SPECIES, ITS
ASSOCIATION WITH COLLOIDAL OR PARTICULATE MATERIALS AND THE SPECIFIC
ASSOCIATED PHYTOPLANKTON ALL CONTRIBUTE TO THE RATE OF UTILIZATION OF
THE PHOSPHATE POOL AND THUS THE RATE OF EUTROPHICATION. (KNAPP-USGS)

FIELD 05C, 05B

ACCESSION NO. W69-09135

TRANSLOCATION OF PHOSPHORUS IN A TROUT STREAM ECOSYSTEM,

MICHIGAN STATE UNIV., EAST LANSING. DEPT. OF FISHERIES AND WILDLIFE; AND
  INSTITUTE FOR FISHERIES RESEARCH, ANN ARBOR, MICH.

ROBERT C. BALL, AND FRANK F. HOOPER.

PROC 1ST NAT SYMP ON RADIOECOLOGY, COLORADO STATE UNIV, FORT COLLINS, SEPT
  10-15, 1961. RADIOECOLOGY, 1963, REINHOLD PUB CORP, NY AND AMER INST OF
  BIOLOGICAL SCIENCES, WASH, DC, SCHULTZ, VINCENT AND KLEMENT JR, ALFRED W
  (EDS), P 217-228. 4 FIG, 7 TAB, 16 REF.

DESCRIPTORS:
    *TRANSLOCATION, *PHOSPHORUS RADIOISOTOPES, *TROUT, *STREAMS,
    *ECOSYSTEMS, RADIOACTIVITY, BIOTA, CYCLES, MICHIGAN, INVERTEBRATES,
    FISH, TEMPERATURE, TRACERS, SAMPLING, PERIPHYTON, AQUATIC PLANTS,
    BACKGROUND RADIATION, GEOCHEMISTRY, SPRING WATERS, PHOSPHORUS,
    ORGANOPHOSPHORUS COMPOUNDS, BACTERIA, DIATOMS, DETRITUS, CYCLING
    NUTRIENTS, LIGHT INTENSITY, FERTILIZATION, GROWTH RATES, EQUILIBRIUM,
    CHARA, ALGAE, AMMOCETES, LAMPREYS, OMNIVORES.

IDENTIFIERS:
    MICHIGAN, CHEBOYGAN COUNTY, STURGEON RIVER, FLUORESCEIN, MACROPHYTES,
    GROUND SOLIDS, UPTAKE, CONSUMER ORGANISMS, SIMULIUM, PHYSA, SALMO
    TRUTTA, ENTOSPHENUS LAMOTTENII, GAMMARUS, CORDULEGASTER.

ABSTRACT:
TO EXPLORE MOVEMENT OF PHOSPHORUS IN A COLD-WATER STREAM AND
RELATIONSHIPS BETWEEN PHOSPHORUS AND STREAM PRODUCTIVITY, SPIKES OF
APPROXIMATELY 23 MILLICURIES OF PHOSPHORUS-32 (P-32) WERE ADDED TO THE
WATER DURING THREE SUMMERS. THE LOSS PATTERN VARIED FROM YEAR TO YEAR,
WITH MUCH UPTAKE APPARENTLY ATTRIBUTED TO PERIPHYTON AND THREE AQUATIC
MACROPHYTES. P-32 RATE OF LOSS FROM PLANTS TO WATER DECREASED WITH TIME
AND CONCENTRATION APPROACHED EQUILIBRIUM LEVEL IN 15 TO 20 DAYS AFTER
SPIKE WAS RELEASED. DECREASE CHANGE RATES SUGGESTED THAT PLANTS WERE
REMOVING RE-CYCLED PHOSPHORUS FROM THE STREAM WATER. HIGHEST UPTAKE BY
PERIPHYTON OCCURRED IN 1958 AND 1960; LOWEST IN 1959. THIS PATTERN WAS
REVERSED BY AQUATIC MACROPHYTES. CONCENTRATION CURVES FOR CONSUMER
ORGANISMS REFLECTED DIFFERENCES IN METABOLIC TURNOVER RATES AND FOOD
RELATIONSHIPS. IN SIMULIUM, CONCENTRATIONS QUICKLY ROSE TO A HIGH
LEVEL, SEEMINGLY DURING PASSAGE OF THE SPIKE. OTHER SMALL FILTER
FEEDERS AND PERIPHYTON SCRAPERS ALSO ACCUMULATED P-32 RAPIDLY, THEN
LOST IT QUICKLY. ACCUMULATING P-32 WAS SLOWER FOR LARGE OMNIVOROUS
STREAM INSECTS AND SMALLER PREDACIOUS FORMS. DIFFERENCES IN CIRCULATION
THROUGH THE ECOSYSTEM, AS NOTED DURING THE 3-YEAR TEST PERIOD, MAY BE
DUE TO DISTRIBUTION OF P-32 BETWEEN SOLUBLE AND PARTICULATE PHASES.
(JONES-WIS)

FIELD 05C

ACCESSION NO. W69-09334

THE PHOSPHORUS PROBLEM,

DEPARTMENT OF THE INTERIOR, WASHINGTON, D.C.

KENNETH M. MACKENTHUN.

J AMER WATER WORKS ASS, VOL 60, NO 9, P 1047-1054, SEPT 1968. 2 TAB, 32 REF.

DESCRIPTORS:
*PHOSPHORUS, *NUTRIENTS, *EUTROPHICATION, *TERTIARY TREATMENT,
NITROGEN, AQUATIC PLANTS, BOTTOM SEDIMENTS, DREDGING, HARVESTING OF
ALGAE, FISH HARVEST, VITAMINS.

IDENTIFIERS:
*WASTE WATER(MUNICIPAL AND INDUSTRIAL), *PHOSPHORUS SOURCES, *NUISANCE
PLANT GROWTHS, *NUTRIENT CONTROL, *PHOSPHORUS CONCENTRATIONS,
MICRONUTRIENTS, AQUATIC ECOSYSTEM, MIDGE LARVAE, LAKE STRATIFICATION,
SURFACE AREA(LAKES), PHAEDACTYLUM, WASTE WATER DIVERSION, INSTREAM
TREATMENT, BOTTOM SEALING, LAKE DILUTION.

ABSTRACT:
REDUCTION OF WASTEWATER PHOSPHORUS INFLOWS TO RECEIVING WATERS IS
NECESSARY TO CHECK ACCELERATING CULTURAL EUTROPHICATION. MINIMIZING
PHOSPHORUS INFLOWS TO SURFACE WATERS IS RECOMMENDED BECAUSE IT IS THE
MOST EASILY REDUCED OF ALL CONSTITUENTS IN WASTE WATERS ESSENTIAL TO
AQUATIC PLANT GROWTH. THE AUTHOR SUGGESTS THAT PREVENTION OF BIOLOGICAL
NUISANCES REQUIRES TOTAL PHOSPHORUS CONCENTRATIONS NOT EXCEEDING 100
MICROGRAMS/LITER AT ANY POINT IN FLOWING WATER; AND 50 MICROGRAMS/LITER
WHERE WATER ENTERS A STANDING BODY (LAKE OR RESERVOIR). THE MOST
EFFECTIVE MEANS OF CONTROLLING PHOSPHORUS LEVELS IN LAKES NOW ARE
TREATMENT OF NUTRIENT POINT SOURCES, WASTE WATER DIVERSIONS AROUND
LAKES, AND DILUTION WITHIN LAKES WHERE FEASIBLE. ONCE NUTRIENTS ARE
COMBINED WITHIN THE ECOSYSTEM OF THE RECEIVING WATER, THEIR REMOVAL IS
TEDIOUS AND EXPENSIVE. RESULTS OF HARVESTING OF AQUATIC CROPS,
DREDGING, AND OTHER MEANS TO REMOVE NUTRIENTS AFTER THEY HAVE REACHED
RECEIVING WATERS SHOULD BE COMPARED WITH INFLOWING NUTRIENT QUANTITIES
TO EVALUATE EFFECTIVENESS. THE PRESERVATION OF SURFACE WATERS IN A
USABLE STATE WILL DEPEND IN THE LONG RUN ON MAXIMUM NUTRIENT REMOVAL AT
THE WASTE SOURCES. (KETELLE-WIS)

FIELD 05C

ACCESSION NO. W69-09340

84

WATER-QUALITY MANAGEMENT AND LAKE EUTROPHICATION: THE LAKE WASHINGTON CASE,

WASHINGTON UNIV., SEATTLE. DEPT. OF ZOOLOGY.

W. THOMAS EDMONDSON.

WATER RESOURCES MANAGEMENT AND PUBLIC POLICY, WASHINGTON UNIV PRESS, SEATTLE,
  P 139-178, 1968. 2 FIG, 1 TAB, 104 REF.

DESCRIPTORS:
   *LAKES, *EUTROPHICATION, *SEWAGE EFFLUENTS, *NUTRIENTS, *NITROGEN,
   PHOSPHORUS, SULFATES, VITAMINS, ALGAE, GROWTH RATES, COPPER SULFATE,
   BOTTOM SEDIMENTS, LAKE MICHIGAN, WASHINGTON, DISSOLVED OXYGEN.

IDENTIFIERS:
   *EFFLUENT DIVERSION, *WATER QUALITY MANAGEMENT, *LAKE WASHINGTON,
   *SEWAGE FERTILIZATION EFFECTS, *BIOLOGICAL PRODUCTIVITY, AQUATIC
   NUISANCES, TRANSPARENCY, AGRICULTURAL RUNOFF, NUTRIENT BUDGETS,
   COMPARATIVE STUDIES, NUTRIENT REMOVAL, LAKE RECOVERY, WISCONSIN, LAKE
   MONONA, LAKE MENDOTA, LAKE WAUBESA, LAKE KEGONSA, LAKE ZURICH,
   OSCILLATORIA RUBESCENS, CLADOPHORA, DINOBRYON DIVERGENS, SWITZERLAND,
   LAKE LUCERNE, LAKE CONSTANCE, WASHINGTON, PUGET SOUND, BARE
   LAKE(ALASKA), LAKE LYNGBY SO(DENMARK), LAKE TAHOE(CALIF), LAKE
   MAGGIORE(ITALY), SPECIES DOMINANCE.

ABSTRACT:
   LAKE WASHINGTON'S CASE PARALLELS OTHER LAKES WORLDWIDE WHERE CULTURAL
   EUTROPHICATION HAS LED TO INCREASED BIOLOGICAL PRODUCTIVITY, LIMITING
   WATER USES. FIRST DEFINITE INDICATION OF ENRICHMENT APPEARED IN 1955
   WITH ABUNDANCE OF OSCILLATORIA RUBESCENS, FOLLOWED BY REDUCTIONS IN
   TRANSPARENCY AND DISSOLVED OXYGEN LEVELS IN WATER. PRESENT SOLUTION TO
   THE PROBLEM HAS BEEN ALMOST TOTAL DIVERSION OF EFFLUENTS FROM THE LAKE,
   COMPLETED IN 1967 AT A COST OF $85 MILLION (INCLUDING NEW TREATMENT
   FACILITIES). EVIDENCE OF THE ROLE OF SEWAGE IN FERTILIZATION AND LAKE
   DETERIORATION COMES FROM MANY SOURCES: DIRECT OBSERVATION; URBANIZATION
   FOLLOWED BY SYMPTOMS OF INCREASED LAKE PRODUCTIVITY; NUTRIENT BUDGET
   STUDIES, INDICATING RELATIVE NUTRIENT PERCENTAGES FROM DIFFERENT
   SOURCES; COMPARISON OF SIMILAR LAKES IN DIFFERENT ENVIRONMENTS; CHANGES
   RESULTING FROM SEWAGE EFFLUENT DIVERSIONS AROUND LAKES; BIOASSAY
   TECHNIQUES OF LAKE WATER SAMPLES; AND PHYSIOLOGICAL STUDIES OF ALGAE
   GROWN IN NUTRIENT SOLUTIONS. THE RELATIVE IMPORTANCE OF VARIOUS
   NUTRIENTS AND MAXIMUM ALLOWABLE CONCENTRATIONS IN LAKES ARE DEPENDENT
   ON SEVERAL INTERRELATED FACTORS INCLUDING DIVERSE EFFECTS OF DIFFERENT
   COMBINATIONS OF NUTRIENTS, AND VARIETY IN REACTIONS BY SPECIES TO
   NUTRIENTS. IMPROVED METHODS OF NUTRIENT REMOVAL FROM SEWAGE WILL
   UNDOUBTEDLY DEVELOP, BUT AT PRESENT, EFFLUENT DIVERSION IS THE ONLY
   SURE METHOD OF CONTROLLING ENRICHMENT OF LAKES. (KETELLE-WIS)

FIELD 05G, 02H

ACCESSION NO. W69-09349

THE ROLE OF ALGAE IN DEGRADING DETERGENT SURFACE ACTIVE AGENTS,

TEXAS UNIV., AUSTIN. DEPT. OF CIVIL ENGINEERING.

ERNST M. DAVIS, AND E. F. GLOYNA.

J WATER POLLUT CONTR FEDERATION, VOL 41, NO 8, PART 1, P 1494-1504, AUG 1969. 11 P, 4 FIG, 5 TAB, 21 REF. GRANT NO NSF-GU-1963.

DESCRIPTORS:
*BIODEGRADATION, *ALGAE, *AQUATIC BACTERIA, DETERGENTS, SURFACTANTS, ALKYLBENZENE SULFONATES, LINEAR ALKYLATE SULFONATES, WATER POLLUTION EFFECTS, STABILIZATION, LAGOONS, TOXICITY, WASTE WATER TREATMENT, WASTE WATER DISPOSAL.

IDENTIFIERS:
*ABS, LAS.

ABSTRACT:
THE RATES OF DEGRADATION OF 3 TYPICAL NONIONIC AND 2 ANIONIC SURFACTANTS BY GREEN AND BLUE-GREEN ALGAE WERE EVALUATED. ADDITIONAL DATA WERE GATHERED FOR ORGANISMS OBTAINED FROM RAW WASTEWATER, ACTIVATED SLUDGE, AND WASTE STABILIZATION POND WATER. RESULTS SHOW THAT ALGAE CONTRIBUTE TO A SMALL EXTENT IN THE DEGRADATION OF COMMON HOUSEHOLD DETERGENTS. MOST DEGRADATION IS ATTRIBUTED TO BACTERIA OR OTHER ORGANISMS. ALGAE DID NOT DEGRADE ALKYL BENZENE SULFONATE WHILE BACTERIA DID (0.45 MG/1/DAY). THE OTHER ANIONIC SURFACTANT, LINEAR ALKYL SULFONATE, WAS DEGRADED IN SMALL AMOUNTS BY SOME OF THE DIFFERENT ALGAL FORMS (0.97 MG/1/DAY BY BACTERIA). NONIONIC DETERGENTS WERE MORE EASILY DEGRADED. (KNAPP-USGS)

FIELD 05D

ACCESSION NO. W69-09454

EFFECT OF LIGHT INTENSITY ON PHOTOSYNTHESIS BY THERMAL ALGAE ADAPTED TO NATURAL AND REDUCED SUNLIGHT,

INDIANA UNIV., INDIANAPOLIS. DEPT. OF MICROBIOLOGY.

THOMAS D. BROCK, AND M. LOUISE BROCK.

LIMNOL AND OCEANOGR, VOL 14, NO 3, P 334-341, MAY 1969. 8 P, 7 TAB, 19 REF.

DESCRIPTORS:
*PHOTOSYNTHESIS, *ALGAE, *THERMAL SPRINGS, *LIGHT INTENSITY, SPRING WATER, WATER CHEMISTRY, WATER TEMPERATURE, CHEMICAL ANALYSIS, ANALYTICAL TECHNIQUES, CLOUD COVER, SPECTROSCOPY, ULTRAVIOLET RADIATION, TRANSMISSIVITY, CHLOROPHYLL.

ABSTRACT:
THE EFFECT OF LIGHT INTENSITY ON THERMAL ALGAE OF HOT SPRINGS OF YELLOWSTONE NATIONAL PARK WAS INVESTIGATED BY MEASURING AUTORADIOGRAPHY OF PHOTOSYNTHETIC ACTIVITY OF CELLS AT DIFFERENT LEVELS IN THE ALGAE MAT. THE RATE OF PHOTOSYNTHESIS DECREASES PROGRESSIVELY WITH DECREASING LIGHT, ALTHOUGH THE MOST EFFICIENT USE IS AT 7-14% OF FULL SUNLIGHT. THE CHLOROPHYLL CONTENT OF THE ALGAE AT THE SURFACE OF THE MAT INCREASES RAPIDLY; HOWEVER, THE CHLOROPHYLL CONTENT OF THE MAT INCREASES SLOWLY OR NOT AT ALL. ALTHOUGH INDIVIDUAL ALGAE CELLS CAN ADAPT RAPIDLY TO CHANGES IN LIGHT, THE ENTIRE POPULATION, BECAUSE OF ITS EXISTENCE IN COMPACT MATS, ADAPTS SLOWLY. AT THE LATITUDE OF YELLOWSTONE THERE IS SUFFICIENT LIGHT THROUGHOUT THE WHOLE YEAR TO STIMULATE THE ALGAE GROWTH EVEN AT CRITICAL TEMPERATURES. (GABRIEL-USGS)

FIELD 02K, 05A

ACCESSION NO. W69-09676

EUTROPHICATION FACTORS IN NORTH CENTRAL FLORIDA LAKES,

FLORIDA UNIV., GAINESVILLE. ENGINEERING AND INDUSTRIAL EXPERIMENT STATION.

P. L. BREZONIK, W. H. MORGAN, E. E. SHANNON, AND H. D. PUTNAM.

ENGINEERING PROGRESS IN FLORIDA. BULL SERIES NO 134 AND WATER RESOURCES
RESEARCH CENTER PUB NO 5, GAINSEVILLE, FLORIDA, AUG 1969. 101 P, 50 FIG, 43
TAB, 124 REF, 4 APPEND. GRANT DON 16, 010(FWPCA). OWRR PROJECT A-002-FLA.

DESCRIPTORS:
WATER QUALITY, *EUTROPHICATION, ALGAL BLOOMS, NUTRIENTS, LIMNOLOGY,
DYSTROPHY, OLIGOTROPHY, *PRIMARY PRODUCTIVITY, *TROPHIC LEVEL, WATER
STORAGE, EVAPORATION, WATER CIRCULATION, SEDIMENTS.

IDENTIFIERS:
SEEPAGE LOSSES, DIURNAL VARIATIONS.

ABSTRACT:
A SMALL LAKE IN AN UNINHABITED DRAINAGE BASIN IN NORTH FLORIDA IS BEING
USED AS A MODEL FOR THE EUTROPHICATION RESEARCH DESCRIBED IN THIS
REPORT. AN ADJACENT LAKE SERVES AS THE CONTROL. BACKGROUND DATA ON THE
CHEMISTRY AND BIOLOGY OF THE LAKES WERE OBTAINED IN ORDER TO BE CERTAIN
OF THEIR SIMILARITY AND TROPHIC STATUS. WATER AND NUTRIENT BUDGETS WERE
DETERMINED AND NUTRIENT ENRICHMENT (TREATED SEWAGE EFFLUENT) HAS BEEN
CONTINUOUS SINCE MARCH 1967. BOTH ROUTINE MONITORING AND SPECIAL (10
DAY) STUDIES WERE CARRIED OUT IN BOTH LAKES COMPRISING THE EXPERIMENTAL
SYSTEM. IN 1968 THE RESEARCH PROGRAM WAS EXTENDED TO INCLUDE OTHER
LAKES OF VARIOUS TYPES AND EXHIBITING VARYING TROPHIC STAGES IN NORTH
CENTRAL FLORIDA. THE EUTROPHIC LAKES IN THE OKLAHOMA CHAIN ARE
INCLUDED. CHANGES IN PARAMETERS HAVE BEEN AND ARE CONTINUING TO BE
DETERMINED AND RELATED TO THE NUTRIENT FLUX. SYSTEMS ANALYSIS
TECHNIQUES HAVE BEEN APPLIED. APPENDICES INCLUDE METHODS OF CHEMICAL
ANALYSES; BIOLOGICAL PROCEDURES, A LISTING OF PROTOZOA, MICROSCOPIC
ALGAE AND SULFUR BACTERIA RECORDED FROM THE EXPERIMENTAL AND CONTROL
LAKES ON FIVE DATES, AND THE MULTIPLE REGRESSION ANALYSES PERFORMED ON
TEN DAY STUDY DATA AND ROUTINE MONTHLY DATA.

FIELD 05C, 02H

ACCESSION NO. W69-09723

NATURAL RADIOACTIVITY IN THE FOOD WEB OF THE BANDED SCULPIN COTTUS CAROLINAE (GILL),

LOUISVILLE UNIV., KY. DEPT. OF BIOLOGY.

W. L. MINCKLEY, J. E. CRADDOCK, AND L. A. KRUMHOLZ.

PROC 1ST NAT SYMP ON RADIOECOLOGY, COLO STATE UNIV, FORT COLLINS, SEPT 10-15, 1961. RADIOECOLOGY, 1963, REINHOLD PUB CORP, NY AND AMER INST OF BIOLOGICAL SCIENCES, WASH DC, SCHULTZ, VINCENT AND KLEMENT JR, ALFRED W (EDS) P229-236. 8 FIG, 4 TAB, 13 REF.

DESCRIPTORS:
*SCULPINS, *RADIOACTIVITY, *FOOD WEBS, KENTUCKY, LIFE HISTORY STUDIES, GAMMA RAYS, SPECTROMETERS, CYANOPHYTA, URANIUM RADIOISOTOPES, STRONTIUM RADIOISOTOPES, CESIUM, FALLOUT, SILTS, DETRITUS, BOTTOM SEDIMENTS, CHUTES, EDDIES, HABITATS, CRAYFISH, GROWTH STAGES, DIATOMS, ALGAE, METABOLISM, ISOPODS, AMPHIPODA, FISH, GROWTH RATES.

IDENTIFIERS:
*COTTUS CAROLINAE, MEAD COUNTY(KY), THORIUM-232, BISMUTH-214, ACTINIUM-228, SWIFT RIFFLES, ZIRCONIUM-95, NIOBIUM-95, RUTHENIUM-106, RHODIUM-106, CERIUM-144, BETA RADIOACTIVITY, GONIOBASIS, DECAPODA, PLECOPTERA, EPHEMEROPTERA, TIPULIDAE, CHIRONOMIDAE, HALF-LIFE.

ABSTRACT:
CONSIDERATION OF LIFE HISTORY AND RADIOECOLOGY OF BANDED SCULPIN (COTTUS CAROLINAE) REVEALS THAT MOST RADIOACTIVITY RESULTS FROM NATURAL DECAY; SOME, FROM FALLOUT. BIOTA OF THE ECOSYSTEM ACCUMULATE MEASURABLE AMOUNTS. IN THE FOOD WEB OF COTTUS CAROLINAE, RADIOACTIVITY CONCENTRATES AT THE PRODUCER LEVEL, WITH DECREASED AMOUNTS AT HIGHER TROPHIC LEVELS INVOLVING ANIMALS. ANIMALS, FED MOSTLY ON DIATOMS AND BLUE-GREEN ALGAE, GENERALLY CONTAINED THE MOST RADIOACTIVITY. VERTEBRATES, AND LARGER INVERTEBRATES CONTAINED RADIOACTIVITY WITH LEVELS VARYING ACCORDING TO SIZE OF INDIVIDUAL ANIMAL. SMALLER ANIMALS CONTAINED MORE RADIOACTIVE MATERIALS PER UNIT WEIGHT THAN LARGER ONES. ACTUAL METABOLIC CHANGES SEEM TO BE AS IMPORTANT IN DETERMINING AMOUNT OF RETENTION AND ACCUMULATION OF RADIOMATERIALS AS ARE AMOUNTS OF THE RADIOMATERIALS INGESTED. SIMILARLY, CUMULATIVE AMOUNTS OF RADIOACTIVITY TAKEN IN AS FOOD, COMPARED WITH EVER-INCREASING BODY BURDEN, ALLOW CALCULATION OF PERCENTAGE OF INGESTED RADIOACTIVITY RETAINED IN THE BODY. PERCENTAGES ARE RELATIVELY HIGH DURING EARLY LIFE OF SCULPINS, DECREASING JUST BEFORE MATURITY, AND INCREASING SLOWLY THROUGHOUT REMAINDER OF LIFE. LACK OF MAJOR VARIATIONS IN THIS LEVEL OF RETENTION INDICATE A CERTAIN METABOLIC NEED, OR SOMEWHAT CONSTANT METABOLIC ACCUMULATION, WHICH COMPRISES APPROXIMATELY 2% OF TOTAL RADIOMATERIALS INGESTED. (JONES-WIS)

FIELD 05C

ACCESSION NO. W69-09742

MODE OF FILAMENTOUS GROWTH OF LEUCOTHRIX MUCOR IN PURE CULTURE AND IN NATURE,
   AS STUDIED BY TRITIATED THYMIDINE AUTORADIOGRAPHY,

INDIANA UNIV., INDIANAPOLIS. DEPT. OF MICROBIOLOGY.

THOMAS D. BROCK.

JOURNAL OF BACTERIOLOGY, VOL 93, NO 3, P 985-990, MARCH 1967. 3 FIG, 5 TAB,
   11 REF.

   DESCRIPTORS:
      *BACTERIA, *CULTURES, *TRITIUM, EPIPHYTOLOGY, MARINE ALGAE, ECOLOGY,
      NUTRIENTS, SEAWATER, ORGANIC COMPOUNDS, HABITATS, STATISTICAL METHODS.

   IDENTIFIERS:
      *LEUCOTHRIX MUCOR, *GROWTH, *FILAMENTS, *AUTORADIOGRAPHY, ANTITHAMNION
      SARNIENSE, TRITIATED GLUCOSE, GONIDIA FORMATION, LONG ISLAND SOUND(NY),
      ICELAND, TRITIATED THYMIDINE, NARRAGANSETT BAY(RI), CAPE
      REYKJANES(ICELAND), SUDURNES(ICELAND), NON-PARAMETRIC TESTS, ONE-SAMPLE
      RUNS TEST, CLUSTER ANALYSIS, REYKJAVIK(ICELAND).

   ABSTRACT:
      GROWTH MODE OF LEUCOTHRIX MUCOR FILAMENTS WAS MEASURED BY
      AUTORADIOGRAPHY WITH TRITIATED THYMIDINE. STUDIES WERE PERFORMED ON L
      MUCOR IN PURE CULTURES IN FREE SUSPENSION, AS AN EPIPHYTE OF PURE
      CULTURE OF THE RED ALGA ANTITHAMNION SARNIENSE, AND AS AN EPIPHYTE OF
      RED ALGAE IN THE SEA. STATISTICAL ANALYSES OF GROWING CELLS
      DISTRIBUTION WAS PERFORMED BY USE OF THE NONPARAMETRIC ONE-SAMPLE RUNS
      TEST AND A CLUSTER ANALYSIS ADAPTED FROM QUADRAT ANALYSES OF PLANT
      ECOLOGISTS. NO EVIDENCE OF PREFERENTIAL GROWTH AT BASE OR TIP OF L
      MUCOR FILAMENTS WAS OBTAINED IN ANY OF THESE STUDIES. HOWEVER, IN
      NATURE, BUT NOT IN THE LABORATORY, THERE WERE REGIONS OF L MUCOR
      FILAMENTS WHICH WERE NONGROWING OR DORMANT. SUCH NONGROWING REGIONS
      COULD INCORPORATE TRITIATED GLUCOSE. IN LABORATORY CULTURE OF
      LEUCOTHRIX ON ANTITHAMNION, LEUCOTHRIX CANNOT GROW ALONE IN THE MEDIUM
      (ASP-6) IN WHICH THE ALGA GROWS, BUT ONLY WHEN ATTACHED TO ALGA. THE
      ALGAE SEEMINGLY RELEASED NUTRIENTS INTO THE WATER WHERE COMPLETE MIXING
      OCCURS. ALL REGIONS OF BACTERIAL FILAMENTS SHOULD HAVE EQUAL ACCESS TO
      THE NUTRIENT SUPPLY AND THERE SHOULD BE NO PREFERENTIAL GROWTH AT THE
      BASE. IN NATURE, EPIPHYTIC BACTERIA PROBABLY DERIVE THEIR NUTRIENTS
      FROM SEAWATER. (JONES-WIS)

   FIELD 05C

   ACCESSION NO. W69-09755

DISTRIBUTION OF PESTICIDES IN SURFACE WATERS,

VIRGINIA POLYTECHNIC INST., BLACKSBURG, VA.

PAUL H. KING, H. H. YEH, PIERRE S. WARREN, AND CLIFFORD W. RANDALL.

J AMER WATER WORKS ASS, VOL 61, NO 9, P 483-486, SEPT 1969. 4 P, 7 FIG, 2
   TAB, 7 REF.

   DESCRIPTORS:
      *PESTICIDE KINETICS, *PESTICIDE REMOVAL, *ADSORPTION, SOILS, COALS,
      DISTRIBUTION PATTERNS, ACTIVATED CARBON, CLAYS, ALGAE, WATER
      PURIFICATION, WATER QUALITY, WATER CHEMISTRY.

   IDENTIFIERS:
      PESTICIDE SORPTION.

   ABSTRACT:
      SEVERAL PAPERS ON PESTICIDE DISTRIBUTION IN WATER, PESTICIDE SORPTION
      IN SOIL, AND PESTICIDE REMOVAL TECHNOLOGY ARE REVIEWED AND SUMMARIZED,
      AND EXPERIMENTAL DATA ARE PRESENTED ON PESTICIDE SORPTION BY SOILS,
      ALGAE, AND ACTIVATED CARBON. THE PESTICIDES STUDIED ARE LINDANE AND
      PARATHION. EXPERIMENTAL RESULTS ARE SHOWN BY SORPTION ISOTHERMS. IN
      GENERAL, HIGH-CLAY SOILS ADSORBED ABOUT TWICE AS MUCH PESTICIDE AS
      SANDY SOILS. ALGAE SORBED ABOUT 10 TIMES AS MUCH AS SOILS. COAL
      ADSORBED ABOUT 2 1/2 TIMES AS MUCH AS SOIL, AND ACTIVATED CARBON
      ADSORBED ABOUT 4 ORDERS OF MAGNITUDE MORE THAN SOIL. (KNAPP-USGS)

   FIELD 05B, 05G

   ACCESSION NO. W69-09884

RADIOECOLOGICAL SURVEILLANCE OF THE WATERWAYS AROUND A NUCLEAR FUELS
REPROCESSING PLANT,

NEW YORK STATE DEPT. OF HEALTH, ALBANY. DIV. OF LABORATORIES AND RESEARCH;
STATE UNIV. OF NEW YORK, ALBANY. DEPT. OF BIOLOGICAL SCIENCES; AND STATE
UNIV. COLL., FREDONIA, N.Y.

N. I. SAX, PAUL C. LEMON, ALLEN H. BENTON, AND JACK J. GABAY.

RADIO HEALTH DATA AND REP, VOL 10, NO 7, P 289-296, JULY 1969. 8 P, 3 FIG, 6
TAB, 26 REF. USPHS SERVICE GRANT, NAT CENTER FOR RADIOLOGICAL HEALTH.

DESCRIPTORS:
*MONITORING, *RADIOISOTOPES, *STREAMS, BIOINDICATORS, *NEW YORK,
NUCLEAR WASTES, WATER POLLUTION SOURCES, PATH OF POLLUTANTS, AQUATIC
LIFE, ALGAE, SILTS, ECOLOGY, AQUATIC ENVIRONMENT.

IDENTIFIERS:
CATTARAUGUS CREEK(NY).

ABSTRACT:
A 3-YR STUDY OF THE AQUATIC ECOSYSTEM AROUND A NUCLEAR FUELS
REPROCESSING PLANT, LOCATED IN WESTERN NEW YORK STATE, WAS CONDUCTED TO
FIND AND EVALUATE NATURAL INDICATORS OF ENVIRONMENTAL CONTAMINATION.
THE STUDY COVERED PREOPERATIONAL AND EARLY POST-OPERATIONAL PHASES. THE
ECOLOGICAL VECTORS SELECTED FOR STUDY WERE ALGAE, SILT, AND FISH, ALL
FROM CATTARAUGUS CREEK AND ITS TRIBUTARIES. SAMPLES WERE COLLECTED AT
SEVERAL POINTS UPSTREAM AND DOWNSTREAM FROM THE PLANT EFFLUENT AND
QUANTITATIVELY ANALYZED BY GAMMA-RAY SPECTROMETRY. ALL VECTORS ANALYZED
INDICATED PROCESS OF UPTAKE AND CONCENTRATION OF RUTHENIUM-RHODIUM-106,
CESIUM-137, CESIUM-134, AND/OR ZIRCONIUM-NIOBIUM-95, AND SOMETIMES
COBALT-60. IN ADDITION, THE CONCEPT OF USING NATURAL INDICATORS NOT
NATIVE TO THE STREAMS UNDER STUDY WAS TESTED BY TRANSLOCATING FRESH
WATER CLAMS FROM CHAUTAUQUA LAKE, 60 MI SOUTHWEST OF THE SITE, AND
PLACING THEM IN THE STREAMS AROUND THE PLANT. THE CLAMS NOT ONLY
THRIVED IN THEIR NEW ENVIRONMENT BUT UPON ANALYSIS SHOWED DEFINITE
INTERACTION BY CONCENTRATION OF THE ABOVE RADIOISOTOPES, WITH THE
SHELLS SHOWING APPROXIMATELY TWICE THE RADIOACTIVITY OF THE SOFT PARTS
OF THE CLAM. THIS CONCEPT MIGHT PROVE TO BE A SENSITIVE INDICATOR OF
ENVIRONMENTAL CONTAMINATION. (KNAPP-USGS)

FIELD 05C, 05A, 05B

ACCESSION NO. W69-10080

THE MEASUREMENT OF CHLOROPHYLL, PRIMARY PRODUCTIVITY, PHOTOPHOSPHORYLATION, AND
MACROMOLECULES IN BENTHIC ALGAL MATS,

INDIANA UNIV., BLOOMINGTON. DEPT. OF MICROBIOLOGY.

THOMAS D. BROCK, AND M. LOUISE BROCK.

LIMNOLOGY AND OCEANOGRAPHY, VOL 12, NO 4, P 600-605, OCT 1967. 1 FIG, 2 TAB,
13 REF.

DESCRIPTORS:
*BENTHIC FLORA, *CHLOROPHYLL, *PRIMARY PRODUCTIVITY, *MEASUREMENT,
PROTEINS, RADIOACTIVITY, CARBON RADIOISOTOPES, PHOSPHORUS
RADIOISOTOPES, ALGAE, BIOCHEMISTRY, ECOLOGY, SAMPLING, THERMAL SPRINGS,
EFFLUENTS, CYANOPHYTA.

IDENTIFIERS:
*PHOTOPHOSPHORYLATION, *MACROMOLECULES, NUCLEIC ACID, RIBONUCLEIC ACID
SYNTHESIS.

ABSTRACT:
THIS IS PART OF A STUDY OF THE BIOCHEMICAL ECOLOGY OF THERMAL
ENVIRONMENTS. ALONG THE GRADIENTS OF HOT-SPRING EFFLUENTS, ALGAE
DEVELOP ON THE BOTTOMS OF SINTER-LINED CHANNELS. THICKNESS AND
CONSISTENCY OF THE MAT IS DETERMINED BY TEMPERATURE, FLOW RATE, AND
TYPE OF FLOW--WHETHER TURBULENT OR LAMINAR. IN CHANNELS WITH LAMINAR
FLOW, THE ALGAL MATS ARE RELATIVELY UNIFORM OVER SEVERAL SQUARE
DECIMETERS. METHODS ARE DESCRIBED FOR MEASURING PRIMARY PRODUCTIVITY,
CHLOROPHYLL, PROTEIN, AND NUCLEIC ACID IN ALGAL CORES. THE CORES ARE
HOMOGENIZED AFTER INCUBATION WITH A RADIOACTIVE ISOTOPE ALLOWING
REPRESENTATIVE SUBSAMPLING. ISOTOPE SELF-ABSORPTION IS NOT A PROBLEM,
BECAUSE ONLY A SMALL FRACTION OF THE CORE IS ACTUALLY FILTERED. THE
METHOD ELIMINATES SAMPLING VARIABILITY SINCE CHLOROPHYLL IS ESTIMATED
IN THE SAME SAMPLES USED TO DETERMINE CARBON-14 (C-14) UPTAKE. ACID OR
ALKALI EXTRACTION WAS USED TO INDICATE THE BIOCHEMICAL FRACTIONS INTO
WHICH THE C-14 WAS INCORPORATED. TOTAL PHOSPHATE CONTENT OF THE SPRING
WATER WAS DETERMINED CHEMICALLY FOR CALCULATION OF SPECIFIC
RADIOACTIVITY OF PHOSPHATE-32 (P-32) ION. RATES OF PHOTOPHOSPHORYLATION
AND RIBONUCLEIC ACID SYNTHESIS WERE MEASURED USING P-32. ALL WORK WAS
DONE WITH BLUE-GREEN ALGAE, AND METHOD SHOULD NOT BE APPLIED TO OTHER
ALGAE WITHOUT PRELIMINARY TESTING. (JONES-WIS)

FIELD 05C, 07B

ACCESSION NO. W69-10151

QUANTITATIVE RELATIONS OF THE FEEDING AND GROWTH OF DAPHNIA PULEX OBTUSA (KURZ) SCOURFIELD,

DEPARTMENT OF WATER TECHNOLOGY, PRAGUE (CZECHOSLOVAKIA).

N. M. KRYUTCHKOVA, AND V. SLADECEK.

HYDROBIOLOGIA, VOL 33, NO 1, P 47-64, 1969. 9 FIG, 4 TAB, 51 REF.

DESCRIPTORS:
   *FEEDING RATES, *GROWTH RATES, *DAPHNIA, PRODUCTIVITY, BIOINDICATORS, SECONDARY PRODUCTIVITY, AQUATIC PRODUCTIVITY, WATER POLLUTION EFFECTS, YEASTS, CRUSTACEA, ALGAE.

IDENTIFIERS:
   *DAPHNIA PULEX OBTUSA, SCENEDESMUS QUADRICAUDA, ECOLOGICAL EFFICIENCIES, TROPHIC ECOLOGY, CLADOCERA.

ABSTRACT:
   AUTHORS FED INDIVIDUAL NEWBORN CLADOCERANS, DAPHNIA PULEX OBTUSA, THREE CONCENTRATIONS OF YEAST (5, 10, 20 MILLIGRAMS/LITER), OR ALGAE, SCENEDESMUS QUADRICAUDA, 5, 10 15 MILLIGRAMS/LITER) IN TEN REPLICATED EXPERIMENTS PER TREATMENT. DURATION OF LIFE, GROWTH, AND FECUNDITY WERE RECORDED FOR EACH DAPHNID. WITH INCREASING CONCENTRATIONS OF YEAST, MEAN TOTAL EGG PRODUCTION INCREASED FROM 101 TO 192; WITH ALGAE, FROM 154 TO 270. MEAN BODY-LENGTH OF DAPHNID FED YEAST ATTAINED 2.4 MILLIMETERS; ALGAE FED, 2.9 MILLIMETERS. OF TOTAL PRODUCTION (BODY WEIGHT + EGG WEIGHT), 66-75% WAS UTILIZED FOR REPRODUCTION IN ENTIRE PERIOD TO DEATH, BEGINNING 5-6 DAYS AFTER INCEPTION OF SEXUAL MATURITY. RELATIVE AVERAGE DAILY DAPHNID GROWTH (ADG) WAS GREATER IN EARLY LIFE, DECREASING WITH AGE; INCREASED WITH INCREASING CONCENTRATIONS OF FOOD, AND WAS GREATER WITH ALGAL FOOD THAN WITH YEAST. AUTHORS CLAIM THAT VALUES FOR ADG (RANGE: 0.23-0.46 FOR ALL TREATMENTS) ARE HIGH AND INDICATE THAT THIS DAPHNIA IS ADAPTED FOR LIFE IN POLLUTED WATER. FOR INCREASING ALGAL CONCENTRATIONS, AVERAGE DAILY RATIO INCREASED (39.1%, 58.8%, 75.9% DAPHNID DRY WEIGHT); EFFICIENCY OF FOOD-UTILIZATION FOR GROWTH DECREASED (56.1%, 47.9%, 35.7%). FILTRATION RATE FOR LIFE SPAN DECREASED WITH INCREASING ALGAL CONCENTRATION FROM 842.3 TO 400 MILLILITERS/MILLIGRAM DRY DAPHNID WEIGHT. (EICHHORN-WIS)

FIELD 05C

ACCESSION NO. W69-10152

FIELD AND EXPERIMENTAL WINTER LIMNOLOGY OF THREE COLORADO MOUNTAIN LAKES,

COLORADO UNIV., BOULDER. DEPT. OF BIOLOGY.

ROBERT W. PENNAK.

ECOLOGY, VOL 49, NO 3, P 505-520, LATE SPRING 1968. 8 FIG, 15 TAB, 36 REF.

DESCRIPTORS:
*LIMNOLOGY, *COLORADO, *LAKES, *MOUNTAINS, *WINTER, CYANOPHYTA, PLANKTON, COPEPODS, EUTROPHICATION, OLIGOTROPHY, TEMPERATURE, PHOTOSYNTHESIS, RAINBOW TROUT, DIATOMS, ANAEROBIC CONDITIONS, ALGAE, ROTIFERS, SESTON, RESPIRATION, ALTITUDE, SAMPLING, ZOOPLANKTON, PHYTOPLANKTON, ICE, BACTERIA, TRIPTON, BOTTOM SEDIMENTS, SOLAR RADIATION, OXYGEN, CARBON DIOXIDE, NANNOPLANKTON.

IDENTIFIERS:
*FIELD STUDIES, *EXPERIMENTAL LIMNOLOGY, CLADOCERAN, MONTANE ZONE, ALPINE ZONE, BOULDER(COLO), TEA LAKE(COLO), BLACK LAKE(COLO), PASS LAKE(COLO), MICRO-ALGAE, FLAGELLATES, CILIATES, COPEPODIDS, NAUPLII, TERRAMYCIN, TETRACYCLINE, STREPTOMYCIN, MILLIPORE FILTRATION, MESOTROPHY.

ABSTRACT:
PHYSICAL, CHEMICAL, AND PLANKTONIC CONDITIONS WERE STUDIED DURING TWO WINTERS IN COLORADO MOUNTAIN LAKES--BLACK LAKE (MESOTROPHIC), PASS LAKE (HIGHLY OLIGOTROPHIC), TEA LAKE (SHALLOW, EUTROPHIC, AND PONDLIKE). WINTER TEMPERATURES OF LOWER WATERS OF BLACK AND PASS LAKES ARE ABOVE 4 DEG C AND AS HIGH AS 5.4 DEG C. SUBSTRATE AND BOTTOM WATERS REACH EQUILIBRIUM IN LATE FEBRUARY OR MARCH. THICK SNOW COVERED BLACK AND PASS LAKES AND NO PHOTOSYNTHESIS OCCURRED FOR 5-7 MONTHS, BUT TEA LAKE HAD LITTLE SNOW AND INTERMITTENT PHOTOSYNTHESIS DURING ALL WINTER. BLACK AND PASS LAKES ARE SUMMER-OLIGOTROPHIC AND WINTER-EUTROPHIC. THEY BECAME SO HIGHLY ANAEROBIC BY MARCH AND APRIL THAT TROUT POPULATIONS DIED. ALL THREE LAKES HAD NEGLIGIBLE WINTER POPULATIONS OF DIATOMS AND GREEN AND BLUE-GREEN ALGAE, BUT POPULATIONS OF MICRO-ALGAE ATTAINED WINTER MAXIMA OF 1-14.5 MILLION CELLS PER LITER, WITH NO CONSISTENT SEASONAL PATTERN. WINTER COPEPOD AND CLADOCERAN POPULATIONS WERE NEGLIGIBLE. ROTIFER POPULATIONS WERE USUALLY DENSE, ESPECIALLY DURING DECEMBER AND JANUARY, BEFORE ONSET OF ANAEROBIOSIS. SESTON VARIED MORE WIDELY DURING MONTHS OF OPEN WATER. WINTER PLANKTON HAD AVERAGE RESPIRATORY RATE IN SITU OF ABOUT TWICE THAT IN CORRESPONDING SAMPLES KEPT IN DARK REFRIGERATOR AT 3 DEG C. (JONES-WIS)

FIELD 02H, 05C

ACCESSION NO. W69-10154

COPPER SULPHATE AIR SPRAY CURES LAKE ALGAE PROBLEM,

SEWER AND WATER ENGINEER, SUDBURY (ONTARIO).

D. D. BROUSE.

WATER AND POLLUTION CONTROL, VOL 104, NO 4, P 25-27, APR 1966. 4 FIG.

DESCRIPTORS:
*COPPER SULFATE, *ALGAL CONTROL, WATER SUPPLY, WATER QUALITY, WATER POLLUTION CONTROL, NUISANCE ALGAE, TASTE, ODOR.

IDENTIFIERS:
*AIR SPRAY, SUDBURY(ONTARIO), LAKE RAMSEY(ONTARIO), ALGAE COUNTS, THRESHOLD ODOR NUMBERS, CROP-DUSTING AIRCRAFT, WATER SAMPLING, ONTARIO WATER RESOURCES COMMISSION, PEST CONTROL ORGANIZATION OF TORONTO, FILAMENTOUS ALGAE, NAVICULA, APHANIZOMENON, ONTARIO(CANADA).

ABSTRACT:
SINCE 1960, SUDBURY, ONTARIO, BORDERING LAKE RAMSEY, HAS TWICE EXPERIENCED SEVERE TASTE AND ODOR PROBLEMS IN ITS MUNICIPAL WATER SUPPLY. LAKE RAMSEY HAS A SURFACE AREA OF 2,023 ACRES, DEPTHS TO 65 FEET, AND LITTLE OUTFLOW. IN OCTOBER 1965, THE ALGAE COUNT PEAKED AT 33,400 AREAL STANDARD UNITS, APHANIZOMENON PREDOMINATING, ACCOMPANIED BY A THRESHOLD ODOR NUMBER OF 200. APHANIZOMENON'S ABILITY TO WITHSTAND WINTER TEMPERATURES NECESSITATED IMMEDIATE ACTION. ON NOVEMBER 6-8, CROPDUSTING AIRCRAFT SPRAYED 20 TONS OF COPPER SULFATE OVER THE LAKE SURFACE RESULTING IN COPPER CONTENT OF 0.20 PARTS PER MILLION (PPM). THE IMMEDIATE AREA OF MUNICIPAL WATER INTAKE WAS AVOIDED. BY NOVEMBER 14TH, THE ALGAE COUNT HAD BEEN REDUCED TO 2,500 AND THE ODOR NUMBER TO 24. ON NOVENBER 22ND, THE ALGAE COUNT HAD REBOUNDED TO 5,000 AND ODOR NUMBER TO 140, WHILE THE COPPER CONCENTRATION REMAINED 0.20 PPM. IN LATE NOVEMBER, A SECOND TREATMENT BROUGHT THE MAXIMUM AVERAGE COPPER CONCENTRATION TO 0.36 PPM. ONE WEEK LATER, AFTER FREEZING OVER, THE LAKE'S ALGAE COUNT WAS BACK DOWN TO 2,000 AND THE ODOR NUMBER TO 17. BY THE END OF DECEMBER, THE WATER WAS SATISFACTORY FOR MUNICIPAL USE WITH ONLY SLIGHT NOTICEABLE ODOR (ALGAE COUNT: 1,300, ODOR NUMBER: 4). (KETELLE-WIS)

FIELD 05G

ACCESSION NO. W69-10155

ALGAE CONTROL WITH COPPER SULFATE,

PORTLAND WATER DISTRICT, MAINE.

W. D. MONIE.

WATER AND SEWAGE WORKS, VOL 103, P 392-397, SEPT 1956. 7 FIG, 8 REF.

DESCRIPTORS:
*ALGAL CONTROL, *COPPER SULFATE, WATER SUPPLY, RESERVOIRS, WATER QUALITY, NEW JERSEY, TEMPERATURE, ORGANIC MATTER, HARDNESS, CARBON DIOXIDE, FISH.

IDENTIFIERS:
*MICROSCOPIC EXAMINATION, *CRITICAL TEMPERATURES, *COPPER DOSAGE, *COPPER APPLICATION METHODS, ALGAE IDENTIFICATION, ALGAE COUNTS, ALGAL GROWTH, SEDGWICK-RAFTER METHOD, TREATMENT TIMING, ALGAE CONTROL REQUIREMENTS, THRESHOLD ODOR TESTS, ASTERIONELLA, APHANIZOMENON, SYNURA, ANABAENA, MELOSIRA, MALLOMONAS, ULOTHRIX, DAVID MONIE TEST, SUMMIT(NJ), CANOE BROOK RESERVOIR(NJ).

ABSTRACT:
ALGAL CONTROL SHOULD BE PRACTICED AS A PREVENTATIVE MEASURE AND NOT A CURE. THE IMPORTANT FACTORS OF CONTROL WITH COPPER SULFATE ARE: (1) ABILITY TO DETERMINE WHEN TREATMENT IS NECESSARY, (2) CORRECT DOSAGE OF COPPER SULFATE, (3) APPLICATION AIMED FOR UNIFORM DISTRIBUTION. THE FIRST STEP IS ALGAE IDENTIFICATION AND COUNTING THROUGH MICROSCOPIC ANALYSIS, COMMONLY USING THE SEDGWICK-RAFTER METHOD. WATER TEMPERATURE DATA ARE USEFUL IN REVEALING CRITICAL TEMPERATURE RANGES IN WHICH GROWTH OF VARIOUS SPECIES IS STIMULATED. MONITORING TEMPERATURE ALLOWS ANTICIPATION OF PERIOD OF RAPID ALGAL GROWTH AND SPECIES COMPOSITION. TIMING OF TREATMENT AND DOSAGE ARE IMPORTANT CONSIDERATIONS, PARTIALLY DEPENDENT ON CHEMICAL CHARACTERISTICS OF WATER, TEMPERATURE, AND TYPE OF ORGANISMS. NO FIXED DOSAGE CAN BE PRESCRIBED, BUT SEVEN FACTORS MAY BE USED AS GUIDELINES: ALGAL SPECIES, AMOUNT OF ORGANIC MATTER, WATER HARDNESS, CARBONIC ACID CONTENT, TEMPERATURE, FISH SPECIES, AND TOTAL VOLUME TO BE TREATED. A SIMPLE, PRACTICAL LABORATORY TEST (DAVID MONIE TEST), DEVELOPED BY THE AUTHOR FOR DETERMINING CORRECT COPPER SULFATE DOSAGE WITHOUT REGARD TO WATER TEMPERATURE AND COMPOSITION AND--WITHIN LIMITS--THE TYPE AND AMOUNT OF ORGANISMS PRESENT, IS DESCRIBED. FINALLY, BOTTOM TOPOGRAPHY IS CRITICAL IN DETERMINING DOSAGES FOR ACHIEVING EQUITABLE DISTRIBUTION. (KETELLE-WIS)

FIELD 05G

ACCESSION NO. W69-10157

EXTRACELLULAR PRODUCTION IN RELATION TO GROWTH OF FOUR PLANKTONIC ALGAE AND OF PHYTOPLANKTON POPULATIONS FROM LAKE ONTARIO,

SCARBOROUGH COLL., TORONTO (ONTARIO).

C. NALEWAJKO, AND L. MARIN.

CANADIAN JOURNAL OF BOTANY, VOL, 47, P 405-413, 1969. 8 FIG, 3 TAB, 9 REF.

DESCRIPTORS:
*PLANKTON, *ALGAE, *PHYTOPLANKTON, *LAKE ONTARIO, *GROWTH RATES, ENVIRONMENTAL EFFECTS, PHOTOSYNTHESIS, DIATOMS, CHLORELLA, CARBON RADIOISOTOPES, LIGHT INTENSITY, CULTURES, RADIOACTIVITY, CARBON DIOXIDE, WATER POLLUTION EFFECTS, WATER POLLUTION SOURCES.

IDENTIFIERS:
*EXTRACELLULAR PRODUCTION, *CARBON FIXATION, *EXCRETION, STEPHANODISCUS TENUIS, ASTERIONELLA FORMOSA, MELOSIRA BINDERANA, GENERATION TIME, CHLORELLA PYRENOIDOSA, DIATOMA ELONGATUM, MELOSIRA ISLANDICA, NITZSCHIA ACICULARIS, NITZSCHIA DISSIPATA, SYNEDRA ACUS, SINKING RATES, GRAZING PRESSURE, POPULATION DENSITY.

ABSTRACT:
AUTHORS PRESENT DATA ON RELATIONSHIP BETWEEN PHOTOSYNTHESIS AND EXTRACELLULAR PRODUCTION AND RELATIVE GROWTH RATE IN FOUR PLANKTONIC ALGAE. THREE ARE DIATOMS ABUNDANT IN LAKE ONTARIO; THE FOURTH SPECIES, CHLORELLA PYRENOIDOSA CHICK, WAS INCLUDED FOR COMPARISON. ALL CULTURES WERE GROWN UNDER LOW LIGHT AND STIRRED CONTINUOUSLY. AT VARIOUS STAGES OF GROWTH THE CARBON-14 METHOD WAS USED TO MEASURE PHOTOSYNTHESIS AND EXCRETION ON EXPOSURE TO HIGH LIGHT INTENSITY. ADDITIONALLY, PHOTOSYNTHESIS AND EXCRETION WERE MEASURED DURING SPRING (1966) INCREASE OF DIATOMS IN LAKE ONTARIO. GROWTH RATES OF PLANKTONIC ALGAL POPULATIONS CANNOT BE MEASURED SINCE FACTORS SUCH AS SINKING AND GRAZING MAY MASK THE TRUE RATE. HOWEVER, SUCH DATA WERE THOUGHT TO BE USEFUL FOR COMPARISON WITH LABORATORY EXPERIMENTS, PARTICULARLY SINCE TWO OF THE SPECIES STUDIED IN CULTURE WERE OF QUANTITATIVE IMPORTANCE IN THE BLOOM. IN THE FOUR SPECIES, BOTH CARBON FIXATION AND EXCRETION (ON BASIS OF ASH-FREE DRY WEIGHT) INCREASE WITH RELATIVE GROWTH RATE OF CULTURES MEASURED IN LOGARITHMIC (TO BASE 10) UNITS. IN NATURAL POPULATIONS, PERCENTAGE EXCRETION VALUES ARE POSTIVELY CORRELATED WITH RELATIVE GROWTH RATES, HOWEVER, ENVIRONMENTAL FACTORS MAY BE OF RELATIVELY GREATER IMPORTANCE AND DETERMINE THE EXTENT OF EXCRETION. (JONES-WISCONSIN)

FIELD 05C, 02H

ACCESSION NO. W69-10158

ALGAL RESPIRATION IN A EUTROPHIC ENVIRONMENT,

PENNSYLVANIA STATE UNIV., UNIVERSITY PARK. DEPT. OF CIVIL ENGINEERING.

ARCHIE J. MCDONNELL, AND R. RUPERT KOUNTZ.

JOURNAL OF WATER POLLUTION CONTROL FEDERATION, VOL 38, NO 5, P 841-847, MAY 1966. 5 FIG, 3 TAB, 9 REF.

DESCRIPTORS:
*ALGAE, *RESPIRATION, *EUTROPHICATION, *DISSOLVED OXYGEN, PENNSYLVANIA, BIOCHEMICAL OXYGEN DEMAND, TEMPERATURE, FLOW, SAMPLING, WASTE WATER DISPOSAL, DIURNAL, EFFLUENTS, NITROGEN, PHOSPHATES, ALKALINITY, FISH, STREAMS, WATER POLLUTION EFFECTS, WATER POLLUTION SOURCES, NUTRIENTS.

IDENTIFIERS:
SPRING CREEK(PA), STATE COLLEGE(PA), FLORA.

ABSTRACT:
PRESENT-DAY COMMUNITY DEVELOPMENT, ESPECIALLY IN SMALL WATERSHEDS, SUCH AS SPRING CREEK, PENNSYLVANIA, HAS COMPOUNDED PROBLEMS OF STREAM POLLUTION CRITERIA. IN MANY STREAM SITUATIONS, DEPLETION OF DISSOLVED OXYGEN RESOURCES IS A RESULT NOT ONLY OF BIOCHEMICAL OXYGEN DEMAND (BOD) BUT ALSO OF THE EFFECT OF COMMUNITY RESPIRATION. EXACTLY HOW MUCH EFFECT EACH OF THESE FACTORS HAS ON THE DISSOLVED OXYGEN LEVEL OF THE STREAM IS DIFFICULT TO EVALUATE. WITH STATISTICAL METHODS OF REGRESSION AND VARIANCE ANALYSIS, IT WAS POSSIBLE TO DEFINE THE BOD PARAMETER AS ONE OF SECONDARY ENGINEERING SIGNIFICANCE AND TO EVALUATE QUANTITATIVELY THE NOCTURNAL OXYGEN DEMANDS OF A HETEROGENEOUS STREAM FLORA. COMMUNITY PLANT RESPIRATION VARIES DIRECTLY WITH STREAM TEMPERATURE AND INVERSELY WITH STREAM FLOW. CHEMICAL ANALYSIS OF CONTRIBUTING SOURCES INDICATED NUTRIENT CONCENTRATIONS SUFFICIENT TO INDUCE AND MAINTAIN A EUTROPHIC ENVIRONMENT. ON THIS BASIS, WHERE POPULATION-TO-RECEIVING WATER RESOURCE RATIO IS HIGH, ANALYSIS FOR SIGNIFICANCE OF PARAMETERS OTHER THAN BOD SHOULD BE MADE AND THE FINDINGS MANIFESTED IN THE TREATMENT PROCESS. THE SOLUTION APPEARS TO LIE IN THE REMOVAL OF NITROGEN AND PHOSPHORUS FROM THE WASTEWATER TREATMENT PLANT EFFLUENT BEFORE IT ENTERS THE CREEK WATERS, WHICH PRACTICALLY IMPLIES TOTAL EXCLUSION. (JONES-WIS)

FIELD 05B

ACCESSION NO. W69-10159

TEMPERATURE OPTIMA FOR ALGAL DEVELOPMENT IN YELLOWSTONE AND ICELAND HOT SPRINGS,

INDIANA UNIV., BLOOMINGTON. DEPT. OF BACTERIOLOGY.

T. D. BROCK, AND M. LOUISE BROCK.

NATURE, VOL 209, NO 5024, P 733-734, FEB 1966. 2 FIG, 7 REF.

DESCRIPTORS:
*ALGAE, *TEMPERATURE, *HOT SPRINGS, HABITATS, ALKALINE WATER, ACIDIC WATER, TRAVERTINE, CALCIUM CARBONATE, BACTERIA, BIOMASS, CHLOROPHYLL, PROTEINS, LIGHT, NUTRIENTS, HYDROSTATIC PRESSURE, CARBON DIOXIDE, CYANOPHYTA, PHYSIOLOGICAL ECOLOGY, THERMAL WATER.

IDENTIFIERS:
*YELLOWSTONE NATIONAL PARK, *ICELAND HOT SPRINGS, SINTER, NUCLEIC ACID, SISJOTHANDI(ICELAND), ENZYME-SUBSTRATE COMPLEXES, RIBONUCLEIC ACID, ORCINOL METHOD.

ABSTRACT:
IN ANALYZING ENVIRONMENTAL FACTORS AFFECTING GROWTH, STUDY OF HABITATS WHERE ONLY A SINGLE FACTOR VARIES IS DESIRABLE. IN HOT SPRINGS TEMPERATURE IS THE ONLY VARIABLE AND ITS RELATION TO BIOLOGICAL DEVELOPMENT CAN BE MEASURED DIRECTLY. IN THE YELLOWSTONE SPRINGS, ALKALINE, ACID, AND TRAVERTINE-DEPOSITING, HIGH IN CALCIUM CARBONATE, EXHIBIT THE GREATEST BIOLOGICAL DEVELOPMENT. OVER SILICEOUS SINTER CONES, WHERE CONSTANT AND UNIFORM THERMAL GRADIENTS MAY EXIST, BLUE-GREEN ALGAE PROLIFERATE. POOLS WERE CHOSEN WHERE INITIAL CHEMICAL COMPOSITION OF WATER FLOWING OVER ALL ORGANISMS ALONG THE THERMAL GRADIENT IS IDENTICAL. THE EXTENT OF ALGAL DEVELOPMENT IS INFLUENCED BY AMOUNT AND RATE OF WATER FLOW AND, SINCE SPRINGS DIFFERED IN THIS FACTOR, THEY ALSO DIFFERED IN ALGAL CONTENT, WHILE EXHIBITING SIMILAR THERMAL GRADIENTS. SAMPLES WERE TAKEN WHENEVER POSSIBLE FROM AREAS HAVING IDENTICAL AMOUNTS AND RATES OF WATER FLOW. CHLOROPHYLL CONTENT WAS ESTIMATED AND RIBONUCLEIC ACID AND PROTEIN ASSAYS MADE. THESE RESULTS WERE EXPRESSED SHOWING RELATIONSHIP BETWEEN TEMPERATURE OF SAMPLE AND BIOCHEMICAL CONTENT PER UNIT AREA. SPECIES NUMBER AND COMPOSITION CHANGED ALONG THE GRADIENT. THE OPTIMUM TEMPERATURE FOR ALGAL DEVELOPMENT IN THESE ALKALINE SPRINGS IS 51-56 DEGREES C. TEMPERATURES WERE LOWER IN ICELAND SPRINGS AND HYDROGEN ION CONCENTRATION HIGHER. (JONES-WIS)

FIELD 05C

ACCESSION NO. W69-10160

THE HABITAT OF LEUCOTHRIX MUCOR, A WIDESPREAD MARINE MICROORGANISM,

INDIANA UNIV., BLOOMINGTON. DEPT. OF BACTERIOLOGY.

THOMAS D. BROCK.

LIMNOLOGY AND OCEANOGRAPHY, VOL 11, NO 2, P 303-307, APRIL 1966. 2 FIG, 7 REF.

DESCRIPTORS:
*MARINE MICROORGANISMS, *HABITATS, EPIPHYTOLOGY, MARINE ALGAE, ECOLOGY, CULTURES, WASHINGTON, WAVES(WATER), CONNECTICUT, RHODE ISLAND, TEMPERATURE, NUTRIENT REQUIREMENTS, PHYSIOCOCHEMICAL PROPERTIES, PHYSIOLOGICAL ECOLOGY, RHODOPHYTA, CHLOROPHYTA.

IDENTIFIERS:
*LEUCOTHRIX MUCOR, MORPHOLOGY, FRIDAY HARBOR(WASH), ANTITHAMNION SARNIENSE, RHODOCHORTON, BANGIA FUSCO-PURPUREA, SPHACELARIA, GONIDIA, CALLOPHYLLIS HAENOPHYLLA, TIDAL CURRENT, PUGET SOUND(WASH), LONG ISLAND SOUND, NARRAGANSETT BAY, CAPE REYKJANES, FAXAFLOI FJORD, ICELAND, DNA BASE COMPOSITION.

ABSTRACT:
LEUCOTHRIX MUCOR IS A LARGE, WIDESPREAD MARINE MICROORGANISM WITH CHARACTERISTIC ORPHOLOGICAL FEATURES RECOGNIZABLE IN NATURAL COLLECTIONS. IT GROWS AS AN EPIPHYTE ON MARINE ALGAE, OCCURRING MOST EXTENSIVELY ON RHODOPHYTES (RED ALGAE) AND ON FILAMENTOUS GREEN ALGAE. AUTHOR DESCRIBES METHOD OF ISOLATION AND SUCCESS IN EXTENSIVE EPIPHYTIC GROWN ON ALGAE IN THE LABORATORY, CONCLUDING THAT THE ALGA NOT ONLY PROVIDES A SUBSTRATUM FOR ATTACHMENT OF LEUCOTHRIX, BUT ALSO NUTRIENTS FOR ITS GROWTH. THE ALGAL CULTURES, WHICH HAVE BEEN MAINTAINED THROUGH SUCCESSIVE TRANSFERS MADE OVER SEVERAL MONTHS, ARE APPARENTLY NOT HARMED IN ANY WAY BY ATTACHMENT AND GROWTH OF L MUCOR. IN NATURE, A WIDE VARIETY OF FILAMENTOUS AND LEAFY RHODOPHYTES ARE COLONIZED WITH L MUCOR. WHERE WATER IS STILL OR SLOW MOVING, L MUCOR IS RARE, BUT IT OCCURS IN EXTREMELY HIGH DENSITIES ON RHODOPHYTES GROWING IN RAPIDLY MOVING WATER. ALL ISOLATES IN PURE CULTURE HAVE BEEN REMARKABLY SIMILAR IN PHYSIOLOGICAL AND MORPHOGENETIC BEHAVIOR, HAVING SIMILAR TEMPERATURE OPTIMA AND NUTRITIONAL REQUIREMENTS. SIX STRAINS HAVE DEOXYRIBONUCLEIC ACID DEMONSTRATING IDENTICAL BASE COMPOSITION. THUS, THE SPECIES, AS DEFINED MORPHOLOGICALLY, COMPRISES A HOMOGENEOUS GROUP OF STRAINS PHYSIOLOGICALLY AND BIOCHEMICALLY. (JONES-WIS)

FIELD 05C, 02L

ACCESSION NO. W69-10161

THE APPLICATION OF MICRO-AUTORADIOGRAPHIC TECHNIQUES TO ECOLOGICAL STUDIES,

INDIANA UNIV., BLOOMINGTON. DEPT. OF MICROBIOLOGY.

M. LOUISE BROCK, AND THOMAS D. BROCK.

INTERNATIONAL ASSOCIATION OF THEORETICAL AND APPLIED LIMNOLOGY,
   COMMUNICATIONS, NO 15, P 1-29, 1968. 6 FIG, 2 TAB, 38 REF.

DESCRIPTORS:
   *ECOLOGY, *RADIOACTIVITY TECHNIQUES, *RADIOISOTOPES, LIMNOLOGY,
   OCEANOGRAPHY, TRACERS, BACTERIA, GROWTH RATES, TRITIUM, PHOTOGRAPHY,
   CARBON RADIOISOTOPES, PHOSPHORUS RADIOISOTOPES, ALGAE, EMULSIONS,
   BACKGROUND RADIATION, IODINE RADIOISOTOPES, NUTRIENT REQUIREMENTS.

IDENTIFIERS:
   *MICRO-AUTORADIOGRAPHY, PHYSIOLOGY, BIOCHEMICAL EXTRACTION, MILLIPORE
   FILTERS, PHASE MICROSCOPY, STAINING, STRIPPING FILM, SYNECHOCOCCUS,
   GRAIN COUNT(AUTORADIOGRAPHY), LEUCOTHRIX MUCOR, ANTITHAMNION SARNIENSE,
   QUANTITATION, GROWTH MODES.

ABSTRACT:
   AUTHORS DESCRIBE AUTORADIOGRAPHY IN DETAIL, NOT ONLY AS AN ADJUNCT TO
   OTHER RADIOACTIVITY EXPERIMENTS, BUT AS A PRINCIPAL TECHNIQUE IN
   LIMNOLOGY, OCEANOGRAPHY, AND RELATED ENVIRONMENTALLY ORIENTED
   DISCIPLINES. TECHNIQUE CAN BE USED AS A SOLE METHOD TO STUDY NUTRITION
   AND PHYSIOLOGY IN THE LABORATORY AS WELL AS IN THE NATURAL ENVIRONMENT.
   SIMPLE BIOCHEMICAL EXTRACTION METHODS ARE DESCRIBED. AUTORADIOGRAPHIC
   RESULTS OBTAINED FROM EXPERIMENTS BASED ON SUCH METHODS ILLUSTRATE THE
   ANALYSIS OF AUTORADIOGRAMS AND DEMONSTRATE THE VALIDITY OF
   AUTORADIOGRAPHIC TECHNIQUES. DETAILED INFORMATION DELINEATES THE USE OF
   AUTORADIOGRAPHY IN EVALUATING MODE AND RATE OF BACTERIA GROWTH.
   DETAILED METHODS FOR PREPARATION OF SPECIMENS, HANDLING OF LIQUID
   AUTORADIOGRAPHIC EMULSION, AND DEVELOPMENT OF AUTORADIOGRAMS ARE GIVEN
   WITH SPECIAL EMPHASIS ON LIMNOLOGICAL APPLICATIONS. COMPARISONS ARE
   MADE WITH CONVENTIONAL METHODS FOR ASSAY OF RADIOACTIVE ISOTOPES.
   TRITIUM WAS THE MOST SUITABLE RADIOACTIVE ISOTOPE FOR THEIR
   EXPERIMENTS. LIQUID EMULSION (NTB-2) AND STRIPPING FILM ARE COMPARED.
   QUALITATIVE EVALUATION CAN BE ACCOMPLISHED WITH TWO-EXPOSURE
   PHOTOGRAPHS BY FOCUSING ON GRAINS TO PHOTOGRAPH THEM AT HALF THE TOTAL
   REQUIRED EXPOSURE, THEN REFOCUSING ON CELLS TO PHOTOGRAPH THEM DURING
   OTHER HALF OF TOTAL REQUIRED EXPOSURE. RELATIVE QUANTITATION AND
   ABSOLUTE QUANTITATION ARE DISCUSSED. (JONES-WISC)

FIELD 07B, 05C

ACCESSION NO. W69-10163

ALGAE AND PHOSPHORUS IN LAKE MINNETONKA,

MINNESOTA UNIV., MINNEAPOLIS. LIMNOLOGICAL RESEARCH CENTER.

ROBERT O. MEGARD.

LIMNOLOGICAL RESEARCH CENTER, MINNESOTA UNIV, MINNEAPOLIS, INTERIM REPORT NO 4, 27 P, DECEMBER 1968. 9 FIG, 3 REF. OWRR PROJECT A-016-MINN.

DESCRIPTORS:
*ALGAE, *PHOSPHORUS, *MINNESOTA, *EUTROPHICATION, POLLUTION ABATEMENT, GROWTH RATES, CHLOROPHYLL, PHOTOSYNTHESIS, CHEMICAL ANALYSIS, BACTERIA, NUTRIENTS, TEMPERATURE, EPILIMNION, NITROGEN COMPOUNDS, SEWAGE EFFLUENTS, LIGHT INTENSITY, LAKES, WATER POLLUTION SOURCES, WATER POLLUTION EFFECTS, WATER POLLUTION CONTROL, NITRATES, NITROGEN FIXATION.

IDENTIFIERS:
*LAKE MINNETONKA(MINN), LAKE MANAGEMENT PROGRAM, ALGAL DENSITY, APHANIZOMENON, MELOSIRA, LYNGBYA.

ABSTRACT:
DATA INDICATE THAT PHOSPHORUS IS THE CRITICAL NUTRIENT FOR ALGAE IN LAKE MINNETONKA, MINNESOTA, DURING SUMMER. THE RELATIONSHIP BETWEEN ALGAE AND NITROGEN WAS NOT STUDIED IN DETAIL BUT NITRATE IS APPARENTLY NOT CRITICAL AT ANY TIME. THIS IS PRESUMABLY DUE TO SYNTHESIS OF NITRATE FROM DISSOLVED GASEOUS NITROGEN BY DOMINANT ALGAS IN SUMMER. PHOSPHORUS PROBABLY WAS NOT A LIMITING NUTRIENT IN SPRING, AT WHICH TIME ANOTHER SUBSTANCE, POSSIBLY NITROGEN, WAS LIMITING. DETAILED ANALYSES OF NUTRIENT BUDGETS AND WATER BALANCES WILL BE REQUIRED TO DETERMINE WHAT MUST BE DONE TO RESTORE LAKE MINNETONKA, AND REQUISITE TIME AND COST OF PROGRAMS. INITIAL OBJECTIVE OF AN INTERIM PROGRAM SHOULD BE TO REDUCE INFLUX OF PHOSPHORUS FROM THE WATERSHED. MOST OF PHOSPHORUS COMES FROM SEWAGE EFFLUENTS, CONTROLLABLE EITHER BY ADVANCED TREATMENT OR BY DIVERSION. DETERIORATION OF THE LAKE WILL BE RETARDED EVEN IF ONLY MODERATE QUANTITIES OF NUTRIENTS ARE PREVENTED FROM ENTERING. ALL MUNICIPALITIES MUST COOPERATE. LAKE MINNETONKA WILL IMPROVE ONLY WHEN ANNUAL NUTRIENT LOSSES EXCEED INFLUX. FURTHER STUDY CAN RESOLVE WHETHER ADVANCED SEWAGE TREATMENT WITH ABATEMENT OF OTHER SOURCES OF NUTRIENTS, OR SEWAGE DIVERSION, WILL BE REQUIRED. (JONES-WIS)

FIELD 05C, 02H

ACCESSION NO. W69-10167

ARTIFICIAL EUTROPHICATION OF LAKE WASHINGTON,.

WASHINGTON UNIV., SEATTLE. DEPT. OF ZOOLOGY.

W. T. EDMONDSON, G. C. ANDERSON, AND DONALD R. PETERSON.

LIMNOLOGY AND OCEANOGRAPHY, VOL 1, P 47-53, 1956. 3 FIG, 1 TAB, 27 REF.

DESCRIPTORS:
*WASHINGTON, *LAKES, *EUTROPHICATION, SEWAGE, BIOTA, PHYTOPLANKTON, DIATOMS, DINOFLAGELLATES, ALGAE, HYPOLIMNION, OXYGEN, PRODUCTIVITY, PHOSPHATES, CHLOROPHYLL, EPILIMNION, EFFLUENT, TEMPERATURE, PHOTOSYNTHESIS, CARBON DIOXIDE, DEPTH, THERMOCLINE, ZOOPLANKTON, TRIPTON, BACTERIA, SEDIMENTATION, COPEPODS, WATER POLLUTION SOURCES, WATER POLLUTION EFFECTS.

IDENTIFIERS:
*LAKE WASHINGTON(WASH), OSCILLATORIA RUBESCENS, SWITZERLAND, PERIDINIUM, OSCILLATORIA AGARDHI, PHORMIDIUM, APHANIZOMENON FLOS-AQUAE, SECCHI DISC TRANSPARENCY, ANABAENA LEMMERMANNI, APHANOCAPSA, BOSMINA LONGISPINA, ZURICHSEE(SWITZERLAND), LINSLEY POND(CONN), BOSMINA COREGONI LONGISPINA.

ABSTRACT:
LAKE WASHINGTON HAS BEEN RECEIVING INCREASING AMOUNTS OF TREATED SEWAGE, AND APPEARS TO BE RESPONDING BY CHANGES IN TYPE AND QUANTITY OF BIOTA. IN 1933 AND 1950, DOMINANT PHYTOPLANKTON ORGANISMS WERE ANABAENA AND VARIOUS DIATOMS AND DINOFLAGELLATES, BUT IN 1955, APPARENTLY FOR THE FIRST TIME, LARGE POPULATIONS OF BLUE-GREEN ALGA, OSCILLATORIA RUBESCENS, OCCURRED, A SPECIES, WHICH CONSTITUTES NUISANCE BLOOMS IN MANY LAKES. AN INTERESTING ECOLOGICAL PROBLEM EXISTS IN CONNECTION WITH THE TWO SPECIES OF OSCILLATORIA IN LAKE WASHINGTON, O AGARDHI AND O RUBESCENS. THE REPLACEMENT OF ONE BY ANOTHER MAY IMPLY A DISTINCT, BUT PERHAPS SUBTLE DIFFERENCE IN ECOLOGICAL REQUIREMENTS. A GREAT INCREASE IN THE HYPOLIMNETIC OXYGEN DEFICIT IS CONSIDERED AS EVIDENCE OF INCREASED PRODUCTIVITY; THE DEFICIT WAS 1.18 MICROGRAMS PER SQUARE CENTIMETER PER MONTH IN 1933, 2.00 IN 1950, AND 3.13 IN 1955. DECREASE IN OXYGEN IS FAIRLY CLOSELY RELATED TO INCREASE IN HYPOLIMNETIC PHOSPHATE CONCENTRATION BETWEEN MEASUREMENTS, MUCH LESS CLOSELY RELATED TO THE CHLOROPHYLL CONCENTRATION IN THE EPILIMNION. FURTHER STUDY OF LAKE WASHINGTON AS ITS EUTROPHICATION PROCEEDS OR, IF EFFLUENTS ARE DIVERTED, OF THE EXTENT TO WHICH THE LAKE REGAINS ITS FORMER, MORE OLIGOTROPHIC, CONDITION WILL BE VALUABLE. (JONES-WIS)

FIELD 05C, 02H

ACCESSION NO. W69-10169

EUTROPHICATION OF LAKES AND RIVERS: ITS ORIGIN AND PREVENTION (IN GERMAN),

KANTONALES LABORATORIUM, ZURICH (SWITZERLAND).

E. A. THOMAS.

VIERTELJAHRSSCHRIFT DER NATURFORSCHENDEN GESELLSCHAFT, ZURICH, VOL 107, NO 3, P 127-140, 1962. 24 REF. IN GERMAN.

DESCRIPTORS:
*LAKES, *RIVERS, *EUTROPHICATION, RUNOFF, FERTILIZERS, NUTRIENTS, NITRATES, PHOSPHATES, PHYTOPLANKTON, IRON COMPOUNDS, ALGAE, PLANTS, WATER POLLUTION EFFECTS, WATER POLLUTION SOURCES, WATER POLLUTION CONTROL, LIMNOLOGY, DIVERSION, STREAMS.

IDENTIFIERS:
ALUMINUM SULFATE, IRON CHLORIDE, IRON SULFATE, CLADOPHORA BLANKETS, HYDRODICTYON, MACROPHYTES.

ABSTRACT:
RUNOFF, EVEN WHEN DEPRIVED OF ITS MECHANICAL AND BIOGENIC CONSTITUENTS BY PURIFICATION, PROVIDES RICH FERTILIZER FOR ALGAE AND HIGHER PLANTS OF STREAMS AND LAKES. NITRATES AND PHOSPHATES ARE PARTICULARLY IMPORTANT STIMULANTS WHICH INDUCE VARIOUS HARMFUL CONSEQUENCES. REMOVAL OF PHOSPHATES FROM THE RUNOFF BY THE USE OF SUITABLE CHEMICAL REACTIONS CONSTITUTES A DIRECT CONTROL MEASURE INHIBITING EUTROPHICATION. THE POSSIBILITY OF DIVERSION IF DISSOLVED PHOSPHATES FOR PRODUCTION OF FARM CROPS IS SUGGESTED. THE METHOD OF EUTROPHICATION CONTROL MUST BE DETERMINED BY THE LIMNOLOGICAL CHARACTERISTICS OF WATER IN QUESTION. (WILDE-WISC)

FIELD 05C

ACCESSION NO. W69-10170

PSEUDOMONAS AERUGINOSA FOR THE EVALUATION OF SWIMMING POOL CHLORINATION AND ALGICIDES,

WISCONSIN UNIV., MADISON. WATER CHEMISTRY LAB.

G. P. FITZGERALD, AND M. E. DERVARTANIAN.

APPLIED MICROBIOLOGY, VOL 17, NO 3, P 415-421, MARCH 1969. 1 FIG, 4 TAB, 19 REF.

DESCRIPTORS:
*PSEUDOMONAS, *SWIMMING POOLS, *BACTERIA, *BACTERICIDES, *CHLORINATION, *AQUATIC BACTERIA, *AQUATIC MICROBIOLOGY, ALGICIDES, BIOASSAY, DISINFECTION, ENVIRONMENTAL EFFECTS, ENVIRONMENTAL SANITATION, INHIBITION, CHLORINE, NITROGEN COMPOUNDS, SEWAGE, SEWAGE BACTERIA, SEWAGE TREATMENT, WATER POLLUTION, WATER POLLUTION EFFECTS, ALGAL CONTROL, ANALYTICAL TECHNIQUES, AQUATIC ALGAE.

IDENTIFIERS:
PSEUDOMONAS AERUGINOSA.

ABSTRACT:
CONCENTRATIONS OF AMMONIA AND THE CHLORINE STABILIZER, CYANURIC ACID, WHICH COULD BE EXPECTED IN SWIMMING POOLS, DECREASED THE RATE OF KILL BY CHLORINE (CL) OF THE POTENTIAL PATHOGEN, PSEUDOMONAS AERUGINOSA. THE EFFECT OF CYANURIC ACID INCREASED AS THE CONCENTRATION OF CL DECREASED, A FACT OF SIGNIFICANCE FROM A PUBLIC HEALTH VIEW. QUATERNARY AMMONIUM ALGICIDES HAD LITTLE EFFECT ON THE KILL RATE OF CL, BUT AN ORGANIC MERCURY ALGICIDE HAD A SYNERGISTIC EFFECT WITH CL WHEN CL-ACTIVITY WAS STRESSED BY THE ADDITION OF AMMONIA OR THE USE OF 100 TIMES THE NORMAL CONCENTRATION OF BACTERIA. THE EFFECT OF NATURAL WATERS, RAIN, BEACHES, AND SWIMMING POOLS ON THE KILL RATE OF 0.5 MILLIGRAMS CL/LITER INDICATED THAT A TREATMENT TIME OF ONE HOUR OR MORE WAS REQUIRED TO KILL 99.9% OF PSEUDOMONAS CELLS AT CONCENTRATION OF A MILLION CELLS PER LITER. THE SYNERGISM OF CL AND THE ORGANIC MERCURY ALGICIDE WAS ALSO DEMONSTRATED WITH THESE WATERS AND WITH SEWAGE TREATMENT PLANT EFFLUENTS. AUTHOR DISCUSSES THE NECESSITY OF DEVELOPING AND USING LABORATORY TESTS WHICH SIMULATE CONDITIONS IN SWIMMING POOLS WITH HEAVY LOADS OF SWIMMERS, AS OPPOSED TO TESTS IN CHLORINE DEMAND-FREE CONDITIONS. (FITZGERALD-WIS)

FIELD 05F

ACCESSION NO. W69-10171

STUDIES ON MORPHOGENESIS IN A BLUE-GREEN ALGA. I. EFFECT OF INORGANIC NITROGEN SOURCES ON DEVELOPMENTAL MORPHOLOGY OF ANABAENA DOLIOLUM,

RAJASTHAN UNIV., JAIPUR (INDIA). DEPT. OF BOTANY.

H. N. SINGH, AND B. S. SRIVASTAVA.

CANADIAN JOURNAL OF MICROBIOLOGY, VOL 14, P 1341-1346, 1968. 5 FIG, 2 TAB, 12 REF.

DESCRIPTORS:
*PHYSIOLOGICAL ECOLOGY, *ENVIRONMENTAL EFFECTS, *MORPHOLOGY, *ALGAE, AQUATIC MICROBIOLOGY, CYANOPHYTA, ESSENTIAL NUTRIENTS, NITROGEN COMPOUNDS, AMMONIA, NITRATES, NITRITES, NUTRIENT REQUIREMENTS, NUTRIENTS, PHYTOPLANKTON, INHIBITION.

IDENTIFIERS:
*ANABAENA DOLIOLUM.

ABSTRACT:
SPORE GERMINATION, HETEROCYST PRODUCTION, HORMOGONE FORMATION, AND SPORULATION ARE THE MORPHOGENETIC STAGES IN DEVELOPMENT OF ANABAENA DOLIOLUM. IN BASAL MEDIUM, SPORULATION IS SIMULTANEOUS WHILE HETEROCYST FORMATION IS SEQUENTIAL. NITRATE-, NITRITE-, AND AMMONIUM-NITROGEN INHIBIT SPORULATION AND HETEROCYST FORMATION, DEGREE OF INHIBITION DEPENDING ON CONCENTRATION AND SOURCE OF INORGANIC NITROGEN. NITRATE AND NITRITE INDUCE LYSIS WHICH IS CONCENTRATION DEPENDENT AND CIRCUMSCRIBED BY TIME, I E, UP TO A CERTAIN STAGE THE LYTIC EVENTS ARE PREVENTABLE BY TRANSFER TO BASAL MEDIUM, BUT BEYOND THIS STAGE THEY BECOME INEVITABLE AND COMPLETE LYSIS OCCURS EVEN IN ABSENCE OF NITRATE OR NITRITE. SEQUENTIAL DIFFERENTIATION AND SPATIAL RELATIONS OF HETEROCYSTS IN A GROWING FILAMENT ARE CHARACTERISTIC OF EACH NITROGEN SOURCE AND INDICATIVE OF THE POLARITY INVOLVED IN GRADIENT OF HETEROCYST FORMATION. (FITZGERALD-WIS)

FIELD 05C

100

ACCESSION NO. W69-10177

EXCESSIVE WATER FERTILIZATION,

WISCONSIN DEPT. OF RESOURCE DEVELOPMENT, MADISON. WATER RESOURCES DIV.

RICHARD B. COREY, ARTHUR D. HASLER, G. FRED LEE, F. H. SCHRAUFNAGEL, AND THOMAS L. WIRTH.

REPORT TO WATER SUBCOMMITTEE, NATURAL RESOURCES COMMITTEE OF STATE AGENCIES, MADISON, WISCONSIN, 50 P, JANUARY 1967. 7 TAB, 131 REF.

DESCRIPTORS:
*WATER, *FERTILIZATION, *EUTROPHICATION, LAKES, STREAMS, ALGAE, FISH, WISCONSIN, SEWAGE, TROUT, NITROGEN, PHOSPHORUS, DETERGENTS, RUNOFF, PERCOLATION, ROADS, ROOFS, PRECIPITATION(ATMOSPHERIC), WETLANDS, SEEPAGE, INDUSTRIAL WASTES, TERTIARY TREATMENT, CHEMICAL PRECIPITATION, HARVESTING, ZONING, DIVERSION, DREDGING, IMPOUNDMENTS, WEEDS, SPORTS, PROPERTY VALUES, TASTE, ODOR, BASE FLOW, LAKE MENDOTA(WIS), BOTTOM SEDIMENTS, MUNICIPAL WASTES, WATER POLLUTION EFFECTS, WATER POLLUTION SOURCES, WATER POLLUTION CONTROL.

IDENTIFIERS:
MANURE, MADISON(WIS), FLUSHING WATERWAYS, MILWAUKEE(WIS), CHICAGO(ILL), ARTIFICIAL CIRCULATION, NUISANCES.

ABSTRACT:
DETERIORATION OF WISCONSIN WATERS WILL PROBABLY INCREASE. EXCESSIVE FERTILIZATION WILL BECOME THE IMPORTANT PUBLIC PROBLEM IN WATER RESOURCES. DOMESTIC SEWAGE CONSTITUTES A MAJOR SOURCE OF NITROGEN AND PHOSPHORUS, APPARENTLY THE MAJOR CAUSATIVE ELEMENTS OF EUTROPHICATION. ACCELERATED USE OF SYNTHETIC DETERGENTS ACCENTUATES A PROBLEM TO WHICH MANY OTHER FACTORS CONTRIBUTE: RUNOFF AND UNDERGROUND PERCOLATION FROM RURAL LANDS; MANURE APPLIED ON FROZEN SOIL; RUNOFF FROM ROOFS AND ROADS; 'WASHING' OF THE ATMOSPHERE BY PRECIPITATION; AND RELEASE OF NITROGEN AND PHOSPHORUS RESULTING FROM DEVELOPMENT OF WETLANDS FOR AGRICULTURAL AND URBAN USE. IMPROVED REMOVAL OF NUTRIENTS FROM SEWAGE APPEARS POSSIBLE. ADDING TERTIARY TREATMENT FOR EFFLUENTS; PERFECTING USE OF CHEMICAL PRECIPITANTS; BIOCHEMICALLY REMOVING NUTRIENTS BY GROWING AND HARVESTING ALGAE IN EFFLUENT LAGOONS; REMOVING NUTRIENTS BY MODIFICATION OF ACTIVATED SLUDGE PROCESSES; LIQUIDIZING AND STORING WINTER MANURE UNTIL SPRING--ALL MAY PROVE EFFECTIVE. ENTIRE DRAINAGE BASINS, SHORES AND FRONTAGES, MAY REQUIRE REZONING. DIVERSION OF EFFLUENTS, FLUSHING WATERWAYS WITH CLEAR WATER, CHEMICALLY CONTROLLING ALGAE AND AQUATIC PLANTS, DREDGING SHALLOW PORTIONS OF LAKES, ARTIFICIALLY CIRCULATING ENTIRE LAKES, DRAWING NUTRIENT-RICH SURPLUS WATER FROM LAKE BOTTOMS, IMPOUNDING STREAMS AND CONTROLLING SUBSEQUENT RELEASES--ALL SHOW PROMISE. (JONES-WIS)

FIELD 05C

ACCESSION NO. W69-10178

THE IMPORTANCE OF EXTRACELLULAR PRODUCTS OF ALGAE IN FRESHWATER,

UNIVERSITY COLL., LONDON (ENGLAND). DEPT. OF BOTANY.

G. E. FOGG, AND D. F. WESTLAKE.

VERH INT VEREIN THEOR ANGEW LIMNOL, VOL 12, P 219-232, 1953. 3 FIG, 2 TAB, 20
    REF.

DESCRIPTORS:
   *ALGAE, *FRESH WATER, *PEPTIDES, COPPER, IRON, IONS, PHOSPHATES,
   ORGANIC COMPOUNDS, TOXICITY, CHLOROPHYTA, LAKES, RESERVOIRS, NITROGEN,
   GROWTH, AMINO ACIDS, CHLAMYDOMONAS, CHLORELLA, WISCONSIN, LAKE
   MICHIGAN, PHYTOPLANKTON, DIATOMS, ECOLOGY, SOIL, CULTURES, CYANOPHYTA,
   WATER POLLUTION SOURCES, WATER POLLUTION EFFECTS.

IDENTIFIERS:
   *EXTRACELLULAR PRODUCTS, ANABAENA CYLINDRICA, ZINC, XANTHOPHYCEAE,
   BACILLARIOPHYCEAE, MYXOPHYCEAE, COMPLEX FORMATION, GROWTH-PROMOTING
   PROPERTIES, CHLAMYDOMONAS MOEWUSII, CHLORELLA PYRENOIDOSA, TRIBONEMA
   AEQUALE, NAVICULA PELLICULOSA, LAKE MENDOTA(WIS), ENGLAND,
   LONDON(ENGLAND), ESTHWAITE(ENGLAND), LOWESWATER(ENGLAND), BARNES SOUTH
   RESERVOIR(ENGLAND), WINDERMERE NORTH BASIN(ENGLAND), LOUGHRIGG
   TARN(ENGLAND), BASSENTHWAITE(ENGLAND), TARN TARN(ENGLAND), SOLWAY
   FIRTH(ENGLAND), LITTLE HAWESWATER(ENGLAND), STAINES RESERVOIR(ENGLAND),
   GLOEOTRICHIA NATANS, OSCILLATORIA, BOTRYOCOCCUS, VOLVOX, UROGLENA,
   APHANIZOMEMON, MICROCYSTIS, COELOSPHAERIUM.

ABSTRACT:
EXPERIMENTS WITH PARTIALLY PURIFIED PREPARATIONS OF THE EXTRACELLULAR
POLYPEPTIDE, PRODUCED DURING NORMAL GROWTH OF THE CYANOPHYTE
(BLUE-GREEN ALGA), ANABAENA CYLINDRICA, SHOW THAT THIS MATERIAL FORMS
COMPLEXES WITH VARIOUS IONS INCLUDING THOSE OF COPPER, ZINC, FERRIC
IRON, PHOSPHATE AND CERTAIN ORGANIC SUBSTANCES. AUTHORS SUGGEST THAT
SUCH COMPLEX FORMATION MAY HAVE BIOLOGICALLY IMPORTANT EFFECTS, AND, AS
AN EXAMPLE, THEY SUGGEST THAT COMPLEX FORMATION BETWEEN EXTRACELLULAR
POLYPEPTIDE AND CUPRIC ION CONSIDERABLY REDUCES TOXICITY OF THE LATTER
TOWARD ANABAENA CYLINDRICA. THIS MAY EXPLAIN THE ERRATIC RESULTS OFTEN
OBTAINED WITH COPPER SULPHATE USED AS AN ALGICIDE. ALGAE REPRESENTING
OTHER CLASSES (CHLOROPHYCEAE, XANTHOPHYCEAE, BACILLARIOPHYCEAE) ALSO
PRODUCE EXTRACELLULAR PEPTIDE. THAT RELATIVELY SUBSTANTIAL AMOUNTS OF
PEPTIDE-NITROGEN OCCUR DISSOLVED IN LAKE WATERS HAS BEEN CONFIRMED FOR
A NUMBER OF ENGLISH LAKES AND RESERVOIRS. AUTHORS SUGGEST THAT THIS
PEPTIDE MAY ORIGINATE PARTIALLY FROM LIVING ALGAE AND THAT BY FORMING
COMPLEXES WITH OTHER DISSOLVED SUBSTANCES, IT MAY EXERT IMPORTANT
EFFECTS ON GROWTH OF AQUATIC ORGANISMS. SPECIFIC SUBSTANCES MAY NOT BE
NECESSARY, AND IN LAKES, COMPLEX-FORMING SUBSTANCES DERIVED FROM
ORGANIC DECAY OR LIVING ALGAE, MAY EFFECTIVELY PRODUCE THE REQUISITE
ENVIRONMENT FOR CYANOPHYTES.(JONES-WIS)

FIELD 05C, 02H

ACCESSION NO. W69-10180

CHANGES IN THE OXYGEN DEFICIT OF LAKE WASHINGTON,

WASHINGTON UNIV., SEATTLE. DEPT. OF ZOOLOGY.

W. T. EDMONDSON.

VERH INT VEREIN THEOR ANGEW LIMNOL, VOL 16, P 153-158, DEC 1966. 3 FIG, 1 TAB, 8 REF.

DESCRIPTORS:
*WASHINGTON, *OXYGEN, *FLUCTUATION, SEWAGE TREATMENT, SOLAR RADIATION, CHLOROPHYLL, PRODUCTIVITY, PHOTOSYNTHESIS, ALGAE, HYPOLIMNION, EPILIMNION, PHYTOPLANKTON, SESTON, CRUSTACEANS, DIATOMS, CHLOROPHYTA, DEPOSITION(SEDIMENTS), WATERSHEDS(BASINS), WATER POLLUTION EFFECTS, CYANOPHYTA, EURTROPHICATION.

IDENTIFIERS:
*LAKE WASHINGTON(WASH), *OXYGEN DEFICIT, SECCHI DISC TRANSPARENCY, MYXOPHYCEA, OSCILLATORIA, ANABAENA, DIAPTOMUS ASHLANDI, EPISCHURA NEVADENSIS, DIAPHANOSOMA LEUCHTENBER GIANUM, CYCLOPS BICUSPIDATUS, REPRODUCTIVE RATE.

ABSTRACT:
RECORDS OF RECENT CHANGES IN OXYGEN REGIME IN LAKE WASHINGTON SHOW THAT, THOUGH INCREASING AMOUNTS OF EFFLUENT FROM SEWAGE TREATMENT PLANTS WERE RECEIVED AND OXYGEN DEFICIT RATE, IN GENERAL WAS HIGHER IN 1956 TO 1964 THAN IN 1933 AND 1950, THE INCREASE FLUCTUATED AND, IN SOME YEARS, RATE WAS SIMILAR TO THAT OF 1950. CHANGES IN DEFICIT DO NOT APPEAR TO BE ATTRIBUTABLE TO CORRESPONDING CHANGES IN PRODUCTIVITY. NO CLEAR RELATION APPEARS BETWEEN DEFICIT AND CHLOROPHYLL OR SOLAR RADIATION INPUT. THUS THE LAKE HAS CONTINUED TO SHOW SIGNS OF ENRICHMENT BY INCREASED PRIMARY PRODUCTIVITY AND ALGAL ABUNDANCE, WHILE OXYGEN DEFICIT AND ITS RATE OF DEVELOPMENT DURING THE SUMMER HAVE NOT CHANGED PROPORTIONATELY. A PARTIAL EXPLANATION MAY LIE IN QUALITATIVE CHANGES IN CHARACTER OF PHYTOPLANKTON. A RELATIVELY SMALL FRACTION OF THE 1950 POPULATION CONSISTED OF COLONIAL MYXOPHYCEAE, WHEREAS IN 1961-1964, THEY DOMINATED THE POPULATION. THE RATE OF DEPOSITION OF MATERIAL TO THE BOTTOM, WHOSE SUBSTANCES SUPPORT THE OXYGEN DEFICIT, APPEARS NOT TO HAVE INCREASED CORRESPONDINGLY TO PRODUCTIVITY OR CROP. THERE IS NO INDICATION THAT CHANGES IN OXYGEN DEFICIT ARE ATTRIBUTABLE TO CHANGES IN WATERSHED OF LAKE. (JONES-WIS)

FIELD 05C, 02H

ACCESSION NO. W69-10182

SEA PLANTS FOR FOOD,

SOUTHAMPTON UNIV. (ENGLAND).

J. E. G. RAYMONT.

THE NEW SCIENTIST, P 10-11, JULY 18, 1957. 1 FIG.

DESCRIPTORS:
*MARINE PLANTS, *ALGAE, CHLORINATION, SODIUM COMPOUNDS, PHOTOSYNTHESIS, CARBON DIOXIDE, NITRATES, PHOSPHATES.

IDENTIFIERS:
*PHAEODACTYLUM, *GAS EFFLUENT, FOOD PRODUCTION, SOUTHAMPTON UNIVERSITY.

ABSTRACT:
A MARINE ALGAE CROP OF PHAEODACTYLUM HAS BEEN EXPERIMENTALLY GROWN BUT NOT SUBJECTED TO THOROUGH CHEMICAL ANALYSIS WHICH WOULD ESTIMATE ITS FOOD VALUE, OR PROTEIN PRODUCED PER NUTRIENT INTAKE. THE DRY WEIGHT OF THE HARVESTED CROP WAS 1/3 GRAM PER LITER CULTURE WHICH WAS A FAVORABLE BEGINNING RESULT. HEAT STERILIZATION WAS THE PREFERRED PROCEDURE IN THE FILTERING AND STERILIZING STEPS. A COMMERCIAL AMMONIUM SULFATE WAS USED AS FERTILIZER. ALGAE GROWTH DEPENDED ON A STRONG ENOUGH LIGHT SOURCE AND WAS THEREFORE GROWN IN SHALLOW AREAS AND IN THE WINTER BY ARTIFICIAL LIGHTING. CARBON DIOXIDE FOR THE PHOTOSYNTHESIS WAS SUPPLIED AS SUCCESSFULLY BY EFFLUENT WASTE GASES OF PARTIALLY PURIFIED WASTE WATER AS BY CYLINDERS. THE EFFLUENT CARBON DIOXIDE IS HOPED TO DECREASE THE NEED FOR NITRATE AND PHOSPHATE. (SHERMAN-VANDERBILT)

FIELD 03C, 05D

ACCESSION NO. W70-00161

LAKE MICHIGAN POLLUTION AND CHICAGO'S SUPPLY,

DEPARTMENT OF WATER AND SEWERS, CHICAGO, ILL.

HYMAN H. GERSTEIN.

J AMERICAN WATER WORKS ASSOCIATION, VOL 57, P 841-857, JULY 1965. 7 FIG, 7 TAB.

DESCRIPTORS:
*LAKE MICHIGAN, *WATER SUPPLY, INTAKES, NUTRIENTS, QUALITY CONTROL, TREATMENT FACILITIES, FILTRATION, SEWAGE, COLIFORMS, EUTROPHICATION, ODOR, AMMONIUM COMPOUNDS, ACTIVATED CARBON, CHLORINE, PHENOLS, OIL WASTES, HYDROGEN ION CONCENTRATION, FLUORIDES, ALKYL BENZENE SULFONATES, WIND VELOCITY, TASTE, COAGULATION, PLANKTON, MONITORING, TEMPERATURE, DETERGENTS, PHOSPHORUS COMPOUNDS, DIATOMS, BIOCHEMICAL OXYGEN DEMAND, DISSOLVED OXYGEN, COLOR, DISSOLVED SOLIDS, BIODEGRADATION, ALGAE, IRON, SULFATES.

IDENTIFIERS:
*POLLUTION, *CHICAGO(ILL), COOK COUNTY(ILL), CALUMET RIVER(ILL), POOLS, WIND DIRECTION, TABELLARIA, FRAGILARIA, ASTERIONELLA, FILTER CLOGGING, STEPHANODISCUS HANTZSCHII, STEPHANODISCUS BINDERANUS, CLADOCERA, WATER MASSES, U S PUBLIC HEALTH SERVICE, GREAT LAKES-ILLINOIS RIVER BASINS PROJECT, INDIANA HARBOR SHIP CANAL.

ABSTRACT:
ALTHOUGH NO SEWAGE AND INDUSTRIAL WASTES ARE DISCHARGED ALONG 30 MILES OF CHICAGO'S LAKE FRONT, GROSS POLLUTION EXISTS IN THE SOUTHERN END OF LAKE MICHIGAN. DATA OVER 40 YEARS REVEAL QUALITY AT THE INTAKES, AND, FROM 1950 TO 1964, ILLUSTRATE POLLUTIONAL TRENDS. A GENERAL INCREASE OF EUTROPHICATION INDICATORS AND INCREASE IN NUMBER AND INTENSITY OF PERIODS WHEN POLLUTANTS SERIOUSLY AFFECTED THE INTAKE AT THE SOUTH PLANT APPEAR IN THE LAST TEN YEARS. SLUGS OF WASTES MAY HAVE ABNORMAL ODORS, LIKE THOSE OF OIL REFINERY WASTES DILUTED WITH LAKE WATER. SINCE 1948, SAMPLES HAVE BEEN COLLECTED ONE DAY EACH WEEK AT ESTABLISHED POINTS. POLLUTIONAL POOLS, INFLUENCED BY WIND-INDUCED LAKE CURRENTS, WERE OBSERVABLE IN A SERIES OF TASTE AND ODOR INCIDENTS. THE PRINCIPAL AGENT IN TREATMENT FOR REMOVAL OF TASTE AND ODOR IS ACTIVATED CARBON. IN TREATING HIGHLY POLLUTED WATERS, APPLICATION OF CHLORINE SUFFICIENT TO DECOMPOSE AMMONIA-NITROGEN MAY NOT ADEQUATELY DESTROY THE BACTERIA, WHEREUPON BACTERIOLOGICALLY UNSAFE WATER MAY RESULT. INCREASING PLANKTON GROWTH NECESSITATES MORE FREQUENT WASHING OF FILTERS; INCREASED GROWTH OF FILAMENTOUS ALGAE TOGETHER WITH APPEARANCE OF NEW SPECIES OF DIATOMS HAVE CAUSED ADDITIONAL PROBLEMS WITH FILTERS. AUTHOR PRESENTS, AS PRACTICAL AND DESIRABLE, A SERIES OF CRITERIA FOR QUALITY OF CHICAGO'S WATER SUPPLY. (JONES-WISCONSIN)

FIELD 05B, 05C, 05G

ACCESSION NO. W70-00263

DEATH OF A LAKE,

DAVID C. FLAHERTY.

QUEST, VOL 6, NO 4, P 7-15, 1968. 9 FIG.

DESCRIPTORS:
*LAKES, *ALGAE, *EUTROPHICATION, OLIGOTROPHY, NUTRIENTS, WATER
TEMPERATURE, DISSOLVED OXYGEN, FISHKILLS, LIMNOLOGY, ALGAL TOXINS,
ALGAL CONTROL, TRACE ELEMENTS, WASHINGTON, NUCLEAR POWERPLANTS, WATER
POLLUTION EFFECTS, PHOSPHORUS, NITROGEN.

IDENTIFIERS:
*CULTURAL EUTROPHY, *NUTRIENT BALANCE, *LAKE AGING, NUTRIENT SOURCES,
WATER SAMPLING, ALGAL GROWTH, OXYGEN DEPLETION, NUCLEAR ACTIVATION
TECHNIQUES, ROCK LAKE(WASH), WILLIAMS LAKE(WASH), CHANNELED
SCABLANDS(WASH), WASHINGTON STATE UNIVERSITY.

ABSTRACT:
ALGAE ARE THE PRIMARY PRODUCERS WHICH INITIATE THE FOOD CHAIN IN LAKES
LEADING UP TO FISH AND MAN. IN NORMAL LAKES, A ROUGH BALANCE EXISTS
BETWEEN INCOMING NUTRIENTS, ALGAL NUMBERS, AND THEIR PREDATORS.
HOWEVER, THIS BALANCE IS UPSET BY MAN'S INFLUENCE, HASTENING CHANGE
FROM OLIGOTROPHIC TO EUTROPHIC CONDITIONS. DETRIMENTAL EFFECTS OF LARGE
NUMBERS OF UNDESIRABLE ALGAL FORMS, INCLUDE OXYGEN DEPLETION RESULTING
IN FISHKILLS; FLAVOR, ODOR, AND COLOR IN WATER SUPPLIES; INDUCED
CORROSION AND SLIME IN INDUSTRIAL WATER SUPPLIES; AND ENDANGERING OF
OTHER AQUATIC AND TERRESTRIAL LIFE FORMS FROM TOXIC SUBSTANCES. STUDIES
CONDUCTED BY DR. WILLIAM FUNK, WASHINGTON STATE UNIVERSITY CIVIL
ENGINEERING PROFESSOR, ON ROCK AND WILLIAMS LAKES IN THE SCABLANDS
BASALTS OF EASTERN WASHINGTON ARE DESCRIBED. THE BASALTS ARE RELATIVELY
LOW IN NITROGEN AND PHOSPHORUS, AND DECOMPOSE SLOWLY, WITH CONSEQUENT
LOW NATURAL NUTRIENT INCOME. HOWEVER, THESE LAKES ARE ENDANGERED BY
HUMAN ACTIVITIES. PRESENT STUDIES MEASURE DISSOLVED OXYGEN,
TEMPERATURE, LIGHT PENETRATION, ALGAL SPECIES, AND CONCENTRATIONS OF
VARIOUS ELEMENTS INCLUDING PHOSPHORUS AND NITROGEN. NUCLEAR ACTIVATION
TECHNIQUES ARE USED TO DETECT TRACE ELEMENTS WHICH DR. FUNK SUSPECTS
MAY HAVE A CRITICAL ROLE IN ALGAL GROWTH. (KETELLE-WISCONSIN)

FIELD 05C, 02H

ACCESSION NO. W70-00264

PRIMARY PRODUCTION IN LABORATORY STREAMS,

OREGON STATE UNIV., CORVALLIS. DEPT. OF FISH AND GAME MANAGEMENT; OREGON
   STATE UNIV., CORVALLIS. DEPT. OF BOTANY; AND OREGON STATE UNIV., CORVALLIS.
   PACIFIC COOPERATIVE WATER POLLUTION AND FISHERIES RESEARCH LABS.

C. DAVID MCINTIRE, ROBERT L. GARRISON, HARRY K. PHINNEY, AND CHARLES E.
   WARREN.

LIMNOLOGY AND OCEANOGRAPHY, VOL 9, NO 1, P 92-102, JAN 1964. 5 FIG, 3 TAB, 20
   REF.

DESCRIPTORS:
   *PRIMARY PRODUCTIVITY, *STREAMS, *LABORATORY TESTS, RESPIRATION,
   ENVIRONMENT, PHOTOSYNTHESIS, OXYGEN, TROPHIC LEVEL, LIGHT INTENSITY,
   ECOLOGY, LOTIC ENVIRONMENT, BIOMASS, FLOW RATES, DEPTH, DISSOLVED
   OXYGEN, ALGAE, VELOCITY, WATER QUALITY, DISSOLVED SOLIDS, HYDROGEN ION
   CONCENTRATION, TEMPERATURE, NITROGEN, NUTRIENTS, ORGANIC MATTER,
   PIGMENTS, CHLOROPHYLL, DIFFUSION, CYANOPHYTA.

IDENTIFIERS:
   COMMUNITY, SPECIFIC CONDUCTANCE, SPECIES, MELOSIRA VARIANS, SYNEDRA
   ULNA, OEDOGONIUM, PHORMIDIUM RETZII, SURFACE DIFFUSION RATE.

ABSTRACT:
   SIX LABORATORY STREAMS WERE USED TO STUDY PRIMARY PRODUCTION AND
   COMMUNITY RESPIRATION IN SIMPLE COMMUNITIES UNDER DIFFERENT
   ENVIRONMENTAL CONDITIONS. A PHOTOSYNTHESIS-RESPIRATION CHAMBER WAS
   DEVELOPED WHICH CAN BE USED TO RELATE PRIMARY PRODUCTION TO VARIOUS
   ENVIRONMENTAL CONDITIONS AND TO VERIFY DIRECT MEASUREMENT OF PRIMARY
   PRODUCTION IN LABORATORY OR NATURAL STREAMS. GROSS PRIMARY PRODUCTION
   IN LABORATORY STREAMS UNDER 6000 LUX OF ILLUMINATION RANGED FROM 2.4 TO
   4.7 GRAMS OXYGEN/SQUARE METER/DAY. COMMUNITY RESPIRATION RANGED FROM
   1.6 TO 4.2 GRAMS OXYGEN/SQUARE METER/DAY. RATIOS OF PHOTOSYNTHESIS TO
   RESPIRATION USUALLY VARIED BETWEEN 1.0 AND 2.0. ILLUMINATION INTENSITY
   WAS APPROXIMATELY LINEARLY RELATED TO PRIMARY PRODUCTION IN RANGE,
   0-4000 LUX, MAXIMUM PHOTOSYNTHETIC RATES OBTAINING AT ABOUT 21,000 LUX.
   COMMUNITIES DEVELOPING ON THE SUBSTRATE WERE SEEDED NATURALLY BY
   SPECIES ENTERING STREAMS THROUGH THE WATER SUPPLY. COMPOSITION OF PLANT
   COMMUNITIES REMAINED SURPRISINGLY CONSTANT, USUALLY VARYING ONLY IN
   RELATIVE ABUNDANCE OF DIFFERENT SPECIES. THE DIATOMS, MELOSIRA VARIANS
   AND SYNEDRA ULNA CONSISTENTLY DOMINATED; OEDOGONIUM SPP WERE SOMETIMES
   VERY ABUNDANT DURING SUMMER. GENERA OF BLUE-GREEN ALGAE WERE
   UBIQUITOUS, FREQUENTLY GROWING ATTACHED TO SIDES OF TROUGHS, NEAR THE
   AIR-WATER INTERFACE. (JONES-WISCONSIN)

FIELD 05C

ACCESSION NO. W70-00265

CHARACTERISTICS OF HYPERTROPHIC LAKES AND CANALS IN CITIES,

STATE ZOOLOGICAL INST. AND MUSEUM, HAMBURG (WEST GERMANY). HYDROBIOLOGICAL
STATION.

HUBERT CASPERS.

INTERNATIONAL ASSOCIATION OF THEORETICAL AND APPLIED LIMNOLOGY, PROCEEDINGS,
VOL 15, P 631-638, FEB 1964. 2 FIG, 2 TAB, 13 REF.

DESCRIPTORS:
*LAKES, *EUTROPHICATION, *CANALS, *CITIES, BIOLOGICAL COMMUNITIES,
SEWAGE, DETRITUS, TUBIFICIDS, PLANKTON, DIATOMS, CYANOPHYTA,
PHYTOPLANKTON, ROTIFERS, CHLOROPHYTA, COPEPODS, CYCLES, CHLAMYDOMONAS,
ALGAE, ZOOPLANKTON, FISH, CHRYSOPHYTA, SESTON, SELF-PURIFICATION.

IDENTIFIERS:
*HYPERTROPHIC WATERS, HAMBURG(GERMANY), CHIRONOMIDS, CLADOCERA,
APHANIZOMENON FLOS-AQUAE, STEPHANODISCUS HANTZSCHII, DIDINIUM,
PARAMECIUM, PHACUS, GLENODINIUM, CRYPTOMONAS, CILIATA, ASTERIONELLA
FORMOSA, RECOLONIZATION, ALSTER LAKE(GERMANY), GERMANY.

ABSTRACT:
ALSTER LAKE IN HAMBURG, GERMANY, IS AN EXAMPLE OF A TYPICAL URBAN LAKE
AND DEMONSTRATES THE SPECIAL CONDITIONS IN CITY WATERS, WHICH ARE
SPECIALIZED HABITATS AND WHERE HYDROBIOLOGICAL STATUS CAN BE MORE
READILY ANALYZED THAN COMPLEX SITES OF 'HEALTHIER' WATERS. URBAN WATERS
MAY CONSTITUTE A RESTRICTED FACIES OF EUTROPIC WATER WHERE DESTRUCTION
OF ORIGINAL COMMUNITIES IS TYPICAL AND NO NEW EQUILIBRIUM DEVELOPS.
RATHER, THE RELATIONSHIP BETWEEN BUILDUP AND BREAKDOWN OF NUTRITIVE
MATERIAL, BETWEEN PRODUCERS AND CONSUMERS, IS ALWAYS IN A STATE OF
TEMPORARY EQUILIBRIUM, AND IS CONSTANTLY IN DANGER OF BECOMING
UNBALANCED. THE ORIGINAL COMMUNITY LOSES ITS MOST IMPORTANT ZONE OF
REGENERATION, THE SUBMERGED PLANTS ALONG BANKS; BOTTOM FAUNA IS
BURDENED WITH ANAEROBIC PROCESSES RESULTING FROM INFLOW OF DETRITUS,
LEAVING ONLY A FEW SPECIES WHOSE MASS DEVELOPMENT IS INHIBITED. THE
SURVIVING PLANKTON TENDS TO INSTABILITY AND EXTREME VARIATIONS FROM
YEAR TO YEAR. NEVERTHELESS, THERE ARE CERTAIN DEFINITE AND REGULAR
EVENTS WHICH MAKE URBANIZED WATERS A UNIQUE EXPERIMENTAL SITUATION,
SINCE THE RESTRICTED FACIES OF EUTROPHIC WATER AFFORDS US LIMNOLOGICAL
INSIGHT INTO FORCED PRODUCTION PROCESSES AND THEIR ECOLOGICAL EFFECTS.
AUTHOR BELIEVES THAT CONDITIONS IN URBAN WATERS IMPROVE WHEN PLANTS
RECOLONIZE THE BANKS. (JONES-WISCONSIN)

FIELD 05C, 02H

ACCESSION NO. W70-00268

THE CONTROL OF WATER WEEDS.

AGRICULTURAL RESEARCH COUNCIL, KIDLINGTON (ENGLAND). WEED RESEARCH
   ORGANIZATION.

E. C. S. LITTLE.

WEED RESEARCH, VOL 8, NO 2, P 79-105, 1968. 363 REF.

DESCRIPTORS:
   *AQUATIC WEEDS, *AQUATIC PLANTS, *AQUATIC WEED CONTROL, WATER
   CONSERVATION, WATER CONTROL, FERTILIZERS, SEWAGE EFFLUENT, NUTRIENTS,
   LAKES, CHANNELS, REVIEWS, BIBLIOGRAPHIES, HERBICIDES, FISH, FRESH
   WATER, ALGAE, PONDS, WATER HYACINTH, CHEMICAL CONTROLS, PARAQUAT,
   DIQUAT, SODIUM ARSENITE, COPPER SULPHATE, MONURON, AMMONIA, DALAPON,
   2-4-5-T UREAS, SEEDS, PROTEINS, MICROORGANISMS, FERMENTATION,
   SURFACTANTS, FLOATING PLANTS, FERNS, ALLIGATORWEED, FORMULATION,
   EMULSIFIERS, SOIL TEXTURE, SILAGE, CHEMICALS, DITCHES, PERSISTENCE,
   IRRIGATION WATER, HARVESTING, MECHANICAL CONTROL, SPRAYING, WATER
   LEVELS, MAMMALS, BIRDS, SNAILS, INSECTS, FUNGI, WATER QUALITY, SOIL
   STERILANTS, EUTROPHICATION, RIVERS, BIOCONTROL, TOXICITY, DRAWDOWN,
   2-4-D, AMINOTRIAZOLE.

IDENTIFIERS:
   TRIAZINE, DICHLOBENIL.

ABSTRACT:
   WATER WEEDS ARE POSING INCREASING PROBLEMS IN MANY COUNTRIES WHICH
   DEPEND ON WATER CONTROL FOR DEVELOPMENT OF AGRICULTURAL, POWER, AND
   TRANSPORT RESOURCES. THE UNITED STATES, BESIDES HAVING ITS SHARE OF
   DIFFICULTIES FROM WATER WEEDS, IS ALSO CONCERNED WITH AQUATIC WEED
   IMPAIRMENT OF INCREASINGLY POPULAR RECREATIONAL ASPECTS OF WATER. HEAVY
   WATER-WEED INFESTATION IS EXPECTED WHEN FERTILE LAND IS SUBMERGED TO
   FORM LAKES, OR WHEN LAKES AND CHANNELS BECOME SILTED. THIS PROBLEM IS
   ACCENTUATED IN DEVELOPED COUNTRIES BY EXTRA PLANT NUTRIENTS REACHING
   WATER SUPPLIES FROM FERTILIZER AND SEWAGE EFFLUENT. AUTHOR PRESENTS A
   COMPREHENSIVE REVIEW OF THE WORLD'S LITERATURE ON AQUATIC WEED CONTROL
   SINCE 1960, TO PROVIDE A GUIDE TO RESEARCH WITH PRIMARY ATTENTION TO
   THOSE PLANTS CAUSING PROBLEMS IN WARM ENVIRONMENTS. THE LITERATURE
   CITATIONS ARE GROUPED AS FOLLOWS: REVIEWS, IDENTIFICATION, GENERAL
   RECOMMENDATIONS, IMPORTANT WATER WEEDS, CHEMICALS USED IN AQUATIC WEED
   CONTROL, CONTROL TECHNIQUES, BIOLOGICAL CONTROL, UTILIZATION OF WATER
   WEEDS, TOXICOLOGY OF HERBICIDES TO FISH, HERBICIDES RESIDUES, AND WATER
   AND ITS EFFECT. AUTHOR INDICATES THE NEED FOR MORE RESEARCH IN
   BIOLOGICAL CONTROL AND UTILIZATION OF WATER WEEDS WHICH MIGHT BE USEFUL
   IN REGIONS WITHOUT THE FINANCIAL RESOURCES TO DEAL WITH THE PROBLEM.
   (SIMSIMAN-WISCONSIN)

FIELD 05G, 04A

ACCESSION NO. W70-00269

CHANGES IN LAKE WASHINGTON FOLLOWING AN INCREASE IN THE NUTRIENT INCOME,

WASHINGTON UNIV., SEATTLE. DEPT. OF ZOOLOGY.

W. T. EDMONDSON.

INTERNATIONAL ASSOCIATION OF THEORETICAL AND APPLIED LIMNOLOGY, PROCEEDINGS,
   VOL 14, P 167-175, JULY 1961. 5 FIG, 1 TAB, 12 REF.

DESCRIPTORS:
   *NUTRIENTS, *PRODUCTIVITY, NUISANCE ALGAE, SEWAGE EFFLUENTS,
   PHOSPHORUS, EUTROPHICATION, WASHINGTON, LAKES, WATER POLLUTION EFFECTS.

IDENTIFIERS:
   *LAKE WASHINGTON(WASH), *LAKE CHANGES, *NUTRIENT CONCENTRATION,
   *NUTRIENT BUDGET, POTENTIAL NUTRIENT CONCENTRATION, AREAL
   INCOME(NUTRIENTS), LAKE VOLUME, MEAN DEPTH, DILUTION, OSCILLATORIA
   RUBESCENS, MADISON LAKES(WIS), CEDAR RIVER(WASH), ZURICHSEE(SWITZ),
   TURLERSEE(SWITZ), GREIFENSEE(SWITZ), PAFFIKERSEE(SWITZ),
   AEGERISEE(SWITZ), FURESO(SWITZ).

ABSTRACT:
   A DENSE BLOOM OF OSCILLATORIA RUBESCENS IN 1955 PROMPTED A STUDY OF
   LAKE WASHINGTON TO DETERMINE EFFECTS OF A GROWING HUMAN POPULATION ON
   LAKE EUTROPHICATION. ONE PHASE WAS AN ENGINEERING SURVEY MADE TO
   ESTABLISH THE MAGNITUDE OF SEWAGE EFFLUENT AND OTHER SOURCES OF
   NUTRIENTS. INCREASES IN NUTRIENTS FROM 1950-1955 WERE SMALL COMPARED
   WITH LIMNOLOGICAL CHANGES. IN ORDER TO COMPARE THE NUTRIENT STATUS OF
   VARIOUS LAKES, A SIMPLE NUMERICAL BASIS OF COMPARISON OF NUTRIENT RATE
   SUPPLY FROM OUTSIDE SOURCES IS DESIRABLE. SYSTEMS WHICH INCORPORATE A
   CONSIDERATION OF THE POTENTIAL CONCENTRATION OF NUTRIENTS IN LAKES ARE
   PREFERABLE TO THOSE RELATING ONLY INCOME TO AREA. POTENTIAL
   CONCENTRATION CAN BE RELATED TO THE AREAL INCOME AND THE MEAN DEPTH OF
   THE LAKE AS FOLLOWS: AREAL INCOME/MEAN DEPTH=POTENTIAL CONCENTRATION
   (GRAMS/SQUARE METER/METER=GRAMS/CUBIC METER). THE PRESENT WORK DEALS
   WITH ONLY A PORTION OF THE PROBLEM--COMPARISON OF INCOME OF PHOSPHORUS
   AMONG LAKES. THE AUTHOR PRESENTS COMPUTATIONS FOR SEVERAL LAKES IN
   WISCONSIN AND EUROPE IN ADDITION TO LAKE WASHINGTON, ALSO SHOWS GRAPHS
   OF AREAL INCOME PLOTTED AGAINST POTENTIAL CONCENTRATION. RESULTS OF
   SUCH CALCULATIONS INDICATE THAT POTENTIAL CONCENTRATION IS MORE CLOSELY
   RELATED TO ALGAL STANDING CROP THAN IS THE AREAL INCOME.
   (KETELLE-WISCONSIN)

FIELD 05C, 06G,

ACCESSION NO. W70-00270

ISOLATION AND CULTURE OF TOXIC STRAINS OF ANABAENA FLOS-AQUAE (LYNGB.) DE BREB.,

NATIONAL RESEARCH COUNCIL OF CANADA, OTTAWA (ONTARIO).

P. R. GORHAM, J. MCLACHLAN, U. T. HAMMER, AND W. K. KIM.

INTERNATIONAL ASSOCIATION OF THEORETICAL AND APPLIED LIMNOLOGY, PROCEEDINGS,
VOL 15, P 796-804, FEBRUARY 1964. 1 FIG, 3 TAB, 22 REF.

DESCRIPTORS:
*ALGAE, *TOXICITY, *CULTURES, *ISOLATION, LIMNOLOGY, PUBLIC HEALTH,
PHYSICOCHEMICAL PROPERTIES, GENETICS, BACTERIA, PEPTIDES, ENVIRONMENTAL
EFFECTS, HETEROGENEITY, IRON, MANGANESE, WATERFOWL, CYANOPHYTA.

IDENTIFIERS:
*ANABAENA FLOS-AQUAE, *STRAINS, PHYCOLOGISTS, VETERINARIANS,
MICROCYSTIS AERUGINOSA, ENDOTOXIN, FAST-DEATH FACTOR, DOSAGE, SPECIES
SUSCEPTIBILITY, ANABAENA LEMMERMANNII, OTTAWA(ONTARIO), BURTON
LAKE(SASKATCHEWAN), HUMBOLDT(SASKATCHEWAN), MICE, MORPHOLOGICAL
CHARACTERS, TRIS.

ABSTRACT:
TOXIC WATERBLOOMS OF PLANKTONIC BLUE-GREEN ALGAE HAVE PUZZLED
PHYCOLOGISTS, LIMNOLOGISTS, VETERINARIANS, AND PUBLIC HEALTH OFFICIALS
FOR MORE THAN 80 YEARS. THE AUTHORS OF THIS PAPER SUCCEEDED IN
ISOLATING TOXIC AND NON-TOXIC STRAINS OF ANABAENA FLOS-AQUAE FROM A
TOXIC BLOOM IN BURTON LAKE, SASKATCHEWAN. OF FOURTEEN STRAINS ISOLATED
IN 1960 AND 1961, EIGHT WERE TOXIC AND SIX NON-TOXIC. THEY SHOWED, FOR
THE FIRST TIME (ALTHOUGH LONG SUSPECTED), THAT THIS ALGA CAN BE
POISONOUS. THE FAST-DEATH FACTOR PRODUCED IS OFTEN SECRETED INTO THE
SURROUNDING WATER. THIS AND OTHER PROPERTIES OF ANABAENA VERY
FAST-DEATH FACTOR INDICATE THAT IT IS DIFFERENT FROM MICROCYSTIS
FAST-DEATH FACTOR. ONE SAMPLE COLLECTED IN 1960 CONTAINED A VERY
FAST-DEATH FACTOR, KILLING MICE IN ONE TO TWO MINUTES AFTER
INTRAPERITONEAL INJECTION WITH A MINIMAL LETHAL DOSE. DEATH WAS
PRECEDED BY SYMPTOMS OF PARALYSIS, TREMORS, AND MILD CONVULSIONS. THE
SALT CONCENTRATIONS GIVING BEST GROWTH OF ONE STRAIN OF ANABAENA
FLOS-AQUAE IN UNIALGAL CULTURE BEAR LITTLE RESEMBLANCE TO THOSE OF THE
LAKE WATER FROM WHICH IT WAS ISOLATED. INTERESTING CHANGES IN GROWTH
HABIT (GRADUAL LOSS OF COLONIAL HABIT AND FILAMENT COILING) WERE
OBSERVED IN CULTURING ANABAENA FLOS-AQUAE, ILLUSTRATING HOW VARIABLE
SOME MORPHOLOGICAL CHARACTERS CAN BE. (JONES-WISCONSIN)

FIELD 05C

ACCESSION NO. W70-00273

CONTRIBUTION TO THE ECOLOGY OF WATER-BLOOM-FORMING BLUE-GREEN ALGAE
   -APHANIZOMENON FLOS AQUAE AND MICROCYSTIS AERUGINOSA,

CESKOSLOVENSKA AKADEMIE VED, PRAGUE. HYDROBIOLOGICAL STATION.

J. HRBACEK.

INTERNATIONAL ASSOCIATION OF THEORETICAL AND APPLIED LIMNOLOGY, PROCEEDINGS,
   VOL 15, P 837-846, FEB 1964. 8 FIG, 2 TAB, 6 REF.

   DESCRIPTORS:
      *ALGAE, *CYANOPHYTA, *ECOLOGY, *EUTROPHICATION, PONDS, RESERVOIRS,
      BACKWATER, FERTILIZATION, NITROGEN COMPOUNDS, PHOSPHORUS COMPOUND,
      PERCHES, FISH, DIATOMS, CARP, PLANKTON, METABOLISM, BIOLOGICAL
      COMMUNITIES, DAPHNIA, NANNOPLANKTON, NUTRIENTS, DEPTH, DOMESTIC
      ANIMALS, SAMPLING, NUISANCE ALGAE.

   IDENTIFIERS:
      *APHANIZOMENON FLOS AQUAE, *MICROCYSTIS AERUGINOSA, BOHEMIA, ELBE
      RIVER, CELAKOVICE(CZECHOSLOVAKIA), BLATNA(CZECHOSLOVAKIA), CYPRINID
      FISHES, CLOROCOCCALES, DAPHNIA PULICARIS, DAPHNIA CUCULLATA, DAPHNIA
      HYALINA, DAPHNIA MAGNA, CZECHOSLOVAKIA.

   ABSTRACT:
      KNOWLEDGE OF INDIVIDUAL FACTORS WHICH EITHER PROMOTE OR INHIBIT
      WATER-BLOOM-FORMING CYANOPHYTES (BLUE-GREEN ALGAE), APHANIZOMENON FLOS
      AQUAE, AND MICROCYSTIS AERUGINOSA IS NOT EXTENSIVE. THIS REPORT IS
      BASED ON A COMPARATIVE STUDY OF WATER-BODIES IN BOHEMIA. TEN BACKWATERS
      IN THE INUNDATION AREA OF THE ELBE RIVER, SIX FISH PONDS, AND THREE
      WATER RESERVOIRS WERE SAMPLED AT MONTHLY INTERVALS FROM APRIL TO
      SEPTEMBER. IN ORDER TO EXAMINE THE RELATIVE IMPORTANCE OF NITROGEN AND
      PHOSPHORUS COMPOUNDS FROM EACH OF THE THREE CATEGORIES OF WATER-BODIES,
      THE FOLLOWING WERE SELECTED: ONE WITH WATER BLOOM CONSISTING OF   .
      APHANIZOMENON WITH OCCASIONALLY SOME MICROCYSTIS; ONE WITHOUT WATER
      BLOOM, AND, INSOFAR AS POSSIBLE, ONE WITH MICROCYSTIS WATER BLOOM. IN
      PONDS, BACKWATERS, AND RESERVOIRS WITH AND WITHOUT WATER BLOOMS, NO
      IMPORTANT DIFFERENCES IN CONTENT OF PHOSPHORUS AND NITROGEN COMPOUNDS
      WERE FOUND. IN ONE BLOOM-CONTAINING RESERVOIR, WATER SHOWED
      SIGNIFICANTLY SMALLER CONCENTRATIONS OF PHOSPHORUS COMPOUNDS THAN IN
      PONDS. AN ASSOCIATION EXISTS BETWEEN CYANOPHYTES AND DAPHNIA. DEPTH HAD
      INCREASING EFFECT ON DEVELOPMENT OF WATER BLOOM OF THESE TWO
      CYANOPHYTES. IN PONDS, INCREASED FISH STOCK ACCOMPANIED DECREASED
      APHANIZOMENON BLOOMS, WHILE IN SOME WATER-BODIES, LARGER FISHSTOCK WAS
      ASSOCIATED WITH DEVELOPMENT OF MICROCYSTIS AERUGINOSA BLOOM.
      (JONES-WISCONSIN)

   FIELD 05C, 02H

   ACCESSION NO. W70-00274

THE USE OF INVERTEBRATES IN A WATER QUALITY INVESTIGATION: A BIOLOGICAL STUDY OF THE MIAMI RIVER, OHIO,

STANFORD UNIV., CALIF. DEPT. OF CIVIL ENGINEERING.

P. JONATHAN YOUNG.

PROC 3RD ANNU AMER WATER RESOURCES CONF, NOV 8-10, 1967, SAN FRANCISCO, CALIF, P 304-313, 1967. 10 P, 2 FIG, 2 TAB, 7 REF.

DESCRIPTORS:
*WATER POLLUTION EFFECTS, *BIOINDICATORS, *WATER QUALITY, *DISSOLVED OXYGEN, *OHIO, WATER TEMPERATURE, SLUDGE WORMS, BLOODWORMS, AQUATIC BACTERIA, ALGAE, MAYFLIES, CADDISFLIES.

IDENTIFIERS:
GREAT MIAMI RIVER(OHIO).

ABSTRACT:
A BIOLOGICAL STUDY WAS MADE OF THE GREAT MIAMI RIVER, OHIO TO DETERMINE THE FACTORS CAUSING VARIATIONS IN AQUATIC LIFE. SAMPLING SITES WERE CHOSEN TO VARY AS LITTLE AS POSSIBLE IN ALL CHARACTERISTICS EXCEPT WATER QUALITY. THE VARIETY OF LIFE WAS MOST RELATED TO DISSOLVED OXYGEN, AND WATER TEMPERATURE WAS THE NEXT MOST IMPORTANT FACTOR. THE WATER POLLUTION CONDITIONS WERE DIVIDED INTO 4 ZONES. VERY POLLUTED: HERE, BACTERIA AND PROTOZOANS ARE ABUNDANT. ONLY A FEW SPECIES OF ANIMALS EXIST (SLUDGE WORMS, BLOOD WORMS, AND MAGGOTS) AND THESE LIVE ON DECAYING MATTER OR BACTERIA. GREEN PLANTS ARE USUALLY ABSENT. SEWAGE FUNGUS MAY OCCUR. POLLUTED: BACTERIA ARE STILL ABUNDANT AND OXYGEN IS LOW. GREEN ALGAE BEGIN TO APPEAR AND BLOOM, AND SOME OF THE MORE TOLERANT INVERTEBRATES AND ROOTED PLANTS ARE FOUND. SEMI-HEALTHY: ALGAE BLOOMS HAVE USUALLY DISAPPEARED. MOST OF THE POLLUTED-FAVORING ORGANISMS ARE DECREASING IN ABUNDANCE WHILE LESS TOLERANT SPECIES BEGIN TO APPEAR. DISSOLVED OXYGEN MAY AVERAGE 5 PPM OR MORE. HEALTHY: THE ANIMAL AND PLANT LIFE IS MORE OR LESS NORMAL. DECOMPOSITION CONTINUES AT A MUCH REDUCED RATE. DISSOLVED OXYGEN IS HIGH AND APPROACHES SATURATION VALUES. THREE INSECTS WERE CHOSEN TO SEPARATE THE CLASSIFICATION ZONES. THE LARVA OF THE DAMSELFLY ISHNURA WAS CHOSEN TO SEPARATE 'POLUTED' FROM 'VERY POLLUTED.' THE LARVA OF THE MAYFLY BAETIS WAS CHOSEN TO SEPARATE 'SEMI-HEALTHY' FROM 'POLLUTED'; AND THE CADDIS FLY LARVA HYDROPSYCHE WAS CHOSEN TO SIGNIFY THE BEGINNING OF THE 'HEALTHY' ZONE. ALONG THE GREAT MIAMI RIVER, THE PROGRESSION OF THESE INVERTEGRATES WAS FOUND TO BE VERY CONSISTENT. (KNAPP-USGS)

FIELD 05C

ACCESSION NO. W70-00475

BACTERIA, CARBON DIOXIDE, AND ALGAL BLOOMS,

WYANDOTTE CHEMICAL CORP., MICH.

L. E. KUENTZEL.

J WATER POLLUT CONTR FEDERATION, VOL 41, NO 10, P 1737-1747, OCT 1969. 11 P, 51 REF.

DESCRIPTORS:
*EUTROPHICATION, *ALGAE, *NUTRIENTS, *CARBON DIOXIDE, *REVIEWS, BIBLIOGRAPHIES, BACTERIA, DECOMPOSING ORGANIC MATTER, WASTE WATER(POLLUTION).

IDENTIFIERS:
ALGAL BLOOMS.

ABSTRACT:
A RECENT LITERATURE SURVEY ON THE SUBJECT OF LAKE EUTROPHICATION, AND ALGAL BLOOMS AND NUTRIENTS AS CAUSATIVE FACTORS, REVEALS THAT THERE IS A DECIDED LACK OF PUBLISHED INFORMATION ON THE ROLE OF THE MAJOR NUTRIENT, CARBON DIOXIDE. GREAT EMPHASIS IS PLACED ON PHOSPHORUS AND NITROGEN AS CAUSATIVE AGENTS IN THE PRODUCTION OF UNWANTED ALGAL GROWTHS, YET A SATISFACTORY ACCOUNTING FOR THE SOURCE OF THE TREMENDOUS AMOUNTS OF CARBON DIOXIDE REQUIRED FOR MASSIVE ALGAL BLOOMS HAS NOT BEEN FOUND IN THE LITERATURE. THE PURPOSES OF THIS PAPER ARE (A) TO PROVIDE SOME LOGISTICS ON CARBON DIOXIDE REQUIREMENTS FOR SUCH BLOOMS, (B) TO ESTABLISH THE FACT THAT SUCH QUANTITIES OF CARBON DIOXIDE ARE NOT AVAILABLE FROM THE ATMOSPHERE AND DISSOLVED SALTS, AND (C) TO POINT OUT THE ROLE PLAYED BY BACTERIA AND DECOMPOSABLE ORGANIC MATTER. (KNAPP-USGS)

FIELD 05C, 10

ACCESSION NO. W70-00664

ECOLOGICAL FACTORS AND THE DISTRIBUTION OF CLADOPHORA GLOMERATA IN THE GREAT
  LAKES,

WISCONSIN UNIV., MILWAUKEE. DEPT. OF BOTANY.

RICHARD P. HERBST.

THE AMER MIDLAND NATUR, VOL 82, NO 1, P 90-98, JULY 1969. 9 P, 3 FIG, 1 TAB,
  15 REF.

DESCRIPTORS:
    *ECOLOGY, *EUTROPHICATION, *GREAT LAKES, *ALGAE, NUTRIENTS, PHOSPHORUS,
    POPULATION, POLLUTANTS, MAPPING, HYDROGEN SULFIDE, CITIES, LAKE HURON,
    LAKE ERIE, LAKE ONTARIO, LAKE MICHIGAN, WATER TEMPERATURE, PHOSPHATES,
    TURBIDITY.

IDENTIFIERS:
    CLADOPHORA GLOMERATA.

ABSTRACT:
    NUTRIENT ENRICHMENT IN THE GREAT LAKES HAS PROVIDED FERTILE AREAS FOR
    GROWTH OF ALGAL NUISANCES. ONE OF THESE SPECIES, CLADOPHORA GLOMERATA,
    HAS BECOME A MAJOR PROBLEM FOR MANY CITIES BORDERING THE GREAT LAKES.
    ECOLOGICAL FACTORS CONCERNING ITS GROWTH IN MILWAUKEE'S HARBOR WERE
    STUDIED, AND ITS DISTRIBUTION IN THE GREAT LAKES DETERMINED. PHOSPHORUS
    LEVELS APPEAR TO BE CLOSELY LINKED WITH CLADOPHORA INCREASES.
    (GABRIEL-USGS)

FIELD 05C, 02H

ACCESSION NO. W70-00667

EFFECT OF PHOTOSYNTHESIS ON OXYGEN SATURATION,

TECHNICAL SCHOOL, NIS (YUGOSLAVIA).

LAZAR R. IGNJATOVIC.

J WATER POLLUT CONTR FEDERATION, VOL 40, NO 5, PART 2, P 151-161, MAY 1968.
  11 P, 3 FIG, 2 TAB, 9 REF.

DESCRIPTORS:
    *PHOTOSYNTHESIS, *OXYGEN, *SATURATION, TEMPERATURE, PRESSURE, MIXING,
    EVAPORATION, SUPERSATURATION, RESERVOIRS, ALGAE, BODIES OF WATER,
    DISSOLVED OXYGEN, NORTH CAROLINA.

IDENTIFIERS:
    *OXYGEN SUPERSATURATION, VERTICAL MIXING, ROANOKE RAPIDS RESERVOIR(NC).

ABSTRACT:
    TEMPERATURE CHANGES, PRESSURE CHANGES, VERTICAL MIXING OF STRATIFIED
    WATER, AND EVAPORATION ARE DISCUSSED AND EVALUATED IN ORDER TO CLARIFY
    THE PHENOMENON OF OXYGEN SUPERSATURATION IN NATURAL BODIES OF WATER.
    DATA FROM ROANOKE RAPIDS RESERVOIR IN NORTH CAROLINA WERE USED.
    SUPERSATURATION WITH DISSOLVED OXYGEN MAY BE AN HIGH AS 25% AT A 5 FT
    DEPTH AND AT AN AVERAGE TEMPERATURE OF 25 DEG C. THESE CHANGES CANNOT
    ACCOUNT FOR OBSERVED SUPERSATURATION. SOME SIGNIFICANCE TO THE
    SUPERSATURATION IS ATTACHED TO THE EFFECTS OF SUNSHINE AND ALGAL
    GROWTH. PHOTOSYNTHESIS IS THE ONLY FORCE CAPABLE OF PRODUCING THE
    SUPERSATURATION OF SUCH A MAGNITUDE. (CARSTEA-USGS)

FIELD 05C, 05B

ACCESSION NO. W70-00683

THE UPTAKE, ACCUMULATION AND EXCHANGE OF STRONTIUM-90 BY OPEN SEA PHYTOPLANKTON,

MIAMI UNIV., FLA. INST. OF MARINE SCIENCE.

E. F. CORCORAN, AND J. F. KIMBALL, JR.

PROC 1ST NAT SYMP ON RADIOECOLOGY, COLORADO STATE UNIV, FT COLLINS, SEPT 10-15, 1961. RADIOECOLOGY, REINHOLD PUB CORP, N Y AND AMER INSTITUTE OF BIOLOGICAL SCIENCES, WASHINGTON, D C, SCHULTZ, VINCENT AND KLEMENT ALFRED W JR (EDS), P 187-191, 1963. 5 FIG, 4 TAB, 7 REF.

DESCRIPTORS:
*STRONTIUM RADIOISOTOPES, *PHYTOPLANKTON, RADIOACTIVITY, WASTE DISPOSAL, NUCLEAR REACTORS, POWERPLANTS, SHIPS, SUBMARINES, EFFLUENTS, ION EXCHANGE, RESINS, LEAKAGE, CESIUM, NUCLEAR EXPLOSIONS, ABSORPTION, ADSORPTION, ATLANTIC OCEAN, DIATOMS, NERITIC ALGAE, IONS, WATER POLLUTION EFFECTS, WATER POLLUTION SOURCES, RADIOECOLOGY.

IDENTIFIERS:
*RADIONUCLIDE UPTAKE, *RADIONUCLIDE ACCUMULATION, *RADIONUCLIDE EXCHANGE, *OPEN SEA, WEAPONS TESTING, NUCLEAR SHIP SAVANNAH, ZIRCONIUM-95, RUBIDIUM-106, NIOBIUM-95, GYMNODINIUM SIMPLEX, FLAGELLATES, ELEUTHERA ISLAND, NITZCHIA SERIATA, NAVICULA, FLORIDA CURRENT, YTTRIUM-90, KATODINIUM ROTUNDATA.

ABSTRACT:
AUTHORS USED FIVE CULTURES OF OCEANIC PHYTOPLANKTON TO DETERMINE THEIR ABILITY TO CONCENTRATE STRONTIUM (SR) FROM SEA WATER, WITH THE OBJECTIVE OF ASCERTAINING WHETHER RESULTS OF STUDIES WITH NERITIC AND FRESHWATER FORMS CAN APPLY TO NERITIC ORGANISMS. THE EXPERIMENTS WERE SHORT-TERMED, AND WERE CONDUCTED DURING ACTIVE GROWTH PHASES TO ELIMINATE POSSIBILITY OF NUTRIENT DEFICIENCIES AND OVER-POPULATION, TO MINIMIZE BACTERIAL INTERFERENCE, AND TO EMULATE NATURAL CONDITIONS AS CLOSELY AS POSSIBLE. ALL CELLS GROWN IN MEDIA CONTAINING STRONTIUM-90-YTTRIUM-90 BECAME RADIOACTIVE. UPTAKE OF SR WAS PROPORTIONAL TO THE CONCENTRATION OF SR IN THE MEDIUM, CELLS BEING ABLE TO CONCENTRATE RADIOACTIVE SR FROM 18.6 TO 381.7 TIMES ITS AMBIENT CONCENTRATION. CELL GROWTH, IN MEDIA CONTAINING UP TO TEN TIMES NORMAL CONCENTRATION OF SR, WAS NOT INHIBITED. MEDIA CONTAINING HIGH CONCENTRATIONS OF SR HAD LITTLE EFFECT ON ABILITY OF CELLS TO ACCUMULATE SR-90. ADSORPTION WAS FOUND TO PLAY AN IMPORTANT ROLE IN ACCUMULATION OF SR-90. WHILE THESE RESULTS ARE PRELIMINARY, IT SHOULD BE NOTED THAT PRIMARY PRODUCERS IN MARINE ENVIRONMENTS CAN ACCUMULATE SR-90 FROM SEA WATER. CONTROLLED LABORATORY EXPERIMENTS HAVE THE ADVANTAGE OF ALLOWING COMPARISONS BETWEEN VARIOUS PHYTOPLANKTERS AND RADIONUCLIDES. (JONES-WISCONSIN)

FIELD 05C

ACCESSION NO. W70-00707

THE ROLE OF PHYTOPLANKTON IN THE CYCLING OF RADIONUCLIDES IN THE MARINE
ENVIRONMENT,

BUREAU OF COMMERCIAL FISHERIES, BEAUFORT, N.C.

T. R. RICE.

PROC 1ST NAT SYMP ON RADIOECOLOGY, COLORADO STATE UNIV, FORT COLLINS, SEPT
10-15, 1961. RADIOECOLOGY, REINHOLD PUB CORP, N Y AND AMER INSTITUTE OF
BIOLOGICAL SCIENCES, WASHINGTON, D C, SCHULTZ, VINCENT AND KLEMENT, ALFRED
W JR (EDS). P 179-185, 1963. 5 FIG, 8 TAB, 23 REF.

DESCRIPTORS:
*PHYTOPLANKTON, *CYCLING, *RADIOISOTOPES, SEA WATER, FOOD CHAINS,
PHOSPHORUS RADIOISOTOPES, COBALT RADIOISOTOPES, URANIUM RADIOISOTOPES,
NUCLEAR EXPLOSIONS, STRONTIUM RADIOISOTOPES, CESIUM, PHYSICOCHEMICAL
PROPERTIES, ADSORPTION, COLLOIDS, ZINC RADIOISOTOPES, CALCIUM,
POTASSIUM, CLAMS, SIZE, ALGAE, DIATOMS, TRACE ELEMENTS, METABOLISM,
IONS, CHLORELLA, WATER POLLUTION EFFECTS, WATER POLLUTION SOURCES,
NICHES, RADIOECOLOGY.

IDENTIFIERS:
*MARINE ENVIRONMENT, CERIUM-144, NITZSCHIA CLOSERIUM, ARTEMIA,
CARTERIA, NORTH TEMPERATE ZONE, PLUTONIUM, HALF-LIFE(RADIONUCLIDE),
FILTER-FEEDING ANIMALS, RUTHENIUM-106, PLATYMONAS, THALASSIOSIRA,
AMPHIDINIUM KLEBSI, PORPHYRIDIUM CRUENTUM, RADIONUCLIDE UPTAKE,
NANNOCHLORIS, THORACOMONAS, YTTRIUM, 'BIOLOGICAL DILUTION',
PHYSIOLOGICAL CONDITION, MERCENARIA MERCENARIA.

ABSTRACT:
THE IMPORTANCE OF PHYTOPLANKTON IN CYCLING RADIONUCLIDES IN OCEANS
DEPENDS UPON NUMBER AND CONCENTRATION OF RADIONUCLIDES ACCUMULATED,
LENGTH OF TIME THEY ARE RETAINED BY THE PHYTOPLANKTON, SIZE OF
PHYTOPLANKTON CELLS, NUMBERS OF CELLS PRESENT IN THE WATER, AND
EFFICIENCY OF DIGESTION OF PHYTOPLANKTERS BY FILTER-FEEDING ANIMALS.
THE AVAILABILITY OF A RADIONUCLIDE IN SEA WATER DEPENDS LARGELY UPON
ITS PHYSICAL STATE (COLLOIDAL OR PARTICULATE) AS WELL AS ITS IONIC
STATE. ACCUMULATION IS DEPENDENT ON AMOUNTS OF THE ELEMENT PRESENT IN
SEA WATER AND THE QUANTITIES OF OTHER METABOLICALLY SIMILAR ELEMENTS.
BINARY FISSION OF RADIOACTIVE PHYTOPLANKTERS IN NONACTIVE WATER RESULTS
IN 50% REDUCTION OF ACTIVITY PER CELLS WITH EACH DIVISION, TERMED AS
'BIOLOGICAL DILUTION.' DIFFERENT SPECIES OF PHYTOPLANKTON, WHICH
ACCUMULATE A RADIONUCLIDE TO THE SAME LEVEL AND ARE REMOVED FROM THE
WATER WITH THE SAME EFFICIENCY BY A FILTER-FEEDING ANIMAL, MAY NOT
RESULT IN THE ANIMAL'S BECOMING EQUALLY RADIOACTIVE, FOR ALL SPECIES OF
PHYTOPLANKTON ARE NOT DIGESTED WITH EQUAL EFFICIENCY. BRINE SHRIMP
(ARTEMIA) CONCENTRATE RADIOCOBALT TO HIGHER LEVELS FED UPON RADIOACTIVE
CARTERIA CELLS, THAN WHEN RADIOACTIVITY IS DERIVED FROM THE WATER.
(JONES-WISCONSIN)

FIELD 05C

ACCESSION NO. W70-00708

OBSERVATIONS ON DEEPWATER PLANTS IN LAKE TAHOE, CALIFORNIA AND NEVADA,

NEVADA FISH AND GAME DEPT., RENO; AND CALIFORNIA STATE DEPT. OF FISH AND
GAME, SACRAMENTO. INLAND FISHERIES BRANCH.

TED C. FRANTZ, AND ALMO J. CORDONE.

ECOLOGY, VOL 48, NO 5, P 709-714, 1967. 3 FIG, 3 TAB, 17 REF.

DESCRIPTORS:
*DEEP WATER, *LAKES, *PLANTS, *CALIFORNIA, *NEVADA, DEPTH, ALGAE,
MOSSES, CHARA, FISH, LIMNOLOGY, LAKE TROUT, CRAYFISH, STONEFLIES,
SNAILS, ECOLOGY, EUTROPHICATION, DOMESTIC WASTES, SHALLOW WATER,
CHLOROPHYTA, CHRYSOPHYTA, CYANOPHYTA, SAMPLING, DENSITY, LIGHT, MUD,
SILTS, SANDS, GRAVELS, SEDIMENTS, WATER POLLUTION SOURCES.

IDENTIFIERS:
*LAKE TAHOE(CALIF-NEV), HYDROPHYTES, LIVERWORTS, MORPHOMETRY,
PACIFASTACUS LENIUSCULUS, ANACHARIS CANADENSIS, MYRIOPHYLLUM,
POTAMOGETON CRISPUS, MICROSPORA WITTROCKII, SPIROGYRA, XANTHOPHYCEAE,
MYXOPHYCEAE, SALVELINUS NAMAYCUSH, SUBSTRATE TYPE.

ABSTRACT:
AQUATIC STUDIES OF LAKE TAHOE HAVE DISCLOSED A REMARKABLE PLANT
ZONATION. WITH SURFACE AREA NOT UNUSUALLY LARGE, IT HAS MAXIMUM DEPTH
OF 1645 FEET, MEAN DEPTH OF 1027 FEET. DEEPWATER MACROSCOPIC
HYDROPHYTES IN LAKE TAHOE, FOUND TO DEPTHS OF 500 FEET, CONSISTED OF
ALGAE, MOSSES, AND LIVERWORTS. MOST WERE CONCENTRATED AT DEPTHS OF
200-350 FEET. ONLY CHARA OCCASIONALLY INVADES AREAS SHALLOW AS 20 FEET;
OTHER DEEPWATER HYDROPHYTES ARE RESTRICTED TO DEPTHS BELOW 50 FEET. THE
LIST OF IDENTIFIED SPECIES, PROBABLY INCOMPLETE, INCLUDES SIX ALGAE,
TEN MOSSES, AND TWO LIVERWORTS. ALTHOUGH LIGHT TRANSMISSION PROBABLY
CONTROLS MAXIMUM DEPTH OF PLANT OCCURRENCE, FACTORS LIMITING INSHORE
DISTRIBUTION ARE LESS CLEAR. SUBSTRATE TYPE AT LAKE'S SOUTH END SEEMS
TO CONTROL SHOREWARD EXTENSION OF PLANTS. THESE DEEPWATER PLANT BEDS
ARE NOT ONLY UNUSUAL SCIENTIFIC INTEREST BUT APPEAR IMPORTANT IN THE
LIFE HISTORY OF TAHOE'S MAJOR GAME FISH, THE LAKE TROUT, WHOSE MAXIMUM
CONCENTRATION COINCIDES WITH THESE PLANT ZONES; PLANTS APPARENTLY
PROVIDE SHELTER FOR CRAYFISH, NONGAME FISHES, AND SMALL LAKE TROUT.
LONGTERM RESEARCH ON ECOLOGY OF THESE DEEPWATER HYDROPHYTE COMMUNITIES
IS NEEDED. LAKE TAHOE IS THREATENED WITH EUTROPHICATION FROM DOMESTIC
WASTES. (JONES-WISCONSIN)

FIELD 02H, 05C

ACCESSION NO. W70-00711

SEASONAL SELECTION OF ESTUARINE BACTERIA BY WATER TEMPERATURE,

RHODE ISLAND UNIV., KINGSTON. NARRAGANSETT MARINE LAB.

JOHN MCN SIEBURTH.

JOURNAL EXPERIMENTAL MARINE BIOLOGY AND ECOLOGY, VOL 1, P 98-121, 1967. 11
FIG, 4 TAB, 39 REF.

DESCRIPTORS:
*BACTERIA, *ESTUARINE ENVIRONMENT, *WATER TEMPERATURE, *SEASONAL, RHODE
ISLAND, SAMPLING, ECOLOGY, AMMONIA, ENZYMES, CLASSIFICATION, GROWTH
RATES, SEAWATER, INCUBATION, PSEUDOMONAS, SUBSURFACE-WATERS, BOTTOM
SEDIMENTS, SURFACE WATERS, DETRITUS, LIGHT INTENSITY, TURBIDITY, ALGAE,
DIATOMS, CYCLING NUTRIENTS, PHYSIOLOGICAL ECOLOGY, WATER POLLUTION
EFFECTS, PHYTOPLANKTON.

IDENTIFIERS:
NARRAGANSETT BAY(RHODE ISLAND), TAXONOMIC TYPES, HETERETROPHY,
BACTERIAL MEDIUM, ARTHROBACTER, FLORA, AMBIENT TEMPERATURE, THERMAL
SENSITIVITY, GENERA, 'FLAVOBACTERIUM-CYTOPHAGA', VIBRIOS,
ACHROMOBACTERS, GRAM-POSITIVE BACTERIA, SARCINE-LIKE, BACILLUS-LIKE,
MICROCOCCUS-LIKE, PSYCHROPHILES, PLEMORPHIC, MYCELOID, CORYNEFORM RODS,
COCCI, ARTHROSPORES, GRAM-NEGATIVE BACTERIA, COLONY DEVELOPMENT.

ABSTRACT:
HETERETROPHIC PLANKTONIC BACTERIA WERE OBSERVED IN TEMPERATURE
NARRAGANSETT BAY, RHODE ISLAND, TO ESTABLISH WHETHER SEASONAL WATER
TEMPERATURE CHANGES ARE SUFFICIENT TO SELECT THERMAL AND TAXONOMIC
TYPES. POPULATION CHANGES WERE EXAMINED IN SEMI-MONTHLY SAMPLES AND
GENERA DETERMINED FOR 2500+ ISOLATES; GROWTH-TEMPERATURE SPECTRA OF 600
REPRESENTATIVE ISOLATES AT 20 TEMPERATURES, 2C (C= DEGREES CELSIUS)
APART, WERE EXAMINED. GROWTH RANGE OF NATURAL BACTERIAL POPULATIONS.
VARIED WITH WATER TEMPERATURE. ALL THERMAL TYPES WERE ISOLATED AND
POPULATION TRENDS OBSERVED. MESOPHILIC ISOLATES (GROWTH ABOVE 10C;
OPTIMUM HAVE ABOVE 30C) WERE ONLY OBTAINED ON PLATES AT 36C WHEN WATER
TEMPERATURE EXCEEDED 10C. SEVERAL MESOPHILES WERE COLD-SENSITIVE AND
LABILE BELOW 18C. OBLIGATE PSYCHROPHILES (GROWTH BELOW 16-20C; OPTIMUM
9-10C) WERE ONLY OBTAINED ON OC PLATES WHERE WATER TEMPERATURES WERE
BELOW 10C. TWO ADDITIONAL THERMAL TYPES WERE OBTAINED ON BOTH O AND 18C
PLATES. ORGANISMS WHICH GREW AT OC (OPTIMUM 10-30C) WERE CONSIDERED
FACULTATIVE PSYCHROPHILES; THOSE WHICH GREW AT OC (NARROWER OPTIMUM,
18-20C), PSYCHROTOLERANT. BESIDES THESE MAJOR THERMAL TYPES, A FEW
ISOLATES OF SEVERAL TAXONOMIC GROUPS EXHIBITED TENDENCY FOR MULTIPLE
TEMPERATURE OPTIMA. SEASONAL SELECTION OF THERMAL TYPES OF WATER
TEMPERATURE OCCURRED IN ALL TAXONOMIC GROUPINGS. (JONES-WISCONSIN)

FIELD 05C

ACCESSION NO. W70-00713

THE ISOLATION, PURIFICATION, AND NUTRIENT SOLUTION REQUIREMENTS OF BLUE-GREEN
ALGAE,

WISCONSIN UNIV., MADISON. DEPT. OF BOTANY.

GERALD C. GERLOFF, GEORGE P. FITZGERALD, AND FOLKE SKOOG.

IN SYMPOSIUM ON THE CULTURING OF ALGAE, CHARLES F. KETTERING FOUNDATION,
DAYTON, OHIO, P 27-44, 1950. 6 TAB.

DESCRIPTORS:
*ISOLATION, *CYANOPHYTES, *NUTRIENT REQUIREMENTS, CULTURES, LAKES,
ALGAE, GROWTH RATES, BACTERIA, ULTRAVIOLET RADIATION, ACTINOMYCETES,
HYDROGEN ION CONCENTRATION, NITRATES, POTASSIUM, CALCIUM, MAGNESIUM,
IRON, ORGANIC MATTER, NITROGEN FIXATION, VOLUME, STREAMS, WATER
POLLUTION EFFECTS, EUTROPHICATION, NUISANCE ALGAE.

IDENTIFIERS:
*PURIFICATION, CULTURE MEDIA, MICROCYSTIS AERUGINOSA, APHANIZOMENON
FLOS-AQUAE, GLOEOTRICHIA ECHINULATA, COCCOCHLORIS PENIOCYSTIS,
MADISON(WIS), CHICAGO NATURAL HISTORY MUSEUM, NOSTOC MUSCORUN, NUTRIENT
CONCENTRATION, WISCONSIN, SULFUR, PHOSPHORUS.

ABSTRACT:
PROCEDURES AND PROGRESS ARE DESCRIBED IN ISOLATING AND CULTURING
REPRESENTATIVE CYANOPHYTES (BLUE-GREEN ALGAE) AND MAKING THEM
BACTERIA-FREE. EACH SPECIES ISOLATED WAS DESIGNATED AS A PERMANENT
SPECIMEN IN CHICAGO NATURAL HISTORY MUSEUM COLLECTION. COCCOCHLORIS
PENIOCYSTIS REQUIRES AN UNUSUALLY HIGH PH FOR MAXIMUM GROWTH.
SYSTEMATIC CONCENTRATION VARIATION OF ONE ELEMENT AT A TIME OF THE
BASIC CULTURE SOLUTION SHOWED THAT HIGHER NITRATE CONCENTRATIONS WERE
SOLELY RESPONSIBLE FOR INCREASED GROWTH. RELATIONSHIP OF PHOSPHORUS,
SULFUR, POTASSIUM, MAGNESIUM, CALCIUM, AND IRON WERE TESTED. THE
DEMONSTRATION INDICATES THAT 24 SPECIES OF CYANOPHYTES CAN BE CULTURED
CONTINUOUSLY IN INORGANIC SOLUTIONS OF KNOWN COMPOSITION, EVIDENCES
STRONGLY THAT, AS A GROUP, THEY DO NOT REQUIRE ORGANIC SUBSTANCES OR
UNKNOWN FACTORS FOR NORMAL GROWTH. IT APPEARS THAT NITROGEN SUPPLY IS
AN IMPORTANT FACTOR IN THEIR GROWTH UNDER NATURAL CONDITIONS.
COCCOCHLORIS PENIOCYSTIS APPARENTLY DOES NOT FIX ATMOSPHERIC NITROGEN.
THE LABORATORY DATA PROBABLY CANNOT BE APPLIED DIRECTLY TO FIELD
CONDITIONS SINCE CONCENTRATIONS OF ESSENTIAL ELEMENTS AS WELL AS
VOLUMES OF SOLUTION FROM WHICH ALGAE EXTRACT THEM, ARE CERTAINLY
IMPORTANT. NEVERTHELESS, LABORATORY DATA SHOULD HELP CONSIDERABLY IN
DETERMINING REQUIREMENTS FOR OPTIMUM GROWTH OF CYANOPHYTES IN LAKES AND
STREAMS. (JONES-WISCONSIN)

FIELD 05C

ACCESSION NO. W70-00719

MODIFIED FILTER MEDIA FROM REMOVAL OF WATER POLLUTANTS,

C. D. AGARWAL, AND A. V. S. PRABHAKARA RAO.

WATER RESEARCH, VOL 2, NO 1, P 43-45, JAN 1968. 6 REF.

DESCRIPTORS:
*WATER POLLUTION TREATMENT, *FILTRATION, *WATER POLLUTION CONTROL,
*FILTERS, SEWAGE EFFLUENTS, ELECTROLYTES, ZETA POTENTIAL, BACTERIA,
PROTOZOA, ALGAE, SOLID WASTES, TERTIARY TREATMENT.

IDENTIFIERS:
POLYELECTROLYTE-COATED FILTERS, BAUXITE.

ABSTRACT:
A SIMPLE FILTER WAS DEVELOPED FOR EFFICIENT REMOVAL OF MICRON-SIZED
PARTICLES. THESE PARTICLES INCLUDE SOME BACTERIA, PROTOZOA, ALGAE,
CYSTS, AND EVEN SOME WORMS. SEWAGE PARTICLES CARRY SMALL ELECTROKINETIC
CHARGES (SURFACE ZETA POTENTIALS), WHICH ARE GENERALLY NEGATIVE. THE
EFFICIENCY OF FILTRATION CAN BE GREATLY IMPROVED IF THE FILTER MEDIA
ARE COATED WITH CATIONIC POLYELECTROLYTES TO RETAIN THE NEGATIVELY
CHARGED PARTICLES. THE EXPERIMENTAL RESULTS OBTAINED BY USING BAUXITE
COATED FILTERS WERE COMPARABLE WITH THOSE USING POLYELECTROLYTES. THE
ECONOMICS OF THE DESIGNED CRITERIA FOR THIS FILTRATION PROGRESS ARE
EVALUATED. (CARSTEA-USGS)

FIELD 05D, 05G

ACCESSION NO. W70-01027

STUDIES ON NATURAL FACTORS AFFECTING PHOSPHATE ABSORPTION AND ITS UTILIZATION
  BY ALGAE,

  JOSEPH SHAPIRO.

  WATER RESEARCH, VOL 2, NO 1, P 21-23, JAN 1968. 2 FIG, 3 REF.

    DESCRIPTORS:
      *EUTROPHICATION, *NUTRIENTS, *WATER POLLUTION SOURCES, *PHOSPHATES,
      ALGAE, WATER POLLUTION TREATMENT, WATER QUALITY CONTROL, SEWAGE
      EFFLUENTS, ABSORPTION, LAKES.

    IDENTIFIERS:
      *PHOSPHATE ABSORPTION.

    ABSTRACT:
      THE ACCUMULATION OF NUTRIENT ELEMENTS IN WATER, PARTICULARLY NITROGEN
      AND PHOSPHORUS, FAVORS THE PROCESS OF EUTROPHICATION. PREVIOUS
      EXPERIMENTS HAVE SHOWN THAT DIFFERENT FACTORS ARE PRESENT IN NATURAL
      WATERS WHICH INHIBIT OR FACILITATE THE USE OF PHOSPHATE BY ALGAE. THE
      'PHOSPHATE SPARING FACTOR' WAS STUDIED BY USING LAKE WATER AND
      MICROCYSTIS AERUGINOSA AS THE TEST ORGANISM. ION EXCHANGE RESINS WERE
      USED TO DETERMINE AN IONIC FACTOR THAT FACILITATED THE USE OF PHOSPHATE
      BY MICROCYSTIS. ADDITION OF SALTS INCREASED THE RATES OF PHOSPHATE
      UPTAKE BY THE ALGAE IN DILUTED LAKES; HOWEVER, NITRATE SEEMED TO BE THE
      MOST IMPORTANT FACTOR. MORE EXPERIMENTAL WORK USING DIFFERENT ALGAE AND
      CHEMICAL COMPONENTS ARE IN PROGRESS. (CARSTEA-USGS)

    FIELD 05C, 05B

    ACCESSION NO. W70-01031

THE INFLUENCE OF ALGAL ANTIBIOSIS ON THE ECOLOGY OF MARINE MICROORGANISMS,

  RHODE ISLAND UNIV., KINGSTON. NARRAGANSETT MARINE LAB.

  JOHN MCNEILL SIEBURTH.

  ADVANCES IN MICROBIOLOGY OF THE SEA, ACADEMIC PRESS, NEW YORK, VOL 1, P
    63-94, 1968. 8 FIG, 122 REF.

    DESCRIPTORS:
      *ALGAE, *ECOLOGY, *MARINE MICROORGANISMS, PHYTOPLANKTON, BACTERICIDES,
      ALGAL TOXINS, LARVAE, PHENOLS, SHELLFISH, OYSTERS, INDUSTRIES,
      CHELATION, BACTERIA, STREPTOCOCCUS, SECONDARY PRODUCTIVITY, SULFIDES,
      SEA WATER, PHAWOPHYTA, CRUSTACEANS, HERBIVORES, CHLOROPHYLL, ENZUMES,
      PEPTIDES, BIOCHEMISTRY, HYDROGEN ION CONCENTRATION, OXIDATION, DIATOMS,
      EPIPHYTOLOGY, METABOLISM, PRIMARY PRODUCTIVITY, FLUORESCENCE, ISOTROPY,
      TRITIUM, SYMBIOSIS, TEMPERATURE, COPEPODS, PROTOZOA, ZOOPLANKTON, SOLAR
      RADIATION, PSEUDOMONAS, PARASITISM, PHYSIOLOGICAL ECOLOGY, WATER
      POLLUTION SOURCES, WATER POLLUTION EFFECTS.

    IDENTIFIERS:
      *ANTIBIOSIS, AUTOLYSIS, EXCRETION, DEPURATION, PHYCOCOLLOIDS, TANNINS,
      FLAGELLATES, PHARMACEUTICALS, EXTRACELLULAR PRODUCTS, CHEMICAL
      RESISTANCE, EUPHAUSIA, PHAEOCYSTIS, SARGASSUM, FRONDS, HYDROCARBONS,
      BACTERIAL ATTACHMENT, ECOTOCARPUS, POLYSIPHONIA, ORGANIC AGGREGATES,
      SEASONAL EFFECTS.

    ABSTRACT:
      AUTHOR REVIEWS THE ROLE IN MARINE ECOLOGY OF ANTIBIOSIS RESULTING FROM
      EXCRETION OF SUBSTANCES SECONDARILY EXCRETED BY ALGAE, A SUBJECT
      CONTRIBUTING TO THE MUCH LARGER PICTURE OF MICROBIAL INTERACTIONS.
      PHOTOSYNTHETIC PLANTS ARE PRIMARILY IMPORTANT TO HETEROTROPHIC
      MICROBIOTA AS PROVIDERS OF PRIMARY NUTRITIONAL SUBSTANCES. CONVERSELY,
      HETEROTROPHIC BACTERIA MAY MARKEDLY INFLUENCE THE NUTRITION AND SUCCESS
      OF MARINE PLANTS, PRESUMABLY RESULTING FROM MODIFICATION OF EXCRETED
      AND AUTOLYTIC PRODUCTS BY DEPOLYMERIZATION AND MINERALIZATION INTO
      USABLE NUTRIENTS, PRODUCTION OF GROWTH AND MORPHOGENETIC FACTORS FOR
      ALGAE, AND BY DETOXIFICATION OF INHIBITORY SUBSTANCES. WHEN POSSIBLE,
      SUCH PROCESSES SHOULD BE STUDIED IN CONCERT, RATHER THAN SEPARATELY.
      REALIZATION THAT ALGAL PRODUCTS MAY BENEFIT, MODIFY, AND EVEN SUPPRESS
      EARLY LIFE STAGES OF MARINE ANIMALS REEMPHASIZES THE IMPORTANCE OF
      INTERDISCIPLINARY ATTACKS ON ECOLOGY. FIELD STUDIES PROMISE TO
      DEMONSTRATE FACTORS WHICH INFLUENCE SUPPRESSION, SELECTION, DOMINANCE,
      AND SUCCESSION OF MARINE MICROORGANISMS. ANY SUCH STUDIES MUST INCLUDE
      CONSIDERATION OF EFFECTS OF EXTRACELLULAR SECRETIONS. AUTHOR SUMMARIZES
      HIS KNOWLEDGE AND UNDERSTANDING OF THE NATURE AND INFLUENCE OF
      INHIBITORY SECONDARY PLANT MATERIALS IN MARINE ENVIRONMENTS TO
      ENCOURAGE CONTINUED INVESTIGATION OF ALGAL ANTIBIOSIS. (JONES-WISCONSIN)

    FIELD 05C

    ACCESSION NO. W70-01068

STUDIES ON ALGAL SUBSTANCES IN THE SEA. III. THE PRODUCTION OF EXTRACELLULAR
ORGANIC MATTER BY LITTORAL MARINE ALGAE,

RHODE ISLAND UNIV., KINGSTON. NARRAGANSETT MARINE LAB.

JOHN MCN SIEBURTH.

J EXPERIMENTAL MARINE BIOLOGY AND ECOLOGY, VOL 3, P 290-309, 1969. 1 FIG, 18
TAB, 26 REF.

DESCRIPTORS:
*OCEANS, *MARINE ALGAE, *LITTORAL, *ORGANIC MATTER, EXUDATION, TIDES,
PHOTOSYNTHESIS, SOLAR RADIATION, SALINITY, RAINFALL, TEMPERATURE,
COLORIMETRY, CARBOHYDRATES, PHENOLS, NITROGEN, RESPIRATION, CARBON,
PRIMARY PRODUCTIVITY, OXYGEN, METABOLISM, WIND VELOCITY, HUMIDITY,
BACTERIA, HYDROGEN ION CONCENTRATION, SULFATES, RESPIRATION, FOOD
CHAINS.

IDENTIFIERS:
*ALGAL SUBSTANCES, *EXTRACELLULAR, FUCUS VESICULOSUS, CHONDRUS CRISPUS,
ASCOPHYLLUM NODOSUM, LAMINARIA DIGITATA, LAMINARIA AGARDHII,
DESICCATION, IMMERSION, NARRAGANSETT MARINE LABORATORY(RHODE ISLAND),
DILUTION RATE, FRUITING BODIES, ULVA LACTUCA VAR LATISSIMA,
POLYSIPHONIA HARVEYI, EPIPHYTIC BACTERIA, SUBLITTORAL, YELLOW MATERIAL,
DUMONTIA INCRASSATA, EMERSION.

ABSTRACT:
SUNLIT AND DARKENED FUCUS VESICULOSUS ACCUMULATE NO DISSOLVED ORGANIC
MATTER IN CLOSED SYSTEMS, IN WHICH PROBLEM OF BACTERIAL GROWTH WAS NOT
ENTIRELY OVERCOME. OPEN SYSTEMS, WITH SEAWATER DILUTIONS, CONSISTENTLY
ACCUMULATED APPRECIABLE QUANTITIES OF EXTRACELLULAR ORGANIC CARBON.
EXUDATION IN FUCUS, COUPLED DIRECTLY WITH PHOTOSYNTHESIS, INCREASED
WITH SOLAR RADIATION. TWO LAMINARIA PRODUCED EXUDATION IN THE DARK.
REDUCING SALINITY OF SEAWATER APPARENTLY DIMINISHED ORGANIC MATTER
EXUDATION RATE BY FUCUS. ORGANIC CARBON IS LOST UPON RE-IMMERSION OF
DISSICATED FUCUS IN SEAWATER. ALGAL CLUMPS ON EXPOSED ROCKS LOST SOME
39 MILLIGRAMS CARBON/100 GRAMS. LIGHT NATURAL RAINFALL MORE EFFECTIVELY
EXTRACTS ORGANIC CARBON THAN ARTIFICIAL RAINFALL. NATURE OF EXUDATES,
DETERMINED BY COLOROMETRIC ASSAY, REVEALED THAT CARBOHYDRATES ARE MOST
ABUNDANT, FOLLOWED BY EQUAL AMOUNTS OF NITROGENUOUS AND POLYPHENOLIC
MATERIAL. DURING SPRING CONDITIONS, FUCUS EXUDES APPROXIMATELY 30% OF
TOTAL OR 40% OF NET CARBON FIXED DAILY. SINCE BIOMASS OF CARBON IN
FUCUS BEDS CAN EXCEED 1000 GRAMS/SQUARE METER AND ALGAE MINIMALLY FIX
6.5 GRAMS/SQUARE METER/DAY, FUCUS EXUDATION OF ORGANIC MATTER EQUAL 5-7
GRAMS CARBON/SQUARE METER/DAY. (JONES-WISCONSIN)

FIELD 05B

ACCESSION NO. W70-01073

STUDIES ON ALGAL SUBSTANCES IN THE SEA. I. GELBSTOFF (HUMIC MATERIAL) IN
   TERRESTRIAL AND MARINE WATERS,

NORWEGIAN INST. OF SEAWEED RESEARCH, TRONDHEIM.

JOHN MCN SIEBURTH, AND ARNE JENSEN.

J EXPERIMENTAL MARINE BIOLOGY AND ECOLOGY, VOL 2, P 174-189, 1968. 8 FIG, 1
   TAB, 24 REF.

   DESCRIPTORS:
      *OCEANS, *MARINE ALGAE, *SEAWATER, *FRESH WATER, *HUMIC ACIDS, *COLOR,
   RIVERS, BOGS, PHENOLS, PIGMENTS, ORGANIC MATTER, BIOMASS, BENTHIC
   FLORA, PHAEOPHYTA, POLAR REGIONS, SPECTROPHOTOMETRY, CHROMATOGRAPHY,
   HYDROGEN ION CONCENTRATION, FLUORESCENCE, CARBOHYDRATES, TEMPERATURE,
   SALINITY, ULTRAVIOLET RADIATION, SULFATES, CELLULOSE, INDUSTRIES, WATER
   POLLUTION SOURCES.

   IDENTIFIERS:
      *GELBSTOFF, *ALGAL SUBSTANCES, SUBPOLAR WATERS, TYNSET(NORWAY), NID
   RIVER(NORWAY), GAULA RIVER(NORWAY), FLAKK(NORWAY), RANHEIM(NORWAY),
   TRONDHEIMSFJORD(NORWAY), SMOLA(NORWAY), HITRA(NORWAY), FROYA(NORWAY),
   FUCUS VESCICULOSUS, ASCOPHYLLUM NODOSUM, ECTOCARPUS CONFERVOIDES, NYLON
   COLUMN TECHNIQUE, PARTICULATE MATTER, BLUE FLUORESCING MATERIAL, YELLOW
   MATERIAL, FRACTIONATION.

   ABSTRACT:
      DISSOLVED YELLOW COLOURING MATTER OF TERRESTRIAL AND MARINE WATERS WAS
   CONCENTRATED IN NYLON COLUMNS. A CONCENTRATION FACTOR OF 10,000 WAS
   OBTAINED WITH A RECOVERY OF APPROXIMATELY 70%. THE METHOD ALLOWED
   ISOLATION OF A REFERENCE 'GELBSTOFF' OF MARINE ORIGIN USED TO ESTIMATE
   GELBSTOFF CONCENTRATIONS IN BOG (17 MILLIGRAMS/LITER (MG/L)), RIVER
   (ABOUT 1 MG/L), AND SEA WATER (0.003-0.8 MG/L). SEASONAL AND
   GEOGRAPHICAL VARIATION OF GELBSTOFF CONCENTRATION IN SEA WATER WAS
   OBSERVED. METHODS BASED ON SPECTRA AND DIFFERENTIAL SPECTRA WERE
   INADEQUATE FOR CHARACTERIZATION OF DIFFERENT TYPES OF YELLOW MATERIAL
   FROM TERRESTRIAL AND SEA WATER; THEY ARE APPARENTLY DERIVED FROM
   POLYPHENOLS. FAIR SEPARATION OF GELBSTOFF WAS EFFECTED FROM DIFFERENT
   SOURCES BY A METHOD OF TWO-DIMENSIONAL PAPER CHROMATOGRAPHY, WHICH THEY
   DEVELOPED. MARINE GELBSTOFF COULD BE DISTINGUISHED FROM BOTH RIVER AND
   BOG WATER PIGMENTS. A CONSIDERABLE FRACTION OF TERRESTRIAL GELBSTOFF
   PRECIPITATED RAPIDLY IN CONTACT WITH SEA WATER. CONTENT AND
   CHARACTERISTICS OF MARINE GELBSTOFF ARE VARIABLE, AND IS UNDETECTABLE
   DURING JANUARY, FEBRUARY, AND MARCH. PART OF THE YELLOW COLOR OF SEA
   WATER IS DUE TO PARTICULATE ORGANIC MATTER, WHICH IS LARGELY AMORPHOUS
   AND SHOWS POLYPHENOL STAINING. PRECIPITATING GELBSTOFF MAY BE A MAJOR
   SOURCE OF THESE ORGANIC AGGREGATES. (JONES-WISCONSIN)

   FIELD 05B

   ACCESSION NO.   W70-01074

RECOVERY OF A SALT MARSH IN PEMBROKESHIRE, SOUTH-WEST WALES, FROM POLLUTION BY
   CRUDE OIL,

   FIELD STUDIES COUNCIL, PEMBROKE (ENGLAND). OIL POLLUTION RESEARCH UNIT.

   E. B. COWELL, AND J. M. BAKER.

   BIOLOGICAL CONSERVATION, VOL 1, NO 4, P 291-296, JULY 1969. 2 FIG, 2 TAB, 8
   REF.

      DESCRIPTORS:
         *ALGAL POISONING, *MARSH PLANTS, *OILY WATER, AQUATIC ALGAE, MARINE
      ALGAE, OIL, MARSH, POLLUTANTS.

      IDENTIFIERS:
         *SALT MARSH, *OIL POLLUTION, *RECOVERY FROM POLLUTION, *KUWAIT CRUDE,
      BENTLASS SALT MARSH.

      ABSTRACT:
         AN ACCOUNT OF THE BIOLOGICAL EFFECTS OF A SPILL OF KUWAIT CRUDE OIL ON
      THE PLANTS AND ALGAE OF THE BENTLASS SALT MARSH NEAR PEMBROKE, S.W.
      WALES. SOME SPECIES OF THE MARSH PLANTS WERE MORE AFFECTED BY THE OIL
      THAN OTHERS. SPILL OCCURRED IN JANUARY 1967 BUT BY JUNE 1968, MOST OF
      THE SPECIES AFFECTED WERE SHOWING SOME SIGNS OF RECOVERY WITH THE
      EXCEPTION OF TRIGLOCHIN MARITIMA. NO ANIMAL OBSERVATIONS.
      (KATZ-WASHINGTON)

      FIELD 05C

      ACCESSION NO.   W70-01231

A COMPOSITE RATING OF ALGAE TOLERATING ORGANIC POLLUTION,

FEDERAL WATER POLLUTION CONTROL ADMINISTRATION, CINCINNATI, OHIO.

C. M. PALMER.

JOURNAL OF PHYCOLOGY, VOL 5, NO 1, P 78-82, 1969. 11 TAB, 29 REF.

DESCRIPTORS:
*ALGAE, *AQUATIC ALGAE, *AQUATIC ENVIRONMENT, POLLUTANTS, BIOLOGICAL COMMUNITIES, WATER POLLUTION EFFECTS, DIATOMS.

IDENTIFIERS:
*POLLUTION-TOLERANT ALGAE, *BLUE GREEN ALGAE, *EUGLENA, PIGMENTED FLAGELLATES.

ABSTRACT:
FROM INFORMATION ON POLLUTION-TOLERANT ALGAE COMPILED FROM REPORTS FROM 165 AUTHORS, THE GENERA AND SPECIES MOST OFTEN REFERRED TO AS SIGNIFICANT FALL INTO A RELATIVELY STABLE SERIES. DIATOMS, PIGMENTED FLAGELLATES, GREEN, AND BLUE-GREEN ALGAE ARE ALL WELL REPRESENTED AMONG THE POLLUTION-TOLERANT GENERA AND SPECIES. THE TOP 8 GENERA ARE EUGLENA, OSCILLATORIA, CHLAMYDOMONAS, SCENEDESMUS, CHLORELLA, NITZSHCIA, NAVICULA, AND STIGEOCLONIUM, AND THE TOP 5 SPECIES, EUGLENA VIRIDIS, NITZSCHIA PALAE, OSCILLATORIA LIMOSA, SCENEDESMUS QUADRICAUDA, AND OSCILLATORIA TENUIS. IN SOME GENERA, E.G., EUGLENA, A SINGLE SPECIES IS FAR MORE SIGNIFICANT THAN ALL OTHERS AS A POLLUTION-TOLERANT FORM. IN OTHER GENERA, E.G., OSCILLATORIA, ONLY A SLIGHT DIFFERENCE DISTINGUISHES THE POLLUTION TOLERANCE OF 2 OR MORE SPECIES. ALGAL GENUS AND SPECIES POLLUTION INDICES ARE PRESENTED FOR USE IN RATING WATER SAMPLES WITH HIGH ORGANIC POLLUTION. (KATZ-WASHINGTON)

FIELD 05C

ACCESSION NO. W70-01233

STUDY OF THE TOXICITY OF CHEMICAL PRODUCTS WITH RESPECT TO THE MARINE FOOD CHAIN (FRENCH),

CENTRE D'ETUDES ET DE RECHERCHES DE BIOLOGIE ET D'OCEANOGRAPHIE MEDICALE, NICE (FRANCE).

M. AUBERT, R. CHARRA, AND G. MALARA.

REVUE INTERNATIONALE D'OCEANOGRAPHIE MEDICALE, VOL XIII-XIV, P 45-72, 1969. 14 FIG, 7 TAB, 7 REF.

DESCRIPTORS:
*DETERGENTS, *INSECTICIDES, *TOXICITY, *WATER POLLUTION, FOOD CHAINS, AQUATIC ALGAE, BRINE SHRIMP, FISH, MUSSELS.

IDENTIFIERS:
*HYDROCARBON, *TOXICITY THRESHOLD.

ABSTRACT:
THE THRESHOLD TOXICITIES OF 5 CHEMICAL POLLUTANTS (3 DETERGENTS, 1 INSECTICIDE, 1 HYDROCARBON) WERE DETERMINED FOR ORGANISMS FROM 3 LEVELS OF THE FOOD CHAIN (PHYTOPLANKTON, ZOOPLANKTON, FISH). THE RESULTS SHOW THAT THE TOXICITY OF THE CHEMICALS VARIES FROM ONE ORGANISM TO ANOTHER AND FROM ONE LIFE STAGE TO ANOTHER IN THE SAME ORGANISM. (TYNER-WASHINGTON)

FIELD 05C

ACCESSION NO. W70-01466

STUDY OF THE BIODEGRADABILITY OF TOXIC CHEMICAL PRODUCTS WITH RESPECT TO THE
   MARINE FOOD CHAIN (FRENCH),

   CENTRE D'ETUDES ET DE RECHERCHES DE BIOLOGIE ET D'OCEANOGRAPHIE MEDICALE,
      NICE (FRANCE).

   M. AUBERT, AND J. P. GAMBAROTTA.

   REVUE INTERNATIONALE D'OCEANOGRAPHIE MEDICALE, VOL XIII-XIV, P 73-105, 1969.
      19 FIG, 3 TAB, 8 REF.

      DESCRIPTORS:
         *BIODEGRADATION, *DETERGENTS, *INSECTICIDES, *MARINE BACTERIA,
         *TOXICITY, *WATER POLLUTION, AQUATIC ALGAE, BRINE SHRIMP, FISH, FOOD
         CHAINS, MUSSELS.

      IDENTIFIERS:
         *HYDROCARBON.

      ABSTRACT:
         TEST SOLUTIONS OF 5 CHEMICAL POLLUTANTS (3 DETERGENTS, 1 INSECTICIDE, 1
         HYDROCARBON) AT CONCENTRATIONS TWICE THE PREVIOUSLY DETERMINED
         THRESHOLD DOSES FOR THE ORGANISMS IN QUESTION WERE SUBJECTED TO
         DEGRADATION BY MARINE MICROORGANISMS FOR 3, 6, OR 9 DAYS PRIOR TO THE
         INTRODUCTION OF THE TEST ORGANISMS. THESE ORGANISMS WERE DRAWN FROM
         DIFFERENT LEVELS OF THE FOOD CHAIN; PRIMARY PRODUCERS, 2 SPECIES OF
         PHYTOPLANKTON; HERBIVORES, ZOOPLANKTON AND MUSSELS; CARNIVORES, FISH.
         IN SOME CASES BIOLOGICAL DEGRADATION OF THE CHEMICAL POLLUTANT CAUSED A
         REDUCTION IN TOXICITY. (TYNER-WASHINGTON)

      FIELD 05C

      ACCESSION NO. W70-01467

EFFECTS OF 'TORREY CANYON' TYPE POLLUTION ON THE UNICELLULAR MARINE ALGA
   PHAEODACTYLUM TRICORNUTUM (FRENCH),

   LABORATOIRE DE PHYSIOLOGIE GENERALE ET COMPAREE DU MUSEUM, PARIS (FRANCE);
      AND LABORATOIRE DE PHYSIOLOGIE DES ETRES MARINS DE L'INSTITUT
      OCEANOGRAPHIQUE, PARIS (FRANCE).

   J. C. LACAZE.

   REVUE INTERNATIONALE D'OCEANOGRAPHIE MEDICALE, VOL XIII-XIV, P 157-179, 1969.
      3 FIG, 2 TAB, 36 REF.

      DESCRIPTORS:
         *DETERGENTS, *DIATOMS, *OIL, *OIL WASTES, *SURFACTANTS, *WATER
         POLLUTION, CLEANING, DISASTERS, OILY WATER, MARINE ALGAE, SHIPS.

      IDENTIFIERS:
         *CRUDE OIL, *GAMOSOL, *OIL SPILLS, *TORREY CANYON, AROMATIC SOLVENTS,
         PHAEODACTYLUM TRICORNUTUM.

      ABSTRACT:
         THE AUTHOR STUDIED THE EFFECTS OF BOTH CRUDE OIL (OF THE SORT CARRIED
         BY THE 'TORREY CANYON') AND GAMOSOL (A DETERGENT USED TO DISPERSE OIL)
         ON PHAEODACTYLUM TRICORNUTUM, A MARINE DIATOM. A 10% REDUCTION IN
         GROWTH WAS OBSERVED IN CULTURES PREPARED WITH 1% PETROLEUM EXTRACTS.
         AFTER AERATION THE REDUCTION IN GROWTH WAS 5%, SOME TOXIC VOLATILE
         COMPOUNDS HAVING BEEN ELIMINATED AND THE OXYGEN CONCENTRATION
         INCREASED. PETROLEUM EXTRACTS WHICH HAD BEEN PREVIOUSLY RAISED TO A
         TEMPERATURE ABOVE 100 DEG C EXHIBITED AN INCREASED TOXICITY (35%
         REDUCTION IN GROWTH). GAMOSOL (14% NONIONIC SURFACTANTS AND 86%
         AROMATIC SOLVENTS) IS MORE TOXIC (50% REDUCTION IN GROWTH AFTER
         EXPOSURE TO 5.6 PPM) THAN THE SURFACE ACTIVE FRACTION (50% REDUCTION IN
         GROWTH AFTER EXPOSURE TO 8 PPM). HOWEVER, THIS TOXICITY VARIES BECAUSE
         OF THE HIGH VOLATILITY OF THE SOLVENT IN DILUTE SOLUTION. A 30 MINUTE
         EXPOSURE TO 40 PPM GAMOSOL KILLED THE CELLS. (TYNER-WASHINGTON)

      FIELD 05C, 05B

      ACCESSION NO. W70-01470

PRACTICAL ASPECTS OF THE DESIGN OF WASTE STABILIZATION PONDS,

W. D. BARLOW.

PROC 9TH STH MUNIC INDUSTR WASTE CONF, P 65-70, 1960.

DESCRIPTORS:
*LAGOONS, *DOMESTIC WASTES, WASTE WATER TREATMENT, INDUSTRIAL WASTES,
NEUTRIENTS, BACTERIA, ALGAE, TREATMENT FACILITIES, TEST PROCEDURES,
METALS.

IDENTIFIERS:
*FINISHING WASTES.

ABSTRACT:
LAGOONS CAN BE DESIGNED TO PROVIDE COMPLETE TREATMENT FOR ANY WASTE
WATER, DOMESTIC AND/OR INDUSTRIAL, PROVIDED IT CONTAINS THE NECESSARY
NUTRIENTS AND IS FREE FROM SUBSTANCES WHICH ARE TOXIC TO BACTERIA OR
ALGAE. THE AUTHOR DISCUSSES THE BASIC PROCESSES INVOLVED; FACTORS
AFFECTING THE DESIGN AND OPERATION OF THE LAGOON (PARTICULARLY UNDER
CONDITIONS IN NORTH AND SOUTH CAROLINA); PERFORMANCE OF THE LAGOON,
INCLUDING METHODS OF MEASUREMENT; AND PRECAUTIONS TO BE TAKEN TO AVOID
HEALTH HAZARDS, POLLUTION AND OTHER NUISANCE PROBLEMS. IN DISCUSSION,
REFERENCE WAS MADE TO LAGOONS USED FOR TREATING MIXED WASTE WATERS AND
20-30 PER CENT OF DOMESTIC SEWAGE, AND ESPECIALLY TO THE PROBLEMS
ENCOUNTERED WHICH APPEAR TO BE CAUSED BY TOXIC METALS IN THE WASTE
WATER. (LIVENGOOD-NORTH CAROLINA STATE UNIV)

FIELD 05F

ACCESSION NO. W70-01519

NUTRIENT LIMITATION OF SUMMER PHYTOPLANKTON GROWTH IN CAYUGA LAKE,

CORNELL UNIV., ITHACA, N.Y.

D. H. HAMILTON, JR.

LIMNOLOGY AND OCEANOGRAPHY, VOL 14, NO 4, P 579-590, JULY 1969. 3 FIG, 9 TAB,
32 REF.

DESCRIPTORS:
*PHYTOPLANKTON, *LIMITING FACTORS, *NUTRIENTS, *PRIMARY PRODUCTIVITY,
PHOSPHATES, SILICATES, CITRATES, BIOASSAY, NEW YORK, LAKES, INHIBITION,
ALGAE, DIATOMS, EUTROPHICATION, CARBON RADIOISOTOPES, CHLOROPHYLL,
HYDROGEN ION CONCENTRATION, MAGNESIUM, CALCIUM.

IDENTIFIERS:
*ENRICHMENT EXPERIMENTS, *NUTRIENT LIMITATION, CAYUGA LAKE(NY),
ASTERIONELLA FORMOSA, SYNEDRA, MELOSIRA, DINOBRYON, FRAGILARIA,
RADIOCARBON METHOD, FINGER LAKES(NY), RODHE'S SOLUTION VIII, NUTRIENT
LIMITATION, CARBON-14.

ABSTRACT:
AUTHOR DESCRIBES EXPERIMENTS PERFORMED TO DETERMINE THE EFFECT OF
DEFINED CULTURE MEDIUM COMPONENTS ON NATURAL PHYTOPLANKTON POPULATIONS.
SAMPLES COLLECTED FROM CAYUGA LAKE (ITHACA, NEW YORK) AT 2-3 DAY
INTERVALS FROM 20 JULY THROUGH 18 AUGUST WERE PLACED IN AN ENRICHMENT
MEDIUM (RODHE'S SOLUTION VIII). ASSIMILATION WAS MEASURED WITH
RADIOCARBON (C-14). MAJOR ALGAL CULTURE COMPONENTS WERE ASTERIONELLA
FORMOSA AND UNIDENTIFIED SPECIES OF SYNEDRA, MELOSIRA, DINOBRYON, AND
FRAGILARIA. RESULTS INDICATED AN ENHANCEMENT EFFECT WITH CALCIUM
NITRATE, DUE APPARENTLY TO THE CALCIUM ELEMENT. MAGNESIUM SULFATE AND
FERRIC CITRATE-CITRIC ACID WERE IRREGULARLY INHIBITORY. PHOSPHATE
FAILED TO STIMULATE CULTURES AND OCASIONALLY PROVED INHIBITING.
NUTRIENT COMBINATIONS YIELDED VARIABLE RESULTS; AUTHOR STATES THIS MAY
BE DUE TO PHENOMENA PECULIAR TO THE SAMPLE BOTTLES OR TO CHANGING
COMPOSITION AND ALTERNATIVELY, DUE TO PHYSIOLOGY OF THE PHYTOPLANKTON
POPULATIONS. TABULAR DATA INCLUDE RESULTS AND SUMMARY OF ENRICHMENT
EXPERIMENTS. (VOIGHTLANDER-WISCONSIN)

FIELD 02H, 05C

ACCESSION NO. W70-01579

EXTENT OF DAMAGE TO COASTAL HABITATS DUE TO THE TORREY CANYON INCIDENT,

FURZEBROOK RESEARCH STATION, WAREHAM (ENGLAND).

D. S. RANWELL.

BIOLOGICAL EFFECTS OF OIL POLLUTION ON LITTORAL COMMUNITIES. SUPPLEMENT TO
VOLUME 2 OF FIELD STUDIES. LONDON, FIELD STUDIES COUNCIL, JULY, 1968. P
39-47.

DESCRIPTORS:
*EMULSIFIERS, *MORTALITY, *MARINE ALGAE, *MARSH PLANTS, *MARINE
ANIMALS, *OILY WATER, LICHENS, TIDAL EFFECTS, DISPERSION, OIL-WATER
INTERFACES, PERSISTENCE, WATER POLLUTION TREATMENT, WATER POLLUTION
EFFECTS.

IDENTIFIERS:
*TORREY CANYON, SUSCEPTIBILITY, REGENERATION, CLIFF VEGETATION, SAND
DUNES.

ABSTRACT:
THE FULL RANGE OF OIL-CONTAMINATED HABITATS RESULTING FROM THE TORREY
CANYON SPILL WERE INVESTIGATED FROM THE AIR AND GROUND. DECONTAMINATION
MEASURES AND TIDAL ACTION EFFECTIVELY CLEARED MOST OIL FROM THE CORNISH
COAST, ALTHOUGH SOME PERSISTED THROUGHOUT 1967. OVER 1 1/2 MILLION
GALLONS OF EMULSIFIERS WERE USED. THEY RAPIDLY CLEARED THE WATER
SURFACE BUT THEY DISTRIBUTED CONTAMINATION BEYOND THE ORIGINAL
CONTAMINATION AREA, BOTH IN DEPTH IN SAND AND WATER. TOXIC ELEMENTS IN
THE EMULSIFIERS DESTROYED SOME MARINE FLORA AND FAUNA. THE GREATEST
DAMAGE OCCURRED ON THE ROCKY SHORELINES WHERE MUCH OIL WAS STRANDED.
INTER-TIDAL ALGAE AND LICHENS SUFFERED THE MOST EXTENSIVELY AMONG
PLANTS BECAUSE OF A CUT-OFF OF LIGHT AND AIR. THE MOST TOXIC FRACTIONS
OF CRUDE OIL ARE VOLATILE AND DISAPPEAR RAPIDLY. MANUAL REMOVAL OF OIL
WAS MUCH PREFERRED OVER CHEMICAL REMOVAL METHODS, AS IT DID NOT ADD TO
POLLUTION. AT LEAST 100 SPECIES OF ALGAE, LICHENS, AND FLOWERING PLANTS
WERE FILLED BY OIL OR EMULSIFIER POLLUTION.
(SJOLSETH-WASHINGTON)

FIELD 05C

ACCESSION NO. W70-01777

EXPERIMENTS WITH SOME DETERGENTS AND CERTAIN INTERTIDAL ALGAE,

UNIVERSITY COLL. OF WALES, ABERYSTWYTH. DEPT. OF BOTANY.

A. D. BONEY.

BIOLOGICAL EFFECTS OF OIL POLLUTION ON LITTORAL COMMUNITIES. SUPPLEMENT TO
VOLUME 2 OF FIELD STUDIES. LONDON, FIELD STUDIES COUNCIL, JULY, 1968. P
55-72.

DESCRIPTORS:
*DETERGENTS, *MARINE ALGAE, WATER POLLUTION EFFECTS, LETHAL LIMIT,
GROWTH RATES, SPORES.

IDENTIFIERS:
*TORREY CANYON, *DETERGENT CONCENTRATIONS, SURVIVAL, SUSCEPTIBILITY,
PHYSIOLOGICAL OBSERVATIONS, CHRONIC EXPOSURE, ACUTE EXPOSURE.

ABSTRACT:
THE DIRECT EFFECTS OF DETERGENT APPLICATION FOLLOWING THE TORREY CANYON
DISASTER ON SEVEN TEST SPECIES OF MARINE ALGAE WERE STUDIED. SIX
DETERGENTS WERE USED EACH AT SEVERAL DIFFERENT CONCENTRATIONS. THE
RECEPTACLE TISSUES OF ASCOPHYLLUM NODOSUM WERE DAMAGED. CELL DAMAGE WAS
EXTENSIVE IN CLADOPHORA RUPESTRIS. BRYOPSIS HYPNOIDES WERE EASILY
KILLED. THE CEPT PHASES OF PRASINOCLADUS MARINUS WERE DESTROYED BY
DETERGENTS. GROWTH AND SPORE PRODUCTION OF ACROHAETIUM INFESTANS WERE
RETARDED. NO IMMEDIATE CELL DAMAGE WAS NOTICED AMONG POLYSIPHONIA
LANOSA AND PORPHYRA UMBILICALIS. BOTH ACUTE AND CHRONIC EXPOSURES WERE
UTILIZED. (SJOLSETH-WASHINGTON)

FIELD 05C

ACCESSION NO. W70-01779

BIOLOGICAL CONSEQUENCES OF OIL POLLUTION AND SHORE CLEANSING,

UNIVERSITY COLL. OF SWANSEA (WALES). DEPT. OF ZOOLOGY.

A. NELSON-SMITH.

BIOLOGICAL EFFECTS OF OIL POLLUTION ON LITTORAL COMMUNITIES. SUPPLEMENT TO
VOLUME 2 OF FIELD STUDIES. LONDON, FIELD STUDIES COUNCIL, JULY, 1968. P
73-80.

DESCRIPTORS:
*OILY WATER, *MARINE ALGAE, *EMULSIFIERS, WATER POLLUTION EFFECTS,
MORTALITY, GASTROPODS, MARINE PLANTS, ON-SITE DATA COLLECTIONS.

IDENTIFIERS:
*TORREY CANYON, REPOPULATION, SUSCEPTIBILITY.

ABSTRACT:
THE EFFECTS OF OIL SPILLS IN DIFFERENT AREAS WERE INVESTIGATED TO
PROVIDE INTERESTING COMPARISONS. OIL SPILLED INTO A SMALL COVE YIELDED
EXTENSIVE SHORE POLLUTION AND MORTALITIES. NO EMULSIFIERS WERE USED,
AND IN LESS THAN A YEAR THE ANIMAL POPULATIONS WERE AGAIN NORMAL.
GASTROPOD MOLLUSCS APPEARED TO BE MOST SERIOUSLY AFFECTED. MARINE ALGAE
POPULATIONS WERE ALSO REDUCED. MISUSE OF EMULSIFIERS EMBEDDED THE OIL
MORE FIRMLY AND RESULTED IN INCREASED MORTALITIES. EMULSIFIERS ENABLE
THE OIL TO WET THE SURFACES OF SHORE ORGANISMS AND TO PENETRATE THEIR
SYSTEMS AND ALSO TO SPREAD THE OIL TO PREVIOUSLY UNREACHED AREAS. TIME
OF YEAR OF OIL SPILL AND MOBILITY OF ORGANISMS REGULATE THE RATE AT
WHICH THE MEMBERS OF THE SHORE COMMUNITY ARE ABLE TO RECOVER FROM OIL
POLLUTION. (SJOLSETH-WASHINGTON)

FIELD 05C

ACCESSION NO. W70-01780

PHYSIOLOGY OF GRANULAR LEUKOCYTES IN FISH BLOOD,

L. I. SMIRNOVA.

TRANSLATION FROM VOPROSY IKHTIOLOGII. PROBLEMS OF ICHTHYOLOGY, VOL 8, NO 5,
1968. P 748-755.

DESCRIPTORS:
*FISH PHYSIOLOGY, *ALGAE, *THERMAL POLLUTION, *TOXICITY.

IDENTIFIERS:
*FISH HEMATOLOGY, *GRANULAR LEUCOCYTES, *MUSCULAR EXERTION,
*2-METHYL-5-VINYLPYRIDINE, *GLUE GREEN ALGAE, BREAM, WHITE BREAM,
LEUKOCYTIC COMPOSITION, ALTERNATING CURRENT.

ABSTRACT:
A CONNECTION WAS ESTABLISHED BETWEEN THE LEUCOCYTIC COMPOSITION OF THE
BLOOD IN FED AND STARVING BREAM. AN INCREASE IN THE NUMBER OF
GRANULOCYTES CIRCULATING IN THE BLOOD WAS OBSERVED IN SOME FAVORABLE
AND SOME STRESS CONDITIONS, SUCH AS INTENSIVE FEEDING, TEMPERATURE
INCREASE, INCREASED ACTIVITY, ELECTRICAL STIMULATION AND EXPOSURE TO
2-METHYL-5-VINYLPYRIDINE, BLUE GREEN ALGAE. WITH THE CHANGES IN THE
WHITE BLOOD CELL COMPOSITION, THERE WERE PARALLEL REDUCTIONS IN THE
RESISTANCE OF RED CELLS AND THEIR SUBSEQUENT DESTRUCTION.
(KATZ-WASHINGTON)

FIELD 05C

ACCESSION NO. W70-01788

AN INVESTIGATION BY RAPID CARBON-14 BIOASSAY OF FACTORS AFFECTING THE CULTURAL
   EUTROPHICATION OF LAKE TAHOE, CALIFORNIA-NEVADA,

CALIFORNIA UNIV., DAVIS. DEPT OF ZOOLOGY; AND ENGINEERING-SCIENCE INC.,
   OAKLAND, CALIF.

CHARLES R. GOLDMAN, AND RALF C. CARTER.

JOURNAL WATER POLLUTION CONTROL FEDERATION, VOL 37, NO 7, P 1044-1059, 1965.
   11 FIG, 5 TAB, 27 REF.

DESCRIPTORS:
   *LAKES, *CARBON RADIOISOTOPES, *BIOASSAY, *EUTROPHICATION, *CALIFORNIA,
   *NEVADA, LIMNOLOGY, PRIMARY PRODUCTIVITY, CHLOROPHYLL, SAMPLING,
   PHYTOPLANKTON, RADIOACTIVITY, PHOTOSYNTHESIS, CHEMICAL ANALYSIS, TRACE
   ELEMENTS, TEMPERATURE, IRON, NUTRIENTS, LIGHT PENETRATION, SEWAGE
   EFFLUENTS, OXYGEN, SOLAR RADIATION, PERIPHYTON, BACTERIA, SPECTROSCOPY,
   COPPER, NITROGEN, PHOSPHORUS, ALGAE, WATER POLLUTION EFFECTS, WATER
   CHEMISTRY.

IDENTIFIERS:
   *LAKE TAHOE(CALIF-NEV), TRANSPARENCY, LIGHT MEASUREMENTS, EUPHOTIC
   ZONE, PLANT CAROTENOIDS, CARBON FIXATION, PELAGIC REGION, MONOMICTIC,
   DEPTH EFFECTS, IONIC CONCENTRATION, SEASONAL EFFECTS.

ABSTRACT:
   ATTEMPTS WERE MADE TO DETERMINE SIGNIFICANT VARIABILITY IN THE PRESENT
   FERTILITY OF LAKE TAHOE, TO ESTIMATE MORE PRECISELY THE LAKE'S
   PRODUCTIVITY BY A MORE COMPREHENSIVE SAMPLING PROGRAM THAN PREVIOUSLY
   POSSIBLE, AND, WITH CARBON-14 BIOASSAY METHODS, TO DETERMINE
   EUTROPHICATION POTENTIAL OF INFLOW, BOTH NATURAL AND THAT ENRICHED WITH
   SEWAGE EFFLUENT. THE EXTREME SENSITIVITY OF THE CARBON-14 METHOD
   PROVIDES A MEANS OF PREDICTING LEVELS OF NUTRIENT ADDITION WHICH MIGHT
   CAUSE OBJECTIONAL EUTROPHICATION. ASSUMPTION THAT THE PHYSIOLOGICAL
   CONDITION OF PLANKTON IS VARIABLE IN DIFFERENT PARTS OF THE LAKE IS
   SUPPORTED BY RESULTS OF STUDY. AREAL ESTIMATES OF PRODUCTIVITY WERE
   DETERMINED BY INTEGRATING PHOTOSYNTHETIC CURVES RESULTING FROM
   CONSIDERATION OF LIGHT EXTINCTION, VARIATION IN TEMPERATURE, AND THE
   DISTRIBUTION OF PLANKTON WITH DEPTH. A SECONDARY SEWAGE EFFLUENT, WHEN
   DILUTED TO 1/3,700 OF ORIGINAL STRENGTH, STIMULATES BIOACTIVITY IN LAKE
   TAHOE SUFFICIENTLY TO PRODUCE A 1% DECREASE IN LIGHT TRANSMITTANCE
   WITHIN 12 DAYS. MEASUREMENTS OF PRIMARY PRODUCTIVITY AROUND THE LAKE'S
   MARGIN SHOW THAT, DESPITE EXTENSIVE MIXING, LOCALIZED SIGNS OF
   EUTROPHICATION ARE EVIDENT. THE EXTREME SENSITIVITY OF LAKE TAHOE
   WATERS TO ANY CHANGE IN NUTRIENT REGIME CLEARLY EVIDENCES THE POTENTIAL
   OF EVEN TREATED EFFLUENT FOR SPEEDING EUTROPHICATION. (JONES-WISCONSIN)

FIELD 05C, 02H

ACCESSION NO. W70-01933

CHANGES IN THE BIOLOGY OF THE LOWER GREAT LAKES,

OHIO STATE UNIV., COLUMBUS. NATURAL RESOURCES INST.

CHARLES A. DAMBACH.

BULLETIN OF THE BUFFALO SOCIETY OF NATURAL SCIENCES, VOL 25, NO 1, P 1-17, 1969. 19 REF.

DESCRIPTORS:
*BIOLOGY, *GREAT LAKES, *LAKE ERIE, *LAKE MICHIGAN, *LAKE ONTARIO, LAKE HURON, AESTHETICS, ECOLOGY, ECONOMICS, OHIO, COMMERCIAL FISHING, PUBLIC HEALTH, WALLEYE, VEGETATION, WILD RICE, SILTS, PLANKTON, MAYFLIES, FAUNA, FISH, OLIGOCHAETES, MIDGES, SNAILS, PHOSPHORUS, NUTRIENTS, ALGAE, DIATOMS, CHLOROPHYTA, DISSOLVED OXYGEN, PIKE, CARP, DRUM(FRESHWATER), CISCO, LAKE TROUT, EUTROPHICATION, STRIPED BASS, CYANOPHYTA, WATER POLLUTION EFFECTS, ELECTRIC POWERPLANTS.

IDENTIFIERS:
HEXAGENIA, CHIRONOMIDAE, PROCLADIUS, CHIRONOMUS PROMOSUS, TRICHOPTERA, LEECHES, FINGERNAIL CLAMS, WHITE FISH, ALEWIFE, GIZZARD SHAD, SEA LAMPHREY, COHO SALMON.

ABSTRACT:
DRAMATIC BIOLOGICAL CHANGES HAVE APPEARED IN BOTTOM FAUNA AND AMONG CERTAIN FISHES OF THE LOWER GREAT LAKES. OF SPECIAL SIGNIFICANCE IS ABUNDANT INCREASE, SINCE 1959, OF THE MIDGE PROCLADIUS, A SUPPOSEDLY MORE POLLUTION-TOLERANT FORM, WHILE CHIRONOMUS PROMOSUS HAS DECREASED, SUGGESTING THAT POLLUTION ZONES HAVE EXTENDED FURTHER INTO THE LAKES. THE MAYFLY IS NOW RARE. BENTHIC FAUNA IS NOW DOMINATED BY OLIGOCHAETES AND MIDGES, WITH SOME FINGERNAIL CLAMS, SNAILS, AND LEECHES ON THE INCREASE. CHEMICAL CONDITIONS PROBABLY PROVIDE A MORE RELIABLE INDEX TO CHANGES THAN PLANKTON DATA, BUT ARE DIFFICULT TO RELATE. SPECIES COMPOSITION, ONCE DOMINATED BY DIATOMS, ARE NOW DOMINATED BY BLUE-GREEN ALGAE. DECLINE OF CERTAIN HIGH QUALITY FISHES, NOTABLY THE BLUE PIKE AND WALLEYE PIKE, IS LARGELY RESPONSIBLE FOR THE ACCELERATED PUBLIC INTEREST IN CORRECTIVE MEASURES. RELATIVE SIGNIFICANCE OF ENVIRONMENT VERSUS OVERFISHING IS DEBATABLE. INCREASE IN EUTROPHICATION RATE OF THE GREAT LAKES, ESPECIALLY LAKE ERIE, IS SIGNIFICANT. HUMAN TECHNOLOGY CAN SO MODIFY THE ENVIRONMENT THAT BIOLOGICAL POPULATIONS ARE SIGNIFICANTLY AFFECTED. BIOLOGISTS WITH REQUISITE KNOWLEDGE, SHOULD DEVELOP BETTER GUIDELINES FOR WEIGHING COSTS OF EACH INCREMENT OF DEGRADATION AND EACH INCREMENT OF IMPROVEMENT. (JONES-WISCONSIN)

FIELD 05C, 02H

ACCESSION NO. W70-01943

SULFIDE PRODUCTION IN WASTE STABILIZATION PONDS,

TEXAS UNIV., AUSTIN. ENVIRONMENTAL HEALTH ENGINEERING RESEARCH LAB.

EARNEST F. GLOYNA, AND ERNESTO ESPINO.

J SANIT ENG DIV, PROC AMER SOC CIVIL ENG, VOL 95, NO SA3, PROC PAPER 6637, P 607-628, JUNE 1969. 19 FIG, 4 TAB, 17 REF.

DESCRIPTORS:
*SULFIDES, *OXIDATION LAGOONS, *DESIGN, *SULFUR BACTERIA, STABILIZATION, WASTE WATER, ALGAE, DIFFUSION, SULFUR.

IDENTIFIERS:
*SULFATE ION, *DESULFOVIBRIO, FACULTATIVE, BOD, DETENTION TIME.

ABSTRACT:
THE PRODUCTION AND RELEASE OF HYDROGEN SULFIDE FROM WASTE STABILIZATION PONDS AND REQUIRED DESIGN MODIFICATIONS, RESULTING FROM INCREASED SULFATE OR HYDROGEN SULFIDE CONTENT ARE DISCUSSED. A 6-FT DEEP FACULTATIVE MODEL POND WAS OPERATED UNDER SEVEN DIFFERENT OPERATING CONDITIONS. THE OBJECTIVE OF THE EXPERIMENTS WAS TO STUDY THE RELATIONSHIPS BETWEEN SULFIDE CONCENTRATION. (LEDBETTER-TEXAS)

FIELD 05D

ACCESSION NO. W70-01971

REMOVAL OF NITROGEN AND PHOSPHORUS FROM WASTE WATER,

STANFORD UNIV., CALIF.

ROLF ELIASSEN, AND GEORGE TCHOBANOGLOUS.

ENVIRON SCI TECHNOL, VOL 3, NO 6, P 536-541, JUNE 1969. 2 TAB, 15 REF.

DESCRIPTORS:
*NUTRIENTS, *NITROGEN, *PHOSPHORUS, EFFICIENCIES, COST COMPARISONS, TREATMENT, BIOLOGICAL TREATMENT, ALGAL CONTROL, AMMONIA, ION EXCHANGE, REVERSE OSMOSIS, DISTILLATION, ELECTRODIALYSIS, SORPTION, CHEMICAL PRECIPITATION, ULTIMATE DISPOSAL.

IDENTIFIERS:
*NUTRIENT REMOVAL, *EUTROPHICATION, ANAEROBIC DENITRIFICATION, LAND APPLICATION, ELECTROCHEMICAL TREATMENT, AMMONIA STRIPPING, ALGAE HARVESTING.

ABSTRACT:
THE PRINCIPAL SOURCES OF NITROGEN AND PHOSPHORUS IN WASTEWATERS ARE TABULATED WITH THE AMOUNTS FROM EACH SOURCE; DOMESTIC AND AGRICULTURAL OPERATIONS ARE ON A PAR WITH INDUSTRIAL OUTPUT FOR NITROGEN DISCHARGES, BUT PHOSPHORUS RELEASES BY INDUSTRY ARE NOT AVAILABLE. NITROGEN REMOVAL MAY BE DONE BY THE FOLLOWING PROCESSES WITH REMOVAL EFFICIENCIES AND COST IN DOLLARS PER MILLION GALLONS IN PARENTHESES: AMMONIA STRIPPING (80-98, 9-25), ANAEROBIC DENITRIFICATION (60-95, 25-30), ALGAE HARVESTING (50-90, 20-35) AND CONVENTIONAL BIOLOGICAL TREATMENT (30-50, 30-100). A SIMILAR LIST FOR PHOSPHORUS REMOVAL FOLLOWS: MODIFIED ACTIVATED SLUDGE (60-80, 30-100), CHEMICAL PRECIPITATION (88-95, 10-70), CHEMICAL PRECIPITATION WITH FILTRATION (95-98, 70-90), AND SORPTION (90-98, 40-70). IN ADDITION, THE CONVENTIONAL BIOLOGICAL TREATMENT REMOVES 10-30 PERCENT OF THE PHOSPHORUS. A LIST OF THE TREATMENT SCHEMES FOR REMOVAL OF BOTH NITROGEN AND PHOSPHORUS IS THE FOLLOWING: ION EXCHANGE (80-92, 86-98, 170-300), ELECTROCHEMICAL TREATMENT (80-85, 4-8 FOR POWER), ELECTRODIALYSIS (30-50, 100-250), REVERSE OSMOSIS (65-95, 250-400), DISTILLATION (90-98, 400-1000), AND LAND APPLICATION (60-90 PHOSPHORUS, 75-150). ULTIMATE DISPOSAL METHODS MAY BE CLASSIFIED INTO 4 GROUPS: DUMPING; SUBSURFACE INJECTION; CONVERSION AND DUMPING; AND CONVERSION, PRODUCT RECOVERY, AND DUMPING. (LEDBETTER-TEXAS)

FIELD 05D

ACCESSION NO. W70-01981

EFFECTS OF INSECTICIDES ON ALGAE,

ONTARIO WATER RESOURCES COMMISSION, TORONTO.

A. E. CHRISTIE.

WATER AND SEWAGE WORKS, VOL 116, NO 5, P 172-176, MAY 1969. 9 TAB, 31 REF.

DESCRIPTORS:
*ALGAE, *TOXICITY, *INSECTICIDES, METABOLISM, RADIOACTIVITY TECHNIQUES, CHLORELLA, OXIDATION LAGOONS.

IDENTIFIERS:
*ALGAE TOXICITY, CARBON-14, DDT, MALATHION, SEVIN.

ABSTRACT:
A STUDY WAS CONDUCTED OF THE TOXICITY AND METABOLISM OF DDT, SEVIN, MALATHION TO CULTURES OF CHLORELLA PYRENOIDOSA AND ALGAE FROM A WASTE STABILIZATION POND. THE INSECTICIDES IN THE APPROPRIATE SOLVENTS WERE ADDED TO CULTURE FLASKS OF ALGAL MEDIUM AND THE NUMBER OF ALGAL CELLS COUNTED AFTER VARIOUS NUMBER OF DAYS INCUBATION. CARBON-14 TAGGED INSECTICIDES WERE USED TO MEASURE UPTAKE AND METABOLISM OF THE INSECTICIDES BY ALGAE. DDT EXHIBITED NO TOXIC PROPERTIES UP TO 100 MG/L AND WAS ONLY SLIGHTLY DEGRADED. SEVIN IS TOXIC AT CONCENTRATIONS ABOVE 0.1 MG/L AND IS NOT ALTERED APPRECIABLY IN ACIDIC MEDIA. MALATHION ALTERS THE MAKEUP OF MIXED CULTURE BUT IS NOT PERSISTENTLY INHIBITORY. (LEDBETTER-TEXAS)

FIELD 05D

ACCESSION NO. W70-02198

QUANTITATIVE CATION REQUIREMENTS OF SEVERAL GREEN AND BLUE-GREEN ALGAE,

WISCONSIN UNIV., MADISON. INST. OF PLANT DEVELOPMENT, AND WISCONSIN UNIV.,
    MADISON. DEPT. OF BOTANY.

G. C. GERLOFF, AND K. A. FISHBECK.

JOURNAL OF PHYCOLOGY, VOL 5, NO 2, P 109-114, 1969. 3 FIG, 3 TAB, 23 REF.

DESCRIPTORS:
    *ALGAE, *CHLOROPHYTA, *CYANOPHYTA, *NUTRIENT REQUIREMENTS, ANALYTICAL
    TECHNIQUES, WATER POLLUTION EFFECTS, PHYTOPLANKTON, PLANKTON, AQUATIC
    MICROBIOLOGY, CYCLING NUTRIENTS, ESSENTIAL NUTRIENTS, CALCIUM,
    MAGNESIUM, POTASSIUM, HARVESTING OF ALGAE, LIMNOLOGY, NUTRIENTS,
    PHYSIOLOGICAL ECOLOGY, CHLORELLA, SCENEDESMUS.

IDENTIFIERS:
    *CATIONS, DRAPARNALDIA, DRAPARNALDIA PLUMOSA, STIGEOCLONIUM,
    MICROCYSTIS, STIGEOCLONIUM TENUE, MICROCYSTIS AERUGINOSA, NOSTOC,
    NOSTOC MUSCORUM, ABSORPTION, SPECTROMETRY.

ABSTRACT:
    AUTHORS EVALUATED REQUIREMENTS OF SIX SPECIES OF GREEN AND BLUE-GREEN
    ALGAE (CHLORELLA PYRENOIDOSA, SCENEDESMUS QUADRICAUDA, DRAPARNALDIA
    PLUMOSA, STIGEOCLONIUM TENUE, MICROCYSTIS AERUGINOSA, NOSTOC MUSCORUM)
    FOR CALCIUM, MAGNESIUM, AND POTASSIUM IN TERMS OF CRITICAL CELL
    CONCENTRATIONS (CCC) OF THE THREE CATIONS. 'CRITICAL CELL
    CONCENTRATION' IS DEFINED AS THE MINIMAL CELL CONTENT OF AN ELEMENT
    WHICH PERMITS MAXIMAL OR NEAR-MAXIMAL GROWTH OF AN ORGANISM. AS
    COMPARED WITH REQUIREMENTS OF ANGIOSPERM CROP PLANTS, CALCIUM NEEDS OF
    ALL SIX SPECIES WERE EXTREMELY LOW (CCC 0.06% OR LESS ON OVEN-DRY
    BASIS); REQUIREMENTS FOR MAGNESIUM WERE EQUAL TO OR SLIGHTLY LESS THAN
    IN AGIOSPERMS (0.15-0.30%) WITH EXCEPTION OF SCENEDESMUS (0.05%); AND
    REQUIREMENTS FOR POTASSIUM VARIED GREATLY, FROM CCC LESS THAN AVERAGE
    VALUES ESTABLISHED FOR ANGIOSPERMS (0.25-0.50%) TO VALUES EQUAL TO OR
    GREATER THAN MEAN VALUES FOR HIGHER PLANTS (0.80-2.40%). THE RESULTS
    SUGGEST DIFFERENCES IN PHYSIOLOGY AND FUNCTION OF CATIONS IN THE ALGAE
    STUDIED AND IN ANGIOSPERMS WHICH SEEM WORTHY OF FURTHER INVESTIGATION.
    (FITZGERALD-WIS)

FIELD 05C

ACCESSION NO. W70-02245

TRISODIUM NITRILOTRIACETATE AND ALGAE,

ONTARIO WATER RESOURCES COMMISSION, TORONTO. DIV. OF RESEARCH.

A. E. CHRISTIE.

ONTARIO WATER RESOURCES COMMISSION, DIV OF RESEARCH PAPER NO 2023, JUNE 1969.
    11 P, 4 TAB, 6 REF.

DESCRIPTORS:
    *ALGAE, ALGAL CONTROL, ALGICIDES, AQUATIC ALGAE, AQUATIC MICROBIOLOGY,
    BIOASSAY, CHLOROPHYTA, CHLORELLA, CYCLING NUTRIENTS, ESSENTIAL
    NUTRIENTS, INHIBITION, NITROGEN COMPOUNDS, NUTRIENT REQUIREMENTS,
    NUTRIENTS, POLLUTION IDENTIFICATION, SEWAGE, SEWAGE EFFLUENTS, SEWAGE
    TREATMENT, WATER POLLUTION CONTROL, WATER POLLUTION EFFECTS, WATER
    POLLUTION TREATMENT, DETERGENTS, TOXICITY, ACTIVATED SLUDGE,
    EUTROPHICATION, BIODEGRADATION, AMINO ACIDS.

IDENTIFIERS:
    *TRISODIUM NITRILOTRIACETATE, NTA, CHLORELLA PYRENOIDOSA, CARBOXYLIC
    ACIDS, AMINO CARBOXYLIC ACIDS.

ABSTRACT:
    AUTHOR REPORTS RESULTS OF PRELIMINARY TESTS TO DETERMINE THE EFFECT OF
    THE POTENTIAL DETERGENT BUILDER, TRISODIUM NITRILOTRIACETATE, ON GROWTH
    OF THE GREEN ALGA, CHLORELLA PYRENOIDOSA (IND 395). INCREASES IN
    NITROGEN (N) FROM 0.14 TO 14 MILLIGRAMS N/LITER RESULTED IN THREEFOLD
    INCREASES IN CELL COUNTS FOR N EITHER FROM TRISODIUM NITRILOTRIACETATE
    OR FROM SODIUM NITRATE. TRISODIUM NITRILOTRIACETATE IN CONCENTRATION OF
    275 MILLIGAMS/LITER WAS NOT TOXIC TO CULTURES OF CHLORELLA AT AN
    INITIAL CONCENTRATION OF 1 MILLION CELLS/MILLILITER. THE FILTRATE OF AN
    ACTIVATED SLUDGE, CAPABLE OF DEGRADING TRISODIUM NITRILOTRIACETATE,
    WOULD SUPPORT THE GROWTH OF CHLORELLA. (FITZGERALD-WIS)

FIELD 05C

ACCESSION NO. W70-02248

DEMONSTRATION OF THE ANTAGONISTIC ACTION OF LARGE AQUATIC PLANTS ON ALGAE AND
  ROTIFERS,

WISCONSIN UNIV., MADISON. DEPT. OF ZOOLOGY.

ARTHUR D. HASLER, AND ELIZABETH JONES.

ECOLOGY, VOL 30, NO 3, P 359-364, JULY 1949. 1 FIG, 4 TAB, 6 REF.

DESCRIPTORS:
  *AQUATIC PLANTS, *ALGAE, *ROTIFERS, PLANKTON, BASS, PHYTOPLANKTON,
  PONDS, ZOOPLANKTON, MINNOWS, PEDALFERS, SCENEDESMUS, CHLAMYDOMONAS,
  TEMPERATURE, CARBON DIOXIDE, HYDROGEN ION CONCENTRATION, BIOCARBONATES,
  DISSOLVED OXYGEN, DETRITUS, BACTERIA, WATER POLLUTION EFFECTS,
  LIMNOLOGY, CYANOPHYTA.

IDENTIFIERS:
  *ANTAGONISTIC ACTION, *AQUATIC MACROPHYTES, ANACHARIS CANADENSIS,
  MACROPHYTES, POTAMOGETON FOLIOSUS, EXPERIMENTAL LIMNOLOGY, BRACHIONUS,
  SYNCHAETA PECTINATA, POLYARTHRA VULGARIS, KARATELLA COHCLEARIS,
  ASPLANCHA, LECANE, BOSMINA, NAVICULA, FRAGILARIA, CYCLOTELLA,
  FLAGELLATES, ANKISTRODESMUS, GOMPHENEMA, SPIROGYRA.

ABSTRACT:
  PLANTS AND ANIMALS VARY GREATLY IN DENSITY AND SPECIES COMPOSITION AND
  ARE CONTROLLED BY EXTREMELY COMPLEX FACTORS. MACROPHYTES MAY HAVE
  CERTAIN COMPETITIVE ADVANTAGES OVER ALGAE WHEN THEIR POPULATIONS ATTAIN
  SPECIFIC DENSITIES. IN EXPERIMENTS REPORTED HERE, AUTHORS AIMED AT
  EVIDENCE CONFIRMING OR CONTRADICTING PREVIOUS CASUAL OBSERVATIONS
  BEARING ON SUCH ANTAGONISMS. 'SILO PONDS' ERECTED WITHIN A LARGE
  HATCHERY POND, WERE DENSELY PLANTED WITH ANACHARIS CANADENSIS
  ASSOCIATED SLIGHTLY WITH POTAMOGETON FOLIOSUS. DURING SUMMER,
  PHYTOPLANKTERS AND ZOOPLANKTERS WERE COUNTED WEEKLY. ZOOPLANKTON OF ALL
  PONDS WERE SUBJECT TO SIMILAR DEGREES OF PREDATION SINCE EACH CONTAINED
  50 FINGERLING LARGEMOUTH BASS. FATHEAD MINNOWS WERE MAINTAINED IN EACH
  POND AS FOOD FOR THE BASS. DENSE GROWTHS OF AQUATIC MACROPHYTES IN THE
  'SILO PONDS' HAD STATISTICALLY SIGNIFICANT INHIBITING EFFECT UPON
  PHYTOPLANKTON AND ROTIFERS, WHEREAS PLANKTONIC CRUSTACEA WERE NOT
  AFFECTED. GROWTH OF BASS VARIED SO GREATLY THAT IT PRECLUDED
  DETERMINATION OF EFFECT OF AQUATIC PLANTS ON FISH GROWTH. LISTS OF
  DOMINANT PLANKTONIC GENERA IN PLANTED AND CONTROL (PLANT-FREE) PONDS
  ARE GIVEN. DATA ON TEMPERATURE, FREE CARBON DIOXIDE, PH, BICARBONATE
  ION, AND DISSOLVED OXYGEN OF EXPERIMENTAL AND CONTROL PONDS ARE
  REPORTED. (JONES-WIS)

FIELD 05C

ACCESSION NO. W70-02249

THE OVERALL PICTURE OF EUTROPHICATION,

TEXAS UNIV., AUSTIN. ENVIRONMENTAL HEALTH ENGINEERING RESEARCH LAB.

E. GUS FRUH.

JOURNAL WATER POLLUTION CONTROL FEDERATION, VOL 39, NO 9, P 1449-1463, 1967.
  1 FIG, 7 TAB, 66 REF.

DESCRIPTORS:
    *EUTROPHICATION, LAKES, ORGANIC WASTES, OXYGEN, STANDING CROP, ALGAE,
    ZOOPLANKTON, FISH, BACTERIA, BENTHIC FAUNA, PHOTOSYNTHESIS, BIOASSAYS,
    AQUATIC WEEDS, THERMAL STRATIFICATION, EPILIMNION, HYPOLIMNION, CYCLING
    NUTRIENTS, TASTE, ODOR, COLOR, TURBIDITY, HYDROGEN SULFIDE, IRON,
    MAGNESIUM, COOLING, PROPERTY VALUES, RECREATION DEMAND, IRRIGATION,
    TRACE ELEMENTS, NITROGEN FIXATION, ATMOSPHERE, GROUNDWATER,
    WATERSHEDS(BASINS), RUNOFF, PHOSPHORUS, NITROGEN, AGRICULTURAL
    WATERSHEDS, FORESTS, WASTE WATER(POLLUTION), LAND MANAGEMENT, COPPER
    SULFATE, SODIUM ARSENITE, HARVESTING, WATER POLLUTION, VITAMINS.

IDENTIFIERS:
    SHORELINE VEGETATION, STORMWATER, DRAINAGE.

ABSTRACT:

NITROGEN AND PHOSPHORUS, IMPORTANT IN PLANT GROWTH, ARE SUBJECT TO
    CONTROL AND MODIFICATION. THE IMMEDIATE QUESTION WHICH CONFRONTS WATER
    QUALITY INVESTIGATORS IS WHETHER NITROGEN AND PHOSPHORUS ARE REALLY THE
    LIMITING FACTORS IN PARTICULAR AQUATIC ENVIRONMENTS WHICH THEY ARE
    INVESTIGATING. SUCH DETERMINATIONS CAN BE ACCOMPLISHED BY VARIOUS
    BIOASSAY TECHNIQUES. IN LABORATORY CULTURES OF ALGAE, TRACE ELEMENTS,
    PREVIOUSLY OVERLOOKED, ARE ESSENTIAL. LIKEWISE, DIFFERENT SOILS HAVE
    BEEN ADDED TO CULTURE MEDIA TO SUPPLY VARIOUS NECESSARY ORGANICS AND
    GROWTH FACTORS. TODAY, KNOWN VITAMINS AND ORGANIC COMPOUNDS ARE USED.
    WATER QUALITY INVESTIGATORS SHOULD DETERMINE WHETHER SUCH
    MICRONUTRIENTS ARE IN SUFFICIENT SUPPLY IN THE HYDROSPHERE UNDER
    VARIOUS ECOLOGICAL CONDITIONS. KNOWLEDGE OF SUCH FACTS PERMITS A
    LOGICAL APPROACH FOR IDENTIFICATION OF PRIMARY NUTRIENT SOURCES. SOME
    NUTRIENT SOURCES, INCLUDING RAINFALL, GROUNDWATER, AND NORMAL WATERSHED
    DRAINAGE, CANNOT BE CONTROLLED. IF WASTEWATER EFFLUENTS CAUSE
    EUTROPHICATION, DIVERSION OR TERTIARY TREATMENT CAN BE APPLIED. IF
    AGRICULTURAL AND/OR URBAN DRAINAGE ARE PRIMARY NUTRIENT SOURCES, LITTLE
    CAN, FOR ALL PRACTICAL PURPOSES, BE ACCOMPLISHED. POSSIBLE ECONOMIC
    APPROACHES UNDER INVESTIGATION IN LAKES ARE SELECTIVE WITHDRAWAL OF
    HYPOLIMNETIC WATERS, DILUTION, AERATION OF THE HYPOLIMNION, AND MIXING
    OF HYPOLIMNETIC WATERS WITH THE EPILIMNION. (JONES-WIS)

FIELD 05C

ACCESSION NO. W70-02251

ALGAL GROWTH AND SEWAGE EFFLUENT IN THE POTOMAC ESTUARY,

JOHNS HOPKINS UNIV., BALTIMORE, MD. DEPT. OF SANITARY ENGINEERING AND WATER RESOURCES.

JOSEPH SHAPIRO, AND ROBERTO RIBEIRO.

JOURNAL WATER POLLUTION CONTROL FEDERATION, VOL 37, NO 7, P 1034-1042, JULY 1965. 6 FIG, 2 TAB, 7 REF.

DESCRIPTORS:
*ALGAE, *SEWAGE EFFLUENTS, *ESTUARIES, FRESH WATER, LAKES, SILTS, PHOSPHATES, NITROGEN, WASTE WATER(POLLUTION), SCENEDESMUS, CYANOPHYTA, CHLOROPHYTA, CHLOROPHYLL, BACTERIA, TRACE ELEMENTS, VITAMINS, AMINO ACIDS, CALCIUM, AMMONIA, NITROGEN FIXATION, IRON COMPOUNDS, LIGHT PENETRATION, WATER POLLUTION EFFECTS, EUTROPHICATION.

IDENTIFIERS:
*ALGAL GROWTH, *POTOMAC RIVER, PLECTONEMA, NOSTOC PALUDOSUM, BLEACHING, WASHINGTON(DC).

ABSTRACT:
ADDITION OF EFFLUENTS FROM SECONDARY WASTEWATER TREATMENT PLANTS TO POTOMAC RIVER GREATLY INCREASES GROWTH OF BOTH GREEN (CHLOROPHYTES) AND BLUE-GREEN ALGAE (CYANOPHYTES) IN PROPORTION TO THE QUANTITY OF EFFLUENT ADDED. AS LITTLE AS 5% EFFLUENT IS EFFECTIVE, AND 40% IS NOT SUPRAOPTIMAL. SINCE CYANOPHYTES ARE ABLE TO PROVIDE THEIR OWN NITROGEN SUPPLY THROUGH FIXATION OF MOLECULAR NITROGEN, PHOSPHATE IS THE SOLE NUTRIENT RESPONSIBLE FOR LIMITING THEIR GROWTH. CHLOROPHYTES REQUIRE BOTH PHOSPHATE AND AMMONIUM-NITROGEN FOR GROWTH. REMOVAL OF AMMONIA FROM EFFLUENTS WILL CONTROL CHLOROPHYTES ONLY, BUT PHOSPHATE REMOVAL WILL LIMIT CYANOPHYTES AS WELL. PARTIAL REMOVAL OF PHOSPHORUS FROM EFFLUENTS WILL CONTROL BOTH ALGAL GROUPS TO A DEGREE COMMENSURATE WITH EXTENT OF REMOVAL. NORMAL PHOSPHORUS CONCENTRATIONS OF THE POTOMAC RIVER ARE SUCH THAT INCREASES IN THESE CONCENTRATIONS WILL STIMULATE FURTHER ALGAL GROWTH. THE HYPOTHESIS THAT TRACE SUBSTANCES ARE NOT LIMITING WAS VERIFIED. SILT, AFFECTING LIGHT PENETRATION AND ADSORPTION OF PHOSPHATES OR AMMONIA ONTO SILT PARTICLES MAY BOTH LEAD TO REDUCED ALGAL GROWTHS. SOLUTION OF THE SILT PROBLEM, THROUGH BUILDING OF DAMS, MUST BE TIED TO PROGRAMS FOR NUTRIENT CONTROL. (JONES-WIS)

FIELD 02L, 05C

ACCESSION NO. W70-02255

COOLING WATER CHEMICAL TREATMENTS,

BETZ LABS., INC., TREVOSE, PA. COOLING WATER SERVICES.

JOHN M. DONOHUE.

IND WATER ENG (IWE) 1ST ANNU AIR/WATER ENG BUYERS GUIDE, P 70, 1969.

DESCRIPTORS:
*WATER TREATMENT, *COOLING WATER, *CORROSION, *CHEMICAL REACTIONS, *WATER REUSE, DISSOLVED OXYGEN, PITTING(CORROSION), RUSTING, METALS, INHIBITORS, SILTS, ORGANIC MATTER, IRON OXIDES, BACTERIA, ALGAE, FUNGI, SULFUR COMPOUNDS, CARBONATES.

IDENTIFIERS:
*METAL CORROSION, COOLING WATER TREATMENT, ON-STREAM DESLUDGING.

ABSTRACT:
CORROSION OF METALS IS PRIMARILY CAUSED BY OXYGEN THAT IS DISSOLVED IN WATER AND THEN RELEASED AS THE COOLING WATER INCREASES IN TEMPERATURE. THE MOST COMMON TYPES OF METAL ATTACK ARE PITTING OF THE METAL SURFACE, GENERAL WASTING AWAY CAUSED BY LOW WATER PH, AND GOUGING OF METAL SURFACES. CHROMATE-BASED TREATMENTS ARE THE MOST EFFECTIVE AND WIDELY USED OF THE INHIBITORS FOR CONTROLLING WATERSIDE ATTACK. COOLING WATER ANTIFOULANTS ARE USED TO PREVENT DEPOSITION OF SILT, IRON OXIDE, ORGANIC MATTER, AND GENERAL COOLING-WATER DEBRIS. THE TECHNIQUES OF ON-STREAM DESLUDGING OF COOLING SYSTEMS WITH SYNTHETIC POLYMERS HAVE BEEN SUCCESSFULLY USED IN COOLING OPERATIONS. BACTERIA, FUNGI, AND ALGAE CAN INTERFERE WITH WATER FLOW, DECREASE HEAT TRANSFER, AND ACCELERATE CORROSION. BIOCIDES DEVELOPED AND USED IN COOLING SYSTEMS INCLUDE ORGANIC SULFUR COMPOUNDS, ORGANO-METALLICS, CARBONATES, DIAMINES, COPPER, CHLORINE, HYPOCHLORITES, ORGANIC CHLORINE DONORS AND BROMINE COMPOUNDS. (CARSTEA-USGS)

FIELD 05D, 08G

ACCESSION NO. W70-02294

FURTHER PLANKTONIC ALGAE OF AUCKLAND SEWAGE TREATMENT PONDS AND OTHER WATERS,

BOLIVAR LABS., SALISBURY (AUSTRALIA).

A. HAUGHEY.

NEW ZEAL J MARINE AND FRESHWATER RES, VOL 3, NO 2, P 245-261, JUNE 1969. 17 P, 6 FIG, 13 REF.

DESCRIPTORS:
*ALGAE, *PLANKTON, *SEWAGE, *SEWAGE TREATMENT, *EUTROPHICATION, PONDS, POLLUTANT IDENTIFICATION, BIOLOGY, WATER POLLUTION, WATER POLLUTION SOURCES, ALGAL CONTROL, BIOINDICATORS.

IDENTIFIERS:
*NEW ZEALAND, SEWAGE TREATMENT POND ALGAE.

ABSTRACT:
SEVENTEEN SPECIES OF ALGAE, NEWLY RECORDED BETWEEN 1965 AND 1968 FROM SEWAGE TREATMENT PONDS AROUND AUCKLAND CITY, ARE DESCRIBED TOGETHER WITH ANOTHER 11 SPECIES FOUND IN EUTROPHIC WATERS OF THE AREA. THESE LATTER NATURAL WATERS ARE OFTEN OF A TRANSIENT NATURE AND USUALLY HEAVILY POLLUTED IN SOME WAY. (GABRIEL-USGS)

FIELD 05C

ACCESSION NO. W70-02304

EFFECT OF HALOGENS ON ALGAE - I. CHLORELLA SOROKINIANA,

TECHNION - ISRAEL INST. OF TECH., HAIFA. SANITARY ENGINEERING LAB.

YEHUDA KOTT, AND J. EDLIS.

SPONSORED BY MEKOROTH WATER CO. WATER RESEARCH, VOL 3, P 251-256, 1969. 4 FIG, 1 TAB, 19 REF.

DESCRIPTORS:
*HALOGENS, *ALGICIDES, CHLORINE, ALGAL CONTROL, CHLORELLA, DISINFECTION, MORTALITY.

IDENTIFIERS:
*KILL RATES, *BROMINE, IODINE, PHOTO REACTIVITY, ALGAL METABOLISM.

ABSTRACT:
BROMINE WITHIN THE CONCENTRATION RANGE OF 0.2 TO 1.0 MG/L CAUSED A DECREASE IN CHLORELLA NUMBER AFTER 2 - 5 HR CONTACT, AFTER WHICH TIME REGROWTH OCCURRED. A SECOND DOSE OF BROMINE AT THE TIME OF REGROWTH BROUGHT ABOUT TOTAL ALGAL KILL. USING CHLORINE AT THE SAME CONCENTRATION HAD AN ALGISTATIC EFFECT, WHILE A SECOND INTRODUCTION OF CHLORINE HAD NO ADDITIONAL EFFECT AT ALL. USING A MIXTURE OF CHLORINE AND BROMINE SOLUTIONS CAUSED RAPID ALGAL DEATH. RESULTS OF HALOGENATION IN THE DARK DEMONSTRATED THE SUPERIORITY OF CHLORINE UNDER THESE CONDITIONS. IT IS ASSUMED THAT PHOTOREACTIVITY OF BROMINE MIGHT BE THE REASON FOR ITS BETTER ALGICIDAL EFFECTS IN THE LIGHT. (SJOLSETH-WASHINGTON)

FIELD 05C

ACCESSION NO. W70-02363

EFFECT OF HALOGENS ON ALGAE - II. CLADOPHORA SP.,

TECHNION - ISRAEL INST. OF TECH., HAIFA. SANITARY ENGINEERING LAB.

NACHUM BETZER, AND YEHUDA KOTT.

SPONSORED BY MEKOROTH WATER CO. WATER RESEARCH, VOL 3, P 257-264, 1969. 7 FIG, 5 TAB, 8 REF.

DESCRIPTORS:
*HALOGENS, ALGICIDES, CYTOLOGICAL STUDIES, CHLORINE, COPPER SULFATE, ALGAL CONTROL, MORTALITY, CHLOROPHYLL, ALGAE, BIOASSAY, VIABILITY.

IDENTIFIERS:
*CLADOPHORA, BROMINE, FILAMENTOUS ALGAE, KILL RATES, PHOTO REACTIVITY, CELL COLOR, CELL DAMAGE.

ABSTRACT:
HALOGENATION OF NATURAL TUFTS GAVE NO FRUITFUL RESULTS. USING ISOLATED FILAMENTS OF CLADOPHORA SP. SHOWED THAT HALOGENS CAUSE CHANGES IN CELL COLOR FROM GREEN TO YELLOW OR EVEN TO COLORLESS, VERY OFTEN DAMAGE TO CELL WALL. LEACHING OF CELL CONTENT ALSO OCCURRED, OR CONTRACTION OF CYTOPLASM WAS OBSERVED. USING DARK FIELD TECHNIQUE IT WAS OBSERVED THAT THE NORMAL RED FLUORESCENCE SEEN IN HEALTHY CELLS DISAPPEARED. VERY HIGH CONCENTRATIONS (50 MG/L AND OVER) CAUSED SEVERE DEFORMATION OF CELLS. IN THE ABSENCE OF A RECOMMENDED PROCEDURE FOR MEASURING VIABILITY OF FILAMENTOUS ALGAE, ONE HUNDRED CELLS WERE COUNTED IN EACH FILAMENT AND A BIOASSAY WAS PERFORMED AT THE END OF EACH EXPERIMENT. ADDITION OF 5 MG/L OF HALOGENS HAS CAUSED SOME EFFECT ON THE ALGAE, WHILE ADDITION OF COPPER SULPHATE AT THIS CONCENTRATION SHOWED NO EFFECT AT ALL. WITH HIGHER CONCENTRATIONS OF HALOGENS OR COPPER SULPHATE, CELL DAMAGE WAS CAUSED. TEN MG/L CHLORINE FOR 2 HR, 10 MG/L BROMINE FOR 10 HR OR 10 MG/L COPPER SULPHATE FOR 4 DAYS OF CONTACT CAUSED COMPLETE ALGAL KILL. (SJOLSETH-WASHINGTON)

FIELD 05C

ACCESSION NO. W70-02364

ALGICIDAL EFFECT OF BROMINE AND CHLORINE ON CHLORELLA PYRENOIDOSA,

TECHNION - ISRAEL INST. OF TECH., HAIFA. SANITARY ENGINEERING LAB.

YEHUDA KOTT, GALILA HERSHKOVITZ, A. SHEMTOB, AND J. B. SLESS.

APPLIED MICROBIOLOGY, 14 (1), P 8-11, 1966. 4 TAB, 3 FIG, 16 REF.

DESCRIPTORS:
*BIOASSAY, *HALOGENS, *GROWTH RATES, *CHLORELLA, CHLORINE, ALGICIDES, ALGAL CONTROL, DISINFECTION, MODE OF ACTION, MORTALITY, INHIBITION, ALGAE.

IDENTIFIERS:
BROMINE, BROMAMINE, CHLORAMINE.

ABSTRACT:
CHLORELLA PYRENOIDOSA WAS FOUND TO GROW RAPIDLY IN TAP WATER. PEAK GROWTH WAS REACHED AFTER 2 TO 3 DAYS. CHLORINE AND BROMINE, ADDED TO SUCH WATER IN CONCENTRATIONS OF .18 TO .42 PPM, WERE SHOWN TO BE EFFECTIVE INHIBITORS OF ALGAL GROWTH. BROMINE AND BROMAMINE WERE PRIMARILY ALGICIDAL, WHEREAS CHLORINE AND CHLORAMINES WERE MAINLY ALGISTATIC. IT IS ASSUMED THAT THE MECHANISMS OF ACTION OF THESE HALOGENS ON CHLORELLA ARE NOT THE SAME. THE NEED TO INVESTIGATE THE MODE OF ACTION OF THESE HALOGENS IS STRESSED. (SJOLSETH-WASHINGTON)

FIELD 05C

ACCESSION NO. W70-02370

CHLORINATION AS A MEANS OF CONTROLLING EARTHY ODORS IN RESERVOIRS,

TECHNION - ISRAEL INST. OF TECH., HAIFA. SANITARY ENGINEERING LAB.

M. REBHUN, M. A. FOX, AND J. B. SLESS.

THE JERUSALEM CONFERENCE FOR WATER QUALITY RESEARCH, P 1-8, 1969. 5 FIG, 10
   REF.

   DESCRIPTORS:
      *ODOR-PRODUCING ALGAE, *TASTE-PRODUCING ALGAE, ALGAE, CHLORINATION,
      WATER QUALITY CONTROL.

   IDENTIFIERS:
      *OSCILLATORIA, NATIONAL WATER CARRIER(ISRAEL), THRESHOLD ODOR NUMBERS.

   ABSTRACT:
      SEASONAL APPEARANCE OF BLUE-GREEN ALGAE (OSCILLATORIA SPP) PRODUCES
      EARTHY ODORS AND TASTES IN THE WATER OF THE TSALMON AND BETH NATUFA
      RESERVOIRS OF THE NATIONAL WATER CARRIER. FIELD TRIALS, IN WHICH WATER
      FROM THESE RESERVOIRS WAS CHLORINATED, SHOWED THAT CHLORINATION WAS NOT
      COMPLETELY EFFECTIVE AS A MEANS OF CONTROLLING THE EARTHY ODOR,
      ALTHOUGH A REDUCTION OF THE ODOR LEVEL WAS ALWAYS OBSERVED, WHEN THE
      EARTHY ODOR WAS THE SOLE ODOR PRESENT IN THE WATER. A MODEL SYSTEM
      EMPLOYING A SPECIES OF OSCILLATORIA, WAS USED TO PROVIDE SAMPLES OF
      WATER HAVING DIFFERENT LEVELS OF THRESHOLD ODOR, AND THE EFFECT OF
      CHLORINATION INVESTIGATED. IT WAS FOUND THAT CHLORINATION REDUCED THE
      INITIAL ODOR TO AN ASYMPTOTIC VALUE WHICH PROVED TO BE DIRECTLY
      PROPORTIONAL TO THE INITIAL ODOR LEVEL. THEY CONCLUDED THAT COMPLETE
      CONTROL OF THE EARTHY ODOR PRODUCED BY OSCILLATORIA IS NOT FEASIBLE BY
      USE OF CHLORINE ALONE, BUT THAT ODOR LEVELS MAY BE REDUCED BY
      CHLORINATION TO AN EXTENT WHERE OTHER METHODS (SUCH AS DILUTION) ARE
      MORE READILY APPLICABLE. (SJOLSETH-WASHINGTON)

   FIELD 05G, 05C

   ACCESSION NO. W70-02373

EXTRACELLULAR PRODUCTS OF PHYTOPLANKTON PHOTOSYNTHESIS,

WESTFIELD COLL., LONDON (ENGLAND). DEPT OF BOTANY; AND UNIVERSITY COLL.,
   LONDON. DEPT. OF BOTANY.

G. E. FOGG, CZESLAWA NALEWAJKO, AND W. D. WATT.

PROCEEDINGS ROYAL SOCIETY (B), VOL 162, P 517-534, 1965. 6 FIG, 4 TAB, 31 REF.

   DESCRIPTORS:
      *PHYTOPLANKTON, *PHOTOSYNTHESIS, BICARBONATES, RADIOACTIVITY, CARBON
      RADIOISOTOPES, LAKES, ORGANIC MATTER, SEA WATER, PRIMARY PRODUCTIVITY,
      LIGHT INTENSITY, TROPHIC LEVEL, SAMPLING TEMPERATURE, TRACERS,
      CHLORELLA, HYDROGEN ION CONCENTRATION, ALKALINITY, DIATOMS,
      TEMPERATURE, WEATHER, EPILIMNION, ALGAE, DENSITY, SEASONAL, MARINE
      PLANTS, WATER POLLUTION SOURCES.

   IDENTIFIERS:
      *EXTRACELLULAR PRODUCTS, PHOTIC ZONE, GLYCOLLIC ACID, CHEMOTROPHY,
      PARTICULATE MATTER, BLELHAM TARN(ENGLAND), CHLORELLA PYRENOIDOSA,
      FIXATION, TRINGFORD(ENGLAND), ANABAENA CYLINDRICA, NORTH SEA, FIRTH OF
      CLYDE(SCOTLAND), SCOTLAND, ENGLAND, LAKE WINDERMERE(ENGLAND), CARBON-14.

   ABSTRACT:
      FOLLOWING EXPOSURE IN SITU (3-24 HOURS) OF LAKE OR SEA WATER SAMPLES,
      WITH ADDED RADIOACTIVE BICARBONATE, RADIOCARBON WAS FOUND IN
      PHYTOPLANKTERS AND IN DISSOLVED ORGANIC MATTER. AMOUNTS IN THE WATER
      VARIED BETWEEN 7 AND 50% OF TOTAL CARBON FIXED IN PHOTIC ZONE OF THE
      WATER COLUMN. THIS PRODUCTION OF LABELLED EXTRACELLULAR ORGANIC MATTER
      OCCURRED UNDER VARIOUS CONDITIONS AND WITH DIVERSE PHYTOPLANKTERS.
      ESTIMATION OF PRIMARY PRODUCTIVITY FROM FIXATION OF CARBON-14 IN
      PARTICULATES ONLY ARE PROBABLY ERRONEOUSLY LOW. LABELLED ORGANIC
      SUBSTANCES IN WATER ARE APPARENTLY LIBERATED BY INTACT
      PHOTOSYNTHESIZING CELLS RATHER THAN FROM CELLS BROKEN DURING
      FILTRATION. GLYCOLLIC ACID IS POSSIBLY A PRINCIPAL SUBSTANCE. OVER A
      WIDE RANGE OF LIGHT INTENSITIES, LIBERATION OF EXTRACELLULAR PRODUCTS
      BY A GIVEN PHYTOPLANKTON POPULATION WAS PROPORTIONAL TO THE CARBON
      FIXED IN CELLS, TENDING TO BE RELATIVELY GREATER AT LOW LIGHT
      INTENSITIES AND AT LIGHT INTENSITIES HIGH ENOUGH TO INHIBIT
      PHOTOSYNTHESIS, WHEN AS MUCH AS 95% OF TOTAL ORGANIC RADIOCARBON MIGHT
      BE EXTRACELLULAR. POPULATION DENSITY, PERIOD OF EXPOSURE TO LABELLED
      BICARBONATE, AND SPECIES DIFFERENCES INFLUENCE SUCH EXCRETION.
      EXTENSIVE EXTRACELLULAR LIBERATION, BY PHYTOPLANKTON, OF PHOTOSYNTHETIC
      PRODUCTS HAS IMPORTANT IMPLICATIONS FOR STUDIES OF AQUATIC TROPHIC
      RELATIONSHIPS. (JONES-WISCONSIN)

   FIELD 05B

   ACCESSION NO. W70-02504

INTERRELATIONS OF DISSOLVED ORGANIC MATTER AND PHYTOPLANKTON,

MICHIGAN UNIV., ANN ARBOR. DEPT OF ZOOLOGY.

GEORGE W. SAUNDERS.

THE BOTANICAL REVIEW, VOL 23, NO 6, P 389-409, 1957. 4 TAB, 131 REF.

DESCRIPTORS:
*ORGANIC MATTER, *PHYTOPLANKTON, LIMNOLOGY, ZOOPLANKTON, ALGAE,
COLLOIDS, TRACE ELEMENTS, DISSOLVED SOLIDS, NUTRIENTS, VITAMINS,
TOXINS, INHIBITORS, SUCCESSION, CHLORELLA, CHELATION, OCHROMONAS,
BACTERIA, FUNGI, AQUATIC PLANTS, ASSAY, DISTRIBUTION, DENSITY,
CHROMATOGRAPHY, RADIOACTIVITY, TRACERS, WATER POLLUTION EFFECTS,
PHYSIOLOGICAL ECOLOGY, ALGICIDES.

IDENTIFIERS:
*DISSOLVED ORGANIC MATTER, ORGANIC DEBRIS, ORGANIC GROWTH FACTORS,
VITAMIN B1, VITAMIN B6, VITAMIN B12, BIOTIN, HISTIDINE, URACIL, REDUCED
SULFUR COMPOUNDS, ANTIBIOTIC EFFECTS, STIMULANTS, GONYAULAX CATENELLA,
GONYAULAX TAMERENSIS, GYMNODINIUM BREVIS, MYCROSYSTIS, ALGASTATIC
SUBSTANCES, ALGADYNAMIC SUBSTANCES, PHORMIDINE, PANDORININE,
SCENEDESMINE, MICROCRUSTACEANS, CYTOTOXIC AGENTS, AUXOTROPHY,
AUTOTROPHY, HETERETROPHY, NUTRITION, PHYSIOLOGY.

ABSTRACT:
LIMNOLOGISTS ATTEMPT TO EXPLAIN DISTRIBUTION AND INTERRELATIONSHIPS OF
FRESHWATER ORGANISMS FROM THE MORE CONVENTIONAL PHYSICAL AND CHEMICAL
POINTS OF VIEW. THE BIOCHEMIST AND PHYSIOLOGIST ARE BEGINNING TO
UNDERSTAND BASIC METABOLIC PATHWAYS OF SYNTHESIS AND DEGRADATION.
MICRO-QUANTITIES OF BOTH ORGANIC AND INORGANIC MATTER ARE SIGNIFICANT
TO GROWTH, BEHAVIOR, AND SUCCESS OF LIVING ORGANISMS. WITH THE NEW
INFORMATION FROM BIOCHEMISTRY AND NEW TOOLS AND TECHNIQUES, SUCH AS
PAPER CHROMATOGRAPHY, CHELATE COMPLEXES, RADIOACTIVE TRACERS, AND
MICROBIOLOGICAL ASSAY, LIMNOLOGISTS MAY COMPLETE THE UNIFIED PICTURE.
THE ATTACK MAY BE INITIATED BY USE OF PURE CULTURES IN THE LABORATORY
TO DETERMINE THE BASIC REQUIREMENTS OF THE MORE SIGNIFICANT ALGAL FORMS
IN TERMS OF THEIR ABILITY TO UTILIZE VARIOUS ORGANIC SUBSTRATES,
REQUIREMENT FOR GROWTH FACTORS, REACTION TOWARD ANTIBIOTICS, AND THEIR
REQUIREMENT FOR TRACE METALS. LIMNOLOGISTS IN THE FIELD MAY DETERMINE
YEARLY CYCLES OF SPECIFIC ORGANIC NUTRIENTS, VITAMINS, AND TRACE
METALS. MANY QUESTIONS MAY BE ANSWERED BY CORRELATING LABORATORY AND
FIELD STUDIES WITH ANALYSES OF POPULATIONS AND THEIR SUCCESSION IN
NATURE. THESE TOOLS SHOULD BE CONSIDERED AND USED TO USHER IN A NEW ERA
IN LIMNOLOGY AND THE INVESTIGATION OF INTERRELATIONS OF DISSOLVED
ORGANIC MATTER, PHYTOPLANKTON, AND ZOOPLANKTON. (JONES-WISCONSIN)

FIELD 05C, 02H

ACCESSION NO. W70-02510

DESIGN PRINCIPLES OF WASTE STABILIZATION PONDS,

NOVA SCOTIA TECHNICAL COLL., HALIFAX.

D. THIRUMURTHI.

J SANIT ENG DIV, PROC AMER SOC CIVIL ENG, VOL 95, NO SA2, PROC PAPER 6515, P
311-330, APR 1969. 8 FIG, 6 TAB, 22 REF.

DESCRIPTORS:
*ALGAE, *DISPERSIONS, *ENVIRONMENTAL ENGINEERING, *SANITARY
ENGINEERING, *SEWAGE TREATMENT, *TRACERS, *WATER POLLUTION, DESIGN,
DESIGN CRITERIA, OXIDATION LAGOONS PHOTOSYNTHESIS.

IDENTIFIERS:
*MODEL TESTS, DYES.

ABSTRACT:
WASTE STABILIZATION PONDS WILL BE MORE EFFECTIVELY EMPLOYED AS
POLLUTION CONTROL SYSTEMS IF SUITABLE DESIGN FORMULA AND CORRESPONDING
DESIGN PARAMETERS ARE MADE AVAILABLE TO DESIGN ENGINEERS. TO ACCOMPLISH
THIS OBJECTIVE, AN ANALOGY HAS BEEN DRAWN BETWEEN A FIRST-ORDER
CHEMICAL REACTOR AND A WASTE STABILIZATION POND. THIS ANALOGY HAS
RESULTED IN THE ADAPTATION OF A CHEMICAL ENGINEERING REACTOR DESIGN
EQUATION FOR DESIGNING STABILIZATION PONDS. TWO DESIGN PARAMETERS,
FIRST-ORDER BOD REMOVAL COEFFICIENT AND DISPERSION INDEX, WERE
CONSIDERED AS THE BASIS OF THE NEW DESIGN APPROACH. SYNTHETIC SEWAGE
WAS USED TO DETERMINE THE BOD REMOVAL KINETICS. A CONTROL POND, WITH
TAP WATER AS THE MEDIUM AND SODIUM CHLORIDE AS A TRACER, WAS USED TO
CALCULATE THE DISPERSION INDEX. THE INTERPRETATION OF THE RESULTS
DEMONSTRATE THE SUCCESSFUL APPLICABILITY OF THE PROPOSED DESIGN
FORMULA. AN EXAMPLE DESIGN PROBLEM HAS BEEN WORKED OUT.
(LEDBETTER-TEXAS)

FIELD 05D

ACCESSION NO. W70-02609

LIMITATION OF ALGAL GROWTH IN SOME CENTRAL AFRICAN WATERS,

MALAWI UNIV., LIMBE.

BRIAN MOSS.

LIMNOLOGY AND OCEANOGRAPHY, VOL 14, NO 4, P 591-601, JULY 1969. 11 P, 3 FIG, 4 TAB, 20 REF.

DESCRIPTORS:
*ALGAE, *PRODUCTIVITY, *EUTROPHICATION, *WATER CHEMISTRY, POLLUTANTS, ALGAL CONTROL, NITRATES, PHOSPHATES, SULFATES, SEASONAL, LAKES, SAMPLING, PONDS, RIVERS, RESERVOIRS, AIR TEMPERATURE, RAIN, GEOLOGY, MAPPING, NUTRIENTS.

IDENTIFIERS:
*AFRICA, MALAWI.

ABSTRACT:
LEVELS OF INORGANIC NUTRIENTS (NITRATE, PHOSPHATE, SULFATE) AVAILABLE FOR ALGAL GROWTH IN NINE BODIES OF WATER IN MALAWI WERE INVESTIGATED, USING ENRICHMENT BIOASSAYS WITH NATURALLY PRESENT ALGAE AS TEST ORGANISMS. NITRATE OR NITRATE PLUS PHOSPHATE WERE POTENTIALLY LIMITING TO ALGAL GROWTH. THERE WAS NO EVIDENCE OF SULFATE LIMITATION. THE RELEVANCE OF ENRICHMENT BIOASSAYS TO WHOLE WATER BODY SITUATIONS IS CONSIDERED. A MODEL, ILLUSTRATING INVALID USE OF ASSAYS IN THE INTERPRETATION OF SEASONAL SUCCESSION IN ALGAL COMMUNITIES, IS DISCUSSED. (GABRIEL-USGS)

FIELD 05C, 05A

ACCESSION NO. W70-02646

BUOYANCY AND SINKING CHARACTERISTICS OF FRESHWATER PHYTOPLANKTON,

RHODE ISLAND UNIV., KINGSTON.

THEODORE J. SMAYDA.

TECHNICAL COMPLETION REPORT, RHODE ISLAND WATER RESOURCES CENTER, 1969. 7 P. OWRR PROJECT B-006-RI.

DESCRIPTORS:
*FLOTATION, *SINKING, *PHYTOPLANKTON, *DIATOMS, *BLUE-GREEN ALGA, COPPER SULFATE.

ABSTRACT:
THE SINKING RATES OF SELECTED FRESHWATER PHYTOPLANKTONS WERE DETERMINED. THE SPECIES WERE THE DIATOMS ASTERIONALLA FORMOSA, TABELLARIA FLOCCULOSA, AND THE BLUE-GREEN ALGA APHANIZOMANON FLOS-AQUAE. THE SINKING RATES OF SOME BRACKISH-COASTAL WATER DIATOMS WHICH HAVE MORPHOLOGICAL HOMOLOGUES IN FRESHWATER WERE ALSO INVESTIGATED TO EXAMINE THE INFLUENCE OF MORPHOLOGY ON SINKING RATE. SINKING RATES OF THE FRESHWATER FORMS USED WERE FOUND TO APPROXIMATE THOSE (APPROXIMATELY 0.2 TO 1 METERS DAY-1) FOR MARINE PHYTOPLANKTERS. INCREASED CULTURE AGE AND COLONY (CHAIN) SIZE FAVORED INCREASED SINKING RATES. SINKING COULD BE HEIGHTENED BY THE ADDITION OF CUSO4, FORMALIN AND KI. KILLING BY HEAT ALSO HEIGHTENED SINKING. THE ADDITION OF 0.1% BUTANOL AND ISOPROPANOL ALSO HEIGHTENED SINKING OF THE DIATOMS USED, BUT ONLY THE BUTANOL TREATED BLUE-GREEN ALGA CELLS RESPONDED THUSLY. THESE ALGAE REMAINED VIABLE DESPITE THE ALCOHOL TREATMENT. THEIR HEIGHTENED SINKING RATE (AND, CONVERSELY, THEREFORE ONE OF THEIR SUSPENSION AIDES) APPEARS TO RELATED IN TOTO (OR PRIMARILY) TO A CHANGE IN TOTAL CELL DENSITY DUE TO OSMOTIC DISRUPTIONS AND/OR DISTURBANCE OF THE CELL MEMBRANE ELECTROKINETICS WHICH CAN INFLUENCE THE HYDROFUGE OR HYDROPHILIC PROPERTIES OF THE CELL. THE RESULTS WITH CUCO4 ADDITION SUGGEST THAT IN ADDITION TO CONTROL OF NOXIOUS WATER BLOOMS THROUGH GROWTH CESSATION BY THIS AGENT, SINKING OF CERTAIN PHYTOPLANKTERS MAY ALSO BE PROMOTED LEADING TO SETTLING OUT OF THE TROPHOGENIC ZONE INTO THE APHOTIC LAYERS WHERE REMINERALIZATION OCCURS.

FIELD 05G

ACCESSION NO. W70-02754

PHYTOPLANKTON FLORA OF NEWFOUND AND WINNISQUAM LAKES, NEW HAMPSHIRE,

NEW HAMPSHIRE UNIV., DURHAM. DEPT. OF BOTANY.

G. K. GRUENDLING, AND A. C. MATHIESON.

RHODORA, VOL 71 (787): P 444-447, 1969. 42 FIG, 117 REF. OWRR PROJECT
A-010-NH.

DESCRIPTORS:
*ALGAE, *TROPHIC LEVELS, AQUATIC ENVIRONMENT, WATER QUALITY, NUTRIENTS,
NORTHEAST U S, ECOLOGICAL DISTRIBUTION, BIOLOGICAL INDICATORS, NEW
HAMPSHIRE.

IDENTIFIERS:
*NEWFOUND LAKE, *WINNISQUAM LAKE, WATER AND PLANTS, WATER POLLUTION
EFFECTS, ECOLOGY, WATER QUALITY CHARACTERIZATION.

ABSTRACT:
A TOTAL OF 185 TAXA OF FRESH WATER ALGAE (PRIMARILY PHYTOPLANKTON) WERE
IDENTIFIED AT WINNISQUAM (PARTIALLY EUTROPHIC) AND NEWFOUND
(OLIGOTROPHIC) LAKES. OF THE 142 TAXA RECORDED AT WINNISQUAM LAKE 85
WERE FOUND EXCLUSIVELY AT THIS LOCATION. THE GREEN ALGAE CONTRIBUTED
THE GREATEST NUMBER OF SPECIES, BUT THE BULK OF THE STANDING CROP WAS
COMPOSED OF BLUE-GREEN ALGAE. A TOTAL OF 100 TAXA WERE RECORDED AT
NEWFOUND LAKE, AND 43 OF THESE WERE FOUND EXCLUSIVELY AT NEWFOUND LAKE.
THE LARGEST NUMBER OF SPECIES WERE GREEN ALGAE. DURING THE SUMMER THE
PHYTOPLANKTON FLORA AT WINNISQUAM LAKE WAS PRIMARILY COMPOSED OF
MEMBERS OF THE CHLOROPHYCEAE AND CYANOPHYCEAE. MEMBERS OF THE
BACILLARIOPHYCEAE, CHRYSOPHYCEAE AND CHLOROPHYCEAE (MAINLY DESMIDS)
CONTRIBUTED THE LARGEST NUMBER OF SPECIES DURING THE SPRING AND FALL.
DIATOMS WERE THE MAJOR COMPONENT OF THE WINTER FLORA. THE SUMMER AND
FALL FLORA AT NEWFOUND LAKE WAS PRIMARILY COMPOSED OF MEMBERS OF THE
CYANOPHYCEAE, CHLOROPHYCEAE (ESPECIALLY DESMIDS) AND CHRYSOPHYCEAE. THE
DIATOMS AND GOLDEN-BROWN ALGAE CONTRIBUTED THE LARGEST NUMBER OF
SPECIES DURING THE WINTER AND SPRING.

FIELD 05C, 02H

ACCESSION NO. W70-02764

ALGAE IN RELATION TO MINE WATER,

WEST VIRGINIA UNIV., MORGANTOWN. WATER RESEARCH INST.

H. D. BENNETT.

PB-188 906. WEST VIRGINIA WATER RESEARCH INSTITUTE, JOURNAL REPORT 1,
(REPRINT FROM CASTANEA VOL 34, NO 3, P 306-328, SEPT 1969). OWRR A-007-WVA.

DESCRIPTORS:
*ALGAE, *ACID MINE WATER, *HABITATS, *ENVIRONMENTAL EFFECTS, *LIMITING
FACTORS, BENTHIC FLORA, BENTHOS, MINE WATER, WATER POLLUTION
EFFECTS, ACID STREAMS, MINE DRAINAGE, CHLOROPHYTA,
CHRYSOPHYTA, EUGLENOPHYTA, CYANOPHYTA, DIATOMS, AQUATIC HABITAT,
POPULATION.

IDENTIFIERS:
*CYTOLOGICAL EFFECTS, ALGAL POPULATION.

ABSTRACT:
AN ANNUAL CYCLE OF BIMONTHLY SAMPLING WAS DONE TO OBTAIN INFORMATION ON
SPECIES OF ALGAE TOLERANT TO ACID MINE WATER AND ON CHEMICAL FACTORS
RELATED TO ECOLOGY OF ALGAE. NITRATE, PHOSPHATE, AND CALCIUM APPEARED
NOT TO BE LIMITING OR MODIFYING. CORRELATION WAS LACKING BETWEEN
FACTORS EXCEPT ACIDITY, IRON, AND PH. A RANGE FOR THESE CHARACTERIZED
EACH HABITAT, AND THEY APPEARED CONTROLLING ON ALGAL POPULATION IN
MORE-ACID STREAMS. THE TOTAL NUMBER OF GENERA AND SPECIES COMPARED
FAVORABLY WITH NUMBERS REPORTED FROM 'UNPOLLUTED' WATERS. NEARLY HALF
THE TOTAL SPECIES FOUND IN MORE-ACID STREAMS EQUALLY WERE IN LESS-ACID
CREEKS, RIVERS, AND PONDS. ALGAE CHARACTERISTIC OF MINE WATER WERE
THOSE COMMON TO THE OTHER HABITATS. REDUCTION IN NUMBERS OF SPECIES AT
HIGHER ACIDITY AND LOWER PH WAS PRIMARILY IN SPECIES LESS COMMON TO A
RANGE OF HABITATS. THE RANGE OF ACIDITY AND PH AT A SITE IS
CHARACTERIZED BY A RANGE IN SPECIES NUMBER AND ABUNDANCE A CHANGE IN
ABUNDANCE BEING DEPENDENT ON THE SPECIES. SOME SUCH AS EUGLENA
MUTABILIS SCH., EUNOTIA TENELLA (GRUN) CLEVE, AND PINNULARIA BRAUNII
(GRUN) CLEVE ARE MOST ABUNDANT IN MINE-POLLUTED WATER. (DODSON-WEST
VIRGINIA)

FIELD 05C

ACCESSION NO. W70-02770

PHYTOPLANKTON POPULATIONS IN RELATION TO TROPHIC LEVELS OF LAKES IN NEW
HAMPSHIRE, U.S.A.

NEW HAMPSHIRE UNIV., DURHAM. DEPT. OF BOTANY.

G. K. GRUENDLING, AND A. C. MATHIESON.

AVAILABLE FROM THE CLEARINGHOUSE AS PB-188 912, $3.00 IN PAPER COPY, $0.65 IN
MICROFICHE. COMPLETION REPORT, UNIVERSITY OF NEW HAMPSHIRE WATER RESOURCES
RESEARCH CENTER, RESEARCH REPORT NO 1, JUNE 1969. 81 P, 15 FIG, 9 TAB. OWRR
PROJECT A-010-NH.

DESCRIPTORS:
*ALGAE, AQUATIC ENVIRONMENT, *AQUATIC ALGAE, WATER QUALITY, *TROPHIC
LEVELS, NUTRIENTS, NORTHEAST U.S., ECOLOGICAL DISTRIBUTION, FREQUENCY
DISTRIBUTION, POPULATION, BIOLOGICAL INDICATORS, NEW HAMPSHIRE.

IDENTIFIERS:
PLANT GROWTH, WATER AND PLANTS, WATER POLLUTION EFFECTS, ECOLOGY,
EFFECTS OF POLLUTION, WATER QUALITY CONTROL, WINNISQUAM LAKE, NEWFOUND
LAKE.

ABSTRACT:
THE RELATIONSHIP BETWEEN COMPOSITION, PERIODICITY AND ABUNDANCE OF
PHYTOPLANKTON SPECIES AND THE TROPHIC STATUS OF TWO NEW HAMPSHIRE LAKES
(WINNISQUAM AND NEWFOUND) WERE INVESTIGATED. NEWFOUND LAKE IS
OLIGOTROPHIC, WHILE WINNISQUAM IS UNDERGOING RAPID EUTROPHICATION. THE
DIFFERENCES IN THE TOTAL STANDING CROP AT THE TWO LAKES WERE RELATED
WITH THEIR DIFFERENCES IN NUTRIENT CONCENTRATIONS. ORTHOPHOSPHATE
APPEARED TO BE THE PRIMARY LIMITING FACTOR OF ALGAL GROWTH AT NEWFOUND
LAKE, WHILE LIGHT, TEMPERATURE AND POSSIBLE EXTRACELLULAR PRODUCTS WERE
THE PRIMARY LIMITING FACTORS AT WINNISQUAM LAKE. A DISCUSSION OF THE
FACTORS INFLUENCING THE PERIODICITY, ABUNDANCE AND DISTRIBUTION OF
PHYTOPLANKTON SPECIES AT THE TWO LAKES IS PRESENTED, AND THE RESULTS
ARE COMPARED WITH PREVIOUS FINDINGS. THE SPECIES COMPOSITION AT THE
LAKES WAS DEPENDENT UPON THEIR TROPHIC STATUS. DOMINANT SPECIES WERE
CONSIDERED TO BE THE BEST INDICATORS OF TROPHIC LEVELS. A SUMMARY OF
POSSIBLE RARE INDICATOR ORGANISMS IS ALSO INCLUDED. THE PHYTOPLANKTON
QUOTIENTS PROPOSED BY VARIOUS AUTHORS WERE APPLIED TO THE LAKES. MOST
OF THE INDICES CORRECTLY DESIGNATED THE EUTROPHIC LAKE, BUT NOT THE
OLIGOTROPHIC ONE.

FIELD 05C, 02H

ACCESSION NO. W70-02772

CALIFORNIA UNIV., BERKELEY. SANITARY ENGINEERING RESEARCH LAB. AND PACIFIC
NORTHWEST WATER LAB. CORVALLIS, OREG.

PROCEEDINGS OF THE EUTROPHICATION-BIOSTIMULATION ASSESSMENT WORKSHOP, JUNE
19-21, 1969, CALIFORNIA UNIV, BERKELEY, SANITARY ENGINEERING RESEARCH LAB
AND NATIONAL EUTROPHICATION RESEARCH PROGRAM, CORVALLIS, OREGON, PACIFIC
NORTHWEST WATER LAB P I-IV + 281 P. 93 FIG, 43 TAB, 341 REF.

DESCRIPTORS:
EUTROPHICATION, *CONFERENCES, WATER POLLUTION EFFECTS, ALGAL CONTROL,
WATER POLLUTION, TOXICITY, LAKES, DESIGN CRITERIA, REMOTE SENSING,
MINNESOTA, SEDIMENT-WATER INTERFACES, CALIFORNIA, BIOASSAY,
SCENEDESMUS, PHYSIOLOGICAL ECOLOGY, KINETICS, FARM WASTES, BENTHOS,
NITRATES, FLUORESCENCE, NITROGEN FIXATION, ENVIRONMENTAL EFFECTS,
NUTRIENT REQUIREMENTS, COMMUNITIES(BIOLOGICAL), CARBON RADIOISOTOPES,
PRODUCTIVITY, RIVERS, TRACE ELEMENTS.

IDENTIFIERS:
*BIOSTIMULATION, ALGAL GROWTH, WATER POLLUTION ASSESSMENT, SANITARY
ENGINEERING RESEARCH LABORATORY(CALIF), NATIONAL EUTROPHICATION
RESEARCH PROGRAM, CONTINUOUS CULTURE, CHEMOSTATS, BATCH CULTURES,
ACETYLENE REDUCTION TECHNIQUE, HETEROTROPHY, FLAGELLATES, LAKE
CLASSIFICATION, SHAGAWA LAKE(MINN).

ABSTRACT:
EUTROPHICATION MAY BE THE MOST SERIOUS AND PRESSING WATER QUALITY
PROBLEM WE FACE, AND NEW SCRUTINY OF RESEARCH APPROACHES IS REQUIRED.
THE DELIBERATIONS ARE PRESENTED OF A GROUP OF WORKERS IN EUTROPHICATION
HELD AT BERKELEY, CALIFORNIA, DURING 19-21 JUNE 1969. THE SESSIONS WERE
SPONSORED BY THE SANITARY ENGINEERING RESEARCH LABORATORY, UNIVERSITY
OF CALIFORNIA, AND THE NATIONAL EUTROPHICATION RESEARCH PROGRAM, US
DEPARTMENT OF THE INTERIOR. SCOPE OF THE REPORT IS INDICATED BY
SUBJECTS OF INDIVIDUAL CONTRIBUTIONS: NEED FOR ASSAYS; CHEMOSTAT
ASSAYS; BATCH ASSAYS FOR DETERMINING ALGAL GROWTH POTENTIAL; ALGAL
GROWTH ASSESSMENT BY FLUORESCENCE TECHNIQUES; KINETIC ASSESSMENT OF
ALGAL GROWTH; FACTORS IN UTILIZATION BY SCENEDESMUS OF NITRATE FROM
AGRICULTURAL WASTES; ENVIRONMENTAL AND NUTRITIONAL REQUIREMENTS FOR
ALGAE; ASSESSMENT OF NITROGEN FIXATION BY ACETYLENE REDUCTION;
HETEROTROPHY IN ALGAL FLAGELLATES; PHYSIOLOGICAL ECOLOGY; PHYSIOLOGICAL
ECOLOGY OF BENTHIC ALGAL COMMUNITIES; MEASUREMENT OF ALGAL GROWTH WITH
RADIOCARBON; LACUSTRINE CLASSIFICATION; PRODUCTIVITY OF RIVERS;
MEASUREMENT OF TRACE ELEMENTS IN THE HYDROSPHERE; DESIGN APPLICATIONS
OF BIOSTIMULATION AND TOXICITY CRITERIA; REMOTE SENSING IN WATER
RESOURCES MANAGEMENT; NUTRIENT EXCHANGE BETWEEN SEDIMENT AND WATER; AND
SHAGAWA LAKE (MINNESOTA) EUTROPHICATION RESEARCH PROJECT. AN INDEX IS
PROVIDED. (EICHHORN-WISCONSIN)

FIELD 05A, 05C

ACCESSION NO. W70-02775

ALGAL GROWTH ASSESSMENTS BY FLUORESCENCE TECHNIQUES,

FEDERAL WATER POLLUTION CONTROL ADMINISTRATION, ALAMEDA, CALIF.

RICHARD C. BAIN, JR.

PROC OF THE EUTROPHICATION-BIOSTIMULATION ASSESSMENT WORKSHOP, JUNE 19-21, 1969, CALIFORNIA UNIV, BERKELEY, SANITARY ENGINEERING RESEARCH LAB AND NAT EUTROPHICATION RESEARCH PROGRAM, CORVALLIS, ORE, PAC NORTHWEST WATER LAB, P 39-55. 9 FIG, 20 REF, APPEND.

DESCRIPTORS:
*ALGAL GROWTH, *BIOASSAYS, *FLUOROMETRY, CALIFORNIA, NUTRIENTS, EUTROPHICATION, CHLOROPHYLL, FLUORESCENCE, LAKES, RIVERS, NITROGEN, BIOMASS, SEWAGE, FARM WASTES, CARBON CYCLE, PHOSPHORUS, PHOSPHATE, NITRATES, INDUSTRIAL WASTES, WATER POLLUTION EFFECTS, DENITRIFICATION, SPECTROPHOTOMETRY.

IDENTIFIERS:
SAN JOAQUIN RIVER(CALIF), MERCED RIVER(CALIF), TUOLUMNE RIVER(CALIF), STANISLAUS RIVER(CALIF), WATER POLLUTION ASSESSMENT, AMERICAN RIVER(CALIF), CLEAR LAKE(CALIF), EEL RIVER(CALIF), JOINT INDUSTRY/GOVERNMENT TASK FORCE ON EUTROPHICATION, PROVISIONAL ALGAL ASSAY PROCEDURE, BATCH CULTURES, CHEMOSTATS, CHLOROPHYLL A, PHOTOBIOLOGY, ALGAL GROWTH POTENTIAL, SAN FRANCISCO BAY.

ABSTRACT:
AUTHOR DESCRIBES PROCEDURES FOR ASSESSING ALGAL GROWTH POTENTIAL, IN JOINT INDUSTRY/GOVERNMENT TASK FORCE'S PROVISIONAL ALGAL ASSAY PROCEDURE, BY ESTIMATING ALGAL CHLOROPHYLL FLUOROMETRICALLY OR SPECTROPHOTOMETRICALLY. WATER SMAPLES ARE FILTERED THROUGH GLASS FIBER FILTERS, PRESERVED WITH MAGNESIUM CARBONATE, AND FILTERS KEPT FROZEN IN DARK REFRIGERATOR. CHLOROPHYLL CONCENTRATION, EXTRACTED INTO SOLUTION BY GRINDING THE FILTER IN ACETONE, IS ESTIMATED BY SPECTROPHOTOMETRY (AT WAVELENGTHS 630, 645 AND 665 NANOMETERS) OR FLUOROMETRY. CORRECTION FOR PHAEOPHYTIN IS ACCOMPLISHED BY A SECOND READING AFTER TREATMENT OF SAMPLE WITH HYDROCHLORIC ACID. ILLUSTRZTIVE DATA FOR VARIOUS CALIFORNIA WATERS RECEIVING SEWAGE, AGRICULTURAL AND INDUSTRIAL WASTES ARE DELINEATED IN THE REPORT. IN BATCH ASSAYS,PEAK FLUORESCENCE DUE TO CHLOROPHYLL, CHARACTERISTICALLY APPEARS WITHINTHE FIRST WEEK, NOT NECESSARILY PARALLELLING PEAK BIOMASSES. DATA ANALYSIS FROM SUCH ASSAYS TOGETHER WITH THOSE FOR NUTRIENT ANALYSIS PERMITS INTERPRETIVE APPLICATION OF THE PROCEDURE. THUS FOR 44 DATA POINTS FROM VARIOUS STREAMS OF SAN JOAQUIN VALLEY, IN WHICH NITROGEN IS THOUGHT TO BE LIMITING, REGRESSION ANALYSIS OF RELATIONSHIP BETWEEN TOTAL N (MILLIGRAMS/LITER) AND CHLOROPHYLL (MICROGRAMS/LITER) YIELDS RATIO, TOTAL N/CHLOROPHYLL, WITH 95% CONFIDENCE INTERVAL OF 4.6-10.8, IN GOOD AGREEMENT WITH LITERATURE VALUES FOR ALGAL CELLS WITH NITROGEN LIMITING. (EICHHORN-WISCONSIN)

FIELD 05A

ACCESSION NO. W70-02777

CONTINUOUS-FLOW (CHEMOSTAT) ASSAYS,

CALIFORNIA UNIV., BERKELEY. SANITARY ENGINEERING RESEARCH LAB.

DONALD B. PORCELLA.

PROCEEDINGS OF THE EUTROPHICATION-BIOSTIMULATION ASSESSMENT WORKSHOP, JUNE
19-21, 1969, CALIFORNIA UNIV, BERKELEY, SANITARY ENGINEERING RESEARCH LAB
AND NATIONAL EUTROPHICATION RESEARCH PROGRAM, CORVALLIS, ORE, PACIFIC
NORTHWEST WATER LAB, P 7-22. 6 FIG, 1 TAB, 30 REF.

DESCRIPTORS:
*BIOASSAYS, NUTRIENTS, PHOSPHORUS, KINETIC THEORY, ENVIRONMENTAL
FACTORS, BIOMASS, GROWTH RATES, ALGAE, OLIGOTROPHY, LIMITING FACTORS,
CHLORELLA, ANALYTICAL TECHNIQUES, CYANOPHYTA, DIATOMS, EUTROPHICATION,
WATER POLLUTION EFFECTS, SYSTEMS ANALYSIS, SUCCESSION, ECOSYSTEMS,
CYCLING NUTRIENTS.

IDENTIFIERS:
*CONTINUOUS CULTURE, *CHEMOSTATS, TURBIDOSTATS, NUTRIENT LIMITED
GROWTH, LUXURY CONSUMPTION, PROVISIONAL ALGAL ASSAY, PROCEDURES,
SELENASTRUM CAPRICORNUTUM, SELENASTRUM GRACILE, GROWTH LIMITATION,
SELENASTRUM, ALGAL GROWTH, ANABAENA, BIOSTIMULATION, WATER POLLUTION
ASSESSMENT.

ABSTRACT:
BECAUSE THEIR DYNAMIC PROPERTIES MORE NEARLY APPROXIMATE ECOLOGICAL
CONDITIONS THAN DO BATCH CULTURES, CONTINUOUS CULTURES OPERATING IN
CHEMOSTAT MODE (CONSTANT FLOW-RATE) HAVE BEEN UTILIZED IN A VARIETY OF
STUDIES OF ALGAL GROWTH, INCLUDING ASSESSMENT OF BIOSTIMULATORY
RESPONSES. REQUIREMENTS FOR STEADY STATE CONTINUOUS FLOW SYSTEMS AND
CHARACTERISTICS OF IDEAL CHEMOSTATS WITH ILLUSTRATIONS OF IDEALIZED
CURVES FROM AN EXPERIMENT CONCERNING RELATIVE CHANGE IN ALGAL CELL
MATERIAL AND LIMITING PHOSPHORUS CONCENTRATIONS ARE INCLUDED. IN
TRANSIENT PHASES OF CHEMOSTAT OPERATION, PHOSPHORUS CONCENTRATIONS ARE
LESS THAN THOSE CALCULATED THEORETICALLY, PROBABLY DUE TO 'LUXURY'
CONSUMPTION BY ALGAE. CONSIDERATIONS OF MASS BALANCES AND KINETICS OF
ALGAL GROWTH, USEFUL IN DERIVATION OF CONTINUOUS CULTURES THEORY, NEED
BE APPLIED TO NATURAL ECOSYSTEMS TO ELICIT USEFUL INFORMATION ON
PRODUCTIVITY, NUTRIENT CYCLING, AND SUCCESSIONAL RELATIONSHIPS. BASED
ON CHEMOSTAT EXPERIMENTS DESCRIBED, UTILIZING BOTH SELENASTRUM GRACILE
AND NATURAL ALGAL POPULATIONS AS 'SEED', ARE AREAS REQUIRING FURTHER
STUDY, INCLUDING THE FOLLOWING PROBLEMS: MULTIPLE LIMITING FACTORS; USE
OF NATURAL 'SEED' POPULATIONS; EXCESSIVE WASHOUT AND EXTINCTION OF
CULTURES IN BIOASSAY OF OLIGOTROPHIC WATERS; DEVELOPMENT OF SYNTHETIC
MEDIA; AND PROFILES OF BIOMASS ANALYSIS AND NUTRIENT CONCENTRATION
DURING TRANSIENT GROWTH IN ORDER TO ESTABLISH APPROPRIATE KINETIC
PARAMETERS. (SEE W70-02775). (EICHHORN-WISCONSIN)

FIELD 05A

ACCESSION NO. W70-02779

STRUCTURAL CHARACTERISTICS OF BENTHIC ALGAL COMMUNITIES IN LABORATORY STREAMS,

OREGON STATE UNIV., CORVALLIS. DEPT. OF BOTANY.

C. DAVID MCINTIRE.

ECOLOGY, VOL 49, NO 3, P 520-537, 1968. 5 FIG, 6 TAB, 31 REF.

DESCRIPTORS:
*BENTHIC FLORA, *ALGAE, *BIOLOGICAL COMMUNITIES, *LABORATORIES,
*STREAMS, LIGHT INTENSITY, CURRENTS(WATER), VELOCITY, DIATOMS,
SEASONAL, BIOMASS, PIGMENTS, PRODUCTIVITY, PERIPHYTON, CHLOROPHYTA,
CHRYSOPHYTA, CYANOPHYTA, SAMPLING, PHYTOPLANKTON, ORGANIC MATTER,
CHLOROPHYLL, DYNAMICS, TEMPERATURE, SILTS, STANDING CROP, ECOLOGY,
PHOTOSYNTHESIS, RESPIRATION, CARBON DIOXIDE, NUTRIENTS, DISSOLVED
OXYGEN, LOTIC ENVIRONMENT.

IDENTIFIERS:
SPECIES DIVERSITY, AUTECOLOGY, MORPHOLOGY, EXPORT LOSSES,
CHEMOSYNTHESIS, ABUNDANCE.

ABSTRACT:
EFFECTS OF LIGHT INTENSITY AND CURRENT VELOCITY ON SPECIES COMPOSITION
AND ECOLOGICAL PROPERTIES OF BENTHIC ALGAL COMMUNITIES WERE
INVESTIGATED IN LABORATORY STREAMS. OF 15 DIATOM TAXA STUDIED, ONLY
MELOSIRA VARIANS, MERIDION CIRCULARE, AND NAVICULA RADIOSA WERE MORE
ABUNDANT IN STREAMS RECEIVING 700-FOOT-CANDLE ILLUMINATION THAN IN
THOSE RECEIVING 150-FOOT-CANDLE. ACHNANTHES EXIGUA, A MINUTISSIMA,
MERIDION CIRCULARE, RHOICOSPHENIA CURVATA, AND NAVICULA RADIOSA WERE
INDIFFERENT TO CURRENT VELOCITY, WHICH HAD POSITIVE EFFECT ON ABUNDANCE
OF NITZSCHIA LINEARIS, ACHNANTHES LANCEOLATA, NAVICULA CRYPTOCEPHALA, N
MINIMA, N SEMINULUM, SYNEDRA ULNA, GOMPHONEMA PARVULUM, G ANGUSTATUM,
COCCONEIS PLACENTULA, AND THE LANCEOLATE NITZSCHIA. MELOSIRA VARIANS
EXHIBITED NEGATIVE RESPONSE TO CURRENT VELOCITY; SIX TAXA, OTHER THAN
DIATOMS, ANABAENA VARIABILIS AND TRIBONEMA MINOR WERE MORE ABUNDANT IN
STREAMS WITH HIGHER LIGHT INTENSITY, BUT ONLY ONE SPECIES, PHORMIDIUM
RETZII, WAS MORE ABUNDANT IN STREAMS WITH CURRENT THAN IN STANDING
WATER. AT A PARTICULAR SEASON, LIGHT INTENSITY, AND CURRENT
VELOCITY--LABORATORY STREAM CONDITIONS--ALLOW ESTABLISHMENT OF ALGAL
COMMUNITIES WITH COMPARATIVELY UNIQUE SPECIES COMPOSITION AND
CHARACTERISTIC BIOMASS, PIGMENT CONCENTRATION, AND INCREMENT OF EXPORT.
TO UNDERSTAND FACTORS INFLUENCING PRODUCTIVITY IN THESE COMMUNITIES,
SOME KNOWLEDGE IS NECESSARY OF AUTECOLOGIES OF COMMUNITY CONSTITUENTS,
AND MECHANISMS REGULATING SPECIES COMPOSITION AND FLORAL DIVERSITY.
(JONES-WISCONSIN)

FIELD 05C

ACCESSION NO. W70-02780

CLADOPHORA GLOMERATA AND CONCOMITANT ALGAE IN THE RIVER SKAWA. DISTRIBUTION AND
CONDITIONS OF APPEARANCE,

WYZSZA SZKOLA ROLNICZA, OLSZTYN-KORTOWA (POLAND). KATEDRA BOTANIKI.

HENRYK CHUDYBA.

ACTA HYDROBIOLOGICA, VOL 10, NO 1-2, P 39-84, 1968. 27 FIG, 12 TAB, 82 REF.

DESCRIPTORS:
*CLADOPHORA, ALGAE, AQUATIC ALGAE, AQUATIC MICROBIOLOGY, AQUATIC
PRODUCTIVITY, CHLOROPHYTA, GONOPHYTA, DIATOMS, ENVIRONMENTAL EFFECTS,
LIMNOLOGY, ECOLOGY, ECOLOGICAL DISTRIBUTION.

IDENTIFIERS:
*SKAWA RIVER(POLAND), MONTANE RIVER COURSE, CARPATHIAN SUBMONTANE
REGION, GRADIENTS(RIVER), THALLI, SCHIZOMYCETES, CYANOPHYTA, FUNGI,
BACILLARIOPHYCEAE, CHAROPHYTA, RHODOPHYTA.

ABSTRACT:
RESULTS OF 137 SAMPLES COLLECTED DURING SEPTEMBER 1-27, 1963 FROM THE
RIVER SKAWA, POLAND, INDICATED THAT TWO GROUPS CAN BE DISTINGUISHED BY
THALLI: C GLOMERATA RHEOBENTHICUM FROM PLACES WITH CURRENT; C GLOMERATA
LIMNOBENTHICUM FROM PLACES WITH NO CURRENT. THE THALLI APPEARED YELLOW
OR BROWN AS A RESULT OF ATTACHED DIATOMS. THE RAPIDITY OF CURRENT DID
NOT PLAY ANY SIGNIFICANT ROLE IN FORMATION OF THE COMMUNITY OF
ACCOMPANYING ALGAE. EPIPHYTES WERE MAINLY DIATOMS, FOLLOWED BY GREEN
AND BLUE-GREEN ALGAS. PENNATAE DIATOMS WERE MOST COMMON; NAVICULA AND
NITZSCHIA WERE THE MOST COMMON GENUS. GROUPS OF ALGAE ACCOMPANYING C
GLOMERATA AT ALL SAMPLING PLACES FORM ONE TYPE OF COMMUNITY AND A
SINGLE PHYTOCENOSE WITH A SIMILAR FLORISTIC COMPOSITION AND STRUCTURE.
(FITZGERALD-WISCONSIN)

FIELD 05C

ACCESSION NO. W70-02784

INTERPRETATION OF RADIONUCLIDE UPTAKE FROM AQUATIC ENVIRONMENTS,

OAK RIDGE NATIONAL LAB., TENN. RADIATION ECOLOGY SECTION.

DANIEL J. NELSON.

NUCLEAR SAFETY, VOL 5, NO 2, P 196-199, 1964. 12 REF.

DESCRIPTORS:
*RADIOISOTOPES, *AQUATIC ENVIRONMENT, CHLORELLA, TRACERS,
PHOTOSYNTHESIS, CARBON RADIOISOTOPES, FISH, CLAMS, STRONTIUM
RADIOISOTOPES, TENNESSEE RIVER, TROPHIC LEVEL, ALGAE, CADDISFLIES, FOOD
CHAINS, STABLE ISOTOPES, RADIOACTIVE WASTE DISPOSAL, OCEANS, FOODS,
ABSORPTION, HAZARDS, CESIUM, HYDROGEN, DEUTERIUM, TRITIUM.

IDENTIFIERS:
*UPTAKE, SPECIFIC ACTIVITY, CHLORELLA PYRENOIDOSA, BIOLOGICAL
HALF-LIFE, CLINCH RIVER(TENNESSEE), PHYSICAL HALF-LIFE, HYDROPSYCHE
COCKERELLI, BODY BURDENS, POMOXIS ANNULARIS, GASTROINTESTINAL TRACT.

ABSTRACT:
SPECIFIC ACTIVITIES IN AQUATIC ORGANISMS HELP EXPLAIN DIFFERENCES IN
RADIONUCLIDE UPTAKE. DATA EVOLVED BY ROUTINE RADIOLOGICAL MONITORING
ESTABLISHES SAFETY STANDARDS FOR RADIONUCLIDE RELEASES TO FRESH-WATER
HABITATS. EQUILIBRIUM LEVELS OF RADIONUCLIDES BETWEEN ORGANISMS AND
ENVIRONMENT ARE EXPECTED ONLY FOR ELEMENTS WITH SHORT BIOLOGICAL
HALF-LIFE IN TISSUE. METHOD OF SPECIFIC ACTIVITIES IS BASED ON
ASSUMPTION THAT IF THE DISTRIBUTION OF A STABLE ELEMENT IS KNOWN
BETWEEN AN ORGANISM AND ITS ENVIRONMENT, IT MAY BE USED TO PREDICT
RADIONUCLIDE CONCENTRATIONS IN WATER ORGANISMS RECEIVING CONSTANT
RADIONUCLIDE RELEASES. THE RELATION OF STRONTIUM AND STRONTIUM-90 IN
FISH MAY BE USED TO CALCULATE EXPECTED BODY BURDENS. SPECIFIC ACTIVITY
RELATION MIGHT BE A CRITERION FOR LIMITING RADIOACTIVE WASTE DISPOSAL
IN OCEANS. WHILE SPECIFIC ACTIVITY OF A CHEMICAL ELEMENT IN ENVIRONMENT
OF HUMAN FOOD ORGANISMS IS BELOW THE PERMISSIBLE SPECIFIC ACTIVITY FOR
THE ELEMENT IN A HUMAN BODY, MAN CANNOT OBTAIN MORE THAN HIS
PERMISSIBLE BODY BURDEN FROM THE SEAS. DIRECT APPLICATION OF THE
SPECIFIC ACTIVITY CONCEPT APPLIES ONLY TO THOSE RADIONUCLIDES WHICH
CONSTITUTE AN ABSORPTION HAZARD. WITH RADIONUCLIDES FOR WHICH
ABSORPTION IS NOT CRITICAL, SPECIFIC ACTIVITIES MAY BE RELATED TO
IRRADIATION OF THE GASTROINTESTINAL TRACT BY ITS CONTENTS.
(JONES-WISCONSIN)

FIELD 05A

ACCESSION NO. W70-02786

FERTILIZATION OF LAKES BY AGRICULTURAL AND URBAN DRAINAGE,

MASSACHUSETTS INST. OF TECH., CAMBRIDGE. DEPT. OF SANITARY CHEMISTRY.

CLAIR N. SAWYER.

NEW ENGLAND WATER WORKS ASSOCIATION, VOL 61, NO 2, P 109-127, JUNE 1947. 5
FIG, 6 TAB, 6 REF.

DESCRIPTORS:
*SEWAGE DISPOSAL, *AGRICULTURAL WATERSHEDS, *RUNOFF, *FERTILIZATION,
LAKES, SEWAGE TREATMENT, SEWAGE BACTERIA, NITROGEN, PHOSPHORUS, ALGAE,
COPPER SULFATE, EFFLUENTS, MICROCYSTIS AERUGINOSA.

IDENTIFIERS:
*URBAN DRAINAGE, *AGRICULTURAL DRAINAGE, BIOLOGICAL RESPONSE,
FERTILIZATION SURVEY, MADISON LAKES(WIS), LAKE MENDOTA(WIS), LAKE
WAUBESA(WIS), LAKE KEGONSA(WIS).

ABSTRACT:
THE REPORTED INVESTIGATIONS DISCLOSE THAT AGRICULTURAL DRAINAGE
CONTRIBUTES ANNUALLY ABOUT 4,500 POUNDS OF NITROGEN AND 255 POUNDS OF
PHOSPHORUS PER SQUARE MILE OF THE WATERSHEDS. BIOLOGICALLY TREATED
SEWAGE SUPPLIES ANNUALLY APPROXIMATELY 6.0 OF NITROGEN AND 1.2 POUNDS
OF PHOSPHORUS PER CAPITA, THUS EQUATING THE ENRICHMENT OF LAKE WATER
FROM 1 SQUARE MILE OF AGRICULTURAL DRAINAGE TO 750 PERSONS FOR NITROGEN
AND TO 212 PERSONS FOR PHOSPHORUS. IN 1946, LAKES MONONA, WAUBESA, AND
KEGONSA, LOCATED ON THE OUTSKIRTS OF THE CITY OF MADISON, RECEIVED
APPROXIMATELY 1,300 TONS OF NITROGEN AND 215 TONS OF PHOSPHORUS IN
ORGANIC AND INORGANIC FORMS. THE LAKES RETAINED FROM 30 TO 60% OF
NITROGEN RECEIVED. THE PHORPHORUS CONCENTRATION IN LAKES WAUBESA AND
KEGONSA ATTAINED THE HIGH LEVEL ABOVE 0.25 PARTS PER MILLION (PPM) AS
COMPARED WITH 0.01-0.02 PPM CONCENTRATION FOUND IN LESS POLLUTED LAKES
OF SOUTHERN WISCONSIN. IMPORTANCE OF THE HYPOLIMNION, THE
NITROGEN:PHOSPHORUS RATIO, AND A 'ROUND-THE-CALENDAR' RECORDS IS
STRESSED. MICROCYSTIS AERUGINOSA WAS A PARTICULARLY OBNOXIOUS
BY-PRODUCT OF EUTROPHICATION, EITHER BECAUSE OF ITS INHERENT
REPRODUCTIVE POTENTIAL OR RESISTANCE TO COPPER SULFATE. OTHER DATA OF
THIS REPORT HAVE CONSIDERABLE IMMEDIATE AS WELL AS HISTORICAL INTEREST.
(WILDE-WISCONSIN)

FIELD 05B

ACCESSION NO. W70-02787

EFFECTS OF ACID MINE WASTES ON PHYTOPLANKTON IN NORTHERN ONTARIO LAKES,

ONTARIO WATER RESOURCES COMMISSION, TORONTO. DIV. OF RESEARCH.

M. G. JOHNSON, M. F. P. MICHALSKI, AND A. E. CHRISTIE.

ONTARIO WATER RESOURCES COMMISSION, DIVISION OF RESEARCH PUBLICATION NO 30,
(1968). 44 P, 6 FIG, 5 TAB, 15 REF, APPEND A, B, C.

DESCRIPTORS:
*ALGAE, *ANALYTICAL TECHNIQUES, *AQUATIC ALGAE, AQUATIC MICROBIOLOGY,
AQUATIC MICROORGANISMS, AQUATIC PLANTS, AQUATIC PRODUCTIVITY, BALANCE
OF NATURE, BIOASSAY, CHLOROPHYTA, CYANOPHYTA, DIATOMS, ENVIRONMENTAL
EFFECTS, ESSENTIAL NUTRIENTS, EUTROPHICATION, BICARBONATES, CARBON,
CARBONATES, CARBON CYCLE.

IDENTIFIERS:
*ACID MINE WATER, *NUTRIENT REQUIREMENTS, *POLLUTANTS, INHIBITION,
LIMNOLOGY, ACIDITY, ACID SOILS, ACIDIC WATER, ACIDS, PHOTOSYNTHESIS,
PHOTOSYNTHETIC OXYGEN, PHYSIOLOGICAL ECOLOGY, PHYTOPLANKTON, PLANKTON,
POLLUTION ABATEMENT, POLLUTANT IDENTIFICATION, WATER POLLUTION CONTROL,
WATER POLLUTION EFFECTS, WATER POLLUTION SOURCES.

ABSTRACT:
STUDIES WERE CARRIED OUT DURING 1965 THROUGH 1967 ON THREE NORTHERN
ONTARIO LAKES; TWO (QUIRKE AND PECORS LAKES) WERE CONTAMINATED BY
QUANTITIES OF FREE MINERAL ACIDITY FROM URANIUM MILLING WASTES AND ONE
(DUNLOP LAKE) WAS UNAFFECTED. DIFFERENCES IN CHEMICAL COMPOSITION IN
THE AFFECTED LAKES, INCLUDING LOW PH VALUES AND INCREASED
CONCENTRATIONS OF SULFATE ION, NITRATE ION, AND CALCIUM ION WERE
DIRECTLY RELATED TO PROCESSES FOR THE EXTRACTION OF URANIUM AND
SUBSEQUENT TREATMENT OF WASTES. NITRATES WERE IN GREATER SUPPLY IN THE
CONTAMINATED LAKES THAN IN DUNLOP LAKE, WHILE PHOSPHORUS AND SILICA
OCCURRED IN SIMILAR CONCENTRATIONS IN ALL THREE LAKES. LOWER
PHYTOPLANKTON POPULATIONS AND INDICES OF DIVERSITY WERE FOUND IN QUIRKE
AND PECORS LAKES THAN IN DUNLOP. AVERAGE PRIMARY PRODUCTIVITIES IN THE
THREE LAKES WERE 126, 71 AND 34 MILLIGRAMS CARBON/SQUARE METER PER DAY,
RESPECTIVELY. IN SITU AREAL AND VOLUMETRIC MEASUREMENTS IN LABORATORY
AND FIELD BIOASSAYS CONFIRMED THE IMPORTANCE OF INORGANIC CARBON IN
LIMITING PRIMARY PRODUCTIVITY. AUTHORS CONCLUDED THAT INORGANIC CARBON
LIMITS PRIMARY PRODUCTIVITY IN THE LAKES CONTAMINATED BY ACID MINE
WASTES. HOWEVER, THEY DID NOT SEPARATE THE EFFECTS OF REDUCED PH ALONE
FROM THE EFFECTS OF REDUCED INORGANIC CARBON IN ARRIVING AT THAT
CONCLUSION. (FITZGERALD-WISCONSIN)

FIELD 05C, 05B

ACCESSION NO. W70-02792

NUTRIENT-PHYTOPLANKTON RELATIONSHIPS IN EIGHT SOUTHERN ONTARIO LAKES,

ONTARIO WATER RESOURCES COMMISSION, TORONTO. DIV. OF RESEARCH.

A. E. CHRISTIE.

ONTARIO WATER RESOURCES COMMISSION, DIV OF RESEARCH PUBLICATION NO 32, OCT
    1968. 37 P, 10 FIG, 7 TAB, 23 REF.

DESCRIPTORS:
    *ALGAE, *PHYTOPLANKTON, *NUTRIENTS, *PHOSPHORUS, *NITROGEN,
    *ALKALINITY, LAKES, PRODUCTIVITY, PHYTOPLANKTON.

IDENTIFIERS:
    *ONTARIO, *CANADA, KUSHOG LAKE(ONTARIO), TWELVE MILE LAKE(ONTARIO),
    GULL LAKE(ONTARIO), BALSAM LAKE(ONTARIO), STURGEON LAKE(ONTARIO),
    BUCKHORN LAKE(ONTARIO), CLEAR LAKE(ONTARIO), RICE LAKE(ONTARIO).

ABSTRACT:
A CHAIN OF EIGHT SOUTHERN ONTARIO LAKES WAS SAMPLED FROM APRIL TO
SEPTEMBER 1967, APPROXIMATELY EVERY TWO WEEKS, TO DEPTHS COMMENSURATE
WITH THE PHOTOTROPHIC ZONE. EACH LAKE WAS SAMPLED AT TWO SITES.
ANALYSIS OF ALL SAMPLES INCLUDED SUSPENDED AND DISSOLVED SOLIDS; TOTAL
ALKALINITY; PH; NITRATE, AMMONIA, ORGANIC NITROGEN; ORTHO AND ACID
HYDROLIZABLE (STANDARD METHOD'S 'TOTAL') PHOSPHORUS; ALGAL GENERA
(REPORTED AS AREAL STANDARD UNITS); AND CHLOROPHYLL A. A FEW SAMPLES
WERE ANALYZED FOR CALCIUM, MAGNESIUM, IRON, SODIUM, POTASSIUM, SULFATE,
AND CHLORINE. ONE LABORATORY GROWTH STUDY ON A SINGLE SAMPLE WAS
REPORTED WHERE GLUCOSE, SODIUM BICARBONATE, AND A COMBINATION OF THE
TWO WAS ADDED TO FLASKS CONTAINING AN 'OLIGOTROPHIC' LAKE WATER SAMPLE.
BASED ON THIS SINGLE LABORATORY EXPERIMENT AND A POSITIVE CORRELATION
BETWEEN ALKALINITY AND ALGAL STANDING CROPS IN THE OTHER LAKE WATER
SAMPLES, IT WAS CONCLUDED THAT THE UPPER RELATIVELY OLIGOTROPHIC LAKES
CONTAINED ADEQUATE NITROGEN AND PHOSPHORUS, BUT COULD SUPPORT GREATER
ALGAL CROPS IF ADDITIONAL INORGANIC OR BIODEGRADABLE ORGANIC CARBON
WERE SUPPLIED. IN THE LOWER LAKES THE CARBON WAS CONCLUDED TO BE IN
AMPLE SUPPLY. IT WAS CONCLUDED FURTHER THAT ENTRY OF NITROGEN AND
PHOSPHORUS INTO ALL EIGHT LAKES SHOULD BE CONTROLLED IN THE FUTURE.
(GERHOLD-WISCONSIN)

FIELD 05C

ACCESSION NO. W70-02795

OBJECTIONABLE ALGAE WITH REFERENCE TO THE KILLING OF FISH AND OTHER ANIMALS,

MICHIGAN STATE UNIV., EAST LANSING. DEPT. OF BOTANY AND PLANT PATHOLOGY.

G. W. PRESCOTT.

HYDROBIOLOGIA, VOL 1, P 1-13, 1948. 6 TAB, 14 REF.

DESCRIPTORS:
*NUISANCE ALGAE, *EUTROPHICATION, CYANOPHYTA, LAKES, RESERVOIRS, SCUM, NITROGEN, PHOSPHORUS, CARBONATES, TEMPERATURE, SHALLOW WATER, HYDROGEN SULFIDE, COPPER SULFATE, CHLOROPHYTA, CRUSTACEANS, SOIL EROSION, SEWAGE EFFLUENTS, FARM WASTES. POPULATION, INDUSTRIES, METHANE, ALKALINITY, HYDROGEN ION CONCENTRATION, TOXINS, OXYGEN, DECOMPOSING ORGANIC MATTER, WATER POLLUTION EFFECTS, PHOTOSYNTHESIS, IOWA, CHLORINE.

IDENTIFIERS:
HYDROXYLAMINE, ANABAENA, MICROCYSTIS, GLOEOTRICHIA ECHINULATA, DINOBRYON, SYNURA UVELLA, FRAGILARIA, TABELLARIA, MELOSIRA, SYNEDRA, COELOSPHAERIUM, SPIROGYRA, CLADOPHORA, DIAPTOMUS, CYCLOPS, GOMPHONEMA, PHORMIDIUM, APHANIZOMENON FLOS-AQUAE, LYNGBYA, STORM LAKE(IOWA), CENTER LAKE(IOWA), DIAMOND LAKE(IOWA).

ABSTRACT:
MOST TROUBLESOME ALGAE IN LAKES AND RESERVOIRS ARE SPECIES OF CYANOPHYTES (BLUE-GREEN ALGAE), WHICH MULTIPLY RAPIDLY AND TEND TO FLOAT HIGH IN THE WATER, FORMING SURFACE SCUMS. UPON DECOMPOSITION, MALODOROUS GASES AND DISAGREEABLE TASTES ARE PRODUCED. CRITICAL CONDITIONS ARISING FROM A SUPERABUNDANT GROWTH OF ALGAE OCCUR IN LAKES WHICH ARE SHALLOW, WARM, AND RICH IN NITROGEN, PHOSPHORUS, AND HALF-BOUND CARBONATES. DENSE 'BLOOMS' ARE PARTICULARLY ANNOYING AND OFTEN CAUSE ECONOMIC LOSS BY RUINING RECREATIONAL SITES, SPOILING DRINKING WATER, CLOGGING SAND FILTERS, KILLING FISH (DIRECTLY OR INDIRECTLY), AND KILLING ANIMALS WHICH DRINK THE HEAVILY INFESTED WATERS. EXPERIMENTS INDICATE THAT FISH MAY BE POISONED BY PRODUCTS OF PROTEIN DECOMPOSITION, INCLUDING HYDROXYLAMINE AND HYDROGEN SULFIDE, WHICH ARE RELEASED WHEN DENSE MASSES OF CYANOPHYTES DECAY. FACTORS WHICH SHOULD BE TAKEN INTO CONSIDERATION FOR EFFECTIVE TREATMENT OF LAKES WITH COPPER SULPHATE ARE LISTED. ADVANTAGES OF APPLYING COPPER SULPHATE BY SPRAY METHOD ARE POINTED OUT. A LABORATORY EXPERIMENT TO TEST THE ALGICIDAL ACTION OF COPPER IS DESCRIBED. DATA INDICATE THAT FISH CAN WITHSTAND MUCH GREATER CONCENTRATIONS OF COPPER SULPHATE THAN THOSE NEEDED TO CONTROL ALGAE. (JONES-WISCONSIN)

FIELD 05C

ACCESSION NO. W70-02803

MINERAL NUTRITION OF PHYTOPLANKTON,

WOODS HOLE OCEANOGRAPHIC INSTITUTION, MASS.

BOSTWICK H. KETCHUM.

ANNUAL REVIEW OF PLANT PHYSIOLOGY, VOL 5, P 55-74, 1954. 2 TAB, 126 REF.

DESCRIPTORS:
*PHYTOPLANKTON, *NUTRIENT REQUIREMENTS, ECOLOGY, CHLORELLA, ALGAE,
ORGANIC MATTER, CULTURES, CARBON, METABOLISM, NITROGEN, PHOSPHORUS,
SULFUR, CALCIUM, MAGNESIUM, POTASSIUM, SODIUM, IRON, MANGANESE, TRACE
ELEMENTS, CHELATION, PHOTOSYNTHESIS, CARBON DIOXIDE, BICARBONATES,
CARBONATES, TEMPERATURE, HYDROGEN ION CONCENTRATION, ALKALINITY, LIGHT
INTENSITY, CYANOPHYTA, IONS, AMMONIA, NITRATES, NITRITES,
DINOFLAGELLATES, CHLOROPHYTA, HUMIC ACIDS.

IDENTIFIERS:
HETEROTROPHIC, CHEMOTROPHIC, AUTOTROPHIC, SILICON, PEDIASTRUM BORYANUM,
STAURASTRUM PARADOXUM, BOTRYOCCUS BRAUNII, FRAGILARIA CROTONENSIS,
NITZSCHIA, ASTERIONELLA GRACILLIMA, GROWTH FACTORS.

ABSTRACT:
PHYTOPLANKTON NUTRITION WAS STUDIED FOR UNDERSTANDING ECOLOGICAL
CONDITIONS FOR GROWTH. USE OF ALGAL ORGANIC MATERIAL AS A POSSIBLE FOOD
SOURCE HAS STIMULATED EXTENSIVE STUDIES OF MASS CULTURES. FOR
COMPARISONS OF RESULTS, NUTRIENT REQUIREMENTS ARE DEFINED AS: ABSOLUTE,
NORMAL, MINIMUM, AND OPTIMUM CONCENTRATIONS. OPTIMUM CONCENTRATION WILL
PERMIT MAXIMUM GROWTH RATE, REPRODUCTION, OR PHOTOSYNTHESIS OF ALGAL
POPULATIONS. THIS IS BASED UPON THE CONCEPT THAT AN INADEQUATE
CONCENTRATION MAY RESULT IN DECREASED ASSIMILATION RATE AND AN
EXCESSIVE CONCENTRATION MAY BE INHIBITORY OR TOXIC. PURE, OR AT LEAST,
UNIALGAL CULTURES, ARE NEEDED FOR THIS STUDY. MOST NUTRITIONAL
INVESTIGATIONS HAVE BEEN MADE ON ALGAE SPECIES EASY TO CULTURE WHICH
PERHAPS DIFFER IN NUTRITION AND PHYSIOLOGY FROM THOSE DIFFICULT TO
CULTURE. MAJOR ABSOLUTE REQUIREMENTS INCLUDE CARBON, NITROGEN,
PHOSPHORUS, SULPHUR, POTASSIUM, AND MAGNESIUM; IRON AND MANGANESE ARE
REQUIRED IN SMALLER AMOUNTS, SODIUM IS UNESSENTIAL AND CALCIUM
REQUIREMENT IS CONTRADICTORY. SILICON IS A MAJOR REQUIREMENT FOR DIATOM
GROWTH BUT UNNECESSARY FOR OTHER FORMS. VARIOUS OTHER ELEMENTS--ZINC,
BORON, COBALT, MOLYBDENUM, AND COPPER MAY BE NECESSARY AS TRACE
ELEMENTS. NORMAL REQUIREMENTS OF VARIOUS ALGAE ARE GIVEN FOR CARBON,
NITROGEN, PHOSPHORUS, UNDER CERTAIN CONDITIONS. (JONES-WISCONSIN)

FIELD 05C

ACCESSION NO. W70-02804

MICRONUTRIENT REQUIREMENTS FOR GREEN PLANTS, ESPECIALLY ALGAE,

CHARLES F. KETTERING RESEARCH LAB., YELLOW SPRINGS, OHIO.

CLYDE EYSTER.

ALGAE AND MAN, JACKSON, DANIEL F, EDITOR, PLENUM PRESS, NY, P 86-119, 1964.
11 FIG, 2 TAB, 174 REF.

DESCRIPTORS:
*NUTRIENT REQUIREMENTS, *PLANT GROWTH, *ALGAE, ELEMENTS(CHEMICAL),
CHLOROPHYTA, CYANOPHYTA, CHLORELLA, MANGANESE, PHOTOSYNTHESIS, OXYGEN,
CARBON DIOXIDE, SCENEDESMUS, CHLOROPHYLL, IRON, CHELATION, ENZYMES,
CHLORINE, LIGHT INTENSITY, NITRATE, EUGLENA, CALCIUM, AMMONIUM SALTS,
NITROGEN FIXATION, BORON, AZOTOBACTER, MOLYBDENUM, COBALT, COPPER,
SODIUM, MAGNESIUM, POTASSIUM, SULFUR, PHOSPHORUS, HYDROGEN.

IDENTIFIERS:
MACRONUTRIENTS, MICRONUTRIENTS, CHLORELLA PYRENOIDOSA, AUTOTROPHIC
GROWTH, HETEROTROPHIC GROWTH, ANKISTRODESMUS, NOSTOC MUSCORUM,
SCENEDESMUS OBLIQUUES, SCENEDESMUS QUADRICAUDA, PORPHYRIDIUM CRUENTUM,
LEMNA MINOR, PHOTOPHOSPHORYLATION, VANADIUM, ZINC, EUGLENA GRACILIS,
ANABAENA CYLINDRICA, BACILLUS SUBTILIS, ASPERGILLUS ORYZAE NIGER,
SILICON, PHANEROGAMS.

ABSTRACT:
FIVE ELEMENTS, MANGANESE, IRON, CHLORINE, ZINC, AND VANADIUM ARE
GENERALLY REQUIRED BY ALGAE AND GREEN PLANTS FOR PHOTOSYNTHESIS.
MANGANESE AND CHLORINE HAVE A ROLE IN OXYGEN EVOLUTION. VANADIUM
ACCELERATES PHOTOSYNTHESIS UNDER HIGH LIGHT INTENSITY BUT THIS EFFECT
HAS BEEN REPORTED ONLY FOR SCENEDESMUS OBLIQUUS AND CHLORELLA. IRON IS
THE METAL PART OF AT LEAST TWO PLANT CYTOCHROMES WHICH OCCUR IN THE
CHLOROPLASTS OF GREEN CELLS AND PROBABLY FACILITATES TRANSFER OF
ELECTRONS DURING PHOTOSYNTHESIS. ROLE OF ZINC IN PHOTOSYNTHESIS IS
UNCERTAIN; IT MAY AID IN HYDROGEN TRANSFER SINCE SINC IS KNOWN TO BE
THE METAL CONSTITUENT OF VARIOUS DEHYDROGENASES. NITROGEN FIXATION IN
ALGAE HAVING THE CAPACITY TO FIX GASEOUS NITROGEN, REQUIRES THE
FOLLOWING ELEMENTS: BORON, CALCIUM, IRON, MOLYBDENUM, AND PERHAPS
COBALT; THESE ELEMENTS MAY BE REQUIRED FOR ONE OR MORE OTHER FUNCTIONS.
CRITICAL CONCENTRATION OF AN ELEMENT FOR ITS TWO OR MORE FUNCTIONS IS
USUALLY MUCH HIGHER FOR NITROGEN FIXATION. ALGAE WHICH ARE FORCED TO
FIX NITROGEN HAVE HIGH BORON AND CALCIUM REQUIREMENTS. PLACE OF COBALT
UTILIZATION IS UNCERTAIN. SOME EVIDENCE INDICATES THAT SMALL
CONCENTRATIONS OF MANGANESE, CALCIUM, BORON, COBALT, COPPER, AND
SILICON ARE REQUIRED FOR OTHER METABOLIC FUNCTIONS. (SEE
W69-07832). (JONES-WISCONSIN)

FIELD 05C

ACCESSION NO. W70-02964

ENVIRONMENTAL CONDITIONS AND THE PATTERN OF METABOLISM IN ALGAE,

WESTFIELD COLL., LONDON (ENGLAND). DEPT. OF BOTANY.

G. E. FOGG.

ALGAE AND MAN, DANIEL F. JACKSON, EDITOR, PLENUM PRESS, NY, P 77-85, 1964. 22 REF.

DESCRIPTORS:
 *ALGAE, *METABOLISM, *ENVIRONMENTS, VARIABILITY, CHLORELLA, CARBON, TEMPERATURE, LIGHT INTENSITY, NUTRIENT REQUIREMENTS, NITROGEN, NITROGEN FIXATION, PHOTOCHEMICAL, CARBON DIOXIDE, SYNTHESIS, ENZYMES, CHLOROPHYTA, RHODOPHYTA, SCENEDESMUS, CHLAMYDOMONAS, ECOLOGY, PHOSPHORUS, SULPHUR, HYDROGEN ION CONCENTRATION, TIME.

IDENTIFIERS:
 BOTRYOCOCCUS BRAUNII, METABOLIC PATTERN, ANABAENA CYLINDRICA, GLYCOLYSIS, TRICARBOXYLIC ACID CYCLE, AUTORADIOGRAPHY, MONODUS SUBTERRANEUS, NAVICULA PELLICULOSA, XANTHOPHYCEAE, BACILLARIOPHYCEAE, NITZSCHIA PALEA, MYXOPHYCEAE, CHLAMYDOMONAS MOEWUSSI, FAT ACCUMULATION.

ABSTRACT:
A SINGLE STRAIN OF ALGAL SPECIE MAY SHOW REMARKABLE VARIATIONS IN INTENSITY AND PATTERN OF ITS METABOLIC ACTIVITIES, DEPENDING ON EXPOSURE CONDITIONS. EFFECTS OF LIGHT INTENSITY, TEMPERATURE, AND HYDROGEN ION CONCENTRATION ON RATES OF INDIVIDUAL METABOLIC PROCESSES AND ON THEIR FINAL GROWTH IN CELL NUMBERS ARE KNOWN, YET THE DIFFERENTIAL EFFECTS THAT THESE CONDITIONS MAY HAVE ON DIFFERENT PROCESSES ARE LARGELY UNKNOWN. TIME FACTOR IS SIGNIFICANT; TEMPERATURE CHANGES MAY PRODUCE MARKED ALTERATIONS IN THE BALANCE OF MAJOR ANABOLIC ALGAL PROCESSES. AVAILABILITY OF METABOLITES MAY AFFECT METABOLIC PATTERN, APART FROM CHANGES ON OVERALL METABOLIC RATE. IN ALGAL POPULATIONS GROWING EXPONENTIALLY, SYNTHESIS OF PROTEINS AND OTHER PROTOPLASMIC CONSTITUENTS PREDOMINATE AND DIRECTLY UTILIZE INTERMEDIATES OF THE PHOTOSYNTHETIC CARBON CYCLE. A SIMILAR EFFECT IS PERHAPS INVOLVED IN EXCRETION OF GLYCOLIC ACID FROM PHOTOSYNTHESIZING CELLS. ALGAE IS ABLE TO PRODUCE ADAPTIVE ENZYMES AND METABOLIC PATTERN IS DETERMINED PRIMARILY BY RELATIVE ACTIVITIES OF THE VARIOUS ENZYME SYSTEMS. IT IS UNCLEAR WHETHER FAT ACCUMULATION IN NITROGEN-DEFICIENT CULTURES RESULTS FROM HALTING OF THE DEVELOPMENTAL CYCLE AT A POINT AT WHICH FAT SYNTHESIS PREDOMINATES. FACTORS AFFECTING FAT ACCUMULATION ARE DISCUSSED. (SEE W69-07832). (JONES-WISCONSIN)

FIELD 05C

ACCESSION NO. W70-02965

INSECTICIDES AND ALGAE: TOXICITY AND DEGRADATION,

ONTARIO WATER RESOURCES COMMISSION, TORONTO. DIV. OF RESEARCH.

A. E. CHRISTIE.

ONTARIO WATER RESOURCES COMMISSION, DIV OF RESEARCH PUBLICATION NO 27, 1968.
   26 P, 9 TAB, 31 REF.

   DESCRIPTORS:
      *ALGAE, *ALGAL CONTROL, *ALGICIDES, ANALYTICAL TECHNIQUES, AQUATIC
      ALGAE, AQUATIC ENVIRONMENT, AQUATIC MICROBIOLOGY, BALANCE OF NATURE,
      BIOASSAY, CHLOROPHYTA, ENVIRONMENTAL EFFECTS, INHIBITION, LIMNOLOGY,
      PESTICIDES, PESTICIDE REMOVAL, PESTICIDE RESIDUES, PHYSIOLOGICAL
      ECOLOGY, PHYTOPLANKTON, PHYTOTOXICITY, PLANKTON, POLLUTANT
      IDENTIFICATION, SEWAGE, SEWAGE DISPOSAL, SEWAGE LAGOONS, SEWAGE
      TREATMENT, STABILIZATION PONDS, WATER POLLUTION CONTROL, WATER
      POLLUTION EFFECTS, DDT.

   IDENTIFIERS:
      CHLORELLA PYRENOIDOSA, PESTICIDE DEGRADATION, SEVIN, MALATHION,
      ACETONE, ETHANOL.

   ABSTRACT:
      RELATIONSHIPS BETWEEN THREE INSECTICIDES (DDT, SEVIN AND MALATHION) AND
      ALGAE WITH RESPECT TO TOXICITY AND DEGRADATION WERE INVESTIGATED USING
      MATERIALS OBTAINED FROM A NEARBY WASTE STABILIZATION POND AND AXENIC
      CULTURES OF CHLORELLA PYRENOIDOSA AT PH 6.0 AND 9.0. DDT EXHIBITED NO
      TOXIC PROPERTIES TO EITHER SYSTEM AT CONCENTRATIONS UP TO 100
      MILLIGRAM/LITER AND WAS BROKEN DOWN ONLY TO A LIMITED EXTENT BY ALGAE.
      MALATHION EXHIBITED NO PERSISTENT INHIBITORY EFFECT TO ALGAE AND
      APPEARED TO BE METABOLIZED FAIRLY READILY IN COMPARISON TO THE OTHER
      TWO INSECTICIDES STUDIED. SEVIN AT 100 MILLIGRAM/LITER WAS MORE TOXIC
      TO THE MIXED POND ALGAE THAN TO CHLORELLA. THE TOXICITY OF SEVIN TO
      CHLORELLA WAS MORE PRONOUNCED AT PH 6 THAN AT PH 9 (30 AND 15%
      INHIBITION, RESPECTIVELY). THE EFFECT OF SEVIN COULD BE DETECTED AT 0.1
      MILLIGRAM/LITER, BUT INCREASING CONCENTRATS HAD LITTLE INCREASED
      EFFECTIVENESS. (FITZGERALD-WISCONSIN)

   FIELD 05G

   ACCESSION NO. W70-02968

THE ORIGIN AND QUANTITIES OF PLANT NUTRIENTS IN LAKE MENDOTA,

WISCONSIN UNIV., MADISON.

GERARD A. ROHLICH, AND WILLIAM L. LEA.

LAKE INVESTIGATIONS COMMITTEE REPORT, WISCONSIN UNIV, MADISON, 1949. 8 P, 6
   TAB.

   DESCRIPTORS:
      *EUTROPHICATION, *AGRICULTURAL WASTES, *ALGAE, *LAKES, *TRIBUTARIES,
      SAMPLING, FLOW, ANALYSIS, FERTILIZATION, PLANT GROWTH SUBSTANCES,
      SEWAGE, SULFATE, MANGANESE, IRON, CARBON DIOXIDE.

   IDENTIFIERS:
      *INLAND, HORIZONTAL SAMPLING, VERTICAL SAMPLING, LAKE MENDOTA(WIS),
      LAKE MONONA(WIS), LAKE WAUBESA(WIS), LAKE KEGONSA(WIS), WEED BEDS,
      PHOSPHORUS, NITROGEN BALANCE.

   ABSTRACT:
      TO OBTAIN A COMPREHENSIVE PICTURE OF EUTROPHICATION DYNAMICS OF LAKE
      MENDOTA, WISCONSIN, THREE SERIES OF STATIONS--INLAND, HORIZONTAL, AND
      VERTICAL-- WERE ESTABLISHED ON THE LAKE AND ITS TRIBUTARIES. ANALYSES
      OF SAMPLES COLLECTED AT INLAND STATIONS ENABLED DETERMINATION OF
      NUTRIENTS RETAINED IN THE LAKE. SAMPLING STATIONS OF THE HORIZONTAL
      SERIES FORMED TRANSECTS FROM UPPER PARTS OF STREAMS TO THEIR MOUTHS AND
      INTO THE LAKE. THESE STATIONS PROVIDED INFORMATION ON THE MODIFICATION
      OF ORIGINAL NUTRIENT CONCENTRATION BY WEED BEDS AND ALGAE. VERTICAL
      SERIES STATIONS WERE LOCATED IN TWO 65-FOOT-DEEP AND ONE 80-FOOT-DEEP
      HOLES. SAMPLING OF THREE WATER HORIZONS--SURFACE, THERMOCLINE AT 10
      METERS, AND ABOUT 0.5 METERS ABOVE THE BOTTOM--AIMED TO REVEAL
      VARIATION IN NUTRIENT CONCENTRATION AND AQUATIC BIOMASS WITH DEPTH.
      LAKE'S INFLOW-OUTFLOW BALANCE CONSTITUTED 130 PLUS OR MINUS 1.2 CUBIC
      FEET/SECOND. ANALYSES OF CHEMICAL FACTORS INCLUDED PH, ALKALINITY,
      DISSOLVED OXYGEN, BIOCHEMICAL OXYGEN DEMAND, SOLUBLE PHOSPHORUS, TOTAL
      PHOSPHORUS, FREE AMMONIA NITROGEN, TOTAL ORGANIC NITROGEN, NITRITE,
      NITRATE, SULFATES, SILICATES, MANGANESE, AND IRON. DURING THE YEAR
      17,460 POUNDS SOLUBLE PHOSPHORUS ENTERED THE LAKE, THE AMOUNT
      CORRESPONDING TO AN APPLICATION OF 1.8 POUNDS/ACRE OF LAKE SURFACE.
      ENRICHMENT IN ORGANIC NITROGEN, LARGELY IN NITRATE FORM, AMOUNTED TO
      259,720 POUNDS, EQUIVALENT TO FERTILIZATION OF 26.7 POUNDS/ACRE.
      (WILDE-WISCONSIN)

   FIELD 05C

   ACCESSION NO. W70-02969

IN VITRO RESPONSES OF SOFT WATER ALGAE TO FERTILIZATION,

ONTARIO WATER RESOURCES COMMISSION, TORONTO. DIV. OF RESEARCH.

A. E. CHRISTIE.

DIVISION OF RESEARCH PAPER NO 2024, AUG 1969. 16 P, 2 FIG, 4 TAB, 9 REF.

DESCRIPTORS:
*BIOASSAY, PHYTOPLANKTON, LAKES, EUTROPHICATION, SAMPLING, SURFACES
WATERS, CARBON, FERTILITY, ALGAE, NITROGEN, PHOSPHORUS, WATER POLLUTION
EFFECTS.

IDENTIFIERS:
*TWELVE MILE LAKE(CANADA), *ONTARIO, *CANADA, NUTRIENT LEVELS, SOFT
WATER.

ABSTRACT:
A 'PRELIMINARY' SINGLE SAMPLE FROM OLIGOTROPHIC TWELVE MILE LAKE,
ONTARIO, WAS BIOASSAYED BY SPIKING WITH SODIUM BICARBONATE, POTASSIUM
NITRATE, AND POTASSIUM DIHYDROGEN PHOSPHATE, OR COMBINATIONS OF THESE
COMPOUNDS. BOTTLES WERE CARRIED 21 DAYS AT 20 C UNDER A DIURNAL CYCLE
OF 14 HOURS LIGHT AND 10 HOURS DARK UNDER THE EXTREMELY HIGH LIGHT
INTENSITY OF 4000-FOOT-CANDLES. THE SAMPLE'S INDIGINOUS POPULATION
SERVED AS THE ASSAY RESPONSE ORGANISMS, AND WERE MEASURED BY AREAL
STANDARD UNITS. THE ONLY RESPONSE BY THE ALGAE IN THE SAMPLE WAS TO A
COMBINATION OF NITROGEN AND PHOSPHORUS. (GERHOLD-WISCONSIN)

FIELD 05C

ACCESSION NO. W70-02973

STUDIES ON CHEMICALS WITH SELECTIVE TOXICITY TO BLUE-GREEN ALGAE,

WISCONSIN UNIV., MADISON. WATER CHEMISTRY LAB.

GEORGE P. FITZGERALD, GERALD C. GERLOFF, AND FOLKE SKOOG.

SEWAGE AND INDUSTRIAL WASTES, VOL 24, NO 7, P 888-896, 1952. 1 FIG, 6 TAB, 7
REF.

DESCRIPTORS:
*TOXICITY, *CYANOPHYTA, *CHEMCONTROL, EUTROPHICATION, BASS, WATER
POLLUTION CONTROL, MINNOWS, NUISANCE ALGAE, SUNFISHES, ALGAE, DAPHNIA,
LAKES, CHLOROPHYTA, STREAMS, DIATOMS, RESERVOIRS, CULTURES, SEWAGE
EFFLUENTS, WISCONSIN, ALGICIDES, HYDROLYSIS, COPPER SULFATE, FISHES,
HERBICIDES, MIDGES, PESTICIDES.

IDENTIFIERS:
*SELECTIVE TOXICITY, CHEMICAL STRUCTURE, 2,3-DICHLORONAPHTHOQUINONE,
MICROCYSTIS AERUGINOSA, APHANIZOMENON FLOS-AQUAE, MINIMUM LETHAL DOSES,
CHU'S MEDIA, QUINONE DERIVATIVES, ALKYL SUBSTITUTED IMIDAZOLINES,
BIOLOGICAL ACTIVITY, QUARTERNARY, AMMONIUM COMPOUNDS, MINIMUM LETHAL
EXPOSURE, PHENANTHROQUINONE, MADISON(WIS), ELODEA, REDOX COMPOUNDS,
CERATOPHYLLUM, MYROPHYLLUM, RHIZOCLONIUM, PHOTOBIOLOGY.

ABSTRACT:
PROCEDURES USED AND RESULTS OF TOXICITY TESTS OF 300 ORGANIC CHEMICALS
TO MICROCYSTIS AERUGINOSA, A BLOOM-PRODUCING CYANOPHYTE (BLUE-GREEN
ALGA) ARE DESCRIBED. MANY COMPOUNDS WERE TOXIC IN CONCENTRATIONS IN THE
RANGE 1-10 PARTS/MILLON (PPM); SEVERAL QUINONE-LIKE COMPOUNDS ARE TOXIC
BELOW 1 PPM, AND ONE, 2,3-DICHLORONAPTHOQUINONE, IS LETHAL AT
CONCENTRATIONS OF 0.002 PPM. APPARENTLY SELECTIVELY TOXIC TO NUISANCE
CYANOPHYTES, IT IS NONTOXIC TO MOST GREEN ALGAE AND HIGHER PLANTS, AND
HAS NO SHORTTERM DELETERIOUS EFFECTS ON SOME FISHES AND INVERTEBRATES
EVEN WHEN PRESENT IN EXCESS OF SATURATION CONCENTRATIONS. FURTHER STUDY
OF THIS COMPOUND TO RULE OUT INJURIOUS EFFECTS TO DESIRABLE SPECIES
AFTER LONG EXPOSURES ARE RECOMMENDED. AGE OF CULTURES OF MICROCYSTIS
INFLUENCE ITS RESPONSE TO THE COMPOUND; CENTRIFUGED CELLS FROM BOTH
6-DAY AND 19-DAY CULTURES WERE MARKEDLY LESS SENSITIVE WHEN RESUSPENDED
IN OLD MEDIUM. THIS EFFECT WAS PARTIALLY COUNTERACTED AFTER
FORTIFICATION OF OLD MEDIUM WITH NUTRIENTS. ESTIMATES OF MINIMUM LETHAL
EXPOSURES INDICATE THAT COMPOUND IS EFFECTIVELY TOXIC AFTER 8-10
MINUTES TREATMENT OF CELLS. LIMITED FIELD TESTS IN BARRELS, TANKS, AND
LAGOONS, CONTAINING DENSE GROWTHS OF NUISANCE ALGAE, PRINCIPALLY
APHANIZOMENON FLOS-AQUAE, IN LAKE WATER OR SEWAGE EFFLUENT, INDICATE
THAT COMPOUND IS EFFECTIVE IN LOW CONCENTRATIONS (0.01-0.1 PPM).
(EICHHORN-WISCONSIN)

FIELD 05G

ACCESSION NO. W70-02982

EVALUATION OF WATER QUALITY TRENDS, SHENANDOAH RIVER, VIRGINIA, FRONT ROYAL TO
BERRYVILLE,

FEDERAL WATER POLLUTION CONTROL ADMINISTRATION, CHARLOTTESVILLE, VA. MIDDLE
ATLANTIC REGION.

JOHN W. BAUMEISTER.

FEDERAL WATER POLLUTION CONTROL ADMINISTRATION, MIDDLE ATLANTIC REGION
TECHNICAL REPORT 2, DEC 1968. 14 P, 5 FIG, 5 TAB, 20 REF.

DESCRIPTORS:
*WATER QUALITY, *WATER POLLUTION, *RIVERS, WATER CHEMISTRY, WATER
POLLUTION SOURCES, PHOSPHORUS, NITROGEN COMPOUNDS, EUTROPHICATION,
PRODUCTIVITY, ALGAE, AQUATIC LIFE.

IDENTIFIERS:
SHENANDOAH RIVER.

ABSTRACT:
PHYSICAL AND CHEMICAL DATA COLLECTED AT THE BERRYVILLE SURVEILLANCE
STATION WERE EVALUATED TO DETERMINE POLLUTION AND EUTROPHICATION TRENDS
IN THE MAIN STEM OF THE SHENANDOAH DOWNSTREAM FROM FRONT ROYAL.
CHEMICAL QUALITY WAS WITHIN THE RAW WATER CRITERIA ESTABLISHED FOR
PUBLIC WATER SUPPLIES; HOWEVER, TOTAL PHOSHORUS AND NITROGEN CONTENT OF
RIVER WATER WAS SUFFICIENTLY HIGH TO INDICATE A FERTILE STREAM CAPABLE
OF PRODUCING ABUNDANT AQUATIC ORGANISMS. THIRTEEN YEARS OF BOTTOM
SAMPLES COLLECTED OVER A 30 YR PERIOD IN THIS REACH SHOW A GRADUAL
INCREASE IN THE PRODUCTIVITY OF AQUATIC INVERTEBRATES. FOR THE MOST
PART, THESE SAMPLES CONTAINED CLEAN WATER AND FACULTATIVE FORMS WHICH
INDICATE THE PRESENCE OF RELATIVELY CLEAN WATER. PHYTOPLANKTON NUMBERS
AND POPULATION COMPOSITION DURING THE PERIOD 1962-1965 INDICATE A
CHANGE FROM A DIATOM DOMINATED POPULATION TO ONE DOMINATED BY GREEN
ALGAE. NUISANCE ALGAE WERE NOT PRESENT IN QUANTITIES SUFFICIENT TO
CAUSE WATER TREATMENT PROBLEMS. (KNAPP-USGS)

FIELD 05A

ACCESSION NO. W70-03068

ALGAL GROWTH AND PRIMARY PRODUCTIVITY IN A THERMAL STREAM,

WASHINGTON UNIV., SEATTLE. DEPT. OF ZOOLOGY.

JOHN G. STOCKNER.

JOURNAL FISHERIES RESEARCH BOARD OF CANADA, VOL 25, NO 10, P 2037-2058, 1968.
13 FIG, 7 TAB, 25 REF.

DESCRIPTORS:
*PRIMARY PRODUCTIVITY, *ALGAE, *THERMAL WATER, PERIPHYTON, BIOMASS,
PHYSIOLOGICAL ECOLOGY, CYANOPHYTA, DIFFUSION, RESPIRATION, PHOSPHATES,
ECOSYSTEMS, ORGANIC MATTER, WATER POLLUTION, THERMAL POLLUTION,
STREAMS, NUTRIENTS, LIGHT, WASHINGTON, WATER CHEMISTRY, OXYGEN,
HYDROGEN ION CONCENTRATION, PHOTOSYNTHESIS, ALKALINITY, NITRATES,
CONDUCTIVITY, CHLORIDES, CARBONATES, SULFATES, BICARBONATES, POTASSIUM,
IRON, SODIUM, CALCIUM, MAGNESIUM, PHOSPHORUS.

IDENTIFIERS:
*ALGAL GROWTH, *THERMOBIOLOGY, *DIURNAL OXYGEN CURVES, CALORIMETRY,
SOLARIMETRY, YELLOWSTONE NATIONAL PARK, MT LASSEN NATIONAL PARK, MT
RAINIER NATIONAL PARK, OHANAPECOSH HOT SPRINGS(WASH), SCHIZOTHRIX,
PHORMIDIUM, WINKLER TECHNIQUE, OXYGEN ELECTRODES, ATOMIC ABSORPTION
SPECTROPHOTOMETRY, TOTAL DISSOLVED SOLIDS, HEDRIODISCUS, CALOPARYPHUS.

ABSTRACT:
TROPHIC RELATIONSHIPS IN FLOWING WATER ARE NOT AS WELL UNDERSTOOD AS
THOSE IN MORE STATIC ENVIRONMENTS. BECAUSE OF THEIR RELATIVELY CONSTANT
CONDITIONS, STREAMS DRAINING THERMAL SPRINGS ARE IDEAL SITES FOR SUCH
STUDIES. ESTIMATES OF PRIMARY PRODUCTION AND RESPIRATION OF BLUE-GREEN
ALGAE IN STREAM DRAINING ONE OF THE OHANAPECOSH HOT SPRINGS,
WASHINGTON, WERE OBTAINED THROUGHOUT 1966 FROM DAILY OXYGEN CURVES.
RATES OF GROSS PRIMARY PRODUCTIVITY WERE SLIGHTLY HIGHER THAN PUBLISHED
VALUES FROM OTHER AQUATIC ECOSYSTEMS, BUT WERE LESS THAN RECENT
ESTIMATES, DERIVED FROM MEASUREMENT OF RADIOCARBON UPTAKE FROM OTHER
THERMAL WATERS. CALCULATED RATIOS, PRODUCTION/RESPIRATION, WERE IN
RANGE 1.1-5.0. AS GROWING SEASON PROGRESSED, PHOTOSYNTHETIC EFFECTS ON
IONIC COMPOSITION OF EFFLUENT WATER, AS WELL AS MORPHOLOGICAL CHANGES
IN ALGAL MAT, WERE OBSERVABLE. CHANGES IN ALGAL BIOMASS DURING
1964-1965, TOGETHER WITH GROWTH RATES IN DENUDED AREAS BY TWO METHODS
OF HARVEST WERE STUDIED. ACCELERATED GROWTH OCCURRED IN MAY AND JUNE,
ASSOCIATED WITH MAXIMAL VALUES FOR PRIMARY PRODUCTION, WHEREAS GROWTH
WAS MINIMAL IN WINTER. SIGNIFICANT DIFFERENCES IN GROWTH BETWEEN YEARS
WERE ATTRIBUTED TO DIFFERENCES IN HARVESTING METHODS, AND TO PRODUCTION
OF AUTO-INHIBITORY TOXINS BY SENESCENT CELLS IN MATURE MAT. ANNUAL
ENERGY BUDGET FOR MAT IS DERIVED FROM 1966 ESTIMATES.
(EICHHORN-WISCONSIN)

FIELD 05C

ACCESSION NO. W70-03309

CONTROL OF BLUE-GREEN ALGAE BLOOMS WITH 2,3-DICHLORONAPHTHOQUINONE,

WISCONSIN UNIV., MADISON. DEPT. OF BOTANY.

GEORGE P. FITZGERALD, AND FOLKE SKOOG.

SEWAGE AND INDUSTRIAL WASTES, VOL 26, NO 9, P 1136-1140, SEPT 1954. 1 FIG, 1 REF.

DESCRIPTORS:
*ALGICIDES, *CYANOPHYTA, *EUTROPHICATION, WISCONSIN, WATER QUALITY CONTROL, LAKES, WATER POLLUTION SOURCES, CHLOROPHYTA, NUISANCE ALGAE, FISH, ZOOPLANKTON, SNAILS, APPLICATION METHODS.

IDENTIFIERS:
*2,3-DICHLORONAPHTHOQUINONE, MICROCYSTIS, APHANIZOMENON, ANABAENA, GLOEOTRICHIA, JANESVILLE(WIS), SPAULDING'S POND(WIS), HYDRODICTYON, NAJAS, ANACHARIS, LAKE WAUBESA(WIS), MACROPHYTES, ALGAL GROWTH.

ABSTRACT:
METHODS FOR APPLICATION OF THE SELECTIVE ALGICIDE, 2,3-DICHLORONAPHTHOQUINONE (2,3-CNQ), AND RESULTS OF A LARGE-SCALE TEST OF ITS EFFECTIVENESS IN CONTROLLING EXCESSIVE GROWTHS OF CYANOPHYTES (BLUE-GREEN ALGAE) IN SPAULDING'S POND, WISCONSIN, ARE REPORTED. THIS HIGHLY EUTROPHIC LAKE OF 27 ACRES NORMALLY SUPPORTS A CONTINUOUS BLOOM OF ALGAE, PRINCIPALLY MICROCYSTIS AND ALPHANIZOMENON, DURING SUMMER. SPRAY APPLICATIONS OF 2,3-CNQ SUSPENSIONS, YIELDING CONCENTRATIONS OF 30-55 PARTS PER BILLION IN THE LAKE, CLUMPED AND KILLED HEAVY GROWTHS OF CYANOPHYTES. SUCH CHEMICAL TREATMENT HAD NO OBSERVABLE HARMFUL EFFECTS ON CHLOROPHYTES (GREEN ALGAE), HIGHER AQUATIC PLANTS, FISH, OR ZOOPLANKTON. SOME INSTANCES OF RAPIDLY RECURRING CYANOPHYTE BLOOMS REQUIRED REPEATED APPLICATIONS. LASTING EFFECTS OF TREATMENT WITH 2,3-CNQ MAY BE INDIRECTLY INCREASED AS A RESULT OF VIGOROUS GROWTH OF CHLOROPHYTES AND MACROPHYTES WHICH FOLLOWS SUPPRESSION OF CYANOPHYTES. (EICHHORN-WISCONSIN)

FIELD 05G, 05C

ACCESSION NO. W70-03310

MICROBIOLOGY OF SEWAGE LAGOONS-EFFECTS OF INDUSTRIAL WASTES ON LAGOON ECOLOGY,

NORTH DAKOTA UNIV., GRAND FORKS. SCHOOL OF MEDICINE.

JOHN W. VENNES.

AVAILABLE FROM THE CLEARINGHOUSE AS PB-189 159, $3.00 IN PAPER COPY $0.65 IN MICROFICHE. COMPLETION REPORT, AUG 1969. 82 P, 3 FIG, 41 TAB, 35 REF. OWRR PROJECT NO A-016-NDAK.

DESCRIPTORS:
*MICROBIOLOGY, *SEWAGE LAGOONS, *INDUSTRIAL WASTES, SEWAGE TREATMENT, NORTH DAKOTA, COLIFORMS, OXIDATION LAGOONS, BIOCHEMICAL OXYGEN DEMAND, PHOTOSYNTHESIS, HYDROGEN SULFIDE, LAKES, ALGAE, ENTERIC BACTERIA, METHANE BACTERIA, CHEMICAL ANALYSIS, WATER POLLUTION, CHLORELLA, SCENEDESMUS.

IDENTIFIERS:
*MICROBIAL ECOLOGY, PURPLE SULFUR BACTERIA, POPULATION EQUIVALENTS(WASTES), RED RIVER VALLEY(NORTH DAKOTA), SULFATE REDUCTION, POTATO PROCESSING WASTES, CHEESE MANUFACTURING WASTES, STARCH MANUFACTURING WASTES, WATER QUALITY STANDARDS, GRAFTON(ND), LAKOTA(ND), HARVEY(ND), CHROMATIUM VINOSUM, THIOCAPSA FLORIDANA, ENTEROCOCCI, PARK RIVER(ND), SHEYENNE RIVER(ND), THIOPEDIA.

ABSTRACT:
STATE AND FEDERAL LEGISLATION, REGULATING EFFLUENT DISCHARGE TO NATURAL WATERS, IS SUCH THAT FOOD PROCESSING INDUSTRIES IN NORTH DAKOTA MUST DISCHARGE WASTES INTO MUNICIPAL LAGOONS RESULTING IN FREQUENT OVERLOADS OF THESE SYSTEMS. LAGOON AT GRAFTON, DESIGNED FOR 5-DAY BIOCHEMICAL OXYGEN DEMAND (BOD-5) OF 30 POUNDS/ACRE PER DAY, RECEIVES 240, LARGELY FROM TWO POTATO PROCESSING PLANTS. INCREASED LOADINGS FAVOR (AS DOMINANT MICROORGANISMS IN THESE LAGOONS) THE PURPLE SULFUR BACTERIA (PSB), WHICH PHOTOSYNTHETICALLY UTILIZE HYDROGEN SULFIDE AND VARIOUS VOLATILE ORGANIC ACIDS FOR GROWTH. MOST REDUCTION IN BOD-5 AT GRAFTON IS ATTRIBUTABLE TO THREE GENERA OF PSB: CHROMATIUM, THIOPEDIA, THIOCAPSA. INCREASED LOADING DOES NOT SIGNIFICANTLY MULTIPLY NUMBERS OF ENTERIC ORGANISMS HERE. A NATURAL LAKE, USED FOR SEWAGE STABILIZATION IN LAKOTA, RECEIVES WHEY; IT RECOVERS TO AN ALGAL PHASE DURING SUMMER, AND SHOWS NO INCREASE IN NUMBERS OF ENTERIC ORGANISMS. AT HARVEY, AN AERATED CONTINUOUS DISCHARGE LAGOON, RECEIVING DOMESTIC WASTES AND INTERMITTENT CREAMERY WASTES, REDUCES BOD-5 TO NEAR STREAM LEVELS, BUT CANNOT REDUCE COLIFORMS TO ACCEPTABLE NUMBERS. IF THE LATTER TREATMENT IS TO SUCCEED IN NORTHERN CLIMATES, CONSIDERATION MUST BE GIVEN TO LONGER DETENTION TIMES, GREATER DELIVERY OF OXYGEN TO CELLS, CHLORINATION, OR REEVALUATION OF STREAM STANDARDS. (EICHHORN-WISCONSIN)

FIELD 05C

ACCESSION NO. W70-03312

A LABORATORY METHOD FOR THE STUDY OF MARINE BENTHIC DIATOMS,

OREGON STATE UNIV., CORVALLIS. DEPT. OF BOTANY.

C. DAVID MCINTIRE, AND BARRY L. WULFF.

LIMNOLOGY AND OCEANOGRAPHY, VOL 14, NO 5, P 667-678, 1969. 6 FIG, 4 TAB, 20 REF.

DESCRIPTORS:
*MARINE ALGAE, *DIATOMS, *BENTHIC FLORA, LIGHT INTENSITY, BIOMASS, DISTRIBUTION, ECOSYSTEMS, POPULATION, OREGON, INTERTIDAL AREAS, PHYSIOLOGICAL ECOLOGY, LITTORAL, ESTUARIES, PHOTOSYNTHESIS, RESPIRATION, PIGMENTS, CHLOROPHYLL, TEMPERATURE, OXYGEN, SALINITY.

IDENTIFIERS:
DESICCATION, YAQUINA BAY(ORE), CAROTENOIDS, SPECIES, DIVERSITY INDEX, ACHNANTHES, ACTINOPTYCHUS, AMPHIPLEURA RUTILANS, AMPHORA, AULACODISCUS, BACILLARIA, BIDDULPHIA, CHAETOCEROS, COCCONEIS, COSCINODISCUS, DIMEROGRAMMA, DIPLONEIS, EUNOTOGRAMMA, FRAGILARIA, GOMPHONEMA, GYROSIGMA, MELOSIRA, NAVICULA, NITZSCHIA, SKELETONEMA, PLAGIOGRAMMA, PLEUROSIGMA, RHAPHONEIS, RHOICOSPHENIA, SURIRELLA, SYNEDRA.

ABSTRACT:
EFFECTS OF LIGHT INTENSITY AND EXPOSURE TO DESICCATION ON VERTICAL DISTRIBUTION AND GROWTH OF MARINE BENTHIC DIATOM POPULATIONS WERE INVESTIGATED IN A LABORATORY MODEL ECOSYSTEM AND RESPIROMETER CHAMBER. DIATOM FLORA DEVELOPED WAS SIMILAR TO THAT FROM FIELD STATIONS IN LOWER YAQUINA BAY, OREGON. VERTICAL DISTRIBUTION OF MANY SPECIES WAS CLOSELY RELATED TO LIGHT INTENSITY AND PERIOD OF EXPOSURE TO DESICCATION. BIOMASS ACCUMULATED RAPIDLY ON SUBSTRATES SUBJECTED TO HIGH LIGHT INTENSITIES WITHOUT EXPOSURE TO DESICCATION. COMMUNITIES ACCLIMATED TO VARIOUS LIGHT INTENSITIES AND PERIOD OF DESICCATION RESPONDED DIFFERENTLY TO CHANGES IN LIGHT INTENSITY IN RESPIROMETER CHAMBER. EXPERIMENTAL RESULTS SHOW THAT LABORATORY APPARATUS DESCRIBED CAN BE USEFUL IN STUDY OF SIMPLIFIED INTERTIDAL COMMUNITIES. SYSTEM PROVIDES SOME LABORATORY CONTROL OVER SALINITY, LIGHT INTENSITY, TIDAL CYCLE, AND PROBABLY TEMPERATURE. MOVEMENTS AND GRAZING ACTIVITIES OF SMALL MARINE INVERTEBRATES COULD BE STUDIED. IT IS EMPHASIZED THAT USE OF SUCH LABORATORY ECOSYSTEMS IS SUBJECT TO CERTAIN LIMITATIONS RESULTING FROM SIMPLIFICATIONS OF NATURE. LABORATORY ECOSYSTEMS ARE BEST USED TO GAIN INFORMATION TO SUPPLEMENT AND HELP UNDERSTAND CONCURRENT OBSERVATIONS IN THE FIELD. (JONES-WISCONSIN)

FIELD 05C

ACCESSION NO. W70-03325

EFFECT OF SURFACE/VOLUME RELATIONSHIP, CO-SUB-2 ADDITION, AERATION, AND MIXING
ON NITRATE UTILIZATION BY SCENEDESMUS CULTURES IN SUBSURFACE AGRICULTURAL
WASTE WATERS,

CALIFORNIA STATE DEPT. OF WATER RESOURCES, FRESNO; AND FEDERAL WATER
    POLLUTION CONTROL ADMINISTRATION, FRESNO, CALIF.

RANDALL BROWN, AND JAMES ARTHUR.

PROCEEDINGS OF THE EUTROPHICATION-BIOSTIMULATION ASSESSMENT WORKSHOP, JUNE
    19-21, 1969, CALIFORNIA UNIV, BERKELEY, SANITARY ENGINEERING RESEARCH LAB
    AND NATIONAL EUTROPHICATION RESEARCH PROGRAM, CORVALLIS, ORE, PAC NORTHWEST
    WATER LAB, P 80-97. 12 FIG, 1 TAB, 1 REF.

DESCRIPTORS:
    *SURFACE, *VOLUME, *CARBON DIOXIDE, *AERATION, *NITRATES, *CULTURES,
    *SCENEDESMUS, *SUBSURFACE WATERS, *AGRICULTURE, *WASTES, ALGAE, ASSAY,
    DISSOLVED SOLIDS, CHEMICAL ANALYSIS, PHOSPHATES, IRON COMPOUNDS,
    HYDROGEN ION CONCENTRATION, ALKALINITY, CALIFORNIA, LIGHT INTENSITY,
    TEMPERATURE, IONS, SPECTROPHOTOMETRY, SUSPENDED LOAD, FLUORESCENCE,
    SUMPS.

IDENTIFIERS:
    STRIPPING, FIREBAUGH CENTER(CALIF), SCENEDESMUS QUADRICAUDA, CELL
    COUNTS, SPECIES DETERMINATION, ABSORBANCE, COENOBIA, SWIRLING, VOLATILE
    SOLIDS.

ABSTRACT:
    A METHOD OF REMOVING NITRATE-NITROGEN FROM SUBSURFACE AGRICULTURAL
    DRAINAGE WATER--ALGAE STRIPPING--ENCOURAGES ALGAL GROWTH, NITRATE IS
    CONVERTED TO CELLULAR MATERIALS, AND ALGAE SEPARATED FROM LIQUID PHASE
    BY CENTRIFUGATION, FLOCCULATION, ETC. THIS WATER, HIGH IN DISSOLVED
    SALTS, IS NOT A COMPLETE MEDIA FOR SCENEDESMUS (THE ALGA USED) GROWTH.
    LABORATORY STUDIES ARE CONDUCTED TO DETERMINE NUTRIENT ADDITIONS
    NECESSARY FOR OPTIMUM GROWTH (AND NITROGEN REMOVAL). ONE EXPERIMENT
    DETERMINED EFFECT OF FLASK SIZE, SURFACE/VOLUME RATIO, AND TYPE OF
    MIXING (INCLUDING AERATION AND CARBON DIOXIDE ADDITION). SURFACE/VOLUME
    RATIO SHOWED MAXIMUM GROWTH OR NITROGEN REMOVAL OCCURRED IN
    250-MILLILITER FLASKS CONTAINING 125 MILLILITERS OF CULTURE. COMPRESSED
    AIR ALONE EFFECTED A SIMILAR GROWTH RATE AS COMPRESSED AIR ENRICHED
    WITH CARBON DIOXIDE, NEITHER ADDITIVE AFFECTING ULTIMATE YIELD. IN THE
    AGRICULTURAL DRAINAGE WATER USED, CARBON WAS PROBABLY A LIMITING
    ELEMENT AND RATE OF EXCHANGE ACROSS THE SURFACE OF THE CULTURE WAS NOT
    SUFFICIENTLY RAPID WITH SWIRLING ONLY. THE 7.3 - 7.6 PH RANGE
    DIFFERENCE AFFECTED NUTRIENT AVAILABILITY. IN DETERMINING OPTIMUM
    AMOUNT OF IRON AND PHOSPHORUS FOR NITROGEN REMOVAL, TYPE OF MIXING WAS
    SIGNIFICANT. MEASUREMENT OF THE NITRATE PARAMETER BY THE SPECIFIC ION
    ELECTRODE METHOD CAN PRODUCE A RAPID AND RELIABLE ESTIMATE OF ALGAL
    ACTIVITY. (SEE W70-02775).
    (JONES-WISCONSIN)

FIELD 05D

ACCESSION NO. W70-03334

ESTABLISHMENT OF TOWN SANITARY DISTRICTS.

WIS STAT ANN SECS 60.305-60.315 (1957), AS AMENDED, (SUPP 1969) 60.316 (SUPP
    1969).

DESCRIPTORS:
    *WISCONSIN, *ADMINISTRATIVE AGENCIES, *WATER WORKS, *WASTE DISPOSAL,
    DOMESTIC WASTES, SEWERS, SEWAGE TREATMENT, SEWAGE DISPOSAL, STORM
    DRAINS, SURFACE DRAINAGE, ALGAL CONTROL, AQUATIC LIFE, WATER
    CONSERVATION, PUBLIC HEALTH, WATER POLLUTION, CONTRACTS, REGULATION,
    RIGHT-OF-WAY, ESTIMATED COSTS, LEGAL ASPECTS, LEGISLATION, SEWAGE
    DISTRICTS.

ABSTRACT:
    THE TOWN SANITARY DISTRICT COMMISSIONS ARE REQUIRED TO PROJECT, PLAN,
    CONSTRUCT, AND MAINTAIN SYSTEMS OF WATER WORKS AND GARBAGE OR REFUSE
    DISPOSAL, INCLUDING SANITARY SEWERS, SURFACE SEWERS, OR STORM WATER
    SEWERS. THEY ARE AUTHORIZED TO PROVIDE FOR SEWAGE COLLECTION AND THE
    CHEMICAL TREATMENT OF WATER FOR CONTROL OF NUISANCE PRODUCING AQUATIC
    GROWTHS. THEY ARE TO UNDERTAKE IMPROVEMENTS FOR THE PROMOTION OF PUBLIC
    HEALTH. THEY MAY REQUIRE THE INSTALLATION OF PRIVATE SEWAGE SYSTEMS.
    CHARGES FOR SERVICES AND METHODS OF FINANCING CONSTRUCTION ARE
    DISCUSSED. THE DEPARTMENT OF RESOURCE DEVELOPMENT MAY ESTABLISH TOWN
    SANITARY DISTRICTS IN AREAS WHERE IT FINDS EXISTING FACILITIES
    INADEQUATE AND WHERE A MENACE TO HEALTH OR POLLUTION OF SURFACE WATERS
    EXISTS. THIS DETERMINATION BY THE DEPARTMENT IS SUBJECT TO JUDICIAL
    REVIEW. (DUSS-FLORIDA)

FIELD 05E

ACCESSION NO. W70-03344

BIOLOGICAL N2 FIXATION IN LAKES,

WISCONSIN UNIV., MADISON. WATER RESOURCES CENTER.

ROBERT H. BURRIS.

AVAILABLE FROM THE CLEARINGHOUSE AS PB-189 163, $3.00 IN PAPER COPY, $0.65 IN
MICROFICHE. WISCONSIN WATER RESOURCES CENTER, TECHNICAL REPORT OWRR
B-020-WIS, 1969. 6 P. OWRR PROJECT B-020-WIS.

DESCRIPTORS:
*NITROGEN-FIXATION, *NUTRIENTS, *LAKES, *ALGAE, *NITROGEN, WISCONSIN.

IDENTIFIERS:
*ANALYTICAL METHODS, *BIOCHEMISTRY METHODS, SEASONS, ACETYLENE
REDUCTION TEST.

ABSTRACT:
ACETYLENE REDUCTION BY NITROGEN-FIXING ORGANISMS WAS USED TO ESTIMATE
POTENTIAL NITROGEN FIXATION BY BLUE-GREEN ALGAE IN LAKES. THE METHOD
GIVES A USEFUL INDEX OF NITROGEN FIXATION WITH REGARD TO THE ROLE OF
NITROGEN-FIXING AGENTS IN THE EUTROPHICATION PROCESS. LAKE MENDOTA
(EUTROPHIC), MADISON, CRYSTAL AND TROUT LAKES (OLIGOTROPHIC), VILAS
COUNTY, WISCONSIN, AND LITTLE ARBOR VITAE LAKE (EUTROPHIC) VILAS
COUNTY, WISCONSIN, WERE SAMPLED AT POINTS 100 FEET FROM SHORE AND AT
MID-LAKE DURING THE SUMMER OF 1968. ANALYSES OF SAMPLES INDICATED THAT
ACETYLENE REDUCTION VARIES DRASTICALLY AMONG LAKES IN A GIVEN AREA
(LITTLE ARBOR VITAE LAKE FIXED 125-380 NANOMOLES C SUB 2 H SUB 4 N/30
MINUTES TO LESS THAN ONE NANOMOLES FOR TROUT LAKE ON JULY 28, 1968),
AND VARIES MARKEDLY WITH TIME IN A SPECIFIC LAKE (FOR LITTLE ARBOR
VITAE LAKE, WHEN SAMPLES WERE COLLECTED ON SEPTEMBER 10, 1969, THE
ACETYLENE REDUCTION PER LITER OF WATER OR PER MG N HAD DECREASED 10 TO
20 FOLD FROM THOSE OF JULY 28, 1968). ACETYLENE REDUCTION IN LAKES WAS
STUDIED WITH RESPECT TO CHANGES IN SEASON, PRESENCE OF
NATURAL-OCCURRING ALGAE (GLOEOTTICHIA COLONIES CHIEFLY), VARIATION IN
DEPTHS, AND DIFFERENCE IN LIGHT. SAMPLES TAKEN FROM GREEN BAY,
WISCONSIN WERE ALSO ANALYSED. (KERRIGAN-WISCONSIN)

FIELD 05C, 02H

ACCESSION NO. W70-03429

CHEMICAL AND DETRITAL FEATURES OF PALOUSE RIVER, IDAHO, RUNOFF FLOWAGE,

EASTERN NEW MEXICO UNIV., PORTALES. DEPT. OF BIOLOGICAL SCIENCES.

PHILIP A. BUSCEMI.

OIKOS, VOL 20, NO 1, P 119-127, 1969. 3 FIG, 3 TAB, 30 REF.

DESCRIPTORS:
*RUNOFF, *GROUNDWATER, *SESTON, *SEDIMENTS, ORGANIC MATTER,
CURRENTS(WATER), WATER POLLUTION SOURCES, SURFACE RUNOFF, RAINFALL,
SNOWMELT, SURFACE WATERS, STREAMS, HYDROGEN ION CONCENTRATION, IRON,
MAGNESIUM, NITRATES, PHOSPHATES, ALGAE, DETRITUS, PULP WASTES,
LIVESTOCK, IDAHO.

IDENTIFIERS:
TREE BARK, CHARCOAL, ALGAL GROWTH, STREAM CONCOURSE, PALOUSE
RIVER(IDAHO).

ABSTRACT:
WATER AND SEDIMENT SAMPLES WERE COLLECTED AT FOUR STATIONS ESTABLISHED
ON PALOUSE RIVER, IDAHO. DETERMINATIONS INCLUDED PH, NITRATE,
PHOSPHATE, IRON, MANGANESE, HARDNESS, SUSPENDED DETRITUS (DRIFTING
SESTON), AND ORGANIC CONTENT OF SEDIMENTARY DEPOSITS. THE RESULTS SHOW
INFLUENCE OF RAINFALL, MELT WATER, STREAM INFLOW, AND MILL POND ON
SESTON CONCENTRATION. VARIATION IN CHEMICAL COMPOSITION OF WATER
INDICATED THE EFFECT OF EITHER MELT WATER RUNOFF OR SUBTERRANEAN
DISCHARGE. THE INCREASED POLLUTION OF WATER AND SEDIMENT ENRICHMENT IN
ORGANIC MATTER WERE CORRELATED WITH LOCATIONS OF LUMBER CAMPS AND
CATTLE GRAZING. (WILDE-WISCONSIN)

FIELD 05B, 05C

ACCESSION NO. W70-03501

THE MINERAL NUTRITION OF MICROCYSTIS AERUGINOSA,

WISCONSIN UNIV., MADISON. DEPT. OF BOTANY.

GERALD C. GERLOFF, GEORGE P. FITZGERALD, AND FOLKE SKOOG.

AMERICAN JOURNAL OF BOTANY, VOL 39, P 26-32, 1952. 7 TAB, 8 REF.

DESCRIPTORS:
*ALGAE, *NUTRIENTS, *NUTRIENT REQUIREMENTS, GROWTH RATES, LABORATORY
TESTS, CULTURES, SODIUM COMPOUNDS, NITROGEN.

IDENTIFIERS:
*MICROCYSTIS AERUGINOSA(KUTZ), CULTURE MEDIA, CHU NO 10 SOLUTION,
MINIMUM NUTRIENTS.

ABSTRACT:
A MICROCYSTIS AERUGINOSA (KUTZ) CULTURE WAS MADE AXENIC BY ULTRAVIOLET
LIGHT AND GROWN IN BASAL MEDIUM OF CHU NO 10 SOLUTION, MODIFIED SO THAT
CATIONS WERE FURNISHED AS CHLORIDE SALTS, AND ANION AS THE SODIUM
SALTS. MINIMUM ELEMENTAL REQUIREMENTS FOR OPTIMUM GROWTH YIELD WAS
DETERMINED BY USE OF THIS MEDIUM. SMALL AMOUNTS OF SODIUM WERE REPORTED
TO BE ESSENTIAL FOR THE GROWTH OF MICROCYSTIS BUT A BENEFICIAL EFFECT
OF EITHER SODIUM CARBONATE OR SODIUM SILICATE WAS FOUND TO BE DUE
PRIMARILY TO PH CONTROL. MICROCYSTIS EXHIBITED AN OPTIMUM GROWTH AT PH
OF APPROXIMATELY 10; AT THIS PH, SODIUM NITRATE, SODIUM NITRITE OR
AMMONIUM CHLORIDE WERE EQUALLY EFFECTIVE NITROGEN SOURCES. WHEN MINIMUM
NUTRIENT CONCENTRATIONS WERE DETERMINED EACH IN TURN, A NEW MEDIUM
COMPOSED OF THESE CONCENTRATIONS WAS FOUND TO SUPPORT MAXIMUM YIELDS AS
COMPARED TO THE SLIGHTLY MODIFIED CHU NO 10 MEDIUM. COMPOSITION OF THIS
MEDIUM, IN MILLIGRAMS/LITER, WAS: NITROGEN 13.6; PHOSPHORUS 0.18;
SULFUR 0.83; POTASSIUM 1.13; MAGNESIUM 0.13; CALCIUM 0.25; IRON 0.06;
AND THE 'USUAL' AMOUNTS OF SODIUM CARBONATE AND SODIUM SILICATE.
DISADVANTAGES OF THIS MEDIUM FOR ROUTINE USE WERE THAT GROWTH LAG, PALE
CELLS, AND REQUIREMENTS FOR A 'LARGER' INOCULUM WAS EXHIBITED.
(GERHOLD-WISCONSIN)

FIELD 05C

ACCESSION NO. W70-03507

THE SHAGAWA LAKE, MINNESOTA, EUTROPHICATION RESEARCH PROJECT,

FEDERAL WATER POLLUTION CONTROL ADMINISTRATION, CORVALLIS, OREG. PACIFIC
NORTHWEST WATER LAB.

ROBERT M. BRICE, AND CHARLES F. POWERS.

PROCEEDINGS OF THE EUTROPHICATION-BIOSTIMULATION ASSESSMENT WORKSHOP, JUNE
19-21, 1969, CALIFORNIA UNIV, BERKELEY, SANITARY ENGINEERING RESEARCH LAB
AND NATIONAL EUTROPHICATION RESEARCH PROGRAM, CORVALLIS, ORE, PAC NORTHWEST
WATER LAB, P 258-269. 5 FIG, 4 TAB.

DESCRIPTORS:
*MINNESOTA, *EUTROPHICATION, SEWAGE EFFLUENTS, TERTIARY TREATMENT,
ALGAE, WATER QUALITY, CONIFEROUS FORESTS, DECIDUOUS FORESTS,
PHOSPHORUS, NITROGEN, CYANOPHYTA, CHLOROPHYTA, DIATOMS, WEIGHT, PILOT
PLANTS, IRON, CHLORELLA, SODIUM, POTASSIUM, SILICATES, DISSOLVED
SOLIDS, CALCIUM, PRODUCTIVITY, STANDING CROP, CHLOROPHYLL, PHOSPHATES,
ADDITIVES.

IDENTIFIERS:
*SHAGAWA LAKE(MINN), ELY(MINN), BLOOMS, CHLORELLA PYRENOIDOSA,
MICROCYSTIS AERUGINOSA, BURNTSIDE RIVER(MINN).

ABSTRACT:
THIS PROJECT SOUGHT TO DETERMINE WHETHER SHAGAWA LAKE WOULD RECOVER
FROM ITS PRESENTLY ADVANCED STATE OF EUTROPHICATION IF SECONDARY SEWAGE
EFFLUENT PRESENTLY ENTERING THE LAKE WERE CHANGED TO NUTRIENT-STRIPPED
TERTIARY. ALTHOUGH DATA ARE INCOMPLETE AND CONCLUSIONS PRELIMINARY,
STUDIES SHOWED LAKE WATER CONTAINING 5% SECONDARY SEWAGE EFFLUENT HAD
APPRECIABLY HIGHER ALGAL PRODUCTIVITY THAN SHAGAWA LAKE WATER ONLY OR
WATER CONTAINING 5% TERTIARY SEWAGE EFFLUENT, THE LATTER ADDITION
MAKING NO APPRECIABLE DIFFERENCE IN THE LAKE WATER. BURNTSIDE RIVER
WATER (80% SURFACE INFLOW) CONTAINING 2% SECONDARY SEWAGE EFFLUENT HAD
APPRECIABLY HIGHER ALGAL PRODUCTIVITY THAN CORRESPONDING MIXTURES
CONTAINING TERTIARY EFFLUENT. FURTHER TESTING OF SHAGAWA LAKE WATER AND
BURNTSIDE RIVER IS CONTINUING, USING OTHER CONCENTRATIONS OF ADDITIVES
TO EXPLORE MORE COMPLETELY THE BIOSTIMULATION ACHIEVED THROUGH SEWAGE
FERTILIZATION. ADDITIONALLY, EXPERIMENTS WILL BE CARRIED OUT IN WHICH
VARIOUS CONCENTRATIONS OF NITROGEN AND PHOSPHORUS ARE ADDED TO
BURNTSIDE RIVER WATER, TO DETERMINE WHETHER EITHER OR BOTH ELEMENTS ACT
TO LIMIT ALGAL PRODUCTION. ANTICIPATED RESULTS OF THESE EXPERIMENTS
WILL PROVIDE INFORMATION NECESSARY TO DECISIONS ON FUTURE MANAGEMENT
POLICIES FOR SHAGAWA LAKE. (SEE W70-02775).
(JONES-WISCONSIN)

FIELD 05C, 02H

ACCESSION NO. W70-03512

EFFECTS OF FIVE HERBICIDES ON THREE GREEN ALGAE,

NORTH TEXAS STATE UNIV., DENTON. DEPT. OF BIOLOGY.

B. DWAIN VANCE, AND DAVID L. SMITH.

TEXAS JOURNAL OF SCIENCE, VOL XX, P 329-337, APR 1969. 1 TAB, 41 REF.

DESCRIPTORS:
*HERBICIDES, *GROWTH RATES, *ALGAE, 2-4-D, AMINOTRIAZOLE, 2-4-5-T,
SCENEDESMUS, CHLAMYDOMONAS, CHLORELLA, MORTALITY, BIOASSAY, INHIBITION,
MODE OF ACTION, PHOTOSYNTHESIS.

IDENTIFIERS:
SIMAZINE 80 W, DACTHAL.

ABSTRACT:
SCREENING TESTS WERE MADE TO DETERMINE IF THE HERBICIDES SIMAZINE 80W,
DACTHAL, AMITROL-T, 2,4-D, AND 2,4,5-T WERE TOXIC TO BACTERIA-FREE
UNIALGAL CULTURES OF SCENEDESMUS QUADRICAUDA, CHLAMYDOMONAS EUGAMETOS,
AND CHLORELLA PYRENOIDOSA TX 7-11-05. DACTHAL, SIMAZINE, 2,4-D AND
2,4,5-T SHOWED NO TOXIC EFFECTS ON ANY OF THE ALGAE AT CONCENTRATIONS
UP TO 200 UG/ML. SIMAZINE INCREASED THE GROWTH OF C. EUGAMETOS IN
CONCENTRATIONS UP TO 200 UG/ML. AMITROL-T WAS INHIBITORY TO ALL THE
ALGAE AT HIGH CONCENTRATIONS BUT SHOWED NO EFFECT AT CONCENTRATIONS
BELOW 150 UG/ML. (SJOLSETH-WASHINGTON)

FIELD 05C

ACCESSION NO. W70-03519

BIOLOGICAL CONCENTRATION OF PESTICIDES BY ALGAE,

NORTH TEXAS STATE UNIV., DENTON. DEPT. OF BIOLOGY.

B. DWAIN VANCE, AND WAYMON DRUMMOND.

JOURNAL AMERICAN WATER WORKS ASSOCIATION, VOL 61, NO 7, P 360-362, JULY 1969.
2 TAB, 11 REF. US PUBLIC HEALTH SERV RESEARCH GRANT CC 00269.

DESCRIPTORS:
*ALGAE, *PESTICIDE TOXICITY, *PESTICIDE REMOVAL, *FOOD WEBS, DDT,
CHLORINATED HYDROCARBON PESTICIDES, ALDRIN, ENDRIN, DIELDRIN,
SCENEDESMUS, GAS CHROMATOGRAPHY, RESISTANCE, BIOASSAY, GROWTH RATES,
PESTICIDE RESIDUES, DETERIORATION.

IDENTIFIERS:
*BIOLOGICAL CONCENTRATION, ANABAENA CYLINDRICA, MICROCYSTIS AERUGINOSA,
OEDOGNIUM.

ABSTRACT:
UNIALGAL CULTURES OF MICROCYSTIS AERUGINOSA, ANABAENA CYLINDRICA,
SCENEDESMUS QUADRICAUDA, AND OEDOGONIUM SP. WERE GROWN IN THE PRESENCE
OF ALDRIN, DIELDRIN, ENDRIN AND DDT IN CONCENTRATIONS OF 5, 10 AND 20
UG/ML. THESE ALGAE ARE EFFECTIVE IN ACCUMULATING PESTICIDES FROM THE
MEDIUM IN WHICH THEY GROW. IN NO CASE WAS CONCENTRATION LESS THAN 100
FOLD. ALGAE WERE SHOWN TO ACCUMULATE DETECTABLE QUANTITIES WITHIN 30
MINUTES AFTER EXPOSURE. THE ALGAE WERE SHOWN TO BE HIGHLY RESISTANT TO
THE PESTICIDE, EXCEPT THAT DIELDRIN, ALDRIN, AND ENDRIN ARE ALGICIDAL
TO M. AERUGINOSA AT CONCENTRATION OF LESS THAN 5 UG/ML. THE AUTHORS
FEEL THAT THE ABILITY OF ALGAE TO CONCENTRATE PESTICIDES WITHOUT
THEMSELVES SUCCUMBING MAY BE A POTENTIAL DANGER TO ORGANISMS HIGHER UP
THE FOOD CHAIN. (SJOLSETH-WASHINGTON)

FIELD 05C

ACCESSION NO. W70-03520

A METHOD FOR PREDICTING THE EFFECTS OF LIGHT INTENSITY ON ALGAL GROWTH AND
PHOSPHOROUS ASSIMILATION,

ABBOTT LABS., NORTH CHICAGO, ILL. DEPT. OF SANITARY ENGINEERING; AND MICHIGAN
UNIV., ANN ARBOR.

H. S. AZAD, AND J. A. BORCHARDT.

JOURNAL WATER POLLUTION CONTROL FEDERATION, VOL 41, NO 11, PART 2, P
R392-R404, NOV 1969. 10 FIG, 14 REF.

DESCRIPTORS:
*ALGAE, *CHLORELLA, *SCENEDESMUS, *LIGHT, PHOSPHORUS, PHOTOSYNTHESIS,
LIGHT INTENSITY, STABILIZATION, PONDS, PENETRATION, ABSORPTION.

IDENTIFIERS:
*BEER-LAMBERT LAW, *FIXED-DENSITY STUDIES, *FIXED-LIGHT TESTS,
*EFFECTIVE-AVERAGE-LIGHT-INTENSITY, PHOTO-ELECTRIC TURBIDOSTAT, LIGHT
EXTINCTION COEFFICIENT, ALGAL CELL DENSITY.

ABSTRACT:
RESEARCH IS DEVOTED TO OBTAINING A BETTER UNDERSTANDING OF THE
INTERACTION OF LIGHT INTENSITY ON THE RATE OF ALGAL GROWTH, AND
PHOSPHOROUS UPTAKE BY ALGAE IN AN AQUEOUS ENVIRONMENT. GREEN ALGAE
CHLORELLA AND SCENEDESMUS WERE GROWN UNDER A CAREFULLY CONTROLLED
ENVIRONMENT. THE LIGHT INTENSITY WAS VARIED BY TWO METHODS: (1)
MAINTENANCE OF A CONSTANT BIOMASS DENSITY BY USING A
PHOTO-ELECTRIC-TURBIDOSTAT APPARATUS PROVIDING FOR VARIABLE LIGHT
INTENSITY CONTROL AT THE SOURCES; (2) MAINTENANCE OF THE INCIDENT
LUMINOUS ENERGY AT A CONSTANT VALUE PROVIDING FOR VARIABLE ALGAE CELL
DENSITY. THE BEER-LAMBERT EXPRESSION (I SUB O/I = E TO THE POWER CAD)
FAILED TO PREDICT THE LIGHT PENETRATION DATA. HOWEVER, AN EMPIRICAL
RELATIONSHIP IN WHICH THE FAMILIAR COEFFICIENT OF EXTINCTION WAS
REPLACED BY AN EXPONENTIAL VARIABLE DID PROVIDE THE NECESSARY
AGREEMENT. ALGAL GROWTH RATE VS. THE CULTURE-FACE-LIGHT INTENSITY
YIELDS A CURVE THAT IMPLIES THE EXISTENCE OF A LIGHT SATURATION POINT.
WHEN GROWTH RATE VS EFFECTIVE AVERAGE LIGHT INTENSITY WAS PLOTTED, A
MORE LOGICAL EFFECT OF LIGHT WAS PRODUCED. (D'AREZZO-TEXAS)

FIELD 05C, 05D

ACCESSION NO. W70-03923

ANIONIC AND NONIONIC SURFACTANT SORPTION AND DEGRADATION BY ALGAE CULTURES,

TEXAS UNIV., AUSTIN. DEPT. ENVIRONMENTAL HEALTH ENGINEERING.

ERNST M. DAVIS, AND EARNEST F. GLOYNA.

JOURNAL OF THE AMERICAN OIL CHEMISTS' SOCIETY, VOL 46, NO 11, P 604-608, NOV
1969. 2 FIG, 3 TAB, 9 REF.

DESCRIPTORS:
*SURFACTANTS, *SORPTION, *ALGAE, CHLOROPHYTA, CYANOPHYTA, DETERGENTS,
SPECTROSCOPY.

IDENTIFIERS:
*LINEAR ALKYL SULFONATES, *AXENIC CULTURES, *ALKYL POLYETHOXYLATE,
*ALKYL PHENOL POLYETHOXYLATE, ORGANIC EXTRACTIONS, INFRARED
SPECTROSCOPY, BLUE-GREEN ALGAE, GREEN ALGAE.

ABSTRACT:
INVESTIGATION WAS MADE TO DETERMINE EFFECTS OF AXENIC CULTURES OF 5
SPECIES OF CYANOPHYTA ALGAE AND 3 SPECIES OF CHLOROPHYTA ALGAE, WHICH
ARE COMMON TO WASTE STABILIZATION PONDS, ON IONIC AND ANIONIC
SURFACTANTS CONTAINED IN THE BULK OF HOUSEHOLD DETERGENT PRODUCTS.
TESTS MADE TO DETERMINE THE SORPTION CHARACTERISTICS OF THE SURFACTANTS
BY THE ALGAE SHOWED RELEASES AND DEGRADATION OF UP TO 99% OF SOME OF
THE COMPONENT PARTS OF THE SURFACTANT MOLECULE. COMPARATIVE TESTS USING
A HETEROGENEOUS MICROCOSM ASSOCIATED WITH LABORATORY ACCLIMATED
STABILIZATION POND WATER SHOWED CAPABILITY OF GREATER REDUCTION OF ALL
SURFACTANT COMPONENTS THAN THE INDIVIDUAL ALGAE SPECIES. THE TESTS
CONFIRMED THE FEASIBILITY OF USING INFRARED SPECTROPHOTOMETRIC ANALYSIS
TECHNIQUES AND THE ADEQUACY OF ORGANIC SOLVENTS FOR EXTRACTION OF
SURFACTANTS PRIOR TO COMPLEXATION IN CONCENTRATIONS OF AND BELOW 50
MG/L. (D'AREZZO-TEXAS)

FIELD 05D

ACCESSION NO. W70-03928

DISTRIBUTION, CHARACTERIZATION, AND NUTRITION OF MARINE MICROORGANISMS FROM THE ALGAE POLYSIPHONIA LANOSA AND ASCOPHYLLUM NODOSUM,

MCGILL UNIV., MONTREAL (QUEBEC). DEPT. OF MICROBIOLOGY AND IMMUNOLOGY; AND NEW BRUNSWICK UNIV., FREDERICTON. DEPT. OF BIOLOGY.

E. C. S. CHAN, AND ELIZABETH A. MCMANUS.

CANADIAN JOURNAL OF MICROBIOLOGY, VOL 15, NO 5, P 409-420, 1969. 5 FIG, 4 TAB, 38 REF.

DESCRIPTORS:
   *MARINE MICROORGANISMS, *ALGAE, *NUTRIENT REQUIREMENTS, *DISTRIBUTION, BACTERIA, SAMPLING, SEA WATER, SEASONAL, PSEUDOMONAS, TEMPERATURE, AMINO ACIDS, INTERTIDAL AREAS, CARBOHYDRATES, YEASTS, ENZYMES, CULTURES, HYDROGEN SULFIDE, ELECTRON MICROSCOPY, BIOCHEMISTRY, VITAMINS, EPIPHYTOLOGY.

IDENTIFIERS:
   *POLYSIPHONIA LANOSA, *ASCOPHYLLUM NODOSUM, VIBRIO, FLAVOBACTER, ESCHERICHIA, SARCINA, STAPHYLOCOCCUS, ACHROMOBACTER(ALKALIGENES), RHODOTORULA, GROWTH FACTORS, COLONIAL MORPHOLOGY, CELLULAR MORPHOLOGY, FLAGELLATION, IODINE TEST, GRAM REACTION, CYTOPHAGA, PROTEUS, SERRATIA, BAY OF FUNDY(CANADA), RHYZOIDS.

ABSTRACT:
   BACTERIAL POPULATIONS OF TWO ASSOCIATED LITTORAL MARINE ALGAE, POLYSIPHONIA LANOSA AND ASCOPHYLLUM NODOSUM, AND THEIR ENVIRONMENTAL SEAWATER WERE STUDIED QUANTITATIVELY. SAMPLINGS SHOWED NUMBERS OF BACTERIA ON P LANOSA AND IN THE SEAWATER REMAINED RELATIVELY CONSTANT WHILE THOSE ON A NODOSUM DECREASED IN MID-SUMMER, AFTER AN APPARENT SPRING MAXIMUM, AND THEN GRADUALLY INCREASED. PURE CULTURES OF 25 ISOLATES WERE STUDIED AND IDENTIFIED TO GENUS LEVEL. THERE WAS A PREPONDERANCE OF THE GENERA VIBRIO AND FLAVOBACTER (8 EACH). THREE OF THE ESCHERICHIA GROUP, 2 OF PSEUDOMONAS, ONE EACH OF GENERA SARCINA, STAPHYLOCOCCUS, AND ACHROMOBACTER (OR ALKALIGENES), AND A PINK YEAST (RHODOTORULA) MADE UP THE OTHER ISOLATES. OPTIMUM TEMPERATURE FOR GROWTH OF MOST ISOLATES WAS AROUND 21C. FIVE ISOLATES FAILED TO GROW AT 30C; 20 DID NOT GROW AT 37C; OF THE 5 GROWING AT 37C, DEVELOPMENT WAS FEEBLE IN EVERY CASE BUT ONE. STUDIES ON GROSS NUTRITIONAL REQUIREMENTS OF THE MARINE ORGANISMS SHOWED THAT 22 OF 25 ISOLATES REQUIRED A SUPPLEMENT OF AMINO ACIDS IN GROWTH MEDIUM. TWO WERE ABLE TO GROW ON BASAL MEDIUM; ONE ISOLATE GREW ONLY IN A MEDIUM SUPPLEMENTED WITH GROWTH FACTORS IN ADDITION TO AMINO ACIDS (JONES-WISCONSIN)

FIELD 05C

ACCESSION NO. W70-03952

AVAILABILITY OF MUD PHOSPHATES FOR THE GROWTH OF ALGAE,

H. L. GOLTERMAN, C. C. BAKELS, AND J. J. JAKOBS-MOGELIN.

VERHANDLUNGEN DER INTERNATIONALEN VEREINIGUNG FUR THEORETISCHE UND ANGEWANDTE LIMNOLOGIE, VOL 17, P 467-479, 1969. 10 FIG, 3 TAB, 10 REF, DISCUSSION.

DESCRIPTORS:
   *EUTROPHICATION, *MUD(LAKE), *PHOSPHORUS, *AQUATIC ALGAE, *BIOASSAY, SCENEDESMUS, PHOSPHATES, WATER POLLUTION SOURCES.

IDENTIFIERS:
   *DUTCH LAKES, *ALLOCHTHONOUS PHOSPHATES, ULTRA-VIOLET LIGHT STERILIZATION, ASTERIONELLA, SCENEDESMUS OBLIQUUS, GROWTH CONSTATS, RIVER RHINE.

ABSTRACT:
   AVILABILITY OF LAKE MUDS PHOSPHORUS TO ALGAE WAS DETERMINED BY EMPLOYING SCENEDESMUS OBLIQUUS AS TEST ORGANISM. CULTURE SOLUTIONS CONTAINED POTASSIUM PHOSPHATE IN DIFFERENT CONCENTRATIONS, CALCIUM PHOSPHATES, IRON PHOSPHATES, AND LAKE MUDS AS THE ONLY SOURCE OF PHOSPHORUS. THE GROWTH CONSTANTS WERE OBTAINED FROM CULTURES WITH INORGANIC PHOSPHATES. BASED ON THE NUMBER OF CELLS OF SCENEDESMUS, A CONSIDERABLE FRACTION OF MUD PHOSPHORUS WAS AVAILABLE TO ALGAE. DEPENDING ON THE ORIGIN OF SEDIMENT, FROM 7 TO 30.5% OF PHOSPHORUS WAS USED FOR CELL GROWTH. STERILIZATION WITH ULTRA-VIOLET LIGHT HAD NO SIGNIFICANT INFLUENCE ON THE GROWTH OF ORGANISMS. RESULTS SUGGESTED THAT SCENEDESMUS IS UTILIZING LARGELY ALLOCHTHONOUS PHOSPHATES, LEAVING SUFFICIENT SUPPLY OF INORGANIC PHOSPHORUS FOR ASTERIONELLA ALGAE. (WILDE-WISCONSIN)

FIELD 05C, 02H

ACCESSION NO. W70-03955

RELATION BETWEEN FILTERING RATE, TEMPERATURE, AND BODY SIZE IN FOUR SPECIES OF DAPHNIA,

YALE UNIV., NEW HAVEN, CONN. OSBORN MEMORIAL LABS.

CAROLYN W. BURNS.

LIMNOLOGY AND OCEANOGRAPHY, VOL 14, NO 5, P 693-700, 1969. 2 FIG, 2 TAB, 18 REF, DISCUSSION.

DESCRIPTORS:
*GRAZING, *TEMPERATURE, *DAPHNIA, PHYSIOLOGICAL ECOLOGY, LAKES, ENVIRONMENTAL EFFECTS, WATER POLLUTION EFFECTS, EPILIMNION, CONNECTICUT, EUGLENA, CHLAMYDOMONAS, PHOSPHORUS RADIOISOTOPES, REGRESSION ANALYSIS, ALGAE, PONDS, PREDATION.

IDENTIFIERS:
*FILTERING RATE, *BODY SIZE, *COMPARATIVE STUDIES, DAPHNIA SCHODLERI, DAPHNIA PULEX, DAPHNIA GALEATA MENDOTAE, DAPHNIA MAGNA, SELENASTRUM, RHODOTORULA, TEMPERATURE COEFFICIENTS, NORTH AMERICA, CENTRAL AMERICA, RATE EQUATIONS, ANALYSIS OF VARIANCE, FOOD CONCENTRATION, MEXICO, PREDICTIVE EQUATIONS, ZOOGEOGRAPHY, THERMOBIOLOGY.

ABSTRACT:
FILTERING RATES (F) OF CLADOCERANS CAN BE MEASURED BY DETERMINING THEIR UPTAKE OF RADIOPHOSPHORUS FROM LABELLED FOOD. F OF FOUR DAPHNID SPECIES, FED ON LABELLED YEAST, WERE MEASURED AT TEMPERATURES OF 15C, 20C, AND 25C. MAXIMUM F INCREASED WITH RISING TEMPERATURE AND BODY SIZE IN ALL SPECIES, BUT A GENERAL EQUATION RELATING F TO BODY LENGTH (L) COULD NOT BE DERIVED. BODY WEIGHT (W) CAN BE DEDUCED FROM L WITH EQUATIONS OF TYPE, LOG W = LOG A + B LOG L, WHERE A AND B ARE EMPIRICALLY DERIVED COEFFICIENTS. DATA ARE TABULATED FOR A AND B OF ALL SPECIES, CONSIDERED COLLECTIVELY AND INDIVIDUALLY. F, EXPRESSED IN MILLILITERS WATER FILTERED PER HOUR/MILLIGRAM DAPHNID DRY WEIGHT, DIFFERED AMONG SPECIES. F FOR ADULT DAPHNIA SCHODLERI (DS) AND DAPHNIA PULEX (DP) WERE SIMILAR, AND AT 20C, WERE SLIGHTLY HIGHER THAN AT 15C OR 25C. F FOR DAPHNIA MAGNA (DM) AND DAPHNIA GALEATA MENDOTAE (DG) INCREASED WITH INCREASING TEMPERATURE. TEMPERATURE COEFFICIENTS (Q-10) FOR F, CHANGING IN RANGE 15-25C WERE, FOR FOUR SPECIES: DG-2.71, DM-2.38, DP-0.94, DS-0.90. THERMAL DETERMINANTS OF MAXIMUM F MAY DETERMINE SUCCESS OF DAPHNIDS OCCUPYING WARM EPILIMNETIC LACUSTRINE WATERS IN SUMMER. (EICHHORN-WISCONSIN)

FIELD 05C

ACCESSION NO. W70-03957

CONCEPTS OF EUTROPHICATION AND TROPHIC BIOLOGY,

OESTERREICHISCHE AKADEMIE DER WISSENSCHAFTEN, VIENNA. LUNZ BIOLOGICAL STATION.

FRANZ RUTTNER.

FUNDAMENTALS OF LIMNOLOGY, UNIVERSITY OF TORONTO PRESS, P 130-131, 159-165, 1963. 2 FIG.

DESCRIPTORS:
*LAKES, *EUTROPHICATION, *BIOMASS, *PRODUCTIVITY, CYANOPHYTA, NUISANCE ALGAE, LIMNOLOGY, CHLOROPHYTA, PLANKTON, PHYTOPLANKTON, WATER POLLUTION EFFECTS, WISCONSIN ZOOPLANKTON, TRIPTON, NANNOPLANKTON, NUTRIENTS, OLIGOTROPHY, TEMPERATURE, LIGHT INTENSITY, DAPHNIA, CONNECTICUT, ROTIFERS.

IDENTIFIERS:
*CONCEPTS, *TROPHIC BIOLOGY, ANABAENA FLOS AQUAE, APHANIZOMENON FLOS AQUAE, GLOEOTRICHIA ECHINULATA, BOTRYOCOCCUS, PLANKTON DISTRIBUTION, WISCONSIN LAKE SURVEY, NITROGEN CONTENT, LAKE ERKEN(SWEDEN), LAKE MENDOTA(WIS), NET PLANKTON, CHLOROPHYLL CONTENT, CENTRIFUGE METHOD, CALCULATED VOLUME METHOD, CYCLOTELLA COMENSIS, RHODOMONAS LACUSTRIS, ENOTOMOSTRACA, DIAPTOMIDS, BOSMINA, CARINTHIA, SUNDA ISLANDS.

ABSTRACT:
THIS STANDARD TEXTBOOK OF LIMNOLOGY INCLUDES A DISCUSSION OF CONCEPTS PERTAINING TO EUTROPHICATION AND TROPHIC ECOLOGY OF LAKES. MECHANICAL FACTORS INFLUENCING PLANKTON DISTRIBUTION INCLUDE DENSITY OF WATER AND PLANKTON, DIFFERENTIAL WATER VISCOSITY, AND TURBULENT EDDY DIFFUSION CURRENTS. THE FOLLOWING DESCRIBES FORMATION OF NUISANCE BLOOMS, 'IN CALM WEATHER THE BLUE-GREEN ALGAE OF THE PLANKTON RISE TO THE SURFACE IN CONSEQUENCE OF THEIR 'GAS VACUOLES' AND THERE AGGREGATE INTO THICK MASSES, OFTEN FORMING A 'WATER BLOOM' RECOGNIZABLE FROM AFAR BY ITS BLUE-GREEN COLORATION. ANABAENA FLOS AQUAE AND APHANIZOMENON FLOS AQUAE REVEAL BY THEIR NAMES THAT THEY BELONG TO THE GROUP OF ALGAE THAT CAN PRODUCE THIS STRIKING PHENOMENON.' BASED UPON A COMPOSITE DEFINITION BY OTHERS, PRODUCTION CAN BE CONSIDERED AS THE TOTAL AMOUNT OF ORGANIC MATTER, REPRESENTING BALANCE BETWEEN ASSIMILATION AND DISSIMILATION, WHICH FORMS IN A DEFINED PERIOD FROM RAW MATERIALS SUPPLIED. STANDING CROP IS DEFINED AS THE INSTANTANEOUS LIVING POPULATION. FERTILITY IS THE AVERAGE BIOMASS OF A UNIT VOLUME OF WATER WITHIN THE TROPHOGENIC ZONE, AND IS GOVERNED BY LIMITING NUTRIENTS, THEIR CONVERSION TO ORGANIC MATTER OCCURRING IN PROPORTION TO TEMPERATURE AND ILLUMINATION. DISCUSSION INCLUDES A CRITIQUE OF TECHNIQUES FOR DETERMINATION OF PLANKTONIC BIOMASS. (EICHHORN-WISCONSIN)

FIELD 02H, 05C

ACCESSION NO.   W70-03959

RX FOR AILING LAKES--A LOW PHOSPHATE DIET,

INTERNATIONAL JOINT COMMISSION-UNITED STATES AND CANADA.

ENVIRONMENTAL SCIENCE AND TECHNOLOGY, VOL 3, NO 12, P 1243-1245, 1969. 2 FIG.

DESCRIPTORS:
*PHOSPHATES, *LAKES, *DETERGENTS, *TERTIARY TREATMENT, *GREAT LAKES, CONTROL, COSTS, EUTROPHICATION, NITRATES, LAKE ERIE, LAKE ONTARIO, POLLUTION ABATEMENT, OLIGOTROPHY, DEPTH, PHYTOPLANKTON, ZOOPLANKTON, PHYSICOCHEMICAL PROPERTIES, DOMESTIC WASTES, SEWAGE, INDUSTRIAL WASTES, AGRICULTURE, ST LAWRENCE RIVER, ALGAE, COLIFORMS, DISSOLVED OXYGEN, DISSOLVED SOLIDS, TEMPERATURE, COLOR, TASTE, HYDROGEN ION CONCENTRATION, IRON, RADIOACTIVITY.

IDENTIFIERS:
CANADA, MESOTROPHY, LAKE NORRVIKEN, LAKE MENDOTA, LAKE FURES, LAKE SEBASTICOOK, LAKE WASHINGTON, LAKE MALAREN, LAKE ANNECY, LAKE VANERN, LAKE CONSTANCE, PFAFFIKERSEE, TURLERSEE, BALDEGGERSEE, GREIFENSEE, ZURICHSEE, MOSES LAKE, HALLWILLERSEE.

ABSTRACT:
STUDY WAS INITIATED IN 1964 WHEN THE INTERNATIONAL JOINT COMMISSION OF THE U S AND CANADA ESTABLISHED ADVISORY GROUPS ON STATUS OF POLLUTION IN LAKES ERIE AND ONTARIO AND SEGMENTS OF THE ST LAWRENCE RIVER. REPORT RECOMMENDS TECHNICAL AND LEGISLATIVE MACHINERY FOR CONTROL MEASURES. DETERGENTS' PHOSPHATE CONTENT SHOULD BE REDUCED IMMEDIATELY TO MINIMUM PRACTICAL LEVELS, WITH COMPLETE REPLACEMENT OF PHOSPHORUS WITH LESS INNOCUOUS SUBSTANCES NO LATER THAN 1972. 80% REMOVAL OF PHOSPHATES FROM ALL EFFLUENTS SHOULD BE EFFECTED BY 1972 IN THE LAKE ERIE BASIN AND BY 1975 IN LAKE ONTARIO. TREATMENT OF WASTE EFFLUENTS FOR PHOSPHATE REMOVAL MUST BE IN ADDITION TO, NOT A SUBSTITUTE FOR DETERGENT REFORMULATION. PHOSPHORUS AND NITROGEN ARE RECOGNIZED AS THE MAJOR NUTRIENTS RESPONSIBLE FOR EUTROPHICATION; IT IS APPARENT THAT PHOSPHATE IS THE CONTROLLING FACTOR IN ENRICHMENT OF LOWER GREAT LAKES. EFFICIENT AND RELATIVELY INEXPENSIVE METHODS ARE AVAILABLE FOR 80-95% REMOVAL OF PHOSPHORUS DURING SEWAGE TREATMENT, WHEREAS COMPARABLE ELIMINATION OF NITROGEN COMPOUNDS IS NOT YET FEASIBLE. COSTS FOR PHOSPHATE REMOVAL AT TREATMENT PLANTS WOULD BE REDUCED BY ONE-HALF TO TWO-THIRDS WITH REPLACEMENT OF PHOSPHATE DETERGENT BUILDERS. (JONES-WISCONSIN)

FIELD 02H, 05C

ACCESSION NO. W70-03964

REPORT OF AN ALGAL BLOOM IN VIET-NAM,

HUNGARIAN UNIV. OF AGRICULTURAL SCIENCES, GODOLLO (HUNGARY). INST. OF BOTANY
AND PLANT PHYSIOLOGY.

T. HORTOBAGYI.

ACTA BIOLOGICA ACADEMIAE SCIENTIARUM HUNGARICAE, VOL 20, NO 1, P 23-34, 1969.
5 FIG, 23 REF.

DESCRIPTORS:
*ALGAE, TEMPERATURE, CYANOPHYTA, CHLOROPHYTA, SCENEDESMUS, BENTHOS,
CONVECTION, WAVES(WATER), FISH, SEWAGE, EUGLENOPHYTA, DIATOMS,
BIOLOGICAL COMMUNITIES, BACKWATER, PONDS, ALASKA.

IDENTIFIERS:
*BLOOM, *VIETNAM, LITTLE LAKE(HANOI), RED RIVER(VIETNAM), CHLOROCOCCUS,
SPINES, SCHIZOMYCOPHYTA, CAULOBACTERIALES, CHLOROCOCCALES,
CONJUGATOPHYCAEA, PLANCTOMYCES, BACILLARIOPHYTA, GLOECAPSA,
MERISMOPEDIA, MICROCYSTIS, ANKISTRODESMUS, PEDIASTRUM, TETRAEDRON,
TETRASTRUM, TISZA(HUNGARY), ZIMONA(HUNGARY), ALGAL TAXONOMY, SPIRULINA,
CHODATELLA, MIRACANTHA, BUZSAK(HUNGARY), JAMUNA(INDIA), HORMOGONALES,
APHANOCAPSA, COELOSPHAERIUM, CHLORONOSTOC, PELODICTYON, SYNECHOCYSTIS,
ANABAENA, LYNGBYA, OSCILLATORIA, ROMERIA, ASTASIA, SCHROEDERIA,
COELASTRUM, DICTYOSPHAERIUM, GLOEOACTINUM, GOLENKINIA, NEPHROCHLAMYS,
NEPHROCYTIUM, OOCYSTIS, SIDEROCELIS, STAURASTRUM.

ABSTRACT:
AN ALGAL BLOOM OF VIVID GREEN COLOR DEVELOPED IN LITTLE LAKE, HANOI
(WATER TEMPERATURE 13-26C). THE LAKE, AN ISOLATED BACKWATER OF THE RED
RIVER, CONTAINS CONTAMINATED WATER. THE BLOOM WAS RICH IN SPECIES AND
NEW TAXONS WITH CYANOPHYTA AND CHLOROPHYTA FROM BOTH SURFACE AND BOTTOM
DOMINATING. SCENEDESMUS VARIABILITY WAS PROMINENT. A SAMPLE INDICATED
115 SPECIES WITHOUT BACILLARIOPHYCEAE. THE NUMBER OF THE SCENEDESMUS
SPECIES WAS LARGER THAN IN THE BLOOM. CYANOPHYTA DOMINATED WITH 82% IN
THE ALGAL BLOOM AND 92% IN THE BOTTOM SAMPLE; CHLOROPHYTA WITH 12% AND
6.8%, RESPECTIVELY. NO DEFINITE LINE OF DEMARCATION EXISTS BETWEEN THE
TWO BIOTOPES BECAUSE THE ALGAL BLOOM DEVELOPED ABOVE THE BOTTOM AND
BOTTOM PLANTS ARE EASILY RAISED NEAR THE SURFACE BY WAVES, CONVECTIONAL
STREAMING AND FISH MOVEMENT. THE ALGAE OF THE TWO BIOTOPES ARE
DIFFERENT, THE PHYTOBIOCOENOSES BEING DISSIMILAR. NUMBER OF COMMON
TAXONS IS 51% OF THE TOTAL; UNDESCRIBED TAXONS IS 33%. PARALLEL
MORPHOSES ARE WELL REPRESENTED BY SOME SCENEDESMUS SPECIES. AMONG SPINY
FORMS OF THE CHLOROCOCCALES, THOSE OF THIN SPINES ARE LESS FREQUENT.
PROTUBERANCES BECOME MORE DEVELOPED IN HANOI AND TOWARD THE TROPICS.
(JONES-WISCONSIN)

FIELD 05C

ACCESSION NO. W70-03969

ALGAE, MAN, AND THE ENVIRONMENT.

PROCEEDINGS OF AN INTERNATIONAL SYMPOSIUM HELD AT SYRACUSE UNIVERSITY, JUNE
18-30, 1967. DANIEL F JACKSON, EDITOR. NEW YORK, SYRACUSE UNIV PRESS, 1968.
554 P.

DESCRIPTORS:
*ALGAE, *ENVIRONMENTAL EFFECTS, *HUMAN POPULATION, *EUTROPHICATION,
*CONFERENCES, NUISANCE ALGAE, PUBLICATIONS, WATER POLLUTION EFFECTS,
NEW YORK, CHLOROPHYTA, NUTRIENTS, ECOSYSTEMS, PHYTOPLANKTON,
PHOTOSYNTHESIS, CYANOPHYTA, CYTOLOGICAL STUDIES, ELECTRON MICROSCOPY,
WASTE TREATMENT, LAKES, WATER QUALITY, VIRUSES, ALGICIDES, FISH,
STREAMS, PRODUCTIVITY.

IDENTIFIERS:
*PHYCOLOGY, NY STATE SCIENCE AND TECHNOLOGY FOUND, SYRACUSE(NY),
SYRACUSE UNIVERSITY, FUNDAMENTAL STUDIES, APPLIED STUDIES, ALGAL
PHYSIOLOGY, ALGAL BIOCHEMISTRY, ALGAL TAXONOMY, ALGAL PHYLOGENY,
CLADOPHORA, ALGAL GENETICS, RUSSIA, BAVARIA, MORICHES BAY(NY), ONONDAGA
LAKE(NY).

ABSTRACT:
PHYCOLOGICAL STUDIES ARE IMPORTANT NOT ONLY BECAUSE OF NUISANCE EFFECTS
OF ALGAE FOR MAN, BUT ALSO FOR THEIR POTENTIAL BENEFITS. TO COMPILE
INFORMATION ON BASIC AND APPLIED PHYCOLOGY, A SYMPOSIUM--WHICH THIS
VOLUME RECORDS--WAS SPONSORED BY NEW YORK STATE SCIENCE AND TECHNOLOGY
FOUNDATION AND SYRACUSE UNIVERSITY, NEW YORK (18-30 JUNE 1967). VOLUME
IS ORGANIZED INTO THREE SECTIONS: FUNDAMENTAL PHYCOLOGY, APPLIED
PHYCOLOGY, AND ALGAL STUDIES IN NEW YORK STATE. THE FOLLOWING TOPICAL
LIST OF 27 PAPERS BY 32 CONTRIBUTORS INDICATES ITS SCOPE: HISTORICAL
PHYCOLOGY (1927-1967); TAXONOMY AND PHYLOGENY IN ALGAL BIOCHEMISTRY AND
PHYSIOLOGY; ALGAL REQUIREMENTS FOR MICRONUTRIENTS; PHOSPHATE METABOLISM
OF GREEN ALGAE; NITROGEN INPUT TO AQUATIC ECOSYSTEMS; LIGHT AND
TEMPERATURE EFFECTS ON ALGAE; PHYTOPLANKTONIC PHOTOSYNTHESIS;
PHOTOSYNTHESIS OF CLADOPHORA; BIOLOGY OF FILAMENTOUS CONJUGATING ALGAE;
APPROACH TO MODERN ALGAL TAXONOMY; TRENDS IN ALGAL GENETICS; CYTOLOGY
OF CYANOPHYTES (BLUE-GREEN ALGAE); ULTRASTRUCTURE OF CYANOPHYTES;
RUSSIAN EXPERIMENTS IN ALGAL CULTURE; PHYCOLOGICAL MEDICINE; RUSSIAN
SOIL ALGOLOGY; HARVESTING WASTE-GROWN MICROALGAE; WASTEWATER TREATMENT
AND ALGAL GROWTH; REGENERATION OF BAVARIAN LAKES; NUTRIENT REMOVAL,
VIRAL DISEASES IN CYANOPHYTES; ALGICIDAL EFFECTS; ALGAE-FISH
RELATIONSHIPS; STREAM ASSIMILATION (NEW YORK); PRODUCTIVITY, MORICHES
BAY (NEW YORK); ALGAL ENVIRONMENT, ONONDAGA LAKE (NEW YORK); REVIEW OF
ALGAL LITERATURE (NEW YORK STATE). AN INDEX IS PROVIDED. (SEE
W70-03974). (EICHHORN-WISCONSIN)

FIELD 05C

ACCESSION NO. W70-03973

ONONDAGA LAKE, NEW YORK--AN UNUSUAL ALGAL ENVIRONMENT,

SYRACUSE UNIV., N. Y. DEPT. OF CIVIL ENGINEERING.

DANIEL F. JACKSON.

ALGAE, MAN, AND THE ENVIRONMENT, SYRACUSE UNIV PRESS, NEW YORK, P 515-524,
   1968.

DESCRIPTORS:
   *ALGAE, *AQUATIC ENVIRONMENTS, *EUTROPHICATION, *MEROMIXIS, *WATER
   POLLUTION EFFECTS, BIOINDICATORS, NUISANCE ALGAE, NEW YORK, SEWAGE
   EFFLUENTS, SEWAGE TREATMENT, INDUSTRIAL WASTES, WATER CHEMISTRY,
   DIATOMS, CHLOROPHYTA, EUGLENOPHYTA, CYANOPHYTA, CHRYSOPHYTA, BIOASSAYS,
   SALINE WATER, CHLORELLA, CHLAMYDOMONAS, EUGLENA.

IDENTIFIERS:
   *ONONDAGA LAKE(NY), EUNOTIA, AMPHORA, GOMPHONEMA, PINNULARIA,
   NITZSCHIA, NAVICULA, SYRACUSE(NY), ENTEROMORPHA INTESTINALIS,
   CLADOPHORA, CYCLOTELLA, HALOCHLOROCOCCUM, SYNECHOCOCCUS, CARTERIA,
   GONIUM, ANKISTRODESMUS, MICROSPORA, PEDIASTRUM, SCENEDESMUS,
   LEPOCINCLIS, ANABAENA, ANACYSTIS, GLOEOTRICHIA, GLOEOCAPSA, STAURONEIS.

ABSTRACT:
   FED BY SALINE SPRINGS, LAKE ONONDAGA HAS HIGHEST KNOWN CHLORIDE CONTENT
   (TYPICALLY 1460 MILLIGRAMS/LITER) OF NEW YORK LAKES AND IS PROBABLY
   MEROMICTIC. ITS MORPHOMETRIC, GEOGRAPHIC AND HISTORICAL FEATURES ARE
   DESCRIBED, AND WATER QUALITY DATA ARE TABULATED TO INCLUDE COMPARISONS
   WITH OTHER LAKES. ALTHOUGH SUBJECT TO EXTENSIVE EUTROPHICATION SINCE
   1863, IT FAILS TO BLOOM WITH PLANKTONIC CYANOPHYTES, BUT ANNUAL BLOOMS
   OF CHLOROPHYTES AND EUGLENOPHYTES HAVE OCCURRED DURING SUMMERS SINCE
   1962. LAKE RECEIVES INDUSTRIAL POLLUTANTS. TWO CHARACTERISTIC AREAS ARE
   OBSERVABLE, BASED ON DISTRIBUTION OF ATTACHED ALGAE, CLADOPHORA AND
   ENTEROMORPHA; VARIATIONS IN THE IONIC RATIO, SODIUM/CALCIUM, ARE
   THOUGHT TO CONTROL THEIR DISTRIBUTION. LAKE SUPPORTS ABUNDANT GROWTH OF
   DIATOMS, DISTRIBUTED AMONG 28 GENERA. IN A STUDY TO EVALUATE EFFECTS ON
   THE LAKE OF REPLACING PRIMARY TREATMENT PLANTS WITH IMPROVED SECONDARY
   TREATMENT, GROWTH OF TEN STRAINS OF CHLOROPHYTES, TWO EUGLENOPHYTES,
   EIGHT DIATOMS, AND FIVE CYANOPHYTES CULTURED IN FOUR CONCENTRATIONS OF
   ENRICHED LAKE WATER WAS COMPARED WITH THEIR GROWTH IN NUTRIENT MEDIA.
   BIOASSAY DATA ARE TABULATED; SUMMARIZED, THEY INDICATE THAT LAKE'S
   WATER WILL NOT SUPPORT GROWTH OF PLANKTONIC CYANOPHYTES, AND THAT SOME
   COMPONENT STIMULATES GROWTH OF DIATOM, CYCLOTELLA, AND THE
   CHLOROPHYTES: CHLAMYDOMONAS, HALOCHLOROCOCCUS, AND CHLORELLA (MILFORD
   STRAIN). (SEE ALSO W70-03973). (EICHHORN-WISCONSIN)

FIELD 05C, 02H

ACCESSION NO. W70-03974

EUTROPHICATION: CAUSES, CONSEQUENCES, CORRECTIVES.

NATIONAL ACADEMY OF SCIENCES, WASHINGTON, D.C.

PROCEEDINGS OF SYMPOSIUM HELD JUNE 11-15, 1967, WISCONSIN UNIV, MADISON.
GERARD A ROHLICH, CHAIRMAN, PLANNING COMMITTEE FOR THE INTERNATIONAL
SYMPOSIUM ON EUTROPHICATION. NAS STANDARD BOOK NO 309-01700-9. PRINTING AND
PUBLISHING OFFICE, NATIONAL ACADEMY OF SCIENCES, WASHINGTON, DC, 1969. 661
P.

DESCRIPTORS:
*EUTROPHICATION, *WATER POLLUTION SOURCES, *WATER POLLUTION EFFECTS,
*WATER POLLUTION CONTROL, *CONFERENCES, STREAMS, ESTUARIES, LAKES,
GREAT LAKES, PHYSIOLOGICAL ECOLOGY, WATER CHEMISTRY, GEOGRAPHICAL
REGIONS, CYCLING NUTRIENTS, ZOOPLANKTON, ALGAE, AQUATIC PLANTS,
PALEOLIMNOLOGY, MATHEMATICAL MODELS, BACTERIA, PHYTOPLANKTON,
GEOCHEMISTRY, BENTHOS, FISH, ECOSYSTEMS.

IDENTIFIERS:
NATIONAL ACADEMY OF SCIENCES(USA), UNIVERSITY OF WISCONSIN, ASIA,
NATIONAL RESEARCH COUNCIL(USA), US ATOMIC ENERGY COMMISSION, NATIONAL
SCIENCE FOUNDATION, OFFICE OF NAVAL RESEARCH, US DEPARTMENT OF
INTERIOR, MADISON(WIS), EUROPE, WATER POLLUTION ASSESSMENT, URBAN
DRAINAGE, AGRICULTURAL DRAINAGE, NORTH AMERICA, FOREST DRAINAGE.

ABSTRACT:
IN 1965, A PLANNING COMMITTEE APPOINTED BY NATIONAL ACADEMY OF
SCIENCES-NATIONAL RESEARCH COUNCIL (NAS-NRC) RECOMMENDED THAT '--AN
INTERNATIONAL SYMPOSIUM ON EUTROPHICATION BE HELD IN ORDER THAT THE
PRESENT WORLDWIDE STATE OF KNOWLEDGE AND UNDERSTANDING OF THIS
PHENOMENON CAN BE DISCUSSED IN OPEN FORUM AND RECOMMENDATIONS DEVELOPED
FOR THE EFFECTIVE MANAGEMENT OF PROBLEMS AND FOR THE COURSE OF FUTURE
RESEARCH.' THAT SYMPOSIUM MET AT UNIVERSITY OF WISCONSIN, MADISON,
11-15 JUNE 1967, SPONSORED BY NAS-NRC, ATOMIC ENERGY COMMISSION,
DEPARTMENT OF INTERIOR, NATIONAL SCIENCE FOUNDATION, AND OFFICE OF
NAVAL RESEARCH. APPROXIMATELY 600 PERSONS FROM 12 COUNTRIES ATTENDED.
THIS REPORT OF PROCEEDINGS INCLUDES AN INTRODUCTORY ADDRESS,
'EUTROPHICATION, PAST AND PRESENT', BY YALE UNIVERSITY PROFESSOR G. E.
HUTCHINSON, AND FOUR SECTIONS OF TECHNICAL PRESENTATIONS AS FOLLOWS
(NUMBERS OF CONTRIBUTIONS TO EACH INDICATED IN PARENTHESES):
GEOGRAPHICAL CONCEPTS (9); DETECTION AND MEASUREMENT (7); PREVENTION
AND CORRECTION (8); SCIENTIFIC CONTRIBUTIONS FROM EUTROPHICATION
RESEARCH (8). TWO PAPERS, PRESENTED AT CONFERENCE, ARE NOT PUBLISHED
HERE. THE REPORT ALSO INCLUDES AN INTRODUCTION, SUMMARY, AND SERIES OF
SPECIFIC RECOMMENDATIONS FORMULATED BY THE ORGANIZING COMMITTEE: 12
RELATING TO EDUCATION AND INFORMATION AND 8 TO RESEARCH. THE COMMITTEE
ALSO RECOMMENDED THAT NAS-NRC FORM A PERMANENT COMMITTEE TO IMPLEMENT
ITS RECOMMENDATIONS. (EICHHORN-WISCONSIN)

FIELD 05C

ACCESSION NO. W70-03975

EUTROPHICATION: CAUSES, CONSEQUENCES, CORRECTIVES.

NATIONAL ACADEMY OF SCIENCES, WASHINGTON, D.C.

PROCEEDINGS OF SYMPOSIUM HELD JUNE 11-15, 1967, WISCONSIN UNIV, MADISON.
GERARD A ROHLICH, CHAIRMAN, PLANNING COMMITTEE FOR THE INTERNATIONAL
SYMPOSIUM ON EUTROPHICATION. NAS STANDARD BOOK NO 309-01700-9. PRINTING AND
PUBLISHING OFFICE, NATIONAL ACADEMY OF SCIENCES, WASHINGTON, DC, 1969. 661
P.

DESCRIPTORS:
*EUTROPHICATION, *WATER POLLUTION SOURCES, *WATER POLLUTION EFFECTS,
*WATER POLLUTION CONTROL, *CONFERENCES, STREAMS, ESTUARIES, LAKES,
GREAT LAKES, PHYSIOLOGICAL ECOLOGY, WATER CHEMISTRY, GEOGRAPHICAL
REGIONS, CYCLING NUTRIENTS, ZOOPLANKTON, ALGAE, AQUATIC PLANTS,
PALEOLIMNOLOGY, MATHEMATICAL MODELS, BACTERIA, PHYTOPLANKTON,
GEOCHEMISTRY, BENTHOS, FISH, ECOSYSTEMS.

IDENTIFIERS:
NATIONAL ACADEMY OF SCIENCES(USA), UNIVERSITY OF WISCONSIN, ASIA,
NATIONAL RESEARCH COUNCIL(USA), US ATOMIC ENERGY COMMISSION, NATIONAL
SCIENCE FOUNDATION, OFFICE OF NAVAL RESEARCH, US DEPARTMENT OF
INTERIOR, MADISON(WIS), EUROPE, WATER POLLUTION ASSESSMENT, URBAN
DRAINAGE, AGRICULTURAL DRAINAGE, NORTH AMERICA, FOREST DRAINAGE.

ABSTRACT:
IN 1965, A PLANNING COMMITTEE APPOINTED BY NATIONAL ACADEMY OF
SCIENCES-NATIONAL RESEARCH COUNCIL (NAS-NRC) RECOMMENDED THAT '--AN
INTERNATIONAL SYMPOSIUM ON EUTROPHICATION BE HELD IN ORDER THAT THE
PRESENT WORLDWIDE STATE OF KNOWLEDGE AND UNDERSTANDING OF THIS
PHENOMENON CAN BE DISCUSSED IN OPEN FORUM AND RECOMMENDATIONS DEVELOPED
FOR THE EFFECTIVE MANAGEMENT OF PROBLEMS AND FOR THE COURSE OF FUTURE
RESEARCH.' THAT SYMPOSIUM MET AT UNIVERSITY OF WISCONSIN, MADISON,
11-15 JUNE 1967, SPONSORED BY NAS-NRC, ATOMIC ENERGY COMMISSION,
DEPARTMENT OF INTERIOR, NATIONAL SCIENCE FOUNDATION, AND OFFICE OF
NAVAL RESEARCH. APPROXIMATELY 600 PERSONS FROM 12 COUNTRIES ATTENDED.
THIS REPORT OF PROCEEDINGS INCLUDES AN INTRODUCTORY ADDRESS,
'EUTROPHICATION, PAST AND PRESENT', BY YALE UNIVERSITY PROFESSOR G. E.
HUTCHINSON, AND FOUR SECTIONS OF TECHNICAL PRESENTATIONS AS FOLLOWS
(NUMBERS OF CONTRIBUTIONS TO EACH INDICATED IN PARENTHESES):
GEOGRAPHICAL CONCEPTS (9); DETECTION AND MEASUREMENT (7); PREVENTION
AND CORRECTION (8); SCIENTIFIC CONTRIBUTIONS FROM EUTROPHICATION
RESEARCH (8). TWO PAPERS, PRESENTED AT CONFERENCE, ARE NOT PUBLISHED
HERE. THE REPORT ALSO INCLUDES AN INTRODUCTION, SUMMARY, AND SERIES OF
SPECIFIC RECOMMENDATIONS FORMULATED BY THE ORGANIZING COMMITTEE: 12
RELATING TO EDUCATION AND INFORMATION AND 8 TO RESEARCH. THE COMMITTEE
ALSO RECOMMENDED THAT NAS-NRC FORM A PERMANENT COMMITTEE TO IMPLEMENT
ITS RECOMMENDATIONS. (EICHHORN-WISCOSNIN)

FIELD 05C

ACCESSION NO. W70-03975

PHYSIOLOGICAL ECOLOGY,

OREGON STATE UNIV., CORVALLIS. DEPT. OF BOTANY.

HARRY K. PHINNEY.

PROCEEDINGS OF THE EUTROPHICATION-BIOSTIMULATION ASSESSMENT WORKSHOP, JUNE
19-21, 1969, CALIFORNIA UNIV, BERKELEY, SANITARY ENGINEERING RESEARCH LAB
AND NATIONAL EUTROPHICATION RESEARCH PROGRAM, CORVALLIS, ORE, PAC NORTHWEST
WATER LAB, P 141-145. 1 FIG, 5 REF.

DESCRIPTORS:
  *PHYSIOLOGICAL ECOLOGY, *AQUATIC ENVIRONMENTS, RIVERS, ENVIRONMENTAL
  EFFECTS, PERIPHYTON, WATER POLLUTION EFFECTS, OREGON, BIOINDICATORS,
  LIGHT INTENSITY, EUTROPHICATION, GRAZING, ALGAE, WATER CHEMISTRY,
  BIOMASS, BENTHOS, SALINITY, SNAILS, SULFATES, NITRATES, PHOSPHATES,
  POTASSIUM, IRON, BICARBONATE, ZINC, COBALT, CALCIUM, MAGNESIUM,
  CURRENTS(WATER).

IDENTIFIERS:
  BERRY CREEK(ORE), OREGON STATE UNIVERSITY, EXPERIMENTAL STREAMS,
  SEASONAL EFFECTS, THERMAL EFFECTS, APHANIZOMENON FLOS AQUAE, METOLIUS
  RIVER(ORE), COMMUNITY STRUCTURE, OXYTREMA, ORGANIC ENRICHMENT,
  INTERACTIONS.

ABSTRACT:
  PHYSIOLOGICAL ECOLOGY MAY BE DEFINED AS THE STUDY OF PHYSIOLOGICAL
  RESPONSES OF ORGANISMS TO VARIATIONS IN ECOLOGICAL CONDITIONS
  EXPERIENCED IN NATURE OR TO THOSE WHICH CAN BE MANIPULATED IN THE
  LABORATORY. BOTH LABORATORY AND FIELD APPROACHES TO ECOLOGY SUFFER FROM
  BASIC PHILOSOPHICAL AND METHODOLOGICAL DEFECTS. PHYSIOLOGICAL
  ECOLOGISTS ATTEMPT TO RETAIN A BALANCED PERSPECTIVE, AND MAINTAIN AN
  AWARENESS OF THESE PITFALLS IN INTERPRETING DATA AND CONSTRUCTING
  GENERALIZATIONS. SPECIFIC STUDIES ARE CITED WHICH ILLUSTRATE THESE
  PRINCIPLES FOR A VARIETY OF AQUATIC ENVIRONMENTS (RIVERS, LABORATORY
  STREAMS, AND ESTUARIES) AND FOR A VARIETY OF INFLUENCING FACTORS
  (CHEMICAL, PHYSICAL, AND BIOTIC). BECAUSE OF COMPLEXITIES IN
  ELUCIDATING CAUSAL FACTORS INFLUENCING SPECIES ABUNDANCE, GREAT CAUTION
  SHOULD BE EXERTED IN IDENTIFYING ALGAL SPECIES AS INDICATORS OF
  POLLUTION. ACCUMULATION OF INFORMATION RELATIVE TO PHYSIOLOGICAL
  ECOLOGY IS SLOW, LABORIOUS, AND FRAUGHT WITH DIFFICULTIES HAMPERING
  CORRECT INTERPRETATION. IT IS UNLIKELY THAT 'CRASH' PROGRAMS DESIGNED
  TO CHARACTERIZE THE PHYSIOLOGICAL ECOLOGY OF CULTURAL EUTROPHICATION
  CAN PROVIDE SOLUTIONS TO OUR PRESENT CRISIS IN TIME TO PREVENT THE
  FORESEEABLE DISASTER. (SEE W70-02775).
  (EICHHORN-WISCONSIN)

FIELD 05C

ACCESSION NO. W70-03978

C-14 UPTAKE AS A SENSITIVE MEASURE OF THE GROWTH OF ALGAL CULTURES,

CALIFORNIA UNIV., DAVIS. DEPT. OF ZOOLOGY; AND CALIFORNIA UNIV., DAVIS. INST. OF ECOLOGY.

CHARLES R. GOLDMAN, MILTON G. TUNZI, AND RICHARD ARMSTRONG.

PROCEEDINGS OF THE EUTROPHICATION-BIOSTIMULATION ASSESSMENT WORKSHOP, JUNE 19-21, 1969, CALIFORNIA UNIV, BERKELEY, SANITARY ENGINEERING RESEARCH LAB AND NATIONAL EUTROPHICATION RESEARCH PROGRAM, CORVALLIS, ORE, PAC NORTHWEST WATER LAB, P 158-170. 6 FIG, 1 TAB, 11 REF.

DESCRIPTORS:
*CARBON RADIOISOTOPES, *ALGAE, *CULTURES, PHYTOPLANKTON, CARBON, RADIOACTIVITY, CHLORELLA, SAMPLING, EFFLUENTS, TRIBUTARIES, ALKALINITY, HYDROGEN ION CONCENTRATION, NUTRIENTS, PHOSPHORUS, NITROGEN, IRON, GROWTH RATES, CHLOROPHYTA, CYANOPHYTA, CHRYSOPHYTA, PHOTOSYNTHESIS.

IDENTIFIERS:
*GROWTH, *UPTAKE, FIXATION, CELL COUNT, OPTICAL DENSITY, PARTICULATE ORGANIC CARBON, LAKE TAHOE(CALIF), CHLORELLA PYRENOIDOSA, SELENASTRUM GRACILE, CARBON ASSIMILATION, BACILLARIOPHYCEAE, DINOPHYCEAE.

ABSTRACT:
RESULTS OF COMPARISONS TO DETERMINE RELIABILITY OF CARBON-14 EXPERIMENTS WITH OTHER METHODS EMPLOYED FOR MEASURING GROWTH IN CULTURES ARE SUMMARIZED. THESE INCLUDE MAKING CELL COUNTS, MEASURING THE INCREASE IN LIGHT EXTINCTION AS THE CELLS MULTIPLY, AND DETERMINING THE BIOMASS OF CELLS PRODUCED. RESULTS OF A CHLORELLA EXPERIMENT PROVIDE A VALUABLE EVALUATION OF THESE METHODS. THE CELL COUNT WAS QUITE SUCCESSFUL; THE OPTICAL DENSITY METHOD WAS FASTEST AND EASIEST BUT WAS MAINLY APPLICABLE TO MEASURING CONCENTRATION OF DENSE CULTURES. THE CARBON-14 METHOD CAN EASILY BE USED FOR MEASURING ALGAL CELL GROWTH IN BOTH NUTRIENT-RICH AND NUTRIENT-POOR WATERS. CARBON-14 CAN DETECT GROWTH IN WATER WITH NATURAL PHYTOPLANKTON WHEN CELL COUNT AND OPTICAL DENSITY METHODS DO NOT GIVE EVIDENCE OF ANY CHANGE IN ALGAL MASS; CARBON-14 DATA GIVE A CARBON FIXATION RATE, WHEREAS CELL COUNT AND OPTICAL DENSITY MEASUREMENTS ARE STATIC MEASURES OF MASS. IN ADDITION, CARBON-14 EXPERIMENTS YIELD VALUES PROPORTIONAL TO THE MASS OF CELLS IN A CULTURE. THE COMBINED EFFECT OF NITROGEN AND PHOSPHORUS WAS TO STIMULATE PHOTOSYNTHESIS, THE GREATEST STIMULATION RESULTING FROM THE MAXIMAL ADDITION OF THE TWO NUTRIENTS. (SEE W70-02775).
(JONES-WISCONSIN)

FIELD 05C

ACCESSION NO. W70-03983

CARBON SOURCES IN ALGAL POPULATIONS AND ALGAL COMMUNITY STRUCTURE,

OKLAHOMA WATER RESOURCES RESEARCH INST., STILLWATER.

TROY C. DORRIS.

AVAILABLE FROM THE CLEARINGHOUSE AS PB-189 523, $3.00 IN PAPER COPY, $0.65 IN MICROFICHE. RESEARCH PROJECT TECHNICAL COMPLETION REPORT, OKLAHOMA WRSIC INSTITUTE (1969). 10 P, 1 FIG, 2 TAB. OWRR PROJECT NO B-005-OKLA.

DESCRIPTORS:
*CARBON, *ALGAE, *POPULATION, *BIOLOGICAL COMMUNITIES, EFFLUENTS, EUTROPHICATION, PHYTOPLANKTON, DECOMPOSING ORGANIC MATTER, OKLAHOMA, RESERVOIRS, SAMPLING, INDUSTRIAL WASTES, MUNICIPAL WASTES, SEWAGE, ADSORPTION, ACTIVATED CARBON, FILTERS, SEASONAL, STRATIFICATION, BIOCHEMICAL OXYGEN DEMAND, CARBON RADIOISOTOPES, DETERGENTS, ANTIFREEZE, OXIDATION, PHOTOSYNTHESIS, OIL WASTES, GARBAGE DUMPS, PIGMENTS, DAMS, PHYSICOCHEMICAL PROPERTIES, CHLOROPHYLL.

IDENTIFIERS:
ARKANSAS RIVER(OKLA), TULSA(OKLA), POLLUTION, FOSSIL, CIMARRON RIVER(OKLA), BIXBY(OKLA), KEYSTONE RESERVOIR(OKLA), CAROTENOIDS.

ABSTRACT:
A PORTION OF A STUDY TO DETERMINE SOURCE AND DISTRIBUTION OF ORGANIC COMPOUNDS IN THE KEYSTONE RESERVOIR, OKLAHOMA, AND THE DEVELOPMENT OF PHYTOPLANKTON POPULATIONS IN RESPONSE TO EUTROPHICATION PROCESSES RESULTING FROM DECOMPOSITION OF ORGANIC COMPOUNDS IN EFFLUENTS ENTERING THE CIMARRON ARM OF THE RESERVOIR IS DESCRIBED. ORGANIC COMPOUNDS FROM INDUSTRIAL AND MUNICIPAL SEWAGE EFFLUENTS WERE SAMPLED BY ADSORPTION ON ACTIVATED CARBON FILTERS AT SIX STATIONS TO DEVELOP INFORMATION ON SEASONAL CHANGES IN QUANTITY AND NATURE OF ORGANIC COMPOUNDS, EFFECT OF STRATIFICATION ON MOVEMENT AND CHANGES OF COMPOUNDS, AND POLLUTION OF THE ARKANSAS RIVER BY EFFLUENTS FROM TULSA. THE EFFORT TO IDENTIFY PARTICULAR COMPOUNDS, TRACE THEIR PROGRESS THROUGH THE RESERVOIR AND DETERMINE EXTENT OF THEIR DEGRADATION BY PASSAGE THROUGH THE RESERVOIR IS EXPECTED TO BE COMPLETED WITH FURTHER STUDY. EXTRACTION OF ADSORBED COMPOUNDS FROM CARBON FILTERS HAS BEEN ACCOMPLISHED BUT DATA WERE INCOMPLETE. PHYTOPLANKTON STUDIES ON THE CIMARRON ARM OF THE RESERVOIR WERE DESIGNED TO STUDY EFFECTS OF EUTROPHICATION OF THE WATER BY DOMESTIC AND INDUSTRIAL EFFLUENTS. THE FULL SPECTRUM OF PHOTOSYNTHETIC AND OTHER PIGMENTS WAS MEASURED AS EACH SAMPLE WAS TAKEN. EFFORTS WERE MADE IN SPECIES IDENTIFICATION. (JONES-WISCONSIN)

FIELD 05B

ACCESSION NO.   W70-04001

PROGRESS AND DEVELOPMENTS IN TREATMENT OF INTEGRATED TEXTILE MILL WASTES,

COLORADO STATE UNIV., FORT COLLINS. NATURAL RESOURCES CENTER.

T. A. ALSPAUGH.

PROCEEDINGS 12TH SOUTHERN MUNICIPAL INDUSTRIAL WASTE CONFERENCE, NORTH
CAROLINA STATE UNIV, RALEIGH, NC, P 46-62, 1964.

DESCRIPTORS:
*ACTIVATED SLUDGE, *BIOCHEMICAL OXYGEN DEMAND, *COLOR, WASTE WATER
TREATMENT, TREATMENT FACILITIES, AERATION, LAGOONS, DOMESTIC WASTES,
SEDIMENTATION, CHLORINATION, SLUDGE, DETERGENTS, SPRAYING, ALGAE,
CENTRIFUGATION, FILTRATION, ANAEROBIC DIGESTION, INCINERATION, IRON
COMPOUNDS, SULFATES.

IDENTIFIERS:
*ANTI-FOAM AGENTS, *TEXTILE WASTES, FIBER WASTES, DYEING WASTES.

ABSTRACT:
A FULL-SCALE PLANT WAS CONSTRUCTED TO TREAT 21 MIL. GAL. PER WEEK BY
THE 24-HOUR HIGH-SOLIDS ACTIVATED-SLUDGE PROCESS IN WHICH THE LONGER
AERATION PERIOD AND THE HIGHER SOLIDS CONCENTRATE IN THE AERATION TANK
BOTH ASSIST IN REMOVING THE HIGHLY-SOLUBLE BIOCHEMICAL OXYGEN DEMAND OF
TEXTILE WASTE WATERS, BUFFERING SHOCK LOADS, AND PROVIDING EFFICIENT
TREATMENT AS REGARDS REMOVAL OF COLOR. THE TEXTILE WASTE WATERS PASS
THROUGH TWO STORAGE LAGOONS IN SERIES (TO PROVIDE EQUALIZATION AND SOME
REMOVAL OF LINT AND DYE PASTE) AND ARE MIXED WITH 10% OF SCREENED AND
DEGRITTED DOMESTIC SEWAGE BEFORE ENTRY INTO THE AERATION TANK. AFTER
FINAL SEDIMENTATION THE EFFLUENT IS CHLORINATED AND REAERATED BEFORE
DISCHARGE TO A CREEK. OPERATING DATA ARE TABULATED SHOWING AN AVERAGE
BIOCHEMICAL OXYGEN DEMAND REMOVAL OF 89-90% AND COLOR REMOVAL OF
50-75%. THE PROBLEM OF FOAMING IS ACCENTUATED BY THE HIGH CONCENTRATE
OF SYNTHETIC DETERGENTS IN TEXTILE WASTE WATERS; THE MOST SUCCESSFUL
TREATMENT METHODS ARE SPRAYING WITH CRUDE WASTE WATER, ALONE OR WITH
THE ADDITION OF ANTI-FOAMING AGENT ON TO THE ROTOR BLADES, AND POSSIBLY
THE REMOVAL AND DESTRUCTION OF FOAM BY A VACUUM PUMP. THE
POORLY-SETTLING SLUDGE WAS FOUND TO CONTAIN HEAVY GROWTHS OF ALGAE. THE
PROBLEM COULD BE OVERCOME BY ALLOWING THE SLUDGE TO BECOME SEPTIC
BEFORE RETURN TO THE AERATION TANK, OR BY INTRODUCING A SHOCK LOAD OF
ORGANIC ALGAE. HOWEVER, THESE ALGAE HAVE GOOD CHARACTERISTICS FOR
REMOVAL OF BIOCHEMICAL OXYGEN DEMAND AND COLOR. EXCESS SLUDGE FROM THE
TREATMENT OF TEXTILE WASTE WATERS IS DIFFICULT TO SETTLE AND
CONCENTRATE. RATES OF VACUUM FILTRATION AND CENTRIFUGATION ARE LOW
COMPARED WITH SEWAGE SLUDGE. AT PRESENT THE SLUDGE IS DISCHARGED TO THE
MUNICIPAL SEWER. OTHER PROMISING METHODS OF DISPOSAL APPEAR TO BE
VACUUM FILTRATION AND CENTRIFUGATION. (LIVENGOOD-NORTH CAROLINA STATE
UNIV)

FIELD 05D

ACCESSION NO. W70-04060

APPLICATION OF RADIOISOTOPE TECHNIQUES TO A CRITICAL WATER RESOURCES PROBLEM AREA - NAMELY NUTRITIONAL POLLUTION,

OKLAHOMA UNIV., NORMAN. DEPT. OF CIVIL ENGINEERING AND SANITARY SCIENCE; AND ARKANSAS UNIV., FAYETTEVILLE. DEPT. OF CIVIL ENGINEERING.

GEORGE W. REID, ROBERT A. GEARHEART, JAMES M. ROBERTSON, AND ROBERT M. SWEAZY.

AVAILABLE FROM THE CLEARINGHOUSE AS PB-189 793, $3.00 IN PAPER COPY, $0.65 IN MICROFICHE. OKLAHOMA STATE UNIVERSITY WATER RESOURCES RESEARCH INSTITUTE PROJECT COMPLETION REPORT, 1969. 14 P, 1 FIG, 1 TAB, 1 APPEND. OWRR PROJ NO A-011-OKLA.

DESCRIPTORS:
*EUTROPHICATION, *NUTRIENTS, *PHOSPHORUS COMPOUNDS, ALGAE, PHOSPHATES, NITRATES, SEWAGE EFFLUENTS, TERTIARY TREATMENT, WATER POLLUTION SOURCES, WATER POLLUTION EFFECTS, LABORATORY TESTS, MODEL STUDIES, RADIOACTIVITY TECHNIQUES, TAGGING.

IDENTIFIERS:
NUTRIENT REQUIREMENTS(ALGAE).

ABSTRACT:
MINIMAL CONCENTRATIONS OF NITROGEN AND PHOSPHORUS NECESSARY FOR THE GROWTH OF ALGAE WERE STUDIED. A RADIOACTIVE PHOSPHORUS METHOD WAS USED AS ONE OF THE PARAMETERS FOR BIOLOGICAL MASS PRODUCED AND PHOSPHORUS UPTAKE RATE IN ALGA CULTURING UNIT. ALL OTHER NECESSARY INORGANIC AND ORGANIC NUTRIENTS, INCLUDING VITAMIN B-12 AND OTHER GROWTH METABOLITES, WERE SUPPLIED IN EXCESS TO ALLOW MAXIMUM ALGAE REACTION TO THE NITROGEN AND PHOSPHORUS. THE RELATIONSHIP BETWEEN NITROGEN, PHOSPHORUS AND ALGAL GROWTH IS EXPRESSED BY A MULTIPLE REGRESSION ANALYSIS PROCEDURE. A BACTERIAL BLOOM OCCURS WITHIN THE SECOND OR THIRD DAY IN CONTRAST TO THE ALGAE GROWTH WHICH USUALLY REQUIRES 6 TO 8 DAYS. THE SOLUBLE PHOSPHATE IS TAKEN UP 80-90% IN THE FIRST DAY. AT A PHOSPHATE CONCENTRATION OF 0.10 MG/L, THE ALGAL BLOOM CONDITION THRESHOLD IS REACHED. REGARDLESS OF THE NITROGEN CONCENTRATION, THIS NARROW RANGE OF SENSITIVITY TO THE PHOSPHORUS CONCENTRATION DEMONSTRATES THE PROBLEM IN ATTEMPTING TO CONTROL NUTRITIONAL POLLUTION. (KNAPP-USGS)

FIELD 05A, 05C

ACCESSION NO. W70-04074

INVESTIGATION OF SOME METHODS FOR INCREASING THE DIGESTIBILITY IN VITRO OF MICROALGAE,

ROYAL INST. OF TECH., STOCKHOLM (SWEDEN). DIV. OF APPLIED BIOCHEMISTRY.

GUDMUND HEDENSKOG, LENNART ENEBO, JITKA VENDLOVA, AND BOHUMIR PROKES.

BIOTECHNOLOGY AND BIOENGINEERING, VOL 11, NO 1, P 37-51, 1969. 6 FIG, 4 TAB, 18 REF.

DESCRIPTORS:
*ALGAE, *MICROORGANISMS, SCENEDESMUS, PROTEINS, WALLS, STRUCTURE, AMINO ACIDS, MECHANICAL EQUIPMENT, ENZYMES, CHEMICAL DEGRADATION, VITAMINS, CELLULOSE, CHLORELLA, PEPTIDES, BACTERIA, SNAILS, WEIGHT, VACUUM DRYING, BOILING.

IDENTIFIERS:
*DIGESTIBILITY, *IN VITRO, IN VIVO, BALL-MILL, MEICELASE, HYDROGEN PEROXIDE, RATS, TRYPSIN, SCENEDESMUS QUADRICAUDA, SCENEDESMUS OBLIQUUS, HELIX POMATIA, CHLORELLA PYRENOIDOSA, NITROGEN UTILIZATION, ROLLER DRYING, SPRAY-DRIED, GLASS BEADS, TRICHODERMA, DISINTEGRATION.

ABSTRACT:
TO INCREASE AVAILABILITY OF CELL BOUND PROTEIN IN SCENEDESMUS ALGAE, MECHANICAL, ENZYMATIC, AND CHEMICAL METHODS OF DEGRADING THE CELL WALL STRUCTURE WERE INVESTIGATED. MECHANICAL TREATMENT INVOLVED THE USE OF A BALL-MILL; ALGAE SUSPENSION TOGETHER WITH GLASS BEADS WAS MILLED IN A WATER-COOLED CHAMBER EQUIPPED WITH ROTATING DISKS. THE CHEMICAL TEST INCLUDED A CELLULOLYTIC ENZYME (MEICELASE) EMPLOYING HYDROGEN PEROXIDE. USING BATCHES IN THE BALL-MILL EXPERIMENTS A COMPLETE DISINTEGRATION WAS ACHIEVED IN A DISINTEGRATOR. TRIALS WERE ALSO PERFORMED WITH A CONTINUOUS DISINTEGRATOR AND DEPENDENCE OF DISINTEGRATION ON BEAD SIZE AND FLOW RATE STUDIED. DISINTEGRATION DETERMINED BY MICROSCOPIC CELL COUNT WAS COMPARED TO INCREASE OF PEPSIN DIGESTIBILITY. THE MEICELASE TREATMENT CAUSED A SLIGHT INCREASE OF PEPSIN DIGESTIBILITY, MEASURED AFTER 3 HOUR PEPSIN INCUBATION. NO INCREASE OF PEPSIN DIGESTIBILITY WAS DETECTED WITH HYDROGEN PEROXIDE TREATMENT. AFTER THE BALL-MILL DISINTEGRATION, 95% OF CONTAMINATING BACTERIA WERE KILLED AND YIELDS OF EXTRACTABLE PROTEINS WERE HIGHER. CAPACITY OF AVAILABLE CONTINUOUS BALL-MILLS IS SUCH THAT THEY COULD BE USED ON A PILOT-PLANT SCALE AND ENERGY COST OF DISINTEGRATION WOULD BE OF THE SAME MAGNITUDE AS THAT OF SEPARATION. (JONES-WISCONSIN)

FIELD 05G

ACCESSION NO. W70-04184

WHAT'S KILLING OUR LAKES,

MINNESOTA UNIV., MINNEAPOLIS. DEPT. SOIL SCIENCE.

LOWELL D. HANSON, AND WILLIAM E. FENSTER.

CROPS AND SOILS, VOL 22, P 13-15, 1969.

DESCRIPTORS:
*LAKES, STREAMS, FISH, LAND, ECOSYSTEMS, EUTROPHICATION, OLIGOTROPHY, SEWAGE, ALGAE, SCUM, WEEDS, NUTRIENTS, SEPTIC TANKS, EFFLUENTS, SWAMPS, FOREST SOILS, MINNESOTA, WISCONSIN, MICHIGAN, PHOSPHORUS, AGRICULTURAL WATERSHEDS, ORGANIC MATTER, SEDIMENTS, FERTILIZERS, WATER POLLUTION, SNOW, RUNOFF, DRAINAGE WATER, GRASSLANDS, SOIL EROSION, GROUNDWATER, POTHOLES.

IDENTIFIERS:
*KILLING, ANIMAL MANURES, LAKE MINNETONKA(MINN), SOIL MINERALS, PLANT RESIDUES, MORRIS(MINN), SOIL PERCOLATION.

ABSTRACT:
PHOSPHORUS SOURCES FROM AGRICULTURAL WATERSHEDS ARE SOIL MINERALS AND ORGANIC MATTER; FERTILIZERS, MANURES, AND PLANT RESIDUES; SEWAGE AND SEDIMENT. IF SOIL REACTS WITH PHOSPHORUS-CONTAINING MATERIALS, SEEPING WATER WILL BE STRIPPED OF PHOSPHORUS. SNOW-MELT CARRIES HIGHER AMOUNTS OF PHOSPHORUS THAN RUNOFF WATERS DURING OTHER SEASONS. SEPTIC TANK WASTE WATER DOES NOT ALWAYS EVAPORATE AS EXPECTED AND SEEPS INTO GROUNDWATER AND LAKES. BOTTOM SEDIMENTS SERVE AS A 'PHOSPHORUS SINK' REDUCING THE EXPLOSIVE BIOLOGICAL EFFECTS OF PHOSPHORUS IN WATER. SOLUTION OF ENVIRONMENTAL PROBLEMS MAY INVOLVE EXTENSIVE LAND-FORMING OPERATIONS TO FORCE SOIL FILTRATION OF RUNOFF WATER. IMMEDIATE IMPROVEMENTS CAN BE EFFECTED BY NOT SPREADING MANURE ON SLOPING FROZEN LAND, DIVERTING FEEDLOT AND GRAIN STORAGE RUNOFF INTO SEEPAGE AREAS, AND STORING MANURE IN AREAS WHERE RUNOFF IS NOT A PROBLEM; LARGE FEED LOTS SHOULD BE LOCATED ON LEVEL, FINE TEXTURED SOILS RATHER THAN SHORELINE SLOPES OF A STREAM OR LAKE. LOCATION DECISIONS IN AGRICULTURE WILL INCREASINGLY BE GUIDED BY POLLUTION CONTROL AGENCY REGULATIONS AND ZONING RESTRICTIONS. LAND USE PLANNING HAS LONG RANGE CONSEQUENCES. COOPERATION OF SOIL SCIENTISTS, AGRICULTURAL ENGINEERS, AND FARMERS' INNOVATIVE TALENTS WILL RESULT IN QUICKER SOLUTIONS. (JONES-WISCONSIN)

FIELD 02H, 05C

ACCESSION NO. W70-04193

THE UPTAKE OF ORGANIC SOLUTES IN LAKE WATER,

UPPSALA UNIV. (SWEDEN). INST. OF LIMNOLOGY.

RICHARD T. WRIGHT, AND JOHN E. HOBBIE.

LIMNOLOGY AND OCEANOGRAPHY, VOL 10, NO 1, P 22-28, JAN 1965. 3 FIG, 3 TAB, 15 REF.

DESCRIPTORS:
*LAKES, *ORGANIC MATTER, *PHYSIOLOGICAL ECOLOGY, RADIOACTIVITY TECHNIQUES, ANALYTICAL TECHNIQUES, WATER POLLUTION EFFECTS, PHYTOPLANKTON, DIFFUSION, ALGAE, BACTERIA, EUTROPHICATION, GLUCOSE, CARBON RADIOISOTOPES, DINOFLAGELLATES, BIOASSAY, BIOINDICATORS.

IDENTIFIERS:
*HETEROTROPHY, *BIOLOGICAL UPTAKE, *MICROBIAL ECOLOGY, MAXIMUM UPTAKE VELOCITY, ACTIVE TRANSPORT, BACTERIAL POPULATIONS, ALGAL POPULATIONS, HETEROTROPHIC POTENTIAL, RELATIVE HETEROTROPHIC POTENTIAL, SWEDEN, ENZYME KINETICS, LINEWEAVER-BURK EQUATION, MICHAELIS KINETICS, FLAGELLATES, LAKE ERKEN(SWEDEN), ACETATE, LAKE NORRVIKEN(SWEDEN), ACETATE BACTERIA, GLUCOSE BACTERIA, CHLAMYDOMONADS.

ABSTRACT:
METHODS EXIST FOR ESTIMATING, WITH RADIOCARBON (C-14) LABELLED ORGANIC COMPOUNDS, THEIR ASSIMILATION BY PLANKTONIC MICROBIOTA. BASED ON LINEWEAVER-BURK MODIFICATION OF MICHAELIS ENZYME KINETICS, A MODIFICATION OF THOSE METHODS IS DESCRIBED WHICH GIVES BETTER ESTIMATES OF HETEROTROPHIC POTENTIAL, PARTICULARLY IN CASES WHERE UPTAKE IS MEDIATED VIA PROCESSES OF ACTIVE TRANSPORT. THE TREATMENT YIELDS A LINEAR RELATIONSHIP SUCH THAT A PLOT OF THE EXPRESSION, (C)(U)(T)/C, VERSUS A HAS A SLOPE WHOSE INVERSE ESTIMATES MAXIMUM UPTAKE VELOCITY (WHERE: C- COUNTS/MINUTE FROM 1 MICROCURIE C-14 IN COUNTING DEVICE; U- MICROCURIES ADDED TO SAMPLE; T- INCUBATION TIME (HOURS); C- COUNTS/MINUTE FROM FILTERED ORGANISMS; A- MILLIGRAMS/LITER OF ADDED SUBSTRATE). WHERE UPTAKE RESULTS FROM DIFFUSION INTO CELL WITH INSTANTANEOUS METABOLISM OF SUBSTRATE, A DIFFUSION CONSTANT CAN BE DEFINED FROM KNOWLEDGE OF UPTAKE VELOCITIES AT TWO DIFFERENT SUBSTRATE CONCENTRATIONS. METHOD WAS TESTED IN TWO SWEDISH LAKES SUBJECT TO EUTROPHICATION. OVER A WIDE RANGE OF SUBSTRATES, ESTIMATES APPARENTLY SHOW EXISTENCE OF TWO UPTAKE MECHANISMS: A TRANSPORT SYSTEM, SATURATED AT LOW SUBSTRATE CONCENTRATIONS, AND A DIFFUSION PROCESS WHOSE VELOCITY INCREASES STEADILY WITH INCREASING SUBSTRATE. EVIDENCE SUGGESTS THAT BACTERIAL POPULATIONS ARE RESPONSIBLE FOR 'TRANSPORT-TYPE' PHENOMENA, AND ALGAL POPULATIONS, FOR UPTAKE BY DIFFUSION. (EICHHORN-WISCONSIN)

FIELD 02H, 05C

ACCESSION NO. W70-04194

LIBERATION OF ORGANIC ACIDS BY GREEN UNICELLULAR ALGAE,

MOSCOW STATE UNIV. (USSR). DEPT. OF MICROBIOLOGY.

I. V. MAKSIMOVA, AND M. N. PIMENOVA.

TRANSLATED FROM MIKROBIOLOGIYA, VOL 37, NO L, P 77-86, JAN-FEB 1969. MICROBIOLOGY, VOL 38, NO 1, P 64-70, 1969. 7 FIG, 4 TAB, 17 REF.

DESCRIPTORS:
*ALGAE, *CHLORELLA, *ORGANIC ACIDS, BIOMASS, ORGANIC COMPOUNDS, CARBOHYDRATES.

IDENTIFIERS:
*GREEN ALGAE, CHLORELLA VULGARIS, CHLORELLA PYRENOIDOSA, VOLATILE ACIDS, NON-VOLATILE ACIDS, KETOACIDS, TAMIYA'S MINERAL MEDIUM, GLYOXYLIC ACID.

ABSTRACT:
AN INVESTIGATION OF THE QUALITATIVE COMPOSITION OF ORGANIC ACIDS LIBERATED TO THE MINERAL MEDIUM BY DIFFERENT STRAINS OF CHLORELLA VULGARIS AND CHLORELLA PYRENOIDOSA. FORMIC, ACETIC, LACTIC, PYRUVIC, A-KETOGLUTARIC, ACETOACETIC IN SOME CASES COMPRISED 25% OF THE EXTRACELLULAR COMPOUND. THE FILTRATES OF THE ALGAE ALWAYS CONTAINED GLYOXIC ACID. THE TOTAL QUANTITY OF ACIDS AND THEIR REPRESENTATION VARIED WITH THE PHASE OF GROWTH AND SIZE OF THE BIOMASS. ANALYSIS OF CELLS INDICATED THAT THE ACIDS WERE NOT DERIVED FROM EXTRACELLULAR COMPOUNDS. SECRETION OF NONVOLATILE ACIDS, GRADUAL ACCUMULATION OF VOLATILE ACIDS, AND THE SHARP INCREASE IN KETOACIDS AT HIGH DENSITY OF THE CULTURE SUGGESTED THAT THE PROCESSES ARE CONTROLLED BY DIFFERENT MACHANISMS. (WILDE-WISCONSIN)

FIELD 05C

ACCESSION NO. W70-04195

GRAVITY FILTRATION OF ALGAL SUSPENSIONS,

ONTARIO WATER RESOURCES COMMISSION, TORONTO. DIV. OF RESEARCH.

R. H. G. ANDREWS.

ONTARIO WATER RESOURCES COMMISSION, DIVISION OF RESEARCH PUBLICATION NO 21,
    AUG 1968. 41 P, 22 FIG, 6 TAB.

DESCRIPTORS:
    *WATER, *FILTERS, *ALGAE, NUISANCE ALGAE, INTAKES, SANDS, FLOW RATES,
    TURBIDITY, CHLORELLA, SCENEDESMUS, EUGLENA.

IDENTIFIERS:
    *WATER SUPPLIES, GREEN ALGAE, HEAD LOSSES, OOCISTUS, FILTRATION MEDIA,
    ANTHRACITE, GRAVITY FILTERS.

ABSTRACT:
    THIS STUDY WAS CONCERNED WITH EFFECTIVENESS OF DIFFERENT FILTER MEDIA
    PREVENTING THE CLOGGING OF WATER SUPPLIES BY ALGAE: CHLORELLA,
    SCENEDESMUS, OOCISTUS, AND EUGLENA. THE DUAL-MEDIA AND ANTHRACITE
    FILTERS PROVED COMPARATIVELY MOST EFFECTIVE, BUT THEIR MAXIMUM
    RETENTION WAS ONLY 50% OF EUGLENA AND 18% OF OTHER ALGAE. THERE WAS NO
    CONSISTENT EFFECT OF THE RATE OF FLOW ON ALGAE REMOVED. THE USE OF
    NEPHELOMETRIC TURBIDIMETER FAILED TO PROVIDE A SIGNIFICANT CORRELATION
    BETWEEN ALGAE COUNTS AND RECORDED TURBIDITIES. THE CONCLUSION IS THAT
    GRAVITY FILTERS ALONE SHOULD NOT BE CONSIDERED FOR REMOVAL OF ALGAE
    FROM RAW WATER SUPPLIES. (WILDE-WISCONSIN)

FIELD 05F

ACCESSION NO. W70-04199

SPECIAL ASPECTS OF NITROGEN FIXATION BY BLUE-GREEN ALGAE,

BRISTOL UNIV. (ENGLAND). DEPT. OF BOTANY; AND WESTFIELD COLL., LONDON
    (ENGLAND). DEPT. OF BOTANY.

ROSALIE M. COX, AND P. FAY.

PROCEEDINGS OF THE ROYAL SOCIETY (B), VOL 172, NO 1029, P 357-366, 1969. 2
    FIG, 3 TAB, 30 REF.

DESCRIPTORS:
    *NITROGEN FIXATION, *ALGAE, LIGHT, CARBON DIOXIDE, INHIBITION, NITROGEN
    FIXING BACTERIA, RADIOACTIVITY TECHNIQUE, PHOTOSYNTHESIS.

IDENTIFIERS:
    *ANABAENA CYLINDRICA, ATP, CMU, NITROGEN-15, PYRUVATE, ACETYLENE
    REDUCTION, REDUCTANT.

ABSTRACT:
    MORE THAN 90% INHIBITION OF CARBON DIOXIDE FIXATION BY
    P-CHLOROPHENYL-1, 1-DIMETHYLUREA (CMU) FAILED TO INFLUENCE NITROGEN
    FIXATION BY NITROGEN-STARVED CELLS OF ANABAENA CYLINDRICA, BUT RETARDED
    THE PROCESS ABOUT 50% IN NORMAL CELLS OF THE ORGANISM. THIS SUGGESTED
    THAT NITROGEN-FIXATION IN ANABAENA IS INDEPENDENT OF REDUCING POTENTIAL
    GENERATED BY THE PHOTO-ELECTRON TRANSPORT, BUT IS STIMULATED BY
    PHOTOSYNTHETICALLY PRODUCED CARBON SKELETONS. INCREASED LIGHT AUGMENTED
    NITROGEN-FIXATION IN THE PRESENCE AND IN ABSENCE OF CMU THUS INDICATING
    THE ESSENTIAL PART OF ATP IN THE PROCESS. THE ESTABLISHED 3:1 RATIO OF
    PYRUVATE DECARBOXYLATION TO NITROGEN-FIXATION IS IN AGREEMENT WITH THE
    HYPOTHESIS THAT PYRUVATE DONATES HYDROGEN FOR NITROGEN REDUCTION, AND
    THAT IN BLUE-GREEN ALGAE THE AVAILABILITY OF THE REDUCTANT IS NOT
    INFLUENCED BY PHOTOSYNTHESIS. (WILDE-WISCONSIN)

FIELD 05C

ACCESSION NO. W70-04249

A STUDY OF THE PROFUNDAL BOTTOM FAUNA OF LAKE WASHINGTON,

WASHINGTON UNIV., SEATTLE. DEPT. OF ZOOLOGY.

RUDOLPH N. THUT.

ECOLOGICAL MONOGRAPHS, VOL 39, NO 1, P 79-110, 1969. 27 FIG, 5 TAB, 34 REF.

DESCRIPTORS:
*BENTHIC FAUNA, NITROGEN, PHOSPHORUS, PHYTOPLANKTON, DIATOMS,
ZOOPLANKTON, DINOFLAGELLATES, ALGAE, DISSOLVED OXYGEN, HYPOLIMNION,
EPILIMNION, LAKE ERIE, INSECTS, DIPTERA, CRUSTACEA, MOLLUSCS,
OLIGOCHAETES, GASTROPODS.

IDENTIFIERS:
*LAKE WASHINGTON(WASH), MACROFAUNA, GYTTJA, OSCILLATORIA RUBESCENS,
ZURICHSEE, FILAMENTOUS ALGAE, FILAMENTOUS BLUE-GREEN ALGAE, OCULAR
MICROMETER, SUCROSE-FLOTATION TECHNIQUE, METALIMNION, BATHYTHERMOGRAPH,
ARTHROPODA, CHIRONOMIDAE, AQUATIC OLIGOCHAETES.

ABSTRACT:
A REVIEW OF THE MACROFAUNA IN THE PROFUNDAL ZONE OF LAKE WASHINGTON
BETWEEN SEPTEMBER 1963 AND SEPTEMBER 1964 AND DISCUSSION. ALTHOUGH
SEWAGE DIVERSION BEGAN IN 1963, PHOSPHATE AND OXYGEN VALUES INDICATED
THE LAKE WAS STILL IN THE EUTROPHIC PHASE. TEN STATIONS WERE CHOSEN AT
5-METER DEPTH INTERVALS FROM 10 TO 55 METERS AND SAMPLED APPROXIMATELY
MONTHLY WITH AN EKMAN DREDGE. 24 SPECIES WERE RECOGNIZED FROM THE
PROFUNDAL ZONE, EACH PRESENTED SEPARATELY, WITH THE EXCEPTION OF
OLIGOCHAETA SPECIES, ALONG WITH THEIR DEPTH AND POPULATION DYNAMICS
THROUGHOUT THE YEAR. THE CHIRONOMIDAE WERE MOST NUMEROUS OF BOTTOM
FAUNA CONSTITUENTS (ABOUT 45% OF THE TOTAL). 13 SPECIES WERE FOUND
DURING THE STUDY, INCLUDING PREDATORS, DEPOSIT-FEEDERS, AND
FILTER-FEEDERS. LARVAE WERE MOST COMMON AT THE SHALLOW-WATER STATIONS
AND BECAME PROGRESSIVELY DIMINISHED WITH INCREASE IN DEPTH. THE
OLIGOCHAETA COMPRISED ABOUT 1/2 OF THE TOTAL NUMBER AND 1/3 OF THE
TOTAL BIOMASS OF THE PROFUNDAL FAUNA. FOUR SPECIES WERE IDENTIFIED. THE
OLIGOCHAETA WERE FOUND IN THE GREATEST NUMBERS AND BIOMASS AT THE
GREATEST DEPTH SAMPLED; THEIR ABUNDANCE PROGRESSIVELY DECLINED WITH
DECREASE IN DEPTH. AMPHIPODA AND SPHAERIDAE WERE PRESENT BUT IN SMALLER
NUMBERS. (HASKINS-WISCONSIN)

FIELD 02H, 05C

ACCESSION NO. W70-04253

WATER POLLUTION BY NUTRIENTS-SOURCES, EFFECTS AND CONTROL, PAPERS PRESENTED AT
1966 ANNUAL MEETING OF MINNESOTA CHAPTER SOIL CONSERVATION SOCIETY OF
AMERICA.

MINNESOTA UNIV., MINNEAPOLIS. WATER RESOURCES RESEARCH CENTER.

AVAILABLE FROM THE CLEARINGHOUSE AS PB-189 794, $3.00 IN PAPER COPY, $0.65 IN
MICROFICHE. WRRC BULLETIN 13, MINNESOTA WATER RESOURCES RESEARCH CENTER,
JUNE 1969. 79 P. OWRR PROJECT A-999-MINN.

DESCRIPTORS:
*WATER POLLUTION SOURCES, *WATER POLLUTION EFFECTS, *EUTROPHICATION,
*NUTRIENTS, ALGAE, FISH POPULATION, *WATER POLLUTION CONTROL, FARM
WASTES, MUNICIPAL WASTES, SEPTIC TANKS, RECREATION WASTES, WATER
QUALITY.

IDENTIFIERS:
*NUTRIENT SOURCES.

ABSTRACT:
THE BULLETIN INCLUDES THE PAPERS PRESENTED AT A CONFERENCE ON 'NUTRIENT
POLLUTION - SOURCES, EFFECTS AND CONTROL' HELD IN MINNEAPOLIS,
MINNESOTA ON JANUARY 8, L969. THE CONFERENCE WAS PLANNED AS THE ANNUAL
MEETING OF THE MINNESOTA CHAPTER, SOIL CONSERVATION SOCIETY OF AMERICA.
THE PAPERS, ALL INDIVIDUALLY ABSTRACTED, INCLUDE THE FOLLOWING TITLES:
NUTRIENTS AND OTHER FORMS OF POLLUTION, DIAGNOSING POLLUTION IN LAKE
MINNETONKA, EFFECT OF EUTROPHICATION ON FISH AND RELATED ORGANISMS,
HEALTH ASPECTS, SURFACE AND GROUNDWATERS, ANIMAL WASTE DISPOSAL
PROBLEMS AND TRENDS IN MINNESOTA, MANAGING LIVESTOCK WASTES TO CONTROL
POLLUTION, RUNOFF AND SEDIMENT AS NUTRIENT SOURCES, CONTROLLING
NUTRIENTS AND ORGANIC TOXICANTS IN RUNOFF, TREATMENT OF MUNICIPAL
WASTES, SEPTIC TANK EFFLUENTS, WATER POLLUTION IN RECREATIONAL AREAS -
SOURCES AND CONTROL, AND SETTING WATER QUALITY STANDARDS AND REGULATING
NUTRIENT SOURCES, IMPLEMENTING POLLUTION CONTROL.

FIELD 05B, 05C, 05G

ACCESSION NO. W70-04266

DIAGNOSING POLLUTION IN LAKE MINNETONKA,

MINNESOTA UNIV., MINNEAPOLIS. LIMNOLOGICAL RESEARCH CENTER.

ROBERT O. MEGARD.

WATER POLLUTION BY NUTRIENTS--SOURCES, EFFECTS AND CONTROL, WATER RESOURCES
RESEARCH CENTER, UNIVERSITY OF MINNESOTA, MINNEAPOLIS, WRRC BULLETIN 13, P
9-14, JUNE 1969. 1 FIG, 1 TAB, 2 REF.

DESCRIPTORS:
*PHOSPHORUS, *NITROGEN, *ALGAE, SEWAGE PLANTS, CHLOROPHYLL,
PHOTOSYNTHESIS, TRIBUTARIES, WATERSHED MANAGEMENT, SEDIMENTS.

IDENTIFIERS:
LAKE MINNETONKA(MINN), NUTRIENT TRAPS, NUTRIENT INFLUX, NUTRIENT
SOURCES, PHOSPHORUS BUDGET, APHANIZOMINON FLOS AQUAE.

ABSTRACT:
INCREASED MONETARY SPENDING ON WASTE DISPOSAL AND WATERSHED MANAGEMENT
IS SUGGESTED AS A RECOURSE TO DETERIORATING QUALITY OF LAKES. AN
EXAMPLE, IS WIDENING THE SCOPE OF SEWAGE TREATMENT TO INCLUDE
OBJECTIVES OF MAINTAINING RECREATIONAL AND AESTHETIC VALUES OF LAKES AS
WELL AS MINIMIZING HEALTH HAZARDS. MEASURING ABUNDANCE AND DAILY GROWTH
OF ALGAE BY CHLOROPHYLL ANALYSIS AND PHOTOSYNTHESIS EXPERIMENTS
RESPECTIVELY, IS A METHOD OF COMPARING LAKES AND ASSESSING
EFFECTIVENESS OF NUTRIENT ABATEMENT PROGRAMS. DATA WAS GATHERED FROM
LOWER LAKE OF LAKE MINNETONKA ADJACENT TO MINNEAPOLIS, MINNESOTA, AND A
DIRECT CORRELATION BETWEEN PHOSPHORUS CONCENTRATION WAS DISCOVERED
DURING JULY AND AUGUST. EACH POUND OF PHOSPHORUS WAS CONSIDERED
RESPONSIBLE FOR 500 LBS OF DRY ALGAE DURING THE TWO MONTHS. DATA ON
INFLUX OF PHOSPHORUS FROM TRIBUTARIES AND CONCENTRATION IN THE LAKE
INDICATED THAT SEDIMENTS WERE THE MAJOR NUTRITION TRAP. THERE WAS DOUBT
WHETHER PHOSPHORUS RELEASED BY SEDIMENTS BETWEEN MAY AND AUGUST WAS
AVAILABLE TO ALGAE. DATA INDICATED THAT REDUCING PHOSPHORUS INFLUX BY
ONE-HALF WOULD SUBSTANTIALLY CONTROL LOWER LAKE'S EUTROPHICATION BASED
ON THE ASSUMPTION THAT A LARGE PORTION OF PHOSPHORUS IS REMOVED BY
BIOGEOCHEMICAL PROCESS. (SEE W70-04266). (BANNERMAN-WISCONSIN)

FIELD 05C, 02H

ACCESSION NO. W70-04268

STUDY OF BACTERIA ASSOCIATED WITH MARINE ALGAE IN CULTURE. I. PRELIMINARY
DETERMINATION OF SPECIES (FRENCH),

STATION MARINE D'ENDOUME, MARSEILLE (FRANCE).

B. R. BERLAND, M. G. BIANCHI, AND S. Y. MAESTRINI.

MARINE BIOLOGY, VOL 2, NO 4, P 350-355, 1969. 3 TAB, 12 REF.

DESCRIPTORS:
*BACTERIA, *MARINE ALGAE, *CULTURES, PSEUDOMONAS, METABOLISM, KINETICS,
NITROGEN COMPOUNDS.

IDENTIFIERS:
SPECIES DETERMINATION, TAXONOMY, FLAVOBACTERIUM, ACHROMOBACTER, GULF OF
MARSEILLES, ASTERIONELLA JAPONICA, CHAETOCEROS LAUNDERI, LAUDERIA
BOREALIS, LEPTOCYLINDRUS DANICUS, PHAEODACTYLUM TRICORNUTUM,
STICHOCHRYSIS IMMOBILIS, VIBRIO, MORPHOLOGY, CULTURAL CHARACTERISTICS,
TERMARY COMPOUNDS, MACROMOLECULES, ANTIBIOTICS, AGARBACTERIUM,
XANTHOMONAS, MICROCOCCUS, STAPHYLOCOCCUS.

ABSTRACT:
BACTERIA STRAINS HAVE BEEN ISOLATED FROM MARINE ALGAE CULTURES AND
ASSIGNED TENTATIVE GENERA AND SPECIES. OBSERVATIONS AND TESTS WERE
CONDUCTED USING BERGEY'S MANUAL AND SPECIFIC PAPERS ON MARINE BACTERIA
AS TAXONOMIC KEYS. THE GENERA PSEUDOMONAS, FLAVOBACTERIUM, AND
ACHROMOBACTER APPEAR MOST IMPORTANT, BASED ON THE NUMBER OF SPECIES.
THIS DOES NOT IMPLY THAT THEY ACTUALLY DOMINATE IN NUMBER OF GERMS AND
IN METABOLIC IMPORTANCE. A LATER STUDY WILL PERMIT DETERMINATION OF THE
PRECISE NATURE AND DIVISION OF SPECIES WHICH ACCOMPANY AN ALGA IN
CULTURE AND SUBSEQUENTLY THE SAME ALGA IN ITS NATURAL ENVIRONMENT. IT
MAY BE NECESSARY TO USE SOME CULTURE MEDIUM MORE DEFINITIVE AND LESS
PLETHORIC, THE MEDIUM CURRENTLY USED BEING TOO RICH IN ORGANIC MATTER
AND NOT CORRESPONDING TO CONDITIONS OF THE NATURAL MEDIUM OF THE
ORGANISMS; THE LITERATURE SHOWS THAT THEIR NUTRITIONAL NEEDS ARE VERY
SIMPLE. THE STUDY IS UNDERTAKEN TO DETERMINE PRECISELY THEIR METABOLIC
NEEDS AND HOW THESE ARE SATISFIED IN A CULTURE OF ALGAE.
(JONES-WISCONSIN)

FIELD 02H, 05C

ACCESSION NO. W70-04280

PHYTOPLANKTON AND RELATED WATER-QUALITY CONDITIONS IN AN ENRICHED ESTUARY,

GEOLOGICAL SURVEY, TACOMA, WASH.

EUGENE B. WELCH.

JOURNAL WATER POLLUTION CONTROL FEDERATION, VOL 40, NO 10, P 1711-1727, 1968.
10 FIG, 27 REF.

DESCRIPTORS:
*ANALYTICAL TECHNIQUES, *AQUATIC ALGAE, *AQUATIC MICROBIOLOGY, *AQUATIC
PRODUCTIVITY, BIOCHEMICAL OXYGEN DEMAND, CYCLING NUTRIENTS, DIATOMS,
ENVIRONMENTAL EFFECTS, ESSENTIAL NUTRIENTS, EUTROPHICATION, LIGHT
INTENSITY, LIGHT PENETRATION, LIMNOLOGY, NITROGEN COMPOUNDS, NUISANCE
ALGAE, NUTRIENT REQUIREMENTS, NUTRIENTS, PHOSPHORUS COMPOUNDS,
PHOTOSYNTHETIC OXYGEN, PHYSIOLOGICAL ECOLOGY, PHYTOPLANKTON, ESTUARIES,
POLLUTANT IDENTIFICATION, SEWAGE, SEWAGE DISPOSAL, SEWAGE EFFLUENTS,
WATER POLLUTION CONTROL, WATER POLLUTION EFFECTS, WATER POLLUTION
SOURCES.

IDENTIFIERS:
*SEATTLE(WASH), RENTON TREATMENT PLANT, DUWAMISH ESTUARY(WASH).

ABSTRACT:
AMMONIA, SOLUBLE PHOSPHATE, AND TOTAL PHOSPHATE CONCENTRATIONS WERE
OBSERVED TO INCREASE IN THE DUWAMISH ESTUARY, WASHINGTON, FOLLOWING
INITIAL DISCHARGE OF EFFLUENT FROM THE RENTON TREATMENT PLANT AT
SEATTLE. A PHYTOPLANKTON BLOOM DOMINATED BY MARINE SPECIES OCCURRED IN
THE LOWER ESTUARY IN AUGUST 1965, ABOUT 1.5 MONTHS FOLLOWING THE
NUTRIENT INCREASE. A BLOOM DID NOT OCCUR IN 1964, PRIOR TO EFFLUENT
DISCHARGE FROM THE TREATMENT PLANT, BUT SOME EVIDENCE SHOWS THAT ALGAL
ACTIVITY WAS GREAT IN AUGUST 1963. THE NUTRIENT INCREASE IN 1965
PROBABLY WAS NOT THE SOLE FACTOR CAUSING THE BLOOM DURING THAT SUMMER
BECAUSE: (A) THE PRE-EFFLUENT NUTRIENT CONCENTRATIONS IN 1963 AND 1964
WERE RELATIVELY HIGH; (B) A BLOOM ALSO OCCURRED AT THE FURTHEST
DOWNSTREAM STATION WHERE NO INCREASE IN NUTRIENTS WAS APPARENT IN 1965
OVER THAT OF THE PRECEDING TWO YEARS; (C) A BLOOM PROBABLY OCCURRED IN
1963, BEFORE THE ADDITION OF NUTRIENTS FROM THE PLANT; AND (D) THE
BLOOM MAXIMUM DID NOT OCCUR IN 1965 UNTIL ABOUT 1.5 MONTHS AFTER THE
NUTRIENT INCREASE, WHEN DISCHARGE AND TIDAL EXCHANGE CONDITIONS WERE
MINIMUM. BLOOM TIMING SEEMED RELATED MOST CLOSELY TO HYDROGRAPHIC
CONDITIONS. (FITZGERALD-WISCONSIN)

FIELD 02L, 05C

ACCESSION NO. W70-04283

COMPETITION BETWEEN PLANKTONIC BACTERIA AND ALGAE FOR ORGANIC SOLUTES,

UPPSALA UNIV. (SWEDEN). INST. OF LIMNOLOGY.

JOHN E. HOBBIE, AND RICHARD T. WRIGHT.

MEMORIE DELL'INSTITUTO ITALIANO DI IDROBIOLOGIA, SUPPL 18, P 175-185, 1965. 4
  FIG, 1 TAB, 5 REF.

DESCRIPTORS:
  *COMPETITION, *PHYTOPLANKTON, *AQUATIC BACTERIA, *ALGAE, ORGANIC
  MATTER, LAKES, DIFFUSION, CARBON RADIOISOTOPES, CHLAMYDOMONAS,
  BIOASSAYS, WATER POLLUTION EFFECTS, ANALYTICAL TECHNIQUES.

IDENTIFIERS:
  *ORGANIC SOLUTES, *HETEROTROPHY, MICROBIAL ECOLOGY, SWEDEN, LAKE
  ERKEN(SWEDEN), ACTIVE TRANSPORT, CARBON-14, ENZYME KINETICS, GLUCOSE,
  MECHAELIS-MENTEN EQUATION, ACETATE, GLYCOLLIC ACID.

ABSTRACT:
  SOPHISTICATED INVESTIGATIVE TECHNIQUES, IN ADDITION TO THE TRADITIONAL
  METHODS OF DIRECT COUNTS AND ENRICHMENT CULTURES, ARE REQUIRED FOR
  BETTER UNDERSTANDING OF ROLES OF MICROORGANISMS IN THE DYNAMICS OF
  LACUSTRINE METABOLISM. A PROBLEM OF CURRENT INTEREST IS THE
  IDENTIFICATION OF ORGANIC SOLUTES UTILIZED BY HETEROTROPHIC BACTERIA
  AND ALGAE, AND ESTIMATION OF THEIR CONCENTRATIONS AND UPTAKE RATES.
  DATA ON ASSIMILATION OF RADIOCARBON-LABELED ORGANIC SUBSTRATES CAN BE
  ANALYZED ON BASES OF THEORIES OF DIFFUSION AND ENZYME KINETICS. SUCH
  METHODS ARE ILLUSTRATED WITH DATA FROM STUDIES ON MODEL CULTURES
  (CHLAMYDOMONAS AND PLANKTONIC BACTERIA) AND SAMPLES FROM LAKE ERKEN
  (SWEDEN). MEASUREMENTS OF VELOCITY UPTAKE OF RADIOISOTOPE OVER A RANGE
  OF SUBSTRATE CONCENTRATIONS REVEAL TWO MECHANISMS: AN ACTIVE TRANSPORT
  SYSTEM, ATTRIBUTABLE TO BACTERIA, AND ANOTHER SYSTEM IN ALGAE WHICH ·
  FOLLOWS DIFFUSION KINETICS. THE LATTER IS EFFECTIVE AT SUBSTRATE
  CONCENTRATIONS ABOVE 500 MICROGRAMS/LITER. IN NATURAL WATERS, BACTERIA
  EFFECTIVELY REDUCE ACETATE AND GLUCOSE CONCENTRATIONS BELOW 20
  MICROGRAMS/LITER, APPARENTLY EFFECTING DRASTIC LIMITATION OF ALGAL
  HETEROTROPHY. IN NATURE, OTHER COMPOUNDS ARE AVAILABLE TO ALGAE AND
  HIGHER TEMPERATURE MAY INCREASE UPTAKE RATES, SO THAT ALGAL
  HETEROTROPHY IS NOT NECESSARILY PRECLUDED. (EICHHORN-WISCONSIN)

FIELD 02H, 05C

ACCESSION NO. W70-04284

A NEW METHOD FOR THE STUDY OF BACTERIA IN LAKES: DESCRIPTION AND RESULTS,

NORTH CAROLINA STATE UNIV., RALEIGH. DEPT. OF ZOOLOGY; AND GORDON COLL., WENHAM, MASS.

JOHN E. HOBBIE, AND RICHARD T. WRIGHT.

MITTEILUNGEN INTERNATIONALE VEREINIGUNG FUR THEORETISCHE UND ANGEWANDTE LIMNOLOGIE, VOL 14, P 64-71, 1968. 5 FIG, 10 REF.

DESCRIPTORS:
*LAKES, *BACTERIA, *ANALYTICAL TECHNIQUES, WATER POLLUTION EFFECTS, ALGAE, ECOSYSTEMS, PHYSIOLOGICAL ECOLOGY, AQUATIC BACTERIA, CARBON RADIOISOTOPES, ENZYMES, KINETICS, PHOTOSYNTHESIS, ORGANIC MATTER, ORGANIC COMPOUNDS, METABOLISM, BIOINDICATORS, BIOASSAYS.

IDENTIFIERS:
*HETEROTROPHY, AUTOTROPHY, CARBON-14, RADIOCARBON UPTAKE TECHNIQUES, GLUCOSE, MICHAELIS-MENTEN EQUATION, SWEDEN, LAKE ERKEN(SWEDEN), LINEWEAVER-BURK EQUATION, MAXIMUM UPTAKE VELOCITY, MICHAELIS CONSTANT, LAKE NORRVIKEN(SWEDEN), LAKE EKOLN(SWEDEN), LAKE LOTSJON(SWEDEN), PERMEASES, CHAMYDOMONADS, LAPPLAND(SWEDEN), ACETATE, TURNOVER TIMES.

ABSTRACT:
ALTHOUGH BACTERIA ARE IMPORTANT IN LACUSTRINE METABOLISM, DETAILS OF THEIR POPULATION DYNAMICS--INCLUDING GENERATION TIMES AND TURNOVER RATES--ARE POORLY UNDERSTOOD. RADIOCARBON-LABELED ORGANIC COMPOUNDS HAVE BEEN USED PREVIOUSLY TO DETERMINE HETEROTROPHIC POTENTIAL OF NATURAL WATERS, BUT CONCENTRATIONS OF SUBSTRATE ADDED HAVE BEEN SUFFICIENTLY LARGE AS TO MAKE INTERPRETATION OF RESULTANT DATA AMBIGUOUS. BASED UPON THEORETICAL TREATMENT ADAPTED FROM LINEWEAVER-BURK MODIFICATION OF MICHAELIS-MENTEN ENZYME KINETICS, AND ON ADDITIONS OF VERY LOW CONCENTRATIONS OF LABELED ORGANIC SUBSTRATES, A NEW METHOD FOR STUDY OF BACTERIAL ASSIMILATION IS DESCRIBED. THAT DERIVATION YIELDS RELATIONSHIP, $T/F = (K + S)/V) + A/V$. HERE, T IS INCUBATION TIME (HOURS); F, PERCENTAGE UPTAKE OF ISOTOPE BY ORGANISMS; K, CONSTANT (SIMILAR TO MICHAELIS CONSTANT); S, SUBSTRATE CONCENTRATION IN NATURE; A, ADDED SUBSTRATE CONCENTRATION; V, MAXIMUM VELOCITY. PLOT OF T/F VERSUS A USUALLY RESULTS IN LINEAR RELATIONSHIP FROM WHICH CAN BE ESTIMATED MAXIMUM UPTAKE VELOCITY (V), TURNOVER TIME, AND APPROXIMATE NATURALLY OCCURRING SUBSTRATE CONCENTRATION. METHOD IS ILLUSTRATED WITH DATA FROM A SERIES OF SWEDISH LAKES. DURING A SINGLE YEAR, V MAY VARY BY 2 ORDERS OF MAGNITUDE WITHIN ONE LAKE; BY 4, BETWEEN LAKES. V IS ALSO A VERY SENSITIVE INDICATOR OF POLLUTION. (EICHHORN-WISCONSIN)

FIELD 02H, 05C

ACCESSION NO. W70-04287

ANNUAL AUTOTROPHIC PRODUCTION OF PHYTOPLANKTON IN LAKE BAIKAL,

K. K. VOTINTSEV, AND G. I. POPOVSKAYA.

TRANSLATED FROM RUSSIAN. ORIGINAL P 205-208. DOKLADY AKADEMII NAUK SSSR, VOL 176, NO 1, P 634-637, 1967. 3 FIG, 1 TAB, 13 REF.

DESCRIPTORS:
*PHYTOPLANKTON, *LAKES, ORGANIC MATTER, SEASONAL, ALGAE, PRIMARY PRODUCTIVITY, HARVESTING OF ALGAE, BIOMASS.

IDENTIFIERS:
SEASONAL DISTRIBUTION, LAKE BAIKAL, PHOTOSYNTHETIC LAYER, MELOSIRA BAICALENSIS, GYMNODINIUM BAICALENSE, CYCLOTELLA MINUTA.

ABSTRACT:
THIS PIONEERING SURVEY OF ORGANIC MATTER PRODUCTION IN LAKE BAIKAL DISCLOSED A WIDE PLACE-AND-TIME VARIATION IN THE BIOMASS OF PHYTOPLANKTON, REPRESENTED LARGELY BY MELOSIRA BAICALENSIS, CYCLOTELLA MINUTA, AND GYMNODINIUM BAICALENSE. ACCORDING TO PRELIMINARY ESTIMATES, THE LAKE HAS AN ANNUAL PRODUCTIVITY OF 157 MILLION TONS (WET WEIGHT) OF HARVESTABLE ALGAE, AND 136 MILLION TONS OF NON-HARVESTABLE ALGAE. THE STUDY OF PRIMARY PRODUCTIVITY BY THE 'FLASK' METHOD, USING OXYGEN, WAS CONFINED TO THE UPPER 25 METERS OF THE PHOTOSYNTHETIC LAYER. THE MEAN ANNUAL BIOMASS PER SURFACE SQUARE METER OF THIS LAYER APPROACHED 7 GRAMS. THE RATIO OF THE AVERAGE ANNUAL PRIMARY PRODUCTION TO THE MEAN ANNUAL INDEX OF PHYTOPLANKTON BIOMASS FOR THE ENTIRE LAKE WAS EQUAL TO 464. PREVIOUS STUDIES HAVE SHOWN THAT VALUES FOR GROSS PHOTOSYNTHESIS ARE CLOSE TO THOSE PERMITTING EFFECTIVE PRODUCTIVITY OF ALGAE. INFORMATION ON FOOD REQUIREMENTS OF MAIN REPRESENTATIVES OF ZOOPLANKTON OF LAKE BAIKAL THUS FAR IS NOT AVAILABLE. (WILDE-WISCONSIN)

FIELD 02H, 05C

ACCESSION NO. W70-04290

DEVELOPMENT OF A METHOD FOR THE TOTAL COUNT OF MARINE BACTERIA ON ALGAE,

MCGILL UNIV., MONTREAL (QUEBEC). DEPT. OF MICROBIOLOGY AND IMMUNOLOGY; AND
NEW BRUNSWICK UNIV., FREDERICTON. DEPT. OF BIOLOGY.

E. C. S. CHAN, AND ELIZABETH A. MCMANUS.

CANADIAN JOURNAL OF MICROBIOLOGY, VOL 13, P 295-301, 1967. 2 FIG, 3 TAB, 10
REF.

DESCRIPTORS:
*MARINE BACTERIA, *ALGAE, *ANALYSIS, *ANALYTICAL TECHNIQUES,
TEMPERATURE, PIPTIDES, INTERTIDAL AREAS, SOIL BACTERIA, LABORATORY
TESTS.

IDENTIFIERS:
*ALGAE HOMOGENIZATION, *BACTERIA COUNT, WARING BLENDOR MORTALITY, CELL
WALL, HEXOSAMINE, ASCOPHYLLUM NODOSUM, POLYSIPHONIA LANOSA, POINT
LEPREAU(CANADA), NEW BRUNSWICK(CANADA), SERVALL OMNI-MIXER HOMOGENIZER.

ABSTRACT:
IN AN INITIAL ATTEMPT TO DEVELOP A TECHNIQUE FOR ENUMERATION OF
BACTERIA ON LITTORAL ALGAE, IT WAS OBSERVED THAT HOMOGENIZATION OF
ALGAL SAMPLES IN THE WARING BLENDOR RESULTED IN MORTALITY OF MARINE
BACTERIA SAMPLES, APPARENTLY DUE TO SINGULAR SUSCEPTIBILITY TO
MANIPULATIONS. FURTHER WORK ON THIS LETHAL EFFECT OF BLENDING SHOWED
TWO DETRIMENTAL FACTORS, HEAT AND MECHANICAL INJURY OF THE CELLS, WERE
IMPLICATED: DUE TO THE INCREASE IN TEMPERATURE DURING HOMOGENIZATION
(TEMPERATURES SLIGHTLY BEYOND 35C RESULT IN RAPID BACTERIAL KILLING)
AND THE HIGH SPEED ROTATION OF THE KNIFE-BLADE ASSEMBLY OF THE WARING
BLENDOR. THESE FINDINGS RESULTED IN A SATISFACTORY TECHNIQUE FOR
OBTAINING HOMOGENATES OF MARINE MATERIALS TO OBTAIN TOTAL BACTERIA
COUNT BY USING A SERVALL OMNI-MIXER HOMOGENIZER AT SPEEDS OF ABOUT 5000
REVOLUTIONS PER MINUTE AND TEMPERATURE OF HOMOGENIZATION NOT EXCEEDING
30C. WITH THIS METHOD, IT WAS POSSIBLE TO MAKE AN EXTENSIVE STUDY OF
ECOLOGY OF MARINE BACTERIA ON POLYSIPHONIA LANOSA AND ASCOPHYLLUM
NODOSUM, THE FORMER ALGA GROWING EPIPHYTICALLY ON ASCOPHYLLUM.
(JONES-WISCONSIN)

FIELD 05A

ACCESSION NO. W70-04365

YEASTS FROM THE NORTH SEA,

MIAMI UNIV., FLA. INST. OF MARINE SCIENCES; MIAMI UNIV., FLA. DEPT. OF
MICROBIOLOGY; AND BIOLOGISCHE ANSTALT HELGOLAND (WEST GERMANY).

S. P. MEYERS, D. G. AHEARN, W. GUNKEL, AND F. J. ROTH, JR.

MARINE BIOLOGY, VOL 1, NO 2, P 118-123, 1967. 6 FIG, 1 TAB, 19 REF.

DESCRIPTORS:
 *YEASTS, POPULATION, MARINE PLANTS, DINOFLAGELLATES, OCEANS,
 BIOINDICATORS, FUNGI, ESTUARIES, BACTERIA, SAMPLING, SEASONAL,
 HYDROGRAPHY, FLORIDA, THERMOCLINE, CURRENTS(WATER), PHYTOPLANKTON,
 ZOOPLANKTON, HYDROGEN ION CONCENTRATION, ALGAE.

IDENTIFIERS:
 *NORTH SEA, MYCOTA, DEBARYOMYCES HANSENII, RHODOTORULA RUBRA, CANDIDA
 DIDDENSII, CANDIDA ZEYLANOIDES, CANDIDA KRUSEI, CANDIDA OBTUSA, CANDIDA
 TROPICALIS, NOCTILUCA MILIARIS, HELGOLAND(GERMANY), AUREOBASIDIUM
 PULLULANS, HANSENIASPORA UVARUM, RADULESPORES, ZYMOLOGICAL,
 ELBE(GERMANY), SPOROBOLOMYCES ROSEUS, AUTOLYSIS, ASCOSPOROGENOUS.

ABSTRACT:
 MYCOLOGICAL EXAMINATION OF ESTUARINE AND OPEN OCEAN ENVIRONMENTS HAVE
 REVEALED DIVERSE POPULATIONS OF YEAST OF VARIOUS TAXA AND PHYSIOLOGICAL
 GROUPS. EVIDENCE INDICATES THAT THESE MICROORGANISMS ARE NORMAL
 COMPONENTS OF OCEAN BIOTA, AND THAT CERTAIN SPECIES MAY SERVE AS
 BIO-INDICATORS FOR SPECIFIC WATER MASSES AND CONDITIONS. DATA ARE
 NEEDED ON THE MYCOTA ASSOCIATED WITH ALGAL 'BLOOMS' AND THEIR EFFECT ON
 GROWTH STIMULATION AND REPRODUCTION OF FUNGI. YEASTS WERE ISOLATED FROM
 TWELVE ESTABLISHED SITES IN THE NORTH SEA FROM 1964 TO 1966. A
 PERCENTAGE FREQUENCY OF 99% WITH POPULATIONS VARYING FROM LESS THAN 10
 TO MORE THAN 3000 VIABLE CELLS/LITER WAS OBSERVED. THIS MYCOTA WAS
 CHARACTERIZED BY CONSIDERABLE SPATIAL AND TEMPORAL FLUCTUATIONS, WITH
 THE DOMINANT YEAST PRESENT THE ASCOSPOROGENOUS SPECIES, DEBARYOMYCES
 HANSENII. THIS TAXON, AS WELL AS OTHER COMMON NORTH SEA YEASTS,
 RHODOTORULA RUBRA AND CANDIDA DIDDENSII, HAVE BEEN REPORTED FREQUENTLY
 FROM OTHER MARINE LOCALES. NOTEWORTHY CONCENTRATIONS OF YEASTS,
 ESPECIALLY D HANSENII, WERE OBSERVED DURING SUMMER, OFTEN IN
 ASSOCIATION WITH VARIOUS STAGES OF DEVELOPMENT OF THE DINOFLAGELLATE,
 NOCTILUCA MILIARIS. POPULATION DYNAMICS OF THE NORTH SEA YEASTS ARE
 DISCUSSED IN RELATION TO SIMILAR STUDIES OF OTHER MARINE ENVIRONMENTS.
 (JONES-WISCONSIN)

FIELD 05C

ACCESSION NO. W70-04368

THE PHYSIOLOGY OF AN ALGAL NUISANCE,

WESTFIELD COLL., LONDON (ENGLAND). DEPT. OF BOTANY.

G. E. FOGG.

PROCEEDINGS ROYAL SOCIETY B, VOL 173, P 175-189, 1969. 4 FIG, 84 REF.

DESCRIPTORS:
    *NUISANCE ALGAE, CYANOPHYTA, VITAMINS, PEPTIDES, ORGANIC MATTER, TRACE
    ELEMENTS, LIGHT INTENSITY, CHLAMYDOMONAS, CHLORELLA, ENZYMES, CARBON
    DIOXIDE, GROWTH RATES, PLANKTON, OXYGEN, NITROGEN FIXATION, NITRATES,
    ELECTRON MICROSCOPY, AMMONIA, SPECTROMETERS, DIFFUSION, TURBULENCE,
    TOXINS, BACTERIA, COPPER SULFATE, CIRCULATION.

IDENTIFIERS:
    *PHYSIOLOGY, SPIROGYRA, MYXOPHYCEAE, MICROCYSTIS, COELOSPHAERIUM,
    OSCILLATORIA, ANABAENA, APHANIZOMENON, GLOEOTRICHIA, POLYPEPTIDES,
    PHOTOASSIMILATION, BERKELSE LAKE(NETHERLANDS), HETEROCYSTS,
    GAS-VACUOLES, LLANGORS LAKE(WALES), HALOBACTERIUM, RESPIROMETER,
    PROTOPLASM, WATER-BLOOMS, LAKE GEORGE(UGANDA), RHINE-MEEUSE
    DELTA(NETHERLANDS), HOHLSPINDELN, MYXOPHYCEAN BLOOMS.

ABSTRACT:
    PHYSIOLOGY OF MICROCYSTIS, COELOSPHAERIUM, OSCILLATORIA, ANABAENA,
    APHANIZOMENON, AND GLOEOTRICHIA ARE DISCUSSED RELATIVE TO DEVELOPMENT
    OF BLOOMS. SINCE THE ALGAE FORM GELATINOUS MASSES, LOCAL EXHAUSTION OF
    AMMONIA MAY FORCE CELLS TO UTILIZE THE NITROGEN MOLECULE, AGGRAVATING
    EUTROPHICATION. REMARKABLE FEATURES OF THE PLANKTONIC BLUE-GREEN ALGAE ·
    ARE CELLS CONTAINING GAS-VACUOLES. THESE GAS-FILLED CAVITIES IN
    PROTOPLASM APPEAR AS INDEFINITE, IRREGULAR, REFRACTIVE BODIES. THE
    RELATIVELY HIGH BUOYANT DENSITY OF THEIR MEMBRANES SHOWS THEM AS
    PROBABLY AT VARIANCE TO OTHER CELLULAR MEMBRANES. THEY ARE SPECIFIC
    STRUCTURES WITH DEFINITE SELECTIVE VALUE. PERHAPS GAS-VACUOLE FORMATION
    IN DIM LIGHT CAUSES A CELL OR COLONY TO RISE; INHIBITION OF GAS-VACUOLE
    FORMATION, AND CONTINUED GROWTH AND CELL DIVISION NEARER THE SURFACE
    SUBSEQUENTLY CAUSING IT TO BECOME HEAVIER AND SINKING. THIS UP AND DOWN
    MOVEMENT WOULD CONTRIBUTE TO MAXIMUM ABSORPTION OF MINERAL NUTRIENTS.
    NEAR THE SURFACE LIGHT INHIBITION OF PHOTOSYNTHESIS MAY RESULT IN
    INABILITY TO 'DILUTE OUT' THE GAS-VACUOLES BY GROWTH, THUS THE ALGA
    REMAINS FLOATING AND BECOMES MORIBUND. COPPER SULPHATE USAGE FOR
    CONTROL OF MYXOPHYCEAN WATER-BLOOMS IS QUESTIONABLE. INHIBITING
    FORMATION OF THESE GAS-VACUOLES OR THEIR DESTRUCTION BY ULTRASONIC
    RADIATION, OR OTHER MEANS, MIGHT EFFECTIVELY CONTROL THE NUISANCE.
    ARTIFICIAL CIRCULATION MAY BREAK DOWN CHEMICAL STRATIFICATION IN THE
    WATER. (JONES-WISCONSIN)

FIELD 05C

ACCESSION NO. W70-04369

PERIPHYTON GROWTH ON ARTIFICIAL SUBSTRATES IN A RADIOACTIVELY CONTAMINATED LAKE,

OAK RIDGE NATIONAL LAB., TENN. RADIATION ECOLOGY SECTION.

ERNEST C. NEAL, BERNARD C. PATTEN, AND CHARLES E. DEPOE.

ECOLOGY, VOL 48, NO 6, P 918-924, 1967. 1 FIG, 1 TAB, 17 REF.

DESCRIPTORS:
*PERIPHYTON, *RADIOACTIVITY, *LAKES, PRODUCTIVITY, RADIOISOTOPES,
TEMPERATURE, DEPTH, BACTERIA, SLIME, ALGAE, CYANOPHYTA, DIATOMS,
CHLOROPHYTA, LIGHT INTENSITY, SUCCESSION, ZINC RADIOISOTOPES, CESIUM,
COBALT RADIOISOTOPES, SEASONAL, RADIOCHEMICAL ANALYSIS, MONITORING,
PIGMENTS, PHYTOPLANKTON, WASTE WATER(POLLUTION), PHOTOSYNTHESIS,
CHLAMYDOMONAS, CHLORELLA, EUGLENA, SCENEDESMUS, PROTOZOA, NEMATODES,
SURFACES.

IDENTIFIERS:
*GROWTH DYNAMICS, *BACTERIAL COLONIZATION, *ARTIFICIAL SUBSTRATES,
*RADIONUCLIDE ACCUMULATION, RUTHENIUM-106, WHITE OAK LAKE(TENN),
ACTIVITY-DENSITY, VERTICAL, MICROCYSTIS, OSCILLATORIA, NAVICULA,
CYMBELLA, LYNGBYA, FRAGILARIA, OEDOGONIUM, SPIROGYRA, CHAETOPHORA,
STIGEOCLONIUM, ANKISTRODESMUS, APHANOCAPSA, CHROOCOCCUS, CLOSTERIUM,
GLEOCAPSA, SPIRULINA, ULOTHRIX, CHIRONOMIDS, MELTON HILL DAM(TENN),
COBALAMIN.

ABSTRACT:
PERIPHYTON COLONIZATION, BIOMASS DEVELOPMENT AND RADIONUCLIDE
ACCUMULATION WERE STUDIED BY SUSPENDING POLYETHYLENE TAPE VERTICALLY
FOR PERIODS TO 9 WEEKS IN A RADIOACTIVELY CONTAMINATED LAKE. BIOMASS
GROWTH WAS NEARLY COMPLETE ON UPPER TAPE SECTIONS AFTER 2 WEEKS, BUT
FULL DEVELOPMENT ON LOWER SECTIONS TOOK LONGER; MAXIMUM BIOMASS
OCCURRED AT 25.4 TO 50.8 CENTIMETERS DEPTH. BLUE-GREENS WERE INITIAL
ALGAL DOMINANTS, SUCCEEDED BY DIATOMS AND FILAMENTOUS GREENS. SPECIES
SUCCESSION CONTINUED BEYOND BIOMASS EQUILIBRIUM, WITH BLUE-GREENS
IMPORTANT IN DEEPER ZONES AND GREENS DEVELOPING BEST IN UPPER, LIGHTED
REGIONS. RADIOISOTOPE CONCENTRATIONS EQUILIBRATED RAPIDLY IN BOTH
ARTIFICIAL SUBSTRATES AND PERIPHYTON BIOMASS. ZINC-65 WAS CONCENTRATED
HIGHLY BY THE POLYETHYLENE TAPE, CESIUM-137 MODERATELY, AND COBALT-60
AND RUTHENIUM-106 NEGLIGIBLY, IF AT ALL. AMBIENT CONCENTRATIONS OF
ZINC-65 IN WATER AND PERIPHYTON WERE BELOW DETECTION. THE OTHER
ISOTOPES CONTRIBUTED ABOUT EQUALLY TO BIOMASS RADIOACTIVITY. BIOMASS
ACTIVITY-DENSITIES OF RUTHENIUM-106 AND CESIUM-137 INCREASED SLIGHTLY
WITH DEPTH; THIS PATTERN WAS PRONOUNCED FOR COBALT-60. FACTORS
CONSIDERED TO ACCOUNT FOR OBSERVED VERTICAL DISTRIBUTIONS INCLUDE:
DIFFERENT RADIOISOTOPE CONCENTRATIONS IN SURFACE AND DEEP WATER; A
HYPERBOLIC RATHER THAN LINEAR RELATIONSHIP BETWEEN ACTIVITY-DENSITY AND
BIOMASS; AND CONCENTRATION OF COBALT-60 BY BLUE-GREEN ALGAE AS AN
ESSENTIAL ELEMENT, OR BY MICROBIOTA FOR USE IN COBALMIN SYNTHESIS.
(JONES-WISCONSIN)

FIELD 05C

ACCESSION NO. W70-04371

BIODEGRADABILITY OF POTENTIAL ORGANIC SUBSTITUTES FOR PHOSPHATES,

ROBERT A. TAFT SANITARY ENGINEERING CENTER, CINCINNATI, OHIO. CINCINNATI
WATER RESEARCH LAB.

ROBERT L. BUNCH, AND M. B. ETTINGER.

PROCEEDINGS 22ND INDUSTRIAL WASTE CONFERENCE, LAFAYETTE, INDIANA, P 393-396,
1967. 1 FIG, 14 REF.

DESCRIPTORS:
*WATER QUALITY, *EUTROPHICATION, *DETERGENTS, *PHOSPHATES, CHELATION,
WASTE TREATMENT, AQUATIC ALGAE, AQUATIC PLANTS, TESTING, BIODEGRADATION.

IDENTIFIERS:
*COMPLEXONES, *ALGAL BLOOMS, NTA, EDTA, HEIDA, LAKE TAHOE(CALIF).

ABSTRACT:
THE UNDESIRABLE EUTROPHICATION OF WATERS MAY BE CONSIDERABLY REDUCED BY
REPLACING PHOSPHATES IN INDUSTRIAL PROCESSES AND DETERGENTS WITH
BIODEGRADABLE CHELATING AGENTS, SUCH AS NITRILOACETIC ACID (NTA) AND
HYDROXYETHYLIMINODIACETIC ACID (HEIDA). THE USE OF DETERGENTS
CONTAINING NON-PHOSPHATE BUILDERS IS SUGGESTED, ESPECIALLY FOR EXTREME
SITUATIONS, SUCH AS LAKE TAHOE. THE USE OF NON-BIODEGRADABLE AGENTS
WHICH COMPLEX HEAVY METALS WILL DEPEND UPON THEIR EFFECT ON WASTE
TREATMENTS, SOFTENING WATER, DRINKING WATER, AND CALCIUM METABOLISM OF
MAN. (WILDE-WISCONSIN)

FIELD 05G

ACCESSION NO. W70-04373

SOME PROBLEMS REMAINING IN ALGAE CULTURING,

STATION D'HYDROBIOLOGIE APPLIQUEE, LE PARACLET PAR BOVES (FRANCE).

A. G. WURTZ.

ALGAE AND MAN, JACKSON, DANIEL F (EDITOR), PLENUM PRESS, NEW YORK, P 120-137, 1964.

DESCRIPTORS:
*ANALYTICAL TECHNIQUES, *ALGAE, *GROWTH RATES, *BIOINDICATORS, *WATER POLLUTION, *WASTE TREATMENT, FOOD CHAINS, CHLORELLA, CHLOROPHYLL, PHYSIOLOGICAL ECOLOGY, CYANOPHYTA, EUTROPHICATION, NUISANCE ALGAE, NITROGEN, PHOSPHATES, HYDROGEN ION CONCENTRATION, CYCLING NUTRIENTS, MUD-WATER INTERFACES, EUGLENOPHYTA, TOXICITY, BIOASSAYS, DIATOMS, EUGLENA, SPHAEROTILUS.

IDENTIFIERS:
*ALGAL CULTURES, *AXENIC CULTURES, PEDIASTRUM, SELF-PURIFICATION, ALGAL ENUMERATION, SCENEDESMUS, GROWTH EQUATIONS, ANKISTRODESMUS, HETEROTROPHY, CONTINUOUS CULTURES, CHLAMYDOMONAS, BATCH CULTURES, OPTICAL DENSITY, SOIL-WATER TECHNIQUES, PHYCOLOGY, MICROCYSTIS, GLOEOTRICHIA, APHANIZOMENON, MICRONUTRIENTS, DINOPHYCEAE, CRYPTOPHYCEAE, CHROMATIUM, OSCILLATORIA, ANTIBIOTIC TREATMENT, CHU'S MEDIUM, ULTRAVIOLET STERILIZATION, CONSTRUCTIVE SYNECOLOGY, TABELLARIA.

ABSTRACT:
NUMEROUS PROBLEMS EXIST AMONG THE INTERFACES BETWEEN PURE-CULTURE STUDY OF ALGAE AND PHYCOLOGICAL ECOLOGY, ESPECIALLY AS REGARDS THE QUALITY OF MAN'S ENVIRONMENT. WHILE ATTAINMENT OF AXENIC CULTURES IS, OF ITSELF, A VALID GOAL, INCREASING USE OF SUCH CULTURES, COUPLED WITH ECOLOGICAL INSIGHTS, MAY YIELD USEFUL INFORMATION PERTAINING TO MANY PERPLEXING PRACTICAL PROBLEMS. EQUATIONS DESCRIBING RELATIONSHIPS AMONG GROWTH, YIELD, AND BIOMASS IN BATCH AND CONTINUOUS CULTURES ARE REVIEWED, AS ARE METHODS FOR EVALUATING ALGAL GROWTH. THERE REMAINS A BASIC NEED FOR STANDARDIZED TECHNIQUES FOR ASSESSING GROWTH IN ALGAL CULTURES. SOME ASPECTS OF TECHNIQUES FOR AXENIC CULTURING ARE REVIEWED. DISCREPANCIES BETWEEN INFORMATION ON MINERAL NUTRITION OF BLUE-GREEN ALGAE, AS DEVELOPED FROM STUDIES OF AXENIC CULTURES, AND FIELD OBSERVATIONS OF CONDITIONS LEADING TO NUISANCE BLOOMS NEED FURTHER STUDY AND CLARIFICATION. THE SOIL-WATER TECHNIQUE, WHICH HAS BEEN INSTRUMENTAL IN SUCCESSFUL CULTURING OF SOME 500 STRAINS OF VARIOUS ALGAE, RAISES FURTHER ECOLOGICAL QUESTIONS, INCLUDING THOSE OF HETEROTROPHY AND CONDITIONS AT THE SOIL-WATER INTERFACE. OTHER SIMILAR QUESTIONS CONSIDERED IN SOME DETAIL IN THIS REPORT INCLUDE INFLUENCE OF INNOCULUM SIZE ON POPULATION GROWTH; ECOLOGICAL IMPLICATIONS OF ALGAL CULTURE; SELF-PURIFICATION OF WASTES BY ALGAE; ALGAL INDICATORS OF WATER QUALITY; AND ROLE OF ALGAE IN FOOD CHAINS. (SEE W69-07832). (EICHHORN-WISCONSIN)

FIELD 05C

ACCESSION NO. W70-04381

ANALYTICAL CHEMISTRY OF PLANT NUTRIENTS,

WISCONSIN, UNIV., MADISON. WATER CHEMISTRY LAB.

G. FRED LEE.

EUTROPHICATION: CAUSES, CONSEQUENCES, CORRECTIVES. PRINTING AND PUBLISHING
OFFICE, NATIONAL ACADEMY OF SCIENCES, WASHINGTON, D C, 1969. P 646-658. 3
FIG, 3 TAB, 22 REF.

DESCRIPTORS:
*ANALYTICAL TECHNIQUES, *AQUATIC WEEDS, *WATER ANALYSIS, *SAMPLING,
PLANT PHYSIOLOGY, EUTROPHICATION, DATA COLLECTION, CHEMICAL ANALYSIS,
RELIABILITY, WATER POLLUTION, WATER POLLUTION EFFECTS, NITROGEN,
PHOSPHORUS, ALGAE, NUTRIENTS, WATER PROPERTIES.

IDENTIFIERS:
*NUTRIENT ELEMENTS, PHOSPHORUS RELEASE, SAMPLE PRESERVATION, WATER
DYNAMICS, WATER CONSTITUENTS, ANALYTICAL CHEMISTRY, WASTE WATER
EFFECTS, LAKE MENDOTA(WIS), BLACK EARTH CREEK(WIS).

ABSTRACT:
VARIOUS STEPS MUST BE TAKEN TO ACHIEVE RATIONAL MANAGEMENT OF WATER
BASINS INVADED OR THREATENED BY NOXIOUS AQUATIC WEEDS. THE FIRST PHASE
OF AN EUTROPHICATION STUDY INVOLVES A DECISION ON THE METHOD, LOCATION,
AND FREQUENCY OF SAMPLING, AND THE MANNER OF SAMPLE PRESERVATION. NEXT,
IS THE SELECTION OF METHODS OF ANALYSES APPROPRIATE TO THE REQUIRED
ACCURACY OF RESULTS. FINALLY, A CRITICAL EVALUATION OF RESULTS OBTAINED
AS TO THEIR GENERAL RELIABILITY AND STATISTICAL SIGNIFICANCE. DETAILED
INFORMATION ON SOURCES, FORMS, AND METHODS OF ANALYSIS APPLICABLE TO
NITROGEN AND PHOSPHORUS ARE GIVEN. ADDITION OF MERCURIC CHLORIDE AND
FREEZING TO -20C, RATHER THAN THE USE OF CHLOROFORM AND HYDROGEN
SULFATE, IS RECOMMENDED FOR THE PRESERVATION OF NITROGEN FORMS AND
PHOSPHORUS COMPOUNDS IN STORED WATER SAMPLES. A WARNING IS GIVEN ABOUT
POSSIBILITY PHOSPHATE LOSS DUE TO SORPTION BY GLASS OR PLASTIC
CONTAINERS. (SEE W70-03975). (WILDE-WISCONSIN)

FIELD 05A, 07B

ACCESSION NO. W70-04382

GEOCHEMISTRY OF EUTROPHICATION,

GENERAL DYNAMICS CORP., GROTON, CONN. ELECTRIC BOAT DIV.

RICHARD J. BENOIT.

EUTROPHICATION: CAUSES, CONSEQUENCES, CORRECTIVES, PRINTING AND PUBLISHING
OFFICE, NATIONAL ACADEMY OF SCIENCES, WASHINGTON, D C, 1969. P 614-630. 3
FIG, 6 TAB, 30 REF.

DESCRIPTORS:
*GEOCHEMISTRY, *EUTROPHICATION, *WATER STRUCTURE, ATMOSPHERE, SEA
WATER, FRESH WATER, SORPTION, SODIUM, ALGAE, PHOSPHORUS, TRACE
ELEMENTS, PLANKTON, LAKES, RIVERS, GREAT LAKES.

IDENTIFIERS:
*EQUILIBRIUM MODEL, LITHOSPHERE, SORPTION REACTIONS, SOLUBILITY
EQUILIBRIUM THEORY.

ABSTRACT:
THE TENDENCY OF THE LAND-BASED HYDROSPHERE TO ATTAIN SIMILAR
COMPOSITION UNDER THE COMBINED INFLUENCES OF THE BUFFERING ACTION OF
ION-EXCHANGE REACTIONS AND INTEGRATION OF TRIBUTARIES IS STRESSED. THE
EQUILIBRIUM MODEL, BASED ON SOLUBILITY EQUILIBRIA OF COMMON
CONSTITUENTS OF THE LITHOSPHERE AND CARBON DIOXIDE CONCENTRATION EQUAL
TO 35,000 ATMOSPHERES, GIVES A REASONABLY SATISFACTORY AGREEMENT WITH
THE ACTUAL COMPOSITION OF WATER OF THE GREAT LAKES. WHILE SORPTION
REACTIONS MAY CAUSE SOME DEVIATIONS FROM THE EQUILIBRIUM MODELS, THE
MAJOR DISCREPANCIES IN THE CHEMICAL COMPOSITION OF WATERS ARE
ATTRIBUTED TO MAN'S ACTIVITY. THIS IS PARTICULARLY TRUE OF THE CONTENTS
OF SODIUM AND PHOSPHORUS, THE TWO ELEMENTS THAT MAY LIMIT OR UNDULY
PROMOTE THE ALGAL GROWTH. TRACE ELEMENTS MAY ALSO ALTER THE TROPHIC
LEVEL OF A LAKE TO CREATE A NUISANCE OR EVEN LEAD TO A DEGRADATION OF A
RESOURCE. (SEE W70-03975). (WILDE-WISCONSIN)

FIELD 05C

192

ACCESSION NO. W70-04385

THE GREAT AND DIRTY LAKES,

GLADWIN HILL.

IN: CONTROLLING POLLUTION: THE ECONOMICS OF A CLEANER AMERICA, ED. MARSHALL
I. GOLDMAN, ENGLEWOOD CLIFFS, N.J.: PRENTICE-HALL, INC., 1967, P. 43-48,
AND SATURDAY REIVEW, OCT. 23, 1965.

    DESCRIPTORS:
    *ALGAE, *INDUSTRIAL WASTES, *MUNICIPAL WASTES, *GREAT LAKES, *WATER
    POLLUTION, SEWAGE, AESTHETICS, AEROBIC CONDITIONS, ANAEROBIC
    CONDITIONS, WATER SUPPLY.

    IDENTIFIERS:
    *DEVOLUTION, *INTERSTATE POLLUTION, *METROPOLITAN AREAS, BLONDIN.

    ABSTRACT:
    THE POLLUTION OF THE GREAT LAKES IS DISCUSSED. FOCUSING ON NIAGRA
    FALLS, THE SOURCES OF POLLUTION FROM DULUTH TO BUFFALO ARE TRACED.
    SPECIAL ATTENTION IS GIVEN TO THE CHICAGO AND DETROIT METROPOLITAN
    AREAS. THE GREAT LAKES ARE TREATED AS ONE CONTINUOUS SYSTEM AND
    SUGGESTS INTERSTATE POLLUTION ABATEMENT AS THE ONLY POSSIBLE RECOURSE,
    REJECTING INTERNATIONAL EFFORTS AS CUMBERSOME AND INTRASTATE EFFORTS AS
    INEFFECTUAL. WHILE RECOGNIZING THE MUNICIPAL WASTE AS A MAJOR POLLUTION
    SOURCE, THE AUTHOR CONCENTRATES ON INDUSTRIAL WASTES FROM THE OIL,
    STEEL, PAPER, SOAP, CHEMICAL AND AUTOMOTIVE INDUSTRIES. HE CITES THE
    CHICAGO AND DETROIT METROPOLITAN AREAS AS THE MAIN DISCHARGERS OF
    AMMONIA NITROGEN, PHENOLS, CYANIDE, OIL, PHOSPHATES, CHLORIDES,
    SUSPENDED AND SETTLEABLE SOLIDS AND NITROGEN COMPOUND WASTES INTO THE
    LAKES. CHICAGO'S EFFORTS TO MAINTAIN A USABLE WATER SUPPLY AND
    DETROIT'S OUTDATED SEWAGE TREATMENT FACILITIES ARE ALSO REVIEWED. THE
    DEATH OF LAKE ERIE IS TREATED ALONG WITH A DISCUSSION OF THE DEVOLUTION
    OF AQUATIC LIFE WHICH ACCOMPANIED IT. THE EFFORTS OF PRIVATE
    INDIVIDUALS AND CONSERVATION GROUPS AND THOSE OF VARIOUS STATE AND
    FEDERAL GOVERNMENTS ARE TREATED. (RICHMOND-CHICAGO)

    FIELD 05G, 05B

    ACCESSION NO. W70-04430

THE FUTURE OF THE LAKE,

WASHINGTON UNIV., SEATTLE. DEPT. OF ZOOLOGY.

W. T. EDMONDSON.

MUNICIPALITY OF METROPOLITAN SEATTLE, METRO QUARTERLY, FALL 1965. 4 P, 1 FIG.

    DESCRIPTORS:
    *ALGAE, *WATER POLLUTION TREATMENT, LAKES, SEWAGE, DRAINAGE DISTRICTS,
    DEBT, COSTS, CITIES.

    IDENTIFIERS:
    *LAKE WASHINGTON(WASH), SEATTLE(WASH), REVENUE BONDS, METRO
    ACT(SEATTLE).

    ABSTRACT:
    ORIGINALLY SEATTLE, WASHINGTON, RELEASED NON-TREATED SEWAGE INTO PUGET
    SOUND AND TREATED SEWAGE INTO LAKE WASHINGTON. SWIMMING IN THE SOUND
    WAS DESTROYED, AND LAKE WASHINGTON BECAME EUTROPHIC. THE METRO ACT
    ENACTED IN 1957 PROVIDED FOR THE ESTABLISHMENT OF A METROPOLITAN
    MUNICIPAL CORPORATION WITH THE POWER TO PERFORM SEWAGE DISPOSAL, WATER,
    PARKS, TRANSPORTATION, COMPREHENSIVE PLANNING AND/OR GARBAGE DISPOSAL.
    IN 1958 THE SEWAGE DISPOSAL PROVISO WAS APPROVED BY 60% VOTE. JULY 1966
    SHOULD SEE THE END OF SEWAGE DISCHARGE INTO LAKE WASHINGTON. $120
    MILLION IN REVENUE BONDS HAVE BEEN SOLD TO DATE, TO BE REPAID BY A $2
    MONTHLY CHARGE PER RESIDENTIAL CONNECTION AND A $2 MONTHLY CHARGE PER
    900 CUBIC FEET OF WATER USED BY INDUSTRIAL AND COMMERCIAL ENTERPRISES.
    PROMPT IMPROVEMENT IN THE LAKE IS EXPECTED WITHIN A FEW YEARS AFTER
    FINAL DIVERSION IN 1966 AS THE WATERSHED IS RELATIVELY POOR IN
    DISSOLVED MINERALS INCLUDING NUTRIENTS NEEDED BY ALGAE, AND AS THERE
    ARE NO MASSIVE DEPOSITS OF ACCUMULATED SOFT SEDIMENTS ON THE LAKE
    BOTTOM RICH IN NUTRIENTS AVAILABLE TO ALGAE, THE PROGNOSIS IS GOOD FOR
    THE RECOVERY OF LAKE WASHINGTON. (POWERS-WISCONSIN)

    FIELD 05C, 02H

    ACCESSION NO. W70-04455

OUTWITTING THE PATIENT ASSASSIN: THE HUMAN USE OF LAKE POLLUTION,

NORTHWESTERN UNIV., EVANSTON, ILL. TECHNOLOGICAL INST.

HAROLD B. GOTAAS.

BULLETIN OF THE ATOMIC SCIENTISTS, P 8-10, MAY 1969.

DESCRIPTORS:
*NUTRIENTS, *BENEFIT-COST ANALYSIS, *LAKE ERIE, PHOSPHORUS, FISH
STOCKING, COMMERCIAL FISH, SALMON, ALGAE, FISH.

IDENTIFIERS:
ANTIPOLLUTION PROGRAMS, BIOLOGICAL BALANCE, OVERFISHING, COHO SALMON,
ALEWIFE, NUTRIENT REMOVAL.

ABSTRACT:
SOME ASPECTS OF THE OCTOBER 1968 DEPARTMENT OF INTERIOR REPORT ON LAKE
ERIE ARE CHALLENGED. THE EMPHASIS ON NUTRIENT REMOVAL WHICH WOULD
RESULT IN IMMEDIATE EXPENDITURE OF $1.1 BILLION TO CONTROL MUNICIPAL
POLLUTION AND $285 MILLION FOR CURBING INDUSTRIAL CONTAMINATION IS
QUESTIONED IN LIGHT OF THE EXPECTED BENEFITS. SEVERAL EFFICACIOUS
ALTERNATIVES FOR REVIVING THE GREAT LAKES ARE SUGGESTED: (1) FURTHER
NUTRIENT RELATIONSHIPS AND COST VERSUS BENEFITS STUDIES SHOULD BE
UNDERTAKEN BEFORE MONEY IS SPENT TO BUILD EXPENSIVE TREATMENT
FACILITIES; (2) HARVEST ALGAE AS A POTENTIAL SOURCE OF FOOD THUS
PREVENTING IT FROM CONTRIBUTING TO THE ORGANIC WASTE LOAD; (3) PREVENT
THE INTRODUCTION OF ALL TOXIC MATERIALS ENABLING DESIRABLE BIOLOGICAL
BALANCES TO BE ESTABLISHED; (4) SEED THE LAKES WITH DESIRABLE FISH TO
ESTABLISH A FOOD CHAIN WHICH WOULD PERMIT THE GROWTH AND REMOVAL OF
NUTRIENTS AS WELL AS FISH FOR FOOD NEEDS AND SPORT; AND (5) ADOPT WATER
QUALITY STANDARDS THAT ARE REALISTIC AS TO THEIR COSTS, BENEFITS, AND
TIME PRIORITIES. (HASKINS-WISCONSIN)

FIELD 05C, 02H

ACCESSION NO. W70-04465

ALGAE FROM WESTERN LAKE ERIE,

OHIO STATE UNIV., COLUMBUS. DEPT. OF BOTANY AND PLANT PATHOLOGY.

CLARENCE E. TAFT, AND W. JACK KISHLER.

OHIO JOURNAL OF SCIENCE, VOL 68, NO 2, P 80-83, 1968. 9 FIG, 7 REF.

DESCRIPTORS:
*ALGAE, *LAKE ERIE, CHLOROPHYTA, CYANOPHYTA, UNITED STATES, HABITATS.

IDENTIFIERS:
*NEW SPECIES, *WESTERN LAKE ERIE(OHIO), GONGROSIRA STAGNALIS,
NEPHROCYTIUM OBESUM W AND G S WEST, RADIOCOCCUS NIMBATUS, URONEMA
ELONGATUM HODGETTS, CHROOCOCCUS PRESCOTTII, CALOTHRIX FUSCA,
MICROCOLEUS LACUSTRIS (RAB) FARLOW, CLADOPHORA, TAXONOMIC DESCRIPTIONS,
GONGROSIRA LACUSTRIS BRAND, GONGROSIRA DEBARYANA RAB, OOCYSTIS,
OSCILLATORIA, ULOTHRIX, EUCAPSIS ALPINA CLEMENTS AND SCHANTZ, SCYTONEMA
ALATUM (CARM) BORZI, SCYTONEMA MYOCHROUS.

ABSTRACT:
FOUR SPECIES OF ALGAE IN THE CHLOROPHYTA (GONGROSIRA STAGNALIS,
NEPHROCYTIUM OBESUM, RADIOCOCCUS NIMBATUS, AND URONEMA ELONGATUM) AND
FIVE IN THE CYANOPHYTA (CHROOCOCCUS PRESCOTTI, CALOTHRIX FUSCA,
MICROCOLEUS LACUSTRIS, SCYTONEMA ALATUM, AND SCYTONEMA MYOCHROUS) ARE
NEWLY REPORTED FOR WESTERN LAKE ERIE. GONGROSIRA STAGNALIS (G S WEST)
SCHMIDLE, COLLECTED FROM THE BASAL FRAGMENTS OF OLD CLADOPHORA, APPEARS
TO BE A NEW RECORD FOR THE UNITED STATES. NEPHROCYTIUM OBESUM W AND G S
WEST, WHICH IS REPORTED AS OFTEN HAVING A SHALLOWLY SCROBICULATE WALL,
IS UNIQUE AND MERITS FURTHER INTENSIVE STUDY. THESE NEW ALGAE
OCCASIONALLY APPEAR IN TEACHING AND RESEARCH COLLECTIONS AT STONE
LABORATORY, PUT-IN-BAY, OHIO. THE HABITAT OF G STAGNALIS ON
CALCIUM-ENCRUSTED CLADOPHORA FILAMENTS IS UNIQUE. (JONES-WISCONSIN)

FIELD 05C

ACCESSION NO. W70-04468

A SIMPLE PIPETTE CONTROL FOR ISOLATING PLANKTONIC ALGAE,

RHODE ISLAND UNIV., KINGSTON. GRADUATE SCHOOL OF OCEANOGRAPHY.

BRENDA J. BOLEYN.

CANADIAN JOURNAL OF MICROBIOLOGY, VOL 13, P 1129-1130, 1967. 1 FIG, 4 REF.

DESCRIPTORS:
*ALGAE, *ANALYTICAL TECHNIQUES, *PLANKTON, *LABORATORY TESTS,
*ISOLATION, *CONTROL, METHODOLOGY, CULTURES, TESTING, CAPILLARY
CONDUCTIVITY, BACTERIA, DIATOMS, TEST PROCEDURES.

IDENTIFIERS:
*PIPETTE, UNIALGAL, NITZSCHIA, ALGAE ISOLATION, CAPILLARITY,
CAPILLARITY CONTROL DEVICE.

ABSTRACT:
A DEVICE FOR SIMPLIFYING THE ISOLATION OF PLANKTON OR OTHER SMALL
PARTICLES HAS BEEN DEVELOPED. IT OPERATES BY CONTROLLING THE
CAPILLARITY OF THE FINE-TIPPED PIPETTE ROUTINELY USED FOR PHYTOPLANKTON
ISOLATION. THE INWARD AND OUTWARD FLOW OF FLUID MAY BE CONTROLLED
SMOOTHLY AND COMPLETELY. THERE ARE A NUMBER OF ADVANTAGES IN THIS
DEVICE OVER THE UNMODIFIED ISOLATING PIPETTE. A STEADY HAND IS EASILY
MAINTAINED SINCE THE HAND GUIDING THE PIPETTE NEED NOT CONTROL
CAPILLARITY ALSO. THE OPERATOR IS ABLE TO NUDGE AND STEER RELUCTANT OR
ADHERING FORMS INTO THE PIPETTE, ESPECIALLY WHERE LONG SLENDER CELLS,
SETOSE FORMS, OR CHAIN-FORMERS MIGHT EASILY CLUSTER AND OCCLUDE THE
TIP. A PIPETTE WITH AN ORIFICE OF SMALL OR LARGE SIZE MAY BE USED,
MAKING IT HELPFUL TO ISOLATE LARGE OR PECULIARLY SHAPED CELLS. THE
EXPLOSIVE NATURE OF PIPETTE DELIVERY IS LESSENED AND BUBBLING IS
PREVENTED. MOTILE ORGANISMS MAY BE FOLLOWED AND CAUGHT READILY. THE
ENTIRE PROCEDURE IS HASTENED, DIMINISHING DANGER OF THERMAL SHOCK,
ESPECIALLY IMPORTANT FOR WINTER FORMS. ALSO THE RISK OF PRODUCING
APPRECIABLE SALINITY CHANGES IS MINIMIZED. (JONES-WISCONSIN)

FIELD 05A

ACCESSION NO. W70-04469

THE EFFECT OF IMPOUNDMENTS ON NATURAL WATERS (IN GERMAN),

AKADEMIYA NAUK SSSR, LENINGRAD (USSR).

W. I. SHADIN.

VERH. INTERNAT. VEREIN, LIMNOL., VOL. IV, P 792-805, JULY 1961.

DESCRIPTORS:
*LAKES, *RESERVOIRS, ZOOPLANKTON, PHYTOPLANKTON, TEMPERATURE,
*EPILIMNION, *HYPOLIMNION, ALGAE, FUNGI, BACTERIA, FISH, LIMNOLOGY,
HYDROLOGIC ASPECTS, AQUATIC ENVIRONMENT, WATER PROPERTIES, BIOLOGICAL
PROPERTIES, ECOLOGY, STRATIFICATION.

IDENTIFIERS:
WATER QUALITY PROPERTIES.

ABSTRACT:
THE BUILDING OF IMPOUNDMENT QUALITATIVELY CHANGES THE BIOLOGICAL AND
ECOLOGICAL SYSTEM OF WATER. A SURVEY OF RESPONSE OF WATER QUALITY TO
THE BUILDING OF RESERVOIRS AND REVIEW OF WATER QUALITY PROPERTIES IN
LAKES AND RESERVOIRS IN DIFFERENT GEOLOGIC AND CLIMATIC CONDITIONS IS
PRESENTED IN THE PAPER. THE WATER QUALITY AND BIOLOGICAL SYSTEM OF
WATER IN IMPOUNDMENT IS DEPENDENT ON MANY FACTORS, SUCH AS HYDROLOGIC
REGIME, CLIMATIC AND REGIONAL CONDITIONS, PURPOSE OF RESERVOIR, ETC.
THE TEMPERATURE DISTRIBUTION INFLUENCES THE DETENTION (THROUGH-FLOW)
TIME AND FORMATION OF DENSITY LAYERS WITH DIFFERENT WATER QUALITY
PARAMETERS. THE CHEMICAL REGIME IS MAINLY DEPENDENT ON SLUDGE
FORMATION. THE AMOUNT OF BACTERIA IN WATER HAS TWO MAXIMUMS DURING A
YEAR. THE FORMATION OF ZOOPLANKTON AND PHYTOPLANKTON IS FURTHER
DISCUSSED. THE PAPER WAS PRESENTED AS A LEADING LECTURE AT XIV-TH
INTERNATIONAL LIMNOLOGICAL CONFERENCE IN 1961. (NOVOTNY-VANDERBILT)

FIELD 02H, 05C

ACCESSION NO. W70-04475

IMPOUNDMENT DESTRATIFICATION FOR RAW WATER QUALITY CONTROL USING EITHER
   MECHANICAL OR DIFFUSED AIR PUMPING,

FEDERAL WATER POLLUTION CONTROL ADMINISTRATION, CINCINNATI, OHIO.

J. M. SYMONS, W. H. IRWIN, E. L. ROBINSON, AND G. G. ROBECK.

J. AMERICAN WATER WORKS ASSOC., VOL 59, NO 10, P 1268-1291, OCTOBER 1967. 20
   FIG., 5 TAB., 19 REF.

DESCRIPTORS:
   *THERMAL STRATIFICATION, *AERATION, *IMPOUNDMENTS, *RESERVOIRS,
   *MIXING, *WATER QUALITY CONTROL, WATER QUALITY, WATER POLLUTION,
   NITROGEN, PHOSPHORUS, AIR, PHYTOPLANKTON, ALGAE, EFFICIENCIES,
   DISSOLVED OXYGEN, OXYGEN, LAKES.

ABSTRACT:
   TWO METHODS ARE COMPARED FOR ARTIFICIALLY DESTRATIFYING LAKES AND
   RESERVOIRS BOTH OF WHICH BROKE UP THE THERMAL STRATIFICATIONS PATTERN
   IN THE LAKES SUCCESSFULLY AND IMPROVED WATER QUALITIES. OTHER
   CONCLUSIONS REACHED WERE THAT MULTIPLE MIXING STARTED IN THE SPRING
   WILL MAINTAIN GOOD QUALITY WATER IN AN IMPOUNDMENT, AND IS SUPERIOR TO
   A SINGLE MIXING IN MIDSUMMER OR EARLY FALL. SOME RISE IN NITROGEN AND
   PHOSPHORUS CONCENTRATIONS OCCURRED IN SURFACE WATER DURING
   DESTRATIFICATION, BUT THE SURFACE POPULATION OF PHYTOPLANKTON AND THE
   PHYTOPLANKTON STANDING CROP DECLINED DURING MIXING. ALGAL POPULATIONS
   INCREASED AGAIN WHEN MIXING STOPPED. DATA TAKEN IN THIS STUDY AND FROM
   5 OTHER STUDIES IN THE LITERATURE SHOWED A DESTRATIFICATION EFFICIENCY
   RANGING FROM 0.1 TO 1.2%, AND AVERAGING 0.5%; AND AN ENERGY INPUT/UNIT
   OF IMPOUNDMENT WATER VOLUME RANGING FROM 0.7 TO 4.9 KW-HR/ACRE-FT. AND
   AVERAGING 2 KW-HR/ACRE-FT. (NOVOTNY-VANDERBILT)

FIELD 05F, 02H

ACCESSION NO. W70-04484

TREATMENT OF ALGAE AND WEEDS IN LAKES AT MADISON,

HEALTH DEPT., MADISON, WIS. LAB.

BERNARD P. DOMOGALLA.

ENGINEERING NEWS-RECORD, VOL 97, NO 24, P 950-954, 1926. 8 FIG, 3 TAB.

DESCRIPTORS:
   *COPPER SULFATE, *EUTROPHICATION, *ALGAE, *AQUATIC WEEDS, LAKES,
   ALGICIDES, AQUATIC WEED CONTROL, MECHANICAL CONTROL, WATER POLLUTION
   CONTROL, CHEMCONTROL, ARSENIC COMPOUNDS, NITROGEN, PHOSPHORUS.

IDENTIFIERS:
   *MADISON LAKES(WIS), *AQUATIC WEED CUTTING, *WEED EXTRACTING STEEL
   CABLES, ALGAL SUCCESSION, ANABAENA, CLADOPHORA, PEDIASTRUM,
   HYDRODICTYON, POTAMOGETON, RANUNCULUS.

ABSTRACT:
   FIELD AND LABORATORY INVESTIGATIONS OF HIGHLY EUTROPHIC LAKES IN THE
   VICINITY OF MADISON YIELDED SEVERAL CONCLUSIONS. THE COPPER SULFATE
   DRAGGING METHOD PROVED TO BE MOST EFFECTIVE FOR SUPPRESSION OF
   BOTTOM-ANCHORED ALGAE ON WIND EXPOSED AREAS. THE SURFACE ALGAE AND
   THOSE INHABITING SHALLOW WATERS WERE BEST CONTROLLED BY SPRAYING WITH
   COPPER SULFATE. WEEDS IN OPEN WATER WERE ELIMINATED BY WEED-CUTTING
   MACHINES. STEMS AND ROOTS OF WEEDS IN SHALLOW WATERS AND NEAR SHORE
   LINES WERE EXTRACTED BY STEEL CABLES WITH CLIPS AND SWIVELS. SEVERAL
   SPECIES OF WATER WEEDS WERE ERADICATED BY SODIUM ARSENITE. THE STUDY
   DISCLOSED A SEASONAL SUCCESSION OF AQUATIC PLANTS AND A PRONOUNCED
   RELATIONSHIP BETWEEN ERADICATION TREATMENTS AND BIOLOGICAL FIXATION OF
   AVAILABLE FORMS OF NITROGEN AND PHOSPHORUS. (WILDE-WISCONSIN)

FIELD 02H, 05C

ACCESSION NO. W70-04494

ALGAL NUTRITION AND EUTROPHICATION,

HASKINS LABS., NEW YORK.

LUIGI PROVASOLI.

EUTROPHICATION: CAUSES, CONSEQUENCES, CORRECTIVES, PRINTING AND PUBLISHING
OFFICE, NATIONAL ACADEMY OF SCIENCES, WASHINGTON, D C, 1969, P 574-593. 2
FIG, 3 TAB, 72 REF.

DESCRIPTORS:
*EUTROPHICATION, *ALGAE, *NUTRIENTS, NITROGEN, PHOSPHORUS, SODIUM,
POTASSIUM, TRACE METALS, VITAMINS, WATER POLLUTION, PHYSIOLOGICAL
ECOLOGY, DIATOMS, OLIGOTROPHY, CYANOPHYTA, CHLOROPHYTA, GREAT LAKES,
BIOINDICATORS, FISH, WASHINGTON, LAKES, INDIANA, OCHROMONAS,
PHYTOPLANKTON, SEAWEEDS.

IDENTIFIERS:
*ALGAL NUTRITION, DITYLUM, CHAETOCEROS, SKELETONEMA, THALASSIOSIRA,
COCCOLITHUS, SARGASSO SEA, SYNURA, GYMNODINIUM, CHEMICAL SYMBIOSIS,
MICROBIAL ECOLOGY, DINOBRYON, VOLVOX, PANDORINA, SPIROGYRA,
SYNECHOCOCCUS, CHROOCOCCUS, OSCILLATORIA, NOSTOC, ANABAENA, ANACYSTIS,
MICROCYSTIS, BOSMINA, ASTERIONELLA, MELOSIRA, SYNEDRA, FRAGILARIA,
CERATIUM, SWEDEN, APHANIZOMENON, RHODOMONAS, CRYPTOMONADS,
CHRYSOMONADS, EUGLENIDS, DINOFLAGELLATES.

ABSTRACT:
PREVALENT EMPHASIS ON NITROGEN AND PHOSPHORUS NOTWITHSTANDING, LITTLE
DOUBT EXISTS THAT OTHER NUTRIENTS PLAY ROLES IN EUTROPHICATION OF
NATURAL WATERS. ALSO, ACCENT IS ON ASPECTS OF BIOLOGICAL OVERPRODUCTION
IN EUTROPHICATION, WHEREAS MORE VISIBLE MANIFESTATIONS ARE CHANGES IN
COMMUNITY STRUCTURE, PRODUCING LESS DESIRABLE SPECIES VIA ALTERATIONS
IN FOOD CHAINS. FROM THESE VIEWPOINTS, UNFORTUNATELY LITTLE IS KNOWN
REGARDING ECOLOGICALLY IMPORTANT SPECIES. SELECTED NUTRITIONAL ASPECTS
OF ALGAE ARE REVIEWED AS REGARDS NITROGEN, PHOSPHORUS, SODIUM,
POTASSIUM, TRACE METALS, VITAMINS, AND CHEMICAL SYMBIOSIS. IN SOME
SITUATIONS, GROWTH OF CHLOROPHYTES (GREEN ALGAE) ARE FAVORED BY HIGH
CONCENTRATIONS OF NITROGEN, PHOSPHORUS, AND POTASSIUM. WHERE
CHLOROPHYTES ARE REPLACED, NITROGEN-FIXING BLUE-GREEN ALGAE MAY TAKE
OVER BECAUSE NITROGEN IS TOO SCARCE TO MEET REQUIREMENTS OF
CHLOROPHYTES. IT IS KNOWN THAT BOTH SODIUM AND POTASSIUM INCREASE
DURING EUTROPHICATION. SUCH CHANGES MAY BE FUNCTIONAL AS WELL AS
INDICATIVE OF EUTROPHICATION. ALTHOUGH TRACE ELEMENTS ARE OF
ACKNOWLEDGED SIGNIFICANCE IN ALGAL NUTRITION, FEW RELIABLE ECOLOGICAL
DATA ARE AVAILABLE, AND ADDITIONAL RESEARCH IS REQUIRED. VITAMINS ARE
LIKELY TO BE RICH IN EUTROPHIC ENVIRONMENTS, A MATTER OF IMPORTANCE
SINCE 197 OF 204 ALGAL SPECIES STUDIED REQUIRE VITAMINS, AND SINCE A
PATTERN OF VITAMIN REQUIREMENTS IS BEGINNING TO EMERGE FOR VARIOUS
TAXONOMIC GROUPS. (SEE W70-03975).
(EICHHORN-WISCONSIN)

FIELD 05C

ACCESSION NO. W70-04503

ALGAE AND METROPOLITAN WASTES.

PUBLIC HEALTH SERVICE, CINCINNATI, OHIO. DIV. OF WATER SUPPLY AND POLLUTION
CONTROL; AND ROBERT A. TAFT SANITARY ENGINEERING CENTER, CINCINNATI, OHIO.

TRANSACTIONS OF SEMINAR ON ALGAE AND METROPOLITAN WASTES HELD AT CINCINNATI,
OHIO, APRIL 27-29, 1960, SPONSORED BY DIV OF WATER SUPPLY AND POLLUTION
CONTROL AND THE ROBERT A TAFT SANITARY ENGINEERING CENTER. BARTSCH, ALFRED
F, CHAIRMAN. ROBERT A TAFT SANITARY ENGINEERING CENTER, SEC TR W61-3, 1961.
162 P.

DESCRIPTORS:
*CONFERENCES, *EUTROPHICATION, *ALGAE, PLANNING, RIVERS, LAKES,
LIMNOLOGY, WISCONSIN, URBANIZATION, CONNECTICUT, OREGON, WASHINGTON,
PRIMARY PRODUCTIVITY, ECOSYSTEMS, ILLINOIS, PHOSPHORUS, DRAINAGE,
AGRICULTURAL WATERSHEDS, NITROGEN FIXATION, CYANOPHYTA, SEWAGE
EFFLUENTS, MINNESOTA, WATER POLLUTION EFFECTS, ALGICIDES, WATER QUALITY.

IDENTIFIERS:
*URBAN WASTES, MADISON(WIS), LAKE ZOAR(CONN), KLAMATH LAKE(ORE), LAKE
WASHINGTON(WASH), ALGAL NUTRITION, ALGAL PHYSIOLOGY, HETEROTROPHY,
MICRONUTRIENTS, BADFISH RIVER(WIS), DETROIT LAKES(MINN), NUTRIENT
REMOVAL, MASS COMMUNICATION.

ABSTRACT:
IN APRIL 1960, A SEMINAR MET IN CINCINNATI, OHIO, TO CONSIDER THE
PROBLEM, ALGAE AND METROPOLITAN WASTES. THE MEETING, OF WHICH THIS
VOLUME CONSTITUTES THE RECORDED TRANSACTIONS, WAS SPONSORED BY DIVISION
OF WATER SUPPLY AND POLLUTION CONTROL AND ROBERT A TAFT SANITARY
ENGINEERING CENTER, BOTH AGENCIES OF THE US PUBLIC HEALTH SERVICE.
DISCUSSIONS AT THE SEMINAR WERE LIMITED TO PREVENTION AND CONTROL OF
OBJECTIONABLE ALGAL BLOOMS ARISING FROM ENRICHMENT OF THE HYDROSPHERE
BY URBAN AND OTHER WASTES. THE CONFERENCE WAS ORGANIZED INTO PANEL
DISCUSSIONS AS FOLLOWS: STATEMENT OF THE PROBLEM, GROWTH
CHARACTERISTICS OF ALGAE, SOURCES OF NUTRIENTS, METHODS OF PREVENTION
OR CONTROL. TOPICAL COVERAGE OF SECTIONS, WITH NUMBER OF CONTRIBUTIONS
TO EACH RECORDED IN PARENTHESES, WILL INDICATE SCOPE AND COVERAGE OF
MEETING: INTRODUCTION (1); SPECIFIC PROBLEMS IN LAKES (4), AND IN
RIVERS (1); ALGAL NUTRITION (3); MEASUREMENT OF PRODUCTIVITY (2); LAND
DRAINAGE (2); WASTES (1); LIMNOLOGICAL RELATIONSHIPS (3); NUTRIENT
LIMITATION (6); ALGICIDE USE (1). THE VOLUME INCLUDES A SUMMATION OF
RESEARCH NEEDS AND TEXT OF A BANQUET ADDRESS BY A JOURNALIST ON THEME
OF COMMUNICATION BETWEEN SCIENTIST AND LAYMAN. (SEE ALSO W70-04507).
(EICHHORN-WISCONSIN)

FIELD 05C

ACCESSION NO. W70-04506

SPECIFIC PROBLEMS IN RIVERS: ALGAE IN RIVERS OF THE UNITED STATES,

PUBLIC HEALTH SERVICE, CINCINNATI, OHIO. DIV. OF WATER SUPPLY AND POLLUTION
CONTROL.

C. MERVIN PALMER.

TRANSACTIONS OF THE SEMINAR ON ALGAE AND METROPOLITAN WASTES, HELD AT
CINCINNATI, OHIO, APRIL 27-29, 1960, SPONSORED BY DIVISION OF WATER SUPPLY
AND POLLUTION CONTROL AND THE ROBERT A TAFT SANITARY ENGINEERING CENTER.
ALGAE AND METROPOLITAN WASTES, U S DEPARTMENT OF HEALTH, EDUCATION, AND
WELFARE, DIVISION OF WATER SUPPLY AND POLLUTION CONTROL; AND ROBERT A TAFT
SANITARY ENGINEERING CENTER, CINCINNATI, OHIO, TECHNICAL REPORT, SEC TR
W61-3, P 34-38, 1961. 3 FIG, 4 REF.

DESCRIPTORS:
*RIVERS, *ALGAE, *SAMPLING, *MISSISSIPPI RIVER, *OHIO RIVER, NUISANCE
ALGAE, WATER QUALITY, SEWAGE EFFLUENTS, DIATOMS, MISSOURI RIVER,
COLUMBIA RIVER, COLORADO RIVER, TENNESSEE RIVER, HUDSON RIVER, DELAWARE
RIVER, WATER POLLUTION EFFECTS.

IDENTIFIERS:
*RIVER STUDIES, *NATIONAL WATER QUALITY NETWORK, ALGAE COUNTS, PLANKTON
ANALYSES, AGRICULTURAL RUNOFF, ARKANSAS RIVER, MERRIMACK RIVER, RED
RIVER, SAVANNAH RIVER, SNAKE RIVER, RIO GRANDE RIVER, POTOMAC RIVER.

ABSTRACT:
INVESTIGATION OF ALGAE IN RIVERS OF THE UNITED STATES HAVE BEEN LACKING
DUE TO THE GENERAL IMPRESSION THAT THEY ARE RELATIVELY SCARCE. HOWEVER,
MORE DATA IS NEEDED TO DETERMINE THEIR SIGNIFICANCE AS POLLUTION
INDICATORS, STREAM PURIFIERS, NUISANCE GROWTHS, AND FOOD FOR FISH.
BASIC INFORMATION SHOULD BE COLLECTED FOR EXTENDED TIME PERIODS DURING
EACH MONTH OF THE YEAR FOR AS MANY RIVERS IN REPRESENTATIVE LOCATIONS
AS POSSIBLE. SUCH DATA SHOULD YIELD NECESSARY BACKGROUND INFORMATION
FOR STUDIES AND AID IN DETERMINING EFFECTS OF PARTICULAR FACTORS ON
RIVER BIOTA. SINCE 1957, THE NATIONAL WATER QUALITY NETWORK PROGRAM OF
THE PUBLIC HEALTH SERVICE HAS CARRIED ON SUCH SAMPLING PROGRAM FOR 16
RIVERS AND THE GREAT LAKES. FOR THE FIRST TWO YEARS THE AVERAGE COUNT
PER MONTHLY SAMPLES WAS 3,625 ALGAE/MILLILITER, LOWEST COUNTS OCCURRING
IN NOVEMBER, DECEMBER, AND MARCH, HIGHEST IN APRIL, SEPTEMBER, AND
OCTOBER. HIGHEST PLANKTON COUNTS OCCURRED IN RIVERS ENRICHED BY
AGRICULTURAL RUNOFF AND SEWAGE EFFLUENTS, PARTICULARLY WHERE TOXIC
INDUSTRIAL WASTES WERE LACKING. THE SAME FEW GENERA OF ALGAE ARE FOUND
TO DOMINATE RIVERS ACROSS THE COUNTRY. TWO COMPARISON STUDIES AT THE
ROBERT A TAFT SANITARY ENGINEERING CENTER DEAL WITH GREEN ALGAE OF THE
MISSISSIPPI AND OHIO RIVERS AND NUMBER OF GENERA OF DIATOMS IN SEVERAL
RIVER SYSTEMS. (SEE W70-04506). (KETELLE-WISCONSIN)

FIELD 05G

ACCESSION NO. W70-04507

ALGAE AS INDICATORS OF POLLUTION,

ACADEMY OF NATURAL SCIENCES OF PHILADELPHIA, PA. DEPT. OF LIMNOLOGY.

RUTH PATRICK.

THIRD SEMINAR, 1962. BIOLOGICAL PROBLEMS IN WATER POLLUTION, ROBERT A TAFT
SANITARY ENGINEERING CENTER, CINCINNATI, OHIO, PUBLIC HEALTH SERVICE
PUBLICATION NO 999-WP-25, P 225-231, 1965. 4 FIG, 1 TAB, 33 REF.

DESCRIPTORS:
*ALGAE, *BIOINDICATORS, *WATER POLLUTION, DIATOMS, MARYLAND,
PENNSYLVANIA, RIVERS, STREAMS, EUGLENA, AQUATIC POPULATIONS,
EUTROPHICATION, CYANOPHYTA, SEWAGE EFFLUENTS, CHLOROPHYTA, RHODOPHYTA,
PHYSIOLOGICAL ECOLOGY, INDUSTRIAL WASTES, HYDROGEN ION CONCENTRATION,
IRON COMPOUNDS, HYDROGEN SULFIDE, LEAD, COPPER, AMMONIA, CONDUCTIVITY,
CHROMIUM, PHENOLS, NITROGEN, CALCIUM, DISSOLVED OXYGEN.

IDENTIFIERS:
COMMUNITY STRUCTURE, RIDLEY CREEK(PA), NOBS CREEK(MD), BACK RIVER(MD),
OSCILLATORIA, CATHERWOOD DIATOMETER, SAVANNAH RIVER, POPULATION
DYNAMICS, ALGAL COMMUNITIES, MICROBIAL ECOLOGY, WATER QUALITY CRITERIA,
CYANIDE, NAPHTHENIC ACID.

ABSTRACT:
EXTENSIVE LITERATURE HAS DEVELOPED REGARDING ALGAL INDICATORS OF
POLLUTION. POLLUTION CONCEPT REFERS TO VARIOUS PHYSICAL AND CHEMICAL
CHANGES IN THE HYDROSPHERE PRODUCED BY HUMAN ACTIVITIES. EARLIEST
RECOGNIZABLE FORM OF POLLUTION RESULTED FROM ORGANIC CONTAMINATION, AND
SPECIES WERE CLASSIFIED ACCORDING TO DEGREES OF ORGANIC POLLUTION THEY
TOLERATED. LITTLE ATTENTION WAS FOCUSED ON POPULATION DYNAMICS OF
VARIOUS SPECIES OR ON HOLISTIC ASPECTS OF ALGAL COMMUNITIES. BECAUSE OF
INCREASING DIVERSITY AND COMPLEXITY OF POLLUTIONAL LOADS AND BECAUSE
ENVIRONMENT STRONGLY AFFECTS THEIR OCCURRENCE AND POPULATION SIZE,
ASSESSING POLLUTION BY PRESENCE OR ABSENCE OF A FEW SPECIES IS INCLINED
TO BE MISLEADING. THUS, WELL ESTABLISHED POPULATIONS OF SEVERAL ALGAL
SPECIES MAY INDICATE PRESENCE OF PARTICULAR FORMS OF POLLUTION, BUT
THEIR ELIMINATION OR DECREASE MAY RESULT FROM CAUSES VERY DIFFERENT
THAN CHANGES IN POLLUTIONAL LOAD. BECAUSE ALGAL COMMUNITIES INTEGRATE
EFFECTS OF INFLUENCING FACTORS, MORE RECENT STUDY OF POLLUTION HAS BEEN
DIRECTED TO CONSIDER STRUCTURE OF THE ENTIRE ALGAL COMMUNITY AS
INDICATIVE OF POLLUTION. EXAMPLES ARE CITED OF SEVERAL SUCH APPROACHES,
CONSIDERING VARIOUS FORMS OF POLLUTION AT A NUMBER OF DIFFERENT
LOCALITIES. DISCUSSION FOLLOWING THE PAPER DEVELOPED INFORMATION THAT
EUGLENA AND OSCILLATORIA ARE OFTEN CITED AS GENERA MOST TOLERANT TO
SEWAGE AND RELATED CONDITIONS. (EICHHORN-WISCONSIN)

FIELD 05C

ACCESSION NO. W70-04510

LAKE EUTROPHICATION -- A NATURAL PROCESS,

GEOLOGICAL SURVEY, ALBANY, N.Y.

PHILLIP E. GREESON.

WATER RESOURCES BULLETIN, VOL 5, NO 4, P 16-30, DECEMBER 1969. 15 P, 6 FIG, 2
TAB, 44 REF.

DESCRIPTORS:
*EUTROPHICATION, *LAKES, NUTRIENTS, WATER POLLUTION EFFECTS,
PRODUCTIVITY, SEDIMENTS, URBANIZATION, WATER POLLUTION TREATMENT,
ALGAE, WATER MANAGEMENT(APPLIED), LIMNOLOGY.

IDENTIFIERS:
EUTROPHICATION TREATMENT.

ABSTRACT:
LAKE EUTROPHICATION IS AN ECONOMIC, RECREATIONAL, AND AESTHETIC PROBLEM
THAT AFFECTS EVERY LAKE OF THE WORLD. EUTROPHICATION IS THE NATURAL
PROCESS OF LAKE AGING, AND PROGRESSES IRRESPECTIVE OF MAN'S ACTIVITIES.
POLLUTION, HOWEVER, CAN HASTEN THE NATURAL RATE OF AGING AND SHORTEN
THE LIFE EXPECTANCY OF A BODY OF WATER. THE EUTROPHICATION OF A LAKE
CONSISTS OF THE GRADUAL PROGRESSION FROM ONE LIFE STAGE TO ANOTHER
BASED ON THE DEGREE OF NOURISHMENT OR PRODUCTIVITY. THE EXTINCTION OF A
LAKE IS ATTRIBUTED TO ENRICHMENT BY NUTRITIVE MATERIALS, BIOLOGICAL
PRODUCTIVITY, DECAY, AND SEDIMENTATION. PRESENTLY USED METHODS FOR
RETARDING EUTROPHICATION ARE THE ABATEMENT OF CULTURAL ENRICHMENT,
TREATMENT OF EUTROPHIC SYMPTOMS, AND CONTROL OF FUNDAMENTAL CAUSES.
(KNAPP-USGS)

FIELD 05C, 02H

ACCESSION NO. W70-04721

URBAN EFFECTS ON QUALITY OF STREAMFLOW,

TEXAS UNIV., AUSTIN. ENVIRONMENTAL HEALTH ENGINEERING RESEARCH LAB.

E. GUS FRUH.

IN: EFFECTS OF WATERSHED CHANGES ON STREAMFLOW, WATER RESOURCES SYMPOSIUM NO
2, AUSTIN, TEXAS, OCTOBER 1968, P 255-282, UNIVERSITY OF TEXAS PRESS,
AUSTIN AND LONDON, 1969. 28 P, 22 FIG, 5 TAB, 15 REF. NSF GRANT GU-1963,
AND TEXAS WATER QUALITY BD CONTRACT NO. 68-69-281.

DESCRIPTORS:
  *RESERVOIRS, *URBANIZATION, *WATER QUALITY, *STRATIFICATION, *TEXAS,
  WATER POLLUTION SOURCES, WATER POLLUTION EFFECTS, DISSOLVED OXYGEN,
  AQUATIC BACTERIA, ALGAE.

IDENTIFIERS:
  AUSTIN(TEX), COLORADO RIVER(TEX).

ABSTRACT:
  THE EFFECTS OF IMPOUNDMENTS AND URBANIZATION ON THE WATER QUALITY OF
  THE COLORADO RIVER OF TEXAS WERE STUDIED IN THE RESERVOIRS NEAR AUSTIN,
  TEXAS. LAKE TRAVIS, UPSTREAM FROM AUSTIN, IS LARGE AND DEEP AND HAS NO
  SIGNIFICANT INPUT OF POLLUTION. LAKE AUSTIN, THE NEXT RESERVOIR
  DOWNSTREAM, IS MUCH SMALLER AND SHALLOWER, AND RECEIVES SOME
  RECREATIONAL AND URBAN RUNOFF POLLUTION. TOWN LAKE, IN AUSTIN, IS SMALL
  AND NARROW WITH PREDOMINANTLY RIVER CHARACTERISTICS, AND RECEIVES SOME
  URBAN RUNOFF POLLUTION. OXYGEN CONCENTRATION-WATER DEPTH DATA FOR THE 3
  RESERVOIRS ARE TABULATED. IN LAKE TRAVIS, EVEN IN THE WINTER, THE
  TEMPERATURE VARIED WITH DEPTH. DURING THE SUMMER, THE OXYGEN FIRST
  BECAME DEPLETED AT THE THERMOCLINE REGION. THROUGHOUT THE FALL, VARIOUS
  DEPTHS OF LAKE TRAVIS BECAME REAERATED, BUT OXYGEN-DEPLETED WATERS WERE
  STILL PASSING THROUGH THE PENSTOCKS IN NOVEMBER. IN WINTER, OXYGEN WAS
  PRESENT AT ALL DEPTHS. IN LAKE AUSTIN, THE SUMMER OXYGEN CONCENTRATION
  IN THE EPILIMNION VARIED AROUND SATURATION. DISSOLVED OXYGEN DECREASED
  STEADILY IN THE HYPOLIMNION DURING THE SUMMER AND WAS CONSISTENTLY
  LOWER ABOVE THE SEDIMENTS. AFTER AUTUMN TURNOVER, OXYGEN REMAINED
  UNIFORM FROM TOP TO BOTTOM. HIGH NUMBERS OF TOTAL AND COLIFORM BACTERIA
  WERE FOUND IN LAKE AUSTIN DURING THE SPRING FOLLOWING PERIODS OF
  INTENSIVE RAINFALL. THE URBAN STREAM, BARTON CREEK, HAD SIGNIFICANTLY
  HIGHER CONCENTRATIONS OF SOLUTES, NUTRIENTS, AND BACTERIA THAN TOWN
  LAKE ABOVE THE STREAM'S ENTRANCE, PARTICULARLY DURING THE SPRING
  RAINFALL PERIOD. ALL OF AUSTIN'S URBAN STREAMS ENTER TOWN LAKE AND LAKE
  AUSTIN, WITH THE STREAMS FROM THE MORE HIGHLY DEVELOPED AREAS ENTERING
  TOWN LAKE. (KNAPP-USGS)

FIELD 04C, 05B

ACCESSION NO. W70-04727

BIOLOGICAL RESPONSES TO NUTRIENTS-EUTROPHICATION: PROBLEMS IN FRESHWATER,

TEXAS UNIV., AUSTIN.

E. GUS FRUH.

ADVANCES IN WATER QUALITY IMPROVEMENT, (EDITORS: GLOYNA, E. F. AND
ECKENFELDER, W. W., JR.), AUSTIN, TEXAS, UNIV OF TEXAS PRESS, 1968. P
49-64, 7 TAB, 82 REF.

DESCRIPTORS:
*EUTROPHICATION, *NITROGEN COMPOUNDS, *PHOSPHORUS COMPOUNDS,
*NUTRIENTS, *BIOASSAY, *RUNOFF, *LAKES, *IMPOUNDMENTS, TERTIARY
TREATMENT, ALGAE, DOMESTIC WASTES.

IDENTIFIERS:
*MICRONUTRIENTS, *NUTRIENT INTERACTIONS, *RADIOCARBON BIOASSAY,
*NUTRIENT TYPES, MICROQUANTITIES, MACROQUANTITIES, ALGAL BLOOMS,
LIEBIG'S LAW OF THE MINIMUM, VOISIN'S LAW OF THE MAXIMUM.

ABSTRACT:
THE PURPOSE OF THE PRESENTATION IS TO PROVIDE A FRAMEWORK OF REFERENCE
ON THE EUTROPHICATION PROBLEMS OF FRESHWATER. IMPORTANT INSIGHTS ON
EUTROPHICATION HAVE COME FROM THE FIELDS OF BOTANY, LIMNOLOGY, AND
SANITARY ENGINEERING. DISCUSSED ARE THE TYPES AND IMPACTS OF CRITICAL
NUTRIENTS. SPECIAL ATTENTION IS FOCUSED ON NITROGEN, AND PHOSPHORUS,
INCLUDING A DETAILED DISCUSSION OF SPECIFIC SOURCES, NAMELY RAINFALL,
GROUNDWATER, FOREST RUNOFF, AGRICULTURAL RUNOFF, URBAN DRAINAGE AND
WASTEWATER EFFLUENTS. ANALYSIS IS MADE OF THE RELATIONSHIP OF NUISANCE
CONDITIONS TO NITROGEN AND PHOSPHORUS. EXAMINATION IS MADE OF THE ROLE
OF ORGANIC AND INORGANIC MICRONUTRIENTS; THE NATURE OF NUTRIENT
INTERACTIONS WITH VARIOUS PHYSICAL, CHEMICAL, AND BIOLOGICAL VARIABLES.
VARIOUS APPROACHES TO CONTROL EUTROPHICATION ARE REPORTED. CONCLUSIONS
ARE REACHED THAT OTHER THAN TERTIARY TREATMENT, DIVERSION OF THE
NUTRIENT-RICH SOURCES APPEARS THE MOST FEASIBLE EUTROPHICATION CONTROL.
HOWEVER, LEACHING OR BACTERIAL DECOMPOSITION OF NUTRIENTS FROM THE
SEDIMENTS SEEMS THE LIMITING FACTOR IN LAKE RECOVERY RATES. MANY
PROCESSES IN AQUATIC ENVIRONMENTS ARE ATTRIBUTED TO BACTERIA, HOWEVER,
CURRENT ANALYTICAL TECHNIQUES ARE INADEQUATE FOR THE QUANTITATIVE
EVALUATION OF BACTERIAL SPECIES AND NUMBER.
(D'AREZZO-TEXAS)

FIELD 05D

ACCESSION NO. W70-04765

BASIS FOR WASTE STABILIZATION POND DESIGNS,

TEXAS UNIV., AUSTIN.

EARNEST F. GLOYNA.

ADVANCES IN WATER QUALITY IMPROVEMENT, (EDITORS: GLOYNA, E. F., AND
ECKENFELDER, W. W., JR.), AUSTIN, TEXAS, UNIV OF TEXAS PRESS, 1968. P
397-408, 1 FIG, 1 TAB, 19 REF.

DESCRIPTORS:
*WASTE TREATMENT, *STABILIZATION, *PONDS, *DESIGN CRITERIA, FACILITIES,
EQUATIONS, AEROBIC CONDITIONS, ANAEROBIC CONDITIONS, ALGAE, ALGAL
CONTROL, ECOLOGICAL DISTRIBUTION, ACTIVATED SLUDGE, MAINTENANCE COSTS,
CONSTRUCTION COSTS.

IDENTIFIERS:
*WASTE STABILIZATION PONDS, *ALGAL PHYSIOLOGY, *DESIGN THEORY AND
CALCULATIONS, PHOTOSYNTHESIS, ILLUMINATION, TEMPERATURE, NUTRIENTS,
FACULTATIVE POND DESIGNS, BOD, SLUDGE.

ABSTRACT:
MANY PROBLEMS ARISING FROM USE OF WASTE STABILIZATION PONDS ARE
TRACEABLE TO INADEQUATE ENGINEERING, POOR MAINTENANCE AND LACK OF
OPERATIONAL SUPERVISION. HOWEVER, THERE IS A MORE FUNDAMENTAL NEED:
DESIGN AND ENGINEERING FOR SYSTEMS USING STABILIZATION PONDS MUST BE
BASED ON AN UNDERSTANDING OF THE PRINCIPLES OF ALGAL PHYSIOLOGY.
GENERALLY, ALGAE GROWING IN A STREAM OR A WASTE STABILIZATION POND ARE
IN A HIGHLY COMPETITIVE ENVIRONMENT. ONE OR MORE OF THE FACTORS
NECESSARY FOR PHOTOSYNTHESIS ARE INTERRELATED, AND UNICELLULAR ALGAE,
IN PARTICULAR WILL REACT RAPIDLY TO ENVIRONMENTAL CHANGES.
ILLUMINATION, TEMPERATURE, AND NUTRIENTS--THE FACTORS WHICH AFFECT
GROWTH RATE ARE EXAMINED CRITICALLY. THE PHENOMENA OF ALGAL PHYSIOLOGY,
TOGETHER WITH ACTUAL TREATMENT OBJECTIVES ARE SYNTHESIZED IN GOOD
DESIGN. IN THE FINAL ANALYSIS, THE SPECIFIC DESIGN OF A WASTE
STABILIZATION POND DEPENDS HEAVILY UPON THE SPECIFICALLY-IMPOSED
TREATMENT OBJECTIVES AND REQUIREMENTS. A POND SYSTEM MAY BE DESIGNED TO
RECEIVE UNTREATED SEWAGE OR INDUSTRIAL WASTES, PRIMARY OR SECONDARY
TREATMENT EFFLUENTS OR EXCESS ACTIVATED SLUDGES AND SETTLEABLE SOLIDS.
CERTAIN CRITICAL FACTORS IN DESIGN DEVELOPMENT ARE DISCUSSED, BASED ON
OPERATIONAL EXPERIENCES AND RESEARCH WHICH PROVIDE VITAL INSIGHTS INTO
ENVIRONMENTAL FACTORS GOVERNING WASTE STABILIZATION PROCESSES.
(D'AREZZO-TEXAS)

FIELD 05D

ACCESSION NO. W70-04786

ADVANCES IN ANAEROBIC POND SYSTEMS DESIGN,

CALIFORNIA UNIV., BERKELEY.

W. J. OSWALD.

ADVANCES IN WATER QUALITY IMPROVEMENT, (EDITORS: GLOYNA, E. F., AND
ECKENFELDER, W. W., JR.), AUSTIN, TEXAS, UNIV OF TEXAS PRESS, 1968. P
409-426, 6 FIG, 2 TAB, 11 REF.

DESCRIPTORS:
*WASTE TREATMENT, *STABILIZATION, *PONDS, *ENVIRONMENTAL EFFECTS,
*BIOLOGICAL TREATMENT, *DESIGN CRITERIA, DISSOLVED OXYGEN, TEMPERATURE,
BIOCHEMICAL OXYGEN DEMAND(BOD), EVAPORATION, FERMENTATION, METHANE,
ALGAL CONTROL, BACTERIA, COLIFORMS.

IDENTIFIERS:
*ANAEROBIC WASTE PONDS, *DEPTH EFFECTS, *HYDRAULIC LOAD EFFECTS,
*ORGANIC LOAD EFFECTS, *DETENTION TIME EFFECTS, *FACULATIVE PONDS,
*AEROBIC PONDS, *ANAEROBIC PONDS, DESIGN FOR ENVIRONMENTAL CONTROL,
RECIRCULATION, BOD REMOVAL, GAS PRODUCTION.

ABSTRACT:
THE BIOLOGICAL ACTIONS AND INTERACTIONS IN WASTE STABILIZATION PONDS
HAVE BEEN COMPLEX ENOUGH TO DETER SCIENTIFIC INVESTIGATION AND THUS TO
FOSTER THE WIDESPREAD USE OF EMPIRICAL DESIGN FORMULATIONS WHICH ARE
BASED ON LIMITED DATA. RECOGNITION OF THE FACT THAT SEDIMENTATION AND
FOUR BASIC BIOLOGICAL REACTIONS INTERACT IN PONDS, MAKES POSSIBLE A
RATIONAL DESIGN OF PONDS. PRECISE FORMULAS HAVE NOT YET BEEN DEVELOPED
GOVERNING ALL VARIABLES. HOWEVER, THE RELATIONSHIPS BETWEEN TEMPERATURE
AND METHANE FERMENTATION, DEPTH AND TEMPERATURE, AND BETWEEN DEPTH AND
OXYGENATION ARE WELL-ESTABLISHED. METHANE FERMENTATION IS THE SINE QUA
NON FOR AEROBIC AND FACULTATIVE PONDS; THEREFORE, DESIGNS MAXIMIZING
TEMPERATURE UNIFORMITY AND MINIMIZING OXYGEN INTRUSION ARE MOST
SUCCESSFUL. IN SERIES PONDS, WHICH RECYCLE OXYGENATED WATER, THE
PREVENTION OF ODOR IN THE INITIAL ANAEROBIC POND IS DONE EASILY BY
OVERLAYING THE PRIMARY ANAEROBIC POND WITH OXYGENATED WATER. FOR
DESTRUCTION OF COLIFORM BACTERIA, BOTH TIME AND TEMPERATURE ARE VITAL
FACTORS. AT NORMAL POND TEMPERATURES, OBTAINING TIME INCREMENTS BY
DEPTH INCREASES IS EASIER THAN GETTING TEMPERATURE INCREMENTS BY DEPTH
DECREASES. WASTES HAVING HIGH CARBON/NITROGEN RATIOS MUST BE OXIDIZED
AND AERATED BEFORE FERMENTATION. (D'AREZZO-TEXAS)

FIELD 05D

ACCESSION NO. W70-04787

WASTE STABILIZATION POND PRACTICES IN THE UNITED STATES,

FEDERAL WATER POLLUTION CONTROL ADMINISTRATION, DALLAS, TEX.

JEROME H. SVORE.

ADVANCES IN WATER QUALITY IMPROVEMENT, (EDITORS: GLOYNA, E. F., AND
ECKENFELDER, W. W., JR.), AUSTIN, TEXAS, UNIV OF TEXAS PRESS, 1968. P
427-434, 2 FIG, 2 TAB, 8 REF, 1 PLATE.

DESCRIPTORS:
*STABILIZATION, *PONDS, *DESIGN, *EFFICIENCIES, *CHLORINATION, *ODOR,
*WASTE TREATMENT, ALGAE, DISINFECTION, BIOCHEMICAL OXYGEN DEMAND(BOD),
AERATION, COLIFORMS, ALGAL CONTROL, DISSOLVED OXYGEN.

IDENTIFIERS:
*AERATED LAGOONS, *MIDWEST(US), *SOUTHWEST(US), *DESIGN CRITERIA, POND
EFFLUENT, COLORADO, WEST(US), STABILIZATION POND LOADINGS, ODOR
PERSISTENCE.

ABSTRACT:
THE DESIGN, OPERATIONS, AND GRADUAL ACCEPTANCE OF WASTE STABILIZATION
PONDS IN THE SOUTHWEST DURING THE PAST 25 YEARS IS TRACED. RECENT
SURVEYS IN COLORADO HAVE SUBSTANTIATED GENERAL EXPERIENCE IN THE
MIDWEST, CLASSIFYING THE PREDOMINANT DEFECTS IN DESIGN, CONSTRUCTION,
AND OPERATION INTO THREE CLASSES: (A) LIQUID DEPTH NOT MAINTAINED, (B)
WATER SURFACE INADEQUATE, AND (C) PRIMARY SOLIDS NOT EVENLY
DISTRIBUTED. A COMPARISON IS MADE OF CURRENT METHODS USED THROUGHOUT
THE SOUTHERN HALF OF THE UNITED STATES REGARDING THE EXCESSIVE ALGAE
PROBLEM. THE DETERMINATION OF LAGOON TREATMENT EFFICIENCY MUST BE MADE
FROM THE VIEWPOINT OF REGIONAL WATER PRIORITIES AND OBJECTIVES. THE
REDUCTION IN REQUIRED CHLORINE DISINFECTION USAGE RESULTING FROM
EFFICIENT POND OPERATION IS DISCUSSED. SOME REASONS FOR THE TREND
TOWARD GREATER USE OF AERATED LAGOONS ARE PRESENTED. WHILE THERE ARE
STILL ISOLATED AREAS WHERE OPINION EXISTS THAT WASTE STABILIZATION
PONDS ARE DANGEROUS, UNDESIRABLE AND, A HEALTH HAZARD, THE MAJOR TREND
IS TOWARD RECOGNITION THAT IF PONDS ARE DESIGNED AND OPERATED PROPERLY
(IN SERIES), THEY WILL PROVIDE MORE TREATMENT PER UNIT COST THAN ANY
OTHER SINGLE PROCESS. (D'AREZZO-TEXAS)

FIELD 05D

ACCESSION NO. W70-04788

EXTRA-DEEP PONDS,

TECHNION - ISRAEL INST. OF TECH., HAIFA; AND WATER PLANNING FOR ISRAEL LTD.,
TEL AVIV.

ALBERTO M. WACHS, AND ANDRE BEREND.

ADVANCES IN WATER QUALITY IMPROVEMENT, (EDITORS GLOYNA, E. F., AND
ECKENFELDER, W. W., JR.), AUSTIN, TEXAS, UNIV OF TEXAS PRESS, 1968. P
450-456, 5 FIG, 1 TAB, 9 REF.

DESCRIPTORS:
*PONDS, *STABILIZATION, *THERMAL STRATIFICATION, *SEASONAL, *SULFIDES,
TEMPERATURE, CLIMATIC DATA, STRATIFICATION, DISSOLVED OXYGEN, ALGAE,
BIOCHEMICAL OXYGEN DEMAND(BOD), MIXING, WINDS.

IDENTIFIERS:
*THERMAL GRADIENT, *BOD REMOVAL, *ISRAEL, *ALGAL CONCENTRATION,
'EXTRA-DEEP PONDS', ASHKELON(ISRAEL), TEMPERATURE PROFILES, VOLATILE
SUSPENDED SOLIDS, MULTI-CELL SERIAL SYSTEM, PH, FACULTATIVE POND,
ANAEROBIC POND.

ABSTRACT:
AN EXPERIMENTAL UNIT WAS BUILT BY TAHAL, LTD, NEAR THE WASTEWATER
TREATMENT FACILITY, ASHKELON, ISRAEL, TO OBTAIN PILOT PLANT PERFORMANCE
DATA ON 'EXTRA-DEEP' WASTE STABILIZATION PONDS, (I.E., DEPTH GREATER
THAN 8 FEET). THE EXPERIENCE GAINED SHOWS THAT WHEN LOADINGS ACCEPTABLE
TO THE ISRAEL CLIMATE WERE USED DURING THE WARM SEASON, A PRONOUNCED
THERMAL GRADIENT CAUSED THE UPPER PART OF THE POND TO BEHAVE AS A
'SEPARATE' FACULTATIVE POND 'FLOATING' ON AN UNDERLYING ANAEROBIC POND.
THIS STRATIFICATION WAS ENHANCED BY THE FACT THAT BOTH INLET AND OUTLET
WERE LOCATED AT THE SURFACE. EFFLUENT WAS RICH IN ALGAE. DURING THE
WARM SEASON, ALGAL CONCENTRATIONS WERE MUCH LOWER AT OR BELOW DEPTHS OF
1 METER; THEREFORE, PLACING OF OUTLET AT THAT DEPTH WOULD DRAW OFF
BETTER EFFLUENT INSOFAR AS ALGAL DENSITY WAS CONCERNED, BUT IT WAS
DEFICIENT IN DISSOLVED OXYGEN AT TIMES. THE THERMAL GRADIENT
DISAPPEARED DURING THE COLD SEASON. MIXING WAS DUE MAINLY TO WIND
ACTION AND INFLUENCED BY POND GEOMETRY. DISSOLVED OXYGEN WAS NOT ALWAYS
PRESENT IN THE UPPER POND LAYERS BUT IT DID CONTAIN CONSIDERABLE ALGAE.
DURING FALL, A POND TURNOVER OCCURRED RESULTING IN MIXING OF LOWER AND
UPPER LAYERS. THIS CAUSED A SURFACE INCREASE OF SULFIDES CONCENTRATION,
A DROP IN PH, BUT NO ODORS. DURING BOTH COLD AND WARM SEASONS, BOD
REMOVAL WAS ABOUT 80%. (D'AREZZO-TEXAS)

FIELD 05D

ACCESSION NO. W70-04790

THE ADAPTATION OF PLANKTON ALGAE. III. WITH SPECIAL CONSIDERATION OF THE
  IMPORTANCE IN NATURE,

ROYAL DANISH SCHOOL OF PHARMACY, COPENHAGEN. DEPT. OF BOTANY.

E. STEEMANN NIELSEN, AND ERIK G. JORGENSEN.

PHYSIOLOGIA PLANTARUM, VOL 21, NO 3, P 647-654, 1968. 2 TAB, 20 REF.

DESCRIPTORS:
  *PLANKTON, *ALGAE, LIGHT INTENSITY, TEMPERATURE, PRIMARY PRODUCTIVITY,
  SURFACES, ARCTIC, ENZYMES, CHLOROPHYLL, PHOTOSYNTHESIS, CHLORELLA,
  GRAZING, ZOOPLANKTON, DIATOMS, RESPIRATION, PIGMENTS, OCEANS, ORGANIC
  MATTER, DIURNAL, LATITUDINAL STUDIES.

IDENTIFIERS:
  PHOTOOXIDATION, CHLORELLA VULGARIS, CHLORELLA VULGARIS(COLUMBIA),
  CHLORELLA PYRENOIDOSA, CHLORELLA SACCHAROPHILA, CHLORELLA ELLIPSOIDEA,
  CHLORELLA LUTEOVIRIDIS, SKELETONEMA COSTATUM, PHYSIOLOGICAL ADJUSTMENT,
  ASTERIONELLA FORMOSA, CYCLOTELLA MENEGHINIANA, PHOTIC ZONE,
  HELSINGOR(DENMARK), SALINO-CLINE.

ABSTRACT:
  ADAPTATION (PHYSIOLOGICAL ADJUSTMANT) OF PLANKTON ALGAE, IN NATURAL
  STATES, TO DIFFERENT LIGHT INTENSITIES AND TEMPERATURES IS DESCRIBED.
  ALTHOUGH DIFFERENT SPECIES VARY IN EXTENT OF ADAPTATION, THEY ALL ADAPT
  TO SOME DEGREE, AND ADAPTATION IS SIGNIFICANT FOR PLANKTON POPULATIONS
  AT ALL DEPTHS. SHADE ADAPTATION OF ALGAE IN THE LOWER PART OF THE
  PHOTIC ZONE IS OFTEN OF MINOR IMPORTANCE TO THE INTEGRAL PRIMARY
  PRODUCTION PER UNIT OF SURFACE, EXCEPT ON OCCASIONS WHEN THE BULK OF
  THE ALGAE IS FOUND IN THIS LOWER PART. IT IS SHOWN THAT THE I-SUB-K (A
  LIGHT INTENSITY DEFINING ONSET OF LIGHT-SATURATED PHOTOSYNTHESIS) OF
  SURFACE PLANKTON DURING SUMMER IN THE ARCTIC IS HIGH DESPITE LOW
  TEMPERATURES DUE TO INCREASE OF THE ENZYME QUANTITIES PER CELL. DAILY
  FLUCTUATIONS IN I-SUB-K ARE PARTIALLY DUE TO PHOTOOXIDATION AND
  PARTIALLY BECAUSE THE PERIODS FOR THE PRODUCTION OF CHLOROPHYLL AND
  PHOTOSYNTHETIC ENZYMES ARE MUTUALLY DISPLACED. DESPITE DAILY
  FLUCTUATIONS IN I-SUB-K, THE DIFFERENCE BETWEEN 'SUN' AND 'SHADE'
  PHYTOPLANKTON IS DISTINCT. THE DIURNAL VARIATION IN PHOTOSYNTHETIC
  ACTIVITY OF PHYTOPLANKTON MAY TAKE PLACE ESPECIALLY AT LOW LATITUDES,
  AND NEAR THE SURFACE; AND TO SOME EXTENT AT HIGHER LATITUDES, AND IN
  LOWER PHOTIC ZONE. (JONES-WISCONSIN)

FIELD 02H, 05C

ACCESSION NO. W70-04809

EUTROPHICATION TESTING PROGRAM FOR MINNESOTA'S 'SKY BLUE WATERS',

SCHOELL AND MADSON, INC., HOPKINS, MINN.

JAMES R. ORR.

WATER AND SEWAGE WORKS, VOL 115, NO 3, P 97-101, 1968. 2 FIG, 2 TAB, 10 REF.

DESCRIPTORS:
  *EUTROPHICATION, *MINNESOTA, RECREATION, SEWAGE DISPOSAL, WATER
POLLUTION, POLLUTION ABATEMENT, COLIFORMS, WATER QUALITY, PHOSPHORUS,
NITROGEN, FLOW, CHEMICAL ANALYSIS, NUTRIENTS, EFFLUENTS, ALGAE, WEEDS,
CARBON DIOXIDE, VITAMINS, NITRATES, INORGANIC COMPOUNDS, SURFACE
RUNOFF, UNDERFLOW, PRECIPITATION(ATMOSPHERIC), DRAINAGE, FERTILIZERS,
FARM WASTES, SOIL EROSION, FROZEN GROUND.

IDENTIFIERS:
  LAKE MINNETONKA(MINN), MINNEAPOLIS-ST PAUL(MINN).

ABSTRACT:
  THE GREATEST SOURCE OF NUTRIENTS TO LAKE MINNETONKA, ONE OF MINNESOTA'S
BEST KNOWN LAKES, IS SEWAGE TREATMENT PLANT EFFLUENT. A MAJOR PART OF
THE PROBLEM WOULD BE SOLVED BY REMOVING NUTRIENTS FROM EFFLUENTS OR
SUPPRESSING SEWAGE EFFLUENT DISCHARGE. LOSS OF LARGE QUANTITIES OF
WASTEWATER OUTSIDE THE WATERSHED COULD AGGRAVATE THE PROBLEM OF
MAINTAINING LAKE LEVEL. ANOTHER APPROACH MAY BE CONTROL OF EXCESSIVE
WEED AND ALGAE PRODUCTION BY CHEMICAL TREATMENT OR MECHANICAL
HARVESTING, HOWEVER, CONTINUED USE OF ALGICIDES MAY EVENTUALLY HAVE A
TOXIC EFFECT ON FISH AND OTHER AQUATIC LIFE; MECHANICAL HARVESTING IS
IMPRACTICAL. ADVANCED WASTE TREATMENT OF SEWAGE DISCHARGED TO LAKE,
WITH EMPHASIS ON REMOVAL OF PHOSPHORUS SHOULD BE CONSIDERED.
MISCELLANEOUS NUTRIENT SOURCES, OTHER THAN SEWAGE PLANT EFFLUENT,
REPRESENT APPROXIMATELY 2/3 TOTAL NITROGEN AND APPROXIMATELY 1/5 TOTAL
PHOSPHORUS INFLOW. WATERSHED DWELLERS CAN MINIMIZE THESE SOURCES BY
PREVENTING DECAYING VEGETATION BEING WASHED INTO THE LAKE BY RAINFALL,
DISCONTINUING COMMERCIAL AND MANURE FERTILIZER USAGE, AND PRACTICING
SOIL EROSION CONTROL. MUNICIPALITIES CAN ASSIST BY KEEPING PUBLIC LANDS
CLEAN, LOCATING WASTE DUMPS TO AVERT DRAINAGE INTO THE LAKE,
ELIMINATING SOIL ABSORPTION SEWAGE DISPOSAL FACILITIES ADJACENT TO
LAKE, AND BY PUBLICIZING THE PROBLEM. (JONES-WISCONSIN)

FIELD 02H, 05C

ACCESSION NO. W70-04810

DISTINCTION BETWEEN BACTERIAL AND ALGAL UTILIZATION OF SOLUBLE SUBSTANCES IN
THE SEA,

MARINE LAB., ABERDEEN (SCOTLAND); AND INDIANA UNIV., INDIANAPOLIS. DEPT. OF
MICROBIOLOGY.

A. L. S. MUNRO, AND T. D. BROCK.

JOURNAL GENERAL MICROBIOLOGY, VOL 51, P 35-42, 1968. 3 FIG, 1 TAB, 14 REF, 3
PLATES.

DESCRIPTORS:
*AQUATIC BACTERIA, *ALGAE, *ANALYTICAL TECHNIQUES, *MARINE
MICROORGANISMS, PHYSIOLOGICAL ECOLOGY, ORGANIC MATTER, WATER POLLUTION
EFFECTS, DIATOMS, CARBON RADIOISOTOPES, KINETICS, TRITIUM, CHLOROPHYLL.

IDENTIFIERS:
*HETEROTROPHY, *ORGANIC SOLUTES, *MARINE ENVIRONMENTS, MICROBIAL
ECOLOGY, FLUORESCENCE MICROSCOPY, ACETATE, CARBON-14, AUTORADIOGRAPHY,
MICHAELIS-MENTEN EQUATION, LOCH EWE(SCOTLAND), SCOTLAND, PHAEOPIGMENTS,
ACRIDINE ORANGE, LINEWEAVER-BURKE EQUATION.

ABSTRACT:
STAINING WITH ACRIDINE ORANGE AND DIRECT OBSERVATION WITH FLUORESCENCE
MICROSCOPY DEMONSTRATES NUMEROUS VIABLE BACTERIA AND DIATOMS ATTACHED
TO SAND GRAINS OF A LITTORAL BEACH TO DEPTHS EXCEEDING 10 CENTIMETERS.
KINETIC ANALYSIS OF UPTAKE OF RADIOCARBON-LABELLED ACETATE OVER RANGE,
10-5000 MICROGRAMS/LITER, DEMONSTRATES TWO UPTAKE MECHANISMS: ONE,
SATURABLE AT LOWER CONCENTRATIONS (MICHAELIS-MENTEN KINETICS), AND
ANOTHER, DESCRIBABLE AS DIFFUSION, THE LATTER HAVING BEEN ATTRIBUTED BY
OTHERS TO ALGAL UPTAKE. AUTORADIOGRAPHY, HOWEVER, REVEALS THAT BACTERIA
ALONE ARE RESPONSIBLE FOR UPTAKE OF TRITIATED ACETATE. IN THOSE
SITUATIONS WHERE MICHAELIS-MENTEN KINETICS APPLIED, MAXIMUM ATTAINABLE
UPTAKE RATES (V) WERE MUCH GREATER THAN ANY PREVIOUSLY REPORTED FOR
WATER, SUCH VALUES FOR V BEING ATTRIBUTABLE TO LARGER MICROBIAL
POPULATIONS IN SAND. SINCE REASONS FOR EXISTENCE OF TWO TYPES OF
KINETICS ARE OBSCURE, IT IS DANGEROUS TO MAKE CONCLUSIONS REGARDING
MICROBIAL HETEROTROPHY BASED ON KINETIC ANALYSIS ALONE. BASED ON THESE
STUDIES, IT SEEMS CLEAR THAT ALGAL HETEROTROPHY IS NEGLIGIBLE IN SEA
WATERS, AND THAT AUTORADIOGRAPHY IS AN EFFECTIVE TOOL IN STUDIES OF
ENERGETICS AND TROPHIC RELATIONSHIPS IN MICROBIAL ECOLOGY.
(EICHHORN-WISCONSIN)

FIELD 05C

ACCESSION NO. W70-04811

EUTROPHICATION AND DETERGENTS,

SOUTHAM BUSINESS PUBLICATIONS LTD., DON MILLS (ONTARIO). READER SERVICE DEPT;
AND TORONTO UNIV. (ONTARIO).

P. H. JONES, AND TOM DAVEY.

WATER AND POLLUTION CONTROL, VOL 106, NO 9, P 22-25, 46, SEPTEMBER 1968.

DESCRIPTORS:
*EUTROPHICATION, PHOSPHORUS, *DETERGENTS, DOMESTIC WASTES, POLLUTANTS,
WATER POLLUTION SOURCES, WATER QUALITY, ALGAE, SEWAGE EFFLUENTS.

IDENTIFIERS:
POLYPHOSPHATES, NUTRIENT BALANCE, FOAMING DETERGENTS, PUBLIC
APPREHENSION.

ABSTRACT:
DETERGENTS ARE ATTACKED AS A MAJOR NUTRIENT SOURCE IN EUTROPHICATION.
SOLUBLE POLYPHOSPHATES SERVING AS FILLER IN DETERGENTS HELP UPSET
NUTRIENT BALANCES IN LAKES AND STIMULATE ALGAL GROWTH. OTHER
SIGNIFICANT SOURCES OF BIOLOGICALLY ACTIVE PHOSPHORUS ARE ANIMAL AND
HUMAN WASTE. BECAUSE MOST DETERGENTS CONTAIN 50% POLYPHOSPHATES, SEWAGE
EFFLUENT PHOSPHATE HAS DOUBLED OR TRIPLED. SUBSTITUTE FILLERS FOR
POLYPHOSPHATES, WHICH CAN ACT AS CHELATING AGENTS AND CONTROL PH, ARE
NOW AVAILABLE. REPLACING PHOSPHORUS IN DETERGENTS WOULD HAVE AN
INSTANTANEOUS BENEFICIAL EFFECT AT A COST OF AN EXTRA 15-20 CENTS A
BOX. PUBLIC PRESSURE COULD FORCE THE DESIRED CHANGE. EXCESS PHOSPHORUS
ACCELERATES A LAKE'S NATURAL AGING PROCESS. (BANNERMAN-WISCONSIN)

FIELD 05B

ACCESSION NO. W70-05084

THE INFLUENCE OF NITROGEN ON HETEROCYST PRODUCTION IN BLUE-GREEN ALGAE,

BUREAU OF COMMERCIAL FISHERIES, ANN ARBOR, MICH. BIOLOGICAL LAB.

ROANN E. OGAWA, AND JOHN F. CARR.

LIMNOLOGY AND OCEANOGRAPHY, VOL 14, NO 3, P 342-351, 1969. 7 FIG, 4 TAB, 26 REF.

DESCRIPTORS:
*ALGAE, *CYANOPHYTA, *NITROGEN FIXATION, NUTRIENTS, LAKE ERIE.

IDENTIFIERS:
*HETEROCYSTS, ANABAENA VARIABILIS, MICROCYSTIS AERUGINOSA, APHANIZOMENON FLOS-AQUAE, OSCILLATORIA, ANABAENA CYLINDRICA B629, ANABAENA INAEQUALIS 381, ANABAENA FLOS AQUAE, TOLYPOTHRIX DISTORTA, GLOEOTRICHIA ECHINULATA LB 1303.

ABSTRACT:
INDIRECT EVIDENCE IS PRESENTED SUGGESTING INVOLVEMENT OF HETEROCYSTS IN UTILIZATION OF ATMOSPHERIC NITROGEN AS SOLE NITROGEN SOURCE FOR ALGAE. SEVEN HETEROCYSTOUS (KNOWN ATMOSPHERIC NITROGEN-FIXING) BLUE-GREEN ALGAE WERE GROWN IN A MODIFIED CHU NO 10 MEDIUM DEVOID OF COMBINED NITROGEN; TWO NON-HETEROCYSTOUS (NON-ATMOSPHERIC-FIXING) BLUE-GREENS DID NOT GROW IN THIS MEDIUM. HETEROCYSTS WERE PRODUCED SHORTLY AFTER INOCULA DEVOID OF HETEROCYSTS WERE PLACED IN NITORGEN-FREE MEDIUM. PRODUCTION WAS GREATEST WHEN ATMOSPHERIC NITORGEN SERVED AS SOLE NITROGEN SOURCE, AND LEAST WHEN AMMONIA-NITROGEN SERVED AS SOLE NITROGEN-SOURCE. NITRATE-NITROGEN PRODUCED AN INTERMEDIATE NUMBER OF HETEROCYSTS. WHEN MEDIUM NITRATE-NITROGEN CONTENT WAS VARIED AS SOLE NITROGEN SOURCE, NUMBERS OF HETEROCYSTS PRODUCED WERE INVERSELY PROPORTIONAL TO THE NITROGEN CONCENTRATION. THEY DID NOT DEVELOP IN THE ABSENCE OF PHOSPHORUS, BUT NO EFFORT WAS MADE TO DETERMINE THE CRITICAL PHOSPHORUS CONCENTRATION FOR THEIR PRODUCTION. PRODUCTION OF HETEROCYSTS IN CIRCUMSTANCES OF LOW COMBINED NITROGEN MAY PROVIDE AN ECOLOGICAL ADVANTAGE. THE RELATIVE NUMBERS OF HETEROCYSTS IN FIELD-COLLECTED SAMPLES INDICATE THE RELATIVE AMOUNT OF AVAILABLE NITROGEN AND A CONTINUING SUPPLY OF AVAILABLE PHOSPHORUS.
(GERHOLD-WISCONSIN)

FIELD 05C

ACCESSION NO. W70-05091

PHYTOPLANKTON AS AN AGENT OF THE SELF-PURIFICATION OF CONTAMINATED WATERS,

FOREIGN TECHNOLOGY DIV., WRIGHT-PATTERSON AFB, OHIO.

G. G. VINBERG, AND T. N. SIVKE.

AVAILABLE FROM THE CLEARINGHOUSE AS TT-67-62992, AND AD-659 310, $3.00 IN
PAPER COPY, $0.65 IN MICROFICHE. TRANSLATION BY FOREIGN TECHNOLOGY
DIVISION, WRIGHT-PATTERSON AIR FORCE BASE, FTD-MT-66-13, AUGUST 18, 1967,
37 P. TRANS. FROM TRUDY VSESOYUZNOGO GIDROBIOLOGICHESKOGO OBSHCHESTVA, VOL
7, P 3-23, 1956.

DESCRIPTORS:
*PHYTOPLANKTON, *SELF-PURIFICATION, *WASTE WATER(POLLUTION), *WATER
PURIFICATION, ALGAE, SEWAGE, ANAEROBIC, OXYGEN, NITRIFICATION,
COLIFORMS, PONDS, CLIMATES, PHOTOSYNTHESIS, PRODUCTIVITY, SYMBIOSIS,
RESERVOIRS, LAKES, REAERATION, SESTON, CHLOROPHYLL, CYCLES, UNITED
STATES, CHLORELLA, BIOCHEMICAL OXYGEN DEMAND, OXIDATION LAGOONS,
CALIFORNIA, TEXAS, DEPTH, RETENTION, NORTH DAKOTA, ICE, SNOW, CARP,
SALMON, IRRIGATION, FILTRATION, ORGANIC MATTER, AMMONIA, HYDROGEN ION
CONCENTRATION, ANTIBIOTICS, SYMBIOSIS, EUGLENA, LIGHT INTENSITY,
NUTRIENTS, TEMPERATURE.

IDENTIFIERS:
PISCICULTURAL PONDS, SVISLOCH' RIVER(USSR), MINSK(USSR), ENGLAND,
OSCILLATORIA, AUSTRALIA, SANTA ROSA(CALIF), LYUBLIN(USSR),
LYUBERETS(USSR), MUNICH(GERMANY), CHLORELLA PYRENOIDOSA, BICHOROMATE,
DRY WEIGHT, B COLI COMMUNAE, EUGLENE GRACILIS, PROTOCOCCALES.

ABSTRACT:
PLANKTONIC ALGAE GROWN ON A LARGE SCALE ON UNDILUTED CITY SEWAGE
SHARPLY ACCELERATES SELF-PURIFICATION PROCESSES; IT IS EXPRESSED IN
FASTER INITIAL LOWERING OF THE BIOCHEMICAL OXYGEN DEMAND, EARLY
TERMINATION OF THE ANAEROBIC PHASE, APPEARANCE OF FREE OXYGEN, AND IN
THE ACCELERATED ONSET OF NITRIFICATION. ACCUMULATION OF A LARGE
QUANTITY OF ORGANIC SUBSTANCES SYNTHESIZED BY THE ALGAE IN THE
COMPOSITION OF THE BODIES OF THE LIVING CELLS IS NOT REFLECTED IN THE
MAGNITUDE OF THE BIOCHEMICAL OXYGEN DEMAND. THE DEATH RATE OF COLIFORM
BACTERIA INCREASED SHARPLY WITH THE LARGE-SCALE DEVELOPMENT OF GREEN
ALGAE IN SELF-CLEANING SEWAGE. EUGLENA AND CHLORELLA ARE COMPARED.
PONDS, FILLED WITH UNDILUTED SEWAGE, ARE THE SIMPLEST METHOD TO USE
GREEN ORGANISMS AS AGENTS OF SELF-PURIFICATION AND IS AN ESPECIALLY
EFFECTIVE METHOD OF WATER RECLAMATION, PARTICULARLY APPLICABLE TO WARM,
DRY, CLIMATIC REGIONS. FURTHER STUDY OF FAVORABLE CONDITIONS FOR
DEVELOPMENT OF PHOTOSYNTHESIZING PLANKTON ORGANISMS WILL ALLEVIATE
PROBLEMS IN LARGE-SCALE CULTIVATION OF ALGAE AS A METHOD OF UTILIZING
SEWAGE AND PARTICULARLY SHOULD CLARIFY THE APPLICABILITY OF USING ALGAE
GROWN ON SEWAGE TO INCREASE THE PRODUCTIVITY OF PISCICULTURAL PONDS.
(JONES-WISCONSIN)

FIELD 05F

ACCESSION NO.   W70-05092

MEASUREMENT AND DETECTION OF EUTROPHICATION,

CENTRAL COASTAL REGIONAL WATER QUALITY CONTROL BOARD, SAN LUIS OBISPO, CALIF.

THOMAS E. BAILEY.

JOURNAL OF THE SANITARY ENGINEERING DIVISION, PROCEEDINGS OF THE AMERICAN
SOCIETY OF CIVIL ENGINEERS, VOL 93, NO SA6, P 121-132, 1967. 6 FIG.

DESCRIPTORS:
*CORRELATION ANALYSIS, *FLUOROMETRY, *FLUORESCENCE, *CHLOROPHYLL,
*EUTROPHICATION, PRIMARY PRODUCTIVITY, PHYTOPLANKTON, PHOTOSYNTHESIS,
BIOMASS, CULTURES, PLANT PHYSIOLOGY, ALGAE, DIURNAL DISTRIBUTION,
OXYGEN, DISSOLVED OXYGEN, MEASUREMENT, ESTUARIES, SANITARY ENGINEERING,
WATER TREATMENT, WASTE TREATMENT.

IDENTIFIERS:
SACRAMENTO-SAN JOAQUIN DELTA, CALIFORNIA WATER PLAN, DELTA WATER
FACILITIES(CALIF), OXYGEN DYNAMICS, SACRAMENTO-SAN JOAQUIN ESTUARY,
ENVIRONMENTAL RELATIONSHIPS, FLUOROMETRIC CHLOROPHYLL TECHNIQUE.

ABSTRACT:
PRIMARY PRODUCTIVITY, AN IMPORTANT INDICATOR OF EUTROPHICATION, WAS
MEASURED INDIRECTLY AS CHLOROPHYLL CONCENTRATION OF PHYTOPLANKTON
STANDING CROP. PIGMENT CONCENTRATIONS WERE DETERMINED
SPECTROPHOTOMETRICALLY AS ABSORBANCE WITH A 90% ACETONE SOLVENT.
FLUORESCENT DYE TRACER TECHNIQUES WERE MODIFIED TO MEASURE CHLOROPHYLL
FLUOROMETRICALLY. LIGHT SCATTERING, QUENCHING, AND VARIATION IN
PHYTOPLANKTON WERE CONSIDERED TO INFLUENCE FLUORESCENCE MEASURED BY THE
TURNER MODEL 111 FLUOROMETER, ESTABLISHING CHLOROPHYLL AS A GOOD
INDICATOR OF EUTROPHICATION BASED ON RELATIONSHIP WITH PHYTOPLANKTON
STANDING CROP AND PHOTOSYNTHETIC ACTIVITY. AN EMPIRICAL EQUATION WAS
DEFINED FOR ESTIMATING PHOTOSYNTHESIS FROM CHLOROPHYLL DATA. DEPENDENCE
OF DISSOLVED OXYGEN SUPERSATURATION ON PLANT METABOLISM RELATED DIURNAL
OXYGEN CONCENTRATION TO CHLOROPHYLL. FLUOROMETRIC TECHNIQUE OF
MEASURING CHLOROPHYLL IS A FAST AND FLEXIBLE METHOD FOR MEASURING
PRIMARY ENVIRONMENTAL RELATIONSHIP AND BIOLOGICAL MONITORING IN SUCH
INSTALLATIONS AS SEWAGE TREATMENT PLANTS. ONCE PRIMARY ENVIRONMENTAL
CHLOROPHYLL RELATIONSHIPS ARE DEFINED, THE FLUOROMETRIC TECHNIQUE CAN
BE USED ALMOST WITHOUT LIMITATION. (BANNERMAN-WISCONSIN)

FIELD 02L, 05C

ACCESSION NO. W70-05094

GLUCOSE AND ACETATE IN FRESHWATER: CONCENTRATIONS AND TURNOVER RATES,

NORTH CAROLINA STATE UNIV., RALEIGH. DEPT. OF ZOOLOGY.

JOHN E. HOBBIE.

PROCEEDINGS OF INTERNATIONAL BIOLOGICAL PROGRAM SYMPOSIUM HELD IN AMSTERDAM
AND NIEUWERSLUIS, OCTOBER 10-16, 1966. CHEMICAL ENVIRONMENT IN THE AQUATIC
HABITAT, GOLTERMAN, H L AND CLYMO, R S, EDITORS, N V NOORD-HOLLANDSCHE
UITGEVERS MAATSCHAPPIJ, AMSTERDAM, 1967. P 245-251, 5 FIG, 9 REF.

DESCRIPTORS:
*CHEMICAL ANALYSIS, ORGANIC MATTER, WATER POLLUTION EFFECTS, ANALYTICAL
TECHNIQUES, KINETICS, AQUATIC BACTERIA, ALGAE, PHYSIOLOGICAL ECOLOGY,
CARBON RADIOISOTOPES, LAKES, PHYTOPLANKTON, BIOASSAYS.

IDENTIFIERS:
*GLUCOSE, *ACETATE, *FRESHWATER ENVIRONMENTS, *TURNOVER RATES,
*HETEROTROPHY, ORGANIC SOLUTES, SWEDEN, LAKE ERKEN(SWEDEN), ANNUAL
CYCLES, MICROBIAL ECOLOGY, RADIOCARBON UPTAKE TECHNIQUES,
MICHAELIS-MENTEN EQUATION, RELATIVE HETEROTROPHIC POTENTIAL,
LINEWEAVER-BURK EQUATION, UPTAKE VELOCITY, MAXIMUM UPTAKE VELOCITY,
SUBSTRATE CONCENTRATIONS.

ABSTRACT:
ORGANIC SOLUTES IN THE HYDROSPHERE ARE EXTREMELY DIVERSE, AND THEIR
DYNAMIC PROPERTIES HAVE GREATER ECOLOGICAL SIGNIFICANCE THAN DO THEIR
INSTANTANEOUS CONCENTRATIONS. USEFUL INFORMATION REGARDING
HETEROTROPHIC POTENTIAL OF LACUSTRINE BACTERIA CAN BE DERIVED FROM
ANALYSIS, ACCORDING TO ENZYME-TYPE KINETICS ORIGINALLY PROPOSED BY
MICHAELIS AND MENTEN, OF THEIR UPTAKE OF RADIOCARBON-LABELED
SUBSTRATES. SUCH TREATMENT YIELDS ESTIMATES OF UPTAKE VELOCITY; MAXIMUM
UPTAKE VELOCITY (V); TURNOVER TIME (T); AND RELATIVE MEASURE (K + S) OF
CONCENTRATIONS OF NATURALLY-OCCURRING SUBSTRATES CONCENTRATION (S),
WHERE K IS ANALOGOUS TO MICHAELIS-MENTEN CONSTANT. THESE KINETIC
PARAMETERS, FOLLOWED OVER ONE ANNUAL CYCLE, ARE DESCRIBED FOR GLUCOSE
AND ACETATE IN LAKE ERKEN, SWEDEN. V, VARYING 40-FOLD, FOLLOWED ALGAL
CYCLES CLOSELY, PEAKING IN JUNE AND SEPTEMBER. THAT CONSTANT REMOVAL
AND SUPPLY OF BOTH COMPOUNDS ARE IN BALANCE, AND THAT THEY DO NOT
ACCUMULATE IS INDICATED BY SEASONALLY CONSTANT VALUE FOR (K + S),
NATURALLY-OCCURRING CONCENTRATIONS (IN MICROGRAMS/LITER) REMAINING AT
APPROXIMATELY 6 FOR GLUCOSE, AND 10 FOR ACETATE. T VARIED FROM 10 HOURS
(SUMMER) TO APPROXIMATELY 1000 HOURS (WINTER). WHEN T = 10, SIZEABLE
FRACTION OF NATURALLY-OCCURRING DISSOLVED ORGANIC CARBON IN LAKE IS
CYCLED, ILLUSTRATING ECOLOGICAL VALUE OF KNOWING T AS COMPARED WITH
KNOWING S. (EICHHORN-WISCONSIN)

FIELD 05C

ACCESSION NO. W70-05109

INVESTIGATION OF THE ODOR NUISANCE OCCURRING IN THE MADISON LAKES PARTICULARLY
LAKES MONONA, WAUBESA AND KEGONSA FROM JULY 1942 TO JULY 1943,

PUBLIC HEALTH SERVICE, MADISON, WIS.; AND WISCONSIN UNIV., MADISON. DEPT. OF
CIVIL ENGINEERING.

C. N. SAWYER, J. B. LACKEY, AND A. T. LENZ.

REPORT TO GOVERNOR'S COMMITTEE, MADISON, WISCONSIN, 1943. 79 P. 15 FIG, 4
TAB, 22 REF, APPENDIX.

DESCRIPTORS:
*ODOR-PRODUCING ALGAE, *NUISANCE ALGAE, *WISCONSIN, *LAKES,
PRODUCTIVITY, NITROGEN, PHOSPHORUS, DRAINAGE, MARSHES, FARMS,
CYANOPHYTA, PLANKTON, ZOOPLANKTON, EUTROPHICATION, STORM DRAINS, SEWAGE
EFFLUENTS, STREAMFLOW, MEASUREMENT, WATER POLLUTION, FERTILIZATION,
INDUSTRIAL WASTES, GARBAGE DUMPS, OIL WASTES, INLETS(WATERWAYS),
ORGANIC MATTER, POTASSIUM, SULFATES, CRUSTACEANS, DIATOMS, COPPER
SULFATE, DINOFLAGELLATES, EUGLENOPHYTA, CHLOROPHYTA, CHLAMYDOMONAS,
FUNGI.

IDENTIFIERS:
*MADISON(WIS), LAKE MONONA(WIS), LAKE WAUBESA(WIS), LAKE KEGONSA(WIS),
BLOOMING, LAKE MENDOTA(WIS), LAKE WINGRA(WIS), ANABAENA SPIROIDES,
MICROCYSTIS AERUGINOSA, CRYPTOMONADIDA, MASTIGOPHORA, CRYPTOPHYCEAE,
CHRYSOPHYCEAE, VOLVOCALES, SARCODINA, CILIATES, CYCLOTELLA,
APHANIZOMENON FLOS AQUAE, MELOSIRA, ASTERIONELLA FORMOSA, FRAGILARIA,
STEPHANODISCUS, PANDORINA.

ABSTRACT:
MADISON LAKES, ESPECIALLY THE SHALLOW DOWNSTREAM LAKES BELOW THE CITY,
HAVE BEEN TROUBLED DURING WARM SEASONS WITH OFFENSIVE ODORS AND
UNSIGHLTY SHORELINES. THE MAJOR CAUSE OF THE OFFENSIVE CONDITIONS ARE A
DIRECT RESULT OF ACCUMULATIONS OF ALGAL GROWTHS IN ADVANCED STAGES OF
DECOMPOSITION. THESE GROWTH STIMULATIONS ARE RELATED TO FERTILIZATION
OF THE WATER WITH MINERALS, SUCH AS NITROGEN, PHOSPHORUS, POTASSIUM,
AND POSSIBLY OTHER ENRICHING ELEMENTS AND COMPOUNDS. DRAINAGE FROM RICH
AGRICULTURAL AREAS AND URBAN COMMUNITIES INTO SHALLOW LAKES ENHANCES
ALGAL PROBLEMS. NITROGEN IS SHOWN TO BE ONE OF THE MAJOR COMPONENTIAL
ELEMENTS, WITH PHOSPHORUS MINOR. NUISANCES IN INLAND LAKES ARE LARGELY
MEMBERS OF THE PLANT KINGDOM. DOMESTIC SEWAGE, ALTHOUGH HIGHLY PURIFIED
BY MODERN TREATMENT METHODS, IS RICH IN PHOSPHORUS, NITROGEN,
POTASSIUM, AND POSSIBLY OTHER ELEMENTS. THE MOST OBNOXIOUS BLOOMS ARE
THOSE OF BLUE-GREEN ALGAE, PARTICULARLY ANABAENA AND MICROCYSTIS.
EXCESSIVE AMOUNTS OF PLANKTON ARE RELATED TO AGRICULTURAL DRAINAGE.
LAKES MONONA, WAUBESA, AND KEGONSA ARE ADDITIONALLY ENRICHED BY STORM
SEWERS, OTHER THAN RAIN WATER, AND BY SEWAGE TREATMENT PLANT EFFLUENT.
(SEE W70-05113). (JONES-WISCONSIN)

FIELD 02H, 05C

ACCESSION NO. W70-05112

INVESTIGATION OF THE ODOR NUISANCE OCCURRING IN THE MADISON LAKES PARTICULARLY
LAKES MONONA, WAUBESA AND KEGONSA FROM JULY 1943 TO JULY 1944,

PUBLIC HEALTH SERVICE, MADISON, WIS.; AND WISCONSIN UNIV., MADISON. DEPT. OF
CIVIL ENGINEERING.

C. N. SAWYER, J. B. LACKEY, AND A. T. LENZ.

REPORT TO GOVERNOR'S COMMITTEE, MADISON, WISCONSIN, 1945. 92 P. 23 FIG, 29
TAB, APPENDIX.

DESCRIPTORS:
*ALGAE, *LAKES, *EUTROPHICATION, *HYDROLOGY, CHEMICAL PROPERTIES,
WISCONSIN, SEWAGE EFFLUENTS, NITROGEN, PHOSPHORUS.

IDENTIFIERS:
*TRUAX FIELD(MADISON) EFFLUENTS, *MADISON LAKES, YAHARA RIVER(WIS),
STORKWEATHER CREEK(WIS), NINE SPRINGS CREEK(WIS), DOOR CREEK(WIS),
DUTCH MILL DRAINAGE DITCH(WIS), LAKE MONONA(WIS), LAKE WAUBESA(WIS),
LAKE KEGONSA(WIS).

ABSTRACT:
THE THREE PARTS OF THIS VOLUMINOUS REPORT PRESENT RESULTS OF DETAILED
HYDROLOGICAL, CHEMICAL, AND BIOLOGICAL INVESTIGATIONS OF THE THREE
MADISON LAKES AND THEIR TRIBUTARIES. THE STUDY CONFIRMED THAT THE
OBJECTIONAL GROWTH OF ALGAE AND LAKE BLOOM ARE PRIMARILY INDUCED BY
INORGANIC FORMS OF NITROGEN AND PHOSPHORUS OF SEWAGE EFFLUENTS.
NUUISANCE BLOOMS ARE ATTRIBUTED CHIEFLY TO BLUE-GREEN ALGAE. THE
RESULTS OF LABORATORY EXPERIMENTS INDICATED THAT REMOVAL OF SOLUBLE
NITROGEN AND PHOSPHORUS FROM SEWAGE WOULD MATERIALLY DECREASE THE ALGAL
DENSITY. THE CRITICAL CONTENTS OF NITROGEN AND PHOSPHORUS IN THE LAKE
WATERS WERE ESTABLISHED AS 0.30 AND 0.015 PPM, RESPECTIVELY. A
SUGGESTION IS MADE TO DUPLICATE LABORATORY TRIALS IN ARTIFICIAL PONDS
TREATED WITH SEWAGE AND SEWAGE FREED FROM ITS NITROGEN AND PHOSPHORUS.
(SEE W70-05112). (WILDE-WISCONSIN)

FIELD 02H, 05C

ACCESSION NO. W70-05113

RAPID POLAROGRAPHIC DETERMINATION OF CHLORINE, BROMINE AND IODINE IN ALGAE,

TRIESTE UNIV. (ITALY). INST. OF THE SCIENCE OF PROPERTIES AND QUALITIES OF
COMMERCIAL PRODUCTS.

C. CALZOLARI, L. GABRIELLI, AND G. PERTOLDI MARLETTA.

ANALYST, VOL 94, NO 1122, P 774-779, SEPTEMBER 1969. 6 P, 1 FIG, 4 TAB, 15
REF.

DESCRIPTORS:
*ALGAE, *CHEMICAL ANALYSIS, *CHLORINE, *ANALYTICAL TECHNIQUES,
*POLAROGRAPHIC ANALYSIS, PHOTOMETRY, HALOGENS.

IDENTIFIERS:
BROMINE, IODINE.

ABSTRACT:
A RAPID METHOD IS DESCRIBED FOR THE POLAROGRAPHIC DETERMINATION OF
CHLORINE, BROMINE AND IODINE IN ALGAE AFTER A SINGLE COMBUSTION
PROCEDURE. FOR THE DETERMINATION OF CHLORINE AND IODINE, THE ANODIC
WAVES OF THEIR IONS WERE RECORDED. FOR BROMINE, ON THE OTHER HAND, THE
BROMIDE IONS WERE OXIDIZED QUANTITATIVELY TO BROMATE IONS, AND THE
CATHODIC WAVE FOR THESE IONS WAS RECORDED. THE METHOD HAS BEEN TESTED
ON SEVERAL SPECIES OF ALGAE FROM THE GULF OF TRIESTE. (KNAPP-USGS)

FIELD 05A

ACCESSION NO. W70-05178

THE EFFECT OF LIGHT ON NITRATE AND NITRITE ASSIMILATION BY CHLORELLA AND
ANKISTRODESMUS,

UNIVERSITY COLL., LONDON (ENGLAND). DEPT. OF BOTANY.

I. MORRIS, AND J. AHMED.

PHYSIOLOGIA PLANTARUM, VOL 22, NO 6, P 1166-1174, 1969. 5 TAB, 20 REF.

DESCRIPTORS:
*ALGAE, *LIGHT, *NITRATES, *NITRITES, AEROBIC CONDITIONS, ANAEROBIC
CONDITIONS, CHLORELLA.

IDENTIFIERS:
CARBON DIOXIDE, CHLORELLA PYRENOIDOSA, ANKISTRODESMUS BRAUNII,
DUNALIELLA TERTIOLECTA, DCMU INHIBITOR.

ABSTRACT:
ASSIMILATION OF NITRATE AND NITRITE NITROGEN BY CHLORELLA PYRENOIDOSA
AND ANKISTRODESMUS BRAUNII WAS OBSERVED UNDER ILLUMINATION IN AEROBIC
AND ANAEROBIC MEDIA IN THE ABSENCE OF CARBON DIOXIDE. THE PRESENCE OF
CARBON DIOXIDE FAILED TO INFLUENCE THE ASSIMILATION OF CHLORELLA, BUT
STIMULATED THAT OF ANKISTRODESMUS. THE RATIOS OF OXYGEN:NITRATE AND
OXYGEN:NITRITE VARIED IN DIFFERENT TRIALS AND WERE HIGHER THAN
CALCULATED VALUES FOR OXYGEN OF 2.0 AND 1.5 EXPECTED IN REDUCTION OF
NITRATE AND NITRITE, RESPECTIVELY. OXYGEN EVOLUTION WAS COMPLETELY
ARRESTED BY 3-(3'-4'-DICHLOROPHENYL)-1-1 DIMETHYL UREA (DCMU) AT A
CONCENTRATION OF 0.000004M. HOWEVER, NITRITE ASSIMILATION BY BOTH ALGAE
AND NITRATE ASSIMILATION BY ANKISTRODESMUS WERE LESS DEPRESSED BY THE
INHIBITOR. (WILDE-WISCONSIN)

FIELD 02K, 05C

ACCESSION NO. W70-05261

SUMMARY OF LITERATURE ON AQUATIC WEED CONTROL,

TORONTO UNIV. (ONTARIO).

J. MURRAY SPEIRS.

CANADIAN FISH CULTURIST, VOL 3, NO 4, P 20-32, 1948. 55 REF.

DESCRIPTORS:
*AQUATIC WEED CONTROL, *BIBLIOGRAPHIES, CATTAILS, BULRUSH, DREDGING,
COPPER SULPHATE, PONDWEEDS, SODIUM ARSENITE, FISH, ANIMALS, ALGAE,
CUTTING MANAGEMENT, POTABLE WATER, HERBICIDES, CHARA, ALKALINITY,
BULLHEADS, EELS, KILLIFISHES, TROUT, SUCKERS, BASS, WATER HYACINTH,
FROGS, CRAYFISH, MOSQUITOES, BACTERIA, IRRIGATION, WILLOW TREES,
FERTILIZERS, BURNING.

IDENTIFIERS:
*LITERATURE, ELEOCHARIS, 2,4-D, AMMATE, CHLORAMINE, DE-K-PRUF-21,
BENOCLORS, NIGROSINE DYE, SODIUM CHLORATE, SANTOBRITE, CLADOPHORA,
POTAMOGETON SPP, LAKE MENDOTA(WIS), ANACHARIS, NAIAS, SCIRPUS,
NYMPHAEA, CRASSIPES EICHORNIA, TYPHA, BRASENIA, DOW CONTACT,
SPARGANIUM, SALIX, GROWTH REGULATORS, MYRIOPHYLLUM, WATER BLOOMS,
CHAINING.

ABSTRACT:
SEVERAL METHODS ARE AVAILABLE TO CONTROL PARTICULAR AQUATICS. TO
CONTROL EMERGENT AQUATICS SUCH AS CATTAILS, BULRUSHES AND SPIKERUSHES,
2,4-D OR AMMATE ARE EFFECTIVE FOR PERMANENT CONTROL; DREDGING AND USE
OF A FLAME THROWER ARE ADEQUATE TEMPORARY CONTROLS. TO CONTROL ALGAE,
COPPER SULPHATE APPEARS TO BE EFFICIENT; CHLORAMINE MAY BE EQUALLY
USEFUL BUT REQUIRES ADDITIONAL TESTING; FOR ALGAE COATING CONCRETE OR
WOODEN WALLS OF FISH TANKS OR RESERVOIRS, DE-K-PRUF-21 IS SUGGESTED.
THE MOST EFFECTIVE CONTROL FOR PONDWEEDS APPEARS TO BE SODIUM ARSENITE
BUT CAUTION MUST BE EXERCISED WHERE THE WATER MIGHT BE A DRINKING
SUPPLY FOR MAMMALS (INCLUDING MAN). BENOCLORS ARE EFFECTIVE CONTROLS
WHERE FISH AND OTHER AQUATIC ANIMALS ARE UNIMPORTANT; NIGROSINE DYE MAY
BE USED TO SHADE OUT THE PONDWEEDS WHERE THE BLACK COLOR GIVEN THE
WATER IS ACCEPTABLE; IN SOME WATERS, PONDWEEDS HAVE BEEN CONTROLLED BY
ADDING ENOUGH FERTILIZER TO PRODUCE A WATER BLOOM SHADING OUT THE
PONDWEEDS; THIS METHOD MAY HAVE MERIT WHERE DENSE ALGAL GROWTHS ARE
PREFERABLE TO PONDWEED GROWTHS. WATER LILIES ARE PARTICULARLY RESISTANT
TO CHEMICAL CONTROL METHODS, AND PERSISTENT CUTTING APPEARS THE ONLY
SATISFACTORY METHOD. (JONES-WISCONSIN)

FIELD 05C, 04A

ACCESSION NO. W70-05263

THE ECOLOGY OF A RESERVOIR,

PAUL A. ERICKSON, AND JOHN T. REYNOLDS.

NATURAL HISTORY, P 48-53, NOVEMBER 1969. 4 FIG.

DESCRIPTORS:
*WATER QUALITY CONTROL, *ECOLOGY, *RESERVOIR, MASSACHUSETTS, POTABLE
WATER, WATER POLLUTION, BACTERIA, ALGAE, NUTRIENTS, INDICATORS, ODOR,
TASTE, FILTERS, WATER TREATMENT, DISSOLVED OXYGEN, EUTROPHICATION,
FISH, LAND, VALUE, LEAVES, FERTILIZERS, PESTICIDES, CELLULOSE, CYCLES,
HARVESTING, DREDGING, MONITORING, MICROBIOLOGY, FOOD CHAINS.

IDENTIFIERS:
QUABBIN RESERVOIR(MASS), BOSTON(MASS).

ABSTRACT:
TECHNIQUES DESIGNED TO SLOW THE EUTROPHICATION PROCESS ARE: REDUCTION
OF NUTRIENTS BY CONTROLING SEWAGE WASTE DISPOSAL, AND CHEMICAL
FERTILIZERS POSSIBLY LEACHING INTO THE WATERSHED, AND BY BACTERIAL
DECOMPOSITION OF BOTTOM SEDIMENTS; BY MECHANICAL HARVESTING OF ALGAL
POPULATIONS, ROOTED AND BLOSSOMING PLANTS; DREDGING; REMOVAL OF
DISSOLVED NUTRIENTS BY PHYSICAL AND CHEMICAL AGENTS; REMOVAL OF ALGAL
POPULATIONS BY CHEMICAL AGENTS AND REDUCTION BY INCREASING FISH
POPULATIONS. THE FINANCIAL AND PHYSICAL PROBLEMS ASSOCIATED WITH THEIR
ACTUAL APPLICATION ARE IMMENSE, AS A RESERVOIR IS THE FOCUS OF
SOCIO-POLITICAL FACTORS WHICH MUST BE WEIGHED IN ANY PROCEDURE. ANOTHER
POSSIBILITY FOR SLOWING DOWN EUTROPHICATION LIES IN THE UNIQUE STUDIES,
CLASSIFIED AS 'MICROBIAL INTERVENTION.' IF THE RESERVOIR ECOSYSTEM
COULD BE MANAGED TO FAVOR BACTERIA GROUPS THAT ALREADY COMPETE WITH
ALGAE FOR DECOMPOSED ORGANIC NUTRIENTS, THE ALGAL FOOD CHAIN WOULD BE
BROKEN AND ALGAL POPULATIONS OF UNDESIRABLE PROPORTIONS COULD NOT
OCCUR, THUS MAXIMIZING THE RESERVOIR'S SELF-CLEANSING CAPABILITY.
MICROBIAL INTERVENTION WOULD TEND TO SOOTHE THE POLITICAL AND SOCIAL
PULSE, FOR IT IS A 'NATURAL PROCESS.' PRECISE INFORMATION DESCRIBING
INTERACTIONS OF ALGAL AND BACTERIAL POPULATIONS MUST BE UNDERSTOOD
BEFORE THEY CAN BE MANIPULATED FOR HUMAN BENEFITS. (JONES-WISCONSIN)

FIELD 05C, 05G

ACCESSION NO. W70-05264

ALGAL FLOCCULATION WITH SYNTHETIC ORGANIC POLYELECTROLYTES,

NOTRE DAME UNIV., IND. DEPT. OF CIVIL ENGINEERING.

MARK W. TENNEY, WAYNE F. ECHELBERGER, JR., RONALD G. SCHUESSLER, AND JOSEPH
L. PAVONI.

APPLIED MICROBIOLOGY, VOL 18, NO 6, P 965-971, 1969. 8 FIG, 11 REF.

DESCRIPTORS:
*ALGAE, *FLOCCULATION, *WASTE WATER TREATMENT, ELECTROLYTES, CHEMICAL
REACTIONS, ELECTROCHEMISTRY, RESISTIVITY, CHLORELLA, SCENEDESMUS.

IDENTIFIERS:
*POLYELECTROLYTES, *SYNTHETIC ORGANIC POLYELECTROLYTES, CATIONIC
ELECTROLYTES, ALGAL SURFACE CHARGE, ELECTROSTATIC REPULSION, DOW C-31,
DOW A-21, NALCO N-670.

ABSTRACT:
CATONIC, ANIONIC, AND NONIONIC SYNTHETIC POLYELECTROLYTES WERE USED IN
ATTEMPTS TO REMOVE ALGAE FROM WASTE WATERS. UNDER CONDITIONS OF THE
STUDY, FLOCCULATION OF ALGAE WAS CAUSED ONLY BY CATIONIC ELECTROLYTES.
THE MECHANISM OF FLOCCULATION IS INTERPRETED AS BRIDGING PHENOMENA
BETWEEN ALGAL CELLS AND THE MATRIX OF LINEARY EXTENDED POLYMER CHAINS.
THE FLOCCULATION REQUIRES A REDUCTION IN THE ALGAL SURFACE CHARGE TO A
LEVEL AT WHICH THE EXTENDED POLYMERS CAN OVERCOME ELECTROSTATIC
REPULSION. THE RATE OF APPLICATION OF CATIONIC FLOCCULANTS DEPENDS UPON
PH OF THE MEDIUM, ALGAL CONCENTRATION, AND ALGAL GROWTH PHASE.
(WILDE-WISCONSIN)

FIELD 05D

ACCESSION NO. W70-05267

ON THE BIOLOGICAL SIGNIFICANCE OF PHOSPHATE ANALYSIS; COMPARISON OF STANDARD
AND NEW METHODS WITH A BIOASSAY,

MINNESOTA UNIV., MINNEAPOLIS. LIMNOLOGICAL RESEARCH CENTER.

WILLIAM CHAMBERLAIN, AND JOSEPH SHAPIRO.

LIMNOLOGY AND OCEANOGRAPHY, VOL 14, NO 6, P 921-927, 1969. 2 FIG, 2 TAB, 13
REF.

DESCRIPTORS:
*WATER CHEMISTRY, *WATER POLLUTION EFFECTS, *PHOSPHATES, *ANALYTICAL
TECHNIQUES, *BIOASSAYS, ORGANOPHOSPHORUS COMPOUNDS, SOLVENT
EXTRACTIONS, PHOSPHORUS COMPOUNDS, LAKES, SEWAGE EFFLUENTS, ARSENIC
COMPOUNDS, SPECTROPHOTOMETRY, ALGAE, MINNESOTA, SODIUM ARSENITE.

IDENTIFIERS:
ARSENATE INTERFERENCE, DISSOLVED INORGANIC PHOSPHATES, SOLUBLE REACTIVE
PHOSPHORUS, PHOSPHATE(BIOLOGICALLY AVAILABLE), MICROCYSTIS AERUGINOSA,
LAKE COMO(MINN), LAKE VADNAIS(MINN), MILLE LAC(MINN), DEMING
LAKE(MINN), ORCHARD LAKE(MINN), LAKE OWASSO(MINN), LAKE
JOSEPHINE(MINN), GRAY'S BAY(MINN), SMITH'S BAY(MINN), PHOSPHATE
DETERMINATION METHODS.

ABSTRACT:
THE POSSIBILITY EXISTS THAT STANDARD ANALYTICAL METHODS MIGHT
OVERESTIMATE DISSOLVED INORGANIC PHOSPHATE, THUS MODIFIED TECHNIQUES
HAVE BEEN INTRODUCED TO OPERATE AN OPERATIONALLY DEFINED FRACTION OF
PHOSPHORUS COMPOUND, 'SOLUBLE REACTIVE PHOSPHORUS.' AVAILABILITY OF
VARIOUS ANALYTICAL PROCEDURES RAISES THE QUESTION OF WHICH ONE MEASURES
THE PHOSPHORUS FRACTION WHICH IS IMPORTANT AS A PLANT NUTRIENT. TWO
SEWAGE EFFLUENTS AND WATER FROM 11 MINNESOTA LAKES WERE ANALYZED BY
FOUR PROCEDURES: HARVEY METHOD, 6-SECOND METHOD, SOLVENT EXTRACTION,
STEPHENS METHOD. ANALYTICAL RESULTS WERE COMPARED WITH MAGNITUDE OF
PHOSPHATE UPTAKE DURING A BIOASSAY EMPLOYING PHOSPHORUS-STARVED CELLS
OF BLUE-GREEN ALGA, MICROCYSTIS AERUGINOSA. WATERS COULD BE ARRANGED
INTO TWO GROUPS BASED UPON MAGNITUDE OF DIFFERENCE BETWEEN RESULTS OF
HARVEY METHOD AND EXTRACTION PROCEDURE. WATERS OF GROUP II, WHERE SUCH
DIFFERENCES WERE LARGE, ALL WERE HIGH IN ARSENATE (16-56 MICROGRAMS
ARSENATE-ARSENIC/LITER), ARSENATE INTERFERENCE FROM SUCH CONCENTRATIONS
BEING SUFFICIENTLY LARGE TO ACCOUNT FOR DISCREPANCIES IN PHOSPHATE
DETERMINATIONS. EXTRACTION PROCEDURE IS LESS SENSITIVE TO ARSENATE
INTERFERENCE THAN ARE OTHER METHODS. IN ANALYZING LAKE WATER FOR
PHOSPHATES, IT IS ADVISABLE TO ASSUME THAT ARSENIC IS PRESENT.
PHOSPHATE ANALYSES RESULTING FROM EXTRACTION PROCEDURE WERE IN GOOD
AGREEMENT WITH BIOLOGICALLY AVAILABLE PHOSPHATES AS DETERMINED BY
BIOASSAY. (EICHHORN-WISCONSIN)

FIELD 05C, 05A

ACCESSION NO. W70-05269

THERMOPHILIC BLUE-GREEN ALGAE AND THE THERMAL ENVIRONMENT,

OREGON UNIV., EUGENE. DEPT. OF BIOLOGY.

RICHARD W. CASTENHOLZ.

BACTERIOLOGICAL REVIEWS, VOL 33, NO 4, P 476-504, 1969. 4 FIG, 4 TAB, 187 REF.

DESCRIPTORS:
*CYANOPHYTA, *ALGAE, *THERMAL SPRINGS, *OREGON, THERMOPHILIC BACTERIA, THERMAO POLLUTION, NUTRIENTS, TEMPERATURE, PHOTOSYNTHESIS, LIGHT INTENSITY.

IDENTIFIERS:
*YELLOWSTONE PARK, *ICELAND, MASTIGOCLADUS LAMINOSUS, PHORMIDIUM LAMINOSUM, OSCILLATORIA TENUIS, OSCILLATORIA TEREBRIFORMIS, MATS, SYNECHOCOCCUS LIVIDUS, BLUE-GREEN ALGAE, FILAMENTS.

ABSTRACT:
THE BLUE-GREEN ALGAE IN GENERAL HAVE HIGHER TEMPERATURE OPTIMA THAN OTHER ALGAE AND SOME OF THEM, THE THERMOPHILES, AND INHABITANTS OF HOT SPRINGS AT TEMPERATURES AS HIGH AS 73-74C. IT IS NO SURPRISE, THEREFORE, THAT THERMAL POLLUTION MAY FAVOR THE GROWTH OF THE BLUE-GREEN ALGAE. WATERS FROM HOT SPRINGS GENERALLY POSSESS MUCH GREATER SOLUTE (NUTRIENT) CONCENTRATIONS THAN SURFACE WATERS AND IT IS PROBABLE THAT MANY THERMAL STREAMS CAUSE EUTROPHICATION IN THE BODIES OF WATER INTO WHICH THEY DRAIN. THE PAPER COVERS THE FOLLOWING TOPICS: DISTRIBUTION OF THERMAL WATERS; DISTRIBUTION OF SPECIES (UPPER AND LOWER TEMPERATURE LIMITS, CLASSIFICATION AND GEOGRAPHICAL DISTRIBUTION, PROBLEMS OF SURVIVAL AND TRANSPORT); STUDIES OF NATURAL POPULATIONS (MAT FORMATION AND STABILITY, MOVEMENTS OF FILAMENTS AND MATS, MEASUREMENTS OF PHOTOSYNTHESIS AND GROWTH); CULTIVATION OF THERMOPHILIC CYANOPHYTES (MEDIUMS AND NUTRITION, ISOLATION AND MAINTENANCE, RATES OF GROWTH, RATES OF PHOTOSYNTHESIS, AND RESPIRATION RATES IN CULTURE); RESPONSES TO TEMPERATURE AND LIGHT INTENSITY (OPTIMAL TEMPERATURE AND LIGHT INTENSITY, EFFECTS OF LIGHT AND TEMPERATURE ON PIGMENTATION, GROWTH AND SURVIVAL AT SUBOPTIMAL TEMPERATURES). (GERHOLD-WISCONSIN)

FIELD 05B

ACCESSION NO. W70-05270

DDT REDUCES PHOTOSYNTHESIS BY MARINE PHYTOPLANKTON,

STATE UNIV. OF NEW YORK, STONY BROOK. DEPT. OF ELECTRICAL ENGINEERING.

CHARLES F. WURSTER, JR.

SCIENCE, VOL 159, P 1474-1475, MARCH 1968. 2 FIG, 21 REF.

DESCRIPTORS:
*WATER POLLUTION EFFECTS, *PHOTOSYNTHESIS, *PHYTOPLANKTON, *DDT, *MARINE ALGAE, TOXICITY, DIATOMS, ESTUARIES, PHYSIOLOGICAL ECOLOGY, MASSACHUSETTS, CHLOROPHYTA, DINOFLAGELLATES, ALGAE, CALIFORNIA, FLORIDA, PESTICIDE RESIDUES, ECOSYSTEMS, FOOD CHAINS, BIOLOGICAL COMMUNITIES, BIOASSAYS, EUTROPHICATION.

IDENTIFIERS:
SKELETONEMA COSTATUM, COCCOLITHUS HUXLEYI, PYRAMIMONAS, PERIDINIUM TROCHOIDEUM, VINEYARD SOUND(MASS), WOODS HOLE(MASS). RADIOCARBON UPTAKE TECHNIQUES, MICROBIAL ECOLOGY, COCCOLITHOPHORES, SEAWATER MEDIUM F.

ABSTRACT:
WIDELY DISTRIBUTED IN THE HYDROSPHERE, THE INSECTICIDE DDT HAS POTENTIALLY GREAT AND DIVERSE ECOLOGICAL EFFECTS. LITTLE IS KNOWN ABOUT ITS INFLUENCE ON PHYTOPLANKTON. OVER RANGE OF CONCENTRATIONS 1-500 PPB, ITS EFFECT ON PHOTOSYNTHESIS, UNDER CONTROLLED CONDITIONS OF LIGHT AND TEMPERATURE, WAS DETERMINED FOR A NATURAL POPULATION OF PHYTOPLANKTON AND FOR FOUR SPECIES OF LABORATORY CULTURED MARINE PHYTOPLANKTERS: SKELETONEMA COSTATUM (DIATOM), COCCOLITHUS HUXLEYI (COCCOLITHOPHORE), PYRAMIMONAS (GREEN ALGA), PERIDINIUM TROCHOIDEUM (DINOFLAGELLATE). WHEN PHOTOSYNTHESIS (MEASURED AS PERCENTAGE OF RADIOCARBON UPTAKE BY UNTREATED CONTROLS) IS PLOTTED AGAINST INCREASING DOSAGE OF DDT, CHARACTERISTIC SIGMOID CURVES RESULT, WITH PHOTOSYNTHESIS DECREASING APPROXIMATELY LINEARLY IN DOSAGE RANGE 1-100 PPB DDT. SIGMOID DOSE-RESPONSE CURVES APPARENTLY INDICATE ABSENCE OF THRESHOLD LEVELS BELOW WHICH NO EFFECTS OCCUR. TESTING RELATIONSHIP IN SKELETONEMA BETWEEN CELL CONCENTRATIONS AND TOXICITY REVEALED THAT PHOTOSYNTHESIS DIMINISHED TO 50% OF CONTROLS WHEN FINAL CELLULAR CONCENTRATIONS WERE REDUCED 100-FOLD. FOR CHEMICAL REASONS, EFFECTIVE DOSES OF DDT IN THESE EXPERIMENTS MAY BE LOWER THAN QUANTITIES ADDED INITIALLY. MOREOVER, BECAUSE DDT BECOMES AVAILABLE NATURALLY AT STEADY RATES, GIVEN CONCENTRATION IN NATURE WOULD HAVE GREATER ECOLOGICAL SIGNIFICANCE THAN SAME INITIAL EXPERIMENTAL CONCENTRATION. SELECTIVE TOXIC STRESS BY DDT MAY CREATE FLORAL IMBALANCE, THEREBY AGGRAVATING LESS DESIRABLE MANIFESTATION OF EUTROPHICATION. (EICHHORN-WISCONSIN)

FIELD 05C

ACCESSION NO. W70-05272

REVIEW ON TOXIGENIC ALGAE,

HEBREW UNIV., JERUSALEM (ISRAEL).

MOSHE SHILO.

VERHANDLUNGEN DER INTERNATIONALEN VEREINIGUNG FUR THEORESTISCHE UND
ANGEWANDTE LIMNOLOGIE, VOL 15, P 782-795, 1964. 90 REF, DISCUSSION.

DESCRIPTORS:
*TOXICITY, *ALGAE, *REVIEWS, HABITATS, CLAMS, MUSSELS, SHELLFISH,
DINOFLAGELLATES, CYANOPHYTA, CHRYSOPHYTA, CALIFORNIA, UNITED STATES,
BRACKISH WATER, ESTUARIES, BACTERIA, MOLLUSKS, VITAMIN B, RED TIDE,
SALINITY, FLORIDA, TEXAS, PHOSPHORUS, MIGRATION, DIURNAL, ALGICIDES,
CLOSTRIDIUM, FISH, AMPHIBIANS, ANABAENA, CYTOLOGICAL STUDIES,
PHYSICOCHEMICAL PROPERTIES.

IDENTIFIERS:
GONYAULAX, GONYAULAX VENIFICUM, PYRODINIUM, GYMNODINIUM BREVE, CANADA,
EUROPE, SOUTH AFRICA, ISRAEL, MICROCYSTIS, PRYMNESIUM PARVUM,
SYNERGISTIC EFFECTS, AXENIC CULTURES, MONOALGAL, PORTUGAL, TAPES,
JAPAN, AMPHIDINIUM, BLOOMS, MYTILUS, TRICHODESMIUM ERYTHREUM,
SAXIDOMUS, CLOSTRIDIUM BOTULINUS, VENERUPIS, OSTREA, ARTHROPODS.

ABSTRACT:
TOXIGENIC ALGAE ARE FOUND IN MARINE, BRACKISH AND FRESH WATER
WORLDWIDE. HIGH CONCENTRATIONS OF TOXIN-FORMING ORGANISMS, STORED IN
CLAMS, MUSSELS, AND SHELLFISH CAUSE FOOD POISONING. DIFFERENT TOXONOMIC
GROUPS CONTAIN TOXIN-FORMING ALGAE--SOME DINOFLAGELLATES, CERTAIN
BLUE-GREEN ALGAE, AND ONE SPECIES OF CHRYSOPHYCEAE. RELATIONSHIP
BETWEEN ALGAL APPEARANCE AND TOXIC PHENOMENA DID NOT PROVE
UNEQUIVOCALLY THAT TOXIN-FORMATION IS CAUSED BY ALGAE THEMSELVES. MANY
FORMS IN TOXIC TIDES AND BLOOMS ARE NOW AVAILABLE IN AXENIC CULTURE.
MASS DEVELOPMENT OF TOXIC ALGAE BEING SPORADIC, INVESTIGATORS SEEK FOR
CRITICAL COMBINATION OF PHYSICOCHEMICAL FACTORS TRIGGERING THEM. BLOOMS
HAVE BEEN ASSOCIATED WITH UPWELLING OF DEEP, COOL, NUTRIENT-RICH WATER,
HEAVY RAINFALL WITH RIVER DISCHARGE, SEWAGE POLLUTION, PHOSPHORUS
CONCENTRATIONS, GROWTH-PROMOTING SUBSTANCES FORMED BY MICROORGANISMS,
SALINITY, ACTIVE OR PASSIVE CONCENTRATIONS OF THE CELLS AFTER GROWTH.
PROPERTIES AND BEHAVIOR OF ALGAL TOXINS ARE DESCRIBED. LACK OF
CORRELATION BETWEEN CONCENTRATION OF POTENTIALLY TOXIC ALGAE AND LEVEL
OF TOXINS IN WATER IS EMPHASIZED. INTRA-AND EXTRA-CELLULAR
CONCENTRATIONS OF TOXIC PRINCIPLES IN AXENIC PRYMNESIUM CULTURES VARIED
WITH STAGES OF GROWTH. OPTIMAL GROWTH CONDITIONS WERE NOT NECESSARILY
OPTIMAL FOR TOXIN BIOSYNTHESIS. SHIFTING CONCENTRATIONS OF CO-FACTORS,
SUCH AS STREPTOMYCIN, SPERMINE, AND OTHERS AS YET UNKNOWN MAY ALTER
TOXIC EFFECT WITHOUT CHANGES IN TOXIN LEVEL. (JONES-WISCONSIN)

FIELD 05B

ACCESSION NO. W70-05372

DISTRIBUTION AND ABUNDANCE OF PHYTOPLANKTON AND ROTIFERS IN A MAIN STEM
MISSOURI RIVER RESERVOIR,

BUREAU OF SPORT FISHERIES AND WILDLIFE, YANKTON, S. DAK. NORTH CENTRAL
RESERVOIR INVESTIGATIONS.

PATRICK L. HUDSON, AND BRUCE C. COWELL.

PROCEEDINGS OF SOUTH DAKOTA ACADEMY OF SCIENCE, VOL 45, P 84-106, 1966. 7
FIG, 5 TAB, 16 REF.

DESCRIPTORS:
*RESERVOIRS, *PHYTOPLANKTON, *ROTIFERS, *DISTRIBUTION, MISSOURI RIVER,
SOUTH DAKOTA, NEBRASKA, ALGAE, NANNOPLANKTON, DIATOMS.

IDENTIFIERS:
*LEWIS AND CLARK LAKE (S D), ABUNDANCE, BLUE-GREEN ALGAE, GREEN ALGAE,
FRAGILARIA, ASTERIONELLA, MELOSIRA, MICROCYSTIS, KERATELLA, POLYARTHRA,
ASPLANCHNA, BRANCHIONUS.

ABSTRACT:
WEEKLY SAMPLES WERE TAKEN AT TEN STATIONS IN LEWIS AND CLARK LAKE
RESERVOIR, BORDERING SOUTH DAKOTA-NEBRASKA, FROM MAY TO OCTOBER;
ADDITIONAL SAMPLES WERE OBTAINED FROM AN AUTOMATIC PLANKTON SAMPLER AT
THE DISCHARGE. MEAN NUMBERS OF NET PHYTOPLANKTON INCREASED FROM
HEADWATERS TO THE DAM; VOLUMES OF NET PHYTOPLANKTON PER LITER OF WATER
VARIED IRREGULARLY THROUGHOUT THE RESERVOIR. DATA FROM THE RESERVOIR
DISCHARGE INDICATE NANNOPLANKTON COMPRISED 84% OF PHYTOPLANKTON VOLUME.
DISCHARGE OF PHYTOPLANKTON TOTALLED 9,058 METRIC TONS FOR ONE YEAR WITH
PEAK DISCHARGES OCCURRING IN MARCH, MAY AND AUGUST. PHYTOPLANKTON
DISCHARGE DOES NOT APPEAR TO BE DIRECTLY RELATED TO VOLUME OF WATER
DISCHARGED. DIATOMS WERE DOMINANT PHYTOPLANKTER IN ALL MONTHS EXCEPT
AUGUST (GREEN ALGAE) AND OCTOBER (BLUE-GREEN ALGAE). ABUNDANCE OF
ROTIFERS INCREASED FROM HEADWATERS DOWNSTREAM. DOMINANT ROTIFERS IN
TERMS OF MONTHS PRESENT AND MEAN NUMBERS PER LITER WERE POLYARTHRA,
ASPLANCHNA AND KERATELLA; TEN ADDITIONAL GENERA WERE NOTED. SEASONAL
ABUNDANCE FIGURES INDICATE A SINGLE PEAK IN LATE SPRING--EARLY SUMMER.
HIGHEST TOTAL MEAN VALUES OCCURRED IN APRIL AND JUNE; LOWEST TOTALS
WERE IN NOVEMBER AND JANUARY. GRAPHICAL AND TABULAR DATA INCLUDE
MONTHLY AVERAGES OF PHYTOPLANKTON AND ROTIFER DENSITIES.
(VOIGTLANDER-WISCONSIN)

FIELD 02H, 05C

ACCESSION NO. W70-05375

AIRBORNE FLUOROMETER APPLICABLE TO MARINE AND ESTUARINE STUDIES,

GEOLOGICAL SURVEY, WASHINGTON, D.C.; AND PERKIN-ELMER CORP., NORWALK, CONN.

GEORGE E. STOERTZ, WILLIAM R. HEMPHILL, AND DAVID A. MARKLE.

MARINE TECHNOLOGY SOCIETY JOURNAL, VOL 3, NO 6, P 11-26, 1969. 24 FIG, 16 REF.

DESCRIPTORS:
*INSTRUMENTATION, *SAMPLING, *MEASUREMENT, *FLUOROMETRY, DYE RELEASES,
CURRENTS(WATER), REMOTE SENSING, SEA WATER, CHLOROPHYLL, OIL, ALGAE,
PACIFIC OCEAN.

IDENTIFIERS:
*FRAUNHOFER LINE DISCRIMINATOR, RHODAMINE WT DYE, AIRBORNE TESTS, OIL
SPILLS, GEOLOGICAL SS POLARIS, CHLOROPHYLL FLUORESCENCE, FABRY-PEROT
FILTERS, SOLAR FRAUNHOFER LINES, ATTENUATION MEASUREMENT APPARATUS,
HYDROGEN F LINE.

ABSTRACT:
THIS STUDY WAS CONCERNED WITH PERFORMANCE AND POSSIBLE IMPROVEMENTS OF
THE FRAUNHOFER LINE DISCRIMINATOR (FLD). THIS FLUOROMETER IS THE FIRST
TO PROVIDE DIRECT MEASUREMENTS OF FLUORESCENCE IN BROAD DAYLIGHT AND
FROM A REMOTE PLATFORM, THUS GREATLY FACILITATING STUDIES OF CURRENT
FLOW AND RATE OF DISPERSION OF RHODAMINE AND OTHER TAGGING DYES. THE
SMALLEST DETECTABLE INCREMENT OF RHODAMINE WT IN 0.5 METER DEPTH WAS 1
PPB. TEMPERATURE SIGNIFICANTLY INFLUENCED THE FLD RESULTS. SHIPBOARD
AND HELICOPTER TESTS OF THE FLD AND PLOTTING OF FLIGHT PATHS BY
TRACKING RADAR WERE SUCCESSFUL. THE DYE WAS DETECTABLE IN AIRBORNE
TESTS AT ALTITUDES EXCEEDING 5000 FEET. A NEW DESIGN HAVING ONLY ONE
PHOTOMULTIPLIER SHOULD ELIMINATE THE NEED FOR MONITORING STANDARD
TARGETS. ADDITIONAL IMPROVEMENTS MAY ENABLE DETECTION OF CHLOROPHYLL
FLUORESCENCE, OIL SPILLS OR LEAKAGES, ALGAL BLOOMS AND INDICATION OF
POLLUTION, AND FACILITATE DISPERSION STUDIES. (WILDE-WISCONSIN)

FIELD 07B, 05A

ACCESSION NO. W70-05377

THE ADAPTATION OF PLANKTON ALGAE. IV. LIGHT ADAPTATION IN DIFFERENT ALGAL
  SPECIES,

ROYAL DANISH SCHOOL OF PHARMACY, COPENHAGEN. DEPT. OF BOTANY.

ERIK G. JORGENSEN.

PHYSIOLOGIA PLANTARUM, VOL 22, NO 6, P 1307-1315, 1969. 4 FIG, 2 TAB, 15 REF.

DESCRIPTORS:
    *PLANKTON, *ALGAE, *LIGHT INTENSITY, CHLORELLA, CHLOROPHYTA, DIATOMS,
    PIGMENTS, CHLOROPHYLL, PHOTOSYNTHESIS, GROWTH RATE, ORGANIC MATTER,
    SCENEDESMUS, CHLAMYDOMONAS, FRESHWATER, MARINE ALGAE, ENZYMES.

IDENTIFIERS:
    *ADAPTATION, *SPECIES, CYCLOTELLA MENEGHINIANA, LIGHT-SATURATED RATE,
    CELL VOLUME, NITZSCHIA CLOSTERIUM, NITZSCHIA PALEA, CHLOROPLASTS,
    SCENEDESMUS OBLIQUUS, SCENEDESMUS QUADRICAUDA, CHLORELLA PYRENOIDOSA,
    CHLORELLA VULGARIS, MONODUS SUBTERRANEUS, ANKISTRODESMUS FALCATUS,
    CHLAMYDOMONAS MOEWUSII, SKELETONEMA COSTATUM, SYNECHOCOCCUS ELONGATUS.

ABSTRACT:
    TWO TYPES OF PLANKTON ALGAE ADAPT TO LIGHT INTENSITIES
    DIFFERENTLY--CHLORELLA AND CYCLOTELLA. THE CHLORELLA TYPE IS MOSTLY
    FOUND AMONG GREEN ALGAE AND ADAPTS TO A NEW LIGHT INTENSITY CHIEFLY BY
    CHANGING THE PIGMENT CONTENT. THE CELLS ADAPTED TO A HIGH LIGHT
    INTENSITY HAVE A LOWER CHLOROPHYLL A CONTENT PER CELL THAN CELLS
    ADAPTED TO LOW LIGHT INTENSITY, THE LATTER WITH RATHER LOW LIGHT
    SATURATION. THE LIGHT-SATURATED RATE OF PHOTOSYNTHESIS IS USUALLY LOWER
    FOR CELLS ADAPTED TO A HIGH LIGHT INTENSITY THAN FOR CELLS ADAPTED TO A
    LOW LIGHT INTENSITY. ACTUAL PHOTOSYNTHESIS (PHOTOSYNTHESIS AT THE LIGHT
    INTENSITY WHERE THE CELLS ARE GROWN) IS NOT MUCH HIGHER AT HIGH LIGHT
    INTENSITY THAN AT LOW. THE CYCLOTELLA, PREVALENT AMONG DIATOMS, ADAPTS
    ONLY BY CHANGING THE LIGHT-SATURATION RATE. CHLOROPHYLL CONTENT IS THE
    SAME IN CELLS GROWN AT LOW AND HIGH INTENSITIES. LIGHT SATURATION FOR
    CELLS GROWN AT A LOW LIGHT INTENSITY IS RATHER HIGH. THE
    LIGHT-SATURATED RATE IS MUCH HIGHER IN THE CASE EXAMINED AT HIGH
    INTENSITY THAN AT LOW. ACTUAL PHOTOSYNTHESIS IS CONSIDERABLY HIGHER FOR
    CELLS GROWN AT HIGH LIGHT INTENSITIES OPPOSED TO LOW. TRANSITIONAL
    TYPES OCCUR. (JONES-WISCONSIN)

FIELD 02K, 05C

ACCESSION NO. W70-05381

ALGAE AND PHOTOSYNTHESIS IN SHAGAWA LAKE, MINNESOTA,

MINNESOTA UNIV., MINNEAPOLIS. LIMNOLOGICAL RESEARCH CENTER.

ROBERT O. MEGARD.

UNIVERSITY OF MINNESOTA, MINNEAPOLIS, LIMNOLOGICAL RESEARCH CENTER, INTERIM REPORT NO 5, APRIL 1969. 5 FIG, 10 TAB, 8 REF.

DESCRIPTORS:
*ALGAE, *PHOTOSYNTHESIS, *MINNESOTA, PHYTOPLANKTON, CYANOPHYTA, DIATOMS, STRATIFICATION, OXYGEN, HYDROGEN ION CONCENTRATION, TEMPERATURE, DEPTH, DENSITY, CHLOROPHYLL, ICE, SEASONAL, DIURNAL, LIGHT INTENSITY, CARBON, SEWAGE TREATMENT, TERTIARY TREATMENT, EFFLUENTS, LIMNOLOGY, WATER CHEMISTRY, RESPIRATION, BATHYMETRY, SALINITY, PHOSPHORUS, PRODUCTIVITY, CHLOROPHYTA, CHRYSOPHYTA, SCENEDESMUS, STANDING CROP.

IDENTIFIERS:
*SHAGAWA LAKE(MINN), APHANIZOMENON FLOS AQUAE, STEPHANODISCUS ASTREA, LAKE MINNETONKA(MINN), MORPHOMETRY, ALGAE BLOOMS, CLEARWATER LAKE(MINN), TROUT LAKE(MINN), KIMBALL LAKE(MINN), CRYPTOPHYTA, PYROPHYTA, CHROOMONAS ACUTA, SCHROEDERIA SETIGERA, MELOSIRA AMBIGUA, CERATIUM HIRUDINELLA, ANKISTRODESMUS, GLENODINIUM PULVISCULUS, SCENEDESMUS QUADRICAUDA.

ABSTRACT:
SHAGAWA LAKE (MINNESOTA) PHYTOPLANKTON WAS DOMINATED BY APHANIZOMENON FLOS AQUAE IN SUMMER AND STEPHANODISCUS ASTREA VAR MINUTULA IN WINTER, MAINLY CONCENTRATED IN DEPTHS ABOVE 4 METERS. CHLOROPHYLL CONTENT OF APHANIZOMENON FILAMENTS IN DIFFERENT PARTS AT THE LAKE SURFACE WERE SIMILAR ON ANY DATE BUT DECREASED 40% IN TWO WEEKS, A POSSIBLE RESPONSE TO RECENT PHOTIC HISTORY OR POPULATION AGING. CHLOROPHYLL CONCENTRATIONS AND ALGAL DENSITIES BENEATH MARCH ICE WERE GREATER THAN IN SOME PARTS OF THE LAKE DURING SUMMER. WHEN ALGAE FROM DIFFERENT LOCALITIES WERE INCUBATED AT 0.5 METER DEPTH (LIGHT INTENSITIES OPTIMAL FOR PHOTOSYNTHESIS ON SUNNY DAYS), AVERAGE DAILY RATE OF PHOTOSYNTHETIC CARBON ASSIMILATION WAS 42 MICROGRAMS CARBON/MICROGRAM CHLOROPHYLL, IDENTICAL TO THE RATE AT THE SAME TEMPERATURE IN LAKE MINNETONKA. THE DENSE ALGAL POPULATIONS IN SURFACE WATERS ABSORBED SO MUCH LIGHT THAT THERE WAS LITTLE PHOTOSYNTHESIS BELOW 3 METERS. ALGAL PHOTOSYNTHETIC CAPACITY BENEATH ALMOST ONE METER CLEAR ICE NEARLY EQUALLED THAT OF CLOUDY SUMMER DAYS. EFFECTS OF ALGAL PHOTOSYNTHETIC CAPACITY WERE MEASURED ON VARYING AMOUNTS OF EFFLUENT FROM SECONDARY AND TERTIARY WASTE TREATMENT PLANTS. PHOTOSYNTHESIS WAS STIMULATED SLIGHTLY BY TERTIARY EFFLUENT, BUT HALTED BY 20% MUNICIPAL WATER. (SEE ALSO W69-00659, W69-10167). (JONES-WISCONSIN)

FIELD 02H, 05C

ACCESSION NO. W70-05387

ALGAE IN WATER SUPPLIES,

ROBERT A. TAFT SANITARY ENGINEERING CENTER, CINCINNATI, OHIO.

C. MERVIN PALMER.

U S PUBLIC HEALTH SERVICE PUBLICATION NO 657, 1959. 88 P. 55 FIG, 13 TAB, 170 REF, 6 PLATES, APPENDIX.

DESCRIPTORS:
*ALGAE, *WATER SUPPLY, *CONTROL, TASTE, ODOR, FILTERS, PLANKTON, WATER POLLUTION, SURFACE WATERS, RESERVOIRS, CONTROL, BIBLIOGRAPHIES, ANALYTICAL TECHNIQUES, DIATOMS, SAMPLING, PHYSICOCHEMICAL PROPERTIES, CENTRIFUGATION, MICROSCOPY, NETS, NANNOPLANKTON, PIGMENTS, CHLOROPHYTA, CYANOPHYTA, CHRYSOPHYTA, RHODOPHYTA, EUGLENOPHYTA, SLIME, COLOR, CORROSION, DISTRIBUTION SYSTEMS, COAGULATION, WATER SOFTENING, TOXICITY, MARINE ALGAE, FRESH WATER, PARASITISM, WATER TREATMENT, INDICATORS, WASTE TREATMENT, COPPER SULFATE, ALGICIDES, PHOTOSYNTHESIS.

IDENTIFIERS:
*IDENTIFICATION, *SIGNIFICANCE, ALGAL KEY, CLUMP COUNT, DESMIDS, FLAGELLATES, MESOPLANKTON, MICROPLANKTON, PSEUDOPLANKTON, CRYPTOPHYCEAE, VOLVOCALES, INDUSTRIAL USES, CLEAN WATER, GENUS, SPECIES, TOUCH SENSATION.

ABSTRACT:
CONTINUED SURVEILLANCE OF ALGAL POPULATIONS UNDER A PLANNED AND UNIFORM APPROACH IS NEEDED TO PUT INFORMATION IN AN UNDERSTANDABLE AND USEFUL FORM. THIS MANUAL IS AN ATTEMPT TO APPRAISE NUISANCE ORGANISM PROBLEMS AND FURNISH INFORMATION FOR REMEDYING SOME DIFFICULTIES. IT DEALS WITH THE ECOLOGY AND LIFE HISTORY OF ALGAE AND GIVES INFORMATION ON FILTER CLOGGING, MAT-FORMING ALGAE, ATTACHED FORMS, ALGICIDES, AND ALGAL CONTROL. LARGE QUANTITIES OF ALGAL MATERIAL CAUSE SERIOUS DIFFICULTIES IN WATER TREATMENT PLANTS. CERTAIN DIATOMS ALMOST INVARIABLY REDUCE THE LENGTH OF FILTER RUNS. SYNURA, IS A TASTE AND ODOR PRODUCER. LOW CONCENTRATIONS OF MOST ALGAE ARE OFTEN AN ASSET IN RAW WATERS. A SIMPLIFIED KEY OF 289 SPECIES WITH DRAWINGS OF MOST IMPORTANT IS GIVEN. THE ALGAE ARE CONSIDERED AND DISPLAYED ACCORDING TO THEIR SIGNIFICANCE TO SANITARY SCIENTISTS AND TECHNICIANS RATHER THAN WITH REGARD TO THEIR EVOLUTIONARY RELATIONSHIP. MOST ALGAE OF IMPORTANCE IN WATER SUPPLIES MAY BE CHARACTERIZED IN FOUR GENERAL GROUPS: BLUE-GREEN ALGAE, GREEN ALGAE, DIATOMS, AND PIGMENTED FLAGELLATES. A FEW MISCELLANEOUS FORMS DO NOT FIT INTO THESE FOUR GROUPS. POLLUTED WATER ALGAE, CLEAN WATER ALGAE, PLANKTON, AND OTHER SURFACE WATER ALGAE ARE INCLUDED.
(JONES-WISCONSIN)

FIELD 05F

ACCESSION NO. W70-05389

ON THE THEORY OF ADDING NUTRIENTS TO LAKES WITH THE OBJECT OF INCREASING TROUT
PRODUCTION,

DALHOUSIE UNIV., HALIFAX (NOVA SCOTIA). ZOOLOGICAL LAB.

F. R. HAYES.

CANADIAN FISH CULTURIST, VOL 10, P 32-37, 1951. 13 REF.

DESCRIPTORS:
 *NUTRIENTS, *LAKES, *TROUT, *FISH MANAGEMENT, FERTILIZATION, ALGAE,
 PLANKTON, NITROGEN, PHOSPHORUS, POTASSIUM, SELECTIVITY, POISONS,
 OXYGEN, TRACERS, PHOSPHORUS RADIOISOTOPES, IONS, HUMIC ACIDS, CARP,
 ICE, IRON COMPOUNDS, HYDROGEN ION CONCENTRATION, ACIDITY, MUD, HYDROGEN
 SULFIDE.

IDENTIFIERS:
 PRINCE EDWARD ISLAND(CANADA), NOVA SCOTIA(CANADA), PENGUINS, LAKE
 ZURICH(SWITZERLAND), WHITEFISH, PUNCHBOWL LAKE(CANADA).

ABSTRACT:
 WHEN NUTRIENT SALTS ARE ADDED TO LAKES FOR FERTILIZATION, 90% MAY
 DISAPPEAR IN AN EXCHANGE REACTION, WITHIN A FEW WEEKS. THE REMAINING
 10% IS LIKELY TO REMAIN IN OR AVAILABLE TO THE WATER CONTRIBUTING TO
 ALGAL GROWTH, LEADING TO ANIMAL PLANKTON GROWTH, AND THENCE TO FISH
 GROWTH. NONE OF THE EXTRA MATERIAL IS PROBABLY SEEN DURING THE SECOND
 SUMMER. WHILE INFORMATION IS LESS COMPLETE REGARDING POTASSIUM AND
 NITROGEN, THAN PHOSPHORUS, THE VIEW IS THAT ABOUT THE SAME PERCENTAGE
 OF EACH WOULD SURVIVE IN WATER AFTER EQUILIBRATION. AS A CONSERVATION
 MEASURE, FERTILIZATION OF LAKES DOES NOT APPEAR TO BE AN ECONOMICALLY
 FEASIBLE PROCEDURE, ALTHOUGH IT CONSTITUTES AN INTERESTING EXPERIMENT
 FOR A PRIVATELY CONTROLLED LAKE. IF THERE ARE NO COARSE FISH TO REPLACE
 TROUT UNDER HIGH FERTILITY CONDITIONS, LARGE TROUT YIELDS MAY BE
 EXPECTED. THIS SEEMS THE SITUATION IN PRINCE EDWARD ISLAND.
 ALTERNATIVELY COARSE FISH MIGHT BE SELECTIVELY POISONED IN CONJUNCTION
 WITH A FERTILIZATION PROGRAM. SELECTIVE POISONING, BY REMOVING COARSE
 FISH FROM SHALLOWS WHILE LEAVING TROUT IN THE DEPTHS, HAS BEEN FOUND
 PROMISING. IT MAY BE THAT TROUT SURVIVE WHERE THEY DO, BECAUSE THEY CAN
 TOLERATE MORE ADVERSE CONDITIONS THAN COMPETING FISH. (JONES-WISCONSIN)

FIELD 02H, 05C

ACCESSION NO. W70-05399

LIMNOLOGICAL RELATIONS OF INDIAN INLAND WATERS WITH SPECIAL REFERENCE TO
WATERBLOOMS,

UNIVERSITY COLL., LONDON (ENGLAND). DEPT. OF BOTANY.

R. N. SINGH.

VERHANDLUNGEN DER INTERNATIONALEN VEREINIGUNG FUR THEORETISCHE UND ANGEWANDTE
LIMNOLOGIE, VOL 12, P 831-836, 1953. 16 REF.

DESCRIPTORS:
*LIMNOLOGY, *WATER, *RIVERS, *LAKES, *PONDS, LIGHT, TEMPERATURE,
FERTILIZATION, FISH, NUISANCE ALGAE, ODOR, TASTE, DEPTH, CIRCULATION,
DIURNAL, WINDS, CURRENTS(WATER), WATER POLLUTION, DRAINAGE, ORGANIC
MATTER, SEWAGE, PHYTOPLANKTON, PRODUCTIVITY, CHLOROPHYTA, HYDROGEN ION
CONCENTRATION, NITRATES, PHOSPHATES, CYANOPHYTA, OXIDATION-REDUCTION
POTENTIAL, BOTTOM SEDIMENTS, EUGLENOPHYTA, TOXICITY.

IDENTIFIERS:
*INDIA, *BLOOMS, RIVER GANGES(INDIA), RAJGHAT(INDIA), BOMBAY(INDIA),
BENARES(INDIA), MACROPHYTES, CLADOPHORA, OEDOGONIA, SPIROGYRA,
MICROCYSTIS AERUGINOSA, ANABAENOPSIS, RACIBORSKII, RAPHIDIOPSIS,
ANABAENA, WOLLEA, OSCILLATORIA, SPIRULINA, VOLVOCALES, GLENODINIUM,
TRACHELOMONAS.

ABSTRACT:
INDIAN WATERS ARE INFESTED WITH WATER BLOOMS. EXPERIMENTS WITH BLOOM
TYPES AND THEIR EFFECT ON ANIMALS PROVED THAT INDIAN WATER BLOOMS ARE
NEVER TOXIC ALTHOUGH THE SAME ALGAL SPECIES ARE POISONOUS IN OTHER
COUNTRIES. BLUE-GREEN ALGAE ENCOURAGE GROWTH OF FISH WHICH FEED UPON
THEM DIRECTLY OR INDIRECTLY, PROVIDING AN EXCELLENT SOURCE OF
FERTILIZATION OF A POND. THE BLOOM CAN ALSO BE UTILIZED AS GREEN
MANURE; DRIED BOTTOM MUDS ARE USED AS A MANURE SOURCE. THE ALGAE BLOOMS
ARE INITAILLY SAPROPHYTIC--SIGNIFICANT IN VIEW OF THEIR SUDDEN
APPEARANCE AFTER A LONG ABSENCE. IN THE GANGES, THE ALGAL FLORA IS
TYPICALLY CHLOROPHYCEAN, BUT A MARKED CHANGE OCCURS WHEN THE RIVER
RECEIVES SEWAGE OR EFFLUENT, THEN A MYXOPHYCEAN BLOOM IS FORMED.
SAMPLES OF ALGAE BLOOM DRIED FOR FERTILIZATION OF HATCHERY PONDS HAVE
BEEN PREPARED WHICH CAN BE INOCULATED AND GROWN SUCCESSFULLY. A
BIOLOGICAL METHOD OF CONTROL BY INTRODUCING FISH AND CRUSTACEANS FOR
FOOD UTILIZATION MIGHT BE ENVISAGED. WHEN BLOOM DISINTEGRATION HAS
REACHED A MAXIMUM AND TEMPERATURE IS HIGH, A POND MAY BE COVERED WITH
DEAD FISHES INVOLVING ECONOMIC LOSS. (JONES-WISCONSIN)

FIELD 02H, 05C

ACCESSION NO. W70-05404

THE EUTROPHICATION OF LAKE KLOPEINER (IN GERMAN),

BIOLOGISCHE STATION LUNZ (AUSTRIA).

I. FINDENEGG.

OESTERREICHISCHE WASSERWIRTSCHAFT, VOL 17, NO 7-8, P 175-181, 1965. 4 FIG, 3 TAB.

DESCRIPTORS:
*EUTROPHICATION, TEMPERATURE STRATIFICATION, DEPTH, PLANKTON, LIGHT PENETRATION, SEASONAL, PRIMARY PRODUCTIVITY, CALCIUM, NITRATES, AMMONIA, PHOSPHATES, HYDROGEN ION CONCENTRATION, PHYTOPLANKTON, ODOR, WINDS, EPILIMNION, HYPOLIMNION, LIGHT INTENSITY, CONDUCTIVITY, ALGAE, SEWAGE, DIATOMS, CYANOPHYTA, CARBON RADIOISOTOPES, POTASSIUM, OXYGEN.

IDENTIFIERS:
*LAKE KLOPEINER(AUSTRIA), LAKE WORTHER(AUSTRIA), VERTICAL, CHALK, METALIMNION, OSCILLATORIA RUBESCENS, DINOBRYON, UROGLENA AMERICANA, CYCLOTELLA, CERATIUM HIRUNDINELLA, PERIDINIUM WILLEI, CHLOROCOCCALES, ANKISTRODESMUS, OOCYSTIS, CRYOMONAS EROSA, COSMARIUM, CHROCOCCUS LIMNETICUS, LYNGBYA LIMNETICA, ANABAENA FLOS AQUAE, TABELLARIA, MELOSIRA, FRAGILARIA CROTONENSIS, APHANIZOMENON.

ABSTRACT:
LAKE KLOPEINER, AUSTRIA, DESPITE HEAVY SEWAGE INFLUX, HAS NOT PROGRESSED FAR IN EUTROPHICATION, ALTHOUGH CONSIDERABLE AMOUNTS OF NUTRIENTS HAVE ACCUMULATED IN THE DEPTHS AS A RESULT OF MEROMIXIS. THIS MEROMICTIC BEHAVIOR IS THE KEY TO UNDERSTANDING THE STRATIFICATION OF OXYGEN AND ITS CHANGES SINCE 1931. THE ZONE FREE OF OXYGEN EXPANDED UPWARD DURING THE COURSE OF THE DECADES AND LAY BETWEEN 20 AND 35 METERS IN 1935 AND IS NOW ELEVATED TO A LEVEL BETWEEN 10-25 METERS. THIS MEROMICTIC BEHAVIOR IN WINTER AND THE STRICT SEPARATION OF BOTH EPI- AND HYPOLIMNION IN SUMMER GIVE A SPECIAL CHARACTER TO THE COURSE OF EUTROPHICATION. THE LAKE HAS AN UNUSUAL THERMIC POSITION: SELDOM DISTURBED BY WINDS AND THEN ONLY FOR SHORT PERIODS, IT HAS NO STRONG CURRENTS; LOCAL BREEZES DISTRIBUTE THE SURFACE WATER, WARMED BY SUNSHINE, ONLY WITHIN A THIN LAYER OF EPILIMNION. AN INCREASE IN AMOUNT OF ALGAE IS DUE TO THE IMMIGRATION OF A SPECIES OF PLANKTON, THE BLOOD ALGA (OSCILLATORIA RUBESCENS), NOT OBSERVED BEFORE. THIS SPECIES IS REGARDED AS AN INDICATOR OF INCREASING EUTROPHICATION ACCELERATED BY DOMESTIC SEWAGE. THERE IS ALSO A DECREASE OF TRANSPARENCY.
(JONES-WISCONSIN)

FIELD 02H, 05C

ACCESSION NO. W70-05405

ENVIRONMENTAL REQUIREMENTS OF FRESH-WATER PLANKTON ALGAE,

UPPSALA UNIV. (SWEDEN). INST. FOR PHYSIOLOGICAL BOTANY.

WILHELM RODHE.

SYMBOLAE BOTANICAE UPSALIENSES, VOL 10, NO 1, P 1-149, 1948. 33 FIG, 30 TAB, 96 REF.

DESCRIPTORS:
*ENVIRONMENT, *FRESH WATER, *PLANKTON, *ALGAE, SCENEDESMUS, LIGHT, CULTURES, IRON COMPOUNDS, DIATOMS, CYANOPHYTA, PHOSPHATES, CHLOROPHYLL, MAGNESIUM, POTASSIUM, TEMPERATURE, EPILIMNION, LIGHT INTENSITY, SEASONAL, PHYSIOLOGICAL ECOLOGY, ECOTYPES, CHLOROPHYTA, PHOTOMETRY, CHLORELLA, NUTRIENTS, TRACE ELEMENTS, CHLAMYDOMONAS, CHRYSOPHYTA, NITRATES, PLANT GROWTH, BICARBONATES, DEPTH.

IDENTIFIERS:
*REQUIREMENTS, SCENEDESMUS, ABSORPTION, CELL CONCENTRATION, UROGLENA, ANKISTRODESMUS, CHLOROCOCCALES, VOLVOCALES, CONJUGATAE, HETEROKONTAE, GLOEOTRICHIA, ASTERIONELLA, DINOBRYON, MELOSIRA, ANEBODA(SWEDEN), UPPSALA(SWEDEN), LAKE ERKEN(SWEDEN), EXTINCTION VALUES, EUDORINA, CULTURE MEDIUM, COELASTRUM, PEDIASTRUM.

ABSTRACT:
THIS WORK ELUCIDATES THE INFLUENCE OF SOME ENVIRONMENTAL FACTORS ON FRESHWATER PLANKTON ALGAE. THE REQUIREMENTS OF SOME ALGAE, ESPECIALLY SCENEDESMUS QUADRICAUDA, WERE STUDIED IN CULTURES UNDER CONTROLLED CONDITIONS AND COMPARED WITH LAKE CONDITIONS. SUBSTANCE PRODUCTION, CELL MULTIPLICATION, AND CHLOROPHYLL FORMATION ARE RATHER INDEPENDENT ASPECTS OF THE DEVELOPMENT OF A CULTURE. PARALLEL DETERMINATIONS SHOULD BE CARRIED OUT FOR THE DIFFERENT PARAMETERS. ALPHA A-DIPYRIDYL AND O-PHENANTHROLINE ARE SENSITIVE AND RELIABLE REAGENTS FOR DETERMINATION OF IRON IN WATER. IN LIGHT, ONE PART CRITRIC ACID TO ONE PART FERRIC CITRATE HAS A STABILIZING EFFECT SUFFICIENT FOR ALGAL CULTURES. A CULTURE SOLUTION WAS COMPOSED SUITABLE FOR UNLIMITED CULTIVATION OF ABOUT 40 SPECIES AND FORMS BELONGING TO CHLOROCOCCALES, VOLVOCALES, CONJUGATAE, HETEROKONTAE, AND DIATOMS. THE PHOSPHATE DEPENDENCE OF GROWTH CAN DIFFER FOR VARIOUS PLANKTON ALGAE. EXPERIMENTAL RESULTS WITH DINOBRYON AND UROGLENA SUGGEST THAT PHOSPHORUS, EVEN IN ORDINARY LAKES, MAY BECOME A MAXIMUM FACTOR FOR SOME ALGAE; ECOLOGICAL OBSERVATIONS CONFIRM THIS SUGGESTION. THE COMPETITION OF ALGAE FOR AVAILABLE NITROGEN IONS IS A FACTOR IN NATURAL CONDITIONS WITH DIFFERENT ALGAL DENSITIES. IN AVERAGE LAKES MAGNESIUM AND POTASSIUM NEVER LIMIT THE DEVELOPMENT OF SCENEDESMUS QUADRICAUDA. MELOSIRA ISLANDICA SSP HELVETICA PREFERS LOW TEMPERATURES; ANKISTRODESMUS FALCATUS HIGHER TEMPERATURES. (JONES-WISCONSIN)

FIELD 02H, 05C

ACCESSION NO. W70-05409

PHOSPHATE FERTILIZATION OF PONDS,

ALABAMA AGRICULTURAL EXPERIMENT STATION, AUBURN.

H. S. SWINGLE, B. C. GOOCH, AND H. R. RABANAL.

PROCEEDINGS OF THE 17TH ANNUAL CONFERENCE, SOUTHEASTERN ASSOCIATION OF GAME AND FISH COMMISSIONERS, HELD AT HOT SPRINGS, ARKANSAS, SEPTEMBER 29-OCTOBER 2, P 213-218, 1963. 1 TAB, 14 REF.

DESCRIPTORS:
*PONDS, *FERTILIZATION, *PHOSPHATE, FISH, PLANKTON, AQUATIC WEED CONTROL, INSECT CONTROL, MOSQUITOS, ALGAL CONTROL, BASS, CARP, CATFISHES, NITROGEN, POTASSIUM.

IDENTIFIERS:
PHOSPHATE FERTILIZERS, BLUEGILLS, MICROCYSTIS SCUMS, PITHIOPHORA ALGAE.

ABSTRACT:
REINFORCEMENT OF PHOSPHATE FERTILIZATION OF FISH PONDS WITH NITROGEN AND POTASSIUM PRODUCED NO SIGNIFICANT INCREASE IN THE BIOMASS OF PLANKTON OR FISH YIELD. IN PONDS PREVIOUSLY TREATED WITH 8-8-2 FERTILIZER, MICROCYSTIS SCUMS CONTINUED TO BE A PROBLEM EVEN WHEN FERTILIZATION WAS LIMITED TO PHOSPHATES. IN SOME INSTANCES, APPLICATION OF PHOSPHATE FERTILIZERS ALONE DELAYED RESPONSE FOR 30 TO 60 DAYS AND PERMITTED THE GROWTH OF PITHOPHORA ALGAE. (WILDE-WISCONSIN)

FIELD 02H, 05C

ACCESSION NO. W70-05417

EFFECTS OF ACID MINE WASTES ON PHYTOPLANKTON IN NORTHERN ONTARIO LAKES,

ONTARIO WATER RESOURCES COMMISSION, TORONTO.

M. G. JOHNSON, M. F. P. MICHALSKI, AND A. E. CHRISTIE.

ONTARIO WATER RESOURCES COMMISSION, DIVISION OF RESEARCH — PUBLICATION NO 30, 1969. 41 P, 5 TAB, 6 FIG, 15 REF, 3 APPEND.

DESCRIPTORS:
*ACID MINE WATER, ACIDIC WATER, HYDROGEN ION CONCENTRATION, WATER CHEMISTRY, *PHYTOPLANKTON, CARBON DIOXIDE, NUTRIENT REQUIREMENTS, POPULATIONS, ENVIRONMENTAL EFFECTS, *PRIMARY PRODUCTIVITY, WATER POLLUTION EFFECTS, BIOASSAYS, ALGAE, COMPARATIVE PRODUCTIVITY, AQUATIC PRODUCTIVITY, DISTRIBUTION, SPECIES DIVERSITY.

IDENTIFIERS:
ONTARIO LAKES, QUIRKE LAKE, PECORS LAKE, DUNLOP LAKE, *INORGANIC CARBON.

ABSTRACT:
STUDIES WERE CARRIED OUT ON THREE NORTHERN ONTARIO LAKES; TWO (QUIRKE AND PECORS LAKES) WERE CONTAMINATED BY FREE MINERAL ACIDITY AND ONE (DUNLOP LAKE) WAS UNAFFECTED. LOWER PHYTOPLANKTON POPULATIONS AND INDICES OF DIVERSITY WERE FOUND IN QUIRKE AND PECORS LAKES THAN IN DUNLOP LAKE. MANY SPECIES OF BACILLARIOPHYCEAE, CHRYSOPHYCEAE AND MYXOPHYCEAE DEVELOPED IN THE REFERENCE LAKE BUT WERE ABSENT OR OCCURRED IN EXTREMELY LOW NUMBERS IN THE CONTAMINATED LAKES. AVERAGE PRIMARY PRODUCTIVITIES IN DUNLOP, QUIRKE AND PECORS LAKES WERE 126, 71 AND 34 MG C M-(2) DAY-(1), RESPECTIVELY. IN SITU AREAL AND VOLUMETRIC MEASUREMENTS IN LABORATORY AND FIELD BIOASSAYS CONFIRMED THE' IMPORTANCE OF INORGANIC CARBON IN LIMITING PRIMARY PRODUCTIVITY. IT IS CONCLUDED THAT INORGANIC CARBON LIMITS PRIMARY PRODUCTIVITY IN THE LAKES CONTAMINATED BY ACID MINE WASTES, WITH THE REDUCED PH AND INORGANIC CARBON DECREASING BOTH THE NUMBER OF SPECIES AND TOTAL NUMBER OF PHYTOPLANKTON. (SJOLSETH AND KATZ-WASHINGTON)

FIELD 05C

ACCESSION NO. W70-05424

THE POTENTIAL ROLE OF FISHERY MANAGEMENT IN THE REDUCTION OF CHAOBORID MIDGE POPULATIONS AND WATER QUALITY ENHANCEMENT,

S. F. COOK, JR.

CALIFORNIA VECTOR VIEWS, VOL 15, NO 7, P 63-70, JULY 1968. 2 FIG, 11 REF.

DESCRIPTORS:
FISH MANAGEMENT, STREAM IMPROVEMENT, *WATER QUALITY CONTROL, *MIDGES, FISH FOOD ORGANISMS, NUTRIENTS, ENERGY TRANSFER, *FOOD CHAINS, FOOD WEBS, FORAGE FISH, PLANKTON, ALGAE, *ALGAL CONTROL, PREDATION, COMPETITION, *INSECT CONTROL.

IDENTIFIERS:
*FISHERY MANAGEMENT, *CHAOBORID MIDGES, CLEAR LAKE(CALIF), NUTRIENT ENERGY PATHWAYS, TROPHIC RELATIONSHIPS.

ABSTRACT:
INFORMATION HAS BEEN ACQUIRED INDICATING THAT AT LEAST PARTIAL BIOLOGICAL CONTROL OF CHAOBORID MIDGES AND PLANKTONIC ALGAE MAY BE POSSIBLE WITH THE USE OF FISH. TWO BASIC CONCEPTS PERTAINING TO THIS RELATIONSHIP HAVE BEEN EXPLORED; CONTROL THROUGH DIRECT INFLUENCE OF PREDATION AND CONTROL THROUGH COMPETITION FOR NUTRIENT ENERGY. THE EMPHASIS IS PLACED ON THE LATTER CONCEPT, AND MEANS FOR ITS IMPLEMENTATION ARE DISCUSSED. THE WORK HAS CENTERED PRIMARILY UPON THE TROPHIC RELATIONSHIPS BETWEEN THE MIDGES AND PLANKTON AND TWO SPECIES OF ATHERINID SILVERSIDES AND THE THREADFIN SHAD. (SJOLSETH-WASHINGTON)

FIELD 05G, 05C

ACCESSION NO. W70-05428

REMOVAL OF SELECTED CONTAMINANTS FROM WATER BY SORPTION OF COAL,

VIRGINIA POLYTECHNIC INST., BLACKSBURG. WATER RESOURCES RESEARCH CENTER.

PAUL H. KING, FRANCIS R. MCNEICE, AND PIERRE S. WARREN.

AVAILABLE FROM THE CLEARINGHOUSE AS PB-190 802, $3.00 IN PAPER COPY, $0.65 IN
MICROFICHE. VIRGINIA POLYTECHNIC INSTITUTE WATER RESOURCES RESEARCH CENTER
BULLETIN 32, NOVEMBER 1969. 75 P, 34 FIG, 3 TAB, 37 REF. OWRR PROJ NO
A-015-VA.

DESCRIPTORS:
*TERTIARY TREATMENT, *PHOSPHORUS COMPOUNDS, *NUTRIENTS, *PESTICIDES,
*WATER POLLUTION CONTROL, EUTROPHICATION, ALGAE, FILTRATION, SORPTION,
FLOCCULATION, COAL, ACTIVATED CARBON.

IDENTIFIERS:
COAL CONTACT PROCESSES, PESTICIDE REMOVAL.

ABSTRACT:
REMOVAL OR LIMITATION OF SOLUBLE PHOSPHORUS IN WASTEWATER EFFLUENTS IS
A KEY FACTOR IN ALLEVIATING THE AESTHETIC AND ECONOMIC PROBLEMS
ASSOCIATED WITH EXCESSIVE ALGAL GROWTH. VERY DILUTE CONCENTRATIONS OF
LONG-LIVED PESTICIDES CAN IN TIME BE LETHAL TO ANIMALS EXPOSED TO THEM.
IF DANGEROUS OR POTENTIALLY DANGEROUS CONCENTRATIONS OF PESTICIDES
EXIST IN WATER UTILIZED FOR HUMAN CONSUMPTION, METHODS MUST BE
DEVELOPED TO REMOVE THEM EASILY AND ECONOMICALLY. ACTIVATED CARBON IS
VERY EFFECTIVE IN ADSORBING ORGANIC COMPOUNDS FROM DILUTE SOLUTION. THE
HIGH COST OF THIS ADSORBENT, HOWEVER, OFTEN PROHIBITS ITS USE IN
LARGE-SCALE TREATMENT PROCESSES. COAL IS RELATIVELY LOW IN COST AND
THUS IS AN ATTRACTIVE FILTER MEDIA IF THE REMOVAL OF DISSOLVED MATERIAL
BY SORPTION CAN BE ACCOMPLISHED CONCURRENTLY WITH THE ELIMINATION OF
SUSPENDED SOLIDS BY FILTRATION. ALTHOUGH COAL IS A POORER SORBENT THAN
IS ACTIVATED CARBON, IT IS CONSIDERABLY MORE ECONOMICAL AND CAN BE
INCINERATED AFTER EXPOSURE TO A WASTEWATER STREAM. THE EXTENT OF UPTAKE
OF SELECTED PHOSPHORUS CONTAINING COMPOUNDS AND SYNTHETIC ORGANIC
PESTICIDES BY COAL WERE EVALUATED IN BOTH BATCH CONTINUOUS FLOW
OPERATIONS. THE KINETICS OF THE UPTAKE PROCESS AS WELL AS THE CAPACITY
OF THE COAL FOR THE SORBATE AT EQUILIBRIUM ARE REPORTED. (KNAPP-USGS)

FIELD 05D

ACCESSION NO. W70-05470

EUTROPHICATION TRENDS IN A CHAIN OF ARTIFICIAL LAKES IN MADRAS (INDIA),

HYDROLOGICAL RESEARCH STATION, KILPAUK (INDIA).

A. SREENIVASAN.

ENVIRONMENTAL HEALTH, VOL 11, P 392-401, 1969. 5 TAB, 22 REF.

DESCRIPTORS:
*RESERVOIRS, *EUTROPHICATION, NITROGEN, OXYGEN, OXIDATION, CARBON
DIOXIDE, ALKALINITY, ALGAE.

IDENTIFIERS:
INDIA, METHYL ORANGE, ALGAL BLOOMS.

ABSTRACT:
EUTROPHICATION OF FOUR CONSECUTIVE ARTIFICIAL LAKES--OOTY, SANDYNULLA,
PYKARA AND GLENMORGAN FOREBAY, INDIA,--IS DISCUSSED. MORPHOMETRIC DATA,
GENERAL CONDITIONS OF THE LAKES FOR THE YEARS 1965-1967, THE
HYDROLOGICAL CONDITIONS OF GLENMORGAN RESERVOIR, AND THE QUALITY OF
INFLUENT WATER INTO OOTY LAKE AND ITS OUTFLOW, ARE FURNISHED. DUE TO
DAMAGE OF A WASTE DISPOSAL PIPE, SEWAGE HAS DISCHARGED INTO OOTY MAKING
IT HYPER-EUTROPHIC. DURING 1962-1963, OXYGEN DEPLETION BELOW 2.0 WAS
NOTED. DURING 1965-1966 STEEP GRADIENTS IN FREE CARBON DIOXIDE AND
METHYL ORANGE ALKALINITY WERE NOTED AND THE ORGANIC NITROGEN CONTENT
REACHED 7.6-18.0 MILLIGRAMS/LITER. OOTY LAKE PLANKTON IS DENSE YEAR
ROUND WITH MICROCYSTIS AERUGINOSA OCCURRING AS A PERMANENT BLOOM.
CARBON DIOXIDE ACCUMULATES IN THE BOTTOM OF SANDYNULLA RESERVOIR, BUT
NOT AS SEVERELY AS IN OOTY. METHYL ORANGE ALKALINITY GRADIENT WAS NOTED
FROM SURFACE TO BOTTOM DURING CERTAIN PERIODS, MORE OR LESS REFLECTING
THE INFLUENCE OF OOTY LAKE. OXYGEN DEPLETION IN THE BOTTOM WAS NOTED
AND ORGANIC NITROGEN CONTENT WAS 2.1-15.0 PPM. PYKARA LAKE SHOWED NO
CHANGE. GLENMORGAN FOREBAY SHOWS THAT IT IS NOT INFLUENCED BY THE
WATERS FROM THE OTHER LAKES. THE REMOVAL OF BOTTOM SEDIMENTS FROM OOTY
IS SUGGESTED TO ALLEVIATE THE PROBLEM. (HASKINS-WISCONSIN)

FIELD 02H, 05C

ACCESSION NO. W70-05550

FACTORS INVOLVED IN DISPOSAL OF SEWAGE EFFLUENTS TO LAKES,

MASSACHUSETTS INST. OF TECH., CAMBRIDGE. DEPT. OF SANITARY CHEMISTRY.

CLAIR N. SAWYER.

SEWAGE AND INDUSTRIAL WASTES, VOL 26, NO 3, P 317-328, 1954. 6 FIG, 3 TAB, 10 REF, DISCUSSION.

DESCRIPTORS:
*LAKES, *SEWAGE EFFLUENTS, *SEWAGE DISPOSAL, ALGAE, AGRICULTURAL WATERSHEDS, NUTRIENTS, FERTILIZATION, PHYSICAL PROPERTIES, CLIMATE, NITROGEN, PHOSPHORUS.

IDENTIFIERS:
ALGAL BLOOMS, ALGAE NUTRITION, MICROCYSTIS AERUGINOSA.

ABSTRACT:
THE SIGNIFICANCE OF FERTILIZATION AS THE CAUSE OF ALGAL BLOOMING AND THE VARIOUS PHYSICAL, CHEMICAL, AND BIOCHEMICAL FACTORS THAT DETERMINE THE SAFE LOADINGS WHICH MAY BE APPLIED ARE DISCUSSED. FUNDAMENTAL CONSIDERATIONS ARE SOURCES OF FERTILIZING MATTER REACHING LAKES INCLUDING GROUNDWATER, SURFACE RUNOFF, AND SEWAGE EFFLUENTS OR INDUSTRIAL WASTES; PERMISSIBLE NUTRIENT OR FERTILIZING LEVELS; AND FACTORS CONSIDERED OF IMPORTANCE TO LAKE BEHAVIOR. CLIMATE; PHYSICAL FACTORS OF THE LAKE INCLUDING AREA, MORPHOLOGY, DEPTH, TEMPERATURE, AND COLOR; BIOLOGICAL FACTORS; AND CHEMICAL FACTORS INCLUDING ESSENTIAL ELEMENTS, CARBON DIOXIDE, PH, AND FEEDBACK FROM BOTTOM DEPOSITS ARE PRESENTED. THE CONTOUR OF THE LAKE DETERMINES TO A GREAT EXTENT WHEN AND HOW ALGAL GROWTH DEVELOPS. FOR LAKES THAT DO NOT STRATIFY, ALL NUTRIENTS ARE AVAILABLE; IN STRATIFIED LAKES, ELEMENTS BELOW THE THERMOCLINE ARE SEALED OFF FROM THE ACTIVE BIOLOGICAL ZONE. ECOLOGICAL RELATIONSHIPS ARE AN IMPORTANT FACTOR IN DETERMINING LAKE BEHAVIOR. NITROGEN IS NEEDED IN THE GREATEST AMOUNT AS A FERTILIZING ELEMENT. DISCUSSIONS OF THE PAPER BY PROFESSOR HAROLD B GOTAAS OF UNIVERSITY OF CALIFORNIA, BERKELEY, AND PROFESSOR JAMES B LACKEY OF THE UNIVERSITY OF FLORIDA, GAINESVILLE, ARE INCLUDED. (HASKINS-WISCONSIN)

FIELD 05C, 02H

ACCESSION NO. W70-05565

THE CHEMICAL CONTROL OF AQUATIC NUISANCES,

WISCONSIN COMMITTEE ON WATER POLLUTION, MADISON.

KENNETH M. MACKENTHUN.

COMMITTEE ON WATER POLLUTION, MADISON, WISCONSIN, JANUARY 1958, 64 P, 22 FIG, 3 TAB.

DESCRIPTORS:
*AQUATIC WEED CONTROL, *HERBICIDES, *COPPER SULFATE, *SODIUM ARSENITE, NUISANCE ALGAE, SNAILS, LEGISLATION, WISCONSIN, CHEMCONTROL, PONDWEEDS, ROOTED AQUATIC PLANTS, RATES OF APPLICATION, APPLICATION TECHNIQUES, SWIMMING, WATERFOWL, BIRDS, COSTS, COST-SHARING.

IDENTIFIERS:
SCHISTOSOME DERMATITIS, COPPER CARBONATE, CERCARIAE, WATER MILFOIL COONTAIL, WILD CELERY, BUR REED, DUCKWEED, WHITE WATER LILY, SWIMMERS' ITCH.

ABSTRACT:
A DETAILED DESCRIPTION OF METHODS OF CONTROL OF AQUATIC ROOTED VEGETATION, ALGAE, AND CERCARIAE, CAUSING SWIMMERS' ITCH, IS GIVEN. THE BIOMASS OF AQUATIC WEEDS WAS REDUCED BY SODIUM ARSENITE TREATMENTS IN SOME 80 LAKES WITH AN APPROXIMATE APPLICATION EFFICIENCY OF 150 GALLONS OF HERBICIDE PER HOUR. ONE OR MORE ALGAE-ERADICATING TREATMENTS WERE GIVEN TO ABOUT 40 LAKES, AN APPLICATION OF 250 POUNDS OF COMMERCIAL COPPER SULFATE CONSUMING ONE HOUR. CONTROL OF CERCARIAE WAS SUCCESSFULLY ACHIEVED IN A LIMITED NUMBER OF LAKES WITH A MIXTURE OF COPPER SULFATE AND COPPER CARBONATE. THE EFFICIENCY OF CONTROL MEASURES WAS MATERIALLY AUGMENTED BY THE USE OF SPECIALLY DESIGNED EQUIPMENT, INCLUDING STEEL BARGES, PORTABLE ALUMINUM ALLOY BARGES, SPRAY BOATS AND ASSEMBLIES, LOADING BOOMS, AND GRAVITY FLOW INJECTORS. (WILDE-WISCONSIN)

FIELD 05C

ACCESSION NO. W70-05568

WASTE WATER RECLAMATION PROJECT FOR ANTELOPE VALLEY AREA,

LOS ANGELES DEPT. OF COUNTY ENGINEER, CALIF.

JOHN A. LAMBIE.

AVAILABLE FROM THE CLEARINGHOUSE AS PB-191 067, $3.00 IN PAPER COPY, $0.65 IN MICROFICHE. LOS ANGELES COUNTY DEPARTMENT OF ENGINEER REPORT, AUGUST 1968. 225 P, 14 FIG, 32 TAB, 71 REF. FWPCA PROJECT 17080---8/68.

DESCRIPTORS:
*WATER REUSE, *RECLAIMED WATER, *CALIFORNIA, WATER QUALITY, AESTHETICS, RECREATION, FISHING, ALGAE, NUTRIENTS, TURBIDITY, TERTIARY TREATMENT, SAMPLING, MONITORING, WATER DEMAND, WATER SUPPLY, ARID LANDS, LIMNOLOGY.

IDENTIFIERS:
LOS ANGELES COUNTY(CALIF).

ABSTRACT:
AN ECONOMICALLY FEASIBLE WASTE WATER RENOVATION PROCESS WAS DEVELOPED IN LOS ANGELES COUNTY. THE TERTIARY TREATED PRODUCT WATER IS PATHOGENICALLY SAFE, ESTHETICALLY PLEASING AND SUITABLE FOR FISH LIFE. PRIMARY USE OF THE WATER WILL BE FOR AN AQUATIC RECREATION PARK, BUT OTHER ANTICIPATED USES INCLUDE SOIL RECLAMATION, IRRIGATION, AND INDUSTRIAL APPLICATIONS. THESE REUSES OF WASTE WATER WILL RESULT IN THE CONSERVATION OF THE PRESENTLY DIMINISHING FRESH WATER SUPPLY IN THIS ARID AREA. ALL RESEARCH AND TESTING WAS DONE WITH THE KNOWLEDGE THAT ULTIMATELY THE PUBLIC MUST ACCEPT THE CONCEPT OF REUSE OF RENOVATED WASTE WATER FOR THE PROJECT TO BE A TOTAL SUCCESS. ESTHETICS OF THE PRODUCT WATER, INCLUDING CLARITY, COLOR AND ODOR WERE IMPORTANT ASPECTS OF THE TREATMENT PROCESS. TEST DATA DEMONSTRATE THAT BACTERIOLOGICAL AND VIRAL REQUIREMENTS CAN BE MET. FISH HAVE SUCCESSFULLY SURVIVED AND PROPAGATED IN THE TEST PONDS. ALGAL GROWTH AND NUTRIENT LEVELS OF WATER IN THE PILOT PONDS ARE CONSIDERED WITHIN ACCEPTABLE LIMITS.

FIELD 05D, 05F

ACCESSION NO. W70-05645

QUANTITATIVE MICRO-DETERMINATION OF LIPID CARBON IN MICROORGANISMS,

CALIFORNIA UNIV., SAN DIEGO, LA JOLLA. INST. OF MARINE RESOURCES; AND SCRIPPS INSTITUTION OF OCEANOGRAPHY, LA JOLLA, CALIF. DIV. OF MARINE BIOLOGY.

O. HOLM-HANSEN, J. COOMBS, B. E. VOLCANI, AND P. M. WILLIAMS.

ANALYTICAL BIOCHEMISTRY, VOL 19, NO 3, P 561-568, 1967. 1 FIG, 1 TAB, 17 REF.

DESCRIPTORS:
*ANALYTICAL TECHNIQUES, *MICROORGANISMS, *LIPIDS, LIGHT, ALGAE, TEMPERATURE, CARBON DIOXIDE, OXIDATION, CHEMICALS, CARBON, GRAVIMETRY.

IDENTIFIERS:
CHLOROFORM-METHANE EXTRACTS, LIPID CARBON, CHLOROPLASTS, INFRARED ANALYSIS.

ABSTRACT:
THE DESCRIBED PROCEDURE FOR THE QUANTITATIVE DETERMINATION OF LIPID IN MICROORGANISMS EMPLOYS COMBUSTION OF THE SOLVENT EXTRACTS TO CARBON DIOXIDE AND MEASUREMENT OF THE LIBERATED CARBON DIOXIDE BY INFRARED ANALYSIS. THE METHOD PERMITS ANALYSES OF SAMPLES CONTAINING AS LITTLE AS 20 MICROGRAMS OF LIPID, AND THE RESULTS ARE COMPARABLE TO THOSE OBTAINED BY GRAVIMETRIC AND COLORIMETRIC METHODS. (WILDE-WISCONSIN)

FIELD 07B, 05A

ACCESSION NO. W70-05651

BIOASSAY OF SEAWATER. IV. THE DETERMINATION OF DISSOLVED BIOTIN IN SEAWATER
USING C-14 UPTAKE BY CELLS OF AMPHIDINIUM CARTERI,

CALIFORNIA UNIV., SAN DIEGO, LA JOLLA. INST. OF MARINE RESOURCES.

A. F. CARLUCCI, AND S. B. SILBERNAGEL.

CANADIAN JOURNAL OF MICROBIOLOGY, VOL 3, P 979-986, 1967. 2 FIG, 3 TAB, 11
REF.

DESCRIPTORS:
*BIOASSAY, *SEA WATER, *VITAMINS, VITAMIN B, ALGAE, CULTURES, CHEMICAL
ANALYSIS, ANALYTICAL TECHNIQUES, DINOFLAGELLATES, MARINE BACTERIA,
WATER CHEMISTRY, CARBON RADIOISOTOPES, PACIFIC OCEAN, CARBON DIOXIDE,
LIGHT, TEMPERATURE.

IDENTIFIERS:
*BIOTIN, *AMPHIDINIUM CARTERI, *RADIOCARBON UPTAKE TECHNIQUES, WATER
POLLUTION IDENTIFICATION, ORGANIC SOLUTES, VITAMIN B-1, VITAMIN B-12,
THIAMINE, SERRATIA MARINORUBRA, GYRODINIUM COHNII, MICROBIAL ECOLOGY,
COULTER COUNTER, GROWTH INHIBITION, CYANOCOBALAMIN, MUTANT BACTERIA,
CYCLOTELLA NANA, GEIGER COUNTER.

ABSTRACT:
THREE VITAMINS OF POTENTIAL IMPORTANCE IN MARINE MICROBIAL ECOLOGY ARE
BIOTIN, CYANOCOBALAMIN (B-12), AND THIAMINE (B-1). MUTANT MARINE
BACTERIA AND HETEROTROPHIC DINOFLAGELLATES HAVE BEEN EMPLOYED IN
BIOASSAYS TO ESTIMATE BIOTIN PREVIOUSLY, BUT SUCH ASSAYS HAVE BEEN
RELATIVELY INSENSITIVE. BIOTIN DISSOLVED IN SEAWATER CAN BE DETERMINED
BY RAPID, SENSITIVE, AND PRECISE TECHNIQUE DESCRIBED. UPTAKE OF
RADIOCARBON-LABELED CARBON DIOXIDE CELLS OF THE MARINE DINOFLAGELLATE,
AMPHIDINIUM CARTERI, DURING 2 HOURS FOLLOWING 94 HOURS OF PREINCUBATION
UNDER CAREFULLY CONTROLLED CONDITIONS IS PROPORTIONAL TO BIOTIN
CONCENTRATION. CELLS CAN BE ENUMERATED IN HEMACYTOMETER OR WITH
ELECTRONIC PARTICLE COUNTER TO EVALUATE GROWTH RESPONSES OF
DINOFLAGELLATE POPULATIONS TO VITAMIN CONCENTRATIONS, BUT INCUBATION
MUST BE EXTENDED TO 144 OR 168 HOURS TO OBTAIN PROPORTIONAL
DOSE-RESPONSE CURVES. BIOTIN CAN BE ASSAYED IN RANGE OF CONCENTRATIONS,
0.2-6.0 NANOGRAMS/LITER, A SENSITIVITY CONSIDERABLY GREATER THAN THOSE
FOR MICROBIOLOGICAL ASSAYS PREVIOUSLY DESCRIBED. SIX REPLICATED
DETERMINATIONS AT 1 AND 3 NANOGRAMS/LITER YIELDED STANDARD DEVIATIONS
OF 0.09 AND 0.28 RESPECTIVLY. FROM THE NORTH PACIFIC OCEAN, 45 SAMPLES
CONTAINED BIOTIN IN CONCENTRATION RANGE, 0.65-3.0 NANOGRAMS/LITER.
AVAILABILITY OF TECHNIQUES FOR ASSAY OF ECOLOGICALLY IMPORTANT
CONCENTRATIONS OF VITAMINS MAKES POSSIBLE A STUDY OF THEIR ROLE IN
MARINE ENVIRONMENTS. (EICHHORN-WISCONSIN)

FIELD 05A

ACCESSION NO. W70-05652

DEVELOPMENT OF A SYMBIOTIC ALGAL-BACTERIAL SYSTEM FOR NUTRIENT REMOVAL FROM
WASTEWATER,

OHIO STATE UNIV., COLUMBUS. DEPT. OF CIVIL ENGINEERING.

F. J. HUMENIK, AND G. P. HANNA, JR.

24TH ANNUAL PURDUE INDUSTRIAL WASTE CONFERENCE, MAY 6-8, PURDUE UNIVERSITY,
LAFAYETTE, INDIANA, 1969. 16 P, 7 FIG, 1 TAB, 15 REF.

DESCRIPTORS:
*WASTE WATER TREATMENT, *NUTRIENTS, ALGAE, BACTERIA, CULTURES, CHEMICAL
OXYGEN DEMAND, NITROGEN, PHOSPHATES, BIOMASS, SYMBIOSIS, OXYDATION
LAGOONS.

IDENTIFIERS:
*CHLORELLA PYRENOIDOSA, SYNTHETIC SEWAGE FEED, SYMBIOTIC RELATIONSHIPS.

ABSTRACT:
AN ALGAL-BACTERIAL CULTURE, DEVELOPED BY INOCULATION OF WASTEWATER WITH
CHLORELLA PYRENOIDOSA, PROVED TO BE EFFICIENT IN REMOVING CERTAIN
NUTRIENTS FROM THE EFFLUENTS. UNDER OPTIMUM CONDITIONS, THE
ALGAL-BACTERIAL FLOC SETTLED VERY RAPIDLY, YIELDING A CLEAR SUPERNAUT.
THE AVERAGE REMOVAL OF PHOSPHATES, HOWEVER, WAS LESS THAN 3%. MAXIMUM
REMOVAL OF CHEMICAL OXYGEN DEMAND AND ORGANIC NITROGEN WAS OBTAINED
DURING UNAERATED OPERATION WITH HARVESTING. THE BIOMASS OXIDATION
REQUIRED 1.27 MILLIGRAMS OF OXYGEN/MILLIGRAM OF SOLIDS. SUPPLEMENTAL
AERATION FAILED TO IMPROVE EITHER REMOVAL OR CONSERVATION OF NUTRIENTS.
THE TERM 'SYMBIOTIC CULTURE' WAS INTRODUCED ON THE BASIS OF A
STEADY-STATE EQUILIBRIUM IN THE OXYGEN CONCENTRATION, MAINTAINED BY
PHOTOSYNTHETIC OXYGENATION AND TOTAL RESPIRATION. (WILDE-WISCONSIN)

FIELD 05D

ACCESSION NO. W70-05655

CULTURAL EUTROPHICATION WITH SPECIAL REFERENCE TO LAKE WASHINGTON,

WASHINGTON UNIV., SEATTLE. DEPT. OF ZOOLOGY.

W. T. EDMONDSON.

MITTEILUNGEN INTERNATIONALE VEREINIGUNG FUR THEORETISCHE UND ANGEWANDTE
LIMNOLOGIE, VOL 17, P 19-32, 1969. 5 FIG, 37 REF, DISCUSSION.

DESCRIPTORS:
*LAKES, *EUTROPHICATION, *WASHINGTON, URBANIZATION, WATER POLLUTION,
SEDIMENTS, PALEOLIMNOLOGY, NUTRIENTS, PRODUCTIVITY, SEDIMENTATION
RATES, STRATIGRAPHY, BIOINDICATORS, ALGAE, OLIGOTROPHY, DIVERSION,
CORES, SEWAGE EFFLUENTS, DIATOMS, PHOSPHORUS, ORGANIC MATTER, NITROGEN,
CHLOROPHYLL.

IDENTIFIERS:
*LAKE WASHINGTON(WASH), *CULTURAL INFLUENCES, MICROCYSTIS, DREDGE
SAMPLES, ANABAENA, SEASONAL EFFECTS, DECOMPOSITION RATES, CHRONOLOGY,
CLADOCERANS, CHYDORID CLADOCERANS, SPECIES COMPOSITION, MICROFOSSILS,
COMPETITIVE EXCLUSION PRINCIPLE, OSCILLATORIA RUBESCENS, PISTON CORER,
FRAGILARIA, MELOSIRA, ASTERIONELLA, ARAPHIDINEAE, CENTRIC DIATOMS,
BOSMINA, DIAPHANOSOMA, OSCILLAXANTHIN.

ABSTRACT:
EUTROPHICATION OF LAKE WASHINGTON BY URBAN SEWAGE AND ITS SUBSEQUENT
RECOVERY AFTER INSTALLATION OF TREATMENT FACILITIES AND EFFLUENT
DIVERSION ARE WELL DOCUMENTED. PALEOLIMNOLOGICAL EVIDENCE BEARING ON
THIS CASE HISTORY IS DESCRIBED. OF MANY DIATOMS OCCURRING IN CORES,
FRAGILARIA CROTONENSIS DOMINATES RECENTLY; MELOSIRA ITALICA, IN THE
DEEPER (OLDER) SEDIMENTS; AND ASTERIONELLA FORMOSA OCCURS AT ALL
DEPTHS. NO SPECIES, ABUNDANT AT ANY DEPTH, IS MISSING FROM SIGNIFICANT
CORE SECTIONS. RELATIVE COMPOSITION OF CENTRIC VERSUS ARAPHIDINATE
GROUPS OF DIATOMS CORRELATE WITH KNOWN LAKE'S EUTROPHICATION HISTORY,
LARGE DECREASES IN ARAPHIDINEAE COINCIDING WITH DIVERSION OF RAW
SEWAGE. USING SPECIES GROUPS MAY OVERCOME INHERENT DISADVANTAGES POISED
BY CONSIDERING SINGLE SPECIES AS BIOINDICATORS. ANIMAL FOSSILS HAVE
BEEN GIVEN LITTLE STUDY, ALTHOUGH BOSMINA LONGIROSTRIS--COMMON
CLADOCERAN IN NATURALLY OR CULTURALLY EUTROPHIC LAKES--IS KNOWN TO BE
ABUNDANT THROUGHOUT SEDIMENTS, WITH INCREASING ABUNDANCE SEEMINGLY
PARALLELING PERIOD OF ENRICHMENT. BASED UPON REASONABLY RELIABLE MARKER
LAYER, SEDIMENTATION SINCE 1916 HAS PROCEEDED AT RATE BETWEEN 2.3 AND
3.1 MILLIMETERS/YEAR. SURFICIAL SEDIMENTS TEND TO BE DISTINCTLY
ENRICHED IN PHOSPHORUS, AFTER INTERPOSED MINIMUM IN 1930-1940, WHEN
SEWAGE WAS DIVERTED. IN GENERAL, EXTENSIVE LIMNOLOGICAL DATA ARE SCANTY
FOR LAKES WITH HISTORY OF EUTROPHICATION, MAKING ELUCIDATION OF
PALEOLIMNOLOGICAL CORRELATIVES DIFFICULT. (EICHHORN-WISCONSIN)

FIELD 05C, 02H

ACCESSION NO. W70-05663

EUTROPHICATION OF LAKES AND STREAMS, CAUSE AND PREVENTION (IN GERMAN),

KANTONALES LABORATORIUM, ZURICH (SWITZERLAND).

E. A. THOMAS.

MONATSBULLETIN DES SCHWEIZERISCHEN VEREINS VON GAS- UND WASSERFACHMANNERN, NO
6/7, P 3-12, 1963. 7 FIG, 24 REF.

DESCRIPTORS:
*EUTROPHICATION, *LAKES, *RIVERS, *ALGAE, WATER POLLUTION CONTROL,
WATER POLLUTION EFFECTS, WATER POLLUTION SOURCES, SEWAGE, PHOSPHORUS,
CHEMCONTROL, INDUSTRIAL WASTES, SEWAGE TREATMENT.

IDENTIFIERS:
RING TUBE, CANALIZATION, ALUMINUM SULFATE, FERRICHLORIDE, IRON SULFATE,
GALVANIZING WASTES, LAKE ZURICH(SWITZERLAND), SWITZERLAND,
KILCHBERG(SWITZERLAND), ZOLLIKON(SWITZERLAND), LAKE TEGERN(BAVARIA),
LAKE SCHLIER(BAVARIA), LAKE ZELL(AUSTRIA), LAKE HALL WILER(SWITZERLAND).

ABSTRACT:
WATER POLLUTION WAS RECOGNIZED IN SOME COUNTRIES A CENTURY AGO WHEN
ZURICH KANTON, SWITZERLAND, ENACTED A LAW IN 1881. WHILE MECHANICAL AND
BIOLOGICAL TREATMENT OF SEWAGE ARE THE MOST GENERALLY USED METHODS OF
WATER PURIFICATION AND EUTROPHICATION REVERSAL, THE RESULTANT EFFLUENTS
STILL PROMOTE ALGAE AND HIGHER WATER PLANTS. LAKES SURROUNDED BY
COMMUNITIES BENEFIT BY SEWAGE CANALIZATION CIRCLING THE LAKE, A PROCESS
USED FOR 50 YEARS FOR A PORTION OF ZURICH AND ADJACENT COMMUNITIES OF
KILCHBERG AND ZOLLIKON, TOTALING A 50,000 POPULATION. LARGE LAKES AND
THOSE WITH IRREGULAR SHORELINES RECEIVING SEWAGE AS WELL AS
NUTRIENT-RICH TRIBUTARY STREAMS REQUIRE PHOSPHATE REMOVAL FROM SEWAGE
BY ALUMINUM SULFATE, BY FERRICHLORIDE, OR BY A COMBINATION OF IRON
CHLORIDE AND IRON SULFATE. TESTS TO PRECIPITATE PHOSPHATE IN SEWAGE
PROVED PROMISING, ESPECIALLY BY A PRECIPITANT PRODUCED BY SWISS SODA
PLANT, ZURZACH, FROM PICKLE WASTES, A PRODUCT OF GALVANIZATION (89.9%
PHOSPHATE ELIMINATED) AT 1/3 TO 1/4 COST OF PURE IRON CHLORIDE. THE
PRODUCT ALSO SERVED TO ELIMINATE INDUSTRIAL WASTES AND TO PRODUCE A
MARKETABLE COMMODITY. CHEMICAL CONTROL OF NUISANCES, CHANNELING THE
HYPOLIMNION FROM A LAKE TO REMOVE NUTRIENTS, AND OXYGENATION EFFECTS
ARE DISCUSSED. (AUEN-WISCONSIN)

FIELD 05C

ACCESSION NO. W70-05668

WASTE TREATMENT FOR THE CONTROL OF HETEROTROPHIC AND AUTOTROPHIC ACTIVITY IN RECEIVING WATERS,

HARVARD UNIV., CAMBRIDGE, MASS. DEPT. OF WATER RESOURCES.

WERNER STUMM, AND MARK W. TENNEY.

PROCEEDINGS 12TH SOUTHERN MUNICIPAL INDUSTRIAL WASTE CONFERENCE, RALEIGH, NORTH CAROLINA, P 97-111, 1963. 5 FIG, 1 TAB, 15 REF.

DESCRIPTORS:
*WASTE TREATMENT, *WATERS, *ORGANISMS, STREAMS, WATER POLLUTION, ANIMALS, BACTERIA, FUNGI, ALGAE, FLOATING PLANTS, ROOTED AQUATIC PLANTS, SELF-PURIFICATION, BIOCHEMICAL OXYGEN DEMAND, TOXICITY, OXYGEN-REDUCTION POTENTIAL, SOLAR RADIATION, ENERGY, CARBON DIOXIDE, PHOTOSYNTHESIS, GROWTH RATE, AEROBIC CONDITIONS, ANAEROBIC CONDITIONS, BIOLOGICAL TREATMENT, FERTILIZATION, RIVERS, SEWAGE, PHOSPHORUS, NITROGEN, CARBON, ALUMINUM.

IDENTIFIERS:
MATHEMATICAL STUDIES, AUTOTROPHIC, HETEROTROPHIC.

ABSTRACT:
THE FEASIBILITY OF AEROBIC BIOLOGICAL WASTE TREATMENT WITH COMPLEMENTARY CHEMICAL TREATMENT FOR FLOCCULATION OF DISPERSED MICROORGANISMS AND REMOVAL OF PHOSPHATES HAS BEEN EXPERIMENTALLY CORROBORATED. RESULTS WERE OBTAINED ON THE EFFICIENCY OF ORGANIC CARBON REMOVED BY BIOLOGICAL AND SUBSEQUENT CHEMICAL TREATMENT WITH ALUM. CHEMICAL TREATMENT ALONE (WITHOUT PRIOR BIOLOGICAL SUBSTRATE UTILIZATION) DOES NOT REMOVE ORGANIC CARBON TO AN APPRECIABLE EXTENT. BACTERIA, IN THE FLOCCULATION PROCESS, ACT AS HYDROPHILIC COLLOIDS. THE STOICHIOMETRIC RELATION BETWEEN COAGULANT DOSE AND NUMBER OF MICROORGANISMS (OR ORGANIC CARBON OF BIOLOGICAL SOLIDS) IS INDICATED. THE REQUIRED TOTAL COAGULANT DEMAND IS A FUNCTION OF BOTH THE CONCENTRATION OF DISPERSED MICROORGANISMS AND THE CONCENTRATION OF PHOSPHORUS PRESENT. RESULTS ILLUSTRATE THAT CHEMICAL COAGULATIVE METHODS CAN SERVE AS A VALUABLE COMPLEMENT TO THE BIOLOGICAL SUBSTRATE UTILIZATION PROCESSES. THE COMPLEMENTARY CHEMICAL AND BIOLOGICAL TREATMENT IS MORE ECONOMICAL THAN THE CONVENTIONAL BIOLOGICAL TREATMENT; IT IS CERTAINLY MORE ECONOMICAL THAN A CONVENTIONAL BIOLOGICAL TREATMENT FOLLOWED BY A THIRD STAGE CHEMICAL TREATMENT. IN THOSE CASES WHERE CHEMICAL TREATMENT IS MORE BENEFICIAL THAN BIOLOGICAL TREATMENT, THE PROPOSED TREATMENT COMBINATION APPEARS TO BE, EVEN FROM AN ECONOMICAL POINT OF VIEW, THE MOST EXPEDIENT WASTE PURIFICATION ARRANGEMENT. (JONES-WISCONSIN)

FIELD 05D

ACCESSION NO. W70-05750

USE OF ALGAL CULTURE METHODS IN EUTROPHICATION RESEARCH (IN NORWEGIAN),

NORSK INSTITUTT FOR VANNFORSKNING, BLINDERN.

OLAV SKULBERG.

ENGLISH SUMMARY. NORDISKE JORDBRUKSFORSKERES FORENING, VOL 47, P 211-215, 1965. 1 FIG, 2 REF.

DESCRIPTORS:
*EUTROPHICATION, ALGAE, CULTURES, CHLOROPHYTA, CYANOPHYTA, BRACKISH WATER, OLIGOTROPHY, TEMPERATURE, LIGHT INTENSITY, CHLOROPHYLL, NUTRIENTS, CONDUCTIVITY, AGRICULTURAL WATERSHEDS, SEWAGE, WATER POLLUTION, CHEMICAL ANALYSIS, PHYSICAL PROPERTIES.

IDENTIFIERS:
BACILLARIOPHYCEAE, SELENASTRUM CAPRICORNUTUM PRINTZ, OSCILLATORIA VAUCHER, SKELETONEMA COSTATUM(GREV) CL, CHLORELLA OVALIS BUTCHER, NJOER LAKE(NORWAY), HOBOL RIVER(NORWAY), VAN LAKE(NORWAY).

ABSTRACT:
UNIALGAL CULTURES ARE USED IN A PROCEDURE TO MEASURE THE FERTILITY OF WATER RECEIVING POLLUTION WITH PLANT NUTRIENTS. THE TECHNIQUE DEVELOPED IS CONSIDERED A SUPPLEMENT TO CHEMICAL AND PHYSICAL METHODS FOR THE HANDLING OF EUTROPHICATION PROBLEMS. SPECIES FROM THE CLASSES OF CHLOROPHYCEAE, CYANOPHYCEAE, AND BACILLARIOPHYCEAE WERE USED AS TEST ALGAE. OF THESE, SELENASTRUM CAPRICORNUTUM, WHICH OCCURS WIDELY IN OLIGOTROPHIC AND EUTROPHIC FRESH WATER LOCALITIES UNDER NATURAL CONDITIONS, WAS USED FOR EXPERIMENTATION MOST EXTENSIVELY. OTHER SUITABLE ALGAE HAVE BEEN OSCILLATORIA VAUCHER SP, SKELETONEMA COSTATUM (GREV) CL, AND CHLORELLA OVALIS BUTCHER. THE METHOD CAN BE USED TO DETERMINE THE EFFICIENCY OF THE WATER TREATMENT PLANTS IN REDUCING PLANT NUTRIENTS IN THE EFFLUENT, AND CAN MEASURE THE TROPHIC LEVEL OF WATER MASS. EXPERIMENTAL METHODS ARE NECESSARY TO GIVE A DETAILED UNDERSTANDING OF THE BIOLOGICAL PHENOMENA WHICH FOLLOW EUTROPHICATION. WITH CAREFUL AND CRITICAL USE OF ALGAL CULTURE EXPERIMENTS, IT WOULD BE POSSIBLE TO DEFINE NUMERICAL RELATIONS BETWEEN THE CONDITIONS OF THE WATER, DEGREE OF POLLUTION AND THE BIOLOGICAL SITUATION. (JONES-WISCONSIN)

FIELD 05C, 07B

ACCESSION NO. W70-05752

FAYETTEVILLE GREEN LAKE, NEW YORK. V. STUDIES OF PRIMARY PRODUCTION AND
ZOOPLANKTON IN A MEROMICTIC MARL LAKE,

CORNELL UNIV., ITHACA, N.Y. DIV. OF BIOLOGICAL SCIENCES.

D. A. CULVER, AND G. J. BRUNSKILL.

LIMNOLOGY AND OCEANOGRAPHY, VOL 14, NO 6, P 862-873, 1969. 7 FIG, 4 TAB, 26
REF.

DESCRIPTORS:
*LAKES, *NEW YORK, *MEROMIXIS, *PRIMARY PRODUCTIVITY, *ZOOPLANKTON,
SULFUR BACTERIA, DIEL MIGRATION, CARBON DIOXIDE, HYDROGEN ION
CONCENTRATION, PHOSPHATES, NITRATES, DAPHNIA, SILICA, THERMAL
STRATIFICATION, SULFIDES, AMMONIUM COMPOUNDS, OXIDATION REDUCTION
POTENTIAL, DISSOLVED OXYGEN, CHEMICAL ANALYSIS, ORGANIC MATTER, ALGAE,
IRON, HYDROGEN SULFIDE, CYANOPHYTA, CHLOROPHYTA, LIGHT PENETRATION.

IDENTIFIERS:
*FAYETTEVILLE GREEN LAKE(NY), *MARL LAKES, CHLOROBIUM, FILINIA,
MONIMOLIMNION, CHEMOCLINE, MIXOLIMNION, RADIOCARBON UPTAKE TECHNIQUES,
BACTERIAL PLATE, BACTERIAL PIGMENTS, DIAPTOMUS, EPISCHURA,
ORTHOCYCLOPS, KERATELLA, POLYARTHRA, CONOCHILUS, PYCNOCLINE,
PHOTOLITHOTROPHIC BACTERIA, DESULFOVIBRIO, MESOTHERMY, MELOSIRA,
SYNEDRA, RHODOMONAS, CYCLOTELLA, CHROMATIUM.

ABSTRACT:
MEROMIXIS OF FAYETTEVILLE GREEN LAKE, NEW YORK, INFLUENCES ITS TROPHIC
STRUCTURE IN THE SENSE THAT BACTERIAL PRIMARY PRODUCTION RESEMBLES
TOTAL PRIMARY PRODUCTION IN EUTROPHIC LAKES, WHILE ALGAL PRODUCTION IS
MORE OLIGOTROPHIC. SEVEN DETERMINATIONS OF RADIOCARBON UPTAKE IN LIGHT
AND DARK BOTTLES WERE MADE BETWEEN OCTOBER 1966 AND THE FOLLOWING JULY.
ANNUAL PRIMARY PRODUCTION WAS ESTIMATED AT 290 GRAMS CARBON/SQUARE
METER, OF WHICH APPROXIMATELY 83% WAS BACTERIAL, PRIMARILY DUE TO WELL
DEVELOPED PLATE OF PHOTOSYNTHETIC SULFUR OXIDIZERS IN THE CHEMOCLINE
(18-20 METERS). ALGAL PRODUCTIVITY MAY BE LIMITED BY UNFAVORABLE RATIO,
MONOVALENT CATIONS:DIVALENT CATIONS, WHICH HAS VALUE OF 0.06 IN THIS
LAKE, WHEREAS HIGHER RATIOS, IN RANGE 2-30, ARE CONSIDERED OPTIMAL.
ZOOPLANKTONIC BIOMASS WAS DOMINATED BY COPEPODS. DENSE CONCENTRATIONS
OF ZOOPLANKTERS OCCUR NEAR THE BACTERIAL PLATE, WHERE THEY PROBABLY
FEED ON BACTERIA OR ON MIXOLIMNETICALLY PRODUCED SESTON, WHICH TENDS TO
ACCUMULATE IN THE CHEMOCLINE. QUANTITATIVE SAMPLING OF ZOOPLANKTON WITH
CLARKE-BUMPUS SAMPLER YIELDED EVIDENCE FOR DIEL MIGRATIONS OF MANY
SPECIES. (EICHHORN-WISCONSIN)

FIELD 02H, 05C

ACCESSION NO. W70-05760

THE EUTROPHICATION OF LAKES AND THE RADICAL CHANGE OF THEIR MATERIAL CONTENT
(IN GERMAN),

WALDEMAR OHLE.

LIMNOLOGISKA FOREINGEN I FINLAND, LIMNOLOGISYMPOSION, HELSINKI, 10-23, P
1-16, 1964. 10 REF.

DESCRIPTORS:
*EUTROPHICATION, *LAKES, NUTRIENTS, FERTILIZATION, BACTERIA, ALGAE,
PLANKTON, TUBIFICIDS, CALCIUM, SILICATES, DIATOMS, WINDS, NITROGEN,
PHOSPHATES, METHANE, MUD, OXYGEN, ANAEROBIC CONDITIONS, HYDROGEN
SULFIDE, IRON, FUNGI, CLAYS, BRACKISH WATER, FISHERIES, LIMNOLOGY.

IDENTIFIERS:
CHIRONOMIDS, LAKE CONSTANCE, WHITEFISH, WURTTEMBERG(GERMANY),
PLON(GERMANY), LAKE SCHLIER(GERMANY), LAKE TEGERN(GERMANY), GERMANY,
FINLAND, SEWAGE RING TUBES.

ABSTRACT:
MOST NORTH GERMAN LAKES HAVE BEEN SUBJECT TO EUTROPHICATION SINCE EARLY
1930'S DUE TO DOMESTIC SEWAGE INFLOWS AND ALLUVIAL SOIL PARTICLES. AT
THE END OF SUMMER STAGNATION, HYPOLIMNETIC OXYGEN IS REPLACED BY
HYDROGEN SULFIDE IN SOLUTION. THE SEDIMENTS, ACCUMULATING LAKE
NUTRIENTS, ARE REGARDED AS RESULTANT OF AUTOCHTHON DYNAMICS OF
METABOLISM AND INFLUENCES OF REGIONAL SURROUNDINGS. 3% TO 5% OF THE
MINERAL FERTILIZER WASHED FROM SLOPING FIELDS ARE EXTREMELY HIGH
AMOUNTS FOR LAKES WITH NO DENSE PLANT GROWTH, ESPECIALLY BECAUSE OF
PHOSPHATE ENRICHMENT. HIGHER AMOUNTS OF PHOSPHORUS ARE ADDED BY SEWAGE,
SUPPLEMENTED BY DETERGENTS. PHOSPHATE IS A DOMINANT PRODUCTION FACTOR
WITH NITROGEN COMPOUNDS OF SECONDARY IMPORTANCE. WHEN LARGE AMOUNTS OF
ORGANIC SUBSTANCES HAVE ACCUMULATED SO THAT DISSOLVED OXYGEN IN DEEP
WATER IS CONSUMED BY BACTERIAL ACTIVITY, LARGER AMOUNTS OF NUTRIENTS
ARE RELEASED FROM MUD INTO HYPOLIMNION. THE MORE BACTERIAL METHANE
OXIDATION DEVELOPS, THE HIGHER UTILIZATION OF OXYGEN, THUS MORE
NUTRIENTS ARE RECIRCULATED FROM THE MUD INTO THE WATER UNDER ANAEROBIC
CONDITIONS. THE INCREASING PRIMARY PRODUCTION IS INTENSIFIED
EXPONENTIALLY. ALL SEWAGE SHOULD BE DIVERTED FROM LAKES BY VARIOUS
MEANS AND CHEMICAL TREATMENT COORDINATED WITH BIOLOGICAL PURIFICATION.
(JONES-WISCONSIN)

FIELD 05C, 02H

ACCESSION NO. W70-05761

INCREASING WASTEWATER FLOW VELOCITY BY USING CHEMICAL ADDITIVES,

WESTERN CO. OF NORTH AMERICA. RICHARDSON, TEX. AND SOUTHERN METHODIST UNIV.,
DALLAS, TEX.

J. L. OVERFIELD, H. R. CRAWFORD, J. K. BAXTER, L. J. HARRINGTON, AND I. W.
SANTRY, JR.

JOURNAL OF THE WATER POLLUTION CONTROL FEDERATION, VOL 41, NO 9, SEPT 1969. P
1570-1585, 17 FIG, 1 TAB.

DESCRIPTORS:
*ADDITIVES, *FLOW RATE, *WASTE WATER, *ALGAE, *SLUDGE, STORMS,
TURBULENCE, PILOT PLANTS, BACTERIA, SEDIMENTATION, TOXICITY,
COST-BENEFIT ANALYSIS.

IDENTIFIERS:
*POLYMERS, *BIOLOGICAL LIFE, *ALGAE BLOOMS, *FISH TOXICITY TEST, SLUDGE
DRYING, SOLIDS-SETTLING, ECONOMIC ADVANTAGE, TEMPERATURE-EFFECT.

ABSTRACT:
PRACTICAL METHODS HAVE BEEN DEVELOPED TO INCREASE WASTE WATER FLOW
VELOCITY BY USING CHEMICAL ADDITIVES WITHOUT INCREASING THE ENERGY
NEEDED TO MOVE THE LIQUID. EXISTING SEWER LINE FLOW CAN BE INCREASED BY
USING POLYMER ADDITIVES THAT WOULD INCREASE THE SOLID REMOVAL
EFFECTIVENESS AT TREATMENT PLANTS. A 100 FT LONG 6 IN SEWER LINE TEST
FACILITY ALONG WITH 25,000-GAL CONTROLLED-TEMPERATURE RESERVOIR IS USED
TO DETERMINE THE INCREASED WATER FLOW WITH ADDITION OF CHEMICAL
ADDITIVES. THERE IS A RAPID INCREASE UP TO 24 TIMES THE NORMAL FLOW.
FLOW COULD BE INCREASED BY A FACTOR OF 2 OR MORE BY USING
CONCENTRATIONS OF ADDITIVES BETWEEN 45 AND 200 ML/L. THE COST/BENEFIT
ANALYSIS FOR A GIVEN EXAMPLE SHOWS THAT THE COST OF RECTIFYING OVERFLOW
FROM A CONSTRICTION BY USING AN ADDITIVE INJECTION SYSTEM, IS LESS THAN
HALF THE PRORATED COST OF CONSTRUCTING A PARALLEL PIPE SYSTEM TO
RELIEVE THE CONDITION. HOWEVER, IF THE OVERFLOWS OCCUR TEN TIMES A
YEAR, THE ADDITIVE INJECTION SYSTEM BECOMES MORE EXPENSIVE. BIOCHEMICAL
STUDIES CONDUCTED INDICATE THAT ADDITIVES DO NOT AFFECT ADVERSELY THE
WASTE WATER BACTERIA, FISH, OR PROMOTE ALGAE GROWTH. FURTHER, THE
ADDITIVES IMPROVE WASTE WATER SETTLING AND SLUDGE DRYING RATES.
ADDITIVE INJECTION SYSTEM, THROUGH PREFERABLE ECONOMICALLY IN MANY
CASES, CANNOT BE APPLIED AS A LONG TERM SOLUTION. (SHANKAR-TEXAS)

FIELD 05D

ACCESSION NO. W70-05819

AN INVESTIGATION INTO UPWARD-FLOW FILTRATION..

WEST HERTFORDSHIRE MAIN DRAINAGE AUTHORITY, RICKMANSWORTH (ENGLAND).

RICHARD WOOD, WILLIAM S. SMITH, AND J. K. MURRAY.

WATER POLLUTION CONTROL, VOL 67, NO 4, 1968. P 421-428, 3 FIG, 7 TAB, 2 REF.

DESCRIPTORS:
*SEWAGE TREATMENT, *FILTRATION, *BIOCHEMICAL OXYGEN DEMAND, *ACTIVATED
SLUDGE, BIOLOGICAL TREATMENT, PILOT PLANTS, HUMUS, REAERATION,
EFFLUENTS, FLOW RATES, FILTERS, ALGAE.

IDENTIFIERS:
*UPWARD FLOW, PUTRESCIBILITY TEST, BODY IMMEDIUM FILTER, BACK-WASHING.

ABSTRACT:
SEWAGE TREATMENT AUTHORITIES ARE INCREASINGLY BEING REQUIRED TO PRODUCE
HIGHER QUALITY EFFLUENTS FROM INCREASING VOLUMES OF STRONGER SEWAGES.
ENLARGED ACTIVATED SLUDGE PLANT IS DESIGNED TO PRODUCE FULLY NITRIFIED
EFFLUENT OF LOW SOLIDS CONTENT, TO WHICH COULD BE ADDED ADDITIONAL
ACTIVATED SLUDGE SOLIDS. THIS MAKES IT POSSIBLE TO CARRY OUT AN
INVESTIGATION OF UPWARD-FLOW SAND FILTRATION. A BODY IMMEDIUM FILTER IS
USED TO DETERMINE THE EFFECTIVENESS IN FILTERING A HIGH-QUALITY
EFFLUENT CONTAINING A WIDE RANGE OF ACTIVATED SLUDGE CONCENTRATIONS.
SATISFACTORY REMOVALS OF COARSER ACTIVATED SLUDGE ARE OBTAINED. BY
DOSING THE FILTER WITH ARTIFICIALLY HIGH PROPORTIONS OF WELL-OXIDIZED
ACTIVATED SLUDGE, HIGH PERCENTAGE REMOVALS ARE OBTAINED. ALTHOUGH THE
UNIT IS MANUALLY CONTROLLED, THIS TYPE OF FILTER LENDS ITSELF TO
AUTOMATED OPERATION AND PROVIDED THAT THE FEED LIQUOR IS OF HIGH
QUALITY AND ONLY CONTAMINATED WITH SOLIDS, VERY HIGH QUALITY EFFLUENTS
CAN BE PRODUCED. (SHANKAR-TEXAS)

FIELD 05D

ACCESSION NO. W70-05821

THE ECOLOGY OF PERIPHYTON IN WESTERN LAKE SUPERIOR: PART I - TAXONOMY AND
DISTRIBUTION,

MINNESOTA UNIV., MINNEAPOLIS. WATER RESOURCES RESEARCH CENTER.

JACKSON L. FOX, THERON O. ODLAUG, AND THEODORE A. OLSON.

AVAILABLE FROM THE CLEARINGHOUSE AS PB-191 214, $3.00 IN PAPER COPY, $0.65 IN
MICROFICHE. MINNESOTA UNIVERSITY WATER RESOURCES RESEARCH CENTER BULLETIN
14, AUGUST 1969. 127 P, 50 FIG, 37 PLATE, 24 TAB, 79 REF, APPEND. OWRR
PROJECT NO A-011-MINN.

DESCRIPTORS:
*LAKE SUPERIOR, *PERIPHYTON, *LIMNOLOGY, *WATER QUALITY, BIOINDICATORS,
AQUATIC LIFE, AQUATIC ALGAE, WATER POLLUTION EFFECTS.

IDENTIFIERS:
PERIPHYTON INVENTORY(LAKE SUPERIOR), WATER QUALITY INDICATORS.

ABSTRACT:
THE PLANT PORTION OF THE EPILITHIC PERIPHYTON OF THE WESTERN ARM OF
LAKE SUPERIOR WAS FOUND TO CONSIST SOLELY OF REPRESENTATIVES FROM THREE
PHYLA OF ALGAE, THE CHRYSOPHYTA, THE CHLOROPHYTA, AND THE CYANOPHYTA.
MEMBERS OF THE PHYLUM CHRYSOPHYTA WERE THE MOST ABUNDANT ORGANISMS.
DIATOMS COMPRISED OVER 90 PER CENT OF THE TOTAL NUMBER OF ORGANISMS.
THE PREDOMINANT GENERA WERE FOUND TO BE SYNEDRA, ACHNANTHES, NAVICULA,
CYMBELLA, AND GOMPHONEMA. THE MEAN TOTAL COUNTS OF ORGANISMS IN THE
NATURALLY OCCURRING PERIPHYTON OF STONY POINT BAY, THE PRIMARY SAMPLING
AREA, RANGED FROM 497,000 PER SQUARE CENTIMETER OF ROCK SURFACE IN 1966
TO 1,470,000 PER SQUARE CENTIMETER IN 1967. THE BIOMASS OF THE
NATURALLY OCCURRING STONY POINT BAY PERIPHYTON, IN TERMS OF DRY WEIGHT,
WAS 153 GRAMS PER SQUARE METER IN 1965. THE ORGANISMS FOUND IN THE
PERIPHYTON OF THE WESTERN ARM OF LAKE SUPERIOR ARE INDICATIVE OF CLEAN
WATER. THE EXTENSIVE SHALLOW WATER AREA OF LAKE SUPERIOR SUPPORTS LARGE
QUANTITIES OF ATTACHED ALGAE, WHICH, AS PRIMARY PRODUCERS, FORM THE
FIRST LINK IN THE FOOD CHAIN. (KNAPP-USGS)

FIELD 02H, 05C

ACCESSION NO. W70-05958

GROWTH OF CHLORELLA PYRENOIDOSA 71105 IN ACTIVATED SLUDGE WASTE EFFLUENT,

GENERAL DYNAMICS CORP., GROTON, CONN. ELECTRIC BOAT DIV.

DONALD E. LEONE.

JOURNAL WATER POLLUTION CONTROL FEDERATION, VOL 41, NO 1, P 51-55, JAN 1969.
5 TAB, 9 REF.

DESCRIPTORS:
*ALGAE, *ACTIVATED SLUDGE, *EFFLUENTS, WASTE WATER RECLAMATION, WASTE
WATER TREATMENT, *NUTRIENTS.

IDENTIFIERS:
MASS CULTURE OF ALGAE, PACKED CELL VOLUME.

ABSTRACT:
PROCESSED WASTE EFFLUENT WAS STUDIED, USING CHLORELLA PYRENOIDOSA
71105, AS A METHOD FOR ALGAE GROWTH AND POSSIBLY WASTE WATER
RECLAMATION. TREATED WASTE EFFLUENT, CENTRIFUGED AND FILTERED TWICE
THROUGH CHARCOAL FOR REMOVAL OF COLOR, WAS SUPPLEMENTED WITH IRON
AND/OR UREA AND COMPARED WITH DL-61 AND EB STANDARD ALGAE MEDIA. WHILE
PRELIMINARY DATA REVEALED DEFICIENT MAGNESIUM, PHOSPHORUS, AND SULFATES
AND SUFFICIENT CALCIUM, NITROGEN, AND IRON FOR NORMAL ALGAL NUTRIENTS;
SUPPLEMENTING WASTE EFFLUENT WITH IRON WAS BENEFICIAL. DATA REVEALED
THAT EFFLUENT ALONE AND EFFLUENT SUPPLEMENTED WITH UREA SHOWED A
SUBSTANTIAL LOWERING OF DOUBLING TIME AND REDUCED PCV (PACKED CELL
VOLUME). THE WASTE EFFLUENT CONTAINING ADDITIONAL IRON AND UREA SHOWED
HIGHER CHLOROPHYLL CONTENT, ALTHOUGH FEWER CELLS WERE PRODUCED, WHEN
COMPARED WITH EFFLUENT SUPPLEMENTED WITH IRON AND DILUTED STANDARD
MEDIA. RESULTS INDICATED THAT PROCESSED WASTE EFFLUENT SUPPLEMENTED
WITH IRON (0.56 MG/L TOTAL) WAS A GOOD MEDIUM FOR CHLORELLA PYRENOIDOSA
71105 AND PRODUCED AS MUCH PCV AS THE DILUTED STANDARD MEDIA.
(MORGAN-TEXAS)

FIELD 05D

ACCESSION NO. W70-06053

THE BIOLOGICAL EFFECT OF COPPER SULPHATE TREATMENT ON LAKE ECOLOGY,

WISCONSIN COMMITTEE ON WATER POLLUTION, MADISON.

KENNETH M. MACKENTHUN, AND HAROLD L. COOLEY.

WISCONSIN ACADEMY OF SCIENCES, ARTS AND LETTERS, VOL 41, P 177-187, 1952. 7
TAB, 18 REF.

DESCRIPTORS:
*BIOLOGY, *EFFECTS, *COPPER SULFATE, *TREATMENT, *LAKES, *ECOLOGY,
BOTTOM SEDIMENTS, MUD, BENTHOS, WISCONSIN, SAMPLING, CONTOURS, DEPTH,
OLIGOCHAETES, TOXICITY, FISH, BLOODWORMS, ALGAL CONTROL.

IDENTIFIERS:
LAKE MENDOTA(WIS), LAKE MONONA(WIS), LAKE NAGAWICKA(WIS), LAKE
PEWAUKEE(WIS), CHIRONOMUS, PROCLADIUS, CORETHRA PUNCTIPENNIS, PISIDIUM
IDAHOENSE, TENDIPES PLUMOSUS, BEAVER DAM LAKE(WIS), DELAVAN LAKE(WIS).

ABSTRACT:
THESE STUDIES WERE DESIGNED TO DETERMINE THE EFFECT OF ACCUMULATED
COPPER IN BOTTOM MUDS, APPLIED AS COPPER SULFATE FOR ALGAL CONTROL, ON
BOTTOM DWELLING ORGANISMS WHICH ARE AVAILABLE AS FISH FOOD. ALTHOUGH
THE TOXIC LIMIT OF COPPER SULFATE, PRECIPITATED AND ACCUMULATED IN
BOTTOM MUDS, UPON CERTAIN TYPES OF BOTTOM-DWELLING ORGANISMS COULD NOT
BE ACCURATELY DETERMINED IN THE TIME ALLOTTED, LABORATORY TESTS
INDICATE THAT IT IS NEAR 9000 PARTS/MILLION COPPER ON A DRY-WEIGHT
BASIS. RESULTS INDICATE THAT COPPER ACCUMULATION IN BOTTOM MUDS, DUE TO
USAGE OF COPPER SULFATE TO CONTROL ALGAE IN HARD WATER LAKES, IS
CONSIDERABLY LOWER IN CONCENTRATION THAN THE AMOUNTS EXPERIMENTALLY
DETERMINED TO HAVE A DELETERIOUS EFFECT ON THE PROFUNDAL
BOTTOM-DWELLING ORGANISMS STUDIED. THESE STUDIES INDICATED THAT
DIFFERENCES OCCURRING IN THE POPULATION DENSITY OF BOTTOM ORGANISMS IN
THE FOUR WISCONSIN LAKES STUDIED ARE DUE TO ECOLOGICAL VARIABLES WITHIN
THESE SEPARATE BODIES OF WATER. (JONES-WISCONSIN)

FIELD 02H, 05C

ACCESSION NO. W70-06217

LIMNOLOGICAL ASPECTS OF RECREATIONAL LAKES,

ROBERT A. TAFT SANITARY ENGINEERING CENTER, CINCINNATI, OHIO. DIV. OF WATER
SUPPLY AND POLLUTION CONTROL.

KENNETH M. MACKENTHUN, WILLIAM M. INGRAM, AND RALPH PORGES.

AVAILABLE FROM THE SUPERINTENDENT OF DOCUMENTS, U.S. GOVERNMENT PRINTING
OFFICE, WASHINGTON, DC, 20402, PRICE $4.25.U S DEP OF HEALTH, EDUCATION,
AND WELFARE, PUBLIC HEALTH SERVICE PUBLICATION NO 1167, 1964. 176 P, 11
FIG, 9 TAB, 52 PLATES, 448 REF.

DESCRIPTORS:
*LIMNOLOGY, *LAKES, *RECREATION, RESERVOIRS, PONDS, TEMPERATURE, LIGHT
PENETRATION, DISSOLVED OXYGEN, BICARBONATES, HYDROGEN ION
CONCENTRATION, NUISANCE ALGAE, AQUATIC PLANTS, BOTTOM SEDIMENTS,
ANIMALS, FISH, INFLOW, OUTLETS, NUTRIENTS, PRODUCTIVITY,
PHOTOSYNTHESIS, EUTROPHICATION, TOXICITY, MIDGES, MOSQUITOES, SAMPLING,
CONTROL, INDUSTRIAL WASTES, TENNESSEE, SHORE-LINE COVER, SNAILS,
WISCONSIN, HUMAN DISEASE, GAME BIRDS, TECHNOLOGY, POPULATION.

IDENTIFIERS:
LEECHES, SCHISTOSOME CERCARIAE, TVA RESERVOIRS, ANOPHELES
QUADRIMACULATUS, MADISON(WIS), PSOROPHORA CILIATA, ARTIFICIAL KEY.

ABSTRACT:
THIS BOOK CONSIDERS THE MANY PROBLEMS ASSOCIATED WITH THE RECREATIONAL
USE OF LAKES, RESERVOIRS, AND PONDS, AND IS DIRECTED TOWARD
INTERPRETATION OF PROBLEMS AND MANAGEMENT OF THE RECREATIONAL LAKE
PHENOMENA. CHAPTERS CONSIDER THE GENERAL PROBLEM OF AQUATIC NUISANCES;
REVIEW THE ECOLOGY OF LAKES, RESERVOIRS AND PONDS AND PRESENT
INFORMATION ON BIOTIC PRODUCTION, LEADING TO AN UNDERSTANDING OF THE
SCOPE AND MAGNITUDE OF BASIC NUISANCE PROBLEMS; DISCUSS NUTRIENTS FROM
SEWAGE AND RUNOFF AND THEIR IMPACT ON BIOLOGICAL GROWTHS; REVIEW PLANT
AND ANIMAL PESTS AFFECTING RECREATIONAL WATER; PRESENT SIMPLE KEYS WITH
ILLUSTRATIONS FOR IDENTIFICATION OF SOME COMMON AQUATIC PLANTS; DISCUSS
THE MECHANICS OF ESTABLISHING A SAMPLING PROGRAM FOR A LAKE, RESERVOIR,
OR POND TO DETERMINE THE PRESENT STATE OF BIOLOGICAL GROWTHS AND
POSSIBLY PREDICT FUTURE TRENDS; AND PRESENT INFORMATION FOR CONTROL OR
ALLEVIATION OF EXCESSIVE PRODUCTION OF BIOLOGICAL NUISANCES. KNOWLEDGE
OF THE MUTUAL RELATIONS BETWEEN ORGANISMS AND THEIR ENVIRONMENT IS
NECESSARY FOR INSTIGATION OF AN EFFECTIVE CONTROL PROGRAM. A GLOSSARY
IS INCLUDED. (JONES-WISCONSIN)

FIELD 02H, 05C

ACCESSION NO. W70-06225

SOME QUANTITATIVE ASPECTS OF ALGAL GROWTH IN LAKE MENDOTA,

WISCONSIN UNIV., MADISON. DEPT. OF ZOOLOGY.

DONALD E. WOHLSCHLAG, AND ARTHUR D. HASLER.

ECOLOGY, VOL 32, NO 4, P 581-593, 1951. 1 FIG, 5 TAB, 27 REF.

DESCRIPTORS:
*ALGAE, *PLANT GROWTH, *WISCONSIN, ECOLOGY, CHEMICAL ANALYSIS,
PHYTOPLANKTON, CYANOPHYTA, SEDIMENTS, PLANT POPULATIONS, WEATHER,
WATER, MOVEMENT, CULTURES, BAYS, STREAMS, WEEDS, LITTORAL, WINDS.

IDENTIFIERS:
*LAKE MENDOTA(WIS), PELAGIC AREAS, HORIZONTAL, BLOOMS, GELATINOUS
MATTER.

ABSTRACT:
DIVERSE AREAS OF LAKE MENDOTA SHOW DIFFERENCES IN QUANTITIES OF ALGAE,
BOTH SEASONALLY AND INTRASEASONALLY. VARIOUS BAY AREAS AND STREAM
MOUTHS, WEEDY LITTORAL, AND PELAGIC STATIONS WITHIN BAYS HAVE
CHARACTERISTIC ALGAL NUMBERS. CERTAIN STATIONS RETAIN CHARACTERISTICS
FROM ONE SEASON TO THE NEXT, OTHERS VARY ON A SEASONAL BASIS. WHERE
STATION INDIVIDUALITY IS INDICATED, IT IS PRESENT OVER AND ABOVE
TEMPORAL VARIATION. ALGAL CULTURES DERIVED FROM SEDIMENTS EXHIBITED
CHARACTERISTIC DIFFERENCES FROM BAY TO BAY AND FROM STATION TO STATION
WITHIN BAYS. GENERALLY, THE SAME STATION INDIVIDUALITIES PRESENT IN
CULTURES WERE PRESENT IN CORRESPONDING COLLECTIONS FROM OVERLYING
WATERS. COARSE, SANDY SEDIMENTS FROM THE SAME LOCALE PRODUCED MORE
ALGAE WHEN OBTAINED PRIOR TO ICE-FREE SPRING CIRCULATION. THE POSSIBLE
SIGNIFICANCE OF BLUE-GREEN ALGAE TO SUBSEQUENT BLOOMS IS DISCUSSED.
ALGAL CONCENTRATIONS APPEAR GREATER AT A LITTORAL AREA AFTER 48-HOUR
ONSHORE WIND THAN AT A SIMILAR UNEXPOSED AREA. CHLOROPHYLL, CHEMICAL,
AND MICROSCOPIC DATA SUGGEST THAT SUDDEN APPEARANCE OF DIPLOCYSTIS
AEROGINOSA IS ACCOMPANIED BY INCREASE IN GELATINOUS MATTER
PROPORTIONATELY GREATER THAN INCREASE IN ACTIVITY REPRESENTED BY
CHLOROPHYLL INCREASE. WIND PRODUCED SEDIMENT AGITATION IS A POSSIBLE
AGENT IN PRODUCTION OF BLUE-GREEN BLOOMS RELATIVE TO SEASONAL
SUCCESSION OF ALGAL BLOOMS. (JONES-WISCONSIN)

FIELD 02H, 05C

ACCESSION NO. W70-06229

KINETICS OF OXYGEN UPTAKE BY DEAD ALGAE,

HOWARD UNIV., WASHINGTON, D.C. DEPT. OF CIVIL ENGINEERING.

MAN M. VARMA, AND FRANCIS DIGIAND.

JOURNAL OF THE WATER POLLUTION CONTROL FEDERATION, VOL 40, NO 4, P 613-626,
APRIL 1968. 11 FIG, 6 TAB, 16 REF.

DESCRIPTORS:
*AGE, *ALGAE, *BIODEGRADATION, *OXYGEN DEMAND, LABORATORY TESTS,
LAGOONS, OXIDATION, PHOTOSYNTHESES, PLANTS, TEMPERATURE, *WASTE WATER
TREATMENT.

IDENTIFIERS:
*UPTAKE, CELLS, CONCENTRATION, COMPOSITION, MATERIAL BALANCES,
ORGANISMS.

ABSTRACT:
THE RATE OF OXYGEN UPTAKE BY DEAD ALGAL CELLS IS INDEPENDENT OF THE
ALGAL CELL CONCENTRATION. THE RATE OF UPTAKE BY YOUNG CELLS IS GREATER
THAN THAT BY OLD CELLS. RATES INCREASE WITH TEMPERATURE UP TO ABOUT 35
DEG C, THEN DECREASE SHARPLY. INCREASED BACTERIAL POPULATION WILL
INCREASE THE BIODEGRADATION OF THE DEAD ALGAE. UPTAKE RATES OBSERVED
RANGED FROM 0.008 MICRO L O SUB 2/MIN/MG ALGAE FOR OLD CELLS AT 20 DEG
C TO 0.294 MICRO L O SUB 2/MIN/MG ALGAE FOR YOUNG CELLS AT 35 DEG C.
MATERIAL BALANCES SHOWED THAT THE TOTAL ACTUAL OXYGEN UPTAKE WAS LESS
THAN THE TOTAL OXYGEN REQUIRED FOR COMPLETE DEGRADATION OF THE
SUBSTRATE BECAUSE OXIDATION WAS INCOMPLETE. (AGUIRRE-TEXAS)

FIELD 05D

ACCESSION NO. W70-06600

OVERLOADED OXIDATION PONDS--TWO CASE STUDIES,

KANSAS UNIV., LAWRENCE. DEPT. OF CIVIL ENGINEERING.

ROSS E. MCKINNEY.

JOURNAL OF THE WATER POLLUTION CONTROL FEDERATION, VOL 40, NO 1, P 49-56, JAN 1968. 9 TAB, 7 REF.

DESCRIPTORS:
*AERATION, *ALGAE, *LAGOONS, *OXIDATION LAGOONS, BIOCHEMICAL OXYGEN DEMAND, CHEMICAL OXYGEN DEMAND, DISSOLVED OXYGEN, ODOR, ORGANIC LOADING, *WASTE WATER TREATMENT.

IDENTIFIERS:
*AERATION TUBES, *DIFFUSERS, *RAYTOWN(MO), LEE'S SUMMIT(MO), OXYGEN TRANSFER, PH, SUSPENDED SOLIDS.

ABSTRACT:
THE USE OF A DIFFUSED AERATION SYSTEM AS A MEANS OF INCREASING THE CAPACITY AND IMPROVING THE PERFORMANCE OF OVERLOADED OXIDATION PONDS WAS EVALUATED IN A ONE YEAR DUAL STUDY OF THE PONDS AT LEE'S SUMMIT AND RAYTOWN, MO. THE SUPPLEMENTAL AERATION SYSTEM CONSISTED OF POLYETHYLENE TUBE DIFFUSERS, A COMMON HEADER, AND AN AIR COMPRESSOR. DATA ACCUMULATED DURING THE TEST PERIOD INCLUDE BOD, COD, SUSPENDED SOLIDS, PH, AND DISSOLVED OXYGEN. BIOCHEMICAL OXYGEN DEMAND (BOD) AND COD REMOVAL AVERAGED 80% AND 54% AT LEE'S SUMMIT, RESPECTIVELY. THE POND COVERED 25 ACRES, WITH AN AVERAGE DEPTH OF 5 FEET AND AN AVERAGE DETENTION PERIOD OF 30 DAYS. THE BOD5 LOAD TO THE POND WAS ESTIMATED TO BE 2550 LBS/DAY. THE OXIDATION POND AT RAYTOWN CONSISTED OF A FIRST-STAGE AERATED LAGOON WITH TWO 15-HP AERATORS FOLLOWED BY A SECOND-STAGE LAGOON, WHERE THE PLASTIC AERATION TUBING WAS INSTALLED, ESTIMATED TO RECEIVE AN ORGANIC LOAD OF 1000 LBS BOD5/DAY. ORGANIC REMOVAL EFFICIENCIES WERE SIMILAR TO THE LEE'S SUMMIT SYSTEM AS A STATISTICAL BASIS. THE PLASTIC TUBING AERATION UNITS CAN PRODUCE AN EFFLUENT UNDER 30 MG/L BOD5 AT A LOADING OF 1000 PEOPLE/D/AC. BUT, THE STUDY INDICATES THAT THE AERATION TUBING SHOULD BE USED WITH CAUTION. IT CAN ASSIST OVERLOADED OXIDATION PONDS BY IMPROVING MIXING, BUT IS NOT A CURE-ALL. CLOGGING OF THE AERATION TUBING WAS A PROBLEM RELIEVED BY TREATMENT WITH HYDROCHLORIC ACID, AND THE BASIC MECHANISM OF TREATMENT WAS INCREASED ALGAL METABOLISM THROUGH BETTER MIXING RATHER THAN DIRECT OXYGEN TRANSFER. (AGUIRRE-TEXAS)

FIELD 05D, 05G

ACCESSION NO. W70-06601

BIOLOGICAL EXTRACTION OF NUTRIENTS,

MICHIGAN UNIV., ANN ARBOR. DEPT. OF CIVIL ENGINEERING.

JACK A. BORCHARDT, AND HARDAM S. AZAD.

JOURNAL OF THE WATER POLLUTION CONTROL FEDERATION, VOL 40, NO 10, P 1739-1754, OCT 1968. 13 FIG, 22 REF.

DESCRIPTORS:
*ALGAE, *PHOSPHORUS, *PHOSPHATES, ENERGY, DI-URNAL DISTRIBUTION, TEMPERATURE, METABOLISM, *WASTE WATER TREATMENT, LIGHT, MICROORGANISMS, PLANTS.

IDENTIFIERS:
*BIOLOGICAL EXTRACTION, GROWTH, SHOCK LOADING.

ABSTRACT:
STUDIES OF PHOSPHATE UPTAKE FROM SYNTHETIC SECONDARY WASTE WATER BY ALGAE IN CONTINUOUS-CULTURE LABORATORY UNITS SHOWED THREE REGIMES OF PHOSPHATE UPTAKE: (1) GROWTH DEPENDENT ON PHOSPHATE CONCENTRATION BELOW 0-1.5 MG/L PHOSPHATE (0-3% OF ALGAL DRY WEIGHT), (2) STORAGE OF PHOSPHATE AT 1.5 TO 4.5 MG/L (3-9%), AND (3) SATURATION AT PHOSPHATE LEVELS GREATER THAN 4.5 MG/L (9%), AT AN ORGANISM DENSITY OF 50 MG/L. AT 15 MG/L THE GROWTH RATE ALSO DECLINED WHEN PHOSPHATE CONCENTRATION WAS LOWERED BEYOND 3% OF THE ALGAL DRY WEIGHT. AS TEMPERATURES BECOME LOWER, HIGHER PHOSPHATE CONCENTRATIONS WERE NEEDED TO GROW THE SAME MASS OF PROTOPLASM. ABOUT TWO-THIRDS OF THE PHOSPHATE UPTAKE ACHIEVED UNDER CONSTANT LIGHT CONDITIONS IS ACHIEVED UNDER A DI-URNAL TYPE LIGHT CYCLE. AFTER SHOCK LOADING PHOSPHATE STARVED CULTURES CAN EXHIBIT PHOSPHATE UPTAKES OF UP TO 20% BY WEIGHT. (HANCUFF-TEXAS)

FIELD 05D, 05C

ACCESSION NO. W70-06604

IMPROVEMENTS IN TREATMENT DESIGN FOR ENHANCING WASTE WATER QUALITY,

WASHINGTON UNIV., SEATTLE. DEPT. OF CIVIL ENGINEERING.

DALE A. CARLSON, AND H. KIRK WILLARD.

WASHINGTON UNIVERSITY WATER RESEARCH CENTER FINAL PROJECT REPORT FOR FY 1968, JULY 11, 1968. 17P, 13 FIG, 20 REF, APPEND. OWRR PROJECT NO A-016-WASH.

DESCRIPTORS:
*TERTIARY TREATMENT, *AEROBIC TREATMENT, OXYGEN DEMAND, BIOCHEMICAL OXYGEN DEMAND, SUSPENDED LOAD, SOLIDS CONTACT PROCESS, SEWAGE TREATMENT, TRICKLING FILTERS, WASTE WATER TREATMENT, PHOSPHATES, NITRATES, ALGAE.

IDENTIFIERS:
SOLIDS REMOVAL.

ABSTRACT:
A NEW TWO-PHASE BIOLOGICAL TERTIARY WASTE TREATMENT SYSTEM INCLUDES AN AEROBIC BACTERIAL GROWTH UNIT AND A PARTICULATE REMOVAL UNIT. IN THE REMOVAL UNIT, THE EFFECTS OF SUSPENDED SOLIDS LOADING ON A COMMUNITY OF ATTACHED PARTICULATE FEEDING ORGANISMS WERE DETERMINED. A BIOMASS YIELD OF 16% OF THE INFLUENT SUSPENDED SOLIDS FOR THE PARTICULATE FEEDERS PRODUCED A LOW EFFLUENT SOLIDS CONCENTRATION OF ONLY 0.069 MG SUSPENDED SOLIDS PER MG B. O. D. REMOVED. THIS DENSE MATERIAL WAS FOUND TO SETTLE RAPIDLY. TOTAL DISSOLVED ORGANIC CARBON WAS OVER 90% DESTROYED. SIMILAR REMOVAL VALUES OF 90 TO 94% WERE MEASURED FOR B. O. D. THE DECREASES OF BOTH ORGANIC CARBON AND B. O. D. WERE HIGHER AT LOWER LOADING RATES. MOST NITROGEN COMPOUNDS WERE OXIDIZED TO NITRATES. AS EXPECTED, THE PERCENT OF NITRIFICATION DECREASED WITH INCREASED LOADING. PHOSPHATES WERE NOT REMOVED EXCEPT FOR THE AMOUNT INCORPORATED INTO WASTED CELL BIOMASS RANGING IN MAGNITUDE FROM 22 TO 36%. (KNAPP-USGS)

FIELD 05D

ACCESSION NO. W70-06792

FUNCTION OF SOLIDS IN ANAEROBIC LAGOON TREATMENT OF WASTE WATER,

MELBOURNE WATER SCIENCE INST. (AUSTRALIA).

C. D. PARKER, AND G. P. SKERRY.

JOURNAL OF THE WATER POLLUTION CONTROL FEDERATION, VOL 40, NO 2, P 192-204, FEBRUARY, 1968. 13 TAB, 4 REF.

DESCRIPTORS:
*ANAEROBIC CONDITIONS, *EFFLICIENCIES, *LAGOONS, *SLUDGE, ALGAE, BIOCHEMICAL OXYGEN DEMAND, MIXING, SEWAGE TREATMENT, WASTE WATER TREATMENT.

IDENTIFIERS:
*ANAEROBIC TREATMENT, *AUSTRALIA, LABORATORY-SCALE STUDIES, PLANT-SCALE STUDIES, REMOVAL.

ABSTRACT:
FIELD STUDIES OF ANAEROBIC LAGOONS IN AUSTRALIA HAVE SHOWN THAT SOLIDS FARTHER FROM THE INLET OF A LAGOON ARE MORE ACTIVE IN PURIFICATION THAN THOSE DEPOSITED CLOSER TO THE INLET, ALTHOUGH IN MANY CASES THEY SHOW LOWER GAS YIELDS. LAGOON SLUDGE PURIFICATION INDEX (LBS BOD/ACRE/DAY/LB VS) VALUES RANGED FROM 0.0053 DURING THE WINTER AT A TYPICAL POND INLET TO 0.00902 DURING THE SUMMER AT THE POND OUTLET. THE ORGANIC LOAD TO THE POND DURING THIS PERIOD WAS ABOUT 482 LBS BOD/DAY/ACRE, AND THE SLUDGE GAS YIELD VARIED FROM 1.2 TO 7.7 ML/DAY/GVS. THE PRESENCE OF ALGAE IN POND SUPERNATANT APPARENTLY DOES NOT REDUCE THE SLUDGE'S BOD REMOVAL CAPACITY. WHERE METHANE FERMENTATION IS INHIBITED, BOD CAN STILL BE REMOVED BY SULFATE REDUCTION. LABORATORY STUDIES SHOWED THAT A SUBSTANTIAL INCREASE (ALMOST 30%) IN BOD REMOVAL CAN BE ACHIEVED BY MIXING SLUDGE WITH THE SUPERNATANT, AND THAT ALGAE PROLIFERATE READILY IN MEDIA WITH HIGH ORGANIC CONTENTS. (AGUIRRE-TEXAS)

FIELD 05D

ACCESSION NO. W70-06899

USE OF UNICELLULAR ALGAE FOR PURIFYING WASTE WATERS FROM MAN-MADE FIBER PLANTS
(IN RUSSIAN),

G. M. P. MORDVINTSEVA, V. V. GRABOVSKAYA, AND V. K. MARINICH.

KHIMICHESKIE VOLOKNA, VOL. 3, P 61-62, 1966.

DESCRIPTORS:
*ALGAE, IRON COMPOUNDS, HYDROGEN SULFIDE, BIOCHEMICAL OXYGEN DEMAND,
NUTRIENTS, WASTE WATER TREATMENT, *FIBER(PLANT).

IDENTIFIERS:
*VISCOSE RAYON PLANT WASTES, *MAN-MADE FIBER PLANTS, NYLON-6, *VISCOSE
RAYON, ZINC COMPOUNDS, CARBON DISULFIDE.

ABSTRACT:
LABORATORY AND PILOT-PLANT EXPERIMENTS ARE REPORTED ON THE USE OF
UNICELLULAR GREEN ALGAE (CHLORELLA AND ANKISTRODESMUS) TO TREAT WASTE
WATERS FROM THE MANUFACTURE OF NYLON-6 AND VISCOSE RAYON. GROWTH OF THE
ALGAE COULD BE STIMULATED BY ADDITION OF NUTRIENTS SUCH AS UREA,
MAGNESIUM SULPHATE, AND POTASSIUM PHOSPHATE. IT WAS FOUND THAT
TREATMENT, WHICH WAS COMPLETE WITHIN 6 DAYS UNDER LABORATORY CONDITIONS
AND 3 DAYS IN THE OPEN AIR, EFFECTED COMPLETE REMOVAL OF ZINC, IRON,
HYDROGEN SULPHIDE, CARBON DISULPHIDE, AND SULPHIDES, REDUCED THE BOD BY
90 PERCENT, AND INCREASED THE DISSOLVED-OXYGEN CONCENTRATION BY 100-400
PER CENT, BUT HAD NO EFFECT ON CALCIUM AND SULPHATES. (LIVENGOOD -
WORK-NORTH CAROLINA STATE UNIV)

FIELD 05D

ACCESSION NO. W70-06925

AN INTRODUCTION TO THE LIMNOLOGY OF HARTBEESPOORT DAM WITH SPECIAL REFERENCE TO
THE EFFECT OF INDUSTRIAL AND DOMESTIC POLLUTION,

NATIONAL INST. FOR WATER RESEARCH, PRETORIA (SOUTH AFRICA).

B. R. ALLANSON, AND J. M. T. M. GIESKES.

HYDROBIOLOGIA, VOL 18, P 77-94, 1961. 11 REF.

DESCRIPTORS:
*RESERVOIRS, *LIMNOLOGY, *INDUSTRIAL WASTES, *WATER POLLUTION EFFECTS,
*EUTROPHICATION, DISSOLVED OXYGEN, HYDROGEN ION CONCENTRATION,
EPILIMNION, PHOTOSYNTHESIS, ALGAE, BOTTOM SEDIMENTS, THERMOCLINE,
HYPOLIMNION, ANIMALS, DENSITY, IRON, PHOSPHORUS, SILICA, SAMPLING,
CHEMICAL ANALYSIS, PHYSICAL PROPERTIES, CARP, TILAPIA, TEMPERATURE,
NUTRIENTS, PHYTOPLANKTON, CYANOPHYTA, NITROGEN, OLIGOCHAETES, DIPTERA.

IDENTIFIERS:
*HARTBEESPOORT DAM(SOUTH AFRICA), TRANSPARENCY, MARGINAL VEGETATION
FAUNA, SANDY BOTTOMS FAUNA, MARGINAL VEGETATION, MICROCYSTIS, ANABAENA,
EUCYCLOPS, NAIS, CRICOTOPUS, AUSTROCLOEON, ENALLAGMA, PSEUDAGRION,
BRACHYTHEMIS, CHYDORUS, PLEUROXUS, TANYPUS, TANYTARSUS, LIMNODRILUS,
EUCYCLOPS, ALONA, CHAOBORUS.

ABSTRACT:
THE CORRELATED DATA OF THIS STUDY OF WATER MASS IN SOUTH AFRICA
CONTRIBUTE TO KNOWLEDGE OF POLLUTION EFFECTS IN A RESERVOIR.
HARTBEESPOORT DAM IS EUTROPHIC WITH A LOW TRANSPARENCY DEVELOPED DURING
THE PAST 25 YEARS. MONOMICTIC WITH COMPLETE CIRCULATION DURING WINTER
MONTHS, LOW DISSOLVED OXYGEN CONCENTRATIONS THROUGHOUT THE RESERVOIR
OFTEN RESULT. NUTRIENTS ARE CONTRIBUTED BY THE JUKSKEI-CROCODILE RIVER
SYSTEM AND TO A LESSER DEGREE FROM THE MAGALIES RIVER. RISES IN THE
EPILIMNETIC PH WERE MAINLY CAUSED BY PHOTOSYNTHETIC ACTIVITY OF ALGAE
IN THE PRESENCE OF MODERATELY BASIC WATER. BOTTOM SEDIMENTS IN THE
DEEPER PORTION OF THE RESERVOIR UNDERWENT RAPID ANAEROBIASIS AFTER THE
ESTABLISHMENT OF A THERMOCLINE AND OXYCLINE IN SUMMER. FAUNA DENSITY IN
THE BOTTOM SEDIMENTS DECREASES RAPIDLY WITH THE ESTABLISHMENT OF THE
HYPOLIMNION. SEDIMENTS ARE RICH IN IRON AND DURING SUMMER THERE WAS AN
INCREASE IN THE FERROUS COMPONENT. THE SOLUBLE AND ACID SOLUBLE
PHOSPHORUS INCREASES IN THE HYPOLIMNION AFTER A WINTER MINIMUM. SILICA
CONTENT IS COMPARABLE WITH PRODUCTIVE WATER MASSES IN OTHER PARTS OF
THE WORLD. FAUNAL ASSOCIATIONS ARE TYPICAL OF OTHER MODERATELY BASIC
AND HIGHLY PRODUCTIVE WATER IN THE SAME GEOGRAPHICAL REGION OF SOUTH
AFRICA. (JONES-WISCONSIN)

FIELD 05C

ACCESSION NO. W70-06975

A COMPOSITE RATING OF ALGAE TOLERATING ORGANIC POLLUTION,

ROBERT A. TAFT WATER RESEARCH CENTER, CINCINNATI, OHIO. ADVANCED WASTE TREATMENT RESEARCH LAB.

C. MERVIN PALMER.

JOURNAL OF PHYCOLOGY, VOL 5, NO 1, P 78-82, 1969. 5 P, 11 TAB, 29 REF.

DESCRIPTORS:
*ALGAE, *BIOINDICATORS, *WATER POLLUTION EFFECTS, *REVIEWS, *WATER QUALITY, WATER POLLUTION, ORGANIC MATTER, BIBLIOGRAPHIES.

IDENTIFIERS:
POLLUTION-TOLERANT ALGAE.

ABSTRACT:
A REVIEW OF POLLUTION-TOLERANT ALGAE USING REPORTS FROM 165 AUTHORS SHOWS THAT THE GENERA AND SPECIES MOST OFTEN REFERRED TO AS SIGNIFICANT FALL INTO A RELATIVELY STABLE SERIES. DIATOMS, PIGMENTED FLAGELLATES, GREEN, AND BLUE-GREEN ALGAE ARE ALL WELL REPRESENTED AMONG THE POLLUTION-TOLERANT GENERA AND SPECIES. EUGLENA, OSCILLATORIA, CHLAMYDOMONAS, SCENEDESMUS, CHLORELLA, NITZSCHIA, NAVICULA, AND STIGEOCLONIUM ARE THE MOST TOLERANT GENERA, AND EUGLENA VIRIDIS, NITZSCHIA PALEA, OSCILLATORIA LIMOSA, SCENEDESMUS QUADRICAUDA, AND OSCILLATORIA TENUIS ARE THE MOST TOLERANT SPECIES. IN SOME GENERA, FOR EXAMPLE EUGLENA, A SINGLE TOLERANT SPECIES IS FAR MORE SIGNIFICANT THAT ALL OTHERS. IN OTHER GENERA, AS IN OSCILLATORIA, ONLY A SLIGHT DIFFERENCE DISTINGUISHES THE POLLUTION TOLERANCE OF TWO OR MORE SPECIES. ALGAL POLLUTION INDICES ARE PRESENTED FOR USE IN RATING WATER SAMPLES WITH HIGH ORGANIC POLLUTION. (KNAPP-USGS)

FIELD 05C

ACCESSION NO. W70-07027

NITRIFICATION AND DENITRIFICATION IN A MODEL WASTE STABILIZATION POND,

TEXAS UNIV., AUSTIN. CENTER FOR RESEARCH IN WATER RESOURCES.

JORGE AGUIRRE, AND EARNEST F. GLOYNA.

ENVIRONMENTAL HEALTH ENGINEERING RESEARCH LABORATORY, REPORT EHE-05-6701, CRWR-19, JUNE 1967. 82 P, 22 FIG, 5 TAB, 38 REF. FWQA CONTRACT WP-00688-38.

DESCRIPTORS:
*DENITRIFICATION, EFFICIENCIES, *NITRIFICATION, NITROGEN COMPOUNDS, NITROGEN CYCLE, ALGAE, BIOLOGICAL TREATMENT, BOTTOM SEDIMENTS, NUTRIENTS, *WASTE WATER TREATMENT, *OXIDATION LAGOONS, MILK, *MODEL STUDIES.

IDENTIFIERS:
*FACULTATIVE LAGOONS, LABORATORY MODEL, COD REMOVAL, MILK SUBSTRATE, NITROGEN REMOVAL, *MILK WASTES.

ABSTRACT:
A LABORATORY MODEL OF A FACULTATIVE WASTE STABILIZATION POND WAS USED TO INVESTIGATE NITRIFICATION AND DENITRIFICATION OF MILK WASTES AS A FUNCTION OF DEPTH. THE MODEL WAS OPERATED ON A SEMI-CONTINUOUS FLOW BASIS USING A DRY MILK PRODUCT AS FEED. THE ORGANIC AND HYDRAULIC LOADING WERE KEPT CONSTANT AT 50 LB BOD5/ACRE/DAY AND 3 INCHES/DAY (5 GAL/D), RESPECTIVELY. GENERAL POND CHARACTERISTICS WERE EVALUATED BY MEASURING D.O., PH, ORP AND TEMPERATURE PROFILES; COD AND THE DIFFERENT FORMS OF NITROGEN WERE MONITORED AT FREQUENT INTERVALS DURING THE TEST PERIOD TO DETERMINE DEPTH PROFILES OF THESE PARAMETERS AS WELL COD REMOVAL EFFICIENCIES AND NITROGEN TRANSPORATIONS. A RAPID REMOVAL OF ORGANICS OCCURRED IN OR NEAR THE BOTTOM SEDIMENT LAYER WITH A CORRESPONDINGLY HIGH DEGREE OF AMMONIFICATION. ABOVE THE SLUDGE LAYER THE CONTENTS REMAINED ESSENTIALLY UNCHANGED EXCEPT FOR MINOR VARIATIONS DUE TO ALGAL DENSITY FLUCTUATIONS. DETECTABLE CONCENTRATIONS OF NITRATES WERE NOT FOUND DURING THE TEST PERIOD, BUT A SUBSTANTIAL AMOUNT OF NITROGEN WAS LOST FROM THE LIQUID FRACTION. COD AND NITROGEN REMOVALS FROM THE WASTE WATER, AS MEASURED BY THE OVERFLOW, WERE 85% AND 44% RESPECTIVELY. (AGUIRRE-TEXAS)

FIELD 05D

ACCESSION NO. W70-07031

SULFIDE PRODUCTION IN WASTE STABILIZATION PONDS,

TEXAS UNIV., AUSTIN. CENTER FOR RESEARCH IN WATER RESOURCES.

ERNESTO ESPINO DE LA O, AND EARNEST F. GLOYNA.

ENVIRONMENTAL HEALTH ENGINEERING RESEARCH LABORATORY, REPORT EHE-04-6802, CRWR NO. 26, MAY 1967. 156 P, 50 FIG, 33 TAB, 75 REF. FWQA CONTRACT WP-00688-03.

DESCRIPTORS:
*HYDROGEN SULFIDE, *ORGANIC LOADING, *SULFATES, *SULFIDES, ALGAE, GASES, ODOR, OXIDATION-REDUCTION POTENTIAL, SULFUR BACTERIA, *WASTE WATER TREATMENT, *OXIDATION LAGOONS, *MODEL STUDIES.

IDENTIFIERS:
DETENTION TIME, LABORATORY MODELS, BOD REMOVAL RATES, DESIGN APPLICATIONS, FACULTATIVE PONDS, INFRARED SPECTROSCOPY, SULFUR COMPOUNDS, WASTE STABILIZATION PONDS.

ABSTRACT:
LABORATORY MODELS, UNDER CONTROLLED ENVIRONMENTAL CONDITIONS, WERE USED TO INVESTIGATE THE PRODUCTION AND RELEASE OF HYDROGEN SULFIDE FROM WASTE STABILIZATION PONDS. A 6 FOOT DEEP MODEL FACULTATIVE POND WAS STUDIED UNDER SEVEN DIFFERENT OPERATING CONDITIONS. THE EXPERIMENTS EVALUATED THE RELATIONSHIP BETWEEN SULFIDE CONCENTRATION IN THE POND AND FOUR OPERATING CHARACTERISTICS: DETENTION, CONCENTRATION OF SULFATE ION IN THE INFLUENT, BOD SURFACE LOAD AND SULFATE ION CONCENTRATION IN TERMS OF SURFACE LOAD. AT CONSTANT DETENTION TIME AND BOD SURFACE LOAD, A LINEAR RELATIONSHIP WAS FOUND TO APPLY BETWEEN SULFIDE CONCENTRATION IN THE POND AND SULFATE CONCENTRATION IN THE INFLUENT. INCREASES IN SULFIDE CONCENTRATIONS IN THE POND FROM ZERO TO 1 AND 2 MG/L WERE ACCOMPANIED WITH DROPS IN OXIDATION-REDUCTION POTENTIALS OF 80 AND 120MV, RESPECTIVELY. FURTHER INCREASES IN SULFIDE CONCENTRATION AFFECTED THE ORP IN A LESSER WAY. SULFIDE CONCENTRATIONS IN THE POND OF ABOUT 7 MG/L WERE ACCOMPANIED BY AN ALMOST COMPLETE DISAPPEARANCE OF ALGAE, BUT THE BOD REMOVAL RATES WERE NOT APPRECIABLY AFFECTED. INFRARED SPECTROSCOPY WAS USED TO IDENTIFY METHANE, HYDROGEN SULFIDE AND CARBON DIOXIDE IN THE GASEOUS EMISSIONS FROM THE POND.
(AGUIRRE-TEXAS)

FIELD 05D

ACCESSION NO. W70-07034

TEMPERATURE AND MANGANESE AS DETERMINING FACTORS IN THE PRESENCE OF DIATOM OR BLUE-GREEN ALGAL FLORAS IN STREAMS,

ACADEMY OF NATURAL SCIENCES OF PHILADELPHIA, PA. DEPT. OF LIMNOLOGY.

RUTH PATRICK, BOWMAN CRUM, AND JOHN COLES.

PROCEEDINGS OF THE NATIONAL ACADEMY OF SCIENCES, VOL 64, NO 2, P 472-478, 1969. 2 TAB, 10 REF.

DESCRIPTORS:
*STREAMS, *ALGAE, *DIATOMS, TEMPERATURE, MANGANESE, CULTURES, RECIRCULATED WATER, NITRATES, PHOSPHATES, SODIUM, WATER POLLUTION EFFECTS, LABORATORY TESTS.

IDENTIFIERS:
GREEN ALGAE, BLUE-GREEN ALGAE, TEMPERATURE EXPERIMENTS, RECYCLED WATER.

ABSTRACT:
TRIALS CONDUCTED IN CULTURE BOXES INDICATED THAT AT AN AVERAGE TEMPERATURE OF 34-38C BLUE-GREEN ALGAE TEND TO REPLACE DIATOMS. BOTH BLUE-GREEN AND GREEN ALGAE, COMMONLY FOUND IN POLLUTED WATERS, WERE FAVORED BY MEDIA WITH MANGANESE CONCENTRATION OF ONLY A FEW PARTS PER BILLION. AN AVERAGE CONCENTRATION OF MANGANESE OF 0.02-0.043 MILLIGRAM/LITER IN NATURAL STREAMS AND 0.04-0.28 MILLIGRAM/LITER IN RECYCLED WATER EXPERIMENT FAVORED DOMINATION OF DIATOM FLORA. INCREASE IN NITRATE AND PHOSPHATE CONTENT OF WATER FAILED TO INFLUENCE THE FLORA OF FREE-FLOWING STREAMS, BUT PROMOTED THE GROWTH OF BLUE-GREEN AND GREEN ALGAE IN RECYCLED WATER OF A LOW MANGANESE CONTENT.
(WILDE-WISCONSIN)

FIELD 05C

ACCESSION NO. W70-07257

EUTROPHICATION IN NORTH AMERICA,

WASHINGTON UNIV., SEATTLE.

W. T. EDMONDSON.

EUTROPHICATION: CAUSES, CONSEQUENCES, CORRECTIVES, P 124-149. PRINTING AND
PUBLISHING OFFICE, NATIONAL ACADEMY OF SCIENCES, WASHINGTON, D C, 1969. 6
FIG, 4 TAB, 70 REF.

DESCRIPTORS:
*EUTROPHICATION, *NUISANCE ALGAE, *SEWAGE, WATER POLLUTION SOURCES,
WATER POLLUTION EFFECTS, PHOSPHORUS, NITROGEN, CHLOROPHYLL,
PRODUCTIVITY, REGIONAL ANALYSIS, RESERVOIRS.

IDENTIFIERS:
NUTRIENT BUDGETS, LAKE SEBASTICOOK(ME), LAKE WINNISQUAM(N H), MADISON
LAKES(WIS), LAKE WASHINGTON(WASH), NUISANCE CONDITIONS.

ABSTRACT:
EXAMPLES OF ARTIFICIAL ENRICHMENT ARE GIVEN. WELL-DOCUMENTED HISTORIES
OF THE EFFECTS OF ARTIFICIAL ENRICHMENT ARE NOT COMMON. CASE HISTORIES
ARE AVAILABLE FOR FOUR LAKES OR LAKE SYSTEMS: LAKE SEBASTICOOK, MAINE;
LAKE WINNISQUAM, NEW HAMPSHIRE; THE MADISON, WISCONSIN, LAKES; AND LAKE
WASHINGTON, WASHINGTON. IN EACH CASE, ARTIFICIAL ENRICHMENT OCCURRED BY
SEWAGE EFFLUENT INPUT; SEBASTICOOK ALSO RECEIVED LARGE INPUTS OF
INDUSTRIAL WASTES HIGH IN NUTRIENTS AND ORGANIC MATTER. ARTIFICIAL
ENRICHMENT IN ALL CASES HAS RESULTED IN LOWERED TRANSPARENCY,
MYXOPHYCEAN BLOOMS AND NUISANCE CONDITIONS OF SURFACE SCUM AND ODOR.
DIVERSION OF SEWAGE EFFLUENTS HAS LED TO IMPROVED CONDITIONS IN LAKE
WASHINGTON AND THE MADISON LAKES MONONA AND WAUBESA, ALTHOUGH LAKE
MENDOTA APPEARS TO BE DETERIORATING. ONE OF THE MAJOR FEATURES OF THE
INVESTIGATIONS OF ARTIFICIAL ENRICHMENT IN NORTH AMERICAN LAKES IS THE
DEVELOPMENT OF NUTRIENT BUDGETS. THIS APPROACH NEEDS FURTHER
DEVELOPMENT AND REFINEMENT, AND WITH CONTINUED DEVELOPMENT OF BIOASSAY
TECHNIQUES CAN BE EXPECTED TO AID FUTURE INVESTIGATIONS. (SEE ALSO
W70-03975). (VOIGTLANDER-WISCONSIN)

FIELD 05C

ACCESSION NO. W70-07261

EFFECTS OF INCREASED TEMPERATURES ON AQUATIC ORGANISMS,

ACADEMY OF NATURAL SCIENCES OF PHILADELPHIA, PA. DEPT. OF LIMNOLOGY.

JOHN CAIRNS, JR.

PROCEEDINGS OF THE 10TH INDUSTRIAL WASTE CONFERENCE, PURDUE UNIVERSITY,
ENGINEERING BULLETIN, VOL. 40, NO. 1, 1956. P 346, 4 FIG, 3 REF.

DESCRIPTORS:
*MICROORGANISMS, *ALGAE, *DOMINANT ORGANISMS, *ACCLIMATIZATION,
DIATOMS, THERMAL POLLUTION, TEMPERATURE, CHEMICAL PROPERTIES, ECOLOGY,
ECOLOGICAL DISTRIBUTION, THERMAL EFFECTS.

IDENTIFIERS:
GREEN ALGAE, BLUE-GREEN ALGAE, HEAT DEATH.

ABSTRACT:
THE POPULATION OF MICROORGANISMS OF A RIVER IS COMPOSED OF THE SPECIES
THAT ARE BEST FITTED TO THE LOCAL ECOLOGICAL CONDITIONS. BECAUSE THERE
IS A FIERCE COMPETITION FOR SPACE, A SPECIES MAY BE ELIMINATED BY AN
UNFAVORABLE, RATHER THAN LETHAL, FACTOR, SUCH AS INCREASED TEMPERATURE.
A MIXED POPULATION OF ALGAE WAS TAKEN FROM A NORMAL HEALTHY STREAM AND
GRADUALLY EXPOSED TO AN INCREASED TEMPERATURE. AT 20 DEG C, A LARGE
NUMBER OF SPECIES OF DIATOMS WERE PRESENT AND DOMINATED THE CULTURES.
AT 30 DEG C, ONLY TWO SPECIES OF DIATOMS WERE NUMEROUS. ABOVE 30 DEG C,
A FEW INDIVIDUAL DIATOMS SURVIVED, BUT GREEN ALGAE PREDOMINATED AT 30
DEG C TO 35 DEG C. IN THE RANGE OF 35 DEG C TO 40 DEG C, BLUE-GREEN
ALGAE WERE DOMINANT. THERE ARE STRONG INDICATIONS THAT THE COMPOSITION
OF THE WATER IS AN IMPORTANT FACTOR IN HEAT DEATH OF AQUATIC ORGANISMS,
SINCE THE CHEMICAL BALANCE IS USUALLY QUITE DIFFERENT IN HEATED WATERS
FROM THAT OF A TYPICAL STREAM. MICROORGANISMS CAN BE ACCLIMATIZED TO
VERY HIGH TEMPERATURES IF THE CHANGES ARE SMALL AND GRADUAL.
(SPEAKMAN-VANDERBILT)

FIELD 05C

ACCESSION NO. W70-07313

RESEARCH TO SAVE AMERICA'S LAKES,

FEDERAL WATER POLLUTION CONTROL ADMINISTRATION, CORVALLIS, OREG. PACIFIC
NORTHWEST WATER LAB.

THOMAS E. MALONEY, AND ALFRED F. BARTSCH.

IN: WATER - 1969, CHEMICAL ENGINEERING PROGRESS SYMPOSIUM SERIES 97, VOL 65,
PUBLISHED BY AMERICAN INSTITUTE OF CHEMICAL ENGINEERS, NEW YORK, P 278-280,
1969. 3 P.

DESCRIPTORS:
*EUTROPHICATION, *LAKES, *WATER POLLUTION CONTROL, *REVIEWS, ALGAE,
WEEDS, ECOLOGY, ENVIRONMENT, CHEMCONTROL, BIOCONTROL, INTEGRATED
CONTROL MEASURES, PHYSICAL CONTROL, WATER QUALITY.

IDENTIFIERS:
*EUTROPHICATION CONTROL.

ABSTRACT:
THE MATURATION PROCESS OF A LAKE IS CALLED EUTROPHICATION AND IS
GREATLY ACCELERATED BY MAN'S ACTIVITIES WHICH INCREASE THE NUTRIENT
INPUT TO THE LAKE. THESE ACTIVITIES INCLUDE AGRICULTURAL DEVELOPMENT,
URBANIZATION, AND THE DISCHARGE OF SEWAGE, INDUSTRIAL WASTE, AND
TREATMENT PLANT EFFLUENTS TO THE LAKES. WITH ADVANCING OF
EUTROPHICATION, SYMPTOMS APPEAR INCLUDING: AN INCREASE IN PRODUCTIVITY
OF ALGAE, WATERWEEDS, AND OTHER LIVING MATTER; THE NITROGEN AND
PHOSPHORUS CONCENTRATIONS INCREASE; DISSOLVED OXYGEN IS DEPLETED IN THE
LOWER DEPTHS; THE MORE DESIRABLE FISH SPECIES BECOME EXTINCT, AND ROUGH
FISH SPECIES APPEAR; AND ROOTED VEGETATION BEGINS TO GROW IN FROM THE
LAKE MARGIN AND EVENTUALLY FILLS THE LAKE.
(KNAPP-USGS)

FIELD 05C, 02H

ACCESSION NO. W70-07393

REMOVAL OF NITROGEN AND PHOSPHORUS,

STANFORD UNIV., CALIF. DEPT. OF CIVIL ENGINEERING.

ROLF ELIASSEN, AND GEORGE TCHOBANOGLOUS.

PROCEEDINGS OF THE INDUSTRIAL WASTE CONFERENCE, 23RD, P 35-48, 1968. 5 TAB,
16 REF.

DESCRIPTORS:
*COST COMPARISONS, *EFFICIENCIES, *NITROGEN, *PHOSPHORUS, ALGAE,
BIOLOGICAL TREATMENT, WASTE WATER TREATMENT.

IDENTIFIERS:
*REMOVAL PROCESSES, CHEMICAL TREATMENT, OVERFERTILIZATION, PHYSICAL
TREATMENT, UNIT OPERATIONS, UNIT PROCESSES.

ABSTRACT:
THE NEED TO REMOVE NITROGEN AND PHOSPHORUS FROM WASTE WATERS STEMS FROM
THE FACT THAT THEY ARE CAPABLE OF STIMULATING AND PROMOTING THE
UNDESIRABLE GROWTH OF ALGAE AND AQUATIC PLANTS. THE IMPORTANT FORMS OF
NITROGEN ARE AMMONIA, NITRITE AND NITRATE, WHILE PHOSPHORUS MAY OCCUR
AS SOLUBLE INORGANIC PHOSPHATES. NITROGEN AND PHOSPHORUS REMOVAL
METHODS MAY BE CLASSIFIED AS BIOLOGICAL, CHEMICAL AND PHYSICAL.
BIOLOGICAL METHODS INCLUDE ACTIVATED SLUDGE, TRICKLING FILTERS AND
ANAEROBIC TREATMENT. CHEMICAL AND PHYSICAL PROCESSES INCLUDE ION
EXCHANGE, ELECTRODIALYSIS, REVERSE OSMOSIS AND DISTILLATION. TYPICAL
REMOVAL EFFICIENCIES RANGE BETWEEN 40 TO 98 PERCENT AND 30 TO 98
PERCENT FOR NITROGEN AND PHOSPHORUS COMPOUNDS, RESPECTIVELY. IN
PLANNING AND DESIGNING FACILITIES TO BE USED FOR THE REMOVAL OF
NITROGEN AND PHOSPHORUS, KEY FACTORS WHICH MUST BE CONSIDERED ARE: (1)
WHETHER COMPOUNDS OF ONE OR BOTH ARE TO BE REMOVED, (2) THE USE TO BE
MADE OF THE TREATED WASTE WATER, (3) THE AVAILABLE MEANS FOR DISPOSING
OF THE ULTIMATE CONTAMINANTS, AND (4) THE ECONOMIC FEASIBILITY OF THE
SELECTED PROCESS OR PROCESSES. THE ESTIMATED REMOVAL COSTS PER MILLION
GALLONS RANGE FROM $4 TO $1000 FOR BOTH NITROGEN AND PHOSPHORUS
DEPENDING ON THE PROCESS USED AND DEGREE OF TREATMENT REQUIRED.
(AGUIRRE-TEXAS)

FIELD 05D

ACCESSION NO. W70-07469

THE MANAGEMENT AND DISPOSAL OF DAIRY MANURE,

WASHINGTON STATE UNIV., PULLMAN. COLL. OF ENGINEERING.

DONALD E. PROCTOR.

PROCEEDINGS OF THE INDUSTRIAL WASTE CONFERENCE, 23RD, 1968. P 554-566, 8 FIG.

DESCRIPTORS:
*ANIMAL WASTES, MANAGEMENT, *FARM MANAGEMENT, *DAIRY INDUSTRY, *ALGAE, *CATTLE, SLURRIES, SPRAYING, ACTIVATED SLUDGE, FARM WASTES.

IDENTIFIERS:
*MANURE, ANAEROBIC LAGOON, AERATED LAGOON.

ABSTRACT:
DAIRY MANURE CAN BE EITHER AN ASSET OR A LIABILITY DEPENDING ON THE FARMER'S MANAGEMENT POLICIES. INCREASED DEMAND FOR LIVESTOCK-DERIVED PRODUCTS, SPECIALIZATION OF FARM OPERATIONS, CONFINEMENT REARING, CHEAPER CHEMICAL FERTILIZERS, URBAN SPRAWL AND FARM AREA ENCROACHMENT, AND HIGHER AESTHETIC STANDARDS ARE ALL FACETS OF THE CHANGING PROBLEM OF MANURE DISPOSAL. OF THESE CHANGES, CONFINEMENT REARING IS MOST SIGNIFICANT. NOT ONLY IS THE MANURE CONCENTRATED INTO A SMALLER AREA, BUT RAINFALL RUNOFF CAN TREBLE THE WASTE VOLUME TO BE HANDLED. TWO WASHINGTON STATE DAIRY FARMS RECEIVED FEDERAL DEMONSTRATION PROJECT GRANTS, THE KNOTT DAIRY FARM OF THE WASHINGTON STATE UNIVERSITY RECEIVING A GRANT FROM THE FWPCA, AND THE MONROE HONOR FARM OF THE STATE OF WASHINGTON INSTITUTIONAL FARM INDUSTRIES RECEIVING A GRANT FROM THE PUBLIC HEALTH SERVICE. THE TWO-YEAR PROJECT AT THE KNOTT DAIRY FARM PROPOSED TO (1) DEMONSTRATE THE CAPABILITIES OF AN ANAEROBIC LAGOON FOR FIRST STAGE TREATMENT OF DAIRY MANURE, (2) DEMONSTRATE THE COMPARATIVE CAPABILITIES AND ECONOMICS OF ACTIVATED SLUDGE AND NATURALLY AERATED LAGOONS FOR SECOND STAGE TREATMENT AND (3) DETERMINE WHETHER IT IS POSSIBLE AND PRACTICAL TO REDUCE THE NITROGEN AND PHOSPHORUS CONTENT OF THE TREATED EFFLUENT BY ALGAE PROPAGATION AND HARVESTING FOR USE AS CATTLE FEED. THE MONROE HONOR FARM PROJECT PROPOSED TO (1) DEMONSTRATE THE CAPABILITIES OF AN ANAEROBIC LAGOON FOR FIRST STAGE TREATMENT OF DAIRY MANURE. (MAKELA-TEXAS)

FIELD 05D

ACCESSION NO. W70-07491

STABILIZATION POND TREATMENT OF SLAUGHTER-HOUSE WASTES,

JAPAN PUBLIC NUISANCE LAB., TOKYO.

ISAMU HORASAWA.

PROCEEDINGS OF THE INDUSTRIAL WASTE CONFERENCE, 23RD, 1968. P 1178-1185, 3 FIG, 6 TAB, 9 REF.

DESCRIPTORS:
*HOGS, *LAGOONS, ALGAE, ORGANIC LOADING, WASTE WATER TREATMENT, WATER UTILIZATION, INDUSTRIAL WASTES, OXIDATION LAGOONS.

IDENTIFIERS:
*JAPAN, *SLAUGHTERHOUSE WASTES, EXPERIMENTAL STUDIES, LAGOON DESIGN, SCALE-UP FACTORS, WASTE COMPOSITION.

ABSTRACT:
WASTES DISCHARGED FROM ABOUT 900 SLAUGHTERHOUSES IN JAPAN, WHICH KILL APPROXIMATELY 6,000,000 ANIMALS PER YEAR, REQUIRE A HIGH DEGREE OF TREATMENT TO PREVENT WATER POLLUTION AND AESTHETIC NUISANCES OF BLOOD COLOR AND OFFENSIVE ODOR. THIS EXPERIMENTAL STUDY WAS DIRECTED TOWARD THE USE OF THE STABILIZATION POND PROCESS BY SMALL SLAUGHTERHOUSES. THE AVERAGE WATER CONSUMPTION BY SLAUGHTERHOUSES LOCATED THROUGHOUT JAPAN IS ABOUT 848 LITERS PER HOG UNIT AND THE AVERAGE BOD LOADING PER HOG UNIT IS 1.255 KG. FOUR EXPERIMENTAL STABILIZATION POND UNITS WERE RUN BOTH IN THE LABORATORY AND EXPOSED TO OUTSIDE WINTER CONDITIONS. THE BOD LOADING ON THE UNITS RANGED FROM 47 TO 239 MG/L. THESE STUDIES SHOWED THAT, WITHIN THE LIMITS OF LABORATORY SCALE TESTING, BOD LOADINGS UP TO 100 MG/L (100 G/M2) CAN BE TOLERATED WITHOUT DISTINCT AESTHETIC NUISANCE. THE DOMINANT ALGAE IN THE PONDS WAS CHLORELLA. THE RESPONSE TO VARIATION IN WEATHER CONDITIONS APPEARED TO BE DEPENDENT TO SOME EXTENT UPON THE BOD LOADING OF THE UNIT. THE MORE HEAVILY LOADED UNITS SHOWED AN INDIFFERENT RESPONSE TO WEATHER CHANGE. BIOCHEMICAL OXYGEN DEMAND (BOD) LOADING FOR STABILIZATION PONDS IS REPORTED TO RANGE FROM LESS THAN 10 MG/L UP TO 150 MG/L DEPENDING ON THE DGREEE OF TREATMENT REQUIRED. (AGUIRRE-TEXAS)

FIELD 05D

ACCESSION NO. W70-07508

THE RESPONSE OF FRESH-WATER PROTOZOAN COMMUNITIES TO HEATED WASTE WATERS,

VIRGINIA POLYTECHNIC INST., BLACKSBURG. DEPT. OF BIOLOGY.

JOHN CAIRNS, JR.

CHESAPEAKE SCIENCE, VOL 10, NO 3 AND 4, P 177-185, 1969. 12 REF, 8 FIG.

DESCRIPTORS:
*WATER POLLUTION EFFECTS, WATER POLLUTION SOURCES, *THERMAL POLLUTION,
THERMAL POWERPLANTS, *THERMAL STRESS, THERMAL WATER, *PROTOZOA, ALGAE,
DIATOMS, ANIMAL BEHAVIOR, *THERMODYNAMIC BEHAVIOR.

IDENTIFIERS:
*TEMPERATURE SHOCK, *PROTOZOAN COMMUNITIES, POTOMAC RIVER, SAVANNAH
RIVER, SHOCK EFFECTS.

ABSTRACT:
THE RESPONSE OF FRESH-WATER PROTOZOAN COMMUNITIES EXPOSED TO BOTH
SEVERE ACUTE TEMPERATURE SHOCKS AS WELL AS SMALL GRADUAL LONG-TERM
INCREASES ARE DISCUSSED. THE FORMER EXPERIMENTS WERE CARRIED OUT IN
PLASTIC TROUGHS WITH A CONSTANT FLOW OF UNFILTERED LAKE WATER. SEVERE
ACUTE SHOCKS (SOME TO NEARLY 50 C) RESULTED IN A MARKED REDUCTION IN
NUMBER OF SPECIES PRESENT. HOWEVER, RECOVERY WAS QUITE RAPID (A MATTER
OF A FEW DAYS) ONCE THE TEMPERATURE STRESS CEASED. OBSERVATIONS OF THE
EFFECTS OF SMALL GRADUAL LONG-TERM INCREASES WERE MADE ON THE PROTOZOAN
COMMUNITIES OF THE SAVANNAH AND POTOMAC RIVERS EACH OF WHICH RECEIVED
HEATED WASTE WATER DISCHARGES. EACH OF THESE STUDIES COVERED A PERIOD
IN EXCESS OF NINE YEARS AND OBSERVATIONS ARE STILL BEING MADE. THERE
WAS NO EVIDENCE THAT INDICATED THE PROTOZOAN COMMUNITIES OF THESE
RIVERS HAD BEEN DEGRADED BY THE SMALL GRADUAL TEMPERATURE INCREASES
RESULTING FROM THE DISCHARGE OF HEATED WASTE WATERS. HOWEVER, THERE IS
EVIDENCE THAT COMPETITIVE EXCLUSION OF ALGAL SPECIES BY OTHER MORE
TOLERANT ALGAL SPECIES MAY CAUSE QUALITATIVE SHIFTS IN THE COMMUNITY
STRUCTURE WHICH MAY BE UNDESIRABLE. (SJOLSETH-WASHINGTON)

FIELD 05C

ACCESSION NO. W70-07847

EFFECTS OF NITROGEN LIMITATION ON THE GROWTH AND COMPOSITION OF UNICELLULAR
ALGAE IN CONTINUOUS CULTURE,

SCHOOL OF AEROSPACE MEDICINE, BROOKS AFB, TEX.

B. RICHARDSON, D. M. ORCUTT, H. A. SCHWERTNER, CARA L. MARTINEZ, AND HAZEL E.
WICKLINE.

APPLIED MICROBIOLOGY, VOL 18, NO 2, P 245-250, AUGUST 1969, 3 FIG, 4 TAB, 19
REF.

DESCRIPTORS:
*GROWTH, *ALGAE, LIPIDS, NUTRIENTS, CHLORELLA, CULTURES.

IDENTIFIERS:
*FAT ACCUMULATION, *LABORATORY STUDIES, *NITROGEN LIMITATION, BATCH
CULTURES, CHLORELLA SOROKINIANA, COMPOSITION OF ALGAE, CONTINUOUS
CULTURES, FATTY ACIDS, OOCYSTIS POLYMORPHA, UNICELLULAR ALGAE.

ABSTRACT:
FAT ACCUMULATION TAKES PLACE IN MANY ALGAE AS A RESPONSE TO EXHAUSTION
OF THE NITROGEN SUPPLY AND MAY PROVIDE A MEANS OF ENCHANCING THE
POTENTIAL FOOD VALUE OF ALGAE. CHEMOSTATIC CONTINUOUS CULTURES OF
CHLORELLA SOROKINIANA AND OOCYSTIS POLYMORPHA WERE SUBJECTED TO
SUCCESSIVE REDUCTIONS IN INFLUENT NITROGEN. AS CELLULAR NITROGEN
CONTENT DECREASED FROM ABOUT 10 TO 4%, OXYGEN EVOLUTION, CARBON DIOXIDE
UPTAKE, CHLOROPHYLL CONTENT, AND TISSUE PRODUCTION WERE DRASTICALLY
REDUCED, BUT TOTAL LIPID CONTENT WAS ESSENTIALLY UNCHANGED. IN
BATCH-CULTURED CELLS, NITROGEN COULD BE REDUCED TO 3% OF DRY WEIGHT,
CAUSING AN INCREASE IN TOTAL FATTY ACIDS AND PRONOUNCED CHANGES IN THE
COMPOSITION OF THE FATTY ACID FRACTION. THESE RESULTS SUGGEST THAT
CELLULAR NITROGEN MUST FALL TO ABOUT 3% OF DRY WEIGHT BEFORE
APPRECIABLE INCREASES IN LIPID SYNTHESIS CAN OCCUR. ALL NITROGEN IS
THEN APPARENTLY COMPLETELY BOUND IN ESSENTIAL CELL CONSTITUENTS, AND
CARBON SUBSEQUENTLY FIXED IS CONVERTED INTO LIPID PRODUCTS. NITROGEN
LIMITATION MAY BE USEFUL IN INCREASING THE FOOD QUALITY OF
BATCH-CULTURED CELLS, BUT THE TECHNIQUE HAS LITTLE VALUE FOR CONTINUOUS
CULTURE SYSTEMS PER SE. (AGUIRRE-TEXAS)

FIELD 05D, 05C

ACCESSION NO. W70-07957

ALGAE HANDLED EFFICIENTLY BY AUGUSTA WATER DISTRICT,

AUGUSTA WATER DISTRICT, MAINE.

SIDNEY S. ANTHONY.

WATER AND SEWAGE WORKS, VOL 116, P 185-189, 1969. 1 FIG, 6 PHOTOS.

DESCRIPTORS:
*ALGAE, *MAINE, *WATER SUPPLY, COPPER SULFATE, RESERVOIRS, WATER
POLLUTION, MICROORGANISMS, HISTORY, TASTE, ODOR, DEEP WELLS, WATER
TREATMENT, MECHANICAL EQUIPMENT, PUMPING PLANTS, MAINE.

IDENTIFIERS:
*AUGUSTA(MAINE), CARLETON POND(MAINE), SYNURA, ASTERIONELLA, TYPHOID
FEVER, KENNEBEC RIVER(MAINE), COBBOSSEECONTEE LAKE(MAINE), BARGE.

ABSTRACT:
THE HISTORY OF THE AUGUSTA, MAINE, WATER SUPPLY AFTER 1870 IS
DESCRIBED. AT THAT TIME THE KENNEBEC RIVER WAS USED AS THE SOURCE OF
WATER; IN THE WINTER OF 1902-1903 WIDESPREAD TYPHOID FEVER OUTBREAKS,
TRACEABLE TO THE WATER SUPPLY, CAUSED THE INAUGURATION OF A WATER
DISTRICT. CARLTON POND, THE PRINCIPAL RESERVOIR, WAS CHOSEN BY THE
AUGUSTA WATER DISTRICT AND PLACED IN OPERATION IN 1906. THE POND,
HOWEVER, IS SUBJECT TO PERIODIC INFESTATIONS OF ALGAE, PRIMARILY
SYNURA, WITH ASTERIONELLA ALSO PRESENT. TO ALLEVIATE THE TASTE AND ODOR
PROBLEM CAUSED BY ALGAE, CONTROL MEASURES ARE TAKEN. A SPECIAL BARGE
AND PUMPS WERE DEVELOPED FOR APPLICATION OF COPPER SULFATE (THE ONLY
CHEMICAL USED) IN DILUTION OF 0.35 MILLIGRAMS/LITER. THE TREATMENT IS
COMPLETED IN ONE DAY, IN THE FALL, WHEN WATER TURBULENCE IS MINIMAL,
AND APPLICATION IS MADE TO THE ENTIRE LAKE AND SHORELINE. ONE TREATMENT
IS USUALLY ADEQUATE TO ELIMINATE THE ALGAE. (JONES-WISCONSIN)

FIELD 05F

ACCESSION NO. W70-08096

EFFECTS OF INSECTICIDES ON ALGAE,

ONTARIO WATER RESOURCES COMMISSION, TORONTO. DIV. OF RESEARCH.

A. E. CHRISTIE.

WATER AND SEWAGE WORKS, VOL 116, P 172-176, 1969. 6 TAB, 31 REF.

DESCRIPTORS:
*ALGAE, *INSECTICIDES, TOXICITY, DEGRADATION(DECOMPOSITION), DDT,
ALKALINITY, STABILIZATION, WASTE STORAGE, PONDS, HYDROLYSIS,
METABOLISM, CHEMICAL REACTIONS, CHLORELLA, TRACERS, RADIOACTIVITY,
CARBON RADIOISOTOPES, HYDROGEN ION CONCENTRATION, SCENEDESMUS, EUGLENA,
BACTERIA, YEASTS, ACTINOMYCETES, PONDING, WASTE TREATMENT.

IDENTIFIERS:
SEVIN, MALATHION, CHLORELLA PYRENOIDOSA, SCHRODERIA, THIMET,
TRICHODERMA.

ABSTRACT:
DDT, SEVIN, AND MALATHION WERE STUDIED TO DETERMINE THEIR EFFECT ON
ALGAL POPULATIONS ASSOCIATED WITH WASTE STABILIZATION PONDS AND THEIR
POSSIBLE DEGRADATION IN AN ALGAL ENVIRONMENT. THE PESTICIDES VARY IN
DEGREE OF TOXICITY TO ALGAE AND IN EXTENT TO WHICH THEY ARE DEGRADED IN
PRESENCE OF ALGAE. DDT EXHIBITED NO TOXIC PROPERTIES UP TO
CONCENTRATIONS OF 100 MILLIGRAMS/LITER AND RECEIVED ONLY SLIGHT
DEGRADATION. SEVIN IS TOXIC AT CONCENTRATIONS ABOVE 0.1
MILLIGRAMS/LITER AND IS NOT ALTERED APPRECIABLY IN ACIDIC MEDIA.
MALATHION APPEARS TO RECEIVE EXTENSIVE CONVERSION AND, ALTHOUGH CAPABLE
OF ALTERING COMPOSITION OF A MIXED ALGAL COMMUNITY, DID NOT DISPLAY
PERSISTENT INHIBITORY EFFECT. EXTENDING THIS TO THE ALKALINE WASTE
STABILIZATION POND AND ITS ALGAL COMMUNITY, ENTRY OF DDT, UP TO 100
MILLIGRAMS/LITEE, IS NOT LIKELY TO BE TOXIC NOR WILL THE MATERIAL BE
DEGRADED BY ALGAE. A SIMILAR QUANTITY OF SEVIN WOULD SERIOUSLY REDUCE
EFFICIENCY OF THE POND WITH CONVERSION MORE LIKELY TO OCCUR BY ALKALINE
HYDROLYSIS THAN ALGAL METABOLIC PROCESSES. AN EQUIVALENT SLUG OF
MALATHION, ALTHOUGH TEMPORARILY INTERRUPTING OPERATION OF THE POND,
COULD BE BROKEN DOWN RAPIDLY BY CHEMICAL AND METABOLIC REACTIONS.
(JONES-WISCONSIN)

FIELD 05C

ACCESSION NO. W70-08097

ALGAL REMOVAL IN UNIT PROCESSES,

ENVIRONMENTAL CONTROL ADMINISTRATION, CINCINNATI, OHIO; AND IOWA UNIV, IOWA
CITY. DEPT. OF ENVIRONMENTAL HEALTH; AND IOWA UNIV., IOWA CITY. DEPT. OF
CIVIL ENGINEERING.

RONALD R. SPEEDY, NEIL B. FISHER, AND DONALD B. MCDONALD.

JOURNAL AMERICAN WATER WORKS ASSOCIATION, VOL 61, P 289-292, 1969. 1 FIG, 5
TAB, 6 REF.

DESCRIPTORS:
*ALGAL CONTROL, *POTABLE WATER, CYANOPHYTA, SULFUR BACTERIA, COLOR,
IRON BACTERIA, TASTE, FILTERS, IOWA, DIATOMS, EUGLENA, CHLAMYDOMONAS,
SAMPLING, SCENEDESMUS, CHLOROPHYTA, CHLORELLA, CHLORINE, LIME,
COAGULATION.

IDENTIFIERS:
*UNIT PROCESSES, GASTROENTERITIS, SYNURA, ANABAENA, IOWA RIVER,
CYCLOTELLA, FRAGILARIA, SYNEDRA, NAVICULA, APHANIZOMENON, OSCILLATORIA,
MELOSIRA, ACTINASTRUM, COELASTRUM, PEDIASTRUM, WESTELLA,
STEPHANODISCUS, ANKISTRODESMUS, CHLOROMONAD, ALUM.

ABSTRACT:
A DETERMINATION OF THE EFFICIENCY OF ALGAL REMOVAL IN VARIOUS UNIT
OPERATIONS OF POTABLE WATER TREATMENT PROCESS WAS MADE. SAMPLES WERE
COLLECTED AT VARIOUS LOCATIONS THROUGHOUT THE TREATMENT PLANT AND
EXAMINED TO DETERMINE KINDS AND NUMBERS OF ALGAL FORMS PRESENT. THE
RESULTS INDICATE A WIDE DISPARITY IN THE EFFICIENCY OF ALGAL REMOVAL
AMONG THE SEVERAL UNIT PROCESSES STUDIED. NUMEROUS ORGANISMS PASS THE
FILTERS INTO FINISHED TAP WATER; COUNTS AS HIGH AS 44
ORGANISMS/MILLILITER WERE OBTAINED. CYCLOTELLA AND CHLORELLA APPEAR
WITH THE GREATEST FREQUENCY IN THE PLANT EFFLUENT. THE USE OF ALUM AS A
COAGULANT SEEMS MODERATELY EFFECTIVE FOR ALGAL REMOVAL, BUT LIME
APPEARS TO BE MUCH MORE EFFICIENT DUE TO FORMATION OF HEAVY FLOC WHICH
ENTRAPS THE CELLS UPON SETTLING. BOTH TYPES OF FILTERS EFFECT A HIGH
DEGREE OF ALGAL REMOVAL AT THE START OF A FILTER RUN; THE SAND
CONTINUES TO SHOW FAIRLY EFFICIENT REMOVAL UNTIL THE HEAD LOSS GOES
ABOVE SIX FEET. THE ANTHRAFILT FILTER SHOWS A HIGHER NUMBER OF ALGAE
PASSING THROUGH THE FILTER. THE IMPORTANCE OF MAINTAINING A CHLORINE
RESIDUAL IS SHOWN BY THE NUMBER OF ALGAE FOUND IN THE FINISHED TAP
WATER. (JONES-WISCONSIN)

FIELD 05F

ACCESSION NO. W70-08107

THE INHIBITORY EFFECTS OF COPPER ON MARINE PHYTOPLANKTON,

DOW CHEMICAL CO., FREEPORT, TEX, TEXAS DIV.

E. F. MANDELLI.

MARINE SCIENCE, VOL 14, P 47-57, 1969. 6 FIG, 1 TAB, 21 REF.

DESCRIPTORS:
*MARINE WATER, *COPPER, *ALGAE, SEA WATER, TEMPERATURE, SALINITY,
BIOASSAY, ANALYTICAL TECHNIQUES, LETHAL LIMITS, DINOFLAGELLATES,
DIATOMS.

IDENTIFIERS:
*COPPER CHLORIDE, UPTAKE, FREUNDLICH ISOTHERM, BLUE-GREEN ALGAE, GREEN
ALGAE.

ABSTRACT:
THE GROWTH OF SEVERAL SPECIES OF DINOFLAGELLATES AND DIATOMS IN BATCH
CULTURES UNDER CONTINUOUS ILLUMINATION WAS INHIBITED AT 20C BY
CONCENTRATION OF COPPER BETWEEN 0.055 AND 0.265 MICROGRAM/MILLILITER.
COCCOCHLORIS ELABANS FAILED TO GROW AT 0.03 MICROGRAM/MILLILITER AND
40C, WHEREAS DUNALIELLA TERTIOLECTA SURVIVED 0.6 MICROGRAM/MILLILITER
AT 30C. SKELETONEMA COSTATUM IN LIGHT-DARK CULTURES WITH MAINTENANCE OF
CONSTANT NUMBER OF CELLS AND COPPER IONS WAS INHIBITED AT 0.05
MICROGRAM/MILLILITER. A POSITIVE CORRELATION WAS FOUND BETWEEN THE LOG
OF COPPER UPTAKE:BIOMASS OF ALGAE RATIO AND THE TEMPERATURE, AND A
NEGATIVE CORRELATION BETWEEN THE SAME RATIO AND SALINITY.
(WILDE-WISCONSIN)

FIELD 05C

ACCESSION NO. W70-08111

GLENODININE, AN ICHTHYOTOXIC SUBSTANCE PRODUCED BY A DINOFLAGELLATE, PERIDINIUM POLONICUM (IN JAPANESE),

TOKYO UNIV. (JAPAN). LAB. OF MARINE BIOCHEMISTRY.

YOSHIRO HASHIMOTO, TOMOTOSHI OKAICHI, LE DUNG DANG, AND TAMAO NOGUCHI.

BULLETIN OF THE JAPANESE SOCIETY OF SCIENTIFIC FISHERIES, VOL 34, NO. 6, 1968. P. 528-534, 7 FIGS, 1 TAB, 8 REFS.

DESCRIPTORS:
WATER POLLUTION SOURCES, WATER POLLUTION EFFECTS, DINOFLAGELLATES, FISHKILL, FISH TOXINS, BIOASSAY, AQUATIC ALGAE, *ALGAL TOXINS, KILLIFISHES, *TOXICITY, ISOLATION.

IDENTIFIERS:
LAKE SAGAMI, JAPAN, *GLENODININE, *PERIDINIUM POLONICUM, PLANKTON BLOOMS.

ABSTRACT:
IN SEPTEMBER 1962, EXTENSIVE WATER BLOOMS OF A DINOFLAGELLATE IDENTIFIED AS PERIDINIUM POLONICUM, ACCOMPANIED BY MASS MORTALITY OF FISH, OCCURRED IN LAKE SAGAMI NEAR TOKYO. AN ATTEMPT WAS MADE TO ISOLATE THE SUSPECTED ICHTHYOTOXIC SUBSTANCE FROM THE PLANKTON COLLECTED FROM THE LAKE. THE SUBSTANCE WAS CONFIRMED TO BE AN ALKALOID AND TO CONTAIN A SULFHYDRYL RADICAL. THE AUTHORS DESIGNATED IT AS GLENODININE. BIOASSAYS SHOWED GLENODININE TO BE EXTREMELY TOXIC TO BOTH FRESHWATER AND MARINE FISHES. (SJOLSETH-WASHINGTON)

FIELD 05C

ACCESSION NO. W70-08372

BIOLOGICAL PARAMETERS FOR THE OPERATION AND CONTROL OF OXIDATION PONDS-II,

DEPARTMENT OF INDUSTRIES, CAPE TOWN (SOUTH AFRICA). DIV. OF SEA FISHERIES.

U. S. TSCHORTNER.

WATER RESEARCH, VOL 2, P 327-346, 1968. 9 FIG, 7 TAB, 24 REF.

DESCRIPTORS:
*ALGAE, *BIOLOGICAL PROPERTIES, CONTROL, *OPERATION AND MAINTENANCE, *OXIDATION LAGOONS, ABSORPTION, BIODEGRADATION, CHEMICAL OXYGEN DEMAND, CHLOROPHYLL, LAGOONS, PONDS, OXIDATION, WASTE WATER TREATMENT.

ABSTRACT:
OXIDATION PONDS ARE AN ECONOMICAL MEANS OF TREATMENT FOR DOMESTIC SEWAGE AND INDUSTRIAL WASTE EFFLUENTS, AND ARE WIDELY USED IN THE LESS POPULATED AREAS OF WARMER CLIMATES. THEIR ACTION RESEMBLES THE NATURAL PURIFICATION PROCESS OF POLLUTED RIVER. THEY REQUIRE LESS ATTENTION AND SKILLED STAFF THAN MECHANICAL SYSTEMS. THE SUCCESSFUL PERFORMANCE OF STABILIZATION, HOWEVER, DEPENDS ON THE BIOLOGICAL BALANCE OF THE SYSTEM. THE PRESENT STUDY DESCRIBES THE DEVELOPMENT OF PARAMETERS FOR THE DETERMINATION OF THE OXYGEN-PRODUCING COMMUNITY (PARAMETERS A AND AP) AND THE EFFICIENCY OF THE POND ON THE BREAKING DOWN OF THE LOAD (PARAMETER E). PARAMETER A REVEALS THE RELATIVE FRACTION OF CHLOROPHYLL-BEARING ORGANISMS OF THE TOTAL LOAD AND ALLOWS A DIRECT MEASUREMENT OF AUTOTROPHIC POND PERFORMANCE. PARAMETER AP REVEALS THE RATE OF PHOTOSYNTHESIS AND DOES NOT DEPEND ON VARIATIONS OF CHLOROPHYLL-A CONTENT OF CELLS. PARAMETER E IS BASED ON THE METABOLISM OF ALGAE AND CRUSTACEOUS; IT ESTIMATES THE CONTRIBUTION OF BOTH TO THE MICROPOPULATION AND INDICATES THE EFFECTIVENESS OF ALGAE IN REMOVING POLLUTION. COMPARISON OF THE COD CONTRIBUTED BY DISSOLVED AND SUSPENDED SOLIDS, WITH THE CALCULATED VALUES FOR ALGAE AND NON-ALGAL MATTER OF THE SUSPENDED FRACTION, REVEALED THAT IN A WELL-BALANCED SYSTEM MORE THAN HALF OF THE TOTAL COD IS CONTRIBUTED BY ALGAL CELLS, THE FIGURE IN THE OVERLOADED PONDS BEING ONLY 20%. (AGUIRRE-TEXAS)

FIELD 05D

ACCESSION NO. W70-08392

EFFECT OF ORGANIC COMPOUNDS ON PHOTOSYNTHETIC OXYGENATION--I. CHLOROPHYLL
DESTRUCTION AND SUPPRESSION OF PHOTOSYNTHETIC OXYGEN PRODUCTION,

TEXAS UNIV., AUSTIN. ENVIRONMENTAL HEALTH ENGINEERING RESEARCH LAB.

JU-CHANG HUANG, AND EARNEST F. GLOYNA.

WATER RESEARCH, VOL 2, P 347-366, 1968. 28 FIG, 3 TAB, 19 REF.

DESCRIPTORS:
   *ALGAE, *CHLOROPHYLL, WATER POLLUTION EFFECTS, *ORGANIC COMPOUNDS,
   *OXYGENATION, *PHOTOSYNTHESIS, CHLORELLA, PESTICIDES, PESTICIDE
   TOXICITY, PHOTOSYNTHETIC OXYGEN, PRODUCTIVITY, *WASTE WATER TREATMENT.

IDENTIFIERS:
   *CHLORELLA PYRENOIDOSA.

ABSTRACT:
   THIS PAPER DESCRIBES THE INHIBITION OF CHLOROPHYLL SYNTHESIS AND THE
   DESTRUCTION OF CHLOROPHYLL IN CHLORELLA PYRENOIDOSA DUE TO THE PRESENCE
   OF SELECTED ORGANIC COMPOUNDS. EMPHASIS IS DIRECTED TOWARD AN
   EXPLANATION OF LABORATORY TECHNIQUES AND AN EVALUATION OF CHANGES IN
   PHOTOSYNTHETIC OXYGEN PRODUCTION DUE TO THE ADDITION OF SELECTED
   COMPOUNDS. CHANGES IN EXTRACTABLE CHLOROPHYLL CONCENTRATIONS ARE
   PRESENTED IN TERMS OF EXPOSURE TIME AND CONCENTRATION OF CHEMICAL
   ADDITIVES. SELECTED ALGAL CULTURES WERE GROWN FOR A TEST PERIOD OF 72
   HOURS IN TEST TUBES UNDER CONSTANT TEMPERATURE CONDITIONS, WITH
   CONTINUOUS ILLUMINATION, AND UNDER CONTROLLED NUTRIENT SUPPLIES. UNDER
   SUCH CONDITIONS THE CHLOROPHYLL CONTENT DECREASES MARKEDLY IN RELATION
   TO THE CONCENTRATION OF THE ORGANIC ADDITIVE. FOR THE COMPOUNDS
   EVALUATED, THE CHLOROPHYLL REDUCTION SEEMS TO BE A FUNCTION OF THE
   SUBSTITUTED GROUPS AND THEIR RELATIVE POSITIONS. ALSO, A DIRECT
   MANOMETRIC TECHNIQUE IS DESCRIBED FOR MEASUREMENT OF PHOTOSYNTHETIC GAS
   EXCHANGE. RESULTS SHOW THAT OXYGEN PRODUCTION IS SUPPRESSED BY
   CHLORINATED AND NITRATED PHENOLS. (AGUIRRE-TEXAS)

FIELD 05D

ACCESSION NO. W70-08416

ADVANCES IN WATER POLLUTION RESEARCH.

PROCEEDINGS 4TH INTERNATIONAL CONFERENCE ON WATER POLLUTION RESEARCH, HELD IN
   PRAGUE, CZECHOSLOVAKIA, APRIL 21-25, 1969: LONDON, PERGAMON PRESS, LTD,
   1969. 936 P.

DESCRIPTORS:
   *WATER RESOURCES, *WATER POLLUTION CONTROL, *RESEARCH EQUIPMENT,
   *ANALYTICAL TECHNIQUES, WATER QUALITY, DISSOLVED OXYGEN, BIOCHEMICAL
   OXYGEN DEMAND, SUSPENDED LOAD, WASTES, SEWAGE, ALGAE, MICROORGANISMS,
   PHOSPHATES, HYDROGENATION, AERATION, FLOCCULATION, FILTRATION, MODEL
   STUDIES, COMPUTERS, NUTRIENTS, EUTROPHICATION.

IDENTIFIERS:
   *WATER POLLUTION RESEARCH.

ABSTRACT:
   THIS VOLUME CONTAINS THE 53 ORIGINAL PAPERS PRESENTED AT THE FOURTH
   INTERNATIONAL CONFERENCE ON WATER POLLUTION RESEARCH HELD IN PRAGUE IN
   APRIL, 1969. AMONG THE TOPICS DISCUSSED ARE THE SELF-PURIFICATION OF
   POLLUTED RIVERS THAT HAVE BEEN ARTIFICIALLY CANALIZED, THE FACTORS THAT
   INFLUENCE POLLUTION UNDER THESE CONDITIONS, AND METHODS FOR PREDICTING
   THE EXTENT OF POLLUTION AND THE DISTRIBUTION OF DISSOLVED OXYGEN IN
   RIVERS. COMPUTERS MAY BE USED TO ANALYZE VARIOUS ALTERNATIVE MEANS OF
   OBTAINING RIVER WATER OF REQUIRED QUALITY. THE MOST IMPORTANT
   CONTRIBUTION TOWARDS POLLUTION CONTROL IS MADE BY PURIFYING SEWAGE AND
   OTHER WASTE WATER. PURIFIED SEWAGE IS AN IMPORTANT SOURCE OF THE
   MINERAL NUTRIENT REQUIREMENTS OF MICROORGANISMS ESPECIALLY ALGAE, AND
   POLLUTION CONTROL MUST INCLUDE NUTRIENT REMOVAL FROM EFFLUENTS.
   (WOODARD-USGS)

FIELD 05G, 05C, 05D

ACCESSION NO. W70-08627

MODIFIED FILTER MEDIA FOR REMOVAL OF WATER POLLUTANTS,

INDIAN INST. OF TECH., KANPUR. DEPT. OF CIVIL ENGINEERING.

G. D. AGRAWAL, AND A. V. S. PRABHAKARA RAO.

IN: ADVANCES IN WATER POLLUTION RESEARCH, PROCEEDINGS 4TH INTERNATIONAL
CONFERENCE ON WATER POLLUTION RESEARCH, HELD IN PRAGUE, CZECHOSLOVAKIA,
APRIL 21-25, 1969: LONDON, PERGAMON PRESS, LTD, P 299-307, 1969. 9 P, 5
FIG, 10 REF.

DESCRIPTORS:
*WATER POLLUTION CONTROL, *FILTRATION, *PARTICLE SIZE, *CLAYS, *SILTS,
SANDS, ALGAE, PLANKTON, MICROORGANISMS, TURBIDITY, ANALYTICAL
TECHNIQUES, LABORATORY TESTS, FOREIGN RESEARCH.

IDENTIFIERS:
*FILTER MEDIA, ALUMINA-COATED SAND, FERRIC OXIDE-COATED SAND, CRUSHED
BAUXITE, ACID-WASHED BAUXITE, KANPUR(INDIA).

ABSTRACT:
EXPERIMENTS WERE MADE TO DEVELOP A SIMPLE FILTER, WITH MODIFIED MEDIA,
FOR EFFICIENT REMOVALS OF MICRO SIZED PARTICLES, WHETHER USED FOR
DISPOSAL TO EXCLUDE PARTICULATE POLLUTANTS FROM EFFLUENTS OR FOR WATER
SUPPLY TO REMOVE PARTICULATE CONTAMINANTS. TWO PYREX GLASS TUBES WERE
USED TO CONTAIN THE FILTER COLUMNS. THE FILTER BED FOR ALL THE RUNS WAS
A 7.6 CM LAYER OF SHINGLE AND COARSE SAND SUPPORTING AN 18 CM LAYER OF
THE MEDIA UNDER STUDY. FILTER MEDIA CONSISTED OF UNTREATED SAND, SAND
COATED WITH ALUMINA, SAND COATED WITH FERRIC OXIDE, CRUSHED LOW-GRADE
BAUXITE AND BAUXITE WASHED WITH ACID. WATER WITH LESS THAN 10 MG/L.
TURBIDITY COULD BE PRODUCED BY FILTRATION OF RAW STREAM WATER (INFLUENT
TURBIDITY 46 MG/L.) OR RAW POND WATER (INFLUENT TURBIDITY 32 MG/L.) AND
REDUCED THE TOTAL BACTERIAL NUMBERS BY ABOUT 50%. (WOODWARD-USGS)

FIELD 05D

ACCESSION NO. W70-08628

FACTORS INFLUENCING PHOSPHATE USE BY ALGAE,

MINNESOTA UNIV., MINNEAPOLIS. LIMNOLOGICAL RESEARCH CENTER.

JOSEPH SHAPIRO, WILLIAM CHAMBERLAIN, AND JUDITH BARRETT.

PARTIALLY SUPPORTED BY FWQA RESEARCH GRANT. IN: ADVANCES IN WATER POLLUTION
RESEARCH, PROCEEDING 4TH INTERNATIONAL CONFERENCE ON WATER POLLUTION
RESEARCH, HELD IN PRAGUE, CZECHOSLOVAKIA, APRIL 21-25, 1969: LONDON,
PERGAMON PRESS, LTD, P 149-160, 1969. 12 P, 10 FIG, 3 TAB, 13 REF.

DESCRIPTORS:
*EUTROPHICATION, *LAKES, *ALGAE, *GROWTH RATES, *PHOSPHATES, LABORATORY
TESTS, MINNESOTA, NUTRIENTS, WATER POLLUTION EFFECTS.

IDENTIFIERS:
*MICROCYSTIS AERUGINOSA, *LAKE WATERS.

ABSTRACT:
MICROCYSTIS AERUGINOSA FROM LAKES IN MINNESOTA, WAS USED TO SHOW THE
RELATIONSHIPS OF PHOSPHORUS TO ALGAE GROWTH IN NATURAL WATER SYSTEMS.
THE RAPID RATE OF ABSORPTION OF PHOSPHATE BY ALGAE SUSPENDED IN LAKE
WATER IS LARGELY DUE TO THE INORGANIC SALTS OF THE WATER AND NOT TO AN
ORGANIC FACTOR. (WOODARD-USGS)

FIELD 05C, 02H

ACCESSION NO. W70-08637

METABOLISM OF LINDANE BY UNICELLULAR ALGAE,

STATE UNIVERSITY COLL., BUFFALO, N.Y. GREAT LAKES LAB.

ROBERT A. SWEENEY.

PROCEEDINGS OF THE 12TH CONFERENCE ON GREAT LAKES RESEARCH, P 98-102, 1969. 3
  FIG, 18 REF. USPHS GRANT 1T-WP-3901-A.

DESCRIPTORS:
  PESTICIDE RESIDUES, ALGAE, GREAT LAKES, METABOLISM, *CHLORELLA,
  *CHLAMYDOMONAS LABORATORY TESTS, GAS CHROMATOGRAPHY, DEGRADATION, WATER
  POLLUTION EFFECTS, AQUATIC ENVIRONMENT, CHLORINATED HYDROCARBON
  PESTICIDES, PLANT PHYSIOLOGY.

IDENTIFIERS:
  *LINDANE, *ALGAE METABOLISM, DETOXIFICATION, LINDANE METABOLISM.

ABSTRACT:
  LINDANE IS AN INSECTICIDE THAT HAS BEEN DETECTED IN THE GREAT LAKES
  BASIN. THE METABOLISM OF THIS PESTICIDE BY CHLORELLA VULGARIS AND
  CHLAMYDOMONAS REINHARDTII IN AXENIC CULTURE WAS ASCERTAINED. EACH
  SPECIES WAS CULTURED IN A SOLUTION INITIALLY CONTAINING 4.5, 2.3, OR
  0.0 PPM LINDANE. USING CHROMATOGRAPHIC TECHNIQUES, A MARKED DECREASE IN
  THE LINDANE CONCENTRATION WITH TIME WAS NOTED. A SUBSTANCE WAS OBSERVED
  IN EXTRACTIONS OF THE SOLUTIONS CONTAINING ALGAE AND LINDANE WHICH WAS
  NOT PRESENT IN EXTRACTS FROM EITHER ALGAE-FREE CULTURES CONTAINING
  LINDANE OR FROM ALGAE IN A LINDANE-FREE MEDIUM. THIS SUBSTANCE WAS
  IDENTIFIED AS 1, 3, 4, 5, 6 --PENTACHLORO-CYCLOHEX-L-ENE, A NON-TOXIC
  LINDANE METABOLITE. THIS INDICATED THAT CHLORELLA AND CHLAMYDOMONAS CAN
  METABOLIZE LINDANE. IT ALSO MAY EXPLAIN IN PART THE RELATIVELY LOW
  CONCENTRATION OF THIS PESTICIDE IN THE GREAT LAKES.
  (SJOLSETH-WASHINGTON)

FIELD 05C

ACCESSION NO. W70-08652

RELATIONSHIPS BETWEEN BACTERIA AND PHYTOPLANKTON,

BYELORUSSIAN STATE UNIV., MINSK (USSR). DEPT. OF INVERTEBRATE ZOOLOGY.

YU S POTAENKO, AND T M MIKHEEVA.

TRANS. FROM MIKROBIOLOGIYA, VOL 38, NO 4, P 722-727. JULY-AUG 1969.
  MICROBIOLOGY, VOL 38, NO 4, P 603-607, 1969. 3 FIG, 37 REF.

DESCRIPTORS:
  *BACTERIA, *PHYTOPLANKTON, *LAKES, ALGAE, MICROORGANISMS, BIOMASS,
  DIATOMS, CYANOPHYTA, IRON, ALGICIDES, ZOOPLANKTON, FISH.

IDENTIFIERS:
  BYELORUSSIA(USSR), POLYCHLOROPINENE.

ABSTRACT:
  SEVERAL FACTORS MAY BE OPERATIONAL SIMULTANEOUSLY IN THE
  CHARACTERISTICS OF THE RELATIONSHIP BETWEEN BACTERIA AND PHYTOPLANKTON.
  STUDY OF BACTERIOPLANKTON AND PHYTOPLANKTON OF EIGHT BYELORUSSIAN LAKES
  IN DIFFERENT YEARS SHOWED THERE WERE EXTREMELY DIVERSE RELATIONSHIPS
  BETWEEN ALGAE AND OTHER MICROORGANISMA: GROWTH OF ALGAE AND BACTERIA
  PROCEEDED IN PARALLEL, MAXIMUM ALGAL DEVELOPMENT CORRESPONDED TO
  MINIMUM BACTERIAL POPULATION AND VICE VERSA, OR THE GROWTH OF THESE
  GROUPS OF ORGANISMS WAS INDEPENDENT. NO DEFINITE RELATIONSHIP WAS
  OBSERVED BETWEEN BACTERIAL GROWTH AND THE PRESENCE OF A GIVEN TOXONOMIC
  ALGAL GROUP. THE USUAL CYCLES FOR INDIVIDUAL SPECIES AND ALGAL GROUPS
  WERE REESTABLISHED WHEN THE DISRUPTED EQUILIBRIUM AMONG DIFFERENT
  ELEMENTS OF THE FOOD CHAINS WAS RESTORED A YEAR AFTER THE FISH KILL.
  FROM A STUDY OF VARIATION IN BACTERIOPLANKTON GROWN IN THE NAROCHANSKII
  LAKES AND FROM DATA IN THE LITERATURE OF THE SEASONAL DYNAMICS OF
  BACTERIAL POPULATION IN A NUMBER OF OTHER LAKES, IT CAN BE CONCLUDED
  THAT THE BACTERIOPLANKTON DISTRIBUTION AND POPULATION CHARACTERISTIC OF
  OTHER EUTROPHIC LAKES IN THE SOVIET UNION IS RESTORED WITHIN 1.5 YEARS
  AFTER TREATMENT WITH PCP COMPOUND. (JONES-WISCONSIN)

FIELD 02H, 05C

ACCESSION NO. W70-08654

BIODEGRADABLE DETERGENTS - SOFT DETERGENTS,

HART CHEMICAL LTD., GUELPH (ONTARIO).

H. ZIMMERMAN.

CANADIAN TEXTILE JOURNAL, VOL. 82, P 43-45, JANUARY 22, 1965. 3 FIG.

DESCRIPTORS:
*ALKYLBENZENE SULFONATES, *BIODEGRADATION, DETERGENTS, WATER POLLUTION, WASTE WATERS, BIOCHEMICAL OXYGEN DEMAND, FISH, ALGAE, FOAMING, SEWAGE TREATMENT.

IDENTIFIERS:
*TEXTILE MILL WASTES, ALFOL ALCOHOLS.

ABSTRACT:
THE ALKYL ARYL SULPHONATES ARE NON-BIODEGRADABLE, WHEREAS SOAPS, THE FATTY ALCOHOL SULPHATES AND OTHER FAT BASED DETERGENTS, ARE READILY BIODEGRADABLE. THESE BIODEGRADABLE SYNTHETIC DETERGENTS ARE AVAILABLE, BUT THEIR USE DOES NOT SOLVE THE OVER-ALL PROBLEMS OF WATER POLLUTION. NEARLY ALL WASTE WATERS FROM TEXTILE OPERATIONS HAVE A HIGH BOD WHICH AFFECTS FISH LIFE. THE DEGRADATION OF AN ALKYL ARYL SULPHONATE IN A SEWAGE TREATMENT PROCESS IS ONLY ABOUT 50 PERCENT, AND IN RIVER WATER IS ONLY ABOUT 75 PERCENT IN 30 DAYS. THE CHEMISTRY OF ALKYL BENZENE SULPHONATES(ABS) AND THE ALFOL ALCOHOL PROCESS IS DISCUSSED. THE MAJOR NUISANCES FROM ABS RESIDUES ARE FOAMING AND ALGAL GROWTHS. THERE IS LITTLE EVIDENCE THAT ABS RESIDUES ARE TOXIC TO HUMANS OR FISH, BUT THE FOAMING ADVERSELY AFFECTS SEWAGE TREATMENT. (LONON-NORTH CAROLINASTATE)

FIELD 05B

ACCESSION NO. W70-08740

HEAT AS A POLLUTANT,

GENERAL ELECTRIC CO., RICHLAND, WASH.

ROBERT T. JASKE.

OREGON STATE UNIV. WATER RESOURCES. RESEARCH INSTITUTE WATER QUALITY CONTROL SEMINAR PROCEEDINGS, SESSION 7, P 61-82, NOVEMBER 1964. 10 FIG.

DESCRIPTORS:
*THERMAL POLLUTION, *TEMPERATURE, HYDROGEN ION CONCENTRATION, NUTRIENTS, ALGAE, FISH, STRATIFICATION, FLOCCULATION, TURBINES, ECONOMIC EFFICIENCY, HEAT BUDGET, SOCIAL ASPECTS.

ABSTRACT:
THE EFFECTS OF INCREASED TEMPERATURE ON PH, NUTRIENTS, ALGAE, TASTE, AND FISH ARE DISCUSSED. THE ADVANTAGES AND DISADVANTAGES OF STRATIFICATION, FLOCCULATION, AND OTHER VARIATIONS IN THE FLUID MECHANICS OF STREAMS ARE CONSIDERED. A SPECIFIC EXAMPLE OF A TYPICAL HEAT-PRODUCING INDUSTRIAL PROCESS OF A TURBINE GENERATOR PLANT IS EXAMINED. THE EFFICIENCY AND THE OVERALL PLANT ECONOMY IS DETERMINED FOR VARYING TEMPERATURES OF THE EFFLUENT. PROBLEMS ARISING FROM STREAM SIZE AND THE ERECTION OF DAMS AND RESERVOIRS ARE MENTIONED. THREE PRIMARY CONCERNS OF FURTHER RESEARCH INCLUDE: (1) IMPROVED ATTENTION TO HEAT BUDGETS AND MATHEMATICAL TREATMENT OF AIR-WATER TRANSPORT PHENOMENA; (2) CLOSER IDENTIFICATION OF THE SPECIFIC CHANGES ASSOCIATED WITH THE TEMPERATURE PARAMETER; AND (3) IMPROVED STANDARDIZATION AND METHODS OF ECONOMIC ANALYSIS FOR IDENTIFICATION OF THE SIGNIFICANT ECONOMIC FACTORS INVOLVED. (OSBORNE-VANDERBILT)

FIELD 05C

ACCESSION NO. W70-08834

NUTRIENT REMOVAL FROM SEWAGE EFFLUENTS BY ALGAL ACTIVITY,

NATIONAL INST. FOR WATER RESEARCH, PRETORIA (SOUTH AFRICA).

J. HEMENS, AND G. J. STANDER.

ADVANCES IN WATER POLLUTION RESEARCH, PROCEEDINGS, FOURTH INTERNATIONAL
  CONFERENCE ON WATER POLLUTION RESEARCH, PRAGUE, APR 21-25, 1969, LONDON,
  PERGAMON PRESS, 1969, P 701-716.

DESCRIPTORS:
  *ALGAE, *HYDROGEN ION CONCENTRATION, *NUTRIENTS, *TERTIARY TREATMENT,
  BIOLOGICAL TREATMENT, NITROGEN, PHOSPHORUS, SEWAGE EFFLUENTS.

IDENTIFIERS:
  *ALGAL ACTIVITY, *NUTRIENT REMOVAL.

ABSTRACT:
  RESULTS OBTAINED IN REMOVING NITROGEN AND PHOSPHORUS FROM SEWAGE
  EFFLUENT BASED ON A MODIFICATION OF THE METHODS REPORTED BY OTHER
  WORKERS IS DESCRIBED. THE RESULTS SUGGEST THAT INSTEAD OF ATTEMPTING TO
  REMOVE NUTRIENTS FROM SEWAGE EFFLUENT BY ASSIMILATION INTO NEW ALGAL
  CELLS IT MAY BE MORE PRACTICAL, UNDER SUITABLE CLIMATIC CONDITIONS, TO
  EXPLOIT THE EFFECTS OF PH INCREASE DUE TO ALGAL PHOTOSYNTHESIS AND TO
  AVOID THE PROBLEMS OF LIGHT UTILIZATION BY DENSE ALGAL CULTURES. THE
  BASIC REQUIREMENTS FOR THIS SYSTEM WOULD BE A LEVEL OF PHOTOSYNTHESIS
  SUFFICIENT TO PRODUCE THE PH INCREASE REQUIRED TO PRECIPITATE PHOSPHATE
  AND TO PROMOTE LOSS OF AMMONIA TO THE ATMOSPHERE. THE DECREASE IN
  AMMONIA AND PHOSPHATE REMOVAL OBSERVED DURING THE NIGHT MIGHT BE
  OVERCOME BY THE PROVISION OF SUFFICIENT FLUORESCENT LIGHT ABOVE THE
  EXPERIMENTAL DISCS TO LESSEN THE PH DECREASE. ALTERNATIVELY, PH CONTROL
  BY THE ADDITION OF ALKAKI MIGHT BE POSSIBLE, PARTICULARLY SINCE IT WAS
  OBSERVED THAT THE REDUCTION IN ALKALINITY DURING DAYLIGHT REDUCED THE
  AMOUNT OF LIME REQUIRED TO REACH A PH OF 10.3 BY AS MUCH AS 75% WHEN
  COMPARED TO THAT NECESSARY TO ACHIEVE THE SAME INCREASE IN THE
  UNTREATED EFFLUENT. (AGUIRRE-TEXAS)

FIELD 05D

ACCESSION NO. W70-08980

WATER QUALITY IN RELATION TO PRODUCTIVITY OF LAKE ASHTABULA RESERVOIR IN
  SOUTHEASTERN NORTH DAKOTA,

NORTH DAKOTA WATER RESOURCES RESEARCH INST., FARGO.

JOHN J. PETERKA.

AVAILABLE FROM NTIS AS PB-193 683, $3.00 IN PAPER COPY, $0.65 IN MICROFICHE.
  RESEARCH PROJECT TECHNICAL COMPLETION REPORT, WI-221-002-69, MARCH 1969. 23
  P, 4 TAB, 7 FIG, 18 REF. OWRR PROJECT A-004-NDAK(1).

DESCRIPTORS:
  RESERVOIR, *PRIMARY PRODUCTIVITY, *EUTROPHICATION, WATER QUALITY,
  *NUTRIENTS, LIMNOLOGY, WATER TEMPERATURE, *ALGAE BLOOMS, CYANOPHYTA,
  NORTH DAKOTA.

IDENTIFIERS:
  LAKE ASHTABULA.

ABSTRACT:
  THIS STUDY ATTEMPTS TO RELATE THE CHEMICAL AND PHYSICAL CHARACTERISTICS
  OF THE LAKE WATER TO STANDING CROPS AND PRODUCTION RATES OF
  PHYTOPLANKTON. THE AVERAGE GROSS PHOTOSYNTHESIS RATE WAS 41 PERCENT OF
  THE OPTIMAL RATE. PHYTOPLANKTON STANDING CROPS AVERAGED $ UG
  CHLOROPHYLL/LITER IN THE EUPHOTIC ZONE. OPTIMAL PHOTOSYNTHESIS PER UNIT
  CHLOROPHYLL AVERAGED 13.5 MG O2/MG CHLOROPHYLL/HOUR. GROSS PRODUCTION
  PER UNIT CHLOROPHYLL DECREASED WITH INCREASE IN PHYTOPLANKTON STANDING
  CROP. ANNUAL GROSS PHOTOSYNTHESIS WAS 1.4 GC/M2/DAY. THREE EMPIRICAL
  EQUATIONS WERE USED TO PREDICT PHOTOSYNTHESIS IN THE EUPHOTIC ZONE.
  GROSS PHOTOSYNTHESIS WAS RELATED TO STANDING CROP AND WATER
  TEMPERATURE, BUT NOT TO CO2 OR MACRO-NUTRIENTS. HIGH NUTRIENT LEVELS
  AND CONTINUAL CIRCULATION OF THE WATER DURING ICE-FREE PERIODS MADE THE
  LAKE EUTROPHIC. (PETERKA-NORTH DAKOTA STATE)

FIELD 5C, 2H

ACCESSION NO. W70-09093

EFFECTIVE PHOSPHORUS REMOVAL BY THE ADDITION OF ALUM TO THE ACTIVATED SLUDGE
   PROCESS,

METROPOLITAN SANITARY DISTRICT OF GREATER CHICAGO, ILL.

DAVID R. ZENZ, AND JOSEPH R. PIVNICKA.

PAPER PRESENTED AT THE 24TH ANNUAL PURDUE INDUSTRIAL WASTES CONFERENCE, MAY
   6-8, 1969. 59 P, 34 FIG, 4 TAB, 21 REF.

DESCRIPTORS:
   *PHOSPHORUS, *PHOSPHATES, *PHOSPHORUS COMPOUNDS, *CHEMICAL
   PRECIPITATION, *CHEMICAL DEGRADATION, *BIOLOGICAL TREATMENT,
   *BIODEGRADATION, *BIOCHEMISTRY, *BIOCHEMICAL OXYGEN DEMAND, *WASTE
   WATER TREATMENT, *WATER POLLUTION CONTROL, *EUTROPHICATION, *ALGAL
   CONTROL, *TERTIARY TREATMENT, SUSPENDED LOAD, PILOT PLANTS, ACTIVATED
   SLUDGE, TREATMENT FACILITIES.

IDENTIFIERS:
   *PHOSPHORUS REMOVAL, *ALUM, *ALUMINUM SULFATE, *ACTIVATED SLUDGE
   PROCESS, HANOVER TREATMENT PLANT.

ABSTRACT:
   THE OBJECTIVE OF THIS EXPERIMENTAL RESEARCH WAS TO EVALUATE, UNDER
   ACTUAL WASTE WATER TREATMENT PLANT OPERATING CONDITIONS, A COMBINED
   CHEMICAL-BIOLOGICAL PROCESS CAPABLE OF PRODUCING AN EFFLUENT LOW IN
   PHOSPHORUS, BOD AND SUSPENDED SOLIDS. TESTS WERE CONDUCTED AT THE
   HANOVER PARK WATER RECLAMATION PLANT IN COOK COUNTY, ILLINOIS. THE
   PLANT CONSISTS OF A 2 MGD CONVENTIONAL ACTIVATED SLUDGE FACILITY AND A
   TERTIARY FACILITY CAPABLE OF TREATING THE ENTIRE FLOW FROM THE
   SECONDARY FACILITY. THE SECONDARY PLANT WAS SEPARATED INTO TWO DISTINCT
   SYSTEMS FOR THE TESTS. ALUM WAS ADDED TO THE INFLUENT OF THE AERATION
   TANK OF ONE SYSTEM BY A FEED PUMP. EFFLUENT SAMPLES, TAKEN
   AUTOMATICALLY AS 24-HOUR COMPOSITES, WERE ANALYZED FOR BOD, SS, VS, PER
   CENT TOTAL SOLIDS, PER CENT VOLATILE SOLIDS, ALKALINITY, S.V.I AND
   NH3-N. AN AUTOANALYZER WAS USED FOR PHOSPHORUS DETERMINATIONS. IT WAS
   FOUND THAT THE PROCESS CAN PRODUCE EFFLUENT PHOSPHORUS CONCENTRATIONS
   OF 2.30 MG/L (TOTAL P) AND 0.83 MG/L (SOLUBLE P) AT A AL:TOTAL P RATIO
   (MOLAR BASIS) OF 1.54 AND A AL:SOLUBLE P RATIO OF 1.85. THIS IS 83 AND
   93 PER CENT REMOVAL, RESPECTIVELY. BOUND PHOSPHORUS WAS NOT RELEASED
   DURING ANAEROBIC DIGESTION. ADDITIONS OF LARGE AMOUNTS OF ALUM DID NOT
   ADVERSELY AFFECT BOD REMOVAL EFFICIENCY, BUT REMOVED HIGHER MICROBIAL
   LIFE FORMS IN THE MIXED LIQUOR. THE SUSPENDED SOLIDS INCREASED AT AN
   ALUM CONCENTRATION OF 252 MG/L. WASTE SLUDGE INCREASED, AND S.V.I
   DECREASED. ALUM DID NOT UPSET THE ANAEROBIC BIOLOGICAL TREATMENT
   PROCESS. (POERTNER)

FIELD 05D

ACCESSION NO. W70-09186

BAFFLED BIOLOGICAL BASIS FOR TREATING POULTRY PLANT WASTES,

SYRACUSE UNIV., N.Y.

NELSON L. NEMEROW.

JOURNAL OF THE WATER POLLUTION CONTROL FEDERATION, VOL 41, NO 9, P 1602-1612,
   SEPTEMBER 1969. 8 FIG, 9 TAB.

DESCRIPTORS:
   *BIOLOGICAL TREATMENT, *OXIDATION, LAGOONS, *BAFFLES, *WASTE WATER
   TREATMENT, *POULTRY, INDUSTRIAL WASTES, LAGOONS, PONDS, PHOTOSYNTHESIS,
   COSTS, EFFICIENCY, BIOCHEMICAL OXYGEN DEMAND, ALGAE, SEDIMENTATION,
   SETTLING BASINS.

IDENTIFIERS:
   *CHICKENS, FOOD WASTES, POULTRY WASTES.

ABSTRACT:
   A POULTRY PLANT IN MILLSBOROUGH, DELAWARE, PROCESSED 10,000 CHICKENS
   PER HOUR WITH A WASTE WATER OF 40,000 GPH, AND AN EFFLUENT OF 2,500 LB
   BOD5/DAY AT AN AVERAGE BOD OF 630 MG/L. BECAUSE THE AREA IS
   COMMERCIALLY AND RECREATIONALLY OF GREAT VALUE A PROGRAM WAS INITIATED
   TO REDUCE THE WASTE WATER CONCENTRATION AT A MAXIMUM COST OF $100,000.
   ADEQUATE SCREENING FOLLOWED BY A TWO-STAGE OXIDATION POND PLANT
   UTILIZING OVER AND UNDER CONTACT BAFFLES IN THE FIRST STAGE FOLLOWED BY
   CHLORINATION PROVIDED A 85 TO 95% BOD REDUCTIONS. THE FIRST STAGE
   CONSISTS OF A BAFFLED HIGH-RATE DEEP POND. THE SECOND STATE IS A
   SHALLOW SYNTHETIC BASIN. LOADINGS OF OVER 200 LB/DAY/ACRE RESULTED IN
   HIGH EFFICIENCY AND COLIFORM COUNTS OF LESS THAN 10/1000 ML. THE FINAL
   COST WAS $90,000. (HANCUFF-TEXAS)

FIELD 05D

ACCESSION NO. W70-09320

ANIONIC AND NONIONIC SURFACTANT SORPTION AND DEGRADATION BY ALGAE CULTURES,

TEXAS UNIV., AUSTIN. ENVIRONMENTAL HEALTH ENGINEERING RESEARCH LAB.

ERNST M. DAVIS, AND EARNEST F. GLOYNA.

JOURNAL OF THE AMERICAN OIL CHEMISTS' SOCIETY, VOL 46, NO 11, P 604-608, 1969. 9 REF, 2 FIG, 3 TAB. NSF GRANT NSF-GU-1963.

DESCRIPTORS:
*LINEAR ALKYLATE SULFONATES, *SURFACTANTS, WATER POLLUTION SOURCES, *DETERGENTS, *DEGRADATION(DECOMPOSITION), CYANOPHYTA, CHLOROPHYTA, *ALGAE, SORPTION, WASTE WATER TREATMENT, CULTURES, CHLORELLA, WATER POLLUTION EFFECTS.

IDENTIFIERS:
*ALKYL POLY ETHOXYLATE, *ALKYL PHENOL POLYETHOXYLATE, ANIONIC SURFACTANTS, NONIONIC SURFACTANTS, INFRARED SPECTRA.

ABSTRACT:
DEGRADATION OF THREE SURFACTANTS HAS BEEN DETERMINED BY ORGANIC EXTRACTION PROCEDURES AND INFRARED SPECTROSCOPY. AXENIC CULTURES OF FIVE SPECIES OF BLUE-GREEN ALGAE AND THREE SPECIES OF GREEN ALGAE WHICH ARE COMMON TO WASTE STABILIZATION PONDS WERE TEST ORGANISMS. ANALYTICAL DATA ARE SHOWN COMPARING THE EFFECTS PRODUCED BY THE ALGAE CULTURES AND A HETEROGENEOUS MICROCOSM. LINEAR ALKYL SULFONATE WAS THE ANIONIC SURFACTANT COMPOUND TESTED. AN ALKYL POLYETHOXYLATE AND AN ALKYL PHENOL POLYETHOXYLATE WERE THE NONIONIC TEST SURFACTANTS. SORPTION OF THE COMPOUNDS BY THE ALGAE USUALLY WAS FOLLOWED BY RELEASE AND DEGRADATION OF UP TO 99% OF SOME OF THE COMPONENT PARTS OF THE SURFACTANT MOLECULE. (SJOLSETH-WASHINGTON)

FIELD 05C

ACCESSION NO. W70-09438

POTOMAC RIVER WATER QUALITY NETWORK, COMPILATION OF DATA-WATER YEAR 1969.

INTERSTATE COMMISSION ON THE POTOMAC RIVER BASIN, WASHINGTON, D.C.

INTERSTATE COMMISSION ON THE POTOMAC RIVER BASIN DATA COMPILATION REPORT, 1969. 92 P, 2 FIG, 3 MAP.

DESCRIPTORS:
*DATA COLLECTIONS, *WATER QUALITY, BIOCHEMICAL OXYGEN DEMAND, SUSPENDED LOAD, TURBIDITY, DATA STORAGE AND RETRIEVAL, COLIFORMS, DISSOLVED OXYGEN, ALGAE, NUTRIENTS, SOLUTES, CHLORIDES, CALCIUM, CARBONATES, PHOSPHATES.

IDENTIFIERS:
POTOMAC RIVER BASIN.

ABSTRACT:
WATER QUALITY DATA FROM THE POTOMAC RIVER BASIN ARE SUMMARIZED. THE WATER STATISTICS ARE OBTAINED FROM SAMPLES TAKEN AT 80 STREAM SAMPLING POINTS IN THE BASIN. THE SAMPLES WERE TAKEN AND ANALYZED BY AGENCIES, INDUSTRIES AND MUNICIPALITIES. AN EXPLANATION OF PARAMETERS IS INCLUDED. THE DATA OBTAINED THIS WATER YEAR WERE STORED IN THE FEDERAL WATER POLLUTION CONTROL ADMINISTRATION COMPUTER- STORET. (KNAPP-USGS)

FIELD 05A, 07A

ACCESSION NO. W70-09557

THE PERSISTENCE OF PESTICIDES IN IMPOUNDED WATERS,

KENTUCKY WATER RESOURCES INST., LEXINGTON.

ROBERT A. LAUDERDALE.

AVAILABLE FROM NTIS AS PB-194 056, $3.00 IN PAPER COPY, $0.65 IN MICROFICHE.
KENTUCKY UNIVERSITY WATER RESOURCES INSTITUTE RESEARCH REPORT NO 17, 1969.
39 P, 8 FIG, 5 TAB, 8 REF. OWRR PROJECT NO A-002-KY(4).

DESCRIPTORS:
*WATER POLLUTION EFFECTS, *PESTICIDES, RESIDUES, *PESTICIDE KINETICS,
*PERSISTENCE, EVAPORATION, BIODEGRADATION, ADSORPTION, ALGAE, CLAYS,
ALDRIN, DIELDRIN, DDT, AERATION, SILTS.

IDENTIFIERS:
CHLORDANE.

ABSTRACT:
THE PERSISTENCE OF THE INSECTICIDES ALDRIN, DIELDRIN, CHLORDANE, AND
DDT IN WATER WERE MEASURED. A SMALL AREA WAS SPRAYED WITH THESE
COMPOUNDS, AND THE INSECTICIDES WERE COLLECTED IN THE RUNOFF WATER FROM
THE AREA. LABORATORY EXPERIMENTS WERE PERFORMED TO DETERMINE THE
EFFECTIVENESS OF AERATION, ADSORPTION ON SILT, AND ADSORPTION ON ALGAE
IN REMOVING THE PESTICIDES FROM WATER. ALL OF THE PESTICIDES WERE FOUND
IN SAMPLES OF WATER COLLECTED FOR THE FULL PERIOD OF THE TESTS. THE
AMOUNTS WHICH WERE FOUND IN A SMALL POND INTO WHICH THE SURFACE WATER
DRAINED WAS CONSISTENTLY LESS THAN THAT FOUND IN THE SURFACE RUNOFF
WATER ITSELF. THE REDUCED CONCENTRATIONS IN THE POND WATER WERE
APPARENTLY THE RESULT OF A COMBINATION OF SEDIMENTATION, ADSORPTION ON
ALGAE OR SILT, OR VOLATILIZATION OF THE INSECTICIDE FROM THE WATER INTO
THE ATMOSPHERE. IN GENERAL, THE EFFICIENCY OF REMOVAL OF INSECTICIDE
FROM WATER BY ANY ONE OF THESE PROCESSES APPEARS TO BE LOW. HOWEVER,
BECAUSE OF THE LARGE RATIO OF ABSORBANT TO INSECTICIDE, THE OVERALL
REMOVAL IS SUBSTANTIAL. (KNAPP-USGS)

FIELD 05B, 05G

ACCESSION NO. W70-09768

PRELIMINARY OBSERVATIONS OF THE HYDROMECHANICS, NUTRIENT CYCLES AND
EUTROPHICATION STATUS OF LAKE KINNERET (LAKE TIBERIAS),

LIMNOLOGICAL LAB., TABIGHA (ISRAEL).

C. SERRUYA, S. SERRUYA, AND T. BERMAN.

VERHANDLUNGEN DER INTERNATIONALEN VEREINIGUNG FUR THEORETISCHE UND ANGEWANDTE
LIMNOLOGIE, VOL 17, P 342-351, 1969. 7 FIG, 1 TAB, 8 REF.

DESCRIPTORS:
*CYCLING NUTRIENTS, *EUTROPHICATION, *HYDRODYNAMICS, LAKES, INTERNAL
WATER, WAVES(WATER), DIURNAL, EPILIMNION, WINDS, THERMOCLINE,
TEMPERATURE, CURRENTS(WATER), HYPOLIMNION, ALGAE, NITROGEN, PHOSPHORUS,
OXYGEN, CARBON, ENZYMES, SAMPLING, BOTTOM SEDIMENTS, SODIUM, POTASSIUM,
IRON, MANGANESE, CALCIUM, PIGMENTS.

IDENTIFIERS:
*LAKE KINNERET(ISRAEL).

ABSTRACT:
EXISTENCE OF INTERNAL WAVES IN LAKE KINNERET (ISRAEL) WAS CONFIRMED. IN
EARLY MAY 1968, WITH WELL DEVELOPED THERMOCLINE, CHARACTERISTIC
INTERNAL WAVES WERE OBSERVED, AVERAGING 11 HOURS, OCCURRING REGULARLY
AT SEMI-DIURNAL INTERVALS, CORRELATED SOMEWHAT WITH PERIODIC WEST WIND.
IN THE HYPOLIMNION, LARGE QUANTITIES OF AMMONIUM-NITROGEN AND
ORTHOPHOSPHATE CONCENTRATIONS INCREASINGLY ACCUMULATED FROM MAY TO
DECEMBER. RATIO OF TOTAL AVAILABLE NITROGEN TO ORTHOPHOSPHATES IS HIGH
IN FEBRUARY AND DECREASES RAPIDLY TO MID-SUMMER. ASSIMILATION RATE AND
REGENERATION IN EPILIMNION IS ABOUT FOUR TIMES GREATER IN SPRING THAN
SUMMER. UNDER PREVAILING LAKE CONDITIONS, DURING THE DECOMPOSITION OF
ORGANIC MATTER IN HYPOLIMNION, FOR EVERY PHOSPHORUS ATOM LIBERATED AS
PHOSPHATE, 580 ATOMS OF OXYGEN ARE USED FOR OXIDATION OF CORRESPONDING
TOTAL QUANTITY OF ORGANIC MATTER. STUDY OF THE EPOLIMNION STRONGLY
SUGGESTS THAT THE HIGH OXYGEN:PHOSPHORUS RATIO REFLECTS THE HIGH
ORGANIC CONTENT OF LAKE ALGAE AND PARTICULARLY THEIR HIGH
NITROGEN:PHOSPHORUS RATIO. HIGH LEVEL OF ORGANIC MATTER INDICATES THAT
SEDIMENTS ARE PROVIDING QUANTITIES OF NUTRIENTS, PARTICULARLY
AMMONIUM-NITROGEN. BIWEEKLY SAMPLES OF ALKALINE MONOPHOSPHOESTERASES
ARE MEASURED. IN BOTTOM DEPOSITS, CONTENT OF ORGANIC CARBON AND
KJELDAHL NITROGEN IS RATHER HIGH. THERE ARE INDICATIONS OF RECENT
ENHANCED EUTROPHICATION, SUPPORTED BY RESULTS OF PIGMENT EXTRACTION
FROM SEDIMENTS. (JONES-WISCONSIN)

FIELD 02H, 05C

ACCESSION NO. W70-09889

A SIMPLE, INEXPENSIVE ALGAL CHEMOSTAT,

NORTH CAROLINA STATE UNIV., RALEIGH. DEPT. OF ZOOLOGY.

EDWARD J. CARPENTER.

LIMNOLOGY AND OCEANOGRAPHY, VOL 13, NO 4, P 720-721, 1968. 1 FIG, 2 REF.

DESCRIPTORS:
*ANALYTICAL TECHNIQUES, *ALGAE, PLANKTON, DIATOMS, MARINE ALGAE, FRESH
WATER, FLOW RATES, PUMPS, ELECTROLYSIS, CULTURES.

IDENTIFIERS:
*CHEMOSTAT, PHYSIOLOGY, FLAGELLATES.

ABSTRACT:
A SIMPLE CHEMOSTAT HAS BEEN DEVELOPED WHICH DOES NOT USE EXPENSIVE
MECHANICAL ACTION PUMPS OR ELABORATE GRAVITY FLOW SYSTEMS. IN ADDITION,
THE ENTIRE APPARATUS CAN BE EASILY STERILIZED AS A UNIT IF THE RUBBER
STOPPERS IN THE CHEMOSTAT ARE LEFT AJAR DURING AUTOCLAVING. AN
ELECTROLYSIS PUMP GENERATES A GAS MIXTURE WHICH DISPLACES THE CULTURE
MEDIUM IN THE MEDIUM RESERVOIR INTO THE ALGAE CULTURE VESSEL. THE FLOW
RATE OF THE CULTURE MEDIUM IS REGULATED BY A VARIABLE RESISTOR IN THE
CIRCUIT AND INDIRECTLY MONITORED WITH A MILLIAMMETER. MIXING AND
AERATION OF ALGAE AND MEDIUM IS ACCOMPLISHED BY PUMPING STERILE,
COTTON-FILTERED AIR THROUGH THE CULTURE VESSEL. AN OVERFLOW VESSEL
EFFECTIVELY PREVENTS BACK CONTAMINATION OF THE CULTURE VESSEL BY
BACTERIA AS A SUSPENSION OF THE ORGANISMS UNDER CULTURE CONTINUALLY
FLOWS OUT. THE VOLUME OF MEDIUM WITHIN THE CULTURE VESSEL IS REGULATED
THROUGH SIPHON ACTION BY ADJUSTING THE HEIGHT OF THE OVERFLOW TUBE.
THIS CHEMOSTAT HAS BEEN USED SUCCESSFULLY IN SEVERAL NUTRIENT LIMITED,
AXENIC EXPERIMENTS ON TWO PLANKTONIC MARINE DIATOMS AND ONE FRESHWATER
FLAGELLATE. (JONES-WISCONSIN)

FIELD 07B, 05C

ACCESSION NO. W70-09904

ON USING INDUSTRIAL AND DOMESTIC WASTES IN AQUACULTURE,

MARYLAND UNIV., SOLOMONS. NATURAL RESOURCES INST.

J. A. MIHURSKY.

AGRICULTURAL ENGINEERING, VOL 50, NO 11, P 667-689, 1969. 1 FIG, 3 TAB, 18
REF.

DESCRIPTORS:
*WASTE DISPOSAL, *AQUACULTURE, *ULTIMATE DISPOSAL, DOMESTIC WASTES,
INDUSTRIAL WASTES, SEWAGE WASTES, BYPRODUCTS, FERTILIZERS, ALGAE,
ZOOPLANKTON, FISH, SHELLFISH, STEAM, ELECTRIC POWER, HEATED WATER,
PROTEINS, AQUATIC MICROORGANISMS, TEMPERATURE, ACTIVATED SLUDGE,
FISHERIES, CARP, CLAMS, PONDS, RUNNING WATERS, FRESH WATER, BRACKISH
WATER, ECONOMICS, CYCLING NUTRIENTS, FOODS.

IDENTIFIERS:
PETROLEUM WASTES, MERCENARIA MERCENARIA, AQUACULTURE, RECYCLING WASTES.

ABSTRACT:
THREE WASTE MATERIALS HAVE THE GREATEST POTENTIAL FOR RECYCLING THROUGH
AQUATIC SYSTEMS INTO USEFUL NUTRIENTS--PETROLEUM WASTES, SEWAGE WASTES,
AND WASTE HEAT. PROTEIN DEVELOPED AND SOLD IN FRANCE IS PRODUCED BY
MICROORGANISMS GROWING ON WASTE LOW-GRADE PETROLEUM. LARGE-SCALE USE OF
MICROBIOLOGICALLY DERIVED PROTEINS SEEMS POSSIBLE. SEWAGE CONVERTED TO
INORGANIC MATERIALS MAY PROVIDE FERTILIZER FOR ALGAL PRODUCTION. WASTE
HEAT MAY ASSIST IN PROVIDING OPTIMAL ENVIRONMENTAL CONDITIONS FOR
BIOLOGICAL GROWTH AT ALL TROPHIC LEVELS. IN VARIOUS WORLD SITUATIONS
SEWAGE OR SEWAGE PRODUCTS ARE USED TO STIMULATE FISHERY PRODUCTION.
ALGAE GROWN IN MASS AIDED BY WATER FROM TREATED SEWAGE AND CONCENTRATED
BY EVAPODRYING TECHNIQUE PRODUCES FOOD FOR CHICKENS. WASTE HEAT FROM
STEAM ELECTRIC STATIONS IS USED TO RAISE CARP IN LATITUDES FARTHER
NORTH THAN USUAL AND TO EXTEND THE GROWING SEASON IN SOME LOCALITIES. A
WASTE FLUE GAS WAS USED TO SUPPLY CARBON DIOXIDE GAS TO MASS-CULTURE AN
ALGAL FOOD SUPPLY. WILD OR NATURAL WATER IS PREFERABLE TO POND CULTURE.
RUNNING WATER SYSTEMS PERMIT DENSE STOCKING, WASTE REMOVAL AND
CONTINUED DISSOLVED OXYGEN RENEWAL. COSTS OF PRODUCING FISH PROTEIN
CONCENTRATE THROUGH AQUACULTURE PLUS WASTE UTILIZATION WILL BE
SUBSIDIZED BY SOCIETY'S NEED TO RECYCLE ITS WASTE MATERIALS INTO USEFUL
FOODSTUFFS. (JONES-WISCONSIN)

FIELD 05E

ACCESSION NO. W70-09905

RECOVERY OF A SALT MARSH IN PEMBROKESHIRE, SOUTH-WEST WALES, FROM POLLUTION BY CRUDE OIL,

K. J. M. BAKER, AND E. B. COWELL.

BIOLOGICAL CONSERVATION, VOL. 1, NO. 4, JULY 1969, P 291-295. 2 FIG, 2 TAB, 8 REF.

DESCRIPTORS:
RESISTANCE, WATER POLLUTION EFFECTS, *SALT MARSHES, OILY WATER, *OIL, *MARINE ALGAE, WATER POLLUTION SOURCES, ON-SITE DATA COLLECTIONS, *ON-SITE INVESTIGATIONS, FREQUENCY, *FREQUENCY ANALYSIS.

IDENTIFIERS:
*RECOVERY, OIL SPILL, *OIL POLLUTION, *FILAMENTOUS GREEN ALGAE, HALIMONE PORTULACOIDES, SUAEDA MARITIMA, ASTER TRIPOLIUM, PEMBROKE, SOUTH-WEST WALES.

ABSTRACT:
BENTLASS SALT MARSH, NEAR PEMBROKE, SOUTH-WEST WALES, WAS POLLUTED BY KUWAIT CRUDE OIL IN JANUARY 1967. FREQUENCY DATA TAKEN IN JUNE 1966, JUNE 1967, AND JUNE 1968, SHOW THAT SOME SPECIES WERE MORE AFFECTED THAN OTHERS BY THE OIL. IN JUNE 1967, SUAEDA MARITIMA, SALICORNIA SPP., HALIMIONE PORTULACOIDES, AND FILAMENTOUS GREEN ALGAE, SHOWED THE GREATEST REDUCTION OF FREQUENCY, WHILE ASTER TRIPOLIUM, COCHLEARIA SPP., TRIGLOCHIN MARITIMA, PUCCINELLIA MARITIMA, JUNCUS GERARDI, LIMONIUM HUMILE, AND SPARTINA TOWNSENDII SHOWED SOME REDUCTION. AT THE TIME FREQUENCIES OF FESTUCA RUBRA, PANTAGO MARITIMA, ARMERIA MARITIMA, ARTEMISIA MARITIMA, GLAUX MARITIMA, AND SPERGULARIA SPP., WERE NOT SIGNIFICANTLY LOWER THAN THE PRE-SPILL LEVELS. BY JUNE 1968 MOST OF THE SPECIES WHICH WERE AFFECTED IN 1967 SHOWED SOME RECOVERY, THE EXCEPTION BEING TRIGLOCHIN MARITIMA, THE ILL-EFFECT ON WHICH HAD INCREASED. THE SPECIES WITH THE GREATEST COVERAGE, FESTUCA RUBRA IN THE UPPER MARSH, PUCCINELLIA MARITIMA IN THE MID-MARSH, AND SPARTINA TOWNSENDII IN THE LOWER MARSH, HAD ALL RECOVERED MORE OR LESS COMPLETELY BY JUNE 1968. (SJOLSETH-WASHINGTON)

FIELD 05C

ACCESSION NO. W70-09976

CONTROL OF AQUATIC PLANT NUISANCES (ESPECIALLY ALGAE) WITH SOME SUBSTITUTED PHENYLUREAS,

IVONNE J. LE COSQUINO DE BUSSY.

VERHANDLUNGEN DER INTERNATIONALEN VEREINIGUNG FUR THEORETISCHE UND ANGEWANDTE LIMNOLOGIE, VOL 17, P 539-545, 1969. 1 FIG, 4 TAB, 9 REF.

DESCRIPTORS:
*AQUATIC PLANTS, *NUISANCE ALGAE, *ALGICIDES, *PHENOLS, *WATER QUALITY CONTROL, CRUSTACEANS, ROTIFERS, SWIMMING POOLS, CHLAMYDOMONAS, DAPHNIA, DIATOMS, CHLORELLA, SCENEDESMUS, OXYGEN, COLIFORMS, SEDIMENTS, MUD, SANDS, SLUDGE, RECREATION.

IDENTIFIERS:
*PHENYLUREAS, DIURON, FLAGELLATES, MONURON, TRANSPARENCY, CYCLOPS, SCENEDESMUS OBLIQUUS, DACTYLOCOCCUS, ESCHERICHIA COLI, OOCYSTIS, LEMNA, SPIRODELA, ELODEA.

ABSTRACT:
THE REQUIREMENT IN THE NETHERLANDS FOR A GOOD MEANS OF CONTROLLING ALGAE NUISANCE IN OPEN-AIR FLOW-THROUGH SWIMMING POOLS OR IN RECREATIONAL PONDS, FOSTERED RESEARCH WITH SUBSTITUTED PHENYLUREAS. THESE HERBICIDES ARE INHIBITORS OF PHOTOSYNTHESIS. HIGHER SUBMERGED WATER PLANTS AND FILAMENTOUS ALGAE CAN BE CONTROLLED VERY WELL BY THE APPLICATION OF DIURON IN EARLY APRIL IN DOSAGES VARYING FROM 0.2 TO 0.4 PPM. IN ORDER TO CONTROL MICROALGAE IN SWIMMING POOLS AND TO OBTAIN A TRANSPARENCY OF 100-250 CENTIMETERS A HIGHER DOSE IS NECESSARY, THE GROWTH OF CRUSTACEA, WHICH EAT SMALL ALGAE BEING DESIRABLE AT THE SAME TIME. THE ADDITION OF DIURON SHOULD BE EFFECTED IN CLEAR WATER PREFERABLY WITH AN INITIAL DOSE OF ONE PPM. IN SMALLER SWIMMING POOLS THE CONCENTRATION OF DIURON MUST BE ABOVE 0.6 PPM IN ORDER TO OBTAIN A TRANSPARENCY OF 120-240 CENTIMETERS. IN LARGE SWIMMING POOLS (6000 CUBIC METERS) A TRANSPARENCY FROM 170-270 CENTIMETERS COULD BE MAINTAINED EVEN DURING THE TIME WHEN THE DIURON IN THE WATER WAS NO LONGER DETECTABLE. (JONES-WISCONSIN)

FIELD 05G

ACCESSION NO. W70-10161

ALGAL RECORDS FOR THREE INDIANA SEWAGE STABILIZATION PONDS,

ROBERT A. TAFT WATER RESEARCH CENTER, CINCINNATI, OHIO. ADVANCED WASTE
    TREATMENT RESEARCH LAB.

C. MERVIN PALMER.

PROCEEDINGS OF THE INDIANA ACADEMY OF SCIENCE, VOL 78, P 139-145, 1968. 3
    TAB, 8 REF.

DESCRIPTORS:
    *SEWAGE LAGOONS, *ALGAE, *SEWAGE TREATMENT, *INDIANA, CHLOROPHYTA,
    CYANOPHYTA, DIATOMS, SEASONAL, EUGLENA, CHLAMYDOMONAS, CHLORELLA,
    SCENEDESMUS, SAMPLING, OXYDATION LAGOONS.

IDENTIFIERS:
    BLUE-GREEN ALGAE, NITZSCHIA, BIOLOGICAL STUDIES, ALGAL IDENTIFICATION.

ABSTRACT:
    THREE SEWAGE STABILIZATION PONDS IN SOUTHEASTERN INDIANA WERE AMONG
    SEVERAL THROUGHOUT THE UNITED STATES SELECTED FOR BIOLOGICAL STUDIES,
    WITH PARTICULAR EMPHASIS ON THE ALGAL FLORA. ALGAL IDENTIFICATIONS HAVE
    BEEN RECORDED FROM 376 SAMPLES COLLECTED FROM THE PONDS DURING A PERIOD
    FROM MAY 1962 TO AUGUST 1968. ALTHOUGH CERTAIN GENERA WERE FOUND
    FREQUENTLY IN ALL THREE PONDS, EACH POND HAD A DISTINCTIVE ALGAL FLORA.
    GREEN ALGAE WERE INVARIABLY THE MOST ABUNDANT; HOWEVER, FLAGELLATES
    WERE ALSO PROMINENT. OF A TOTAL OF 64 GENERA OF THE MOST SIGNIFICANT
    AND ABUNDANT ALGAE THERE WERE 29 GREEN ALGAE, 19 FLAGELLATES, 10
    BLUE-GREEN ALGAE, AND 6 DIATOMS. SOME GENERA WERE LIMITED TO THE SUMMER
    SEASON, WHILE OTHERS WERE MOST PROMINENT IN SPRING AND FALL OR IN
    WINTER. THE POLLUTION-TOLERANT ALGAE, EUGLENA AND NITZSCHIA, WERE
    ABUNDANT AND PERSISTENT IN ALL THREE PONDS. THE BIOCHEMICAL OXYGEN
    DEMAND AT THE INTAKES WAS ABOUT 500 PPM AND FOR THE EFFLUENTS LESS THAN
    50 PPM. THE FIRST VIRUS INFECTING A BLUE-GREEN ALGA WAS ISOLATED FROM
    ONE OF THE PONDS IN RIPLEY COUNTY. (JONES-WISCONSIN)

FIELD 05D

ACCESSION NO. W70-10173

MANAGEMENT OF AQUATIC VASCULAR PLANTS AND ALGAE,

CORNELL UNIV., ITHACA, N.Y. DEPT. OF BOTANY.

HUGH F. MULLIGAN.

EUTROPHICATION: CAUSES, CONSEQUENCES, CORRECTIVES, P 464-482. PRINTING AND
PUBLISHING OFFICE, NATIONAL ACADEMY OF SCIENCES, WASHINGTON, D.C., 1969. 2
TAB, 110 REF.

DESCRIPTORS:
*AQUATIC WEED CONTROL, *AQUATIC PLANTS, *HERBICIDES, *VASCULAR TISSUES,
*ALGAE, ECOSYSTEMS, CHARA, BENTHIC FLORA, FLOATING PLANTS,
PHYTOPLANKTON, AQUATIC INSECTS, FISH, GAME BIRDS, ANIMALS, FOOD CHAINS,
OXYGEN, ROTIFERS, WEEDS, WATER HYACINTH, ALLIGATORWEED, CHEMCONTROL,
MECHANICAL CONTROL, BIOCONTROL, COPPER SULFATE, SODIUM ARSENITE,
DIQUAT, TOXICITY, SNAILS, MAMMALS, TILAPIA, CARP, FERTILIZATION,
NORTHEAST US, NEW YORK, TENNESSEE VALLEY, 2-4-D.

IDENTIFIERS:
CERATOPHYLLUM, MYRIOPHYLLUM, EICHHORNIA, POTAMOGETON, TRAPA, PISTIA,
ALTERNANTHERA, HETERANTHERA, NAJAS, SILVEX, ENDOTHAL, POTASSIUM
PERMANGANATE, FENAC, SILVER NITRATE, SIMAZIN, NUPHAR, AGASICLES,
MANATEE, MARISA.

ABSTRACT:
CHEMICAL TREATMENTS OFFER, AT BEST, TEMPORARY RELIEF FROM OVERABUNDANCE
OF AQUATIC VEGETATION. EVEN WHEN EXCELLENT RESULTS ARE OBTAINED,
TREATMENTS MUST BE REPEATED ANNUALLY. PRESENTLY, AQUATIC HERBICIDES ARE
NONSELECTIVE, USUALLY KILLING ALL PLANTS BY CONTACT. IT WOULD BE MORE
EFFICACIOUS TO USE AQUATIC HERBICIDES POSSESSING SELECTIVE TOXICITY
AGAINST SPECIFIC AQUATIC WEEDS, THUS PERMITTING THE REESTABLISHMENT OF
DESIRABLE PLANTS. THE COST OF DEVELOPING AQUATIC HERBICIDES WOULD BE
GREATER AS APPOSED TO AGRICULTURAL HERBICIDES BECAUSE OF THE
SPECIALIZED TESTING REQUIREMENTS, AND GENERALLY, THE POTENTIAL MARKET
WOULD NOT BE AS GREAT. MECHANICAL CONTROL IS VERY PROMISING: IT
INVOLVES HAND-PULLING OR CUTTING AND HARVESTING PLANTS FROM RESTRICTED
AREAS. BIOLOGICAL CONTROL HAS MET WITH ONLY LIMITED SUCCESS, BUT IS AN
AVENUE OF PROMISING POTENTIAL. THIS APPROACH WOULD INCLUDE INTRODUCING
ORGANISMS THAT ARE INIMICAL TO THE TARGET ORGANISMS AND MANIPULATING
THE EXISTING AQUATIC ENVIRONMENT. IN A CONCEPT, THIS TYPE OF CONTROL
USES BIOLOGY AGAINST BIOLOGY. AMONG THE ORGANISMS UTILIZED IN CONTROL
OF AQUATIC PLANTS ARE FISH, SNAILS, MAMMALS, AND BEETLES. BIOLOGICAL
CONTROL IS THE ONLY METHOD THAT COULD PROVIDE A PERMANENT SOLUTION TO
THE PROBLEM OF EXCESSIVE AQUATIC VEGETATION. (SEE ALSO W70-03975)
(JONES-WISCONSIN)

FIELD 02I, 05C

ACCESSION NO. W70-10175

TREATMENT OF WASTE FROM POLYESTER MANUFACTURING. OPERATIONS,

FIBER INDUSTRIES, INC., CHARLOTTE, N.C.; AND DAVIS AND FLOYD ENGINEERS, INC., GREENWOOD, S.C.

C. E. STEINMETZ, AND W. J. DAY.

CHAMICAL ENGINEERING PROGRESS, SYMPOSIUM SERIES VOL 65, NO 97, 1969. P 188-190.

DESCRIPTORS:
*WATER REUSE, *RECLAIMED WATER, ORGANIC COMPOUNDS, CHEMICAL WASTES, ION EXCHANGE, BIOCHEMICAL OXYGEN DEMAND, CHEMICAL OXYGEN DEMAND, COLOR, ACTIVATED SLUDGE, ALGAE, WASTE WATER TREATMENT.

IDENTIFIERS:
*POLYESTER FIBERS, MAN-MADE FIBERS, METHYL ALCOHOL, ETHYLENE GLYCOL, POLYMERS, DIMETHYL TEREPHTHALATE.

ABSTRACT:
THIS IS A PROGRESS REPORT TO MAY 1969, ON A WATER REUSE PROJECT, JOINTLY UNDERTAKEN BY THE SHELBY PLANT OF FIBER INDUSTRIES, AND THE FWPCA. THIS PLANT APPLIES A LUBRICANT, OR SO-CALLED 'FIBER FINISH', TO THE POLYESTER FIBER MANUFACTURED THERE. IT IS OF A PROPRIETARY NATURE BUT FOR PURPOSES OF WASTE TREATMENT SCHEMES MAY BE CONSIDERED A LONG CHAIN FATTY ACID OR DERIVATIVE THEREOF. THIS MATERIAL, BLOW DOWN WATER FROM THE POWER PLANT, AND SMALL AMOUNTS OF CHEMICAL WASTES, INCLUDING METHYL ALCOHOL, ETHYLENE GLYCOL, DIMETHYL TEREPHTHALATE, LOW MOLECULAR WEIGHT POLYMER AND TITANIUM DIOXIDE, ION EXCHANGE RECHARGE CHEMICALS, HOUSEKEEPING AND LABORATORY WASTES ARE THE ESSENTIAL WASTE PRODUCTS. IN A PILOT PLANT STUDY THE WASTE WATER PREVIOUSLY TREATED IN AN EXTENDED AERATION ACTIVATED SLUDGE SYSTEM IS FURTHER TREATED. AS RECEIVED IT HAS A COLOR OF 50 UNITS, C.O.D. OF 280-300 PPM AND B.O.D. OF 8-24 PPM. ALGAE ARE REMOVED BY SCREENING AND A CARBON UNIT REDUCES THE COLOR TO 10 UNITS, C.O.D. TO 125 PPM AND B.O.D. TO 4 PPM. ONE HALF OF THIS WATER IS DISCHARGED AND THE REST RETURNED TO THE SYSTEM AFTER COOLING AND REMOVAL OF CHROMIUM AND ZINC SALTS. (WORK-NORTH CAROLINA STATE)

FIELD 05D

ACCESSION NO. W70-10433

A REVIEW OF THE LITERATURE OF 1966 ON WASTE WATER AND WATER POLLUTION CONTROL,

C. M. WEISS.

JOURNAL WATER POLLUTION CONTROL FEDERATION, VOL 39, NO 7, P 1049-1154, JULY 1967. 917 REF.

DESCRIPTORS:
*WASTE WATER TREATMENT, *BIOLOGICAL TREATMENT, *WATER POLLUTION CONTROL, FISH, TOXICITY, FRESH WATER, PULP WASTES, PESTICIDE RESIDUES, HERBICIDES, ESTUARIES, WATER QUALITY, DETERGENTS, ALGAL CONTROL, INDUSTRIAL WASTES, THERMAL POLLUTION, METHODOLOGY, INSTRUMENTATION, SEA WATER.

IDENTIFIERS:
METAL TOXICITY.

ABSTRACT:
A REVIEW OF PAPERS PUBLISHED DURING 1966 IS PRESENTED. MATERIAL COVERED INCLUDES: BIOLOGICAL STUDIES EVALUATING WATER POLLUTION, CAUSES OF FISH MORTALITY, TOXIC NATURE OF SUBSTANCES IN FRESH WATER, WATER QUALITY, DETERGENTS AS RELATED TO ALGAL GROWTH AND FISH DAMAGE, DOMESTIC AND INDUSTRIAL WASTE TREATMENT, METAL TOXICITY ASSAY, EFFECT OF WASTES FROM THE PULPING AND FOREST INDUSTRIES, THE POTENTIAL HAZARDS FROM PESTICIDE AND HERBICIDE RESIDUES, EFFECT OF THERMAL DISCHARGES, PHYSIOLOGICAL AND PATHOLOGICAL CONDITIONS IN FISH CAUSED BY TOXIC COMPOUNDS, METHODOLOGY AND INSTRUMENTATION, AND ESTUARINE AND MARINE POLLUTION.
(LIVENGOOD-NORTH CAROLINA STATE)

FIELD 05D, 10

ACCESSION NO. W70-10437

BEVERLY HILLS WATER DEPARTMENT SOLVES ALGAE PROBLEM WITH KMNO-SUB-4,

BEVERLY HILLS MUNICIPAL WATER DEPT., CALIF.

MARY JO WITT.

WATER AND SEWAGE WORKS, VOL 115, P 400-403, 1968. 2 FIG, 6 PHOTOS.

DESCRIPTORS:
*POTABLE WATER, *WATER QUALITY CONTROL, *ALGAE, COPPER SULFATE,
CHLORINE, WELLS, CALIFORNIA.

IDENTIFIERS:
*POTASSIUM PERMANGANATE, BEVERLY HILLS(CALIF).

ABSTRACT:
BEVERLY HILLS (CALIFORNIA) OPERATES TWO LIME SOFTENING AND FILTRATION
PLANTS FOR THEIR WELL WATER. PLANTS NO 1 AND NO 2 TREAT AN AVERAGE OF
5.5 MGD AND 2.5 MGD, RESPECTIVELY. IN MARCH 1963, A 900,000 GAL
SEDIMENTATION BASIN IN PLANT NO 1 WAS REPLACED BY A 250,000 GAL BASIN.
COPPER SULFATE AND CHLORINE HAD CONTROLLED ALGAE AND BACTERIAL GROWTHS
IN THE OLD BASIN BUT CONVERTED ALGAE AND BACTERIA TO A GREENISH-BLACK
TAR-LIKE SCUM IN THE NEW BASIN. IN JULY 1964, A 1.8 MILLIGRAM/LITER
DOSE OF POTASSIUM PERMANGANATE WAS ADDED TO THE BASIN ALONG WITH
CHLORINE AND COPPER SULFATE RESULTING IN NOTICEABLE IMPROVEMENT WITHIN
A DAY OR SO; CONTINUED TREATMENT ELIMINATED THE GROWTH. FOR MAXIMUM
CONTACT, POTASSIUM PERMANGANATE WAS FED INTO THE PUMP DISCHARGE LINE AT
THE FURTHEST WELL. SINCE STARTING POTASSIUM PERMANGANATE TREATMENT,
COPPER SULFATE APPLICATION WAS ELIMINATED; SHUTDOWNS FOR CLEAN-OUT ARE
NECESSARY ONLY EVERY 9-12 MONTHS, AND THERE IS PRACTICALLY NO ALGAE IN
THE SETTLING BASIN. DURING THE SUMMER AND FALL LARGE PIECES OF ALGAE
WOULD BREAK LOOSE FROM THE CLARIFIER AND BASIN WALLS IN PLANT NO 2. IN
OCTOBER 1965, A 1.2-1.5 MILLIGRAM/LITER CONTINUING DOSE OF POTASSIUM
PERMANGANATE, FED TO THE WELL FARTHEST FROM THE PLANT, FREED THE GROWTH
FROM THE BASIN WALLS. (HASKINGS-WISCONSIN)

FIELD 05F

ACCESSION NO. W71-00110

SEASONAL DYNAMICS OF THE REDUCING SUGAR CONTENT OF WATER IN THE KIEV RESERVOIR
AND INFLOWING RIVERS,

AKADEMIYA NAUK URSR, KIEV. INSTYTUT HIDROBIOLOGII.

G. A. YENAKI.

TRANSLATION OF GIDROBIOLOGICHESKIY ZHURNAL, VOL 5, NO 3, 1969.
HYDROBIOLOGICAL JOURNAL, VOL 5, NO 3, P 47-50, 1969. 4 P, 2 FIG, 1 TAB, 13
REF.

DESCRIPTORS:
*RESERVOIRS, *EUTROPHICATION, *CARBOHYDRATES, *NUTRIENTS,
*BIODEGRADATION, ALGAE, BACTERIA, AQUATIC MICROORGANISMS, SAMPLING,
WATER ANALYSIS, WATER QUALITY, POLLUTANT IDENTIFICATION.

IDENTIFIERS:
KIEV RESERVOIR(USSR).

ABSTRACT:
CARBOHYDRATE-LIKE COMPOUNDS IN NATURAL WATERS PLAY AN IMPORTANT ROLE IN
PROCESSES WHICH OCCUR WITHIN RESERVOIRS AND IN PARTICULAR IN PROCESSES
OF FORMATION AND TRANSFORMATION OF ORGANIC SUBSTANCES IN NATURAL
WATERS. MONTHLY DETERMINATIONS WERE MADE IN 1965 AND 1966 OF THE
REDUCING SUGAR CONTENT OF WATER IN THE KIEV RESERVOIR AND THE RIVERS
FLOWING INTO IT--THE DNIEPER, PRIPYAT', AND TETEREV. DETERMINATION OF
SUGARS WAS CONDUCTED ON FILTERED WATER BY A SPECTROPHOTOMETRICAL
METHOD. THE HIGH CONTENT OF REDUCING SUGARS IN WATER OF THE KIEV
RESERVOIR IS ONE OF THE MAIN FACTORS RESPONSIBLE FOR THE INTENSIVE
DEVELOPMENT OF MICROORGANISMS, THE INTENSIFICATION OF BIOCHEMICAL
PROCESSES, AND THE TRANSFORMATION OF BIOGENIC ELEMENTS. THIS CONDITION
INVOLVES AN INCREASE IN THE NUMBER OF PLANKTONIC ORGANISMS AND THE
OCCURRENCE OF 'BLOOMING' OF THE WATER. THE CONTENT OF REDUCING SUGARS
IN WATER OF THE KIEV RESERVOIR IN 1965 AND 1966 VARIED FROM 0.26-1.33
MG/LITER, AND IN THE RIVERS FLOWING INTO THE RESERVOIR THIS VALUE WAS
EQUIVALENT TO 0.17-1.63 MG/LITER. THE GLUCOSE CARBON WAS EQUIVALENT TO
0.7-6.7% AND 0.9-4.1% OF THE TOTAL ORGANIC CARBON, RESPECTIVELY. THE
MAXIMUM QUANTITY OF SUGARS WAS FOUND IN SPRING AND SUMMER, AND THE
MINIMUM QUANTITY WAS FOUND IN WINTER. THE INCREASE IN CONTENT OF
REDUCING SUGARS IN THE WATER OF THE KIEV RESERVOIR WAS ONE OF THE
FACTORS WHICH CAUSED 'BLOOMING' OF THE WATER IN THE SPRING-AUTUMN
PERIOD OF 1966. (KNAPP-USGS)

FIELD 05B

ACCESSION NO. W71-00221

BOTTOM FAUNA AND EUTROPHICATION,

COPENHAGEN UNIV., HILLEROD (DENMARK). FRESHWATER-BIOLOGICAL LAB.

PETUR M. JONASSON.

EUTROPHICATION: CAUSES, CONSEQUENCES, CORRECTIVES, P 274-305. PRINTING AND
  PUBLISHING OFFICE, NATIONAL ACADEMY OF SCIENCES, WASHINGTON, DC, 1969. 10
  FIG, 3 TAB, 47 REF.

DESCRIPTORS:
  *BENTHIC FAUNA, *EUTROPHICATION, *NUTRIENTS, LAKES, ECOLOGY, PRIMARY
  PRODUCTIVITY, ALGAE, PLANKTON, SUBMERGED PLANTS, PHYSICOCHEMICAL
  PROPERTIES, PHYTOPLANKTON, DISTRIBUTION, DEPTH, OXYGEN, ALKALINITY,
  HYDROGEN ION CONCENTRATION, STANDING CROP, LIGHT, BIOMASS, RESPIRATION,
  LIFE CYCLES, OLIGOCHAETES, MOLLUSKS, STRATIFICATION, DIATOMS,
  CHLOROPHYTA, CYANOPHYTA, DAPHNIA, SNAILS.

IDENTIFIERS:
  LAKE ERSOM(DENMARK).

ABSTRACT:
  BOTTOM FAUNA FITS INTO AN ECOLOGICAL PATTERN SET BY PRIMARY PRODUCTION
  OF ALGAE, SUBMERGED MACROPHYTES, AND PHYSICAL AND CHEMICAL FACTORS. THE
  PHYSICOCHEMICAL FACTORS AFFECT SIZE AND SEASONAL TREND OF PRIMARY
  PRODUCTION. IN TURN, PRIMARY PRODUCTION DETERMINES RANGE OF VARIOUS
  PHYSICAL AND CHEMICAL FACTORS AT THE BOTTOM. PHYTOPLANKTON PRODUCTION,
  VERTICAL DISTRIBUTION AND ABUNDANCE OF MACROPHYTES ARE IMPORTANT
  DETERMINANTS OF OXYGEN REGIME, ALKALINITY, PH, AND OTHER FACTORS
  ESPECIALLY IMPORTANT IN RELATION TO VERTICAL AND TEMPORAL VARIATION.
  BOTTOM FAUNA IS CORRESPONDINGLY AFFECTED IN QUALITY, QUANTITY, AND
  DEPTH DISTRIBUTION. ABOVE THE THERMOCLINE IS A RICH FAUNA WITH HIGH
  OXYGEN DEMANDS; BELOW THE THERMOCLINE ARE A FEW SPECIALISTS TOLERATING
  LOW OXYGEN TENSION. THESE QUALITATIVE CONTRASTS ARE PARALLELED BY
  QUANTITATIVE CONTRASTS. INCREASING NUTRIENTS TO EPILIMNION RESULTS IN
  INCREASED STANDING CROP AND PHYTOPLANKTON PRODUCTION. MACROPHYTES CLIMB
  TO COMPENSATE FOR DECREASED ILLUMINATION. LAKE CHANGES INTO ONE DEVOID
  OF SUBMERGED VEGETATION, DOMINATED BY NYMPHOIDS SHOWING ONLY THE LOWER
  PEAK OF ANIMAL BIOMASS. INCREASING NUTRIENTS TO HYPOLIMNION CAUSES
  RAPIDLY DEVELOPING OXYGEN DEFICIT, LOWERING RESPIRATORY ACTIVITY,
  REDUCING GROWTH PERIODS, AND OTHER ADVERSE EFFECTS ON BOTTOM FAUNA.
  DURING FINAL STAGE OF EUTROPHICATION, SESSILE MUD-DWELLERS MAY DROP TO
  ALMOST ZERO OR BECOME RESTRICTED TO OLIGOCHAETES. (JONES-WISCONSIN)

FIELD 05C

ACCESSION NO. W71-00665

PHYSIOLOGICAL-ECOLOGICAL STUDIES OF BENTHIC ALGAE IN LABORATORY STREAMS,

OREGON STATE UNIV., CORVALLIS. DEPT. OF BOTANY.

C. DAVID MCINTIRE.

JOURNAL WATER POLLUTION CONTROL FEDERATION, VOL 40, NO 11, PART 1, P
    1940-1952, 1968. 3 FIG, 2 TAB, 18 REF.

DESCRIPTORS:
    *PHYSIOLOGICAL ECOLOGY, *BENTHIC FLORA, *ALGAE, *HYDRAULIC MODELS,
    LABORATORY TESTS, PERIPHYTON, STREAMS, PHOTOSYNTHESIS, RESPIRATION,
    MICROORGANISMS, LIGHT INTENSITY, CARBON DIOXIDE, DISSOLVED OXYGEN,
    TEMPERATURE, VELOCITY, CYANOPHYTA, BIOMASS, INSECTS, DIATOMS,
    CHLOROPHYTA, CHRYSOPHYTA, GRAZING, SNAILS, AGE, TURBIDITY, DIPTERA,
    MAYFLIES, DETRITUS.

IDENTIFIERS:
    *RESPIROMETER CHAMBER.

ABSTRACT:
    PHYSIOLOGICAL ECOLOGY OF BENTHIC ALGAL COMMUNITIES WERE STUDIED IN
    LABORATORY STREAMS. RELATIVE ABUNDANCE OF ALGAL SPECIES DEPENDED ON
    SEASONAL FLUCTUATIONS IN CHEMICAL AND PHYSICAL PROPERTIES OF THE CREEK
    WATER IN THE EXPERIMENTAL STREAMS AND COMBINATIONS OF LIGHT INTENSITY
    AND CURRENT VELOCITY. PERIPHYTON ACCUMULATED ON RUBBLE AND GRAVEL
    GROWING SURFACES MOST RAPIDLY AT RELATIVELY HIGH LIGHT INTENSITIES AND
    CURRENT VELOCITIES, BUT WAS RETARDED AT HIGH DENSITIES OF GRAZING
    SNAILS. EXPORT RATE OF COMMUNITY MATERIAL DEPENDED ON FACTORS
    INFLUENCING BIOMASS ACCUMULATION, COMMUNITY AGE, CURRENT VELOCITY AND
    SILT LOAD. WITH COMMUNITIES DEVELOPED AT A LIGHT INTENSITY OF 550-FOOT
    CANDLE, INCREASE IN CONCENTRATION OF MOLECULAR CARBON DIOXIDE MARKEDLY
    INCREASED PHOTOSYNTHETIC RATE AT ILLUMINATION INTENSITIES NEAR AND
    ABOVE 1000-FOOT CANDLE. AT LIGHT INTENSITIES BELOW 500-FT C, A
    TEMPERATURE INCREASE OF 10C HAD NO SIGNIFICANT EFFECT ON PHOTOSYNTHETIC
    RATE, BUT AT INTENSITIES ABOVE 1000-FT C, THE TEMPERATURE COEFFICIENT
    VARIED BETWEEN 1.3 AND 1.6. COMMUNITY RESPIRATION INCREASED
    EXPONENTIALLY WITH TEMPERATURE INCREASES IN AIR-SATURATED WATER.
    RESPIRATORY RATES DECREASED RAPIDLY AS DISSOLVED OXYGEN WAS REDUCED
    BELOW AIR-SATURATION. EXPERIMENTS WITH LABORATORY COMMUNITIES SUGGESTED
    THAT, IN CERTAIN ECOLOGICAL INVESTIGATIONS, THE BENTHIC PERIPHYTON
    COMMUNITY MAY BE VIEWED AS A SINGLE FUNCTIONAL UNIT OR QUASI-ORGANISM.
    (JONES-WISCONSIN)

FIELD 07B, 05C

ACCESSION NO. W71-00668

PRIMARY PRODUCTIVITY, CHEMO-ORGANOTHROPHY, AND NUTRITIONAL INTERACTIONS OF
    EPIPHYTIC ALGAE AND BACTERIA ON MACROPHYTES IN THE LITTORAL OF A LAKE,

MICHIGAN STATE UNIV., EAST LANSING. DEPT. OF BOTANY AND PLANT PATHOLOGY.

HAROLD LEROY ALLEN.

AVAILABLE FROM NTIS AS COO-1599-25(PT. 2), $3.00 IN PAPER COPY, $0.95 IN
    MICROFICHE. 1969. 186 P. PH D THESIS.

DESCRIPTORS:
    *LAKES, *PRODUCTIVITY, MICHIGAN, *EPIPHYTOLOGY, NUTRIENTS, PLANT
    PATHOLOGY, *ALGAE.

IDENTIFIERS:
    LAWRENCE LAKE(MICHIGAN), *MACROPHYTES.

ABSTRACT:
    ASSESSMENT OF EPIPHYTIC ALGAL AND BACTERIAL IN SITU COMMUNITY
    METABOLISM, AND PHYSIOLOGICAL-NUTRITIONAL INTERRELATIONSHIPS OF
    MACROPHYTE-EPIPHYTE SYSTEMS, WERE INVESTIGATED IN THE LITTORAL ZONE OF
    A SMALL TEMPERATE LAKE FROM APRIL 1968 THROUGH MAY 1969. ANNUAL PRIMARY
    PRODUCTIVITY, CHEMO-ORGANOTROPHIC UTILIZATION OF DISSOLVED ORGANIC
    COMPOUNDS, AND FIELD AND LABORATORY STUDIES OF MACROPHYTE-EPIPHYTE
    INTERACTIONS WERE MONITORED BY CARBON-14 TECHNIQUES. QUALITATIVE AND
    QUANTITATIVE PHOTOSYNTHETIC PIGMENT COMPOSITION, AND A BRIEF TAXONOMIC
    EXAMINATION OF THE SESSILE COMPLEX, ACCOMPANIED MEASUREMENT OF FIELD
    PARAMETERS. PRODUCTIVITY MEASUREMENTS OF EPIPHYTIC ALGAE ON ARTIFICIAL
    SUBSTRATA COLONIZED IN EMERGENT (SCIRPUS ACUTUS MUHL.) AND SUBMERGENT
    (NAJAS FLEXILIS L. AND CHARA SPP.) MACROPHYTIC VEGETATION SITES WERE
    COMPARED OVER AN ANNUAL PERIOD WITH PIGMENT (CHLOROPHYLL A AND TOTAL
    PLANT CAROTENOIDS) ESTIMATES OF BIOMASS.

FIELD 02H, 05C

ACCESSION NO. W71-00832

TREATMENT OF DAIRY MANURE BY LAGOONING,

WASHINGTON STATE UNIV., PULLMAN.

SURINDER K. BHAGAT, AND DONALD E. PROCTOR.

JOURNAL WATER POLLUTION CONTROL FEDERATION, VOL 41, NO 5, 1969, P 785-795. 9
  FIG, 7 TAB, 6 REF.

DESCRIPTORS:
  *FARM WASTES, *FARM LAGOONS, *BIODEGRADATION, ALGAE, BIOCHEMICAL OXYGEN
  DEMAND, ANAEROBIC DIGESTION, AEROBIC TREATMENT, CHEMICAL OXYGEN DEMAND,
  STORAGE CAPACITY, EFFLUENTS, CONSTRUCTION.

IDENTIFIERS:
  *DAIRY MANURE, TOTAL SOLIDS, VOLITILE SOLIDS, NON-DEGRADABLE SOLID.

ABSTRACT:
  BECAUSE OF HIGH SOLIDS CONTENT OF DAIRY MANURE WASTE, ANAEROBIC LAGOONS
  CAN BE USED SATISFACTORILY AS A PRIMARY WASTE TREATMENT. AVERAGE
  REMOVALS OF BOD, COD, TS, AND VS ABOVE 86 PERCENT CAN BE ACCOMPLISHED
  WITH AN APPLIED LOADING OF 70 LB VS/DAY/1000 CU. FT. (1120 G/DAY/CU.
  M.). AN ANAEROBIC LAGOON CAN ACT AS A SEDIMENTATION, FLOTATION, AND
  ANAEROBIC DIGESTION PROCESS UNIT WHILE SIMULTANEOUSLY PROVIDING
  LONG-TERM STORAGE FOR NON-DEGRADABLE SOLID RESIDUE. THE EFFLUENT FROM
  THE ANAEROBIC LAGOON RETAINS MOST OF THE NUTRIENTS PRESENT IN THE RAW
  MANURE WASTE AND THUS HAS FERTILIZER VALUE. THE EFFLUENT HAS ORGANIC
  MATTER WHICH CAN BE OXIDIZED. THE EFFLUENT CAN BE APPLIED TO A FIELD OR
  SUBJECTED TO FURTHER TREATMENT. THE SECONDARY TREATMENT CAN BE AN
  AERATED LAGOON, OXIDATION DITCH, OR AN OXIDATION POND. THE RESULTS OF
  THE BATCH TYPE AEROBIC TREATMENT INDICATED THAT AN EFFLUENT BOD AT 20
  MG/L CAN BE ACHIEVED BY A 24-HR. AERATION PERIOD. (CHRISTENBURY-IOWA
  STATE)

FIELD 05D

ACCESSION NO. W71-00936

NATURAL FILTERS FOR AGRICULTURAL WASTES,

SOIL CONSERVATION SERVICE, WASHINGTON, D.C.

W. E. BULLARD, JR.

SOIL CONSERVATION, VOL 34, NO 4, NOVEMBER 1968, P 75-77. 2 FIG.

DESCRIPTORS:
  *FARM WASTES, *SPRINKLER IRRIGATION, *ORGANIC WASTES, *WASTE WATER
  DISPOSAL, AIR POLLUTION, WATER POLLUTION, ODOR, EFFLUENTS, SEWAGE
  EFFLUENTS, DILUTION, DECOMPOSING ORGANIC MATTER, INSECTS, MITES,
  BACTERIS, FUNGI, ALGAE, NITRITES, NITRATES, DETERGENTS, PHOSPHATES.

IDENTIFIERS:
  *BIOLOGIC 'DISPOSERS', MICROSCOPIC ORGANISMS, PAPERMILL WASTE EFFLUENT,
  CHEESE FACTORY WASTE WATERS.

ABSTRACT:
  FARMERS AND PROCESSORS OF FARM PRODUCTS ARE FINDING THAT THE WASTE
  PRODUCTS OF THEIR OPERATIONS GENERALLY CAN BE RETURNED TO THE LAND WITH
  LESS HAZARD TO THE ENVIRONMENT THAN WHEN DISCHARGED INTO STREAMS. BY
  COMPLETING THE NATURAL CYCLE OF GROWTH, DEATH, AND DECAY ON THE LAND
  WHERE CROPS ARE PRODUCED THEY MAKE USE OF A LEGION OF 'DISPOSER'
  ORGANISMS IN THE SOIL CAPABLE OF DECOMPOSING ORGANIC WASTES ON SITE.
  WHEN SPRAYED ON GRASS OR CROPS, THE EFFLUENT SERVES THE DUAL PURPOSE OF
  IRRIGATING AND FERTILIZING THE FIELD, THUS, AGRICULTURE HAS THE
  POTENTIAL MEANS OF DISPOSING OF ITS OWN WASTES AND PREVENTING OR
  REDUCING ENVIRONMENTAL POLLUTION. NUMEROUS EXAMPLES ARE GIVEN OF
  RETURNING SEWAGE TREATMENT EFFLUENTS, PULP AND PAPERMILL EFFLUENTS, AND
  CHEESE FACTORY WASTE WATERS, AMONG OTHERS, TO THE LAND BY SPRINKLER
  IRRIGATION. THE PRINCIPLE BEHIND THESE SUCCESSFUL OPERATIONS IS THAT OF
  GETTING MATERIAL PRODUCED FROM THE LAND BACK ONTO THE LAND WHERE THEY
  CAN BE USED AGAIN IN PRODUCTION. (WHITE-IOWA STATE)

FIELD 05D

ACCESSION NO. W71-00940

REQUIREMENTS FOR MICROBIAL REDUCTION OF FARM ANIMAL WASTES,

SOUTH DAKOTA STATE UNIV., BROOKINGS. DEPT. OF BACTERIOLOGY.

EDWARD C. BERRY.

PROCEEDINGS NATIONAL SYMPOSIUM ON ANIMAL WASTE MANAGEMENT, ASAE PUBLICATION NO SP-0366, MICHIGAN STATE UNIVERSITY, MAY 1966, P 56-58. 2 TAB, 1 FIG, 13 REF.

DESCRIPTORS:
*FARM WASTES, *LAGOONS, *BIODEGRADATION, *SEWAGE BACTERIA, ODOR, ANAEROBIC CONDITIONS, AEROBIC CONDITIONS, E. COLI, BACTERIA, FUNGI, ACTINOMYCETES, PROTOZOA, ALGAE.

IDENTIFIERS:
*CHEMICAL ENVIRONMENT, *PHYSICAL ENVIRONMENT, PHAGE, FERMENTOR, FACULTATIVE, MICROAEROPHILIC, INOCULATION.

ABSTRACT:
THE CARDINAL PRINCIPLE ON WHICH ALL SANITATION REDUCTION WORK IS BASED IS TO PROVIDE AN ENVIRONMENT IN WHICH THE MICRO-ORGANISMS CAN BRING ABOUT CONVERSION OF UNDESIRABLE MATERIAL TO A NON-OFFENSIVE AND STABLE STATE IN THE SHORTEST POSSIBLE TIME. TO BRING THIS ABOUT IT IS NECESSARY TO CONSIDER (A) THE WASTES WE WANT REDUCED AND THEIR END PRODUCTS AND (B) THE ORGANISMS THAT WE WANT TO PERFORM THIS CHORE FOR US. THE MICRO-ORGANISMS INVOLVED IN MANURE REDUCTION ARE TO BE FOUND IN THE FOLLOWING GROUPS: (A) BACTERIA (AEROBIC, ANAEROBIC, MICROAEROPHILIC, FACULTATIVE OR OBLIGATE), (B) FUNGI, (C) ACTINOMYCETE, (D) PROTOZOA, (E) ALGAE, AND (F) PHAGE. EACH ORGANISM FINDS ITS OPTIMUM ENVIRONMENT UNDER FAIRLY RESTRICTED ENVIRONMENTAL CONDITIONS. THIS PAPER GIVES EXAMPLES OF THE MOST COMMON MICRO-ORGANISMS AND THE EFFECT THE ENVIRONMENT HAS ON THEIR ACTIVITY.
(CHRISTENBURY-IOWA STATE)

FIELD 05D

ACCESSION NO. W71-02009

ECOLOGICAL CHANGES OF APPLIED SIGNIFICANCE INDUCED BY THE DISCHARGE OF HEATED WATERS,

ASTON UNIV., BIRMINGHAM (ENGLAND). DEPT. OF BIOLOGICAL SCIENCES.

H. A. HAWKES.

IN: ENGINEERING ASPECTS OF THERMAL POLLUTION, CHAPTER 2, P 15-57, 1969. 4 TAB, 3 FIG, 84 REF.

DESCRIPTORS:
*TEMPERATURE EFFECTS, *THERMAL POLLUTION, *ECOLOGY, ALGAE, BACTERIA, COLIFORMS, FISH, FUNGI, HEAT, WATER POLLUTION, STREAMS, RIVERS, LAKES, PHYSIOLOGICAL ECOLOGY, LIMITING FACTORS, ECOSYSTEMS, MICROORGANISMS, AQUATIC MICROORGANISMS, ENZYMES.

ABSTRACT:
THE GENERAL REVIEW OF ECOLOGICAL ASPECTS OF THERMAL POLLUTION IS DISCUSSED. TEMPERATURE OF WATER IS AN ECOLOGICAL FACTOR WHICH INFLUENCES ALL FORMS OF BIOLOGICAL LIFE IN WATER. THE MINIMUM, OPTIMUM, AND HARMFUL TEMPERATURE RANGES FOR ALL KINDS OF BIOTA IN WATERS IS DISCUSSED. THE TEMPERATURE INFLUENCES THE MICROBIOLOGICAL ORGANISMS BY ITS EFFECTS ON ENZYMES, WHICH ARE INACTIVATED AT HIGH TEMPERATURES. THE MOST OBVIOUS EFFECT OF THE RAISED WATER TEMPERATURE ON FISH IS THE DIRECT AUTECOLOGICAL ONE, AFFECTING THE DISTRIBUTION OF EACH SPECIES ACCORDING TO THEIR TEMPERATURE TOLERANCE RANGES. THE OPTIMUM TEMPERATURE RANGES FOR FISH ARE PRESENTED, AND INDIRECT EFFECTS OF RAISED TEMPERATURE ON FISH ARE DISCUSSED (DECREASE OF OXYGEN LEVEL, INCREASE OF TOXICITY EFFECTS, ETC.). THE PAPER ALSO DISCUSSES THE TEMPERATURE EFFECTS ON MACRO-INVERTEBRATES, ALGAE, BACTERIA, AND FUNGI.
(NOVOTNY-VANDERBILT)

FIELD 05C

ACCESSION NO. W71-02479

BIOLOGICAL ASPECTS OF THERMAL POLLUTION.

VANDERBILT UNIV., NASHVILLE, TENN. DEPT. OF ENVIRONMENTAL AND WATER RESOURCES
ENGINEERING.

PROCEEDINGS OF THE NATIONAL SYMPOSIUM ON THERMAL POLLUTION, SPONSORED BY THE
FEDERAL WATER POLLUTION CONTROL ADMINISTRATION AND VANDERBILT UNIVERSITY,
PORTLAND, OREGON, JUNE 3-5, 1968, PRICE: $7.95. PETER A. KRENKEL AND FRANK
L. PARKER, EDITORS, VANDERBILT UNIVERSITY PRESS, NASHVILLE, 1969. 351 P.

DESCRIPTORS:
*THERMAL POLLUTION, *THERMAL POWERPLANTS, *ECOLOGY, *BIOTA,
*TEMPERATURE, *AQUATIC ENVIRONMENT, *AQUATIC LIFE, *AQUATIC
POPULATIONS, ALGAE, BENTHOS, ZOOPLANKTON, FRESHWATER FISH, ESTUARINE
ENVIRONMENT, IMPOUNDED WATERS, FRESHWATER ALGAE, ANADROMOUS FISHES.

ABSTRACT:
THE PROCEEDINGS OF THE NATIONAL SYMPOSIUM ON THERMAL POLLUTION -
BIOLOGICAL CONSIDERATIONS - ARE INCLUDED IN THIS BOOK. A BRIEF
DISCUSSION OF THE ENGINEERING PROBLEMS OF THERMAL POLLUTION INCLUDES
STRATIFICATION OF IMPOUNDED WATERS, THE CONTRIBUTION OF HEAT FROM POWER
PLANTS, COOLING WATER REQUIREMENTS, NATURAL HEAT DISSIPATION
MECHANISMS, AND ALTERNATIVE METHODS FOR COOLING WATER. REPORTS AND
DISCUSSIONS CONCERNING THE EFFECTS OF THERMAL POLLUTION ON FRESHWATER
ALGAE, MARINE ZOOPLANKTON, BRITISH FRESHWATER FISHES, ANADROMOUS
FISHES, AND FRESHWATER BENTHOS ARE PRESENTED. THE DEVELOPMENT OF
THERMAL REQUIREMENTS OF FRESHWATER FISHES, THEORETICAL CONSIDERATIONS
OF THERMAL POLLUTION ON MARINE FISHES AT VARIOUS LIFE STAGES, AND THE
POTENTIAL EFFECT OF THERMAL ALTERATION ON MARINE AND ESTUARINE BENTHOS
ARE ALSO DISCUSSED. MOST PRESENTATIONS INCLUDE REVIEWS OF PAST AND
CURRENT LITERATURE AND RECOMMENDATIONS, AND THE FINAL PRESENTATION
DEALS EXCLUSIVELY WITH RESEARCH NEEDS FOR THERMAL POLLUTION. (SEE ALSO
W71-02494 AND W71-02496) (SPEAKMAN-VANDERBILT)

FIELD 05C, 05B, 05D

ACCESSION NO. W71-02491

ASPECTS OF THE POTENTIAL EFFECT OF THERMAL ALTERATION ON MARINE AND ESTUARINE
BENTHOS,

OREGON STATE UNIV., NEWPORT. YAQUINA BILOGICAL LAB.

JOEL W. HEDGPETH, AND JEFFERSON J. GONOR.

IN: BIOLOGICAL ASPECTS OF THERMAL POLLUTION, CHAPTER 4, P 80-118, 1969. 7
FIG, 1 TAB, 91 REF.

DESCRIPTORS:
*BENTHOS, *TEMPERATURE, *THERMAL POLLUTION, LABORATORY ANIMALS,
ACCLIMATIZATION, MARINE ANIMALS, AQUATIC LIFE, AQUATIC ANIMALS, BENTHIC
FAUNA, AQUATIC PLANTS, MARINE ALGAE, BENTHIC FLORA, ESTUARIES,
*ESTUARINE ENVIRONMENT, BAYS, *FLUCTUATION, RESISTANCE, LIFE CYCLES,
GROWTH STAGES.

IDENTIFIERS:
TEMPERATURE RANGES, TEMPERATURE REQUIREMENTS.

ABSTRACT:
LABORATORY EXPERIMENTS ON THERMAL TOLERANCES, DEATH POINTS, AND THE
LIFE, WITHOUT REFERENCE TO THE NATURAL CONDITIONS, INCLUDING PREVIOUS
TEMPERATURE EXPERIENCE AND STATE OF TIDE OR SEASON AT WHICH
EXPERIMENTAL MATERIAL WAS GATHERED, HAVE QUESTIONABLE UTILITY IN
REFERENCE TO WHAT THE ORGANISMS MAY ACTUALLY DO IN NATURE. INTENSIVE
FIELD STUDIES WITH IN SITU MEASUREMENTS OF THE ENVIRONMENT AND THE
ORGANISMS, COMBINED WITH CONTINUOUS MONITORING, ESPECIALLY OF THERMAL
GRADIENTS, ARE NEEDED. FIELD EVIDENCE INDICATES THAT SOME INTERTIDAL
HERBIVORES ARE WELL ADJUSTED TO THE TEMPERATURE RANGES FROM SEAWATER TO
RATHER HIGH AIR TEMPERATURES FOR VARYING PERIODS OF TIME, AND THAT
INDEED SUCH A TEMPERATURE RANGE MAY BE AN ECOLOGICAL REQUIREMENT,
RATHER THAN AN ENVIRONMENT TO BE ENDURED. HOW EXTENSIVE THIS MAY BE,
AND TO WHAT DEGREE IT MAY APPLY ALSO TO SUBTIDAL ORGANISMS, REMAINS TO
BE DETERMINED. THERE IS EVIDENCE INDICATING THAT SOME MARINE ORGANISMS
DO NOT FLOURISH IN A STABLE TEMPERATURE REGIME BUT REQUIRE THE
VARIATION AROUND THE STATISTICAL MEAN. HOWEVER, SUCH TEMPERATURE
REQUIREMENTS ARE NOT YET ESTABLISHED FOR MANY ORGANISMS, SINCE MOST
EXPERIMENTS INVOLVING LABORATORY CULTURE ARE OF COMPARATIVELY SHORT
DURATION. IN CONTRAST TO THE OBSERVED ABILITY OF MANY MARINE ORGANISMS
TO WITHSTAND WIDE RANGES OF ENVIRONMENTAL TEMPERATURES AT SOME STAGES
OF THEIR LIFE CYCLES, THERE IS THE GROWING BODY OF EVIDENCE THAT
COMPARATIVELY SMALL FLUCTUATIONS IN OCEANIC TEMPERATURES MAY INFLUENCE
THE DISTRIBUTION AND ABUNDANCE OF MANY SPECIES, ESPECIALLY THOSE THAT
OCCUR IN LARGE POPULATIONS. (SEE ALSO W71-02491) (SPEAKMAN-VANDERBILT)

FIELD 05C

ACCESSION NO. W71-02494

SOME EFFECTS OF TEMPERATURE ON FRESHWATER ALGAE,

ACADEMY OF NATURAL SCIENCES OF PHILADELPHIA, PA. DEPT. OF LIMNOLOGY.

RUTH PATRICK.

IN: BIOLOGICAL ASPECTS OF THERMAL POLLUTION, CHAPTER 7, P 161-185, 1969. 53
REF.

DESCRIPTORS:
*ALGAE, *TEMPERATURE, DIATOMS, PHOTOSYNTHESIS.

IDENTIFIERS:
*BLUE-GREEN ALGAE, *GREEN ALGAE, *RED ALGAE, *TEMPERATURE TOLERANCE,
OPTIMUM GROWTH TEMPERATURE, CONDENSER PAGGAGE.

ABSTRACT:
MOST OF THE COMMON FRESHWATER ALGAE BELONG TO THE GREEN ALGAE, DIATOMS,
RED ALGAE, AND BLUE-GREEN ALGAE GROUPS. IN EACH OF THESE MAJOR GROUPS
THERE ARE MANY SPECIES. THESE SPECIES COVER A WIDE RANGE OF
TEMPERATURES FROM THOSE THAT PREFER COOL-WATER CONDITIONS TO THOSE THAT
PREFER VERY WARM WATER CONDITIONS. EACH SPECIES HAS A RANGE OF
TEMPERATURE THAT IT CAN TOLERATE AND A RANGE IN WHICH ITS GROWTH IS
OPTIMUM. IN GENERAL, THE BLUE-GREEN ALGAE HAVE MORE SPECIES THAT PREFER
TEMPERATURES FROM 35C UPWARD, WHEREAS THE GREEN ALGAE HAVE A RELATIVELY
LARGE NUMBER OF SPECIES THAT GROW BEST IN TEMPERATURES RANGING UP TO
35C. MOST DIATOM SPECIES PREFER TEMPERATURES BELOW 30C. THE EFFECT OF
ARTIFICIALLY INCREASING THE TEMPERATURE REGIME OF A SPECIES TENDS TO
INCREASE GROWTH AND PHOTOSYNTHESIS SO LONG AS LIGHT IS SUFFICIENT FOR
THESE FUNCTIONS AND THE LIMITS OF TEMPERATURE TOLERANCE ARE NOT
REACHED. AS ONE APPROACHES THE LIMITS OF TEMPERATURE TOLERANCE FOR A
SPECIES, CELL DIVISION IS REPRESSED, AS IS PHOTOSYNTHESIS, AND THE
FORMATION OF REPRODUCTIVE CELLS MAY BE REPRESSED. THE PATTERN OF GROWTH
MAY BE GREATLY ALTERED. THE TEMPERATURES AT WHICH SPECIES MAY
SUCCESSFULLY GROW IN THE LABORATORY MAY BE HIGHER THAN THOSE WHICH
OCCUR IN NATURE BECAUSE OF LACK OF COMPETITION WITH OTHER SPECIES UNDER
NATURAL CONDITIONS AND THE EFFECT OF PREDATOR PRESSURE. STUDIES MADE
CONCERNING THE EFFECT ON ALGAE OF PASSING THEM THROUGH A CONDENSER
INDICATE THAT IF THE TEMPERATURE DOES NOT EXCEED 34-34.5C, LITTLE, IF
ANY, HARM IS DONE. (SEE ALSO W71-02491) (SPEAKMAN-VANDERBILT)

FIELD 05C

ACCESSION NO. W71-02496

AUTOECOLOGICAL AND SAPROBIOLOGICAL INVESTIGATIONS OF FRESH WATER CILIATES (IN
GERMAN),

BONN UNIV. (GERMANY). ZOOLOGICAL INST.

HARTMUT BICK.

ENGLISH SUMMARY. HYDROBIOLOGIA, VOL 31, NO 1, P 17-36, 1968. 13 REF.

DESCRIPTORS:
*INDICATORS, *PROTOZOA, *ENVIRONMENT, *ORGANIC MATTER, WATER QUALITY,
TEMPERATURE, HYDROGEN ION CONCENTRATION, OXYGEN, CARBON DIOXIDE,
AMMONIA, NITRITES, HYDROGEN SULFIDE, SALINITY, LAKES, NUTRIENTS,
BACTERIA, ALGAE, BRACKISH WATER, ALKALINE WATER, SODIUM.

IDENTIFIERS:
*CILIATES, FLAGELLATES, SAPROBIEN SYSTEM, CILIATE AUTOECOLOGY.

ABSTRACT:
ECOLOGICAL VALENCIES OF 31 SP OF FRESH WATER CILIATES ARE PRESENTED IN
TABULATED FORM. ENVIRONMENTAL FACTORS INCLUDE TEMPERTURE, PH, OXYGEN,
CARBON DIOXIDE, FREE AND BOUND AMMONIA, NITRITES, HYDROGEN SULFIDE, AND
THE TOTAL CONCENTRATION OF SALTS IN BRACKISH WATERS AND INLAND ALKALINE
LAKES ENRICHED IN SODIUM. THIS RECORD IS SUPPLEMENTED BY A NUMBER OF
BACTERIA, ACCOMPANYING DIFFERENT CILIATES. AS FAR AS POSSIBLE, THE
OPTIMUM RANGES OF SITE FACTORS ARE GIVEN FOR DIFFERENT SPECIES. THE
AUTOECOLOGICAL RESULTS ARE RELATED TO THE INDICATOR VALUE OF SPECIES
FOR THE SAPROBIEN SYSTEM OF MONITORING WATER QUALITY. (WILDE-WISCONSIN)

FIELD 05G, 02H

ACCESSION NO. W71-03031

RAPIDLY LABELED POLYPHOSPHATES AND METAPHOSPHATES IN THE BLUE-GREEN ALGA ANACYSTIS NIDULANS (IN GERMAN),

TECHNISCHE HOCHSCHULE, HANOVER (WEST GERMANY). INST. FOR BOTANY.

R. NIEMEYER, AND G. RICHTER.

ENGLISH SUMMARY. ARCHIV FUR MIKROBIOLOGIE, VOL 69, P 54-59, 1969. 3 FIG, 10 REF.

DESCRIPTORS:
*ANALYTICAL TECHNIQUES, *PHOSPHATES, *ALGAE, CHROMOTOGRAPHY, RADIOCHEMICAL ANALYSIS, TRACERS.

IDENTIFIERS:
*ANACYSTIS NIDULANS, OLIGOPHOSPHATES, METAPHOSPHATES, PULSE-LABELING, POLYPHOSPHATES, FRACTIONATION.

ABSTRACT:
THE CELLS OF THE BLUE-GREEN ALGA ANACYSTIS NIDULANS WERE PULSE-LABELED WITH PHOSPHORUS-32-ORTHOPHOSPHATE AND GROWN FOR 20 HOURS IN PHOSPHORUS-FREE MEDIUM. MOST OF THE PHOSPHORUS-32 WAS INCORPORATED INTO INORGANIC PHOSPHATES WHICH WERE ISOLATED BY JOINING EXTRACTION WITH THE NUCLEIC ACID AND FRACTIONATION ON METHYLATED SERUM ALBUMIN AND KIESELGUR. THE COLUMN CHROMTOGRAPHY OF THE ISOLATE REVEALED 8 PHOSPHATE RESIDUES INCLUDING OLIGOPHOSPHATES AND CIRCULAR METAPHOSPHATES OF THUS FAR UNKNOWN ORIGIN AND FUNCTION. THE TRIMETAPHOSPHATE INCORPORATED CONSIDERABLY MORE RADIOACTIVITY THAN OTHER LOW-MOLECULAR INORGANIC PHOSPHATES. (WILDE-WISCONSIN)

FIELD 07B, 05C

ACCESSION NO. W71-03033

ELECTROKINETIC PHENOMENA IN THE FILTRATION OF ALGAL SUSPENSIONS,

CORNELL, HOWLAND, HAYES AND MERRYFIELD, SEATTLE, WASH.; AND MICHIGAN UNIV., ANN ARBOR. DEPT. OF CIVIL ENGINEERING.

GERALD W. FOESS, AND JACK A. BORCHARDT.

JOURNAL AMERICAN WATER WORKS ASSOCIATION, VOL 61, P 333-338, 1969. 10 FIG, 1 TAB, 18 REF.

DESCRIPTORS:
*ALGAE, *SANDS, *FILTRATION, FILTERS, ELECTROPHORESIS, CHLORELLA, SCENEDESMUS, ADSORPTION, SEWAGE TREATMENT, PARTICLE SIZE.

IDENTIFIERS:
*ALGAL SUSPENSIONS, ELECTROKINETIC EFFECTS, ANN ARBOR(MICH), SAND BEDS.

ABSTRACT:
THIS INVESTIGATION REVEALED THAT THE ATTACHMENT MECHANISM FOR DISCRETE ALGAE IN SAND FILTRATION INVOLVES A SURFACE AFFINITY BETWEEN THE ALGAL CELLS AND SAND GRAINS, AN INTERACTION WHICH CAN BE CHEMICALLY CONTROLLED. A LOWERING OF THE SUSPENSION PH VALUE INCREASES ADSORPTION OF THE PARTICLES AND THEREBY SIGNIFICANTLY AUGMENTS THE REMOVAL OF SUSPENDED MATTER. THE REMOVAL IS ALSO INCREASED DUE TO A REDUCTION OF THE REPULSIVE ENERGY OF SAND BY ITS COATING WITH POSITIVELY CHARGED SUBSTANCES. THESE MODIFICATIONS OF SAND AND SUSPENDED PARTICLE PROPERTIES PROMISE TO INCREASE FILTER REMOVAL EFFICIENCIES BEYOND THOSE POSSIBLE WITH NORMAL DOSES OF COAGULANTS AND WITH MINIMAL HEAD LOSS BUILDUP. (WILDE-WISCONSIN)

FIELD 05D

ACCESSION NO. W71-03035

CULTURAL EUTROPHICATION: THE NATURE OF ·THE PROBLEM AND SOME RECOMMENDATIONS FOR
ITS STUDY AND CONTROL,

FLORIDA UNIV., GAINESVILLE. DEPT. OF ENVIRONMENTAL ENGINEERING.

PATRICK L. BREZONIK.

DEPARTMENT OF ENVIRONMENTAL ENGINEERING, UNIVERSITY OF FLORIDA, GAINESVILLE,
UNDATED. 26 P, 5 TAB, 51 REF. OWRR PROJECT B-004-FLA(2).

DESCRIPTORS:
*LAKES, *EUTROPHICATION, *WATER POLLUTION CONTROL, *WATER POLLUTION
SOURCES, *WATER POLLUTION EFFECTS, WATER UTILIZATION, ANALYTICAL
TECHNIQUES, NUTRIENTS, NITROGEN, PHOSPHORUS, AGRICULTURE, WATER
MANAGEMENT(APPLIED), GEOCHEMISTRY, SOIL TYPES, HYDROLOGY, CLIMATES,
TROPHIC LEVEL, NITROGEN FIXATION, FOREST, WINDS, LIGHT, TEMPERATURE,
HYDROGEN ION CONCENTRATION, DISSOLVED OXYGEN, CHLOROPHYLL, ALGAE,
CONDUCTIVITY, SEDIMENTS, BENTHIC FAUNA, RECREATION, BIOASSAY, FUNGI,
VIRUSES.

IDENTIFIERS:
*CULTURAL EUTROPHICATION, TROPHIC INDICATORS, NUTRIENT BUDGET.

ABSTRACT:
THE CAUSAL CHANGES IN WATER QUALITY FROM PRE-INDUSTRIAL MAN TO PRESENT
HIGH-DENSITY POPULATIONS ARE DESCRIBED. OF DESIRABLE TECHNIQUES,
MATHEMATICAL MODELS AND BIOASSAYS MAKE POSSIBLE COMPARISONS OF THE
NUTRITIONAL STATUS OF THE DIFFERENT SPECIES OF ALGAE MICROSCOPICALLY,
BY CYTOCHEMICAL STAINING OF ENZYMES. THE FIELD OF PALEOLIMNOLOGY POINTS
OUT THE DIFFERENTIATION OF NATURAL AND CULTURAL EFFECTS ON LAKES. STUDY
OF SEDIMENT CORES GIVES A BASIS FOR COMPARISON IN EUTROPHICATION
DEVELOPMENT. BLOOM-FORMING BLUE-GREEN ALGAE CHARACTERISTICS POINT TO
NEED OF KNOWLEDGE OF ALGAL PHYSIOLOGY. THE BEST SOLUTION TO THE PROBLEM
OF CULTURAL EUTROPHICATION IS ELIMINATION OF NUTRIENTS FROM SOURCE
WATERS. IN RESTORING LAKES, METHODS OF PREVENTING NUTRIENT RELEASE FROM
SEDIMENTS, AND RESEARCH ON EFFECTIVENESS OF PROVIDING GAME FISH
SPAWNING AND FEEDING AREAS IS NEEDED. REFINING THE TOOLS FOR LAKE
MANAGEMENT, INCLUDING LEGAL DEVELOPMENTS IN SHORELAND CORRIDOR ZONING
ARE PREREQUISITES. GREATER ACCURACY IS NEEDED IN PREDICTING RESPONSE OF
A LAKE TO A GIVEN NUTRIENT INPUT OR TO CHANGES IN LAND UTILIZATION TO
MAKE POSSIBLE LAKE BASIN ZONING AND DEVELOPMENT OF LAND USE PATTERNS ON
A RATIONAL BASIS. (JONES-WISCONSIN)

FIELD 05C

ACCESSION NO. W71-03048

UTILIZATION OF ALKOXYSILANES FOR REDUCTION OF THE ABUNDANCE OF PHYTOPLANKTON
DURING 'BLOOMS' OF WATER BASINS (IN RUSSIAN),

MOSCOW STATE UNIV. (USSR).

N. S. STROGANOV, V. G. KHOBOTIEV, L. V. KOLOSOVA, AND M. A. KADINA.

DOKLADY AKADEMII NAUK SSSR, VOL 181, NO 5, P 1257-1259, 1968. 2 FIG, 2 TAB,
13 REF.

DESCRIPTORS:
*HERBICIDES, *ALGAE, TOXICITY, SCENEDESMUS, CHLORELLA, DAPHNIA,
DETERIORATION, WATER TREATMENT, DEGRADATION(DECOMPOSITION).

IDENTIFIERS:
*ALGAL BLOOMS, *ORGANO-SILICATES, *ALKOXYSILANES, SCENEDESMUS
QUADRICAUDA, CHLORELLA VULGARIS, DAPHNIA MAGNA, LIMNAEA STAGNALIS,
OUSPENSKII'S MEDIUM.

ABSTRACT:
A SEARCH FOR HERBICIDES SUPPRESSING ALGAL BLOOMS, BUT HARMLESS TO
ZOOPLANKTERS AND OTHER HYDROBIONTS LEAD TO INVESTIGATION OF
ORGANO-SILICATES: VENYLTRIETHOXYSILANE, TETRAETHOXYSILANE, AND
TRIFLUOROPROPENYL-(METHYL)-DIETHOXYSILANE. THE TEST ORGANISIMS INCLUDED
SCENEDESMUS QUADRICAUDE, CHLORELLA VULGARIS, DAPHNIA MAGNA, AND LIMNAEA
STAGNALIS. THE COUNT OF SURVIVED SPECIMENS SHOWED THAT THE
CONCENTRATION OF 0.01 MG/LITER OF ALL INVESTIGATED COMPOUNDS ELIMINATED
93% TO 96% OF ALGAL CELLS WITHOUT DAMAGING EITHER DAPHNIA OR MOLLUSKS.
AMONG THE ADVANTAGES OF THE ALKOXYSILANES IS THEIR RELATIVELY RAPID
DISINTEGRATION IN WATER. (WILDE-WISCONSIN)

FIELD 05F, 05C

ACCESSION NO. W71-03056

THE EFFECTS OF DIELDRIN ON DIATOMS,

KANSAS UNIV., LAWRENCE. DEPT. OF ZOOLOGY.

JOHN CAIRNS, JR.

MOSQUITO NEWS, VOL 28, P 172-179, 1968. 1 TAB, 1 FIG, 6 REF. USPHS GRANT
EF-00266.

DESCRIPTORS:
*DIELDRIN, *DIATOMS, LABORATORY TESTS, INSECTICIDES, WATER POLLUTION
EFFECTS, ALGAE.

IDENTIFIERS:
*EFFECTS, NAVICULA SEMINULUM VAR.HUSTEDTII.

ABSTRACT:
THE EFFECTS OF DIELDRIN ON NAVICULA SEMINULUM WAS DESCRIBED. AT A
DIELDRIN CONCENTRATION OF 12.8 PPM THE NUMBER OF DIATOMS WAS ONLY HALF
THAT IN THE CONTROLS AT THE END OF THE 5-DAY GROWTH PERIOD. THERE WAS
NO INCREASE IN THE NUMBER OF CELLS AT 32.0 PPM, AND A CONCENTRATION OF
1.8 PPM WAS ONLY 10.6 PERCENT DIFFERENT FROM THE CONTROLS.
(WAHTOLA-WASHINGTON)

FIELD 05C

ACCESSION NO. W71-03183

THE ISOLATION OF PROTOZOA AND ALGAE FROM THEIR BACTERIAL CONTAMINANTS,

MASSACHUSETTS UNIV., WALTHAM. DEPT. OF ENVIRONMENTAL SCIENCES.

ROBERT A. COLER, AND HAIM B. GUNNER.

PHYTON, VOL 26, NO 2, P 191-194, NOVEMBER 1969. 3 FIG, 11 REF. OWRR PROJECT
A-020-MASS(3).

DESCRIPTORS:
*ISOLATION, *BACTERIA, *PROTOZOA, *ALGAE, LETHAL LIMIT,
ELECTROPHORESIS, MICROORGANISMS, ELECTRIC CURRENTS, *WASTE TREATMENT,
WASTE WATER TREATMENT.

ABSTRACT:
A TECHNIQUE IS OUTLINED FOR THE ISOLATION OF CILIATED PROTOZOA AND
GREEN AND BLUE-GREEN ALGAE FROM BACTERIA. THE PROCEDURE IS DESIGNED TO
EXPLOIT BOTH THE ELECTROPHORETIC MOBILITY OF SELECTED CILIATED PROTOZOA
AND FLAGELLATED ALGAE AND THE APPARENTLY GREATER VULNERABILITY OF
BACTERIA OVER PROTOZOA AND ALGAE TO THE LETHAL AFFECTS OF AN ELECTRIC
FIELD.

FIELD 05F, 05D

ACCESSION NO. W71-03775

MOVEMENT OF ALGAL- AND FUNGAL-BOUND RADIOSTRONTUM AS CHELATE COMPLEXES IN A
CALCAREOUS SOIL,

ARIZONA UNIV., TUCSON.

WALLACE H. FULLER, AND MICHAEL F. L'ANNUNZIATA.

ATOMIC ENERGY COMMISSION CONTRACT AT (11-1)-947. SOIL SCIENCE, VOL 107, NO 3,
MARCH 1969, P 223-230. 3 TAB, 5 FIG, 7 REF.

DESCRIPTORS:
*CHELATION, *STRONTIUM RADIOISOTOPES, *ALGAE, *FUNGI, *CALCAREOUS SOIL,
SOIL WATER, SOIL WATER MOVEMENT, LEACHING, ARID LANDS, DEPTH,
CYANOPHYTA, CHLOROPHYTA, SIEROZEMS, SOIL MICROBIOLOGY, DISTRIBUTION
PATTERNS, SOIL PROFILES.

IDENTIFIERS:
*RADIOSTRONTIUM, *ALGAL-BOUND RADIOSTRONTIUM, *FUNGAL-BOUND
RADIOSTRONTIUM, *MICROBIAL ACTIVITY, *RADIOSTRONTIUM COMPLEXES, DTPA,
SOIL WATER PROFILE.

ABSTRACT:
THIS REPORT EVALUATES THE EFFECT OF MICROBIAL ACTIVITY ON MOVEMENT OF
ALGAL- AND FUNGAL-BOUND RAIOSTRONTIUM, THE EFFECT OF DTPA ON SR SUPER
89 MOBILITY ADSORBED BY ALGAL AND FUNGAL CELLS IN A CALCAREOUS SOIL,
AND EXTRA-CELLULAR ALGAL EXCRETIONS AS SR SUPER 89 COMPLEXING AGENTS.
LEACHING WITH A WATER-FORMALDEHYDE AND WATER SOLUTION CAUSED LITTLE SR
SUPER 89-BOUND MOVEMENT BY ASPERGILLUS NIGER (FUNGI), RHISOPUS ARRHIZUS
(FUNGI), OSCILLATORIA BIJUGA (BLUE-GREEN ALGA) AND CHLOROCOCCUM SP. (A
GREEN ALGA) IN MOHAVE SANDY LOAM. LEACHING WITH 0.067 M DTPA SOLUTION
DISPLACED FROM 0.2 TO 1.6 PERCENT OF THE ALGAL- AND FUNGAL-BOUND SR
SUPER 89 BELOW A 20-CM SOIL DEPTH. THE SR SUPER 89 PROFILE DISTRIBUTION
REMAINING IN THE LEACHED SOIL CONTRASTED WITH THE DTPA-TREATED AND THE
WATER-TREATED SOIL COLUMNS. WHEREVER DTPA WAS ADDED SR SUPER 89 MOVED
DOWNWARD, WHEREAS IN THE WATER-TREATED CONTROL SR SUPER 89 REMAINED IN
THE UPPER 2.5 CM. NEGATIVELY CHARGED SR SUPER 89 COMPLEXES FORMED FROM
EXCRETIONS OF THE DESERT SOIL ALGA ANACYSTIS MENEGH ARE SHOWN. TABLES
INDICATE SR SUPER 89 MOVEMENT WITH RESPECT TO ALGAL AND FUNGAL TISSUE,
BIOLOGICALLY BOUND SR SUPER 89 FROM DTPA TREATMENT AND TOTAL
EXTRACTABLE SR SUPER 89. FIGURES ARE INCLUDED OF EXCHANGEABLE AND FIXED
SR SUPER 89 WITH SOIL DEPTH FOR VARIOUS BIOLOGICAL SOIL TREATMENTS, AND
EVIDENCE OF SR SUPER 89 IN A CELL-FREE ANACYSTIS MEDIUM AS NEGATIVE
COMPLEXES. (POPKIN-ARIZONA)

FIELD 05B, 02G

ACCESSION NO. W71-04067

BIODEGRADABILITY OF POTENTIAL ORGANIC SUBSTITUTES FOR PHOSPHATES,

ROBERT A. TAFT SANITARY ENGINEERING CENTER, CINCINNATI, OHIO. CINCINNATI
WATER RESEARCH LAB.

ROBERT L. BUNCH, AND M. B. ETTINGER.

INDUSTRIAL WASTE CONFERENCE, 22ND, MAY 2, 3, 4, 1967, PURDUE UNIVERSITY, VOL
52, NO 3, P 393-396, JULY 1968. 1 FIG, 14 REF.

DESCRIPTORS:
*ALGAE, *BIODEGRADATION, *PHOSPHATES, *EUTROPHICATION, NUTRIENTS,
NITROGEN, WASTE WATER TREATMENT, PHOSPHOROUS, RUNOFF, DOMESTIC WASTES,
DETERGENTS, INDUSTRIAL WASTES, *CHELATION.

IDENTIFIERS:
URBAN RUNOFF.

ABSTRACT:
EUTROPHICATION WILL BE AN INCREASINGLY IMPORTANT PROBLEM IN THE FUTURE
AS MORE WATERS ARE IMPOUNDED AND REUSED. CONSIDERATION SHOULD BE GIVEN
NOT ONLY TO IMPROVING METHODS OF NUTRIENT REMOVAL FROM SEWAGE
INFLUENTS, BUT ALSO TO FINDING ACCEPTABLE SUBSTITUTES FOR PHOSPHATES IN
INDUSTRIAL PROCESSES AND DETERGENT FORMULATIONS. PRIVATE INDUSTRY CAN
MAKE A CONTRIBUTION BY DEVELOPING ECONOMICALLY PRACTICAL MATERIALS
WHICH WILL BE CONVERTED INTO INNOCUOUS PRODUCTS AT A RATE SUFFICIENTLY
RAPID TO PREVENT SIGNIFICANT ACCUMULATION OF POLLUTANTS. IT APPEARS
THAT BIODEGRADABLE CHELATING AGENTS MAY BE CONSIDERED AS A PARTIAL
SUBSTITUTE FOR PHOSPHATES IN CERTAIN TYPES OF DETERGENT FORMULATIONS.
THE RELATIONSHIP BETWEEN THE STRUCTURE OF CHELATING AGENTS AND THE LOSS
OF CHELATING ABILITY DUE TO BIOLOGICAL ACTION REMAINS TO BE ANSWERED.
RESULTS ON COMPOUNDS NOW UNDER STUDY SHOULD GIVE A BETTER UNDERSTANDING
OF THE TYPE OF CHELATING COMPOUNDS THAT WOULD MINIMIZE ENVIRONMENTAL
CONTAMINATION. SO FAR, ONLY COMPOUNDS CONTAINING ONLY ORE NITROGEN
MOLECULE HAVE SHOW BIODEGRADABILITY. IT DOES NOT SEEM DESIRABLE TO USE
NON-BIODEGRADABLE CHELATING AGENTS THAT STRONGLY COMPLEX HEAVY METALS
UNLESS IT HAS BEEN DEMONSTRATED THAT THEY DO NOT INTERFERE WITH WASTE
TREATMENT, COAGULATION, WATER SOFTENING, CALCIUM METABOLISM OF MAN, OR
CONVEY METALS INTO THE DRINKING WATER. (ELLIS-TEXAS)

FIELD 05D, 05C

ACCESSION NO. W71-04072

PERIPHYTON BIOMASS-CHLOROPHYLL RATIO AS AN INDEX OF WATER QUALITY,

FEDERAL WATER POLLUTION CONTROL ADMINISTRATION, CINCINNATI, OHIO. ANALYTICAL
QUALITY CONTROL LAB.

CORNELIUS I. WEBER, AND BEN H. MCFARLAND.

AVAILABLE FROM: EPA WQO, ANALYTICAL QUALITY CONTROL LABORATORY, 1014
BROADWAY, CINCINNATI, OHIO 45202. MARCH 1969. 19 P, 8 TAB, 20 REF.

DESCRIPTORS:
*WATER QUALITY, *BIOINDICATORS, *CHLOROPHYLL, *OHIO RIVER, *ANALYTICAL
TECHNIQUES, BIOMASS, WATER POLLUTION, ALGAE, PERIPHYTON, PHOSPHORUS,
HYDROGEN ION CONCENTRATION, ALKALINITY, HARDNESS(WATER)), CHLORIDES,
ORGANIC MATTER, CARBON, WATER POLLUTION EFFECTS, BACTERIA, PERIPHYTON,
OHIO.

IDENTIFIERS:
*AUTOTROPHIC INDEX, CHLOROPHYLL A, SHAYLER RUN, ORGANIC CARBON,
CINCINNATI(OHIO).

ABSTRACT:
A REVIEW WAS MADE OF THE LITERATURE REGARDING THE RELATIONSHIP OF
BIOMASS AND CHLOROPHYLL AS A FIRST STEP IN DETERMINING THE POTENTIAL
USE OF THE BIOMASS-CHLOROPHYLL RATIO AS AN INDEX OF WATER QUALITY.
LABORATORY STUDIES WERE THEN BEGUN IN 1967 ON SAMPLES TAKEN AT TWO
STATIONS ABOVE AND BELOW THE MAJOR WASTE OUTFALL IN THE OHIO RIVER AT
CINCINNATI. IN 1968, SAMPLES TAKEN BY THE FWPCA NEWTOWN FISH TOXICOLOGY
LABORATORY FROM A SMALL CREEK (SHAYLER RUN) WERE INCLUDED IN THE STUDY.
SINCE CHLOROPHYLL A IS THE ONLY FORM OF CHLOROPHYLL FOUND IN ALL ALGAE,
THE BIOMASS-CHLOROPHYLL WHICH IS TERMED THE AUTOTROPHIC INDEX SHOULD BE
BASED EXCLUSIVELY ON THIS PIGMENT. THE AUTOTROPHIC INDICES FOR THE
SAMPLES TAKEN AT THE TWO OHIO RIVER STATIONS WERE SIGNIFICANTLY
DIFFERENT. THE DATA OBTAINED AT THE UPSTREAM STATION WERE SIMILAR TO
THOSE FROM SHAYLER RUN, AND INDICATED A PREDOMINANTLY AUTOTROPHIC
PERIPHYTON. IN CONTRAST TO THE UPSTREAM STATION, THE HIGH INDICES AT
THE STATION BELOW THE WASTE OUTFALL SUGGESTED A 'CONSUMER-TYPE'
COMMUNITY. THE AUTOTROPHIC INDEX SHOWS CONSIDERABLE PROMISE AS A TOOL
IN WATER QUALITY STUDIES. DATA ON WATER CHEMISTRY, AUTOTROPHIC INDICES,
COMPOSITION OF PERIPHYTON, AND TEST METHODS ARE INCLUDED.
(LITTLE-BATTELLE)

FIELD 05A, 05C

ACCESSION NO. W71-04206

ALGAL NUTRIENT RESPONSES IN AGRICULTURAL WASTE WATER,

FEDERAL WATER QUALITY ADMINISTRATION, FRESNO, CALIF.; AND CALIFORNIA STATE
DEPT. OF WATER RESOURCES, FRESNO.

JAMES F. ARTHUR, RANDALL L. BROWN, BRUCE A. BUTTERFIELD, AND JOEL C. GOLDMAN.

IN: COLLECTED PAPERS REGARDING NITRATES IN AGRICULTURAL WASTE WATERS, FEDERAL
WATER QUALITY ADMINISTRATION WATER POLLUTION CONTROL RESEARCH SERIES 13030
ELY, 12/69, P 123-141, DECEMBER 1969. 19 P, 8 FIG, 1 TAB, 12 REF. FWQA
PROJECT 13030 ELY.

DESCRIPTORS:
*ALGAE, *WASTE WATER TREATMENT, *WATER POLLUTION CONTROL,
*DENITRIFICATION, *BIODEGRADATION, *DRAINAGE WATER, CALIFORNIA,
DRAINAGE PROGRAMS, ANAEROBIC BACTERIA, ANEROBIC CONDITIONS, LAGOONS.

IDENTIFIERS:
*CENTRAL VALLEY(CALIFORNIA), *BACTERIAL DENITRIFICATION.

ABSTRACT:
ALGAL ASSIMILATION OF NUTRIENTS INTO CELLULAR MATERIAL WITH SUBSEQUENT
REMOVAL FROM THE GROWTH MEDIUM IS A FEASIBLE PROCESS TO REMOVE NITRATES
FROM IRRIGATION WASTE WATER. THE EFFICIENCY OF THE PROPOSED SYSTEM IS
GREATLY ENHANCED IF AS MANY VARIABLES AS POSSIBLE ARE OPTIMIZED,
LEAVING ONLY NITROGEN THE LIMITING NUTRIENT. ORTHOPHOSPHATE ADDITIONS
OF 2.0-3.0 MG/LITER P ARE REQUIRED THE YEAR ROUND TO REMOVE 20.0
MG/LITER NITRATE-NITROGEN FROM THE GROWTH MEDIUM. IRON AND CARBON ALSO
HAVE BEEN FOUND TO BE LIMITING ALGAL GROWTH AND NITROGEN ASSIMILATION
DURING PART OF THE YEAR. (KNAPP-USGS)

FIELD 05D, 05B

ACCESSION NO. W71-04554

THE EFFECTS OF NITROGEN REMOVAL ON THE ALGAL GROWTH POTENTIAL OF SAN JOAQUIN
  VALLEY AGRICULTURAL TILE DRAINAGE EFFLUENTS,

  CALIFORNIA STATE DEPT. OF WATER RESOURCES, FRESNO; AND FEDERAL WATER QUALITY
    ADMINISTRATION, ALAMEDA, CALIF.

  RANDALL L. BROWN, RICHARD C. BAIN, JR., AND MILTON G. TUNZI.

  IN: COLLECTED PAPERS REGARDING NITRATES IN AGRICULTURAL WASTE WATERS, FEDERAL
    WATER QUALITY ADMINISTRATION WATER POLLUTION CONTROL RESEARCH SERIES 13030
    ELY, 12/69, P 143-155, DECEMBER 1969. 13 P, 4 FIG, 4 TAB, 3 REF. FWQA
    PROJECT 13030 ELY.

    DESCRIPTORS:
      *ALGAE, *WASTE WATER TREATMENT, *WATER POLLUTION CONTROL,
      *DENITRIFICATION, *BIODEGRADATION, *DRAINAGE WATER, CALIFORNIA,
      DRAINAGE PROGRAMS, ANAEROBIC BACTERIA, ANAEROBIC CONDITIONS, LAGOONS.

    IDENTIFIERS:
      *CENTRAL VALLEY(CALIFORNIA), *BACTERIAL DENITRIFICATION.

    ABSTRACT:
      LABORATORY CULTURE EXPERIMENTS WERE MADE TO DETERMINE THE EFFECTIVENESS
      OF THE TWO BIOLOGICAL PROCESSES UNDER INVESTIGATION, ALGAL STRIPPING
      AND BACTERIAL DENITRIFICATION, FOR REMOVING THE ALGAL GROWTH POTENTIAL
      OF THE TILE DRAINAGE WATER WHEN ADDED TO POTENTIAL RECEIVING WATERS IN
      THE SACRAMENTO-SAN JOAQUIN DELTA. ALGAL GROWTH POTENTIAL TESTS IN TWO
      DIFFERENT LABORATORIES INDICATE THAT NITRATE-RICH AGRICULTURAL
      DRAINAGE, WHEN MIXED WITH SAN JOAQUIN RIVER DELTA WATER, STIMULATES
      ALGAL GROWTH. SELECTIVE REMOVAL OF NITRATE-NITROGEN BY ANAEROBIC
      DENITRIFICATION OR REMOVAL OF NUTRIENTS BY ALGAL CELLS GROWN IN SHALLOW
      PONDS YIELDED COMPARABLE BIOASSAY RESULTS. EUTROPHICATION DUE TO
      AGRICULTURAL WASTE WATERS CAN BE CONTROLLED BY TREATMENT.
      (KNAPP-USGS)

    FIELD 05D, 05B

    ACCESSION NO. W71-04555

HARVESTING OF ALGAE GROWN IN AGRICULTURAL WASTE WATERS,

  CALIFORNIA STATE DEPT. OF WATER RESOURCES, FRESNO; AND BUREAU OF RECLAMATION,
    FRESNO, CALIF.

  BRUCE A. BUTTERFIELD, AND JAMES R. JONES.

  IN: COLLECTED PAPERS REGARDING NITRATES IN AGRICULTURAL WASTE WATERS, FEDERAL
    WATER QUALITY ADMINISTRATION WATER POLLUTION CONTROL RESEARCH SERIES 13030
    ELY, 12/69, P 157-163, DECEMBER 1969. 7 P, 1 TAB. FWQA PROJECT 13030 ELY.

    DESCRIPTORS:
      *ALGAE, *WASTE WATER TREATMENT, *WATER POLLUTION CONTROL,
      *DENITRIFICATION, *BIODEGRADATION, *DRAINAGE WATER, CALIFORNIA,
      DRAINAGE PROGRAMS, ANAEROBIC BACTERIA, ANAEROBIC CONDITIONS, LAGOONS.

    IDENTIFIERS:
      *ALGAE HARVESTING, *CENTRAL VALLEY(CALIF.).

    ABSTRACT:
      THROUGH LABORATORY AND FIELD TESTING OF ALGAE HARVESTING METHODS FOR
      DENITRIFICATION, IT WAS SHOWN THAT EFFECTIVE CONCENTRATION CAN BE
      ACCOMPLISHED USING THE FLOCCULATION-SEDIMENTATION PROCESS TO REMOVE
      90-95 PERCENT OF THE SUSPENDED SOLIDS FROM ALGAE-LADEN AGRICULTURAL
      WASTE WATER. DEWATERING AND DRYING CAN BE ACCOMPLISHED BUT THE
      EFFICIENCIES OF THE UNITS TESTED WERE LOW. THE NEED TO RECIRCULATE THE
      WATER WOULD INCREASE THE OVERALL COST; HOWEVER, IT IS BELIEVED BETTER
      RESULTS WILL BE ACHIEVED IN LARGER CAPACITY UNITS.
      (KNAPP-USGS)

    FIELD 05D, 05B

    ACCESSION NO. W71-04556

COMBINED NUTRIENT REMOVAL AND TRANSPORT SYSTEM FOR TILE DRAINAGE FROM THE SAN JOAQUIN VALLEY,

CALIFORNIA STATE DEPT. OF WATER RESOURCES, FRESNO; AND FEDERAL WATER QUALITY ADMINISTRATION, FRESNO, CALIF.; AND CALIFORNIA UNIV., BERKELEY. DEPT. OF SANITARY ENGINEERING AND PUBLIC HEALTH.

JOEL C. GOLDMAN, JAMES F. ARTHUR, WILLIAM J. OSWALD, AND LOUIS A. BECK.

IN: COLLECTED PAPERS REGARDING NITRATES IN AGRICULTURAL WASTE WATERS, FEDERAL WATER QUALITY ADMINISTRATION WATER POLLUTION CONTROL RESEARCH SERIES 13030 ELY, 12/69, P 165-186, DECEMBER 1969. 22 P, 4 FIG, 1 TAB, APPEND. FWQA PROJECT 13030 ELY.

DESCRIPTORS:
*ALGAE, *WASTE WATER TREATMENT, *WATER POLLUTION CONTROL, *DENITRIFICATION, *BIODEGRADATION, *DRAINAGE WATER, CALIFORNIA, DRAINAGE PROGRAMS, ANAEROBIC BACTERIA, ANAEROBIC CONDITIONS, LAGOONS.

IDENTIFIERS:
*ALGAE HARVESTING, *CENTRAL VALLEY(CALIF.).

ABSTRACT:
CURRENT PLANS CALL FOR TREATMENT OF AGRICULTURAL WASTE WATER FOR NUTRIENT (NITROGEN) REMOVAL FROM THE PROPOSED SAN LUIS AND MASTER DRAINS PRIOR TO DISCHARGE INTO THE BAY-DELTA AREA. OF THE SEVERAL TREATMENT PROCESSES BEING INVESTIGATED, THE ALGAE STRIPPING PROCESS IS ESTIMATED TO REQUIRE BETWEEN 6,000 AND 12,000 ACRES OF LAND TO ACCOMPLISH THIS TASK. BECAUSE EVERY BODY OF WATER IS A POTENTIAL ALGAL GROWTH SYSTEM, THERE ARE SEVERAL ALTERNATIVES WHICH WILL GREATLY REDUCE THE TOTAL COST OF TREATMENT AND PERHAPS IMPROVE THE OVERALL EFFICIENCY OF NUTRIENT REMOVAL. IN-LINE TREATMENT USING THE DRAIN AND THE DRAINAGE RESERVOIRS FOR ALGAE GROWTH MAY BE AN ECONOMICAL AND PRACTICAL METHOD FOR NUTRIENT REMOVAL. BECAUSE KESTERSON RESERVOIR IS AN INTEGRAL PART OF THE PROPOSED DRAINAGE SYSTEM AND CONTAINS THE REQUIRED AREA NEEDED FOR TREATMENT BY ALGAE STRIPPING, IT SEEMS LOGICAL TO USE IT AS A DUAL-PURPOSE TREATMENT AND STORAGE RESERVOIR. KESTERSON RESERVOIR, IF MODIFIED AS SUGGESTED, HAS THE POTENTIAL TO PROVIDE NITROGEN REMOVALS IN EXCESS OF 90%. (KNAPP-USGS)

FIELD 05D, 05G, 05B

ACCESSION NO. W71-04557

SOME MICROBIAL-CHEMICAL INTERACTIONS AS SYSTEMS PARAMETERS IN LAKE ERIE,

OHIO STATE UNIV., COLUMBUS. MICROBIAL AND CELLULAR BIOLOGY.

P. R. DUGAN, J. I. FREA, AND R. M. PFISTER.

IN: SYSTEMS ANALYSIS FOR GREAT LAKES WATER RESOURCES, P 21-28, OCTOBER 1969. 8 P, 2 FIG, 16 REF. OWRR PROJECT A-999-OHIO(3).

DESCRIPTORS:
*SYSTEMS ANALYSIS, DATA COLLECTIONS, *ALGAE CONTROL, *LAKE ERIE, BIOCHEMICAL OXYGEN DEMAND, BACTERIA, ECOLOGY, WATER POLLUTION EFFECTS, LAKES, *CYANOPHYTA, *EUTROPHICATION, WATER QUALITY.

ABSTRACT:
SOME MICROBIAL-CHEMICAL INTERACTIONS AS SYSTEMS PARAMETERS IN LAKE ERIE WERE PRESENTED. ONE OF THE MAJOR PROBLEMS CITED WAS THE INCREASED GROWTH RATE OF BLUE GREEN ALGAE AND OTHER MICROORGANISMS IN THE LAKE WHICH HAVE LED TO OBJECTIONS BECAUSE OF DECREASED RECREATIONAL VALUE, MORTALITY OF FISH AND DOMESTIC ANIMALS, CLOGGING OF WATER SUPPLY INTAKE FILTERS, AND DEPLETION OF OXYGEN IN THE WATER. FOUR GENERAL PARAMETERS WERE CONSIDERED IN RELATIONSHIP TO ACCELERATED GROWTH OF BLUE GREEN ALGAE: (1) AMOUNT OF LIGHT (ENERGY); (2) NITROGEN SUPPLY; (3) CO2 OR CO3; AND (4) MINERALS. SEVERAL OBSERVATIONS WERE MADE FROM DATA COLLECTED IN THE WESTERN BASIN OF LAKE ERIE DURING THE SPRING AND SUMMER OF 1969. USING BOD AS AN EXAMPLE OF RECYCLING, ROLE OF BACTERIA IN MAKING NUTRIENTS AVAILABLE FOR ALGAE GROWTH WAS SHOWN SIGNIFICANT; AND ALGAE, ONCE ABOVE A CRITICAL CONCENTRATION SIGNIFICANTLY ENRICH THEIR OWN ENVIRONMENTS WITH ORGANICS, WHICH INDICATED A SPIRALING INCREASE IN RATE OF EUTROPHICATION. SUGGESTIONS FOR DECREASING ALGAE AND BACTERIA POPULATION IN THE LAKE WERE: PREVENTING ORGANIC AND MINERAL NUTRIENTS FROM ENTERING THE WATER COLUMN AND REMOVING SEDIMENTS OF RELATIVELY HIGH ORGANIC CONTENT PHYSICALLY OR ALLOWING THEM TO DECREASE NATURALLY. (KRISS-CORNELL)

FIELD 06A, 05C, 02H

ACCESSION NO. W71-04758

POULTRY PROCESSING WASTE TREATMENT AT MILLSBORO, DELAWARE,

SYRACUSE UNIV., N.Y. DEPT. OF CIVIL ENGINEERING.

NELSON L. NEMEROW.

PROCEEDINGS, INDUSTRIAL WASTE CONFERENCE, 22ND, MAY 2, 3, 4, 1967, PURDUE
UNIVERSITY, VOL LII, NO 3, JULY 1968, P 526-536. 4 FIG, 4 TAB, 1 REF.

DESCRIPTORS:
*POULTRY, *OXIDATION LAGOONS, *ORGANIC LOADING, ALGAE, AMMONIA,
SEDIMENTATION, FLOTATION, BIODEGRADATION, CHLORINATION, COLIFORM,
BACTERIA, BIOCHEMICAL OXYGEN DEMAND, DELAWARE, *WASTE WATER TREATMENT.

IDENTIFIERS:
MILLSBORO(DELAWARE), SWAN CREEK, INDIAN RIVER.

ABSTRACT:
9,000 TO 10,000 CHICKENS WITH AN AVERAGE WEIGHT OF 3.75 LBS ARE
PROCESSED DURING AN 8-11 HOUR WORKING DAY AT MILLSBORO, DELAWARE. THE
WASTES FROM THIS OPERATION ARE TREATED IN A TWO STAGE OXIDATION POND
WITH EFFLUENT CHLORINATION. SAMPLES WERE COLLECTED IN APRIL, JUNE, AND
DECEMBER OF 1966, AND MARCH OF 1967. THE BOD LOADING WAS FOUND TO BE
1390 LBS/DAY, WITH A LOADING ON THE FIRST POND OF 935 LB BOD/ACRE. THIS
POND EFFECTED A 72.5 PERCENT REDUCTION IN BOD. THE SECOND POND HAD A
LOADING OF 79 LB BOD/ACRE, WITH AN ADDITIONAL BOD REDUCTION OF 9.5%
WHEN ALGAE ARE NOT REMOVED FROM THE EFFLUENT, AND 70% WHEN ALGAE ARE
REMOVED. THE OVERALL REMOVAL, BASED ON THE 4 SAMPLES PER YEAR, WAS 92%
FOR THE FILTERED EFFLUENT, AND 75% WHEN THE ALGAE ARE NOT FILTERED OUT.
(LOWRY-TEXAS)

FIELD 05D

ACCESSION NO. W71-05013

THE EFFECT OF VARYING AMOUNTS AND RATIOS OF NITROGEN AND PHOSPHATES ON ALGAE
BLOOMS,

ACADEMY OF NATURAL SCIENCES OF PHILADELPHIA, PA. DEPT. OF LIMNOLOGY.

RUTH PATRICK.

PROCEEDINGS, INDUSTRIAL WASTE CONFERENCE, 21ST, MAY 3, 4, 5, 1966, PURDUE
UNIVERSITY, VOL L, NO 2, MARCH 1966, P 41-51. 6 TAB, 6 REF.

DESCRIPTORS:
*NITROGEN, *PHOSPHATES, *ALGAE, *EUTROPHICATION, DIATOMS, MICROBIOLOGY,
NITRATES, AMMONIA, WATER POLLUTION EFFECTS, WASTE WATER TREATMENT,
PENNSYLVANIA.

IDENTIFIERS:
*GLUCOSE, WATERLOO MILLS(PA).

ABSTRACT:
A SERIES OF EXPERIMENTS WAS UNDERTAKEN AT THE WATERLOO MILLS FIELD
RESEARCH STATION TO TRY TO DETERMINE THE EFFECT OF NITROGEN AS NITRATES
OR AMMONIUM AND PHOSPHATES IN VARYING AMOUNTS AND RATIOS ON NATURAL
COMMUNITIES OF DIATOMS. SOME EXPERIMENTS WERE CARRIED OUT TO DETERMINE
THE EFFECTS OF ADDING SMALL AMOUNTS OF GLUCOSE TO THESE ENRICHED
WATERS. CERTAIN SPECIES INCREASED IN ABUNDANCE WITH INCREASES IN
NITROGEN AS NITRATES AND/OR A AMMONIUM AND WITH INCREASES IN PHOSPHATES
AND SOME OF THEM WITH THE ADDITION OF SMALL AMOUNTS OF GLUCOSE.
HOWEVER, THE HEAVY DIATOM GROWTHS WHICH OFTEN OCCUR IN STREAMS ENRICHED
WITH EFFLUENTS FROM SEWAGE TREATMENT PLANTS DID NOT OCCUR WHICH
INDICATES OTHER FACTORS ARE IMPORTANT TO THE GROWTH OF DIATOMS AND
HENCE, ALGAL BLOOMS. (ELLIS-TEXAS)

FIELD 05C, 05D

ACCESSION NO. W71-05026

APPLICATION OF INDUSTRIAL PHOTOSYNTHESIS PROCESS TO WASTE WATER RENOVATION SYSTEMS,

NORTH AMERICAN AVIATION, INC., EL SEGUNDO, CALIF. S AND ID LIFE SCIENCES.

R. H. T. MATTONI, H. NUGENT MYRICK, AND E. C. KELLER, JR.

PROCEEDINGS, INDUSTRIAL WASTE CONFERENCE, 20TH, MAY 4, 5, 6, 1965. ENGINEERING BULLETIN OF PURDUE UNIVERSITY, VOL XLIX, NO 4, P 684-705. 17 FIG, 1 TAB.

DESCRIPTORS:
*ALGAE, *BIOMASS, WATER PURIFICATION, *PROTOZOA, *FUNGI, *BACTERIA, HYDROGEN ION CONCENTRATION, TEMPERATURE, ALKALINITY, BICARBONATES, BACTERICIDE, KINETICS, HARDNESS, COLIFORMS, CYTOLOGICAL STUDIES, EVAPORATION, MICROORGANISMS, CALIFORNIA, *PHOTOSYNTHESIS, WASTE WATER TREATMENT.

IDENTIFIERS:
INDUSTRIAL PHOTOSYNTHESIS, MEAN DOUBLING TIME, PACKED CELL VOLUMES, BACTERIAL PLATE COUNTS, LANCASTER(CALIF).

ABSTRACT:
INDUSTRIAL PHOTOSYNTHESIS REFERS TO THE CONTROLLED GROWTH OF COMPLEX POPULATIONS OF MICROORGANISMS, MAINLY UNICELLULAR ALGAE, ON COMPLEX ORGANIC SUBSTRATES SUCH AS PRIMARY AND SECONDARY WASTE WATERS. THE GOAL OF THIS PROCESS IS TO PRODUCE SALABLE BIOMASS IN ADDITION TO INCREASING THE WATER QUALITY TO LEVELS ACCEPTABLE FOR IRRIGATIONAL OR RECREATIONAL USES. THREE FULL-SCALE TANK REACTORS WERE USED TO PERFORM EXPERIMENTS ON SELECTION PROCESSES, PH DEPENDENCE, AND TEMPERATURE DEPENDENCE. THE NATURAL SELECTION OF A SIMPLE ARRAY OF SPECIES FROM A COMPLEX SEED WAS DEMONSTRATED. PH, HOWEVER, WAS SHOWN TO HAVE LITTLE EFFECT ON THE RELATIVE SPECIES FREQUENCY PROFILES. IT WAS ALSO DETERMINED THAT WHEN REACTORS ARE OPERATED IN AN IDENTICAL FASHION, THEY DO PRODUCE IDENTICAL POPULATIONS OF MICROORGANISMS. THE EXPERIMENT WAS SITUATED IN AN AREA KNOWN FOR HIGH TEMPERATURE AND CLOUD FREE DAYS. DURING THE EXPERIMENT, HOWEVER, THE REGION WAS EXPERIENCING ITS ANNUAL TEMPERATURE MINIMUMS, WITH ICE FORMING ON THE TOPS OF THE REACTORS. IN SPITE OF THIS, GROWTH RATES OBTAINED WERE COMPARABLE TO GROWTH RATES DURING THE SPRING, SUMMER, AND AUTUMN, DEMONSTRATING THESE SYSTEMS TO BE APPLICABLE EVEN WHERE FREEZING TEMPERATURES ARE ENCOUNTERED DURING PART OF THE DIURNAL CYCLE. FINALLY, THE FINAL REDUCTIONS OF BACTERIA WERE SHOWN TO BE A FUNCTION OF ALGAE GROWTH. THIS WAS ATTRIBUTED TO BACTERIOCIDAL ENZYMES EXCRETED BY THE ALGAE. (LOWRY-TEXAS)

FIELD 05D

ACCESSION NO. W71-05267

A TECHNICAL ASSESSMENT OF CURRENT WATER QUALITY CONDITIONS AND FACTORS AFFECTING WATER QUALITY IN THE UPPER POTOMAC ESTUARY,

ENVIRONMENTAL PROTECTION AGENCY, ANNAPOLIS, MD. CHESAPEAKE TECHNICAL SUPPORT LAB.

NORBERT A. JAWORSKI, DONALD W. LEAR, JR., AND JOHAN A. AALTO.

WATER QUALITY STATUS REPORT. CTSL TECHNICAL REPORT NO 5, MAR, MARCH 1969. 55 P, 17 FIG, 6 TAB, 21 REF.

DESCRIPTORS:
*BIOCHEMICAL OXYGEN DEMAND, *NUTRIENTS, *DISSOLVED OXYGEN, *ALGAE, COLIFORMS, EUTROPHICATION(PHYTOPLANKTON), ESTUARIES, WATER POLLUTION SOURCES, WATER POLLUTION EFFECTS, NITROGEN, COLIFORMS, WATER QUALITY.

IDENTIFIERS:
*POTOMAC RIVER ESTUARY, NUTRIENT-PHYTOPLANKTON RELATIONSHIPS.

ABSTRACT:
WATER QUALITY CONDITIONS IN THE UPPER POTOMAC ESTUARY INCLUDING BOD, DISSOLVED OXYGEN, PHOSPHORUS, NITROGEN AND COLIFORM DENSITIES, ARE PRESENTED AND DISCUSSED. THE SOURCES AND RELATIONSHIPS OF NUTRIENTS TO ALGAL STANDING CROP ARE STUDIED. MAJOR SOURCES OF CARBONACEOUS AND NITROGENOUS OXYGEN DEMAND AS WELL AS THE DEMAND FROM THE ALGAL STANDING CROP ON DISSOLVED OXYGEN ARE EVALUATED. (AALTO-CHESAPEAKE TECHNICAL SUPPORT LABORATORY)

FIELD 05B, 05C, 02L

ACCESSION NO. W71-05407

WATER QUALITY AND WASTE WATER LOADINGS UPPER PO.TOMAC ESTUARY DURING 1969,

ENVIRONMENTAL PROTECTION AGENCY, ANNAPOLIS, MD. CHESAPEAKE TECHNICAL SUPPORT
    LAB.

NORBERT A. JAWORSKI.

WATER QUALITY STATUS REPORT. CTSL TECHNICAL REPORT NO 27, MAR, NOV 1969. 62
    P, 14 FIG, 9 TAB, 4 REF, APPEND.

    DESCRIPTORS:
        *BIOCHEMICAL OXYGEN DEMAND, *PHOSPHORUS, *NITROGEN, *COLIFORMS,
        *DISSOLVED OXYGEN, ESTUARIES, ALGAE, WATER POLLUTION SOURCES, WATER
        POLLUTION EFFECTS, WATER QUALITY.

    IDENTIFIERS:
        *POTOMAC RIVER ESTUARY.

    ABSTRACT:
        FOR THE FIRST EIGHT MONTHS OF 1969, ABOUT 55% OF THE BOD ENTERING THE
        POTOMAC ESTUARY FROM ALL MAJOR SOURCES WAS FROM WASTE WATER DISCHARGES
        IN THE WASHINGTON AREA. ALSO THESE DISCHARGES CONTRIBUTED 86% OF THE
        TOTAL PHOSPHORUS AND 66% OF THE TOTAL NITROGEN. TOTAL CARBON ENTERING
        THE ESTUARY FROM ALL SOURCES WAS MAINLY IN THE FORM OF INORGANIC
        CARBON. DISSOLVED OXYGEN LEVELS WERE MEASURED. AS A RESULT OF INCREASED
        CHLORINATION OF THE WASTE WATER TREATMENT FACILITY EFFLUENTS, COLIFORM
        DENSITIES WERE SIGNIFICANTLY LOWER. TO FACILITATE THE DETERMINATION OF
        WASTE WATER DISCHARGE LOADINGS, THE ESTUARY WAS ZONED INTO FIFTEEN MILE
        REACHES. MAXIMUM WASTE DISCHARGE LOADINGS FOR BOD (ORGANIC CARBON),
        NITROGEN AND PHOSPHORUS HAVE BEEN DETERMINED FOR ZONE I IN THE
        WASHINGTON AREA. PRELIMINARY ESTIMATES OF LOADINGS HAVE BEEN
        ESTABLISHED FOR ZONE II DOWNSTREAM FROM THE WASHINGTON AREA.
        (AALTO-CHESAPEAKE TECHNICAL SUPPORT LABORATORY)

    FIELD 05B, 05C, 02L

    ACCESSION NO. W71-05409

PARAMETERS AFFECTING THE GROWTH OF ULVA LATISSIMA IN A POLLUTED ESTUARY,

NORTHEASTERN UNIV., BOSTON, MASS.

THOMAS D. WAITE, AND CONSTANTINE GREGORY.

NORTH EASTERN REGIONAL ANTIPOLLUTION CONFERENCE, ANNUAL, PROCEEDINGS,
    KINGSTON, RHODE ISLAND, JULY 22-25, 1969, P 12-15.

    DESCRIPTORS:
        *ESTUARIES, *HARBORS, *SEWAGE, WATER POLLUTION EFFECTS, *ALGAE.

    IDENTIFIERS:
        *ULVA LATISSIMA, *AMMONIA NITROGEN.

    ABSTRACT:
        ULVA LATISSIMA, A MARINE FLORA THAT HAS PRODUCED SEVERE POLLUTION
        PROBLEMS DUE TO ITS UNRESTRAINED GROWTH IN ESTUARIES AND HARBORS AS A
        RESULT OF DOMESTIC SEWAGE FERTILIZING THE MARINE ENVIRONMENT WAS
        STUDIED. THE CORRELATION BETWEEN AMMONIA NITROGEN CONCENTRATION IN
        SEAWATER AND RATE OF GROWTH OF ULVA LATISSIMA WAS INVESTIGATED. THE
        GREATEST GROWTH RATE IS REALIZED AT 0.7 MG/L AMMONIA NITROGEN
        CONCENTRATION. REDUCTION OF AMMONIA CONCENTRATION TO LESS THAN 0.5 MG/L
        IS NECESSARY TO REGULATE EXCESSIVE GROWTH OF ULVA LATISSIMA.
        (ENSIGN-PAI)

    FIELD 05C

    ACCESSION NO. W71-05500

SOME BIOGEOCHEMICAL CONSIDERATIONS ON THE RADIOACTIVE CONTAMINATION OF MARINE
BIOTA AND ENVIRONMENTS,

INTERNATIONAL LABORATORY OF MARINE RADIOACTIVITY, MONTE CARLO (MONACO).

RINNOSUKE FUKAI.

SOME BIOGEOCHEMICAL CONSIDERATIONS ON THE RADIOACTIVE CONTAMINATION OF MARINE
BIOTA AND ENVIRONMENTS, PAPER, PERGAMON PRESS, NEW YORK, 1969, P 391-394. 1
FIG, 1 TAB, 7 REF.

DESCRIPTORS:
*RADIOACTIVITY, *TRACE ELEMENTS, *AQUATIC LIFE, WATER POLLUTION
EFFECTS, MARINE ANIMALS, ALGAE, MOLLUSKS, FISH, RADIOACTIVE WASTES.

IDENTIFIERS:
*BIOGEOCHEMICAL DATA, *ABUNDANCE RATIOS.

ABSTRACT:
RADIOACTIVE CONTAMINATION OF MARINE ENVIRONMENTS AND BIOTA IS A LINK IN
THE BIOGEOCHEMICAL PROCESS IN THE SEA. THE IMPORTANCE OF THE
ACCUMULATION OF BIOGEOCHEMICAL DATA ON THE DISTRIBUTIONS OF TRACE
ELEMENTS IN MARINE BIOTA IN RELATION TO THE PREDICTION OF RADIOACTIVE
CONTAMINATION IS DISCUSSED. THE REASON FOR OBTAINING THE 'STANDARD
ABUNDANCE' OF TRACE ELEMENTS FOR CERTAIN GROUPS OF MARINE ORGANISMS IS
EXPLAINED AND THE IMPORTANCE OF THE INTERPRETATION OF THE ANALYTICAL
RESULTS AND COMPUTATIONS OF THE ABUNDANCE RATIO IS MENTIONED.
(ENSIGN-PAI)

FIELD 05A, 05C

ACCESSION NO. W71-05531

UTILIZATION OF ALKOXYSILANES FOR REDUCTION OF THE ABUNDANCE OF PHYTOPLANKTON
DURING 'BLOOMS' OF WATER BASINS (IN RUSSIAN),

MOSCOW STATE UNIV. (USSR).

N. S. STROGANOV, V. G. KHOBOTIEV, L. V. KOLOSOVA, AND M. A. KADINA.

DOKLADY ADADEMII NAUK SSSR, VOL 181, NO 5, P 1257-1259, 1968. 2 FIG, 2 TAB,
13 REF.

DESCRIPTORS:
*HERBICIDES, *ALGAE, TOXICITY, SCENEDESMUS, CHLORELLA, DAPHNIA,
DETERIORATION, WATER TREATMENT, DEGRADATION(DECOMPOSITION).

IDENTIFIERS:
*ALGAL BLOOMS, *ORGANO-SILICATES, *ALKOXYSILANES, SCENEDESMUS
QUADRICAUDA, CHLORELLA VULGARIS, DAPHNIA MAGNA, LIMNAEA STAGNALIS,
OUSPENSKII'S MEDIUM.

ABSTRACT:
A SEARCH FOR HERBICIDES SUPPRESSING ALGAL BLOOMS, BUT HARMLESS TO
ZOOPLANKTERS AND OTHER HYDROBIONTS LEAD TO INVESTIGATION OF
ORGANO-SILICATES: VENYLTRIETHOXYSILANE, TETRAETHOXYSILANE, AND
TRIFLUOROPROPENYL-(METHYL)-DIETHOXYSILANE. THE TEST ORGANISM INCLUDED
SCENEDESMUS QUADRICAUDE, CHLORELLA VULGARIS, DAPHNIA MAGNA, AND LIMNAEA
STAGNALIS. THE COUNT OF SURVIVED SPECIMENS SHOWED THAT THE
CONCENTRATION OF 0.01 MG/LITER OF ALL INVESTIGATED COMPOUNDS ELIMINATED
93% TO 96% OF ALGAL CELLS WITHOUT DAMAGING EITHER DAPHNIA OR MOLLUSKS.
AMONG THE ADVANTAGES OF THE ALKOXYSILANES IS THEIR RELATIVELY RAPID
DISINTEGRATION IN WATER. (WILDE-WISCONSIN)

FIELD 05F, 05C

ACCESSION NO. W71-05719

CHEMICAL AND BIOLOGICAL REACTIONS FROM LAGOONS USED FOR CATTLE,

WISCONSIN UNIV., MADISON.

S. A. WITZEL, ELIZABETH MCCOY, AND RICHARD LEHNER.

ASAE PAPER NO 64-417. TRANSACTIONS OF THE AMERICAN SOCIETY OF AGRICULTURAL
ENGINEERS, VOL 8, P 449-451, 1965. 1 FIG, 16 REF.

DESCRIPTORS:
*OXIDATION LAGOONS, *BIOCHEMICAL OXYGEN DEMAND, *ANAEROBICS, ALGAE,
BACTERIA, WATER POLLUTION SOURCES, ODOR, FARM WASTE, SLUDGE, SOLID
WASTES, DECOMPOSTING ORGANIC MATTER, LAGOONS, CATTLE.

IDENTIFIERS:
ORGANIC NITROGEN, BACTERIOLOGICAL STUDY, AEROBICITY.

ABSTRACT:
AN EXPERIMENTAL LAGOON TO RECEIVE THE WASTES AS LIQUID MANURE FROM SIX
BULLS WAS CONSTRUCTED. IN AN EXPERIMENTAL BARN THE MANURE FROM THE
BULLS WAS WASHED DAILY INTO TWO GUTTERS 24 IN. WIDE AT THE TOP. THE
GUTTERS HAD A CAPACITY OF 2000 GALLONS. THE GUTTERS WERE FLUSHED EVERY
SEVEN DAYS INTO A LAGOON. THE CIRCULAR LAGOON HAD A 60 FT DIAMETER AT
THE TOP, A 40 FT DIAMETER AT THE BOTTOM AND WAS 5 FT DEEP. LIQUID
MANURE SAMPLES WERE TAKEN FROM THE BARN GUTTER AND FROM THE LAGOON
MANURE LIQUID AND SLUDGE. BOD TESTS WERE CONDUCTED ON ALL SAMPLES AND
THE PERCENT BOD REDUCTION FROM GUTTER TO LAGOON WAS CALCULATED. OTHER
TESTS WERE MADE TO DETERMINE PERCENT TOTAL SOLIDS REMOVAL, ORGANIC
NITROGEN CONTENT, PH VALUES, AND TEMPERATURE. EXTENSIVE TESTS WERE MADE
AND REPORTED ON BACTERIA CONTENT, BOTH QUANTITATIVE AND QUALITATIVE.
FIVE CONCLUSIONS OF THE STUDY ARE MENTIONED. (PARKER-IOWA STATE)

FIELD 05D

ACCESSION NO. W71-05742

REMOVAL OF CONTAMINANTS FROM LAKE ONTARIO BY NATURAL PROCESSES,

CANADIAN OCEANOGRAPHIC DATA CENTRE, OTTAWA (ONTARIO).

H. E. SWEERS.

IN: PROCEEDINGS TWELFTH CONFERENCE ON GREAT LAKES RESEARCH, MAY 5-7, 1969,
UNIVERSITY OF MICHIGAN, ANN ARBOR: INTERNATIONAL ASSOCIATION FOR GREAT
LAKES RESEARCH, P 734-741, 1969. 8 P, 4 FIG, 7 REF.

DESCRIPTORS:
*SELF-PURIFICATION, *WATER POLLUTION CONTROL, *LAKE ONTARIO, *WATER
CIRCULATION, PATH OF POLLUTANTS, NUTRIENTS, ALGAE, STRATIFIED FLOW,
MIXING, FLOW, MODEL STUDIES, MATHEMATICAL MODELS, GREAT LAKES, THERMAL
STRATIFICATION.

IDENTIFIERS:
REMOVAL TIME(POLLUTANTS).

ABSTRACT:
A MODEL IS DEVELOPED TO CALCULATE THE REMOVAL TIME OF A CONSERVATIVE
CONTAMINANT FROM A LAKE, TAKING SUMMER STRATIFICATION INTO ACCOUNT. THE
BASIC ASSUMPTIONS OF THE MODEL WERE COMPARED WITH CONDITIONS ACTUALLY
OCCURRING IN LAKE ONTARIO, AND IT IS SHOWN THAT STRATIFICATION HAS
LITTLE EFFECT ON THE CALCULATED REMOVAL TIME. THE RESULTS ARE
EXTRAPOLATED TO THE BEHAVIOR OF NON-CONSERVATIVE PARAMETERS. A SHARP
REDUCTION IN THE RATE OF INPUT OF NUTRIENTS COULD RESULT IN A MARKED
DECREASE IN ALGAL GROWTH WITHIN A YEAR AFTER SUCH MEASURES BECOME
EFFECTIVE. (KNAPP-USGS)

FIELD 05G, 05B, 02H

ACCESSION NO. W71-05878

INVESTIGATIONS OF DAILY VARIATIONS IN CHEMICAL BACTERIOLOGICAL AND BIOLOGICAL
PARAMETERS AT TWO LAKE ONTARIO LOCATIONS NEAR TORONTO, PART I--CHEMISTRY,

ONTARIO WATER RESOURCES COMMISSION, REXDALE.

T. G. BRYDGES.

IN: PROCEEDINGS TWELFTH CONFERENCE ON GREAT LAKES RESEARCH, MAY 5-7, 1969,
UNIVERSITY OF MICHIGAN, ANN ARBOR: INTERNATIONAL ASSOCIATION FOR GREAT
LAKES RESEARCH, P 750-759, 1969. 10 P, 4 FIG, 1 TAB, 5 REF.

DESCRIPTORS:
*LAKE ONTARIO, *WATER QUALITY, *MONITORING, BACTERIA, WATER CHEMISTRY,
PHOSPHORUS, NITROGEN, AMMONIA, ALGAE, NUTRIENTS, SEWAGE, WASTE
WATER(POLLUTION), WATER POLLUTION SOURCES, MUNICIPAL WASTES, GREAT
LAKES, *POLLUTANT IDENTIFICATION.

IDENTIFIERS:
*TORONTO(ONTARIO).

ABSTRACT:
AN INVESTIGATION OF DAILY VARIATIONS IN CHEMICAL, BACTERIOLOGICAL AND
ALGAL PARAMETERS WAS CARRIED OUT USING SAMPLES COLLECTED FIVE DAYS A
WEEK FROM TORONTO HARBOUR AND FROM THE INTAKE OF THE R. C. HARRIS
FILTRATION PLANT. THE INTAKE EXTENDS 2500 M FROM SHORE AND IS 12 M DEEP
IN 23 M OF WATER. THE STUDY BEGAN IN JULY 1968. WEEKLY DEPTH AND
SEDIMENT SAMPLES WERE COLLECTED TO SUPPLEMENT THE DAILY DATA. FOURTEEN
CHEMICAL PARAMETERS WERE MONITORED. TOTAL PHOSPHORUS CONCENTRATION WAS
THE BEST INDICATOR OF THE PRESENCE OF URBAN DRAINAGE. AMMONIA AND
ORGANIC NITROGEN CONCENTRATIONS IN THE HARBOR VARY INVERSELY;
CONSEQUENTLY KJELDAHL NITROGEN IS A BETTER MEASURE OF WATER QUALITY
THAN EITHER OF THEM. SURFACE NITRATE CONCENTRATIONS VARY WITH WIND
SPEED AND BIOLOGICAL ACTIVITY AND DO NOT ALWAYS REFLECT THE EFFECTS OF
RUNOFF. CHLOROPHYLL AND SOLUBLE PHOSPHORUS CONCENTRATIONS IN THE HARBOR
VARY INVERSELY. PHOSPHORUS WAS APPARENTLY NOT LIMITING ALGAL GROWTH,
BUT THERE WERE INSUFFICIENT DATA TO TEST THE RELATIONSHIP FOR THE OPEN
LAKE. DAILY DATA WERE REQUIRED TO DEFINE SOME INTER-PARAMETER
RELATIONSHIPS WHICH WERE NOT DEFINED BY DATA COLLECTED AT WEEKLY
INTERVALS. (KNAPP-USGS)

FIELD 05A, 05B, 02H

ACCESSION NO. W71-05879

75 YEARS OF IMPROVEMENT IN WATER SUPPLY QUALITY,

AMERICAN WATER WORKS ASSOCIATION, NEW YORK.

ABEL WOLMAN.

IN: WATER, HEALTH AND SOCIETY, BLOOMINGTON, INDIANA UNIVERSITY PRESS, P
93-103, 1969.

DESCRIPTORS:
*WATER SUPPLY, *WATER QUALITY, *WATER TREATMENT, FILTRATION,
CHLORINATION, COAGULATION, WATERSHEDS, SEWERAGE ALGAE, POTABLE WATER.

IDENTIFIERS:
*AMERICAN WATER WORKS ASSOCIATION, CHEMICAL INDUSTRY, ODOR, TASTE,
CONSUMER.

ABSTRACT:
THIS ARTICLE DISCUSSES THE VARIOUS EFFORTS OF THE AMERICAL WATER WORKS
ASSOCIATION, IN BOTH TECHNOLOGICAL AND SCIENTIFIC FIELDS, TO PROVIDE
WATER OF MAXIMUM USEFULNESS AND SATISFACTION TO THE CONSUMER OVER THE
LAST 75 YEARS. ADVANCES IN WATER SUPPLY SUCH AS EFFECTIVE COAGULATION,
SEDIMENTATION, FILTRATION, DISINFECTION AND THE USE OF CHLORINATION ARE
SURVEYED AS MILESTONES IN WATER SUPPLY QUALITY. SPECIFIC WATER QUALITY
ANALYSIS DIFFICULTIES ARE ASSESSED FROM THE POINTS OF VIEW OF BOTH THE
WATER DIAGNOSTICIAN AND THE CONSUMER. PARTICULAR ATTENTION IS GIVEN TO
THE IDENTIFICATION OF SOURCES AND CAUSES OF TASTE AND ODOR, AND THE
RELATIONSHIP OF THESE TO ALGAE AND TO THE BURGEONING CHEMICAL INDUSTRY.
THE USE OF SUCH REMEDIES AS POWDERED ACTIVATED CARBON, OZONE,
CHLORINATION, AND CHLORINE DIOXIDE IS DISCUSSED. ADVANCES IN THE WATER
SUPPLY QUALITY ARE ATTRIBUTED TO THE DEVELOPMENT OF STANDARD METHODS OF
ANALYSIS AND TO THE JUDICIOUS USE OF COMMERCIAL EQUIPMENT. A RAPID
TREND IS HYPOTHESIZED TOWARD THE LITERAL TAILORING OF WATER FOR THE
DOMESTIC, INDUSTRIAL, OR FARMING CONSUMER AS REQUIREMENTS OF WATER
SUPPLY QUALITY BECOME MORE NUMEROUS AND COMPLEX. (MURPHY-RUTGERS)

FIELD 06E, 05G

ACCESSION NO. W71-05943

INVESTIGATIONS OF THE TECHNIQUES OF PURIFICATION OF ALGAL CULTURES: ELABORATION OF METHODS APPLICABLE TO DESMIDS, FILAMENTOUŠ CYANOPHYTES, AND DIATOMS, (IN FRENCH),

CENTRE NATIONAL DE LA RECHERCHE SCIENTIFIQUE, GIF-SUR-YVETTE (FRANCE). CENTRE DE RECHERCHES HYDROBIOLOGIQUES.

M. TASSIGNY, G. LAPORTE, AND R. POURRIOT.

ENGLISH SUMMARY. ANNALES DE L'ISTITUT PASTEUR, VOL 117, NO 1, P 64-75, 1969. 2 TAB, 22 REF.

DESCRIPTORS:
*LABORATORY TESTS, *ALGAE, CULTURES, DIATOMS, CYANOPHYTA, BIOASSAY, TEST PROCEDURES.

IDENTIFIERS:
*PURE ALGAL CULTURES.

ABSTRACT:
SEVERAL METHODS OF OBTAINING PURE CULTURES OF ALGAE WERE SUBJECTED TO CRITICAL REVIEW: DILUTION AND USE OF PIPETTES, DILUTION FOLLOWING AN ULTRASONIC TREATMENT, USE OF ANTIBIOTICS, AND GERMINATION OF STERILIZED SPORES. EACH OF THESE METHODS OFFERED A POSSIBILITY OF OBTAINING PURE CULTURES OF A LIMITED NUMBER OF SPECIES. THE SIMPLE TECHNIQUE OF DILUTION AND USE OF PIPETTES PROVIDED THE GREATEST CHANCE OF SUCCESS. A PROPER CHOICE OF THE METHOD AND CERTAIN MODIFICATIONS PERMITTED ISOLATION OF MICROSTERIAE, CLOSTERIUM, DRAPARNALDIA, NITZSCHIA, AND SEVERAL FILAMENTOUS CYANOPHYCEAE. (WILDE-WISCONSIN)

FIELD 07B, 05A

ACCESSION NO. W71-06002

PRACTICAL ASPECTS OF THE DESIGN OF WASTE STABILIZATION PONDS,

WOOTON (L.E.) AND CO., RALEIGH, N.C.

W. D. BARLOW.

NINTH SOUTHERN MUNICIPAL AND INDUSTRIAL WASTE CONFERENCE PROCEEDINGS (NORTH CAROLINA STATE UNIVERSITY, RALEIGH, NORTH CAROLINA, APRIL 7-8 1960) P 65-74, TECHNICAL PRESS, RALEIGH, NORTH CAROLINA, 1960.

DESCRIPTORS:
*LAGOONS, *DOMESTIC WASTES, WASTE WATER TREATMENT, INDUSTRIAL WASTES, NUTRIENTS, BACTERIA, ALGAE, TREATMENT FACILITIES, TEST PROCEDURES, METALS.

IDENTIFIERS:
*FINISHING WASTES.

ABSTRACT:
LAGOONS CAN BE DESIGNED TO PROVIDE COMPLETE TREATMENT FOR ANY WASTE WATER, DOMESTIC AND/OR INDUSTRIAL, PROVIDED IT CONTAINS THE NECESSARY NUTRIENTS AND IS FREE FROM SUBSTANCES WHICH ARE TOXIC TO BACTERIA OR ALGAE. THE AUTHOR DISCUSSES THE BASIC PROCESSES INVOLVED; FACTORS AFFECTING THE DESIGN AND OPERATION OF THE LAGOON (PARTICULARLY UNDER CONDITIONS IN NORTH AND SOUTH CAROLINA); PERFORMANCE OF THE LAGOON, INCLUDING METHODS OF MEASUREMENT; AND PRECAUTIONS TO BE TAKEN TO AVOID HEALTH HAZARDS, POLLUTION AND OTHER NUISANCE PROBLEMS. IN DISCUSSION, REFERENCE WAS MADE TO LAGOONS USED FOR TREATING MIXED WASTE WATERS AND 20-30 PERCENT OF DOMESTIC SEWAGE, AND ESPECIALLY TO THE PROBLEMS ENCOUNTERED WHICH APPEAR TO BE CAUSED BY TOXIC METALS IN THE WASTE WATER. (LIVENGOOD — NORTH CAROLINA STATE UNIVERSITY)

FIELD 05D

ACCESSION NO. W71-06737

NATION-WIDE RESEARCH ON ANIMAL WASTE DISPOSAL,

FEDERAL WATER POLLUTION CONTROL ADMINISTRATION, CHICAGO, ILL. LAKE MICHIGAN
BASIN OFFICE.

JACOB O. DUMELLE.

IN: PROCEEDINGS OF FARM ANIMAL WASTE AND BY-PRODUCT MANAGEMENT CONFERENCE,
UNIVERSITY EXTENSION, UNIVERSITY OF WISCONSIN, P 80-81, NOVEMBER 6-7, 1969.

DESCRIPTORS:
*FARM WASTES, *RESEARCH AND DEVELOPMENT, GRANTS, ALGAE, AIR POLLUTION,
WATER POLLUTION, SOIL CONTAMINATION, NUTRIENTS.

IDENTIFIERS:
*FWPCA, ACTIVATED ALGAE, FEEDLOTS, OXIDATION DITCH.

ABSTRACT:
THE ARTICLE GIVES BRIEF DESCRIPTIONS OF RESEARCH PROJECTS WHICH THE
FEDERAL WATER POLLUTION CONTROL ADMINISTRATION IS HELPING TO FUND. ONE
SUCH PROJECT UNDER WAY IN CALIFORNIA IS TRYING TO DETERMINE THE
PRACTICABILITY OF PRODUCING AND HARVESTING ALGAE TO REMOVE NUTRIENTS
FROM AGRICULTURAL DRAINAGE WATERS. OTHER PROJECTS INVOLVE CATTLE
FEEDLOT RUNOFF, AND DAIRY WASTE WATERS. BESIDES RESEARCH ON TREATMENT
METHODS, SOME PROJECTS ARE TRYING TO FIND OUT HOW MUCH NUTRIENT RUNS
OFF, AND HOW MUCH GETS INTO WATER. (WHITE-IOWA STATE)

FIELD 05G, 05D

ACCESSION NO. W71-06821

SOME PROBLEMS IN THE TESTING OF MATERIALS WITH ALGAE,

CENTRAL LAB. TNO, DELFT (NETHERLANDS).

H. J. HUECK, AND D. M. M. ADEMA.

SUMMARIES IN GERMAN, ENGLISH, FRENCH AND SPANISH. MATERIAL UND ORGANISMEN,
VOL 2, NO 2, P 141-152, 1967. 1 FIG, 4 TAB, 15 REF.

DESCRIPTORS:
*ALGICIDES, *PAINTS, *TEST PROCEDURES, ANTIFOULING MATERIALS, TESTING,
BIODEGRADATION, ALGAE, MARINE ALGAE, GROWTH RATES, ANALYTICAL
TECHNIQUES, LABORATORY EQUIPMENT, ASBESTOS CEMENT, CHLORELLA, CARBON
DIOXIDE.

IDENTIFIERS:
*TEST DEVELOPMENT, *TEST PROBLEMS, HORMIDIUM SP.

ABSTRACT:
A FEW CONSIDERATIONS ARE GIVEN ABOUT THE TESTING OF ALGICIDES
INCORPORATED IN PAINTS AND STONY MATERIALS. TWO EXPERIMENTAL DESIGNS
ARE DESCRIBED: ONE FOR TESTING ALGICIDAL PAINTS WITH AQUATIC ALGAE, AND
ONE FOR TESTING ALGICIDAL ASBESTOS-CEMENT WITH AERIAL ALGAE. FOR THE
FIRST PURPOSE A STATIC TEST IS UNDESIRABLE BECAUSE OF THE BUILDING UP
OF TOXIC CONCENTRATIONS IN THE MEDIUM WHEREAS THE ACTUAL TOXICITY AT
THE SURFACE OF THE MATERIAL SHOULD BE MEASURED. A DEVICE IS PROPOSED
THAT USES A CONTINUOUSLY INNOCULATED MEDIUM FLOWING SLOWLY THROUGH A
TEST AQUARIUM. FOR THE TESTING OF ASBESTOS-CEMENT AND OTHER ALKALINE
STONY MATERIALS WITH AERIAL ALGAE IT IS FOUND THAT THE AVAILABILITY OF
CO2 NEAR THE SURFACE MAY BE A LIMITING FACTOR. ANOTHER DEVICE IS
DESCRIBED TO TEST SAMPLES IN A FLOW OF AIR ENRICHED WITH 5% CO2.
(SJOLSETH-WASHINGTON)

FIELD 05C

ACCESSION NO. W71-07339

EXISTING AND POTENTIAL PROBLEMS OF EXCESSIVE EUTROPHICATION IN THE TENNESSEE VALLEY,

TENNESSEE VALLEY AUTHORITY, CHATTANOOGA. WATER QUALITY BRANCH.

EUGENE B. WELCH.

PROCEEDINGS OF THE 7TH ANNUAL SANITARY AND WATER RESOURCES ENGINEERING CONFERENCE, VANDERBILT UNIVERSITY, NASHVILLE, TENNESSEE, P 45-73, 1968. 7 FIG, 8 REF.

DESCRIPTORS:
*EUTROPHICATION, *TENNESSEE VALLEY AUTHORITY PROJECT, *RESERVOIRS, TEMPERATURE, DISSOLVED OXYGEN, DENSITY, ROOTED AQUATIC PLANTS, ALGAE, CHLOROPHYLL, HYPOLIMNION, PHOSPHORUS, NITROGEN, PHYTOPLANKTON, PRODUCTIVITY, TURBIDITY, SURFACES, DEPTH, DISTRIBUTION, TIME, EPILIMNION, THERMOCLINE, LIGHT PENETRATION, STANDING CROP, INDUSTRIAL WASTES, RAINFALL, ALABAMA.

IDENTIFIERS:
CHEROKEE RESERVOIR(TENN), HOLSTON RIVER(TENN), POTOMOGETON PECTINATUS, POTOMOGETON CRISPUS, HETERAMTHERA, PICKWICK RESERVOIR(TENN), NAJAS, PHOTIC ZONE, DOUGLAS RESERVOIR(TENN), NORRIS RESERVOIR(TENN), FRENCH BROAD RIVER(TENN), CHICKAMAUGA RESERVOIR(TENN), FLORENCE(ALA).

ABSTRACT:
PATTERN OF HYPOLIMNETIC OXYGEN DEPLETION IN THREE TVA SYSTEM RESERVOIRS APPEARS RELATED TO PHYTOPLANKTON PRODUCTIVITY, WHICH IS AFFECTED BY SEVERAL ENVIRONMENTAL FACTORS. IN RESERVOIRS STUDIED, NUTRIENT (NITROGEN AND PHOSPHORUS) CONTENT POTENTIAL LIMITS PRODUCTION ONLY DURING INTENSE SUMMER STRATIFICATION. THERMAL STRATIFICATION AFFECTING NUTRIENT DISTRIBUTION AND LIGHT AVAILABILITY, TURBIDITY AFFECTING LIGHT PENETRATION, AND WATER DETENTION TIME MAY AT VARIOUS PERIODS AND PLACES CONTROL PRODUCTION REGARDLESS OF NUTRIENT CONTENT. FIELD AND LABORATORY EVIDENCE SHOWS THAT GROWTH OF TWO NUISANCE MACROPHYTES IN TENNESSEE VALLEY IS PROBABLY NOT CONTROLLED BY NITROGEN AND PHOSPHORUS CONTENT OF WATER. NUTRIENT REQUIREMENTS ARE APPARENTLY LESS THAN AMOUNTS MEASURED AND AVAILABLE. ORGANIC SEDIMENT IS RELATIVELY MORE IMPORTANT AS NUTRITION SOURCES FOR MAXIMUM GROWTH THAN SURROUNDING WATER. AMOUNT OF AVAILABLE LIGHT AT LEAST PARTLY EXPLAINS ANNUAL AND SEASONAL VARIABILITY IN NUISANCE MACROPHYTE GROWTH. LIGHT REACHING STREAM OR LAKE BOTTOM AND AVAILABLE FOR PLANT GROWTH DURING GROWING SEASON FROM YEAR TO YEAR IS AFFECTED BY INTERACTION OF INCIDENT LIGHT, WATER DEPTH, AND TURBIDITY OR RELATIVE PENETRATION. THESE, OR RELATED FACTORS, WERE MOST FAVORABLE DURING THE TWO YEARS WHEN NUISANCE MACROPHYTE PROBLEMS WERE GREATEST IN PICKWICK RESERVOIR, AND WERE RELATED TO PLANT GROWTH IN HOLSTON RIVER DURING 1967. (JONES-WISCONSIN)

FIELD 05C, 02H

ACCESSION NO. W71-07698

GLENODININE, AN ICHTHYOTOXIC SUBSTANCE PRODUCED BY A DINOFLAGELLATE, PERIDINIUM POLONICUM,

TOKYO UNIV. (JAPAN). LAB. OF MARINE BIOCHEMISTRY.

YOSHIRO HASHIMOTO, TOMOTOSHI OKAICHI, LE DUNG DANG, AND TAMAO NOGUCHI.

BULLETIN OF THE JAPANESE SOCIETY OF SCIENTIFIC FISHERIES, VOL 34, NO 6, P 528-534, JUNE 1968. 7 FIG, 1 TAB, 8 REF.

DESCRIPTORS:
*FISHKILL, *TOXINS, *ALGAL POISONING, *ALGAL TOXINS, *PHYTOTOXICITY, *FISH DISEASES, *FISH TOXINS, TOXICITY, HYDROGEN ION CONCENTRATION, DISSOLVED OXYGEN, ALGAE, PHYTOPLANKTON, ALKALINE WATER, ALKALINITY, FISH FARMING, FISH FOOD ORGANISMS, FISH PHYSIOLOGY, RED TIDE, ORGANIC COMPOUNDS, DINOFLAGELLATES.

IDENTIFIERS:
*PERIDINIUM SP., *GLENODININE, ALKALOID TOXINS, IBOGAALKALOID, ORGANIC TOXINS.

ABSTRACT:
MASS MORTALITIES ASSOCIATED WITH PHYTOPLANKTON BLOOMS SOMETIMES OCCUR IN JAPANESE FISH CULTURE OPERATIONS. THE AUTHORS OPINE THAT THIS MAY BE THE FIRST REPORT IDENTIFYING THE TOXIC SPECIES AND ITS ASSOCIATED TOXIN. THE OFFENDER, GLENODININE, WAS ISOLATED FROM THE DINOFLAGELLATE, PERIDINIUM POLONICUM. THE TOXIN IS AN ALKALOID SHOWING SOME RESEMBLANCE TO IBOGAALKALOID, AND HAS THE SULFHYDRYL FUNCTION IN THE MOLECULE. FISHKILLS OCCURRED IN THE AFTERNOON OR EVENING WHEN DISSOLVED OXYGEN WAS HIGH AND PH LEVELS OF 8.7-9.2 WERE PRESENT. ALTHOUGH MORTALITIES SOMETIMES OCCURRED AFTER BLOOMS RATHER THAN CONCURRENTLY, LETHAL LEVELS OF GLENODININE WERE FOUND IN THE WATER DURING THE MORTALITIES. THE SUBSTANCE IS APPARENTLY TOXIC ONLY IN ALKALINE MEDIUM, CREATED BY HIGH PHYTOPLANKTONIC PHOTOSYNTHETIC ACTIVITY. (LEGORE-WASHINGTON)

FIELD 05C, 05A

ACCESSION NO. W71-07731

MANURE LAGOONS......DESIGN CRITERIA AND MANAGEMENT,

MARYLAND UNIV., COLLEGE PARK. DEPT. OF AGRICULTURAL ENGINEERING.

HARRY J. EBY.

ASAE PAPER NO 61-935. AGRICULTURAL ENGINEERING JOURNAL, VOL 43, P 698-701,
   714-715, DEC 1962. 6 FIG, 1 TAB, 19 REF.

   DESCRIPTORS:
    *FARM LAGOONS, *DESIGN CRITERIA, WATER TEMPERATURE, SEWAGE TREATMENT,
    AEROBIC BACTERIA, AQUATIC PLANTS, ANAEROBIC BACTERIA, ALGAE,
    BIOCHEMICAL OXYGEN DEMAND, OXIDATION LAGOONS, SLUDGE, PHOTOSYNTHETIC
    OXYGEN, FARM WASTES, WASTE WATER TREATMENT.

   IDENTIFIERS:
    *SITE SELECTION, LOADING.

   ABSTRACT:
    CRITERIA TO BE CONSIDERED WHEN DESIGNING A LAGOON FOR TREATMENT OF
    WASTES PRODUCED BY ANIMALS IN CONFINEMENT IS DISCUSSED. IT MENTIONS
    SITUATIONS WHERE LAGOONS WOULD NOT BE FEASIBLE. SEVEN CRITERIA FOR SITE
    SELECTION ARE GIVEN. THE PHYSICAL, CHEMICAL AND BIOLOGICAL FACTORS
    DISCUSSED INCLUDE TEMPERATURE, LIGHT, SPECIFIC GRAVITY, MIXING,
    NUTRITIONAL EFFECTS, PH EFFECTS, TOXIC EFFECTS, AND INTERRELATIONSHIP
    OF BIOLOGICAL SPECIES. ALSO MENTIONED IS THE ALGAL-BACTERIAL
    RELATIONSHIP. DESIGN FACTORS FOR SIZE AND VOLUME ARE GIVEN. THE ARTICLE
    CONCLUDES WITH MANAGEMENT PROBLEMS ENCOUNTERED SUCH AS FLOATING DEBRIS,
    OVERLOADING, INTERMITTENT LOADING, AQUATIC WEEDS AND SLUDGE BUILD-UP.
    (PARKER-IOWA STATE)

   FIELD 05D

   ACCESSION NO. W71-08221

TREATMENT OF POTATO PROCESSING WASTES AT SALADA FOODS LTD., ALLISTON,

ONTARIO WATER RESOURCES COMMISSION, TORONTO.

T. D. ARMSTRONG, AND B. I. BOYKO.

PROCEEDINGS, ONTARIO INDUSTRIAL WASTE CONFERENCE, 16TH, JUNE 1969, NIAGARA
   FALLS, ONTARIO, P 188-208.

   DESCRIPTORS:
    *POTATOES, *INDUSTRIAL WASTES, SCREENS, SEDIMENTATION, ORGANIC LOADING,
    ACTIVATED SLUDGE, AERATION, DISSOLVED OXYGEN, SLUDGE, OXIDATION
    LAGOONS, ALGAE, ADMINISTRATION, MAINTENANCE, *WASTE WATER TREATMENT.

   IDENTIFIERS:
    *POTATO PROCESSING WASTES, AEROBIC LAGOONS, ANAEROBIC LAGOONS.

   ABSTRACT:
    THE SALADA PLANT AT ALLISTON, ONTARIO, WAS CONSTRUCTED IN 1959. AT THIS
    TIME, 5000 LBS/HR OF POTATOES WERE PROCESSED INTO POTATO FLAKES DURING
    A 24 HOUR DAY. SINCE THAT TIME NEW PRODUCTS HAVE BEEN ADDED, AND WATER
    CONSUMPTION, INITIALLY 200,000 GPD NOW RANGES FROM 630,000 TO 750,000
    GPD DURING THE AUGUST TO MAY PROCESSING SEASON. THE COMPANY AND THE
    TOWN WERE TO CONSTRUCT SECONDARY TREATMENT FACILITIES JOINTLY. THE
    INITIAL INSTALLATION WAS DESIGNED FOR A WASTE FLOW OF 150,000 GPD AND
    INCLUDED SCREENING, SEDIMENTATION, AN ANAEROBIC LAGOON, AND TWO AEROBIC
    LAGOONS. THE LAGOON FILLED FOR 5 1/2 MONTHS, AND EFFLUENT ON THE FIRST
    DAY OF OVERFLOW WAS 376,000 GALLONS WITH CORRESPONDING 6010 LBS BOD.
    THE PLANT WAS OVERLOADED BOTH HYDRAULICALLY AND ORGANICALLY, THE
    ANAEROBIC LAGOON WAS CHANGED TO AN AEROBIC LAGOON, TWO CONCRETE
    CLARIFIERS WERE ADDED, EACH WITH 2.75 HOURS DETENTION TIME, AND BUBBLE
    GUN AERATION DEVICES WERE INSTALLED. SINCE THE PLANT WAS STILL
    OVER-LOADED, A COMPREHENSIVE SURVEY WAS ORDERED, AND FROM THE RESULTS
    OF THIS STUDY, A 6.2 HOUR DETENTION TIME ACTIVATED SLUDGE BASIN WAS
    BUILT. TOTAL COST OF THE TREATMENT PLANT HAD RISEN TO $350,000.
    ALTHOUGH PROBLEMS WERE STILL ENCOUNTERED, THIS TREATMENT FACILITY
    PROVIDED ADEQUATE TREATMENT. HAD THE NECESSARY LONG RANGE PLANNING BEEN
    DONE EARLIER, A SERIOUS POLLUTIONAL PROBLEM MAY HAVE BEEN AVOIDED.
    (LOWRY-TEXAS)

   FIELD 05D

ACCESSION NO. W71-08970

STUDIES ON ZINC IN FRESHWATER ECOSYSTEMS,

OBERLIN COLL., OHIO.

EDWARD J. KORMONDY.

AVAILABLE FROM THE NATIONAL TECHNICAL INFORMATION SERVICE AS COO-1499-3,
$3.00 IN PAPER COPY, $0.95 IN MICROFICHE. REPORT COO-1499-3, JAN 1969. 32
P. CONTRACT AT(11-1)-1499.

DESCRIPTORS:
*ZINC RADIOISOTOPES, *AQUATIC INSECTS, *AQUATIC PLANTS, AQUATIC
ENVIRONMENT, ECOSYSTEMS, AQUATIC HABITATS, DRAGONFLIES, ALGAE,
SEDIMENTS, SOIL-WATER-PLANT RELATIONSHIPS, NUCLEAR WASTES, ABSORPTION,
CHELATION, METABOLISM, AQUARIA, WATER POLLUTION EFFECTS, ADSORPTION.

IDENTIFIERS:
*ZINC.

ABSTRACT:
ZINC-65 IS SIGNIFICANTLY CONCENTRATED IN THE AQUATIC ENVIRONMENT FROM
NUCLEAR WASTES. THE PRESENT EXPERIMENTS WITH MICROCOSMS (DRAGONFLY
LARVAE, ALGAE, WATER, AND SOIL) SHOW A DEPENDENCE OF UPTAKE ON
TEMPERATURE, VOLUME OF THE AQUARIUM AND LIGHT; WHICH SUGGESTS THAT
METABOLIC ACTIVITY PLAYS A ROLE. IN PARTICULAR, UPTAKE BY AQUATIC
PLANTS WAS AFFECTED BY PHOTOSYNTHESIS. ONLY ABOUT 1% AS MUCH UPTAKE BY
PLANTS OCCURRED IN THE PRESENCE OF THE CHELATING AGENT EDTA.
PRELIMINARY STUDIES SHOWED VARIATIONS IN CHELATION BY SOIL BETWEEN
VARIOUS SOIL TYPES. THE BINDING OF ZINC BY SOIL WAS REDUCED BY THE
PRESENCE OF PLANTS. (BOPP-NSIC)

FIELD 05C

ACCESSION NO. W71-09005

HEALTH PHYSICS DIVISION ANNUAL PROGRESS REPORT FOR PERIOD ENDING JULY 31, 1969:
SYSTEMS ECOLOGY, WATERSHED AQUATIC HABITAT INTERACTIONS.

OAK RIDGE NATIONAL LAB., TENN.

AVAILABLE FROM THE NATIONAL TECHNICAL INFORMATION SERVICE AS ORNL-4446, $3.00
IN PAPER COPY, $0.95 IN MICROFICHE. REPORT ORNL-4446, (P 137-162), JULY
1969.

DESCRIPTORS:
*ECOLOGY, *MATHEMATICAL MODELS, *WATERSHED MANAGEMENT, GEOCHEMISTRY,
LAKES, CONIFEROUS FORESTS, ALGAE, INSECTS, PHOSPHORUS RADIOISOTOPES,
TECHNOLOGY, NATURAL RESOURCES, ECOLOGICAL DISTRIBUTION, STOCHASTIC
PROCESSES, ECOSYSTEMS, AQUATIC HABITATS, WATER POLLUTION EFFECTS.

ABSTRACT:
THE PURPOSE OF THE ECOLOGICAL SYSTEMS ANALYSIS PROGRAM IS TO DEVELOP A
METHODOLOGY FOR MAKING USEFUL PREDICTIONS ABOUT THE OUTCOME OF
ENVIRONMENTAL INTERACTIONS. PROGRESS IS REPORTED ON THE FOLLOWING
PROJECTS: PERTIENCE OF RECENT ECOLOGICAL LITERATURE TO THE MODELING OF
ECOSYSTEMS, BIOGEOCHEMICAL ECOLOGY RESEARCH COLLECTION, PARAMETER
IDENTIFICATION IN SYSTEMS ECOLOGY, NUMERICAL METHODS IN SYSTEMS
ECOLOGY, A VERSATILE COMPARTMENT SYSTEM SIMULATION PROGRAM, A
STOCHASTIC MODEL OF FEEDING IN A FOREST CENTIPEDE, A MODEL FOR CEDAR
BOG LAKE (MINNESOTA), ANALYSIS OF THE CHANGE-IN-RATIO MODEL FOR
ESTIMATING POPULATION ABUNDANCE, AND SYSTEMS ANALYSIS OF BALSAM
FORESTS. THE WALKER BRANCH WATERSHED PROJECT WAS ORGANIZED TO ASSESS
THE POTENTIAL ECOLOGICAL IMPACT OF NEW TECHNOLOGY BEING APPLIED TO THE
FIELD OF NATURAL RESOURCE MANAGEMENT. A COMPARTMENT MODEL IS
ILLUSTRATED FOR EVALUATION OF TRANSFER RATES AND CHARACTERISTICS OF THE
TRANSFER FUNCTIONS AMONG THE PRINCIPAL COMPONENTS OF THE WATERSHED
ECOSYSTEM. A MATERIAL BALANCE STUDY USING P-32 AND PERIPHYTIC ALGAE IN
A CLOSED SYSTEM IN THE LABORATORY WAS CONDUCTED TO OBTAIN DATA ON
UPTAKE AND TURNOVER OF PHOSPHORUS BY STREAM DIATOMS.
(BOPP-NSIC)

FIELD 05C

ACCESSION NO. W71-09012

RADIOLOGICAL PHYSICS DIVISION. ANNUAL REPORT, JULY 1968-JUNE 1969.

ARGONNE NATIONAL LAB., ILL.

AVAILABLE FROM THE NATIONAL TECHNICAL INFORMATION SERVICE AS ANL-7615, $3.00
IN PAPER COPY, $0.95 IN MICROFICHE. REPORT ANL-7615, 1969. 233 P.

DESCRIPTORS:
*WATER POLLUTION EFFECTS, *FALLOUT, *ECOSYSTEMS, GREAT LAKES, TRACE
ELEMENTS, RADIUM RADIOISOTOPES, THORIUM RADIOISOTOPES, URANIUM
RADIOISOTOPES, MARINE ALGAE, WATER POLLUTION SOURCES.

IDENTIFIERS:
*CESIUM RADIOISOTOPES.

ABSTRACT:
INCLUDES 46 ARTICLES OF WHICH THREE CONCERN RADIOISOTOPES IN AQUATIC
ECOSYSTEMS: CONCENTRATIONS OF TRACE ELEMENTS IN GREAT LAKES FISH;
CONCENTRATION OF RADIUM, THORIUM, AND URANIUM BY TROPICAL ALGAE; AND
BEHAVIOR OF FALLOUT CESIUM-137 IN AQUATIC AND TERRESTRIAL SYSTEMS.
(BOPP-NSIC)

FIELD 05G, 05C

ACCESSION NO. W71-09013

THE USE OF AERO-HYDRAULIC GUNS IN THE BIOLOGICAL TREATMENT OF ORGANIC WASTES,

PROCTOR AND REDFERN, TORONTO (ONTARIO).

C. S. DUTTON, AND C. P. FISHER.

PROCEEDINGS, INDUSTRIAL WASTE CONFERENCE, 21, VOL L, NO 2, MARCH 1966. P
403-423, 7 FIG.

DESCRIPTORS:
*OXIDATION LAGOONS, *ORGANIC WASTES, *OXYGENATION, AEROBIC TREATMENT,
ANAEROBIC CONDITIONS, CANNERIES, ACTIVATED SLUDGE, BIOCHEMICAL OXYGEN
DEMAND, OXIDATION-REDUCTION POTENTIAL, AQUATIC ALGAE, *BIOLOGICAL
TREATMENT, *WASTE WATER TREATMENT.

IDENTIFIERS:
AERO-HYDRAULIC GUN, ALLISTON(ONTARIO).

ABSTRACT:
THE AERATED LAGOON TREATMENT OF ORGANIC WASTES IS AN ECONOMICAL AND
INCREASINGLY POPULAR PROCESS. HOWEVER, THE DEGREE OF TREATMENT HAS BEEN
LIMITED BY THE NATURE OF DEVICES AVAILABLE FOR ACCOMPLISHING
OXYGENATION. RECENT EXPERIMENTS UTILIZED AERO-HYDRAULIC GUNS IN SIMPLE
AERATED LAGOONS FOR TREATMENT OF ORGANIC WASTES. THE GUNS CONSISTED OF
A POLYETHYLENE STACK PIPE, BUBBLE GENERATOR, AND AN AIR SUPPLY THAT
PERIODICALLY RELEASED NUMEROUS SMALL BUBBLES FROM THE LAGOON BOTTOM.
THE RESULTING FLOW PATTERN OF LIQUID FROM THE LAGOON BOTTOM INVOLVED A
STRONG RADIAL CURRENT AT THE SURFACE. OXYGENATION CAPACITY OF THE GUNS
WERE DERIVED IN A SERIES OF CURVES THAT PLOTTED LBS O2/HR AGAINST FREE
AIR CONSUMPTION. CORRECTION FACTORS PERMITTED APPLICATION OF THE CURVES
TO ACTUAL DESIGNS. A GUN CAN ALSO FUNCTION AS A POSITIVE SLUDGE RETURN
PUMP BY SEALING UP ONE OF THE PORTS AND CONNECTING A SUCTION PIPE. AT A
POTATO WASTE TREATMENT FACILITY IN ALLISTON, ONTARIO, SOME 790,000 GPD
OF POTATO WASTE WAS DISCHARGED DAILY ALONG WITH A LOAD OF 5,000 LBS BOD
AND A LIKE AMOUNT OF SUSPENDED SOLIDS. RECOMMENDATIONS FOR THIS
FACILITY WERE: (1) TWO CONVENTIONAL SETTLING BASINS, EACH 24 FT SQUARE
EQUIPPED WITH A SLUDGE SCRAPER, (2) A SIMPLE AERATED LAGOON EQUIPPED
WITH 127 GUNS, AND (3) TWO ALGAE LAGOONS. DR. C. P. FISHER STUDIED THE
SYSTEM AND FOUND BOD REMOVAL ACHIEVED BY THE AERATED LAGOON WAS
APPROXIMATELY 90% WITH LITTLE OR NO MEASURABLE RESIDUAL DISSOLVED
OXYGEN. THERE WAS NO DOUBT THAT THE THOROUGH MIXING PRODUCED BY THE
GUNS WAS RESPONSIBLE TO A GREATER OR LESSER DEGREE FOR THE HIGH
REMOVALS. OTHER STUDIES WERE CONDUCTED ON POULTRY AND FRUIT CANNING
WASTES. (BURDETTE-TEXAS)

FIELD 05D

ACCESSION NO. W71-09546

FOOD CANNERY WASTE TREATMENT BY LAGOONS AND DITCHES AT SHEPPARTON, VICTORIA, AUSTRALIA,

MELBOURNE WATER SCIENCE INST. (AUSTRALIA).

C. D. PARKER.

PROCEEDINGS, INDUSTRIAL WASTE CONFERENCE, 21, VOL L, NO 2, MARCH 1966, P 284-302, 5 FIG, 13 TAB, 6 REF.

DESCRIPTORS:
*CANNERIES, *INDUSTRIAL WASTES, PEACHES, APRICOTS, PEARS, CITRUS FRUITS, TOMATOES, *AEROBIC TREATMENT, *ANAEROBIC DIGESTION, BIOCHEMICAL OXYGEN DEMAND, ACTIVATED SLUDGE, COST COMPARISONS, SEDIMENTS, ALGAE, HYDROGEN ION CONCENTRATION, ELECTRIC POWER, SEWAGE DISPOSAL, *WASTE WATER TREATMENT, *OXIDATION LAGOONS.

IDENTIFIERS:
SHEPPARTION(AUSTRALIA), OXIDATION DITCHES, SOUP, FOOD WASTES.

ABSTRACT:
CANNERY WASTE IS CHARACTERIZED BY HIGH BOD AND HIGH ACIDITY. THE USUAL METHOD OF TREATMENT IS SPRAY OR FLOOD IRRIGATION OR AEROBIC TYPE LAGOONS WITH HEAVY DOSAGE OF SODIUM NITRATE FOR ODOR CONTROL. IN SHEPPARTON, VICTORIA, EXPANSIONS IN THE FACILITIES OF SHEPPARTON PRESERVING COMPANY, THE LARGEST CANNERY IN THE SOUTHERN HEMISPHERE, AND THE BUILDING OF A SOUP CANNERY IN THE AREA CAUSED PROBLEMS IN THE TREATMENT OF THE RESULTING SEWAGE. THERE WAS A PRESSING NEED FOR AN EFFICIENT MEANS OF PURIFYING, WITHOUT CAUSING NUISANCE, THE LARGE SEASONAL FLOWS OF HIGHLY POLLUTED FOOD WASTES FROM TWO CANNERIES ON AREAS CLOSELY ADJACENT TO A PROSPEROUS AND RAPIDLY DEVELOPING URBAN CENTER. THE SHORT SEASONAL NATURE OF CANNERY OPERATIONS AND CONSEQUENT HIGH CAPITAL COST MADE THE PROBLEM EVEN MORE DIFFICULT. IT WAS DECIDED TO EXPERIMENT WITH ANAEROBIC AND AEROBIC TYPE LAGOONS AND OXIDATION DITCHES. THE WASTE FROM EACH FIRM IS MIXED WITH SEWAGE PLANT EFFLUENT AND TREATED IN THE NEW PROCESSING UNIT WITH VERY GOOD RESULTS WHICH ARE DESCRIBED IN GREAT DETAIL. THE TOTAL COST OF THE NEW UNIT, EXCLUDING THE COST OF INFLOW AND OUTFLOW PIPELINES AND LABOR AND MAINTAINENCE, WAS $155,000 TO THE PRESERVING CO., AND $143,000 (AMERICAN DOLLARS) TO THE SOUP CANNERY PLUS ANNUAL COSTS OF $5,364 AND $8,192 RESPECTIVELY, EXCLUDING PIPELINES, PUMPING STATIONS, AND MAINTAINENCE LABOR. THE NET RESULT WAS ALMOST COMPLETE PURIFICATION AT COSTS THE INDUSTRIES COULD ACCEPT. (LOWRY-TEXAS)

FIELD 05D

ACCESSION NO. W71-09548

REPORT TO THE CONTRA COSTA COUNTY WATER AGENCY ON A STUDY OF THE EFFECTS OF THE PROPOSED FEDERAL SAN LUIS INTERCEPTOR DRAIN AND THE STATE SAN JOAQUIN VALLEY MASTER DRAIN.

METCALF AND EDDY, INC., PALO ALTO, CALIF.

OCT 30, 1964. 93 P, 4 FIG, 32 TAB, 37 REF.

DESCRIPTORS:
*AGRICULTURAL CHEMICALS, *DELTAS, *DRAINAGE EFFECTS, *WATER QUALITY, *FARM WASTES, *INDUSTRIAL WASTES, MUNICIPAL WASTES, NUTRIENTS, PESTICIDES, RECREATION, ALGAE, OXYGEN DEMAND, WATER POLLUTION, MUNICIPAL WATER, WATER REQUIREMENTS, WATER CONVEYANCE, WATER COSTS, ARTIFICIAL WATER COURSES, CALIFORNIA.

IDENTIFIERS:
*CONTRA COSTA COUNTY, WATER DETERIORATION, DRAIN, SAN JOAQUIN VALLEY MASTER DRAIN, SAN FRANCISCO BAY-DELTA.

ABSTRACT:
THE PROBABLE EFFECTS ON THE OFFSHORE WATERS OF CONTRA COSTA COUNTY, CALIFORNIA BY THE PROPOSED DRAIN WHICH WILL CONVEY AGRICULTURAL SUBSURFACE WASTE WATERS FROM THE SAN JOAQUIN VALLEY AND WASTES FROM THE SAN LUIS DIVISION OF THE CENTRAL VALLEY PROJECT ARE DISCUSSED. THE POINT OF DISCHARGE WOULD BE THE SAN JOAQUIN RIVER IN THE VICINITY OF THE ANTIOCH BRIDGE. DISCHARGE IS EXPECTED TO BEGIN IN 1969 AT A RATE OF 75 CFS AND INCREASE TO 900 CFS BY 1995. SERIOUS DETERIORATION OF THE QUALITY OF OFFSHORE WATERS CAN BE EXPECTED BY 1975. ACQUATIC LIFE AND RECREATIONAL USE OF THE OFFSHORE WATERS COULD BE SERIOUSLY AFFECTED BY THE DISCHARGE OF NUTRIENTS AND PESTICIDES FROM THE DRAIN. NUTRIENTS ARE EXPECTED TO HAVE A GREATER EFFECT THAN PESTICIDES. THE NUTRIENTS IN THE DRAIN WILL PRODUCE GROWTHS OF ALGAE IN THE DRAIN WHICH BY THEIR DEATH AND DECAY WILL EXERT AN OXYGEN DEMAND IN THE RIVER AND BAY VARYING FROM 7500 LBS PER DAY IN 1975 TO 27,000 LBS PER DAY IN 1995. INDUSTRY AND THE CITY OF ANTIOCH WILL BE ABLE TO USE RIVER WATER A LESSER PROPORTION OF THE TIME TO SATISFY HIGH QUALITY REQUIREMENTS. THE DISCHARGE OF THE DRAIN WILL HAVE NO SIGNIFICANT EFFECT ON WATER FOR AGRICULTURAL USE. (POERTNER)

FIELD 06G, 05B

ACCESSION NO. W71-09577

SPECIFIC COMPOSITION OF PHYTOPLANKTON IN A LAKE WARMED UP BY WASTEWATER FROM A
THERMAL POWER STATION AND IN LAKES WITH NORMAL TEMPERATURES,

INSTYTUT RYBACTWA SRODLADOWEGO, OLSZTYN-KORTOWO (POLAND).

J. POLTORACKA.

TRANSLATION FROM POLISH AEC-TR-7194. ACTA SOCIETATIS BOTANICORUM POLONIAE,
POLISH, VOL. 37, NO. 2, 1968, P. 297-325, 5 FIG., 1 TAB., 5 REF.

DESCRIPTORS:
*THERMAL POWERPLANTS, *LAKES, TEMPERATURE, PHYTOPLANKTON, ALGAE, HEATED
WATER, POPULATION, WATER POLLUTION EFFECTS.

IDENTIFIERS:
SPECIES, TAXONS.

ABSTRACT:
A PRELIMINARY EVALUATION OF THE SPECIFIC COMPOSITION OF PHYTOPLANKTON
WAS CARRIED OUT IN LAKES LICHEN, MIKORZYN AND SLESIN, CHARACTERIZED BY
DIFFERENT TEMPERATURES CAUSED BY THE ACTIVITY OF THE POWERPLANT AT
KONIN. THE HIGHEST WATER TEMPERATURES WERE DISPLAYED BY LAKE LICHEN
(7.4-27.5C), AND THE LOWEST TEMPERATURES PREVAILED IN LAKE SLESIN
(0.8-20.7C). DURING THE ENTIRE PERIOD OF INVESTIGATION IN ALL THREE
LAKES, 305 SPECIES WERE FOUND AND 414 TAXONS WERE IDENTIFIED.
PREDOMINANT AMONG THEM WERE BACILLARIOPHYCEAE (44 PERCENT),
CHLOROPHYCEAE (34 PERCENT) AND CYNONPHYCEAE (11 PERCENT). THE HIGHEST
NUMBER OF PHYTOPLANKTON COMPONENTS (285) WERE FOUND IN LAKE LICHEN, AND
THEIR LOWEST NUMBER (197) WAS ASCERTAINED IN LAKE SLESIN, WHICH WAS
ASSOCIATED WITH GREAT DIFFERENCES IN THE SHARE OF CHLOROPHYCEAE WHOSE
HIGHEST NUMBER WAS FOUND IN THE WARMEST, LAKE LICHEN, WHEREAS THEIR
LOWEST NUMBER WAS DISCOVERED IN LAKE SLESIN, THE COLDEST LAKE. THE
NUMBER OF SPECIES OF ALGAE IN LAKE LICHEN WAS RATHER STABLE·THROUGHOUT
THE YEAR BUT FLUCTUATED SEASONALLY IN OTHER LAKES. A MOST MARKED
CORRELATION BETWEEN THE NUMBER OF THE TAXONS OF ALGAE AND SURFACE WATER
TEMPERATURE WAS OBSERVED IN LAKE LICHEN. WITH RESPECT TO THE SIZE OF
INDIVIDUAL POPULATION, 14 PLANKTONIC SPECIES WERE DOMINANT AT VARIOUS
PERIODS IN DIFFERENT LAKES. (UPADHYAYA-VANDERBILT)

FIELD 05C, 02H

ACCESSION NO. W71-09767

UPTAKE OF RADIOACTIVE NUCLIDES BY AQUATIC ORGANISMS: THE APPLICATION OF THE
EXPONENTIAL MODEL,

TOKYO UNIV. (JAPAN). DEPT. OF FISHERIES.

Y. HIYAMA, AND M. SHIMIZU.

ENVIRONMENTAL CONTAMINATION BY RADIOACTIVE MATERIALS, VIENNA, INTERNATIONAL
ATOMIC ENERGY (1969), P 463-476.

DESCRIPTORS:
*MARINE ANIMALS, *RADIOISOTOPES, ABSORPTION, MARINE ALGAE, MARINE FISH,
INVERTEBRATES, CRUSTACEANS, MOLLUSKS, STRONTIUM RADIOISOTOPES, ZINC
RADIOISOTOPES, IODINE RADIOISOTOPES, COBALT RADIOISOTOPES, WATER
POLLUTION EFFECTS.

IDENTIFIERS:
CESIUM RADIOISOTOPES, CADMIUM RADIOISOTOPES, CERIUM RADIOISOTOPES,
ECHINODERMS.

ABSTRACT:
A COMPARTMENTAL MODEL WAS USED TO CORRELATE EXPERIMENTAL DATA ON
RADIONUCLIDES (CS, SR, ZN, CD, CE, I, AND CO) UPTAKE BY MARINE ALGAE,
FISH AND INVERTEBRATES. FOR SOME SYSTEMS THE CONCENTRATION FACTORS
VARIED WITH TIME AS MUCH AS 100-FOLD. (BOPP-NSIC)

FIELD 05C

ACCESSION NO. W71-09850

INFLUENCE OF THE PHYSIOCOCHEMICAL FORM OF RUTHENIUM ON CONTAMINATION OF MARINE ORGANISMS, (IN FRENCH),

DEPARTEMENT DE LA PROTECTION SANITAIRE, CHERBOURG (FRANCE). SECTION DE RADIO-ECOLOGIE.

PIERRE GUEGUENIAT, PIERRE BOVARD, AND JACQUES ANCELLIN.

COMPTES RENDUS, SERIES D, VOL 268, FEB 10, 1969, P 976-979.

DESCRIPTORS:
*MARINE ANIMALS, *MARINE PLANTS, RADIOISOTOPES, PHYSICOCHEMICAL PROPERTIES, ANION ADSORPTION, COLLOIDS, AQUARIA, ABSORPTION, WATER POLLUTION EFFECTS, CATION EXCHANGE, CATION ADSORPTION, HYDROLYSIS, MARINE ALGAE, INVERTEBRATES, NUCLEAR WASTES, WATER POLLUTION SOURCES.

IDENTIFIERS:
*RUTHENIUM RADIOISOTOPES.

ABSTRACT:
THE DERIVATIVES OF RUTHENIUM NITROSYLS FROM EFFLUENTS OF FUEL REPROCESSING PLANTS WERE FOUND IN VARIOUS PHYSICOCHEMICAL FORMS IN SEA WATER - AS PRECIPITATES, IN SOLUTION, OR IN COLLOIDAL FORM. THE FORM IN WHICH THE RUTHENIUM WAS PRESENT DEPENDED UPON THE AMOUNTS OF OTHER SUSPENDED MATERIAL IN THE SEA WATER. EXPERIMENTS IN AQUARIA SHOWED THAT RUTHENIUM PRESENT IN PRECIPITATES CAUSED MORE CONTAMINATION OF MARINE ALGAE AND INVERTEBRATES AS COMPARED WITH SOLUBLE (CATIONIC) RUTHENIUM. (BOPP-NSIC)

FIELD 05B, 05C

ACCESSION NO. W71-09863

WATER QUALITY INFLUENCES ON OUTDOOR RECREATION IN THE LAKE ONTARIO BASIN,

BUREAU OF OUTDOOR RECREATION, ANN ARBOR, MICH.

ROBERT J. HENLEY.

IN: PROCEEDINGS, TENTH CONFERENCE ON GREAT LAKES RESEARCH, ANN ARBOR, BRAUN-BRUMFIELD, INC, 1967, P 427-440.

DESCRIPTORS:
*WATER QUALITY, *RECREATION, *SWIMMING, *WATER POLLUTION EFFECTS, BACTERIA, ALGAE, FISH, SEWERS, RUNOFF, WASTES, WATER SUPPLY, EFFLUENTS, STREAMS, LAKES, RIVERS, RESERVOIRS, BOATING, EUTHROPHICATION.

IDENTIFIERS:
*LAKE ONTARIO, WATER TURBIDITY, INDUSTRIAL PLANTS, WATER SPORTS.

ABSTRACT:
THE RELATIONSHIPS BETWEEN WATER QUALITY AND THE USES OF RECREATIONAL WATER RESOURCES IN THE LAKE ONTARIO BASIN ARE SURVEYED. THERE IS A DISCUSSION OF WATER QUALITY INDICES (HIGH BACTERIA COUNTS, ALGAE MASSES, FISH MORTALITY AND WATER TURPIDITY). RECREATION VALUE OF WATER IS ANALYZED IN LIGHT OF THE INFLUENCE OF POOR WATER QUALITY ON SWIMMING, WITH PARTICULAR ATTENTION TO SITUATIONS IN WHICH SWIMMING HAS BEEN BANNED BY PUBLIC OFFICIALS. THE VALUE OF POLLUTION CONTROL TO SUCH WATER RECREATION IS STRESSED. PREDICTIONS ARE MADE FOR THE FUTURE DEMAND OF THE PUBLIC FOR SWIMMING AND OTHER WATER-RELATED RECREATION ACTIVITIES AND INDICATIONS POINT TO AN INCREASE IN THIS DEMAND. (MURPHY-RUTGERS)

FIELD 05C

ACCESSION NO. W71-09880

DISSOLVED ORGANIC MATTER AND PHYTOPLANKTONIC PRODUCTIVITY IN MARL LAKES,

MICHIGAN STATE UNIV., HICKORY CORNERS. W. K. KELLOGG BIOLOGICAL STATION.

ROBERT G. WETZEL.

MITTEILUNGEN INTERNATIONALE VEREINIGUNG FUR THEORETISCHE UND ANGEWANDTE
LIMNOLOGIE, VOL 14, P 261-270, 1968. 1 FIG, 1 TAB, 40 REF. NSF GRANT
GB-1452.

DESCRIPTORS:
*PHYTOPLANKTON, *ALGAE, *PRIMARY PRODUCTIVITY, *ORGANIC MATTER,
NUTRIENTS, LAKES, WATER PROPERTIES, BACTERIA, BIOASSAY, METABOLISM,
ABSORPTION, COMPETITION, OLIGOTROPHY, VITAMINS, HEAVY METALS, TOXICITY,
DIATOMS, MARL.

IDENTIFIERS:
*MARL LAKES, DISSOLVED ORGANIC MATTER, ARCTIC ALGAE, AUTOTROPHIC
METABOLISM, HETEROTROPHY, CROOKED LAKE(IND), LITTLE CROOKED LAKE(IND).

ABSTRACT:
THE HIGH EFFICIENCY OF BACTERIA IN UTILIZATION OF DISSOLVED ORGANIC
MATTER DEPRIVES ALGAE OF ENERGY MATERIAL AND DEPRESSES THEIR GROWTH. AN
IMPORTANT PART IN THE COMPETITION OF THE TWO GROUPS OF ORGANISMS IS
PLAYED BY CHELATION INFLUENCING THE AVAILABILITY OF NUTRIENTS, THEIR
RATIO, AND POTENCY OF TOXICANTS. THESE RELATIONSHIPS WERE ELUCIDATED BY
BIOASSAYS CONDUCTED SIMULTANEOUSLY IN TWO CONNECTED HARD WATER LAKES
CONTAINING DIFFERENT CONCENTRATION OF DISSOLVED ORGANIC MATTER.
(WILDE-WISCONSIN)

FIELD 05C, 02H

ACCESSION NO. W71-10079

EFFECTS OF ZOOPLANKTON ON PHOTOSYNTHESIS BY ALGAE IN LAKES,

RICHMOND UNIV., VA. DEPT. OF BIOLOGY.

JOHN W. BISHOP.

VIRGINIA POLYTECHNIC INSTITUTE, BLACKSBURG, VIRGINIA, WATER RESOURCES
RESEARCH CENTER, BULLETIN NO 31, P 89-93, 1969. OWRR PROJECT A-002-VA(1).

DESCRIPTORS:
*ZOOPLANKTON, *PHOTOSYNTHESIS, *ALGAE, LAKES, OXYGEN, EUTROPHICATION,
ALGAL CONTROL, BIOCONTROL, TURBULENCE, COPEPODS, ROTIFERS, CHLOROPHYLL,
PIGMENTS, IMPOUNDMENTS, RESERVOIRS, ANALYTICAL TECHNIQUES, ON-SITE
TESTS.

IDENTIFIERS:
ASTERIONELLA SPIROIDES, CLADOCERANS, WESTHAMPTON LAKE(VA).

ABSTRACT:
POSSIBILITY OF BIOLOGICAL CONTROL OF ALGAE IN NUTRIENT ENRICHED WATERS
PROMPTED STUDY OF ZOOPLANKTON AS POTENTIAL REGULATORS OF ALGAE.
EXPERIMENTS WERE CONDUCTED IN AQUARIA WHERE ASSEMBLAGES OF ALGAE WERE
CONFINED (ONE IN PAIRED AQUARIA IN THE LAKE WITH AND ONE WITHOUT
MIGRATORY ZOOPLANKTON). EFFECTS OF THE AQUARIA ON THREE PHYSICAL
FACTORS OF LAKE WATER, LIGHT INTENSITY, TEMPERATURE, AND TURBULENCE
WERE COMPARED. DIFFERENCES BETWEEN QUANTITIES AND RATES OF ALGAL
PHOTOSYNTHESIS IN THE AQUARIA WERE USED AS ESTIMATES OF ZOOPLANKTON
EFFECT ON ALGAE. THE CHLOROPHYLL METHOD FOR MEASUREMENT OF ALGAL
CONCENTRATION CORRECTED FOR PHEO-PIGMENTS. RATE OF RADIOACTIVE CARBON
UPTAKE AS MEASURE OF PHOTOSYNTHESIS WAS STUDIED, USING A LIQUID
SCINTILLATION COUNTER AND BY THE OXYGEN EVOLUTION METHOD WHICH WAS
SUFFICIENTLY ACCURATE AND EASIER FOR ROUTINE EXPERIMENTS. PHEO-PIGMENTS
CONSISTED OF NEARLY 52% OF THE ALGAL PIGMENTS ANALYZED AND INCREASED
FROM THE SURFACE TOWARD LAKE BOTTOM; ROTIFERS WERE ABOUT 60% OF
ZOOPLANKTON DURING MID=SUMMER; CLADOCERANS WERE NOT MORE THAN &%.
EXPERIMENTS TO ASCERTAIN EFFECTS OF AQUARIA AND ZOOPLANKTON ON
PHOTOSYNTHESIS ARE TO BE CONTINUED.
(JONES-WISCONSIN)

FIELD 05C

ACCESSION NO. W71-10098

EFFECT OF AGRICULTURE ON WATER QUALITY,

FEDERAL WATER POLLUTION CONTROL ADMINISTRATION, EVANSVILLE, IND. LOWER OHIO
BASIN OFFICE.

T. R. SMITH.

IN: 2ND COMPENDIUM OF ANIMAL WASTE MANAGEMENT, JUNE 1969, PAPER NO. 11, 11 P,
9 REF.

DESCRIPTORS:
WATER QUALITY CONTROL, *NUTRIENTS, *PESTICIDES, FARM WASTES, SILT,
EROSION, SEDIMENTATION, RUNOFF, FERTILIZERS, NITROGEN, PHOSPHORUS,
ALGAE, LIVESTOCK, WATER POLLUTION EFFECTS.

IDENTIFIERS:
FEEDLOTS, WABASH RIVER BASIN.

ABSTRACT:
THE MAIN SOURCES OF AGRICULTURAL ASSOCIATED WATER POLLUTION IN HUMID
REGIONS ARE: (1) SILT FROM SOIL EROSION; (2) FERTILIZERS, MAINLY
PHOSPHORUS AND NITROGEN COMPOUNDS (3) PESTICIDES; AND (4) ORGANIC
WASTES FROM FEEDLOTS. IN ARID REGIONS, IRRIGATION RETURN FLOWS ARE A
PROBLEM. THIS PAPER DISCUSSES THE EFFECT OF EACH OF THESE SOURCES CAN
HAVE ON WATER QUALITY AND SUGGESTS SOME PREVENTATIVE MEASURES. IT IS
NECESSARY THAT AGRICULTURALISTS PLAN TO CONTROL POLLUTIONAL EFFECTS ON
THEIR ACTIVITY. (CHRISTENBURY-ISU)

FIELD 05G, 05C

ACCESSION NO. W71-10376

DEMONSTRATE FEASIBILITY OF THE USE OF ULTRASONIC FILTRATION IN TREATING THE
OVERFLOWS FROM COMBINED AND/OR STORM SEWERS.

ACOUSTICA ASSOCIATES, INC., LOS ANGELES, CALIF.

AVAILABLE FROM THE NATIONAL TECHNICAL INFORMATION SERVICE AS PB-201 745,
$3.00 IN PAPER COPY, $0.95 IN MICROFICHE. FINAL REPORT SEPTEMBER 22, 1969,
86 P, 10 FIG, 1 TAB, 3 APPEND. FWPCA PROGRAM 11020--09/67, CONTRACT NO.
14-12-23.

DESCRIPTORS:
*ULTRASONICS, *SUSPENDED LOAD, *FILTRATION, ALGAE, COLLOIDS,
FLOCCULATION, MIXING, ECONOMIC FEASIBILITY, TECHNICAL FEASIBILITY, COST
ANALYSIS, SLUDGE, INCINERATION, STORM RUN-OFF, SEDIMENTATION,
BIOCHEMICAL OXYGEN DEMAND, WASTE WATER TREATMENT, OVERFLOW, SEWERS,
STORM DRAINS, SEWAGE EFFLUENTS, *SEWAGE TREATMENT, STORM RUNOFF.

IDENTIFIERS:
BACKWASHING, *SUSPENDED SOLIDS, *COMBINED SEWERS, STORMWATER OVERFLOWS.

ABSTRACT:
THE FEASIBILITY, BOTH ECONOMIC AND TECHNICAL, OF USING ULTRASONIC
FILTRATION AS AN AID TO REMOVING SUSPENDED IMPURITIES WAS INVESTIGATED.
INPUTS TO THE TEST APPARATUS INCLUDED DRINKING WATER, SIMULATED SEWAGE,
PRIMARY AND SECONDARY SEWAGE EFFLUENTS, ALGAE POND EFFLUENT, COARSELY
SCREENED, RAW SEWAGE DILUTED 2:1, 5:1, AND 10:1 WITH SIMULATED STORM
DRAINAGE, AND DELIBERATELY CONCENTRATED RAW SEWAGE. DIRT, SAND, AND
LEAVES WERE ADDED IN SOME TEST RUNS WITH NO APPARENT EFFECT ON
PERFORMANCE. 20 AND 50 MICRON FILTER ELEMENTS, OPERATED AT
APPROXIMATELY 10 GPM/FT2 WITH HEAD LOSSES FROM 1 TO 4 PSI, REDUCED THE
BOD AND SUSPENDED SOLIDS OF RAW SEWAGE DILUTED IN VARYING DEGREES WITH
WATER BY AN AVERAGE OF 40 AND 70% RESPECTIVELY. SYSTEM PERFORMANCE WAS
INTERMEDIATE BETWEEN THAT OF PRIMARY AND SECONDARY TREATMENT. OTHER
BENEFITS INCLUDED: (1) APPLICATION OF ULTRASONIC ENERGY INCREASED THE
TOTAL QUANTITY OF EFFLUENT FILTERED BETWEEN BACKWASH CYCLES FROM 4 TO
18 TIMES THAT TREATED WITH SIMILAR EQUIPMENT WITHOUT ULTRASONICS
APPLIED; AND (2) APPLICATION OF ULTRASONIC ENERGY RESTORED THE FILTER
ELEMENTS TO 'LIKE-NEW' CONDITION WITHOUT NECESSITATING FREQUENT FILTER
REPLACEMENT. COST FIGURES FOR AN ULTRASONIC FILTRATION SYSTEM CAPABLE
OF HANDLING 1.4 MGD, EXHIBITING PERFORMANCE MIDWAY BETWEEN PRIMARY AND
SECONDARY TREATMENT, AND OCCUPYING ONLY SEVERAL HUNDRED SQUARE FEET,
RANGED BETWEEN 2.8 AND 3.5 CENTS/1000 GALLONS. BY COMPARISON, COMBINED
PRIMARY-SECONDARY TREATMENT COST FIGURES WERE REPORTED AS 10 CENTS/1000
GALLONS EXCLUDING THE COST OF THE CONSIDERABLE AMOUNT OF LAND INVOLVED.
(LOWRY-TEXAS)

FIELD 05D

ACCESSION NO. W71-10654

BIOLOGICAL PROBLEMS IN WATER POLLUTION,

ROBERT A TAFT SANITARY ENGINEERING CENTER, CINCINNATI, OHIO.

CLARENCE M. TARZWELL.

AVAILABLE FROM THE NATIONAL TECHNICAL INFORMATION SERVICE AS PB-196 627,
$3.00 IN PAPER COPY, $0.95 IN MICROFICHE. TRANSACTIONS, SEMINAR ON
BIOLOGICAL PROBLEMS IN WATER POLLUTION, TAFT ENGINEERING CENTER, APR 23-25,
1956. 272 P.

DESCRIPTORS:
*WATER POLLUTION, *AQUATIC BIOLOGY, *BIOINDICATORS, *BIOASSAY,
TOXICITY, AQUATIC MAMMALS, ALGAE, FISH, DIATOMS, FUNGI, INSECTS,
BACTERIA, PROTOZOA, SEWAGE, LAKES, NUTRIENTS.

IDENTIFIERS:
*INDICATOR SPECIES, MICROORGANISMS, FRESH WATER BIOLOGY, SUSPENDED
SEDIMENTS, MARINE BIOLOGY, STREAM POLLUTION, *WATER POLLUTION
DETECTION.

ABSTRACT:
BIOLOGISTS ENGAGED IN POLLUTION INVESTIGATIONS AND RESEARCH OFTEN WORK
ALONE AND ARE SOMEWHAT ISOLATED. FOR SOME TIME, THEREFORE, THERE HAD
BEEN RECOGNIZED A NEED FOR A CONFERENCE OF THOSE ENGAGED IN THE STUDY
OF BIOLOGICAL PROBLEMS IN WATER POLLUTION CONTROL, TO ACQUAINT THEM
WITH CURRENT DEVELOPMENTS AND NEW METHODS OF APPROACH, AND TO ENABLE
THEM TO BECOME ACQUAINTED WITH OTHER WORKERS IN THE FIELD. THE FIRST
SUCH GATHERING WAS HELD AS A SEMINAR AT THE ROBERT A. TAFT SANITARY
ENGINEERING CENTER, APRIL 23 - 27, 1956. THE SEMINAR CONSISTED OF PANEL
DISCUSSIONS AND WAS PLANNED SO THAT MOST OF THE TIME WAS DEVOTED TO
COMMENTARY FROM THE FLOOR WITH ONLY SHORT PRESENTATIONS BY PANEL
MEMBERS. SUBJECTS DISCUSSED WERE (1) USE AND VALUE OF BIOASSAYS; (2)
USE AND VALUE OF BIOLOGICAL INDICATORS OF POLLUTION; (3) CURRENT
INVESTIGATIONS IN WATER POLLUTION BIOLOGY; (4) WATER QUALITY CRITERIA
FOR AQUATIC LIFE; AND (5) TRAINING OF SANITARY AQUATIC BIOLOGISTS.

FIELD 05C

ACCESSION NO. W71-10786

THE DEPENDENCE OF PRIMARY PRODUCTION ON THE COMPOSITION OF PHYTOPLANKTON
(ZAVISIMOST PERVICHNOI PRODUKTSH OT SOSTAVA FITOPLANKTONA),

NAVAL OCEANOGRAPHIC OFFICE WASHINGTON, D.C.

I. L. PYRINA.

AVAILABLE FROM THE NATIONAL TECHNICAL INFORMATION SERVICE AS AD-718 176,
$3.00 IN PAPER COPY, $0.95 IN MICROFICHE. NAVAL OCEANOGRAPHIC TRANSLATION
201, 1963. TRANSLATION: PERVICHNAYA PRODUKTSIYA MORAY I VNUTVENNIKH, P
308-313, 1961.

DESCRIPTORS:
*PLANKTON, *PHOTOSYNTHESIS, PHYTOPLANKTON, ALGAE, TEMPERATURE, SOLAR
RADIATION.

IDENTIFIERS:
MARINE BIOLOGY, *PRIMARY BIOLOGICAL PRODUCTIVITY.

ABSTRACT:
THE DEPENDENCE OF PRIMARY PRODUCTION ON THE COMPOSITION OF
PHYTOPLANKTON IS BASED ON SPECIAL OBSERVATIONS CARRIED OUT IN THE
SUMMER OF 1958 IN THE IVAN'KOVSKOYE, RYBINSKOYE AND KUYBYSHEVSKOYA
WATER BASINS, AS WELL AS IN CERTAIN SECTORS OF THE VOLGA RIVER. IN
ADDITION TO THE PRIMARY PRODUCTION, THE SPECIES' COMPOSITION AND THE
BIOMASS OF PHYTOPLANKTON WAS DETERMINED, THE ENERGY OF SOLAR LIGHT
STRIKING THE WATER SURFACE, THE TEMPERATURE AND TRANSPARENCY OF WATER.
THE INSTRUMENTS AND METHODS USED ARE BRIEFLY DESCRIBED, ILLUSTRATING
GRAPHICALLY THE VARIATIONS OF PHOTOSYNTHESIS IN VARIOUS ALGAL GROUPS
AND THE UTILIZATION OF LIGHT ENERGY BY VARIOUS PHYTOPLANKTON SPECIES IN
THE ABOVE-MENTIONED WATER BASINS.

FIELD 05C, 02I

ACCESSION NO. W71-10788

PHOTOSYNTHETIC RECLAMATION OF AGRICULTURAL SOLID AND LIQUID WASTES--SECOND
    PROGRESS REPORT,

CALIFORNIA UNIV., BERKELEY. SANITARY ENGINEERING RESEARCH LAB.

GORDON L. DUGAN, CLARENCE G. GOLUEKE, WILLIAM J. OSWALD, AND CHARLES E.
    RIXFORD.

CALIFORNIA UNIVERSITY, SANITARY ENGINEERING RESEARCH LABORATORY, REPORT NO.
    70-1, 165 P, 24 FIG, 55 TAB, 51 REF. US PUBLIC HEALTH SERVICE 5R01 UI
    00566-03.

    DESCRIPTORS:
      *FARM WASTES, *POULTRY, WASTE WATER TREATMENT, SEDIMENTATION, OXIDATION
      LAGOONS, ANAEROBIC DIGESTION, ALGAE, SLUDGE, METHANE, ORGANIC LOADING,
      HYDROGEN ION CONCENTRATION, TEMPERATURE, CENTRIFUGATION, COAGULATION,
      DEWATERING, *COST ANALYSIS.

    IDENTIFIERS:
      VOLATILE SOLIDS, GRIT.

    ABSTRACT:
    A 36-WEEK STUDY WAS INITIATED TO PROVIDE INFORMATION ON THE ECONOMICS
    OF TREATING ANIMAL WASTES. CHICKENS WERE CHOSEN, MAINLY FOR CONVENIENCE
    SAKE, AS THE WASTE PRODUCERS TO BE STUDIED, AND 113-TWENTY WEEK OLD
    LEGHORN PULLETS WERE PLACED IN CAGES. THE TREATMENT SYSTEM USED
    CONSISTED OF INCLINED TROUGHS UNDER THE CAGES WHICH WERE FLUSHED WITH
    WATER FROM A FLUSHING BUCKET, GRIT REMOVAL, SEDIMENTATION, AN OXIDATION
    LAGOON FOR THE SEDIMENTATION TANK SUPERNATANT, AND AN ANAEROBIC
    DIGESTER FOR THE SOLIDS. THE SYSTEMS APPROACH WAS USED, AND SYSTEM
    BALANCES WERE PERFORMED FOR TOTAL SOLIDS, VOLATILE SOLIDS, TOTAL
    UNOXIDIZED NITROGEN AND ENERGY FOR THE CHICKENS, SEDIMENTATION TANK,
    DIGESTER, AND ALGAE. ALL BALANCES WERE PERFORMED FROM WEEK 5 THROUGH
    WEEK 36 EXCEPT FOR THE DIGESTER, WHERE OPERATION WAS TERMINATED AT WEEK
    24 AND THE SOLIDS DEWATERED, DRIED, AND STUDIED FOR POSSIBLE FURTHER
    REUSE. AN ECONOMIC ANALYSIS OF AN INTEGRATED SYSTEM OF 100,000 LAYING
    HENS REVEALED A COST OF APPROXIMATELY 2 CENTS/DOZEN EGGS FOR A SYSTEM
    BASED ON THE ONE TESTED. HOWEVER, CONSIDERATION OF AN EXTREMELY
    CONSERVATIVE ALGAE HARVESTING RATE OF 12 TONS/ACRE/YEAR AT A PRICE OF 5
    CENTS/LB (DRY WEIGHT) DROPPED THE OVERALL WASTE HANDLING OUTLAY TO
    APPROXIMATELY 1 CENT/DOZEN. ON THIS BASIS, ADDITIONAL STUDIES WERE
    RECOMMENDED TO AID IN THE IMPLEMENTATION OF TREATMENT FACILITIES FOR
    ANIMAL WASTES AS SOON AS POSSIBLE. (LOWRY-TEXAS)

    FIELD 05D

    ACCESSION NO. W71-11375

SULFIDE PRODUCTION IN WASTE STABILIZATION PONDS,

TEXAS UNIV., AUSTIN. CENTER FOR RESEARCH IN WATER RESOURCES.

ERNESTO ESPINO DE LA O, AND E. F. GLOYNA.

TECHNICAL REPORT NO. EHE-04-6802, CRWR-26, MAY 1967, 156 P, 50 FIG, 33 TAB, 75 REF. FWPCA CONTRACT WP-00688-03.

DESCRIPTORS:
*OXIDATION PONDS, *HYDROGEN SULFIDE, *SULFIDES, ALGAE, ORGANIC LOADING, BIOCHEMICAL OXYGEN DEMAND, OXIDATION-REDUCTION POTENTIAL, DISSOLVED OXYGEN, ODORS, CORROSION, OXIDATION, PHOTOSYNTHESIS, ANALYTICAL TECHNIQUES, SPECTROSCOPY, COLORIMETRY, SAMPLING, METHANE, CARBON DIOXIDE, DIFFUSION, *WASTE WATER TREATMENT.

ABSTRACT:
EXPERIMENTS WERE CONDUCTED ON A 6 FT DEEP MODEL FACULTATIVE POND TO DETERMINE THE RELATIONSHIP BETWEEN SULFIDE CONCENTRATION IN THE POND AND FOUR OPERATING PARAMETERS: (1) DETENTION TIME, (2) CONCENTRATION OF SULFATE ION IN THE INFLUENT, (3) BOD SURFACE LOAD, AND (4) SULFATE ION CONCENTRATION IN TERMS OF SURFACE LOAD. ALL EXPERIMENTS WERE CONDUCTED UNDER CONTROLLED ENVIRONMENTAL CONDITIONS, ENCOMPASSING 7 DIFFERENT OPERATIONAL CONDITIONS. AT CONSTANT DETENTION TIME AND CONSTANT BOD SURFACE LOADING, A LINEAR RELATIONSHIP EXISTED BETWEEN SULFIDE CONCENTRATION IN THE POND AND SULFATE CONCENTRATION IN THE INFLUENT. INCREASES IN POND SULFIDE CONCENTRATION FROM NEAR ZERO TO 1 AND 2 MG/L, RESPECTIVELY, WERE ACCOMPANIED BY DROPS IN THE OXIDATION-REDUCTION POTENTIAL OF 80 AND 120 MICRON. FURTHER INCREASES IN SULFIDE CONCENTRATION AFFECTED THE ORP TO A MUCH LESSER EXTENT. THIS PHENOMENON ESTABLISHED ORP MEASUREMENT AS AN INDICATOR OF POTENTIAL ODOR TROUBLES AND/OR POND SYSTEM UPSET. ALGAE DISAPPEARED ALMOST ENTIRELY WHEN SULFIDE CONCENTRATIONS IN THE POND ROSE ABOVE 7.0 MG/L BUT BOD REMOVAL RATES WERE NOT DETECTABLY DIMINSIHED. INFRARED SPECTROSCOPIC ANALYSIS OF GASEOUS EMISSIONS FROM THE POND IDENTIFIED THREE OF THE COMPONENT GASES AS METHATE, HYDROGEN SULFIDE, AND CARBON DIOXIDE. INTERFERENCE FROM MOISTURE IN THE AIR PROHIBITED FURTHER IDENTIFICATION. (LOWRY-TEXAS)

FIELD 05D

ACCESSION NO. W71-11377

RADIOACTIVITY TRANSPORT IN WATER--EFFECTS OF ORGANIC POLLUTION ON RADIONUCLIDE TRANSPORT,

TEXAS UNIV., AUSTIN. CENTER FOR RESEARCH IN WATER RESOURCES.

KAZUHIRO FUTAGAWA, AND ERNEST F. GLOYNA.

TECHNICAL REPORT NO. 16, EHE-04-6803, CRWR-27, ORO-16, MAY 1968, 61 P, 14 FIG, 4 TAB, 31 REF. U.S. ATOMIC ENERGY COMMISSION CONTRACT AT-(11-1)-490.

DESCRIPTORS:
*RADIONUCLIDE, *DISSOLVED OXYGEN, *RIVERS, OXIDATION, PHOTOSYNTHESIS, RESPIRATION, SCOUR, SATURATION, MIXING, DIFFUSION, SEDIMENTATION, ABSORPTION, BIOCHEMICAL OXYGEN DEMAND, AERATION, TEMPERATURE, FLOW, ALGAE, BENTHOS, AEROBIC CONDITIONS, ANAEROBIC CONDITIONS, WATER POLLUTION SOURCES, PATH OF POLLUTANT.

ABSTRACT:
RADIONUCLIDE TRANSPORT IN WATER CAN BE AFFECTED BY THE PRESENCE OF ORGANIC POLLUTION IN TWO WAYS. ONE OF THESE WAYS IS RELATED TO THE LACK OF DISSOLVED OXYGEN CAUSED BY BIOLOGICAL DECOMPOSITION OF ORGANIC POLLUTANTS. THE OTHER MECHANISM IS DETERMINED BY THE EFFECTS OF THE INTERACTIONS OF THE RADIONUCLIDES WITH THE ORGANIC MATERIAL ITSELF. AS A PRELIMINARY STEP TOWARDS A MORE COMPLETE EXPLANATION OF THE DISSOLVED OXYGEN DEFICIT EFFECT, THE ECOLOGICAL ENVIRONMENT OF A RIVER SYSTEM WAS SIMULATED IN A SPECIALLY ADAPTED RESEARCH FLUME. DISTRIBUTION OF DISSOLVED OXYGEN AND RE-AERATION RATES WERE THEN MATHEMATICALLY MODELED USING DATA TAKEN FROM THE MODEL RIVER. USING A WATER DEPTH OF 1.0 FT, REAERATION RATE CONSTANTS WERE DETERMINED AT A FLOW OF 1.7 IN/MIN AND 28C. THE VALUES FOR THE RE-AERATION RATE CONSTANTS WERE 0.18 AND 0.23 RESPECTIVELY. WITH A 200 MG/L CONCENTRATION OF DEXTRINE IN THE INFLUENT, THE BIOLOGICAL OXIDATION RATE CONSTANT IN THE MODEL RIVER WAS DETERMINED TO BE 0.006, WHICH WAS QUITE LOW IN COMPARISON WITH VALUES DETERMINED ON FULL SCALE RIVERS. IT WAS ALSO DETERMINED THE VARIABILITY OF THE ENVIRONMENTAL FACTORS WHICH DETERMINE PHOTOSYNTHETIC OXYGEN PRODUCTION PRECLUDED THE USE OF ALGAL PRODUCED OXYGEN AS A RELIABLE OXYGEN SOURCE. HOWEVER, SINCE THE VARIOUS PHOTOSYNTHETIC-RESPIRATIONSHIPS CAN BE MEASURED IN A MODEL WITH SUITABLE ACCURACY, THE FUTURE, MORE INTENSIVE, RADIONUCLIDE TRANSPORT STUDIES CAN PROCEED. (LOWRY-TEXAS)

FIELD 05B

ACCESSION NO. W71-11381

FACTORS AFFECTING THE USE OF PURIFIED SEWAGE EFFLUENTS FOR COOLING PURPOSES,

JOHANNESBURG MUNICIPALITY (SOUTH AFRICA).

D. W. OSBORN.

WATER POLLUTION CONTROL, VOL 69, NO. 4, P 456-464, 5 FIG, 1 TAB, 1 REF.

DESCRIPTORS:
*COOLING WATER, *ELECTRIC POWER PLANTS, *WATER REUSE, COOLING TOWERS,
DAMS, ALGAE, HYDROGEN ION CONCENTRATION, TEMPERATURE, ALKALINITY,
NITRIFICATION, *CORROSION, CARBON DIOXIDE, SOLUBILITY, *PHOSPHATES,
CHEMICAL PRECIPITATION, FILTRATION, AMMONIA, WATER PURIFICATION, WASTE
WATER TREATMENT.

IDENTIFIERS:
*NITRIC ACID, *LANGELIER INDEX.

ABSTRACT:
ONE OF THE MORE IMPORTANT PROBLEMS INVOLVED IN USING PURIFIED SEWAGE
EFFLUENT AS COOLING WATER FOR POWER STATIONS HAS BEEN THE OXIDATION OF
AMMONIA TO NITRIC ACID. SERIOUS CORROSION DAMAGE TO CONCRETE AND METAL
FIXTURES HAS BEEN THE RESULT IN SEVERAL INSTANCES. IN ADDITION TO
CORROSION, PHOSPHATE PRECIPITATION FROM LOW ALKALINITY, LOW PH WATERS
HAS ALSO CAUSED MANY PROBLEMS IN LINES AND PUMPS. THEREFORE,
CONSTRUCTION OF THE NEW ORLANDO POWER STATION INCLUDED A LARGE DAM. THE
65 ACRE AREA, 180 X 10 6 GAL. VOLUME FORMED BY THE DAM SERVED AS A
MATURATION POND IN WHICH THE AMMONIA PRESENT IN THE WATER IS SLOWLY
DISTILLED OFF AND NOT CONVERTED TO NITRIC ACID. THE POND IS MAINTAINED
AT 24C THE YEAR ROUND, SO THAT AMMONIA AND BOD REMOVALS CAN CONTINUE
DURING BOTH SUMMER AND WINTER CONDITIONS. DURING THIS DETENTION PERIOD,
THE ALGAE ALSO RAISE THE PH LEVELS OF THE WATER TO THE POINT WHERE
PHOSPHATE IS PRECIPITATED OUT BEFORE IT CAN FOUL THE PIPES AND VALVES
OF THE COOLING TOWERS. THE OPERATIONAL EXPERIENCE NEEDED TO CORRECT
THESE PROBLEMS WAS GAINED OVER A 30 YEAR PERIOD. SUCCESSFUL UTILIZATION
OF WASTE WATER EFFLUENTS TO AUGMENT SCARCE WATER SUPPLIES HAS LONG BEEN
A WAY OF LIFE IN JOHANNESBURG, AND THE EXPERIENCE GAINED THERE HAS
CONTRIBUTED MUCH TO THE NEW TECHNOLOGY OF WATER RE-USE. (LOWRY-TEXAS)

FIELD 05D

ACCESSION NO.   W71-11393

THE TOXICITY OF COPPER COMPLEXES TOWARDS SCENEDESMUS QUADRICAUDA BREB.,

MOSCOW STATE UNIV. (USSR). DEPT. OF HYDROBIOLOGY.

V. G. KHOBOT'EV, V. I. KAPKOV, AND E. G. RUKHADZE.

MICROBIOLOGY (USSR), VOL 38, NO 5, P 729-731, SEPTEMBER-OCTOBER 1969. 2 TAB,
8 REF.

DESCRIPTORS:
*COPPER, *TOXICITY, *ALGAE, *WATER POLLUTION EFFECTS, CHELATION,
SPECTROPHOTOMETRY, NUTRIENTS, HEAVY METALS.

IDENTIFIERS:
PROTOCOCCUS, SCENEDESMUS QUADRICAUDA.

ABSTRACT:
THE TOXIC EFFECTS OF COPPER ON PROTOCOCCAL ALGAE WERE STUDIED USING TWO
TYPES OF COPPER COMPOUNDS: CHELATES, THE MAJORITY OF WHICH ARE SOLUBLE
IN WATER (COPPER BIS-O-HYDROXYQUINOLINE, COPPER BIS-O-VANILLAL, AND
COPPER BIS-O-VANILLALMETHYLIMINE), AND A COMPLEX OF COPPER WITH
PYRIDINE (COPPER DICHLORIDE PYRIDINATE). CONCENTRATIONS OF 0.64-32 MG
COPPER/LITER OF MEDIUM FREE FROM K SUB 2 CO SUB 3 AND KH SUB 2 PO SUB 4
WERE USED. THE RESULTS SHOWED THE COPPER COMPLEXES TO HAVE A TOXIC
ACTION ON THE ALGAE. THE GREATEST TOXIC ACTION WAS SHOWN BY COPPER
DICHLORIDE PYRIDINATE, WHICH AT CONCENTRATIONS OF 0.4 MG COPPER/LITER
REDUCED THE NUMBER OF ALGA CELLS BY FOUR TIMES AFTER THE SEVENTH DAY AS
COMPARED WITH THE CONTROL. COPPER CHELATES ALSO SHOWED TOXIC ACTION BUT
AT SOMEWHAT LARGER CONCENTRATION. THE LEAST TOXIC WAS COPPER
BIS-O-HYDROXYQUINOLINE, WHICH AT ALL CONCENTRATIONS TESTED ONLY
INHIBITED GROWTH. (MORTLAND-BATTELLE)

FIELD 05C

ACCESSION NO.   W71-11561

A COMBINATION OF RING-OVEN AND CIRCULAR CHROMATOGRAPHY OF TRACE METAL ANALYSIS,

DALHOUSIE UNIV., HALIFAX (NOVA SCOTIA). DEPT. OF CHEMISTRY.

R. W. FREI, AND C. A. STOCKTON.

MIKROCHIMICA ACTA, P 1196-1203, 1969. 1 FIG, 2 TAB, 20 REF.

DESCRIPTORS:
*CHROMATOGRAPHY, *COBALT, *COPPER, *IRON, TRACE ELEMENTS, ALGAE, SEA WATER, FRESH WATER.

IDENTIFIERS:
*NICKEL, DETECTION LIMITS, CHEMICAL INTERFERENCES.

ABSTRACT:
A CHROMATOGRAPHIC METHOD FOR METAL CHELATES OF PYRIDINE-2-ALDEHYDE-2QUINOLYLHYDRAZONE(PAQH) HAS BEEN ADAPTED TO THE RING-OVEN TECHNIQUE. ADVANTAGE HAS BEEN TAKEN OF THE LARGE DIFFERENCE IN RF VALUES OF THE COBALT, COPPER, NICKEL AND IRON(III) CHELATES ON ALUMINA, THE ACID STABILITY OF THE COBALT PAQH COMPLEX, SELECTIVE FORMATION OF CHLORIDE COMPLEXES OF THESE METALS AND THE USE OF SPECIFIC SPRAY REAGENTS FOR COPPER AND IRON, SUCH AS RUBEANIC ACID AND KSCN TO DEVELOP A SIMPLE AND RAPID FIELD METHOD FOR TRACE METAL ANALYSIS. QUANTITATIVE ESTIMATION OF THESE FOUR METALS WITH AN ACCURACY RANGING FROM 5.0 TO 12 PERCENT HAS BEEN CARRIED OUT IN LESS THAN 30 MINUTES. DETECTION LIMITS WERE FOUND TO BE ABOUT 0.01 MICROGRAM FOR NICKEL, 0.008 MICROGRAM FOR COPPER AND IRON AND 0.005 MICROGRAM FOR COBALT. OTHER IONS DID NOT INTERFERE IN UP TO 10-FOLD EXCESS. THE SUITABILITY OF THIS METHOD FOR FIELD WORK WAS TESTED BY MONITORING TRACE METALS IN FRESH AND SEA WATER SUPPLIES AND IN ALGAE. (LITTLE-BATTELLE)

FIELD 05A

ACCESSION NO. W71-11687

BIOLOGICAL HANDBOOK FOR ENGINEERS.

NATIONAL AERONAUTICS AND SPACE ADMINISTRATION, HUNTSVILLE, ALA. GEORGE C. MARSHALL SPACE FLIGHT CENTER.

NASA CONTRACTOR REPORT 61237, JUNE 1968. 225 P, 64 FIG, 74 TAB, 146 REF.

DESCRIPTORS:
*MICROBIOLOGY, *CLASSIFICATION, *CULTURES, *BIOASSAYS, PROTOZOA, BACTERIA, FUNGI, VIRUSES, ALGAE, LICHENS, ACTIMOMYCETES, CYTOLOGICAL STUDIES, FILTRATION, ULTRASONICS, DISINFECTION, SAMPLING, CENTRIFUGATION, SEDIMENTATION, CHEMICAL PRECIPITATION, LIFE CYCLES, GENETICS, E. COLI, ISOLATION.

IDENTIFIERS:
RICKETTSIAE, STERILIZATION, BACILLUS SUBTILIS, BACILLUS STEAROTHERMOPHILUS, PROTEUS VULGARIS, PSEUDOMONAS AERUGINOSA, STAPHYLOCOCCUS AUREUS, LACTOBACILLUS CASEI, BACILLUS THURINGIENSIS, AGARS, STAPHYLOCOCCUS EPIDERMIDIS, MICROCOCCUS, STREPTOCOCCUS, BREVIBACTERIUM, CORYNEBACTERIUM, ELECTROSTATIC PRECIPITATION, THERMAL PRECIPITATION.

ABSTRACT:
USEFUL MICROBIOLOGICAL BACKGROUND INFORMATION HAS BEEN COMPILED IN A HANDBOOK FOR ENGINEERS AND SCIENTISTS WORKING ON BIO-RELATED PROJECTS. IT IS INTENDED AS AN AID IN: (1) EVALUATING THE EFFECTS OF ENGINEERING PROCEDURES ON MICROBIAL LIFE; (2) DETERMINING THE EFFECTS OF DECONTAMINATION OF STERILIZATION PROCEDURES ON THE PERFORMANCE OF AN OVERALL SYSTEM; AND (3) UNDERSTANDING THE LANGUAGE OF MICROBIOLOGISTS. THE INFORMATION PRESENTED COVERS SUCH SUBJECTS AS: (1) THE VARIOUS CATEGORIES OF MICROORGANISMS, THEIR PHYSICAL CHARACTERISTICS, THEIR RELATIVE ABUNDANCES IN VARIOUS LOCATIONS, THEIR LIFE CYCLES, AND THEIR GENETICS; (2) THE METHODS AND APPLICATIONS OF DECONTAMINATION AND STERILIZATION BY WET OR DRY HEAT, CHEMICAL TREATMENT, RADIATION EXPOSURE, AND PRESSURE TREATMENT; (3) THE RELATIVE SUCCESS OF VARIOUS METHODS OF CONTAMINATION CONTROL, SUCH AS FILTRATION, CLEAN ROOMS, LAMINAR AIR FLOWS, AND SHIELDING; AND (4) THE THEORY AND TECHNIQUES OF BIOASSAY--THE DETECTING, IDENTIFYING, AND COUNTING OF MICROBES IN SPECIFIC LOCATIONS. THE FINAL SECTION COMPRISES A GLOSSARY OF COMMON MICROBIOLOGICAL TERMS. THE MANY REFERENCES CITED THROUGHOUT THE WORK GUIDE THE INTERESTED READER TO ADDITIONAL INFORMATION. (MORTLAND-BATTELLE)

FIELD 05A

ACCESSION NO. W71-11823

THERMAL EFFECTS AND NUCLEAR POWER STATIONS IN THE U.S.A.,

ARGONNE NATIONAL LAB., ILL. CENTER FOR ENVIRONMENTAL STUDIES; AND BATTELLE
MEMORIAL INST., RICHLAND, WASH. PACIFIC NORTHWEST LABS.

D. MILLER, J. V. TOKAR, AND R. E. NAKATANI.

AVAILABLE FROM THE NATIONAL TECHNICAL INFORMATION SERVICE AS CONF-700810-30,
$3.00 IN PAPER COPY, $0.95 IN MICROFICHE. REPORT NO IAEA-SM-146/30 (NO
DATE). 14 P, 1 TAB, 21 REF.

DESCRIPTORS:
*THERMAL POLLUTION, *NUCLEAR POWERPLANTS, *WATER QUALITY CONTROL,
*BIODEGRADATION, WATER UTILIZATION, FISH, ALGAE, MODEL STUDIES,
BIOASSAY, AQUATIC LIFE, COOLING WATER, SITES.

ABSTRACT:
OF 94 NUCLEAR POWER PLANTS IN THE U.S., 60.6 PERCENT ARE COOLED BY
ONCE-THROUGH SYSTEMS IN WHICH THE COOLING WATER IS RETURNED DIRECTLY TO
A NATURAL WATER SOURCE. THESE THERMAL DISCHARGES UNDOUBTEDLY MODIFY THE
BIOLOGICAL COMMUNITIES THAT ARE EXPOSED. ALTHOUGH GENERAL EFFECTS ARE
KNOWN, THERE ARE INSUFFICIENT DATA AVAILABLE AT PRESENT TO PERMIT
EFFECTIVE EVALUATION OF PLANT SITES AND DISCHARGE DETAILS ON THE BIOTIC
COMMUNITIES. IT IS THEREFORE NECESSARY THAT STUDIES BE UNDERTAKEN
RELATIVE TO POWER PLANT SITING AND OPERATION TO ENSURE COMPLIANCE WITH
WATER QUALITY STANDARDS, TO DEFINE MIXING ZONES, TO PREDICT TEMPERATURE
DISTRIBUTIONS IN RECEIVING WATER, TO ASSESS BIOLOGICAL PERTURBATIONS,
TO DESIGN INTAKE AND OUTFALL STRUCTURES TO MINIMIZE BIOLOGICAL DAMAGE,
TO CONTROL NUISANCE GROWTHS OF PLANTS AND ALGAE, AND TO DETERMINE
SUBLETHAL EFFECTS OF TEMPERATURE ON AQUATIC LIFE. (LITTLE-BATTELLE)

FIELD 05C, 08A

ACCESSION NO. W71-11881

A SURVEY OF THE BOTTOM FAUNA IN LAKES OF THE OKANAGAN VALLEY, BRITISH COLUMBIA,

FISHERIES RESEARCH BOARD OF CANADA, WINNIPEG (MANITOBA). FRESHWATER INST.

OLE A. SEATHER.

FISHERIES RESEARCH BOARD OF CANADA (MS) TECHNICAL REPORT NO 196 (UNDATED). 31
P, 3 FIG, 7 TAB, 15 REF.

DESCRIPTORS:
*LAKES, *BENTHIC FAUNA, *TROPHIC LEVELS, *SURVEYS, BIOLOGICAL
COMMUNITIES, BIOINDICATORS, LAKE MORPHOLOGY, ECOLOGICAL DISTRIBUTION,
MIDGES, WATER POLLUTION EFFECTS, OLIGOCHAETES, DIPTERA, ALGAE, INSECTS.

IDENTIFIERS:
TROPHIC COMMUNITIES, OKANAGAN LAKE(CANADA), SKAHA LAKE(CANADA), OSOYOOS
LAKE(CANADA), DEFORMED CHIRONOMIDS.

ABSTRACT:
BOTTOM FAUNA WERE STUDIED TO DETERMINE POLLUTION EFFECTS IN THREE LARGE
LAKES IN OKANAGAN BASIN, BRITISH COLUMBIA. A COMPARISON WITH EARLIER
DATA INDICATES LAKE OKANAGAN IS NOT ONLY POLLUTED IN CERTAIN AREAS, BUT
BOTTOM FAUNA OF WHOLE LAKE HAS CHANGED. THERE HAS BEEN SIGNIFICANT
INCREASE IN TOTAL NUMBERS OF BOTTOM ORGANISMS FOR STATIONS SHALLOWER
THAN 50 METERS AND SIGNIFICANT INCREASE IN TOTAL NUMBERS OF
CHIRONOMIDS, OLIGOCHAETES, PISIDIUM AND PERHAPS OTHERS. MOST STRIKING
CHANGE IS IN PERCENTAGE DISTRIBUTION BETWEEN GROUPS. OLIGOCHAETES HAVE
INCREASED FROM 15% OF MACROBENTHOS TO 50% TO 60% OF FAUNA, SUGGESTIVE,
TOGETHER WITH DEFORMED CHIRONOMIDS OCCURRENCE, OF INSECTICIDE
POLLUTION. THE LAKE AS A WHOLE IS STILL OLIGOTROPHIC ALTHOUGH NOT
ULTRAOLIGOTROPHIC. THE FEW SKAHA LAKE STATIONS WERE INSUFFICIENT FOR
DETAILED CHARACTERIZATION; IT HAS PROBABLY BEEN MODERATELY EUTROPHIC A
LONG TIME AND A DRAMATIC CHANGE IN BOTTOM FAUNA MAY HAVE OCCURRED QUITE
RECENTLY. OSOYOOS LAKE'S NORTHERN AND CENTRAL BASINS ARE, ACCORDING TO
COMPOSITION OF BOTTOM FAUNA, MODERATELY AND STRONGLY EUTROPHIC,
RESPECTIVELY. IN OKANAGAN AND SKAHA LAKES CHIRONOMIDS WITH DEFORMED
MOUTH-PARTS AND THICKENED BODY WALLS WERE FOUND. DEFORMED CHIRONOMIDS
MAY BE USEFUL IN DETECTING SOME FORMS OF POLLUTION. (JONES-WISCONSIN)

FIELD 05C

ACCESSION NO. W71-12077

EFFECTS OF TOXIC ORGANICS ON PHOTOSYNTHETIC REOXYGENATION,

TEXAS UNIV., AUSTIN. CENTER FOR RESEARCH IN WATER RESOURCES.

JU-CHANG HUANG, AND EARNEST F. GLOYNA.

TECHNICAL REPORT NO. EHE-07-6701, CRWR-20, AUGUST 1967, 163 P, 64 FIG, 24 TAB, 78 REF. EPA GRANT WP-00688-03.

DESCRIPTORS:
*REAERATION, *CHLOROPHYLL, *OXYGENATION, *TOXICITY, ORGANIC COMPOUNDS, *PHENOLS, PESTICIDES, OXIDATION LAGOONS, ALGAE, PHOTOSYNTHESIS, DISSOLVED OXYGEN, LABORATORY TESTS, TEMPERATURE, WASTE WATER TREATMENT.

ABSTRACT:
THE RELATIVE TOXICITY OF 33 PHENOLIC-TYPE COMPOUNDS AND 8 PESTICIDES WAS INVESTIGATED TO DETERMINE THE EXTENT TO WHICH CHLOROPHYLL INHIBITION BY CERTAIN INDUSTRIAL WASTES DECREASES STREAM DISSOLVED OXYGEN CONTENTS. PHENOLIC-TYPE GROUPS INCLUDED THOSE WITH MONO-AND POLY-SUBSTITUTED GROUPS SUCH AS -BR, -CL, -CH3, -NO2, -NH2, AND -OH. PESTICIDES STUDIED INCLUDED DDT, LINDANE, 2,4-D, SODIUM SALT OF 2, 4-D, 2, 45-T, DOW SODIUM TCA, DOW ESTERON 99, AND DOW FORMULA 40. ORGANIC COMPOUND TOXICITY WAS EVALUATED IN TWO WAYS. FIRST, TEST TUBE ALGAL CULTURES WERE GROWN AND CHEMOSTATED AT 25C UNDER CONTINUOUS ILLUMINATION FOR 72 HOURS. KNOP'S SOLUTION, INCLUDING THE HUTNER-EDTA MICRO ELEMENT SYSTEM, WAS USED AS CULTURE MEDIUM, AND A 5% CARBON DIOXIDE IN AIR GAS WAS BLOWN THROUGH TO SUPPLY THE INORGANIC CARBON SOURCE AND THE MIXING. SECONDLY, WARBURG MANOMETRIC TECHNIQUES WERE USED TO MEASURE GAS EXCHANGE OCCURING WHEN CARBONATE-BICARBONATE BUFFER IS USED AS A SUSPENDING FLUID. RESULTS SHOWED THE DECREASE IN CHLOROPHYLL CONTENT TO BE GENERALLY LOGARITHMIC TO TOXIC ORGANIC CONCENTRATION, WITH SUBSEQUENT SUPPRESSED OXYGEN PRODUCTION. FULL SCALE TESTS SUBSTANTIATED THESE CONCLUSIONS, THEREFORE A TOXICITY COMPENSATION FACTOR MUST BE INTRODUCED INTO DESIGN EQUATIONS FOR WASTE STABILIZATION PONDS RECEIVING TOXIC ORGANICS. (LOWRY-TEXAS)

FIELD 05C, 05D

ACCESSION NO. W71-12183

METHODS FOR DETERMINING THE PRIMARY PRODUCTIVITY OF EPIPELIC AND EPIPSAMMIC ALGAL ASSOCIATIONS,

BRISTOL UNIV. (ENGLAND). DEPT. OF BOTANY.

M. HICKMAN.

LIMNOLOGY AND OCEANOGRAPHY, VOL 14, NO 6, P 936-941, 1969. 7 TAB, 13 REF.

DESCRIPTORS:
PRIMARY PRODUCTIVITY, *BIOLOGICAL COMMUNITIES, *ALGAE, *METHODOLOGY, CARBON RADIOISOTOPES, DIATOMS, MEASUREMENT, SEDIMENTS, SAMPLING, DEPTH, SEA WATER, FRES WATER, LITTORAL, DISSOLVED OXYGEN, DIURNAL, MIGRATION PATTERNS.

IDENTIFIERS:
*EPIPSAMMIC ALGAE, *EPIPELIC ALGAE, ALGAL ASSOCIATIONS, SHEAR WATER(ENGLAND), AMPHORA OVALIS, OPEPHORA MARTYI, ABBOT'S POND(SOMERSET, ENGLAND), SAND GRAINS, TIDAL CYCLES.

ABSTRACT:
QUANTITATIVE TECHNIQUES OF DEFINED ACCURACY WERE DEVELOPED FOR ESTIMATING PRIMARY PRODUCTIVITY OF BOTH EPIPELIC AND EPIPSAMMIC ALGAL ASSOCIATIONS OF BOTH FRESH AND MARINE SHALLOW AND DEEP-WATER SITES. FOR THE EPIPSAMMIC ALGAL ASSOCIATION SAMPLES WERE TAKEN FROM A RICH DIATOMAL FLORA ATTACHED TO SAND GRAIN SURFACES IN THE LITTORAL ZONE OF A SMALL LAKE. THE ALGAE WERE SEPARATED FROM THE EPIPELIC ALGAE AND DETRITAL MATERIAL AND A C-14 SOURCE ADDED. THE ALGAE WERE REMOVED FROM THE SAND GRAINS BEFORE ASSAY FOR C-14 BY SONIFICATION. THE EPIPELIC ALGAL ASSOCIATION WAS STUDIED BY TRAPPING THE ALGAE WITH A DOUBLE LAYER OF LENS TISSUE AND C-14 METHOD FOR ASSAY. HARVESTING TIME IS CRITICAL, SINCE EPIPELIC POPULATIONS SHOW MARKED DIURNAL MIGRATION RHYTHMS. AFTER INCUBATION, ALGAE WERE REMOVED FROM LENS TISSUE. SINCE 87.5% OF THE EPIPELIC POPULATION WAS TRAPPED IN THE TISSUE AND 85.8% OF THE TRAPPED POPULATION REMOVED FROM IT, PRODUCTIVITY ESTIMATES WERE ATTRIBUTABLE TO 73.3% OF TOTAL POPULATION. THE PRIMARY PRODUCTIVITY OF BOTH EPIPSAMMIC AND EPIPELIC ALGAE CAN ALSO BE STUDIED WITH OXYGEN TECHNIQUES. THE ROUTINE METHODS FOR BOTH ASSOCIATIONS, THE EPIPELIC AND EPIPSAMMIC, WERE STATISTICALLY DEFINED. (JONES-WISCONSIN)

FIELD 05C, 05A

ACCESSION NO. W71-12870

REMOVAL OF PHOSPHOROUS FROM A SEWAGE LAGOON BY HARVESTING ALGA,

MISSISSIPPI STATE UNIV., STATE COLLEGE, DEPT. OF CIVIL ENGINEERING.

J. E. LITTLE.

MASTER'S THESIS, AUGUST 1969. 41 P, 11 FIG, 2 TAB, 30 REF.

DESCRIPTORS:
*EUTROPHICATION, *PHOSPHOROUS, ALGAE, *HARVESTING OF ALGAE, EFFLUENT,
QUALITY CONTROL, OXIDATION LAGOON, NITROGEN, NUTRIENTS, FILTRATION,
MICROORGANISMS, ANALYTICAL TECHNIQUES, *WASTE WATER TREATMENT,
TREATMENT FACILITIES, MISSISSIPPI.

ABSTRACT:
EUTROPHICATION, DEFINED AS THE PROCESS IN WHICH SURFACE BODIES OF WATER
ARE ENRICHED BY THE EXCESSIVE ADDITION OF NUTRIENTS, AFFECTS AS MUCH AS
56% OF TOTAL MUNICIPAL SURFACE WATER SUPPLIES IN THE UNITED STATES.
EXCESSIVE ALGAL GROWTH IN EUTROPHIED WATER CAUSED DIFFICULTIES IN
TREATING SEWAGE AND REDUCED THE RECREATION VALUE OF LAKES AND STREAMS.
BECAUSE EUTROPHICATION IS A PRODUCT OF NUTRIENT ENRICHMENT, IT WAS
DECIDED THAT CONTROL MEASURES SHOULD BE BASED ON EITHER REMOVAL,
REDUCTION, OR PREVENTION OF THE ADDITION OF NUTRIENTS. MAKING USE OF
THE SEWAGE LAGOON RESEARCH FACILITY AT MISSISSIPPI STATE UNIVERSITY, AN
ATTEMPT WAS MADE TO SEE WHETHER THE PHOSPHOROUS CONTENT OF AN OXIDATION
POND EFFLUENT COULD BE BROUGHT TO ACCEPTABLE LEVELS THROUGH HARVESTING
THE ALGAE CONTAINED IN THE POND EFFLUENT. A SIX MONTH STUDY REVEALED
THAT: (1) HARVESTING ALGAE REDUCED THE PHOSPHATE CONCENTRATION TO 91%
MAXIMUM REMOVAL, (2) HARVESTING ALGAE WILL NOT REMOVE ENOUGH
PHOSPHOROUS FROM OXIDATION POND EFFLUENT TO CONTROL ALGAL GROWTHS IN
RECEIVING STREAMS AND LAKES UNLESS A LARGE QUANTITY OF ALMOST
PHOSPHOROUS-FREE WATER IS AVAILABLE FOR DILUTION. (3) A FILTER WITH A
PORE SIZE OF 45 MICRON OR LESS APPEARS TO REMOVE ALL INSOLUBLE
PHOSPHATE FROM OXIDATION POND EFFLUENT, BUT ADEQUATE FLOW IS DIFFICULT
TO ATTAIN, (4) IT DOES NOT APPEAR THAT THE METHOD STUDIED IS PRACTICAL
BRCAUSE A SIGNIFICANT DEGREE OF PHOSPHOROUS REMOVAL IS ACHIEVED ONLY
WHEN BOD REMOVAL EFFICIENCY IS LOW, THEREBY DEFEATING THE ORIGINAL
PURPOSE OF OXIDATION PONDS. (ADKINS-TEXAS)

FIELD 05D, 05C

ACCESSION NO. W71-13313

THE MECHANISM OF PHOSPHOROUS REMOVAL IN ACTIVATED SLUDGE,

MAINE UNIV., ORONO. DEPT. OF CIVIL ENGINEERING.

RALPH E. OULTON.

MASTER'S THESIS, UNIVERSITY OF MAINE, JANUARY 1969. 87 PAGES, 31 FIG, 20 REF.

DESCRIPTORS:
*EUTROPHICATION, *ALGAE, NUTRIENTS, NITROGEN, PHOSPHOROUS, NITROGEN
FIXING BACTERIA, *ACTIVATED SLUDGE, STORAGE, TEMPERATURE,
MICROORGANISMS, NITROGEN COMPOUNDS, *PHOSPHORUS COMPOUNDS, ADSORPTION,
REMOVAL, DEGRADATION, CHEMICAL OXYGEN DEMAND, *WASTE WATER TREATMENT.

IDENTIFIERS:
MIXED LIQUOR, SUSPENDED SOLIDS, *PHOSPHOROUS UPTAKE RATE.

ABSTRACT:
COMPOUNDS OF NITROGEN AND PHOSPHOROUS ARE THE MAJOR NUTRIENTS IN THE
EUTROPHICATION PROCESS. IN ORDER TO CONTROL EUTROPHICATION, THE SUPPLY
OF AT LEAST ONE OF THESE NUTRIENTS MUST BE CONTROLLED. NITROGEN-FIXING
BACTERIA OFFER A VIRTUALLY UNLIMITED SUPPLY OF NITROGEN TO ALGAE;
THEREFORE, PHOSPHORUS IS THE NUTRIENT WHICH CAN BEST BE LIMITED. TESTS
WERE MADE TO DETERMINE WHETHER THE UPTAKE OF SOLUBLE PHOSPHOROUS BY
ACTIVATED SLUDGE MIXED LIQUOR SUSPENDED SOLIDS IS RELATED TO THE
VARIATION IN STORED ENERGY PRODUCTS WITHIN THE CELLS COMPRISING THE
MLSS. RESULTS OF THE ANALYTICAL TESTS PERFORMED REVEALED SEVERAL FACTS,
(1) AN ACCIDENTAL INCREASE IN TEMPERATURE FROM 25 DEG C TO 55 DEG C
CAUSED AN INCREASE IN THE SOLUBLE PHOSPHOROUS UPTAKE RATE FROM 40 MG/L
TO 80 MG/L, (2) REMOVAL OF SOLUBLE PHOSPHORUS FROM SOLUTION IS RELATED
TO THE CONTINUED METABOLISM OF THE ACTIVATED SLUDGE CELLS EVEN AFTER
ALL AVAILABLE EXTERNAL ENERGY SOURCES ARE EXHAUSTED, (3) THE FIRST
TOTAL SOLUBLE PHOSPHORUS CONCENTRATION HAD NO EFFECT ON THE PHOSPHORUS
UPTAKE RATE IN THE SYSTEM STUDIED, (4) THE PHOSPHORUS REMOVAL AND USE
OF STORAGE PRODUCTS AFTER THE EXHAUSTION OF EXTERNAL COD WAS
INDEPENDENT OF THE FIRST MLSS CONCENTRATION AND THE TEMPERATURE
VARIATIONS STUDIED. (ATKINS-TEXAS)

FIELD 05D, 05C

ACCESSION NO. W71-13335

THE RATE AND EXTENT OF ALGAL DECOMPOSITION IN ANAEROBIC WATERS,

KENTUCKY UNIV., LEXINGTON.

EDWARD G. FOREE, AND PERRY L. MCCARTY.

PROCEEDINGS, INDUSTRIAL WASTE CONFERENCE, 24TH, MAY 6-8, 1969, P 13-36, 13
FIG, 8 TAB, 9 REF.

DESCRIPTORS:
*ALGAE, *LABORATORY TESTS, *ANAEROBIC CONDITIONS, LIGHT INTENSITY,
EUTROPHICATION, NUTRIENTS, *FERMENTATION, TOXICITY, CHEMICAL OXYGEN
DEMAND, HYDROGEN ION CONCENTRATION, TEMPERATURE, PROTEINS,
CARBOHYDRATES, LIPIDS, KINETICS, MATHEMATICAL MODELS, NEUTRALIZATION,
*DEGRADATION(DECOMPOSITION), METHANE, WATER POLLUTION EFFECTS.

IDENTIFIERS:
*METHANE FERMENTATION.

ABSTRACT:
ALGAE CULTURES DECOMPOSING UNDER DARK, ANAEROBIC CONSTANT TEMPERATURE
CONDITIONS WERE STUDIED IN THE LABORATORY TO EVALUATE THE EFFECTS OF
CELL COMPOSITION, PH, TEMPERATURE, AND TO DETECT ANY CHANGES IN METHANE
FERMENTATION AND SULGATE REDUCTION. IT WAS DETERMINED THAT
STABILIZATION OF ALGAE OCCURRED BY SULFATE REDUCTION AS LONG AS SULFATE
REMAINED IN SOLUTION. ONCE THE SULFATE SUPPLY WAS EXHAUSTED,
STABILIZATION CONTINUED THROUGH METHANE FERMENTATION. DARK, ANAEROBIC
DECAY OF ALGAE WAS DETERMINED AS: M= (MO - FMO) E-KT+ F MO, WHERE F=
REFRACTORY ORGANIC PORTION HAVING AN AVERAGE VALUE OF 0.40 + 0.15 FROM
15 TO 25C, ALGAL DECOMPOSITION RATE IS TEMPERATURE DEPENDENT: K/KO=ECK
(T-TO), WITH AN EXPECTED CK OF 0.055/C. PROTEIN, CARBOHYDRATE AND LIPID
CONTENTS OF ALGAE CAN BE RELIABLY CALCULATED AS A FUNCTION OF NITROGEN
CONTENT AND COD TO VOLATILE SOLIDS RATIO OF THE CELLS. (LOWRY-TEXAS)

FIELD 05C

ACCESSION NO. W71-13413

EXPERIENCE IN THE TREATMENT AND RE-USE OF INDUSTRIAL WASTE WATERS,

JOHNS HOPKINS UNIV., BALTIMORE, MD. DEPT. OF GEOGRAPHY AND ENVIRONMENTAL
ENGINEERING.

C. E. RENN.

PROCEEDINGS, INDUSTRIAL WASTE CONFERENCE, 24TH, MAY 6, 7, AND 8, 1969. P
962-968.

DESCRIPTORS:
*WATER REUSE, *WASTE WATER TREATMENT, *INDUSTRIAL WASTES, OXIDATION
PONDS, ALGAE, BENTHIC FAUNA, WATERSHEDS, WATER SHORTAGE FILTERS,
SNAILS, ODOR, COOLING TOWERS, FILTERS, EFFLUENT SAMPLING.

IDENTIFIERS:
HYDROGEN SULFIDE, IMHOFF TANKS, PRECIPITATES.

ABSTRACT:
THE WORLDS LARGEST ELECTRIC MACHINE TOOL PLANT REUSES IT'S WATER ON A
REGULAR BASIS. INITIALLY, THE PLANT HAD AN ADEQUATE WATER SUPPLY FROM 3
TO 5 WELLS PUMPING AT A RATE OF 100 TO 150 GPM. PLANT EXPANSION AND NEW
PROCESSES FORCED THE DEVELOPMENT OF A WATER REUSE SCHEME. WASTE WATER
FROM THE PLANT IS TREATED IN 2 IMHOFF TANKS AND A HIGH RATE TRICKLING
FILTER AND THEN STORED IN A LARGE POND. WATER IS DRAWN FROM THE POND,
FILTERED AND FED BACK TO THOSE PARTS OF THE PLANT THAT CAN USE IT OR BE
ADAPTED TO USE IT. THERE WERE NUMEROUS PROBLEMS ASSOCIATED WITH
PERFECTION THIS SYSTEM, THE POND GOES ANAEROBIC IN THE SUMMER, HYDROGEN
SULFIDE PRODUCED IN THE POND ESCAPED INTO THE PLANT, SNAILS COLLECTED
IN DEAD ENDS, ALGAL FILAMENTS CLOGGED HEAT EXCHANGERS, COOLING WATER
TUNNELS WERE STOPPED UP AND ODORS WERE EMITTED FROM THE EVAPORATIVE
COOLERS. APPROACHES TO THE SOLUTION OF THESE PROBLEMS, SUCCESSFUL AND
UNSUCCESSFUL, ARE PRESENTED AND THE RESULTS DISCUSSED. A UNIQUE PROFILE
SAMPLER WAS DEVELOPED AND USED TO DETERMINE THE WATER QUALITY. THE
CHANGES THAT OCCUR IN THE POND ARE DISCUSSED AND AN INDICATION OF THE
PROCESSES BRINGING ABOUT THESE CHANGES ARE PRESENTED. PREDICTABLY, THE
POND IS GRADUALLY FILLING WITH PRECIPITATES. IN THE PROCESS, THE
COMPANY HAS MADE A LOT OF USE OF A LITTLE WATER AND IS DISCHARGING A
BETTER EFFLUENT TO THE RECEIVING STREAM. (GOESSLING-TEXAS)

FIELD 05D

ACCESSION NO. W72-00027

WHAT IS POLLUTION,

M. I. GOLDMAN, AND R. SHOOP.

IN: CONTROLLING POLLUTION: THE ECONOMICS OF A CLEANER, AMERICA, P 59-70,
  PRENTICE HALL INC., ENGLEWOOD CLIFFS, NEW JERSEY, 1967. 12 P.

DESCRIPTORS:
  *WATER POLLUTION SOURCES, *WATER POLLUTION EFFECTS, *CLASSIFICATION,
  *WATER POLLUTION CONTROL, WATERCOURSES(LEGAL), ORGANIC WASTES, PULP
  WASTES, INDUSTRIAL WASTES, TOXINS, SOLID WASTES, RADIOACTIVE WASTES,
  HEATED WATER, THERMAL POLLUTION, BACTERIA, EFFLUENTS, OXYGEN DEMAND,
  AEROBIC TREATMENT, PHOTOSYNTHESIS, ALGAE, SEWAGE, MUNICIPAL WASTES,
  SEWAGE TREATMENT, TREATMENT FACILITIES, LEGAL ASPECTS.

ABSTRACT:
  A BRIEF SURVEY IS MADE OF THE VARIOUS KINDS OF AIR AND WATER POLLUTION,
  THE MEANS OF MEASURING THESE POLLUTANTS, AND THE METHODS EMPLOYED TO
  CONTROL POLLUTION. POLLUTION IS A RELATIVE CONCEPT, AS POLLUTION OCCURS
  ONLY WHEN IMPURITIES RISE ABOVE A CERTAIN SPECIFIED LEVEL. WATER
  POLLUTION CAN BE CLASSIFIED IN FIVE ASPECTS: (1) PUTRESCIBLE MATERIALS
  ARE ORDINARY DOMESTIC AND INDUSTRIAL ORGANIC WASTES; (2) HEATED
  EFFLUENTS THAT CAUSE THERMAL POLLUTION; (3) TOXINS WHICH ARE NOT EASILY
  BROKEN DOWN BY BIOLOGICAL MEANS AND WHICH ARE NOT READILY NEUTRALIZED
  BY WATERCOURSES; (4) INERT WASTES THAT ENTER AS SOLIDS INTO WATER BUT
  ARE NOT INVOLVED IN CHEMICAL REACTIONS; AND (5) RADIOACTIVE SUBSTANCES.
  THE ARTICLE DESCRIBES THE PROCESS OF OXYGEN DEPLETION IN WATERS AND THE
  NATURAL AND ARTIFICIAL MEANS TO RESTORE AN OXYGEN BALANCE. PHYSICAL AND
  BIOLOGICAL MEANS TO MEASURE WATER POLLUTION ARE EXAMINED, ALONG WITH
  THE METHODS EMPLOYED TO TREAT THESE VARIOUS POLLUTANTS. THE PROCESS AND
  EFFECTIVENESS OF PRIMARY, SECONDARY, AND TERTIARY SEWAGE TREATMENT ARE
  DISCUSSED. (REES-FLORIDA)

FIELD 05G, 05B

ACCESSION NO. W72-00462

AQUATIC PLANT COMMUNITIES IN LAKE BUTTE DES MORTS: PHASE 2. EFFECTS OF HIGHER
  AQUATIC PLANTS, MARSH WATER AND MARSH SEDIMENTS ON PHYTOPLANKTON,

WISCONSIN STATE UNIV., OSHKOSH. DEPT. OF BIOLOGY.

WILLIAM E. SLOEY.

WISCONSIN DEPT OF NATURAL RESOURCES, MADISON, PROJECT COMPLETION REPORT,
  SUMMER 1969. 20 P, 3 FIG, 6 TAB, 4 REF.

DESCRIPTORS:
  *AQUATIC PLANTS, *INHIBITORS, *MARSHES, *LAKES, *PHYTOPLANKTON,
  BIOASSAY, LABORATORY TESTS, PRODUCTIVITY, ABSORPTION, SEDIMENTS,
  ORGANIC MATTER, ALGAE, DIATOMS, CYANOPHYTA, CHLOROPHYTA, BULRUSH, PLANT
  GROWTH SUBSTANCES, CHLORELLA, CHLAMYDOMONAS, SCENEDESMUS, WISCONSIN.

IDENTIFIERS:
  *SCIRPUS FLUVIATILIS, *LAKE BUTTE DES MORTS(WISC), GROWTH STIMULATION,
  SURFACE LITTER.

ABSTRACT:
  TO VERIFY DIFFERENCES BETWEEN MARSH AND OPEN LAKE WATER IN LAKE BUTTE
  DES MORTS (WISCONSIN) AND TO DETERMINE IF SOME SUBSTANCE OR SUBSTANCES
  ARE PRODUCED IN MARSHES WHICH INHIBIT OR LIMIT GROWTH OF PLANKTONIC
  ALGAE, THE EFFECTS OF MARSH WATER, BOTTOM MUD, AND AQUATIC PLANT
  EXTRACTS UPON THE PHOTOSYNTHETIC RATE OF NATURAL PHYTOPLANKTON BY USE
  OF THE C-14 PRIMARY PRODUCTION TECHNIQUE WERE STUDIED. PARTIALLY
  DECAYED LITTER AND SURFACE ORGANIC PARTICLES INHIBITED PHYTOPLANKTON
  C-14 PRODUCTIVITY WHILE SUB-SURFACE LITTER AND DECAYED ORGANIC MATTER
  STIMULATED PHYTOPLANKTON. MILLIPORE FILTERED MARSH WATER SOMETIMES
  STIMULATED AND SOMETIMES INHIBITED C-14 UPTAKE. A FEW OF THE 16 NATIVE
  MARSH PLANTS TESTED INHIBITED C-14 UPTAKE WHEN MACERATED EXTRACTS OF
  THE PLANTS WERE EMPLOYED; THE MOST PRONOUNCED INHIBITOR WAS SCIRPUS
  FLUVIATILIS. THE PRESENT STANDS OF S FLUVIATILIS IN LAKE BUTTE DES
  MORTS SHOULD BE PROTECTED AND ENCOURAGED TO PROLIFERATE. SOME OTHER
  PLANT EXTRACTS GREATLY STIMULATED C-14 UPTAKE OF PHYTOPLANKTON, PERHAPS
  DUE TO THEIR PRODUCTION OF STIMULATORY SUBSTANCES OR TO EXTRACTION OF
  MINERALS AND/OR PLANT GROWTH SUBSTANCES FROM THE TISSUES DURING
  MACERATION. (JONES-WISCONSIN)

FIELD 05C

ACCESSION NO. W72-00845

MASS CULTURING OF ALGAE IN CONTROLLED LIGHT-DARK DETENTION SYSTEMS: I. EFFECTS
OF CULTURE DENSITY,

NORWICH UNIV., NORTHFIELD, VT. DEPT. OF CIVIL ENGINEERING.

G. R. PYPER, W. J. WEBER, JR., AND J. A. BORCHARDT.

PROCEEDINGS, INDUSTRIAL WASTE CONFERENCE, 21ST, MAY 3, 4, AND 5, 1966, P
1003-1020, 8 FIG, 2 TAB, 20 REF. USPRS RES. GRANT WP-00155.

DESCRIPTORS:
*ALGAE, NITROGEN, PHOSPHORUS, *EUTROPHICATION, PHOTOSYNTHESIS, LAGOONS,
LABORATORY TESTS, LIGHT INTENSITY, PHOTOPERIODISM, LIGHT PENETRATION,
WATER POLLUTION EFFECTS, WATER POLLUTION CONTROL, *NUTRIENTS.

IDENTIFIERS:
*NUTRIENT REMOVAL, CONTINUOUS FLOW SYSTEM.

ABSTRACT:
THE POLLUTION-ACCELERATED AGING OF SURFACE WATERS HAS BECOME A MATTER
OF CONCERN OVER THE PAST SEVERAL YEARS. CONTINUING INCREASES IN THE
VOLUMES OF WASTEWATER INDICATE THAT MORE THOROUGH PURIFICATION OF THESE
DISCHARGES MAY BE THE ONLY FEASIBLE MEASURE FOR THE PROTECTION OF OUR
NATURAL WATERS. AQUATIC PLANTS CAN BE UTILIZED IN LAGOONS TO EXTRACT
PHOSPHORUS AND NITROGEN FROM WASTEWATERS PRIOR TO DISCHARGE TO THE
RECEIVING WATER. LABORATORY STUDIES HAVE BEEN MADE ON THE ALGAL GROWTH
RATES AND YIELDS AS A FUNCTION OF DARK TO LIGHT RATIOS AND AS A
FUNCTION OF ALGAL CONCENTRATION. THE GENERAL CONCLUSIONS ARE: (1) MASS
CULTURES OF ALGAE CAN BE GROWN IN A CONTINUOUS SYSTEM CONSISTING OF A
DARK RESERVOIR FROM WHICH THE CULTURE CAN BE CIRCULATED THROUGH A LIGHT
CELL; (2) THERE IS A MAXIMUM LIGHT UTILIZATION RATE WHICH GIVES A
MAXIMUM YIELD WITH RESPECT TO DARK TO LIGHT DETENTION TIMES; (3) GROWTH
RATE VARIES LINEARLY WITH DARK TO LIGHT RATIOS OVER A BROAD RANGE OF
VALUES; (4) NUTRIENT REMOVAL CHARACTERISTICS ARE NOT ADVERSELY AFFECTED
BY LIGHT AND DARK CULTURING SYSTEMS; (5) NUTRIENT REMOVAL APPEARS TO BE
A FUNCTION OF CULTURE DENSITY; (6) CURRENT WORK IS ONLY A BEGINNING AND
MUST BE EXTENDED; AND (7) A MAJOR FACTOR TO BE STUDIED IS THE EFFECT OF
FLOW VELOCITY THROUGH THE LIGHT CELL. (GOESSLING-TEXAS)

FIELD 05C, 05G

ACCESSION NO. W72-00917

ENVIRONMENTAL MONITORING ASSOCIATED WITH DISCHARGES OF RADIOACTIVE WASTE DURING 1969 FROM UKAEA ESTABLISHMENTS,

UNITED KINGDOM ATOMIC ENERGY AUTHORITY, HERWELL (ENGLAND). AUTHORITY HEALTH AND SAFETY BRANCH.

E. T. WRAY.

AVAILABLE FROM THE NATIONAL TECHNICAL INFORMATION SERVICE AS AHSB(RP)R-105. PRICE - $3.00, MICROFICHE $0.95. ATOMIC HEALTH AND SAFETY BRANCH REPORT NO. AHSB(RP)R-105.

DESCRIPTORS:
*MONITORING, *DISCHARGE MEASUREMENT, *EFFLUENTS, *RADIOISOTOPES, *WASTE DISPOSAL, *WASTE STORAGE, *WATER POLLUTION, WATER POLLUTION SOURCES, WATER POLLUTION CONTROL, RADIOACTIVE WASTE, WATER QUALITY ACT, PATH OF POLLUTANT, FOOD CHAINS, SAMPLING, SURVEYS, MARINE ALGAE, AQUATIC LIFE, MILK, ATLANTIC OCEAN, LANDFILLS, DISPOSAL, REGULATION, WATER POLICY, GOVERNMENTS, GASES, NUCLEAR WASTES.

IDENTIFIERS:
CONCENTRATION, ENVIRONMENTAL MONITORING.

ABSTRACT:
THIS REPORT SUMMARISES THE RESULTS OF THE MONITORING PROGRAMS CARRIED OUT IN CONNECTION WITH RADIOACTIVE WASTE DISCHARGES FROM THE PRINCIPAL ESTABLISHMENTS OF THE UKAEA IN 1969. THE INFORMATION GIVEN INCLUDES THE ESSENTIAL FREATURES OF THE AUTHORIZATIONS FOR DISCHARGES WHICH WERE MADE THE RESULTS OF ENVIRONMENTAL MONITORING. DERIVED WORKING LIMITS ARE QUOTED TO PROVIDE A BASIS FOR COMPARISON OF THE ENVIRONMENTAL MONITORING RESULTS. MOST ESTABLISHMENTS CARRIED OUT SAMPLING AND ANALYSIS ADDITIONAL TO THE MINIMUM REQUIRED BY STATUTE, AND SOME REFERENCE TO THIS HAS BEEN MADE IN THE REPORT. ONE OF THE PRINCIPAL METHODS OF ESTIMATING THE EFFECT OF GASEOUS DISCHARGES TO THE ENVIRONMENT IS BY THE SAMPLING AND ANALYSIS OF MILK. THE ACTIVITY FOUND IN THE VICINITY OF AUTHORITY ESTABLISHMENTS IS ALMOST ALWAYS DUE TO FALLOUT AND NOT TO CONTAMINATION OF THE ENVIRONMENT BY THE ESTABLISHMENT ITSELF. METHODS OF TREATMENT OF RADIOACTIVE WASTE IN THE UKAEA HAVE BEEN UNDER INVESTIGATION WITH THE OBJECT OF IMPROVING THE SYSTEM OF HANDLING AND DISCHARGE OF THE WASTE. THE APPENDIX TO THE REPORT CONTAINS A SUMMARY OF THE GENERAL PRINCIPLES GOVERNING THE NATURE OF THE POSSIBLE HAZARDS WHICH MIGHT ARISE AS A CONSEQUENCE OF THESE DISCHARGES. (HOUSER-NSIC)

FIELD 05B

ACCESSION NO. W72-00941

EVALUATION OF EFFECT OF IMPOUNDMENT ON WATER QUALITY IN CHENEY RESERVOIR,

COLORADO STATE UNIV., FORT COLLINS.

J. C. WARD, AND S. KARAKI.

RESEARCH REPORT NO 25, SEPT 1969. 67 P, 38 FIG, 19 TAB, 23 REF.

DESCRIPTORS:
EFFECTS, *IMPOUNDED WATERS, *WATER QUALITY, *DISSOLVED SOLIDS, DISSOLVED OXYGEN, EVAPORATION CONTROL, WATER BUDGET, WATER CHEMISTRY, WATER TEMPERATURE, HEAT BUDGET, TURBIDITY, *RESERVOIR EVAPORATION, SALINITY, ALGAE, STRATIFICATION, SALT BALANCE, BIBLIOGRAPHIES.

IDENTIFIERS:
USBR RESEARCH CONTRACTS, CHENEY DAM, KANS.

ABSTRACT:
A STUDY WAS CONDUCTED TO DETERMINE THE EFFECT OF IMPOUNDMENT ON THE QUALITY OF WATER IN CHENEY RESERVOIR NEAR WICHITA, KANS. THE RESULTS SHOWED THAT AN INCREASE IN THE DISSOLVED SOLIDS CONCENTRATION WAS DIRECTLY RELATED TO EVAPORATION. ON AN ANNUAL BASIS, 42% OF THE TOTAL INFLOW WAS EVAPORATED FROM THE RESERVOIR. EVAPORATION CONTROL WAS SUGGESTED FOR CONTROLLING DISSOLVED SOLIDS CONCENTRATION; HOWEVER, THIS WOULD CAUSE A POSSIBLY UNDESIRABLE INCREASE IN RESERVOIR TEMPERATURE OF 12 TO 19 DEG F. THE BIOLOGICAL ACTIVITY WITHIN THE RESERVOIR DID NOT SEEM TO AFFECT THE WATER QUALITY MATERIALLY, ALTHOUGH SOME ODORS WERE DETECTED. THE ODOR APPEARS TO HAVE STABILIZED AT A THRESHOLD NUMBER OF ABOUT 5. THE EFFECT OF THE INTERACTION BETWEEN MICRO-ORGANISMS AND NUTRIENTS WAS CHARACTERIZED IN THE ANALYSIS OF PHOSPHATES, NITRATES, AND SILICA CONCENTRATIONS IN THE RESERVOIR. THE DISSOLVED OXYGEN SATURATION DECREASED FROM 100% AT THE WATER SURFACE TO 82% AT 25 FT DEPTH. NO RESERVOIR STRATIFICATION WAS DETECTED DURING THE PERIOD OF DATA COLLECTION. (USBA)

FIELD 05F

ACCESSION NO. W72-01773

BIOLOGY OF WATER POLLUTION: A COLLECTION OF SELECTED PAPERS ON STREAM
POLLUTION, WASTE WATER, AND WATER TREATMENT.

FEDERAL WATER POLLUTION CONTROL ADMINISTRATION, WASHINGTON, D.C.

FEDERAL WATER POLLUTION CONTROL ADMINISTRATION PUBLICATION CWA-3, 1967. 290
P, 107 FIG, 58 TAB, 2 PHOTO, 530 REF.

DESCRIPTORS:
*WATER POLLUTION, WASTE DISPOSAL, *SEWAGE TREATMENT, *WASTE WATER
TREATMENT, *WASTE WATER DISPOSAL, *INDUSTRIAL WASTES, *WATER POLLUTION
CONTROL, FISH CONSERVATION, POLLUTION ABATEMENT, BIBLIOGRAPHIES,
ECOLOGY, BIOLOGY, FISH, PLANKTON, WATER QUALITY, ALGAE, AQUATIC LIFE,
STREAM POLLUTION, BACTERIA, IMPURITIES, WASTE TREATMENT.

ABSTRACT:
THIS BOOK OF SELECTED PUBLICATIONS ON BIOLOGY OF WATER POLLUTION, WATER
TREATMENT AND SEWAGE AND INDUSTRIAL WASTE TREATMENT CONTAINS SOME OF
THE MANY EXCELLENT AND BASIC PERTINENT BIOLOGICAL PAPERS THAT HAVE BEEN
COMMONLY INACCESSIBLE TO THE CONTEMPORARY INVESTIGATOR. THESE PAPERS
OFTEN QUOTED AND ARE A PORTION OF THE FOUNDATION UPON WHICH MODERN
AQUATIC ECOLOGICAL SCIENTIFIC THOUGHT AND DECISIONS ARE OFTEN BASED IN
SUMMATING WATER POLLUTION CONTROL INVESTIGATIONS. THIS COMPILED
COLLECTION WILL BE OF ASSISTANCE IN 3 PHASES OF WATER POLLUTION
ABATEMENT: (1) IT WILL PROVIDE A TECHNICAL SERVICE TO THE AQUATIC
ECOLOGIST THROUGH ASSEMBLAGE OF INFORMATIVE LITERATURE; (2) IT WILL
ILLUSTRATE MANY OF THE CONCEPTS UPON WHICH REGULATIONS HAVE BEEN
FORMULATED FOR THE PROTECTION OF AQUATIC LIFE; AND (3) IT WILL AID IN
THE TRAINING OF NEW ENVIRONMENTAL SCIENTISTS TO MEET TODAY'S AND
TOMORROW'S PERSONNEL NEEDS IN THE CONSERVATION OF OUR NATION'S NATURAL
RESOURCES. (SEE ALSO W72-01788 THRU W72-01819)

FIELD 05C, 05G

ACCESSION NO. W72-01786

SEWAGE, ALGAE AND FISH,

PUBLIC HEALTH SERVICE, CINCINNATI, OHIO.

FLOYD J. BRINLEY.

IN: BIOLOGY OF WATER POLLUTION, P 10-13. 1 FIG, 6 REF. COMPILED BY L. E.
KEUP, W. M. INGRAM, AND K. M. MACKENTHUN, FEDERAL WATER POLLUTION CONTROL
ADMINISTRATION, WASHINGTON, D C, 1967.

DESCRIPTORS:
*SEWAGE, *STREAMS, *WATER POLLUTION EFFECTS, *FISH, ALGAE, TROPHIC
LEVEL, PRODUCTIVITY, FOOD CHAINS, TOXICITY, WATER POLLUTION CONTROL,
DECOMPOSING ORGANIC MATTER, CARBON DIOXIDE.

IDENTIFIERS:
*OHIO RIVER BASIN.

ABSTRACT:
DISCHARGE OF RAW SEWAGE INTO STREAMS IMPARTS TOXICITY TO THE AREA
SURROUNDING THE OUTLET. AT A DISTANCE DOWNSTREAM, HOWEVER, SEWAGE
UNDERGOES BACTERIOLOGICAL DECOMPOSITION, LOSES ITS TOXICITY, AND ACTS
AS A FERTILIZER, STIMULATING THE GROWTH OF PLANKTON AND CONSEQUENT
INCREASE OF THE FISH POPULATION. SECONDARY SEWAGE TREATMENT ELIMINATES
TOXIC INGREDIENTS AND THE ENTIRE STREAM IS BENEFITED BY THE INFLOW OF
AVAILABLE NUTRIENTS. (SEE ALSO W72-01786) (WILDE-WISCONSIN)

FIELD 05C

ACCESSION NO. W72-01788

A HEAVY MORTALITY OF FISHES RESULTING FROM THE DECOMPOSITION OF ALGAE IN THE
YAHARA RIVER, WISCONSIN,

WISCONSIN DEPT. OF CONSERVATION, MADISON; AND WISCONSIN BOARD OF HEALTH,
   MADISON.

KENNETH M. MACKENTHUN, AND ALFRED F. BARTSCH.

IN: BIOLOGY OF WATER POLLUTION, P 75-78. 3 FIG, 1 TAB, 2 REF. COMPILED BY L E
   KEUP, W M INGRAM, AND K M MACKENTHUN, FEDERAL WATER CONTROL ADMINISTRATION,
   WASHINGTON, D C, 1967.

   DESCRIPTORS:
      *FISHKILL, *CYANOPHYTA, *DECOMPOSING ORGANIC MATTER, TOXICITY,
      DISSOLVED OXYGEN, ALGAE, WISCONSIN.

   IDENTIFIERS:
      *YAHARA RIVER(WIS), APHANIZOMENON FLOS AQUAE, OXYGEN DEPLETION.

   ABSTRACT:
      EARLY IN OCTOBER, 1946, A HEAVY MORTALITY OF ALL FISH SPECIES OCCURRED
      IN THE YAHARA RIVER, WISCONSIN. BEFORE EXPIRING, THE FISH, INCLUDING
      CARP, NORTHERN PIKE, WALLEYE PIKE, CRAPPIES, SUCKERS, AND EEL,
      CONGREGATED NEAR THE SHORE BREATHING AT THE SURFACE. THE MORTALITY WAS
      CORRELATED WITH THE CIRTICALLY LOW CONCENTRATION OF DISSOLVED
      OXYGEN--UNDER 1.0 PPM; THIS DEFICIENCY RESULTED FROM THE DECOMPOSITION
      OF AN ALGAL MASS, LARGELY APHANIZOMENON FLOS AQUAE. AS REVEALED BY
      LABORATORY TRIALS, TOXIC SUBSTANCES RELEASED BY DECOMPOSING ALGAL
      TISSUES WERE A CONTRIBUTING FACTOR. (SEE ALSO W72-01786)
      (WILDE-WISCONSIN)

   FIELD 05C

   ACCESSION NO. W72-01797

SUGGESTED CLASSIFICATION OF ALGAE AND PROTOZOA IN SANITARY SCIENCE,

   ROBERT A. TAFT SANITARY ENGINEERING CENTER, CINCINNATI, OHIO. INTERFERENCE
      ORGANISMS STUDIES AND WATER SUPPLY AND WATER POLLUTION CONTROL RESEARCH.

   C. MERVIN, PALMER, AND WILLIAM M. INGRAM.

   IN: BIOLOGY OF WATER POLLUTION, P 79-83. 1 FIG, 1 TAB, 24 REF. COMPILED BY L.
      E. KEUP, W. M. INGRAM, AND K. M. MACKENTHUN, FEDERAL WATER POLLUTION
      CONTROL ADMINISTRATION, WASHINGTON, D.C., 1967.

      DESCRIPTORS:
         *ALGAE, *PROTOZOA, *SYSTEMATICS, *SANITARY ENGINEERING, MICROORGANISMS,
         NUISANCE ALGAE, SEWAGE TREATMENT, SELF-PURIFICATION, DISSOLVED OXYGEN,
         PHOTOSYNTHESIS.

      IDENTIFIERS:
         *PIGMENTED FLAGELLATES, *NON-PIGMENTED FLAGELLATES.

      ABSTRACT:
         TO MEET THE TAXONOMIC REQUIREMENTS OF SANITARY SCIENTISTS REGARDING THE
         DEMARKATION LINE BETWEEN ALGAE AND PROTOZOA, IT IS RECOMMENDED TO
         SEPARATE THE FLAGELLATES ON THE BASIS OF THEIR ABILITY TO PRODUCE
         OXYGEN. ACCORDINGLY, THE FLAGELLATES WITH PHOTOSYNTHETIC PIGMENTS WOULD
         BE ALGAE, AND WITHOUT THE PIGMENTS--PROTOZOA. A LIST OF FLAGELLATES OF
         ALGAL AND PROTOZOAN TYPES IS INCLUDED. (SEE ALSO W72-01786)
         (WILDE-WISCONSIN)

      FIELD 05C, 02H

      ACCESSION NO. W72-01798

AQUATIC ORGANISMS AS AN AID IN SOLVING WASTE DISPOSAL PROBLEMS,

ACADEMY OF NATURAL SCIENCES, PHILADELPHIA, PA. DEPT. OF LIMNOLOGY.

RUTH PATRICK.

IN: BIOLOGY OF WATER POLLUTION, P 108-110. 6 REF, DISCUSSION. COMPILED BY L E
KEUP, W M INGRAM, AND K M MACKENTHUN, FEDERAL WATER POLLUTION CONTROL
ADMINISTRATION, WASHINGTON, D C, 1967.

DESCRIPTORS:
*BIOASSAY, *AQUATIC LIFE, *WASTE DISPOSAL, *WATER POLLUTION EFFECTS,
FOOD CHAINS, BACTERIA, DEGRADATION(DECOMPOSITION), ALGAE, OXYGENATION,
SELF-PURIFICATION, LABORATORY TESTS, INVERTEBRATES, FISH, WATER
ANALYSIS, INSECTS, SNAILS, TOXICITY, RIVERS, HISTOGRAMS.

IDENTIFIERS:
*RIVER SURVEYS.

ABSTRACT:
SEVERAL LABORATORY TESTS OR BIOASSAYS, SUPPLEMENTING DETERMINATIONS OF
OXYGEN DEMANDS, ARE DESCRIBED AS AIDS IN SOLVING WASTE DISPOSAL
PROBLEMS. THE SUGGESTED TEST ORGANISMS INCLUDE FISH, INSECTS, SNAILS,
AND ALGAE, PARTICULARLY NITZSCHIA LINEARIS. THESE TESTS ARE INTENDED
PRIMARILY TO DETECT THE TOXICITY OF POLLUTANTS. THE SECOND APPROACH,
THE BIOLOGICAL SURVEY OF RIVERS, COMPRISES IDENTIFICATION OF ALL
AQUATIC ORGANISMS, TOTAL BACTERIAL AND COLIFORM COUNTS, AND THE
DETERMINATION OF BOD AND OTHER CHEMICAL CHARACTERISTICS OF WATER. (SEE
ALSO W72-01786) (WILDE-WISCONSIN)

FIELD 05C

ACCESSION NO. W72-01801

CHEMICAL COMPOSITION OF ALGAE AND ITS RELATIONSHIP TO TASTE AND ODOR,

MICHIGAN UNIV., ANN ARBOR. SCHOOL OF PUBLIC HEALTH.

GERARD A. ROHLICH, AND WILLIAM B. SARLES.

IN: BIOLOGY OF WATER POLLUTION, P 232-235. COMPILED BY W. M. INGRAM, L. E.
KEUP, AND K. M. MACKENTHUN, FEDERAL WATER POLLUTION CONTROL ADMINISTRATION,
WASHINGTON, D. C., 1967. 4 TAB, 12 REF.

DESCRIPTORS:
*WATER POLLUTION SOURCES, *ALGAE, *TASTE, *ODOR, *AMINO ACIDS, POTABLE
WATER, ACTINOMYCETES, PLANT GROWTH, DECOMPOSING ORGANIC MATTER,
CHEMICAL ANALYSIS, PROTEINS, BACTERIA.

IDENTIFIERS:
ODORIFEROUS PRODUCTS, ORGANISMS' ODORS.

ABSTRACT:
THE OBJECTIONABLE TASTE AND ODOR OF DRINKING WATER IS LARGELY DUE TO
DECOMPOSITION OF NUCLEO-PROTEINS, PHOSPHO-PROTEINS, LIPIDS, AND OTHER
TISSUE CONSTITUENTS OF AQUATIC BIOTA, PARTICULARLY ALGAE IN ASSOCIATION
WITH BACTERIA AND ACTINOMYCETES. THE ODOR OF DRINKING WATER, RANGING
FROM VILE STENCH AND FISHY OR COD-LIVER OIL TO EARTHY, SWEET GRASSY AND
AROMATIC, IS IMPARTED BY DIATOMECEAE, CYANOPHYCEAE, CHLOROPHYCEAE,
ACTINOMYCES, AND PROTOZOA. THE ODOR MAY ALSO ORIGINATE FROM BACTERIAL
ACTION LIBERATING AMMONIA FROM AMINO-ACIDS AND PRODUCING VARIOUS
ALCOHOLS. ODOR MAY BE INTENSIFIED BY DECOMPOSITION OF ALGAE, ESPECIALLY
FOLLOWING THEIR DESTRUCTION BY CHLORINATION, COPPER SULFATE, AND
SIMILAR ERADICATING TREATMENTS. (SEE ALSO W72-01786) (WILDE-WISCONSIN)

FIELD 05A, 05C

ACCESSION NO. W72-01812

AQUATIC BIOLOGY AND THE WATER WORKS ENGINEER,

JAMES B. LACKEY.

IN: BIOLOGY OF WATER POLLUTION, P 236-239. COMPILED BY W. M. INGRAM, L. E.
KEUP, AND K. M. MACKENTHUN, FEDERAL WATER POLLUTION CONTROL ADMINISTRATION,
WASHINGTON, D. C., 1967. 9 FIG.

DESCRIPTORS:
*SANITARY ENGINEERING, *WATER SUPPLY, *BIOLOGICAL PROPERTIES, *WATER
POLLUTION SOURCES, TASTE, ODOR, ALGAE, RESERVOIRS, WATERSHEDS(BASINS),
INSECTICIDES, HARDNESS(WATER), DIATOMS, FOULING, PLANKTON, SEWAGE,
DOMESTIC WASTE, INDUSTRIAL WASTES, FARM WASTES.

IDENTIFIERS:
*BIOLOGICAL PROBLEMS, SCIOTO RIVER(OHIO), LAKE MENDOTA(WIS), LAKE
MONONA(WIS), LAKE WAUBESA(WIS), LAKE KEGONSA(WIS).

ABSTRACT:
AN ENGINEER PROPOSING TO USE RIVER OR LAKE WATER FOR THE NEEDS OF A
CITY OR AN INDUSTRY SHOULD CONSIDER THE PERTINENT PHYSICAL, CHEMICAL,
AND BIOLOGICAL PROPERTIES OF THE WATER. THE USE OF WATER FOR EITHER
HUMAN OR INDUSTRIAL CONSUMPTION MAY BE PRECLUDED BY A NUMBER OF
UNFAVORABLE CHARACTERISTICS, SUCH AS HIGH CONCENTRATION OF NITRATES,
PHOSPHATES, SEWAGE, AND SUBSEQUENT ALGAL BLOOMS, ABUNDANT PLANKTON WITH
ITS HIGH BOD AND FILTER CLOGGING POTENTIAL, THE PRESENCE OF PESTICIDES,
TOXIC INDUSTRIAL WASTES, OR PATHOGENIC ORGANISMS. (SEE ALSO W72-01786)
(WILDE-WISCONSIN)

FIELD 05C

ACCESSION NO. W72-01813

PRE-TREATMENT BASIN FOR ALGAE REMOVAL,

MENASHA ELECTRIC AND WATER UTILITIES, WIS.

ANDREW J. MARX.

IN: BIOLOGY OF WATER POLLUTION, P 239-244. COMPILED BY W. M. INGRAM, L. E.
KEUP, AND K. M. MACKENTHUN, FEDERAL WATER POLLUTION CONTROL ADMINISTRATION,
WASHINGTON, D. C., 1967. 1 FIG, 2 PHOTOS.

DESCRIPTORS:
*WATER PURIFICATION, *ALGAL CONTROL, *WATER SUPPLY, *BASINS, WATER
POLLUTION TREATMENT, COPPER SULFATE, ODOR, COLIFORMS, ECONOMICS.

IDENTIFIERS:
LAKE WINNEBAGO(WIS), MENASHA(WIS).

ABSTRACT:
FOLLOWING TRIALS WITH A SMALL PILOT PLANT, THE ELECTRIC AND WATER
UTILITY COMPANY OF MENASHA, WISCONSIN CONSTRUCTED A 600 X 700 FT
PRETREATMENT BASIN AT THE MOUTH OF FOX RIVER. THIS BASIN PERMITTED
ALGAE CONTROL BY A SOLUTION OF COPPER SULFATE, PREPARED BY DISSOLVING
500 LBS OF COPPER SULFATE AND 1 QT OF SULFURIC ACID IN 1,000 GALLONS OF
WATER IN A WOODEN TANK. THE TANK IS CONNECTED TO A SMALL LIQUID FEEDER
WITH A METERING DEVICE TO DELIVER 2 PPM OF COPPER SULFATE. THIS WATER
PRETREATMENT ERADICATED ALGAE AND WEEDS, DECREASED THE TURBIDITY,
REDUCED ODOR AND COLIFORM ORGANISMS, REDUCED THE EXPENSE ON CHEMICALS,
ELIMINATED THE ANCHOR ICE, AND GREATLY SIMPLIFIED THE MANAGEMENT. IN
1951 THE AGE OF THE INSTALLATION WAS 5 YEARS. (SEE ALSO W72-01786)
(WILDE-WISCONSIN)

FIELD 05F, 05C

ACCESSION NO. W72-01814

PROTOZOA AND ACTIVATED SLUDGE,

MASSACHUSETTS INST. OF TECH., CAMBRIDGE. DEPT. OF SANITARY ENGINEERING.

ROSS E. MCKINNEY, AND ANDREW GRAM.

IN: BIOLOGY OF WATER POLLUTION, P 252-262. COMPILED BY W. M. INGRAM, L. E. KEUP, AND K. M. MACKENTHUN, FEDERAL WATER POLLUTION CONTROL ADMINISTRATION, WASHINGTON, D. C., 1967. 7 FIG, 1 TAB, 18 REF.

DESCRIPTORS:
*SEWAGE TREATMENT, *PROTOZOA, *ACTIVATED SLUDGE, *BACTERIA, METABOLISM, GROWTH RATES, MICROORGANISMS, ALGAE, FUNGI, LABORATORY TESTS, BIOCHEMICAL OXYGEN DEMAND, SEWAGE EFFLUENTS, SANITARY ENGINEERING.

IDENTIFIERS:
SARCODINA, MASTIGOPHORA, HOLOZOIC MASTIGOPHORA, HOLOPHYTIC MASTIGOPHORA, CILIATA.

ABSTRACT:
TO DELINEATE THE ROLE OF PROTOZOA IN SLUDGE ACTIVATION, EXPERIMENTS WERE CONDUCTED UNDER THE CONCEPT OF 'MINIMUM CONDITIONS.' THE LATTER WERE ESTABLISHED BY A COMBINATION OF A BALANCED SOLUBLE SUBSTRATE, BACTERIA FROM ACTIVATED SLUDGE, AND A PURE CULTURE OF PROTOZOA. THE FLOC FORMED WAS THE RESULT OF METABOLIC ACTIVITY OF MICROORGANISMS IN THE MIXED LIQUOR. THE FLAGELLATE AND CILIATE PROTOZOA INCLUDED CHILOMONAS PARAMECIUM, EUGLENA GRACILIS, TETRAHYMENA FILII, AND GLAUCOMA SCINTILLANS. THE TRIALS DEMONSTRATED THE GENERAL PRINCIPLES OF COMPETITION OF PROTOZOA WITH BACTERIA AND THE IMPORTANCE OF PROTOZOA IN SLUDGE ACTIVATION. THE SUCCESSION OF PROTOZOA--SARCODINA, THE HOLOPHYTIC AND HOLOZOIC MASTIGOPHORA, AND THE FREE-SWIMMING AND STALKED CILIATA--WAS ESTABLISHED TO SERVE AS A GUIDE TO BETTER OPERATIONS. THE BACTERIAL FLOCCULATION WAS CONFIRMED IN THE CONTROL UNIT. (SEE ALSO W72-01786) (WILDE-WISCONSIN)

FIELD 05D, 05C

ACCESSION NO. W72-01817

BIOLOGICAL FACTORS IN TREATMENT OF RAW SEWAGE IN ARTIFICIAL PONDS,

ROBERT A. TAFT SANITARY ENGINEERING CENTER, CINCINNATI, OHIO; AND SOUTH DAKOTA STATE COLL., BROOKINGS. DEPT. OF ZOOLOGY.

A. F. BARTSCH, AND M. O. ALLUM.

IN: BIOLOGY OF WATER POLLUTION, P 262-269. COMPILED BY L. E. KEUP, W. M. INGRAM, AND K. M. MACKENTHUN, FEDERAL WATER POLLUTION CONTROL ADMINISTRATION, WASHINGTON, D. C., 1967. 6 FIG, 7 TAB, 18 REF.

DESCRIPTORS:
*SEWAGE TREATMENT, *OXIDATION LAGOONS, *ALGAE, *BACTERIA, *DISSOLVED OXYGEN, SEWAGE LAGOONS, PLANT PHYSIOLOGY, ANIMAL PHYSIOLOGY, NORTH DAKOTA, SOUTH DAKOTA, PHOTOSYNTHESIS, LIGHT PENETRATION, ON-SITE TESTS, BIOCHEMICAL OXYGEN DEMAND, DIURNAL, CHLOROPHYLL.

IDENTIFIERS:
*PHOTOSYNTHETIC OXYGEN, LEMMON(SD), KADOKA(SD).

ABSTRACT:
FREEDOM FROM NUISANCE CONDITIONS AND THE ACCEPTABILITY OF RAW SEWAGE PONDS TO MUNICIPALITIES AS TREATMENT FACILITIES ARE INTIMATELY DEPENDENT UPON THE DISSOLVED OXYGEN SUPPLY IN SEWAGE. LIGHT AND DARK BOTTLE ANALYSES AND THE ESTIMATED OXYGEN PRODUCTION BASED ON VERDUIN'S FACTOR PERMIT ESTABLISHMENT OF SUITABLE LOADING AND DEPTH OF SEWAGE LAGOONS. THE USE OF POND DESIGN EQUATIONS, HOWEVER, REQUIRES CAUTION BECAUSE PROFILE VARIATIONS, WEATHER CONDITIONS, DIURNAL CHANGES, AND OTHER FACTORS MAY IMPOSE CONSIDERABLE DEVIATION IN DESIGN. (SEE ALSO W72-01786) (WILDE-WISCONSIN)

FIELD 05D, 05C

ACCESSION NO. W72-01818

TRICKLING FILTER ECOLOGY,

ROBERT A. TAFT SANITARY ENGINEERING CENTER, CINCINNATI, OHIO.

WILLIAM B. COOKE.

IN: BIOLOGY OF WATER POLLUTION, P 269-287. COMPILED BY L. E. KEUP, W. M.
INGRAM, AND K. M. MACKENTHUN, FEDERAL WATER POLLUTION CONTROL
ADMINISTRATION, WASHINGTON, D. C., 1967. 7 FIG, 3 TAB, 45 REF.

DESCRIPTORS:
*BIOLOGICAL COMMUNITIES, *SEWAGE TREATMENT, *TRICKLING FILTERS, *BIOTA,
SEWAGE BACTERIA, SYSTEMATICS, FUNGI, ALGAE, DIATOMS, PROTOZOA,
PLANKTON, WORMS, NEMATODES, ROTIFERS, LARVAE, SNAILS, MITES, MOLDS,
GROWTH RATES, HABITATS.

IDENTIFIERS:
*ZOOGLEAL SLIME, DAYTON(OHIO), VANDALIA(OHIO), RHIZOPODS, SPIDERS, MILK
WASTE TREATMENT, COLONY COUNTS, ARTHROPODS.

ABSTRACT:
THE POPULATIONS OF SEVERAL TRICKLING FILTERS IN THE USA AND ENGLAND ARE
REVIEWED. THE COMPILED LIST OF THE COMMON MEMBERS OF THIS MAN-MADE
ECOSYSTEM, EXCEEDING 200 SPECIES, INCLUDES HERBIVORES, CARNIVORES, AND
SAPROBES, SERVING AS PRIMARY OR SECONDARY DECOMPOSERS. THIS POPULATION
EFFECTIVELY REMOVES SOLIDS, COLLOIDS, AND DISSOLVED MATTER, THUS
REDUCING BOD OF THE SEWAGE. NO CORRELATION BETWEEN THE SURFACE GROWTH
AND TEMPERATURE OR PRECIPITATION WAS OBSERVED. THE FUNCTIONING OF THE
POPULATION DEPENDS LARGELY ON PREFORMED ORGANIC MATTER. AT DEATH THE
ORGANISMS ARE ATTACKED BY SCAVENGERS AND THUS REENTER THE METABOLIC
SYSTEM. (SEE ALSO W72-01786) (WILDE-WISCONSIN)

FIELD 05D, 05C

ACCESSION NO. W72-01819

CHEMISTRY OF NITROGEN AND PHOSPHORUS IN WATER.

AMERICAN WATER WORKS ASSOCIATION, NEW YORK. WATER QUALITY DIV. COMMITTEE ON
NUTRIENTS IN WATER.

DESCRIPTORS:
WASTE WATER TREATMENT, WATER TREATMENT, *ALGAE, *EUTROPHICATION,
*NITROGEN, NITROGEN CYCLE, *PHOSPHORUS, PHOSPHORUS COMPOUNDS, WATER
SUPPLY, WATER MANAGEMENT.

ABSTRACT:
OF THE MAJOR ELEMENTS ESSENTIAL TO ALGAL GROWTH, NITROGEN AND
PHOSPHORUS ARE THE ONES MOST LIKELY TO BE FOUND IN LIMITED QUANTITIES
IN NATURAL WATERS. SINCE THEY THEREFORE REPRESENT PROMISING WEAK LINKS
IN ALGAL LIFE CYCLES, THEIR CHEMICAL STATES AND BEHAVIOR IN WATER ARE
EXAMINED TO SEE HOW WATER TREATMENT MIGHT BENEFIT. LARGE SUPPLIES OF
NITROGEN AND PHOSPHORUS ARE PRESENT IN MANY BODIES OF WATER EITHER IN
THE SEDIMENTS, THE ATMOSPHERE ABOVE, OR IN THE FORM OF DISSOLVED GAS.
THESE FORMS MAY BE AVAILABLE FOR THE GROWTH OF ALGAE AND OTHER AQUATIC
PLANTS, BUT THE RATES AT WHICH THEY BECOME AVAILABLE IS SLOW. THESE
RATES ARE IMPORTANT, HOWEVER, AS THEY TEND TO CONTROL THE AMOUNT OF
VEGETABLE GROWTH WHICH CAN BE SUPPORTED. SOLUBLE NITROGEN AND
PHOSPHORUS CONTAINED IN WASTEWATER EFFLUENTS ARE IN A READILY AVAILABLE
FORM. IF DISCHARGED TO NATURAL WATERS, THEY CAN STIMULATE GROWTH FAR IN
EXCESS OF THAT WHICH WOULD OCCUR NATURALLY. THUS, IN ASSESSING THE
EXTENT OF NUTRIENT-RELATED PROBLEMS AND THEIR CONTROL, THE WATER
MANAGER MUST EVALUATE THE SIGNIFICANCE OF THE READILY AVAILABLE FORMS
AND ALSO THE FORMS WHICH MAY BE RELEASED SLOWLY FROM THE SUSPENDED
PARTICLES AND THE SEDIMENTS. TO DO THIS, THE BODY OF WATER MUST BE
CONSIDERED A CHEMICAL REACTOR WHICH IS CONTROLLED BY THE KINETICS OF A
NUMBER OF PROCESSES. (GOESSLING-TEXAS)

FIELD 05C, 05D, 02K

ACCESSION NO. W72-01867

317

EUTROPHICATION: SMALL FLORIDA LAKES AS MODELS TO STUDY THE PROCESS,

FLORIDA UNIV., GAINSVILLE. DEPT. OF ENVIRONMENTAL ENGINEERING.

PATRICK L. BREZONIK, AND HUGH D. PUTNAM.

PROCEEDINGS OF THE SEVENTEENTH SOUTHERN WATER RESOURCES AND POLLUTION CONTROL
CONFERENCE, APRIL 16-18, 1968. P 315-333, 3 FIG, 8 TAB, 9 REF. GRANT NO.
16010DON.

DESCRIPTORS:
*EUTROPHICATION, *LIMNOLOGY, *ESSENTIAL NUTRIENTS, *PRIMARY
PRODUCTIVITY, AQUATIC ALGAE, TROPHIC LEVEL, WATER QUALITY, FLORIDA.

IDENTIFIERS:
*ANDERSON-CUE LAKE, MELROSE(FLORIDA).

ABSTRACT:
EUTROPHICATION IS A PROCESS OF LAKE AGING WHICH CAN BE ACCELERATED BY
NUTRIENT ENRICHMENT. THIS ENRICHMENT MANIFESTS ITSELF IN A VARIETY OF
LARGELY DELETERIOUS EFFECTS ON THE LACUSTRINE WATER QUALITY AND BIOTA.
RESPONSE OF LAKES TO INCREASED NUTRIENT LOADS IS HIGHLY INDIVIDUALISTIC
AND DEPENDENT ON NUMEROUS PHYSICAL, CHEMICAL AND BIOLOGICAL FACTORS
WHICH DEFINE THE LAKE'S ORIGINAL TROPHIC STRUCTURE. THE ASSIMILATIVE
CAPACITY OF A LAKE, I.E. THE NUTRIENT LOAD IT CAN ACCEPT WITHOUT
DEVIATING FROM CERTAIN IMPOSED WATER QUALITY STANDARDS AND CONDITIONS,
MUST BE KNOWN FOR INTELLIGENT WATER QUALITY MANAGEMENT. A SMALL
ISOLATED FLORIDA LAKE HAS RECEIVED A CONTROLLED AMOUNT OF NUTRIENT
LOADING. ANDERSON-CUE IS AN OLIGOTROPHIC LAKE OF 19.1 ACRES. A
PRELIMINARY INVESTIGATION INDICATED A SMALL NUTRIENT SUPPLY IN THE
LAKE, WHICH REFLECTED THE IMPOVERISHED SANDY SOILS, SPARSE LAKE BIOTA,
AND LOW PRIMARY PRODUCTIVITY. SEWAGE EFFLUENT ENRICHED WITH NITROGEN
AND PHOSPHOROUS COMPOUNDS HAS BEEN ADDED TO THE EXPERIMENTAL LAKE SINCE
MARCH 1967. THE LAKE'S RESPONSE TO THE INCREASED NUTRIENT FLUX IS BEING
DETERMINED BY MONITORING A VARIETY OF CHEMICAL AND BIOLOGICAL
PARAMETERS, INCLUDING PHYTOPLANKTON SPECIES COMPOSITION, PRIMARY
PRODUCTION, CHLOROPHYLL, NITROGEN AND PHOSPHOROUS FORMS, DISSOLVED
OXYGEN AND TRACE METALS. BIOASSAY METHODS HAVE INDICATED THAT
PHOSPHOROUS IS STILL THE LIMITING NUTRIENT. IN THE FIRST YEAR OF STUDY,
EFFECTS ON WATER QUALITY AND TROPHIC STATUS HAVE BEEN SMALL. (EPA
ABSTRACT)

FIELD 05B, 05C

ACCESSION NO. W72-01990

A PARTIAL CHECKLIST OF FLORIDA FRESH-WATER ALGAE AND PROTOZOA WITH REFERENCE TO
MCCLOUD AND CUE LAKES,

FLORIDA UNIV., GAINESVILLE. DEPT. OF ENVIRONMENTAL ENGINEERING.

JAMES B. LACKEY, AND ELSIE W. LACKEY.

AVAILABLE FROM THE NATIONAL TECHNICAL INFORMATION SERVICE AS PB-179 071,
$3.00 IN PAPER COPY, $0.95 IN MICROFICHE. WATER RESOURCES RESEARCH CENTER,
PUBLICATION NO. 3, BULLETIN SERIES NO. 131, FLORIDA ENGINEERING AND
INDUSTRIAL EXPERIMENT STATION, VOL 21, NO 11, NOV 1967, P 1-28. 16 TAB, 24
REF. OWRR B-004-FLA(7). FWPCA GRANT 16010 DON.

DESCRIPTORS:
*AQUATIC ALGAE, *PROTOZOA, *FLORIDA, *EUTROPHICATION, AQUATIC
MICROORGANISMS, TROPHIC LEVEL.

IDENTIFIERS:
CUE LAKE(FLORIDA), MCCLOUD LAKE(FLORIDA).

ABSTRACT:
FLORIDA IS A LAND WHERE THERE ARE MANY KINDS OF FRESH WATER AND MANY
KINDS OF CLIMATE. SINCE THERE IS USUALLY A DISTINCT RAINY SEASON,
BLOOMS TEND TO BE PREVALENT DURING OR JUST AFTER THIS SEASON. IN
EUTROPHIC WATERS, BLOOMS MAY OCCUR ANYTIME. ALL THESE MATTERS MAKE
FLORIDA WATERS AN EXTREMELY VALUABLE RESEARCH AREA FOR THE
MICROBIOLOGIST. THE MOST INTENSIVE WORK HAS BEEN DONE ON TWO LAKES NEAR
MELROSE, FLORIDA: CUE LAKE AND MCCLOUD LAKE. CUE LAKE BECAME SUBJECT TO
ROUTINE FERTILIZATION SINCE 1967. IN ADDITION TO THESE TWO LAKES, 33
OTHER LOCATIONS WERE SAMPLED. THE REPORT CONTAINS A LIST OF OCCURRING
SPECIES. A WIDE SPECTRUM OF ALGAE AND PROTOZOA WERE IDENTIFIED
EMPHASIZING THAT MOST OF THESE ORGANISMS ARE COSMOPOLITAN. MANY OF THEM
DO NOT CROSS BROAD ECOLOGICAL BOUNDARIES BUT NARROW BOUNDARIES AND EACH
SPECIES HAS A MAXIMUM SET OF CONDITIONS UNDER WHICH IT ATTAINS MAXIMUM
NUMBERS. IT IS USEFUL TO RECOGNIZE THE SPECIES WE SEE BECAUSE BY
CORRELATING NUMBERS OF INDIVIDUALS WITH KNOWN CONDITIONS WE MAY REACH
CONCLUSIONS REGARDING TAXONOMIC AND PHYSIOLOGICAL RELATIONSHIPS.

FIELD 05A, 05C

ACCESSION NO. W72-01993

PRELIMINARY RESEARCH IN THE LABORATORY ON EXPERIMENTAL BRACKISH ECOSYSTEMS,

J. C. LACAZE, C. HALLOPEAU, AND M. VOIGT.

BULL MUS NAT HIST NATUR (PARIS). 41(5): 1278-1289. ILLUS. 1969.

DESCRIPTORS:
ALGAE, BRACKISH, ECOSYSTEMS, FAUNA, FLORA, LABORATORY, MINERALS, POLLUTION.

ABSTRACT:
FOUR AQUATIC ECOSYSTEMS WERE STUDIED STARTING FROM ALGAE, SEDIMENT, AND WATER TAKEN FROM A BRACKISH MARSH OF THE ARCACHON REGION. THE IMPORTANT THING IS TO OBTAIN SYSTEMS POSSESSING CHARACTERISTIC STRUCTURE, BEHAVIOR AND REPLICABILITY. THE EXPERIMENT LASTED A YEAR (JULY 1966 - AUG. 1967) DURING WHICH TIME FAUNA, FLORA, NITRATES, PHOSPHATES, AND ALKALINITY WERE ANALYZED. SPECIAL IMPORTANCE WAS GIVEN TO THE MICRO-FLORA OF DIATOMS DEVELOPING ON THE IMMERSED SUBSTRATA. THE EXPERIMENTAL ECOSYSTEMS WERE STUDIED AS BIOLOGICAL MATERIAL FOR POLLUTION STUDIES.--COPYRIGHT 1971, BIOLOGICAL ABSTRACTS, INC.

FIELD 05A

ACCESSION NO. W72-02203

UTILIZATION OF ALGAE BY DAPHINIA AS INFLUENCED BY CELL SENESCENCE AND UV IRRADIATION,

MARYLAND UNIV., COLLEGE PARK. DEPT. OF ZOOLOGY.

R. G. STROSS, J. C. JONES, F. M. UNGER, AND J. M. VAIL.

PROCEEDINGS, INDUSTRIAL WASTE CONFERENCE, 20TH, MAY 4, 5, AND 6, 1965. P 706-714, 4 FIG, 5 TAB, 17 REF.

DESCRIPTORS:
WATER QUALITY, *ALGAL CONTROL, EUTROPHICATION, DAPHNIA, CHLAMYDOMONAS, LABORATORY TESTS, PARTICLE SIZE, POPULATION, GROWTH RATES, WATER POLLUTION CONTROL.

IDENTIFIERS:
ALGAL SENESCENCE, GROWTH INHIBITIONS.

ABSTRACT:
SUSPENSIONS OF ALGAE, BACTERIA AND DETRITUS ARE INGESTED BY MANY GROUPS OF ANIMALS. DAPHINIA ARE EQUIPPED WITH THORACIC LIMBS WHICH COLLECT PARTICLES FROM DILUTE SUSPENSIONS. THE NUMBER OF PARTICLES INGESTED DEPENDS ON SIZE, DENSITY AND OTHER FACTORS SUCH AS NUTRITIONAL DEFICIENCIES OR INDEGESTIBILITY. A STUDY HAS BEEN COMPLETED IN WHICH THE UTILIZATION OF CHLAMYDOMONAS REINHARDI (DANGEARD I.U. STRAINS NO. 89 AND 90) BY DAPHNIA PULEX DE GEER WAS MEASURED WHICH ALLOWED AN EVALUATION OF THE EFFECT OF SENESCENT AND UV IRRADIATED ALGAE ON THE INTRINSIC RATES OF INCREASE OF DAPHNIA POPULATIONS. THE RESULTS OF THESE TESTS WERE: (1) THE INTAKE OF ALGAL CELLS BY DAPHNIA DECREASED WITH THE AGE OF THE ALGAL CULTURE; (2) INHIBITION WAS PRODUCED BY THE CELLS AND THE PARTICLE-FREE MEDIUM OF THE CULTURE; (3) THE FEEDING OF SENESCENT CELLS REDUCED THE INTRINSIC RATE OF POPULATION INCREASE TO NEARLY ONE THIRD THAT OF CONTROL POPULATIONS; (4) BIRTH RATE OF DAPHNIA WAS MORE STRONGLY INFLUENCED THAN DEATH RATE; (5) IRRADIATION OF LOG-PHASE CELLS WITH FAR UV STIMULATES THE EFFECT OF CELL SENESCENCE ON GROWTH OF THE DAPHNIA; (6) LIPID PEROXIDES IN THE CELLS INCREASED WITH THE DOSE OF UV BUT MOST OF THE GROWTH INHIBITION RESULTED FROM THE MINIMUM DOSE; AND, (7) A HYPOTHETICAL POPULATION WAS CONSTRUCTED TO SHOW THE INFLUENCE OF ALGAL SENESCENCE ON DENSITY AND PERMISSABLE WASHOUT RATE OF DAPHNIA POPULATION IN EQUILIBRIUM WITH ITS FOOD SUPPLY. (GOESSLING-TEXAS)

FIELD 05G, 05C

ACCESSION NO. W72-02417

THE EFFECT OF LAS VEGAS WASH EFFLUENT UPON THE WATER QUALITY IN LAKE MEAD,

BUREAU OF RECLAMATION, DENVER, COLO. ENGINEERING AND RESEARCH CENTER.

D. A. HOFFMAN, P. R. TRAMUTT, AND F. C. HELLER.

AVAILABLE FROM THE NATIONAL TECHNICAL INFORMATION SERVICE AS PB-200 137.
$3.00 IN PAPER COPY, $0.95 IN MICROFICHE. REC-ERC-71-11, JAN 71, 32P.

DESCRIPTORS:
*NEVADA, *EUTROPHICATION, *LAKES, *WATER POLLUTION, DEGRADATION, SEWAGE
TREATMENT, NITROGEN COMPOUNDS, PHOSPHORUS COMPOUNDS, RESERVOIRS, ALGAE,
GROUND WATER, NITRATES, PHOSPHATES, LAS VEGAS BAY.

IDENTIFIERS:
DISSOLVED ORGANIC MATTER, LAS VEGAS, NEVADA, *LAKE MEAD, LAS VEGAS BAY.

ABSTRACT:
EFFLUENTS FROM THE CLARK COUNTY AND LAS VEGAS, NEV SEWAGE TREATMENT
PLANTS ARE THE PRINCIPAL SOURCES OF WATER FLOWING INTO LAS VEGAS WASH.
WATER IN THE WASH FLOWS INTO LAS VEGAS BAY, AN ARM OF THE BOULDER BASIN
REACH OF LAKE MEAD. A STUDY WAS CONDUCTED TO DETERMINE THE QUALITY OF
WATER IN LAS VEGAS WASH AND THE EFFECT OF THAT WATER ON THE WATER
QUALITY OF THE LAKE. LARGE AMOUNTS OF TOTAL DISSOLVED SOLIDS AND ALGAE
NUTRIENTS ENTER THE LAKE FROM THE WASH. CHLOROPHYLL A VALUES,
INDICATORS OF ALGAE BLOOMS, WERE 20-25 TIMES HIGHER IN LAS VEGAS BAY
THAN IN THE CONTROL STATIONS ELSEWHERE IN LAKE MEAD. AN UNACCOUNTABLE
INCREASE IN CATIONS AND ANIONS OCCURS IN THE WASH. THE SOURCE OF THE
ADDITION IS NOT KNOWN BUT MAY BE CAUSED BY UNMEASURED GROUND-WATER
INFLOW. METHODS OF COLLECTIONS AND ANALYSIS FOR NITROGEN AND PHOSPHORUS
ARE DESCRIBED.

FIELD 02H, 05C

ACCESSION NO. W72-02817

STUDIES ON ALGAE FROM EUTROPHIC AND OLIGOTROPHIC WATERS AND THEIR USE IN THE
BIO-ASSAY OF WATER QUALITY,

MICHIGAN STATE UNIV., EAST LANSING. DEPT. OF BOTANY AND PLANT PATHOLOGY.

BRIAN MOSS.

AVAILABLE FROM THE NATIONAL TECHNICAL INFORMATION SERVICE AS PB-205 805,
$3.00 IN PAPER COPY, $0.95 IN MICROFICHE. PROJECT COMPLETION REPORT 11 P.
OWRR A-034-MICH(1).

DESCRIPTORS:
ALGAE, *NUTRIENT REQUIREMENTS, BIOASSAY, EUTROPHICATION, OLIGOTROPHY,
LAKES, MICHIGAN, BICARBONATES, IRON, PHOSPHORUS, NITROGEN, WATER
QUALITY.

ABSTRACT:
THE NUTRITION OF SEVERAL SPECIES OF ALGAE FOUND IN EUTROPHIC AND
OLIGOTROPHIC LAKES AND THE USEFULNESS OF THESE ALGAE IN BIO-ASSAYS FOR
DETECTING LIMITING NUTRIENTS IN NATURAL WATERS WERE INVESTIGATED. THE
ROLES OF BICARBONATE, PH, MONOVALENT AND DIVALENT CATIONS, PHOSPHORUS,
NITROGEN AND IRON ARE DISCUSSED. (BAHR-MICHIGAN)

FIELD 05A, 05C

320

ACCESSION NO. W72-03183

PETROCHEMICAL EFFLUENTS TREATMENT PRACTICES, DETAILED REPORT,

ENGINEERING-SCIENCE, INC., TEXAS, AUSTIN.

EARNEST F. GLOYNA, AND DAVIS L. FORD.

AVAILABLE FROM NTIS AS PB-205 824, $3.00 IN PAPER COPY, $0.95 IN MICROFICHE.
WATER POLLUTION CONTROL RESEARCH SERIES, FEB 70, 249 P, 31 FIG, 46 TAB, 301
REF, 2 APPEND. FWPCA PROGRAM NO 12020 02/70, CONTRACT NO 14-12-461.

DESCRIPTORS:
*OIL, *OIL WASTES, *CHEMICAL WASTES, INDUSTRIAL WASTES, WATER ANALYSIS,
ANALYTICAL TECHNIQUES, SAMPLING, ORGANIC MATTER, TOXICITY, PHENOLS,
ODOR, COLOR, TURBIDITY, BIOCHEMICAL OXYGEN DEMAND, HYDROGEN ION
CONCENTRATION, ORGANIC LOADING, *WATER REUSE, RESEARCH AND DEVELOPMENT,
WASTE IDENTIFICATION, ALGAE, *WASTE WATER TREATMENT.

ABSTRACT:
500 NEW PETROLEUM PRODUCTS ARE INTRODUCED ON THE MARKET EACH YEAR,
MAKING THE PETROLEUM INDUSTRY AND ITS ALLIED PETROCHEMICL INDUSTRIES
SOME OF THE NATION'S FASTEST GROWING INDUSTRIES. THE PETROCHEMICAL
INDUSTRY IS PROJECTED AS INCREASING BY 9% PER YEAR THROUGH 1975. MOST
OF THE POLLUTIONAL LOAD FROM PETROCHEMICAL INDUSTRIES EMANATES FROM
PROCESS WASTE STREAMS. THE PROCESSES FOR REDUCING ITS POTENTIAL HARM TO
THE ENVIRONMENT ARE WELL DEVELOPED AND UNDERSTOOD. HOWEVER, CONSTANT
RESEARCH AND DEVELOPMENT EFFORTS ARE CONTINUALLY INTRODUCING NEW
POLLUTANTS INTO THE ENVIRONMENT. THEREFORE, DEVELOPMENT OF NEW PRODUCTS
SHOULD NOW INCLUDE DEVELOPMENT OF TREATMENT SCHEMES BY WHICH THE WASTES
FROM THE NEW PRODUCTS MAY BE ALLEVIATED. CARE MUST BE EXERCISED EVEN IN
THE SAMPLING PROCEDURES, SINCE MANY PETROLEUM-BASED CHEMICALS INTERFERE
WITH THE ACCEPTED ANALYTICAL TEST PROCEDURES. FINALLY, FEASIBILITY
STUDIES ARE NEEDED TO DETERMINE THE CAPABILITY OF THE PETROCHEMICAL
INDUSTRIES TO RE-USE MORE OF THEIR WASTEWATERS IN THE PLANT, AND
PHYSICAL, CHEMICAL, AND BIOLOGICAL TREATMENT ARE DISCUSSED. (SEE ALSO
W70-07511) (LOWRY-TEXAS)

FIELD 05D

ACCESSION NO. W72-03299

TASTE AND ODOR PROBLEMS IN NEW RESERVOIRS IN WOODED AREAS,

SEATTLE DEPT. OF WATER, WASH.

E. J. ALLEN.

JOURNAL OF THE AMERICAN WATER WORKS ASSOCIATION, VOL 52, NO 8, P 1027-1032,
AUGUST 1960. 1 FIG, 8 REF.

DESCRIPTORS:
*WATER TREATMENT, *TASTE, *ODOR, *RESERVOIRS, *DECIDUOUS TREES,
*DECOMPOSING ORGANIC MATTER, *ALGAE, *PHENOLS.

ABSTRACT:
THE NEARLY UNIVERSAL APPEARANCE OF TASTE AND ODORS IN RESERVOIRS DUE TO
MANY DIFFERENT CAUSES, ARE DISCUSSED WHICH INCLUDE DECAYING VEGETATION
ORIGINALLY PRESENT ON THE SITE, DECIDUOUS TREES ADJACENT TO THE
RESERVOIRS WITH DECAY OF FALLEN LEAVES, ALGAE FED BY THE LEAF
INFUSIONS, PHENOLS PRODUCED BY DECAYING LEAVES, (BEAVERS) BUILDING DAMS
AND DRAGGING VEGETABLE MATERIALS INTO THE RESERVOIR, WEED DECAY, WATER
COLLECTING IN ROAD DITCHES PROVIDING ENVIRONMENT FOR ALGAE GROWTHS.
TASTE AND ODOR PROBLEMS IN NEW RESERVOIRS IN WOODED AREAS CAN BE
AVOIDED BY RECOGNITION OF THE ENVIRONMENTAL CONDITIONS CONDUCIVE TO
THEIR PRODUCTION AND BY TAKING THE PROPER MEASURES TO ELIMINATE SUCH
CONDITIONS. (BEAN-AWWA)

FIELD 05F

ACCESSION NO. W72-03475

ALGAE AND OTHER INTERFERENCE ORGANISMS IN WATER OF THE SOUTH CENTRAL UNITED STATES,

ROBERT A. TAFT SANITARY ENGINEERING CENTER, CINCINNATI, OHIO.

C. M. PALMER.

JOURNAL OF THE AMERICAN WATER WORKS ASSOCIATION, VOL. 3, NO. 7, P 897-914, JULY 1960. 2 TAB, 91 REF.

DESCRIPTORS:
*WATER TREATMENT, *ALGAE, *ALGAL CONTROL, *TASTE, *ODOR, *NATURAL STREAMS, *DISTRIBUTION SYSTEMS, CHLORINATION, COPPER SULFATE, ACTIVATED CARBON, WATER SUPPLY.

IDENTIFIERS:
SOUTH - CENTRAL STATES.

ABSTRACT:
COVERING THE 6 SOUTH-CENTRAL STATES OF ALABAMA, ARKANSAS, LOUISIANA, MISSISSIPPI, OKLAHOMA, AND TEXAS, THE DIFFERENCES IN DIFFERENT WATERSHEDS ARE DISCUSSED ALONG WITH THE TYPES OF ORGANISMS MOST PREVALENT, THE DIFFICULTIES CAUSED IN TREATMENT PLANTS, IN IMPOUNDMENTS, AND BY INFESTATIONS IN DISTRIBUTION SYSTEMS. TASTES AND ODORS CAUSED ARE DISCUSSED, ALSO TOXIC ALGAE. CONTROL BY CHEMICALS SUCH AS COPPER SULFATE, ALGICIDES, CHLORINE AND CHLORINE-AMMONIA ARE DISCUSSED. THE COMMON USE OF ACTIVATED CARBON IS ELABORATED, ALSO MECHANICAL CONTROLS FOR REDUCING TASTE AND ODORS IN A FILTER PLANT. NINETY - ONE REFERENCES ARE NOTED. (BEAN-AWWA)

FIELD 05F

ACCESSION NO. W72-03562

USE OF ACTIVATED CARBON TO PREVENT WATER SUPPLY CONTAMINATION,

WEST VIRGINIA PULP AND PAPER COMPANY, COVINGTON, VA. CARBON TECHNICAL SERVICE LAB.

A. Y. HYNDSHAW.

WATER AND WASTES ENGINEERING, VOL 6, NO 2, FEBRUARY 1969, P 42-44.

DESCRIPTORS:
*ACTIVATED CARBON, *ADSORPTION, *WATER TREATMENT, *WASTE WATER TREATMENT, ALGAE, FILTRATION, NEW HAMPSHIRE, NEW YORK, VIRGINIA, ORGANIC COMPOUNDS, LAGOON, FILTERS, PESTICIDES, SLUDGE, DIGESTION, INDIANA.

IDENTIFIERS:
*CARBON FILTERS, LEBANON, BATAVIA, CLARKSBURG, BUFFALO, DUNREITH, OIL SPILLS.

ABSTRACT:
MOST ORGANICS THAT MIGHT BE POTENTIAL CONTAMINANTS OF POTABLE WATER SUPPLIES CAN BE EFFECTIVELY REMOVED OR INACTIVATED BY APPLICATION OF ACTIVATED CARBON. IN 1929, AFTER PHENOL-CONTAMINATED WATER CAUSED SPOILAGE OF SOME MEAT AT A CHICAGO MEAT PACKING COMPANY, IT WAS FOUND THAT POWDERED CARBON COULD BE USED TO PREVENT THIS TYPE OF CONTAMINATION. SINCE THAT TIME, USE OF ACTIVATED CARBON HAS EXPANDED GREATLY. TODAY MOST WATER TREATMENT PLANTS USE ACTIVATED CARBON TO REMOVE TASTE AND ODOR-CAUSING SUBSTANCES THAT ARE NOT REMOVED BY SUCH PROCESSES AS COAGULATION, FLOCCULATION, AND FILTRATION. APPLICATIONS OF CARBON TO IMPOUNDED SUPPLIES WILL ALSO CONTROL ALGAE GROWTHS. POWDERED CARBON HAS BEEN USED VERY EFFECTIVELY TO CORRECT CONDITIONS RESULTING FROM OIL SPILLS. THE VERY GREAT SURFACE AREAS ATTRACT AND HOLD LARGE QUANTITIES OF OIL. AFTER A FIRE IN A WAREHOUSE RELEASED LARGE AMOUNTS OF ETHYLENE GLYCOL INTO A RIVER IN CLARKSBURG, VIRGINIA, ACTIVATED CARBON WAS FED AT A RATE OF 10 MG/L UNTIL THE CONTAMINATION PASSED. THIS AMOUNT WAS SUFFICIENT TO ELIMINATE TASTES AND ODORS. IN BUFFALO, AN INDUSTRY USED 55 GALLON BARRELS WITH CARBON AS MAKESHIFT FILTERS TO ELIMINATE TRICRESYL PHOSPHATE FROM WATER BEFORE DISCHARGING IT INTO A STREAM. CARBON TREATMENT HAS ALSO BEEN USED TO ADSORB DDT AND SIMILAR PESTICIDES. ADDITION OF POWDERED CARBON AT THE GRIT CHAMBER IN WATER TREATMENT PLANTS HAS BEEN SHOWN TO REDUCE ATMOSPHERIC ODORS, ACCELERATE SETTLING, IMPROVE DIGESTION, AND REDUCE FOAM FORMATION. (BIGGS-TEXAS)

FIELD 05G, 05F

ACCESSION NO. W72-03641

SAND FILTRATION OF ALGAL SUSPENSIONS,

MICHIGAN UNIV., ANN ARBOR. DEPT. OF SANITARY ENGINEERING.

J. A. BORCHARDT, AND C. R. O'MELIA.

JOURNAL OF THE AMERICAN WATER WORKS ASSOCIATION, VOL 53, NO 12, P 1493-1502, DECEMBER 1961. 6 FIG, 3 TAB, 10 REF.

DESCRIPTORS:
*WATER TREATMENT, *ALGAE, *FILTRATION, HEAD LOSS, FILTERS, *SCENEDESMUS, FLOCCULATION, WATER POLLUTION TREATMENT.

ABSTRACT:
FILTRATION OF BOTH ALGAL AND FLOCCULENT PARTICLES THROUGH SAND BEDS WAS TESTED. EIGHT-FOOT PLEXIGLASS FILTERS WITH INSIDE DIAMETER OF 3 3/8 IN WERE UTILIZED WITH OTTAWA SAND DIVIDED IN THREE FRACTIONS, GRADED BY MICROSCOPIC MEASUREMENTS. ALL RUNS WERE MADE AFTER TAPPING THE FILTERS TO PRODUCE POROSITY OF 40%. ALGAE OF GENERA SCENEDESMUS, ANKISTRODESMUS AND ANABAENA WERE CULTURED WITH INORGANIC SALTS, AERATED AND TEMPERATURE CONTROLLED. HEAD LOSSES THROUGHOUT THE BEDS WERE DETERMINED BY 7 PIEZOMETER TUBES, AND ALGAE SAMPLES WERE COUNTED WITH AN ELECTRONIC PARTICLE COUNTER. MINIMUM REMOVAL EFFICIENCIES OF 33, 22 AND 10% WERE PRODUCED BY THE THREE SAND SIZES 0.316, 0.397, AND 0.524 MM RESPECTIVELY. REMOVAL OF THE STRINGY ANABAENA CONSISTENTLY WAS HIGHER THAN OF THE SMALLER MORE COMPACT ANKISTRODESMUS. ALGAL REMOVAL IN EVERY CASE DECREASED WITH TIME TO A CONSTANT MINIMUM VALUE. SIGNIFICANT NUMBERS OF ALGAE WERE FOUND IN EVERY SAMPLE THROUGHOUT THE RESEARCH. HEAD LOSS INCREASE WAS LINEAR WITH TIME THROUGHOUT THE GREATER PART OF EACH RUN. FUNCTIONALLY, THE BEHAVIOR OF THE FILTER BED IS THE SAME IN THE REMOVAL OF EITHER FLOCCULANT OR NON-FLOCCULANT PARTICLES. SOME SURFACE STRAINING TAKES PLACE FOLLOWED BY CONTINUOUS UNIFORM REMOVAL WITH DEPTH. THE PRESENCE OF FLOCCULENT MATERIAL ASSISTS IN THE ENTRAPMENT OF ALGAE CELLS, THEREFORE MULTIPLE ADDITIONS OF CHEMICALS AND CONTROLLED FLOCCULATION MAY BE MORE IMPORTANT THAN SEDIMENTATION CAPACITY. ADDITION OF CHEMICALS TO THE FILTER INFLUENT MIGHT BE BEST. (BEAN-AWWA)

FIELD 05F

ACCESSION NO.   W72-03703

ALGAE AND OTHER INTERFERENCE ORGANISMS IN WATER SUPPLIES OF CALIFORNIA,

ROBERT A. TAFT SANITARY ENGINEERING CENTER, CINCINNATI, OHIO.

C. M. PALMER.

JOURNAL OF THE AMERICAN WATER WORKS ASSOCIATION, VOL. 53, NO. 10, P.1297-1312, OCTOBER 1961. 100 REF.

DESCRIPTORS:
*WATER TREATMENT, *ALGAE, *TASTE, *ODOR, *ALGICIDES, ACTIVATED CARBON, FILTRATION, *TOXICITY, *DISTRIBUTION SYSTEMS, WATER POLLUTION, WATER QUALITY.

ABSTRACT:
70% OF ALL PRECIPITATION IN CALIFORNIA OCCURS DURING THREE WINTER MONTHS. RAINFALL IN NORTHERN CALIFORNIA EXCEEDS 100 IN. PER YEAR; IN THE EXTREME SOUTH, 3 IN. OR LESS PER YEAR. WATER OF WINTER FLOWS MUST BE STORED AND SOME FROM WET YEARS MUST BE HELD FOR USE IN DRY YEARS, PARTICULARLY AS DRY PERIODS SOMETIMES EXTEND OVER A FEW YEARS. WATER IN VARIOUS AREAS VARIES GREATLY IN PHYSICO-CHEMICAL PROPERTIES, IN GENERAL BEING SOFT IN THE NORTH AND HARD TOWARD THE SOUTH. NEARLY 40% OF WATER UTILIZED IS DRAWN FROM WELLS AND SPRINGS. THE MANY TYPES OF ALGAE FOUND IN CALIFORNIA WATERS ARE LISTED. RESERVOIRS PROBLEMS, AQUATIC ANIMALS, PHYSICO-CHEMICAL CHANGES, TASTE AND ODOR CAUSES, TREATMENT PLANT PROBLEMS, INFESTATIONS IN DISTRIBUTION SYSTEMS, ALGICIDES FOR RESERVOIRS, ACTIVATED CARBON, SELECTIVE DRAFT, RESERVOIR AND CANAL COVERS, CONTROL OF WEEDS, WATERSHED CONTROL, CONTROL BY FILTRATION, CONTROL OF ORGANISMS IN DISTRIBUTION SYSTEMS AND TOXIC ALGAE ARE DISCUSSED. (BEAN - AWWA)

FIELD 05A, 05F

ACCESSION NO.   W72-03746

REMOVAL OF ALGAE BY MICROSTRAINERS,

ONTARIO WATER RESOURCES COMMISSION, TORONTO (ONTARIO).

A. E. BERRY.

JOURNAL OF THE AMERICAN WATER WORKS ASSOCIATION, VOL 53, NO 12, P 1503-1508,
  DECEMBER 1961. 2 REF.

    DESCRIPTORS:
      *WATER TREATMENT, *ALGAE, FILTRATION, *SLIME, *ULTRAVIOLET RADIATION,
      *TREATMENT FACILITIES, FILTERS.

    IDENTIFIERS:
      *MICROSTRAINERS, HYPOCHLORITES.

    ABSTRACT:
      MICROSTRAINERS MAY SERVE AS THE SOLE FILTRATION PROCESS, AS TREATMENT
      AHEAD OF BOTH SLOW AND RAPID SAND FILTERS, OR AS FILTERS OF SEWAGE AND
      INDUSTRIAL EFFLUENTS AFTER TREATMENT. THEY ARE USED ON DRINKING WATER
      AND INDUSTRIAL WATER. SELECTION OF MICROSTRAINER MUST BE RELATED TO
      EFFICIENCIES IN REMOVAL OF ALGAE AND ECONOMICS IN COMPARISON WITH OTHER
      METHODS OF TREATMENT. SELECTION OF APPROPRIATE UNITS AND DESIGN OF THE
      INSTALLATION MUST BE RELATED TO THE FILTERABILITY INDEX OF THE WATER
      FOR SELECTION OF FABRIC, AND MAXIMUM CAPACITY IS DEPENDENT ON MAXIMUM
      SUBMERGENCE. IN THE UNITED STATES 12 INSTALLATIONS ARE REPORTED IN USE.
      IN CANADA 15 INSTALLATIONS ARE IN USE OR WILL BE SHORTLY. OF THESE, 13
      ARE SERVING AS THE ONLY TREATMENT APART FROM DISINFECTION AND TWO ARE
      USED AHEAD OF RAPID SAND FILTERS. SOME DIFFICULTIES HAVE ARISEN. SOME
      UNITS HAD INADEQUATE MEANS OF CONTROLLING THE HEAD LOSS THROUGH THE
      FABRIC, WHICH WAS DAMAGED. WASHWATER PRESSURES WERE NOT EFFECTIVELY
      CONTROLLED, SLIMING AND BINDING OF THE FABRICS REQUIRED TREATMENT WITH
      SODIUM HYPOCHLORITE OR OTHER CHEMICALS AND IT HAS BEEN FOUND DESIRABLE
      TO EQUIP ALL UNITS WITH ULTRA-VIOLET LIGHTS AND TO OPERATE THEM
      CONTINUOUSLY. BLINDING OF THE FABRIC HAS BEEN ENCOUNTERED FROM SOME
      CAUSE OTHER THAN SLIMES. THE CAUSE REMAINS TO BE DETERMINED. (BEAN-AWWA)

    FIELD 05F

    ACCESSION NO. W72-03824

EXPERIENCES WITH MUNICIPAL DIATOMITE FILTERS EXPERIENCES IN NEW YORK,

NEW YORK STATE DEPT. OF HEALTH, LATHAN.

G. W. MOORE.

JOURNAL OF THE AMERICAN WATER WORKS ASSOCIATION, VOL. 54, NO. 12, P
  1500-1504, DECEMBER 1962. 3 TAB.

    DESCRIPTORS:
      *WATER TREATMENT, *TURBIDITY, *IRON, *MANGANESE, *COLOR, *ALGAE,
      *COLLOIDS, FLOW RATES, *FILTERS, *NEW YORK, TREATMENT FACILITIES.

    IDENTIFIERS:
      *DIATOMITE FILTERS, FILTER AIDS.

    ABSTRACT:
      EXPERIENCES AT THE VARIOUS DIATOMITE FILTER INSTALLATIONS IN N.Y. ARE
      REVIEWED, AND THE N.Y. STATE HEALTH DEPARTMENT'S POSITION REGARDING
      DIATOMITE FILTRATION IS SUMMARIZED. IN GENERAL DIATOMITE FILTERS SHOULD
      BE CONSIDERED AS FINISHING FILTERS WHERE THE WATER IS NOT HEAVILY
      POLLUTED, THE AVERAGE TURBIDITY IS 10 OR LESS, AND MAXIMUM TURBIDITY
      DOES NOT EXCEED 30 FOR ANY APPRECIABLE TIME. THEY SHOULD BE APPROVED
      ONLY IF PRETREATMENT IS INCLUDED WHERE THERE IS A PROBLEM OF IRON
      REMOVAL, FIXED COLOR, HEAVY ALGAE LAODING, OR TURBIDITY IS IN FINE
      SUSPENSION OR COLLOIDAL FORM. AT LEAST TWO UNITS SHOULD BE PROVIDED.
      PERMISSIBLE RATE OF FILTRATION SHOULD DEPEND UPON THE CHARACTER OF THE
      RAW WATER AND THE DEGREE OR TYPE OF SUPERVISION AND LABORATORY CONTROL.
      THE NORMAL FILTER RATE SHOULD BE 1 GPM/SQ FT. EITHER HORIZONTAL OR
      VERTICAL ELEMENTS ARE ACCEPTABLE, ALSO PRESSURE OR VACUUM TYPES. THE
      DIATOMITE USED SHOULD BE ACCEPTABLE TO THE REVIEWING AUTHORITY.
      (BEAN-AWWA)

    FIELD 05F

    ACCESSION NO. W72-04145

USE OF POTASSIUM PERMANGANATE IN WATER TREATMENT,

CEDAR RAPIDS WATER WORKS, IOWA.

A. K. CHERRY.

JOURNAL OF THE AMERICAN WATER WORKS ASSOCIATION, VOL. 54, NO. 4, P 417-424,
APRIL 1962. 1 FIG, 1 TAB.

DESCRIPTORS:
*WATER TREATMENT, *ALGAE, *ACTINOMYCETES, *ODOR, ACTIVATED CARBON,
AMMONIA, CHLORINE.

IDENTIFIERS:
*POTASSIUM PERMANGANATE.

ABSTRACT:
THRESHOLD ODORS OF RAW WATER, NORMALLY UNDER 200, ROSE TO AN AVERAGE OF
896 IN JULY WITH A MAXIMUM ONE DAY OF 4,000. ALGAE COUNT AVERAGED
48,600 PER ML. AVERAGE PERMANGANATE DOSAGE WAS 10.1 PPM, AND CARBON
APPLICATION 51 PPM. FINISHED WATER ODORS WERE AS HIGH AS 14-18. THE
MOST SUCCESSFUL TREATMENT WAS DISCONTINUANCE OF SOFTENING AND THE
APPLICATION OF PERMANGANATE AT THE INTAKE, PERMITTING IT TO PASS
THROUGH THE PRIMARY MIXERS AND CLARIFIERS WITHOUT ADDING OTHER
CHEMICALS. ALUM WAS APPLIED IN THE SECONDARY MIXERS, WITH LIME FOR PH
ADJUSTMENT. CARBON WAS APPLIED AS THE WATER PASSED TO THE FINAL
CLARIFIERS. AMMONIA AND CHLORINE APPLICATION WAS CHANGED TO THE CLEAR
WELL AFTER FILTRATION. BY THIS TREATMENT, ODOR WAS REDUCED FROM
400-2,000 TO 3-5 IN FINISHED WATER. THE ABNORMALLY LARGE NUMBER OF
KNOWN ODOR PRODUCING ALGAE WAS ASSUMED TO BE THE PRINCIPAL CAUSE OF THE
UNDESIRABLE TASTES AND ODORS, HOWEVER, PART OF THE ODORS COULD BE
TRACED TO ACTINOMYCETES. (BEAN-AWWA)

FIELD 05F

ACCESSION NO. W72-04155

ODOR CONTROL WITH CARBON AND PERMANGANATE AT DES MOINES,

DES MOINES WATER WORKS, IOWA.

JOURNAL OF THE AMERICAN WATER WORKS ASSOCIATION, VOL 60, NO 10, P 1195-1198,
OCTOBER 1968. 2 TAB.

DESCRIPTORS:
*WATER TREATMENT, ODOR, *ALGAE, *DECOMPOSING ORGANIC MATTER, *POTASSIUM
COMPOUNDS, *ACTIVATED CARBON, ACTINOMYCETES, IOWA.

IDENTIFIERS:
*POTASSIUM PERMANGANATE, *DES MOINES(IOWA).

ABSTRACT:
AT PERIODS DURING THE YEAR THE THRESHOLD ODOR OF THE WATER SUPPLY AT
DES MOINES ROSE TO LEVELS CAUSING WIDESPREAD COMPLAINTS OF CONSUMERS.
THESE ODORS ARE TYPICALLY EARTHY OR MUSTY IN CHARACTER. INDUSTRIAL
WASTE OR DOMESTIC SEWAGE CONTAMINATION ARE NOT PROBLEMS IN THE RACCOON
RIVER AN OCCASIONAL SOURCE OF RAW WATER FOR DES MOINES. THE ODORS
RESULT FROM GROWTH OF ALGAE OR WASHING OF EXTRACTS FROM DECAYED
VEGETATION. ACTINOMYCETES HAVE BEEN ISOLATED WITH THE SAME ODOR
CHARACTERISTICS AS THE RIVER WATER. POTASSIUM PERMANGANATE IS A
VALUABLE ADJUNCT TO THE NORMAL FEED OF ACTIVATED CARBON DURING PERIODS
OF RIVER INTAKE; WHEN THE THRESHOLD ODOR OF THE RACCOON RIVER EXCEEDS
25, POTASSIUM PERMANGANATE IS FED. AT THRESHOLD ODOR LEVELS BELOW 25,
EITHER MATERIAL WILL PRODUCE SATISFACTORY WATER, AND THE CHOICE IS
PRIMARILY ONE OF CONVENIENCE. (BEAN-AWWA)

FIELD 05F

ACCESSION NO. W72-04242

ION-EXCHANGE FOR RECLAMATION OF REUSABLE SUPPLIES,

STANFORD UNIV., CALIF. DEPT. OF CIVIL ENGINEERING.

R. ELIASSEN, B. M. WYCKOFF, AND C. D. TONKIN.

JOURNAL OF THE AMERICAN WATER WORKS ASSOCIATION, VOL 57, NO 9, P 1113-1122, SEPTEMBER 1965. 6 FIG, 3 TAB, 4 REF.

DESCRIPTORS:
*WATER TREATMENT, *WATER REUSE, *RECLAMATION, *GROUNDWATER, RECHARGE, *ION EXCHANGE, *ORGANIC MATTER, *FILTRATION, DIATOMACEOUS EARTH, ALGAE, TREATMENT FACILITIES.

IDENTIFIERS:
FILTER AIDS, DIATOMITE FILTERS.

ABSTRACT:
PILOT PLANT STUDIES ON SEWAGE AT PALO ALTO, CALIF., INDICATE THE MOST ECONOMIC TREATMENT TO RECOVER REUSABLE WATER (POTABILITY NOT MENTIONED) IS THE ION-EXCHANGE PROCESS. INORGANIC MATERIALS, SUCH AS NITROGEN AND PHOSPHORUS COMPOUNDS, AND ORGANIC MATERIALS MEASURED BY THE CHEMICAL OXYGEN DEMAND, ALKYL BENZENE SULFURATE AND COLOR TESTS, CAN BE REMOVED IN ONE OPERATION, PROVIDING FILTRATION PRECEDES THE ION-EXCHANGE PROCESS. WATER SUITABLE FOR MANY PURPOSES, INCLUDING SURFACE STORAGE, CAN BE PRODUCED AT A COST COMPETITIVE WITH EXISTING SUPPLIES IN HIGH COST AREAS. DIATOMACEOUS-EARTH FILTRATION, FOLLOWED BY ANION-EXCHANGE, CAN BE ACCOMPLISHED AT A COST OF ABOUT 21 CENTS PER 1,000 GAL. ALTHOUGH WATER RECLAIMED WITH ION-EXCHANGE WILL STILL CONTAIN PHOSPHOROUS AND NITROGEN COMPOUNDS, THE PROBLEM OF ALGAE GROWTH IN RESERVOIRS WILL BE MINIMIZED. A LESSER DEGREE OF NUTRIENT REMOVAL, AND CONSEQUENT LOWER-COST TREATMENT, WOULD SUFFICE FOR RECHARGE OF GROUND WATER. (BEAN-AWWA)

FIELD 05F

ACCESSION NO. W72-04318

ORGANIC MATTER AND FINISHED WATER QUALITY,

EAST BAY MUNICIPAL UTILITY DISTRICT, OAKLAND, CALIF.

J. J. CONNORS, AND R. B. BAKER.

JOURNAL OF THE AMERICAN WATER WORKS ASSOCIATION, VOL 61, NO 3, P 107-113, MARCH 1969. 6 FIG, 2 TAB, 10 REF.

DESCRIPTORS:
*WATER TREATMENT, WATER QUALITY, *ORGANIC MATTER, ALGAE, COPPER SULFATE, *CHEMICAL OXYGEN DEMAND.

IDENTIFIERS:
TOTAL ORGANIC CARBON.

ABSTRACT:
WATER QUALITY DETERIORATION CAN OCCUR BECAUSE OF ORGANIC MATTER AFTER WATER HAS LEFT THE TREATMENT PLANT. SIGNIFICANT DIFFERENCES IN THE AMOUNT OF ORGANIC MATTER AND ALGAL BIOMASS CAN OCCUR BUT NOT BE INDICATED BY CHEMICAL ANALYSIS. COPPER SULFATE MAY NOT CONTROL TOTAL ORGANIC MATTER. IN SOME RESERVOIRS THE TOTAL ORGANIC LOAD AS MEASURED BY THE CHEMICAL OXYGEN DEMAND (COD) IS ESTABLISHED BY THE CHARACTER OF THE RUNOFF. ORGANIC REMOVALS BY FILTRATION ARE SLIGHT COMPARED WITH THOSE BY COAGULATION AND SEDIMENTATION. THE TOC APPARATUS FOR MEASURING 'TOTAL ORGANIC CARBON' WILL BE USEFUL FOR CONTINUED STUDIES. THE COD TEST, WHATEVER IT MEASURES, CAN BE MADE TO YIELD GOOD PRECISION WITH POTABLE WATERS. (BEAN-AWWA)

FIELD 05F

ACCESSION NO. W72-04495

BIOLOGICAL CONCENTRATION OF PESTICIDES BY ALGAE,

NORTH TEXAS STATE UNIV., DENTON. DEPARTMENT OF BIOLOGY.

B. D. VANCE, AND W. DRUMMOND.

JOURNAL OF THE AMERICAN WATER WORKS ASSOCIATION, VOL 61, NO 7, P 360-362,
JULY 1969. 2 TAB, 11 REF.

DESCRIPTORS:
*WATER QUALITY, *ALGAE, *PESTICIDES, *FISH, *FISH KILLS, *ROTIFERS,
*ECOLOGY, *ECOSYSTEMS.

IDENTIFIERS:
*POTENTIATION, *ORGANIC DEGRADATION.

ABSTRACT:
ALGAE ARE EXTREMELY EFFICIENT POTENTIATORS OF PESTICIDE RESIDUES.
CONCENTRATION OF MANY PESTICIDES BY CERTAIN ALGAE OCCURS QUITE RAPIDLY,
AND THERE IS APPARENTLY VERY LITTLE, IF ANY, DEGRADATION BY THE ALGAE.
THEREFORE THESE HIGHLY CONCENTRATED RESIDUES CAN BE BIOLOGICALLY
TRANSFERRED TO HIGHER MEMBERS OF THE FOOD CHAIN SUCH AS ROTIFERS AND
SMALL FISH. MOST HIGHER MEMBERS DO NOT APPEAR TO BE NEARLY AS RESISTANT
AS THE ALGAE TO THESE CHLORINATED COMPOUNDS. CONSEQUENTLY, FISH KILLS
AND SIMILAR CATASTROPHIES MAY BE BROUGHT ABOUT IN WATER CONTAINING AS
LITTLE AS 1 PPB OF A PESTICIDE. SUCH POTENTIATION AND BIOLOGICAL
TRANSFER HAS BEEN AMPLY DEMONSTRATED, BEGINNING WITH FISH AND
CONTINUING UP THE FOOD CHAIN TO MAN. IT IS IMPORTANT TO KNOW THE
EFFECTS OF SUCH CHEMICALS UPON ECOSYSTEMS AND HOW SUCH CHLORINATED
HYDROCARBON MOLECULES MAY INFLUENCE THE BASIC ECOLOGY OF OUR
ENVIRONMENT. IT WAS THOUGHT AT ONE TIME THAT INSECTICIDES DID NOT CAUSE
DEATH IN PLANTS, EVEN IN GROSS OVER-DOSAGES. HOWEVER, TESTS SHOW THAT
DIELDRIN, ALDRIN, AND ENDRIN ARE ALGICIDAL TO M. AERUGINOSA AT A
CONCENTRATION OF LESS THAN 5 NG/ML. (BEAN-AWWA)

FIELD 05B, 05C

ACCESSION NO. W72-04506

EUTROPHICATION OF SMALL RESERVOIRS IN THE GREAT PLAINS,

NEBRASKA UNIV., LINCOLN.

GARY L. HERGENRADER, AND MARK J. HAMMER.

AVAILABLE FROM THE NATIONAL TECHNICAL INFORMATION SERVICE AS PB-206 953,
$3.00 IN PAPER COPY, $0.95 IN MICROFICHE. MIMEO (UNDATED). 26 P, 4 FIG, 4
TAB, 6 REF. OWRR A-014-NEB(5).

DESCRIPTORS:
*EUTROPHICATION, *RESERVOIRS, *GREAT PLAINS, ALGAE, NEBRASKA,
TURBIDITY, STRATIFICATION, TEMPERATURE, CHLOROPHYLL, CYANOPHYTA,
AQUATIC PLANTS, PHOSPHORUS, DENSITY, WATER POLLUTION SOURCES, RUNOFF,
LIGHT PENETRATION, SAMPLING, PRIMARY PRODUCTIVITY, PHYTOPLANKTON,
EUGLENA, CHLOROPHYTA, NITRATES, AMMONIA, CARBON, RECREATION, WATER
POLLUTION CONTROL, ALKALINITY, DISSOLVED SOLIDS.

IDENTIFIERS:
*SALT VALLEY RESERVOIRS(NEB), *CLEAR WATER RESERVOIRS, *TURBID
RESERVOIRS, CYCLOTELLA, TRACHELOMONAS, MELOSIRA, STEPHANODISCUS,
ANABAENA, APHANIZOMENON, MICROCYSTIS, POTAMOGETON, POLYGONUM,
SAGITTARIA, NAJAS.

ABSTRACT:
LIMNOLOGICAL STUDIES OF FIVE RESERVOIRS IN THE SALT VALLEY WATERSHED
DISTRICT (EASTERN NEBRASKA) WERE INITIATED TO DETERMINE EXISTING
TROPHIC CONDITIONS, ESTIMATE EUTROPHICATION RATES, IF POSSIBLE, BY
MEASURING CHANGES IN SEVERAL PARAMETERS, IDENTIFY SOURCES OF NUTRIENT
INPUTS, AND EVALUATE PREVENTATIVE AND REMEDIAL MEASURES. DURING JUNE,
JULY, AND AUGUST, EACH OF THE STUDY RESERVOIRS WAS SAMPLED AT WEEKLY
INTERVALS; ONE LAKE WAS SAMPLED THROUGHOUT THE YEAR, BUT LESS
FREQUENTLY DURING ICE-COVER. WATER SAMPLES WERE ANALYZED FOR DISSOLVED
OXYGEN, TEMPERATURE, UNDERWATER LIGHT INTENSITY, ALKALINITY, PH,
HARDNESS, DISSOLVED AND SUSPENDED SOLIDS, IRON, COD, PHOSPHATES,
AMMONIA, NITRATE AND ORGANIC NITROGEN, CHLORIDE, SULFATE, AND
TURBIDITY; ALGAE IDENTIFIED, AND PRIMARY PRODUCTION DETERMINED. RUNOFF
WATERS IMPOUNDED IN THE RESERVOIRS CONTAIN SUFFICIENT NUTRIENT SALTS TO
SUPPORT ABUNDANT GROWTHS OF AQUATIC PLANTS. RESERVOIRS, LIGHT-LIMITED
BY SOIL TURBIDITY, SUPPORT NEITHER ABUNDANT GROWTHS OF AQUATIC PLANTS
NOR DENSE BLUE-GREEN ALGAL BLOOMS; CLEAR WATER RESERVOIRS ARE VERY
EUTROPHIC. EUTROPHICATION RATE IS VERY RAPID AND APPARENTLY DIRECTLY
RELATED TO AGE. CONTROL OF PHOTOSYNTHESIS THROUGH INHIBITION OF
SUNLIGHT PENETRATION BY THE ADDITION OF VARIOUS SUBSTANCES INTO THE
RESERVOIRS DIRECTLY OR TO THE WATER SURFACE SHOULD BE INVESTIGATED.
(JONES-WISCONSIN)

FIELD 05C, 02H

ACCESSION NO. W72-04761

BIOLOGICAL AND BACTERIOLOGICAL EVALUATION OF PILOT PLANT ARTIFICIAL RECHARGE
EXPERIMENTS,

PRAGUE DEPT. OF WATER TECHNOLOGY (CZECHOSLOVAKIA).

A. MORAVCOVA, L. MASINOVA, AND V. BERNATOVA.

WATER RESEARCH, VOL 2, NO 4, P 265-276, 1968. 8 FIG, 5 TAB, 13 REF.

DESCRIPTORS:
*ARTIFICIAL RECHARGE, COPPER, *ALGICIDES, ALGAE, *EVALUATION,
FILTRATION.

IDENTIFIERS:
SILVER.

ABSTRACT:
THE EFFECT OF THE ALGICIDE CA-350 (A COMBINATION OF COPPER AND SILVER)
WAS INVESTIGATED ON FOUR SAND FILTERS, MODELS OF AN ARTIFICIAL RECHARGE
BASIN. USING 300 MICROGRAM/1 (PPB) CU IN A PREVENTIVE TREATMENT THE
DEVELOPMENT OF FILAMENTOUS ALGAL FORMS WAS FULLY CONTROLLED WITH THE
CONSERVATION OF A NORMAL BACTERIAL MICROFLORA. THE PROCESSES OF
MINERALIZATION OF ORGANIC MATTER CONTINUED UNDER AEROBIC CONDITIONS.
COPPER WAS GRADUALLY ELUTED FROM THE SAND LAYERS OF THE MODEL FILTERS,
NEVER REACHING ITS MAXIMUM ALLOWABLE CONCENTRATION IN THE OUTFLOWING
WATER. (SKOGERBOE-COLORADO STATE)

FIELD 04B, 05F

ACCESSION NO. W72-05143

ECOLOGY OF CLADOPHORA FRACTA AND CLADOPHORA GLOMERATA,

SYRACUSE UNIV., N.Y.

DANIEL F. JACKSON, AND SHUNN-DAR LIN.

FEDERAL WATER POLLUTION CONTROL ADMINISTRATION, NO 16010--05/68, FINAL
REPORT, MAY 1968. 133 P, 10 FIG, 19 TAB, 93 REF, APPEND. FWPCA-WP 00782.

DESCRIPTORS:
*ALGAE, *EUTROPHICATION, *PHOTOSYNTHESIS, *RESPIRATION, GREAT LAKES,
NUISANCE ALGAE, WATER QUALITY, NUTRIENTS, PHOSPHORUS, NITROGEN, CALCIUM
CHLORIDE, TEMPERATURE, METABOLISM, EPIPHYTOLOGY, ON-SITE TESTS,
LABORATORY TESTS, LOTIC ENVIRONMENTS, LENTIC ENVIRONMENTS.

IDENTIFIERS:
*CLADOPHORA FRACTA, CLADOPHORA GLOMERATA, BUTTERNUT CREEK(NY).

ABSTRACT:
DETERMINATIONS OF RESPIRATION AND PHOTOSYNTHETIC RATES OF CLADOPHORA
FRACTA AND CLADOPHORA GLOMERATA UNDER NATURAL AND LABORATORY
CONDITIONS, AND NITRATE AND PHOSPHATE REQUIREMENTS UNDER LABORATORY
CONDITIONS WERE MADE. IN THE LABORATORY, CLADOPHORA FRACTA RESPIRATION
AND PHOTOSYNTHETIC RATES WERE GREATER WHEN OSCILLATED AT 120 STROKES
PER MINUTE THAN WITH NO OSCILLATION. THE HIGHEST CLADOPHORA FRACTA
METABOLIC RATES WERE OBSERVED IN THE MEDIUM LACKING CALCIUM CHLORIDE.
MARKEDLY LOW RATES WERE OBTAINED IN MEDIA LACKING NITRATE AND
PHOSPHATE. WITH SMALL INCREASES (0.00 TO 0.01 MG/L) OF EITHER NITRATE
OR PHOSPHATE CONCENTRATION, CLADOPHORA GLOMERATE PHOTOSYNTHETIC RATE
INCREASED SIGNIFICANTLY. WITH NO PHOSPHATE, CLADOPHORA GLOMERATE
RESPIRATION RATE WAS ERRATIC. PHOTOSYNTHETIC RATE WAS NOT AFFECTED BY
NITRATE CONCENTRATION VARYING FROM 0.01 TO 100.00 MG/L. IT IS CONCLUDED
THAT PHOSPHORUS IS A LIMITING FACTOR OF CLADOPHORA GROWTH. CLADOPHORA
METABOLIC RATES WERE NOT INFLUENCED BY PHOSPHATE CONCENTRATION LEVELS
0.01 TO 5.00G/L IN NITRATE-FREE MEDIA. GENERALLY, CLADOPHORA GLOMERATE
RESPIRATION AND PHOTOSYNTHETIC RATES INCREASED AS THE COMBINATION OF
NITRATE AND PHOSPHATE DOSES INCREASED. CLADOPHORA GLOMERATE RESPIRATION
AND PHOTOSYNTHETIC RATES UNDER NATURAL CONDITIONS WERE ABOUT THREE AND
SIX TIMES, RESPECTIVELY, GREATER THAN UNDER LABORATORY CONDITIONS.
(JONES-WISCONSIN)

FIELD 05C

ACCESSION NO. W72-05453

POWERS OF LAKE AUTHORITIES.

PUBLIC ACT NO. 416, CONNECTICUT LEGISLATIVE SERVICE, P 430 (1969). 1 P.

DESCRIPTORS:
*CONNECTICUT, *LAKES, *LOCAL GOVERNMENTS, *WATER QUALITY CONTROL,
ALGAE, AQUATIC WEEDS, ADMINISTRATION, SUPERVISORY CONTROL(POWER),
ADMINISTRATIVE AGENCIES.

ABSTRACT:
BY AMENDMENT TO CONNECTICUT STATUES, LEGISLATURES OF LOCAL GOVERNMENTS
ARE AUTHORIZED TO GRANT LAKE AUTHORITIES POWER TO CONTROL AND ABATE
ALGAE AND AQUATIC WEEDS, IN COOPERATION WITH THE STATE WATER RESOURCES
COMMISSION, AND STUDY WATER MANAGEMENT FOR RECOMMENDATIONS TO LOCAL
GOVERNMENTS. LAKE AUTHORITIES SHALL HAVE NO JURISDICTION IN MATTERS
SUBJECT TO REGULATION BY THE CONNECTICUT BOARD OF FISHERIES AND GAME.
(HART-FLORIDA)

FIELD 06E, 05G

ACCESSION NO. W72-05774

FORMAL DISCUSSION OF, 'THE RECLAMATION OF SEWAGE EFFLUENT FOR DOMESTIC USE,

METROPOLITAN WATER BOARD, LONDON (ENGLAND).

N. P. BURMAN.

PROCEEDINGS, INTERNATIONAL ASSOCIATION ON WATER POLLUTION RESEARCH
CONFERENCE, 3RD, MUNICH, GERMANY, 1966, P 19-23, 2 FIG, 1 TAB, 2 REF.

DESCRIPTORS:
*WATER PURIFICATION, *FLOW AUGMENTATION, *LAGOONS, ACTIVATED SLUDGE,
ALGAE, NUTRIENTS, VIRUSES, BACTERIA, E. COLI, PUBLIC HEALTH, PATHOGENIC
BACTERIA, TEMPERATURE, NITRATES, AMMONIA, *WATER REUSE, WASTE WATER
TREATMENT.

ABSTRACT:
SECONDARY EFFLUENT FROM A CONVENTIONAL ACTIVATED SLUDGE PLANT WAS
PASSED THROUGH A SERIES OF 3 LAGOONS. EFFLUENT FROM THE LAGOONS FLOWED
INTO THE RIVER LEE AT RYE MEADS. CHEMICAL, BACTERIOLOGICAL, AND
BIOLOGICAL QUALITY WERE OVER A PERIOD OF 2.5 YEARS. BACTERIAL
REDUCTIONS DEMONSTRATED A WIDE VARIABILITY WITH TEMPERATURE AND TYPE OF
ORGANISMS. IN GENERAL, E. COLI, TOTAL COLIFORM, AND FECAL STREPTOCOCCI
WERE THE MOST COMPLETELY REMOVED (IN EXCESS OF 90% ALWAYS). THE DEGREE
OF BACTERIAL IMPROVEMENT WAS NOT LINEAR, CHANGING WITH TEMPERATURE.
ENTEROVIRUSES WERE FOUND IN HALF THE SAMPLES OF THE LAGOON INFLUENT AND
ONLY TWICE IN THE LAGOON EFFLUENT, FULLY JUSTIFYING THE PLACE OF
MATURATION PONDS IN WATER RECLAMATION. QUESTIONS WERE ALSO RAISED ABOUT
THE FLOTATION PROCESS, AND TEMPERATURE OF THE ALGAE NUTRIENT STRIPPING
PROCESS, AS WELL AS ABS REMOVALS. (LOWRY-TEXAS)

FIELD 05D

ACCESSION NO. W72-06017

THE EFFECT OF ALGAL CONCENTRATION, LUMINOUS INTENSITY, TEMPERATURE, AND DIURNAL
CYCLE OR PERIODICITY UPON GROWTH OF MIXED ALGAL CULTURES FROM WASTE
STABILIZATION LAGOONS AS DETERMINED ON THE WARBURG APPARATUS,

MISSOURI UNIV., COLUMBIA. DEPT. OF CHEMICAL ENGINEERING.

R. H. LUEBBERS, AND D. N. PARIKH.

PROCEEDINGS, INDUSTRIAL WASTE CONFERENCE, 21ST, MAY 3-5, 1966, PART I, P
348-367. 13 FIG, 21 REF.

DESCRIPTORS:
*WASTE WATER TREATMENT, *OXIDATION LAGOONS, *ALGAE, LIGHT INTENSITY,
CARBON DIOXIDE, GROWTH RATES, TEMPERATURE, PILOT PLANTS, HYDROGEN ION
CONCENTRATION, ORGANIC LOADING, CENTRIFUGATION, RESPIRATION,
BIOCHEMICAL OXYGEN DEMAND, NITROGEN FIXATION.

IDENTIFIERS:
*OXYGEN PRODUCTION.

ABSTRACT:
THE NORMAL PERFORMANCE AND EFFECT OF VARIOUS OPERATING VARIABLES ON
WASTE STABILIZATION LAGOONS WERE INVESTIGATED IN A 2 STAGE EXPERIMENTAL
STUDY. THE RESULTS OF THE FIRST STAGE PILOT PLANT STUDY, PARTICULARLY
THE EFFECT OF SUPPLEMENTARY LIGHT ON ALGAL GROWTH PROMPTED THE USE OF
THE WARBURG APPARATUS. USING SUBSTRATE FROM THE PILOT PLANT LAGOONS,
TESTS RUNS WERE MADE BOTH WITH AND WITHOUT A SOURCE OF ACTINIC LIGHT,
AND WITH KOH, A 'CARBON DIOXIDE BUFFER', OR NO MATERIAL IN THE CENTER
WELL OF THE WARBURG FLASKS. THE STANDARD 5 DAY BOD OCCURRED IN 50 HOURS
AND THE NITROGEN PLATEAU AT 350 HOURS. O2 PRODUCTION IS ALMOST A LINEAR
FUNCTION OF ALGAL AND OTHER BIOTA CONCENTRATION. O2 PRODUCTION
INCREASED AS A LOGARITHMIC FUNCTION OF LIGHT CONCENTRATION UP TO 720
FT-CANDLES, REMAINED CONSTANT UNTIL 4500 FT-CANDLES, AND THEN DECREASED
WITH THE ALGAE DYING WITHIN 24 HOURS. O2 PRODUCTION VARIED WITH
TEMPERATURE, VERY LOW AT 10 DEG C, INCREASING RATE AT AN INCREASING
RATE TO ABOVE 20 DEG C, AND CONTINUES TO INCREASE AT A DECREASING RATE.
MAXIMUM RATE OF O2 PRODUCTION IS ABOVE 35 DEG C. O2 CONSUMPTION WITHOUT
LIGHT INCREASED LINEARLY WITH TEMPERATURE. BOTH DIURNAL CYCLE AND
PERIODICITY EFFECTED O2 PRODUCTION WITH A MAXIMUM RATE OF PRODUCTION
OBTAINED WITH A 55 TO 60% PERIODICITY. (MORGAN-TEXAS)

FIELD 05D, 05C

ACCESSION NO. W72-06022

FIELD STUDIES ON SEDIMENT-WATER ALGAL NUTRIENT INTERCHANGE PROCESSES AND WATER
   QUALITY OF UPPER KLAMATH AND AGENCY LAKES, JULY 1967-MARCH 1969,

   PACIFIC NORTHWEST WATER LAB., CORVALLIS, OREG.

   A. R. GAHLER.

   AVAILABLE FROM THE NATIONAL TECHNICAL INFORMATION SERVICE AS PB-207 643,
      $3.00 IN PAPER COPY, $0.95 IN MICROFICHE. WORKING PAPER NO. 66, OCTOBER
      1969, 33 P, 1 FIG, 9 TAB, 6 REF, APPEND. FWPCA PROGRAM 16010---10/69.

      DESCRIPTORS:
         *EUTROPHICATION, LAKE, *LAKE SEDIMENTS, *NUTRIENTS, NITROGEN COMPOUNDS,
         PHOSPHOROUS COMPOUNDS, HYDROGEN ION CONCENTRATION, CONDUCTIVITY,
         TEMPERATURE, DISSOLVED OXYGEN, PHYTOPLANKTON, WATER ANALYSIS, ALGAL
         CONTROL, ALGAE, WATER QUALITY, SEDIMENT-WATER INTERFACES, OREGON,
         SESSILE ALGAE, CYANOPHYTA.

      IDENTIFIERS:
         *UPPER KLAMATH LAKE(ORE.), AGENCY LAKE(ORE.), *OSCILLATORA.

      ABSTRACT:
         STUDIES OF ALGAL NUTRIENT INTERCHANGE BETWEEN SEDIMENT AND WATER UNDER
         ENVIRONMENTAL CONDITIONS WERE CARRIED OUT IN UPPER KLAMATH LAKE,
         OREGON, FROM JULY 1967 TO MARCH 1969. EXPERIMENTAL 'POOLS' OF LAKE
         WATER IN CONTACT WITH THE SEDIMENT AND EXPERIMENTAL POOLS OF WATER NOT
         EXPOSED TO THE SEDIMENT WERE COMPARED WITH THE OPEN LAKE FROM NOVEMBER
         1967 TO JUNE 1968. WATER QUALITY MEASUREMENTS IN AGENCY AND UPPER
         KLAMATH LAKES WERE MADE TO DETERMINE WHETHER INTERCHANGE PROCESSES
         COULD BE OBSERVED DIRECTLY IN THE WATER, TO ESTABLISH CONDITIONS FOR
         LABORATORY INTERCHANGE TESTS, AND TO COMPARE LAKE CONDITIONS WITH THE
         EXPERIMENTAL POOLS. DATA WERE OBTAINED FROM JULY 1967 TO MARCH 1969 ON
         PH, CONDUCTIVITY, TEMPERATURE, DISSOLVED OXYGEN, CHEMICAL COMPOSITION,
         AND PHYTOPLANKTON. INTERCHANGE DEFINITELY OCCURRED WHEN OSCILLATORA
         FLOATED TO THE LAKE SURFACE WITH ATTACHED SEDIMENT WHICH CONTAINED
         SOLUBLE NITROGEN AND PHOSPHORUS COMPOUNDS. A PLASTIC-BOTTOMED POOL OF
         WATER NOT EXPOSED TO SEDIMENT EXHIBITED HIGHER OXYGEN CONTENT UNDER THE
         ICE AND HAD LESS PHYTOPLANKTON GROWTH IN SPRING THAN THE POOLS EXPOSED
         TO SEDIMENTS. THE EFFECTS OF GAS EVOLUTION, WIND, CURRENTS, FISH,
         BOATING, BENTHOS, DIFFUSION, ETC., ON THE SHALLOW LAKES WAS NOT
         QUANTITATIVELY DETERMINED, BUT IT SEEMS QUITE PROBABLE THAT ANYTHING
         THAT STIRS THE SEDIMENT CAUSES INTERCHANGE OF NUTRIENTS. (ALSO SEE
         W72-06052)

      FIELD 05C, 02H

      ACCESSION NO. W72-06051

INTERIM REPORT, UPPER KLAMATH LAKE STUDIES OREGON,

   FEDERAL WATER POLLUTION CONTROL ADMINISTRATION, CORVALLIS, OREG. PACIFIC
      NORTHWEST WATER LAB.

   W. E. MILLER, AND J. C. TASH.

   AVAILABLE FROM THE NATIONAL TECHNICAL INFORMATION SERVICE AS PB-207 640,
      $3.00 IN PAPER COPY, $0.95 IN MICROFICHE. WATER POLLUTION CONTROL RESEARCH
      SERIES, PUBLICATION WP-20-8, SEPTEMBER 1967, 37 P, 3 REF, 14 TAB, 30 REF.
      FWPCA PROGRAM 16010---09/67.

      DESCRIPTORS:
         *EUTROPHICATION, LAKES, LAKE SEDIMENTS, ALGAL CONTROL, *NUTRIENTS,
         ALGAE, BENTHIC FAUNA, CHEMICAL ANALYSIS, SESSILE ALGAE, *PRIMARY
         PRODUCTIVITY, OREGON, CYANOPHYTA.

      IDENTIFIERS:
         *UPPER KLAMATH LAKE(ORE.), *APHANIZOMENON FLOS-AQUAE, *CARBON-14
         MEASUREMENTS.

      ABSTRACT:
         THE WATER QUALITY IN UPPER KLAMATH LAKE WATERSHED DURING THE PERIOD
         MARCH 1965 TO APRIL 1966 IS DEFINED. THE SOURCES OF ALGAL NUTRIENTS AND
         OTHER CONSTITUENTS IN WATER FLOWING INTO AND OUT OF UPPER KLAMATH LAKE
         ARE IDENTIFIED AND THE QUANTITY OF THESE CONSTITUENTS IS COMPARED IN
         PRISTINE STREAMS, CANALS, RIVERS, AGRICULTURAL DRAINAGE, AND SPRINGS.
         CHEMICAL ANALYSES WERE MADE OF ALGAE, LAKE SEDIMENT AND BOTTOM FAUNA
         SAMPLES. ALGAL SPECIES WERE IDENTIFIED AND MEASURED; DURING THE PERIOD
         OF THE STUDY APHANIZOMENON FLOS-AQUAE REPRESENTED ABOUT 90-99 PERCENT
         OF THE TOTAL ALGAL CROP DURING THE SUMMER. LIMITING NUTRIENT STUDIES
         USING THE ISOTOPE CARBON-14 TO MEASURE VARIATIONS IN PRIMARY
         PRODUCTIVITY WERE CONDUCTED IN SITU AND IN THE LABORATORY; DATA FROM
         THESE STUDIES WERE DIFFICULT TO INTERPRET. (SEE ALSO W72-06051)

      FIELD 05C, 02H

      ACCESSION NO. W72-06052

MECHANISMS OF ANAEROBIC WASTE TREATMENT,

CALIFORNIA UNIV., BERKELEY.

GORDON L. DUGAN, AND WILLIAM J. OSWALD.

IN: POTATO WASTE TREATMENT, PROCEEDINGS OF A SYMPOSIUM SPONSORED BY FWPCA AND
IDAHO UNIVERSITY, MARCH 1968. P 5-17, 5 FIG, 1 TAB.

DESCRIPTORS:
*OXIDATION LAGOONS, *ANAEROBIC DIGESTION, *MUNICIPAL WASTES, METHANE
BACTERIA, ACID BACTERIA, SULFUR BACTERIA, ALGAE, MIXING, HYDROGEN ION
CONCENTRATION, TEMPERATURE, TOXICITY, DISSOLVED OXYGEN, LIGHT
INTENSITY, SLUDGE, DEPTH, ORGANIC LOADING, DESIGN CRITERIA, ODORS,
EFFICIENCIES, *WASTE WATER TREATMENT.

IDENTIFIERS:
*DETENTION TIME.

ABSTRACT:
ANAEROBIC TREATMENT OF VARIOUS MUNICIPAL AND INDUSTRIAL WASTES IS
PRACTICED MAINLY IN ANAEROBIC DIGESTERS, FACULTATIVE LAGOONS, AND
ANAEROBIC LAGOONS. DECOMPOSITION TAKES PLACE IN TWO PHASES, DURING
ANAEROBIC DIGESTION. THE FIRST, OR 'ACID' PHASE INVOLVES THE BREAKDOWN
OF SETTLED SLUDGE TO ORGANIC VOLATILE ACIDS AND ENERGY, AND THE SECOND
OR 'GAS' PHASE CONSISTS OF THE CONVERSION OF THE ORGANIC VOLATILE ACIDS
TO CO_2, CH_4, H_2, N_2, AND MINOR AMOUNTS OF OTHER GASES. DIGESTION
TEMPERATURES USUALLY LIE SOMEWHERE BETWEEN 15 AND 35 DEG C, WITH GAS
COLLECTION AND RE-USE AS A HEAT SOURCE BEING PRACTICED IN CONVENTIONAL
ANAEROBIC DIGESTION. DIGESTER LOADING IS EXPRESSED AS LBS VOLATILE
SOLIDS/FT3/DAY AND RANGE FROM 0.08 TO 0.1. FACULTATIVE PONDS ARE
INITIALLY LOADED AT FROM 20 TO 50 LBS BOD5/ACRE/DAY, BUT ONCE
ESTABLISHED THEY MAY BE LOADED AS HIGH AS 1000 LBS BOD5/ACRE/DAY.
DETENTION TIMES OF 3, 30, 90 AND 8-10 DAYS ARE COMMON FOR AEROBIC
PONDS, CONVENTIONAL DIGESTION, FACULTATIVE PONDS, AND ANAEROBIC PONDS
RESPECTIVELY, WITH EFFICIENCIES FOR SAME BEING 80% FOR AEROBIC PONDS,
85 TO 95% FOR FACULTATIVE PONDS, AND 70% FOR ANAEROBIC PONDS. LACK OF
CONTROL OF ODOR PROBLEMS HAS HINDERED APPLICATION OF ANAEROBIC PONDS,
AND RESEARCH WORK TO DEVELOP SUCH CONTROLS IS NEEDED.
(LOWRY-TEXAS)

FIELD 05D

ACCESSION NO. W72-06854

SAND FILTRATION OF PARTICULATE MATTER,

WYOMING UNIV., LARAMIE. DEPT. OF CIVIL ENGINEERING.

E. DAVIS, AND J. A. BORCHARDT.

JOURNAL SANITARY ENGINEERING DIVISION, AMERICAN SOCIETY OF CIVIL ENGINEERS,
VOL. 92, NO. SA5, P 47-60, PAPER 4940, OCTOBER 1966. 8 FIG, 2 TAB. USPHS
GRANT WP-00329.

DESCRIPTORS:
*WATER TREATMENT, *FILTRATION, SANITARY ENGINEERING, *FILTERS, *DESIGN,
COAGULATION, ALGAE, ACTIVATED CARBON, *FLOCCULATION, *HEAD LOSS.

IDENTIFIERS:
*FILTER MEDIA.

ABSTRACT:
THE PENETRATION OF PARTICULATE MATTER THROUGH SAND FILTERS CAN BE
ELIMINATED OR GREATLY REDUCED BY THE APPLICATION OF A COAGULATING
CHEMICAL SUPPLIED AS A SOLUTION DIRECTLY TO THE FILTER INFLUENT. AT
HIGH DENSITIES OF ORGANISMS, THE INABILITY TO REMOVE ALL CELLS OF
SELENASTOUM AND THE DIMINISHING RATE OF REMOVAL ASSOCIATED WITH
INCREASED COAGULANT DOSAGES INDICATES THE NECESSITY OF PRETREATMENT.
THE MODIFICATION OF THE PATTERN OF COAGULANT DEPOSIT, THE HEAD LOSS
INCREASE PATTERN AND THE INABILITY TO REMOVE ALL ALGAL CELLS AT HIGH
DENSITIES SEEM TO INDICATE THE IMPORTANCE OF THE CHEMICAL NATURE OF THE
FLOC IN ANY FILTRATION MECHANISM. THE LOCATION OF POWDERED ACTIVATED
CARBON DEPOSITED UNDER PLAIN FILTRATION WAS SHOWN TO BE LOGARITHMIC
WITH DEPTH. THE LOCATION OF DEPOSITED COAGULANT AND ACTIVATED CARBON
DECLINED LOGARITHMICALLY WITH DEPTH IN THE UPPER PART OF THE UNIFORM
SIZE FILTER MEDIA. (BEAN-AWWARF)

FIELD 05F

ACCESSION NO. W72-07334

ALGAE CONTROL,

WISCONSIN COMMITTEE ON WATER POLLUTION, MADISON.

K. M. MACKENTHUN.

PUBLIC WORKS, VOL 91, NO 9, P 114-116 AND 158, SEPTEMBER 1960. 2 FIG, 8 REF.

DESCRIPTORS:
WATER TREATMENT, *ALGAE, *COPPER SULFATE, *ALKALINITY, WATER QUALITY
CONTROL, *CONTROL, *ALGAL CONTROL, TOXICITY.

ABSTRACT:
DAMAGE CAUSED BY ALGAL GROWTHS IS DESCRIBED. THE DIFFERENCE IN
TREATMENT OF WATERS WITH LOW ALKALINITY AND THOSE WITH OVER 40 MG/L
METHYL ORANGE ALKALINITY IS DISCUSSED. THE APPLICATION OF COPPER
SULFATE, EQUIPMENT REQUIRED, CONTROL OF THE APPLICATION, AND THE
TOXICITY OF THE APPLICATION TO AQUATIC LIFE ARE DESCRIBED. MORE THAN
HALF THE STATES REPORT COMPLETE SUPERVISION OF FIELD APPLICATION OF
CHEMICALS, AND IN ADDITION ONE-FIFTH REPORT THAT THEY SPOT CHECK FIELD
APPLICATIONS. (BEAN-AWWARF)

FIELD 04A, 05F, 05G, 05C

ACCESSION NO. W72-07442

FIXATION OF ATMOSPHERIC NITROGEN BY NONLEGUMES IN WET MOUNTAIN MEADOWS,

AGRICULTURAL RESEARCH SERVICE, FORT COLLINS, COLO. SOIL AND WATER
CONSERVATION RESEARCH DIV.

L. K. PORTER, AND A. R. GRABLE.

AGRONOMY JOURNAL, VOL 61, NO 4, P 521-523, 1969, 1 FIG, 5 TAB, 8 REF.
JULY-AUGUST.

DESCRIPTORS:
*NITROGEN FIXATION, IRRIGATION, ORGANIC MATTER, *GRASSLANDS, MOUNTAINS,
*SOIL MICROORGANISMS, SOIL ALGAE, TURF, SOIL MICROBIOLOGY, WATER
POLLUTION SOURCES.

IDENTIFIERS:
*MOUNTAIN MEADOW, NITROGEN-15, SOD MATS, NON-LEGUMES, *NITROGEN FIXING
ORGANISMS.

ABSTRACT:
MOUNTAIN SOILS SUBJECTED TO EXCESSIVE IRRIGATION AND HIGH WATER TABLES
ACCUMULATE ORGANIC MATTER IN SOD MATS. MEADOW SOILS OFTEN CONTAIN TWO
OR THREE TIMES MORE N THAN EQUAL AREAS OF ADJACENT DRY SOILS. SYMBIOTIC
NITROGEN FIXATION BY LEGUMES IS ONE POSSIBLE SOURCE OF N FOR SOD MAT
FORMATION. FIXATION OF N2 BY FREE-LIVING ORGANISMS HAS ALSO BEEN
POSTULATED AND WAS VERIFIED BY THE LABORATORY STUDIES REPORTED HERE.
NITROGEN FIXATION BY SOD MATS CONTAINING NO LEGUMES WAS DETERMINED AT
18C IN ATMOSPHERES CONTAINING N2 15. BOTH PHOTOSYNTHETIC AND
NONPHOTOSYNTHETIC ORGANISMS APPEARED TO FIX N2. IN 10 DAYS, MATS IN THE
DARK FIXED FROM 0.76 TO 1.90 KG N2 HA-1 AND ILLUMINATED MATS FIXED FROM
3.72 TO 6.86 KG N HA-1. (SKOGERBOE-COLORADO STATE)

FIELD 05B, 03F

ACCESSION NO. W72-08110

SEWAGE NUTRIENT REMOVAL BY A SHALLOW ALGAL STREAM,

NATIONAL INST. FOR WATER RESEARCH, PRETORIA (SOUTH AFRICA).

J. HEMENS, AND M. H. MASON.

WATER RESEARCH, VOL. 2, NO. 4, P 277-287, 1968. 5 FIG, 1 TAB, 28 REF.

DESCRIPTORS:
*ALGAE, *NITROGEN, *PHOSPHORUS, *SEWAGE EFFLUENT, *SEWAGE TREATMENT,
*NUTRIENT REMOVAL, *WATER REUSE, SEASONAL, WATER TREATMENT.

IDENTIFIERS:
*SEWAGE NUTRIENT REMOVAL.

ABSTRACT:
THE USE OF ALGAE PONDS FOR REMOVING NITROGEN AND PHOSPHORUS BY
ASSIMILATION FROM SECONDARY SEWAGE EFFLUENT SUFFERS FROM THE
DISADVANTAGE OF INEFFICIENT LIGHT UTILIZATION BY THE DENSE ALGAL
CULTURES REQUIRED AND FROM THE PRACTICAL PROBLEM OF ECONOMICALLY
REMOVING THE ALGAE FROM SUSPENSION. AN EXPERIMENT IS DESCRIBED IN WHICH
THE PH INCREASE RESULTING FROM ALGAL PHOTOSYNTHESIS IN A SHALLOW STREAM
WAS USED TO REMOVE PHOSPHATE BY PRECIPITATION AND NITROGEN BY
ASSIMILATION AND LOSS TO THE ATMOSPHERE AS AMMONIA GAS. ASSOCIATION OF
PRECIPITATED PHOSPHATE WITH THE ALGAL CELLS PRODUCED A GRANULAR ALGAL
SEDIMENT EASILY REMOVED BY GRAVITY SETTLEMENT. REMOVAL OF NITROGEN AND
PHOSPHATE EXCEEDED 90 PER CENT IN THE WARMER SEASONS BUT AT WINTER
TEMPERATURES THE EFFICIENCY WAS LESS, DUE TO DECREASE IN THE ELEVATION
OF PH VALUE. THE METHOD APPEARS TO HAVE POSSIBLE APPLICATION FOR RURAL
COMMUNITIES IN EQUABLE CLIMATES WHERE WATER RE-USE IS DESIRABLE.
(SKOGERBOE-COLORADO STATE)

FIELD 05D

ACCESSION NO. W72-08133

DIATOMITE FILTRATION OF GOOD QUALITY SURFACE SUPPLIES,

JOHNS-MANVILLE PRODUCTS CORP., MANVILLE, N.J.

D. W. DAVIS, T. S. BROWN, S. SYROTYNSKI, AND E. HENDERSON.

CHEMICAL ENGINEERING PROGRESS SYMPOSIUM SERIES NO. 97, VOL 65, 1969, P
133-139, 3 FIG, 2 TAB, 10 REF.

DESCRIPTORS:
*DIATOMACEOUS EARTH, *FILTRATION, *SEPARATION TECHNIQUES, TURBIDITY,
SUSPENDED SOLIDS, ALGAE, PILOT PLANTS, HEAD LOSS, PERMEABILITY,
TEMPERATURE, OPERATION AND MAINTENANCE, DESIGN CRITERIA, *WATER
TREATMENT, *NEW YORK, WATER PURIFICATION.

IDENTIFIERS:
*MASSENA(NY).

ABSTRACT:
THE DESIGN OF DIATOMACEOUS EARTH FILTERS FOR WATER PURIFICATION STILL
REQUIRES PILOT PLANT DATA ON THE ACTUAL FILTRATION CHARACTERISTICS OF
EACH WATER TO BE FILTERED. IN ADDITION, THE TRADITIONAL APPROACH TO
DIATOMITE FILTRATION USES BODY FEED LEVELS EXPRESSED AS A RATIO OF
MILLIGRAMS/LITER OF BODY FEED TO THE MEASURED TURBIDITY. HOWEVER, FULL
SCALE OPERATIONAL DATA OBTAINED AT THE MASSENA, NEW YORK WATER
TREATMENT PLANT HAVE DEMONSTRATED THAT THE BODY FEED TO TURBIDITY
REQUIRED FOR ECONOMIC OPERATION IS MUCH HIGHER THAN THE RANGE OF 2-5
COMMONLY FOUND TO BE DESIRABLE AT OTHER LOCATIONS. SINCE THESE PROBLEMS
WERE ANTICIPATED, MEASUREMENTS OF THE WATER QUALITY BY DIRECT
MICROSCOPIC COUNTS WAS ALSO EMPLOYED, AND SUBSEQUENT INVESTIGATIONS
REVEALED THAT THE TOTAL COUNT VARIATIONS EXERTED A GREATER INFLUENCE ON
THE FILTER PERFORMANCE THAN THE TURBIDITY (OPTICALLY MEASURED). FURTHER
ITEMS FOR CONSIDERATION WHEN PLANNING AN EXPERIMENTAL PROGRAM TO OBTAIN
DESIGN DATA INCLUDE: (1) REVIEW OF HISTORICAL DATA; (2) USE OF A WIDE
ENOUGH VARIATION OF BODY FEED LEVELS AND FILTER AIDS TO PRECLUDE
EXCESSIVE EXTRAPOLATION; AND (3) TESTS SHOULD BE RUN AT TIMES OF
POOREST FEED WATER QUALITY. (LOWRY-TEXAS)

FIELD 05F

ACCESSION NO. W72-08686

ALGAL REMOVAL WITH A DIATOMACEOUS EARTH REVERSIBLE CAKE FILTER,

GRANGER FILTER CO., WAKEFIELD, MASS.

J. G. BROWN.

CHEMICAL ENGINEERING PROGRESS SYMPOSIUM SERIES NO. 97, VOL 65, 1969, P
128-132, 1 TAB, 7 REF.

DESCRIPTORS:
*ALGAE, *FILTRATION, *DIATOMACEOUS EARTH, SEPARATION TECHNIQUES,
PRESSURE, FLOW RATES, SLUDGE, SLURRIES, AUTOMATIC CONTROL, PIPING
SYSTEMS(MECHANICAL), CLEANING, COST ANALYSIS, WATER PURIFICATION,
*WATER TREATMENT.

IDENTIFIERS:
*ALGAE HARVESTING, *DIATOMITE FILTRATION.

ABSTRACT:
LAGOON WATERS CONTAINING VARYING CONCENTRATIONS OF GREEN ALGAE WERE
FILTERED THROUGH A STANDARD GRANGER C-33 REVERSIBLE FILTER WITH
DIATOMITE PRECOAT AT HIGH VELOCITY DURING A 10 WEEK PILOT PLANT TEST.
THE FILTER CLOTH COMPRISED 22 SQUARE FEET AND WAS MADE OF MONOFILAMENT
POLYPROPYLENE. THE PILOT PLANT INCLUDED AN AUTOMATIC CYCLE CONTROLLER
WITH AIR OPERATED VALVES, AND A DIATOMITE SLURRY HANDLING SYSTEM
INCLUDING A SLUDGE PUMP, LIFT PUMP, AND AN AGITATED DIATOMITE SETTLING
AND DECANTING TANK. THE SUCCESS OF THE SELF-PLUGGING BACKWASH OF THE
REVERSIBLE FILTRATION PROCESS WAS DEMONSTRATED, AS WAS THE
EFFECTIVENESS OF USING HIGH INTERNAL FILTER VELOCITIES, LOW PRESSURE
DROPS ACROSS THE CAKE, AND SMOOTH MONOFILAMENT POLYPROPYLENE CLOTH.
FILTRATE AVERAGED 10 PPM OF ALGAE, WHILE THE RECYCLED CONCENTRATED
STREAM CONTAINED 500 PPM OF ALGAE. TOTAL COSTS FOR REMOVAL OF 0.54 LB
ALGAE/1000 GALLONS USING 2.2 LB OF DIATOMITE/LB OF ALGAE WERE ESTIMATED
AT 8.6 TO 13.6 CENTS/1000 GALLONS PROCESSED. AFTER 10 WEEKS OF NEARLY
CONTINUOUS OPERATION, THE FILTER DID NOT REQUIRE CLEANING. THE
EXPERIENCE IN OPERATION ACQUIRED FROM THE PILOT PLANT TESTS WILL
HOPEFULLY MAKE THE ECONOMICS OF THE PROCESS EVEN MORE ATTRACTIVE.
(LOWRY-TEXAS)

FIELD 05F

ACCESSION NO. W72-08688

PRE-TREATMENT LICKS ALGAE PROBLEM,

OSHKOSH DEPT. OF PUBLIC WORKS, WIS.

J. STRAUSS.

WATER AND SEWAGE WORKS, VOL 110, NO 7, P 267-268, JULY 1963. 2 FIG.

DESCRIPTORS:
*WATER TREATMENT, SANITARY ENGINEERING, *FILTRATION, DESIGN, *ALGAE,
COPPER SULFATE, SEDIMENTATION, ACTIVATED CARBON, CHLORINE, COSTS,
TREATMENT FACILITIES, CLEANING, *ALGAL CONTROL, WISCONSIN.

IDENTIFIERS:
ALUM, *OSHKOSH(WIS), BACKWASHING.

ABSTRACT:
ON AN 8.8 MGD FILTER PLANT AT OSHKOSH, 9.8 PERCENT WASHWATER WAS
SOMETIMES REQUIRED, BECAUSE ALGAE CLOGGED THE FILTERS. CHLORINE, ALUM,
AND ACTIVATED CARBON APPLIED INSIDE THE PLANT REDUCED UNPLEASANT TASTES
AND ODORS BUT DID NOT ELIMINATE SUFFICIENT QUANTITIES OF ALGAE. A 30
ACRE PRESETTLING BASIN WAS CONSTRUCTED IN A BAY NEAR THE PLANT, WITH
BAFFLE WALLS TO INSURE MAXIMUM RETENTION. WATER ENTERS THE BASIN
THROUGH A 36 IN. STEEL LINE TO 2,000 FT. OUT IN THE LAKE. COPPER
SULFATE SOLUTION IS PUMPED INTO THE INTAKE LINE AS WATER ENTERS THE
BASIN. AFTER THE PRE-SETTLING CHLORINE, ACTIVATED CARBON AND ALUM ARE
APPLIED IN THE MIXING TANK. TREATMENT IN CONVENTIONAL SETTLING BASINS
AND FILTRATION FOLLOW. DURING THE FIRST FIVE MONTHS OF OPERATION THE
AVERAGE NUMBER OF FILTERS WASHED DAILY WAS 3.8 RATHER THAN THE PREVIOUS
18.6. BACKWASHING WATER WAS REDUCED TO 2.6 PERCENT AND COST OF
CHEMICALS WAS REDUCED TO 70 PERCENT OF THE PREVIOUS COST. (BEAN-AWWARF)

FIELD 05F

ACCESSION NO. W72-08859

MEASUREMENT AND CONTROL OF PARTICULATE MATTER IN FILTER EFFLUENTS,

FLINT WATER UTILITY, MICH.

R. M. HARWOOD.

JOURNAL OF THE AMERICAN WATER WORKS ASSOCIATION, VOL. 55, NO. 4, P 487-489, APRIL 1963. 1 REF.

DESCRIPTORS:
*WATER TREATMENT, *FILTRATION, *ALGAE, *WATER QUALITY, *ACTIVATED CARBON, *ELECTROLYTES, FILTERS, *ANALYTICAL TECHNIQUES.

IDENTIFIERS:
*MILLIPORE FILTERS.

ABSTRACT:
THE SEARCH FOR IMPROVED QUALITY PUTS MORE AND MORE EMPHASIS ON MEASURMENT AND CONTROL. TYPE AA MILIPORE FILTERS, PORE SIZE 0.8 MICRONS WITH 9.6-SQ CM FILTER AREA AND SUCTION EQUAL TO 12 IN. MERCURY COLUMN ARE USED, TO MEASURE MATERIALS IN THE EFFLUENT OF FILTERS. THE MOST RAPID METHOD FOR DETECTION AND ANALYSIS IS THE SIMPLE PATCH TEST. THE FILTER CAKE IS MOUNTED AND COMPARED WITH PREVIOUS SAMPLES COLLECTED FROM THE SAME SYSTEM. THE PARTICULATE DISCOLORATION OF THE FILTER IS A ROUGH INDICATION OF THE CLEANLINESS OF THE SAMPLE. FOR ALGAE COUNT THE FILTER IS WASHED AND SUBJECTED TO MICROSCOPIC COUNTING, REPORTED IN ORGANISMS PER GALLON. THE DRY FILTER MAY BE PLACED ON THE MICROSCOPE STAGE FOR DIRECT COUNTING OF PARTICLES BY REFLECTED LIGHT. CARBON AND FLOC PARTICLES CAN BE VISUALLY IDENTIFIED. TIME OF FLOW THROUGH THE MEMBRANE FILTER WILL INDICATE THE FILTERABILITY. MOST OF THE ALGAE FORMS IN THE RAW WATER MAY APPEAR ON THE MEMBRANE FILTER, ALSO FINES OF THE ACTIVATED CARBON. POLYELECTROLYTES MAY BE USEFUL IN REDUCING PENETRATION IN THE PLANT FILTERS. (BEAN-AWWARF)

FIELD 05F, 05D

ACCESSION NO. W72-09693

RESEARCH ON ALGAL ODOR,

ROBERT A. TAFT SANITARY ENGINEERING CENTER, CINCINNATI, OHIO.

T. E. MALONEY.

JOURNAL OF THE AMERICAN WATER WORK ASSOCIATION, VOL. 55, NO. 4, P 481-486, APRIL 1963. 1 FIG, 2 TAB, 3 REF.

DESCRIPTORS:
*WATER TREATMENT, *WATER QUALITY, *ALGAE, *ODOR, HYDROGEN ION CONCENTRATION.

ABSTRACT:
PRELIMINARY STUDIES IN RESEARCH ON THE ROLE OF ALGAE IN CAUSING ODORS IN WATER SUPPLIES ARE PRESENTED. INVESTIGATIONS CARRIED OUT WITH CHLOROCOCCUM, A UNICELLULAR GREEN ALGA, IN PURE CULTURE INDICATE THAT THE ODORIFEROUS MATERIAL IS, FOR THE MOST PART, RETAINED WITHIN THE CELLS UNTIL THEIR DEATH OR DISINTEGRATION. THE ODOR PRODUCED BY THIS ALGA INCREASED NOTABLY UNDER EXPERIMENTAL CONDITIONS IN THE ACIDIC PH RANGE. FRACTIONIZATION OF DISINTEGRATED ALGAL CELLS REVEALED THAT A NUMBER OF CHEMICAL COMPOUNDS ARE INVOLVED IN THE ODORIFEROUS MATERIAL. (BEAN-AWWARF)

FIELD 05F, 05C

ACCESSION NO. W72-09694

EFFECTS OF DEPTH OF WASTE STABILIZATION POND PERFORMANCE,

VIRGINIA POLYTECHNIC INST., BLACKSBURG.

S. K. ANDERSON.

MASTER'S THESIS, 1969, 75 P, 17 FIG, 4 TAB, 37 REF.

DESCRIPTORS:
*OXIDATION LAGOONS, *WASTE WATER TREATMENT, ALGAE, PHOTOSYNTHESIS,
DEPTH, THERMOCLINE, CLIMATIC ZONES, ORGANIC LOADING, DISSOLVED OXYGEN,
HYDROGEN ION CONCENTRATION, BIOCHEMICAL OXYGEN DEMAND, VIRGINIA.

IDENTIFIERS:
*SUSPENDED SOLIDS, CHASE CITY(VA).

ABSTRACT:
AN 8 ACRE POND WITH A 3 FT DEPTH AND A 4 ACRE POND WITH A 6 FT DEPTH
WERE BOTH LOADED AT 26 LB BOD/ACRE/DAY. BOTH PONDS HAD BEEN OPERATING
FOR TWO YEARS ON DOMESTIC WASTES. FOLLOWING A 45 DAY ACCLIMATION
PERIOD, DISSOLVED OXYGEN, TEMPERATURE, AND PH MEASUREMENTS WERE TAKEN
DURING 24 HOUR TESTING PERIODS SPACED AT 2 WEEK INTERVALS FOR A 12 WEEK
PERIOD. BOD AND SUSPENDED SOLIDS CONCENTRATIONS WERE ALSO MEASURED
DURING EACH TESTING PERIOD. ALL POND LEVELS CONTAINED DISSOLVED OXYGEN
AT ALL TIMES, WITH THE DEEPER LEVELS SOMETIMES APPROACHING THE
ANAEROBIC STATE. MEANWHILE, ALGAL PHOTOSYNTHESIS PRODUCED LEVELS OF 20
MG/1 DISSOLVED OXYGEN IN THE UPPER LEVELS OF THE POND, WHILE RAISING TO
PH ABOVE 10. BECAUSE OF THE FREQUENT D.O. INTRUSIONS TO THE BOTTOM
LEVEL, METHANE FERMENTATION WAS NEVER ESTABLISHED. THE 6 FT POND DEPTH
IN THE DEEPER POND WAS SUFFICIENT TO INSURE THE ESTABLISHMENT OF A
DEFINITE THERMOCLINE UNDER EXISTING CLIMATIC CONDITIONS IN VIRGINIA.
BOTH THE BOD AND THE SUSPENDED SOLIDS IN THE EFFLUENT PRESENT AT
CONCENTRATIONS OF 30-40 PPM AND 350 PPM RESPECTIVELY, WERE ATTRIBUTED
TO ALGAE CARRIED OVER IN THE EFFLUENT. IN ACCORDANCE WITH OTHER RECENT
STUDIES AND WITH THEORY, IT WAS DEMONSTRATED THAT THE DEEPER POND
CONSISTENTLY PRODUCED A SUPERIOR EFFLUENT IN TERMS OF BOTH BOD AND SS.
(LOWRY-TEXAS)

FIELD 05D

ACCESSION NO. W72-10573

ALGAE OF THE LITORAL OF WEST COAST OF SAKHALIN,

T. F. SHCHAPOVA, AND V. B. VOZZHINSKAYA.

AVAILABLE FROM THE NATIONAL TECHNICAL INFORMATION SERVICE AS AD-724 266,
$3.00 IN PAPER COPY, $0.95 IN MICROFICHE. NAVAL OCEANOGRAPHIC OFFICE
TRANSLATION 339. TRANS. OF VODOROSLI LITORAL: ZAPADNOGO POBERZH'YA
SAKHALINA, TRUDY INSTITUTA OKEANOLOGII, VOL. 34, P 123-146, 1960. 5 FIG, 2
TAB, 33 REF.

DESCRIPTORS:
*ALGAE, *LITTORAL, *AQUATIC PLANTS, ISLANDS, ECOLOGICAL DISTRIBUTION,
BIOLOGICAL COMMUNITIES, SYSTEMATICS, MARINE PLANTS.

IDENTIFIERS:
*SAKHALIN ISLAND(USSR).

ABSTRACT:
THE WESTERN COAST OF SAKHALIN ISLAND WAS STUDIED ON AN EXPEDITION OF
THE INSTITUTE OF OCEANOLOGY USSR IN 1954. THE COASTAL BELT INCLUDING
THE LITTORAL AND SUBLITTORAL FROM KHOE IN THE NORTH TO IVANOVKA IN THE
SOUTH, ABOUT 600 KM, IS DESCRIBED POINTING OUT TYPE OF BOTTOM,
AMPLITUDE OF TIDES, EXPOSURE TO WAVES AND EFFECTS OF WARM OR COLD
CURRENTS. ABOUT 500 QUALITATIVE AND QUANTITATIVE SAMPLES WERE OBTAINED
AND A HERBARIUM OF 700 PLANTS WAS PREPARED. FROM 1860 TO 1926 VARIOUS
EXPEDITIONS AND INVESTIGATIONS SAMPLED ALGAE AT SEVERAL POINTS AND THE
BOTANICAL HISTORY IS GIVEN. AFTER THE RETURN OF SOUTH SAKHALIN TO THE
SOVIET UNION FROM JAPAN COMPREHENSIVE STUDIES OF ALGAE IN THE AREA WERE
UNDERTAKEN, CONCENTRATING ON THE ECOLOGY AND PRACTICAL APPLICATION OF
MACROPHYTES. A CONSIDERABLE PART OF THESE STUDIES HAS REMAINED
UNPUBLISHED. IN THIS WORK ALGAE ARE DEFINED TO THE SPECIES, ALSO THEIR
QUANTITY AND DISTRIBUTION ARE EVALUATED AND DESCRIBED. A COMPREHENSIVE
LIST OF ALGAE IN THE APPENDIX SUMS UP THE LOCATION OF SAMPLES, THEIR
SIZE, FORM, QUANTITY, AND DISTRIBUTION. (JONES-WISCONSIN)

FIELD 05C, 05B

ACCESSION NO. W72-10623

TOXICITY OF PETROCHEMICALS IN THE AQUATIC ENVIRONMENT,

TEXAS UNIV., AUSTIN. DEPT. OF CIVIL ENGINEERING.

J. F. MALINA, JR.

WATER AND SEWAGE WORKS, VOL 111, NO 10, OCTOBER 1964, P 456-460. 1 FIG, 2
    TAB, 20 REF.

DESCRIPTORS:
    *POLLUTANT IDENTIFICATION, *TOXICITY, *BIOASSAY, FISH, AQUATIC
    ENVIRONMENT, ORGANIC PESTICIDES, HYDROGEN ION CONCENTRATION,
    TEMPERATURE, CHLORINATED HYDROCARBON PESTICIDES, ALGAE, LABORATORY
    TESTS, WATER POLLUTION EFFECTS.

IDENTIFIERS:
    *PETROCHEMICAL COMPOUNDS, MEAN TOLERANCE LIMIT, BLUEGILLS, CHLORELLA
    PYRENOIDOSA, BENZENE COMPOUNDS.

ABSTRACT:
    THE ORGANIC CHEMICAL CONCENTRATION CARRIED IN MANY WATER COURSES HAS
    BEEN INCREASING AS A RESULT OF THE GROWING INDUSTRIAL, AGRICULTURAL,
    AND DOMESTIC APPLICATIONS OF PETROCHEMICAL PRODUCTS. THE TOXIC EFFECTS
    OF VARIOUS PETROCHEMICALS ON AQUATIC ORGANISMS WERE SUMMARIZED. THE
    RESULTS OF BIOASSAY TESTS INDICATING THE ACUTE TOXICITY OF A SPECIFIC
    COMPOUND ON ACCLIMATIZED TEST ANIMALS, NAMELY FISH, WERE ARRANGED IN
    TABULAR FORM. THE TOXICITY DATA WERE REPORTED AS THE 96 HOUR MEDIAN
    TOLERANCE LIMIT. IT WAS NOTED THAT IN CONDUCTING A TOXICITY BIOASSAY IT
    IS NECESSARY TO USE THE SAME SPECIES AND SIZE OF TEST ANIMAL AND THAT
    THE WATER MUST BE OF UNIFORM QUALITY AND THE PH AND TEMPERATURE MUST BE
    RELATIVELY CONSTANT. THE CHLORINATED AND PHOSPHORUS CONTAINING ORGANIC
    PESTICIDES WERE MORE TOXIC TO FISH AND INVERTEBRATES THAN ANY OF THE
    COMPOUNDS TESTED. ALKYL ARYL SULFONATE, LACTONITRIDE, SODIUM BUTYL
    MECAPTIDE, SODIUM CYANIDE AND TETRAETHYL LEAD WERE TOXIC TO BLUEGILLS
    IN CONCENTRATIONS LESS THAN 10 MG/L. ALSO, THE TOXICITY OF VARIOUS
    ORGANIC CHEMICALS TO A SPECIES OF GREEN ALGAE CHLORELLA PYRENOIDOSA
    WHICH HAD BEEN EVALUATED IN LABORATORY STUDIES WERE REPORTED. COMPOUNDS
    OF THE BENZENE SERIES EXHIBITED THE HIGHEST TOXICITY WITH NITROBENZENE
    BEING TOXIC AT CONCENTRATIONS OF 0.01 MG/L OR LESS. (GALWARDI-TEXAS)

FIELD 05C, 05B

ACCESSION NO. W72-11105

INFLUENCE OF SITE CHARACTERISTICS ON QUALITY OF IMPOUNDED WATER,

WASHINGTON UNIV., SEATTLE. DEPT. OF CIVIL ENGINEERING.

R. O. SYLVESTER, AND R. W. SEABLOOM.

JOURNAL OF THE AMERICAN WATER WORKS ASSOCIATION, VOL 57, NO 12, P 1528-1546,
    DEC. 1965. 11 FIG, 4 TAB, 24 REF.

DESCRIPTORS:
    *RESERVOIR SITES, WATER RESOURCES DEVELOPMENT, WATER QUALITY, SOIL
    INVESTIGATIONS, *IMPOUNDED WATERS, IMPOUNDMENTS, *DISSOLVED OXYGEN,
    BIOCHEMICAL OXYGEN DEMAND, ALGAL CONTROL, *COLOR, *ALKALINITY,
    *NITRATES, NITROGEN, *CONDUCTIVITY, *LEACHING, IRON, TEMPERATURE, SOIL
    ANALYSIS, TASTE-PRODUCING ALGAE, HYDROGEN ION CONCENTRATION, *ORGANIC
    SOILS, MINERALOGY.

IDENTIFIERS:
    WOOD, ALGAL COUNT, GREEN RIVER, WASHINGTON, HOWARD A. HANSON RESERVOIR,
    TACOMA(WASH).

ABSTRACT:
    SOILS AT IMPOUNDMENT SITES SHOULD BE TESTED FOR THEIR EFFECT ON
    OVERLYING WATER. ORGANIC SOILS CREATE THE MOST SEVERE PROBLEMS BUT
    LEACHING THEM OR COVERING THEM WITH AS LITTLE AS 2 INCHES OF MINERAL
    SOIL MAY BE EFFECTIVE. TESTS WERE PERFORMED ON HOWARD A. HANSON
    RESERVOIR, NEAR TACOMA, WASHINGTON, TO ASCERTAIN PHYSICAL, CHEMICAL AND
    BIOLOGIC EFFECTS. BASED ON THESE STUDIES, RECOMMENDED WATER QUALITY
    TESTS OVER SOILS OF IMPOUNDMENTS SHOULD INCLUDE DO, COLOR, NITRATE,
    AMMONIA, ALGAL COUNTS AND PH MEASUREMENTS OF AT LEAST ONE MONTH
    DURATION. LEACHING AND EXCHANGE STUDIES WILL REVEAL IF SITE SHOULD BE
    ABANDONED OR IF COVER SOIL IS FEASIBLE. SHORELINES SHOULD BE
    INVESTIGATED FOR EFFECT ON TURBIDITY. COMPLETE WATER QUALITY ANALYSES
    UNDER VARIOUS FLOW CONDITIONS ARE NEEDED TO EVALUATE FUTURE RESERVOIR
    QUALITY AS RELATED TO THE UNDERLYING SOIL. (FLACK-AWWARF)

FIELD 05G, 05F, 02G

ACCESSION NO. W72-11906

MODIFICATION OF SOME PHYSIOLOGICAL METHODS FOR ESTIMATION OF WATER QUALITY AND
THEIR RELATIONSHIP TO DEGREES OF SAPROBITY (MODIFIKACE VYBRANYCH
FYZIOLOGICKYCH METOD STANOVENI KVALITY VODY A JEJICH VZTAH K STUPNUM SAPROB
ITY),

VYSOKA SKOLA CHEMICO-TECHNOLOGICKA, PRAGUE (CZECHOSLOVAKIA). DEPT. OF WATER
TECHNOLOGY.

D. MATULOVA.

VODNI HOSPODARSTVI, VOL 9, P 393-395, 1967. 1 TAB, 18 REF. ENGLISH SUMMARY.

DESCRIPTORS:
*WATER QUALITY, *BIOINDICATORS, *WATER ANALYSIS, *BIOASSAY,
*COMPARATIVE PRODUCTIVITY, AQUATIC MICROBIOLOGY, EUTROPHICATION,
MUNICIPAL WASTES, FRESHWATER ALGAE, WATER POLLUTION EFFECTS, DOMESTIC
WASTES, INDUSTRIAL WASTES, ENVIRONMENTAL EFFECTS, ANALYTICAL
TECHNIQUES, PHYSIOLOGICAL ECOLOGY, PLANT PHYSIOLOGY.

IDENTIFIERS:
*ALGAL CULTURE, *BIOMASS TITER, *BIOLOGICAL INDEX, *SAPROBITY, PRAGUE,
CZECHOSLOVAKIA, LABORATORY PROCEDURES, STATISTICAL PROCEDURES, BMT.

ABSTRACT:
STUDIES WERE MADE WITH SEWAGE AND INDUSTRIAL WASTES COLLECTED FROM
SEVERAL AREAS NEAR PRAGUE, ESPECIALLY WHERE THE BOTIC EMPTIES INTO THE
RIVER VLTAVA, TO DETERMINE ITS EFFECT ON ALGAE. A TECHNIQUE WAS
DEVELOPED TO EVALUATE ITS TOXICITY TO ALGAE. ON THE BASIS OF RESULTS
OBTAINED USING THE BIOMASS TITER (BMT) AS A CRITERION, A BIOLOGICAL
INDEX OF WATER QUALITY WAS OBTAINED. THE INDEX ENABLED THE CALCULATION
OF DIFFERENT DEGREES OF SAPROBITY. THE VALUES OBTAINED BY THE
LABORATORY EXPERIMENTS WERE COMPARED WITH FINDINGS IN THE FIELD. A GOOD
CORRELATION OF BETA SAPROBIC TO POLYSAPROBIC WATERS WAS FOUND.
(KATZ-WASHINGTON)

FIELD 05C

ACCESSION NO. W72-12240

THE BIO-FOULING MENACE IN COOLING SYSTEMS,

A. MENNIE, AND E. TEHLE.

EFFLUENT AND WATER TREATMENT JOURNAL, VOL 9, NO. 9, P 493-499, SEPTEMBER
1969, 4 FIG.

DESCRIPTORS:
*COOLING WATER, *BIOMASS, *SLIME, ALGAE, FUNGI, BACTERIA, BACTERICIDES,
CHLORINE, COSTS, *WASTE WATER TREATMENT.

IDENTIFIERS:
BIOCIDES, PRODUCTION LOSS, QUATERNARY AMMONIUM COMPOUNDS, PHENOLIC
COMPOUNDS, ORGANO-METALLIC COMPOUNDS, AMINES, ENGLAND.

ABSTRACT:
THE DEVELOPMENT OF BIOLOGICAL GROWTHS IN INDUSTRIAL COOLING SYSTEMS HAS
PRODUCED DIFFICULT AND SERIOUS PROBLEMS, AND STEPS MUST BE TAKEN TO
CONTROL OR ELIMINATE BIOFOULING. EXCESSIVE GROWTH AND DEVELOPMENT OF
ALGAE, FUNGI AND BACTERIA HAVE RESULTED IN BIOLOGICAL FOULING IN
COOLING WATER SYSTEMS. TWO CASE-HISTORIES ILLUSTRATING HOW BIOLOLOGICAL
FOULING CAN ACCUMULATE TO SHUT DOWN ESSENTIAL PRODUCTION PROCESSES WERE
PRESENTED. ONE CONCERNING A RECIRCULATING COOLING SYSTEM IN A
PHARMACEUTICAL PLANT WHERE THE COST OF EACH SHUTDOWN WAS ABOUT 1,000
POUNDS PER HOUR, AND THE ANNUAL LOSS OF PRODUCTION ABOUT 12 HOURS.
TREATMENT INCLUDED THE USE OF A SYNTHETIC ANTIFOULANT WHICH BROKE UP
THE GROWTH DEBRIS, DISPERSING IT INTO A HARMLESS AND EASILY REMOVED
CLOUD OF FINE PARTICLES. THE SECOND CASE INVOLVED A RECIRCULATING
COOLING SYSTEM OF A CHEMICAL BY-PRODUCTS COMPLEX OF A STEELWORKS COKE
PRODUCING PLANT. THE COST OF A COMPLETE CLEAN-UP TO RESTORE FULL
OPERATING EFFICIENCIES WAS ASSESSED AT 3,000 POUNDS, OF WHICH HALF WAS
LOSS OF PRODUCTION. TYPES OF BIOCIDES USED INCLUDE QUARTERNARY AMMONIUM
COMPOUNDS, PHENOLIC COMPOUNDS, ORGANO-METALLIC COMPOUNDS AND AMINES.
ONE COMMON METHOD OF INHIBITING ALGAL AND FUNGAL GROWTH IN LARGE
COOLING SYSTEMS MAKES EFFECTIVE USE OF CHLORINE AS A OXIDIZING BIOCIDE.
HOWEVER, THE USE OF CHLORINE AND RELATED CONTROL EQUIPMENT WAS SHOWN TO
BE EXPENSIVE WHEREAS NON-OXIDIZING BIOCIDES CAN BE INTRODUCED TO THE
LARGEST OF SYSTEMS AT MUCH LOWER COST BY DIRECT ADDITION TO THE COOLING
TOWER SUMP. (GALWARDI-TEXAS)

FIELD 05D, 05F

ACCESSION NO. W72-12987

ADVANCES IN WATER TREATMENT,

OMAHA METROPOLITAN UTILITIES DISTRICT, NEBR.

J. F. ERDEI.

JOURNAL AMERICAN WATER WORKS ASSOCIATION, VOL 55, NO 7, P 845-856, JULY 1963.
1 TAB, 16 REF.

DESCRIPTORS:
*WATER TREATMENT, SANITARY ENGINEERING, *PATHOGENIC BACTERIA, ALGAE,
ORGANIC MATTER, RADIOACTIVITY, PESTICIDES, TASTE, ODOR, COAGULATION,
ELECTROLYTES, POTABLE WATER.

ABSTRACT:
REVIEWS WERE CONDUCTED OF THE STATUS OF INCREASING WATER POLLUTION
PROBLEMS. USE OF CHEMICALS, NEW SCIENTIFIC TOOLS, ANALYTICAL
TECHNIQUES, TREATMENT PROCESSES, POLYELECTROLYTE USAGE, COAGULATION,
JAR TEST, PATHOGENIC ORGANISMS, ALGAE, TRACE ELEMENTS, PESTICIDES,
RADIOSOTOPES, AND TASTES AND ODORS, ARE DISCUSSED. MORE STUDY IS NEEDED
ON METHODS OF TREATING DRINKING WATER, A GOOD SHARE OF WHICH SHOULD
INVOLVE ORGANIC CONTAMINANTS. THERE IS NO ASSURANCE THAT SUCH HARMFUL
SUBSTANCES IN SOLUTION WILL SIGNIFY THEIR PRESENCE BY PRODUCING TASTE
OR ODOR, AND MANY INTERESTING OPPORTUNITIES FOR RESEARCH EXIST IN THESE
AREAS. (BEAN-AWWARF)

FIELD 05F

ACCESSION NO. W72-13601

CHEMICAL ASPECTS OF ACTINOMYCETE METABOLITES AS CONTRIBUTORS OF TASTE AND ODOR,

IOWA UNIV., IOWA CITY. STATE HYGIENIC LAB.

R. L. MORRIS, J. D. DOUGHERTY, AND G. W. RONALD.

JOURNAL OF THE AMERICAN WATER WORKS ASSOCIATION, VOL 55, NO 10, P 1380-1390,
OCTOBER 1963. 7 FIG, 1 TAB, 8 REF.

DESCRIPTORS:
*WATER TREATMENT, *ACTINOMYCETES, *ALGAE, *TASTE, *ODOR, WATER QUALITY
CONTROL, POLLUTANT IDENTIFICATION, IOWA.

IDENTIFIERS:
*METABOLITES, *CEDAR RIVER(IOWA).

ABSTRACT:
THE EXTREMELY SEVERE MUSTY TASTES AND ODORS THAT OCCURRED IN THE CEDAR
RIVER WERE DUE TO CERTAIN METABOLITES OF THE ACTINOMYCETES,
PARTICULARLY A 'NEUTRAL FRACTION'. NO CORRELATION EXISTED BETWEEN THE
INTENSITY OF TASTE PROBLEMS AND THRESHOLD ODOR VALUES, AS THE BURNING,
MUSTY TASTE OF CRITICAL FRACTIONS OF ACTIMYCETE CULTURES WERE FAR MORE
PERSISTENT THAN THE ASSOCIATED ODOR PHENOMENA. METABOLITES WERE
EXTRACTED, PURIFIED AND DETERMINATION MADE AS TO WHICH PRODUCED THE
TYPICAL TASTES AND ODORS. THESE WERE PURIFIED BY ELUTRIATION WITH
VARIOUS SOLVENTS. WHEN THE REMAINING ETHER SOLUTION WAS DRIED WITH
SODIUM SULFATE A PALE YELLOW OIL CALLED THE 'NEUTRAL FRACTION' WAS
RECLAIMED; IT HAD A DECIDEDLY MUSTY TASTE AND ODOR. VARIOUS 'NEUTRAL
FRACTIONS' IN SPECTROGRAMS, SHOWED CORRELATIONS OF BOUND HYDROXYL AND
CARBONYL GROUPS AT WAVELENGTHS 2.9-2.95 AND 5.8-5.85 MICRONS
RESPECTIVELY. (BEAN-AWWARF)

FIELD 05F

ACCESSION NO. W72-13605

EFFECTS OF RIVER PHYSICAL AND CHEMICAL CHARACTERISTICS ON AQUATIC LIFE,

ACADEMY OF NATURAL SCIENCES OF PHILADELPHIA, PA.

R. PATRICK.

JOURNAL OF THE AMERICAN WATER WORKS ASSOCIATION, VOL 54, NO 5, P 544-550, MAY
1962. 2 REF.

DESCRIPTORS:
*AQUATIC LIFE, *AQUATIC POPULATIONS, *RIVERS, ALGAE, PROTOZOA, AQUATIC
INSECTS, FISH REPRODUCTION, INDUSTRIAL WASTES, ECOLOGY, DISSOLVED
OXYGEN, HARDNESS, TIDAL WATERS, ORGANIC MATTER, CONDUCTIVITY,
TEMPERATURE, BACTERIA, NUTRIENTS, ESTUARINE ENVIRONMENT, PONDS,
BREEDINGS, SALINE WATER-FRESHWATER INTERFACES, COLLOIDS.

IDENTIFIERS:
*RIVER REGIONS, *PHYSICAL CHARACTERISTICS, *CHEMICAL CHARACTERISTICS,
ORGANISMS, SUSPENDED SOLIDS, SALTWATER TONGUE, BENTHIC FORMS, EPIPHYTIC
FORMS, COLLOIDAL FRACTION, MULTIPLE-USES, SAVANNAH RIVER, GULF COAST.

ABSTRACT:
THE PHYSICAL AND CHEMICAL CHARACTERISTICS OF A RIVER HAVE IMPORTANT
EFFECTS ON ITS AQUATIC LIFE AND, IN TURN, AQUATIC LIFE CAN INDICATE
POLLUTION AFFECTS ON THE RIVER. HEAD WATER REGIONS HAVE RELATIVELY LOW
CAPACITY FOR ORGANIC ASSIMILATION, SERVE AS EXCELLENT BREEDING ZONES,
AND MUST BE MAINTAINED IN THEIR NATURAL STATE. THE MAIN TRUNK REGION
CAN ACCOMMODATE WASTES IF THEY ARE DISPERSED THROUGHOUT THE STREAM.
THIS SECTION IS RICH IN HABITATS AND POPULATIONS MAY BE LARGE. THE
TIDAL SECTION IS COMPLEX, EXHIBITING RETENTION, LIMITED MIXING,
SALTWATER INTERFACE EFFECTS, ALL OF WHICH AFFECT THE AQUATIC LIFE.
CHEMICAL AND PHYSICAL ASPECTS INCLUDE DISSOLVED OXYGEN, HARDNESS,
ORGANIC MATTER, TOTAL CONDUCTIVITY, TEMPERATURE, SUSPENDED SOLIDS,
WHICH AFFECT THE VARIETY AND NUMBERS OF SPECIES OF AQUATIC LIFE. IN
PLANNING THE USES OF A RIVER THE EFFECTS OF WASTES ON AQUATIC LIFE MUST
BE CONSIDERED. (FLACK-AWWARF)

FIELD 05C, 02L, 05B

ACCESSION NO. W73-00748

RELATIONSHIP OF IMPOUNDMENT TO WATER QUALITY,

GEOLOGICAL SURVEY, WASHINGTON, D.C. WATER QUALITY BRANCH.

S. K. LOVE.

JOURNAL OF THE AMERICAN WATER WORKS ASSOCIATION, VOL 53, NO 5, P 559-568, MAY
1961. 5 FIG, 1 TAB, 13 REF.

DESCRIPTORS:
*RESERVOIR OPERATIONS, DISSOLVED OXYGEN, DISSOLVED SOLIDS,
SEDIMENTATION RATES, MANGANESE, LOW-FLOW AUGMENTATION, TENNESSEE RIVER,
ALGAE, IRON, ALKALINITY, HARDNESS, RADIOACTIVITY, TEMPERATURE,
SALINITY, REAERATION, FISH, TASTE, ODOR, OXYGEN SAG, POTOMAC RIVER.

IDENTIFIERS:
*RESERVOIR EFFECTS, LAKE MEAD, DILUTION, ALGAL GROWTHS, OXYGEN
DEPLETION.

ABSTRACT:
EFFECTS OF IMPOUNDMENT ON WATER INCLUDE REDUCTION OF POLLUTANTS,
LEVELING OF VARIATIONS IN DISSOLVED SOLIDS, HARDNESS, PH AND
ALKALINITY, TEMPERATURE REDUCTION, SEDIMENT ENTRAPMENT, LOW FLOW
AUGMENTATION AND DILUTION, INCREASED ALGAE GROWTHS, REDUCED DISSOLVED
OXYGEN, INCREASE IN CARBON DIOXIDE, IRON, MANGANESE, AND ALKALINITY,
AND INCREASED DISSOLVED SOLIDS DUE TO EVAPORATION OR DISSOLVING OF
ROCKS. CASES OF BENEFICIAL AND DETRIMENTAL EFFECTS INCLUDE LAKE MEAD,
THE TENNESSEE VALLEY, SWITZERLAND LAKES, CALIFORNIA AND TEXAS. THE
POTOMAC RIVER IS ANALYZED IN SOME DETAIL. YET, MOST STUDIES OF
IMPOUNDMENT EFFECTS ARE MORE QUANTITATIVE THAN QUALITATIVE. THERE IS
NEED FOR BETTER STUDIES OF STREAM IMPOUNDMENTS SO THAT PROBLEMS WITH
PROPOSED IMPOUNDMENTS CAN BE RESOLVED IN ADVANCE. (FLACK-AWWARF)

FIELD 05C, 02H

ACCESSION NO. W73-00750

PREVENTION OF WATER SOURCE CONTAMINATION,

HENNINGSON, DURHAM AND RICHARDSON, INC., OMAHA, NEBR.

P. BOLTON.

JOURNAL OF THE AMERICAN WATER WORKS ASSOCIATION, VOL. 53, NO 10, P 1243-1250, OCTOBER 1961, 1 TAB.

DESCRIPTORS:
*WATER POLLUTION SOURCES, *WATER POLLUTION EFFECTS, *WATER SOURCES, *GROUNDWATER, DDT, MISSISSIPPI RIVER BASIN, URBANIZATION, ALGAL POISONING, GASOLINE, FLOODS, AQUIFERS, WATER LAW, NITRATES, TURBIDITY, FLOODWATER, REGULATION, WELL REGULATIONS.

IDENTIFIERS:
*MASTER PLANNING, *SOURCE CONTAMINATION, CYANIDE, NEW PRODUCT CONTROL, PLUMBING CODES, WELL DRILLING CODES.

ABSTRACT:
SAFE WATER SUPPLIES REQUIRE CONSIDERATIONS OF LOCATION, CONSTRUCTION AND OPERATION. INCREASED COMPLEXITY OF INDUSTRIALIZED CIVILIZATION HAS INTRODUCED NEW SOURCES OF WATER SUPPLY CONTAMINANTS. SOURCES OF CONTAMINATION OF SURFACE AND GROUND WATER ARE LISTED. CAUSES OF GROUNDWATER POLLUTION ARE OFTEN THE PROMISCUOUS STORAGE OF WASTES ON THE GROUND SURFACE NEAR A WELL FIELD OR ABOVE A WATER BEARING FORMATION. SUDDEN INTRODUCTION OF CONTAMINANTS IS A TYPICAL CAUSE OF SURFACE WATER POLLUTION. A NATIONAL MASTER PLAN FOR PREVENTING CONTAMINATION OF WATER RESOURCES IS PROPOSED. (FLACK-AWWARF)

FIELD 05B, 05G

ACCESSION NO. W73-00758

THE NEED FOR NUTRIENT CONTROL,

METCALF AND EDDY, INC., BOSTON, MASS.

C. N. SAWYER.

JOURNAL WATER POLLUTION CONTROL FEDERATION, VOL 40, NO 3, P 363-370, MARCH 1968. 5 FIG, 1 TAB, 11 REF.

DESCRIPTORS:
*WATER QUALITY, *NUTRIENTS, *EUTROPHICATION, NITROGEN, PHOSPHORUS, *NUTRIENT REMOVAL, SOIL EROSION, ALGAE, *WATER POLLUTION CONTROL, WATER QUALITY CONTROL.

ABSTRACT:
NUTRIENT CONTROL IS NEEDED TO ARREST EUTROPHICATION OF LAKES AND STREAMS. CULTURAL DRAINAGE IS A MAJOR CONTRIBUTOR OF NUTRIENTS AND A LOGICAL POINT OF ATTACK ON THE PROBLEM. SOIL EROSION CONTROL AND PREVENTION OF LEACHING FROM ANIMAL MANURES ARE HELPFUL AGRICULTURAL PRACTICES. PHOSPHORUS REMOVAL OFFERS THE MOST PROMISE FOREUTROPHICATION CONTROL, SINCE SOME BLUE-GREEN ALGAE CAN FIX NITROGEN AND OTHER NUTRIENTS ARE PRESENT IN NATURAL WATERS OR EXIST IN QUANTITIES TOO SMALL FOR PRACTICE REMOVAL. CHEMICAL ANALYSIS OF THE LAKE WATER FREQUENTLY DURING THE YEAR CAN DETERMINE WHICH NUTRIENTS ARE CRITICAL, I.E., WHICH NUTRIENTS DECLINE IN CONCENTRATION AS PRIMARY PRODUCTIVITY INCREASES. (BEAN-AWWARF)

FIELD 05G, 05C

ACCESSION NO. W73-01122

ECOLOGY OF A EUTROPHIC ENVIRONMENT: MATHEMATICAL ANALYSIS OF DATA, (ECOLOGIE
D'UN MILIEU EUTROPHIQUE: TRAITEMENT MATHEMATIQUE DES DONNEES).

CENTRE UNIVERSITAIRE DE LUMINY, MARSEILLE (FRANCE). LABORATOIRE
D'HYDROBIOLOGIE MARINE.

DESCRIPTORS:
*PHYTOPLANKTON, *STATISTICAL METHODS, *ZOOPLANKTON, *BIOMASS,
*SUCESSION, *ISOLATION, DIATOMS, ECOLOGY, SALINITY, ORGANIC MATTER,
RIVERS, ESTUARINE ENVIRONMENT, WATER TEMPERATURE, *EUTROPHICATION,
AQUATIC LIFE, SESTON, SCENEDESMUS, MATHEMATICAL STUDIES, DATA
COLLECTIONS, MARINE ANIMALS, PRIMARY PRODUCTIVITY, CHLOROPHYLL,
PIGMENTS, MAGNESIUM COMPOUNDS, DISTRIBUTION PATTERNS, PERSISTENCE,
WATER PROPERTIES, PHYSICAL PROPERTIES, AQUATIC PRODUCTIVITY, MARINE
ALGAE, SALT TOLERANCE, CORRELATION ANALYSIS, CHRYSOPHYTA, CHLOROPHYTA,
FISH EGGS, LARVAE, CRUSTACEANS, COPEPODS, PHOSPHATES, NITRATES.

IDENTIFIERS:
*CHLOROPHYLL A, DATA INTERPRETATION, *RHONE RIVER, NAVICULA, MELOSIRA,
THALASSIOSIRA, COSCINDISCUS, BACTERIOSTRUM, LEPTOCYLINDRUS, CORYCAEUS,
OIKOPLEURA, FRITILLARIA, OBELIA, CLUSTER ANALYSIS, MULTIVARIATE
ANALYSIS, PRINCIPAL-COMPONENT ANALYSIS, PART CORRELATION, DECAPODS,
SPECIES DIVERSITY INDEX.

ABSTRACT:
NUMEROUS DATA ON PHYSICAL, CHEMICAL AND BIOLOGICAL PARAMETERS IN THE
DILUTION LAYER OF THE RHONE MOUTH HAVE BEEN STUDIED BY MULTIVARIATE
TECHNIQUES: PRINCIPAL-COMPONENT ANALYSIS, PART CORRELATION. A NEW
TECHNIQUE OF CLUSTER ANALYSIS IS ALSO PROPOSED. BY THESE MEANS, A VERY
EURYHALINE GROUP OF ZOOPLANKTON SPECIES HAS BEEN ISOLATED AND THE
EXTREMELY LOW SENSIBILITY OF PHYTOPLANKTON TOWARDS, SALINITY HAS BEEN
SHOWN. HOWEVER, TEMPERATURE SEEMS TO BE THE MOST IMPORTANT ECOLOGICAL
FACTOR. INSTABILITY AND EUTROPHY OF THIS AREA DO NOT APPEAR TO DISTURB
THE PHYTOPLANKTON CYCLE, WHICH OCCURS WITH ITS USUAL SUCCESSIONS.
CHLOROPHYLL A AND ORGANIC MATTER DO NOT SEEM OF VALUE FOR ESTIMATION OF
THE BIOMASS IN THE AREA STUDIED. (LONG-BATTELLE)

FIELD 05C, 05A

ACCESSION NO. W73-01446

THE INTERRELATIONSHIPS AMONG PLANKTON, ATTACHED ALGAE AND THE PHOSPHORUS CYCLE
IN ARTIFICIAL OPEN SYSTEMS,

TORONTO UNIV. (ONTARIO).

J. L. CONFER.

AVAILABLE FROM UNIV. MICROFILMS, INC., ANN ARBOR, MICH., 48106. PH. D.
THESIS, 1969.

DESCRIPTORS:
*PLANKTON, *ALGAE, *SESSILE ALGAE, *CYCLING NUTRIENTS, ECOSYSTEMS,
PHOSPHORUS, SCENEDESMUS, CHLOROPHYTA, BACTERIA, PHOSPHORUS
RADIOISOTOPES, INFLOW, CYANOPHYTA, LITTORAL, NUTRIENTS, MODEL STUDIES,
BIOASSAY, RADIOACTIVITY TECHNIQUES, ABSORPTION.

IDENTIFIERS:
*PHOSPHORUS CYCLE, GLOEOTRICHA, P-32.

ABSTRACT:
THE DISTRIBUTION AND RATES OF CIRCULATION OF PHOSPHORUS WERE STUDIED IN
200 LITER MICROECOSYSTEMS. THE SYSTEMS WERE MAINTAINED AS OPEN SYSTEMS
WITH A CONSTANT INFLUX OF WATER CONTAINING PHOSPHORUS. THE EFFECTS OF
DIFFERENT INFLOWING PHOSPHORUS CONCENTRATIONS WERE DETERMINED.
FILAMENTOUS ALGAL GROWTH ATTACHED TO THE SIDES OF THE TANKS HAD A
DOMINANT INFLUENCE ON THE PHOSPHORUS CONCENTRATION IN THE OPEN WATER.
THERE WAS A NEGATIVE CORRELATION BETWEEN THE AMOUNT OF ALGAE ATTACHED
TO THE SIDES AND PHOSPHORUS CONCENTRATION IN THE OPEN WATER. THE
ATTACHED GROWTH REMOVED PARTICLES AS SMALL AS BACTERIA AND AS LARGE AS
SCENEDESMUS FROM THE OPEN WATER. THE ESTIMATED RATE OF MOVEMENT FOR
VARIOUS PARTICLES RANGED FROM 14 TO 385 PERCENT PER DAY. WHEN THE
INFLOWING WATER WAS ENRICHED WITH PHOSPHORUS, THE SPECIES COMPOSITION
OF THE LITTORAL ALGAE WAS CHANGED. BLUE-GREEN ALGAE, PREDOMINANTLY
GLOEOTRICHA, REPLACED THE GREEN ALGAL CHARACTERISTIC OF UNFERTILIZED
PONDS. PHOSPHORUS CONTENT OF THE OPEN WATER WAS NOT INCREASED BY
ENRICHMENT BECAUSE THE BLUE-GREEN ALGAE ACCUMULATED A LARGER FRACTION
OF THE INFLOWING PHOSPHORUS THAN DID THE GREEN ALGAE. RESULTS OF
EXPERIMENTS IN WHICH RADIOACTIVE PHOSPHORUS WAS ADDED TO THE OPEN WATER
WERE SIMILAR TO THOSE OBTAINED PREVIOUSLY IN LAKES. THE KINETICS OF
P-32 LOSS FROM THE OPEN WATER ARE THEORETICALLY COMPATIBLE EITHER WITH
AN EQUILIBRIUM DISTRIBUTION BETWEEN THE OPEN WATER AND THE ATTACHED
ALGAE OR OTHER LITTORAL SOLIDS, OR WITH A STEADY STATE SYSTEM IN WHICH
THERE IS A NET MOVEMENT FROM THE OPEN WATER TO THE ATTACHED ALGAE OR
LITTORAL SOLIDS. (HOLOMAN-BATTELLE)

FIELD 05C, 05B

ACCESSION NO. W73-01454

KINETICS OF ALGAL BIOMASS PRODUCTION SYSTEMS WITH RESPECT TO INTENSITY AND
  NITROGEN CONCENTRATION,

CALIFORNIA UNIV., BERKELEY.

G. SHELEF.

AVAILABLE FROM UNIVERSITY MICROFILMS, INC., ANN ARBOR, MICHIGAN 48106, PH D.
  DISSERTATION, 1968, 265 P.

  DESCRIPTORS:
    *NITROGEN, *IRRADIATION, *ALGAE, WASTE TREATMENT, EUTHROPHICATION,
    MODEL STUDIES, BIOMASS, *LIGHT INTENSITY.

  IDENTIFIERS:
    KINETIC MODELS, CHLORELLA PYRENOIDOSA, *ALGAE BIOMASS.

  ABSTRACT:
    THREE KINETIC MODELS TO DESCRIBE THE RELATIONSHIP BETWEEN THE
    IRRADIANCE AND ALGAL GROWTH AND PRODUCTION WERE DEVELOPED AND
    EVALUATED. THESE THREE WERE BASED UPON: (A) ENERGY BALANCE AND/OR A
    'ZERO ORDER-FIRST ORDER' KINETIC RELATIONSHIP; (B) THE RECTANGULAR
    HYPERBOLA FUNCTION; AND, (C) THE EXPONENTIAL FUNCTION. THE MODELS
    INCORPORATED THE SPECTRAL DISTRIBUTION OF THE LIGHT SOURCE AS IT
    AFFECTED THE ESSENTIAL PARAMETERS - LIGHT CONVERSION EFFICIENCY AND
    LIGHT TRANSMISSION IN AN OPTICALLY DENSE ALGAL SUSPENSION. OF THE
    THREE, THE MODEL BASED ON THE EXPONENTIAL RELATIONSHIP PROVED MOST
    COMPATIBLE WITH THE EXPERIMENTAL RESULTS UNDER THE CONDITIONS OF A
    CHEMOSTATIC ALGATRON REACTION. THE MODIFIED EXPONENTIAL FUNCTION BEST
    DESCRIBED THE RELATIONSHIP BETWEEN NITRATE-NITROGEN CONCENTRATION AND
    ALGAL SPECIFIC GROWTH RATE UNDER CONDITIONS OF LIGHT SATURATION IN A
    CONTINUOUS CULTURE OF CHLORELLA PYRENOIDOSA. THE USE OF THE MEF WAS
    APPLICABLE IN OPTICALLY DENSE ALGAL CULTURES IN DESCRIBING THE
    RELATIONSHIP BETWEEN NITRATE-NITROGEN CONCENTRATION AND THE AVERAGE NET
    ALGAL SPECIFIC GROWTH RATE OR NET ALGAL PRODUCTION. (ANDERSON-TEXAS)

  FIELD 05C, 05B

  ACCESSION NO. W73-02218

KENOSHA INCREASES PLANT CAPACITY WITH MICROSTRAINERS,

KENOSHA WATER DEPT., WIS.

O. F. NELSON.

WATER WORKS AND WASTES ENGINEERING, VOL 2, NO 7, P 43-46, JULY 1965, 3 FIG, 1
  TAB.

  DESCRIPTORS:
    *WATER TREATMENT, *ALGAE, *FILTERS, SANDS, LAKE MICHIGAN, WATER QUALITY
    CONTROL, FILTRATION, WATER PURIFICATION, MEMBRANES, SCREENS, TREATMENT
    FACILITIES, *WISCONSIN.

  IDENTIFIERS:
    *SCREENING, *KENOSHA(WIS), *STRAINING, *MICROSTRAINERS.

  ABSTRACT:
    THE INSTALLATION OF MICROSTRAINERS AHEAD OF ALL OTHER TREATMENT AT THE
    KENOSHA, WISCONSIN WATER TREATMENT PLANT HAS RESULTED IN EXTENDED
    FILTER RUNS, 46 TO 97 PERCENT REMOVAL OF ALGAE, 25 PERCENT INCREASE IN
    PLANT CAPACITY AND ELIMINATION OF TASTE AND ODOR PROBLEMS. THE PROCESS
    COSTS ARE APPROXIMATELY $1.50 PER MILLION GALLONS OF WATER TREATED WHEN
    THE PLANT IS PROCESSING 12 MILLION GALLONS PER DAY. (FLACK-AWWARF)

  FIELD 05F

ACCESSION NO. W73-02426

SIGNIFICANT DESCRIPTOR INDEX

The starred descriptors indexed here have been selected as best describing the content of each paper. If additional references to these or other terms are desired, consult the *General Index.* The numbers at the right are the accession numbers appearing at the end of the abstracts.

, 'BREAKING' OF / *MADISON(WIS),
RID CLIMATES, SEMIARID CLIMATES,
        *BIOGEOCHEMICAL DATA,
MS, *ALGAE, *DOMINANT ORGANISMS,
S, *TURNOVER RATES, */ *GLUCOSE,
YDROGEN ION CONCENTRATION, WATE/
ORGANISMS, *ALGAE, *AQUATIC BAC/
REMENTS, *POLLUTANTS, INHIBITIO/
IRONMENTAL EFFECTS, *LI/ *ALGAE,
ODOR, WATER Q/ *WATER TREATMENT,
CARBO/ *WATER TREATMENT, *ALGAE,
IC MATTER, *POTASSIUM COMPOUNDS,
WATER TREATMENT, *WASTE WATER T/
TRATION, *ALGAE, *WATER QUALITY,
MOVAL, *ALUM, *ALUMINUM SULFATE,
STE WATER RECLAMATION, / *ALGAE,
ION, *BIOCHEMICAL OXYGEN DEMAND,
A/ *SEWAGE TREATMENT, *PROTOZOA,
OROUS, NITROGEN FIXING BACTERIA,
ICATION, LIME, *NITROGEN, HYDRO/
XYGEN DEMAND, *COLOR, WASTE WAT/
 MENEGHINIANA, LIGHT-SATURATED /
TER, *ALGAE, *SLUDGE, STORMS, T/
WORKS, *WASTE DISPO/ *WISCONSIN,
S, *LAKES, *SEDIMENTS, *EUTROPH/
DE KINETICS, *PESTICIDE REMOVAL,
ASTE WATER T/ *ACTIVATED CARBON,
ENT R/ *AESTHETIC DETERIORATION,
SOUTHWEST(US), *DESIGN CRITERIA/
TOWN(MO), LEE'S SUMMIT(MO), OXY/
DATION LAGOONS, BIOCHEMICAL OXY/
OIRS, / *THERMAL STRATIFICATION,
RFACE, *VOLUME, *CARBON DIOXIDE,
TIME EFFECTS, *FACULATIVE PONDS,
, BIOCHEMI/ *TERTIARY TREATMENT,
PEARS, CITRUS FRUITS, TOMATOES,
ED WASTE TREATMENT, *NUTRIENT R/

XYGEN DEMAND, LABORATORY TESTS,/
, *FERTILIZAT/ *SEWAGE DISPOSAL,
L RESPONSE, FE/ *URBAN DRAINAGE,
KES, *TRIBUTAR/ *EUTROPHICATION,
 *DRAINAGE EFFECTS, *WATER QUAL/
SCENEDESMUS, *SUBSURFACE WATERS,
E RAMSEY(ONTARIO), ALGAE COUNTS/
ND RADIATION, *AQUATIC HABITATS,
        *BLUE-GREEN ALGAE,
C MODELS, CHLORELLA PYRENOIDOSA,
T,/ *POLYMERS, *BIOLOGICAL LIFE,
S, LIMNOLOGY, WATER TEMPERATURE,
EW HAMPSHIRE, ALGAE, ODOR, TAST/
TEMS ANALYSIS, DATA COLLECTIONS,
Y(CALIF.).:
Y(CALIF.).:
TRATION.:
COUNT, WARING BLENDOR MORTALITY/
, LINDANE METABOLISM./ *LINDANE,
MALATHION, SEVIN.:

*'WORKING' OF LAKES, *WATER-BLOOM	W69-08283
*ABS, LAS.:	W69-09454
*ABSORPTION, *CALCIUM, SOIL CONTA	W69-03491
*ABUNDANCE RATIOS.:	W71-05531
*ACCLIMATIZATION, DIATOMS, THERMA	W70-07313
*ACETATE, *FRESHWATER ENVIRONMENT	W70-05109
*ACID MINE WATER, ACIDIC WATER, H	W70-05424
*ACID MINE WATER, *AQUAITIC MICRO	W69-00096
*ACID MINE WATER, *NUTRIENT REQUI	W70-02792
*ACID MINE WATER, *HABITATS, *ENV	W70-02770
*ACTINOMYCETES, *ALGAE, *TASTE, *	W72-13605
*ACTINOMYCETES, *ODOR, ACTIVATED	W72-04155
*ACTIVATED CARBON, ACTINOMYCETES,	W72-04242
*ACTIVATED CARBON, *ADSORPTION, *	W72-03641
*ACTIVATED CARBON, *ELECTROLYTES,	W72-09693
*ACTIVATED SLUDGE PROCESS, HANOVE	W70-09186
*ACTIVATED SLUDGE, *EFFLUENTS, WA	W70-06053
*ACTIVATED SLUDGE, BIOLOGICAL TRE	W70-05821
*ACTIVATED SLUDGE, *BACTERIA, MET	W72-01817
*ACTIVATED SLUDGE, STORAGE, TEMPE	W71-13335
*ACTIVATED SLUDGE, ALGAE, EUTROPH	W68-00012
*ACTIVATED SLUDGE, *BIOCHEMICAL O	W70-04060
*ADAPTATION, *SPECIES, CYCLOTELLA	W70-05381
*ADDITIVES, *FLOW RATE, *WASTE WA	W70-05819
*ADMINISTRATIVE AGENCIES, *WATER	W70-03344
*ADSORPTION, *PHOSPHORUS COMPOUND	W69-03075
*ADSORPTION, SOILS, COALS, DISTRI	W69-09884
*ADSORPTION, *WATER TREATMENT, *W	W72-03641
*ADVANCED WASTE TREATMENT, *NUTRI	W69-07818
*AERATED LAGOONS, *MIDWEST(US), *	W70-04788
*AERATION TUBES, *DIFFUSERS, *RAY	W70-06601
*AERATION, *ALGAE, *LAGOONS, *OXI	W70-06601
*AERATION, *IMPOUNDMENTS, *RESERV	W70-04484
*AERATION, *NITRATES, *CULTURES,	W70-03334
*AEROBIC PONDS, *ANAEROBIC PONDS,	W70-04787
*AEROBIC TREATMENT, OXYGEN DEMAND	W70-06792
*AEROBIC TREATMENT, *ANAEROBIC DI	W71-09548
*AESTHETIC DETERIORATION, *ADVANC	W69-07818
*AFRICA, MALAWI.:	W70-02646
*AGE, *ALGAE, *BIODEGRADATION, *O	W70-06600
*AGRICULTURAL WATERSHEDS, *RUNOFF	W70-02787
*AGRICULTURAL DRAINAGE, BIOLOGICA	W70-02787
*AGRICULTURAL WASTES, *ALGAE, *LA	W70-02969
*AGRICULTURAL CHEMICALS, *DELTAS,	W71-09577
*AGRICULTURE, *WASTES, ALGAE, ASS	W70-03334
*AIR SPRAY, SUDBURY(ONTARIO), LAK	W69-10155
*ALASKA, LAKES, RIVERS, HOT SPRIN	W69-08272
*ALGA BLOOMS, ALGAE PHYSIOLOGY.:	W69-03512
*ALGAE BIOMASS.: KINETI	W73-02218
*ALGAE BLOOMS, *FISH TOXICITY TES	W70-05819
*ALGAE BLOOMS, CYANOPHYTA, NORTH	W70-09093
*ALGAE CONTROL, *WATER SUPPLY, *N	W69-08282
*ALGAE CONTROL, *LAKE ERIE, BIOCH	W71-04758
*ALGAE HARVESTING, *CENTRAL VALLE	W71-04556
*ALGAE HARVESTING, *CENTRAL VALLE	W71-04557
*ALGAE HARVESTING, *DIATOMITE FIL	W72-08688
*ALGAE HOMOGENIZATION, *BACTERIA	W70-04365
*ALGAE METABOLISM, DETOXIFICATION	W70-08652
*ALGAE TOXICITY, CARBON-14, DDT,	W70-02198

345

SEWAGE, WATER POLLUTION EFFECTS,   *ALGAE.: *ESTUARIES, *HARBORS, *   W71-05500
OGY, NUTRIENTS, PLANT PATHOLOGY,   *ALGAE.: /Y, MICHIGAN, *EPIPHYTOL   W71-00832
L, TEST PROCEDURES, SCENEDESMUS,   *ALGAE.: /ER TEMPERATURE, SEASONA   W68-00472
    *PHOSPHORUS, *IRON, *LIMNOLOGY,   *ALGAE.: /HMENT TESTS, *NITROGEN,   W69-06004
                            BERLIN,   *ALGAE-TITRE.:
AQUATIC WEEDS, *BIOASSAY, *ESSE/   *ALGAE, *ANALYTICAL TECHNIQUES, *   W69-03364
CYCLING NUTRIENTS, *EUTROPHICAT/   *ALGAE, *ANALYTICAL TECHNIQUES, *   W69-03358
WATER, *AQUATIC MICROORGANISMS,   *ALGAE, *AQUATIC BACTERIA, *AQUAT   W69-00096
*AQUATIC WEEDS, *INHIBITION, *P/   *ALGAE, *ALGAL TOXINS, *AQUARIA,   W69-03188
*LIMNOLOGY, FISHKILL, NUTRIENTS,   *ALGAE, *ALGAL CONTROL, ALGAL POI   W69-05844
L NUTRIENTS, *EUTROPHICATION, */   *ALGAE, *ALGAL CONTROL, *ESSENTIA   W69-04798
BIOASSAY, *ESSENTIAL NUTRIENTS, /   *ALGAE, *ANALYTICAL TECHNIQUES, *   W69-04802
CHNIQUES, TEM/ *MARINE BACTERIA,   *ALGAE, *ANALYSIS, *ANALYTICAL TE   W70-04365
UTROPHICATION, *MEROMIXIS, *WAT/   *ALGAE, *AQUATIC ENVIRONMENTS, *E   W70-03974
MARINE MICRO/ *AQUATIC BACTERIA,   *ALGAE, *ANALYTICAL TECHNIQUES, *   W70-04811
PLANKTON, *LABORATORY TESTS, *I/   *ALGAE, *ANALYTICAL TECHNIQUES, *   W70-04469
COPPER SULFATE, *EUTROPHICATION,   *ALGAE, *AQUATIC WEEDS, LAKES, AL   W70-04494
ENTS, WASTE WATER RECLAMATION, /   *ALGAE, *ACTIVATED SLUDGE, *EFFLU   W70-06053
AQUATIC ALGAE, AQUATIC MICROBIO/   *ALGAE, *ANALYTICAL TECHNIQUES, *   W70-02792
S, ANALYTICAL TECHNIQUES, AQUAT/   *ALGAE, *ALGAL CONTROL, *ALGICIDE   W70-02968
TS, *ENVIRONMENTAL EFFECTS, *LI/   *ALGAE, *ACID MINE WATER, *HABITA   W70-02770
ENVIRONMENT, POLLUTANTS, BIOLOG/   *ALGAE, *AQUATIC ALGAE, *AQUATIC   W70-01233
ENTS, SURFACTA/ *BIODEGRADATION,   *ALGAE, *AQUATIC BACTERIA, DETERG   W69-09454
ODOR, *NATURA/ *WATER TREATMENT,   *ALGAE, *ALGAL CONTROL, *TASTE, *   W72-03562
TIVATED CARBO/ *WATER TREATMENT,   *ALGAE, *ACTINOMYCETES, *ODOR, AC   W72-04155
*LABORATORIES, / *BENTHIC FLORA,   *ALGAE, *BIOLOGICAL COMMUNITIES,   W70-02780
DEMAND, LABORATORY TESTS,/ *AGE/   *ALGAE, *BIODEGRADATION, *OXYGEN   W70-06600
ONTROL, *OPERATION AND MAINTENA/   *ALGAE, *BIOLOGICAL PROPERTIES, C   W70-08392
LLUTION EFFECTS, *REVIEWS, *WAT/   *ALGAE, *BIOINDICATORS, *WATER PO   W70-07027
ION, *PROTOZOA, *FUNGI, *BACTER/   *ALGAE, *BIOMASS, WATER PURIFICAT   W71-05267
TES, *EUTROPHICATION, NUTRIENTS/   *ALGAE, *BIODEGRADATION, *PHOSPHA   W71-04072
E TREATMENT, *OXIDATION LAGOONS,   *ALGAE, *BACTERIA, *DISSOLVED OXY   W72-01818
LLUTION, DIATOMS, MARYLAND, PEN/   *ALGAE, *BIOINDICATORS, *WATER PO   W70-04510
ICATION, *PLANT GROWTH, ANALYTI/   *ALGAE, *BIBLIOGRAPHIES, *EUTROPH   W69-04801
ROPHICATION, WATER POLLUTION EF/   *ALGAE, *BIOASSAY, NUTRIENTS, EUT   W69-06864
N, *FERTILIZATION, *INLAND WATE/   *ALGAE, *BIOASSAY, *EUTROPHICATIO   W69-07833
ATER TREATMENT, *POLLUTANT IDEN/   *ALGAE, *BIOLOGICAL TREATMENT, *W   W69-07389
CYTOLOGICAL STUDIES, CULTURES, /   *ALGAE, *CARBON, *CARBON CYCLE, *   W69-08278
RINE, *ANALYTICAL TECHNIQUES, */   *ALGAE, *CHEMICAL ANALYSIS, *CHLO   W70-05178
XATION, NUTRIENTS, LAKE ERIE.:   *ALGAE, *CYANOPHYTA, *NITROGEN FI   W70-05091
    *LIGHT, PHOSPHORUS, PHOTOSYNTH/   *ALGAE, *CHLORELLA, *SCENEDESMUS,   W70-03923
S, BIOMASS, ORGANIC COMPOUNDS, /   *ALGAE, *CHLORELLA, *ORGANIC ACID   W70-04195
CARBON,/ *CARBON RADIOISOTOPES,   *ALGAE, *CULTURES, PHYTOPLANKTON,   W70-03983
MES, *ESSENTIAL NUTRIENTS, *NIT/   *ALGAE, *CYCLING NUTRIENTS, *ENZY   W69-03185
, PHYTOPLANKTON, BIBLIOGRAPHIES,   *ALGAE, *CHEMICAL PROPERTIES, NUT   W68-00856
ARM MANAGEMENT, *DAIRY INDUSTRY,   *ALGAE, *CATTLE, SLURRIES, SPRAYI   W70-07491
TION EFFECTS, *ORGANIC COMPOUND/   *ALGAE, *CHLOROPHYLL, WATER POLLU   W70-08416
, *NUTRIENT REQUIREMENTS, ANALY/   *ALGAE, *CHLOROPHYTA, *CYANOPHYTA   W70-02245
UTROPHICATION, PONDS, RESERVOIR/   *ALGAE, *CYANOPHYTA, *ECOLOGY, *E   W70-00274
DITY, *IRON, *MANGANESE, *COLOR,   *ALGAE, *COLLOIDS, FLOW RATES, *F   W72-04145
ITY, WATER QUA/ WATER TREATMENT,   *ALGAE, *COPPER SULFATE, *ALKALIN   W72-07442
ER, *PO/ *WATER TREATMENT, ODOR,   *ALGAE, *DECOMPOSING ORGANIC MATT   W72-04242
TAL ENGINEERING, *SANITARY ENGI/   *ALGAE, *DISPERSIONS, *ENVIRONMEN   W70-02609
L IMATIZATION, / *MICROORGANISMS,   *ALGAE, *DOMINANT ORGANISMS, *ACC   W70-07313
NGANESE, CULTURES, RE/ *STREAMS,   *ALGAE, *DIATOMS, TEMPERATURE, MA   W70-07257
HYTOPLANKTON, *DISSOLVED OXYGEN,   *ALGAE, *DIURNAL DISTRIBUTION, *C   W69-03611
HUMAN POPULATION, *EUTROPHICATI/   *ALGAE, *ENVIRONMENTAL EFFECTS, *   W70-03973
GANISMS, PHYTOPLANKTON, BACTERI/   *ALGAE, *ECOLOGY, *MARINE MICROOR   W70-01068
PHY, NUTRIENTS, WATER T/ *LAKES,   *ALGAE, *EUTROPHICATION, OLIGOTRO   W70-00264
ATER TREATMENT, WATER TREATMENT,   *ALGAE, *EUTROPHICATION, *NITROGE   W72-01867
MICROB/ *NITROGEN, *PHOSPHATES,   *ALGAE, *EUTROPHICATION, DIATOMS,   W71-05026
NTHESIS, *RESPIRATION, GREAT LA/   *ALGAE, *EUTROPHICATION, *PHOTOSY   W72-05453
ILTERS, *SCEN/ *WATER TREATMENT,   *ALGAE, *FILTRATION, HEAD LOSS, F   W72-03703
S EARTH, SEPARATION TECHNIQUES,/   *ALGAE, *FILTRATION, *DIATOMACEOU   W72-08688
HIGAN, WATER / *WATER TREATMENT,   *ALGAE, *FILTERS, SANDS, LAKE MIC   W73-02426
ATION, *STRONTIUM RADIOISOTOPES,   *ALGAE, *FUNGI, *CALCAREOUS SOIL,   W71-04067
COPPER, IRON, IONS, PHOSPHATES,/   *ALGAE, *FRESH WATER, *PEPTIDES,   W69-10180
ER TREATMENT, ELECTROLYTES, CHE/   *ALGAE, *FLOCCULATION, *WASTE WAT   W70-05267
TORS, */ *ANALYTICAL TECHNIQUES,   *ALGAE, *GROWTH RATES, *BIOINDICA   W70-04381
S, LAB/ *EUTROPHICATION, *LAKES,   *ALGAE, *GROWTH RATES, *PHOSPHATE   W70-08637
ON, *NUTRIENTS, *TERTIARY TREAT/   *ALGAE, *HYDROGEN ION CONCENTRATI   W70-08980
LOGICAL ECOLOGY, *BENTHIC FLORA,   *ALGAE, *HYDRAULIC MODELS, LABORA   W71-00668
COBALT, MANGANESE, ATLANTIC OCE/   *ALGAE, *HERRINGS, COPPER, IRON,   W69-08525
WAGE, EFFLUENTS, CARBON DIOXIDE,   *ALGAE, *HYDROGEN ION CONCENTRATI   W68-00855
CIPAL WASTES, *GREAT LAKES, *WA/   *ALGAE, *INDUSTRIAL WASTES, *MUNI   W70-04430
DEGRADATION(DECOMPOSITION), DDT/   *ALGAE, *INSECTICIDES, TOXICITY,   W70-08097
ONS, BIOCHEMICAL OXY/ *AERATION,   *ALGAE, *LAGOONS, *OXIDATION LAGO   W70-06601
OBIC CONDITIONS, LIGHT INTENSIT/   *ALGAE, *LABORATORY TESTS, *ANAER   W71-13413
PHICATION, *AGRICULTURAL WASTES,   *ALGAE, *LAKES, *TRIBUTARIES, SAM   W70-02969
LA, CHLOROPHYTA, DIA/ *PLANKTON,   *ALGAE, *LIGHT INTENSITY, CHLOREL   W70-05381
*HYDROLOGY, CHEMICAL PROPERTIES/   *ALGAE, *LAKES, *EUTROPHICATION,   W70-05113
TES, AEROBIC CONDITIONS, ANAERO/   *ALGAE, *LIGHT, *NITRATES, *NITRI   W70-05261
CYANOPHYTA, UNITED STATES, HABI/   *ALGAE, *LAKE ERIE, CHLOROPHYTA,   W70-04468
S, ISLANDS, ECOLOGICAL DISTRIBU/   *ALGAE, *LITTORAL, *AQUATIC PLANT   W72-10623
MUS, PROTEINS, WALLS, STRUCTURE/   *ALGAE, *MICROORGANISMS, SCENEDES   W70-04184
*NITROGEN COMPOUNDS, *NUTRIENTS,   *ALGAE, *MICROORGANISMS, BIODEGRA   W69-06970
S, VARIABILITY, CHLORELLA, CARB/   *ALGAE, *METABOLISM, *ENVIRONMENT   W70-02965
TIVITY, *BIOLOGICAL COMMUNITIES,   *ALGAE, *METHODOLOGY, CARBON RADI   W71-12870
PPER SULFATE, RESERVOIRS, WATER/   *ALGAE, *MAINE, *WATER SUPPLY, CO   W70-08096
DE, *REVIEWS, / *EUTROPHICATION,   *ALGAE, *NUTRIENTS, *CARBON DIOXI   W70-00664

```
DISTRIB/ *MARINE MICROORGANISMS,          *ALGAE, *NUTRIENT REQUIREMENTS, *    W70-03952
EN-FIXATION, *NUTRIENTS, *LAKES,          *ALGAE, *NITROGEN, WISCONSIN.: /G      W70-03429
UIREMENTS, GROWTH RATES, LABORA/          *ALGAE, *NUTRIENTS, *NUTRIENT REQ      W70-03507
SPHORUS, SODIU/ *EUTROPHICATION,          *ALGAE, *NUTRIENTS, NITROGEN, PHO      W70-04503
SEWAGE EFFLUENT, *SEWAGE TREATM/          *ALGAE, *NITROGEN, *PHOSPHORUS,        W72-08133
WATER TREATMENT, *WATER QUALITY,          *ALGAE, *ODOR, HYDROGEN ION CONCE      W72-09694
ES, *DECOMPOSING ORGANIC MATTER,          *ALGAE, *PHENOLS.: /DECIDUOUS TRE      W72-03475
KILLS, *ROTIFE/ *WATER QUALITY,           *ALGAE, *PESTICIDES, *FISH, *FISH      W72-04506
TA, PHYTOPLANKTON, CYANOPHYTA, /           *ALGAE, *PHOTOSYNTHESIS, *MINNESO     W70-05387
TICIDE REMOVAL, *FOOD WEBS, DDT/          *ALGAE, *PESTICIDE TOXICITY, *PES      W70-03520
COMMUNITIES, EFFLUENTS/ *CARBON,          *ALGAE, *POPULATION, *BIOLOGICAL       W70-04001
RATES, *TURBULENCE, CHLORRELA, /          *ALGAE, *PHOTOSYNTHESIS, *GROWTH       W69-03730
*EUTROPHICATION, POLLUTION ABAT/          *ALGAE, *PHOSPHORUS, *MINNESOTA,       W69-10167
ARIO, *GROWTH RATES,/ *PLANKTON,          *ALGAE, *PHYTOPLANKTON, *LAKE ONT      W69-10158
, ECOLOGY, CHEMICAL ANALYSIS, P/          *ALGAE, *PLANT GROWTH, *WISCONSIN      W70-06229
ENERGY, DI-URNAL DISTRIBUTION,/           *ALGAE, *PHOSPHORUS, *PHOSPHATES,      W70-06604
*SANITARY ENGINEERING, MICROORG/          *ALGAE, *PROTOZOA, *SYSTEMATICS,       W72-01798
RGANIC MATTER, / *PHYTOPLANKTON,          *ALGAE, *PRIMARY PRODUCTIVITY, *O      W71-10079
S, *PHOSPHORUS, *NITROGEN, *ALK/          *ALGAE, *PHYTOPLANKTON, *NUTRIENT      W70-02795
ATION, *WATER CHEMISTRY, POLLUT/          *ALGAE, *PRODUCTIVITY, *EUTROPHIC      W70-02646
GE TREATMENT, *EUTROPHICATION, /          *ALGAE, *PLANKTON, *SEWAGE, *SEWA      W70-02304
, PHYTOPLANKTO/ *AQUATIC PLANTS,          *ALGAE, *ROTIFERS, PLANKTON, BASS      W70-02249
TION, *DISSOLVED OXYGEN, PENNSY/          *ALGAE, *RESPIRATION, *EUTROPHICA      W69-10159
, BIBLIOGRAPHIES, CHLORELLA, CH/          *ALGAE, *RESEARCH AND DEVELOPMENT      W69-06865
*EUTROPHICA/ *DISSOLVED OXYGEN,           *ALGAE, *RESPIRATION, *ESTUARIES,      W69-07520
, *WATER POLLUTION SO/ *ECOLOGY,          *ALGAE, *RIVERS, *WATER POLLUTION      W69-07096
, *TEMPERATURE, SOUTH CAROLINA,/          *ALGAE, *RADIOISOTOPES, *SORPTION      W69-07845
, MUSSELS, SHELLFISH/ *TOXICITY,          *ALGAE, *REVIEWS, HABITATS, CLAMS      W70-05372
IVER, *OHIO RIVER, NUI/ *RIVERS,          *ALGAE, *SAMPLING, *MISSISSIPPI R      W70-04507
ECTS, DRAINAGE / *LAKES, PLANTS,          *ALGAE, *SALINITY, IRRIGATION EFF      W70-00360
LF OF MEXICO, A/ *RADIOISOTOPES,          *ALGAE, *SORPTION, ABSORPTION, GU      W69-08524
RIES, FRESH WATER, LAKES, SILTS/          *ALGAE, *SEWAGE EFFLUENTS, *ESTUA      W70-02255
TIVES, *FLOW RATE, *WASTE WATER,          *ALGAE, *SLUDGE, STORMS, TURBULEN      W70-05819
ERS, ELECTROPHORESIS, CHLORELLA/          *ALGAE, *SANDS, *FILTRATION, FILT      W71-03035
NA, CHLOROPHYT/ *SEWAGE LAGOONS,          *ALGAE, *SEWAGE TREATMENT, *INDIA      W70-10173
NUTRIENTS, ECOSYSTEM/ *PLANKTON,          *ALGAE, *SESSILE ALGAE, *CYCLING       W73-01454
, ACTIVATED C/ *WATER TREATMENT,          *ALGAE, *TASTE, *ODOR, *ALGICIDES      W72-03746
WATER TREATMENT, *ACTINOMYCETES,          *ALGAE, *TASTE, *ODOR, WATER QUAL      W72-13605
OTOSYNTHESIS.:                            *ALGAE, *TEMPERATURE, DIATOMS, PH      W71-02496
    METABOLISM, RADIOACTIVITY TECH/       *ALGAE, *TOXICITY, *INSECTICIDES,      W70-02198
CITY.:           *FISH PHYSIOLOGY,        *ALGAE, *THERMAL POLLUTION, *TOXI      W70-01788
ENVIRONMENT, WATER QUALITY, NUT/          *ALGAE, *TROPHIC LEVELS, AQUATIC       W70-02764
S, HABITATS, ALKALINE WATER, AC/          *ALGAE, *TEMPERATURE, *HOT SPRING      W69-10160
INTENSITY, SPR/ *PHOTOSYNTHESIS,          *ALGAE, *THERMAL SPRINGS, *LIGHT       W69-09676
OLATION, LIMNOLOGY, PUBLIC HEAL/          *ALGAE, *TOXICITY, *CULTURES, *IS      W70-00273
DS, P/ *WATER POLLUTION SOURCES,          *ALGAE, *TASTE, *ODOR, *AMINO ACI      W72-01812
, THERMOPHILIC BAC/ *CYANOPHYTA,          *ALGAE, *THERMAL SPRINGS, *OREGON      W70-05270
N, BIOMA/ *PRIMARY PRODUCTIVITY,          *ALGAE, *THERMAL WATER, PERIPHYTO      W70-03309
TASTE, ODOR, FILTERS, PLANKTON,/          *ALGAE, *WATER SUPPLY, *CONTROL,       W70-05389
T, LAKES, SEWAGE, DRAINAGE DIST/          *ALGAE, *WATER POLLUTION TREATMEN      W70-04455
CATION, COLOR, TASTE, ODOR, TOX/          *ALGAE, *WATER QUALITY, *EUTROPHI      W69-05697
CHELATION,/ *COPPER, *TOXICITY,           *ALGAE, *WATER POLLUTION EFFECTS,      W71-11561
WATER POLLUTION CONTROL, *DENIT/          *ALGAE, *WASTE WATER TREATMENT, *      W71-04555
WATER POLLUTION CONTROL, *DENIT/          *ALGAE, *WASTE WATER TREATMENT, *      W71-04556
WATER POLLUTION CONTROL, *DENIT/          *ALGAE, *WASTE WATER TREATMENT, *      W71-04554
WATER POLLUTION CONTROL, *DENIT/          *ALGAE, *WASTE WATER TREATMENT, *      W71-04557
    *WATER TREATMENT, *FILTRATION,        *ALGAE, *WATER QUALITY, *ACTIVATE      W72-09693
AQUATIC ALGAE, AQUATIC MICROBI/           *ALGAE, ALGAL CONTROL, ALGICIDES,     W70-02248
ATIC ALGAE, WATER QUALITY, *TRO/          *ALGAE, AQUATIC ENVIRONMENT, *AQU      W70-02772
IRONMENTAL EFFECTS, *MORPHOLOGY,          *ALGAE, AQUATIC MICROBIOLOGY, CYA      W69-10177
FISHKILL, / *PESTICIDE RESIDUES,          *ALGAE, BIOASSAY, BIOINDICATORS,       W69-08565
UNITIES, CULTU/ *EUTROPHICATION,          *ALGAE, BIOASSAY, BIOLOGICAL COMM      W69-07832
DETERG/ *SURFACTANTS, *SORPTION,          *ALGAE, CHLOROPHYTA, CYANOPHYTA/       W70-03928
ERIA/ *EUTROPHICATION, PLANKTON,          *ALGAE, CHEMICAL PROPERTIES, BACT      W68-00483
, NITROGEN, PHOSPHORUS, DIATOMS,          *ALGAE, CHLOROPHYTA, CYANOPHYTA,       W68-00172
OUNDS, PHOTOSYN/ *MARINE PLANTS,          *ALGAE, CHLORINATION, SODIUM COMP      W70-00161
LYTICAL TECHNIQUES, *PHOSPHATES,          *ALGAE, CHROMOTOGRAPHY, RADIOCHEM      W71-03033
E WATER, *WATER QUALITY CONTROL,          *ALGAE, COPPER SULFATE, CHLORINE,      W71-00110
, *NUTRIENTS, *DISSOLVED OXYGEN,          *ALGAE, COLIFORMS, EUTROPHICATION      W71-05407
NGINEERING, *FILTRATION, DESIGN,          *ALGAE, COPPER SULFATE, SEDIMENTA      W72-08859
HYTA, BIOASS/ *LABORATORY TESTS,          *ALGAE, CULTURES, DIATOMS, CYANOP      W71-06002
CULTURES, DIAGENESIS, SEDIMENTS/          *ALGAE, CYANOPHYTA, CHRYSOPHYTA,       W69-08284
    *HERBICIDES, *VASCULAR TISSUES,       *ALGAE, ECOSYSTEMS, CHARA, BENTHI      W70-10175
S, *MINNESOTA, *LAKES, *DIATOMS,          *ALGAE, ECOLOGY, SOUTH DAKOTA, FI      W69-00659
ENT REQUIREMENTS, *PLANT GROWTH,          *ALGAE, ELEMENTS(CHEMICAL), CHLOR      W70-02964
AVIOLET RADIA/ *WATER TREATMENT,          *ALGAE, FILTRATION, *SLIME, *ULTR      W72-03824
SULFIDE, BIOCHEMICAL OXYGEN DEM/          *ALGAE, IRON COMPOUNDS, HYDROGEN       W70-06925
    *ZOOPLANKTON, *PHOTOSYNTHESIS,        *ALGAE, LAKES, OXYGEN, EUTROPHICA      W71-10098
ISOLATION, *BACTERIA, *PROTOZOA,          *ALGAE, LETHAL LIMIT, ELECTROPHOR      W71-03775
LLA, CULTURES.:          *GROWTH,         *ALGAE, LIPIDS, NUTRIENTS, CHLORE      W70-07957
R TREATMENT, *OXIDATION LAGOONS,          *ALGAE, LIGHT INTENSITY, CARBON D      W72-06022
HIBITION, N/ *NITROGEN FIXATION,          *ALGAE, LIGHT, CARBON DIOXIDE, IN      W70-04249
URE, PRIMARY PRODUCT/ *PLANKTON,          *ALGAE, LIGHT INTENSITY, TEMPERAT      W70-04809
TROPHICATION, PHOTOSYNTHESIS, L/          *ALGAE, NITROGEN, PHOSPHORUS, *EU      W72-00917
PHORUS, NITRO/ *EUTROPHICATION,           *ALGAE, NUTRIENTS, NITROGEN, PHOS      W71-13335
, *EUTROPHICATION, *GREAT LAKES,          *ALGAE, NUTRIENTS, PHOSPHORUS, PO      W70-00667
SANDS, FLOW R/ *WATER, *FILTERS,          *ALGAE, NUISANCE ALGAE, INTAKES,       W70-04199
S, *SEWAGE TRE/ *EUTROPHICATION,          *ALGAE, NUTRIENTS, DOMESTIC WASTE      W68-00248
HYTOPLANKTON, *AQUATIC BACTERIA,          *ALGAE, ORGANIC MATTER, LAKES, DI      W70-04284
ATION, LIGHT, TE/ *GROWTH RATES,          *ALGAE, PHOTOSYNTHESIS, EUTROPHIC      W69-07442
```

ONS, *BIOCHEMICAL OXYGEN DEMAND,   *ANAEROBICS, ALGAE, BACTERIA, WAT   W71-05742
S, TOMATOES, *AEROBIC TREATMENT,   *ANAEROBIC DIGESTION, BIOCHEMICAL   W71-09548
NSIT/ *ALGAE, *LABORATORY TESTS,   *ANAEROBIC CONDITIONS, LIGHT INTE   W71-13413
FECTS, *HYDRAULIC LOAD EFFECTS,/   *ANAEROBIC WASTE PONDS, *DEPTH EF   W70-04787
ACULATIVE PONDS, *AEROBIC PONDS,   *ANAEROBIC PONDS, DESIGN FOR ENVI   W70-04787
WASTES, MET/ *OXIDATION LAGOONS,   *ANAEROBIC DIGESTION, *MUNICIPAL   W72-06854
, TEM/ *MARINE BACTERIA, *ALGAE,   *ANALYSIS, *ANALYTICAL TECHNIQUES   W70-04365
ALGAE, *AQUATIC MICROBIOLOGY, */   *ANALYTICAL TECHNIQUES, *AQUATIC   W70-04283
GROWTH RATES, *BIOINDICATORS, */   *ANALYTICAL TECHNIQUES, *ALGAE, *   W70-04381
INE BACTERIA, *ALGAE, *ANALYSIS,   *ANALYTICAL TECHNIQUES, TEMPERATU   W70-04365
LUTION EFFEC/ *LAKES, *BACTERIA,   *ANALYTICAL TECHNIQUES, WATER POL   W70-04287
WEEDS, *WATER ANALYSIS, *SAMPLI/   *ANALYTICAL TECHNIQUES, *AQUATIC   W70-04382
Y METHODS, SEASONS, ACETYLENE R/   *ANALYTICAL METHODS, *BIOCHEMISTR   W70-03429
, *LABORATORY TESTS, *I/ *ALGAE,   *ANALYTICAL TECHNIQUES, *PLANKTON   W70-04469
ANISMS, *LIPIDS, LIGHT, ALGAE, /   *ANALYTICAL TECHNIQUES, *MICROORG   W70-05651
ICRO/ *AQUATIC BACTERIA, *ALGAE,   *ANALYTICAL TECHNIQUES, *MARINE M   W70-04811
, *CHEMICAL ANALYSIS, *CHLORINE,   *ANALYTICAL TECHNIQUES, *POLAROGR   W70-05178
POLLUTION EFFECTS, *PHOSPHATES,   *ANALYTICAL TECHNIQUES, *BIOASSAY   W70-05269
, *ESSENTIAL NUTRIENTS,/ *ALGAE,   *ANALYTICAL TECHNIQUES, *BIOASSAY   W69-04802
NUTRIENTS, *EUTROPHICAT/ *ALGAE,   *ANALYTICAL TECHNIQUES, *CYCLING   W69-03358
SOURCES, SEPARATION TECHNIQUES,   *ANALYTICAL TECHNIQUES, CLASSIFIC   W69-01273
, *ENZYMES, *NUTRIENT REQUIREME/   *ANALYTICAL TECHNIQUES, *BIOASSAY   W69-03373
NUTRIENTS, *EUTROPHICATION, *PH/   *ANALYTICAL TECHNIQUES, *CYCLING   W69-03370
WEEDS, *BIOASSAY, *ESSE/ *ALGAE,   *ANALYTICAL TECHNIQUES, *AQUATIC   W69-03364
CARBON, *ELECTROLYTES, FILTERS,   *ANALYTICAL TECHNIQUES.: /TIVATED   W72-09693
TORS, *CHLOROPHYLL, *OHIO RIVER,   *ANALYTICAL TECHNIQUES, BIOMASS,   W71-04206
ES, *ALGAE, CHROMOTOGRAPHY, RAD/   *ANALYTICAL TECHNIQUES, *PHOSPHAT   W71-03033
ON CONTROL, *RESEARCH EQUIPMENT,   *ANALYTICAL TECHNIQUES, WATER QUA   W70-08627
LANKTON, DIATOMS, MARINE ALGAE,/   *ANALYTICAL TECHNIQUES, *ALGAE, P   W70-09904
ALGAE, AQUATIC MICROBIO/ *ALGAE,   *ANALYTICAL TECHNIQUES, *AQUATIC   W70-02792
A OGLIMUN, UDO/ *TRACE ELEMENTS,   *ANCHOA LAMPROTAENIA, *OPISTHONEM   W69-08525
DA).:   *ANDERSON-CUE LAKE, MELROSE(FLORI   W72-01990
MANAGEMENT, *DAIRY INDUSTRY, */   *ANIMAL WASTES, MANAGEMENT, *FARM   W70-07491
CROPHYTES, ANACHARIS CANADENSIS/   *ANTAGONISTIC ACTION, *AQUATIC MA   W70-02249
S, FIBER WASTES, DYEING WASTES./   *ANTI-FOAM AGENTS, *TEXTILE WASTE   W70-04060
, DEPURATION, PHYCOCOLLOIDS, TA/   *ANTIBIOSIS, AUTOLYSIS, EXCRETION   W70-01068
CYSTIS AERUGINOSA, BOHEMIA, ELB/   *APHANIZOMENON FLOS AQUAE, *MICRO   W70-00274
N-14/ *UPPER KLAMATH LAKE(ORE.),   *APHANIZOMENON FLOS-AQUAE, *CARBO   W72-06052
DOMESTIC WAST/ *WASTE DISPOSAL,   *AQUACULTURE, *ULTIMATE DISPOSAL,   W70-09905
TION, *P/ *ALGAE, *ALGAL TOXINS,   *AQUARIA, *AQUATIC WEEDS, *INHIBI   W70-03188
LIBRIUM, BIOLOGICAL COMMUNITIES,   *AQUATIC ALGAE, HALOPHYTES.: /QUI   W69-00360
TES, SEMIARID CLIM/ *SOIL ALGAE,   *AQUATIC ALGAE, ALGAE, ARID CLIMA   W69-03491
LOGY, */ *ANALYTICAL TECHNIQUES,   *AQUATIC ALGAE, *AQUATIC MICROBIO   W70-04283
CATION, *MUD(LAKE), *PHOSPHORUS,   *AQUATIC ALGAE, *BIOASSAY, SCENED   W70-03955
ENT, POLLUTANTS, BIOLOG/ *ALGAE,   *AQUATIC ALGAE, *AQUATIC ENVIRONM   W70-01233
*ALGAE, *ANALYTICAL TECHNIQUES,   *AQUATIC ALGAE, AQUATIC MICROBIOL   W70-02792
RO/ *ALGAE, AQUATIC ENVIRONMENT,   *AQUATIC ALGAE, WATER QUALITY, *T   W70-02772
DA, *EUTROPHICATION, AQUATIC MI/   *AQUATIC ALGAE, *PROTOZOA, *FLORI   W72-01993
*BIOASSAY, T/ *WATER POLLUTION,   *AQUATIC BIOLOGY, *BIOINDICATORS,   W71-10786
A, *BACTERICIDES, *CHLORINATION,   *AQUATIC BACTERIA, *AQUATIC MICRO   W69-10171
RFACTA/ *BIODEGRADATION, *ALGAE,   *AQUATIC BACTERIA, DETERGENTS, SU   W69-09454
C/ *COMPETITION, *PHYTOPLANKTON,   *AQUATIC BACTERIA, *ALGAE, ORGANI   W70-04284
AQUIATIC MICROORGANISMS, *ALGAE,   *AQUATIC BACTERIA, *AQUATIC FUNGI   W69-00096
TICAL TECHNIQUES, *MARINE MICRO/   *AQUATIC BACTERIA, *ALGAE, *ANALY   W70-04811
GAE, PROTOZOA, *AQUATIC INSECTS,   *AQUATIC DRIFT, FISH FOOD ORGANIS   W69-02782
ATION, *MEROMIXIS, *WAT/ *ALGAE,   *AQUATIC ENVIRONMENTS, *EUTROPHIC   W70-03974
VIRONME/ *PHYSIOLOGICAL ECOLOGY,   *AQUATIC ENVIRONMENTS, RIVERS, EN   W70-03978
S, *RADIUM RADIOISOTOPES, *URAN/   *AQUATIC ENVIRONMENTS, *ECOSYSTEM   W69-07846
TRACERS, PHOTOS/ *RADIOISOTOPES,   *AQUATIC ENVIRONMENT, CHLORELLA,   W70-02786
BIOLOG/ *ALGAE, *AQUATIC ALGAE,   *AQUATIC ENVIRONMENT, POLLUTANTS,   W70-01233
*ECOLOGY, *BIOTA, *TEMPERATURE,   *AQUATIC ENVIRONMENT, *AQUATIC LI   W71-02491
ISMS, *ALGAE, *AQUATIC BACTERIA,   *AQUATIC FUNGI, YEASTS, HYDROGEN   W69-00096
, RIVERS/ *BACKGROUND RADIATION,   *AQUATIC HABITATS, *ALASKA, LAKES   W69-08272
FISH FOOD ORG/ ALGAE, PROTOZOA,   *AQUATIC INSECTS, *AQUATIC DRIFT,   W69-02782
, AQUATIC / *ZINC RADIOISOTOPES,   *AQUATIC INSECTS, *AQUATIC PLANTS   W71-09005
WATER POLLUTION EFFE/ *BIOASSAY,   *AQUATIC LIFE, *WASTE DISPOSAL, *   W72-01801
*RADIOACTIVITY, *TRACE ELEMENTS,   *AQUATIC LIFE, WATER POLLUTION EF   W71-05531
MPERATURE, *AQUATIC ENVIRONMENT,   *AQUATIC LIFE, *AQUATIC POPULATIO   W71-02491
NS, *RIVERS, ALGAE, PROTOZOA, A/   *AQUATIC LIFE, *AQUATIC POPULATIO   W73-00748
ANADENSIS/ *ANTAGONISTIC ACTION,   *AQUATIC MACROPHYTES, ANACHARIS C   W70-02249
CHLORINATION, *AQUATIC BACTERIA,   *AQUATIC MICROBIOLOGY, ALGICIDES,   W69-10171
ICAL ECOLOGY, *CYANOPHYTA, ALGA/   *AQUATIC MICROBIOLOGY, *PHYSIOLOG   W69-03518
, *MARINE MICROO/ *MICROBIOLOGY,   *AQUATIC MICROBIOLOGY, *ESTUARIES   W69-03752
ICAL TECHNIQUES, *AQUATIC ALGAE,   *AQUATIC MICROBIOLOGY, *AQUATIC P   W70-04283
IC ALGAE, *AQUATIC MICROBIOLOGY,   *AQUATIC PRODUCTIVITY, BIOCHEMICA   W70-04283
QUATIC ALGAE, AGING(BROLOGICAL),   *AQUATIC PRODUCTIVITY, LIMNOLOGY,   W68-00912
ROGEN, PHOSPHORUS, BICARBONATES,   *AQUATIC POPULATIONS.: /LGAE, NIT   W68-00468
NTROL, WATER CO/ *AQUATIC WEEDS,   *AQUATIC PLANTS, *AQUATIC WEED CO   W70-00269
S, PLANKTON, BASS, PHYTOPLANKTO/   *AQUATIC PLANTS, *ALGAE, *ROTIFER   W70-02249
SCULAR T/ *AQUATIC WEED CONTROL,   *AQUATIC PLANTS, *HERBICIDES, *VA   W70-10175
ATIC ENVIRONMENT, *AQUATIC LIFE,   *AQUATIC POPULATIONS, ALGAE, BENT   W71-02491
*ALGICIDES, *PHENOLS, *WATER Q/   *AQUATIC PLANTS, *NUISANCE ALGAE,   W70-10161
RSHES, *LAKES, *PHYTOPLANKTON, /   *AQUATIC PLANTS, *INHIBITORS, *MA   W72-00845
RADIOISOTOPES, *AQUATIC INSECTS,   *AQUATIC PLANTS, AQUATIC ENVIRONM   W71-09005
GAE, PROTOZOA, A/ *AQUATIC LIFE,   *AQUATIC POPULATIONS, *RIVERS, AL   W73-00748
CAL DISTRIBU/ *ALGAE, *LITTORAL,   *AQUATIC PLANTS, ISLANDS, ECOLOGI   W72-10623
LANTS, *HERBICIDES, *VASCULAR T/   *AQUATIC WEED CONTROL, *AQUATIC P   W70-10175
*AQUATIC WEEDS, *AQUATIC PLANTS,   *AQUATIC WEED CONTROL, WATER CONS   W70-00269
*AQUATIC WEED CONTROL, WATER CO/   *AQUATIC WEEDS, *AQUATIC PLANTS,   W70-00269
*ALGAE, *ANALYTICAL TECHNIQUES,   *AQUATIC WEEDS, *BIOASSAY, *ESSEN   W69-03364

AL/ *FARM WASTES, *FARM LAGOONS,
, ODOR,/ *FARM WASTES, *LAGOONS,
ION, *CARBOHYDRATES, *NUTRIENTS,
UTION CONTROL, *DENITRIFICATION,
TROPHICATION, NUTRIENTS/ *ALGAE,
UTION CONTROL, *DENITRIFICATION,
UTION CONTROL, *DENITRIFICATION,
UTION CONTROL, *DENITRIFICATION,
RPLANTS, *WATER QUALITY CONTROL,
SECTICIDES, *MARINE BACTERIA, */
BACTERIA, DETERGENTS, SURFACTA/
RATIOS.:
IO RIVER, *ANAL/ *WATER QUALITY,
TER POLLUTION, *AQUATIC BIOLOGY,
DISSO/ *WATER POLLUTION EFFECTS,
EFFECTS, *REVIEWS, *WAT/ *ALGAE,
CHNIQUES, *ALGAE, *GROWTH RATES,
DIATOMS, MARYLAND, PEN/ *ALGAE,
*BIOASSAY, *COM/ *WATER QUALITY,
C ORGANISMS, PAPERMILL WASTE EF/
LUTION / *WASTE WATER TREATMENT,
HOCK LOADING.:
*FISH TOXICITY TEST,/ *POLYMERS,
*OPERATION AND MAINTENA/ *ALGAE,
, LAGOONS, *BAFFLES, *WASTE WAT/
PITATION, *CHEMICAL DEGRADATION,
, *SEWAGE FERTILIZATION EFFECTS,
*METHODOL/ PRIMARY PRODUCTIVITY,
R(OHIO), LAKE MENDOTA(WIS), LAK/
TREATMENT, *TRICKLING FILTERS, /
TARY ENGINEERING, *WATER SUPPLY,
UCTION POTENTIAL, AQUATIC ALGAE,
ORIES, / *BENTHIC FLORA, *ALGAE,
*ALGAL CULTURE, *BIOMASS TITER,
*PONDS, *ENVIRONMENTAL EFFECTS,
NA CYLINDRICA, MICROCYSTIS AERU/
S/ *CARBON, *ALGAE, *POPULATION,
OLOGY, MAXIMUM U/ *HETEROTROPHY,
ATMENT, *POLLUTANT IDEN/ *ALGAE,
CTIONS, BLACK SEA, CONTACT ZONE/
*WASTE DISCHARGES, FLORA, FAUNA,
E, *LAKE MICHIGAN, *LAKE ONTARI/
TE, *TREATMENT, *LAKES, *ECOLOG/
, *SAPROBITY, P/ *ALGAL CULTURE,
ACTERIA, BACTER/ *COOLING WATER,
ATISTICAL METHODS, *ZOOPLANKTON,
OTOZOA, *FUNGI, *BACTER/ *ALGAE,
TA, NU/ *LAKES, *EUTROPHICATION,
TER POLLUTION ASSESSMENT, SANIT/
, PHOSPHORUS RADIOISOTOPES, TRO/
*THERMAL POWERPLANTS, *ECOLOGY,
E TREATMENT, *TRICKLING FILTERS,
Y, LAKE WAUBESA, LYNGB/ *BLOOMS,
ADIOCARBON UPTAKE TECHNIQUES, W/
OI), RED RIVER(VIETNAM), CHLORO/
DIVERSITY, LAKE WAUBESA, LYNGB/
GHAT(INDIA), BOMBAY(IND/ *INDIA,
ENTE/ *POLLUTION-TOLERANT ALGAE,
NKING, *PHYTOPLANKTON, *DIATOMS,
*RED ALGAE, *TEMPERATURE TOLERA/
ALGAE PHYSIOLOGY.:
CENTRATION, / *THERMAL GRADIENT,
DAPHNIA SCHOD/ *FILTERING RATE,
MONTANA, *HYDROCARBONS, *OLEFI/
LLUTION, *NUTRIENT REQUIREMENTS,

Y, ALGAL METABOLIS/ *KILL RATES,
M RADIOISOTOPES, *ALGAE, *FUNGI,
ONMENT, AQUATI/ *PHOTOSYNTHESIS,
SEMIARID CLIMATES, *ABSORPTION,
, *ALGAE, *DIURNAL DISTRIBUTION,
*WATER REUSE, *RECLAIMED WATER,
ON, *ESTUARIES, *EUTROPHICATION,
PES, *BIOASSAY, *EUTROPHICATION,
E/ *DEEP WATER, *LAKES, *PLANTS,
ELVE MILE LAKE(ONTARI/ *ONTARIO,
LVE MILE LAKE(CANADA), *ONTARIO,
UNITIE/ *LAKES, *EUTROPHICATION,
EACHES, APRICOTS, PEARS, CITRUS/
E/ *RESERVOIRS, *EUTROPHICATION,
ES, CULTURES, / *ALGAE, *CARBON,
ATES, *CULTU/ *SURFACE, *VOLUME,
ROPHICATION, *ALGAE, *NUTRIENTS,
PHAN/ *EXTRACELLULAR PRODUCTION,
, CLARKSBURG, BUFFALO, DUNREITH/
*EUTROPHICATION, *CALI/ *LAKES,
ULTURES, PHYTOPLANKTON, CARBON,/
RE.), *APHANIZOMENON FLOS-AQUAE,
OLOGICAL COMMUNITIES, EFFLUENTS/
CAL STUDIES, CULTURES, / *ALGAE,

*BIODEGRADATION, ALGAE, BIOCHEMIC   W71-00936
*BIODEGRADATION, *SEWAGE BACTERIA   W71-02009
*BIODEGRADATION, ALGAE, BACTERIA,   W71-00221
*BIODEGRADATION, *DRAINAGE WATER,   W71-04557
*BIODEGRADATION, *PHOSPHATES, *EU   W71-04072
*BIODEGRADATION, *DRAINAGE WATER,   W71-04554
*BIODEGRADATION, *DRAINAGE WATER,   W71-04556
*BIODEGRADATION, *DRAINAGE WATER,   W71-04555
*BIODEGRADATION, WATER UTILIZATIO   W71-11881
*BIODEGRADATION, *DETERGENTS, *IN   W70-01467
*BIODEGRADATION, *ALGAE, *AQUATIC   W69-09454
*BIOGEOCHEMICAL DATA, *ABUNDANCE   W71-05531
*BIOINDICATORS, *CHLOROPHYLL, *OH   W71-04206
*BIOINDICATORS, *BIOASSAY, TOXIC   W71-10786
*BIOINDICATORS, *WATER QUALITY, *   W70-00475
*BIOINDICATORS, *WATER POLLUTION   W70-07027
*BIOINDICATORS, *WATER POLLUTION,   W70-04381
*BIOINDICATORS, *WATER POLLUTION,   W70-04510
*BIOINDICATORS, *WATER ANALYSIS,   W72-12240
*BIOLOGIC 'DISPOSERS', MICROSCOPI   W71-00940
*BIOLOGICAL TREATMENT, *WATER POL   W70-10437
*BIOLOGICAL EXTRACTION, GROWTH, S   W70-06604
*BIOLOGICAL LIFE, *ALGAE BLOOMS,   W70-05819
*BIOLOGICAL PROPERTIES, CONTROL,   W70-08392
*BIOLOGICAL TREATMENT, *OXIDATION   W70-09320
*BIOLOGICAL TREATMENT, *BIODEGRAD   W70-09186
*BIOLOGICAL PRODUCTIVITY, AQUATIC   W69-09349
*BIOLOGICAL COMMUNITIES, *ALGAE,   W71-12870
*BIOLOGICAL PROBLEMS, SCIOTO RIVE   W72-01813
*BIOLOGICAL COMMUNITIES, *SEWAGE   W72-01819
*BIOLOGICAL PROPERTIES, *WATER PO   W72-01813
*BIOLOGICAL TREATMENT, *WASTE WAT   W71-09546
*BIOLOGICAL COMMUNITIES, *LABORAT   W70-02780
*BIOLOGICAL INDEX, *SAPROBITY, PR   W72-12240
*BIOLOGICAL TREATMENT, *DESIGN CR   W70-04787
*BIOLOGICAL CONCENTRATION, ANABAE   W70-03520
*BIOLOGICAL COMMUNITIES, EFFLUENT   W70-04001
*BIOLOGICAL UPTAKE, *MICROBIAL EC   W70-04194
*BIOLOGICAL TREATMENT, *WATER TRE   W69-07389
*BIOLOGICAL OCEANOGRAPHY, *TRANSA   W69-07440
*BIOLOGICAL COMMUNITIES, INTERTID   W68-00010
*BIOLOGY, *GREAT LAKES, *LAKE ERI   W70-01943
*BIOLOGY, *EFFECTS, *COPPER SULFA   W70-06217
*BIOMASS TITER, *BIOLOGICAL INDEX   W72-12240
*BIOMASS, *SLIME, ALGAE, FUNGI, B   W72-12987
*BIOMASS, *SUCESSION, *ISOLATION,   W73-01446
*BIOMASS, WATER PURIFICATION, *PR   W71-05267
*BIOMASS, *PRODUCTIVITY, CYANOPHY   W70-03959
*BIOSTIMULATION, ALGAL GROWTH, WA   W70-02775
*BIOTA, *COLUMBIA RIVER, CHROMIUM   W69-07853
*BIOTA, *TEMPERATURE, *AQUATIC EN   W71-02491
*BIOTA, SEWAGE BACTERIA, SYSTEMAT   W72-01819
*BIOTIC FACTORS, SPECIES DIVERSIT   W69-01977
*BIOTIN, *AMPHIDINIUM CARTERI, *R   W70-05652
*BLOOM, *VIETNAM, LITTLE LAKE(HAN   W70-03969
*BLOOMS, *BIOTIC FACTORS, SPECIES   W69-01977
*BLOOMS, RIVER GANGES(INDIA), RAJ   W70-05404
*BLUE GREEN ALGAE, *EUGLENA, PIGM   W70-01233
*BLUE-GREEN ALGA, COPPER SULFATE.   W70-02754
*BLUE-GREEN ALGAE, *GREEN ALGAE,   W71-02496
*BLUE-GREEN ALGAE, *ALGA BLOOMS,   W69-03512
*BOD REMOVAL, *ISRAEL, *ALGAL CON   W70-04790
*BODY SIZE, *COMPARATIVE STUDIES,   W70-03957
*BOTRYOCOCCUS BRAUNII, *ANACYSTIS   W69-08284
*BOTTOM SEDIMENTS, ON-SITE DATA C   W68-00511
*BROMINE, *IODINE.:   W69-08562
*BROMINE, IODINE, PHOTO REACTIVIT   W70-02363
*CALCAREOUS SOIL, SOIL WATER, SOI   W71-04067
*CALCIUM CARBONATE, AQUATIC ENVIR   W69-00446
*CALCIUM, SOIL CONTAMINATION, *ST   W69-03491
*CALIFORNIA, ESTIMATING EQUATIONS   W69-03611
*CALIFORNIA, WATER QUALITY, AESTH   W70-05645
*CALIFORNIA, WATER POLLUTION, REA   W69-07520
*CALIFORNIA, *NEVADA, LIMNOLOGY,   W70-01933
*CALIFORNIA, *NEVADA, DEPTH, ALGA   W70-00711
*CANADA, KUSHOG LAKE(ONTARIO), TW   W70-02795
*CANADA, NUTRIENT LEVELS, SOFT WA   W70-02973
*CANALS, *CITIES, BIOLOGICAL COMM   W70-00268
*CANNERIES, *INDUSTRIAL WASTES, P   W71-09548
*CARBOHYDRATES, *NUTRIENTS, *BIOD   W71-00221
*CARBON CYCLE, *CYTOLOGICAL STUDI   W69-08278
*CARBON DIOXIDE, *AERATION, *NITR   W70-03334
*CARBON DIOXIDE, *REVIEWS, BIBLIO   W70-00664
*CARBON FIXATION, *EXCRETION, STE   W69-10158
*CARBON FILTERS, LEBANON, BATAVIA   W72-03641
*CARBON RADIOISOTOPES, *BIOASSAY,   W70-01933
*CARBON RADIOISOTOPES, *ALGAE, *C   W70-03983
*CARBON-14 MEASUREMENTS.: /LAKE(O   W72-06052
*CARBON, *ALGAE, *POPULATION, *BI   W70-04001
*CARBON, *CARBON CYCLE, *CYTOLOGI   W69-08278

DIA PLUMOSA, STIGEOCLONIUM, MIC/
GEMENT, *DAIRY INDUSTRY, *ALGAE,
 *METABOLITES,
IS LUTHE/ *DNA, *ORGANIC CARBON,
A, ALGAE, PHYTOPLANKTON, *LAKES,
TERIAL DENITRIFICATION.:
TERIAL DENITRIFICATION.:
 *ALGAE HARVESTING,
 *ALGAE HARVESTING,

IF), NUTRI/ *FISHERY MANAGEMENT,
PES, *ALGAE, *FUNGI, *CALCAREOU/
, DETERGENTS, INDUSTRIAL WASTES,
S, WATE/ *TOXICITY, *CYANOPHYTA,
ENT, *COOLING WATER, *CORROSION,
ENVIRONMENT, PHAGE, FERMENTOR, /
OSPHATES, *PHOSPHORUS COMPOUNDS,
POUNDS, *CHEMICAL PRECIPITATION,
R, WATER POLLUTION EFFECTS, ANA/
NALYTICAL TECHNIQUES, */ *ALGAE,
WATER SAMPLING, *WATER ANALYSIS,
LANKTON, BIBLIOGRAPHIES, *ALGAE,
IRCULATION, NUTRIENT ABSORPTION,
HOSPHORUS, ALGAE, CYANOPHYTA, N/
IS AERUGINOSA, *NUTRIENT AVAILA/
IONS, *PHYSICAL CHARACTERISTICS,
ES, WATER AN/ *OIL, *OIL WASTES,
C MATTER, ALGAE, COPPER SULFATE,
ES.:
NT LIMITED/ *CONTINUOUS CULTURE,
CALUMET RIVER(ILL),/ *POLLUTION,
ASTES.:
T LAKES, METABOLISM, *CHLORELLA,
ALGAE, GREAT LAKES, METABOLISM,

ASSAY, *HALOGENS, *GROWTH RATES,
N, / *BIOCHEMICAL OXYGEN DEMAND,
DEMAND, *PHOTOSYNTHETIC OXYGEN,
SEWAGE FEED, SYMBIOTIC RELATIO/
SS, ORGANIC COMPOUNDS, / *ALGAE,
PHOSPHORUS, PHOTOSYNTH/ *ALGAE,
*PONDS, *DESIGN, *EFFICIENCIES,
POOLS, *BACTERIA, *BACTERICIDES,
, */ *ALGAE, *CHEMICAL ANALYSIS,
IS, *FLUOROMETRY, *FLUORESCENCE,
*ENZYMES, *ESSENTIAL NUTRIENTS/
TY, *MEASUREMEN/ *BENTHIC FLORA,
ENT REQUIREMENTS, ANALY/ *ALGAE,
ECTS, *ORGANIC COMPOUND/ *ALGAE,
*WATER QUALITY, *BIOINDICATORS,
CITY, ORGANIC COMP/ *REAERATION,
ON, *RHONE RIVER, NAVICULA, MEL/
, *IRON, TRACE ELEMENTS, ALGAE,/
SYSTEM, CILIATE AUTOECOLOGY.:
LAKES, *EUTROPHICATION, *CANALS,
ALGAE, KILL RATES, PHOTO REACT/
, AQUATIC MICROBIOLOGY, AQUATIC/
OMERATA, BUTTERNUT CREEK(NY).:
TIAL NUTRIENTS, *EUTROPHICATION,
HICATION, AQUATIC ALGAE, *LAKES,
SSAYS, PROTOZOA,/ *MICROBIOLOGY,
URCES, *WATER POLLUTION EFFECTS,
OL, *FILTRATION, *PARTICLE SIZE,
R/ *SALT VALLEY RESERVOIRS(NEB),
EMENTS, ALGAE,/ *CHROMATOGRAPHY,
NUTRIENTS, *TRACE ELEMENTS, ALG/
DEMAND, *PHOSPHORUS, *NITROGEN,
RON, *MANGANESE, *COLOR, *ALGAE,
, *TURBIDITY, *IRON, *MANGANESE,
AL OXYGEN DEMAND, ALGAL CONTROL,
TER, *FRESH WATER, *HUMIC ACIDS,
DGE, *BIOCHEMICAL OXYGEN DEMAND,
INTER, CYANOPHYTA, / *LIMNOLOGY,
ORUS RADIOISOTOPES, TRO/ *BIOTA,
BACKWASHING, *SUSPENDED SOLIDS,
OD/ *FILTERING RATE, *BODY SIZE,
ORS, *WATER ANALYSIS, *BIOASSAY,
UATIC BACTERIA, *ALGAE, ORGANIC/
EDTA, HEIDA, LAKE TAHOE(CALIF)/
DIANS, PIVERS ISLAND(N C), DETE/
AENA FLOS AQUAE, APHANIZOMENON /
ALKALINITY, *NITRATES, NITROGEN,
MAN POPULATION, *EUTROPHICATION,
FECTS, *WATER POLLUTION CONTROL,
FECTS, *WATER POLLUTION CONTROL,
NSLATIONS, ALGAE, CRUSTACEANS, /
LGAE, PLANNING, RIVERS, LAKES, /
ECTS, ALGAL CON/ EUTROPHICATION,
RNMENTS, *WATER QUALITY CONTROL/
TURBIDOSTATS, NUTRIENT LIMITED/

*CATIONS, DRAPARNALDIA, DRAPARNAL   W70-02245
*CATTLE, SLURRIES, SPRAYING, ACTI   W70-07491
*CEDAR RIVER(IOWA).:   W72-13605
*CELL SIZE, EUKARYOTES, MONOCHRYS   W69-08278
*CELLULOSE, DISTRIBUTION PATTERNS   W69-00976
*CENTRAL VALLEY(CALIFORNIA), *BAC   W71-04555
*CENTRAL VALLEY(CALIFORNIA), *BAC   W71-04554
*CENTRAL VALLEY(CALIF.).:   W71-04556
*CENTRAL VALLEY(CALIF.).:   W71-04557
*CESIUM RADIOISOTOPES.:   W71-09013
*CHAOBORID MIDGES, CLEAR LAKE(CAL   W70-05428
*CHELATION, *STRONTIUM RADIOISOTO   W71-04067
*CHELATION.: /FF, DOMESTIC WASTES   W71-04072
*CHEMCONTROL, EUTROPHICATION, BAS   W70-02982
*CHEMICAL REACTIONS, *WATER REUSE   W70-02294
*CHEMICAL ENVIRONMENT, *PHYSICAL   W71-02009
*CHEMICAL PRECIPITATION, *CHEMICA   W70-09186
*CHEMICAL DEGRADATION, *BIOLOGICA   W70-09186
*CHEMICAL ANALYSIS, ORGANIC MATTE   W70-05109
*CHEMICAL ANALYSIS, *CHLORINE, *A   W70-05178
*CHEMICAL ANALYSIS, PH, SAMPLING,   W69-00659
*CHEMICAL PROPERTIES, NUTRIENTS,   W68-00856
*CHEMICAL PROPERTIES, EPILMNION,   W68-00857
*CHEMICAL ANALYSIS, *NITROGEN, *P   W69-05868
*CHEMICAL COMPOSITION, *MICROCYST   W69-05868
*CHEMICAL CHARACTERISTICS, ORGANI   W73-00748
*CHEMICAL WASTES, INDUSTRIAL WAST   W72-03299
*CHEMICAL OXYGEN DEMAND.: /ORGANI   W72-04495
*CHEMOSTAT, PHYSIOLOGY, FLAGELLAT   W70-09904
*CHEMOSTATS, TURBIDOSTATS, NUTRIE   W70-02779
*CHICAGO(ILL), COOK COUNTY(ILL),   W70-00263
*CHICKENS, FOOD WASTES, POULTRY W   W70-09320
*CHLAMYDOMONAS LABORATORY TESTS,   W70-08652
*CHLORELLA, *CHLAMYDOMONAS LABORA   W70-08652
*CHLORELLA PYRENOIDOSA.:   W70-08416
*CHLORELLA, CHLORINE, ALGICIDES,   W70-02370
*CHLORELLA, *PHOTOSYNTHETIC OXYGE   W69-03362
*CHLORELLA, ALGAE, ANALYTICAL TEC   W69-03371
*CHLORELLA PYRENOIDOSA, SYNTHETIC   W70-05065
*CHLORELLA, *ORGANIC ACIDS, BIOMA   W70-04195
*CHLORELLA, *SCENEDESMUS, *LIGHT,   W70-03923
*CHLORINATION, *ODOR, *WASTE TREA   W70-04788
*CHLORINATION, *AQUATIC BACTERIA,   W69-10171
*CHLORINE, *ANALYTICAL TECHNIQUES   W70-05178
*CHLOROPHYLL, *EUTROPHICATION, PR   W70-05094
*CHLOROPHYTA, *CYCLING NUTRIENTS,   W69-03186
*CHLOROPHYLL, *PRIMARY PRODUCTIVI   W69-10151
*CHLOROPHYTA, *CYANOPHYTA, *NUTRI   W70-02245
*CHLOROPHYLL, WATER POLLUTION EFF   W70-08416
*CHLOROPHYLL, *OHIO RIVER, *ANALY   W71-04206
*CHLOROPHYLL, *OXYGENATION, *TOXI   W71-12183
*CHLOROPHYLL A, DATA INTERPRETATI   W73-01446
*CHROMATOGRAPHY, *COBALT, *COPPER   W71-11687
*CILIATES, FLAGELLATES, SAPROBIEN   W71-03031
*CITIES, BIOLOGICAL COMMUNITIES,   W70-00268
*CLADOPHORA, BROMINE, FILAMENTOUS   W70-02364
*CLADOPHORA, ALGAE, AQUATIC ALGAE   W70-02784
*CLADOPHORA FRACTA, CLADOPHORA GL   W72-05453
*CLADOPHORA, *NUISANCE ALGAE, *SE   W69-04798
*CLASSIFICATION, DESMIDS, OLIGOTR   W68-00255
*CLASSIFICATION, *CULTURES, *BIOA   W71-11823
*CLASSIFICATION, *WATER POLLUTION   W72-00462
*CLAYS, *SILTS, SANDS, ALGAE, PLA   W70-08628
*CLEAR WATER RESERVOIRS, *TURBID   W72-04761
*COBALT, *COPPER, *IRON, TRACE EL   W71-11687
*COBALT, *CYANOPHYTA, *ESSENTIAL   W69-06277
*COLIFORMS, *DISSOLVED OXYGEN, ES   W71-05409
*COLLOIDS, FLOW RATES, *FILTERS,   W72-04145
*COLOR, *ALGAE, *COLLOIDS, FLOW R   W72-04145
*COLOR, *ALKALINITY, *NITRATES, N   W72-11906
*COLOR, RIVERS, BOGS, PHENOLS, PI   W70-01074
*COLOR, WASTE WATER TREATMENT, TR   W70-04060
*COLORADO, *LAKES, *MOUNTAINS, *W   W69-10154
*COLUMBIA RIVER, CHROMIUM, PHOSPH   W69-07853
*COMBINED SEWERS, STORMWATER OVER   W71-10654
*COMPARATIVE STUDIES, DAPHNIA SCH   W70-03957
*COMPARATIVE PRODUCTIVITY, AQUATI   W72-12240
*COMPETITION, *PHYTOPLANKTON, *AQ   W70-04284
*COMPLEXONES, *ALGAL BLOOMS, NTA,   W70-04373
*CONCENTRATIONS, AEQUIPECTEN IRRA   W69-08267
*CONCEPTS, *TROPHIC BIOLOGY, ANAB   W70-03959
*CONDUCTIVITY, *LEACHING, IRON, T   W72-11906
*CONFERENCES, NUISANCE ALGAE, PUB   W70-03973
*CONFERENCES, STREAMS, ESTUARIES,   W70-03975
*CONFERENCES, STREAMS, ESTUARIES,   W70-03975
*CONFERENCES, *OCEANOGRAPHY, *TRA   W69-07440
*CONFERENCES, *EUTROPHICATION, *A   W70-04506
*CONFERENCES, WATER POLLUTION EFF   W70-02775
*CONNECTICUT, *LAKES, *LOCAL GOVE   W72-05774
*CONTINUOUS CULTURE, *CHEMOSTATS,   W70-02779

IORATION, DRAIN, SAN JOAQUIN VA/
KALINITY, WATER QUALITY CONTROL,
, *LABORATORY TESTS, *ISOLATION,
LANKTON,/ *ALGAE, *WATER SUPPLY,
THERMAL POLLUTION, *TEMPERATURE,
ALGAE, FUNGI, BACTERIA, BACTER-
LANTS, *WATER REUSE, COOLING TO/
ICAL REACTION/ *WATER TREATMENT,
AL TEMPERATURES, *COPPER DOSAGE,
ICH ISOTHERM, BLUE-GREEN ALGAE,/
INATION, *CRITICAL TEMPERATURES,
SERVOIRS, WATER/ *ALGAL CONTROL,
WATER SUPPLY, WATER QUALITY, WA/
 *ALGAL CONTROL, *NEW HAMPSHIRE,
ES, *ECOLOG/ *BIOLOGY, *EFFECTS,
ER QUA/ WATER TREATMENT, *ALGAE,
ON/ *NUISANCE ALGAE, *ALGICIDES,
 *ALGAE, *AQUATIC WEEDS, LAKES,/
UATIC WEED CONTROL, *HERBICIDES,
*TOXIN/ *EUTROPHICATION, *LAKES,
RATURE, SALINITY/ *MARINE WATER,
LGAE,/ *CHROMATOGRAPHY, *COBALT,
R POLLUTION EFFECTS, CHELATION,/
NTROL, ALGICIDES,/ *LAKES, *MUD,
FISH, ECOSYSTEMS, / *PHOSPHORUS,
IVE TREATMENT, ALGAE NUISANCE, /
RY, *FLUORESCENCE, *CHLOROPHYLL/
TURE, ALKALINITY, NITRIFICATION,
WATER TREATMENT, *COOLING WATER,
GATION, COAGULATION, DEWATERING,
 *NITROGEN, *PHOSPHORUS, ALGAE,/
), THORIUM-232, BISMUTH-214, AC/
AMICS, CELLULAR SIZE, CHLORELLA/
OSAGE/ *MICROSCOPIC EXAMINATION,
, *TORREY CANYON, AROMATIC SOLV/
ANCE, *LAKE AGING, NUTRIENT SOU/
, DREDG/ *LAKE WASHINGTON(WASH),
INDICATORS, NUTRIENT BUDGET.:
NONA(WIS), LAKE TAHOE(CALIF), L/
IONAL USE, WASTE EFFLUENTS.:
LOGICAL ECOLOGY, *PHYTOPLANKTON,
 *CARBON RADIOISOTOPES, *ALGAE,
M, KI/ *BACTERIA, *MARINE ALGAE,
N DIOXIDE, *AERATION, *NITRATES,
PUBLIC HEAL/ *ALGAE, *TOXICITY,
, MARINE ALGAE, ECOL/ *BACTERIA,
 *MICROBIOLOGY, *CLASSIFICATION,
GAL NUTRITION, ALGAL PHYSIOLOGY/
, *TRACE ELEMENTS, ALG/ *COBALT,
CONSIN, WATER QUALI/ *ALGICIDES,
GY, *SALINE WATER, *THERMAL WAT/
 *NUTRIENT REQUIREMENTS, *PHYSI/
BIOLOGY, *PHYSIOLOGICAL ECOLOGY,
GY, *PHYTOPLANKTON, *CULTURES, /
ANKTON, ALGAE, *NUTRIENT REQUIR/
YSIOLOGICAL ECOLOGY, *WATER POL/
GY, ALGAE, ANA/ *EUTROPHICATION,
AGE, FISH POPULATIONS, PLANKTON,
NUTRIENTS, LAKE ERIE.:   *ALGAE,
INGS, *OREGON, THERMOPHILIC BAC/
WATER POLLUTION EFFECTS, LAKES,
MATTER, TOXICITY, D/ *FISHKILL,
ATION, PONDS, RESERVOIR/ *ALGAE,
NTS, CULTURES, LAKE/ *ISOLATION,
TS, ANALY/ *ALGAE, *CHLOROPHYTA,
HICATION, BASS, WATE/ *TOXICITY,
ON, *HYDRODYNAMICS, LAKES, INTE/
REAMS, *WATER POLLUTION EFFECTS,
ON, *PH/ *ANALYTICAL TECHNIQUES,
SENTIAL NUTRIENTS/ *CHLOROPHYTA,
 *ALGAE, *ANALYTICAL TECHNIQUES,
SENTIAL NUTRIENTS, *NIT/ *ALGAE,
ER INFERFAC/ *PHOSPHATE, *LAKES,
L EFFECTS, *ESSENTIAL NUTRIENTS/
NKTON, *WATER POLLUTION SOURCES,
OMPOUNDS, *PHOSPHORUS COMPOUNDS,
N, ALGAE, PLANT G/ *FERTILIZERS,
OTOPES, STREAMFLOW/ *PERIPHYTON,
US RADIOISOTOPES, *SALT MARSHES,
LANKTON, *ALGAE, *SESSILE ALGAE,
ER, FOOD CHAINS/ *PHYTOPLANKTON,
ATION.:
 *ALGAE, *CARBON, *CARBON CYCLE,
S, MANAGEMENT, *FARM MANAGEMENT,
TILE SOLIDS, NON-DEGRADABLE SOL/
S QUADRICAUDA, ECOLOGICAL EFFIC/
 *FEEDING RATES, *GROWTH RATES,
LAKES, / *GRAZING, *TEMPERATURE,
R QUALITY, *ARTIFICIAL RECHARGE,
, BIOCHEMICAL OXYGEN DEMAND, SU/

*CONTRA COSTA COUNTY, WATER DETER   W71-09577
*CONTROL, ALGAL CONTROL, TOXICIT   W72-07442
*CONTROL, METHODOLOGY, CULTURES,   W70-04469
*CONTROL, TASTE, ODOR, FILTERS, P   W70-05389
*COOLING WATER, CHEMICAL PROPERTI   W69-05023
*COOLING WATER, *BIOMASS, *SLIME,   W72-12987
*COOLING WATER, *ELECTRIC POWER P   W71-11393
*COOLING WATER, *CORROSION, *CHEM   W70-02294
*COPPER APPLICATION METHODS, ALGA   W69-10157
*COPPER CHLORIDE, UPTAKE, FREUNDL   W70-08111
*COPPER DOSAGE, *COPPER APPLICATI   W69-10157
*COPPER SULFATE, WATER SUPPLY, RE   W69-10157
*COPPER SULFATE, *ALGAL CONTROL,   W69-10155
*COPPER SULFATE, SEWAGE EFFLUENTS   W69-08674
*COPPER SULFATE, *TREATMENT, *LAK   W70-06217
*COPPER SULFATE, ALKALINITY, WAT   W72-07442
*COPPER SULFATE, *WATER QUALITY C   W69-05704
*COPPER SULFATE, *EUTROPHICATION,   W70-04494
*COPPER SULFATE, *SODIUM ARSENITE   W70-05568
*COPPER SULFATE, *ALGAL CONTROL,   W68-00859
*COPPER, *ALGAE, SEA WATER, TEMPE   W70-08111
*COPPER, *IRON, TRACE ELEMENTS, A   W71-11687
*COPPER, *TOXICITY, *ALGAE, *WATE   W71-11561
*COPPER, COPPER SULFATE, ALGAL CO   W69-03366
*CORAL, *REEFS, *MARINE ANIMALS,   W69-08526
*CORRECTIVE TREATMENT, *PREVENTAT   W69-08282
*CORRELATION ANALYSIS, *FLUOROMET   W70-05094
*CORROSION, CARBON DIOXIDE, SOLUB   W71-11393
*CORROSION, *CHEMICAL REACTIONS,   W70-02294
*COST ANALYSIS.: /ATURE, CENTRIFU   W71-11375
*COST COMPARISONS, *EFFICIENCIES,   W70-07469
*COTTUS CAROLINAE, MEAD COUNTY(KY   W69-09742
*COULTER COUNTER, *POPULATION DYN   W69-06540
*CRITICAL TEMPERATURES, *COPPER D   W69-10157
*CRUDE OIL, *GAMOSOL, *OIL SPILLS   W70-01470
*CULTURAL EUTROPHY, *NUTRIENT BAL   W70-00264
*CULTURAL INFLUENCES, MICROCYSTIS   W70-05663
*CULTURAL EUTROPHICATION, TROPHIC   W71-03048
*CULTURAL EUTROPHICATION, LAKE MO   W69-06858
*CULTURAL EUTROPHICATION, RECREAT   W69-06535
*CULTURES, ALGAE, AQUATIC PRODUCT   W69-03514
*CULTURES, PHYTOPLANKTON, CARBON,   W70-03983
*CULTURES, PSEUDOMONAS, METABOLIS   W70-04280
*CULTURES, *SCENEDESMUS, *SUBSURF   W70-03334
*CULTURES, *ISOLATION, LIMNOLOGY,   W70-00273
*CULTURES, *TRITIUM, EPIPHYTOLOGY   W69-09755
*CULTURES, *BIOASSAYS, PROTOZOA,   W71-11823
*CYANOCOBALAMIN, ALGAL GROWTH, AL   W69-06277
*CYANOPHYTA, *ESSENTIAL NUTRIENTS   W69-06277
*CYANOPHYTA, *EUTROPHICATION, WIS   W70-03310
*CYANOPHYTA, *PHYSIOLOGICAL ECOLO   W69-03516
*CYANOPHYTA, *NITROGEN COMPOUNDS,   W69-03515
*CYANOPHYTA, ALGAE, ANALYTICAL TE   W69-03518
*CYANOPHYTA, *PHYSIOLOGICAL ECOLO   W69-03514
*CYANOPHYTA, *EUTROPHICATION, *PL   W69-03512
*CYANOPHYTA, *EUTROPHICATION, *PH   W69-03513
*CYANOPHYTA, *PHYSIOLOGICAL ECOLO   W69-03517
*CYANOPHYTA, WASHINGTON, DC, *EST   W68-00461
*CYANOPHYTA, *NITROGEN FIXATION,   W70-05091
*CYANOPHYTA, *ALGAE, *THERMAL SPR   W70-05270
*CYANOPHYTA, *EUTROPHICATION, WAT   W71-04758
*CYANOPHYTA, *DECOMPOSING ORGANIC   W72-01797
*CYANOPHYTA, *ECOLOGY, *EUTROPHIC   W70-00274
*CYANOPHYTES, *NUTRIENT REQUIREME   W70-00719
*CYANOPHYTA, *NUTRIENT REQUIREMEN   W70-02245
*CYANOPHYTA, *CHEMCONTROL, EUTROP   W70-02982
*CYCLING NUTRIENTS, *EUTROPHICATI   W70-09889
*CYCLING NUTRIENTS, *EUTROPHICATI   W69-03369
*CYCLING NUTRIENTS, *EUTROPHICATI   W69-03370
*CYCLING NUTRIENTS, *ENZYMES, *ES   W69-03186
*CYCLING NUTRIENTS, *EUTROPHICATI   W69-03358
*CYCLING NUTRIENTS, *ENZYMES, *ES   W69-03185
*CYCLING NUTRIENTS, *SEDIMENT-WAT   W69-06859
*CYCLING NUTRIENTS, *ENVIRONMENTA   W69-04804
*CYCLING NUTRIENTS, ALGAE, ANALYT   W69-04800
*CYCLING NUTRIENTS, WATER CHEMIST   W69-05142
*CYCLING NUTRIENTS, EUTROPHICATIO   W69-07826
*CYCLING NUTRIENTS, *ZINC RADIOIS   W69-07862
*CYCLING NUTRIENTS, GEORGIA, SEDI   W69-08274
*CYCLING NUTRIENTS, ECOSYSTEMS, P   W73-01454
*CYCLING, *RADIOISOTOPES, SEA WAT   W70-00708
*CYTOLOGICAL EFFECTS, ALGAL POPUL   W70-02770
*CYTOLOGICAL STUDIES, CULTURES, F   W69-08278
*DAIRY INDUSTRY, *ALGAE, *CATTLE,   W70-07491
*DAIRY MANURE, TOTAL SOLIDS, VOLI   W71-00936
*DAPHNIA PULEX OBTUSA, SCENEDESMU   W69-10152
*DAPHNIA, PRODUCTIVITY, BIOINDICA   W69-10152
*DAPHNIA, PHYSIOLOGICAL ECOLOGY,   W70-03957
*DATA COLLECTIONS, PIT RECHARGE,   W69-05894
*DATA COLLECTIONS, *WATER QUALITY   W70-09557

353

ION EFFECTS, *CYCLING NUTRIENTS,
TECHNIQUES, *CYCLING NUTRIENTS,
YSIOLOGICAL ECOLOGY, ALGAE, ANA/
, *NUTRIENT REQUIR/ *CYANOPHYTA,
COLOGY, *WATER POL/ *CYANOPHYTA,
ALGAE, NUISANCE ALGAE, PLANKTON/
IBLIOGRAPHIES, *ALGAE, *CHEMICA/
NKTON, ALGAE, SEASONAL, CYANOPH/
, BIOLOGICAL TREATMENT, BIBLIOG/
CUT, WATER POLLUTION, *NUTRIENT/
TER BLOOM, AQUATIC ALGAE, AGING/
LANTS, BIBLIOGRAPHIES, CYANOPHY/
S, UNITED STATES, PHOSPHORUS, N/
, CHEMICAL PROPERTIES, BACTERIA/
SULFATE, *ALGAL CONTROL, *TOXIN/
ES, BIBLIOGRAPHIES, *PHOSPHORUS/
N, ALGAE, *NUISANCE ALGAE, SPRA/
S, OXYGEN REQUIREMENTS, ICED LA/
AKES, *PHOSPHORUS RADIOISOTOPE,/
MENT, DRAINAGE SYSTEMS, FLOW RA/
SOURCES, *LAKES, NUTRIENTS, FE/
ALGAE, *RESPIRATION, *ESTUARIES,
*INLAND WATE/ *ALGAE, *BIOASSAY,
, BIOLOGICAL COMMUNITIES, CULTU/
ON, URBANIZATION, DOMESTIC WAST/
QUATIC WEEDS, CYANOPHYTA, FISHK/
CHLOROP/ *PHYTOPLANKTON, *PONDS,
ISH, *LIMNOLOGY, FISHKI/ *LAKES,
MANGANESE, *WISCONSIN, ALGAE, C/
*LIMNOLOGY, *PHYTOPLANKTON, FIS/
ATIC PLA/ *AQUATIC WEED CONTROL,
OR, TOX/ *ALGAE, *WATER QUALITY,
NALYTI/ *ALGAE, *BIBLIOGRAPHIES,
L CONTROL, *ESSENTIAL NUTRIENTS,
CONTROL, *AQUATIC WEED CONTROL,
CONTROL, *AQUATIC WEED CONTROL,
NDS, *NUTRIENT/ *BIBLIOGRAPHIES,
LOR, FISHKILL, LAKES, ODOR, TOX/
, ATMOSPHERE, SE/ *GEOCHEMISTRY,
OSPHATES, CHELA/ *WATER QUALITY,
SOURCES, *WATER POLLUTION EFFE/
CTIVITY, CYANOPHYTA, NU/ *LAKES,
*ALGAE, *AQUATIC ENVIRONMENTS,
NTAL EFFECTS, *HUMAN POPULATION,
SPHORUS, *AQUATIC ALGAE, *BIOAS/
SOURCES, *WATER POLLUTION EFFE/
, TERTIARY TREATMEN/ *MINNESOTA,
QUALI/ *ALGICIDES, *CYANOPHYTA,
URCES, *WATER POLLUTION EFFECTS,
SPHORUS COMPOUNDS, ALGAE, PHOSP/
S, WATER POLLUTION EFFECTS, PRO/
NDS, *PHOSPHORUS COMPOUNDS, *NU/
TS, NITROGEN, PHOSPHORUS, SODIU/
WEEDS, LAKES,/ *COPPER SULFATE,
, RIVERS, LAKES, / *CONFERENCES,
ANIZATION, WATER POLLUT/ *LAKES,
, OXIDATION, CARBO/ *RESERVOIRS,
TIFICATION, DEPTH, PLANKTON, LI/
ICAL PROPERTIES/ *ALGAE, *LAKES,
RY, *FLUORESCENCE, *CHLOROPHYLL,
ERGENTS, DOMESTIC WASTES, POLLU/
EATION, SEWAGE DISPOSAL, WATER /
ER POLLUTION SOURCES, *PHOSPHAT/
TON, *SEWAGE, *SEWAGE TREATMENT,
CARBON RADIOISOTOPES, *BIOASSAY,
ASTES, OXYGEN, STANDING CROP, A/
IFICATION, L/ *NUTRIENT REMOVAL,
S, RESERVOIRS,/ *NUISANCE ALGAE,
STES, *ALGAE, *LAKES, *TRIBUTAR/
, POLLUT/ *ALGAE, *PRODUCTIVITY,
S, ECOLOGY, PRI/ *BENTHIC FAUNA,
CONTROL, *WATER POLLUT/ *LAKES,
*NUTRIENTS, *BIODE/ *RESERVOIRS,
*ALGAE, WATER POLLUTION CONTRO/
S, FERTILIZATION, BACTERIA, ALG/
CHLOROPHYTA, CYANOPHYTA, BRACK/
ASTES, *WATER POLLUTION EFFECTS,
OLLUTION CONTROL, *REVIEWS, ALG/

*SEWAGE, WATER POLLUTION SOURC/
*GROWTH RATES, *PHOSPHATES, LAB/
LAKES, INTE/ *CYCLING NUTRIENTS,
TMENT, *WATER POLLUTION CONTROL,
ESERVOIR, *PRIMARY PRODUCTIVITY,
*NITROGEN, *PHOSPHATES, *ALGAE,
ION EFFECTS, LAKES, *CYANOPHYTA,
E, *BIODEGRADATION, *PHOSPHATES,
ENTIAL NUTRIENTS, *PRIMARY PROD/
ATIC ALGAE, *PROTOZOA, *FLORIDA,
ATMENT, WATER TREATMENT, *ALGAE,

*EUTROPHICATION, ALGAE, AQUATIC A	W69-03369
*EUTROPHICATION, *PHOSPHORUS COMP	W69-03370
*EUTROPHICATION, *CYANOPHYTA, *PH	W69-03517
*EUTROPHICATION, *PLANKTON, ALGAE	W69-03512
*EUTROPHICATION, *PHYSIOLOGICAL E	W69-03513
*EUTROPHICATION, *LAKE MICHIGAN,	W68-01244
*EUTROPHICATION, PHYTOPLANKTON, B	W68-00856
*EUTROPHICATION, *LAKES, PHYTOPLA	W68-00860
*EUTROPHICATION, LABORATORY TESTS	W68-00855
*EUTROPHICATION, *LAKES, CONNECTI	W68-00511
*EUTROPHICATION, ALGAL BLOOMS, WA	W68-00912
*EUTROPHICATION, ALGAE, AQUATIC P	W68-00680
*EUTROPHICATION, LAKES, *NUTRIENT	W68-00478
*EUTROPHICATION, PLANKTON, *ALGAE	W68-00483
*EUTROPHICATION, *LAKES, *COPPER	W68-00859
*EUTROPHICATION, *LAKES, ACID LAK	W68-00857
*EUTROPHICATION, *LAKES, WISCONSI	W68-00488
*EUTROPHICATION, *LAKES, NUTRIENT	W68-00487
*EUTROPHICATION, OLIGOTROPHIC, *L	W68-00858
*EUTROPHICATION, *POLLUTION ABATE	W68-00479
*EUTROPHICATION, *WATER POLLUTION	W69-07084
*EUTROPHICATION, *CALIFORNIA, WAT	W69-07520
*EUTROPHICATION, *FERTILIZATION,	W69-07833
*EUTROPHICATION, *ALGAE, BIOASSAY	W69-07832
*EUTROPHICATION, *LAKES, GLACIATI	W69-07818
*EUTROPHICATION, *LAKES, ALGAE, A	W69-06535
*EUTROPHICATION, ALGAE, BIOMASS,	W69-06540
*EUTROPHICATION, FERTILIZATION, F	W69-05844
*EUTROPHICATION, *IRON, *LAKES, *	W69-05867
*EUTROPHICATION, *FERTILIZATION,	W69-06858
*EUTROPHICATION, *HARVESTING, AQU	W69-06276
*EUTROPHICATION, COLOR, TASTE, OD	W69-05697
*EUTROPHICATION, *PLANT GROWTH, A	W69-04801
*EUTROPHICATION, *CLADOPHORA, *NU	W69-04798
*EUTROPHICATION, AQUATIC PLANTS,	W69-05706
*EUTROPHICATION, AQUATIC PLANTS,	W69-05705
*EUTROPHICATION, *NITROGEN COMPOU	W69-04805
*EUTROPHICATION, ALGAE, DUCKS, CO	W69-08279
*EUTROPHICATION, *WATER STRUCTURE	W70-04385
*EUTROPHICATION, *DETERGENTS, *PH	W70-04373
*EUTROPHICATION, *WATER POLLUTION	W70-03975
*EUTROPHICATION, *BIOMASS, *PRODU	W70-03959
*EUTROPHICATION, *MEROMIXIS, *WAT	W70-03974
*EUTROPHICATION, *CONFERENCES, NU	W70-03973
*EUTROPHICATION, *MUD(LAKE), *PHO	W70-03955
*EUTROPHICATION, *WATER POLLUTION	W70-03975
*EUTROPHICATION, SEWAGE EFFLUENTS	W70-03512
*EUTROPHICATION, WISCONSIN, WATER	W70-03310
*EUTROPHICATION, *NUTRIENTS, ALGA	W70-04266
*EUTROPHICATION, *NUTRIENTS, *PHO	W70-04074
*EUTROPHICATION, *LAKES, NUTRIENT	W70-04721
*EUTROPHICATION, *NITROGEN COMPOU	W70-04765
*EUTROPHICATION, *ALGAE, *NUTRIEN	W70-04503
*EUTROPHICATION, *ALGAE, *AQUATIC	W70-04494
*EUTROPHICATION, *ALGAE, PLANNING	W70-04506
*EUTROPHICATION, *WASHINGTON, URB	W70-05663
*EUTROPHICATION, NITROGEN, OXYGEN	W70-05550
*EUTROPHICATION, TEMPERATURE STRA	W70-05405
*EUTROPHICATION, *HYDROLOGY, CHEM	W70-05113
*EUTROPHICATION, PRIMARY PRODUCTI	W70-05094
*EUTROPHICATION, PHOSPHORUS, *DET	W70-05084
*EUTROPHICATION, *MINNESOTA, RECR	W70-04810
*EUTROPHICATION, *NUTRIENTS, *WAT	W70-01031
*EUTROPHICATION, PONDS, POLLUTANT	W70-02304
*EUTROPHICATION, *CALIFORNIA, *NE	W70-01933
*EUTROPHICATION, LAKES, ORGANIC W	W70-02251
*EUTROPHICATION, ANAEROBIC DENITR	W70-01981
*EUTROPHICATION, CYANOPHYTA, LAKE	W70-02803
*EUTROPHICATION, *AGRICULTURAL WA	W70-02969
*EUTROPHICATION, *WATER CHEMISTRY	W70-02646
*EUTROPHICATION, *NUTRIENTS, LAKE	W71-00665
*EUTROPHICATION, *WATER POLLUTION	W71-03048
*EUTROPHICATION, *CARBOHYDRATES,	W71-00221
*EUTROPHICATION, *LAKES, *RIVERS,	W70-05668
*EUTROPHICATION, *LAKES, NUTRIENT	W70-05761
*EUTROPHICATION, ALGAE, CULTURES,	W70-05752
*EUTROPHICATION, DISSOLVED OXYGEN	W70-06975
*EUTROPHICATION, *LAKES, *WATER P	W70-07393
*EUTROPHICATION CONTROL.:	W70-07393
*EUTROPHICATION, *NUISANCE ALGAE,	W70-07261
*EUTROPHICATION, *LAKES, *ALGAE,	W70-08637
*EUTROPHICATION, *HYDRODYNAMICS,	W70-09889
*EUTROPHICATION, *ALGAL CONTROL,	W70-09186
*EUTROPHICATION, WATER QUALITY, *	W70-09093
*EUTROPHICATION, DIATOMS, MICROBI	W71-05026
*EUTROPHICATION, WATER QUALITY.: /	W71-04758
*EUTROPHICATION, NUTRIENTS, NITRO	W71-04072
*EUTROPHICATION, *LIMNOLOGY, *ESS	W72-01990
*EUTROPHICATION, AQUATIC MICROORG	W72-01993
*EUTROPHICATION, *NITROGEN, NITRO	W72-01867

OLLUTION, DEGRADATION,/ *NEVADA,
L/ *ALGAE, NITROGEN, PHOSPHORUS,
GAE, *HARVESTING OF ALGAE, EFFL/
S, NITROGEN, PHOSPHOROUS, NITRO/
Y AUTHORITY PROJECT, *RESERVOIR/
*ALGAE, *PHOSPHORUS, *MINNESOTA,
HYTOPLANKT/ *WASHINGTON, *LAKES,
RIENTS, WATER T/ *LAKES, *ALGAE,
ALGAE, / *WATER, *FERTILIZATION,
ERS, NUTRIENTS/ *LAKES, *RIVERS,
ENT, N/ *PHOSPHORUS, *NUTRIENTS,
TES, NUTRIENTS, ALGAE, WATER PO/
NTATION, BIBLIOGRAPHIES, PHOSPH/
S, *NUTRIENTS, *NITROGE/ *LAKES,
, SCUM, FISHING, WISCONSIN, BIB/
ES, PONDS, FERTILIZERS, ALGAE, /
, BIOLOGICAL COMMUNITIE/ *LAKES,
TS, *CARBON DIOXIDE, *REVIEWS, /
*ALGAE, *CYANOPHYTA, *ECOLOGY,
LGAE, NUTRIENTS, PHOS/ *ECOLOGY,
N, PENNSY/ *ALGAE, *RESPIRATION,
TRIENTS, LIMNOLO/ WATER QUALITY,
ENVIRONMENT, WATER TEMPERATURE,
ORU/ *WATER QUALITY, *NUTRIENTS/
MENTS, ALGAL CONTROL, *NUTRIENT/
MENTS, *NUTRIENTS, NITROGEN COM/
EAT PLAINS, ALGAE, NEBRASKA, TU/
*RESPIRATION, GREAT LA/ *ALGAE,
ARGE, COPPER, *ALGICIDES, ALGAE,
AR PRODUCTION, *CARBON FIXATION,
AN, MONTANE ZON/ *FIELD STUDIES,
N FIXATION, *EXCRETION, STEPHAN/
CYLINDRICA, ZINC, XANTHOPHYCEA/
ONE, GLYCOLLIC ACID, CHEMOTROPH/
, CHONDRUS C/ *ALGAL SUBSTANCES,
FFECTS, *DETENTION TIME EFFECTS,
MODEL, COD REMOVAL, MILK SUBSTR/
S, TR/ *WATER POLLUTION EFFECTS,
WATER TEMPERATURE, SEWAGE TREAT/
LGAE, BIOCHEMICAL/ *FARM WASTES,
, */ *ANIMAL WASTES, MANAGEMENT,
DEGRADATION, ALGAE, BIOCHEMICAL/
ON, *ORGANIC WASTES, *WASTE WAT/
DATION, *SEWAGE BACTERIA, ODOR,/
RAINAGE EFFECTS, *WATER QUALITY,
OPMENT, GRANTS, ALGAE, AIR POLL/
ER TREATMENT, SEDIMENTATION, OX/
TES, EUTROPHIC/ *NITROGEN CYCLE,
UDIES, *NITROGEN LIMITATION, BA/
RL LAKES, CHLOROBIUM, FILINIA, /
APHNIA, PRODUCTIVITY, BIOINDICA/
SITY, EUTROPHICATION, NUTRIENTS,
LAKES, STREAMS, ALGAE, / *WATER,
GRICULTURAL WATERSHEDS, *RUNOFF,
OPLANKTON, FIS/ *EUTROPHICATION,
GAE, *BIOASSAY, *EUTROPHICATION,
PLANKTON, AQUATIC WEED/ *PONDS,
OXYGEN REQUIREMENTS, ICED LAKES,
EUTROPHICATION, ALGAE, PLANT G/
UTRIENTS, WASTE WATER TREATMENT,
NOLOGY, CLADOCERAN, MONTANE ZON/
VERY, OIL SPILL, *OIL POLLUTION,
ITH/ *LEUCOTHRIX MUCOR, *GROWTH,
D, FERRIC OXIDE-COATED SAND, CR/

PARATIVE STUDIES, DAPHNIA SCHOD/
INTAKES, SANDS, FLOW R/ *WATER,
ILTRATION, SANITARY ENGINEERING,
, *ALGAE, *COLLOIDS, FLOW RATES,
ATER / *WATER TREATMENT, *ALGAE,
ATION, *WATER POLLUTION CONTROL,
TRO/ *WATER POLLUTION TREATMENT,
YS, */ *WATER POLLUTION CONTROL,
DEMAND, *ACT/ *SEWAGE TREATMENT,
ESIS, CHLORELLA/ *ALGAE, *SANDS,
*ULTRASONICS, *SUSPENDED LOAD,
*ION EXCHANGE, *ORGANIC MATTER,
, *FILTERS, */ *WATER TREATMENT,
*SCEN/ *WATER TREATMENT, *ALGAE,
TY, *ACTIVATE/ *WATER TREATMENT,
TREATMENT, SANITARY ENGINEERING,
SEPARATION TECHNIQUES,/ *ALGAE,
ES, TURBID/ *DIATOMACEOUS EARTH,

, *ALGAL TOXINS, *PHYTOTOXICITY,
CYTES, *MUSCULAR EXERTION, *2-M/
ITY, *ALGAE, *PESTICIDES, *FISH,
ALG/ *NUTRIENTS, *LAKES, *TROUT,
L POLLUTION, *TOXICITY.:

*EUTROPHICATION, *LAKES, *WATER P   W72-02817
*EUTROPHICATION, PHOTOSYNTHESIS,   W72-00917
*EUTROPHICATION, *PHOSPHOROUS, AL   W71-13313
*EUTROPHICATION, *ALGAE, NUTRIENT   W71-13335
*EUTROPHICATION, *TENNESSEE VALLE   W71-07698
*EUTROPHICATION, POLLUTION ABATEM   W69-10167
*EUTROPHICATION, SEWAGE, BIOTA, P   W69-10169
*EUTROPHICATION, OLIGOTROPHY, NUT   W70-00264
*EUTROPHICATION, LAKES, STREAMS,   W69-10178
*EUTROPHICATION, RUNOFF, FERTILIZ   W69-10170
*EUTROPHICATION, *TERTIARY TREATM   W69-09340
*EUTROPHICATION, *LAKES, *PHOSPHA   W69-09135
*EUTROPHICATION, *REVIEWS, DOCUME   W69-08518
*EUTROPHICATION, *SEWAGE EFFLUENT   W69-09349
*EUTROPHICATION, CYANOPHYTA, ODOR   W69-08283
*EUTROPHICATION, *NUTRIENTS, *LAK   W69-08668
*EUTROPHICATION, *CANALS, *CITIES   W70-00268
*EUTROPHICATION, *ALGAE, *NUTRIEN   W70-00664
*EUTROPHICATION, PONDS, RESERVOIR   W70-00274
*EUTROPHICATION, *GREAT LAKES, *A   W70-00667
*EUTROPHICATION, *DISSOLVED OXYGE   W69-10159
*EUTROPHICATION, ALGAL BLOOMS, NU   W69-09723
*EUTROPHICATION, AQUATIC LIFE, SE   W73-01446
*EUTROPHICATION, NITROGEN, PHOSPH   W73-01122
*EUTROPHICATION, LAKES, LAKE SEDI   W72-06052
*EUTROPHICATION, LAKE, *LAKE SEDI   W72-06051
*EUTROPHICATION, *RESERVOIRS, *GR   W72-04761
*EUTROPHICATION, *PHOTOSYNTHESIS,   W72-05453
*EVALUATION, FILTRATION.: /L RECH   W72-05143
*EXCRETION, STEPHANODISCUS TENUIS   W69-10158
*EXPERIMENTAL LIMNOLOGY, CLADOCER   W69-10154
*EXTRACELLULAR PRODUCTION, *CARBO   W69-10158
*EXTRACELLULAR PRODUCTS, ANABAENA   W69-10180
*EXTRACELLULAR PRODUCTS, PHOTIC Z   W70-02504
*EXTRACELLULAR PRODUCTS, FUCUS VESICULOSUS   W70-01073
*FACULATIVE PONDS, *AEROBIC PONDS   W70-04787
*FACULTATIVE LAGOONS, LABORATORY   W70-07031
*FALLOUT, *ECOSYSTEMS, GREAT LAKE   W71-09013
*FARM LAGOONS, *DESIGN CRITERIA,   W71-08221
*FARM LAGOONS, *BIODEGRADATION, A   W71-00936
*FARM MANAGEMENT, *DAIRY INDUSTRY   W70-07491
*FARM WASTES, *FARM LAGOONS, *BIO   W71-00936
*FARM WASTES, *SPRINKLER IRRIGATI   W71-00940
*FARM WASTES, *LAGOONS, *BIODEGRA   W71-02009
*FARM WASTES, *INDUSTRIAL WASTES,   W71-09577
*FARM WASTES, *RESEARCH AND DEVEL   W71-06821
*FARM WASTES, *POULTRY, WASTE WAT   W71-11375
*FARM WASTE, GROUND WATER, *NITRA   W69-05323
*FAT ACCUMULATION, *LABORATORY ST   W70-07957
*FAYETTEVILLE GREEN LAKE(NY), *MA   W70-05760
*FEEDING RATES, *GROWTH RATES, *D   W69-10152
*FERMENTATION, TOXICITY, CHEMICAL   W71-13413
*FERTILIZATION, *EUTROPHICATION,   W69-10154
*FERTILIZATION, LAKES, SEWAGE TRE   W70-02787
*FERTILIZATION, *LIMNOLOGY, *PHYT   W69-06858
*FERTILIZATION, *INLAND WATERWAYS   W69-07833
*FERTILIZATION, *PHOSPHATE, FISH,   W70-05417
*FERTILIZATION, WATER FLEAS, LARV   W68-00487
*FERTILIZERS, *CYCLING NUTRIENTS,   W69-07826
*FIBER(PLANT).: /OXYGEN DEMAND, N   W70-06925
*FIELD STUDIES, *EXPERIMENTAL LIM   W69-10154
*FILAMENTOUS GREEN ALGAE, HALIMON   W70-09976
*FILAMENTS, *AUTORADIOGRAPHY, ANT   W69-09755
*FILTER MEDIA, ALUMINA-COATED SAN   W70-08628
*FILTER MEDIA.:   W72-07334
*FILTERING RATE, *BODY SIZE, *COM   W70-03957
*FILTERS, *ALGAE, NUISANCE ALGAE,   W70-04199
*FILTERS, *DESIGN, COAGULATION, A   W72-07334
*FILTERS, *NEW YORK, TREATMENT FA   W72-04145
*FILTERS, SANDS, LAKE MICHIGAN, W   W73-02426
*FILTERS, SEWAGE EFFLUENTS, ELECT   W70-01027
*FILTRATION, *WATER POLLUTION CON   W70-01027
*FILTRATION, *PARTICLE SIZE, *CLA   W70-08628
*FILTRATION, *BIOCHEMICAL OXYGEN   W70-05821
*FILTRATION, FILTERS, ELECTROPHOR   W71-03035
*FILTRATION, ALGAE, COLLOIDS, FLO   W71-10654
*FILTRATION, DIATOMACEOUS EARTH,   W72-04318
*FILTRATION, SANITARY ENGINEERING   W72-07334
*FILTRATION, HEAD LOSS, FILTERS,   W72-03703
*FILTRATION, *ALGAE, *WATER QUALI   W72-09693
*FILTRATION, DESIGN, *ALGAE, COPP   W72-08859
*FILTRATION, *DIATOMACEOUS EARTH,   W72-08688
*FILTRATION, *SEPARATION TECHNIQU   W72-08686
*FINISHING WASTES.:   W71-06737
*FINISHING WASTES.:   W70-01519
*FISH DISEASES, *FISH TOXINS, TOX   W71-07731
*FISH HEMATOLOGY, *GRANULAR LEUCO   W70-01788
*FISH KILLS, *ROTIFERS, *ECOLOGY,   W72-04506
*FISH MANAGEMENT, FERTILIZATION,   W70-05399
*FISH PHYSIOLOGY, *ALGAE, *THERMA   W70-01788

357

QUATIC POPULATIONS, OLIGOTROPHY,
*PHYTOTOXICITY, *FISH DISEASES,
*BIOLOGICAL LIFE, *ALGAE BLOOMS,
L EFFECTS, *ESSENTIAL NUTRIENTS,
ER QUALITY, *ALGAE, *PESTICIDES,
REAMS, *WATER POLLUTION EFFECTS,
TS, *ESSENTIAL NUTRIENTS, *FISH,
IDGES, CLEAR LAKE(CALIF), NUTRI/
ING ORGANIC MATTER, TOXICITY, D/
ING, *ALGAL TOXINS, *PHYTOTOXIC/
GHT TESTS, */ *BEER-LAMBERT LAW,
ERT LAW, *FIXED-DENSITY STUDIES,
MENT, ELECTROLYTES, CHE/ *ALGAE,
LATION, ALGAE, ACTIVATED CARBON,
C MI/ *AQUATIC ALGAE, *PROTOZOA,
TON, *DIATOMS, *BLUE-GREEN ALGA/
IVATED SLU/ *WATER PURIFICATION,
*SLUDGE, STORMS, T/ *ADDITIVES,
S, *ESTUARINE ENVIRONMENT, BAYS,
OLAR RADI/ *WASHINGTON, *OXYGEN,
RELATION ANALYSIS, *FLUOROMETRY,
OROPHYLL/ *CORRELATION ANALYSIS,
TATION, *SAMPLING, *MEASUREMENT,
TS, / *ALGAL GROWTH, *BIOASSAYS,
SMS, NUTRIENTS, ENERGY TRANSFER,
DE TOXICITY, *PESTICIDE REMOVAL,
Y ST/ *SCULPINS, *RADIOACTIVITY,
HODAMINE WT DYE, AIRBORNE TESTS/
-SITE INVESTIGATIONS, FREQUENCY,
IRON, IONS, PHOSPHATES,/ *ALGAE,
CEANS, *MARINE ALGAE, *SEAWATER,
SCENEDESMUS, LIGH/ *ENVIRONMENT,
OMPOUNDS, *NITROGEN UTILIZATION,
TER FLEAS, LARVAE, GROWTH RATES,
ER RATES, */ *GLUCOSE, *ACETATE,
UM, *ALGAL-BOUND RADIOSTRONTIUM,
, WATER PURIFICATION, *PROTOZOA,
STRONTIUM RADIOISOTOPES, *ALGAE,
, OXIDATION DITCH.:
NYON, AROMATIC SOLV/ *CRUDE OIL,
OUTHAMPTON UNIV/ *PHAEODACTYLUM,
BPOLAR WATERS, TYNSET(NORWAY), /
WATER STRUCTURE, ATMOSPHERE, SE/
OGAALKALOID, O/ *PERIDINIUM SP.,
UM, PLANKTO/ LAKE SAGAMI, JAPAN,

NVIRONMENTS, *TURNOVER RATES, */
TION, *2-METHYL-5-VINYLPYRIDINE,
XERTION, *2-M/ *FISH HEMATOLOGY,
ION, IRRIGATION, ORGANIC MATTER,
 PHYSIOLOGICAL ECOLOGY, LAKES, /
DETERGENTS, *TERTIARY TREATMENT,
TRIAL WASTES, *MUNICIPAL WASTES,
WATER POLLUTION CONTROL, WATER /
ICHIGAN, *LAKE ONTARI/ *BIOLOGY,
PHOS/ *ECOLOGY, *EUTROPHICATION,
ATER POLLUTION / *WATER QUALITY,
U/ *EUTROPHICATION, *RESERVOIRS,
ATURE TOLERA/ *BLUE-GREEN ALGAE,
 CHLORELLA PYRENOIDOSA, VOLATIL/
, ORGANIC MATTER, CURR/ *RUNOFF,
ENT, *WATER REUSE, *RECLAMATION,
LLUTION EFFECTS, *WATER SOURCES,
NIZATION, *ARTIFICIAL SUBSTRATE/
*ANALYTICAL TECHNIQUES, *ALGAE,
NOTRIAZOLE, 2-4-5-/ *HERBICIDES,
RELA, / *ALGAE, *PHOTOSYNTHESIS,
ESIS, EUTROPHICATION, LIGHT, TE/
URES, NUTRIENTS, RIVERS, LAKES,
*EUTROPHICATION, *LAKES, *ALGAE,
, *PHYTOPLANKTON, *LAKE ONTARIO,
VITY, BIOINDICA/ *FEEDING RATES,
NE, ALGIC/ *BIOASSAY, *HALOGENS,
APHY, ANTITH/ *LEUCOTHRIX MUCOR,
S, CHLORELLA, CULTURES.:
COUNT, OPTICAL DENSITY, PARTICU/
LGAE, E/ *MARINE MICROORGANISMS,
, *LI/ *ALGAE, *ACID MINE WATER,
STUDIES, CHLORINE, COPPER SULF/
ALGAL CONTROL, CHLORELLA, DISIN/
LLA, CHLORINE, ALGIC/ *BIOASSAY,
CLIDES, ASSIMILATION, BA-LA, HA/
N EFFECTS, *ALGAE.: *ESTUARIES,
 TRANSPARENCY, MARGINAL VEGETAT/
OPHICATION, *PHOSPHOROUS, ALGAE,
C WEED CONTROL, *EUTROPHICATION,
ACTIVATED CARBON, *FLOCCULATION,
E, 2-4-D, AMINOTRIAZOLE, 2-4-5-/
TOTOXICITY, ALGAL POISO/ *2-4 D,
DIUM ARS/ *AQUATIC WEED CONTROL,

*FISH POPULATIONS, PLANT POPULATI   W68-00172
*FISH TOXINS, TOXICITY, HYDROGEN   W71-07731
*FISH TOXICITY TEST, SLUDGE DRYIN   W70-05819
*FISH, *FISHERIES, ALGAE, AQUATIC   W69-04804
*FISH, *FISH KILLS, *ROTIFERS, *E   W72-04506
*FISH, ALGAE, TROPHIC LEVEL, PROD   W72-01788
*FISHERIES, ALGAE, AQUATIC ALGAE,   W69-04804
*FISHERY MANAGEMENT, *CHAOBORID M   W70-05428
*FISHKILL, *CYANOPHYTA, *DECOMPOS   W72-01797
*FISHKILL, *TOXINS, ALGAL POISON   W71-07731
*FIXED-DENSITY STUDIES, *FIXED-LI   W70-03923
*FIXED-LIGHT TESTS, *EFFECTIVE-AV   W70-03923
*FLOCCULATION, *WASTE WATER TREAT   W70-05267
*FLOCCULATION, *HEAD LOSS.: /OAGU   W72-07334
*FLORIDA, *EUTROPHICATION, AQUATI   W72-01993
*FLOTATION, *SINKING, *PHYTOPLANK   W70-02754
*FLOW AUGMENTATION, *LAGOONS, ACT   W72-06017
*FLOW RATE, *WASTE WATER, *ALGAE,   W70-05819
*FLUCTUATION, RESISTANCE, LIFE CY   W71-02494
*FLUCTUATION, SEWAGE TREATMENT, S   W69-10182
*FLUORESCENCE, *CHLOROPHYLL, *EUT   W70-05094
*FLUOROMETRY, *FLUORESCENCE, *CHL   W70-05094
*FLUOROMETRY, DYE RELEASES, CURRE   W70-05377
*FLUOROMETRY, CALIFORNIA, NUTRIEN   W70-02777
*FOOD CHAINS, FOOD WEBS, FORAGE F   W70-05428
*FOOD WEBS, DDT, CHLORINATED HYDR   W70-03520
*FOOD WEBS, KENTUCKY, LIFE HISTOR   W69-09742
*FRAUNHOFER LINE DISCRIMINATOR, R   W70-05377
*FREQUENCY ANALYSIS.: /TIONS, *ON   W70-09976
*FRESH WATER, *PEPTIDES, COPPER,.   W69-10180
*FRESH WATER, *HUMIC ACIDS, *COLO   W70-01074
*FRESH WATER, *PLANKTON, *ALGAE,   W70-05409
*FRESH WATER, ALGAE, NITRATE AMMO   W69-04521
*FRESH WATER FISH, MICHIGAN, TROU   W68-00487
*FRESHWATER ENVIRONMENTS, *TURNOV   W70-05109
*FUNGAL-BOUND RADIOSTRONTIUM, *MI   W71-04067
*FUNGI, *BACTERIA, HYDROGEN ION C   W71-05267
*FUNGI, *CALCAREOUS SOIL, SOIL WA   W71-04067
*FWPCA, ACTIVATED ALGAE, FEEDLOTS   W71-06821
*GAMOSOL, *OIL SPILLS, *TORREY CA   W70-01470
*GAS EFFLUENT, FOOD PRODUCTION, S   W70-00161
*GELBSTOFF, *ALGAL SUBSTANCES, SU   W70-01074
*GEOCHEMISTRY, *EUTROPHICATION, *   W70-04385
*GLENODININE, ALKALOID TOXINS, IB   W71-07731
*GLENODININE, *PERIDINIUM POLONIC   W70-08372
*GLUCOSE, WATERLOO MILLS(PA).:   W71-05026
*GLUCOSE, *ACETATE, *FRESHWATER E   W70-05109
*GLUE GREEN ALGAE, BREAM, WHITE B   W70-01788
*GRANULAR LEUCOCYTES, *MUSCULAR E   W70-01788
*GRASSLANDS, MOUNTAINS, *SOIL MIC   W72-08110
*GRAZING, *TEMPERATURE, *DAPHNIA,   W70-03957
*GREAT LAKES, CONTROL, COSTS, EUT   W70-03964
*GREAT LAKES, *WATER POLLUTION, S   W70-04430
*GREAT LAKES, *WATER POLLUTION, *   W69-03948
*GREAT LAKES, *LAKE ERIE, *LAKE M   W70-01943
*GREAT LAKES, *ALGAE, NUTRIENTS,   W70-00667
*GREAT LAKES, *TRACE ELEMENTS, *W   W69-08562
*GREAT PLAINS, ALGAE, NEBRASKA, T   W72-04761
*GREEN ALGAE, *RED ALGAE, *TEMPER   W71-02496
*GREEN ALGAE, CHLORELLA VULGARIS,   W70-04195
*GROUNDWATER, *SESTON, *SEDIMENTS   W70-03501
*GROUNDWATER, RECHARGE, *ION EXCH   W72-04318
*GROUNDWATER, DDT, MISSISSIPPI RI   W73-00758
*GROWTH DYNAMICS, *BACTERIAL COLO   W70-04371
*GROWTH RATES, *BIOINDICATORS, *W   W70-04381
*GROWTH RATES, *ALGAE, 2-4-D, AMI   W70-03519
*GROWTH RATES, *TURBULENCE, CHLOR   W69-03730
*GROWTH RATES, *ALGAE, PHOTOSYNTH   W69-07442
*GROWTH RATES, *WATER TEMPERATURE   W68-00472
*GROWTH RATES, *PHOSPHATES, LABOR   W70-08637
*GROWTH RATES, ENVIRONMENTAL EFFE   W69-10158
*GROWTH RATES, *DAPHNIA, PRODUCTI   W69-10152
*GROWTH RATES, *CHLORELLA, CHLORI   W70-02370
*GROWTH, *FILAMENTS, *AUTORADIOGR   W69-09755
*GROWTH, *ALGAE, LIPIDS, NUTRIENT   W70-07957
*GROWTH, *UPTAKE, FIXATION, CELL   W70-03983
*HABITATS, EPIPHYTOLOGY, MARINE A   W69-10161
*HABITATS, *ENVIRONMENTAL EFFECTS   W70-02770
*HALOGENS, ALGICIDES, CYTOLOGICAL   W70-02364
*HALOGENS, *ALGICIDES, CHLORINE,   W70-02363
*HALOGENS, *GROWTH RATES, *CHLORE   W70-02370
*HANFORD(WASH), *REACTOR, RADIONU   W69-07853
*HARBORS, *SEWAGE, WATER POLLUTIO   W71-05500
*HARTBEESPOORT DAM(SOUTH AFRICA),   W70-06975
*HARVESTING OF ALGAE, EFFLUENT, Q   W71-13313
*HARVESTING, AQUATIC PLANTS, AQUA   W69-06276
*HEAD LOSS.: /OAGULATION, ALGAE,   W72-07334
*HERBICIDES, *GROWTH RATES, *ALGA   W70-03519
*HERBICIDES, *PHYTOPLANKTON, *PHY   W69-00994
*HERBICIDES, *COPPER SULFATE, *SO   W70-05568

ENEDESMUS, CHLORELLA, DAPHNIA, /
C WEED CONTROL, *AQUATIC PLANTS,
ENEDESMUS, CHLORELLA, DAPHNIA, /
MANGANESE, ATLANTIC OCE/ *ALGAE,
, MICROCYSTIS AERUGINOSA, APHAN/
*MARINE ENVIRONMENTS, MICROBIAL/
R ENVIRONMENTS, *TURNOVER RATES,
, *MICROBIAL ECOLOGY, MAXIMUM U/
SWEDEN, LAKE/ *ORGANIC SOLUTES,
-14, RADIOCARBON UPTAKE TECHNIQ/
OADING, WASTE WATER TREATMENT, /
WATER, AC/ *ALGAE, *TEMPERATURE,
*ALGAE, *ENVIRONMENTAL EFFECTS,
ALGAE, *SEAWATER, *FRESH WATER,
ECOLOGY, *BENTHIC FLORA, *ALGAE,
BIC WASTE PONDS, *DEPTH EFFECTS,
.:

CUS BRAUNII, *ANACYSTIS MONTANA,
LING NUTRIENTS, *EUTROPHICATION,
RIENTS, *TERTIARY TREAT/ *ALGAE,
NG, *SULFATES, *SULFIDES, ALGAE/
AE, ORGANIC L/ *OXIDATION PONDS,
FLUENTS, CARBON DIOXIDE, *ALGAE,
*ALGAE, *LAKES, *EUTROPHICATION,
MANY), CHIRONOMIDS, CLADOCERA, /
NKTON, TEMPERATURE, *EPILIMNION,
LEI/ *YELLOWSTONE NATIONAL PARK,
, PHORMIDIUM/ *YELLOWSTONE PARK,
LGAL KEY, CLUMP COUNT, DESMIDS,/
, *DISSOLVED SOLIDS, D/ EFFECTS,
ER QUALITY, SOIL INVESTIGATIONS,
ERMAL STRATIFICATION, *AERATION,
NTS, *BIOASSAY, *RUNOFF, *LAKES,
ICELASE, HYDROG/ *DIGESTIBILITY,
IA), RAJGHAT(INDIA), BOMBAY(IND/
UDY.:
OONS, *ALGAE, *SEWAGE TREATMENT,
ENT, *ORGANIC MATTER, WATER QUA/
S, FRESH WATER BIOLOGY, SUSPEND/
R REUSE, *WASTE WATER TREATMENT,
REATMENT, *WASTE WATER DISPOSAL,
S, *WATER QUALITY, *FARM WASTES,
MENTATION, ORGANIC L/ *POTATOES,
COTS, PEARS, CITRUS/ *CANNERIES,
ION EF/ *RESERVOIRS, *LIMNOLOGY,
STES, *GREAT LAKES, *WA/ *ALGAE,
*MICROBIOLOGY, *SEWAGE LAGOONS,
UNDWATER BASINS, WATER SUPPLY, /
OXINS, *AQUARIA, *AQUATIC WEEDS,
HYTOPLANKTON, / *AQUATIC PLANTS,
*EUTROPHICATION, *FERTILIZATION,
TICAL SAMPLING, LAKE MENDOTA(WI/
ON, ALGAE, SEASONAL, CYANOPHYTA,
LAKE, PECORS LAKE, DUNLOP LAKE,
CONTROL, PREDATION, COMPETITION,
*/ *BIODEGRADATION, *DETERGENTS,
POLLUTION, FOOD CH/ *DETERGENTS,
CTIVITY TECH/ *ALGAE, *TOXICITY,
ION(DECOMPOSITION), DDT/ *ALGAE,
SUREMENT, *FLUOROMETRY, DYE REL/
TUARIES, *ESTUARINE ENVIRONMENT,
TAN AREAS, BLONDIN/ *DEVOLUTION,
ISCHARGES, FLORA, FAUNA, *BIOLO/
E, *ALGAL CONTROL, WEED CONTROL,
*BROMINE,
OSSES, HYPOLIMNION, THERMOCLINE,
AMATION, *GROUNDWATER, RECHARGE,
NSIN, ALGAE, C/ *EUTROPHICATION,
T TESTS, *NITROGEN, *PHOSPHORUS,
,/ *WATER TREATMENT, *TURBIDITY,
HROMATOGRAPHY, *COBALT, *COPPER,
MENT, EUTHROPHICATIO/ *NITROGEN,
OPLANKTON, *BIOMASS, *SUCESSION,
*ALGAE, LETHAL LIMIT, ELECTROP/
L/ *ALGAE, *TOXICITY, *CULTURES,
NT REQUIREMENTS, CULTURES, LAKE/
S, *PLANKTON, *LABORATORY TESTS,
*THERMAL GRADIENT, *BOD REMOVAL,
XPERIMENTAL STUDIES, LAGOON DES/
STRAINERS.: *SCREENING,
STRAINERS.:
OTO REACTIVITY, ALGAL METABOLIS/
NNETONKA(MINN), SOIL MINERALS, /
UTION, *RECOVERY FROM POLLUTION,
*ALGAE, *BIOLOGICAL COMMUNITIES,
MITATION, BA/ *FAT ACCUMULATION,
*PRIMARY PRODUCTIVITY, *STREAMS,
ES, DIATOMS, CYANOPHYTA, BIOASS/
DITIONS, LIGHT INTENSIT/ *ALGAE,
NALYTICAL TECHNIQUES, *PLANKTON,

*HERBICIDES, *ALGAE, TOXICITY, SC   W71-05719
*HERBICIDES, *VASCULAR TISSUES, *   W70-10175
*HERBICIDES, *ALGAE, TOXICITY, SC   W71-03056
*HERRINGS, COPPER, IRON, COBALT,   W69-08525
*HETEROCYSTS, ANABAENA VARIABILIS   W70-05091
*HETEROTROPHY, *ORGANIC SOLUTES,   W70-04811
*HETEROTROPHY, ORGANIC SOLUTES, S   W70-05109
*HETEROTROPHY, *BIOLOGICAL UPTAKE   W70-04194
*HETEROTROPHY, MICROBIAL ECOLOGY,   W70-04284
*HETEROTROPHY, AUTOTROPHY, CARBON   W70-04287
*HOGS, *LAGOONS, ALGAE, ORGANIC L   W70-07508
*HOT SPRINGS, HABITATS, ALKALINE   W69-10160
*HUMAN POPULATION, *EUTROPHICATIO   W70-03973
*HUMIC ACIDS, *COLOR, RIVERS, BOG   W70-01074
*HYDRAULIC MODELS, LABORATORY TES   W71-00668
*HYDRAULIC LOAD EFFECTS, *ORGANIC   W70-04787
*HYDROCARBON, *TOXICITY THRESHOLD   W70-01466
*HYDROCARBON.:   W70-01467
*HYDROCARBONS, *OLEFINS, GREEN RI   W69-08284
*HYDRODYNAMICS, LAKES, INTERNAL W   W70-09889
*HYDROGEN ION CONCENTRATION, *NUT   W70-08980
*HYDROGEN SULFIDE, *ORGANIC LOADI   W70-07034
*HYDROGEN SULFIDE, *SULFIDES, ALG   W71-11377
*HYDROGEN ION CONCENTRATION, PHOS   W68-00855
*HYDROLOGY, CHEMICAL PROPERTIES,   W70-05113
*HYPERTROPHIC WATERS, HAMBURG(GER   W70-00268
*HYPOLIMNION, ALGAE, FUNGI, BACTE   W70-04475
*ICELAND HOT SPRINGS, SINTER, NUC   W69-10160
*ICELAND, MASTIGOCLADUS LAMINOSUS   W70-05270
*IDENTIFICATION, *SIGNIFICANCE, A   W70-05389
*IMPOUNDED WATERS, *WATER QUALITY   W72-01773
*IMPOUNDED WATERS, IMPOUNDMENTS,   W72-11906
*IMPOUNDMENTS, *RESERVOIRS, *MIXI   W70-04484
*IMPOUNDMENTS, TERTIARY TREATMENT   W70-04765
*IN VITRO, IN VIVO, BALL-MILL, ME   W70-04184
*INDIA, *BLOOMS, RIVER GANGES(IND   W70-05404
*INDIA, MOOSI RIVER ECOLOGICAL ST   W69-07096
*INDIANA, CHLOROPHYTA, CYANOPHYTA   W70-10173
*INDICATORS, *PROTOZOA, *ENVIRONM   W71-03031
*INDICATOR SPECIES, MICROORGANISM   W71-10786
*INDUSTRIAL WASTES, OXIDATION PON   W72-00027
*INDUSTRIAL WASTES, *WATER POLLUT   W72-01786
*INDUSTRIAL WASTES, MUNICIPAL WAS   W71-09577
*INDUSTRIAL WASTES, SCREENS, SEDI   W71-08970
*INDUSTRIAL WASTES, PEACHES, APRI   W71-09548
*INDUSTRIAL WASTES, *WATER POLLUT   W70-06975
*INDUSTRIAL WASTES, *MUNICIPAL WA   W70-04430
*INDUSTRIAL WASTES, SEWAGE TREATM   W70-03312
*INFILTRATION, *PONDS, *SOIL, GRO   W69-07838
*INHIBITION, *PHYSIOLOGICAL ECOLO   W69-03188
*INHIBITORS, *MARSHES, *LAKES, *P   W72-00845
*INLAND WATERWAYS, *WATER POLLUTI   W69-07833
*INLAND, HORIZONTAL SAMPLING, VER   W70-02969
*INORGANIC COMPOUNDS, PHOSPHORUS,   W68-00860
*INORGANIC CARBON.: /AKES, QUIRKE   W70-05424
*INSECT CONTROL.: /ALGAE, *ALGAL   W70-05428
*INSECTICIDES, *MARINE BACTERIA,   W70-01467
*INSECTICIDES, *TOXICITY, *WATER   W70-01466
*INSECTICIDES, METABOLISM, RADIOA   W70-02198
*INSECTICIDES, TOXICITY, DEGRADAT   W70-08097
*INSTRUMENTATION, *SAMPLING, *MEA   W70-05377
*INSTRUMENTATION, *POLLUTION IDEN   W69-06203
*INTERSTATE POLLUTION, *METROPOLI   W70-04430
*INTERTIDAL COMMUNITIES, *WASTE D   W68-00010
*INVESTIGATIONS, ZOO-PLANKTON, SP   W68-00468
*IODINE.:   W69-08562
*ION EXCHANGE, WATER CIRCULATION,   W68-00857
*ION EXCHANGE, *ORGANIC MATTER, *   W72-04318
*IRON, *LAKES, *MANGANESE, *WISCO   W69-05867
*IRON, *LIMNOLOGY, *ALGAE.: /HMEN   W69-06004
*IRON, *MANGANESE, *COLOR, *ALGAE   W72-04145
*IRON, TRACE ELEMENTS, ALGAE, SEA   W71-11687
*IRRADIATION, *ALGAE, WASTE TREAT   W73-02218
*ISOLATION, DIATOMS, ECOLOGY, SAL   W73-01446
*ISOLATION, *BACTERIA, *PROTOZOA,   W71-03775
*ISOLATION, LIMNOLOGY, PUBLIC HEA   W70-00273
*ISOLATION, *CYANOPHYTES, *NUTRIE   W70-00719
*ISOLATION, *CONTROL, METHODOLOGY   W70-04469
*ISRAEL, *ALGAL CONCENTRATION, 'E   W70-04790
*JAPAN, *SLAUGHTERHOUSE WASTES, E   W70-07508
*KENOSHA(WIS), *STRAINING, *MICRO   W73-02426
*KILL RATES, *BROMINE, IODINE, PH   W70-02363
*KILLING, ANIMAL MANURES, LAKE MI   W70-04193
*KUWAIT CRUDE, BENTLASS SALT MARS   W70-01231
*LABORATORIES, *STREAMS, LIGHT IN   W70-02780
*LABORATORY STUDIES, *NITROGEN LI   W70-07957
*LABORATORY TESTS, RESPIRATION, E   W70-00265
*LABORATORY TESTS, *ALGAE, CULTUR   W71-06002
*LABORATORY TESTS, *ANAEROBIC CON   W71-13413
*LABORATORY TESTS, *ISOLATION, *C   W70-04469

E, *PILOT-PLANT/ *ALGAL CULTURE, *LABORATORY STUDIES, *MASS CULTUR W69-06865
 *EUTROPHICATION, TEST ANALYSIS, *LABORATORY TESTS, TEST PROCEDURE W68-00472
ALYSIS, LIMNOLOGY, GROWTH RATES, *LABORATORY TESTS, *SCENEDESMUS, W68-00475
 WATER TREATMENT, INDUSTRIAL WA/ *LAGOONS, *DOMESTIC WASTES, WASTE W71-06737
 WASTE WATER TREATMENT, / *HOGS, *LAGOONS, ALGAE, ORGANIC LOADING, W70-07508
E BACTERIA, ODOR,/ *FARM WASTES, *LAGOONS, *BIODEGRADATION, *SEWAG W71-02009
OBIC CONDITIONS, *EFFLICIENCIES, *LAGOONS, *SLUDGE, ALGAE, BIOCHEM W70-06899
CHEMICAL OXY/ *AERATION, *ALGAE, *LAGOONS, *OXIDATION LAGOONS, BIO W70-06601
 WATER TREATMENT, INDUSTRIAL WA/ *LAGOONS, *DOMESTIC WASTES, WASTE W70-01519
URIFICATION, *FLOW AUGMENTATION, *LAGOONS, ACTIVATED SLUDGE, ALGAE W72-06017
RAL EUTROPHY, *NUTRIENT BALANCE, *LAKE AGING, NUTRIENT SOURCES, WA W70-00264
TH STIMUL/ *SCIRPUS FLUVIATILIS, *LAKE BUTTE DES MORTS(WISC), GROW W72-00845
ATION, / *LAKE WASHINGTON(WASH), *LAKE CHANGES, *NUTRIENT CONCENTR W70-00270
ATA COLLECTIONS, *ALGAE CONTROL, *LAKE ERIE, BIOCHEMICAL OXYGEN DE W71-04758
 ONTARI/ *BIOLOGY, *GREAT LAKES, *LAKE ERIE, *LAKE MICHIGAN, *LAKE W70-01943
TER POLLUTION / *EUTROPHICATION, *LAKE ERIE, *NUTRIENTS, ALGAE, WA W69-01445
ROL, *POLLUTION ABATEMEN/ *OHIO, *LAKE ERIE, *WATER POLLUTION CONT W69-06305
TA, UNITED STATES, HABI/ *ALGAE, *LAKE ERIE, CHLOROPHYTA, CYANOPHY W70-04468
TRIENTS, *BENEFIT-COST ANALYSIS, *LAKE ERIE, PHOSPHORUS, FISH STOC W70-04465
 *LAKE KINNERET(ISRAEL).: W70-09889
RTHER(AUSTRIA), VERTICAL, CHALK/ *LAKE KLOPEINER(AUSTRIA), LAKE WO W70-05405
GANIC MATTER, LAS VEGAS, NEVADA, *LAKE MEAD, LAS VEGAS BAY.: /D OR W72-02817
, HORIZONTAL, BLOOMS, GELATINOU/ *LAKE MENDOTA(WIS), PELAGIC AREAS W70-06229
OLOGY, *GREAT LAKES, *LAKE ERIE, *LAKE MICHIGAN, *LAKE ONTARIO, LA W70-01943
TAKES, NUTRIENTS, QUALITY CONTR/ *LAKE MICHIGAN, *WATER SUPPLY, IN W70-00263
N, NUISANCE ALG/ *PHYTOPLANKTON, *LAKE MICHIGAN, *WATER CIRCULATIO W69-05763
LGAE, PLANKTON/ *EUTROPHICATION, *LAKE MICHIGAN, ALGAE, NUISANCE A W68-01244
GEMENT PROGRAM, ALGAL DENSITY, / *LAKE MINNETONKA(MINN), LAKE MANA W69-10167
LANKTON, *ALGAE, *PHYTOPLANKTON, *LAKE ONTARIO, *GROWTH RATES, ENV W69-10158
KES, *LAKE ERIE, *LAKE MICHIGAN, *LAKE ONTARIO, LAKE HURON, AESTHE W70-01943
ATION, *WATER POLLUTION CONTROL, *LAKE ONTARIO, *WATER CIRCULATION W71-05878
ONITORING, BACTERIA, WATER CHEM/ *LAKE ONTARIO, *WATER QUALITY, *M W71-05879
NDUSTRIAL PLANTS, WATER SPORTS./ *LAKE ONTARIO, WATER TURBIDITY, I W71-09880
OGEN COM/ *EUTROPHICATION, LAKE, *LAKE SEDIMENTS, *NUTRIENTS, NITR W72-06051
NOLOGY, *WATER QUALITY, BIOINDI/ *LAKE SUPERIOR, *PERIPHYTON, *LIM W70-05958
NCY, LIGHT MEASUREMENTS, EUPHOT/ *LAKE TAHOE(CALIF-NEV), TRANSPARE W70-01933
ES, LIVERWORTS, MORPHOMETRY, PA/ *LAKE TAHOE(CALIF-NEV), HYDROPHYT W70-00711
NGES, *NUTRIENT CONCENTRATION, / *LAKE WASHINGTON(WASH), *LAKE CHA W70-00270
EFICIT, SECCHI DISC TRANSPARENC/ *LAKE WASHINGTON(WASH), *OXYGEN D W69-10182
RIA RUBESCENS, SWITZERLAND, PER/ *LAKE WASHINGTON(WASH), OSCILLATO W69-10169
SION, *WATER QUALITY MANAGEMENT, *LAKE WASHINGTON, *SEWAGE FERTILI W69-09349
 INFLUENCES, MICROCYSTIS, DREDG/ *LAKE WASHINGTON(WASH), *CULTURAL W70-05663
ASH), REVENUE BONDS, METRO ACT(/ *LAKE WASHINGTON(WASH), SEATTLE(W W70-04455
A, GYTTJA, OSCILLATORIA RUBESCE/ *LAKE WASHINGTON(WASH), MACROFAUN W70-04253
 *MICROCYSTIS AERUGINOSA, *LAKE WATERS.: W70-08637
OPHYTA, FRESH WATER FISH, LAKES, *LAKE ZURICH, *MUNICIPAL WASTES, W68-00680
 *NITROGEN-FIXATION, *NUTRIENTS, *LAKES, *ALGAE, *NITROGEN, WISCON W70-03429
HOSPHATES, LAB/ *EUTROPHICATION, *LAKES, *ALGAE, *GROWTH RATES, *P W70-08637
OLIGOTROPHY, NUTRIENTS, WATER T/ *LAKES, *ALGAE, *EUTROPHICATION, W70-00264
LEVELS, *SURVEYS, BIOLOGICAL CO/ *LAKES, *BENTHIC FAUNA, *TROPHIC W71-12077
CHNIQUES, WATER POLLUTION EFFEC/ *LAKES, *BACTERIA, *ANALYTICAL TE W70-04287
ONTROL, *TOXIN/ *EUTROPHICATION, *LAKES, *COPPER SULFATE, *ALGAL C W68-00859
 BACTERIA, ALGAE, PHYTOPLANKTON, *LAKES, *CELLULOSE, DISTRIBUTION W69-00976
 *EUTROPHICATION, AQUATIC ALGAE, *LAKES, *CLASSIFICATION, DESMIDS, W68-00255
MENT-WATER INFERFAC/ *PHOSPHATE, *LAKES, *CYCLING NUTRIENTS, *SEDI W69-06859
IOASSAY, *EUTROPHICATION, *CALI/ *LAKES, *CARBON RADIOISOTOPES, *B W70-01933
, S/ BIBLIOGRAPHIES, *MINNESOTA, *LAKES, *DIATOMS, *ALGAE, ECOLOGY W69-00659
EATMENT, *GREAT LA/ *PHOSPHATES, *LAKES, *DETERGENTS, *TERTIARY TR W70-03964
, *PRODUCTIVITY, CYANOPHYTA, NU/ *LAKES, *EUTROPHICATION, *BIOMASS W70-03959
ATION, FISH, *LIMNOLOGY, FISHKI/ *LAKES, *EUTROPHICATION, FERTILIZ W69-05844
TON, URBANIZATION, WATER POLLUT/ *LAKES, *EUTROPHICATION, *WASHING W70-05663
GY, CHEMICAL PROPERTIES/ *ALGAE, *LAKES, *EUTROPHICATION, *HYDROLO W70-05113
BIOTA, PHYTOPLANKT/ *WASHINGTON, *LAKES, *EUTROPHICATION, SEWAGE, W69-10169
EFFLUENTS, *NUTRIENTS, *NITROGE/ *LAKES, *EUTROPHICATION, *SEWAGE W69-09349
 *CITIES, BIOLOGICAL COMMUNITIE/ *LAKES, *EUTROPHICATION, *CANALS, W70-00268
TS, *COPPER SULFATE, *TREATMENT, *LAKES, *ECOLOGY, BOTTOM SEDIMENT W70-06217
OLLUTION CONTROL, *WATER POLLUT/ *LAKES, *EUTROPHICATION, *WATER P W71-03048
 *NUTRIENTS, *BIOASSAY, *RUNOFF, *LAKES, *IMPOUNDMENTS, TERTIARY T W70-04765
R QUALITY CONTROL/ *CONNECTICUT, *LAKES, *LOCAL GOVERNMENTS, *WATE W72-05774
LGAE, C/ *EUTROPHICATION, *IRON, *LAKES, *MANGANESE, *WISCONSIN, A W69-05867
FATE, ALGAL CONTROL, ALGICIDES,/ *LAKES, *MUD, *COPPER, COPPER SUL W69-03366
OPHYTA, / *LIMNOLOGY, *COLORADO, *LAKES, *MOUNTAINS, *WINTER, CYAN W69-10154
RIMARY PRODUCTIVITY, *ZOOPLANKT/ *LAKES, *NEW YORK, *MEROMIXIS, *P W70-05760
OGICAL ECOLOGY, RADIOACTIVITY T/ *LAKES, *ORGANIC MATTER, *PHYSIOL W70-04194
ENTATION RATES, *EUTROPHICATION, *LAKES, *PRIMARY PRODUCTIVITY, *P W69-00632
 *EUTROPHICATION, OLIGOTROPHIC, *LAKES, *PHOSPHORUS RADIOISOTOPE, W68-00858
E,/ *LIMNOLOGY, *WATER, *RIVERS, *LAKES, *PONDS, LIGHT, TEMPERATUR W70-05404
*EPIPHYTOLOGY, NUTRIENTS, PLANT/ *LAKES, *PRODUCTIVITY, MICHIGAN, W71-00832
VADA, DEPTH, ALGAE/ *DEEP WATER, *LAKES, *PLANTS, *CALIFORNIA, *NE W70-00711
LGAE, WATER PO/ *EUTROPHICATION, *LAKES, *PHOSPHATES, NUTRIENTS, A W69-09135
C PLANTS, *INHIBITORS, *MARSHES, *LAKES, *PHYTOPLANKTON, BIOASSAY, W72-00845
 RUNOFF, FERTILIZERS, NUTRIENTS/ *LAKES, *RIVERS, *EUTROPHICATION, W69-10170
LLUTION CONTRO/ *EUTROPHICATION, *LAKES, *RIVERS, *ALGAE, WATER PO W70-05668
 PONDS, TEMPERATURE,/ *LIMNOLOGY, *LAKES, *RECREATION, RESERVOIRS, W70-06225
 PHYTOPLANKTON, TEMPERATURE, *E/ *LAKES, *RESERVOIRS, ZOOPLANKTON, W70-04475
E DISPOSAL, ALGAE, AGRICULTURAL/ *LAKES, *SEWAGE EFFLUENTS, *SEWAG W70-05565
SORPTION, *PHOSPHORUS COMPOUNDS, *LAKES, *SEDIMENTS, *EUTROPHICATI W69-03075
 FERTILIZATION, ALG/ *NUTRIENTS, *LAKES, *TROUT, *FISH MANAGEMENT, W70-05399
N, *AGRICULTURAL WASTES, *ALGAE, *LAKES, *TRIBUTARIES, SAMPLING, F W70-02969
 *REVIEWS, ALG/ *EUTROPHICATION, *LAKES, *WATER POLLUTION CONTROL, W70-07393

TION,/ *NEVADA, *EUTROPHICATION,   *LAKES, *WATER POLLUTION, DEGRADA   W72-02817
S, *PHOSPHORUS/ *EUTROPHICATION,   *LAKES, ACID LAKES, BIBLIOGRAPHIE   W68-00857
NOPHYTA, FISHK/ *EUTROPHICATION,   *LAKES, ALGAE, AQUATIC WEEDS, CYA   W69-06535
OMAS/ *BACTERIA, *PHYTOPLANKTON,   *LAKES, ALGAE, MICROORGANISMS, BI   W70-08654
ION, *NUTRIENT/ *EUTROPHICATION,   *LAKES, CONNECTICUT, WATER POLLUT   W68-00511
DOMESTIC WAST/ *EUTROPHICATION,   *LAKES, GLACIATION, URBANIZATION,   W69-07818
IC PRODUCTIVITY, EUTROPHICATION,   *LAKES, LIMNOLOGY, *PLANKTON, ALG   W69-00446
MENTS, ICED LA/ *EUTROPHICATION,   *LAKES, NUTRIENTS, OXYGEN REQUIRE   W68-00487
ATION, *WATER POLLUTION SOURCES,   *LAKES, NUTRIENTS, FERTILIZERS, P   W69-07084
N EFFECTS, PRO/ *EUTROPHICATION,   *LAKES, NUTRIENTS, WATER POLLUTIO   W70-04721
BACTERIA, ALG/ *EUTROPHICATION,   *LAKES, NUTRIENTS, FERTILIZATION,   W70-05761
ALGAE, PRIMARY/ *PHYTOPLANKTON,   *LAKES, ORGANIC MATTER, SEASONAL,   W70-04290
SONAL, CYANOPH/ *EUTROPHICATION,   *LAKES, PHYTOPLANKTON, ALGAE, SEA   W68-00860
, IRRIGATION EFFECTS, DRAINAGE /   *LAKES, PLANTS, *ALGAE, *SALINITY   W69-00360
, / *EUTROPHICATION, *NUTRIENTS,   *LAKES, PONDS, FERTILIZERS, ALGAE   W69-08668
ES/ *PERIPHYTON, *RADIOACTIVITY,   *LAKES, PRODUCTIVITY, RADIOISOTOP   W70-04371
AE, *NUISANCE ALGAE, *WISCONSIN,   *LAKES, PRODUCTIVITY, NITROGEN, P   W70-05112
YSTEMS, EUTROPHICATION, OLIGOTR/   *LAKES, STREAMS, FISH, LAND, ECOS   W70-04193
N, ALGAE,/ *THERMAL POWERPLANTS,   *LAKES, TEMPERATURE, PHYTOPLANKTO   W71-09767
CE ALGAE, SPRA/ *EUTRIPHICATION,   *LAKES, WISCONSIN, ALGAE, *NUISAN   W68-00488
E, NUISANCE AL/ *EUTRIPHICATION,   *LAKES, WISCONSIN, COPPER SULPHAT   W68-00468
                   *NITRIC ACID,   *LANGELIER INDEX.:   W71-11393
TRATES, NITROGEN, *CONDUCTIVITY,   *LEACHING, IRON, TEMPERATURE, SOI   W72-11906
IDAY HARBOR(WASH), ANTITHAMNION,   *LEUCOTHRIX MUCOR, MORPHOLOGY, FR   W69-10161
MENTS, *AUTORADIOGRAPHY, ANTITH/   *LEUCOTHRIX MUCOR, *GROWTH, *FILA   W69-09755
DANCE, BLUE-GREEN ALGAE, GREEN /   *LEWIS AND CLARK LAKE (S D), ABUN   W70-05375
ROPHYTA, DIA/ *PLANKTON, *ALGAE,   *LIGHT INTENSITY, CHLORELLA, CHLO   W70-05381
HESIS, *ALGAE, *THERMAL SPRINGS,   *LIGHT INTENSITY, SPRING WATER, W   W69-09676
ICATION, MODEL STUDIES, BIOMASS,   *LIGHT INTENSITY.: /ENT, EUTROPH   W73-02218
OBIC CONDITIONS, ANAERO/ *ALGAE,   *LIGHT, *NITRATES, *NITRITES, AER   W70-05261
ALGAE, *CHLORELLA, *SCENEDESMUS,   *LIGHT, PHOSPHORUS, PHOTOSYNTHESI   W70-03923
RIMARY PRODUCTI/ *PHYTOPLANKTON,   *LIMITING FACTORS, *NUTRIENTS, *P   W70-01579
ABITATS, *ENVIRONMENTAL EFFECTS,   *LIMITING FACTORS, BENTHIC FLORA,   W70-02770
OUNTAINS, *WINTER, CYANOPHYTA, /   *LIMNOLOGY, *COLORADO, *LAKES, *M   W69-10154
    *PRIMARY PROD/ *EUTROPHICATION,   *LIMNOLOGY, *ESSENTIAL NUTRIENTS,   W72-01990
DI/ *LAKE SUPERIOR, *PERIPHYTON,   *LIMNOLOGY, *WATER QUALITY, BIOIN   W70-05958
RESERVOIRS, PONDS, TEMPERATURE,/   *LIMNOLOGY, *LAKES, *RECREATION,   W70-06225
WATER POLLUTION EF/ *RESERVOIRS,   *LIMNOLOGY, *INDUSTRIAL WASTES, *   W70-06975
ES, *PONDS, LIGHT, TEMPERATURE,/   *LIMNOLOGY, *WATER, *RIVERS, *LAK   W70-05404
IS, PRODUCTIVITY, WATER QUALITY,   *LIMNOLOGY, MINNESOTA, WATER ANAL   W69-00632
*EUTROPHICATION, *FERTILIZATION,   *LIMNOLOGY, *PHYTOPLANKTON, FISH,   W69-06858
, *NITROGEN, *PHOSPHORUS, *IRON,   *LIMNOLOGY, *ALGAE.: /HMENT TESTS   W69-06004
OPHICATION, FERTILIZATION, FISH,   *LIMNOLOGY, FISHKILL, NUTRIENTS,   W69-05844
XIFICATION, LINDANE METABOLISM./   *LINDANE, *ALGAE METABOLISM, DETO   W70-08652
FACTANTS, WATER POLLUTION SOURC/   *LINEAR ALKYLATE SULFONATES, *SUR   W70-09438
CULTURES, *ALKYL POLYETHOXYLAT/   *LINEAR ALKYL SULFONATES, *AXENIC   W70-03928
CAL TECHNIQUES, *MICROORGANISMS,   *LIPIDS, LIGHT, ALGAE, TEMPERATUR   W70-05651
MMATE, CHLORAMINE, DE-K-PRUF-21/   *LITERATURE, ELEOCHARIS, 2,4-D, A   W70-05263
TION, T/ *OCEANS, *MARINE ALGAE,   *LITTORAL, *ORGANIC MATTER, EXUDA   W70-01073
DS, ECOLOGICAL DISTRIBU/ *ALGAE,   *LITTORAL, *AQUATIC PLANTS, ISLAN   W72-10623
Y CONTROL/ *CONNECTICUT, *LAKES,   *LOCAL GOVERNMENTS, *WATER QUALIT   W72-05774
UENTS, *WATER POLLUTION CONTROL,   *LOCAL GOVERNMENTS, STATE GOVERNM   W69-06909
BONUCLEI/ *PHOTOPHOSPHORYLATION,   *MACROMOLECULES, NUCLEIC ACID, RI   W69-10151
         LAWRENCE LAKE(MICHIGAN),   *MACROPHYTES.:   W71-00832
*TRUAX FIELD(MADISON) EFFLUENTS,   *MADISON LAKES, YAHARA RIVER(WIS)   W70-05113
D CUTTING, *WEED EXTRACTING STE/   *MADISON LAKES(WIS), *AQUATIC WEE   W70-04494
LAKE WAUBESA(WIS), LAKE KEGONSA/   *MADISON(WIS), LAKE MONONA(WIS),   W70-05112
KE MENDOTA(WIS), LAKE MONONA(WI/   *MADISON(WIS), NUISANCE ODORS, LA   W69-03366
S, *WATER-BLOOM, 'BREAKING' OF /   *MADISON(WIS), *'WORKING' OF LAKE   W69-08283
FATE, RESERVOIRS, WATER/ *ALGAE,   *MAINE, *WATER SUPPLY, COPPER SUL   W70-08096
*V/ *VISCOSE RAYON PLANT WASTES,   *MAN-MADE FIBER PLANTS, NYLON-6,   W70-06925
*EUTROPHICATION, *IRON, *LAKES,   *MANGANESE, *WISCONSIN, ALGAE, CA   W69-05867
ER TREATMENT, *TURBIDITY, *IRON,   *MANGANESE, *COLOR, *ALGAE, *COLL   W72-04145
D LAGOON.:   *MANURE, ANAEROBIC LAGOON, AERATE   W70-07491
*SALT MARSHES, OILY WATER, *OIL/   *MARINE ALGAE, WATER POLLUTION SO   W70-09976
RINE / *EMULSIFIERS, *MORTALITY,   *MARINE ALGAE, *MARSH PLANTS, *MA   W70-01777
FECTS, LETHAL LIMI/ *DETERGENTS,   *MARINE ALGAE, WATER POLLUTION EF   W70-01779
C MATTER, EXUDATION, T/ *OCEANS,   *MARINE ALGAE, *LITTORAL, *ORGANI   W70-01073
WATER, *HUMIC ACIDS, */ *OCEANS,   *MARINE ALGAE, *SEAWATER, *FRESH   W70-01074
R POLLUTION EFFECT/ *OILY WATER,   *MARINE ALGAE, *EMULSIFIERS, WATE   W70-01780
, PHYSICOC/ *ZINC RADIOISOTOPES,   *MARINE ALGAE, *RESINS, SEA WATER   W69-08275
SYNTHESIS, *PHYTOPLANKTON, *DDT,   *MARINE ALGAE, TOXICITY, DIATOMS,   W70-05272
FLORA, LIGHT INTENSITY, BIOMAS/   *MARINE ALGAE, *DIATOMS, *BENTHIC   W70-03325
ONAS, METABOLISM, KI/ *BACTERIA,   *MARINE ALGAE, *CULTURES, PSEUDOM   W70-04280
Y, *MARINE ALGAE, *MARSH PLANTS,   *MARINE ANIMALS, *OILY WATER, LIC   W70-01777
, / *PHOSPHORUS, *CORAL, *REEFS,   *MARINE ANIMALS, FISH, ECOSYSTEMS   W69-08526
ABSORPTION, MARINE ALGAE, MARIN/   *MARINE ANIMALS, *RADIOISOTOPES,   W71-09850
RADIOISOTOPES, PHYSICOCHEMICAL /   *MARINE ANIMALS, *MARINE PLANTS,   W71-09863
ION, *DETERGENTS, *INSECTICIDES,   *MARINE BACTERIA, *TOXICITY, *WAT   W70-01467
IS, *ANALYTICAL TECHNIQUES, TEM/   *MARINE BACTERIA, *ALGAE, *ANALYS   W70-04365
*HETEROTROPHY, *ORGANIC SOLUTES,   *MARINE ENVIRONMENTS, MICROBIAL E   W70-04811
NITZSCHIA CLOSERIUM, ARTEMIA, C/   *MARINE ENVIRONMENT, CERIUM-144,   W70-00708
, EPIPHYTOLOGY, MARINE ALGAE, E/   *MARINE MICROORGANISMS, *HABITATS   W69-10161
KTON, BACTERI/ *ALGAE, *ECOLOGY,   *MARINE MICROORGANISMS, PHYTOPLAN   W70-01068
*ALGAE, *ANALYTICAL TECHNIQUES,   *MARINE MICROORGANISMS, PHYSIOLOG   W70-04811
NUTRIENT REQUIREMENTS, *DISTRIB/   *MARINE MICROORGANISMS, *ALGAE, *   W70-03952
TS, RECREATION, WAT/ *NUTRIENTS,   *MARINE MICROORGANISMS, *POLLUTAN   W69-04276
QUATIC MICROBIOLOGY, *ESTUARIES,   *MARINE MICROORGANISMS, MARINE AL   W69-03752
ION, SODIUM COMPOUNDS, PHOTOSYN/   *MARINE PLANTS, *ALGAE, CHLORINAT   W70-00161
YSICOCHEMICAL / *MARINE ANIMALS,   *MARINE PLANTS, RADIOISOTOPES, PH   W71-09863

EA WATER, TEMPERATURE, SALINITY/
/ *FAYETTEVILLE GREEN LAKE(NY),
TTER, ARCTIC ALGAE, AUTOTROPHIC/
IC ALGAE, MAR/ *ALGAL POISONING,
IERS, *MORTALITY, *MARINE ALGAE,
/ *AQUATIC PLANTS, *INHIBITORS,
ORELLA PYRENOIDOSA(EMERSON STRA/
AL CULTURE, *LABORATORY STUDIES,

NATION, CYANIDE, NEW PRODUCT CO/
MANAGEMENT, GEOCHEMIS/ *ECOLOGY,
OROPHYLL, *PRIMARY PRODUCTIVITY,
EL/ *INSTRUMENTATION, *SAMPLING,
C ENVIRONMENTS, *EUTROPHICATION,
, *ZOOPLANKT/ *LAKES, *NEW YORK,
BILITY, CHLORELLA, CARB/ *ALGAE,
:
REATMENT, ON-STREAM DESLUDGING./

*BIOLOGICAL COMMUNITIES, *ALGAE,
VOLUTION, *INTERSTATE POLLUTION,
Y, BIOCHEMICAL EXTRACTION, MILL/
M, *FUNGAL-BOUND RADIOSTRONTIUM,
ETEROTROPHY, *BIOLOGICAL UPTAKE,
BACTERIA, POPULATION EQUIVALEN/
INDUSTRIAL WASTES, SEWAGE TREAT/
OORGANISMS, *WATER POLLUTION, */
OGY, *ESTUARIES, *MARINE MICROO/
CULTURES, *BIOASSAYS, PROTOZOA,/
ELB/ *APHANIZOMENON FLOS AQUAE,
TERS.:
T AVAILA/ *CHEMICAL COMPOSITION,
ENT AVAILABILITY, *ALGAL GROWTH,
LTURE MEDIA, CHU NO 10 SOLUTION/
CTIONS, *RADIOCARBON BIOASSAY, /
ALGAE, / *ANALYTICAL TECHNIQUES,
TEINS, WALLS, STRUCTURE/ *ALGAE,
, */ *MICROBIOLOGY, *MINERALOGY,
N COMPOUNDS, *NUTRIENTS, *ALGAE,
T ORGANISMS, *ACCLIMATIZATION, /
AL TEMPERATURES, *COPPER DOSAGE/

NING, *KENOSHA(WIS), *STRAINING,
OVEMENT, *WATER QUALITY CONTROL,
SIGN CRITERIA/ *AERATED LAGOONS,
ILK SUBSTRATE, NITROGEN REMOVAL,

, *MUNICIPAL / *WATER POLLUTION,
RADIOLARIA, FORAMINIFERA, THIO/
TER POLLUTION, */ *MICROBIOLOGY,
GAE, ECOLOGY, S/ BIBLIOGRAPHIES,
HYTA, / *ALGAE, *PHOTOSYNTHESIS,
SPOSAL, WATER / *EUTROPHICATION,
GE EFFLUENTS, TERTIARY TREATMEN/
UTION ABAT/ *ALGAE, *PHOSPHORUS,
NUI/ *RIVERS, *ALGAE, *SAMPLING,
ION, *IMPOUNDMENTS, *RESERVOIRS,
ICAL TECHNIQUES, CLASSIFICATION,
TMENT, *OXIDATION LAGOONS, MILK,
R TREATMENT, *OXIDATION LAGOONS,

EAMS, BIOINDICATORS, *NEW YORK,/
*LAKE ONTARIO, *WATER QUALITY,
NT, *EFFLUENTS, *RADIOISOTOPES,
ECOLOGY, *ENVIRONMENTAL EFFECTS,
PLANTS, *MARINE / *EMULSIFIERS,
D MATS, NON-LEGUMES, *NITROGEN /
*LIMNOLOGY, *COLORADO, *ALGAE,
ALGAE, *BIOAS/ *EUTROPHICATION,
GAL CONTROL, ALGICIDES,/ *LAKES,
INE WASTES, *RADIOACTIVE WASTES,
WATER FISH, LAKES, *LAKE ZURICH,
*WA/ *ALGAE, *INDUSTRIAL WASTES,
N LAGOONS, *ANAEROBIC DIGESTION,
EMATOLOGY/ *GRANULAR LEUCOCYTES,
ALGAE COUNTS, P/ *RIVER STUDIES,
, *ALGAL CONTROL, *TASTE, *ODOR,
*WATER POLLUTION, DEGRADATION,/
R, *LAKES, *PLANTS, *CALIFORNIA,
Y, *EUTROPHICATION, *CALIFORNIA,
MMISSION, ALGAL BLOOMS, NUTRIEN/
*NUISANCE ALGAE, *ALGAL CONTROL,
*ALGAE CONTROL, *WATER SUPPLY,
OHIO), GONGROSIRA STAGNALIS, NE/
OTOPES, *STREAMS, BIOINDICATORS,
RODUCTIVITY/ *ZOOPLANKT/ *LAKES,
*COLLOIDS, FLOW RATES, *FILTERS,
SIGN CRITERIA, *WATER TREATMENT,
ND ALGAE.:
WATER AND PLANTS, WATER POLLUT/

*MARINE WATER, *COPPER, *ALGAE, S   W70-08111
*MARL LAKES, CHLOROBIUM, FILINIA,   W70-05760
*MARL LAKES, DISSOLVED ORGANIC MA   W71-10079
*MARSH PLANTS, *OILY WATER, AQUAT   W70-01231
*MARSH PLANTS, *MARINE ANIMALS, *   W70-01777
*MARSHES, *LAKES, *PHYTOPLANKTON,   W72-00845
*MASS CULTURE, *ALGAL GROWTH, CHL   W69-07442
*MASS CULTURE, *PILOT-PLANT STUDI   W69-06865
*MASSENA(NY).:   W72-08686
*MASTER PLANNING, *SOURCE CONTAMI   W73-00758
*MATHEMATICAL MODELS, *WATERSHED   W71-09012
*MEASUREMENT, PROTEINS, RADIOACTI   W69-10151
*MEASUREMENT, *FLUOROMETRY, DYE R   W70-05377
*MEROMIXIS, *WATER POLLUTION EFFE   W70-03974
*MEROMIXIS, *PRIMARY PRODUCTIVITY   W70-05760
*METABOLISM, *ENVIRONMENTS, VARIA   W70-02965
*METABOLITES, *CEDAR RIVER(IOWA).   W72-13605
*METAL CORROSION, COOLING WATER T   W70-02294
*METHANE FERMENTATION.:   W71-13413
*METHODOLOGY, CARBON RADIOISOTOPE   W71-12870
*METROPOLITAN AREAS, BLONDIN.: /E   W70-04430
*MICRO-AUTORADIOGRAPHY, PHYSIOLOG   W69-10163
*MICROBIAL ACTIVITY, *RADIOSTRONT   W71-04067
*MICROBIAL ECOLOGY, MAXIMUM UPTAK   W70-04194
*MICROBIAL ECOLOGY, PURPLE SULFUR   W70-03312
*MICROBIOLOGY, *SEWAGE LAGOONS, *   W70-03312
*MICROBIOLOGY, *MINERALOGY, *MICR   W69-07428
*MICROBIOLOGY, *AQUATIC MICROBIOL   W69-03752
*MICROBIOLOGY, *CLASSIFICATION, *   W71-11823
*MICROCYSTIS AERUGINOSA, BOHEMIA,   W70-00274
*MICROCYSTIS AERUGINOSA, *LAKE WA   W70-08637
*MICROCYSTIS AERUGINOSA, *NUTRIEN   W69-05868
*MICROCYSTIS AERUGINOSA, ANTAGONI   W69-05867
*MICROCYSTIS AERUGINOSA(KUTZ), CU   W70-03507
*MICRONUTRIENTS, *NUTRIENT INTERA   W70-04765
*MICROORGANISMS, *LIPIDS, LIGHT,   W70-05651
*MICROORGANISMS, SCENEDESMUS, PRO   W70-04184
*MICROORGANISMS, *WATER POLLUTION   W69-07428
*MICROORGANISMS, BIODEGRADATION,   W69-06970
*MICROORGANISMS, *ALGAE, *DOMINAN   W70-07313
*MICROSCOPIC EXAMINATION, *CRITIC   W69-10157
*MICROSTRAINERS, HYPOCHLORITES.:   W72-03824
*MICROSTRAINERS.:   *SCREE W73-02426
*MIDGES, FISH FOOD ORGANISMS, NUT   W70-05428
*MIDWEST(US), *SOUTHWEST(US), *DE   W70-04788
*MILK WASTES.: /L, COD REMOVAL, M   W70-07031
*MILLIPORE FILTERS.:   W72-09693
*MINE WASTES, *RADIOACTIVE WASTES   W69-01165
*MINERAL TRANSFORMATION, METAZOA,   W69-07428
*MINERALOGY, *MICROORGANISMS, *WA   W69-07428
*MINNESOTA, *LAKES, *DIATOMS, *AL   W69-00659
*MINNESOTA, PHYTOPLANKTON, CYANOP   W70-05387
*MINNESOTA, RECREATION, SEWAGE DI   W70-04810
*MINNESOTA, *EUTROPHICATION, SEWA   W70-03512
*MINNESOTA, *EUTROPHICATION, POLL   W69-10167
*MISSISSIPPI RIVER, *OHIO RIVER,   W70-04507
*MIXING, *WATER QUALITY CONTROL,   W70-04484
*MODEL STUDIES, WASTE IDENTIFICAT   W69-01273
*MODEL STUDIES.: /ASTE WATER TREA   W70-07031
*MODEL STUDIES.: /IA, *WASTE WATE   W70-07034
*MODEL TESTS, DYES.:   W70-02609
*MONITORING, *RADIOISOTOPES, *STR   W69-10080
*MONITORING, BACTERIA, WATER CHEM   W71-05879
*MONITORING, *DISCHARGE MEASUREME   W72-00941
*MORPHOLOGY, *ALGAE, AQUATIC MICR   W69-10177
*MORTALITY, *MARINE ALGAE, *MARSH   W70-01777
*MOUNTAIN MEADOW, NITROGEN-15, SO   W72-08110
*MOUNTAINS, *WINTER, CYANOPHYTA,   W69-10154
*MUD(LAKE), *PHOSPHORUS, *AQUATIC   W70-03955
*MUD, *COPPER, COPPER SULFATE, AL   W69-03366
*MUNICIPAL WASTES, AQUATIC ALGAE,   W69-01165
*MUNICIPAL WASTES, NUTRIENTS, NUI   W68-00680
*MUNICIPAL WASTES, *GREAT LAKES,   W70-04430
*MUNICIPAL WASTES, METHANE BACTER   W72-06854
*MUSCULAR EXERTION, *2-METHYL-5-V   W70-01788
*NATIONAL WATER QUALITY NETWORK,   W70-04507
*NATURAL STREAMS, *DISTRIBUTION S   W72-03562
*NEVADA, *EUTROPHICATION, *LAKES,   W72-02817
*NEVADA, DEPTH, ALGAE, MOSSES, CH   W70-00711
*NEVADA, LIMNOLOGY, PRIMARY PRODU   W70-01933
*NEW HAMPSHIRE WATER POLLUTION CO   W69-08674
*NEW HAMPSHIRE, *COPPER SULFATE,   W69-08674
*NEW HAMPSHIRE, ALGAE, ODOR, TAST   W69-08282
*NEW SPECIES, *WESTERN LAKE ERIE(W70-04468
*NEW YORK, NUCLEAR WASTES, WATER   W69-10080
*NEW YORK, *MEROMIXIS, *PRIMARY P   W70-05760
*NEW YORK, TREATMENT FACILITIES.:   W72-04145
*NEW YORK, WATER PURIFICATION.: /   W72-08686
*NEW ZEALAND, SEWAGE TREATMENT PO   W70-02304
*NEWFOUND LAKE, *WINNISQUAM LAKE,   W70-02764

, LAKE SEDIMENTS, ALGAL CONTROL,
OGEN, PHOSPHORU/ *WATER QUALITY,
FFECTS, WATER POLLUTION CONTROL,
LGA/ *BIOCHEMICAL OXYGEN DEMAND,
TES, SIL/ WATER QUALITY CONTROL,
CE ALGAE, SEWAGE EFFLUENTS, PHO/
IEWS, / *EUTROPHICATION, *ALGAE,
TIARY TREATMENT, N/ *PHOSPHORUS,
TROPHICATION, *SEWAGE EFFLUENTS,
IZERS, ALGAE, / *EUTROPHICATION,
CES, *PHOSPHAT/ *EUTROPHICATION,
HYTOPLANKTON, *LIMITING FACTORS,
S, EFFICIENCIES, COST COMPARISO/
N, *ALK/ *ALGAE, *PHYTOPLANKTON,
*EUTROPHICATION, WATER QUALITY,
AE, *HYDROGEN ION CONCENTRATION,
LAMATION, WASTE WATER TREATMENT,
*BENTHIC FAUNA, *EUTROPHICATION,
*EUTROPHICATION, *CARBOHYDRATES,
S, GROWTH RATES, LABORA/ *ALGAE,
OGEN, WISCO/ *NITROGEN-FIXATION,
LUTION EFFECTS, *EUTROPHICATION,
, ALGAE, PHOSP/ *EUTROPHICATION,
S, *LAKE ERIE, PHOSPHORUS, FISH/
MS, BIODEG/ *NITROGEN COMPOUNDS,
S, *POLLUTANTS, RECREATION, WAT/
OPHICATION, *NITROGEN COMPOUNDS,
N / *EUTROPHICATION, *LAKE ERIE,
ABATEMENT, NU/ *EUTROPHICATION,
OMPOUNDS, PH/ *WATER PROPERTIES,
ORUS, N/ *EUTROPHICATION, LAKES,
AQUATIC PRODUCTIVITY, LIMNOLOGY,
ENT, SECONDARY TREATMENT, STABI/
SODIU/ *EUTROPHICATION, *ALGAE,
OMPOUNDS, *PHOSPHORUS COMPOUNDS,
MANAGEMENT, FERTILIZATION, ALG/
URES, C/ *WASTE WATER TREATMENT,
REATMENT, *PHOSPHORUS COMPOUNDS,
AE, CRUSTACEANS, / *CONFERENCES,
, *ORGANIC MATTER, EXUDATION, T/
, *FRESH WATER, *HUMIC ACIDS, */
TION, / *WATER ANALYSIS, *ODORS,
ALGAE, *WISCONSIN, *LAKES, PROD/
DUCING ALGAE, ALGAE, CHLORINATI/
WATER TREATMENT, *ALGAE, *TASTE,
LLUTION SOURCES, *ALGAE, *TASTE,
*ALGAE, *ALGAL CONTROL, *TASTE,
EES, / *WATER TREATMENT, *TASTE,
N, *EFFICIENCIES, *CHLORINATION,
EATMENT, *ALGAE, *ACTINOMYCETES,
EATMENT, *WATER QUALITY, *ALGAE,
*ACTINOMYCETES, *ALGAE, *TASTE,
TROPHICATION, / *WATER ANALYSIS,
, *SAMPLING, *MISSISSIPPI RIVER,

Y, *BIOINDICATORS, *CHLOROPHYLL,
ON CONTROL, *POLLUTION ABATEMEN/
ATER QUALITY, *DISSOLVED OXYGEN,
LLUTION, *KUWAIT C/ *SALT MARSH,
N ALGAE, / *RECOVERY, OIL SPILL,
ATIC SOLV/ *CRUDE OIL, *GAMOSOL,
:
P/ *DETERGENTS, *DIATOMS, *OIL,
DUSTRIAL WASTES, WATER AN/ *OIL,
ECTS, *SALT MARSHES, OILY WATER,
*WATER P/ *DETERGENTS, *DIATOMS,
ES, INDUSTRIAL WASTES, WATER AN/
*MARSH PLANTS, *MARINE ANIMALS,
*ALGAL POISONING, *MARSH PLANTS,
SIFIERS, WATER POLLUTION EFFECT/
NACYSTIS MONTANA, *HYDROCARBONS,
URCES, ON-SITE DATA COLLECTIONS,
SYRACUSE, WARBURG RESPIROMETER,/
ORA, GOMPHONEMA, PINNULARIA, NI/
S, S/ *TWELVE MILE LAKE(CANADA),
TARIO), TWELVE MILE LAKE(ONTARI/
ULATION, *RADIONUCLIDE EXCHANGE,
*BIOLOGICAL PROPERTIES, CONTROL,
ELEMENTS, *ANCHOA LAMPROTAENIA,
PHYTA, *ALGAE, *THERMAL SPRINGS,
GAE, HYDROGEN ION CONCENTRATION,
COMPOUNDS, / *ALGAE, *CHLORELLA,
RYOTES, MONOCHRYSIS LUTHE/ *DNA,
OPHYLL, WATER POLLUTION EFFECTS,
*POTENTIATION,
FIDES, ALGAE/ *HYDROGEN SULFIDE,
/ *POULTRY, *OXIDATION LAGOONS,
FFECTS, *HYDRAULIC LOAD EFFECTS,
COLOGY, RADIOACTIVITY T/ *LAKES,
, *ALGAE, *PRIMARY PRODUCTIVITY,

*NUTRIENTS, ALGAE, BENTHIC FAUNA,   W72-06052
*NUTRIENTS, *EUTROPHICATION, NITR   W73-01122
*NUTRIENTS.: /, WATER POLLUTION E   W72-00917
*NUTRIENTS, *DISSOLVED OXYGEN, *A   W71-05407
*NUTRIENTS, *PESTICIDES, FARM WAS   W71-10376
*NUTRIENTS, *PRODUCTIVITY, NUISAN   W70-00270
*NUTRIENTS, *CARBON DIOXIDE, *REV   W70-00664
*NUTRIENTS, *EUTROPHICATION, *TER   W69-09340
*NUTRIENTS, *NITROGEN, PHOSPHORUS   W69-09349
*NUTRIENTS, *LAKES, PONDS, FERTIL   W69-08668
*NUTRIENTS, *WATER POLLUTION SOUR   W70-01031
*NUTRIENTS, *PRIMARY PRODUCTIVITY   W70-01579
*NUTRIENTS, *NITROGEN, *PHOSPHORU   W70-01981
*NUTRIENTS, *PHOSPHORUS, *NITROGE   W70-02795
*NUTRIENTS, LIMNOLOGY, WATER TEMP   W70-09093
*NUTRIENTS, *TERTIARY TREATMENT,   W70-08980
*NUTRIENTS.: /TS, WASTE WATER REC   W70-06053
*NUTRIENTS, LAKES, ECOLOGY, PRIMA   W71-00665
*NUTRIENTS, *BIODEGRADATION, ALGA   W71-00221
*NUTRIENTS, *NUTRIENT REQUIREMENT   W70-03507
*NUTRIENTS, *LAKES, *ALGAE, *NITR   W70-03429
*NUTRIENTS, ALGAE, FISH POPULATIO   W70-04266
*NUTRIENTS, *PHOSPHORUS COMPOUNDS   W70-04074
*NUTRIENTS, *BENEFIT-COST ANALYSI   W70-04465
*NUTRIENTS, *ALGAE, *MICROORGANIS   W69-06970
*NUTRIENTS, *MARINE MICROORGANISM   W69-04276
*NUTRIENTS, *PHOSPHORUS COMPOUNDS   W69-04805
*NUTRIENTS, ALGAE, WATER POLLUTIO   W69-01445
*NUTRIENTS, POLLUTANTS, POLLUTION   W69-01453
*NUTRIENTS, *VIRGINIA, NITROGEN C   W69-03695
*NUTRIENTS, UNITED STATES, PHOSPH   W68-00478
*NUTRIENTS, BIO MASS, OLIGOTROPHY   W68-00912
*NUTRIENTS, PONDS, PRIMARY TREATM   W68-00012
*NUTRIENTS, NITROGEN, PHOSPHORUS,   W70-04503
*NUTRIENTS, *BIOASSAY, *RUNOFF, *   W70-04765
*NUTRIENTS, *LAKES, *TROUT, *FISH   W70-05399
*NUTRIENTS, ALGAE, BACTERIA, CULT   W70-05655
*NUTRIENTS, *PESTICIDES, *WATER P   W70-05470
*OCEANOGRAPHY, *TRANSLATIONS, ALG   W69-07440
*OCEANS, *MARINE ALGAE, *LITTORAL   W70-01073
*OCEANS, *MARINE ALGAE, *SEAWATER   W70-01074
*ODOR PRODUCING ALGAE, EUTROPHICA   W69-00387
*ODOR-PRODUCING ALGAE, *NUISANCE   W70-05112
*ODOR-PRODUCING ALGAE, *TASTE-PRO   W70-02373
*ODOR, *ALGICIDES, ACTIVATED CARB   W72-03746
*ODOR, *AMINO ACIDS, POTABLE WATE   W72-01812
*ODOR, *NATURAL STREAMS, *DISTRIB   W72-03562
*ODOR, *RESERVOIRS, *DECIDUOUS TR   W72-03475
*ODOR, *WASTE TREATMENT, ALGAE, D   W70-04788
*ODOR, ACTIVATED CARBON, AMMONIA,   W72-04155
*ODOR, HYDROGEN ION CONCENTRATION   W72-09694
*ODOR, WATER QUALITY CONTROL, POL   W72-13605
*ODORS, *ODOR PRODUCING ALGAE, EU   W69-00387
*OHIO RIVER, NUISANCE ALGAE, WATE   W70-04507
*OHIO RIVER BASIN.:   W72-01788
*OHIO RIVER, *ANALYTICAL TECHNIQU   W71-04206
*OHIO, *LAKE ERIE, *WATER POLLUTI   W69-06305
*OHIO, WATER TEMPERATURE, SLUDGE   W70-00475
*OIL POLLUTION, *RECOVERY FROM PO   W70-01231
*OIL POLLUTION, *FILAMENTOUS GREE   W70-09976
*OIL SPILLS, *TORREY CANYON, AROM   W70-01470
*OIL SPILLS, PROTEIN AQUICULTURE.   W69-05891
*OIL WASTES, *SURFACTANTS, *WATER   W70-01470
*OIL WASTES, *CHEMICAL WASTES, IN   W72-03299
*OIL, *MARINE ALGAE, WATER POLLUT   W70-09976
*OIL, *OIL WASTES, *SURFACTANTS,   W70-01470
*OIL, *OIL WASTES, *CHEMICAL WAST   W72-03299
*OILY WATER, LICHENS, TIDAL EFFEC   W70-01777
*OILY WATER, AQUATIC ALGAE, MARIN   W70-01231
*OILY WATER, *MARINE ALGAE, *EMUL   W70-01780
*OLEFINS, GREEN RIVER FORMATION,   W69-08284
*ON-SITE INVESTIGATIONS, FREQUENC   W70-09976
*ONONDAGA LAKE, ONONDAGA COUNTY,   W69-02959
*ONONDAGA LAKE(NY), EUNOTIA, AMPH   W70-03974
*ONTARIO, *CANADA, NUTRIENT LEVEL   W70-02973
*ONTARIO, *CANADA, KUSHOG LAKE(ON   W70-02795
*OPEN SEA, WEAPONS TESTING, NUCLE   W70-00707
*OPERATION AND MAINTENANCE, *OXID   W70-08392
*OPISTHONEMA OGLIMUN, UDOTEA FLAB   W69-08525
*OREGON, THERMOPHILIC BACTERIA, T   W70-05270
*ORGANI: /SIS, TASTE-PRODUCING AL   W72-11906
*ORGANIC ACIDS, BIOMASS, ORGANIC   W70-04195
*ORGANIC CARBON, *CELL SIZE, EUKA   W69-08278
*ORGANIC COMPOUNDS, *OXYGENATION,   W70-08416
*ORGANIC DEGRADATION.:   W72-04506
*ORGANIC LOADING, *SULFATES, *SUL   W70-07034
*ORGANIC LOADING, ALGAE, AMMONIA,   W71-05013
*ORGANIC LOAD EFFECTS, *DETENTION   W70-04787
*ORGANIC MATTER, *PHYSIOLOGICAL E   W70-04194
*ORGANIC MATTER, NUTRIENTS, LAKES   W71-10079

CATORS, *PROTOZOA, *ENVIRONMENT,
CEANS, *MARINE ALGAE, *LITTORAL,
LIMNOLOGY, ZOOPLANKTON, ALGAE, /
*WATER TREATMENT, WATER QUALITY,
DWATER, RECHARGE, *ION EXCHANGE,
MICROBIAL ECOLOGY, SWEDEN, LAKE/
MENTS, MICROBIAL/ *HETEROTROPHY,
M WASTES, *SPRINKLER IRRIGATION,
ROBIC TREAT/ *OXIDATION LAGOONS,
ION,/ *WASTE TREATMENT, *WATERS,
, SCENEDESMUS QU/ *ALGAL BLOOMS,
, SCENEDESMUS QU/ *ALGAL BLOOMS,
RIER(ISRAEL), THRESHOLD ODOR NU/
H LAKE(ORE.), AGENCY LAKE(ORE.),
                                ALUM,
TREATMENT, ALGAE, PHOTOSYNTHESI/
GESTION, *MUNICIPAL WASTES, MET/
 INTENS/ *WASTE WATER TREATMENT,
FUR BACTERIA, STABIL/ *SULFIDES,
ING, ALGAE, AMMONIA, / *POULTRY,
OXYGEN DEMAND, *ANAEROBICS, ALG/
ISPOSAL, *WASTE WATER TREATMENT,
ES, *OXYGENATION, AEROBIC TREAT/
ERIA, *DISSO/ *SEWAGE TREATMENT,
DE, *SULFIDES, ALGAE, ORGANIC L/
XY/ *AERATION, *ALGAE, *LAGOONS,
ROL, *OPERATION AND MAINTENANCE,
TRIENTS, *WASTE WATER TREATMENT,
ACTERIA, *WASTE WATER TREATMENT,
ASTE WAT/ *BIOLOGICAL TREATMENT,
, *PHOSPHORUS, SEWAGE TREATMENT,
  *AGE, *ALGAE, *BIODEGRADATION,
SPARENC/ *LAKE WASHINGTON(WASH),

MIXING, ROANOKE RAPIDS RESERVO/
, PRESSURE, MI/ *PHOTOSYNTHESIS,
ATMENT, SOLAR RADI/ *WASHINGTON,
ION EFFECTS, *ORGANIC COMPOUNDS,
COMP/ *REAERATION, *CHLOROPHYLL,
DATION LAGOONS, *ORGANIC WASTES,
ULING MATERIALS, TE/ *ALGICIDES,
 POLLUTION CONTROL, *FILTRATION,
TREATMENT, SANITARY ENGINEERING,
OSPHATES,/ *ALGAE, *FRESH WATER,
AKE SAGAMI, JAPAN, *GLENODININE,
KALOID TOXINS, IBOGAALKALOID, O/
UALITY, BIOINDI/ *LAKE SUPERIOR,
ES, PRODUCTIVITY, RADIOISOTOPES,
*ZINC RADIOISOTOPES, STREAMFLOW/
, RESIDUES, *PESTICIDE KINETICS,
 KINE/ *WATER POLLUTION EFFECTS,
 EFFECTS, *PESTICIDES, RESIDUES,
TER QUALITY CONTROL, *NUTRIENTS,
SSAY, BIOINDICATORS, FISHKILL, /
SOILS, COA/ *PESTICIDE KINETICS,
EMOVAL, *ADSORPTION, SOILS, COA/
EMOVAL, *FOOD WEBS, DDT/ *ALGAE,
DT/ *ALGAE, *PESTICIDE TOXICITY,
LIZATION, *PESTICIDE METABOLISM,
 SOUTH CAROLINA, LYOPHILIZATION,
OTHIOATE PESTICIDES, HERBICIDES,
HOSPHORUS COMPOUNDS, *NUTRIENTS,
*ROTIFE/ *WATER QUALITY, *ALGAE,
LERANCE LIMIT, BLUEGILLS, CHLOR/
 LIGURIAN/ *RADIONUCLIDE UPTAKE,
OD PRODUCTION, SOUTHAMPTON UNIV/
OMPOSING ORGANIC MATTER, *ALGAE/
N, *TOXICITY, ORGANIC COMPOUNDS,
TS, *NUISANCE ALGAE, *ALGICIDES,
, MONURON, TRANSPARENCY, CYCLOP/
S, *CHEMICAL PRECI/ *PHOSPHORUS,
, *LAKES, *ALGAE, *GROWTH RATES,
TRIBUTION,/ *ALGAE, *PHOSPHORUS,
HY, RAD/ *ANALYTICAL TECHNIQUES,
ION, CARBON DIOXIDE, SOLUBILITY,
RIENTS/ *ALGAE, *BIODEGRADATION,
ION, DIATOMS, MICROB/ *NITROGEN,
TER PO/ *EUTROPHICATION, *LAKES,

IENTS, *WATER POLLUTION SOURCES,
IENTS, *SEDIMENT-WATER INFERFAC/
IC WEED/ *PONDS, *FERTILIZATION,
ISTRY, *WATER POLLUTION EFFECTS,
LING NUTRIENTS, *EUTROPHICATION,
Y, *EUTROPHICATION, *DETERGENTS,
 *TERTIARY TREATMENT, *GREAT LA/
MIXED LIQUOR, SUSPENDED SOLIDS,
OF ALGAE, EFFL/ *EUTROPHICATION,
ROORGANISMS, NITROGEN COMPOUNDS,
S, / *BIOCHEMICAL OXYGEN DEMAND,

*ORGANIC MATTER, WATER QUALITY, T   W71-03031
*ORGANIC MATTER, EXUDATION, TIDES   W70-01073
*ORGANIC MATTER, *PHYTOPLANKTON,   W70-02510
*ORGANIC MATTER, ALGAE, COPPER SU   W72-04495
*ORGANIC MATTER, *FILTRATION, DIA   W72-04318
*ORGANIC SOLUTES, *HETEROTROPHY,   W70-04284
*ORGANIC SOLUTES, *MARINE ENVIRON   W70-04811
*ORGANIC WASTES, *WASTE WATER DIS   W71-00940
*ORGANIC WASTES, *OXYGENATION, AE   W71-09546
*ORGANISMS, STREAMS, WATER POLLUT   W70-05750
*ORGANO-SILICATES, *ALKOXYSILANES   W71-03056
*ORGANO-SILICATES, *ALKOXYSILANES   W71-05719
*OSCILLATORIA, NATIONAL WATER CAR   W70-02373
*OSCILLATORA.:   *UPPER KLAMAT   W72-06051
*OSHKOSH(WIS), BACKWASHING.:   W72-08859
*OXIDATION LAGOONS, *WASTE WATER   W72-10573
*OXIDATION LAGOONS, *ANAEROBIC DI   W72-06854
*OXIDATION LAGOONS, *ALGAE, LIGHT   W72-06022
*OXIDATION LAGOONS, *DESIGN, *SUL   W71-01971
*OXIDATION LAGOONS, *ORGANIC LOAD   W71-05013
*OXIDATION LAGOONS, *BIOCHEMICAL   W71-05742
*OXIDATION LAGOONS.: /R, SEWAGE D   W71-09548
*OXIDATION LAGOONS, *ORGANIC WAST   W71-09546
*OXIDATION LAGOONS, *ALGAE, *BACT   W72-01818
*OXIDATION PONDS, *HYDROGEN SULFI   W71-11377
*OXIDATION LAGOONS, BIOCHEMICAL O   W70-06601
*OXIDATION LAGOONS, ABSORPTION, B   W70-08392
*OXIDATION LAGOONS, MILK, *MODEL   W70-07031
*OXIDATION LAGOONS, *MODEL STUDIE   W70-07034
*OXIDATION, LAGOONS, *BAFFLES, *W   W70-09320
*OXIDATION LAGOONS.: /NCENTRATION   W68-00012
*OXYGEN DEMAND, LABORATORY TESTS,   W70-06600
*OXYGEN DEFICIT, SECCHI DISC TRAN   W69-10182
*OXYGEN PRODUCTION.:   W72-06022
*OXYGEN SUPERSATURATION, VERTICAL   W70-00683
*OXYGEN, *SATURATION, TEMPERATURE   W70-00683
*OXYGEN, *FLUCTUATION, SEWAGE TRE   W69-10182
*OXYGENATION, *PHOTOSYNTHESIS, CH   W70-08416
*OXYGENATION, *TOXICITY, ORGANIC   W71-12183
*OXYGENATION, AEROBIC TREATMENT,   W71-09546
*PAINTS, *TEST PROCEDURES, ANTIFO   W71-07339
*PARTICLE SIZE, *CLAYS, *SILTS, S   W70-08628
*PATHOGENIC BACTERIA, ALGAE, ORGA   W72-13601
*PEPTIDES, COPPER, IRON, IONS, PH   W69-10180
*PERIDINIUM POLONICUM, PLANKTON B   W70-08372
*PERIDINIUM SP., *GLENODININE, AL   W71-07731
*PERIPHYTON, *LIMNOLOGY, *WATER Q   W70-05958
*PERIPHYTON, *RADIOACTIVITY, *LAK   W70-04371
*PERIPHYTON, *CYCLING NUTRIENTS,   W69-07862
*PERSISTENCE, EVAPORATION, BIODEG   W70-09768
*PESTICIDES, RESIDUES, *PESTICIDE   W70-09768
*PESTICIDE KINETICS, *PERSISTENCE   W70-09768
*PESTICIDES, FARM WASTES, SILT, E   W71-10376
*PESTICIDE RESIDUES, *ALGAE, BIOA   W69-08565
*PESTICIDE REMOVAL, *ADSORPTION,   W69-09884
*PESTICIDE KINETICS, *PESTICIDE R   W69-09884
*PESTICIDE TOXICITY, *PESTICIDE R   W70-03520
*PESTICIDE REMOVAL, *FOOD WEBS, D   W70-03520
*PESTICIDE ACCUMULATION.: /LYOPHI   W69-02782
*PESTICIDE METABOLISM, *PESTICIDE   W69-02782
*PESTICIDE RESIDUES, DIATOMS, CRA   W69-02782
*PESTICIDES, *WATER POLLUTION CON   W70-05470
*PESTICIDES, *FISH, *FISH KILLS,   W72-04506
*PETROCHEMICAL COMPOUNDS, MEAN TO   W72-11105
*PHAEODACTYLUM TRICORNUTUM, EDTA,   W69-08275
*PHAEODACTYLUM, *GAS EFFLUENT, FO   W70-00161
*PHENOLS.: /DECIDUOUS TREES, *DEC   W72-03475
*PHENOLS, PESTICIDES, OXIDATION L   W71-12183
*PHENOLS, *WATER QUALITY CONTROL,   W70-10161
*PHENYLUREAS, DIURON, FLAGELLATES   W70-10161
*PHOSPHATES, *PHOSPHORUS COMPOUND   W70-09186
*PHOSPHATES, LABORATORY TESTS, MI   W70-08637
*PHOSPHATES, ENERGY, DI-URNAL DIS   W70-06604
*PHOSPHATES, *ALGAE, CHROMOTOGRAP   W71-03033
*PHOSPHATES, CHEMICAL PRECIPITATI   W71-11393
*PHOSPHATES, *EUTROPHICATION, NUT   W71-04072
*PHOSPHATES, *ALGAE, *EUTROPHICAT   W71-05026
*PHOSPHATES, NUTRIENTS, ALGAE, WA   W69-09135
*PHOSPHATE ABSORPTION.:   W70-01031
*PHOSPHATES, ALGAE, WATER POLLUTI   W70-01031
*PHOSPHATE, *LAKES, *CYCLING NUTR   W69-06859
*PHOSPHATE, FISH, PLANKTON, AQUAT   W70-05417
*PHOSPHATES, *ANALYTICAL TECHNIQU   W70-05269
*PHOSPHATES, *PHYSIOLOGICAL ECOLO   W69-03358
*PHOSPHATES, CHELATION, WASTE TRE   W70-04373
*PHOSPHATES, *LAKES, *DETERGENTS,   W70-03964
*PHOSPHOROUS UPTAKE RATE.:   W71-13335
*PHOSPHOROUS, ALGAE, *HARVESTING   W71-13313
*PHOSPHORUS COMPOUNDS, ADSORPTION   W71-13335
*PHOSPHORUS, *NITROGEN, *COLIFORM   W71-05409

NKTON, ALGAE, / *ORGANIC MATTER,
BICARBONATES, RADIOACTIVITY, CA/
REEN ALGA/ *FLOTATION, *SINKING,
PHORUS, *NITROGEN, *ALK/ *ALGAE,
MENTS, ECOLOGY, CHLORELLA, ALGA/
, *NUTRIENTS, *PRIMARY PRODUCTI/
ODS, *ZOOPLANKTON, *BIOMASS, *S/
*ALGAL POISONING, *ALGAL TOXINS,
D, *HERBICIDES, *PHYTOPLANKTON,
IBITION, *PHYSIOLOGICAL ECOLOGY,
ENTED FLAGELLATES.:
RRENTS(WATER), LIMNOLOGY, LAKES,
BORATORY STUDIES, *MASS CULTURE,
GAE ISOLATION, CAPILLARITY, CAP/
*ALGAE, *ANALYTICAL TECHNIQUES,
Y, TEMPERATURE, PRIMARY PRODUCT/
IGH/ *ENVIRONMENT, *FRESH WATER,
TY, CHLORELLA, CHLOROPHYTA, DIA/
TROPHICATION, *LAKES, LIMNOLOGY,
, *LAKES, *PRIMARY PRODUCTIVITY,
R/ *CYANOPHYTA, *EUTROPHICATION,
PLANKTON, ALGAE, TEMPERATURE, S/
MENT, *EUTROPHICATION, / *ALGAE,
, *LAKE ONTARIO, *GROWTH RATES,/
, *CYCLING NUTRIENTS, ECOSYSTEM/
HEMICAL/ *NUTRIENT REQUIREMENTS,
Y, CHEMICAL ANALYSIS, P/ *ALGAE,
UM, SILICA, DIATOMS, RESISTANCE,
BIBLIOGRAPHIES, *EUTROPHICATION,
SPHATES, SILICATES, IRON OXIDES,
PTH, ALGAE/ *DEEP WATER, *LAKES,
HLORINE, *ANALYTICAL TECHNIQUES,
ATERS, ECOSYSTEMS, ESTUARINE EN/
TRIENTS, *MARINE MICROORGANISMS,
CAL TREATMENT, *WATER TREATMENT,
MENT, *POLLUTANT IDENTIFICATION,
E WATER, *NUTRIENT REQUIREMENTS,
, MUNICIPAL WASTES, GREAT LAKES,
ITY, *BIOASSAY, FISH, AQUATIC E/
GREEN ALGAE, *EUGLENA, PIGMENTE/
OUNTY(ILL), CALUMET RIVER(ILL),/
TION, *POLLUTION IDENTIFICATION,
 ERIE, *WATER POLLUTION CONTROL,
E ENVIRONMENT, *INSTRUMENTATION,
TER POLLUTION SOURCES, DREDGING,
STEMS, FLOW RA/ *EUTROPHICATION,
ANIC POLYELECTROLYTES, CATIONIC/
S, METHYL ALCOHOL, ETHYLENE GLY/
AE BLOOMS, *FISH TOXICITY TEST,/
M NODOSUM, VIBRIO, FLAVOBACTER,/
CHLORINATION, */ *STABILIZATION,
WASTE TREATMENT, *STABILIZATION,
WASTE TREATMENT, *STABILIZATION,
IOMASS, CHLOROP/ *PHYTOPLANKTON,
E, FISH, PLANKTON, AQUATIC WEED/
STRATIFICATION, *SEASONAL, *SUL/
, WATER SUPPLY, */ *INFILTRATION,
NOLOGY, *WATER, *RIVERS, *LAKES,
ZE, CHLORELLA/ *COULTER COUNTER,
IES, EFFLUENTS/ *CARBON, *ALGAE,
NTROL, *ALGAE, COPPER SULFATE, /
R BACTERIA, COL/ *ALGAL CONTROL,
HILLS(CALIF).:
L TESTS, ALGISTATIC TESTS, SLIM/
AE, *DECOMPOSING ORGANIC MATTER,
NES(IOWA).:

C LAGOONS, ANAEROBIC LAGOONS.:
REENS, SEDIMENTATION, ORGANIC L/
ON.:
PHYTOPLANKTON RELATIONSHIPS.:

C PALUDOSUM, BLE/ *ALGAL GROWTH,
GANIC LOADING, ALGAE, AMMONIA, /
SEDIMENTATION, OX/ *FARM WASTES,
BAFFLES, *WASTE WATER TREATMENT,
*TEMPERATURE, *COOLING WATER, /
ISANCE, / *CORRECTIVE TREATMENT,
:   MARINE BIOLOGY,
ATTER, / *PHYTOPLANKTON, *ALGAE,
LIMNOLOGY, *ESSENTIAL NUTRIENTS,
ATION, WATER QUALITY/ RESERVOIR,
*LAKES, *NEW YORK, *MEROMIXIS,
, *LIMITING FACTORS, *NUTRIENTS,
*LABORATORY TESTS, RESPIRATION,/
N/ *BENTHIC FLORA, *CHLOROPHYLL,
MNOLOGY, DYSTROPHY, OLIGOTROPHY,
HERMAL WATER, PERIPHYTON, BIOMA/
ULATIONS, ENVIRONMENTAL EFFECTS,
UTRIENTS, BIO MASS, OLIGOTROPHY,

*PHYTOPLANKTON, LIMNOLOGY, ZOOPLA   W70-02510
*PHYTOPLANKTON, *PHOTOSYNTHESIS,   W70-02504
*PHYTOPLANKTON, *DIATOMS, *BLUE-G   W70-02754
*PHYTOPLANKTON, *NUTRIENTS, *PHOS   W70-02795
*PHYTOPLANKTON, *NUTRIENT REQUIRE   W70-02804
*PHYTOPLANKTON, *LIMITING FACTORS   W70-01579
*PHYTOPLANKTON, *STATISTICAL METH   W73-01446
*PHYTOTOXICITY, *FISH DISEASES, *   W71-07731
*PHYTOTOXICITY, ALGAL POISONING,   W69-00994
*PHYTOTOXICITY, ALGAL CONTROL, AL   W69-03188
*PIGMENTED FLAGELLATES, *NON-PIGM   W72-01798
*PILOT PLANTS, *NUISANCE ALGAE, R   W68-00479
*PILOT-PLANT STUDIES, ALGAL GROWT   W69-06865
*PIPETTE, UNIALGAL, NITZSCHIA, AL   W70-04469
*PLANKTON, *LABORATORY TESTS, *IS   W70-04469
*PLANKTON, *ALGAE, LIGHT INTENSIT   W70-04809
*PLANKTON, *ALGAE, SCENEDESMUS, L   W70-05409
*PLANKTON, *ALGAE, *LIGHT INTENSI   W70-05381
*PLANKTON, ALGAE, BIBLIOGRAPHIES,   W69-00446
*PLANKTON, ALGAE, *PHOTOSYNTHESIS   W69-00632
*PLANKTON, ALGAE, *NUTRIENT REQUI   W69-03512
*PLANKTON, *PHOTOSYNTHESIS, PHYTO   W71-10788
*PLANKTON, *SEWAGE, *SEWAGE TREAT   W70-02304
*PLANKTON, *ALGAE, *PHYTOPLANKTON   W69-10158
*PLANKTON, *ALGAE, *SESSILE ALGAE   W73-01454
*PLANT GROWTH, *ALGAE, ELEMENTS(C   W70-02964
*PLANT GROWTH, *WISCONSIN, ECOLOG   W70-06229
*PLANT GROWTH REGULATORS, FRESH W   W68-00856
*PLANT GROWTH, ANALYTICAL TECHNIQ   W69-04801
*PLANT POPULATIONS, REVIEWS, LIGH   W68-00860
*PLANTS, CALIFORNIA, *NEVADA, DE   W70-00711
*POLAROGRAPHIC ANALYSIS, PHOTOMET   W70-05178
*POLLUTANTS, *ESTUARIES, *TIDAL W   W68-00010
*POLLUTANTS, RECREATION, WATER SP   W69-04276
*POLLUTANT IDENTIFICATION, *POLLU   W69-07389
*POLLUTANT ABATEMENT, EFFLUENTS,   W69-07389
*POLLUTANTS, INHIBITION, LIMNOLOG   W70-02792
*POLLUTANT IDENTIFICATION.: /RCES   W71-05879
*POLLUTANT IDENTIFICATION, *TOXIC   W72-11105
*POLLUTION-TOLERANT ALGAE, *BLUE   W70-01233
*POLLUTION, *CHICAGO(ILL), COOK C   W70-00263
*POLLUTION ABATEMENT, TIDAL WATER   W69-06203
*POLLUTION ABATEMENT, POLLUTANTS,   W69-06305
*POLLUTION IDENTIFICATION, *POLLU   W69-06203
*POLLUTION ABATEMENT, DIKES, SOIL   W68-00172
*POLLUTION ABATEMENT, DRAINAGE SY   W68-00479
*POLYELECTROLYTES, *SYNTHETIC ORG   W70-05267
*POLYESTER FIBERS, MAN-MADE FIBER   W70-10433
*POLYMERS, *BIOLOGICAL LIFE, *ALG   W70-05819
*POLYSIPHONIA LANOSA, *ASCOPHYLLU   W70-03952
*PONDS, *DESIGN, *EFFICIENCIES, *   W70-04788
*PONDS, *DESIGN CRITERIA, FACILIT   W70-04786
*PONDS, *ENVIRONMENTAL EFFECTS, *   W70-04787
*PONDS, *EUTROPHICATION, ALGAE, B   W69-06540
*PONDS, *FERTILIZATION, *PHOSPHAT   W70-05417
*PONDS, *STABILIZATION, *THERMAL   W70-04790
*PONDS, *SOIL, GROUNDWATER BASINS   W69-07838
*PONDS, LIGHT, TEMPERATURE, FERTI   W70-05404
*POPULATION DYNAMICS, CELLULAR SI   W69-06540
*POPULATION, *BIOLOGICAL COMMUNIT   W74-04001
*POTABLE WATER, *WATER QUALITY CO   W71-00110
*POTABLE WATER, CYANOPHYTA, SULFU   W70-08107
*POTASSIUM PERMANGANATE, BEVERLY   W71-00110
*POTASSIUM PERMANGANATE, ALGICIDA   W69-05704
*POTASSIUM COMPOUNDS, *ACTIVATED   W72-04242
*POTASSIUM PERMANGANATE, *DES MOI   W72-04242
*POTASSIUM PERMANGANATE.:   W72-04155
*POTATO PROCESSING WASTES, AEROBI   W71-08970
*POTATOES, *INDUSTRIAL WASTES, SC   W71-08970
*POTENTIATION, *ORGANIC DEGRADATI   W72-04506
*POTOMAC RIVER ESTUARY, NUTRIENT-   W71-05407
*POTOMAC RIVER ESTUARY.:   W71-05409
*POTOMAC RIVER, PLECTONEMA, NOSTO   W70-02255
*POULTRY, *OXIDATION LAGOONS, *OR   W71-05013
*POULTRY, WASTE WATER TREATMENT,   W71-11375
*POULTRY, INDUSTRIAL WASTES, LAGO   W70-09320
*POWERPLANTS, *THERMAL POLLUTION,   W69-05023
*PREVENTATIVE TREATMENT, ALGAE NU   W69-08282
*PRIMARY BIOLOGICAL PRODUCTIVITY.   W71-10788
*PRIMARY PRODUCTIVITY, *ORGANIC M   W71-10079
*PRIMARY PRODUCTIVITY, AQUATIC AL   W72-01990
*PRIMARY PRODUCTIVITY, *EUTROPHIC   W70-09093
*PRIMARY PRODUCTIVITY, *ZOOPLANKT   W70-05760
*PRIMARY PRODUCTIVITY, PHOSPHATES   W70-01579
*PRIMARY PRODUCTIVITY, *STREAMS,   W70-00265
*PRIMARY PRODUCTIVITY, *MEASUREME   W69-10151
*PRIMARY PRODUCTIVITY, *TROPHIC L   W69-09723
*PRIMARY PRODUCTIVITY, *ALGAE, *T   W70-03309
*PRIMARY PRODUCTIVITY, WATER POLL   W70-05424
*PRIMARY PRODUCTIVITY, TROPIC LEV   W68-00912

OROPHYLL A, DATA INTERPRETATION,
ERISTICS, *CHEMICAL CHARACTERIS/
UALITY NETWORK, ALGAE COUNTS, P/

FERTILIZERS, NUTRIENTS/ *LAKES,
ISSIPPI RIVER, *OHIO RIVER, NUI/
EMPERATURE,/ *LIMNOLOGY, *WATER,
CONTRO/ *EUTROPHICATION, *LAKES,
POLLUTION SO/ *ECOLOGY, *ALGAE,
ATIC LIFE, *AQUATIC POPULATIONS,
RADIONUCLIDE, *DISSOLVED OXYGEN,
WATER QUALITY, *WATER POLLUTION,
I / *RESERVOIRS, *PHYTOPLANKTON,
LANKTO/ *AQUATIC PLANTS, *ALGAE,
*PESTICIDES, *FISH, *FISH KILLS,
POSAL, *AGRICULTURAL WATERSHEDS,
OMPOUNDS, *NUTRIENTS, *BIOASSAY,
SEDIMENTS, ORGANIC MATTER, CURR/

NOPHYTA, *PHYSIOLOGICAL ECOLOGY,
AINAGE / *LAKES, PLANTS, *ALGAE,
OPES, *PHOSPHORUS RADIOISOTOPES,
OVERY FROM POLLUTION, *KUWAIT C/
STANCE, WATER POLLUTION EFFECTS,
EAR WATER RESERVOIRS, *TURBID R/
ATION EFFECTS, DRAINAGE EFFECTS,
*AQUATIC WEEDS, *WATER ANALYSIS,
HIO RIVER, NUI/ *RIVERS, *ALGAE,
ETRY, DYE REL/ *INSTRUMENTATION,
CTROPHORESIS, CHLORELLA/ *ALGAE,
ONS, *ENVIRONMENTAL ENGINEERING,
*ALGAE, *PROTOZOA, *SYSTEMATICS,
PLY, *BIOLOGICAL PROPERTIES, *W/
IOMASS TITER, *BIOLOGICAL INDEX,
ATANS, SEAWEED, ZIRCONIUM-95-NI/
NIUM-95-NI/ *SARGASSUM FLUITANS,
E, MI/ *PHOTOSYNTHESIS, *OXYGEN,
*FILTRATION, HEAD LOSS, FILTERS,
*AERATION, *NITRATES, *CULTURES,
PHOTOSYNTH/ *ALGAE, *CHLORELLA,
GROWTH RATES, *LABORATORY TESTS,
DES MORTS(WISC), GROWTH STIMUL/
NING, *MICROSTRAINERS.:
WEBS, KENTUCKY, LIFE HISTORY ST/
ALGAE, CULTURES, CH/ *BIOASSAY,
ZATION, *THERMAL STRATIFICATION,
ENVIRONMENT, *WATER TEMPERATURE,
PLANT, DUWAMISH ESTUARY(WASH).:
CIDS, */ *OCEANS, *MARINE ALGAE,
DOSUM, ALGINIC ACID, FUCUS VESI/
ATE, *LAKES, *CYCLING NUTRIENTS,
*RUNOFF, *GROUNDWATER, *SESTON,
LING NUTRIENTS.:
, *PHOSPHORUS COMPOUNDS, *LAKES,
UCTURE, 2,3-DICHLORONAPHTHOQUIN/
ION CONTROL, *LAKE ONTARIO, *WA/
POLLUTION), *WA/ *PHYTOPLANKTON,
DIATOMACEOUS EARTH, *FILTRATION,
S, ECOSYSTEM/ *PLANKTON, *ALGAE,
ER, CURR/ *RUNOFF, *GROUNDWATER,
STES, *LAGOONS, *BIODEGRADATION,
ATERSHEDS, *RUNOFF, *FERTILIZAT/
URAL/ *LAKES, *SEWAGE EFFLUENTS,
AL, ALGAE, AGRICULTURAL/ *LAKES,
ON CONTROL, *LOCAL / *WISCONSIN,
ESH WATER, LAKES, SILTS/ *ALGAE,
ITROGE/ *LAKES, *EUTROPHICATION,
*ALGAE, *NITROGEN, *PHOSPHORUS,
TY MANAGEMENT, *LAKE WASHINGTON,
TREATMENT, *INDIANA, CHLOROPHYT/
ES, SEWAGE TREAT/ *MICROBIOLOGY,

, *PHOSPHORUS, *SEWAGE EFFLUENT,
GAE, NUTRIENTS, DOMESTIC WASTES,
SEWAGE, *PHOSPHORUS, PHOSPHATES,
ON CONTROL, PACIFI/ *WASHINGTON,
ROPHYT/ *SEWAGE LAGOONS, *ALGAE,
BIOCHEMICAL OXYGEN DEMAND, *ACT/
N, / *ALGAE, *PLANKTON, *SEWAGE,
INEERING, *SANITARY ENGINEERING,
WATER POLLUTION, WASTE DISPOSAL,
TIVATED SLUDGE, *BACTERIA, META/
OONS, *ALGAE, *BACTERIA, *DISSO/
TERS, / *BIOLOGICAL COMMUNITIES,
STORM DRAINS, SEWAGE EFFLUENTS,
ON EFFECTS, FISH, ALGAE, TROPH/
OPHICATION, / *ALGAE, *PLANKTON,
N, *CLADOPHORA, *NUISANCE ALGAE,
MENT, BIBLIOGRAPHIES, CHLORELLA,

*RHONE RIVER, NAVICULA, MELOSIRA,   W73-01446
*RIVER REGIONS, *PHYSICAL CHARACT   W73-00748
*RIVER STUDIES, *NATIONAL WATER Q   W70-04507
*RIVER SURVEYS.:   W72-01801
*RIVERS, *EUTROPHICATION, RUNOFF,   W69-10170
*RIVERS, *ALGAE, *SAMPLING, *MISS   W70-04507
*RIVERS, *LAKES, *PONDS, LIGHT, T   W70-05404
*RIVERS, *ALGAE, WATER POLLUTION   W70-05668
*RIVERS, *WATER POLLUTION, *WATER   W69-07096
*RIVERS, ALGAE, PROTOZOA, AQUATIC   W73-00748
*RIVERS, OXIDATION, PHOTOSYNTHESI   W71-11381
*RIVERS, WATER CHEMISTRY, WATER P   W70-03068
*ROTIFERS, *DISTRIBUTION, MISSOUR   W70-05375
*ROTIFERS, PLANKTON, BASS, PHYTOP   W70-02249
*ROTIFERS, *ECOLOGY, *ECOSYSTEMS.   W72-04506
*RUNOFF, *FERTILIZATION, LAKES, S   W70-02787
*RUNOFF, *LAKES, *IMPOUNDMENTS, T   W70-04765
*RUNOFF, *GROUNDWATER, *SESTON, *   W70-03501
*RUTHENIUM RADIOISOTOPES.:   W71-09863
*SAKHALIN ISLAND(USSR).:   W72-10623
*SALINE WATER, *THERMAL WATER, AL   W69-03516
*SALINITY, IRRIGATION EFFECTS, DR   W69-00360
*SALT MARSHES, *CYCLING NUTRIENTS   W69-08274
*SALT MARSH, *OIL POLLUTION, *REC   W70-01231
*SALT MARSHES, OILY WATER, *OIL,   W70-09976
*SALT VALLEY RESERVOIRS(NEB), *CL   W72-04761
*SAMPLING, ALGAE TOXINS, SALINE W   W69-00360
*SAMPLING, PLANT PHYSIOLOGY, EUTR   W70-04382
*SAMPLING, *MISSISSIPPI RIVER, *O   W70-04507
*SAMPLING, *MEASUREMENT, *FLUOROM   W70-05377
*SANDS, *FILTRATION, FILTERS, ELE   W71-03035
*SANITARY ENGINEERING, *SEWAGE TR   W70-02609
*SANITARY ENGINEERING, MICROORGAN   W72-01798
*SANITARY ENGINEERING, *WATER SUP   W72-01813
*SAPROBITY, PRAGUE, CZECHOSLOVAKI   W72-12240
*SARGASSUM FLUITANS, *SARGASSUM N   W69-08524
*SARGASSUM NATANS, SEAWEED, ZIRCO   W69-08524
*SATURATION, TEMPERATURE, PRESSUR   W70-00683
*SCENEDESMUS, FLOCCULATION, WATER   W72-03703
*SCENEDESMUS, *SUBSURFACE WATERS,   W70-03334
*SCENEDESMUS, *LIGHT, PHOSPHORUS,   W70-03923
*SCENEDESMUS, DIATOMACEOUS EARTH,   W68-00475
*SCIRPUS FLUVIATILIS, *LAKE BUTTE   W72-00845
*SCREENING, *KENOSHA(WIS), *STRAI   W73-02426
*SCULPINS, *RADIOACTIVITY, *FOOD   W69-09742
*SEA WATER, *VITAMINS, VITAMIN B,   W70-05652
*SEASONAL, *SULFIDES, TEMPERATURE   W70-04790
*SEASONAL, RHODE ISLAND, SAMPLING   W70-00713
*SEATTLE(WASH), RENTON TREATMENT   W70-04283
*SEAWATER, *FRESH WATER, *HUMIC A   W70-01074
*SEAWEED EXTRACTS, ASCOPHYLLUM NO   W69-07826
*SEDIMENT-WATER INERFACE, MUD-WA   W69-06859
*SEDIMENTS, ORGANIC MATTER, CURRE   W70-03501
*SEDIMENTS, *NUTRIENT BUDGET, CYC   W68-00511
*SEDIMENTS, *EUTROPHICATION, ALGA   W69-03075
*SELECTIVE TOXICITY, CHEMICAL STR   W70-02982
*SELF-PURIFICATION, *WATER POLLUT   W71-05878
*SELF-PURIFICATION, *WASTE WATER(W70-05092
*SEPARATION TECHNIQUES, TURBIDITY   W72-08686
*SESSILE ALGAE, *CYCLING NUTRIENT   W73-01454
*SESTON, *SEDIMENTS, ORGANIC MATT   W70-03501
*SEWAGE BACTERIA, ODOR, ANAEROBIC   W71-02009
*SEWAGE DISPOSAL, *AGRICULTURAL W   W70-02787
*SEWAGE DISPOSAL, ALGAE, AGRICULT   W70-05565
*SEWAGE EFFLUENTS, *SEWAGE DISPOS   W70-05565
*SEWAGE EFFLUENTS, *WATER POLLUTI   W69-06909
*SEWAGE EFFLUENTS, *ESTUARIES, FR   W70-02255
*SEWAGE EFFLUENTS, *NUTRIENTS, *N   W69-09349
*SEWAGE EFFLUENT, *SEWAGE TREATME   W72-08133
*SEWAGE FERTILIZATION EFFECTS, *B   W69-09349
*SEWAGE LAGOONS, *ALGAE, *SEWAGE   W70-10173
*SEWAGE LAGOONS, *INDUSTRIAL WAST   W70-03312
*SEWAGE NUTRIENT REMOVAL.:   W72-08133
*SEWAGE TREATMENT, *NUTRIENT REMO   W72-08133
*SEWAGE TREATMENT, SEWAGE, BIOLOG   W68-00248
*SEWAGE TREATMENT, DOMESTIC WASTE   W68-00256
*SEWAGE TREATMENT, *WATER POLLUTI   W69-03683
*SEWAGE TREATMENT, *INDIANA, CHLO   W70-10173
*SEWAGE TREATMENT, *FILTRATION, *   W70-05821
*SEWAGE TREATMENT, *EUTROPHICATIO   W70-02304
*SEWAGE TREATMENT, *TRACERS, *WAT   W70-02609
*SEWAGE TREATMENT, *WASTE WATER T   W72-01786
*SEWAGE TREATMENT, *PROTOZOA, *AC   W72-01817
*SEWAGE TREATMENT, *OXIDATION LAG   W72-01818
*SEWAGE TREATMENT, *TRICKLING FIL   W72-01819
*SEWAGE TREATMENT, STORM RUNOFF.:   W71-10654
*SEWAGE, *STREAMS, *WATER POLLUTI   W72-01788
*SEWAGE, *SEWAGE TREATMENT, *EUTR   W70-02304
*SEWAGE, ALGICIDES, AQUATIC ALGAE   W69-04798
*SEWAGE, EFFLUENTS, CARBON DIOXID   W68-00855

*ALGAE.: *ESTUARIES, *HARBORS,
EUTROPHICATION, *NUISANCE ALGAE,
LOOMS, CHLORELLA PYRENOIDOSA, M/
N FLOS AQUAE, STEPHANODISCUS AS/
OUNT, DESMIDS,/ *IDENTIFICATION,
TRATION, *PARTICLE SIZE, *CLAYS,
*SYSTEMS ANALYSIS, *ECOSYSTEMS,
S, *BLUE-GREEN ALGA/ *FLOTATION,

ER COURSE, CARPATHIAN SUBMONTAN/
TAL STUDIES, LAGOON DES/ *JAPAN,
R TREATMENT, *ALGAE, FILTRATION,
ACTER/ *COOLING WATER, *BIOMASS,
TIONS, *EFFLICIENCIES, *LAGOONS,
FLOW RATE, *WASTE WATER, *ALGAE,
L, *HERBICIDES, *COPPER SULFATE,
E, ARID CLIMATES, SEMIARID CLIM/
MATTER, *GRASSLANDS, MOUNTAINS,
SUPPLY, / *INFILTRATION, *PONDS,
ROLINA,/ *ALGAE, *RADIOISOTOPES,
YANOPHYTA, DETERG/ *SURFACTANTS,
XICO, A/ *RADIOISOTOPES, *ALGAE,
EW PRODUCT CO/ *MASTER PLANNING,
*AERATED LAGOONS, *MIDWEST(US),
, LIGHT-SATURATED / *ADAPTATION,
ASTES, *WASTE WAT/ *FARM WASTES,
RITERIA, FACI/ *WASTE TREATMENT,
CATION, *SEASONAL, *SUL/ *PONDS,
ENTAL EFFECTS/ *WASTE TREATMENT,
*EFFICIENCIES, *CHLORINATION, */
N, *BIOMASS, *S/ *PHYTOPLANKTON,
*SCREENING, *KENOSHA(WIS),
IANS, MIC/ *ANABAENA FLOS-AQUAE,
, *URBANIZATION, *WATER QUALITY,
ION, DECOMPOSING ORGANIC MATTER,
S, BACTERIA, ORGANIC POLLUTANTS,
S, *CYCLING NUTRIENTS, *EUTROPH/
HOSPHORUS RADIOISOTOPES, *TROUT,
K,/ *MONITORING, *RADIOISOTOPES,
IRATION,/ *PRIMARY PRODUCTIVITY,
RATURE, MANGANESE, CULTURES, RE/
ICAL COMMUNITIES, *LABORATORIES,
S, *FISH, ALGAE, TROPH/ *SEWAGE,
*FUNGI, *CALCAREOU/ *CHELATION,
LANKTON, RADIOACTIVITY, WASTE D/
N, *CALCIUM, SOIL CONTAMINATION,
TRATES, *CULTURES, *SCENEDESMUS,
METHODS, *ZOOPLANKTON, *BIOMASS,
ULTATIVE, BOD, DETENTION TIME.:
ROGEN SULFIDE, *ORGANIC LOADING,
DE, *ORGANIC LOADING, *SULFATES,
ESIGN, *SULFUR BACTERIA, STABIL/
DATION PONDS, *HYDROGEN SULFIDE,
ERMAL STRATIFICATION, *SEASONAL,
ES, *OXIDATION LAGOONS, *DESIGN,
E, *AERATION, *NITRATES, *CULTU/
CHLOROPHYTA, CYANOPHYTA, DETERG/
TS, *DIATOMS, *OIL, *OIL WASTES,
RC/ *LINEAR ALKYLATE SULFONATES,
*BENTHIC FAUNA, *TROPHIC LEVELS,
RS, STORMWATER OVE/ BACKWASHING,
AE, COLLOIDS, FLO/ *ULTRASONICS,
.:
ERICIDES, *CHLORI/ *PSEUDOMONAS,
TS/ *WATER QUALITY, *RECREATION,
ES, CATIONIC/ *POLYELECTROLYTES,
NG, MICROORG/ *ALGAE, *PROTOZOA,
NS, *ALGAE CONTROL, *LAKE ERIE,/
SIMULATION ANALYSIS, WATER POLL/
LORINATI/ *ODOR-PRODUCING ALGAE,
WATER POLLUTION SOURCES, *ALGAE,
EATMENT, *ACTINOMYCETES, *ALGAE,
EATMENT, *ALGAE, *ALGAL CONTROL,
TED C/ *WATER TREATMENT, *ALGAE,
DUOUS TREES, / *WATER TREATMENT,
ATS, ALKALINE WATER, AC/ *ALGAE,
TRATION, NU/ *THERMAL POLLUTION,
MMUNITIES, POTOMAC RIVER, SAVAN/
ESIS.: *ALGAE,
ALGAE, *GREEN ALGAE, *RED ALGAE,
L POWERPLANTS, *ECOLOGY, *BIOTA,
LLUTION, *ECOLOGY, ALGAE, BACTE/
LABORATORY ANIMALS, / *BENTHOS,
LGAE, *RADIOISOTOPES, *SORPTION,
POWERPLANTS, *THERMAL POLLUTION,
ICAL ECOLOGY, LAKES, / *GRAZING,
CT, *RESERVOIR/ *EUTROPHICATION,
N ION CONCENTRATION, *NUTRIENTS,
*EUTROPHICATION, *ALGAL CONTROL,
ATMENT, OXYGEN DEMAND, BIOCHEMI/

*SEWAGE, WATER POLLUTION EFFECTS,   W71-05500
*SEWAGE, WATER POLLUTION SOURCES,   W70-07261
*SHAGAWA LAKE(MINN), ELY(MINN), B   W70-03512
*SHAGAWA LAKE(MINN), APHANIZOMENO   W70-05387
*SIGNIFICANCE, ALGAL KEY, CLUMP C   W70-05389
*SILTS, SANDS, ALGAE, PLANKTON, M   W70-08628
*SIMULATION ANALYSIS, WATER POLLU   W69-06536
*SINKING, *PHYTOPLANKTON, *DIATOM   W70-02754
*SITE SELECTION, LOADING.:   W71-08221
*SKAWA RIVER(POLAND), MONTANE RIV   W70-02784
*SLAUGHTERHOUSE WASTES, EXPERIMEN   W70-07508
*SLIME, *ULTRAVIOLET RADIATION, *   W72-03824
*SLIME, ALGAE, FUNGI, BACTERIA, B   W72-12987
*SLUDGE, ALGAE, BIOCHEMICAL OXYGE   W70-06899
*SLUDGE, STORMS, TURBULENCE, PILO   W70-05819
*SODIUM ARSENITE, NUISANCE ALGAE,   W70-05568
*SOIL ALGAE, *AQUATIC ALGAE, ALGA   W69-03491
*SOIL MICROORGANISMS, SOIL ALGAE,   W72-08110
*SOIL, GROUNDWATER BASINS, WATER   W69-07838
*SORPTION, *TEMPERATURE, SOUTH CA   W69-07845
*SORPTION, *ALGAE, CHLOROPHYTA, C   W70-03928
*SORPTION, ABSORPTION, GULF OF ME   W69-08524
*SOURCE CONTAMINATION, CYANIDE, N   W73-00758
*SOUTHWEST(US), *DESIGN CRITERIA,   W70-04788
*SPECIES, CYCLOTELLA MENEGHINIANA   W70-05381
*SPRINKLER IRRIGATION, *ORGANIC W   W71-00940
*STABILIZATION, *PONDS, *DESIGN C   W70-04786
*STABILIZATION, *THERMAL STRATIFI   W70-04790
*STABILIZATION, *PONDS, *ENVIRONM   W70-04787
*STABILIZATION, *PONDS, *DESIGN,·   W70-04788
*STATISTICAL METHODS, *ZOOPLANKTO   W73-01446
*STRAINING, *MICROSTRAINERS.:   W73-02426
*STRAINS, PHYCOLOGISTS, VETERINAR   W70-00273
*STRATIFICATION, *TEXAS, WATER PO   W70-04727
*STREAM POLLUTION.: /TER UTILIZAT   W68-00483
*STREAMS, EUGLENA, NUTRIENTS, PLA   W68-00483
*STREAMS, *WATER POLLUTION EFFECT   W69-03369
*STREAMS, *ECOSYSTEMS, RADIOACTIV   W69-09334
*STREAMS, BIOINDICATORS, *NEW YOR   W69-10080
*STREAMS, *LABORATORY TESTS, RESP   W70-00265
*STREAMS, *ALGAE, *DIATOMS, TEMPE   W70-07257
*STREAMS, LIGHT INTENSITY, CURREN   W70-02780
*STREAMS, *WATER POLLUTION EFFECT   W72-01788
*STRONTIUM RADIOISOTOPES, *ALGAE,   W71-04067
*STRONTIUM RADIOISOTOPES, *PHYTOP   W70-00707
*STRONTIUM RADIOSOTOPES, CYTOLOGI   W69-03491
*SUBSURFACE WATERS, *AGRICULTURE,   W70-03334
*SUCESSION, *ISOLATION, DIATOMS,   W73-01446
*SULFATE ION, *DESULFOVIBRIO, FAC   W70-01971
*SULFATES, *SULFIDES, ALGAE, GASE   W70-07034
*SULFIDES, ALGAE, GASES, ODOR, OX   W70-07034
*SULFIDES, *OXIDATION LAGOONS, *D   W70-01971
*SULFIDES, ALGAE, ORGANIC LOADING   W71-11377
*SULFIDES, TEMPERATURE, CLIMATIC   W70-04790
*SULFUR BACTERIA, STABILIZATION,   W70-01971
*SURFACE, *VOLUME, *CARBON DIOXID   W70-03334
*SURFACTANTS, *SORPTION, *ALGAE,   W70-03928
*SURFACTANTS, *WATER POLLUTION, C   W70-01470
*SURFACTANTS, WATER POLLUTION SOU   W70-09438
*SURVEYS, BIOLOGICAL COMMUNITIES,   W71-12077
*SUSPENDED SOLIDS, *COMBINED SEWE   W71-10654
*SUSPENDED LOAD, *FILTRATION, ALG   W71-10654
*SUSPENDED SOLIDS, CHASE CITY(VA)   W72-10573
*SWIMMING POOLS, *BACTERIA, *BACT   W69-10171
*SWIMMING, *WATER POLLUTION EFFEC   W71-09880
*SYNTHETIC ORGANIC POLYELECTROLYT   W70-05267
*SYSTEMATICS, *SANITARY ENGINEERI   W72-01798
*SYSTEMS ANALYSIS, DATA COLLECTIO   W71-04758
*SYSTEMS ANALYSIS, *ECOSYSTEMS, *   W69-06536
*TASTE-PRODUCING ALGAE, ALGAE, CH   W70-02373
*TASTE, *ODOR, *AMINO ACIDS, POTA   W72-01812
*TASTE, *ODOR, WATER QUALITY CONT   W72-13560
*TASTE, *ODOR, *NATURAL STREAMS,   W72-03562
*TASTE, *ODOR, *ALGICIDES, ACTIVA   W72-03746
*TASTE, *ODOR, *RESERVOIRS, *DECI   W72-03475
*TEMPERATURE, *HOT SPRINGS, HABIT   W69-10160
*TEMPERATURE, HYDROGEN ION CONCEN   W70-08834
*TEMPERATURE SHOCK, *PROTOZOAN CO   W70-07847
*TEMPERATURE, DIATOMS, PHOTOSYNTH   W71-02496
*TEMPERATURE TOLERANCE, OPTIMUM G   W71-02496
*TEMPERATURE, *AQUATIC ENVIRONMEN   W71-02491
*TEMPERATURE EFFECTS, *THERMAL PO   W71-02479
*TEMPERATURE, *THERMAL POLLUTION,   W71-02494
*TEMPERATURE, SOUTH CAROLINA, CES   W69-07845
*TEMPERATURE, *COOLING WATER, CHE   W69-05023
*TEMPERATURE, *DAPHNIA, PHYSIOLOG   W70-03957
*TENNESSEE VALLEY AUTHORITY PROJE   W71-07698
*TERTIARY TREATMENT, BIOLOGICAL T   W70-08980
*TERTIARY TREATMENT, SUSPENDED LO   W70-09186
*TERTIARY TREATMENT, *AEROBIC TRE   W70-06792

US, *NUTRIENTS, *EUTROPHICATION,
PHOSPHATES, *LAKES, *DETERGENTS,
COMPOUNDS, *NUTRIENTS, *PESTICI/
, HORMIDIUM SP.:
                    *TEST DEVELOPMENT,
ERIALS, TE/ *ALGICIDES, *PAINTS,
*WATER QUALITY, *STRATIFICATION,
OLS.:
EING WASTES./ *ANTI-FOAM AGENTS,

*ISRAEL, *ALGAL CONCENTRATION, /
*COOLING WATER, / *POWERPLANTS,
HYDROGEN ION CONCENTRATION, NU/
BIOTA, *TEM/ *THERMAL POLLUTION,
RPLANTS, *ECOLOGY, *BIOTA, *TEM/
IMALS, / *BENTHOS, *TEMPERATURE,
AE, BACTE/ *TEMPERATURE EFFECTS,
FFECTS, WATER POLLUTION SOURCES,
PERATURE, PHYTOPLANKTON, ALGAE,/
RPLANTS, *WATER QUALITY CONTROL/
          *FISH PHYSIOLOGY, *ALGAE,
POLLUTION, THERMAL POWERPLANTS,
Y, SPR/ *PHOTOSYNTHESIS, *ALGAE,
N, *IMPOUNDMENTS, *RESERVOIRS, /
L, *SUL/ *PONDS, *STABILIZATION,
PHILIC BAC/ *CYANOPHYTA, *ALGAE,
  *PRIMARY PRODUCTIVITY, *ALGAE,
OLOGICAL ECOLOGY, *SALINE WATER,
URVES, CALORIMET/ *ALGAL GROWTH,
ALGAE, DIATOMS, ANIMAL BEHAVIOR,
INE EN/ *POLLUTANTS, *ESTUARIES,

CEPTIBILITY.:
RUDE OIL, *GAMOSOL, *OIL SPILLS,
TRATIONS, SURVIVAL, SUSCEPTIBIL/
EGENERATION, CLIFF VEGETATION, /
                    *HYDROCARBON,
OGY, *ALGAE, *THERMAL POLLUTION,
ISM, RADIOACTIVITY TECH/ *ALGAE,
*INSECTICIDES, *MARINE BACTERIA,
CH/ *DETERGENTS, *INSECTICIDES,
ROL, EUTROPHICATION, BASS, WATE/
ION, *CHLOROPHYLL, *OXYGENATION,
ON EFFECTS, CHELATION,/ *COPPER,
GAE, *ALGAL TOXINS, KILLIFISHES,
LIMNOLOGY, PUBLIC HEAL/ *ALGAE,
TATS, CLAMS, MUSSELS, SHELLFISH/
AQUATIC ALGAE, WATER POLLUTION,
S, ACTIVATED CARBON, FILTRATION,
IC E/ *POLLUTANT IDENTIFICATION,
TOXINS, *PHYTOTOXIC/ *FISHKILL/
*COPPER SULFATE, *ALGAL CONTROL,
YANOPHYTA, *ESSENTIAL NUTRIENTS,
ATER POLLUTION / *RADIOACTIVITY,
ENIA, *OPISTHONEMA OGLIMUN, UDO/
/ *WATER QUALITY, *GREAT LAKES,
ENGINEERING, *SEWAGE TREATMENT,
ZONE/ *BIOLOGICAL OCEANOGRAPHY,
TEMS, *PHOSPHORUS RADIOISOTOPES,
LEM, LEYDEN, THE HAGUE, MINERAL/
, / *CONFERENCES, *OCEANOGRAPHY,
ISOTOPES, *TROUT, *STREAMS, *EC/
LOGY, *EFFECTS, *COPPER SULFATE,
*SLIME, *ULTRAVIOLET RADIATION,
CULTURAL WASTES, *ALGAE, *LAKES,
COMMUNITIES, *SEWAGE TREATMENT,
, CHLORELLA PYRENOIDOSA, CARBOX/
GAE, ECOL/ *BACTERIA, *CULTURES,
QUAE, APHANIZOMENON / *CONCEPTS,
GOTROPHY, *PRIMARY PRODUCTIVITY,
, *AQUATIC ALGAE, WATER QUALITY,
ENT, WATER QUALITY, NUT/ *ALGAE,
ICAL CO/ *LAKES, *BENTHIC FAUNA,
ZATION, ALG/ *NUTRIENTS, *LAKES,
TION, *PHOSPHORUS RADIOISOTOPES,
*MADISON LAKES, YAHARA RIVER(WI/
ORUS RADIOISOTOPES, *TRANSFER, /
S(NEB), *CLEAR WATER RESERVOIRS,
OLOR, *ALGAE,/ *WATER TREATMENT,
*PHOTOSYNTHESIS, *GROWTH RATES,
ETATE, *FRESHWATER ENVIRONMENTS,
LAKES, *PHOSPHORUS RADIOISOTOPE,
IO, *CANADA, NUTRIENT LEVELS, S/
   *WASTE DISPOSAL, *AQUACULTURE,
ILTRATION, ALGAE, COLLOIDS, FLO/
ENT, *ALGAE, FILTRATION, *SLIME,
N.:
SYNURA, ANABAENA, IOWA RIVER, /
IZOMENON FLOS-AQUAE, *CARBON-14/
LAKE(ORE.), *OSCILLATORA.:

*TERTIARY TREATMENT, NITROGEN, AQ W69-09340
*TERTIARY TREATMENT, *GREAT LAKES W70-03964
*TERTIARY TREATMENT, *PHOSPHORUS   W70-05470
*TEST DEVELOPMENT, *TEST PROBLEMS W71-07339
*TEST PROBLEMS, HORMIDIUM SP.:   W71-07339
*TEST PROCEDURES, ANTIFOULING MAT W71-07339
*TEXAS, WATER POLLUTION SOURCES,   W70-04727
*TEXTILE MILL WASTES, ALFOL ALCOH W70-08740
*TEXTILE WASTES, FIBER WASTES, DY W70-04060
*THERMAL BAR.:   W69-05763
*THERMAL GRADIENT, *BOD REMOVAL,   W70-04790
*THERMAL POLLUTION, *TEMPERATURE, W69-05023
*THERMAL POLLUTION, *TEMPERATURE, W70-08834
*THERMAL POWERPLANTS, *ECOLOGY, * W71-02491
*THERMAL POLLUTION, *THERMAL POWE W71-02491
*THERMAL POLLUTION, LABORATORY AN W71-02494
*THERMAL POLLUTION, *ECOLOGY, ALG W71-02479
*THERMAL POLLUTION, THERMAL POWER W70-07847
*THERMAL POWERPLANTS, *LAKES, TEM W71-09767
*THERMAL POLLUTION, *NUCLEAR POWE W71-11881
*THERMAL POLLUTION, *TOXICITY.:   W70-01788
*THERMAL STRESS, THERMAL WATER, * W70-07847
*THERMAL SPRINGS, *LIGHT INTENSIT W69-09676
*THERMAL STRATIFICATION, *AERATIO W70-04484
*THERMAL STRATIFICATION, *SEASONA W70-04790
*THERMAL SPRINGS, *OREGON, THERMO W70-05270
*THERMAL WATER, PERIPHYTON, BIOMA W70-03309
*THERMAL WATER, ALGAE, AQUATIC AL W69-03516
*THERMOBIOLOGY, *DIURNAL OXYGEN C W70-03309
*THERMODYNAMIC BEHAVIOR.: /OZOA,   W70-07847
*TIDAL WATERS, ECOSYSTEMS, ESTUAR W68-00010
*TORONTO(ONTARIO).:   W71-05879
*TORREY CANYON, REPOPULATION, SUS W70-01780
*TORREY CANYON, AROMATIC SOLVENTS W70-01470
*TORREY CANYON, *DETERGENT CONCEN W70-01779
*TORREY CANYON, SUSCEPTIBILITY, R W70-01777
*TOXICITY THRESHOLD.:   W70-01466
*TOXICITY.:   *FISH PHYSIOL W70-01788
*TOXICITY, *INSECTICIDES, METABOL W70-02198
*TOXICITY, *WATER POLLUTION, AQUA W70-01467
*TOXICITY, *WATER POLLUTION, FOOD W70-01466
*TOXICITY, *CYANOPHYTA, *CHEMCONT W70-02982
*TOXICITY, ORGANIC COMPOUNDS, *PH W71-12183
*TOXICITY, *ALGAE, *WATER POLLUTI W71-11561
*TOXICITY, ISOLATION.: /QUATIC AL W70-08372
*TOXICITY, *CULTURES, *ISOLATION, W70-00273
*TOXICITY, *ALGAE, *REVIEWS, HABI W70-05372
*TOXICITY, BIOLOGICAL MEMBRANES,   W69-05306
*TOXICITY, *DISTRIBUTION SYSTEMS, W72-03746
*TOXICITY, *BIOASSAY, FISH, AQUAT W72-11105
*TOXINS, *ALGAL POISONING, *ALGAL W71-07731
*TOXINS, ALGAE, RESISTANCE, FISH, W68-00859
*TRACE ELEMENTS, ALGAE, EUTROPHIC W69-06277
*TRACE ELEMENTS, *AQUATIC LIFE, W W71-05531
*TRACE ELEMENTS, *ANCHOA LAMPROTA W69-08525
*TRACE ELEMENTS, *WATER POLLUTION W69-08562
*TRACERS, *WATER POLLUTION, DESIG W70-02609
*TRANSACTIONS, BLACK SEA, CONTACT W69-07440
*TRANSFER, WATER, BACTERIA, SEDIM W69-07861
*TRANSFORMATIONS, AMSTERDAM, HAAR W69-07838
*TRANSLATIONS, ALGAE, CRUSTACEANS W69-07440
*TRANSLOCATION, *PHOSPHORUS RADIO W69-09334
*TREATMENT, *LAKES, *ECOLOGY, BOT W70-06217
*TREATMENT FACILITIES, FILTERS.: / W72-03824
*TRIBUTARIES, SAMPLING, FLOW, ANA W70-02969
*TRICKLING FILTERS, *BIOTA, SEWAG W71-01819
*TRISODIUM NITRILOTRIACETATE, NTA W70-02248
*TRITIUM, EPIPHYTOLOGY, MARINE AL W69-09755
*TROPHIC BIOLOGY, ANABAENA FLOS A W70-03959
*TROPHIC LEVEL, WATER STORAGE, EV W69-09723
*TROPHIC LEVELS, NUTRIENTS, NORTH W70-02772
*TROPHIC LEVELS, AQUATIC ENVIRONM W70-02764
*TROPHIC LEVELS, *SURVEYS, BIOLOG W71-12077
*TROUT, *FISH MANAGEMENT, FERTILI W70-05399
*TROUT, *STREAMS, *ECOSYSTEMS, RA W69-09334
*TRUAX FIELD(MADISON) EFFLUENTS,   W70-05113
*TUBIFICIDS, *ECOSYSTEMS, *PHOSPH W69-07861
*TURBID RESERVOIRS, CYCLOTELLA, T W72-04761
*TURBIDITY, *IRON, *MANGANESE, *C W72-04145
*TURBULENCE, CHLORRELA, SCENEDESM W69-03730
*TURNOVER RATES, *HETEROTROPHY, O W70-05109
*TURNOVERS, NUTRIENTS, PHOSPHORUS W68-00858
*TWELVE MILE LAKE(CANADA), *ONTAR W70-02973
*ULTIMATE DISPOSAL, DOMESTIC WAST W70-09905
*ULTRASONICS, *SUSPENDED LOAD, *F W71-10654
*ULTRAVIOLET RADIATION, *TREATMEN W72-03824
*ULVA LATISSIMA, *AMMONIA NITROGE W71-05500
*UNIT PROCESSES, GASTROENTERITIS, W70-08107
*UPPER KLAMATH LAKE(ORE.), *APHAN W72-06052
*UPPER KLAMATH LAKE(ORE.), AGENCY W72-06051

371

CHLOROPHYLL, *OHIO RIVER, *ANAL/
 FILTRATION, CHL/ *WATER SUPPLY,
ERIA, WATER CHEM/ *LAKE ONTARIO,
OPPER SULFATE, / *POTABLE WATER,
PERIOR, *PERIPHYTON, *LIMNOLOGY,
NCE ALGAE, *ALGICIDES, *PHENOLS,
N DEMAND, SU/ *DATA COLLECTIONS,
TER POLLUTION EFFECTS, *REVIEWS,
LLUTION EFFECTS, *BIOINDICATORS,
ACE ELEMENTS, *WATER POLLUTION /
WASHINGTON/ *EFFLUENT DIVERSION,
ES, *FISH, *FISH KILLS, *ROTIFE/
CUT, *LAKES, *LOCAL GOVERNMENTS,
OPHICATION, NITROGEN, PHOSPHORU/
WATER ANALYSIS, *BIOASSAY, *COM/
DROGEN ION CO/ *WATER TREATMENT,
 TREATMENT, *FILTRATION, *ALGAE,
RGE, *DATA COLLECTIONS, PIT REC/
AE, *ALGICIDES, *COPPER SULFATE,
COLOR, TASTE, ODOR, TOX/ *ALGAE,
 *RESERVOIR, MASSACHUSETTS, POT/
OUNDMENTS, *RESERVOIRS, *MIXING,
*TE/ *RESERVOIRS, *URBANIZATION,
MANAGEMENT, STREAM IMPROVEMENT,
GE INPUT, BIOLOGICAL PROPERTIES,
 *RIVERS, WATER CHEMISTRY, WATE/
*DETERGENTS, *PHOSPHATES, CHELA/
N CONTROL, *RESEARCH EQUIPMENT,/
RGANIC COMPOUNDS, CHEMICAL WAST/
G WATER, *ELECTRIC POWER PLANTS,
ENT, *INDUSTRIAL WASTES, OXIDAT/
CONCENTRATION, ORGANIC LOADING,
*CORROSION, *CHEMICAL REACTIONS,
CALIFORNIA, WATER QUALITY, AEST/
COMPOU/ *WASTE WATER TREATMENT,
ENITRIFICATION, *BIODEGRADATION,
NDWATER, RECH/ *WATER TREATMENT,
 TEMPERATURE, NITRATES, AMMONIA,
GE TREATMENT, *NUTRIENT REMOVAL,
URCES, *WATER POLLUTION EFFECTS,
 *GEOCHEMISTRY, *EUTROPHICATION,
D LOSSES, OOCISTUS, FILTRATION /
GAE, ODOR, TAST/ *ALGAE CONTROL,
RWAYS, *WATER POLLUTION EFFECTS,
DOR, FILTERS, PLANKTON,/ *ALGAE,
TIES, *W/ *SANITARY ENGINEERING,
ER PURIFICATION, *ALGAL CONTROL,
ATER TREATMENT, FILTRATION, CHL/
SERVOIRS, WATER/ *ALGAE, *MAINE,
, QUALITY CONTR/ *LAKE MICHIGAN,
ACTERIA, *ESTUARINE ENVIRONMENT,
S, RIVERS, LAKES, *GROWTH RATES,
 *ALGAE, *BIOLOGICAL TREATMENT,
 *WATER SUPPLY, *WATER QUALITY,
*CORROSION, *CHEMICAL REACTION/
*ALGAE, *TASTE, *ODOR, WATER Q/
*ALGAE, *ODOR, HYDROGEN ION CO/
ERING, *FILTRATION, DESIGN, *AL/
ERING, *PATHOGENIC BACTERIA, AL/
LGAE, *WATER QUALITY, *ACTIVATE/
S, SANDS, LAKE MICHIGAN, WATER /
ST ANALYSIS, WATER PURIFICATION,
ND MAINTENANCE, DESIGN CRITERIA,
CONTROL, *TASTE, *ODOR, *NATURA/
*RESERVOIRS, *DECIDUOUS TREES, /
TION, HEAD LOSS, FILTERS, *SCEN/
*ORGANIC MATTER, ALGAE, COPPER /
NITARY ENGINEERING, *FILTERS, */
RECLAMATION, *GROUNDWATER, RECH/
ON, *MANGANESE, *COLOR, *ALGAE,/
ION, *SLIME, *ULTRAVIOLET RADIA/
*ODOR, *ALGICIDES, ACTIVATED C/
DECOMPOSING ORGANIC MATTER, *PO/
*ACTIVATED CARBON, *ADSORPTION,
MYCETES, *ODOR, ACTIVATED CARBO/
NS, HISTORY, WATER PURIFICATION,
ONSIN, *ADMINISTRATIVE AGENCIES,
DISON(WIS), *'WORKING' OF LAKES,
CATION, LAKES, STREAMS, ALGAE, /
E ALGAE, INTAKES, SANDS, FLOW R/
LIGHT, TEMPERATURE,/ *LIMNOLOGY,
ER POLLUTION,/ *WASTE TREATMENT,
 *ECOLOGY, *MATHEMATICAL MODELS,
CONTROL, ALGICIDES, BIOCONTROL,
KES(WIS), *AQUATIC WEED CUTTING,
IRA STAGNALIS, NE/ *NEW SPECIES,
S, WATER POLLUT/ *NEWFOUND LAKE,
, *COLORADO, *LAKES, *MOUNTAINS,
LYSIS, P/ *ALGAE, *PLANT GROWTH,
RODUCING ALGAE, *NUISANCE ALGAE,

*WATER QUALITY, *BIOINDICATORS, *	W71-04206
*WATER QUALITY, *WATER TREATMENT,	W71-05943
*WATER QUALITY, *MONITORING, BACT	W71-05879
*WATER QUALITY CONTROL, *ALGAE, C	W71-00110
*WATER QUALITY, BIOINDICATORS, AQ	W70-05958
*WATER QUALITY CONTROL, CRUSTACEA	W70-10161
*WATER QUALITY, BIOCHEMICAL OXYGE	W70-09557
*WATER QUALITY, WATER POLLUTION,	W70-07027
*WATER QUALITY, *DISSOLVED OXYGEN	W70-00475
*WATER QUALITY, *GREAT LAKES, *TR	W69-08562
*WATER QUALITY MANAGEMENT, *LAKE	W69-09349
*WATER QUALITY, *ALGAE, *PESTICID	W72-04506
*WATER QUALITY CONTROL, ALGAE, AQ	W72-05774
*WATER QUALITY, *NUTRIENTS, *EUTR	W73-01122
*WATER QUALITY, *BIOINDICATORS, *	W72-12240
*WATER QUALITY, *ALGAE, *ODOR, HY	W72-09694
*WATER QUALITY, *ACTIVATED CARBON	W72-09693
*WATER QUALITY, *ARTIFICIAL RECHA	W69-05894
*WATER QUALITY CONTROL, ALGAL CON	W69-05704
*WATER QUALITY, *EUTROPHICATION,	W69-05697
*WATER QUALITY CONTROL, *ECOLOGY,	W70-05264
*WATER QUALITY CONTROL, WATER QUA	W70-04484
*WATER QUALITY, *STRATIFICATION,	W70-04727
*WATER QUALITY CONTROL, *MIDGES,	W70-05428
*WATER QUALITY, TURBIDITY, PLANT	W68-00475
*WATER QUALITY, *WATER POLLUTION,	W70-03068
*WATER QUALITY, *EUTROPHICATION,	W70-04373
*WATER RESOURCES, *WATER POLLUTIO	W70-08627
*WATER REUSE, *RECLAIMED WATER, O	W70-10433
*WATER REUSE, COOLING TOWERS, DAM	W71-11393
*WATER REUSE, *WASTE WATER TREATM	W72-00027
*WATER REUSE, RESEARCH AND DEVELO	W72-03299
*WATER REUSE, DISSOLVED OXYGEN, P	W70-02294
*WATER REUSE, *RECLAIMED WATER, *	W70-05645
*WATER REUSE, *NITROGEN, NITROGEN	W69-08053
*WATER REUSE, *RECLAIMED WATER, N	W69-08054
*WATER REUSE, *RECLAMATION, *GROU	W72-04318
*WATER REUSE, WASTE WATER TREATME	W72-06017
*WATER REUSE, SEASONAL, WATER TRE	W72-08133
*WATER SOURCES, *GROUNDWATER, DDT	W73-00758
*WATER STRUCTURE, ATMOSPHERE, SEA	W70-04385
*WATER SUPPLIES, GREEN ALGAE, HEA	W70-04199
*WATER SUPPLY, *NEW HAMPSHIRE, AL	W69-08282
*WATER SUPPLY, BIOINDICATORS, CAL	W69-07833
*WATER SUPPLY, *CONTROL, TASTE, O	W70-05389
*WATER SUPPLY, *BIOLOGICAL PROPER	W72-01813
*WATER SUPPLY, *BASINS, WATER POL	W72-01814
*WATER SUPPLY, *WATER QUALITY, *W	W71-05943
*WATER SUPPLY, COPPER SULFATE, RE	W70-08096
*WATER SUPPLY, INTAKES, NUTRIENTS	W70-00263
*WATER TEMPERATURE, *SEASONAL, RH	W70-00713
*WATER TEMPERATURE, SEASONAL, TES	W68-00472
*WATER TREATMENT, *POLLUTANT IDEN	W69-07389
*WATER TREATMENT, FILTRATION, CHL	W71-05943
*WATER TREATMENT, *COOLING WATER,	W70-02294
*WATER TREATMENT, *ACTINOMYCETES,	W72-13605
*WATER TREATMENT, *WATER QUALITY,	W72-09694
*WATER TREATMENT, SANITARY ENGINE	W72-08859
*WATER TREATMENT, SANITARY ENGINE	W72-13601
*WATER TREATMENT, *FILTRATION, *A	W72-09693
*WATER TREATMENT, *ALGAE, *FILTER	W73-02426
*WATER TREATMENT.: / CLEANING, CO	W72-08688
*WATER TREATMENT, *NEW YORK, WATE	W72-08686
*WATER TREATMENT, *ALGAE, *ALGAL	W72-03562
*WATER TREATMENT, *TASTE, *ODOR,	W72-03475
*WATER TREATMENT, *ALGAE, *FILTRA	W72-03703
*WATER TREATMENT, WATER QUALITY,	W72-04495
*WATER TREATMENT, *FILTRATION, SA	W72-07334
*WATER TREATMENT, *WATER REUSE, *	W72-04318
*WATER TREATMENT, *TURBIDITY, *IR	W72-04145
*WATER TREATMENT, *ALGAE, FILTRAT	W72-03824
*WATER TREATMENT, *ALGAE, *TASTE,	W72-03746
*WATER TREATMENT, ODOR, *ALGAE, *	W72-04242
*WATER TREATMENT, *WASTE WATER TR	W72-03641
*WATER TREATMENT, *ALGAE, *ACTINO	W72-04155
*WATER UTILIZATION, DECOMPOSING O	W68-00483
*WATER WORKS, *WASTE DISPOSAL, DO	W70-03344
*WATER-BLOOM, 'BREAKING' OF LAKES	W69-08283
*WATER, *FERTILIZATION, *EUTROPHI	W69-10178
*WATER, *FILTERS, *ALGAE, NUISANC	W70-04199
*WATER, *RIVERS, *LAKES, *PONDS,	W70-05404
*WATERS, *ORGANISMS, STREAMS, WAT	W70-05750
*WATERSHED MANAGEMENT, GEOCHEMIST	W71-09012
*WATERSHED MANAGEMENT, *ENVIRONME	W68-00478
*WEED EXTRACTING STEEL CABLES, AL	W70-04494
*WESTERN LAKE ERIE(OHIO), GONGROS	W70-04468
*WINNISQUAM LAKE, WATER AND PLANT	W70-02764
*WINTER, CYANOPHYTA, PLANKTON, CO	W69-10154
*WISCONSIN, ECOLOGY, CHEMICAL ANA	W70-06229
*WISCONSIN, *LAKES, PRODUCTIVITY,	W70-05112

GENERAL INDEX

The middle rank of this index lists in alphabetical order each word of each descriptor except those included in the *Significant Descriptor Index,* which should be consulted first. The numbers at the right are the accession numbers appearing at the end of the abstract.

KE ERIE, LAKE ONTARIO, POLLUTION	ABATEMENT, OLIGOTROPHY, DEPTH, PH	W70-03964
NUTRIENTS, POLLUTANTS, POLLUTION	ABATEMENT, NUISANCE ALGAE, PHOSPH	W69-01453
RA/ *EUTROPHICATION, *POLLUTION	ABATEMENT, DRAINAGE SYSTEMS, FLOW	W68-00479
D), OYSTERS, CHANNELS, POLLUTION	ABATEMENT, HUMAN POPULATION.: /IL	W68-00461
ON SOURCES, DREDGING, *POLLUTION	ABATEMENT, DIKES, SOIL CONSE: /TI	W68-00172
S, CONTROL, POLLUTION ABATEMENT,	ABATEMENT, PACIFIC COAST REGION,	W69-03683
LGAE, PLANTS, CONTROL, POLLUTION	ABATEMENT, ABATEMENT, PACIFIC COA	W69-03683
UTANT IDENTIFICATION, *POLLUTANT	ABATEMENT, EFFLUENTS, STREAMFLOW,	W69-07389
ODOR-PRODUCING ALGAE, POLLUTION	ABATEMENT, SEWAGE TREATMENT, SEWA	W69-06909
UTION IDENTIFICATION, *POLLUTION	ABATEMENT, TIDAL WATERS, MIXING,	W69-06203
ER POLLUTION CONTROL, *POLLUTION	ABATEMENT, POLLUTANTS, WASTES, TH	W69-06305
OSAL, WATER POLLUTION, POLLUTION	ABATEMENT, COLIFORMS, WATER QUALI	W70-04810
YTOPLANKTON, PLANKTON, POLLUTION	ABATEMENT, POLLUTANT IDENTIFICATI	W70-02792
OL, FISH CONSERVATION, POLLUTION	ABATEMENT, BIBLIOGRAPHIES, ECOLOG	W72-01786
SOTA, *EUTROPHICATION, POLLUTION	ABATEMENT, GROWTH RATES, CHLOROPH	W69-10167
AMPHORA OVALIS, OPEPHORA MARTYI,	ABBOT'S POND(SOMERSET, ENGLAND),	W71-12870
L COUNTS, SPECIES DETERMINATION,	ABSORBANCE, COENOBIA, SWIRLING, V	W70-03334
ABILIZATION, PONDS, PENETRATION,	ABSORPTION.: /LIGHT INTENSITY, ST	W70-03923
NIQUE, OXYGEN ELECTRODES, ATOMIC	ABSORPTION SPECTROPHOTOMETRY, TOT	W70-03309
ROG/ *REQUIREMENTS, SCENEDESMUS,	ABSORPTION, CELL CONCENTRATION, U	W70-05409
TREATMENT, LAGOONS, OXYGENATION,	ABSORPTION, ECOLOGY, WATER PURIFI	W69-07389
NGE, WATER CIRCULATION, NUTRIENT	ABSORPTION, *CHEMICAL PROPERTIES,	W68-00857
OTTOM SEDIMENTS, BIBLIOGRAPHIES,	ABSORPTION, FISHES, ALGAE.: /S, B	W68-00858
LABORATORY TESTS, PRODUCTIVITY,	ABSORPTION, SEDIMENTS, ORGANIC MA	W72-00845
IXING, DIFFUSION, SEDIMENTATION,	ABSORPTION, BIOCHEMICAL OXYGEN DE	W71-11381
*MARINE ANIMALS, *RADIOISOTOPES,	ABSORPTION, MARINE ALGAE, MARINE	W71-09850
BACTERIA, BIOASSAY, METABOLISM,	ABSORPTION, COMPETITION, OLIGOTRO	W71-10079
T RELATIONSHIPS, NUCLEAR WASTES,	ABSORPTION, CHELATION, METABOLISM	W71-09005
N ADSORPTION, COLLOIDS, AQUARIA,	ABSORPTION, WATER POLLUTION EFFEC	W71-09863
ADIOISOTOPES, *ALGAE, *SORPTION,	ABSORPTION, GULF OF MEXICO, ATLAN	W69-08524
AGE, CESIUM, NUCLEAR EXPLOSIONS,	ABSORPTION, ADSORPTION, ATLANTIC	W70-00707
GINOSA, NOSTOC, NOSTOC MUSCORUM,	ABSORPTION, SPECTROMETRY.: / AERU	W70-02245
E WASTE DISPOSAL, OCEANS, FOODS,	ABSORPTION, HAZARDS, CESIUM, HYDR	W70-02786
*PHOSPHATE	ABSORPTION.:	W70-01031
ALITY CONTROL, SEWAGE EFFLUENTS,	ABSORPTION, LAKES.: /NT, WATER QU	W70-01031
MAINTENANCE, *OXIDATION LAGOONS,	ABSORPTION, BIODEGRADATION, CHEMI	W70-08392
ASSAY, RADIOACTIVITY TECHNIQUES,	ABSORPTION.: / MODEL STUDIES, BIO	W73-01454
, EXPORT LOSSES, CHEMOSYNTHESIS,	ABUNDANCE.: /TECOLOGY, MORPHOLOGY	W70-02780
N / *LEWIS AND CLARK LAKE (S D),	ABUNDANCE, BLUE-GREEN ALGAE, GREE	W70-05375
ITY OF WISCONSIN, ASIA/ NATIONAL	ACADEMY OF SCIENCES(USA), UNIVERS	W70-03975
ITY OF WISCONSIN, ASIA/ NATIONAL	ACADEMY OF SCIENCES(USA), UNIVERS	W70-03975
FUNGIA FUNGITES, ZOOXANTHELLAE,	ACANTHURUS TRIOSTEGUS, CONUS EBRA	W69-08526
	ACCELERATED EUTROPHICATION.:	W68-01244
L POLLUTION, LABORATORY ANIMALS,	ACCLIMATIZATION, MARINE ANIMALS,	W71-02494
, *NITROGEN LIMITATION, BA/ *FAT	ACCUMULATION, *LABORATORY STUDIES	W70-07957
DIONUCLIDE UPTAKE, *RADIONUCLIDE	ACCUMULATION, *RADIONUCLIDE EXCHA	W70-00707
EAE, CHLAMYDOMONAS MOEWUSSI, FAT	ACCUMULATION.: /A PALEA, MYXOPHYC	W70-02965
PESTICIDE METABOLISM, *PESTICIDE	ACCUMULATION.: /LYOPHILIZATION, *	W69-02782
FICIAL SUBSTRATES, *RADIONUCLIDE	ACCUMULATION, RUTHENIUM-106, WHIT	W70-04371
COLOGY, FLUORESCENCE MICROSCOPY,	ACETATE, CARBON-14, AUTORADIOGRAP	W70-04811
COSE, MECHAELIS-MENTEN EQUATION,	ACETATE, GLYCOLLIC ACID.: /S, GLU	W70-04284
FLAGELLATES, LAKE ERKEN(SWEDEN),	ACETATE, LAKE NORRVIKEN(SWEDEN),	W70-04194
CHAMYDOMONADS, LAPPLAND(SWEDEN),	ACETATE, TURNOVER TIMES: /EASES,	W70-04287
E DEGRADATION, SEVIN, MALATHION,	ACETONE, ETHANOL.: /OSA, PESTICID	W70-02968
URE, CHEMOSTATS, BATCH CULTURES,	ACETYLENE REDUCTION TECHNIQUE, HE	W70-02775
ATP, CMU, NITROGEN-15, PYRUVATE,	ACETYLENE REDUCTION, REDUCTANT.: /	W70-04249
*BIOCHEMISTRY METHODS, SEASONS,	ACETYLENE REDUCTION TEST.: /HODS,	W70-03429
NOIDS, SPECIES, DIVERSITY INDEX,	ACHNANTHES, ACTINOPTYCHUS, AMPHIP	W70-03325
ATION, TAXONOMY, FLAVOBACTERIUM,	ACHROMOBACTER, GULF OF MARSEILLES	W70-04280
RICHIA, SARCINA, STAPHYLOCOCCUS,	ACHROMOBACTER(ALKALIGENES), RHODO	W70-03952
VOBACTERIUM-CYTOPHAGA', VIBRIOS,	ACHROMOBACTERS, GRAM-POSITIVE BAC	W70-00713
M, MELOSIRA ISLANDICA, NITZSCHIA	ACICULARIS, NITZSCHIA DISSIPATA,	W69-10158
ICIPAL WASTES, METHANE BACTERIA,	ACID BACTERIA, SULFUR BACTERIA, A	W72-06854
DRICA, GLYCOLYSIS, TRICARBOXYLIC	ACID CYCLE, AUTORADIOGRAPHY, MONO	W70-02965
LAX POLYEDRA, 3,5-DIAMINOBENZOIC	ACID DIHYDROCHLORIDE, INFRARED GA	W69-08278
PHORUS/ *EUTROPHICATION, *LAKES,	ACID LAKES, BIBLIOGRAPHIES, *PHOS	W68-00857
INHIBITION, LIMNOLOGY, ACIDITY,	ACID SOILS, ACIDIC WATER, ACIDS,	W70-02792
LUTION, WATER POLLUTION EFFECTS,	ACID STREAMS, MINE DRAINAGE, CHLO	W70-02770
GEOGRAPHICAL REGIONS, KENTUCKY,	ACID STREAMS, MICROFLORA.: /HIPS,	W68-00891

377

CULES, NUCLEIC ACID, RIBONUCLEIC
OUNTY, KENTUCKY.:
TY CRITERIA, CYANIDE, NAPHTHENIC
TEN EQUATION, ACETATE, GLYCOLLIC
MIYA'S MINERAL MEDIUM, GLYOXYLIC
DE-COATED SAND, CRUSHED BAUXITE,
*NITRIC
PRODUCTS, PHOTIC ZONE, GLYCOLLIC
TS, ASCOPHYLLUM NODOSUM, ALGINIC
SUBSTRATE COMPLEXES, RIBONUCLEIC
LATION, *MACROMOLECULES, NUCLEIC
AND HOT SPRINGS, SINTER, NUCLEIC
RINGS, HABITATS, ALKALINE WATER,
LIMNOLOGY, ACIDITY, ACID SOILS,
TRATION, WATE/ *ACID MINE WATER,
LLUTANTS, INHIBITION, LIMNOLOGY,
NDS, HYDROGEN ION CONCENTRATION,
LANTS, AMMONIA, HARDNESS(WATER),
OLLUSKS, MINE ACIDS, COAL MINES,
RBOXYLIC ACIDS, AMINO CARBOXYLIC
DETERGENTS, RADIOACTIVITY, AMINO
*SEAWATER, *FRESH WATER, *HUMIC
HLORELLA PYRENOIDOSA, CARBOXYLIC
, / *ALGAE, *CHLORELLA, *ORGANIC
TRACE ELEMENTS, VITAMINS, AMINO
HORUS RADIOISOTOPES, IONS, HUMIC
ERVOIRS, NITROGEN, GROWTH, AMINO
, DIATOMS, CORAL, MOLLUSKS, MINE
PSEUDOMONAS, TEMPERATURE, AMINO
SA, VOLATILE ACIDS, NON-VOLATILE
ROTEINS, WALLS, STRUCTURE, AMINO
CHLORELLA PYRENOIDOSA, VOLATILE
LGAE, CONTINUOUS CULTURES, FATTY
IDITY, ACID SOILS, ACIDIC WATER,
S, *ALGAE, *TASTE, *ODOR, *AMINO
TLAND), SCOTLAND, PHAEOPIGMENTS,
LIFE, PTYCHOCHEILUS OREGONENSIS,
HOLE OCEANOGRAPHIC INSTITUTION,
TTLE(WASH), REVENUE BONDS, METRO
RADIOACTIVE WASTE, WATER QUALITY
FUNGI, VIRUSES, ALGAE, LICHENS,
ZOMENON, OSCILLATORIA, MELOSIRA,
Y(KY), THORIUM-232, BISMUTH-214,
SMUS, EUGLENA, BACTERIA, YEASTS,
TIONS, E. COLI, BACTERIA, FUNGI,
OR, *AMINO ACIDS, POTABLE WATER,
BACTERIA, ULTRAVIOLET RADIATION,
ALGAE, EUTROPHICATION, ALCOHOLS,
UM COMPOUNDS, *ACTIVATED CARBON,
ES, DIVERSITY INDEX, ACHNANTHES,
CHARIS CANADENSIS/ *ANTAGONISTIC
L CONTROL, DISINFECTION, MODE OF
Y, BIOASSAY, INHIBITION, MODE OF
IPAL WASTES, SEWAGE, ADSORPTION,
, ELECTRODIALYSIS, ION EXCHANGE,
EMICAL OXYGEN DEMAND, *BACTERIA,
ENT, SOLID WASTES, MICROBIOLOGY,
IARY TREATMENT, ALGAE, BACTERIA,
N, SORPTION, FLOCCULATION, COAL,
ONTROL, ECOLOGICAL DISTRIBUTION,
TION, BIOCHEMICAL OXYGEN DEMAND,
SEDIMENTATION, ORGANIC LOADING,
ANAEROBIC CONDITIONS, CANNERIES,
ION DITCH.:   *FWPCA,
, CHEMICAL OXYGEN DEMAND, COLOR,
AE, *CATTLE, SLURRIES, SPRAYING,
T, SUSPENDED LOAD, PILOT PLANTS,
TIC MICROORGANISMS, TEMPERATURE,
S, COALS, DISTRIBUTION PATTERNS,
ATION, ODOR, AMMONIUM COMPOUNDS,
RS, *DESIGN, COAGULATION, ALGAE,
LGAE, *TASTE, *ODOR, *ALGICIDES,
S, CHLORINATION, COPPER SULFATE,
, *ALGAE, *ACTINOMYCETES, *ODOR,
N, *FLOW AUGMENTATION, *LAGOONS,
, COPPER SULFATE, SEDIMENTATION,
ROWTH, OXYGEN DEPLETION, NUCLEAR
COLOGY, MAXIMUM UPTAKE VELOCITY,
OGY, SWEDEN, LAKE ERKEN(SWEDEN),
ENIUM-106, WHITE OAK LAKE(TENN),
*ALGAL
BOUND RADIOSTRONTIUM, *MICROBIAL
TITUTED IMIDAZOLINES, BIOLOGICAL
BIOLOGICAL HA/ *UPTAKE, SPECIFIC
IS, NITZSCHIA DISSIPATA, SYNEDRA
YPTOPHYTA, PYROPHYTA, CHROOMONAS
OBSERVATIONS, CHRONIC EXPOSURE,
G CROP, CHLOROPHYLL, PHOSPHATES,
LETONEMA COSTATUM, PHYSIOLOGICAL
LUDGE, OXIDATION LAGOONS, ALGAE,

ACID SYNTHESIS.: /ION, *MACROMOLE   W69-10151
ACID WASTE MICROFLORA, MCCREARY C   W69-00096
ACID.: /BIAL ECOLOGY, WATER QUALI   W70-04510
ACID.: /S, GLUCOSE, MECHAELIS-MEN   W70-04284
ACID.: /TILE ACIDS, KETOACIDS, TA   W70-04195
ACID-WASHED BAUXITE, KANPUR(INDIA   W70-08628
ACID, *LANGELIER INDEX.:   W71-11393
ACID, CHEMOTROPHY, PARTICULATE MA   W70-02504
ACID, FUCUS VESICULOSUS, GROWTH F   W69-07826
ACID, ORCINOL METHOD.: /, ENZYME-   W69-10160
ACID, RIBONUCLEIC ACID SYNTHESIS.   W69-10151
ACID, SISJOTHANDI(ICELAND), ENZYM   W69-10160
ACIDIC WATER, TRAVERTINE, CALCIUM   W69-10160
ACIDIC WATER, ACIDS, PHOTOSYNTHES   W70-02792
ACIDIC WATER, HYDROGEN ION CONCEN   W70-05424
ACIDITY, ACID SOILS, ACIDIC WATER   W70-02792
ACIDITY, MUD, HYDROGEN SULFIDE.: /   W70-05399
ACIDITY, OXYGEN, ALGAE, SELF-PURI   W69-07838
ACIDITY, PYRITE, SULFUR COMPOUNDS   W69-07428
ACIDS.: /HLORELLA PYRENOIDOSA, CA   W70-02248
ACIDS.: /ISTRIBUTION, CHELATION,   W69-08275
ACIDS, *COLOR, RIVERS, BOGS, PHEN   W70-01074
ACIDS, AMINO CARBOXYLIC ACIDS.: /   W70-02248
ACIDS, BIOMASS, ORGANIC COMPOUNDS   W70-04195
ACIDS, CALCIUM, AMMONIA, NITROGEN   W70-02255
ACIDS, CARP, ICE, IRON COMPOUNDS,   W70-05399
ACIDS, CHLAMYDOMONAS, CHLORELLA,   W69-10180
ACIDS, COAL MINES, ACIDITY, PYRIT   W69-07428
ACIDS, INTERTIDAL AREAS, CARBOHYD   W70-03952
ACIDS, KETOACIDS, TAMIYA'S MINERA   W70-04195
ACIDS, MECHANICAL EQUIPMENT, ENZY   W70-04184
ACIDS, NON-VOLATILE ACIDS, KETOAC   W70-04195
ACIDS, OOCYSTIS POLYMORPHA, UNICE   W70-07957
ACIDS, PHOTOSYNTHESIS, PHOTOSYNTH   W70-02792
ACIDS, POTABLE WATER, ACTINOMYCET   W72-01812
ACRIDINE ORANGE, LINEWEAVER-BURKE   W70-04811
ACROCHEILUS ALUTACEUS, ULOTHRIX,   W69-07853
ACROPORA HYACINTHUS, POCILLOPORA   W69-08526
ACT(SEATTLE).: /INGTON(WASH), SEA   W70-04455
ACT, PATH OF POLLUTANT, FOOD CHAI   W72-00941
ACTIMOMYCETES, CYTOLOGICAL STUDIE   W71-11823
ACTINASTRUM, COELASTRUM, PEDIASTR   W70-08107
ACTINIUM-228, SWIFT RIFFLES, ZIRC   W69-09742
ACTINOMYCETES, PONDING, WASTE TRE   W70-08097
ACTINOMYCETES, PROTOZOA, ALGAE.: /   W71-02009
ACTINOMYCETES, PLANT GROWTH, DECO   W72-01812
ACTINOMYCETES, HYDROGEN ION CONCE   W70-00719
ACTINOMYCETES, CYANOPHYTA, SPECTR   W69-00387
ACTINOMYCETES, IOWA.: /, *POTASSI   W72-04242
ACTINOPTYCHUS, AMPHIPLEURA RUTILA   W70-03325
ACTION, *AQUATIC MACROPHYTES, ANA   W70-02249
ACTION, MORTALITY, INHIBITION, AL   W70-02370
ACTION, PHOTOSYNTHESIS.: /ORTALIT   W70-03519
ACTIVATED CARBON, FILTERS, SEASON   W70-04001
ACTIVATED SLUDGE, DENITRIFICATION   W69-01453
ACTIVATED SLUDGE, IRON COMPOUNDS,   W68-00256
ACTIVATED SLUDGE, RATES, CHLORELL   W68-00248
ACTIVATED SLUDGE.: / DEMAND, TERT   W69-08054
ACTIVATED CARBON.: /AE, FILTRATIO   W70-05470
ACTIVATED SLUDGE, MAINTENANCE COS   W70-04786
ACTIVATED SLUDGE, COST COMPARISON   W71-09548
ACTIVATED SLUDGE, AERATION, DISSO   W71-08970
ACTIVATED SLUDGE, BIOCHEMICAL OXY   W71-09546
ACTIVATED ALGAE, FEEDLOTS, OXIDAT   W71-06821
ACTIVATED SLUDGE, ALGAE, WASTE WA   W70-10433
ACTIVATED SLUDGE, FARM WASTES.: /   W70-07491
ACTIVATED SLUDGE, TREAT: /REATMEN   W70-09186
ACTIVATED SLUDGE, FISHERIES, CARP   W70-09905
ACTIVATED CARBON, CLAYS, ALGAE, W   W69-09884
ACTIVATED CARBON, CHLORINE, PHENO   W70-00263
ACTIVATED CARBON, *FLOCCULATION,   W72-07334
ACTIVATED CARBON, FILTRATION, *TO   W72-03746
ACTIVATED CARBON, WATER SUPPLY.: /   W72-03562
ACTIVATED CARBON, AMMONIA, CHLORI   W72-04155
ACTIVATED SLUDGE, ALGAE, NUTRIENT   W72-06017
ACTIVATED CARBON, CHLORINE, COSTS   W72-08859
ACTIVATION TECHNIQUES, ROCK LAKE(W70-00264
ACTIVE TRANSPORT, BACTERIAL POPUL   W70-04194
ACTIVE TRANSPORT, CARBON-14, ENZY   W70-04284
ACTIVITY-DENSITY, VERTICAL, MICRO   W70-04371
ACTIVITY, *NUTRIENT REMOVAL.:   W70-08980
ACTIVITY, *RADIOSTRONTIUM COMPLEX   W71-04067
ACTIVITY, QUARTERNARY, AMMONIUM C   W70-02982
ACTIVITY, CHLORELLA PYRENOIDOSA,   W70-02786
ACUS, SINKING RATES, GRAZING PRES   W69-10158
ACUTA, SCHROEDERIA SETIGERA, MELO   W70-05387
ACUTE EXPOSURE.: /, PHYSIOLOGICAL   W70-01779
ADD: /CIUM, PRODUCTIVITY, STANDIN   W70-03512
ADJUSTMENT, ASTERIONELLA FORMOSA,   W70-04809
ADMINISTRATION, MAINTENANCE, *WAS   W71-08970

ION, SUPERVISORY CONTROL(POWER), Y CONTROL, ALGAE, AQUATIC WEEDS, EFFECTS, CATION EXCHANGE, CATION HYSICOCHEMICAL PROPERTIES, ANION QUARIA, WATER POLLUTION EFFECTS, OMPOUNDS, *PHOSPHORUS COMPOUNDS, IUM, PHYSICOCHEMICAL PROPERTIES, NUCLEAR EXPLOSIONS, ABSORPTION, CE, EVAPORATION, BIODEGRADATION, HORESIS, CHLORELLA, SCENEDESMUS, ASTES, MUNICIPAL WASTES, SEWAGE, LGAE, PERIPHYTON, INVERTEBRATES, MUD-WATER INTERFACE, PHOSPHORUS, RYING, SOLIDS-SETTLING, ECONOMIC NSEE(SWITZ), PAFFIKERSEE(SWITZ), TRATION FACTOR, LIMNODRILUS SPP, CHLORELLA PYRENOIDOSA, TRIBONEMA AND(N C), DETE/ *CONCENTRATIONS, *MANURE, ANAEROBIC LAGOON, ISMS, PHOSPHATES, HYDROGENATION, E, CLAYS, ALDRIN, DIELDRIN, DDT, TION, BIOCHEMICAL OXYGEN DEMAND, GANIC LOADING, ACTIVATED SLUDGE, AERATION, NUTRIENTS, OXYGEN SAG, TREATMENT, TREATMENT FACILITIES, BIOCHEMICAL OXYGEN DEMAND(BOD), LINITY, TEMPERATURE, MONITORING, RIO).:
R TEMPERATURE, SEWAGE TREATMENT, UD, BENTHOS, HYDROGEN, BACTERIA, ORGANIC MATTER, AIR TEMPERATURE, CRITERIA, FACILITIES, EQUATIONS, E, *LIGHT, *NITRATES, *NITRITES, R POLLUTION, SEWAGE, AESTHETICS, MPERATURE, FLOW, ALGAE, BENTHOS, DE, PHOTOSYNTHESIS, GROWTH RATE, RIA, ODOR, ANAEROBIC CONDITIONS, S.:   *POTATO PROCESSING WASTES, , *ORGANIC WASTES, *OXYGENATION, TERIA, EFFLUENTS, OXYGEN DEMAND, GEN DEMAND, ANAEROBIC DIGESTION, NITROGEN, BACTERIOLOGICAL STUDY, S, PROTEUS VULGARIS, PSEUDOMONAS *MICROCYSTIS
PSEUDOMONAS
ISTS, VETERINARIANS, MICROCYSTIS ZOMENON FLOS AQUAE, *MICROCYSTIS EAE, NOSTOCHINEAE, CLATHROCYSTIS TION, CULTURE MEDIA, MICROCYSTIS SULFATE, EFFLUENTS, MICROCYSTIS HLORONAPHTHOQUINONE, MICROCYSTIS STIGEOCLONIUM TENUE, MICROCYSTIS LORELLA PYRENOIDOSA, MICROCYSTIS ANABAENA CYLINDRICA, MICROCYSTIS CHU NO 10 SOLUTION/ *MICROCYSTIS ANABAENA SPIROIDES, MICROCYSTIS ANABAENA VARIABILIS, MICROCYSTIS OGICALLY AVAILABLE), MICROCYSTIS EDOGONIA, SPIROGYRA, MICROCYSTIS MS, ALGAE NUTRITION, MICROCYSTIS CHLORIS PENIOCYSTIS, DIPLOCYSTIS ANABAENA FLOS-AQUAE, MICROCYSTIS EMICAL COMPOSITION, *MICROCYSTIS ITY, *ALGAL GROWTH, *MICROCYSTIS E, PHOTOMICROGRAPHS, MICROCYSTIS TER, *CALIFORNIA, WATER QUALITY, LAKES, *WATER POLLUTION, SEWAGE, IGAN, *LAKE ONTARIO, LAKE HURON, EGETAT/ *HARTBEESPOORT DAM(SOUTH IUM BREVE, CANADA, EUROPE, SOUTH , INDUSTRIAL PLANTS, STRUCTURES, DS, MACROMOLECULES, ANTIBIOTICS, ERLAND, PERIDINIUM, OSCILLATORIA M, LAMINARIA DIGITATA, LAMINARIA S CASEI, BACILLUS THURINGIENSIS, SILVER NITRATE, SIMAZIN, NUPHAR, A, CHRYSOPHYTA, GRAZING, SNAILS, Y CONTROL(POWER), ADMINISTRATIVE SPO/ *WISCONSIN, *ADMINISTRATIVE *UPPER KLAMATH LAKE(ORE.), STES, DYEING WASTES./ *ANTI-FOAM OTOCARPUS, POLYSIPHONIA, ORGANIC OMS, WATER BLOOM, AQUATIC ALGAE, TROPHY, *NUTRIENT BALANCE, *LAKE AQUATIC NUISANCES, TRANSPARENCY, NUTRIENTS, NITROGEN, PHOSPHORUS, OPHYLL, NUTRIENTS, CONDUCTIVITY, WISCONSIN, MICHIGAN, PHOSPHORUS, TION ASSESSMENT, URBAN DRAINAGE, STES, SEWAGE, INDUSTRIAL WASTES,

ADMINISTRATIVE AGENCIES.: /ISTRAT   W72-05774
ADMINISTRATION; SUPERVISORY CONTR   W72-05774
ADSORPTION, HYDROLYSIS, MARINE AL   W71-09863
ADSORPTION, COLLOIDS, AQUARIA, AB   W71-09863
ADSORPTION.: /TION, METABOLISM, A   W71-09005
ADSORPTION, REMOVAL, DEGRADATION,   W71-13335
ADSORPTION, COLLOIDS, ZINC RADIOI   W70-00708
ADSORPTION, ATLANTIC OCEAN, DIATO   W70-00707
ADSORPTION, ALGAE, CLAYS, ALDRIN,   W70-09768
ADSORPTION, SEWAGE TREATMENT, PAR   W71-03035
ADSORPTION, ACTIVATED CARBON, FIL   W70-04001
ADSORPTION, PHYTOPLANKTON, SUCKER   W69-07853
ADSORPTION, EUTROPHICATION, ALGAE   W69-06859
ADVANTAGE, TEMPERATURE-EFFECT.: /   W70-05819
AEGERISEE(SWITZ), FURES: / GREIFE   W70-00270
AEOLOSOMA HEMPRICHI, ESCHERICHIA   W69-07861
AEQUALE, NAVICULA PELLICULOSA, LA   W69-10180
AEQUIPECTEN IRRADIANS, PIVERS ISL   W69-08267
AERATED LAGOON.:   W70-07491
AERATION, FLOCCULATION, FILTRATIO   W70-08627
AERATION, SILTS.: /SORPTION, ALGA   W70-09768
AERATION, TEMPERATURE, FLOW, ALGA   W71-11381
AERATION, DISSOLVED OXYGEN, SLUDG   W71-08970
AERATION, BIOCHEMICAL OXYGEN DEMA   W69-07520
AERATION, LAGOONS, DOMESTIC WASTE   W70-04060
AERATION, COLIFORMS, ALGAL CONTRO   W70-04788
AERIAL PHOTOGRAPHY, INFRARED RADI   W68-00010
AERO-HYDRAULIC GUN, ALLISTON(ONTA   W71-09546
AEROBIC BACTERIA, AQUATIC PLANTS,   W71-08221
AEROBIC BACTERIA, METABOLISM, CAN   W69-07838
AEROBIC BACTERIA, WATER TEMPERATU   W69-07096
AEROBIC CONDITIONS, ANAEROBIC CON   W70-04786
AEROBIC CONDITIONS, ANAEROBIC CON   W70-05261
AEROBIC CONDITIONS, ANAEROBIC CON   W70-04430
AEROBIC CONDITIONS, ANAEROBIC CON   W71-11381
AEROBIC CONDITIONS, ANAEROBIC CON   W70-05750
AEROBIC CONDITIONS, E. COLI, BACT   W71-02009
AEROBIC LAGOONS, ANAEROBIC LAGOON   W71-08970
AEROBIC TREATMENT, ANAEROBIC COND   W71-09546
AEROBIC TREATMENT, PHOTOSYNTHESIS   W72-00462
AEROBIC TREATMENT, CHEMICAL OXYGE   W71-00936
AEROBICITY.:   ORGANIC   W71-05742
AERUGINOSA, STAPHYLOCOCCUS AUREUS   W71-11823
AERUGINOSA, *LAKE WATERS.:   W70-08637
AERUGINOSA.:   W69-10171
AERUGINOSA, ENDOTOXIN, FAST-DEATH   W70-00273
AERUGINOSA, BOHEMIA, ELBE RIVER,   W70-00274
AERUGINOSA, COELOSPHAERIUM KUTZIN   W69-08283
AERUGINOSA, APHANIZOMENON FLOS-AQ   W70-00719
AERUGINOSA.: /ORUS, ALGAE, COPPER   W70-02787
AERUGINOSA, APHANIZOMENON FLOS-AQ   W70-02982
AERUGINOSA, NOSTOC, NOSTOC MUSCOR   W70-02245
AERUGINOSA, BURNTSIDE RIVER(MINN)   W70-03512
AERUGINOSA, OEDOGNIUM.: /RATION,   W70-03520
AERUGINOSA(KUTZ), CULTURE MEDIA,   W70-03507
AERUGINOSA, CRYPTOMONADIDA, MASTI   W70-05112
AERUGINOSA, APHANIZOMENON FLOS-AQ   W70-05091
AERUGINOSA, LAKE COMO(MINN), LAKE   W70-05269
AERUGINOSA, ANABAENOPSIS, RACIBOR   W70-05404
AERUGINOSA.:   ALGAL BLOO   W70-05565
AERUGINOSA, NOSTOC MUSCORUM, VANA   W69-06277
AERUGINOSA, ALGAL CULTURES, CHEMO   W69-06864
AERUGINOSA, *NUTRIENT AVAILABILIT   W69-05868
AERUGINOSA, ANTAGONISMS, ALGAL NU   W69-05867
AERUGINOSA, ANABAENA CIRCINALIS,   W69-05704
AESTHETICS, RECREATION, FISHING,   W70-05645
AESTHETICS, AEROBIC CONDITIONS, A   W70-04430
AESTHETICS, ECOLOGY, ECONOMICS, O   W70-01943
AFRICA), TRANSPARENCY, MARGINAL V   W70-06975
AFRICA, ISRAEL, MICROCYSTIS, PRYM   W70-05372
AFTERBAYS, WATER TYPES, INTAKES.:   W69-05023
AGARBACTERIUM, XANT: /ARY COMPOUN   W70-04280
AGARDHI, PHORMIDIUM, APHANIZOMENO   W69-10169
AGARDHII, DESICCATION, IMMERSION,   W70-01073
AGARS, STAPHYLOCOCCUS EPIDERMIDIS   W71-11823
AGASICLES, MANATEE, MARISA.: /C,   W70-10175
AGE, TURBIDITY, DIPTERA, MAYFL: /   W71-00668
AGENCIES.: /ISTRATION, SUPERVISOR   W72-05774
AGENCIES, *WATER WORKS, *WASTE DI   W70-03344
AGENCY LAKE(ORE.), *OSCILLATORA.:   W72-06051
AGENTS, *TEXTILE WASTES, FIBER WA   W70-04060
AGGREGATES, SEASONAL EFFECTS.: /C   W70-01068
AGING(BROLOGICAL), *AQUATIC PRODU   W68-00912
AGING, NUTRIENT SOURCES, WATER SA   W70-00264
AGRICULTURAL RUNOFF, NUTRIENT BUD   W69-09349
AGRICULTURE, WATER MANAGEMENT(APP   W71-03048
AGRICULTURAL WATERSHEDS, SEWAGE,   W70-05752
AGRICULTURAL WATERSHEDS, ORGANIC   W70-04193
AGRICULTURAL DRAINAGE, NORTH AMER   W70-03975
AGRICULTURE, ST LAWRENCE RIVER, A   W70-03964

AL RECHARGE, COPPER, *ALGICIDES,   ALGAE, *EVALUATION, FILTRATION.: /   W72-05143
*TEMPERATURE TOLERA/ *BLUE-GREEN   ALGAE, *GREEN ALGAE, *RED ALGAE,   W71-02496
  *EUTROPHICATION, *PHOSPHOROUS,   ALGAE, *HARVESTING OF ALGAE, EFFL   W71-13313
ESMIDS/ *EUTROPHICATION, AQUATIC   ALGAE, *LAKES, *CLASSIFICATION, D   W68-00255
, EXUDATION, T/ *OCEANS, *MARINE   ALGAE, *LITTORAL, *ORGANIC MATTER   W70-01073
EMULSIFIERS, *MORTALITY, *MARINE   ALGAE, *MARSH PLANTS, *MARINE ANI   W70-01777
OASSAY, EUTROPHICATION, OLIGOTR/   ALGAE, *NUTRIENT REQUIREMENTS, BI   W72-03183
YTA, *EUTROPHICATION, *PLANKTON,   ALGAE, *NUTRIENT REQUIREMENTS.: /   W69-03512
TROPHICATION, *LAKES, WISCONSIN,   ALGAE, *NUISANCE ALGAE, SPRAYING,   W68-00488
N, *LAKES, PROD/ *ODOR-PRODUCING   ALGAE, *NUISANCE ALGAE, *WISCONSI   W70-05112
PRIMARY PRODUCTIVITY, *PLANKTON,   ALGAE, *PHOTOSYNTHESIS, PRODUCTIV   W69-00632
OPHICATION, AQUATIC MI/ *AQUATIC   ALGAE, *PROTOZOA, *FLORIDA, *EUTR   W72-01993
AUNA, CHEMICAL ANALYSIS, SESSILE   ALGAE, *PRIMARY PRODUCTIVITY, ORE   W72-06052
OLERA/ *BLUE-GREEN ALGAE, *GREEN   ALGAE, *RED ALGAE, *TEMPERATURE T   W71-02496
OC/ *ZINC RADIOISOTOPES, *MARINE   ALGAE, *RESINS, SEA WATER, PHYSIC   W69-08275
HICATION, *CLADOPHORA, *NUISANCE   ALGAE, *SEWAGE, ALGICIDES, AQUATI   W69-04798
OURC/ *EUTROPHICATION, *NUISANCE   ALGAE, *SEWAGE, WATER POLLUTION S   W70-07261
HUMIC ACIDS, */ *OCEANS, *MARINE   ALGAE, *SEAWATER, *FRESH WATER, *   W70-01074
GAE, CHLORINATI/ *ODOR-PRODUCING   ALGAE, *TASTE-PRODUCING ALGAE, AL   W70-02373
-GREEN ALGAE, *GREEN ALGAE, *RED   ALGAE, *TEMPERATURE TOLERANCE, OP   W71-02496
VELOPMENT, WASTE IDENTIFICATION,   ALGAE, *WASTE WATER TREATMENT.: /   W72-03299
*NUTRIENT REMOVAL, SOIL EROSION,   ALGAE, *WATER POLLUTION CONTROL,   W73-01122
*ODOR-PRODUCING ALGAE, *NUISANCE   ALGAE, *WISCONSIN, *LAKES, PRODUC   W70-05112
*FILTERS, *DESIGN, COAGULATION,   ALGAE, ACTIVATED CARBON, *FLOCCUL   W72-07334
YGEN, SLUDGE, OXIDATION LAGOONS,   ALGAE, ADMINISTRATION, MAINTENANC   W71-08970
GAL BLOOMS, WATER BLOOM, AQUATIC   ALGAE, AGING(BROLOGICAL), *AQUATI   W68-00912
AGE EFFLUENTS, *SEWAGE DISPOSAL,   ALGAE, AGRICULTURAL WATERSHEDS, N   W70-05565
ESEARCH AND DEVELOPMENT, GRANTS,   ALGAE, AIR POLLUTION, WATER POLLU   W71-06821
WA/ *EPIPSAMMIC ALGAE, *EPIPELIC   ALGAE, ALGAL ASSOCIATIONS, SHEAR   W71-12870
RODUCING ALGAE, *TASTE-PRODUCING   ALGAE, ALGAE, CHLORINATION, WATER   W70-02373
TS, COST COMPARISONS, ECONOMICS,   ALGAE, ALGAL CONTROL, ALGICIDES,   W69-08518
ONDITIONS, ANAEROBIC CONDITIONS,   ALGAE, ALGAL CONTROL, ECOLOGICAL   W70-04786
INTERKILLING, FISHKILL, NUISANCE   ALGAE, ALGAE, SEASONAL, LIMNOLOGY   W68-00487
OUNDS, *WATER POLLUTION SOURCES,   ALGAE, ALGAL CONTROL, ALGAL POISO   W69-03370
ARID CLIM/ *SOIL ALGAE, *AQUATIC   ALGAE, ALGAE, ARID CLIMATES, SEMI   W69-03491
ROL, *WATER POLLUTION TREATMENT,   ALGAE, ALGAL CONTROL, ANALYTICAL   W69-03374
OLOGY, *WATER POLLUTION EFFECTS,   ALGAE, ALGAL CONTROL, HARVESTING,   W69-03513
C WASTES, NUTRIENTS, PHOSPHORUS,   ALGAE, ALGICIDES, HARVESTING OF A   W69-07818
, NUTRIENTS, TERTIARY TREATMENT,   ALGAE, AMMONIA.: /DENITRIFICATION   W69-08053
ATION LAGOONS, *ORGANIC LOADING,   ALGAE, AMMONIA, SEDIMENTATION, FL   W71-05013
NT, IMPOUNDED WATERS, FRESHWATER   ALGAE, ANADROMOUS FISHES.: /RONME   W71-02491
ION SOURCES, *CYCLING NUTRIENTS,   ALGAE, ANALYTICAL TECHNIQUES, CLA   W69-04800
OTOSYNTHETIC OXYGEN, *CHLORELLA,   ALGAE, ANALYTICAL TECHNIQUES, AQU   W69-03371
REMENTS, *PHYSIOLOGICAL ECOLOGY,   ALGAE, ANALYTICAL TECHNIQUES, AQU   W69-03515
SIOLOGICAL ECOLOGY, *CYANOPHYTA,   ALGAE, ANALYTICAL TECHNIQUES, AQU   W69-03518
NOPHYTA, *PHYSIOLOGICAL ECOLOGY,   ALGAE, ANALYTICAL TECHNIQUES, AQU   W69-03517
N CYCLE, *NUTRIENT REQUIREMENTS,   ALGAE, ANALYTICAL TECHNIQUES, AQU   W69-03186
LORELLA, *PHOTOSYNTHETIC OXYGEN,   ALGAE, ANALYTICAL TECHNIQUES, AQU   W69-03362
OUNDS, *WATER POLLUTION EFFECTS,   ALGAE, ANIMALS, AQUATIC PLANTS, B   W69-04805
GAE, *SEWAGE, ALGICIDES, AQUATIC   ALGAE, AQUATIC PRODUCTIVITY, BIOA   W69-04798
FISH, *FISHERIES, ALGAE, AQUATIC   ALGAE, AQUATIC MICROBIOLOGY, AQUA   W69-04804
REMENTS, *PHYTOPLANKTON, AQUATIC   ALGAE, AQUATIC MICROBIOLOGY, AQUA   W69-04802
AL POISONING, ALGICIDES, AQUATIC   ALGAE, AQUATIC ENVIRONMENT, AQUAT   W69-05844
IFICATION, PHOSPHORUS COMPOUNDS,   ALGAE, AQUATIC PLANTS.: / *DENITR   W69-05323
AL NUTRIENTS, *FISH, *FISHERIES,   ALGAE, AQUATIC ALGAE, AQUATIC MIC   W69-04804
FISHK/ *EUTROPHICATION, *LAKES,   ALGAE, AQUATIC WEEDS, CYANOPHYTA,   W69-06535
, ANALYTICAL TECHNIQUES, AQUATIC   ALGAE, AQUATIC MICROBIOLOGY, AQUA   W69-03186
, ANALYTICAL TECHNIQUES, AQUATIC   ALGAE, AQUATIC MICROBIOLOGY, AQUA   W69-03185
, ANALYTICAL TECHNIQUES, AQUATIC   ALGAE, AQUATIC BACTERIA, AQUATIC   W69-03362
, ANALYTICAL TECHNIQUES, AQUATIC   ALGAE, AQUATIC BACTERIA, AQUATIC   W69-03188
TS, *NITROGEN COMPOUNDS, AQUATIC   ALGAE, AQUATIC MICROBIOLOGY, CHLA   W69-03364
*PHYSIOLOGICAL ECOLOGY, AQUATIC   ALGAE, AQUATIC MICROBIOLOGY, CHLO   W69-03358
, ANALYTICAL TECHNIQUES, AQUATIC   ALGAE, AQUATIC MICROBIOLOGY, BALA   W69-03374
*EUTROPHICATION, ALGAE, AQUATIC   ALGAE, AQUATIC BACTERIA, AQUATIC   W69-03369
, ANALYTICAL TECHNIQUES, AQUATIC   ALGAE, AQUATIC MICROBIOLOGY, BALA   W69-03517
LING NUTRIENTS, *EUTROPHICATION,   ALGAE, AQUATIC ALGAE, AQUATIC BAC   W69-03369
, ANALYTICAL TECHNIQUES, AQUATIC   ALGAE, AQUATIC BACTERIA, AQUATIC   W69-03371
, *SALINE WATER, *THERMAL WATER,   ALGAE, AQUATIC ALGAE, AQUATIC MIC   W69-03516
, *THERMAL WATER, ALGAE, AQUATIC   ALGAE, AQUATIC MICROBIOLOGY, BALA   W69-03516
LOGY, *PHYTOPLANKTON, *CULTURES,   ALGAE, AQUATIC PRODUCTIVITY, CYCL   W69-03514
GAL CONTROL, HARVESTING, AQUATIC   ALGAE, AQUATIC ANIMALS, AQUATIC M   W69-03513
IREMENTS, *PHOSPHORUS COMPOUNDS,   ALGAE, AQUATIC ALGAE, AQUATIC MIC   W69-03373
, ANALYTICAL TECHNIQUES, AQUATIC   ALGAE, AQUATIC MICROBIOLOGY, AQUA   W69-03515
PHORUS COMPOUNDS, ALGAE, AQUATIC   ALGAE, AQUATIC MICROBIOLOGY, AQUA   W69-03373
IONS, AQUATIC PLANTS, WATERSHED,   ALGAE, AQUATIC VEGETATION, NITROG   W68-00511
HIES, CYANOPHY/ *EUTROPHICATION,   ALGAE, AQUATIC PLANTS, BIBLIOGRAP   W68-00680
S, FLOW RATES, CONN STREAM FLOW,   ALGAE, AQUATIC ALGAE, CURRENTS(WA   W68-00479
S, EUTROPHICATION, PRODUCTIVITY,   ALGAE, AQUATIC LIFE.: /N COMPOUND   W70-03068
CYCLING NUTRIENTS, ZOOPLANKTON,   ALGAE, AQUATIC PLANTS, PALEOLIMNO   W70-03975
CYCLING NUTRIENTS, ZOOPLANKTON,   ALGAE, AQUATIC PLANTS, PALEOLIMNO   W70-03975
LATION, WASTE TREATMENT, AQUATIC   ALGAE, AQUATIC PLANTS, TESTING, B   W70-04373
OGEN ION CONCENTRATION, NUISANCE   ALGAE, AQUATIC PLANTS, BOTTOM SED   W70-06225
, FISH, PLANKTON, WATER QUALITY,   ALGAE, AQUATIC LIFE, STREAM POLLU   W72-01786
HAINS, SAMPLING, SURVEYS, MARINE   ALGAE, AQUATIC LIFE, MILK, ATLANT   W72-00941
LGAL CONTROL, ALGICIDES, AQUATIC   ALGAE, AQUATIC MICROBIOLOGY, BIOA   W70-02248
*ANALYTICAL TECHNIQUES, *AQUATIC   ALGAE, AQUATIC MICROBIOLOGY, AQUA   W70-02792
, ANALYTICAL TECHNIQUES, AQUATIC   ALGAE, AQUATIC ENVIRONMENT, AQUAT   W70-02968
TIC/ *CLADOPHORA, ALGAE, AQUATIC   ALGAE, AQUATIC MICROBIOLOGY, AQUA   W70-02784
ROBIOLOGY, AQUATIC/ *CLADOPHORA,   ALGAE, AQUATIC ALGAE, AQUATIC MIC   W70-02784
RNMENTS, *WATER QUALITY CONTROL,   ALGAE, AQUATIC WEEDS, ADMINISTRAT   W72-05774

IM/ *SOIL ALGAE, *AQUATIC ALGAE,   ALGAE, ARID CLIMATES, SEMIARID CL W69-03491
E WATERS, *AGRICULTURE, *WASTES,   ALGAE, ASSAY, DISSOLVED SOLIDS, C W70-03334
DISSOLVED ORGANIC MATTER, ARCTIC   ALGAE, AUTOTROPHIC METABOLISM, HE W71-10079
ICAL OXYGEN DEMAND, *ANAEROBICS,   ALGAE, BACTERIA, WATER POLLUTION W71-05742
S, *THERMAL POLLUTION, *ECOLOGY,   ALGAE, BACTERIA, COLIFORMS, FISH, W71-02479
ES, *NUTRIENTS, *BIODEGRADATION,   ALGAE, BACTERIA, AQUATIC MICROORG W71-00221
FECTS, PHYTOPLANKTON, DIFFUSION,   ALGAE, BACTERIA, EUTROPHICATION, W70-04194
BIBLIOGRAPHIES, LAKES, NUISANCE   ALGAE, BACTERIA, WATER POLLUTION W69-08283
YGEN DEMAND, TERTIARY TREATMENT,   ALGAE, BACTERIA, ACTIVATED SLUDGE W69-08054
STE WATER TREATMENT, *NUTRIENTS,   ALGAE, BACTERIA, CULTURES, CHEMIC W70-05655
ATIC LIFE, *AQUATIC POPULATIONS,   ALGAE, BENTHOS, ZOOPLANKTON, FRES W71-02491
IC FAUNA, AQUATIC PLANTS, MARINE   ALGAE, BENTHIC FLORA, ESTUARIES, W71-02494
USTRIAL WASTES, OXIDATION PONDS,   ALGAE, BENTHIC FAUNA, WATERSHEDS, W72-00027
ND, AERATION, TEMPERATURE, FLOW,   ALGAE, BENTHOS, AEROBIC CONDITION W71-11381
ENTS, ALGAL CONTROL, *NUTRIENTS,   ALGAE, BENTHIC FAUNA, CHEMICAL AN W72-06052
N, *LAKES, LIMNOLOGY, *PLANKTON,   ALGAE, BIBLIOGRAPHIES, CHEMICAL P W69-00446
EFFECTS, *PRIMARY PRODUCTIVITY,   ALGAE, BIOMASS, CALCIUM COMPOUNDS W69-02959
, ANALYTICAL TECHNIQUES, AQUATIC   ALGAE, BIOASSAY, CYCLING NUTRIENT W69-03518
AL POISONING, ALGICIDES, AQUATIC   ALGAE, BIOASSAY, CHLOROPHYTA, CYA W69-03370
TRATIFICATION, DISSOLVED OXYGEN,   ALGAE, BIOCHEMICAL OXYGEN DEMAND(W70-04790
HLAMYDOMONAS,/ *RAIN, *VITAMINS,   ALGAE, BIOASSAY, BIOINDICATORS, C W69-06273
ANKTON, *PONDS, *EUTROPHICATION,   ALGAE, BIOMASS, CHLOROPHYLL, NEW W69-06540
MARY PRODUCTIVITY, HARVESTING OF   ALGAE, BIOMASS.: /NAL, ALGAE, PRI W70-04290
ATIC PLANTS, ANAEROBIC BACTERIA,   ALGAE, BIOCHEMICAL OXYGEN DEMAND, W71-08221
*FARM LAGOONS, *BIODEGRADATION,   ALGAE, BIOCHEMICAL OXYGEN DEMAND, W71-00936
FFLICIENCIES, *LAGOONS, *SLUDGE,   ALGAE, BIOCHEMICAL OXYGEN DEMAND, W70-06899
IENCIES, *NITROGEN, *PHOSPHORUS,   ALGAE, BIOLOGICAL TREATMENT, WAST W70-07469
ROGEN COMPOUNDS, NITROGEN CYCLE,   ALGAE, BIOLOGICAL TREATMENT, BOTT W70-07031
CONTROL, MORTALITY, CHLOROPHYLL,   ALGAE, BIOASSAY, VIABILITY.: /AL W70-02364
TOPES, PHOSPHORUS RADIOISOTOPES,   ALGAE, BIOCHEMISTRY, ECOLOGY, SAM W69-10151
TOLOGY, ENVIRONMENTAL C/ BENTHIC   ALGAE, BLOOMS, CLOSED SYSTEMS, CY W69-07832
TH.:   GREEN   ALGAE, BLUE-GREEN ALGAE, HEAT DEA W70-07313
URE EXPERIMENTS, RECYCLED/ GREEN   ALGAE, BLUE-GREEN ALGAE, TEMPERAT W70-07257
ON, SUPERSATURATION, RESERVOIRS,   ALGAE, BODIES OF WATER, DISSOLVED W70-00683
ION, EPILIMNION, PHOTOSYNTHESIS,   ALGAE, BOTTOM SEDIMENTS, THERMOCL W70-06975
SEWAGE TREATMENT, HARVESTING OF   ALGAE, BOTTOM SEDIMENTS, WEED CON W68-00478
ITY, LAKES, NUTRIENTS, BACTERIA,   ALGAE, BRACKISH WATER, ALKALINE W W71-03031
A, FLORA, LABORATORY, MINERALS,/   ALGAE, BRACKISH, ECOSYSTEMS, FAUN W72-02203
HYL-5-VINYLPYRIDINE, *GLUE GREEN   ALGAE, BREAM, WHITE BREAM, LEUKOC W70-01788
ICITY, *WATER POLLUTION, AQUATIC   ALGAE, BRINE SHRIMP, FISH, FOOD C W70-01467
POLLUTION, FOOD CHAINS, AQUATIC   ALGAE, BRINE SHRIMP, FISH, MUSSEL W70-01466
TENNESSEE RIVER, TROPHIC LEVEL,   ALGAE, CADDISFLIES, FOOD CHAINS, W70-02786
*LAKES, *MANGANESE, *WISCONSIN,   ALGAE, CALCIUM, CHELATION, CYANOP W69-05867
S, CHLOROPHYTA, DINOFLAGELLATES,   ALGAE, CALIFORNIA, FLORIDA, PESTI W70-05272
N, SLUDGE, DETERGENTS, SPRAYING,   ALGAE, CENTRIFUGATION, FILTRATION W70-04060
LLA PYRENOIDOSA, VOLATIL/ *GREEN   ALGAE, CHLORELLA VULGARIS, CHLORE W70-04195
, CORES, DIATOMS, ODOR-PRODUCING   ALGAE, CHLOROPHYLL, CHEMTROL, COP W68-00470
G ALGAE, *TASTE-PRODUCING ALGAE,   ALGAE, CHLORINATION, WATER QUALIT W70-02373
DENSITY, ROOTED AQUATIC PLANTS,   ALGAE, CHLOROPHYLL, HYPOLIMNION, W71-07698
ION, BIODEGRADATION, ADSORPTION,   ALGAE, CLAYS, ALDRIN, DIELDRIN, D W70-09768
ES, *SEDIMENTS, *EUTROPHICATION,   ALGAE, CLAYS, SILICATES.: /, *LAK W69-03075
AGRICULTURE, ST LAWRENCE RIVER,   ALGAE, COLIFORMS, DISSOLVED: /ES, W70-03964
ED SOILDS, PHOSPHATES, NITROGEN,   ALGAE, COLIFORMS.: /IRON, DISSOLV W69-05894
S, *SUSPENDED LOAD, *FILTRATION,   ALGAE, COLLOIDS, FLOCCULATION, MI W71-10654
LANKTON, LIMNOLOGY, ZOOPLANKTON,   ALGAE, COLLOIDS, TRACE ELEMENTS, W70-02510
ER POLLUTION EFFECTS, BIOASSAYS,   ALGAE, COMPARATIVE PRODUCTIVITY, W70-05424
ELLA SOROKINIANA, COMPOSITION OF   ALGAE, CONTINUOUS CULTURES, FATTY W70-07957
BACTERIA, NITROGEN, PHOSPHORUS,   ALGAE, COPPER SULFATE, EFFLUENTS, W70-02787
WATER QUALITY, *ORGANIC MATTER,   ALGAE, COPPER SULFATE, *CHEMICAL W72-04495
S, *OCEANOGRAPHY, *TRANSLATIONS,   ALGAE, CRUSTACEANS, BENTHOS, BRIN W69-07440
SEA WATER, *VITAMINS, VITAMIN B,   ALGAE, CULTURES, CHEMICAL ANALYSI W70-05652
NOPHYTA, BRACK/ *EUTROPHICATION,   ALGAE, CULTURES, CHLOROPHYTA, CYA W70-05752
CONN STREAM FLOW, ALGAE, AQUATIC   ALGAE, CURRENTS(WATER), LIMNOLOGY W68-00479
SODIUM ARSENITE, FISH, ANIMALS,   ALGAE, CUTTING MANAGEMENT, POTABL W70-05263
NTRATION, FUNGI, YEASTS, PYRITE,   ALGAE, CYANOPHYTA, DOMINANT ORGAN W68-00891
NALYSIS, *NITROGEN, *PHOSPHORUS,   ALGAE, CYANOPHYTA, NITRATES, NITR W69-05868
ERATURE, DEPTH, BACTERIA, SLIME,   ALGAE, CYANOPHYTA, DIATOMS, CHLOR W70-04371
IDES, ORGANIC MATTER,/ *NUISANCE   ALGAE, CYANOPHYTA, VITAMINS, PEPT W70-04369
ATER INTERFACES, OREGON, SESSILE   ALGAE, CYANOPHYTA.: /, SEDIMENT-W W72-06051
AQUATIC ENVIRONMENT, PHOSPHATES,   ALGAE, DAPHNIA, GROWTH RATES, PRE W69-06536
NOWS, NUISANCE ALGAE, SUNFISHES,   ALGAE, DAPHNIA, LAKES, CHLOROPHYT W70-02982
EMPERATURE, WEATHER, EPILIMNION,   ALGAE, DENSITY, SEASONAL, MARINE W70-02504
MAGNESIUM, NITRATES, PHOSPHATES,   ALGAE, DETRITUS, PULP WASTES, LIV W70-03501
-PRODUCING ALGAE, ODOR-PRODUCING   ALGAE, DIATOMS, CYANOPHYTA, CHLOR W69-05704
, SNAILS, PHOSPHORUS, NUTRIENTS,   ALGAE, DIATOMS, CHLOROPHYTA, DISS W70-01943
AY, NEW YORK, LAKES, INHIBITION,   ALGAE, DIATOMS, EUTROPHICATION, C W70-01579
TRESS, THERMAL WATER, *PROTOZOA,   ALGAE, DIATOMS, ANIMAL BEHAVIOR, W70-07847
GE BACTERIA, SYSTEMATICS, FUNGI,   ALGAE, DIATOMS, PROTOZOA, PLANKTO W72-01819
TION, SEDIMENTS, ORGANIC MATTER,   ALGAE, DIATOMS, CYANOPHYTA, CHLOR W72-00845
TUS, LIGHT INTENSITY, TURBIDITY,   ALGAE, DIATOMS, CYCLING NUTRIENTS W70-00713
CALCIUM, POTASSIUM, CLAMS, SIZE,   ALGAE, DIATOMS, TRACE ELEMENTS, M W70-00708
RIA, STABILIZATION, WASTE WATER,   ALGAE, DIFFUSION, SULFUR.: /BACTE W70-01971
S, ZOOPLANKTON, DINOFLAGELLATES,   ALGAE, DISSOLVED OXYGEN, HYPOLIMN W70-04253
NATION, *ODOR, *WASTE TREATMENT,   ALGAE, DISINFECTION, BIOCHEMICAL W70-04788
ASS, CULTURES, PLANT PHYSIOLOGY,   ALGAE, DIURNAL DISTRIBUTION, OXYG W70-05094
MPOUNDMENTS, TERTIARY TREATMENT,   ALGAE, DOMESTIC WASTES.: /KES, *I W70-04765
KES, ODOR, TOX/ *EUTROPHICATION,   ALGAE, DUCKS, COLOR, FISHKILL, LA W69-08279
S, AQUATIC POPULATIONS, NUISANCE   ALGAE, DUCKS(WILD), OYSTERS, CHAN W68-00461
OISOTOPES, STREAMFLOW, SAMPLING,   ALGAE, ECOSYSTEMS, ENVIRONMENT, R W69-07862
NIQUES, WATER POLLUTION EFFECTS,   ALGAE, ECOSYSTEMS, PHYSIOLOGICAL W70-04287

LLING, FISHKILL, NUISANCE ALGAE,
HICATION, *LAKES, PHYTOPLANKTON,
 *COPPER, *IRON, TRACE ELEMENTS,
, AQUATIC HABITATS, DRAGONFLIES,
ENCY, BIOCHEMICAL OXYGEN DEMAND,
ARDNESS(WATER), ACIDITY, OXYGEN,
 PESTICIDE RESIDUES, PHOSPHATES,
POLLUTION), *WATER PURIFICATION,
OLLUTION SOURCES, WATER QUALITY,
, LIGHT INTENSITY, CONDUCTIVITY,
ROBIC TREATMENT, PHOTOSYNTHESIS,
EERING, MICROORGANISMS, NUISANCE
TRIENTS, *PRODUCTIVITY, NUISANCE
G, DISASTERS, OILY WATER, MARINE
ATH OF POLLUTANTS, AQUATIC LIFE,
ON LAGOONS, ANAEROBIC DIGESTION,
FATE, *SODIUM ARSENITE, NUISANCE
A POTENTIAL, BACTERIA, PROTOZOA,
KES, WISCONSIN, ALGAE, *NUISANCE
RESERVOIR EVAPORATION, SALINITY,
, PATH OF POLLUTANTS, NUTRIENTS,
UTION CONTROL, MINNOWS, NUISANCE
ATER POLLUTION CONTROL, NUISANCE
 *PHOTOSYNTHESIS, PHYTOPLANKTON,
ECYCLED/ GREEN ALGAE, BLUE-GREEN
*MICROORGANISMS, *LIPIDS, LIGHT,
AN, *WATER CIRCULATION, NUISANCE
S, *PHYTOPLANKTON, *DDT, *MARINE
IAL WASTES, NUTRIENTS, BACTERIA,
AL WASTES, NEUTRIENTS, BACTERIA,
*FILTRATION, DIATOMACEOUS EARTH,
, *PRIMARY PRODUCTIVITY, AQUATIC
*WATER POLLUTION EFFECTS, *FISH,
AINS, *SOIL MICROORGANISMS, SOIL
STES, *MUNICIPAL WASTES, AQUATIC
RATES, DEPTH, DISSOLVED OXYGEN,
DEMAND, COLOR, ACTIVATED SLUDGE,
RSHES, OILY WATER, *OIL, *MARINE
NDICATORS, AQUATIC LIFE, AQUATIC
TTERNS, ACTIVATED CARBON, CLAYS,
*LAKES, *PHOSPHATES, NUTRIENTS,
ON EFFECTS, CHLORIDES, HALOGENS,
S, *DISSOLVED OXYGEN, ESTUARIES,
S, URANIUM RADIOISOTOPES, MARINE
ETHAL LIMI/ *DETERGENTS, *MARINE
 POLLUTION SOURCES, *PHOSPHATES,
E, AQUATIC ENVIRONMENT, *AQUATIC
ICATION, *LAKE ERIE, *NUTRIENTS,
ERSHED MANAGEMENT, HARVESTING OF
TION, WATER POLLUTION TREATMENT,
PPI RIVER, *OHIO RIVER, NUISANCE
, BIOLOGICAL MEMBRANES,/ AQUATIC
MIC FEASIBILITY, LAKES, NUISANCE
ION SOURCES, IRRIGATION, AQUATIC
ENSITY, EUTROPHICATION, GRAZING,
E EFFLUENTS, TERTIARY TREATMENT,
, WATER ANALYSIS, ALGAL CONTROL,
SPIRATION, GREAT LAKES, NUISANCE
ON, MUNICIPAL WASTES, FRESHWATER
 ANALYSIS, NUTRIENTS, EFFLUENTS,
TER POLLUTION CONTROL, *REVIEWS,
TER, TOXICITY, DISSOLVED OXYGEN,
HLOROPHYTA, CYANOPHYTA, NUISANCE
*WASTE TREATMENT, *AQUICULTURE,
WASTES, BYPRODUCTS, FERTILIZERS,
C WASTES, OXYGEN, STANDING CROP,
COPEPODS, CYCLES, CHLAMYDOMONAS,
E ON EUTROPHICATION, PROVISIONAL
 LUXURY CONSUMPTION, PROVISIONAL
IPSAMMIC ALGAE, *EPIPELIC ALGAE,
PLIED STUDIES, ALGAL PHYSIOLOGY,
C ALGAE, AGING/ *EUTROPHICATION,
ICROQUANTITIES, MACROQUANTITIES,
CROCYSTIS AERUGINOSA.:
            INDIA, METHYL ORANGE,

 WATER QUALITY, *EUTROPHICATION,
HIRE WATER POLLUTION COMMISSION,
T, LIGHT EXTINCTION COEFFICIENT,
YNAMICS, EXTRACELLULAR PRODUCTS,
NNAH RIVER, POPULATION DYNAMICS,
EMAND(BOD), AERATION, COLIFORMS,
NS, ANAEROBIC CONDITIONS, ALGAE,
PORATION, FERMENTATION, METHANE,
TROL, INSECT CONTROL, MOSQUITOS,
RANE, NORTHEAST U.S/ CYANAPHYTA,
UATIC PRODUCTIVITY, HYPOLIMNION,
SULFATE, *WATER QUALITY CONTROL,
 STORM DRAINS, SURFACE DRAINAGE,
TION, SUCCESSION, WATER QUALITY,

ALGAE, SEASONAL, LIMNOLOGY, LIGHT   W68-00487
ALGAE, SEASONAL, CYANOPHYTA, *INO   W68-00860
ALGAE, SEA WATER, FRESH WATER.: /   W71-11687
ALGAE, SEDIMENTS, SOIL-WATER-PLAN   W71-09005
ALGAE, SEDIMENTATION, SETTLING BA   W70-09320
ALGAE, SELF-PURIFICATION, BACTERI   W69-07838
ALGAE, SEWAGE, WATER POLLUTION, W   W69-06305
ALGAE, SEWAGE, ANAEROBIC, OXYGEN,   W70-05092
ALGAE, SEWAGE EFFLUENTS.: /ATER P   W70-05084
ALGAE, SEWAGE, DIATOMS, CYANOPHYT   W70-05405
ALGAE, SEWAGE, MUNICIPAL WA: / AE   W72-00462
ALGAE, SEWAGE TREATMENT, SELF-PUR   W72-01798
ALGAE, SEWAGE EFFLUENTS, PHOSPHOR   W70-00270
ALGAE, SHIPS.: /OLLUTION, CLEANIN   W70-01470
ALGAE, SILTS, ECOLOGY, AQUATIC EN   W69-10080
ALGAE, SLUDGE, METHANE, ORGANIC L   W71-11375
ALGAE, SNAILS, LEGISLATION, WISCO   W70-05568
ALGAE, SOLID WASTES, TERTIARY TRE   W70-01027
ALGAE, SPRAYING, SEASONAL, *ALGAL   W68-00488
ALGAE, STRATIFICATION, SALT BALAN   W72-01773
ALGAE, STRATIFIED FLOW, MIXING, F   W71-05878
ALGAE, SUNFISHES, ALGAE, DAPHNIA,   W70-02982
ALGAE, TASTE, ODOR.: / QUALITY, W   W69-10155
ALGAE, TEMPERATURE, SOLAR RADIATI   W71-10788
ALGAE, TEMPERATURE EXPERIMENTS, R   W70-07257
ALGAE, TEMPERATURE, CARBON DIOXID   W70-05651
ALGAE, THERMAL POLLUTION, DIFFUSI   W69-05763
ALGAE, TOXICITY, DIATOMS, ESTUARI   W70-05272
ALGAE, TREATMENT FACILITIES, TEST   W71-06737
ALGAE, TREATMENT FACILITIES, TEST   W70-01519
ALGAE, TREATMENT FACILITIES.: /,   W72-04318
ALGAE, TROPHIC LEVEL, WATER QUALI   W72-01990
ALGAE, TROPHIC LEVEL, PRODUCTIVIT   W72-01788
ALGAE, TURF, SOIL MICROBIOLOGY, W   W72-08110
ALGAE, URANIUM RADIOISOTOPES, RAD   W69-01165
ALGAE, VELOCITY, WATER QUALITY, D   W70-00265
ALGAE, WASTE WATER TREATMENT.: /   W70-10433
ALGAE, WATER POLLUTION SOURCES, O   W70-09976
ALGAE, WATER POLLUTION EFFECTS.: /   W70-05958
ALGAE, WATER PURIFICATION, WATER   W69-09884
ALGAE, WATER POLLUTION EFFECTS, W   W69-09135
ALGAE, WATER POLLUTION SOURCES, W   W69-08562
ALGAE, WATER POLLUTION SOURCES, W   W71-05409
ALGAE, WATER POLLUTION SOURCES.: /   W71-09013
ALGAE, WATER POLLUTION EFFECTS, L   W70-01779
ALGAE, WATER POLLUTION TREATMENT,   W70-01031
ALGAE, WATER QUALITY, *TROPHIC LE   W70-02772
ALGAE, WATER POLLUTION SOURCES, N   W69-01445
ALGAE, WATER ZONING.: /MENTS, WAT   W68-00859
ALGAE, WATER MANAGEMENT(APPLIED),   W70-04721
ALGAE, WATER QUALITY, SEWAGE EFFL   W70-04507
ALGAE, WATER POLLUTION, *TOXICITY   W69-05306
ALGAE, WATER POLLUTION CONTROL, N   W69-06276
ALGAE, WATER ANALYSIS, CARBONATES   W69-07096
ALGAE, WATER CHEMISTRY, BIOMASS,   W70-03978
ALGAE, WATER QUALITY, CONIFEROUS   W70-03512
ALGAE, WATER QUALITY, SEDIMENT-WA   W72-06051
ALGAE, WATER QUALITY, NUTRIENTS,   W72-05453
ALGAE, WATER POLLUTION EFFECTS, D   W72-12240
ALGAE, WEEDS, CARBON DIOXIDE, VIT   W70-04810
ALGAE, WEEDS, ECOLOGY, ENVIRONMEN   W70-07393
ALGAE, WISCONSIN.: /G ORGANIC MAT   W72-01797
ALGAE, WISCONSIN, NUTRIENTS, INDI   W69-01977
ALGAE, YEASTS, BACTERIA, FISH, PR   W69-05891
ALGAE, ZOOPLANKTON, FISH, SHELLFI   W70-09905
ALGAE, ZOOPLANKTON, FISH, BACTERI   W70-02251
ALGAE, ZOOPLANKTON, FISH, CHRYSOP   W70-00268
ALGAL ASSAY PROCEDURE, BATCH CULT   W70-02777
ALGAL ASSAY, PROCEDURES, SELENAST   W70-02779
ALGAL ASSOCIATIONS, SHEAR WATER(E   W71-12870
ALGAL BIOCHEMISTRY, ALGAL TAXONOM   W70-03973
ALGAL BLOOMS, WATER BLOOM, AQUATI   W68-00912
ALGAL BLOOMS, LIEBIG'S LAW OF THE   W70-04765
ALGAL BLOOMS, ALGAE NUTRITION, MI   W70-05565
ALGAL BLOOMS.:   W70-05550
ALGAL BLOOMS.:   W70-00664
ALGAL BLOOMS, NUTRIENTS, LIMNOLOG   W69-09723
ALGAL BLOOMS, NUTRIENT LOAD, SYNT   W69-08674
ALGAL CELL DENSITY.: / TURBIDOSTA   W70-03923
ALGAL CELLS.: /E, PHYTOPLANKTON D   W69-05697
ALGAL COMMUNITIES, MICROBIAL ECOL   W70-04510
ALGAL CONTROL, DISSOLVED OXYGEN.:   W70-04788
ALGAL CONTROL, ECOLOGICAL DISTRIB   W70-04786
ALGAL CONTROL, BACTERIA, COLIFORM   W70-04787
ALGAL CONTROL, BASS, CARP, CATFIS   W70-05417
ALGAL CONTROL, *ALGAL TOXINS MEMB   W69-05306
ALGAL CONTROL, SEASONAL, ANNUAL T   W69-05142
ALGAL CONTROL, SLIME, TASTE-PRODU   W70-05704
ALGAL CONTROL, AQUATIC LIFE, WATE   W70-03344
ALGAL CONTROL, ALGICIDES, DOMINAN   W69-01977

CENEDESMUS, CHLORELLA, *EUGLENA,
, *MUD, *COPPER, COPPER SULFATE,
LOGICAL ECOLOGY, *PHYTOTOXICITY,
*WATER POLLUTION EFFECTS, ALGAE,
ATER POLLUTION TREATMENT, ALGAE,
*WATER POLLUTION SOURCES, ALGAE,
T COMPARISONS, ECONOMICS, ALGAE,
LUTION, WATER POLLUTION EFFECTS,
HKILLS, LIMNOLOGY, ALGAL TOXINS,
, LAKES, OXYGEN, EUTROPHICATION,
N, *WATER CHEMISTRY, POLLUTANTS,
RENCES, WATER POLLUTION EFFECTS,
TREATMENT, BIOLOGICAL TREATMENT,
UDIES, CHLORINE, COPPER SULFATE,
*HALOGENS, *ALGICIDES, CHLORINE,
LUTION, WATER POLLUTION SOURCES,
 ALGAE, AQUATIC MICROBI/ *ALGAE,
*CHLORELLA, CHLORINE, ALGICIDES,
TES, TOXICITY, FISH, BLOODWORMS,
RIES, WATER QUALITY, DETERGENTS,
YGEN, BIOCHEMICAL OXYGEN DEMAND,
, PHYTOPLANKTON, WATER ANALYSIS,
HICATION, LAKES, LAKE SEDIMENTS,
TON, HOWARD A. HANSON RES/ WOOD,
                             *PURE
S-AQUAE, MICROCYSTIS AERUGINOSA,
(MINN), LAKE MANAGEMENT PROGRAM,
, PEDIASTRUM, SELF-PURIFICATION,
MY, ALGAL PHYLOGENY, CLADOPHORA,
ALOUSE RIV/ TREE BARK, CHARCOAL,
 LAKE WAUBESA(WIS), MACROPHYTES,
UGINOSA, *NUTRIENT AVAILABILITY,
L GROWTH, NUTRIENT AVAILABILITY,
LENA GRACILIS Z,/ *VITAMIN B-12,
S CULTURE, *PILOT-PLANT STUDIES,
GAL PHYSIOLOGY/ *CYANOCOBALAMIN,
ILLARIOPHYCEAE, CHLORO/ *EUROPE,
TH EQUATION, INCIDENT LIGHT UTI/
UTRIENT SOURCES, WATER SAMPLING,
AE IDENTIFICATION, ALGAE COUNTS,
ESSMENT, SANIT/ *BIOSTIMULATION,
 GROWTH LIMITATION, SELENASTRUM,
IR EFFECTS, LAKE MEAD, DILUTION,
, NITZSCHIA, BIOLOGICAL STUDIES,
*IDENTIFICATION, *SIGNIFICANCE,
OMINE, IODINE, PHOTO REACTIVITY,
AKE(ORE), LAKE WASHINGTON(WASH),
IENT AVAILABILITY, ALGAL GROWTH,
LOT-PLANT STUDIES, ALGAL GROWTH,
  *CYANOCOBALAMIN, ALGAL GROWTH,
OCYSTIS AERUGINOSA, ANTAGONISMS,

AL BIOCHEMISTRY, ALGAL TAXONOMY,
MENTAL STUDIES, APPLIED STUDIES,
A, ANTAGONISMS, ALGAL NUTRITION,
, ALGAL GROWTH, ALGAL NUTRITION,
, ALGAL GROWTH, ALGAL NUTRITION,
, ALGAL GROWTH, ALGAL NUTRITION,
SHINGTON(WASH), ALGAL NUTRITION,
TRIENTS, *ALGAE, *ALGAL CONTROL,
N SOURCES, ALGAE, ALGAL CONTROL,
   *PHYTOPLANKTON, *PHYTOTOXICITY,
, *PHYTOTOXICITY, ALGAL CONTROL,
SIPPI RIVER BASIN, URBANIZATION,
RANSPORT, BACTERIAL POPULATIONS,
          *CYTOLOGICAL EFFECTS,
ONS.:
, *WEED EXTRACTING STEEL CABLES,
TROLYTES, CATIONIC ELECTROLYTES,
TISZA(HUNGARY), ZIMONA(HUNGARY),
 PHYSIOLOGY, ALGAL BIOCHEMISTRY,
ION.:
LUTA/ BIOCHEMICAL OXYGEN DEMAND,
MS, PHYTOPLANKTON, BACTERICIDES,
ED OXYGEN, FISHKILLS, LIMNOLOGY,
TS, CHLOROPHYLL A, PHOTOBIOLOGY,
GYMNODINIUM BREVIS, MYCROSYSTIS,
, SLIM/ *POTASSIUM PERMANGANATE,
PHORA, *NUISANCE ALGAE, *SEWAGE,
*ALGAL CONTROL, ALGAL POISONING,
S, NUTRIENTS, PHOSPHORUS, ALGAE,
, COPPER SULFATE, ALGAL CONTROL,
N, WATER QUALITY, ALGAL CONTROL,
 ALGAL CONTROL, ALGAL POISONING,
ATION, *WATER POLLUTION CONTROL,
, LAKES, WATER QUALITY, VIRUSES,
 PHOSPHORUS, MIGRATION, DIURNAL,
NESOTA, WATER POLLUTION EFFECTS,
, *ALGAE, *AQUATIC WEEDS, LAKES,
WTH RATES, *CHLORELLA, CHLORINE,

ALGAL CONTROL, BACTERICIDES, GROW   W69-00994
ALGAL CONTROL, ALGICIDES, TRACE E   W69-03366
ALGAL CONTROL, ALGAL POISONING, A   W69-03188
ALGAL CONTROL, HARVESTING, AQUATI   W69-03513
ALGAL CONTROL, ANALYTICAL TECHNIQ   W69-03374
ALGAL CONTROL, ALGAL POISONING, A   W69-03370
ALGAL CONTROL, ALGICIDES, SEDIMEN   W69-08518
ALGAL CONTRO: /EATMENT, WATER POL   W69-10171
ALGAL CONTROL, TRACE ELEMENTS, WA   W70-00264
ALGAL CONTROL, BIOCONTROL, TURBUL   W71-10098
ALGAL CONTROL, NITRATES, PHOSPHAT   W70-02646
ALGAL CONTROL, WATER POLLUTION, T   W70-02775
ALGAL CONTROL, AMMONIA, ION .EXCHA   W70-01981
ALGAL CONTROL, MORTALITY, CHLOROP   W70-02364
ALGAL CONTROL, CHLORELLA, DISINFE   W70-02363
ALGAL CONTROL, BIOINDICATORS.: /L   W70-02304
ALGAL CONTROL, ALGICIDES, AQUATIC   W70-02248
ALGAL CONTROL, DISINFECTION, MODE   W70-02370
ALGAL CONTROL.: /DEPTH, OLIGOCHAE   W70-06217
ALGAL CONTROL, INDUSTRIAL WASTES,   W70-10437
ALGAL CONTROL, *COLOR, *ALKALINIT   W72-11906
ALGAL CONTROL, ALGAE, WATER QUALI   W72-06051
ALGAL CONTROL, *NUTRIENTS, ALGAE,   W72-06052
ALGAL COUNT, GREEN RIVER, WASHING   W72-11906
ALGAL CULTURES.:   W71-06002
ALGAL CULTURES, CHEMOSTATS, CARBO   W69-06864
ALGAL DENSITY, APHANIZOMENON, MEL   W69-10167
ALGAL ENUMERATION, SCENEDESMUS, G   W70-04381
ALGAL GENETICS, RUSSIA, BAVARIA,   W70-03973
ALGAL GROWTH, STREAM CONCOURSE, P   W70-03501
ALGAL GROWTH.: /NAJAS, ANACHARIS,   W70-03310
ALGAL GROWTH, ALGAL NUTRITION, AL   W69-05868
ALGAL GROWTH POTENTIAL, SELENASTR   W69-06864
ALGAL GROWTH, CYANOCOBALAMIN, EUG   W69-06273
ALGAL GROWTH, ALGAL NUTRITION, AL   W69-06865
ALGAL GROWTH, ALGAL NUTRITION, AL   W69-06277
ALGAL GROWTH, ANKISTRODESMUS, BAC   W69-07833
ALGAL GROWTH KINETICS, MONOD GROW   W69-03730
ALGAL GROWTH, OXYGEN DEPLETION, N   W70-00264
ALGAL GROWTH, SEDGWICK-RAFTER MET   W69-10157
ALGAL GROWTH, WATER POLLUTION ASS   W70-02775
ALGAL GROWTH, ANABAENA, BIOSTIMUL   W70-02779
ALGAL GROWTHS, OXYGEN DEPLETION.:   W73-00750
ALGAL IDENTIFICATION.: /EEN ALGAE   W70-10173
ALGAL KEY, CLUMP COUNT, DESMIDS,   W70-05389
ALGAL METABOLISM.: /LL RATES, *BR   W70-02363
ALGAL NUTRITION, ALGAL PHYSIOLOGY   W70-04506
ALGAL NUTRITION, ALGAL PHYSIOLOGY   W69-05868
ALGAL NUTRITION, ALGAL PHYSIOLOGY   W69-06865
ALGAL NUTRITION, ALGAL PHYSIOLOGY   W69-06277
ALGAL NUTRITION, ALGAL PHYSIOLOGY   W69-05867
ALGAL NUTRIENTS.:   W69-05844
ALGAL PHYLOGENY, CLADOPHORA, ALGA   W70-03973
ALGAL PHYSIOLOGY, ALGAL BIOCHEMIS   W70-03973
ALGAL PHYSIOLOGY, LAKE MENDOTA(WI   W69-05867
ALGAL PHYSIOLOGY, KJELDAHL PROCED   W69-05868
ALGAL PHYSIOLOGY, AMERICAN RESEAR   W69-06865
ALGAL PHYSIOLOGY, CALOTHRIX PARIE   W69-06277
ALGAL PHYSIOLOGY, HETEROTROPHY, M   W70-04506
ALGAL POISONING, ALGICIDES, AQUAT   W69-05844
ALGAL POISONING, ALGICIDES, AQUAT   W69-03370
ALGAL POISONING, SCENEDESMUS, CHL   W69-00994
ALGAL POISONING, ANALYTICAL TECHN   W69-03188
ALGAL POISONING, GASOLINE, FLOODS   W73-00758
ALGAL POPULATIONS, HETEROTROPHIC   W70-04194
ALGAL POPULATION.:   W70-02770
ALGAL SENESCENCE, GROWTH INHIBITI   W72-02417
ALGAL SUCCESSION, ANABAENA, CLADO   W70-04494
ALGAL SURFACE CHARGE, ELECTROSTAT   W70-05267
ALGAL TAXONOMY, SPIRULINA, CHODAT   W70-03969
ALGAL TAXONOMY, ALGAL PHYLOGENY,   W70-03973
ALGAL TAXONOMY, ALGAE IDENTIFICAT   W69-03514
ALGAL TOXINS, ORGANIC WASTES, POL   W69-01273
ALGAL TOXINS, LARVAE, PHENOLS, SH   W70-01068
ALGAL TOXINS, ALGAL CONTROL, TRAC   W70-00264
ALGAL: / BATCH CULTURES, CHEMOSTA   W70-02777
ALGASTATIC SUBSTANCES, ALGADYNAMI   W70-02510
ALGICIDAL TESTS, ALGISTATIC TESTS   W69-05704
ALGICIDES, AQUATIC ALGAE, AQUATIC   W69-04798
ALGICIDES, AQUATIC ALGAE, AQUATIC   W69-05844
ALGICIDES, HARVESTING OF ALGAE, E   W69-07818
ALGICIDES, TRACE ELEMENTS, CORES,   W69-03366
ALGICIDES, DOMINANT ORGANISMS, GR   W69-01977
ALGICIDES, AQUATIC ALGAE, BIOASSA   W69-03370
ALGICIDES, BIOCONTROL, *WATERSHED   W68-00478
ALGICIDES, FISH, STRE: /TREATMENT   W70-03973
ALGICIDES, CLOSTRIDIUM, FISH, AMP   W70-05372
ALGICIDES, WATER QUALITY.: /, MIN   W70-04506
ALGICIDES, AQUATIC WEED CONTROL,   W70-04494
ALGICIDES, ALGAL CONTROL, DISINFE   W70-02370

MICROBI/ *ALGAE, ALGAL CONTROL,   ALGICIDES, AQUATIC ALGAE, AQUATIC   W70-02248
HLORINE, COPPER SULF/ *HALOGENS,   ALGICIDES, CYTOLOGICAL STUDIES, C   W70-02364
ES, SEWAGE EFFLUENTS, WISCONSIN,   ALGICIDES, HYDROLYSIS, COPPER SUL   W70-02982
BACTERIA, *AQUATIC MICROBIOLOGY,   ALGICIDES, BIOASSAY, DISINFECTION   W69-10171
ECONOMICS, ALGAE, ALGAL CONTROL,   ALGICIDES, SEDIMENTS, DISSOLVED O   W69-08518
MASS, DIATOMS, CYANOPHYTA, IRON,   ALGICIDES, ZOOPLANKTON, FISH.: /O   W70-08654
D EXTRACTS, ASCOPHYLLUM NODOSUM,   ALGINIC ACID, FUCUS VESICULOSUS,   W69-07826
M PERMANGANATE, ALGICIDAL TESTS,   ALGISTATIC TESTS, SLIMEPRODUCING   W69-05704
DOSA, FUCALES, ALKANES, ALKENES,   ALIPHATIC HYDROCARBONS, PETROLEUM   W69-08284
A, STAPHYLOCOCCUS, ACHROMOBACTER(ALKALIGENES), RHODOTORULA, GROWTH   W70-03952
ERATURE, *HOT SPRINGS, HABITATS,   ALKALINE WATER, ACIDIC WATER, TRA   W69-10160
BACTERIA, ALGAE, BRACKISH WATER,   ALKALINE WATER, SODIUM.: /IENTS,   W71-03031
ED OXYGEN, ALGAE, PHYTOPLANKTON,   ALKALINE WATER, ALKALINITY, FISH   W71-07731
, PHYTOPLANKTON, ALKALINE WATER,   ALKALINITY, FISH FARMING, FISH FO   W71-07731
RUS, HYDROGEN ION CONCENTRATION,   ALKALINITY, HARDNESS(WATER), CHLO   W71-04206
ION CONCENTRATION, TEMPERATURE,   ALKALINITY, BICARBONATES, BACTERI   W71-05267
ION CONCENTRATION, TEMPERATURE,   ALKALINITY, NITRIFICATION, *CORRO   W71-11393
ON, DISTRIBUTION, DEPTH, OXYGEN,   ALKALINITY, HYDROGEN ION CONCENTR   W71-00665
DEGRADATION(DECOMPOSITION), DDT,   ALKALINITY, STABILIZATION, WASTE   W70-08097
EFFLUENTS, NITROGEN, PHOSPHATES,   ALKALINITY, FISH, STREAMS, WATER   W69-10159
URE, HYDROGEN ION CONCENTRATION,   ALKALINITY,: /ARBONATES, TEMPERAT   W70-02804
POPULATION, INDUSTRIES, METHANE,   ALKALINITY, HYDROGEN ION CONCENTR   W70-02803
LLA, HYDROGEN ION CONCENTRATION,   ALKALINITY, DIATOMS, TEMPERATURE,   W70-02504
N CONCENTRATION, PHOTOSYNTHESIS,   ALKALINITY, NITRATES, COND: /N IO   W70-03309
NDS, HYDROGEN ION CONCENTRATION,   ALKALINITY, CALIFORNIA, LIGHT INT   W70-03334
AMPLING, EFFLUENTS, TRIBUTARIES,   ALKALINITY, HYDROGEN ION CONCENTR   W70-03983
ONDUCTIVITY, CALCIUM, MAGNESIUM,   ALKALINITY, SULFATES, CHLORIDES,   W69-05894
YGEN, OXIDATION, CARBON DIOXIDE,   ALKALINITY, ALGAE.: /NITROGEN, OX   W70-05550
OTABLE WATER, HERBICIDES, CHARA,   ALKALINITY, BULLHEADS, EELS, KILL   W70-05263
R PROPERTIES, WATER TEMPERATURE,   ALKALINITY, *PHOTOSYNTHESIS, LIGH   W68-00248
N, TENNESSEE RIVER, ALGAE, IRON,   ALKALINITY, HARDNESS, RADIOACTIVI   W73-00750
*PERIDINIUM SP., *GLENODININE,   ALKALOID TOXINS, IBOGAALKALOID, O   W71-07731
CHLORELLA PYRENOIDOSA, FUCALES,   ALKANES, ALKENES, ALIPHATIC HYDRO   W69-08284
A PYRENOIDOSA, FUCALES, ALKANES,   ALKENES, ALIPHATIC HYDROCARBONS,   W69-08284
EN ION CONCENTRATION, FLUORIDES,   ALKYL BENZENE SULFONATES, WIND VE   W70-00263
HU'S MEDIA, QUINONE DERIVATIVES,   ALKYL SUBSTITUTED IMIDAZOLINES, B   W70-02982
S, *ALKYL POLYETHOXYLAT/ *LINEAR   ALKYL SULFONATES, *AXENIC CULTURE   W70-03928
ALKYLBENZENE SULFONATES, LINEAR   ALKYLATE SULFONATES, WATER POLLUT   W69-09454
, WATER POLLUTION SOURC/ *LINEAR   ALKYLATE SULFONATES, *SURFACTANTS   W70-09438
CTERIA, DETERGENTS, SURFACTANTS,   ALKYLBENZENE SULFONATES, LINEAR A   W69-09454
ROTIFERS, WEEDS, WATER HYACINTH,   ALLIGATORWEED, CHEMCONTROL, MECHA   W70-10175
AERO-HYDRAULIC GUN,   ALLISTON(ONTARIO).:   W71-09546
YTARSUS, LIMNODRILUS, EUCYCLOPS,   ALONA, CHAOBORUS.: / TANYPUS, TAN   W70-06975
OLOGY, CLADOCERAN, MONTANE ZONE,   ALPINE ZONE, BOULDER(COLO), TEA L   W69-10154
IONELLA FORMOSA, RECOLONIZATION,   ALSTER LAKE(GERMANY), GERMANY.: /   W70-00268
TRUN/ DUPLIN RIVER(GA), SPARTINA   ALTEMIFLORA, ORGANISMS, TURSIOPS   W69-08274
NIA, POTAMOGETON, TRAPA, PISTIA,   ALTERNANTHERA, HETERANTHERA, NAJA   W70-01788
E BREAM, LEUKOCYTIC COMPOSITION,   ALTERNATING CURRENT.: /REAM, WHIT   W70-01788
, ROTIFERS, SESTON, RESPIRATION,   ALTITUDE, SAMPLING, ZOOPLANKTON,   W69-10154
US, ANKISTRODESMUS, CHLOROMONAD,   ALUM.: /M, WESTELLA, STEPHANODISC   W70-08107
:   ALUM, *OSHKOSH(WIS), BACKWASHING.   W72-08859
-COATED SAND, CR/ *FILTER MEDIA,   ALUMINA-COATED SAND, FERRIC OXIDE   W70-08628
IRON SULFATE, CLADOPHORA BLANKE/   ALUMINUM SULFATE, IRON CHLORIDE,   W69-10170
IRON S/ RING TUBE, CANALIZATION,   ALUMINUM SULFATE, FERRICHLORIDE,   W70-05668
CHEILUS OREGONENSIS, ACROCHEILUS   ALUTACEUS, ULOTHRIX, CLADOPHORA,   W69-07853
IAL MEDIUM, ARTHROBACTER, FLORA,   AMBIENT TEMPERATURE, THERMAL SENS   W70-00713
, SCHROEDERIA SETIGERA, MELOSIRA   AMBIGUA, CERATIUM HIRUDINELLA, AN   W70-05387
TEMPERATURE COEFFICIENTS, NORTH   AMERICA, CENTRAL AMERICA, RATE EQ   W70-03957
GE, AGRICULTURAL DRAINAGE, NORTH   AMERICA, FOREST DRAINAGE.: /RAINA   W70-03975
GE, AGRICULTURAL DRAINAGE, NORTH   AMERICA, FOREST DRAINAGE.: /RAINA   W70-03975
FICIENTS, NORTH AMERICA, CENTRAL   AMERICA, RATE EQUATIONS, ANALYSIS   W70-03957
GAL NUTRITION, ALGAL PHYSIOLOGY,   AMERICAN RESEARCH AND DEVELOPMENT   W69-06865
IF), WATER POLLUTION ASSESSMENT,   AMERICAN RIVER(CALIF), CLEAR LAKE   W70-02777
A RUBESCENS, DINOBRYON, UROGLENA   AMERICANA, CYCLOTELLA, CERATIUM H   W70-05405
UNDS, ORGANO-METALLIC COMPOUNDS,   AMINES, ENGLAND.: /PHENOLIC COMPO   W72-12987
TION, DETERGENTS, RADIOACTIVITY,   AMINO ACIDS.: /ISTRIBUTION, CHELA   W69-08275
SONAL, PSEUDOMONAS, TEMPERATURE,   AMINO ACIDS, INTERTIDAL AREAS, CA   W70-03952
MUS, PROTEINS, WALLS, STRUCTURE,   AMINO ACIDS, MECHANICAL EQUIPMENT   W70-04184
TERIA, TRACE ELEMENTS, VITAMINS,   AMINO ACIDS, CALCIUM, AMMONIA, NI   W70-02255
S, RESERVOIRS, NITROGEN, GROWTH,   AMINO ACIDS, CHLAMYDOMONAS, CHLOR   W69-10180
A PYRENOIDOSA, CARBOXYLIC ACIDS,   AMINO CARBOXYLIC ACIDS.: /HLORELL   W70-02248
S, *GROWTH RATES, *ALGAE, 2-4-D,   AMINOTRIAZOLE, 2-4-5-T, SCENEDESM   W70-03519
*LITERATURE, ELEOCHARIS, 2,4-D,   AMMATE, CHLORAMINE, DE-K-PRUF-21,   W70-05263
TION, ELECTROCHEMICAL TREATMENT,   AMMATE, CHLORAMINE, DE-K-PRUF-21,   W70-05263
ENTS, TERTIARY TREATMENT, ALGAE,   AMMONIA STRIPPING, ALGAE HARVESTI   W70-01981
BACTERIA, TEMPERATURE, NITRATES,   AMMONIA.: /DENITRIFICATION, NUTRI   W69-08053
UTRIENTS, ENVIRONMENTAL EFFECTS,   AMMONIA, *WATER REUSE, WASTE WATE   W72-06017
CHEMISTRY, PHOSPHORUS, NITROGEN,   AMMONIA, AMMONIUM COMPOUNDS, NITR   W69-03364
HYDROGEN SULFIDE, LEAD, COPPER,   AMMONIA, ALGAE, NUTRIENTS, SEWAGE   W71-05879
EUGLENA, CHLOROPHYTA, NITRATES,   AMMONIA, CONDUCTIVITY, CHR: /NDS,   W70-04510
YCETES, *ODOR, ACTIVATED CARBON,   AMMONIA, CARBON, RECREATION, WATE   W72-04761
ENITE, COPPER SULPHATE, MONURON,   AMMONIA, CHLORINE.: /AE, *ACTINOM   W72-04155
RHODE ISLAND, SAMPLING, ECOLOGY,   AMMONIA, DALAPON, 2-4-: /DIUM ARS   W70-00269
, TEMPERATURE, NITRATES, PLANTS,   AMMONIA, ENZYMES, CLASSIFICATION,   W70-00713
OGICAL TREATMENT, ALGAL CONTROL,   AMMONIA, HARDNESS(WATER), ACIDITY   W69-07838
VITAMINS, AMINO ACIDS, CALCIUM,   AMMONIA, ION EXCHANGE, REVERSE OS   W70-01981
L NUTRIENTS, NITROGEN COMPOUNDS,   AMMONIA, NITROGEN FIXATION, IRON   W70-02255
TRATION, OXYGEN, CARBON DIOXIDE,   AMMONIA, NITRATES, NITRITES, NUTR   W69-10177
ITRIFICATION, ORGANIC COMPOUNDS,   AMMONIA, NITRITES, HYDROGEN SULFI   W71-03031
R, NITROGEN COMPOUNDS, NITROGEN,   AMMONIA, NITRITES, NITRATES, YEAS   W69-06970
                                     AMMONIA, NUTRIENTS, BIOCHEMICAL O   W69-08054

```
*OXYGENATION, AEROBIC TREATMENT,      ANAEROBIC CONDITIONS, CANNERIES,     W71-09546
, EQUATIONS, AEROBIC CONDITIONS,      ANAEROBIC CONDITIONS, ALGAE, ALGA    W70-04786
AL SYSTEM, PH, FACULTATIVE POND,      ANAEROBIC POND.: /MULTI-CELL SERI    W70-04790
, *NITRITES, AEROBIC CONDITIONS,      ANAEROBIC CONDITIONS, CHLORELLA.:    W70-05261
AESTHETICS, AEROBIC CONDITIONS,       ANAEROBIC CONDITIONS, WATER SUPPL    W70-04430
TER PURIFICATION, ALGAE, SEWAGE,      ANAEROBIC, OXYGEN, NITRIFICATION     W70-05092
NETWORK, ALGAE COUNTS, PLANKTON       ANALYSES, AGRICULTURAL RUNOFF, AR    W70-04507
CENTRAL AMERICA, RATE EQUATIONS,      ANALYSIS OF VARIANCE, FOOD CONCEN    W70-03957
D CARBON, CHLOROPLASTS, INFRARED      ANALYSIS.: /ETHANE EXTRACTS, LIPI    W70-05651
PROPERTIES, PHOSPHORUS, CHEMICAL      ANALYSIS.: / WEED CONTROL, WATER     W68-00479
, COAGULATION, DEWATERING, *COST      ANALYSIS.: /ATURE, CENTRIFUGATION    W71-11375
SOURCES, WATER CHEMISTRY, WATER       ANALYSIS.: /LGAE, WATER POLLUTION    W69-08562
ENTATION, TOXICITY, COST-BENEFIT      ANALYSIS.: /ANTS, BACTERIA, SEDIM    W70-05819
TIGATIONS, FREQUENCY, *FREQUENCY      ANALYSIS.: /TIONS, *ON-SITE INVES    W70-09976
TIAL, DISSOLVED OXYGEN, CHEMICAL      ANALYSIS, ORGANIC MATTER, ALGAE,     W70-05760
ORUS, SILICA, SAMPLING, CHEMICAL      ANALYSIS, PHYSICAL PROPERTIES, CA    W70-06975
H, *WISCONSIN, ECOLOGY, CHEMICAL      ANALYSIS, PHYTOPLANKTON, CYANOPHY    W70-06229
EWAGE, WATER POLLUTION, CHEMICAL      ANALYSIS, PHYSICAL PROPERTIES.: /    W70-05752
OROPHYLL, PRODUCTIVITY, REGIONAL      ANALYSIS, RESERVOIRS.: /OGEN, CHL    W70-07261
MICROORGANISMS, SAMPLING, WATER       ANALYSIS, WATER QUALITY, POLLUTAN    W71-00221
E, CHROMOTOGRAPHY, RADIOCHEMICAL      ANALYSIS, TRACERS.: /HATES, *ALGA    W71-03033
TRY, WATER TEMPERATURE, CHEMICAL      ANALYSIS, ANALYTICAL TECHNIQUES,     W69-09676
S, ONE-SAMPLE RUNS TEST, CLUSTER      ANALYSIS, REYKJAVIK(ICELAND).: /T    W69-09755
OPHYLL, PHOTOSYNTHESIS, CHEMICAL      ANALYSIS, BACTERIA, NUTRIENTS, TE    W69-10167
ITY, TECHNICAL FEASIBILITY, COST      ANALYSIS, SLUDGE, INCINERATION, S    W71-10654
E CONTROL, *LAKE ERIE,/ *SYSTEMS      ANALYSIS, DATA COLLECTIONS, *ALGA    W71-04758
ESTS, INVERTEBRATES, FISH, WATER      ANALYSIS, INSECTS, SNAILS, TOXICI    W72-01801
MPOSING ORGANIC MATTER, CHEMICAL      ANALYSIS, PROTEINS, BACTERIA.: /O    W72-01812
WASTES, INDUSTRIAL WASTES, WATER      ANALYSIS, ANALYTICAL TECHNIQUES,     W72-03299
TIVITY, PHOTOSYNTHESIS, CHEMICAL      ANALYSIS, TRACE ELEMENTS, TEMPERA    W70-01933
WATER POLLUTION EFFECTS, SYSTEMS      ANALYSIS, SUCCESSION, ECOSYSTEMS,    W70-02779
S, *TRIBUTARIES, SAMPLING, FLOW,      ANALYSIS, FERTILIZATION, PLANT GR    W70-02969
ALGAE, EUTROPHICATION, / *WATER       ANALYSIS, *ODORS, *ODOR PRODUCING    W69-00387
OLLUTION, WATER SAMPLING, *WATER      ANALYSIS, *CHEMICAL ANALYSIS, PH,    W69-00659
LING, *WATER ANALYSIS, *CHEMICAL      ANALYSIS, PH, SAMPLING, ON-SITE D    W69-00659
TY, *LIMNOLOGY, MINNESOTA, WATER      ANALYSIS, CALCIUM CARBONATE, HYPO    W69-00632
PROCEDUR/ *EUTROPHICATION, TEST       ANALYSIS, *LABORATORY TESTS, TEST    W68-00472
TION, ALGAE, GERMANY, VOLUMETRIC      ANALYSIS, LIMNOLOGY, GROWTH RATES    W68-00475
MIN B, ALGAE, CULTURES, CHEMICAL      ANALYSIS, ANALYTICAL TECHNIQUES,     W70-05652
TECHNIQUES, */ *ALGAE, *CHEMICAL      ANALYSIS, *CHLORINE, *ANALYTICAL     W70-05178
ENCE, *CHLOROPHYLL/ *CORRELATION      ANALYSIS, *FLUOROMETRY, *FLUORESC    W70-05094
PHORUS, NITROGEN, FLOW, CHEMICAL      ANALYSIS, NUTRIENTS, EFFLUENTS, A    W70-04810
TICAL TECHNIQUES, *POLAROGRAPHIC      ANALYSIS, PHOTOMETRY, HALOGENS.: /    W70-05178
OLLUTION EFFECTS, ANA/ *CHEMICAL      ANALYSIS, ORGANIC MATTER, WATER P    W70-05109
PHORUS RADIOISOTOPES, REGRESSION      ANALYSIS, ALGAE, PONDS, PREDATION    W70-03957
ATION, DATA COLLECTION, CHEMICAL      ANALYSIS, RELIABILITY, WATER POLL    W70-04382
CHNIQUES, *AQUATIC WEEDS, *WATER      ANALYSIS, *SAMPLING, PLANT PHYSIO    W70-04382
FISH/ *NUTRIENTS, *BENEFIT-COST       ANALYSIS, *LAKE ERIE, PHOSPHORUS,    W70-04465
SOTOPES, SEASONAL, RADIOCHEMICAL      ANALYSIS, MONITORING, PIGMENTS, P    W70-04371
ERIA, METHANE BACTERIA, CHEMICAL      ANALYSIS, WATER POLLUTION, CHLORE    W70-03312
SSAY, DISSOLVED SOLIDS, CHEMICAL      ANALYSIS, PHOSPHATES, IRON COMPOU    W70-03334
VELOCITY, TEMPERATURE, CHEMICAL       ANALYSIS, PHYSIOLOGICAL ECOLOGY,     W69-07862
WATER, GAMMA RAYS, RADIOCHEMICAL      ANALYSIS, TRITIUM, CARBON RADIOIS    W69-08269
ID DIHYDROCHLORIDE, INFRARED GAS      ANALYSIS, COULTER COUNTER, EUGLEN    W69-08278
N ANALYSIS, WATER POLL/ *SYSTEMS      ANALYSIS, *ECOSYSTEMS, *SIMULATIO    W69-06536
ALGAE, CYANOPHYTA, N/ *CHEMICAL       ANALYSIS, *NITROGEN, *PHOSPHORUS,    W69-05868
ALYSIS, *ECOSYSTEMS, *SIMULATION      ANALYSIS, WATER POLLUTION EFFECTS    W69-06536
UTRIENTS, WATER CHEMISTRY, WATER      ANALYSIS, PHOSPHORUS, PONDS, PHYT    W69-05142
NATION, MICROORGANISMS, CHEMICAL      ANALYSIS, TEMPERATURE, NITRATES,     W69-07838
IRRIGATION, AQUATIC ALGAE, WATER      ANALYSIS, CARBONATES, NITRATES, C    W69-07096
TEMS(MECHANICAL), CLEANING, COST      ANALYSIS, WATER PURIFICATION, *WA    W72-08688
VED OXYGEN, PHYTOPLANKTON, WATER      ANALYSIS, ALGAL CONTROL, ALGAE, W    W72-06051
, ALGAE, BENTHIC FAUNA, CHEMICAL      ANALYSIS, SESSILE ALGAE, *PRIMARY    W72-06052
QUALITY, *BIOINDICATORS, *WATER       ANALYSIS, *BIOASSAY, *COMPARATIVE    W72-12240
, CLUSTER ANALYSIS, MULTIVARIATE      ANALYSIS, PRINCIPAL-COMPONENT ANA    W73-01446
EACHING, IRON, TEMPERATURE, SOIL      ANALYSIS, TASTE-PRODUCING ALGAE,     W72-11906
TE ANALYSIS, PRINCIPAL-COMPONENT      ANALYSIS, PART CORRELATION, DECAP    W73-01446
RA, FRITILLARIA, OBELIA, CLUSTER      ANALYSIS, MULTIVARIATE ANALYSIS,     W73-01446
L WASTES, ENVIRONMENTAL EFFECTS,      ANALYTICAL TECHNIQUES, PHYSIOLOGI    W72-12240
*EUTROPHICATION, *PLANT GROWTH,       ANALYTICAL TECHNIQUES, ANTIBIOTIC    W69-04801
RCES, *CYCLING NUTRIENTS, ALGAE,      ANALYTICAL TECHNIQUES, CLAY LOAM,    W69-04800
ER DYNAMICS, WATER CONSTITUENTS,      ANALYTICAL CHEMISTRY, WASTE WATER    W70-04382
ASSAYS, WATER POLLUTION EFFECTS,      ANALYTICAL TECHNIQUES.: /NAS, BIO    W70-04284
OLOGY, RADIOACTIVITY TECHNIQUES,      ANALYTICAL TECHNIQUES, WATER POLL    W70-04194
MATTER, WATER POLLUTION EFFECTS,      ANALYTICAL TECHNIQUES, KINETICS,     W70-05109
AE, CULTURES, CHEMICAL ANALYSIS,      ANALYTICAL TECHNIQUES, DINOFLAGEL    W70-05652
RVOIRS, CONTROL, BIBLIOGRAPHIES,      ANALYTICAL TECHNIQUES, DIATOMS, S    W70-05389
ALGAL CONTROL, ALGAL POISONING,      ANALYTICAL TECHNIQUES, AQUATIC AL    W69-03188
, *NUTRIENT REQUIREMENTS, ALGAE,      ANALYTICAL TECHNIQUES, AQUATIC AL    W69-03186
, *PHOTOSYNTHETIC OXYGEN, ALGAE,      ANALYTICAL TECHNIQUES, AQUATIC AL    W69-03362
TIAL NUTRIENTS, *NITROGEN CYCLE,      ANALYTICAL TECHNIQUES, AQUATIC AL    W69-03185
PECTROSCOPY, GAS CHROMATOGRAPHY,      ANALYTICAL TECHNIQUES, WATER POLL    W69-00387
CAL ECOLOGY, *CYANOPHYTA, ALGAE,      ANALYTICAL TECHNIQUES, AQUATIC AL    W69-03518
, *PHYSIOLOGICAL ECOLOGY, ALGAE,      ANALYTICAL TECHNIQUES, AQUATIC AL    W69-03517
, *PHYSIOLOGICAL ECOLOGY, ALGAE,      ANALYTICAL TECHNIQUES, AQUATIC AL    W69-03515
HETIC OXYGEN, *CHLORELLA, ALGAE,      ANALYTICAL TECHNIQUES, AQUATIC AL    W69-03371
TREATMENT, ALGAE, ALGAL CONTROL,      ANALYTICAL TECHNIQUES, AQUATIC AL    W69-03374
GAE, *ALGAL CONTROL, *ALGICIDES,      ANALYTICAL TECHNIQUES, AQUATIC AL    W70-02968
HY, LIMITING FACTORS, CHLORELLA,      ANALYTICAL TECHNIQUES, CYANOPHYTA    W70-02779
NOPHYTA, *NUTRIENT REQUIREMENTS,      ANALYTICAL TECHNIQUES, WATER POLL    W70-02245
```

ASONAL EFFECTS, THERMAL EFFECTS,	APHANIZOMENON FLOS AQUAE, METOLIU	W70-03978
LORONAPHTHOQUINONE, MICROCYSTIS,	APHANIZOMENON, ANABAENA, GLOEOTRI	W70-03310
E WAUBESA, LYNGBYA, MICROCYSTIS,	APHANIZOMENON, ANABAENA, MELOSIRA	W69-01977
IABILIS, MICROCYSTIS AERUGINOSA,	APHANIZOMENON FLOS-AQUAE, OSCILLA	W70-05091
SARCODINA, CILIATES, CYCLOTELLA,	APHANIZOMENON FLOS AQUAE, MELOSIR	W70-05112
ODISCUS AS/ *SHAGAWA LAKE(MINN),	APHANIZOMENON FLOS AQUAE, STEPHAN	W70-05387
OSIRA, STEPHANODISCUS, ANABAENA,	APHANIZOMENON, MICROCYSTIS, POTAM	W72-04761
IENT SOURCES, PHOSPHORUS BUDGET,	APHANIZOMINON FLOS AQUAE.: / NUTR	W70-04268
, STIGEOCLONIUM, ANKISTRODESMUS,	APHANOCAPSA, CHROOCOCCUS, CLOSTER	W70-04371
NSPARENCY, ANABAENA LEMMERMANNI,	APHANOCAPSA, BOSMINA LONGISPINA,	W69-10169
NKTON, AQUATIC ANIMALS, ANIMALS,	APPALACHIAN MOUNTAIN REGION.: /LA	W69-03695
REMOVAL, AGRICULTURAL DRAINAGE,	APPALACHIAN MOUNTAIN CLUB HUTS, T	W69-07818
R LINES, ATTENUATION MEASUREMENT	APPARATUS, HYDROGEN F LINE.: /OFE	W70-05377
ROOTED AQUATIC PLANTS, RATES OF	APPLICATION, APPLICATION TECHNIQU	W70-05568
IC PLANTS, RATES OF APPLICATION,	APPLICATION TECHNIQUES, SWIMMING,	W70-05568
LGAE, FISH, ZOOPLANKTON, SNAILS,	APPLICATION METHODS.: /NUISANCE A	W70-03310
RATURES, *COPPER DOSAGE, *COPPER	APPLICATION METHODS, ALGAE IDENTI	W69-10157
ODELS, BOD REMOVAL RATES, DESIGN	APPLICATIONS, FACULTATIVE PONDS,	W70-07034
ANAEROBIC DENITRIFICATION, LAND	APPLICATION, ELECTROCHEMICAL TREA	W70-01981
UNIVERSITY, FUNDAMENTAL STUDIES,	APPLIED STUDIES, ALGAL PHYSIOLOGY	W70-03973
EATMENT, ALGAE, WATER MANAGEMENT(	APPLIED), LIMNOLOGY.: /LLUTION TR	W70-04721
S, AGRICULTURE, WATER MANAGEMENT(	APPLIED), GEOCHEMISTRY, SOIL TYPE	W71-03048
ANCE, FOAMING DETERGENTS, PUBLIC	APPREHENSION.: /TES, NUTRIENT BAL	W70-05084
ES, *INDUSTRIAL WASTES, PEACHES,	APRICOTS, PEARS, CITRUS FRUITS, T	W71-09548
M WASTES, MERCENARIA MERCENARIA,	AQUACULTURE, RECYCLING WASTES.: /	W70-09905
HORUS BUDGET, APHANIZOMINON FLOS	AQUAE.: / NUTRIENT SOURCES, PHOSP	W70-04268
LAKE(ORE.), *APHANIZOMENON FLOS-	AQUAE, *CARBON-14 MEASUREMENTS.: /	W72-06052
OHEMIA, ELB/ *APHANIZOMENON FLOS	AQUAE, *MICROCYSTIS AERUGINOSA, B	W70-00274
TERINARIANS, MIC/ *ANABAENA FLOS-	AQUAE, *STRAINS, PHYCOLOGISTS, VE	W70-00273
RIUM KUTZINGIANUM, ANABAENA FLOS	AQUAE, ANABAENA MENDOTAE, ANABAEN	W69-08283
*TROPHIC BIOLOGY, ANABAENA FLOS	AQUAE, APHANIZOMENON FLOS AQUAE,	W70-03959
A FLOS AQUAE, APHANIZOMENON FLOS-	AQUAE, GLOEOTRICHIA ECHINULATA, B	W70-03959
S AERUGINOSA, APHANIZOMENON FLOS-	AQUAE, GLOEOTRICHIA ECHINULATA, C	W70-00719
, PHORMIDIUM, APHANIZOMENON FLOS-	AQUAE, LYNGBYA, STORM LAKE(IOWA),	W70-02803
, CYCLOTELLA, APHANIZOMENON FLOS	AQUAE, MELOSIRA, ASTERIONELLA FOR	W70-05112
RMAL BIOLOGY, ANABAENA FLOS	AQUAE, METOLIUS RIVER(ORE), COMMU	W70-03978
RUM CAPRICORNUTUM, ANABAENA FLOS-	AQUAE, MICROCYSTIS AERUGINOSA, AL	W69-06864
S AERUGINOSA, APHANIZOMENON FLOS-	AQUAE, MINIMUM LETHAL DOSES, CHU'	W70-02982
S AERUGINOSA, APHANIZOMENON FLOS-	AQUAE, OSCILLATORIA, ANABAENA CYL	W70-05091
A RIVER(WIS), APHANIZOMENON FLOS	AQUAE, OXYGEN DEPLETION.: *YAHAR	W72-01797
, PHORMIDIUM, APHANIZOMENON FLOS-	AQUAE, SECCHI DISC TRANSPARENCY,	W69-10169
S, CLADOCERA, APHANIZOMENON FLOS-	AQUAE, STEPHANODISCUS HANTZSCHII,	W70-00268
A LAKE(MINN), APHANIZOMENON FLOS	AQUAE, STEPHANODISCUS ASTREA, LAK	W70-05387
LYNGBYA LIMNETICA, ANABAENA FLOS	AQUAE, TABELLARIA, MELO: /TICUS,	W70-05405
NA INAEQUALIS 381, ANABAENA FLOS	AQUAE, TOLYPOTHRIX DISTORTA, GLOE	W70-05091
IES, ANION ADSORPTION, COLLOIDS,	AQUARIA, ABSORPTION, WATER POLLUT	W71-09863
SORPTION, CHELATION, METABOLISM,	AQUARIA, WATER POLLUTION EFFECTS,	W71-09005
, OXIDATION-REDUCTION POTENTIAL,	AQUATIC ALGAE, *BIOLOGICAL TREATM	W71-09546
UTRIENTS, *PRIMARY PRODUCTIVITY,	AQUATIC ALGAE, TROPHIC LEVEL, WAT	W72-01990
LGICIDES, ANALYTICAL TECHNIQUES,	AQUATIC ALGAE, AQUATIC ENVIRONMEN	W70-02968
IA, *TOXICITY, *WATER POLLUTION,	AQUATIC ALGAE, BRINE SHRIMP, FISH	W70-01467
ING, *MARSH PLANTS, *OILY WATER,	AQUATIC ALGAE, MARINE ALGAE, OIL,	W70-01231
, *WATER POLLUTION, FOOD CHAINS,	AQUATIC ALGAE, BRINE SHRIMP, FISH	W70-01466
ALGAE, ALGAL CONTROL, ALGICIDES,	AQUATIC ALGAE, AQUATIC MICROBIOLO	W70-02248
GY, AQUATIC/ *CLADOPHORA, ALGAE,	AQUATIC ALGAE, AQUATIC MICROBIOLO	W70-02784
FISHKILL, FISH TOXINS, BIOASSAY,	AQUATIC ALGAE, *ALGAL TOXINS, KIL	W70-08372
TY, BIOINDICATORS, AQUATIC LIFE,	AQUATIC ALGAE, WATER POLLUTION EF	W70-05958
R POLLUTION SOURCES, IRRIGATION,	AQUATIC ALGAE, WATER ANALYSIS, CA	W69-07096
SANCE ALGAE, *SEWAGE, ALGICIDES,	AQUATIC ALGAE, AQUATIC PRODUCTIVI	W69-04798
TOXICITY, BIOLOGICAL MEMBRANES,/	AQUATIC ALGAE, WATER POLLUTION, *	W69-05306
ROL, ALGAL POISONING, ALGICIDES,	AQUATIC ALGAE, AQUATIC ENVIRONMEN	W69-05844
IENTS, *FISH, *FISHERIES, ALGAE,	AQUATIC ALGAE, AQUATIC MICROBIOLO	W69-04804
NT REQUIREMENTS, *PHYTOPLANKTON,	AQUATIC ALGAE, AQUATIC MICROBIOLO	W69-04802
TES, CHELATION, WASTE TREATMENT,	AQUATIC ALGAE, AQUATIC PLANTS, TE	W70-04373
A, ALGAE, ANALYTICAL TECHNIQUES,	AQUATIC ALGAE, AQUATIC BACTERIA,	W69-03371
Y, ALGAE, ANALYTICAL TECHNIQUES,	AQUATIC ALGAE, AQUATIC MICROBIOLO	W69-03517
LGAE, ALGAL CONTROL, HARVESTING,	AQUATIC ALGAE, AQUATIC ANIMALS, A	W69-03513
NE WATER, *THERMAL WATER, ALGAE,	AQUATIC ALGAE, AQUATIC MICROBIOLO	W69-03516
Y, ALGAE, ANALYTICAL TECHNIQUES,	AQUATIC ALGAE, AQUATIC MICROBIOLO	W69-03515
ROL, ALGAL POISONING, ALGICIDES,	AQUATIC ALGAE, BIOASSAY, CHLOROPH	W69-03370
A, ALGAE, ANALYTICAL TECHNIQUES,	AQUATIC ALGAE, BIOASSAY, CYCLING	W69-03518
CONTROL, ANALYTICAL TECHNIQUES,	AQUATIC ALGAE, AQUATIC MICROBIOLO	W69-03374
S, *PHOSPHORUS COMPOUNDS, ALGAE,	AQUATIC ALGAE, AQUATIC MICROBIOLO	W69-03373
CTIVE WASTES, *MUNICIPAL WASTES,	AQUATIC ALGAE, URANIUM RADIOISOTO	W69-01165
TRIENTS, *EUTROPHICATION, ALGAE,	AQUATIC ALGAE, AQUATIC BACTERIA,	W69-03369
SPHATES, *PHYSIOLOGICAL ECOLOGY,	AQUATIC ALGAE, AQUATIC MICROBIOLO	W69-03358
EN CYCLE, ANALYTICAL TECHNIQUES,	AQUATIC ALGAE, AQUATIC MICROBIOLO	W69-03185
S, ALGAE, ANALYTICAL TECHNIQUES,	AQUATIC ALGAE, AQUATIC MICROBIOLO	W69-03186
N, ALGAE, ANALYTICAL TECHNIQUES,	AQUATIC ALGAE, AQUATIC BACTERIA,	W69-03362
NUTRIENTS, *NITROGEN COMPOUNDS,	AQUATIC ALGAE, AQUATIC MICROBIOLO	W69-03364
OISONING, ANALYTICAL TECHNIQUES,	AQUATIC ALGAE, AQUATIC BACTERIA,	W69-03188
RATES, CONN STREAM FLOW, ALGAE,	AQUATIC ALGAE, CURRENTS(WATER), L	W68-00479
TION, ALGAL BLOOMS, WATER BLOOM,	AQUATIC ALGAE, AGING(BROLOGICAL),	W68-00912
ATION, DESMIDS/ *EUTROPHICATION,	AQUATIC ALGAE, *LAKES, *CLASSIFIC	W68-00255
TROL, HARVESTING/ AQUATIC ALGAE,	AQUATIC ANIMALS, AQUATIC MICROBIO	W69-03513
KTON, AQUATIC LIFE, ZOOPLANKTON,	AQUATIC ANIMALS, ANIMALS, APPALAC	W69-03695
N, MARINE ANIMALS, AQUATIC LIFE,	AQUATIC ANIMALS, BENTHIC FAUNA, A	W71-02494
ATURE, SLUDGE WORMS, BLOODWORMS,	AQUATIC BACTERIA, ALGAE, MAYFLIES	W70-00475
TICAL TECHNIQUES, AQUATIC ALGAE,	AQUATIC BACTERIA, AQUATIC MICROBI	W69-03371
PHICATION, ALGAE, AQUATIC ALGAE,	AQUATIC BACTERIA, AQUATIC MICROBI	W69-03369

```
TICAL TECHNIQUES, AQUATIC ALGAE,        AQUATIC BACTERIA, AQUATIC MICROBI    W69-03188
TICAL TECHNIQUES, AQUATIC ALGAE,        AQUATIC BACTERIA, AQUATIC MICROBI    W69-03362
OSYSTEMS, PHYSIOLOGICAL ECOLOGY,        AQUATIC BACTERIA, CARBON RADIOISO    W70-04287
TES, NITRATES, YEASTS, BACTERIA,        AQUATIC BACTERIA, WATER POLLUTION    W69-06970
ANALYTICAL TECHNIQUES, KINETICS,        AQUATIC BACTERIA, ALGAE, PHYSIOLO    W70-05109
UTION EFFECTS, DISSOLVED OXYGEN,        AQUATIC BACTERIA, ALGAE.: /R POLL    W70-04727
LAKE, BRITISH COLUMBIA, CANADA,         AQUATIC ECOSYSTEM MODEL, SIMULATI    W69-06536
CONCENTRATIONS, MICRONUTRIENTS,         AQUATIC ECOSYSTEM, MIDGE LARVAE,     W69-09340
TIC LIFE, ALGAE, SILTS, ECOLOGY,        AQUATIC ENVIRONMENT.: /ANTS, AQUA    W69-10080
DATION, WATER POLLUTION EFFECTS,        AQUATIC ENVIRONMENT, CHLORINATED     W70-08652
Y, NUT/ *ALGAE, *TROPHIC LEVELS,        AQUATIC ENVIRONMENT, WATER QUALIT    W70-02764
AE, WATER QUALITY, *TRO/ *ALGAE,        AQUATIC ENVIRONMENT, *AQUATIC ALG    W70-02772
TICAL TECHNIQUES, AQUATIC ALGAE,        AQUATIC ENVIRONMENT, AQUATIC MICR    W70-02968
QUATIC INSECTS, *AQUATIC PLANTS,        AQUATIC ENVIRONMENT, ECOSYSTEMS,     W71-09005
UCTIVITY, MODEL STUDIES, SALMON,        AQUATIC ENVIRONMENT, PHOSPHATES,     W69-06536
ONING, ALGICIDES, AQUATIC ALGAE,        AQUATIC ENVIRONMENT, AQUATIC PLAN    W69-05844
UTROPHICATION, LAKES, LIMNOLOGY,        AQUATIC ENVIRONMENT, AQUATIC PROD    W69-05142
, LIMNOLOGY, HYDROLOGIC ASPECTS,        AQUATIC ENVIRONMENT, WATER PROPER    W70-04475
TION, WATER SPORTS, ENVIRONMENT,        AQUATIC ENVIRONMENT, ESTUARINE EN    W69-04276
TOSYNTHESIS, *CALCIUM CARBONATE,        AQUATIC ENVIRONMENT, AQUATIC PROD    W69-00446
ION, *TOXICITY, *BIOASSAY, FISH,        AQUATIC ENVIRONMENT, ORGANIC PEST    W72-11105
WATER POLLUTION EFFECTS, FUNGI,         AQUATIC FUNGI.: /LLUTION CONTROL,    W69-05706
TOCHASTIC PROCESSES, ECOSYSTEMS,        AQUATIC HABITATS, WATER POLLUTION    W71-09012
AQUATIC ENVIRONMENT, ECOSYSTEMS,        AQUATIC HABITATS, DRAGONFLIES, AL    W71-09005
GLENOPHYTA, CYANOPHYTA, DIATOMS,        AQUATIC HABITAT, POPULATION.: /EU    W70-02770
FLOATING PLANTS, PHYTOPLANKTON,         AQUATIC INSECTS, FISH, GAME BIRDS    W70-10175
TIONS, *RIVERS, ALGAE, PROTOZOA,        AQUATIC INSECTS, FISH REPRODUCTIO    W73-00748
ER TEMPERATURE, *EUTROPHICATION,        AQUATIC LIFE, SESTON, SCENEDESMUS    W73-01446
ACCLIMATIZATION, MARINE ANIMALS,        AQUATIC LIFE, AQUATIC ANIMALS, BE    W71-02494
, *WATER QUALITY, BIOINDICATORS,        AQUATIC LIFE, AQUATIC ALGAE, WATE    W70-05958
OPHICATION, PRODUCTIVITY, ALGAE,        AQUATIC LIFE.: /N COMPOUNDS, EUTR    W70-03068
PLANKTON, WATER QUALITY, ALGAE,         AQUATIC LIFE, STREAM POLLUTION, B    W72-01786
SAMPLING, SURVEYS, MARINE ALGAE,        AQUATIC LIFE, MILK, ATLANTIC OCEA    W72-00941
ALGAE, MODEL STUDIES, BIOASSAY,         AQUATIC LIFE, COOLING WATER, SITE    W71-11881
ION SOURCES, PATH OF POLLUTANTS,        AQUATIC LIFE, ALGAE, SILTS, ECOLO    W69-10080
IES, ECOLOGY, SALINITY, BENTHOS,        AQUATIC LIFE, ALGAE, PLANTS, WATE    W69-05023
STANDING CROP, MICROENVIRONMENT,        AQUATIC LIFE, ENVIRONMENT, MICROO    W69-03611
BACTERIA, ENVIRONMENTAL EFFECTS,        AQUATIC LIFE, AQUATIC MICROORGANI    W69-04276
NIC COMPOUNDS, METALS, PLANKTON,        AQUATIC LIFE, ZOOPLANKTON, AQUATI    W69-03695
SURFACE DRAINAGE, ALGAL CONTROL,        AQUATIC LIFE, WATER CONSERVATION,    W70-03344
INDICATORS, *BIOASSAY, TOXICITY,        AQUATIC MAMMALS, ALGAE, FISH, DIA    W71-10786
OZOA, *FLORIDA, *EUTROPHICATION,        AQUATIC MICROORGANISMS, TROPHIC L    W72-01993
AL EFFECTS, *MORPHOLOGY, *ALGAE,        AQUATIC MICROBIOLOGY, CYANOPHYTA,    W69-10177
ICAL TECHNIQUES, *AQUATIC ALGAE,        AQUATIC MICROBIOLOGY, AQUATIC MIC    W70-02792
ATIC ALGAE, AQUATIC ENVIRONMENT,        AQUATIC MICROBIOLOGY, BALANCE OF     W70-02968
TIC ALGAE, AQUATIC MICROBIOLOGY,        AQUATIC MICROORGANISMS, AQUATIC P    W70-02792
LADOPHORA, ALGAE, AQUATIC ALGAE,        AQUATIC MICROBIOLOGY, AQUATIC PRO    W70-02784
FFECTS, PHYTOPLANKTON, PLANKTON,        AQUATIC MICROBIOLOGY, CYCLING NUT    W70-02245
NTROL, ALGICIDES, AQUATIC ALGAE,        AQUATIC MICROBIOLOGY, BIOASSAY, C    W70-02248
BIODEGRADATION, ALGAE, BACTERIA,        AQUATIC MICROORGANISMS, SAMPLING,    W71-00221
ORS, ECOSYSTEMS, MICROORGANISMS,        AQUATIC MICROORGANISMS, ENZYMES.:    W71-02479
C POWER, HEATED WATER, PROTEINS,        AQUATIC MICROORGANISMS, TEMPERATU    W70-09905
TICAL TECHNIQUES, AQUATIC ALGAE,        AQUATIC MICROBIOLOGY, AQUATIC WEE    W69-03515
RONMENTAL EFFECTS, AQUATIC LIFE,        AQUATIC MICROORGANISMS, MICROORGA    W69-04276
AQUATIC ALGAE, AQUATIC ANIMALS,         AQUATIC MICROBIOLOGY, AQUATIC PRO    W69-03513
AQUATIC ALGAE, AQUATIC BACTERIA,        AQUATIC MICROBIOLOGY, BIOASSAY, C    W69-03371
TICAL TECHNIQUES, AQUATIC ALGAE,        AQUATIC MICROBIOLOGY, BALANCE OF     W69-03374
TICAL TECHNIQUES, AQUATIC ALGAE,        AQUATIC MICROBIOLOGY, BALANCE OF     W69-03517
COMPOUNDS, ALGAE, AQUATIC ALGAE,        AQUATIC MICROBIOLOGY, AQUATIC PRO    W69-03373
MAL WATER, ALGAE, AQUATIC ALGAE,        AQUATIC MICROBIOLOGY, BALANCE OF     W69-03516
AQUATIC ALGAE, AQUATIC BACTERIA,        AQUATIC MICROBIOLOGY, BALANCE OF     W69-03369
OLOGICAL ECOLOGY, AQUATIC ALGAE,        AQUATIC MICROBIOLOGY, CHLOROPHYTA    W69-03358
AQUATIC ALGAE, AQUATIC BACTERIA,        AQUATIC MICROBIOLOGY, BALANCE OF     W69-03188
TROGEN COMPOUNDS, AQUATIC ALGAE,        AQUATIC MICROBIOLOGY, CHLAMYDOMON    W69-03364
AQUATIC ALGAE, AQUATIC BACTERIA,        AQUATIC MICROBIOLOGY, BIOASSAY, C    W69-03362
TICAL TECHNIQUES, AQUATIC ALGAE,        AQUATIC MICROBIOLOGY, AQUATIC PRO    W69-03185
TICAL TECHNIQUES, AQUATIC ALGAE,        AQUATIC MICROBIOLOGY, AQUATIC PRO    W69-03186
, *PHYTOPLANKTON, AQUATIC ALGAE,        AQUATIC MICROBIOLOGY, AQUATIC PRO    W69-04802
FISHERIES, ALGAE, AQUATIC ALGAE,        AQUATIC MICROBIOLOGY, AQUATIC WEE    W69-04804
SSAY, *COMPARATIVE PRODUCTIVITY,        AQUATIC MICROBIOLOGY, EUTROPHICAT    W72-12240
FECTS, *BIOLOGICAL PRODUCTIVITY,        AQUATIC NUISANCES, TRANSPARENCY,     W69-09349
GRAPH, ARTHROPODA, CHIRONOMIDAE,        AQUATIC OLIGOCHAETES.: /THYTHERMO    W70-04253
WASTE TREATMENT, AQUATIC ALGAE,         AQUATIC PLANTS, TESTING, BIODEGRA    W70-04373
G NUTRIENTS, ZOOPLANKTON, ALGAE,        AQUATIC PLANTS, PALEOLIMNOLOGY, M    W70-03975
G NUTRIENTS, ZOOPLANKTON, ALGAE,        AQUATIC PLANTS, PALEOLIMNOLOGY, M    W70-03975
ATIC ALGAE, AQUATIC ENVIRONMENT,        AQUATIC PLANTS, AQUATIC POPULATIO    W69-05844
ON, PHOSPHORUS COMPOUNDS, ALGAE,        AQUATIC PLANTS.: / *DENITRIFICATI    W69-05323
C WEED CONTROL, *EUTROPHICATION,        AQUATIC PLANTS, HERBICIDES, PESTI    W69-05705
C WEED CONTROL, *EUTROPHICATION,        AQUATIC PLANTS, HERBICIDES, PESTI    W69-05706
LLUTION EFFECTS, ALGAE, ANIMALS,        AQUATIC PLANTS, BIOMASS, FERTILIZ    W69-04805
L, *EUTROPHICATION, *HARVESTING,        AQUATIC PLANTS, AQUATIC WEEDS, EC    W69-06276
*ALGAL CONTROL, FISHKILL, ROOTED        AQUATIC PLANTS, GLACIAL LAKES, PL    W68-00488
YANOPHY/ *EUTROPHICATION, ALGAE,        AQUATIC PLANTS, BIBLIOGRAPHIES, C    W68-00680
PLANTS, *NUISANCE ALGAE, ROOTED         AQUATIC PLANTS, *RESERVOIRS, HYDR    W68-00479
MENTS, ON-SITE DATA COLLECTIONS,        AQUATIC PLANTS, WATERSHED, ALGAE,    W68-00511
, CHEMCONTROL, PONDWEEDS, ROOTED        AQUATIC PLANTS, RATES OF APPLICAT    W70-05568
, TRACERS, SAMPLING, PERIPHYTON,        AQUATIC PLANTS, BACKGROUND RADIAT    W69-09334
, *TERTIARY TREATMENT, NITROGEN,        AQUATIC PLANTS, BOTTOM SEDIMENTS,    W69-09340
AQUATIC ANIMALS, BENTHIC FAUNA,         AQUATIC PLANTS, MARINE ALGAE, BEN    W71-02494
N CONCENTRATION, NUISANCE ALGAE,        AQUATIC PLANTS, BOTTOM SEDIMENTS,    W70-06225
```

```
, ALGAE, FLOATING PLANTS, ROOTED        AQUATIC PLANTS, SELF-PURIFICATION    W70-05750
ON, OCHROMONAS, BACTERIA, FUNGI,        AQUATIC PLANTS, ASSAY, DISTRIBUTI    W70-02510
BIOLOGY, AQUATIC MICROORGANISMS,        AQUATIC PLANTS, AQUATIC PRODUCTIV    W70-02792
AGE TREATMENT, AEROBIC BACTERIA,        AQUATIC PLANTS, ANAEROBIC BACTERI    W71-08221
ISSOLVED OXYGEN, DENSITY, ROOTED        AQUATIC PLANTS, ALGAE, CHLOROPHYL    W71-07698
RATURE, CHLOROPHYLL, CYANOPHYTA,        AQUATIC PLANTS, PHOSPHORUS, DENSI    W72-04761
VANIA, RIVERS, STREAMS, EUGLENA,        AQUATIC POPULATIONS, EUTROPHICATI    W70-04510
FISH, AMPHIBIANS, INVERTEBRATES,        AQUATIC POPULATIONS, BIBLIOGRAPHI    W68-00859
UTROPHICATION, MAINE, NUTRIENTS,        AQUATIC POPULATIONS, WATER QUALIT    W68-00470
AN POPULATION, SEWAGE EFFLUENTS,        AQUATIC POPULATIONS, NUISANCE ALG    W68-00461
NG, RECREATION, SILT, NUTRIENTS,        AQUATIC POPULATIONS, OLIGOTROPHY,    W68-00172
TIC ENVIRONMENT, AQUATIC PLANTS,        AQUATIC POPULATIONS, CONSERVATION    W69-05844
 LIMNOLOGY, AQUATIC ENVIRONMENT,        AQUATIC PRODUCTIVITY, HYPOLIMNION    W69-05142
TIC ALGAE, AQUATIC MICROBIOLOGY,        AQUATIC PRODUCTIVITY, CYANOPHYTA,    W69-04802
EWAGE, ALGICIDES, AQUATIC ALGAE,        AQUATIC PRODUCTIVITY, BIOASSAY, C    W69-04798
 CARBONATE, AQUATIC ENVIRONMENT,        AQUATIC PRODUCTIVITY, EUTROPHICAT    W69-00446
TIC ALGAE, AQUATIC MICROBIOLOGY,        AQUATIC PRODUCTIVITY, AQUATIC WEE    W69-03185
TIC ALGAE, AQUATIC MICROBIOLOGY,        AQUATIC PRODUCTIVITY, AQUATIC WEE    W69-03186
PHYTOPLANKTON, *CULTURES, ALGAE,        AQUATIC PRODUCTIVITY, CYCLING NUT    W69-03514
TIC ALGAE, AQUATIC MICROBIOLOGY,        AQUATIC PRODUCTIVITY, AQUATIC WEE    W69-03373
C ANIMALS, AQUATIC MICROBIOLOGY,        AQUATIC PRODUCTIVITY, BALANCE OF     W69-03513
ALGAE, COMPARATIVE PRODUCTIVITY,        AQUATIC PRODUCTIVITY, DISTRIBUTIO    W70-05424
 MICROORGANISMS, AQUATIC PLANTS,        AQUATIC PRODUCTIVITY, BALANCE OF     W70-02792
TIC ALGAE, AQUATIC MICROBIOLOGY,        AQUATIC PRODUCTIVITY, CHLOROPHYTA    W70-02784
FFECTS, WATER POLLUTION SOURCES,        AQUATIC PRODUCTIVITY, LIMNOLOGY,     W69-09135
ICATORS, SECONDARY PRODUCTIVITY,        AQUATIC PRODUCTIVITY, WATER POLLU    W69-10152
QUATIC PLANTS, WATERSHED, ALGAE,        AQUATIC VEGETATION, NITROGEN, PHO    W68-00511
ROBIOLOGY, AQUATIC PRODUCTIVITY,        AQUATIC WEEDS, CHLOROPHYTA, CYANO    W69-03373
TIC ALGAE, AQUATIC MICROBIOLOGY,        AQUATIC WEEDS, BALANCE OF NATURE,    W69-03515
ROBIOLOGY, AQUATIC PRODUCTIVITY,        AQUATIC WEEDS, BALANCE OF NATURE,    W69-03185
ROBIOLOGY, AQUATIC PRODUCTIVITY,        AQUATIC WEEDS, BIOASSAY, CYANOPHY    W69-03186
ION, *PHOSPHATE, FISH, PLANKTON,        AQUATIC WEED CONTROL, INSECT CONT    W70-05417
AQUATIC WEEDS, LAKES, ALGICIDES,        AQUATIC WEED CONTROL, MECHANICAL     W70-04494
TIC ALGAE, AQUATIC MICROBIOLOGY,        AQUATIC WEEDS, BALANCE OF NATURE,    W69-04804
 *EUTROPHICATION, *LAKES, ALGAE,        AQUATIC WEEDS, CYANOPHYTA, FISHKI    W69-06535
ON, *HARVESTING, AQUATIC PLANTS,        AQUATIC WEEDS, ECONOMIC FEASIBILI    W69-06276
AUNA, PHOTOSYNTHESIS, BIOASSAYS,        AQUATIC WEEDS, THERMAL STRATIFICA    W70-02251
, *WATER QUALITY CONTROL, ALGAE,        AQUATIC WEEDS, ADMINISTRATION, SU    W72-05774
                *OIL SPILLS, PROTEIN    AQUICULTURE.:                        W69-05891
GAL POISONING, GASOLINE, FLOODS,        AQUIFERS, WATER LAW, NITRATES, TU    W73-00758
GILARIA, MELOSIRA, ASTERIONELLA,        ARAPHIDINEAE, CENTRIC DI: /R, FRA    W70-05663
ONS, ELECTROKINETIC EFFECTS, ANN        ARBOR(MICH), SAND BEDS.: /USPENSI    W71-03035
LAKES, DISSOLVED ORGANIC MATTER,        ARCTIC ALGAE, AUTOTROPHIC METABOL    W71-10079
VER, NEBESNA RIVER, MOOSE CREEK,        ARCTIC CIRCLE HOT SPRINGS, RADON,    W69-08272
 PRIMARY PRODUCTIVITY, SURFACES,        ARCTIC, ENZYMES, CHLOROPHYLL, PHO    W70-04809
 SEDIMENTS, GEOLOGIC FORMATIONS,        ARCTIC, TUNDRA, SCALING, GRAVELS,    W69-08272
ITY, BIOLOGICAL MEMBRANES, HUMID        AREA.: /, WATER POLLUTION, *TOXIC    W69-05306
AE, LAKE STRATIFICATION, SURFACE        AREA(LAKES), PHAEDACTYLUM, WASTE     W69-09340
OTENTIAL NUTRIENT CONCENTRATION,        AREAL INCOME(NUTRIENTS), LAKE VOL    W70-00270
ERSTATE POLLUTION, *METROPOLITAN        AREAS, BLONDIN.: /EVOLUTION, *INT    W70-04430
ERATURE, AMINO ACIDS, INTERTIDAL        AREAS, CARBOHYDRATES, YEASTS, ENZ    W70-03952
NOU/ *LAKE MENDOTA(WIS), PELAGIC        AREAS, HORIZONTAL, BLOOMS, GELATI    W70-06229
, POPULATION, OREGON, INTERTIDAL        AREAS, PHYSIOLOGICAL ECOLOGY, LIT    W70-03325
EMPERATURE, PIPTIDES, INTERTIDAL        AREAS, SOIL BACTERIA, LABORATORY     W70-04365
IL ALGAE, *AQUATIC ALGAE, ALGAE,        ARID CLIMATES, SEMIARID CLIMATES,    W69-03491
ING, WATER DEMAND, WATER SUPPLY,        ARID LANDS, LIMNOLOGY.: / MONITOR    W70-05645
, SOIL WATER MOVEMENT, LEACHING,        ARID LANDS, DEPTH, CYANOPHYTA, CH    W71-04067
N ANALYSES, AGRICULTURAL RUNOFF,        ARKANSAS RIVER, MERRIMACK RIVER,     W70-04507
, POLLUTION, FOSSIL, CIMARRON R/        ARKANSAS RIVER(OKLA), TULSA(OKLA)    W70-04001
OL, *OIL SPILLS, *TORREY CANYON,        AROMATIC SOLVENTS, PHAEODACTYLUM     W70-01470
INORGANIC PHOSPHATES, SOLUBLE R/        ARSENATE INTERFERENCE, DISSOLVED     W70-05269
POUNDS, LAKES, SEWAGE EFFLUENTS,        ARSENIC COMPOUNDS, SPECTROPHOTOME    W70-05269
 POLLUTION CONTROL, CHEMCONTROL,        ARSENIC COMPOUNDS, NITROGEN, PHOS    W70-04494
OMETRY, ALGAE, MINNESOTA, SODIUM        ARSENITE.: /OMPOUNDS, SPECTROPHOT    W70-05269
ICIDES, *COPPER SULFATE, *SODIUM        ARSENITE, NUISANCE ALGAE, SNAILS,    W70-05568
PPER SULPHATE, PONDWEEDS, SODIUM        ARSENITE, FISH, ANIMALS, ALGAE, C    W70-05263
NTROLS, PARAQUAT, DIQUAT, SODIUM        ARSENITE, COPPER SULPHATE, MONURO    W70-00269
OCONTROL, COPPER SULFATE, SODIUM        ARSENITE.: /ECHANICAL CONTROL, BI    W70-10175
TIONS, BLACK SEA, CONTACT ZONES,        ARTEMIA, BALANUS, DESORPTION, EMB    W69-07440
CERIUM-144, NITZSCHIA CLOSERIUM,        ARTEMIA, CARTERIA, NORTH TEMPERAT    W70-00708
 HETERETROPHY, BACTERIAL MEDIUM,        ARTHROBACTER, FLORA, AMBIENT TEMP    W70-00713
TE, CHEMOSYNTHESIS, CONCRETIONS,        ARTHROBACTER, PYROPHOSPHATE, ORTH    W69-07428
, METALIMNION, BATHYTHERMOGRAPH,        ARTHROPODA, CHIRONOMIDAE, AQUATIC    W70-04253
UM BOTULINUS, VENERUPIS, OSTREA,        ARTHROPODS.: /AXIDOMUS, CLOSTRIDI    W70-05372
 WASTE TREATMENT, COLONY COUNTS,        ARTHROPODS.: /PODS, SPIDERS, MILK    W72-01819
ARCH AND DEVELOPMENT CORP(MASS),        ARTHUR D LITTLE INC(MASS), CARNEG    W69-06865
IAL FERTILIZERS.:                       ARTIFICIAL FERTILIZATION, COMMERC    W68-00487
, WATER CONVEYANCE, WATER COSTS,        ARTIFICIAL WATER COURSES, CALIFOR    W71-09577
S, MILWAUKEE(WIS), CHICAGO(ILL),        ARTIFICIAL CIRCULATION, NUISANCES    W69-10178
ADISON(WIS), PSOROPHORA CILIATA,        ARTIFICIAL KEY.: /DRIMACULATUS, M    W70-06225
ECHNIQUES, LABORATORY EQUIPMENT,        ASBESTOS CEMENT, CHLORELLA, CARBO    W71-07339
S VESICULOSUS, CHONDRUS CRISPUS,        ASCOPHYLLUM NODOSUM, LAMINARIA DI    W70-01073
OYA(NORWAY), FUCUS VESCICULOSUS,        ASCOPHYLLUM NODOSUM, ECTOCARPUS C    W70-01074
, FUCUS VESI/ *SEAWEED EXTRACTS,        ASCOPHYLLUM NODOSUM, ALGINIC ACID    W69-07826
ORTALITY, CELL WALL, HEXOSAMINE,        ASCOPHYLLUM NODOSUM, POLYSIPHONIA    W70-04365
POROBOLOMYCES ROSEUS, AUTOLYSIS,        ASCOSPOROGENOUS.: /BE(GERMANY), S    W70-04368
NCENTRATION, 'EXTRA-DEEP PONDS',        ASHKELON(ISRAEL), TEMPERATURE PRO    W70-04790
SCILLATORIA, ANABAENA, DIAPTOMUS        ASHLANDI, EPISCHURA NEVADENSIS, D    W69-10182
                            LAKE        ASHTABULA.:                          W70-09093
S(USA), UNIVERSITY OF WISCONSIN,        ASIA, NATIONAL RESEARCH COUNCIL(U    W70-03975
```

S(USA), UNIVERSITY OF WISCONSIN,
EDIES, JUDICIAL DECISIONS, LEGAL
EFFICIENCY, HEAT BUDGET, SOCIAL
RIA, FISH, LIMNOLOGY, HYDROLOGIC
T-OF-WAY, ESTIMATED COSTS, LEGAL
A CYLINDRICA, BACILLUS SUBTILIS,
VULGARIS, KARATELLA COHCLEARIS,
ROCYSTIS, KERATELLA, POLYARTHRA,
UTROPHICATION, PROVISIONAL ALGAL
BACTERIA, FUNGI, AQUATIC PLANTS,
S, *AGRICULTURE, *WASTES, ALGAE,
Y CONSUMPTION, PROVISIONAL ALGAL
LENA GRACILIS Z, MICROBIOLOGICAL
ON(WIS), EUROPE, WATER POLLUTION
ON(WIS), EUROPE, WATER POLLUTION
N, ALGAL GROWTH, WATER POLLUTION
BIOSTIMULATION, WATER POLLUTION
US RIVER(CALIF), WATER POLLUTION
OSA, SELENASTRUM GRACILE, CARBON
NITROGEN
(WASH), *REACTOR, RADIONUCLIDES,
OPIC LEVEL, WATER QUALITY, WASTE
LUTION, MARINE ORGANISMS, MARINE
DOR, TAST/ *AMERICAN WATER WORKS
IC ALGAE, *EPIPELIC ALGAE, ALGAL
PORTULACOIDES, SUAEDA MARITIMA,
), CARLETON POND(MAINE), SYNURA,
NS, WESTHAMPTON LAKE(VA).:
AGILARIA CROTONENSIS, NITZSCHIA,
ENT LIMITATION, CAYUGA LAKE(NY),
RECTION, TABELLARIA, FRAGILARIA,
ENODINIUM, CRYPTOMONAS, CILIATA,
UIREMENTS, THRESHOLD ODOR TESTS,
XCRETION, STEPHANODISCUS TENUIS,
HROMOBACTER, GULF OF MARSEILLES,
LTRA-VIOLET LIGHT STERILIZATION,
ALGAE, GREEN ALGAE, FRAGILARIA,
ANACYSTIS, MICROCYSTIS, BOSMINA,
TATUM, PHYSIOLOGICAL ADJUSTMENT,
ANIZOMENON FLOS AQUAE, MELOSIRA,
TAE, HETEROKONTAE, GLOEOTRICHIA,
TON CORER, FRAGILARIA, MELOSIRA,
MENON FLOS AQUAE, STEPHANODISCUS
APHICAL REGIONS, COASTAL PLAINS,
LOSIONS, ABSORPTION, ADSORPTION,
COPPER, IRON, COBALT, MANGANESE,
ION, ABSORPTION, GULF OF MEXICO,
ARINE ALGAE, AQUATIC LIFE, MILK,
UTROPHICATION, *WATER STRUCTURE,
RUNOFF, UNDERFLOW, PRECIPITATION(
ION, ROADS, ROOFS, PRECIPITATION(
06, ANTIMONY-125, GRO/ *ENIWETOK
5, GRO/ *ENIWETOK ATOLL, *BIKINI
OCEA, WOODS HOLE OCEAN/ ENIWETOK
ER TECHNIQUE, OXYGEN ELECTRODES,
TIONAL RESEARCH COUNCIL(USA), US
TIONAL RESEARCH COUNCIL(USA), US
ACETYLENE/ *ANABAENA CYLINDRICA,
FRONDS, HYDROCARBONS, BACTERIAL
FILTERS, SOLAR FRAUNHOFER LINES,
ATION RATES, MANGANESE, LOW-FLOW
SLU/ *WATER PURIFICATION, *FLOW
, AMPHIPLEURA RUTILANS, AMPHORA,
CA MILIARIS, HELGOLAND(GERMANY),
MONAS AERUGINOSA, STAPHYLOCOCCUS
:
UP, FOOD WASTES.:   SHEPPARTION(
SK(USSR), ENGLAND, OSCILLATORIA,
KLOPEINER(AUSTRIA), LAKE WORTHER(
VERTICAL, CHALK/ *LAKE KLOPEINER(
LAKE SCHLIER(BAVARIA), LAKE ZELL(
NA, EUCYCLOPS, NAIS, CRICOTOPUS,
SSES, CHEMOS/ SPECIES DIVERSITY,
UTROPHICATION, *TENNESSEE VALLEY
LATES, SAPROBIEN SYSTEM, CILIATE
LLA PYRENOIDOSA(EMERSON STRAIN),
DUCTS, HORMONES, AUXINS, METABO/
GERMANY), SPOROBOLOMYCES ROSEUS,
PHYCOCOLLOIDS, TA/ *ANTIBIOSIS,
E, FLOW RATES, SLUDGE, SLURRIES,
LYSIS, TRICARBOXYLIC ACID CYCLE,
HEMICAL EXTRACTION, MILL/ *MICRO-
FILM, SYNECHOCOCCUS, GRAIN COUNT(
MICROSCOPY, ACETATE, CARBON-14,
BO/ HETEROTROPHIC, CHEMOTROPHIC,
UTRIENTS, CHLORELLA PYRENOIDOSA,
MATHEMATICAL STUDIES,
ED ORGANIC MATTER, ARCTIC ALGAE,
N UPTAKE TECHNIQ/ *HETEROTROPHY,
XTRACELLULAR PRODUCTS, HORMONES,

ASIA, NATIONAL RESEARCH COUNCIL(U W70-03975
ASPECTS.: /, STREAMS, SEWAGE, REM W69-06909
ASPECTS.: /ON, TURBINES, ECONOMIC W70-08834
ASPECTS, AQUATIC ENVIRONMENT, WAT W70-04475
ASPECTS, LEGISLATION, SEWAGE DIST W70-03344
ASPERGILLUS ORYZAE NIGER, S: /AEN W70-02964
ASPLANCHA, LECANE, BOSMINA, NAVIC W70-02249
ASPLANCHNA, BRANCHIONUS.: /A, MIC W70-05375
ASSAY PROCEDURE, BATCH CULTURES,   W70-02777
ASSAY, DISTRIBUTION, DENSITY, CHR W70-02510
ASSAY, DISSOLVED SOLIDS, CHEMICAL W70-03334
ASSAY, PROCEDURES, SELENASTRUM CA W70-02779
ASSAY, SPECIES COMPOSITION.: /EUG W69-06273
ASSESSMENT, URBAN DRAINAGE, AGRIC W70-03975
ASSESSMENT, URBAN DRAINAGE, AGRIC W70-03975
ASSESSMENT, SANITARY ENGINEERING   W70-02775
ASSESSMENT.: /L GROWTH, ANABAENA, W70-02779
ASSESSMENT, AMERICAN RIVER(CALIF) W70-02777
ASSIMILATION, BACILLARIOPHYCEAE,   W70-03983
ASSIMILATION.:   W69-06970
ASSIMILATION, BA-LA, HALF-LIFE, P W69-07853
ASSIMILATIVE CAPACITY, STANDING C W68-00912
ASSOCIATIONS, MOSCOW, PETROLEUM P W69-07440
ASSOCIATION, CHEMICAL INDUSTRY, O W71-05943
ASSOCIATIONS, SHEAR WATER(ENGLAND W71-12870
ASTER TRIPOLIUM, PEMBROKE, SOUTH- W70-09976
ASTERIONELLA, TYPHOID FEVER, KENN W70-08096
ASTERIONELLA SPIROIDES, CLADOCERA W71-10098
ASTERIONELLA GRACILLIMA, GROWTH F W70-02804
ASTERIONELLA FORMOSA, SYNEDRA, ME W70-01579
ASTERIONELLA, FILTER CLOGGING, ST W70-00263
ASTERIONELLA FORMOSA, RECOLONIZAT W70-00268
ASTERIONELLA, APHANIZOMENON, SYNU W69-10157
ASTERIONELLA FORMOSA, MELOSIRA BI W69-10158
ASTERIONELLA JAPONICA, CHAETOCERO W70-04280
ASTERIONELLA, SCENEDESMUS OBLIQUU W70-03955
ASTERIONELLA, MELOSIRA, MICROCYST W70-05375
ASTERIONELLA, MELOSIRA, SYNEDRA,   W70-04503
ASTERIONELLA FORMOSA, CYCLOTELLA   W70-04809
ASTERIONELLA FORMOSA, FRAGILARIA, W70-05112
ASTERIONELLA, DINOBRYON, MELOSIRA W70-05409
ASTERIONELLA, ARAPHIDINEAE, CENTR W70-05663
ASTREA, LAKE MINNETONKA(MINN), MO W70-05387
ATLANTIC COASTAL PLAIN, INORGANIC W69-03695
ATLANTIC OCEAN, DIATOMS, NERITIC   W70-00707
ATLANTIC OCEAN, PUERTO RICO, SPEC W69-08525
ATLANTIC OCEAN, RADIOACTIVITY, GA W69-08524
ATLANTIC OCEAN, LANDFILLS, DISPOS W72-00941
ATMOSPHERE, SEA WATER, FRESH WATE W70-04385
ATMOSPHERIC), DRAINAGE, FERTILIZE W70-04810
ATMOSPHERIC), WETLANDS, SEEPAGE,   W69-10178
ATOLL, *BIKINI ATOLL, RUTHENIUM-1 W69-08269
ATOLL, RUTHENIUM-106, ANTIMONY-12 W69-08269
ATOLL, TURNOVER TIME, TRIDACNA CR W69-08526
ATOMIC ABSORPTION SPECTROPHOTOMET W70-03309
ATOMIC ENERGY COMMISSION, NATIONA W70-03975
ATOMIC ENERGY COMMISSION, NATIONA W70-03975
ATP, CMU, NITROGEN-15, PYRUVATE,   W70-04249
ATTACHMENT, ECOTOCARPUS, POLYSIPH W70-01068
ATTENUATION MEASUREMENT APPARATUS W70-05377
AUGMENTATION, TENNESSEE RIVER, AL W73-00750
AUGMENTATION, *LAGOONS, ACTIVATED W72-06017
AULACODISCUS, BACILLARIA, BIDDULP W70-03325
AUREOBASIDIUM PULLULANS, HANSENIA W70-04368
AUREUS, LACTOBACILLUS CASEI, BACI W71-11823
AUSTIN(TEX), COLORADO RIVER(TEX). W70-04727
AUSTRALIA), OXIDATION DITCHES, SO W71-09548
AUSTRALIA, SANTA ROSA(CALIF), LYU W70-05092
AUSTRIA), VERTICAL, CHALK, METALI W70-05405
AUSTRIA), LAKE WORTHER(AUSTRIA),   W70-05405
AUSTRIA), LAKE HALL WILER(SWITZER W70-05668
AUSTROCLOEON, ENALLAGMA, PSEUDAGR W70-06975
AUTECOLOGY, MORPHOLOGY, EXPORT LO W70-02780
AUTHORITY PROJECT, *RESERVOIRS, T W71-07698
AUTOECOLOGY.:   *CILIATES, FLAGEL W71-03031
AUTOINHIBITORS.: / GROWTH, CHLORE W69-07442
AUTOINHIBITORS, EXTRACELLULAR PRO W69-04801
AUTOLYSIS, ASCOSPOROGENOUS.: /BE(W70-04368
AUTOLYSIS, EXCRETION, DEPURATION, W70-01068
AUTOMATIC CONTROL, PIPING SYSTEMS W72-08688
AUTORADIOGRAPHY, MONODUS SUBTERRA W70-02965
AUTORADIOGRAPHY, PHYSIOLOGY, BIOC W69-10163
AUTORADIOGRAPHY), LEUCOTHRIX MUCO W69-10163
AUTORADIOGRAPHY, MICHAELIS-MENTEN W70-04811
AUTOTROPHIC, SILICON, PEDIASTRUM   W70-02804
AUTOTROPHIC GROWTH, HETEROTROPHIC W70-02964
AUTOTROPHIC, HETEROTROPHIC.:   W70-05750
AUTOTROPHIC METABOLISM, HETEROTRO W71-10079
AUTOTROPHY, CARBON-14, RADIOCARBO W70-04287
AUXINS, METABOLITES, INHIBITION,   W69-04801

ROCYSTIS AERUGINOSA, / *NUTRIENT   AVAILABILITY, *ALGAL GROWTH, *MIC   W69-05867
ICROCYSTIS AERUGINOSA, *NUTRIENT   AVAILABILITY, ALGAL GROWTH, ALGAL   W69-05868
IAL, SE/ *ALGAL GROWTH, NUTRIENT   AVAILABILITY, ALGAL GROWTH POTENT   W69-06864
GETATION, NITROGEN, PHOSPHOROUS,   AVAILABLE NUTRIENTS, SEDIMENT COR   W68-00511
OSPHORUS, PHOSPHATE(BIOLOGICALLY   AVAILABLE), MICROCYSTIS AERUGINOS   W70-05269
, *FIXED-LIGHT TESTS, *EFFECTIVE-   AVERAGE-LIGHT-INTENSITY, PHOTO-EL   W70-03923
IUM PARVUM, SYNERGISTIC EFFECTS,   AXENIC CULTURES, MONOALGAL, PORTU   W70-05372
SALTS, NITROGEN FIXATION, BORON,   AZOTOBACTER, MOLYBDENUM, COBALT,   W70-02964
OIDOSA, BICHROMATE, DRY WEIGHT,   B COLI COMMUNAE, EUGLENE GRACILIS   W70-05092
WATER POLLUTION SOURCES, VITAMIN   B.: /S, WATER POLLUTION EFFECTS,   W69-06277
CATION, ORGANIC SOLUTES, VITAMIN   B-1, VITAMIN B-12, THIAMINE, SERR   W70-05652
STOC MUSCORUM, VANADIUM, VITAMIN   B-12.: /IPLOCYSTIS AERUGINOSA, NO   W69-06277
N, EUGLENA GRACILIS Z,/ *VITAMIN   B-12, ALGAL GROWTH, CYANOCOBALAMI   W69-06273
IC SOLUTES, VITAMIN B-1, VITAMIN   B-12, THIAMINE, SERRATIA MARINORU   W70-05652
, *SEA WATER, *VITAMINS, VITAMIN   B, ALGAE, CULTURES, CHEMICAL ANAL   W70-05652
IES, BACTERIA, MOLLUSKS, VITAMIN   B, RED TIDE, SALINITY, FLORIDA, T   W70-05372
OR, RADIONUCLIDES, ASSIMILATION,   BA-LA, HALF-LIFE, PTYCHOCHEILUS O   W69-07853
RUTILANS, AMPHORA, AULACODISCUS,   BACILLARIA, BIDDULPHIA, CHAETOCER   W70-03325
UM GRACILE, CARBON ASSIMILATION,   BACILLARIOPHYCEAE, DINOPHYCEAE.: /   W70-03983
CONJUGATOPHYCAEA, PLANCTOMYCES,   BACILLARIOPHYTA, GLOECAPSA, MERIS   W70-03969
E, ALGAL GROWTH, ANKISTRODESMUS,   BACILLARIOPHYCEAE, CHLOROPHYCEAE,   W69-07833
CULA PELLICULOSA, XANTHOPHYCEAE,   BACILLARIOPHYCEAE, NITZSCHIA PALE   W70-02965
CHIZOMYCETES, CYANOPHYTA, FUNGI,   BACILLARIOPHYCEAE, CHAROPHYTA, RH   W70-02784
PRICORNUTUM PRINTZ, OSCILLATORI/   BACILLARIOPHYCEAE, SELENASTRUM CA   W70-05752
CYLINDRICA, ZINC, XANTHOPHYCEAE,   BACILLARIOPHYCEAE, MYXOPHYCEAE, C   W69-10180
YCOBACTERIUM, BACILLUS SUBTILIS,   BACILLUS MYCOIDES, BACILLUS MESEN   W69-07838
LUS SUBTILIS, BACILLUS MYCOIDES,   BACILLUS MESENTERICUS, DIATOMEA,   W69-07838
ION, PSEUDOMONAS, MYCOBACTERIUM,   BACILLUS SUBTILIS, BACILLUS MYCOI   W69-07838
A GRACILIS, ANABAENA CYLINDRICA,   BACILLUS SUBTILIS, ASPERGILLUS OR   W70-02964
OTH/ RICKETTSIAE, STERILIZATION,   BACILLUS SUBTILIS, BACILLUS STEAR   W71-11823
TERILIZATION, BACILLUS SUBTILIS,   BACILLUS STEAROTHERMOPHILUS, PROT   W71-11823
CUS AUREUS, LACTOBACILLUS CASEI,   BACILLUS THURINGIENSIS, AGARS, ST   W71-11823
POSITIVE BACTERIA, SARCINE-LIKE,   BACILLUS-LIKE, MICROCOCCUS-LIKE,   W70-00713
IDLEY CREEK(PA), NOBS CREEK(MD),   BACK RIVER(MD), OSCILLATORIA, CAT   W70-04510
LITY TEST, BODY IMMEDIUM FILTER,   BACK-WASHING.: / FLOW, PUTRESCIBL   W70-05821
ING, PERIPHYTON, AQUATIC PLANTS,   BACKGROUND RADIATION, GEOCHEMISTR   W69-09334
RADIOISOTOPES, ALGAE, EMULSIONS,   BACKGROUND RADIATION, IODINE RADI   W69-10163
, ALGAE, INSECTS, INVERTEBRATES,   BACKGROUND RADIATION, FLOCCULATIO   W69-07846
COMBINED SEWERS, STORMWATER OVE/   BACKWASHING, *SUSPENDED SOLIDS, *   W71-10654
ALUM, *OSHKOSH(WIS),   BACKWASHING.:   W72-08859
TROPHICATION, PONDS, RESERVOIRS,   BACKWATER, FERTILIZATION, NITROGE   W70-00274
DIATOMS, BIOLOGICAL COMMUNITIES,   BACKWATER, PONDS, ALASKA.: /YTA,   W70-03969
ER, CHEMICAL ANALYSIS, PROTEINS,   BACTERIA.: /OMPOSING ORGANIC MATT   W72-01812
AQUATIC LIFE, STREAM POLLUTION,   BACTERIA, IMPURITIES, WASTE TREAT   W72-01786
RICKLING FILTERS, *BIOTA, SEWAGE   BACTERIA, SYSTEMATICS, FUNGI, ALG   W72-01819
POLLUTION EFFECTS, FOOD CHAINS,   BACTERIA, DEGRADATION(DECOMPOSITI   W72-01801
TERIA, AQUATIC PLANTS, ANAEROBIC   BACTERIA, ALGAE, BIOCHEMICAL OXYG   W71-08221
, FISH, DIATOMS, FUNGI, INSECTS,   BACTERIA, PROTOZOA, SEWAGE, LAKES   W71-10786
MMING, *WATER POLLUTION EFFECTS,   BACTERIA, ALGAE, FISH, SEWERS, RU   W71-09880
ATURE, SEWAGE TREATMENT, AEROBIC   BACTERIA, AQUATIC PLANTS, ANAEROB   W71-08221
RIENTS, LAKES, WATER PROPERTIES,   BACTERIA, BIOASSAY, METABOLISM, A   W71-10079
HEATED WATER, THERMAL POLLUTION,   BACTERIA, EFFLUENTS, OXYGEN DEMAN   W72-00462
EN, PHOSPHOROUS, NITROGEN FIXING   BACTERIA, *ACTIVATED SLUDGE, STOR   W71-13335
*CULTURES, *BIOASSAYS, PROTOZOA,   BACTERIA, FUNGI, VIRUSES, ALGAE,   W71-11823
IA, DRAINAGE PROGRAMS, ANAEROBIC   BACTERIA, ANAEROBIC CONDITIONS, L   W71-04556
IO, *WATER QUALITY, *MONITORING,   BACTERIA, WATER CHEMISTRY, PHOSPH   W71-05879
ADATION, CHLORINATION, COLIFORM,   BACTERIA, BIOCHEMICAL OXYGEN DEMA   W71-05013
YGEN DEMAND, *ANAEROBICS, ALGAE,   BACTERIA, WATER POLLUTION SOURCES   W71-05742
IA, DRAINAGE PROGRAMS, ANAEROBIC   BACTERIA, ANEROBIC CONDITIONS, LA   W71-04554
CARBON, WATER POLLUTION EFFECTS,   BACTERIA, PERIPHYTON, OHIO.: /R,   W71-04206
IA, DRAINAGE PROGRAMS, ANAEROBIC   BACTERIA, ANAEROBIC CONDITIONS, L   W71-04557
T, INDUSTRIAL WASTES, NUTRIENTS,   BACTERIA, ALGAE, TREATMENT FACILI   W71-06737
ERIE, BIOCHEMICAL OXYGEN DEMAND,   BACTERIA, ECOLOGY, WATER POLLUTIO   W71-04758
IA, DRAINAGE PROGRAMS, ANAEROBIC   BACTERIA, ANAEROBIC CONDITIONS, L   W71-04555
S, ACHROMOBACTERS, GRAM-POSITIVE   BACTERIA, SARCINE-LIKE, BACILLUS-   W70-00713
COCHEMICAL PROPERTIES, GENETICS,   BACTERIA, PEPTIDES, ENVIRONMENTAL   W70-00273
OXIDE, *REVIEWS, BIBLIOGRAPHIES,   BACTERIA, DECOMPOSING ORGANIC MAT   W70-00664
LUDGE WORMS, BLOODWORMS, AQUATIC   BACTERIA, ALGAE, MAYFLIES, CADDIS   W70-00475
ZOOPLANKTON, PHYTOPLANKTON, ICE,   BACTERIA, TRIPTON, BOTTOM SEDIMEN   W69-10154
, TRAVERTINE, CALCIUM CARBONATE,   BACTERIA, BIOMASS, CHLOROPHYLL, P   W69-10160
IMNOLOGY, OCEANOGRAPHY, TRACERS,   BACTERIA, GROWTH RATES, TRITIUM,   W69-10163
BIODEGRADATION, *ALGAE, *AQUATIC   BACTERIA, DETERGENTS, SURFACTANTS   W69-09454
RUS, ORGANOPHOSPHORUS COMPOUNDS,   BACTERIA, DIATOMS, DETRITUS, CYCL   W69-09334
ERMOCLINE, ZOOPLANKTON, TRIPTON,   BACTERIA, SEDIMENTATION, COPEPODS   W69-10169
OTOSYNTHESIS, CHEMICAL ANALYSIS,   BACTERIA, NUTRIENTS, TEMPERATURE,   W69-10167
RICIDES, *CHLORINATION, *AQUATIC   BACTERIA, *AQUATIC MICROBIOLOGY,   W69-10171
TROGEN COMPOUNDS, SEWAGE, SEWAGE   BACTERIA, SEWAGE TREATMENT, WATER   W69-10171
ODUCTIVITY, *ZOOPLANKTON, SULFUR   BACTERIA, DIEL MIGRATION, CARBON   W70-05760
S, PYCNOCLINE, PHOTOLITHOTROPHIC   BACTERIA, DESULFOVIBRIO, MESOTHER   W70-05760
TORMS, TURBULENCE, PILOT PLANTS,   BACTERIA, SEDIMENTATION, TOXICITY   W70-05819
LAKES, NUTRIENTS, FERTILIZATION,   BACTERIA, ALGAE, PLANKTON, TUBIFI   W70-05761
AGOONS, *BIODEGRADATION, *SEWAGE   BACTERIA, ODOR, ANAEROBIC CONDITI   W71-02009
IDE, SALINITY, LAKES, NUTRIENTS,   BACTERIA, ALGAE, BRACKISH WATER,   W71-03031
NS, AEROBIC CONDITIONS, E. COLI,   BACTERIA, FUNGI, ACTINOMYCETES, P   W71-02009
RMAL POLLUTION, *ECOLOGY, ALGAE,   BACTERIA, COLIFORMS, FISH, FUNGI,   W71-02479
TRIENTS, *BIODEGRADATION, ALGAE,   BACTERIA, AQUATIC MICROORGANISMS,   W71-00221
ENTRATION, SCENEDESMUS, EUGLENA,   BACTERIA, YEASTS, ACTINOMYCETES,   W70-08097
OTABLE WATER, CYANOPHYTA, SULFUR   BACTERIA, COLOR, IRON BACTERIA, T   W70-08107
TION-REDUCTION POTENTIAL, SULFUR   BACTERIA, *WASTE WATER TREATMENT,   W70-07034
TA, SULFUR BACTERIA, COLOR, IRON   BACTERIA, TASTE, FILTERS, IOWA, D   W70-08107

HLORELLA, CHELATION, OCHROMONAS,
LAKES, SEWAGE TREATMENT, SEWAGE
POLYSIPHONIA HARVEYI, EPIPHYTIC
RES, LAKES, ALGAE, GROWTH RATES,
, INDUSTRIAL WASTES, NEUTRIENTS,
BOLISM, WIND VELOCITY, HUMIDITY,
S, ELECTROLYTES, ZETA POTENTIAL,
OYSTERS, INDUSTRIES, CHELATION,
TERGENTS, *INSECTICIDES, *MARINE
TES, DISSOLVED OXYGEN, DETRITUS,
EN, SOLAR RADIATION, PERIPHYTON,
TS, ORGANIC MATTER, IRON OXIDES,
PHYTA, CHLOROPHYTA, CHLOROPHYLL,
CROP, ALGAE, ZOOPLANKTON, FISH,
DATION LAGOONS, *DESIGN, *SULFUR
OLIMNOLOGY, MATHEMATICAL MODELS,
OLIMNOLOGY, MATHEMATICAL MODELS,
, PHYSIOLOGICAL ECOLOGY, AQUATIC
PHYTOPLANKTON, DIFFUSION, ALGAE,
TITION, *PHYTOPLANKTON, *AQUATIC
IDE, INHIBITION, NITROGEN FIXING
CELLULOSE, CHLORELLA, PEPTIDES,
ENT REQUIREMENTS, *DISTRIBUTION,
MICROBIAL ECOLOGY, PURPLE SULFUR
ALGAE, ENTERIC BACTERIA, METHANE
N SULFIDE, LAKES, ALGAE, ENTERIC
, DIFFUSION, TURBULENCE, TOXINS,
BIOINDICATORS, FUNGI, ESTUARIES,
PIPTIDES, INTERTIDAL AREAS, SOIL
TESTING, CAPILLARY CONDUCTIVITY,
LYTICAL TECHNIQUES, TEM/ *MARINE
DIOISOTOPES, TEMPERATURE, DEPTH,
HOS, HYDROGEN, BACTERIA, AEROBIC
SUPPLY, MUD, BENTHOS, HYDROGEN,
NIA, NITRITES, NITRATES, YEASTS,
MATTER, AIR TEMPERATURE, AEROBIC
RATES, YEASTS, BACTERIA, AQUATIC
XPLOSIONS, BIOTA, SPECTROMETERS,
MAND, TERTIARY TREATMENT, ALGAE,
RADIOISOTOPES, *TRANSFER, WATER,
G NUTRIENTS, GEORGIA, SEDIMENTS,
GRAPHIES, LAKES, NUISANCE ALGAE,
NT, *AQUICULTURE, ALGAE, YEASTS,
NTATION, METHANE, ALGAL CONTROL,
FECTS, DISSOLVED OXYGEN, AQUATIC
ION, *HYPOLIMNION, ALGAE, FUNGI,
ATES, BRACKISH WATER, ESTUARIES,
L SPRINGS, *OREGON, THERMOPHILIC
ER TREATMENT, *NUTRIENTS, ALGAE,
HIBITION, CYANOCOBALAMIN, MUTANT
HNIQUES, DINOFLAGELLATES, MARINE
REAMS, WATER POLLUTION, ANIMALS,
TH, FROGS, CRAYFISH, MOSQUITOES,
POTABLE WATER, WATER POLLUTION,
AL TECHNIQUES, KINETICS, AQUATIC
HNIQUES, *MARINE MICRO/ *AQUATIC
MICROORGANISMS, *ALGAE, *AQUATIC
ON, *ALGAE, CHEMICAL PROPERTIES,
VIRONMENT, WATER QUALITY, SEWAGE
CHNIQUES, AQUATIC ALGAE, AQUATIC
NE ALGAE, MARINE ANIMALS, MARINE
CHNIQUES, AQUATIC ALGAE, AQUATIC
LAKES, *CELLULOSE, DI/ *ECOLOGY,
CHNIQUES, AQUATIC ALGAE, AQUATIC
N, ALGAE, AQUATIC ALGAE, AQUATIC
TTER, CONDUCTIVITY, TEMPERATURE,
ANITARY ENGINEERING, *PATHOGENIC
HORUS, SCENEDESMUS, CHLOROPHYLL,
*BIOMASS, *SLIME, ALGAE, FUNGI,
COLI, PUBLIC HEALTH, PATHOGENIC
UDGE, ALGAE, NUTRIENTS, VIRUSES,
BACTERIA, ACID BACTERIA, SULFUR
L WASTES, METHANE BACTERIA, ACID
TION, *MUNICIPAL WASTES, METHANE
GREAT LAKES BASIN,
TAKE VELOCITY, ACTIVE TRANSPORT,
SARGASSUM, FRONDS, HYDROCARBONS,
, RADIOCARBON UPTAKE TECHNIQUES,
AKE TECHNIQUES, BACTERIAL PLATE,
, TAXONOMIC TYPES, HETERETROPHY,
BLING TIME, PACKED CELL VOLUMES,
ATURE, ALKALINITY, BICARBONATES,
E MICROORGANISMS, PHYTOPLANKTON,
ORELLA, *EUGLENA, ALGAL CONTROL,
*SLIME, ALGAE, FUNGI, BACTERIA,
.: ORGANIC NITROGEN,
XYGEN, ALGAE, SELF-PURIFICATION,
RA, THALASSIOSIRA, COSCINDISCUS,
ORGANIC MATTER, INSECTS, MITES,
Y, HETEROTROPHY, MICRONUTRIENTS,

BACTERIA, FUNGI, AQUATIC PLANTS,	W70-02510
BACTERIA, NITROGEN, PHOSPHORUS, A	W70-02787
BACTERIA, SUBLITTORAL, YELLOW MAT	W70-01073
BACTERIA, ULTRAVIOLET RADIATION,	W70-00719
BACTERIA, ALGAE, TREATMENT FACILI	W70-01519
BACTERIA, HYDROGEN ION CONCENTRAT	W70-01073
BACTERIA, PROTOZOA, ALGAE, SOLID	W70-01027
BACTERIA, STREPTOCOCCUS, SECONDAR	W70-01068
BACTERIA, *TOXICITY, *WATER POLLU	W70-01467
BACTERIA, WATER POLLUTION EFFECTS	W70-02249
BACTERIA, SPECT: /EFFLUENTS, OXYG	W70-01933
BACTERIA, ALGAE, FUNGI, SULFUR CO	W70-02294
BACTERIA, TRACE ELEMENTS, VITAMIN	W70-02255
BACTERIA, BENTHIC FAUNA, PHOTOSYN	W70-02251
BACTERIA, STABILIZATION, WASTE WA	W70-01971
BACTERIA, PHYTOPLANKTON, GEOCHEMI	W70-03975
BACTERIA, PHYTOPLANKTON, GEOCHEMI	W70-03975
BACTERIA, CARBON RADIOISOTOPES, E	W70-04287
BACTERIA, EUTROPHICATION, GLUCOSE	W70-04194
BACTERIA, *ALGAE, ORGANIC MATTER,	W70-04284
BACTERIA, RADIOACTIVITY TECHNIQUE	W70-04249
BACTERIA, SNAILS, WEIGHT, VACUUM	W70-04184
BACTERIA, SAMPLING, SEA WATER, SE	W70-03952
BACTERIA, POPULATION EQUIVALENTS(	W70-03312
BACTERIA, CHEMICAL ANALYSIS, WATE	W70-03312
BACTERIA, METHANE BACTERIA, CHEMI	W70-03312
BACTERIA, COPPER SULFATE, CIRCULA	W70-04369
BACTERIA, SAMPLING, SEASONAL, HYD	W70-04368
BACTERIA, LABORATORY TESTS.: /E,	W70-04365
BACTERIA, DIATOMS, TEST PROCEDURE	W70-04469
BACTERIA, *ALGAE, *ANALYSIS, *ANA	W70-04365
BACTERIA, SLIME, ALGAE, CYANOPHYT	W70-04371
BACTERIA, METABOLISM, CANALS, SAN	W69-07838
BACTERIA, AEROBIC BACTERIA, METAB	W69-07838
BACTERIA, AQUATIC BACTERIA, WATER	W69-06970
BACTERIA, WATER TEMPERATURE, IRON	W69-07096
BACTERIA, WATER POLLUTION EFFECTS	W69-06970
BACTERIA, ALGAE, SANDS, HYDROGEN	W69-08267
BACTERIA, ACTIVATED SLUDGE.: / DE	W69-08054
BACTERIA, SEDIMENTS, MINNOWS, DET	W69-07861
BACTERIA, DETRITUS, ESTUARIES, PL	W69-08274
BACTERIA, WATER POLLUTION EFFECTS	W69-08283
BACTERIA, FISH, PROTEINS, OIL WAS	W69-05891
BACTERIA, COLIFORMS.: /ION, FERME	W70-04787
BACTERIA, ALGAE.: /R POLLUTION EF	W70-04727
BACTERIA, FISH, LIMNOLOGY, HYDROL	W70-04475
BACTERIA, MOLLUSKS, VITAMIN B, RE	W70-05372
BACTERIA, THERMAO POLLUTION, NUTR	W70-05270
BACTERIA, CULTURES, CHEMICAL OXYG	W70-05655
BACTERIA, CYCLOTELLA NANA, GEIGER	W70-05652
BACTERIA, WATER CHEMISTRY, CARBON	W70-05652
BACTERIA, FUNGI, ALGAE, FLOATING	W70-05750
BACTERIA, IRRIGATION, WILLOW TREE	W70-05263
BACTERIA, ALGAE, NUTRIENTS, INDIC	W70-05264
BACTERIA, ALGAE, PHYSIOLOGICAL EC	W70-05109
BACTERIA, *ALGAE, *ANALYTICAL TEC	W70-04811
BACTERIA, *AQUATIC FUNGI, YEASTS,	W69-00096
BACTERIA, ORGANIC POLLUTANTS, *ST	W68-00483
BACTERIA, ENVIRONMENTAL EFFECTS,	W69-04276
BACTERIA, AQUATIC MICROBIOLOGY, B	W69-03371
BACTERIA, MARINE FUNGI.: /S, MARI	W69-03752
BACTERIA, AQUATIC MICROBIOLOGY, B	W69-03362
BACTERIA, ALGAE, PHYTOPLANKTON, *	W69-00976
BACTERIA, AQUATIC MICROBIOLOGY, B	W69-03188
BACTERIA, AQUATIC MICROBIOLOGY, B	W69-03369
BACTERIA, NUTRIENTS, ESTUARINE EN	W73-00748
BACTERIA, ALGAE, ORGANIC MATTER,	W72-13601
BACTERIA, PHOSPHORUS RADIOISOTOPE	W73-01454
BACTERIA, BACTERICIDES, CHLORINE,	W72-12987
BACTERIA, TEMPERATURE, NITRATES,	W72-06017
BACTERIA, E. COLI, PUBLIC HEALTH,	W72-06017
BACTERIA, ALGAE, MIXING, HYDROGEN	W72-06854
BACTERIA, SULFUR BACTERIA, ALGAE,	W72-06854
BACTERIA, ACID BACTERIA, SULFUR B	W72-06854
BACTERIAL POLLUTION.:	W69-03948
BACTERIAL POPULATIONS, ALGAL POPU	W70-04194
BACTERIAL ATTACHMENT, ECOTOCARPUS	W70-01068
BACTERIAL PLATE, BACTERIAL PIGMEN	W70-05760
BACTERIAL PIGMENTS, DIAPTOMUS, EP	W70-05760
BACTERIAL MEDIUM, ARTHROBACTER, F	W70-00713
BACTERIAL PLATE COUNTS, LANCASTER	W71-05267
BACTERICIDE, KINETICS, HARDNESS,	W71-05267
BACTERICIDES, ALGAL TOXINS, LARVA	W70-01068
BACTERICIDES, GROWTH RATES.: /CHL	W69-00994
BACTERICIDES, CHLORINE, COSTS, *W	W72-12987
BACTERIOLOGICAL STUDY, AEROBICITY	W71-05742
BACTERIOPHAGE, RESERVOIRS, SPORES	W69-07838
BACTERIOSTRUM, LEPTOCYLINDRUS, CO	W73-01446
BACTERIS, FUNGI, ALGAE, NITRITES,	W71-00940
BADFISH RIVER(WIS), DETROIT LAKES	W70-04506

LGAL CONTROL, *NUTRIENTS, ALGAE,
GMENTS, ORGANIC MATTER, BIOMASS,
NTAL EFFECTS, *LIMITING FACTORS,
A, AQUATIC PLANTS, MARINE ALGAE,
SUES, *ALGAE, ECOSYSTEMS, CHARA,
RADIOISOTOPES, IRON, MANGANESE,
PENDED SOLIDS, SALTWATER TONGUE,
L PROPERTIES, ECOLOGY, SALINITY,
ATION, TEMPERATURE, FLOW, ALGAE,
RANSLATIONS, ALGAE, CRUSTACEANS,
PHYTA, CHLOROPHYTA, SCENEDESMUS,
WATER BASINS, WATER SUPPLY, MUD,
LIMITING FACTORS, BENTHIC FLORA,
ECOLOGY, KINETICS, FARM WASTES,
ALGAE, WATER CHEMISTRY, BIOMASS,
*ECOLOGY, BOTTOM SEDIMENTS, MUD,
FE, *AQUATIC POPULATIONS, ALGAE,
Y FROM POLLUTION, *KUWAIT CRUDE,
LUEGILLS, CHLORELLA PYRENOIDOSA,
CONCENTRATION, FLUORIDES, ALKYL
POLYPEPTIDES, PHOTOASSIMILATION,

IVERSITY, EXPERIMENTAL STREAMS,/
UM-106, RHODIUM-106, CERIUM-144,
N, CENTRAL CITY(COLO), MUTATION,
        *POTASSIUM PERMANGANATE,
WATER POLLUTION, ORGANIC MATTER,
IENTS, LAKES, CHANNELS, REVIEWS,
ENTS, *CARBON DIOXIDE, *REVIEWS,
CATION, *REVIEWS, DOCUMENTATION,
NSERVATION, POLLUTION ABATEMENT,
E, STRATIFICATION, SALT BALANCE,
ODOR, SCUM, FISHING, WISCONSIN,
LGAE, *RESEARCH AND DEVELOPMENT,
S, *DIATOMS, *ALGAE, ECOLOGY, S/
HICATION, ALGAE, AQUATIC PLANTS,
ROPHICATION, *LAKES, ACID LAKES,
ERTEBRATES, AQUATIC POPULATIONS,
A COLLECTIONS, BOTTOM SEDIMENTS,
INORGANIC COMPOUNDS, PHOSPHORUS,
*EUTROPHICATION, PHYTOPLANKTON,
ES, LIMNOLOGY, *PLANKTON, ALGAE,
ORY TESTS, BIOLOGICAL TREATMENT,
ACE WATERS, RESERVOIRS, CONTROL,
TS, ALGAE, NITROGEN, PHOSPHORUS,
OGY, WATER PURIFICATION, CARBON,
ES, PHOSPHATES, POTASSIUM, IRON,
N, OLIGOTROPHY, LAKES, MICHIGAN,
RATION, TEMPERATURE, ALKALINITY,
T PENETRATION, DISSOLVED OXYGEN,
*PHYTOPLANKTON, *PHOTOSYNTHESIS,
PHOTOSYNTHESIS, CARBON DIOXIDE,
NTIAL NUTRIENTS, EUTROPHICATION,
GERMANY), CHLORELLA PYRENOIDOSA,
SOMA LEUCHTENBER GIANUM, CYCLOPS
PHORA, AULACODISCUS, BACILLARIA,
, ASTERIONELLA FORMOSA, MELOSIRA
ISCUS HANTZSCHII, STEPHANODISCUS
UCTIVITY, LIMNOLOGY, *NUTRIENTS,
ONDITIONS, ANAEROBIC CONDITIONS,
                          PESTICIDE
, / *PESTICIDE RESIDUES, *ALGAE,
AQUATIC MICROBIOLOGY, ALGICIDES,
EA WATER, TEMPERATURE, SALINITY,
GELLATES, FISHKILL, FISH TOXINS,
PRODUCTIVITY, BALANCE OF NATURE,
MICROBIOLOGY, BALANCE OF NATURE,
, MORTALITY, CHLOROPHYLL, ALGAE,
TIC ALGAE, AQUATIC MICROBIOLOGY,
PHOSPHATES, SILICATES, CITRATES,
NT-WATER INTERFACES, CALIFORNIA,
, CULTURES, DIATOMS, CYANOPHYTA,
ALGAE, *NUTRIENT REQUIREMENTS,
MARSHES, *LAKES, *PHYTOPLANKTON,
ION, FISH, ALGAE, MODEL STUDIES,
KES, WATER PROPERTIES, BACTERIA,
TIC PRODUCTIVITY, AQUATIC WEEDS,
QUATIC WEEDS, BALANCE OF NATURE,
BACTERIA, AQUATIC MICROBIOLOGY,
ONING, ALGICIDES, AQUATIC ALGAE,
Y, BALANCE OF NATURE, WISCONSIN,
MICROBIOLOGY, BALANCE OF NATURE,
QUATIC WEEDS, BALANCE OF NATURE,
TIC ALGAE, AQUATIC PRODUCTIVITY,
MICROBIOLOGY, BALANCE OF NATURE,
MICROBIOLOGY, BALANCE OF NATURE,
TICAL TECHNIQUES, AQUATIC ALGAE,
BACTERIA, AQUATIC MICROBIOLOGY,
GAS CHROMATOGRAPHY, RESISTANCE,

BENTHIC FAUNA, CHEMICAL ANALYSIS,   W72-06052
BENTHIC FLORA, PHAEOPHYTA, POLAR   W70-01074
BENTHIC FLORA, BENTHOS, MINE WATE   W70-02770
BENTHIC FLORA, ESTUARIES, *ESTUAR   W71-02494
BENTHIC FLORA, FLOATING PLANTS, P   W70-10175
BENTHIC FLORA, STREAMS, CULTURES,   W69-07845
BENTHIC FORMS, EPIPHYTIC FORMS, C   W73-00748
BENTHOS, AQUATIC LIFE, ALGAE, PLA   W69-05023
BENTHOS, AEROBIC CONDITIONS, ANAE   W71-11381
BENTHOS, BRINE SHRIMP, ECOSYSTEMS   W69-07440
BENTHOS, CONVECTION, WAVES(WATER)   W70-03969
BENTHOS, HYDROGEN, BACTERIA, AERO   W69-07838
BENTHOS, MINE WATER, WATER POLLUT   W70-02770
BENTHOS, NITRATES, FLUORESCENCE,   W70-02775
BENTHOS, SALINITY, SNAILS, SULFAT   W70-03978
BENTHOS, WISCONSIN, SAMPLING, CON   W70-06217
BENTHOS, ZOOPLANKTON, FRESHWATER   W71-02491
BENTLASS SALT MARSH.: /, *RECOVER   W70-01231
BENZENE COMPOUNDS.: /NCE LIMIT, B   W72-11105
BENZENE SULFONATES, WIND VELOCITY   W70-00263
BERKELSE LAKE(NETHERLANDS), HETER   W70-04369
BERLIN(WEST), GERMANY.:   W68-00472
BERLIN, *ALGAE-TITRE.:   W68-00475
BERRY CREEK(ORE), OREGON STATE UN   W70-03978
BETA RADIOACTIVITY, GONIOBASIS, D   W69-09742
BETA RAYS, COSMIC RADIATION, S: /   W69-08272
BEVERLY HILLS(CALIF).:   W71-00110
BIBLIOGRAPHIES.: /WATER QUALITY,   W70-07027
BIBLIOGRAPHIES, HERBICIDES, FISH,   W70-00269
BIBLIOGRAPHIES, BACTERIA, DECOMPO   W70-00664
BIBLIOGRAPHIES, PHOSPHORUS COMPOU   W69-08518
BIBLIOGRAPHIES, ECOLOGY, BIOLOGY,   W72-01786
BIBLIOGRAPHIES.: / SALINITY, ALGA   W72-01773
BIBLIOGRAPHIES, LAKES, NUISANCE A   W69-08283
BIBLIOGRAPHIES, CHLORELLA, CHLORO   W69-06865
BIBLIOGRAPHIES, *MINNESOTA, *LAKE   W69-00659
BIBLIOGRAPHIES, CYANOPHYTA, FRESH   W68-00680
BIBLIOGRAPHIES, *PHOSPHORUS RADIO   W68-00857
BIBLIOGRAPHIES, BOTTOM SEDIMENTS,   W68-00859
BIBLIOGRAPHIES, ABSORPTION, FISHE   W68-00858
BIBLIOGRAPHIES, NITRATES, NUTRIEN   W68-00860
BIBLIOGRAPHIES, *ALGAE, *CHEMICAL   W68-00856
BIBLIOGRAPHIES, CHEMICAL PRECIPIT   W69-00446
BIBLIOGRAPHIES, CHLORELLA, *SEWAG   W68-00855
BIBLIOGRAPHIES, ANALYTICAL TECHNI   W70-05389
BICARBONATES, *AQUATIC POPULATION   W68-00468
BICARBONATES, PHOTOSYNTHESIS, NUT   W69-07389
BICARBONATE, ZINC, COBALT, CALCIU   W70-03978
BICARBONATES, IRON, PHOSPHORUS, N   W72-03183
BICARBONATES, BACTERICIDE, KINETI   W71-05267
BICARBONATES, HYDROGEN ION CONCEN   W70-06225
BICARBONATES, RADIOACTIVITY, CARB   W70-02504
BICARBONATES, CARBONATES, TEMPERA   W70-02804
BICARBONATES, CARBON, CARBONATES,   W70-02792
BICHOROMATE, DRY WEIGHT, B COLI C   W70-05092
BICUSPIDATUS, REPRODUCTIVE RATE.:   W69-10182
BIDDULPHIA, CHAETOCEROS, COCCONEI   W70-03325
BINDERANA, GENERATION TIME, CHLOR   W69-10158
BINDERANUS, CLADOCERA, WATER MASS   W70-00263
BIO MASS, OLIGOTROPHY, *PRIMARY P   W68-00912
BIO: /SIS, GROWTH RATE, AEROBIC C   W70-05750
BIOACCUMULATION.:   W69-08565
BIOASSAY, BIOINDICATORS, FISHKILL   W69-08565
BIOASSAY, DISINFECTION, ENVIRONME   W69-10171
BIOASSAY, ANALYTICAL TECHNIQUES,   W70-08111
BIOASSAY, AQUATIC ALGAE, *ALGAL T   W70-08372
BIOASSAY, CHLOROPHYTA, CYANOPHYTA   W70-02792
BIOASSAY, CHLOROPHYTA, ENVIRONMEN   W70-02968
BIOASSAY, VIABILITY.: /AL CONTROL   W70-02364
BIOASSAY, CHLOROPHYTA, CHLORELLA,   W70-02248
BIOASSAY, NEW YORK, LAKES, INHIBI   W70-01579
BIOASSAY, SCENEDESMUS, PHYSIOLOGI   W70-02775
BIOASSAY, TEST PROCEDURES.: /LGAE   W71-06002
BIOASSAY, EUTROPHICATION, OLIGOTR   W72-03183
BIOASSAY, LABORATORY TESTS, PRODU   W72-00845
BIOASSAY, AQUATIC LIFE, COOLING W   W71-11881
BIOASSAY, METABOLISM, ABSORPTION,   W71-10079
BIOASSAY, CYANOPHYTA, ENVIRONMENT   W69-03186
BIOASSAY, CHLOROPHYTA, CYANOPHYTA   W69-03185
BIOASSAY, CHLOROPHYTA, CYANOPHYTA   W69-03362
BIOASSAY, CHLOROPHYTA, CYANOPHYTA   W69-03370
BIOASSAY, BIOCHEMICAL OXYGEN DEMA   W69-03369
BIOASSAY, CHLOROPHYTA, CYANOPHYTA   W69-03188
BIOASSAY, CHLOROPHYTA, DIATOMS, C   W69-03515
BIOASSAY, CYCLING NUTRIENTS, ENVI   W69-04798
BIOASSAY, CHLOROPHYTA, DIATOMS, E   W69-03516
BIOASSAY, BIOCHEMICAL OXYGEN DEMA   W69-03374
BIOASSAY, CYCLING NUTRIENTS, ENVI   W69-03518
BIOASSAY, CHLOROPHYTA, CYANOPHYTA   W69-03371
BIOASSAY, GROWTH RATES, PESTICIDE   W70-03520

MYDOMONAS, CHLORELLA, MORTALITY,
RADIOISOTOPES, DINOFLAGELLATES,
RIENT INTERACTIONS, *RADIOCARBON
CULTU/ *EUTROPHICATION, *ALGAE,
MONAS,/ *RAIN, *VITAMINS, ALGAE,
LYTICAL TECHNIQUES, ANTIBIOTICS,
TORAL, NUTRIENTS, MODEL STUDIES,
OISOTOPES, LAKES, PHYTOPLANKTON,
UNDS, METABOLISM, BIOINDICATORS,
ERFACES, EUGLENOPHYTA, TOXICITY,
ON RADIOISOTOPES, CHLAMYDOMONAS,
OPHYTA, CYANOPHYTA, CHRYSOPHYTA,
CHAINS, BIOLOGICAL COMMUNITIES,
TIVITY, WATER POLLUTION EFFECTS,
, BENTHIC FAUNA, PHOTOSYNTHESIS,
IDE, HYDROGEN ION CONCENTRATION,
NS, CANNERIES, ACTIVATED SLUDGE,
ANTS, ANAEROBIC BACTERIA, ALGAE,
N, STORM RUN-OFF, SEDIMENTATION,
TREATMENT, *ANAEROBIC DIGESTION,
SION, SEDIMENTATION, ABSORPTION,
ULFIDES, ALGAE, ORGANIC LOADING,
IGHT PENETRATION, ON-SITE TESTS,
PHENOLS, ODOR, COLOR, TURBIDITY,
ALGAE, FUNGI, LABORATORY TESTS,
ONS, *ALGAE CONTROL, *LAKE ERIE,
HLORINATION, COLIFORM, BACTERIA,
WATER QUALITY, DISSOLVED OXYGEN,
E, *LAGOONS, *OXIDATION LAGOONS,
EROBIC TREATMENT, OXYGEN DEMAND,
RON COMPOUNDS, HYDROGEN SULFIDE,
NCIES, *LAGOONS, *SLUDGE, ALGAE,
LAGOONS, *BIODEGRADATION, ALGAE,
, WATER POLLUTION, WASTE WATERS,
OTOSYNTHESIS, COSTS, EFFICIENCY,
, CHEMICAL WASTES, ION EXCHANGE,
ATA COLLECTIONS, *WATER QUALITY,
CRO-AUTORADIOGRAPHY, PHYSIOLOGY,
*DISSOLVED OXYGEN, PENNSYLVANIA,
UATIC PLANTS, SELF-PURIFICATION,
YCLES, UNITED STATES, CHLORELLA,
CATION, DISSOLVED OXYGEN, ALGAE,
TREATMENT, ALGAE, DISINFECTION,
, DISSOLVED OXYGEN, TEMPERATURE,
OBIOLOGY, *AQUATIC PRODUCTIVITY,
LTERS, SEASONAL, STRATIFICATION,
A, COLIFORMS, OXIDATION LAGOONS,
GOVERNMENTS, STATE GOVERNMENTS,
NUTRIENTS, OXYGEN SAG, AERATION,
S, NITROGEN, AMMONIA, NUTRIENTS,
GY, BALANCE OF NATURE, BIOASSAY,
OF NATURE, WISCONSIN, BIOASSAY,
TOXINS, ORGANIC WASTES, POLLUTA/
S, CHEMICAL PRECIPITATION, IONS,
ASTES, *ALGAE, WATER PROPERTIES,
IMPOUNDMENTS, *DISSOLVED OXYGEN,
GEN, HYDROGEN ION CONCENTRATION,
NG, CENTRIFUGATION, RESPIRATION,
EN SULFIDE, ELECTRON MICROSCOPY,
STUDIES, ALGAL PHYSIOLOGY, ALGAL
PHOSPHORUS RADIOISOTOPES, ALGAE,
CHLOROPHYLL, ENZUMES, PEPTIDES,

NARY AMMONIUM COMPOUNDS, PHENOL/
GATIONS, ZOO-PLANKTON, SPRAYING,
ER POLLUTION CONTROL, ALGICIDES,
CHEMCONTROL, MECHANICAL CONTROL,
OLOGY, ENVIRONMENT, CHEMCONTROL,
, EUTROPHICATION, ALGAL CONTROL,
MONIA, SEDIMENTATION, FLOTATION,
ANTIFOULING MATERIALS, TESTING,
*OXIDATION LAGOONS, ABSORPTION,
TICS, *PERSISTENCE, EVAPORATION,
AND, MINNESOTA, WATER CHEMISTRY,
, NITROGEN COMPOUNDS, OXIDATION,
RIENTS, *ALGAE, *MICROORGANISMS,
ALGAE, AQUATIC PLANTS, TESTING,
PLANTS, DINOFLAGELLATES, OCEANS,
WATER POLLUTION EFFECTS, OREGON,
MIXIS, *WATER POLLUTION EFFECTS,
, ORGANIC COMPOUNDS, METABOLISM,
OPES, DINOFLAGELLATES, BIOASSAY,
AIN, *VITAMINS, ALGAE, BIOASSAY,
FFECTS, WATER POLLUTION SOURCES,
OLLUTION EFFECTS, *WATER SUPPLY,
, *PHYTOPLANKTON, BRITISH ISLES,
PHYTA, CHLOROPHYTA, GREAT LAKES,
DIMENTATION RATES, STRATIGRAPHY,
TON, *LIMNOLOGY, *WATER QUALITY,

BIOASSAY, INHIBITION, MODE OF ACT   W70-03519
BIOASSAY, BIOINDICATORS.: /CARBON   W70-04194
BIOASSAY, *NUTRIENT TYPES, MICROQ   W70-04765
BIOASSAY, BIOLOGICAL COMMUNITIES,   W69-07832
BIOASSAY, BIOINDICATORS, CHLAMYDO   W69-06273
BIOASSAY, NORON, CALCIUM, CHLORIN   W69-04801
BIOASSAY, RADIOACTIVITY TECHNIQUE   W73-01454
BIOASSAYS.: /ECOLOGY, CARBON RADI   W70-05109
BIOASSAYS.: /ATTER, ORGANIC COMPO   W70-04287
BIOASSAYS, DIATOMS,: /D-WATER INT   W70-04381
BIOASSAYS, WATER POLLUTION EFFECT   W70-04284
BIOASSAYS, SALINE WATER, CHLORELL   W70-03974
BIOASSAYS, EUTROPHICATION.: /FOOD   W70-05272
BIOASSAYS, ALGAE, COMPARATIVE PRO   W70-05424
BIOASSAYS, AQUATIC WEEDS, THERMAL   W70-02251
BIOCARBONATES, DISSOLVED OXYGEN,   W70-02249
BIOCHEMICAL OXYGEN DEMAND, OXIDAT   W71-09546
BIOCHEMICAL OXYGEN DEMAND, OXIDAT   W71-08221
BIOCHEMICAL OXYGEN DEMAND, WASTE   W71-10654
BIOCHEMICAL OXYGEN DEMAND, ACTIVA   W71-09548
BIOCHEMICAL OXYGEN DEMAND, AERATI   W71-11381
BIOCHEMICAL OXYGEN DEMAND, OXIDAT   W71-11377
BIOCHEMICAL OXYGEN DEMAND, DIURNA   W72-01818
BIOCHEMICAL OXYGEN DEMAND, HYDROG   W72-03299
BIOCHEMICAL OXYGEN DEMAND, SEWAGE   W72-01817
BIOCHEMICAL OXYGEN DEMAND, BACTER   W71-04758
BIOCHEMICAL OXYGEN DEMAND, DELAWA   W71-05013
BIOCHEMICAL OXYGEN DEMAND, SUSPEN   W70-08627
BIOCHEMICAL OXYGEN DEMAND, CHEMIC   W70-06601
BIOCHEMICAL OXYGEN DEMAND, SUSPEN   W70-06792
BIOCHEMICAL OXYGEN DEMAND, NUTRIE   W70-06925
BIOCHEMICAL OXYGEN DEMAND, MIXING   W70-06899
BIOCHEMICAL OXYGEN DEMAND, ANAERO   W71-00936
BIOCHEMICAL OXYGEN DEMAND, FISH,   W70-08740
BIOCHEMICAL OXYGEN DEMAND, ALGAE,   W70-09320
BIOCHEMICAL OXYGEN DEMAND, CHEMIC   W70-10433
BIOCHEMICAL OXYGEN DEMAND, SUSPEN   W70-09557
BIOCHEMICAL EXTRACTION, MILLIPORE   W69-10163
BIOCHEMICAL OXYGEN DEMAND, TEMPER   W69-10159
BIOCHEMICAL OXYGEN DEMAND, TOXICI   W70-05750
BIOCHEMICAL OXYGEN DEMAND, OXIDAT   W70-05092
BIOCHEMICAL OXYGEN DEMAND(BOD), M   W70-04790
BIOCHEMICAL OXYGEN DEMAND(BOD), A   W70-04788
BIOCHEMICAL OXYGEN DEMAND(BOD), E   W70-04787
BIOCHEMICAL OXYGEN DEMAND, CYCLIN   W70-04283
BIOCHEMICAL OXYGEN DEMAND, CARBON   W70-04001
BIOCHEMICAL OXYGEN DEMAND, PHOTOS   W70-03312
BIOCHEMICAL OXYGEN DEMAND, ODOR-P   W69-06909
BIOCHEMICAL OXYGEN DEMAND, PHOTOS   W69-07520
BIOCHEMICAL OXYGEN DEMAND, TERTIA   W69-08054
BIOCHEMICAL OXYGEN DEMAND, CHLORO   W69-03374
BIOCHEMICAL OXYGEN DEMAND, ENVIRO   W69-03369
BIOCHEMICAL OXYGEN DEMAND, ALGAL   W69-01273
BIOCHEMICAL OXYGEN DEMAND, MINNES   W69-00446
BIOCHEMICAL OXYGEN DEMAND, *BACTE   W68-00256
BIOCHEMICAL OXYGEN DEMAND, ALGAL   W72-11906
BIOCHEMICAL OXYGEN DEMAND, VIRGIN   W72-10573
BIOCHEMICAL OXYGEN DEMAND, NITROG   W72-06022
BIOCHEMISTRY, VITAMINS, EPIPHYTOL   W70-03952
BIOCHEMISTRY, ALGAL TAXONOMY, ALG   W70-03973
BIOCHEMISTRY, ECOLOGY, SAMPLING,   W69-10151
BIOCHEMISTRY, HYDROGEN ION CONCEN   W70-01068
BIOCIDES.:   W69-05706
BIOCIDES.:   W69-05705
BIOCIDES, PRODUCTION LOSS, QUATER   W72-12987
BIOCONTROL, INFECTION, NUTRIENTS,   W68-00468
BIOCONTROL, *WATERSHED MANAGEMENT   W68-00478
BIOCONTROL, COPPER SULFATE, SODIU   W70-10175
BIOCONTROL, INTEGRATED CONTROL ME   W70-07393
BIOCONTROL, TURBULENCE, COPEPODS,   W71-10098
BIODEGRADATION, CHLORINATION, COL   W71-05013
BIODEGRADATION, ALGAE, MARINE ALG   W71-07339
BIODEGRADATION, CHEMICAL OXYGEN D   W70-08392
BIODEGRADATION, ADSORPTION, ALGAE   W70-09768
BIODEGRADATION.: /ICAL OXYGEN DEM   W69-00446
BIODEGRADATION, NITRIFICATION, DE   W69-08053
BIODEGRADATION, NITRIFICATION, DE   W69-06970
BIODEGRADATION.: /ATMENT, AQUATIC   W70-04373
BIOINDICATORS, FUNGI, ESTUARIES,   W70-04368
BIOINDICATORS, LIGHT INTENSITY, E   W70-03978
BIOINDICATORS, NUISANCE ALGAE, NE   W70-03974
BIOINDICATORS, BIOASSAYS.: /ATTER   W70-04287
BIOINDICATORS.: /CARBON RADIOISOT   W70-04194
BIOINDICATORS, CHLAMYDOMONAS, CAR   W69-06273
BIOINDICATORS, NITROGEN FIXATION.   W69-06864
BIOINDICATORS, CALCIUM, CHLORIDES   W69-07833
BIOINDICATORS, *ESTIMATING EQUATI   W68-00255
BIOINDICATORS, FISH, WASHINGTON,   W70-04503
BIOINDICATORS, ALGAE, OLIGOTROPHY   W70-05663
BIOINDICATORS, AQUATIC LIFE, AQUA   W70-05958

R, PLECTONEMA, NOSTOC PALUDOSUM,　BLEACHING, WASHINGTON(DC).: /RIVE　W70-02255
CHEMOTROPHY, PARTICULATE MATTER,　BLELHAM TARN(ENGLAND), CHLORELLA　W70-02504
IZATION, *BACTERIA COUNT, WARING　BLENDOR MORTALITY, CELL WALL, HEX　W70-04365
RAS CHODATI, DINOPHYCEAE, FILTER　BLOCKING, HETERETROPHS, ITALY, LA　W69-07833
POLLUTION, *METROPOLITAN AREAS,　BLONDIN.: /EVOLUTION, *INTERSTATE　W70-04430
ON, BENTHIC FAUNA, SLUDGE WORMS,　BLOODWORMS, WATER POLLUTION.: /KT　W68-01244
WATER TEMPERATURE, SLUDGE WORMS,　BLOODWORMS, AQUATIC BACTERIA, ALG　W70-00475
H, OLIGOCHAETES, TOXICITY, FISH,　BLOODWORMS, ALGAL CONTROL.: /DEPT　W70-06217
IS), *'WORKING' OF LAKES, *WATER-　BLOOM, 'BREAKING' OF LAKES, CHROO　W69-08283
TREATMENT, ALGAE NUISANCE, ALGAE　BLOOM, ALGAE GROWTH, WATER DENSIT　W69-08282
ROPHICATION, ALGAL BLOOMS, WATER　BLOOM, AQUATIC ALGAE, AGING(BROLO　W68-00912
USE.:　　　　　*DEFINITIONS, *WATER　BLOOM, TOXIC ALGAE, RECREATIONAL　W69-08279
WAUBESA(WIS), LAKE KEGONSA(WIS),　BLOOMING, LAKE MENDOTA(WIS), LAKE　W70-05112
　　INDIA, METHYL ORANGE, ALGAL　BLOOMS.:　　　　　　　　　　　W70-05550
　　　　　　　　　　　　　　ALGAL　BLOOMS.:　　　　　　　　　　　W70-00664
*PERIDINIUM POLONICUM, PLANKTON　BLOOMS.: /I, JAPAN, *GLENODININE,　W70-08372
LYMERS, *BIOLOGICAL LIFE, *ALGAE　BLOOMS, *FISH TOXICITY TEST, SLUD　W70-05819
YSILANES, SCENEDESMUS QU/ *ALGAL　BLOOMS, *ORGANO-SILICATES, *ALKOX　W71-03056
YSILANES, SCENEDESMUS QU/ *ALGAL　BLOOMS, *ORGANO-SILICATES, *ALKOX　W71-05719
TIS AERUGINOSA.:　　　　　ALGAL　BLOOMS, ALGAE NUTRITION, MICROCYS　W70-05565
　　*BLUE-GREEN ALGAE, *ALGA　BLOOMS, ALGAE PHYSIOLOGY.:　　　W69-03512
REGULATORS, MYRIOPHYLLUM, WATER　BLOOMS, CHAINING.: /SALIX, GROWTH　W70-05263
*SHAGAWA LAKE(MINN), ELY(MINN),　BLOOMS, CHLORELLA PYRENOIDOSA, MI　W70-03512
ETONKA(MINN), MORPHOMETRY, ALGAE　BLOOMS, CLEARWATER LAKE(MINN), TR　W70-05387
ENVIRONMENTAL C/ BENTHIC ALGAE,　BLOOMS, CLOSED SYSTEMS, CYTOLOGY,　W69-07832
OLOGY, WATER TEMPERATURE, *ALGAE　BLOOMS, CYANOPHYTA, NORTH DAKOTA.　W70-09093
WIS), PELAGIC AREAS, HORIZONTAL,　BLOOMS, GELATINOUS MATTER.: /OTA(　W70-06229
RESPIROMETER, PROTOPLASM, WATER-　BLOOMS, LAKE GEORGE(UGANDA), RHIN　W70-04369
ANTITIES, MACROQUANTITIES, ALGAL　BLOOMS, LIEBIG'S LAW OF THE MINIM　W70-04765
UGAL, TAPES, JAPAN, AMPHIDINIUM,　BLOOMS, MYTILUS, TRICHODESMIUM ER　W70-05372
HOE(CALIF)/ *COMPLEXONES, *ALGAL　BLOOMS, NTA, EDTA, HEIDA, LAKE TA　W70-04373
ATER POLLUTION COMMISSION, ALGAL　BLOOMS, NUTRIENT LOAD, SYNTHETIC　W69-08674
QUALITY, *EUTROPHICATION, ALGAL　BLOOMS, NUTRIENTS, LIMNOLOGY, DYS　W69-09723
E, AGING/ *EUTROPHICATION, ALGAL　BLOOMS, WATER BLOOM, AQUATIC ALGA　W68-00912
N TECHNIQUE, PARTICULATE MATTER,　BLUE FLUORESCING MATE: /LON COLUM　W70-01074
OGICAL STUDIES, ALGAL IDENTIFIC/　BLUE-GREEN ALGAE, NITZSCHIA, BIOL　W70-10173
ERIMENTS, RECYCLED/ GREEN ALGAE,　BLUE-GREEN ALGAE, TEMPERATURE EXP　W70-07257
DE, UPTAKE, FREUNDLICH ISOTHERM,　BLUE-GREEN ALGAE, GREEN ALGAE.: /　W70-08111
　　　　　　　　　　GREEN ALGAE,　BLUE-GREEN ALGAE, HEAT DEATH.:　　W70-0.7313
RACTIONS, INFRARED SPECTROSCOPY,　BLUE-GREEN ALGAE, GREEN ALGAE.: /　W70-03928
, FILAMENTOUS ALGAE, FILAMENTOUS　BLUE-GREEN ALGAE, OCULAR MICROMET　W70-04253
IS, MATS, SYNECHOCOCCUS LIVIDUS,　BLUE-GREEN ALGAE, FILAMENTS.: /RM　W70-05270
AND CLARK LAKE (S D), ABUNDANCE,　BLUE-GREEN ALGAE, GREEN ALGAE, FR　W70-05375
N, FLUORESCENCE, PHOTOSYNTHESIS,　BLUE-GREEN ALGAE, PHYTOPLANKTON,　W69-05697
HIOPHORA/ PHOSPHATE FERTILIZERS,　BLUEGILLS, MICROCYSTIS SCUMS, PIT　W70-05417
COMPOUNDS, MEAN TOLERANCE LIMIT,　BLUEGILLS, CHLORELLA PYRENOIDOSA,　W72-11105
CEDURES, STATISTICAL PROCEDURES,　BMT.: /HOSLOVAKIA, LABORATORY PRO　W72-12240
EAMS, LAKES, RIVERS, RESERVOIRS,　BOATING, EUTROPHICATION.: /, STR　W71-09880
TENTION TIME, LABORATORY MODELS,　BOD REMOVAL RATES, DESIGN APPLICA　W70-07034
ONMENTAL CONTROL, RECIRCULATION,　BOD REMOVAL, GAS PRODUCTION.: /IR　W70-04787
CTION, BIOCHEMICAL OXYGEN DEMAND(　BOD), AERATION, COLIFORMS, ALGAL　W70-04788
ATURE, BIOCHEMICAL OXYGEN DEMAND(　BOD), EVAPORATION, FERMENTATION,　W70-04787
ALGAE, BIOCHEMICAL OXYGEN DEMAND(　BOD), MIXING, WINDS.: /D OXYGEN,　W70-04790
ON, *DESULFOVIBRIO, FACULTATIVE,　BOD, DETENTION TIME.: *SULFATE I　W70-01971
IENTS, FACULTATIVE POND DESIGNS,　BOD, SLUDGE.: / TEMPERATURE, NUTR　W70-04786
ONS, GEOGRAPHICAL REGIONS, BAYS,　BODIES OF WATER.: /T REGION, REGI　W69-03683
ERSATURATION, RESERVOIRS, ALGAE,　BODIES OF WATER, DISSOLVED OXYGEN　W70-00683
ISLAND), DILUTION RATE, FRUITING　BODIES, ULVA LACTUCA VAR LATISSIM　W70-01073
LF-LIFE, HYDROPSYCHE COCKERELLI,　BODY BURDENS, POMOXIS ANNULARIS,　W70-02786
PWARD FLOW, PUTRESCIBILITY TEST,　BODY IMMEDIUM FILTER, BACK-WASHIN　W70-05821
R, *HUMIC ACIDS, *COLOR, RIVERS,　BOGS, PHENOLS, PIGMENTS, ORGANIC　W70-01074
AQUAE, *MICROCYSTIS AERUGINOSA,　BOHEMIA, ELBE RIVER, CELAKOVICE(C　W70-00274
, SNAILS, WEIGHT, VACUUM DRYING,　BOILING.: /LA, PEPTIDES, BACTERIA　W70-04184
ED, ZIRCONIUM-95-NIOBIUM-95, PRE-　BOMB MATERIAL, RUTHENIUM-106-RHOD　W69-08524
R GANGES(INDIA), RAJGHAT(INDIA),　BOMBAY(INDIA), BENARES(INDIA), MA　W70-05404
ON(WASH), SEATTLE(WASH), REVENUE　BONDS, METRO ACT(SEATTLE).: /INGT　W70-04455
MOEA, PISONIA, PANDANUS, MUSCLE,　BONE, LEAVES, SHELL, RHODIUM-102.　W69-08269
, CHAETOCEROS LAUNDERI, LAUDERIA　BOREALIS, LEPTOCYLINDRUS DANICUS,　W70-04280
LLERITE, ORPIMENT, TETRAHEDRITE,　BORNITE, CHALCOPYRITE, CHEMOSYNTH　W69-07428
MONIUM SALTS, NITROGEN FIXATION,　BORON, AZOTOBACTER, MOLYBDENUM, C　W70-02964
CKING, HETERETROPHS, ITALY, LAKE　BORREVANNET NORWAY, LAKE LUGANO I　W69-07833
RIA, PANDORINA MORUM, PEDIASTRUM　BORYA: / PUSILLA, OOCYSTIS SOLITA　W69-06540
NAH RIVER LABORATORY, PLECTONEMA　BORYANUM, REACTOR.:　　　　SAVAN　W69-07845
AUTOTROPHIC, SILICON, PEDIASTRUM　BORYANUM, STAURASTRUM PARADOXUM,　W70-02804
WITZERLAND), LINSLEY POND(CONN),　BOSMINA COREGONI LONGISPINA.: /(S　W69-10169
ABAENA LEMMERMANNI, APHANOCAPSA,　BOSMINA LONGISPINA, ZURICHSEE(SWI　W69-10169
NABAENA, ANACYSTIS, MICROCYSTIS,　BOSMINA, ASTERIONELLA, MELOSIRA,　W70-04503
A COHCLEARIS, ASPLANCHA, LECANE,　BOSMINA, NAVICULA, FRAGILARIA, CY　W70-02249
WINNISQUAM, LONG POND, CONCORD,　BOSTON LOT RESERVOIR, LEBANON, WA　W69-08674
　　　QUABBIN RESERVOIR(MASS),　BOSTON(MASS).:　　　　　　　　W70-05264
BORYANUM, STAURASTRUM PARADOXUM,　BOTRYOCCUS BRAUNII, FRAGILARIA CR　W70-02804
ATTERN, ANABAENA CYLINDRICA, GL/　BOTRYOCOCCUS BRAUNII, METABOLIC P　W70-02965
AQUAE, GLOEOTRICHIA ECHINULATA,　BOTRYOCOCCUS, PLANKTON DISTRIBUTI　W70-03959
R DIVERSION, INSTREAM TREATMENT,　BOTTOM SEALING, LAKE DILUTION.: /　W69-09340
, SMELTS, FISHKILL, CRUSTACEANS,　BOTTOM SEDIMENTS.: /AGE EFFLUENTS　W69-08674
TMENT, NITROGEN, AQUATIC PLANTS,　BOTTOM SEDIMENTS, DREDGING, HARVE　W69-09340
ESIUM, FALLOUT, SILTS, DETRITUS,　BOTTOM SEDIMENTS, CHUTES, EDDIES,　W69-09742
E, GROWTH RATES, COPPER SULFATE,　BOTTOM SEDIMENTS, LAKE MICHIGAN,　W69-09349
PSEUDOMONAS, SUBSURFACE-WATERS,　BOTTOM SEDIMENTS, SURFACE WATERS,　W70-00713
LANKTON, ICE, BACTERIA, TRIPTON,　BOTTOM SEDIMENTS, SOLAR RADIATION　W69-10154

ILIMNION, PHOTOSYNTHESIS, ALGAE,
NUISANCE ALGAE, AQUATIC PLANTS,
E, *TREATMENT, *LAKES, *ECOLOGY,
LE, ALGAE, BIOLOGICAL TREATMENT,
YGEN, CARBON, ENZYMES, SAMPLING,
*DISTRIBUTION, ANIMALS, PLANTS,
NG, GRAVELS, MUD, SANDS, SHALES,
*CHEMICAL PROPERTIES, EPILMNION,
ODELS, ON-SITE DATA COLLECTIONS,
TIC POPULATIONS, BIBLIOGRAPHIES,
TREATMENT, HARVESTING OF ALGAE,
RACE ELEMENTS, CORES, WISCONSIN,
MARGINAL VEGETATION FAUNA, SANDY
RYTHREUM, SAXIDOMUS, CLOSTRIDIUM
ERAN, MONTANE ZONE, ALPINE ZONE,
AL-BOUND RADIOSTRONTIUM, *FUNGAL-
ND RADI/ *RADIOSTRONTIUM, *ALGAL-
OLIOSUS, EXPERIMENTAL LIMNOLOGY,
OCLOEON, ENALLAGMA, PSEUDAGRION,
GEN SULFIDE, IRON, FUNGI, CLAYS,
LTURES, CHLOROPHYTA, CYANOPHYTA,
KES, NUTRIENTS, BACTERIA, ALGAE,
DS, RUNNING WATERS, FRESH WATER,
HYTA, CALIFORNIA, UNITED STATES,
A, LABORATORY, MINERALS,/ ALGAE,
RATELLA, POLYARTHRA, ASPLANCHNA,
SCRIPTIONS, GONGROSIRA LACUSTRIS
E, PINGO LAKE, TWELVE MILE LAKE,
AEA, CRASSIPES EICHORNIA, TYPHA,
ROCARBONS, *OLEFI/ *BOTRYOCOCCUS
ELLA PYRENOIDOSA, ANKISTRODESMUS
TAURASTRUM PARADOXUM, BOTRYOCOCCUS
ENA CYLINDRICA, GL/ BOTRYOCOCCUS
ORKING' OF LAKES, *WATER-BLOOM, '
*GLUE GREEN ALGAE, BREAM, WHITE
INYLPYRIDINE, *GLUE GREEN ALGAE,
S, ESTUARINE ENVIRONMENT, PONDS,
NIFICUM, PYRODINIUM, GYMNODINIUM
DIS, MICROCOCCUS, STREPTOCOCCUS,
ONYAULAX TAMERENSIS, GYMNODINIUM
, ORGANISMS, TURSIOPS TRUNCATUS,
NS, ALGAE, CRUSTACEANS, BENTHOS,
ION, FOOD CHAINS, AQUATIC ALGAE,
*WATER POLLUTION, AQUATIC ALGAE,
ECOSYSTEM MODEL/ KOOTENAY LAKE,
S, OLIGOTROPHIC, *PHYTOPLANKTON,
, NORRIS RESERVOIR(TENN), FRENCH
ATER BLOOM, AQUATIC ALGAE, AGING(
                            BROMINE,

RATES, PHOTO REACT/ *CLADOPHORA,

NOSA, POINT LEPREAU(CANADA), NEW
NTARIO), STURGEON LAKE(ONTARIO),
ON, TROPHIC INDICATORS, NUTRIENT
ANNUAL BUDGET, PHOSPHORUS ANNUAL
UX, NUTRIENT SOURCES, PHOSPHORUS
                    *SEDIMENTS, *NUTRIENT
IC, TROPHOLYTIC, NITROGEN ANNUAL
UTRIENT CONCENTRATION, *NUTRIENT
BINES, ECONOMIC EFFICIENCY, HEAT
EMISTRY, WATER TEMPERATURE, HEAT
YGEN, EVAPORATION CONTROL, WATER
Y, AGRICULTURAL RUNOFF, NUTRIENT
KE WINNISQUAM(N H), MA/ NUTRIENT
S, LEBANON, BATAVIA, CLARKSBURG,
, HERBICIDES, CHARA, ALKALINITY,
TROL, *BIBLIOGRAPHIES, CATTAILS,
IATOMS, CYANOPHYTA, CHLOROPHYTA,
R MILFOIL COONTAIL, WILD CELERY,
FE, HYDROPSYCHE COCKERELLI, BODY
EROTROPHIC POTENTIAL, LINEWEAVER-
DEN, ENZYME KINETICS, LINEWEAVER-
, LAKE ERKEN(SWEDEN), LINEWEAVER-
NTS, ACRIDINE ORANGE, LINEWEAVER-
NOIDOSA, MICROCYSTIS AERUGINOSA,
A LEMMERMANNII, OTTAWA(ONTARIO),
TATUM(GREV) CL, CHLORELLA OVALIS
MUL/ *SCIRPUS FLUVIATILIS, *LAKE
RA FRACTA, CLADOPHORA GLOMERATA,
NE.:
NDUSTRIAL WASTES, SEWAGE WASTES,
ORGANIC GROWTH FACTORS, VITAMIN
VITAMIN B1, VITAMIN B6, VITAMIN
WTH FACTORS, VITAMIN B1, VITAMIN
SCILLATORIA, ANABAENA CYLINDRICA
SLAND(N C), DETECTOR, BEAUFORT(N
ECTEN IRRADIANS, PIVERS ISLAND(N
GE, ELECTROSTATIC REPULSION, DOW
LLUSKS, STRATIFICATION, DIATOMS,

BOTTOM SEDIMENTS, THERMOCLINE, HY   W70-06975
BOTTOM SEDIMENTS, ANIMALS, FISH,   W70-06225
BOTTOM SEDIMENTS, MUD, BENTHOS, W   W70-06217
BOTTOM SEDIMENTS, NUTRIENTS, *WAS   W70-07031
BOTTOM SEDIMENTS, SODIUM, POTASSI   W70-09889
BOTTOM SEDIMENTS, SOILS, NUCLEAR   W69-08269
BOTTOM SEDIMENTS, CATTAILS, RADIO   W69-08272
BOTTOM SEDIMENTS.: / ABSORPTION,   W68-00857
BOTTOM SEDIMENTS, BIBLIOGRAPHIES,   W68-00858
BOTTOM SEDIMENTS, WATERSHED MANAG   W68-00859
BOTTOM SEDIMENTS, WEED CONTROL, S   W68-00478
BOTTOM SEDIMENT, BENTHIC FAUNA.: /   W69-03366
BOTTOMS FAUNA, MARGINAL VEGETATIO   W70-06975
BOTULINUS, VENERUPIS, OSTREA, ART   W70-05372
BOULDER(COLO), TEA LAKE(COLO), BL   W69-10154
BOUND RADIOSTRONTIUM, *MICROBIAL   W71-04067
BOUND RADIOSTRONTIUM, *FUNGAL-BOU   W71-04067
BRACHIONUS, SYNCHAETA PECTINATA,   W70-02249
BRACHYTHEMIS, CHYDORUS, PLEUROXUS   W70-06975
BRACKISH WATER, FISHERIES, LIMNOL   W70-05761
BRACKISH WATER, OLIGOTROPHY, TEMP   W70-05752
BRACKISH WATER, ALKALINE WATER, S   W71-03031
BRACKISH WA: /S, CARP, CLAMS, PON   W70-09905
BRACKISH WATER, ESTUARIES, BACTER   W70-05372
BRACKISH, ECOSYSTEMS, FAUNA, FLOR   W72-02203
BRANCHIONUS.: /A, MICROCYSTIS, KE   W70-05375
BRAND, GONGROSIRA DEBARYANA RAB,   W70-04468
BRANT LAKE, IMIKPUK LAKE, NORTH M   W69-08272
BRASENIA, DOW CONTACT, SPARGANIUM   W70-05263
BRAUNII, *ANACYSTIS MONTANA, *HYD   W69-08284
BRAUNII, DUNALIELLA TERTIOLECTA, '   W70-05261
BRAUNII, FRAGILARIA CROTONENSIS,   W70-02804
BRAUNII, METABOLIC PATTERN, ANABA   W70-02965
BREAKING' OF LAKES, CHROOCOCCACEA   W69-08283
BREAM, LEUKOCYTIC COMPOSITION, AL   W70-01788
BREAM, WHITE BREAM, LEUKOCYTIC CO   W70-01788
BREEDINGS, SALINE WATER-FRESHWATE   W73-00748
BREVE, CANADA, EUROPE, SOUTH AFRI   W70-05372
BREVIBACTERIUM, CORYNEBACTERIUM,   W71-11823
BREVIS, MYCROSYSTIS, ALGASTATIC S   W70-02510
BREVOORTIA TYRRANUS, CONCENTRATIO   W69-08274
BRINE SHRIMP, ECOSYSTEMS, FISH, P   W69-07440
BRINE SHRIMP, FISH, MUSSELS.: /UT   W70-01466
BRINE SHRIMP, FISH, FOOD CHAINS,   W70-01467
BRITISH COLUMBIA, CANADA, AQUATIC   W69-06536
BRITISH ISLES, BIOINDICATORS, *ES   W68-00255
BROAD RIVER(TENN), CHICKAMAUGA RE   W71-07698
BROLOGICAL), *AQUATIC PRODUCTIVIT   W68-00912
BROMAMINE, CHLORAMINE.:   W70-02370
BROMINE, BROMAMINE, CHLORAMINE.:   W70-02370
BROMINE, FILAMENTOUS ALGAE, KILL   W70-02364
BROMINE, IODINE.:   W70-05178
BRUNSWICK(CANADA), SERVALL OMNI-M   W70-04365
BUCKHORN LAKE(ONTARIO), CLEAR LAK   W70-02795
BUDGET.:   *CULTURAL EUTROPHICATI   W71-03048
BUDGET.: / TROPHOLYTIC, NITROGEN   W69-05142
BUDGET, APHANIZOMINON FLOS AQUAE.   W70-04268
BUDGET, CYCLING NUTRIENTS.:   W68-00511
BUDGET, PHOSPHORUS ANNUAL BUDGET.   W69-05142
BUDGET, POTENTIAL NUTRIENT CONCEN   W70-00270
BUDGET, SOCIAL ASPECTS.: /ON, TUR   W70-08834
BUDGET, TURBIDITY, *RESERVOIR EVA   W72-01773
BUDGET, WATER CHEMISTRY, WATER TE   W72-01773
BUDGETS, COMPARATIVE STUDIES, NUT   W69-09349
BUDGETS, LAKE SEBASTICOOK(ME), LA   W70-07261
BUFFALO, DUNREITH, OIL SPILLS.: /   W72-03641
BULLHEADS, EELS, KILLIFISHES, TRO   W70-05263
BULRUSH, DREDGING, COPPER SULPHAT   W70-05263
BULRUSH, PLANT GROWTH SUBSTANCES,   W72-00845
BUR REED, DUCKWEED, WHITE WATER L   W70-05568
BURDENS, POMOXIS ANNULARIS, GASTR   W70-02786
BURK EQUATION, UPTAKE VELOCITY, M   W70-05109
BURK EQUATION, MICHAELIS KINETICS   W70-04194
BURK EQUATION, MAXIMUM UPTAKE VEL   W70-04287
BURKE EQUATION.: /AND, PHAEOPIGME   W70-04811
BURNTSIDE RIVER(MINN).: /LLA PYRE   W70-03512
BURTON LAKE(SASKATCHEWAN), HUMBOL   W70-00273
BUTCHER, NJOER LAKE(NORWAY), HOBO   W70-05752
BUTTE DES MORTS(WISC), GROWTH STI   W72-00845
BUTTERNUT CREEK(NY).:   *CLADOPHO   W72-05453
BYELORUSSIA(USSR), POLYCHLOROPINE   W70-08654
BYPRODUCTS, FERTILIZERS, ALGAE, Z   W70-09905
B1, VITAMIN B6, VITAMIN B12, BIOT   W70-02510
B12, BIOTIN, HISTIDINE, URACIL, R   W70-02510
B6, VITAMIN B12, BIOTIN, HISTIDIN   W70-02510
B629, ANABAENA INAEQUALIS 381, AN   W70-05091
C), CRASSOSTREA VIRGINICA, MERCEN   W69-08267
C), DETECTOR, BEAUFORT(N C), CRAS   W69-08267
C-31, DOW A-21, NALCO N-670.: /AR   W70-05267
C: /LIFE CYCLES, OLIGOCHAETES, MO   W71-00665

```
INC, COBALT, CALCIUM, MAGNESIUM,     C: /TASSIUM, IRON, BICARBONATE, Z     W70-03978
CUTTING, *WEED EXTRACTING STEEL      CABLES, ALGAL SUCCESSION, ANABAEN     W70-04494
ULTER COUNTER, EUGLENA GRACILIS,     CACHONINA NIEI, SKELETONEMA COSTA     W69-08278
PHYTOPLANKTON, SUCKERS, SHINERS,     CADDISFLIES, TEMPERATURE, FLOW, D     W69-07853
SEE RIVER, TROPHIC LEVEL, ALGAE,     CADDISFLIES, FOOD CHAINS, STABLE      W70-02786
UATIC BACTERIA, ALGAE, MAYFLIES,     CADDISFLIES.: /MS, BLOODWORMS, AQ     W70-00475
IOISOTOPE/ CESIUM RADIOISOTOPES,     CADMIUM RADIOISOTOPES, CERIUM RAD     W71-09850
WATER, ACIDIC WATER, TRAVERTINE,     CALCIUM CARBONATE, BACTERIA, BIOM     W69-10160
LOGY, MINNESOTA, WATER ANALYSIS,     CALCIUM CARBONATE, HYPOLIMNION.: /    W69-00632
NUTRIENTS, PHOSPHORUS, NITROGEN,     CALCIUM CHLORIDE, TEMPERATURE, ME     W72-05453
RY PRODUCTIVITY, ALGAE, BIOMASS,     CALCIUM COMPOUNDS, CHLORIDES, SAL     W69-02959
EN ION CONCENTRATION, MAGNESIUM,     CALCIUM.: /S, CHLOROPHYLL, HYDROG     W70-01579
GHT INTENSITY, NITRATE, EUGLENA,     CALCIUM, AMMONIUM SALTS, NITROGEN     W70-02964
ELEMENTS, VITAMINS, AMINO ACIDS,     CALCIUM, AMMONIA, NITROGEN FIXATI     W70-02255
, NUTRIENTS, SOLUTES, CHLORIDES,     CALCIUM, CARBONATES, PHOSPHATES.:     W70-09557
, *MANGANESE, *WISCONSIN, ALGAE,     CALCIUM, CHELATION, CYANOPHYTA, N     W69-05867
S, ANTIBIOTICS, BIOASSAY, NORON,     CALCIUM, CHLORINE, COBALT, COPPER     W69-04801
S, *WATER SUPPLY, BIOINDICATORS,     CALCIUM, CHLORIDES, CONDUCTIVITY,     W69-07833
ER, SEDIMENT LOAD, CONDUCTIVITY,     CALCIUM, MAGNESIUM, ALKALINITY, S     W69-05894
ES, TESTS, NITROGEN, PHOSPHORUS,     CALCIUM, MAGNESIUM, SODIUM, SILIC     W68-00856
IRON, BICARBONATE, ZINC, COBALT,     CALCIUM, MAGNESIUM, C: /TASSIUM,      W70-03978
NUTRIENTS, ESSENTIAL NUTRIENTS,     CALCIUM, MAGNESIUM, POTASSIUM, HA     W70-02245
M, NITROGEN, PHOSPHORUS, SULFUR,     CALCIUM, MAGNESIUM, POTASSIUM, SO     W70-02804
CENTRATION, NITRATES, POTASSIUM,     CALCIUM, MAGNESIUM, IRON, ORGANIC     W70-00719
SEASONAL, PRIMARY PRODUCTIVITY,     CALCIUM, NITRATES, AMMONIA, PHOSP     W70-05405
UM, SILICATES, DISSOLVED SOLIDS,     CALCIUM, PRODUCTIVITY, STANDING C     W70-03512
N, COLLOIDS, ZINC RADIOISOTOPES,     CALCIUM, POTASSIUM, CLAMS, SIZE,      W70-00708
IUM, POTASSIUM, IRON, MANGANESE,     CALCIUM, PIGMENTS.: /DIMENTS, SOD     W70-09889
IA, ALGAE, PLANKTON, TUBIFICIDS,     CALCIUM, SILICATES, DIATOMS, WIND     W70-05761
ICO, SPECTROSCOPY, DISTRIBUTION,     CALCIUM, TROPHIC LEVELS, FOOD WEB     W69-08525
HYLL CONTENT, CENTRIFUGE METHOD,     CALCULATED VOLUME METHOD, CYCLOTE     W70-03959
L PHYSIOLOGY, *DESIGN THEORY AND     CALCULATIONS, PHOTOSYNTHESIS, ILL     W70-04786
LGAE HARVESTING, *CENTRAL VALLEY(    CALIF.).:                        *A     W71-04556
LGAE HARVESTING, *CENTRAL VALLEY(    CALIF.).:                        *A     W71-04557
SIUM PERMANGANATE, BEVERLY HILLS(    CALIF.).:                     *POTAS     W71-00110
          LOS ANGELES COUNTY(    CALIF).:                              W70-05645
MS, NTA, EDTA, HEIDA, LAKE TAHOE(    CALIF.).: /OMPLEXONES, *ALGAL BLOO     W70-04373
ACTERIAL PLATE COUNTS, LANCASTER(    CALIF.).: / PACKED CELL VOLUMES, B     W71-05267
ULATE ORGANIC CARBON, LAKE TAHOE(    CALIF), CHLORELLA PYRENOIDOSA, SE     W70-03983
UTION ASSESSMENT, AMERICAN RIVER(    CALIF), CLEAR LAKE(CALIF), EEL RI     W70-02777
MERICAN RIVER(CALIF), CLEAR LAKE(    CALIF), EEL RIVER(CALIF), JOINT I     W70-02777
F), CLEAR LAKE(CALIF), EEL RIVER(    CALIF), JOINT INDUSTRY/GOVERNMENT     W70-02777
ON, LAKE MONONA(WIS), LAKE TAHOE(    CALIF), LAKE WASHINGTON(WASH), ZU     W69-06858
ILLATORIA, AUSTRALIA, SANTA ROSA(    CALIF), LYUBLIN(USSR), LYUBERETS(     W70-05092
UMNE RIVER(CA/ SAN JOAQUIN RIVER(    CALIF), MERCED RIVER(CALIF), TUOL     W70-02777
ENGINEERING RESEARCH LABORATORY(    CALIF), NATIONAL EUTROPHICATION R     W70-02775
T, *CHAOBORID MIDGES, CLEAR LAKE(    CALIF), NUTRIENT ENERGY PATHWAYS,     W70-05428
TER PLAN, DELTA WATER FACILITIES(    CALIF), OXYGEN DYNAMICS, SACRAMEN     W70-05094
CEL/ STRIPPING, FIREBAUGH CENTER(    CALIF), SCENEDESMUS QUADRICAUDA,      W70-03334
CED RIVER(CALIF), TUOLUMNE RIVER(    CALIF), STANISLAUS RIVER(CALIF),      W70-02777
AQUIN RIVER(CALIF), MERCED RIVER(    CALIF), TUOLUMNE RIVER(CALIF), ST     W70-02777
E RIVER(CALIF), STANISLAUS RIVER(    CALIF), WATER POLLUTION ASSESSMEN     W70-02777
EASUREMENTS, EUPHOT/ *LAKE TAHOE(    CALIF-NEV), TRANSPARENCY, LIGHT M     W70-01933
TS, MORPHOMETRY, PA/ *LAKE TAHOE(    CALIF-NEV), HYDROPHYTES, LIVERWOR     W70-00711
XYGEN DEMAND, OXIDATION LAGOONS,     CALIF: / CHLORELLA, BIOCHEMICAL O     W70-05092
R/ SACRAMENTO-SAN JOAQUIN DELTA,     CALIFORNIA WATER PLAN, DELTA WATE     W70-05094
ROPHYTA, DINOFLAGELLATES, ALGAE,     CALIFORNIA, FLORIDA, PESTICIDE RE     W70-05272
LLATES, CYANOPHYTA, CHRYSOPHYTA,     CALIFORNIA, UNITED STATES, BRACKI     W70-05372
N ION CONCENTRATION, ALKALINITY,     CALIFORNIA, LIGHT INTENSITY, TEMP     W70-03334
OSPHORUS, WASHINGTON, WISCONSIN,     CALIFORNIA.: / NUISANCE ALGAE, PH     W69-06858
SOTA, SEDIMENT-WATER INTERFACES,     CALIFORNIA, BIOASSAY, SCENEDESMUS     W70-02775
ROWTH, *BIOASSAYS, *FLUOROMETRY,     CALIFORNIA, NUTRIENTS, EUTROPHICA     W70-02777
BIODEGRADATION, *DRAINAGE WATER,     CALIFORNIA, DRAINAGE PROGRAMS, AN     W71-04555
ES, EVAPORATION, MICROORGANISMS,     CALIFORNIA, *PHOTOSYNTHESIS, WAST     W71-05267
CATION.:            *CENTRAL VALLEY(    CALIFORNIA), *BACTERIAL DENITRIFI     W71-04554
BIODEGRADATION, *DRAINAGE WATER,     CALIFORNIA, DRAINAGE PROGRAMS, AN     W71-04554
BIODEGRADATION, *DRAINAGE WATER,     CALIFORNIA, DRAINAGE PROGRAMS, AN     W71-04556
BIODEGRADATION, *DRAINAGE WATER,     CALIFORNIA, DRAINAGE PROGRAMS, AN     W71-04557
CATION.:            *CENTRAL VALLEY(    CALIFORNIA), *BACTERIAL DENITRIFI     W71-04555
COSTS, ARTIFICIAL WATER COURSES,     CALIFORNIA.: / CONVEYANCE, WATER     W71-09577
COPPER SULFATE, CHLORINE, WELLS,     CALIFORNIA.: /Y CONTROL, *ALGAE,      W71-00110
-PURPUREA, SPHACELARIA, GONIDIA,     CALLOPHYLLIS HAENOPHYLLA, TIDAL C     W69-10161
DISSOLVED SOLIDS, HEDRIODISCUS,     CALOPARYP: /CTROPHOTOMETRY, TOTAL     W70-03309
BIOLOGY, *DIURNAL OXYGEN CURVES,     CALORIMETRY, SOLARIMETRY, YELLOWS     W70-03309
ODGETTS, CHROOCOCCUS PRESCOTTII,     CALOTHRIX FUSCA, MICROCOLEUS LACU     W70-04468
GAL NUTRITION, ALGAL PHYSIOLOGY,     CALOTHRIX PARIETINA, COCCOCHLORIS     W69-06277
*CHICAGO(ILL), COOK COUNTY(ILL),     CALUMET RIVER(ILL), POOLS, WIND D     W70-00263
NAVICULA, APHANIZOMENON, ONTARIO(    CANADA).: /O, FILAMENTOUS ALGAE,      W69-10155
LAND), WHITEFISH, PUNCHBOWL LAKE(    CANADA).: /S, LAKE ZURICH(SWITZER     W70-05399
ENT LEVELS, S/ *TWELVE MILE LAKE(    CANADA), *ONTARIO, *CANADA, NUTRI     W70-02973
SKAHA LAKE(CANADA), OSOYOOS LAKE(    CANADA), DEFORMED CHIRONOMIDS.: /     W71-12077
GUINS, LAK/ PRINCE EDWARD ISLAND(    CANADA), NOVA SCOTIA(CANADA), PEN     W70-05399
LYSIPHONIA LANOSA, POINT LEPREAU(    CANADA), NEW BRUNSWICK(CANADA), S     W70-04365
KANAGAN LAKE(CANADA), SKAHA LAKE(    CANADA), OSOYOOS LAKE(CANADA), DE     W71-12077
WARD ISLAND(CANADA), NOVA SCOTIA(    CANADA), PENGUINS, LAKE ZURICH(SW     W70-05399
PROTEUS, SERRATIA, BAY OF FUNDY/    CANADA), RHYZOIDS.: /, CYTOPHAGA,     W70-03952
T LEPREAU(CANADA), NEW BRUNSWICK(    CANADA), SERVALL OMNI-MIXER HOMOG     W70-04365
OPHIC COMMUNITIES, OKANAGAN LAKE(    CANADA), SKAHA LAKE(CANADA), OSOY     W71-12077
KOOTENAY LAKE, BRITISH COLUMBIA,     CANADA, AQUATIC ECOSYSTEM MODEL,      W69-06536
, PYRODINIUM, GYMNODINIUM BREVE,     CANADA, EUROPE, SOUTH AFRICA, ISR     W70-05372
```

N, LAKE MENDOTA, LAKE FURES, LA/
*AQUATIC MACROPHYTES, ANACHARIS
IFASTACUS LENIUSCULUS, ANACHARIS
INS PROJECT, INDIANA HARBOR SHIP
ERRICHLORIDE, IRON S/ RING TUBE,
A, AEROBIC BACTERIA, METABOLISM,
CES HANSENII, RHODOTORULA RUBRA,
DIDDENSII, CANDIDA ZEYLANOIDES,
IDA ZEYLANOIDES, CANDIDA KRUSEI,
CANDIDA KRUSEI, CANDIDA OBTUSA,
TORULA RUBRA, CANDIDA DIDDENSII,
TREATMENT, ANAEROBIC CONDITIONS,
, SURVIVAL, SUSCEPTIBIL/ *TORREY
, *GAMOSOL, *OIL SPILLS,   *TORREY
ITY.:   *TORREY
ION, CLIFF VEGETATION, / *TORREY
CHEMICAL OXYGEN DEMAND, STORAGE
ATER QUALITY, WASTE ASSIMILATIVE
THYMIDINE, NARRAGANSETT BAY(RI),
ISLAND SOUND, NARRAGANSETT BAY,
A, ALGAE ISOLATION, CAPILLARITY,
GAL, NITZSCHIA, ALGAE ISOLATION,
METHODOLOGY, CULTURES, TESTING,
AL GROWTH POTENTIAL, SELENASTRUM
BACILLARIOPHYCEAE, SELENASTRUM
L ASSAY, PROCEDURES, SELENASTRUM
ION CONCENTRATION, FLUORESCENCE,
NFALL, TEMPERATURE, COLORIMETRY,
NTRATION, TEMPERATURE, PROTEINS,
GROWTH SUBSTANCES, PLANT GROWTH,
, AMINO ACIDS, INTERTIDAL AREAS,
IDS, BIOMASS, ORGANIC COMPOUNDS,
YRENOIDOSA, SELENASTRUM GRACILE,
N, BIOMASS, SEWAGE, FARM WASTES,
ICARBONATES, CARBON, CARBONATES,
ITROGEN FIXATION, PHOTOCHEMICAL,
ENTS, CHELATION, PHOTOSYNTHESIS,
EWAGE, SULFATE, MANGANESE, IRON,
NGANESE, PHOTOSYNTHESIS, OXYGEN,
MUS, CHLAMYDOMONAS, TEMPERATURE,
NITY, NITRIFICATION, *CORROSION,
COLORIMETRY, SAMPLING, METHANE,
ENT, ASBESTOS CEMENT, CHLORELLA,
ROL, DECOMPOSING ORGANIC MATTER,
SULFUR BACTERIA, DIEL MIGRATION,
MICROORGANISMS, LIGHT INTENSITY,
ROGEN ION CONCENTRATION, OXYGEN,
NUTRIENTS, HYDROSTATIC PRESSURE,
ENSITY, CULTURES, RADIOACTIVITY,
ATURE, ORGANIC MATTER, HARDNESS,
NT, TEMPERATURE, PHOTOSYNTHESIS,
ODIUM COMPOUNDS, PHOTOSYNTHESIS,
ITROGEN FIXATION, *ALGAE, LIGHT,
LAMYDOMONAS, CHLORELLA, ENZYMES,
ROPHICATION, LIGHT, TEMPERATURE,
, PHYSIOLOGICAL ECOLOGY, OXYGEN,
, CHLORELLA, *SEWAGE, EFFLUENTS,
WATER CHEMISTRY, *PHYTOPLANKTON,
ON RADIOISOTOPES, PACIFIC OCEAN,
ENTIAL, SOLAR RADIATION, ENERGY,
ON, NITROGEN, OXYGEN, OXIDATION,
PIDS, LIGHT, ALGAE, TEMPERATURE,
DOSA, ANKISTRODESMUS BRAUNII, D/
RIENTS, EFFLUENTS, ALGAE, WEEDS,
AGOONS, *ALGAE, LIGHT INTENSITY,
*VISCOSE RAYON, ZINC COMPOUNDS,
UPHOTIC ZONE, PLANT CAROTENOIDS,
ALGAE, DIATOMS, EUTROPHICATION,
IS, BICARBONATES, RADIOACTIVITY,
ORELLA, TRACERS, PHOTOSYNTHESIS,
LORELLA, TRACERS, RADIOACTIVITY,
REMENT, PROTEINS, RADIOACTIVITY,
TOSYNTHESIS, DIATOMS, CHLORELLA,
WTH RATES, TRITIUM, PHOTOGRAPHY,
MMUNITIES, *ALGAE, *METHODOLOGY,
ATER POLLUTION EFFECTS, DIATOMS,
A, ALGAE, PHYSIOLOGICAL ECOLOGY,
ARINE BACTERIA, WATER CHEMISTRY,
AE, SEWAGE, DIATOMS, CYANOPHYTA,
RADIOCHEMICAL ANALYSIS, TRITIUM,
ECOLOGY, OXYGEN, CARBON DIOXIDE,
CTERIA, EUTROPHICATION, GLUCOSE,
GICAL ECOLOGY, AQUATIC BACTERIA,
TION, BIOCHEMICAL OXYGEN DEMAND,
RGANIC MATTER, LAKES, DIFFUSION,
                      TOTAL ORGANIC
N, FLOCCULATION, COAL, ACTIVATED
RS LAKE, DUNLOP LAKE, *INORGANIC
OSA, ALGAL CULTURES, CHEMOSTATS,
GLAND, LAKE WINDERMERE(ENGLAND),

CANADA, MESOTROPHY, LAKE NORRVIKE	W70-03964
CANADENSIS, MACROPHYTES, POTAMOGE	W70-02249
CANADENSIS, MYRIOPHYLLUM, POTAMOG	W70-00711
CANAL.: /LAKES-ILLINOIS RIVER BAS	W70-00263
CANALIZATION, ALUMINUM SULFATE, F	W70-05668
CANALS, SANDS, CHLORINATION, MICR	W69-07838
CANDIDA DIDDENSII, CANDIDA ZEYLAN	W70-04368
CANDIDA KRUSEI, CANDIDA OBTUSA, C	W70-04368
CANDIDA OBTUSA, CANDIDA TROPICALI	W70-04368
CANDIDA TROPICALIS, NOCTILUCA MIL	W70-04368
CANDIDA ZEYLANOIDES, CANDIDA KRUS	W70-04368
CANNERIES, ACTIVATED SLUDGE, BIOC	W71-09546
CANYON, *DETERGENT CONCENTRATIONS	W70-01779
CANYON, AROMATIC SOLVENTS, PHAEOD	W70-01470
CANYON, REPOPULATION, SUSCEPTIBIL	W70-01780
CANYON, SUSCEPTIBILITY, REGENERAT	W70-01777
CAPACITY, EFFLUENTS, CONSTRUCTION	W71-00936
CAPACITY, STANDING CROP, ANNUAL T	W68-00912
CAPE REYKJANES(ICELAND), SUDURNES	W69-09755
CAPE REYKJANES, FAXAFLOI FJORD, I	W69-10161
CAPILLARITY CONTROL DEVICE.: /CHI	W70-04469
CAPILLARITY, CAPILLARITY CONTROL	W70-04469
CAPILLARY CONDUCTIVITY, BACTERIA,	W70-04469
CAPRICORNUTUM, ANABAENA FLOS-AQUA	W69-06864
CAPRICORNUTUM PRINTZ, OSCILLATORI	W70-05752
CAPRICORNUTUM, SELENASTRUM GRACIL	W70-02779
CARBOHYDRATES, TEMPERATURE, SALIN	W70-01074
CARBOHYDRATES, PHENOLS, NITROGEN,	W70-01073
CARBOHYDRATES, LIPIDS, KINETICS,	W71-13413
CARBOHYDRATES.: /N, ALGAE, PLANT	W69-07826
CARBOHYDRATES, YEASTS, ENZYMES, C	W70-03952
CARBOHYDRATES.: /LLA, *ORGANIC AC	W70-04195
CARBON ASSIMILATION, BACILLARIOPH	W70-03983
CARBON CYCLE, PHOSPHORUS, PHOSPHA	W70-02777
CARBON CYCLE.: /EUTROPHICATION, B	W70-02792
CARBON DIOXIDE, SYNTHESIS, ENZYME	W70-02965
CARBON DIOXIDE, BICARBONATES, CAR	W70-02804
CARBON DIOXIDE.: /H SUBSTANCES, S	W70-02969
CARBON DIOXIDE, SCENEDESMUS, CHLO	W70-02964
CARBON DIOXIDE, HYDROGEN ION CONC	W70-02249
CARBON DIOXIDE, SOLUBILITY, *PHOS	W71-11393
CARBON DIOXIDE, DIFFUSION, *WASTE	W71-11377
CARBON DIOXIDE.: /BORATORY EQUIPM	W71-07339
CARBON DIOXIDE.: / POLLUTION CONT	W72-01788
CARBON DIOXIDE, HYDROGEN ION CONC	W70-05760
CARBON DIOXIDE, DISSOLVED OXYGEN,	W71-00668
CARBON DIOXIDE, AMMONIA, NITRITES	W71-03031
CARBON DIOXIDE, CYANOPHYTA, PHYSI	W69-10160
CARBON DIOXIDE, WATER POLLUTION E	W69-10158
CARBON DIOXIDE, FISH.: /Y, TEMPER	W69-10157
CARBON DIOXIDE, DEPTH, THERMOCLIN	W69-10169
CARBON DIOXIDE, NITRATES, PHOSPHA	W70-00161
CARBON DIOXIDE, INHIBITION, NITRO	W70-04249
CARBON DIOXIDE, GROWTH RATES, PLA	W70-04369
CARBON DIOXIDE, NUTRIENTS, BIOLOG	W69-07442
CARBON DIOXIDE, CARBON RADIOISOTO	W69-07862
CARBON DIOXIDE, *ALGAE, *HYDROGEN	W68-00855
CARBON DIOXIDE, NUTRIENT REQUIREM	W70-05424
CARBON DIOXIDE, LIGHT, TEMPERATUR	W70-05652
CARBON DIOXIDE, PHOTOSYNTHESIS, G	W70-05750
CARBON DIOXIDE, ALKALINITY, ALGAE	W70-05550
CARBON DIOXIDE, OXIDATION, CHEMIC	W70-05651
CARBON DIOXIDE, CHLORELLA PYRENOI	W70-05261
CARBON DIOXIDE, VITAMINS, NITRATE	W70-04810
CARBON DIOXIDE, GROWTH RATES, TEM	W72-06022
CARBON DISULFIDE.: /NTS, NYLON-6,	W70-06925
CARBON FIXATION, PELAGIC REGION,	W70-01933
CARBON RADIOISOTOPES, CHLOROPHYLL	W70-01579
CARBON RADIOISOTOPES, LAKES, ORGA	W70-02504
CARBON RADIOISOTOPES, FISH, CLAMS	W70-02786
CARBON RADIOISOTOPES, HYDROGEN IO	W70-08097
CARBON RADIOISOTOPES, PHOSPHORUS	W69-10151
CARBON RADIOISOTOPES, LIGHT INTEN	W69-10158
CARBON RADIOISOTOPES, PHOSPHORUS	W69-10163
CARBON RADIOISOTOPES, DIATOMS, ME	W71-12870
CARBON RADIOISOTOPES, KINETICS, T	W70-04811
CARBON RADIOISOTOPES, LAKES, PHYT	W70-05109
CARBON RADIOISOTOPES, PACIFIC OCE	W70-05652
CARBON RADIOISOTOPES, POTASSIUM,	W70-05405
CARBON RADIOISOTOPES, BIOTA, CLAM	W69-08269
CARBON RADIOISOTOPES, DETRITUS, C	W69-07862
CARBON RADIOISOTOPES, DINOFLAGELL	W70-04194
CARBON RADIOISOTOPES, ENZYMES, KI	W70-04287
CARBON RADIOISOTOPES, DETERGENTS,	W70-04001
CARBON RADIOISOTOPES, CHLAMYDOMON	W70-04284
CARBON.:	W72-04495
CARBON.: /AE, FILTRATION, SORPTIO	W70-05470
CARBON.: /AKES, QUIRKE LAKE, PECO	W70-05424
CARBON-14.: / MICROCYSTIS AERUGIN	W69-06864
CARBON-14.: /TLAND), SCOTLAND, EN	W70-02504

UTION VIII, NUTRIENT LIMITATION, : *ALGAE TOXICITY, LUORESCENCE MICROSCOPY, ACETATE, HNIO/ *HETEROTROPHY, AUTOTROPHY, ERKEN(SWEDEN), ACTIVE TRANSPORT, MENT, *WASTE WATER T/ *ACTIVATED ONOCHRYSIS LUTHE/ *DNA, *ORGANIC LGAE, *WATER QUALITY, *ACTIVATED N, COAGULATION, ALGAE, ACTIVATED *POTASSIUM COMPOUNDS, *ACTIVATED *ACTINOMYCETES, *ODOR, ACTIVATED ON, ECOLOGY, WATER PURIFICATION, S, EUTROPHICATION, BICARBONATES, R, AMMONIUM COMPOUNDS, ACTIVATED LOROFORM-METHANE EXTRACTS, LIPID ULFATE, SEDIMENTATION, ACTIVATED OROPHYLL A, SHAYLER RUN, ORGANIC DISTRIBUTION PATTERNS, ACTIVATED Y, BIOINDICATORS, CHLAMYDOMONAS, E, NITROGEN, PHOSPHORUS, OXYGEN, TION, SAMPLING, SURFACES WATERS, S, SEWAGE, ADSORPTION, ACTIVATED TE, *ODOR, *ALGICIDES, ACTIVATED N DIOXIDE, OXIDATION, CHEMICALS, CAL DENSITY, PARTICULATE ORGANIC ALGAE, ORGANIC MATTER, CULTURES, PHENOLS, NITROGEN, RESPIRATION, ALGAE, *CULTURES, PHYTOPLANKTON, CHLOROPHYTA, NITRATES, AMMONIA, SONAL, DIURNAL, LIGHT INTENSITY, ONMENTS, VARIABILITY, CHLORELLA, TER), CHLORIDES, ORGANIC MATTER, ATION, COPPER SULFATE, ACTIVATED CIDIC WATER, TRAVERTINE, CALCIUM SCHISTOSOME DERMATITIS, COPPER QUATI/ *PHOTOSYNTHESIS, *CALCIUM NNESOTA, WATER ANALYSIS, CALCIUM , AQUATIC ALGAE, WATER ANALYSIS, TS, SOLUTES, CHLORIDES, CALCIUM, PHICATION, BICARBONATES, CARBON, IRS, SCUM, NITROGEN, PHOSPHORUS, S, CARBON DIOXIDE, BICARBONATES, ALGAE, FUNGI, SULFUR COMPOUNDS, NOIDOSA, CARBOXYLIC ACIDS, AMINO ATE, NTA, CHLORELLA PYRENOIDOSA, ERIONELLA, TYP/ *AUGUSTA(MAINE), ASS), ARTHUR D LITTLE INC(MASS), CARNEGIE DEPT OF PLANT BIOLOGY, F WATER, DISSOLVED OXYGEN, NORTH , *SORPTION, *TEMPERATURE, SOUTH ER, OYSTERS, CLAMS, CRABS, NORTH IDE METABOLISM, *PESTICID/ SOUTH UM-232, BISMUTH-214, AC/ *COTTUS ASUREMENTS, EUPHOTIC ZONE, PLANT OKLA), KEYSTONE RESERVOIR(OKLA), DESICCATION, YAQUINA BAY(ORE), MOSQUITOS, ALGAL CONTROL, BASS, RE, ACTIVATED SLUDGE, FISHERIES, ROPHYTA, DISSOLVED OXYGEN, PIKE, SYNTHESIS, PHYTOPLANKTON, CISCO, ADIOISOTOPES, IONS, HUMIC ACIDS, OMPOUND, PERCHES, FISH, DIATOMS, L ANALYSIS, PHYSICAL PROPERTIES, R(POLAND), MONTANE RIVER COURSE, U/ *OSCILLATORIA, NATIONAL WATER NIQUES, W/ *BIOTIN, *AMPHIDINIUM ALIELLA TERTIOLECTA, AMPHIDINIUM E, CHLOROPHYCEAE, CHRYSOPHYCEAE, HALOCHLOROCOCCUM, SYNECHOCOCCUS, 4, NITZSCHIA CLOSERIUM, ARTEMIA, HYLOCOCCUS AUREUS, LACTOBACILLUS C EFFECTS, STIMULANTS, GONYAULAX ITOS, ALGAL CONTROL, BASS, CARP,), BACK RIVER(MD), OSCILLATORIA, LUTION EFFECTS, CATION EXCHANGE, RPTION, WATER POLLUTION EFFECTS, THETIC ORGANIC POLYELECTROLYTES, C WEED CONTROL, *BIBLIOGRAPHIES, SANDS, SHALES, BOTTOM SEDIMENTS,

POSTING ORGANIC MATTER, LAGOONS, OLLUTION EFFECTS, WATER QUALITY, COCCUS, SPINES, SCHIZOMYCOPHYTA, PERIMENTS, *NUTRIENT LIMITATION, A RUBESCENS, MADISON LAKES(WIS), AERUGINOSA, BOHEMIA, ELBE RIVER, AE, WATER MILFOIL COONTAIL, WILD E, KILL RATES, PHOTO REACTIVITY, EMENTS, SCENEDESMUS, ABSORPTION, ICU/ *GROWTH, *UPTAKE, FIXATION,

CARBON-14.: /KES(NY), RODHE'S SOL   W70-01579
CARBON-14, DDT, MALATHION, SEVIN.   W70-02198
CARBON-14, AUTORADIOGRAPHY, MICHA   W70-04811
CARBON-14, RADIOCARBON UPTAKE TEC   W70-04287
CARBON-14, ENZYME KINETICS, GLUCO   W70-04284
CARBON, *ADSORPTION, *WATER TREAT   W72-03641
CARBON, *CELL SIZE, EUKARYOTES, M   W69-08278
CARBON, *ELECTROLYTES, FILTERS, *   W72-09693
CARBON, *FLOCCULATION, *HEAD LOSS   W72-07334
CARBON, ACTINOMYCETES, IOWA.: /,   W72-04242
CARBON, AMMONIA, CHLORINE.: /AE,   W72-04155
CARBON, BICARBONATES, PHOTOSYNTHE   W69-07389
CARBON, CARBONATES, CARBON CYCLE.   W70-02792
CARBON, CHLORINE, PHENOLS, OIL WA   W70-00263
CARBON, CHLOROPLASTS, INFRARED AN   W70-05651
CARBON, CHLORINE, COSTS, TREATMEN   W72-08859
CARBON, CINCINNATI(OHIO).: /, CHL   W71-04206
CARBON, CLAYS, ALGAE, WATER PURIF   W69-09884
CARBON, ECOSYSTEMS, EUGLENA, EUTR   W69-06273
CARBON, ENZYMES, SAMPLING, BOTTOM   W70-09889
CARBON, FERTILITY, ALGAE, NITROGE   W70-02973
CARBON, FILTERS, SEASONAL, STRATI   W70-04001
CARBON, FILTRATION, *TOXICITY, *D   W72-03746
CARBON, GRAVIMETRY.: /TURE, CARBO   W70-05651
CARBON, LAKE TAHOE(CALIF), CHLORE   W70-03983
CARBON, METABOLISM, NITROGEN, PHO   W70-02804
CARBON, PRIMARY PRODUCTIVITY, OXY   W70-01073
CARBON, RADIOACTIVITY, CHLORELLA,   W70-03983
CARBON, RECREATION, WATER POLLUTI   W72-04761
CARBON, SEWAGE TREATMENT, TERTIAR   W70-05387
CARBON, TEMPERATURE, LIGHT INTENS   W70-02965
CARBON, WATER POLLUTION EFFECTS,   W71-04206
CARBON, WATER SUPPLY.: /, CHLORIN   W72-03562
CARBONATE, BACTERIA, BIOMASS, CHL   W69-10160
CARBONATE, CERCARIAE, WATER MILFO   W70-05568
CARBONATE, AQUATIC ENVIRONMENT, A   W69-00446
CARBONATE, HYPOLIMNION.: /OGY, MI   W69-00632
CARBONATES, NITRATES, CHLORIDES,   W69-07096
CARBONATES, PHOSPHATES.: /NUTRIEN   W70-09557
CARBONATES, CARBON CYCLE.: /EUTRO   W70-02792
CARBONATES, TEMPERATURE, SHALLOW   W70-02803
CARBONATES, TEMPERATURE, HYDROGEN   W70-02804
CARBONATES.: /N OXIDES, BACTERIA,   W70-02294
CARBOXYLIC ACIDS.: /HLORELLA PYRE   W70-02248
CARBOXYLIC ACIDS, AMINO CARBOXYLI   W70-02248
CARLETON POND(MAINE), SYNURA, AST   W70-08096
CARNEGIE DEPT OF PLANT BIOLOGY, C   W69-06865
CARNEGIE INSTITUTION OF WASHINGTO   W69-06865
CAROLINA.: /OIRS, ALGAE, BODIES O   W70-00683
CAROLINA, CESIUM, STRONTIUM RADIO   W69-07845
CAROLINA, SEDIMENTS, RADIOACTIVIT   W69-08267
CAROLINA, LYOPHILIZATION, *PESTIC   W69-02782
CAROLINAE, MEAD COUNTY(KY), THORI   W69-09742
CAROTENOIDS, CARBON FIXATION, PEL   W70-01933
CAROTENOIDS.: /IVER(OKLA), BIXBY(W70-04001
CAROTENOIDS, SPECIES, DIVERSITY I   W70-03325
CARP, CATFISHES, NITROGEN, POTASS   W70-05417
CARP, CLAMS, PONDS, RUNNING WATER   W70-09905
CARP, D: /S, ALGAE, DIATOMS, CHLO   W70-01943
CARP, DIATOMS, CHLOROPHYTA, CHLAM   W69-02959
CARP, ICE, IRON COMPOUNDS, HYDROG   W70-05399
CARP, PLANKTON, METABOLISM, BIOLO   W70-00274
CARP, TILAPIA, TEMPERATURE, NU: /   W70-06975
CARPATHIAN SUBMONTANE REGION, GRA   W70-02784
CARRIER(ISRAEL), THRESHOLD ODOR N   W70-02373
CARTERI, *RADIOCARBON UPTAKE TECH   W70-05652
CARTERI, SYRACOSPHAERA ELONGATA,   W69-08278
CARTERIA, COMPARATIVE STUDIES, CR   W69-07833
CARTERIA, GONIUM, ANKISTRODESMUS,   W70-03974
CARTERIA, NORTH TEMPERATE ZONE, P   W70-00708
CASEI, BACILLUS THURINGIENSIS, AG   W71-11823
CATENELLA, GONYAULAX TAMERENSIS,   W70-02510
CATFISHES, NITROGEN, POTASSIUM.: /   W70-05417
CATHERWOOD DIATOMETER, SAVANNAH R   W70-04510
CATION ADSORPTION, HYDROLYSIS, MA   W71-09863
CATION EXCHANGE, CATION ADSORPTIO   W71-09863
CATIONIC ELECTROLYTES, ALGAL SURF   W70-05267
CATTAILS, BULRUSH, DREDGING, COPP   W70-05263
CATTAILS, RADIOACTIVITY, INTERFAC   W69-08272
CATTARAUGUS CREEK(NY).:   W69-10080
CATTLE.: /GE, SOLID WASTES, DECOM   W71-05742
CATTLE.: /ODOR, TOXICITY, WATER P   W69-08279
CAULOBACTERIALES, CHLOROCOCCALES,   W70-03969
CAYUGA LAKE(NY), ASTERIONELLA FOR   W70-01579
CEDAR RIVER(WASH), ZURICHSEE(SWIT   W70-00270
CELAKOVICE(CZECHOSLOVAKIA), BLATN   W70-00274
CELERY, BUR REED, DUCKWEED, WHITE   W70-05568
CELL COLOR, CELL DAMAGE.: /S ALGA   W70-02364
CELL CONCENTRATION, UROGLENA, ANK   W70-05409
CELL COUNT, OPTICAL DENSITY, PART   W70-03983

CALIF), SCENEDESMUS QUADRICAUDA,
S, PHOTO REACTIVITY, CELL COLOR,
HT EXTINCTION COEFFICIENT, ALGAL
VOLATILE SUSPENDED SOLIDS, MULTI-
EGHINIANA, LIGHT-SATURATED RATE,
ESIS, MEAN DOUBLING TIME, PACKED
   MASS CULTURE OF ALGAE, PACKED
COUNT, WARING BLENDOR MORTALITY,
S, EXTRACELLULAR PRODUCTS, ALGAL
, MATERIAL BALANCES, O/ *UPTAKE,
TH FACTORS, COLONIAL MORPHOLOGY,
R COUNTER, *POPULATION DYNAMICS,
 CHEMICAL DEGRADATION, VITAMINS,
LEAVES, FERTILIZERS, PESTICIDES,
ULTRAVIOLET RADIATION, SULFATES,
, LABORATORY EQUIPMENT, ASBESTOS
QUAE, LYNGBYA, STORM LAKE(IOWA),
CAUDA, CEL/ STRIPPING, FIREBAUGH
URE COEFFICIENTS, NORTH AMERICA,
RCTIC CIRCLE HOT SPRINGS, RADON,
            SOUTH -
IRA, ASTERIONELLA, ARAPHIDINEAE,
ING, PHYSICOCHEMICAL PROPERTIES,
GE, DETERGENTS, SPRAYING, ALGAE,
 ION CONCENTRATION, TEMPERATURE,
ASONICS, DISINFECTION, SAMPLING,
CONCENTRATION, ORGANIC LOADING,
T PLANKTON, CHLOROPHYLL CONTENT,
LEGANS, EUGLENA GRACILIS, FOREST
, MELOSIRA, SYNEDRA, FRAGILARIA,
ERIA SETIGERA, MELOSIRA AMBIGUA,
UROGLENA AMERICANA, CYCLOTELLA,
HANIZOMENON, ANABAENA, MELOSIRA,
ENA.:   HYDRODICTYON, ELODEA,
HORNIA, POTAMOGETON, TRAPA, PIS/
LES QUADRI/ LEECHES, SCHISTOSOME
ME DERMATITIS, COPPER CARBONATE,
ISOTOPES, CADMIUM RADIOISOTOPES,
ARTEMIA, C/ *MARINE ENVIRONMENT,
-95, RUTHENIUM-106, RHODIUM-106,
-106, RUTHENIUM-103-RHODIUM-103,
BIUM, EUROPIUM-155, SILVER-110M,
IOISOTOPES, CERIUM RADIOISOTOPE/
 SUCCESSION, ZINC RADIOISOTOPES,
OTOPES, STRONTIUM RADIOISOTOPES,
ANS, FOODS, ABSORPTION, HAZARDS,
, ION EXCHANGE, RESINS, LEAKAGE,
OSIONS, STRONTIUM RADIOISOTOPES,
ASSIUM RADIOISOTOPES, MANGANESE,
XPLOSIONS, COBALT RADIOISOTOPES,
N, *TEMPERATURE, SOUTH CAROLINA,
NTS, DINOPHYCEAE, CRYPTOPHYCEAE,
RSEILLES, ASTERIONELLA JAPONICA,
ODISCUS, BACILLARIA, BIDDULPHIA,
IOSI/ *ALGAL NUTRITION, DITYLUM,
AGILARIA, OEDOGONIUM, SPIROGYRA,
ORS, MYRIOPHYLLUM, WATER BLOOMS,
, MONITORING, MICROBIOLOGY, FOOD
TOXICITY, *WATER POLLUTION, FOOD
, *WATER POLLUTION EFFECTS, FOOD
ICIDE RESIDUES, ECOSYSTEMS, FOOD
OLLUTION, *WASTE TREATMENT, FOOD
UTRIENTS, ENERGY TRANSFER, *FOOD
ALGAE, BRINE SHRIMP, FISH, FOOD
FISH, GAME BIRDS, ANIMALS, FOOD
*RADIOISOTOPES, SEA WATER, FOOD
ITY ACT, PATH OF POLLUTANT, FOOD
LEVEL, ALGAE, CADDISFLIES, FOOD
ROPHIC LEVEL, PRODUCTIVITY, FOOD
RGINIA, PENNSYLVANIA, MARCASITE,
ORPIMENT, TETRAHEDRITE, BORNITE,
LAKE WORTHER(AUSTRIA), VERTICAL,
LLATORIA RUBESCENS, OSCILLATORIA
           *RESPIROMETER
LAKE LOTSJON(SWEDEN), PERMEASES,
/ *LAKE WASHINGTON(WASH), *LAKE
EARTH, MUD, LAKES, CULTIVATION,
LAKE(WASH), WILLIAMS LAKE(WASH),
WAGE EFFLUENT, NUTRIENTS, LAKES,
NCE ALGAE, DUCKS(WILD), OYSTERS,
, LIMNODRILUS, EUCYCLOPS, ALONA,
MENT, POTABLE WATER, HERBICIDES,
LAR TISSUES, *ALGAE, ECOSYSTEMS,
, *NEVADA, DEPTH, ALGAE, MOSSES,
EFFECTS, ECOLOGY, WATER QUALITY
IS, VIBRIO, MORPHOLOGY, CULTURAL
SICAL CHARACTERISTICS, *CHEMICAL
TERIS/ *RIVER REGIONS, *PHYSICAL
SKATCHEWAN), MICE, MORPHOLOGICAL
NCOURSE, PALOUSE RIV/ TREE BARK,

CELL COUNTS, SPECIES DETERMINATIO   W70-03334
CELL DAMAGE.: /S ALGAE, KILL RATE   W70-02364
CELL DENSITY.: / TURBIDOSTAT, LIG   W70-03923
CELL SERIAL SYSTEM, PH, FACULTATI   W70-04790
CELL VOLUME, NITZSCHIA CLOSTERIUM   W70-05381
CELL VOLUMES, BACTERIAL PLATE COU   W71-05267
CELL VOLUME.:   W70-06053
CELL WALL, HEXOSAMINE, ASCOPHYLLU   W70-04365
CELLS.: /E, PHYTOPLANKTON DYNAMIC   W69-05697
CELLS, CONCENTRATION, COMPOSITION   W70-06600
CELLULAR MORPHOLOGY, FLAGELLATION   W70-03952
CELLULAR SIZE, CHLORELLA VULGARIS   W69-06540
CELLULOSE, CHLORELLA, PEPTIDES, B   W70-04184
CELLULOSE, CYCLES, HARVESTING, DR   W70-05264
CELLULOSE, INDUSTR: /, SALINITY,   W70-01074
CEMENT, CHLORELLA, CARBON DIOXIDE   W71-07339
CENTER LAKE(IOWA), DIAMOND LAKE(I   W70-02803
CENTER(CALIF), SCENEDESMUS QUADRI   W70-03334
CENTRAL AMERICA, RATE EQUATIONS,   W70-03957
CENTRAL CITY(COLO), MUTATION, BET   W69-08272
CENTRAL STATES.:   W72-03562
CENTRIC DI: /R, FRAGILARIA, MELOS   W70-05663
CENTRIFUGATION, MICROSCOPY, NETS,   W70-05389
CENTRIFUGATION, FILTRATION, ANERO   W70-04060
CENTRIFUGATION, COAGULATION, DEWA   W71-11375
CENTRIFUGATION, SEDIMENTATION, CH   W71-11823
CENTRIFUGATION, RESPIRATION, BIOC   W72-06022
CENTRIFUGE METHOD, CALCULATED VOL   W70-03959
CENTRIFUGE, GLENODINIUM PULVISCUL   W69-06540
CER: /STIS, BOSMINA, ASTERIONELLA   W70-04503
CERATIUM HIRUDINELLA, ANKISTRODES   W70-05387
CERATIUM HIRUNDINELLA, PERIDINIUM   W70-05405
CERATIUM, OSCILLATORIA, EUDORINA.   W69-01977
CERATOPHYLLUM, MICROCYSTIS, ANABA   W68-00479
CERATOPHYLLUM, MYRIOPHYLLUM, EICH   W70-10175
CERCARIAE, TVA RESERVOIRS, ANOPHE   W70-06225
CERCARIAE, WATER MILFOIL COONTAIL   W70-05568
CERIUM RADIOISOTOPES, ECHINODERMS   W71-09850
CERIUM-144, NITZSCHIA CLOSERIUM,   W70-00708
CERIUM-144, BETA RADIOACTIVITY, G   W69-09742
CERIUM-144-PRASEODYMIUM-144.: /UM   W69-08524
CERIUM, GUETTARDA, IPOMOEA, PISON   W69-08269
CESIUM RADIOISOTOPES, CADMIUM RAD   W71-09850
CESIUM, COBALT RADIOISOTOPES, SEA   W70-04371
CESIUM, FALLOUT, SILTS, DETRITUS,   W69-09742
CESIUM, HYDROGEN, DEUTERIUM, TRIT   W70-02786
CESIUM, NUCLEAR EXPLOSIONS, ABSOR   W70-00707
CESIUM, PHYSICOCHEMICAL PROPERTIE   W70-00708
CESIUM, RADIUM RADIOISOTOPES, URA   W69-08524
CESIUM, STRONTIUM RADIOISOTOPES,   W69-08269
CESIUM, STRONTIUM RADIOISOTOPES,   W69-07845
CH: /, APHANIZOMENON, MICRONUTRIE   W70-04381
CHAETOCEROS LAUNDERI, LAUDERIA BO   W70-04280
CHAETOCEROS, COCCONEIS, COSCINODI   W70-03325
CHAETOCEROS, SKELETONEMA, THALASS   W70-04503
CHAETOPHORA, STIGEOCLONIUM, ANKIS   W70-04371
CHAINING.: /SALIX, GROWTH REGULAT   W70-05263
CHAINS.: /S, HARVESTING, DREDGING   W70-05264
CHAINS, AQUATIC ALGAE, BRINE SHRI   W70-01466
CHAINS, BACTERIA, DEGRADATION(DEC   W72-01801
CHAINS, BIOLOGICAL COMMUNITIES, B   W70-05272
CHAINS, CHLORELLA, CHLOROPHYLL, P   W70-04381
CHAINS, FOOD WEBS, FORAGE FISH, P   W70-05428
CHAINS, MUSSELS.: /UTION, AQUATIC   W70-01467
CHAINS, OXYGEN, ROTIFERS, WEEDS,   W70-10175
CHAINS, PHOSPHORUS RADIOISOTOPES,   W70-00708
CHAINS, SAMPLING, SURVEYS, MARINE   W72-00941
CHAINS, STABLE ISOTOPES, RADIOACT   W70-02786
CHAINS, TOXICITY, WATER POLLUTION   W72-01788
CHALCOCITE, COVELLITE, SPHALERITE   W69-07428
CHALCOPYRITE, CHEMOSYNTHESIS, CON   W69-07428
CHALK, METALIMNION, OSCILLATORIA   W70-05405
CHALYBIA, HYDRODICTYON RETICULATU   W69-05704
CHAMBER.:   W71-00668
CHAMYDOMONADS, LAPPLAND(SWEDEN),   W70-04287
CHANGES, *NUTRIENT CONCENTRATION,   W70-00270
CHANNEL MORPHOLOGY, SEWAGE INPUT,   W68-00475
CHANNELED SCABLANDS(WASH), WASHIN   W70-00264
CHANNELS, REVIEWS, BIBLIOGRAPHIES   W70-00269
CHANNELS, POLLUTION ABATEMENT, HU   W68-00461
CHAOBORUS.: / TANYPUS, TANYTARSUS   W70-06975
CHARA, ALKALINITY, BULLHEADS, EEL   W70-05263
CHARA, BENTHIC FLORA, FLOATING PL   W70-10175
CHARA, FISH, LIMNOLOGY, LAKE TROU   W70-00711
CHARACTERIZATION.: /TER POLLUTION   W70-02764
CHARACTERISTICS, TERMARY COMPOUND   W70-04280
CHARACTERISTICS, ORGANISMS, SUSPE   W73-00748
CHARACTERISTICS, *CHEMICAL CHARAC   W73-00748
CHARACTERS, TRIS.: /, HUMBOLDT(SA   W70-00273
CHARCOAL, ALGAL GROWTH, STREAM CO   W70-03501

ONIC ELECTROLYTES, ALGAL SURFACE	CHARGE, ELECTROSTATIC REPULSION,	W70-05267
PHYTA, FUNGI, BACILLARIOPHYCEAE,	CHAROPHYTA, RHODOPHYTA.: /, CYANO	W70-02784
*SUSPENDED SOLIDS,	CHASE CITY(VA).:	W72-10573
FLUORESCEIN, MACROPH/ MICHIGAN,	CHEBOYGAN COUNTY, STURGEON RIVER,	W69-09334
NISMS, PAPERMILL WASTE EFFLUENT,	CHEESE FACTORY WASTE WATERS.: /GA	W71-00940
CTION, POTATO PROCESSING WASTES,	CHEESE MANUFACTURING WASTES, STAR	W70-03312
ATION, *DETERGENTS, *PHOSPHATES,	CHELATION, WASTE TREATMENT, AQUAT	W70-04373
ESE, *WISCONSIN, ALGAE, CALCIUM,	CHELATION, CYANOPHYTA, NUISANCE A	W69-05867
REDUCTION(CHEMICAL), HYDROLYSIS,	CHELATION, FUNGI, ALGAE, PROTOZOA	W69-07428
OPERTIES, TRACERS, DISTRIBUTION,	CHELATION, DETERGENTS, RADIOACTIV	W69-08275
SHELLFISH, OYSTERS, INDUSTRIES,	CHELATION, BACTERIA, STREPTOCOCCU	W70-01068
SCENEDESMUS, CHLOROPHYLL, IRON,	CHELATION, ENZYMES, CHLORINE, LIG	W70-02964
IRON, MANGANESE, TRACE ELEMENTS,	CHELATION, PHOTOSYNTHESIS, CARBON	W70-02804
HIBITORS, SUCCESSION, CHLORELLA,	CHELATION, OCHROMONAS, BACTERIA,	W70-02510
ALGAE, *WATER POLLUTION EFFECTS,	CHELATION, SPECTROPHOTOMETRY, NUT	W71-11561
IPS, NUCLEAR WASTES, ABSORPTION,	CHELATION, METABOLISM, AQUARIA, W	W71-09005
NTATION, SEDIMENTS, TASTE, WATER	CHEM: /PLANKTON, SULFATES, SEDIME	W69-07833
SNAILS, LEGISLATION, WISCONSIN,	CHEMCONTROL, PONDWEEDS, ROOTED AQ	W70-05568
ION SOURCES, SEWAGE, PHOSPHORUS,	CHEMCONTROL, INDUSTRIAL WASTES, S	W70-05668
ONTROL, WATER POLLUTION CONTROL,	CHEMCONTROL, ARSENIC COMPOUNDS, N	W70-04494
, WATER HYACINTH, ALLIGATORWEED,	CHEMCONTROL, MECHANICAL CONTROL,	W70-10175
AE, WEEDS, ECOLOGY, ENVIRONMENT,	CHEMCONTROL, BIOCONTROL, INTEGRAT	W70-07393
ANT GROWTH, *WISCONSIN, ECOLOGY,	CHEMICAL ANALYSIS, PHYTOPLANKTON,	W70-06229
ION POTENTIAL, DISSOLVED OXYGEN,	CHEMICAL ANALYSIS, ORGANIC MATTER	W70-05760
N, PHOSPHORUS, SILICA, SAMPLING,	CHEMICAL ANALYSIS, PHYSICAL PROPE	W70-06975
WTH, DECOMPOSING ORGANIC MATTER,	CHEMICAL ANALYSIS, PROTEINS, BACT	W72-01812
, RADIOACTIVITY, PHOTOSYNTHESIS,	CHEMICAL ANALYSIS, TRACE ELEMENTS	W70-01933
ER CHEMISTRY, WATER TEMPERATURE,	CHEMICAL ANALYSIS, ANALYTICAL TEC	W69-09676
ES, CHLOROPHYLL, PHOTOSYNTHESIS,	CHEMICAL ANALYSIS, BACTERIA, NUTR	W69-10167
ITY, PHOSPHORUS, NITROGEN, FLOW,	CHEMICAL ANALYSIS, NUTRIENTS, EFF	W70-04810
INS, VITAMIN B, ALGAE, CULTURES,	CHEMICAL ANALYSIS, ANALYTICAL TEC	W70-05652
RSHEDS, SEWAGE, WATER POLLUTION,	CHEMICAL ANALYSIS, PHYSICAL PROPE	W70-05752
S, CHLORINATION, MICROORGANISMS,	CHEMICAL ANALYSIS, TEMPERATURE, N	W69-07838
NTENSITY, VELOCITY, TEMPERATURE,	CHEMICAL ANALYSIS, PHYSIOLOGICAL	W69-07862
EUTROPHICATION, DATA COLLECTION,	CHEMICAL ANALYSIS, RELIABILITY, W	W70-04382
ERIC BACTERIA, METHANE BACTERIA,	CHEMICAL ANALYSIS, WATER POLLUTIO	W70-03312
ALGAE, ASSAY, DISSOLVED SOLIDS,	CHEMICAL ANALYSIS, PHOSPHATES, IR	W70-03334
L, WATER PROPERTIES, PHOSPHORUS,	CHEMICAL ANALYSIS.: / WEED CONTRO	W68-00479
NUTRIENTS, ALGAE, BENTHIC FAUNA,	CHEMICAL ANALYSIS, SESSILE ALGAE,	W72-06052
NEGIE INSTITUTION OF WASHINGTON,	CHEMICAL COMPOSITION(ENGLAND), FO	W69-06865
R, ALGAE, PONDS, WATER HYACINTH,	CHEMICAL CONTROLS, PARAQUAT, DIQU	W70-00269
, MECHANICAL EQUIPMENT, ENZYMES,	CHEMICAL DEGRADATION, VITAMINS, C	W70-04184
*NICKEL, DETECTION LIMITS,	CHEMICAL INTERFERENCES.:	W71-11687
MERICAN WATER WORKS ASSOCIATION,	CHEMICAL INDUSTRY, ODOR, TASTE, C	W71-05943
RIENTS, *FERMENTATION, TOXICITY,	CHEMICAL OXYGEN DEMAND, HYDROGEN	W71-13413
DSORPTION, REMOVAL, DEGRADATION,	CHEMICAL OXYGEN DEMAND, *WASTE WA	W71-13335
OONS, BIOCHEMICAL OXYGEN DEMAND,	CHEMICAL OXYGEN DEMAND, DISSOLVED	W70-06601
ONS, ABSORPTION, BIODEGRADATION,	CHEMICAL OXYGEN DEMAND, CHLOROPHY	W70-08392
ANGE, BIOCHEMICAL OXYGEN DEMAND,	CHEMICAL OXYGEN DEMAND, COLOR, AC	W70-10433
IC DIGESTION, AEROBIC TREATMENT,	CHEMICAL OXYGEN DEMAND, STORAGE C	W71-00936
ENTS, ALGAE, BACTERIA, CULTURES,	CHEMICAL OXYGEN DEMAND, NITROGEN,	W70-05655
ES, *EUTROPHICATION, *HYDROLOGY,	CHEMICAL PROPERTIES, WISCONSIN, S	W70-05113
N, *TEMPERATURE, *COOLING WATER,	CHEMICAL PROPERTIES, ECOLOGY, SAL	W69-05023
PLANKTON, ALGAE, BIBLIOGRAPHIES,	CHEMICAL PRECIPITATION, IONS, BIO	W69-00446
UTROPHICATION, PLANKTON, *ALGAE,	CHEMICAL PROPERTIES, BACTERIA, OR	W68-00483
THERMAL POLLUTION, TEMPERATURE,	CHEMICAL PROPERTIES, ECOLOGY, ECO	W70-07313
IOXIDE, SOLUBILITY, *PHOSPHATES,	CHEMICAL PRECIPITATION, FILTRATIO	W71-11393
, CENTRIFUGATION, SEDIMENTATION,	CHEMICAL PRECIPITATION, LIFE CYCL	W71-11823
TROPHICATION, SPECIES DIVERSITY,	CHEMICAL PRECIPITATION.: /URAL EU	W69-08518
RIAL WASTES, TERTIARY TREATMENT,	CHEMICAL PRECIPITATION, HARVESTIN	W69-10178
TION, ELECTRODIALYSIS, SORPTION,	CHEMICAL PRECIPITATION, ULTIMATE	W70-01981
UTICALS, EXTRACELLULAR PRODUCTS,	CHEMICAL RESISTANCE, EUPHAUSIA, P	W70-01068
, PONDS, HYDROLYSIS, METABOLISM,	CHEMICAL REACTIONS, CHLORELLA, TR	W70-08097
APHICAL REGIONS, PHOTOSYNTHESIS,	CHEMICAL REACTIONS, ZOOPLANKTON,	W69-03611
E WATER TREATMENT, ELECTROLYTES,	CHEMICAL REACTIONS, ELECTROCHEMIS	W70-05267
RGASSO SEA, SYNURA, GYMNODINIUM,	CHEMICAL SYMBIOSIS, MICROBIAL ECO	W70-04503
APHTHOQUIN/ *SELECTIVE TOXICITY,	CHEMICAL STRUCTURE, 2,3-DICHLORON	W70-02982
TION, PHYSI/ *REMOVAL PROCESSES,	CHEMICAL TREATMENT, OVERFERTILIZA	W70-07469
LAIMED WATER, ORGANIC COMPOUNDS,	CHEMICAL WASTES, ION EXCHANGE, BI	W70-10433
*PLANT GROWTH, *ALGAE, ELEMENTS(	CHEMICAL), CHLOROPHYTA, CYANOPHYT	W70-02964
CORROSION, OXIDATION, REDUCTION(	CHEMICAL), HYDROLYSIS, CHELATION,	W69-07428
TURE, CARBON DIOXIDE, OXIDATION,	CHEMICALS, CARBON, GRAVIMETRY.: /	W70-05651
EUTROPHICATION, SEQUENCE, LAKES,	CHEMICALS, FISHING, SPORT FISHING	W68-00172
ECTS, *WATER QUAL/ *AGRICULTURAL	CHEMICALS, *DELTAS, *DRAINAGE EFF	W71-09577
RIFICATION, WATER QUALITY, WATER	CHEMISTRY.: /AYS, ALGAE, WATER PU	W69-09884
T INTENSITY, SPRING WATER, WATER	CHEMISTRY, WATER TEMPERATURE, CHE	W69-09676
, WATER POLLUTION SOURCES, WATER	CHEMISTRY, WATER ANALYSIS.: /LGAE	W69-08562
TY, *MONITORING, BACTERIA, WATER	CHEMISTRY, PHOSPHORUS, NITROGEN,	W71-05879
ION CONTROL, WATER BUDGET, WATER	CHEMISTRY, WATER TEMPERATURE, HEA	W72-01773
RIENTS, LIGHT, WASHINGTON, WATER	CHEMISTRY, OXYGEN, HYDROGEN ION C	W70-03309
*WATER POLLUTION, *RIVERS, WATER	CHEMISTRY, WATER POLLUTION SOURCE	W70-03068
CTIVITY, *EUTROPHICATION, *WATER	CHEMISTRY, POLLUTANTS, ALGAL CONT	W70-02646
VIRONMENT, MICROORGANISMS, WATER	CHEMISTRY, PLANTS, SOUTHWEST U. S	W69-03611
DISPOSAL, SEWAGE DISPOSAL, WATER	CHEMISTRY, ECOLOGY, WAT: / WASTE	W69-04276
OXYGEN DEMAND, MINNESOTA, WATER	CHEMISTRY, BIODEGRADATION.: /ICAL	W69-00446
GELLATES, MARINE BACTERIA, WATER	CHEMISTRY, CARBON RADIOISOTOPES,	W70-05652
YDROGEN ION CONCENTRATION, WATER	CHEMISTRY, *PHYTOPLANKTON, CARBON	W70-05424
TS, *PHOSPHATES, *ANALYT/ *WATER	CHEMISTRY, *WATER POLLUTION EFFEC	W70-05269
ENT, EFFLUENTS, LIMNOLOGY/ WATER	CHEMISTRY, RESPIRATION, BATHYMETR	W70-05387
SODIUM, VIRUSES, VITAMINS, WATER	CHEMISTRY, WATER POLLUT: /MENTS,	W69-04801

ENTS, NUTRIENTS, TOXICITY, WATER
NTS, PHOSPHORUS COMPOUNDS, WATER
OUNDS, *CYCLING NUTRIENTS, WATER
Y, STANDING CROP, STREAMS, WATER
DUCTIVITY, SOIL CHEMISTRY, WATER
KTON, PRIMARY PRODUCTIVITY, SOIL
, WATER CONSTITUENTS, ANALYTICAL
ES, PHYSIOLOGICAL ECOLOGY, WATER
ES, PHYSIOLOGICAL ECOLOGY, WATER
ATMENT, INDUSTRIAL WASTES, WATER
PHICATION, GRAZING, ALGAE, WATER
OROBIUM, FILINIA, MONIMOLIMNION,
ASSAY PROCEDURE, BATCH CULTURES,
RCH PROGRAM, CONTINUOUS CULTURE,
STIS AERUGINOSA, ALGAL CULTURES,
AHEDRITE, BORNITE, CHALCOPYRITE,
ONOMIC GROUPS, MOLECULAR FILTER,
LOGY, MORPHOLOGY, EXPORT LOSSES,
N, PEDIASTRUM BO/ HETEROTROPHIC,
TS, PHOTIC ZONE, GLYCOLLIC ACID,
AGEMENT, *ENVIRONMENTAL EFFECTS,
OR-PRODUCING ALGAE, CHLOROPHYLL,
    USBR RESEARCH CONTRACTS,
RIVER(TENN), POTOMOGETON PECTI/
LORIS PENIOCYSTIS, MADISON(WIS),
SHING WATERWAYS, MILWAUKEE(WIS),
TENN), FRENCH BROAD RIVER(TENN),
, OSOYOOS LAKE(CANADA), DEFORMED
PTERA, EPHEMEROPTERA, TIPULIDAE,
ROPHIC WATERS, HAMBURG(GERMANY),
MUS PROMOSUS, TRICHO/ HEXAGENIA,
EFISH, WURTTEMBERG(GERMANY), PL/
N, BATHYTHERMOGRAPH, ARTHROPODA,
AWICKA(WIS), LAKE PEWAUKEE(WIS),
GENIA, CHIRONOMIDAE, PROCLADIUS,
ATER(POLLUTION), PHOTOSYNTHESIS,
TRACE ELEMENTS, LIGHT INTENSITY,
TEROTROPHY, CONTINUOUS CULTURES,
DIFFUSION, CARBON RADIOISOTOPES,
ASSAYS, SALINE WATER, CHLORELLA,
PILIMNION, CONNECTICUT, EUGLENA,
OTRIAZOLE, 2-4-5-T, SCENEDESMUS,
SCO, CARP, DIATOMS, CHLOROPHYTA,
NTS, INDICATOR SPECIES, EUGLENA,
TIC ALGAE, AQUATIC MICROBIOLOGY,
ALGAE, BIOASSAY, BIOINDICATORS,
RANEUS, ANKISTRODESMUS FALCATUS,
TE, ORGANIC MATTER, SCENEDESMUS,
ELLA, NUTRIENTS, TRACE ELEMENTS,
MINNOWS, PEDALFERS, SCENEDESMUS,
E, NITZSCHIA PALEA, MYXOPHYCEAE,
OPHYTA, RHODOPHYTA, SCENEDESMUS,
CEANS, ROTIFERS, SWIMMING POOLS,
YTA, DIATOMS, SEASONAL, EUGLENA,
FILTERS, IOWA, DIATOMS, EUGLENA,
ON, GROWTH-PROMOTING PROPERTIES,
, NITROGEN, GROWTH, AMINO ACIDS,
, CHLOROPHYTA, COPEPODS, CYCLES,
NT GROWTH SUBSTANCES, CHLORELLA,
ONTROL, EUTROPHICATION, DAPHNIA,
    BROMINE, BROMAMINE,
TURE, ELEOCHARIS, 2,4-D, AMMATE,
BENOCLORS, NIGROSINE DYE, SODIUM

OGEN LIMITATION, BATCH CULTURES,
, THIMET, TRI/ SEVIN, MALATHION,
, SKELETONEMA COSTATUM(GREV) CL,
ILANES, SCENEDESMUS QUADRICAUDA,
RISODIUM NITRILOTRIACETATE, NTA,
E MATTER, BLELHAM TARN(ENGLAND),
DEGRADATION, SEVIN, MALATHION, /
MACRONUTRIENTS, MICRONUTRIENTS,
HA/ *UPTAKE, SPECIFIC ACTIVITY,
ILANES, SCENEDESMUS QUADRICAUDA,
PERTIES, CHLAMYDOMONAS MOEWUSII,
SIRA BINDERANA, GENERATION TIME,
SMUS BRAUNII, D/ CARBON DIOXIDE,
YUBERETS(USSR), MUNICH(GERMANY),
LIQUUS, SCENEDESMUS QUADRICAUDA,
DRICAUDA, CHLORELLA PYRENOIDOSA,
OLUMBIA), CHLORELLA PYRENOIDOSA,
AROPHILA, CHLORELLA ELLIPSOIDEA,
TOOXIDATION, CHLORELLA VULGARIS,
GARIS(COLUMBIA)/ PHOTOOXIDATION,
OIDOSA, CHLORELLA SACCHAROPHILA,
S, CHLORELLA VULGARIS(COLUMBIA),
ULATION DYNAMICS, CELLULAR SIZE,
A/ *MASS CULTURE, *ALGAL GROWTH,
EDIMENTS, PRECAMBRIAN SEDIMENTS,
A LAKE(MINN), ELY(MINN), BLOOMS,

CHEMISTRY, WATER POLLUTION EFFECT W69-05867
CHEMISTRY, WATER POLLUTION EFFECT W69-05868
CHEMISTRY, WATER ANALYSIS, PHOSPH W69-05142
CHEMISTRY, WATER: /S, PRODUCTIVIT W69-04805
CHEMISTRY, WATER POLLUTION EFFECT W69-06273
CHEMISTRY, WATER CHEMISTRY, WATER W69-06273
CHEMISTRY, WASTE WATER EFFECTS, L W70-04382
CHEMISTRY, GEOGRAPHICAL REGIONS, W70-03975
CHEMISTRY, GEOGRAPHICAL REGIONS, W70-03975
CHEMISTRY, DIATOMS, CHLOROPHYTA, W70-03974
CHEMISTRY, BIOMASS, BENTHOS, SALI W70-03978
CHEMOCLINE, MIXOLIMNION, RADIOCAR W70-05760
CHEMOSTATS, CHLOROPHYLL A, PHOTOB W70-02777
CHEMOSTATS, BATCH CULTURES, ACETY W70-02775
CHEMOSTATS, CARBON-14.: / MICROCY W69-06864
CHEMOSYNTHESIS, CONCRETIONS, ARTH W69-07428
CHEMOSYNTHETIC, SAPROPHYTIC, PLAT W68-00891
CHEMOSYNTHESIS, ABUNDANCE.: /TECO W70-02780
CHEMOTROPHIC, AUTOTROPHIC, SILICO W70-02804
CHEMOTROPHY, PARTICULATE MATTER, W70-02504
CHEMTROL, ALGAE.: /*WATERSHED MAN W68-00478
CHEMTROL, COPPER SULFATE, LAKES, W68-00470
CHENEY DAM, KANS.: W72-01773
CHEROKEE RESERVOIR(TENN), HOLSTON W71-07698
CHICAGO NATURAL HISTORY MUSEUM, N W70-00719
CHICAGO(ILL), ARTIFICIAL CIRCULAT W69-10178
CHICKAMAUGA RESERVOIR(TENN), FLOR W71-07698
CHIRONOMIDS.: /SKAHA LAKE(CANADA) W71-12077
CHIRONOMIDAE, HALF-LIFE.: / PLECO W69-09742
CHIRONOMIDS, CLADOCERA, APHANIZOM W70-00268
CHIRONOMIDAE, PROCLADIUS, CHIRONO W70-01943
CHIRONOMIDS, LAKE CONSTANCE, WHIT W70-05761
CHIRONOMIDAE, AQUATIC OLIGOCHAETE W70-04253
CHIRONOMUS, PROCLADIUS, CORETHRA W70-06217
CHIRONOMUS PROMOSUS, TRICHOPTERA, W70-01943
CHL: /NTS, PHYTOPLANKTON, WASTE W W70-04371
CHLAMYDOMONAS, CHLORELLA, ENZYMES W70-04369
CHLAMYDOMONAS, BATCH CULTURES, OP W70-04381
CHLAMYDOMONAS, BIOASSAYS, WATER P W70-04284
CHLAMYDOMONAS, EUGLENA.: /TA, BIO W70-03974
CHLAMYDOMONAS, PHOSPHORUS RADIOIS W70-03957
CHLAMYDOMONAS, CHLORELLA, MORTALI W70-03519
CHLAMYDOMONAS, CHLORELLA.: /N, CI W69-02959
CHLAMYDOMONAS.: /ISCONSIN, NUTRIE W69-01977
CHLAMYDOMONAS, CHLOROPHYTA, CYANO W69-03364
CHLAMYDOMONAS, CARBON, ECOSYSTEMS W69-06273
CHLAMYDOMONAS MOEWUSII, SKELETONE W70-05381
CHLAMYDOMONAS, FRESHWATER, MARINE W70-05381
CHLAMYDOMONAS, CHRYSOPHYTA.: /LOR W70-05409
CHLAMYDOMONAS, TEMPERATURE, CARBO W70-02249
CHLAMYDOMONAS MOEWUSSI, FAT ACCUM W70-02965
CHLAMYDOMONAS, ECOLOGY, PHOSPHORU W70-02965
CHLAMYDOMONAS, DAPHNIA, DIATOMS, W70-10161
CHLAMYDOMONAS, CHLORELLA, SCENEDE W70-10173
CHLAMYDOMONAS, SAMPLING, SCENEDES W70-08107
CHLAMYDOMONAS MOEWUSII, CHLORELLA W69-10180
CHLAMYDOMONAS, CHLORELLA, WISCONS W69-10180
CHLAMYDOMONAS, ALGAE, ZOOPLANKTON W70-00268
CHLAMYDOMONAS, SCENEDESMUS, WISCO W72-00845
CHLAMYDOMONAS, LABORATORY TESTS, W72-02417
CHLORAMINE.: W70-02370
CHLORAMINE, DE-K-PRUF-21, BENOCLO W70-05263
CHLORATE, SANTOBRITE, CLADOPHORA, W70-05263
CHLORDANE.: W70-09768
CHLORELLA SOROKINIANA, COMPOSITIO W70-07957
CHLORELLA PYRENOIDOSA, SCHRODERIA W70-08097
CHLORELLA OVALIS BUTCHER, NJOER L W70-05752
CHLORELLA VULGARIS, DAPHNIA MAGNA W71-03056
CHLORELLA PYRENOIDOSA, CARBOXYLIC W70-02248
CHLORELLA PYRENOIDOSA, FIXATION, W70-02504
CHLORELLA PYRENOIDOSA, PESTICIDE W70-02968
CHLORELLA PYRENOIDOSA, AUTOTROPHI W70-02964
CHLORELLA PYRENOIDOSA, BIOLOGICAL W70-02786
CHLORELLA VULGARIS, DAPHNIA MAGNA W71-05719
CHLORELLA PYRENOIDOSA, TRIBONEMA W69-10180
CHLORELLA PYRENOIDOSA, DIATOMA EL W69-10158
CHLORELLA PYRENOIDOSA, ANKISTRODE W70-05261
CHLORELLA PYRENOIDOSA, BICHOROMAT W70-05092
CHLORELLA PYRENOIDOSA, CHLORELLA W70-05381
CHLORELLA VULGARIS, MONODUS SUBTE W70-05381
CHLORELLA SACCHAROPHILA, CHLORELL W70-04809
CHLORELLA LUTEOVIRIDIS, SKELETONE W70-04809
CHLORELLA VULGARIS(COLUMBIA), CHL W70-04809
CHLORELLA VULGARIS, CHLORELLA VUL W70-04809
CHLORELLA ELLIPSOIDEA, CHLORELLA W70-04809
CHLORELLA PYRENOIDOSA, CHLORELLA W70-04809
CHLORELLA VULGARIS, CORNELL UNIVE W69-06540
CHLORELLA PYRENOIDOSA(EMERSON STR W69-07442
CHLORELLA PYRENOIDOSA, FUCALES, A W69-08284
CHLORELLA PYRENOIDOSA, MICROCYSTI W70-03512

```
GANIC CARBON, LAKE TAHOE(CALIF),        CHLORELLA PYRENOIDOSA, SELENASTRU    W70-03983
EDESMUS OBLIQUUS, HELIX POMATIA,        CHLORELLA PYRENOIDOSA, NITROGEN U    W70-04184
GREEN ALGAE, CHLORELLA VULGARIS,        CHLORELLA PYRENOIDOSA, VOLATILE A    W70-04195
ENOIDOSA, VOLATIL/ *GREEN ALGAE,        CHLORELLA VULGARIS, CHLORELLA PYR    W70-04195
MASS.:            KINETIC MODELS,       CHLORELLA PYRENOIDOSA, *ALGAE BIO    W73-02218
MEAN TOLERANCE LIMIT, BLUEGILLS,        CHLORELLA PYRENOIDOSA, BENZENE CO    W72-11105
ONDITIONS, ANAEROBIC CONDITIONS,        CHLORELLA.: /*NITRITES, AEROBIC C    W70-05261
OMS, CHLOROPHYTA, CHLAMYDOMONAS,        CHLORELLA.: /N, CISCO, CARP, DIAT    W69-02959
Y, ALGAL POISONING, SCENEDESMUS,        CHLORELLA, *EUGLENA, ALGAL CONTRO    W69-00994
IOLOGY, ACTIVATED SLUDGE, RATES,        CHLORELLA, HYDROGEN ION CONCENTRA    W68-00248
GICAL TREATMENT, BIBLIOGRAPHIES,        CHLORELLA, *SEWAGE, EFFLUENTS, CA    W68-00855
OROPHYLL, CYCLES, UNITED STATES,        CHLORELLA, BIOCHEMICAL OXYGEN DEM    W70-05092
, ELECTROCHEMISTRY, RESISTIVITY,        CHLORELLA, SCENEDESMUS.: /ACTIONS    W70-05267
ES, CHLOROPHYLL, PHOTOSYNTHESIS,        CHLORELLA, GRAZING, ZOOPLANKTON,     W70-04809
NKTON, *ALGAE, *LIGHT INTENSITY,        CHLORELLA, CHLOROPHYTA, DIATOMS,     W70-05381
OTYPES, CHLOROPHYTA, PHOTOMETRY,        CHLORELLA, NUTRIENTS, TRACE ELEME    W70-05409
EGRADATION, VITAMINS, CELLULOSE,        CHLORELLA, PEPTIDES, BACTERIA, SN    W70-04184
S, SANDS, FLOW RATES, TURBIDITY,        CHLORELLA, SCENEDESMUS, EUGLENA.:    W70-04199
OPHYTA, BIOASSAYS, SALINE WATER,        CHLORELLA, CHLAMYDOMONAS, EUGLENA    W70-03974
PLANKTON, CARBON, RADIOACTIVITY,        CHLORELLA, SAMPLING, EFFLUENTS, T    W70-03983
OMS, WEIGHT, PILOT PLANTS, IRON,        CHLORELLA, SODIUM, POTASSIUM, SIL    W70-03512
MICAL ANALYSIS, WATER POLLUTION,        CHLORELLA, SCENEDESMUS.: /IA, CHE    W70-03312
5-T, SCENEDESMUS, CHLAMYDOMONAS,        CHLORELLA, MORTALITY, BIOASSAY, I    W70-03519
, *WASTE TREATMENT, FOOD CHAINS,        CHLORELLA, CHLOROPHYLL, PHYSIOLOG    W70-04381
 LIGHT INTENSITY, CHLAMYDOMONAS,        CHLORELLA, ENZYMES, CARBON DIOXID    W70-04369
AND DEVELOPMENT, BIBLIOGRAPHIES,        CHLORELLA, CHLOROPHYTA, DIATOMS,     W69-06865
FFECTS, PHOTOSYNTHESIS, DIATOMS,        CHLORELLA, CARBON RADIOISOTOPES,     W69-10158
WTH, AMINO ACIDS, CHLAMYDOMONAS,        CHLORELLA, WISCONSIN, LAKE MICHIG    W69-10180
, *ALGAE, TOXICITY, SCENEDESMUS,        CHLORELLA, DAPHNIA, DETERIORATION    W71-05719
TORY EQUIPMENT, ASBESTOS CEMENT,        CHLORELLA, CARBON DIOXIDE.: /BORA    W71-07339
ULRUSH, PLANT GROWTH SUBSTANCES,        CHLORELLA, CHLAMYDOMONAS, SCENEDE    W72-00845
, OLIGOTROPHY, LIMITING FACTORS,        CHLORELLA, ANALYTICAL TECHNIQUES,    W70-02779
OISOTOPES, *AQUATIC ENVIRONMENT,        CHLORELLA, TRACERS, PHOTOSYNTHESI    W70-02786
*NUTRIENT REQUIREMENTS, ECOLOGY,        CHLORELLA, ALGAE, ORGANIC MATTER,    W70-02804
MICAL), CHLOROPHYTA, CYANOPHYTA,        CHLORELLA, MANGANESE, PHOTOSYNTHE    W70-02964
ISM, *ENVIRONMENTS, VARIABILITY,        CHLORELLA, CARBON, TEMPERATURE, L    W70-02965
UTRIENTS, PHYSIOLOGICAL ECOLOGY,        CHLORELLA, SCENEDESMUS.: /LOGY, N    W70-02245
, SAMPLING TEMPERATURE, TRACERS,        CHLORELLA, HYDROGEN ION CONCENTRA    W70-02504
ICIDES, CHLORINE, ALGAL CONTROL,        CHLORELLA, DISINFECTION, MORTALIT    W70-02363
OLISM, RADIOACTIVITY TECHNIQUES,        CHLORELLA, OXIDATION LAGOONS.: /B    W70-02198
 TOXINS, INHIBITORS, SUCCESSION,        CHLORELLA, CHELATION, OCHROMONAS,    W70-02510
OBIOLOGY, BIOASSAY, CHLOROPHYTA,        CHLORELLA, CYCLING NUTRIENTS, ESS    W70-02248
ATION, FILTERS, ELECTROPHORESIS,        CHLORELLA, SCENEDESMUS, ADSORPTIO    W71-03035
, *ALGAE, TOXICITY, SCENEDESMUS,        CHLORELLA, DAPHNIA, DETERIORATION    W71-03056
, *OXYGENATION, *PHOTOSYNTHESIS,        CHLORELLA, PESTICIDES, PESTICIDE     W70-08416
OWTH, *ALGAE, LIPIDS, NUTRIENTS,        CHLORELLA, CULTURES.:           *GR    W70-07957
PLING, SCENEDESMUS, CHLOROPHYTA,        CHLORELLA, CHLORINE, LIME, COAGUL    W70-08107
 METABOLISM, CHEMICAL REACTIONS,        CHLORELLA, TRACERS, RADIOACTIVITY    W70-08097
EASONAL, EUGLENA, CHLAMYDOMONAS,        CHLORELLA, SCENEDESMUS, SAMPLING,    W70-10173
WASTE WATER TREATMENT, CULTURES,        CHLORELLA, WATER POLLUTION EFFECT    W70-09438
CHLAMYDOMONAS, DAPHNIA, DIATOMS,        CHLORELLA, SCENEDESMUS, OXYGEN, C    W70-10161
HERM, BLUE-GREEN ALGAE,/ *COPPER        CHLORIDE, UPTAKE, FREUNDLICH ISOT    W70-08111
A BLANKE/ ALUMINUM SULFATE, IRON        CHLORIDE, IRON SULFATE, CLADOPHOR    W69-10170
S, PHOSPHORUS, NITROGEN, CALCIUM        CHLORIDE, TEMPERATURE, METABOLISM    W72-05453
MENTS, *WATER POLLUTION EFFECTS,        CHLORIDES, HALOGENS, ALGAE, WATER    W69-08562
YGEN, ALGAE, NUTRIENTS, SOLUTES,        CHLORIDES, CALCIUM, CARBONATES, P    W69-09557
ON, ALKALINITY, HARDNESS(WATER),        CHLORIDES, ORGANIC MATTER, CARBON    W71-04206
MAGNESIUM, ALKALINITY, SULFATES,        CHLORIDES, SODIUM, IRON, DISSOLVE    W69-05894
 ANALYSIS, CARBONATES, NITRATES,        CHLORIDES, AMMONIUM COMPOUNDS, OX    W69-07096
SUPPLY, BIOINDICATORS, CALCIUM,         CHLORIDES, CONDUCTIVITY, CULTURES    W69-07833
GAE, BIOMASS, CALCIUM COMPOUNDS,        CHLORIDES, SALINE LAKES, HEAVY ME    W69-02959
LGAE, OXYGEN SAG, WATER QUALITY,        CHLORIDES, PHOSPHORUS, SULFATES,     W69-03948
TICIDE REMOVAL, DDT, *FOOD WEBS,        CHLORINATED HYDROCARBON PESTICIDE    W70-03520
ON EFFECTS, AQUATIC ENVIRONMENT,        CHLORINATED HYDROCARBON PESTICIDE    W70-08652
 ION CONCENTRATION, TEMPERATURE,        CHLORINATED HYDROCARBON PESTICIDE    W72-11105
HOTOSYN/ *MARINE PLANTS, *ALGAE,        CHLORINATION, SODIUM COMPOUNDS, P    W70-00161
, *TASTE-PRODUCING ALGAE, ALGAE,        CHLORINATION, WATER QUALITY CONTR    W70-02373
Y, *WATER TREATMENT, FILTRATION,        CHLORINATION, COAGULATION, WATERS    W71-05943
TION, FLOTATION, BIODEGRADATION,        CHLORINATION, COLIFORM, BACTERIA,    W71-05013
STREAMS, *DISTRIBUTION SYSTEMS,         CHLORINATION, COPPER SULFATE, ACT    W72-03562
DOMESTIC WASTES, SEDIMENTATION,         CHLORINATION, SLUDGE, DETERGENTS,    W70-04060
ERIA, METABOLISM, CANALS, SANDS,        CHLORINATION, MICROORGANISMS, CHE    W69-07838
ODOR, ACTIVATED CARBON, AMMONIA,        CHLORINE.: /AE, *ACTINOMYCETES, *    W72-04155
SEDIMENTATION, ACTIVATED CARBON,        CHLORINE, COSTS, TREATMENT FACILI    W72-08859
, FUNGI, BACTERIA, BACTERICIDES,        CHLORINE, COSTS, *WASTE WATER TRE    W72-12987
OTICS, BIOASSAY, NORON, CALCIUM,        CHLORINE, COBALT, COPPER, ECOLOGY    W69-04801
GENS, *GROWTH RATES, *CHLORELLA,        CHLORINE, ALGICIDES, ALGAL CONTRO    W70-02370
A, DISIN/ *HALOGENS, *ALGICIDES,        CHLORINE, ALGAL CONTROL, CHLORELL    W70-02363
 ALGICIDES, CYTOLOGICAL STUDIES,        CHLORINE, COPPER SULFATE, ALGAL C    W70-02364
PHYLL, IRON, CHELATION, ENZYMES,        CHLORINE, LIGHT INTENSITY, NITRAT    W70-02964
ONMENTAL SANITATION, INHIBITION,        CHLORINE, NITROGEN COMPOUNDS, SEW    W69-10171
IUM COMPOUNDS, ACTIVATED CARBON,        CHLORINE, PHENOLS, OIL WASTES, HY    W70-00263
CONTROL, *ALGAE, COPPER SULFATE,        CHLORINE, WELLS, CALIFORNIA.: /Y     W71-00110
EDESMUS, CHLOROPHYTA, CHLORELLA,        CHLORINE, LIME, COAGULATION.: /EN    W70-08107
LLE GREEN LAKE(NY), *MARL LAKES,        CHLOROBIUM, FILINIA, MONIMOLIMNIO    W70-05760
LAKE(HANOI), RED RIVER(VIETNAM),        CHLOROCOCCUS, SPINES, SCHIZOMYCOP    W70-03969
HIZOMYCOPHYTA, CAULOBACTERIALES,        CHLOROCOCCALES, CONJUGATOPHYCAEA,    W70-03969
ATION, UROGLENA, ANKISTRODESMUS,        CHLOROCOCCALES, VOLVOCALES, CONJU    W70-05409
HIRUNDINELLA, PERIDINIUM WILLEI,        CHLOROCOCCALES, ANKISTRODESMUS, O    W70-05405
D CARBON, CHLOROPLASTS, INFRARE/        CHLOROFORM-METHANE EXTRACTS, LIPI    W70-05651
```

409

```
AQUATIC MICROBIOLOGY, BIOASSAY,        CHLOROPHYTA, CYANOPHYTA, DIATOMS,    W69-03371
ATER POLLUTION EFFECTS, DIATOMS,       CHLOROPHYTA, CYANOPHYTA, NUISANCE    W69-01977
TIC ALGAE, AQUATIC MICROBIOLOGY,       CHLOROPHYTA, CYANOPHYTA, ESSENTIA    W69-03358
GY, BALANCE OF NATURE, BIOASSAY,       CHLOROPHYTA, CYANOPHYTA, DIATOMS,    W69-03188
DS, BALANCE OF NATURE, BIOASSAY,       CHLOROPHYTA, CYANOPHYTA, ENVIRONM    W69-03185
OPLANKTON, CISCO, CARP, DIATOMS,       CHLOROPHYTA, CHLAMYDOMONAS, CHLOR    W69-02959
   AQUATIC MICROBIOLOGY, BIOASSAY,     CHLOROPHYTA, CYANOPHYTA, ENVIRONM    W69-03362
ICIDES, AQUATIC ALGAE, BIOASSAY,       CHLOROPHYTA, CYANOPHYTA, DIATOMS,    W69-03370
DS, BALANCE OF NATURE, BIOASSAY,       CHLOROPHYTA, DIATOMS, CYCLING NUT    W69-03515
GY, BALANCE OF NATURE, BIOASSAY,       CHLOROPHYTA, DIATOMS, ENVIRONMENT    W69-03516
AL EFFECTS, ESSENTIAL NUTRIENTS,       CHLOROPHYTA, DIATOMS, INHIBITION,    W69-03517
TIC PRODUCTIVITY, AQUATIC WEEDS,       CHLOROPHYTA, CYANOPHYTA, DIATOMS,    W69-03373
SSAY, BIOCHEMICAL OXYGEN DEMAND,       CHLOROPHYTA, CYCLING NUTRIENTS, E    W69-03374
ITY, WATER TEMPERATURE, DIATOMS,       CHLOROPHYTA.: /IEWS, LIGHT INTENS    W68-00860
MENT, BIBLIOGRAPHIES, CHLORELLA,       CHLOROPHYTA, DIATOMS, LIGHT, LIPI    W69-06865
CING ALGAE, DIATOMS, CYANOPHYTA,       CHLOROPHYTA, DIATOMS, LIGHT, LIPI    W69-06865
L POLLUTION, DIFFUSION, DIATOMS,       CHLOROPHYTA, IRON, MANGANESE, EUT    W69-05704
-METHANE EXTRACTS, LIPID CARBON,       CHLOROPHYTA, CHRYSOPHYTA, DINOFLA    W69-05763
HIA CLOSTERIUM, NITZSCHIA PALEA,       CHLOROPLASTS, INFRARED ANALYSIS.:    W70-05651
SIS, *GROWTH RATES, *TURBULENCE,       CHLOROPLASTS, SCENEDESMUS OBLIQUU    W70-05381
ARY), ALGAL TAXONOMY, SPIRULINA,       CHLORRELA, SCENEDESMUS, KINETICS,    W69-03730
RUCIGENIA RECTANGULARIS, DICERAS       CHODATE: /A(HUNGARY), ZIMONA(HUNG    W70-03969
XTRACELLULAR, FUCUS VESICULOSUS,       CHODATI, DINOPHYCEAE, FILTER BLOC    W69-07833
, COPPER, AMMONIA, CONDUCTIVITY,       CHONDRUS CRISPUS, ASCOPHYLLUM NOD    W70-01073
TIS, CRYOMONAS EROSA, COSMARIUM,       CHR: /NDS, HYDROGEN SULFIDE, LEAD    W70-04510
TON(ND), LAKOTA(ND), HARVEY(ND),       CHROCOCCUS LIMNETICUS, LYNGBYA LI    W70-05405
DRIN, DIELDRIN, SCENEDESMUS, GAS       CHROMATIUM VINOSUM, THIOCAPSA FLO    W70-03312
URES, DIAGENESIS, SEDIMENTS, GAS       CHROMATOGRAPHY, RESISTANCE, BIOAS    W70-03520
S, CYANOPHYTA, SPECTROSCOPY, GAS       CHROMATOGRAPHY, LIPIDS, OIL, OIL     W69-08284
N, *WATER POLLUTION EFFECTS, GAS       CHROMATOGRAPHY, ANALYTICAL TECHNI    W69-00387
OLAR REGIONS, SPECTROPHOTOMETRY,       CHROMATOGRAPHY.: /T IDENTIFICATIO    W69-02782
S, ASSAY, DISTRIBUTION, DENSITY,       CHROMATOGRAPHY, HYDROGEN ION CONC    W70-01074
AMYDOMONAS LABORATORY TESTS, GAS       CHROMATOGRAPHY, RADIOACTIVITY, TR    W70-02510
S, TRO/ *BIOTA, *COLUMBIA RIVER,       CHROMATOGRAPHY, DEGRADATION, WATE    W70-08652
TECHNIQUES, *PHOSPHATES, *ALGAE,       CHROMIUM, PHOSPHORUS RADIOISOTOPE    W69-07853
ITY, PHYSIOLOGICAL OBSERVATIONS,       CHROMOTOGRAPHY, RADIOCHEMICAL ANA    W71-03033
AL EFFECTS, DECOMPOSITION RATES,       CHRONIC EXPOSURE, ACUTE EXPOSURE.    W70-01779
ATER-BLOOM, 'BREAKING' OF LAKES,       CHRONOLOGY, CLADOCERANS, CHYDORID    W70-05663
ORINA, SPIROGYRA, SYNECHOCOCCUS,       CHROOCOCCACEAE, NOSTOCHINEAE, CLA    W69-08283
TUS, URONEMA ELONGATUM HODGETTS,       CHROOCOCCUS, OSCILLATORIA, NOSTOC    W70-04503
UM, ANKISTRODESMUS, APHANOCAPSA,       CHROOCOCCUS PRESCOTTII, CALOTHRIX    W70-04468
E(MINN), CRYPTOPHYTA, PYROPHYTA,       CHROOCOCCUS, CLOSTERIUM, GLEO: /I    W70-04371
DA, MASTIGOPHORA, CRYPTOPHYCEAE,       CHROOMONAS ACUTA, SCHROEDERIA SET    W70-05387
ACILLARIOPHYCEAE, CHLOROPHYCEAE,       CHRYSOPHYCEAE, VOLVOCALES, SARCOD    W70-05112
, SEDIMENTS/ *ALGAE, CYANOPHYTA,       CHRYSOPHYCEAE, CARTERIA, COMPARAT    W69-07833
DIFFUSION, DIATOMS, CHLOROPHYTA,       CHRYSOPHYTA, CULTURES, DIAGENESIS    W69-08284
SH, DINOFLAGELLATES, CYANOPHYTA,       CHRYSOPHYTA, DINOFLAGELLATES, CYA    W69-05763
GMENTS, CHLOROPHYTA, CYANOPHYTA,       CHRYSOPHYTA, CALIFORNIA, UNITED S    W70-05372
, TRACE ELEMENTS, CHLAMYDOMONAS,       CHRYSOPHYTA, RHODOPHYTA, EUGLENOP    W70-05389
RATES, CHLOROPHYTA, CYANOPHYTA,       CHRYSOPHYTA,: /LORELLA, NUTRIENTS    W70-05409
PHYTA, EUGLENOPHYTA, CYANOPHYTA,       CHRYSOPHYTA, PHOTOSYNTHESIS.: /TH    W70-03983
AMS, MINE DRAINAGE, CHLOROPHYTA,       CHRYSOPHYTA, BIOASSAYS, SALINE WA    W70-03974
TIVITY, PERIPHYTON, CHLOROPHYTA,       CHRYSOPHYTA, EUGLENOPHYTA, CYANOP    W70-02770
, INSECTS, DIATOMS, CHLOROPHYTA,       CHRYSOPHYTA, CYANOPHYTA, SAMPLING    W70-02780
TES, SHALLOW WATER, CHLOROPHYTA,       CHRYSOPHYTA, GRAZING, SNAILS, AGE    W71-00668
MONAS, ALGAE, ZOOPLANKTON, FISH,       CHRYSOPHYTA, CYANOPHYTA, SAMPLING    W70-00711
AERUGINOSA(KUTZ), CULTURE MEDIA,       CHRYSOPHYTA, SESTON, SELF-PURIFIC    W70-00268
LOS-AQUAE, MINIMUM LETHAL DOSES,       CHU NO 10 SOLUTION, MINIMUM NUTRI    W70-03507
LTS, DETRITUS, BOTTOM SEDIMENTS,       CHU'S MEDIA, QUINONE DERIVATIVES,    W70-02982
RATES, CHRONOLOGY, CLADOCERANS,       CHUTES, EDDIES, HABITATS, CRAYFIS    W69-09742
AGMA, PSEUDAGRION, BRACHYTHEMIS,       CHYDORID CLADOCERANS, SPECIES COM    W70-05663
OPHORA, HOLOPHYTIC MASTIGOPHORA,       CHYDORUS, PLEUROXUS, TANYPUS, TAN    W70-06975
ULATUS, MADISON(WIS), PSOROPHORA       CILIATA.: /PHORA, HOLOZOIC MASTIG    W72-01817
HACUS, GLENODINIUM, CRYPTOMONAS,       CILIATA, ARTIFICIAL KEY.: /DRIMAC    W70-06225
, FLAGELLATES, SAPROBIEN SYSTEM,       CILIATA, ASTERIONELLA FORMOSA, RE    W70-00268
COLO), MICRO-ALGAE, FLAGELLATES,       CILIATE AUTOECOLOGY.:  *CILIATES     W71-03031
OPHYCEAE, VOLVOCALES, SARCODINA,       CILIATES, COPEPODIDS, NAUPLII, TE    W69-10154
IATOMEA, CLADOPHORA, FLAGELLATA,       CILIATES, CYCLOTELLA, APHANIZOMEN    W70-05112
TULSA(OKLA), POLLUTION, FOSSIL,       CILIATES, RANUNCULUS CIRCINATUS,    W69-07838
A, SHAYLER RUN, ORGANIC CARBON,       CIMARRON RIVER(OKLA), BIXBY(OKLA)    W70-04001
IS), LAKE WAUBESA(WIS), ANABAENA       CINCINNATI(OHIO).: /, CHLOROPHYLL    W71-04206
MICROCYSTIS AERUGINOSA, ANABAENA       CIR: /MENDOTA(WIS), LAKE MONONA(W    W69-08283
FLAGELLATA, CILIATES, RANUNCULUS       CIRCINALIS, GLOEOTRICHIA ECHINULA    W69-05704
BESNA RIVER, MOOSE CREEK, ARCTIC       CIRCINATUS, POTAMOGETON PUSILLUS.    W69-07838
PLANKTON, *LAKE MICHIGAN, *WATER       CIRCLE HOT SPRINGS, RADON, CENTRA    W69-08272
ATER, FRESH WATER, SHORES, WATER       CIRCULATION, NUISANCE ALGAE, THER    W69-05763
OXINS, BACTERIA, COPPER SULFATE,       CIRCULATION, FLOODS, DENSITY DURR    W69-06203
SANCE ALGAE, ODOR, TASTE, DEPTH,       CIRCULATION.: /ION, TURBULENCE, T    W70-04369
HERMOCLINE, *ION EXCHANGE, WATER       CIRCULATION, DIURNAL, WINDS, CURR    W70-05404
E(WIS), CHICAGO(ILL), ARTIFICIAL       CIRCULATION, NUTRIENT ABSORPTION,    W68-00857
ATER STORAGE, EVAPORATION, WATER       CIRCULATION, NUISANCES.: /ILWAUKE    W69-10178
N CONTROL, *LAKE ONTARIO, *WATER       CIRCULATION, SEDIMENTS.: /EVEL, W    W69-09723
, PHOTOSYNTHESIS, PHYTOPLANKTON,       CIRCULATION, PATH OF POLLUTANTS,    W71-05878
DRAINAGE DISTRICTS, DEBT, COSTS,       CISCO, CARP, DIATOMS, CHLOROPHYTA    W69-02959
ANTS, MAPPING, HYDROGEN SULFIDE,       CITIES.: /ATMENT, LAKES, SEWAGE,     W70-04455
UCTIVITY, PHOSPHATES, SILICATES,       CITIES, LAKE HURON, LAKE ERIE, LA    W70-00667
ASTES, PEACHES, APRICOTS, PEARS,       CITRATES, BIOASSAY, NEW YORK, LAK    W70-01579
RCLE HOT SPRINGS, RADON, CENTRAL       CITRUS FRUITS, TOMATOES, *AEROBIC    W71-09548
           *SUSPENDED SOLIDS, CHASE    CITY(COLO), MUTATION, BETA RAYS,     W69-08272
ONTROL, PACIFIC NORTHWEST U. S.,       CITY(VA).:                           W72-10573
                                       CIVIL ENGINEERING, WASTE TREATMEN    W69-03683
```

CHER, SKELETONEMA COSTATUM(GREV)
LATION MODEL, PHOSPHORUS EFFECT,
L EFFICIENCIES, TROPHIC ECOLOGY,
, HAMBURG(GERMANY), CHIRONOMIDS,
CHII, STEPHANODISCUS BINDERANUS,
TUDIES, *EXPERIMENTAL LIMNOLOGY,
.:   ASTERIONELLA SPIROIDES,
HRONOLOGY, CLADOCERANS, CHYDORID
DECOMPOSITION RATES, CHRONOLOGY,
A), BENARES(INDIA), MACROPHYTES,
YE, SODIUM CHLORATE, SANTOBRITE,
LES, ALGAL SUCCESSION, ANABAENA,
ACROCHEILUS ALUTACEUS, ULOTHRIX,
BACILLUS MESENTERICUS, DIATOMEA,
SYRACUSE, WARBURG RESPIROMETER,
ROCOLEUS LACUSTRIS (RAB) FARLOW,
ALGAL TAXONOMY, ALGAL PHYLOGENY,
(NY), ENTEROMORPHA INTESTINALIS,
TE, IRON CHLORIDE, IRON SULFATE,

EDRA, COELOSPHAERIUM, SPIROGYRA,
REEK(NY).: *CLADOPHORA FRACTA,
RAISIN RIVER, DETROIT R/ LAKE ST
TEMPERATURE, SEA WATER, OYSTERS,
UM, CARBON RADIOISOTOPES, BIOTA,
ITY, *ALGAE, *REVIEWS, HABITATS,
TIVATED SLUDGE, FISHERIES, CARP,
DIOISOTOPES, CALCIUM, POTASSIUM,
SIS, CARBON RADIOISOTOPES, FISH,
TRICHOPTERA, LEECHES, FINGERNAIL
-GREEN ALGAE, GREEN / *LEWIS AND
ARBON FILTERS, LEBANON, BATAVIA,
 HETEROTROPHY, FLAGELLATES, LAKE
L ING, ECOLOGY, AMMONIA, ENZYMES,
HNIQUES, *ANALYTICAL TECHNIQUES,
S, CHROOCOCCACEAE, NOSTOCHINEAE,
YDEN, THE HAGUE, MINERALIZATION,
S, ALGAE, ANALYTICAL TECHNIQUES,
ODEGRADATION, ADSORPTION, ALGAE,
TION PATTERNS, ACTIVATED CARBON,
, HYDROGEN SULFIDE, IRON, FUNGI,
NALYTICAL TECHNIQUES, CLAY LOAM,
DIMENTS, *EUTROPHICATION, ALGAE,
AE, VOLVOCALES, INDUSTRIAL USES,
 *SURFACTANTS, *WATER POLLUTION,
ROL, PIPING SYSTEMS(MECHANICAL),
NE, COSTS, TREATMENT FACILITIES,
SESSMENT, AMERICAN RIVER(CALIF),
NTARIO), BUCKHORN LAKE(ONTARIO),
Y MANAGEMENT, *CHAOBORID MIDGES,
INN), MORPHOMETRY, ALGAE BLOOMS,
N, SUSCEPTIBILITY, REGENERATION,
TILIZATION, PHYSICAL PROPERTIES,
NITRIFICATION, COLIFORMS, PONDS,
GAE, *AQUATIC ALGAE, ALGAE, ARID
, ALGAE, ARID CLIMATES, SEMIARID
HEMISTRY, SOIL TYPES, HYDROLOGY,
EASONAL, *SULFIDES, TEMPERATURE,
TOSYNTHESIS, DEPTH, THERMOCLINE,
RENOIDOSA, BIOLOGICAL HALF-LIFE,
ZONE, HELSINGOR(DENMARK), SALINO-
FRAGILARIA, ASTERIONELLA, FILTER
ZECHOSLOVAKIA), CYPRINID FISHES,
MENTAL C/ BENTHIC ALGAE, BLOOMS,
VIRONMENT, CERIUM-144, NITZSCHIA
TED RATE, CELL VOLUME, NITZSCHIA
ESMUS, APHANOCAPSA, CHROOCOCCUS,
CHODESMIUM ERYTHREUM, SAXIDOMUS,
, MIGRATION, DIURNAL, ALGICIDES,
ANALYSIS, ANALYTICAL TECHNIQUES,
L DRAINAGE, APPALACHIAN MOUNTAIN
ATION, *SIGNIFICANCE, ALGAL KEY,
RIC TESTS, ONE-SAMPLE RUNS TEST,
OIKOPLEURA, FRITILLARIA, OBELIA,
 CYLINDRICA, NORTH SEA, FIRTH OF
LENE/ *ANABAENA CYLINDRICA, ATP,
LOGICAL DILUTION', PHYSIOLOGICAL
AZOTOBACTER, MOLYBDENUM, COBALT,
RHODOPHYTA, EUGLENOPHYTA, SLIME,
ULFONATES, WIND VELOCITY, TASTE,
HYTA, CHLORELLA, CHLORINE, LIME,
ON, TEMPERATURE, CENTRIFUGATION,
TMENT, FILTRATION, CHLORINATION,
TIVITY, PESTICIDES, TASTE, ODOR,
ENGINEERING, *FILTERS, *DESIGN,
 REMOVAL.:
MS, CORAL, MOLLUSKS, MINE ACIDS,
TRATION, SORPTION, FLOCCULATION,
IDE REMOVAL, *ADSORPTION, SOILS,
ON ABATEMENT, ABATEMENT, PACIFIC

CL, CHLORELLA OVALIS BUTCHER, NJO	W70-05752
CLADECERA.: /COSYSTEM MODEL, SIMU	W69-06536
CLADOCERA.: /ADRICAUDA, ECOLOGICA	W69-10152
CLADOCERA, APHANIZOMENON FLOS-AQU	W70-00268
CLADOCERA, WATER MASSES, U S PUBL	W70-00263
CLADOCERAN, MONTANE ZONE, ALPINE	W69-10154
CLADOCERANS, WESTHAMPTON LAKE(VA)	W71-10098
CLADOCERANS, SPECIES COMPOSITION,	W70-05663
CLADOCERANS, CHYDORID CLADOCERANS	W70-05663
CLADOPHORA, OEDOGONIA, SPIROGYRA,	W70-05404
CLADOPHORA, POTAMOGETON SPP, LAKE	W70-05263
CLADOPHORA, PEDIASTRUM, HYDRODICT	W70-04494
CLADOPHORA, PROSOPIUM WILLIAMSONI	W69-07853
CLADOPHORA, FLAGELLATA, CILIATES,	W69-07838
CLADOPHORA, ENTEROMORPHA, LEPOCIN	W70-02959
CLADOPHORA, TAXONOMIC DESCRIPTION	W70-04468
CLADOPHORA, ALGAL GENETICS, RUSSI	W70-03973
CLADOPHORA, CYCLOTELLA, HALOCHLOR	W70-03974
CLADOPHORA BLANKETS, HYDRODICTYON	W69-10170
CLADOPHORA GLOMERATA.:	W70-00667
CLADOPHORA, DIAPTOMUS, CYCLOPS, G	W70-02803
CLADOPHORA GLOMERATA, BUTTERNUT C	W72-05453
CLAIR, ROUGE RIVER, HURON RIVER,	W69-01445
CLAMS, CRABS, NORTH CAROLINA, SED	W69-08267
CLAMS, FALLOUT, PACIFIC OCE: /ITI	W69-08269
CLAMS, MUSSELS, SHELLFISH, DINOFL	W70-05372
CLAMS, PONDS, RUNNING WATERS, FRE	W70-09905
CLAMS, SIZE, ALGAE, DIATOMS, TRAC	W70-00708
CLAMS, STRONTIUM RADIOISOTOPES, T	W70-02786
CLAMS, WHITE FISH, ALEWIFE, GIZZA	W70-01943
CLARK LAKE (S D), ABUNDANCE, BLUE	W70-05375
CLARKSBURG, BUFFALO, DUNREITH, OI	W72-03641
CLASSIFICATION, SHAGAWA LAKE(MINN	W70-02775
CLASSIFICATION, GROWTH RATES, SEA	W70-00713
CLASSIFICATION, *MODEL STUDIES, W	W69-01273
CLATHROCYSTIS AERUGINOSA, COELOSP	W69-08283
CLAY LENSES, RHINE RIVER, AMSTERD	W69-07838
CLAY LOAM, CLAYS, COLLOIDS, EUTRO	W69-04800
CLAYS, ALDRIN, DIELDRIN, DDT, AER	W70-09768
CLAYS, ALGAE, WATER PURIFICATION,	W69-09884
CLAYS, BRACKISH WATER, FISHERIES,	W70-05761
CLAYS, COLLOIDS, EUTROPHICATION,	W69-04800
CLAYS, SILICATES.: /, *LAKES, *SE	W69-03075
CLEAN WATER, GENUS, SPECIES, TOUC	W70-05389
CLEANING, DISASTERS, OILY WATER,	W70-01470
CLEANING, COST ANALYSIS, WATER PU	W72-08688
CLEANING, *ALGAL CONTROL, WISCONS	W72-08859
CLEAR LAKE(CALIF), EEL RIVER(CALI	W70-02777
CLEAR LAKE(ONTARIO), RICE LAKE(ON	W70-02795
CLEAR LAKE(CALIF), NUTRIENT ENERG	W70-05428
CLEARWATER LAKE(MINN), TROUT LAKE	W70-05387
CLIFF VEGETATION, SAND DUNES.: /O	W70-01777
CLIMATE, NITROGEN, PHOSPHORUS.: /	W70-05565
CLIMATES, PHOTOSYNTHESIS, PRODUCT	W70-05092
CLIMATES, SEMIARID CLIMATES, *ABS	W69-03491
CLIMATES, *ABSORPTION, *CALCIUM,	W69-03491
CLIMATES, TROPHIC LEVEL, NITROGEN	W71-03048
CLIMATIC DATA, STRATIFICATION, DI	W70-04790
CLIMATIC ZONES, ORGANIC LOADING,	W72-10573
CLINCH RIVER(TENNESSEE), PHYSICAL	W70-02786
CLINE.: /LA MENEGHINIANA, PHOTIC	W70-04809
CLOGGING, STEPHANODISCUS HANTZSCH	W70-00263
CLOROCOCCALES, DAPHNIA PULICARIS,	W70-00274
CLOSED SYSTEMS, CYTOLOGY, ENVIRON	W69-07832
CLOSERIUM, ARTEMIA, CARTERIA, NOR	W70-00708
CLOSTERIUM, NITZSCHIA PALEA, CHLO	W70-05381
CLOSTERIUM, GLEO: /IUM, ANKISTROD	W70-04371
CLOSTRIDIUM BOTULINUS, VENERUPIS,	W70-05372
CLOSTRIDIUM, FISH, AMPHIBIANS, AN	W70-05372
CLOUD COVER, SPECTROSCOPY, ULTRAV	W69-09676
CLUB HUTS, TEGERNSEE, GERMANY, EX	W69-07818
CLUMP COUNT, DESMIDS, FLAGELLATES	W70-05389
CLUSTER ANALYSIS, REYKJAVIK(ICELA	W69-09755
CLUSTER ANALYSIS, MULTIVARIATE AN	W73-01446
CLYDE(SCOTLAND), SCOTLAND, ENGLAN	W70-02504
CMU, NITROGEN-15, PYRUVATE, ACETY	W70-04249
CO: / THORACOMONAS, YTTRIUM, 'BIO	W70-00708
CO: /, NITROGEN FIXATION, BORON,	W70-02964
CO: /A, CYANOPHYTA, CHRYSOPHYTA,	W70-05389
COAGULATION, PLANKTON, MONITORING	W70-00263
COAGULATION.: /ENEDESMUS, CHLOROP	W70-08107
COAGULATION, DEWATERING, *COST AN	W71-11375
COAGULATION, WATERSHEDS, SEWERAGE	W71-05943
COAGULATION, ELECTROLYTES, POTABL	W72-13601
COAGULATION, ALGAE, ACTIVATED CAR	W72-07334
COAL CONTACT PROCESSES, PESTICIDE	W70-05470
COAL MINES, ACIDITY, PYRITE, SULF	W69-07428
COAL, ACTIVATED CARBON.: /AE, FIL	W70-05470
COALS, DISTRIBUTION PATTERNS, ACT	W69-09884
COAST REGION, REGIONS, GEOGRAPHIC	W69-03683

OUTHWEST U. S., REGIONS, PACIFIC COAST REGIONS, GEOGRAPHICAL REGIO W69-03611
TIPLE-USES, SAVANNAH RIVER, GULF COAST.: / COLLOIDAL FRACTION, MUL W73-00748
EGIONS, COASTAL PLAINS, ATLANTIC COASTAL PLAIN, INORGANIC COMPOUND W69-03695
, REGIONS, GEOGRAPHICAL REGIONS, COASTAL PLAINS, ATLANTIC COASTAL W69-03695
POLYELECTROLYTE- COATED FILTERS, BAUXITE.: W70-01027
SAND, CR/ *FILTER MEDIA, ALUMINA- COATED SAND, FERRIC OXIDE-COATED W70-08628
LUMINA-COATED SAND, FERRIC OXIDE- COATED SAND, CRUSHED BAUXITE, ACI W70-08628
HAINS, PHOSPHORUS RADIOISOTOPES, COBALT RADIOISOTOPES, URANIUM RAD W70-00708
OISOTOPES, IODINE RADIOISOTOPES, COBALT RADIOISOTOPES, WATER POLLU W71-09850
ENTHIC FLORA, STREAMS, CULTURES, COBALT RADIOISOTOPES.: /GANESE, B W69-07845
ENTS, SOILS, NUCLEAR EXPLOSIONS, COBALT RADIOISOTOPES, CESIUM, STR W69-08269
ION, ZINC RADIOISOTOPES, CESIUM, COBALT RADIOISOTOPES, SEASONAL, R W70-04371
ASSIUM, IRON, BICARBONATE, ZINC, COBALT, CALCIUM, MAGNESIUM, C: /T W70-03978
ASSAY, NORON, CALCIUM, CHLORINE, COBALT, COPPER, ECOLOGY, ENVIRONM W69-04801
BORON, AZOTOBACTER, MOLYBDENUM, COBALT, CO: /, NITROGEN FIXATION, W70-02964
*ALGAE, *HERRINGS, COPPER, IRON, COBALT, MANGANESE, ATLANTIC OCEAN W69-08525
MEADOW LAKE, GYTTJA, DUFF, OOZE, COBBLES, EQUISETUM, THORIUM-232, W69-08272
ID FEVER, KENNEBEC RIVER(MAINE), COBBOSSEECONTEE LAKE(MAINE), BARG W70-08096
PHIC, MYCELOID, CORYNEFORM RODS, COCCI,: /E, PSYCHROPHILES, PLEMOR W70-00713
-AQUAE, GLOEOTRICHIA ECHINULATA, COCCOCHLORIS PENIOCYSTIS, MADISON W70-00719
PHYSIOLOGY, CALOTHRIX PARIETINA, COCCOCHLORIS PENIOCYSTIS, DIPLOCY W69-06277
E TECHNIQUES, MICROBIAL ECOLOGY, COCCOLITHOPHORES, SEAWATER MEDIUM W70-05272
PERIDINI/ SKELETONEMA COSTATUM, COCCOLITHUS HUXLEYI, PYRAMIMONAS, W70-05272
ROS, SKELETONEMA, THALASSIOSIRA, COCCOLITHUS, SARGASSO SEA, SYNURA W70-04503
LLARIA, BIDDULPHIA, CHAETOCEROS, COCCONEIS, COSCINODISCUS, DIMEROG W70-03325
PHYSICAL HALF-LIFE, HYDROPSYCHE COCKERELLI, BODY BURDENS, POMOXIS W70-02786
ATIVE LAGOONS, LABORATORY MODEL, COD REMOVAL, MILK SUBSTRATE, NITR W70-07031
L, PLUMBING CODES, WELL DRILLING CODES.: /NIDE, NEW PRODUCT CONTRO W73-00758
E, NEW PRODUCT CONTROL, PLUMBING CODES, WELL DRILLING CODES.: /NID W73-00758
IC TURBIDOSTAT, LIGHT EXTINCTION COEFFICIENT, ALGAL CELL DENSITY.: W70-03923
ASTRUM, RHODOTORULA, TEMPERATURE COEFFICIENTS, NORTH AMERICA, CENT W70-03957
ALUES, EUDORINA, CULTURE MEDIUM, COELASTRUM, PEDIASTRUM.: /CTION V W70-05409
LLATORIA, MELOSIRA, ACTINASTRUM, COELASTRUM, PEDIASTRUM, WESTELLA, W70-08107
, TABELLARIA, MELOSIRA, SYNEDRA, COELOSPHAERIUM, SPIROGYRA, CLADOP W70-02803
OGYRA, MYXOPHYCEAE, MICROCYSTIS, COELOSPHAERIUM, OSCILLATORIA, ANA W70-04369
INEAE, CLATHROCYSTIS AERUGINOSA, COELOSPHAERIUM KUTZINGIANUM, ANAB W69-08283
ECIES DETERMINATION, ABSORBANCE, COENOBIA, SWIRLING, VOLATILE SOLI W70-03334
, POLYARTHRA VULGARIS, KARATELLA COHCLEARIS, ASPLANCHA, LECANE, BO W70-02249
SERRATIA MARINORUBRA, GYRODINIUM COHNII, MICROBIAL ECOLOGY, COULTE W70-05652
BIOLOGICAL BALANCE, OVERFISHING, COHO SALMON, ALEWIFE, NUTRIENT RE W70-04465
IFE, GIZZARD SHAD, SEA LAMPHREY, COHO SALMON.: /, WHITE FISH, ALEW W70-01943
EMEROPTERA, PLECOPTERA, DIPTERA, COLEOPTERA, HEMIPTERA, MOLLUSCA, W69-07846
DOSA, BICHOROMATE, DRY WEIGHT, B COLI COMMUNEA, EUGLENE GRACILIS, W70-05092
NDITIONS, AEROBIC CONDITIONS, E. COLI, BACTERIA, FUNGI, ACTINOMYCE W71-02009
ATION, LIFE CYCLES, GENETICS, E. COLI, ISOLATION.: /MICAL PRECIPIT W71-11823
AEOLOSOMA HEMPRICHI, ESCHERICHIA COLI, MYRIOPHYLLUM, LEPOMIS MACRO W69-07861
QUUS, DACTYLOCOCCUS, ESCHERICHIA COLI, OOCYSTIS, LEMNA, SPIRODELA, W70-10161
NUTRIENTS, VIRUSES, BACTERIA, E. COLI, PUBLIC HEALTH, PATHOGENIC B W72-06017
N, BIODEGRADATION, CHLORINATION, COLIFORM, BACTERIA, BIOCHEMICAL O W71-05013
DS, PHOSPHATES, NITROGEN, ALGAE, COLIFORMS.: /IRON, DISSOLVED SOIL W69-05894
ETHANE, ALGAL CONTROL, BACTERIA, COLIFORMS.: /ION, FERMENTATION, M W70-04787
POLLUTION, POLLUTION ABATEMENT, COLIFORMS, WATER QUALITY, PHOSPHO W70-04810
AL OXYGEN DEMAND(BOD), AERATION, COLIFORMS, ALGAL CONTROL, DISSOLV W70-04788
NAEROBIC, OXYGEN, NITRIFICATION, COLIFORMS, PONDS, CLIMATES, PHOTO W70-05092
SEWAGE TREATMENT, NORTH DAKOTA, COLIFORMS, OXIDATION LAGOONS, BIO W70-03312
LTURE, ST LAWRENCE RIVER, ALGAE, COLIFORMS, DISSOLVED: /ES, AGRICU W70-03964
TER POLLUTION EFFECTS, NITROGEN, COLIFORMS, WATER QUALITY.: /S, WA W71-05407
ENTS, *DISSOLVED OXYGEN, *ALGAE, COLIFORMS, EUTROPHICATION(PHYTOPL W71-05407
BACTERICIDE, KINETICS, HARDNESS, COLIFORMS, CYTOLOGICAL STUDIES, E W71-05267
TREATMENT, COPPER SULFATE, ODOR, COLIFORMS, ECONOMICS.: /OLLUTION W72-01814
CHLORELLA, SCENEDESMUS, OXYGEN, COLIFORMS, SEDIMENTS, MUD, SANDS, W70-10161
ITY, DATA STORAGE AND RETRIEVAL, COLIFORMS, DISSOLVED OXYGEN, ALGA W70-09557
TION, *ECOLOGY, ALGAE, BACTERIA, COLIFORMS, FISH, FUNGI, HEAT, WAT W71-02479
FACILITIES, FILTRATION, SEWAGE, COLIFORMS, EUTROPHICATION, ODOR, W70-00263
HEMICAL OXYGEN DEMAND, SU/ *DATA COLLECTIONS, *WATER QUALITY, BIOC W70-09557
POLLUTION SOURCES, ON-SITE DATA COLLECTIONS, *ON-SITE INVESTIGATI W70-09976
E ERIE,/ *SYSTEMS ANALYSIS, DATA COLLECTIONS, *ALGAL CONTROL, *LAK W71-04758
ODS, MARINE PLANTS, ON-SITE DATA COLLECTIONS.: /MORTALITY, GASTROP W70-01780
PHYSIOLOGY, EUTROPHICATION, DATA COLLECTION, CHEMICAL ANALYSIS, RE W70-04382
ITY, *ARTIFICIAL RECHARGE, *DATA COLLECTIONS, PIT RECHARGE, WATER W69-05894
HORUS RADIOISOTOPE, ON-SITE DATA COLLECTIONS, FISHES, PLANKTON, AL W68-00857
*BOTTOM SEDIMENTS, ON-SITE DATA COLLECTIONS, AQUATIC PLANTS, WATE W68-00511
YSIS, PH, SAMPLING, ON-SITE DATA COLLECTIONS, LIMNOLOGY.: /AL ANAL W69-00659
ATHEMATICAL MODELS, ON-SITE DATA COLLECTIONS, BOTTOM SEDIMENTS, BI W68-00858
SMUS, MATHEMATICAL STUDIES, DATA COLLECTIONS, MARINE ANIMALS, PRIM W73-01446
SPRING CREEK(PA), STATE COLLEGE(PA), FLORA.: W69-10159
BENTHIC FORMS, EPIPHYTIC FORMS, COLLOIDAL FRACTION, MULTIPLE-USES W73-00748
INE WATER-FRESHWATER INTERFACES, COLLOIDS.: /PONDS, BREEDINGS, SAL W73-00748
CHEMICAL PROPERTIES, ADSORPTION, COLLOIDS, ZINC RADIOISOTOPES, CAL W70-00708
, LIMNOLOGY, ZOOPLANKTON, ALGAE, COLLOIDS, TRACE ELEMENTS, DISSOLV W70-02510
PENDED LOAD, *FILTRATION, ALGAE, COLLOIDS, FLOCCULATION, MIXING, E W71-10654
AL PROPERTIES, ANION ADSORPTION, COLLOIDS, AQUARIA, ABSORPTION, WA W71-09863
AL TECHNIQUES, CLAY LOAM, CLAYS, COLLOIDS, EUTROPHICATION, IMPOUND W69-04800
NE ZONE, BOULDER(COLO), TEA LAKE(COLO), BLACK LAKE(COLO), PASS LAK W69-10154
LO), BLACK LAKE(COLO), PASS LAKE(COLO), MICRO-ALGAE, FLAGELLATES, W69-10154
HOT SPRINGS, RADON, CENTRAL CITY(COLO), MUTATION, BETA RAYS, COSMI W69-08272
OLO), TEA LAKE(COLO), BLACK LAKE(COLO), PASS LAKE(COLO), MICRO-ALG W69-10154
NTANE ZONE, ALPINE ZONE, BOULDER(COLO), TEA LAKE(COLO), BLACK LAKE W69-10154
S), RHODOTORULA, GROWTH FACTORS, COLONIAL MORPHOLOGY, CELLULAR MOR W70-03952
TE/ *GROWTH DYNAMICS, *BACTERIAL COLONIZATION, *ARTIFICIAL SUBSTRA W70-04371

415

N CYCLE, *PHOSPHORUS, PHOSPHORUS   COMPOUNDS, WATER SUPPLY, WATER MA   W72-01867
IRE, NEW YORK, VIRGINIA, ORGANIC   COMPOUNDS, LAGOON, FILTERS, PESTI   W72-03641
RATURE, MICROORGANISMS, NITROGEN   COMPOUNDS, *PHOSPHORUS COMPOUNDS,   W71-13335
NITROGEN COMPOUNDS, *PHOSPHORUS   COMPOUNDS, ADSORPTION, REMOVAL, D   W71-13335
*OXYGENATION, *TOXICITY, ORGANIC   COMPOUNDS, *PHENOLS, PESTICIDES,   W71-12183
TION, FILTRATION, MODEL STUDIES,   COMPUTERS, NUTRIENTS, EUTROPHICAT   W70-08627
RBON RADIOISOTOPES, HYDROGEN ION   CONCENTRATION, SCENEDESMUS, EUGLE   W70-08097
ARY TREAT/ *ALGAE, *HYDROGEN ION   CONCENTRATION, *NUTRIENTS, *TERTI   W70-08980
TION, *TEMPERATURE, HYDROGEN ION   CONCENTRATION, NUTRIENTS, ALGAE,   W70-08834
, DISSOLVED OXYGEN, HYDROGEN ION   CONCENTRATION, EPILIMNION, PHOTOS   W70-06975
IAL BALANCES, O/ *UPTAKE, CELLS,   CONCENTRATION, COMPOSITION, MATER   W70-06600
YGEN, BICARBONATES, HYDROGEN ION   CONCENTRATION, NUISANCE ALGAE, AQ   W70-06225
PHYTON, PHOSPHORUS, HYDROGEN ION   CONCENTRATION, ALKALINITY, HARDNE   W71-04206
OXYGEN, ALKALINITY, HYDROGEN ION   CONCENTRATION, STANDING CROP, LIG   W71-00665
ALITY, TEMPERATURE, HYDROGEN ION   CONCENTRATION, OXYGEN, CARBON DIO   W71-03031
OWERS, DAMS, ALGAE, HYDROGEN ION   CONCENTRATION, TEMPERATURE, ALKAL   W71-11393
TORING.:   CONCENTRATION, ENVIRONMENTAL MONI   W72-00941
ICAL OXYGEN DEMAND, HYDROGEN ION   CONCENTRATION, TEMPERATURE, PROTE   W71-13413
ICAL OXYGEN DEMAND, HYDROGEN ION   CONCENTRATION, ORGANIC LOADING, *   W72-03299
, SEDIMENTS, ALGAE, HYDROGEN ION   CONCENTRATION, ELECTRIC POWER, SE   W71-09548
E, ORGANIC LOADING, HYDROGEN ION   CONCENTRATION, TEMPERATURE, CENTR   W71-11375
H TOXINS, TOXICITY, HYDROGEN ION   CONCENTRATION, DISSOLVED OXYGEN,   W71-07731
*FUNGI, *BACTERIA, HYDROGEN ION   CONCENTRATION, TEMPERATURE, ALKAL   W71-05267
HOSPHORUS, SULPHUR, HYDROGEN ION   CONCENTRATION, TIME.: /ECOLOGY, P   W70-02965
CHEMISTRY, OXYGEN, HYDROGEN ION   CONCENTRATION, PHOTOSYNTHESIS, AL   W70-03309
NATES, TEMPERATURE, HYDROGEN ION   CONCENTRATION, ALKALINITY,: /ARBO   W70-02804
ETHANE, ALKALINITY, HYDROGEN ION   CONCENTRATION, TOXINS, OXYGEN, DE   W70-02803
RE, CARBON DIOXIDE, HYDROGEN ION   CONCENTRATION, BIOCARBONATES, DIS   W70-02249
TRACERS, CHLORELLA, HYDROGEN ION   CONCENTRATION, ALKALINITY, DIATOM   W70-02504
IDES, BIOCHEMISTRY, HYDROGEN ION   CONCENTRATION, OXIDATION,: / PEPT   W70-01068
TOPES, CHLOROPHYLL, HYDROGEN ION   CONCENTRATION, MAGNESIUM, CALCIUM   W70-01579
IBIL/ *TORREY CANYON, *DETERGENT   CONCENTRATIONS, SURVIVAL, SUSCEPT   W70-01779
RY, CHROMATOGRAPHY, HYDROGEN ION   CONCENTRATION, FLUORESCENCE, CARB   W70-01074
HUMIDITY, BACTERIA, HYDROGEN ION   CONCENTRATION, SULFATES, RESPIRAT   W70-01073
MONOMICTIC, DEPTH EFFECTS, IONIC   CONCENTRATION, SEASONAL EFFECTS.:   W70-01933
, *NUTRIENT CONTROL, *PHOSPHORUS   CONCENTRATIONS, MICRONUTRIENTS, A   W69-09340
HENOLS, OIL WASTES, HYDROGEN ION   CONCENTRATION, FLUORIDES, ALKYL B   W70-00263
, DISSOLVED SOLIDS, HYDROGEN ION   CONCENTRATION, TEMPERATURE, NITRO   W70-00265
USEUM, NOSTOC MUSCORUN, NUTRIENT   CONCENTRATION, WISCONSIN, SULFUR,   W70-00719
ION, ACTINOMYCETES, HYDROGEN ION   CONCENTRATION, NITRATES, POTASSIU   W70-00719
RIENT BUDGET, POTENTIAL NUTRIENT   CONCENTRATION, AREAL INCOME(NUTRI   W70-00270
(WASH), *LAKE CHANGES, *NUTRIENT   CONCENTRATION, *NUTRIENT BUDGET,   W70-00270
ATIC FUNGI, YEASTS, HYDROGEN ION   CONCENTRATION, IRON, SULFATE, TEM   W69-00096
WATER TEMPERATURE, HYDROGEN ION   CONCENTRATION, FUNGI, YEASTS, PYR   W68-00891
N DIOXIDE, *ALGAE, *HYDROGEN ION   CONCENTRATION, PHOSPHORUS, NITROG   W68-00855
N, LIME, *NITROGEN, HYDROGEN ION   CONCENTRATION, *PHOSPHORUS, SEWAG   W68-00012
, RATES, CHLORELLA, HYDROGEN ION   CONCENTRATION, PLANT TISSUES, SCE   W68-00248
MMONIA, PHOSPHATES, HYDROGEN ION   CONCENTRATION, PHYTOPLANKTON, ODO   W70-05405
IVITY, CHLOROPHYTA, HYDROGEN ION   CONCENTRATION, NITRATES, PHOSPHAT   W70-05404
TIFICATION, OXYGEN, HYDROGEN ION   CONCENTRATION, TEMPERATURE, DEPTH   W70-05387
CE, IRON COMPOUNDS, HYDROGEN ION   CONCENTRATION, ACIDITY, MUD, HYDR   W70-05399
T, *BOD REMOVAL, *ISRAEL, *ALGAL   CONCENTRATION, 'EXTRA-DEEP PONDS'   W70-04790
INDUSTRIAL WASTES, HYDROGEN ION   CONCENTRATION, IRON COMPOUNDS, HY   W70-04510
XIMUM UPTAKE VELOCITY, SUBSTRATE   CONCENTRAT: / UPTAKE VELOCITY, MA   W70-05109
S, SCENEDESMUS, ABSORPTION, CELL   CONCENTRATION, UROGLENA, ANKISTRO   W70-05409
ATER, ACIDIC WATER, HYDROGEN ION   CONCENTRATION, WATER CHEMISTRY, *   W70-05424
ON, CARBON DIOXIDE, HYDROGEN ION   CONCENTRATION, PHOSPHATES, NITRAT   W70-05760
ERIA, ALGAE, SANDS, HYDROGEN ION   CONCENTRATION.: /CTROMETERS, BACT   W69-08267
SP/ RADIOAUTOGRAPHS, HALF-LIFE,   CONCENTRATION FACTOR, LIMNODRILUS   W69-07861
TRUNCATUS, BREVOORTIA TYRRANUS,   CONCENTRATION PROCESSES.: /RSIOPS   W69-08274
UCTIVITY, CULTURES, HYDROGEN ION   CONCENTRATION, IRON, LAKES, MAGNE   W69-07833
A, MICROCYSTIS AERU/ *BIOLOGICAL   CONCENTRATION, ANABAENA CYLINDRIC   W70-03520
CE WATERS, STREAMS, HYDROGEN ION   CONCENTRATION, IRON, MAGNESIUM, N   W70-03501
ES, IRON COMPOUNDS, HYDROGEN ION   CONCENTRATION, ALKALINITY, CALIFO   W70-03334
IONS, ANALYSIS OF VARIANCE, FOOD   CONCENTRATION, MEXICO, PREDICTIVE   W70-03957
TARIES, ALKALINITY, HYDROGEN ION   CONCENTRATION, NUTRIENTS, PHOSPHO   W70-03983
NKTON, ZOOPLANKTON, HYDROGEN ION   CONCENTRATION, ALGAE.: / PHYTOPLA   W70-04368
TROGEN, PHOSPHATES, HYDROGEN ION   CONCENTRATION, CYCLING NUTRIENTS,   W70-04381
TURE, PILOT PLANTS, HYDROGEN ION   CONCENTRATION, ORGANIC LOADING, C   W72-06022
SPHOROUS COMPOUNDS, HYDROGEN ION   CONCENTRATION, CONDUCTIVITY, TEMP   W72-06051
ITY, *ALGAE, *ODOR, HYDROGEN ION   CONCENTRATION.: /ENT, *WATER QUAL   W72-09694
RIA, ALGAE, MIXING, HYDROGEN ION   CONCENTRATION, TEMPERATURE, TOXIC   W72-06854
, DISSOLVED OXYGEN, HYDROGEN ION   CONCENTRATION, BIOCHEMICAL OXYGEN   W72-10573
ORGANIC PESTICIDES, HYDROGEN ION   CONCENTRATION, TEMPERATURE, CHLOR   W72-11105
TE-PRODUCING ALGAE, HYDROGEN ION   CONCENTRATION, *ORGANI: /SIS, TAS   W72-11906
LGAE TREATMENT COSTS, DINOBRYON,   CONCORD(N H), PEANACOOK LAKE(N H)   W69-08282
IRE, LAKE WINNISQUAM, LONG POND,   CONCORD, BOSTON LOT RESERVOIR, LE   W69-08674
, CHARCOAL, ALGAL GROWTH, STREAM   CONCOURSE, PALOUSE RIVER(IDAHO).:   W70-03501
E, CHALCOPYRITE, CHEMOSYNTHESIS,   CONCRETIONS, ARTHROBACTER, PYROPH   W69-07428
SYNTHESIS, ALKALINITY, NITRATES,   COND: /N ION CONCENTRATION, PHOTO   W70-03309
NCE, OPTIMUM GROWTH TEMPERATURE,   CONDENSER PAGGAGE.: /ATURE TOLERA   W71-02496
R, ANAEROBIC CONDITIONS, AEROBIC   CONDITIONS, E. COLI, BACTERIA, FU   W71-02009
SEWAGE BACTERIA, ODOR, ANAEROBIC   CONDITIONS, AEROBIC CONDITIONS, E   W71-02009
MS, ANAEROBIC BACTERIA, ANEROBIC   CONDITIONS, LAGOONS.: /AGE PROGRA   W71-04554
METHANE, MUD, OXYGEN, ANAEROBIC   CONDITIONS, HYDROGEN SULFIDE, IRO   W70-05761
ONS, *SLUDGE, ALGAE,/ *ANAEROBIC   CONDITIONS, *EFFLICIENCIES, *LAGO   W70-06899
LAKE WASHINGTON(WASH), NUISANCE   CONDITIONS.: /MADISON LAKES(WIS)   W70-07261
AINBOW TROUT, DIATOMS, ANAEROBIC   CONDITIONS, ALGAE, ROTIFERS, SEST   W69-10154
S, ANAEROBIC BACTERIA, ANAEROBIC   CONDITIONS, LAGOONS.: /GE PROGRAM   W71-04557
S, ANAEROBIC BACTERIA, ANAEROBIC   CONDITIONS, LAGOONS.: /GE PROGRAM   W71-04556

NORTHEAST U.S/ CYANAPHYTA, ALGAL   CONTROL, *ALGAL TOXINS MEMBRANE,   W69-05306
EUTROPH/ *BIBLIOGRAPHIES, *ALGAL   CONTROL, *AQUATIC WEED CONTROL, *   W69-05705
TING, AQUATIC PLA/ *AQUATIC WEED   CONTROL, *EUTROPHICATION, *HARVES   W69-06276
WAGE EFFLUENTS, *WATER POLLUTION   CONTROL, *LOCAL GOVERNMENTS, STAT   W69-06909
IO, *LAKE ERIE, *WATER POLLUTION   CONTROL, *POLLUTION ABATEMENT, PO   W69-06305
SHIRE, ALGAE, ODOR, TAST/ *ALGAE   CONTROL, *WATER SUPPLY, *NEW HAMP   W69-08282
FATE, *SODIUM ARS/ *AQUATIC WEED   CONTROL, *HERBICIDES, *COPPER SUL   W70-05568
REAM IMPROVEMENT, *WATER QUALITY   CONTROL, *MIDGES, FISH FOOD ORGAN   W70-05428
S, BULRUSH, DREDG/ *AQUATIC WEED   CONTROL, *BIBLIOGRAPHIES, CATTAIL   W70-05263
SSACHUSETTS, POT/ *WATER QUALITY   CONTROL, *ECOLOGY, *RESERVOIR, MA   W70-05264
 *LAKES, *COPPER SULFATE, *ALGAL   CONTROL, *TOXINS, ALGAE, RESISTAN   W68-00859
EUTROPHICATION, *WATER POLLUTION   CONTROL, *WATER POLLUTION TREATME   W69-03374
UTROPHICATION, */ *ALGAE, *ALGAL   CONTROL, *ESSENTIAL NUTRIENTS, *E   W69-04798
ANCE ALGAE, *ALGAL CONTROL, WEED   CONTROL, *INVESTIGATIONS, ZOO-PLA   W68-00468
LUTION EFFECTS, *WATER POLLUTION   CONTROL, *CONFERENCES, STREAMS, E   W70-03975
LUTION EFFECTS, *WATER POLLUTION   CONTROL, *CONFERENCES, STREAMS, E   W70-03975
T, *FILTRATION, *WATER POLLUTION   CONTROL, *FILTERS, SEWAGE EFFLUEN   W70-01027
ECHNIQUES, AQUAT/ *ALGAE, *ALGAL   CONTROL, *ALGICIDES, ANALYTICAL T   W70-02968
WAT/ *WATER PURIFICATION, *ALGAL   CONTROL, *WATER SUPPLY, *BASINS,   W72-01814
 *WATER TREATMENT, *ALGAE, *ALGAL   CONTROL, *TASTE, *ODOR, *NATURAL   W72-03562
LLUTION EFFECTS, WATER POLLUTION   CONTROL, *NUTRIENTS.: /, WATER PO   W72-00917
LEAR POWERPLANTS, *WATER QUALITY   CONTROL, *BIODEGRADATION, WATER U   W71-11881
ALYSIS, DATA COLLECTIONS, *ALGAE   CONTROL, *LAKE ERIE, BIOCHEMICAL   W71-04758
F-PURIFICATION, *WATER POLLUTION   CONTROL, *LAKE ONTARIO, *WATER CI   W71-05878
ATER TREATMENT, *WATER POLLUTION   CONTROL, *DENITRIFICATION, *BIODE   W71-04556
ATER TREATMENT, *WATER POLLUTION   CONTROL, *DENITRIFICATION, *BIODE   W71-04557
FARM WASTES, SIL/ WATER POLLUTION   CONTROL, *NUTRIENTS, *PESTICIDES,   W71-10376
TA, SULFUR BACTERIA, COL/ *ALGAL   CONTROL, POTABLE WATER, CYANOPHY   W70-08107
IZE, *CLAYS, */ *WATER POLLUTION   CONTROL, *FILTRATION, *PARTICLE S   W70-08628
CATION, *LAKES, *WATER POLLUTION   CONTROL, *REVIEWS, ALGAE, WEEDS,   W70-07393
 *ALGAE, *BIOLOGICAL PROPERTIES,   CONTROL, *OPERATION AND MAINTENAN   W70-08392
ATER RESOURCES, *WATER POLLUTION   CONTROL, *RESEARCH EQUIPMENT, *AN   W70-08627
ATER TREATMENT, *WATER POLLUTION   CONTROL, *DENITRIFICATION, *BIODE   W71-04554
EUTROPHICATION, *WATER POLLUTION   CONTROL, *WATER POLLUTION SOURCES   W71-03048
ATER TREATMENT, *WATER POLLUTION   CONTROL, *DENITRIFICATION, *BIODE   W71-04555
IDES, *VASCULAR T/ *AQUATIC WEED   CONTROL, *AQUATIC PLANTS, *HERBIC   W70-10175
 *POTABLE WATER, *WATER QUALITY   CONTROL, *ALGAE, COPPER SULFATE,   W71-00110
CONTROL, *EUTROPHICATION, *ALGAL   CONTROL, *TERTIARY TREATMENT, SUS   W70-09186
ATER TREATMENT, *WATER POLLUTION   CONTROL, *EUTROPHICATION, *ALGAL   W70-09186
UPPLY, RESERVOIRS, WATER/ *ALGAL   CONTROL, *COPPER SULFATE, WATER S   W69-10157
SULFATE/ *NUISANCE ALGAE, *ALGAL   CONTROL, *NEW HAMPSHIRE, *COPPER   W69-08674
UISANCE PLANT GROWTHS, *NUTRIENT   CONTROL, *PHOSPHORUS CONCENTRATIO   W69-09340
ENT, BIOLOGICAL TREATMENT, ALGAL   CONTROL, AMMONIA, ION EXCHANGE, R   W70-01981
, AQUATIC MICROBI/ *ALGAE, ALGAL   CONTROL, ALGICIDES, AQUATIC ALGAE   W70-02248
DRAINS, SURFACE DRAINAGE, ALGAL   CONTROL, AQUATIC LIFE, WATER CONS   W70-03344
OLAR RADIATION, *WATER POLLUTION   CONTROL, ALGICIDES, BIOCONTROL, *   W68-00478
OLLUTION TREATMENT, ALGAE, ALGAL   CONTROL, ANALYTICAL TECHNIQUES, A   W69-03374
SUCCESSION, WATER QUALITY, ALGAL   CONTROL, ALGICIDES, DOMINANT ORGA   W69-01977
L ECOLOGY, *PHYTOTOXICITY, ALGAL   CONTROL, ALGAL POISONING, ANALYTI   W69-03188
, *COPPER, COPPER SULFATE, ALGAL   CONTROL, ALGICIDES, TRACE ELEMENT   W69-03366
POLLUTION SOURCES, ALGAE, ALGAL   CONTROL, ALGAL POISONING, ALGICID   W69-03370
ARISONS, ECONOMICS, ALGAE, ALGAL   CONTROL, ALGICIDES, SEDIMENTS, DI   W69-08518
 *COPPER SULFATE, *WATER QUALITY   CONTROL, ALGAL CONTROL, SLIME, TA   W69-05704
HKILL, NUTRIENTS, *ALGAE, *ALGAL   CONTROL, ALGAL POISONING, ALGICID   W69-05844
OCAL GOVERNMENTS, *WATER QUALITY   CONTROL, ALGAE, AQUATIC WEEDS, AD   W72-05774
OPLANKTON, WATER ANALYSIS, ALGAL   CONTROL, ALGAE, WATER QUALITY, SE   W72-06051
SMUS, CHLORELLA, *EUGLENA, ALGAL   CONTROL, BACTERICIDES, GROWTH RAT   W69-00994
INSECT CONTROL, MOSQUITOS, ALGAL   CONTROL, BASS, CARP, CATFISHES, N   W70-05417
ON, FERMENTATION, METHANE, ALGAL   CONTROL, BACTERIA, COLIFORMS.: /I   W70-04787
ION, SURFACE WATERS, RESERVOIRS,   CONTROL, BIBLIOGRAPHIES, ANALYTIC   W70-05389
, WATER POLLUTION SOURCES, ALGAL   CONTROL, BIOINDICATORS.: /LLUTION   W70-02304
TORWEED, CHEMCONTROL, MECHANICAL   CONTROL, BIOCONTROL, COPPER SULFA   W70-10175
S, OXYGEN, EUTROPHICATION, ALGAL   CONTROL, BIOCONTROL, TURBULENCE,   W71-10098
ICIDES, *PHENOLS, *WATER QUALITY   CONTROL, CRUSTACEANS, ROTIFERS, S   W70-10161
ENS, *ALGICIDES, CHLORINE, ALGAL   CONTROL, CHLORELLA, DISINFECTION,   W70-02363
HANICAL CONTROL, WATER POLLUTION   CONTROL, CHEMCONTROL, ARSENIC COM   W70-04494
ERTIARY TREATMENT, *GREAT LAKES,   CONTROL, COSTS, EUTROPHICATION, N   W70-03964
BOD), AERATION, COLIFORMS, ALGAL   CONTROL, DISSOLVED OXYGEN.: /AND(W70-04788
ELLA, CHLORINE, ALGICIDES, ALGAL   CONTROL, DISINFECTION, MODE OF AC   W70-02370
HAINS, TOXICITY, WATER POLLUTION   CONTROL, DECOMPOSING ORGANIC MATT   W72-01788
CHLAMYDO/ WATER QUALITY, *ALGAL   CONTROL, EUTROPHICATION, DAPHNIA,   W72-02417
AEROBIC CONDITIONS, ALGAE, ALGAL   CONTROL, ECOLOGICAL DISTRIBUTION,   W70-04786
S, *PESTICIDES, *WATER POLLUTION   CONTROL, EUTROPHICATION, ALGAE, F   W70-05470
ISH POPULATION, *WATER POLLUTION   CONTROL, FARM WASTES, MUNICIPAL W   W70-04266
LGAE, SPRAYING, SEASONAL, *ALGAL   CONTROL, FISHKILL, ROOTED AQUATIC   W68-00488
USTRIAL WASTES, *WATER POLLUTION   CONTROL, FISH CONSERVATION, POLLU   W72-01786
ICAL TREATMENT, *WATER POLLUTION   CONTROL, FISH, TOXICITY, FRESH WA   W70-10437
NTROL, WATER CONSERVATION, WATER   CONTROL, FERTILIZERS, SEWAGE EFFL   W70-00269
DS, LAKE MICHIGAN, WATER QUALITY   CONTROL, FILTRATION, WATER PURIFI   W73-02426
POLLUTION EFFECTS, ALGAE, ALGAL   CONTROL, HARVESTING, AQUATIC ALGA   W69-03513
TE, FISH, PLANKTON, AQUATIC WEED   CONTROL, INSECT CONTROL, MOSQUITO   W70-05417
WATER QUALITY, DETERGENTS, ALGAL   CONTROL, INDUSTRIAL WASTES, THERM   W70-10437
Y, MIDGES, MOSQUITOES, SAMPLING,   CONTROL, INDUS: /ICATION, TOXICIT   W70-06225
LLUTION SOURCES, WATER POLLUTION   CONTROL, LIMNOLOGY, DIVERSION, ST   W69-10170
CATION, WISCONSIN, WATER QUALITY   CONTROL, LAKES, WATER POLLUTION S   W70-03310
PHICATION, BASS, NUISANCE ALGAE,   CONTROL, MINNOWS, NUISANCE ALGAE,   W70-02982
CHLORINE, COPPER SULFATE, ALGAL   CONTROL, MORTALITY, CHLOROPHYLL,   W70-02364
ON, AQUATIC WEED CONTROL, INSECT   CONTROL, MOSQUITOS, ALGAL CONTROL   W70-05417
, LAKES, ALGICIDES, AQUATIC WEED   CONTROL, MECHANICAL CONTROL, WATE   W70-04494
NUISANCE ALGAE, WATER POLLUTION   CONTROL, NUTRIENTS, WATER POLLUTI   W69-06276

TER CHEMISTRY, POLLUTANTS, ALGAL CONTROL, NITRATES, PHOSPHATES, SU W70-02646
LLUTION EFFECTS, WATER POLLUTION CONTROL, NITRATES,. NITROGEN FI: / W69-10167
, WATER QUALITY, WATER POLLUTION CONTROL, NUISANCE ALGAE, TASTE, O W69-10155
LLUTION SOURCES, WATER POLLUTION CONTROL, OXYGEN BALANCE, SEWAGE, W69-08668
TING OF ALGAE, EFFLUENT, QUALITY CONTROL, OXIDATION LAGOON, NITROG W71-13313
GE FISH, PLANKTON, ALGAE, *ALGAL CONTROL, PREDATION, COMPETITION, W70-05428
WAGE TREATMENT, *WATER POLLUTION CONTROL, PACIFIC NORTHWEST U. S., W69-03683
ENT, ENGINEERING, ALGAE, PLANTS, CONTROL, POLLUTION ABATEMENT, ABA W69-03683
TAMINATION, CYANIDE, NEW PRODUCT CONTROL, PLUMBING CODES, WELL DRI W73-00758
AE, *TASTE, *ODOR, WATER QUALITY CONTROL, POLLUTANT IDENTIFICATION W72-13605
TES, SLUDGE, SLURRIES, AUTOMATIC CONTROL, PIPING SYSTEMS(MECHANICA W72-08688
PONDS, DESIGN FOR ENVIRONMENTAL CONTROL, RECIRCULATION, BOD REMOV W70-04787
LLUTION SOURCES, WATER POLLUTION CONTROL, RADIOACTIVE WASTE, WATER W72-00941
LLUTION TREATMENT, WATER QUALITY CONTROL, SEWAGE EFFLUENTS, ABSORP W70-01031
OF ALGAE, BOTTOM SEDIMENTS, WEED CONTROL, SOLAR RADIATION, *WATER W68-00478
E, *WATER QUALITY CONTROL, ALGAL CONTROL, SLIME, TASTE-PRODUCING A W69-05704
PRODUCTIVITY, HYPOLIMNION, ALGAL CONTROL, SEASONAL, ANNUAL TURNOVE W69-05142
PLY, INTAKES, NUTRIENTS, QUALITY CONTROL, TREATMENT FACILITIES, FI W70-00263
, LIMNOLOGY, ALGAL TOXINS, ALGAL CONTROL, TRACE ELEMENTS, WASHINGT W70-00264
UALITY CONTROL, *CONTROL, *ALGAL CONTROL, TOXICITY.: /ITY, WATER Q W72-07442
ENT FACILITIES, CLEANING, ALGAL CONTROL, WISCONSIN.: /STS, TREATM W72-08859
EROSION, ALGAE, *WATER POLLUTION CONTROL, WATER QUALITY CONTROL.: / W73-01122
, *AQUATIC PLANTS, *AQUATIC WEED CONTROL, WATER CONSERVATION, WATE W70-00269
ITY, WA/ *COPPER SULFATE, *ALGAL CONTROL, WATER SUPPLY, WATER QUAL W69-10155
, WATER POLLUTION EFFECTS, ALGAL CONTROL, WATER POLLUTION, TOXICIT W70-02775
ECTS OF POLLUTION, WATER QUALITY CONTROL, WINNISQUAM LAKE, NEWFOUN W70-02772
EWAGE TREATMENT, WATER POLLUTION CONTROL, WATER POLLUTION EFFECTS, W70-02248
IDENTIFICATION, WATER POLLUTION CONTROL, WATER POLLUTION EFFECTS, W70-02792
CLASSIFICATION, WATER POLLUTION CONTROL, WATERCOURSES(LEGAL), ORG W72-00462
S, DISSOLVED OXYGEN, EVAPORATION CONTROL, WATER BUDGET, WATER CHEM W72-01773
RATED CONTROL MEASURES, PHYSICAL CONTROL, WATER QUALITY.: /, INTEG W70-07393
CIDES, HARVESTING, WATER QUALITY CONTROL, WATER POLLUTION CONTROL, W69-05706
QUALITY CONTROL, WATER POLLUTION CONTROL, WATER POLLUTION EFFECTS, W69-05706
QUALITY CONTROL, WATER POLLUTION CONTROL, WATER POLLUTION EFFECTS. W69-05705
CIDES, HARVESTING, WATER QUALITY CONTROL, WATER POLLUTION CONTROL, W69-05705
, WATER QUALITY, WATER POLLUTION CONTROL, WATER PO: /STE TREATMENT W69-06535
SULPHATE, NUISANCE ALGAE, *ALGAL CONTROL, WEED CONTROL, *INVESTIGA W68-00468
ATER POLLUTION, *WATER POLLUTION CONTROL, WATER POLLUTION EFFECTS, W69-03948
OIRS, HYDROELECTRIC PLANTS, WEED CONTROL, WATER PROPERTIES, PHOSPH W68-00479
NTIFICATION, POLLUTANTS, QUALITY CONTROL, WATER POLLUTION, *WATER W69-01273
*RIVERS, *ALGAE, WATER POLLUTION CONTROL, WATER POLLUTION EFFECTS, W70-05668
AQUATIC WEED CONTROL, MECHANICAL CONTROL, WATER POLLUTION CONTROL, W70-04494
ERVOIRS, *MIXING, *WATER QUALITY CONTROL, WATER QUALITY, WATER POL W70-04484
PONDS, WATER HYACINTH, CHEMICAL CONTROLS, PARAQUAT, DIQUAT, SODIU W70-00269
NTHELLAE, ACANTHURUS TRIOSTEGUS, CONUS EBRAEUS.: / FUNGITES, ZOOXA W69-08526
LOROPHYTA, SCENEDESMUS, BENTHOS, CONVECTION, WAVES(WATER), FISH, S W70-03969
OM, ALGAE GROWTH, WATER DENSITY, CONVECTION CURRENTS, ALGAE TREATM W69-08282
WATER, WATER REQUIREMENTS, WATER CONVEYANCE, WATER COSTS, ARTIFICI W71-09577
LL),/ *POLLUTION, *CHICAGO(ILL), COOK COUNTY(ILL), CALUMET RIVER(I W70-00263
TRIC POWER PLANTS, *WATER REUSE, COOLING TOWERS, DAMS, ALGAE, HYDR W71-11393
SHORTAGE FILTERS, SNAILS, ODOR, COOLING TOWERS, FILTERS, EFFLUENT W72-00027
STUDIES, BIOASSAY, AQUATIC LIFE, COOLING WATER, SITES.: /E, MODEL W71-11881
M DESLUDGING./ *METAL CORROSION, COOLING WATER TREATMENT, ON-STREA W70-02294
DROGEN SULFIDE, IRON, MAGNESIUM, COOLING, PROPERTY VALUES, RECREAT W70-02251
BONATE, CERCARIAE, WATER MILFOIL COONTAIL, WILD CELERY, BUR REED, W70-05568
RO-ALGAE, FLAGELLATES, CILIATES, COPEPODIDS, NAUPLII, TERRAMYCIN, W69-10154
, *WINTER, CYANOPHYTA, PLANKTON, COPEPODS, EUTROPHICATION, OLIGOTR W69-10154
RIPTON, BACTERIA, SEDIMENTATION, COPEPODS, WATER POLLUTION SOURCES W69-10169
PLANKTON, ROTIFERS, CHLOROPHYTA, COPEPODS, CYCLES, CHLAMYDOMONAS, W70-00268
CONTROL, BIOCONTROL, TURBULENCE, COPEPODS, ROTIFERS, CHLOROPHYLL, W71-10098
R MILFO/ SCHISTOSOME DERMATITIS, COPPER CARBONATE, CERCARIAE, WATE W70-05568
ES, CATTAILS, BULRUSH, DREDGING, COPPER SULPHATE, PONDWEEDS, SODIU W70-05263
E, LIGHT TEMPERATURE, NUTRIENTS, COPPER SULFATE, STRATIFICATION.: / W69-08282
N, TURBULENCE, TOXINS, BACTERIA, COPPER SULFATE, CIRCULATION.: /IO W70-04369
GICIDES,/ *LAKES, *MUD, *COPPER, COPPER SULFATE, ALGAL CONTROL, AL W69-03366
TRIPHICATION, *LAKES, WISCONSIN, COPPER SULPHATE, NUISANCE ALGAE, W68-00468
NG ALGAE, CHLOROPHYLL, CHEMTROL, COPPER SULFATE, LAKES, NITROGEN.: W68-00470
SINS, WATER POLLUTION TREATMENT, COPPER SULFATE, ODOR, COLIFORMS, W72-01814
TRIBUTION SYSTEMS, CHLORINATION, COPPER SULFATE, ACTIVATED CARBON, W72-03562
, VITAMINS, ALGAE, GROWTH RATES, COPPER SULFATE, BOTTOM SEDIMENTS, W69-09349
RAQUAT, DIQUAT, SODIUM ARSENITE, COPPER SULPHATE, MONURON, AMMONIA W70-00269
, CYTOLOGICAL STUDIES, CHLORINE, COPPER SULFATE, ALGAL CONTROL, MO W70-02364
ISCONSIN, ALGICIDES, HYDROLYSIS, COPPER SULFATE, FISHES, HERBICIDE W70-02982
SHALLOW WATER, HYDROGEN SULFIDE, COPPER SULFATE, CHLOROPHYTA, CRUS W70-02803
IA, NITROGEN, PHOSPHORUS, ALGAE, COPPER SULFATE, EFFLUENTS, MICROC W70-02787
TON, *DIATOMS, *BLUE-GREEN ALGA, COPPER SULFATE.: /NG, *PHYTOPLANK W70-02754
*ALGAE, *MAINE, *WATER SUPPLY, COPPER SULFATE, RESERVOIRS, WATER W70-08096
MECHANICAL CONTROL, BIOCONTROL, COPPER SULFATE, SODIUM ARSENITE,: W70-10175
*WATER QUALITY CONTROL, *ALGAE, COPPER SULFATE, CHLORINE, WELLS, W71-00110
QUALITY, *ORGANIC MATTER, ALGAE, COPPER SULFATE, *CHEMICAL OXYGEN W72-04495
NG, *FILTRATION, DESIGN, *ALGAE, COPPER SULFATE, SEDIMENTATION, AC W72-08859
ATION, FI/ *ARTIFICIAL RECHARGE, COPPER, *ALGICIDES, ALGAE, *EVALU W72-05143
MPOUNDS, HYDROGEN SULFIDE, LEAD, COPPER, AMMONIA, CONDUCTIVITY, CH W70-04510
ORON, CALCIUM, CHLORINE, COBALT, COPPER, ECOLOGY, ENVIRONMENTAL EF W69-04801
ATLANTIC OCE/ *ALGAE, *HERRINGS, COPPER, IRON, COBALT, MANGANESE, W69-08525
*ALGAE, *FRESH WATER, *PEPTIDES, COPPER, IRON, IONS, PHOSPHATES, O W69-10180
TOZOA, LICHENS, PLANTS, DIATOMS, CORAL, MOLLUSKS, MINE ACIDS, COAL W69-07428
NTOSPHENUS LAMOTTENII, GAMMARUS, CORDULEGASTER.: / SALMO TRUTTA, E W69-09334
ND), LINSLEY POND(CONN), BOSMINA COREGONI LONGISPINA.: /(SWITZERLA W69-10169
, OSCILLATORIA RUBESCENS, PISTON CORER, FRAGILARIA, MELOSIRA, ASTE W70-05663

S, AVAILABLE NUTRIENTS, SEDIMENT
RRENTS(WATER), INDUSTRIAL WASTE,
, ALGAE, OLIGOTROPHY, DIVERSION,
TROL, ALGICIDES, TRACE ELEMENTS,
EE(WIS), CHIRONOMUS, PROCLADIUS,
LLULAR SIZE, CHLORELLA VULGARIS,
MERICAN RESEARCH AND DEVELOPMENT
INCIPAL-COMPONENT ANALYSIS, PART
REUSE, DISSOLVED OXYGEN, PITTING(
T, ON-STREAM DESLUDGING./ *METAL
ENTIAL, DISSOLVED OXYGEN, ODORS,
RIFICATION, *BACTERIA, *ENZYMES,
, BACTERIOSTRUM, LEPTOCYLINDRUS,
, STREPTOCOCCUS, BREVIBACTERIUM,
HROPHILES, PLEMORPHIC, MYCELOID,
VICULA, MELOSIRA, THALASSIOSIRA,
DULPHIA, CHAETOCEROS, COCCONEIS,
SMUS, OOCYSTIS, CRYOMONAS EROSA,
CITY(COLO), MUTATION, BETA RAYS,
ORUS, FISH/ *NUTRIENTS, *BENEFIT-
G SYSTEMS(MECHANICAL), CLEANING,
SIBILITY, TECHNICAL FEASIBILITY,
OXYGEN DEMAND, ACTIVATED SLUDGE,
OGEN, *PHOSPHORUS, EFFICIENCIES,
TION SOURCES, CYCLING NUTRIENTS,
IME, OXYGEN, TERTIARY TREATMENT,
CTERIA, SEDIMENTATION, TOXICITY,
IMMING, WATERFOWL, BIRDS, COSTS,
, DRAIN, SAN JOAQUIN VA/ *CONTRA
SCILLATORIA VAUCHER, SKELETONEMA
RAMIMONAS, PERIDINI/ SKELETONEMA
AMYDOMONAS MOEWUSII, SKELETONEMA
ORELLA LUTEOVIRIDIS, SKELETONEMA
LIS, CACHONINA NIEI, SKELETONEMA
MAINTENANCE COSTS, CONSTRUCTION
ACTERIA, BACTERICIDES, CHLORINE,
REMENTS, WATER CONVEYANCE, WATER
EWAGE, DRAINAGE DISTRICTS, DEBT,
N, ACTIVATED SLUDGE, MAINTENANCE
UES, SWIMMING, WATERFOWL, BIRDS,
ECTION CURRENTS, ALGAE TREATMENT
LAGOONS, PONDS, PHOTOSYNTHESIS,
REATMENT, *GREAT LAKES, CONTROL,
ULATION, RIGHT-OF-WAY, ESTIMATED
ION, ACTIVATED CARBON, CHLORINE,
EL RIVER, DOLORES RIVER, URAVAN,
CHLORIDE, INFRARED GAS ANALYSIS,
INIUM COHNII, MICROBIAL ECOLOGY,
SCONSIN, ASIA, NATIONAL RESEARCH
SCONSIN, ASIA, NATIONAL RESEARCH
PPING FILM, SYNECHOCOCCUS, GRAIN
*SIGNIFICANCE, ALGAL KEY, CLUMP
EMOSYNTHETIC, SAPROPHYTIC, PLATE
OWARD A. HANSON RES/ WOOD, ALGAL
*GROWTH, *UPTAKE, FIXATION, CELL
*ALGAE HOMOGENIZATION, *BACTERIA
ACTERIA, CYCLOTELLA NANA, GEIGER
LLULAR SIZE, CHLORELLA/ *COULTER
, INFRARED GAS ANALYSIS, COULTER
HNII, MICROBIAL ECOLOGY, COULTER
UNIVERSITY, ELECTRONIC PARTICLE
NIUM PULVISCULUS, HAEMOCYTOMETER
ODS, ALGAE IDENTIFICATION, ALGAE
RS, MILK WASTE TREATMENT, COLONY
ED CELL VOLUMES, BACTERIAL PLATE
NAL WATER QUALITY NETWORK, ALGAE
), SCENEDESMUS QUADRICAUDA, CELL
IO), LAKE RAMSEY(ONTARIO), ALGAE
                         LOS ANGELES
*POLLUTION, *CHICAGO(ILL), COOK
214, AC/ *COTTUS CAROLINAE, MEAD
ACID WASTE MICROFLORA, MCCREARY
IN, MACROPH/ MICHIGAN, CHEBOYGAN
METER,/ *ONONDAGA LAKE, ONONDAGA
NIC PHOSPHATES.:   HARRIS
N, SAN JOAQUIN VA/ *CONTRA COSTA
AWA RIVER(POLAND), MONTANE RIVER
E, WATER COSTS, ARTIFICIAL WATER
SYLVANIA, MARCASITE, CHALCOCITE,
IS, ANALYTICAL TECHNIQUES, CLOUD
TURE, SEA WATER, OYSTERS, CLAMS,
TON, ECOSYSTEMS, SINKS, TRACERS,
HARIS, NAIAS, SCIRPUS, NYMPHAEA,
D(N C), DETECTOR, BEAUFORT(N C),
ATER PROPERTIES.:
S, *PESTICIDE RESIDUES, DIATOMS,
RS, BASS, WATER HYACINTH, FROGS,
MENTS, CHUTES, EDDIES, HABITATS,
RA, FISH, LIMNOLOGY, LAKE TROUT,
TRUCTURE, RIDLEY CREEK(PA), NOBS

CORES.: /ON, NITROGEN, PHOSPHOROU   W68-00511
CORES, DIATOMS, ODOR-PRODUCING AL   W68-00470
CORES, SEWAGE EFFLUENTS, DIATOMS,   W70-05663
CORES, WISCONSIN, BOTTOM SEDIMENT   W69-03366
CORETHRA PUNCTIPENNIS, PISIDIUM I   W70-06217
CORNELL UNIVERSITY, ELECTRONIC PA   W69-06540
CORP(MASS), ARTHUR D LITTLE INC(M   W69-06865
CORRELATION, DECAPODS, SPECIES DI   W73-01446
CORROSION)), RUSTING, METALS, INHI   W70-02294
CORROSION, COOLING WATER TREATMEN   W70-02294
CORROSION, OXIDATION, PHOTOSYNTHE   W71-11377
CORROSION, OXIDATION, REDUCTION(C   W69-07428
CORYCAEUS, OIKOPLEURA, FRITILLARI   W73-01446
CORYNEBACTERIUM, ELECTROSTATIC PR   W71-11823
CORYNEFORM RODS, COCCI,: /E, PSYC   W70-00713
COSCINDISCUS, BACTERIOSTRUM, LEPT   W73-01446
COSCINODISCUS, DIMEROGRAMMA, DIPL   W70-03325
COSMARIUM, CHROCOCCUS LIMNETICUS,   W70-05405
COSMIC RADIATION, S: /N, CENTRAL   W69-08272
COST ANALYSIS, *LAKE ERIE, PHOSPH   W70-04465
COST ANALYSIS, WATER PURIFICATION   W72-08688
COST ANALYSIS, SLUDGE, INCINERATI   W71-10654
COST COMPARISONS, SEDIMENTS, ALGA   W71-09548
COST COMPARISONS, TREATMENT, BIOL   W70-01981
COST COMPARISONS, ECONOMICS, ALGA   W69-08518
COST COMPARISONS.: / COMPOUNDS, L   W68-00256
COST-BENEFIT ANALYSIS.: /ANTS, BA   W70-05819
COST-SHARING.: /ON TECHNIQUES, SW   W70-05568
COSTA COUNTY, WATER DETERIORATION   W71-09577
COSTATUM(GREV) CL, CHLORELLA OVAL   W70-05752
COSTATUM, COCCOLITHUS HUXLEYI, PY   W70-05272
COSTATUM, SYNECHOCOCCUS ELON: /HL   W70-05381
COSTATUM, PHYSIOLOGICAL ADJUSTMEN   W70-04809
COSTATUM, DUNALIELLA TERTIOLECTA,   W69-08278
COSTS.: /UTION, ACTIVATED SLUDGE,   W70-04786
COSTS, *WASTE WATER TREATMENT.: /   W72-12987
COSTS, ARTIFICIAL WATER COURSES,   W71-09577
COSTS, CITIES.: /ATMENT, LAKES, S   W70-04455
COSTS, CONSTRUCTION COSTS.: /UTIO   W70-04786
COSTS, COST-SHARING.: /ON TECHNIQ   W70-05568
COSTS, DINOBRYON, CONCORD(N H), P   W69-08282
COSTS, EFFICIENCY, BIOCHEMICAL OX   W70-09320
COSTS, EUTROPHICATION, NITRATES,   W70-03964
COSTS, LEGAL ASPECTS, LEGISLATION   W70-03344
COSTS, TREATMENT FACILITIES, CLEA   W72-08859
COTTUS SPP, RHINICHTHYS OSCULUS,   W69-07846
COULTER COUNTER, EUGLENA GRACILIS   W69-08278
COULTER COUNTER, GROWTH INHIBITIO   W70-05652
COUNCIL(USA), US ATOMIC ENERGY CO   W70-03975
COUNCIL(USA), US ATOMIC ENERGY CO   W70-03975
COUNT(AUTORADIOGRAPHY), LEUCOTHRI   W69-10163
COUNT, DESMIDS, FLAGELLATES, MESO   W70-05389
COUNT, FILAMENTOUS, UNICELLULAR,   W68-00891
COUNT, GREEN RIVER, WASHINGTON, H   W72-11906
COUNT, OPTICAL DENSITY, PARTICULA   W70-03983
COUNT, WARING BLENDOR MORTALITY,   W70-04365
COUNTER.: /ANOCOBALAMIN, MUTANT B   W70-05652
COUNTER, *POPULATION DYNAMICS, CE   W69-06540
COUNTER, EUGLENA GRACILIS, CACHON   W69-08278
COUNTER, GROWTH INHIBITION, CYANO   W70-05652
COUNTING, EUDORINA ELEGANS, EUGLE   W69-06540
COUNTING, ITHACA(NY), GLENODINIUM   W69-06540
COUNTS, ALGAL GROWTH, SEDGWICK-RA   W69-10157
COUNTS, ARTHROPODS.: /PODS, SPIDE   W72-01819
COUNTS, LANCASTER(CALIF).: / PACK   W71-05267
COUNTS, PLANKTON ANALYSES, AGRICU   W70-04507
COUNTS, SPECIES DETERMINATION, AB   W70-03334
COUNTS, THRESHOLD ODOR NUMBERS, C   W69-10155
COUNTY(CALIF).:   W70-05645
COUNTY(ILL), CALUMET RIVER(ILL),   W70-00263
COUNTY(KY), THORIUM-232, BISMUTH-   W69-09742
COUNTY, KENTUCKY.:   W69-00096
COUNTY, STURGEON RIVER, FLUORESCE   W69-09334
COUNTY, SYRACUSE, WARBURG RESPIRO   W69-02959
COUNTY, TEX, POLYPHOSPHATES, ORGA   W69-04800
COUNTY, WATER DETERIORATION, DRAI   W71-09577
COURSE, CARPATHIAN SUBMONTANE REG   W70-02784
COURSES, CALIFORNIA.: / CONVEYANC   W71-09577
COVELLITE, SPHALERITE, MILLERITE,   W69-07428
COVER, SPECTROSCOPY, ULTRAVIOLET   W69-09676
CRABS, NORTH CAROLINA, SEDIMENTS,   W69-08267
CRABS, SHRIMP, MULLETS, FISH, ALG   W69-08274
CRASSIPES EICHORNIA, TYPHA, BRASE   W70-05263
CRASSOSTREA VIRGINICA, MERCENARIA   W69-08267
CRATER LAKE, ORE, GROWTH RATES, W   W68-00478
CRAYFISH, POLLUTANT IDENTIFICATIO   W69-02782
CRAYFISH, MOSQUITOES, BACTERIA, I   W70-05263
CRAYFISH, GROWTH STAGES, DIATOMS,   W69-09742
CRAYFISH, STONEFLIES, SNAILS, ECO   W70-00711
CREEK(MD), BACK RIVER(MD), OSCILL   W70-04510

421

NAL ALGAL ASSAY PROCEDURE, BATCH
A, STREAMS, DIATOMS, RESERVOIRS,
HLORELLA, ALGAE, ORGANIC MATTER,
OPHYTES, *NUTRIENT REQUIREMENTS,
LANKTON, DIATOMS, ECOLOGY, SOIL,
HYTOLOGY, MARINE ALGAE, ECOLOGY,
RADIOISOTOPES, LIGHT INTENSITY,
NGANESE, BENTHIC FLORA, STREAMS,
ALCIUM, CHLORIDES, CONDUCTIVITY,
IOASSAY, BIOLOGICAL COMMUNITIES,
E, MICROCYSTIS AERUGINOSA, ALGAL
*ALGAE, CYANOPHYTA, CHRYSOPHYTA,
BON CYCLE, *CYTOLOGICAL STUDIES,
, BRACK/ *EUTROPHICATION, ALGAE,
NT, *NUTRIENTS, ALGAE, BACTERIA,
ER, *VITAMINS, VITAMIN B, ALGAE,
VUM, SYNERGISTIC EFFECTS, AXENIC
TON, *ALGAE, SCENEDESMUS, LIGHT,
ANKTON, PHOTOSYNTHESIS, BIOMASS,
CARBOHYDRATES, YEASTS, ENZYMES,
GROWTH RATES, LABORATORY TESTS,
LINEAR ALKYL SULFONATES, *AXENIC
DESMUS, HETEROTROPHY, CONTINUOUS
SOLATION, *CONTROL, METHODOLOGY,
S CULTURES, CHLAMYDOMONAS, BATCH
CATION/ *ALGAL CULTURES, *AXENIC
STRUM, SELF-PURIFICATION/ *ALGAL
KOCYTIC COMPOSITION, ALTERNATING
CALLOPHYLLIS ·HAENOPHYLLA, TIDAL
ZCHIA SERIATA, NAVICULA, FLORIDA
RIES, *STREAMS, LIGHT INTENSITY,
WINDS, THERMOCLINE, TEMPERATURE,
DROGRAPHY, FLORIDA, THERMOCLINE,
TON, *SEDIMENTS, ORGANIC MATTER,
ENT, *FLUOROMETRY, DYE RELEASES,
TH, CIRCULATION, DIURNAL, WINDS,
ANOPHYTA, ENVIRONMENTAL EFFECTS,
, BIOTA, WASTE WATER(POLLUTION),
REAM FLOW, ALGAE, AQUATIC ALGAE,
ROWTH, WATER DENSITY, CONVECTION
ORESIS, MICROORGANISMS, ELECTRIC
*THERMOBIOLOGY, *DIURNAL OXYGEN
ARSENITE, FISH, ANIMALS, ALGAE,
ADISON LAKES(WIS), *AQUATIC WEED
TOXINS MEMBRANE, NORTHEAST U.S/
ECOLOGY, WATER QUALITY CRITERIA,
PLANNING, *SOURCE CONTAMINATION,
LTER COUNTER, GROWTH INHIBITION,
Z,/ *VITAMIN B-12, ALGAL GROWTH,
N, *LAKES, ALGAE, AQUATIC WEEDS,
A, CHRYSOPHYTA, DINOFLAGELLATES,
, ODOR-PRODUCING ALGAE, DIATOMS,
ROBIOLOGY, AQUATIC PRODUCTIVITY,
NSIN, ALGAE, CALCIUM, CHELATION,
, *NITROGEN, *PHOSPHORUS, ALGAE,
, DIAGENESIS, SEDIMENTS/ *ALGAE,
WISCONSIN, BIB/ *EUTROPHICATION,
DISSOLVED OXYGEN, OXYGEN DEMAND,
ARVESTING OF ALGAE, ENVIRONMENT,
N, ALGAE, CULTURES, CHLOROPHYTA,
L ECOLOGY, DIATOMS, OLIGOTROPHY,
TIC POPULATIONS, EUTROPHICATION,
L WATERSHEDS, NITROGEN FIXATION,
HORUS, DRAINAGE, MARSHES, FARMS,
ELS, SHELLFISH, DINOFLAGELLATES,
LTURES, IRON COMPOUNDS, DIATOMS,
ESIS, *MINNESOTA, PHYTOPLANKTON,
ENTRATION, NITRATES, PHOSPHATES,
CTIVITY, ALGAE, SEWAGE, DIATOMS,
PLANKTON, PIGMENTS, CHLOROPHYTA,
, DEPTH, BACTERIA, SLIME, ALGAE,
*ALGAE, *LAKE ERIE, CHLOROPHYTA,
OROPHYLL, PHYSIOLOGICAL ECOLOGY,
RGANIC MATTER,/ *NUISANCE ALGAE,
*SORPTION, *ALGAE, CHLOROPHYTA,
CATION, *BIOMASS, *PRODUCTIVITY,
S FORESTS, PHOSPHORUS, NITROGEN,
MUS, BENTH/ *ALGAE, TEMPERATURE,
TOMS, CHLOROPHYTA, EUGLENOPHYTA,
, PHYTOPLANKTON, PHOTOSYNTHESIS,
IRON, GROWTH RATES, CHLOROPHYTA,
S, DIATOMS, *ALGAE, CHLOROPHYTA,
NOLOGY, PLANT GROWTH REGULATORS,
AQUATIC PLANTS, BIBLIOGRAPHIES,
CATION, ALCOHOLS, ACTINOMYCETES,
N, FUNGI, YEASTS, PYRITE, ALGAE,
PHYTOPLANKTON, ALGAE, SEASONAL,
F NATURE, BIOASSAY, CHLOROPHYTA,
IC ALGAE, BIOASSAY, CHLOROPHYTA,
N EFFECTS, DIATOMS, CHLOROPHYTA,

CULTURES, CHEMOSTATS, CHLOROPHYLL   W70-02777
CULTURES, SEWAGE EFFLUENTS, WISCO   W70-02982
CULTURES, CARBON, METABOLISM, NIT   W70-02804
CULTURES, LAKES, ALGAE, GROWTH RA   W70-00719
CULTURES, CYANOPHYTA, WATER POLLU   W69-10180
CULTURES, WASHINGTON, WAVES(WATER   W69-10161
CULTURES, RADIOACTIVITY, CARBON D   W69-10158
CULTURES, COBALT RADIOISOTOPES.: /   W69-07845
CULTURES, HYDROGEN ION CONCENTRAT   W69-07833
CULTURES, DIATOMS, ECOLOGY, NUISA   W69-07832
CULTURES, CHEMOSTATS, CARBON-14.:   W69-06864
CULTURES, DIAGENESIS, SEDIMENTS,   W69-08284
CULTURES, FLUOROMETRY, BIOMASS, O   W69-08278
CULTURES, CHLOROPHYTA, CYANOPHYTA   W70-05752
CULTURES, CHEMICAL OXYGEN DEMAND,   W70-05655
CULTURES, CHEMICAL ANALYSIS, ANAL   W70-05652
CULTURES, MONOALGAL, PORTUGAL, TA   W70-05372
CULTURES, IRON COMPOUNDS, DIATOMS   W70-05409
CULTURES, PLANT PHYSIOLOGY, ALGAE   W70-05094
CULTURES, HYDROGEN SULFIDE, ELECT   W70-03952
CULTURES, SODIUM COMPOUNDS, NITRO   W70-03507
CULTURES, *ALKYL POLYETHOXYLATE,   W70-03928
CULTURES, CHLAMYDOMONAS, BATCH CU   W70-04381
CULTURES, TESTING, CAPILLARY COND   W70-04469
CULTURES, OPTICAL DENSITY, SOIL-W   W70-04381
CULTURES, PEDIASTRUM, SELF-PURIFI   W70-04381
CULTURES, *AXENIC CULTURES, PEDIA   W70-04381
CURRENT.: /REAM, WHITE BREAM, LEU   W70-01788
CURRENT, PUGET SOUND(WASH), LONG   W69-10161
CURRENT, YTTRIUM-90, KATODINIUM· R   W70-00707
CURRENTS(WATER), VELOCITY, DIATOM   W70-02780
CURRENTS(WATER), HYPOLIMNION, ALG   W70-09889
CURRENTS(WATER), PHYTOPLANKTON, Z   W70-04368
CURRENTS(WATER), WATER POLLUTION   W70-03501
CURRENTS(WATER), REMOTE SENSING,   W70-05377
CURRENTS(WATER), WATER POLLUTION,   W70-05404
CURRENTS(WATER), INDUSTRIAL WASTE   W68-00470
CURRENTS(WATER).: /BEACHES, SANDS   W68-00010
CURRENTS(WATER), LIMNOLOGY, LAKES   W68-00479
CURRENTS, ALGAE TREATMENT COSTS,   W69-08282
CURRENTS, *WASTE TREATMENT, WASTE   W71-03775
CURVES, CALORIMETRY, SOLARIMETRY,   W70-03309
CUTTING MANAGEMENT, POTABLE WATER   W70-05263
CUTTING, *WEED EXTRACTING STEEL C   W70-04494
CYANAPHYTA, ALGAL CONTROL, *ALGAL   W69-05306
CYANIDE, NAPHTHENIC ACID.: /BIAL   W70-04510
CYANIDE, NEW PRODUCT CONTROL, PLU   W73-00758
CYANOCOBALAMIN, MUTANT BACTERIA,   W70-05652
CYANOCOBALAMIN, EUGLENA GRACILIS   W69-06273
CYANOPHYTA, FISHKILL, FERTILIZATI   W69-06535
CYANOPHYTA.: /DIATOMS, CHLOROPHYT   W69-05763
CYANOPHYTA, CHLOROPHYTA, IRON, MA   W69-05704
CYANOPHYTA, ENVIRONMENTAL EFFECTS   W69-04802
CYANOPHYTA, NUISANCE ALGAE, NUTRI   W69-05867
CYANOPHYTA, NITRATES, NITROGEN CO   W69-05868
CYANOPHYTA, CHRYSOPHYTA, CULTURES   W69-08284
CYANOPHYTA, ODOR, SCUM, FISHING,   W69-08283
CYANOPHYTA, DIAT: /S, SEDIMENTS,   W69-08518
CYANOPHYTA.: /ALGAE, ALGICIDES, H   W69-07818
CYANOPHYTA, BRACKISH WATER, OLIGO   W70-05752
CYANOPHYTA, CHLOROPHYTA, GREAT LA   W70-04503
CYANOPHYTA, SEWAGE EFFLUENTS, CHL   W70-04510
CYANOPHYTA, SEWAGE EFFLUENTS, MIN   W70-04506
CYANOPHYTA, PLANKTON, ZOOPLANKTON   W70-05112
CYANOPHYTA, CHRYSOPHYTA, CALIFORN   W70-05372
CYANOPHYTA, PHOSPHATES, CHLOROPHY   W70-05405
CYANOPHYTA, DIATOMS, STRATIFICATI   W70-05387
CYANOPHYTA, OXIDATION-REDU: /CONC   W70-05404
CYANOPHYTA, CARBON RADIOISOTOPES,   W70-05405
CYANOPHYTA, CHRYSOPHYTA, RHODOPHY   W70-05389
CYANOPHYTA, DIATOMS, CHLOROPHYTA,   W70-04371
CYANOPHYTA, UNITED STATES, HABITA   W70-04468
CYANOPHYTA, EUTROPHICATION, NUISA   W70-04381
CYANOPHYTA, VITAMINS, PEPTIDES, O   W70-04369
CYANOPHYTA, DETERGENTS, SPECTROSC   W70-03928
CYANOPHYTA, NUISANCE ALGAE, LIMNO   W70-03959
CYANOPHYTA, CHLOROPHYTA, DIATOMS,   W70-03512
CYANOPHYTA, CHLOROPHYTA, SCENEDES   W70-03969
CYANOPHYTA, CHRYSOPHYTA, BIOASSAY   W70-03974
CYANOPHYTA, CYTOLOGICAL STUDIES,   W70-03973
CYANOPHYTA, CHRYSOPHYTA, PHOTOSYN   W70-03983
CYANOPHYTA, LAND MANAGEMENT, WATE   W68-00172
CYANOPHYTA, ENVIRONMENTAL EFFECTS   W68-00470
CYANOPHYTA, FRESH WATER FISH, LAK   W68-00680
CYANOPHYTA, SPECTROSCOPY, GAS CHR   W69-00387
CYANOPHYTA, DOMINANT ORGANISM, NU   W68-00891
CYANOPHYTA, *INORGANIC COMPOUNDS,   W68-00860
CYANOPHYTA, ENVIRONMENTAL EFFECTS   W69-03185
CYANOPHYTA, DIATOMS, DISINFECTION   W69-03370
CYANOPHYTA, NUISANCE ALGAE, WISCO   W69-01977

S, SYNURA, ANABAENA, IOWA RIVER,
, BOSMINA, NAVICULA, FRAGILARIA,
ALENSIS, GYMNODINIUM BAICALENSE,
ETHOD, CALCULATED VOLUME METHOD,
MORPHA INTESTINALIS, CLADOPHORA,
CLIS, CHLOROGONIUM, SCENEDESMUS,
VOLVOCALES, SARCODINA, CILIATES,
, DINOBRYON, UROGLENA AMERICANA,
TURATED / *ADAPTATION, *SPECIES,
DJUSTMENT, ASTERIONELLA FORMOSA,
CYANOCOBALAMIN, MUTANT BACTERIA,
RESERVOIRS, *TURBID RESERVOIRS,
OS-AQUAE, OSCILLATORIA, ANABAENA
, PYRUVATE, ACETYLENE/ *ANABAENA
OLOGICAL CONCENTRATION, ANABAENA
ON, TRINGFORD(ENGLAND), ANABAENA
NII, METABOLIC PATTERN, ANABAENA
ZINC, EUGLENA GRACILIS, ANABAENA
EXTRACELLULAR PRODUCTS, ANABAENA
OCYSTIS, OSCILLATORIA, NAVICULA,
OVAKIA), BLATNA(CZECHOSLOVAKIA),
PPER SULF/ *HALOGENS, ALGICIDES,
, ALGAE, LICHENS, ACTIMOMYCETES,
, KINETICS, HARDNESS, COLIFORMS,
TON, PHOTOSYNTHESIS, CYANOPHYTA,
IUM, FISH, AMPHIBIANS, ANABAENA,
NATION, *STRONTIUM RADIOSOTOPES,
C ALGAE, BLOOMS, CLOSED SYSTEMS,
ION, IODINE TEST, GRAM REACTION,
ITIVITY, GENERA, 'FLAVOBACTERIUM-
DAPHNIA HYALINA, DAPHNIA MAGNA,
BOHEMIA, ELBE RIVER, CELAKOVICE(
LAKOVICE(CZECHOSLOVAKIA), BLATNA(
GICAL INDEX, *SAPROBITY, PRAGUE,
D DEVELOPMENT CORP(MASS), ARTHUR
GREEN / *LEWIS AND CLARK LAKE (S
PHYTOTOXICITY, ALGAL POISO/ *2-4
IDES, *GROWTH RATES, *ALGAE, 2-4-
21/ *LITERATURE, ELEOCHARIS, 2,4-
A, DISSOLVED OXYGEN, PIKE, CARP,
                           SIMAZINE 80 W,
, CYCLOPS, SCENEDESMUS OBLIQUUS,
*ALGAE BLOOMS, CYANOPHYTA, NORTH
(WASTES), RED RIVER VALLEY(NORTH
WASTES), SEWAGE TREATMENT, NORTH
*DIATOMS, *ALGAE, ECOLOGY, SOUTH
TRIBUTION, MISSOURI RIVER, SOUTH
PHYSIOLOGY, NORTH DAKOTA, SOUTH
IOLOGY, ANIMAL PHYSIOLOGY, NORTH
PPER SULPHATE, MONURON, AMMONIA,
OENSE, TENDIPES PLUMOSUS, BEAVER
MARGINAL VEGETAT/ *HARTBEESPOORT
USBR RESEARCH CONTRACTS, CHENEY
OTO REACTIVITY, CELL COLOR, CELL
S, *WATER REUSE, COOLING TOWERS,
AUDERIA BOREALIS, LEPTOCYLINDRUS
LOROCOCCALES, DAPHNIA PULICARIS,
APHNIA SCHODLERI, DAPHNIA PULEX,
IA PULICARIS, DÁPHNIA CUCULLATA,
HNIA CUCULLATA, DAPHNIA HYALINA,
QUADRICAUDA, CHLORELLA VULGARIS,
QUADRICAUDA, CHLORELLA VULGARIS,
PULEX, DAPHNIA GALEATA MENDOTAE,
TIVE STUDIES, DAPHNIA SCHODLERI,
CYPRINID FISHES, CLOROCOCCALES,
BODY SIZE, *COMPARATIVE STUDIES,
Y, TEMPERATURE, LIGHT INTENSITY,
*ALGAL CONTROL, EUTROPHICATION,
OXICITY, SCENEDESMUS, CHLORELLA,
OXICITY, SCENEDESMUS, CHLORELLA,
, SWIMMING POOLS, CHLAMYDOMONAS,
ENVIRONMENT, PHOSPHATES, ALGAE,
UISANCE ALGAE, SUNFISHES, ALGAE,
ABOLISM, BIOLOGICAL COMMUNITIES,
ENTRATION, PHOSPHATES, NITRATES,
LANT PHYSIOLOGY, EUTROPHICATION,
ANALYSIS, PH, SAMPLING, ON-SITE
ENTS, *BOTTOM SEDIMENTS, ON-SITE
UM, MATHEMATICAL MODELS, ON-SITE
PHOSPHORUS RADIOISOTOPE, ON-SITE
STROPODS, MARINE PLANTS, ON-SITE
WATER POLLUTION SOURCES, ON-SITE
*LAKE ERIE,/ *SYSTEMS ANALYSIS,
ENEDESMUS, MATHEMATICAL STUDIES,
, NAVICULA, MEL/ *CHLOROPHYLL A,
MAND, SUSPENDED LOAD, TURBIDITY,
                          *BIOGEOCHEMICAL
*SULFIDES, TEMPERATURE, CLIMATIC
MELOSIRA, MALLOMONAS, ULOTHRIX,
ZOPODS, SPIDER/ *ZOOGLEAL SLIME,

CYCLOTELLA, FRAGILARIA, SYNEDRA,	W70-08107
CYCLOTELLA, FLAGELLATES, ANKISTRO	W70-02249
CYCLOTELLA MINUTA.: /ELOSIRA BAIC	W70-04290
CYCLOTELLA COMENSIS, RHODOMONAS L	W70-03959
CYCLOTELLA, HALOCHLOROCOCCUM, SYN	W70-03974
CYCLOTELLA, STEPHANODSICUS.: /CIN	W69-02959
CYCLOTELLA, APHANIZOMENON FLOS AQ	W70-05112
CYCLOTELLA, CERATIUM HIRUNDINELLA	W70-05405
CYCLOTELLA MENEGHINIANA, LIGHT-SA	W70-05381
CYCLOTELLA MENEGHINIANA, PHOTIC Z	W70-04809
CYCLOTELLA NANA, GEIGER COUNTER.:	W70-05652
CYCLOTELLA, TRACHELOMONAS, MELOSI	W72-04761
CYLINDRICA B629, ANABAENA INAEQUA	W70-05091
CYLINDRICA, ATP, CMU, NITROGEN-15	W70-04249
CYLINDRICA, MICROCYSTIS AERUGINOS	W70-03520
CYLINDRICA, NORTH SEA, FIRTH OF C	W70-02504
CYLINDRICA, GLYCOLYSIS, TRICARBOX	W70-02965
CYLINDRICA, BACILLUS SUBTILIS, AS	W70-02964
CYLINDRICA, ZINC, XANTHOPHYCEAE,	W69-10180
CYMBELLA, LYNGBYA, FRAGILARIA, OE	W70-04371
CYPRINID FISHES, CLOROCOCCALES, D	W70-00274
CYTOLOGICAL STUDIES, CHLORINE, CO	W70-02364
CYTOLOGICAL STUDIES, FILTRATION,	W71-11823
CYTOLOGICAL STUDIES, EVAPORATION,	W71-05267
CYTOLOGICAL STUDIES, ELECTRON MIC	W70-03973
CYTOLOGICAL STUDIES, PHY: /OSTRID	W70-05372
CYTOLOGICAL STUDIES.: /IL CONTAMI	W69-03491
CYTOLOGY, ENVIRONMENTAL CONDITION	W69-07832
CYTOPHAGA, PROTEUS, SERRATIA, BAY	W70-03952
CYTOPHAGA', VIBRIOS, ACHROMOBACTE	W70-00713
CZECHOSLOVAKIA.: /HNIA CUCULLATA,	W70-00274
CZECHOSLOVAKIA), BLATNA(CZECHOSLO	W70-00274
CZECHOSLOVAKIA), CYPRINID FISHES,	W70-00274
CZECHOSLOVAKIA, LABORATORY PROCED	W72-12240
D LITTLE INC(MASS), CARNEGIE DEPT	W69-06865
D), ABUNDANCE, BLUE-GREEN ALGAE,	W70-05375
D, *HERBICIDES, *PHYTOPLANKTON, *	W69-00994
D, AMINOTRIAZOLE, 2-4-5-T, SCENED	W70-03519
D, AMMATE, CHLORAMINE, DE-K-PRUF-	W70-05263
D: /S, ALGAE, DIATOMS, CHLOROPHYT	W70-01943
DACTHAL.:	W70-03519
DACTYLOCOCCUS, ESCHERICHIA COLI,	W70-10161
DAKOTA.: /GY, WATER TEMPERATURE,	W70-09093
DAKOTA), SULFATE REDUCTION, POTAT	W70-03312
DAKOTA, COLIFORMS, OXIDATION LAGO	W70-03312
DAKOTA, FIELD INVESTIGATIONS, WAT	W69-00659
DAKOTA, NEBRASKA, ALGAE, NANNOPLA	W70-05375
DAKOTA, PHOTOSYNTHESIS, LIGHT PEN	W72-01818
DAKOTA, SOUTH DAKOTA, PHOTOSYNTHE	W72-01818
DALAPON, 2-4-: /DIUM ARSENITE, CO	W70-00269
DAM LAKE(WIS), DELAVAN LAKE(WIS).	W70-06217
DAM(SOUTH AFRICA), TRANSPARENCY,	W70-06975
DAM, KANS.:	W72-01773
DAMAGE.: /S ALGAE, KILL RATES, PH	W70-02364
DAMS, ALGAE, HYDROGEN ION CONCENT	W71-11393
DANICUS, PHAEODACTYLUM TRICORNUTU	W70-04280
DAPHNIA CUCULLATA, DAPHNIA HYALIN	W70-00274
DAPHNIA GALEATA MENDOTAE, DAPHNIA	W70-03957
DAPHNIA HYALINA, DAPHNIA MAGNA, C	W70-00274
DAPHNIA MAGNA, CZECHOSLOVAKIA.: /	W70-00274
DAPHNIA MAGNA, LIMNAEA STAGNALIS,	W71-05719
DAPHNIA MAGNA, LIMNAEA STAGNALIS,	W71-03056
DAPHNIA MAGNA, SELENASTRUM, RHODO	W70-03957
DAPHNIA PULEX, DAPHNIA GALEATA ME	W70-03957
DAPHNIA PULICARIS, DAPHNIA CUCULL	W70-00274
DAPHNIA SCHODLERI, DAPHNIA PULEX,	W70-03957
DAPHNIA, CONNECTICUT, ROTIFERS.: /	W70-03959
DAPHNIA, CHLAMYDOMONAS, LABORATOR	W72-02417
DAPHNIA, DETERIORATION, WATER TRE	W71-05719
DAPHNIA, DETERIORATION, WATER TRE	W71-03056
DAPHNIA, DIATOMS, CHLORELLA, SCEN	W70-10161
DAPHNIA, GROWTH RATES, PREDATION.	W69-06536
DAPHNIA, LAKES, CHLOROPHYTA, STRE	W70-02982
DAPHNIA, NANNOPLANKTON, NUTRIENTS	W70-00274
DAPHNIA, SILICA, THERMAL STRATIFI	W70-05760
DATA COLLECTION, CHEMICAL ANALYSI	W70-04382
DATA COLLECTIONS, LIMNOLOGY.: /AL	W69-00659
DATA COLLECTIONS, AQUATIC PLANTS,	W68-00511
DATA COLLECTIONS, BOTTOM SEDIMENT	W68-00858
DATA COLLECTIONS, FISHES, PLANKTO	W68-00857
DATA COLLECTIONS.: /MORTALITY, GA	W70-01780
DATA COLLECTIONS, *ON-SITE INVEST	W70-09976
DATA COLLECTIONS, *ALGAE CONTROL,	W71-04758
DATA COLLECTIONS, MARINE ANIMALS,	W73-01446
DATA INTERPRETATION, *RHONE RIVER	W73-01446
DATA STORAGE AND RETRIEVAL, COLIF	W70-09557
DATA, *ABUNDANCE RATIOS.:	W71-05531
DATA, STRATIFICATION, DISSOLVED O	W70-04790
DAVID MONIE TEST, SUM: /ANABAENA,	W69-10157
DAYTON(OHIO), VANDALIA(OHIO), RHI	W72-01819

426

THAL DOSES, CHU'S MEDIA, QUINONE
EEN RIVER FORMATION, CHLOROPHYLL
CARIAE, WATER MILFO/ SCHISTOSOME
SCIRPUS FLUVIATILIS, *LAKE BUTTE
BATEMENT, EFFLUENTS, STREAMFLOW,
B) FARLOW, CLADOPHORA, TAXONOMIC
ROTENOIDS, SPECIES, DIVERSITY I/
IA DIGITATA, LAMINARIA AGARDHII,
ATORY MODELS, BOD REMOVAL RATES,
ACERS, *WATER POLLUTION, DESIGN,
ATER POLLUTION, TOXICITY, LAKES,
SLUDGE, DEPTH, ORGANIC LOADING,
TURE, OPERATION AND MAINTENANCE,
AEROBIC PONDS, *ANAEROBIC PONDS,
NITARY ENGINEERING, *FILTRATION,
ENT, *TRACERS, *WATER POLLUTION,
ES, EXPERIMENTAL STUDIES, LAGOON
URE, NUTRIENTS, FACULTATIVE POND
OLING WATER TREATMENT, ON-STREAM

FICANCE, ALGAL KEY, CLUMP COUNT,
ALGAE, *LAKES, *CLASSIFICATION,

CONTACT ZONES, ARTEMIA, BALANUS,
INE, PHOTOLITHOTROPHIC BACTERIA,
ERENCES.:   *NICKEL,
REAM POLLUTION, *WATER POLLUTION
DE, ULOTHRIX, UPTAKE, HALF-LIFE,
N IRRADIANS, PIVERS ISLAND(N C),
LLIN LAKE, GRAPHITE LAKE, OLD J/
, BOD REMOVAL RATES, DESIGN APP/
DESULFOVIBRIO, FACULTATIVE, BOD,
ENE SULFONATES, *BIODEGRADATION,
CIDES, ESTUARIES, WATER QUALITY,
HOROUS, RUNOFF, DOMESTIC WASTES,
UNGI, ALGAE, NITRITES, NITRATES,
GE, TROUT, NITROGEN, PHOSPHORUS,
BLOOMS, NUTRIENT LOAD, SYNTHETIC
TION, *ALGAE, *AQUATIC BACTERIA,
RACERS, DISTRIBUTION, CHELATION,
HATES, NUTRIENT BALANCE, FOAMING
*ALGAE, CHLOROPHYTA, CYANOPHYTA,
MENTATION, CHLORINATION, SLUDGE,
EN DEMAND, CARBON RADIOISOTOPES,
ROWTH RATES, PESTICIDE RESIDUES,
EATMENT, *NUTRIENT R/ *AESTHETIC
SCENEDESMUS, CHLORELLA, DAPHNIA,
VA/ *CONTRA COSTA COUNTY, WATER
SCENEDESMUS, CHLORELLA, DAPHNIA,
UADRICAUDA, CELL COUNTS, SPECIES
TERIUM, ACHROMOBACTER, / SPECIES
N), SMITH'S BAY(MINN), PHOSPHATE
M./ *LINDANE, *ALGAE METABOLISM,
US COMPOUNDS, BACTERIA, DIATOMS,
SOTOPES, CESIUM, FALLOUT, SILTS,
BIOLOGICAL COMMUNITIES, SEWAGE,
OTTOM SEDIMENTS, SURFACE WATERS,
BIOCARBONATES, DISSOLVED OXYGEN,
UM, NITRATES, PHOSPHATES, ALGAE,
R, BACTERIA, SEDIMENTS, MINNOWS,
S, GEORGIA, SEDIMENTS, BACTERIA,
N DIOXIDE, CARBON RADIOISOTOPES,
RONUTRIENTS, BADFISH RIVER(WIS),
IVER, HURON RIVER, RAISIN RIVER,
TION, HAZARDS, CESIUM, HYDROGEN,
IDIUM SP.:   *TEST
OLL/ *FARM WASTES, *RESEARCH AND
DING, *WATER REUSE, RESEARCH AND
RELLA, CH/ *ALGAE, *RESEARCH AND
HYSIOLOGY, AMERICAN RESEARCH AND
RESERVOIR SITES, WATER RESOURCES
CAPILLARITY, CAPILLARITY CONTROL
RE, CENTRIFUGATION, COAGULATION,
PHOSPHORUS, *PHOSPHATES, ENERGY,
ERIONELLA, ARAPHIDINEAE, CENTRIC
ANOPHYTA, CHRYSOPHYTA, CULTURES,
ICULOSA, GONYAULAX POLYEDRA, 3,5-
M LAKE(IOWA), CENTER LAKE(IOWA),
ASHLANDI, EPISCHURA NEVADENSIS,
OPHYCEA, OSCILLATORIA, ANABAENA,
PHAERIUM, SPIROGYRA, CLADOPHORA,
ERIAL PLATE, BACTERIAL PIGMENTS,
YGEN, OXYGEN DEMAND, CYANOPHYTA,
ION TIME, CHLORELLA PYRENOIDOSA,
*LABORATORY TESTS, *SCENEDESMUS,
E, *ORGANIC MATTER, *FILTRATION,
MYCOIDES, BACILLUS MESENTERICUS,
ER(MD), OSCILLATORIA, CATHERWOOD
                     FILTER AIDS,
NEBRASKA, ALGAE, NANNOPLANKTON,

DERIVATIVES, ALKYL SUBSTITUTED IM   W70-02982
DERIVATIVES, RECENT SEDIMENTS, PR   W69-08284
DERMATITIS, COPPER CARBONATE, CER   W70-05568
DES MORTS(WISC), GROWTH STIMULATI   W72-00845
DESALINATION, TERTIARY TREATMENT,   W69-07389
DESCRIPTIONS, GONGROSIRA LACUSTRI   W70-04468
DESICCATION, YAQUINA BAY(ORE), CA   W70-03325
DESICCATION, IMMERSION, NARRAGANS   W70-01073
DESIGN APPLICATIONS, FACULTATIVE   W70-07034
DESIGN CRITERIA, OXIDATION LAGOON   W70-02609
DESIGN CRITERIA, REMOTE SENSING,   W70-02775
DESIGN CRITERIA, ODORS, EFFICIENC   W72-06854
DESIGN CRITERIA, *WATER TREATMENT   W72-08686
DESIGN FOR ENVIRONMENTAL CONTROL,   W70-04787
DESIGN, *ALGAE, COPPER SULFATE, S   W72-08859
DESIGN, DESIGN CRITERIA, OXIDATIO   W70-02609
DESIGN, SCALE-UP FACTORS, WASTE C   W70-07508
DESIGNS, BOD, SLUDGE.: / TEMPERAT   W70-04786
DESLUDGING.: /METAL CORROSION, CO   W70-02294
DESMIDS.:   W68-00860
DESMIDS, FLAGELLATES, MESOPLANKTO   W70-05389
DESMIDS, OLIGOTROPHIC, *PHYTOPLAN   W68-00255
DESMINS, FLAGELLATES.:   W68-00172
DESORPTION, EMBRYOLOGY, ENERGY TR   W69-07440
DESULFOVIBRIO, MESOTHERMY, MELOSI   W70-05760
DETECTION LIMITS, CHEMICAL INTERF   W71-11687
DETECTION.: /, MARINE BIOLOGY, ST   W71-10786
DETECTOR.:   *RADIONUCLI   W69-07862
DETECTOR, BEAUFORT(N C), CRASSOST   W69-08267
DETECTOR, ALASKA, HUDEUC LAKE, PU   W69-08272
DETENTION TIME, LABORATORY MODELS   W70-07034
DETENTION TIME.: *SULFATE ION, *   W70-01971
DETERGENTS, WATER POLLUTION, WAST   W70-08740
DETERGENTS, ALGAL CONTROL, INDUST   W70-10437
DETERGENTS, INDUSTRIAL WASTES, *C   W71-04072
DETERGENTS, PHOSPHATES.: /ERIS, F   W71-00940
DETERGENTS, RUNOFF, PERCOLATION,   W69-10178
DETERGENTS, FISH MANAGERS, FISH Y   W69-08674
DETERGENTS, SURFACTANTS, ALKYLBEN   W69-09454
DETERGENTS, RADIOACTIVITY, AMINO   W69-08275
DETERGENTS, PUBLIC APPREHENSION.:   W70-05084
DETERGENTS, SPECTROSCOPY.: /ION,   W70-03928
DETERGENTS, SPRAYING, ALGAE, CENT   W70-04060
DETERGENTS, ANTIFREEZ: /ICAL OXYG   W70-04001
DETERIORATION.: /NCE, BIOASSAY, G   W70-03520
DETERIORATION, *ADVANCED WASTE TR   W69-07818
DETERIORATION, WATER TREATMENT, D   W71-03056
DETERIORATION, DRAIN, SAN JOAQUIN   W71-09577
DETERIORATION, WATER TREATMENT, D   W71-05719
DETERMINATION, ABSORBANCE, COENOB   W70-03334
DETERMINATION, TAXONOMY, FLAVOBAC   W70-04280
DETERMINATION: /), GRAY'S BAY(MIN   W70-05269
DETOXIFICATION, LINDANE METABOLIS   W70-08652
DETRITUS, CYCLING NUTRIENTS, LIG: W69-09334
DETRITUS, BOTTOM SEDIMENTS, CHUTE   W69-09742
DETRITUS, TUBIFICIDS, PLANKTON, D   W70-00268
DETRITUS, LIGHT INTENSITY, TURBID   W70-00713
DETRITUS, BACTERIA, WATER POLLUTI   W70-02249
DETRITUS, PULP WASTES, LIVESTOCK,   W70-03501
DETRITUS, CYCLING NUTRIENTS, PLAN   W69-07861
DETRITUS, ESTUARIES, PLANKTON, EC   W69-08274
DETRITUS, COLUMBIA RIVER, ELECTRO   W69-07862
DETROIT LAKES(MINN), NUTRIENT REM   W70-04506
DETROIT RIVER, MAUMEE RIVER.: / R   W69-01445
DEUTERIUM, TRITIUM.: /ODS, ABSORP   W70-02786
DEVELOPMENT, *TEST PROBLEMS, HORM   W71-07339
DEVELOPMENT, GRANTS, ALGAE, AIR P   W71-06821
DEVELOPMENT, WASTE IDENTIFICATION   W72-03299
DEVELOPMENT, BIBLIOGRAPHIES, CHLO   W69-06865
DEVELOPMENT CORP(MASS), ARTHUR D   W69-06865
DEVELOPMENT, WATER QUALITY, SOIL   W72-11906
DEVICE.: /CHIA, ALGAE ISOLATION,   W70-04469
DEWATERING, *COST ANALYSIS.: /ATU   W71-11375
DI-URNAL DISTRIBUTION, TEMPERATUR   W70-06604
DI: /R, FRAGILARIA, MELOSIRA, AST   W70-05663
DIAGENESIS, SEDIMENTS, GAS CHROMA   W69-08284
DIAMINOBENZOIC ACID DIHYDROCHLORI   W69-08278
DIAMOND LAKE(IOWA).: /NGBYA, STOR   W70-02803
DIAPHANOSOMA LEUCHTENBER GIANUM,   W69-10182
DIAPTOMUS ASHLANDI, EPISCHURA NEV   W69-10182
DIAPTOMUS, CYCLOPS, GOMPHONEMA, P   W70-02803
DIAPTOMUS, EPISCHURA, ORTHOCYCLOP   W70-05760
DIAT: /S, SEDIMENTS, DISSOLVED OX   W69-08518
DIATOMA ELONGATUM, MELOSIRA ISLAN   W69-10158
DIATOMACEOUS EARTH, MUD, LAKES, C   W68-00475
DIATOMACEOUS EARTH, ALGAE, TREATM   W72-04318
DIATOMEA, CLADOPHORA, FLAGELLATA,   W69-07838
DIATOMETER, SAVANNAH RIVER, POPUL   W70-04510
DIATOMITE FILTERS.:   W72-04318
DIATOMS.: /I RIVER, SOUTH DAKOTA,   W70-05375

ENII, RHODOTORULA RUBRA, CANDIDA	DIDDENSII, CANDIDA ZEYLANOIDES, C	W70-04368
OUAE, STEPHANODISCUS HANTZSCHII,	DIDINIUM, PARAMECIUM, PHACUS, GLE	W70-00268
, *ZOOPLANKTON, SULFUR BACTERIA,	DIEL MIGRATION, CARBON DIOXIDE, H	W70-05760
RBON PESTICIDES, ALDRIN, ENDRIN,	DIELDRIN, SCENEDESMUS, GAS CHROMA	W70-03520
WATER POLLUTION EFFECTS, ALDRIN,	DIELDRIN, ENDRIN, DDT.: /UTANTS,	W69-08565
DSORPTION, ALGAE, CLAYS, ALDRIN,	DIELDRIN, DDT, AERATION, SILTS.: /	W70-09768
OMETRY, N-HEPTADECANE, MONOENES,	DIENES, TRIENES.: /S, MASS SPECTR	W69-08284
NIUM, PHORMIDIUM RETZII, SURFACE	DIFFUSION RATE.: /RA ULNA, OEDOGO	W70-00265
YSIOLOGICAL ECOLOGY, CYANOPHYTA,	DIFFUSION, RESPIRATION, PHOSPHATE	W70-03309
ABILIZATION, WASTE WATER, ALGAE,	DIFFUSION, SULFUR.: /BACTERIA, ST	W70-01971
TION, SCOUR, SATURATION, MIXING,	DIFFUSION, SEDIMENTATION, ABSORPT	W71-11381
MPLING, METHANE, CARBON DIOXIDE,	DIFFUSION, *WASTE WATER TREATMENT	W71-11377
ISANCE ALGAE, THERMAL POLLUTION,	DIFFUSION, DIATOMS, CHLOROPHYTA,	W69-05763
ROSCOPY, AMMONIA, SPECTROMETERS,	DIFFUSION, TURBULENCE, TOXINS, BA	W70-04369
, *ALGAE, ORGANIC MATTER, LAKES,	DIFFUSION, CARBON RADIOISOTOPES,	W70-04284
OLLUTION EFFECTS, PHYTOPLANKTON,	DIFFUSION, ALGAE, BACTERIA, EUTRO	W70-04194
RIFUGATION, FILTRATION, ANEROBIC	DIGESTION, INCINERATION, IRON COM	W70-04060
, *AEROBIC TREATMENT, *ANAEROBIC	DIGESTION, BIOCHEMICAL OXYGEN DEM	W71-09548
ON, OXIDATION LAGOONS, ANAEROBIC	DIGESTION, ALGAE, SLUDGE, METHANE	W71-11375
ON, FILTERS, PESTICIDES, SLUDGE,	DIGESTION, INDIANA.: /OUNDS, LAGO	W72-03641
HEMICAL OXYGEN DEMAND, ANAEROBIC	DIGESTION, AEROBIC TREATMENT, CHE	W71-00936
*OXIDATION LAGOONS, *ANAEROBIC	DIGESTION, *MUNICIPAL WASTES, MET	W72-06854
, ASCOPHYLLUM NODOSUM, LAMINARIA	DIGITATA, LAMINARIA AGARDHII, DES	W70-01073
OLYEDRA, 3,5-DIAMINOBENZOIC ACID	DIHYDROCHLORIDE, INFRARED GAS ANA	W69-08278
DREDGING, *POLLUTION ABATEMENT,	DIKES, SOIL CONSE: /TION SOURCES,	W68-00172
MARINE LABORATORY(RHODE ISLAND),	DILUTION RATE, FRUITING BODIES, U	W70-01073
TREATMENT, BOTTOM SEALING, LAKE	DILUTION.: /R DIVERSION, INSTREAM	W69-09340
IENTS), LAKE VOLUME, MEAN DEPTH,	DILUTION, OSCILLATORIA RUBESCENS,	W70-00270
OR, EFFLUENTS, SEWAGE EFFLUENTS,	DILUTION, DECOMPOSING ORGANIC MAT	W71-00940
*RESERVOIR EFFECTS, LAKE MEAD,	DILUTION, ALGAL GROWTHS, OXYGEN D	W73-00750
ORACOMONAS, YTTRIUM, 'BIOLOGICAL	DILUTION', PHYSIOLOGICAL CO: / TH	W70-00708
CEROS, COCCONEIS, COSCINODISCUS,	DIMEROGRAMMA, DIPLONEIS, EUNOTOGR	W70-03325
OHOL, ETHYLENE GLYCOL, POLYMERS,	DIMETHYL TEREPHTHALATE.: /HYL ALC	W70-10433
PHAERIUM PULCHELLUM, GOMPHONEMA,	DINOBRYON.: /RETICULATUM, DICTYOS	W69-05704
CURRENTS, ALGAE TREATMENT COSTS,	DINOBRYON, CONCORD(N H), PEANACOO	W69-08282
TAE, GLOEOTRICHIA, ASTERIONELLA,	DINOBRYON, MELOSIRA, ANEBODA(SWED	W70-05409
LIMNION, OSCILLATORIA RUBESCENS,	DINOBRYON, UROGLENA AMERICANA, CY	W70-05405
AL SYMBIOSIS, MICROBIAL ECOLOGY,	DINOBRYON, VOLVOX, PANDORINA, SPI	W70-04503
ELLA FORMOSA, SYNEDRA, MELOSIRA,	DINOBRYON, FRAGILARIA, RADIOCARBO	W70-01579
CYSTIS, GLOEOTRICHIA ECHINULATA,	DINOBRYON, SYNURA UVELLA, FRAGILA	W70-02803
TICAL TECHNIQUES, LETHAL LIMITS,	DINOFLAGELLATES, DIATOMS.: /ANALY	W70-08111
OURCES, WATER POLLUTION EFFECTS,	DINOFLAGELLATES, FISHKILL, FISH T	W70-08372
, BIOTA, PHYTOPLANKTON, DIATOMS,	DINOFLAGELLATES, ALGAE, HYPOLIMNI	W69-10169
GY, RED TIDE, ORGANIC COMPOUNDS,	DINOFLAGELLATES.: / FISH PHYSIOLO	W71-07731
TATS, CLAMS, MUSSELS, SHELLFISH,	DINOFLAGELLATES, CYANOPHYTA, CHRY	W70-05372
OGY, MASSACHUSETTS, CHLOROPHYTA,	DINOFLAGELLATES, ALGAE, CALIFORNI	W70-05272
ANALYSIS, ANALYTICAL TECHNIQUES,	DINOFLAGELLATES, MARINE BACTERIA,	W70-05652
ATOMS, CHLOROPHYTA, CHRYSOPHYTA,	DINOFLAGELLATES, CYANOPHYTA.: /DI	W69-05763
, GLUCOSE, CARBON RADIOISOTOPES,	DINOFLAGELLATES, BIOASSAY, BIOIND	W70-04194
OPLANKTON, DIATOMS, ZOOPLANKTON,	DINOFLAGELLATES, ALGAE, DISSOLVED	W70-04253
ASTS, POPULATION, MARINE PLANTS,	DINOFLAGELLATES, OCEANS, BIOINDIC	W70-04368
, APHANIZOMENON, MICRONUTRIENTS,	DINOPHYCEAE, CRYPTOPHYCEAE, CH: /	W70-04381
ASSIMILATION, BACILLARIOPHYCEAE,	DINOPHYCEAE.: /M GRACILE, CARBON	W70-03983
RECTANGULARIS, DICERAS CHODATI,	DINOPHYCEAE, FILTER BLOCKING, HET	W69-07833
BESTOS CEMENT, CHLORELLA, CARBON	DIOXIDE.: /BORATORY EQUIPMENT, AS	W71-07339
COMPOSING ORGANIC MATTER, CARBON	DIOXIDE.: / POLLUTION CONTROL, DE	W72-01788
SULFATE, MANGANESE, IRON, CARBON	DIOXIDE.: /H SUBSTANCES, SEWAGE,	W70-02969
ION, *ALGAE, *NUTRIENTS, *CARBON	DIOXIDE, *REVIEWS, BIBLIOGRAPHIES	W70-00664
ULTU/ *SURFACE, *VOLUME, *CARBON	DIOXIDE, *AERATION, *NITRATES, *C	W70-03334
ELLA, *SEWAGE, EFFLUENTS, CARBON	DIOXIDE, *ALGAE, *HYDROGEN ION CO	W68-00855
ROGEN, OXYGEN, OXIDATION, CARBON	DIOXIDE, ALKALINITY, ALGAE.: /NIT	W70-05550
ON CONCENTRATION, OXYGEN, CARBON	DIOXIDE, AMMONIA, NITRITES, HYDRO	W71-03031
HELATION, PHOTOSYNTHESIS, CARBON	DIOXIDE, BICARBONATES, CARBONATES	W70-02804
TS, HYDROSTATIC PRESSURE, CARBON	DIOXIDE, CYANOPHYTA, PHYSIOLOGICA	W69-10160
NKISTRODESMUS BRAUNII, D/ CARBON	DIOXIDE, CHLORELLA PYRENOIDOSA, A	W70-05261
OLOGICAL ECOLOGY, OXYGEN, CARBON	DIOXIDE, CARBON RADIOISOTOPES, DE	W69-07862
PERATURE, PHOTOSYNTHESIS, CARBON	DIOXIDE, DEPTH, THERMOCLINE, ZOOP	W69-10169
GANISMS, LIGHT INTENSITY, CARBON	DIOXIDE, DISSOLVED OXYGEN, TEMPER	W71-00668
METRY, SAMPLING, METHANE, CARBON	DIOXIDE, DIFFUSION, *WASTE WATER	W71-11377
ORGANIC MATTER, HARDNESS, CARBON	DIOXIDE, FISH.: /Y, TEMPERATURE,	W69-10157
ONAS, CHLORELLA, ENZYMES, CARBON	DIOXIDE, GROWTH RATES, PLANKTON,	W70-04369
*ALGAE, LIGHT INTENSITY, CARBON	DIOXIDE, GROWTH RATES, TEMPERATUR	W72-06022
BACTERIA, DIEL MIGRATION, CARBON	DIOXIDE, HYDROGEN ION CONCENTRATI	W70-05760
LAMYDOMONAS, TEMPERATURE, CARBON	DIOXIDE, HYDROGEN ION CONCENTRATI	W70-02249
FIXATION, *ALGAE, LIGHT, CARBON	DIOXIDE, INHIBITION, NITROGEN FIX	W70-04249
OISOTOPES, PACIFIC OCEAN, CARBON	DIOXIDE, LIGHT, TEMPERATURE.: /DI	W70-05652
HEMISTRY, *PHYTOPLANKTON, CARBON	DIOXIDE, NUTRIENT REQUIREMENTS, P	W70-05424
TION, LIGHT, TEMPERATURE, CARBON	DIOXIDE, NUTRIENTS, BIOLOGICAL PR	W69-07442
OMPOUNDS, PHOTOSYNTHESIS, CARBON	DIOXIDE, NITRATES, PHOSPHATES.: /	W70-00161
IGHT, ALGAE, TEMPERATURE, CARBON	DIOXIDE, OXIDATION, CHEMICALS, CA	W70-05651
SOLAR RADIATION, ENERGY, CARBON	DIOXIDE, PHOTOSYNTHESIS, GROWTH R	W70-05750
, PHOTOSYNTHESIS, OXYGEN, CARBON	DIOXIDE, SCENEDESMUS, CHLOROPHYLL	W70-02964
FIXATION, PHOTOCHEMICAL, CARBON	DIOXIDE, SYNTHESIS, ENZYMES, CHLO	W70-02965
ITRIFICATION, *CORROSION, CARBON	DIOXIDE, SOLUBILITY, *PHOSPHATES,	W71-11393
EFFLUENTS, ALGAE, WEEDS, CARBON	DIOXIDE, VITAMINS, NITRATES, INOR	W70-04810
CULTURES, RADIOACTIVITY, CARBON	DIOXIDE, WATER POLLUTION EFFECTS,	W69-10158
ETINA, COCCOCHLORIS PENIOCYSTIS,	DIPLOCYSTIS AERUGINOSA, NOSTOC MU	W69-06277
IS, COSCINODISCUS, DIMEROGRAMMA,	DIPLONEIS, EUNOTOGRAMMA, FRAGILAR	W70-03325
POLLUTION EFFECTS, OLIGOCHAETES,	DIPTERA, ALGAE, INSECTS.: /WATER	W71-12077
EPILIMNION, LAKE ERIE, INSECTS,	DIPTERA, CRUSTACEA, MOLLUSCS, OLI	W70-04253

NITROGEN, PHOSPHORUS, PLANKTON,
NTRATION, TEMPERATURE, TOXICITY,
CLIMATIC ZONES, ORGANIC LOADING,
TION, CONDUCTIVITY, TEMPERATURE,
IR OPERATIONS, DISSOLVED OXYGEN,
S, SEDIM/ *RESERVOIR OPERATIONS,
ION, INDUSTRIAL WASTES, ECOLOGY,
AWRENCE RIVER, ALGAE, COLIFORMS,
, ION EXCHANGE, REVERSE OSMOSIS,
ANABAENA FLOS AQUAE, TOLYPOTHRIX
GAL ASPECTS, LEGISLATION, SEWAGE
FLORA, LIGHT INTENSITY, BIOMASS,
HINULATA, BOTRYOCOCCUS, PLANKTON
YNTHETIC LAYER, MELOSI/ SEASONAL
PLANT PHYSIOLOGY, ALGAE, DIURNAL
ALGAE, ALGAL CONTROL, ECOLOGICAL
DUCTIVITY, AQUATIC PRODUCTIVITY,
SSOLVED OXYGEN, *ALGAE, *DIURNAL
YTOPLANKTON, *LAKES, *CELLULOSE,
ICOCHEMICAL PROPERTIES, TRACERS,
A, FUNGI, AQUATIC PLANTS, ASSAY,
ENTS, NORTHEAST U.S., ECOLOGICAL
IENTS, NORTHEAST U S, ECOLOGICAL
OLOGICAL DISTRIBUTION, FREQUENCY
, LIMNOLOGY, ECOLOGY, ECOLOGICAL
ITY, TURBIDITY, SURFACES, DEPTH,
ORS, LAKE MORPHOLOGY, ECOLOGICAL
Y, NATURAL RESOURCES, ECOLOGICAL
S, *PHOSPHATES, ENERGY, DI-URNAL
PROPERTIES, ECOLOGY, ECOLOGICAL
MICAL PROPERTIES, PHYTOPLANKTON,
A, SIEROZEMS, SOIL MICROBIOLOGY,
CEAN, PUERTO RICO, SPECTROSCOPY,
OVAL, *ADSORPTION, SOILS, COALS,
ATIC PLANTS, ISLANDS, ECOLOGICAL
EATMENT, LAKES, SEWAGE, DRAINAGE
SE RAYON, ZINC COMPOUNDS, CARBON
VATED ALGAE, FEEDLOTS, OXIDATION
CREEK(WIS), DUTCH MILL DRAINAGE
HEPPARTION(AUSTRALIA), OXIDATION
, THALASSIOSI/ *ALGAL NUTRITION,
LTURES, PLANT PHYSIOLOGY, ALGAE,
                       SEEPAGE LOSSES,
A, TEXAS, PHOSPHORUS, MIGRATION,
ESTS, BIOCHEMICAL OXYGEN DEMAND,
SAMPLING, WASTE WATER DISPOSAL,
S, INTERNAL WATER, WAVES(WATER),
ITY, CHLOROPHYLL, ICE, SEASONAL,
IGMENTS, OCEANS, ORGANIC MATTER,
TER, LITTORAL, DISSOLVED OXYGEN,
ODOR, TASTE, DEPTH, CIRCULATION,
NSPARENCY, CYCLOP/ *PHENYLUREAS,
ENT, *LAKE WASHINGTON/ *EFFLUENT
AKES), PHAEDACTYLUM, WASTE WATER
ECIPITATION, HARVESTING, ZONING,
ER POLLUTION CONTROL, LIMNOLOGY,
OINDICATORS, ALGAE, OLIGOTROPHY,
BAY(ORE), CAROTENOIDS, SPECIES,
T CORRELATION, DECAPODS, SPECIES
DUCTIVITY, DISTRIBUTION, SPECIES
CULTURAL EUTROPHICATION, SPECIES
BLOOMS, *BIOTIC FACTORS, SPECIES
, EXPORT LOSSES, CHEMOS/ SPECIES
KJANES, FAXAFLOI FJORD, ICELAND,
OSPH/ *EUTROPHICATION, *REVIEWS,
                       *ANABAENA
IVER, DURANGO, SAN MIGUEL RIVER,
NANNOPLANKTON, NUTRIENTS, DEPTH,
SNAILS, ECOLOGY, EUTROPHICATION,
AQUACULTURE, *ULTIMATE DISPOSAL,
TREATMENT, PHOSPHOROUS, RUNOFF,
TOMS, FOULING, PLANKTON, SEWAGE,
LAKES, GLACIATION, URBANIZATION,
WATER QUALITY, LAKE MORPHOMETRY,
TROPHICATION, *ALGAE, NUTRIENTS,
, PHOSPHATES, *SEWAGE TREATMENT,
OGEN, SEWAGE, INDUSTRIAL WASTES,
ENTS, TERTIARY TREATMENT, ALGAE,
CATION, PHOSPHORUS, *DETERGENTS,
, *WATER WORKS, *WASTE DISPOSAL,
T FACILITIES, AERATION, LAGOONS,
TON, PHYSICOCHEMICAL PROPERTIES,
ALGAE, WATER POLLUTION EFFECTS,
ALITY, ALGAL CONTROL, ALGICIDES,
ASTS, PYRITE, ALGAE, CYANOPHYTA,
K(WIS), NINE SPRINGS CREEK(WIS),
*CRITICAL TEMPERATURES, *COPPER
A, ENDOTOXIN, FAST-DEATH FACTOR,
MENON FLOS-AQUAE, MINIMUM LETHAL
INDUSTRIAL PHOTOSYNTHESIS, MEAN

DISSOLVED OXYGEN, HYPOLIMNION, GR   W68-00253
DISSOLVED OXYGEN, LIGHT INTENSITY   W72-06854
DISSOLVED OXYGEN, HYDROGEN ION CO   W72-10573
DISSOLVED OXYGEN, PHYTOPLANKTON,   W72-06051
DISSOLVED SOLIDS, SEDIMENTATION R   W73-00750
DISSOLVED OXYGEN, DISSOLVED SOLID   W73-00750
DISSOLVED OXYGEN, HARDNESS, TIDAL   W73-00748
DISSOLVED: /ES, AGRICULTURE, ST L   W70-03964
DISTILLATION, ELECTRODIALYSIS, SO   W70-01981
DISTORTA, GLOEOTRICHIA ECHINULATA   W70-05091
DISTRI: /WAY, ESTIMATED COSTS, LE   W70-03344
DISTRIBUTION, ECOSYSTEMS, POPULAT   W70-03325
DISTRIBUTION, WISCONSIN LAKE SURV   W70-03959
DISTRIBUTION, LAKE BAIKAL, PHOTOS   W70-04290
DISTRIBUTION, OXYGEN, DISSOLVED O   W70-05094
DISTRIBUTION, ACTIVATED SLUDGE, M   W70-04786
DISTRIBUTION, SPECIES DIVERSITY.:   W70-05424
DISTRIBUTION, *CALIFORNIA, ESTIMA   W69-03611
DISTRIBUTION PATTERNS, MEMBRANE P   W69-00976
DISTRIBUTION, CHELATION, DETERGEN   W69-08275
DISTRIBUTION, DENSITY, CHROMATOGR   W70-02510
DISTRIBUTION, FREQUENCY DISTRIBUT   W70-02772
DISTRIBUTION, BIOLOGICAL INDICATO   W70-02764
DISTRIBUTION, POPULATION, BIOLOGI   W70-02772
DISTRIBUTION.: /RONMENTAL EFFECTS   W70-02784
DISTRIBUTION, TIME, EPILIMNION, T   W71-07698
DISTRIBUTION, MIDGES, WATER POLLU   W71-12077
DISTRIBUTION, STOCHASTIC PROCESSE   W71-09012
DISTRIBUTION, TEMPERATURE, METABO   W70-06604
DISTRIBUTION, THERMAL EFFECTS.: /   W70-07313
DISTRIBUTION, DEPTH, OXYGEN, ALKA   W71-00665
DISTRIBUTION PATTERNS, SOIL PROFI   W71-04067
DISTRIBUTION, CALCIUM, TROPHIC LE   W69-08525
DISTRIBUTION PATTERNS, ACTIVATED   W69-09884
DISTRIBUTION, BIOLOGICAL COMMUNIT   W72-10623
DISTRICTS, DEBT, COSTS, CITIES.: /   W70-04455
DISULFIDE.: /NTS, NYLON-6, *VISCO   W70-06925
DITCH.:   *FWPCA, ACTI   W71-06821
DITCH(WIS), LAKE MONONA(WIS), LAK   W70-05113
DITCHES, SOUP, FOOD WASTES.:   S   W71-09548
DITYLUM, CHAETOCEROS, SKELETONEMA   W70-04503
DIURNAL DISTRIBUTION, OXYGEN, DIS   W70-05094
DIURNAL VARIATIONS.:   W69-09723
DIURNAL, ALGICIDES, CLOSTRIDIUM,   W70-05372
DIURNAL, CHLOROPHYLL.: /ON-SITE T   W72-01818
DIURNAL, EFFLUENTS, NITROGEN, PHO   W69-10159
DIURNAL, EPILIMNION, WINDS, THERM   W70-09889
DIURNAL, LIGHT INTENSITY, CARBON,   W70-05387
DIURNAL, LATITUDINAL STUDIES.: /P   W70-04809
DIURNAL, MIGRATION PATTERNS.: /WA   W71-12870
DIURNAL, WINDS, CURRENTS(WATER),   W70-05404
DIURON, FLAGELLATES, MONURON, TRA   W70-10161
DIVERSION, *WATER QUALITY MANAGEM   W69-09349
DIVERSION, INSTREAM TREATMENT, BO   W69-09340
DIVERSION, DREDGING, IMPOUNDMENTS   W69-10178
DIVERSION, STREAMS.: /OURCES, WAT   W69-10170
DIVERSION, CORES, SEWAGE EFFLUENT   W70-05663
DIVERSITY INDEX, ACHNANTHES, ACTI   W70-03325
DIVERSITY INDEX.: / ANALYSIS, PAR   W73-01446
DIVERSITY.: /CTIVITY, AQUATIC PRO   W70-05424
DIVERSITY, CHEMICAL PRECIPITATION   W69-08518
DIVERSITY, LAKE WAUBESA, LYNGBYA,   W69-01977
DIVERSITY, AUTECOLOGY, MORPHOLOGY   W70-02780
DNA BASE COMPOSITION.: / CAPE REY   W69-10161
DOCUMENTATION, BIBLIOGRAPHIES, PH   W69-08518
DOLIOLUM.:   W69-10177
DOLORES RIVER, URAVAN, COTTUS SPP   W69-07846
DOMESTIC ANIMALS, SAMPLING, NUISA   W70-00274
DOMESTIC WASTES, SHALLOW WATER, C   W70-00711
DOMESTIC WASTES, INDUSTRIAL WASTE   W70-09905
DOMESTIC WASTES, DETERGENTS, INDU   W71-04072
DOMESTIC WASTE, INDUSTRIAL WASTES   W72-01813
DOMESTIC WASTES, NUTRIENTS, PHOSP   W69-07818
DOMESTIC WASTES, PLANT GROWTH SUB   W68-00470
DOMESTIC WASTES, *SEWAGE TREATMEN   W68-00248
DOMESTIC WASTES, *ALGAE, WATER PR   W68-00256
DOMESTIC WASTES, SEWAGE TREATMENT   W68-00478
DOMESTIC WASTES.: /KES, *IMPOUNDM   W70-04765
DOMESTIC WASTES, POLLUTANTS, WATE   W70-05084
DOMESTIC WASTES, SEWERS, SEWAGE T   W70-03344
DOMESTIC WASTES, SEDIMENTATION, C   W70-04060
DOMESTIC WASTES, SEWAGE, INDUSTRI   W70-03964
DOMESTIC WASTES, INDUSTRIAL WASTE   W72-12240
DOMINANT ORGANISMS, GRAZING, WATE   W69-01977
DOMINANT ORGANISM, NUTRIENTS, RAI   W68-00891
DOOR CREEK(WIS), DUTCH MILL DRAIN   W70-05113
DOSAGE, *COPPER APPLICATION METHO   W69-10157
DOSAGE, SPECIES SUSCEPTIBILITY, A   W70-00273
DOSES, CHU'S MEDIA, QUINONE DERIV   W70-02982
DOUBLING TIME, PACKED CELL VOLUME   W71-05267

PITATION, LIFE CYCLES, GENETICS,
E, NUTRIENTS, VIRUSES, BACTERIA,
IENT REQUIREMENTS, PHYSIOLOGICAL
FFECTS, LAKE MENDOTA(WIS), BLACK
ECHNIQUES, TURBID/ *DIATOMACEOUS
ATTER, *FILTRATION, DIATOMACEOUS
ESTS, *SCENEDESMUS, DIATOMACEOUS
LGAE, *FILTRATION, *DIATOMACEOUS
AE, ACANTHURUS TRIOSTEGUS, CONUS
POLYCHAETA, CRUSTACEA, MOLLUSCA,
OISOTOPES, CERIUM RADIOISOTOPES,
ZOMENON FLOS-AQUAE, GLOEOTRICHIA
BAENA, MICROCYSTIS, GLOEOTRICHIA
ZOMENON FLOS AQUAE, GLOEOTRICHIA
LYPOTHRIX DISTORTA, GLOEOTRICHIA
NABAENA CIRCINALIS, GLOEOTRICHIA
    EUTROPHICATION PREVENTION,
        *INDIA, MOOSI RIVER
ONDITIONS, ALGAE, ALGAL CONTROL,
ALITY, NUTRIENTS, NORTHEAST U S,
VELS, NUTRIENTS, NORTHEAST U.S.,
TAL EFFECTS, LIMNOLOGY, ECOLOGY,
OBTUSA, SCENEDESMUS QUADRICAUDA,
, TECHNOLOGY, NATURAL RESOURCES,
BIOINDICATORS, LAKE MORPHOLOGY,
E, CHEMICAL PROPERTIES, ECOLOGY,
TORAL, *AQUATIC PLANTS, ISLANDS,
*HYDRAULIC MODEL/ *PHYSIOLOGICAL
*MORPHOLOGY, *AL/ *PHYSIOLOGICAL
IVERS, ENVIRONME/ *PHYSIOLOGICAL
EDS, *INHIBITION, *PHYSIOLOGICAL
WAT/ *CYANOPHYTA, *PHYSIOLOGICAL
TIC MICROBIOLOGY, *PHYSIOLOGICAL
S, / *CYANOPHYTA, *PHYSIOLOGICAL
*EUTROPHICATION, *PHYSIOLOGICAL
ION, *CYANOPHYTA, *PHYSIOLOGICAL
ENT REQUIREMENTS, *PHYSIOLOGICAL
ION, *PHOSPHATES, *PHYSIOLOGICAL
ALGAE, ECOSYSTEMS, PHYSIOLOGICAL
NTS, AQUATIC LIFE, ALGAE, SILTS,
EASONAL, RHODE ISLAND, SAMPLING,
UTION ABATEMENT, BIBLIOGRAPHIES,
ECOLOGICAL EFFICIENCIES, TROPHIC
ATS, EPIPHYTOLOGY, MARINE ALGAE,
LGAE, *PLANT GROWTH, *WISCONSIN,
RIPHYTON, BIOMASS, PHYSIOLOGICAL
LANKTON, *NUTRIENT REQUIREMENTS,
NOLOGY, NUTRIENTS, PHYSIOLOGICAL
ELLA, CHLOROPHYLL, PHYSIOLOGICAL
C BACTERIA, ALGAE, PHYSIOLOGICAL
BON UPTAKE TECHNIQUES, MICROBIAL
RA, GYRODINIUM COHNII, MICROBIAL
M, CHEMICAL SYMBIOSIS, MICROBIAL
, WATER POLLUTION, PHYSIOLOGICAL
REPRODUCTION, INDUSTRIAL WASTES,
TENSITY, SEASONAL, PHYSIOLOGICAL
LCIUM, CHLORINE, COBALT, COPPER,
PLANTS, WATER POLLUTION EFFECTS,
NVIRONMENTAL EFFECTS, LIMNOLOGY,
ONTARIO, LAKE HURON, AESTHETICS,
EMPERATURE, CHEMICAL PROPERTIES,
CONTROL, *REVIEWS, ALGAE, WEEDS,
T, CRAYFISH, STONEFLIES, SNAILS,
*MARINE ENVIRONMENTS, MICROBIAL
PHYTA, RHODOPHYTA, PHYSIOLOGICAL
SSAY, SCENEDESMUS, PHYSIOLOGICAL
TROPHIC LEVEL, LIGHT INTENSITY,
MS, RIVERS, LAKES, PHYSIOLOGICAL
INTERTIDAL AREAS, PHYSIOLOGICAL
ERATURE, *DAPHNIA, PHYSIOLOGICAL
, *BIOLOGICAL UPTAKE, *MICROBIAL
IATOMS, ESTUARIES, PHYSIOLOGICAL
COMMUNITIES, CULTURES, DIATOMS,
IUM, EPIPHYTOLOGY, MARINE ALGAE,
CHEMICAL ANALYSIS, PHYSIOLOGICAL
NE MICROORGANISMS, PHYSIOLOGICAL
IENT REQUIREMENTS, PHYSIOLOGICAL
EMENTS, NUTRIENTS, PHYSIOLOGICAL
HORUS, PHOSPHATES, PHYSIOLOGICAL
TROPHICATION, *NUTRIENTS, LAKES,
MPERATURE, SILTS, STANDING CROP,
OSYNTHETIC OXYGEN, PHYSIOLOGICAL
ESTICIDE RESIDUES, PHYSIOLOGICAL
POPULATION EQUIVALEN/ *MICROBIAL
YTA, SCENEDESMUS, CHLAMYDOMONAS,
YTICAL TECHNIQUES, PHYSIOLOGICAL
EMICAL PROPERTIES, PHYSIOLOGICAL
WEDEN), ANNUAL CYCLES, MICROBIAL
*ORGANIC MATTER, *PHYSIOLOGICAL
OLUTES, *HETEROTROPHY, MICROBIAL

E. COLI, ISOLATION.: /MICAL PRECI   W71-11823
E. COLI, PUBLIC HEALTH, PATHOGENI   W72-06017
E: /F ALGAE, NUISANCE ALGAE, NUTR   W69-03362
EARTH CREEK(WIS).: /WASTE WATER E   W70-04382
EARTH, *FILTRATION, *SEPARATION T   W72-08686
EARTH, ALGAE, TREATMENT FACILITIE   W72-04318
EARTH, MUD, LAKES, CULTIVATION, C   W68-00475
EARTH, SEPARATION TECHNIQUES, PRE   W72-08688
EBRAEUS.: / FUNGITES, ZOOXANTHELL   W69-08526
ECHINODERMATA, PACIFIC OCEAN.: /   W69-08526
ECHINODERMS.: /OPES, CADMIUM RADI   W71-09850
ECHINULATA, COCCOCHLORIS PENIOCYS   W70-00719
ECHINULATA, DINOBRYON, SYNURA UVE   W70-02803
ECHINULATA, BOTRYOCOCCUS, PLANKTO   W70-03959
ECHINULATA LB 1303.: /S AQUAE, TO   W70-05091
ECHINULATE, OSCILLATORIA RUBESCEN   W69-05704
ECOLOGIC IMBALANCES.:   W69-07084
ECOLOGICAL STUDY.:   W69-07096
ECOLOGICAL DISTRIBUTION, ACTIVATE   W70-04786
ECOLOGICAL DISTRIBUTION, BIOLOGIC   W70-02764
ECOLOGICAL DISTRIBUTION, FREQUENC   W70-02772
ECOLOGICAL DISTRIBUTION.: /RONMEN   W70-02784
ECOLOGICAL EFFICIENCIES, TROPHIC   W69-10152
ECOLOGICAL DISTRIBUTION, STOCHAST   W71-09012
ECOLOGICAL DISTRIBUTION, MIDGES,   W71-12077
ECOLOGICAL DISTRIBUTION, THERMAL   W70-07313
ECOLOGICAL DISTRIBUTION, BIOLOGIC   W72-10623
ECOLOGY, *BENTHIC FLORA, *ALGAE,   W71-00668
ECOLOGY, *ENVIRONMENTAL EFFECTS,   W69-10177
ECOLOGY, *AQUATIC ENVIRONMENTS, R   W70-03978
ECOLOGY, *PHYTOTOXICITY, ALGAL CO   W69-03188
ECOLOGY, *SALINE WATER, *THERMAL   W69-03516
ECOLOGY, *CYANOPHYTA, ALGAE, ANAL   W69-03518
ECOLOGY, *PHYTOPLANKTON, *CULTURE   W69-03514
ECOLOGY, *WATER POLLUTION EFFECTS   W69-03513
ECOLOGY, ALGAE, ANALYTICAL TECHNI   W69-03517
ECOLOGY, ALGAE, ANALYTICAL TECHNI   W69-03515
ECOLOGY, AQUATIC ALGAE, AQUATIC M   W69-03358
ECOLOGY, AQUATIC BACTERIA, CARBON   W70-04287
ECOLOGY, AQUATIC ENVIRONMENT.: /A   W69-10080
ECOLOGY, AMMONIA, ENZYMES, CLASSI   W70-00713
ECOLOGY, BIOLOGY, FISH, PLANKTON,   W72-01786
ECOLOGY, CLADOCERA.: /ADRICAUDA,   W69-10152
ECOLOGY, CULTURES, WASHINGTON, WA   W69-10161
ECOLOGY, CHEMICAL ANALYSIS, PHYTO   W70-06229
ECOLOGY, CYANOPHYTA, DIFFUSION, R   W70-03309
ECOLOGY, CHLORELLA, ALGAE, ORGANI   W70-02804
ECOLOGY, CHLORELLA, SCENEDESMUS.:   W70-02245
ECOLOGY, CYANOPHYTA, EUTROPHICATI   W70-04381
ECOLOGY, CARBON RADIOISOTOPES, LA   W70-05109
ECOLOGY, COCCOLITHOPHORES, SEAWAT   W70-05272
ECOLOGY, COULTER COUNTER, GROWTH   W70-05652
ECOLOGY, DINOBRYON, VOLVOX, PANDO   W70-04503
ECOLOGY, DIATOMS, OLIGOTROPHY, CY   W70-04503
ECOLOGY, DISSOLVED OXYGEN, HARDNE   W73-00748
ECOLOGY, ECOTYPES, CHLOROPHYTA, P   W70-05409
ECOLOGY, ENVIRONMENTAL EFFECTS, G   W69-04801
ECOLOGY, EFFECTS OF POLLUTION, WA   W70-02772
ECOLOGY, ECOLOGICAL DISTRIBUTION.   W70-02784
ECOLOGY, ECONOMICS, OHIO, COMMERC   W70-01943
ECOLOGY, ECOLOGICAL DISTRIBUTION,   W70-07313
ECOLOGY, ENVIRONMENT, CHEMCONTROL   W70-07393
ECOLOGY, EUTROPHICATION, DOMESTIC   W70-00711
ECOLOGY, FLUORESCENCE MICROSCOPY,   W70-04811
ECOLOGY, INDUSTRIAL WASTES, HYDRO   W70-04510
ECOLOGY, KINETICS, FARM WASTES, B   W70-02775
ECOLOGY, LOTIC ENVIRONMENT, BIOMA   W70-00265
ECOLOGY, LIMITING FACTORS, ECOSYS   W71-02479
ECOLOGY, LITTORAL, ESTUARIES, PHO   W70-03325
ECOLOGY, LAKES, ENVIRONMENTAL EFF   W70-03957
ECOLOGY, MAXIMUM UPTAKE VELOCITY,   W70-04194
ECOLOGY, MASSACHUSETTS, CHLOROPHY   W70-05272
ECOLOGY, NUISANCE ALGAE, PHAEOPHY   W69-07832
ECOLOGY, NUTRIENTS, SEAWATER, ORG   W69-09755
ECOLOGY, OXYGEN, CARBON DIOXIDE,   W69-07862
ECOLOGY, ORGANIC MATTER, WATER PO   W70-04811
ECOLOGY, PHYTOPLANKTON, SE: /NUTR   W69-03371
ECOLOGY, PH: /ON, NUTRIENT REQUIR   W69-03364
ECOLOGY, PHYTOPLAN: /IENTS, PHOSP   W69-03373
ECOLOGY, PRIMARY PRODUCTIVITY, AL   W71-00665
ECOLOGY, PHOTOSYNT: /DYNAMICS, TE   W70-02780
ECOLOGY, PHYTOPLANKTON, PLANKTON,   W70-02792
ECOLOGY, PHYTOPLANKTON, PHYTOTOXI   W70-02968
ECOLOGY, PURPLE SULFUR BACTERIA,   W70-03312
ECOLOGY, PHOSPHORUS, SULPHUR, HYD   W70-02965
ECOLOGY, PLANT PHYSIOLOGY.: /ANAL   W72-12240
ECOLOGY, RHODOPHYTA, CHLOROPHYA.   W69-10161
ECOLOGY, RADIOCARBON UPTAKE TECHN   W70-05109
ECOLOGY, RADIOACTIVITY TECHNIQUES   W70-04194
ECOLOGY, SWEDEN, LAKE ERKEN(SWEDE   W70-04284

OPERTIES, BIOLOGICAL PROPERTIES,
ESOTA, *LAKES, *DIATOMS, *ALGAE,
LING WATER, CHEMICAL PROPERTIES,
IOISOTOPES, ALGAE, BIOCHEMISTRY,
ICHIGAN, PHYTOPLANKTON, DIATOMS,
*SUCESSION, *ISOLATION, DIATOMS,
OXIDE, CYANOPHYTA, PHYSIOLOGICAL
CYCLING NUTRIENTS, PHYSIOLOGICAL
PLANTS, WATER POLLUTION EFFECTS,
HEMICAL OXYGEN DEMAND, BACTERIA,
AGOONS, OXYGENATION, ABSORPTION,
EWAGE DISPOSAL, WATER CHEMISTRY,
AKES, GREAT LAKES, PHYSIOLOGICAL
AKES, GREAT LAKES, PHYSIOLOGICAL
CS, ALGAL COMMUNITIES, MICROBIAL
SPHORUS COMPOUNDS, PHYSIOLOGICAL
SLUDGE DRYING, SOLIDS-SETTLING,
ICATION, FLOCCULATION, TURBINES,
COLLOIDS, FLOCCULATION, MIXING,
, AQUATIC PLANTS, AQUATIC WEEDS,
COPPER SULFATE, ODOR, COLIFORMS,
LAKE HURON, AESTHETICS, ECOLOGY,
ING NUTRIENTS, COST COMPARISONS,
RITISH COLUMBIA, CANADA, AQUATIC
RATIONS, MICRONUTRIENTS, AQUATIC
, *REEFS, *MARINE ANIMALS, FISH,
S, SYSTEMS ANALYSIS, SUCCESSION,
FUSION, RESPIRATION, PHOSPHATES,
ORY, MINERALS,/ ALGAE, BRACKISH,
TIC PLANTS, AQUATIC ENVIRONMENT,
TRIBUTION, STOCHASTIC PROCESSES,
IDES, *VASCULAR TISSUES, *ALGAE,
GICAL ECOLOGY, LIMITING FACTORS,
DICATORS, CHLAMYDOMONAS, CARBON,
, DETRITUS, ESTUARIES, PLANKTON,
ES, STREAMFLOW, SAMPLING, ALGAE,
STACEANS, BENTHOS, BRINE SHRIMP,
COMPOUNDS, MOLYBDENUM, ROTIFERS,
, FISH, RADIOECOLOGY, EFFLUENTS,
ASHINGTON, PRIMARY PRODUCTIVITY,
IA, FLORIDA, PESTICIDE RESIDUES,
EW YORK, CHLOROPHYTA, NUTRIENTS,
WATER POLLUTION EFFECTS, ALGAE,
TR/ *LAKES, STREAMS, FISH, LAND,
NTENSITY, BIOMASS, DISTRIBUTION,
HLORRELA, SCENEDESMUS, KINETICS,
ANTS, *ESTUARIES, *TIDAL WATERS,
SSILE ALGAE, *CYCLING NUTRIENTS,
ROCARBONS, BACTERIAL ATTACHMENT,
SEASONAL, PHYSIOLOGICAL ECOLOGY,
SCICULOSUS, ASCOPHYLLUM NODOSUM,
RITUS, BOTTOM SEDIMENTS, CHUTES,
COMPLEXONES, *ALGAL BLOOMS, NTA,
AKE, *PHAEODACTYLUM TRICORNUTUM,
A(CANADA), PENGUINS, LAK/ PRINCE
RIVER(CALIF), CLEAR LAKE(CALIF),
S, CHARA, ALKALINITY, BULLHEADS,
ECONOMIC ADVANTAGE, TEMPERATURE-
EL, SIMULATION MODEL, PHOSPHORUS
ATER POLLUTION EFFECTS, ECOLOGY,
UTION TREATMENT, WATER POLLUTION
IA, ORGANIC AGGREGATES, SEASONAL
S, IONIC CONCENTRATION, SEASONAL
GEN, PHOSPHORUS, WATER POLLUTION
, AQUATIC ALGAE, WATER POLLUTION
URES, CHLORELLA, WATER POLLUTION
SOTA, NUTRIENTS, WATER POLLUTION
MAC RIVER, SAVANNAH RIVER, SHOCK
ECOLOGICAL DISTRIBUTION, THERMAL
LLUTION SOURCES, WATER POLLUTION
SHINGTON, LAKES, WATER POLLUTION
T RADIOISOTOPES, WATER POLLUTION
LGAE, LIVESTOCK, WATER POLLUTION
TER, POPULATION, WATER POLLUTION
QUATIC HABITATS, WATER POLLUTION
ITION), METHANE, WATER POLLUTION
QUATIC BACTERIA, WATER POLLUTION
ALGAE, BACTERIA, WATER POLLUTION
ICAL PROPERTIES, WATER POLLUTION
WATER CHEMISTRY, WATER POLLUTION
LLUTION CONTROL, WATER POLLUTION
DELAWARE RIVER, WATER POLLUTION
XYGEN, PLANKTON, WATER POLLUTION
KTON, *REVIEWS, *WATER POLLUTION
                       PHYSIOLOGICAL
ABORATORY TESTS, WATER POLLUTION
LUTION SOURCES, *WATER POLLUTION
TY, IRRIGATION EFFECTS, DRAINAGE
ROPH/ *STREAMS, *WATER POLLUTION
LUTION SOURCES, *WATER POLLUTION

ECOLOGY, STRATIFICATION.: /TER PR   W70-04475
ECOLOGY, SOUTH DAKOTA, FIELD INVE   W69-00659
ECOLOGY, SALINITY, BENTHOS, AQUAT   W69-05023
ECOLOGY, SAMPLING, THERMAL SPRING   W69-10151
ECOLOGY, SOIL, CULTURES, CYANOPHY   W69-10180
ECOLOGY, SALINITY, ORGANIC MATTER   W73-01446
ECOLOGY, THERMAL WATER.: /RBON DI   W69-10160
ECOLOGY, WATER: /ALGAE, DIATOMS,   W70-00713
ECOLOGY, WATER QUALITY CHARACTERI   W70-02764
ECOLOGY, WATER POLLUTION EFFECTS,   W71-04758
ECOLOGY, WATER PURIFICATION, CARB   W69-07389
ECOLOGY, WAT: / WASTE DISPOSAL, S   W69-04276
ECOLOGY, WATER CHEMISTRY, GEOGRAP   W70-03975
ECOLOGY, WATER CHEMISTRY, GEOGRAP   W70-03975
ECOLOGY, WATER QUALITY CRITERIA,   W70-04510
ECOLOGY,: / ALGAE, NUTRIENTS, PHO   W69-04802
ECONOMIC ADVANTAGE, TEMPERATURE-E   W70-05819
ECONOMIC EFFICIENCY, HEAT BUDGET,   W70-08834
ECONOMIC FEASIBILITY, TECHNICAL F   W71-10654
ECONOMIC FEASIBILITY, LAKES, NUIS   W69-06276
ECONOMICS.: /OLLUTION TREATMENT,   W72-01814
ECONOMICS, OHIO, COMMERCIAL FISHI   W70-01943
ECONOMICS, ALGAE, ALGAL CONTROL,   W69-08518
ECOSYSTEM MODEL, SIMULATION MODEL   W69-06536
ECOSYSTEM, MIDGE LARVAE, LAKE STR   W69-09340
ECOSYSTEMS, CYCLING NUTRIENTS, TE   W69-08526
ECOSYSTEMS, CYCLING NUTRIENTS.: /   W70-02779
ECOSYSTEMS, ORGANIC MATTER, WATER   W70-03309
ECOSYSTEMS, FAUNA, FLORA, LABORAT   W72-02203
ECOSYSTEMS, AQUATIC HABITATS, DRA   W71-09005
ECOSYSTEMS, AQUATIC HABITATS, WAT   W71-09012
ECOSYSTEMS, CHARA, BENTHIC FLORA,   W70-10175
ECOSYSTEMS, MICROORGANISMS, AQUAT   W71-02479
ECOSYSTEMS, EUGLENA, EUTROPHICATI   W69-06273
ECOSYSTEMS, SINKS, TRACERS, CRABS   W69-08274
ECOSYSTEMS, ENVIRONMENT, RETENTIO   W69-07862
ECOSYSTEMS, FISH, PHYTOPLANKTON,   W69-07440
ECOSYSTEMS, PHOTOS: /ITE, SULFUR   W69-07428
ECOSYSTEMS, PLANKTON, ALGAE, PERI   W69-07853
ECOSYSTEMS, ILLINOIS, PHOSPHORUS,   W70-04506
ECOSYSTEMS, FOOD CHAINS, BIOLOGIC   W70-05272
ECOSYSTEMS, PHYTOPLANKTON, PHOTOS   W70-03973
ECOSYSTEMS, PHYSIOLOGICAL ECOLOGY   W70-04287
ECOSYSTEMS, EUTROPHICATION, OLIGO   W70-04193
ECOSYSTEMS, POPULATION, OREGON, I   W70-03325
ECOSYSTEMS.: /TES, *TURBULENCE, C   W69-03730
ECOSYSTEMS, ESTUARINE ENVIRONMENT   W68-00010
ECOSYSTEMS, PHOSPHORUS, SCENEDESM   W73-01454
ECOTOCARPUS, POLYSIPHONIA, ORGANI   W70-01068
ECOTYPES, CHLOROPHYTA, PHOTOMETRY   W70-05409
ECTOCARPUS CONFERVOIDES, NYLON CO   W70-01074
EDDIES, HABITATS, CRAYFISH, GROWT   W69-09742
EDTA, HEIDA, LAKE TAHOE(CALIF).: /   W70-04373
EDTA, LIGURIAN SEA, EUROPEAN MEDI   W69-08275
EDWARD ISLAND(CANADA), NOVA SCOTI   W70-05399
EEL RIVER(CALIF), JOINT INDUSTRY/   W70-02777
EELS, KILLIFISHES, TROUT, SUCKERS   W70-05263
EFFECT.: /RYING, SOLIDS-SETTLING,   W70-05819
EFFECT, CLADECERA.: /COSYSTEM MOD   W69-06536
EFFECTS OF POLLUTION, WATER QUALI   W70-02772
EFFECTS.: /ERSISTENCE, WATER POLL   W70-01777
EFFECTS.: /COTOCARPUS, POLYSIPHON   W70-01068
EFFECTS.: /NOMICTIC, DEPTH EFFECT   W70-01933
EFFECTS.: /ERTILITY, ALGAE, NITRO   W70-02973
EFFECTS.: /DICATORS, AQUATIC LIFE   W70-05958
EFFECTS.: / WATER TREATMENT, CULT   W70-09438
EFFECTS.: /ABORATORY TESTS, MINNE   W70-08637
EFFECTS.: /ZOAN COMMUNITIES, POTO   W70-07847
EFFECTS.: / PROPERTIES, ECOLOGY,   W70-07313
EFFECTS.: /, CYANOPHYTA, WATER PO   W69-10180
EFFECTS.: /US, EUTROPHICATION, WA   W70-00270
EFFECTS.: /E RADIOISOTOPES, COBAL   W71-09850
EFFECTS.: /ITROGEN, PHOSPHORUS, A   W71-10376
EFFECTS.: /KTON, ALGAE, HEATED WA   W71-09767
EFFECTS.: /OCESSES, ECOSYSTEMS, A   W71-09012
EFFECTS.: / *DEGRADATION(DECOMPOS   W71-13413
EFFECTS.: /S, YEASTS, BACTERIA, A   W69-06970
EFFECTS.: /HIES, LAKES, NUISANCE   W69-08283
EFFECTS.: /IDE, NUTRIENTS, BIOLOG   W69-07442
EFFECTS.: / NUTRIENTS, TOXICITY,   W69-05867
EFFECTS.: /LITY CONTROL, WATER PO   W69-05705
EFFECTS.: /E RIVER, HUDSON RIVER,   W70-04507
EFFECTS.: /NTS, PHOTOSYNTHETIC, O   W69-03514
EFFECTS.: /OLIGOTROPHY, PHYTOPLAN   W68-00680
EFFECTS.:   W68-00859
EFFECTS.: /N PESTICIDES, ALGAE, L   W72-11105
EFFECTS, *WATER SOURCES, *GROUNDW   W73-00758
EFFECTS, *SAMPLING, ALGAE TOXINS,   W69-00360
EFFECTS, *CYCLING NUTRIENTS, *EUT   W69-03369
EFFECTS, *PRIMARY PRODUCTIVITY, A   W69-02959

434

```
     *ANAEROBIC WASTE PONDS, *DEPTH   EFFECTS, *HYDRAULIC LOAD EFFECTS,    W70-04787
IC LOAD EFFECTS, *DETENTION TIME      EFFECTS, *FACULATIVE PONDS, *AERO    W70-04787
ULIC LOAD EFFECTS, *ORGANIC LOAD      EFFECTS, *DETENTION TIME EFFECTS,    W70-04787
   *DEPTH EFFECTS, *HYDRAULIC LOAD     EFFECTS, *ORGANIC LOAD EFFECTS, *   W70-04787
LIZATION, *PONDS, *ENVIRONMENTAL      EFFECTS, *BIOLOGICAL TREATMENT, *    W70-04787
LANKTON, *DDT,/ *WATER POLLUTION      EFFECTS, *PHOTOSYNTHESIS, *PHYTOP    W70-05272
ATER CHEMISTRY, *WATER POLLUTION      EFFECTS, *PHOSPHATES, *ANALYTICAL    W70-05269
ENTS, POPULATIONS, ENVIRONMENTAL      EFFECTS, *PRIMARY PRODUCTIVITY, W    W70-05424
YCLING NUTRIENTS, *ENVIRONMENTAL      EFFECTS, *ESSENTIAL NUTRIENTS, *F    W69-04804
LAND WATERWAYS, *WATER POLLUTION      EFFECTS, *WATER SUPPLY, BIOINDICA    W69-07833
LUTION SOURCES, *WATER POLLUTION      EFFECTS, *EUTROPHICATION, *NUTRIE    W70-04266
LUTION SOURCES, *WATER POLLUTION      EFFECTS, *WATER POLLUTION CONTROL    W70-03975
OPHICATI/ *ALGAE, *ENVIRONMENTAL      EFFECTS, *HUMAN POPULATION, *EUTR    W70-03973
LUTION SOURCES, *WATER POLLUTION      EFFECTS, *WATER POLLUTION CONTROL    W70-03975
LUTION SOURCES, *WATER POLLUTION      EFFECTS, *CLASSIFICATION, *WATER     W72-00462
R QUALITY, *DISSOLVED SOLIDS, D/      EFFECTS, *IMPOUNDED WATERS, *WATE    W72-01773
REAT LAKES, TR/ *WATER POLLUTION      EFFECTS, *FALLOUT, *ECOSYSTEMS, G    W71-09013
AL CHEMICALS, *DELTAS, *DRAINAGE      EFFECTS, *WATER QUALITY, *FARM WA    W71-09577
WAGE, *STREAMS, *WATER POLLUTION      EFFECTS, *FISH, ALGAE, TROPHIC LE    W72-01788
ARBORS, *SEWAGE, WATER POLLUTION      EFFECTS, *ALGAE.: *ESTUARIES, *H     W71-05500
OLOGICAL ECOLOGY, *ENVIRONMENTAL      EFFECTS, *MORPHOLOGY, *ALGAE, AQU    W69-10177
ASHINGTON, *SEWAGE FERTILIZATION      EFFECTS, *BIOLOGICAL PRODUCTIVITY    W69-09349
UALITY, *DISSO/ *WATER POLLUTION      EFFECTS, *BIOINDICATORS, *WATER Q    W70-00475
E, *CHLOROPHYLL, WATER POLLUTION      EFFECTS, *ORGANIC COMPOUNDS, *OXY    W70-08416
PESTICIDE KINE/ *WATER POLLUTION      EFFECTS, *PESTICIDES, RESIDUES, *    W70-09768
R, / RESISTANCE, WATER POLLUTION      EFFECTS, *SALT MARSHES, OILY WATE    W70-09976
 *BIOINDICATORS, *WATER POLLUTION     EFFECTS, *REVIEWS, *WATER QUALITY    W70-07027
USTRIAL WASTES, *WATER POLLUTION      EFFECTS, *EUTROPHICATION, DISSOLV    W70-06975
LOGY, ALGAE, BACTE/ *TEMPERATURE      EFFECTS, *THERMAL POLLUTION, *ECO    W71-02479
WATER, *HABITATS, *ENVIRONMENTAL      EFFECTS, *LIMITING FACTORS, BENTH    W70-02770
               *CYTOLOGICAL          EFFECTS, ALGAL POPULATION.:          W70-02770
WATER POLLUTION, WATER POLLUTION      EFFECTS, ACID STREAMS, MINE DRAIN    W70-02770
N, *CONFERENCES, WATER POLLUTION      EFFECTS, ALGAL CONTROL, WATER POL    W70-02775
LGAL SUSPENSIONS, ELECTROKINETIC      EFFECTS, ANN ARBOR(MICH), SAND BE    W71-03035
S, INSECTICIDES, WATER POLLUTION      EFFECTS, ALGAE.: /LABORATORY TEST    W71-03183
HY, DEGRADATION, WATER POLLUTION      EFFECTS, AQUATIC ENVIRONMENT, CHL    W70-08652
ITY, POLLUTANTS, WATER POLLUTION      EFFECTS, ALDRIN, DIELDRIN, ENDRIN    W69-08565
WATER POLLUTION, WATER POLLUTION      EFFECTS, ALGAL CONTRO: /EATMENT,     W69-10171
BOLISM, AQUARIA, WATER POLLUTION      EFFECTS, ADSORPTION.: /TION, META    W71-09005
REAMS, SEASONAL EFFECTS, THERMAL      EFFECTS, APHANIZOMENON FLOS AQUAE    W70-03978
ONAS, BIOASSAYS, WATER POLLUTION      EFFECTS, ANALYTICAL TECHNIQUES.: /   W70-04284
ICAL TECHNIQUES, WATER POLLUTION      EFFECTS, ALGAE, ECOSYSTEMS, PHYSI    W70-04287
ORUS COMPOUNDS, *WATER POLLUTION      EFFECTS, ALGAE, ANIMALS, AQUATIC     W69-04805
 ORGANIC MATTER, WATER POLLUTION      EFFECTS, ANALYTICAL TECHNIQUES, K    W70-05109
ENTS, MINNESOTA, WATER POLLUTION      EFFECTS, ALGICIDES, WATER QUALITY    W70-04506
, PRYMNESIUM PARVUM, SYNERGISTIC      EFFECTS, AXENIC CULTURES, MONOALG    W70-05372
CYCLING NUTRIENTS, ENVIRONMENTAL      EFFECTS, AMMONIA, AMMONIUM COMPOU    W69-03364
OGICAL ECOLOGY, *WATER POLLUTION      EFFECTS, ALGAE, ALGAL CONTROL, HA    W69-03513
, SEWAGE BACTERIA, ENVIRONMENTAL      EFFECTS, AQUATIC LIFE, AQUATIC MI    W69-04276
INDUSTRIAL WASTES, ENVIRONMENTAL      EFFECTS, ANALYTICAL TECHNIQUES, P    W72-12240
RY PRODUCTIVITY, WATER POLLUTION      EFFECTS, BIOASSAYS, ALGAE, COMPAR    W70-05424
ON, *MEROMIXIS, *WATER POLLUTION      EFFECTS, BIOINDICATORS, NUISANCE     W70-03974
ION, *SWIMMING, *WATER POLLUTION      EFFECTS, BACTERIA, ALGAE, FISH, S    W71-09880
 MATTER, CARBON, WATER POLLUTION      EFFECTS, BACTERIA, PERIPHYTON, OH    W71-04206
RIA, ABSORPTION, WATER POLLUTION      EFFECTS, CATION EXCHANGE, CATION     W71-09863
XICITY, *ALGAE, *WATER POLLUTION      EFFECTS, CHELATION, SPECTROPHOTOM    W71-11561
ERSHEDS(BASINS), WATER POLLUTION      EFFECTS, CYANOPHYTA, EURTROPHICAT    W69-10182
TRACE ELEMENTS, *WATER POLLUTION      EFFECTS, CHLORIDES, HALOGENS, ALG    W69-08562
ATORS, CYANOPHYTA, ENVIRONMENTAL      EFFECTS, CURRENTS(WATER), INDUSTR    W68-00470
RSHED MANAGEMENT, *ENVIRONMENTAL      EFFECTS, CHEMTROL, ALGAE.: /*WATE    W68-00478
ANISMS, GRAZING, WATER POLLUTION      EFFECTS, DIATOMS, CHLOROPHYTA, CY    W69-01977
S, *ALGAE, *SALINITY, IRRIGATION      EFFECTS, DRAINAGE EFFECTS, *SAMPL    W69-00360
EDGE SAMPLES, ANABAENA, SEASONAL      EFFECTS, DECOMPOSITION RATES, CHR    W70-05663
LLUTION SOURCES, WATER POLLUTION      EFFECTS, DISSOLVED OXYGEN, AQUATI    W70-04727
 ORGANIC MATTER, WATER POLLUTION      EFFECTS, DIATOMS, CARBON RADIOISO    W70-04811
LLUTION SOURCES, WATER POLLUTION      EFFECTS, DINOFLAGELLATES, FISHKIL    W70-08372
DUSTRIAL WASTES, WATER POLLUTION      EFFECTS, DENITRIFICATION, SPECTRO    W70-02777
CAL COMMUNITIES, WATER POLLUTION      EFFECTS, DIATOMS.: /ANTS, BIOLOGI    W70-01233
ALS, *OILY WATER, LICHENS, TIDAL      EFFECTS, DISPERSION, OIL-WATER IN    W70-01777
RESHWATER ALGAE, WATER POLLUTION      EFFECTS, DOMESTIC WASTES, INDUSTR    W72-12240
ATER AND PLANTS, WATER POLLUTION      EFFECTS, ECOLOGY, EFFECTS OF POLL    W70-02772
ATER AND PLANTS, WATER POLLUTION      EFFECTS, ECOLOGY, WATER QUALITY C    W70-02764
ANOPHYTA, DIATOMS, ENVIRONMENTAL      EFFECTS, ESSENTIAL NUTRIENTS, EUT    W70-02792
GHT PENETRATION, WATER POLLUTION      EFFECTS, EUTROPHICATION.: /DS, LI    W70-02255
SAY, DISINFECTION, ENVIRONMENTAL      EFFECTS, ENVIRONMENTAL SANITATION    W69-10171
VOLUME, STREAMS, WATER POLLUTION      EFFECTS, EUTROPHICATION, NUISANCE    W70-00719
PHYTA, CYANOPHYTA, ENVIRONMENTAL      EFFECTS, ESSENTIAL NUTRIENTS, FRE    W69-03362
PHYTA, CYANOPHYTA, ENVIRONMENTAL      EFFECTS, EUTROPHICATION, LIMNOLOG    W69-03185
CAL OXYGEN DEMAND, ENVIRONMENTAL      EFFECTS, ENVIRONMENTAL SANITATION    W69-03369
CYCLING NUTRIENTS, ENVIRONMENTAL      EFFECTS, ESSENTIAL NUTRIENTS, ENV    W69-03374
OMS, DISINFECTION, ENVIRONMENTAL      EFFECTS, ESSENTIAL NUTRIENTS, FRE    W69-03370
CYCLING NUTRIENTS, ENVIRONMENTAL      EFFECTS, ESSENTIAL NUTRIENTS, NUT    W69-03373
CYCLING NUTRIENTS, ENVIRONMENTAL      EFFECTS, ENZYMES, ESSENTIAL N: /     W69-03188
ASSAY, CYANOPHYTA, ENVIRONMENTAL      EFFECTS, EUTROPHICATION, LIMNOLOG    W69-03186
ANOPHYTA, DIATOMS, ENVIRONMENTAL      EFFECTS, ESSENTIAL NUTRIENTS, HAR    W69-03371
CYCLING NUTRIENTS, ENVIRONMENTAL      EFFECTS, ESSENTIAL NUTRIENTS, EUT    W69-03518
CYCLING NUTRIENTS, ENVIRONMENTAL      EFFECTS, ESSENTIAL NUTRIENTS, CHL    W69-03517
LLUTION CONTROL, WATER POLLUTION      EFFECTS, EUTROPHICATION, FISHKILL    W69-03948
CYCLING NUTRIENTS, ENVIRONMENTAL      EFFECTS, ESSENTIAL NUTRIENTS, EUT    W69-03514
CYCLING NUTRIENTS, ENVIRONMENTAL      EFFECTS, ENVIRONMENTAL SANITATION    W69-03513
```

NT, NITROGEN, PHOSPHORUS, SEWAGE   EFFLUENTS.: /, BIOLOGICAL TREATME   W70-08980
OLLUTION, WATER POLLUTION, ODOR,   EFFLUENTS, SEWAGE EFFLUENTS, DILU   W71-00940
LLUTION, ODOR, EFFLUENTS, SEWAGE   EFFLUENTS, DILUTION, DECOMPOSING   W71-00940
OXYGEN DEMAND, STORAGE CAPACITY,   EFFLUENTS, CONSTRUCTION.: /MICAL   W71-00936
PILOT PLANTS, HUMUS, REAERATION,   EFFLUENTS, FLOW RATES, FILTERS, A   W70-05821
UCTIVITY, NUISANCE ALGAE, SEWAGE   EFFLUENTS, PHOSPHORUS, EUTROPHICA   W70-00270
LOGY, SAMPLING, THERMAL SPRINGS,   EFFLUENTS, CYANOPHYTA.: /TRY, ECO   W69-10151
, WASTE WATER DISPOSAL, DIURNAL,   EFFLUENTS, NITROGEN, PHOSPHATES,   W69-10159
NION, NITROGEN COMPOUNDS, SEWAGE   EFFLUENTS, LIGHT INTENSITY, LAKES   W69-10167
T, WATER QUALITY CONTROL, SEWAGE   EFFLUENTS, ABSORPTION, LAKES.: /N   W70-01031
LUTION CONTROL, *FILTERS, SEWAGE   EFFLUENTS, ELECTROLYTES, ZETA POT   W70-01027
POWERPLANTS, SHIPS, SUBMARINES,   EFFLUENTS, ION EXCHANGE, RESINS,   W70-00707
MPSHIRE, *COPPER SULFATE, SEWAGE   EFFLUENTS, SMELTS, FISHKILL, CRUS   W69-08674
*LAKES, *EUTROPHICATION, *SEWAGE   EFFLUENTS, *NUTRIENTS, *NITROGEN,   W69-09349
ER, THERMAL POLLUTION, BACTERIA,   EFFLUENTS, OXYGEN DEMAND, AEROBIC   W72-00462
IOCHEMICAL OXYGEN DEMAND, SEWAGE   EFFLUENTS, SANITARY ENGINEERING.:   W72-01817
OW, SEWERS, STORM DRAINS, SEWAGE   EFFLUENTS, *SEWAGE TREATMENT, STO   W71-10654
S, RUNOFF, WASTES, WATER SUPPLY,   EFFLUENTS, STREAMS, LAKES, RIVERS   W71-09880
R, LAKES, SILTS/ *ALGAE, *SEWAGE   EFFLUENTS, *ESTUARIES, FRESH WATE   W70-02255
N IDENTIFICATION, SEWAGE, SEWAGE   EFFLUENTS, SEWAGE TREATMENT, WATE   W70-02248
OSPHORUS, ALGAE, COPPER SULFATE,   EFFLUENTS, MICROCYSTIS AERUGINOSA   W70-02787
IENTS, LIGHT PENETRATION, SEWAGE   EFFLUENTS, OXYGEN, SOLAR RADIATIO   W70-01933
MS, RESERVOIRS, CULTURES, SEWAGE   EFFLUENTS, WISCONSIN, ALGICIDES,   W70-02982
RUSTACEANS, SOIL EROSION, SEWAGE   EFFLUENTS, FARM WASTES. POPULATIO   W70-02803
OL, *LOCAL / *WISCONSIN, *SEWAGE   EFFLUENTS, *WATER POLLUTION CONTR   W69-06909
IFICATION, *POLLUTANT ABATEMENT,   EFFLUENTS, STREAMFLOW, DESALINATI   W69-07389
Y DURRENTS, MEASUREMENTS, SEWAGE   EFFLUENTS, TEMPERATURE, OXYGEN, S   W69-06203
, SEDIMENTS, FISH, RADIOECOLOGY,   EFFLUENTS, ECOSYSTEMS, PLANKTON,   W69-07853
TROPHICATION, CYANOPHYTA, SEWAGE   EFFLUENTS, CHLOROPHYTA, RHODOPHYT   W70-04510
GEN FIXATION, CYANOPHYTA, SEWAGE   EFFLUENTS, MINNESOTA, WATER POLLU   W70-04506
W, CHEMICAL ANALYSIS, NUTRIENTS,   EFFLUENTS, ALGAE, WEEDS, CARBON D   W70-04810
NCE ALGAE, WATER QUALITY, SEWAGE   EFFLUENTS, DIATOMS, MISSOURI RIVE   W70-04507
E, AGRICULTURAL/ *LAKES, *SEWAGE   EFFLUENTS, *SEWAGE DISPOSAL, ALGA   W70-05565
TROPHY, DIVERSION, CORES, SEWAGE   EFFLUENTS, DIATOMS, PHOSPHORUS, O   W70-05663
E TREATMENT, TERTIARY TREATMENT,   EFFLUENTS, LIMNOLOGY, WATER CHEMI   W70-05387
OPHICATION, STORM DRAINS, SEWAGE   EFFLUENTS, STREAMFLOW, MEASUREMEN   W70-05112
AL PROPERTIES, WISCONSIN, SEWAGE   EFFLUENTS, NITROGEN, PHOSPHORUS.:   W70-05113
RIVER(WI/ *TRUAX FIELD(MADISON)   EFFLUENTS, *MADISON LAKES, YAHARA   W70-05113
SPHORUS COMPOUNDS, LAKES, SEWAGE   EFFLUENTS, ARSENIC COMPOUNDS, SPE   W70-05269
NNESOTA, *EUTROPHICATION, SEWAGE   EFFLUENTS, TERTIARY TREATMENT, AL   W70-03512
AE, PHOSPHATES, NITRATES, SEWAGE   EFFLUENTS, TERTIARY TREATMENT, WA   W70-04074
LATION, *BIOLOGICAL COMMUNITIES,   EFFLUENTS, EUTROPHICATION, PHYTOP   W70-04001
IOACTIVITY, CHLORELLA, SAMPLING,   EFFLUENTS, TRIBUTARIES, ALKALINIT   W70-03983
NUISANCE ALGAE, NEW YORK, SEWAGE   EFFLUENTS, SEWAGE TREATMENT, INDU   W70-03974
WEEDS, NUTRIENTS, SEPTIC TANKS,   EFFLUENTS, SWAMPS, FOREST SOILS,   W70-04193
LIOGRAPHIES, CHLORELLA, *SEWAGE,   EFFLUENTS, CARBON DIOXIDE, *ALGAE   W68-00855
UARIES, HUMAN POPULATION, SEWAGE   EFFLUENTS, AQUATIC POPULATIONS, N   W68-00461
IS/ CERATOPHYLLUM, MYRIOPHYLLUM,   EICHHORNIA, POTAMOGETON, TRAPA, P   W70-10175
AS, SCIRPUS, NYMPHAEA, CRASSIPES   EICHORNIA, TYPHA, BRASENIA, DOW C   W70-05263
NT, LAKE NORRVIKEN(SWEDEN), LAKE   EKOLN(SWEDEN), LAKE LOTSJON(SWEDE   W70-04287
MICROCYSTIS AERUGINOSA, BOHEMIA,   ELBE RIVER, CELAKOVICE(CZECHOSLOV   W70-00274
ARUM, RADULESPORES, ZYMOLOGICAL,   ELBE(GERMANY), SPOROBOLOMYCES ROS   W70-04368
ELECTROPHORESIS, MICROORGANISMS,   ELECTRIC CURRENTS, *WASTE TREATME   W71-03775
LANKTON, FISH, SHELLFISH, STEAM,   ELECTRIC POWER, HEATED WATER, PRO   W69-09905
GAE, HYDROGEN ION CONCENTRATION,   ELECTRIC POWER, SEWAGE DISPOSAL,   W71-09548
, OUTLETS, HYDROELECTRIC PLANTS,   ELECTRIC POWERPLANTS, ENGINEERING   W69-05023
E-AVERAGE-LIGHT-INTENSITY, PHOTO-   ELECTRIC TURBIDOSTAT, LIGHT EXTIN   W70-03923
LECTROLYTES, CHEMICAL REACTIONS,   ELECTROCHEMISTRY, RESISTIVITY, CH   W70-05267
NITRIFICATION, LAND APPLICATION,   ELECTROCHEMICAL TREATMENT, AMMONI   W70-01981
IDIUM, WINKLER TECHNIQUE, OXYGEN   ELECTRODES, ATOMIC ABSORPTION SPE   W70-03309
, REVERSE OSMOSIS, DISTILLATION,   ELECTRODIALYSIS, SORPTION, CHEMIC   W70-01981
Y TREATMENT, TERTIARY TREATMENT,   ELECTRODIALYSIS, ION EXCHANGE, AC   W69-01453
(MICH), SAN/ *ALGAL SUSPENSIONS,   ELECTROKINETIC EFFECTS, ANN ARBOR   W71-03035
FRESH WATER, FLOW RATES, PUMPS,   ELECTROLYSIS, CULTURES.: / ALGAE,   W70-09904
ROL, *FILTERS, SEWAGE EFFLUENTS,   ELECTROLYTES, ZETA POTENTIAL, BAC   W70-01027
GANIC POLYELECTROLYTES, CATIONIC   ELECTROLYTES, ALGAL SURFACE CHARG   W70-05267
ULATION, *WASTE WATER TREATMENT,   ELECTROLYTES, CHEMICAL REACTIONS,   W70-05267
CIDES, TASTE, ODOR, COAGULATION,   ELECTROLYTES, POTABLE WATER.: /TI   W72-13601
MES, CULTURES, HYDROGEN SULFIDE,   ELECTRON MICROSCOPY, BIOCHEMISTRY   W70-03952
EN, NITROGEN FIXATION, NITRATES,   ELECTRON MICROSCOPY, AMMONIA, SPE   W70-04369
CYANOPHYTA, CYTOLOGICAL STUDIES,   ELECTRON MICROSCOPY, WASTE TREATM   W70-03973
LA VULGARIS, CORNELL UNIVERSITY,   ELECTRONIC PARTICLE COUNTING, EUD   W69-06540
TOPES, DETRITUS, COLUMBIA RIVER,   ELECTRONICS, SCALING.: / RADIOISO   W69-07862
*PROTOZOA, *ALGAE, LETHAL LIMIT,   ELECTROPHORESIS, MICROORGANISMS,   W71-03775
E, *SANDS, *FILTRATION, FILTERS,   ELECTROPHORESIS, CHLORELLA, SCENE   W71-03035
BREVIBACTERIUM, CORYNEBACTERIUM,   ELECTROSTATIC PRECIPITATION, THER   W71-11823
CTROLYTES, ALGAL SURFACE CHARGE,   ELECTROSTATIC REPULSION, DOW C-31   W70-05267
ONIC PARTICLE COUNTING, EUDORINA   ELEGANS, EUGLENA GRACILIS, FOREST   W69-06540
IREMENTS, *PLANT GROWTH, *ALGAE,   ELEMENTS(CHEMICAL), CHLOROPHYTA,   W70-02964
, SODIUM, IRON, MANGANESE, TRACE   ELEMENTS, CHELATION, PHOTOSYNTHES   W70-02804
PLANKTON, ALGAE, COLLOIDS, TRACE   ELEMENTS, DISSOLVED SOLIDS, NUTRI   W70-02510
TA, CHLOROPHYLL, BACTERIA, TRACE   ELEMENTS, VITAMINS, AMINO ACIDS,   W70-02255
THESIS, CHEMICAL ANALYSIS, TRACE   ELEMENTS, TEMPERATURE, IRON, NUTR   W70-01933
, *COBALT, *COPPER, *IRON, TRACE   ELEMENTS, ALGAE, SEA WATER, FRESH   W71-11687
*ECOSYSTEMS, GREAT LAKES, TRACE   ELEMENTS, RADIUM RADIOISOTOPES, T   W71-09013
LLUTION / *RADIOACTIVITY, *TRACE   ELEMENTS, *AQUATIC LIFE, WATER PO   W71-05531
AMS, SIZE, ALGAE, DIATOMS, TRACE   ELEMENTS, METABOLISM.: /SSIUM, CL   W70-00708
ER QUALITY, *GREAT LAKES, *TRACE   ELEMENTS, *WATER POLLUTION EFFECT   W69-08562
OPISTHONEMA OGLIMUN, UDO/ *TRACE   ELEMENTS, *ANCHOA LAMPROTAENIA, *   W69-08525
GAL TOXINS, ALGAL CONTROL, TRACE   ELEMENTS, WASHINGTON, NUCLEAR POW   W70-00264

437

TA, *ESSENTIAL NUTRIENTS, *TRACE   ELEMENTS, ALGAE, EUTROPHICATION, W69-06277
TRY, CHLORELLA, NUTRIENTS, TRACE   ELEMENTS, CHLAMYDOMONAS, CHRYSOPH W70-05409
PLE PRESERVATION, WAT/ *NUTRIENT   ELEMENTS, PHOSPHORUS RELEASE, SAM W70-04382
SODIUM, ALGAE, PHOSPHORUS, TRACE   ELEMENTS, PLANKTON, LAKES, RIVERS W70-04385
  PEPTIDES, ORGANIC MATTER, TRACE   ELEMENTS, LIGHT INTENSITY, CHLAMY W70-04369
  ALGAL CONTROL, ALGICIDES, TRACE   ELEMENTS, CORES, WISCONSIN, BOTTO W69-03366
, PREDATION, PRODUCTIVITY, TRACE   ELEMENTS, SODIUM, VIRUSES, VITAMI W69-04801
MINE, DE-K-PRUF-21/ *LITERATURE,   ELEOCHARIS, 2,4-D, AMMATE, CHLORA W70-05263
YMNODINIUM SIMPLEX, FLAGELLATES,   ELEUTHERA ISLAND, NITZCHIA SERIAT W70-00707
                        ELLIOT LAKE, ONTARIO.: W69-01165
LORELLA SACCHAROPHILA, CHLORELLA   ELLIPSOIDEA, CHLORELLA LUTEOVIRID W70-04809
OLI, OOCYSTIS, LEMNA, SPIRODELA,   ELODEA.: /LOCOCCUS, ESCHERICHIA C W70-10161
S, ANABAENA.:   HYDRODICTYON,   ELODEA, CERATOPHYLLUM, MICROCYSTI W68-00479
PHENANTHROQUINONE, MADISON(WIS),   ELODEA, REDOX COMP: /L EXPOSURE, W70-02982
LETONEMA COSTATUM, SYNECHOCOCCUS   ELON: /HLAMYDOMONAS MOEWUSII, SKE W70-05381
PHIDINIUM CARTERI, SYRACOSPHAERA   ELONGATA, THALASSIOSIRA FLUVIAT: / W69-08278
T, RADIOCOCCUS NIMBATUS, URONEMA   ELONGATUM HODGETTS, CHROOCOCCUS P W70-04468
, CHLORELLA PYRENOIDOSA, DIATOMA   ELONGATUM, MELOSIRA ISLANDICA, NI W69-10158
NOIDOSA, M/ *SHAGAWA LAKE(MINN),   ELY(MINN), BLOOMS, CHLORELLA PYRE W70-03512
S, ARTEMIA, BALANUS, DESORPTION,   EMBRYOLOGY, ENERGY TRANSFORMATION W69-07440
AL GROWTH, CHLORELLA PYRENOIDOSA(EMERSON STRAIN), AUTOINHIBITORS.: W69-07442
PHOSPHORUS RADIOISOTOPES, ALGAE,   EMULSIONS, BACKGROUND RADIATION, W69-10163
  NAIS, CRICOTOPUS, AUSTROCLOEON,   ENALLAGMA, PSEUDAGRION, BRACHYTHE W70-06975
RA, HETERANTHERA, NAJAS, SILVEX,   ENDOTHAL, POTASSIUM PERMANGANATE, W70-10175
NARIANS, MICROCYSTIS AERUGINOSA,   ENDOTOXIN, FAST-DEATH FACTOR, DOS W70-00273
UTION EFFECTS, ALDRIN, DIELDRIN,   ENDRIN, DDT.: /UTANTS, WATER POLL W69-08565
  HYDROCARBON PESTICIDES, ALDRIN,   ENDRIN, DIELDRIN, SCENEDESMUS, GA W70-03520
RESEARCH COUNCIL(USA), US ATOMIC   ENERGY COMMISSION, NATIONAL SCIEN W70-03975
RESEARCH COUNCIL(USA), US ATOMIC   ENERGY COMMISSION, NATIONAL SCIEN W70-03975
GES, CLEAR LAKE(CALIF), NUTRIENT   ENERGY PATHWAYS, TROPHIC RELATION W70-05428
  FISH FOOD ORGANISMS, NUTRIENTS,   ENERGY TRANSFER, *FOOD CHAINS, FO W70-05428
BALANUS, DESORPTION, EMBRYOLOGY,   ENERGY TRANSFORMATIONS, HETEROTRO W69-07440
TION POTENTIAL, SOLAR RADIATION,   ENERGY, CARBON DIOXIDE, PHOTOSYNT W70-05750
ALGAE, *PHOSPHORUS, *PHOSPHATES,   ENERGY, DI-URNAL DISTRIBUTION, TE W70-06604
RONMENTAL ENGINEERING, *SANITARY   ENGINEERING, *SEWAGE TREATMENT, * W70-02609
AE, *DISPERSIONS, *ENVIRONMENTAL   ENGINEERING, *SANITARY ENGINEERIN W70-02609
R POLLUTION ASSESSMENT, SANITARY   ENGINEERING RESEARCH LABORATORY(C W70-02775
ROTOZOA, *SYSTEMATICS, *SANITARY   ENGINEERING, MICROORGANISMS, NUIS W72-01798
OGICAL PROPERTIES, *W/ *SANITARY   ENGINEERING, *WATER SUPPLY, *BIOL W72-01813
MAND, SEWAGE EFFLUENTS, SANITARY   ENGINEERING.: /CHEMICAL OXYGEN DE W72-01817
MEASUREMENT, ESTUARIES, SANITARY   ENGINEERING, WATER TREATMENT, WAS W70-05094
IC PLANTS, ELECTRIC POWERPLANTS,   ENGINEERING STRUCTURES, INDUSTRIA W69-05023
IL ENGINEERING, WASTE TREATMENT,   ENGINEERING, ALGAE, PLANTS, CONTR W69-03683
, PACIFIC NORTHWEST U. S., CIVIL   ENGINEERING, WASTE TREATMENT, ENG W69-03683
SENTIAL NUTRIENTS, ENVIRONMENTAL   ENGINEERING, H: /NTAL EFFECTS, ES W69-03374
, AL/ *WATER TREATMENT, SANITARY   ENGINEERING, *PATHOGENIC BACTERIA W72-13601
*AL/ *WATER TREATMENT, SANITARY   ENGINEERING, *FILTRATION, DESIGN, W72-08859
TREATMENT, *FILTRATION, SANITARY   ENGINEERING, *FILTERS, *DESIGN, C W72-07334
GANO-METALLIC COMPOUNDS, AMINES,   ENGLAND.: /PHENOLIC COMPOUNDS, OR W72-12987
WASHINGTON, CHEMICAL COMPOSITION(ENGLAND), FOOD SOURCES, GREENHOUS W69-06865
A MARTYI, ABBOT'S POND(SOMERSET,   ENGLAND), SAND GRAINS, TIDAL CYCL W71-12870
ALGAL ASSOCIATIONS, SHEAR WATER(ENGLAND), AMPHORA OVALIS, OPEPHOR W71-12870
OTLAND, ENGLAND, LAKE WINDERMERE(ENGLAND), CARBON-14.: /TLAND), SC W70-02504
PYRENOIDOSA, FIXATION, TRINGFORD(ENGLAND), ANABAENA CYLINDRICA, NO W70-02504
PARTICULATE MATTER, BLELHAM TARN(ENGLAND), CHLORELLA PYRENOIDOSA, W70-02504
KE MENDOTA(WIS), ENGLAND, LONDON(ENGLAND), ESTHWAITE(ENGLAND), LOW W69-10180
LAND, LONDON(ENGLAND), ESTHWAITE(ENGLAND), LOWESWATER(ENGLAND), BA W69-10180
, ESTHWAITE(ENGLAND), LOWESWATER(ENGLAND), BARNES SOUTH RES: /AND) W69-10180
PELLICULOSA, LAKE MENDOTA(WIS),   ENGLAND, LONDON(ENGLAND), ESTHWAI W69-10180
TH OF CLYDE(SCOTLAND), SCOTLAND,   ENGLAND, LAKE WINDERMERE(ENGLAND) W70-02504
SLOCH' RIVER(USSR), MINSK(USSR),   ENGLAND, OSCILLATORIA, AUSTRALIA, W70-05092
IDACNA CROCEA, WOODS HOLE OCEAN/   ENIWETOK ATOLL, TURNOVER TIME, TR W69-08526
COLORADO RIVER, TEXAS, *NUTRIENT   ENRICHMENT TESTS, *NITROGEN, *PHO W69-06004
ITY STRUCTURE, OXYTREMA, ORGANIC   ENRICHMENT, INTERACTIONS.: /OMMUN W70-03978
  HYDROGEN SULFIDE, LAKES, ALGAE,   ENTERIC BACTERIA, METHANE BACTERI W70-03312
UM VINOSUM, THIOCAPSA FLORIDANA,   ENTEROCOCCI, PARK RIVER: /HROMATI W70-03312
TZSCHIA, NAVICULA, SYRACUSE(NY),   ENTEROMORPHA INTESTINALIS, CLADOP W70-03974
ARBURG RESPIROMETER, CLADOPHORA,   ENTEROMORPHA, LEPOCINCLIS, CHLORO W69-02959
, SIMULIUM, PHYSA, SALMO TRUTTA,   ENTOSPHENUS LAMOTTENII, GAMMARUS, W69-09334
ASTRUM, SELF-PURIFICATION, ALGAL   ENUMERATION, SCENEDESMUS, GROWTH W70-04381
OGY, HYDROLOGIC ASPECTS, AQUATIC   ENVIRONMENT, WATER PROPERTIES, BI W70-04475
EROMIXIS, *WAT/ *ALGAE, *AQUATIC   ENVIRONMENTS, *EUTROPHICATION, *M W70-03974
*PHYSIOLOGICAL ECOLOGY, *AQUATIC   ENVIRONMENTS, RIVERS, ENVIRONMENT W70-03978
, *AQUATIC ENVIRONMENTS, RIVERS,   ENVIRONMENTAL EFFECTS, PERIPHYTON W70-03978
A, PHYSIOLOGICAL ECOLOGY, LAKES,   ENVIRONMENTAL EFFECTS, WATER POLL W70-03957
AND, CYCLING NUTRIENTS, DIATOMS,   ENVIRONMENTAL EFFECTS, ESSENTIAL W70-04283
AL EFFECTS, ESSENTIAL NUTRIENTS,   ENVIRONMENTAL ENGINEERING, H: /NT W69-03374
OASSAY, CHLOROPHYTA, CYANOPHYTA,   ENVIRONMENTAL EFFECTS, EUTROPHICA W69-03185
YTA, DIATOMS, CYCLING NUTRIENTS,   ENVIRONMENTAL EFFECTS, ESSENTIAL W69-03373
YTA, DIATOMS, CYCLING NUTRIENTS,   ENVIRONMENTAL EFFECTS, ENZYMES, E W69-03188
N DEMAND, ENVIRONMENTAL EFFECTS,   ENVIRONMENTAL SANITATION, ESSENTI W69-03369
  CHLOROPHYTA, CYCLING NUTRIENTS,   ENVIRONMENTAL EFFECTS, ESSENTIAL W69-03374
SSAY, BIOCHEMICAL OXYGEN DEMAND,   ENVIRONMENTAL EFFECTS, ENVIRONMEN W69-03369
ANOPHYTA, DIATOMS, DISINFECTION,   ENVIRONMENTAL EFFECTS, ESSENTIAL W69-03370
HLOROPHYTA, CYANOPHYTA, DIATOMS,   ENVIRONMENTAL EFFECTS, ESSENTIAL W69-03371
TIC WEEDS, BIOASSAY, CYANOPHYTA,   ENVIRONMENTAL EFFECTS, EUTROPHICA W69-03186
CYANOPHYTA, ESSENTIAL NUTRIENTS,   ENVIRONMENTAL EFFECTS, NITROGEN C W69-03358
OASSAY, CHLOROPHYTA, CYANOPHYTA,   ENVIRONMENTAL EFFECTS, ESSENTIAL W69-03362
YTA, DIATOMS, CYCLING NUTRIENTS,   ENVIRONMENTAL EFFECTS, AMMONIA, A W69-03364
TER SPORTS, ENVIRONMENT, AQUATIC   ENVIRONMENT, ESTUARINE ENVIRONMEN W69-04276

WATER QUALITY, SEWAGE BACTERIA,
CE OF NATURE, CYCLING NUTRIENTS,
UTRIENTS, ENVIRONMENTAL EFFECTS,
YTA, DIATOMS, CYCLING NUTRIENTS,
, AQUATIC ENVIRONMENT, ESTUARINE
TANTS, RECREATION, WATER SPORTS,
AE, BIOASSAY, CYCLING NUTRIENTS,
TY, BIOASSAY, CYCLING NUTRIENTS,
CE OF NATURE, CYCLING NUTRIENTS,
 MICROENVIRONMENT, AQUATIC LIFE,
 BIOASSAY, CHLOROPHYTA, DIATOMS,
LORINE, COBALT, COPPER, ECOLOGY,
PRODUCTIVITY, CYCLING NUTRIENTS,
T GROWTH REGULATORS, CYANOPHYTA,
AL WATERS, ECOSYSTEMS, ESTUARINE
SIS, *CALCIUM CARBONATE, AQUATIC
OLLUTION/ *ESTUARIES, *ESTUARINE
ATION, LAKES, LIMNOLOGY, AQUATIC
QUATIC PRODUCTIVITY, CYANOPHYTA,
LGICIDES, AQUATIC ALGAE, AQUATIC
, MODEL STUDIES, SALMON, AQUATIC
OW, SAMPLING, ALGAE, ECOSYSTEMS,
LOOMS, CLOSED SYSTEMS, CYTOLOGY,
 ALGICIDES, HARVESTING OF ALGAE,
M RADIOISOTOPES, *URAN/ *AQUATIC
IVITY, MARINE ANIMALS, ESTUARINE
 *GLUCOSE, *ACETATE, *FRESHWATER
SACRAMENTO-SAN JOAQUIN ESTUARY,
RIENT REQUIREMENTS, POPULATIONS,
DS, *ANAEROBIC PONDS, DESIGN FOR
ROPHY, *ORGANIC SOLUTES, *MARINE
 *LABORATORY TESTS, RESPIRATION,
NFECTION, ENVIRONMENTAL EFFECTS,
GICIDES, BIOASSAY, DISINFECTION,
 LIGHT INTENSITY, ECOLOGY, LOTIC
S, GENETICS, BACTERIA, PEPTIDES,
 *SEASONAL/ *BACTERIA, *ESTUARINE
A CLOSERIUM, ARTEMIA, C/ *MARINE
, ALGAE, SILTS, ECOLOGY, AQUATIC
N, *LAKE ONTARIO, *GROWTH RATES,
CHNIQUES, AQUATIC ALGAE, AQUATIC
F NATURE, BIOASSAY, CHLOROPHYTA,
 *ALGAE, *TROPHIC LEVELS, AQUATIC
FLUORESCENCE, NITROGEN FIXATION,
NTS, PHOSPHORUS, KINETIC THEORY,
R QUALITY, *TRO/ *ALGAE, AQUATIC
CHLOROPHYTA, GONOPHYTA, DIATOMS,
PHOTOS/ *RADIOISOTOPES, *AQUATIC
HLOROPHYTA, CYANOPHYTA, DIATOMS,
 *ALGAE, *AQUATIC ALGAE, *AQUATIC
                     CONCENTRATION,
NSECTS, *AQUATIC PLANTS, AQUATIC
HIC FLORA, ESTUARIES, *ESTUARINE
KTON, FRESHWATER FISH, ESTUARINE
 *CHEMICAL ENVIRONMENT, *PHYSICAL
, *BIOTA, *TEMPERATURE, *AQUATIC
T, PHAGE, FERMENTOR, / *CHEMICAL
WATER POLLUTION EFFECTS, AQUATIC
 *REVIEWS, ALGAE, WEEDS, ECOLOGY,
ESTIC WASTES, INDUSTRIAL WASTES,
, BACTERIA, NUTRIENTS, ESTUARINE
RGANIC MATTER, RIVERS, ESTUARINE
E TESTS, LABORATORY TESTS, LOTIC
ESTS, LOTIC ENVIRONMENTS, LENTIC
XICITY, *BIOASSAY, FISH, AQUATIC
ACEANS, HERBIVORES, CHLOROPHYLL,
HETEROTROPHIC POTENTIAL, SWEDEN,
N), ACTIVE TRANSPORT, CARBON-14,
LEIC ACID, SISJOTHANDI(ICELAND),
GANISMS, AQUATIC MICROORGANISMS,
MONAS, FRESHWATER, MARINE ALGAE,
 PRODUCTIVITY, SURFACES, ARCTIC,
INO ACIDS, MECHANICAL EQUIPMENT,
AL AREAS, CARBOHYDRATES, YEASTS,
NSITY, CHLAMYDOMONAS, CHLORELLA,
AND, SAMPLING, ECOLOGY, AMMONIA,
ICAL, CARBON DIOXIDE, SYNTHESIS,
S, CHLOROPHYLL, IRON, CHELATION,
UTRIENTS, ENVIRONMENTAL EFFECTS,
 BACTERIA, CARBON RADIOISOTOPES,
GEN, PHOSPHORUS, OXYGEN, CARBON,
ONIOBASIS, DECAPODA, PLECOPTERA,
PUNCTATUS, TRICHOPTERA, ODONATA,
PLANTS, GLACIAL LAKES, PLANKTON,
INGIENSIS, AGARS, STAPHYLOCOCCUS
ACES, DEPTH, DISTRIBUTION, TIME,
ACTERIA, NUTRIENTS, TEMPERATURE,
TIVITY, PHOSPHATES, CHLOROPHYLL,
TOSYNTHESIS, ALGAE, HYPOLIMNION,
AL WATER, WAVES(WATER), DIURNAL,

ENVIRONMENTAL EFFECTS, AQUATIC LI   W69-04276
ENVIRONMENTAL EFFECTS, ESSENTIAL   W69-03517
ENVIRONMENTAL SANITATION, ESSENTI   W69-03513
ENVIRONMENTAL EFFECTS, ESSENTIAL   W69-03515
ENVIRONMENT, WATER QUALITY, SEWAG   W69-04276
ENVIRONMENT, AQUATIC ENVIRONMENT,   W69-04276
ENVIRONMENTAL EFFECTS, ESSENTIAL   W69-03518
ENVIRONMENTAL EFFECTS, HARVESTING   W69-04798
ENVIRONMENTAL EFFECTS, ENVIRONMEN   W69-03513
ENVIRONMENT, MICROORGANISMS, WATE   W69-03611
ENVIRONMENTAL EFFECTS, INHIBITION   W69-03516
ENVIRONMENTAL EFFECTS, GRAZING, I   W69-04801
ENVIRONMENTAL EFFECTS, ESSENTIAL   W69-03514
ENVIRONMENTAL EFFECTS, CURRENTS(W   W68-00470
ENVIRONMENT, MARINE ALGAE, SALINI   W68-00010
ENVIRONMENT, AQUATIC PRODUCTIVITY   W69-00446
ENVIRONMENT, *INSTRUMENTATION, *P   W69-06203
ENVIRONMENT, AQUATIC PRODUCTIVITY   W69-05142
ENVIRONMENTAL EFFECTS, EUTROPHICA   W69-04802
ENVIRONMENT, AQUATIC PLANTS, AQUA   W69-05844
ENVIRONMENT, PHOSPHATES, ALGAE, D   W69-06536
ENVIRONMENT, RETENTION, PHOTOPERI   W69-07862
ENVIRONMENTAL CONDITIONS, EXTRACE   W69-07832
ENVIRONMENT, CYANOPHYTA.: /ALGAE,   W69-07818
ENVIRONMENTS, *ECOSYSTEMS, *RADIU   W69-07846
ENVIRONMENT, FALLOUT, NUCLEAR EXP   W69-08267
ENVIRONMENTS, *TURNOVER RATES, *H   W70-05109
ENVIRONMENTAL RELATIONSHIPS, FLUO   W70-05094
ENVIRONMENTAL EFFECTS, *PRIMARY P   W70-05424
ENVIRONMENTAL CONTROL, RECIRCULAT   W70-04787
ENVIRONMENTS, MICROBIAL ECOLOGY,   W70-04811
ENVIRONMENT, PHOTOSYNTHESIS, OXYG   W70-00265
ENVIRONMENTAL SANITATION, INHIBIT   W69-10171
ENVIRONMENTAL EFFECTS, ENVIRONMEN   W69-10171
ENVIRONMENT, BIOMASS, FLOW RATES,   W70-00265
ENVIRONMENTAL EFFECTS, HETEROGENE   W70-00273
ENVIRONMENT, *WATER TEMPERATURE,   W70-00713
ENVIRONMENT, CERIUM-144, NITZSCHI   W70-00708
ENVIRONMENT.: /ANTS, AQUATIC LIFE   W69-10080
ENVIRONMENTAL EFFECTS, PHOTOSYNTH   W69-10158
ENVIRONMENT, AQUATIC MICROBIOLOGY   W70-02968
ENVIRONMENTAL EFFECTS, INHIBITION   W70-02968
ENVIRONMENT, WATER QUALITY, NUTRI   W70-02764
ENVIRONMENTAL EFFECTS, NUTRIE: /   W70-02775
ENVIRONMENTAL FACTORS, BIOMASS, G   W70-02779
ENVIRONMENT, *AQUATIC ALGAE, WATE   W70-02772
ENVIRONMENTAL EFFECTS, LIMNOLOGY,   W70-02784
ENVIRONMENT, CHLORELLA, TRACERS,   W70-02786
ENVIRONMENTAL EFFECTS, ESSENTIAL   W70-02792
ENVIRONMENT, POLLUTANTS, BIOLOGIC   W70-01233
ENVIRONMENTAL MONITORING.:   W72-00941
ENVIRONMENT, ECOSYSTEMS, AQUATIC   W71-09005
ENVIRONMENT, BAYS, *FLUCTUATION,   W71-02494
ENVIRONMENT, IMPOUNDED WATERS, FR   W71-02491
ENVIRONMENT, PHAGE, FERMENTOR, FA   W71-02009
ENVIRONMENT, *AQUATIC LIFE, *AQUA   W71-02491
ENVIRONMENT, *PHYSICAL ENVIRONMEN   W71-02009
ENVIRONMENT, CHLORINATED HYDROCAR   W70-08652
ENVIRONMENT, CHEMCONTROL, BIOCONT   W70-07393
ENVIRONMENTAL EFFECTS, ANALYTICAL   W72-12240
ENVIRONMENT, PONDS, BREEDINGS, SA   W73-00748
ENVIRONMENT, WATER TEMPERATURE, *   W73-01446
ENVIRONMENTS, LENTIC ENVIRONMENTS   W72-05453
ENVIRONMENTS.: /STS, LABORATORY T   W72-05453
ENVIRONMENT, ORGANIC PESTICIDES,   W72-11105
ENZUMES, PEPTIDES, BIOCHEMISTRY,   W70-01068
ENZYME KINETICS, LINEWEAVER-BURK   W70-04194
ENZYME KINETICS, GLUCOSE, MECHAEL   W70-04284
ENZYME-SUBSTRATE COMPLEXES, RIBON   W69-10160
ENZYMES.: /S, ECOSYSTEMS, MICROOR   W71-02479
ENZYMES.: / SCENEDESMUS, CHLAMYDO   W70-05381
ENZYMES, CHLOROPHYLL, PHOTOSYNTHE   W70-04809
ENZYMES, CHEMICAL DEGRADATION, VI   W70-04184
ENZYMES, CULTURES, HYDROGEN SULFI   W70-03952
ENZYMES, CARBON DIOXIDE, GROWTH R   W70-04369
ENZYMES, CLASSIFICATION, GROWTH R   W70-00713
ENZYMES, CHLOROPHYTA, RHODOPHYTA,   W70-02965
ENZYMES, CHLORINE, LIGHT INTENSIT   W70-02964
ENZYMES, ESSENTIAL N: / CYCLING N   W69-03188
ENZYMES, KINETICS, PHOTOSYNTHESIS   W70-04287
ENZYMES, SAMPLING, BOTTOM SEDIMEN   W70-09889
EPHEMEROPTERA, TIPULIDAE, CHIRONO   W69-09742
EPHEMEROPTERA, PLECOPTERA, DIPTER   W69-07846
EPIDEMICS.: /ILL, ROOTED AQUATIC   W60-00488
EPIDERMIDIS, MICROCOCCUS, STREPTO   W71-11823
EPILIMNION, THERMOCLINE, LIGHT PE   W71-07698
EPILIMNION, NITROGEN COMPOUNDS, S   W69-10167
EPILIMNION, EFFLUENT, TEMPERATURE   W69-10169
EPILIMNION, PHYTOPLANKTON, SESTON   W69-10182
EPILIMNION, WINDS, THERMOCLINE, T   W70-09889

RACTS, REGULATION, RIGHT-OF-WAY,
URNAL DISTRIBUTION, *CALIFORNIA,
PHYSIOLOGICAL ECOLOGY, LITTORAL,
S, OCEANS, BIOINDICATORS, FUNGI,
CONTROL, *CONFERENCES, STREAMS,
CONTROL, *CONFERENCES, STREAMS,
, SEDIMENTS, BACTERIA, DETRITUS,
, UNITED STATES, BRACKISH WATER,
MARINE ALGAE, TOXICITY, DIATOMS,
, DISSOLVED OXYGEN, MEASUREMENT,
PESTICIDE RESIDUES, HERBICIDES,
TS, MARINE ALGAE, BENTHIC FLORA,
, EUTROPHICATION(PHYTOPLANKTON),
, *COLIFORMS, *DISSOLVED OXYGEN,
S, ZOOPLANKTON, FRESHWATER FISH,
, RADIOACTIVITY, MARINE ANIMALS,

NVIRONMENT, AQUATIC ENVIRONMENT,
RIES, *TIDAL WATERS, ECOSYSTEMS,
EMPERATURE, BACTERIA, NUTRIENTS,
ALINITY, ORGANIC MATTER, RIVERS,
                    *POTOMAC RIVER
RENTON TREATMENT PLANT, DUWAMISH
DYNAMICS, SACRAMENTO-SAN JOAQUIN
ELATIONSHIPS.: *POTOMAC RIVER
IONS, SAN JOAQUIN RIVER, POTOMAC
TION, SEVIN, MALATHION, ACETONE,
OXYLATE, ANIONIC SU/ *ALKYL POLY
MAN-MADE FIBERS, METHYL ALCOHOL,
ANYPUS, TANYTARSUS, LIMNODRILUS,
GETATION, MICROCYSTIS, ANABAENA,
Y, ELECTRONIC PARTICLE COUNTING,
ELOSIRA, CERATIUM, OSCILLATORIA,
RKEN(SWEDEN), EXTINCTION VALUES,
ICLE COUNTING, EUDORINA ELEGANS,
2, ALGAL GROWTH, CYANOCOBALAMIN,
D GAS ANALYSIS, COULTER COUNTER,
PHOSPHORYLATION, VANADIUM, ZINC,
RBIDITY, CHLORELLA, SCENEDESMUS,
WATER, CHLORELLA, CHLAMYDOMONAS,
, PENNSYLVANIA, RIVERS, STREAMS,
ION CONCENTRATION, SCENEDESMUS,
, TASTE, FILTERS, IOWA, DIATOMS,
, CYANOPHYTA, DIATOMS, SEASONAL,
ORINE, LIGHT INTENSITY, NITRATE,
FFECTS, EPILIMNION, CONNECTICUT,
N, NUTRIENTS, INDICATOR SPECIES,
ARY PRODUCTIVITY, PHYTOPLANKTON,
LAMYDOMONAS, CARBON, ECOSYSTEMS,
A, ORGANIC POLLUTANTS, *STREAMS,
TE, DRY WEIGHT, B COLI COMMUNAE,
OPHYTA, CHRYSOPHYTA, RHODOPHYTA,
ION, WAVES(WATER), FISH, SEWAGE,
CHEMISTRY, DIATOMS, CHLOROPHYTA,
NUTRIENTS, MUD-WATER INTERFACES,
INAGE, CHLOROPHYTA, CHRYSOPHYTA,
NA, *ORGANIC CARBON, *CELL SIZE,
NULARIA, NI/ *ONONDAGA LAKE(NY),
DISCUS, DIMEROGRAMMA, DIPLONEIS,
R PRODUCTS, CHEMICAL RESISTANCE,
RANSPARENCY, LIGHT MEASUREMENTS,

NIUM, GYMNODINIUM BREVE, CANADA,
TMENT OF INTERIOR, MADISON(WIS),
TMENT OF INTERIOR, MADISON(WIS),
TRICORNUTUM, EDTA, LIGURIAN SEA,
239, RATS, ZIRCONIUM-95-NIOBIUM,
R POLLUTION EFFECTS, CYANOPHYTA,
ES, RIVERS, RESERVOIRS, BOATING,
IATION, *ALGAE, WASTE TREATMENT,
DUCTIVITY, AQUATIC MICROBIOLOGY,
YNTHESIS, *ALGAE, LAKES, OXYGEN,
OLVED OXYGEN, *ALGAE, COLIFORMS,
BIC CONDITIONS, LIGHT INTENSITY,
WATER QUALITY, *ALGAL CONTROL,
NUTRIENT REQUIREMENTS, BIOASSAY,
E, SEWAGE EFFLUENTS, PHOSPHORUS,
, FILTRATION, SEWAGE, COLIFORMS,
SH, STONEFLIES, SNAILS, ECOLOGY,
TREAMS, WATER POLLUTION EFFECTS,
CYANOPHYTA, PLANKTON, COPEPODS,
KES, INHIBITION, ALGAE, DIATOMS,
ER POLLUTION EFFECTS, ALGAL CON/
TECHNIQUES, CYANOPHYTA, DIATOMS,
ARCH LABORATORY(CALIF), NATIONAL
NDUSTRY/GOVERNMENT TASK FORCE ON
OROMETRY, CALIFORNIA, NUTRIENTS,
AL EFFECTS, ESSENTIAL NUTRIENTS,
*BIOASSAY, PHYTOPLANKTON, LAKES,
PHOSPHORUS, NITROGEN COMPOUNDS,

ESTIMATED COSTS, LEGAL ASPECTS, L	W70-03344
ESTIMATING EQUATIONS, OXYGENATION	W69-03611
ESTUARIES, PHOTOSYNTHESIS, RESPIR	W70-03325
ESTUARIES, BACTERIA, SAMPLING, SE	W70-04368
ESTUARIES, LAKES, GREAT LAKES, PH	W70-03975
ESTUARIES, LAKES, GREAT LAKES, PH	W70-03975
ESTUARIES, PLANKTON, ECOSYSTEMS,	W69-08274
ESTUARIES, BACTERIA, MOLLUSKS, VI	W70-05372
ESTUARIES, PHYSIOLOGICAL ECOLOGY,	W70-05272
ESTUARIES, SANITARY ENGINEERING,	W70-05094
ESTUARIES, WATER QUALITY, DETERGE	W70-10437
ESTUARIES, *ESTUARINE ENVIRONMENT	W71-02494
ESTUARIES, WATER POLLUTION SOURCE	W71-05407
ESTUARIES, ALGAE, WATER POLLUTION	W71-05409
ESTUARINE ENVIRONMENT, IMPOUNDED	W71-02491
ESTUARINE ENVIRONMENT, FALLOUT, N	W69-08267
ESTUARINE MICROBIOLOGY.:	W69-03752
ESTUARINE ENVIRONMENT, WATER QUAL	W69-04276
ESTUARINE ENVIRONMENT, MARINE ALG	W68-00010
ESTUARINE ENVIRONMENT, PONDS, BRE	W73-00748
ESTUARINE ENVIRONMENT, WATER TEMP	W73-01446
ESTUARY.:	W71-05409
ESTUARY(WASH).: *SEATTLE(WASH),	W70-04283
ESTUARY, ENVIRONMENTAL RELATIONSH	W70-05094
ESTUARY, NUTRIENT-PHYTOPLANKTON R	W71-05407
ESTUARY, SAN FRANCISCO BAY.: /IAT	W69-07520
ETHANOL.: /OSA, PESTICIDE DEGRADA	W70-02968
ETHOXYLATE, *ALKYL PHENOL POLYETH	W70-09438
ETHYLENE GLYCOL, POLYMERS, DIMETH	W70-10433
EUCYCLOPS, ALONA, CHAOBORUS.: / T	W70-06975
EUCYCLOPS, NAIS, CRICOTOPUS, AUST	W70-06975
EUDORINA ELEGANS, EUGLENA GRACILI	W69-06540
EUDORINA.: /IZOMENON, ANABAENA, M	W69-01977
EUDORINA, CULTURE MEDIUM, COELAST	W70-05409
EUGLENA GRACILIS, FOREST CENTRIFU	W69-06540
EUGLENA GRACILIS Z, MICROBIOLOGIC	W69-06273
EUGLENA GRACILIS, CACHONINA NIEI,	W69-08278
EUGLENA GRACILIS, ANABAENA CYLIND	W70-02964
EUGLENA.: / SANDS, FLOW RATES, TU	W70-04199
EUGLENA.: /TA, BIOASSAYS, SALINE	W70-03974
EUGLENA, AQUATIC POPULATIONS, EUT	W70-04510
EUGLENA, BACTERIA, YEASTS, ACTINO	W70-08097
EUGLENA, CHLAMYDOMONAS, SAMPLING,	W70-08107
EUGLENA, CHLAMYDOMONAS, CHLORELLA	W70-10173
EUGLENA, CALCIUM, AMMONIUM SALTS,	W70-02964
EUGLENA, CHLAMYDOMONAS, PHOSPHORU	W70-03957
EUGLENA, CHLAMYDOMONAS.: /ISCONSI	W69-01977
EUGLENA, CHLOROPHYTA, NITRATES, A	W72-04761
EUGLENA, EUTROPHICATION, ORGANIC	W69-06273
EUGLENA, NUTRIENTS, PLANT POPULAT	W68-00483
EUGLENE GRACILIS, PROTOCOCCALES.:	W70-05092
EUGLENOPHYTA, SLIME, CO: /A, CYAN	W70-05389
EUGLENOPHYTA, DIATOMS, BIOLOGICAL	W70-03969
EUGLENOPHYTA, CYANOPHYTA, CHRYSOP	W70-03974
EUGLENOPHYTA, TOXICITY, BIOASSAYS	W70-04381
EUGLENOPHYTA, CYANOPHYTA, DIATOMS	W70-02770
EUKARYOTES, MONOCHRYSIS LUTHERI,	W69-08278
EUNOTIA, AMPHORA, GOMPHONEMA, PIN	W70-03974
EUNOTOGRAMMA, FRAGILARIA, GOMPHON	W70-03325
EUPHAUSIA, PHAEOCYSTIS, SARGASSUM	W70-01068
EUPHOTIC ZONE, PLANT CAROTENOIDS,	W70-01933
EUROPE.:	W69-04804
EUROPE, SOUTH AFRICA, ISRAEL, MIC	W70-05372
EUROPE, WATER POLLUTION ASSESSMEN	W70-03975
EUROPE, WATER POLLUTION ASSESSMEN	W70-03975
EUROPEAN MEDITERRANEAN.: /CTYLUM	W69-08275
EUROPIUM-155, SILVER-110M, CERIUM	W69-08269
EURTROPHICATION.: /(BASINS), WATE	W69-10182
EUTHROPHICATION.: /, STREAMS, LAK	W71-09880
EUTHROPHICATION, MODEL STUDIES, B	W73-02218
EUTROPHICATION, MUNICIPAL WASTES,	W72-12240
EUTROPHICATION, *ALGAL CONTROL, BI	W71-10098
EUTROPHICATION(PHYTOPLANKTON), ES	W71-05407
EUTROPHICATION, NUTRIENTS, *FERME	W71-13413
EUTROPHICATION, DAPHNIA, CHLAMYDO	W72-02417
EUTROPHICATION, OLIGOTROPHY, LAKE	W72-03183
EUTROPHICATION, WASHINGTON, LAKES	W70-00270
EUTROPHICATION, ODOR, AMMONIUM CO	W70-00263
EUTROPHICATION, DOMESTIC WASTES,	W70-00711
EUTROPHICATION, NUISANCE ALGAE.: /	W70-00719
EUTROPHICATION, OLIGOTROPHY, TEMP	W69-10154
EUTROPHICATION, CARBON RADIOISOTO	W70-01579
EUTROPHICATION, *CONFERENCES, WAT	W70-02779
EUTROPHICATION, WATER POLLUTION E	W70-02779
EUTROPHICATION RESEARCH PROGRAM,	W70-02775
EUTROPHICATION, PROVISIONAL ALGAL	W70-02777
EUTROPHICATION, CHLOROPHYLL, FLUO	W70-02777
EUTROPHICATION, BICARBONATES, CAR	W70-02792
EUTROPHICATION, SAMPLING, SURFACE	W70-02973
EUTROPHICATION, PRODUCTIVITY, ALG	W70-03068

```
CITY, *CYANOPHYTA, *CHEMCONTROL,          EUTROPHICATION, BASS, WATER POLLU     W70-02982
RATION, WATER POLLUTION EFFECTS,          EUTROPHICATION.: /DS, LIGHT PENET     W70-02255
L STUDIES, COMPUTERS, NUTRIENTS,          EUTROPHICATION.: /ILTRATION, MODE     W70-08627
S, PRODUCTIVITY, PHOTOSYNTHESIS,          EUTROPHICATION, TOXICITY, MIDGES,     W70-06225
S, NUTRIENT BUDGET.:    *CULTURAL         EUTROPHICATION, TROPHIC INDICATOR     W71-03048
, CHEMICAL PRECIPITATI/ CULTURAL          EUTROPHICATION, SPECIES DIVERSITY     W69-08518
WATER QUALITY, WISCONSIN, LAKES,          EUTROPHICATION.: /UTION EFFECTS,      W69-05868
 WASTE, GROUND WATER, *NITRATES,          EUTROPHICATION, *WATER POLLUTION      W69-05323
HOSPHORUS, PONDS, PHYTOPLANKTON,          EUTROPHICATION, LAKES, LIMNOLOGY,     W69-05142
QUATIC WEEDS, BALANCE OF NATURE,          EUTROPHICATION, FISH FARMING, FIS     W69-04804
A, CHLOROPHYTA, IRON, MANGANESE,          EUTROPHICATION, WATER POLLUTION C     W69-05704
TERFACE, PHOSPHORUS, ADSORPTION,          EUTROPHICATION, ALGAE, MACROAQUAT     W69-06859
 WASTE EFFLUENTS.:      *CULTURAL         EUTROPHICATION, RECREATIONAL USE,     W69-06535
LAKE TAHOE(CALIF), L/ *CULTURAL           EUTROPHICATION, LAKE MONONA(WIS),     W69-06858
AS, CARBON, ECOSYSTEMS, EUGLENA,          EUTROPHICATION, ORGANIC MATTER, P     W69-06273
F/ *ALGAE, *BIOASSAY, NUTRIENTS,          EUTROPHICATION, WATER POLLUTION E     W69-06864
IC IMBALANCES.:                           EUTROPHICATION PREVENTION, ECOLOG     W69-07084
TRIENTS, *TRACE ELEMENTS, ALGAE,          EUTROPHICATION, MOLYBDENUM, NITRO     W69-06277
FERTILIZERS, *CYCLING NUTRIENTS,          EUTROPHICATION, ALGAE, PLANT GROW     W69-07826
H RATES, *ALGAE, PHOTOSYNTHESIS,          EUTROPHICATION, LIGHT, TEMPERATUR     W69-07442
BIOINDICATORS, LIGHT INTENSITY,           EUTROPHICATION, GRAZING, ALGAE, W     W70-03978
T, *GREAT LAKES, CONTROL, COSTS,          EUTROPHICATION, NITRATES, LAKE ER     W70-03964
OLOGICAL COMMUNITIES, EFFLUENTS,          EUTROPHICATION, PHYTOPLANKTON, DE     W70-04001
IS, *SAMPLING, PLANT PHYSIOLOGY,          EUTROPHICATION, DATA COLLECTION,      W70-04382
YSIOLOGICAL ECOLOGY, CYANOPHYTA,          EUTROPHICATION, NUISANCE ALGAE, N     W70-04381
AL EFFECTS, ESSENTIAL NUTRIENTS,          EUTROPHICATION, LIGHT INTENSITY,      W70-04283
STREAMS, FISH, LAND, ECOSYSTEMS,          EUTROPHICATION, OLIGOTROPHY, SEWA     W70-04193
TON, DIFFUSION, ALGAE, BACTERIA,          EUTROPHICATION, GLUCOSE, CARBON R     W70-04194
TER TREATMENT, DISSOLVED OXYGEN,          EUTROPHICATION, FISH, LAND, VALUE     W70-05264
ANOPHYTA, PLANKTON, ZOOPLANKTON,          EUTROPHICATION, STORM DRAINS, SEW     W70-05112
OLOGICAL COMMUNITIES, BIOASSAYS,          EUTROPHICATION.: /FOOD CHAINS, BI     W70-05272
                                          EUTROPHICATION TREATMENT.:            W70-04721
S, EUGLENA, AQUATIC POPULATIONS,          EUTROPHICATION, CYANOPHYTA, SEWAG     W70-04510
CIDES, *WATER POLLUTION CONTROL,          EUTROPHICATION, ALGAE, FILTRATION     W70-05470
, *ODORS, *ODOR PRODUCING ALGAE,          EUTROPHICATION, ALCOHOLS, ACTINOM     W69-00387
VIRONMENT, AQUATIC PRODUCTIVITY,          EUTROPHICATION, *LAKES, LIMNOLOGY     W69-00446
                   ACCELERATED            EUTROPHICATION.:                      W68-01244
ANOPHYTA, ENVIRONMENTAL EFFECTS,          EUTROPHICATION, LIMNOLOGY, NITROG     W69-03186
ANOPHYTA, ENVIRONMENTAL EFFECTS,          EUTROPHICATION, LIMNOLOGY, NITROG     W69-03185
ISMS, NITROGEN, NUTRIENT/ ALGAE,          EUTROPHICATION, LAKES, MICROORGAN     W68-00481
HYDRO/ *ACTIVATED SLUDGE, ALGAE,          EUTROPHICATION, LIME, *NITROGEN,      W68-00012
UES, CLAY LOAM, CLAYS, COLLOIDS,          EUTROPHICATION, IMPOUNDMENTS, LAK     W69-04800
ONTROL, WATER POLLUTION CONTROL,          EUTROPHICATION, FISHKILL, ALGAE,      W69-03948
AL EFFECTS, ESSENTIAL NUTRIENTS,          EUTROPHICATION, INHIBITION, LIGHT     W69-03514
ON POTENTIAL, CYCLING NUTRIENTS,          EUTROPHICATION.: /IDATION-REDUCTI     W69-04521
ANOPHYTA, ENVIRONMENTAL EFFECTS,          EUTROPHICATION, LIMNOLOGY, NITROG     W69-04802
AL EFFECTS, ESSENTIAL NUTRIENTS,          EUTROPHICATION, FREEZE DRYING, FR     W69-03518
AL EFFECTS, ESSENTIAL NUTRIENTS,          EUTROPHICATION, INHIBITION, LIGHT     W69-03515
E AGING, NUTRIENT SOU/ *CULTURAL          EUTROPHY, *NUTRIENT BALANCE, *LAK     W70-00264
, TEMPERATURE, PRESSURE, MIXING,          EVAPORATION, SUPERSATURATION, RES     W70-00683
, *TROPHIC LEVEL, WATER STORAGE,          EVAPORATION, WATER CIRCULATION, S     W69-09723
ESTICIDE KINETICS, *PERSISTENCE,          EVAPORATION, BIODEGRADATION, ADSO     W70-09768
SOLVED SOLIDS, DISSOLVED OXYGEN,          EVAPORATION CONTROL, WATER BUDGET     W72-01773
AT BUDGET, TURBIDITY, *RESERVOIR         EVAPORATION, SALINITY, ALGAE, STR     W72-01773
COLIFORMS, CYTOLOGICAL STUDIES,           EVAPORATION, MICROORGANISMS, CALI     W71-05267
BIOCHEMICAL OXYGEN DEMAND(BOD),           EVAPORATION, FERMENTATION, METHAN     W70-04787
MICHAELIS-MENTEN EQUATION, LOCH           EWE(SCOTLAND), SCOTLAND, PHAEOPIG     W70-04811
ES, *COPPER DOSAGE/ *MICROSCOPIC          EXAMINATION, *CRITICAL TEMPERATUR     W69-10157
LIDE ACCUMULATION, *RADIONUCLIDE          EXCHANGE, *OPEN SEA, WEAPONS TEST     W70-00707
HIPS, SUBMARINES, EFFLUENTS, ION          EXCHANGE, RESINS, LEAKAGE, CESIUM     W70-00707
 WATER POLLUTION EFFECTS, CATION          EXCHANGE, CATION ADSORPTION, HYDR     W71-09863
COMPOUNDS, CHEMICAL WASTES, ION           EXCHANGE, BIOCHEMICAL OXYGEN DEMA     W70-10433
ENT, ALGAL CONTROL, AMMONIA, ION          EXCHANGE, REVERSE OSMOSIS, DISTIL     W70-01981
TREATMENT, ELECTRODIALYSIS, ION           EXCHANGE, ACTIVATED SLUDGE, DENIT     W69-01453
, HYPOLIMNION, THERMOCLINE, *ION          EXCHANGE, WATER CIRCULATION, MASS     W68-00857
ONS, EXTRACELLULAR PRODUCTS, GAS          EXCHANGE, GREEN PLANTS, MASS CULT     W69-07832
ON, *GROUNDWATER, RECHARGE, *ION          EXCHANGE, *ORGANIC MATTER, *FILTR     W72-04318
ITION, MICROFOSSILS, COMPETITIVE          EXCLUSION PRINCIPLE, OSCILLATORIA     W70-05663
IDS, TA/ *ANTIBIOSIS, AUTOLYSIS,          EXCRETION, DEPURATION, PHYCOCOLLO     W71-01068
*GRANULAR LEUCOCYTES, *MUSCULAR           EXERTION, *2-METHYL-5-VINYLPYRIDI     W70-01788
N CLUB HUTS, TEGERNSEE, GERMANY,          EXPANDING TECHNOLOGY, OVERPOPULAT     W69-07818
K(ORE), OREGON STATE UNIVERSITY,          EXPERIMENTAL STREAMS, SEASONAL EF     W70-03978
, CAYUGA LAKE(NY), / *ENRICHMENT          EXPERIMENTS, *NUTRIENT LIMITATION     W70-01579
CROPHYTES, POTAMOGETON FOLIOSUS,          EXPERIMENTAL LIMNOLOGY, BRACHIONU     W70-02249
E, BLUE-GREEN ALGAE, TEMPERATURE          EXPERIMENTS, RECYCLED WATER.: /GA     W70-07257
*JAPAN, *SLAUGHTERHOUSE WASTES,           EXPERIMENTAL STUDIES, LAGOON DESI     W70-07508
, URANIUM RADIOISOTOPES, NUCLEAR          EXPLOSIONS, STRONTIUM RADIOISOTOP     W70-00708
RESINS, LEAKAGE, CESIUM, NUCLEAR          EXPLOSIONS, ABSORPTION, ADSORPTIO     W70-00707
BOTTOM SEDIMENTS, SOILS, NUCLEAR          EXPLOSIONS, COBALT RADIOISOTOPES,     W69-08269
NE ENVIRONMENT, FALLOUT, NUCLEAR          EXPLOSIONS, BIOTA, SPECTROMETERS,     W69-08267
VERSITY, AUTECOLOGY, MORPHOLOGY,          EXPORT LOSSES, CHEMOSYNTHESIS, AB     W70-02780
VATIONS, CHRONIC EXPOSURE, ACUTE          EXPOSURE.: /, PHYSIOLOGICAL OBSER     W70-01779
SIOLOGICAL OBSERVATIONS, CHRONIC          EXPOSURE, ACUTE EXPOSURE.: /, PHY     W70-01779
MONIUM COMPOUNDS, MINIMUM LETHAL          EXPOSURE, PHENANTHROQUINONE, MADI     W70-02982
HOTO-ELECTRIC TURBIDOSTAT, LIGHT          EXTINCTION COEFFICIENT, ALGAL CEL     W70-03923
ALA(SWEDEN), LAKE ERKEN(SWEDEN),          EXTINCTION VALUES, EUDORINA, CULT     W70-05409
 *ISRAEL, *ALGAL CONCENTRATION, '         EXTRA-DEEP PONDS', ASHKELON(ISRAE     W70-04790
OLOGY, ENVIRONMENTAL CONDITIONS,          EXTRACELLULAR PRODUCTS, GAS EXCHA     W69-07832
IC RATE, PHYTOPLANKTON DYNAMICS,          EXTRACELLULAR PRODUCTS, ALGAL CEL     W69-05697
AUXINS, METABO/ AUTOINHIBITORS,           EXTRACELLULAR PRODUCTS, HORMONES,     W69-04801
```

S, FLAGELLATES, PHARMACEUTICALS,

S), *AQUATIC WEED CUTTING, *WEED
L PHENOL POLYETHOXYLATE, ORGANIC
ANOPHOSPHORUS COMPOUNDS, SOLVENT
OGRAPHY, PHYSIOLOGY, BIOCHEMICAL
.:   *BIOLOGICAL
STS, INFRARE/ CHLOROFORM-METHANE
GINIC ACID, FUCUS VESI/ *SEAWEED
GAE, *LITTORAL, *ORGANIC MATTER,
ACROPORA HYACINTHUS, POCILLOPORA
MEASUREMENT APPARATUS, HYDROGEN
OCCOLITHOPHORES, SEAWATER MEDIUM
LARIS, CHLOROPHYLL FLUORESCENCE,
LIFORNIA WATER PLAN, DELTA WATER
ATION, *PONDS, *DESIGN CRITERIA,
MS, FISH GENETICS, FISH HANDLING
WASTE WATER TREATMENT, TREATMENT
ENTS, QUALITY CONTROL, TREATMENT
ENTS, BACTERIA, ALGAE, TREATMENT
ENTS, BACTERIA, ALGAE, TREATMENT
WASTE WATER TREATMENT, TREATMENT
, *FILTERS, *NEW YORK, TREATMENT
LTRAVIOLET RADIATION, *TREATMENT
RBON, CHLORINE, COSTS, TREATMENT
OMACEOUS EARTH, ALGAE, TREATMENT
N, MEMBRANES, SCREENS, TREATMENT
ERUGINOSA, ENDOTOXIN, FAST-DEATH
GRAPHS, HALF-LIFE, CONCENTRATION
ASTERIONELLA GRACILLIMA, GROWTH
DUCTI/ *PHYTOPLANKTON, *LIMITING
ENVIRONMENTAL EFFECTS, *LIMITING
S, KINETIC THEORY, ENVIRONMENTAL
ES, ALGAE, OLIGOTROPHY, LIMITING
LKALIGENES), RHODOTORULA, GROWTH
PHYSIOLOGICAL ECOLOGY, LIMITING
ACID, FUCUS VESICULOSUS, GROWTH
ISHKILL, FERTILIZATION, LIMITING
TARD GROWTH TEST, ORGANIC GROWTH
WAUBESA, LYNGB/ *BLOOMS, *BIOTIC
, ORGANIC DEBRIS, ORGANIC GROWTH
STUDIES, LAGOON DESIGN, SCALE-UP
PAPERMILL WASTE EFFLUENT, CHEESE
L ENVIRONMENT, PHAGE, FERMENTOR,
OVAL RATES, DESIGN APPLICATIONS,
: *SULFATE ION, *DESULFOVIBRIO,
INATION, TEMPERATURE, NUTRIENTS,
S, MULTI-CELL SERIAL SYSTEM, PH,
DUS SUBTERRANEUS, ANKISTRODESMUS
ANIMALS, ESTUARINE ENVIRONMENT,
BON RADIOISOTOPES, BIOTA, CLAMS,
STRONTIUM RADIOISOTOPES, CESIUM,
SCA, MICROCOLEUS LACUSTRIS (RAB)
ATION, *WATER POLLUTION CONTROL,
PHYSIOLOGICAL ECOLOGY, KINETICS,
SOIL EROSION, SEWAGE EFFLUENTS,
VERS, NITROGEN, BIOMASS, SEWAGE,
IES, SPRAYING, ACTIVATED SLUDGE,
MESTIC WASTE, INDUSTRIAL WASTES,
, WATER POLLUTION SOURCES, ODOR,
, SLUDGE, PHOTOSYNTHETIC OXYGEN,
ONTROL, *NUTRIENTS, *PESTICIDES,
MATE DISPOSAL, WATER REUSE, FISH
OF NATURE, EUTROPHICATION, FISH
ALKALINE WATER, ALKALINITY, FISH
, PHOSPHORUS, DRAINAGE, MARSHES,
CROCYSTIS AERUGINOSA, ENDOTOXIN,
PHYCEAE, CHLAMYDOMONAS MOEWUSSI,
N OF ALGAE, CONTINUOUS CULTURES,
CONSIN, BOTTOM SEDIMENT, BENTHIC
ITIES, *WASTE DISCHARGES, FLORA,
S, LAKES, ECOLOGY, PRI/ *BENTHIC
BIOLOGICAL CO/ *LAKES, *BENTHIC
C LIFE, AQUATIC ANIMALS, BENTHIC
TROL, *NUTRIENTS, ALGAE, BENTHIC
RICE, SILTS, PLANKTON, MAYFLIES,
S,/ ALGAE, BRACKISH, ECOSYSTEMS,
VEGETATION FAUNA, SANDY BOTTOMS
OPLANKTON, DIATOMS, ZO/ *BENTHIC
LANKTON, FISH, BACTERIA, BENTHIC
RANSPARENCY, MARGINAL VEGETATION
UISANCE ALGAE, PLANKTON, BENTHIC
OXIDATION PONDS, ALGAE, BENTHIC
ARRAGANSETT BAY, CAPE REYKJANES,
TERIA, IRRIGATION, WILLOW TREES,
PLANTS, AQUATIC WEEDS, ECONOMIC
, FLOCCULATION, MIXING, ECONOMIC
ECONOMIC FEASIBILITY, TECHNICAL
ON TREATMENT, FINANCING, GRANTS,
LA PYRENOIDOSA, SYNTHETIC SEWAGE

EXTRACELLULAR PRODUCTS, CHEMICAL	W70-01068
EXTRACTABLE PHOSPHORUS.:	W69-03075
EXTRACTING STEEL CABLES, ALGAL SU	W70-04494
EXTRACTIONS, INFRARED SPECTROSCOP	W70-03928
EXTRACTIONS, PHOSPHORUS COMPOUNDS	W70-05269
EXTRACTION, MILLIPORE FILTERS, PH	W69-10163
EXTRACTION, GROWTH, SHOCK LOADING	W70-06604
EXTRACTS, LIPID CARBON, CHLOROPLA	W70-05651
EXTRACTS, ASCOPHYLLUM NODOSUM, AL	W69-07826
EXUDATION, TIDES, PHOTOSYNTHESIS,	W70-01073
EYDOUXI, PORITES LOBATA, LEPTASTR	W69-08526
F LINE.: /OFER LINES, ATTENUATION	W70-05377
F.: /NIQUES, MICROBIAL ECOLOGY, C	W70-05272
FABRY-PEROT FILTERS, SOLAR FRAUNH	W70-05377
FACILITIES(CALIF), OXYGEN DYNAMIC	W70-05094
FACILITIES, EQUATIONS, AEROBIC CO	W70-04786
FACILITIES, FISH MANAGEMENT, LIMN	W69-04804
FACILITIES, AERATION, LAGOONS, DO	W70-04060
FACILITIES, FILTRATION, SEWAGE, C	W70-00263
FACILITIES, TEST PROCEDURES, META	W70-01519
FACILITIES, TEST PROCEDURES, META	W71-06737
FACILITIES, MISSISSIPPI.: /UES, *	W71-13313
FACILITIES.: /OLLOIDS, FLOW RATES	W72-04145
FACILITIES, FILTERS.: /*SLIME, *U	W72-03824
FACILITIES, CLEANING, *ALGAL CONT	W72-08859
FACILITIES.: /, *FILTRATION, DIAT	W72-04318
FACILITIES, *WISCONSIN.: /FICATIO	W73-02426
FACTOR, DOSAGE, SPECIES SUSCEPTIB	W70-00273
FACTOR, LIMNODRILUS SPP, AEOLOSOM	W69-07861
FACTORS.: /ROTONENSIS, NITZSCHIA.	W70-02804
FACTORS, *NUTRIENTS, *PRIMARY PRO	W70-01579
FACTORS, BENTHIC FLORA, BENTHOS,	W70-02770
FACTORS, BIOMASS, GROWTH RATES, A	W70-02779
FACTORS, CHLORELLA, ANALYTICAL TE	W70-02779
FACTORS, COLONIAL MORPHOLOGY, CEL	W70-03952
FACTORS, ECOSYSTEMS, MICROORGANIS	W71-02479
FACTORS, LAMINARIA SACCHARINA, MA	W69-07826
FACTORS, NITROGEN COMPOUNDS, NUTR	W69-06535
FACTORS, POLYSACCHARIDES.: /, MUS	W69-07826
FACTORS, SPECIES DIVERSITY, LAKE	W69-01977
FACTORS, VITAMIN B1, VITAMIN B6,	W70-02510
FACTORS, WASTE COMPOSITION.: /AL	W70-07508
FACTORY WASTE WATERS.: /GANISMS,	W71-00940
FACULTATIVE, MICROAEROPHILIC, INO	W71-02009
FACULTATIVE PONDS, INFRARED SPECT	W70-07034
FACULTATIVE, BOD, DETENTION TIME.	W70-01971
FACULTATIVE POND DESIGNS, BOD, SL	W70-04786
FACULTATIVE, ANAEROBIC POND.	W70-04790
FALCATUS, CHLAMYDOMONAS MOEWUSII,	W70-05381
FALLOUT, NUCLEAR EXPLOSIONS, BIOT	W69-08267
FALLOUT, PACIFIC OCE: /ITIUM, CAR	W69-08269
FALLOUT, SILTS, DETRITUS, BOTTOM	W69-09742
FARLOW, CLADOPHORA, TAXONOMIC DES	W70-04468
FARM WASTES, MUNICIPAL WASTES, SE	W70-04266
FARM WASTES, BENTHOS, NITRATES, F	W70-02775
FARM WASTES. POPULATION, INDUSTRI	W70-02803
FARM WASTES, CARBON CYCLE, PHOSPH	W70-02777
FARM WASTES.: /AE, *CATTLE, SLURR	W70-07491
FARM WASTES.: /ANKTON, SEWAGE, DO	W72-01813
FARM WASTE, SLUDGE, SOLID WASTES,	W71-05742
FARM WASTES, WASTE WATER TREATMEN	W71-08221
FARM WASTES, SILT, EROSION, SEDIM	W71-10376
FARMING.: / SULFITE LIQUORS, ULTI	W69-05891
FARMING, FISH ROOD ORGANISMS, FIS	W69-04804
FARMING, FISH FOOD ORGANISMS, FIS	W71-07731
FARMS, CYANOPHYTA, PLANKTON, ZOOP	W70-05112
FAST-DEATH FACTOR, DOSAGE, SPECIE	W70-00273
FAT ACCUMULATION.: /A PALEA, MYXO	W70-02965
FATTY ACIDS, OOCYSTIS POLYMORPHA,	W70-07957
FAUNA.: /ACE ELEMENTS, CORES, WIS	W69-03366
FAUNA, *BIOLOGICAL COMMUNITIES, I	W68-00010
FAUNA, *EUTROPHICATION, *NUTRIENT	W71-00665
FAUNA, *TROPHIC LEVELS, *SURVEYS,	W71-12077
FAUNA, AQUATIC PLANTS, MARINE ALG	W71-02494
FAUNA, CHEMICAL ANALYSIS, SESSILE	W72-06052
FAUNA, FISH, OLIGOCHAETES, MIDGES	W70-01943
FAUNA, FLORA, LABORATORY, MINERAL	W72-02203
FAUNA, MARGINAL VEGETATION, MICRO	W70-06975
FAUNA, NITROGEN, PHOSPHORUS, PHYT	W70-04253
FAUNA, PHOTOSYNTHESIS, BIOASSAYS,	W70-02251
FAUNA, SANDY BOTTOMS FAUNA, MARGI	W70-06975
FAUNA, SLUDGE WORMS, BLOODWORMS,	W68-01244
FAUNA, WATERSHEDS, WATER SHORTAGE	W72-00027
FAXAFLOI FJORD, ICELAND, DNA BASE	W69-10161
FE: /S, CRAYFISH, MOSQUITOES, BAC	W70-05263
FEASIBILITY, LAKES, NUISANCE ALGA	W69-06276
FEASIBILITY, TECHNICAL FEASIBILIT	W71-10654
FEASIBILITY, COST ANALYSIS, SLUDG	W71-10654
FEDERAL GOVERNMENT, STATE GOVERNM	W69-06305
FEED, SYMBIOTIC RELATIONSHIPS.: /	W70-05655

HALF-LIFE(RADIONUCLIDE), FILTER-

*FWPCA, ACTIVATED ALGAE,
NDOTHAL, POTASSIUM PERMANGANATE,
*METHANE
OXYGEN DEMAND(BOD), EVAPORATION,
T, *PHYSICAL ENVIRONMENT, PHAGE,
LTER MEDIA, ALUMINA-COATED SAND,
CANALIZATION, ALUMINUM SULFATE,
ARIA, FORAMINIFERA, THIOBACILLUS-
HYDROGE/ STRIP MINES, DRAINAGE,
MPLING, SURFACES WATERS, CARBON,
ARIES, SAMPLING, FLOW, ANALYSIS,
L DRAINAGE, BIOLOGICAL RESPONSE,
N, PONDS, RESERVOIRS, BACKWATER,
EMENT, *LAKE WASHINGTON, *SEWAGE
ZERS.:   ARTIFICIAL
ION, ALGAE, MACROAQUATIC PLANTS,
TIC WEEDS, CYANOPHYTA, FISHKILL,
FISHKI/ *LAKES, *EUTROPHICATION,
NIMALS, AQUATIC PLANTS, BIOMASS,
TROPHICATION, *LAKES, NUTRIENTS,
ICULTURAL WATERSHEDS, NUTRIENTS,
W, MEASUREMENT, WATER POLLUTION,
LAKES, *TROUT, *FISH MANAGEMENT,
KES, *PONDS, LIGHT, TEMPERATURE,
RAINAGE, MECHANIC/ LAKE LUCERNE,
TION, *NUTRIENTS, *LAKES, PONDS,
RIVERS, *EUTROPHICATION, RUNOFF,
TER CONSERVATION, WATER CONTROL,
STES, SEWAGE WASTES, BYPRODUCTS,
EROSION, SEDIMENTATION, RUNOFF,
TION, FISH, LAND, VALUE, LEAVES,
IS SCUMS, PITHIOPHORA/ PHOSPHATE
PITATION(ATMOSPHERIC), DRAINAGE,
, */ *DOMESTIC ANIMALS, *WASTES,
TION SOURCES, *LAKES, NUTRIENTS,
FICIAL FERTILIZATION, COMMERCIAL
HEDS, ORGANIC MATTER, SEDIMENTS,
), SYNURA, ASTERIONELLA, TYPHOID
TION CONTROL, NITRATES, NITROGEN
SE RAYON PLANT WASTES, *MAN-MADE
TI-FOAM AGENTS, *TEXTILE WASTES,
LCOHOL, ETHYLENE GLY/ *POLYESTER
GLY/ *POLYESTER FIBERS, MAN-MADE
, *ALGAE, ECOLOGY, SOUTH DAKOTA,
N LAKES, YAHARA RIVER(WI/ *TRUAX
CHANIC/ LAKE LUCERNE, FERTILIZED
CONTROL ORGANIZATION OF TORONTO,
OTO REACT/ *CLADOPHORA, BROMINE,
HETIC, SAPROPHYTIC, PLATE COUNT,
CILLATORIA RUBESCENS, ZURICHSEE,
S, ZURICHSEE, FILAMENTOUS ALGAE,
OCCUS LIVIDUS, BLUE-GREEN ALGAE,
KE(NY), *MARL LAKES, CHLOROBIUM,
MICROSCOPY, STAINING, STRIPPING
*DIATOMITE FILTERS,

S, DICERAS CHODATI, DINOPHYCEAE,
LARIA, FRAGILARIA, ASTERIONELLA,
TONIUM, HALF-LIFE(RADIONUCLIDE),
TRESCIBILITY TEST, BODY IMMEDIUM
IC,/ TAXONOMIC GROUPS, MOLECULAR
ATION MEDIA, ANTHRACITE, GRAVITY
ADIATION, *TREATMENT FACILITIES,
*MILLIPORE
FILTER AIDS, DIATOMITE
ACTIVATED CARBON, *ELECTROLYTES,
*ALGAE, *FILTRATION, HEAD LOSS,
S, *SEWAGE TREATMENT, *TRICKLING
AERATION, EFFLUENTS, FLOW RATES,
POLYELECTROLYTE-COATED
LA/ *ALGAE, *SANDS, *FILTRATION,
S, SNAILS, ODOR, COOLING TOWERS,
*DIATOMITE
IA, COLOR, IRON BACTERIA, TASTE,
BURG, BUFFALO, DUNREITH/ *CARBON
INIA, ORGANIC COMPOUNDS, LAGOON,
IOCHEMICAL EXTRACTION, MILLIPORE
R SUPPLY, *CONTROL, TASTE, ODOR,
OPHYLL FLUORESCENCE, FABRY-PEROT
E, ADSORPTION, ACTIVATED CARBON,
AUNA, WATERSHEDS, WATER SHORTAGE
ESS, SEWAGE TREATMENT, TRICKLING
RIENTS, INDICATORS, ODOR, TASTE,
CONTROL, EUTROPHICATION, ALGAE,
SPRAYING, ALGAE, CENTRIFUGATION,
EN ALGAE, HEAD LOSSES, OOCISTUS,
ENATION, AERATION, FLOCCULATION,
SPHATES, CHEMICAL PRECIPITATION,

FEEDING ANIMALS, RUTHENIUM-106, P W70-00708
FEEDLOTS, WABASH RIVER BASIN.: W71-10376
FEEDLOTS, OXIDATION DITCH.: W71-06821
FENAC, SILVER NITRATE, SIMAZIN, N W70-10175
FERMENTATION.: W71-13413
FERMENTATION, METHANE, ALGAL CONT W70-04787
FERMENTOR, FACULTATIVE, MICROAERO W71-02009
FERRIC OXIDE-COATED SAND, CRUSHED W70-08628
FERRICHLORIDE, IRON SULFATE, GALV W70-05668
FERROBACILLUS, THIOBACILLUS DENIT W69-07428
FERROBACILLUS, WATER TEMPERATURE, W68-00891
FERTILITY, ALGAE, NITROGEN, PHOSP W70-02973
FERTILIZATION, PLANT GROWTH SUBST W70-02969
FERTILIZATION SURVEY, MADISON LAK W70-02787
FERTILIZATION, NITROGEN COMPOUNDS W70-00274
FERTILIZATION EFFECTS, *BIOLOGICA W69-09349
FERTILIZATION, COMMERCIAL FERTILI W68-00487
FERTILIZATION.: /ION, EUTROPHICAT W69-06859
FERTILIZATION, LIMITING FACTORS, W69-06535
FERTILIZATION, FISH, *LIMNOLOGY, W69-05844
FERTILIZATION, LAKES, LIMNOLOGY, W69-04805
FERTILIZATION, BACTERIA, ALGAE, P W70-05761
FERTILIZATION, PHYSICAL PROPERTIE W70-05565
FERTILIZATION, INDUSTRIAL WASTES, W70-05112
FERTILIZATION, ALGAE, PLANKTON, N W70-05399
FERTILIZATION, FISH, NUISANCE ALG W70-05404
FERTILIZED FIELDS, SUBTERRANEAN D W69-08668
FERTILIZERS, ALGAE, NITROGEN, PHO W69-08668
FERTILIZERS, NUTRIENTS, NITRATES, W69-10170
FERTILIZERS, SEWAGE EFFLUENT, NUT W70-00269
FERTILIZERS, ALGAE, ZOOPLANKTON, W70-09905
FERTILIZERS, NITROGEN, PHOSPHORUS W71-10376
FERTILIZERS, PESTICIDES, CELLULOS W70-05264
FERTILIZERS, BLUEGILLS, MICROCYST W70-05417
FERTILIZER: /FF, UNDERFLOW, PRECI W70-04810
FERTILIZERS, *ESSENTIAL NUTRIENTS W69-05323
FERTILIZERS, PHOSPHATES, NITRATES W69-07084
FERTILIZERS.:   ARTI W68-00487
FERTILIZERS, WATER POLLUTION, SNO W70-04193
FEVER, KENNEBEC RIVER(MAINE), COB W70-08096
FI: /LLUTION EFFECTS, WATER POLLU W69-10167
FIBER PLANTS, NYLON-6, *VISCOSE R W70-06925
FIBER WASTES, DYEING WASTES.: /AN W70-04060
FIBERS, MAN-MADE FIBERS, METHYL A W70-10433
FIBERS, METHYL ALCOHOL, ETHYLENE W70-10433
FIELD INVESTIGATIONS, WATER POLLU W69-00659
FIELD(MADISON) EFFLUENTS, *MADISO W70-05113
FIELDS, SUBTERRANEAN DRAINAGE, ME W69-08668
FILAMENTOUS ALGAE, NAVICULA, APHA W69-10155
FILAMENTOUS ALGAE, KILL RATES, PH W70-02364
FILAMENTOUS, UNICELLULAR, NATURAL W68-00891
FILAMENTOUS ALGAE, FILAMENTOUS BL W70-04253
FILAMENTOUS BLUE-GREEN ALGAE, OCU W70-04253
FILAMENTS.: /RMIS, MATS, SYNECHOC W70-05270
FILINIA, MONIMOLIMNION, CHEMOCLIN W70-05760
FILM, SYNECHOCOCCUS, GRAIN COUNT(W69-10163
FILTER AIDS.: W72-04145
FILTER AIDS, DIATOMITE FILTERS.: W72-04318
FILTER BLOCKING, HETERETROPHS, IT W69-07833
FILTER CLOGGING, STEPHANODISCUS H W70-00263
FILTER-FEEDING ANIMALS, RUTHENIUM W70-00708
FILTER, BACK-WASHING.: / FLOW, PU W68-00891
FILTER, CHEMOSYNTHETIC, SAPROPHYT W68-00891
FILTERS.: /OSSES, OOCISTUS, FILTR W70-04199
FILTERS.: /*SLIME, *ULTRAVIOLET R W72-03824
FILTERS.: W72-09693
FILTERS.: W72-04318
FILTERS, *ANALYTICAL TECHNIQUES.: W72-09693
FILTERS, *SCENEDESMUS, FLOCCULATI W72-03703
FILTERS, *BIOTA, SEWAGE BACTERIA, W72-01819
FILTERS, ALGAE.: /ANTS, HUMUS, RE W70-05821
FILTERS, BAUXITE.: W70-01027
FILTERS, ELECTROPHORESIS, CHLOREL W71-03035
FILTERS, EFFLUENT SAMPLING.: /TER W72-00027
FILTERS, FILTER AIDS.: W72-04145
FILTERS, IOWA, DIATOMS, EUGLENA, W70-08107
FILTERS, LEBANON, BATAVIA, CLARKS W72-03641
FILTERS, PESTICIDES, SLUDGE, DIGE W72-03641
FILTERS, PHASE MICROSCOPY, STAINI W69-10163
FILTERS, PLANKTON, WATER POLLUTIO W70-05389
FILTERS, SOLAR FRAUNHOFER LINES, W70-05377
FILTERS, SEASONAL, STRATIFICATION W70-04001
FILTERS, SNAILS, ODOR, COOLING TO W72-00027
FILTERS, WASTE WATER TREATMENT, P W70-06792
FILTERS, WATER TREATMENT, DISSOLV W70-05264
FILTRATION, SORPTION, FLOCCULATIO W70-05470
FILTRATION, ANEROBIC DIGESTION, I W70-04060
FILTRATION MEDIA, ANTHRACITE, GRA W70-04199
FILTRATION, MODEL STUDIES, COMPUT W70-08627
FILTRATION, AMMONIA, WATER PURIFI W71-11393

444

ION LAGOON, NITROGEN, NUTRIENTS,
IMOMYCETES, CYTOLOGICAL STUDIES,
RADIA/ *WATER TREATMENT, *ALGAE,
, *WASTE WATER TREATMENT, ALGAE,
R, *ALGICIDES, ACTIVATED CARBON,
WATER QUALITY, *WATER TREATMENT,
CYCLINE, STREPTOMYCIN, MILLIPORE
Y CONTROL, TREATMENT FACILITIES,
    *ALGAE HARVESTING, *DIATOMITE
*ALGICIDES, ALGAE, *EVALUATION,
MICHIGAN, WATER QUALITY CONTROL,
RCES, WATER POLLUTION TREATMENT,
FRAGILARIA, RADIOCARBON METHOD,
PROMOSUS, TRICHOPTERA, LEECHES,
, LAKE TEGERN(GERMANY), GERMANY,
MUS QUADRICAUDA, CEL/ STRIPPING,
ANABAENA CYLINDRICA, NORTH SEA,
ASTES, *WATER POLLUTION CONTROL,
TON, ALKALINE WATER, ALKALINITY,
ULTIMATE DISPOSAL, WATER REUSE,
LANCE OF NATURE, EUTROPHICATION,
*WATER QUALITY CONTROL, *MIDGES,
AQUATIC INSECTS, *AQUATIC DRIFT,
WATER, ALKALINITY, FISH FARMING,
SH FARMING, FISH ROOD ORGANISMS,
H ROOD ORGANISMS, FISH GENETICS,
, DREDGING, HARVESTING OF ALGAE,
IENT LOAD, SYNTHETIC DETERGENTS,
ETICS, FISH HANDLING FACILITIES,
NT, *WATER QUALITY CONTROL, *MI/
ATIC DRIFT, FISH FOOD ORGANISMS,
SH FARMING, FISH FOOD ORGANISMS,
STANDING CROP, ANNUAL TURNOVER,
OPHYTA/ *EUTROPHICATION, SEWAGE,
TROPHICATION, *NUTRIENTS, ALGAE,
LGAE, PROTOZOA, AQUATIC INSECTS,
E, EUTROPHICATION, FISH FARMING,
NALYSIS, *LAKE ERIE, PHOSPHORUS,
FOOD ORGANISMS, FISH PHYSIOLOGY,
ECTS, DINOFLAGELLATES, FISHKILL,
HETIC DETERGENTS, FISH MANAGERS,
GROWTH RATES, ALGAE, FRESH WATER
COMMERCIAL FISH, SALMON, ALGAE,
A, IRON, ALGICIDES, ZOOPLANKTON,
ATTER, HARDNESS, CARBON DIOXIDE,
*EUTROPHICATION, FERTILIZATION,
LEECHES, FINGERNAIL CLAMS, WHITE
TERS, BIOCHEMICAL OXYGEN DEMAND,
ODEGRADATION, WATER UTILIZATION,
TRACERS, CRABS, SHRIMP, MULLETS,
ROL, *TOXINS, ALGAE, RESISTANCE,
DIURNAL, ALGICIDES, CLOSTRIDIUM,
ATE, PONDWEEDS, SODIUM ARSENITE,
IFICATION, *TOXICITY, *BIOASSAY,
ANDING CROP, ALGAE, ZOOPLANKTON,
, DEPTH, OLIGOCHAETES, TOXICITY,
LAMYDOMONAS, ALGAE, ZOOPLANKTON,
SYNTHESIS, CARBON RADIOISOTOPES,
S, PHOSPHORUS COMPOUND, PERCHES,
OXICITY, AQUATIC MAMMALS, ALGAE,
*CORAL, *REEFS, *MARINE ANIMALS,
BENTHOS, ZOOPLANKTON, FRESHWATER
ON, AQUATIC ALGAE, BRINE SHRIMP,
EWS, BIBLIOGRAPHIES, HERBICIDES,
OGY, ALGAE, BACTERIA, COLIFORMS,
PHYTOPLANKTON, AQUATIC INSECTS,
ION, *LIMNOLOGY, *PHYTOPLANKTON,
METABOLISM, ISOPODS, AMPHIPODA,
ANTS, BOTTOM SEDIMENTS, ANIMALS,
ABSORPTION, MARINE ALGAE, MARINE
RAPHIES, CYANOPHYTA, FRESH WATER
ISSOLVED OXYGEN, EUTROPHICATION,
ATION, OLIGOTR/ *LAKES, STREAMS,
LIMNION, ALGAE, FUNGI, BACTERIA,
DA, DEPTH, ALGAE, MOSSES, CHARA,
RVAE, GROWTH RATES, *FRESH WATER
NS, AQUATIC ALGAE, BRINE SHRIMP,
ATIONS, PLANT POPULATIONS, ROUGH
GHT, TEMPERATURE, FERTILIZATION,
ILTS, PLANKTON, MAYFLIES, FAUNA,
NTHOS, BRINE SHRIMP, ECOSYSTEMS,
NDS, *FERTILIZATION, *PHOSPHATE,
*FOOD CHAINS, FOOD WEBS, FORAGE
IBLIOGRAPHIES, ECOLOGY, BIOLOGY,
ULTURE, ALGAE, YEASTS, BACTERIA,
RADIOISOTOPES, IRON, SEDIMENTS,
MARINE ANIMALS, ALGAE, MOLLUSKS,
HORUS, FISH STOCKING, COMMERCIAL
TIVITY, INVERTEBRATES, PLANKTON,
NTHOS, CONVECTION, WAVES(WATER),

FILTRATION, MICROORGANISMS, ANALY   W71-13313
FILTRATION, ULTRASONICS, DISINFEC   W71-11823
FILTRATION, *SLIME, *ULTRAVIOLET   W72-03824
FILTRATION, NEW HAMPSHIRE, NEW YO   W72-03641
FILTRATION, *TOXICITY, *DISTRIBUT   W72-03746
FILTRATION, CHLORINATION, COAGULA   W71-05943
FILTRATION, MESOTROPHY.: /, TETRA   W69-10154
FILTRATION, SEWAGE, COLIFORMS, EU   W70-00263
FILTRATION.:   W72-08688
FILTRATION.: /L RECHARGE, COPPER,   W72-05143
FILTRATION, WATER PURIFICATION, M   W73-02426
FINANCING, GRANTS, FEDERAL GOVERN   W69-06305
FINGER LAKES(NY), RODHE'S SOLUTIO   W70-01579
FINGERNAIL CLAMS, WHITE FISH, ALE   W70-01943
FINLAND, SEWAGE RING TUBES.: /NY)   W70-05761
FIREBAUGH CENTER(CALIF), SCENEDES   W70-03334
FIRTH OF CLYDE(SCOTLAND), SCOTLAN   W70-02504
FISH CONSERVATION, POLLUTION ABAT   W72-01786
FISH FARMING, FISH FOOD ORGANISMS   W71-07731
FISH FARMING.: / SULFITE LIQUORS,   W69-05891
FISH FARMING, FISH ROOD ORGANISMS   W69-04804
FISH FOOD ORGANISMS, NUTRIENTS, E   W70-05428
FISH FOOD ORGANISMS, FISH PHYSIOL   W69-02782
FISH FOOD ORGANISMS, FISH PHYSIOL   W71-07731
FISH GENETICS, FISH HANDLING FACI   W69-04804
FISH HANDLING FACILITIES, FISH MA   W69-04804
FISH HARVEST, VITAMINS.: /DIMENTS   W69-09340
FISH MANAGERS, FISH YIELDS, UNDER   W69-08674
FISH MANAGEMENT, LIMNOLOGY, NITRO   W69-04804
FISH MANAGEMENT, STREAM IMPROVEME   W70-05428
FISH PHYSIOLOGY, FISH TAXONOMY, D   W69-02782
FISH PHYSIOLOGY, RED TIDE, ORGANI   W71-07731
FISH POPULATIONS.: /IVE CAPACITY,   W68-00912
FISH POPULATIONS, PLANKTON, *CYAN   W68-00461
FISH POPULATION, *WATER POLLUTION   W70-04266
FISH REPRODUCTION, INDUSTRIAL WAS   W73-00748
FISH ROOD ORGANISMS, FISH GENETIC   W69-04804
FISH STOCKING, COMMERCIAL FISH, S   W70-04465
FISH TAXONOMY, DDT, PHOSPHOTHIOAT   W69-02782
FISH TOXINS, BIOASSAY, AQUATIC AL   W70-08372
FISH YIELDS, UNDERWATER SHOCK WAV   W69-08674
FISH.: /, LAKE ERIE, STATISTICS,   W68-00253
FISH.: /HOSPHORUS, FISH STOCKING,   W70-04465
FISH.: /OMASS, DIATOMS, CYANOPHYT   W70-08654
FISH.: /Y, TEMPERATURE, ORGANIC M   W69-10157
FISH, *LIMNOLOGY, FISHKILL, NUTRI   W69-05844
FISH, ALEWIFE, GIZZARD SHAD, SEA   W70-01943
FISH, ALGAE, FOAMING, SEWAGE TREA   W70-08740
FISH, ALGAE, MODEL STUDIES, BIOAS   W71-11881
FISH, ALGAE.: /COSYSTEMS, SINKS,   W69-08274
FISH, AMPHIBIANS, INVERTEBRATES,   W68-00859
FISH, AMPHIBIANS, ANABAENA, CYTOL   W70-05372
FISH, ANIMALS, ALGAE, CUTTING MAN   W70-05263
FISH, AQUATIC ENVIRONMENT, ORGANI   W72-11105
FISH, BACTERIA, BENTHIC FAUNA, PH   W70-02251
FISH, BLOODWORMS, ALGAL CONTROL.:   W70-06217
FISH, CHRYSOPHYTA, SESTON, SELF-P   W70-00268
FISH, CLAMS, STRONTIUM RADIOISOTO   W70-02786
FISH, DIATOMS, CARP, PLANKTON, ME   W70-00274
FISH, DIATOMS, FUNGI, INSECTS, BA   W71-10786
FISH, ECOSYSTEMS, CYCLING NUTRIEN   W69-08526
FISH, ESTUARINE ENVIRONMENT, IMPO   W71-02491
FISH, FOOD CHAINS, MUSSELS.: /UTI   W70-01467
FISH, FRESH WATER, ALGAE, PONDS,   W70-00269
FISH, FUNGI, HEAT, WATER POLLUTIO   W71-02479
FISH, GAME BIRDS, ANIMALS, FOOD C   W70-10175
FISH, GREAT LAKES, LEGISLATION, N   W69-06858
FISH, GROWTH RATES.: /OMS, ALGAE,   W69-09742
FISH, INFLOW, OUTLETS, NUTRIENTS,   W70-06225
FISH, INVERTEBRATES, CRUSTACEANS,   W71-09850
FISH, LAKES, *LAKE ZURICH, *MUNIC   W68-00680
FISH, LAND, VALUE, LEAVES, FERTIL   W70-05264
FISH, LAND, ECOSYSTEMS, EUTROPHIC   W70-04193
FISH, LIMNOLOGY, HYDROLOGIC ASPEC   W70-04475
FISH, LIMNOLOGY, LAKE TROUT, CRAY   W70-00711
FISH, MICHIGAN, TROUT, WINTERKILL   W68-00487
FISH, MUSSELS.: /UTION, FOOD CHAI   W70-01466
FISH, NITROGEN, PHOSPHORUS, DIATO   W68-00172
FISH, NUISANCE ALGAE, ODOR, TASTE   W70-05404
FISH, OLIGOCHAETES, MIDGES, SNAIL   W70-01943
FISH, PHYTOPLANKTON, PLANKTON, RA   W69-07440
FISH, PLANKTON, AQUATIC WEED CONT   W70-05417
FISH, PLANKTON, ALGAE, *ALGAL CON   W70-05428
FISH, PLANKTON, WATER QUALITY, AL   W72-01786
FISH, PROTEINS, OIL WASTES, OILY   W69-05891
FISH, RADIOECOLOGY, EFFLUENTS, EC   W69-07853
FISH, RADIOACTIVE WASTES.: /CTS,   W71-05531
FISH, SALMON, ALGAE, FISH.: /HOSP   W70-04465
FISH, SEA WATER, GAMMA RAYS, RADI   W69-08269
FISH, SEWAGE, EUGLENOPHYTA, DIATO   W70-03969

445

LUTION EFFECTS, BACTERIA, ALGAE,
FERTILIZERS, ALGAE, ZOOPLANKTON,
CONCENTRATION, NUTRIENTS, ALGAE,
ITROGEN, PHOSPHATES, ALKALINITY,
TER QUALITY, VIRUSES, ALGICIDES,
MPERATURE, SALINITY, REAERATION,
CYCLES, MICHIGAN, INVERTEBRATES,
TMENT, *WATER POLLUTION CONTROL,
L COMMUNITIES, SEDIMENTS, BIOTA,
YTA, GREAT LAKES, BIOINDICATORS,
LABORATORY TESTS, INVERTEBRATES,
HICATION, LAKES, STREAMS, ALGAE,
ES, CHLOROPHYTA, NUISANCE ALGAE,
, TEMPERATURE, ACTIVATED SLUDGE
N, FUNGI, CLAYS, BRACKISH WATER,
RS, FRESHWATER ALGAE, ANADROMOUS
NTS, BIBLIOGRAPHIES, ABSORPTION,
BLATNA(CZECHOSLOVAKIA), CYPRINID
DES, HYDROLYSIS, COPPER SULFATE,
OTOPE, ON-SITE DATA COLLECTIONS,
QUALITY, AESTHETICS, RECREATION,
LIGHT PENETRATION, OXYGEN, SPORT
OGY, ECONOMICS, OHIO, COMMERCIAL
LAKES, CHEMICALS, FISHING, SPORT
ION, SEQUENCE, LAKES, CHEMICALS,
ICATION, CYANOPHYTA, ODOR, SCUM,
OPHICATION, ALGAE, DUCKS, COLOR,
FERTILIZATION, FISH, *LIMNOLOGY,
LGAE, AQUATIC WEEDS, CYANOPHYTA,
AYING, SEASONAL, *ALGAL CONTROL,
 MICHIGAN, TROUT, WINTERKILLING,
LLUTION EFFECTS, EUTROPHICATION,
*ALGAE, BIOASSAY, BIOINDICATORS,
LFATE, SEWAGE EFFLUENTS, SMELTS,
LUTION EFFECTS, DINOFLAGELLATES,
R TEMPERATURE, DISSOLVED OXYGEN,
SOURCES, BIOINDICATORS, NITROGEN
CHEMICAL OXYGEN DEMAND, NITROGEN
TER, *GRASSLANDS, MOU/ *NITROGEN
OPHICATION, MOLYBDENUM, NUTRIEN
TROGEN, NITROGEN CYCLE, NITROGEN
GY, NITROGEN COMPOUNDS, NITROGEN
GY, NITROGEN COMPOUNDS, NITROGEN
GY, NITROGEN COMPOUNDS, NITROGEN-
GY, NITROGEN COMPOUNDS, NITROGEN
IA, AMMONIUM COMPOUNDS, NITROGEN
 *ALGAE, *CYANOPHYTA, *NITROGEN
GRICULTURAL WATERSHEDS, NITROGEN
ATES, PLANKTON, OXYGEN, NITROGEN
SITY, PARTICU/ *GROWTH, *UPTAKE,
IOXIDE, INHIBITION, N/ *NITROGEN
GAE, *NITROGEN, WISCO/ *NITROGEN-
, IRON, ORGANIC MATTER, NITROGEN
XTRACELLULAR PRODUCTION, *CARBON
LIMATES, TROPHIC LEVEL, NITROGEN
ZONE, PLANT CAROTENOIDS, CARBON
REQUIREMENTS, NITROGEN, NITROGEN
ALCIUM, AMMONIUM SALTS, NITROGEN
NITRATES, FLUORESCENCE, NITROGEN
CIDS, CALCIUM, AMMONIA, NITROGEN
ENGLAND), CHLORELLA PYRENOIDOSA,
NITROGEN, PHOSPHOROUS, NITROGEN
ON DIOXIDE, INHIBITION, NITROGEN
SOD MATS, NON-LEGUMES, *NITROGEN
TT BAY, CAPE REYKJANES, FAXAFLOI
IA, *OPISTHONEMA OGLIMUN, UDOTEA
ENTERICUS, DIATOMEA, CLADOPHORA,
RK EQUATION, MICHAELIS KINETICS,
ALGAL KEY, CLUMP COUNT, DESMIDS,
                          DESMINS,
), PASS LAKE(COLO), MICRO-ALGAE,
NIOBIUM-95, GYMNODINIUM SIMPLEX,
URATION, PHYCOCOLLOIDS, TANNINS,
NTED FLAGELLATES, *NON-PIGMENTED
LLATES.:   *PIGMENTED
AVICULA, FRAGILARIA, CYCLOTELLA,
DUCTION TECHNIQUE, HETEROTROPHY,
GREEN ALGAE, *EUGLENA, PIGMENTED
LIATE AUTOECOLOGY.:   *CILIATES,
     *CHEMOSTAT, PHYSIOLOGY,
Y, CYCLOP/ *PHENYLUREAS, DIURON,
MORPHOLOGY, CELLULAR MORPHOLOGY,
ER(NORWAY), GAULA RIVER(NORWAY),
E, THERMAL SENSITIVITY, GENERA, '
A, *ASCOPHYLLUM NODOSUM, VIBRIO,
SPECIES DETERMINATION, TAXONOMY,
CED LAKES, *FERTILIZATION, WATER
ANIMALS, BACTERIA, FUNGI, ALGAE,
COSYSTEMS, CHARA, BENTHIC FLORA,
PHATES, HYDROGENATION, AERATION,

FISH, SEWERS, RUNOFF, WASTES, WAT   W71-09880
FISH, SHELLFISH, STEAM, ELECTRIC   W70-09905
FISH, STRATIFICATION, FLOCCULATIO   W70-08834
FISH, STREAMS, WATER POLLUTION EF   W69-10159
FISH, STRE: /TREATMENT, LAKES, WA   W70-03973
FISH, TASTE, ODOR, OXYGEN SAG, PO   W73-00750
FISH, TEMPERATURE, TRACERS, SAMPL   W69-09334
FISH, TOXICITY, FRESH WATER, PULP   W70-10437
FISH, UTAH, ALGAE, INSECTS, INVER   W69-07846
FISH, WASHINGTON, LAKES, INDIANA,   W70-04503
FISH, WATER ANALYSIS, INSECTS, SN   W72-01801
FISH, WISCONSIN, SEWAGE, TROUT, N   W69-10178
FISH, ZOOPLANKTON, SNAILS, APPLIC   W70-03310
FISHERIES, CARP, CLAMS, PONDS, RU   W70-09905
FISHERIES, LIMNOLOGY.: /FIDE, IRO   W70-05761
FISHES.: /RONMENT, IMPOUNDED WATE   W71-02491
FISHES, ALGAE.: /S, BOTTOM SEDIME   W68-00858
FISHES, CLOROCOCCALES, DAPHNIA PU   W70-00274
FISHES, HERBICIDES, MIDGES, PESTI   W70-02982
FISHES, PLANKTON, ALGAE, FROGS, M   W68-00857
FISHING, ALGAE, NUTRIENTS, TURBID   W70-05645
FISHING, ON-SITE INVESTIGATIONS.:   W68-00487
FISHING, PUBLIC HEALTH, WALLEYE,   W70-01943
FISHING, RECREATION, SILT, NUTRIE   W68-00172
FISHING, SPORT FISHING, RECREATIO   W68-00172
FISHING, WISCONSIN, BIBLIOGRAPHIE   W69-08283
FISHKILL, LAKES, ODOR, TOXICITY,   W69-08279
FISHKILL, NUTRIENTS, *ALGAE, *ALG   W69-05844
FISHKILL, FERTILIZATION, LIMITING   W69-06535
FISHKILL, ROOTED AQUATIC PLANTS,   W68-00488
FISHKILL, NUISANCE ALGAE, ALGAE,   W68-00487
FISHKILL, ALGAE, OXYGEN SAG, WATE   W69-03948
FISHKILL, HAZARDS, PESTICIDE KINE   W69-08565
FISHKILL, CRUSTACEANS, BOTTOM SED   W69-08674
FISHKILL, FISH TOXINS, BIOASSAY,   W70-08372
FISHKILLS, LIMNOLOGY, ALGAL TOXIN   W70-00264
FIXATION.: /CTS, WATER POLLUTION   W69-06864
FIXATION.: /ION, RESPIRATION, BIO   W72-06022
FIXATION, IRRIGATION, ORGANIC MAT   W72-08110
FIXATION, NUISANCE ALGAE, NUTRIEN   W69-06277
FIXATION, PHOSPHATES, PHOSPHORUS,   W69-04805
FIXATION, NUTRIENTS, NU: /LIMNOLO   W69-03518
FIXATION, NUISANCE ALGAE, NUTRIEN   W69-04802
FIXATION, PHOSPHORUS COMP: /MNOLO   W69-03186
FIXATION, NUTRIENT REQUIREMENT: /   W69-03185
FIXATION, NUTRIENT REQUIREMENTS,   W69-03364
FIXATION, NUTRIENTS, LAKE ERIE.:   W70-05091
FIXATION, CYANOPHYTA, SEWAGE EFFL   W70-04506
FIXATION, NITRATES, ELECTRON MICR   W70-04369
FIXATION, CELL COUNT, OPTICAL DEN   W70-03983
FIXATION, *ALGAE, LIGHT, CARBON D   W70-04249
FIXATION, *NUTRIENTS, *LAKES, *AL   W70-03429
FIXATION, VOLUME, STREAMS, WATER   W70-00719
FIXATION, *EXCRETION, STEPHANODIS   W69-10158
FIXATION, FOREST, WINDS, LIGHT, T   W71-03048
FIXATION, PELAGIC REGION, MONOMIC   W70-01933
FIXATION, PHOTOCHEMICAL, CARBON D   W70-02965
FIXATION, BORON, AZOTOBACTER, MOL   W70-02964
FIXATION, ENVIRONMENTAL EFFECTS,   W70-02775
FIXATION, IRON COMPOUNDS, LIGHT P   W70-02255
FIXATION, TRINGFORD(ENGLAND), ANA   W70-02504
FIXING BACTERIA, *ACTIVATED SLUDG   W71-13335
FIXING BACTERIA, RADIOACTIVITY TE   W70-04249
FIXING ORGANISMS.: /NITROGEN-15,   W72-08110
FJORD, ICELAND, DNA BASE COMPOSIT   W69-10161
FLABELLUM, ZINC, NICKEL, PUERTO R   W69-08525
FLAGELLATA, CILIATES, RANUNCULUS   W69-07838
FLAGELLATES, LAKE ERKEN(SWEDEN),   W70-04194
FLAGELLATES, MESOPLANKTON, MICROP   W70-05389
FLAGELLATES.:   W68-00172
FLAGELLATES, CILIATES, COPEPODIDS   W69-10154
FLAGELLATES, ELEUTHERA ISLAND, NI   W70-00707
FLAGELLATES, PHARMACEUTICALS, EXT   W70-01068
FLAGELLATES.:   *PIGME   W72-01798
FLAGELLATES, *NON-PIGMENTED FLAGE   W72-01798
FLAGELLATES, ANKISTRODESMUS, GOMP   W70-02249
FLAGELLATES, LAKE CLASSIFICATION,   W70-02775
FLAGELLATES.: /RANT ALGAE, *BLUE   W70-01233
FLAGELLATES, SAPROBIEN SYSTEM, CI   W71-03031
FLAGELLATES.:   W70-09904
FLAGELLATES, MONURON, TRANSPARENC   W70-10161
FLAGELLATION, IODINE TEST, GRAM R   W70-03952
FLAKK(NORWAY), RANHEIM(NORWAY), T   W70-01074
FLAVOBACTERIUM-CYTOPHAGA', VIBRIO   W70-00713
FLAVOBACTERIUM, ESCHERICHIA, SARCINA   W70-03952
FLAVOBACTERIUM, ACHROMOBACTER, GU   W70-04280
FLEAS, LARVAE, GROWTH RATES, *FRE   W68-00487
FLOATING PLANTS, ROOTED AQUATIC P   W70-05750
FLOATING PLANTS, PHYTOPLANKTON, A   W70-10175
FLOCCULATION, FILTRATION, MODEL S   W70-08627

446

TS, ALGAE, FISH, STRATIFICATION,   FLOCCULATION, TURBINES, ECONOMIC   W70-08834
EAD LOSS, FILTERS, *SCENEDESMUS,   FLOCCULATION, WATER POLLUTION TRE   W72-03703
D, *FILTRATION, ALGAE, COLLOIDS,   FLOCCULATION, MIXING, ECONOMIC FE   W71-10654
ON, ALGAE, FILTRATION, SORPTION,   FLOCCULATION, COAL, ACTIVATED CAR   W70-05470
RTEBRATES, BACKGROUND RADIATION,   FLOCCULATION, POLLUTANTS, SAMPLIN   W69-07846
TION, ALGAL POISONING, GASOLINE,   FLOODS, AQUIFERS, WATER LAW, NITR   W73-00758
ATER, SHORES, WATER CIRCULATION,   FLOODS, DENSITY DURRENTS, MEASURE   W69-06203
WATER LAW, NITRATES, TURBIDITY,   FLOODWATER, REGULATION, WELL REGU   W73-00758
NG CREEK(PA), STATE COLLEGE(PA),   FLORA.:   SPRI W69-10159
*PHYSIOLOGICAL ECOLOGY, *BENTHIC   FLORA, *ALGAE, *HYDRAULIC MODELS,   W71-00668
ITIES, *LABORATORIES, / *BENTHIC   FLORA, *ALGAE, *BIOLOGICAL COMMUN   W70-02780
DUCTIVITY, *MEASUREMEN/ *BENTHIC   FLORA, *CHLOROPHYLL, *PRIMARY PRO   W69-10151
BACTERIAL MEDIUM, ARTHROBACTER,   FLORA, AMBIENT TEMPERATURE, THERM   W70-00713
ECTS, *LIMITING FACTORS, BENTHIC   FLORA, BENTHOS, MINE WATER, WATER   W70-02770
IC PLANTS, MARINE ALGAE, BENTHIC   FLORA, ESTUARIES, *ESTUARINE ENVI   W71-02494
COMMUNITIES, *WASTE DISCHARGES,   FLORA, FAUNA, *BIOLOGICAL COMMUNI   W68-00010
LGAE, ECOSYSTEMS, CHARA, BENTHIC   FLORA, FLOATING PLANTS, PHYTOPLAN   W70-10175
AE, BRACKISH, ECOSYSTEMS, FAUNA,   FLORA, LABORATORY, MINERALS, POLL   W72-02203
MARINE ALGAE, *DIATOMS, *BENTHIC   FLORA, LIGHT INTENSITY, BIOMASS,   W70-03325
ORGANIC MATTER, BIOMASS, BENTHIC   FLORA, PHAEOPHYTA, POLAR REGIONS,   W70-01074
OTOPES, IRON, MANGANESE, BENTHIC   FLORA, STREAMS, CULTURES, COBALT   W69-07845
N), CHICKAMAUGA RESERVOIR(TENN),   FLORENCE(ALA).: / BROAD RIVER(TEN   W71-07698
AND, NITZCHIA SERIATA, NAVICULA,   FLORIDA CURRENT, YTTRIUM-90, KATO   W70-00707
E, TROPHIC LEVEL, WATER QUALITY,   FLORIDA.: /UCTIVITY, AQUATIC ALGA   W72-01990
*ANDERSON-CUE LAKE, MELROSE(FLORIDA).:   W72-01990
CUE LAKE(FLORIDA), MCCLOUD LAKE(FLORIDA).:   W72-01993
CUE LAKE(FLORIDA), MCCLOUD LAKE(FLORIDA).:   W72-01993
OFLAGELLATES, ALGAE, CALIFORNIA,   FLORIDA, PESTICIDE RESIDUES, ECOS   W70-05272
, VITAMIN B, RED TIDE, SALINITY,   FLORIDA, TEXAS, PHOSPHORUS, MIGRA   W70-05372
SAMPLING, SEASONAL, HYDROGRAPHY,   FLORIDA, THERMOCLINE, CURRENTS(WA   W70-04368
), CHROMATIUM VINOSUM, THIOCAPSA   FLORIDANA, ENTEROCOCCI, PARK RIVE   W70-03312
EPTS, *TROPHIC BIOLOGY, ANABAENA   FLOS AQUAE, APHANIZOMENON FLOS AQ   W70-03959
ABAENA FLOS AQUAE, APHANIZOMENON   FLOS AQUAE, GLOEOTRICHIA ECHINULA   W70-03959
PHOSPHORUS BUDGET, APHANIZOMINON   FLOS AQUAE.: / NUTRIENT SOURCES,   W70-04268
, THERMAL EFFECTS, APHANIZOMENON   FLOS AQUAE, METOLIUS RIVER(ORE),   W70-03978
HAGAWA LAKE(MINN), APHANIZOMENON   FLOS AQUAE, STEPHANODISCUS ASTREA   W70-05387
CUS, LYNGBYA LIMNETICA, ANABAENA   FLOS AQUAE, TABELLARIA, MELO: /TI   W70-05405
IATES, CYCLOTELLA, APHANIZOMENON   FLOS AQUAE, MELOSIRA, ASTERIONELL   W70-05112
NABAENA INAEQUALIS 381, ANABAENA   FLOS AQUAE, TOLYPOTHRIX DISTORTA,   W70-05091
SPHAERIUM KUTZINGIANUM, ANABAENA   FLOS AQUAE, ANABAENA MENDOTAE, AN   W69-08283
YAHARA RIVER(WIS), APHANIZOMENON   FLOS AQUAE, OXYGEN DEPLETION.:   *   W72-01797
SA, BOHEMIA, ELB/ *APHANIZOMENON   FLOS AQUAE, *MICROCYSTIS AERUGINO   W70-00274
S, VETERINARIANS, MIC/ *ANABAENA   FLOS-AQUAE, *STRAINS, PHYCOLOGIST   W70-00273
CYSTIS AERUGINOSA, APHANIZOMENON   FLOS-AQUAE, GLOEOTRICHIA ECHINULA   W70-00719
ARDHI, PHORMIDIUM, APHANIZOMENON   FLOS-AQUAE, SECCHI DISC TRANSPARE   W69-10169
NOMIDS, CLADOCERA, APHANIZOMENON   FLOS-AQUAE, STEPHANODISCUS HANTZS   W70-00268
CYSTIS AERUGINOSA, APHANIZOMENON   FLOS-AQUAE, MINIMUM LETHAL DOSES,   W70-02982
ONEMA, PHORMIDIUM, APHANIZOMENON   FLOS-AQUAE, LYNGBYA, STORM LAKE(I   W70-02803
ENASTRUM CAPRICORNUTUM, ANABAENA   FLOS-AQUAE, MICROCYSTIS AERUGINOS   W69-06864
CYSTIS AERUGINOSA, APHANIZOMENON   FLOS-AQUAE, OSCILLATORIA, ANABAEN   W70-05091
AMATH LAKE(ORE.), *APHANIZOMENON   FLOS-AQUAE, *CARBON-14 MEASUREMEN   W72-06052
LGAE, OCULAR MICROMETER, SUCROSE-   FLOTATION TECHNIQUE, METALIMNION,   W70-04253
, ALGAE, AMMONIA, SEDIMENTATION,   FLOTATION, BIODEGRADATION, CHLORI   W71-05013
, INTERTIDAL ALGAE, PUGET SOUND,   FLOURESCEIN DYE.: /AL COMMUNITIES   W68-00010
IMENTATION RATES, MANGANESE, LOW-   FLOW AUGMENTATION, TENNESSEE RIVE   W73-00750
SEPARATION TECHNIQUES, PRESSURE,   FLOW RATES, SLUDGE, SLURRIES, AUT   W72-08688
ION ABATEMENT, DRAINAGE SYSTEMS,   FLOW RATES, CONN STREAM FLOW, ALG   W68-00479
NUISANCE ALGAE, INTAKES, SANDS,   FLOW RATES, TURBIDITY, CHLORELLA,   W70-04199
NESE, *COLOR, *ALGAE, *COLLOIDS,   FLOW RATES, *FILTERS, *NEW YORK,   W72-04145
OGY, LOTIC ENVIRONMENT, BIOMASS,   FLOW RATES, DEPTH, DISSOLVED OXYG   W70-00265
TOMS, MARINE ALGAE, FRESH WATER,   FLOW RATES, PUMPS, ELECTROLYSIS,   W70-09904
S, HUMUS, REAERATION, EFFLUENTS,   FLOW RATES, FILTERS, ALGAE.: /ANT   W70-05821
*NUTRIENT REMOVAL, CONTINUOUS   FLOW SYSTEM.:   W72-00917
N DEMAND, AERATION, TEMPERATURE,   FLOW, ALGAE, BENTHOS, AEROBIC CON   W71-11381
SYSTEMS, FLOW RATES, CONN STREAM   FLOW, ALGAE, AQUATIC ALGAE, CURRE   W68-00479
*LAKES, *TRIBUTARIES, SAMPLING,   FLOW, ANALYSIS, FERTILIZATION, PL   W70-02969
R QUALITY, PHOSPHORUS, NITROGEN,   FLOW, CHEMICAL ANALYSIS, NUTRIENT   W70-04810
INERS, CADDISFLIES, TEMPERATURE,   FLOW, DIATOMS, MANGANESE.: /S, SH   W69-07853
TS, NUTRIENTS, ALGAE, STRATIFIED   FLOW, MIXING, FLOW, MODEL STUDIES   W71-05878
ALGAE, STRATIFIED FLOW, MIXING,   FLOW, MODEL STUDIES, MATHEMATICAL   W71-05878
MMEDIUM FILTER, BACK-WA/ *UPWARD   FLOW, PUTRESCIBILITY TEST, BODY I   W70-05821
ICAL OXYGEN DEMAND, TEMPERATURE,   FLOW, SAMPLING, WASTE WATER DISPO   W69-10159
EED, ZIRCONIUM-95-NI/ *SARGASSUM   FLUITANS, SARGASSUM NATANS, SEAW   W69-08524
ECTROPHOTOMETRY, SUSPENDED LOAD,   FLUORESCE: /TEMPERATURE, IONS, SP   W70-03334
HEBOYGAN COUNTY, STURGEON RIVER,   FLUORESCEIN, MACROPHYTES, GROUND   W69-09334
PHY, HYDROGEN ION CONCENTRATION,   FLUORESCENCE, CARBOHYDRATES, TEMP   W70-01074
FARM WASTES, BENTHOS, NITRATES,   FLUORESCENCE, NITROGEN FIXATION,   W70-02775
TS, EUTROPHICATION, CHLOROPHYLL,   FLUORESCENCE, LAKES, RIVERS, NITR   W70-02777
AKES, RIVERS, LIGHT PENETRATION,   FLUORESCENCE, PHOTOSYNTHESIS, BLU   W69-05697
ENVIRONMENTS, MICROBIAL ECOLOGY,   FLUORESCENCE MICROSCOPY, ACETATE,   W70-04811
OLOGICAL SS POLARIS, CHLOROPHYLL   FLUORESCENCE, FABRY-PEROT FILTERS   W70-05377
HNIQUE, PARTICULATE MATTER, BLUE   FLUORESCING MATE: /LON COLUMN TEC   W70-01074
TES, HYDROGEN ION CONCENTRATION,   FLUORIDES, ALKYL BENZENE SULFONAT   W70-00263
RY, ENVIRONMENTAL RELATIONSHIPS,   FLUOROMETRIC CHLOROPHYLL TECHNIQU   W70-05094
*CYTOLOGICAL STUDIES, CULTURES,   FLUOROMETRY, BIOMASS, OCEANOGRAPH   W69-08278
), CHICAG/ MANURE, MADISON(WIS),   FLUSHING WATERWAYS, MILWAUKEE(WIS   W69-10178
OSPHAERA ELONGATA, THALASSIOSIRA   FLUVIAT: /HIDINIUM CARTERI, SYRAC   W69-08278
S(WISC), GROWTH STIMUL/ *SCIRPUS   FLUVIATILIS, *LAKE BUTTE DES MORT   W72-00845
ER WASTES, DYEING WASTES./ *ANTI-   FOAM AGENTS, *TEXTILE WASTES, FIB   W70-04060
OLYPHOSPHATES, NUTRIENT BALANCE,   FOAMING DETERGENTS, PUBLIC APPREH   W70-05084

447

TANCE, *PLANT GROWTH REGULATORS,
NTS, BIBLIOGRAPHIES, CYANOPHYTA,
TIDAL WATERS, MIXING, SEA WATER,
TRUCTURE, ATMOSPHERE, SEA WATER,
TER, SCENEDESMUS, CHLAMYDOMONAS,
NS, ALGAE, BENTHOS, ZOOPLANKTON,
E ENVIRONMENT, IMPOUNDED WATERS,
, PONDS, BREEDINGS, SALINE WATER-
UTROPHICATION, MUNICIPAL WASTES,
LGAE,/ *COPPER CHLORIDE, UPTAKE,
    *LEUCOTHRIX MUCOR, MORPHOLOGY,
YLINDRUS, CORYCAEUS, OIKOPLEURA,
, SUCKERS, BASS, WATER HYACINTH,
CTIONS, FISHES, PLANKTON, ALGAE,
PHAUSIA, PHAEOCYSTIS, SARGASSUM,
), SMOLA(NORWAY), HITRA(NORWAY),
RRIGATION, GROUND WATER, RUNOFF,
RY(RHODE ISLAND), DILUTION RATE,
PEACHES, APRICOTS, PEARS, CITRUS
EDIMENTS, CHLORELLA PYRENOIDOSA,
), HITRA(NORWAY), FROYA(NORWAY),
LGAL SUBSTANCES, *EXTRACELLULAR,
COPHYLLUM NODOSUM, ALGINIC ACID,
RACUSE(NY), SYRACUSE UNIVERSITY,
PHAGA, PROTEUS, SERRATIA, BAY OF
OLLUTION EFFECTS, FUNGI, AQUATIC
ANIMALS, MARINE BACTERIA, MARINE
C CONDITIONS, E. COLI, BACTERIA,
ATTER, INSECTS, MITES, BACTERIS,
A, SEWAGE BACTERIA, SYSTEMATICS,
HEMICAL), HYDROLYSIS, CHELATION,
ER POLLUTION, ANIMALS, BACTERIA,
ONTROL, WATER POLLUTION EFFECTS,
CHELATION, OCHROMONAS, BACTERIA,
ALLI, SCHIZOMYCETES, CYANOPHYTA,
EPILIMNION, *HYPOLIMNION, ALGAE,
WATER, *BIOMASS, *SLIME, ALGAE,
DITIONS, HYDROGEN SULFIDE, IRON,
GELLATES, OCEANS, BIOINDICATORS,
LGAE, BACTERIA, COLIFORMS, FISH,
C MAMMALS, ALGAE, FISH, DIATOMS,
TH RATES, MICROORGANISMS, ALGAE,
R, IRON OXIDES, BACTERIA, ALGAE,
*BIOASSAYS, PROTOZOA, BACTERIA,
GAE, *AQUATIC BACTERIA, *AQUATIC
URE, HYDROGEN ION CONCENTRATION,
TES LOBATA, LEPTASTREA PURPUREA,
ATA, LEPTASTREA PURPUREA, FUNGIA
KE NORRVIKEN, LAKE MENDOTA, LAKE
KERSEE(SWITZ), AEGERISEE(SWITZ),
HROOCOCCUS PRESCOTTII, CALOTHRIX
SARNIENSE, RHODOCHORTON, BANGIA
NALIS, NEPHROCYTIUM OBESUM W AND
SMS, TURSIOPS TRUN/ DUPLIN RIVER(
CHODLERI, DAPHNIA PULEX, DAPHNIA
BITION, SILICON, VANADIUM, ZINC,
TE, FERRICHLORIDE, IRON SULFATE,
PLANKTON, AQUATIC INSECTS, FISH,
KENTUCKY, LIFE HISTORY STUDIES,
, ATLANTIC OCEAN, RADIOACTIVITY,
ATES, PLANKTON, FISH, SEA WATER,
, ORGANIC MATTER, RADIOISOTOPES,
TRUTTA, ENTOSPHENUS LAMOTTENII,
MBAY(IND/ *INDIA, *BLOOMS, RIVER
ERTILIZATION, INDUSTRIAL WASTES,
C ACID DIHYDROCHLORIDE, INFRARED
CULTURES, DIAGENESIS, SEDIMENTS,
CETES, CYANOPHYTA, SPECTROSCOPY,
ATION, *WATER POLLUTION EFFECTS,
, ENDRIN, DIELDRIN, SCENEDESMUS,
*CHLAMYDOMONAS LABORATORY TESTS,
DITIONS, EXTRACELLULAR PRODUCTS,
ROL, RECIRCULATION, BOD REMOVAL,
LAKE(NETHERLANDS), HETEROCYSTS,
NG, *SULFATES, *SULFIDES, ALGAE,
, URBANIZATION, ALGAL POISONING,
, IOWA RIVER, / *UNIT PROCESSES,
BODY BURDENS, POMOXIS ANNULARIS,
ER POLLUTION EFFECTS, MORTALITY,
USTACEA, MOLLUSCS, OLIGOCHAETES,
NSET(NORWAY), NID RIVER(NORWAY),
UTANT BACTERIA, CYCLOTELLA NANA,
LAGIC AREAS, HORIZONTAL, BLOOMS,
EMPERATURE, THERMAL SENSITIVITY,
LLA FORMOSA, MELOSIRA BINDERANA,
LTH, PHYSICOCHEMICAL PROPERTIES,
ICAL PRECIPITATION, LIFE CYCLES,
GAL PHYLOGENY, CLADOPHORA, ALGAL
RMING, FISH ROOD ORGANISMS, FISH
S, INDUSTRIAL USES, CLEAN WATER,

FRESH WATER LAKES, LABORATORY TES   W68-00856
FRESH WATER FISH, LAKES, *LAKE ZU   W68-00680
FRESH WATER, SHORES, WATER CIRCUL   W69-06203
FRESH WATER, SORPTION, SODIUM, AL   W70-04385
FRESHWATER, MARINE ALGAE, ENZYMES   W70-05381
FRESHWATER FISH, ESTUARINE ENVIRO   W71-02491
FRESHWATER ALGAE, ANADROMOUS FISH   W71-02491
FRESHWATER INTERFACES, COLLOIDS.:   W73-00748
FRESHWATER ALGAE, WATER POLLUTION   W72-12240
FREUNDLICH ISOTHERM, BLUE-GREEN A   W70-08111
FRIDAY HARBOR(WASH), ANTITHAMNION   W69-10161
FRITILLARIA, OBELIA, CLUSTER ANAL   W73-01446
FROGS, CRAYFISH, MOSQUITOES, BACT   W70-05263
FROGS, MOSSES, HYPOLIMNION, THERM   W68-00857
FRONDS, HYDROCARBONS, BACTERIAL A   W70-01068
FROYA(NORWAY), FUCUS VESCICULOSUS   W70-01074
FROZEN GROUND, SOIL EROSION, WATE   W69-05323
FRUITING BODIES, ULVA LACTUCA VAR   W70-01073
FRUITS, TOMATOES, *AEROBIC TREATM   W71-09548
FUCALES, ALKANES, ALKENES, ALIPHA   W69-08284
FUCUS VESCICULOSUS, ASCOPHYLLUM N   W70-01074
FUCUS VESICULOSUS, CHONDRUS CRISP   W70-01073
FUCUS VESICULOSUS, GROWTH FACTORS   W69-07826
FUNDAMENTAL STUDIES, APPLIED STUD   W70-03973
FUNDY(CANADA), RHYZOIDS.: /, CYTO   W70-03952
FUNGI.: /LLUTION CONTROL, WATER P   W69-05706
FUNGI.: /S, MARINE ALGAE, MARINE   W69-03752
FUNGI, ACTINOMYCETES, PROTOZOA, A   W71-02009
FUNGI, ALGAE, NITRITES, NITRATES,   W71-00940
FUNGI, ALGAE, DIATOMS, PROTOZOA,   W72-01819
FUNGI, ALGAE, PROTOZOA, LICHENS,   W69-07428
FUNGI, ALGAE, FLOATING PLANTS, RO   W70-05750
FUNGI, AQUATIC FUNGI.: /LLUTION C   W69-05706
FUNGI, AQUATIC PLANTS, ASSAY, DIS   W70-02510
FUNGI, BACILLARIOPHYCEAE, CHAROPH   W70-02784
FUNGI, BACTERIA, FISH, LIMNOLOGY,   W70-04475
FUNGI, BACTERIA, BACTERICIDES, CH   W72-12987
FUNGI, CLAYS, BRACKISH WATER, FIS   W70-05761
FUNGI, ESTUARIES, BACTERIA, SAMPL   W70-04368
FUNGI, HEAT, WATER POLLUTION, STR   W71-02479
FUNGI, INSECTS, BACTERIA, PROTOZO   W71-10786
FUNGI, LABORATORY TESTS, BIOCHEMI   W72-01817
FUNGI, SULFUR COMPOUNDS, CARBONAT   W70-02294
FUNGI, VIRUSES, ALGAE, LICHENS, A   W71-11823
FUNGI, YEASTS, HYDROGEN ION CONCE   W69-00096
FUNGI, YEASTS, PYRITE, ALGAE, CYA   W68-00891
FUNGIA FUNGITES, ZOOXANTHELLAE, A   W69-08526
FUNGITES, ZOOXANTHELLAE, ACANTHUR   W69-08526
FURES, LAKE SEBASTICOOK, LAKE WAS   W70-03964
FURES: / GREIFENSEE(SWITZ), PAFFI   W70-00270
FUSCA, MICROCOLEUS LACUSTRIS (RAB   W70-04468
FUSCO-PURPUREA, SPHACELARIA, GONI   W69-10161
G S WEST, RADIOCOCCUS NIMBATUS, U   W70-04468
GA), SPARTINA ALTEMIFLORA, ORGANI   W69-08274
GALEATA MENDOTAE, DAPHNIA MAGNA,   W70-03957
GALLIUM.: /INS, METABOLITES, INHI   W69-04801
GALVANIZING WASTES, LAKE ZURICH(S   W70-05668
GAME BIRDS, ANIMALS, FOOD CHAINS,   W70-10175
GAMMA RAYS, SPECTROMETERS, CYANOP   W69-09742
GAMMA RAYS, POTASSIUM RADIOISOTOP   W69-08524
GAMMA RAYS, RADIOCHEMICAL ANALYSI   W69-08269
GAMMA RAYS, URANIUM RADIOISOTOPES   W69-08272
GAMMARUS, CORDULEGASTER.: / SALMO   W69-09334
GANGES(INDIA), RAJGHAT(INDIA), BO   W70-05404
GARBAGE DUMPS, OIL WASTES, INLETS   W70-05112
GAS ANALYSIS, COULTER COUNTER, EU   W69-08278
GAS CHROMATOGRAPHY, LIPIDS, OIL,   W69-08284
GAS CHROMATOGRAPHY, ANALYTICAL TE   W69-00387
GAS CHROMATOGRAPHY.: /T IDENTIFIC   W69-02782
GAS CHROMATOGRAPHY, RESISTANCE, B   W70-03520
GAS CHROMATOGRAPHY, DEGRADATION,   W70-08652
GAS EXCHANGE, GREEN PLANTS, MASS   W69-07832
GAS PRODUCTION.: /IRONMENTAL CONT   W70-04787
GAS-VACUOLES, LLANGORS LAKE(WALES   W70-04369
GASES, ODOR, OXIDATION-REDUCTION   W70-07034
GASOLINE, FLOODS, AQUIFERS, WATER   W73-00758
GASTROENTERITIS, SYNURA, ANABAENA   W70-08107
GASTROINTESTINAL TRACT.: /RELLI,   W70-02786
GASTROPODS, MARINE PLANTS, ON-SIT   W70-01780
GASTROPODS.: /NSECTS, DIPTERA, CR   W70-04253
GAULA RIVER(NORWAY), FLAKK(NORWAY   W70-01074
GEIGER COUNTER.: /ANOCOBALAMIN, M   W70-05652
GELATINOUS MATTER.: /OTA(WIS), PE   W70-06229
GENERA, 'FLAVOBACTERIUM-CYTOPHAGA   W70-00713
GENERATION TIME, CHLORELLA PYRENO   W69-10158
GENETICS, BACTERIA, PEPTIDES, ENV   W70-00273
GENETICS, E. COLI, ISOLATION.: /M   W71-11823
GENETICS, RUSSIA, BAVARIA, MORICH   W70-03973
GENETICS, FISH HANDLING FACILITIE   W69-04804
GENUS, SPECIES, TOUCH SENSATION.:   W70-05389

MODELS, BACTERIA, PHYTOPLANKTON,   GEOCHEMIS: /NOLOGY, MATHEMATICAL   W70-03975
MODELS, BACTERIA, PHYTOPLANKTON,   GEOCHEMIS: /NOLOGY, MATHEMATICAL   W70-03975
L MODELS, *WATERSHED MANAGEMENT,   GEOCHEMISTRY, LAKES, CONIFEROUS F   W71-09012
IC PLANTS, BACKGROUND RADIATION,   GEOCHEMISTRY, SPRING WATERS, PHOS   W69-09334
TURE, WATER MANAGEMENT(APPLIED),   GEOCHEMISTRY, SOIL TYPES, HYDROLO   W71-03048
OGICAL ECOLOGY, WATER CHEMISTRY,   GEOGRAPHICAL REGIONS, CYCLING NUT   W70-03975
OGICAL ECOLOGY, WATER CHEMISTRY,   GEOGRAPHICAL REGIONS, CYCLING NUT   W70-03975
, RAINFALL-RUNOFF RELATIONSHIPS,   GEOGRAPHICAL REGIONS, KENTUCKY, A   W68-00891
, PACIFIC COAST REGION, REGIONS,   GEOGRAPHICAL REGIONS, BAYS, BODIE   W69-03683
REGIONS, PACIFIC COAST REGIONS,   GEOGRAPHICAL REGIONS, PHOTOSYNTHE   W69-03611
TIDES, SOUTHEAST U. S., REGIONS,   GEOGRAPHICAL REGIONS, COASTAL PLA   W69-03695
HABITATS, DEPTH, AIR, SEDIMENTS,   GEOLOGIC FORMATIONS, ARCTIC, TUND   W69-08272
DYE, AIRBORNE TESTS, OIL SPILLS,   GEOLOGICAL SS POLARIS, CHLOROPHYL   W70-05377
SERVOIRS, AIR TEMPERATURE, RAIN,   GEOLOGY, MAPPING, NUTRIENTS.: /RE   W70-02646
, PROTOPLASM, WATER-BLOOMS, LAKE   GEORGE(UGANDA), RHINE-MEEUSE DELT   W70-04369
ALT MARSHES, *CYCLING NUTRIENTS,   GEORGIA, SEDIMENTS, BACTERIA, DET   W69-08274
                    BERLIN(WEST),   GERMANY.:   W68-00472
ONIZATION, ALSTER LAKE(GERMANY),   GERMANY.: /IONELLA FORMOSA, RECOL   W70-00268
/ *HYPERTROPHIC WATERS, HAMBURG(GERMANY), CHIRONOMIDS, CLADOCERA,   W70-00268
OSA, RECOLONIZATION, ALSTER LAKE(GERMANY), GERMANY.: /IONELLA FORM   W70-00268
RADULESPORES, ZYMOLOGICAL, ELBE(GERMANY), SPOROBOLOMYCES ROSEUS,   W70-04368
S, NOCTILUCA MILIARIS, HELGOLAND(GERMANY), AUREOBASIDIUM PULLULANS   W70-04368
KE SCHLIER(GERMANY), LAKE TEGERN(GERMANY), GERMANY, FINLAND, SEWAG   W70-05761
ONSTANCE, WHITEFISH, WURTTEMBERG(GERMANY), PLON(GERMANY), LAKE SCH   W70-05761
NY), PLON(GERMANY), LAKE SCHLIER(GERMANY), LAKE TEGERN(GERMANY), G   W70-05761
FISH, WURTTEMBERG(GERMANY), PLON(GERMANY), LAKE SCHLIER(GERMANY),   W70-05761
N(USSR), LYUBERETS(USSR), MUNICH(GERMANY), CHLORELLA PYRENOIDOSA,   W70-05092
N MOUNTAIN CLUB HUTS, TEGERNSEE,   GERMANY, EXPANDING TECHNOLOGY, OV   W69-07818
(GERMANY), LAKE TEGERN(GERMANY),   GERMANY, FINLAND, SEWAGE RING TUB   W70-05761
NOLOGY,/ *EUTROPHICATION, ALGAE,   GERMANY, VOLUMETRIC ANALYSIS, LIM   W68-00475
DENSIS, DIAPHANOSOMA LEUCHTENBER   GIANUM, CYCLOPS BICUSPIDATUS, REP   W69-10182
NAIL CLAMS, WHITE FISH, ALEWIFE,   GIZZARD SHAD, SEA LAMPHREY, COHO   W70-01943
FISHKILL, ROOTED AQUATIC PLANTS,   GLACIAL LAKES, PLANKTON, EPIDEMIC   W68-00488
C WAST/ *EUTROPHICATION, *LAKES,   GLACIATION, URBANIZATION, DOMESTI   W69-07818
ION, ROLLER DRYING, SPRAY-DRIED,   GLASS BEADS, TRICHODERMA, DISINTE   W70-04184
, LAKE SHATUTAKEE, MASCOMA LAKE,   GLEN LAKE.: / WADLEIGH STATE PARK   W69-08674
I, DIDINIUM, PARAMECIUM, PHACUS,   GLENODINIUM, CRYPTOMONAS, CILIATA   W70-00268
OCYTOMETER COUNTING, ITHACA(NY),   GLENODINIUM QUADRIDENS, OOCYSTIS   W69-06540
ENA GRACILIS, FOREST CENTRIFUGE,   GLENODINIUM PULVISCULUS, HAEMOCYT   W69-06540
IUM HIRUDINELLA, ANKISTRODESMUS,   GLENODINIUM PULVISCULUS, SCENEDES   W70-05387
LLATORIA, SPIRULINA, VOLVOCALES,   GLENODINIUM, TRACHELOMONAS.: /SCI   W70-05404
OCAPSA, CHROOCOCCUS, CLOSTERIUM,   GLEO: /IUM, ANKISTRODESMUS, APHAN   W70-04371
NABAENA HASSALI, LYNGBYA NOLLEI,   GLOCOTRICHIA PISUM, NOSTOC VERRUC   W69-08283
, PLANCTOMYCES, BACILLARIOPHYTA,   GLOECAPSA, MERISMOPEDIA, MICROCYS   W70-03969
ABAENA, ANACYSTIS, GLOEOTRICHIA,   GLOEOCAPSA, STAURONEIS.: /LIS, AN   W70-03974
EPOCINCLIS, ANABAENA, ANACYSTIS,   GLOEOTRICHIA, GLOEOCAPSA, STAURON   W70-03974
ATORIA, ANABAENA, APHANIZOMENON,   GLOEOTRICHIA, POLYPEPTIDES, PHOTO   W70-04369
HNIQUES, PHYCOLOGY, MICROCYSTIS,   GLOEOTRICHIA, APHANIZOMENON, MICR   W70-04381
AQUAE, APHANIZOMENON FLOS AQUAE,   GLOEOTRICHIA ECHINULATA, BOTRYOCO   W70-03959
AERUGINOSA, ANABAENA CIRCINALIS,   GLOEOTRICHIA ECHINULATE, OSCILLAT   W69-05704
CALES, CONJUGATAE, HETEROKONTAE,   GLOEOTRICHIA, ASTERIONELLA, DINOB   W70-05409
LOS AQUAE, TOLYPOTHRIX DISTORTA,   GLOEOTRICHIA ECHINULATA LB 1303.:   W70-05091
INOSA, APHANIZOMENON FLOS-AQUAE,   GLOEOTRICHIA ECHINULATA, COCCOCHL   W70-00719
XYLAMINE, ANABAENA, MICROCYSTIS,   GLOEOTRICHIA ECHINULATA, DINOBRYO   W70-02803
CYSTIS, APHANIZOMENON, ANABAENA,   GLOEOTRICHIA, JANESVILLE(WIS), SP   W70-03310
              *PHOSPHORUS CYCLE,   GLOEOTRICHA, P-32.:   W73-01454
                     CLADOPHORA   GLOMERATA.:   W70-00667
*CLADOPHORA FRACTA, CLADOPHORA   GLOMERATA, BUTTERNUT CREEK(NY).:   W72-05453
ALGAE, BACTERIA, EUTROPHICATION,   GLUCOSE, CARBON RADIOISOTOPES, DI   W70-04194
NTITHAMNION SARNIENSE, TRITIATED   GLUCOSE, GONIDIA FORMATION, LONG   W69-09755
ORT, CARBON-14, ENZYME KINETICS,   GLUCOSE, MECHAELIS-MENTEN EQUATIO   W70-04284
, RADIOCARBON UPTAKE TECHNIQUES,   GLUCOSE, MICHAELIS-MENTEN EQUATIO   W70-04287
FIBERS, METHYL ALCOHOL, ETHYLENE   GLYCOL, POLYMERS, DIMETHYL TEREPH   W70-10433
ACELLULAR PRODUCTS, PHOTIC ZONE,   GLYCOLLIC ACID, CHEMOTROPHY, PART   W70-02504
HAELIS-MENTEN EQUATION, ACETATE,   GLYCOLLIC ACID.: /S, GLUCOSE, MEC   W70-04284
IC PATTERN, ANABAENA CYLINDRICA,   GLYCOLYSIS, TRICARBOXYLIC ACID CY   W70-02965
OACIDS, TAMIYA'S MINERAL MEDIUM,   GLYOXYLIC ACID.: /TILE ACIDS, KET   W70-04195
LA, FLAGELLATES, ANKISTRODESMUS,   GOMPHENEMA, SPIROGYRA.: /CYCLOTEL   W70-02249
CLADOPHORA, DIAPTOMUS, CYCLOPS,   GOMPHONEMA, PHORMIDIUM, APHANIZOM   W70-02803
ONEIS, EUNOTOGRAMMA, FRAGILARIA,   GOMPHONEMA, GYROSIGMA, MELOSIRA,   W70-03325
DAGA LAKE(NY), EUNOTIA, AMPHORA,   GOMPHONEMA, PINNULARIA, NITZSCHIA   W70-03974
TUM, DICTYOSPHAERIUM PULCHELLUM,   GOMPHONEMA, DINOBRYON.: /RETICULA   W69-05704
DOPHORA, TAXONOMIC DESCRIPTIONS,   GONGROSIRA LACUSTRIS BRAND, GONGR   W70-04468
ONS, GONGROSIRA LACUSTRIS BRAND,   GONGROSIRA DEBARYANA RAB, OOCYSTI   W70-04468
ECIES, *WESTERN LAKE ERIE(OHIO),   GONGROSIRA STAGNALIS, NEPHROCYTIU   W70-04468
ON SARNIENSE, TRITIATED GLUCOSE,   GONIDIA FORMATION, LONG ISLAND SO   W69-09755
GIA FUSCO-PURPUREA, SPHACELARIA,   GONIDIA, CALLOPHYLLIS HAENOPHYLLA   W69-10161
CERIUM-144, BETA RADIOACTIVITY,   GONIOBASIS, DECAPODA, PLECOPTERA,   W69-09742
COCCUM, SYNECHOCOCCUS, CARTERIA,   GONIUM, ANKISTRODESMUS, MICROSPOR   W70-03974
UATIC PRODUCTIVITY, CHLOROPHYTA,   GONOPHYTA, DIATOMS, ENVIRONMENTAL   W70-02784
ANTIBIOTIC EFFECTS, STIMULANTS,   GONYAULAX CATENELLA, GONYAULAX TA   W70-02510
STIMULANTS, GONYAULAX CATENELLA,   GONYAULAX TAMERENSIS, GYMNODINIUM   W70-02510
S LUTHERI, NAVICULA PELLICULOSA,   GONYAULAX POLYEDRA, 3,5-DIAMINOBE   W69-08278
GYMNODINIUM BREVE, C/ GONYAULAX,   GONYAULAX VENIFICUM, PYRODINIUM,   W70-05372
YRODINIUM, GYMNODINIUM BREVE, C/   GONYAULAX, GONYAULAX VENIFICUM, P   W70-05372
RANTS, FEDERAL GOVERNMENT, STATE   GOVERNMENTS, WATER QUALITY: /G, G   W69-06305
NTROL, *LOCAL GOVERNMENTS, STATE   GOVERNMENTS, BIOCHEMICAL OXYGEN D   W69-06909
*WATER POLLUTION CONTROL, *LOCAL   GOVERNMENTS, STATE GOVERNMENTS, B   W69-06909
MENT, FINANCING, GRANTS, FEDERAL   GOVERNMENT, STATE GOVERNMENTS, WA   W69-06305
OL/ *CONNECTICUT, *LAKES, *LOCAL   GOVERNMENTS, *WATER QUALITY CONTR   W72-05774

450

LORELLA PYRENOIDOSA, SELENASTRUM
STRUM CAPRICORNUTUM, SELENASTRUM
 GROWTH, CYANOCOBALAMIN, EUGLENA
NTING, EUDORINA ELEGANS, EUGLENA
ALYSIS, COULTER COUNTER, EUGLENA
WEIGHT, B COLI COMMUNAE, EUGLENE
YLATION, VANADIUM, ZINC, EUGLENA
ONENSIS, NITZSCHIA, ASTERIONELLA
*ALGAL CONCENTRATION, / *THERMAL
E, CARPATHIAN SUBMONTANE REGION,
WASTES, WATER QUALITY STANDARDS,
, STRIPPING FILM, SYNECHOCOCCUS,
'S POND(SOMERSET, ENGLAND), SAND
LOGY, FLAGELLATION, IODINE TEST,
PHAGA', VIBRIOS, ACHROMOBACTERS,
SAVANNAH RIVER, SNAKE RIVER, RIO
STES, *RESEARCH AND DEVELOPMENT,
POLLUTION TREATMENT, FINANCING,
LASKA, HUDEUC LAKE, PULLIN LAKE,
N, SNOW, RUNOFF, DRAINAGE WATER,
ATIONS, ARCTIC, TUNDRA, SCALING,
NSITY, LIGHT, MUD, SILTS, SANDS,
E, OXIDATION, CHEMICALS, CARBON,
S, FILTRATION MEDIA, ANTHRACITE,
SSO(MINN), LAKE JOSEPHINE(MINN),
TA, SYNEDRA ACUS, SINKING RATES,
LIGHT INTENSITY, EUTROPHICATION,
 ECOLOGY, ENVIRONMENTAL EFFECTS,
ATOMS, CHLOROPHYTA, CHRYSOPHYTA,
, ALGICIDES, DOMINANT ORGANISMS,
HYLL, PHOTOSYNTHESIS, CHLORELLA,
UTION.:
NITROGEN COMPOUNDS, PHOSPHATES,
, DISSOLVED OXYGEN, HYPOLIMNION,
NCES, STREAMS, ESTUARIES, LAKES,
NCES, STREAMS, ESTUARIES, LAKES,
TROPHY, CYANOPHYTA, CHLOROPHYTA,
EMENTS, PLANKTON, LAKES, RIVERS,
RS, PHOSPHATES, NITRATES, ALGAE,
LIMNOLOGY, *PHYTOPLANKTON, FISH,
LA, / PESTICIDE RESIDUES, ALGAE,
SSES, U S PUBLIC HEALTH SERVICE,
EL STUDIES, MATHEMATICAL MODELS,
UTION SOURCES, MUNICIPAL WASTES,
 EFFECTS, *FALLOUT, *ECOSYSTEMS,
, *PHOTOSYNTHESIS, *RESPIRATION,

LICH ISOTHERM, BLUE-GREEN ALGAE,
                GREEN ALGAE, BLUE-
AT DEATH.:
PTAKE, FREUNDLICH ISOTHERM, BLUE-
L STUDIES, ALGAL IDENTIFIC/ BLUE-
LL, *OIL POLLUTION, *FILAMENTOUS
LGAE, *TEMPERATURE TOLERA/ *BLUE-
MPERATURE EXPERIMENTS, RECYCLED/
NTS, RECYCLED/ GREEN ALGAE, BLUE-
 *PHYTOPLANKTON, *DIATOMS, *BLUE-
*POLLUTION-TOLERANT ALGAE, *BLUE
*2-METHYL-5-VINYLPYRIDINE, *GLUE
UORESCENCE, PHOTOSYNTHESIS, BLUE-
AMENTOUS ALGAE, FILAMENTOUS BLUE-
S, FILTRATION / *WATER SUPPLIES,
 SPECTROSCOPY, BLUE-GREEN ALGAE,
ONS, INFRARED SPECTROSCOPY, BLUE-
PHYSIOLOGY.:   *BLUE-
ATS, SYNECHOCOCCUS LIVIDUS, BLUE-
LARK LAKE (S D), ABUNDANCE, BLUE-
D), ABUNDANCE, BLUE-GREEN ALGAE,
ROBIUM, FILINIA, / *FAYETTEVILLE
'S POND(WIS), LAKE WAUBESA(WIS),
CELLULAR PRODUCTS, GAS EXCHANGE,
ONTANA, *HYDROCARBONS, *OLEFINS,
. HANSON RES/ WOOD, ALGAL COUNT,
POSITION(ENGLAND), FOOD SOURCES,
KERSEE, TURLERSEE, BALDEGGERSEE,
ICHSEE(SWITZ), TURLERSEE(SWITZ),
IA VAUCHER, SKELETONEMA COSTATUM(
                VOLATILE SOLIDS,
RIVER, FLUORESCEIN, MACROPHYTES,
US COMPOUNDS, RESERVOIRS, ALGAE,
CITY, SOIL POROSITY, IRRIGATION,
C/ *NITROGEN CYCLE, *FARM WASTE,
ON, GROUND WATER, RUNOFF, FROZEN
LL, RUTHENIUM-106, ANTIMONY-125,
/ *INFILTRATION, *PONDS, *SOIL,
NTHETIC, SAPROPHYTIC,/ TAXONOMIC
ERIONELLA, SCENEDESMUS OBLIQUUS,
ALGAL ENUMERATION, SCENEDESMUS,
TI/ ALGAL GROWTH KINETICS, MONOD
ACTER(ALKALIGENES), RHODOTORULA,

GRACILE, CARBON ASSIMILATION, BAC W70-03983
GRACILE, GROWTH LIMITATION, SELEN W70-02779
GRACILIS Z, MICROBIOLOGICAL ASSAY W69-06273
GRACILIS, FOREST CENTRIFUGE, GLEN W69-06540
GRACILIS, CACHONINA NIEI, SKELETO W69-08278
GRACILIS, PROTOCOCCALES.: /, DRY W70-05092
GRACILIS, ANABAENA CYLINDRICA, BA W70-02964
GRACILLIMA, GROWTH FACTORS.:/ROT W70-02804
GRADIENT, *BOD REMOVAL, *ISRAEL, W70-04790
GRADIENTS(RIVER), THALLI, SCHIZOM W70-02784
GRAFTON(ND), LAKOTA(ND), HARVEY(N W70-03312
GRAIN COUNT(AUTORADIOGRAPHY), LEU W69-10163
GRAINS, TIDAL CYCLES.: /YI, ABBOT W71-12870
GRAM REACTION, CYTOPHAGA, PROTEUS W70-03952
GRAM-POSITIVE BACTERIA, SARCINE-L W70-00713
GRANDE RIVER, POTOMAC RIVER.: /, W70-04507
GRANTS, ALGAE, AIR POLLUTION, WAT W71-06821
GRANTS, FEDERAL GOVERNMENT, STATE W69-06305
GRAPHITE LAKE, OLD JOHN LAKE, PIN W69-08272
GRASSLANDS, SOIL EROSION,: /LUTIO W70-04193
GRAVELS, MUD, SANDS, SHALES, BOTT W69-08272
GRAVELS, SEDIMENTS, WATER POLLUTI W70-00711
GRAVIMETRY.: /TURE, CARBON DIOXID W70-05651
GRAVITY FILTERS.: /OSSES, OOCISTU W70-04199
GRAY'S BAY(MINN), SMITH'S BAY(MIN W70-05269
GRAZING PRESSURE, POPULATION DENS W69-10158
GRAZING, ALGAE, WATER CHEMISTRY, W70-03978
GRAZING, IRON, MANGANESE, MOLYBDE W69-04801
GRAZING, SNAILS, AGE, TURBIDITY, W71-00668
GRAZING, WATER POLLUTION EFFECTS, W69-01977
GRAZING, ZOOPLANKTON, DIATOMS, RE W70-04809
GREAT LAKES BASIN, BACTERIAL POLL W69-03948
GREAT LAKES, MICHIGAN.: /SOURCES, W69-01445
GREAT LAKES, LAKE ERIE, STATISTIC W68-00253
GREAT LAKES, PHYSIOLOGICAL ECOLOG W70-03975
GREAT LAKES, PHYSIOLOGICAL ECOLOG W70-03975
GREAT LAKES, BIOINDICATORS, FISH, W70-04503
GREAT LAKES.: /OSPHORUS, TRACE EL W70-04385
GREAT LAKES.: /TRIENTS, FERTILIZE W69-07084
GREAT LAKES, LEGISLATION, NITROGE W69-06858
GREAT LAKES, METABOLISM, *CHLOREL W70-08652
GREAT LAKES-ILLINOIS RIVER BASINS W70-00263
GREAT LAKES, THERMAL STRATIFICATI W71-05878
GREAT LAKES, *POLLUTANT IDENTIFIC W71-05879
GREAT LAKES, TRACE ELEMENTS, RADI W71-09013
GREAT LAKES, NUISANCE ALGAE, WATE W72-05453
GREAT MIAMI RIVER(OHIO).:   W70-00475
GREEN ALGAE.: /DE, UPTAKE, FREUND W70-08111
GREEN ALGAE, HEAT DEATH.:   W70-07313
GREEN ALGAE, BLUE-GREEN ALGAE, HE W70-07313
GREEN ALGAE, GREEN ALGAE.: /DE, U W70-08111
GREEN ALGAE, NITZSCHIA, BIOLOGICA W70-10173
GREEN ALGAE, HALIMONE PORTULACOID W70-09976
GREEN ALGAE, *GREEN ALGAE, *RED A W71-02496
GREEN ALGAE, BLUE-GREEN ALGAE, TE W70-07257
GREEN ALGAE, TEMPERATURE EXPERIME W70-07257
GREEN ALGA, COPPER SULFATE.: /NG, W70-02754
GREEN ALGAE, *EUGLENA, PIGMENTED W70-01233
GREEN ALGAE, BREAM, WHITE BREAM, W70-01788
GREEN ALGAE, PHYTOPLANKTON, OHIO W69-05697
GREEN ALGAE, OCULAR MICROMETER, S W70-04253
GREEN ALGAE, HEAD LOSSES, OOCISTU W70-04199
GREEN ALGAE.: /RACTIONS, INFRARED W70-03928
GREEN ALGAE, GREEN ALGAE.: /RACTI W70-03928
GREEN ALGAE, *ALGA BLOOMS, ALGAE W69-03512
GREEN ALGAE, FILAMENTS.: /RMIS, M W70-05270
GREEN ALGAE, GREEN ALGAE, FRAGILA W70-05375
GREEN ALGAE, FRAGILARIA, ASTERION W70-05375
GREEN LAKE(NY), MARL LAKES, CHLO W70-05760
GREEN LAKE(WIS).: /VAL, SPAULDING W69-05868
GREEN PLANTS, MASS CULTURE, MEDIC W69-07832
GREEN RIVER FORMATION, CHLOROPHYL W69-08284
GREEN RIVER, WASHINGTON, HOWARD A W72-11906
GREENHOUSE STUDIES, GROWTH K: /OM W69-06865
GREIFENSEE, ZURICHSEE, MOSES LAKE W70-03964
GREIFENSEE(SWITZ), PAFFIKERSEE(SW W70-00270
GREV) CL, CHLORELLA OVALIS BUTCHE W70-05752
GRIT.:   W71-11375
GROUND SOLIDS, UPTAKE, CONSUMER O W69-09334
GROUND WATER, NITRATES, PHOSPHATE W72-02817
GROUND WATER, RUNOFF, FROZEN GROU W69-05323
GROUND WATER, *NITRATES, EUTROPHI W69-05323
GROUND, SOIL EROSION, WATER POLLU W69-05323
GROUNDWATER, BISMUTH-207, VERTEBR W69-08269
GROUNDWATER BASINS, WATER SUPPLY, W69-07838
GROUPS, MOLECULAR FILTER, CHEMOSY W68-00891
GROWTH CONSTATS, RIVER RHINE.: /T W70-03955
GROWTH EQUATIONS, ANKISTRODESMUS, W70-04381
GROWTH EQUATION, INCIDENT LIGHT U W69-03730
GROWTH FACTORS, COLONIAL MORPHOLO W70-03952

OL, MUSTARD GROWTH TEST, ORGANIC
ALGINIC ACID, FUCUS VESICULOSUS,
MATTER, ORGANIC DEBRIS, ORGANIC
ZSCHIA, ASTERIONELLA GRACILLIMA,
     ALGAL SENESCENCE,
ROBIAL ECOLOGY, COULTER COUNTER,
OOD SOURCES, GREENHOUSE STUDIES,
ATION, INCIDENT LIGHT UTI/ ALGAL
RICORNUTUM, SELENASTRUM GRACILE,
HAMNION SARNIENSE, QUANTITATION,
TH, NUTRIENT AVAILABILITY, ALGAL
ENT, PHOSPHATES, ALGAE, DAPHNIA,
ENTRATION, PHOSPHORUS, NITROGEN,
NA, ALGAL CONTROL, BACTERICIDES,
TILIZATION, WATER FLEAS, LARVAE,
AT LAKES, LAKE ERIE, STATISTICS,
     CRATER LAKE, ORE,
VOLUMETRIC ANALYSIS, LIMNOLOGY,
CARBON DIOXIDE, PHOTOSYNTHESIS,
TS, CHLOROPHYLL, PHOTOSYNTHESIS,
TRIENTS, *NUTRIENT REQUIREMENTS,
ATOGRAPHY, RESISTANCE, BIOASSAY,
ORELLA, ENZYMES, CARBON DIOXIDE,
NTS, PHOSPHORUS, NITROGEN, IRON,
OCEANOGRAPHY, TRACERS, BACTERIA,
OPHICATION, POLLUTION ABATEMENT,
OLISM, ISOPODS, AMPHIPODA, FISH,
ORUS, SULFATES, VITAMINS, ALGAE,
MMONIA, ENZYMES, CLASSIFICATION,
REMENTS, CULTURES, LAKES, ALGAE,
ENVIRONMENTAL FACTORS, BIOMASS,
POLLUTION EFFECTS, LETHAL LIMIT,
ESTS, PARTICLE SIZE, POPULATION,
S, LARVAE, SNAILS, MITES, MOLDS,
D SLUDGE, *BACTERIA, METABOLISM,
EGRADATION, ALGAE, MARINE ALGAE,
LIGHT INTENSITY, CARBON DIOXIDE,
DOW CONTACT, SPARGANIUM, SALIX,
WTH SUBSTANCES, LIMNOLOGY, PLANT
ICA, DIATOMS, RESISTANCE, *PLANT
ON, MANGANESE, MOLYBDENUM, PLANT
TES, EDDIES, HABITATS, CRAYFISH,
UATION, RESISTANCE, LIFE CYCLES,
IS, *LAKE BUTTE DES MORTS(WISC),
YTA, CHLOROPHYTA, BULRUSH, PLANT
, ANALYSIS, FERTILIZATION, PLANT
PHOMETRY, DOMESTIC WASTES, PLANT
TS, EUTROPHICATION, ALGAE, PLANT
*TEMPERATURE TOLERANCE, OPTIMUM
IA SACCHARINA, MANNITOL, MUSTARD
*ESTIMATING EQUATIONS, PERIOD OF
WAUBESA(WIS), MACROPHYTES, ALGAL
MYXOPHYCEAE, COMPLEX FORMATION,
*NUTRIENT REQUIREMENTS, *PLANT
CALIFORNIA, NUTRIENTS, / *ALGAL
*NUTRIENT AVAILABILITY, *ALGAL
A, NOSTOC PALUDOSUM, BLE/ *ALGAL
OXYGEN CURVES, CALORIMET/ *ALGAL
ICAL ANALYSIS, P/ *ALGAE, *PLANT
A, *NUTRIENT AVAILABILITY, ALGAL
URE, *PILOT-PLANT STUDIES, ALGAL
YSIOLOGY/ *CYANOCOBALAMIN, ALGAL
TA, LAKES, RESERVOIRS, NITROGEN,
UTOTROPHIC GROWTH, HETEROTROPHIC
H LIMITATION, SELENASTRUM, ALGAL
OPHYCEAE, CHLORO/ *EUROPE, ALGAL
RAPHIES, *EUTROPHICATION, *PLANT
, PLANT GROWTH SUBSTANCES, PLANT
RSON STRA/ *MASS CULTURE, *ALGAL
RACILIS Z,/ *VITAMIN B-12, ALGAL
ABLE WATER, ACTINOMYCETES, PLANT
LORELLA PYRENOIDOSA, AUTOTROPHIC
, TURBIDOSTATS, NUTRIENT LIMITED
GAL GROWTH POTENTIAL, SE/ *ALGAL
T SOURCES, WATER SAMPLING, ALGAL
AND, PHOSPHORUS COMPOUNDS, PLANT
NTIFICATION, ALGAE COUNTS, ALGAL
     *BIOLOGICAL EXTRACTION,
RIV/ TREE BARK, CHARCOAL, ALGAL
GAE NUISANCE, ALGAE BLOOM, ALGAE
OLLUTION EFFECTS, ECOLOGY/ PLANT
T, SANIT/ *BIOSTIMULATION, ALGAL
SPHORUS SOURCES, *NUISANCE PLANT
ECTS, LAKE MEAD, DILUTION, ALGAL
ROPIUM-155, SILVER-110M, CERIUM,
, MULTIPLE-USES, SAVANNAH RIVER,
, FLAVOBACTERIUM, ACHROMOBACTER,
, *ALGAE, *SORPTION, ABSORPTION,
RIO), TWELVE MILE LAKE(ONTARIO),
     AERO-HYDRAULIC

GROWTH FACTORS, POLYSACCHARIDES.: W69-07826
GROWTH FACTORS, LAMINARIA SACCHAR W69-07826
GROWTH FACTORS, VITAMIN B1, VITAM W70-02510
GROWTH FACTORS.: /ROTONENSIS, NIT W70-02804
GROWTH INHIBITIONS.:   W72-02417
GROWTH INHIBITION, CYANOCOBALAMIN W70-05652
GROWTH K: /OMPOSITION(ENGLAND), F W69-06865
GROWTH KINETICS, MONOD GROWTH EQU W69-03730
GROWTH LIMITATION, SELENASTRUM, A W70-02779
GROWTH MODES.: /HRIX MUCOR, ANTIT W69-10163
GROWTH POTENTIAL, SELENASTRUM CAP W69-06864
GROWTH RATES, PREDATION.: /VIRONM W69-06536
GROWTH RATES, STABILIZATION PONDS W68-00855
GROWTH RATES.: /CHLORELLA, *EUGLE W69-00994
GROWTH RATES, *FRESH WATER FISH,   W68-00487
GROWTH RATES, ALGAE, FRESH WATER   W68-00253
GROWTH RATES, WATER PROPERTIES.:   W68-00478
GROWTH RATES, *LABORATORY TESTS,   W68-00475
GROWTH RATE, AEROBIC CONDITIONS,   W70-05750
GROWTH RATE, ORGANIC MATTER, SCEN W70-05381
GROWTH RATES, LABORATORY TESTS, C W70-03507
GROWTH RATES, PESTICIDE RESIDUES, W70-03520
GROWTH RATES, PLANKTON, OXYGEN, N W70-04369
GROWTH RATES, CHLOROPHYTA, CYANOP W70-03983
GROWTH RATES, TRITIUM, PHOTOGRAPH W69-10163
GROWTH RATES, CHLOROPHYLL, PHOTOS W69-10167
GROWTH RATES.: /OMS, ALGAE, METAB W69-09742
GROWTH RATES, COPPER SULFATE, BOT W69-09349
GROWTH RATES, SEAWATER, INCUBATIO W70-00713
GROWTH RATES, BACTERIA, ULTRAVIOL W70-00719
GROWTH RATES, ALGAE, OLIGOTROPHY, W70-02779
GROWTH RATES, SPORES.: /E, WATER   W70-01779
GROWTH RATES, WATER POLLUTION CON W72-02417
GROWTH RATES, HABITATS.: /ROTIFER W72-01819
GROWTH RATES, MICROORGANISMS, ALG W72-01817
GROWTH RATES, ANALYTICAL TECHNIQU W71-07339
GROWTH RATES, TEMPERATURE, PILOT   W72-06022
GROWTH REGULATORS, MYRIOPHYLLUM,   W70-05263
GROWTH REGULATORS, CYANOPHYTA, EN W68-00470
GROWTH REGULATORS, FRESH WATER LA W68-00856
GROWTH REGULATORS, PREDATION, PRO W69-04801
GROWTH STAGES, DIATOMS, ALGAE, ME W69-09742
GROWTH STAGES.: /NT, BAYS, *FLUCT W71-02494
GROWTH STIMULATION, SURFACE LITTE W72-00845
GROWTH SUBSTANCES, CHLORELLA, CHL W72-00845
GROWTH SUBSTANCES, SEWAGE, SULFAT W70-02969
GROWTH SUBSTANCES, LIMNOLOGY, PLA W68-00470
GROWTH SUBSTANCES, PLANT GROWTH,   W69-07826
GROWTH TEMPERATURE, CONDENSER PAG W71-02496
GROWTH TEST, ORGANIC GROWTH FACTO W69-07826
GROWTH.: / ISLES, BIOINDICATORS,   W68-00255
GROWTH.: /NAJAS, ANACHARIS, LAKE   W70-03310
GROWTH-PROMOTING PROPERTIES, CHLA W69-10180
GROWTH, *ALGAE, ELEMENTS(CHEMICAL W70-02964
GROWTH, *BIOASSAYS, *FLUOROMETRY, W70-02777
GROWTH, *MICROCYSTIS AERUGINOSA,   W69-05867
GROWTH, *POTOMAC RIVER, PLECTONEM W70-02255
GROWTH, *THERMOBIOLOGY, *DIURNAL   W70-03309
GROWTH, *WISCONSIN, ECOLOGY, CHEM W70-06229
GROWTH, ALGAL NUTRITION, ALGAL PH W69-05868
GROWTH, ALGAL NUTRITION, ALGAL PH W69-06865
GROWTH, ALGAL NUTRITION, ALGAL PH W69-06277
GROWTH, AMINO ACIDS, CHLAMYDOMONA W69-10180
GROWTH, ANKISTRODESMUS, NOSTOC MU W70-02964
GROWTH, ANABAENA, BIOSTIMULATION, W70-02779
GROWTH, ANKISTRODESMUS, BACILLARI W69-07833
GROWTH, ANALYTICAL TECHNIQUES, AN W69-04801
GROWTH, CARBOHYDRATES.: /N, ALGAE W69-07826
GROWTH, CHLORELLA PYRENOIDOSA(EME W69-07442
GROWTH, CYANOCOBALAMIN, EUGLENA G W69-06273
GROWTH, DECOMPOSING ORGANIC MATTE W72-01812
GROWTH, HETEROTROPHIC GROWTH, ANK W70-02964
GROWTH, LUXURY CONSUMPTION, PROVI W70-02779
GROWTH, NUTRIENT AVAILABILITY, AL W69-06864
GROWTH, OXYGEN DEPLETION, NUCLEAR W70-00264
GROWTH, PRIMARY PRODUCTIVITY, RIV W69-06535
GROWTH, SEDGWICK-RAFTER METHOD, T W69-10157
GROWTH, SHOCK LOADING.:   W70-06604
GROWTH, STREAM CONCOURSE, PALOUSE W70-03501
GROWTH, WATER DENSITY, CONVECTION W69-08282
GROWTH, WATER AND PLANTS, WATER P W70-02772
GROWTH, WATER POLLUTION ASSESSMEN W70-02775
GROWTHS, *NUTRIENT CONTROL, *PHOS W69-09340
GROWTHS, OXYGEN DEPLETION.: / EFF W73-00750
GUETTARDA, IPOMOEA, PISONIA, PAND W69-08269
GULF COAST.: / COLLOIDAL FRACTION W73-00748
GULF OF MARSEILLES, ASTERIONELLA   W70-04280
GULF OF MEXICO, ATLANTIC OCEAN, R W69-08524
GULL LAKE(ONTARIO), BALSAM LAKE(O W70-02795
GUN, ALLISTON(ONTARIO).:   W71-09546

CATENELLA, GONYAULAX TAMERENSIS, GYMNODINIUM BREVIS, MYCROSYSTIS, W70-02510
UM-95, RUBIDIUM-106, NIOBIUM-95, GYMNODINIUM SIMPLEX, FLAGELLATES, W70-00707
TIC LAYER, MELOSIRA BAICALENSIS, GYMNODINIUM BAICALENSE, CYCLOTELL W70-04290
CCOLITHUS, SARGASSO SEA, SYNURA, GYMNODINIUM, CHEMICAL SYMBIOSIS, W70-04503
GONYAULAX VENIFICUM, PYRODINIUM, GYMNODINIUM BREVE, CANADA, EUROPE W70-05372
THIAMINE, SERRATIA MARINORUBRA, GYRODINIUM COHNII, MICROBIAL ECOL W70-05652
OGRAMMA, FRAGILARIA, GOMPHONEMA, GYROSIGMA, MELOSIRA, NAVICULA, NI W70-03325
IMIKPUK LAKE, NORTH MEADOW LAKE, GYTTJA, DUFF, OOZE, COBBLES, EQUI W69-08272
KE WASHINGTON(WASH), MACROFAUNA, GYTTJA, OSCILLATORIA RUBESCENS, Z W70-04253
, CONCORD(N H), PEANACOOK LAKE(N H).: / TREATMENT COSTS, DINOBRYON W69-08282
BASTICOOK(ME), LAKE WINNISQUAM(IS H), MADISON LAKES(WIS), LAKE WASH W70-07261
MENT COSTS, DINOBRYON, CONCORD(N H), PEANACOOK LAKE(N H).: / TREAT W69-08282
ENTS, ENVIRONMENTAL ENGINEERING, H: /NTAL EFFECTS, ESSENTIAL NUTRI W69-03374
AL/ *TRANSFORMATIONS, AMSTERDAM, HAARLEM, LEYDEN, THE HAGUE, MINER W69-07838
TA, CYANOPHYTA, DIATOMS, AQUATIC HABITAT, POPULATION.: /EUGLENOPHY W70-02770
ILS, MITES, MOLDS, GROWTH RATES, HABITATS.: /ROTIFERS, LARVAE, SNA W72-01819
HYTA, CYANOPHYTA, UNITED STATES, HABITATS.: /, *LAKE ERIE, CHLOROP W70-04468
RIVERS, HOT SPRINGS, TERRESTRIAL HABITATS, DEPTH, AIR, SEDIMENTS, W69-08272
*BACKGROUND RADIATION, *AQUATIC HABITATS, *ALASKA, LAKES, RIVERS, W69-08272
SH/ *TOXICITY, *ALGAE, *REVIEWS, HABITATS, CLAMS, MUSSELS, SHELLFI W70-05372
C PROCESSES, ECOSYSTEMS, AQUATIC HABITATS, WATER POLLUTION EFFECTS W71-09012
ENVIRONMENT, ECOSYSTEMS, AQUATIC HABITATS, DRAGONFLIES, ALGAE, SED W71-09005
OTTOM SEDIMENTS, CHUTES, EDDIES, HABITATS, CRAYFISH, GROWTH STAGES W69-09742
TS, SEAWATER, ORGANIC COMPOUNDS, HABITATS, STATISTICAL METHODS.: / W69-09755
GAE, *TEMPERATURE, *HOT SPRINGS, HABITATS, ALKALINE WATER, ACIDIC W69-10160
RIFUGE, GLENODINIUM PULVISCULUS, HAEMOCYTOMETER COUNTING, ITHACA(N W69-06540
HACELARIA, GONIDIA, CALLOPHYLLIS HAENOPHYLLA, TIDAL CURRENT, PUGET W69-10161
AMSTERDAM, HAARLEM, LEYDEN, THE HAGUE, MINERALIZATION, CLAY LENSE W69-07838
OPTERA, TIPULIDAE, CHIRONOMIDAE, HALF-LIFE.: / PLECOPTERA, EPHEMER W69-09742
NORTH TEMPERATE ZONE, PLUTONIUM, HALF-LIFE(RADIONUCLIDE), FILTER-F W70-00708
HLORELLA PYRENOIDOSA, BIOLOGICAL HALF-LIFE, CLINCH RIVER(TENNESSEE W70-02786
LINCH RIVER(TENNESSEE), PHYSICAL HALF-LIFE, HYDROPSYCHE COCKERELLI W70-02786
*RADIONUCLIDE, ULOTHRIX, UPTAKE, HALF-LIFE, DETECTOR.: W69-07862
IONUCLIDES, ASSIMILATION, BA-LA, HALF-LIFE, PTYCHOCHEILUS OREGONEN W69-07853
LIMNODRILUS SP/ RADIOAUTOGRAPHS, HALF-LIFE, CONCENTRATION FACTOR, W69-07861
UTION, *FILAMENTOUS GREEN ALGAE, HALIMONE PORTULACOIDES, SUAEDA MA W70-09976
VARIA, LAKE ZELL(AUSTRIA), LAKE HALL WILER(SWITZERLAND).: /IER(BA W70-05668
EIFENSEE, ZURICHSEE, MOSES LAKE, HALLWILLERSEE.: /BALDEGGERSEE, GR W70-03964
-VACUOLES, LLANGORS LAKE(WALES), HALOBACTERIUM, RESPIROMETER, PROT W70-04369
TINALIS, CLADOPHORA, CYCLOTELLA, HALOCHLOROCOCCUM, SYNECHOCOCCUS, W70-03974
AROGRAPHIC ANALYSIS, PHOTOMETRY, HALOGENS.: /ICAL TECHNIQUES, *POL W70-05178
ER POLLUTION EFFECTS, CHLORIDES, HALOGENS, ALGAE, WATER POLLUTION W69-08562
CAL COMMUNITIES, *AQUATIC ALGAE, HALOPHYTES.: /QUILIBRIUM, BIOLOGI W69-00360
ADOCERA, / *HYPERTROPHIC WATERS, HAMBURG(GERMANY), CHIRONOMIDS, CL W70-00268
ION, ALGAL BLOOMS, NUTRIEN/ *NEW HAMPSHIRE WATER POLLUTION COMMISS W69-08674
TION, BIOLOGICAL INDICATORS, NEW HAMPSHIRE.: /DISTRIBUTION, POPULA W70-02772
TION, BIOLOGICAL INDICATORS, NEW HAMPSHIRE.: / ECOLOGICAL DISTRIBU W70-02764
ANCE ALGAE, *ALGAL CONTROL, *NEW HAMPSHIRE, *COPPER SULFATE, SEWAG W69-08674
ABAENA, PHYGON XL(DICHLONE), NEW HAMPSHIRE, LAKE WINNISQUAM, LONG W69-08674
REATMENT, ALGAE, FILTRATION, NEW HAMPSHIRE, NEW YORK, VIRGINIA, OR W72-03641
GAE CONTROL, *WATER SUPPLY, *NEW HAMPSHIRE, ALGAE, ODOR, TASTE, LI W69-08282
D ORGANISMS, FISH GENETICS, FISH HANDLING FACILITIES, FISH MANAGEM W69-04804
O/ *BLOOM, *VIETNAM, LITTLE LAKE(HANOI), RED RIVER(VIETNAM), CHLOR W70-03969
FATE, *ACTIVATED SLUDGE PROCESS, HANOVER TREATMENT PLANT.: /UM SUL W70-09186
RMANY), AUREOBASIDIUM PULLULANS, HANSENIASPORA UVARUM, RADULESPORE W70-04368
*NORTH SEA, MYCOTA, DEBARYOMYCES HANSENII, RHODOTORULA RUBRA, CAND W70-04368
EEN RIVER, WASHINGTON, HOWARD A. HANSON RESERVOIR, TACOMA(WASH).: / W72-11906
MENON FLOS-AQUAE, STEPHANODISCUS HANTZSCHII, DIDINIUM, PARAMECIUM, W70-00268
FILTER CLOGGING, STEPHANODISCUS HANTZSCHII, STEPHANODISCUS BINDER W70-00263
IS RIVER BASINS PROJECT, INDIANA HARBOR SHIP CANAL.: /LAKES-ILLINO W70-00263
OTHRIX MUCOR, MORPHOLOGY, FRIDAY HARBOR(WASH), ANTITHAMNION SARNIE W69-10161
N ION CONCENTRATION, ALKALINITY, HARDNESS(WATER), CHLORIDES, ORGAN W71-04206
ATERSHEDS(BASINS), INSECTICIDES, HARDNESS(WATER), DIATOMS, FOULING W72-01813
TURE, NITRATES, PLANTS, AMMONIA, HARDNESS(WATER), ACIDITY, OXYGEN, W69-07838
CHLORIDES, PHOSPHORUS, SULFATES, HARDNESS(WATER).: /ATER QUALITY, W69-03948
RBONATES, BACTERICIDE, KINETICS, HARDNESS, COLIFORMS, CYTOLOGICAL W71-05267
EY, TEMPERATURE, ORGANIC MATTER, HARDNESS, CARBON DIOXIDE, FISH.: / W69-10157
STES, ECOLOGY, DISSOLVED OXYGEN, HARDNESS, TIDAL WATERS, ORGANIC M W73-00748
RIVER, ALGAE, IRON, ALKALINITY, HARDNESS, RADIOACTIVITY, TEMPERAT W73-00750
S, ORGANIC PHOSPHATES.: HARRIS COUNTY, TEX, POLYPHOSPHATE W69-04800
DGING, HARVESTING OF ALGAE, FISH HARVEST, VITAMINS.: /DIMENTS, DRE W69-09340
NTS, BOTTOM SEDIMENTS, DREDGING, HARVESTING OF ALGAE, FISH HARVEST W69-09340
EATMENT, CHEMICAL PRECIPITATION, HARVESTING, ZONING, DIVERSION, DR W69-10178
.).:   *ALGAE HARVESTING, *CENTRAL VALLEY(CALIF W71-04557
.).:   *ALGAE HARVESTING, *CENTRAL VALLEY(CALIF W71-04556
, CALCIUM, MAGNESIUM, POTASSIUM, HARVESTING OF ALGAE, LIMNOLOGY, N W70-02245
ATMENT, AMMONIA STRIPPING, ALGAE HARVESTING.: /ELECTROCHEMICAL TRE W70-01981
UTRIENTS, ENVIRONMENTAL EFFECTS, HARVESTING ALGAE, INHIBITION, LIG W69-04798
SEDIMENTS, WATERSHED MANAGEMENT, HARVESTING OF ALGAE, WATER ZONING W68-00859
AL EFFECTS, ESSENTIAL NUTRIENTS, HARVESTING OF ALGAE, NUISANCE ALG W69-03371
N EFFECTS, ALGAE, ALGAL CONTROL, HARVESTING, AQUATIC ALGAE, AQUATI W69-03513
SANITATION, ESSENTIAL NUTRIENTS, HARVESTING OF ALGAE, INHIBITION, W69-03369
REEZE DRYING, FREEZE-THAW TESTS, HARVESTING OF ALGAE, NUISANCE ALG W69-03362
MESTIC WASTES, SEWAGE TREATMENT, HARVESTING OF ALGAE, BOTTOM SEDIM W68-00478
S, PHOSPHORUS, ALGAE, ALGICIDES, HARVESTING OF ALGAE, ENVIRONMENT, W69-07818
PLANTS, HERBICIDES, PESTICIDES, HARVESTING, WATER QUALITY CONTROL W69-05705
PLANTS, HERBICIDES, PESTICIDES, HARVESTING, WATER QUALITY CONTROL W69-05706
AL, ALGAE, PRIMARY PRODUCTIVITY, HARVESTING OF ALGAE, BIOMASS.: /N W70-04290
, PESTICIDES, CELLULOSE, CYCLES, HARVESTING, DREDGING, MONITORING, W70-05264
.:   *ALGAE HARVESTING, *DIATOMITE FILTRATION W72-08688

NDARDS, GRAFTON(ND), LAKOTA(ND), HARVEY(ND), CHROMATIUM VINOSUM, T W70-03312
TUCA VAR LATISSIMA, POLYSIPHONIA HARVEYI, EPIPHYTIC BACTERIA, SUBL W70-01073
UAE, ANABAENA MENDOTAE, ANABAENA HASSALI, LYNGBYA NOLLEI, GLOCOTRI W69-08283
OSAL, OCEANS, FOODS, ABSORPTION, HAZARDS, CESIUM, HYDROGEN, DEUTER W70-02786
OASSAY, BIOINDICATORS, FISHKILL, HAZARDS, PESTICIDE KINETICS, PEST W69-08565
TREATMENT, *ALGAE, *FILTRATION, HEAD LOSS, FILTERS, *SCENEDESMUS, W72-03703
DED SOLIDS, ALGAE, PILOT PLANTS, HEAD LOSS, PERMEABILITY, TEMPERAT W72-08686
/ *WATER SUPPLIES, GREEN ALGAE, HEAD LOSSES, OOCISTUS, FILTRATION W70-04199
DOCERA, WATER MASSES, U S PUBLIC HEALTH SERVICE, GREAT LAKES-ILLIN W70-00263
RUSES, BACTERIA, E. COLI, PUBLIC HEALTH, PATHOGENIC BACTERIA, TEMP W72-06017
S, *ISOLATION, LIMNOLOGY, PUBLIC HEALTH, PHYSICOCHEMICAL PROPERTIE W70-00273
OHIO, COMMERCIAL FISHING, PUBLIC HEALTH, WALLEYE, VEGETATION, WILD W70-01943
LIFE, WATER CONSERVATION, PUBLIC HEALTH, WATER POLLUTION, CONTRACT W70-03344
ER CHEMISTRY, WATER TEMPERATURE, HEAT BUDGET, TURBIDITY, *RESERVOI W72-01773
, TURBINES, ECONOMIC EFFICIENCY, HEAT BUDGET, SOCIAL ASPECTS.: /ON W70-08834
GREEN ALGAE, BLUE-GREEN ALGAE, HEAT DEATH.: W70-07313
ACTERIA, COLIFORMS, FISH, FUNGI, HEAT, WATER POLLUTION, STREAMS, R W71-02479
HELLFISH, STEAM, ELECTRIC POWER, HEATED WATER, PROTEINS, AQUATIC M W70-09905
OLID WASTES, RADIOACTIVE WASTES, HEATED WATER, THERMAL POLLUTION, W72-00462
MPERATURE, PHYTOPLANKTON, ALGAE, HEATED WATER, POPULATION, WATER P W71-09767
PETITION, OLIGOTROPHY, VITAMINS, HEAVY METALS, TOXICITY, DIATOMS, W71-10079
N, SPECTROPHOTOMETRY, NUTRIENTS, HEAVY METALS.: /EFFECTS, CHELATIO W71-11561
POUNDS, CHLORIDES, SALINE LAKES, HEAVY METALS, INDUSTRIAL WASTES, W69-02959
TOMETRY, TOTAL DISSOLVED SOLIDS, HEDRIODISCUS, CALOPARYP: /CTROPHO W70-03309
XONES, *ALGAL BLOOMS, NTA, EDTA, HEIDA, LAKE TAHOE(CALIF).: /OMPLE W70-04373
TROPICALIS, NOCTILUCA MILIARIS, HELGOLAND(GERMANY), AUREOBASIDIUM W70-04368
ADRICAUDA, SCENEDESMUS OBLIQUUS, HELIX POMATIA, CHLORELLA PYRENOID W70-04184
TELLA MENEGHINIANA, PHOTIC ZONE, HELSINGOR(DENMARK), SALINO-CLINE. W70-04809
*MUSCULAR EXERTION, *2-M/ *FISH HEMATOLOGY, *GRANULAR LEUCOCYTES, W70-01788
PLECOPTERA, DIPTERA, COLEOPTERA, HEMIPTERA, MOLLUSCA, RAFFINATE, T W69-07846
CTOR, LIMNODRILUS SPP, AEOLOSOMA HEMPRICHI, ESCHERICHIA COLI, MYRI W69-07861
EUM CRUDES, MASS SPECTROMETRY, N- HEPTADECANE, MONOENES, DIENES, TR W69-08284
*EUTROPHICATION, AQUATIC PLANTS, HERBICIDES, PESTICIDES, HARVESTIN W69-05705
*EUTROPHICATION, AQUATIC PLANTS, HERBICIDES, PESTICIDES, HARVESTIN W69-05706
TTING MANAGEMENT, POTABLE WATER, HERBICIDES, CHARA, ALKALINITY, BU W70-05263
DDT, PHOSPHOTHIOATE PESTICIDES, HERBICIDES, *PESTICIDE RESIDUES, W69-02782
ROLYSIS, COPPER SULFATE, FISHES, HERBICIDES, MIDGES, PESTICIDES.: / W70-02982
PULP WASTES, PESTICIDE RESIDUES, HERBICIDES, ESTUARIES, WATER QUAL W70-10437
ANNELS, REVIEWS, BIBLIOGRAPHIES, HERBICIDES, FISH, FRESH WATER, AL W70-00269
WATER, PHAWOPHYTA, CRUSTACEANS, HERBIVORES, CHLOROPHYLL, ENZUMES, W70-01068
MERCENARIA MERCENARIA, PANOPEUS HERBSTII.: /RASSOSTREA VIRGINICA, W69-08267
PECTINATUS, POTOMOGETON CRISPUS, HETERAMTHERA, PICKWICK RESERVOIR(W71-07698
N, TRAPA, PISTIA, ALTERNANTHERA, HETERANTHERA, NAJAS, SILVEX, ENDO W70-10175
(RHODE ISLAND), TAXONOMIC TYPES, HETERETROPHY, BACTERIAL MEDIUM, A W70-00713
I, DINOPHYCEAE, FILTER BLOCKING, HETERETROPHS, ITALY, LAKE BORREVA W69-07833
ION, BERKELSE LAKE(NETHERLANDS), HETEROCYSTS, GAS-VACUOLES, LLANGO W70-04369
PEPTIDES, ENVIRONMENTAL EFFECTS, HETEROGENEITY, IRON, MANGANESE, W W70-00273
OCCALES, VOLVOCALES, CONJUGATAE, HETEROKONTAE, GLOEOTRICHIA, ASTER W70-05409
HAELIS-MENTEN EQUATION, RELATIVE HETEROTROPHIC POTENTIAL, LINEWEAV W70-05109
GAL NUTRITION, ALGAL PHYSIOLOGY, HETEROTROPHY, MICRONUTRIENTS, BAD W70-04506
THEMATICAL STUDIES, AUTOTROPHIC, HETEROTROPHIC.: MA W70-05750
ROWTH EQUATIONS, ANKISTRODESMUS, HETEROTROPHY, CONTINUOUS CULTURES W70-04381
POPULATIONS, ALGAL POPULATIONS, HETEROTROPHIC POTENTIAL, RELATIVE W70-04194
ETEROTROPHIC POTENTIAL, RELATIVE HETEROTROPHIC POTENTIAL, SWEDEN, W70-04194
RYOLOGY, ENERGY TRANSFORMATIONS, HETEROTROPHY, ORGANIC POLLUTION, W69-07440
C ALGAE, AUTOTROPHIC METABOLISM, HETEROTROPHY, CROOKED LAKE(IND), W71-10079
TROPHIC, SILICON, PEDIASTRUM BO/ HETEROTROPHIC, CHEMOTROPHIC, AUTO W70-02804
PYRENOIDOSA, AUTOTROPHIC GROWTH, HETEROTROPHIC GROWTH, ANKISTRODES W70-02964
, ACETYLENE REDUCTION TECHNIQUE, HETEROTROPHY, FLAGELLATES, LAKE C W70-02775
US, CHIRONOMUS PROMOSUS, TRICHO/ HEXAGENIA, CHIRONOMIDAE, PROCLADI W70-01943
NG BLENDOR MORTALITY, CELL WALL, HEXOSAMINE, ASCOPHYLLUM NODOSUM, W70-04365
EXAS, *NUTRIENT ENRICHMENT TEST/ HIGHLAND LAKES, COLORADO RIVER, T W69-06004
*POTASSIUM PERMANGANATE, BEVERLY HILLS(CALIF).: W71-00110
GERA, MELOSIRA AMBIGUA, CERATIUM HIRUDINELLA, ANKISTRODESMUS, GLEN W70-05387
AMERICANA, CYCLOTELLA, CERATIUM HIRUNDINELLA, PERIDINIUM WILLEI, W70-05405
VITAMIN B6, VITAMIN B12, BIOTIN, HISTIDINE, URACIL, REDUCED SULFUR W70-02510
SECTS, SNAILS, TOXICITY, RIVERS, HISTOGRAMS.: / WATER ANALYSIS, IN W72-01801
S, MADISON(WIS), CHICAGO NATURAL HISTORY MUSEUM, NOSTOC MUSCORUN, W70-00719
VITY, *FOOD WEBS, KENTUCKY, LIFE HISTORY STUDIES, GAMMA RAYS, SPEC W69-09742
WATER POLLUTION, MICROORGANISMS, HISTORY, TASTE, ODOR, DEEP WELLS, W70-08096
A, NUTRIENTS, PLANT POPULATIONS, HISTORY, WATER PURIFICATION, *WAT W68-00483
IMSFJORD(NORWAY), SMOLA(NORWAY), HITRA(NORWAY), FROYA(NORWAY), FUC W70-01074
LIS BUTCHER, NJOER LAKE(NORWAY), HOBOL RIVER(NORWAY), VAN.LAKE(NOR W70-05752
CCUS NIMBATUS, URONEMA ELONGATUM HODGETTS, CHROOCOCCUS PRESCOTTII, W70-04468
VER TIME, TRIDACNA CROCEA, WOODS HOLE OCEANOGRAPHIC INSTITUTION, A W69-08526
EUM, VINEYARD SOUND(MASS), WOODS HOLE(MASS). RADIOCARBON UPTAKE TE W70-05272
IGOPHORA, HOLOZOIC MASTIGOPHORA, HOLOPHYTIC MASTIGOPHORA, CILIATA. W72-01817
MASTI/ SARCODINA, MASTIGOPHORA, HOLOZOIC MASTIGOPHORA, HOLOPHYTIC W72-01817
PECTI/ CHEROKEE RESERVOIR(TENN), HOLSTON RIVER(TENN), POTOMOGETON W71-07698
WARING BLENDOR MORTALITY/ *ALGAE HOMOGENIZATION, *BACTERIA COUNT, W70-04365
WICK(CANADA), SERVALL OMNI-MIXER HOMOGENIZER.: /CANADA), NEW BRUNS W70-04365
PLING, LAKE MENDOTA(WI/ *INLAND, HORIZONTAL SAMPLING, VERTICAL SAM W70-02969
AKE MENDOTA(WIS), PELAGIC AREAS, HORIZONTAL, BLOOMS, GELATINOUS MA W70-06229
EST DEVELOPMENT, *TEST PROBLEMS, HORMIDIUM SP.: *T W71-07339
IBITORS, EXTRACELLULAR PRODUCTS, HORMONES, AUXINS, METABOLITES, IN W69-04801
ABITATS, *ALASKA, LAKES, RIVERS, HOT SPRINGS, TERRESTRIAL HABITATS W69-08272
IVER, MOOSE CREEK, ARCTIC CIRCLE HOT SPRINGS, RADON, CENTRAL CITY(W69-08272
INIER NATIONAL PARK, OHANAPECOSH HOT SPRINGS(WASH), SCHIZOTHRIX, P W70-03309
LOWSTONE NATIONAL PARK, *ICELAND HOT SPRINGS, SINTER, NUCLEIC ACID W69-10160
COUNT, GREEN RIVER, WASHINGTON, HOWARD A. HANSON RESERVOIR, TACOM W72-11906

E LAKE, OLD J/ DETECTOR, ALASKA,   HUDEUC LAKE, PULLIN LAKE, GRAPHIT   W69-08272
COLORADO RIVER, TENNESSEE RIVER,   HUDSON RIVER, DELAWARE RIVER, WAT   W70-04507
YTA, WASHINGTON, DC, *ESTUARIES,   HUMAN POPULATION, SEWAGE EFFLUENT   W68-00461
, CHANNELS, POLLUTION ABATEMENT,   HUMAN POPULATION.: /ILD) OYSTERS   W68-00461
RIO), BURTON LAKE(SASKATCHEWAN),   HUMBOLDT(SASKATCHEWAN), MICE, MOR   W70-00273
 PHOSPHORUS RADIOISOTOPES, IONS,   HUMIC ACIDS, CARP, ICE, IRON COMP   W70-05399
*TOXICITY, BIOLOGICAL MEMBRANES,   HUMID AREA.: /, WATER POLLUTION,   W69-05306
YGEN, METABOLISM, WIND VELOCITY,   HUMIDITY, BACTERIA, HYDROGEN ION   W70-01073
LOGICAL TREATMENT, PILOT PLANTS,   HUMUS, REAERATION, EFFLUENTS, FLO   W70-05821
TRASTRUM, TISZA(HUNGARY), ZIMONA(HUNGARY), ALGAL TAXONOMY, SPIRULI   W70-03969
M, TETRAEDRON, TETRASTRUM, TISZA(HUNGARY), ZIMONA(HUNGARY), ALGAL   W70-03969
T R/ LAKE ST CLAIR, ROUGE RIVER,   HURON RIVER, RAISIN RIVER, DETROI   W69-01445
KE MICHIGAN, *LAKE ONTARIO, LAKE   HURON, AESTHETICS, ECOLOGY, ECONO   W70-01943
, HYDROGEN SULFIDE, CITIES, LAKE   HURON, LAKE ERIE, LAKE ONTARIO, L   W70-00667
*EFFECTS, NAVICULA SEMINULUM VAR.   HUSTEDTII.:   W71-03183
INAGE, APPALACHIAN MOUNTAIN CLUB   HUTS, TEGERNSEE, GERMANY, EXPANDI   W69-07818
KELETONEMA COSTATUM, COCCOLITHUS   HUXLEYI, PYRAMIMONAS, PERIDINIUM   W70-05272
IS, ORGANIC MATTER, ALGAE, IRON,   HY: /LVED OXYGEN, CHEMICAL ANALYS   W70-05760
HES, TROUT, SUCKERS, BASS, WATER   HYACINTH, FROGS, CRAYFISH, MOSQUI   W70-05263
, OXYGEN, ROTIFERS, WEEDS, WATER   HYACINTH, ALLIGATORWEED, CHEMCONT   W70-10175
FRESH WATER, ALGAE, PONDS, WATER   HYACINTH, CHEMICAL CONTROLS, PARA   W70-00269
ANOGRAPHIC INSTITUTION, ACROPORA   HYACINTHUS, POCILLOPORA EYDOUXI,   W69-08526
ARIS, DAPHNIA CUCULLATA, DAPHNIA   HYALINA, DAPHNIA MAGNA, CZECHOSLO   W70-00274
REST, WINDS, LIGHT, TEMPERATURE,   HYD: /EVEL, NITROGEN FIXATION, FO   W71-03048
:   AERO-   HYDRAULIC GUN, ALLISTON(ONTARIO),   W71-09546
AQUATIC ENVIRONMENT, CHLORINATED   HYDROCARBON PESTICIDES, PLANT PHY   W70-08652
 PHAEOCYSTIS, SARGASSUM, FRONDS,   HYDROCARBONS, BACTERIAL ATTACHMEN   W70-01068
LES, ALKANES, ALKENES, ALIPHATIC   HYDROCARBONS, PETROLEUM CRUDES, M   W69-08284
AL, *FOOD WEBS, DDT, CHLORINATED   HYDROCARBON PESTICIDES, ALDRIN, E   W70-03520
RATION, TEMPERATURE, CHLORINATED   HYDROCARBON PESTICIDES, ALGAE, LA   W72-11105
NABAENA, CLADOPHORA, PEDIASTRUM,   HYDRODICTYON, POTAMOGETON, RANUNC   W70-04494
UBESCENS, OSCILLATORIA CHALYBIA,   HYDRODICTYON RETICULATUM, DICTYOS   W69-05704
UM, MICROCYSTIS, ANABAENA.:   HYDRODICTYON, ELODEA, CERATOPHYLL   W68-00479
ON SULFATE, CLADOPHORA BLANKETS,   HYDRODICTYON, MACROPHYTES.: /, IR   W69-10170
LLE(WIS), SPAULDING'S POND(WIS),   HYDRODICTYON, NAJAS, ANACHARIS, L   W70-03310
TED AQUATIC PLANTS, *RESERVOIRS,   HYDROELECTRIC PLANTS, WEED CONTRO   W68-00479
IES, WATER TEMPERATURE, OUTLETS,   HYDROELECTRIC PLANTS, ELECTRIC PO   W69-05023
TENUATION MEASUREMENT APPARATUS,   HYDROGEN F LINE.: /OFER LINES, AT   W70-05377
DIATOMS, STRATIFICATION, OXYGEN,   HYDROGEN ION CONCENTRATION, TEMPE   W70-05387
*ACID MINE WATER, ACIDIC WATER,   HYDROGEN ION CONCENTRATION, WATER   W70-05424
KTON, PRODUCTIVITY, CHLOROPHYTA,   HYDROGEN ION CONCENTRATION, NITRA   W70-05404
, NITRATES, AMMONIA, PHOSPHATES,   HYDROGEN ION CONCENTRATION, PHYTO   W70-05405
CIDS, CARP, ICE, IRON COMPOUNDS,   HYDROGEN ION CONCENTRATION, ACIDI   W70-05399
 DIEL MIGRATION, CARBON DIOXIDE,   HYDROGEN ION CONCENTRATION, PHOSP   W70-05760
ICAL ECOLOGY, INDUSTRIAL WASTES,   HYDROGEN ION CONCENTRATION, IRON   W70-04510
OMETERS, BACTERIA, ALGAE, SANDS,   HYDROGEN ION CONCENTRATION.: /CTR   W69-08267
LORIDES, CONDUCTIVITY, CULTURES,   HYDROGEN ION CONCENTRATION, IRON,   W69-07833
EUTROPHICATION, LIME, *NITROGEN,   HYDROGEN ION CONCENTRATION, *PHOS   W68-00012
IVATED SLUDGE, RATES, CHLORELLA,   HYDROGEN ION CONCENTRATION, PLANT   W68-00248
ERROBACILLUS, WATER TEMPERATURE,   HYDROGEN ION CONCENTRATION, FUNGI   W68-00891
ACTERIA, *AQUATIC FUNGI, YEASTS,   HYDROGEN ION CONCENTRATION, IRON,   W69-00096
NCE ALGAE, NITROGEN, PHOSPHATES,   HYDROGEN ION CONCENTRATION, CYCLI   W70-04381
SIS, PHOSPHATES, IRON COMPOUNDS,   HYDROGEN ION CONCENTRATION, ALKAL   W70-03334
OWMELT, SURFACE WATERS, STREAMS,   HYDROGEN ION CONCENTRATION, IRON,   W70-03501
LUENTS, TRIBUTARIES, ALKALINITY,   HYDROGEN ION CONCENTRATION, NUTRI   W70-03983
ER), PHYTOPLANKTON, ZOOPLANKTON,   HYDROGEN ION CONCENTRATION, ALGAE   W70-04368
ONATES, CARBONATES, TEMPERATURE,   HYDROGEN ION CONCENTRATION, ALKAL   W70-02804
S, ECOLOGY, PHOSPHORUS, SULPHUR,   HYDROGEN ION CONCENTRATION, TIME.   W70-02965
INGTON, WATER CHEMISTRY, OXYGEN,   HYDROGEN ION CONCENTRATION, PHOTO   W70-03309
ND VELOCITY, HUMIDITY, BACTERIA,   HYDROGEN ION CONCENTRATION, SULFA   W70-01073
ECTROPHOTOMETRY, CHROMATOGRAPHY,   HYDROGEN ION CONCENTRATION, FLUOR   W70-01074
RBON RADIOISOTOPES, CHLOROPHYLL,   HYDROGEN ION CONCENTRATION, MAGNE   W70-01579
TEMPERATURE, TRACERS, CHLORELLA,   HYDROGEN ION CONCENTRATION, ALKAL   W70-02504
AS, TEMPERATURE, CARBON DIOXIDE,   HYDROGEN ION CONCENTRATION, BIOCA   W70-02249
INDUSTRIES, METHANE, ALKALINITY,   HYDROGEN ION CONCENTRATION, TOXIN   W70-02803
WATER QUALITY, DISSOLVED SOLIDS,   HYDROGEN ION CONCENTRATION, TEMPE   W70-00265
, CHLORINE, PHENOLS, OIL WASTES,   HYDROGEN ION CONCENTRATION, FLUOR   W70-00263
VIOLET RADIATION, ACTINOMYCETES,   HYDROGEN ION CONCENTRATION, NITRA   W70-00719
ENZUMES, PEPTIDES, BIOCHEMISTRY,   HYDROGEN ION CONCENTRATION, OXIDA   W70-01068
OACTIVITY, CARBON RADIOISOTOPES,   HYDROGEN ION CONCENTRATION, SCENE   W70-08097
THERMAL POLLUTION, *TEMPERATURE,   HYDROGEN ION CONCENTRATION, NUTRI   W70-08834
TER, WATER QUALITY, TEMPERATURE,   HYDROGEN ION CONCENTRATION, OXYGE   W71-03031
, ALGAE, PERIPHYTON, PHOSPHORUS,   HYDROGEN ION CONCENTRATION, ALKAL   W71-04206
TION, DEPTH, OXYGEN, ALKALINITY,   HYDROGEN ION CONCENTRATION, STAND   W71-00665
UTROPHICATION, DISSOLVED OXYGEN,   HYDROGEN ION CONCENTRATION, EPILI   W70-06975
DISSOLVED OXYGEN, BICARBONATES,   HYDROGEN ION CONCENTRATION, NUISA   W70-06225
T COMPARISONS, SEDIMENTS, ALGAE,   HYDROGEN ION CONCENTRATION, ELECT   W71-09548
LUDGE, METHANE, ORGANIC LOADING,   HYDROGEN ION CONCENTRATION, TEMPE   W71-11375
SE, COOLING TOWERS, DAMS, ALGAE,   HYDROGEN ION CONCENTRATION, TEMPE   W71-11393
N, *PROTOZOA, *FUNGI, *BACTERIA,   HYDROGEN ION CONCENTRATION, TEMPE   W71-05267
ISEASES, *FISH TOXINS, TOXICITY,   HYDROGEN ION CONCENTRATION, DISSO   W71-07731
DITY, BIOCHEMICAL OXYGEN DEMAND,   HYDROGEN ION CONCENTRATION, ORGAN   W72-03299
OXICITY, CHEMICAL OXYGEN DEMAND,   HYDROGEN ION CONCENTRATION, TEMPE   W71-13413
GANIC LOADING, DISSOLVED OXYGEN,   HYDROGEN ION CONCENTRATION, BIOCH   W72-10573
OMPOUNDS, PHOSPHOROUS COMPOUNDS,   HYDROGEN ION CONCENTRATION, CONDU   W72-06051
, *WATER QUALITY, *ALGAE, *ODOR,   HYDROGEN ION CONCENTRATION.: /ENT   W72-09694
ATES, TEMPERATURE, PILOT PLANTS,   HYDROGEN ION CONCENTRATION, ORGAN   W72-06022
ENVIRONMENT, ORGANIC PESTICIDES,   HYDROGEN ION CONCENTRATION, TEMPE   W72-11105
 SULFUR BACTERIA, ALGAE, MIXING,   HYDROGEN ION CONCENTRATION, TEMPE   W72-06854
ANALYSIS, TASTE-PRODUCING ALGAE,   HYDROGEN ION CONCENTRATION, *ORGA   W72-11906

CLAYS, COLLOIDS, EUTROPHICATION,
VESTIGATIONS, *IMPOUNDED WATERS,
OL, *MI/ FISH MANAGEMENT, STREAM
IFE, STREAM POLLUTION, BACTERIA,
ABAENA CYLINDRICA B629, ANABAENA
MENT CORP(MASS), ARTHUR D LITTLE
KINETICS, MONOD GROWTH EQUATION,
FILTRATION, ANEROBIC DIGESTION,
SIBILITY, COST ANALYSIS, SLUDGE,
AL NUTRIENT CONCENTRATION, AREAL
ICATION, GROWTH RATES, SEAWATER,
D LAKE(IND), LITTLE CROOKED LAKE(
LISM, HETEROTROPHY, CROOKED LAKE(
    *NITRIC ACID, *LANGELIER
ION, DECAPODS, SPECIES DIVERSITY
URE, *BIOMASS TITER, *BIOLOGICAL
CAROTENOIDS, SPECIES, DIVERSITY
, ORGANIC CARBON, / *AUTOTROPHIC
ITE, ACID-WASHED BAUXITE, KANPUR(
S(INDIA), RAJGHAT(INDIA), BOMBAY(
MS, RIVER GANGES(INDIA), RAJGHAT(
T(INDIA), BOMBAY(INDIA), BENARES(
D/ *INDIA, *BLOOMS, RIVER GANGES(
S.:
MILLSBORO(DELAWARE), SWAN CREEK,
S-ILLINOIS RIVER BASINS PROJECT,
, PESTICIDES, SLUDGE, DIGESTION,
CATORS, FISH, WASHINGTON, LAKES,
NCE ALGAE, WISCONSIN, NUTRIENTS,
ION, BACTERIA, ALGAE, NUTRIENTS,
CULTURAL EUTROPHICATION, TROPHIC
RY(LAKE SUPERIOR), WATER QUALITY
RIBUTION, POPULATION, BIOLOGICAL
LOGICAL DISTRIBUTION, BIOLOGICAL
, MOSQUITOES, SAMPLING, CONTROL,
RADIATION, SULFATES, CELLULOSE,
C WASTES, WASTE WATER TREATMENT,
PHOSPHORUS, PHOSPHATE, NITRATES,
FF, DOMESTIC WASTES, DETERGENTS,
ER TREATMENT, WATER UTILIZATION,
WASTE WATER TREATMENT, *POULTRY,
LITY, DETERGENTS, ALGAL CONTROL,
IMATE DISPOSAL, DOMESTIC WASTES,
, *OIL WASTES, *CHEMICAL WASTES,
LANKTON, SEWAGE, DOMESTIC WASTE,
C WASTES, WASTE WATER TREATMENT,
OUBLING TIME, PACKED CELL VOLUM/
IGHT PENETRATION, STANDING CROP,
*LAKE ONTARIO, WATER TURBIDITY,
L), ORGANIC WASTES, PULP WASTES,
ATMOSPHERIC), WETLANDS, SEEPAGE,
*NU/ *WASTE WATER(MUNICIPAL AND
WATER POLLUTION, FERTILIZATION,
SEWAGE, PHOSPHORUS, CHEMCONTROL,
KTON, CRYPTOPHYCEAE, VOLVOCALES,
ODOPHYTA, PHYSIOLOGICAL ECOLOGY,
DES, SALINE LAKES, HEAVY METALS,
MENTAL EFFECTS, CURRENTS(WATER),
S, PHOSPHORUS, NITROGEN, SEWAGE,
ERTIES, DOMESTIC WASTES, SEWAGE,
OKLAHOMA, RESERVOIRS, SAMPLING,
AGE EFFLUENTS, SEWAGE TREATMENT,
RPLANTS, ENGINEERING STRUCTURES,
ATIC INSECTS, FISH REPRODUCTION,
LUTION EFFECTS, DOMESTIC WASTES,
AE, PHENOLS, SHELLFISH, OYSTERS,
LUENTS, FARM WASTES. POPULATION,
(CALIF), EEL RIVER(CALIF), JOINT
ATER WORKS ASSOCIATION, CHEMICAL
GEMENT, *FARM MANAGEMENT, *DAIRY
-PLANKTON, SPRAYING, BIOCONTROL,
CLING NUTRIENTS, *SEDIMENT-WATER
TERIA, PHOSPHORUS RADIOISOTOPES,
BOTTOM SEDIMENTS, ANIMALS, FISH,
LAKE WASHINGTON(WASH), *CULTURAL
URNOVERS, NUTRIENTS, PHOSPHORUS,
(MINN), NUTRIENT TRAPS, NUTRIENT
CTS, LIPID CARBON, CHLOROPLASTS,
INOBENZOIC ACID DIHYDROCHLORIDE,
MONITORING, AERIAL PHOTOGRAPHY,
ETHOXYLATE, ORGANIC EXTRACTIONS,
APPLICATIONS, FACULTATIVE PONDS,
RFACTANTS, NONIONIC SURFACTANTS,
   ALGAL SENESCENCE, GROWTH
RIENT REQUIREMENTS, *POLLUTANTS,
ATES, BIOASSAY, NEW YORK, LAKES,
TION, MODE OF ACTION, MORTALITY,
NUTRIENTS, ESSENTIAL NUTRIENTS,
OROPHYTA, ENVIRONMENTAL EFFECTS,
MENTS, NUTRIENTS, PHYTOPLANKTON,

IMPOUNDMENTS, LAKES, LOAMS, RESER W69-04800
IMPOUNDMENTS, *DISSOLVED OXYGEN, W72-11906
IMPROVEMENT, *WATER QUALITY CONTR W70-05428
IMPURITIES, WASTE TREATMENT.: / L W72-01786
INAEQUALIS 381, ANABAENA FLOS AQU W70-05091
INC(MASS), CARNEGIE DEPT OF PLANT W69-06865
INCIDENT LIGHT UTILIZATION.: /TH W69-03730
INCINERATION, IRON COMPOUNDS, SUL W70-04060
INCINERATION, STORM RUN-OFF, SEDI W71-10654
INCOME(NUTRIENTS), LAKE VOLUME, M W70-00270
INCUBATION, PSEUDOMONAS, SUBSURFA W70-00713
IND).: /ISM, HETEROTROPHY, CROOKE W71-10079
IND), LITTLE CROOKED LAKE(IND).: / W71-10079
INDEX.: W71-11393
INDEX.: / ANALYSIS, PART CORRELAT W73-01446
INDEX, *SAPROBITY, PRAGUE, CZECHO W72-12240
INDEX, ACHNANTHES, ACTINOPTYCHUS, W70-03325
INDEX, CHLOROPHYLL A, SHAYLER RUN W71-04206
INDIA).: /ATED SAND, CRUSHED BAUX W70-08628
INDIA), BENARES(INDIA), MACROPHYT W70-05404
INDIA), BOMBAY(INDIA), BENARES(IN W70-05404
INDIA), MACROPHYTES, CLADOPHORA, W70-05404
INDIA), RAJGHAT(INDIA), BOMBAY(IN W70-05404
INDIA, METHYL ORANGE, ALGAL BLOOM W70-05550
INDIAN RIVER.: W71-05013
INDIANA HARBOR SHIP CANAL.: /LAKE W70-00263
INDIANA.: /OUNDS, LAGOON, FILTERS W72-03641
INDIANA, OCHROMONAS, PHYTOPLANKTO W70-04503
INDICATOR SPECIES, EUGLENA, CHLAM W69-01977
INDICATORS, ODOR, TASTE, FILTERS, W70-05264
INDICATORS, NUTRIENT BUDGET.:   * W71-03048
INDICATORS.:   PERIPHYTON INVENTO W70-05958
INDICATORS, NEW HAMPSHIRE.: /DIST W70-02772
INDICATORS, NEW HAMPSHIRE.: / ECO W70-02764
INDUS: /ICATION, TOXICITY, MIDGES W70-06225
INDUSTR: /, SALINITY, ULTRAVIOLET W70-01074
INDUSTRIAL WASTES, NEUTRIENTS, BA W70-01519
INDUSTRIAL WASTES, WATER POLLUTIO W70-02777
INDUSTRIAL WASTES, *CHELATION.: / W71-04072
INDUSTRIAL WASTES, OXIDATION LAGO W70-07508
INDUSTRIAL WASTES, LAGOONS, PONDS W70-09320
INDUSTRIAL WASTES, THERMAL POLLUT W70-10437
INDUSTRIAL WASTES, SEWAGE WASTES, W70-09905
INDUSTRIAL WASTES, WATER ANALYSIS W72-03299
INDUSTRIAL WASTES, FARM WASTES.: / W72-01813
INDUSTRIAL WASTES, NUTRIENTS, BAC W71-06737
INDUSTRIAL PHOTOSYNTHESIS, MEAN D W71-05267
INDUSTRIAL WASTE: /THERMOCLINE, L W71-07698
INDUSTRIAL PLANTS, WATER SPORTS.: W71-09880
INDUSTRIAL WASTES, TOXINS, SOLID W72-00462
INDUSTRIAL WASTES, TERTIARY TREAT W69-10178
INDUSTRIAL), *PHOSPHORUS SOURCES, W69-09340
INDUSTRIAL WASTES, GARBAGE DUMPS, W70-05112
INDUSTRIAL WASTES, SEWAGE TREATME W70-05668
INDUSTRIAL USES, CLEAN WATER, GEN W70-05389
INDUSTRIAL WASTES, HYDROGEN ION C W70-04510
INDUSTRIAL WASTES, LAKES, PHOTOSY W69-02959
INDUSTRIAL WASTE, CORES, DIATOMS, W68-00470
INDUSTRIAL WASTES, DOMESTIC WASTE W68-00478
INDUSTRIAL WASTES, AGRICULTURE, S W70-03964
INDUSTRIAL WASTES, MUNICIPAL WAST W70-04001
INDUSTRIAL WASTES, WATER CHEMISTR W70-03974
INDUSTRIAL PLANTS, STRUCTURES, AF W69-05023
INDUSTRIAL WASTES, ECOLOGY, DISSO W73-00748
INDUSTRIAL WASTES, ENVIRONMENTAL W72-12240
INDUSTRIES, CHELATION, BACTERIA, W70-01068
INDUSTRIES, METHANE, ALKALINITY, W70-02803
INDUSTRY/GOVERNMENT TASK FORCE ON W70-02777
INDUSTRY, ODOR, TASTE, CONSUMER.: W71-05943
INDUSTRY, *ALGAE, *CATTLE, SLURRI W70-07491
INFECTION, NUTRIENTS, ALGAE, NITR W68-00468
INFERFACE, MUD-WATER INTERFACE, P W69-06859
INFLOW, CYANOPHYTA, LITTORAL, NUT W73-01454
INFLOW, OUTLETS, NUTRIENTS, PRODU W70-06225
INFLUENCES, MICROCYSTIS, DREDGE S W70-05663
INFLUENT SEWAGE, EQUILIBRIUM, MAT W68-00858
INFLUX, NUTRIENT SOURCES, PHOSPHO W70-04268
INFRARED ANALYSIS.: /ETHANE EXTRA W70-05651
INFRARED GAS ANALYSIS, COULTER CO W69-08278
INFRARED RADIATION, DYE RELEASES, W68-00010
INFRARED SPECTROSCOPY, BLUE-GREEN W70-03928
INFRARED SPECTROSCOPY, SULFUR COM W70-07034
INFRARED SPECTRA.: /E, ANIONIC SU W70-09438
INHIBITIONS.: W72-02417
INHIBITION, LIMNOLOGY, ACIDITY, A W70-02792
INHIBITION, ALGAE, DIATOMS, EUTRO W70-01579
INHIBITION, ALGAE.: /OL, DISINFEC W70-02370
INHIBITION, NITROGEN COMPOUNDS, N W70-02248
INHIBITION, LIMNOLOGY, PESTICIDES W70-02968
INHIBITION.: /S, NUTRIENT REQUIRE W69-10177

457

FECTS, ENVIRONMENTAL SANITATION,
CHLORELLA, MORTALITY, BIOASSAY,
, *ALGAE, LIGHT, CARBON DIOXIDE,
REEZE DRYING, FREEZE-THAW TESTS,
SANITATION, ESSENTIAL NUTRIENTS,
NUTRIENTS, HARVESTING OF ALGAE,
, HORMONES, AUXINS, METABOLITES,
NTIAL NUTRIENTS, EUTROPHICATION,
DIATOMS, ENVIRONMENTAL EFFECTS,
NTIAL NUTRIENTS, EUTROPHICATION,
ENTAL EFFECTS, HARVESTING ALGAE,
NUTRIENTS, CHLOROPHYTA, DIATOMS,
ECOLOGY, COULTER COUNTER, GROWTH
II, DUNALIELLA TERTIOLECTA, DCMU
DS, NUTRIENTS, VITAMINS, TOXINS,
ING(CORROSION), RUSTING, METALS,
STES, GARBAGE DUMPS, OIL WASTES,
R, FACULTATIVE, MICROAEROPHILIC,
ARSENATE INTERFERENCE, DISSOLVED
BON DIOXIDE, VITAMINS, NITRATES,
PLAINS, ATLANTIC COASTAL PLAIN,
TION, CHANNEL MORPHOLOGY, SEWAGE
PLANKTON, AQUATIC WEED CONTROL,
RIN, *DIATOMS, LABORATORY TESTS,
RESERVOIRS, WATERSHEDS(BASINS),
S, OLIGOCHAETES, DIPTERA, ALGAE,
/ *ZINC RADIOISOTOPES, *AQUATIC
D ORG/ ALGAE, PROTOZOA, *AQUATIC
LS, ALGAE, FISH, DIATOMS, FUNGI,
, VELOCITY, CYANOPHYTA, BIOMASS,
OLIMNION, EPILIMNION, LAKE ERIE,
G PLANTS, PHYTOPLANKTON, AQUATIC
RIVERS, ALGAE, PROTOZOA, AQUATIC
MENTS, BIOTA, FISH, UTAH, ALGAE,
ION, DECOMPOSING ORGANIC MATTER,
AKES, CONIFEROUS FORESTS, ALGAE,
RTEBRATES, FISH, WATER ANALYSIS,
STIGLEOCLONIUS,
CROCEA, WOODS HOLE OCEANOGRAPHIC
DEPT OF PLANT BIOLOGY, CARNEGIE
DACTYLUM, WASTE WATER DIVERSION,
THERMAL POLLUTION, METHODOLOGY,
UCTURES, AFTERBAYS, WATER TYPES,
*LAKE MICHIGAN, *WATER SUPPLY,
FILTERS, *ALGAE, NUISANCE ALGAE,
ONMENT, CHEMCONTROL, BIOCONTROL,
MPERATURE, PHOTOSYNTHESIS, LIGHT
, MODEL STUDIES, BIOMASS, *LIGHT
OXIDATION LAGOONS, *ALGAE, LIGHT
OXICITY, DISSOLVED OXYGEN, LIGHT
L, ICE, SEASONAL, DIURNAL, LIGHT
, DIA/ *PLANKTON, *ALGAE, *LIGHT
, EPILIMNION, HYPOLIMNION, LIGHT
, TEMPERATURE, EPILIMNION, LIGHT
RODUCT/ *PLANKTON, *ALGAE, LIGHT
OLIGOTROPHY, TEMPERATURE, LIGHT
NUTRIENTS, EUTROPHICATION, LIGHT
HOSPHORUS, PHOTOSYNTHESIS, LIGHT
TESTS, *EFFECTIVE-AVERAGE-LIGHT-
N, ALKALINITY, CALIFORNIA, LIGHT
OLIGOTROPHY, TEMPERATURE, LIGHT
TS, OREGON, BIOINDICATORS, LIGHT
YTA, DIATOMS, CHLOROPHYTA, LIGHT
IC MATTER, TRACE ELEMENTS, LIGHT
RETENTION, PHOTOPERIODISM, LIGHT
IAL NUTRIENTS, INHIBITION, LIGHT
UTROPHICATION, INHIBITION, LIGHT
FFECTS, INHIBITION, LIGHT, LIGHT
IATOMS, INHIBITION, LIGHT, LIGHT
EEZE-THAW TESTS, FREEZING, LIGHT
UTROPHICATION, INHIBITION, LIGHT
LANT POPULATIONS, REVIEWS, LIGHT
SPIRATION, MICROORGANISMS, LIGHT
IS, OXYGEN, TROPHIC LEVEL, LIGHT
*ALGAE, *THERMAL SPRINGS, *LIGHT
SURFACE WATERS, DETRITUS, LIGHT
MPOUNDS, SEWAGE EFFLUENTS, LIGHT
LLA, CARBON RADIOISOTOPES, LIGHT
TS, *ANAEROBIC CONDITIONS, LIGHT
LAGOONS, LABORATORY TESTS, LIGHT
TER, PRIMARY PRODUCTIVITY, LIGHT
ELLA, CARBON, TEMPERATURE, LIGHT
LATION, ENZYMES, CHLORINE, LIGHT
*DIATOMS, *BENTHIC FLORA, LIGHT
, *LABORATORIES, *STREAMS, LIGHT
E, OXYTREMA, ORGANIC ENRICHMENT,
AY, / *MICRONUTRIENTS, *NUTRIENT
IMENT-WATER INFERFACE, MUD-WATER
IMENTS, CATTAILS, RADIOACTIVITY,
ON, CYCLING NUTRIENTS, MUD-WATER

INHIBITION, CHLORINE, NITROGEN CO   W69-10171
INHIBITION, MODE OF ACTION, PHOTO   W70-03519
INHIBITION, NITROGEN FIXING BACTE   W70-04249
INHIBITION, NUTRIENT REQUI: /S, F   W69-03370
INHIBITION, LIGHT INTENSITY, LIMN   W69-03513
INHIBITION, LIGHT PENETRATION, NI   W69-03369
INHIBITION, SILICON, VANADIUM, ZI   W69-04801
INHIBITION, LIGHT INTENSITY, LIGH   W69-03514
INHIBITION, LIGHT, LIGHT INTENSIT   W69-03516
INHIBITION, LIGHT INTENSITY, LIGH   W69-03515
INHIBITION, LIGHT PENETRATION, LI   W69-04798
INHIBITION, LIGHT, LIGHT INTENSIT   W69-03517
INHIBITION, CYANOCOBALAMIN, MUTAN   W70-05652
INHIBITOR.: /ANKISTRODESMUS BRAUN   W70-05261
INHIBITORS, SUCCESSION, CHLORELLA   W70-02510
INHIBITORS, SILTS, ORGANIC MATTER   W70-02294
INLETS(WATERWAYS), ORGANIC MA: /A   W70-05112
INOCULATION.: /T, PHAGE, FERMENTO   W71-02009
INORGANIC PHOSPHATES, SOLUBLE REA   W70-05269
INORGANIC COMPOUNDS, SURFACE RUNO   W70-04810
INORGANIC COMPOUNDS, METALS, PLAN   W69-03695
INPUT, BIOLOGICAL PROPERTIES, *WA   W68-00475
INSECT CONTROL, MOSQUITOS, ALGAL   W70-05417
INSECTICIDES, WATER POLLUTION EFF   W71-03183
INSECTICIDES, HARDNESS(WATER), DI   W72-01813
INSECTS.: /WATER POLLUTION EFFECT   W71-12077
INSECTS, *AQUATIC PLANTS, AQUATIC   W71-09005
INSECTS, *AQUATIC DRIFT, FISH FOO   W69-02782
INSECTS, BACTERIA, PROTOZOA, SEWA   W71-10786
INSECTS, DIATOMS, CHLOROPHYTA, CH   W71-00668
INSECTS, DIPTERA, CRUSTACEA, MOLL   W70-04253
INSECTS, FISH, GAME BIRDS, ANIMAL   W70-10175
INSECTS, FISH REPRODUCTION, INDUS   W73-00748
INSECTS, INVERTEBRATES, BACKGROUN   W69-07846
INSECTS, MITES, BACTERIS, FUNGI,   W71-00940
INSECTS, PHOSPHORUS RADIOISOTOPES   W71-09012
INSECTS, SNAILS, TOXICITY, RIVERS   W72-01801
INSOLUBLE WASTES, CULTURE MEDIA.:   W68-00248
INSTITUTION, ACROPORA HYACINTHUS,   W69-08526
INSTITUTION OF WASHINGTON, CHEMIC   W69-06865
INSTREAM TREATMENT, BOTTOM SEALIN   W69-09340
INSTRUMENTATION, SEA WATER.: /ES,   W70-10437
INTAKES.: /INDUSTRIAL PLANTS, STR   W69-05023
INTAKES, NUTRIENTS, QUALITY CONTR   W70-00263
INTAKES, SANDS, FLOW RATES, TURBI   W70-04199
INTEGRATED CONTROL MEASURES, PHYS   W70-07393
INTENSITY.: /UTION, NUTRIENTS, TE   W70-05270
INTENSITY.: /ENT, EUTHROPHICATION   W73-02218
INTENSITY, CARBON DIOXIDE, GROWTH   W72-06022
INTENSITY, SLUDGE, DEPTH, ORGANIC   W72-06854
INTENSITY, CARBON, SEWAGE TREATME   W70-05387
INTENSITY, CHLORELLA, CHLOROPHYTA   W70-05381
INTENSITY, CONDUCTIVITY, ALGAE, S   W70-05405
INTENSITY, SEASONAL, PHYSIOLOGICA   W70-05409
INTENSITY, TEMPERATURE, PRIMARY P   W70-04809
INTENSITY, CHLOROPHYLL, NUTRIENTS   W70-05752
INTENSITY, LIGHT PENETRATION, LIM   W70-04283
INTENSITY, STABILIZATION, PONDS,   W70-03923
INTENSITY, PHOTO-ELECTRIC TURBIDO   W70-03923
INTENSITY, TEMPERATURE, IONS, SPE   W70-03334
INTENSITY, DAPHNIA, CONNECTICUT,   W70-03959
INTENSITY, EUTROPHICATION, GRAZIN   W70-03978
INTENSITY, SUCCESSION, ZINC RADIO   W70-04371
INTENSITY, CHLAMYDOMONAS, CHLOREL   W70-04369
INTENSITY, VELOCITY, TEMPERATURE,   W69-07862
INTENSITY, LIMNOLOGY,: /N, ESSENT   W69-03513
INTENSITY, LIGHT PENETRAT: /TS, E   W69-03515
INTENSITY, LIGHT PENETRATION, LIM   W69-03516
INTENSITY, LIGHT PENETRATION, LIM   W69-03517
INTENSITY, LIGHT PENETRATION, LIM   W69-03518
INTENSITY, LIGHT PENETRATION, LIM   W69-03514
INTENSITY, WATER TEMPERATURE, DIA   W68-00860
INTENSITY, CARBON DIOXIDE, DISSOL   W71-00668
INTENSITY, ECOLOGY, LOTIC ENVIRON   W70-00265
INTENSITY, SPRING WATER, WATER CH   W69-09676
INTENSITY, TURBIDITY, ALGAE, DIAT   W70-00713
INTENSITY, LAKES, WATER POLLUTION   W69-10167
INTENSITY, CULTURES, RADIOACTIVIT   W69-10158
INTENSITY, EUTROPHICATION, NUTRIE   W71-13413
INTENSITY, PHOTOPERIODISM, LIGHT   W72-00917
INTENSITY, TROPHIC LEVEL, SAMPLIN   W70-02504
INTENSITY, NUTRIENT REQUIREMENTS,   W70-02965
INTENSITY, NITRATE, EUGLENA, CALC   W70-02964
INTENSITY, BIOMASS, DISTRIBUTION,   W70-03325
INTENSITY, CURRENTS(WATER), VELOC   W70-02780
INTERACTIONS.: /OMMUNITY STRUCTUR   W70-03978
INTERACTIONS, *RADIOCARBON BIOASS   W70-04765
INTERFACE, PHOSPHORUS, ADSORPTION   W69-06859
INTERFACES, ORGANIC MATTER, RADIO   W69-08272
INTERFACES, EUGLENOPHYTA, TOXICIT   W70-04381

NSING, MINNESOTA, SEDIMENT-WATER   INTERFACES, CALIFORNIA, BIOASSAY,   W70-02775
L EFFECTS, DISPERSION, OIL-WATER   INTERFACES, PERSISTENCE, WATER PO   W70-01777
E, WATER QUALITY, SEDIMENT-WATER   INTERFACES, OREGON, SESSILE ALGAE   W72-06051
EEDINGS, SALINE WATER-FRESHWATER   INTERFACES, COLLOIDS.: /PONDS, BR   W73-00748
CKEL, DETECTION LIMITS, CHEMICAL   INTERFERENCES.:   *NI   W71-11687
  PHOSPHATES, SOLUBLE R/ ARSENATE   INTERFERENCE, DISSOLVED INORGANIC   W70-05269
NAVAL RESEARCH, US DEPARTMENT OF   INTERIOR, MADISON(WIS), EUROPE, W   W70-03975
NAVAL RESEARCH, US DEPARTMENT OF   INTERIOR, MADISON(WIS), EUROPE, W   W70-03975
HICATION, *HYDRODYNAMICS, LAKES,   INTERNAL WATER, WAVES(WATER), DIU   W70-09889
ODUCTION BIOLOGY, RUSSIA, SECOND   INTERNATIONAL OCEANOGRAPHIC CONGR   W69-07440
ICULA, MEL/ *CHLOROPHYLL A, DATA   INTERPRETATION, *RHONE RIVER, NAV   W73-01446
MONAS, TEMPERATURE, AMINO ACIDS,   INTERTIDAL AREAS, CARBOHYDRATES,   W70-03952
CHNIQUES, TEMPERATURE, PIPTIDES,   INTERTIDAL AREAS, SOIL BACTERIA,   W70-04365
  FAUNA, *BIOLOGICAL COMMUNITIES,   INTERTIDAL ALGAE, PUGET SOUND, FL   W68-00010
  ECOSYSTEMS, POPULATION, OREGON,   INTERTIDAL AREAS, PHYSIOLOGICAL E   W70-03325
CULA, SYRACUSE(NY), ENTEROMORPHA   INTESTINALIS, CLADOPHORA, CYCLOTE   W70-03974
UALITY INDICATORS.:   PERIPHYTON   INVENTORY(LAKE SUPERIOR), WATER Q   W70-05958
-PURIFICATION, LABORATORY TESTS,   INVERTEBRATES, FISH, WATER ANALYS   W72-01801
PTION, HYDROLYSIS, MARINE ALGAE,   INVERTEBRATES, NUCLEAR WASTES, WA   W71-09863
TION, MARINE ALGAE, MARINE FISH,   INVERTEBRATES, CRUSTACEANS, MOLLU   W71-09850
TIVITY, BIOTA, CYCLES, MICHIGAN,   INVERTEBRATES, FISH, TEMPERATURE,   W69-09334
E, RESISTANCE, FISH, AMPHIBIANS,   INVERTEBRATES, AQUATIC POPULATION   W68-00859
GANISMS, MICROORGANISMS, SESTON,   INVERTEBRATES, SHELLFISH, WASTE D   W69-04276
OTA, FISH, UTAH, ALGAE, INSECTS,   INVERTEBRATES, BACKGROUND RADIATI   W69-07846
MS, PLANKTON, ALGAE, PERIPHYTON,   INVERTEBRATES, ADSORPTION, PHYTOP   W69-07853
LING NUTRIENTS, PLANKTON, ALGAE,   INVERTEBRATES, TURBIDITY, WORMS,   W69-07861
GAE, IRON, BIRDS, RADIOACTIVITY,   INVERTEBRATES, PLANKTON, FISH, SE   W69-08269
AE, ECOLOGY, SOUTH DAKOTA, FIELD   INVESTIGATIONS, WATER POLLUTION,   W69-00659
, OXYGEN, SPORT FISHING, ON-SITE   INVESTIGATIONS.: /GHT PENETRATION   W68-00487
-SITE DATA COLLECTIONS, *ON-SITE   INVESTIGATIONS, FREQUENCY, *FREQU   W70-09976
DEVELOPMENT, WATER QUALITY, SOIL   INVESTIGATIONS, *IMPOUNDED WATERS   W72-11906
EMULSIONS, BACKGROUND RADIATION,   IODINE RADIOISOTOPES, NUTRIENT RE   W69-10163
DIOISOTOPES, ZINC RADIOISOTOPES,   IODINE RADIOISOTOPES, COBALT RADI   W71-09850
LLULAR MORPHOLOGY, FLAGELLATION,   IODINE TEST, GRAM REACTION, CYTOP   W70-03952
                         BROMINE,   IODINE.:   W70-05178
ETABOLIS/ *KILL RATES, *BROMINE,   IODINE, PHOTO REACTIVITY, ALGAL M   W70-02363
RE, TRACERS, CHLORELLA, HYDROGEN   ION CONCENTRATION, ALKALINITY, DI   W70-02504
RATURE, CARBON DIOXIDE, HYDROGEN   ION CONCENTRATION, BIOCARBONATES,   W70-02249
Y, PHOSPHORUS, SULPHUR, HYDROGEN   ION CONCENTRATION, TIME.: /ECOLOG   W70-02965
ATER CHEMISTRY, OXYGEN, HYDROGEN   ION CONCENTRATION, PHOTOSYNTHESIS   W70-03309
ARBONATES, TEMPERATURE, HYDROGEN   ION CONCENTRATION, ALKALINITY,: /   W70-02804
TY, HUMIDITY, BACTERIA, HYDROGEN   ION CONCENTRATION, SULFATES, RESP   W70-01073
OMETRY, CHROMATOGRAPHY, HYDROGEN   ION CONCENTRATION, FLUORESCENCE,   W70-01074
OISOTOPES, CHLOROPHYLL, HYDROGEN   ION CONCENTRATION, MAGNESIUM, CAL   W70-01579
S, METHANE, ALKALINITY, HYDROGEN   ION CONCENTRATION, TOXINS, OXYGEN   W70-02803
THANE, ORGANIC LOADING, HYDROGEN   ION CONCENTRATION, TEMPERATURE, C   W71-11375
SONS, SEDIMENTS, ALGAE, HYDROGEN   ION CONCENTRATION, ELECTRIC POWER   W71-09548
NG TOWERS, DAMS, ALGAE, HYDROGEN   ION CONCENTRATION, TEMPERATURE, A   W71-11393
CHEMICAL OXYGEN DEMAND, HYDROGEN   ION CONCENTRATION, ORGANIC LOADIN   W72-03299
CHEMICAL OXYGEN DEMAND, HYDROGEN   ION CONCENTRATION, TEMPERATURE, P   W71-13413
*FISH TOXINS, TOXICITY, HYDROGEN   ION CONCENTRATION, DISSOLVED OXYG   W71-07731
ZOA, *FUNGI, *BACTERIA, HYDROGEN   ION CONCENTRATION, TEMPERATURE, A   W71-05267
DIATION, ACTINOMYCETES, HYDROGEN   ION CONCENTRATION, NITRATES, POTA   W70-00719
PEPTIDES, BIOCHEMISTRY, HYDROGEN   ION CONCENTRATION, OXIDATION,: /   W70-01068
LITY, DISSOLVED SOLIDS, HYDROGEN   ION CONCENTRATION, TEMPERATURE, N   W70-00265
E, PHENOLS, OIL WASTES, HYDROGEN   ION CONCENTRATION, FLUORIDES, ALK   W70-00263
TH, OXYGEN, ALKALINITY, HYDROGEN   ION CONCENTRATION, STANDING CROP,   W71-00665
TION, DISSOLVED OXYGEN, HYDROGEN   ION CONCENTRATION, EPILIMNION, PH   W70-06975
D OXYGEN, BICARBONATES, HYDROGEN   ION CONCENTRATION, NUISANCE ALGAE   W70-06225
OLLUTION, *TEMPERATURE, HYDROGEN   ION CONCENTRATION, NUTRIENTS, ALG   W70-08834
ERTIARY TREAT/ *ALGAE, *HYDROGEN   ION CONCENTRATION, *NUTRIENTS, *T   W70-08980
, CARBON RADIOISOTOPES, HYDROGEN   ION CONCENTRATION, SCENEDESMUS, E   W70-08097
PERIPHYTON, PHOSPHORUS, HYDROGEN   ION CONCENTRATION, ALKALINITY, HA   W71-04206
R QUALITY, TEMPERATURE, HYDROGEN   ION CONCENTRATION, OXYGEN, CARBON   W71-03031
OGY, INDUSTRIAL WASTES, HYDROGEN   ION CONCENTRATION, IRON COMPOUNDS   W70-04510
RATION, CARBON DIOXIDE, HYDROGEN   ION CONCENTRATION, PHOSPHATES, NI   W70-05760
NE WATER, ACIDIC WASTES, HYDROGEN   ION CONCENTRATION, WATER CHEMISTR   W70-05424
P, ICE, IRON COMPOUNDS, HYDROGEN   ION CONCENTRATION, ACIDITY, MUD,   W70-05399
DUCTIVITY, CHLOROPHYTA, HYDROGEN   ION CONCENTRATION, NITRATES, PHOS   W70-05404
S, AMMONIA, PHOSPHATES, HYDROGEN   ION CONCENTRATION, PHYTOPLANKTON,   W70-05405
STRATIFICATION, OXYGEN, HYDROGEN   ION CONCENTRATION, TEMPERATURE, D   W70-05387
PHATES, IRON COMPOUNDS, HYDROGEN   ION CONCENTRATION, ALKALINITY, CA   W70-03334
URFACE WATERS, STREAMS, HYDROGEN   ION CONCENTRATION, IRON, MAGNESIU   W70-03501
RIBUTARIES, ALKALINITY, HYDROGEN   ION CONCENTRATION, NUTRIENTS, PHO   W70-03983
OPLANKTON, ZOOPLANKTON, HYDROGEN   ION CONCENTRATION, ALGAE.: / PHYT   W70-04368
, NITROGEN, PHOSPHATES, HYDROGEN   ION CONCENTRATION, CYCLING NUTRIE   W70-04381
LUS, WATER TEMPERATURE, HYDROGEN   ION CONCENTRATION, FUNGI, YEASTS,   W68-00891
*AQUATIC FUNGI, YEASTS, HYDROGEN   ION CONCENTRATION, IRON, SULFATE,   W69-00096
ARBON DIOXIDE, *ALGAE, *HYDROGEN   ION CONCENTRATION, PHOSPHORUS, NI   W68-00855
ATION, LIME, *NITROGEN, HYDROGEN   ION CONCENTRATION, *PHOSPHORUS, S   W68-00012
UDGE, RATES, CHLORELLA, HYDROGEN   ION CONCENTRATION, PLANT TISSUES,   W68-00248
BACTERIA, ALGAE, SANDS, HYDROGEN   ION CONCENTRATION.: /CTROMETERS,   W69-08267
CONDUCTIVITY, CULTURES, HYDROGEN   ION CONCENTRATION, IRON, LAKES, M   W69-07833
ACTERIA, ALGAE, MIXING, HYDROGEN   ION CONCENTRATION, TEMPERATURE, T   W72-06854
  PHOSPHOROUS COMPOUNDS, HYDROGEN   ION CONCENTRATION, CONDUCTIVITY,   W72-06051
NT, ORGANIC PESTICIDES, HYDROGEN   ION CONCENTRATION, TEMPERATURE, C   W72-11105
  TASTE-PRODUCING ALGAE, HYDROGEN   ION CONCENTRATION, *ORGANI: /SIS,   W72-11906
QUALITY, *ALGAE, *ODOR, HYDROGEN   ION CONCENTRATION.: /ENT, *WATER   W72-09694
DING, DISSOLVED OXYGEN, HYDROGEN   ION CONCENTRATION, BIOCHEMICAL OX   W72-10573
PERATURE, PILOT PLANTS, HYDROGEN   ION CONCENTRATION, ORGANIC LOADIN   W72-06022

                                                                         459

IARY TREATMENT, ELECTRODIALYSIS,
ANIC COMPOUNDS, CHEMICAL WASTES,
S, SHIPS, SUBMARINES, EFFLUENTS,
EATMENT, ALGAL CONTROL, AMMONIA,
  BOD, DETENTION TIME.: *SULFATE
GION, MONOMICTIC, DEPTH EFFECTS,
RAPHIES, CHEMICAL PRECIPITATION,
ACERS, PHOSPHORUS RADIOISOTOPES,
  WATER, *PEPTIDES, COPPER, IRON,
A, LIGHT INTENSITY, TEMPERATURE,
C OCEAN, DIATOMS, NERITIC ALGAE,
STROENTERITIS, SYNURA, ANABAENA,
ACTIVATED CARBON, ACTINOMYCETES,
NTROL, POLLUTANT IDENTIFICATION,
      *METABOLITES, *CEDAR RIVER(
ASSIUM PERMANGANATE, *DES MOINES(
  CENTER LAKE(IOWA), DIAMOND LAKE(
  FLOS-AQUAE, LYNGBYA, STORM LAKE(
A, STORM LAKE(IOWA), CENTER LAKE(
  , IRON BACTERIA, TASTE, FILTERS,
  SILVER-110M, CERIUM, GUETTARDA,
PERTY VALUES, RECREATION DEMAND,
NOPHYTA, SULFUR BACTERIA, COLOR,
OPHORA BLANKE/ ALUMINUM SULFATE,
ATES, PHOSPHATES, PHYTOPLANKTON,
  BIOCHEMICAL OXYGEN DEM/ *ALGAE,
IUM, AMMONIA, NITROGEN FIXATION,
  , CHEMICAL ANALYSIS, PHOSPHATES,
NEROBIC DIGESTION, INCINERATION,
S, IONS, HUMIC ACIDS, CARP, ICE,
E, SCENEDESMUS, LIGHT, CULTURES,
TES, HYDROGEN ION CONCENTRATION,
ND, *BACTERIA, ACTIVATED SLUDGE,
UTRIENTS, PHOSPHATES, SILICATES,
HIBITORS, SILTS, ORGANIC MATTER,
ALUMINUM SULFATE, IRON CHLORIDE,
ALUMINUM SULFATE, FERRICHLORIDE,
BIC BACTERIA, WATER TEMPERATURE,
S, BIOMASS, DIATOMS, CYANOPHYTA,
NTATION, TENNESSEE RIVER, ALGAE,
NITRATES, PHOSPHATES, POTASSIUM,
RADIOISOTOPES, MANGANESE, ALGAE,
CES, SEWAGE, SULFATE, MANGANESE,
OXIDE, SCENEDESMUS, CHLOROPHYLL,
  , DIATOMS, WEIGHT, PILOT PLANTS,
OCE/ *ALGAE, *HERRINGS, COPPER,
TY, SULFATES, CHLORIDES, SODIUM,
IC CONDITIONS, HYDROGEN SULFIDE,
NUTRIENTS, PHOSPHORUS, NITROGEN,
ANALYSIS, ORGANIC MATTER, ALGAE,
*FRESH WATER, *PEPTIDES, COPPER,
RES, HYDROGEN ION CONCENTRATION,
AMS, HYDROGEN ION CONCENTRATION,
OR, TURBIDITY, HYDROGEN SULFIDE,
M, MAGNESIUM, POTASSIUM, SODIUM,
ONMENTAL EFFECTS, HETEROGENEITY,
OM SEDIMENTS, SODIUM, POTASSIUM,
DIOISOTOPES, ZINC RADIOISOTOPES,
IATOMS, CYANOPHYTA, CHLOROPHYTA,
  ENVIRONMENTAL EFFECTS, GRAZING,
IS, TRACE ELEMENTS, TEMPERATURE,
  , POTASSIUM, CALCIUM, MAGNESIUM,
  , HYPOLIMNION, ANIMALS, DENSITY,
  , LAKES, MICHIGAN, BICARBONATES,
OPHIC LEVEL, ZINC RADIOISOTOPES,
STS, HYDROGEN ION CONCENTRATION,
ROGEN, *CONDUCTIVITY, *LEACHING,
TE/ *CONCENTRATIONS, AEQUIPECTEN
  CROPS, TOXICITY, SOIL POROSITY,
UTION, *WATER POLLUTION SOURCES,
AKES, PLANTS, *ALGAE, *SALINITY,
  CRAYFISH, MOSQUITOES, BACTERIA,
SLANDS, MOU/ *NITROGEN FIXATION,
TE WAT/ *FARM WASTES, *SPRINKLER
  CURRENT, PUGET SOUND(WASH), LONG
GLUCOSE, GONIDIA FORMATION, LONG
A), PENGUINS, LAK/ PRINCE EDWARD
S, AEQUIPECTEN IRRADIANS, PIVERS
                        *SAKHALIN
AGANSETT MARINE LABORATORY(RHODE
ROPHY, B/ NARRAGANSETT BAY(RHODE
  SIMPLEX, FLAGELLATES, ELEUTHERA
ER TEMPERATURE, *SEASONAL, RHODE
WAVES(WATER), CONNECTICUT, RHODE
OSA, DIATOMA ELONGATUM, MELOSIRA
GAE, *LITTORAL, *AQUATIC PLANTS,
TROPHIC, *PHYTOPLANKTON, BRITISH
  TOXINS, KILLIFISHES, *TOXICITY,
  LIFE CYCLES, GENETICS, E. COLI,
ETTE, UNIALGAL, NITZSCHIA, ALGAE

ION EXCHANGE, ACTIVATED SLUDGE, D   W69-01453
ION EXCHANGE, BIOCHEMICAL OXYGEN   W70-10433
ION EXCHANGE, RESINS, LEAKAGE, CE   W70-00707
ION EXCHANGE, REVERSE OSMOSIS, DI   W70-01981
ION, *DESULFOVIBRIO, FACULTATIVE,   W70-01971
IONIC CONCENTRATION, SEASONAL EFF   W70-01933
IONS, BIOCHEMICAL OXYGEN DEMAND,   W69-00446
IONS, HUMIC ACIDS, CARP, ICE, IRO   W70-05399
IONS, PHOSPHATES, ORGANIC COMPOUN   W69-10180
IONS, SPECTROPHOTOMETRY, SUSPENDE   W70-03334
IONS, WATER POLLUTION EFFECTS, WA   W70-00707
IOWA RIVER, CYCLOTELLA, FRAGILARI   W70-08107
IOWA.: /, *POTASSIUM COMPOUNDS, *   W72-04242
IOWA.: /, *ODOR, WATER QUALITY CO   W72-13605
IOWA).:   W72-13605
IOWA).:   *POT   W72-04242
IOWA).: /NGBYA, STORM LAKE(IOWA),   W70-02803
IOWA), CENTER LAKE(IOWA), DIAMOND   W70-02803
IOWA), DIAMOND LAKE(IOWA).: /NGBY   W70-02803
IOWA, DIATOMS, EUGLENA, CHLAMYDOM   W70-08107
IPOMOEA, PISONIA, PANDANUS, MUSCL   W69-08269
IR: /RON, MAGNESIUM, COOLING, PRO   W70-02251
IRON BACTERIA, TASTE, FILTERS, IO   W70-08107
IRON CHLORIDE, IRON SULFATE, CLAD   W69-10170
IRON COMPOUNDS, ALGAE, PLANTS, WA   W69-10170
IRON COMPOUNDS, HYDROGEN SULFIDE,   W70-06925
IRON COMPOUNDS, LIGHT PENETRATION   W70-02255
IRON COMPOUNDS, HYDROGEN ION CONC   W70-03334
IRON COMPOUNDS, SULFATES.: /ON, A   W70-04060
IRON COMPOUNDS, HYDROGEN ION CONC   W70-05399
IRON COMPOUNDS, DIATOMS, CYANOPHY   W70-05409
IRON COMPOUNDS, HYDROGEN SULFIDE,   W70-04510
IRON COMPOUNDS, LIME, OXYGEN, TER   W68-00256
IRON OXIDES, *PLANT POPULATIONS,   W68-00860
IRON OXIDES, BACTERIA, ALGAE, FUN   W70-02294
IRON SULFATE, CLADOPHORA BLANKETS   W69-10170
IRON SULFATE, GALVANIZING WASTES,   W70-05668
IRON.: /ER, AIR TEMPERATURE, AERO   W69-07096
IRON, ALGICIDES, ZOOPLANKTON, FIS   W70-08654
IRON, ALKALINITY, HARDNESS, RADIO   W73-00750
IRON, BICARBONATE, ZINC, COBALT,   W70-03978
IRON, BIRDS, RADIOACTIVITY, INVER   W69-08269
IRON, CARBON DIOXIDE.: /H SUBSTAN   W70-02969
IRON, CHELATION, ENZYMES, CHLORIN   W70-02964
IRON, CHLORELLA, SODIUM, POTASSIU   W70-03512
IRON, COBALT, MANGANESE, ATLANTIC   W69-08525
IRON, DISSOLVED SOILDS, PHOSPHATE   W69-05894
IRON, FUNGI, CLAYS, BRACKISH WATE   W70-05761
IRON, GROWTH RATES, CHLOROPHYTA,   W70-03983
IRON, HY: /LVED OXYGEN, CHEMICAL   W70-05760
IRON, IONS, PHOSPHATES, ORGANIC C   W69-10180
IRON, LAKES, MAGNESIUM, NUTRIENTS   W69-07833
IRON, MAGNESIUM, NITRATES, PHOSPH   W70-03501
IRON, MAGNESIUM, COOLING, PROPERT   W70-02251
IRON, MANGANESE, TRACE ELEMENTS,   W70-02804
IRON, MANGANESE, WATERFOWL, CYANO   W70-00273
IRON, MANGANESE, CALCIUM, PIGMENT   W70-09889
IRON, MANGANESE, BENTHIC FLORA, S   W69-07845
IRON, MANGANESE, EUTROPHICATION,   W69-05704
IRON, MANGANESE, MOLYBDENUM, PLAN   W69-04801
IRON, NUTRIENTS, LIGHT PENETRATIO   W70-01933
IRON, ORGANIC MATTER, NITROGEN FI   W70-00719
IRON, PHOSPHORUS, SILICA, SAMPLIN   W70-06975
IRON, PHOSPHORUS, NITROGEN, WATER   W72-03183
IRON, SEDIMENTS, FISH, RADIOECOLO   W69-07853
IRON, SULFATE, TEMPERATURE, KENTU   W69-00096
IRON, TEMPERATURE, SOIL ANALYSIS,   W72-11906
IRRADIANS, PIVERS ISLAND(N C), DE   W69-08267
IRRIGATION, GROUND WATER, RUNOFF,   W69-05323
IRRIGATION, AQUATIC ALGAE, WATER   W69-07096
IRRIGATION EFFECTS, DRAINAGE EFFE   W69-00360
IRRIGATION, WILLOW TREES, FE: /S,   W70-05263
IRRIGATION, ORGANIC MATTER, *GRAS   W72-08110
IRRIGATION, *ORGANIC WASTES, *WAS   W71-00940
ISLAND SOUND, NARRAGANSETT BAY, C   W69-10161
ISLAND SOUND(NY), ICELAND, TRITIA   W69-09755
ISLAND(CANADA), NOVA SCOTIA(CANAD   W70-05399
ISLAND(N C), DETECTOR, BEAUFORT(N   W69-08267
ISLAND(USSR).:   W72-10623
ISLAND), DILUTION RATE, FRUITING   W70-01073
ISLAND), TAXONOMIC TYPES, HETERET   W70-00713
ISLAND, NITZCHIA SERIATA, NAVICUL   W70-00707
ISLAND, SAMPLING, ECOLOGY, AMMONI   W70-00713
ISLAND, TEMPERATURE, NUTRIENT REQ   W69-10161
ISLANDICA, NITZSCHIA ACICULARIS,   W69-10158
ISLANDS, ECOLOGICAL DISTRIBUTION,   W72-10623
ISLES, BIOINDICATORS, *ESTIMATING   W68-00255
ISOLATION.: /QUATIC ALGAE, *ALGAL   W70-08372
ISOLATION.: /MICAL PRECIPITATION,   W71-11823
ISOLATION, CAPILLARITY, CAPILLARI   W70-04469

TERIA, *DISSOLVED OXYGEN, SEWAGE
HEMICAL OXYGEN DEMAND, OXIDATION
OLLUTION EFFECTS, STABILIZATION,
, LIGHT PENETRATION, PHOSPHORUS,
ALGAE, GREEN / *LEWIS AND CLARK
, LAKE WASHINGTON, LAKE MALAREN,

, MELOSI/ SEASONAL DISTRIBUTION,
R BLOCKING, HETERETROPHS, ITALY,
IQUE, HETEROTROPHY, FLAGELLATES,
ILABLE), MICROCYSTIS AERUGINOSA,
MBERG(GERMANY), PL/ CHIRONOMIDS,
LAREN, LAKE ANNECY, LAKE VANERN,
UBESA(WIS), SPAULDING POND(WIS),
TREAM TREATMENT, BOTTOM SEALING,
ONSTANT, LAKE NORRVIKEN(SWEDEN),
, *NITROGEN FIXATION, NUTRIENTS,
ALIS, NE/ *NEW SPECIES, *WESTERN
OXYGEN, HYPOLIMNION, EPILIMNION,
COSTS, EUTROPHICATION, NITRATES,
XYGEN, HYPOLIMNION, GREAT LAKES,
GEN SULFIDE, CITIES, LAKE HURON,
N LAKE SURVEY, NITROGEN CONTENT,
OPHY, MICROBIAL ECOLOGY, SWEDEN,
CHAELIS-MENTEN EQUATION, SWEDEN,
MICHAELIS KINETICS, FLAGELLATES,
TROPHY, ORGANIC SOLUTES, SWEDEN,
NEBODA(SWEDEN), UPPSALA(SWEDEN),

Y, LAKE NORRVIKEN, LAKE MENDOTA,
METER, PROTOPLASM, WATER-BLOOMS,
ER(BAVARIA), LAKE ZELL(AUSTRIA),
PPING, HYDROGEN SULFIDE, CITIES,
, *LAKE MICHIGAN, *LAKE ONTARIO,
D LAKE(MINN), LAKE OWASSO(MINN),
MONONA(WIS), LAKE WAUBESA(WIS),
MONONA(WIS), LAKE WAUBESA(WIS),
E MONONA(WIS), LAKE WINGRA(WIS),
MENDOTA(WIS), LAKE MONONA(WIS),
MENDOTA(WIS), LAKE WAUBESA(WIS),
MONONA(WIS), LAKE WAUBESA(WIS),
ONA, LAKE MENDOTA, LAKE WAUBESA,
MONONA(WIS), LAKE WAUBESA(WIS),
KEN(SWEDEN), LAKE EKOLN(SWEDEN),
SUBTERRANEAN DRAINAGE, MECHANIC/
ITALY, LAKE BORREVANNET NORWAY,
KE SEBASTICOOK, LAKE WASHINGTON,
NSITY, / *LAKE MINNETONKA(MINN),
ANNET NORWAY, LAKE LUGANO ITALY,
S, OXYGEN D/ *RESERVOIR EFFECTS,
S), LAKE KEGONSA(WIS), LAKE WAU/
CROPS, TR/ TAKASUKA POND(JAPAN),
GAL NUTRITION, ALGAL PHYSIOLOGY,
RICHIA PISUM, NOSTOC VERRUCOSUM,
ADA, MESOTROPHY, LAKE NORRVIKEN,
GEN CONTENT, LAKE ERKEN(SWEDEN),
CHEMISTRY, WASTE WATER EFFECTS,

*MADISON(WIS), NUISANCE ODORS,
TE, CLADOPHORA, POTAMOGETON SPP,
S), LAKE KEGONSA(WIS), BLOOMING,
ECOVERY, WISCONSIN, LAKE MONONA,
A AEQUALE, NAVICULA PELLICULOSA,
AL PROBLEMS, SCIOTO RIVER(OHIO),
TAL SAMPLING, VERTICAL SAMPLING,
TION SURVEY, MADISON LAKES(WIS),
S), LAKE NAGAWICKA(WIS), LAKE P/
MYDOMONAS, CHLORELLA, WISCONSIN,
OPPER SULFATE, BOTTOM SEDIMENTS,
HURON, LAKE ERIE, LAKE ONTARIO,
ATMENT, *ALGAE, *FILTERS, SANDS,
OS AQUAE, STEPHANODISCUS ASTREA,
S-ST PAUL(MINN).:
RAPS, NUTRIENT INFLUX, NUTRIENT/
ALS, / *KILLING, ANIMAL MANURES,
DUTCH MILL DRAINAGE DITCH(WIS),
S), LAKE KEGONSA/ *MADISON(WIS),
ISANCE ODORS, LAKE MENDOTA(WIS),
C VERRUCOSUM, LAKE MENDOTA(WIS),
L PHYSIOLOGY, LAKE MENDOTA(WIS),
F), L/ *CULTURAL EUTROPHICATION,
S), LAKE WAU/ LAKE MENDOTA(WIS),
MOVAL, LAKE RECOVERY, WISCONSIN,
WIS), LAKE P/ LAKE MENDOTA(WIS),
CAL SAMPLING, LAKE MENDOTA(WIS),
RIVER(OHIO), LAKE MENDOTA(WIS),
ICAL COMMUNITIES, BIOINDICATORS,
ATIC POPULATIONS, WATER QUALITY,
UKEE LAKE(WIS), NORTH LAKE(WIS),
MENDOTA(WIS), LAKE MONONA(WIS),

LAGOONS, PLANT PHYSIOLOGY, ANIMAL   W72-01818
LAGOONS, SLUDGE, PHOTOSYNTHETIC O   W71-08221
LAGOONS, TOXICITY, WASTE WATER TR   W69-09454
LAGOONS: /LINITY, *PHOTOSYNTHESIS   W68-00248
LAKE (S D), ABUNDANCE, BLUE-GREEN   W70-05375
LAKE ANNECY, LAKE VANERN, LAKE CO   W70-03964
LAKE ASHTABULA.:   W70-09093
LAKE BAIKAL, PHOTOSYNTHETIC LAYER   W70-04290
LAKE BORREVANNET NORWAY, LAKE LUG   W69-07833
LAKE CLASSIFICATION, SHAGAWA LAKE   W70-02775
LAKE COMO(MINN), LAKE VADNAIS(MIN   W70-05269
LAKE CONSTANCE, WHITEFISH, WURTTE   W70-05761
LAKE CONSTANCE, PFAFFIKERSEE, TUR   W70-03964
LAKE DELAVAN(WIS), PEWAUKEE LAKE(W69-05867
LAKE DILUTION.: /R DIVERSION, INS   W69-09340
LAKE EKOLN(SWEDEN), LAKE LOTSJON(W70-04287
LAKE ERIE.:   *ALGAE, *CYANOPHYTA   W70-05091
LAKE ERIE(OHIO), GONGROSIRA STAGN   W70-04468
LAKE ERIE, INSECTS, DIPTERA, CRUS   W70-04253
LAKE ERIE, LAKE ONTARIO, POLLUTIO   W70-03964
LAKE ERIE, STATISTICS, GROWTH RAT   W68-00253
LAKE ERIE, LAKE ONTARIO, LAKE MIC   W70-00667
LAKE ERKEN(SWEDEN), LAKE MENDOTA(W70-03959
LAKE ERKEN(SWEDEN), ACTIVE TRANSP   W70-04284
LAKE ERKEN(SWEDEN), LINEWEAVER-BU   W70-04287
LAKE ERKEN(SWEDEN), ACETATE, LAKE   W70-04194
LAKE ERKEN(SWEDEN), ANNUAL CYCLES   W70-05109
LAKE ERKEN(SWEDEN), EXTINCTION VA   W70-05409
LAKE ERSOM(DENMARK).:   W71-00665
LAKE FURES, LAKE SEBASTICOOK, LAK   W70-03964
LAKE GEORGE(UGANDA), RHINE-MEEUSE   W70-04369
LAKE HALL WILER(SWITZERLAND).: /I   W70-05668
LAKE HURON, LAKE ERIE, LAKE ONTAR   W70-00667
LAKE HURON, AESTHETICS, ECOLOGY,   W70-01943
LAKE JOSEPHINE(MINN), GRAY'S BAY(W70-05269
LAKE KEGONSA(WIS).: /H(WIS), LAKE   W70-05113
LAKE KEGONSA(WIS), BLOOMING, LAKE   W70-05112
LAKE KEGONSA(WIS), LAKE WAUBESA(W   W69-03366
LAKE KEGONSA(WIS), LAKE WAUBESA(W   W69-06276
LAKE KEGONSA(WIS).: /(WIS), LAKE   W70-02787
LAKE KEGONSA(WIS), WEED BEDS, PHO   W70-02969
LAKE KEGONSA, LAKE ZURICH, OSCILL   W69-09349
LAKE KEGONSA(WIS).: /A(WIS), LAKE   W72-01813
LAKE LOTSJON(SWEDEN), PERMEASES,   W70-04287
LAKE LUCERNE, FERTILIZED FIELDS,   W69-08668
LAKE LUGANO ITALY, LAKE MARIDALSV   W69-07833
LAKE MALAREN, LAKE ANNECY, LAKE V   W70-03964
LAKE MANAGEMENT PROGRAM, ALGAL DE   W69-10167
LAKE MARIDALSVANNET NORWAY, LAKE   W69-07833
LAKE MEAD, DILUTION, ALGAL GROWTH   W73-00750
LAKE MENDOTA(WIS), LAKE MONONA(WI   W69-06276
LAKE MENDOTA(WIS), PHYTOPLANKTON   W69-05142
LAKE MENDOTA(WIS), LAKE MONONA(WI   W69-05867
LAKE MENDOTA(WIS), LAKE MONONA(WI   W69-08283
LAKE MENDOTA, LAKE FURES, LAKE SE   W70-03964
LAKE MENDOTA(WIS), NET PLANKTON,   W70-03959
LAKE MENDOTA(WIS), BLACK EARTH CR   W70-04382
LAKE MENDOTA(WIS).:   W69-03185
LAKE MENDOTA(WIS), LAKE MONONA(WI   W69-03366
LAKE MENDOTA(WIS), ANACHARIS, NAI   W70-05263
LAKE MENDOTA(WIS), LAKE WINGRA(WI   W70-05112
LAKE MENDOTA, LAKE WAUBESA, LAKE   W69-09349
LAKE MENDOTA(WIS), ENGLAND, LONDO   W69-10180
LAKE MENDOTA(WIS), LAKE MONONA(WI   W72-01813
LAKE MENDOTA(WIS), LAKE MONONA(WI   W70-02969
LAKE MENDOTA(WIS), LAKE WAUBESA(W   W70-02787
LAKE MENDOTA(WIS), LAKE MONONA(WI   W70-06217
LAKE MICHIGAN, PHYTOPLANKTON, DIA   W69-10180
LAKE MICHIGAN, WASHINGTON, DISSOL   W69-09349
LAKE MICHIGAN, WATER TEMPERATURE,   W70-00667
LAKE MICHIGAN, WATER QUALITY CONT   W73-02426
LAKE MINNETONKA(MINN), MORPHOMETR   W70-05387
LAKE MINNETONKA(MINN), MINNEAPOLI   W70-04810
LAKE MINNETONKA(MINN), NUTRIENT T   W70-04268
LAKE MINNETONKA(MINN), SOIL MINER   W70-04193
LAKE MONONA(WIS), LAKE WAUBESA(WI   W70-05113
LAKE MONONA(WIS), LAKE WAUBESA(WI   W70-05112
LAKE MONONA(WIS), LAKE WINGRA(WIS   W69-03366
LAKE MONONA(WIS), LAKE WAUBESA(WI   W69-08283
LAKE MONONA(WIS), LAKE WAUBESA(WI   W69-05867
LAKE MONONA(WIS), LAKE TAHOE(CALI   W69-06858
LAKE MONONA(WIS), LAKE KEGONSA(WI   W69-06276
LAKE MONONA, LAKE MENDOTA, LAKE W   W69-09349
LAKE MONONA(WIS), LAKE NAGAWICKA(W70-06217
LAKE MONONA(WIS), LAKE WAUBESA(WI   W70-02969
LAKE MONONA(WIS), LAKE WAUBESA(WI   W72-01813
LAKE MORPHOLOGY, ECOLOGICAL DISTR   W71-12077
LAKE MORPHOMETRY, DOMESTIC WASTES   W68-00470
LAKE NAGAWICKA(WIS).: /WIS), PEWA   W69-05867
LAKE NAGAWICKA(WIS), LAKE PEWAUKE   W70-06217

464

OMETRY, ALGAE BLOOMS, CLEARWATER
MS, CLEARWATER LAKE(MINN), TROUT
MINN), TROUT LAKE(MINN), KIMBALL
S(MINN), MILLE LAC(MINN), DEMING
INN), DEMING LAKE(MINN), ORCHARD
NOBRYON, CONCORD(N H), PEANACOOK
DES, PHOTOASSIMILATION, BERKELSE
CHLORELLA OVALIS BUTCHER, NJOER
ORWAY), HOBOL RIVER(NORWAY), VAN
ARIA, MORICHES BAY(NY), ONONDAGA
ONEMA, PINNULARIA, NI/ *ONONDAGA
, FILINIA, / *FAYETTEVILLE GREEN
TS, *NUTRIENT LIMITATION, CAYUGA
), BUCKHORN LAKE(ONTARIO), CLEAR
, BALSAM LAKE(ONTARIO), STURGEON
TWELVE MILE LAKE(ONTARIO), GULL
NTARI/ *ONTARIO, *CANADA, KUSHOG
USHOG LAKE(ONTARIO), TWELVE MILE
STURGEON LAKE(ONTARIO), BUCKHORN
RIO), GULL LAKE(ONTARIO), BALSAM
ARIO), CLEAR LAKE(ONTARIO), RICE
SCILLATORA.:   *UPPER KLAMATH
UPPER KLAMATH LAKE(ORE.), AGENCY
QUAE, *CARBON-14/ *UPPER KLAMATH
N(WIS), LAKE ZOAR(CONN), KLAMATH
RMANNII, OTTAWA(ONTARIO), BURTON
LATION, RUTHENIUM-106, WHITE OAK
ROIDES, CLADOCERANS, WESTHAMPTON
EROCYSTS, GAS-VACUOLES, LLANGORS
LEAR ACTIVATION TECHNIQUES, ROCK
IQUES, ROCK LAKE(WASH), WILLIAMS
S, BEAVER DAM LAKE(WIS), DELAVAN
D(WIS), LAKE WAUBESA(WIS), GREEN
IS), LAKE DELAVAN(WIS), PEWAUKEE
(WIS), PEWAUKEE LAKE(WIS), NORTH
E, TENDIPES PLUMOSUS, BEAVER DAM
E, *BIOAS/ *EUTROPHICATION, *MUD(
QUIRKE LAKE, PECORS LAKE, DUNLOP
, NITROGEN COM/ *EUTROPHICATION,
PLANTS, WATER POLLUT/ *NEWFOUND
HN LAKE, PINGO LAKE, TWELVE MILE
QUATIC ECOSYSTEM MODEL/ KOOTENAY
TARIO LAKES, QUIRKE LAKE, PECORS
E PARK, LAKE SHATUTAKEE, MASCOMA
TOR, ALASKA, HUDEUC LAKE, PULLIN
LAKE, IMIKPUK LAKE, NORTH MEADOW
EE, GREIFENSEE, ZURICHSEE, MOSES
GO LAKE, TWELVE MILE LAKE, BRANT
                          *ANDERSON-CUE
ATER QUALITY CONTROL, WINNISQUAM
E MILE LAKE, BRANT LAKE, IMIKPUK
DEUC LAKE, PULLIN LAKE, GRAPHITE
WARBURG RESPIROMETER,/ *ONONDAGA
                                ELLIOT
OPERTIES.:   CRATER
INORGANIC/ ONTARIO LAKES, QUIRKE
IN LAKE, GRAPHITE LAKE, OLD JOHN
OLD J/ DETECTOR, ALASKA, HUDEUC
PHITE LAKE, OLD JOHN LAKE, PINGO
LUT/ *NEWFOUND LAKE, *WINNISQUAM
:   GREAT
L, SEWAGE EFFLUENTS, ABSORPTION,
, PLANKTON, LAKES, RIVERS, GREAT
NCIES, DISSOLVED OXYGEN, OXYGEN,
OSPHATES, NITRATES, ALGAE, GREAT
NTS, BADFISH RIVER(WIS), DETROIT
ARIA, RADIOCARBON METHOD, FINGER
E, FERTILIZATION SURVEY, MADISON
OSCILLATORIA RUBESCENS, MADISON
), LAKE WINNISQUAM(N H), MADISON
, *WEED EXTRACTING STE/ *MADISON
AKE STRATIFICATION, SURFACE AREA(
U S PUBLIC HEALTH SERVICE, GREAT
ECOLOGY, *EUTROPHICATION, *GREAT
ULTRA-VIOLET LIGHT STER/ *DUTCH
COLOGY, WATER POLLUTION EFFECTS,
IENTS, OXYGEN REQUIREMENTS, ICED
PROCEDURES, NURTRIENTS, RIVERS,
S, CYANOPHYTA, FRESH WATER FISH,
, *LAKE ONTARI/ *BIOLOGY, *GREAT
PHOSPHORUS, N/ *EUTROPHICATION,
GAE, CURRENTS(WATER), LIMNOLOGY,
SOURCES, MUNICIPAL WASTES, GREAT
LLUTION / *WATER QUALITY, *GREAT
F / *MADISON(WIS), *'WORKING' OF
OLLUTION CONTROL, WATER / *GREAT
ASTES, *MUNICIPAL WASTES, *GREAT
ICATION, *ALGAE, *AQUATIC WEEDS,
HOTOSYNTHESIS, HYDROGEN SULFIDE,
NUTRIENT REQUIREMENTS, CULTURES,

LAKE(MINN), TROUT LAKE(MINN), KIM   W70-05387
LAKE(MINN), KIMBALL LAKE(MINN), C   W70-05387
LAKE(MINN), CRYPTOPHYTA, PYROPHYT   W70-05387
LAKE(MINN), ORCHARD LAKE(MINN), L   W70-05269
LAKE(MINN), LAKE OWASSO(MINN), LA   W70-05269
LAKE(N H).: / TREATMENT COSTS, DI   W69-08282
LAKE(NETHERLANDS), HETEROCYSTS, G   W70-04369
LAKE(NORWAY), HOBOL RIVER(NORWAY)   W70-05752
LAKE(NORWAY).: /HER, NJOER LAKE(N   W70-05752
LAKE(NY).: /GENETICS, RUSSIA, BAV   W70-03973
LAKE(NY), EUNOTIA, AMPHORA, GOMPH   W70-03974
LAKE(NY), *MARL LAKES, CHLOROBIUM   W70-05760
LAKE(NY), ASTERIONELLA FORMOSA, S   W70-01579
LAKE(ONTARIO), RICE LAKE(ONTARIO)   W70-02795
LAKE(ONTARIO), BUCKHORN LAKE(ONTA   W70-02795
LAKE(ONTARIO), BALSAM LAKE(ONTARI   W70-02795
LAKE(ONTARIO), TWELVE MILE LAKE(O   W70-02795
LAKE(ONTARIO), GULL LAKE(ONTARIO)   W70-02795
LAKE(ONTARIO), CLEAR LAKE(ONTARIO   W70-02795
LAKE(ONTARIO), STURGEON LAKE(ONTA   W70-02795
LAKE(ONTARIO).: /UCKHORN LAKE(ONT   W70-02795
LAKE(ORE.), AGENCY LAKE(ORE.), *O   W72-06051
LAKE(ORE.), *OSCILLATORA.:   *   W72-06051
LAKE(ORE.), *APHANIZOMENON FLOS-A   W72-06052
LAKE(ORE), LAKE WASHINGTON(WASH),   W70-04506
LAKE(SASKATCHEWAN), HUMBOLDT(SASK   W70-00273
LAKE(TENN), ACTIVITY-DENSITY, VER   W70-04371
LAKE(VA).:   ASTERIONELLA SPI   W71-10098
LAKE(WALES), HALOBACTERIUM, RESPI   W70-04369
LAKE(WASH), WILLIAMS LAKE(WASH),   W70-00264
LAKE(WASH), CHANNELED SCABLANDS(W   W70-00264
LAKE(WIS).: /SE, TENDIPES PLUMOSU   W70-06217
LAKE(WIS).: /VAL, SPAULDING'S PON   W69-05868
LAKE(WIS), NORTH LAKE(WIS), LAKE   W69-05867
LAKE(WIS), LAKE NAGAWICKA(WIS).: /   W69-05867
LAKE(WIS), DELAVAN LAKE(WIS).: /S   W70-06217
LAKE), *PHOSPHORUS, *AQUATIC ALGA   W70-03955
LAKE, *INORGANIC CARBON.: /AKES,   W70-05424
LAKE, *LAKE SEDIMENTS, *NUTRIENTS   W72-06051
LAKE, *WINNISQUAM LAKE, WATER AND   W70-02764
LAKE, BRANT LAKE, IMIKPUK LAKE, N   W69-08272
LAKE, BRITISH COLUMBIA, CANADA, A   W69-06536
LAKE, DUNLOP LAKE, *INORGANIC CAR   W70-05424
LAKE, GLEN LAKE.: / WADLEIGH STAT   W69-08674
LAKE, GRAPHITE LAKE, OLD JOHN LAK   W69-08272
LAKE, GYTTJA, DUFF, OOZE, COBBLES   W69-08272
LAKE, HALLWILLERSEE.: /BALDEGGERS   W70-03964
LAKE, IMIKPUK LAKE, NORTH MEADOW   W69-08272
LAKE, MELROSE(FLORIDA).:   W72-01990
LAKE, NEWFOUND LAKE.: /LLUTION, W   W70-02772
LAKE, NORTH MEADOW LAKE, GYTTJA,   W69-08272
LAKE, OLD JOHN LAKE, PINGO LAKE,   W69-08272
LAKE, ONONDAGA COUNTY, SYRACUSE,   W69-02959
LAKE, ONTARIO.:   W69-01165
LAKE, ORE, GROWTH RATES, WATER PR   W68-00478
LAKE, PECORS LAKE, DUNLOP LAKE, *   W70-05424
LAKE, PINGO LAKE, TWELVE MILE LAK   W69-08272
LAKE, PULLIN LAKE, GRAPHITE LAKE,   W69-08272
LAKE, TWELVE MILE LAKE, BRANT LAK   W69-08272
LAKE, WATER AND PLANTS, WATER POL   W70-02764
LAKES BASIN, BACTERIAL POLLUTION.   W69-03948
LAKES.: /NT, WATER QUALITY CONTRO   W70-01031
LAKES.: /OSPHORUS, TRACE ELEMENTS   W70-04385
LAKES.: /PLANKTON, ALGAE, EFFICIE   W70-04484
LAKES.: /TRIENTS, FERTILIZERS, PH   W69-07084
LAKES(MINN), NUTRIENT REMOVAL, MA   W70-04506
LAKES(NY), RODHE'S SOLUTION VIII,   W70-01579
LAKES(WIS), LAKE MENDOTA(WIS), LA   W70-02787
LAKES(WIS), CEDAR RIVER(WASH), ZU   W70-00270
LAKES(WIS), LAKE WASHINGTON(WASH)   W70-07261
LAKES(WIS), *AQUATIC WEED CUTTING   W70-04494
LAKES), PHAEDACTYLUM, WASTE WATER   W69-09340
LAKES-ILLINOIS RIVER BASINS PROJE   W70-00263
LAKES, *ALGAE, NUTRIENTS, PHOSPHO   W70-00667
LAKES, *ALLOCHTHONOUS PHOSPHATES,   W70-03955
LAKES, *CYANOPHYTA, *EUTROPHICATI   W71-04758
LAKES, *FERTILIZATION, WATER FLEA   W68-00487
LAKES, *GROWTH RATES, *WATER TEMP   W68-00472
LAKES, *LAKE ZURICH, *MUNICIPAL W   W68-00680
LAKES, *LAKE ERIE, *LAKE MICHIGAN   W70-01943
LAKES, *NUTRIENTS, UNITED STATES,   W68-00478
LAKES, *PILOT PLANTS, *NUISANCE A   W68-00479
LAKES, *POLLUTANT IDENTIFICATION.   W71-05879
LAKES, *TRACE ELEMENTS, *WATER PO   W69-08562
LAKES, *WATER-BLOOM, 'BREAKING' O   W69-08283
LAKES, *WATER POLLUTION, *WATER P   W69-03948
LAKES, *WATER POLLUTION, SEWAGE,   W70-04430
LAKES, ALGICIDES, AQUATIC WEED CO   W70-04494
LAKES, ALGAE, ENTERIC BACTERIA, M   W70-03312
LAKES, ALGAE, GROWTH RATES, BACTE   W70-00719

S/ *EUTROPHICATION, *LAKES, ACID
, CYANOPHYTA, CHLOROPHYTA, GREAT
ERS, SEWAGE EFFLUENT, NUTRIENTS,
FISH/ *EUTROPHICATION, SEQUENCE,
YETTEVILLE GREEN LAKE(NY), *MARL
LGAE, SUNFISHES, ALGAE, DAPHNIA,
KES, *WATER-BLOOM, 'BREAKING' OF
TRIENT ENRICHMENT TEST/ HIGHLAND
NTS, *TERTIARY TREATMENT, *GREAT
ERSHED MANAGEMENT, GEOCHEMISTRY,
DESMUS, DIATOMACEOUS EARTH, MUD,
TROL, WATER POLLUTION, TOXICITY,
ACTERIA, *ALGAE, ORGANIC MATTER,
ARCTIC ALGAE, AUTOTROPHIC/ *MARL
NA, *EUTROPHICATION, *NUTRIENTS,
*DAPHNIA, PHYSIOLOGICAL ECOLOGY,
FECTS, WATER QUALITY, WISCONSIN,
SURFA/ *BIOASSAY, PHYTOPLANKTON,
CONFERENCES, STREAMS, ESTUARIES,
CONFERENCES, STREAMS, ESTUARIES,
IUM COMPOUNDS, CHLORIDES, SALINE
BIOINDICATORS, FISH, WASHINGTON,
S, CITRATES, BIOASSAY, NEW YORK,
*EUTROPHICATION, *HYDRODYNAMICS,
T GROWTH REGULATORS, FRESH WATER
OLVED OXYGEN, HYPOLIMNION, GREAT
ROL, *NUTRIENT/ *EUTROPHICATION,
OGY, *PHYTOPLANKTON, FISH, GREAT
, PHYTOPLANKTON, EUTROPHICATION,
PLANTS, BIOMASS, FERTILIZATION,
ATION, *ALGAE, PLANNING, RIVERS,
S, EUTROPHICATION, IMPOUNDMENTS,
YDROGEN ION CONCENTRATION, IRON,
PESTICIDE RESIDUES, ALGAE, GREAT
AY, EUTROPHICATION, OLIGOTROPHY,
NUTRIENT/ ALGAE, EUTROPHICATION,
GEN COMPOUNDS, PHOSPHATES, GREAT
PHYLL, CHEMTROL, COPPER SULFATE,
TIC WEEDS, ECONOMIC FEASIBILITY,
HING, WISCONSIN, BIBLIOGRAPHIES,
TOSYNTHESIS, *RESPIRATION, GREAT
CTS, BACTERIA, PROTOZOA, SEWAGE,
TES, HYDROGEN SULFIDE, SALINITY,
, ALGAE, DUCKS, COLOR, FISHKILL,
ANDING CROP, A/ *EUTROPHICATION,
OACTIVITY, CARBON RADIOISOTOPES,
ANKTON, *PHOTOSYNTHESIS, *ALGAE,
ALYSIS, WATER POLLUTION EFFECTS,
HEAVY METALS, INDUSTRIAL WASTES,
STREAMS, ESTUARIES, LAKES, GREAT
STREAMS, ESTUARIES, LAKES, GREAT
L ECOLOGY, CARBON RADIOISOTOPES,
ATER POLLUTION, STREAMS, RIVERS,
, ROOTED AQUATIC PLANTS, GLACIAL
SPHORUS, *NITROGEN, *ALKALINITY,
DUNLOP LAKE, *INORGANIC/ ONTARIO
UCTIVITY, SYMBIOSIS, RESERVOIRS,
AE, *EUTROPHICATION, CYANOPHYTA,
OMPOUNDS, TOXICITY, CHLOROPHYTA,
TION, CHLOROPHYLL, FLUORESCENCE,
ATER SUPPLY, EFFLUENTS, STREAMS,
HORUS, TRACE ELEMENTS, PLANKTON,
ION, *AQUATIC HABITATS, *ALASKA,
ODOR, TOXINS, DISSOLVED OXYGEN,
PHOSPHATES, SULFATES, SEASONAL,
RSHEDS, *RUNOFF, *FERTILIZATION,
GAE, *WATER POLLUTION TREATMENT,
TRACTIONS, PHOSPHORUS COMPOUNDS,
LUENTS, *ESTUARIES, FRESH WATER,
*FERTILIZATION, *EUTROPHICATION,
WASTE DISPOSAL, WATER POLLUTION,
*CHEMICAL PROPERTIES, NUTRIENTS,
DIES, MATHEMATICAL MODELS, GREAT
TS, *FALLOUT, *ECOSYSTEMS, GREAT
ITY, *ORGANIC MATTER, NUTRIENTS,
RUS, EUTROPHICATION, WASHINGTON,
WAGE EFFLUENTS, LIGHT INTENSITY,
ISCONSIN, WATER QUALITY CONTROL,
RON MICROSCOPY, WASTE TREATMENT,
ELD(MADISON) EFFLUENTS, *MADISON
QUALITY STANDARDS, GRAFTON(ND),
ES, *FIXED-LIGHT TESTS, */ *BEER-
CUS VESICULOSUS, GROWTH FACTORS,
US CRISPUS, ASCOPHYLLUM NODOSUM,
LUM NODOSUM, LAMINARIA DIGITATA,
TIGOCLADUS LAMINOSUS, PHORMIDIUM
NE PARK, *ICELAND, MASTIGOCLADUS
PHYSA, SALMO TRUTTA, ENTOSPHENUS
FISH, ALEWIFE, GIZZARD SHAD, SEA
N, UDO/ *TRACE ELEMENTS, *ANCHOA

LAKES, BIBLIOGRAPHIES, *PHOSPHORU   W68-00857
LAKES, BIOINDICATORS, FISH, WASHI   W70-04503
LAKES, CHANNELS, REVIEWS, BIBLIOG   W70-00269
LAKES, CHEMICALS, FISHING, SPORT   W68-00172
LAKES, CHLOROBIUM, FILINIA, MONIM   W70-05760
LAKES, CHLOROPHYTA, STREAMS, DIAT   W70-02982
LAKES, CHROOCOCCACEAE, NOSTOCHINE   W69-08283
LAKES, COLORADO RIVER, TEXAS, *NU   W69-06004
LAKES, CONTROL, COSTS, EUTROPHICA   W70-03964
LAKES, CONIFEROUS FORESTS, ALGAE,   W71-09012
LAKES, CULTIVATION, CHANNEL MORPH   W68-00475
LAKES, DESIGN CRITERIA, REMOTE SE   W70-02775
LAKES, DIFFUSION, CARBON RADIOISO   W70-04284
LAKES, DISSOLVED ORGANIC MATTER,   W71-10079
LAKES, ECOLOGY, PRIMARY PRODUCTIV   W71-00665
LAKES, ENVIRONMENTAL EFFECTS, WAT   W70-03957
LAKES, EUTROPHICATION.: /UTION EF   W69-05868
LAKES, EUTROPHICATION, SAMPLING,   W70-02973
LAKES, GREAT LAKES, PHYSIOLOGICAL   W70-03975
LAKES, GREAT LAKES, PHYSIOLOGICAL   W70-03975
LAKES, HEAVY METALS, INDUSTRIAL W   W69-02959
LAKES, INDIANA, OCHROMONAS, PHYTO   W70-04503
LAKES, INHIBITION, ALGAE, DIATOMS   W70-01579
LAKES, INTERNAL WATER, WAVES(WATE   W70-09889
LAKES, LABORATORY TESTS.: / *PLAN   W68-00856
LAKES, LAKE ERIE, STATISTICS, GRO   W68-00253
LAKES, LAKE SEDIMENTS, ALGAL CONT   W72-06052
LAKES, LEGISLATION, NITROGEN, NUI   W69-06858
LAKES, LIMNOLOGY, AQUATIC ENVIRON   W69-05142
LAKES, LIMNOLOGY, MIDGES, NITRATE   W69-04805
LAKES, LIMNOLOGY, WISCONSIN, URBA   W70-04506
LAKES, LOAMS, RESERVOIRS, TEXAS.:   W69-04800
LAKES, MAGNESIUM, NUTRIENTS, NUIS   W69-07833
LAKES, METABOLISM, *CHLORELLA, *C   W70-08652
LAKES, MICHIGAN, BICARBONATES, IR   W72-03183
LAKES, MICROORGANISMS, NITROGEN,   W68-00481
LAKES, MICHIGAN.: /SOURCES, NITRO   W69-01445
LAKES, NITROGEN.: / ALGAE, CHLORO   W68-00470
LAKES, NUISANCE ALGAE, WATER POLL   W69-06276
LAKES, NUISANCE ALGAE, BACTERIA,   W69-08283
LAKES, NUISANCE ALGAE, WATER QUAL   W72-05453
LAKES, NUTRIENTS.: /, FUNGI, INSE   W71-10786
LAKES, NUTRIENTS, BACTERIA, ALGAE   W71-03031
LAKES, ODOR, TOXICITY, WATER POLL   W69-08279
LAKES, ORGANIC WASTES, OXYGEN, ST   W70-02251
LAKES, ORGANIC MATTER, SEA WATER,   W70-02504
LAKES, OXYGEN, EUTROPHICATION, AL   W71-10098
LAKES, PHOSPHORUS, TROPHIC LEVEL,   W69-06536
LAKES, PHOTOSYNTHESIS, PHYTOPLANK   W69-02959
LAKES, PHYSIOLOGICAL ECOLOGY, WAT   W70-03975
LAKES, PHYSIOLOGICAL ECOLOGY, WAT   W70-03975
LAKES, PHYTOPLANKTON, BIOASSAYS.:   W70-05109
LAKES, PHYSIOLOGICAL ECOLOGY, LIM   W71-02479
LAKES, PLANKTON, EPIDEMICS.: /ILL   W68-00488
LAKES, PRODUCTIVITY, PHYTOPLANKTO   W70-02795
LAKES, QUIRKE LAKE, PECORS LAKE,   W70-05424
LAKES, REAERATION, SESTON, CHLORO   W70-05092
LAKES, RESERVOIRS, SCUM, NITROGEN   W70-02803
LAKES, RESERVOIRS, NITROGEN, GROW   W69-10180
LAKES, RIVERS, NITROGEN, BIOMASS,   W70-02777
LAKES, RIVERS, RESERVOIRS, BOATIN   W71-09880
LAKES, RIVERS, GREAT LAKES.: /OSP   W70-04385
LAKES, RIVERS, HOT SPRINGS, TERRE   W69-08272
LAKES, RIVERS, LIGHT PENETRATION,   W69-05697
LAKES, SAMPLING, PONDS, RIVERS, R   W70-02646
LAKES, SEWAGE TREATMENT, SEWAGE B   W70-02787
LAKES, SEWAGE, DRAINAGE DISTRICTS   W70-04455
LAKES, SEWAGE EFFLUENTS, ARSENIC   W70-05269
LAKES, SILTS, PHOSPHATES, NITROGE   W70-02255
LAKES, STREAMS, ALGAE, FISH, WISC   W69-10178
LAKES, STREAMS, SEWAGE, REMEDIES,   W69-06909
LAKES, TESTS, NITROGEN, PHOSPHORU   W68-00856
LAKES, THERMAL STRATIFICATION.: /   W71-05878
LAKES, TRACE ELEMENTS, RADIUM RAD   W71-09013
LAKES, WATER PROPERTIES, BACTERIA   W71-10079
LAKES, WATER POLLUTION EFFECTS.: /   W70-00270
LAKES, WATER POLLUTION SOURCES, W   W69-10167
LAKES, WATER POLLUTION SOURCES, C   W70-03310
LAKES, WATER QUALITY, VIRUSES, AL   W70-03973
LAKES, YAHARA RIVER(WIS), STORKWE   W70-05113
LAKOTA(ND), HARVEY(ND), CHROMATIU   W70-03312
LAMBERT LAW, *FIXED-DENSITY STUDI   W70-03923
LAMINARIA SACCHARINA, MANNITOL, M   W69-07826
LAMINARIA DIGITATA, LAMINARIA AGA   W70-01073
LAMINARIA AGARDHII, DESICCATION,   W70-01073
LAMINOSUM, OSCILLATORIA TENUIS, O   W70-05270
LAMINOSUS, PHORMIDIUM LAMINOSUM,   W70-05270
LAMOTTENII, GAMMARUS, CORDULEGAST   W69-09334
LAMPHREY, COHO SALMON.: /, WHITE   W70-01943
LAMPROTAENIA, *OPISTHONEMA OGLIMU   W69-08525

VOLUMES, BACTERIAL PLATE COUNTS,
TION, ANAEROBIC DENITRIFICATION,
*ALGAE, CHLOROPHYTA, CYANOPHYTA,
OLIGOTR/ *LAKES, STREAMS, FISH,
ED OXYGEN, EUTROPHICATION, FISH,
ATIC LIFE, MILK, ATLANTIC OCEAN,
L WATER MOVEMENT, LEACHING, ARID
WATER DEMAND, WATER SUPPLY, ARID
RIO, FLAVOBACTER,/ *POLYSIPHONIA
SCOPHYLLUM NODOSUM, POLYSIPHONIA
EDEN), PERMEASES, CHAMYDOMONADS,
ES, *FERTILIZATION, WATER FLEAS,
RIENTS, AQUATIC ECOSYSTEM, MIDGE
TON, BACTERICIDES, ALGAL TOXINS,
TON, WORMS, NEMATODES, ROTIFERS,
UND WATER, NITRATES, PHOSPHATES,
, LAS VEGAS, NEVADA, *LAKE MEAD,
S VEG/ DISSOLVED ORGANIC MATTER,
    *ABS,
Y, YELLOWSTONE NATIONAL PARK, MT
RUITING BODIES, ULVA LACTUCA VAR
    *ULVA
OCEANS, ORGANIC MATTER, DIURNAL,
JAPONICA, CHAETOCEROS LAUNDERI,
TERIONELLA JAPONICA, CHAETOCEROS
ANTITIES, ALGAL BLOOMS, LIEBIG'S
G'S LAW OF THE MINIMUM, VOISIN'S
ED-LIGHT TESTS, */ *BEER-LAMBERT
ASOLINE, FLOODS, AQUIFERS, WATER
, SEWAGE, WATER POLLUTION, WATER
YTES.:
DUSTRIAL WASTES, AGRICULTURE, ST
ION, LAKE BAIKAL, PHOTOSYNTHETIC
ISTORTA, GLOEOTRICHIA ECHINULATA
SOIL WATER, SOIL WATER MOVEMENT,
RON COMPOUNDS, HYDROGEN SULFIDE,
EFFLUENTS, ION EXCHANGE, RESINS,
TROPHICATION, FISH, LAND, VALUE,
PISONIA, PANDANUS, MUSCLE, BONE,
FALO, DUNREITH/ *CARBON FILTERS,
, CONCORD, BOSTON LOT RESERVOIR,
KARATELLA COHCLEARIS, ASPLANCHA,
TUBES, *DIFFUSERS, *RAYTOWN(MO),
HIRONOMUS PROMOSUS, TRICHOPTERA,
VA RESERVOIRS, ANOPHELES QUADRI/
E, REMEDIES, JUDICIAL DECISIONS,
, RIGHT-OF-WAY, ESTIMATED COSTS,
POLLUTION CONTROL, WATERCOURSES(
ESTIMATED COSTS, LEGAL ASPECTS,
HYTOPLANKTON, FISH, GREAT LAKES,
RSENITE, NUISANCE ALGAE, SNAILS,
ADOW, NITROGEN-15, SOD MATS, NON-
CCHI DISC TRANSPARENCY, ANABAENA
SPECIES SUSCEPTIBILITY, ANABAENA
    *PHOTOSYNTHETIC OXYGEN
DRICAUDA, PORPHYRIDIUM CRUENTUM,
CUS, ESCHERICHIA COLI, OOCYSTIS,
WORTS, MORPHOMETRY, PACIFASTACUS
THE HAGUE, MINERALIZATION, CLAY
ATORY TESTS, LOTIC ENVIRONMENTS,
OSPORA, PEDIASTRUM, SCENEDESMUS,
METER, CLADOPHORA, ENTEROMORPHA,
ESCHERICHIA COLI, MYRIOPHYLLUM,
OSUM, POLYSIPHONIA LANOSA, POINT
LLOPORA EYDOUXI, PORITES LOBATA,
ROS LAUNDERI, LAUDERIA BOREALIS,
RA, COSCINDISCUS, BACTERIOSTRUM,
PHANIZOMENON FLOS-AQUAE, MINIMUM
ARY, AMMONIUM COMPOUNDS, MINIMUM
ALGAE, WATER POLLUTION EFFECTS,
N, *BACTERIA, *PROTOZOA, *ALGAE,
BIOASSAY, ANALYTICAL TECHNIQUES,
ISCHURA NEVADENSIS, DIAPHANOSOMA
2-M/ *FISH HEMATOLOGY, *GRANULAR
S, GRAIN COUNT(AUTORADIOGRAPHY),
GREEN ALGAE, BREAM, WHITE BREAM,
AQUATIC MICROORGANISMS, TROPHIC
N, PRIMARY PRODUCTIVITY, TROPHIC
OTOPES, TENNESSEE RIVER, TROPHIC
PHOTOSYNTHESIS, OXYGEN, TROPHIC
ES, HYDROLOGY, CLIMATES, TROPHIC
N EFFECTS, *FISH, ALGAE, TROPHIC
ECTS, LAKES, PHOSPHORUS, TROPHIC
TIVITY, LIGHT INTENSITY, TROPHIC
UCTIVITY, AQUATIC ALGAE, TROPHIC
   *PRIMARY PRODUCTIVITY, *TROPHIC
Y, *PRIMARY PRODUCTIVITY, TROPIC
HOSPHORUS RADIOISOTOPES, TROPHIC
*LAKES, *BENTHIC FAUNA, *TROPHIC
R QUALITY, NUT/ *ALGAE, *TROPHIC

LANCASTER(CALIF).: / PACKED CELL   W71-05267
LAND APPLICATION, ELECTROCHEMICAL   W70-01981
LAND MANAGEMENT, WATER POLLUTION   W68-00172
LAND, ECOSYSTEMS, EUTROPHICATION,   W70-04193
LAND, VALUE, LEAVES, FERTILIZERS,   W70-05264
LANDFILLS, DISPOSAL, REGULATION,:   W72-00941
LANDS, DEPTH, CYANOPHYTA, CHLOROP   W71-04067
LANDS, LIMNOLOGY.: / MONITORING,   W70-05645
LANOSA, *ASCOPHYLLUM NODOSUM, VIB   W70-03952
LANOSA, POINT LEPREAU(CANADA), NE   W70-04365
LAPPLAND(SWEDEN), ACETATE, TURNOV   W70-04287
LARVAE, GROWTH RATES, *FRESH WATE   W68-00487
LARVAE, LAKE STRATIFICATION, SURF   W69-09340
LARVAE, PHENOLS, SHELLFISH, OYSTE   W70-01068
LARVAE, SNAILS, MITES, MOLDS, GRO   W72-01819
LAS VEGAS BAY.: /OIRS, ALGAE, GRO   W72-02817
LAS VEGAS BAY.: /D ORGANIC MATTER   W72-02817
LAS VEGAS, NEVADA, *LAKE MEAD, LA   W72-02817
LAS.:   W69-09454
LASSEN NATIONAL PARK, MT RAINIER   W70-03309
LATISSIMA, POLYSIPHONIA HARVEYI,   W70-01073
LATISSIMA, *AMMONIA NITROGEN.:   W71-05500
LATITUDINAL STUDIES.: /PIGMENTS,   W70-04809
LAUDERIA BOREALIS, LEPTOCYLINDRUS   W70-04280
LAUNDERI, LAUDERIA BOREALIS, LEPT   W70-04280
LAW OF THE MINIMUM, VOISIN'S LAW   W70-04765
LAW OF THE MAXIMUM.: /OOMS, LIEBI   W70-04765
LAW, *FIXED-DENSITY STUDIES, *FIX   W70-03923
LAW, NITRATES, TURBIDITY, FLOODWA   W73-00758
LAW, WATER POLLUTION EFFECTS, WAT   W69-06305
LAWRENCE LAKE(MICHIGAN), *MACROPH   W71-00832
LAWRENCE RIVER, ALGAE, COLIFORMS,   W70-03964
LAYER, MELOSIRA BAICALENSIS, GYMN   W70-04290
LB 1303.: /S AQUAE, TOLYPOTHRIX D   W70-05091
LEACHING, ARID LANDS, DEPTH, CYAN   W71-04067
LEAD, COPPER, AMMONIA, CONDUCTIVI   W70-04510
LEAKAGE, CESIUM, NUCLEAR EXPLOSIO   W70-00707
LEAVES, FERTILIZERS, PESTICIDES,   W70-05264
LEAVES, SHELL, RHODIUM-102.: /A,   W69-08269
LEBANON, BATAVIA, CLARKSBURG, BUF   W72-03641
LEBANON, WADLEIGH STATE PARK, LAK   W69-08674
LECANE, BOSMINA, NAVICULA, FRAGIL   W70-02249
LEE'S SUMMIT(MO), OXYGEN TRANSFER   W70-06601
LEECHES, FINGERNAIL CLAMS, WHITE   W70-01943
LEECHES, SCHISTOSOME CERCARIAE, T   W70-06225
LEGAL ASPECTS.: /, STREAMS, SEWAG   W69-06909
LEGAL ASPECTS, LEGISLATION, SEWAG   W70-03344
LEGAL), ORGANIC WASTES, PULP WAST   W72-00462
LEGISLATION, SEWAGE DISTRI: /WAY,   W70-03344
LEGISLATION, NITROGEN, NUISANCE A   W69-06858
LEGISLATION, WISCONSIN, CHEMCONTR   W70-05568
LEGUMES, *NITROGEN FIXING ORGANIS   W72-08110
LEMMERMANNI, APHANOCAPSA, BOSMINA   W69-10169
LEMMERMANNII, OTTAWA(ONTARIO), BU   W70-00273
LEMMON(SD), KADOKA(SD).:   W72-01818
LEMNA MINOR, PHOTOPHOSPHORYLATION   W70-02964
LEMNA, SPIRODELA, ELODEA.: /LOCOC   W70-10161
LENIUSCULUS, ANACHARIS CANADENSIS   W70-00711
LENSES, RHINE RIVER, AMSTERDAM RH   W69-07838
LENTIC ENVIRONMENTS.: /STS, LABOR   W72-05453
LEPOCINCLIS, ANABAENA, ANACYSTIS,   W70-03974
LEPOCINCLIS, CHLOROGONIUM, SCENED   W69-02959
LEPOMIS MACROCHIRUS, PIMEPHALES N   W69-07861
LEPREAU(CANADA), NEW BRUNSWICK(CA   W70-04365
LEPTASTREA PURPUREA, FUNGIA FUNGI   W69-08526
LEPTOCYLINDRUS DANICUS, PHAEODACT   W70-04280
LEPTOCYLINDRUS, CORYCAEUS, OIKOPL   W73-01446
LETHAL DOSES, CHU'S MEDIA, QUINON   W70-02982
LETHAL EXPOSURE, PHENANTHROQUINON   W70-02982
LETHAL LIMIT, GROWTH RATES, SPORE   W70-01779
LETHAL LIMIT, ELECTROPHORESIS, MI   W71-03775
LETHAL LIMITS, DINOFLAGELLATES, D   W70-08111
LEUCHTENBER GIANUM, CYCLOPS BICUS   W69-10182
LEUCOCYTES, *MUSCULAR EXERTION, *   W70-01788
LEUCOTHRIX MUCOR, ANTITHAMNION SA   W69-10163
LEUKOCYTIC COMPOSITION, ALTERNATI   W70-01788
LEVEL.: /LORIDA, *EUTROPHICATION,   W72-01993
LEVEL.: /VITY, LIMNOLOGY, PLANKTO   W69-09135
LEVEL, ALGAE, CADDISFLIES, FOOD C   W70-02786
LEVEL, LIGHT INTENSITY, ECOLOGY,   W70-00265
LEVEL, NITROGEN FIXATION, FOREST,   W71-03048
LEVEL, PRODUCTIVITY, FOOD CHAINS,   W72-01788
LEVEL, PRODUCTIVITY, MODEL STUDIE   W69-06536
LEVEL, SAMPLING TEMPERATURE, TRAC   W70-02504
LEVEL, WATER QUALITY, FLORIDA.: /   W72-01990
LEVEL, WATER STORAGE, EVAPORATION   W69-09723
LEVEL, WATER QUALITY, WASTE ASSIM   W68-00912
LEVEL, ZINC RADIOISOTOPES, IRON,   W69-07853
LEVELS, *SURVEYS, BIOLOGICAL COMM   W71-12077
LEVELS, AQUATIC ENVIRONMENT, WATE   W70-02764

, DISTRIBUTION, CALCIUM, TROPHIC | LEVELS, FOOD WEBS.: /SPECTROSCOPY | W69-08525
C ALGAE, WATER QUALITY, *TROPHIC | LEVELS, NUTRIENTS, NORTHEAST U.S. | W70-02772
DA), *ONTARIO), *CANADA, NUTRIENT | LEVELS, SOFT WATER.: /E LAKE(CANA | W70-02973
SFORMATIONS, AMSTERDAM, HAARLEM, | LEYDEN, THE HAGUE, MINERALIZATION | W69-07838
BACTERIA, FUNGI, VIRUSES, ALGAE, | LICHENS, ACTIMOMYCETES, CYTOLOGIC | W71-11823
ELATION, FUNGI, ALGAE, PROTOZOA, | LICHENS, PLANTS, DIATOMS, CORAL, | W69-07428
S, *MARINE ANIMALS, *OILY WATER, | LICHENS, TIDAL EFFECTS, DISPERSIO | W70-01777
, MACROQUANTITIES, ALGAL BLOOMS, | LIEBIG'S LAW OF THE MINIMUM, VOIS | W70-04765
NTATION, CHEMICAL PRECIPITATION, | LIFE CYCLES, GENETICS, E. COLI, I | W71-11823
BAYS, *FLUCTUATION, RESISTANCE, | LIFE CYCLES, GROWTH STAGES.: /NT, | W71-02494
OP, LIGHT, BIOMASS, RESPIRATION, | LIFE CYCLES, OLIGOCHAETES, MOLLUS | W71-00665
OACTIVITY, *FOOD WEBS, KENTUCKY, | LIFE HISTORY STUDIES, GAMMA RAYS, | W69-09742
A, TIPULIDAE, CHIRONOMIDAE, HALF- | LIFE.: / PLECOPTERA, EPHEMEROPTER | W69-09742
ON, PRODUCTIVITY, ALGAE, AQUATIC | LIFE.: /N COMPOUNDS, EUTROPHICATI | W70-03068
TEMPERATE ZONE, PLUTONIUM, HALF- | LIFE(RADIONUCLIDE), FILTER-FEEDIN | W70-00708
TY TEST,/ *POLYMERS, *BIOLOGICAL | LIFE, *ALGAE BLOOMS, *FISH TOXICI | W70-05819
, *AQUATIC ENVIRONMENT, *AQUATIC | LIFE, *AQUATIC POPULATIONS, ALGAE | W71-02491
RS, ALGAE, PROTOZOA, A/ *AQUATIC | LIFE, *AQUATIC POPULATIONS, *RIVE | W73-00748
LUTION EFFE/ *BIOASSAY, *AQUATIC | LIFE, *WASTE DISPOSAL, *WATER POL | W72-01801
CES, PATH OF POLLUTANTS, AQUATIC | LIFE, ALGAE, SILTS, ECOLOGY, AQUA | W69-10080
LOGY, SALINITY, BENTHOS, AQUATIC | LIFE, ALGAE, PLANTS, WATER PROPER | W69-05023
, ENVIRONMENTAL EFFECTS, AQUATIC | LIFE, AQUATIC MICROORGANISMS, MIC | W69-04276
IZATION, MARINE ANIMALS, AQUATIC | LIFE, AQUATIC ANIMALS, BENTHIC FA | W71-02494
QUALITY, BIOINDICATORS, AQUATIC | LIFE, AQUATIC ALGAE, WATER POLLUT | W69-05958
LLA PYRENOIDOSA, BIOLOGICAL HALF- | LIFE, CLINCH RIVER(TENNESSEE), PH | W70-02786
DRILUS SP/ RADIOAUTOGRAPHS, HALF- | LIFE, CONCENTRATION FACTOR, LIMNO | W69-07861
MODEL STUDIES, BIOASSAY, AQUATIC | LIFE, COOLING WATER, SITES.: /E, | W71-11881
ONUCLIDE, ULOTHRIX, UPTAKE, HALF- | LIFE, DETECTOR.: *RADI | W69-07862
CROP, MICROENVIRONMENT, AQUATIC | LIFE, ENVIRONMENT, MICROORGANISMS | W69-03611
RIVER(TENNESSEE), PHYSICAL HALF- | LIFE, HYDROPSYCHE COCKERELLI, BOD | W70-02786
, SURVEYS, MARINE ALGAE, AQUATIC | LIFE, MILK, ATLANTIC OCEAN, LANDF | W72-00941
LIDES, ASSIMILATION, BA-LA, HALF- | LIFE, PTYCHOCHEILUS OREGONENSIS, | W69-07853
RATURE, *EUTROPHICATION, AQUATIC | LIFE, SESTON, SCENEDESMUS, MATHEM | W73-01446
N, WATER QUALITY, ALGAE, AQUATIC | LIFE, STREAM POLLUTION, BACTERIA, | W72-01786
IVITY, *TRACE ELEMENTS, *AQUATIC | LIFE, WATER POLLUTION EFFECTS, MA | W71-05531
DRAINAGE, ALGAL CONTROL, AQUATIC | LIFE, WATER CONSERVATION, PUBLIC | W70-03344
OUNDS, METALS, PLANKTON, AQUATIC | LIFE, ZOOPLANKTON, AQUATIC ANIMAL | W69-03695
MS, DETRITUS, CYCLING NUTRIENTS, | LIG: / COMPOUNDS, BACTERIA, DIATO | W69-09334
ITY, PHOTO-ELECTRIC TURBIDOSTAT, | LIGHT EXTINCTION COEFFICIENT, ALG | W70-03923
GHT, PHOSPHORUS, PHOTOSYNTHESIS, | LIGHT INTENSITY, STABILIZATION, P | W70-03923
TRATION, ALKALINITY, CALIFORNIA, | LIGHT INTENSITY, TEMPERATURE, ION | W70-03334
IENTS, OLIGOTROPHY, TEMPERATURE, | LIGHT INTENSITY, DAPHNIA, CONNECT | W70-03959
ORGANIC MATTER, TRACE ELEMENTS, | LIGHT INTENSITY, CHLAMYDOMONAS, C | W70-04369
NTIAL NUTRIENTS, EUTROPHICATION, | LIGHT INTENSITY, LIGHT PENETRATIO | W70-04283
EFFECTS, OREGON, BIOINDICATORS, | LIGHT INTENSITY, EUTROPHICATION, | W70-03978
YANOPHYTA, DIATOMS, CHLOROPHYTA, | LIGHT INTENSITY, SUCCESSION, ZINC | W70-04371
NTS, EUTROPHICATION, INHIBITION, | LIGHT INTENSITY, LIGHT PENETRAT: / | W69-03515
NTAL EFFECTS, INHIBITION, LIGHT, | LIGHT INTENSITY, LIGHT PENETRATIO | W69-03516
NG, FREEZE-THAW TESTS, FREEZING, | LIGHT INTENSITY, LIGHT PENETRATIO | W69-03518
YTA, DIATOMS, INHIBITION, LIGHT, | LIGHT INTENSITY, LIGHT PENETRATIO | W69-03517
ES, *PLANT POPULATIONS, REVIEWS, | LIGHT INTENSITY, WATER TEMPERATUR | W68-00860
NTS, EUTROPHICATION, INHIBITION, | LIGHT INTENSITY, LIGHT PENETRATIO | W69-03514
ESSENTIAL NUTRIENTS, INHIBITION, | LIGHT INTENSITY, LIMNOLOGY,: /N, | W69-03513
MENT, RETENTION, PHOTOPERIODISM, | LIGHT INTENSITY, VELOCITY, TEMPER | W69-07862
WATER, OLIGOTROPHY, TEMPERATURE, | LIGHT INTENSITY, CHLOROPHYLL, NUT | W70-05752
MARY PRODUCT/ *PLANKTON, *ALGAE, | LIGHT INTENSITY, TEMPERATURE, PRI | W70-04809
TS, TEMPERATURE, PHOTOSYNTHESIS, | LIGHT INTENSITY.: /UTION, NUTRIEN | W70-05270
ASSIUM, TEMPERATURE, EPILIMNION, | LIGHT INTENSITY, SEASONAL, PHYSIO | W70-05409
ROPHYLL, ICE, SEASONAL, DIURNAL, | LIGHT INTENSITY, CARBON, SEWAGE T | W70-05387
WINDS, EPILIMNION, HYPOLIMNION, | LIGHT INTENSITY, CONDUCTIVITY, AL | W70-05405
MENTS, SURFACE WATERS, DETRITUS, | LIGHT INTENSITY, TURBIDITY, ALGAE | W70-00713
YNTHESIS, OXYGEN, TROPHIC LEVEL, | LIGHT INTENSITY, ECOLOGY, LOTIC E | W70-00265
CHLORELLA, CARBON RADIOISOTOPES, | LIGHT INTENSITY, CULTURES, RADIOA | W69-10158
GEN COMPOUNDS, SEWAGE EFFLUENTS, | LIGHT INTENSITY, LAKES, WATER POL | W69-10167
ESIS, LAGOONS, LABORATORY TESTS, | LIGHT INTENSITY, PHOTOPERIODISM, | W72-00917
RY TESTS, *ANAEROBIC CONDITIONS, | LIGHT INTENSITY, EUTROPHICATION, | W71-13413
NITIES, *LABORATORIES, *STREAMS, | LIGHT INTENSITY, CURRENTS(WATER), | W70-02780
CHLORELLA, CARBON, TEMPERATURE, | LIGHT INTENSITY, NUTRIENT REQUIRE | W70-02965
N, CHELATION, ENZYMES, CHLORINE, | LIGHT INTENSITY, NITRATE, EUGLENA | W70-02964
ALGAE, *DIATOMS, *BENTHIC FLORA, | LIGHT INTENSITY, BIOMASS, DISTRIB | W70-03325
SEA WATER, PRIMARY PRODUCTIVITY, | LIGHT INTENSITY, TROPHIC LEVEL, S | W70-02504
IS, RESPIRATION, MICROORGANISMS, | LIGHT INTENSITY, CARBON DIOXIDE, | W71-00668
URE, TOXICITY, DISSOLVED OXYGEN, | LIGHT INTENSITY, SLUDGE, DEPTH, O | W72-06854
ENT, *OXIDATION LAGOONS, *ALGAE, | LIGHT INTENSITY, CARBON DIOXIDE, | W72-06022
TAHOE(CALIF-NEV), TRANSPARENCY, | LIGHT MEASUREMENTS, EUPHOTIC ZONE | W70-01933
S, TEMPERATURE, IRON, NUTRIENTS, | LIGHT PENETRATION, SEWAGE EFFLUEN | W70-01933
TROGEN FIXATION, IRON COMPOUNDS, | LIGHT PENETRATION, WATER POLLUTIO | W70-02255
RESERVOIRS, PONDS, TEMPERATURE, | LIGHT PENETRATION, DISSOLVED OXYG | W70-06225
LIGHT INTENSITY, PHOTOPERIODISM, | LIGHT PENETRATION, WATER POLLUTIO | W72-00917
, TIME, EPILIMNION, THERMOCLINE, | LIGHT PENETRATION, STANDING CROP, | W71-07698
A, SOUTH DAKOTA, PHOTOSYNTHESIS, | LIGHT PENETRATION, ON-SITE TESTS, | W72-01818
WATER POLLUTION SOURCES, RUNOFF, | LIGHT PENETRATION, SAMPLING, PRIM | W72-04761
STRATIFICATION, DEPTH, PLANKTON, | LIGHT PENETRATION, SEASONAL, PRIM | W70-05405
DISSOLVED OXYGEN, LAKES, RIVERS, | LIGHT PENETRATION, FLUORESCENCE, | W69-05697
HARVESTING OF ALGAE, INHIBITION, | LIGHT PENETRATION, NITROGEN COMP: | W69-03369
ON, INHIBITION, LIGHT INTENSITY, | LIGHT PENETRATION, LIMNOLOGY, NIT | W69-03514
S, HARVESTING ALGAE, INHIBITION, | LIGHT PENETRATION, LIMNOLOGY, NIT | W69-04798
IBITION, LIGHT, LIGHT INTENSITY, | LIGHT PENETRATION, LIMNOLOGY, PHO | W69-03516
ON, INHIBITION, LIGHT INTENSITY, | LIGHT PENETRAT: /TS, EUTROPHICATI | W69-03515
ESTS, FREEZING, LIGHT INTENSITY, | LIGHT PENETRATION, LIMNOLOGY, NIT | W69-03518

IBITION, LIGHT, LIGHT INTENSITY, LIGHT PENETRATION, LIMNOLOGY, NUT W69-03517
RE, ALKALINITY, *PHOTOSYNTHESIS, LIGHT PENETRATION, PHOSPHORUS, LA W68-00248
GAE, ALGAE, SEASONAL, LIMNOLOGY, LIGHT PENETRATION, OXYGEN, SPORT W68-00487
EUTROPHICATION, LIGHT INTENSITY, LIGHT PENETRATION, LIMNOLOGY, NIT W70-04283
 CYCLING NUTRIENTS, TEMPERATURE, LIGHT QUALITY, CHLOROPHYLL, ZOOPL W69-08526
THONOUS PHOSPHATES, ULTRA-VIOLET LIGHT STERILIZATION, ASTERIONELLA W70-03955
W HAMPSHIRE, ALGAE, ODOR, TASTE, LIGHT TEMPERATURE, NUTRIENTS, COP W69-08282
, *FIXED-DENSITY STUDIES, *FIXED- LIGHT TESTS, *EFFECTIVE-AVERAGE-L W70-03923
MONOD GROWTH EQUATION, INCIDENT LIGHT UTILIZATION.: /TH KINETICS W69-03730
-LIGHT TESTS, *EFFECTIVE-AVERAGE- LIGHT-INTENSITY, PHOTO-ELECTRIC T W70-03923
PECIES, CYCLOTELLA MENEGHINIANA, LIGHT-SATURATED RATE, CELL VOLUME W70-05381
IQUES, *MICROORGANISMS, *LIPIDS, LIGHT, ALGAE, TEMPERATURE, CARBON W70-05651
ON CONCENTRATION, STANDING CROP, LIGHT, BIOMASS, RESPIRATION, LIFE W71-00665
, N/ *NITROGEN FIXATION, *ALGAE, LIGHT, CARBON DIOXIDE, INHIBITION W70-04249
 *PLANKTON, *ALGAE, SCENEDESMUS, LIGHT, CULTURES, IRON COMPOUNDS, W70-05409
HLOROPHYTA, DIATOMS, INHIBITION, LIGHT, LIGHT INTENSITY, LIGHT PEN W69-03517
VIRONMENTAL EFFECTS, INHIBITION, LIGHT, LIGHT INTENSITY, LIGHT PEN W69-03516
CHLORELLA, CHLOROPHYTA, DIATOMS, LIGHT, LIPIDS, NUTRIENT REQUIREME W69-06865
ABOLISM, *WASTE WATER TREATMENT, LIGHT, MICROORGANISMS, PLANTS.: / W70-06604
, CYANOPHYTA, SAMPLING, DENSITY, LIGHT, MUD, SILTS, SANDS, GRAVELS W70-00711
BIOMASS, CHLOROPHYLL, PROTEINS, LIGHT, NUTRIENTS, HYDROSTATIC PRE W69-10160
ITROGEN FIXATION, FOREST, WINDS, LIGHT, TEMPERATURE, HYD: /EVEL, N W71-03048
PHOTOSYNTHESIS, EUTROPHICATION, LIGHT, TEMPERATURE, CARBON DIOXID W69-07442
*WATER, *RIVERS, *LAKES, *PONDS, LIGHT, TEMPERATURE, FERTILIZATION W70-05404
, PACIFIC OCEAN, CARBON DIOXIDE, LIGHT, TEMPERATURE.: /DIOISOTOPES W70-05652
L POLLUTION, STREAMS, NUTRIENTS, LIGHT, WASHINGTON, WATER CHEMISTR W70-03309
PHAEODACTYLUM TRICORNUTUM, EDTA, LIGURIAN SEA, EUROPEAN MEDITERRAN W69-08275
GRAM-POSITIVE BACTERIA, SARCINE- LIKE, BACILLUS-LIKE, MICROCOCCUS- W70-00713
BACTERIA, SARCINE-LIKE, BACILLUS- LIKE, MICROCOCCUS-LIKE, PSYCHROPH W70-00713
LIKE, BACILLUS-LIKE, MICROCOCCUS- LIKE, PSYCHROPHILES, PLEMORPHIC, W70-00713
BUR REED, DUCKWEED, WHITE WATER LILY, SWIMMERS' ITCH.: /D CELERY, W70-05568
D SLUDGE, ALGAE, EUTROPHICATION, LIME, *NITROGEN, HYDROGEN ION CON W68-00012
HLOROPHYTA, CHLORELLA, CHLORINE, LIME, COAGULATION.: /ENEDESMUS, C W70-08107
CTIVATED SLUDGE, IRON COMPOUNDS, LIME, OXYGEN, TERTIARY TREATMENT, W68-00256
EMICAL COMPOUNDS, MEAN TOLERANCE LIMIT, BLUEGILLS, CHLORELLA PYREN W72-11105
TERIA, *PROTOZOA, *ALGAE, LETHAL LIMIT, ELECTROPHORESIS, MICROORGA W71-03775
WATER POLLUTION EFFECTS, LETHAL LIMIT, GROWTH RATES, SPORES.: /E, W70-01779
NRICHMENT EXPERIMENTS, *NUTRIENT LIMITATION, CAYUGA LAKE(NY), ASTE W70-01579
RODHE'S SOLUTION VIII, NUTRIENT LIMITATION, CARBON-14.: /KES(NY), W70-01579
TUM, SELENASTRUM GRACILE, GROWTH LIMITATION, SELENASTRUM, ALGAL GR W70-02779
, *LABORATORY STUDIES, *NITROGEN LIMITATION, BATCH CULTURES, CHLOR W70-07957
EMOSTATS, TURBIDOSTATS, NUTRIENT LIMITED GROWTH, LUXURY CONSUMPTIO W70-02779
ROWTH RATES, ALGAE, OLIGOTROPHY, LIMITING FACTORS, CHLORELLA, ANAL W70-02779
S, LAKES, PHYSIOLOGICAL ECOLOGY, LIMITING FACTORS, ECOSYSTEMS, MIC W71-02479
OPHYTA, FISHKILL, FERTILIZATION, LIMITING FACTORS, NITROGEN COMPOU W69-06535
   *NICKEL, DETECTION LIMITS, CHEMICAL INTERFERENCES.: W71-11687
Y, ANALYTICAL TECHNIQUES, LETHAL LIMITS, DINOFLAGELLATES, DIATOMS. W70-08111
LORELLA VULGARIS, DAPHNIA MAGNA, LIMNAEA STAGNALIS, OUSPENSKII'S M W71-03056
LORELLA VULGARIS, DAPHNIA MAGNA, LIMNAEA STAGNALIS, OUSPENSKII'S M W71-05719
, CHROCOCCUS LIMNETICUS, LYNGBYA LIMNETICA, ANABAENA FLOS AQUAE, T W70-05405
NAS EROSA, COSMARIUM, CHROCOCCUS LIMNETICUS, LYNGBYA LIMNETICA, AN W70-05405
HALF-LIFE, CONCENTRATION FACTOR, LIMNODRILUS SPP, AEOLOSOMA HEMPRI W69-07861
PLEUROXUS, TANYPUS, TANYTARSUS, LIMNODRILUS, EUCYCLOPS, ALONA, CH W70-06975
EMAND, WATER SUPPLY, ARID LANDS, LIMNOLOGY.: / MONITORING, WATER D W70-05645
LAYS, BRACKISH WATER, FISHERIES, LIMNOLOGY.: /FIDE, IRON, FUNGI, C W70-05761
LGAE, WATER MANAGEMENT(APPLIED), LIMNOLOGY.: /LLUTION TREATMENT, A W70-04721
PLING, ON-SITE DATA COLLECTIONS, LIMNOLOGY.: /AL ANALYSIS, PH, SAM W69-00659
CTIVITY, EUTROPHICATION, *LAKES, LIMNOLOGY, *PLANKTON, ALGAE, BIBL W69-00446
LOGICAL), *AQUATIC PRODUCTIVITY, LIMNOLOGY, *NUTRIENTS, BIO MASS, W68-00912
NUISANCE ALGAE, ALGAE, SEASONAL, LIMNOLOGY, LIGHT PENETRATION, OXY W68-00487
WASTES, PLANT GROWTH SUBSTANCES, LIMNOLOGY, PLANT GROWTH REGULATOR W68-00470
E, GERMANY, VOLUMETRIC ANALYSIS, LIMNOLOGY, GROWTH RATES, *LABORAT W68-00475
ION, PLANT TISSUES, SCENEDESMUS, LIMNOLOGY, WATER PROPERTIES, WATE W68-00248
AQUATIC ALGAE, CURRENTS(WATER), LIMNOLOGY, LAKES, *PILOT PLANTS, W68-00479
HT INTENSITY, LIGHT PENETRATION, LIMNOLOGY, NITROGEN COMPOUNDS, NI W69-03518
HT INTENSITY, LIGHT PENETRATION, LIMNOLOGY, NUTRIENT REQUIREMENTS, W69-03517
, BIOMASS, FERTILIZATION, LAKES, LIMNOLOGY, MIDGES, NITRATES, NITR W69-04805
HT INTENSITY, LIGHT PENETRATION, LIMNOLOGY, PHOTOSYNTHETIC OXYGEN, W69-03516
, INHIBITION, LIGHT PENETRATION, LIMNOLOGY, NITROGEN COMPOUNDS, NU W69-04798
NMENTAL EFFECTS, EUTROPHICATION, LIMNOLOGY, NITROGEN COMPOUNDS, NI W69-04802
ING FACILITIES, FISH MANAGEMENT, LIMNOLOGY, NITROGEN COMPOUNDS, NI W69-04804
HT INTENSITY, LIGHT PENETRATION, LIMNOLOGY, NITROGEN COMPOUNDS, NU W69-03514
NMENTAL EFFECTS, EUTROPHICATION, LIMNOLOGY, NITROGEN COMPOUNDS, NI W69-03186
NMENTAL EFFECTS, EUTROPHICATION, LIMNOLOGY, NITROGEN COMPOUNDS, NI W69-03185
TS, INHIBITION, LIGHT INTENSITY, LIMNOLOGY,: /N, ESSENTIAL NUTRIEN W69-03513
, TERTIARY TREATMENT, EFFLUENTS, LIMNOLOGY, WATER CHEMISTRY, RESPI W70-05387
PLANKTON, EUTROPHICATION, LAKES, LIMNOLOGY, AQUATIC ENVIRONMENT, A W69-05142
HT INTENSITY, LIGHT PENETRATION, LIMNOLOGY, NITROGEN COMPOUNDS, NU W70-04283
ITY, CYANOPHYTA, NUISANCE ALGAE, LIMNOLOGY, CHLOROPHYTA, PLANKTON, W70-03959
N, ALGAE, FUNGI, BACTERIA, FISH, LIMNOLOGY, HYDROLOGIC ASPECTS, AQ W70-04475
*ALGAE, PLANNING, RIVERS, LAKES, LIMNOLOGY, WISCONSIN, URBANIZATIO W70-04506
TION, WATER QUALITY, *NUTRIENTS, LIMNOLOGY, WATER TEMPERATURE, *AL W70-09093
EMENTS, *POLLUTANTS, INHIBITION, LIMNOLOGY, ACIDITY, ACID SOILS, A W70-02792
DIATOMS, ENVIRONMENTAL EFFECTS, LIMNOLOGY, ECOLOGY, ECOLOGICAL DI W70-02784
PHICATION, *CALIFORNIA, *NEVADA, LIMNOLOGY, PRIMARY PRODUCTIVITY, W70-01933
POTASSIUM, HARVESTING OF ALGAE, LIMNOLOGY, NUTRIENTS, PHYSIOLOGIC W70-02245
VIRONMENTAL EFFECTS, INHIBITION, LIMNOLOGY, PESTICIDES, PESTICIDE W70-02968
*ORGANIC MATTER, *PHYTOPLANKTON, LIMNOLOGY, ZOOPLANKTON, ALGAE, CO W70-02510
CTERIA, WATER POLLUTION EFFECTS, LIMNOLOGY, CYANOPHYTA.: /ITUS, BA W70-02249
TAMOGETON FOLIOSUS, EXPERIMENTAL LIMNOLOGY, BRACHIONUS, SYNCHAETA W70-02249

PTH, ALGAE, MOSSES, CHARA, FISH,   LIMNOLOGY, LAKE TROUT, CRAYFISH,   W70-00711
OURCES, WATER POLLUTION CONTROL,   LIMNOLOGY, DIVERSION, STREAMS.: /   W69-10170
N/ *FIELD STUDIES, *EXPERIMENTAL   LIMNOLOGY, CLADOCERAN, MONTANE ZO   W69-10154
VITY TECHNIQUES, *RADIOISOTOPES,   LIMNOLOGY, OCEANOGRAPHY, TRACERS,   W69-10163
TOXICITY, *CULTURES, *ISOLATION,   LIMNOLOGY, PUBLIC HEALTH, PHYSICO   W70-00273
RE, DISSOLVED OXYGEN, FISHKILLS,   LIMNOLOGY, ALGAL TOXINS, ALGAL CO   W70-00264
CATION, ALGAL BLOOMS, NUTRIENTS,   LIMNOLOGY, DYSTROPHY, OLIGOTROPHY   W69-09723
N SOURCES, AQUATIC PRODUCTIVITY,   LIMNOLOGY, PLANKTON, PRIMARY PROD   W69-09135
PLATTE RIVER,   LINCOLN(NEBR).:   W69-05894
LGAE METABOLISM, DETOXIFICATION,   LINDANE METABOLISM.: /LINDANE, *A   W70-08652
DYE, AIRBORNE TESTS/ *FRAUNHOFER   LINE DISCRIMINATOR, RHODAMINE WT   W70-05377
EASUREMENT APPARATUS, HYDROGEN F   LINE.: /OFER LINES, ATTENUATION M   W70-05377
CTANTS, ALKYLBENZENE SULFONATES,   LINEAR ALKYLATE SULFONATES, WATER   W69-09454
-PEROT FILTERS, SOLAR FRAUNHOFER   LINES, ATTENUATION MEASUREMENT AP   W70-05377
PHAEOPIGMENTS, ACRIDINE ORANGE,   LINEWEAVER-BURKE EQUATION.: /AND,   W70-04811
ELATIVE HETEROTROPHIC POTENTIAL,   LINEWEAVER-BURKE EQUATION, UPTAKE   W70-05109
ION, SWEDEN, LAKE ERKEN(SWEDEN),   LINEWEAVER-BURK EQUATION, MAXIMUM   W70-04287
ENTIAL, SWEDEN, ENZYME KINETICS,   LINEWEAVER-BURK EQUATION, MICHAEL   W70-04194
GISPINA, ZURICHSEE(SWITZERLAND),   LINSLEY POND(CONN), BOSMINA COREG   W69-10169
RE/ CHLOROFORM-METHANE EXTRACTS,   LIPID CARBON, CHLOROPLASTS, INFRA   W70-05651
RATURE, PROTEINS, CARBOHYDRATES,   LIPIDS, KINETICS, MATHEMATICAL MO   W71-13413
TURES.:   *GROWTH, *ALGAE,   LIPIDS, NUTRIENTS, CHLORELLA, CUL   W70-07957
LA, CHLOROPHYTA, DIATOMS, LIGHT,   LIPIDS, NUTRIENT REQUIREMENTS, OR   W69-06865
, SEDIMENTS, GAS CHROMATOGRAPHY,   LIPIDS, OIL, OIL SHALE.: /GENESIS   W69-08284
OROUS UPTAKE RATE.:   MIXED   LIQUOR, SUSPENDED SOLIDS, *PHOSPH   W71-13335
OIL WASTES, OILY WATER, SULFITE   LIQUORS, ULTIMATE DISPOSAL, WATER   W69-05891
SOLUBILITY / *EQUILIBRIUM MODEL,   LITHOSPHERE, SORPTION REACTIONS,   W70-04385
SC), GROWTH STIMULATION, SURFACE   LITTER.: /LAKE BUTTE DES MORTS(WI   W72-00845
HETEROTROPHY, CROOKED LAKE(IND),   LITTLE CROOKED LAKE(IND).: /ISM,   W71-10079
DEVELOPMENT CORP(MASS), ARTHUR D   LITTLE INC(MASS), CARNEGIE DEPT O   W69-06865
TNAM), CHLORO/ *BLOOM, *VIETNAM,   LITTLE LAKE(HANOI), RED RIVER(VIE   W70-03969
G, DEPTH, SEA WATER, FRES WATER,   LITTORAL, DISSOLVED OXYGEN, DIURN   W71-12870
CULTURES, BAYS, STREAMS, WEEDS,   LITTORAL, WINDS.: /TER, MOVEMENT,   W70-06229
AL AREAS, PHYSIOLOGICAL ECOLOGY,   LITTORAL, ESTUARIES, PHOTOSYNTHES   W70-03325
DIOISOTOPES, INFLOW, CYANOPHYTA,   LITTORAL, NUTRIENTS, MODEL STUDIE   W73-01454
E TAHOE(CALIF-NEV), HYDROPHYTES,   LIVERWORTS, MORPHOMETRY, PACIFAST   W70-00711
RS, NITROGEN, PHOSPHORUS, ALGAE,   LIVESTOCK, WATER POLLUTION EFFECT   W71-10376
S, ALGAE, DETRITUS, PULP WASTES,   LIVESTOCK, IDAHO.: /ES, PHOSPHATE   W70-03501
REBRIFORMIS, MATS, SYNECHOCOCCUS   LIVIDUS, BLUE-GREEN ALGAE, FILAME   W70-05270
NDS), HETEROCYSTS, GAS-VACUOLES,   LLANGORS LAKE(WALES), HALOBACTERI   W70-04369
ONDS), *DEPTH EFFECTS, *HYDRAULIC   LOAD EFFECTS, *ORGANIC LOAD EFFEC   W70-04787
HYDRAULIC LOAD EFFECTS, *ORGANIC   LOAD EFFECTS, *DETENTION TIME EFF   W70-04787
S, FLO/ *ULTRASONICS, *SUSPENDED   LOAD, *FILTRATION, ALGAE, COLLOID   W71-10654
UPPLY, MUNICIPAL WATER, SEDIMENT   LOAD, CONDUCTIVITY, CALCIUM, MAGN   W69-05894
NS, SPECTROPHOTOMETRY, SUSPENDED   LOAD, FLUORESCE: /TEMPERATURE, IO   W70-03334
, *TERTIARY TREATMENT, SUSPENDED   LOAD, PILOT PLANTS, ACTIVATED SLU   W70-09186
HEMICAL OXYGEN DEMAND, SUSPENDED   LOAD, SOLIDS CONTACT PROCESS, SEW   W70-06792
MMISSION, ALGAL BLOOMS, NUTRIENT   LOAD, SYNTHETIC DETERGENTS, FISH   W69-08674
HEMICAL OXYGEN DEMAND, SUSPENDED   LOAD, TURBIDITY, DATA STORAGE AND   W70-09557
HEMICAL OXYGEN DEMAND, SUSPENDED   LOAD, WASTES, SEWAGE, ALGAE, MICR   W70-08627
OGICAL EXTRACTION, GROWTH, SHOCK   LOADING.:   *BIOL   W70-06604
*SITE SELECTION,   LOADING.:   W71-08221
ROGEN ION CONCENTRATION, ORGANIC   LOADING, *WATER REUSE, RESEARCH A   W72-03299
GAE/ *HYDROGEN SULFIDE, *ORGANIC   LOADING, *SULFATES, *SULFIDES, AL   W70-07034
DISSOLVED OXYGEN, ODOR, ORGANIC   LOADING, *WASTE WATER TREATMENT.:   W70-06601
RY, *OXIDATION LAGOONS, *ORGANIC   LOADING, ALGAE, AMMONIA, SEDIMENT   W71-05013
SCREENS, SEDIMENTATION, ORGANIC   LOADING, ACTIVATED SLUDGE, AERATI   W71-08970
LFIDE, *SULFIDES, ALGAE, ORGANIC   LOADING, BIOCHEMICAL OXYGEN DEMAN   W71-11377
ROGEN ION CONCENTRATION, ORGANIC   LOADING, CENTRIFUGATION, RESPIRAT   W72-06022
MOCLINE, CLIMATIC ZONES, ORGANIC   LOADING, DISSOLVED OXYGEN, HYDROG   W72-10573
NTENSITY, SLUDGE, DEPTH, ORGANIC   LOADING, DESIGN CRITERIA, ODORS,   W72-06854
ALGAE, SLUDGE, METHANE, ORGANIC   LOADING, HYDROGEN ION CONCENTRATI   W71-11375
*HOGS, *LAGOONS, ALGAE, ORGANIC   LOADING, WASTE WATER TREATMENT, W   W70-07508
DO, WEST(US), STABILIZATION POND   LOADINGS, ODOR PERSISTENCE.: /ORA   W70-04788
GAE, ANALYTICAL TECHNIQUES, CLAY   LOAM, CLAYS, COLLOIDS, EUTROPHICA   W69-04800
OPHICATION, IMPOUNDMENTS, LAKES,   LOAMS, RESERVOIRS, TEXAS.: / EUTR   W69-04800
US, POCILLOPORA EYDOUXI, PORITES   LOBATA, LEPTASTREA PURPUREA, FUNG   W69-08526
APHY, MICHAELIS-MENTEN EQUATION,   LOCH EWE(SCOTLAND), SCOTLAND, PHA   W70-04811
OSA, LAKE MENDOTA(WIS), ENGLAND,   LONDON(ENGLAND), ESTHWAITE(ENGLAN   W69-10180
ATED GLUCOSE, GONIDIA FORMATION,   LONG ISLAND SOUND(NY), ICELAND, T   W69-09755
IDAL CURRENT, PUGET SOUND(WASH),   LONG ISLAND SOUND, NARRAGANSETT B   W69-10161
NEW HAMPSHIRE, LAKE WINNISQUAM,   LONG POND, CONCORD, BOSTON LOT RE   W69-08674
EMMERMANNI, APHANOCAPSA, BOSMINA   LONGISPINA, ZURICHSEE(SWITZERLAND   W69-10169
LEY POND(CONN), BOSMINA COREGONI   LONGISPINA.: /(SWITZERLAND), LINS   W69-10169
    LOS ANGELES COUNTY(CALIF).:   W70-05645
TED CARBON, *FLOCCULATION, *HEAD   LOSS.: /OAGULATION, ALGAE, ACTIVA   W72-07334
TMENT, *ALGAE, *FILTRATION, HEAD   LOSS, FILTERS, *SCENEDESMUS, FLOC   W72-03703
OLIDS, ALGAE, PILOT PLANTS, HEAD   LOSS, PERMEABILITY, TEMPERATURE,   W72-08686
DS, PHENOL/ BIOCIDES, PRODUCTION   LOSS, QUATERNARY AMMONIUM COMPOUN   W72-12987
, AUTECOLOGY, MORPHOLOGY, EXPORT   LOSSES, CHEMOSYNTHESIS, ABUNDANCE   W70-02780
SEEPAGE   LOSSES, DIURNAL VARIATIONS.:   W69-09723
ATER SUPPLIES, GREEN ALGAE, HEAD   LOSSES, OOCISTUS, FILTRATION MEDI   W70-04199
QUAM, LONG POND, CONCORD, BOSTON   LOT RESERVOIR, LEBANON, WADLEIGH   W69-08674
LEVEL, LIGHT INTENSITY, ECOLOGY,   LOTIC ENVIRONMENT, BIOMASS, FLOW   W70-00265
ON-SITE TESTS, LABORATORY TESTS,   LOTIC ENVIRONMENTS, LENTIC ENVIRO   W72-05453
WEDEN), LAKE EKOLN(SWEDEN), LAKE   LOTSJON(SWEDEN), PERMEASES, CHAMY   W70-04287
SEDIMENTATION RATES, MANGANESE,   LOW-FLOW AUGMENTATION, TENNESSEE   W73-00750
ON(ENGLAND), ESTHWAITE(ENGLAND),   LOWESWATER(ENGLAND), BARNES SOUTH   W69-10180
RRANEAN DRAINAGE, MECHANIC/ LAKE   LUCERNE, FERTILIZED FIELDS, SUBTE   W69-08668
Y, LAKE BORREVANNET NORWAY, LAKE   LUGANO ITALY, LAKE MARIDALSVANNET   W69-07833

CHLORELLA ELLIPSOIDEA, CHLORELLA   LUTEOVIRIDIS, SKELETONEMA COSTATU W70-04809
LL SIZE, EUKARYOTES, MONOCHRYSIS   LUTHERI, NAVICULA PELLICULOSA, GO W69-08278
PHYSIOLOGY, KJELDAHL PROCEDURE,   LUXURY CONSUMPTION, NUTRIENT REMO W69-05868
OSTATS, NUTRIENT LIMITED GROWTH,   LUXURY CONSUMPTION, PROVISIONAL A W70-02779
OSMARIUM, CHROCOCCUS LIMNETICUS,   LYNGBYA LIMNETICA, ANABAENA FLOS   W70-05405
AENA MENDOTAE, ANABAENA HASSALI,   LYNGBYA NOLLEI, GLOCOTRICHIA PISU W69-08283
ENSITY, APHANIZOMENON, MELOSIRA,   LYNGBYA.: /EMENT PROGRAM, ALGAL D W69-10167
SCILLATORIA, NAVICULA, CYMBELLA,   LYNGBYA, FRAGILARIA, OEDOGONIUM,   W70-04371
SPECIES DIVERSITY, LAKE WAUBESA,   LYNGBYA, MICROCYSTIS, APHANIZOMEN W69-01977
IDIUM, APHANIZOMENON FLOS-AQUAE,   LYNGBYA, STORM LAKE(IOWA), CENTER W70-02803
LISM, *PESTICID/ SOUTH CAROLINA,   LYOPHILIZATION, *PESTICIDE METABO W69-02782
ANTA ROSA(CALIF), LYUBLIN(USSR),   LYUBERETS(USSR), MUNICH(GERMANY), W70-05092
A, AUSTRALIA, SANTA ROSA(CALIF),   LYUBLIN(USSR), LYUBERETS(USSR), M W70-05092
STES, INLETS(WATERWAYS), ORGANIC   MA: /ASTES, GARBAGE DUMPS, OIL WA W70-05112
SORPTION, EUTROPHICATION, ALGAE,   MACROAQUATIC PLANTS, FERTILIZATIO W69-06859
CHIA COLI, MYRIOPHYLLUM, LEPOMIS   MACROCHIRUS, PIMEPHALES NOTATUS.: W69-07861
RUBESCE/ *LAKE WASHINGTON(WASH),   MACROFAUNA, GYTTJA, OSCILLATORIA   W70-04253
RACTERISTICS, TERMARY COMPOUNDS,   MACROMOLECULES, ANTIBIOTICS, AGAR W70-04280
HLORELLA PYRENOIDOSA, AUTOTROPH/   MACRONUTRIENTS, MICRONUTRIENTS, C W70-02964
S, ANACHARIS, LAKE WAUBESA(WIS),   MACROPHYTES, ALGAL GROWTH.: /NAJA W70-03310
CROPHYTES, ANACHARIS CANADENSIS,   MACROPHYTES, POTAMOGETON FOLIOSUS W70-02249
*ANTAGONISTIC ACTION, *AQUATIC   MACROPHYTES, ANACHARIS CANADENSIS W70-02249
ADOPHORA BLANKETS, HYDRODICTYON,   MACROPHYTES.: /, IRON SULFATE, CL W69-10170
TY, STURGEON RIVER, FLUORESCEIN,   MACROPHYTES, GROUND SOLIDS, UPTAK W69-09334
, BOMBAY(INDIA), BENARES(INDIA),   MACROPHYTES, CLADOPHORA, OEDOGONI W70-05404
NUTRIENT TYPES, MICROQUANTITIES,   MACROQUANTITIES, ALGAL BLOOMS, LI W70-04765
VISCOSE RAYON PLANT WASTES, *MAN- MADE FIBER PLANTS, NYLON-6, *VISC W70-06925
LENE GLY/ *POLYESTER FIBERS, MAN- MADE FIBERS, METHYL ALCOHOL, ETHY W70-10433
ICOOK(ME), LAKE WINNISQUAM(N H),   MADISON LAKES(WIS), LAKE WASHINGT W70-07261
ILUTION, OSCILLATORIA RUBESCENS,   MADISON LAKES(WIS), CEDAR RIVER(W W70-00270
RESPONSE, FERTILIZATION SURVEY,   MADISON LAKES(WIS), LAKE MENDOTA(W70-02787
HAL EXPOSURE, PHENANTHROQUINONE,   MADISON(WIS), ELODEA, REDOX COMP: W70-02982
MILWAUKEE(WIS), CHICAG/ MANURE,   MADISON(WIS), FLUSHING WATERWAYS, W69-10178
ULATA, COCCOCHLORIS PENIOCYSTIS,   MADISON(WIS), CHICAGO NATURAL HIS W70-00719
OIRS, ANOPHELES QUADRIMACULATUS,   MADISON(WIS), PSOROPHORA CILIATA, W70-06225
AMATH LAKE(ORE),/ *URBAN WASTES,   MADISON(WIS), LAKE ZOAR(CONN), KL W70-04506
ARCH, US DEPARTMENT OF INTERIOR,   MADISON(WIS), EUROPE, WATER POLLU W70-03975
ARCH, US DEPARTMENT OF INTERIOR,   MADISON(WIS), EUROPE, WATER POLLU W70-03975
KEGONSA(WIS), LAKE WAUBESA(WIS),   MADISON(WIS), UNIVERSITY OF WISCO W69-06276
   MADISON(WIS).: W69-03369
S, YAHARA RIVER(WI/ *TRUAX FIELD(MADISON) EFFLUENTS, *MADISON LAKE W70-05113
UCTIVITY, CHLOROPHYLL, PIGMENTS,   MAGN: /RINE ANIMALS, PRIMARY PROD W73-01446
ULLATA, DAPHNIA HYALINA, DAPHNIA   MAGNA, CZECHOSLOVAKIA.: /HNIA CUC W70-00274
UDA, CHLORELLA VULGARIS, DAPHNIA   MAGNA, LIMNAEA STAGNALIS, OUSPENS W71-03056
UDA, CHLORELLA VULGARIS, DAPHNIA   MAGNA, LIMNAEA STAGNALIS, OUSPENS W71-05719
APHNIA GALEATA MENDOTAE, DAPHNIA   MAGNA, SELENASTRUM, RHODOTORULA,   W70-03957
YDROGEN ION CONCENTRATION, IRON,   MAGNESIUM, NITRATES, PHOSPHATES,   W70-03501
ARBONATE, ZINC, COBALT, CALCIUM,   MAGNESIUM, C: /TASSIUM, IRON, BIC W70-03978
OPHYTA, PHOSPHATES, CHLOROPHYLL,   MAGNESIUM, POTASSIUM, TEMPERATURE W70-05409
, NITROGEN, PHOSPHORUS, CALCIUM,   MAGNESIUM, SODIUM, SILICA, DIATOM W68-00856
ION CONCENTRATION, IRON, LAKES,   MAGNESIUM, NUTRIENTS, NUISANCE AL W69-07833
ENT LOAD, CONDUCTIVITY, CALCIUM,   MAGNESIUM, ALKALINITY, SULFATES,   W69-05894
N, NITRATES, POTASSIUM, CALCIUM,   MAGNESIUM, IRON, ORGANIC MATTER,   W70-00719
EN, PHOSPHORUS, SULFUR, CALCIUM,   MAGNESIUM, POTASSIUM, SODIUM, IRO W70-02804
RBIDITY, HYDROGEN SULFIDE, IRON,   MAGNESIUM, COOLING, PROPERTY VALU W70-02251
YLL, HYDROGEN ION CONCENTRATION,   MAGNESIUM, CALCIUM.: /S, CHLOROPH W70-01579
S, ESSENTIAL NUTRIENTS, CALCIUM,   MAGNESIUM, POTASSIUM, HARVESTING   W70-02245
OSCILLATORIA TENUIS, RAM/ PROTON   MAGNETIC RESONANCE SPECTROMETRY,   W69-00387
NICAL EQUIPMENT, PUMPING PLANTS,   MAINE.: /, WATER TREATMENT, MECHA W70-08096
VER(MAINE), COBBOSSEECONTEE LAKE(MAINE), BARGE.: /VER, KENNEBEC RI W70-08096
URA, ASTERIONELLA, TYP/ *AUGUSTA(MAINE), CARLETON POND(MAINE), SYN W70-08096
A, TYPHOID FEVER, KENNEBEC RIVER(MAINE), COBBOSSEECONTEE LAKE(MAIN W70-08096
*AUGUSTA(MAINE), CARLETON POND(MAINE), SYNURA, ASTERIONELLA, TYP W70-08096
IONS, WATER QU/ *EUTROPHICATION,   MAINE, NUTRIENTS, AQUATIC POPULAT W68-00470
NAL VARIA/ LAKE SEBASTICOOK, ME,   MAINE, PERIODIC VARIATIONS, SEASO W68-00481
DISTRIBUTION, ACTIVATED SLUDGE,   MAINTENANCE COSTS, CONSTRUCTION C W70-04786
PERTIES, CONTROL, *OPERATION AND   MAINTENANCE, *OXIDATION LAGOONS,   W70-08392
LAGOONS, ALGAE, ADMINISTRATION,   MAINTENANCE, *WASTE WATER TREATME W71-08970
LITY, TEMPERATURE, OPERATION AND   MAINTENANCE, DESIGN CRITERIA, *WA W72-08686
BASTICOOK, LAKE WASHINGTON, LAKE   MALAREN, LAKE ANNECY, LAKE VANERN W70-03964
SCHRODERIA, THIMET, TRI/ SEVIN,   MALATHION, CHLORELLA PYRENOIDOSA, W70-08097
*ALGAE TOXICITY, CARBON-14, DDT,   MALATHION, SEVIN.: W70-02198
A, PESTICIDE DEGRADATION, SEVIN,   MALATHION, ACETONE, ETHANOL.: /OS W70-02968
*AFRICA,   MALAWI.: W70-02646
NON, SYNURA, ANABAENA, MELOSIRA,   MALLOMONAS, ULOTHRIX, DAVID MONIE W69-10157
RS, *BIOASSAY, TOXICITY, AQUATIC   MAMMALS, ALGAE, FISH, DIATOMS, FU W71-10786
ETHYLENE GLY/ *POLYESTER FIBERS,   MAN-MADE FIBERS, METHYL ALCOHOL,   W70-10433
IRY INDUSTRY, */ *ANIMAL WASTES,   MANAGEMENT, *FARM MANAGEMENT, *DA W70-07491
ANIMAL WASTES, MANAGEMENT, *FARM   MANAGEMENT, *DAIRY INDUSTRY, *ALG W70-07491
, PHOSPHORUS, AGRICULTURE, WATER   MANAGEMENT(APPLIED), GEOCHEMISTRY W71-03048
*MATHEMATICAL MODELS, *WATERSHED   MANAGEMENT, GEOCHEMISTRY, LAKES,   W71-09012
S COMPOUNDS, WATER SUPPLY, WATER   MANAGEMENT.: /OSPHORUS, PHOSPHORU W72-01867
, / *LAKE MINNETONKA(MINN), LAKE   MANAGEMENT PROGRAM, ALGAL DENSITY W69-10167
FLUENT DIVERSION, *WATER QUALITY   MANAGEMENT, *LAKE WASHINGTON, *SE W69-09349
YNTHESIS, TRIBUTARIES, WATERSHED   MANAGEMENT, SEDIMENTS.: /, PHOTOS W70-04268
OLLUTION TREATMENT, ALGAE, WATER   MANAGEMENT(APPLIED), LIMNOLOGY.: / W70-04721
NUTRIENTS, *LAKES, *TROUT, *FISH   MANAGEMENT, FERTILIZATION, ALGAE,   W70-05399
EAR LAKE(CALIF), NUTRI/ *FISHERY   MANAGEMENT, *CHAOBORID MIDGES, CL W70-05428
WATER QUALITY CONTROL, *MI/ FISH   MANAGEMENT, STREAM IMPROVEMENT, * W70-05428
E, FISH, ANIMALS, ALGAE, CUTTING   MANAGEMENT, POTABLE WATER, HERBIC W70-05263

ON PATTERNS, MEMBRANE PROCESSES,
STUARIES, PHYSIOLOGICAL ECOLOGY,
Y CONTROL, *ECOLOGY, *RESERVOIR,
CUS BINDERANUS, CLADOCERA, WATER
ATION, DRAIN, SAN JOAQUIN VALLEY
UM/ *YELLOWSTONE PARK, *ICELAND,
STIS AERUGINOSA, CRYPTOMONADIDA,
ARCODINA, MASTIGOPHORA, HOLOZOIC
OLOZOIC MASTIGOPHORA, HOLOPHYTIC
RA, HOLOPHYTIC MASTI/ SARCODINA,
ICULATE MATTER, BLUE FLUORESCING
LLS, CONCENTRATION, COMPOSITION,
IC BACTERIA, SUBLITTORAL, YELLOW
IRCONIUM-95-NIOBIUM-95, PRE-BOMB
S, *TEST PROCEDURES, ANTIFOULING
OW, MIXING, FLOW, MODEL STUDIES,
CARBOHYDRATES, LIPIDS, KINETICS,
, HETEROTROPHIC.:
S, INFLUENT SEWAGE, EQUILIBRIUM,
AQUATIC PLANTS, PALEOLIMNOLOGY,
AQUATIC PLANTS, PALEOLIMNOLOGY,
UATIC LIFE, SESTON, SCENEDESMUS,
OUNTAIN MEADOW, NITROGEN-15, SOD
UIS, OSCILLATORIA TEREBRIFORMIS,
, HORIZONTAL, BLOOMS, GELATINOUS
UOUS TREES, *DECOMPOSING ORGANIC
ECHARGE, *ION EXCHANGE, *ORGANIC
EN FIXATION, IRRIGATION, ORGANIC
OR, *ALGAE, *DECOMPOSING ORGANIC
, ZOOPLANKTON, ALGAE, / *ORGANIC
ADIOACTIVITY T/ *LAKES, *ORGANIC
UTILIZATION, DECOMPOSING ORGANIC
UNDS, OXYGEN, SILICATES, ORGANIC
YGEN, CHEMICAL ANALYSIS, ORGANIC
EATMENT, WATER QUALITY, *ORGANIC
, ABSORPTION, SEDIMENTS, ORGANIC
*MARL LAKES, DISSOLVED ORGANIC
BOGS, PHENOLS, PIGMENTS, ORGANIC
UALITY, WATER POLLUTION, ORGANIC
ON COLUMN TECHNIQUE, PARTICULATE
C ACID, CHEMOTROPHY, PARTICULATE
DNESS(WATER), CHLORIDES, ORGANIC
ION CONTROL, DECOMPOSING ORGANIC
LANT GROWTH, DECOMPOSING ORGANIC
SAMPLING, PHYTOPLANKTON, ORGANIC
HARDNESS, TIDAL WATERS, ORGANIC
OLOGY, CHLORELLA, ALGAE, ORGANIC
ER, *SESTON, *SEDIMENTS, ORGANIC
ATION, PIGMENTS, OCEANS, ORGANIC
ARINE ALGAE, *LITTORAL, *ORGANIC
NEW JERSEY, TEMPERATURE, ORGANIC
S, DILUTION, DECOMPOSING ORGANIC
TALS, INHIBITORS, SILTS, ORGANIC
MEAD, LAS VEG/ DISSOLVED ORGANIC
LID WASTES, DECOMPOSTING ORGANIC
QUATIC BACTERIA, *ALGAE, ORGANIC
TS, DIATOMS, PHOSPHORUS, ORGANIC
ALCIUM, MAGNESIUM, IRON, ORGANIC
*PRIMARY PRODUCTIVITY, *ORGANIC
YTOPLANKTON, DECOMPOSING ORGANIC
INETICS, PHOTOSYNTHESIS, ORGANIC
ROWTH FACTOR/ *DISSOLVED ORGANIC
RE, NITROGEN, NUTRIENTS, ORGANIC
, NUTRIENT REQUIREMENTS, ORGANIC
EUGLENA, EUTROPHICATION, ORGANIC
DIOACTIVITY, INTERFACES, ORGANIC
HOGENIC BACTERIA, ALGAE, ORGANIC
TOMS, ECOLOGY, SALINITY, ORGANIC
OSYNTHESIS, GROWTH RATE, ORGANIC
TER POLLUTION, DRAINAGE, ORGANIC
*PHYTOPLANKTON, *LAKES, ORGANIC
AGRICULTURAL WATERSHEDS, ORGANIC
ON RADIOISOTOPES, LAKES, ORGANIC
AL TECHNIQUES, SAMPLING, ORGANIC
CYANOPHYTA, *DECOMPOSING ORGANIC
YTA, VITAMINS, PEPTIDES, ORGANIC
, PHYSIOLOGICAL ECOLOGY, ORGANIC
ANA/ *CHEMICAL ANALYSIS, ORGANIC
PHOSPHATES, ECOSYSTEMS, ORGANIC
S, BACTERIA, DECOMPOSING ORGANIC
PROTOZOA, *ENVIRONMENT, *ORGANIC
ER, RAISIN RIVER, DETROIT RIVER,
-BURK EQUATION, UPTAKE VELOCITY,
EDEN), LINEWEAVER-BURK EQUATION,
ICAL UPTAKE, *MICROBIAL ECOLOGY,
THE MINIMUM, VOISIN'S LAW OF THE
SNAILS, AGE, TURBIDITY, DIPTERA,
DWORMS, AQUATIC BACTERIA, ALGAE,
ION, WILD RICE, SILTS, PLANKTON,
CUE LAKE(FLORIDA),

MASSACHUSETTS.: /LOSE, DISTRIBUTI   W69-00976
MASSACHUSETTS, CHLOROPHYTA, DINOF   W70-05272
MASSACHUSETTS, POTABLE WATER, WAT   W70-05264
MASSES, U S PUBLIC HEALTH SERVICE   W70-00263
MASTER DRAIN, SAN FRANCISCO BAY-D   W71-09577
MASTIGOCLADUS LAMINOSUS, PHORMIDI   W70-05270
MASTIGOPHORA, CRYPTOPHYCEAE, CHRY   W70-05112
MASTIGOPHORA, HOLOPHYTIC MASTIGOP   W72-01817
MASTIGOPHORA, CILIATA.: /PHORA, H   W72-01817
MASTIGOPHORA, HOLOZOIC MASTIGOPHO   W72-01817
MATE: /LON COLUMN TECHNIQUE, PART   W70-01074
MATERIAL BALANCES, ORGANISMS.: /E   W70-06600
MATERIAL, DU: /A HARVEYI, EPIPHYT   W70-01073
MATERIAL, RUTHENIUM-106-RHODIUM-1   W69-08524
MATERIALS, TESTING, BIODEGRADATIO   W71-07339
MATHEMATICAL MODELS, GREAT LAKES,   W71-05878
MATHEMATICAL MODELS, NEUTRALIZATI   W71-13413
MATHEMATICAL STUDIES, AUTOTROPHIC   W70-05750
MATHEMATICAL MODELS, ON-SITE DATA   W68-00858
MATHEMATICAL MODELS, BACTERIA, PH   W70-03975
MATHEMATICAL MODELS, BACTERIA, PH   W70-03975
MATHEMATICAL STUDIES, DATA COLLEC   W73-01446
MATS, NON-LEGUMES, *NITROGEN FIXI   W72-08110
MATS, SYNECHOCOCCUS LIVIDUS, BLUE   W70-05270
MATTER.: /OTA(WIS), PELAGIC AREAS   W70-06229
MATTER, *ALGAE, *PHENOLS.: /DECID   W72-03475
MATTER, *FILTRATION, DIATOMACEOUS   W72-04318
MATTER, *GRASSLANDS, MOUNTAINS, *   W72-08110
MATTER, *POTASSIUM COMPOUNDS, *AC   W72-04242
MATTER, *PHYTOPLANKTON, LIMNOLOGY   W70-02510
MATTER, *PHYSIOLOGICAL ECOLOGY, R   W70-04194
MATTER, *STREAM POLLUTION.: /TER   W68-00483
MATTER, AIR TEMPERATURE, AEROBIC   W69-07096
MATTER, ALGAE, IRON, HY: /LVED OX   W70-05760
MATTER, ALGAE, COPPER SULFATE, *C   W72-04495
MATTER, ALGAE, DIATOMS, CYANOPHYT   W72-00845
MATTER, ARCTIC ALGAE, AUTOTROPHIC   W71-10079
MATTER, BIOMASS, BENTHIC FLORA, P   W70-01074
MATTER, BIBLIOGRAPHIES.: /WATER Q   W70-07027
MATTER, BLUE FLUORESCING MATE: /L   W70-01074
MATTER, BLELHAM TARN(ENGLAND), CH   W70-02504
MATTER, CARBON, WATER POLLUTION E   W71-04206
MATTER, CARBON DIOXIDE.: / POLLUT   W72-01788
MATTER, CHEMICAL ANALYSIS, PROTEI   W72-01812
MATTER, CHLOROPHYLL, DYNAMICS, TE   W70-02780
MATTER, CONDUCTIVITY, TEMPERATURE   W73-00748
MATTER, CULTURES, CARBON, METABOL   W70-02804
MATTER, CURRENTS(WATER), WATER PO   W70-03501
MATTER, DIURNAL, LATITUDINAL STUD   W70-04809
MATTER, EXUDATION, TIDES, PHOTOSY   W70-01073
MATTER, HARDNESS, CARBON DIOXIDE,   W69-10157
MATTER, INSECTS, MITES, BACTERIS,   W71-00940
MATTER, IRON OXIDES, BACTERIA, AL   W70-02294
MATTER, LAS VEGAS, NEVADA, *LAKE   W72-02817
MATTER, LAGOONS, CATTLE.: /GE, SO   W71-05742
MATTER, LAKES, DIFFUSION, CARBON   W70-04284
MATTER, NITROGEN, CHLOROPHYLL.: /   W70-05663
MATTER, NITROGEN FIXATION, VOLUME   W70-00719
MATTER, NUTRIENTS, LAKES, WATER P   W71-10079
MATTER, OKLAHOMA, RESERVOIRS, SAM   W70-04001
MATTER, ORGANIC COMPOUNDS, METABO   W70-04287
MATTER, ORGANIC DEBRIS, ORGANIC G   W70-02510
MATTER, P: /CENTRATION, TEMPERATU   W70-00265
MATTER, PHOTOSYNTHESIS, TEMPERATU   W69-06865
MATTER, PHYTOPLANKTON, PRIMARY PR   W69-06273
MATTER, RADIOISOTOPES, GAMMA RAYS   W69-08272
MATTER, RADIOACTIVITY, PESTICIDES   W72-13601
MATTER, RIVERS, ESTUARINE ENVIRON   W73-01446
MATTER, SCENEDESMUS, CHLAMYDOMONA   W70-05381
MATTER, SEWAGE, PHYTOPLANKTON, PR   W70-05404
MATTER, SEASONAL, ALGAE, PRIMARY   W70-04290
MATTER, SEDIMENTS, FERTILIZERS, W   W70-04193
MATTER, SEA WATER, PRIMARY PRODUC   W70-02504
MATTER, TOXICITY, PHENOLS, ODOR,   W72-03299
MATTER, TOXICITY, DISSOLVED OXYGE   W72-01797
MATTER, TRACE ELEMENTS, LIGHT INT   W70-04369
MATTER, WATER POLLUTION EFFECTS,   W70-04811
MATTER, WATER POLLUTION EFFECTS,   W70-05109
MATTER, WATER POLLUTION, THERMAL   W70-03309
MATTER, WASTE WATER(POLLUTION).: /   W70-00664
MATTER, WATER QUALITY, TEMPERATUR   W71-03031
MAUMEE RIVER.: / RIVER, HURON RIV   W69-01445
MAXIMUM UPTAKE VELOCITY, SUBSTRAT   W70-05109
MAXIMUM UPTAKE VELOCITY, MICHAELI   W70-04287
MAXIMUM UPTAKE VELOCITY, ACTIVE T   W70-04194
MAXIMUM.: /OOMS, LIEBIG'S LAW OF   W70-04765
MAYFL: /A, CHRYSOPHYTA, GRAZING,   W71-00668
MAYFLIES, CADDISFLIES.: /MS, BLOO   W70-00475
MAYFLIES, FAUNA, FISH, OLIGOCHAET   W70-01943
MCCLOUD LAKE(FLORIDA).:   W72-01993

473

ONTENT, LAKE ERKEN(SWEDEN), LAKE | MENDOTA(WIS), NET PLANKTON, CHLOR | W70-03959
MESOTROPHY, LAKE NORRVIKEN, LAKE | MENDOTA, LAKE FURES, LAKE SEBASTI | W70-03964
RY, WISCONSIN, LAKE MONONA, LAKE | MENDOTA, LAKE WAUBESA, LAKE KEGON | W69-09349
, DAPHNIA PULEX, DAPHNIA GALEATA | MENDOTAE, DAPHNIA MAGNA, SELENAST | W70-03957
M, ANABAENA FLOS AQUAE, ANABAENA | MENDOTAE, ANABAENA HASSALI, LYNGB | W69-08283
ADAPTATION, *SPECIES, CYCLOTELLA | MENEGHINIANA, LIGHT-SATURATED RAT | W70-05381
ASTERIONELLA FORMOSA, CYCLOTELLA | MENEGHINIANA, PHOTIC ZONE, HELSIN | W70-04809
N-14, AUTORADIOGRAPHY, MICHAELIS- | MENTEN EQUATION, LOCH EWE(SCOTLAN | W70-04811
BON UPTAKE TECHNIQUES, MICHAELIS- | MENTEN EQUATION, RELATIVE HETEROT | W70-05109
YME KINETICS, GLUCOSE, MECHAELIS- | MENTEN EQUATION, ACETATE, GLYCOLL | W70-04284
E TECHNIQUES, GLUCOSE, MICHAELIS- | MENTEN EQUATION, SWEDEN, LAKE ERK | W70-04287
ER(CA/ SAN JOAQUIN RIVER(CALIF), | MERCED RIVER(CALIF), TUOLUMNE RIV | W70-02777
E, RECYCLING / PETROLEUM WASTES, | MERCENARIA MERCENARIA, AQUACULTUR | W70-09905
G / PETROLEUM WASTES, MERCENARIA | MERCENARIA, AQUACULTURE, RECYCLIN | W70-09905
ORT(N C), CRASSOSTREA VIRGINICA, | MERCENARIA MERCENARIA, PANOPEUS H | W69-08267
RASSOSTREA VIRGINICA, MERCENARIA | MERCENARIA, PANOPEUS HERBSTII.: / | W69-08267
CES, BACILLARIOPHYTA, GLOECAPSA, | MERISMOPEDIA, MICROCYSTIS, ANKIST | W70-03969
CULTURAL RUNOFF, ARKANSAS RIVER, | MERRIMACK RIVER, RED RIVER, SAVAN | W70-04507
LIS, BACILLUS MYCOIDES, BACILLUS | MESENTERICUS, DIATOMEA, CLADOPHOR | W69-07838
UMP COUNT, DESMIDS, FLAGELLATES, | MESOPLANKTON, MICROPLANKTON, PSEU | W70-05389
TROPHIC BACTERIA, DESULFOVIBRIO, | MESOTHERMY, MELOSIRA, SYNEDRA, RH | W70-05760
MENDOTA, LAKE FURES, LA/ CANADA, | MESOTROPHY, LAKE NORRVIKEN, LAKE | W70-03964
EPTOMYCIN, MILLIPORE FILTRATION, | MESOTROPHY.: /, TETRACYCLINE, STR | W69-10154
DRICA, GL/ BOTRYOCOCCUS BRAUNII, | METABOLIC PATTERN, ANABAENA CYLIN | W70-02965
PLANTS, MASS CULTURE, MEDICINE, | METABOLIC PATTERNS, MICROALGAE, M | W69-07832
GEN, BACTERIA, AEROBIC BACTERIA, | METABOLISM, CANALS, SANDS, CHLORI | W69-07838
E ALGAE, *CULTURES, PSEUDOMONAS, | METABOLISM, KINETICS, NITROGEN CO | W70-04280
GANIC MATTER, ORGANIC COMPOUNDS, | METABOLISM, BIOINDICATORS, BIOASS | W70-04287
LINA, LYOPHILIZATION, *PESTICIDE | METABOLISM, *PESTICIDE ACCUMULATI | W69-02782
RGANIC MATTER, CULTURES, CARBON, | METABOLISM, NITROGEN, PHOSPHORUS, | W70-02804
ALGAE, *TOXICITY, *INSECTICIDES, | METABOLISM, RADIOACTIVITY TECHNIQ | W70-02198
IODINE, PHOTO REACTIVITY, ALGAL | METABOLISM.: /LL RATES, *BROMINE, | W70-02363
, GROWTH STAGES, DIATOMS, ALGAE, | METABOLISM, ISOPODS, AMPHIPODA, F | W69-09742
, FISH, DIATOMS, CARP, PLANKTON, | METABOLISM, BIOLOGICAL COMMUNITIE | W70-00274
ALGAE, DIATOMS, TRACE ELEMENTS, | METABOLISM.: /SSIUM, CLAMS, SIZE, | W70-00708
N, PRIMARY PRODUCTIVITY, OXYGEN, | METABOLISM, WIND VELOCITY, HUMIDI | W70-01073
URNAL DISTRIBUTION, TEMPERATURE, | METABOLISM, *WASTE WATER TREATMEN | W70-06604
ASTE STORAGE, PONDS, HYDROLYSIS, | METABOLISM, CHEMICAL REACTIONS, C | W70-08097
NE METABOLISM./ *LINDANE, *ALGAE | METABOLISM, DETOXIFICATION, LINDA | W70-08652
ABOLISM, DETOXIFICATION, LINDANE | METABOLISM.: /LINDANE, *ALGAE MET | W70-08652
DE RESIDUES, ALGAE, GREAT LAKES, | METABOLISM, *CHLORELLA, *CHLAMYDO | W70-08652
A, *ACTIVATED SLUDGE, *BACTERIA, | METABOLISM, GROWTH RATES, MICROOR | W72-01817
R WASTES, ABSORPTION, CHELATION, | METABOLISM, AQUARIA, WATER POLLUT | W71-09005
PROPERTIES, BACTERIA, BIOASSAY, | METABOLISM, ABSORPTION, COMPETITI | W71-10079
ATTER, ARCTIC ALGAE, AUTOTROPHIC | METABOLISM, HETEROTROPHY, CROOKED | W71-10079
, CALCIUM CHLORIDE, TEMPERATURE, | METABOLISM, EPIPHYTOLOGY, ON-SITE | W72-05453
ULAR PRODUCTS, HORMONES, AUXINS, | METABOLITES, INHIBITION, SILICON, | W69-04801
 | METAL TOXICITY.: | W70-10437
ER, SUCROSE-FLOTATION TECHNIQUE, | METALIMNION, BATHYTHERMOGRAPH, AR | W70-04253
RTHER(AUSTRIA)), VERTICAL, CHALK, | METALIMNION, OSCILLATORIA RUBESCE | W70-05405
UNDS, PHENOLIC COMPOUNDS, ORGANO- | METALLIC COMPOUNDS, AMINES, ENGLA | W72-12987
CTROPHOTOMETRY, NUTRIENTS, HEAVY | METALS.: /EFFECTS, CHELATION, SPE | W71-11561
ENT FACILITIES, TEST PROCEDURES, | METALS.: /BACTERIA, ALGAE, TREATM | W71-06737
ENT FACILITIES, TEST PROCEDURES, | METALS.: /BACTERIA, ALGAE, TREATM | W70-01519
EN, PITTING(CORROSION), RUSTING, | METALS, INHIBITORS, SILTS, ORGANI | W70-02294
, CHLORIDES, SALINE LAKES, HEAVY | METALS, INDUSTRIAL WASTES, LAKES, | W69-02959
STAL PLAIN, INORGANIC COMPOUNDS, | METALS, PLANKTON, AQUATIC LIFE, Z | W69-03695
ON, OLIGOTROPHY, VITAMINS, HEAVY | METALS, TOXICITY, DIATOMS, MARL.: | W71-10079
PHORUS, SODIUM, POTASSIUM, 'TRACE | METALS, VITAMINS, WATER POLLUTION | W70-04503
YSTIS NIDULANS, OLIGOPHOSPHATES, | METAPHOSPHATES, PULSE-LABELING, P | W71-03033
, THIO/ *MINERAL TRANSFORMATION, | METAZOA, RADIOLARIA, FORAMINIFERA | W69-07428
LAKES, ALGAE, ENTERIC BACTERIA, | METHANE BACTERIA, CHEMICAL ANALYS | W70-03312
IC DIGESTION, *MUNICIPAL WASTES, | METHANE BACTERIA, ACID BACTERIA, | W72-06854
HLOROPLASTS, INFRARE/ CHLOROFORM- | METHANE EXTRACTS, LIPID CARBON, C | W70-05651
BOD), EVAPORATION, FERMENTATION, | METHANE, ALGAL CONTROL, BACTERIA, | W70-04787
WASTES, POPULATION, INDUSTRIES, | METHANE, ALKALINITY, HYDROGEN ION | W70-02803
TROSCOPY, COLORIMETRY, SAMPLING, | METHANE, CARBON DIOXIDE, DIFFUSIO | W71-11377
MS, WINDS, NITROGEN, PHOSPHATES, | METHANE, MUD, OXYGEN, ANAEROBIC C | W70-05761
EROBIC DIGESTION, ALGAE, SLUDGE, | METHANE, ORGANIC LOADING, HYDROGE | W71-11375
ON, *DEGRADATION(DECOMPOSITION), | METHANE, WATER POLLUTION EFFECTS. | W71-13413
LEXES, RIBONUCLEIC ACID, ORCINOL | METHOD.: /, ENZYME-SUBSTRATE COMP | W69-10160
CHLOROPHYLL CONTENT, CENTRIFUGE | METHOD, CALCULATED VOLUME METHOD, | W70-03959
RIFUGE METHOD, CALCULATED VOLUME | METHOD, CYCLOTELLA COMENSIS, RHOD | W70-03959
NOBRYON, FRAGILARIA, RADIOCARBON | METHOD, FINGER LAKES(NY), RODHE'S | W70-01579
S, ALGAL GROWTH, SEDGWICK-RAFTER | METHOD, TREATMENT TIMING, ALGAE C | W69-10157
TRIAL WASTES, THERMAL POLLUTION, | METHODOLOGY, INSTRUMENTATION, SEA | W70-10437
ORY TESTS, *ISOLATION, *CONTROL, | METHODOLOGY, CULTURES, TESTING, C | W70-04469
DISC, VISIBILITY TESTS, CONTROL | METHODS.: SECCHI | W68-00488
COMPOUNDS, HABITATS, STATISTICAL | METHODS.: /TS, SEAWATER, ORGANIC | W69-09755
ZOOPLANKTON, SNAILS, APPLICATION | METHODS.: /NUISANCE ALGAE, FISH, | W70-03310
EASONS, ACETYLENE R/ *ANALYTICAL | METHODS, *BIOCHEMISTRY METHODS, S | W70-03429
*S/ *PHYTOPLANKTON, *STATISTICAL | METHODS, *ZOOPLANKTON, *BIOMASS, | W73-01446
PPER DOSAGE, *COPPER APPLICATION | METHODS, ALGAE IDENTIFICATION, AL | W69-10157
NALYTICAL METHODS, *BIOCHEMISTRY | METHODS, SEASONS, ACETYLENE REDUC | W70-03429
LYESTER FIBERS, MAN-MADE FIBERS, | METHYL ALCOHOL, ETHYLENE GLYCOL, | W70-10433
INDIA, | METHYL ORANGE, ALGAL BLOOMS.: | W70-05550
UCOCYTES, *MUSCULAR EXERTION, *2- | METHYL-5-VINYLPYRIDINE, *GLUE GRE | W70-01788
FECTS, APHANIZOMENON FLOS AQUAE, | METOLIUS RIVER(ORE), COMMUNITY ST | W70-03978
), SEATTLE(WASH), REVENUE BONDS, | METRO ACT(SEATTLE).: /INGTON(WASH | W70-04455
, *SORPTION, ABSORPTION, GULF OF | MEXICO, ATLANTIC OCEAN, RADIOACTI | W69-08524

ALGAL BLOOMS, ALGAE NUTRITION, MICROCYSTIS AERUGINOSA.: W70-05565
ETEROCYSTS, ANABAENA VARIABILIS, MICROCYSTIS AERUGINOSA, APHANIZOM W70-05091
HOSPHATE FERTILIZERS, BLUEGILLS, MICROCYSTIS SCUMS, PITHIOPHORA AL W70-05417
LADOPHORA, OEDOGONIA, SPIROGYRA, MICROCYSTIS AERUGINOSA, ANABAENOP W70-05404
E/ *PURIFICATION, CULTURE MEDIA, MICROCYSTIS AERUGINOSA, APHANIZOM W70-00719
NS, PHYCOLOGISTS, VETERINARIANS, MICROCYSTIS AERUGINOSA, ENDOTOXIN W70-00273
ATA, D/ HYDROXYLAMINE, ANABAENA, MICROCYSTIS, GLOEOTRICHIA ECHINUL W70-02803
LGAE, COPPER SULFATE, EFFLUENTS, MICROCYSTIS AERUGINOSA.: /ORUS, A W70-02787
EN/ *2,3-DICHLORONAPHTHOQUINONE, MICROCYSTIS, APHANIZOMENON, ANABA W70-03310
URE, 2,3-DICHLORONAPHTHOQUINONE, MICROCYSTIS AERUGINOSA, APHANIZOM W70-02982
ARNALDIA PLUMOSA, STIGEOCLONIUM, MICROCYSTIS, STIGEOCLONIUM TENUE, W70-02245
ICROCYSTIS, STIGEOCLONIUM TENUE, MICROCYSTIS AERUGINOSA, NOSTOC, N W70-02245
TOMS FAUNA, MARGINAL VEGETATION, MICROCYSTIS, ANABAENA, EUCYCLOPS, W70-06975
DISCUS, ANABAENA, APHANIZOMENON, MICROCYSTIS, POTAMOGETON, POLYGON W72-04761
ONS, OXYGENATION, STANDING CROP, MICROENVIRONMENT, AQUATIC LIFE, E W69-03611
REGIONS, KENTUCKY, ACID STREAMS, MICROFLORA.: /HIPS, GEOGRAPHICAL W68-00891
UCKY.:   ACID WASTE MICROFLORA, MCCREARY COUNTY, KENT W69-00096
LADOCERANS, SPECIES COMPOSITION, MICROFOSSILS, COMPETITIVE EXCLUSI W70-05663
MENTOUS BLUE-GREEN ALGAE, OCULAR MICROMETER, SUCROSE-FLOTATION TEC W70-04253
IS, GLOEOTRICHIA, APHANIZOMENON, MICRONUTRIENTS, DINOPHYCEAE, CRYP W70-04381
 ALGAL PHYSIOLOGY, HETEROTROPHY, MICRONUTRIENTS, BADFISH RIVER(WIS W70-04506
METABOLIC PATTERNS, MICROALGAE, MICRONUTRIENTS, PHYCOLOGY, SPACE W69-07832
DOSA, AUTOTROPH/ MACRONUTRIENTS, MICRONUTRIENTS, CHLORELLA PYRENOI W70-02964
ROL, *PHOSPHORUS CONCENTRATIONS, MICRONUTRIENTS, AQUATIC ECOSYSTEM W69-09340
CTERI/ *ALGAE, *ECOLOGY, *MARINE MICROORGANISMS, PHYTOPLANKTON, BA W70-01068
TOLOGY, MARINE ALGAE, E/ *MARINE MICROORGANISMS, *HABITATS, EPIPHY W69-10161
E, AQUATIC MICROBIOLOGY, AQUATIC MICROORGANISMS, AQUATIC PLANTS, A W70-02792
, *WASTE WATER TREATMENT, LIGHT, MICROORGANISMS, PLANTS.: /ABOLISM W70-06604
, LETHAL LIMIT, ELECTROPHORESIS, MICROORGANISMS, ELECTRIC CURRENTS W71-03775
SYSTEMS, MICROORGANISMS, AQUATIC MICROORGANISMS, ENZYMES.: /S, ECO W71-02479
Y, LIMITING FACTORS, ECOSYSTEMS, MICROORGANISMS, AQUATIC MICROORGA W71-02479
MS, PHOTOSYNTHESIS, RESPIRATION, MICROORGANISMS, LIGHT INTENSITY, W71-00668
 HEATED WATER, PROTEINS, AQUATIC MICROORGANISMS, TEMPERATURE, ACTI W70-09905
DATION, ALGAE, BACTERIA, AQUATIC MICROORGANISMS, SAMPLING, WATER A W71-00221
, *PHYTOPLANKTON, *LAKES, ALGAE, MICROORGANISMS, BIOMASS, DIATOMS, W70-08654
TE, RESERVOIRS, WATER POLLUTION, MICROORGANISMS, HISTORY, TASTE, O W70-08096
DED LOAD, WASTES, SEWAGE, ALGAE, MICROORGANISMS, PHOSPHATES, HYDRO W70-08627
*SILTS, SANDS, ALGAE, PLANKTON, MICROORGANISMS, TURBIDITY, ANALYT W70-08628
ED SLUDGE, STORAGE, TEMPERATURE, MICROORGANISMS, NITROGEN COMPOUND W71-13335
NITROGEN, NUTRIENTS, FILTRATION, MICROORGANISMS, ANALYTICAL TECHNI W71-13313
TEMATICS, *SANITARY ENGINEERING, MICROORGANISMS, NUISANCE ALGAE, S W72-01798
TERIA, METABOLISM, GROWTH RATES, MICROORGANISMS, ALGAE, FUNGI, LAB W72-01817
LORIDA, *EUTROPHICATION, AQUATIC MICROORGANISMS, TROPHIC LEVEL.: / W72-01993
GY, SUSPEND/ *INDICATOR SPECIES, MICROORGANISMS, FRESH WATER BIOLO W71-10786
YTOLOGICAL STUDIES, EVAPORATION, MICROORGANISMS, CALIFORNIA, *PHOT W71-05267
SM, CANALS, SANDS, CHLORINATION, MICROORGANISMS, CHEMICAL ANALYSIS W69-07838
REQUIREMENTS, *DISTRIB/ *MARINE MICROORGANISMS, *ALGAE, *NUTRIENT W70-03952
*ANALYTICAL TECHNIQUES, *MARINE MICROORGANISMS, PHYSIOLOGICAL ECO W70-04811
BAC/ *ACID MINE WATER, *AQUATIC MICROORGANISMS, *ALGAE, *AQUATIC W69-00096
IC LIFE, AQUATIC MICROORGANISMS, MICROORGANISMS, SESTON, INVERTEBR W69-04276
L EFFECTS, AQUATIC LIFE, AQUATIC MICROORGANISMS, MICROORGANISMS, S W69-04276
ICROBIOLOGY, *ESTUARIES, *MARINE MICROORGANISMS, MARINE ALGAE, MAR W69-03752
EATION, WAT/ *NUTRIENTS, *MARINE MICROORGANISMS, *POLLUTANTS, RECR W69-04276
MENT, AQUATIC LIFE, ENVIRONMENT, MICROORGANISMS, WATER CHEMISTRY, W69-03611
T/ ALGAE, EUTROPHICATION, LAKES, MICROORGANISMS, NITROGEN, NUTRIEN W68-00481
R, *GRASSLANDS, MOUNTAINS, *SOIL MICROORGANISMS, SOIL ALGAE, TURF, W72-08110
MIDS, FLAGELLATES, MESOPLANKTON, MICROPLANKTON, PSEUDOPLANKTON, CR W70-05389
ARBON BIOASSAY, *NUTRIENT TYPES, MICROQUANTITIES, MACROQUANTITIES, W70-04765
WASTE EF/ *BIOLOGIC 'DISPOSERS', MICROSCOPIC ORGANISMS, PAPERMILL W71-00940
ACTION, MILLIPORE FILTERS, PHASE MICROSCOPY, STAINING, STRIPPING F W69-10163
 MICROBIAL ECOLOGY, FLUORESCENCE MICROSCOPY, ACETATE, CARBON-14, A W70-04811
ICAL PROPERTIES, CENTRIFUGATION, MICROSCOPY, NETS, NANNOPLANKTON, W70-05389
URES, HYDROGEN SULFIDE, ELECTRON MICROSCOPY, BIOCHEMISTRY, VITAMIN W70-03952
GEN FIXATION, NITRATES, ELECTRON MICROSCOPY, AMMONIA, SPECTROMETER W70-04369
A, CYTOLOGICAL STUDIES, ELECTRON MICROSCOPY, WASTE TREATMENT, LAKE W70-03973
ARTERIA, GONIUM, ANKISTRODESMUS, MICROSPORA, PEDIASTRUM, SCENEDESM W70-03974
RIOPHYLLUM, POTAMOGETON CRISPUS, MICROSPORA WITTROCKII, SPIROGYRA, W70-00711
CRONUTRIENTS, AQUATIC ECOSYSTEM, MIDGE LARVAE, LAKE STRATIFICATION W69-09340
 *FISHERY MANAGEMENT, *CHAOBORID MIDGES, CLEAR LAKE(CALIF), NUTRIE W70-05428
HESIS, EUTROPHICATION, TOXICITY, MIDGES, MOSQUITOES, SAMPLING, CON W70-06225
FERTILIZATION, LAKES, LIMNOLOGY, MIDGES, NITRATES, NITROGEN, NITRO W69-04805
PER SULFATE, FISHES, HERBICIDES, MIDGES, PESTICIDES.: /OLYSIS, COP W70-02982
LIES, FAUNA, FISH, OLIGOCHAETES, MIDGES, SNAILS, PHOSPHORUS, NUTRI W70-01943
HOLOGY, ECOLOGICAL DISTRIBUTION, MIDGES, WATER POLLUTION EFFECTS, W71-12077
ORAL, DISSOLVED OXYGEN, DIURNAL, MIGRATION PATTERNS.: /WATER, LITT W71-12870
OPLANKTON, SULFUR BACTERIA, DIEL MIGRATION, CARBON DIOXIDE, HYDROG W70-05760
ITY, FLORIDA, TEXAS, PHOSPHORUS, MIGRATION, DIURNAL, ALGICIDES, CL W70-05372
AN, / ANIMAS RIVER, DURANGO, SAN MIGUEL RIVER, DOLORES RIVER, URAV W69-07846
ADA, NUTRIENT LEVELS, S/ *TWELVE MILE LAKE(CANADA), *ONTARIO, *CAN W70-02973
DA, KUSHOG LAKE(ONTARIO), TWELVE MILE LAKE(ONTARIO), GULL LAKE(ONT W70-02795
LD JOHN LAKE, PINGO LAKE, TWELVE MILE LAKE, BRANT LAKE, IMIKPUK LA W69-08272
PPER CARBONATE, CERCARIAE, WATER MILFOIL COONTAIL, WILD CELERY, BU W70-05568
A, CANDIDA TROPICALIS, NOCTILUCA MILIARIS, HELGOLAND(GERMANY), AUR W70-04368
, LABORATORY MODEL, COD REMOVAL, MILK SUBSTRATE, NITROGEN REMOVAL, W70-07031
DALIA(OHIO), RHIZOPODS, SPIDERS, MILK WASTE TREATMENT, COLONY COUN W72-01819
R TREATMENT, *OXIDATION LAGOONS, MILK, *MODEL STUDIES.: /ASTE WATE W70-07031
EYS, MARINE ALGAE, AQUATIC LIFE, MILK, ATLANTIC OCEAN, LANDFILLS, W72-00941
EEK(WIS), DOOR CREEK(WIS), DUTCH MILL DRAINAGE DITCH(WIS), LAKE MO W70-05113
              *TEXTILE MILL WASTES, ALFOL ALCOHOLS.: W70-08740
BILITY, *IN VITRO, IN VIVO, BALL- MILL, MEICELASE, HYDROGEN PEROXID W70-04184

477

MORTALITY, BIOASSAY, INHIBITION,
UTION EFFECTS, LABORATORY TESTS,
US, TROPHIC LEVEL, PRODUCTIVITY,
ATION, FLOCCULATION, FILTRATION,
, STRATIFIED FLOW, MIXING, FLOW,
WATER UTILIZATION, FISH, ALGAE,
CYANOPHYTA, LITTORAL, NUTRIENTS,
ASTE TREATMENT, EUTROPHICATION,
*FACULTATIVE LAGOONS, LABORATORY
TIONS, SOLUBILITY / *EQUILIBRIUM
ATIC ECOSYSTEM MODEL, SIMULATION
UMBIA, CANADA, AQUATIC ECOSYSTEM
OCHEMIS/ *ECOLOGY, *MATHEMATICAL
TS, PALEOLIMNOLOGY, MATHEMATICAL
TS, PALEOLIMNOLOGY, MATHEMATICAL
APP/ DETENTION TIME, LABORATORY
LGAE BIOMASS.:   KINETIC
LOW, MODEL STUDIES, MATHEMATICAL
ENTHIC FLORA, *ALGAE, *HYDRAULIC
, LIPIDS, KINETICS, MATHEMATICAL
EWAGE, EQUILIBRIUM, MATHEMATICAL
SARNIENSE, QUANTITATION, GROWTH
MOTING PROPERTIES, CHLAMYDOMONAS
RODESMUS FALCATUS, CHLAMYDOMONAS
ALEA, MYXOPHYCEAE, CHLAMYDOMONAS
      *POTASSIUM PERMANGANATE, *DES
ROTIFERS, LARVAE, SNAILS, MITES,
SAPROPHYTIC,/ TAXONOMIC GROUPS,
DIPTERA, COLEOPTERA, HEMIPTERA,
N, ALGAE, POLYCHAETA, CRUSTACEA,
IE, INSECTS, DIPTERA, CRUSTACEA,
LICHENS, PLANTS, DIATOMS, CORAL,
KISH WATER, ESTUARIES, BACTERIA,
EFFECTS, MARINE ANIMALS, ALGAE,
ISH, INVERTEBRATES, CRUSTACEANS,
TION, LIFE CYCLES, OLIGOCHAETES,
EN FIXATION, BORON, AZOTOBACTER,
IDITY, PYRITE, SULFUR COMPOUNDS,
ELEMENTS, ALGAE, EUTROPHICATION,
FECTS, GRAZING, IRON, MANGANESE,
IRA, MALLOMONAS, ULOTHRIX, DAVID
MARL LAKES, CHLOROBIUM, FILINIA,
Y, TERTIARY TREATMENT, SAMPLING,
E, CYCLES, HARVESTING, DREDGING,
NE ALGAE, SALINITY, TEMPERATURE,
EASONAL, RADIOCHEMICAL ANALYSIS,
Y, TASTE, COAGULATION, PLANKTON,
      CONCENTRATION, ENVIRONMENTAL
GISTIC EFFECTS, AXENIC CULTURES,
CARBON, *CELL SIZE, EUKARYOTES,
IGHT UTI/ ALGAL GROWTH KINETICS,
PYRENOIDOSA, CHLORELLA VULGARIS,
LIC ACID CYCLE, AUTORADIOGRAPHY,
ASS SPECTROMETRY, N-HEPTADECANE,
CARBON FIXATION, PELAGIC REGION,
AMPLING, LAKE MENDOTA(WIS), LAKE
R(OHIO), LAKE MENDOTA(WIS), LAKE
LAKE P/ LAKE MENDOTA(WIS), LAKE
RUCOSUM, LAKE MENDOTA(WIS), LAKE
     *CULTURAL EUTROPHICATION, LAKE
SIOLOGY, LAKE MENDOTA(WIS), LAKE
AKE WAU/ LAKE MENDOTA(WIS), LAKE
H MILL DRAINAGE DITCH(WIS), LAKE
AKE KEGONSA/ *MADISON(WIS), LAKE
E ODORS, LAKE MENDOTA(WIS), LAKE
, LAKE RECOVERY, WISCONSIN, LAKE
BOTRYOCOCCUS BRAUNII, *ANACYSTIS
SUBMONTAN/ *SKAWA RIVER(POLAND),
ERIMENTAL LIMNOLOGY, CLADOCERAN,
ODIUM ARSENITE, COPPER SULPHATE,
HENYLUREAS, DIURON, FLAGELLATES,
232, YUKON RIVER, NEBESNA RIVER,
                            *INDIA,
ALGAL GENETICS, RUSSIA, BAVARIA,
), HUMBOLDT(SASKATCHEWAN), MICE,
ANTITHAMNION/ *LEUCOTHRIX MUCOR,
      SPECIES DIVERSITY, AUTECOLOGY,
COMMUNITIES, BIOINDICATORS, LAKE
STICHOCHRYSIS IMMOBILIS, VIBRIO,
S, COLONIAL MORPHOLOGY, CELLULAR
TORULA, GROWTH FACTORS, COLONIAL
MUD, LAKES, CULTIVATION, CHANNEL
POPULATIONS, WATER QUALITY, LAKE
S ASTREA, LAKE MINNETONKA(MINN),
F-NEV), HYDROPHYTES, LIVERWORTS,
, SOIL MINERALS, PLANT RESIDUES,
ONTROL, CHLORELLA, DISINFECTION,
, COPPER SULFATE, ALGAL CONTROL,
L, DISINFECTION, MODE OF ACTION,
IFIERS, WATER POLLUTION EFFECTS,

MODE OF ACTION, PHOTOSYNTHESIS.: /   W70-03519
MODEL STUDIES, RADIOACTIVITY TECH   W70-04074
MODEL STUDIES, SALMON, AQUATIC EN   W69-06536
MODEL STUDIES, COMPUTERS, NUTRIEN   W70-08627
MODEL STUDIES, MATHEMATICAL MODEL   W71-05878
MODEL STUDIES, BIOASSAY, AQUATIC   W71-11881
MODEL STUDIES, BIOASSAY, RADIOACT   W73-01454
MODEL STUDIES, BIOMASS, *LIGHT IN   W73-02218
MODEL, COD REMOVAL, MILK SUBSTRAT   W70-07031
MODEL, LITHOSPHERE, SORPTION REAC   W70-04385
MODEL, PHOSPHORUS EFFECT, CLADECE   W69-06536
MODEL, SIMULATION MODEL, PHOSPHOR   W69-06536
MODELS, *WATERSHED MANAGEMENT, GE   W71-09012
MODELS, BACTERIA, PHYTOPLANKTON,   W70-03975
MODELS, BACTERIA, PHYTOPLANKTON,   W70-03975
MODELS, BOD REMOVAL RATES, DESIGN   W70-07034
MODELS, CHLORELLA PYRENOIDOSA, *A   W73-02218
MODELS, GREAT LAKES, THERMAL STRA   W71-05878
MODELS, LABORATORY TESTS, PERIPHY   W71-00668
MODELS, NEUTRALIZATION, *DEGRADAT   W71-13413
MODELS, ON-SITE DATA COLLECTIONS,   W68-00858
MODES.: /HRIX MUCOR, ANTITHAMNION   W69-10163
MOEWUSII, CHLORELLA PYRENOIDOSA,   W69-10180
MOEWUSII, SKELETONEMA COSTATUM, S   W70-05381
MOEWUSSI, FAT ACCUMULATION.: /A P   W70-02965
MOINES(IOWA).:   W72-04242
MOLDS, GROWTH RATES, HABITATS.: /   W72-01819
MOLECULAR FILTER, CHEMOSYNTHETIC,   W68-00891
MOLLUSCA, RAFFINATE, THORIUM, TAI   W69-07846
MOLLUSCA, ECHINODERMATA, PACIFIC   W69-08526
MOLLUSCS, OLIGOCHAETES, GASTROPOD   W70-04253
MOLLUSKS, MINE ACIDS, COAL MINES,   W69-07428
MOLLUSKS, VITAMIN B, RED TIDE, SA   W70-05372
MOLLUSKS, FISH, RADIOACTIVE WASTE   W71-05531
MOLLUSKS, STRONTIUM RADIOISOTOPES   W71-09850
MOLLUSKS, STRATIFICATION, DIATOMS   W71-00665
MOLYBDENUM, COBALT, CO: /, NITROG   W70-02964
MOLYBDENUM, ROTIFERS, ECOSYSTEMS,   W69-07428
MOLYBDENUM, NITROGEN FIXATION, NU   W69-06277
MOLYBDENUM, PLANT GROWTH REGULATO   W69-04801
MONIE TEST, SUM: /ANABAENA, MELOS   W69-10157
MONIMOLIMNION, CHEMOCLINE, MIXOLI   W70-05760
MONITORING, WATER DEMAND, WATER S   W70-05645
MONITORING, MICROBIOLOGY, FOOD CH   W70-05264
MONITORING, AERIAL PHOTOGRAPHY, I   W68-00010
MONITORING, PIGMENTS, PHYTOPLANKT   W70-04371
MONITORING, TEM: /S, WIND VELOCIT   W70-00263
MONITORING.:   W72-00941
MONOALGAL, PORTUGAL, TAPES, JAPAN   W70-05372
MONOCHRYSIS LUTHERI, NAVICULA PEL   W69-08278
MONOD GROWTH EQUATION, INCIDENT L   W69-03730
MONODUS SUBTERRANEUS, ANKISTRODES   W70-05381
MONODUS SUBTERRANEUS, NAVICULA PE   W70-02965
MONOENES, DIENES, TRIENES.: /S, M   W69-08284
MONOMICTIC, DEPTH EFFECTS, IONIC   W70-01933
MONONA(WIS), LAKE WAUBESA(WIS), L   W70-02969
MONONA(WIS), LAKE WAUBESA(WIS), L   W72-01813
MONONA(WIS), LAKE NAGAWICKA(WIS),   W70-06217
MONONA(WIS), LAKE WAUBESA(WIS), A   W69-08283
MONONA(WIS), LAKE TAHOE(CALIF), L   W69-06858
MONONA(WIS), LAKE WAUBESA(WIS), S   W69-05867
MONONA(WIS), LAKE KEGONSA(WIS), L   W69-06276
MONONA(WIS), LAKE WAUBESA(WIS), L   W70-05113
MONONA(WIS), LAKE WAUBESA(WIS), L   W70-05112
MONONA(WIS), LAKE WINGRA(WIS), LA   W69-03366
MONONA, LAKE MENDOTA, LAKE WAUBES   W69-09349
MONTANA, *HYDROCARBONS, *OLEFINS,   W69-08284
MONTANE RIVER COURSE, CARPATHIAN   W70-02784
MONTANE ZONE, ALPINE ZONE, BOULDE   W69-10154
MONURON, AMMONIA, DALAPON, 2-4-: /   W70-00269
MONURON, TRANSPARENCY, CYCLOPS, S   W70-10161
MOOSE CREEK, ARCTIC CIRCLE HOT SP   W69-08272
MOOSI RIVER ECOLOGICAL STUDY.:   W69-07096
MORICHES BAY(NY), ONONDAGA LAKE(N   W70-03973
MORPHOLOGICAL CHARACTERS, TRIS.: /   W70-00273
MORPHOLOGY, FRIDAY HARBOR(WASH),   W69-10161
MORPHOLOGY, EXPORT LOSSES, CHEMOS   W70-02780
MORPHOLOGY, ECOLOGICAL DISTRIBUTI   W71-12077
MORPHOLOGY, CULTURAL CHARACTERIST   W70-04280
MORPHOLOGY, FLAGELLATION, IODINE   W70-03952
MORPHOLOGY, CELLULAR MORPHOLOGY,   W70-03952
MORPHOLOGY, SEWAGE INPUT, BIOLOGI   W68-00475
MORPHOMETRY, DOMESTIC WASTES, PLA   W68-00470
MORPHOMETRY, ALGAE BLOOMS, CLEARW   W70-05387
MORPHOMETRY, PACIFASTACUS LENIUSC   W70-00711
MORRIS(MINN), SOIL PERCOLATION.: /   W70-04193
MORTALITY.: /S, CHLORINE, ALGAL C   W70-02363
MORTALITY, CHLOROPHYLL, ALGAE, BI   W70-02364
MORTALITY, INHIBITION, ALGAE.: /O   W70-02370
MORTALITY, GASTROPODS, MARINE PLA   W70-01780

ESMUS, CHLAMYDOMONAS, CHLORELLA,
*BACTERIA COUNT, WARING BLENDOR
PUS FLUVIATILIS, *LAKE BUTTE DES
A, OOCYSTIS SOLITARIA, PANDORINA
ORGANISMS, MARINE ASSOCIATIONS,
EGGERSEE, GREIFENSEE, ZURICHSEE,
WATER HYACINTH, FROGS, CRAYFISH,
UTROPHICATION, TOXICITY, MIDGES,
IC WEED CONTROL, INSECT CONTROL,
LIFORNIA, *NEVADA, DEPTH, ALGAE,
FISHES, PLANKTON, ALGAE, FROGS,
RICULTURAL DRAINAGE, APPALACHIAN
IC ANIMALS, ANIMALS, APPALACHIAN
ON, ORGANIC MATTER, *GRASSLANDS,
ANT POPULATIONS, WEATHER, WATER,
OUS SOIL, SOIL WATER, SOIL WATER
ETRY, YELLOWSTONE NATIONAL PARK,
L PARK, MT LASSEN NATIONAL PARK,
RADIOGRAPHY, ANTITH/ *LEUCOTHRIX
UNT(AUTORADIOGRAPHY), LEUCOTHRIX
WASH), ANTITHAMNION/ *LEUCOTHRIX
ENTS, *SEDIMENT-WATER INFERFACE,
ONCENTRATION, CYCLING NUTRIENTS,
ROUNDWATER BASINS, WATER SUPPLY,
KES, *ECOLOGY, BOTTOM SEDIMENTS,
OGEN ION CONCENTRATION, ACIDITY,
SCENEDESMUS, DIATOMACEOUS EARTH,
, NITROGEN, PHOSPHATES, METHANE,
RCTIC, TUNDRA, SCALING, GRAVELS,
S, OXYGEN, COLIFORMS, SEDIMENTS,
PHYTA, SAMPLING, DENSITY, LIGHT,
, SINKS, TRACERS, CRABS, SHRIMP,
ILES, VOLATILE SUSPENDED SOLIDS,
HYTIC FORMS, COLLOIDAL FRACTION,
LARIA, OBELIA, CLUSTER ANALYSIS,
LYUBLIN(USSR), LYUBERETS(USSR),
ONS, PIT RECHARGE, WATER SUPPLY,
POLLUTION CONTROL, FARM WASTES,
RS, SAMPLING, INDUSTRIAL WASTES,
IC MICROBIOLOGY, EUTROPHICATION,
HORUS SOURCES, *NU/ *WASTE WATER(
, PHOTOSYNTHESIS, ALGAE, SEWAGE,
FARM WASTES, *INDUSTRIAL WASTES,
OXYGEN DEMAND, WATER POLLUTION,
UTION), WATER POLLUTION SOURCES,
RDA, IPOMOEA, PISONIA, PANDANUS,
OPY, MASS SPECTROMETRY, SYMPLOCA
, DIPLOCYSTIS AERUGINOSA, NOSTOC
C GROWTH, ANKISTRODESMUS, NOSTOC
YSTIS AERUGINOSA, NOSTOC, NOSTOC
O NATURAL HISTORY MUSEUM, NOSTOC
ON(WIS), CHICAGO NATURAL HISTORY
BRINE SHRIMP, FISH, FOOD CHAINS,
UATIC ALGAE, BRINE SHRIMP, FISH,
LGAE, *REVIEWS, HABITATS, CLAMS,
LAMINARIA SACCHARINA, MANNITOL,
OWTH INHIBITION, CYANOCOBALAMIN,
INGS, RADON, CENTRAL CITY(COLO),
LIKE, PSYCHROPHILES, PLEMORPHIC,
LDERS, COMPOSITION, PSEUDOMONAS,
IUM, BACILLUS SUBTILIS, BACILLUS
ODOTORULA RUBRA, CA/ *NORTH SEA,
TAMERENSIS, GYMNODINIUM BREVIS,
NIUSCULUS, ANACHARIS CANADENSIS,
ETON, TRAPA, PIS/ CERATOPHYLLUM,
OMA HEMPRICHI, ESCHERICHIA COLI,
ANIUM, SALIX, GROWTH REGULATORS,
PES, JAPAN, AMPHIDINIUM, BLOOMS,
HAERIUM/ *PHYSIOLOGY, SPIROGYRA,
OCKII, SPIROGYRA, XANTHOPHYCEAE,
ANTHOPHYCEAE, BACILLARIOPHYCEAE,
FICIT, SECCHI DISC TRANSPARENCY,
ILLARIOPHYCEAE, NITZSCHIA PALEA,
ISLAND(N C), DETECTOR, BEAUFORT(
IPECTEN IRRADIANS, PIVERS ISLAND(
ON, CONCORD(N H), PEANACOOK LAKE(
SEBASTICOOK(ME), LAKE WINNISQUAM(
ATMENT COSTS, DINOBRYON, CONCORD(
OLEUM CRUDES, MASS SPECTROMETRY,
LSION, DOW C-31, DOW A-21, NALCO
NTAL EFFECTS, ENZYMES, ESSENTIAL
LAKE(WIS), NORTH LAKE(WIS), LAKE
OTA(WIS), LAKE MONONA(WIS), LAKE
P, LAKE MENDOTA(WIS), ANACHARIS,
ICROCYSTIS, ANABAENA, EUCYCLOPS,
AMOGETON, POLYGONUM, SAGITTARIA,
LDING'S POND(WIS), HYDRODICTYON,
THERA, PICKWICK RESERVOIR(TENN),
IA, ALTERNANTHERA, HETERANTHERA,
C REPULSION, DOW C-31, DOW A-21,

MORTALITY, BIOASSAY, INHIBITION,   W70-03519
MORTALITY, CELL WALL, HEXOSAMINE,   W70-04365
MORTS(WISC), GROWTH STIMULATION,   W72-00845
MORUM, PEDIASTRUM BORYA: / PUSILL   W69-06540
MOSCOW, PETROLEUM POLLUTION, PHOT   W69-07440
MOSES LAKE, HALLWILLERSEE.: /BALD   W70-03964
MOSQUITOES, BACTERIA, IRRIGATION,   W70-05263
MOSQUITOES, SAMPLING, CONTROL, IN   W70-06225
MOSQUITOS, ALGAL CONTROL, BASS, C   W70-05417
MOSSES, CHARA, FISH, LIMNOLOGY, L   W70-00711
MOSSES, HYPOLIMNION, THERMOCLINE,   W68-00857
MOUNTAIN CLUB HUTS, TEGERNSEE, GE   W69-07818
MOUNTAIN REGION.: /LANKTON, AQUAT   W69-03695
MOUNTAINS, *SOIL MICROORGANISMS,   W72-08110
MOVEMENT, CULTURES, BAYS, STREAMS   W70-06229
MOVEMENT, LEACHING, ARID LANDS, D   W71-04067
MT LASSEN NATIONAL PARK, MT RAINI   W70-03309
MT RAINIER NATIONAL PARK, OHANAPE   W70-03309
MUCOR, *GROWTH, *FILAMENTS, *AUTO   W69-09755
MUCOR, ANTITHAMNION SARNIENSE, QU   W69-10163
MUCOR, MORPHOLOGY, FRIDAY HARBOR(W69-10161
MUD-WATER INTERFACE, PHOSPHORUS,   W69-06859
MUD-WATER INTERFACES, EUGLENOPHYT   W70-04381
MUD, BENTHOS, HYDROGEN, BACTERIA,   W69-07838
MUD, BENTHOS, WISCONSIN, SAMPLING   W70-06217
MUD, HYDROGEN SULFIDE.: /DS, HYDR   W70-05399
MUD, LAKES, CULTIVATION, CHANNEL   W68-00475
MUD, OXYGEN, ANAEROBIC CONDITIONS   W70-05761
MUD, SANDS, SHALES, BOTTOM SEDIME   W69-08272
MUD, SANDS, SLUDGE, RECREATION.: /   W70-10161
MUD, SILTS, SANDS, GRAVELS, SEDIM   W70-00711
MULLETS, FISH, ALGAE.: /COSYSTEMS   W69-08274
MULTI-CELL SERIAL SYSTEM, PH, FAC   W70-04790
MULTIPLE-USES, SAVANNAH RIVER, GU   W73-00748
MULTIVARIATE ANALYSIS, PRINCIPAL-   W73-01446
MUNICH(GERMANY), CHLORELLA PYRENO   W70-05092
MUNICIPAL WATER, SEDIMENT LOAD, C   W69-05894
MUNICIPAL WASTES, SEPTIC TANKS, R   W70-04266
MUNICIPAL WASTES, SEWAGE, ADSORPT   W70-04001
MUNICIPAL WASTES, FRESHWATER ALGA   W72-12240
MUNICIPAL AND INDUSTRIAL), *PHOSP   W69-09340
MUNICIPAL WA: / AEROBIC TREATMENT   W72-00462
MUNICIPAL WASTES, NUTRIENTS, PEST   W71-09577
MUNICIPAL WATER, WATER REQUIREMEN   W71-09577
MUNICIPAL WASTES, GREAT LAKES, *P   W71-05879
MUSCLE, BONE, LEAVES, SHELL, RHOD   W69-08269
MUSCORUM.: /UIS, RAMMAN SPECTROSC   W69-00387
MUSCORUM, VANADIUM, VITAMIN B-12.   W69-06277
MUSCORUM, SCENEDESMUS OBLIQUUES,   W70-02964
MUSCORUM, ABSORPTION, SPECTROMETR   W70-02245
MUSCORUN, NUTRIENT CONCENTRATION,   W70-00719
MUSEUM, NOSTOC MUSCORUN, NUTRIENT   W70-00719
MUSSELS.: /UTION, AQUATIC ALGAE,   W70-01467
MUSSELS.: /UTION, FOOD CHAINS, AQ   W70-01466
MUSSELS, SHELLFISH, DINOFLAGELLAT   W70-05372
MUSTARD GROWTH TEST, ORGANIC GROW   W69-07826
MUTANT BACTERIA, CYCLOTELLA NANA,   W70-05652
MUTATION, BETA RAYS, COSMIC RADIA   W69-08272
MYCELOID, CORYNEFORM RODS, COCCI,   W70-00713
MYCOBACTERIUM, BACILLUS SUBTILIS,   W69-07838
MYCOIDES, BACILLUS MESENTERICUS,   W69-07838
MYCOTA, DEBARYOMYCES HANSENII, RH   W70-04368
MYCROSYSTIS, ALGASTATIC SUBSTANCE   W70-02510
MYRIOPHYLLUM, POTAMOGETON CRISPUS   W70-00711
MYRIOPHYLLUM, EICHHORNIA, POTAMOG   W70-10175
MYRIOPHYLLUM, LEPOMIS MACROCHIRUS   W69-07861
MYRIOPHYLLUM, WATER BLOOMS, CHAIN   W70-05263
MYTILUS, TRICHODESMIUM ERYTHREUM,   W70-05372
MYXOPHYCEAE, MICROCYSTIS, COELOSP   W70-04369
MYXOPHYCEAE, SALVELINUS NAMAYCUSH   W70-00711
MYXOPHYCEAE, COMPLEX FORMATION, G   W69-10180
MYXOPHYCEAE, OSCILLATORIA, ANABAEN   W69-10182
MYXOPHYCEAE, CHLAMYDOMONAS MOEWUS   W70-02965
N C), CRASSOSTREA VIRGINICA, MERC   W69-08267
N C), DETECTOR, BEAUFORT(N C), CR   W69-08267
N H).: / TREATMENT COSTS, DINOBRY   W69-08282
N H), MADISON LAKES(WIS), LAKE WA   W70-07261
N H), PEANACOOK LAKE(N H).: / TRE   W69-08282
N-HEPTADECANE, MONOENES, DIENES,   W69-08284
N-670.: /ARGE, ELECTROSTATIC REPU   W70-05267
N: / CYCLING NUTRIENTS, ENVIRONME   W69-03188
NAGAWICKA(WIS).: /WIS), PEWAUKEE   W69-05867
NAGAWICKA(WIS), LAKE PEWAUKEE(WIS   W70-06217
NAIAS, SCIRPUS, NYMPHAEA, CRASSIP   W70-05263
NAIS, CRICOTOPUS, AUSTROCLOEON, E   W70-06975
NAJAS.: /OMENON, MICROCYSTIS, POT   W72-04761
NAJAS, ANACHARIS, LAKE WAUBESA(WI   W70-03310
NAJAS, PHOTIC ZONE, DOUGLAS RESER   W71-07698
NAJAS, SILVEX, ENDOTHAL, POTASSIU   W70-10175
NALCO N-670.: /ARGE, ELECTROSTATI   W70-05267

OGLIMUN, UDOTEA FLABELLUM, ZINC,   NICKEL, PUERTO RICO, LA PARGUERA,   W69-08525
SUBPOLAR WATERS, TYNSET(NORWAY),   NID RIVER(NORWAY), GAULA RIVER(NO   W70-01074
OSPHATES, PULSE-LABE/ *ANACYSTIS   NIDULANS, OLIGOPHOSPHATES, METAPH   W71-03033
TER, EUGLENA GRACILIS, CACHONINA   NIEI, SKELETONEMA COSTATUM, DUNAL   W69-08278
LUS SUBTILIS, ASPERGILLUS ORYZAE   NIGER, S: /AENA CYLINDRICA, BACIL   W70-02964
RAMINE, DE-K-PRUF-21, BENOCLORS,   NIGROSINE DYE, SODIUM CHLORATE, S   W70-05263
ESUM W AND G S WEST, RADIOCOCCUS   NIMBATUS, URONEMA ELONGATUM HODGE   W70-04468
R(WIS), STORKWEATHER CREEK(WIS),   NINE SPRINGS CREEK(WIS), DOOR CRE   W70-05113
UM NATANS, SEAWEED, ZIRCONIUM-95-   NIOBIUM-95, PRE-BOMB MATERIAL, RU   W69-08524
28, SWIFT RIFFLES, ZIRCONIUM-95,   NIOBIUM-95, RUTHENIUM-106, RHODIU   W69-09742
NAH, ZIRCONIUM-95, RUBIDIUM-106,   NIOBIUM-95, GYMNODINIUM SIMPLEX,   W70-00707
LUTONIUM-239, RATS, ZIRCONIUM-95-   NIOBIUM, EUROPIUM-155, SILVER-110   W69-08269
ORWAY, LAKE ZURICH, SWITZERLAND,   NITELV RIVER NORWAY, NORWA: /ET N   W69-07833
TILIZATION, *FRESH WATER, ALGAE,   NITRATE AMMONIUM, NITRITE, OXIDAT   W69-04521
                           DUCKS,   NITRATE POISONING.:   W69-04805
YMES, CHLORINE, LIGHT INTENSITY,   NITRATE, EUGLENA, CALCIUM, AMMONI   W70-02964
SIUM PERMANGANATE, FENAC, SILVER   NITRATE, SIMAZIN, NUPHAR, AGASICL   W70-10175
ACTERIS, FUNGI, ALGAE, NITRITES,   NITRATES, DETERGENTS, PHOSPHATES.   W71-00940
HICATION, DIATOMS, MICROBIOLOGY,   NITRATES, AMMONIA, WATER POLLUTIO   W71-05026
E, CULTURES, RECIRCULATED WATER,   NITRATES, PHOSPHATES, SODIUM, WAT   W70-07257
STE WATER TREATMENT, PHOSPHATES,   NITRATES, ALGAE.: /NG FILTERS, WA   W70-06792
KINETICS, FARM WASTES, BENTHOS,   NITRATES, FLUORESCENCE, NITROGEN   W70-02775
ON CYCLE, PHOSPHORUS, PHOSPHATE,   NITRATES, INDUSTRIAL WASTES, WATE   W70-02777
ION, PHOTOSYNTHESIS, ALKALINITY,   NITRATES, COND: /N ION CONCENTRAT   W70-03309
STRY, POLLUTANTS, ALGAL CONTROL,   NITRATES, PHOSPHATES, SULFATES, S   W70-02646
TES, HYDROGEN ION CONCENTRATION,   NITRATES, POTASSIUM, CALCIUM, MAG   W70-00719
FFECTS, WATER POLLUTION CONTROL,   NITRATES, NITROGEN FI: /LLUTION E   W69-10167
RUNOFF, FERTILIZERS, NUTRIENTS,   NITRATES, PHOSPHATES, PHYTOPLANKT   W69-10170
TS, NITROGEN COMPOUNDS, AMMONIA,   NITRATES, NITRITES, NUTRIENT REQU   W69-10177
PHOTOSYNTHESIS, CARBON DIOXIDE,   NITRATES, PHOSPHATES.: /OMPOUNDS,   W70-00161
OPLANKTON, EUGLENA, CHLOROPHYTA,   NITRATES, AMMONIA, CARBON, RECREA   W72-04761
RESERVOIRS, ALGAE, GROUND WATER,   NITRATES, PHOSPHATES, LAS VEGAS B   W72-02817
ATION, LAKES, LIMNOLOGY, MIDGES,   NITRATES, NITROGEN, NITROGEN CYCL   W69-04805
ENT, NUISANCE ALGAE, PHOSPHATES,   NITRATES, RIVERS, SEWAGE TREATMEN   W69-01453
NDS, PHOSPHORUS, BIBLIOGRAPHIES,   NITRATES, NUTRIENTS, PHOSPHATES,   W68-00860
CHEMICAL ANALYSIS, TEMPERATURE,   NITRATES, PLANTS, AMMONIA, HARDNE   W69-07838
IC COMPOUNDS, AMMONIA, NITRITES,   NITRATES, YEASTS, BACTERIA, AQUAT   W69-06970
GAE, WATER ANALYSIS, CARBONATES,   NITRATES, CHLORIDES, AMMONIUM COM   W69-07096
RIENTS, FERTILIZERS, PHOSPHATES,   NITRATES, ALGAE, GREAT LAKES.: /T   W69-07084
*PHOSPHORUS, ALGAE, CYANOPHYTA,   NITRATES, NITROGEN COMPOUNDS, NUI   W69-05868
WEEDS, CARBON DIOXIDE, VITAMINS,   NITRATES, INORGANIC COMPOUNDS, SU   W70-04810
, PRIMARY PRODUCTIVITY, CALCIUM,   NITRATES, AMMONIA, PHOSPHATES, HY   W70-05405
YTA, HYDROGEN ION CONCENTRATION,   NITRATES, PHOSPHATES, CYANOPHYTA,   W70-05404
N ION CONCENTRATION, PHOSPHATES,   NITRATES, DAPHNIA, SILICA, THERMA   W70-05760
US COMPOUNDS, ALGAE, PHOSPHATES,   NITRATES, SEWAGE EFFLUENTS, TERTI   W70-04074
HOS, SALINITY, SNAILS, SULFATES,   NITRATES, PHOSPHATES, POTASSIUM,   W70-03978
CONCENTRATION, IRON, MAGNESIUM,   NITRATES, PHOSPHATES, ALGAE, DETR   W70-03501
CONTROL, COSTS, EUTROPHICATION,   NITRATES, LAKE ERIE, LAKE ONTARIO   W70-03964
KTON, OXYGEN, NITROGEN FIXATION,   NITRATES, ELECTRON MICROSCOPY, AM   W70-04369
NE, FLOODS, AQUIFERS, WATER LAW,   NITRATES, TURBIDITY, FLOODWATER,   W73-00758
ATHOGENIC BACTERIA, TEMPERATURE,   NITRATES, AMMONIA, *WATER REUSE,   W72-06017
LGAE, SEWAGE, ANAEROBIC, OXYGEN,   NITRIFICATION, COLIFORMS, PONDS,   W70-05092
*MICROORGANISMS, BIODEGRADATION,   NITRIFICATION, DENITRIFICATION, O   W69-06970
UNDS, OXIDATION, BIODEGRADATION,   NITRIFICATION, DENITRIFICATION, N   W69-08053
RATION, TEMPERATURE, ALKALINITY,   NITRIFICATION, *CORROSION, CARBON   W71-11393
PYRENOIDOSA, CARBOX/ *TRISODIUM   NITRILOTRIACETATE, NTA, CHLORELLA   W70-02248
WATER, ALGAE, NITRATE AMMONIUM,   NITRITE, OXIDATION-REDUCTION POTE   W69-04521
ION, ORGANIC COMPOUNDS, AMMONIA,   NITRITES, NITRATES, YEASTS, BACTE   W69-06970
EN COMPOUNDS, AMMONIA, NITRATES,   NITRITES, NUTRIENT REQUIREMENTS,   W69-10177
OXYGEN, CARBON DIOXIDE, AMMONIA,   NITRITES, HYDROGEN SULFIDE, SALIN   W71-03031
, MITES, BACTERIS, FUNGI, ALGAE,   NITRITES, NITRATES, DETERGENTS, P   W71-00940
CTERIOPHAGE, RESERVOIRS, SPORES,   NITRITES,: /SELF-PURIFICATION, BA   W69-07838
                                     NITROGEN ASSIMILATION.:   W69-06970
TON CROPS, TROPHIC, TROPHOLYTIC,   NITROGEN ANNUAL BUDGET, PHOSPHORU   W69-05142
NSA(WIS), WEED BEDS, PHOSPHORUS,   NITROGEN BALANCE.: /S), LAKE KEGO   W70-02969
R POLLUTION SOURCES, PHOSPHORUS,   NITROGEN COMPOUNDS, EUTROPHICATIO   W70-03068
ESSENTIAL NUTRIENTS, INHIBITION,   NITROGEN COMPOUNDS, NUTRIENT REQU   W70-02248
N, EFFICIENCIES, *NITRIFICATION,   NITROGEN COMPOUNDS, NITROGEN CYCL   W70-07031
TRIFICATION, NITROGEN COMPOUNDS,   NITROGEN CYCLE, ALGAE, BIOLOGICAL   W70-07031
RIENTS, TEMPERATURE, EPILIMNION,   NITROGEN COMPOUNDS, SEWAGE EFFLUE   W69-10167
CYANOPHYTA, ESSENTIAL NUTRIENTS,   NITROGEN COMPOUNDS, AMMONIA, NITR   W69-10177
ANITATION, INHIBITION, CHLORINE,   NITROGEN COMPOUNDS, SEWAGE, SEWAG   W69-10171
VOIRS, BACKWATER, FERTILIZATION,   NITROGEN COMPOUNDS, PHOSPHORUS CO   W70-00274
GAE, *EUTROPHICATION, *NITROGEN,   NITROGEN CYCLE, *PHOSPHORUS, PHOS   W72-01867
, DEGRADATION, SEWAGE TREATMENT,   NITROGEN COMPOUNDS, PHOSPHORUS CO   W72-02817
GE, TEMPERATURE, MICROORGANISMS,   NITROGEN COMPOUNDS, *PHOSPHORUS C   W71-13335
US, ALGAE, CYANOPHYTA, NITRATES,   NITROGEN COMPOUNDS, NUISANCE ALGA   W69-05868
FERTILIZATION, LIMITING FACTORS,   NITROGEN COMPOUNDS, NUTRIENTS, OD   W69-06535
ATMENT, *WATER REUSE, *NITROGEN,   NITROGEN COMPOUNDS, OXIDATION, BI   W69-08053
*WATER REUSE, *RECLAIMED WATER,   NITROGEN COMPOUNDS, NITROGEN, AMM   W69-08054
OGRAPHIES, PHOSPHORUS COMPOUNDS,   NITROGEN COMPOUNDS, BIOINDICATORS   W69-08518
OPERTIES, *NUTRIENTS, *VIRGINIA,   NITROGEN COMPOUNDS, PHOSPHORUS, N   W69-03695
Y, LIGHT PENETRATION, LIMNOLOGY,   NITROGEN COMPOUNDS, NITROGEN FIXA   W69-03518
N, LIGHT PENETRATION, LIMNOLOGY,   NITROGEN COMPOUNDS, NUTRIENT REQU   W69-04798
ECTS, EUTROPHICATION, LIMNOLOGY,   NITROGEN COMPOUNDS, NITROGEN FIXA   W69-04802
IES, FISH MANAGEMENT, LIMNOLOGY,   NITROGEN COMPOUNDS, NITROGEN CYCL   W69-04804
, LIMNOLOGY, NITROGEN COMPOUNDS,   NITROGEN CYCLE, NUTRIENT REQUIREM   W69-04804
OGY, MIDGES, NITRATES, NITROGEN,   NITROGEN CYCLE, NITROGEN FIXATION   W69-04805
ALGAE, WATER POLLUTION SOURCES,   NITROGEN COMPOUNDS, PHOSPHATES, G   W69-01445
UTRIENTS, ENVIRONMENTAL EFFECTS,   NITROGEN COMPOUNDS, NUTRIENT REQU   W69-03358

ECTS, EUTROPHICATION, LIMNOLOGY,   NITROGEN COMPOUNDS, NITROGEN FIXA W69-03185
Y, LIGHT PENETRATION, LIMNOLOGY,   NITROGEN COMPOUNDS, NUTRIENT REQU W69-03514
ECTS, EUTROPHICATION, LIMNOLOGY,   NITROGEN COMPOUNDS, NITROGEN-FIXA W69-03186
, INHIBITION, LIGHT PENETRATION,   NITROGEN COMP: /RVESTING OF ALGAE W69-03369
EUDOMONAS, METABOLISM, KINETICS,   NITROGEN COMPOUNDS.: /ULTURES, PS W70-04280
Y, LIGHT PENETRATION, LIMNOLOGY,   NITROGEN COMPOUNDS, NUISANCE ALGA W70-04283
RIBUTION, WISCONSIN LAKE SURVEY,   NITROGEN CONTENT, LAKE ERKEN(SWED W70-03959
KE, *LAKE SEDIMENTS, *NUTRIENTS,   NITROGEN COMPOUNDS, PHOSPHOROUS C W72-06051
TION, BIOCHEMICAL OXYGEN DEMAND,   NITROGEN FIXATION.: /ION, RESPIRA W72-06022
GHT, CARBON DIOXIDE, INHIBITION,   NITROGEN FIXING BACTERIA, RADIOAC W70-04249
GROWTH RATES, PLANKTON, OXYGEN,   NITROGEN FIXATION, NITRATES, ELEC W70-04369
AINAGE, AGRICULTURAL WATERSHEDS,   NITROGEN FIXATION, CYANOPHYTA, SE W70-04506
TS, AMMONIA, AMMONIUM COMPOUNDS,   NITROGEN FIXATION, NUTRIENT REQUI W69-03364
, LIMNOLOGY, NITROGEN COMPOUNDS,   NITROGEN FIXATION, NUTRIENT REQUI W69-03185
RATES, NITROGEN, NITROGEN CYCLE,   NITROGEN FIXATION, PHOSPHATES, PH W69-04805
, LIMNOLOGY, NITROGEN COMPOUNDS,   NITROGEN FIXATION, NUISANCE ALGAE W69-04802
, LIMNOLOGY, NITROGEN COMPOUNDS,   NITROGEN FIXATION, NUTRIENTS, NU: W69-03518
GAE, EUTROPHICATION, MOLYBDENUM,   NITROGEN FIXATION, NUISANCE ALGAE W69-06277
OLLUTION SOURCES, BIOINDICATORS,   NITROGEN FIXATION.: /CTS, WATER P W69-06864
UTRIENTS, NITROGEN, PHOSPHOROUS,   NITROGEN FIXING BACTERIA, *ACTIVA W71-13335
TER POLLUTION CONTROL, NITRATES,   NITROGEN FI: /LLUTION EFFECTS, WA W69-10167
MAGNESIUM, IRON, ORGANIC MATTER,   NITROGEN FIXATION, VOLUME, STREAM W70-00719
ROLOGY, CLIMATES, TROPHIC LEVEL,   NITROGEN FIXATION, FOREST, WINDS, W71-03048
NUTRIENT REQUIREMENTS, NITROGEN,   NITROGEN FIXATION, PHOTOCHEMICAL, W70-02965
, AMINO ACIDS, CALCIUM, AMMONIA,   NITROGEN FIXATION, IRON COMPOUNDS W70-02255
UGLENA, CALCIUM, AMMONIUM SALTS,   NITROGEN FIXATION, BORON, AZOTOBA W70-02964
BENTHOS, NITRATES, FLUORESCENCE,   NITROGEN FIXATION, ENVIRONMENTAL W70-02775
EL, COD REMOVAL, MILK SUBSTRATE,   NITROGEN REMOVAL, *MILK WASTES.: / W70-07031
POMATIA, CHLORELLA PYRENOIDOSA,   NITROGEN UTILIZATION, ROLLER DRYI W70-04184
STS, CULTURES, SODIUM COMPOUNDS,   NITROGEN.: / RATES, LABORATORY TE W70-03507
CHEMTROL, COPPER SULFATE, LAKES,   NITROGEN.: / ALGAE, CHLOROPHYLL, W68-00470
R POLLUTION EFFECTS, PHOSPHORUS,   NITROGEN.: /EAR POWERPLANTS, WATE W70-00264
*ULVA LATISSIMA, *AMMONIA   NITROGEN: W71-05500
, LIMNOLOGY, NITROGEN COMPOUNDS,   NITROGEN-FIXATION, PHOSPHORUS COM W69-03186
*ANABAENA CYLINDRICA, ATP, CMU,   NITROGEN-15, PYRUVATE, ACETYLENE W70-04249
S, *NITROGEN / *MOUNTAIN MEADOW,   NITROGEN-15, SOD MATS, NON-LEGUME W72-08110
QUALITY, NUTRIENTS, PHOSPHORUS,   NITROGEN, CALCIUM CHLORIDE, TEMPE W72-05453
*COLOR, *ALKALINITY, *NITRATES,   NITROGEN, *CONDUCTIVITY, *LEACHIN W72-11906
TY, *NUTRIENTS, *EUTROPHICATION,   NITROGEN, PHOSPHORUS, *NUTRIENT R W73-01122
ON, DIATOMS, ZO/ *BENTHIC FAUNA,   NITROGEN, PHOSPHORUS, PHYTOPLANKT W70-04253
, DECIDUOUS FORESTS, PHOSPHORUS,   NITROGEN, CYANOPHYTA, CHLOROPHYTA W70-03512
NTRATION, NUTRIENTS, PHOSPHORUS,   NITROGEN, IRON, GROWTH RATES, CHL W70-03983
EUTROPHICATION, NUISANCE ALGAE,   NITROGEN, PHOSPHATES, HYDROGEN IO W70-04381
ROPHICATION, *ALGAE, *NUTRIENTS,   NITROGEN, PHOSPHORUS, SODIUM, POT W70-04503
CHEMCONTROL, ARSENIC COMPOUNDS,   NITROGEN, PHOSPHORUS.: / CONTROL, W70-04494
WATER QUALITY, WATER POLLUTION,   NITROGEN, PHOSPHORUS, AIR, PHYTOP W70-04484
LUTION, WATER POLLUTION EFFECTS,   NITROGEN, PHOSPHORUS, ALGAE, NUTR W70-04382
OL, INFECTION, NUTRIENTS, ALGAE,   NITROGEN, PHOSPHORUS, BICARBONATE W68-00468
ENTS, UNITED STATES, PHOSPHORUS,   NITROGEN, SEWAGE, INDUSTRIAL WAST W68-00478
, PLANT POPULATIONS, ROUGH FISH,   NITROGEN, PHOSPHORUS, DIATOMS, *A W68-00172
HICATION, LAKES, MICROORGANISMS,   NITROGEN, NUTRIENTS, PHOSPHORUS, W68-00481
ISS/ *EUTROPHICATION, NUTRIENTS,   NITROGEN, PHOSPHORUS, PLANKTON, D W68-00253
ES, LIMNOLOGY, MIDGES, NITRATES,   NITROGEN, NITROGEN CYCLE, NITROGE W69-04805
ERTIES, NUTRIENTS, LAKES, TESTS,   NITROGEN, PHOSPHORUS, CALCIUM, MA W68-00856
N ION CONCENTRATION, PHOSPHORUS,   NITROGEN, GROWTH RATES, STABILIZA W68-00855
SHED, ALGAE, AQUATIC VEGETATION,   NITROGEN, PHOSPHOROUS, AVAILABLE W68-00511
AIMED WATER, NITROGEN COMPOUNDS,   NITROGEN, AMMONIA, NUTRIENTS, BIO W69-08054
FISH, GREAT LAKES, LEGISLATION,   NITROGEN, NUISANCE ALGAE, PHOSPHO W69-06858
AKES, PONDS, FERTILIZERS, ALGAE,   NITROGEN, PHOSPHORUS, WATER POLLU W69-08668
N, DISSOLVED SOILDS, PHOSPHATES,   NITROGEN, ALGAE, COLIFORMS.: /IRO W69-05894
ORMS, WATER QUALITY, PHOSPHORUS,   NITROGEN, FLOW, CHEMICAL ANALYSIS W70-04810
OMS, PHOSPHORUS, ORGANIC MATTER,   NITROGEN, CHLOROPHYLL.: /TS, DIAT W70-05663
CIUM, SILICATES, DIATOMS, WINDS,   NITROGEN, PHOSPHATES, METHANE, MU W70-05761
ULTURES, CHEMICAL OXYGEN DEMAND,   NITROGEN, PHOSPHATES, BIOMASS, SY W70-05655
N, PHYSICAL PROPERTIES, CLIMATE,   NITROGEN, PHOSPHORUS.: /TILIZATIO W70-05565
O/ *RESERVOIRS, *EUTROPHICATION,   NITROGEN, OXYGEN, OXIDATION, CARB W70-05550
CONTROL, BASS, CARP, CATFISHES,   NITROGEN, POTASSIUM.: /TOS, ALGAL W70-05417
FERTILIZATION, ALGAE, PLANKTON,   NITROGEN, PHOSPHORUS, POTASSIUM, W70-05399
ES, WISCONSIN, SEWAGE EFFLUENTS,   NITROGEN, PHOSPHORUS.: / PROPERTI W70-05113
WISCONSIN, *LAKES, PRODUCTIVITY,   NITROGEN, PHOSPHORUS, DRAINAGE, M W70-05112
AEROBICITY.:   ORGANIC   NITROGEN, BACTERIOLOGICAL STUDY, W71-05742
IA, WATER CHEMISTRY, PHOSPHORUS,   NITROGEN, AMMONIA, ALGAE, NUTRIEN W71-05879
OURCES, WATER POLLUTION EFFECTS,   NITROGEN, COLIFORMS, WATER QUALIT W71-05407
OPHYLL, HYPOLIMNION, PHOSPHORUS,   NITROGEN, PHYTOPLANKTON, PRODUCTI W71-07698
TION, PHOTOSYNTHESIS, L/ *ALGAE,   NITROGEN, PHOSPHORUS, *EUTROPHICA W72-00917
TROPHICATION, *ALGAE, NUTRIENTS,   NITROGEN, PHOSPHOROUS, NITROGEN F W71-13335
ALITY CONTROL, OXIDATION LAGOON,   NITROGEN, NUTRIENTS, FILTRATION, W71-13313
BICARBONATES, IRON, PHOSPHORUS,   NITROGEN, WATER QUALITY.: /HIGAN, W72-03183
IMENTATION, RUNOFF, FERTILIZERS,   NITROGEN, PHOSPHORUS, ALGAE, LIVE W71-10376
ION CONCENTRATION, TEMPERATURE,   NITROGEN, NUTRIENTS, ORGANIC MATT W70-00265
CHLOROPHYTA, LAKES, RESERVOIRS,   NITROGEN, GROWTH, AMINO ACIDS, CH W69-10180
RIMETRY, CARBOHYDRATES, PHENOLS,   NITROGEN, RESPIRATION, CARBON, PR W70-01073
ER DISPOSAL, DIURNAL, EFFLUENTS,   NITROGEN, PHOSPHATES, ALKALINITY, W69-10159
FISH, WISCONSIN, SEWAGE, TROUT,   NITROGEN, PHOSPHORUS, DETERGENTS, W69-10178
OPHICATION, *TERTIARY TREATMENT,   NITROGEN, AQUATIC PLANTS, BOTTOM W69-09340
R POLLUTION EFFECTS, PHOSPHORUS,   NITROGEN, CHLOROPHYLL, PRODUCTIVI W70-07261
TES, *EUTROPHICATION, NUTRIENTS,   NITROGEN, WASTE WATER TREATMENT, W71-04072
NALYTICAL TECHNIQUES, NUTRIENTS,   NITROGEN, PHOSPHORUS, AGRICULTURE W71-03048

ENTS(WATER), HYPOLIMNION, ALGAE,   NITROGEN, PHOSPHORUS, OXYGEN, CAR   W70-09889
TREATMENT, BIOLOGICAL TREATMENT,   NITROGEN, PHOSPHORUS, SEWAGE EFFL   W70-08980
OPHYTA, LAKES, RESERVOIRS, SCUM,   NITROGEN, PHOSPHORUS, CARBONATES,   W70-02803
R, CULTURES, CARBON, METABOLISM,   NITROGEN, PHOSPHORUS, SULFUR, CAL   W70-02804
WAGE TREATMENT, SEWAGE BACTERIA,   NITROGEN, PHOSPHORUS, ALGAE, COPP   W70-02787
LL, FLUORESCENCE, LAKES, RIVERS,   NITROGEN, BIOMASS, SEWAGE, FARM W   W70-02777
WATER, LAKES, SILTS, PHOSPHATES,   NITROGEN, WASTE WATER(POLLUTION),   W70-02255
NTENSITY, NUTRIENT REQUIREMENTS,   NITROGEN, NITROGEN FIXATION, PHOT   W70-02965
ATERS, CARBON, FERTILITY, ALGAE,   NITROGEN, PHOSPHORUS, WATER POLLU   W70-02973
, FLAGELLATES, ELEUTHERA ISLAND,   NITZCHIA SERIATA, NAVICULA, FLORI   W70-00707
*MARINE ENVIRONMENT, CERIUM-144,   NITZSCHIA CLOSERIUM, ARTEMIA, CAR   W70-00708
A ELONGATUM, MELOSIRA ISLANDICA,   NITZSCHIA ACICULARIS, NITZSCHIA D   W69-10158
ISLANDICA, NITZSCHIA ACICULARIS,   NITZSCHIA DISSIPATA, SYNEDRA ACUS   W69-10158
ANTHOPHYCEAE, BACILLARIOPHYCEAE,   NITZSCHIA PALEA, MYXOPHYCEAE, CHL   W70-02965
GHT-SATURATED RATE, CELL VOLUME,   NITZSCHIA CLOSTERIUM, NITZSCHIA P   W70-05381
LL VOLUME, NITZSCHIA CLOSTERIUM,   NITZSCHIA PALEA, CHLOROPLASTS, SC   W70-05381
LARITY, CAP/ *PIPETTE, UNIALGAL,   NITZSCHIA, ALGAE ISOLATION, CAPIL   W70-04469
AMPHORA, GOMPHONEMA, PINNULARIA,   NITZSCHIA, NAVICULA, SYRACUSE(NY)   W70-03974
, GYROSIGMA, MELOSIRA, NAVICULA,   NITZSCHIA, SKELETONEMA, PLAGIOGRA   W70-03325
BRAUNII, FRAGILARIA CROTONENSIS,   NITZSCHIA, ASTERIONELLA GRACILLIM   W70-02804
GAL IDENTIFIC/ BLUE-GREEN ALGAE,   NITZSCHIA, BIOLOGICAL STUDIES, AL   W70-10173
V) CL, CHLORELLA OVALIS BUTCHER,   NJOER LAKE(NORWAY), HOBOL RIVER(N   W70-05752
GINOSA(KUTZ), CULTURE MEDIA, CHU   NO 10 SOLUTION, MINIMUM NUTRIENTS   W70-03507
ITY STRUCTURE, RIDLEY CREEK(PA),   NOBS CREEK(MD), BACK RIVER(MD), O   W70-04510
DIDA OBTUSA, CANDIDA TROPICALIS,   NOCTILUCA MILIARIS, HELGOLAND(GER   W70-04368
*SEAWEED EXTRACTS, ASCOPHYLLUM   NODOSUM, ALGINIC ACID, FUCUS VESI   W69-07826
FUCUS VESCICULOSUS, ASCOPHYLLUM   NODOSUM, ECTOCARPUS CONFERVOIDES,   W70-01074
S, CHONDRUS CRISPUS, ASCOPHYLLUM   NODOSUM, LAMINARIA DIGITATA, LAMI   W70-01073
LL WALL, HEXOSAMINE, ASCOPHYLLUM   NODOSUM, POLYSIPHONIA LANOSA, POI   W70-04365
OLYSIPHONIA LANOSA, *ASCOPHYLLUM   NODOSUM, VIBRIO, FLAVOBACTER, ESC   W70-03952
DOTAE, ANABAENA HASSALI, LYNGBYA   NOLLEI, GLOCOTRICHIA PISUM, NOSTO   W69-08283
, TOTAL SOLIDS, VOLITILE SOLIDS,   NON-DEGRADABLE SOLID.: /RY MANURE   W71-00936
N MEADOW, NITROGEN-15, SOD MATS,   NON-LEGUMES, *NITROGEN FIXING ORG   W72-08110
NES(ICELAND), SUDURNES(ICELAND),   NON-PARAMETRIC TESTS, ONE-SAMPLE   W69-09755
LLA PYRENOIDOSA, VOLATILE ACIDS,   NON-VOLATILE ACIDS, KETOACIDS, TA   W70-04195
ETHOXYLATE, ANIONIC SURFACTANTS,   NONIONIC SURFACTANTS, INFRARED SP   W70-09438
CHNIQUES, ANTIBIOTICS, BIOASSAY,   NORON, CALCIUM, CHLORINE, COBALT,   W69-04801
C ZONE, DOUGLAS RESERVOIR(TENN),   NORRIS RESERVOIR(TENN), FRENCH BR   W71-07698
AKE ERKEN(SWEDEN), ACETATE, LAKE   NORRVIKEN(SWEDEN), A: /ELLATES, L   W70-04194
LOCITY, MICHAELIS CONSTANT, LAKE   NORRVIKEN(SWEDEN), LAKE EKOLN(SWE   W70-04287
ES, LA/ CANADA, MESOTROPHY, LAKE   NORRVIKEN, LAKE MENDOTA, LAKE FUR   W70-03964
ORULA, TEMPERATURE COEFFICIENTS,   NORTH AMERICA, CENTRAL AMERICA, R   W70-03957
DRAINAGE, AGRICULTURAL DRAINAGE,   NORTH AMERICA, FOREST DRAINAGE.: /   W70-03975
DRAINAGE, AGRICULTURAL DRAINAGE,   NORTH AMERICA, FOREST DRAINAGE.: /   W70-03975
EA WATER, OYSTERS, CLAMS, CRABS,   NORTH CAROLINA, SEDIMENTS, RADIOA   W69-08267
DIES OF WATER, DISSOLVED OXYGEN,   NORTH CAROLINA.: /OIRS, ALGAE, BO   W70-00683
T PHYSIOLOGY, ANIMAL PHYSIOLOGY,   NORTH DAKOTA, SOUTH DAKOTA, PHOTO   W72-01818
TURE, *ALGAE BLOOMS, CYANOPHYTA,   NORTH DAKOTA.: /GY, WATER TEMPERA   W70-09093
STRIAL WASTES, SEWAGE TREATMENT,   NORTH DAKOTA, COLIFORMS, OXIDATIO   W70-03312
ALENTS(WASTES), RED RIVER VALLEY(NORTH DAKOTA), SULFATE REDUCTION,   W70-03312
ELAVAN(WIS), PEWAUKEE LAKE(WIS),   NORTH LAKE(WIS), LAKE NAGAWICKA(W   W69-05867
LAKE, BRANT LAKE, IMIKPUK LAKE,   NORTH MEADOW LAKE, GYTTJA, DUFF,   W69-08272
D(ENGLAND), ANABAENA CYLINDRICA,   NORTH SEA, FIRTH OF CLYDE(SCOTLAN   W70-02504
IA CLOSERIUM, ARTEMIA, CARTERIA,   NORTH TEMPERATE ZONE, PLUTONIUM,   W70-00708
ITY, *TROPHIC LEVELS, NUTRIENTS,   NORTHEAST U.S., ECOLOGICAL DISTRI   W70-02772
NMENT, WATER QUALITY, NUTRIENTS,   NORTHEAST U S, ECOLOGICAL DISTRIB   W70-02764
CONTROL, *ALGAL TOXINS MEMBRANE,   NORTHEAST U.S.: /ANAPHYTA, ALGAL   W69-05306
WATER POLLUTION CONTROL, PACIFIC   NORTHWEST U. S., CIVIL ENGINEERIN   W69-03683
WITZERLAND, NITELV RIVER NORWAY,   NORWA: /ET NORWAY, LAKE ZURICH, S   W69-07833
), HOBOL RIVER(NORWAY), VAN LAKE(NORWAY).: /HER, NJOER LAKE(NORWAY   W70-05752
, NID RIVER(NORWAY), GAULA RIVER(NORWAY), FLAKK(NORWAY), RANHEIM(N   W70-01074
LA(NORWAY), HITRA(NORWAY), FROYA(NORWAY), FUCUS VESCICULOSUS, ASCO   W70-01074
RD(NORWAY), SMOLA(NORWAY), HITRA(NORWAY), FROYA(NORWAY), FUCUS VES   W70-01074
ATERS, TYNSET(NORWAY), NID RIVER(NORWAY), GAULA RIVER(NORWAY), FLA   W70-01074
, TRONDHEIMSFJORD(NORWAY), SMOLA(NORWAY), HITRA(NORWAY), FROYA(NOR   W70-01074
RELLA OVALIS BUTCHER, NJOER LAKE(NORWAY), HOBOL RIVER(NORWAY), VAN   W70-05752
STANCES, SUBPOLAR WATERS, TYNSET(NORWAY), NID RIVER(NORWAY), GAULA   W70-01074
WAY), GAULA RIVER(NORWAY), FLAKK(NORWAY), RANHEIM(NORWAY), TRONDHE   W70-01074
RANHEIM(NORWAY), TRONDHEIMSFJORD(NORWAY), SMOLA(NORWAY), HITRA(NOR   W70-01074
(NORWAY), FLAKK(NORWAY), RANHEIM(NORWAY), TRONDHEIMSFJORD(NORWAY),   W70-01074
NJOER LAKE(NORWAY), HOBOL RIVER(NORWAY), VAN LAKE(NORWAY).: /HER,   W70-05752
ETROPHS, ITALY, LAKE BORREVANNET   NORWAY, LAKE LUGANO ITALY, LAKE M   W69-07833
UGANO ITALY, LAKE MARIDALSVANNET   NORWAY, LAKE ZURICH, SWITZERLAND,   W69-07833
URICH, SWITZERLAND, NITELV RIVER   NORWAY, NORWA: /ET NORWAY, LAKE Z   W69-07833
OCYSTIS, DIPLOCYSTIS AERUGINOSA,   NOSTOC MUSCORUM, VANADIUM, VITAMI   W69-06277
MICROCYSTIS AERUGINOSA, NOSTOC,   NOSTOC MUSCORUM, ABSORPTION, SPEC   W70-02245
OTROPHIC GROWTH, ANKISTRODESMUS,   NOSTOC MUSCORUM, SCENEDESMUS OBLI   W70-02964
CHICAGO NATURAL HISTORY MUSEUM,   NOSTOC MUSCORUN, NUTRIENT CONCENT   W70-00719
WTH, *POTOMAC RIVER, PLECTONEMA,   NOSTOC PALUDOSUM, BLEACHING, WASH   W70-02255
GBYA NOLLEI, GLOCOTRICHIA PISUM,   NOSTOC VERRUCOSUM, LAKE MENDOTA(W   W69-08283
CCUS, CHROOCOCCUS, OSCILLATORIA,   NOSTOC, ANABAENA, ANACYSTIS, MICR   W70-04503
M TENUE, MICROCYSTIS AERUGINOSA,   NOSTOC, NOSTOC MUSCORUM, ABSORPTI   W70-02245
AKING' OF LAKES, CHROOCOCCACEAE,   NOSTOCHINEAE, CLATHROCYSTIS AERUG   W69-08283
LEPOMIS MACROCHIRUS, PIMEPHALES   NOTATUS.: /IA COLI, MYRIOPHYLLUM,   W69-07861
K/ PRINCE EDWARD ISLAND(CANADA),   NOVA SCOTIA(CANADA), PENGUINS, LA   W70-05399
X/ *TRISODIUM NITRILOTRIACETATE,   NTA, CHLORELLA PYRENOIDOSA, CARBO   W70-02248
F)/ *COMPLEXONES, *ALGAL BLOOMS,   NTA, EDTA, HEIDA, LAKE TAHOE(CALI   W70-04373
IES, CARP, TILAPIA, TEMPERATURE,   NU: /L ANALYSIS, PHYSICAL PROPERT   W70-06975
S, NITROGEN FIXATION, NUTRIENTS,   NU: /LIMNOLOGY, NITROGEN COMPOUND   W69-03518
ALGAL GROWTH, OXYGEN DEPLETION,   NUCLEAR ACTIVATION TECHNIQUES, RO   W70-00264

ISOTOPES, URANIUM RADIOISOTOPES,   NUCLEAR EXPLOSIONS, STRONTIUM RAD   W70-00708
CHANGE, RESINS, LEAKAGE, CESIUM,   NUCLEAR EXPLOSIONS, ABSORPTION, A   W70-00707
ESTUARINE ENVIRONMENT, FALLOUT,   NUCLEAR EXPLOSIONS, BIOTA, SPECTR   W69-08267
PLANTS, BOTTOM SEDIMENTS, SOILS,   NUCLEAR EXPLOSIONS, COBALT RADIOI   W69-08269
ROL, TRACE ELEMENTS, WASHINGTON,   NUCLEAR POWERPLANTS, WATER POLLUT   W70-00264
, RADIOACTIVITY, WASTE DISPOSAL,   NUCLEAR REACTORS, POWERPLANTS, SH   W70-00707
NGE, *OPEN SEA, WEAPONS TESTING,   NUCLEAR SHIP SAVANNAH, ZIRCONIUM-   W70-00707
REAMS, BIOINDICATORS, *NEW YORK,   NUCLEAR WASTES, WATER POLLUTION S   W69-10080
SOIL-WATER-PLANT RELATIONSHIPS,   NUCLEAR WASTES, ABSORPTION, CHELA   W71-09005
IS, MARINE ALGAE, INVERTEBRATES,   NUCLEAR WASTES, WATER POLLUTION S   W71-09863
HOSPHORYLATION, *MACROMOLECULES,   NUCLEIC ACID, RIBONUCLEIC ACID SY   W69-10151
K, *ICELAND HOT SPRINGS, SINTER,   NUCLEIC ACID, SISJOTHANDI(ICELAND   W69-10160
UALITY, WATER POLLUTION CONTROL,   NUISANCE ALGAE, TASTE, ODOR.: / Q   W69-10155
LLUTION EFFECTS, EUTROPHICATION,   NUISANCE ALGAE.: /REAMS, WATER PO   W70-00719
PTH, DOMESTIC ANIMALS, SAMPLING,   NUISANCE ALGAE.: /, NUTRIENTS, DE   W70-00274
PHO/ *NUTRIENTS, *PRODUCTIVITY,   NUISANCE ALGAE, SEWAGE EFFLUENTS,   W70-00270
ARY ENGINEERING, MICROORGANISMS,   NUISANCE ALGAE, SEWAGE TREATMENT,   W72-01798
TES, HYDROGEN ION CONCENTRATION,   NUISANCE ALGAE, AQUATIC PLANTS, B   W70-06225
POLLUTION SOURCES, CHLOROPHYTA,   NUISANCE ALGAE, FISH, ZOOPLANKTON   W70-03310
ATER POLLUTION CONTROL, MINNOWS,   NUISANCE ALGAE, SUNFISHES, ALGAE,   W70-02982
ON, LAKES, MAGNESIUM, NUTRIENTS,   NUISANCE ALGAE, OLIGOTROPHY, ODOR   W69-07833
IES, CULTURES, DIATOMS, ECOLOGY,   NUISANCE ALGAE, PHAEOPHYTA, PHOTO   W69-07832
ISCONSIN, BIBLIOGRAPHIES, LAKES,   NUISANCE ALGAE, BACTERIA, WATER P   W69-08283
AT LAKES, LEGISLATION, NITROGEN,   NUISANCE ALGAE, PHOSPHORUS, WASHI   W69-06858
, MOLYBDENUM, NITROGEN FIXATION,   NUISANCE ALGAE, NUTRIENTS, VITAMI   W69-06277
CALCIUM, CHELATION, CYANOPHYTA,   NUISANCE ALGAE, NUTRIENT REQUIREM   W69-05867
KE MICHIGAN, *WATER CIRCULATION,   NUISANCE ALGAE, THERMAL POLLUTION   W69-05763
DS, ECONOMIC FEASIBILITY, LAKES,   NUISANCE ALGAE, WATER POLLUTION C   W69-06276
A, NITRATES, NITROGEN COMPOUNDS,   NUISANCE ALGAE, NUTRIENT REQUIREM   W69-05868
NITROGEN COMPOUNDS, PHOSPHORUS,   NUISANCE ALGAE, SALTS, TIDES, SOU   W69-03695
EN COMPOUNDS, NITROGEN FIXATION,   NUISANCE ALGAE, NUTRIENTS, PHOSPH   W69-04802
IATOMS, CHLOROPHYTA, CYANOPHYTA,   NUISANCE ALGAE, WISCONSIN, NUTRIE   W69-01977
HICATION, *LAKE MICHIGAN, ALGAE,   NUISANCE ALGAE, PLANKTON, BENTHIC   W68-01244
H, *MUNICIPAL WASTES, NUTRIENTS,   NUISANCE ALGAE, OLIGOTROPHY, PHYT   W68-00680
POLLUTANTS, POLLUTION ABATEMENT,   NUISANCE ALGAE, PHOSPHATES, NITRA   W69-01453
EFFLUENTS, AQUATIC POPULATIONS,   NUISANCE ALGAE, DUCKS(WILD), OYST   W68-00461
KES, WISCONSIN, COPPER SULPHATE,   NUISANCE ALGAE, *ALGAL CONTROL, W   W68-00468
TROUT, WINTERKILLING, FISHKILL,   NUISANCE ALGAE, ALGAE, SEASONAL,   W68-00487
THAW TESTS, HARVESTING OF ALGAE,   NUISANCE ALGAE, NUTRIENT REQUIREM   W69-03362
NUTRIENTS, HARVESTING OF ALGAE,   NUISANCE ALGAE, NUTRIENT REQUIREM   W69-03371
*MISSISSIPPI RIVER, *OHIO RIVER,   NUISANCE ALGAE, WATER QUALITY, SE   W70-04507
OGY, CYANOPHYTA, EUTROPHICATION,   NUISANCE ALGAE, NITROGEN, PHOSPHA   W70-04381
OLLUTION EFFECTS, BIOINDICATORS,   NUISANCE ALGAE, NEW YORK, SEWAGE   W70-03974
, *EUTROPHICATION, *CONFERENCES,   NUISANCE ALGAE, PUBLICATIONS, WAT   W70-03973
MASS, *PRODUCTIVITY, CYANOPHYTA,   NUISANCE ALGAE, LIMNOLOGY, CHLORO   W70-03959
, LIMNOLOGY, NITROGEN COMPOUNDS,   NUISANCE ALGAE, NUTRIENT REQUIREM   W70-04283
LOW R/ *WATER, *FILTERS, *ALGAE,   NUISANCE ALGAE, INTAKES, SANDS, F   W70-04199
EMPERATURE, FERTILIZATION, FISH,   NUISANCE ALGAE, ODOR, TASTE, DEPT   W70-05404
OPPER SULFATE, *SODIUM ARSENITE,   NUISANCE ALGAE, SNAILS, LEGISLATI   W70-05568
ESIS, *RESPIRATION, GREAT LAKES,   NUISANCE ALGAE, WATER QUALITY, NU   W72-05453
KES(WIS), LAKE WASHINGTON(WASH),   NUISANCE CONDITIONS.: /MADISON LA   W70-07261
, LAKE MONONA(WI/ *MADISON(WIS),   NUISANCE ODORS, LAKE MENDOTA(WIS)   W69-03366
, *PREVENTATIVE TREATMENT, ALGAE   NUISANCE, ALGAE BLOOM, ALGAE GROW   W69-08282
GO(ILL), ARTIFICIAL CIRCULATION,   NUISANCES.: /ILWAUKEE(WIS), CHICA   W69-10178
BIOLOGICAL PRODUCTIVITY, AQUATIC   NUISANCES, TRANSPARENCY, AGRICULT   W69-09349
CARRIER(ISRAEL), THRESHOLD ODOR   NUMBERS.: /ATORIA, NATIONAL WATER   W70-02373
O), ALGAE COUNTS, THRESHOLD ODOR   NUMBERS, CROP-DUSTING AIRCRAFT, W   W69-10155
FENAC, SILVER NITRATE, SIMAZIN,   NUPHAR, AGASICLES, MANATEE, MARIS   W70-10175
BORATORY TESTS, TEST PROCEDURES,   NURTRIENTS, RIVERS, LAKES, *GROWT   W68-00472
FIXATION, ENVIRONMENTAL EFFECTS,   NUTRIE: / FLUORESCENCE, NITROGEN   W70-02775
ION EXCHANGE, WATER CIRCULATION,   NUTRIENT ABSORPTION, *CHEMICAL PR   W68-00857
TH POTENTIAL, SE/ *ALGAL GROWTH,   NUTRIENT AVAILABILITY, ALGAL GROW   W69-06864
NTS, PUBLIC APP/ POLYPHOSPHATES,   NUTRIENT BALANCE, FOAMING DETERGE   W70-05084
K(ME), LAKE WINNISQUAM(N H), MA/   NUTRIENT BUDGETS, LAKE SEBASTICOO   W70-07261
ROPHICATION, TROPHIC INDICATORS,   NUTRIENT BUDGET.:   *CULTURAL EUT   W71-03048
ANSPARENCY, AGRICULTURAL RUNOFF,   NUTRIENT BUDGETS, COMPARATIVE STU   W69-09349
ION, *NUTRIENT BUDGET, POTENTIAL   NUTRIENT CONCENTRATION, AREAL INC   W70-00270
HISTORY MUSEUM, NOSTOC MUSCORUN,   NUTRIENT CONCENTRATION, WISCONSIN   W70-00719
BORID MIDGES, CLEAR LAKE(CALIF),   NUTRIENT ENERGY PATHWAYS, TROPHIC   W70-05428
INNETONKA(MINN), NUTRIENT TRAPS,   NUTRIENT INFLUX, NUTRIENT SOURCES   W70-04268
LUTION COMMISSION, ALGAL BLOOMS,   NUTRIENT LOAD, SYNTHETIC DETERGEN   W69-08674
LAKE(CANADA), *ONTARIO, *CANADA,   NUTRIENT LEVELS, SOFT WATER.: /E   W70-02973
AKES(NY), RODHE'S SOLUTION VIII,   NUTRIENT LIMITATION, CARBON-14.: /   W70-01579
TURE, *CHEMOSTATS, TURBIDOSTATS,   NUTRIENT LIMITED GROWTH, LUXURY C   W70-02779
                                          NUTRIENT MEDIA.:   W69-03491
ESTING OF ALGAE, NUISANCE ALGAE,   NUTRIENT REQUIREMENTS, PHYSIOLOGI   W69-03362
ESTING OF ALGAE, NUISANCE ALGAE,   NUTRIENT REQUIREMENTS, PHYSIOLOGI   W69-03371
TAL EFFECTS, NITROGEN COMPOUNDS,   NUTRIENT REQUIREMENTS, PHOSPHORUS   W69-03358
, FREEZE-THAW TESTS, INHIBITION,   NUTRIENT REQUI: /S, FREEZE DRYING   W69-03370
, LIMNOLOGY, NITROGEN COMPOUNDS,   NUTRIENT REQUIREMENTS, PHOTOSYNTH   W69-03514
UM COMPOUNDS, NITROGEN FIXATION,   NUTRIENT REQUIREMENTS, NUTRIENTS,   W69-03364
EN COMPOUNDS, NITROGEN FIXATION,   NUTRIENT REQUIREMENT: /GY, NITROG   W69-03185
Y, LIGHT PENETRATION, LIMNOLOGY,   NUTRIENT REQUIREMENTS, NUTRIENTS,   W69-03517
ROGEN COMPOUNDS, NITROGEN CYCLE,   NUTRIENT REQUIREME: /MNOLOGY, NIT   W69-04804
, LIMNOLOGY, NITROGEN COMPOUNDS,   NUTRIENT REQUIREMENTS, NUTRIENTS,   W69-04798
ROGEN COMPOUNDS, NUISANCE ALGAE,   NUTRIENT REQUIREMENTS, NUTRIENTS,   W70-04283
                                          NUTRIENT REQUIREMENTS(ALGAE).:   W70-04074
RIVER(WIS), DETROIT LAKES(MINN),   NUTRIENT REMOVAL, MASS COMMUNICAT   W70-04506
ERFISHING, COHO SALMON, ALEWIFE,   NUTRIENT REMOVAL.: /L BALANCE, OV   W70-04465
*PHYTOPLANKTON, CARBON DIOXIDE,   NUTRIENT REQUIREMENTS, POPULATION   W70-05424

ROPHYTA, DIATOMS, LIGHT, LIPIDS,
L PROCEDURE, LUXURY CONSUMPTION,
ROGEN COMPOUNDS, NUISANCE ALGAE,
ION, CYANOPHYTA, NUISANCE ALGAE,
OF WISCONSIN, YAHARA RIVER(WIS),
 INHIBITION, NITROGEN COMPOUNDS,
N, TEMPERATURE, LIGHT INTENSITY,
NT BUDGETS, COMPARATIVE STUDIES,
RADIATION, IODINE RADIOISOTOPES,
ICUT, RHODE ISLAND, TEMPERATURE,
DS, AMMONIA, NITRATES, NITRITES,
                           *SEWAGE
 *NUTRIENT BALANCE, *LAKE AGING,
NUTRIENT TRAPS, NUTRIENT INFLUX,
NUTRIENT/ LAKE MINNETONKA(MINN),
IPS.:   *POTOMAC RIVER ESTUARY,
R POLLUTION, SOIL CONTAMINATION,
CTERIA, PROTOZOA, SEWAGE, LAKES,
FFECTS, WATER POLLUTION SOURCES,
 SUCCESSION, ECOSYSTEMS, CYCLING
ERATURE, RAIN, GEOLOGY, MAPPING,
DIA, CHU NO 10 SOLUTION, MINIMUM
                             ALGAL
MENTS, *NUTRIENT BUDGET, CYCLING
IENT CONCENTRATION, AREAL INCOME(
L, FERTILIZERS, SEWAGE EFFLUENT,
E, *EUTROPHICATION, OLIGOTROPHY,
UNITIES, DAPHNIA, NANNOPLANKTON,
NTRATION, TEMPERATURE, NITROGEN,
ICHIGAN, *WATER SUPPLY, INTAKES,
ROBIOLOGY, CYANOPHYTA, ESSENTIAL
S, CHLOROPHYLL, PROTEINS, LIGHT,
IS, CHEMICAL ANALYSIS, BACTERIA,
NITRITES, NUTRIENT REQUIREMENTS,
OPHICATION, RUNOFF, FERTILIZERS,
HYTOLOGY, MARINE ALGAE, ECOLOGY,
, *EUTROPHICATION, ALGAL BLOOMS,
OPHICATION, *LAKES, *PHOSPHATES,
ERIA, DIATOMS, DETRITUS, CYCLING
RBIDITY, ALGAE, DIATOMS, CYCLING
PHICATION, *GREAT LAKES, *ALGAE,
EPILIMNION, HYPOLIMNION, CYCLING
ATIC ENVIRONMENT, WATER QUALITY,
RACE ELEMENTS, DISSOLVED SOLIDS,
 WATER QUALITY, *TROPHIC LEVELS,
ENVIRONMENTAL EFFECTS, ESSENTIAL
SSAYS, *FLUOROMETRY, CALIFORNIA,
EORY, ENVIRONMENTAL/ *BIOASSAYS,
ION, THERMAL POLLUTION, STREAMS,
N, AQUATIC MICROBIOLOGY, CYCLING
ACE ELEMENTS, TEMPERATURE, IRON,
OMPOUNDS, NUTRIENT REQUIREMENTS,
GY, CYCLING NUTRIENTS, ESSENTIAL
 HARVESTING OF ALGAE, LIMNOLOGY,
TES, MIDGES, SNAILS, PHOSPHORUS,
CHLOROPHYTA, CHLORELLA, CYCLING
LA, CYCLING NUTRIENTS, .ESSENTIAL
S, CHELATION, SPECTROPHOTOMETRY,
Y PRODUCTIVITY, *ORGANIC MATTER,
STRIAL WASTES, MUNICIPAL WASTES,
ER TREATMENT, INDUSTRIAL WASTES,
HORUS, NITROGEN, AMMONIA, ALGAE,
CIRCULATION, PATH OF POLLUTANTS,
 NITRO/ *EUTROPHICATION, *ALGAE,
ROL, OXIDATION LAGOON, NITROGEN,
LIGHT INTENSITY, EUTROPHICATION,
HICATION, *LIMNOLOGY, *ESSENTIAL
N, *PHOSPHATES, *EUTROPHICATION,
LIZATION, ANALYTICAL TECHNIQUES,
DROGEN SULFIDE, SALINITY, LAKES,
 ANIMALS, FISH, INFLOW, OUTLETS,
CAL TREATMENT, BOTTOM SEDIMENTS,
FIDE, BIOCHEMICAL OXYGEN DEMAND,
ODYNAMICS, LAKES, INTE/ *CYCLING
TIVITY, MICHIGAN, *EPIPHYTOLOGY,
IFORMS, DISSOLVED OXYGEN, ALGAE,
        *GROWTH, *ALGAE, LIPIDS,
ES, LABORATORY TESTS, MINNESOTA,
ATION, MODEL STUDIES, COMPUTERS,
URE, HYDROGEN ION CONCENTRATION,
*LAKE ZURICH, *MUNICIPAL WASTES,
PHORUS RADIOISOTOPE, *TURNOVERS,
, CYANOPHYTA, DOMINANT ORGANISM,
HYTA, NUISANCE ALGAE, WISCONSIN,
S, *ALGAE, *CHEMICAL PROPERTIES,
HORUS, BIBLIOGRAPHIES, NITRATES,
ECHNIQUES, *BIOASSAY, *ESSENTIAL
LGAE, *ALGAL CONTROL, *ESSENTIAL
ENVIRONMENTAL EFFECTS, ESSENTIAL
OMPOUNDS, NUTRIENT REQUIREMENTS,

NUTRIENT REQUIREMENTS, ORGANIC MA   W69-06865
NUTRIENT REMOVAL, SPAULDING'S PON   W69-05868
NUTRIENT REQUIREMENTS, NUTRIENTS,   W69-05868
NUTRIENT REQUIREMENTS, NUTRIENTS,   W69-05867
NUTRIENT REMOVAL.: /, UNIVERSITY   W69-06276
NUTRIENT REQUIREMENTS, NUTRIENTS,   W70-02248
NUTRIENT REQUIREMENTS, NITROGEN,   W70-02965
NUTRIENT REMOVAL, LAKE RECOVERY,   W69-09349
NUTRIENT REQUIREMENTS.:/KGROUND   W69-10163
NUTRIENT REQUIREMENTS, PHYSIOCOCH   W69-10161
NUTRIENT REQUIREMENTS, NUTRIENTS,   W69-10177
NUTRIENT REMOVAL.:   W72-08133
NUTRIENT SOURCES, WATER SAMPLING,   W70-00264
NUTRIENT SOURCES, PHOSPHORUS BUDG   W70-04268
NUTRIENT TRAPS, NUTRIENT INFLUX,   W70-04268
NUTRIENT-PHYTOPLANKTON RELATIONSH   W71-05407
NUTRIENTS.: / AIR POLLUTION, WATE   W71-06821
NUTRIENTS.: /, FUNGI, INSECTS, BA   W71-10786
NUTRIENTS.: /S, WATER POLLUTION E   W69-10159
NUTRIENTS.: /S, SYSTEMS ANALYSIS,   W70-02779
NUTRIENTS.: /RESERVOIRS, AIR TEMP   W70-02646
NUTRIENTS.: /SA(KUTZ), CULTURE ME   W70-03507
NUTRIENTS.:   W69-05844
NUTRIENTS.:   *SEDI W68-00511
NUTRIENTS), LAKE VOLUME, MEAN DEP   W70-00270
NUTRIENTS, LAKES, CHANNELS, REVIE   W70-00269
NUTRIENTS, WATER TEMPERATURE, DIS   W70-00264
NUTRIENTS, DEPTH, DOMESTIC ANIMAL   W70-00274
NUTRIENTS, ORGANIC MATTER, P: /CE   W70-00265
NUTRIENTS, QUALITY CONTROL, TREAT   W70-00263
NUTRIENTS, NITROGEN COMPOUNDS, AM   W69-10177
NUTRIENTS, HYDROSTATIC PRESSURE,   W69-10160
NUTRIENTS, TEMPERATURE, EPILIMNIO   W69-10167
NUTRIENTS, PHYTOPLANKTON, INHIBIT   W69-10177
NUTRIENTS, NITRATES, PHOSPHATES,   W69-10170
NUTRIENTS, SEAWATER, ORGANIC COMP   W69-09755
NUTRIENTS, LIMNOLOGY, DYSTROPHY,   W69-09723
NUTRIENTS, ALGAE, WATER POLLUTION   W69-09135
NUTRIENTS, LIG: / COMPOUNDS, BACT   W69-09334
NUTRIENTS, PHYSIOLOGICAL ECOLOGY,   W70-00713
NUTRIENTS, PHOSPHORUS, POPULATION   W70-00667
NUTRIENTS, TASTE, ODOR, COLOR, TU   W70-02251
NUTRIENTS, NORTHEAST U S, ECOLOGI   W70-02764
NUTRIENTS, VITAMINS, TOXINS, INHI   W70-02510
NUTRIENTS, NORTHEAST U.S., ECOLOG   W70-02772
NUTRIENTS, EUTROPHICATION, BICARB   W70-02792
NUTRIENTS, EUTROPHICATION, CHLORO   W70-02777
NUTRIENTS, PHOSPHORUS, KINETIC TH   W70-02779
NUTRIENTS, LIGHT, WASHINGTON, WAT   W70-03309
NUTRIENTS, ESSENTIAL NUTRIENTS, C   W70-02245
NUTRIENTS, LIGHT PENETRATION, SEW   W70-01933
NUTRIENTS, POLLUTION IDENTIFICATI   W70-02248
NUTRIENTS, CALCIUM, MAGNESIUM, PO   W70-02245
NUTRIENTS, PHYSIOLOGICAL ECOLOGY,   W70-02245
NUTRIENTS, ALGAE, DIATOMS, CHLORO   W70-01943
NUTRIENTS, ESSENTIAL NUTRIENTS, I   W70-02248
NUTRIENTS, INHIBITION, NITROGEN C   W70-02248
NUTRIENTS, HEAVY METALS.: /EFFECT   W71-11561
NUTRIENTS, LAKES, WATER PROPERTIE   W71-10079
NUTRIENTS, PESTICIDES, RECREATION   W71-09577
NUTRIENTS, BACTERIA, ALGAE, TREAT   W71-06737
NUTRIENTS, SEWAGE, WASTE WATER(PO   W71-05879
NUTRIENTS, ALGAE, STRATIFIED FLOW   W71-05878
NUTRIENTS, NITROGEN, PHOSPHOROUS,   W71-13335
NUTRIENTS, FILTRATION, MICROORGAN   W71-13313
NUTRIENTS, *FERMENTATION, TOXICIT   W71-13413
NUTRIENTS, *PRIMARY PRODUCTIVITY,   W72-01990
NUTRIENTS, NITROGEN, WASTE WATER   W71-04072
NUTRIENTS, NITROGEN, PHOSPHORUS,   W71-03048
NUTRIENTS, BACTERIA, ALGAE, BRACK   W71-03031
NUTRIENTS, PRODUCTIVITY, PHOTOSYN   W70-06225
NUTRIENTS, *WASTE WATER TREATMENT   W70-07031
NUTRIENTS, WASTE WATER TREATMENT,   W70-06925
NUTRIENTS, *EUTROPHICATION, *HYDR   W70-09889
NUTRIENTS, PLANT PATHOLOGY, *ALGA   W71-00832
NUTRIENTS, SOLUTES, CHLORIDES, CA   W70-09557
NUTRIENTS, CHLORELLA, CULTURES.:   W70-07957
NUTRIENTS, WATER POLLUTION EFFECT   W70-08637
NUTRIENTS, EUTROPHICATION.: /ILTR   W70-08627
NUTRIENTS, ALGAE, FISH, STRATIFIC   W70-08834
NUTRIENTS, NUISANCE ALGAE, OLIGOT   W68-00680
NUTRIENTS, PHOSPHORUS, INFLUENT S   W68-00858
NUTRIENTS, RAINFALL-RUNOFF RELATI   W68-00891
NUTRIENTS, INDICATOR SPECIES, EUG   W69-01977
NUTRIENTS, LAKES, TESTS, NITROGEN   W68-00856
NUTRIENTS, PHOSPHATES, SILICATES,   W68-00860
NUTRIENTS, *NUTRIENT REQUIREMENTS   W69-04802
NUTRIENTS, *EUTROPHICATION, *CLAD   W69-04798
NUTRIENTS, EUTROPHICATION, FREEZE   W69-03518
NUTRIENTS, PHOSPHORUS COMPOUNDS,:   W69-04798

PRODUCTIVITY, BIOASSAY, CYCLING   NUTRIENTS, ENVIRONMENTAL EFFECTS,   W69-04798
*PHOSPHORUS COMPOUNDS, *CYCLING   NUTRIENTS, WATER CHEMISTRY, WATER   W69-05142
ION-REDUCTION POTENTIAL, CYCLING   NUTRIENTS, EUTROPHICATION.: /IDAT   W69-04521
ATER POLLUTION SOURCES, *CYCLING   NUTRIENTS, ALGAE, ANALYTICAL TECH   W69-04800
IMNOLOGY, NUTRIENT REQUIREMENTS,   NUTRIENTS, PHOTOSYNTHET: /TION, L   W69-03517
LOGY, BALANCE OF NATURE, CYCLING   NUTRIENTS, ENVIRONMENTAL EFFECTS,   W69-03517
TROGEN FIXATION, NUISANCE ALGAE,   NUTRIENTS, PHOSPHORUS COMPOUNDS,   W69-04802
EN COMPOUNDS, NITROGEN FIXATION,   NUTRIENTS, NU: /LIMNOLOGY, NITROG   W69-03518
ENVIRONMENTAL EFFECTS, ESSENTIAL   NUTRIENTS, CHLOROPHYTA, DIATOMS,   W69-03517
NVIRONMENTAL EFFECTS, *ESSENTIAL   NUTRIENTS, *FISH, *FISHERIES, ALG   W69-04804
AQUATIC ALGAE, BIOASSAY, CYCLING   NUTRIENTS, ENVIRONMENTAL EFFECTS,   W69-03518
, *ESSENTIAL NUTRIENTS/ *CYCLING   NUTRIENTS, *ENVIRONMENTAL EFFECTS   W69-04804
ENVIRONMENTAL EFFECTS, ESSENTIAL   NUTRIENTS, FREEZE DRYING, FREEZE-   W69-03370
TA, CYANOPHYTA, DIATOMS, CYCLING   NUTRIENTS, ENVIRONMENTAL EFFECTS,   W69-03188
FIXATION, NUTRIENT REQUIREMENTS,   NUTRIENTS, PHYSIOLOGICAL ECOLOGY,   W69-03364
Y, CHLOROPHYTA, DIATOMS, CYCLING   NUTRIENTS, ENVIRONMENTAL EFFECTS,   W69-03515
AL EFFECTS, ESSENTIAL NUTRIENTS,   NUTRIENTS, PHOSPHORUS, PHOSPHATES   W69-03373
E, AQUATIC PRODUCTIVITY, CYCLING   NUTRIENTS, ENVIRONMENTAL EFFECTS,   W69-03514
UTRIENTS/ *CHLOROPHYTA, *CYCLING   NUTRIENTS, *ENZYMES, *ESSENTIAL N   W69-03186
UTRIENTS, *NIT/ *ALGAE, *CYCLING   NUTRIENTS, *ENZYMES, *ESSENTIAL N   W69-03185
NUTRIENTS, *ENZYMES, *ESSENTIAL   NUTRIENTS, *NITROGEN CYCLE, *NUTR   W69-03186
ENVIRONMENTAL EFFECTS, ESSENTIAL   NUTRIENTS, NUTRIENTS, PHOSPHORUS,   W69-03373
VITY, BALANCE OF NATURE, CYCLING   NUTRIENTS, ENVIRONMENTAL EFFECTS,   W69-03513
TA, CYANOPHYTA, DIATOMS, CYCLING   NUTRIENTS, ENVIRONMENTAL EFFECTS,   W69-03364
LOROPHYTA, CYANOPHYTA, ESSENTIAL   NUTRIENTS, ENVIRONMENTAL EFFECTS,   W69-03358
ENVIRONMENTAL EFFECTS, ESSENTIAL   NUTRIENTS, EUTROPHICATION, INHIBI   W69-03515
ENVIRONMENTAL EFFECTS, ESSENTIAL   NUTRIENTS, FREEZE DRYING, FREEZE-   W69-03362
IRONMENTAL SANITATION, ESSENTIAL   NUTRIENTS, HARVESTING OF ALGAE, I   W69-03369
ENVIRONMENTAL EFFECTS, ESSENTIAL   NUTRIENTS, EUTROPHICATION, INHIBI   W69-03514
IRONMENTAL SANITATION, ESSENTIAL   NUTRIENTS, INHIBITION, LIGHT INTE   W69-03513
NUTRIENTS, *ENZYMES, *ESSENTIAL   NUTRIENTS, *NITROGEN CYCLE, ANALY   W69-03185
TIC WEEDS, *BIOASSAY, *ESSENTIAL   NUTRIENTS, *NITROGEN COMPOUNDS, A   W69-03364
TA, CYANOPHYTA, DIATOMS, CYCLING   NUTRIENTS, ENVIRONMENTAL EFFECTS,   W69-03373
ENVIRONMENTAL EFFECTS, ESSENTIAL   NUTRIENTS, HARVESTING OF ALGAE, N   W69-03371
*ANALYTICAL TECHNIQUES, *CYCLING   NUTRIENTS, *EUTROPHICATION, *PHOS   W69-03358
GEN DEMAND, CHLOROPHYTA, CYCLING   NUTRIENTS, ENVIRONMENTAL EFFECTS,   W69-03374
ATER POLLUTION EFFECTS, *CYCLING   NUTRIENTS, *EUTROPHICATION, ALGAE   W69-03369
*ANALYTICAL TECHNIQUES, *CYCLING   NUTRIENTS, *EUTROPHICATION, *PHOS   W69-03370
ENVIRONMENTAL EFFECTS, ESSENTIAL   NUTRIENTS, ENVIRONMENTAL ENGINEER   W69-03374
ATER QU/ *EUTROPHICATION, MAINE,   NUTRIENTS, AQUATIC POPULATIONS, W   W68-00470
SPORT FISHING, RECREATION, SILT,   NUTRIENTS, AQUATIC POPULATIONS, O   W68-00172
GE TRE/ *EUTROPHICATION, *ALGAE,   NUTRIENTS, DOMESTIC WASTES, *SEWA   W68-00248
C POLLUTANTS, *STREAMS, EUGLENA,   NUTRIENTS, PLANT POPULATIONS, HIS   W68-00483
SPRAYING, BIOCONTROL, INFECTION,   NUTRIENTS, ALGAE, NITROGEN, PHOSP   W68-00468
CED LA/ *EUTROPHICATION, *LAKES,   NUTRIENTS, OXYGEN REQUIREMENTS, I   W68-00487
PLANKTON, DISS/ *EUTROPHICATION,   NUTRIENTS, NITROGEN, PHOSPHORUS,   W68-00253
LAKES, MICROORGANISMS, NITROGEN,   NUTRIENTS, PHOSPHORUS, PLANKTON,   W68-00481
NITROGEN, PHOSPHOROUS, AVAILABLE   NUTRIENTS, SEDIMENT CORES: /ON,   W68-00511
CE ALGAE, NUTRIENT REQUIREMENTS,   NUTRIENTS, TOXICITY, WATER CHEMIS   W69-05867
ION, FISH, *LIMNOLOGY, FISHKILL,   NUTRIENTS, *ALGAE, *ALGAL CONTROL   W69-05844
*WASTES, FERTILIZERS, *ESSENTIAL   NUTRIENTS, *NITRIFICATION, CROPS,   W69-05323
CE ALGAE, NUTRIENT REQUIREMENTS,   NUTRIENTS, PHOSPHORUS COMPOUNDS,   W69-05868
*COBALT, *CYANOPHYTA, *ESSENTIAL   NUTRIENTS, *TRACE ELEMENTS, ALGAE   W69-06277
TROGEN FIXATION, NUISANCE ALGAE,   NUTRIENTS, VITAMINS, WATER POLLUT   W69-06277
ALGAE, WATER POLLUTION CONTROL,   NUTRIENTS, WATER POLLUTION EFFECT   W69-06276
HT, TEMPERATURE, CARBON DIOXIDE,   NUTRIENTS, BIOLOGICAL PROPERTIES,   W69-07442
ING FACTORS, NITROGEN COMPOUNDS,   NUTRIENTS, ODOR, OLIGOTROPHY, OXY   W69-06535
AC/ *PHOSPHATE, *LAKES, *CYCLING   NUTRIENTS, *SEDIMENT-WATER INTERF   W69-06859
WATER POLLUTION SOURCES, *LAKES,   NUTRIENTS, FERTILIZERS, PHOSPHATE   W69-07084
N, BICARBONATES, PHOTOSYNTHESIS,   NUTRIENTS, MICROBIOLOGY.: / CARBO   W69-07389
POLLUTION EF/ *ALGAE, *BIOASSAY,   NUTRIENTS, EUTROPHICATION, WATER   W69-06864
SOTOPES, *SALT MARSHES, *CYCLING   NUTRIENTS, GEORGIA, SEDIMENTS, BA   W69-08274
IMALS, FISH, ECOSYSTEMS, CYCLING   NUTRIENTS, TEMPERATURE, LIGHT QUA   W69-08526
WATER POLLUTION SOURCES, CYCLING   NUTRIENTS, COST COMPARISONS, ECON   W69-08518
ODOR, TASTE, LIGHT TEMPERATURE,   NUTRIENTS, COPPER SULFATE, STRATI   W69-08282
ENTS, MINNOWS, DETRITUS, CYCLING   NUTRIENTS, PLANKTON, ALGAE, INVER   W69-07861
PLANT G/ *FERTILIZERS, CYCLING   NUTRIENTS, EUTROPHICATION, ALGAE,   W69-07826
TRATION, IRON, LAKES, MAGNESIUM,   NUTRIENTS, NUISANCE ALGAE, OLIGOT   W69-07833
TREAMFLOW/ *PERIPHYTON, *CYCLING   NUTRIENTS, *ZINC RADIOISOTOPES, S   W69-07862
IA, WATER POLLUTION, REAERATION,   NUTRIENTS, OXYGEN SAG, AERATION,   W69-07520
NITRIFICATION, DENITRIFICATION,   NUTRIENTS, TERTIARY TREATMENT, AL   W69-08053
, URBANIZATION, DOMESTIC WASTES,   NUTRIENTS, PHOSPHORUS, ALGAE, ALG   W69-07818
EN COMPOUNDS, NITROGEN, AMMONIA,   NUTRIENTS, BIOCHEMICAL OXYGEN DEM   W69-08054
EFFECTS, NEW YORK, CHLOROPHYTA,   NUTRIENTS, ECOSYSTEMS, PHYTOPLANK   W70-03973
LANKTON, TRIPTON, NANNOPLANKTON,   NUTRIENTS, OLIGOTROPHY, TEMPERATU   W70-03959
ENVIRONMENTAL EFFECTS, ESSENTIAL   NUTRIENTS, EUTROPHICATION, LIGHT   W70-04283
CE ALGAE, NUTRIENT REQUIREMENTS,   NUTRIENTS, PHOSPHORUS COMPOUND: /   W70-04283
OCHEMICAL OXYGEN DEMAND, CYCLING   NUTRIENTS, DIATOMS, ENVIRONMENTAL   W70-04283
TS, NITROGEN, PHOSPHORUS, ALGAE,   NUTRIENTS, WATER PROPERTIES.: /EC   W70-04382
ROGEN ION CONCENTRATION, CYCLING   NUTRIENTS, MUD-WATER INTERFACES,   W70-04381
Y, GEOGRAPHICAL REGIONS, CYCLING   NUTRIENTS, ZOOPLANKTON, ALGAE, AQ   W70-03975
ITY, HYDROGEN ION CONCENTRATION,   NUTRIENTS, PHOSPHORUS, NITROGEN,   W70-03983
Y, GEOGRAPHICAL REGIONS, CYCLING   NUTRIENTS, ZOOPLANKTON, ALGAE, AQ   W70-03975
PHY, SEWAGE, ALGAE, SCUM, WEEDS,   NUTRIENTS, SEPTIC TANKS, EFFLUENT   W70-04193
OROPHYTA, PHOTOMETRY, CHLORELLA,   NUTRIENTS, TRACE ELEMENTS, CHLAMY   W70-05409
L, *MIDGES, FISH FOOD ORGANISMS,   NUTRIENTS, ENERGY TRANSFER, *FOOD   W70-05428
*CYANOPHYTA, *NITROGEN FIXATION,   NUTRIENTS, LAKE ERIE.: *ALGAE,   W70-05091
S, PRO/ *EUTROPHICATION, *LAKES,   NUTRIENTS, WATER POLLUTION EFFECT   W70-04721
ESIS, ILLUMINATION, TEMPERATURE,   NUTRIENTS, FACULTATIVE POND DESIG   W70-04786
TROGEN, FLOW, CHEMICAL ANALYSIS,   NUTRIENTS, EFFLUENTS, ALGAE, WEED   W70-04810

A, ALG/ *EUTROPHICATION, *LAKES,
TION, SEDIMENTS, PALEOLIMNOLOGY,
ICS, RECREATION, FISHING, ALGAE,
ALGAE, AGRICULTURAL WATERSHEDS,
E, LIGHT INTENSITY, CHLOROPHYLL,
LIC BACTERIA, THERMAO POLLUTION,
ATER POLLUTION, BACTERIA, ALGAE,
AGOONS, ACTIVATED SLUDGE, ALGAE,
, NUISANCE ALGAE, WATER QUALITY,
*ALGAE, *SESSILE ALGAE, *CYCLING
UCTIVITY, TEMPERATURE, BACTERIA,
S, INFLOW, CYANOPHYTA, LITTORAL,
.:   ALGAL BLOOMS, ALGAE
E), LAKE WASHINGTON(WASH), ALGAL
SKELETONEMA, THALASSIOSI/ *ALGAL
NOCOBALAMIN, ALGAL GROWTH, ALGAL
ANT STUDIES, ALGAL GROWTH, ALGAL
VAILABILITY, ALGAL GROWTH, ALGAL
S AERUGINOSA, ANTAGONISMS, ALGAL
OUND, SYRACUSE(NY),/ *PHYCOLOGY,
                        *MASSENA(
            CATTARAUGUS CREEK(
PHORA GLOMERATA, BUTTERNUT CREEK(
MORICHES BAY(NY), ONONDAGA LAKE(
INIA, / *FAYETTEVILLE GREEN LAKE(
NUTRIENT LIMITATION, CAYUGA LAKE(
A, NITZSCHIA, NAVICULA, SYRACUSE(
, PINNULARIA, NI/ *ONONDAGA LAKE(
HAEMOCYTOMETER COUNTING, ITHACA(
DIA FORMATION, LONG ISLAND SOUND(
S, RUSSIA, BAVARIA, MORICHES BAY(
RADIOCARBON METHOD, FINGER LAKES(
E AND TECHNOLOGY FOUND, SYRACUSE(
ODOSUM, ECTOCARPUS CONFERVOIDES,
WASTES, *MAN-MADE FIBER PLANTS,
WIS), ANACHARIS, NAIAS, SCIRPUS,
CUMULATION, RUTHENIUM-106, WHITE
YCAEUS, OIKOPLEURA, FRITILLARIA,
NGROSIRA STAGNALIS, NEPHROCYTIUM
US, NOSTOC MUSCORUM, SCENEDESMUS
ANSPARENCY, CYCLOPS, SCENEDESMUS
EDESMUS QUADRICAUDA, SCENEDESMUS
ATION, ASTERIONELLA, SCENEDESMUS
PALEA, CHLOROPLASTS, SCENEDESMUS
L, SUSCEPTIBILITY, PHYSIOLOGICAL
ANOIDES, CANDIDA KRUSEI, CANDIDA
ECOLOGICAL EFFIC/ *DAPHNIA PULEX
, BIOTA, CLAMS, FALLOUT, PACIFIC
MOLLUSCA, ECHINODERMATA, PACIFIC
CHLOROPHYLL, OIL, ALGAE, PACIFIC
Y, CARBON RADIOISOTOPES, PACIFIC
ABSORPTION, ADSORPTION, ATLANTIC
AE, AQUATIC LIFE, MILK, ATLANTIC
RON, COBALT, MANGANESE, ATLANTIC
RPTION, GULF OF MEXICO, ATLANTIC
IME, TRIDACNA CROCEA, WOODS HOLE
CULTURES, FLUOROMETRY, BIOMASS,
GY, RUSSIA, SECOND INTERNATIONAL
K SEA, CONTACT ZONE/ *BIOLOGICAL
QUES, *RADIOISOTOPES, LIMNOLOGY,
MARINE PLANTS, DINOFLAGELLATES,
PES, RADIOACTIVE WASTE DISPOSAL,
DIATOMS, RESPIRATION, PIGMENTS,
ISH, WASHINGTON, LAKES, INDIANA,
UCCESSION, CHLORELLA, CHELATION,
E, FILAMENTOUS BLUE-GREEN ALGAE,
CTALURUS PUNCTATUS, TRICHOPTERA,
WATER CARRIER(ISRAEL), THRESHOLD
NTARIO), ALGAE COUNTS, THRESHOLD
S), STABILIZATION POND LOADINGS,
CONTROL REQUIREMENTS, THRESHOLD
CONTROL, NUISANCE ALGAE, TASTE,
ENTS, BIOCHEMICAL OXYGEN DEMAND,
L, SLIME, TASTE-PRODUCING ALGAE,
NDUSTRIAL WASTE, CORES, DIATOMS,
C MATTER, *PO/ *WATER TREATMENT,
*WATER POLLUTION SOURCES, TASTE,
WAGE, COLIFORMS, EUTROPHICATION,
IODEGRADATION, *SEWAGE BACTERIA,
ADIOACTIVITY, PESTICIDES, TASTE,
UTION TREATMENT, COPPER SULFATE,
GANIC MATTER, TOXICITY, PHENOLS,
MNION, CYCLING NUTRIENTS, TASTE,
WATER SHORTAGE FILTERS, SNAILS,
MICROORGANISMS, HISTORY, TASTE,
AIR POLLUTION, WATER POLLUTION,
CTERIA, WATER POLLUTION SOURCES,
*WATER SUPPLY, *CONTROL, TASTE,
, NITROGEN COMPOUNDS, NUTRIENTS,
OXYGEN DEMAND, DISSOLVED OXYGEN,

NUTRIENTS, FERTILIZATION, BACTERI   W70-05761
NUTRIENTS, PRODUCTIVITY, SEDIMENT   W70-05663
NUTRIENTS, TURBIDITY, TERTIARY TR   W70-05645
NUTRIENTS, FERTILIZATION, PHYSICA   W70-05565
NUTRIENTS, CONDUCTIVITY, AGRICULT   W70-05752
NUTRIENTS, TEMPERATURE, PHOTOSYNT   W70-05270
NUTRIENTS, INDICATORS, ODOR, TAST   W70-05264
NUTRIENTS, VIRUSES, BACTERIA, E.   W72-06017
NUTRIENTS, PHOSPHORUS, NITROGEN,   W72-05453
NUTRIENTS, ECOSYSTEMS, PHOSPHORUS   W73-01454
NUTRIENTS, ESTUARINE ENVIRONMENT,   W73-00748
NUTRIENTS, MODEL STUDIES, BIOASSA   W73-01454
NUTRITION, MICROCYSTIS AERUGINOSA   W70-05565
NUTRITION, ALGAL PHYSIOLOGY, HETE   W70-04506
NUTRITION, DITYLUM, CHAETOCEROS,   W70-04503
NUTRITION, ALGAL PHYSIOLOGY, CALO   W69-06277
NUTRITION, ALGAL PHYSIOLOGY, AMER   W69-06865
NUTRITION, ALGAL PHYSIOLOGY, KJEL   W69-05868
NUTRITION, ALGAL PHYSIOLOGY, LAKE   W69-05867
NY STATE SCIENCE AND TECHNOLOGY F   W70-03973
NY).:   W72-08686
NY).:   W69-10080
NY).:   *CLADOPHORA FRACTA, CLADO W72-05453
NY).: /GENETICS, RUSSIA, BAVARIA,   W70-03973
NY), *MARL LAKES, CHLOROBIUM, FIL   W70-05760
NY), ASTERIONELLA FORMOSA, SYNEDR   W70-01579
NY), ENTEROMORPHA INTESTINALIS, C   W70-03974
NY), EUNOTIA, AMPHORA, GOMPHONEMA   W70-03974
NY), GLENODINIUM QUADRIDENS, OOCY   W69-06540
NY), ICELAND, TRITIATED THYMIDINE   W69-09755
NY), ONONDAGA LAKE(NY).: /GENETIC   W70-03973
NY), RODHE'S SOLUTION VIII, NUTRI   W70-01579
NY), SYRACUSE UNIVERSITY, FUNDAME   W70-03973
NYLON COLUMN TECHNIQUE, PARTICULA   W70-01074
NYLON-6, *VISCOSE RAYON, ZINC COM   W70-06925
NYMPHAEA, CRASSIPES EICHORNIA, TY   W70-05263
OAK LAKE(TENN), ACTIVITY-DENSITY,   W70-04371
OBELIA, CLUSTER ANALYSIS, MULTIVA   W73-01446
OBESUM W AND G S WEST, RADIOCOCCU   W70-04468
OBLIQUUES, SCENEDESMUS QUADRICAUD   W70-02964
OBLIQUUS, DACTYLOCOCCUS, ESCHERIC   W70-10161
OBLIQUUS, HELIX POMATIA, CHLORELL   W70-04184
OBLIQUUS, GROWTH CONSTATS, RIVER   W70-03955
OBLIQUUS, SCENEDESMUS QUADRICAUDA   W70-05381
OBSERVATIONS, CHRONIC EXPOSURE, A   W70-01779
OBTUSA, CANDIDA TROPICALIS, NOCTI   W70-04368
OBTUSA, SCENEDESMUS QUADRICAUDA,   W69-10152
OCE: /ITIUM, CARBON RADIOISOTOPES   W69-08269
OCEAN.: / POLYCHAETA, CRUSTACEA,   W69-08526
OCEAN.: /OTE SENSING, SEA WATER,   W70-05377
OCEAN, CARBON DIOXIDE, LIGHT, TEM   W70-05652
OCEAN, DIATOMS, NERITIC ALGAE, IO   W70-00707
OCEAN, LANDFILLS, DISPOSAL, REGUL   W72-00941
OCEAN, PUERTO RICO, SPECTROSCOPY,   W69-08525
OCEAN, RADIOACTIVITY, GAMMA RAYS,   W69-08524
OCEANOGRAPHIC INSTITUTION, ACROPO   W69-08526
OCEANOGRAPHY.: /OLOGICAL STUDIES,   W69-08278
OCEANOGRAPHIC CONGRESS, HYPON: /O   W69-07440
OCEANOGRAPHY, *TRANSACTIONS, BLAC   W69-07440
OCEANOGRAPHY, TRACERS, BACTERIA,   W69-10163
OCEANS, BIOINDICATORS, FUNGI, EST   W70-04368
OCEANS, FOODS, ABSORPTION, HAZARD   W70-02786
OCEANS, ORGANIC MATTER, DIURNAL,   W70-04809
OCHROMONAS, PHYTOPLANKTON, SEAWEE   W70-04503
OCHROMONAS, BACTERIA, FUNGI, AQUA   W70-02510
OCULAR MICROMETER, SUCROSE-FLOTAT   W70-04253
ODONATA, EPHEMEROPTERA, PLECOPTER   W69-07846
ODOR NUMBERS.: /ATORIA, NATIONAL   W70-02373
ODOR NUMBERS, CROP-DUSTING AIRCRA   W69-10155
ODOR PERSISTENCE.: /ORADO, WEST(U   W70-04788
ODOR TESTS, ASTERIONELLA, APHANIZ   W69-10157
ODOR.: / QUALITY, WATER POLLUTION   W69-10155
ODOR-PRODUCING ALGAE, POLLUTION A   W69-06909
ODOR-PRODUCING ALGAE, DIATOMS, CY   W69-05704
ODOR-PRODUCING ALGAE, CHLOROPHYLL   W68-00470
ODOR, *ALGAE, *DECOMPOSING ORGANI   W72-04242
ODOR, ALGAE, RESERVOIRS, WATERSHE   W72-01813
ODOR, AMMONIUM COMPOUNDS, ACTIVAT   W70-00263
ODOR, ANAEROBIC CONDITIONS, AEROB   W71-02009
ODOR, COAGULATION, ELECTROLYTES,   W72-13601
ODOR, COLIFORMS, ECONOMICS.: /OLL   W72-01814
ODOR, COLOR, TURBIDITY, BIOCHEMIC   W72-03299
ODOR, COLOR, TURBIDITY, HYDROGEN   W70-02251
ODOR, COOLING TOWERS, FILTERS, EF   W72-00027
ODOR, DEEP WELLS, WATER TREATMENT   W70-08096
ODOR, EFFLUENTS, SEWAGE EFFLUENTS   W71-00940
ODOR, FARM WASTE, SLUDGE, SOLID W   W71-05742
ODOR, FILTERS, PLANKTON, WATER PO   W70-05389
ODOR, OLIGOTROPHY, OXYGEN, OXYGEN   W69-06535
ODOR, ORGANIC LOADING, *WASTE WAT   W70-06601

LFATES, *SULFIDES, ALGAE, GASES, ODOR, OXIDATION-REDUCTION POTENTI W70-07034
LINITY, REAERATION, FISH, TASTE, ODOR, OXYGEN SAG, POTOMAC RIVER.: W73-00750
TS, NUISANCE ALGAE, OLIGOTROPHY, ODOR, PHYTOPLANKTON, SULFATES, SE W69-07833
IB/ *EUTROPHICATION, CYANOPHYTA, ODOR, SCUM, FISHING, WISCONSIN, B W69-08283
R SUPPLY, *NEW HAMPSHIRE, ALGAE, ODOR, TASTE, LIGHT TEMPERATURE, N W69-08282
ILIZATION, FISH, NUISANCE ALGAE, ODOR, TASTE, DEPTH, CIRCULATION, W70-05404
A, ALGAE, NUTRIENTS, INDICATORS, ODOR, TASTE, FILTERS, WATER TREAT W70-05264
ASSOCIATION, CHEMICAL INDUSTRY, ODOR, TASTE, CONSUMER.: /ER WORKS W71-05943
, DUCKS, COLOR, FISHKILL, LAKES, ODOR, TOXICITY, WATER POLLUTION E W69-08279
, *EUTROPHICATION, COLOR, TASTE, ODOR, TOXINS, DISSOLVED OXYGEN, L W69-05697
ON CONCENTRATION, PHYTOPLANKTON, ODOR, WINDS, EPILIMNION, HYPOLIMN W70-05405
ODORS.: ODORIFEROUS PRODUCTS, ORGANISMS' W72-01812
ODORIFEROUS PRODUCTS, ORGANISMS' ODORS.: W72-01812
ION POTENTIAL, DISSOLVED OXYGEN, ODORS, CORROSION, OXIDATION, PHOT W71-11377
RGANIC LOADING, DESIGN CRITERIA, ODORS, EFFICIENCIES, *WASTE WATER W72-06854
NONA(WI/ *MADISON(WIS), NUISANCE ODORS, LAKE MENDOTA(WIS), LAKE MO W69-03366
INDRICA, MICROCYSTIS AERUGINOSA, OEDOGNIUM.: /RATION, ANABAENA CYL W70-03520
INDIA), MACROPHYTES, CLADOPHORA, OEDOGONIA, SPIROGYRA, MICROCYSTIS W70-05404
, CYMBELLA, LYNGBYA, FRAGILARIA, OEDOGONIUM, SPIROGYRA, CHAETOPHOR W70-04371
MELOSIRA VARIANS, SYNEDRA ULNA, OEDOGONIUM, PHORMIDIUM RETZII, SU W70-00265
SLUDGE, INCINERATION, STORM RUN- OFF, SEDIMENTATION, BIOCHEMICAL O W71-10654
ON, NATIONAL SCIENCE FOUNDATION, OFFICE OF NAVAL RESEARCH, US DEPA W70-03975
ON, NATIONAL SCIENCE FOUNDATION, OFFICE OF NAVAL RESEARCH, US DEPA W70-03975
NCHOA LAMPROTAENIA, *OPISTHONEMA OGLIMUN, UDOTEA FLABELLUM, ZINC, W69-08525
PARK, MT RAINIER NATIONAL PARK, OHANAPECOSH HOT SPRINGS(WASH), SC W70-03309
BLUE-GREEN ALGAE, PHYTOPLANKTON, OHIO RIVER, PENNSYLVANIA, WATER S W69-05697
N EFFECTS, BACTERIA, PERIPHYTON, OHIO.: /R, CARBON, WATER POLLUTIO W71-04206
   GREAT MIAMI RIVER(OHIO).: W70-00475
RUN, ORGANIC CARBON, CINCINNATI(OHIO).: /, CHLOROPHYLL A, SHAYLER W71-04206
*NEW SPECIES, *WESTERN LAKE ERIE(OHIO), GONGROSIRA STAGNALIS, NEPH W70-04468
IOLOGICAL PROBLEMS, SCIOTO RIVER(OHIO), LAKE MENDOTA(WIS), LAKE MO W72-01813
AL SLIME, DAYTON(OHIO), VANDALIA(OHIO), RHIZOPODS, SPIDERS, MILK W W72-01819
SPIDER/ *ZOOGLEAL SLIME, DAYTON(OHIO), VANDALIA(OHIO), RHIZOPODS, W72-01819
AESTHETICS, ECOLOGY, ECONOMICS, OHIO, COMMERCIAL FISHING, PUBLIC W70-01943
TRUM, LEPTOCYLINDRUS, CORYCAEUS, OIKOPLEURA, FRITILLARIA, OBELIA, W73-01446
GAS CHROMATOGRAPHY, LIPIDS, OIL, OIL SHALE.: /GENESIS, SEDIMENTS, W69-08284
ENTOUS GREEN ALGAE, / *RECOVERY, OIL SPILL, *OIL POLLUTION, *FILAM W70-09976
, CLARKSBURG, BUFFALO, DUNREITH, OIL SPILLS.: /S, LEBANON, BATAVIA W72-03641
HODAMINE WT DYE, AIRBORNE TESTS, OIL SPILLS, GEOLOGICAL SS POLARIS W70-05377
NDUSTRIAL WASTES, GARBAGE DUMPS, OIL WASTES, INLETS(WATERWAYS), OR W70-05112
EASTS, BACTERIA, FISH, PROTEINS, OIL WASTES, OILY WATER, SULFITE L W69-05891
VATED CARBON, CHLORINE, PHENOLS, OIL WASTES, HYDROGEN ION CONCENTR W70-00263
HENS, TIDAL EFFECTS, DISPERSION, OIL-WATER INTERFACES, PERSISTENCE W70-01777
EY CANYON, AROMATIC SOLV/ *CRUDE OIL, *GAMOSOL, *OIL SPILLS, *TORR W70-01470
SENSING, SEA WATER, CHLOROPHYLL, OIL, ALGAE, PACIFIC OCEAN.: /OTE W70-05377
ER, AQUATIC ALGAE, MARINE ALGAE, OIL, MARSH, POLLUTANTS.: /ILY WAT W70-01231
NTS, GAS CHROMATOGRAPHY, LIPIDS, OIL, OIL SHALE.: /GENESIS, SEDIME W69-08284
RIA, FISH, PROTEINS, OIL WASTES, OILY WATER, SULFITE LIQUORS, ULTI W69-05891
POLLUTION, CLEANING, DISASTERS, OILY WATER, MARINE ALGAE, SHIPS.: W70-01470
OLLUTION EFFECTS, *SALT MARSHES, OILY WATER, *OIL, *MARINE ALGAE, W70-09976
(CANADA), / TROPHIC COMMUNITIES, OKANAGAN LAKE(CANADA), SKAHA LAKE W71-12077
OLLUTION, FOSSIL, CIMARRON RIVER(OKLA), BIXBY(OKLA), KEYSTONE RESE W70-04001
BIXBY(OKLA), KEYSTONE RESERVOIR(OKLA), CAROTENOIDS.: /IVER(OKLA), W70-04001
SIL, CIMARRON RIVER(OKLA), BIXBY(OKLA), KEYSTONE RESERVOIR(OKLA), W70-04001
N R/ ARKANSAS RIVER(OKLA), TULSA(OKLA), POLLUTION, FOSSIL, CIMARRO W70-04001
SSIL, CIMARRON R/ ARKANSAS RIVER(OKLA), TULSA(OKLA), POLLUTION, FO W70-04001
TON, DECOMPOSING ORGANIC MATTER, OKLAHOMA, RESERVOIRS, SAMPLING, I W70-04001
AKE, PULLIN LAKE, GRAPHITE LAKE, OLD JOHN LAKE, PINGO LAKE, TWELVE W69-08272
RTHROPODA, CHIRONOMIDAE, AQUATIC OLIGOCHAETES.: /THYTHERMOGRAPH, A W70-04253
S, DIPTERA, CRUSTACEA, MOLLUSCS, OLIGOCHAETES, GASTROPODS.: /NSECT W70-04253
MIDGES, WATER POLLUTION EFFECTS, OLIGOCHAETES, DIPTERA, ALGAE, INS W71-12077
OMASS, RESPIRATION, LIFE CYCLES, OLIGOCHAETES, MOLLUSKS, STRATIFIC W71-00665
NSIN, SAMPLING, CONTOURS, DEPTH, OLIGOCHAETES, TOXICITY, FISH, BLO W70-06217
PLANKTON, MAYFLIES, FAUNA, FISH, OLIGOCHAETES, MIDGES, SNAILS, PHO W70-01943
PULSE-LABE/ *ANACYSTIS NIDULANS, OLIGOPHOSPHATES, METAPHOSPHATES, W71-03033
S, BIOMASS, GROWTH RATES, ALGAE, OLIGOTROPHY, LIMITING FACTORS, CH W70-02779
MENTS, BIOASSAY, EUTROPHICATION, OLIGOTROPHY, LAKES, MICHIGAN, BIC W72-03183
BOLISM, ABSORPTION, COMPETITION, OLIGOTROPHY, VITAMINS, HEAVY META W71-10079
*LAKES, *ALGAE, *EUTROPHICATION, OLIGOTROPHY, NUTRIENTS, WATER TEM W70-00264
NKTON, COPEPODS, EUTROPHICATION, OLIGOTROPHY, TEMPERATURE, PHOTOSY W69-10154
NUTRIENTS, LIMNOLOGY, DYSTROPHY, OLIGOTROPHY, *PRIMARY PRODUCTIVIT W69-09723
AND, ECOSYSTEMS, EUTROPHICATION, OLIGOTROPHY, SEWAGE, ALGAE, SCUM, W70-04193
PHYSIOLOGICAL ECOLOGY, DIATOMS, OLIGOTROPHY, CYANOPHYTA, CHLOROPH W70-04503
KE ONTARIO, POLLUTION ABATEMENT, OLIGOTROPHY, DEPTH, PHYTOPLANKTON W70-03964
IPTON, NANNOPLANKTON, NUTRIENTS, OLIGOTROPHY, TEMPERATURE, LIGHT I W70-03959
SIUM, NUTRIENTS, NUISANCE ALGAE, OLIGOTROPHY, ODOR, PHYTOPLANKTON, W69-07833
OGEN COMPOUNDS, NUTRIENTS, ODOR, OLIGOTROPHY, OXYGEN, OXYGEN DEMAN W69-06535
ATIGRAPHY, BIOINDICATORS, ALGAE, OLIGOTROPHY, DIVERSION, CORES, SE W70-05663
YTA, CYANOPHYTA, BRACKISH WATER, OLIGOTROPHY, TEMPERATURE, LIGHT I W70-05752
LAKES, *CLASSIFICATION, DESMIDS, OLIGOTROPHIC, *PHYTOPLANKTON, BRI W68-00255
NUTRIENTS, AQUATIC POPULATIONS, OLIGOTROPHY, *FISH POPULATIONS, P W68-00172
STES, NUTRIENTS, NUISANCE ALGAE, OLIGOTROPHY, PHYTOPLANKTON, *REVI W68-00680
LIMNOLOGY, *NUTRIENTS, BIO MASS, OLIGOTROPHY, *PRIMARY PRODUCTIVIT W68-00912
RADIOISOTOPE,/ *EUTROPHICATION, OLIGOTROPHIC, *LAKES, *PHOSPHORUS W68-00858
, NEW BRUNSWICK(CANADA), SERVALL OMNI-MIXER HOMOGENIZER.: /CANADA) W70-04365
(ICELAND), NON-PARAMETRIC TESTS, ONE-SAMPLE RUNS TEST, CLUSTER ANA W69-09755
G RESPIROMETER,/ *ONONDAGA LAKE, ONONDAGA COUNTY, SYRACUSE, WARBUR W72-02959
SSIA, BAVARIA, MORICHES BAY(NY), ONONDAGA LAKE(NY).: /GENETICS, RU W70-03973
S LAKE, DUNLOP LAKE, *INORGANIC/ ONTARIO LAKES, QUIRKE LAKE, PECOR W70-05424
USTING AIRCRAFT, WATER SAMPLING, ONTARIO WATER RESOURCES COMMISSIO W69-10155

MENT, ODOR, *ALGAE, *DECOMPOSING   ORGANIC MATTER, *POTASSIUM COMPOU   W72-04242
, *LAKE MEAD, LAS VEG/ DISSOLVED   ORGANIC MATTER, LAS VEGAS, NEVADA   W72-02817
UDGE, SOLID WASTES, DECOMPOSTING   ORGANIC MATTER, LAGOONS, CATTLE.:   W71-05742
R POLLUTION CONTROL, DECOMPOSING   ORGANIC MATTER, CARBON DIOXIDE.: /   W72-01788
CETES, PLANT GROWTH, DECOMPOSING   ORGANIC MATTER, CHEMICAL ANALYSIS   W72-01812
UCTIVITY, ABSORPTION, SEDIMENTS,   ORGANIC MATTER, ALGAE, DIATOMS, C   W72-00845
HKILL, *CYANOPHYTA, *DECOMPOS/   ORGANIC MATTER, TOXICITY, DISSOLV   W72-01797
*WATER QUALITY, WATER POLLUTION,   ORGANIC MATTER, BIBLIOGRAPHIES.:   W70-07027
EFFLUENTS, DILUTION, DECOMPOS   ORGANIC MATTER, INSECTS, MITES, B   W71-00940
ITY, HARDNESS(WATER), CHLORIDES,   ORGANIC MATTER, CARBON, WATER POL   W71-04206
TY, CARBON RADIOISOTOPES, LAKES,   ORGANIC MATTER, SEA WATER, PRIMAR   W70-02504
RGANIC GROWTH FACTOR/ *DISSOLVED   ORGANIC MATTER, ORGANIC DEBRIS, O   W70-02510
TING, METALS, INHIBITORS, SILTS,   ORGANIC MATTER, IRON OXIDES, BACT   W70-02294
IRATION, PHOSPHATES, ECOSYSTEMS,   ORGANIC MATTER, WATER POLLUTION,   W70-03309
ENTS, ECOLOGY, CHLORELLA, ALGAE,   ORGANIC MATTER, CULTURES, CARBON,   W70-02804
OPHYTA, SAMPLING, PHYTOPLANKTON,   ORGANIC MATTER, CHLOROPHYLL, DYNA   W70-02780
EMPERATURE, NITROGEN, NUTRIENTS,   ORGANIC MATTER, P: /CENTRATION, T   W70-00265
OGRAPHIES, BACTERIA, DECOMPOSING   ORGANIC MATTER, WASTE WATER(POLLU   W70-00664
SSIUM, CALCIUM, MAGNESIUM, IRON,   ORGANIC MATTER, NITROGEN FIXATION   W70-00719
RIVERS, BOGS, PHENOLS, PIGMENTS,   ORGANIC MATTER, BIOMASS, BENTHIC   W70-01074
UALITY, NEW JERSEY, TEMPERATURE,   ORGANIC MATTER, HARDNESS, CARBON   W69-10157
, LIPIDS, NUTRIENT REQUIREMENTS,   ORGANIC MATTER, PHOTOSYNTHESIS, T   W69-06865
UM COMPOUNDS, OXYGEN, SILICATES,   ORGANIC MATTER, AIR TEMPERATURE,   W69-07096
AILS, RADIOACTIVITY, INTERFACES,   ORGANIC MATTER, RADIOISOTOPES, GA   W69-08272
YSTEMS, EUGLENA, EUTROPHICATION,   ORGANIC MATTER, PHYTOPLANKTON, PR   W69-06273
ROUNDWATER, *SESTON, *SEDIMENTS,   ORGANIC MATTER, CURRENTS(WATER),   W70-03501
PHORUS, AGRICULTURAL WATERSHEDS,   ORGANIC MATTER, SEDIMENTS, FERTIL   W70-04193
TION, PHYTOPLANKTON, DECOMPOSING   ORGANIC MATTER, OKLAHOMA, RESERVO   W70-04001
CYANOPHYTA, VITAMINS, PEPTIDES,   ORGANIC MATTER, TRACE ELEMENTS, L   W70-04369
ZYMES, KINETICS, PHOTOSYNTHESIS,   ORGANIC MATTER, ORGANIC COMPOUNDS   W70-04287
PRIMARY/ *PHYTOPLANKTON, *LAKES,   ORGANIC MATTER, SEASONAL, ALGAE,   W70-04290
KTON, *AQUATIC BACTERIA, *ALGAE,   ORGANIC MATTER, LAKES, DIFFUSION,   W70-04284
*WATER UTILIZATION, DECOMPOSING   ORGANIC MATTER, *STREAM POLLUTION   W68-00483
, OIL WASTES, INLETS(WATERWAYS),   ORGANIC MA: /ASTES, GARBAGE DUMPS   W70-05112
RGANISMS, PHYSIOLOGICAL ECOLOGY,   ORGANIC MATTER, WATER POLLUTION E   W70-04811
FFECTS, ANA/ *CHEMICAL ANALYSIS,   ORGANIC MATTER, WATER POLLUTION E   W70-05109
, RESPIRATION, PIGMENTS, OCEANS,   ORGANIC MATTER, DIURNAL, LATITUDI   W70-04809
TER), WATER POLLUTION, DRAINAGE,   ORGANIC MATTER, SEWAGE, PHYTOPLAN   W70-05404
LL, PHOTOSYNTHESIS, GROWTH RATE,   ORGANIC MATTER, SCENEDESMUS, CHLA   W70-05381
OLVED OXYGEN, CHEMICAL ANALYSIS,   ORGANIC MATTER, ALGAE, IRON, HY: /   W70-05760
EFFLUENTS, DIATOMS, PHOSPHORUS,   ORGANIC MATTER, NITROGEN, CHLOROP   W70-05663
STUDY, AEROBICITY.:   ORGANIC NITROGEN, BACTERIOLOGICAL   W71-05742
SSAY, FISH, AQUATIC ENVIRONMENT,   ORGANIC PESTICIDES, HYDROGEN ION   W72-11105
RIS COUNTY, TEX, POLYPHOSPHATES,   ORGANIC PHOSPHATES.:   HAR   W69-04800
, CHEMICAL PROPERTIES, BACTERIA,   ORGANIC POLLUTANTS, *STREAMS, EUG   W68-00483
C/ *POLYELECTROLYTES, *SYNTHETIC   ORGANIC POLYELECTROLYTES, CATIONI   W70-05267
Y TRANSFORMATIONS, HETEROTROPHY,   ORGANIC POLLUTION, MARINE ORGANIS   W69-07440
*TURNOVER RATES, *HETEROTROPHY,   ORGANIC SOLUTES, SWEDEN, LAKE ERK   W70-05109
WATER POLLUTION IDENTIFICATION,   ORGANIC SOLUTES, VITAMIN B-1, VIT   W70-05652
ALKALOID TOXINS, IBOGAALKALOID,   ORGANIC TOXINS.: /, *GLENODININE,   W71-07731
ON CONTROL, WATERCOURSES(LEGAL),   ORGANIC WASTES, PULP WASTES, INDU   W72-00462
CROP, A/ *EUTROPHICATION, LAKES,   ORGANIC WASTES, OXYGEN, STANDING   W70-02251
CAL OXYGEN DEMAND, ALGAL TOXINS,   ORGANIC WASTES, POLLUTANT IDENTIF   W69-01273
ITE, ALGAE, CYANOPHYTA, DOMINANT   ORGANISM, NUTRIENTS, RAINFALL-RUN   W68-00891
COMPOSITION, MATERIAL BALANCES,   ORGANISMS.: /ELLS, CONCENTRATION,   W70-06600
S, NON-LEGUMES, *NITROGEN FIXING   ORGANISMS.: /NITROGEN-15, SOD MAT   W72-08110
TS, PHOSPHORUS, PLANKTON, PLANTS(ORGANISMS).: /, NITROGEN, NUTRIEN   W68-00481
GAL CONTROL, ALGICIDES, DOMINANT   ORGANISMS, GRAZING, WATER POLLUTI   W69-01977
SECTS, *AQUATIC DRIFT, FISH FOOD   ORGANISMS, FISH PHYSIOLOGY, FISH   W69-02782
ICATION, FISH FARMING, FISH ROOD   ORGANISMS, FISH GENETICS, FISH HA   W69-04804
LITY CONTROL, *MIDGES, FISH FOOD   ORGANISMS, NUTRIENTS, ENERGY TRAN   W70-05428
ROPHY, ORGANIC POLLUTION, MARINE   ORGANISMS, MARINE ASSOCIATIONS, M   W69-07440
RIVER(GA), SPARTINA ALTEMIFLORA,   ORGANISMS, TURSIOPS TRUNCATUS, BR   W69-08274
TICS, *CHEMICAL CHARACTERISTICS,   ORGANISMS, SUSPENDED SOLIDS, SALT   W73-00748
ICROORGANISMS, *ALGAE, *DOMINANT   ORGANISMS, *ACCLIMATIZATION, DIAT   W70-07313
IOLOGIC 'DISPOSERS', MICROSCOPIC   ORGANISMS, PAPERMILL WASTE EFFLUE   W71-00940
ALINITY, FISH FARMING, FISH FOOD   ORGANISMS, FISH PHYSIOLOGY, RED T   W71-07731
GROUND SOLIDS, UPTAKE, CONSUMER   ORGANISMS, SIMULIUM, PHYSA, SALMO   W69-09334
            ODORIFEROUS PRODUCTS,   ORGANISMS' ODORS.:   W72-01812
SOURCES COMMISSION, PEST CONTROL   ORGANIZATION OF TORONTO, FILAMENT   W69-10155
M COMPOUNDS, PHENOLIC COMPOUNDS,   ORGANO-METALLIC COMPOUNDS, AMINES   W72-12987
STRY, SPRING WATERS, PHOSPHORUS,   ORGANOPHOSPHORUS COMPOUNDS, BACTE   W69-09334
ALYTICAL TECHNIQUES, *BIOASSAYS,   ORGANOPHOSPHORUS COMPOUNDS, SOLVE   W70-05269
OVELLITE, SPHALERITE, MILLERITE,   ORPIMENT, TETRAHEDRITE, BORNITE,   W69-07428
PIGMENTS, DIAPTOMUS, EPISCHURA,   ORTHOCYCLOPS, KERATELLA, POLYARTH   W70-05760
NS, ARTHROBACTER, PYROPHOSPHATE,   ORTHOPHOSPHATE, DISMUTATION, TANN   W69-07428
, BACILLUS SUBTILIS, ASPERGILLUS   ORYZAE NIGER, S: /AENA CYLINDRICA   W70-02964
BESA, LAKE KEGONSA, LAKE ZURICH,   OSCILLATO: /AKE MENDOTA, LAKE WAU   W69-09349
ND, PER/ *LAKE WASHINGTON(WASH),   OSCILLATORIA RUBESCENS, SWITZERLA   W69-10169
ESCENS, SWITZERLAND, PERIDINIUM,   OSCILLATORIA AGARDHI, PHORMIDIUM,   W69-10169
KE VOLUME, MEAN DEPTH, DILUTION,   OSCILLATORIA RUBESCENS, MADISON L   W70-00270
I DISC TRANSPARENCY, MYXOPHYCEA,   OSCILLATORIA, ANABAENA, DIAPTOMUS   W69-10182
YNEDRA, NAVICULA, APHANIZOMENON,   OSCILLATORIA, MELOSIRA, ACTINASTR   W70-08107
INULATE, OSCILLATORIA RUBESCENS,   OSCILLATORIA CHALYBIA, HYDRODICTY   W69-05704
INALIS, GLOEOTRICHIA ECHINULATE,   OSCILLATORIA RUBESCENS, OSCILLATO   W69-05704
ELENASTRUM CAPRICORNUTUM PRINTZ,   OSCILLATORIA VAUCHER, SKELETONEMA   W70-05752
COMPETITIVE EXCLUSION PRINCIPLE,   OSCILLATORIA RUBESCENS, PISTON CO   W70-05663
LAMINOSUM, OSCILLATORIA TENUIS,   OSCILLATORIA TEREBRIFORMIS, MATS,   W70-05270
LAMINOSUS, PHORMIDIUM LAMINOSUM,   OSCILLATORIA TENUIS, OSCILLATORIA   W70-05270
), VERTICAL, CHALK, METALIMNION,   OSCILLATORIA RUBESCENS, DINOBRYON   W70-05405

S, ACTIVATED SLUDGE, BIOCHEMICAL | OXYGEN DEMAND, OXIDATION-REDUCTIO | W71-09546
LVED OXYGEN, *ALGA/ *BIOCHEMICAL | OXYGEN DEMAND, *NUTRIENTS, *DISSO | W71-05407
*OXIDATION LAGOONS, *BIOCHEMICAL | OXYGEN DEMAND, *ANAEROBICS, ALGAE | W71-05742
BIC BACTERIA, ALGAE, BIOCHEMICAL | OXYGEN DEMAND, OXIDATION LAGOONS, | W71-08221
OGEN, *COLIFORMS, / *BIOCHEMICAL | OXYGEN DEMAND, *PHOSPHORUS, *NITR | W71-05409
POLLUTION, BACTERIA, EFFLUENTS, | OXYGEN DEMAND, AEROBIC TREATMENT, | W72-00462
FERMENTATION, TOXICITY, CHEMICAL | OXYGEN DEMAND, HYDROGEN ION CONCE | W71-13413
, REMOVAL, DEGRADATION, CHEMICAL | OXYGEN DEMAND, *WASTE WATER TREAT | W71-13335
R, COLOR, TURBIDITY, BIOCHEMICAL | OXYGEN DEMAND, HYDROGEN ION CONCE | W72-03299
TION, ON-SITE TESTS, BIOCHEMICAL | OXYGEN DEMAND, DIURNAL, CHLOROPHY | W72-01818
I, LABORATORY TESTS, BIOCHEMICAL | OXYGEN DEMAND, SEWAGE EFFLUENTS, | W72-01817
ALGAE, COPPER SULFATE, *CHEMICAL | OXYGEN DEMAND.: /ORGANIC MATTER, | W72-04495
ES, SEDIMENTS, DISSOLVED OXYGEN, | OXYGEN DEMAND, CYANOPHYTA, DIAT: / | W69-08518
ENTS, ODOR, OLIGOTROPHY, OXYGEN, | OXYGEN DEMAND, PHOSPHORUS COMPOUN | W69-06535
, STATE GOVERNMENTS, BIOCHEMICAL | OXYGEN DEMAND, ODOR-PRODUCING ALG | W69-06909
AMMONIA, NUTRIENTS, BIOCHEMICAL | OXYGEN DEMAND, TERTIARY TREATMENT | W69-08054
XYGEN SAG, AERATION, BIOCHEMICAL | OXYGEN DEMAND, PHOTOSYNTHETIC OXY | W69-07520
TMENT, *FILTRATION, *BIOCHEMICAL | OXYGEN DEMAND, *ACTIVATED SLUDGE, | W70-05821
AE, BACTERIA, CULTURES, CHEMICAL | OXYGEN DEMAND, NITROGEN, PHOSPHAT | W70-05655
, SELF-PURIFICATION, BIOCHEMICAL | OXYGEN DEMAND, TOXICITY, OXYGEN-R | W70-05750
OLVED OXYGEN, ALGAE, BIOCHEMICAL | OXYGEN DEMAND(BOD), MIXING, WINDS | W70-04790
ALGAE, DISINFECTION, BIOCHEMICAL | OXYGEN DEMAND(BOD), AERATION, COL | W70-04788
OXYGEN, TEMPERATURE, BIOCHEMICAL | OXYGEN DEMAND(BOD), EVAPORATION, | W70-04787
D STATES, CHLORELLA, BIOCHEMICAL | OXYGEN DEMAND, OXIDATION LAGOONS, | W70-05092
PRECIPITATION, IONS, BIOCHEMICAL | OXYGEN DEMAND, MINNESOTA, WATER C | W69-00446
NIC WASTES, POLLUTA/ BIOCHEMICAL | OXYGEN DEMAND, ALGAL TOXINS, ORGA | W69-01273
E, WATER PROPERTIES, BIOCHEMICAL | OXYGEN DEMAND, *BACTERIA, ACTIVAT | W68-00256
YGEN, *CHLORELLA, / *BIOCHEMICAL | OXYGEN DEMAND, *PHOTOSYNTHETIC OX | W69-03371
OF NATURE, BIOASSAY, BIOCHEMICAL | OXYGEN DEMAND, CHLOROPHYTA, CYCLI | W69-03374
WISCONSIN, BIOASSAY, BIOCHEMICAL | OXYGEN DEMAND, ENVIRONMENTAL EFFE | W69-03369
SYNTHETIC OXYGEN, / *BIOCHEMICAL | OXYGEN DEMAND, *CHLORELLA, *PHOTO | W69-03362
QUATIC PRODUCTIVITY, BIOCHEMICAL | OXYGEN DEMAND, CYCLING NUTRIENTS, | W70-04283
NAL, STRATIFICATION, BIOCHEMICAL | OXYGEN DEMAND, CARBON RADIOISOTOP | W70-04001
*ACTIVATED SLUDGE, *BIOCHEMICAL | OXYGEN DEMAND, *COLOR, WASTE WATE | W70-04060
GATION, RESPIRATION, BIOCHEMICAL | OXYGEN DEMAND, NITROGEN FIXATION. | W72-06022
, *DISSOLVED OXYGEN, BIOCHEMICAL | OXYGEN DEMAND, ALGAL CONTROL, *CO | W72-11906
N ION CONCENTRATION, BIOCHEMICAL | OXYGEN DEMAND, VIRGINIA.: /YDROGE | W72-10573
E MEAD, DILUTION, ALGAL GROWTHS, | OXYGEN DEPLETION.: / EFFECTS, LAK | W73-00750
(WIS), APHANIZOMENON FLOS AQUAE, | OXYGEN DEPLETION.: *YAHARA RIVER | W72-01797
S, WATER SAMPLING, ALGAL GROWTH, | OXYGEN DEPLETION, NUCLEAR ACTIVAT | W70-00264
, DELTA WATER FACILITIES(CALIF), | OXYGEN DYNAMICS, SACRAMENTO-SAN J | W70-05094
, PHORMIDIUM, WINKLER TECHNIQUE, | OXYGEN ELECTRODES, ATOMIC ABSORPT | W70-03309
TROPHICATION, *LAKES, NUTRIENTS, | OXYGEN REQUIREMENTS, ICED LAKES, | W68-00487
EUTROPHICATION, FISHKILL, ALGAE, | OXYGEN SAG, WATER QUALITY, CHLORI | W69-03948
OLLUTION, REAERATION, NUTRIENTS, | OXYGEN SAG, AERATION, BIOCHEMICAL | W69-07520
, REAERATION, FISH, TASTE, ODOR, | OXYGEN SAG, POTOMAC RIVER.: /NITY | W73-00750
*RAYTOWN(MO), LEE'S SUMMIT(MO), | OXYGEN TRANSFER, PH, SUSPENDED SO | W70-06601
EETER-PHELPS EQUATION, DISSOLVED | OXYGEN VARIATIONS, SAN JOAQUIN RI | W69-07520
AL OXYGEN DEMAND, PHOTOSYNTHETIC | OXYGEN.: /AG, AERATION, BIOCHEMIC | W69-07520
IFORMS, ALGAL CONTROL, DISSOLVED | OXYGEN.: /AND(BOD), AERATION, COL | W70-04788
MICHIGAN, WASHINGTON, DISSOLVED | OXYGEN.: / BOTTOM SEDIMENTS, LAKE | W69-09349
HEMICAL OXYGEN DEMAND, TOXICITY, | OXYGEN-REDUCTION POTENTIAL, SOLAR | W70-05750
TUARIES, *EUTROPHICA/ *DISSOLVED | OXYGEN, *ALGAE, *RESPIRATION, *ES | W69-07520
TION/ *PHYTOPLANKTON, *DISSOLVED | OXYGEN, *ALGAE, *DIURNAL DISTRIBU | W69-03611
N DEMAND, *NUTRIENTS, *DISSOLVED | OXYGEN, *ALGAE, COLIFORMS, EUTROP | W71-05407
L OXYGEN DEMAND, *PHOTOSYNTHETIC | OXYGEN, *CHLORELLA, ALGAE, ANALYT | W69-03371
TORS, *WATER QUALITY, *DISSOLVED | OXYGEN, *OHIO, WATER TEMPERATURE, | W70-00475
SYNTH/ *RADIONUCLIDE, *DISSOLVED | OXYGEN, *RIVERS, OXIDATION, PHOTO | W71-11381
GEN ION CONCENTRATION, DISSOLVED | OXYGEN, ALGAE, PHYTOPLANKTON, ALK | W71-07731
ANIC MATTER, TOXICITY, DISSOLVED | OXYGEN, ALGAE, WISCONSIN.: /G ORG | W72-01797
SS, FLOW RATES, DEPTH, DISSOLVED | OXYGEN, ALGAE, VELOCITY, WATER QU | W70-00265
TOPLANKTON, DISTRIBUTION, DEPTH, | OXYGEN, ALKALINITY, HYDROGEN ION | W71-00665
RETRIEVAL, COLIFORMS, DISSOLVED | OXYGEN, ALGAE, NUTRIENTS, SOLUTES | W70-09557
AND, *CHLORELLA, *PHOTOSYNTHETIC | OXYGEN, ALGAE, ANALYTICAL TECHNIQ | W69-03362
MONIA, HARDNESS(WATER), ACIDITY, | OXYGEN, ALGAE, SELF-PURIFICATION, | W69-07838
DATA, STRATIFICATION, DISSOLVED | OXYGEN, ALGAE, BIOCHEMICAL OXYGEN | W70-04790
ROGEN, PHOSPHATES, METHANE, MUD, | OXYGEN, ANAEROBIC CONDITIONS, HYD | W70-05761
TER POLLUTION EFFECTS, DISSOLVED | OXYGEN, AQUATIC BACTERIA, ALGAE.: | W70-04727
RE, LIGHT PENETRATION, DISSOLVED | OXYGEN, BICARBONATES, HYDROGEN IO | W70-06225
NIQUES, WATER QUALITY, DISSOLVED | OXYGEN, BIOCHEMICAL OXYGEN DEMAND | W70-08627
WATERS, IMPOUNDMENTS, *DISSOLVED | OXYGEN, BIOCHEMICAL OXYGEN DEMAND | W72-11906
ON, ALGAE, NITROGEN, PHOSPHORUS, | OXYGEN, CARBON, ENZYMES, SAMPLING | W70-09889
URE, HYDROGEN ION CONCENTRATION, | OXYGEN, CARBON DIOXIDE, AMMONIA, | W71-03031
ELLA, MANGANESE, PHOTOSYNTHESIS, | OXYGEN, CARBON DIOXIDE, SCENEDESM | W70-02964
ANALYSIS, PHYSIOLOGICAL ECOLOGY, | OXYGEN, CARBON DIOXIDE, CARBON RA | W69-07862
N REDUCTION POTENTIAL, DISSOLVED | OXYGEN, CHEMICAL ANALYSIS, ORGANI | W70-05760
DIATOMS, CHLORELLA, SCENEDESMUS, | OXYGEN, COLIFORMS, SEDIMENTS, MUD | W70-10161
ROGEN ION CONCENTRATION, TOXINS, | OXYGEN, DECOMPOSIN: /ALINITY, HYD | W70-02803
RATION, BICARBONATES, DISSOLVED | OXYGEN, DETRITUS, BACTERIA, WATER | W70-02249
SERVOIRS, TEMPERATURE, DISSOLVED | OXYGEN, DENSITY, ROOTED AQUATIC P | W71-07698
FRES WATER, LITTORAL, DISSOLVED | OXYGEN, DIURNAL, MIGRATION PATTER | W71-12870
GY, ALGAE, DIURNAL DISTRIBUTION, | OXYGEN, DISSOLVED OXYGEN, MEASURE | W70-05094
*RESERVOIR OPERATIONS, DISSOLVED | OXYGEN, DISSOLVED SOLIDS, SEDIMEN | W73-00750
NITROGEN, *COLIFORMS, *DISSOLVED | OXYGEN, ESTUARIES, ALGAE, WATER P | W71-05409
*PHOTOSYNTHESIS, *ALGAE, LAKES, | OXYGEN, EUTROPHICATION, ALGAL CON | W71-10098
TERS, WATER TREATMENT, DISSOLVED | OXYGEN, EUTROPHICATION, FISH, LAN | W70-05264
TY, *DISSOLVED SOLIDS, DISSOLVED | OXYGEN, EVAPORATION CONTROL, WATE | W72-01773
LAGOONS, SLUDGE, PHOTOSYNTHETIC | OXYGEN, FARM WASTES, WASTE WATER | W71-08221
TS, WATER TEMPERATURE, DISSOLVED | OXYGEN, FISHKILLS, LIMNOLOGY, ALG | W70-00264
TRIAL WASTES, ECOLOGY, DISSOLVED | OXYGEN, HARDNESS, TIDAL WATERS, O | W73-00748

ION, ALGAL PHYSIOLOGY, CALOTHRIX	PARIETINA, COCCOCHLORIS PENIOCYST	W69-06277
HIOCAPSA FLORIDANA, ENTEROCOCCI,	PARK RIVER: /HROMATIUM VINOSUM, T	W70-03312
R, NUCLEI/ *YELLOWSTONE NATIONAL	PARK, *ICELAND HOT SPRINGS, SINTE	W69-10160
INOSUS, PHORMIDIUM/ *YELLOWSTONE	PARK, *ICELAND, MASTIGOCLADUS LAM	W70-05270
SERVOIR, LEBANON, WADLEIGH STATE	PARK, LAKE SHATUTAKEE, MASCOMA LA	W69-08674
OLARIMETRY, YELLOWSTONE NATIONAL	PARK, MT LASSEN NATIONAL PARK, MT	W70-03309
ATIONAL PARK, MT LASSEN NATIONAL	PARK, MT RAINIER NATIONAL PARK, O	W70-03309
TIONAL PARK, MT RAINIER NATIONAL	PARK, OHANAPECOSH HOT SPRINGS(WAS	W70-03309
S, PRINCIPAL-COMPONENT ANALYSIS,	PART CORRELATION, DECAPODS, SPECI	W73-01446
, CORNELL UNIVERSITY, ELECTRONIC	PARTICLE COUNTING, EUDORINA ELEGA	W69-06540
S, ADSORPTION, SEWAGE TREATMENT,	PARTICLE SIZE.: /ELLA, SCENEDESMU	W71-03035
CHLAMYDOMONAS, LABORATORY TESTS,	PARTICLE SIZE, POPULATION, GROWTH	W72-02417
NE, GLYCOLLIC ACID, CHEMOTROPHY,	PARTICULATE MATTER, BLELHAM TARN(	W70-02504
RVOIDES, NYLON COLUMN TECHNIQUE,	PARTICULATE MATTER, BLUE FLUORESC	W70-01074
ON, CELL COUNT, OPTICAL DENSITY,	PARTICULATE ORGANIC CARBON, LAKE	W70-03983
ISRAEL, MICROCYSTIS, PRYMNESIUM	PARVUM, SYNERGISTIC EFFECTS, AXEN	W70-05372
EA LAKE(COLO), BLACK LAKE(COLO),	PASS LAKE(COLO), MICRO-ALGAE, FLA	W69-10154
WASTES, WATER POLLUTION SOURCES,	PATH OF POLLUTANTS, AQUATIC LIFE,	W69-10080
ACTIVE WASTE, WATER QUALITY ACT,	PATH OF POLLUTANT, FOOD CHAINS, S	W72-00941
AKE ONTARIO, *WATER CIRCULATION,	PATH OF POLLUTANTS, NUTRIENTS, AL	W71-05878
ITIONS, WATER POLLUTION SOURCES,	PATH OF POLLUTANT.: /AEROBIC COND	W71-11381
ACTERIA, E. COLI, PUBLIC HEALTH,	PATHOGENIC BACTERIA, TEMPERATURE,	W72-06017
*EPIPHYTOLOGY, NUTRIENTS, PLANT	PATHOLOGY, *ALGAE.: /Y, MICHIGAN,	W71-00832
EAR LAKE(CALIF), NUTRIENT ENERGY	PATHWAYS, TROPHIC RELATIONSHIPS.:	W70-05428
BOTRYOCOCCUS BRAUNII, METABOLIC	PATTERN, ANABAENA CYLINDRICA, GLY	W70-02965
OLVED OXYGEN, DIURNAL, MIGRATION	PATTERNS.: /WATER, LITTORAL, DISS	W71-12870
SOIL MICROBIOLOGY, DISTRIBUTION	PATTERNS, SOIL PROFILES.: /OZEMS,	W71-04067
TION, SOILS, COALS, DISTRIBUTION	PATTERNS, ACTIVATED CARBON, CLAYS	W69-09884
ASS CULTURE, MEDICINE, METABOLIC	PATTERNS, MICROALGAE, MICRONUTRIE	W69-07832
*LAKES, *CELLULOSE, DISTRIBUTION	PATTERNS, MEMBRANE PROCESSES, MAS	W69-00976
MINNETONKA(MINN), MINNEAPOLIS-ST	PAUL(MINN).: LAKE	W70-04810
*CANNERIES, *INDUSTRIAL WASTES,	PEACHES, APRICOTS, PEARS, CITRUS	W71-09548
COSTS, DINOBRYON, CONCORD(N H),	PEANACOOK LAKE(N H).: / TREATMENT	W69-08282
TRIAL WASTES, PEACHES, APRICOTS,	PEARS, CITRUS FRUITS, TOMATOES, *	W71-09548
NIC/ ONTARIO LAKES, QUIRKE LAKE,	PECORS LAKE, DUNLOP LAKE, *INORGA	W70-05424
LIMNOLOGY, BRACHIONUS, SYNCHAETA	PECTINATA, POLYARTHRA VULGARIS, K	W70-02249
HOLSTON RIVER(TENN), POTOMOGETON	PECTINATUS, POTOMOGETON CRISPUS,	W71-07698
ON, PONDS, ZOOPLANKTON, MINNOWS,	PEDALFERS, SCENEDESMUS, CHLAMYDOM	W70-02249
MOTROPHIC, AUTOTROPHIC, SILICON,	PEDIASTRUM BORYANUM, STAURASTRUM	W70-02804
LOSIRA, ACTINASTRUM, COELASTRUM,	PEDIASTRUM, WESTELLA, STEPHANODIS	W70-08107
INA, CULTURE MEDIUM, COELASTRUM,	PEDIASTRUM.: /CTION VALUES, EUDOR	W70-05409
STIS SOLITARIA, PANDORINA MORUM,	PEDIASTRUM BORYA: / PUSILLA, OOCY	W69-06540
IUM, ANKISTRODESMUS, MICROSPORA,	PEDIASTRUM, SCENEDESMUS, LEPOCINC	W70-03974
UCCESSION, ANABAENA, CLADOPHORA,	PEDIASTRUM, HYDRODICTYON, POTAMOG	W70-04494
LGAL CULTURES, *AXENIC CULTURES,	PEDIASTRUM, SELF-PURIFICATION, AL	W70-04381
IA, MICROCYSTIS, ANKISTRODESMUS,	PEDIASTRUM, TETRAEDRON, TETRASTRU	W70-03969
, GELATINOU/ *LAKE MENDOTA(WIS),	PELAGIC AREAS, HORIZONTAL, BLOOMS	W70-06229
NT CAROTENOIDS, CARBON FIXATION,	PELAGIC REGION, MONOMICTIC, DEPTH	W70-01933
, MONODUS SUBTERRANEUS, NAVICULA	PELLICULOSA, XANTHOPHYCEAE, BACIL	W70-02965
OSA, TRIBONEMA AEQUALE, NAVICULA	PELLICULOSA, LAKE MENDOTA(WIS), E	W69-10180
S, MONOCHRYSIS LUTHERI, NAVICULA	PELLICULOSA, GONYAULAX POLYEDRA,	W69-08278
UAEDA MARITIMA, ASTER TRIPOLIUM,	PEMBROKE, SOUTH-WEST WALES.: /, S	W70-09976
HIBITION, LIGHT, LIGHT INTENSITY, LIGHT	PENETRAT: /TS, EUTROPHICATION, IN	W69-03515
N, LIGHT, LIGHT INTENSITY, LIGHT	PENETRATION, LIMNOLOGY, PHOTOSYNT	W69-03516
HIBITION, LIGHT INTENSITY, LIGHT	PENETRATION, LIMNOLOGY, NITROGEN	W69-03514
TING OF ALGAE, INHIBITION, LIGHT	PENETRATION, NITROGEN COMP: /RVES	W69-03369
FREEZING, LIGHT INTENSITY, LIGHT	PENETRATION, LIMNOLOGY, NITROGEN	W69-03518
VESTING ALGAE, INHIBITION, LIGHT	PENETRATION, LIMNOLOGY, NITROGEN	W69-04798
N, LIGHT, LIGHT INTENSITY, LIGHT	PENETRATION, LIMNOLOGY, NUTRIENT	W69-03517
KALINITY, *PHOTOSYNTHESIS, LIGHT	PENETRATION, PHOSPHORUS, LAGOONS	W68-00248
LGAE, SEASONAL, LIMNOLOGY, LIGHT	PENETRATION, OXYGEN, SPORT FISHIN	W68-00487
VED OXYGEN, LAKES, RIVERS, LIGHT	PENETRATION, FLUORESCENCE, PHOTOS	W69-05697
INTENSITY, STABILIZATION, PONDS,	PENETRATION, ABSORPTION.: /LIGHT	W70-03923
HICATION, LIGHT INTENSITY, LIGHT	PENETRATION, LIMNOLOGY, NITROGEN	W70-04283
FICATION, DEPTH, PLANKTON, LIGHT	PENETRATION, SEASONAL, PRIMARY PR	W70-05405
VOIRS, PONDS, TEMPERATURE, LIGHT	PENETRATION, DISSOLVED OXYGEN, BI	W70-06225
PERATURE, IRON, NUTRIENTS, LIGHT	PENETRATION, SEWAGE EFFLUENTS, OX	W70-01933
FIXATION, IRON COMPOUNDS, LIGHT	PENETRATION, WATER POLLUTION EFFE	W70-02255
, EPILIMNION, THERMOCLINE, LIGHT	PENETRATION, STANDING CROP, INDUS	W71-07698
INTENSITY, PHOTOPERIODISM, LIGHT	PENETRATION, WATER POLLUTION EFFE	W72-00917
POLLUTION SOURCES, RUNOFF, LIGHT	PENETRATION, SAMPLING, PRIMARY PR	W72-04761
TH DAKOTA, PHOTOSYNTHESIS, LIGHT	PENETRATION, ON-SITE TESTS, BIOCH	W72-01818
ND(CANADA), NOVA SCOTIA(CANADA),	PENGUINS, LAKE ZURICH(SWITZERLAND	W70-05399
ALOTHRIX PARIETINA, COCCOCHLORIS	PENIOCYSTIS, DIPLOCYSTIS AERUGINO	W69-06277
TRICHIA ECHINULATA, COCCOCHLORIS	PENIOCYSTIS, MADISON(WIS), CHICAG	W70-00719
TROPHICATION, *DISSOLVED OXYGEN,	PENNSYLVANIA, BIOCHEMICAL OXYGEN	W69-10159
EFFECTS, WASTE WATER TREATMENT,	PENNSYLVANIA.: /, WATER POLLUTION	W71-05026
ILLUS DENITRIFICANS, W VIRGINIA,	PENNSYLVANIA, MARCASITE, CHALCOCI	W69-07428
LGAE, PHYTOPLANKTON, OHIO RIVER,	PENNSYLVANIA, WATER SUPPLY.: /N A	W69-05697
ER POLLUTION, DIATOMS, MARYLAND,	PENNSYLVANIA, RIVERS, STREAMS, EU	W70-04510
NCE ALGAE, CYANOPHYTA, VITAMINS,	PEPTIDES, ORGANIC MATTER, TRACE E	W70-04369
VITAMINS, CELLULOSE, CHLORELLA,	PEPTIDES, BACTERIA, SNAILS, WEIGH	W70-04184
ERBIVORES, CHLOROPHYLL, ENZUMES,	PEPTIDES, BIOCHEMISTRY, HYDROGEN	W70-01068
PROPERTIES, GENETICS, BACTERIA,	PEPTIDES, ENVIRONMENTAL EFFECTS,	W70-00273
COMPOUNDS, PHOSPHORUS COMPOUND,	PERCHES, FISH, DIATOMS, CARP, PLA	W70-00274
PHOSPHORUS, DETERGENTS, RUNOFF,	PERCOLATION, ROADS, ROOFS, PRECIP	W69-10178
ANT RESIDUES, MORRIS(MINN), SOIL	PERCOLATION.: / SOIL MINERALS, PL	W70-04193
CLOTELLA, CERATIUM HIRUNDINELLA,	PERIDINIUM WILLEI, CHLOROCOCCALES	W70-05405
OCCOLITHUS HUXLEYI, PYRAMIMONAS,	PERIDINIUM TROCHOIDEUM, VINEYARD	W70-05272
LLATORIA RUBESCENS, SWITZERLAND,	PERIDINIUM, OSCILLATORIA AGARDHI,	W69-10169

DICATORS, *ESTIMATING EQUATIONS,
IA/ LAKE SEBASTICOOK, ME, MAINE,
R), WATER QUALITY INDICATORS.:
, RIVERS, ENVIRONMENTAL EFFECTS,
TS, ECOSYSTEMS, PLANKTON, ALGAE,
TEMPERATURE, TRACERS, SAMPLING,
BIOMASS, WATER POLLUTION, ALGAE,
TER POLLUTION EFFECTS, BACTERIA,
RAULIC MODELS, LABORATORY TESTS,
LUENTS, OXYGEN, SOLAR RADIATION,
CTIVITY, *ALGAE, *THERMAL WATER,
BIOMASS, PIGMENTS, PRODUCTIVITY,
).:   *POTASSIUM
JAS, SILVEX, ENDOTHAL, POTASSIUM
                          *POTASSIUM
                          *POTASSIUM
GISTATIC TESTS, SLIM/ *POTASSIUM
ALGAE, PILOT PLANTS, HEAD LOSS,
N(SWEDEN), LAKE LOTSJON(SWEDEN),
CHLOROPHYLL FLUORESCENCE, FABRY-
, BALL-MILL, MEICELASE, HYDROGEN
TABILIZATION POND LOADINGS, ODOR
ISPERSION, OIL-WATER INTERFACES,
ARIO WATER RESOURCES COMMISSION,

ATHION, / CHLORELLA PYRENOIDOSA,
HIBITION, LIMNOLOGY, PESTICIDES,
, PESTICIDES, PESTICIDE REMOVAL,
ICITY, FRESH WATER, PULP WASTES,
LAKES, METABOLISM, *CHLORELLA, /
YNTHESIS, CHLORELLA, PESTICIDES,
TES, ALGAE, CALIFORNIA, FLORIDA,
          COAL CONTACT PROCESSES,
ISTANCE, BIOASSAY, GROWTH RATES,
STES, THERMAL POLLUTION, TOXINS,
IOINDICATORS, FISHKILL, HAZARDS,
LL, HAZARDS, PESTICIDE KINETICS,

ION, AQUATIC PLANTS, HERBICIDES,
ION, AQUATIC PLANTS, HERBICIDES,
BS, DDT, CHLORINATED HYDROCARBON
AND, VALUE, LEAVES, FERTILIZERS,
SH TAXONOMY, DDT, PHOSPHOTHIOATE
RONMENT, CHLORINATED HYDROCARBON
ION, *PHOTOSYNTHESIS, CHLORELLA,
ATE, FISHES, HERBICIDES, MIDGES,
EFFECTS, INHIBITION, LIMNOLOGY,
ANIC COMPOUNDS, LAGOON, FILTERS,
TY, ORGANIC COMPOUNDS, *PHENOLS,
ES, MUNICIPAL WASTES, NUTRIENTS,
SH, AQUATIC ENVIRONMENT, ORGANIC
ERATURE, CHLORINATED HYDROCARBON
, ORGANIC MATTER, RADIOACTIVITY,
ENARIA, AQUACULTURE, RECYCLING /
ALKENES, ALIPHATIC HYDROCARBONS,
MS, MARINE ASSOCIATIONS, MOSCOW,
NG POND(WIS), LAKE DELAVAN(WIS),
(WIS), LAKE NAGAWICKA(WIS), LAKE
CY, LAKE VANERN, LAKE CONSTANCE,
OLIDS, MULTI-CELL SERIAL SYSTEM,
ER ANALYSIS, *CHEMICAL ANALYSIS,
E'S SUMMIT(MO), OXYGEN TRANSFER,
UTRIENTS, PHYSIOLOGICAL ECOLOGY,
ANTZSCHII, DIDINIUM, PARAMECIUM,
TIFICATION, SURFACE AREA(LAKES),
CHEMICAL RESISTANCE, EUPHAUSIA,
ORREY CANYON, AROMATIC SOLVENTS,
OREALIS, LEPTOCYLINDRUS DANICUS,
IATOMS, ECOLOGY, NUISANCE ALGAE,
MATTER, BIOMASS, BENTHIC FLORA,
N, LOCH EWE(SCOTLAND), SCOTLAND,
IRONMENT, *PHYSICAL ENVIRONMENT,
OCOLLOIDS, TANNINS, FLAGELLATES,
L EXTRACTION, MILLIPORE FILTERS,
ODUCTIVITY, SULFIDES, SEA WATER,
VARIATIONS, SAN JOAQU/ STREETER-
POUNDS, MINIMUM LETHAL EXPOSURE,
*ALKYL POLY ETHOXYLATE, *ALKYL
S, *ALKYL POLYETHOXYLATE, *ALKYL
, QUATERNARY AMMONIUM COMPOUNDS,
URE, COLORIMETRY, CARBOHYDRATES,
NDS, ACTIVATED CARBON, CHLORINE,
PLING, ORGANIC MATTER, TOXICITY,
MIC ACIDS, *COLOR, RIVERS, BOGS,
TERICIDES, ALGAL TOXINS, LARVAE,
STANCES, ALGADYNAMIC SUBSTANCES,
HOT SPRINGS(WASH), SCHIZOTHRIX,
DIAPTOMUS, CYCLOPS, GOMPHONEMA,
RIANS, SYNEDRA ULNA, OEDOGONIUM,
ERIDINIUM, OSCILLATORIA AGARDHI,

PERIOD OF GROWTH.: / ISLES, BIOIN   W68-00255
PERIODIC VARIATIONS, SEASONAL VAR   W68-00481
PERIPHYTON INVENTORY(LAKE SUPERIO   W70-05958
PERIPHYTON, WATER POLLUTION EFFEC   W70-03978
PERIPHYTON, INVERTEBRATES, ADSORP   W69-07853
PERIPHYTON, AQUATIC PLANTS, BACKG   W69-09334
PERIPHYTON, PHOSPHORUS, HYDROGEN   W71-04206
PERIPHYTON, OHIO.: /R, CARBON, WA   W71-04206
PERIPHYTON, STREAMS, PHOTOSYNTHES   W71-00668
PERIPHYTON, BACTERIA, SPECT: /EFF   W70-01933
PERIPHYTON, BIOMASS, PHYSIOLOGICA   W70-03309
PERIPHYTON, CHLOROPHYTA, CHRYSOPH   W70-02780
PERMANGANATE, BEVERLY HILLS(CALIF   W71-00110
PERMANGANATE, FENAC, SILVER NITRA   W70-10175
PERMANGANATE, *DES MOINES(IOWA).:   W72-04242
PERMANGANATE.:   W72-04155
PERMANGANATE, ALGICIDAL TESTS, AL   W69-05704
PERMEABILITY, TEMPERATURE, OPERAT   W72-08686
PERMEASES, CHAMYDOMONADS, LAPPLAN   W70-04287
PEROT FILTERS, SOLAR FRAUNHOFER L   W70-05377
PEROXIDE, RATS, TRYPSIN, SCENEDES   W70-04184
PERSISTENCE.: /ORADO, WEST(US), S   W70-04788
PERSISTENCE, WATER POLLUTION TREA   W70-01777
PEST CONTROL ORGANIZATION OF TORO   W69-10155
PESTICIDE SORPTION.:   W69-09884
PESTICIDE DEGRADATION, SEVIN, MAL   W70-02968
PESTICIDE REMOVAL, PESTICIDE RESI   W70-02968
PESTICIDE RESIDUES, PHYSIOLOGICAL   W70-02968
PESTICIDE RESIDUES, HERBICIDES, E   W70-10437
PESTICIDE RESIDUES, ALGAE, GREAT   W70-08652
PESTICIDE TOXICITY, PHOTOSYNTHETI   W70-08416
PESTICIDE RESIDUES, ECOSYSTEMS, F   W70-05272
PESTICIDE REMOVAL.:   W70-05470
PESTICIDE RESIDUES, DETERIORATION   W70-03520
PESTICIDE RESIDUES, PHOSPHATES, A   W69-06305
PESTICIDE KINETICS, PESTICIDE TOX   W69-08565
PESTICIDE TOXICITY, POLLUTANTS, W   W69-08565
PESTICIDE BIOACCUMULATION.:   W69-08565
PESTICIDES, HARVESTING, WATER QUA   W69-05706
PESTICIDES, HARVESTING, WATER QUA   W69-05705
PESTICIDES, ALDRIN, ENDRIN, DIELD   W70-03520
PESTICIDES, CELLULOSE, CYCLES, HA   W70-05264
PESTICIDES, HERBICIDES, *PESTICID   W69-02782
PESTICIDES, PLANT PHYSIOLOGY.: /I   W70-08652
PESTICIDES, PESTICIDE TOXICITY, P   W70-08416
PESTICIDES.: /OLYSIS, COPPER SULF   W70-02982
PESTICIDES, PESTICIDE REMOVAL, PE   W70-02968
PESTICIDES, SLUDGE, DIGESTION, IN   W72-03641
PESTICIDES, OXIDATION LAGOONS, AL   W71-12183
PESTICIDES, RECREATION, ALGAE, OX   W71-09577
PESTICIDES, HYDROGEN ION CONCENTR   W72-11105
PESTICIDES, ALGAE, LABORATORY TES   W72-11105
PESTICIDES, TASTE, ODOR, COAGULAT   W72-13601
PETROLEUM WASTES, MERCENARIA MERC   W70-09905
PETROLEUM CRUDES, MASS SPECTROMET   W69-08284
PETROLEUM POLLUTION, PHOTOSYNTHET   W69-07440
PEWAUKEE LAKE(WIS), NORTH LAKE(WI   W69-05867
PEWAUKEE(WIS), CHIRONOMUS, PROCLA   W70-06217
PFAFFIKERSEE, TURLERSEE, BALDEGGE   W70-03964
PH, FACULTATIVE POND, ANAEROBIC P   W70-04790
PH, SAMPLING, ON-SITE DATA COLLEC   W69-00659
PH, SUSPENDED SOLIDS.: /N(MO), LE   W70-06601
PH: /ON, NUTRIENT REQUIREMENTS, N   W69-03364
PHACUS, GLENODINIUM, CRYPTOMONAS,   W70-00268
PHAEDACTYLUM, WASTE WATER DIVERSI   W69-09340
PHAEOCYSTIS, SARGASSUM, FRONDS, H   W70-01068
PHAEODACTYLUM TRICORNUTUM.: /, *T   W70-01470
PHAEODACTYLUM TRICORNUTUM, STICHO   W70-04280
PHAEOPHYTA, PHOTOSYNTHESIS, PHYTO   W69-07832
PHAEOPHYTA, POLAR REGIONS, SPECTR   W70-01074
PHAEOPIGMENTS, ACRIDINE ORANGE, L   W70-04811
PHAGE, FERMENTOR, FACULTATIVE, MI   W71-02009
PHARMACEUTICALS, EXTRACELLULAR PR   W70-01068
PHASE MICROSCOPY, STAINING, STRIP   W69-10163
PHAWOPHYTA, CRUSTACEANS, HERBIVOR   W70-01068
PHELPS EQUATION, DISSOLVED OXYGEN   W69-07520
PHENANTHROQUINONE, MADISON(WIS),   W70-02982
PHENOL POLYETHOXYLATE, ANIONIC SU   W70-09438
PHENOL POLYETHOXYLATE, ORGANIC EX   W70-03928
PHENOLIC COMPOUNDS, ORGANO-METALL   W72-12987
PHENOLS, NITROGEN, RESPIRATION, C   W70-01073
PHENOLS, OIL WASTES, HYDROGEN ION   W70-00263
PHENOLS, ODOR, COLOR, TURBIDITY,   W72-03299
PHENOLS, PIGMENTS, ORGANIC MATTER   W70-01074
PHENOLS, SHELLFISH, OYSTERS, INDU   W70-01068
PHORMIDINE, PANDORININE,: /IC SUB   W70-02510
PHORMIDIUM, WINKLER TECHNIQUE, OX   W70-03309
PHORMIDIUM, APHANIZOMENON FLOS-AQ   W70-02803
PHORMIDIUM RETZII, SURFACE DIFFUS   W70-00265
PHORMIDIUM, APHANIZOMENON FLOS-AQ   W69-10169

497

OSYSTEM MODEL, SIMULATION MODEL,
 INFERFACE, MUD-WATER INTERFACE,
GOTROPHY, OXYGEN, OXYGEN DEMAND,
 WATER POLLUTION EFFECTS, LAKES,
ION, DOMESTIC WASTES, NUTRIENTS,
IOTA, *COLUMBIA RIVER, CHROMIUM,
, DOCUMENTATION, BIBLIOGRAPHIES,
S, FERTILIZERS, ALGAE, NITROGEN,
LITY, WATER POLLUTION, NITROGEN,
 WATER, SORPTION, SODIUM, ALGAE,
OL, ARSENIC COMPOUNDS, NITROGEN,
VATION, WAT/ *NUTRIENT ELEMENTS,
N, *ALGAE, *NUTRIENTS, NITROGEN,
DUCTIVITY, ECOSYSTEMS, ILLINOIS,
TER POLLUTION EFFECTS, NITROGEN,
NEFIT-COST ANALYSIS, *LAKE ERIE,
EN ION CONCENTRATION, NUTRIENTS,
 MINNESOTA, WISCONSIN, MICHIGAN,
ROUS FORESTS, DECIDUOUS FORESTS,
ECTICUT, EUGLENA, CHLAMYDOMONAS,
CHLORELLA, *SCENEDESMUS, *LIGHT,
UTRIENT REQUIREMENTS, NUTRIENTS,
TRIENT INFLUX, NUTRIENT SOURCES,
S, ZO/ *BENTHIC FAUNA, NITROGEN,
ENTS, *EUTROPHICATION, NITROGEN,
 *CYCLING NUTRIENTS, ECOSYSTEMS,
NEDESMUS, CHLOROPHYTA, BACTERIA,
OSYNTHESIS, L/ *ALGAE, NITROGEN,
S, MICHIGAN, BICARBONATES, IRON,
EN, NITROGEN CYCLE, *PHOSPHORUS,
E TREATMENT, NITROGEN COMPOUNDS,
YLL, CYANOPHYTA, AQUATIC PLANTS,
ALGAE, WATER QUALITY, NUTRIENTS,
, RUNOFF, FERTILIZERS, NITROGEN,
IFEROUS FORESTS, ALGAE, INSECTS,
RING, BACTERIA, WATER CHEMISTRY,
ALGAE, CHLOROPHYLL, HYPOLIMNION,
TECHNIQUES, NUTRIENTS, NITROGEN,
ER POLLUTION, ALGAE, PERIPHYTON,
LIMNION, ANIMALS, DENSITY, IRON,
OURCES, WATER POLLUTION EFFECTS,
), HYPOLIMNION, ALGAE, NITROGEN,
 BIOLOGICAL TREATMENT, NITROGEN,
SOTOPES, SEA WATER, FOOD CHAINS,
ONCENTRATION, WISCONSIN, SULFUR,
*GREAT LAKES, *ALGAE, NUTRIENTS,
RTILIZATION, NITROGEN COMPOUNDS,
UISANCE ALGAE, SEWAGE EFFLUENTS,
PLANTS, WATER POLLUTION EFFECTS,
OTOGRAPHY, CARBON RADIOISOTOPES,
CONSIN, SEWAGE, TROUT, NITROGEN,
OACTIVITY, CARBON RADIOISOTOPES,
FFLUENTS, *NUTRIENTS, *NITROGEN,
ON, GEOCHEMISTRY, SPRING WATERS,
H, OLIGOCHAETES, MIDGES, SNAILS,
BON, FERTILITY, ALGAE, NITROGEN,
MISTRY, WATER POLLUTION SOURCES,
), LAKE KEGONSA(WIS), WEED BEDS,
EDESMUS, CHLAMYDOMONAS, ECOLOGY,
S, CARBON, METABOLISM, NITROGEN,
WAGE, FARM WASTES, CARBON CYCLE,
MENT, SEWAGE BACTERIA, NITROGEN,
KES, RESERVOIRS, SCUM, NITROGEN,
ONMENTAL/ *BIOASSAYS, NUTRIENTS,
 PHYSIOLOGY, FISH TAXONOMY, DDT,
ORMOSA, CYCLOTELLA MENEGHINIANA,
OTROPH/ *EXTRACELLULAR PRODUCTS,
PICKWICK RESERVOIR(TENN), NAJAS,
, FILAMENTOUS ALGAE, KILL RATES,
 *KILL RATES, *BROMINE, IODINE,
FECTIVE-AVERAGE-LIGHT-INTENSITY,
NON, GLOEOTRICHIA, POLYPEPTIDES,
URES, CHEMOSTATS, CHLOROPHYLL A,
TS, NITROGEN, NITROGEN FIXATION,
BACTERIA, GROWTH RATES, TRITIUM,
TEMPERATURE, MONITORING, AERIAL
YARTHRA, CONOCHILUS, PYCNOCLINE,
NIQUES, *POLAROGRAPHIC ANALYSIS,
 ECOLOGY, ECOTYPES, CHLOROPHYTA,
TIC TESTS, SLIMEPRODUCING ALGAE,
S, CHLORELLA VULGARIS(COLUMBIA)/
SYSTEMS, ENVIRONMENT, RETENTION,
BORATORY TESTS, LIGHT INTENSITY,
PHYRIDIUM CRUENTUM, LEMNA MINOR,
OLYBDENUM, ROTIFERS, ECOSYSTEMS,
, SILTS, STANDING CROP, ECOLOGY,
ACID SOILS, ACIDIC WATER, ACIDS,
YANOPHYTA, CHLORELLA, MANGANESE,
NESE, TRACE ELEMENTS, CHELATION,
IC WATER, ACIDS, PHOTOSYNTHESIS,

PHOSPHORUS EFFECT, CLADECERA.: /C   W69-06536
PHOSPHORUS, ADSORPTION, EUTROPHIC   W69-06859
PHOSPHORUS COMPOUNDS, PLANT GROWT   W69-06535
PHOSPHORUS, TROPHIC LEVEL, PRODU?   W69-06536
PHOSPHORUS, ALGAE, ALGICIDES, HAR   W69-07818
PHOSPHORUS RADIOISOTOPES, TROPHIC   W69-07853
PHOSPHORUS COMPOUNDS, NITROGEN CO   W69-08518
PHOSPHORUS, WATER POLLUTION EFFEC   W69-08668
PHOSPHORUS, AIR, PHYTOPLANKTON, A   W70-04484
PHOSPHORUS, TRACE ELEMENTS, PLANK   W70-04385
PHOSPHORUS.: / CONTROL, CHEMCONTR   W70-04494
PHOSPHORUS RELEASE, SAMPLE PRESER   W70-04382
PHOSPHORUS, SODIUM, POTASSIUM, TR   W70-04503
PHOSPHORUS, DRAINAGE, AGRICULTURA   W70-04506
PHOSPHORUS, ALGAE, NUTRIENTS, WAT   W70-04382
PHOSPHORUS, FISH STOCKING, COMMER   W70-04465
PHOSPHORUS, NITROGEN, IRON, GROWT   W70-03983
PHOSPHORUS, AGRICULTURAL WATERSHE   W70-04193
PHOSPHORUS, NITROGEN, CYANOPHYTA,   W70-03512
PHOSPHORUS RADIOISOTOPES, REGRESS   W70-03957
PHOSPHORUS, PHOTOSYNTHESIS, LIGHT   W70-03923
PHOSPHORUS COMPOUND: /CE ALGAE, N   W70-04283
PHOSPHORUS BUDGET, APHANIZOMINON   W70-04268
PHOSPHORUS, PHYTOPLANKTON, DIATOM   W70-04253
PHOSPHORUS, *NUTRIENT REMOVAL, SO   W73-01122
PHOSPHORUS, SCENEDESMUS, CHLOROPH   W73-01454
PHOSPHORUS RADIOISOTOPES, INFLOW,   W73-01454
PHOSPHORUS, *EUTROPHICATION, PHOT   W72-00917
PHOSPHORUS, NITROGEN, WATER QUALI   W72-03183
PHOSPHORUS COMPOUNDS, WATER SUPPL   W72-01867
PHOSPHORUS COMPOUNDS, RESERVOIRS,   W72-02817
PHOSPHORUS, DENSITY, WATER POLLUT   W72-04761
PHOSPHORUS, NITROGEN, CALCIUM CHL   W72-05453
PHOSPHORUS, ALGAE, LIVESTOCK, WAT   W71-10376
PHOSPHORUS RADIOISOTOPES, TECHNOL   W71-09012
PHOSPHORUS, NITROGEN, AMMONIA, AL   W71-05879
PHOSPHORUS, NITROGEN, PHYTOPLANKT   W71-07698
PHOSPHORUS, AGRICULTURE, WATER MA   W71-03048
PHOSPHORUS, HYDROGEN ION CONCENTR   W71-04206
PHOSPHORUS, SILICA, SAMPLING, CHE   W70-06975
PHOSPHORUS, NITROGEN, CHLOROPHYLL   W70-07261
PHOSPHORUS, OXYGEN, CARBON, ENZYM   W70-09889
PHOSPHORUS, SEWAGE EFFLUENTS.: /,   W70-08980
PHOSPHORUS RADIOISOTOPES, COBALT   W70-00708
PHOSPHORUS.: /USCORUN, NUTRIENT C   W70-00719
PHOSPHORUS, POPULATION, POLLUTANT   W70-00667
PHOSPHORUS COMPOUND, PERCHES, FIS   W70-00274
PHOSPHORUS, EUTROPHICATION, WASHI   W70-00270
PHOSPHORUS, NITROGEN.: /EAR POWER   W70-00264
PHOSPHORUS RADIOISOTOPES, ALGAE,   W69-10163
PHOSPHORUS, DETERGENTS, RUNOFF, P   W69-10178
PHOSPHORUS RADIOISOTOPES, ALGAE,   W69-10151
PHOSPHORUS, SULFATES, VITAMINS, A   W69-09349
PHOSPHORUS, ORGANOPHOSPHORUS COMP   W69-09334
PHOSPHORUS, NUTRIENTS, ALGAE, DIA   W70-01943
PHOSPHORUS, WATER POLLUTION EFFEC   W70-02973
PHOSPHORUS, NITROGEN COMPOUNDS, E   W70-03068
PHOSPHORUS, NITROGEN BALANCE.: /S   W70-02969
PHOSPHORUS, SULPHUR, HYDROGEN ION   W70-02965
PHOSPHORUS, SULFUR, CALCIUM, MAGN   W70-02804
PHOSPHORUS, PHOSPHATE, NITRATES,   W70-02777
PHOSPHORUS, ALGAE, COPPER SULFATE   W70-02787
PHOSPHORUS, CARBONATES, TEMPERATU   W70-02803
PHOSPHORUS, KINETIC THEORY, ENVIR   W70-02779
PHOSPHOTHIOATE PESTICIDES, HERBIC   W69-02782
PHOTIC ZONE, HELSINGOR(DENMARK),   W70-04809
PHOTIC ZONE, GLYCOLLIC ACID, CHEM   W70-02504
PHOTIC ZONE, DOUGLAS RESERVOIR(TE   W71-07698
PHOTO REACTIVITY, CELL COLOR, CEL   W70-02364
PHOTO REACTIVITY, ALGAL METABOLIS   W70-02363
PHOTO-ELECTRIC TURBIDOSTAT, LIGHT   W70-03923
PHOTOASSIMILATION, BERKELSE LAKE(W70-04369
PHOTOBIOLOGY, ALGAL: / BATCH CULT   W70-02777
PHOTOCHEMICAL, CARBON DIOXIDE, SY   W70-02965
PHOTOGRAPHY, CARBON RADIOISOTOPES   W69-10163
PHOTOGRAPHY, INFRARED RADIATION,   W68-00010
PHOTOLITHOTROPHIC BACTERIA, DESUL   W70-05760
PHOTOMETRY, HALOGENS.: /ICAL TECH   W70-05178
PHOTOMETRY, CHLORELLA, NUTRIENTS,   W70-05409
PHOTOMICROGRAPHS, MICROCYSTIS AER   W69-05704
PHOTOOXIDATION, CHLORELLA VULGARI   W70-04809
PHOTOPERIODISM, LIGHT INTENSITY,   W69-07862
PHOTOPERIODISM, LIGHT PENETRATION   W72-00917
PHOTOPHOSPHORYLATION, VANADIUM, Z   W70-02964
PHOTOS: /ITE, SULFUR COMPOUNDS, M   W69-07428
PHOTOSYNT: /DYNAMICS, TEMPERATURE   W70-02780
PHOTOSYNTHESIS, PHOTOSYNTHETIC OX   W70-02792
PHOTOSYNTHESIS, OXYGEN, CARBON DI   W70-02964
PHOTOSYNTHESIS, CARBON DIOXIDE, B   W70-02804
PHOTOSYNTHETIC OXYGEN, PHYSIOLOGI   W70-02792

ENVIRONMENT, CHLORELLA, TRACERS,
SIGN CRITERIA, OXIDATION LAGOONS
, FISH, BACTERIA, BENTHIC FAUNA,
OONS, BIOCHEMICAL OXYGEN DEMAND,
AL ECOLOGY, LITTORAL, ESTUARIES,
GEN, HYDROGEN ION CONCENTRATION,
G, PHYTOPLANKTON, RADIOACTIVITY,
CIDES, OXIDATION LAGOONS, ALGAE,
-PURIFICATION, DISSOLVED OXYGEN,
EN, PHOSPHORUS, *EUTROPHICATION,
XYGEN DEMAND, AEROBIC TREATMENT,
MAND, OXIDATION LAGOONS, SLUDGE,
LVED OXYGEN, *RIVERS, OXIDATION,
EN, ODORS, CORROSION, OXIDATION,
OGY, NORTH DAKOTA, SOUTH DAKOTA,
TH RATES, ENVIRONMENTAL EFFECTS,
ILIMNION, EFFLUENT, TEMPERATURE,
MENT, GROWTH RATES, CHLOROPHYLL,
ATION, OLIGOTROPHY, TEMPERATURE,
 CHLORINATION, SODIUM COMPOUNDS,
TION, CHLOROPHYLL, PRODUCTIVITY,
TESTS, RESPIRATION, ENVIRONMENT,
RGANIC MATTER, EXUDATION, TIDES,
DUSTRIAL WASTES, LAGOONS, PONDS,
 PESTICIDES, PESTICIDE TOXICITY,
TORY TESTS, PERIPHYTON, STREAMS,
UTLETS, NUTRIENTS, PRODUCTIVITY,
N ION CONCENTRATION, EPILIMNION,
ATORY TESTS, LAGOONS, OXIDATION,
 *ALGAE, *TEMPERATURE, DIATOMS,
E, PACKED CELL VOLUM/ INDUSTRIAL
IFICATION, CARBON, BICARBONATES,
IGHT, TE/ *GROWTH RATES, *ALGAE,
TION, BIOCHEMICAL OXYGEN DEMAND,
NT REQUIREMENTS, ORGANIC MATTER,
NS, MOSCOW, PETROLEUM POLLUTION,
OGY, NUISANCE ALGAE, PHAEOPHYTA,
N DYNAMICS, EXTRACELLULAR PRODU/
LIGHT PENETRATION, FLUORESCENCE,
S, ARCTIC, ENZYMES, CHLOROPHYLL,
*DESIGN THEORY AND CALCULATIONS,
ARY PRODUCTIVITY, PHYTOPLANKTON,
ION, COLIFORMS, PONDS, CLIMATES,
 DIATOMS, PIGMENTS, CHLOROPHYLL,
LLUTION, NUTRIENTS, TEMPERATURE,
DIATION, ENERGY, CARBON DIOXIDE,
ETALS, INDUSTRIAL WASTES, LAKES,
Y, LIGHT PENETRATION, LIMNOLOGY,
OMPOUNDS, NUTRIENT REQUIREMENTS,
T REGIONS, GEOGRAPHICAL REGIONS,
UTRIENT REQUIREMENTS, NUTRIENTS,
SONAL DISTRIBUTION, LAKE BAIKAL,
GAE, SEWAGE PLANTS, CHLOROPHYLL,
ADIOISOTOPES, ENZYMES, KINETICS,
CTERIA, RADIOACTIVITY TECHNIQUE,
LANKTON, WASTE WATER(POLLUTION),
ENTS, ECOSYSTEMS, PHYTOPLANKTON,
SAY, INHIBITION, MODE OF ACTION,
SCENEDESMUS, *LIGHT, PHOSPHORUS,
OPHYTA, CYANOPHYTA, CHRYSOPHYTA,
, *WASTE WATER TREATMENT, ALGAE,
, ANABAENA, CYTOLOGICAL STUDIES,
UTOLYSIS, EXCRETION, DEPURATION,
 *ANABAENA FLOS-AQUAE, *STRAINS,
 DENSITY, SOIL-WATER TECHNIQUES,
RNS, MICROALGAE, MICRONUTRIENTS,
UNDERWATER SHOCK WAVE, ANABAENA,
CHEMISTRY, ALGAL TAXONOMY, ALGAL
E, CONSUMER ORGANISMS, SIMULIUM,
OL, INTEGRATED CONTROL MEASURES,
F-LIFE, CLINCH RIVER(TENNESSEE),
CA, SAMPLING, CHEMICAL ANALYSIS,
ER POLLUTION, CHEMICAL ANALYSIS,
SHEDS, NUTRIENTS, FERTILIZATION,
FICATION, BIOLOGICAL PROPERTIES,
AL TREATMENT, OVERFERTILIZATION,
GAE, PLANKTON, SUBMERGED PLANTS,
ATION, LIMNOLOGY, PUBLIC HEALTH,
STRONTIUM RADIOISOTOPES, CESIUM,
, *MARINE PLANTS, RADIOISOTOPES,
L TECHNIQUES, DIATOMS, SAMPLING,
PTH, PHYTOPLANKTON, ZOOPLANKTON,
ARINE ALGAE, *RESINS, SEA WATER,
PERATURE, NUTRIENT REQUIREMENTS,
TS, PHYSIOCOCHEMICAL PROPERTIES,
URE, CARBON DIOXIDE, CYANOPHYTA,
GAE, DIATOMS, CYCLING NUTRIENTS,
YTTRIUM, 'BIOLOGICAL DILUTION',
LLUTION, STREAMS, RIVERS, LAKES,
YNTHESIS, PHOTOSYNTHETIC OXYGEN,

PHOTOSYNTHESIS, CARBON RADIOISOTO   W70-02786
PHOTOSYNTHESIS.: /ION, DESIGN, DE   W70-02609
PHOTOSYNTHESIS, BIOASSAYS, AQUATI   W70-02251
PHOTOSYNTHESIS, HYDROGEN SULFIDE,   W70-03312
PHOTOSYNTHESIS, RESPIRATION, PIGM   W70-03325
PHOTOSYNTHESIS, ALKALINITY, NITRA   W70-03309
PHOTOSYNTHESIS, CHEMICAL ANALYSIS   W70-01933
PHOTOSYNTHESIS, DISSOLVED OXYGEN,   W71-12183
PHOTOSYNTHESIS.: /TREATMENT, SELF   W72-01798
PHOTOSYNTHESIS, LAGOONS, LABORATO   W72-00917
PHOTOSYNTHESIS, ALGAE, SEWAGE, MU   W72-00462
PHOTOSYNTHETIC OXYGEN, FARM WASTE   W71-08221
PHOTOSYNTHESIS, RESPIRATION, SCOU   W71-11381
PHOTOSYNTHESIS, ANALYTICAL TECHNI   W71-11377
PHOTOSYNTHESIS, LIGHT PENETRATION   W72-01818
PHOTOSYNTHESIS, DIATOMS, CHLORELL   W69-10158
PHOTOSYNTHESIS, CARBON DIOXIDE, D   W69-10169
PHOTOSYNTHESIS, CHEMICAL ANALYSIS   W69-10167
PHOTOSYNTHESIS, RAINBOW TROUT, DI   W69-10154
PHOTOSYNTHESIS, CARBON DIOXIDE, N   W70-00161
PHOTOSYNTHESIS, ALGAE, HYPOLIMNIO   W69-10182
PHOTOSYNTHESIS, OXYGEN, TROPHIC L   W70-00265
PHOTOSYNTHESIS, SOLAR RADIATION,   W70-01073
PHOTOSYNTHESIS, COSTS, EFFICIENCY   W70-09320
PHOTOSYNTHETIC OXYGEN, PRODUCTIVI   W70-08416
PHOTOSYNTHESIS, RESPIRATION, MICR   W71-00668
PHOTOSYNTHESIS, EUTROPHICATION, T   W70-06225
PHOTOSYNTHESIS, ALGAE, BOTTOM SED   W70-06975
PHOTOSYNTHESES, PLANTS, TEMPERATU   W70-06600
PHOTOSYNTHESIS.:   W71-02496
PHOTOSYNTHESIS, MEAN DOUBLING TIM   W71-05267
PHOTOSYNTHESIS, NUTRIENTS, MICROB   W69-07389
PHOTOSYNTHESIS, EUTROPHICATION, L   W69-07442
PHOTOSYNTHETIC OXYGEN.: /AG, AERA   W69-07520
PHOTOSYNTHESIS, TEMPERATURE, TROP   W69-06865
PHOTOSYNTHETIC PIGMENTS, PRODUCTI   W69-07440
PHOTOSYNTHESIS, PHYTOPLANKTON, PR   W69-07832
PHOTOSYNTHETIC RATE, PHYTOPLANKTO   W69-05697
PHOTOSYNTHESIS, BLUE-GREEN ALGAE,   W69-05697
PHOTOSYNTHESIS, CHLORELLA, GRAZIN   W70-04809
PHOTOSYNTHESIS, ILLUMINATION, TEM   W70-04786
PHOTOSYNTHESIS, BIOMASS, CULTURES   W70-05094
PHOTOSYNTHESIS, PRODUCTIVITY, SYM   W70-05092
PHOTOSYNTHESIS, GROWTH RATE, ORGA   W70-05381
PHOTOSYNTHESIS, LIGHT INTENSITY.:   W70-05270
PHOTOSYNTHESIS, GROWTH RATE, AERO   W70-05750
PHOTOSYNTHESIS, PHYTOPLANKTON, CI   W69-02959
PHOTOSYNTHETIC OXYGEN, TEMPERATUR   W69-03516
PHOTOSYNTHETIC, OXYGEN, PLANKTON,   W69-03514
PHOTOSYNTHESIS, CHEMICAL REACTION   W69-03611
PHOTOSYNTHET: /TION, LIMNOLOGY, N   W69-03517
PHOTOSYNTHETIC LAYER, MELOSIRA BA   W70-04290
PHOTOSYNTHESIS, TRIBUTARIES, WATE   W70-04268
PHOTOSYNTHESIS, ORGANIC MATTER, O   W70-04287
PHOTOSYNTHESIS.: /ROGEN FIXING BA   W70-04249
PHOTOSYNTHESIS, CHL: /NTS, PHYTOP   W70-04371
PHOTOSYNTHESIS, CYANOPHYTA, CYTOL   W70-03973
PHOTOSYNTHESIS.: /ORTALITY, BIOAS   W70-03519
PHOTOSYNTHESIS, LIGHT INTENSITY,   W70-03923
PHOTOSYNTHESIS.: /TH RATES, CHLOR   W70-03983
PHOTOSYNTHESIS, DEPTH, THERMOCLIN   W72-10573
PHY: /OSTRIDIUM, FISH, AMPHIBIANS   W70-01068
PHYCOCOLLOIDS, TANNINS, FLAGELLAT   W70-01068
PHYCOLOGISTS, VETERINARIANS, MICR   W70-00273
PHYCOLOGY, MICROCYSTIS, GLOEOTRIC   W70-04381
PHYCOLOGY, SPACE RESEARCH, TOXIC   W69-07832
PHYGON XL(DICHLONE), NEW HAMPSHIR   W69-08674
PHYLOGENY, CLADOPHORA, ALGAL GENE   W70-03973
PHYSA, SALMO TRUTTA, ENTOSPHENUS   W69-09334
PHYSICAL CONTROL, WATER QUALITY.:   W70-07393
PHYSICAL HALF-LIFE, HYDROPSYCHE C   W70-02786
PHYSICAL PROPERTIES, CARP, TILAPI   W70-06975
PHYSICAL PROPERTIES.: /EWAGE, WAT   W70-05752
PHYSICAL PROPERTIES, CLIMATE, NIT   W70-05565
PHYSICAL PROPERTIES, WASTE WATER(W69-01273
PHYSICAL TREATMENT, UNIT OPERATIO   W70-07469
PHYSICOCHEMICAL PROPERTIES, PHYTO   W71-00665
PHYSICOCHEMICAL PROPERTIES, GENET   W70-00273
PHYSICOCHEMICAL PROPERTIES, ADSOR   W70-00708
PHYSICOCHEMICAL PROPERTIES, ANION   W71-09863
PHYSICOCHEMICAL PROPERTIES, CENTR   W70-05389
PHYSICOCHEMICAL PROPERTIES, DOMES   W70-03964
PHYSICOCHEMICAL PROPERTIES, TRACE   W69-08275
PHYSIOCOCHEMICAL PROPERTIES, PHYS   W69-10161
PHYSIOLOGICAL ECOLOGY, RHODOPHYTA   W69-10161
PHYSIOLOGICAL ECOLOGY, THERMAL WA   W69-10160
PHYSIOLOGICAL ECOLOGY, WATER: /AL   W70-00713
PHYSIOLOGICAL CO: / THORACOMONAS,   W70-00708
PHYSIOLOGICAL ECOLOGY, LIMITING F   W71-02479
PHYSIOLOGICAL ECOLOGY, PHYTOPLANK   W70-02792

TIONS, SURVIVAL, SUSCEPTIBILITY,   PHYSIOLOGICAL OBSERVATIONS, CHRON   W70-01779
OF ALGAE, LIMNOLOGY, NUTRIENTS,   PHYSIOLOGICAL ECOLOGY, CHLORELLA,   W70-02245
ATION, OREGON, INTERTIDAL AREAS,   PHYSIOLOGICAL ECOLOGY, LITTORAL,   W70-03325
IDE REMOVAL, PESTICIDE RESIDUES,   PHYSIOLOGICAL ECOLOGY, PHYTOPLANK   W70-02968
RMAL WATER, PERIPHYTON, BIOMASS,   PHYSIOLOGICAL ECOLOGY, CYANOPHYTA   W70-03309
LIFORNIA, BIOASSAY, SCENEDESMUS,   PHYSIOLOGICAL ECOLOGY, KINETICS,   W70-02775
RACERS, WATER POLLUTION EFFECTS,   PHYSIOLOGICAL: / RADIOACTIVITY, T   W70-02510
TEMPERATURE, CHEMICAL ANALYSIS,   PHYSIOLOGICAL ECOLOGY, OXYGEN, CA   W69-07862
GRAZING, *TEMPERATURE, *DAPHNIA,   PHYSIOLOGICAL ECOLOGY, LAKES, ENV   W70-03957
, ESTUARIES, LAKES, GREAT LAKES,   PHYSIOLOGICAL ECOLOGY, WATER CHEM   W70-03975
, ESTUARIES, LAKES, GREAT LAKES,   PHYSIOLOGICAL ECOLOGY, WATER CHEM   W70-03975
TALS, VITAMINS, WATER POLLUTION,   PHYSIOLOGICAL ECOLOGY, DIATOMS, O   W70-04503
CHAINS, CHLORELLA, CHLOROPHYLL,   PHYSIOLOGICAL ECOLOGY, CYANOPHYTA   W70-04381
TION EFFECTS, ALGAE, ECOSYSTEMS,   PHYSIOLOGICAL ECOLOGY, AQUATIC BA   W70-04287
NION, LIGHT INTENSITY, SEASONAL,   PHYSIOLOGICAL ECOLOGY, ECOTYPES,   W70-05409
E, TOXICITY, DIATOMS, ESTUARIES,   PHYSIOLOGICAL ECOLOGY, MASSACHUSE   W70-05272
NETICS, AQUATIC BACTERIA, ALGAE,   PHYSIOLOGICAL ECOLOGY, CARBON RAD   W70-05109
LUENTS, CHLOROPHYTA, RHODOPHYTA,   PHYSIOLOGICAL ECOLOGY, INDUSTRIAL   W70-04510
HNIQUES, *MARINE MICROORGANISMS,   PHYSIOLOGICAL ECOLOGY, ORGANIC MA   W70-04811
EOVIRIDIS, SKELETONEMA COSTATUM,   PHYSIOLOGICAL ADJUSTMENT, ASTERIO   W70-04809
                                          PHYSIOLOGICAL EFFECTS.:   W68-00859
NUTRIENTS, PHOSPHORUS COMPOUNDS,   PHYSIOLOGICAL ECOLOGY,: / ALGAE,   W69-04802
TRIENTS, PHOSPHORUS, PHOSPHATES,   PHYSIOLOGICAL ECOLOGY, PHYTOPLAN: W69-03373
UTRIENT REQUIREMENTS, NUTRIENTS,   PHYSIOLOGICAL ECOLOGY, PH: /ON, N   W69-03364
CE ALGAE, NUTRIENT REQUIREMENTS,   PHYSIOLOGICAL E: /F ALGAE, NUISAN   W69-03362
CE ALGAE, NUTRIENT REQUIREMENTS,   PHYSIOLOGICAL ECOLOGY, PHYTOPLANK   W69-03371
EFFECTS, ANALYTICAL TECHNIQUES,   PHYSIOLOGICAL ECOLOGY, PLANT PHYS   W72-12240
ES, PHYSIOLOGICAL ECOLOGY, PLANT   PHYSIOLOGY.: /ANALYTICAL TECHNIQU   W72-12240
GREEN ALGAE, *ALGA BLOOMS, ALGAE   PHYSIOLOGY.:   *BLUE-   W69-03512
DRIFT, FISH FOOD ORGANISMS, FISH   PHYSIOLOGY, FISH TAXONOMY, DDT, P   W69-02782
THESIS, BIOMASS, CULTURES, PLANT   PHYSIOLOGY, ALGAE, DIURNAL DISTRI   W70-05094
ASTE STABILIZATION PONDS, *ALGAL   PHYSIOLOGY, *DESIGN THEORY AND CA   W70-04786
WATER ANALYSIS, *SAMPLING, PLANT   PHYSIOLOGY, EUTROPHICATION, DATA   W70-04382
ON(WASH), ALGAL NUTRITION, ALGAL   PHYSIOLOGY, HETEROTROPHY, MICRONU   W70-04506
STUDIES, APPLIED STUDIES, ALGAL   PHYSIOLOGY, ALGAL BIOCHEMISTRY, A   W70-03973
AGONISMS, ALGAL NUTRITION, ALGAL   PHYSIOLOGY, LAKE MENDOTA(WIS), LA   W69-05867
L GROWTH, ALGAL NUTRITION, ALGAL   PHYSIOLOGY, CALOTHRIX PARIETINA,   W69-06277
L GROWTH, ALGAL NUTRITION, ALGAL   PHYSIOLOGY, KJELDAHL PROCEDURE, L   W69-05868
L GROWTH, ALGAL NUTRITION, ALGAL   PHYSIOLOGY, AMERICAN RESEARCH AND   W69-06865
UTION, *TOXICITY.:   *FISH   PHYSIOLOGY, *ALGAE, *THERMAL POLL   W70-01788
                         *CHEMOSTAT,   PHYSIOLOGY, FLAGELLATES.:   W70-09904
ED HYDROCARBON PESTICIDES, PLANT   PHYSIOLOGY.: /IRONMENT, CHLORINAT   W70-08652
N, MILL/ *MICRO-AUTORADIOGRAPHY,   PHYSIOLOGY, BIOCHEMICAL EXTRACTIO   W69-10163
AGOONS, PLANT PHYSIOLOGY, ANIMAL   PHYSIOLOGY, NORTH DAKOTA, SOUTH D   W72-01818
ED OXYGEN, SEWAGE LAGOONS, PLANT   PHYSIOLOGY, ANIMAL PHYSIOLOGY, NO   W72-01818
RMING, FISH FOOD ORGANISMS, FISH   PHYSIOLOGY, RED TIDE, ORGANIC COM   W71-07731
OSPHATES, PHYSIOLOGICAL ECOLOGY,   PHYTOPLAN: /IENTS, PHOSPHORUS, PH   W69-03373
IREMENTS, PHYSIOLOGICAL ECOLOGY,   PHYTOPLANKTON, SE: /NUTRIENT REQU   W69-03371
L WASTES, LAKES, PHOTOSYNTHESIS,   PHYTOPLANKTON, CISCO, CARP, DIATO   W69-02959
HOSPHORUS COMPOUNDS, PHOSPHORUS,   PHYTOPLANKTON.: / REQUIREMENTS, P   W69-03358
YANOPH/ *EUTROPHICATION, *LAKES,   PHYTOPLANKTON, ALGAE, SEASONAL, C   W68-00860
LGAE, *CHEMICA/ *EUTROPHICATION,   PHYTOPLANKTON, BIBLIOGRAPHIES, *A   W68-00856
, DI/ *ECOLOGY, BACTERIA, ALGAE,   PHYTOPLANKTON, *LAKES, *CELLULOSE   W69-00976
TER ANALYSIS, PHOSPHORUS, PONDS,   PHYTOPLANKTON, EUTROPHICATION, LA   W69-05142
POND(JAPAN), LAKE MENDOTA(WIS),   PHYTOPLANKTON CROPS, TROPHIC, TRO   W69-05142
TS, NUISANCE ALGAE, OLIGOTROPHY,   PHYTOPLANKTON, *REVIEWS, *WATER P   W68-00680
BRINE SHRIMP, ECOSYSTEMS, FISH,   PHYTOPLANKTON, PLANKTON, RADIOACT   W69-07440
EUTROPHICATION, ORGANIC MATTER,   PHYTOPLANKTON, PRIMARY PRODUCTIVI   W69-06273
ULAR PRODU/ PHOTOSYNTHETIC RATE,   PHYTOPLANKTON DYNAMICS, EXTRACELL   W69-05697
HOTOSYNTHESIS, BLUE-GREEN ALGAE,   PHYTOPLANKTON, OHIO RIVER, PENNSY   W69-05697
ISANCE ALGAE, OLIGOTROPHY, ODOR,   PHYTOPLANKTON, SULFATES, SEDIMENT   W69-07833
YTON, INVERTEBRATES, ADSORPTION,   PHYTOPLANKTON, SUCKERS, SHINERS,   W69-07853
GAE, PHAEOPHYTA, PHOTOSYNTHESIS,   PHYTOPLANKTON, PRIMARY PRODUCTIVI   W69-07832
N ABATEMENT, OLIGOTROPHY, DEPTH,   PHYTOPLANKTON, ZOOPLANKTON, PHYSI   W70-03964
OROPHYTA, NUTRIENTS, ECOSYSTEMS,   PHYTOPLANKTON, PHOTOSYNTHESIS, CY   W70-03973
IMNOLOGY, CHLOROPHYTA, PLANKTON,   PHYTOPLANKTON, WATER POLLUTION EF   W70-03959
LAKES, *RESERVOIRS, ZOOPLANKTON,   PHYTOPLANKTON, TEMPERATURE, *EPIL   W70-04475
TON, LAKES, INDIANA, OCHROMONAS,   PHYTOPLANKTON, SEAWEEDS.: /ASHING   W70-04503
TION, NITROGEN, PHOSPHORUS, AIR,   PHYTOPLANKTON, ALGAE, EFFICIENCIE   W70-04484
HIC FAUNA, NITROGEN, PHOSPHORUS,   PHYTOPLANKTON, DIATOMS, ZOOPLANKT   W70-04253
ANALYSIS, MONITORING, PIGMENTS,   PHYTOPLANKTON, WASTE WATER(POLLUT   W70-04371
A, THERMOCLINE, CURRENTS(WATER),   PHYTOPLANKTON, ZOOPLANKTON, HYDRO   W70-04368
ADIOISOTOPES, *ALGAE, *CULTURES,   PHYTOPLANKTON, CARBON, RADIOACTIV   W70-03983
, MATHEMATICAL MODELS, BACTERIA,   PHYTOPLANKTON, GEOCHEMIS: /NOLOGY   W70-03975
, MATHEMATICAL MODELS, BACTERIA,   PHYTOPLANKTON, GEOCHEMIS: /NOLOGY   W70-03975
TIES, EFFLUENTS, EUTROPHICATION,   PHYTOPLANKTON, DECOMPOSING ORGANI   W70-04001
NIQUES, WATER POLLUTION EFFECTS,   PHYTOPLANKTON, DIFFUSION, ALGAE,   W70-04194
PHICATION, PRIMARY PRODUCTIVITY,   PHYTOPLANKTON, PHOTOSYNTHESIS, BI   W70-05094
GY, CARBON RADIOISOTOPES, LAKES,   PHYTOPLANKTON, BIOASSAYS.: /ECOLO   W70-05109
RAINAGE, ORGANIC MATTER, SEWAGE,   PHYTOPLANKTON, PRODUCTIVITY, CHLO   W70-05404
TES, HYDROGEN ION CONCENTRATION,   PHYTOPLANKTON, ODOR, WINDS, EPILI   W70-05405
AE, *PHOTOSYNTHESIS, *MINNESOTA,   PHYTOPLANKTON, CYANOPHYTA, DIATOM   W70-05387
*POTOMAC RIVER ESTUARY, NUTRIENT-   PHYTOPLANKTON RELATIONSHIPS.:   W71-05407
POLIMNION, PHOSPHORUS, NITROGEN,   PHYTOPLANKTON, PRODUCTIVITY, TURB   W71-07698
ALGAE, COLIFORMS, EUTROPHICATION(PHYTOPLANKTON), ESTUARIES, WATER   W71-05407
RATION, DISSOLVED OXYGEN, ALGAE,   PHYTOPLANKTON, ALKALINE WATER, AL   W71-07731
SAMPLING, PRIMARY PRODUCTIVITY,   PHYTOPLANKTON, EUGLENA, CHLOROPHY   W72-04761
, S/ *PLANKTON, *PHOTOSYNTHESIS,   PHYTOPLANKTON, ALGAE, TEMPERATURE   W71-10788
OWERPLANTS, *LAKES, TEMPERATURE,   PHYTOPLANKTON, ALGAE, HEATED WATE   W71-09767
*EUTROPHICATION, SEWAGE, BIOTA,   PHYTOPLANKTON, DIATOMS, DINOFLAGE   W69-10169

500

RELLA, WISCONSIN, LAKE MICHIGAN,   PHYTOPLANKTON, ·DIATOMS, ECOLOGY,   W69-10180
NUTRIENTS, NITRATES, PHOSPHATES,   PHYTOPLANKTON, IRON COMPOUNDS, AL   W69-10170
UTRIENT REQUIREMENTS, NUTRIENTS,   PHYTOPLANKTON, INHIBITION.: /S, N   W69-10177
ECOLOGY, *MARINE MICROORGANISMS,   PHYTOPLANKTON, BACTERICIDES, ALGA   W70-01068
 ALGAE, HYPOLIMNION, EPILIMNION,   PHYTOPLANKTON, SESTON, CRUSTACEAN   W69-10182
, PLANKTON, DIATOMS, CYANOPHYTA,   PHYTOPLANKTON, ROTIFERS, CHLOROPH   W70-00268
ALTITUDE, SAMPLING, ZOOPLANKTON,   PHYTOPLANKTON, ICE, BACTERIA, TRI   W69-10154
 BENTHIC FLORA, FLOATING PLANTS,   PHYTOPLANKTON, AQUATIC INSECTS, F   W70-10175
NTS, PHYSICOCHEMICAL PROPERTIES,   PHYTOPLANKTON, DISTRIBUTION, DEPT   W71-00665
SIN, ECOLOGY, CHEMICAL ANALYSIS,   PHYTOPLANKTON, CYANOPHYTA, SEDIME   W70-06229
UCTIVITY, CHLOROPHYLL, SAMPLING,   PHYTOPLANKTON, RADIOACTIVITY, PHO   W70-01933
NIQUES, WATER POLLUTION EFFECTS,   PHYTOPLANKTON, PLANKTON, AQUATIC   W70-02245
LGAE, *ROTIFERS, PLANKTON, BASS,   PHYTOPLANKTON, PONDS, ZOOPLANKTON   W70-02249
ION, SAMPLING, SURFA/ *BIOASSAY,   PHYTOPLANKTON, LAKES, EUTROPHICAT   W70-02973
RESIDUES, PHYSIOLOGICAL ECOLOGY,   PHYTOPLANKTON, PHYTOTOXICITY, PLA   W70-02968
C OXYGEN, PHYSIOLOGICAL ECOLOGY,   PHYTOPLANKTON, PLANKTON, POLLUTIO   W70-02792
RYSOPHYTA, CYANOPHYTA, SAMPLING,   PHYTOPLANKTON, ORGANIC MATTER, CH   W70-02780
ALKALINITY, LAKES, PRODUCTIVITY,   PHYTOPLANKTON.: /US, *NITROGEN, *   W70-02795
, TEMPERATURE, DISSOLVED OXYGEN,   PHYTOPLANKTON, WATER ANALYSIS, AL   W72-06051
OLOGICAL ECOLOGY, PHYTOPLANKTON,   PHYTOTOXICITY, PLANKTON, POLLUTAN   W70-02968
TOMOGETON CRISPUS, HETERAMTHERA,   PICKWICK RESERVOIR(TENN), NAJAS,   W71-07698
     *PIGMENTED FLAGELLATES, *NON-   PIGMENTED FLAGELLATES.:   W72-01798
AE, *BLUE GREEN ALGAE, *EUGLENA,   PIGMENTED FLAGELLATES.: /RANT ALG   W70-01233
SSIUM, IRON, MANGANESE, CALCIUM,   PIGMENTS.: /DIMENTS, SODIUM, POTA   W70-09889
ES, PHOTOSYNTHESIS, RESPIRATION,   PIGMENTS, CHLOROPHYLL, TEMPERATUR   W70-03325
ITY, DIATOMS, SEASONAL, BIOMASS,   PIGMENTS, PRODUCTIVITY, PERIPHYTO   W70-02780
COPEPODS, ROTIFERS, CHLOROPHYLL,   PIGMENTS, IMPOUNDMENTS, RESERVOIR   W71-10098
, *COLOR, RIVERS, BOGS, PHENOLS,   PIGMENTS, ORGANIC MATTER, BIOMASS   W70-01074
IMARY PRODUCTIVITY, CHLOROPHYLL,   PIGMENTS, MAGN: /RINE ANIMALS, PR   W73-01446
MICROSCOPY, NETS, NANNOPLANKTON,   PIGMENTS, CHLOROPHYLL, CYANOPHYTA   W70-05389
OPLANKTON, DIATOMS, RESPIRATION,   PIGMENTS, OCEANS, ORGANIC MATTER,   W70-04809
CHLORELLA, CHLOROPHYTA, DIATOMS,   PIGMENTS, CHLOROPHYLL, PHOTOSYNTH   W70-05381
QUES, BACTERIAL PLATE, BACTERIAL   PIGMENTS, DIAPTOMUS, EPISCHURA, O   W70-05760
IOCHEMICAL ANALYSIS, MONITORING,   PIGMENTS, PHYTOPLANKTON, WASTE WA   W70-04371
ROLEUM POLLUTION, PHOTOSYNTHETIC   PIGMENTS, PRODUCTION BIOLOGY, RUS   W69-07440
, CHLOROPHYTA, DISSOLVED OXYGEN,   PIKE, CARP, D: /S, ALGAE, DIATOMS   W70-01943
TIARY TREATMENT, SUSPENDED LOAD,   PILOT PLANTS, ACTIVATED SLUDGE, T   W70-09186
A, CHLOROPHYTA, DIATOMS, WEIGHT,   PILOT PLANTS, IRON, CHLORELLA, SO   W70-03512
ED SLUDGE, BIOLOGICAL TREATMENT,   PILOT PLANTS, HUMUS, REAERATION,   W70-05821
AE, *SLUDGE, STORMS, TURBULENCE,   PILOT PLANTS, BACTERIA, SEDIMENTA   W70-05819
XIDE, GROWTH RATES, TEMPERATURE,   PILOT PLANTS, HYDROGEN ION CONCEN   W72-06022
BIDITY, SUSPENDED SOLIDS, ALGAE,   PILOT PLANTS, HEAD LOSS, PERMEABI   W72-08686
RIOPHYLLUM, LEPOMIS MACROCHIRUS,   PIMEPHALES NOTATUS.: /IA COLI, MY   W69-07861
E, GRAPHITE LAKE, OLD JOHN LAKE,   PINGO LAKE, TWELVE MILE LAKE, BRA   W69-08272
), EUNOTIA, AMPHORA, GOMPHONEMA,   PINNULARIA, NITZSCHIA, NAVICULA,   W70-03974
GE, SLURRIES, AUTOMATIC CONTROL,   PIPING SYSTEMS(MECHANICAL), CLEAN   W72-08688
LYTICAL TECHNIQUES, TEMPERATURE,   PIPTIDES, INTERTIDAL AREAS, SOIL   W70-04365
VER(USSR), MINSK(USSR), ENGLAND/   PISCICULTURAL PONDS, SVISLOCH' RI   W70-05092
OCLADIUS, CORETHRA PUNCTIPENNIS,   PISIDIUM IDAHOENSE, TENDIPES PLUM   W70-06217
10M, CERIUM, GUETTARDA, IPOMOEA,   PISONIA, PANDANUS, MUSCLE, BONE,   W69-08269
 EICHHORNIA, POTAMOGETON, TRAPA,   PISTIA, ALTERNANTHERA, HETERANTHE   W70-10175
INCIPLE, OSCILLATORIA RUBESCENS,   PISTON CORER, FRAGILARIA, MELOSIR   W70-05663
LI, LYNGBYA NOLLEI, GLOCOTRICHIA,   PISUM, NOSTOC VERRUCOSUM, LAKE ME   W69-08283
IAL RECHARGE, *DATA COLLECTIONS,   PIT RECHARGE, WATER SUPPLY, MUNIC   W69-05894
S, BLUEGILLS, MICROCYSTIS SCUMS,   PITHIOPHORA ALGAE.: /E FERTILIZER   W70-05417
*WATER REUSE, DISSOLVED OXYGEN,   PITTING(CORROSION), RUSTING, META   W70-02294
TRATIONS, AEQUIPECTEN IRRADIANS,   PIVERS ISLAND(N C), DETECTOR, BEA   W69-08267
AVICULA, NITZSCHIA, SKELETONEMA,   PLAGIOGRAMMA,: /IGMA, MELOSIRA, N   W70-03325
COASTAL PLAINS, ATLANTIC COASTAL   PLAIN, INORGANIC COMPOUNDS, METAL   W69-03695
ROPHICATION, *RESERVOIRS, *GREAT   PLAINS, ALGAE, NEBRASKA, TURBIDIT   W72-04761
S, GEOGRAPHICAL REGIONS, COASTAL   PLAINS, ATLANTIC COASTAL PLAIN, I   W69-03695
 JOAQUIN DELTA, CALIFORNIA WATER   PLAN, DELTA WATER FACILITIES(CALI   W70-05094
HLOROCOCCALES, CONJUGATOPHYCAEA,   PLANCTOMYCES, BACILLARIOPHYTA, GL   W70-03969
R QUALITY NETWORK, ALGAE COUNTS,   PLANKTON ANALYSES, AGRICULTURAL R   W70-04507
NODININE, *PERIDINIUM POLONICUM,   PLANKTON BLOOMS.: /I, JAPAN, *GLE   W70-08372
RICHIA ECHINULATA, BOTRYOCOCCUS,   PLANKTON DISTRIBUTION, WISCONSIN   W70-03959
E ALGAE, LIMNOLOGY, CHLOROPHYTA,   PLANKTON, PHYTOPLANKTON, WATER PO   W70-03959
(SWEDEN), LAKE MENDOTA(WIS), NET   PLANKTON, CHLOROPHYLL CONTENT, CE   W70-03959
GAE, PHOSPHORUS, TRACE ELEMENTS,   PLANKTON, LAKES, RIVERS, GREAT LA   W70-04385
S, CARBON DIOXIDE, GROWTH RATES,   PLANKTON, OXYGEN, NITROGEN FIXATI   W70-04369
CHAINS, FOOD WEBS, FORAGE FISH,   PLANKTON, ALGAE, *ALGAL CONTROL,   W70-05428
FERTILIZATION, *PHOSPHATE, FISH,   PLANKTON, AQUATIC WEED CONTROL, I   W70-05417
MPERATURE STRATIFICATION, DEPTH,   PLANKTON, LIGHT PENETRATION, SEAS   W70-05405
ANAGEMENT, FERTILIZATION, ALGAE,   PLANKTON, NITROGEN, PHOSPHORUS, P   W70-05399
*CONTROL, TASTE, ODOR, FILTERS,   PLANKTON, WATER POLLUTION, SURFAC   W70-05389
 FERTILIZATION, BACTERIA, ALGAE,   PLANKTON, TUBIFICIDS, CALCIUM, SI   W70-05761
AGE, MARSHES, FARMS, CYANOPHYTA,   PLANKTON, ZOOPLANKTON, EUTROPHICA   W70-05112
IN, INORGANIC COMPOUNDS, METALS,   PLANKTON, AQUATIC LIFE, ZOOPLANKT   W69-03695
D AQUATIC PLANTS, GLACIAL LAKES,   PLANKTON, EPIDEMICS.: /ILL, ROOTE   W68-00488
TIES, BACTERIA/ *EUTROPHICATION,   PLANKTON, *ALGAE, CHEMICAL PROPER   W68-00483
NUTRIENTS, NITROGEN, PHOSPHORUS,   PLANKTON, DISSOLVED OXYGEN, HYPOL   W68-00253
NITROGEN, NUTRIENTS, PHOSPHORUS,   PLANKTON, PLANTS(ORGANISMS).: /,   W68-00481
ATION, SEWAGE, FISH POPULATIONS,   PLANKTON, *CYANOPHYTA, WASHINGTON   W68-00461
ED CONTROL, *INVESTIGATIONS, ZOO-   PLANKTON, SPRAYING, BIOCONTROL, I   W68-00468
MICHIGAN, ALGAE, NUISANCE ALGAE,   PLANKTON, BENTHIC FAUNA, SLUDGE W   W68-01244
N-SITE DATA COLLECTIONS, FISHES,   PLANKTON, ALGAE, FROGS, MOSSES, H   W68-00857
REMENTS, PHOTOSYNTHETIC, OXYGEN,   PLANKTON, WATER POLLUTION EFFECTS   W69-03514
S, RADIOACTIVITY, INVERTEBRATES,   PLANKTON, FISH, SEA WATER, GAMMA   W69-08269
OECOLOGY, EFFLUENTS, ECOSYSTEMS,   PLANKTON, ALGAE, PERIPHYTON, INVE   W69-07853
WS, DETRITUS, CYCLING NUTRIENTS,   PLANKTON, ALGAE, INVERTEBRATES, T   W69-07861

ONTROL, FISHKILL, ROOTED AQUATIC PLANTS, GLACIAL LAKES, PLANKTON, W68-00488
ONTROL, *EUTROPHICATION, AQUATIC PLANTS, HERBICIDES, PESTICIDES, H W69-05705
ONTROL, *EUTROPHICATION, AQUATIC PLANTS, HERBICIDES, PESTICIDES, H W69-05706
, SUSPENDED SOLIDS, ALGAE, PILOT PLANTS, HEAD LOSS, PERMEABILITY, W72-08686
DGE, BIOLOGICAL TREATMENT, PILOT PLANTS, HUMUS, REAERATION, EFFLUE W70-05821
GROWTH RATES, TEMPERATURE, PILOT PLANTS, HYDROGEN ION CONCENTRATIO W72-06022
OROPHYTA, DIATOMS, WEIGHT, PILOT PLANTS, IRON, CHLORELLA, SODIUM, W70-03512
IBU/ *ALGAE, *LITTORAL, *AQUATIC PLANTS, ISLANDS, ECOLOGICAL DISTR W72-10623
AR PRODUCTS, GAS EXCHANGE, GREEN PLANTS, MASS CULTURE, MEDICINE, M W69-07832
T, MECHANICAL EQUIPMENT, PUMPING PLANTS, MAINE.: /, WATER TREATMEN W70-08096
ANIMALS, BENTHIC FAUNA, AQUATIC PLANTS, MARINE ALGAE, BENTHIC FLO W71-02494
ON PLANT WASTES, *MAN-MADE FIBER PLANTS, NYLON-6, *VISCOSE RAYON, W70-06925
S, MORTALITY, GASTROPODS, MARINE PLANTS, ON-SITE DATA COLLECTIONS. W70-01780
NTS, ZOOPLANKTON, ALGAE, AQUATIC PLANTS, PALEOLIMNOLOGY, MATHEMATI W70-03975
NTS, ZOOPLANKTON, ALGAE, AQUATIC PLANTS, PALEOLIMNOLOGY, MATHEMATI W70-03975
VITY, ALGAE, PLANKTON, SUBMERGED PLANTS, PHYSICOCHEMICAL PROPERTIE W71-00665
, CHARA, BENTHIC FLORA, FLOATING PLANTS, PHYTOPLANKTON, AQUATIC IN W70-10175
CHLOROPHYLL, CYANOPHYTA, AQUATIC PLANTS, PHOSPHORUS, DENSITY, WATE W72-04761
MICAL / *MARINE ANIMALS, *MARINE PLANTS, RADIOISOTOPES, PHYSICOCHE W71-09863
NTROL, PONDWEEDS, ROOTED AQUATIC PLANTS, RATES OF APPLICATION, APP W70-05568
BACTERIA, FUNGI, ALGAE, FLOATING PLANTS, ROOTED AQUATIC PLANTS, SE W70-05750
FLOATING PLANTS, ROOTED AQUATIC PLANTS, SELF-PURIFICATION, BIOCHE W70-05750
MICROORGANISMS, WATER CHEMISTRY, PLANTS, SOUTHWEST U. S., REGIONS, W69-03611
GINEERING STRUCTURES, INDUSTRIAL PLANTS, STRUCTURES, AFTERBAYS, WA W69-05023
REATMENT, AQUATIC ALGAE, AQUATIC PLANTS, TESTING, BIODEGRADATION.: W70-04373
OONS, OXIDATION, PHOTOSYNTHESES, PLANTS, TEMPERATURE, *WASTE WATER W70-06600
RIO, WATER TURBIDITY, INDUSTRIAL PLANTS, WATER SPORTS.: /LAKE ONTA W71-09880
ECOLOGY/ PLANT GROWTH, WATER AND PLANTS, WATER POLLUTION EFFECTS, W70-02772
AKE, *WINNISQUAM LAKE, WATER AND PLANTS, WATER POLLUTION EFFECTS, W70-02764
PLANKTON, IRON COMPOUNDS, ALGAE, PLANTS, WATER POLLUTION EFFECTS, W69-10170
Y, BENTHOS, AQUATIC LIFE, ALGAE, PLANTS, WATER PROPERTIES, WATER T W69-05023
N-SITE DATA COLLECTIONS, AQUATIC PLANTS, WATERSHED, ALGAE, AQUATIC W68-00511
ANTS, *RESERVOIRS, HYDROELECTRIC PLANTS, WEED CONTROL, WATER PROPE W68-Q0479
ALGAE, DENSITY, SEASONAL, MARINE PLANTS,: /, WEATHER, EPILIMNION, W70-02504
, PACKED CELL VOLUMES, BACTERIAL PLATE COUNTS, LANCASTER(CALIF).: / W71-05267
ER, CHEMOSYNTHETIC, SAPROPHYTIC, PLATE COUNT, FILAMENTOUS, UNICELL W68-00891
BON UPTAKE TECHNIQUES, BACTERIAL PLATE, BACTERIAL PIGMENTS, DIAPTO W70-05760
                                  PLATTE RIVER, LINCOLN(NEBR).:   W69-05894
-FEEDING ANIMALS, RUTHENIUM-106, PLATYMONAS, THALASSIOSIRA, AMPHID W70-00708
OACTIVITY, GONIOBASIS, DECAPODA, PLECOPTERA, EPHEMEROPTERA, TIPULI W69-09742
HOPTERA, ODONATA, EPHEMEROPTERA, PLECOPTERA, COLEOPTERA, W69-07846
SAVANNAH RIVER LABORATORY, PLECTONEMA BORYANUM, REACTOR.: W69-07845
*ALGAL GROWTH, *POTOMAC RIVER, PLECTONEMA, NOSTOC PALUDOSUM, BLE W70-02255
MICROCOCCUS-LIKE, PSYCHROPHILES, PLEMORPHIC, MYCELOID, CORYNEFORM W70-00713
DAGRION, BRACHYTHEMIS, CHYDORUS, PLEUROXUS, TANYPUS, TANYTARSUS, L W70-06975
WHITEFISH, WURTTEMBERG(GERMANY), PLON(GERMANY), LAKE SCHLIER(GERMA W70-05761
N, CYANIDE, NEW PRODUCT CONTROL, PLUMBING CODES, WELL DRILLING COD W73-00758
IONS, DRAPARNALDIA, DRAPARNALDIA PLUMOSA, STIGEOCLONIUM, MICROCYST W70-02245
IS, PISIDIUM IDAHOENSE, TENDIPES PLUMOSUS, BEAVER DAM LAKE(WIS), D W70-06217
WATER, BISMUTH-207, VERTEBRATES, PLUTONIUM-239, RATS, ZIRCONIUM-95 W69-08269
CARTERIA, NORTH TEMPERATE ZONE, PLUTONIUM, HALF-LIFE(RADIONUCLIDE W70-00708
, WATER POLLUTION CONTROL, WATER PO: /STE TREATMENT, WATER QUALITY W69-06535
NSTITUTION, ACROPORA HYACINTHUS, POCILLOPORA EYDOUXI, PORITES LOBA W69-08526
UM NODOSUM, POLYSIPHONIA LANOSA, POINT LEPREAU(CANADA), NEW BRUNSW W70-04365
DUCKS, NITRATE POISONING.: W69-04805
OPLANKTON, *PHYTOTOXICITY, ALGAL POISONING, SCENEDESMUS, CHLORELLA W69-00994
CES, ALGAE, ALGAL CONTROL, ALGAL POISONING, ALGICIDES, AQUATIC ALG W69-03370
TOTOXICITY, ALGAL CONTROL, ALGAL POISONING, ANALYTICAL TECHNIQUES, W69-03188
S, *ALGAE, *ALGAL CONTROL, ALGAL POISONING, ALGICIDES, AQUATIC ALG W69-05844
ATER, AQUATIC ALGAE, MAR/ *ALGAL POISONING, *MARSH PLANTS, *OILY W W70-01231
OXIC/ *FISHKILL, *TOXINS, *ALGAL POISONING, *ALGAL TOXINS, *PHYTOT W71-07731
RIVER BASIN, URBANIZATION, ALGAL POISONING, GASOLINE, FLOODS, AQUI W73-00758
SPHORUS, POTASSIUM, SELECTIVITY, POISONS, OXYGEN, TRACERS, PHOSPHO W70-05399
RPATHIAN SUBMONTAN/ *SKAWA RIVER(POLAND), MONTANE RIVER COURSE, CA W70-02784
MASS, BENTHIC FLORA, PHAEOPHYTA, POLAR REGIONS, SPECTROPHOTOMETRY, W70-01074
TESTS, OIL SPILLS, GEOLOGICAL SS POLARIS, CHLOROPHYLL FLUORESCENCE W70-05377
S, RHINE RIVER, AMSTERDAM RHINE, POLDERS, COMPOSITION, PSEUDOMONAS W69-07838
VITAMINS, WATER CHEMISTRY, WATER POLLUT: /MENTS, SODIUM, VIRUSES, W69-04801
D, ALGAL TOXINS, ORGANIC WASTES, POLLUTANT IDENTIFICATION, POLLUTA W69-01273
IDE RESIDUES, DIATOMS, CRAYFISH, POLLUTANT IDENTIFICATION, *WATER W69-02782
, PLANKTON, POLLUTION ABATEMENT, POLLUTANT IDENTIFICATION, WATER P W70-02792
EATMENT, *EUTROPHICATION, PONDS, POLLUTANT IDENTIFICATION, BIOLOGY W70-02304
ANKTON, PHYTOTOXICITY, PLANKTON, POLLUTANT IDENTIFI: /OGY, PHYTOPL W70-02968
, WATER ANALYSIS, WATER QUALITY, POLLUTANT IDENTIFICATION.: /PLING W71-00221
E, *ODOR, WATER QUALITY CONTROL, POLLUTANT IDENTIFICATION, IOWA.: / W72-13605
WATER POLLUTION SOURCES, PATH OF POLLUTANT.: /AEROBIC CONDITIONS, W71-11381
ASTE, WATER QUALITY ACT, PATH OF POLLUTANT, FOOD CHAINS, SAMPLING, W72-00941
REMOVAL TIME(POLLUTANTS).: W71-05878
RIO, *WATER CIRCULATION, PATH OF POLLUTANTS, NUTRIENTS, ALGAE, STR W71-05878
UTROPHICATION, *WATER CHEMISTRY, POLLUTANTS, ALGAL CONTROL, NITRAT W70-02646
ALGAE, MARINE ALGAE, OIL, MARSH, POLLUTANTS.: /ILY WATER, AQUATIC W70-01231
TIC ALGAE, *AQUATIC ENVIRONMENT, POLLUTANTS, BIOLOGICAL COMMUNITIE W70-01233
WATER POLLUTION SOURCES, PATH OF POLLUTANTS, AQUATIC LIFE, ALGAE, W69-10080
TRIENTS, PHOSPHORUS, POPULATION, POLLUTANTS, MAPPING, HYDROGEN SUL W70-00667
NU/ *EUTROPHICATION, *NUTRIENTS, POLLUTANTS, POLLUTION ABATEMENT, W69-01453
ASTES, POLLUTANT IDENTIFICATION, POLLUTANTS, QUALITY CONTROL, WATE W69-01273
AL PROPERTIES, BACTERIA, ORGANIC POLLUTANTS, *STREAMS, EUGLENA, NU W68-00483
KGROUND RADIATION, FLOCCULATION, POLLUTANTS, SAMPLING.: /ATES, BAC W69-07846
N CONTROL, *POLLUTION ABATEMENT, POLLUTANTS, WASTES, THERMAL POLLU W69-06305
DE KINETICS, PESTICIDE TOXICITY, POLLUTANTS, WATER POLLUTION EFFEC W69-08565

S, *DETERGENTS, DOMESTIC WASTES,   POLLUTANTS, WATER POLLUTION SOURC   W70-05084
MONIA, CARBON, RECREATION, WATER   POLLUTI: /LOROPHYTA, NITRATES, AM   W72-04761
L, *WATER SUPPLY, *BASINS, WATER   POLLUTION TREATMENT, COPPER SULFA   W72-01814
SCENEDESMUS, FLOCCULATION, WATER   POLLUTION TREATMENT.: /FILTERS, *   W72-03703
POPULATION, GROWTH RATES, WATER   POLLUTION CONTROL.: /RTICLE SIZE, *   W72-02417
ANTS, PHOSPHORUS, DENSITY, WATER   POLLUTION SOURCES, RUNOFF, LIGHT   W72-04761
SYSTEMS, AQUATIC HABITATS, WATER   POLLUTION EFFECTS.: /OCESSES, ECO   W71-09012
, WATER POLLUTION SOURCES, WATER   POLLUTION EFFECTS, WATER QUALITY.   W71-05409
AEROBICS, ALGAE, BACTERIA, WATER   POLLUTION SOURCES, ODOR, FARM WAS   W71-05742
TION, METABOLISM, AQUARIA, WATER   POLLUTION EFFECTS, ADSORPTION.: /   W71-09005
SYSTEMS, GREAT LAKES, TR/ *WATER   POLLUTION EFFECTS, *FALLOUT, *ECO   W71-09013
OXYGEN, ESTUARIES, ALGAE, WATER   POLLUTION SOURCES, WATER POLLUTIO   W71-05409
DIOISOTOPES, MARINE ALGAE, WATER   POLLUTION SOURCES.: /, URANIUM RA   W71-09013
UARIES, *HARBORS, *SEWAGE, WATER   POLLUTION EFFECTS, *ALGAE.: *EST   W71-05500
E ELEMENTS, *AQUATIC LIFE, WATER   POLLUTION EFFECTS, MARINE ANIMALS   W71-05531
E, WASTE WATER(POLLUTION), WATER   POLLUTION SOURCES, MUNICIPAL WAST   W71-05879
*WA/ *SELF-PURIFICATION, *WATER   POLLUTION CONTROL, *LAKE ONTARIO,   W71-05878
IC LIFE, *WASTE DISPOSAL, *WATER   POLLUTION EFFECTS, FOOD CHAINS, B   W72-01801
, *ODOR, *AMINO ACIDS, P/ *WATER   POLLUTION SOURCES, *ALGAE, *TASTE   W72-01812
TION CONTROL, FISH CONSERVATION,   POLLUTION ABATEMENT, BIBLIOGRAPHI   W72-01786
, WATER POLLUTION EFFECTS, WATER   POLLUTION CONTROL, *NUTRIENTS.: /   W72-00917
ON EFFECTS, *CLASSIFICAT/ *WATER   POLLUTION SOURCES, *WATER POLLUTI   W72-00462
OSAL, *INDUSTRIAL WASTES, *WATER   POLLUTION CONTROL, FISH CONSERVAT   W72-01786
*WATER POLLUTION SOURCES, *WATER   POLLUTION EFFECTS, *CLASSIFICATIO   W72-00462
TY, FOOD CHAINS, TOXICITY, WATER   POLLUTION CONTROL, DECOMPOSING OR   W72-01788
ICAL DISTRIBUTION, MIDGES, WATER   POLLUTION EFFECTS, OLIGOCHAETES,   W71-12077
IODISM, LIGHT PENETRATION, WATER   POLLUTION EFFECTS, WATER POLLUTIO   W72-00917
TROPH/ *SEWAGE, *STREAMS, *WATER   POLLUTION EFFECTS, *FISH, ALGAE,   W72-01788
STORAGE, *WATER POLLUTION, WATER   POLLUTION SOURCES, WATER POLLUTIO   W72-00941
, *BIOLOGICAL PROPERTIES, *WATER   POLLUTION SOURCES, TASTE, ODOR, A   W72-01813
, WATER POLLUTION SOURCES, WATER   POLLUTION CONTROL, RADIOACTIVE WA   W72-00941
N(DECOMPOSITION), METHANE, WATER   POLLUTION EFFECTS.: / *DEGRADATIO   W71-13413
EFFECTS, *CLASSIFICATION, *WATER   POLLUTION CONTROL, WATERCOURSES(L   W72-00462
OIDS, AQUARIA, ABSORPTION, WATER   POLLUTION EFFECTS, CATION EXCHANG   W71-09863
HEATED WATER, POPULATION, WATER   POLLUTION EFFECTS.: /KTON, ALGAE,   W71-09767
, *RECREATION, *SWIMMING, *WATER   POLLUTION EFFECTS, BACTERIA, ALGA   W71-09880
OPPER, *TOXICITY, *ALGAE, *WATER   POLLUTION EFFECTS, CHELATION, SPE   W71-11561
ONS, ANAEROBIC CONDITIONS, WATER   POLLUTION SOURCES, PATH OF POLLUT   W71-11381
PES, COBALT RADIOISOTOPES, WATER   POLLUTION EFFECTS.: /E RADIOISOTO   W71-09850
SPHORUS, ALGAE, LIVESTOCK, WATER   POLLUTION EFFECTS.: /ITROGEN, PHO   W71-10376
RTEBRATES, NUCLEAR WASTES, WATER   POLLUTION SOURCES.: / ALGAE, INVE   W71-09863
IOLOGY, STREAM POLLUTION, *WATER   POLLUTION DETECTION.: /, MARINE B   W71-10786
CATION, WASHINGTON, LAKES, WATER   POLLUTION EFFECTS.: /US, EUTROPHI   W70-00270
ENTS), WATERSHEDS(BASINS), WATER   POLLUTION EFFECTS, CYANOPHYTA, EU   W69-10182
, *WATER QUALITY, *DISSO/ *WATER   POLLUTION EFFECTS, *BIOINDICATORS   W70-00475
GTON, NUCLEAR POWERPLANTS, WATER   POLLUTION EFFECTS, PHOSPHORUS, NI   W70-00264
ITY, AQUATIC PRODUCTIVITY, WATER   POLLUTION EFFECTS, YEASTS, CRUSTA   W69-10152
TER SUPPLY, WATER QUALITY, WATER   POLLUTION CONTROL, NUISANCE ALGAE   W69-10155
INEAR ALKYLATE SULFONATES, WATER   POLLUTION EFFECTS, STABILIZATION,   W69-09454
, WATER POLLUTION EFFECTS, WATER   POLLUTION SOURCES, AQUATIC PRODUC   W69-09135
SPHATES, NUTRIENTS, ALGAE, WATER   POLLUTION EFFECTS, WATER POLLUTIO   W69-09135
*NEW YORK, NUCLEAR WASTES, WATER   POLLUTION SOURCES, PATH OF POLLUT   W69-10080
, WATER POLLUTION EFFECTS, WATER   POLLUTION SOURCES, RADIOECOLOGY.:   W70-00707
FIXATION, VOLUME, STREAMS, WATER   POLLUTION EFFECTS, EUTROPHICATION   W70-00719
TOMS, NERITIC ALGAE, IONS, WATER   POLLUTION EFFECTS, WATER POLLUTIO   W70-00707
TROPHICATION, *NUTRIENTS, *WATER   POLLUTION SOURCES, *PHOSPHATES, A   W70-01031
URCES, *PHOSPHATES, ALGAE, WATER   POLLUTION TREATMENT, WATER QUALIT   W70-01031
SANDS, GRAVELS, SEDIMENTS, WATER   POLLUTION SOURCES.: /MUD, SILTS,   W70-00711
*WATER POLLUTION CONTRO/ *WATER   POLLUTION TREATMENT, *FILTRATION,   W70-01027
N TREATMENT, *FILTRATION, *WATER   POLLUTION CONTROL, *FILTERS, SEWA   W70-01027
, WATER POLLUTION EFFECTS, WATER   POLLUTION CONTROL, NITRATES, NITR   W69-10167
S, LIGHT INTENSITY, LAKES, WATER   POLLUTION SOURCES, WATER POLLUTIO   W69-10167
, WATER POLLUTION SOURCES, WATER   POLLUTION EFFECTS.: /, CYANOPHYTA   W69-10180
OIL, CULTURES, CYANOPHYTA, WATER   POLLUTION SOURCES, WATER POLLUTIO   W69-10180
COMPOUNDS, ALGAE, PLANTS, WATER   POLLUTION EFFECTS, WATER POLLUTIO   W69-10170
, WATER POLLUTION EFFECTS, WATER   POLLUTION SOURCES.: /RBON DIOXIDE   W69-10158
, SEDIMENTATION, COPEPODS, WATER   POLLUTION SOURCES, WATER P: /ERIA   W69-10169
, WATER POLLUTION EFFECTS, WATER   POLLUTION SOURCES, NUTRIENTS.: /S   W69-10159
, WATER POLLUTION EFFECTS, WATER   POLLUTION SOURCES, WATER POLLUTIO   W69-10170
US, *MINNESOTA, *EUTROPHICATION,   POLLUTION ABATEMENT, GROWTH RATES   W69-10167
, WATER POLLUTION SOURCES, WATER   POLLUTION CONTROL, LIMNOLOGY, DIV   W69-10170
REATMENT, WATER POLLUTION, WATER   POLLUTION EFFECTS, ALGAL CONTRO: /   W69-10171
, WATER POLLUTION SOURCES, WATER   POLLUTION EFFECTS, WATER POLLUTIO   W69-10167
ALKALINITY, FISH, STREAMS, WATER   POLLUTION EFFECTS, WATER POLLUTIO   W69-10159
OACTIVITY, CARBON DIOXIDE, WATER   POLLUTION EFFECTS, WATER POLLUTIO   W69-10158
WATER POLLUTION TREATMENT, WATER   POLLUTION EFFECTS.: /ERSISTENCE,   W70-01777
XYGEN, DETRITUS, BACTERIA, WATER   POLLUTION EFFECTS, LIMNOLOGY, CYA   W70-02249
TS, ANALYTICAL TECHNIQUES, WATER   POLLUTION EFFECTS, PHYTOPLANKTON,   W70-02245
DETERGENTS, *MARINE ALGAE, WATER   POLLUTION EFFECTS, LETHAL LIMIT,   W70-01779
, WATER POLLUTION CONTROL, WATER   POLLUTION EFFECTS, WATER: /ATMENT   W70-02248
S, BIOLOGICAL PROCESSES, WATER   POLLUTION EFFECTS, DIATOMS.: /ANT   W70-01233
ARINE ALGAE, *EMULSIFIERS, WATER   POLLUTION EFFECTS, MORTALITY, GAS   W70-01780
FLUENTS, SEWAGE TREATMENT, WATER   POLLUTION CONTROL, WATER POLLUTIO   W70-02248
UTRIENT REQUIREMENTS, NUTRIENTS,   POLLUTION IDENTIFICATION, SEWAGE,   W70-02248
R INTERFACES, PERSISTENCE, WATER   POLLUTION TREATMENT, WATER POLLUT   W70-01777
ROPHICATION, *CONFERENCES, WATER   POLLUTION EFFECTS, ALGAL CONTROL,   W70-02775
NE WATER, WATER POLLUTION, WATER   POLLUTION EFFECTS, ACID STREAMS,   W70-02770
STIMULATION, ALGAL GROWTH, WATER   POLLUTION ASSESSMENT, SANITARY EN   W70-02775
GROWTH, WATER AND PLANTS, WATER   POLLUTION EFFECTS, ECOLOGY, EFFEC   W70-02772
BIOLOGY, WATER POLLUTION, WATER   POLLUTION SOURCES, ALGAL CONTROL,   W70-02304

```
Y, RADIOACTIVITY, TRACERS, WATER    POLLUTION EFFECTS, PHYSIOLOGICAL:     W70-02510
AM LAKE, WATER AND PLANTS, WATER    POLLUTION EFFECTS, ECOLOGY, WATER     W70-02764
POUNDS, LIGHT PENETRATION, WATER    POLLUTION EFFECTS, EUTROPHICATION     W70-02255
ER QUALITY CONTROL, LAKES, WATER    POLLUTION SOURCES, CHLOROPHYTA, N     W70-03310
     *RIVERS, WATER CHEMISTRY, WATER    POLLUTION SOURCES, PHOSPHORUS, NI     W70-03068
GAE, NITROGEN, PHOSPHORUS, WATER    POLLUTION EFFECTS.: /ERTILITY, AL     W70-02973
ROL, EUTROPHICATION, BASS, WATER    POLLUTION CONTROL, MINNOWS, NUISA     W70-02982
C MATTER, CURRENTS(WATER), WATER    POLLUTION SOURCES, SURFACE RUNOFF     W70-03501
, STANISLAUS RIVER(CALIF), WATER    POLLUTION ASSESSMENT, AMERICAN RI     W70-02777
, DIATOMS, EUTROPHICATION, WATER    POLLUTION EFFECTS, SYSTEMS ANALYS     W70-02779
, WATER POLLUTION CONTROL, WATER    POLLUTION EFFECTS, WATER POLLUTIO     W70-02792
POLLUTANT IDENTIFICATION, WATER     POLLUTION CONTROL, WATER POLLUTIO     W70-02792
TRATES, INDUSTRIAL WASTES, WATER    POLLUTION EFFECTS, DENITRIFICATIO     W70-02777
COLOGY, PHYTOPLANKTON, PLANKTON,    POLLUTION ABATEMENT, POLLUTANT ID     W70-02792
, WATER POLLUTION EFFECTS, WATER    POLLUTION SOURCES.: /TION CONTROL     W70-02792
ANABAENA, BIOSTIMULATION, WATER     POLLUTION ASSESSMENT.: /L GROWTH,     W70-02779
T, *BIOLOGICAL TREATMENT, *WATER    POLLUTION CONTROL, FISH, TOXICITY     W70-10437
ATER, *OIL, *MARINE ALGAE, WATER    POLLUTION SOURCES, ON-SITE DATA C     W70-09976
OILY WATER, / RESISTANCE, WATER     POLLUTION EFFECTS, *SALT MARSHES,     W70-09976
ESIDUES, *PESTICIDE KINE/ *WATER    POLLUTION EFFECTS, *PESTICIDES, R     W70-09768
, WATER POLLUTION SOURCES, WATER    POLLUTION EFFECTS, PHOSPHORUS, NI     W70-07261
*EUTROPHICATION, *LAKES, *WATER     POLLUTION CONTROL, *REVIEWS, ALGA     W70-07393
N SOURCES, *THERMAL POLL/ *WATER    POLLUTION EFFECTS, WATER POLLUTIO     W70-07847
*ALGAE, *BIOINDICATORS, *WATER      POLLUTION EFFECTS, *REVIEWS, *WAT     W70-07027
LOGY, *INDUSTRIAL WASTES, *WATER    POLLUTION EFFECTS, *EUTROPHICATIO     W70-06975
*WATER POLLUTION EFFECTS, WATER     POLLUTION SOURCES, *THERMAL POLLU     W70-07847
RATES, PHOSPHATES, SODIUM, WATER    POLLUTION EFFECTS, LABORATORY TES     W70-07257
*NUISANCE ALGAE, *SEWAGE, WATER     POLLUTION SOURCES, WATER POLLUTIO     W70-07261
ATORY TESTS, INSECTICIDES, WATER    POLLUTION EFFECTS, ALGAE.: /LABOR     W71-03183
S, ORGANIC MATTER, CARBON, WATER    POLLUTION EFFECTS, BACTERIA, PERI     W71-04206
PHYTOPLANKTON), ESTUARIES, WATER    POLLUTION SOURCES, WATER POLLUTIO     W71-05407
*LAKES, *EUTROPHICATION, *WATER     POLLUTION CONTROL, *WATER POLLUTI     W71-03048
*WATER POLLUTION CONTROL, *WATER    POLLUTION SOURCES, *WATER POLLUTI     W71-03048
, WATER POLLUTION SOURCES, WATER    POLLUTION EFFECTS, NITROGEN, COLI     W71-05407
, *WASTE WATER TREATMENT, *WATER    POLLUTION CONTROL, *DENITRIFICATI     W71-04555
DEMAND, BACTERIA, ECOLOGY, WATER    POLLUTION EFFECTS, LAKES, *CYANOP     W71-04758
, *WASTE WATER TREATMENT, *WATER    POLLUTION CONTROL, *DENITRIFICATI     W71-04556
*WATER POLLUTION SOURCES, WATER     POLLUTION EFFECTS, WATER UTILIZAT     W71-03048
IOLOGY, NITRATES, AMMONIA, WATER    POLLUTION EFFECTS, WASTE WATER TR     W71-05026
, *WASTE WATER TREATMENT, *WATER    POLLUTION CONTROL, *DENITRIFICATI     W71-04554
, *WASTE WATER TREATMENT, *WATER    POLLUTION CONTROL, *DENITRIFICATI     W71-04557
N EFFECTS, DINOFLAGELLATE/ WATER    POLLUTION SOURCES, WATER POLLUTIO     W70-08372
                          *WATER     POLLUTION RESEARCH.:                  W70-08627
PARTICLE SIZE, *CLAYS, */ *WATER    POLLUTION CONTROL, *FILTRATION, *     W70-08628
, *WASTE WATER TREATMENT, *WATER    POLLUTION CONTROL, *EUTROPHICATIO     W70-09186
  WATER POLLUTION SOURCES, WATER    POLLUTION EFFECTS, DINOFLAGELLATE     W70-08372
UND/ *ALGAE, *CHLOROPHYLL, WATER    POLLUTION EFFECTS, *ORGANIC COMPO     W70-08416
PMENT,/ *WATER RESOURCES, *WATER    POLLUTION CONTROL, *RESEARCH EQUI     W70-08627
ROMATOGRAPHY, DEGRADATION, WATER    POLLUTION EFFECTS, AQUATIC ENVIRO     W70-08652
MENT, CULTURES, CHLORELLA, WATER    POLLUTION EFFECTS.: / WATER TREAT     W70-09438
STS, MINNESOTA, NUTRIENTS, WATER    POLLUTION EFFECTS.: /ABORATORY TE     W70-08637
SULFONATES, *SURFACTANTS, *WATER    POLLUTION SOURCES, *DETERGENTS, *     W70-09438
, WATER POLLUTION SOURCES, WATER    POLLUTION EFFECTS, DISSOLVED OXYG     W70-04727
, SEDIMENTS, URBANIZATION, WATER    POLLUTION TREATMENT, ALGAE, WATER     W70-04721
ANALYSIS, ORGANIC MATTER, WATER     POLLUTION EFFECTS, ANALYTICAL TEC     W70-05109
L ECOLOGY, ORGANIC MATTER, WATER    POLLUTION EFFECTS, DIATOMS, CARBO     W70-04811
CATION, *LAKES, NUTRIENTS, WATER    POLLUTION EFFECTS, PRODUCTIVITY,      W70-04721
MESTIC WASTES, POLLUTANTS, WATER    POLLUTION SOURCES, WATER QUALITY,     W70-05084
, *STRATIFICATION, *TEXAS, WATER    POLLUTION SOURCES, WATER POLLUTIO     W70-04727
EWAGE DISPOSAL, WATER POLLUTION,    POLLUTION ABATEMENT, COLIFORMS, W     W70-04810
S, *PHYTOPLANKTON, *DDT,/ *WATER    POLLUTION EFFECTS, *PHOTOSYNTHESI     W70-05272
ANALYT/ *WATER CHEMISTRY, *WATER    POLLUTION EFFECTS, *PHOSPHATES, *     W70-05269
*NUTRIENTS, *PESTICIDES, *WATER     POLLUTION CONTROL, EUTROPHICATION     W70-05470
TS, *PRIMARY PRODUCTIVITY, WATER    POLLUTION EFFECTS, BIOASSAYS, ALG     W70-05424
, *LAKES, *RIVERS, *ALGAE, WATER    POLLUTION CONTROL, WATER POLLUTIO     W70-05668
, WATER POLLUTION CONTROL, WATER    POLLUTION EFFECTS, WATER POLLUTIO     W70-05668
, WATER POLLUTION EFFECTS, WATER    POLLUTION SOURCES, SEWAGE, PHOSPH     W70-05668
UATIC LIFE, AQUATIC ALGAE, WATER    POLLUTION EFFECTS.: /DICATORS, AQ     W70-05958
OCARBON UPTAKE TECHNIQUES, WATER    POLLUTION IDENTIFICATION, ORGANIC     W70-05652
UCTIVITY, WASTE TREATMENT, WATER    POLLUTION EFFECTS, WATER POLLUTIO     W69-08518
GAE, NITROGEN, PHOSPHORUS, WATER    POLLUTION EFFECTS, WATER POLLUTIO     W69-08668
LL, LAKES, ODOR, TOXICITY, WATER    POLLUTION EFFECTS, WATER QUALITY,     W69-08279
T LAKES, *TRACE ELEMENTS, *WATER    POLLUTION EFFECTS, CHLORIDES, HAL     W69-08562
CIDE TOXICITY, POLLUTANTS, WATER    POLLUTION EFFECTS, ALDRIN, DIELDR     W69-08565
HLORIDES, HALOGENS, ALGAE, WATER    POLLUTION SOURCES, WATER CHEMISTR     W69-08562
S, NUTRIEN/ *NEW HAMPSHIRE WATER    POLLUTION COMMISSION, ALGAL BLOOM     W69-08674
, WATER POLLUTION EFFECTS, WATER    POLLUTION SOURCES, CYCLING NUTRIE     W69-08518
NUISANCE ALGAE, BACTERIA, WATER     POLLUTION EFFECTS.: /HIES, LAKES,     W69-08283
, WATER POLLUTION SOURCES, WATER    POLLUTION CONTROL, OXYGEN BALANCE     W69-08668·
, WATER POLLUTION EFFECTS, WATER    POLLUTION SOURCES, WATER POLLUTIO     W69-08668·
, WATER POLLUTION EFFECTS, WATER    POLLUTION SOURCES.: /ER CHEMISTRY     W69-06273
HEMISTRY, WATER CHEMISTRY, WATER    POLLUTION EFFECTS, WATER POLLUTIO     W69-06273
NITRATES, EUTROPHICATION, *WATER    POLLUTION SOURCES, AGRICULTURAL W     W69-05323
, WATER POLLUTION EFFECTS, WATER    POLLUTION SOURCES, WISCONSIN.: /S     W69-06276
LUTION CONTROL, NUTRIENTS, WATER    POLLUTION EFFECTS, WATER POLLUTIO     W69-06276
TEMEN/ *OHIO, *LAKE ERIE, *WATER    POLLUTION CONTROL, *POLLUTION ABA     W69-06305
NG, WATER QUALITY CONTROL, WATER    POLLUTION CONTROL, WATER POLLUTIO     W69-05706
NG, WATER QUALITY CONTROL, WATER    POLLUTION CONTROL, WATER POLLUTIO     W69-05705
, WATER POLLUTION CONTROL, WATER    POLLUTION EFFECTS.: /LITY CONTROL     W69-05705
, WATER POLLUTION CONTROL, WATER    POLLUTION EFFECTS, FUNGI, AQUATIC     W69-05706
```

```
LGAE, NUTRIENTS, VITAMINS, WATER    POLLUTION EFFECTS, WATER POLLUTIO    W69-06277
ATER POLLUTION, WATER LAW, WATER    POLLUTION EFFECTS, WATER POLLUTIO    W69-06305
OMPOUNDS, WATER CHEMISTRY, WATER    POLLUTION EFFECTS, WATER QUALITY,    W69-05868
TY, LAKES, NUISANCE ALGAE, WATER    POLLUTION CONTROL, NUTRIENTS, WAT    W69-06276
, WATER POLLUTION EFFECTS, WATER    POLLUTION SOURCES, WATER POLLUTIO    W69-06305
MANGANESE, EUTROPHICATION, WATER    POLLUTION CONTROL.: /HYTA, IRON,     W69-05704
TOXICITY, WATER CHEMISTRY, WATER    POLLUTION EFFECTS.: / NUTRIENTS,     W69-05867
, WATER POLLUTION SOURCES, WATER    POLLUTION TREATMENT, FINANCING, G    W69-06305
, WATER POLLUTION EFFECTS, WATER    POLLUTION SOURCES, VITAMIN B.: /S    W69-06277
PRODUCTIVITY, SYSTEMATICS, WATER    POLLUTION EFFECTS, WATER POLLUTIO    W69-07832
, WATER POLLUTION EFFECTS, WATER    POLLUTION SOURCES, WATER SUPPLY.:    W69-07832
ATION, *INLAND WATERWAYS, *WATER    POLLUTION EFFECTS, *WATER SUPPLY,    W69-07833
TREATMENT, WATER QUALITY, WATER    POLLUTION CONTROL, WATER PO: /STE     W69-06535
EMS, *SIMULATION ANALYSIS, WATER    POLLUTION EFFECTS, LAKES, PHOSPHO    W69-06536
NUTRIENTS, EUTROPHICATION, WATER    POLLUTION EFFECTS, WATER POLLUTIO    W69-06864
TS, BIOLOGICAL PROPERTIES, WATER    POLLUTION EFFECTS.: /IDE, NUTRIEN    W69-07442
NTS, FE/ *EUTROPHICATION, *WATER    POLLUTION SOURCES, *LAKES, NUTRIE    W69-07084
, WATER POLLUTION EFFECTS, WATER    POLLUTION SOURCES, BIOINDICATORS,    W69-06864
ONSIN, *SEWAGE EFFLUENTS, *WATER    POLLUTION CONTROL, *LOCAL GOVERNM    W69-06909
ACTERIA, AQUATIC BACTERIA, WATER    POLLUTION EFFECTS.: /S, YEASTS, B    W69-06970
S, RADIOECOLOGY, SORPTION, WATER    POLLUTION EFFECTS, WATER POLLUTIO    W69-07440
, WATER POLLUTION EFFECTS, WATER    POLLUTION SOURCES.: /GY, SORPTION    W69-07440
EN DEMAND, ODOR-PRODUCING ALGAE,    POLLUTION ABATEMENT, SEWAGE TREAT    W69-06909
RIVERS, *WATER POLLUTION, *WATER    POLLUTION SOURCES, IRRIGATION, AQ    W69-07096
PHYTOPLANKTON, *REVIEWS, *WATER    POLLUTION EFFECTS.: /OLIGOTROPHY,    W68-00680
ANOPHYTA, LAND MANAGEMENT, *POL    POLLUTION SOURCES, DREDGING, *POL    W68-00172
CONTROL, SOLAR RADIATION, *WATER    POLLUTION CONTROL, ALGICIDES, BIO    W68-00478
DUCKS(WILD), OYSTERS, CHANNELS,    POLLUTION ABATEMENT, HUMAN POPULA    W68-00461
E ERIE, *NUTRIENTS, ALGAE, WATER    POLLUTION SOURCES, NITROGEN COMPO    W69-01445
CONTROL, WATER POLLUTION, WATER    POLLUTION SOURCES, SEPARATION TEC    W69-01273
RIVERS, SEWAGE TREATMENT, WATER    POLLUTION SOURCES, PRIMARY TREATM    W69-01453
ICATION, *NUTRIENTS, POLLUTANTS,    POLLUTION ABATEMENT, NUISANCE ALG    W69-01453
*WATER POLLUTION SOURCES, *WATER    POLLUTION EFFECTS, *PRIMARY PRODU    W69-02959
POLLUTANT IDENTIFICATION, WATER    POLLUTION EFFECTS, GAS CHROMATOGR    W69-02782
HY, ANALYTICAL TECHNIQUES, WATER    POLLUTION SOURCES.: /CHROMATOGRAP    W69-00387
ON EFFE/ *EUTROPHICATION, *WATER    POLLUTION SOURCES, *WATER POLLUTI    W69-02959
MINANT ORGANISMS, GRAZING, WATER    POLLUTION EFFECTS, DIATOMS, CHLOR    W69-01977
LAKES, *WATER POLLUTION, *WATER    POLLUTION CONTROL, WATER POLLUTIO    W69-03948
S, *PHOSPHORUS COMPOUNDS, *WATER    POLLUTION EFFECTS, ALGAE, ANIMALS    W69-04805
NEERING, ALGAE, PLANTS, CONTROL,    POLLUTION ABATEMENT, ABATEMENT, P    W69-03683
*WATER POLLUTION CONTROL, WATER    POLLUTION EFFECTS, EUTROPHICATION    W69-03948
OMPOUNDS, *PHYTOPLANKTON, *WATER    POLLUTION SOURCES, *CYCLING NUTRI    W69-04800
NGTON, *SEWAGE TREATMENT, *WATER    POLLUTION CONTROL, PACIFIC NORTHW    W69-03683
, *PHYSIOLOGICAL ECOLOGY, *WATER    POLLUTION EFFECTS, ALGAE, ALGAL C    W69-03513
NTHETIC, OXYGEN, PLANKTON, WATER    POLLUTION EFFECTS.: /NTS, PHOTOSY    W69-03514
*WATER POLLUTION CONTROL, *WATER    POLLUTION TREATMENT, ALGAE, ALGAL    W69-03374
N, *PHOSPHORUS COMPOUNDS, *WATER    POLLUTION SOURCES, ALGAE, ALGAL C    W69-03370
REMENTS, *EUTROPHICATION, *WATER    POLLUTION CONTROL, *WATER POLLUTI    W69-03374
ENTS, *EUTROPH/ *STREAMS, *WATER    POLLUTION EFFECTS, *CYCLING NUTRI    W69-03369
ON EFFECTS, *EUTROPHICAT/ *WATER    POLLUTION SOURCES, *WATER POLLUTI    W70-04266
CHLAMYDOMONAS, BIOASSAYS, WATER    POLLUTION EFFECTS, ANALYTICAL TEC    W70-04284
A, *ANALYTICAL TECHNIQUES, WATER    POLLUTION EFFECTS, ALGAE, ECOSYST    W70-04287
*WATER POLLUTION SOURCES, *WATER    POLLUTION EFFECTS, *EUTROPHICATIO    W70-04266
, ALGAE, FISH POPULATION, *WATER    POLLUTION CONTROL, FARM WASTES, M    W70-04266
ES, ANALYTICAL TECHNIQUES, WATER    POLLUTION EFFECTS, PHYTOPLANKTON,    W70-04194
*WATER POLLUTION EFFECTS, *WATER    POLLUTION CONTROL, *CONFERENCES,     W70-03975
ENTAL EFFECTS, PERIPHYTON, WATER    POLLUTION EFFECTS, OREGON, BIOIND    W70-03978
TROPHICATION, *MEROMIXIS, *WATER    POLLUTION EFFECTS, BIOINDICATORS,    W70-03974
*WATER POLLUTION SOURCES, *WATER    POLLUTION EFFECTS, *WATER POLLUTI    W70-03975
ON EFFE/ *EUTROPHICATION, *WATER    POLLUTION SOURCES, *WATER POLLUTI    W70-03975
, WATER POLLUTION SOURCES, WATER    POLLUTION EFFECTS, LABORATORY TES    W70-04074
IOR, MADISON(WIS), EUROPE, WATER    POLLUTION ASSESSMENT, URBAN DRAIN    W70-03975
*WATER POLLUTION EFFECTS, *WATER    POLLUTION CONTROL, *CONFERENCES,     W70-03975
ON EFFE/ *EUTROPHICATION, *WATER    POLLUTION SOURCES, *WATER POLLUTI    W70-03975
*WATER POLLUTION SOURCES, *WATER    POLLUTION EFFECTS, *WATER POLLUTI    W70-03975
UENTS, TERTIARY TREATMENT, WATER    POLLUTION SOURCES, WATER POLLUTIO    W70-04074
IOR, MADISON(WIS), EUROPE, WATER    POLLUTION ASSESSMENT, URBAN DRAIN    W70-03975
ES, ENVIRONMENTAL EFFECTS, WATER    POLLUTION EFFECTS, EPILIMNION, CO    W70-03957
TRATES, LAKE ERIE, LAKE ONTARIO,    POLLUTION ABATEMENT, OLIGOTROPHY,    W70-03964
SANCE ALGAE, PUBLICATIONS, WATER    POLLUTION EFFECTS, NEW YORK, CHLO    W70-03973
, PLANKTON, PHYTOPLANKTON, WATER    POLLUTION EFFECTS, WISCONSIN ZOOP    W70-03959
, SCENEDESMUS, PHOSPHATES, WATER    POLLUTION SOURCES.: /E, *BIOASSAY    W70-03955
IABILITY, WATER POLLUTION, WATER    POLLUTION EFFECTS, NITROGEN, PHOS    W70-04382
NTROL, MECHANICAL CONTROL, WATER    POLLUTION CONTROL, CHEMCONTROL, A    W70-04494
SON RIVER, DELAWARE RIVER, WATER    POLLUTION EFFECTS.: /E RIVER, HUD    W70-04507
E, DRAINAGE DIST/ *ALGAE, *WATER    POLLUTION TREATMENT, LAKES, SEWAG    W70-04455
WAGE EFFLUENTS, MINNESOTA, WATER    POLLUTION EFFECTS, ALGICIDES, WAT    W70-04506
ON EFFECTS, *WATER SOURC/ *WATER    POLLUTION SOURCES, *WATER POLLUTI    W73-00758
*WATER POLLUTION SOURCES, *WATER    POLLUTION EFFECTS, *WATER SOURCES    W73-00758
VAL, SOIL EROSION, ALGAE, *WATER    POLLUTION CONTROL, WATER QUALITY    W73-01122
WASTES, FRESHWATER ALGAE, WATER    POLLUTION EFFECTS, DOMESTIC WASTE    W72-12240
, TURF, SOIL MICROBIOLOGY, WATER    POLLUTION SOURCES.: /, SOIL ALGAE    W72-08110
, ALGAE, LABORATORY TESTS, WATER    POLLUTION EFFECTS.: /N PESTICIDES    W72-11105
GREAT LAKES BASIN, BACTERIAL    POLLUTION.:                          W69-03948
SLUDGE WORMS, BLOODWORMS, WATER    POLLUTION.: /KTON, BENTHIC FAUNA,    W68-01244
OMPOSING ORGANIC MATTER, *STREAM    POLLUTION.: /TER UTILIZATION, DEC    W68-00483
*ESTUARINE    POLLUTION.:                          W69-06203
NA, FLORA, LABORATORY, MINERALS,    POLLUTION.: /ISH, ECOSYSTEMS, FAU    W72-02203
, NUTRIENTS, SEWAGE, WASTE WATER(    POLLUTION), WATER POLLUTION SOURC    W71-05879
HOSPHATES, NITROGEN, WASTE WATER(    POLLUTION), SCENEDESMUS, CYANOPHY    W70-02255
```

506

```
SING ORGANIC MATTER, WASTE WATER(   POLLUTION).: /, BACTERIA, DECOMPO     W70-00664
ACHES, SANDS, BIOTA, WASTE WATER(    POLLUTION), CURRENTS(WATER).: /BE     W68-00010
PHYSICAL PROPERTIES, WASTE WATER(    POLLUTION), WATER QUALITY: /IES,      W69-01273
ENTS, PHYTOPLANKTON, WASTE WATER(    POLLUTION), PHOTOSYNTHESIS, CHL: /    W70-04371
*SELF-PURIFICATION, *WASTE WATER(    POLLUTION), *WATER PURIFICATION,      W70-05092
                                     POLLUTION-TOLERANT ALGAE.:            W70-07027
ATER POLLUTION SOURCES, *THERMAL     POLLUTION, THERMAL POWERPLANTS, *     W70-07847
CCLIMATIZATION, DIATOMS, THERMAL     POLLUTION, TEMPERATURE, CHEMICAL      W70-07313
*REVIEWS, *WATER QUALITY, WATER      POLLUTION, ORGANIC MATTER, BIBLIO     W70-07027
IODEGRADATION, DETERGENTS, WATER     POLLUTION, WASTE WATERS, BIOCHEMI     W70-08740
ION CONCENTRATION, NU/ *THERMAL      POLLUTION, *TEMPERATURE, HYDROGEN     W70-08834
OPPER SULFATE, RESERVOIRS, WATER     POLLUTION, MICROORGANISMS, HISTOR     W70-08096
TICAL TECHNIQUES, BIOMASS, WATER     POLLUTION, ALGAE, PERIPHYTON, PHO     W71-04206
*BENTHOS, *TEMPERATURE, *THERMAL     POLLUTION, LABORATORY ANIMALS, AC     W71-02494
AE, / *RECOVERY, OIL SPILL, *OIL     POLLUTION, *FILAMENTOUS GREEN ALG     W70-09976
TROL, INDUSTRIAL WASTES, THERMAL     POLLUTION, METHODOLOGY, INSTRUMEN     W70-10437
STES, *WASTE WATER DISPOSAL, AIR     POLLUTION, WATER POLLUTION, ODOR,     W71-00940
R DISPOSAL, AIR POLLUTION, WATER     POLLUTION, ODOR, EFFLUENTS, SEWAG     W71-00940
*TEMPERATURE EFFECTS, *THERMAL       POLLUTION, *ECOLOGY, ALGAE, BACTE     W71-02479
*ECOLOGY, *BIOTA, *TEM/ *THERMAL     POLLUTION, *THERMAL POWERPLANTS,      W71-02491
IFORMS, FISH, FUNGI, HEAT, WATER     POLLUTION, STREAMS, RIVERS, LAKES     W71-02479
ACTERIA, SEWAGE TREATMENT, WATER     POLLUTION, WATER POLLUTION EFFECT     W69-10171
ON EFFECTS, ALGAL CONTROL, WATER     POLLUTION, TOXICITY, LAKES, DESIG     W70-02775
WAGE TREATMENT, *TRACERS, *WATER     POLLUTION, DESIGN, DESIGN CRITERI     W70-02609
T IDENTIFICATION, BIOLOGY, WATER     POLLUTION, WATER POLLUTION SOURCE     W70-02304
ION EFFECTS, ECOLOGY, EFFECTS OF     POLLUTION, WATER QUALITY CONTROL,     W70-02772
LORA, BENTHOS, MINE WATER, WATER     POLLUTION, WATER POLLUTION EFFECT     W70-02770
RY, WATE/ *WATER QUALITY, *WATER     POLLUTION, *RIVERS, WATER CHEMIST     W70-03068
MATTER, WATER POLLUTION, THERMAL     POLLUTION, STREAMS, NUTRIENTS, LI     W70-03309
CTERIA, CHEMICAL ANALYSIS, WATER     POLLUTION, CHLORELLA, SCENEDESMUS     W70-03312
NSERVATION, PUBLIC HEALTH, WATER     POLLUTION, CONTRACTS, REGULATION,     W70-03344
COSYSTEMS, ORGANIC MATTER, WATER     POLLUTION, THERMAL POLLUTION, STR     W70-03309
RINE BACTERIA, *TOXICITY/ *WATER     POLLUTION, AQUATIC ALGAE, BRINE S     W70-01467
OIL WASTES, *SURFACTANTS, *WATER     POLLUTION, CLEANING, DISASTERS, O     W70-01470
ON, *KUWAIT C/ *SALT MARSH, *OIL     POLLUTION, *RECOVERY FROM POLLUTI     W70-01231
*INSECTICIDES, *TOXICITY, *WATER     POLLUTION, FOOD CHAINS, AQUATIC A     W70-01466
ISH PHYSIOLOGY, *ALGAE, *THERMAL     POLLUTION, *TOXICITY.:          *F     W70-01788
, *OIL POLLUTION, *RECOVERY FROM     POLLUTION, *KUWAIT CRUDE, BENTLAS     W70-01231
DEVELOPMENT, GRANTS, ALGAE, AIR      POLLUTION, WATER POLLUTION, SOIL      W71-06821
NTS, ALGAE, AIR POLLUTION, WATER     POLLUTION, SOIL CONTAMINATION, NU     W71-06821
TY, *DISTRIBUTION SYSTEMS, WATER     POLLUTION, WATER QUALITY.: /OXICI     W72-03746
*EUTROPHICATION, *LAKES, *WATER      POLLUTION, DEGRADATION, SEWAGE TR     W72-02817
*WATER QUALITY CONTROL/ *THERMAL     POLLUTION, *NUCLEAR POWERPLANTS,      W71-11881
INDICATORS, *BIOASSAY, T/ *WATER     POLLUTION, *AQUATIC BIOLOGY, *BIO     W71-10786
EDIMENTS, MARINE BIOLOGY, STREAM     POLLUTION, *WATER POLLUTION DETEC     W71-10786
ION, ALGAE, OXYGEN DEMAND, WATER     POLLUTION, MUNICIPAL WATER, WATER     W71-09577
ITY, ALGAE, AQUATIC LIFE, STREAM     POLLUTION, BACTERIA, IMPURITIES,      W72-01786
DISPOSAL, *WASTE STORAGE, *WATER     POLLUTION, WATER POLLUTION SOURCE     W72-00941
VE WASTES, HEATED WATER, THERMAL     POLLUTION, BACTERIA, EFFLUENTS, O     W72-00462
E TREATMENT, *WASTE WATE/ *WATER     POLLUTION, WASTE DISPOSAL, *SEWAG     W72-01786
CREATION, SEWAGE DISPOSAL, WATER     POLLUTION, POLLUTION ABATEMENT, C     W70-04810
*WASHINGTON, URBANIZATION, WATER     POLLUTION, SEDIMENTS, PALEOLIMNOL     W70-05663
TERS, *ORGANISMS, STREAMS, WATER     POLLUTION, ANIMALS, BACTERIA, FUN     W70-05750
LTURAL WATERSHEDS, SEWAGE, WATER     POLLUTION, CHEMICAL ANALYSIS, PHY     W70-05752
, ODOR, FILTERS, PLANKTON, WATER     POLLUTION, SURFACE WATERS, RESERV     W70-05389
L, WINDS, CURRENTS(WATER), WATER     POLLUTION, DRAINAGE, ORGANIC MATT     W70-05404
SACHUSETTS, POTABLE WATER, WATER     POLLUTION, BACTERIA, ALGAE, NUTRI     W70-05264
, STREAMFLOW, MEASUREMENT, WATER     POLLUTION, FERTILIZATION, INDUSTR     W70-05112
, THERMOPHILIC BACTERIA, THERMAO     POLLUTION, NUTRIENTS, TEMPERATURE     W70-05270
LONDIN/ *DEVOLUTION, *INTERSTATE     POLLUTION, *METROPOLITAN AREAS, B     W70-04430
CAL ANALYSIS, RELIABILITY, WATER     POLLUTION, WATER POLLUTION EFFECT     W70-04382
TY CONTROL, WATER QUALITY, WATER     POLLUTION, NITROGEN, PHOSPHORUS,      W70-04484
PAL WASTES, *GREAT LAKES, *WATER     POLLUTION, SEWAGE, AESTHETICS, AE     W70-04430
M, TRACE METALS, VITAMINS, WATER     POLLUTION, PHYSIOLOGICAL ECOLOGY,     W70-04503
*ALGAE, *BIOINDICATORS, *WATER       POLLUTION, DIATOMS, MARYLAND, PEN     W70-04510
TH RATES, *BIOINDICATORS, *WATER     POLLUTION, *WASTE TREATMENT, FOOD     W70-04381
KANSAS RIVER(OKLA), TULSA(OKLA),     POLLUTION, FOSSIL, CIMARRON RIVER     W70-04001
R, SEDIMENTS, FERTILIZERS, WATER     POLLUTION, SNOW, RUNOFF, DRAINAGE     W70-04193
LLUTANTS, QUALITY CONTROL, WATER     POLLUTION, *WATER POLLUTION SOURC     W69-01273
TIVE WASTES, *MUNICIPAL / *WATER     POLLUTION, *MINE WASTES, *RADIOAC     W69-01165
OTA, FIELD INVESTIGATIONS, WATER     POLLUTION, WATER SAMPLING, *WATER     W69-00659
TION, *LAKES, CONNECTICUT, WATER     POLLUTION, *NUTRIENT REQUIREMENTS     W68-00511
WATER, / *POWERPLANTS, *THERMAL      POLLUTION, *TEMPERATURE, *COOLING     W69-05023
OL, WATER / *GREAT LAKES, *WATER     POLLUTION, *WATER POLLUTION CONTR     W69-03948
MEMBRANES,/ AQUATIC ALGAE, WATER     POLLUTION, *TOXICITY, BIOLOGICAL      W69-05306
TIC OXYGEN, TEMPERATURE, THERMAL     POLLUTION, THERMAL SPRINGS, THERM     W69-03516
OZEN GROUND, SOIL EROSION, WATER     POLLUTION, *DENITRIFICATION, PHOS     W69-05323
NSERVATION, WATER QUALITY, WATER     POLLUTION, WATER RESOURCES.: / CO     W69-05844
PHOSPHATES, ALGAE, SEWAGE, WATER     POLLUTION, WATER LAW, WATER POLLU     W69-06305
ENT, POLLUTANTS, WASTES, WATER       POLLUTION, TOXINS, PESTICIDE RESI     W69-06305
ULATION, NUISANCE ALGAE, THERMAL     POLLUTION, DIFFUSION, DIATOMS, CH     W69-05763
ERALOGY, *MICROORGANISMS, *WATER     POLLUTION, *WATER PURIFICATION, *     W69-07428
ASSOCIATIONS, MOSCOW, PETROLEUM      POLLUTION, PHOTOSYNTHETIC PIGMENT     W69-07440
ECOLOGY, *ALGAE, *RIVERS, *WATER     POLLUTION, *WATER POLLUTION SOURC     W69-07096
TROPHICATION, *CALIFORNIA, WATER     POLLUTION, REAERATION, NUTRIENTS,     W69-07520
ORMATIONS, HETEROTROPHY, ORGANIC     POLLUTION, MARINE ORGANISMS, MARI     W69-07440
DISPOSAL, WASTE DISPOSAL, WATER      POLLUTION, LAKES, STREAMS, SEWAGE     W69-06909
JAPAN, *GLENODININE, *PERIDINIUM     POLONICUM, PLANKTON BLOOMS.: /I,      W70-08372
LYETHOXYLATE, ANIONIC SU/ *ALKYL     POLY ETHOXYLATE, *ALKYL PHENOL PO     W70-09438
BRACHIONUS, SYNCHAETA PECTINATA,     POLYARTHRA VULGARIS, KARATELLA CO     W70-02249
```

ELOSIRA, MICROCYSTIS, KERATELLA,
SCHURA, ORTHOCYCLOPS, KERATELLA,
CHLOROPHYLL, ZOOPLANKTON, ALGAE,
                    BYELORUSSIA(USSR),
NAVICULA PELLICULOSA, GONYAULAX
ELECTROLYTES, *SYNTHETIC ORGANIC
AUXITE.:
L POLY ETHOXYLATE, *ALKYL PHENOL
YL POLYETHOXYLATE, *ALKYL PHENOL
ONATES, *AXENIC CULTURES, *ALKYL
MENON, MICROCYSTIS, POTAMOGETON,
METHYL ALCOHOL, ETHYLENE GLYCOL,
  CULTURES, FATTY ACIDS, OOCYSTIS
NA, APHANIZOMENON, GLOEOTRICHIA,
  FOAMING DETERGENTS, PUBLIC APP/
S.:   HARRIS COUNTY, TEX,
METAPHOSPHATES, PULSE-LABELING,
TH TEST, ORGANIC GROWTH FACTORS,
HEXOSAMINE, ASCOPHYLLUM NODOSUM,
IES, ULVA LACTUCA VAR LATISSIMA,
CTERIAL ATTACHMENT, ECTOCARPUS,
UDA, SCENEDESMUS OBLIQUUS, HELIX
PSYCHE COCKERELLI, BODY BURDENS,
  *NEW ZEALAND, SEWAGE TREATMENT
PERATURE, NUTRIENTS, FACULTATIVE
SOUTHWEST(US), *DESIGN CRITERIA,
OLORADO, WEST(US), STABILIZATION
  PH, FACULTATIVE POND, ANAEROBIC
ZURICHSEE(SWITZERLAND), LINSLEY
HYTOPLANKTON CROPS, TR/ TAKASUKA
, TYP/ *AUGUSTA(MAINE), CARLETON
OVALIS, OPEPHORA MARTYI, ABBOT'S
IA, JANESVILLE(WIS), SPAULDING'S
S), LAKE WAUBESA(WIS), SPAULDING
N, NUTRIENT REMOVAL, SPAULDING'S
L SERIAL SYSTEM, PH, FACULTATIVE
HAMPSHIRE, LAKE WINNISQUAM, LONG
BACTERIA, YEASTS, ACTINOMYCETES,
R COMPOUNDS, WASTE STABILIZATION
ENTION TIME EFFECTS, *FACULATIVE
  THEORY AN/ *WASTE STABILIZATION
CTS, *FACULATIVE PONDS, *AEROBIC
  LOAD EFFECTS,/ *ANAEROBIC WASTE
ES, ALGAE, ORGANIC L/ *OXIDATION
OLOGICAL COMMUNITIES, BACKWATER,
T, *INDUSTRIAL WASTES, OXIDATION
UTRIENTS, ESTUARINE ENVIRONMENT,
XYGEN, NITRIFICATION, COLIFORMS,
ONDS, *AEROBIC PONDS, *ANAEROBIC
ROPHICATION, *NUTRIENTS, *LAKES,
Y, STABILIZATION, WASTE STORAGE,
DESIGN APPLICATIONS, FACULTATIVE
EN DEMAND, CHLOROPHYLL, LAGOONS,
  LIGHT INTENSITY, STABILIZATION,
TRY, INDUSTRIAL WASTES, LAGOONS,
TRY, WATER ANALYSIS, PHOSPHORUS,
WAGE TREATMENT, *EUTROPHICATION,
PES, REGRESSION ANALYSIS, ALGAE,
RY TREATMENT, STABI/ *NUTRIENTS,
HYTA, *ECOLOGY, *EUTROPHICATION,
ATES, SEASONAL, LAKES, SAMPLING,
  SLUDGE, FISHERIES, CARP, CLAMS,
CONDARY TREATMENT, STABILIZATION
GEN, GROWTH RATES, STABILIZATION
SK(USSR), ENGLAND/ PISCICULTURAL
*LAKES, *RECREATION, RESERVOIRS,
CIDES, FISH, FRESH WATER, ALGAE,
, PLANKTON, BASS, PHYTOPLANKTON,
ALGAL CONCENTRATION, 'EXTRA-DEEP
RUSH, DREDGING, COPPER SULPHATE,
SLATION, WISCONSIN, CHEMCONTROL,
*CHLORI/ *PSEUDOMONAS, *SWIMMING
CRUSTACEANS, ROTIFERS, SWIMMING
COUNTY(ILL), CALUMET RIVER(ILL),
, *ALGAE, NUTRIENTS, PHOSPHORUS,
SINKING RATES, GRAZING PRESSURE,
RONMENT, *AQUATIC LIFE, *AQUATIC
ON, CYANOPHYTA, SEDIMENTS, PLANT
PHYTA, DIATOMS, AQUATIC HABITAT,
  *CYTOLOGICAL EFFECTS, ALGAL
IBUTION, FREQUENCY DISTRIBUTION,
OMASS, DISTRIBUTION, ECOSYSTEMS,
ECOLOGY, PURPLE SULFUR BACTERIA,
, SEWAGE EFFLUENTS, FARM WASTES.
TOPLANKTON, ALGAE, HEATED WATER,
LABORATORY TESTS, PARTICLE SIZE,
DIOXIDE, NUTRIENT REQUIREMENTS,
WOOD DIATOMETER, SAVANNAH RIVER,
TROPHY, *FISH POPULATIONS, PLANT
  POPULATIONS, OLIGOTROPHY, *FISH

POLYARTHRA, ASPLANCHNA, BRANCHION   W70-05375
POLYARTHRA, CONOCHILUS, PYCNOCLIN   W70-05760
POLYCHAETA, CRUSTACEA, MOLLUSCA,   W69-08526
POLYCHLOROPINENE.:   W70-08654
POLYEDRA, 3,5-DIAMINOBENZOIC ACID   W69-08278
POLYELECTROLYTES, CATIONIC ELECTR   W70-05267
POLYELECTROLYTE-COATED FILTERS, B   W71-01027
POLYETHOXYLATE, ANIONIC SURFACTAN   W70-09438
POLYETHOXYLATE, ORGANIC EXTRACTIO   W70-03928
POLYETHOXYLATE, *ALKYL PHENOL POL   W70-03928
POLYGONUM, SAGITTARIA, NAJAS.: /O   W72-04761
POLYMERS, DIMETHYL TEREPHTHALATE.   W70-10433
POLYMORPHA, UNICELLULAR ALGAE.: /   W70-07957
POLYPEPTIDES, PHOTOASSIMILATION,   W70-04369
POLYPHOSPHATES, NUTRIENT BALANCE,   W70-05084
POLYPHOSPHATES, ORGANIC PHOSPHATE   W69-04800
POLYPHOSPHATES, FRACTIONATION.: /   W71-03033
POLYSACCHARIDES.: /, MUSTARD GROW   W69-07826
POLYSIPHONIA LANOSA, POINT LEPREA   W70-04365
POLYSIPHONIA HARVEYI, EPIPHYTIC B   W70-01073
POLYSIPHONIA, ORGANIC AGGREGATES,   W70-01068
POMATIA, CHLORELLA PYRENOIDOSA, N   W70-04184
POMOXIS ANNULARIS, GASTROINTESTIN   W70-02786
POND ALGAE.:   W70-02304
POND DESIGNS, BOD, SLUDGE.: / TEM   W70-04786
POND EFFLUENT, COLORADO, WEST(US)   W70-04788
POND LOADINGS, ODOR PERSISTENCE.:   W70-04788
POND.: /MULTI-CELL SERIAL SYSTEM,   W70-04790
POND(CONN), BOSMINA COREGONI LONG   W69-10169
POND(JAPAN), LAKE MENDOTA(WIS), P   W69-05142
POND(MAINE), SYNURA, ASTERIONELLA   W70-08096
POND(SOMERSET, ENGLAND), SAND GRA   W71-12870
POND(WIS), HYDRODICTYON, NAJAS, A   W70-03310
POND(WIS), LAKE DELAVAN(WIS), PEW   W69-05867
POND(WIS), LAKE WAUBESA(WIS), GRE   W69-05868
POND, ANAEROBIC POND.: /MULTI-CEL   W70-04790
POND, CONCORD, BOSTON LOT RESERVO   W69-08674
PONDING, WASTE TREATMENT.: /ENA,   W70-08097
PONDS.: /ARED SPECTROSCOPY, SULFU   W70-07034
PONDS, *AEROBIC PONDS, *ANAEROBIC   W70-04787
PONDS, *ALGAL PHYSIOLOGY, *DESIGN   W70-04786
PONDS, *ANAEROBIC PONDS, DESIGN F   W70-04787
PONDS, *DEPTH EFFECTS, *HYDRAULIC   W70-04787
PONDS, *HYDROGEN SULFIDE, *SULFID   W71-11377
PONDS, ALASKA.: /YTA, DIATOMS, BI   W70-03969
PONDS, ALGAE, BENTHIC FAUNA, WATE   W72-00027
PONDS, BREEDINGS, SALINE WATER-FR   W73-00748
PONDS, CLIMATES, PHOTOSYNTHESIS,   W70-05092
PONDS, DESIGN FOR ENVIRONMENTAL C   W70-04787
PONDS, FERTILIZERS, ALGAE, NITROG   W69-08668
PONDS, HYDROLYSIS, METABOLISM, CH   W70-08097
PONDS, INFRARED SPECTROSCOPY, SUL   W70-07034
PONDS, OXIDATION, WASTE WATER TRE   W70-08392
PONDS, PENETRATION, ABSORPTION.: /   W70-03923
PONDS, PHOTOSYNTHESIS, COSTS, EFF   W70-09320
PONDS, PHYTOPLANKTON, EUTROPHICAT   W69-05142
PONDS, POLLUTANT IDENTIFICATION,   W70-02304
PONDS, PREDATION.: /US RADIOISOTO   W70-03957
PONDS, PRIMARY TREATMENT, SECONDA   W68-00012
PONDS, RESERVOIRS, BACKWATER, FER   W70-00274
PONDS, RIVERS, RESERVOIRS, AIR TE   W70-02646
PONDS, RUNNING WATERS, FRESH WATE   W70-09905
PONDS, SEWAGE LAGOONS.: /MENT, SE   W68-00012
PONDS, SEWAGE DISPOSAL.: /, NITRO   W68-00855
PONDS, SVISLOCH' RIVER(USSR), MIN   W70-05092
PONDS, TEMPERATURE, LIGHT PENETRA   W70-06225
PONDS, WATER HYACINTH, CHEMICAL C   W70-00269
PONDS, ZOOPLANKTON, MINNOWS, PEDA   W70-02249
PONDS', ASHKELON(ISRAEL), TEMPERA   W70-04790
PONDWEEDS, SODIUM ARSENITE, FISH,   W70-05263
PONDWEEDS, ROOTED AQUATIC PLANTS,   W70-05568
POOLS, *BACTERIA, *BACTERICIDES,   W69-10171
POOLS, CHLAMYDOMONAS, DAPHNIA, DI   W70-10161
POOLS, WIND DIRECTION, TABELLARIA   W70-00263
POPULATION, POLLUTANTS, MAPPING,   W70-00667
POPULATION DENSITY.: /EDRA ACUS,   W69-10158
POPULATIONS, ALGAE, BENTHOS, ZOOP   W71-02491
POPULATIONS, WEATHER, WATER, MOVE   W70-06229
POPULATION.: /EUGLENOPHYTA, CYANO   W70-02770
POPULATION.:   W70-02770
POPULATION, BIOLOGICAL INDICATORS   W70-02772
POPULATION, OREGON, INTERTIDAL AR   W70-03325
POPULATION EQUIVALENTS(WASTES), R   W70-03312
POPULATION, INDUSTRIES, METHANE,   W70-02803
POPULATION, WATER POLLUTION EFFEC   W71-09767
POPULATION, GROWTH RATES, WATER P   W72-02417
POPULATIONS, ENVIRONMENTAL EFFECT   W70-05424
POPULATION DYNAMICS, ALGAL COMMUN   W70-04510
POPULATIONS, ROUGH FISH, NITROGEN   W68-00172
POPULATIONS, PLANT POPULATIONS, R   W68-00172

EATION, SILT, NUTRIENTS, AQUATIC POPULATIONS, OLIGOTROPHY, *FISH P W68-00172
*WATER QUALITY, TURBIDITY, PLANT POPULATIONS.: /GICAL PROPERTIES, W68-00475
ATION, MAINE, NUTRIENTS, AQUATIC POPULATIONS, WATER QUALITY, LAKE W68-00470
OSPHORUS, BICARBONATES, *AQUATIC POPULATIONS.: /LGAE, NITROGEN, PH W68-00468
REAMS, EUGLENA, NUTRIENTS, PLANT POPULATIONS, HISTORY, WATER PURIF W68-00483
ATION, SEWAGE EFFLUENTS, AQUATIC POPULATIONS, NUISANCE ALGAE, DUCK W68-00461
NELS, POLLUTION ABATEMENT, HUMAN POPULATION.: /ILD), OYSTERS, CHAN W68-00461
A/ *EUTROPHICATION, SEWAGE, FISH POPULATIONS, PLANKTON, *CYANOPHYT W68-00461
ASHINGTON, DC, *ESTUARIES, HUMAN POPULATION, SEWAGE EFFLUENTS, AQU W68-00461
, SILICATES, IRON OXIDES, *PLANT POPULATIONS, REVIEWS, LIGHT INTEN W68-00860
PHIBIANS, INVERTEBRATES, AQUATIC POPULATIONS, BIBLIOGRAPHIES, BOTT W68-00859
DING CROP, ANNUAL TURNOVER, FISH POPULATIONS.: /IVE CAPACITY, STAN W68-00912
, *ENVIRONMENTAL EFFECTS, *HUMAN POPULATION, *EUTROPHICATION, *CON W70-03973
RT, BACTERIAL POPULATIONS, ALGAL POPULATIONS, HETEROTROPHIC POTENT W70-04194
ITY, ACTIVE TRANSPORT, BACTERIAL POPULATIONS, ALGAL POPULATIONS, H W70-04194
ICATION, *NUTRIENTS, ALGAE, FISH POPULATION, *WATER POLLUTION CONT W70-04266
AGELLATES, OCEANS, BIO/ *YEASTS, POPULATION, MARINE PLANTS, DINOFL W70-04368
IVERS, STREAMS, EUGLENA, AQUATIC POPULATIONS, EUTROPHICATION, CYAN W70-04510
RONMENT, AQUATIC PLANTS, AQUATIC POPULATIONS, CONSERVATION, WATER W69-05844
OZOA, A/ *AQUATIC LIFE, *AQUATIC POPULATIONS, *RIVERS, ALGAE, PROT W73-00748
HYACINTHUS, POCILLOPORA EYDOUXI, PORITES LOBATA, LEPTASTREA PURPUR W69-08526
IFICATION, CROPS, TOXICITY, SOIL POROSITY, IRRIGATION, GROUND WATE W69-05323
IQUUES, SCENEDESMUS QUADRICAUDA, PORPHYRIDIUM CRUENTUM, LEMNA MINO W70-02964
ALASSIOSIRA, AMPHIDINIUM KLEBSI, PORPHYRIDIUM CRUENTUM, RADIONUCLI W70-00708
CTS, AXENIC CULTURES, MONOALGAL, PORTUGAL, TAPES, JAPAN, AMPHIDINI W70-05372
ILAMENTOUS GREEN ALGAE, HALIMONE PORTULACOIDES, SUAEDA MARITIMA, A W70-09976
', VIBRIOS, ACHROMOBACTERS, GRAM- POSITIVE BACTERIA, SARCINE-LIKE, W70-00713
AE, *TASTE, *ODOR, *AMINO ACIDS, POTABLE WATER, ACTINOMYCETES, PLA W72-01812
ION, WATERSHEDS, SEWERAGE ALGAE, POTABLE WATER.: /NATION, COAGULAT W71-05943
LOGY, *RESERVOIR, MASSACHUSETTS, POTABLE WATER, WATER POLLUTION, B W70-05264
MALS, ALGAE, CUTTING MANAGEMENT, POTABLE WATER, HERBICIDES, CHARA, W70-05263
ODOR, COAGULATION, ELECTROLYTES, POTABLE WATER.: /TICIDES, TASTE, W72-13601
HLORATE, SANTOBRITE, CLADOPHORA, POTAMOGETON SPP, LAKE MENDOTA(WIS W70-05263
CILIATES, RANUNCULUS CIRCINATUS, POTAMOGETON PUSILLUS.: /GELLATA, W69-07838
PHORA, PEDIASTRUM, HYDRODICTYON, POTAMOGETON, RANUNCULUS.: / CLADO W70-04494
ENA, APHANIZOMENON, MICROCYSTIS, POTAMOGETON, POLYGONUM, SAGITTARI W72-04761
CHARIS CANADENSIS, MYRIOPHYLLUM, POTAMOGETON CRISPUS, MICROSPORA W W70-00711
YLLUM, MYRIOPHYLLUM, EICHHORNIA, POTAMOGETON, TRAPA, PISTIA, ALTER W70-10175
ACHARIS CANADENSIS, MACROPHYTES, POTAMOGETON FOLIOSUS, EXPERIMENTA W70-02249
NTHERA, NAJAS, SILVEX, ENDOTHAL, POTASSIUM PERMANGANATE, FENAC, SI W70-10175
CEAN, RADIOACTIVITY, GAMMA RAYS, POTASSIUM RADIOISOTOPES, MANGANES W69-08524
BASS, CARP, CATFISHES, NITROGEN, POTASSIUM.: /TOS, ALGAL CONTROL, W70-05417
YANOPHYTA, CARBON RADIOISOTOPES, POTASSIUM, OXY: /WAGE, DIATOMS, C W70-05405
PLANKTON, NITROGEN, PHOSPHORUS, POTASSIUM, SELECTIVITY, POISONS, W70-05399
SPHATES, CHLOROPHYLL, MAGNESIUM, POTASSIUM, TEMPERATURE, EPILIMNIO W70-05409
S, NITROGEN, PHOSPHORUS, SODIUM, POTASSIUM, TRACE METALS, VITAMINS W70-04503
SULFATES, NITRATES, PHOSPHATES, POTASSIUM, IRON, BICARBONATE, ZIN W70-03978
PLANTS, IRON, CHLORELLA, SODIUM, POTASSIUM, SILICATES, DISSOLVED S W70-03512
PLING, BOTTOM SEDIMENTS, SODIUM, POTASSIUM, IRON, MANGANESE, CALCI W70-09889
L NUTRIENTS, CALCIUM, MAGNESIUM, POTASSIUM, HARVESTING OF ALGAE, L W70-02245
RUS, SULFUR, CALCIUM, MAGNESIUM, POTASSIUM, SODIUM, IRON, MANGANES W70-02804
GEN ION CONCENTRATION, NITRATES, POTASSIUM, CALCIUM, MAGNESIUM, IR W70-00719
DS, ZINC RADIOISOTOPES, CALCIUM, POTASSIUM, CLAMS, SIZE, ALGAE, DI W70-00708
ORTH DAKOTA), SULFATE REDUCTION, POTATO PROCESSING WASTES, CHEESE W70-03312
CONCENTRATION, *NUTRIENT BUDGET, POTENTIAL NUTRIENT CONCENTRATION, W70-00270
GE EFFLUENTS, ELECTROLYTES, ZETA POTENTIAL, BACTERIA, PROTOZOA, AL W70-01027
GASES, ODOR, OXIDATION-REDUCTION POTENTIAL, SULFUR BACTERIA, *WAST W70-07034
YGEN DEMAND, OXIDATION-REDUCTION POTENTIAL, AQUATIC ALGAE, *BIOLOG W71-09546
YGEN DEMAND, OXIDATION-REDUCTION POTENTIAL, DISSOLVED OXYGEN, ODOR W71-11377
ALGAL POPULATIONS, HETEROTROPHIC POTENTIAL, RELATIVE HETEROTROPHIC W70-04194
OTENTIAL, RELATIVE HETEROTROPHIC POTENTIAL, SWEDEN, ENZYME KINETIC W70-04194
EQUATION, RELATIVE HETEROTROPHIC POTENTIAL, LINEWEAVER-BURK EQUATI W70-05109
M COMPOUNDS, OXIDATION REDUCTION POTENTIAL, DISSOLVED OXYGEN, CHEM W70-05760
MAND, TOXICITY, OXYGEN-REDUCTION POTENTIAL, SOLAR RADIATION, ENERG W70-05750
RIENT AVAILABILITY, ALGAL GROWTH POTENTIAL, SELENASTRUM CAPRICORNU W69-06864
UM, NITRITE, OXIDATION-REDUCTION POTENTIAL, CYCLING NUTRIENTS, EUT W69-04521
N VARIATIONS, SAN JOAQUIN RIVER, POTOMAC ESTUARY, SAN FRANCISCO BA W69-07520
, SNAKE RIVER, RIO GRANDE RIVER, POTOMAC RIVER.: /, SAVANNAH RIVER W70-04507
 POTOMAC RIVER BASIN.: W70-09557

E SHOCK, *PROTOZOAN COMMUNITIES, POTOMAC RIVER, SAVANNAH RIVER, SH W70-07847
, FISH, TASTE, ODOR, OXYGEN SAG, POTOMAC RIVER.: /NITY, REAERATION W73-00750
VOIR(TENN), HOLSTON RIVER(TENN), POTOMOGETON PECTINATUS, POTOMOGET W71-07698
R(TENN), POTOMOGETON PECTINATUS, POTOMOGETON CRISPUS, HETERAMTHERA W71-07698
*CHICKENS, FOOD WASTES, POULTRY WASTES.: W70-09320
NG TO/ *COOLING WATER, *ELECTRIC POWER PLANTS, *WATER REUSE, COOLI W71-11393
INISTRATION, SUPERVISORY CONTROL(POWER), ADMINISTRATIVE AGENCIES.: W72-05774
FISH, SHELLFISH, STEAM, ELECTRIC POWER, HEATED WATER, PROTEINS, AQ W70-09905
OGEN ION CONCENTRATION, ELECTRIC POWER, SEWAGE DISPOSAL, *WASTE WA W71-09548
PHYTOPLANKTON, ALGAE,/ *THERMAL POWERPLANTS, *LAKES, TEMPERATURE, W71-09767
OL/ *THERMAL POLLUTION, *NUCLEAR POWERPLANTS, *WATER QUALITY CONTR W71-11881
EM/ *THERMAL POLLUTION, *NUCLEAR POWERPLANTS, *ECOLOGY, *BIOTA, *T W71-02491
CES, *THERMAL POLLUTION, THERMAL POWERPLANTS, *THERMAL STRESS, THE W70-07847
ASTE DISPOSAL, NUCLEAR REACTORS, POWERPLANTS, SHIPS, SUBMARINES, E W70-00707
CE ELEMENTS, WASHINGTON, NUCLEAR POWERPLANTS, WATER POLLUTION EFFE W70-00264
, HYDROELECTRIC PLANTS, ELECTRIC POWERPLANTS, ENGINEERING STRUCTUR W69-05023
, *BIOLOGICAL INDEX, *SAPROBITY, PRAGUE, CZECHOSLOVAKIA, LABORATOR W72-12240
NIUM-103-RHODIUM-103, CERIUM-144- PRASEODYMIUM-144.: /UM-106, RUTHE W69-08524
EAWEED, ZIRCONIUM-95-NIOBIUM-95, PRE-BOMB MATERIAL, RUTHENIUM-106- W69-08524
L DERIVATIVES, RECENT SEDIMENTS, PRECAMBRIAN SEDIMENTS, CHLORELLA W69-08284
ION, SPECIES DIVERSITY, CHEMICAL PRECIPITATION.: /URAL EUTROPHICAT W69-08518

ALGAE, BIBLIOGRAPHIES, CHEMICAL
KEGONSA(WIS), LAKE WAUBESA(WIS),
UNDS, SURFACE RUNOFF, UNDERFLOW,
NOFF, PERCOLATION, ROADS, ROOFS,
ES, TERTIARY TREATMENT, CHEMICAL
*PHOSPHORUS COMPOUNDS, *CHEMICAL
OLUBILITY, *PHOSPHATES, CHEMICAL
TROSTATIC PRECIPITATION, THERMAL
UGATION, SEDIMENTATION, CHEMICAL
, CORYNEBACTERIUM, ELECTROSTATIC
HYDROGEN SULFIDE, IMHOFF TANKS,
CTRODIALYSIS, SORPTION, CHEMICAL
S, ALGAE, DAPHNIA, GROWTH RATES,
GRESSION ANALYSIS, ALGAE, PONDS,
PLANKTON, ALGAE, *ALGAL CONTROL,
BDENUM, PLANT GROWTH REGULATORS,
NCE, FOOD CONCENTRATION, MEXICO,
ELONGATUM HODGETTS, CHROOCOCCUS
ENTS, PHOSPHORUS RELEASE, SAMPLE
S, LIGHT, NUTRIENTS, HYDROSTATIC
DRA ACUS, SINKING RATES, GRAZING
XYGEN, *SATURATION, TEMPERATURE,
US EARTH, SEPARATION TECHNIQUES,
                   EUTROPHICATION
SPHORUS COMPOUNDS, PLANT GROWTH,
, PHOTOSYNTHESIS, PHYTOPLANKTON,
, ORGANIC MATTER, PHYTOPLANKTON,
CONNECTICUT, OREGON, WASHINGTON,
ORGANIC MATTER, SEASONAL, ALGAE,
ON, LIGHT PENETRATION, SEASONAL,
, *CHLOROPHYLL, *EUTROPHICATION,
E, LIGHT INTENSITY, TEMPERATURE,
ATA COLLECTIONS, MARINE ANIMALS,
, NITROGEN, RESPIRATION, CARBON,
ODUCTIVITY, LIMNOLOGY, PLANKTON,
*CALIFORNIA, *NEVADA, LIMNOLOGY,
AKES, ORGANIC MATTER, SEA WATER,
 COMMUNITIES, *ALGAE, *METHODOL/
FF, LIGHT PENETRATION, SAMPLING,
ION, *NUTRIENTS, LAKES, ECOLOGY,
ATMENT, WATER POLLUTION SOURCES,
TMENT, STABI/ *NUTRIENTS, PONDS,
A SCOTIA(CANADA), PENGUINS, LAK/
ANALYSIS, MULTIVARIATE ANALYSIS,
ROFOSSILS, COMPETITIVE EXCLUSION
YCEAE, SELENASTRUM CAPRICORNUTUM
E MENDOTA(WIS), LAK/ *BIOLOGICAL
            *TEST DEVELOPMENT, *TEST
ICATION, PROVISIONAL ALGAL ASSAY
TION, ALGAL PHYSIOLOGY, KJELDAHL
ATER TEMPERATURE, SEASONAL, TEST
NALYSIS, *LABORATORY TESTS, TEST
CTIVITY, BACTERIA, DIATOMS, TEST
MPTION, PROVISIONAL ALGAL ASSAY,
LGAE, TREATMENT FACILITIES, TEST
TOMS, CYANOPHYTA, BIOASSAY, TEST
, TE/ *ALGICIDES, *PAINTS, *TEST
LGAE, TREATMENT FACILITIES, TEST
AGUE, CZECHOSLOVAKIA, LABORATORY
BORATORY PROCEDURES, STATISTICAL
MINUM SULFATE, *ACTIVATED SLUDGE
, SUSPENDED LOAD, SOLIDS CONTACT
TREATMENT, UNIT OPERATIONS, UNIT
EVOORTIA TYRRANUS, CONCENTRATION
DISTRIBUTION PATTERNS, MEMBRANE
                   COAL CONTACT
ERFERTILIZATION, PHYSI/ *REMOVAL
A, ANABAENA, IOWA RIVER, / *UNIT
LOGICAL DISTRIBUTION, STOCHASTIC
S, ANAEROBIC LAGOONS.:   *POTATO
KOTA), SULFATE REDUCTION, POTATO
TRICHO/ HEXAGENIA, CHIRONOMIDAE,
LAKE PEWAUKEE(WIS), CHIRONOMUS,
*WISCONSIN, *LAKES, PROD/ *ODOR-
 *WATER ANALYSIS, *ODORS, *ODOR
RIAL WASTE, CORES, DIATOMS, ODOR
ROL, ALGAL CONTROL, SLIME, TASTE-
IME, TASTE-PRODUCING ALGAE, ODOR-
BIOCHEMICAL OXYGEN DEMAND, TASTE-
ALGAE, ALGAE, CHLORINATI/ *ODOR-
I/ *ODOR-PRODUCING ALGAE, *TASTE-
EMPERATURE, SOIL ANALYSIS, TASTE-
URCE CONTAMINATION, CYANIDE, NEW
                        *OXYGEN
IUM COMPOUNDS, PHENOL/ BIOCIDES,
AEODACTYLUM, *GAS EFFLUENT, FOOD
CRETION, STEPHAN/ *EXTRACELLULAR
LUTION, PHOTOSYNTHETIC PIGMENTS,
RECIRCULATION, BOD REMOVAL, GAS
PONDS, CLIMATES, PHOTOSYNTHESIS,

PRECIPITATION, IONS, BIOCHEMICAL   W69-00446
PRECIPITATION.: /NGRA(WIS), LAKE   W69-03366
PRECIPITATION(ATMOSPHERIC), DRAIN   W70-04810
PRECIPITATION(ATMOSPHERIC), WETLA   W69-10178
PRECIPITATION, HARVESTING, ZONING   W69-10178
PRECIPITATION, *CHEMICAL DEGRADAT   W70-09186
PRECIPITATION, FILTRATION, AMMONI   W71-11393
PRECIPITATION.: /EBACTERIUM, ELEC   W71-11823
PRECIPITATION, LIFE CYCLES, GENET   W71-11823
PRECIPITATION, THERMAL PRECIPITAT   W71-11823
PRECIPITATES.:   W72-00027
PRECIPITATION, ULTIMATE DISPOSAL.   W70-01981
PREDATION.: /VIRONMENT, PHOSPHATE   W69-06536
PREDATION.: /US RADIOISOTOPES, RE   W70-03957
PREDATION, COMPETITION, *INSECT C   W70-05428
PREDATION, PRODUCTIVITY, TRACE EL   W69-04801
PREDICTIVE EQUATIONS, ZOOGEOGRAPH   W70-03957
PRESCOTTII, CALOTHRIX FUSCA, MICR   W70-04468
PRESERVATION, WATER DYNAMICS, WAT   W70-04382
PRESSURE, CARBON DIOXIDE, CYANOPH   W69-10160
PRESSURE, POPULATION DENSITY.: /E   W69-10158
PRESSURE, MIXING, EVAPORATION, SU   W70-00683
PRESSURE, FLOW RATES, SLUDGE, SLU   W72-08688
PREVENTION, ECOLOGIC IMBALANCES.:   W69-07084
PRIMARY PRODUCTIVITY, RIVERS, TAS   W69-06535
PRIMARY PRODUCTIVITY, SYSTEMATICS   W69-07832
PRIMARY PRODUCTIVITY, SOIL CHEMIS   W69-06273
PRIMARY PRODUCTIVITY, ECOSYSTEMS,   W70-04506
PRIMARY PRODUCTIVITY, HARVESTING   W70-04290
PRIMARY PRODUCTIVITY, CALCIUM, NI   W70-05405
PRIMARY PRODUCTIVITY, PHYTOPLANKT   W70-05094
PRIMARY PRODUCTIVITY, SURFACES, A   W70-04809
PRIMARY PRODUCTIVITY, CHLOROPHYLL   W73-01446
PRIMARY PRODUCTIVITY, OXYGEN, MET   W70-01073
PRIMARY PRODUCTIVITY, TROPHIC LEV   W69-09135
PRIMARY PRODUCTIVITY, CHLOROPHYLL   W70-01933
PRIMARY PRODUCTIVITY, LIGHT INTEN   W70-02504
PRIMARY PRODUCTIVITY, *BIOLOGICAL   W71-12870
PRIMARY PRODUCTIVITY, PHYTOPLANKT   W72-04761
PRIMARY PRODUCTIVITY, ALGAE, PLAN   W71-00665
PRIMARY TREATMENT, SECONDARY TREA   W69-01453
PRIMARY TREATMENT, SECONDARY TREA   W68-00012
PRINCE EDWARD ISLAND(CANADA), NOV   W70-05399
PRINCIPAL-COMPONENT ANALYSIS, PAR   W73-01446
PRINCIPLE, OSCILLATORIA RUBESCENS   W70-05663
PRINTZ, OSCILLATORIA VAUCHER, SKE   W70-05752
PROBLEMS, SCIOTO RIVER(OHIO), LAK   W72-01813
PROBLEMS, HORMIDIUM SP.:   W71-07339
PROCEDURE, BATCH CULTURES, CHEMOS   W70-02777
PROCEDURE, LUXURY CONSUMPTION, NU   W69-05868
PROCEDURES, SCENEDESMUS, *ALGAE.:   W68-00472
PROCEDURES, NURTRIENTS, RIVERS, L   W68-00472
PROCEDURES.: /NG, CAPILLARY CONDU   W70-04469
PROCEDURES, SELENASTRUM CAPRICORN   W70-02779
PROCEDURES, METALS.: /BACTERIA, A   W70-01519
PROCEDURES.: /LGAE, CULTURES, DIA   W71-06002
PROCEDURES, ANTIFOULING MATERIALS   W71-07339
PROCEDURES, METALS.: /BACTERIA, A   W71-06737
PROCEDURES, STATISTICAL PROCEDURE   W72-12240
PROCEDURES, BMT.: /HOSLOVAKIA, LA   W72-12240
PROCESS, HANOVER TREATMENT PLANT.   W70-09186
PROCESS, SEWAGE TREATMENT, TRICKL   W70-06792
PROCESSES.: /ILIZATION, PHYSICAL   W70-07469
PROCESSES.: /RSIOPS TRUNCATUS, BR   W69-08274
PROCESSES, MASSACHUSETTS.: /LOSE,   W69-00976
PROCESSES, PESTICIDE REMOVAL.:   W70-05470
PROCESSES, CHEMICAL TREATMENT, OV   W70-07469
PROCESSES, GASTROENTERITIS, SYNUR   W70-08107
PROCESSES, ECOSYSTEMS, AQUATIC HA   W71-09012
PROCESSING WASTES, AEROBIC LAGOON   W71-08970
PROCESSING WASTES, CHEESE MANUFAC   W70-03312
PROCLADIUS, CHIRONOMUS PROMOSUS,   W70-01943
PROCLADIUS, CORETHRA PUNCTIPENNIS   W70-06217
PRODUCING ALGAE, *NUISANCE ALGAE,   W70-05112
PRODUCING ALGAE, EUTROPHICATION,   W69-00387
PRODUCING ALGAE, CHLOROPHYLL, CHE   W68-00470
PRODUCING ALGAE, ODOR-PRODUCING A   W69-05704
PRODUCING ALGAE, DIATOMS, CYANOPH   W69-05704
PRODUCING ALGAE, POLLUTION ABATEM   W69-06909
PRODUCING ALGAE, *TASTE-PRODUCING   W70-02373
PRODUCING ALGAE, ALGAE, CHLORINAT   W70-02373
PRODUCING ALGAE, HYDROGEN ION CON   W72-11906
PRODUCT CONTROL, PLUMBING CODES,   W73-00758
PRODUCTION.:   W72-06022
PRODUCTION LOSS, QUATERNARY AMMON   W72-12987
PRODUCTION, SOUTHAMPTON UNIVERSIT   W70-00161
PRODUCTION, *CARBON FIXATION, *EX   W69-10158
PRODUCTION BIOLOGY, RUSSIA, SECON   W69-07440
PRODUCTION.: /IRONMENTAL CONTROL,   W70-04787
PRODUCTIVITY, SYMBIOSIS, RESERVOI   W70-05092

RIENTS, WATER POLLUTION EFFECTS,   PRODUCTIVITY, SEDIMENTS, URBANIZA   W70-04721
OPHYLL, *EUTROPHICATION, PRIMARY   PRODUCTIVITY, PHYTOPLANKTON, PHOT   W70-05094
INTENSITY, TEMPERATURE, PRIMARY   PRODUCTIVITY, SURFACES, ARCTIC, E   W70-04809
SANCE ALGAE, *WISCONSIN, *LAKES,   PRODUCTIVITY, NITROGEN, PHOSPHORU   W70-05112
*NEW YORK, *MEROMIXIS, *PRIMARY   PRODUCTIVITY, *ZOOPLANKTON, SULFU   W70-05760
ENTS, PALEOLIMNOLOGY, NUTRIENTS,   PRODUCTIVITY, SEDIMENTATION RATES   W70-05663
C MATTER, SEWAGE, PHYTOPLANKTON,   PRODUCTIVITY, CHLOROPHYTA, HYDROG   W70-05404
S, BIOASSAYS, ALGAE, COMPARATIVE   PRODUCTIVITY, AQUATIC PRODUCTIVIT   W70-05424
ENVIRONMENTAL EFFECTS, *PRIMARY   PRODUCTIVITY, WATER POLLUTION EFF   W70-05424
OMPARATIVE PRODUCTIVITY, AQUATIC   PRODUCTIVITY, DISTRIBUTION, SPECI   W70-05424
T PENETRATION, SEASONAL, PRIMARY   PRODUCTIVITY, CALCIUM, NITRATES,   W70-05405
COMPOUNDS, PLANT GROWTH, PRIMARY   PRODUCTIVITY, RIVERS, TASTE, THER   W69-06535
AKES, PHOSPHORUS, TROPHIC LEVEL,   PRODUCTIVITY, MODEL STUDIES, SALM   W69-06536
C MATTER, PHYTOPLANKTON, PRIMARY   PRODUCTIVITY, SOIL CHEMISTRY, WAT   W69-06273
WATER POLLUTION SOURCES, AQUATIC   PRODUCTIVITY, LIMNOLOGY, PLANKTON   W69-09135
TROGEN COMPOUNDS, BIOINDICATORS,   PRODUCTIVITY, WASTE TREATMENT, WA   W69-08518
YNTHESIS, PHYTOPLANKTON, PRIMARY   PRODUCTIVITY, SYSTEMATICS, WATER   W69-07832
ATER POLLUTION EFFECTS, *PRIMARY   PRODUCTIVITY, ALGAE, BIOMASS, CAL   W69-02959
TE, AQUATIC ENVIRONMENT, AQUATIC   PRODUCTIVITY, EUTROPHICATION, *LA   W69-00446
GAE, AGING(BROLOGICAL), *AQUATIC   PRODUCTIVITY, LIMNOLOGY, *NUTRIEN   W68-00912
EUTROPHICATION, *LAKES, *PRIMARY   PRODUCTIVITY, *PLANKTON, ALGAE, *   W69-00632
LANKTON, ALGAE, *PHOTOSYNTHESIS,   PRODUCTIVITY, WATER QUALITY, *LIM   W69-00632
BIO MASS, OLIGOTROPHY, *PRIMARY   PRODUCTIVITY, TROPIC LEVEL, WATER   W68-00912
E, AQUATIC MICROBIOLOGY, AQUATIC   PRODUCTIVITY, CYANOPHYTA, ENVIRON   W69-04802
GY, AQUATIC ENVIRONMENT, AQUATIC   PRODUCTIVITY, HYPOLIMNION, ALGAL   W69-05142
LGICIDES, AQUATIC ALGAE, AQUATIC   PRODUCTIVITY, BIOASSAY, CYCLING N   W69-04798
IXATION, PHOSPHATES, PHOSPHORUS,   PRODUCTIVITY, STANDING CROP, STRE   W69-04805
NT GROWTH REGULATORS, PREDATION,   PRODUCTIVITY, TRACE ELEMENTS, SOD   W69-04801
E, AQUATIC MICROBIOLOGY, AQUATIC   PRODUCTIVITY, AQUATIC WEEDS, BIOA   W69-03186
E, AQUATIC MICROBIOLOGY, AQUATIC   PRODUCTIVITY, AQUATIC WEEDS, BALA   W69-03185
E, AQUATIC MICROBIOLOGY, AQUATIC   PRODUCTIVITY, AQUATIC WEEDS, CHLO   W69-03373
NKTON, *CULTURES, ALGAE, AQUATIC   PRODUCTIVITY, CYCLING NUTRIENTS,   W69-03514
S, AQUATIC MICROBIOLOGY, AQUATIC   PRODUCTIVITY, BALANCE OF NATURE,   W69-03513
CUT, OREGON, WASHINGTON, PRIMARY   PRODUCTIVITY, ECOSYSTEMS, ILLINOI   W70-04506
IPHYTON, *RADIOACTIVITY, *LAKES,   PRODUCTIVITY, RADIOISOTOPES, TEMP   W70-04371
MATTER, SEASONAL, ALGAE, PRIMARY   PRODUCTIVITY, HARVESTING OF ALGAE   W70-04290
*AQUATIC MICROBIOLOGY, *AQUATIC   PRODUCTIVITY, BIOCHEMICAL OXYGEN   W70-04283
ATES, DISSOLVED SOLIDS, CALCIUM,   PRODUCTIVITY, STANDING CROP, CHLO   W70-03512
TES, ALGAE, HYPOLIMNION, OXYGEN,   PRODUCTIVITY, PHOSPHATES, CHLOROP   W69-10169
T, SOLAR RADIATION, CHLOROPHYLL,   PRODUCTIVITY, PHOTOSYNTHESIS, ALG   W69-10182
RY TESTS, RESPIRATION,/ *PRIMARY   PRODUCTIVITY, *STREAMS, *LABORATO   W70-00265
SECONDARY PRODUCTIVITY, AQUATIC   PRODUCTIVITY, WATER POLLUTION EFF   W69-10152
TY, LIMNOLOGY, PLANKTON, PRIMARY   PRODUCTIVITY, TROPHIC LEVEL.: /VI   W69-09135
IC FLORA, *CHLOROPHYLL, *PRIMARY   PRODUCTIVITY, *MEASUREMENT, PROTE   W69-10151
RATES, *GROWTH RATES, *DAPHNIA,   PRODUCTIVITY, BIOINDICATORS, SECO   W69-10152
DYSTROPHY, OLIGOTROPHY, *PRIMARY   PRODUCTIVITY, *TROPHIC LEVEL, WAT   W69-09723
TIVITY, BIOINDICATORS, SECONDARY   PRODUCTIVITY, AQUATIC PRODUCTIVIT   W69-10152
RTILIZATION EFFECTS, *BIOLOGICAL   PRODUCTIVITY, AQUATIC NUISANCES,   W69-09349
EN, RESPIRATION, CARBON, PRIMARY   PRODUCTIVITY, OXYGEN, METABOLISM,   W70-01073
CTERIA, STREPTOCOCCUS, SECONDARY   PRODUCTIVITY, SULFIDES, SEA WATER   W70-01068
GANIC MATTER, SEA WATER, PRIMARY   PRODUCTIVITY, LIGHT INTENSITY, TR   W70-02504
NIA, *NEVADA, LIMNOLOGY, PRIMARY   PRODUCTIVITY, CHLOROPHYLL, SAMPLI   W70-01933
NG FACTORS, *NUTRIENTS, *PRIMARY   PRODUCTIVITY, PHOSPHATES, SILICAT   W70-01579
ROGEN COMPOUNDS, EUTROPHICATION,   PRODUCTIVITY, ALGAE, AQUATIC LIFE   W70-03068
TER, PERIPHYTON, BIOMA/ *PRIMARY   PRODUCTIVITY, *ALGAE, *THERMAL WA   W70-03309
, *NITROGEN, *ALKALINITY, LAKES,   PRODUCTIVITY, PHYTOPLANKTON.: /US   W70-02795
GANISMS, AQUATIC PLANTS, AQUATIC   PRODUCTIVITY, BALANCE OF NATURE,   W70-02792
E, AQUATIC MICROBIOLOGY, AQUATIC   PRODUCTIVITY, CHLOROPHYTA, GONOPH   W70-02784
MS, SEASONAL, BIOMASS, PIGMENTS,   PRODUCTIVITY, PERIPHYTON, CHLOROP   W70-02780
PHORUS, NITROGEN, PHYTOPLANKTON,   .PRODUCTIVITY, TURBIDITY, SURFACES   W71-07698
TS, *FISH, ALGAE, TROPHIC LEVEL,   PRODUCTIVITY, FOOD CHAINS, TOXICI   W72-01788
TON, BIOASSAY, LABORATORY TESTS,   PRODUCTIVITY, ABSORPTION, SEDIMEN   W72-00845
TIES, *ALGAE, *METHODOL/ PRIMARY   PRODUCTIVITY, *BIOLOGICAL COMMUNI   W71-12870
, *ESSENTIAL NUTRIENTS, *PRIMARY   PRODUCTIVITY, AQUATIC ALGAE, TROP   W72-01990
T PENETRATION, SAMPLING, PRIMARY   PRODUCTIVITY, PHYTOPLANKTON, EUGL   W72-04761
*PHYTOPLANKTON, *ALGAE, *PRIMARY   PRODUCTIVITY, *ORGANIC MATTER, NU   W71-10079
INE BIOLOGY, *PRIMARY BIOLOGICAL   PRODUCTIVITY.:   MAR   W71-10788
TOXICITY, PHOTOSYNTHETIC OXYGEN,   PRODUCTIVITY, *WASTE WATER TREATM   W70-08416
TER QUALITY/ RESERVOIR, *PRIMARY   PRODUCTIVITY, *EUTROPHICATION, WA   W70-09093
OSPHORUS, NITROGEN, CHLOROPHYLL,   PRODUCTIVITY, REGIONAL ANALYSIS,   W70-07261
ISH, INFLOW, OUTLETS, NUTRIENTS,   PRODUCTIVITY, PHOTOSYNTHESIS, EUT   W70-06225
TRIENTS, LAKES, ECOLOGY, PRIMARY   PRODUCTIVITY, ALGAE, PLANKTON, SU   W71-00665
NALYSIS, *BIOASSAY, *COMPARATIVE   PRODUCTIVITY, AQUATIC MICROBIOLOG   W72-12240
NALYSIS, SESSILE ALGAE, *PRIMARY   PRODUCTIVITY, OREGON, CYANOPHYTA.   W72-06052
ECTIONS, MARINE ANIMALS, PRIMARY   PRODUCTIVITY, CHLOROPHYLL, PIGMEN   W73-01446
                     ODORIFEROUS   PRODUCTS, ORGANISMS' ODORS.:   W72-01812
ACID, CHEMOTROPH/ *EXTRACELLULAR   PRODUCTS, PHOTIC ZONE, GLYCOLLIC   W70-02504
, PHARMACEUTICALS, EXTRACELLULAR   PRODUCTS, CHEMICAL RESISTANCE, EU   W70-01068
NC, XANTHOPHYCEA/ *EXTRACELLULAR   PRODUCTS, ANABAENA CYLINDRICA, ZI   W69-10180
O/ AUTOINHIBITORS, EXTRACELLULAR   PRODUCTS, HORMONES, AUXINS, METAB   W69-04801
MENTAL CONDITIONS, EXTRACELLULAR   PRODUCTS, GAS EXCHANGE, GREEN PLA   W69-07832
PLANKTON DYNAMICS, EXTRACELLULAR   PRODUCTS, ALGAL CELLS.: /E, PHYTO   W69-05697
TIUM COMPLEXES, DTPA, SOIL WATER   PROFILE.: / ACTIVITY, *RADIOSTRON   W71-04067
OGY, DISTRIBUTION PATTERNS, SOIL   PROFILES.: /OZEMS, SOIL MICROBIOL   W71-04067
', ASHKELON(ISRAEL), TEMPERATURE   PROFILES, VOLATILE SUSPENDED SOLI   W70-04790
INNETONKA(MINN), LAKE MANAGEMENT   PROGRAM, ALGAL DENSITY, APHANIZOM   W69-10167
NATIONAL EUTROPHICATION RESEARCH   PROGRAM, CONTINUOUS CULTURE, CHEM   W70-02775
NAGE WATER, CALIFORNIA, DRAINAGE   PROGRAMS, ANAEROBIC BACTERIA, ANA   W71-04556
NAGE WATER, CALIFORNIA, DRAINAGE   PROGRAMS, ANAEROBIC BACTERIA, ANE   W71-04554
NAGE WATER, CALIFORNIA, DRAINAGE   PROGRAMS, ANAEROBIC BACTERIA, ANA   W71-04555

NAGE WATER, CALIFORNIA, DRAINAGE
RFISHING, COHO SA/ ANTIPOLLUTION
ION, *TENNESSEE VALLEY AUTHORITY
REAT LAKES-ILLINOIS RIVER BASINS
ONOMIDAE, PROCLADIUS, CHIRONOMUS
YCEAE, COMPLEX FORMATION, GROWTH-
T REQUIREMENTS, PHYSIOCOCHEMICAL
PLEX FORMATION, GROWTH-PROMOTING
, PUBLIC HEALTH, PHYSICOCHEMICAL
SOTOPES, CESIUM, PHYSICOCHEMICAL
RING, *WATER SUPPLY, *BIOLOGICAL
MATTER, NUTRIENTS, LAKES, WATER
, RADIOISOTOPES, PHYSICOCHEMICAL
UBMERGED PLANTS, PHYSICOCHEMICAL
ING, CHEMICAL ANALYSIS, PHYSICAL
POLLUTION, TEMPERATURE, CHEMICAL
ND MAINTENA/ *ALGAE, *BIOLOGICAL
SPHORUS, ALGAE, NUTRIENTS, WATER
ECTS, AQUATIC ENVIRONMENT, WATER
NT, WATER PROPERTIES, BIOLOGICAL
                        WATER QUALITY
ON, ZOOPLANKTON, PHYSICOCHEMICAL
ATOMS, SAMPLING, PHYSICOCHEMICAL
TRIENTS, FERTILIZATION, PHYSICAL
ION, CHEMICAL ANALYSIS, PHYSICAL
OPHICATION, *HYDROLOGY, CHEMICAL
SINS, SEA WATER, PHYSICOCHEMICAL
N DIOXIDE, NUTRIENTS, BIOLOGICAL
UATIC LIFE, ALGAE, PLANTS, WATER
RATURE, *COOLING WATER, CHEMICAL
, NITROGEN COMPOUNDS, PH/ *WATER
, NUTRIENT ABSORPTION, *CHEMICAL
WASTE IDENTIFICATION, BIOLOGICAL
BIOLOGICAL PROPERTIES, PHYSICAL
TION, PLANKTON, *ALGAE, CHEMICAL
TRIC PLANTS, WEED CONTROL, WATER
R LAKE, ORE, GROWTH RATES, WATER
, DOMESTIC WASTES, *ALGAE, WATER
IBLIOGRAPHIES, *ALGAE, *CHEMICAL
S, SCENEDESMUS, LIMNOLOGY, WATER
HOLOGY, SEWAGE INPUT, BIOLOGICAL
LFIDE, IRON, MAGNESIUM, COOLING,
ALUTACEUS, ULOTHRIX, CLADOPHORA,
                        *OIL SPILLS,
, ALGAE, YEASTS, BACTERIA, FISH,
E, *MICROORGANISMS, SCENEDESMUS,
M, ELECTRIC POWER, HEATED WATER,
ION CONCENTRATION, TEMPERATURE,
GANIC MATTER, CHEMICAL ANALYSIS,
BACTERIA, BIOMASS, CHLOROPHYLL,
MARY PRODUCTIVITY, *MEASUREMENT,
IS, BACILLUS STEAROTHERMOPHILUS,
TEST, GRAM REACTION, CYTOPHAGA,
COLI COMMUNAE, EUGLENE GRACILIS,
UDA.:
METRY, OSCILLATORIA TENUIS, RAM/
S), HALOBACTERIUM, RESPIROMETER,
TIC DRIFT, FISH FOOD ORG/ ALGAE,
OLYSIS, CHELATION, FUNGI, ALGAE,
ATOMS, FUNGI, INSECTS, BACTERIA,
FICATION, *CULTURES, *BIOASSAYS,
TEMATICS, FUNGI, ALGAE, DIATOMS,
LYTES, ZETA POTENTIAL, BACTERIA,
BACTERIA, FUNGI, ACTINOMYCETES,
TIC POPULATIONS, *RIVERS, ALGAE,
ITED GROWTH, LUXURY CONSUMPTION,
NT TASK FORCE ON EUTROPHICATION,
2,4-D, AMMATE, CHLORAMINE, DE-K-
UTH AFRICA, ISRAEL, MICROCYSTIS,
OTOPUS, AUSTROCLOEON, ENALLAGMA,
WTH RATES, SEAWATER, INCUBATION,

OTHERMOPHILUS, PROTEUS VULGARIS,
DAM RHINE, POLDERS, COMPOSITION,
TERIA, *MARINE ALGAE, *CULTURES,
, SAMPLING, SEA WATER, SEASONAL,
ES, MESOPLANKTON, MICROPLANKTON,
S QUADRIMACULATUS, MADISON(WIS),
BACILLUS-LIKE, MICROCOCCUS-LIKE,
ASSIMILATION, BA-LA, HALF-LIFE,
ENT BALANCE, FOAMING DETERGENTS,
CULTURES, *ISOLATION, LIMNOLOGY,
US, CLADOCERA, WATER MASSES, U S
OMICS, OHIO, COMMERCIAL FISHING,
QUATIC LIFE, WATER CONSERVATION,
NTS, VIRUSES, BACTERIA, E. COLI,
N, *CONFERENCES, NUISANCE ALGAE,
UDOTEA FLABELLUM, ZINC, NICKEL,
BALT, MANGANESE, ATLANTIC OCEAN,
L COMMUNITIES, INTERTIDAL ALGAE,

PROGRAMS, ANAEROBIC BACTERIA, ANA   W71-04557
PROGRAMS, BIOLOGICAL BALANCE, OVE   W70-04465
PROJECT, *RESERVOIRS, TEMPERATURE   W71-07698
PROJECT, INDIANA HARBOR SHIP CANA   W70-00263
PROMOSUS, TRICHOPTERA, LEECHES, F   W70-01943
PROMOTING PROPERTIES, CHLAMYDOMON   W69-10180
PROPERTIES, PHYSIOLOGICAL ECOLOGY   W69-10161
PROPERTIES, CHLAMYDOMONAS MOEWUSI   W69-10180
PROPERTIES, GENETICS, BACTERIA, P   W70-00273
PROPERTIES, ADSORPTION, COLLOIDS,   W70-00708
PROPERTIES, *WATER POLLUTION SOUR   W72-01813
PROPERTIES, BACTERIA, BIOASSAY, M   W71-10079
PROPERTIES, ANION ADSORPTION, COL   W71-09863
PROPERTIES, PHYTOPLANKTON, DISTRI   W71-00665
PROPERTIES, CARP, TILAPIA, TEMPER   W70-06975
PROPERTIES, ECOLOGY, ECOLOGICAL D   W70-07313
PROPERTIES, CONTROL, *OPERATION A   W70-08392
PROPERTIES.: /ECTS, NITROGEN, PHO   W70-04382
PROPERTIES, BIOLOGICAL PROPERTIES   W70-04475
PROPERTIES, ECOLOGY, STRATIFICATI   W70-04475
PROPERTIES.:   W70-04475
PROPERTIES, DOMESTIC WASTES, SEWA   W70-03964
PROPERTIES, CENTRIFUGATION, MICRO   W70-05389
PROPERTIES, CLIMATE, NITROGEN, PH   W70-05565
PROPERTIES.: /EWAGE, WATER POLLUT   W70-05752
PROPERTIES, WISCONSIN, SEWAGE EFF   W70-05113
PROPERTIES, TRACERS, DISTRIBUTION   W69-08275
PROPERTIES, WATER POLLUTION EFFEC   W69-07442
PROPERTIES, WATER TEMPERATURE, OU   W69-05023
PROPERTIES, ECOLOGY, SALINITY, BE   W69-05023
PROPERTIES, *NUTRIENTS, *VIRGINIA   W69-03695
PROPERTIES, EPILMNION, BOTTOM SED   W68-00857
PROPERTIES, PHYSICAL PROPERTIES,   W69-01273
PROPERTIES, WASTE WATER(POLLUTION   W69-01273
PROPERTIES, BACTERIA, ORGANIC POL   W68-00483
PROPERTIES, PHOSPHORUS, CHEMICAL   W68-00479
PROPERTIES.:   CRATE   W68-00478
PROPERTIES, BIOCHEMICAL OXYGEN DE   W68-00256
PROPERTIES, NUTRIENTS, LAKES, TES   W68-00856
PROPERTIES, WATER TEMPERATURE, AL   W68-00248
PROPERTIES, *WATER QUALITY, TURBI   W68-00475
PROPERTY VALUES, RECREATION DEMAN   W70-02251
PROSOPIUM WILLIAMSONI.: /CHEILUS   W69-07853
PROTEIN AQUICULTURE.:   W69-05891
PROTEINS, OIL WASTES, OILY WATER,   W69-05891
PROTEINS, WALLS, STRUCTURE, AMINO   W70-04184
PROTEINS, AQUATIC MICROORGANISMS,   W70-09905
PROTEINS, CARBOHYDRATES, LIPIDS,   W71-13413
PROTEINS, BACTERIA.: /OMPOSING OR   W72-01812
PROTEINS, LIGHT, NUTRIENTS, HYDRO   W69-10160
PROTEINS, RADIOACTIVITY, CARBON R   W69-10151
PROTEUS VULGARIS, PSEUDOMONAS AER   W71-11823
PROTEUS, SERRATIA, BAY OF FUNDY(C   W70-03952
PROTOCOCCALES.: /, DRY WEIGHT, B   W70-05092
PROTOCOCCUS, SCENEDESMUS QUADRICA   W71-11561
PROTON MAGNETIC RESONANCE SPECTRO   W69-00387
PROTOPLASM, WATER-BLOOMS, LAKE GE   W70-04369
PROTOZOA, *AQUATIC INSECTS, *AQUA   W69-02782
PROTOZOA, LICHENS, PLANTS, DIATOM   W69-07428
PROTOZOA, SEWAGE, LAKES, NUTRIENT   W71-10786
PROTOZOA, BACTERIA, FUNGI, VIRUSE   W71-11823
PROTOZOA, PLANKTON, WORMS, NEMATO   W72-01819
PROTOZOA, ALGAE, SOLID WASTES, TE   W70-01027
PROTOZOA, ALGAE.: /IONS, E. COLI,   W71-02009
PROTOZOA, AQUATIC INSECTS, FISH R   W73-00748
PROVISIONAL ALGAL ASSAY, PROCEDUR   W70-02779
PROVISIONAL ALGAL ASSAY PROCEDURE   W70-02777
PRUF-21, BENOCLORS, NIGROSINE DYE   W70-05263
PRYMNESIUM PARVUM, SYNERGISTIC EF   W70-05372
PSEUDAGRION, BRACHYTHEMIS, CHYDOR   W70-06975
PSEUDOMONAS, SUBSURFACE-WATERS, B   W70-00713
PSEUDOMONAS AERUGINOSA.:   W69-10171
PSEUDOMONAS AERUGINOSA, STAPHYLOC   W71-11823
PSEUDOMONAS, MYCOBACTERIUM, BACIL   W69-07838
PSEUDOMONAS, METABOLISM, KINETICS   W70-04280
PSEUDOMONAS, TEMPERATURE, AMINO A   W70-03952
PSEUDOPLANKTON, CRYPTOPHYCEAE, VO   W70-05389
PSOROPHORA CILIATA, ARTIFICIAL KE   W70-06225
PSYCHROPHILES, PLEMORPHIC, MYCELO   W70-00713
PTYCHOCHEILUS OREGONENSIS, ACROCH   W69-07853
PUBLIC APPREHENSION.: /TES, NUTRI   W70-05084
PUBLIC HEALTH, PHYSICOCHEMICAL PR   W70-00273
PUBLIC HEALTH SERVICE, GREAT LAKE   W70-00263
PUBLIC HEALTH, WALLEYE, VEGETATIO   W70-01943
PUBLIC HEALTH, WATER POLLUTION, C   W70-03344
PUBLIC HEALTH, PATHOGENIC BACTERI   W72-06017
PUBLICATIONS, WATER POLLUTION EFF   W70-03973
PUERTO RICO, LA PARGUERA, JOYUDA.   W69-08525
PUERTO RICO, SPECTROSCOPY, DISTRI   W69-08525
PUGET SOUND, FLOURESCEIN DYE.: /A   W68-00010

FICATION, WASTE TREATMENT, WATER | QUALITY, WATER POLLUTION CONTROL, | W69-06535
LUTION EFFECTS, *REVIEWS, *WATER | QUALITY, WATER POLLUTION, ORGANIC | W70-07027
GAL CONTROL, WATER SUPPLY, WATER | QUALITY, WATER POLLUTION, | W69-10155
ALGAE, WATER PURIFICATION, WATER | QUALITY, WATER POLLUTION CONTROL, | W69-10155
RNMENT, STATE GOVERNMENTS, WATER | QUALITY, WATER CHEMISTRY.: /AYS, | W69-09884
S, WASTE WATER(POLLUTION), WATER | QUALITY: /G, GRANTS, FEDERAL GOVE | W69-06305
X MUCOR, ANTITHAMNION SARNIENSE, | QUALITY: /IES, PHYSICAL PROPERTIE | W69-01273
IDAZOLINES, BIOLOGICAL ACTIVITY, | QUANTITATION, GROWTH MODES.: /HRI | W69-10163
ENOL/ BIOCIDES, PRODUCTION LOSS, | QUARTERNARY, AMMONIUM COMPOUNDS, | W70-02982
NIMUM LETHAL DOSES, CHU'S MEDIA, | QUATERNARY AMMONIUM COMPOUNDS, PH | W72-12987
LAKE, *INORGANIC/ ONTARIO LAKES, | QUINONE DERIVATIVES, ALKYL SUBSTI | W70-02982
IX FUSCA, MICROCOLEUS LACUSTRIS (| QUIRKE LAKE, PECORS LAKE, DUNLOP | W70-05424
TRIS BRAND, GONGROSIRA DEBARYANA | RAB) FARLOW, CLADOPHORA, TAXONOMI | W70-04468
CYSTIS AERUGINOSA, ANABAENOPSIS, | RAB, OOCYSTIS, OSCILLATORIA: /CUS | W70-04468
NKTON, ALGAE, TEMPERATURE, SOLAR | RACIBORSKII, RAPHIDIOPSIS, ANABAE | W70-05404
FILTRATION, *SLIME, *ULTRAVIOLET | RADIATION.: /OSYNTHESIS, PHYTOPLA | W71-10788
SEWAGE EFFLUENTS, OXYGEN, SOLAR | RADIATION, *TREATMENT FACILITIES, | W72-03824
ES, ALGAE, EMULSIONS, BACKGROUND | RADIATION, PERIPHYTON, BACTERIA, | W70-01933
COVER, SPECTROSCOPY, ULTRAVIOLET | RADIATION, IODINE RADIOISOTOPES, | W69-10163
TRIPTON, BOTTOM SEDIMENTS, SOLAR | RADIATION, TRANSMISSIVITY, CHLORO | W69-09676
YTON, AQUATIC PLANTS, BACKGROUND | RADIATION, OXYG: /ICE, BACTERIA, | W69-10154
TUATION, SEWAGE TREATMENT, SOLAR | RADIATION, GEOCHEMISTRY, SPRING W | W69-09334
ON, TIDES, PHOTOSYNTHESIS, SOLAR | RADIATION, CHLOROPHYLL, PRODUCTIV | W69-10182
MPERATURE, SALINITY, ULTRAVIOLET | RADIATION, SALINITY, RAINFALL, TE | W70-01073
WTH RATES, BACTERIA, ULTRAVIOLET | RADIATION, SULFATES, CELLULOSE, I | W70-01074
XYGEN-REDUCTION POTENTIAL, SOLAR | RADIATION, ACTINOMYCETES, HYDROGE | W70-00719
NG, AERIAL PHOTOGRAPHY, INFRARED | RADIATION, ENERGY, CARBON DIOXIDE | W70-05750
M SEDIMENTS, WEED CONTROL, SOLAR | RADIATION, DYE RELEASES, BEACHES, | W68-00010
LO), MUTATION, BETA RAYS, COSMIC | RADIATION, *WATER POLLUTION CONTR | W68-00478
ASKA, LAKES, RIVERS/ *BACKGROUND | RADIATION, S: /N, CENTRAL CITY(CO | W69-08272
SECTS, INVERTEBRATES, BACKGROUND | RADIATION, *AQUATIC HABITATS, *AL | W69-08272
, MANGANESE, ALGAE, IRON, BIRDS, | RADIATION, FLOCCULATION, POLLUTAN | W69-07846
LES, BOTTOM SEDIMENTS, CATTAILS, | RADIOACTIVITY, INVERTEBRATES, PLA | W69-08269
RABS, NORTH CAROLINA, SEDIMENTS, | RADIOACTIVITY, INTERFACES, ORGANI | W69-08272
RIBUTION, CHELATION, DETERGENTS, | RADIOACTIVITY, MARINE ANIMALS, ES | W69-08267
, *TROUT, *STREAMS, *ECOSYSTEMS, | RADIOACTIVITY, AMINO ACIDS.: /IST | W69-08275
GULF OF MEXICO, ATLANTIC OCEAN, | RADIOACTIVITY, BIOTA, CYCLES, MIC | W69-09334
, FISH, PHYTOPLANKTON, PLANKTON, | RADIOACTIVITY, GAMMA RAYS, POTASS | W69-08524
ANKTON, PLANKTON, RADIOACTIVITY, | RADIOACTIVITY, RADIOACTIVITY EFFE | W69-07440
ITION, NITROGEN FIXING BACTERIA, | RADIOACTIVITY EFFECTS, RADIOECOLO | W69-07440
LABORATORY TESTS, MODEL STUDIES, | RADIOACTIVITY TECHNIQUE, PHOTOSYN | W70-04249
MATTER, *PHYSIOLOGICAL ECOLOGY, | RADIOACTIVITY TECHNIQUES, TAGGING | W70-04074
CULTURES, PHYTOPLANKTON, CARBON, | RADIOACTIVITY TECHNIQUES, ANALYTI | W70-04194
M RADIOISOTOPES, *PHYTOPLANKTON, | RADIOACTIVITY, CHLORELLA, SAMPLIN | W70-03983
CTIVITY, *MEASUREMENT, PROTEINS, | RADIOACTIVITY, WASTE DISPOSAL, NU | W70-00707
6, RHODIUM-106, CERIUM-144, BETA | RADIOACTIVITY, CARBON RADIOISOTOP | W69-10151
OPES, LIGHT INTENSITY, CULTURES, | RADIOACTIVITY, GONIOBASIS, DECAPO | W69-09742
OPHYLL, SAMPLING, PHYTOPLANKTON, | RADIOACTIVITY, CARBON DIOXIDE, WA | W69-10158
CITY, *INSECTICIDES, METABOLISM, | RADIOACTIVITY, PHOTOSYNTHESIS, CH | W70-01933
BUTION, DENSITY, CHROMATOGRAPHY, | RADIOACTIVITY TECHNIQUES, CHLOREL | W70-02198
, *PHOTOSYNTHESIS, BICARBONATES, | RADIOACTIVITY, TRACERS, WATER POL | W70-02510
S, FOOD CHAINS, STABLE ISOTOPES, | RADIOACTIVITY, CARBON RADIOISOTOP | W70-02504
AL WASTES, TOXINS, SOLID WASTES, | RADIOACTIVE WASTE DISPOSAL, OCEAN | W70-02786
OURCES, WATER POLLUTION CONTROL, | RADIOACTIVE WASTES, HEATED WATER, | W72-00462
ANIMALS, ALGAE, MOLLUSKS, FISH, | RADIOACTIVE WASTE, WATER QUALITY | W72-00941
L REACTIONS, CHLORELLA, TRACERS, | RADIOACTIVE WASTES.: /CTS, MARINE | W71-05531
BACTERIA, ALGAE, ORGANIC MATTER, | RADIOACTIVITY, CARBON RADIOISOTOP | W70-08097
RIENTS, MODEL STUDIES, BIOASSAY, | RADIOACTIVITY, PESTICIDES, TASTE, | W72-13601
GAE, IRON, ALKALINITY, HARDNESS, | RADIOACTIVITY TECHNIQUES, ABSORPT | W73-01454
NTRATION FACTOR, LIMNODRILUS SP/ | RADIOACTIVITY, TEMPERATURE, SALIN | W73-00750
ROTROPHY, AUTOTROPHY, CARBON-14, | RADIOAUTOGRAPHS, HALF-LIFE, CONCE | W69-07861
IMNION, CHEMOCLINE, MIXOLIMNION, | RADIOCARBON UPTAKE TECHNIQUES, GL | W70-04287
NNUAL CYCLES, MICROBIAL ECOLOGY, | RADIOCARBON UPTAKE TECHNIQUES, BA | W70-05760
D SOUND(MASS), WOODS HOLE(MASS). | RADIOCARBON UPTAKE TECHNIQUES, MI | W70-05109
MELOSIRA, DINOBRYON, FRAGILARIA, | RADIOCARBON UPTAKE TECHNIQUES, MI | W70-05272
SPHATES, *ALGAE, CHROMOTOGRAPHY, | RADIOCARBON METHOD, FINGER LAKES(| W70-01579
COBALT RADIOISOTOPES, SEASONAL, | RADIOCHEMICAL ANALYSIS, TRACERS.: | W71-03033
ON, FISH, SEA WATER, GAMMA RAYS, | RADIOCHEMICAL ANALYSIS, MONITORIN | W70-04371
HROCYTIUM OBESUM W AND G S WEST, | RADIOCHEMICAL ANALYSIS, TRITIUM, | W69-08269
ISOTOPES, IRON, SEDIMENTS, FISH, | RADIOCOCCUS NIMBATUS, URONEMA ELO | W70-04468
ACTIVITY, RADIOACTIVITY EFFECTS, | RADIOECOLOGY, EFFLUENTS, ECOSYSTE | W69-07853
FFECTS, WATER POLLUTION SOURCES, | RADIOECOLOGY, SORPTION, WATER POL | W69-07440
A WATER, FOOD CHAINS, PHOSPHORUS | RADIOECOLOGY.: /WATER POLLUTION E | W70-00707
S, COBALT RADIOISOTOPES, URANIUM | RADIOISOTOPES, COBALT RADIOISOTOP | W70-00708
TIES, ADSORPTION, COLLOIDS, ZINC | RADIOISOTOPES, NUCLEAR EXPLOSIONS | W70-00708
PHOSPHORUS RADIOISOTOPES, COBALT | RADIOISOTOPES, CALCIUM, POTASSIUM | W70-00708
S, NUCLEAR EXPLOSIONS, STRONTIUM | RADIOISOTOPES, URANIUM RADIOISOTO | W70-00708
ESIS, DIATOMS, CHLORELLA, CARBON | RADIOISOTOPES, CESIUM, PHYSICOCHE | W70-00708
NS, BACKGROUND RADIATION, IODINE | RADIOISOTOPES, LIGHT INTENSITY, C | W69-10158
ES, TRITIUM, PHOTOGRAPHY, CARBON | RADIOISOTOPES, NUTRIENT REQUIREME | W69-10163
CARBON RADIOISOTOPES, PHOSPHORUS | RADIOISOTOPES, PHOSPHORUS RADIOIS | W69-10163
PROTEINS, RADIOACTIVITY, CARBON | RADIOISOTOPES, ALGAE, EMULSIONS, | W69-10163
CARBON RADIOISOTOPES, PHOSPHORUS | RADIOISOTOPES, PHOSPHORUS RADIOIS | W69-10151
ECTROMETERS, CYANOPHYTA, URANIUM | RADIOISOTOPES, ALGAE, BIOCHEMISTR | W69-10151
URANIUM RADIOISOTOPES, STRONTIUM | RADIOISOTOPES, STRONTIUM RADIOISO | W69-09742
DIOACTIVITY, WASTE D/ *STRONTIUM | RADIOISOTOPES, CESIUM, FALLOUT, S | W69-09742
ALCAREOU/ *CHELATION, *STRONTIUM | RADIOISOTOPES, *PHYTOPLANKTON, RA | W70-00707
, TRACERS, RADIOACTIVITY, CARBON | RADIOISOTOPES, *ALGAE, *FUNGI, *C | W71-04067
DIATOMS, EUTROPHICATION, CARBON | RADIOISOTOPES, HYDROGEN ION CONCE | W70-08097
HICATION, *CALI/ *LAKES, *CARBON | RADIOISOTOPES, CHLOROPHYLL, HYDRO | W70-01579
                                | RADIOISOTOPES, *BIOASSAY, *EUTROP | W70-01933

ISOTOPES, FISH, CLAMS, STRONTIUM
TRACERS, PHOTOSYNTHESIS, CARBON
ARBONATES, RADIOACTIVITY, CARBON
*CESIUM
S, RADIUM RADIOISOTOPES, THORIUM
*AQUATIC PLANTS, AQUATIC / *ZINC
, THORIUM RADIOISOTOPES, URANIUM
ESTS, ALGAE, INSECTS, PHOSPHORUS
AT LAKES, TRACE ELEMENTS, RADIUM
ES, *ALGAE, *METHODOLOGY, CARBON
S, CADMIUM RADIOISOTOPES, CERIUM
PES, CERIUM RADIOISOTOPE/ CESIUM
S, STRONTIUM RADIOISOTOPES, ZINC
CRUSTACEANS, MOLLUSKS, STRONTIUM
ES, IODINE RADIOISOTOPES, COBALT
*RUTHENIUM
OPES, ZINC RADIOISOTOPES, IODINE
*MARINE ANIMALS, *MARINE PLANTS,
E/ CESIUM RADIOISOTOPES, CADMIUM
DIOISOTOPES, TROPHIC LEVEL, ZINC
S, URANIUM RADIOISOTOPES, RADIUM
VIRONMENTS, *ECOSYSTEMS, *RADIUM
OILS, NUCLEAR EXPLOSIONS, COBALT
HYTON, *CYCLING NUTRIENTS, *ZINC
DIOISOTOPES, GAMMA RAYS, URANIUM
, OXYGEN, CARBON DIOXIDE, CARBON
ECTS, SALINITY, TEMPERATU/ *ZINC
ITY, INTERFACES, ORGANIC MATTER,
EMICAL ANALYSIS, TRITIUM, CARBON
M, STRONTIUM RADIOISOTOPES, ZINC
FICIDS, *ECOSYSTEMS, *PHOSPHORUS
OUTH CAROLINA, CESIUM, STRONTIUM
RADIOISOTOPES, CESIUM, STRONTIUM
*RADIUM RADIOISOTOPES, *URANIUM
FLORA, STREAMS, CULTURES, COBALT
MBIA RIVER, CHROMIUM, PHOSPHORUS
*EC/ *TRANSLOCATION, *PHOSPHORUS
*ZINC RADIOISOTOPES, *PHOSPHORUS
M, RADIUM RADIOISOTOPES, URANIUM
SOTOPES, *SALT MARSHES, */ *ZINC
OACTIVITY, GAMMA RAYS, POTASSIUM
TOPES, MANGANESE, CESIUM, RADIUM
SINS, SEA WATER, PHYSICOC/ *ZINC
NC RADIOISOTOPES, CESIUM, COBALT
OACTIVITY, *LAKES, PRODUCTIVITY,
COLOGY, AQUATIC BACTERIA, CARBON
IGHT INTENSITY, SUCCESSION, ZINC
MATTER, LAKES, DIFFUSION, CARBON
EUTROPHICATION, GLUCOSE, CARBON
PHYTOPLANKTON, CARBON,/ *CARBON
IOCHEMICAL OXYGEN DEMAND, CARBON
GLENA, CHLAMYDOMONAS, PHOSPHORUS
LLUTION EFFECTS, DIATOMS, CARBON
E, PHYSIOLOGICAL ECOLOGY, CARBON
ACTERIA, WATER CHEMISTRY, CARBON
AGE, DIATOMS, CYANOPHYTA, CARBON
ONS, OXYGEN, TRACERS, PHOSPHORUS
KES, BIBLIOGRAPHIES, *PHOSPHORUS
E, URANIUM RADIOISOTOPES, RADIUM
LIGOTROPHIC, *LAKES, *PHOSPHORUS
L WASTES, AQUATIC ALGAE, URANIUM
HLOROPHYTA, BACTERIA, PHOSPHORUS
MINERAL TRANSFORMATION, METAZOA,
A, HA/ *HANFORD(WASH), *REACTOR,
ERATE ZONE, PLUTONIUM, HALF-LIFE(
M KLEBSI, PORPHYRIDIUM CRUENTUM,
, SOIL CONTAMINATION, *STRONTIUM
I/ *RADIOSTRONTIUM, *ALGAL-BOUND
ND RADIOSTRONTIUM, *FUNGAL-BOUND
MS, GREAT LAKES, TRACE ELEMENTS,
IC ALGAE, URANIUM RADIOISOTOPES,
MMA RAYS, URANIUM RADIOISOTOPES,
ADIOISOTOPES, MANGANESE, CESIUM,
REEK, ARCTIC CIRCLE HOT SPRINGS,
PULLULANS, HANSENIASPORA UVARUM,
COLEOPTERA, HEMIPTERA, MOLLUSCA,
E COUNTS, ALGAL GROWTH, SEDGWICK-
RS, RESERVOIRS, AIR TEMPERATURE,
HY, TEMPERATURE, PHOTOSYNTHESIS,
A, DOMINANT ORGANISM, NUTRIENTS,
ESIS, SOLAR RADIATION, SALINITY,
LLUTION SOURCES, SURFACE RUNOFF,
ARK, MT LASSEN NATIONAL PARK, MT
CLAIR, ROUGE RIVER, HURON RIVER,
A, *BLOOMS, RIVER GANGES(INDIA),
ECTROMETRY, OSCILLATORIA TENUIS,
IR SPRAY, SUDBURY(ONTARIO), LAKE
: TEMPERATURE
LA RIVER(NORWAY), FLAKK(NORWAY),
LADOPHORA, FLAGELLATA, CILIATES,

RADIOISOTOPES, TENNESSEE RIVER, T   W70-02786
RADIOISOTOPES, FISH, CLAMS, STRON   W70-02786
RADIOISOTOPES, LAKES, ORGANIC MAT   W70-02504
RADIOISOTOPES.:   W71-09013
RADIOISOTOPES, URANIUM RADIOISOTO   W71-09013
RADIOISOTOPES, *AQUATIC INSECTS,   W71-09005
RADIOISOTOPES, MARINE ALGAE, WATE   W71-09013
RADIOISOTOPES, TECHNOLOGY, NATURA   W71-09012
RADIOISOTOPES, THORIUM RADIOISOTO   W71-09013
RADIOISOTOPES, DIATOMS, MEASUREME   W71-12870
RADIOISOTOPES, ECHINODERMS.: /OPE   W71-09850
RADIOISOTOPES, CADMIUM RADIOISOTO   W71-09850
RADIOISOTOPES, IODINE RADIOISOTOP   W71-09850
RADIOISOTOPES, ZINC RADIOISOTOPES   W71-09850
RADIOISOTOPES, WATER POLLUTION EF   W71-09850
RADIOISOTOPES.:   W71-09863
RADIOISOTOPES, COBALT RADIOISOTOP   W71-09850
RADIOISOTOPES, PHYSICOCHEMICAL PR   W71-09863
RADIOISOTOPES, CERIUM RADIOISOTOP   W71-09850
RADIOISOTOPES, IRON, SEDIMENTS, F   W69-07853
RADIOISOTO: /OISOTOPES, GAMMA RAY   W69-08272
RADIOISOTOPES, *URANIUM RADIOISOT   W69-07846
RADIOISOTOPES, CESIUM, STRONTIUM   W69-08269
RADIOISOTOPES, STREAMFLOW, SAMPLI   W69-07862
RADIOISOTOPES, RADIUM RADIOISOTO: W69-08272
RADIOISOTOPES, DETRITUS, COLUMBIA   W69-07862
RADIOISOTOPES, *ENVIRONMENTAL EFF   W69-08267
RADIOISOTOPES, GAMMA RAYS, URANIU   W69-08272
RADIOISOTOPES, BIOTA, CLAMS, FALL   W69-08269
RADIOISOTOPES, IRON, MANGANESE, B   W69-07845
RADIOISOTOPES, *TRANSFER, WATER,   W69-07861
RADIOISOTOPES, ZINC RADIOISOTOPES   W69-07845
RADIOISOTOPES, MANGANESE, ALGAE,   W69-08269
RADIOISOTOPES, RIVERS, MILLS, COL   W69-07846
RADIOISOTOPES.: /GANESE, BENTHIC   W69-07845
RADIOISOTOPES, TROPHIC LEVEL, ZIN   W69-07853
RADIOISOTOPES, *TROUT, *STREAMS,   W69-09334
RADIOISOTOPES, *SALT MARSHES, *CY   W69-08274
RADIOISOTOPES.: /MANGANESE, CESIU   W69-08524
RADIOISOTOPES, *PHOSPHORUS RADIOI   W69-08274
RADIOISOTOPES, MANGANESE, CESIUM,   W69-08524
RADIOISOTOPES, URANIUM RADIOISOTO   W69-08524
RADIOISOTOPES, *MARINE ALGAE, *RE   W69-08275
RADIOISOTOPES, SEASONAL, RADIOCHE   W70-04371
RADIOISOTOPES, TEMPERATURE, DEPTH   W70-04371
RADIOISOTOPES, ENZYMES, KINETICS,   W70-04287
RADIOISOTOPES, CESIUM, COBALT RAD   W70-04371
RADIOISOTOPES, CHLAMYDOMONAS, BIO   W70-04284
RADIOISOTOPES, DINOFLAGELLATES, B   W70-04194
RADIOISOTOPES, *ALGAE, *CULTURES,   W70-03983
RADIOISOTOPES, DETERGENTS, ANTIFR   W70-04001
RADIOISOTOPES, REGRESSION ANALYSI   W70-03957
RADIOISOTOPES, KINETICS, TRITIUM,   W70-04811
RADIOISOTOPES, LAKES, PHYTOPLANKT   W70-05109
RADIOISOTOPES, PACIFIC OCEAN, CAR   W70-05652
RADIOISOTOPES, POTASSIUM, OXY: /W   W70-05405
RADIOISOTOPES, IONS, HUMIC ACIDS,   W70-05399
RADIOISOTOPES, ON-SITE DATA COLLEC   W70-00857
RADIOISOTOPES.: /ES, AQUATIC ALGA   W69-01165
RADIOISOTOPE, *TURNOVERS, NUTRIEN   W68-00858
RADIOISOTOPE, RADIUM RADIOISOTOP   W69-01165
RADIOISOTOPES, INFLOW, CYANOPHYTA   W73-01454
RADIOLARIA, FORAMINIFERA, THIOBAC   W69-07428
RADIONUCLIDES, ASSIMILATION, BA-L   W69-07853
RADIONUCLIDE, FILTER-FEEDING ANI   W70-00708
RADIONUCLIDE UPTAKE, NANNOCHLORIS   W70-00708
RADIOSOTOPES, CYTOLOGICAL STUDIES   W69-03491
RADIOSTRONTIUM, *FUNGAL-BOUND RAD   W71-04067
RADIOSTRONTIUM, *MICROBIAL ACTIVI   W71-04067
RADIUM RADIOISOTOPES, THORIUM RAD   W71-09013
RADIUM RADIOISOTOPES.: /ES, AQUAT   W69-01165
RADIUM RADIOISOTO: /OISOTOPES, GA   W69-08272
RADIUM RADIOISOTOPES, URANIUM RAD   W69-08524
RADON, CENTRAL CITY(COLO), MUTATI   W69-08272
RADULESPORES, ZYMOLOGICAL, ELBE(G   W70-04368
RAFFINATE, THORIUM, TAILINGS.: /   W69-07846
RAFTER METHOD, TREATMENT TIMING,   W69-10157
RAIN, GEOLOGY, MAPPING, NUTRIENTS   W70-02646
RAINBOW TROUT, DIATOMS, ANAEROBIC   W69-10154
RAINFALL-RUNOFF RELATIONSHIPS, GE   W68-00891
RAINFALL, TEMPERATURE, COLORIMETR   W70-01073
RAINFALL, SNOWMELT, SURFACE WATER   W70-03501
RAINIER NATIONAL PARK, OHANAPECOS   W70-03309
RAISIN RIVER, DETROIT RIVER, MAUM   W69-01445
RAJGHAT(INDIA), BOMBAY(INDIA), BE   W70-05404
RAMMAN SPECTROSCOPY, MASS SPECTRO   W69-00387
RAMSEY(ONTARIO), ALGAE COUNTS, TH   W69-10155
RANGES, TEMPERATURE REQUIREMENTS.   W71-02494
RANHEIM(NORWAY), TRONDHEIMSFJORD(W70-01074
RANUNCULUS CIRCINATUS, POTAMOGETO   W69-07838

516

TRUM, HYDRODICTYON, POTAMOGETON,
NOSA, ANABAENOPSIS, RACIBORSKII,
RATION, VERTICAL MIXING, ROANOKE
NORTH AMERICA, CENTRAL AMERICA,
NDED SOLIDS, *PHOSPHOROUS UPTAKE
CLOPS BICUSPIDATUS, REPRODUCTIVE
MIDIUM RETZII, SURFACE DIFFUSION
UDIES, DAPHNIA SCHOD/ *FILTERING
GE, STORMS, T/ *ADDITIVES, *FLOW
DIOXIDE, PHOTOSYNTHESIS, GROWTH
LA MENEGHINIANA, LIGHT-SATURATED
BORATORY(RHODE ISLAND), DILUTION
OROPHYLL, PHOTOSYNTHESIS, GROWTH
RACELLULAR PRODU/ *PHOTOSYNTHETIC
ONDWEEDS, ROOTED AQUATIC PLANTS,
AL CONTROL, BACTERICIDES, GROWTH
ISOPODS, AMPHIPODA, FISH, GROWTH
TROPHICATION, LIGHT, TE/ *GROWTH
LE, 2-4-5-/ *HERBICIDES, *GROWTH
ICAL TECHNIQUES, *ALGAE, *GROWTH
ACTIVITY, ALGAL METABOLIS/ *KILL
C/ *BIOASSAY, *HALOGENS, *GROWTH
OINDICA/ *FEEDING RATES, *GROWTH
PRIMARY PRODUCTIV/ SEDIMENTATION
*COLOR, *ALGAE, *COLLOIDS, FLOW
ION, WATER FLEAS, LARVAE, GROWTH
RODUCTIVITY, BIOINDICA/ *FEEDING
ESHWATER ENVIRONMENTS, *TURNOVER
TRIC ANALYSIS, LIMNOLOGY, GROWTH
ICATION, *LAKES, *ALGAE, *GROWTH
*ALGAE, *PHOTOSYNTHESIS, *GROWTH
RTRIENTS, RIVERS, LAKES, *GROWTH
S, LAKE ERIE, STATISTICS, GROWTH
NMENTAL FACTORS, BIOMASS, GROWTH
ION, ALGAE, MARINE ALGAE, GROWTH
, CULTURES, LAKES, ALGAE, GROWTH
ION, POLLUTION ABATEMENT, GROWTH
MICROBIOLOGY, ACTIVATED SLUDGE,
OSPHORUS, NITROGEN, IRON, GROWTH
SEASONAL EFFECTS, DECOMPOSITION
BATEMENT, DRAINAGE SYSTEMS, FLOW
ULFATES, VITAMINS, ALGAE, GROWTH
LOTIC ENVIRONMENT, BIOMASS, FLOW
, LABORATORY MODELS, BOD REMOVAL
PLANKTON, *LAKE ONTARIO, *GROWTH
MUS, REAERATION, EFFLUENTS, FLOW
DISSIPATA, SYNEDRA ACUS, SINKING
AE, SNAILS, MITES, MOLDS, GROWTH
, *NUTRIENT REQUIREMENTS, GROWTH
DISSOLVED SOLIDS, SEDIMENTATION
E, *BACTERIA, METABOLISM, GROWTH
HY, RESISTANCE, BIOASSAY, GROWTH
BROMINE, FILAMENTOUS ALGAE, KILL
ENZYMES, CARBON DIOXIDE, GROWTH
OSPHATES, ALGAE, DAPHNIA, GROWTH
MARINE ALGAE, FRESH WATER, FLOW
ENZYMES, CLASSIFICATION, GROWTH
ATION TECHNIQUES, PRESSURE, FLOW
ON EFFECTS, LETHAL LIMIT, GROWTH
ON, PHOSPHORUS, NITROGEN, GROWTH
NTS, PRODUCTIVITY, SEDIMENTATION
NTENSITY, CARBON DIOXIDE, GROWTH
RAPHY, TRACERS, BACTERIA, GROWTH
ANCE ALGAE, INTAKES, SANDS, FLOW
        CRATER LAKE, ORE, GROWTH
ARTICLE SIZE, POPULATION, GROWTH
*BIOGEOCHEMICAL DATA, *ABUNDANCE
L, MEICELASE, HYDROGEN PEROXIDE,
207, VERTEBRATES, PLUTONIUM-239,
ER PLANTS, NYLON-6, *V/ *VISCOSE
FIBER PLANTS, NYLON-6, *VISCOSE
NTRAL CITY(COLO), MUTATION, BETA
NTIC OCEAN, RADIOACTIVITY, GAMMA
PLANKTON, FISH, SEA WATER, GAMMA
CKY, LIFE HISTORY STUDIES, GAMMA
NIC MATTER, RADIOISOTOPES, GAMMA
FLAGELLATION, IODINE TEST, GRAM
IUM MODEL, LITHOSPHERE, SORPTION
EGIONS, PHOTOSYNTHESIS, CHEMICAL
REATMENT, ELECTROLYTES, CHEMICAL
HYDROLYSIS, METABOLISM, CHEMICAL
ING WATER, *CORROSION, *CHEMICAL
ED INORGANIC PHOSPHATES, SOLUBLE
MENTOUS ALGAE, KILL RATES, PHOTO
L RATES, *BROMINE, IODINE, PHOTO
LABORATORY, PLECTONEMA BORYANUM,
CTIVITY, WASTE DISPOSAL, NUCLEAR
N, *CALIFORNIA, WATER POLLUTION
TREATMENT, PILOT PLANTS, HUMUS,
Y, SYMBIOSIS, RESERVOIRS, LAKES,

RANUNCULUS.: / CLADOPHORA, PEDIAS   W70-04494
RAPHIDIOPSIS, ANABAENA, WOLLEA, O   W70-05404
RAPIDS RESERVOIR(NC).: /SUPERSATU   W70-00683
RATE EQUATIONS, ANALYSIS OF VARIA   W70-03957
RATE.:   MIXED LIQUOR, SUSPE   W71-13335
RATE.: /MA LEUCHTENBER GIANUM, CY   W69-10182
RATE.: /RA ULNA, OEDOGONIUM, PHOR   W70-00265
RATE, *BODY SIZE, *COMPARATIVE ST   W70-03957
RATE, *WASTE WATER, *ALGAE, *SLUD   W70-05819
RATE, AEROBIC CONDITIONS, ANAEROB   W70-05750
RATE, CELL VOLUME, NITZSCHIA CLOS   W70-05381
RATE, FRUITING BODIES, ULVA LACTU   W70-01073
RATE, ORGANIC MATTER, SCENEDESMUS   W70-05381
RATE, PHYTOPLANKTON DYNAMICS, EXT   W69-05697
RATES OF APPLICATION, APPLICATION   W70-05568
RATES.: /CHLORELLA, *EUGLENA, ALG   W69-00994
RATES.: /OMS, ALGAE, METABOLISM,   W69-09742
RATES, *ALGAE, PHOTOSYNTHESIS, EU   W69-07442
RATES, *ALGAE, 2-4-D, AMINOTRIAZO   W70-03519
RATES, *BIOINDICATORS, *WATER POL   W70-04381
RATES, *BROMINE, IODINE, PHOTO RE   W70-02363
RATES, *CHLORELLA, CHLORINE, ALGI   W70-02370
RATES, *DAPHNIA, PRODUCTIVITY, BI   W69-10152
RATES, *EUTROPHICATION, *LAKES, *   W69-00632
RATES, *FILTERS, *NEW YORK, TREAT   W72-04145
RATES, *FRESH WATER FISH, MICHIGA   W68-00487
RATES, *GROWTH RATES, *DAPHNIA, P   W69-10152
RATES, *HETEROTROPHY, ORGANIC SOL   W70-05109
RATES, *LABORATORY TESTS, *SCENED   W68-00475
RATES, *PHOSPHATES, LABORATORY TE   W70-08637
RATES, *TURBULENCE, CHLORRELA, SC   W69-03730
RATES, *WATER TEMPERATURE, SEASON   W68-00472
RATES, ALGAE, FRESH WATER FISH.: /   W68-00253
RATES, ALGAE, OLIGOTROPHY, LIMITI   W70-02779
RATES, ANALYTICAL TECHNIQUES, LAB   W71-07339
RATES, BACTERIA, ULTRAVIOLET RADI   W70-00719
RATES, CHLOROPHYLL, PHOTOSYNTHESI   W69-10167
RATES, CHLORELLA, HYDROGEN ION CO   W68-00248
RATES, CHLOROPHYTA, CYANOPHYTA, C   W70-03983
RATES, CHRONOLOGY, CLADOCERANS, C   W70-05663
RATES, CONN STREAM FLOW, ALGAE, A   W68-00479
RATES, COPPER SULFATE, BOTTOM SED   W69-09349
RATES, DEPTH, DISSOLVED OXYGEN, A   W70-00265
RATES, DESIGN APPLICATIONS, FACUL   W70-07034
RATES, ENVIRONMENTAL EFFECTS, PHO   W69-10158
RATES, FILTERS, ALGAE.: /ANTS, HU   W70-05821
RATES, GRAZING PRESSURE, POPULATI   W69-10158
RATES, HABITATS.: /ROTIFERS, LARV   W72-01819
RATES, LABORATORY TESTS, CULTURES   W70-03507
RATES, MANGANESE, LOW-FLOW AUGMEN   W73-00750
RATES, MICROORGANISMS, ALGAE, FUN   W72-01817
RATES, PESTICIDE RESIDUES, DETERI   W70-03520
RATES, PHOTO REACTIVITY, CELL COL   W70-02364
RATES, PLANKTON, OXYGEN, NITROGEN   W70-04369
RATES, PREDATION.: /VIRONMENT, PH   W69-06536
RATES, PUMPS, ELECTROLYSIS, CULTU   W70-09904
RATES, SEAWATER, INCUBATION, PSEU   W70-00713
RATES, SLUDGE, SLURRIES, AUTOMATI   W72-08688
RATES, SPORES.: /E, WATER POLLUTI   W70-01779
RATES, STABILIZATION PONDS, SEWAG   W68-00855
RATES, STRATIGRAPHY, BIOINDICATOR   W70-05663
RATES, TEMPERATURE, PILOT PLANTS,   W72-06022
RATES, TRITIUM, PHOTOGRAPHY, CARB   W69-10163
RATES, TURBIDITY, CHLORELLA, SCEN   W70-04199
RATES, WATER PROPERTIES.:   W68-00478
RATES, WATER POLLUTION CONTROL.: /   W72-02417
RATIOS.:   W71-05531
RATS, TRYPSIN, SCENEDESMUS QUADRI   W70-04184
RATS, ZIRCONIUM-95-NIOBIUM, EUROP   W69-08269
RAYON PLANT WASTES, *MAN-MADE FIB   W70-06925
RAYON, ZINC COMPOUNDS, CARBON DIS   W70-06925
RAYS, COSMIC RADIATION, S: /N, CE   W69-08272
RAYS, POTASSIUM RADIOISOTOPES, MA   W69-08524
RAYS, RADIOCHEMICAL ANALYSIS, TRI   W69-08269
RAYS, SPECTROMETERS, CYANOPHYTA,   W69-09742
RAYS, URANIUM RADIOISOTOPES, RADI   W69-08272
REACTION, CYTOPHAGA, PROTEUS, SER   W70-03952
REACTIONS, SOLUBILITY EQUILIBRIUM   W70-04385
REACTIONS, ZOOPLANKTON, AN: /AL R   W69-03611
REACTIONS, ELECTROCHEMISTRY, RESI   W70-05267
REACTIONS, CHLORELLA, TRACERS, RA   W70-08097
REACTIONS, *WATER REUSE, DISSOLVE   W70-02294
REACTIVE PHOSPHORUS, PHOSPHATE(BI   W70-05269
REACTIVITY, CELL COLOR, CELL DAMA   W70-02364
REACTIVITY, ALGAL METABOLISM.: /L   W70-02363
REACTOR.:   SAVANNAH RIVER   W69-07845
REACTORS, POWERPLANTS, SHIPS, SUB   W70-00707
REAERATION, NUTRIENTS, OXYGEN SAG   W69-07520
REAERATION, EFFLUENTS, FLOW RATES   W70-05821
REAERATION, SESTON, CHLOROPHYLL,   W70-05092

517

ACTIVITY, TEMPERATURE, SALINITY,
MATION, CHLOROPHYLL DERIVATIVES,
REC/ *WATER QUALITY, *ARTIFICIAL
RECHARGE, *DATA COLLECTIONS, PIT
AE, *EVALUATION, FI/ *ARTIFICIAL
USE, *RECLAMATION, *GROUNDWATER,
EMPERATURE, MANGANESE, CULTURES,
ESIGN FOR ENVIRONMENTAL CONTROL,
SLUDGE, *EFFLUENTS, WASTE WATER
, CILIATA, ASTERIONELLA FORMOSA,
ILAMENTOUS, UNICELLULAR, NATURAL
STUDIES, NUTRIENT REMOVAL, LAKE
, SEDIMENTS, MUD, SANDS, SLUDGE,
HYTA, NITRATES, AMMONIA, CARBON,
L WASTES, NUTRIENTS, PESTICIDES,
ESIUM, COOLING, PROPERTY VALUES,
INE MICROORGANISMS, *POLLUTANTS,
EMICALS, FISHING, SPORT FISHING,
R / *EUTROPHICATION, *MINNESOTA,
RNIA, WATER QUALITY, AESTHETICS,
IONS, *WATER BLOOM, TOXIC ALGAE,
.: *CULTURAL EUTROPHICATION,
MUNICIPAL WASTES, SEPTIC TANKS,
COMPARATIVE STUDIES, CRUCIGENIA
ALGAE, TEMPERATURE EXPERIMENTS,
CENARIA MERCENARIA, AQUACULTURE,
POPULATION EQUIVALENTS(WASTES),
M, *VIETNAM, LITTLE LAKE(HANOI),
ARKANSAS RIVER, MERRIMACK RIVER,
, BACTERIA, MOLLUSKS, VITAMIN B,
FOOD ORGANISMS, FISH PHYSIOLOGY,
ROQUINONE, MADISON(WIS), ELODEA,
HOSPHATES, CYANOPHYTA, OXIDATION-
B12, BIOTIN, HISTIDINE, URACIL,
, PYRUVATE, ACETYLENE REDUCTION,
OXYGEN DEMAND, TOXICITY, OXYGEN-
S, AMMONIUM COMPOUNDS, OXIDATION
ATE AMMONIUM, NITRITE, OXIDATION-
STATS, BATCH CULTURES, ACETYLENE
STRY METHODS, SEASONS, ACETYLENE
HEMICAL OXYGEN DEMAND, OXIDATION-
HEMICAL OXYGEN DEMAND, OXIDATION-
S, ALGAE, GASES, ODOR, OXIDATION-
*ENZYMES, CORROSION, OXIDATION,
NITROGEN-15, PYRUVATE, ACETYLENE
ER VALLEY(NORTH DAKOTA), SULFATE
LFOIL COONTAIL, WILD CELERY, BUR
*TORREY CANYON, SUSCEPTIBILITY,
S, ANIMALS, APPALACHIAN MOUNTAIN
ER COURSE, CARPATHIAN SUBMONTANE
ENOIDS, CARBON FIXATION, PELAGIC
TEMENT, ABATEMENT, PACIFIC COAST
OGEN, CHLOROPHYLL, PRODUCTIVITY,
SYNTHESIS, TEMPERATURE, TROPICAL
S, *CHEMICAL CHARACTERIS/ *RIVER
ST REGION, REGIONS, GEOGRAPHICAL
AST U. S., REGIONS, GEOGRAPHICAL
Y, WATER CHEMISTRY, GEOGRAPHICAL
Y, WATER CHEMISTRY, GEOGRAPHICAL
, SALTS, TIDES, SOUTHEAST U. S.,
ABATEMENT, PACIFIC COAST REGION,
ST U. S., REGIONS, PACIFIC COAST
NOFF RELATIONSHIPS, GEOGRAPHICAL
IFIC COAST REGIONS, GEOGRAPHICAL
MISTRY, PLANTS, SOUTHWEST U. S.,
BENTHIC FLORA, PHAEOPHYTA, POLAR
MONAS, PHOSPHORUS RADIOISOTOPES,
LTH, WATER POLLUTION, CONTRACTS,
NTIC OCEAN, LANDFILLS, DISPOSAL,
TY, FLOODWATER, REGULATION, WELL
NITRATES, TURBIDITY, FLOODWATER,
GANESE, MOLYBDENUM, PLANT GROWTH
ATOMS, RESISTANCE, *PLANT GROWTH
STANCES, LIMNOLOGY, PLANT GROWTH
NTACT, SPARGANIUM, SALIX, GROWTH
UTRIENT ENERGY PATHWAYS, TROPHIC
SYNTHETIC SEWAGE FEED, SYMBIOTIC
N JOAQUIN ESTUARY, ENVIRONMENTAL
NISM, NUTRIENTS, RAINFALL-RUNOFF
GAE, SEDIMENTS, SOIL-WATER-PLANT
ESTUARY, NUTRIENT-PHYTOPLANKTON
QUES, MICHAELIS-MENTEN EQUATION,
ATIONS, HETEROTROPHIC POTENTIAL,
*NUTRIENT ELEMENTS, PHOSPHORUS
*MEASUREMENT, *FLUOROMETRY, DYE
OGRAPHY, INFRARED RADIATION, DYE
A COLLECTION, CHEMICAL ANALYSIS,
LLUTION, LAKES, STREAMS, SEWAGE,
, DYE RELEASES, CURRENTS(WATER),
OXICITY, LAKES, DESIGN CRITERIA,

REAERATION, FISH, TASTE, ODOR, OX   W73-00750
RECENT SEDIMENTS, PRECAMBRIAN SED   W69-08284
RECHARGE, *DATA COLLECTIONS, PIT   W69-05894
RECHARGE, WATER SUPPLY, MUNICIPAL   W69-05894
RECHARGE, COPPER, *ALGICIDES, ALG   W72-05143
RECHARGE, *ION EXCHANGE, *ORGANIC   W72-04318
RECIRCULATED WATER, NITRATES, PHO   W70-07257
RECIRCULATION, BOD REMOVAL, GAS P   W70-04787
RECLAMATION, WASTE WATER TREATMEN   W70-06053
RECOLONIZATION, ALSTER LAKE(GERMA   W70-00268
RECOVERY.: /HYTIC, PLATE COUNT, F   W68-00891
RECOVERY, WISCONSIN, LAKE MONONA   W69-09349
RECREATION.: /, OXYGEN, COLIFORMS   W70-10161
RECREATION, WATER POLLUTI: /LOROP   W72-04761
RECREATION, ALGAE, OXYGEN DEMAND,   W71-09577
RECREATION DEMAND, IR: /RON, MAGN   W70-02251
RECREATION, WATER SPORTS, ENVIRON   W69-04276
RECREATION, SILT, NUTRIENTS, AQUA   W68-00172
RECREATION, SEWAGE DISPOSAL, WATE   W70-04810
RECREATION, FISHING, ALGAE, NUTRI   W70-05645
RECREATIONAL USE.:   *DEFINIT   W69-08279
RECREATIONAL USE, WASTE EFFLUENTS   W69-06535
RECREATION WASTES, WATER QUALITY.   W70-04266
RECTANGULARIS, DICERAS CHODATI, D   W69-07833
RECYCLED WATER.: /GAE, BLUE-GREEN   W70-07257
RECYCLING WASTES.: /M WASTES, MER   W70-09905
RED RIVER VALLEY(NORTH DAKOTA), S   W70-03312
RED RIVER(VIETNAM), CHLOROCOCCUS,   W70-03969
RED RIVER, SAVANNAH RIVER, SNAKE   W70-04507
RED TIDE, SALINITY, FLORIDA, TEXA   W70-05372
RED TIDE, ORGANIC COMPOUNDS, DINO   W71-07731
REDOX COMP: /L EXPOSURE, PHENANTH   W70-02982
REDU: /CONCENTRATION, NITRATES, P   W70-05404
REDUCED SULFUR COMPOUNDS, ANTIBIO   W70-02510
REDUCTANT.: /TP, CMU, NITROGEN-15   W70-04249
REDUCTION POTENTIAL, SOLAR RADIAT   W70-05750
REDUCTION POTENTIAL, DISSOLVED OX   W70-05760
REDUCTION POTENTIAL, CYCLING NUTR   W69-04521
REDUCTION TECHNIQUE, HETEROTROPHY   W70-02775
REDUCTION TEST.: /HODS, *BIOCHEMI   W70-03429
REDUCTION POTENTIAL, AQUATIC ALGA   W71-09546
REDUCTION POTENTIAL, DISSOLVED OX   W71-11377
REDUCTION POTENTIAL, SULFUR BACTE   W70-07034
REDUCTION(CHEMICAL), HYDROLYSIS,   W69-07428
REDUCTION, REDUCTANT.: /TP, CMU,   W70-04249
REDUCTION, POTATO PROCESSING WAST   W70-03312
REED, DUCKWEED, WHITE WATER LILY,   W70-05568
REGENERATION, CLIFF VEGETATION, S   W70-01777
REGION.: /LANKTON, AQUATIC ANIMAL   W69-03695
REGION, GRADIENTS(RIVER), THALLI,   W70-02784
REGION, MONOMICTIC, DEPTH EFFECTS   W70-01933
REGION, REGIONS, GEOGRAPHICAL REG   W69-03683
REGIONAL ANALYSIS, RESERVOIRS.: /   W70-07261
REGIONS.: / ORGANIC MATTER, PHOTO   W69-06865
REGIONS.: /PHYSICAL CHARACTERISTIC   W73-00748
REGIONS, BAYS, BODIES OF WATER.: /   W69-03683
REGIONS, COASTAL PLAINS, ATLANTIC   W69-03695
REGIONS, CYCLING NUTRIENTS, ZOOPL   W70-03975
REGIONS, CYCLING NUTRIENTS, ZOOPL   W70-03975
REGIONS, GEOGRAPHICAL REGIONS, CO   W69-03695
REGIONS, GEOGRAPHICAL REGIONS, BA   W69-03683
REGIONS, GEOGRAPHICAL REGIONS, PH   W69-03611
REGIONS, KENTUCKY, ACID STREAMS,   W68-00891
REGIONS, PHOTOSYNTHESIS, CHEMICAL   W69-03611
REGIONS, PACIFIC COAST REGIONS, G   W69-03611
REGIONS, SPECTROPHOTOMETRY, CHROM   W70-01074
REGRESSION ANALYSIS, ALGAE, PONDS   W70-03957
REGULATION, RIGHT-OF-WAY, ESTIMAT   W70-03344
REGULATION.: /IC LIFE, MILK, ATLA   W72-00941
REGULATIONS.: / NITRATES, TURBIDI   W73-00758
REGULATION, WELL REGULATIONS.: /   W73-00758
REGULATORS, PREDATION, PRODUCTIVI   W69-04801
REGULATORS, FRESH WATER LAKES, LA   W68-00856
REGULATORS, CYANOPHYTA, ENVIRONME   W68-00470
REGULATORS, MYRIOPHYLLUM, WATER B   W70-05263
RELATIONSHIPS.: /R LAKE(CALIF), N   W70-05428
RELATIONSHIPS.: /LA PYRENOIDOSA,   W70-05655
RELATIONSHIPS, FLUOROMETRIC CHLOR   W70-05094
RELATIONSHIPS, GEOGRAPHICAL REGIO   W68-00891
RELATIONSHIPS, NUCLEAR WASTES, AB   W71-09005
RELATIONSHIPS.:   *POTOMAC RIVER   W71-05407
RELATIVE HETEROTROPHIC POTENTIAL,   W70-05109
RELATIVE HETEROTROPHIC POTENTIAL,   W70-04194
RELEASE, SAMPLE PRESERVATION, WAT   W70-04382
RELEASES, CURRENTS(WATER), REMOTE   W70-05377
RELEASES, BEACHES, SANDS, BIOTA,   W68-00010
RELIABILITY, WATER POLLUTION, WAT   W70-04382
REMEDIES, JUDICIAL DECISIONS, LEG   W69-06909
REMOTE SENSING, SEA WATER, CHLORO   W70-05377
REMOTE SENSING, MINNESOTA, SEDIME   W70-02775

519

DINE, NARRAGANSETT BAY(RI), CAPE   REYKJANES(ICELAND), SUDURNES(ICEL   W69-09755
ND SOUND, NARRAGANSETT BAY, CAPE   REYKJANES, FAXAFLOI FJORD, ICELAN   W69-10161
PLE RUNS TEST, CLUSTER ANALYSIS,   REYKJAVIK(ICELAND).: /TS, ONE-SAM   W69-09755
UE, MINERALIZATION, CLAY LENSES,   RHINE RIVER, AMSTERDAM RHINE, POL   W69-07838
OBLIQUUS, GROWTH CONSTATS, RIVER   RHINE.: /TERIONELLA, SCENEDESMUS   W70-03955
TER-BLOOMS, LAKE GEORGE(UGANDA),   RHINE-MEEUSE DELTA(NETHERLANDS),:   W70-04369
Y LENSES, RHINE RIVER, AMSTERDAM   RHINE, POLDERS, COMPOSITION, PSEU   W69-07838
LORES RIVER, URAVAN, COTTUS SPP,   RHINICHTHYS OSCULUS, SUCKERS, ICT   W69-07846
E, DAYTON(OHIO), VANDALIA(OHIO),   RHIZOPODS, SPIDERS, MILK WASTE TR   W72-01819
*FRAUNHOFER LINE DISCRIMINATOR,   RHODAMINE WT DYE, AIRBORNE TESTS,   W70-05377
GTON, WAVES(WATER), CONNECTICUT,   RHODE ISLAND, TEMPERATURE, NUTRIE   W69-10161
, NARRAGANSETT MARINE LABORATORY(RHODE ISLAND), DILUTION RATE, FRU   W70-01073
, *WATER TEMPERATURE, *SEASONAL,   RHODE ISLAND, SAMPLING, ECOLOGY,   W70-00713
ETERETROPHY, B/ NARRAGANSETT BAY(RHODE ISLAND), TAXONOMIC TYPES, H   W70-00713
M-95, NIOBIUM-95, RUTHENIUM-106,   RHODIUM-106, CERIUM-144, BETA RAD   W69-09742
US, MUSCLE, BONE, LEAVES, SHELL,   RHODIUM-102.: /A, PISONIA, PANDAN   W69-08269
PRE-BOMB MATERIAL, RUTHENIUM-106-   RHODIUM-106, RUTHENIUM-103-RHODIU   W69-08524
M-106-RHODIUM-106, RUTHENIUM-103-   RHODIUM-106, CERIUM-144-PRASEODYM   W69-08524
R(WASH), ANTITHAMNION SARNIENSE,   RHODOCHORTON, BANGIA FUSCO-PURPUR   W69-10161
, MESOTHERMY, MELOSIRA, SYNEDRA,   RHODOMON: /ACTERIA, DESULFOVIBRIO   W70-05760
UME METHOD, CYCLOTELLA COMENSIS,   RHODOMONAS LACUSTR: /LCULATED VOL   W70-03959
, SEWAGE EFFLUENTS, CHLOROPHYTA,   RHODOPHYTA, PHYSIOLOGICAL ECOLOGY   W70-04510
OPHYTA, CYANOPHYTA, CHRYSOPHYTA,   RHODOPHYTA, EUGLENOPHYTA, SLIME,   W70-05389
OPERTIES, PHYSIOLOGICAL ECOLOGY,   RHODOPHYTA, CHLOROPHYTA.: /CAL PR   W69-10161
, BACILLARIOPHYCEAE, CHAROPHYTA,   RHODOPHYTA.: /, CYANOPHYTA, FUNGI   W70-02784
SYNTHESIS, ENZYMES, CHLOROPHYTA,   RHODOPHYTA, SCENEDESMUS, CHLAMYDO   W70-02965
CUS, ACHROMOBACTER(ALKALIGENES),   RHODOTORULA, GROWTH FACTORS, COLO   W70-03952
TAE, DAPHNIA MAGNA, SELENASTRUM,   RHODOTORULA, TEMPERATURE COEFFICI   W70-03957
, MYCOTA, DEBARYOMYCES HANSENII,   RHODOTORULA RUBRA, CANDIDA DIDDEN   W70-04368
SERRATIA, BAY OF FUNDY(CANADA),   RHYZOIDS.: /, CYTOPHAGA, PROTEUS,   W70-03952
ATED THYMIDINE, NARRAGANSETT BAY(RI), CAPE REYKJANES(ICELAND), SUD   W69-09755
, *MACROMOLECULES, NUCLEIC ACID,   RIBONUCLEIC ACID SYNTHESIS.: /ION   W69-10151
ND), ENZYME-SUBSTRATE COMPLEXES,   RIBONUCLEIC ACID, ORCINOL METHOD.   W69-10160
E(ONTARIO), CLEAR LAKE(ONTARIO),   RICE LAKE(ONTARIO).: /UCKHORN LAK   W70-02795
EALTH, WALLEYE, VEGETATION, WILD   RICE, SILTS, PLANKTON, MAYFLIES,   W70-01943
LUS SUBTILIS, BACILLUS STEAROTH/   RICKETTSIAE, STERILIZATION, BACIL   W71-11823
FLABELLUM, ZINC, NICKEL, PUERTO   RICO, LA PARGUERA, JOYUDA.: /OTEA   W69-08525
ANGANESE, ATLANTIC OCEAN, PUERTO   RICO, SPECTROSCOPY, DISTRIBUTION,   W69-08525
BACK RIVE/ COMMUNITY STRUCTURE,   RIDLEY CREEK(PA), NOBS CREEK(MD),   W70-04510
BISMUTH-214, ACTINIUM-228, SWIFT   RIFFLES, ZIRCONIUM-95, NIOBIUM-95   W69-09742
OLLUTION, CONTRACTS, REGULATION,   RIGHT-OF-WAY, ESTIMATED COSTS, LE   W70-03344
SULFATE, FERRICHLORIDE, IRON S/   RING TUBE, CANALIZATION, ALUMINUM   W70-05668
RMANY), GERMANY, FINLAND, SEWAGE   RING TUBES.: /NY), LAKE TEGERN(GE   W70-05761
ER, SAVANNAH RIVER, SNAKE RIVER,   RIO GRANDE RIVER, POTOMAC RIVER.:   W70-04507
TH SERVICE, GREAT LAKES-ILLINOIS   RIVER BASINS PROJECT, INDIANA HAR   W70-00263
FEEDLOTS, WABASH   RIVER BASIN.:   W71-10376
*OHIO   RIVER BASIN.:   W72-01788
POTOMAC   RIVER BASIN.:   W70-09557
, *GROUNDWATER, DDT, MISSISSIPPI   RIVER BASIN, URBANIZATION, ALGAL   W73-00758
N/ *SKAWA RIVER(POLAND), MONTANE   RIVER COURSE, CARPATHIAN SUBMONTA   W70-02784
*INDIA, MOOSI   RIVER ECOLOGICAL STUDY.:   W69-07096
KTON RELATIONSHIPS.:   *POTOMAC   RIVER ESTUARY, NUTRIENT-PHYTOPLAN   W71-05407
*POTOMAC   RIVER ESTUARY.:   W71-05409
, *HYDROCARBONS, *OLEFINS, GREEN   RIVER FORMATION, CHLOROPHYLL DERI   W69-08284
A), BOMBAY(IND/ *INDIA, *BLOOMS,   RIVER GANGES(INDIA), RAJGHAT(INDI   W70-05404
ANUM, REACTOR.:   SAVANNAH   RIVER LABORATORY, PLECTONEMA BORY   W69-07845
LAKE ZURICH, SWITZERLAND, NITELV'   RIVER NORWAY, NORWA: /ET NORWAY,   W69-07833
ESMUS OBLIQUUS, GROWTH CONSTATS,   RIVER RHINE.: /TERIONELLA, SCENED   W70-03955
ULATION EQUIVALENTS(WASTES), RED   RIVER VALLEY(NORTH DAKOTA), SULFA   W70-03312
SHENANDOAH   RIVER.:   W70-03068
RO(DELAWARE), SWAN CREEK, INDIAN   RIVER.:   MILLSBO   W71-05013
SIN RIVER, DETROIT RIVER, MAUMEE   RIVER.: / RIVER, HURON RIVER, RAI   W69-01445
RIVER, RIO GRANDE RIVER, POTOMAC   RIVER.: /, SAVANNAH RIVER, SNAKE   W70-04507
TASTE, ODOR, OXYGEN SAG, POTOMAC   RIVER.: /NITY, REAERATION, FISH,   W73-00750
UOLUMNE RIVER(CALIF), STANISLAUS   RIVER(CALIF), WATER POLLUTION ASS   W70-02777
SAN JOAQUIN RIVER(CALIF), MERCED   RIVER(CALIF), TUOLUMNE RIVER(CALI   W70-02777
R(CALIF), CLEAR LAKE(CALIF), EEL   RIVER(CALIF), JOINT INDUSTRY/GOVE   W70-02777
), MERCED RIVER(CALIF), TUOLUMNE   RIVER(CALIF), STANISLAUS RIVER(CA   W70-02777
, TUOLUMNE RIVER(CA/ SAN JOAQUIN   RIVER(CALIF), MERCED RIVER(CALIF)   W70-02777
R POLLUTION ASSESSMENT, AMERICAN   RIVER(CALIF), CLEAR LAKE(CALIF),   W70-02777
ORGANISMS, TURSIOPS TRUN/ DUPLIN   RIVER(GA), SPARTINA ALTEMIFLORA,   W69-08274
ROWTH, STREAM CONCOURSE, PALOUSE   RIVER(IDAHO).: /CHARCOAL, ALGAL G   W70-03501
(ILL), COOK COUNTY(ILL), CALUMET   RIVER(ILL), POOLS, WIND DIRECTION   W70-00263
*METABOLITES, *CEDAR   RIVER(IOWA).:   W72-13605
IONELLA, TYPHOID FEVER, KENNEBEC   RIVER(MAINE), COBBOSSEECONTEE LAK   W70-08096
CREEK(PA), NOBS CREEK(MD), BACK   RIVER(MD), OSCILLATORIA, CATHERWO   W70-04510
ICROCYSTIS AERUGINOSA, BURNTSIDE   RIVER(MINN).: /LLA PYRENOIDOSA, M   W70-03512
TCHER, NJOER LAKE(NORWAY), HOBOL   RIVER(NORWAY), VAN LAKE(NORWAY).:   W70-05752
OLAR WATERS, TYNSET(NORWAY), NID   RIVER(NORWAY), GAULA RIVER(NORWAY   W70-01074
ORWAY), NID RIVER(NORWAY), GAULA   RIVER(NORWAY), FLAKK(NORWAY), RAN   W70-01074
GREAT MIAMI   RIVER(OHIO).:   W70-00475
AK/ *BIOLOGICAL PROBLEMS, SCIOTO   RIVER(OHIO), LAKE MENDOTA(WIS), L   W72-01813
ON, FOSSIL, CIMARRON R/ ARKANSAS   RIVER(OKLA), TULSA(OKLA), POLLUTI   W70-04001
LA), POLLUTION, FOSSIL, CIMARRON   RIVER(OKLA), BIXBY(OKLA), KEYSTON   W70-04001
HANIZOMENON FLOS AQUAE, METOLIUS   RIVER(ORE), COMMUNITY STRUCTURE,   W70-03978
SE, CARPATHIAN SUBMONTAN/ *SKAWA   RIVER(POLAND), MONTANE RIVER COUR   W70-02784
SA, BIOLOGICAL HALF-LIFE, CLINCH   RIVER(TENNESSEE), PHYSICAL HALF-L   W70-02786
HEROKEE RESERVOIR(TENN), HOLSTON   RIVER(TENN), POTOMOGETON PECTINAT   W71-07698
IS RESERVOIR(TENN), FRENCH BROAD   RIVER(TENN), CHICKAMAUGA RESERVOI   W71-07698
AUSTIN(TEX), COLORADO   RIVER(TEX).:   W70-04727

PISCICULTURAL PONDS, SVISLOCH'
VIETNAM, LITTLE LAKE(HANOI), RED
SCENS, MADISON LAKES(WIS), CEDAR
UAE, OXYGEN DEPLETION.: *YAHARA
OTROPHY, MICRONUTRIENTS, BADFISH
FFLUENTS, *MADISON LAKES, YAHARA
UNIVERSITY OF WISCONSIN, YAHARA
IAN SUBMONTANE REGION, GRADIENTS(
OINDICATORS, *CHLOROPHYLL, *OHIO
*ALGAE, *SAMPLING, *MISSISSIPPI
WASTES, AGRICULTURE, ST LAWRENCE
LOW-FLOW AUGMENTATION, TENNESSEE
NERALIZATION, CLAY LENSES, RHINE
CYSTIS AERUGINOSA, BOHEMIA, ELBE
ISOTOPES, TRO/ *BIOTA, *COLUMBIA
AGE EFFLUENTS, DIATOMS, MISSOURI
IATOMS, MISSOURI RIVER, COLUMBIA
NTERITIS, SYNURA, ANABAENA, IOWA
O RIVER, TENNESSEE RIVER, HUDSON
ROUGE RIVER, HURON RIVER, RAISIN
NIMAS RIVER, DURANGO, SAN MIGUEL
DOLORES RIVER, URAVAN, / ANIMAS
ADIOISOTOPES, DETRITUS, COLUMBIA
IGAN, CHEBOYGAN COUNTY, STURGEON
RACTION, MULTIPLE-USES, SAVANNAH
RIVER, COLORADO RIVER, TENNESSEE
DETROIT R/ LAKE ST CLAIR, ROUGE
                              PLATTE
RON RIVER, RAISIN RIVER, DETROIT
S, AGRICULTURAL RUNOFF, ARKANSAS
HORIUM-232, YUKON RIVER, NEBESNA
L A, DATA INTERPRETATION, *RHONE
S, EQUISETUM, THORIUM-232, YUKON
PLING, *MISSISSIPPI RIVER, *OHIO
GREEN ALGAE, PHYTOPLANKTON, OHIO
UM, BLE/ *ALGAL GROWTH, *POTOMAC
CATHERWOOD DIATOMETER, SAVANNAH
H RIVER, SNAKE RIVER, RIO GRANDE
D OXYGEN VARIATIONS, SAN JOAQUIN
AKE ST CLAIR, ROUGE RIVER, HURON
UNOFF, ARKANSAS RIVER, MERRIMACK
RED RIVER, SAVANNAH RIVER, SNAKE
NSAS RIVER, MERRIMACK RIVER, RED
*PROTOZOAN COMMUNITIES, POTOMAC
UNITIES, POTOMAC RIVER, SAVANNAH
IMACK RIVER, RED RIVER, SAVANNAH
OTIFERS, *DISTRIBUTION, MISSOURI
RIVER, COLUMBIA RIVER, COLORADO
T TEST/ HIGHLAND LAKES, COLORADO
RONTIUM RADIOISOTOPES, TENNESSEE
RANGO, SAN MIGUEL RIVER, DOLORES
ON RES/ WOOD, ALGAL COUNT, GREEN
EE RIVER, HUDSON RIVER, DELAWARE
PSA FLORIDANA, ENTEROCOCCI, PARK
ESH WATER, *HUMIC ACIDS, *COLOR,
ECOLOGY, *AQUATIC ENVIRONMENTS,
OLOGY, SALINITY, ORGANIC MATTER,
TRACE ELEMENTS, PLANKTON, LAKES,
YSIS, INSECTS, SNAILS, TOXICITY,
QUATIC HABITATS, *ALASKA, LAKES,
UTROPHICATION, *ALGAE, PLANNING,
TS, TEST PROCEDURES, NUTRIENTS,
HEAT, WATER POLLUTION, STREAMS,
TOXINS, DISSOLVED OXYGEN, LAKES,
SOTOPES, *URANIUM RADIOISOTOPES,
HLOROPHYLL, FLUORESCENCE, LAKES,
EASONAL, LAKES, SAMPLING, PONDS,
PPLY, EFFLUENTS, STREAMS, LAKES,
NCE ALGAE, PHOSPHATES, NITRATES,
DIATOMS, MARYLAND, PENNSYLVANIA,
NT GROWTH, PRIMARY PRODUCTIVITY,
DETERGENTS, RUNOFF, PERCOLATION,
UPERSATURATION, VERTICAL MIXING,
, NUCLEAR ACTIVATION TECHNIQUES,
CARBON METHOD, FINGER LAKES(NY),
PLEMORPHIC, MYCELOID, CORYNEFORM
RENOIDOSA, NITROGEN UTILIZATION,
TROPHICATION, FISH FARMING, FISH
NTS, RUNOFF, PERCOLATION, ROADS,
TURE, DISSOLVED OXYGEN, DENSITY,
*PILOT PLANTS, *NUISANCE ALGAE,
SONAL, *ALGAL CONTROL, FISHKILL,
, FUNGI, ALGAE, FLOATING PLANTS,
SCONSIN, CHEMCONTROL, PONDWEEDS,
, OSCILLATORIA, AUSTRALIA, SANTA
L, ELBE(GERMANY), SPOROBOLOMYCES
INTENSITY, DAPHNIA, CONNECTICUT,
E, SULFUR COMPOUNDS, MOLYBDENUM,
IOCONTROL, TURBULENCE, COPEPODS,
ZOA, PLANKTON, WORMS, NEMATODES,

RIVER(USSR), MINSK(USSR), ENGLAND   W70-05092
RIVER(VIETNAM), CHLOROCOCCUS, SPI   W70-03969
RIVER(WASH), ZURICHSEE(SWITZ), TU   W70-00270
RIVER(WIS), APHANIZOMENON FLOS AQ   W72-01797
RIVER(WIS), DETROIT LAKES(MINN),   W70-04506
RIVER(WIS), STORKWEATHER CREEK(WI   W70-05113
RIVER), NUTRIENT REMOVAL.: /,   W69-06276
RIVER), THALLI, SCHIZOMYCETES, CY   W70-02784
RIVER, *ANALYTICAL TECHNIQUES, BI   W71-04206
RIVER, *OHIO RIVER, NUISANCE ALGA   W70-04507
RIVER, ALGAE, COLIFORMS, DISSOLVE   W70-03964
RIVER, ALGAE, IRON, ALKALINITY, H   W73-00750
RIVER, AMSTERDAM RHINE, POLDERS,   W69-07838
RIVER, CELAKOVICE(CZECHOSLOVAKIA)   W70-00274
RIVER, CHROMIUM, PHOSPHORUS RADIO   W69-07853
RIVER, COLUMBIA RIVER, COLORADO R   W70-04507
RIVER, COLORADO RIVER, TENNESSEE   W70-04507
RIVER, CYCLOTELLA, FRAGILARIA, SY   W70-08107
RIVER, DELAWARE RIVER, WATER POLL   W70-04507
RIVER, DETROIT RIVER, MAUMEE RIVE   W69-01445
RIVER, DOLORES RIVER, URAVAN, COT   W69-07846
RIVER, DURANGO, SAN MIGUEL RIVER,   W69-07846
RIVER, ELECTRONICS, SCALING.: / R   W69-07862
RIVER, FLUORESCEIN, MACROPHYTES,   W69-09334
RIVER, GULF COAST.: / COLLOIDAL F   W73-00748
RIVER, HUDSON RIVER, DELAWARE RIV   W70-04507
RIVER, HURON RIVER, RAISIN RIVER,   W69-01445
RIVER, LINCOLN(NEBR).:   W69-05894
RIVER, MAUMEE RIVER.: / RIVER, HU   W69-01445
RIVER, MERRIMACK RIVER, RED RIVER   W70-04507
RIVER, MOOSE CREEK, ARCTIC CIRCLE   W69-08272
RIVER, NAVICULA, MELOSIRA, THALAS   W73-01446
RIVER, NEBESNA RIVER, MOOSE CREEK   W69-08272
RIVER, NUISANCE ALGAE, WATER QUAL   W70-04507
RIVER, PENNSYLVANIA, WATER SUPPLY   W69-05697
RIVER, PLECTONEMA, NOSTOC PALUDOS   W70-02255
RIVER, POPULATION DYNAMICS, ALGAL   W70-04510
RIVER, POTOMAC RIVER.: /, SAVANNA   W70-04507
RIVER, POTOMAC ESTUARY, SAN FRANC   W69-07520
RIVER, RAISIN RIVER, DETROIT RIVE   W69-01445
RIVER, RED RIVER, SAVANNAH RIVER,   W70-04507
RIVER, RIO GRANDE RIVER, POTOMAC   W70-04507
RIVER, SAVANNAH RIVER, SNAKE RIVE   W70-04507
RIVER, SAVANNAH RIVER, SHOCK EFFE   W70-07847
RIVER, SHOCK EFFECTS.: /ZOAN COMM   W70-07847
RIVER, SNAKE RIVER, RIO GRANDE RI   W70-04507
RIVER, SOUTH DAKOTA, NEBRASKA, AL   W70-05375
RIVER, TENNESSEE RIVER, HUDSON RI   W70-04507
RIVER, TEXAS, *NUTRIENT ENRICHMEN   W69-06004
RIVER, TROPHIC LEVEL, ALGAE, CADD   W70-02786
RIVER, URAVAN, COTTUS SPP, RHINIC   W69-07846
RIVER, WASHINGTON, HOWARD A. HANS   W72-11906
RIVER, WATER POLLUTION EFFECTS.: /   W70-04507
RIVER: /HROMATIUM VINOSUM, THIOCA   W70-03312
RIVERS, BOGS, PHENOLS, PIGMENTS,   W70-01074
RIVERS, ENVIRONMENTAL EFFECTS, PE   W70-03978
RIVERS, ESTUARINE ENVIRONMENT, WA   W73-01446
RIVERS, GREAT LAKES.: /OSPHORUS,   W70-04385
RIVERS, HISTOGRAMS.: / WATER ANAL   W72-01801
RIVERS, HOT SPRINGS, TERRESTRIAL   W69-08272
RIVERS, LAKES, LIMNOLOGY, WISCONS   W70-04506
RIVERS, LAKES, *GROWTH RATES, *WA   W68-00472
RIVERS, LAKES, PHYSIOLOGICAL ECOL   W71-02479
RIVERS, LIGHT PENETRATION, FLUORE   W69-05697
RIVERS, MILLS, COLORADO, WASTES,   W69-07846
RIVERS, NITROGEN, BIOMASS, SEWAGE   W70-02777
RIVERS, RESERVOIRS, AIR TEMPERATU   W70-02646
RIVERS, RESERVOIRS, BOATING, EUTH   W71-09880
RIVERS, SEWAGE TREATMENT, WATER P   W69-01453
RIVERS, STREAMS, EUGLENA, AQUATIC   W70-04510
RIVERS, TASTE, THERMAL STRATIFICA   W69-06535
ROADS, ROOFS, PRECIPITATION(ATMOS   W69-10178
ROANOKE RAPIDS RESERVOIR(NC).: /S   W70-00683
ROCK LAKE(WASH), WILLIAMS LAKE(WA   W70-00264
RODHE'S SOLUTION VIII, NUTRIENT L   W70-01579
RODS, COCCI,: /E, PSYCHROPHILES,   W70-00713
ROLLER DRYING, SPRAY-DRIED, GLASS   W70-04184
ROOD ORGANISMS, FISH GENETICS, FI   W69-04804
ROOFS, PRECIPITATION(ATMOSPHERIC)   W69-10178
ROOTED AQUATIC PLANTS, ALGAE, CHL   W71-07698
ROOTED AQUATIC PLANTS, *RESERVOIR   W68-00479
ROOTED AQUATIC PLANTS, GLACIAL LA   W68-00488
ROOTED AQUATIC PLANTS, SELF-PURIF   W70-05750
ROOTED AQUATIC PLANTS, RATES OF A   W70-05568
ROSA(CALIF), LYUBLIN(USSR), LYUBE   W70-05092
ROSEUS, AUTOLYSIS, ASCOSPOROGENOU   W70-04368
ROTIFERS.: /, TEMPERATURE, LIGHT   W70-03959
ROTIFERS, ECOSYSTEMS, PHOTOS: /IT   W69-07428
ROTIFERS, CHLOROPHYLL, PIGMENTS,   W71-10098
ROTIFERS, LARVAE, SNAILS, MITES,   W72-01819

TOMS, CYANOPHYTA, PHYTOPLANKTON,
MS, ANAEROBIC CONDITIONS, ALGAE,
S, ANIMALS, FOOD CHAINS, OXYGEN
ER QUALITY CONTROL, CRUSTACEANS,
  CURRENT, YTTRIUM-90, KATODINIUM
RIVER, DETROIT R/ LAKE ST CLAIR,
  POPULATIONS, PLANT POPULATIONS,
TRICHIA ECHINULATE, OSCILLATORIA
MACROFAUNA, GYTTJA, OSCILLATORIA
CHALK, METALIMNION, OSCILLATORIA
XCLUSION PRINCIPLE, OSCILLATORIA
AN DEPTH, DILUTION, OSCILLATORIA
E WASHINGTON(WASH), OSCILLATORIA
EAR SHIP SAVANNAH, ZIRCONIUM-95,
BARYOMYCES HANSENII, RHODOTORULA
SIS, SLUDGE, INCINERATION, STORM
IC INDEX, CHLOROPHYLL A, SHAYLER
, FISHERIES, CARP, CLAMS, PONDS,
NT ORGANISM, NUTRIENTS, RAINFALL-
                    URBAN
LUENTS, *SEWAGE TREATMENT, STORM
CONTROL, OXYGEN BALANCE, SEWAGE,
  PLANKTON ANALYSES, AGRICULTURAL
TE WATER TREATMENT, PHOSPHOROUS,
TILIZERS, WATER POLLUTION, SNOW,
S, SILT, EROSION, SEDIMENTATION,
LAKES, *RIVERS, *EUTROPHICATION,
OSITY, IRRIGATION, GROUND WATER,
ENSITY, WATER POLLUTION SOURCES,
NCES, TRANSPARENCY, AGRICULTURAL
ITROGEN, PHOSPHORUS, DETERGENTS,
WATER POLLUTION SOURCES, SURFACE
ES, INORGANIC COMPOUNDS, SURFACE
, BACTERIA, ALGAE, FISH, SEWERS,
NON-PARAMETRIC TESTS, ONE-SAMPLE
ENY, CLADOPHORA, ALGAL GENETICS,
IC PIGMENTS, PRODUCTION BIOLOGY,
LVED OXYGEN, PITTING(CORROSION),
FFLES, ZIRCONIUM-95, NIOBIUM-95,
UCLIDE), FILTER-FEEDING ANIMALS,
5-NIOBIUM-95, PRE-BOMB MATERIAL,
RIAL, RUTHENIUM-106-RHODIUM-106,
  *ENIWETOK ATOLL, *BIKINI ATOLL,
TES, *RADIONUCLIDE ACCUMULATION,
THES, ACTINOPTYCHUS, AMPHIPLEURA
E(MINN), GRAY'S BAY(MINN), SMITH'
INN), LAKE JOSEPHINE(MINN), GRAY'
, GREEN / *LEWIS AND CLARK LAKE (
QUANTITIES, ALGAL BLOOMS, LIEBIG'
BIG'S LAW OF THE MINIMUM, VOISIN'
AQUAE, MINIMUM LETHAL DOSES, CHU'
A, LIMNAEA STAGNALIS, OUSPENSKII'
A, LIMNAEA STAGNALIS, OUSPENSKII'
OLATILE ACIDS, KETOACIDS, TAMIYA'
A OVALIS, OPEPHORA MARTYI, ABBOT'
CHIA, JANESVILLE(WIS), SPAULDING'
ION, NUTRIENT REMOVAL, SPAULDING'
ANUS, CLADOCERA, WATER MASSES, U
  METHOD, FINGER LAKES(NY), RODHE'
S, *DIFFUSERS, *RAYTOWN(MO), LEE'
LIS, NEPHROCYTIUM OBESUM W AND G
ON CONTROL, PACIFIC NORTHWEST U.
C LEVELS, NUTRIENTS, NORTHEAST U.
  CHEMISTRY, PLANTS, SOUTHWEST U.
LGAE, SALTS, TIDES, SOUTHEAST U.
GAL TOXINS MEMBRANE, NORTHEAST U.
  QUALITY, NUTRIENTS, NORTHEAST U
TILIS, ASPERGILLUS ORYZAE NIGER,
ON, BETA RAYS, COSMIC RADIATION,
LOSUS, GROWTH FACTORS, LAMINARIA
CHLORELLA PYRENOIDOSA, CHLORELLA
IFORNIA WATER PLAN, DELTA WATER/
ILITIES(CALIF), OXYGEN DYNAMICS,
N, REAERATION, NUTRIENTS, OXYGEN
ATION, FISH, TASTE, ODOR, OXYGEN
ICATION, FISHKILL, ALGAE, OXYGEN
IDINIUM POLONICUM, PLANKTO/ LAKE
OCYSTIS, POTAMOGETON, POLYGONUM,
S, CALCIUM COMPOUNDS, CHLORIDES,
FFECTS, *SAMPLING, ALGAE TOXINS,
NOPHYTA, CHRYSOPHYTA, BIOASSAYS,
E ENVIRONMENT, PONDS, BREEDINGS,
HLOROPHYLL, TEMPERATURE, OXYGEN,
RBIDITY, *RESERVOIR EVAPORATION,
  *ALGAE, SEA WATER, TEMPERATURE,
NIA, NITRITES, HYDROGEN SULFIDE,
PHOTOSYNTHESIS, SOLAR RADIATION,
NCE, CARBOHYDRATES, TEMPERATURE,
ESS, RADIOACTIVITY, TEMPERATURE,
N, *ISOLATION, DIATOMS, ECOLOGY,

ROTIFERS, CHLOROPHYTA, COPEPODS,   W70-00268
ROTIFERS, SESTON, RESPIRATION, AL   W69-10154
ROTIFERS, WEEDS, WATER HYACINTH,   W70-10175
ROTIFERS, SWIMMING POOLS, CHLAMYD   W70-10161
ROTUNDATA.: /A, NAVICULA, FLORIDA   W70-00707
ROUGE RIVER, HURON RIVER, RAISIN   W69-01445
ROUGH FISH, NITROGEN, PHOSPHORUS,   W68-00172
RUBESCENS, OSCILLATORIA CHALYBIA,   W69-05704
RUBESCENS, ZURICHSEE, FILAMENTOUS   W70-04253
RUBESCENS, DINOBRYON, UROGLENA AM   W70-05405
RUBESCENS, PISTON CORER, FRAGILAR   W70-05663
RUBESCENS, MADISON LAKES(WIS), CE   W70-00270
RUBESCENS, SWITZERLAND, PERIDINIU   W69-10169
RUBIDIUM-106, NIOBIUM-95, GYMNODI   W70-00707
RUBRA, CANDIDA DIDDENSII, CANDIDA   W70-04368
RUN-OFF, SEDIMENTATION, BIOCHEMIC   W71-10654
RUN, ORGANIC CARBON, CINCINNATI(O   W71-04206
RUNNING WATERS, FRESH WATER, BRAC   W70-09905
RUNOFF RELATIONSHIPS, GEOGRAPHICA   W68-00891
RUNOFF.:   W71-04072
RUNOFF.: /TORM DRAINS, SEWAGE EFF   W71-10654
RUNOFF.: /URCES, WATER POLLUTION   W69-08668
RUNOFF, ARKANSAS RIVER, MERRIMACK   W70-04507
RUNOFF, DOMESTIC WASTES, DETERGEN   W71-04072
RUNOFF, DRAINAGE WATER, GRASSLAND   W70-04193
RUNOFF, FERTILIZERS, NITROGEN, PH   W71-10376
RUNOFF, FERTILIZERS, NUTRIENTS, N   W69-10170
RUNOFF, FROZEN GROUND, SOIL EROSI   W69-05323
RUNOFF, LIGHT PENETRATION, SAMPLI   W72-04761
RUNOFF, NUTRIENT BUDGETS, COMPARA   W69-09349
RUNOFF, PERCOLATION, ROADS, ROOFS   W69-10178
RUNOFF, RAINFALL, SNOWMELT, SURFA   W70-03501
RUNOFF, UNDERFLOW, PRECIPITATION(W70-04810
RUNOFF, WASTES, WATER SUPPLY, EFF   W71-09880
RUNS TEST, CLUSTER ANALYSIS, REYK   W69-09755
RUSSIA, BAVARIA, MORICHES BAY(NY)   W70-03973
RUSSIA, SECOND INTERNATIONAL OCEA   W69-07440
RUSTING, METALS, INHIBITORS, SILT   W70-02294
RUTHENIUM-106, RHODIUM-106, CERIU   W69-09742
RUTHENIUM-106, PLATYMONAS, THALAS   W70-00708
RUTHENIUM-106-RHODIUM-106, RUTHEN   W69-08524
RUTHENIUM-103-RHODIUM-103, CERIUM   W69-08524
RUTHENIUM-106, ANTIMONY-125, GROU   W69-08269
RUTHENIUM-106, WHITE OAK LAKE(TEN   W70-04371
RUTILANS, AMPHORA, AULACODISCUS,   W70-03325
S BAY(MINN), PHOSPHATE DETERMINAT   W70-05269
S BAY(MINN), SMITH'S BAY(MINN), P   W70-05269
S D), ABUNDANCE, BLUE-GREEN ALGAE   W70-05375
S LAW OF THE MINIMUM, VOISIN'S LA   W70-04765
S LAW OF THE MAXIMUM.: /OOMS, LIE   W70-04765
S MEDIA, QUINONE DERIVATIVES, ALK   W70-02982
S MEDIUM.: /ULGARIS, DAPHNIA MAGN   W71-05719
S MEDIUM.: /ULGARIS, DAPHNIA MAGN   W71-03056
S MINERAL MEDIUM, GLYOXYLIC ACID.   W70-04195
S POND(SOMERSET, ENGLAND), SAND G   W71-12870
S POND(WIS), HYDRODICTYON, NAJAS,   W70-03310
S POND(WIS), LAKE WAUBESA(WIS), G   W69-05868
S PUBLIC HEALTH SERVICE, GREAT LA   W70-00263
S SOLUTION VIII, NUTRIENT LIMITAT   W70-01579
S SUMMIT(MO), OXYGEN TRANSFER, PH   W70-06601
S WEST, RADIOCOCCUS NIMBATUS, URO   W70-04468
S., CIVIL ENGINEERING, WASTE TREA   W69-03683
S., ECOLOGICAL DISTRIBUTION, FREQ   W70-02772
S., REGIONS, PACIFIC COAST REGION   W69-03611
S., REGIONS, GEOGRAPHICAL REGIONS   W69-03695
S.: /ANAPHYTA, ALGAL CONTROL, *AL   W69-05306
S, ECOLOGICAL DISTRIBUTION, BIOLO   W70-02764
S: /AENA CYLINDRICA, BACILLUS SUB   W70-02964
S: /N, CENTRAL CITY(COLO), MUTATI   W68-08272
SACCHARINA, MANNITOL, MUSTARD GRO   W69-07826
SACCHAROPHILA, CHLORELLA ELLIPSOI   W70-04809
SACRAMENTO-SAN JOAQUIN DELTA, CAL   W70-05094
SACRAMENTO-SAN JOAQUIN ESTUARY, E   W70-05094
SAG, AERATION, BIOCHEMICAL OXYGEN   W69-07520
SAG, POTOMAC RIVER.: /NITY, REAER   W73-00750
SAG, WATER QUALITY, CHLORIDES, PH   W69-03948
SAGAMI, JAPAN, *GLENODININE, *PER   W70-08372
SAGITTARIA, NAJAS.: /OMENON, MICR   W72-04761
SALINE LAKES, HEAVY METALS, INDUS   W69-02959
SALINE WATER, EQUILIBRIUM, BIOLOG   W70-00360
SALINE WATER, CHLORELLA, CHLAMYDO   W70-03974
SALINE WATER-FRESHWATER INTERFACE   W73-00748
SALINITY.: /PIRATION, PIGMENTS, C   W70-03325
SALINITY, ALGAE, STRATIFICATION,   W72-01773
SALINITY, BIOASSAY, ANALYTICAL TE   W70-08111
SALINITY, LAKES, NUTRIENTS, BACTE   W71-03031
SALINITY, RAINFALL, TEMPERATURE,   W70-01073
SALINITY, ULTRAVIOLET RADIATION,   W70-01074
SALINITY, REAERATION, FISH, TASTE   W73-00750
SALINITY, ORGANIC MATTER, RIVERS,   W73-01446

TER CHEMISTRY, BIOMASS, BENTHOS,
R, CHEMICAL PROPERTIES, ECOLOGY,
ARINE ENVIRONMENT, MARINE ALGAE,
EFFLUENTS, TEMPERATURE, OXYGEN,
SOTOPES, *ENVIRONMENTAL EFFECTS,
, MOLLUSKS, VITAMIN B, RED TIDE,
MISTRY, RESPIRATION, BATHYMETRY,
PHOTIC ZONE, HELSINGOR(DENMARK),
ASENIA, DOW CONTACT, SPARGANIUM,
UMER ORGANISMS, SIMULIUM, PHYSA,
GIZZARD SHAD, SEA LAMPHREY, COHO
GICAL BALANCE, OVERFISHING, COHO
FISH STOCKING, COMMERCIAL FISH,
EL, PRODUCTIVITY, MODEL STUDIES,
SALINITY, ALGAE, STRATIFICATION,
LLUTION, *KUWAIT CRUDE, BENTLASS
RATE, EUGLENA, CALCIUM, AMMONIUM
NDS, PHOSPHORUS, NUISANCE ALGAE,
CS, ORGANISMS, SUSPENDED SOLIDS,
YRA, XANTHOPHYCEAE, MYXOPHYCEAE,
NT ELEMENTS, PHOSPHORUS RELEASE,
LAND), NON-PARAMETRIC TESTS, ONE-
INFLUENCES, MICROCYSTIS, DREDGE
LIGHT INTENSITY, TROPHIC LEVEL,
OOLING TOWERS, FILTERS, EFFLUENT
ATION, FLOCCULATION, POLLUTANTS,
*ZINC RADIOISOTOPES, STREAMFLOW,
TES, FISH, TEMPERATURE, TRACERS,
N, TOXICITY, MIDGES, MOSQUITOES,
IMENTS, MUD, BENTHOS, WISCONSIN,
, TURBIDITY, TERTIARY TREATMENT,
ANALYTICAL TECHNIQUES, DIATOMS,
IC MATTER, OKLAHOMA, RESERVOIRS,
ARBON, RADIOACTIVITY, CHLORELLA,
EMENTS, *DISTRIBUTION, BACTERIA,
ORS, FUNGI, ESTUARIES, BACTERIA,
IGATIONS, WATER POLLUTION, WATER
NALYSIS, *CHEMICAL ANALYSIS, PH,
PATH OF POLLUTANT, FOOD CHAINS,
RCES, RUNOFF, LIGHT PENETRATION,
ANALYSIS, ANALYTICAL TECHNIQUES,
DIATOMS, MEASUREMENT, SEDIMENTS,
TION, ULTRASONICS, DISINFECTION,
QUES, SPECTROSCOPY, COLORIMETRY,
ATES, SULFATES, SEASONAL, LAKES,
OPHYTA, CHRYSOPHYTA, CYANOPHYTA,
IMARY PRODUCTIVITY, CHLOROPHYLL,
S, *ALGAE, *LAKES, *TRIBUTARIES,
MENDOTA(WI/ *INLAND, HORIZONTAL
D, HORIZONTAL SAMPLING, VERTICAL
PLANKTON, LAKES, EUTROPHICATION,
, SESTON, RESPIRATION, ALTITUDE,
RS, CROP-DUSTING AIRCRAFT, WATER
S, ALGAE, BIOCHEMISTRY, ECOLOGY,
RATURE, *SEASONAL, RHODE ISLAND,
OPHYTA, CHRYSOPHYTA, CYANOPHYTA,
E AGING, NUTRIENT SOURCES, WATER
RIENTS, DEPTH, DOMESTIC ANIMALS,
XYGEN DEMAND, TEMPERATURE, FLOW,
PHORUS, OXYGEN, CARBON, ENZYMES,
DIATOMS, EUGLENA, CHLAMYDOMONAS,
NSITY, IRON, PHOSPHORUS, SILICA,
ACTERIA, AQUATIC MICROORGANISMS,
DOMONAS, CHLORELLA, SCENEDESMUS,
SAN JOAQUIN VALLEY MASTER DRAIN,
JOAQUIN RIVER, POTOMAC ESTUARY,
ON, DISSOLVED OXYGEN VARIATIONS,
ER PLAN, DELTA WATER/ SACRAMENTO-
IF), OXYGEN DYNAMICS, SACRAMENTO-
NTY, WATER DETERIORATION, DRAIN,
RIVER(CALIF), TUOLUMNE RIVER(CA/
URAVAN, / ANIMAS RIVER, DURANGO,
INETIC EFFECTS, ANN ARBOR(MICH),
REGENERATION, CLIFF VEGETATION,
ABBOT'S POND(SOMERSET, ENGLAND),
COATED SAND, FERRIC OXIDE-COATED
R/ *FILTER MEDIA, ALUMINA-COATED
*PARTICLE SIZE, *CLAYS, *SILTS,
ADIATION, DYE RELEASES, BEACHES,
IC BACTERIA, METABOLISM, CANALS,
*ALGAE, NUISANCE ALGAE, INTAKES,
ING, DENSITY, LIGHT, MUD, SILTS,
SPECTROMETERS, BACTERIA, ALGAE,
TER TREATMENT, *ALGAE, *FILTERS,
, TUNDRA, SCALING, GRAVELS, MUD,
YGEN, COLIFORMS, SEDIMENTS, MUD,
ENCY, MARGINAL VEGETATION FAUNA,
OXYGEN DEMAND, SEWAGE EFFLUENTS,
WTH, WATER POLLUTION ASSESSMENT,
OXYGEN, MEASUREMENT, ESTUARIES,

SALINITY, SNAILS, SULFATES, NITRA   W70-03978
SALINITY, BENTHOS, AQUATIC LIFE,   W69-05023
SALINITY, TEMPERATURE, MONITORING   W68-00010
SALINITY, ALGAE, MECHANICAL EQUIP   W69-06203
SALINITY, TEMPERATURE, SEA WATER,   W69-08267
SALINITY, FLORIDA, TEXAS, PHOSPHO   W70-05372
SALINITY, PHOSPHORU: /, WATER CHE   W70-05387
SALINO-CLINE.: /LA MENEGHINIANA,   W70-04809
SALIX, GROWTH REGULATORS, MYRIOPH   W70-05263
SALMO TRUTTA, ENTOSPHENUS LAMOTTE   W69-09334
SALMON.: /, WHITE FISH, ALEWIFE,   W70-01943
SALMON, ALEWIFE, NUTRIENT REMOVAL   W70-04465
SALMON, ALGAE, FISH.: /HOSPHORUS,   W70-04465
SALMON, AQUATIC ENVIRONMENT, PHOS   W69-06536
SALT BALANCE, BIBLIOGRAPHIES.: /   W72-01773
SALT MARSH.: /, *RECOVERY FROM PO   W70-01231
SALTS, NITROGEN FIXATION, BORON,   W70-02964
SALTS, TIDES, SOUTHEAST U. S., RE   W69-03695
SALTWATER TONGUE, BENTHIC FORMS,   W73-00748
SALVELINUS NAMAYCUSH, SUBSTRATE T   W70-00711
SAMPLE PRESERVATION, WATER DYNAMI   W70-04382
SAMPLE RUNS TEST, CLUSTER ANALYSI   W69-09755
SAMPLES, ANABAENA, SEASONAL EFFEC   W70-05663
SAMPLING TEMPERATURE, TRACERS, CH   W70-02504
SAMPLING.: /TERS, SNAILS, ODOR, C   W72-00027
SAMPLING.: /ATES, BACKGROUND RADI   W69-07846
SAMPLING, ALGAE, ECOSYSTEMS, ENVI   W69-07862
SAMPLING, PERIPHYTON, AQUATIC PLA   W69-09334
SAMPLING, CONTROL, INDUS: /ICATIO   W70-06225
SAMPLING, CONTOURS, DEPTH, OLIGOC   W70-06217
SAMPLING, MONITORING, WATER DEMAN   W70-05645
SAMPLING, PHYSICOCHEMICAL PROPERT   W70-05389
SAMPLING, INDUSTRIAL WASTES, MUNI   W70-04001
SAMPLING, EFFLUENTS, TRIBUTARIES,   W70-03983
SAMPLING, SEA WATER, SEASONAL, PS   W70-03952
SAMPLING, SEASONAL, HYDROGRAPHY,   W70-04368
SAMPLING, *WATER ANALYSIS, *CHEMI   W69-00659
SAMPLING, ON-SITE DATA COLLECTION   W69-00659
SAMPLING, SURVEYS, MARINE ALGAE,   W72-00941
SAMPLING, PRIMARY PRODUCTIVITY, P   W72-04761
SAMPLING, ORGANIC MATTER, TOXICIT   W72-03299
SAMPLING, DEPTH, SEA WATER, FRES   W71-12870
SAMPLING, CENTRIFUGATION, SEDIMEN   W71-11823
SAMPLING, METHANE, CARBON DIOXIDE   W71-11377
SAMPLING, PONDS, RIVERS, RESERVOI   W70-02646
SAMPLING, PHYTOPLANKTON, ORGANIC   W70-02780
SAMPLING, PHYTOPLANKTON, RADIOACT   W70-01933
SAMPLING, FLOW, ANALYSIS, FERTILI   W70-02969
SAMPLING, VERTICAL SAMPLING, LAKE   W70-02969
SAMPLING, LAKE MENDOTA(WIS), LAKE   W70-02969
SAMPLING, SURFACES WATERS, CARBON   W70-02973
SAMPLING, ZOOPLANKTON, PHYTOPLANK   W69-10154
SAMPLING, ONTARIO WATER RESOURCES   W69-10155
SAMPLING, THERMAL SPRINGS, EFFLUE   W69-10151
SAMPLING, ECOLOGY, AMMONIA, ENZYM   W70-00713
SAMPLING, DENSITY, LIGHT, MUD, SI   W70-00711
SAMPLING, ALGAL GROWTH, OXYGEN DE   W70-00264
SAMPLING, NUISANCE ALGAE.: /, NUT   W70-00274
SAMPLING, WASTE WATER DISPOSAL, D   W69-10159
SAMPLING, BOTTOM SEDIMENTS, SODIU   W70-09889
SAMPLING, SCENEDESMUS, CHLOROPHYT   W70-08107
SAMPLING, CHEMICAL ANALYSIS, PHYS   W70-06975
SAMPLING, WATER ANALYSIS, WATER Q   W71-00221
SAMPLING, OXYDATION LAGOONS.: /MY   W70-10173
SAN FRANCISCO BAY-DELTA.: /RAIN,   W71-09577
SAN FRANCISCO BAY.: /IATIONS, SAN   W69-07520
SAN JOAQUIN RIVER, POTOMAC ESTUAR   W69-07520
SAN JOAQUIN DELTA, CALIFORNIA WAT   W70-05094
SAN JOAQUIN ESTUARY, ENVIRONMENTA   W70-05094
SAN JOAQUIN VALLEY MASTER DRAIN,   W71-09577
SAN JOAQUIN RIVER(CALIF), MERCED   W70-02777
SAN MIGUEL RIVER, DOLORES RIVER,   W69-07846
SAND BEDS.: /USPENSIONS, ELECTROK   W71-03035
SAND DUNES.: /ON, SUSCEPTIBILITY,   W70-01777
SAND GRAINS, TIDAL CYCLES.: /YI,   W71-12870
SAND, CRUSHED BAUXITE, ACID-WASHE   W70-08628
SAND, FERRIC OXIDE-COATED SAND, C   W70-08628
SANDS, ALGAE, PLANKTON, MICROORGA   W70-08628
SANDS, BIOTA, WASTE WATER(POLLUTI   W68-00010
SANDS, CHLORINATION, MICROORGANIS   W69-07838
SANDS, FLOW RATES, TURBIDITY, CHL   W70-04199
SANDS, GRAVELS, SEDIMENTS, WATER   W70-00711
SANDS, HYDROGEN ION CONCENTRATION   W69-08267
SANDS, LAKE MICHIGAN, WATER QUALI   W73-02426
SANDS, SHALES, BOTTOM SEDIMENTS,   W69-08272
SANDS, SLUDGE, RECREATION.: /, OX   W70-10161
SANDY BOTTOMS FAUNA, MARGINAL VEG   W70-06975
SANITARY ENGINEERING.: /CHEMICAL   W72-01817
SANITARY ENGINEERING RESEARCH LAB   W70-02775
SANITARY ENGINEERING, WATER TREAT   W70-05094

BACTERIA, AL/ *WATER TREATMENT,   SANITARY ENGINEERING, *PATHOGENIC   W72-13601
*WATER TREATMENT, *FILTRATION,   SANITARY ENGINEERING, *FILTERS, *   W72-07334
, DESIGN, *AL/ *WATER TREATMENT,   SANITARY ENGINEERING, *FILTRATION   W72-08859
RONMENTAL EFFECTS, ENVIRONMENTAL   SANITATION, ESSENTIAL NUTRIENTS,   W69-03513
RONMENTAL EFFECTS, ENVIRONMENTAL   SANITATION, ESSENTIAL NUTRIENTS,   W69-03369
RONMENTAL EFFECTS, ENVIRONMENTAL   SANITATION, INHIBITION, CHLORINE,   W69-10171
NGLAND, OSCILLATORIA, AUSTRALIA,   SANTA ROSA(CALIF), LYUBLIN(USSR),   W70-05092
NIGROSINE DYE, SODIUM CHLORATE,   SANTOBRITE, CLADOPHORA, POTAMOGET   W70-05263
LOGY.:   *CILIATES, FLAGELLATES,   SAPROBIEN SYSTEM, CILIATE AUTOECO   W71-03031
                                         SAPROLEGNIA.:   W68-00468
OLECULAR FILTER, CHEMOSYNTHETIC,   SAPROPHYTIC, PLATE COUNT, FILAMEN   W68-00891
IBRIO, FLAVOBACTER, ESCHERICHIA,   SARCINA, STAPHYLOCOCCUS, ACHROMOB   W70-03952
BACTERS, GRAM-POSITIVE BACTERIA,   SARCINE-LIKE, BACILLUS-LIKE, MICR   W70-00713
MASTIGOPHORA, HOLOPHYTIC MASTI/   SARCODINA, MASTIGOPHORA, HOLOZOIC   W72-01817
CEAE, CHRYSOPHYCEAE, VOLVOCALES,   SARCODINA, CILIATES, CYCLOTELLA,   W70-05112
EMA, THALASSIOSIRA, COCCOLITHUS,   SARGASSO SEA, SYNURA, GYMNODINIUM   W70-04503
ISTANCE, EUPHAUSIA, PHAEOCYSTIS,   SARGASSUM, FRONDS, HYDROCARBONS,   W70-01068
RIDAY HARBOR(WASH), ANTITHAMNION   SARNIENSE, RHODOCHORTON, BANGIA F   W69-10161
, LEUCOTHRIX MUCOR, ANTITHAMNION   SARNIENSE, QUANTITATION, GROWTH M   W69-10163
, *AUTORADIOGRAPHY, ANTITHAMNION   SARNIENSE, TRITIATED GLUCOSE, GON   W69-09755
TON LAKE(SASKATCHEWAN), HUMBOLDT(SASKATCHEWAN), MICE, MORPHOLOGICA   W70-00273
II, OTTAWA(ONTARIO), BURTON LAKE(SASKATCHEWAN), HUMBOLDT(SASKATCHE   W70-00273
, CYCLOTELLA MENEGHINIANA, LIGHT-   SATURATED RATE, CELL VOLUME, NITZ   W70-05381
TOSYNTHESIS, RESPIRATION, SCOUR,   SATURATION, MIXING, DIFFUSION, SE   W71-11381
ZOAN COMMUNITIES, POTOMAC RIVER,   SAVANNAH RIVER, SHOCK EFFECTS.: /   W70-07847
LLATORIA, CATHERWOOD DIATOMETER,   SAVANNAH RIVER, POPULATION DYNAMI   W70-04510
VER, MERRIMACK RIVER, RED RIVER,   SAVANNAH RIVER, SNAKE RIVER, RIO   W70-04507
NEMA BORYANUM, REACTOR.:   SAVANNAH RIVER LABORATORY, PLECTO   W69-07845
LLOIDAL FRACTION, MULTIPLE-USES,   SAVANNAH RIVER, GULF COAST.: / CO   W73-00748
A, WEAPONS TESTING, NUCLEAR SHIP   SAVANNAH, ZIRCONIUM-95, RUBIDIUM-   W70-00707
YTILUS, TRICHODESMIUM ERYTHREUM,   SAXIDOMUS, CLOSTRIDIUM BOTULINUS,   W70-05372
, WILLIAMS LAKE(WASH), CHANNELED   SCABLANDS(WASH), WASHINGTON STATE   W70-00264
REATMENT, *AUSTRALIA, LABORATORY-   SCALE STUDIES, PLANT-SCALE STUDIE   W70-06899
LABORATORY-SCALE STUDIES, PLANT-   SCALE STUDIES, REMOVAL.: /TRALIA,   W70-06899
RIMENTAL STUDIES, LAGOON DESIGN,   SCALE-UP FACTORS, WASTE COMPOSITI   W70-07508
US, COLUMBIA RIVER, ELECTRONICS,   SCALING.: / RADIOISOTOPES, DETRIT   W69-07862
OGIC FORMATIONS, ARCTIC, TUNDRA,   SCALING, GRAVELS, MUD, SANDS, SHA   W69-08272
EMISTRY, RESISTIVITY, CHLORELLA,   SCENEDESMUS.: /ACTIONS, ELECTROCH   W70-05267
IS, GROWTH RATE, ORGANIC MATTER,   SCENEDESMUS, CHLAMYDOMONAS, FRESH   W70-05381
OROPLASTS, SCENEDESMUS OBLIQUUS,   SCENEDESMUS QUADRICAUDA, CHLORELL   W70-05381
, NITZSCHIA PALEA, CHLOROPLASTS,   SCENEDESMUS OBLIQUUS, SCENEDESMUS   W70-05381
CENTRATION, UROG/ *REQUIREMENTS,   SCENEDESMUS, ABSORPTION, CELL CON   W70-05409
*FRESH WATER, *PLANKTON, *ALGAE,   SCENEDESMUS, LIGHT, CULTURES, IRO   W70-05409
DESMUS, GLENODINIUM PULVISCULUS,   SCENEDESMUS: /RUDINELLA, ANKISTRO   W70-05387
PURIFICATION, ALGAL ENUMERATION,   SCENEDESMUS, GROWTH EQUATIONS, AN   W70-04381
ORUS, *AQUATIC ALGAE, *BIOASSAY,   SCENEDESMUS, PHOSPHATES, WATER PO   W70-03955
, 2-4-D, AMINOTRIAZOLE, 2-4-5-T,   SCENEDESMUS, CHLAMYDOMONAS, CHLOR   W70-03519
GHT STERILIZATION, ASTERIONELLA,   SCENEDESMUS OBLIQUUS, GROWTH CONS   W70-03955
ODESMUS, MICROSPORA, PEDIASTRUM,   SCENEDESMUS, LEPOCINCLIS, ANABAEN   W70-03974
CIDES, ALDRIN, ENDRIN, DIELDRIN,   SCENEDESMUS, GAS CHROMATOGRAPHY,   W70-03520
RATURE, CYANOPHYTA, CHLOROPHYTA,   SCENEDESMUS, BENTHOS, CONVECTION,   W70-03969
UCTURE/ *ALGAE, *MICROORGANISMS,   SCENEDESMUS, PROTEINS, WALLS, STR   W70-04184
YDROGEN PEROXIDE, RATS, TRYPSIN,   SCENEDESMUS QUADRICAUDA, SCENEDES   W70-04184
RYPSIN, SCENEDESMUS QUADRICAUDA,   SCENEDESMUS OBLIQUUS, HELIX POMAT   W70-04184
LOW RATES, TURBIDITY, CHLORELLA,   SCENEDESMUS, EUGLENA.: / SANDS, F   W70-04199
RPHA, LEPOCINCLIS, CHLOROGONIUM,   SCENEDESMUS, CYCLOTELLA, STEPHANO   W69-02959
*PHYTOTOXICITY, ALGAL POISONING,   SCENEDESMUS, CHLORELLA, *EUGLENA,   W69-00994
TURE, SEASONAL, TEST PROCEDURES,   SCENEDESMUS, *ALGAE.: /ER TEMPERA   W68-00472
ON CONCENTRATION, PLANT TISSUES,   SCENEDESMUS, LIMNOLOGY, WATER PRO   W68-00248
H RATES, *TURBULENCE, CHLORRELA,   SCENEDESMUS, KINETICS, ECOSYSTEMS   W69-03730
*HERBICIDES, *ALGAE, TOXICITY,   SCENEDESMUS, CHLORELLA, DAPHNIA,   W71-03056
RGANO-SILICATES, *ALKOXYSILANES,   SCENEDESMUS QUADRICAUDA, CHLORELL   W71-03056
ERS, ELECTROPHORESIS, CHLORELLA,   SCENEDESMUS, ADSORPTION, SEWAGE T   W71-03035
MONURON, TRANSPARENCY, CYCLOPS,   SCENEDESMUS OBLIQUUS, DACTYLOCOCC   W70-10161
GLENA, CHLAMYDOMONAS, CHLORELLA,   SCENEDESMUS, SAMPLING, OXYDATION   W70-10173
AS, DAPHNIA, DIATOMS, CHLORELLA,   SCENEDESMUS, OXYGEN, COLIFORMS, S   W70-10161
UGLENA, CHLAMYDOMONAS, SAMPLING,   SCENEDESMUS, CHLOROPHYTA, CHLOREL   W70-08107
PES, HYDROGEN ION CONCENTRATION,   SCENEDESMUS, EUGLENA, BACTERIA, Y   W70-08097
AL EFFIC/ *DAPHNIA PULEX OBTUSA,   SCENEDESMUS QUADRICAUDA, ECOLOGIC   W69-10152
                     PROTOCOCCUS,   SCENEDESMUS QUADRICAUDA.:   W71-11561
ANCES, CHLORELLA, CHLAMYDOMONAS,   SCENEDESMUS, WISCONSIN.: /H SUBST   W72-00845
RGANO-SILICATES, *ALKOXYSILANES,   SCENEDESMUS QUADRICAUDA, CHLORELL   W71-05719
*HERBICIDES, *ALGAE, TOXICITY,   SCENEDESMUS, CHLORELLA, DAPHNIA,   W71-05719
NTERFACES, CALIFORNIA, BIOASSAY,   SCENEDESMUS, PHYSIOLOGICAL ECOLOG   W70-02775
ITROGEN, WASTE WATER(POLLUTION),   SCENEDESMUS, CYANOPHYTA, CHLOROPH   W70-02255
HYSIOLOGICAL ECOLOGY, CHLORELLA,   SCENEDESMUS.:/LOGY, NUTRIENTS, P   W70-02245
ZOOPLANKTON, MINNOWS, PEDALFERS,   SCENEDESMUS, CHLAMYDOMONAS, TEMPE   W70-02249
IPPING, FIREBAUGH CENTER(CALIF),   SCENEDESMUS QUADRICAUDA, CELL COU   W70-03334
SIS, WATER POLLUTION, CHLORELLA,   SCENEDESMUS.: /IA, CHEMICAL ANALY   W70-03312
ANKISTRODESMUS, NOSTOC MUSCORUM,   SCENEDESMUS OBLIQUUES, SCENEDESMU   W70-02964
MUSCORUM, SCENEDESMUS OBLIQUUES,   SCENEDESMUS QUADRICAUDA, PORPHYRI   W70-02964
NTHESIS, OXYGEN, CARBON DIOXIDE,   SCENEDESMUS, CHLOROPHYLL, IRON, C   W70-02964
NZYMES, CHLOROPHYTA, RHODOPHYTA,   SCENEDESMUS, CHLAMYDOMONAS, ECOLO   W70-02965
PHICATION, AQUATIC LIFE, SESTON,   SCENEDESMUS, MATHEMATICAL STUDIES   W73-01466
TRIENTS, ECOSYSTEMS, PHOSPHORUS,   SCENEDESMUS, CHLOROPHYTA, BACTERI   W73-01454
OIRS, ANOPHELES QUADRI/ LEECHES,   SCHISTOSOME CERCARIAE, TVA RESERV   W70-06225
RBONATE, CERCARIAE, WATER MILFO/   SCHISTOSOME DERMATITIS, COPPER CA   W70-05568
EGION, GRADIENTS(RIVER), THALLI,   SCHIZOMYCETES, CYANOPHYTA, FUNGI,   W70-02784
(VIETNAM), CHLOROCOCCUS, SPINES,   SCHIZOMYCOPHYTA, CAULOBACTERIALES   W70-03969
, OHANAPECOSH HOT SPRINGS(WASH),   SCHIZOTHRIX, PHORMIDIUM, WINKLER   W70-03309

AND), LAKE TEGERN(BAVARIA), LAKE SCHLIER(BAVARIA), LAKE ZELL(AUSTR W70-05668
RG(GERMANY), PLON(GERMANY), LAKE SCHLIER(GERMANY), LAKE TEGERN(GER W70-05761
E, *COMPARATIVE STUDIES, DAPHNIA SCHODLERI, DAPHNIA PULEX, DAPHNIA W70-03957
ALATHION, CHLORELLA PYRENOIDOSA, SCHRODERIA, THIMET, TRICHODERMA.: W70-08097
TA, PYROPHYTA, CHROOMONAS ACUTA, SCHROEDERIA SETIGERA, MELOSIRA AM W70-05387
ACUSE(NY),/ *PHYCOLOGY, NY STATE SCIENCE AND TECHNOLOGY FOUND, SYR W70-03973
OMIC ENERGY COMMISSION, NATIONAL SCIENCE FOUNDATION, OFFICE OF NAV W70-03975
OMIC ENERGY COMMISSION, NATIONAL SCIENCE FOUNDATION, OFFICE OF NAV W70-03975
ONSIN, ASIA/ NATIONAL ACADEMY OF SCIENCES(USA), UNIVERSITY OF WISC W70-03975
ONSIN, ASIA/ NATIONAL ACADEMY OF SCIENCES(USA), UNIVERSITY OF WISC W70-03975
WIS/ *BIOLOGICAL PROBLEMS, SCIOTO RIVER(OHIO), LAKE MENDOTA(W72-01813
MENDOTA(WIS), ANACHARIS, NAIAS, SCIRPUS, NYMPHAEA, CRASSIPES EICH W70-05263
INCE EDWARD ISLAND(CANADA), NOVA SCOTIA(CANADA), PENGUINS, LAKE ZU W70-05399
HAELIS-MENTEN EQUATION, LOCH EWE(SCOTLAND), SCOTLAND, PHAEOPIGMENT W70-04811
DRICA, NORTH SEA, FIRTH OF CLYDE(SCOTLAND), SCOTLAND, ENGLAND, LAK W70-02504
H SEA, FIRTH OF CLYDE(SCOTLAND), SCOTLAND, ENGLAND, LAKE WINDERMER W70-02504
EN EQUATION, LOCH EWE(SCOTLAND), SCOTLAND, PHAEOPIGMENTS, ACRIDINE W70-04811
ON, PHOTOSYNTHESIS, RESPIRATION, SCOUR, SATURATION, MIXING, DIFFUS W71-11381
*POTATOES, *INDUSTRIAL WASTES, SCREENS, SEDIMENTATION, ORGANIC L W71-08970
, WATER PURIFICATION, MEMBRANES, SCREENS, TREATMENT FACILITIES, *W W73-02426
UTROPHICATION, CYANOPHYTA, ODOR, SCUM, FISHING, WISCONSIN, BIBLIOG W69-08283
, CYANOPHYTA, LAKES, RESERVOIRS, SCUM, NITROGEN, PHOSPHORUS, CARBO W70-02803
ION, OLIGOTROPHY, SEWAGE, ALGAE, SCUM, WEEDS, NUTRIENTS, SEPTIC TA W70-04193
TILIZERS, BLUEGILLS, MICROCYSTIS SCUMS, `PITHIOPHORA ALGAE.: /E FER W70-05417
HETIC OXYGEN, LEMMON(SD), KADOKA(SD).:   *PHOTOSYNT W72-01818
*PHOTOSYNTHETIC OXYGEN, LEMMON(SD), KADOKA(SD).: W72-01818
OLOGICAL ECOLOGY, PHYTOPLANKTON, SE: /NUTRIENT REQUIREMENTS, PHYSI W69-03371
ITE FISH, ALEWIFE, GIZZARD SHAD, SEA LAMPHREY, COHO SALMON.: /, WH W70-01943
N, METHODOLOGY, INSTRUMENTATION, SEA WATER.: /ES, THERMAL POLLUTIO W70-10437
*MARINE WATER, *COPPER, *ALGAE, SEA WATER, TEMPERATURE, SALINITY, W70-08111
ISOTOPES, LAKES, ORGANIC MATTER, SEA WATER, PRIMARY PRODUCTIVITY, W70-02504
ENT, SEDIMENTS, SAMPLING, DEPTH, SEA WATER, FRES WATER, LITTORAL, W71-12870
R, *IRON, TRACE ELEMENTS, ALGAE, SEA WATER, FRESH WATER.: / *COPPE W71-11687
ECONDARY PRODUCTIVITY, SULFIDES, SEA WATER, PHAWOPHYTA, CRUSTACEAN W70-01068
NKTON, *CYCLING, *RADIOISOTOPES, SEA WATER, FOOD CHAINS, PHOSPHORU W70-00708
CURRENTS(WATER), REMOTE SENSING, SEA WATER, CHLOROPHYLL, OIL, ALGA W70-05377
ISTRIBUTION, BACTERIA, SAMPLING, SEA WATER, SEASONAL, PSEUDOMONAS, W70-03952
N, *WATER STRUCTURE, ATMOSPHERE, SEA WATER, FRESH WATER, SORPTION, W70-04385
SOTOPES, *MARINE ALGAE, *RESINS, SEA WATER, PHYSICOCHEMICAL PROPER W69-08275
EFFECTS, SALINITY, TEMPERATURE, SEA WATER, OYSTERS, CLAMS, CRABS, W69-08267
, INVERTEBRATES, PLANKTON, FISH, SEA WATER, GAMMA RAYS, RADIOCHEMI W69-08269
ABATEMENT, TIDAL WATERS, MIXING, SEA WATER, FRESH WATER, SHORES, W W69-06203
EANOGRAPHY, *TRANSACTIONS, BLACK SEA, CONTACT ZONES, ARTEMIA, BALA W69-07440
YLUM TRICORNUTUM, EDTA, LIGURIAN SEA, EUROPEAN MEDITERRANEAN.: /CT W69-08275
AND), ANABAENA CYLINDRICA, NORTH SEA, FIRTH OF CLYDE(SCOTLAND), SC W70-02504
I, RHODOTORULA RUBRA, CA/ *NORTH SEA, MYCOTA, DEBARYOMYCES HANSENI W70-04368
ASSIOSIRA, COCCOLITHUS, SARGASSO SEA, SYNURA, GYMNODINIUM, CHEMICA W70-04503
N, *RADIONUCLIDE EXCHANGE, *OPEN SEA, WEAPONS TESTING, NUCLEAR SHI W70-00707
SION, INSTREAM TREATMENT, BOTTOM SEALING, LAKE DILUTION.: /R DIVER W69-09340
L, PHOTOSYNTHETIC LAYER, MELOSI/ SEASONAL DISTRIBUTION, LAKE BAIKA W70-04290
NIVERSITY, EXPERIMENTAL STREAMS, SEASONAL EFFECTS, THERMAL EFFECTS W70-03978
YSTIS, DREDGE SAMPLES, ANABAENA, SEASONAL EFFECTS, DECOMPOSITION R W70-05663
OLYSIPHONIA, ORGANIC AGGREGATES, SEASONAL EFFECTS.: /COTOCARPUS, P W70-01068
TH EFFECTS, IONIC CONCENTRATION, SEASONAL EFFECTS.: /NOMICTIC, DEP W70-01933
ME, MAINE, PERIODIC VARIATIONS, SEASONAL VARIATIONS, SOURCES.: /, W68-00481
LGAE, *NUISANCE ALGAE, SPRAYING, SEASONAL, *ALGAL CONTROL, FISHKIL W68-00488
ROWTH RATES, *WATER TEMPERATURE, SEASONAL, TEST PROCEDURES, SCENED W68-00472
FISHKILL, NUISANCE ALGAE, ALGAE, SEASONAL, LIMNOLOGY, LIGHT PENETR W68-00487
ITY, HYPOLIMNION, ALGAL CONTROL, SEASONAL, ANNUAL TURNOVER.: /CTIV W69-05142
N, *LAKES, PHYTOPLANKTON, ALGAE, SEASONAL, CYANOPHYTA, *INORGANIC W68-00860
EPTH, DENSITY, CHLOROPHYLL, ICE, SEASONAL, DIURNAL, LIGHT INTENSIT W70-05387
TH, PLANKTON, LIGHT PENETRATION, SEASONAL, PRIMARY PRODUCTIVITY, C W70-05405
RE, EPILIMNION, LIGHT INTENSITY, SEASONAL, PHYSIOLOGICAL ECOLOGY, W70-05409
TION, ACTIVATED CARBON, FILTERS, SEASONAL, STRATIFICATION, BIOCHEM W70-04001
S, CESIUM, COBALT RADIOISOTOPES, SEASONAL, RADIOCHEMICAL ANALYSIS, W70-04371
LANKTON, *LAKES, ORGANIC MATTER, SEASONAL, ALGAE, PRIMARY PRODUCTI W70-04290
, ESTUARIES, BACTERIA, SAMPLING, SEASONAL, HYDROGRAPHY, FLORIDA, T W70-04368
, BACTERIA, SAMPLING, SEA WATER, SEASONAL, PSEUDOMONAS, TEMPERATUR W70-03952
NITRATES, PHOSPHATES, SULFATES, SEASONAL, LAKES, SAMPLING, PONDS, W70-02646
HER, EPILIMNION, ALGAE, DENSITY, SEASONAL, MARINE PLANTS,: /, WEAT W70-02504
RENTS(WATER), VELOCITY, DIATOMS, SEASONAL, BIOMASS, PIGMENTS, PROD W70-02780
HLOROPHYTA, CYANOPHYTA, DIATOMS, SEASONAL, EUGLENA, CHLAMYDOMONAS, W70-10173
*NUTRIENT REMOVAL, *WATER REUSE, SEASONAL, WATER TREATMENT.: /NT, W72-08133
METHODS, *BIOCHEMISTRY METHODS, SEASONS, ACETYLENE REDUCTION TEST W70-03429
RO ACT(/ *LAKE WASHINGTON(WASH), SEATTLE(WASH), REVENUE BONDS, MET W70-04455
(WASH), REVENUE BONDS, METRO ACT(SEATTLE).: /INGTON(WASH), SEATTLE W70-04455
OBIAL ECOLOGY, COCCOLITHOPHORES, SEAWATER MEDIUM F.: /NIQUES, MICR W70-05272
S, CLASSIFICATION, GROWTH RATES, SEAWATER, INCUBATION, PSEUDOMONAS W70-00713
ARINE ALGAE, ECOLOGY, NUTRIENTS, SEAWATER, ORGANIC COMPOUNDS, HABI W69-09755
SUM FLUITANS, *SARGASSUM NATANS, SEAWEED, ZIRCONIUM-95-NIOBIUM-95, W69-08524
IANA, OCHROMONAS, PHYTOPLANKTON, SEAWEEDS.: /ASHINGTON, LAKES, IND W70-04503
, LAKE MENDOTA, LAKE FURES, LAKE SEBASTICOOK, LAKE WASHINGTON, LAK W70-03964
VARIATIONS, SEASONAL VARIA/ LAKE SEBASTICOOK, ME, MAINE, PERIODIC W68-00481
N H), MA/ NUTRIENT BUDGETS, LAKE SEBASTICOOK(ME), LAKE WINNISQUAM(W70-07261
SHINGTON(WASH), *OXYGEN DEFICIT, SECCHI DISC TRANSPARENCY, MYXOPHY W69-10182
IDIUM, APHANIZOMENON FLOS-AQUAE, SECCHI DISC TRANSPARENCY, ANABAEN W69-10169
NTROL METHODS.: SECCHI DISC, VISIBILITY TESTS, CO W68-00488
NTS, PRODUCTION BIOLOGY, RUSSIA, SECOND INTERNATIONAL OCEANOGRAPHI W69-07440
IENTS, PONDS, PRIMARY TREATMENT, SECONDARY TREATMENT, STABILIZATIO W68-00012
TION SOURCES, PRIMARY TREATMENT, SECONDARY TREATMENT, TERTIARY TRE W69-01453

IA, PRODUCTIVITY, BIOINDICATORS,   SECONDARY PRODUCTIVITY, AQUATIC P   W69-10152
LATION, BACTERIA, STREPTOCOCCUS,   SECONDARY PRODUCTIVITY, SULFIDES,   W70-01068
ION, ALGAE COUNTS, ALGAL GROWTH,   SEDGWICK-RAFTER METHOD, TREATMENT   W69-10157
HOSPHOROUS, AVAILABLE NUTRIENTS,   SEDIMENT CORES.: /ON, NITROGEN, P   W68-00511
, WATER SUPPLY, MUNICIPAL WATER,   SEDIMENT LOAD, CONDUCTIVITY, CALC   W69-05894
ERIA, REMOTE SENSING, MINNESOTA,   SEDIMENT-WATER INTERFACES, CALIFO   W70-02775
L CONTROL, ALGAE, WATER QUALITY,   SEDIMENT-WATER INTERFACES, OREGON   W72-06051
EMENTS, CORES, WISCONSIN, BOTTOM   SEDIMENT, BENTHIC FAUNA.: /ACE EL   W69-03366
ION, *LAKES, *PRIMARY PRODUCTIV/   SEDIMENTATION RATES, *EUTROPHICAT   W69-00632
, ODOR, PHYTOPLANKTON, SULFATES,   SEDIMENTATION, SEDIMENTS, TASTE,   W69-07833
ATION, LAGOONS, DOMESTIC WASTES,   SEDIMENTATION, CHLORINATION, SLUD   W70-04060
BULENCE, PILOT PLANTS, BACTERIA,   SEDIMENTATION, TOXICITY, COST-BEN   W70-05819
NOLOGY, NUTRIENTS, PRODUCTIVITY,   SEDIMENTATION RATES, STRATIGRAPHY   W70-05663
DESIGN, *ALGAE, COPPER SULFATE,   SEDIMENTATION, ACTIVATED CARBON,   W72-08859
SOLVED OXYGEN, DISSOLVED SOLIDS,   SEDIMENTATION RATES, MANGANESE, L   W73-00750
ZOOPLANKTON, TRIPTON, BACTERIA,   SEDIMENTATION, COPEPODS, WATER PO   W69-10169
IOCHEMICAL OXYGEN DEMAND, ALGAE,   SEDIMENTATION, SETTLING BASINS.: /   W70-09320
ORGANIC LOADING, ALGAE, AMMONIA,   SEDIMENTATION, FLOTATION, BIODEGR   W71-05013
*POULTRY, WASTE WATER TREATMENT,   SEDIMENTATION, OXIDATION LAGOONS,   W71-11375
GE, INCINERATION, STORM RUN-OFF,   SEDIMENTATION, BIOCHEMICAL OXYGEN   W71-10654
DES, FARM WASTES, SILT, EROSION,   SEDIMENTATION, RUNOFF, FERTILIZER   W71-10376
CTION, SAMPLING, CENTRIFUGATION,   SEDIMENTATION, CHEMICAL PRECIPITA   W71-11823
, SATURATION, MIXING, DIFFUSION,   SEDIMENTATION, ABSORPTION, BIOCHE   W71-11381
ES, *INDUSTRIAL WASTES, SCREENS,   SEDIMENTATION, ORGANIC LOADING, A   W71-08970
EVAPORATION, WATER CIRCULATION,   SEDIMENTS.: /EVEL, WATER STORAGE,   W69-09723
IBUTARIES, WATERSHED MANAGEMENT,   SEDIMENTS.: /, PHOTOSYNTHESIS, TR   W70-04268
S, FISHKILL, CRUSTACEANS, BOTTOM   SEDIMENTS.: /AGE EFFLUENTS, SMELT   W69-08674
AL PROPERTIES, EPILMNION, BOTTOM   SEDIMENTS.: / ABSORPTION, *CHEMIC   W68-00857
DIATOMS, CHLOROPHYTA, DEPOSITION(SEDIMENTS), WATERSHEDS(BASINS), W   W69-10182
, ICE, BACTERIA, TRIPTON, BOTTOM   SEDIMENTS, SOLAR RADIATION, OXYG: W69-10154
FALLOUT, SILTS, DETRITUS, BOTTOM   SEDIMENTS, CHUTES, EDDIES, HABITA   W69-09742
NITROGEN, AQUATIC PLANTS, BOTTOM   SEDIMENTS, DREDGING, HARVESTING O   W69-09340
TH RATES, COPPER SULFATE, BOTTOM   SEDIMENTS, LAKE MICHIGAN, WASHING   W69-09349
MONAS, SUBSURFACE-WATERS, BOTTOM   SEDIMENTS, SURFACE WATERS, DETRIT   W70-00713
GHT, MUD, SILTS, SANDS, GRAVELS,   SEDIMENTS, WATER POLLUTION SOURCE   W70-00711
IC HABITATS, DRAGONFLIES, ALGAE,   SEDIMENTS, SOIL-WATER-PLANT RELAT   W71-09005
IVATED SLUDGE, COST COMPARISONS,   SEDIMENTS, ALGAE, HYDROGEN ION CO   W71-09548
OISOTOPES, DIATOMS, MEASUREMENT,   SEDIMENTS, SAMPLING, DEPTH, SEA W   W71-12870
, FRESH WATER BIOLOGY, SUSPENDED   SEDIMENTS, MARINE BIOLOGY, STREAM   W71-10786
TESTS, PRODUCTIVITY, ABSORPTION,   SEDIMENTS, ORGANIC MATTER, ALGAE,   W72-00845
ARBON, ENZYMES, SAMPLING, BOTTOM   SEDIMENTS, SODIUM, POTASSIUM, IRO   W70-09889
N, PHOTOSYNTHESIS, ALGAE, BOTTOM   SEDIMENTS, THERMOCLINE, HYPOLIMNI   W70-06975
AE, BIOLOGICAL TREATMENT, BOTTOM   SEDIMENTS, NUTRIENTS, *WASTE WATE   W70-07031
YSIS, PHYTOPLANKTON, CYANOPHYTA,   SEDIMENTS, PLANT POPULATIONS, WEA   W70-06229
SCENEDESMUS, OXYGEN, COLIFORMS,   SEDIMENTS, MUD, SANDS, SLUDGE, RE   W70-10161
ENT, HARVESTING OF ALGAE, BOTTOM   SEDIMENTS, WEED CONTROL, SOLAR RA   W68-00478
*NUTRIENT REQUIREMENTS, *BOTTOM   SEDIMENTS, ON-SITE DATA COLLECTIO   W68-00511
ON-SITE DATA COLLECTIONS, BOTTOM   SEDIMENTS, BIBLIOGRAPHIES, ABSORP   W68-00858
ULATIONS, BIBLIOGRAPHIES, BOTTOM   SEDIMENTS, WATERSHED MANAGEMENT,   W68-00859
S, RECENT SEDIMENTS, PRECAMBRIAN   SEDIMENTS, CHLORELLA PYRENOIDOSA,   W69-08284
ALGAE, ALGAL CONTROL, ALGICIDES,   SEDIMENTS, DISSOLVED OXYGEN, OXYG   W69-08518
CHLOROPHYLL DERIVATIVES, RECENT   SEDIMENTS, PRECAMBRIAN SEDIMENTS,   W69-08284
RYSOPHYTA, CULTURES, DIAGENESIS,   SEDIMENTS, GAS CHROMATOGRAPHY, LI   W69-08284
ANKTON, SULFATES, SEDIMENTATION,   SEDIMENTS, TASTE, WATER CHEM: /PL   W69-07833
WASTES, BIOLOGICAL COMMUNITIES,   SEDIMENTS, BIOTA, FISH, UTAH, ALG   W69-07846
LEVEL, ZINC RADIOISOTOPES, IRON,   SEDIMENTS, FISH, RADIOECOLOGY, EF   W69-07853
ES, *CYCLING NUTRIENTS, GEORGIA,   SEDIMENTS, BACTERIA, DETRITUS, ES   W69-08274
IBUTION, ANIMALS, PLANTS, BOTTOM   SEDIMENTS, SOILS, NUCLEAR EXPLOSI   W69-08269
ERRESTRIAL HABITATS, DEPTH, AIR,   SEDIMENTS, GEOLOGIC FORMATIONS, A   W69-08272
PES, *TRANSFER, WATER, BACTERIA,   SEDIMENTS, MINNOWS, DETRITUS, CYC   W69-07861
VELS, MUD, SANDS, SHALES, BOTTOM   SEDIMENTS, CATTAILS, RADIOACTIVIT   W69-08272
S, CLAMS, CRABS, NORTH CAROLINA,   SEDIMENTS, RADIOACTIVITY, MARINE   W69-08267
URAL WATERSHEDS, ORGANIC MATTER,   SEDIMENTS, FERTILIZERS, WATER POL   W70-04193
POLLUTION EFFECTS, PRODUCTIVITY,   SEDIMENTS, URBANIZATION, WATER PO   W70-04721
CE ALGAE, AQUATIC PLANTS, BOTTOM   SEDIMENTS, ANIMALS, FISH, INFLOW,   W70-06225
ATMENT, *LAKES, *ECOLOGY, BOTTOM   SEDIMENTS, MUD, BENTHOS, WISCONSI   W70-06217
, URBANIZATION, WATER POLLUTION,   SEDIMENTS, PALEOLIMNOLOGY, NUTRIE   W70-05663
NT/ *EUTROPHICATION, LAKES, LAKE   SEDIMENTS, ALGAL CONTROL, *NUTRIE   W72-06052
OM/ *EUTROPHICATION, LAKE, *LAKE   SEDIMENTS, *NUTRIENTS, NITROGEN C   W72-06051
S.:   SEEPAGE LOSSES, DIURNAL VARIATION   W69-09723
PITATION(ATMOSPHERIC), WETLANDS,   SEEPAGE, INDUSTRIAL WASTES, TERTI   W69-10178
*SITE   SELECTION, LOADING.:   W71-08221
NITROGEN, PHOSPHORUS, POTASSIUM,   SELECTIVITY, POISONS, OXYGEN, TRA   W70-05399
OSCILLATORI/ BACILLARIOPHYCEAE,   SELENASTRUM CAPRICORNUTUM PRINTZ,   W70-05752
E(CALIF), CHLORELLA PYRENOIDOSA,   SELENASTRUM GRACILE, CARBON ASSIM   W70-03983
GALEATA MENDOTAE, DAPHNIA MAGNA,   SELENASTRUM, RHODOTORULA, TEMPERA   W70-03957
ABILITY, ALGAL GROWTH POTENTIAL,   SELENASTRUM CAPRICORNUTUM, ANABAE   W69-06864
URES, SELENASTRUM CAPRICORNUTUM,   SELENASTRUM GRACILE, GROWTH LIMIT   W70-02779
TRUM GRACILE, GROWTH LIMITATION,   SELENASTRUM, ALGAL GROWTH, ANABAE   W70-02779
ISIONAL ALGAL ASSAY, PROCEDURES,   SELENASTRUM CAPRICORNUTUM, SELENA   W70-02779
UISANCE ALGAE, SEWAGE TREATMENT,   SELF-PURIFICATION, DISSOLVED OXYG   W72-01798
OMPOSITION), ALGAE, OXYGENATION,   SELF-PURIFICATION, LABORATORY TES   W72-01801
KTON, FISH, CHRYSOPHYTA, SESTON,   SELF-PURIFICATION.: /GAE, ZOOPLAN   W70-00268
(WATER), ACIDITY, OXYGEN, ALGAE,   SELF-PURIFICATION, BACTERIOPHAGE,   W69-07838
S, *AXENIC CULTURES, PEDIASTRUM,   SELF-PURIFICATION, ALGAL ENUMERAT   W70-04381
G PLANTS, ROOTED AQUATIC PLANTS,   SELF-PURIFICATION, BIOCHEMICAL OX   W70-05750
TIC ALGAE, ALGAE, ARID CLIMATES,   SEMIARID CLIMATES, *ABSORPTION, *   W69-03491
*EFFECTS, NAVICULA   SEMINULUM VAR.HUSTEDTII.:   W71-03183
ALGAL   SENESCENCE, GROWTH INHIBITIONS.:   W72-02417
EAN WATER, GENUS, SPECIES, TOUCH   SENSATION.: / INDUSTRIAL USES, CL   W70-05389

, LAKES, DESIGN CRITERIA, REMOTE SENSING, MINNESOTA, SEDIMENT-WATE W70-02775
ELEASES, CURRENTS(WATER), REMOTE SENSING, SEA WATER, CHLOROPHYLL, W70-05377
RA, AMBIENT TEMPERATURE, THERMAL SENSITIVITY, GENERA, 'FLAVOBACTER W70-00713
UTION, *WATER POLLUTION SOURCES, SEPARATION TECHNIQUES, *ANALYTICA W69-01273
FILTRATION, *DIATOMACEOUS EARTH, SEPARATION TECHNIQUES, PRESSURE, W72-08688
, FARM WASTES, MUNICIPAL WASTES, SEPTIC TANKS, RECREATION WASTES, W70-04266
, ALGAE, SCUM, WEEDS, NUTRIENTS, SEPTIC TANKS, EFFLUENTS, SWAMPS, W70-04193
NG, SPORT FISH/ *EUTROPHICATION, SEQUENCE, LAKES, CHEMICALS, FISHI W68-00172
ILE SUSPENDED SOLIDS, MULTI-CELL SERIAL SYSTEM, PH, FACULTATIVE PO W70-04790
ATES, ELEUTHERA ISLAND, NITZCHIA SERIATA, NAVICULA, FLORIDA CURREN W70-00707
MIN B-1, VITAMIN B-12, THIAMINE, SERRATIA MARINORUBRA, GYRODINIUM W70-05652
AM REACTION, CYTOPHAGA, PROTEUS, SERRATIA, BAY OF FUNDY(CANADA), R W70-03952
(CANADA), NEW BRUNSWICK(CANADA), SERVALL OMNI-MIXER HOMOGENIZER.: / W70-04365
WATER MASSES, U S PUBLIC HEALTH SERVICE, GREAT LAKES-ILLINOIS RIV W70-00263
ENTHIC FAUNA, CHEMICAL ANALYSIS, SESSILE ALGAE, *PRIMARY PRODUCTIV W72-06052
DIMENT-WATER INTERFACES, OREGON, SESSILE ALGAE, CYANOPHYTA.: /, SE W72-06051
, RESERVOIRS, LAKES, REAERATION SESTON, CHLOROPHYLL, CYCLES, UNIT W70-05092
NION, EPILIMNION, PHYTOPLANKTON, SESTON, CRUSTACEANS, DIATOMS, CHL W69-10182
MICROORGANISMS, MICROORGANISMS, SESTON, INVERTEBRATES, SHELLFISH, W69-04276
BIC CONDITIONS, ALGAE, ROTIFERS, SESTON, RESPIRATION, ALTITUDE, SA W69-10154
, *EUTROPHICATION, AQUATIC LIFE, SESTON, SCENEDESMUS, MATHEMATICAL W73-01446
ZOOPLANKTON, FISH, CHRYSOPHYTA, SESTON, SELF-PURIFICATION.: /GAE, W70-00268
A, CHROOMONAS ACUTA, SCHROEDERIA SETIGERA, MELOSIRA AMBIGUA, CERAT W70-05387
EN DEMAND, ALGAE, SEDIMENTATION, SETTLING BASINS.: /OCHEMICAL OXYG W70-09320
CITY TEST, SLUDGE DRYING, SOLIDS- SETTLING, ECONOMIC ADVANTAGE, TEM W70-05819
CITY, CARBON-14, DDT, MALATHION, SEVIN.:   *ALGAE TOXI W70-02198
ENOIDOSA, PESTICIDE DEGRADATION, SEVIN, MALATHION, ACETONE, ETHANO W70-02968
OIDOSA, SCHRODERIA, THIMET, TRI/ SEVIN, MALATHION, CHLORELLA PYREN W70-08097
ZATION, LAKES, SEWAGE TREATMENT, SEWAGE BACTERIA, NITROGEN, PHOSPH W70-02787
INE, NITROGEN COMPOUNDS, SEWAGE, SEWAGE BACTERIA, SEWAGE TREATMENT W69-10171
ENT, *TRICKLING FILTERS, *BIOTA, SEWAGE BACTERIA, SYSTEMATICS, FUN W72-01819
RINE ENVIRONMENT, WATER QUALITY, SEWAGE BACTERIA, ENVIRONMENTAL EF W69-04276
ATES, SHELLFISH, WASTE DISPOSAL, SEWAGE DISPOSAL, WATER CHEMISTRY, W69-04276
OWTH RATES, STABILIZATION PONDS, SEWAGE DISPOSAL.: /, NITROGEN, GR W68-00855
ICATION, *MINNESOTA, RECREATION, SEWAGE DISPOSAL, WATER POLLUTION, W70-04810
ION ABATEMENT, SEWAGE TREATMENT, SEWAGE DISPOSAL, WASTE DISPOSAL, W69-06909
N CONCENTRATION, ELECTRIC POWER, SEWAGE DISPOSAL, *WASTE WATER TRE W71-09548
STS, LEGAL ASPECTS, LEGISLATION, SEWAGE DISTRI: /WAY, ESTIMATED CO W70-03344
ASTES, SEWERS, SEWAGE TREATMENT, SEWAGE DISPOSAL, STORM DRAINS, SU W70-03344
, DIATOMS, RESERVOIRS, CULTURES, SEWAGE EFFLUENTS, WISCONSIN, ALGI W70-02982
HYTA, CRUSTACEANS, SOIL EROSION, SEWAGE EFFLUENTS, FARM WASTES. PO W70-02803
OLLUTION IDENTIFICATION, SEWAGE, SEWAGE EFFLUENTS, SEWAGE TREATMEN W70-02248
N, NUTRIENTS, LIGHT PENETRATION, SEWAGE EFFLUENTS, OXYGEN, SOLAR R W70-01933
ESTS, BIOCHEMICAL OXYGEN DEMAND, SEWAGE EFFLUENTS, SANITARY ENGINE W72-01817
OVERFLOW, SEWERS, STORM DRAINS, SEWAGE EFFLUENTS, *SEWAGE TREATME W71-10654
EPILIMNION, NITROGEN COMPOUNDS, SEWAGE EFFLUENTS, LIGHT INTENSITY W69-10167
, *PRODUCTIVITY, NUISANCE ALGAE, SEWAGE EFFLUENTS, PHOSPHORUS, EUT W70-00270
ION, WATER CONTROL, FERTILIZERS, SEWAGE EFFLUENT, NUTRIENTS, LAKES W70-00269
TER POLLUTION CONTROL, *FILTERS, SEWAGE EFFLUENTS, ELECTROLYTES, Z W70-01027
REATMENT, WATER QUALITY CONTROL, SEWAGE EFFLUENTS, ABSORPTION, LAK W70-01031
TREATMENT, NITROGEN, PHOSPHORUS, SEWAGE EFFLUENTS.: /, BIOLOGICAL W70-08980
ATER POLLUTION, ODOR, EFFLUENTS, SEWAGE EFFLUENTS, DILUTION, DECOM W71-00940
*NEW HAMPSHIRE, *COPPER SULFATE, SEWAGE EFFLUENTS, SMELTS, FISHKIL W69-08674
DENSITY DURRENTS, MEASUREMENTS, SEWAGE EFFLUENTS, TEMPERATURE, OX W69-06203
N, EUTROPHICATION, STORM DRAINS, SEWAGE EFFLUENTS, STREAMFLOW, MEA W70-05112
N SOURCES, WATER QUALITY, ALGAE, SEWAGE EFFLUENTS.: /ATER POLLUTIO W70-05084
, OLIGOTROPHY, DIVERSION, CORES, SEWAGE EFFLUENTS, DIATOMS, PHOSPH W70-05663
NS, PHOSPHORUS COMPOUNDS, LAKES, SEWAGE EFFLUENTS, ARSENIC COMPOUN W70-05269
CHEMICAL PROPERTIES, WISCONSIN, SEWAGE EFFLUENTS, NITROGEN, PHOSP W70-05113
C, *ESTUARIES, HUMAN POPULATION, SEWAGE EFFLUENTS, AQUATIC POPULAT W68-00461
ATORS, NUISANCE ALGAE, NEW YORK, SEWAGE EFFLUENTS, SEWAGE TREATMEN W70-03974
EN/ *MINNESOTA, *EUTROPHICATION, SEWAGE EFFLUENTS, TERTIARY TREATM W70-03512
DS, ALGAE, PHOSPHATES, NITRATES, SEWAGE EFFLUENTS, TERTIARY TREATM W70-04074
, NUISANCE ALGAE, WATER QUALITY, SEWAGE EFFLUENTS, DIATOMS, MISSOU W70-04507
ONS, EUTROPHICATION, CYANOPHYTA, SEWAGE EFFLUENTS, CHLOROPHYTA, RH W70-04510
, NITROGEN FIXATION, CYANOPHYTA, SEWAGE EFFLUENTS, MINNESOTA, WATE W70-04506
CHLORELLA PYRENOIDOSA, SYNTHETIC SEWAGE FEED, SYMBIOTIC RELATIONSH W70-05655
CULTIVATION, CHANNEL MORPHOLOGY, SEWAGE INPUT, BIOLOGICAL PROPERTI W68-00475
TREATMENT, STABILIZATION PONDS, SEWAGE LAGOONS.: /MENT, SECONDARY W68-00012
E, *BACTERIA, *DISSOLVED OXYGEN, SEWAGE LAGOONS, PLANT PHYSIOLOGY, W72-01818
*PHOSPHORUS, *NITROGEN, *OXYGEN, SEWAGE PLANTS, CHLOROPHYLL, PHOTO W70-04268
GERN(GERMANY), GERMANY, FINLAND, SEWAGE RING TUBES.: /NY), LAKE TE W70-05761
CHEMCONTROL, INDUSTRIAL WASTES, SEWAGE TREATMENT.: /, PHOSPHORUS, W70-05668
IURNAL, LIGHT INTENSITY, CARBON, SEWAGE TREATMENT, TERTIARY TREATM W70-05387
GAE, NEW YORK, SEWAGE EFFLUENTS, SEWAGE TREATMENT, INDUSTRIAL WAST W70-03974
USTRIAL WASTES, DOMESTIC WASTES, SEWAGE TREATMENT, HARVESTING OF A W68-00478
ION CONCENTRATION, *PHOSPHORUS, SEWAGE TREATMENT, *OXIDATION LAGO W68-00012
E, PHOSPHATES, NITRATES, RIVERS, SEWAGE TREATMENT, WATER POLLUTION W69-01453
CING ALGAE, POLLUTION ABATEMENT, SEWAGE TREATMENT, SEWAGE DISPOSAL W69-06909
MICROORGANISMS, NUISANCE ALGAE, SEWAGE TREATMENT, SELF-PURIFICATI W72-01798
IGN CRITERIA, WATER TEMPERATURE, SEWAGE TREATMENT, AEROBIC BACTERI W71-08221
, *WATER POLLUTION, DEGRADATION, SEWAGE TREATMENT, NITROGEN COMPOU W72-02817
EN DEMAND, FISH, ALGAE, FOAMING, SEWAGE TREATMENT.: /CHEMICAL OXYG W70-08740
ORELLA, SCENEDESMUS, ADSORPTION, SEWAGE TREATMENT, PARTICLE SIZE.: W71-03035
OCHEMICAL OXYGEN DEMAND, MIXING, SEWAGE TREATMENT, WASTE WATER TRE W70-06899
ED LOAD, SOLIDS CONTACT PROCESS, SEWAGE TREATMENT, TRICKLING FILTE W70-06792
POUNDS, SEWAGE, SEWAGE BACTERIA, SEWAGE TREATMENT, WATER POLLUTION W69-10171
SHINGTON, *OXYGEN, *FLUCTUATION, SEWAGE TREATMENT, SOLAR RADIATION W69-10182
ATION, SEWAGE, SEWAGE EFFLUENTS, SEWAGE TREATMENT, WATER POLLUTION W70-02248
*RUNOFF, *FERTILIZATION, LAKES, SEWAGE TREATMENT, SEWAGE BACTERIA W70-02787

528

AGE LAGOONS, *INDUSTRIAL WASTES,　　SEWAGE TREATMENT, NORTH DAKOTA, C　　W70-03312
SPOSAL, DOMESTIC WASTES, SEWERS,　　SEWAGE TREATMENT, SEWAGE DISPOSAL　　W70-03344
　　　　　　　　　　　　　*NEW ZEALAND,　　SEWAGE TREATMENT POND ALGAE.:　　W70-02304
ESTIC WASTES, INDUSTRIAL WASTES,　　SEWAGE WASTES, BYPRODUCTS, FERTIL　　W70-09905
*SEWAGE TREATM/ *EUTROPHICATION,　　SEWAGE, *PHOSPHORUS, PHOSPHATES,　　W68-00256
STRIAL WASTES, MUNICIPAL WASTES,　　SEWAGE, ADSORPTION, ACTIVATED CAR　　W70-04001
*GREAT LAKES, *WATER POLLUTION,　　SEWAGE, AESTHETICS, AEROBIC CONDI　　W70-04430
MS, EUTROPHICATION, OLIGOTROPHY,　　SEWAGE, ALGAE, SCUM, WEEDS, NUTRI　　W70-04193
DEMAND, SUSPENDED LOAD, WASTES,　　SEWAGE, ALGAE, MICROORGANISMS, PH　　W70-08627
ON), *WATER PURIFICATION, ALGAE,　　SEWAGE, ANAEROBIC, OXYGEN, NITRIF　　W70-05092
ESTIC WASTES, *SEWAGE TREATMENT,　　SEWAGE, BIOLOGICAL TREATMENT, SOL　　W68-00248
INGTON, *LAKES, *EUTROPHICATION,　　SEWAGE, BIOTA, PHYTOPLANKTON, DIA　　W69-10169
REATMENT FACILITIES, FILTRATION,　　SEWAGE, COLIFORMS, EUTROPHICATION　　W70-00263
*CITIES, BIOLOGICAL COMMUNITIES,　　SEWAGE, DETRITUS, TUBIFICIDS, PLA　　W70-00268
INTENSITY, CONDUCTIVITY, ALGAE,　　SEWAGE, DIATOMS, CYANOPHYTA, CARB　　W70-05405
ER), DIATOMS, FOULING, PLANKTON,　　SEWAGE, DOMESTIC WASTE, INDUSTRIA　　W72-01813
ATER POLLUTION TREATMENT, LAKES,　　SEWAGE, DRAINAGE DISTRICTS, DEBT,　　W70-04455
NUTRIENTS, PHOSPHORUS, INFLUENT　　SEWAGE, EQUILIBRIUM, MATHEMATICAL　　W68-00858
CONVECTION, WAVES(WATER), FISH,　　SEWAGE, EUGLENOPHYTA, DIATOMS, BI　　W70-03969
AKES, RIVERS, NITROGEN, BIOMASS,　　SEWAGE, FARM WASTES, CARBON CYCLE　　W70-02777
N, *CYANOPHYTA/ *EUTROPHICATION,　　SEWAGE, FISH POPULATIONS, PLANKTO　　W68-00461
ED STATES, PHOSPHORUS, NITROGEN,　　SEWAGE, INDUSTRIAL WASTES, DOMEST　　W68-00478
CAL PROPERTIES, DOMESTIC WASTES,　　SEWAGE, INDUSTRIAL WASTES, AGRICU　　W70-03964
GI, INSECTS, BACTERIA, PROTOZOA,　　SEWAGE, LAKES, NUTRIENTS.: /, FUN　　W71-10786
REATMENT, PHOTOSYNTHESIS, ALGAE,　　SEWAGE, MUNICIPAL WA: / AEROBIC T　　W72-00462
UTION, DRAINAGE, ORGANIC MATTER,　　SEWAGE, PHYTOPLANKTON, PRODUCTIVI　　W70-05404
FFECTS, WATER POLLUTION SOURCES,　　SEWAGE, PHOSPHORUS, CHEMCONTROL,　　W70-05668
WATER POLLUTION, LAKES, STREAMS,　　SEWAGE, REMEDIES, JUDICIAL DECISI　　W69-06909
LLUTION CONTROL, OXYGEN BALANCE,　　SEWAGE, RUNOFF.: /URCES, WATER PO　　W69-08668
IENTS, POLLUTION IDENTIFICATION,　　SEWAGE, SEWAGE EFFLUENTS, SEWAGE　　W70-02248
N, CHLORINE, NITROGEN COMPOUNDS,　　SEWAGE, SEWAGE BACTERIA, SEWAGE T　　W69-10171
ZATION, PLANT GROWTH SUBSTANCES,　　SEWAGE, SULFATE, MANGANESE, IRON,　　W70-02969
STREAMS, ALGAE, FISH, WISCONSIN,　　SEWAGE, TROUT, NITROGEN, PHOSPHOR　　W69-10178
OGEN, AMMONIA, ALGAE, NUTRIENTS,　　SEWAGE, WASTE WATER(POLLUTION), W　　W71-05879
IDE RESIDUES, PHOSPHATES, ALGAE,　　SEWAGE, WATER POLLUTION, WATER LA　　W69-06305
TIVITY, AGRICULTURAL WATERSHEDS,　　SEWAGE, WATER POLLUTION, CHEMICAL　　W70-05752
NATION, COAGULATION, WATERSHEDS,　　SEWERAGE ALGAE, POTABLE WATER.: /　　W71-05943
EFFECTS, BACTERIA, ALGAE, FISH,　　SEWERS, RUNOFF, WASTES, WATER SUP　　W71-09880
WASTE DISPOSAL, DOMESTIC WASTES,　　SEWERS, SEWAGE TREATMENT, SEWAGE　　W70-03344
WASTE WATER TREATMENT, OVERFLOW,　　SEWERS, STORM DRAINS, SEWAGE EFFL　　W71-10654
NG, *SUSPENDED SOLIDS, *COMBINED　　SEWERS, STORMWATER OVERFLOWS.: /I　　W71-10654
MS, WHITE FISH, ALEWIFE, GIZZARD　　SHAD, SEA LAMPHREY, COHO SALMON.:　　W70-01943
LAGELLATES, LAKE CLASSIFICATION,　　SHAGAWA LAKE(MINN).: /ROTROPHY, F　　W70-02775
CHROMATOGRAPHY, LIPIDS, OIL, OIL　　SHALE.: /GENESIS, SEDIMENTS, GAS　　W69-08284
A, SCALING, GRAVELS, MUD, SANDS,　　SHALES, BOTTOM SEDIMENTS, CATTAIL　　W69-08272
PHORUS, CARBONATES, TEMPERATURE,　　SHALLOW WATER, HYDROGEN SULFIDE,　　W70-02803
EUTROPHICATION, DOMESTIC WASTES,　　SHALLOW WATER, CHLOROPHYTA, CHRYS　　W70-00711
G, WATERFOWL, BIRDS, COSTS, COST-　　SHARING.: /ON TECHNIQUES, SWIMMIN　　W70-05568
BANON, WADLEIGH STATE PARK, LAKE　　SHATUTAKEE, MASCOMA LAKE, GLEN LA　　W69-08674
UTOTROPHIC INDEX, CHLOROPHYLL A,　　SHAYLER RUN, ORGANIC CARBON, CINC　　W71-04206
PELIC ALGAE, ALGAL ASSOCIATIONS,　　SHEAR WATER(ENGLAND), AMPHORA OVA　　W71-12870
PANDANUS, MUSCLE, BONE, LEAVES,　　SHELL, RHODIUM-102.: /A, PISONIA,　　W69-08269
VIEWS, HABITATS, CLAMS, MUSSELS,　　SHELLFISH, DINOFLAGELLATES, CYANO　　W70-05372
RGANISMS, SESTON, INVERTEBRATES,　　SHELLFISH, WASTE DISPOSAL, SEWAGE　　W70-04276
IZERS, ALGAE, ZOOPLANKTON, FISH,　　SHELLFISH, STEAM, ELECTRIC POWER,　　W70-09905
, ALGAL TOXINS, LARVAE, PHENOLS,　　SHELLFISH, OYSTERS, INDUSTRIES, C　　W70-01068
　　　　　　　　　　　　　　　　　　　　SHENANDOAH RIVER.:　　　　　　　　　　W70-03068
DITCHES, SOUP, FOOD WASTES.:　　SHEPPARTION(AUSTRALIA), OXIDATION　　W71-09548
ORPTION, PHYTOPLANKTON, SUCKERS,　　SHINERS, CADDISFLIES, TEMPERATURE　　W69-07853
R BASINS PROJECT, INDIANA HARBOR　　SHIP CANAL.: /LAKES-ILLINOIS RIVE　　W70-00263
EN SEA, WEAPONS TESTING, NUCLEAR　　SHIP SAVANNAH, ZIRCONIUM-95, RUBI　　W70-00707
STERS, OILY WATER, MARINE ALGAE,　　SHIPS.: /OLLUTION, CLEANING, DISA　　W70-01470
, NUCLEAR REACTORS, POWERPLANTS,　　SHIPS, SUBMARINES, EFFLUENTS, ION　　W70-00707
, POTOMAC RIVER, SAVANNAH RIVER,　　SHOCK EFFECTS.: /ZOAN COMMUNITIES　　W70-07847
*BIOLOGICAL EXTRACTION, GROWTH,　　SHOCK LOADING.:　　　　　　　　　　　　W70-06604
ANAGERS, FISH YIELDS, UNDERWATER　　SHOCK WAVE, ANABAENA, PHYGON XL(D　　W69-08674
TOMAC RIVER, SAVAN/ *TEMPERATURE　　SHOCK, *PROTOZOAN COMMUNITIES, PO　　W70-07847
DRAINAGE.:　　　　　　　　　　　　　　SHORELINE VEGETATION, STORMWATER,　　W70-02251
MIXING, SEA WATER, FRESH WATER,　　SHORES, WATER CIRCULATION, FLOODS　　W69-06203
BENTHIC FAUNA, WATERSHEDS, WATER　　SHORTAGE FILTERS, SNAILS, ODOR, C　　W72-00027
GAE, CRUSTACEANS, BENTHOS, BRINE　　SHRIMP, ECOSYSTEMS, FISH, PHYTOPL　　W69-07440
POLLUTION, AQUATIC ALGAE, BRINE　　SHRIMP, FISH, FOOD CHAINS, MUSSEL　　W70-01467
OOD CHAINS, AQUATIC ALGAE, BRINE　　SHRIMP, FISH, MUSSELS.: /UTION, F　　W70-01466
OSYSTEMS, SINKS, TRACERS, CRABS,　　SHRIMP, MULLETS, FISH, ALGAE.: /C　　W69-08274
DEPTH, CYANOPHYTA, CHLOROPHYTA,　　SIEROZEMS, SOIL MICROBIOLOGY, DIS　　W71-04067
RUS, CALCIUM, MAGNESIUM, SODIUM,　　SILICA, DIATOMS, RESISTANCE, *PLA　　W68-00856
MALS, DENSITY, IRON, PHOSPHORUS,　　SILICA, SAMPLING, CHEMICAL ANALYS　　W70-06975
, PHOSPHATES, NITRATES, DAPHNIA,　　SILICA, THERMAL STRATIFICATION, S　　W70-05760
, *EUTROPHICATION, ALGAE, CLAYS,　　SILICATES.: /, *LAKES, *SEDIMENTS　　W69-03075
NITRATES, NUTRIENTS, PHOSPHATES,　　SILICATES, IRON OXIDES, *PLANT PO　　W68-00860
, PLANKTON, TUBIFICIDS, CALCIUM,　　SILICATES, DIATOMS, WINDS, NITROG　　W70-05761
DES, AMMONIUM COMPOUNDS, OXYGEN,　　SILICATES, ORGANIC MATTER, AIR TE　　W69-07096
N, CHLORELLA, SODIUM, POTASSIUM,　　SILICATES, DISSOLVED SOLIDS, CALC　　W70-03512
ESMUS QU/ *ALGAL BLOOMS, *ORGANO-　　SILICATES, *ALKOXYSILANES, SCENED　　W73-03056
RIMARY PRODUCTIVITY, PHOSPHATES,　　SILICATES, CITRATES, BIOASSAY, NE　　W70-01579
ESMUS QU/ *ALGAL BLOOMS, *ORGANO-　　SILICATES, *ALKOXYSILANES, SCENED　　W71-05719
PHIC, CHEMOTROPHIC, AUTOTROPHIC,　　SILICON, PEDIASTRUM BORYANUM, STA　　W72-02804
AUXINS, METABOLITES, INHIBITION,　　SILICON, VANADIUM, ZINC, GALLIUM.　　W69-04801
IENTS, *PESTICIDES, FARM WASTES,　　SILT, EROSION, SEDIMENTATION, RUN　　W71-10376
HING, SPORT FISHING, RECREATION,　　SILT, NUTRIENTS, AQUATIC POPULATI　　W68-00172

ALDRIN, DIELDRIN, DDT, AERATION,
RADIOISOTOPES, CESIUM, FALLOUT,
POLLUTANTS, AQUATIC LIFE, ALGAE,
N), RUSTING, METALS, INHIBITORS,
*ESTUARIES, FRESH WATER, LAKES,
WALLEYE, VEGETATION, WILD RICE,
, SAMPLING, DENSITY, LIGHT, MUD,
OROPHYLL, DYNAMICS, TEMPERATURE,
, POTASSIUM PERMANGANATE, FENAC,

CONIUM-95-NIOBIUM, EUROPIUM-155,
ERNANTHERA, HETERANTHERA, NAJAS,
ANGANATE, FENAC, SILVER NITRATE,

IUM-106, NIOBIUM-95, GYMNODINIUM
CANADA, AQUATIC ECOSYSTEM MODEL,
IDS, UPTAKE, CONSUMER ORGANISMS,
TZSCHIA DISSIPATA, SYNEDRA ACUS,
ESTUARIES, PLANKTON, ECOSYSTEMS,
ONAL PARK, *ICELAND HOT SPRINGS,
T SPRINGS, SINTER, NUCLEIC ACID,
GAE, WATER POLLUTION SOURCES, ON-
Y, GASTROPODS, MARINE PLANTS, ON-
LIBRIUM, MATHEMATICAL MODELS, ON-
UIREMENTS, *BOTTOM SEDIMENTS, ON-
ES, *PHOSPHORUS RADIOISOTOPE, ON-
MICAL ANALYSIS, PH, SAMPLING, ON-
ATION, OXYGEN, SPORT FISHING, ON-
S, ON-SITE DATA COLLECTIONS, *ON-
RE, METABOLISM, EPIPHYTOLOGY, ON-
SYNTHESIS, LIGHT PENETRATION, ON-
VOIRS, ANALYTICAL TECHNIQUES, ON-
AY, AQUATIC LIFE, COOLING WATER,
T, WATER QUALITY, SO/ *RESERVOIR
TION, SEWAGE TREATMENT, PARTICLE
CONTROL, *FILTRATION, *PARTICLE
IA SCHOD/ *FILTERING RATE, *BODY
OPES, CALCIUM, POTASSIUM, CLAMS,
, *POPULATION DYNAMICS, CELLULAR
HE/ *DNA, *ORGANIC CARBON, *CELL
ONAS, LABORATORY TESTS, PARTICLE
MUNITIES, OKANAGAN LAKE(CANADA),
, MELOSIRA, NAVICULA, NITZSCHIA,
UGLENA GRACILIS, CACHONINA NIEI,
NUTRITION, DITYLUM, CHAETOCEROS,
UM PRINTZ, OSCILLATORIA VAUCHER,
HUXLEYI, PYRAMIMONAS, PERIDINI/
ALCATUS, CHLAMYDOMONAS MOEWUSII,
PSOIDEA, CHLORELLA LUTEOVIRIDIS,
S, TEMPERATURE, DEPTH, BACTERIA,
PHYTA, RHODOPHYTA, EUGLENOPHYTA,
O), RHIZOPODS, SPIDER/ *ZOOGLEAL
QUALITY CONTROL, ALGAL CONTROL,
GICIDAL TESTS, ALGISTATIC TESTS,
GAE BLOOMS, *FISH TOXICITY TEST,
M, *ALUMINUM SULFATE, *ACTIVATED
XYGEN, *OHIO, WATER TEMPERATURE,
ALGAE, PLANKTON, BENTHIC FAUNA,
, FACULTATIVE POND DESIGNS, BOD,
MENT, ALGAE, BACTERIA, ACTIVATED
D, *COLOR, WASTE WAT/ *ACTIVATED
TREATMENT, *PROTOZOA, *ACTIVATED
ECLAMATION, / *ALGAE, *ACTIVATED
TION, ORGANIC LOADING, ACTIVATED
UGMENTATION, *LAGOONS, ACTIVATED
OXYGEN DEMAND, COLOR, ACTIVATED
ME, *NITROGEN, HYDRO/ *ACTIVATED
EMICAL OXYGEN DEMAND, *ACTIVATED
CONDITIONS, CANNERIES, ACTIVATED
HEMICAL OXYGEN DEMAND, ACTIVATED
IALYSIS, ION EXCHANGE, ACTIVATED
ES, SEDIMENTATION, CHLORINATION,
SSOLVED OXYGEN, LIGHT INTENSITY,
DS, LAGOON, FILTERS, PESTICIDES,
E, SLURRIES, SPRAYING, ACTIVATED
RGANISMS, TEMPERATURE, ACTIVATED
ICAL FEASIBILITY, COST ANALYSIS,
GEN DEMAND, *BACTERIA, ACTIVATED
OLOGICAL DISTRIBUTION, ACTIVATED
ONS, ANAEROBIC DIGESTION, ALGAE,
DGE, AERATION, DISSOLVED OXYGEN,
XYGEN DEMAND, OXIDATION LAGOONS,
WASTES, MICROBIOLOGY, ACTIVATED
OLIFORMS, SEDIMENTS, MUD, SANDS,
ECHNIQUES, PRESSURE, FLOW RATES,
UTION SOURCES, ODOR, FARM WASTE,
OGEN FIXING BACTERIA, *ACTIVATED
ED LOAD, PILOT PLANTS, ACTIVATED
DAIRY INDUSTRY, *ALGAE, *CATTLE,
S, PRESSURE, FLOW RATES, SLUDGE,

SILTS.: /SORPTION, ALGAE, CLAYS,   W70-09768
SILTS, DETRITUS, BOTTOM SEDIMENTS   W69-09742
SILTS, ECOLOGY, AQUATIC ENVIRONME   W69-10080
SILTS, ORGANIC MATTER, IRON OXIDE   W70-02294
SILTS, PHOSPHATES, NITROGEN, WAST   W70-02255
SILTS, PLANKTON, MAYFLIES, FAUNA,   W70-01943
SILTS, SANDS, GRAVELS, SEDIMENTS,   W70-00711
SILTS, STANDING CROP, ECOLOGY, PH   W70-02780
SILVER NITRATE, SIMAZIN, NUPHAR,   W70-10175
SILVER.:   W72-05143
SILVER-110M, CERIUM, GUETTARDA, I   W69-08269
SILVEX, ENDOTHAL, POTASSIUM PERMA   W70-10175
SIMAZIN, NUPHAR, AGASICLES, MANAT   W70-10175
SIMAZINE 80 W, DACTHAL.:   W70-03519
SIMPLEX, FLAGELLATES, ELEUTHERA I   W70-00707
SIMULATION MODEL, PHOSPHORUS EFFE   W69-06536
SIMULIUM, PHYSA, SALMO TRUTTA, EN   W69-09334
SINKING RATES, GRAZING PRESSURE,   W69-10158
SINKS, TRACERS, CRABS, SHRIMP, MU   W69-08274
SINTER, NUCLEIC ACID, SISJOTHANDI   W69-10160
SISJOTHANDI(ICELAND), ENZYME-SUBS   W69-10160
SITE DATA COLLECTIONS, *ON-SITE I   W70-09976
SITE DATA COLLECTIONS.: /MORTALIT   W70-01780
SITE DATA COLLECTIONS, BOTTOM SED   W68-00858
SITE DATA COLLECTIONS, AQUATIC PL   W68-00511
SITE DATA COLLECTIONS, FISHES, PL   W68-00857
SITE DATA COLLECTIONS, LIMNOLOGY.   W69-00659
SITE INVESTIGATIONS.: /GHT PENETR   W68-00487
SITE INVESTIGATIONS, FREQUENCY, *   W70-09976
SITE TESTS, LABORATORY TESTS, LOT   W72-05453
SITE TESTS, BIOCHEMICAL OXYGEN DE   W72-01818
SITE TESTS.: /IMPOUNDMENTS, RESER   W71-10098
SITES.: /E, MODEL STUDIES, BIOASS   W71-11881
SITES, WATER RESOURCES DEVELOPMEN   W72-11906
SIZE.: /ELLA, SCENEDESMUS, ADSORP   W71-03035
SIZE, *CLAYS, *SILTS, SANDS, ALGA   W70-08628
SIZE, *COMPARATIVE STUDIES, DAPHN   W70-03957
SIZE, ALGAE, DIATOMS, TRACE ELEME   W70-00708
SIZE, CHLORELLA VULGARIS, CORNELL   W69-06540
SIZE, EUKARYOTES, MONOCHRYSIS LUT   W69-08278
SIZE, POPULATION, GROWTH RATES, W   W72-02417
SKAHA LAKE(CANADA), OSOYOOS LAKE(W71-12077
SKELETONEMA, PLAGIOGRAMMA,: /IGMA   W70-03325
SKELETONEMA COSTATUM, DUNALIELLA   W69-08278
SKELETONEMA, THALASSIOSIRA, COCCO   W70-04503
SKELETONEMA COSTATUM(GREV) CL, CH   W70-05752
SKELETONEMA COSTATUM, COCCOLITHUS   W70-05272
SKELETONEMA COSTATUM, SYNECHOCOCC   W70-05381
SKELETONEMA COSTATUM, PHYSIOLOGIC   W70-04809
SLIME, ALGAE, CYANOPHYTA, DIATOMS   W70-04371
SLIME, CO: /A, CYANOPHYTA, CHRYSO   W70-05389
SLIME, DAYTON(OHIO), VANDALIA(OHI   W72-01819
SLIME, TASTE-PRODUCING ALGAE, ODO   W69-05704
SLIMEPRODUCING ALGAE, PHOTOMICROG   W69-05704
SLUDGE DRYING, SOLIDS-SETTLING, E   W70-05819
SLUDGE PROCESS, HANOVER TREATMENT   W70-09186
SLUDGE WORMS, BLOODWORMS, AQUATIC   W70-00475
SLUDGE WORMS, BLOODWORMS, WATER P   W68-01244
SLUDGE.: / TEMPERATURE, NUTRIENTS   W70-04786
SLUDGE.: / DEMAND, TERTIARY TREAT   W69-08054
SLUDGE, *BIOCHEMICAL OXYGEN DEMAN   W70-04060
SLUDGE, *BACTERIA, METABOLISM, GR   W72-01817
SLUDGE, *EFFLUENTS, WASTE WATER R   W70-06053
SLUDGE, AERATION, DISSOLVED OXYGE   W71-08970
SLUDGE, ALGAE, NUTRIENTS, VIRUSES   W72-06017
SLUDGE, ALGAE, WASTE WATER TREATM   W70-10433
SLUDGE, ALGAE, EUTROPHICATION, LI   W68-00012
SLUDGE, BIOLOGICAL TREATMENT, PIL   W70-05821
SLUDGE, BIOCHEMICAL OXYGEN DEMAND   W71-09546
SLUDGE, COST COMPARISONS, SEDIMEN   W71-09548
SLUDGE, DENITRIFICATION.: /ECTROD   W69-01453
SLUDGE, DETERGENTS, SPRAYING, ALG   W70-04060
SLUDGE, DEPTH, ORGANIC LOADING, D   W72-06854
SLUDGE, DIGESTION, INDIANA.: /OUN   W72-03641
SLUDGE, FARM WASTES.: /AE, *CATTL   W70-07491
SLUDGE, FISHERIES, CARP, CLAMS, P   W70-09905
SLUDGE, INCINERATION, STORM RUN-O   W71-10654
SLUDGE, IRON COMPOUNDS, LIME, OXY   W68-00256
SLUDGE, MAINTENANCE COSTS, CONSTR   W70-04786
SLUDGE, METHANE, ORGANIC LOADING,   W71-11375
SLUDGE, OXIDATION LAGOONS, ALGAE,   W71-08970
SLUDGE, PHOTOSYNTHETIC OXYGEN, FA   W71-08221
SLUDGE, RATES, CHLORELLA, HYDROGE   W68-00248
SLUDGE, RECREATION.: /, OXYGEN, C   W70-10161
SLUDGE, SLURRIES, AUTOMATIC CONTR   W72-08688
SLUDGE, SOLID WASTES, DECOMPOSTIN   W71-05742
SLUDGE, STORAGE, TEMPERATURE, MIC   W71-13335
SLUDGE, TREAT: /REATMENT, SUSPEND   W70-09186
SLURRIES, SPRAYING, ACTIVATED SLU   W70-07491
SLURRIES, AUTOMATIC CONTROL, PIPI   W72-08688

OPPER SULFATE, SEWAGE EFFLUENTS,

SEPHINE(MINN), GRAY'S BAY(MINN),
ORWAY), TRONDHEIMSFJORD(NORWAY),
LOROPHYTA, CHRYSOPHYTA, GRAZING,
ISANCE ALGAE, FISH, ZOOPLANKTON,
AKE TROUT, CRAYFISH, STONEFLIES,
SODIUM ARSENITE, NUISANCE ALGAE,
MS, NEMATODES, ROTIFERS, LARVAE,
ERSHEDS, WATER SHORTAGE FILTERS,
UNA, FISH, OLIGOCHAETES, MIDGES,
TRY, BIOMASS, BENTHOS, SALINITY,
, FISH, WATER ANALYSIS, INSECTS,
, CHLORELLA, PEPTIDES, BACTERIA,
IVER, RED RIVER, SAVANNAH RIVER,
S, FERTILIZERS, WATER POLLUTION,
URCES, SURFACE RUNOFF, RAINFALL,
CONOMIC EFFICIENCY, HEAT BUDGET,
  *MOUNTAIN MEADOW, NITROGEN-15,
ROL, BIOCONTROL, COPPER SULFATE,
ICAL CONTROLS, PARAQUAT, DIQUAT,
TROPHOTOMETRY, ALGAE, MINNESOTA,
ING, COPPER SULPHATE, PONDWEEDS,
UF-21, BENOCLORS, NIGROSINE DYE,
NE PLANTS, *ALGAE, CHLORINATION,
TES, LABORATORY TESTS, CULTURES,
  BRACKISH WATER, ALKALINE WATER,
EA WATER, FRESH WATER, SORPTION,
ALKALINITY, SULFATES, CHLORIDES,
, CALCIUM, MAGNESIUM, POTASSIUM,
MES, SAMPLING, BOTTOM SEDIMENTS,
NUTRIENTS, NITROGEN, PHOSPHORUS,
, PILOT PLANTS, IRON, CHLORELLA,
PHOSPHORUS, CALCIUM, MAGNESIUM,
N, PRODUCTIVITY, TRACE ELEMENTS,
TED WATER, NITRATES, PHOSPHATES,
TARIO, *CANADA, NUTRIENT LEVELS,
MOUNTAINS, *SOIL MICROORGANISMS,
Y, *LEACHING, IRON, TEMPERATURE,
URE, PIPTIDES, INTERTIDAL AREAS,
OPLANKTON, PRIMARY PRODUCTIVITY,
NG, *POLLUTION ABATEMENT, DIKES,
CLIMATES, *ABSORPTION, *CALCIUM,
  AIR POLLUTION, WATER POLLUTION,
LFATE, CHLOROPHYTA, CRUSTACEANS,
ND WATER, RUNOFF, FROZEN GROUND,
OFF, DRAINAGE WATER, GRASSLANDS,
, PHOSPHORUS, *NUTRIENT REMOVAL,
RCES DEVELOPMENT, WATER QUALITY,
ICROORGANISMS, SOIL ALGAE, TURF,
NOPHYTA, CHLOROPHYTA, SIEROZEMS,
MANURES, LAKE MINNETONKA(MINN),
S, PLANT RESIDUES, MORRIS(MINN),
*NITRIFICATION, CROPS, TOXICITY,
OBIOLOGY, DISTRIBUTION PATTERNS,
NAGEMENT(APPLIED), GEOCHEMISTRY,
ALGAE, *FUNGI, *CALCAREOUS SOIL,
I, *CALCAREOUS SOIL, SOIL WATER,
*RADIOSTRONTIUM COMPLEXES, DTPA,
, DRAGONFLIES, ALGAE, SEDIMENTS,
BATCH CULTURES, OPTICAL DENSITY,
PHYTOPLANKTON, DIATOMS, ECOLOGY,
PES, *ALGAE, *FUNGI, *CALCAREOUS
LORIDES, SODIUM, IRON, DISSOLVED
BITION, LIMNOLOGY, ACIDITY, ACID
*PESTICIDE REMOVAL, *ADSORPTION,
TANKS, EFFLUENTS, SWAMPS, FOREST
IMALS, PLANTS, BOTTOM SEDIMENTS,
UORESCENCE, FABRY-PEROT FILTERS,
ITY, OXYGEN-REDUCTION POTENTIAL,
  BOTTOM SEDIMENTS, WEED CONTROL,
ERIA, TRIPTON, BOTTOM SEDIMENTS,
  *FLUCTUATION, SEWAGE TREATMENT,
XUDATION, TIDES, PHOTOSYNTHESIS,
ATION, SEWAGE EFFLUENTS, OXYGEN,
YTOPLANKTON, ALGAE, TEMPERATURE,
RNAL OXYGEN CURVES, CALORIMETRY,
URCES, ODOR, FARM WASTE, SLUDGE,
STES, INDUSTRIAL WASTES, TOXINS,
TIAL, BACTERIA, PROTOZOA, ALGAE,
T, SEWAGE, BIOLOGICAL TREATMENT,
  VOLITILE SOLIDS, NON-DEGRADABLE
L OXYGEN DEMAND, SUSPENDED LOAD,

, OXYGEN TRANSFER, PH, SUSPENDED
CE, COENOBIA, SWIRLING, VOLATILE
SH TOXICITY TEST, SLUDGE DRYING,
TER OVE/ BACKWASHING, *SUSPENDED
:   MIXED LIQUOR, SUSPENDED
TECHNIQUES, TURBIDITY, SUSPENDED

SMELTS, FISHKILL, CRUSTACEANS, BO	W69-08674
SMITH LAKE(ALASKA).:	W69-03515
SMITH'S BAY(MINN), PHOSPHATE DETE	W70-05269
SMOLA(NORWAY), HITRA(NORWAY), FRO	W70-01074
SNAILS, AGE, TURBIDITY, DIPTERA,	W71-00668
SNAILS, APPLICATION METHODS.: /NU	W70-03310
SNAILS, ECOLOGY, EUTROPHICATION,	W70-00711
SNAILS, LEGISLATION, WISCONSIN, C	W70-05568
SNAILS, MITES, MOLDS, GROWTH RATE	W72-01819
SNAILS, ODOR, COOLING TOWERS, FIL	W72-00027
SNAILS, PHOSPHORUS, NUTRIENTS, AL	W70-01943
SNAILS, SULFATES, NITRATES, PHOSP	W70-03978
SNAILS, TOXICITY, RIVERS, HISTOGR	W72-01801
SNAILS, WEIGHT, VACUUM DRYING, BO	W70-04184
SNAKE RIVER, RIO GRANDE RIVER, PO	W70-04507
SNOW, RUNOFF, DRAINAGE WATER, GRA	W70-04193
SNOWMELT, SURFACE WATERS, STREAMS	W70-03501
SOCIAL ASPECTS.: /ON, TURBINES, E	W70-08834
SOD MATS, NON-LEGUMES, *NITROGEN	W72-08110
SODIUM ARSENITE.: /ECHANICAL CONT	W70-10175
SODIUM ARSENITE, COPPER SULPHATE,	W70-00269
SODIUM ARSENITE.: /OMPOUNDS, SPEC	W70-05269
SODIUM ARSENITE, FISH, ANIMALS, A	W70-05263
SODIUM CHLORATE, SANTOBRITE, CLAD	W70-05263
SODIUM COMPOUNDS, PHOTOSYNTHESIS,	W70-00161
SODIUM COMPOUNDS, NITROGEN.: / RA	W70-03507
SODIUM.: /IENTS, BACTERIA, ALGAE,	W71-03031
SODIUM, ALGAE, PHOSPHORUS, TRACE	W70-04385
SODIUM, IRON, DISSOLVED SOILDS, P	W69-05894
SODIUM, IRON, MANGANESE, TRACE EL	W70-02804
SODIUM, POTASSIUM, IRON, MANGANES	W70-09889
SODIUM, POTASSIUM, TRACE METALS,	W70-04503
SODIUM, POTASSIUM, SILICATES, DIS	W70-03512
SODIUM, SILICA, DIATOMS, RESISTAN	W68-00856
SODIUM, VIRUSES, VITAMINS, WATER	W69-04801
SODIUM, WATER POLLUTION EFFECTS,	W70-07257
SOFT WATER.: /E LAKE(CANADA), *ON	W70-02973
SOIL ALGAE, TURF, SOIL MICROBIOLO	W72-08110
SOIL ANALYSIS, TASTE-PRODUCING AL	W72-11906
SOIL BACTERIA, LABORATORY TESTS.:	W70-04365
SOIL CHEMISTRY, WATER CHEMISTRY,	W69-06273
SOIL CONSE: /TION SOURCES, DREDGI	W68-00172
SOIL CONTAMINATION, *STRONTIUM RA	W69-03491
SOIL CONTAMINATION, NUTRIENTS.: /	W71-06821
SOIL EROSION, SEWAGE EFFLUENTS, F	W70-02803
SOIL EROSION, WATER POLLUTION, *D	W69-05323
SOIL EROSION,: /LUTION, SNOW, RUN	W70-04193
SOIL EROSION, ALGAE, *WATER POLLU	W73-01122
SOIL INVESTIGATIONS, *IMPOUNDED W	W72-11906
SOIL MICROBIOLOGY, WATER POLLUTIO	W72-08110
SOIL MICROBIOLOGY, DISTRIBUTION P	W71-04067
SOIL MINERALS, PLANT RESIDUES, MO	W70-04193
SOIL PERCOLATION.: / SOIL MINERAL	W70-04193
SOIL POROSITY, IRRIGATION, GROUND	W69-05323
SOIL PROFILES.: /OZEMS, SOIL MICR	W71-04067
SOIL TYPES, HYDROLOGY, CLIMATES,	W71-03048
SOIL WATER, SOIL WATER MOVEMENT,	W71-04067
SOIL WATER MOVEMENT, LEACHING, AR	W71-04067
SOIL WATER PROFILE.: / ACTIVITY,	W71-04067
SOIL-WATER-PLANT RELATIONSHIPS, N	W71-09005
SOIL-WATER TECHNIQUES, PHYCOLOGY,	W70-04381
SOIL, CULTURES, CYANOPHYTA, WATER	W69-10180
SOIL, SOIL WATER, SOIL WATER MOVE	W71-04067
SOILDS, PHOSPHATES, NITROGEN, ALG	W69-05894
SOILS, ACIDIC WATER, ACIDS, PHOTO	W70-02792
SOILS, COALS, DISTRIBUTION PATTER	W69-09884
SOILS, MINNESOTA, WISCONSIN, MICH	W70-04193
SOILS, NUCLEAR EXPLOSIONS, COBALT	W69-08269
SOLAR FRAUNHOFER LINES, ATTENUATI	W70-05377
SOLAR RADIATION, ENERGY, CARBON D	W70-05750
SOLAR RADIATION, *WATER POLLUTION	W68-00478
SOLAR RADIATION, OXYG: /ICE, BACT	W69-10154
SOLAR RADIATION, CHLOROPHYLL, PRO	W69-10182
SOLAR RADIATION, SALINITY, RAINFA	W70-01073
SOLAR RADIATION, PERIPHYTON, BACT	W70-01933
SOLAR RADIATION.: /OSYNTHESIS, PH	W71-10788
SOLARIMETRY, YELLOWSTONE NATIONAL	W73-03309
SOLID WASTES, DECOMPOSTING ORGANI	W71-05742
SOLID WASTES, RADIOACTIVE WASTES,	W72-00462
SOLID WASTES, TERTIARY TREATMENT.	W70-01027
SOLID WASTES, MICROBIOLOGY, ACTIV	W68-00248
SOLID.: /RY MANURE, TOTAL SOLIDS,	W71-00936
SOLIDS CONTACT PROCESS, SEWAGE TR	W70-06792
SOLIDS REMOVAL.:	W70-06792
SOLIDS.: /N(MO), LEE'S SUMMIT(MO)	W70-06601
SOLIDS.: /DETERMINATION, ABSORBAN	W70-03334
SOLIDS-SETTLING, ECONOMIC ADVANTA	W70-05819
SOLIDS, *COMBINED SEWERS, STORMWA	W71-10654
SOLIDS, *PHOSPHOROUS UPTAKE RATE.	W71-13335
SOLIDS, ALGAE, PILOT PLANTS, HEAD	W72-08686

POTASSIUM, SILICATES, DISSOLVED   SOLIDS, CALCIUM, PRODUCTIVITY, ST   W70-03512
                    *SUSPENDED   SOLIDS, CHASE CITY(VA).:   W72-10573
*WASTES, ALGAE, ASSAY, DISSOLVED   SOLIDS, CHEMICAL ANALYSIS, PHOSPH   W70-03334
TERS, *WATER QUALITY, *DISSOLVED   SOLIDS, DISSOLVED OXYGEN, EVAPORA   W72-01773
                         VOLATILE   SOLIDS, GRIT.:   W71-11375
ECTROPHOTOMETRY, TOTAL DISSOLVED   SOLIDS, HEDRIODISCUS, CALOPARYP: /   W70-03309
LOCITY, WATER QUALITY, DISSOLVED   SOLIDS, HYDROGEN ION CONCENTRATIO   W70-00265
URE PROFILES, VOLATILE SUSPENDED   SOLIDS, MULTI-CELL SERIAL SYSTEM,   W70-04790
Y MANURE, TOTAL SOLIDS, VOLITILE   SOLIDS, NON-DEGRADABLE SOLID.: /R   W71-00936
LOIDS, TRACE ELEMENTS, DISSOLVED   SOLIDS, NUTRIENTS, VITAMINS, TOXI   W70-02510
CTERISTICS, ORGANISMS, SUSPENDED   SOLIDS, SALTWATER TONGUE, BENTHIC   W73-00748
ONS, DISSOLVED OXYGEN, DISSOLVED   SOLIDS, SEDIMENTATION RATES, MANG   W73-00750
FLUORESCEIN, MACROPHYTES, GROUND   SOLIDS, UPTAKE, CONSUMER ORGANISM   W69-09334
ADABLE SOL/ *DAIRY MANURE, TOTAL   SOLIDS, VOLITILE SOLIDS, NON-DEGR   W71-00936
DENS, OOCYSTIS PUSILLA, OOCYSTIS   SOLITARIA, PANDORINA MORUM, PEDIA   W69-06540
LITHOSPHERE, SORPTION REACTIONS,   SOLUBILITY EQUILIBRIUM THEORY.: /   W70-04385
ION, *CORROSION, CARBON DIOXIDE,   SOLUBILITY, *PHOSPHATES, CHEMICAL   W71-11393
DISSOLVED INORGANIC PHOSPHATES,   SOLUBLE REACTIVE PHOSPHORUS, PHOS   W70-05269
CROBIAL/ *HETEROTROPHY, *ORGANIC   SOLUTES, *MARINE ENVIRONMENTS, MI   W70-04811
ECOLOGY, SWEDEN, LAKE/ *ORGANIC   SOLUTES, *HETEROTROPHY, MICROBIAL   W70-04284
SOLVED OXYGEN, ALGAE, NUTRIENTS,   SOLUTES, CHLORIDES, CALCIUM, CARB   W70-09557
ER RATES, *HETEROTROPHY, ORGANIC   SOLUTES, SWEDEN, LAKE ERKEN(SWEDE   W70-05109
OLLUTION IDENTIFICATION, ORGANIC   SOLUTES, VITAMIN B-1, VITAMIN B-1   W70-05652
ETHOD, FINGER LAKES(NY), RODHE'S   SOLUTION VIII, NUTRIENT LIMITATIO   W70-01579
(KUTZ), CULTURE MEDIA, CHU NO 10   SOLUTION, MINIMUM NUTRIENTS.: /SA   W70-03507
AYS, ORGANOPHOSPHORUS COMPOUNDS,   SOLVENT EXTRACTIONS, PHOSPHORUS C   W70-05269
SPILLS, *TORREY CANYON, AROMATIC   SOLVENTS, PHAEODACTYLUM TRICORNUT   W70-01470
S, OPEPHORA MARTYI, ABBOT'S POND(SOMERSET, ENGLAND), SAND GRAINS,   W71-12870
ATION, BATCH CULTURES, CHLORELLA   SOROKINIANA, COMPOSITION OF ALGAE   W70-07957
*EQUILIBRIUM MODEL, LITHOSPHERE,   SORPTION REACTIONS, SOLUBILITY EQ   W70-04385
                        PESTICIDE   SORPTION.:   W69-09884
CYANOPHYTA, CHLOROPHYTA, *ALGAE,   SORPTION, WASTE WATER TREATMENT,   W70-09438
, DISTILLATION, ELECTRODIALYSIS,   SORPTION, CHEMICAL PRECIPITATION,   W70-01981
OSPHERE, SEA WATER, FRESH WATER,   SORPTION, SODIUM, ALGAE, PHOSPHOR   W70-04385
TROPHICATION, ALGAE, FILTRATION,   SORPTION, FLOCCULATION, COAL, ACT   W70-05470
OACTIVITY EFFECTS, RADIOECOLOGY,   SORPTION, WATER POLLUTION EFFECTS   W69-07440
PERIDINIUM TROCHOIDEUM, VINEYARD   SOUND(MASS), WOODS HOLE(MASS). RA   W70-05272
, GONIDIA FORMATION, LONG ISLAND   SOUND(NY), ICELAND, TRITIATED THY   W69-09755
AENOPHYLLA, TIDAL CURRENT, PUGET   SOUND(WASH), LONG ISLAND SOUND, N   W69-10161
UNITIES, INTERTIDAL ALGAE, PUGET   SOUND, FLOURESCEIN DYE.: /AL COMM   W68-00010
, PUGET SOUND(WASH), LONG ISLAND   SOUND, NARRAGANSETT BAY, CAPE REY   W69-10161
N(AUSTRALIA), OXIDATION DITCHES,   SOUP, FOOD WASTES.:   SHEPPARTIO   W71-09548
S, MARINE ALGAE, WATER POLLUTION   SOURCES.: /, URANIUM RADIOISOTOPE   W71-09013
NUCLEAR WASTES, WATER POLLUTION   SOURCES.: / ALGAE, INVERTEBRATES,   W71-09863
LLUTION EFFECTS, WATER POLLUTION   SOURCES.: /RBON DIOXIDE, WATER PO   W69-10158
VELS, SEDIMENTS, WATER POLLUTION   SOURCES.: /MUD, SILTS, SANDS, GRA   W70-00711
LLUTION EFFECTS, WATER POLLUTION   SOURCES.: /TION CONTROL, WATER PO   W70-02792
VARIATIONS, SEASONAL VARIATIONS,   SOURCES.: /, ME, MAINE, PERIODIC   W68-00481
ICAL TECHNIQUES, WATER POLLUTION   SOURCES.: /CHROMATOGRAPHY, ANALYT   W69-00387
LLUTION EFFECTS, WATER POLLUTION   SOURCES.: /GY, SORPTION, WATER PO   W69-07440
LLUTION EFFECTS, WATER POLLUTION   SOURCES.: /ER CHEMISTRY, WATER PO   W69-06273
                        *NUTRIENT   SOURCES.:   W70-04266
MUS, PHOSPHATES, WATER POLLUTION   SOURCES.: /E, *BIOASSAY, SCENEDES   W70-03955
IL MICROBIOLOGY, WATER POLLUTION   SOURCES.: /, SOIL ALGAE, TURF, SO   W72-08110
, *WATER SOURC/ *WATER POLLUTION   SOURCES, *WATER POLLUTION EFFECTS   W73-00758
*WATER POLLUTION EFFECTS, *WATER   SOURCES, *GROUNDWATER, DDT, MISSI   W73-00758
EUTROPHICATION, *WATER POLLUTION   SOURCES, *WATER POLLUTION EFFECTS   W70-03975
, *EUTROPHICAT/ *WATER POLLUTION   SOURCES, *WATER POLLUTION EFFECTS   W70-04266
EUTROPHICATION, *WATER POLLUTION   SOURCES, *WATER POLLUTION EFFECTS   W70-03975
EUTROPHICATION, *WATER POLLUTION   SOURCES, *LAKES, NUTRIENTS, FERTI   W69-07084
*PHYTOPLANKTON, *WATER POLLUTION   SOURCES, *WATER POLLUTION EFFECTS   W69-02959
ON, *NUTRIENTS, *WATER POLLUTION   SOURCES, *CYCLING NUTRIENTS, ALGA   W69-04800
PAL AND INDUSTRIAL), *PHOSPHORUS   SOURCES, *PHOSPHATES, ALGAE, WATE   W70-01031
, *CLASSIFICAT/ *WATER POLLUTION   SOURCES, *NUISANCE PLANT GROWTHS,   W69-09340
AMINO ACIDS, P/ *WATER POLLUTION   SOURCES, *WATER POLLUTION EFFECTS   W72-00462
S, *SURFACTANTS, WATER POLLUTION   SOURCES, *ALGAE, *TASTE, *ODOR, *   W72-01812
LLUTION EFFECTS, WATER POLLUTION   SOURCES, *DETERGENTS, *DEGRADATIO   W70-09438
LUTION CONTROL, *WATER POLLUTION   SOURCES, *THERMAL POLLUTION, THER   W70-07847
WATER POLLUTION, WATER POLLUTION   SOURCES, *WATER POLLUTION EFFECTS   W71-03048
EUTROPHICATION, *WATER POLLUTION   SOURCES, ALGAL CONTROL, BIOINDICA   W70-02304
ORUS COMPOUNDS, *WATER POLLUTION   SOURCES, AGRICULTURAL WATERSHEDS.   W69-05323
LLUTION EFFECTS, WATER POLLUTION   SOURCES, ALGAE, ALGAL CONTROL, AL   W69-03370
LLUTION EFFECTS, WATER POLLUTION   SOURCES, AQUATIC PRODUCTIVITY, LI   W69-09135
LLUTION EFFECTS, WATER POLLUTION   SOURCES, BIOINDICATORS, NITROGEN   W69-06864
CONTROL, LAKES, WATER POLLUTION   SOURCES, CYCLING NUTRIENTS, COST   W69-08518
LAND MANAGEMENT, WATER POLLUTION   SOURCES, CHLOROPHYTA, NUISANCE AL   W70-03310
MICAL COMPOSITION(ENGLAND), FOOD   SOURCES, DREDGING, *POLLUTION ABA   W68-00172
ATER POLLUTION, *WATER POLLUTION   SOURCES, GREENHOUSE STUDIES, GROW   W69-06865
ATER(POLLUTION), WATER POLLUTION   SOURCES, IRRIGATION, AQUATIC ALGA   W69-07096
LLUTION EFFECTS, WATER POLLUTION   SOURCES, MUNICIPAL WASTES, GREAT   W71-05879
UTRIENTS, ALGAE, WATER POLLUTION   SOURCES, NUTRIENTS.: /S, WATER PO   W69-10159
ALGAE, BACTERIA, WATER POLLUTION   SOURCES, NITROGEN COMPOUNDS, PHOS   W69-01445
, *MARINE ALGAE, WATER POLLUTION   SOURCES, ODOR, FARM WASTE, SLUDGE   W71-05742
OBIC CONDITIONS, WATER POLLUTION   SOURCES, ON-SITE DATA COLLECTIONS   W70-09976
NUCLEAR WASTES, WATER POLLUTION   SOURCES, PATH OF POLLUTANT.: /AER   W71-11381
WATER CHEMISTRY, WATER POLLUTION   SOURCES, PATH OF POLLUTANTS, AQUA   W69-10080
EWAGE TREATMENT, WATER POLLUTION   SOURCES, PHOSPHORUS, NITROGEN COM   W70-03068
TRAPS, NUTRIENT INFLUX, NUTRIENT   SOURCES, PRIMARY TREATMENT, SECON   W69-01453
LLUTION EFFECTS, WATER POLLUTION   SOURCES, PHOSPHORUS BUDGET, APHAN   W70-04268
                                               SOURCES, RADIOECOLOGY.: /WATER PO   W70-00707

US GREEN ALGAE, / *RECOVERY, OIL
ARKSBURG, BUFFALO, DUNREITH, OIL
SOLV/ *CRUDE OIL, *GAMOSOL, *OIL
MINE WT DYE, AIRBORNE TESTS, OIL
*OIL
ED RIVER(VIETNAM), CHLOROCOCCUS,
CHERICHIA COLI, OOCYSTIS, LEMNA,
TES, ANKISTRODESMUS, GOMPHENEMA,
LOSIRA, SYNEDRA, COELOSPHAERIUM,
CRISPUS, MICROSPORA WITTROCKII,
IS, COELOSPHAERIUM/ *PHYSIOLOGY,
LYNGBYA, FRAGILARIA, OEDOGONIUM,
Y, DINOBRYON, VOLVOX, PANDORINA,
ROPHYTES, CLADOPHORA, OEDOGONIA,
WIS), LAKE WINGRA(WIS), ANABAENA
ON LAKE(VA).: ASTERIONELLA
ANABAENA, WOLLEA, OSCILLATORIA,
ZIMONA(HUNGARY), ALGAL TAXONOMY,
CTS, LETHAL LIMIT, GROWTH RATES,
TION, BACTERIOPHAGE, RESERVOIRS,
RES, ZYMOLOGICAL, ELBE(GERMANY),
LOGY, LIGHT PENETRATION, OXYGEN,
ENCE, LAKES, CHEMICALS, FISHING,
BIDITY, INDUSTRIAL PLANTS, WATER
, *POLLUTANTS, RECREATION, WATER
ONCENTRATION FACTOR, LIMNODRILUS
TOBRITE, CLADOPHORA, POTAMOGETON
R, DOLORES RIVER, URAVAN, COTTUS
OGEN UTILIZATION, ROLLER DRYING,
SEY(ONTARIO), ALGAE COUNTS/ *AIR
STRY, *ALGAE, *CATTLE, SLURRIES,
HLORINATION, SLUDGE, DETERGENTS,
SCONSIN, ALGAE, *NUISANCE ALGAE,
, *INVESTIGATIONS, ZOO-PLANKTON,
A), FLORA.:
ERMAL SPRINGS, *LIGHT INTENSITY,
KGROUND RADIATION, GEOCHEMISTRY,
), STORKWEATHER CREEK(WIS), NINE
R NATIONAL PARK, OHANAPECOSH HOT
PHOTOSYNTHESIS, *ALGAE, *THERMAL
C/ *CYANOPHYTA, *ALGAE, *THERMAL
STRY, ECOLOGY, SAMPLING, THERMAL
, AC/ *ALGAE, *TEMPERATURE, *HOT
, MOOSE CREEK, ARCTIC CIRCLE HOT
TONE NATIONAL PARK, *ICELAND HOT
ATS, *ALASKA, LAKES, RIVERS, HOT
TURE, THERMAL POLLUTION, THERMAL
NE TESTS, OIL SPILLS, GEOLOGICAL
R, RAISIN RIVER, DETROIT R/ LAKE
INDUSTRIAL WASTES, AGRICULTURE,
KE MINNETONKA(MINN), MINNEAPOLIS-
ND EFFLUENT, COLORADO, WEST(US),
OLOGY, *DESIGN THEORY AN/ *WASTE
PHOTOSYNTHESIS, LIGHT INTENSITY,
SPHORUS, NITROGEN, GROWTH RATES,
TREATMENT, SECONDARY TREATMENT,
ONATES, WATER POLLUTION EFFECTS,
OONS, *DESIGN, *SULFUR BACTERIA,
ROSCOPY, SULFUR COMPOUNDS, WASTE
DECOMPOSITION), DDT, ALKALINITY,
ALGAE, CADDISFLIES, FOOD CHAINS,
RESISTANCE, LIFE CYCLES, GROWTH
DIES, HABITATS, CRAYFISH, GROWTH
VULGARIS, DAPHNIA MAGNA, LIMNAEA
VULGARIS, DAPHNIA MAGNA, LIMNAEA
TERN LAKE ERIE(OHIO), GONGROSIRA
IPORE FILTERS, PHASE MICROSCOPY,
UFACTURING WASTES, WATER QUALITY
L, DYNAMICS, TEMPERATURE, SILTS,
, LAKES, ORGANIC WASTES, OXYGEN,
THERMOCLINE, LIGHT PENETRATION,
ITY, HYDROGEN ION CONCENTRATION,
D SOLIDS, CALCIUM, PRODUCTIVITY,
TY, WASTE ASSIMILATIVE CAPACITY,
TIMATING EQUATIONS, OXYGENATION,
HATES, PHOSPHORUS, PRODUCTIVITY,
R(CALIF), TUOLUMNE RIVER(CALIF),
ULGARIS, PSEUDOMONAS AERUGINOSA,
, BACILLUS THURINGIENSIS, AGARS,
AVOBACTER, ESCHERICHIA, SARCINA,
ES, CHEESE MANUFACTURING WASTES,
SPRING CREEK(PA),
ING, GRANTS, FEDERAL GOVERNMENT,
ION CONTROL, *LOCAL GOVERNMENTS,
LOT RESERVOIR, LEBANON, WADLEIGH
D, SYRACUSE(NY),/ *PHYCOLOGY, NY
REAMS,/ BERRY CREEK(ORE), OREGON
ELED SCABLANDS(WASH), WASHINGTON
SOUTH - CENTRAL
CHRYSOPHYTA, CALIFORNIA, UNITED

SPILL, *OIL POLLUTION, *FILAMENTO   W70-09976
SPILLS.: /S, LEBANON, BATAVIA, CL   W72-03641
SPILLS, *TORREY CANYON, AROMATIC   W70-01470
SPILLS, GEOLOGICAL SS POLARIS, CH   W70-05377
SPILLS, PROTEIN AQUICULTURE.:   W69-05891
SPINES, SCHIZOMYCOPHYTA, CAULOBAC   W70-03969
SPIRODELA, ELODEA.: /LOCOCCUS, ES   W70-10161
SPIROGYRA.: /CYCLOTELLA, FLAGELLA   W70-02249
SPIROGYRA, CLADOPHORA, DIAPTOMUS,   W70-02803
SPIROGYRA, XANTHOPHYCEAE, MYXOPHY   W70-00711
SPIROGYRA, MYXOPHYCEAE, MICROCYST   W70-04369
SPIROGYRA, CHAETOPHORA, STIGEOCLO   W70-04371
SPIROGYRA, SYNECHOCOCCUS, CHROOCO   W70-04503
SPIROGYRA, MICROCYSTIS AERUGINOSA   W70-05404
SPIROIDES, MICROCYSTIS AERUGINOSA   W70-05112
SPIROIDES, CLADOCERANS, WESTHAMPT   W71-10098
SPIRULINA, VOLVOCALES, GLENODINIU   W70-05404
SPIRULINA, CHODATE: /A(HUNGARY),   W70-03969
SPORES.: /E, WATER POLLUTION EFFE   W70-01779
SPORES, NITRITES,: /SELF-PURIFICA   W69-07838
SPOROBOLOMYCES ROSEUS, AUTOLYSIS,   W70-04368
SPORT FISHING, ON-SITE INVESTIGAT   W68-00487
SPORT FISHING, RECREATION, SILT,   W68-00172
SPORTS.: /LAKE ONTARIO, WATER TUR   W71-09880
SPORTS, ENVIRONMENT, AQUATIC ENVI   W69-04276
SPP, AEOLOSOMA HEMPRICHI, ESCHERI   W69-07861
SPP, LAKE MENDOTA(WIS), ANACHARIS   W70-05263
SPP, RHINICHTHYS OSCULUS, SUCKERS   W69-07846
SPRAY-DRIED, GLASS BEADS, TRICHOD   W70-04184
SPRAY, SUDBURY(ONTARIO), LAKE RAM   W69-10155
SPRAYING, ACTIVATED SLUDGE, FARM   W70-07491
SPRAYING, ALGAE, CENTRIFUGATION,   W70-04060
SPRAYING, SEASONAL, *ALGAL CONTRO   W68-00488
SPRAYING, BIOCONTROL, INFECTION,   W68-00468
SPRING CREEK(PA), STATE COLLEGE(P   W69-10159
SPRING WATER, WATER CHEMISTRY, WA   W69-09676
SPRING WATERS, PHOSPHORUS, ORGANO   W69-09334
SPRINGS CREEK(WIS), DOOR CREEK(WI   W70-05113
SPRINGS(WASH), SCHIZOTHRIX, PHORM   W70-03309
SPRINGS, *LIGHT INTENSITY, SPRING   W69-09676
SPRINGS, *OREGON, THERMOPHILIC BA   W70-05270
SPRINGS, EFFLUENTS, CYANOPHYTA.: /   W69-10151
SPRINGS, HABITATS, ALKALINE WATER   W69-10160
SPRINGS, RADON, CENTRAL CITY(COLO   W69-08272
SPRINGS, SINTER, NUCLEIC ACID, SI   W69-10160
SPRINGS, TERRESTRIAL HABITATS, DE   W69-08272
SPRINGS, THERMAL STRES: / TEMPERA   W69-03516
SS POLARIS, CHLOROPHYLL FLUORESCE   W70-05377
ST CLAIR, ROUGE RIVER, HURON RIVE   W69-01445
ST LAWRENCE RIVER, ALGAE, COLIFOR   W70-03964
ST PAUL(MINN).:   LA   W70-04810
STABILIZATION POND LOADINGS, ODOR   W70-04788
STABILIZATION PONDS, *ALGAL PHYSI   W70-04786
STABILIZATION PONDS, PENETRATION   W70-03923
STABILIZATION PONDS, SEWAGE DISPO   W68-00855
STABILIZATION PONDS, SEWAGE LAGOO   W68-00012
STABILIZATION, LAGOONS, TOXICITY,   W69-09454
STABILIZATION, WASTE WATER, ALGAE   W70-01971
STABILIZATION PONDS.: /ARED SPECT   W70-07034
STABILIZATION, WASTE STORAGE, PON   W70-08097
STABLE ISOTOPES, RADIOACTIVE WAST   W70-02786
STAGES.: /NT, BAYS, *FLUCTUATION,   W71-02494
STAGES, DIATOMS, ALGAE, METABOLIS   W69-09742
STAGNALIS, OUSPENSKII'S MEDIUM.: /   W71-03056
STAGNALIS, OUSPENSKII'S MEDIUM.: /   W71-05719
STAGNALIS, NEPHROCYTIUM OBESUM W   W70-04468
STAINING, STRIPPING FILM, SYNECHO   W69-10163
STANDARDS, GRAFTON(ND), LAKOTA(ND   W70-03312
STANDING CROP, ECOLOGY, PHOTOSYNT   W70-02780
STANDING CROP, ALGAE, ZOOPLANKTON   W70-02251
STANDING CROP, INDUSTRIAL WASTE: /   W71-07698
STANDING CROP, LIGHT, BIOMASS, RE   W71-00665
STANDING CROP, CHLOROPHYLL, PHOSP   W70-03512
STANDING CROP, ANNUAL TURNOVER, F   W68-00912
STANDING CROP, MICROENVIRONMENT,   W69-03611
STANDING CROP, STREAMS, WATER CHE   W69-04805
STANISLAUS RIVER(CALIF), WATER PO   W70-02777
STAPHYLOCOCCUS AUREUS, LACTOBACIL   W71-11823
STAPHYLOCOCCUS EPIDERMIDIS, MICRO   W71-11823
STAPHYLOCOCCUS, ACHROMOBACTER(ALK   W70-03952
STARCH MANUFACTURING WASTES, WATE   W70-03312
STATE COLLEGE(PA), FLORA.:   W69-10159
STATE GOVERNMENTS, WATER QUALITY: W69-06305
STATE GOVERNMENTS, BIOCHEMICAL OX   W69-06909
STATE PARK, LAKE SHATUTAKEE, MASC   W69-08674
STATE SCIENCE AND TECHNOLOGY FOUN   W70-03973
STATE UNIVERSITY, EXPERIMENTAL ST   W70-03978
STATE UNIVERSITY.: /(WASH), CHANN   W70-00264
STATES.:   W72-03562
STATES, BRACKISH WATER, ESTUARIES   W70-05372

534

TON, CHLOROPHYLL, CYCLES, UNITED   STATES, CHLORELLA, BIOCHEMICAL OX   W70-05092
 CHLOROPHYTA, CYANOPHYTA, UNITED   STATES, HABITATS.: /, *LAKE ERIE,   W70-04468
ATION, LAKES, *NUTRIENTS, UNITED   STATES, PHOSPHORUS, NITROGEN, SEW   W68-00478
ER, ORGANIC COMPOUNDS, HABITATS,   STATISTICAL METHODS.: /TS, SEAWAT   W69-09755
SLOVAKIA, LABORATORY PROCEDURES,   STATISTICAL PROCEDURES, BMT.: /HO   W72-12240
LIMNION, GREAT LAKES, LAKE ERIE,   STATISTICS, GROWTH RATES, ALGAE,   W68-00253
C, SILICON, PEDIASTRUM BORYANUM,   STAURASTRUM PARADOXUM, BOTRYOCCUS   W70-02804
YSTIS, GLOEOTRICHIA, GLOEOCAPSA,   STAURONEIS.: /LIS, ANABAENA, ANAC   W70-03974
E, ZOOPLANKTON, FISH, SHELLFISH,   STEAM, ELECTRIC POWER, HEATED WAT   W70-09905
ION, BACILLUS SUBTILIS, BACILLUS   STEAROTHERMOPHILUS, PROTEUS VULGA   W71-11823
C WEED CUTTING, *WEED EXTRACTING   STEEL CABLES, ALGAL SUCCESSION, A   W70-04494
MINN), APHANIZOMENON FLOS AQUAE,   STEPHANODISCUS ASTREA, LAKE MINNE   W70-05387
OTELLA, TRACHELOMONAS, MELOSIRA,   STEPHANODISCUS, ANABAENA, APHANIZ   W72-04761
OELASTRUM, PEDIASTRUM, WESTELLA,   STEPHANODISCUS, ANKISTRODESMUS, C   W70-08107
GING, STEPHANODISCUS HANTZSCHII,   STEPHANODISCUS BINDERANUS, CLADOP   W70-00263
OCERA, APHANIZOMENON FLOS-AQUAE,   STEPHANODISCUS HANTZSCHII, DIDINI   W70-00268
, ASTERIONELLA, FILTER CLOGGING,   STEPHANODISCUS HANTZSCHII, STEPHA   W70-00263
N, *CARBON FIXATION, *EXCRETION,   STEPHANODISCUS TENUIS, ASTERIONEL   W69-10158
GONIUM, SCENEDESMUS, CYCLOTELLA,   STEPHANODSICUS.: /CINCLIS, CHLORO   W69-02959
S PHOSPHATES, ULTRA-VIOLET LIGHT   STERILIZATION, ASTERIONELLA, SCEN   W70-03955
BACILLUS STEAROTH/ RICKETTSIAE,   STERILIZATION, BACILLUS SUBTILIS,   W71-11823
ICUS, PHAEODACTYLUM TRICORNUTUM,   STICHOCHRYSIS IMMOBILIS, VIBRIO,   W70-04280
OGONIUM, SPIROGYRA, CHAETOPHORA,   STIGEOCLONIUM, ANKISTRODESMUS, AP   W70-04371
PARNALDIA, DRAPARNALDIA PLUMOSA,   STIGEOCLONIUM, MICROCYSTIS, STIGE   W70-02245
OSA, STIGEOCLONIUM, MICROCYSTIS,   STIGEOCLONIUM TENUE, MICROCYSTIS,   W70-02245
 CULTURE MEDIA.:   STIGLEOCLONIUS, INSOLUBLE WASTES,   W68-00248
R COMPOUNDS, ANTIBIOTIC EFFECTS,   STIMULANTS, GONYAULAX CATENELLA,   W70-02510
KE BUTTE DES MORTS(WISC), GROWTH   STIMULATION, SURFACE LITTER.: /LA   W72-00845
OURCES, ECOLOGICAL DISTRIBUTION,   STOCHASTIC PROCESSES, ECOSYSTEMS,   W71-09012
IS, *LAKE ERIE, PHOSPHORUS, FISH   STOCKING, COMMERCIAL FISH, SALMON   W70-04465
LIMNOLOGY, LAKE TROUT, CRAYFISH,   STONEFLIES, SNAILS, ECOLOGY, EUTR   W70-00711
SUSPENDED LOAD, TURBIDITY, DATA   STORAGE AND RETRIEVAL, COLIFORMS,   W70-09557
EATMENT, CHEMICAL OXYGEN DEMAND,   STORAGE CAPACITY, EFFLUENTS, CONS   W71-00936
SOTOPES, *WASTE DISPOSAL, *WASTE   STORAGE, *WATER POLLUTION, WATER   W72-00941
DUCTIVITY, *TROPHIC LEVEL, WATER   STORAGE, EVAPORATION, WATER CIRCU   W69-09723
ALKALINITY, STABILIZATION, WASTE   STORAGE, PONDS, HYDROLYSIS, METAB   W70-08097
ING BACTERIA, *ACTIVATED SLUDGE,   STORAGE, TEMPERATURE, MICROORGANI   W71-13335
ADISON LAKES, YAHARA RIVER(WIS),   STORKWEATHER CREEK(WIS), NINE SPR   W70-05113
ON, ZOOPLANKTON, EUTROPHICATION,   STORM DRAINS, SEWAGE EFFLUENTS, S   W70-05112
TER TREATMENT, OVERFLOW, SEWERS,   STORM DRAINS, SEWAGE EFFLUENTS, *   W71-10654
WAGE TREATMENT, SEWAGE DISPOSAL,   STORM DRAINS, SURFACE DRAINAGE, A   W70-03344
HANIZOMENON FLOS-AQUAE, LYNGBYA,   STORM LAKE(IOWA), CENTER LAKE(IOW   W70-02803
ANALYSIS, SLUDGE, INCINERATION,   STORM RUN-OFF, SEDIMENTATION, BIO   W71-10654
GE EFFLUENTS, *SEWAGE TREATMENT,   STORM RUNOFF.: /TORM DRAINS, SEWA   W71-10654
, *WASTE WATER, *ALGAE, *SLUDGE,   STORMS, TURBULENCE, PILOT PLANTS,   W70-05819
PENDED SOLIDS, *COMBINED SEWERS,   STORMWATER OVERFLOWS.: /ING, *SUS   W71-10654
          SHORELINE VEGETATION,   STORMWATER, DRAINAGE.:   W70-02251
H, CHLORELLA PYRENOIDOSA(EMERSON   STRAIN), AUTOINHIBITORS.: / GROWT   W69-07442
TURE, NUTRIENTS, COPPER SULFATE,   STRATIFICATION.: /, LIGHT TEMPERA   W69-08282
UCTIVITY, RIVERS, TASTE, THERMAL   STRATIFICATION, WASTE TREATMENT,   W69-06535
TRATES, DAPHNIA, SILICA, THERMAL   STRATIFICATION, SULFIDES, AMMONIU   W70-05760
*PONDS, *STABILIZATION, *THERMAL   STRATIFICATION, *SEASONAL, *SULFI   W70-04790
DES, TEMPERATURE, CLIMATIC DATA,   STRATIFICATION, DISSOLVED OXYGEN,   W70-04790
TOPLANKTON, CYANOPHYTA, DIATOMS,   STRATIFICATION, OXYGEN, HYDROGEN   W70-05387
LI/ *EUTROPHICATION, TEMPERATURE   STRATIFICATION, DEPTH, PLANKTON,   W70-05405
 BIOLOGICAL PROPERTIES, ECOLOGY,   STRATIFICATION.: /TER PROPERTIES,   W70-04475
NDMENTS, *RESERVOIRS, / *THERMAL   STRATIFICATION, *AERATION, *IMPOU   W70-04484
VATED CARBON, FILTERS, SEASONAL,   STRATIFICATION, BIOCHEMICAL OXYGE   W70-04001
IOASSAYS, AQUATIC WEEDS, THERMAL   STRATIFICATION, EPILIMNION, HYPOL   W70-02251
IR EVAPORATION, SALINITY, ALGAE,   STRATIFICATION, SALT BALANCE, BIB   W72-01773
CAL MODELS, GREAT LAKES, THERMAL   STRATIFICATION.: /DIES, MATHEMATI   W71-05878
INS, ALGAE, NEBRASKA, TURBIDITY,   STRATIFICATION, TEMPERATURE, CHLO   W72-04761
TRATION, NUTRIENTS, ALGAE, FISH,   STRATIFICATION, FLOCCULATION, TUR   W70-08834
CYCLES, OLIGOCHAETES, MOLLUSKS,   STRATIFICATION, DIATOMS, C: /LIFE   W71-00665
IC ECOSYSTEM, MIDGE LARVAE, LAKE   STRATIFICATION, SURFACE AREA(LAKE   W69-09340
OF POLLUTANTS, NUTRIENTS, ALGAE,   STRATIFIED FLOW, MIXING, FLOW, MO   W71-05878
ODUCTIVITY, SEDIMENTATION RATES,   STRATIGRAPHY, BIOINDICATORS, ALGA   W70-05663
ALITY, VIRUSES, ALGICIDES, FISH,   STRE: /TREATMENT, LAKES, WATER QU   W70-03973
EE BARK, CHARCOAL, ALGAL GROWTH,   STREAM CONCOURSE, PALOUSE RIVER(I   W70-03501
ION, COOLING WATER TREATMENT, ON-   STREAM DESLUDGING.: /METAL CORROS   W70-02294
AINAGE SYSTEMS, FLOW RATES, CONN   STREAM FLOW, ALGAE, AQUATIC ALGAE   W68-00479
Y CONTROL, *MI/ FISH MANAGEMENT,   STREAM IMPROVEMENT, *WATER QUALIT   W70-05428
ER QUALITY, ALGAE, AQUATIC LIFE,   STREAM POLLUTION, BACTERIA, IMPUR   W72-01786
ENDED SEDIMENTS, MARINE BIOLOGY,   STREAM POLLUTION, *WATER POLLUTIO   W71-10786
 STORM DRAINS, SEWAGE EFFLUENTS,   STREAMFLOW, MEASUREMENT, WATER PO   W70-05112
*POLLUTANT ABATEMENT, EFFLUENTS,   STREAMFLOW, DESALINATION, TERTIAR   W69-07389
 NUTRIENTS, *ZINC RADIOISOTOPES,   STREAMFLOW, SAMPLING, ALGAE, ECOS   W69-07862
N CONTROL, LIMNOLOGY, DIVERSION,   STREAMS.: /OURCES, WATER POLLUTIO   W69-10170
CONTROL, *TASTE, *ODOR, *NATURAL   STREAMS, *DISTRIBUTION SYSTEMS, C   W72-03562
IZATION, *EUTROPHICATION, LAKES,   STREAMS, ALGAE, FISH, WISCONSIN,   W69-10178
 IRON, MANGANESE, BENTHIC FLORA,   STREAMS, CULTURES, COBALT RADIOIS   W69-07845
AE, DAPHNIA, LAKES, CHLOROPHYTA,   STREAMS, DIATOMS, RESERVOIRS, CUL   W70-02982
POLLUTION CONTROL, *CONFERENCES,   STREAMS, ESTUARIES, LAKES, GREAT   W70-03975
POLLUTION CONTROL, *CONFERENCES,   STREAMS, ESTUARIES, LAKES, GREAT   W70-03975
 MARYLAND, PENNSYLVANIA, RIVERS,   STREAMS, EUGLENA, AQUATIC POPULAT   W70-04510
EUTROPHICATION, OLIGOTR/ *LAKES,   STREAMS, FISH, LAND, ECOSYSTEMS,   W70-04193
NFALL, SNOWMELT, SURFACE WATERS,   STREAMS, HYDROGEN ION CONCENTRATI   W70-03501
WASTES, WATER SUPPLY, EFFLUENTS,   STREAMS, LAKES, RIVERS, RESERVOIR   W71-09880
N, WATER POLLUTION EFFECTS, ACID   STREAMS, MINE DRAINAGE, CHLOROPHY   W70-02770
RAPHICAL REGIONS, KENTUCKY, ACID   STREAMS, MICROFLORA.: /HIPS, GEOG   W68-00891

ER POLLUTION, THERMAL POLLUTION,
S, LABORATORY TESTS, PERIPHYTON,
H, FUNGI, HEAT, WATER POLLUTION,
N STATE UNIVERSITY, EXPERIMENTAL
ISPOSAL, WATER POLLUTION, LAKES,
US, PRODUCTIVITY, STANDING CROP,
TREATMENT, *WATERS, *ORGANISMS,
WATER, MOVEMENT, CULTURES, BAYS,
N, PHOSPHATES, ALKALINITY, FISH,
TTER, NITROGEN FIXATION, VOLUME,
ED OXYGEN VARIATIONS, SAN JOAQU/
INDUSTRIES, CHELATION, BACTERIA,
COCCUS EPIDERMIDIS, MICROCOCCUS,
UPLII, TERRAMYCIN, TETRACYCLINE,
LUTION, THERMAL SPRINGS, THERMAL
N, THERMAL POWERPLANTS, *THERMAL
LUS, WATER TEMPERATURE, HYDROGE/
ERS, PHASE MICROSCOPY, STAINING,
), SCENEDESMUS QUADRICAUDA, CEL/
ECTROCHEMICAL TREATMENT, AMMONIA
RBON RADIOISOTOPES, FISH, CLAMS,
ANOPHYTA, URANIUM RADIOISOTOPES,
DIOISOTOPES, NUCLEAR EXPLOSIONS,
TEBRATES, CRUSTACEANS, MOLLUSKS,
S, COBALT RADIOISOTOPES, CESIUM,
ERATURE, SOUTH CAROLINA, CESIUM,
S, SCENEDESMUS, PROTEINS, WALLS,
, METOLIUS RIVER(ORE), COMMUNITY
EMISTRY, *EUTROPHICATION, *WATER
CREEK(MD), BACK RIVE/ COMMUNITY
N/ *SELECTIVE TOXICITY, CHEMICAL
G STRUCTURES, INDUSTRIAL PLANTS,
LECTRIC POWERPLANTS, ENGINEERING
ONTIUM RADIOSOTOPES, CYTOLOGICAL
NIC MATTER, DIURNAL, LATITUDINAL
MENT, *OXIDATION LAGOONS, *MODEL
*OXIDATION LAGOONS, MILK, *MODEL
*FAT ACCUMULATION, *LABORATORY
CLADOCERAN, MONTANE ZON/ *FIELD
NETWORK, ALGAE COUNTS, P/ *RIVER
BEER-LAMBERT LAW, *FIXED-DENSITY
ANT/ *ALGAL CULTURE, *LABORATORY
IES, *MASS CULTURE, *PILOT-PLANT
TY, FUNDAMENTAL STUDIES, APPLIED
SYRACUSE UNIVERSITY, FUNDAMENTAL
IC.:   MATHEMATICAL
EEN ALGAE, NITZSCHIA, BIOLOGICAL
UTILIZATION, FISH, ALGAE, MODEL
REATMENT, EUTROPHICATION, MODEL
HYTA, LITTORAL, NUTRIENTS, MODEL
FLOCCULATION, FILTRATION, MODEL
HALOGENS, ALGICIDES, CYTOLOGICAL
SOPHYCEAE, CARTERIA, COMPARATIVE
BON, *CARBON CYCLE, *CYTOLOGICAL
G RATE, *BODY SIZE, *COMPARATIVE
ESTON, SCENEDESMUS, MATHEMATICAL
NTHESIS, CYANOPHYTA, CYTOLOGICAL
HARDNESS, COLIFORMS, CYTOLOGICAL
HENS, ACTIMOMYCETES, CYTOLOGICAL
OOD WEBS, KENTUCKY, LIFE HISTORY
GLAND), FOOD SOURCES, GREENHOUSE
UGHTERHOUSE WASTES, EXPERIMENTAL
TIFIED FLOW, MIXING, FLOW, MODEL
F, NUTRIENT BUDGETS, COMPARATIVE
NT, *AUSTRALIA, LABORATORY-SCALE
MPHIBIANS, ANABAENA, CYTOLOGICAL
EFFECTS, LABORATORY TESTS, MODEL
ATORY-SCALE STUDIES, PLANT-SCALE
OPHIC LEVEL, PRODUCTIVITY, MODEL
CHNIQUES, CLASSIFICATION, *MODEL
*INDIA, MOOSI RIVER ECOLOGICAL
RGANIC NITROGEN, BACTERIOLOGICAL
(ONTARIO), BALSAM LAKE(ONTARIO),
OPH/ MICHIGAN, CHEBOYGAN COUNTY,
N ALGAE, HALIMONE PORTULACOIDES,
NIA HARVEYI, EPIPHYTIC BACTERIA,
AR REACTORS, POWERPLANTS, SHIPS,
Y PRODUCTIVITY, ALGAE, PLANKTON,
MONTANE RIVER COURSE, CARPATHIAN
*GELBSTOFF, *ALGAL SUBSTANCES,
ET(NORWAY), / *GELBSTOFF, *ALGAL
VESICULOSUS, CHONDRUS C/ *ALGAL
SIS, FERTILIZATION, PLANT GROWTH
BREVIS, MYCROSYSTIS, ALGASTATIC
GASTATIC SUBSTANCES, ALGADYNAMIC
LOROPHYTA, BULRUSH, PLANT GROWTH
ROPHICATION, ALGAE, PLANT GROWTH
Y, DOMESTIC WASTES, PLANT GROWTH
EDIA, QUINONE DERIVATIVES, ALKYL
XOPHYCEAE, SALVELINUS NAMAYCUSH,

STREAMS, NUTRIENTS, LIGHT, WASHIN   W70-03309
STREAMS, PHOTOSYNTHESIS, RESPIRAT   W71-00668
STREAMS, RIVERS, LAKES, PHYSIOLOG   W71-02479
STREAMS, SEASONAL EFFECTS, THERMA   W70-03978
STREAMS, SEWAGE, REMEDIES, JUDICI   W69-06909
STREAMS, WATER CHEMISTRY, WATER: /   W69-04805
STREAMS, WATER POLLUTION, ANIMALS   W70-05750
STREAMS, WEEDS, LITTORAL, WINDS.:   W70-06229
STREAMS, WATER POLLUTION EFFECTS,   W69-10159
STREAMS, WATER POLLUTION EFFECTS,   W70-00719
STREETER-PHELPS EQUATION, DISSOLV   W69-07520
STREPTOCOCCUS, SECONDARY PRODUCTI   W70-01068
STREPTOCOCCUS, BREVIBACTERIUM, CO   W71-11823
STREPTOMYCIN, MILLIPORE FILTRATIO   W69-10154
STRES: / TEMPERATURE, THERMAL POL   W69-03516
STRESS, THERMAL WATER, *PROTOZOA,   W70-07847
STRIP MINES, DRAINAGE, FERROBACIL   W68-00891
STRIPPING FILM, SYNECHOCOCCUS, GR   W69-10163
STRIPPING, FIREBAUGH CENTER(CALIF   W70-03334
STRIPPING, ALGAE HARVESTING.: /EL   W70-01981
STRONTIUM RADIOISOTOPES, TENNESSE   W70-02786
STRONTIUM RADIOISOTOPES, CESIUM,   W69-09742
STRONTIUM RADIOISOTOPES, CESIUM,   W70-00708
STRONTIUM RADIOISOTOPES, ZINC RAD   W71-09850
STRONTIUM RADIOISOTOPES, MANGANES   W69-08269
STRONTIUM RADIOISOTOPES, ZINC· RAD   W69-07845
STRUCTURE, AMINO ACIDS, MECHANICA   W70-04184
STRUCTURE, OXYTREMA, ORGANIC ENRI   W70-03978
STRUCTURE, ATMOSPHERE, SEA WATER,   W70-04385
STRUCTURE, RIDLEY CREEK(PA), NOBS   W70-04510
STRUCTURE, 2,3-DICHLORONAPHTHOQUI   W70-02982
STRUCTURES, AFTERBAYS, WATER TYPE   W69-05023
STRUCTURES, INDUSTRIAL PLANTS, ST   W69-05023
STUDIES.: /IL CONTAMINATION, *STR   W69-03491
STUDIES.: /PIGMENTS, OCEANS, ORGA   W70-04809
STUDIES.: /IA, *WASTE WATER TREAT   W70-07034
STUDIES.: /ASTE WATER TREATMENT,   W70-07031
STUDIES, *NITROGEN LIMITATION, BA   W70-07957
STUDIES, *EXPERIMENTAL LIMNOLOGY,   W69-10154
STUDIES, *NATIONAL WATER QUALITY   W70-04507
STUDIES, *FIXED-LIGHT TESTS, *EFF   W70-03923
STUDIES, *MASS CULTURE, *PILOT-PL   W69-06865
STUDIES, ALGAL GROWTH, ALGAL NUTR   W69-06865
STUDIES, ALGAL PHYSIOLOGY, ALGAL   W70-03973
STUDIES, APPLIED STUDIES, ALGAL P   W70-03973
STUDIES, AUTOTROPHIC, HETEROTROPH   W70-05750
STUDIES, ALGAL IDENTIFICATION.: /   W70-10173
STUDIES, BIOASSAY, AQUATIC LIFE,   W71-11881
STUDIES, BIOMASS, *LIGHT INTENSIT   W73-02218
STUDIES, BIOASSAY, RADIOACTIVITY   W73-01454
STUDIES, COMPUTERS, NUTRIENTS, EU   W70-08627
STUDIES, CHLORINE, COPPER SULFATE   W70-02364
STUDIES, CRUCIGENIA RECTANGULARIS   W69-07833
STUDIES, CULTURES, FLUOROMETRY, B   W69-08278
STUDIES, DAPHNIA SCHODLERI, DAPHN   W70-03957
STUDIES, DATA COLLECTIONS, MARINE   W73-01446
STUDIES, ELECTRON MICROSCOPY, WAS   W70-03973
STUDIES, EVAPORATION, MICROORGANI   W71-05267
STUDIES, FILTRATION, ULTRASONICS,   W71-11823
STUDIES, GAMMA RAYS, SPECTROMETER   W69-09742
STUDIES, GROWTH K: /OMPOSITION(EN   W69-06865
STUDIES, LAGOON DESIGN, SCALE-UP   W70-07508
STUDIES, MATHEMATICAL MODELS, GRE   W71-05878
STUDIES, NUTRIENT REMOVAL, LAKE R   W69-09349
STUDIES, PLANT-SCALE STUDIES, REM   W70-06899
STUDIES, PHY: /OSTRIDIUM, FISH, A   W70-05372
STUDIES, RADIOACTIVITY TECHNIQUES   W70-04074
STUDIES, REMOVAL.: /TRALIA, LABOR   W70-06899
STUDIES, SALMON, AQUATIC ENVIRONM   W69-06536
STUDIES, WASTE IDENTIFICATION, BI   W69-01273
STUDY.:   W69-07096
STUDY, AEROBICITY.:   O   W71-05742
STURGEON LAKE(ONTARIO), BUCKHORN   W70-02795
STURGEON RIVER, FLUORESCEIN, MACR   W69-09334
SUAEDA MARITIMA, ASTER TRIPOLIUM,   W70-09976
SUBLITTORAL, YELLOW MATERIAL, DU: W70-01073
SUBMARINES, EFFLUENTS, ION EXCHAN   W70-00707
SUBMERGED PLANTS, PHYSICOCHEMICAL   W71-00665
SUBMONTANE REGION, GRADIENTS(RIVE   W70-02784
SUBPOLAR WATERS, TYNSET(NORWAY),   W70-01074
SUBSTANCES, SUBPOLAR WATERS, TYNS   W70-01074
SUBSTANCES, *EXTRACELLULAR, FUCUS   W70-01073
SUBSTANCES, SEWAGE, SULFATE, MANG   W70-02969
SUBSTANCES, ALGADYNAMIC SUBSTANCE   W70-02510
SUBSTANCES, PHORMIDINE, PANDORINI   W70-02510
SUBSTANCES, CHLORELLA, CHLAMYDOMO   W72-00845
SUBSTANCES, PLANT GROWTH, CARBOHY   W69-07826
SUBSTANCES, LIMNOLOGY, PLANT GROW   W68-00470
SUBSTITUTED IMIDAZOLINES, BIOLOGI   W70-02982
SUBSTRATE TYPE.: /NTHOPHYCEAE, MY   W70-00711

ID, SISJOTHANDI(ICELAND), ENZYME- | SUBSTRATE COMPLEXES, RIBONUCLEIC | W69-10160
LOCITY, MAXIMUM UPTAKE VELOCITY, | SUBSTRATE CONCENTRAT: / UPTAKE VE | W70-05109
ORATORY MODEL, COD REMOVAL, MILK | SUBSTRATE, NITROGEN REMOVAL, *MIL | W70-07031
TERIAL COLONIZATION, *ARTIFICIAL | SUBSTRATES, *RADIONUCLIDE ACCUMUL | W70-04371
AWATER, INCUBATION, PSEUDOMONAS, | SUBSURFACE-WATERS, BOTTOM SEDIMEN | W70-00713
CYCLE, AUTORADIOGRAPHY, MONODUS | SUBTERRANEUS, NAVICULA PELLICULOS | W70-02965
OSA, CHLORELLA VULGARIS, MONODUS | SUBTERRANEUS, ANKISTRODESMUS FALC | W70-05381
LAKE LUCERNE, FERTILIZED FIELDS, | SUBTERRANEAN DRAINAGE, MECHANICAL | W69-08668
DOMONAS, MYCOBACTERIUM, BACILLUS | SUBTILIS, BACILLUS MYCOIDES, BACI | W69-07838
S, ANABAENA CYLINDRICA, BACILLUS | SUBTILIS, ASPERGILLUS ORYZAE NIGE | W70-02964
ETTSIAE, STERILIZATION, BACILLUS | SUBTILIS, BACILLUS STEAROTHERMOPH | W71-11823
S, VITAMINS, TOXINS, INHIBITORS, | SUCCESSION, CHLORELLA, CHELATION, | W70-02510
UTION EFFECTS, SYSTEMS ANALYSIS, | SUCCESSION, ECOSYSTEMS, CYCLING N | W70-02779
S, CHLOROPHYTA, LIGHT INTENSITY, | SUCCESSION, ZINC RADIOISOTOPES, C | W70-04371
D EXTRACTING STEEL CABLES, ALGAL | SUCCESSION, ANABAENA, CLADOPHORA, | W70-04494
CONTROL, ALGIC/ *EUTROPHICATION, | SUCCESSION, WATER QUALITY, ALGAL | W69-01977
HEADS, EELS, KILLIFISHES, TROUT, | SUCKERS, BASS, WATER HYACINTH, FR | W70-05263
COTTUS SPP, RHINICHTHYS OSCULUS, | SUCKERS, ICTALURUS PUNCTATUS, TRI | W69-07846
ATES, ADSORPTION, PHYTOPLANKTON, | SUCKERS, SHINERS, CADDISFLIES, TE | W69-07853
-GREEN ALGAE, OCULAR MICROMETER, | SUCROSE-FLOTATION TECHNIQUE, META | W70-04253
ARIO), ALGAE COUNTS/ *AIR SPRAY, | SUDBURY(ONTARIO), LAKE RAMSEY(ONT | W69-10155
AY(RI), CAPE REYKJANES(ICELAND), | SUDURNES(ICELAND), NON-PARAMETRIC | W69-09755
RED RIVER VALLEY(NORTH DAKOTA), | SULFATE REDUCTION, POTATO PROCESS | W70-03312
IATOMS, *BLUE-GREEN ALGA, COPPER | SULFATE.: /NG, *PHYTOPLANKTON, *D | W70-02754
PPLY, WATER QUALITY, WA/ *COPPER | SULFATE, *ALGAL CONTROL, WATER SU | W69-10155
, *ORGANIC MATTER, ALGAE, COPPER | SULFATE, *CHEMICAL OXYGEN DEMAND. | W72-04495
PHORUS REMOVAL, *ALUM, *ALUMINUM | SULFATE, *ACTIVATED SLUDGE PROCES | W70-09186
*AQUATIC WEEDS, LAKES,/ *COPPER | SULFATE, *EUTROPHICATION, *ALGAE, | W70-04494
LOG/ *BIOLOGY, *EFFECTS, *COPPER | SULFATE, *TREATMENT, *LAKES, *ECO | W70-06217
ED CONTROL, *HERBICIDES, *COPPER | SULFATE, *SODIUM ARSENITE, NUISAN | W70-05568
*EUTROPHICATION, *LAKES, *COPPER | SULFATE, *ALGAL CONTROL, *TOXINS, | W68-00859
SANCE ALGAE, *ALGICIDES, *COPPER | SULFATE, *WATER QUALITY CONTROL, | W69-05704
WATER TREATMENT, *ALGAE, *COPPER | SULFATE, *ALKALINITY, WATER QUALI | W72-07442
,/ *LAKES, *MUD, *COPPER, COPPER | SULFATE, ALGAL CONTROL, ALGICIDES | W69-03366
ON SYSTEMS, CHLORINATION, COPPER | SULFATE, ACTIVATED CARBON, WATER | W72-03562
OGICAL STUDIES, CHLORINE, COPPER | SULFATE, ALGAL CONTROL, MORTALITY | W70-02364
INS, ALGAE, GROWTH RATES, COPPER | SULFATE, BOTTOM SEDIMENTS, LAKE M | W69-09349
NUM SULFATE, IRON CHLORIDE, IRON | SULFATE, CLADOPHORA BLANKETS, HYD | W69-10170
WATER, HYDROGEN SULFIDE, COPPER | SULFATE, CHLOROPHYTA, CRUSTACEANS | W70-02803
QUALITY CONTROL, *ALGAE, COPPER | SULFATE, CHLORINE, WELLS, CALIFOR | W71-00110
ULENCE, TOXINS, BACTERIA, COPPER | SULFATE, CIRCULATION.: /ION, TURB | W70-04369
ROGEN, PHOSPHORUS, ALGAE, COPPER | SULFATE, EFFLUENTS, MICROCYSTIS A | W70-02787
N, ALGICIDES, HYDROLYSIS, COPPER | SULFATE, FISHES, HERBICIDES, MIDG | W70-02982
ING TUBE, CANALIZATION, ALUMINUM | SULFATE, FERRICHLORIDE, IRON SULF | W70-05668
NUM SULFATE, FERRICHLORIDE, IRON | SULFATE, GALVANIZING WASTES, LAKE | W70-05668
ATE, CLADOPHORA BLANKE/ ALUMINUM | SULFATE, IRON CHLORIDE, IRON SULF | W69-10170
E, CHLOROPHYLL, CHEMTROL, COPPER | SULFATE, LAKES, NITROGEN.: / ALGA | W68-00470
PLANT GROWTH SUBSTANCES, SEWAGE, | SULFATE, MANGANESE, IRON, CARBON | W70-02969
ATER POLLUTION TREATMENT, COPPER | SULFATE, ODOR, COLIFORMS, ECONOMI | W72-01814
E, *MAINE, *WATER SUPPLY, COPPER | SULFATE, RESERVOIRS, WATER POLLUT | W70-08096
ICAL CONTROL, BIOCONTROL, COPPER | SULFATE, SODIUM ARSENITE,: /ECHAN | W70-10175
CONTROL, *NEW HAMPSHIRE, *COPPER | SULFATE, SEWAGE EFFLUENTS, SMELTS | W69-08674
T TEMPERATURE, NUTRIENTS, COPPER | SULFATE, STRATIFICATION.: /, LIGH | W69-08282
LTRATION, DESIGN, *ALGAE, COPPER | SULFATE, SEDIMENTATION, ACTIVATED | W72-08859
YDROGEN ION CONCENTRATION, IRON, | SULFATE, TEMPERATURE, KENTUCKY.: / | W69-00096
, WATER/ *ALGAL CONTROL, *COPPER | SULFATE, WATER SUPPLY, RESERVOIRS | W69-10157
N, INCINERATION, IRON COMPOUNDS, | SULFATES.: /ON, ANAEROBIC DIGESTIO | W70-04060
MASS, BENTHOS, SALINITY, SNAILS, | SULFATES, NITRATES, PHOSPHATES, P | W70-03978
QUALITY, CHLORIDES, PHOSPHORUS, | SULFATES, HARDNESS(WATER).: /ATER | W69-03948
LIGOTROPHY, ODOR, PHYTOPLANKTON, | SULFATES, SEDIMENTATION, SEDIMENT | W69-07833
CALCIUM, MAGNESIUM, ALKALINITY, | SULFATES, CHLORIDES, SODIUM, IRON | W69-05894
UTRIENTS, *NITROGEN, PHOSPHORUS, | SULFATES, VITAMINS, ALGAE, GROWTH | W69-09349
SALINITY, ULTRAVIOLET RADIATION, | SULFATES, CELLULOSE, INDUSTR: /, | W70-01074
RIA, HYDROGEN ION CONCENTRATION, | SULFATES, RESPIRATION, FOOD: /CTE | W70-01073
L CONTROL, NITRATES, PHOSPHATES, | SULFATES, SEASONAL, LAKES, SAMPLI | W70-02646
NTRATION, ACIDITY, MUD, HYDROGEN | SULFIDE.: /DS, HYDROGEN ION CONCE | W70-05399
TES, *SULFIDES, ALGAE/ *HYDROGEN | SULFIDE, *ORGANIC LOADING, *SULFA | W70-07034
C L/ *OXIDATION PONDS, *HYDROGEN | SULFIDE, *SULFIDES, ALGAE, ORGANI | W71-11377
*ALGAE, IRON COMPOUNDS, HYDROGEN | SULFIDE, BIOCHEMICAL OXYGEN DEMAN | W70-06925
ERATURE, SHALLOW WATER, HYDROGEN | SULFIDE, COPPER SULFATE, CHLOROPH | W70-02803
N, POLLUTANTS, MAPPING, HYDROGEN | SULFIDE, CITIES, LAKE HURON, LAKE | W70-00667
STS, ENZYMES, CULTURES, HYDROGEN | SULFIDE, ELECTRON MICROSCOPY, BIO | W70-03952
, ANAEROBIC CONDITIONS, HYDROGEN | SULFIDE, IRON, FUNGI, CLAYS, BRAC | W70-05761
ODOR, COLOR, TURBIDITY, HYDROGEN | SULFIDE, IRON, MAGNESIUM, COOLING | W70-02251
ES.: HYDROGEN | SULFIDE, IMHOFF TANKS, PRECIPITAT | W72-00027
DEMAND, PHOTOSYNTHESIS, HYDROGEN | SULFIDE, LAKES, ALGAE, ENTERIC BA | W70-03312
RATION, IRON COMPOUNDS, HYDROGEN | SULFIDE, LEAD, COPPER, AMMONIA, C | W70-04510
IDE, AMMONIA, NITRITES, HYDROGEN | SULFIDE, SALINITY, LAKES, NUTRIEN | W71-03031
OCOCCUS, SECONDARY PRODUCTIVITY, | SULFIDES, SEA WATER, PHAWOPHYTA, | W70-01068
SILICA, THERMAL STRATIFICATION, | SULFIDES, AMMONIUM COMPOUNDS, OXI | W70-05760
ROTEINS, OIL WASTES, OILY WATER, | SULFITE LIQUORS, ULTIMATE DISPOSA | W69-05891
KYL POLYETHOXYLAT/ *LINEAR ALKYL | SULFONATES, *AXENIC CULTURES, *AL | W70-03928
RATION, FLUORIDES, ALKYL BENZENE | SULFONATES, WIND VELOCITY, TASTE, | W70-00263
GENTS, SURFACTANTS, ALKYLBENZENE | SULFONATES, LINEAR ALKYLATE SULFO | W69-09454
ZENE SULFONATES, LINEAR ALKYLATE | SULFONATES, WATER POLLUTION EFFEC | W69-09454
RGENTS, WATER POL/ *ALKYLBENZENE | SULFONATES, *BIODEGRADATION, DETE | W70-08740
OLLUTION SOURC/ *LINEAR ALKYLATE | SULFONATES, *SURFACTANTS, WATER P | W70-09438
ROL, *POTABLE WATER, CYANOPHYTA, | SULFUR BACTERIA, COLOR, IRON BACT | W70-08107
, OXIDATION-REDUCTION POTENTIAL, | SULFUR BACTERIA, *WASTE WATER TRE | W70-07034
ALEN/ *MICROBIAL ECOLOGY, PURPLE | SULFUR BACTERIA, POPULATION EQUIV | W70-03312

```
MARY PRODUCTIVITY, *ZOOPLANKTON,          SULFUR BACTERIA, DIEL MIGRATION,       W70-05760
METHANE BACTERIA, ACID BACTERIA,          SULFUR BACTERIA, ALGAE, MIXING, H      W72-06854
DS, COAL MINES, ACIDITY, PYRITE,          SULFUR COMPOUNDS, MOLYBDENUM, ROT      W69-07428
   OXIDES, BACTERIA, ALGAE, FUNGI,        SULFUR COMPOUNDS, CARBONATES.: /N     W70-02294
OTIN, HISTIDINE, URACIL, REDUCED          SULFUR COMPOUNDS, ANTIBIOTIC EFFE     W70-02510
VE PONDS, INFRARED SPECTROSCOPY,          SULFUR COMPOUNDS, WASTE STABILIZA     W70-07034
   , WASTE WATER, ALGAE, DIFFUSION,       SULFUR.: /BACTERIA, STABILIZATION     W70-01971
ETABOLISM, NITROGEN, PHOSPHORUS,          SULFUR, CALCIUM, MAGNESIUM, POTAS     W70-02804
TRIENT CONCENTRATION, WISCONSIN,          SULFUR, PHOSPHORUS.: /USCORUN, NU     W70-00719
   DIQUAT, SODIUM ARSENITE, COPPER        SULPHATE, MONURON, AMMONIA, DALAP     W70-00269
TAILS, BULRUSH, DREDGING, COPPER          SULPHATE, PONDWEEDS, SODIUM ARSEN     W70-05263
ATION, *LAKES, WISCONSIN, COPPER          SULPHATE, NUISANCE ALGAE, *ALGAL      W68-00468
AMYDOMONAS, ECOLOGY, PHOSPHORUS,          SULPHUR, HYDROGEN ION CONCENTRATI     W70-02965
NAS, ULOTHRIX, DAVID MONIE TEST,          SUM: /ANABAENA, MELOSIRA, MALLOMO     W69-10157
   *DIFFUSERS, *RAYTOWN(MO), LEE'S        SUMMIT(MO), OXYGEN TRANSFER, PH,      W70-06601
INVERTEBRATES, TURBIDITY, WORMS,          SUNFISHES.: /S, PLANKTON, ALGAE,      W69-07861
ONTROL, MINNOWS, NUISANCE ALGAE,          SUNFISHES, ALGAE, DAPHNIA, LAKES,     W70-02982
RS.:    PERIPHYTON INVENTORY(LAKE         SUPERIOR), WATER QUALITY INDICATO     W70-05958
   , *WATER QUALITY, BIOINDI/ *LAKE       SUPERIOR, *PERIPHYTON, *LIMNOLOGY     W70-05958
   , PRESSURE, MIXING, EVAPORATION,       SUPERSATURATION, RESERVOIRS, ALGA     W70-00683
ROANOKE RAPIDS RESERVO/ *OXYGEN           SUPERSATURATION, VERTICAL MIXING,     W70-00683
   , AQUATIC WEEDS, ADMINISTRATION,       SUPERVISORY CONTROL(POWER), ADMIN     W72-05774
S, OOCISTUS, FILTRATION / *WATER          SUPPLIES, GREEN ALGAE, HEAD LOSSE     W70-04199
ONS, ANAEROBIC CONDITIONS, WATER          SUPPLY.: /HETICS, AEROBIC CONDITI     W70-04430
   , WATER POLLUTION SOURCES, WATER       SUPPLY.: /WATER POLLUTION EFFECTS     W69-07832
   OHIO RIVER, PENNSYLVANIA, WATER        SUPPLY.: /N ALGAE, PHYTOPLANKTON,     W69-05697
SULFATE, ACTIVATED CARBON, WATER          SUPPLY.: /, CHLORINATION, COPPER      W72-03562
W/ *SANITARY ENGINEERING, *WATER          SUPPLY, *BIOLOGICAL PROPERTIES, *     W72-01813
FICATION, *ALGAL CONTROL, *WATER          SUPPLY, *BASINS, WATER POLLUTION      W72-01814
LTERS, PLANKTON,/ *ALGAE, *WATER          SUPPLY, *CONTROL, TASTE, ODOR, FI     W70-05389
OR, TAST/ *ALGAE CONTROL, *WATER          SUPPLY, *NEW HAMPSHIRE, ALGAE, OD     W69-08282
EATMENT, FILTRATION, CHL/ *WATER          SUPPLY, *WATER QUALITY, *WATER TR     W71-05943
MONITORING, WATER DEMAND, WATER           SUPPLY, ARID LANDS, LIMNOLOGY.: /     W70-05645
*WATER POLLUTION EFFECTS, *WATER          SUPPLY, BIOINDICATORS, CALCIUM, C     W69-07833
S, WATER/ *ALGAE, *MAINE, *WATER          SUPPLY, COPPER SULFATE, RESERVOIR     W70-08096
H, SEWERS, RUNOFF, WASTES, WATER          SUPPLY, EFFLUENTS, STREAMS, LAKES     W71-09880
TY CONTR/ *LAKE MICHIGAN, *WATER          SUPPLY, INTAKES, NUTRIENTS, QUALI     W70-00263
*SOIL, GROUNDWATER BASINS, WATER          SUPPLY, MUD, BENTHOS, HYDROGEN, B     W69-07838
COLLECTIONS, PIT RECHARGE, WATER          SUPPLY, MUNICIPAL WATER, SEDIMENT     W69-05894
   CONTROL, *COPPER SULFATE, WATER        SUPPLY, RESERVOIRS, WATER QUALITY     W69-10157
R SULFATE, *ALGAL CONTROL, WATER          SUPPLY, WATER QUALITY, WATER POLL     W69-10155
RUS, PHOSPHORUS COMPOUNDS, WATER          SUPPLY, WATER MANAGEMENT.: /OSPHO     W72-01867
DGE LARVAE, LAKE STRATIFICATION,          SURFACE AREA(LAKES), PHAEDACTYLUM     W69-09340
ES, CATIONIC ELECTROLYTES, ALGAL          SURFACE CHARGE, ELECTROSTATIC REP     W70-05267
   , OEDOGONIUM, PHORMIDIUM RETZII,       SURFACE DIFFUSION RATE.: /RA ULNA     W70-00265
   , SEWAGE DISPOSAL, STORM DRAINS,       SURFACE DRAINAGE, ALGAL CONTROL,      W70-03344
MORTS(WISC), GROWTH STIMULATION,          SURFACE LITTER.: /LAKE BUTTE DES      W72-00845
WATER), WATER POLLUTION SOURCES,          SURFACE RUNOFF, RAINFALL, SNOWMEL     W70-03501
   , NITRATES, INORGANIC COMPOUNDS,       SURFACE RUNOFF, UNDERFLOW, PRECIP     W70-04810
TERS, PLANKTON, WATER POLLUTION,          SURFACE WATERS, RESERVOIRS, CONTR     W70-05389
FACE RUNOFF, RAINFALL, SNOWMELT,          SURFACE WATERS, STREAMS, HYDROGEN     W70-03501
URFACE-WATERS, BOTTOM SEDIMENTS,          SURFACE WATERS, DETRITUS, LIGHT I     W70-00713
LAKES, EUTROPHICATION, SAMPLING,          SURFACES WATERS, CARBON, FERTILIT     W70-02973
ANKTON, PRODUCTIVITY, TURBIDITY,          SURFACES, DEPTH, DISTRIBUTION, TI     W71-07698
MPERATURE, PRIMARY PRODUCTIVITY,          SURFACES, ARCTIC, ENZYMES, CHLORO     W70-04809
   , *AQUATIC BACTERIA, DETERGENTS,       SURFACTANTS, ALKYLBENZENE SULFONA     W69-09454
L PHENOL POLYETHOXYLATE, ANIONIC          SURFACTANTS, NONIONIC SURFACTANTS     W70-09438
E, ANIONIC SURFACTANTS, NONIONIC          SURFACTANTS, INFRARED SPECTRA.: /     W70-09438
OLOGICAL RESPONSE, FERTILIZATION          SURVEY, MADISON LAKES(WIS), LAKE      W70-02787
TON DISTRIBUTION, WISCONSIN LAKE          SURVEY, NITROGEN CONTENT, LAKE ER     W70-03959
                            *RIVER        SURVEYS.:                             W72-01801
OLLUTANT, FOOD CHAINS, SAMPLING,          SURVEYS, MARINE ALGAE, AQUATIC LI     W72-00941
NYON, *DETERGENT CONCENTRATIONS,          SURVIVAL, SUSCEPTIBILITY, PHYSIOL     W70-01779
FF VEGETATION, / *TORREY CANYON,          SUSCEPTIBILITY, REGENERATION, CLI     W70-01777
   *TORREY CANYON, REPOPULATION,          SUSCEPTIBILITY.:                      W70-01780
ERGENT CONCENTRATIONS, SURVIVAL,          SUSCEPTIBILITY, PHYSIOLOGICAL OBS     W70-01779
ST-DEATH FACTOR, DOSAGE, SPECIES          SUSCEPTIBILITY, ANABAENA LEMMERMA     W70-00273
RATURE, IONS, SPECTROPHOTOMETRY,          SUSPENDED LOAD, FLUORESCE: /TEMPE     W70-03334
TAKE RATE.:         MIXED LIQUOR,         SUSPENDED SOLIDS, *PHOSPHOROUS UP     W71-13335
OORGANISMS, FRESH WATER BIOLOGY,          SUSPENDED SEDIMENTS, MARINE BIOLO     W71-10786
AL CONTROL, *TERTIARY TREATMENT,          SUSPENDED LOAD, PILOT PLANTS, ACT     W70-09186
LITY, BIOCHEMICAL OXYGEN DEMAND,          SUSPENDED LOAD, TURBIDITY, DATA S     W70-09557
YGEN, BIOCHEMICAL OXYGEN DEMAND,          SUSPENDED LOAD, WASTES, SEWAGE, A     W70-08627
MAND, BIOCHEMICAL OXYGEN DEMAND,          SUSPENDED LOAD, SOLIDS CONTACT PR     W70-06792
SUMMIT(MO), OXYGEN TRANSFER, PH,          SUSPENDED SOLIDS.: /N(MO), LEE'S      W70-06601
   , TEMPERATURE PROFILES, VOLATILE       SUSPENDED SOLIDS, MULTI-CELL SERI     W70-04790
EPARATION TECHNIQUES, TURBIDITY,          SUSPENDED SOLIDS, ALGAE, PILOT PL     W72-08686
ICAL CHARACTERISTICS, ORGANISMS,          SUSPENDED SOLIDS, SALTWATER TONGU     W73-00748
TS, ANN ARBOR(MICH), SAN/ *ALGAL          SUSPENSIONS, ELECTROKINETIC EFFEC     W71-03035
   ), ENGLAND/ PISCICULTURAL PONDS,       SVISLOCH' RIVER(USSR), MINSK(USSR     W70-05092
RIENTS, SEPTIC TANKS, EFFLUENTS,          SWAMPS, FOREST SOILS, MINNESOTA,      W70-04193
              MILLSBORO(DELAWARE),        SWAN CREEK, INDIAN RIVER.:            W71-05013
   SWEDEN), ACETATE, LAKE NORRVIKEN(      SWEDEN), A: /ELLATES, LAKE ERKEN(     W70-04194
INETICS, FLAGELLATES, LAKE ERKEN(         SWEDEN), ACETATE, LAKE NORRVIKEN(     W70-04194
RMEASES, CHAMYDOMONADS, LAPPLAND(         SWEDEN), ACETATE, TURNOVER TIMES:     W70-04287
BIAL ECOLOGY, SWEDEN, LAKE ERKEN(         SWEDEN), ACTIVE TRANSPORT, CARBON     W70-04284
ANIC SOLUTES, SWEDEN, LAKE ERKEN(         SWEDEN), ANNUAL CYCLES, MICROBIAL     W70-05109
EN), UPPSALA(SWEDEN), LAKE ERKEN(         SWEDEN), EXTINCTION VALUES, EUDOR     W70-05409
LOSIRA, ANEBODA(SWEDEN), UPPSALA(         SWEDEN), LAKE ERKEN(SWEDEN), EXTI     W70-05409
TEN EQUATION, SWEDEN, LAKE ERKEN(         SWEDEN), LINEWEAVER-BURK EQUATION     W70-04287
```

KE NORRVIKEN(SWEDEN), LAKE EKOLN(SWEDEN), LAKE LOTSJON(SWEDEN), PE W70-04287
CHAELIS CONSTANT, LAKE NORRVIKEN(SWEDEN), LAKE EKOLN(SWEDEN), LAKE W70-04287
EY, NITROGEN CONTENT, LAKE ERKEN(SWEDEN), LAKE MENDOTA(WIS), NET P W70-03959
LAKE EKOLN(SWEDEN), LAKE LOTSJON(SWEDEN), PERMEASES, CHAMYDOMONADS W70-04287
LA, DINOBRYON, MELOSIRA, ANEBODA(SWEDEN), UPPSALA(SWEDEN), LAKE ER W70-05409
ELATIVE HETEROTROPHIC POTENTIAL, SWEDEN, ENZYME KINETICS, LINEWEAV W70-04194
HETEROTROPHY, MICROBIAL ECOLOGY, SWEDEN, LAKE ERKEN(SWEDEN), ACTIV W70-04284
COSE, MICHAELIS-MENTEN EQUATION, SWEDEN, LAKE ERKEN(SWEDEN), LINEW W70-04287
 *HETEROTROPHY, ORGANIC SOLUTES, SWEDEN, LAKE ERKEN(SWEDEN), ANNUA W70-05109
-232, BISMUTH-214, ACTINIUM-228, SWIFT RIFFLES, ZIRCONIUM-95, NIOB W69-09742
EED, DUCKWEED, WHITE WATER LILY, SWIMMERS' ITCH.: /D CELERY, BUR R W70-05568
CONTROL, CRUSTACEANS, ROTIFERS, SWIMMING POOLS, CHLAMYDOMONAS, DA W70-10161
ICATION, APPLICATION TECHNIQUES, SWIMMING, WATERFOWL, BIRDS, COSTS W70-05568
RMINATION, ABSORBANCE, COENOBIA, SWIRLING, VOLATILE SOLIDS.: /DETE W70-03334
, GREIFENSEE(SWITZ), PAFFIKERSEE(SWITZ), AEGERISEE(SWITZ), FURES: / W70-00270
), PAFFIKERSEE(SWITZ), AEGERISEE(SWITZ), FURES: / GREIFENSEE(SWITZ W70-00270
SH), ZURICHSEE(SWITZ), TURLERSEE(SWITZ), GREIFENSEE(SWITZ), PAFFIK W70-00270
Z), TURLERSEE(SWITZ), GREIFENSEE(SWITZ), PAFFIKERSEE(SWITZ), AEGER W70-00270
S), CEDAR RIVER(WASH), ZURICHSEE(SWITZ), TURLERSEE(SWITZ), GREIFEN W70-00270
N(WASH), OSCILLATORIA RUBESCENS, SWITZERLAND, PERIDINIUM, OSCILLAT W69-10169
A, BOSMINA LONGISPINA, ZURICHSEE(SWITZERLAND), LINSLEY POND(CONN), W69-10169
A(CANADA), PENGUINS, LAKE ZURICH(SWITZERLAND), WHITEFISH, PUNCHBOW W70-05399
E ZELL(AUSTRIA), LAKE HALL WILER(SWITZERLAND).: /IER(BAVARIA), LAK W70-05668
ASTES, LAKE ZURICH(SWITZERLAND), SWITZERLAND, KILCHBERG(SWITZERLAN W70-05668
KILCHBERG(SWITZERLAND), ZOLLIKON(SWITZERLAND), LAKE TEGERN(BAVARIA W70-05668
GALVANIZING WASTES, LAKE ZURICH(SWITZERLAND), SWITZERLAND, KILCHB W70-05668
ZERLAND), SWITZERLAND, KILCHBERG(SWITZERLAND), ZOLLIKON(SWITZERLAN W70-05668
IDALSVANNET NORWAY, LAKE ZURICH, SWITZERLAND, NITELV RIVER NORWAY, W69-07833
CAL AND BIOLOGICAL PURIFICATION, SWITZERLAND.: / DRAINAGE, MECHANI W69-08668
LAKE WASHINGTON(WASH), ZURICHSEE(SWITZERLAND.: /KE TAHOE(CALIF), W69-06858
, NITROGEN, PHOSPHATES, BIOMASS, SYMBIOSIS, OXYDATION LAGOONS.: /D W70-05655
S, PHOTOSYNTHESIS, PRODUCTIVITY, SYMBIOSIS, RESERVOIRS, LAKES, REA W70-05092
A, SYNURA, GYMNODINIUM, CHEMICAL SYMBIOSIS, MICROBIAL ECOLOGY, DIN W70-04503
ENOIDOSA, SYNTHETIC SEWAGE FEED, SYMBIOTIC RELATIONSHIPS.: /LA PYR W70-05655
SPECTROSCOPY, MASS SPECTROMETRY, SYMPLOCA MUSCORUM.: /UIS, RAMMAN W69-00387
ERIMENTAL LIMNOLOGY, BRACHIONUS, SYNCHAETA PECTINATA, POLYARTHRA V W70-02249
SCOPY, STAINING, STRIPPING FILM, SYNECHOCOCCUS, GRAIN COUNT(AUTORA W69-10163
SCILLATORIA TEREBRIFORMIS, MATS, SYNECHOCOCCUS LIVIDUS, BLUE-GREEN W70-05270
MOEWUSII, SKELETONEMA COSTATUM, SYNECHOCOCCUS ELON: /HLAMYDOMONAS W70-05381
N, VOLVOX, PANDORINA, SPIROGYRA, SYNECHOCOCCUS, CHROOCOCCUS, OSCIL W70-04503
A, CYCLOTELLA, HALOCHLOROCOCCUM, SYNECHOCOCCUS, CARTERIA, GONIUM, W70-03974
ACICULARIS, NITZSCHIA DISSIPATA, SYNEDRA ACUS, SINKING RATES, GRAZ W69-10158
ANCE, SPECIES, MELOSIRA VARIANS, SYNEDRA ULNA, OEDOGONIUM, PHORMID W70-00265
RAGILARIA, TABELLARIA, MELOSIRA, SYNEDRA, COELOSPHAERIUM, SPIROGYR W70-02803
BOSMINA, ASTERIONELLA, MELOSIRA, SYNEDRA, FRAGILARIA, CER: /STIS, W70-04503
LAKE(NY), ASTERIONELLA FORMOSA, SYNEDRA, MELOSIRA, DINOBRYON, FRA W70-01579
A RIVER, CYCLOTELLA, FRAGILARIA, SYNEDRA, NAVICULA, APHANIZOMENON, W70-08107
LFOVIBRIO, MESOTHERMY, MELOSIRA, SYNEDRA, RHODOMON: /ACTERIA, DESU W70-05760
MICROCYSTIS, PRYMNESIUM PARVUM, SYNERGISTIC EFFECTS, AXENIC CULTU W70-05372
, NUCLEIC ACID, RIBONUCLEIC ACID SYNTHESIS.: /ION, *MACROMOLECULES W69-10151
, PHOTOCHEMICAL, CARBON DIOXIDE, SYNTHESIS, ENZYMES, CHLOROPHYTA, W70-02965
RELATIO/ *CHLORELLA PYRENOIDOSA, SYNTHETIC SEWAGE FEED, SYMBIOTIC W70-05655
ON, ALGAL BLOOMS, NUTRIENT LOAD, SYNTHETIC DETERGENTS, FISH MANAGE W69-08674
EOTRICHIA ECHINULATA, DINOBRYON, SYNURA UVELLA, FRAGILARIA, TABELL W70-02803
TS, ASTERIONELLA, APHANIZOMENON, SYNURA, ANABAENA, MELOSIRA, MALLO W69-10157
UNIT PROCESSES, GASTROENTERITIS, SYNURA, ANABAENA, IOWA RIVER, CYC W70-08107
TA(MAINE), CARLETON POND(MAINE), SYNURA, ASTERIONELLA, TYPHOID FEV W70-08096
SIRA, COCCOLITHUS, SARGASSO SEA, SYNURA, GYMNODINIUM, CHEMICAL SYM W70-04503
ERTIOLECTA, AMPHIDINIUM CARTERI, SYRACOSPHAERA ELONGATA, THALASSIO W69-08278
TECHNOLOGY FOUND, SYRACUSE(NY), SYRACUSE UNIVERSITY, FUNDAMENTAL W70-03973
TE SCIENCE AND TECHNOLOGY FOUND, SYRACUSE(NY), SYRACUSE UNIVERSITY W70-03973
PINNULARIA, NITZSCHIA, NAVICULA, SYRACUSE(NY), ENTEROMORPHA INTEST W70-03974
*ONONDAGA LAKE, ONONDAGA COUNTY, SYRACUSE, WARBURG RESPIROMETER, C W69-02959
UTRIENT REMOVAL, CONTINUOUS FLOW SYSTEM.:   *N W72-00917
CILIATES, FLAGELLATES, SAPROBIEN SYSTEM, CILIATE AUTOECOLOGY.:   * W71-03031
PENDED SOLIDS, MULTI-CELL SERIAL SYSTEM, PH, FACULTATIVE POND, ANA W70-04790
OPLANKTON, PRIMARY PRODUCTIVITY, SYSTEMATICS, WATER POLLUTION EFFE W69-07832
ILTERS, *BIOTA, SEWAGE BACTERIA, SYSTEMATICS, FUNGI, ALGAE, DIATOM W72-01819
IBUTION, BIOLOGICAL COMMUNITIES, SYSTEMATICS, MARINE PLANTS.: /STR W72-10623
CATION, WATER POLLUTION EFFECTS, SYSTEMS ANALYSIS, SUCCESSION, ECO W70-02779
RRIES, AUTOMATIC CONTROL, PIPING SYSTEMS(MECHANICAL), CLEANING, CO W72-08688
 *NATURAL STREAMS, *DISTRIBUTION SYSTEMS, CHLORINATION, COPPER SUL W72-03562
C/ BENTHIC ALGAE, BLOOMS, CLOSED SYSTEMS, CYTOLOGY, ENVIRONMENTAL W69-07832
, *POLLUTION ABATEMENT, DRAINAGE SYSTEMS, FLOW RATES, CONN STREAM W68-00479
RATION, *TOXICITY, *DISTRIBUTION SYSTEMS, WATER POLLUTION, WATER Q W72-03746
GAE, 2-4-D, AMINOTRIAZOLE, 2-4-5- T, SCENEDESMUS, CHLAMYDOMONAS, CH W70-03519
LIMNETICA, ANABAENA FLOS AQUAE, TABELLARIA, MELO: /TICUS, LYNGBYA W70-05405
RYON, SYNURA UVELLA, FRAGILARIA, TABELLARIA, MELOSIRA, SYNEDRA, CO W70-02803
VER(ILL), POOLS, WIND DIRECTION, TABELLARIA, FRAGILARIA, ASTERIONE W70-00263
TON, HOWARD A. HANSON RESERVOIR, TACOMA(WASH).: /EN RIVER, WASHING W72-11906
UDIES, RADIOACTIVITY TECHNIQUES, TAGGING.: /RATORY TESTS, MODEL ST W70-04074
                            LAKE TAHOE.:   W69-09135
HICATION, LAKE MONONA(WIS), LAKE TAHOE(CALIF), LAKE WASHINGTON(WAS W69-06858
PARTICULATE ORGANIC CARBON, LAKE TAHOE(CALIF), CHLORELLA PYRENOIDO W70-03983
L BLOOMS, NTA, EDTA, HEIDA, LAKE TAHOE(CALIF).: /OMPLEXONES, *ALGA W70-04373
VERWORTS, MORPHOMETRY, PA/ *LAKE TAHOE(CALIF-NEV), HYDROPHYTES, LI W70-00711
IGHT MEASUREMENTS, EUPHOT/ *LAKE TAHOE(CALIF-NEV), TRANSPARENCY, L W70-01933
A, MOLLUSCA, RAFFINATE, THORIUM, TAILINGS.: / COLEOPTERA, HEMIPTER W69-07846
A(WIS), PHYTOPLANKTON CROPS, TR/ TAKASUKA POND(JAPAN), LAKE MENDOT W69-05142
, GONYAULAX CATENELLA, GONYAULAX TAMERENSIS, GYMNODINIUM BREVIS, M W70-02510

, NON-VOLATILE ACIDS, KETOACIDS, TAMIYA'S MINERAL MEDIUM, GLYOXYLI W70-04195
, SCUM, WEEDS, NUTRIENTS, SEPTIC TANKS, EFFLUENTS, SWAMPS, FOREST W70-04193
HYDROGEN SULFIDE, IMHOFF TANKS, PRECIPITATES.: W72-00027
WASTES, MUNICIPAL WASTES, SEPTIC TANKS, RECREATION WASTES, WATER Q W70-04266
TE, ORTHOPHOSPHATE, DISMUTATION, TANN: / ARTHROBACTER, PYROPHOSPHA W69-07428
TION, DEPURATION, PHYCOCOLLOIDS, TANNINS, FLAGELLATES, PHARMACEUTI W70-01068
ACHYTHEMIS, CHYDORUS, PLEUROXUS, TANYPUS, TANYTARSUS, LIMNODRILUS, W70-06975
S, CHYDORUS, PLEUROXUS, TANYPUS, TANYTARSUS, LIMNODRILUS, EUCYCLOP W70-06975
C CULTURES, MONOALGAL, PORTUGAL, TAPES, JAPAN, AMPHIDINIUM, BLOOMS W70-05372
PHY, PARTICULATE MATTER, BLELHAM TARN(ENGLAND), CHLORELLA PYRENOID W70-02504
ALIF), JOINT INDUSTRY/GOVERNMENT TASK FORCE ON EUTROPHICATION, PRO W70-02777
Y CONTROL, ALGAL CONTROL, SLIME, TASTE-PRODUCING ALGAE, ODOR-PRODU W69-05704
RON, TEMPERATURE, SOIL ANALYSIS, TASTE-PRODUCING ALGAE, HYDROGEN I W72-11906
NZENE SULFONATES, WIND VELOCITY, TASTE, COAGULATION, PLANKTON, MON W70-00263
IATION, CHEMICAL INDUSTRY, ODOR, TASTE, CONSUMER.: /ER WORKS ASSOC W71-05943
ION, FISH, NUISANCE ALGAE, ODOR, TASTE, DEPTH, CIRCULATION, DIURNA W70-05404
AE, NUTRIENTS, INDICATORS, ODOR, TASTE, FILTERS, WATER TREATMENT, W70-05264
BACTERIA, COLOR, IRON BACTERIA, TASTE, FILTERS, IOWA, DIATOMS, EU W70-08107
LY, *NEW HAMPSHIRE, ALGAE, ODOR, TASTE, LIGHT TEMPERATURE, NUTRIEN W69-08282
*ALGAE, *WATER SUPPLY, *CONTROL, TASTE, ODOR, FILTERS, PLANKTON, W W70-05389
QUALITY, *EUTROPHICATION, COLOR, TASTE, ODOR, TOXINS, DISSOLVED OX W69-05697
LUTION, MICROORGANISMS, HISTORY, TASTE, ODOR, DEEP WELLS, WATER TR W70-08096
RTIES, *WATER POLLUTION SOURCES, TASTE, ODOR, ALGAE, RESERVOIRS, W W72-01813
LLUTION CONTROL, NUISANCE ALGAE, TASTE, ODOR.: / QUALITY, WATER PO W69-10155
HYPOLIMNION, CYCLING NUTRIENTS, TASTE, ODOR, COLOR, TURBIDITY, HY W70-02251
TTER, RADIOACTIVITY, PESTICIDES, TASTE, ODOR, COAGULATION, ELECTRO W72-13601
URE, SALINITY, REAERATION, FISH, TASTE, ODOR, OXYGEN SAG, POTOMAC W73-00750
H, PRIMARY PRODUCTIVITY, RIVERS, TASTE, THERMAL STRATIFICATION, WA W69-06535
FATES, SEDIMENTATION, SEDIMENTS, TASTE, WATER CHEM: /PLANKTON, SUL W69-07833
R, CHEMOSYNTHETIC, SAPROPHYTIC,/ TAXONOMIC GROUPS, MOLECULAR FILTE W68-00891
USTRIS (RAB) FARLOW, CLADOPHORA, TAXONOMIC DESCRIPTIONS, GONGROSIR W70-04468
NARRAGANSETT BAY(RHODE ISLAND), TAXONOMIC TYPES, HETERETROPHY, AL W70-00713
BACTER, / SPECIES DETERMINATION, TAXONOMY, FLAVOBACTERIUM, ACHROMO W70-04280
OLOGY, ALGAL BIOCHEMISTRY, ALGAL TAXONOMY, ALGAL PHYLOGENY, CLADOP W70-03973
HUNGARY), ZIMONA(HUNGARY), ALGAL TAXONOMY, SPIRULINA, CHODATE: /A(W70-03969
ORGANISMS, FISH PHYSIOLOGY, FISH TAXONOMY, DDT, PHOSPHOTHIOATE PES W69-02782
ALGAL TAXONOMY, ALGAE IDENTIFICATION.: W69-03514
SPECIES, TAXONS.: W71-09767
ONE, ALPINE ZONE, BOULDER(COLO), TEA LAKE(COLO), BLACK LAKE(COLO), W69-10154
N, MIXING, ECONOMIC FEASIBILITY, TECHNICAL FEASIBILITY, COST ANALY W71-10654
NSHIPS, FLUOROMETRIC CHLOROPHYLL TECHNIQUE.: /NVIRONMENTAL RELATIO W70-05094
N FIXING BACTERIA, RADIOACTIVITY TECHNIQUE, PHOTOSYNTHESIS.: /ROGE W70-04249
AR MICROMETER, SUCROSE-FLOTATION TECHNIQUE, METALIMNION, BATHYTHER W70-04253
ARPUS CONFERVOIDES, NYLON COLUMN TECHNIQUE, PARTICULATE MATTER, BL W70-01074
CH CULTURES, ACETYLENE REDUCTION TECHNIQUE, HETEROTROPHY, FLAGELLA W70-02775
SCHIZOTHRIX, PHORMIDIUM, WINKLER TECHNIQUE, OXYGEN ELECTRODES, ATO W70-03309
G FACTORS, CHLORELLA, ANALYTICAL TECHNIQUES, CYANOPHYTA, DIATOMS, W70-02779
UTRIENT REQUIREMENTS, ANALYTICAL TECHNIQUES, WATER POLLUTION EFFEC W70-02245
CIDES, METABOLISM, RADIOACTIVITY TECHNIQUES, CHLORELLA, OXIDATION W70-02198
CONTROL, *ALGICIDES, ANALYTICAL TECHNIQUES, AQUATIC ALGAE, AQUATI W70-02968
IC MICROBIO/ *ALGAE, *ANALYTICAL TECHNIQUES, *AQUATIC ALGAE, AQUAT W70-02792
E, CHEMICAL ANALYSIS, ANALYTICAL TECHNIQUES, CLOUD COVER, SPECTROS W69-09676
EN DEPLETION, NUCLEAR ACTIVATION TECHNIQUES, ROCK LAKE(WASH), WILL W70-00264
LOGY, / *ECOLOGY, *RADIOACTIVITY TECHNIQUES, *RADIOISOTOPES, LIMNO W69-10163
UNDMENTS, RESERVOIRS, ANALYTICAL TECHNIQUES, ON-SITE TESTS.: /IMPO W71-10098
TION, PHOTOSYNTHESIS, ANALYTICAL TECHNIQUES, SPECTROSCOPY, COLORIM W71-11377
TION, MICROORGANISMS, ANALYTICAL TECHNIQUES, *WASTE WATER TREATMEN W71-13313
ALGAE, GROWTH RATES, ANALYTICAL TECHNIQUES, LABORATORY EQUIPMENT, W71-07339
STES, WATER ANALYSIS, ANALYTICAL TECHNIQUES, SAMPLING, ORGANIC MAT W72-03299
*RESEARCH EQUIPMENT, *ANALYTICAL TECHNIQUES, WATER QUALITY, DISSOL W70-08627
, SALINITY, BIOASSAY, ANALYTICAL TECHNIQUES, LETHAL LIMITS, DINOFL W70-08111
TOMS, MARINE ALGAE,/ *ANALYTICAL TECHNIQUES, *ALGAE, PLANKTON, DIA W70-09904
ORGANISMS, TURBIDITY, ANALYTICAL TECHNIQUES, LABORATORY TESTS, FOR W70-08628
CHROMOTOGRAPHY, RAD/ *ANALYTICAL TECHNIQUES, *PHOSPHATES, *ALGAE, W71-03033
S, WATER UTILIZATION, ANALYTICAL TECHNIQUES, NUTRIENTS, NITROGEN, W71-03048
OPHYLL, *OHIO RIVER, *ANALYTICAL TECHNIQUES, BIOMASS, WATER POLLUT W71-04206
OACTIVITY TECHNIQUES, ANALYTICAL TECHNIQUES, WATER POLLUTION EFFEC W70-04194
IOLOGICAL ECOLOGY, RADIOACTIVITY TECHNIQUES, ANALYTICAL TECHNIQUES W70-04194
TS, MODEL STUDIES, RADIOACTIVITY TECHNIQUES, TAGGING.: /RATORY TES W70-04074
TIC MICROBIOLOGY, */ *ANALYTICAL TECHNIQUES, *AQUATIC ALGAE, *AQUA W70-04283
*LAKES, *BACTERIA, *ANALYTICAL TECHNIQUES, WATER POLLUTION EFFEC W70-04287
, *BIOINDICATORS, */ *ANALYTICAL TECHNIQUES, *ALGAE, *GROWTH RATES W70-04381
ER POLLUTION EFFECTS, ANALYTICAL TECHNIQUES.: /NAS, BIOASSAYS, WAT W70-04284
R ANALYSIS, *SAMPLI/ *ANALYTICAL TECHNIQUES, *AQUATIC WEEDS, *WATE W70-04382
Y, CARBON-14, RADIOCARBON UPTAKE TECHNIQUES, GLUCOSE, MICHAELIS-ME W70-04287
, *ALGAE, *ANALYSIS, *ANALYTICAL TECHNIQUES, TEMPERATURE, PIPTIDES W70-04365
RES, OPTICAL DENSITY, SOIL-WATER TECHNIQUES, PHYCOLOGY, MICROCYSTI W70-04381
Y TESTS, *I/ *ALGAE, *ANALYTICAL TECHNIQUES, *PLANKTON, *LABORATOR W70-04469
BIAL ECOLOGY, RADIOCARBON UPTAKE TECHNIQUES, MICHAELIS-MENTEN EQUA W70-05109
IC BACTERIA, *ALGAE, *ANALYTICAL TECHNIQUES, *MARINE MICROORGANISM W70-04811
ER POLLUTION EFFECTS, ANALYTICAL TECHNIQUES, KINETICS, AQUATIC BAC W70-05109
TROL, BIBLIOGRAPHIES, ANALYTICAL TECHNIQUES, DIATOMS, SAMPLING, PH W70-05389
IUM CARTERI, *RADIOCARBON UPTAKE TECHNIQUES, WATER POLLUTION IDENT W70-05652
S, CHEMICAL ANALYSIS, ANALYTICAL TECHNIQUES, DINOFLAGELLATES, MARI W70-05652
IDS, LIGHT, ALGAE, / *ANALYTICAL TECHNIQUES, *MICROORGANISMS, *LIP W70-05651
ATES OF APPLICATION, APPLICATION TECHNIQUES, SWIMMING, WATERFOWL, W70-05568
ANALYSIS, *CHLORINE, *ANALYTICAL TECHNIQUES, *POLAROGRAPHIC ANALYS W70-05178
FFECTS, *PHOSPHATES, *ANALYTICAL TECHNIQUES, *BIOASSAYS, ORGANOPHO W70-05269
S HOLE(MASS). RADIOCARBON UPTAKE TECHNIQUES, MICROBIAL ECOLOGY, CO W70-05272
MIXOLIMNION, RADIOCARBON UPTAKE TECHNIQUES, BACTERIAL PLATE, BACT W70-05760

SSAY, *ESSE/ *ALGAE, *ANALYTICAL   TECHNIQUES, *AQUATIC WEEDS, *BIOA   W69-03364
GICAL ECOLOGY, ALGAE, ANALYTICAL   TECHNIQUES, AQUATIC ALGAE, AQUATI   W69-03517
*NUTRIENT REQUIREME/ *ANALYTICAL   TECHNIQUES, *BIOASSAY, *ENZYMES,   W69-03373
N, *CHLORELLA, ALGAE, ANALYTICAL   TECHNIQUES, AQUATIC ALGAE, AQUATI   W69-03371
ALGAE, ALGAL CONTROL, ANALYTICAL   TECHNIQUES, AQUATIC ALGAE, AQUATI   W69-03374
THETIC OXYGEN, ALGAE, ANALYTICAL   TECHNIQUES, AQUATIC ALGAE, AQUATI   W69-03362
, *CYANOPHYTA, ALGAE, ANALYTICAL   TECHNIQUES, AQUATIC ALGAE, BIOASS   W69-03518
ROL, ALGAL POISONING, ANALYTICAL   TECHNIQUES, AQUATIC ALGAE, AQUATI   W69-03188
GICAL ECOLOGY, ALGAE, ANALYTICAL   TECHNIQUES, AQUATIC ALGAE, AQUATI   W69-03515
EUTROPHICAT/ *ALGAE, *ANALYTICAL   TECHNIQUES, *CYCLING NUTRIENTS, *   W69-03358
EUTROPHICATION, *PH/ *ANALYTICAL   TECHNIQUES, *CYCLING NUTRIENTS, *   W69-03370
REQUIREMENTS, ALGAE, ANALYTICAL   TECHNIQUES, AQUATIC ALGAE, AQUATI   W69-03186
, GAS CHROMATOGRAPHY, ANALYTICAL   TECHNIQUES, WATER POLLUTION SOURC   W69-00387
NTS, *NITROGEN CYCLE, ANALYTICAL   TECHNIQUES, AQUATIC ALGAE, AQUATI   W69-03185
ER POLLUTION SOURCES, SEPARATION   TECHNIQUES, *ANALYTICAL TECHNIQUE   W69-01273
PARATION TECHNIQUES, *ANALYTICAL   TECHNIQUES, CLASSIFICATION, *MODE   W69-01273
ING NUTRIENTS, ALGAE, ANALYTICAL   TECHNIQUES, CLAY LOAM, CLAYS, COL   W69-04800
NUTRIENTS,/ *ALGAE, *ANALYTICAL   TECHNIQUES, *BIOASSAY, *ESSENTIAL   W69-04802
ATION, *PLANT GROWTH, ANALYTICAL   TECHNIQUES, ANTIBIOTICS, BIOASSAY   W69-04801
STUDIES, BIOASSAY, RADIOACTIVITY   TECHNIQUES, ABSORPTION.: / MODEL   W73-01454
*DIATOMACEOUS EARTH, SEPARATION   TECHNIQUES, PRESSURE, FLOW RATES,   W72-08688
EARTH, *FILTRATION, *SEPARATION   TECHNIQUES, TURBIDITY, SUSPENDED   W72-08686
ECTROLYTES, FILTERS, *ANALYTICAL   TECHNIQUES.: /TIVATED CARBON, *EL   W72-09693
NVIRONMENTAL EFFECTS, ANALYTICAL   TECHNIQUES, PHYSIOLOGICAL ECOLOGY   W72-12240
*PHYCOLOGY, NY STATE SCIENCE AND   TECHNOLOGY FOUND, SYRACUSE(NY), S   W70-03973
S, TEGERNSEE, GERMANY, EXPANDING   TECHNOLOGY, OVERPOPULATION.: /HUT   W69-07818
SECTS, PHOSPHORUS RADIOISOTOPES,   TECHNOLOGY, NATURAL RESOURCES, EC   W71-09012
ND), ZOLLIKON(SWITZERLAND), LAKE   TEGERN(BAVARIA), LAKE SCHLIER(BAV   W70-05668
NY), LAKE SCHLIER(GERMANY), LAKE   TEGERN(GERMANY), GERMANY, FINLAND   W70-05761
APPALACHIAN MOUNTAIN CLUB HUTS,   TEGERNSEE, GERMANY, EXPANDING TEC   W69-07818
AGULATION, PLANKTON, MONITORING,   TEM: /S, WIND VELOCITY, TASTE, CO   W70-00263
SERIUM, ARTEMIA, CARTERIA, NORTH   TEMPERATE ZONE, PLUTONIUM, HALF-L   W70-00708
UM, ARTHROBACTER, FLORA, AMBIENT   TEMPERATURE, THERMAL SENSITIVITY,   W70-00713
, *ESTUARINE ENVIRONMENT, *WATER   TEMPERATURE, *SEASONAL, RHODE ISL   W70-00713
ON, FLUORESCENCE, CARBOHYDRATES,   TEMPERATURE, SALINITY, ULTRAVIOLE   W70-01074
R RADIATION, SALINITY, RAINFALL,   TEMPERATURE, COLORIMETRY, CARBOHY   W70-01073
N, OLIGOTROPHY, NUTRIENTS, WATER   TEMPERATURE, DISSOLVED OXYGEN, FI   W70-00264
KE ONTARIO, LAKE MICHIGAN, WATER   TEMPERATURE, PHOSPHATES, TURBIDIT   W70-00667
SYNTHESIS, *OXYGEN, *SATURATION,   TEMPERATURE, PRESSURE, MIXING, EV   W70-00683
*DISSOLVED OXYGEN, *OHIO, WATER   TEMPERATURE, SLUDGE WORMS, BLOODW   W70-00475
IDS, HYDROGEN ION CONCENTRATION,   TEMPERATURE, NITROGEN, NUTRIENTS,   W70-00265
TER), CONNECTICUT, RHODE ISLAND,   TEMPERATURE, NUTRIENT REQUIREMENT   W69-10161
L ANALYSIS, BACTERIA, NUTRIENTS,   TEMPERATURE, EPILIMNION, NITROGEN   W69-10167
ANIA, BIOCHEMICAL OXYGEN DEMAND,   TEMPERATURE, FLOW, SAMPLING, WAST   W69-10159
LOROPHYLL, EPILIMNION, EFFLUENT,   TEMPERATURE, PHOTOSYNTHESIS, CARB   W69-10169
NG WATER, WATER CHEMISTRY, WATER   TEMPERATURE, CHEMICAL ANALYSIS, A   W69-09676
DS, EUTROPHICATION, OLIGOTROPHY,   TEMPERATURE, PHOTOSYNTHESIS, RAIN   W69-10154
OIRS, WATER QUALITY, NEW JERSEY,   TEMPERATURE, ORGANIC MATTER, HARD   W69-10157
CROSCOPIC EXAMINATION, *CRITICAL   TEMPERATURES, *COPPER DOSAGE, *CO   W69-10157
AUTHORITY PROJECT, *RESERVOIRS,   TEMPERATURE, DISSOLVED OXYGEN, DE   W71-07698
,/ *THERMAL POWERPLANTS, *LAKES,   TEMPERATURE, PHYTOPLANKTON, ALGAE   W71-09767
LAGOONS, *DESIGN CRITERIA, WATER   TEMPERATURE, SEWAGE TREATMENT, AE   W71-08221
LIC HEALTH, PATHOGENIC BACTERIA,   TEMPERATURE, NITRATES, AMMONIA, *   W72-06017
Y, CARBON DIOXIDE, GROWTH RATES,   TEMPERATURE, PILOT PLANTS, HYDROG   W72-06022
RUS, NITROGEN, CALCIUM CHLORIDE,   TEMPERATURE, METABOLISM, EPIPHYTO   W72-05453
ASKA, TURBIDITY, STRATIFICATION,   TEMPERATURE, CHLOROPHYLL, CYANOPH   W72-04761
SOLVED OXYGEN, LABORATORY TESTS,   TEMPERATURE, WASTE WATER TREATMEN   W71-12183
GAE, HYDROGEN ION CONCENTRATION,   TEMPERATURE, ALKALINITY, NITRIFIC   W71-11393
ING, HYDROGEN ION CONCENTRATION,   TEMPERATURE, CENTRIFUGATION, COAG   W71-11375
SYNTHESIS, PHYTOPLANKTON, ALGAE,   TEMPERATURE, SOLAR RADIATION.: /O   W71-10788
RIA, *ACTIVATED SLUDGE, STORAGE,   TEMPERATURE, MICROORGANISMS, NITR   W71-13335
HEMICAL OXYGEN DEMAND, AERATION,   TEMPERATURE, FLOW, ALGAE, BENTHOS   W71-11381
R BUDGET, WATER CHEMISTRY, WATER   TEMPERATURE, HEAT BUDGET, TURBIDI   W72-01773
AND, HYDROGEN ION CONCENTRATION,   TEMPERATURE, PROTEINS, CARBOHYDRA   W71-13413
FIXATION, FOREST, WINDS, LIGHT,   TEMPERATURE, HYD: /EVEL, NITROGEN   W71-03048
RIA, HYDROGEN ION CONCENTRATION,   TEMPERATURE, ALKALINITY, BICARBON   W71-05267
ARBON DIOXIDE, DISSOLVED OXYGEN,   TEMPERATURE, VELOCITY, CYANOPHYTA   W71-00668
          TEMPERATURE RANGES,   TEMPERATURE REQUIREMENTS.:   W71-02494
*ORGANIC MATTER, WATER QUALITY,   TEMPERATURE, HYDROGEN ION CONCENT   W71-03031
RATURE TOLERANCE, OPTIMUM GROWTH   TEMPERATURE, CONDENSER PAGGAGE.: /   W71-02496
EQUIREMENTS.:   TEMPERATURE RANGES, TEMPERATURE R   W71-02494
ROTEINS, AQUATIC MICROORGANISMS,   TEMPERATURE, ACTIVATED SLUDGE, FI   W70-09905
EPILIMNION, WINDS, THERMOCLINE,   TEMPERATURE, CURRENTS(WATER), HYP   W70-09889
TER, *COPPER, *ALGAE, SEA WATER,   TEMPERATURE, SALINITY, BIOASSAY,   W70-08111
TY, *NUTRIENTS, LIMNOLOGY, WATER   TEMPERATURE, *ALGAE BLOOMS, CYANO   W70-09093
SICAL PROPERTIES, CARP, TILAPIA,   TEMPERATURE, NU: /L ANALYSIS, PHY   W70-06975
, ENERGY, DI-URNAL DISTRIBUTION,   TEMPERATURE, METABOLISM, *WASTE W   W70-06604
RE/ *STREAMS, *ALGAE, *DIATOMS,   TEMPERATURE, MANGANESE, CULTURES,   W70-07257
ION, DIATOMS, THERMAL POLLUTION,   TEMPERATURE, CHEMICAL PROPERTIES,   W70-07313
GREEN ALGAE, BLUE-GREEN ALGAE,   TEMPERATURE EXPERIMENTS, RECYCLED   W70-07257
VARIABILITY, CHLORELLA, CARBON,   TEMPERATURE, LIGHT INTENSITY, NUT   W70-02965
OXIDE, BICARBONATES, CARBONATES,   TEMPERATURE, HYDROGEN ION CONCENT   W70-02804
C MATTER, CHLOROPHYLL, DYNAMICS,   TEMPERATURE, SILTS, STANDING CROP   W70-02780
ITROGEN, PHOSPHORUS, CARBONATES,   TEMPERATURE, SHALLOW WATER, HYDRO   W70-02803
ERS, SCENEDESMUS, CHLAMYDOMONAS,   TEMPERATURE, CARBON DIOXIDE, HYDR   W70-02249
EMICAL ANALYSIS, TRACE ELEMENTS,   TEMPERATURE, IRON, NUTRIENTS, LIG   W70-01933
TENSITY, TROPHIC LEVEL, SAMPLING   TEMPERATURE, TRACERS, CHLORELLA,   W70-02504
CENTRATION, ALKALINITY, DIATOMS,   TEMPERATURE, WEATHER, EPILIMNION,   W70-02504
, PONDS, RIVERS, RESERVOIRS, AIR   TEMPERATURE, RAIN, GEOLOGY, MAPPI   W70-02646
PIRATION, PIGMENTS, CHLOROPHYLL,   TEMPERATURE, OXYGEN, SALINITY.: /   W70-03325

TY, CALIFORNIA, LIGHT INTENSITY,
ERATURE, AEROBIC BACTERIA, WATER
ORGANIC MATTER, PHOTOSYNTHESIS,
YNTHESIS, EUTROPHICATION, LIGHT,
, SILICATES, ORGANIC MATTER, AIR
CROORGANISMS, CHEMICAL ANALYSIS,
ENVIRONMENTAL EFFECTS, SALINITY,
, SUCKERS, SHINERS, CADDISFLIES,
DISM, LIGHT INTENSITY, VELOCITY,
MEASUREMENTS, SEWAGE EFFLUENTS,
SHIRE, ALGAE, ODOR, TASTE, LIGHT
, MICHIGAN, INVERTEBRATES, FISH,
, ECOSYSTEMS, CYCLING NUTRIENTS,
DS-SETTLING, ECONOMIC ADVANTAGE,
IDATION, PHOTOSYNTHESES, PLANTS,
*RECREATION, RESERVOIRS, PONDS,
TA, BRACKISH WATER, OLIGOTROPHY,
GEN, HYDROGEN ION CONCENTRATION,
A, THERMAO POLLUTION, NUTRIENTS,
IC OCEAN, CARBON DIOXIDE, LIGHT,
RGANISMS, *LIPIDS, LIGHT, ALGAE,
LOROPHYLL, MAGNESIUM, POTASSIUM,
, PLANKTON, LI/ *EUTROPHICATION,
*RIVERS, *LAKES, *PONDS, LIGHT,
IFICATION, *SEASONAL, *SULFIDES,
ANKTON, *ALGAE, LIGHT INTENSITY,
SIGN CRITERIA, DISSOLVED OXYGEN,
A-DEEP PONDS', ASHKELON(ISRAEL),
S, PHOTOSYNTHESIS, ILLUMINATION,
LANKTON, NUTRIENTS, OLIGOTROPHY,
EA WATER, SEASONAL, PSEUDOMONAS,
MAGNA, SELENASTRUM, RHODOTORULA,
YTA, SCENEDESMUS, BENTH/ *ALGAE,
IRS, ZOOPLANKTON, PHYTOPLANKTON,
ES, PRODUCTIVITY, RADIOISOTOPES,
NALYSIS, *ANALYTICAL TECHNIQUES,
PLANTS, WATER PROPERTIES, WATER
ON CONCENTRATION, IRON, SULFATE,
REVIEWS, LIGHT INTENSITY, WATER
, DRAINAGE, FERROBACILLUS, WATER
IMNOLOGY, PHOTOSYNTHETIC OXYGEN,
RONMENT, MARINE ALGAE, SALINITY,
RS, LAKES, *GROWTH RATES, *WATER
MNOLOGY, WATER PROPERTIES, WATER
ING, HYDROGEN ION CONCENTRATION,
*CONDUCTIVITY, *LEACHING, IRON,
ION CONCENTRATION, CONDUCTIVITY,
LINITY, HARDNESS, RADIOACTIVITY,
S, ORGANIC MATTER, CONDUCTIVITY,
DES, HYDROGEN ION CONCENTRATION,
PLANTS, HEAD LOSS, PERMEABILITY,
RS, ESTUARINE ENVIRONMENT, WATER
UNCTIPENNIS, PISIDIUM IDAHOENSE,
N, RUTHENIUM-106, WHITE OAK LAKE(
ERVOIR(TENN), FRENCH BROAD RIVER(
VER(TENN), CHICKAMAUGA RESERVOIR(
ESERVOIR(TENN), NORRIS RESERVOIR(
OGETON PECTI/ CHEROKEE RESERVOIR(
HETERAMTHERA, PICKWICK RESERVOIR(
, PHOTIC ZONE, DOUGLAS RESERVOIR(
E RESERVOIR(TENN), HOLSTON RIVER(
CLAMS, STRONTIUM RADIOISOTOPES,
COLUMBIA RIVER, COLORADO RIVER,
ANGANESE, LOW-FLOW AUGMENTATION,
OLOGICAL HALF-LIFE, CLINCH RIVER(
NIUM, MICROCYSTIS, STIGEOCLONIUM
TION, *EXCRETION, STEPHANODISCUS
ORMIDIUM LAMINOSUM, OSCILLATORIA
NANCE SPECTROMETRY, OSCILLATORIA
SCILLATORIA TENUIS, OSCILLATORIA
YLENE GLYCOL, POLYMERS, DIMETHYL
OLOGY, CULTURAL CHARACTERISTICS,
, CILIATES, COPEPODIDS, NAUPLII,
SKA, LAKES, RIVERS, HOT SPRINGS,
ION, DENITRIFICATION, NUTRIENTS,
ENTS, BIOCHEMICAL OXYGEN DEMAND,
UENTS, STREAMFLOW, DESALINATION,
*RUNOFF, *LAKES, *IMPOUNDMENTS,
TES, NITRATES, SEWAGE EFFLUENTS,
NSITY, CARBON, SEWAGE TREATMENT,
NG, ALGAE, NUTRIENTS, TURBIDITY,
TREATMENT, SECONDARY TREATMENT,
E, IRON COMPOUNDS, LIME, OXYGEN,
NDS, SEEPAGE, INDUSTRIAL WASTES,
, PROTOZOA, ALGAE, SOLID WASTES,
UTROPHICATION, SEWAGE EFFLUENTS,
KISTRODESMUS BRAUNII, DUNALIELLA
SKELETONEMA COSTATUM, DUNALIELLA
TEST PROCEDUR/ *EUTROPHICATION,
EST ANALYSIS, *LABORATORY TESTS,

TEMPERATURE, IONS, SPECTROPHOTOME	W70-03334
TEMPERATURE, IRON.: /ER, AIR TEMP	W69-07096
TEMPERATURE, TROPICAL REGIONS.: /	W69-06865
TEMPERATURE, CARBON DIOXIDE, NUTR	W69-07442
TEMPERATURE, AEROBIC BACTERIA, WA	W69-07096
TEMPERATURE, NITRATES, PLANTS, AM	W69-07838
TEMPERATURE, SEA WATER, OYSTERS,	W69-08267
TEMPERATURE, FLOW, DIATOMS, MANGA	W69-07853
TEMPERATURE, CHEMICAL ANALYSIS, P	W69-07862
TEMPERATURE, OXYGEN, SALINITY, AL	W69-06203
TEMPERATURE, NUTRIENTS, COPPER SU	W69-08282
TEMPERATURE, TRACERS, SAMPLING, P	W69-09334
TEMPERATURE, LIGHT QUALITY, CHLOR	W69-08526
TEMPERATURE-EFFECT.: /RYING, SOLI	W70-05819
TEMPERATURE, *WASTE WATER TREATME	W70-06600
TEMPERATURE, LIGHT PENETRATION, D	W70-06225
TEMPERATURE, LIGHT INTENSITY, CHL	W70-05752
TEMPERATURE, DEPTH, DENSITY, CHLO	W70-05387
TEMPERATURE, PHOTOSYNTHESIS, LIGH	W70-05270
TEMPERATURE.: /DIOISOTOPES, PACIF	W70-05652
TEMPERATURE, CARBON DIOXIDE, OXID	W70-05651
TEMPERATURE, EPILIMNION, LIGHT IN	W70-05409
TEMPERATURE STRATIFICATION, DEPTH	W70-05405
TEMPERATURE, FERTILIZATION, FISH,	W70-05404
TEMPERATURE, CLIMATIC DATA, STRAT	W70-04790
TEMPERATURE, PRIMARY PRODUCTIVITY	W70-04809
TEMPERATURE, BIOCHEMICAL OXYGEN D	W70-04787
TEMPERATURE PROFILES, VOLATILE SU	W70-04790
TEMPERATURE, NUTRIENTS, FACULTATI	W70-04786
TEMPERATURE, LIGHT INTENSITY, DAP	W70-03959
TEMPERATURE, AMINO ACIDS, INTERTI	W70-03952
TEMPERATURE COEFFICIENTS, NORTH A	W70-03957
TEMPERATURE, CYANOPHYTA, CHLOROPH	W70-03969
TEMPERATURE, *EPILIMNION, *HYPOLI	W70-04475
TEMPERATURE, DEPTH, BACTERIA, SLI	W70-04371
TEMPERATURE, PIPTIDES, INTERTIDAL	W70-04365
TEMPERATURE, OUTLETS, HYDROELECTR	W69-05023
TEMPERATURE, KENTUCKY.: /DROGEN I	W69-00096
TEMPERATURE, DIATOMS, CHLOROPHYTA	W68-00860
TEMPERATURE, HYDROGEN ION CONCENT	W68-00891
TEMPERATURE, THERMAL POLLUTION, T	W69-03516
TEMPERATURE, MONITORING, AERIAL P	W68-00010
TEMPERATURE, SEASONAL, TEST PROCE	W68-00472
TEMPERATURE, ALKALINITY, *PHOTOSY	W68-00248
TEMPERATURE, TOXICITY, DISSOLVED	W72-06854
TEMPERATURE, SOIL ANALYSIS, TASTE	W72-11906
TEMPERATURE, DISSOLVED OXYGEN, PH	W72-06051
TEMPERATURE, SALINITY, REAERATION	W73-00750
TEMPERATURE, BACTERIA, NUTRIENTS,	W73-00748
TEMPERATURE, CHLORINATED HYDROCAR	W72-11105
TEMPERATURE, OPERATION AND MAINTE	W72-08686
TEMPERATURE, *EUTROPHICATION, AQU	W73-01446
TENDIPES PLUMOSUS, BEAVER DAM LAK	W70-06217
TENN), ACTIVITY-DENSITY, VERTICAL	W70-04371
TENN), CHICKAMAUGA RESERVOIR(TENN	W71-07698
TENN), FLORENCE(ALA).: / BROAD RI	W71-07698
TENN), FRENCH BROAD RIVER(TENN),	W71-07698
TENN), HOLSTON RIVER(TENN), POTOM	W71-07698
TENN), NAJAS, PHOTIC ZONE, DOUGLA	W71-07698
TENN), NORRIS RESERVOIR(TENN), FR	W71-07698
TENN), POTOMOGETON PECTINATUS, PO	W71-07698
TENNESSEE RIVER, TROPHIC LEVEL, A	W70-02786
TENNESSEE RIVER, HUDSON RIVER, DE	W70-04507
TENNESSEE RIVER, ALGAE, IRON, ALK	W73-00750
(TENNESSEE); PHYSICAL HALF-LIFE, H	W70-02786
TENUE, MICROCYSTIS AERUGINOSA, NO	W70-02245
TENUIS, ASTERIONELLA FORMOSA, MEL	W69-10158
TENUIS, OSCILLATORIA TEREBRIFORMI	W70-05270
TENUIS, RAMMAN SPECTROSCOPY, MASS	W69-00387
TEREBRIFORMIS, MATS, SYNECHOCOCCU	W70-05270
TEREPHTHALATE.: /HYL ALCOHOL, ETH	W70-10433
TERMARY COMPOUNDS, MACROMOLECULES	W70-04280
TERRAMYCIN, TETRACYCLINE, STREPTO	W69-10154
TERRESTRIAL HABITATS, DEPTH, AIR,	W69-08272
TERTIARY TREATMENT, ALGAE, AMMONI	W69-08053
TERTIARY TREATMENT, ALGAE, BACTER	W69-08054
TERTIARY TREATMENT, LAGOONS, OXYG	W69-07389
TERTIARY TREATMENT, ALGAE, DOMEST	W70-04765
TERTIARY TREATMENT, WATER POLLUTI	W70-04074
TERTIARY TREATMENT, EFFLUENTS, LI	W70-05387
TERTIARY TREATMENT, SAMPLING, MON	W70-05645
TERTIARY TREATMENT, ELECTRODIALYS	W69-01453
TERTIARY TREATMENT, COST COMPARIS	W68-00256
TERTIARY TREATMENT, CHEMICAL PREC	W69-10178
TERTIARY TREATMENT.: /L, BACTERIA	W70-01027
TERTIARY TREATMENT, ALGAE, WATER	W70-03512
TERTIOLECTA, DCMU INHIBITOR.: /AN	W70-05261
TERTIOLECTA, AMPHIDINIUM CARTERI,	W69-08278
TEST ANALYSIS, *LABORATORY TESTS,	W68-00472
TEST PROCEDURES, NURTRIENTS, RIVE	W68-00472

S, *WATER TEMPERATURE, SEASONAL,   TEST PROCEDURES, SCENEDESMUS, *AL   W68-00472
CONDUCTIVITY, BACTERIA, DIATOMS,   TEST PROCEDURES.: /NG, CAPILLARY   W70-04469
IA, ALGAE, TREATMENT FACILITIES,   TEST PROCEDURES, METALS.: /BACTER   W70-01519
IA, ALGAE, TREATMENT FACILITIES,   TEST PROCEDURES, METALS.: /BACTER   W71-06737
, DIATOMS, CYANOPHYTA, BIOASSAY,   TEST PROCEDURES.: /LGAE, CULTURES   W71-06002
DS, SEASONS, ACETYLENE REDUCTION   TEST.: /HODS, *BIOCHEMISTRY METHO   W70-03429
WA/ *UPWARD FLOW, PUTRESCIBILITY   TEST, BODY IMMEDIUM FILTER, BACK-   W70-05821
ARAMETRIC TESTS, ONE-SAMPLE RUNS   TEST, CLUSTER ANALYSIS, REYKJAVIK   W69-09755
MORPHOLOGY, FLAGELLATION, IODINE   TEST, GRAM REACTION, CYTOPHAGA, P   W70-03952
HARINA, MANNITOL, MUSTARD GROWTH   TEST, ORGANIC GROWTH FACTORS, POL   W69-07826
E, *ALGAE BLOOMS, *FISH TOXICITY   TEST, SLUDGE DRYING, SOLIDS-SETTL   W70-05819
ALLOMONAS, ULOTHRIX, DAVID MONIE   TEST, SUM: /ANABAENA, MELOSIRA, M   W69-10157
OCEDURES, ANTIFOULING MATERIALS,   TESTING, BIODEGRADATION, ALGAE, M   W71-07339
, AQUATIC ALGAE, AQUATIC PLANTS,   TESTING, BIODEGRADATION.: /ATMENT   W70-04373
*CONTROL, METHODOLOGY, CULTURES,   TESTING, CAPILLARY CONDUCTIVITY,   W70-04469
IDE EXCHANGE, *OPEN SEA, WEAPONS   TESTING, NUCLEAR SHIP SAVANNAH, Z   W70-00707
ER POLLUTION EFFECTS, LABORATORY   TESTS.: / PHOSPHATES, SODIUM, WAT   W70-07257
S, FRESH WATER LAKES, LABORATORY   TESTS.: / *PLANT GROWTH REGULATOR   W68-00856
AREAS, SOIL BACTERIA, LABORATORY   TESTS.: /E, PIPTIDES, INTERTIDAL   W70-04365
, ANALYTICAL TECHNIQUES, ON-SITE   TESTS.: /IMPOUNDMENTS, RESERVOIRS   W71-10098
CYANOPHYTA, BIOASS/ *LABORATORY   TESTS, *ALGAE, CULTURES, DIATOMS,   W71-06002
HT INTENSIT/ *ALGAE, *LABORATORY   TESTS, *ANAEROBIC CONDITIONS, LIG   W71-13413
ED-DENSITY STUDIES, *FIXED-LIGHT   TESTS, *EFFECTIVE-AVERAGE-LIGHT-I   W70-03923
CHNIQUES, *PLANKTON, *LABORATORY   TESTS, *ISOLATION, *CONTROL, METH   W70-04469
VER, TEXAS, *NUTRIENT ENRICHMENT   TESTS, *NITROGEN, *PHOSPHORUS, *I   W69-06004
OLOGY, GROWTH RATES, *LABORATORY   TESTS, *SCENEDESMUS, DIATOMACEOUS   W68-00475
OTASSIUM PERMANGANATE, ALGICIDAL   TESTS, ALGISTATIC TESTS, SLIMEPRO   W69-05704
ROL REQUIREMENTS, THRESHOLD ODOR   TESTS, ASTERIONELLA, APHANIZOMENO   W69-10157
ESIS, LIGHT PENETRATION, ON-SITE   TESTS, BIOCHEMICAL OXYGEN DEMAND,   W72-01818
ANISMS, ALGAE, FUNGI, LABORATORY   TESTS, BIOCHEMICAL OXYGEN DEMAND,   W72-01817
IOG/ *EUTROPHICATION, LABORATORY   TESTS, BIOLOGICAL TREATMENT, BIBL   W68-00855
       SECCHI DISC, VISIBILITY   TESTS, CONTROL METHODS.:   W68-00488
EMENTS, GROWTH RATES, LABORATORY   TESTS, CULTURES, SODIUM COMPOUNDS   W70-03507
                          *MODEL   TESTS, DYES.:   W70-02609
NALYTICAL TECHNIQUES, LABORATORY   TESTS, FOREIGN RESEARCH.: /ITY, A   W70-08628
TION, FREEZE DRYING, FREEZE-THAW   TESTS, FREEZING, LIGHT INTENSITY,   W69-03518
RELLA, *CHLAMYDOMONAS LABORATORY   TESTS, GAS CHROMATOGRAPHY, DEGRAD   W70-08652
ENTS, FREEZE DRYING, FREEZE-THAW   TESTS, HARVESTING OF ALGAE, NUISA   W69-03362
ENTS, FREEZE DRYING, FREEZE-THAW   TESTS, INHIBITION, NUTRIENT REQUI   W69-03370
*DIELDRIN, *DIATOMS, LABORATORY   TESTS, INSECTICIDES, WATER POLLUT   W71-03183
N, SELF-PURIFICATION, LABORATORY   TESTS, INVERTEBRATES, FISH, WATER   W72-01801
ETABOLISM, EPIPHYTOLOGY, ON-SITE   TESTS, LABORATORY TESTS, LOTIC EN   W72-05453
TION, *OXYGEN DEMAND, LABORATORY   TESTS, LAGOONS, OXIDATION, PHOTOS   W70-06600
TOSYNTHESIS, LAGOONS, LABORATORY   TESTS, LIGHT INTENSITY, PHOTOPERI   W72-00917
OLOGY, ON-SITE TESTS, LABORATORY   TESTS, LOTIC ENVIRONMENTS, LENTIC   W72-05453
H RATES, *PHOSPHATES, LABORATORY   TESTS, MINNESOTA, NUTRIENTS, WATE   W70-08637
ER POLLUTION EFFECTS, LABORATORY   TESTS, MODEL STUDIES, RADIOACTIVI   W70-04074
AL PROPERTIES, NUTRIENTS, LAKES,   TESTS, NITROGEN, PHOSPHORUS, CALC   W68-00856
ATOR, RHODAMINE WT DYE, AIRBORNE   TESTS, OIL SPILLS, GEOLOGICAL SS   W70-05377
UDURNES(ICELAND), NON-PARAMETRIC   TESTS, ONE-SAMPLE RUNS TEST, CLUS   W69-09755
PHNIA, CHLAMYDOMONAS, LABORATORY   TESTS, PARTICLE SIZE, POPULATION,   W72-02417
E, *HYDRAULIC MODELS, LABORATORY   TESTS, PERIPHYTON, STREAMS, PHOTO   W71-00668
TOPLANKTON, BIOASSAY, LABORATORY   TESTS, PRODUCTIVITY, ABSORPTION,   W72-00845
DUCTIVITY, *STREAMS, *LABORATORY   TESTS, RESPIRATION, ENVIRONMENT,   W70-00265
ATE, ALGICIDAL TESTS, ALGISTATIC   TESTS, SLIMEPRODUCING ALGAE, PHOT   W69-05704
IS, DISSOLVED OXYGEN, LABORATORY   TESTS, TEMPERATURE, WASTE WATER T   W71-12183
TION, TEST ANALYSIS, *LABORATORY   TESTS, TEST PROCEDURES, NURTRIENT   W68-00472
ON PESTICIDES, ALGAE, LABORATORY   TESTS, WATER POLLUTION EFFECTS.: /   W72-11105
COPEPODIDS, NAUPLII, TERRAMYCIN,   TETRACYCLINE, STREPTOMYCIN, MILLI   W69-10154
TIS, ANKISTRODESMUS, PEDIASTRUM,   TETRAEDRON, TETRASTRUM, TISZA(HUN   W70-03969
SPHALERITE, MILLERITE, ORPIMENT,   TETRAHEDRITE, BORNITE, CHALCOPYRI   W69-07428
ODESMUS, PEDIASTRUM, TETRAEDRON,   TETRASTRUM, TISZA(HUNGARY), ZIMON   W70-03969
       AUSTIN(TEX), COLORADO RIVER(TEX).:   W70-04727
                          AUSTIN(TEX), COLORADO RIVER(TEX).:   W70-04727
PHATES.:   HARRIS COUNTY,   TEX, POLYPHOSPHATES, ORGANIC PHOS   W69-04800
MENTS, LAKES, LOAMS, RESERVOIRS,   TEXAS.: / EUTROPHICATION, IMPOUND   W69-04800
HIGHLAND LAKES, COLORADO RIVER,   TEXAS, *NUTRIENT ENRICHMENT TESTS   W69-06004
B, RED TIDE, SALINITY, FLORIDA,   TEXAS, PHOSPHORUS, MIGRATION, DIU   W70-05372
CARTERI, SYRACOSPHAERA ELONGATA,   THALASSIOSIRA FLUVIAT: /HIDINIUM   W69-08278
TYLUM, CHAETOCEROS, SKELETONEMA,   THALASSIOSIRA, COCCOLITHUS, SARGA   W70-04503
MALS, RUTHENIUM-106, PLATYMONAS,   THALASSIOSIRA, AMPHIDINIUM KLEBSI   W70-00708
RHONE RIVER, NAVICULA, MELOSIRA,   THALASSIOSIRA, COSCINDISCUS, BACT   W73-01446
ONTANE REGION, GRADIENTS(RIVER),   THALLI, SCHIZOMYCETES, CYANOPHYTA   W70-02784
NUTRIENTS, FREEZE DRYING, FREEZE-   THAW TESTS, HARVESTING OF ALGAE,   W69-03362
NUTRIENTS, FREEZE DRYING, FREEZE-   THAW TESTS, INHIBITION, NUTRIENT   W69-03370
PHICATION, FREEZE DRYING, FREEZE-   THAW TESTS, FREEZING, LIGHT INTEN   W69-03518
ONDS, *ALGAL PHYSIOLOGY, *DESIGN   THEORY AND CALCULATIONS, PHOTOSYN   W70-04786
EACTIONS, SOLUBILITY EQUILIBRIUM   THEORY.: /LITHOSPHERE, SORPTION R   W70-04385
, NUTRIENTS, PHOSPHORUS, KINETIC   THEORY, ENVIRONMENTAL FACTORS, BI   W70-02779
COLOGY, ECOLOGICAL DISTRIBUTION,   THERMAL EFFECTS.: / PROPERTIES, E   W70-07313
ENTAL STREAMS, SEASONAL EFFECTS,   THERMAL EFFECTS, APHANIZOMENON FL   W70-03978
TOSYNTHETIC OXYGEN, TEMPERATURE,   THERMAL POLLUTION, THERMAL SPRING   W69-03516
N ABATEMENT, POLLUTANTS, WASTES,   THERMAL POLLUTION, TOXINS, PESTIC   W69-06305
TER CIRCULATION, NUISANCE ALGAE,   THERMAL POLLUTION, DIFFUSION, DIA   W69-05763
ION SOURCES, *THERMAL POLLUTION,   THERMAL POWERPLANTS, *THERMAL STR   W70-07847
ISMS, *ACCLIMATIZATION, DIATOMS,   THERMAL POLLUTION, TEMPERATURE, C   W70-07313
LGAL CONTROL, INDUSTRIAL WASTES,   THERMAL POLLUTION, METHODOLOGY, I   W70-10437
ORGANIC MATTER, WATER POLLUTION,   THERMAL POLLUTION, STREAMS, NUTRI   W70-03309
ADIOACTIVE WASTES, HEATED WATER,   THERMAL POLLUTION, BACTERIA, EFFL   W72-00462
UM, ELECTROSTATIC PRECIPITATION,   THERMAL PRECIPITATION.: /EBACTERI   W71-11823

TER, FLORA, AMBIENT TEMPERATURE,
BIOCHEMISTRY, ECOLOGY, SAMPLING,
TEMPERATURE, THERMAL POLLUTION,
RMAL POLLUTION, THERMAL SPRINGS,
ARY PRODUCTIVITY, RIVERS, TASTE,
ATES, NITRATES, DAPHNIA, SILICA,
ATHEMATICAL MODELS, GREAT LAKES,
HESIS, BIOASSAYS, AQUATIC WEEDS,
ANOPHYTA, PHYSIOLOGICAL ECOLOGY,
AL POWERPLANTS, *THERMAL STRESS,
*OREGON, THERMOPHILIC BACTERIA,
DICTIVE EQUATIONS, ZOOGEOGRAPHY,
SEASONAL, HYDROGRAPHY, FLORIDA,
GAE, FROGS, MOSSES, HYPOLIMNION,
THESIS, ALGAE, BOTTOM SEDIMENTS,
ER), DIURNAL, EPILIMNION, WINDS,
YNTHESIS, CARBON DIOXIDE, DEPTH,
DISTRIBUTION, TIME, EPILIMNION,
T, ALGAE, PHOTOSYNTHESIS, DEPTH,
LGAE, *THERMAL SPRINGS, *OREGON,
UTES, VITAMIN B-1, VITAMIN B-12,
LORELLA PYRENOIDOSA, SCHRODERIA,
TAZOA, RADIOLARIA, FORAMINIFERA,
ERA, THIOBACILLUS-FERROBACILLUS,
HARVEY(ND), CHROMATIUM VINOSUM,
DIONUCLIDE UPTAKE, NANNOCHLORIS,
ELEMENTS, RADIUM RADIOISOTOPES,
TTUS CAROLINAE, MEAD COUNTY(KY),
DUFF, OOZE, COBBLES, EQUISETUM,
HEMIPTERA, MOLLUSCA, RAFFINATE,
ING, ALGAE CONTROL REQUIREMENTS,
E RAMSEY(ONTARIO), ALGAE COUNTS,
NATIONAL WATER CARRIER(ISRAEL),
 *HYDROCARBON, *TOXICITY
S, LACTOBACILLUS CASEI, BACILLUS
ND SOUND(NY), ICELAND, TRITIATED
NIDIA, CALLOPHYLLIS HAENOPHYLLA,
SOMERSET, ENGLAND), SAND GRAINS,
E ANIMALS, *OILY WATER, LICHENS,
IFICATION, *POLLUTION ABATEMENT,
OGY, DISSOLVED OXYGEN, HARDNESS,
ORGANISMS, FISH PHYSIOLOGY, RED
CTERIA, MOLLUSKS, VITAMIN B, RED
RAL, *ORGANIC MATTER, EXUDATION,
OSPHORUS, NUISANCE ALGAE, SALTS,
YSIS, PHYSICAL PROPERTIES, CARP,
ORGANIC LOAD EFFECTS, *DETENTION
 *DETENTION
RIO, FACULTATIVE, BOD, DETENTION
HUR, HYDROGEN ION CONCENTRATION,
 REMOVAL
, MELOSIRA BINDERANA, GENERATION
, SURFACES, DEPTH, DISTRIBUTION,
VAL RATES, DESIGN APP/ DETENTION
AL PHOTOSYNTHESIS, MEAN DOUBLING
OCEAN/ ENIWETOK ATOLL, TURNOVER
PLAND(SWEDEN), ACETATE, TURNOVER
EDGWICK-RAFTER METHOD, TREATMENT
PODA, PLECOPTERA, EPHEMEROPTERA,
C PLANTS, *HERBICIDES, *VASCULAR
YDROGEN ION CONCENTRATION, PLANT
IASTRUM, TETRAEDRON, TETRASTRUM,
ITY, P/ *ALGAL CULTURE, *BIOMASS
 BERLIN, *ALGAE-
 *PETROCHEMICAL COMPOUNDS, MEAN
ALGAE, *RED ALGAE, *TEMPERATURE
 POLLUTION-
, *EUGLENA, PIGMENTE/ *POLLUTION-
QUALIS 381, ANABAENA FLOS AQUAE,
APRICOTS, PEARS, CITRUS FRUITS,
SMS, SUSPENDED SOLIDS, SALTWATER
ON, PEST CONTROL ORGANIZATION OF
IC ABSORPTION SPECTROPHOTOMETRY,

N-DEGRADABLE SOL/ *DAIRY MANURE,
ES, CLEAN WATER, GENUS, SPECIES,
ER PLANTS, *WATER REUSE, COOLING
E FILTERS, SNAILS, ODOR, COOLING
 *DEFINITIONS, *WATER BLOOM,
ENTS, PHYCOLOGY, SPACE RESEARCH,
GICAL LIFE, *ALGAE BLOOMS, *FISH
 METAL
NTROL, *CONTROL, *ALGAL CONTROL,
ION CONCENTRATION, TEMPERATURE,
 *WATER POLLUTION CONTROL, FISH,
ON), DDT/ *ALGAE, *INSECTICIDES,
DAPHNIA, / *HERBICIDES, *ALGAE,
DAPHNIA, / *HERBICIDES, *ALGAE,
CHLORELLA, PESTICIDES, PESTICIDE
ATION, NUTRIENTS, *FERMENTATION,

THERMAL SENSITIVITY, GENERA, 'FLA   W70-00713
THERMAL SPRINGS, EFFLUENTS, CYANO   W69-10151
THERMAL SPRINGS, THERMAL STRES: /   W69-03516
THERMAL STRES: / TEMPERATURE, THE   W69-03516
THERMAL STRATIFICATION, WASTE TRE   W69-06535
THERMAL STRATIFICATION, SULFIDES,   W70-05760
THERMAL STRATIFICATION.: /DIES, M   W71-05878
THERMAL STRATIFICATION, EPILIMNIO   W70-02251
THERMAL WATER.: /RBON DIOXIDE, CY   W69-10160
THERMAL WATER, *PROTOZOA, ALGAE,   W70-07847
THERMAO POLLUTION, NUTRIENTS, TEM   W70-05270
THERMOBIOLOGY.: /ION, MEXICO, PRE   W70-03957
THERMOCLINE, CURRENTS(WATER), PHY   W70-04368
THERMOCLINE, *ION EXCHANGE, WATER   W68-00857
THERMOCLINE, HYPOLIMNION, ANIMALS   W70-06975
THERMOCLINE, TEMPERATURE, CURRENT   W70-09889
THERMOCLINE, ZOOPLANKTON, TRIPTON   W69-10169
THERMOCLINE, LIGHT PENETRATION, S   W71-07698
THERMOCLINE, CLIMATIC ZONES, ORGA   W72-10573
THERMOPHILIC BACTERIA, THERMAO PO   W70-05270
THIAMINE, SERRATIA MARINORUBRA, G   W70-05652
THIMET, TRICHODERMA.: /ATHION, CH   W70-08097
THIOBACILLUS-FERROBACILLUS, THIOB   W69-07428
THIOBACILLUS DENITRIFICANS, W VIR   W69-07428
THIOCAPSA FLORIDANA, ENTEROCOCCI,   W70-03312
THORACOMONAS, YTTRIUM, 'BIOLOGICA   W70-00708
THORIUM RADIOISOTOPES, URANIUM RA   W71-09013
THORIUM-232, BISMUTH-214, ACTINIU   W69-09742
THORIUM-232, YUKON RIVER, NEBESNA   W69-08272
THORIUM, TAILINGS.: / COLEOPTERA,   W69-07846
THRESHOLD ODOR TESTS, ASTERIONELL   W69-10157
THRESHOLD ODOR NUMBERS, CROP-DUST   W69-10155
THRESHOLD ODOR NUMBERS.: /ATORIA,   W70-02373
THRESHOLD.:   W70-01466
THURINGIENSIS, AGARS, STAPHYLOCOC   W71-11823
THYMIDINE, NARRAGANSETT BAY(RI),   W69-09755
TIDAL CURRENT, PUGET SOUND(WASH)   W69-10161
TIDAL CYCLES.: /YI, ABBOT'S POND(W71-12870
TIDAL EFFECTS, DISPERSION, OIL-WA   W70-01777
TIDAL WATERS, MIXING, SEA WATER,   W69-06203
TIDAL WATERS, ORGANIC MATTER, CON   W73-00748
TIDE, ORGANIC COMPOUNDS, DINOFLAG   W71-07731
TIDE, SALINITY, FLORIDA, TEXAS, P   W70-05372
TIDES, PHOTOSYNTHESIS, SOLAR RADI   W70-01073
TIDES, SOUTHEAST U. S., REGIONS,   W69-03695
TILAPIA, TEMPERATURE, NU: /L ANAL   W70-06975
TIME EFFECTS, *FACULATIVE PONDS,   W70-04787
TIME.:   W72-06854
TIME.:   *SULFATE ION, *DESULFOVIB   W70-01971
TIME.: /ECOLOGY, PHOSPHORUS, SULP   W70-02965
TIME(POLLUTANTS).:   W71-05878
TIME, CHLORELLA PYRENOIDOSA, DIAT   W69-10158
TIME, EPILIMNION, THERMOCLINE, LI   W71-07698
TIME, LABORATORY MODELS, BOD REMO   W70-07034
TIME, PACKED CELL VOLUMES, BACTER   W71-05267
TIME, TRIDACNA CROCEA, WOODS HOLE   W69-08526
TIMES: /EASES, CHAMYDOMONADS, LAP   W70-04287
TIMING, ALGAE CONTROL REQUIREMENT   W69-10157
TIPULIDAE, CHIRONOMIDAE, HALF-LIF   W69-09742
TISSUES, *ALGAE, ECOSYSTEMS, CHAR   W70-10175
TISSUES, SCENEDESMUS, LIMNOLOGY,   W68-00248
TISZA(HUNGARY), ZIMONA(HUNGARY),   W70-03969
TITER, *BIOLOGICAL INDEX, *SAPROB   W72-12240
TITRE.:   W68-00475
TOLERANCE LIMIT, BLUEGILLS, CHLOR   W72-11105
TOLERANCE, OPTIMUM GROWTH TEMPERA   W71-02496
TOLERANT ALGAE.:   W70-07027
TOLERANT ALGAE, *BLUE GREEN ALGAE   W70-01233
TOLYPOTHRIX DISTORTA, GLOEOTRICHI   W70-05091
TOMATOES, *AEROBIC TREATMENT, *AN   W71-09548
TONGUE, BENTHIC FORMS, EPIPHYTIC   W73-00748
TORONTO, FILAMENTOUS ALGAE, NAVIC   W69-10155
TOTAL DISSOLVED SOLIDS, HEDRIODIS   W70-03309
TOTAL ORGANIC CARBON.:   W72-04495
TOTAL SOLIDS, VOLITILE SOLIDS, NO   W71-00936
TOUCH SENSATION.: / INDUSTRIAL US   W70-05389
TOWERS, DAMS, ALGAE, HYDROGEN ION   W71-11393
TOWERS, FILTERS, EFFLUENT SAMPLIN   W72-00027
TOXIC ALGAE, RECREATIONAL USE.:   W69-08279
TOXIC ALGAE.: /OALGAE, MICRONUTRI   W69-07832
TOXICITY TEST, SLUDGE DRYING, SOL   W70-05819
TOXICITY.:   W70-10437
TOXICITY.: /ITY, WATER QUALITY CO   W72-07442
TOXICITY, DISSOLVED OXYGEN, LIGHT   W72-06854
TOXICITY, FRESH WATER, PULP WASTE   W70-10437
TOXICITY, DEGRADATION(DECOMPOSITI   W70-08097
TOXICITY, SCENEDESMUS, CHLORELLA,   W71-05719
TOXICITY, SCENEDESMUS, CHLORELLA,   W71-03056
TOXICITY, PHOTOSYNTHETIC OXYGEN,   W70-08416
TOXICITY, CHEMICAL OXYGEN DEMAND,   W71-13413

544

WATER ANALYSIS, INSECTS, SNAILS,
TA, *DECOMPOSING ORGANIC MATTER,
EVEL, PRODUCTIVITY, FOOD CHAINS,
OTROPHY, VITAMINS, HEAVY METALS,
LOGY, *BIOINDICATORS, *BIOASSAY,
IQUES, SAMPLING, ORGANIC MATTER,
Y, *FISH DISEASES, *FISH TOXINS,
-DICHLORONAPHTHOQUIN/ *SELECTIVE
ON, SEVIN.:   *ALGAE
ALGAL CONTROL, WATER POLLUTION,
EFFECTS, STABILIZATION, LAGOONS,
, PHOSPHATES, ORGANIC COMPOUNDS,
TION, BIOCHEMICAL OXYGEN DEMAND,
PLANTS, BACTERIA, SEDIMENTATION,
, CONTOURS, DEPTH, OLIGOCHAETES,
PHOTOSYNTHESIS, EUTROPHICATION,
TOPLANKTON, *DDT, *MARINE ALGAE,
S, COLOR, FISHKILL, LAKES, ODOR,
S, PESTICIDE KINETICS, PESTICIDE
UTRIENT REQUIREMENTS, NUTRIENTS,
UTRIENTS, *NITRIFICATION, CROPS,
OD WEBS, DDT/ *ALGAE, *PESTICIDE
-WATER INTERFACES, EUGLENOPHYTA,
YANAPHYTA, ALGAL CONTROL, *ALGAL
D TOXINS, IBOGAALKALOID, ORGANIC
*INHIBITION, *P/ *ALGAE, *ALGAL
TOXINS, *ALGAL POISONING, *ALGAL
GEN, FISHKILLS, LIMNOLOGY, ALGAL
ROMETERS, DIFFUSION, TURBULENCE,
DINOFLAGELLATES, FISHKILL, FISH
ROPHICATION, COLOR, TASTE, ODOR,
NIUM SP., *GLENODININE, ALKALOID
VED SOLIDS, NUTRIENTS, VITAMINS,
BIOASSAY, AQUATIC ALGAE, *ALGAL
YTOPLANKTON, BACTERICIDES, ALGAL
BIOCHEMICAL OXYGEN DEMAND, ALGAL
ITY, HYDROGEN ION CONCENTRATION,
ANTS, WASTES, THERMAL POLLUTION,
AINAGE EFFECTS, *SAMPLING, ALGAE
PULP WASTES, INDUSTRIAL WASTES,
OTOXICITY, *FISH DISEASES, *FISH
LLOUT, *ECOSYSTEMS, GREAT LAKES,
GRAPHY, *COBALT, *COPPER, *IRON,
ASSIUM, SODIUM, IRON, MANGANESE,
Y, ZOOPLANKTON, ALGAE, COLLOIDS,
OROPHYTA, CHLOROPHYLL, BACTERIA,
OTOSYNTHESIS, CHEMICAL ANALYSIS,
UM, CLAMS, SIZE, ALGAE, DIATOMS,
GY, ALGAL TOXINS, ALGAL CONTROL,
LATORS, PREDATION, PRODUCTIVITY,
LFATE, ALGAL CONTROL, ALGICIDES,
AMINS, PEPTIDES, ORGANIC MATTER,
TION, SODIUM, ALGAE, PHOSPHORUS,
HOTOMETRY, CHLORELLA, NUTRIENTS,
, PHOSPHORUS, SODIUM, POTASSIUM,
OGRAPHY, RADIOCHEMICAL ANALYSIS,
OTOPES, LIMNOLOGY, OCEANOGRAPHY,
HIC LEVEL, SAMPLING TEMPERATURE,
ES, PLANKTON, ECOSYSTEMS, SINKS,
TER, PHYSICOCHEMICAL PROPERTIES,
M, SELECTIVITY, POISONS, OXYGEN,
*AQUATIC ENVIRONMENT, CHLORELLA,
, CHEMICAL REACTIONS, CHLORELLA,
NVERTEBRATES, FISH, TEMPERATURE,
, CHROMATOGRAPHY, RADIOACTIVITY,
*TURBID RESERVOIRS, CYCLOTELLA,
RULINA, VOLVOCALES, GLENODINIUM,
OXIS ANNULARIS, GASTROINTESTINAL
WN(MO), LEE'S SUMMIT(MO), OXYGEN
OOD ORGANISMS, NUTRIENTS, ENERGY
IA, FORAMINIFERA, THIO/ *MINERAL
, DESORPTION, EMBRYOLOGY, ENERGY
TROSCOPY, ULTRAVIOLET RADIATION,
PRODUCTIVITY, AQUATIC NUISANCES,
IZOMENON FLOS-AQUAE, SECCHI DISC
H), *OXYGEN DEFICIT, SECCHI DISC
HARTBEESPOORT DAM(SOUTH AFRICA),
S, DIURON, FLAGELLATES, MONURON,
EUPHOT/ *LAKE TAHOE(CALIF-NEV),
EDEN, LAKE ERKEN(SWEDEN), ACTIVE
MAXIMUM UPTAKE VELOCITY, ACTIVE
HYLLUM, EICHHORNIA, POTAMOGETON,
LAKE MINNETONKA(MINN), NUTRIENT
S, ALKALINE WATER, ACIDIC WATER,
PILOT PLANTS, ACTIVATED SLUDGE,
CTIVATED SLUDGE PROCESS, HANOVER
KES, NUTRIENTS, QUALITY CONTROL,
GROWTH, SEDGWICK-RAFTER METHOD,
ES, NEUTRIENTS, BACTERIA, ALGAE,
*NEW ZEALAND, SEWAGE

```
SLUDGE, *BACTERIA, META/ *SEWAGE     TREATMENT, *PROTOZOA, *ACTIVATED      W72-01817
LLUTION, WASTE DISPOSAL, *SEWAGE     TREATMENT, *WASTE WATER TREATMENT    W72-01786
  *SEWAGE TREATMENT, *WASTE WATER    TREATMENT, *WASTE WATER DISPOSAL,    W72-01786
, RHIZOPODS, SPIDERS, MILK WASTE     TREATMENT, COLONY COUNTS, ARTHROP    W72-01819
SUPPLY, *BASINS, WATER POLLUTION     TREATMENT, COPPER SULFATE, ODOR,     W72-01814
LGAE, *BACTERIA, *DISSO/ *SEWAGE     TREATMENT, *OXIDATION LAGOONS, *A    W72-01818
EAD LOSS, FILTERS, *SCEN/ *WATER     TREATMENT, *ALGAE, *FILTRATION, H    W72-03703
LGAE, LIGHT INTENS/ *WASTE WATER     TREATMENT, *OXIDATION LAGOONS, *A    W72-06022
NGANESE, *COLOR, *ALGAE,/ *WATER     TREATMENT, *TURBIDITY, *IRON, *MA    W72-04145
SING ORGANIC MATTER, *PO/ *WATER     TREATMENT, ODOR, *ALGAE, *DECOMPO    W72-04242
E, *EUTROPHICATION,/ WASTE WATER     TREATMENT, WATER TREATMENT, *ALGA    W72-01867
N,/ WASTE WATER TREATMENT, WATER     TREATMENT, *ALGAE, *EUTROPHICATIO    W72-01867
ATED CARBON, *ADSORPTION, *WATER     TREATMENT, *WASTE WATER TREATMENT    W72-03641
C MATTER, ALGAE, COPPER / *WATER     TREATMENT, WATER QUALITY, *ORGANI    W72-04495
 *ALGICIDES, ACTIVATED C/ *WATER     TREATMENT, *ALGAE, *TASTE, *ODOR,    W72-03746
TION, *GROUNDWATER, RECH/ *WATER     TREATMENT, *WATER REUSE, *RECLAMA    W72-04318
LIME, *ULTRAVIOLET RADIA/ *WATER     TREATMENT, *ALGAE, FILTRATION, *S    W72-03824
, *WATER TREATMENT, *WASTE WATER     TREATMENT, ALGAE, FILTRATION, NEW    W72-03641
OIRS, *DECIDUOUS TREES, / *WATER     TREATMENT, *TASTE, *ODOR, *RESERV    W72-03475
, *TASTE, *ODOR, *NATURA/ *WATER     TREATMENT, *ALGAE, *ALGAL CONTROL    W72-03562
R POLLUTION, DEGRADATION, SEWAGE     TREATMENT, NITROGEN COMPOUNDS, PH    W72-02817
, *ODOR, ACTIVATED CARBO/ *WATER     TREATMENT, *ALGAE, *ACTINOMYCETES    W72-04155
LYTICAL TECHNIQUES, *WASTE WATER     TREATMENT, TREATMENT FACILITIES,     W71-13313
RAINS, SEWAGE EFFLUENTS, *SEWAGE     TREATMENT, STORM RUNOFF.: /TORM D    W71-10654
RM WASTES, *POULTRY, WASTE WATER     TREATMENT, SEDIMENTATION, OXIDATI    W71-11375
MICAL OXYGEN DEMAND, WASTE WATER     TREATMENT, OVERFLOW, SEWERS, STOR    W71-10654
TIAL, AQUATIC ALGAE, *BIOLOGICAL     TREATMENT, *WASTE WATER TREATMENT    W71-09546
TERIA, WATER TEMPERATURE, SEWAGE     TREATMENT, AEROBIC BACTERIA, AQUA    W71-08221
IC WASTES, *OXYGENATION, AEROBIC     TREATMENT, ANAEROBIC CONDITIONS,     W71-09546
R, SEWAGE DISPOSAL, *WASTE WATER     TREATMENT, *OXIDATION LAGOONS.: /    W71-09548
S, *DOMESTIC WASTES, WASTE WATER     TREATMENT, INDUSTRIAL WASTES, NUT    W71-06737
ITRUS FRUITS, TOMATOES, *AEROBIC     TREATMENT, *ANAEROBIC DIGESTION,     W71-09548
R SUPPLY, *WATER QUALITY, *WATER     TREATMENT, FILTRATION, CHLORINATI    W71-05943
GAE, *PLANKTON, *SEWAGE, *SEWAGE     TREATMENT, *EUTROPHICATION, PONDS    W70-02304
*METAL CORROSION, COOLING WATER      TREATMENT, ON-STREAM DESLUDGING.:    W70-02294
, *SANITARY ENGINEERING, *SEWAGE     TREATMENT, *TRACERS, *WATER POLLU    W70-02609
MPARISONS, TREATMENT, BIOLOGICAL     TREATMENT, ALGAL CONTROL, AMMONIA    W70-01981
AND APPLICATION, ELECTROCHEMICAL     TREATMENT, AMMONIA STRIPPING, ALG    W70-01981
EFFICIENCIES, COST COMPARISONS,      TREATMENT, BIOLOGICAL TREATMENT,     W70-01981
SION, *CHEMICAL REACTION/ *WATER     TREATMENT, *COOLING WATER, *CORRO    W70-02294
SEWAGE, SEWAGE EFFLUENTS, SEWAGE     TREATMENT, WATER POLLUTION CONTRO    W70-02248
ES, PERSISTENCE, WATER POLLUTION     TREATMENT, WATER POLLUTION EFFECT    W70-01777
F, *FERTILIZATION, LAKES, SEWAGE     TREATMENT, SEWAGE BACTERIA, NITRO    W70-02787
OONS, *INDUSTRIAL WASTES, SEWAGE     TREATMENT, NORTH DAKOTA, COLIFORM    W70-03312
 DOMESTIC WASTES, SEWERS, SEWAGE     TREATMENT, SEWAGE DISPOSAL, STORM    W70-03344
TION, SEWAGE EFFLUENTS, TERTIARY     TREATMENT, ALGAE, WATER QUALITY,     W70-03512
VIRONMENTAL EFFECTS, *BIOLOGICAL     TREATMENT, *DESIGN CRITERIA, DISS    W70-04787
ES, *CHLORINATION, *ODOR, *WASTE     TREATMENT, ALGAE, DISINFECTION, B    W70-04788
IES, SANITARY ENGINEERING, WATER     TREATMENT, WASTE TREATMENT.: /UAR    W70-05094
, *ENVIRONMENTAL EFFECTS/ *WASTE     TREATMENT, *STABILIZATION, *PONDS    W70-04787
TREAMS, WATER POLLUTION,/ *WASTE     TREATMENT, *WATERS, *ORGANISMS, S    W70-05750
CAL OXYGEN DEMAND, *ACT/ *SEWAGE     TREATMENT, *FILTRATION, *BIOCHEMI    W70-05821
TERIA, CULTURES, C/ *WASTE WATER     TREATMENT, *NUTRIENTS, ALGAE, BAC    W70-05655
E WATER RECLAMATION, WASTE WATER     TREATMENT, *NUTRIENTS.: /TS, WAST    W70-06053
D, *ACTIVATED SLUDGE, BIOLOGICAL     TREATMENT, PILOT PLANTS, HUMUS, R    W70-05821
, NUTRIENTS, TURBIDITY, TERTIARY     TREATMENT, SAMPLING, MONITORING,     W70-05645
*NUTRIENTS, *PESTICI/ *TERTIARY      TREATMENT, *PHOSPHORUS COMPOUNDS,    W70-05470
LIGHT INTENSITY, CARBON, SEWAGE      TREATMENT, TERTIARY TREATMENT, EF    W70-05387
RBON, SEWAGE TREATMENT, TERTIARY     TREATMENT, EFFLUENTS, LIMNOLOGY,     W70-05387
ORS, ODOR, TASTE, FILTERS, WATER     TREATMENT, DISSOLVED OXYGEN, EUTR    W70-05264
GAE, *FLOCCULATION, *WASTE WATER     TREATMENT, ELECTROLYTES, CHEMICAL    W70-05267
, *DESIGN CRITERIA, FACI/ *WASTE     TREATMENT, *STABILIZATION, *PONDS    W70-04786
*LAKES, *IMPOUNDMENTS, TERTIARY      TREATMENT, ALGAE, DOMESTIC WASTES    W70-04765
E DIST/ *ALGAE, *WATER POLLUTION     TREATMENT, LAKES, SEWAGE, DRAINAG    W70-04455
S, URBANIZATION, WATER POLLUTION     TREATMENT, ALGAE, WATER MANAGEMEN    W70-04721
ATES, SEWAGE EFFLUENTS, TERTIARY     TREATMENT, WATER POLLUTION SOURCE    W70-04074
YGEN DEMAND, *COLOR, WASTE WATER     TREATMENT, TREATMENT FACILITIES,     W70-04060
CATORS, *WATER POLLUTION, *WASTE     TREATMENT, FOOD CHAINS, CHLORELLA    W70-04381
S, *PHOSPHATES, CHELATION, WASTE     TREATMENT, AQUATIC ALGAE, AQUATIC    W70-04373
, *LAKES, *DETERGENTS, *TERTIARY     TREATMENT, *GREAT LAKES, CONTROL,    W70-03964
W YORK, SEWAGE EFFLUENTS, SEWAGE     TREATMENT, INDUSTRIAL WASTES, WAT    W70-03974
DIES, ELECTRON MICROSCOPY, WASTE     TREATMENT, LAKES, WATER QUALITY,     W70-03973
RECTIVE TREATMENT, *PREVENTATIVE     TREATMENT, ALGAE NUISANCE, ALGAE     W69-08282
OINDICATORS, PRODUCTIVITY, WASTE     TREATMENT, WATER POLLUTION EFFECT    W69-08518
ENTS, *EUTROPHICATION, *TERTIARY     TREATMENT, NITROGEN, AQUATIC PLAN    W69-09340
T, ALGAE NUISANCE, / *CORRECTIVE     TREATMENT, *PREVENTATIVE TREATMEN    W69-08282
REAMFLOW, DESALINATION, TERTIARY     TREATMENT, LAGOONS, OXYGENATION,     W69-07389
E, *BIOLOGICAL TREATMENT, *WATER     TREATMENT, *POLLUTANT IDENTIFICAT    W69-07389
LUTANT IDEN/ *ALGAE, *BIOLOGICAL     TREATMENT, *WATER TREATMENT, *POL    W69-07389
C DETERIORATION, *ADVANCED WASTE     TREATMENT, *NUTRIENT REMOVAL, AGR    W69-07818
GAE, POLLUTION ABATEMENT, SEWAGE     TREATMENT, SEWAGE DISPOSAL, WASTE    W69-06909
TRIFICATION, NUTRIENTS, TERTIARY     TREATMENT, ALGAE, AMMONIA.: /DENI    W69-08053
N, NITROGEN COMPOU/ *WASTE WATER     TREATMENT, *WATER REUSE, *NITROGE    W69-08053
CHEMICAL OXYGEN DEMAND, TERTIARY     TREATMENT, ALGAE, BACTERIA, ACTIV    W69-08054
LLUTION SOURCES, WATER POLLUTION     TREATMENT, FINANCING, GRANTS, FED    W69-06305
EASTS, BACTERIA, FISH, P/ *WASTE     TREATMENT, *AQUICULTURE, ALGAE, Y    W69-05891
E, THERMAL STRATIFICATION, WASTE     TREATMENT, WATER QUALITY, WATER P    W69-06535
LUTION CONTROL, *WATER POLLUTION     TREATMENT, ALGAE, ALGAL CONTROL,     W69-03374
OL, PACIFI/ *WASHINGTON, *SEWAGE     TREATMENT, *WATER POLLUTION CONTR    W69-03683
U. S., CIVIL ENGINEERING, WASTE      TREATMENT, ENGINEERING, ALGAE, PL    W69-03683
```

RYON, MELOSIRA, ANEBODA(SWEDEN), UPPSALA(SWEDEN), LAKE ERKEN(SWEDE W70-05409
, SUSPENDED SOLIDS, *PHOSPHOROUS UPTAKE RATE.: MIXED LIQUOR W71-13335
MPHIDINIUM CARTERI, *RADIOCARBON UPTAKE TECHNIQUES, WATER POLLUTIO W70-05652
), WOODS HOLE(MASS). RADIOCARBON UPTAKE TECHNIQUES, MICROBIAL ECOL W70-05272
, MICROBIAL ECOLOGY, RADIOCARBON UPTAKE TECHNIQUES, MICHAELIS-MENT W70-05109
OCLINE, MIXOLIMNION, RADIOCARBON UPTAKE TECHNIQUES, BACTERIAL PLAT W70-05760
TOTROPHY, CARBON-14, RADIOCARBON UPTAKE TECHNIQUES, GLUCOSE, MICHA W70-04287
INEWEAVER-BURK EQUATION, MAXIMUM UPTAKE VELOCITY, MICHAELIS CONSTA W70-04287
AKE, *MICROBIAL ECOLOGY, MAXIMUM UPTAKE VELOCITY, ACTIVE TRANSPORT W70-04194
UATION, UPTAKE VELOCITY, MAXIMUM UPTAKE VELOCITY, SUBSTRATE CONCEN W70-05109
NTIAL, LINEWEAVER-BURK EQUATION, UPTAKE VELOCITY, MAXIMUM UPTAKE V W70-05109
UM U/ *HETEROTROPHY, *BIOLOGICAL UPTAKE, *MICROBIAL ECOLOGY, MAXIM W70-04194
M, EDTA, LIGURIAN/ *RADIONUCLIDE UPTAKE, *PHAEODACTYLUM TRICORNUTU W69-08275
N, *RADIONUCLIDE / *RADIONUCLIDE UPTAKE, *RADIONUCLIDE ACCUMULATIO W70-00707
EIN, MACROPHYTES, GROUND SOLIDS, UPTAKE, CONSUMER ORGANISMS, SIMUL W69-09334
-GREEN ALGAE,/ *COPPER CHLORIDE, UPTAKE, FREUNDLICH ISOTHERM, BLUE W70-08111
      *RADIONUCLIDE, ULOTHRIX, UPTAKE, HALF-LIFE, DETECTOR.: W69-07862
PHYRIDIUM CRUENTUM, RADIONUCLIDE UPTAKE, NANNOCHLORIS, THORACOMONA W70-00708
VITAMIN B12, BIOTIN, HISTIDINE, URACIL, REDUCED SULFUR COMPOUNDS, W70-02510
OISOTOPES, COBALT RADIOISOTOPES, URANIUM RADIOISOTOPES, NUCLEAR EX W70-00708
RAYS, SPECTROMETERS, CYANOPHYTA, URANIUM RADIOISOTOPES, STRONTIUM W69-09742
ISOTOPES, THORIUM RADIOISOTOPES, URANIUM RADIOISOTOPES, MARINE ALG W71-09013
TTER, RADIOISOTOPES, GAMMA RAYS, URANIUM RADIOISOTOPES, RADIUM RAD W69-08272
E, CESIUM, RADIUM RADIOISOTOPES, URANIUM RADIOISOTOPES.: /MANGANES W69-08524
MUNICIPAL WASTES, AQUATIC ALGAE, URANIUM RADIOISOTOPES, RADIUM RAD W69-01165
SAN MIGUEL RIVER, DOLORES RIVER, URAVAN, COTTUS SPP, RHINICHTHYS O W69-07846
OPE, WATER POLLUTION ASSESSMENT, URBAN DRAINAGE, AGRICULTURAL DRAI W70-03975
OPE, WATER POLLUTION ASSESSMENT, URBAN DRAINAGE, AGRICULTURAL DRAI W70-03975
                                 URBAN RUNOFF.: W71-04072
FFECTS, PRODUCTIVITY, SEDIMENTS, URBANIZATION, WATER POLLUTION TRE W70-04721
RS, LAKES, LIMNOLOGY, WISCONSIN, URBANIZATION, CONNECTICUT, OREGON W70-04506
ROPHICATION, *LAKES, GLACIATION, URBANIZATION, DOMESTIC WASTES, NU W69-07818
S, *EUTROPHICATION, *WASHINGTON, URBANIZATION, WATER POLLUTION, SE W70-05663
R, DDT, MISSISSIPPI RIVER BASIN, URBANIZATION, ALGAL POISONING, GA W73-00758
SPHORUS, *PHOSPHATES, ENERGY, DI- URNAL DISTRIBUTION, TEMPERATURE, W70-06604
CILLATORIA RUBESCENS, DINOBRYON, UROGLENA AMERICANA, CYCLOTELLA, C W70-05405
ABSORPTION, CELL CONCENTRATION, UROGLENA, ANKISTRODESMUS, CHLOROC W70-05409
G S WEST, RADIOCOCCUS NIMBATUS, URONEMA ELONGATUM HODGETTS, CHROO W70-04468
NATIONAL RESEARCH COUNCIL(USA), US ATOMIC ENERGY COMMISSION, NATI W70-03975
NATIONAL RESEARCH COUNCIL(USA), US ATOMIC ENERGY COMMISSION, NATI W70-03975
ATION, OFFICE OF NAVAL RESEARCH, US DEPARTMENT OF INTERIOR, MADISO W70-03975
ATION, OFFICE OF NAVAL RESEARCH, US DEPARTMENT OF INTERIOR, MADISO W70-03975
AGOONS, *MIDWEST(US), *SOUTHWEST(US), *DESIGN CRITERIA, POND EFFLU W70-04788
ERIA/ *AERATED LAGOONS, *MIDWEST(US), *SOUTHWEST(US), *DESIGN CRIT W70-04788
A, POND EFFLUENT, COLORADO, WEST(US), STABILIZATION POND LOADINGS, W70-04788
IA/ NATIONAL ACADEMY OF SCIENCES(USA), UNIVERSITY OF WISCONSIN, AS W70-03975
IA/ NATIONAL ACADEMY OF SCIENCES(USA), UNIVERSITY OF WISCONSIN, AS W70-03975
ASIA, NATIONAL RESEARCH COUNCIL(USA), US ATOMIC ENERGY COMMISSION W70-03975
ASIA, NATIONAL RESEARCH COUNCIL(USA), US ATOMIC ENERGY COMMISSION W70-03975
AM, KANS.: USBR RESEARCH CONTRACTS, CHENEY D W72-01773
BLOOM, TOXIC ALGAE, RECREATIONAL USE.: *DEFINITIONS, *WATER W69-08279
          *DOMESTIC WATER USE.: W68-00483
RAL EUTROPHICATION, RECREATIONAL USE, WASTE EFFLUENTS.: *CULTU W69-06535
OPHYCEAE, VOLVOCALES, INDUSTRIAL USES, CLEAN WATER, GENUS, SPECIES W70-05389
MS, COLLOIDAL FRACTION, MULTIPLE- USES, SAVANNAH RIVER, GULF COAST. W73-00748
          *SAKHALIN ISLAND(USSR).: W72-10623
          KIEV RESERVOIR(USSR).: W71-00221
DS, SVISLOCH' RIVER(USSR), MINSK(USSR), ENGLAND, OSCILLATORIA, AUS W70-05092
ALIA, SANTA ROSA(CALIF), LYUBLIN(USSR), LYUBERETS(USSR), MUNICH(GE W70-05092
ICULTURAL PONDS, SVISLOCH' RIVER(USSR), MINSK(USSR), ENGLAND, OSCI W70-05092
CALIF), LYUBLIN(USSR), LYUBERETS(USSR), MUNICH(GERMANY), CHLORELLA W70-05092
          BYELORUSSIA(USSR), POLYCHLOROPINENE.: W70-08654
UNITIES, SEDIMENTS, BIOTA, FISH, UTAH, ALGAE, INSECTS, INVERTEBRAT W69-07846
TORY, WATER PURIFICATION, *WATER UTILIZATION, DECOMPOSING ORGANIC W68-00483
GROWTH EQUATION, INCIDENT LIGHT UTILIZATION.: /TH KINETICS, MONOD W69-03730
   *NITROGEN COMPOUNDS, *NITROGEN UTILIZATION, *FRESH WATER, ALGAE, W69-04521
CHLORELLA PYRENOIDOSA, NITROGEN UTILIZATION, ROLLER DRYING, SPRAY W70-04184
NG, WASTE WATER TREATMENT, WATER UTILIZATION, INDUSTRIAL WASTES, O W70-07508
   *WATER POLLUTION EFFECTS, WATER UTILIZATION, ANALYTICAL TECHNIQUE W71-03048
CONTROL, *BIODEGRADATION, WATER UTILIZATION, FISH, ALGAE, MODEL S W71-11881
ASIDIUM PULLULANS, HANSENIASPORA UVARUM, RADULESPORES, ZYMOLOGICAL W70-04368
IA ECHINULATA, DINOBRYON, SYNURA UVELLA, FRAGILARIA, TABELLARIA, M W70-02803
S, CLADOCERANS, WESTHAMPTON LAKE(VA).: ASTERIONELLA SPIROIDE W71-10098
   *SUSPENDED SOLIDS, CHASE CITY(VA).: W72-10573
E(NETHERLANDS), HETEROCYSTS, GAS- VACUOLES, LLANGORS LAKE(WALES), H W70-04369
TIDES, BACTERIA, SNAILS, WEIGHT, VACUUM DRYING, BOILING.: /LA, PEP W70-04184
ERUGINOSA, LAKE COMO(MINN), LAKE VADNAIS(MINN), MILLE LAC(MINN), D W70-05269
OIR/ *EUTROPHICATION, *TENNESSEE VALLEY AUTHORITY PROJECT, *RESERV W71-07698
ETERIORATION, DRAIN, SAN JOAQUIN VALLEY MASTER DRAIN, SAN FRANCISC W71-09577
TER RESERVOIRS, *TURBID R/ *SALT VALLEY RESERVOIRS(NEB), *CLEAR WA W72-04761
NITRIFICATION.: *CENTRAL VALLEY(CALIFORNIA), *BACTERIAL DE W71-04555
   *ALGAE HARVESTING, *CENTRAL VALLEY(CALIF.).: W71-04556
   *ALGAE HARVESTING, *CENTRAL VALLEY(CALIF.).: W71-04557
NITRIFICATION.: *CENTRAL VALLEY(CALIFORNIA), *BACTERIAL DE W71-04554
N EQUIVALENTS(WASTES), RED RIVER VALLEY(NORTH DAKOTA), SULFATE RED W70-03312
GEN, EUTROPHICATION, FISH, LAND, VALUE, LEAVES, FERTILIZERS, PESTI W70-05264
, LAKE ERKEN(SWEDEN), EXTINCTION VALUES, EUDORINA, CULTURE MEDIUM, W70-05409
ON, MAGNESIUM, COOLING, PROPERTY VALUES, RECREATION DEMAND, IR: /R W70-02251
KE(NORWAY), HOBOL RIVER(NORWAY), VAN LAKE(NORWAY).: /HER, NJOER LA W70-05752
ETABOLITES, INHIBITION, SILICON, VANADIUM, ZINC, GALLIUM.: /INS, M W69-04801

```
             LAKES, *WA/ *ALGAE, *INDUSTRIAL     WASTES, *MUNICIPAL WASTES, *GREAT     W70-04430
     TION, *MINE WASTES, *RADIOACTIVE          WASTES, *MUNICIPAL WASTES, AQUATI    W69-01165
     AT/ *OXIDATION LAGOONS, *ORGANIC          WASTES, *OXYGENATION, AEROBIC TRE    W71-09546
     ATMENT, SEDIMENTATION, OX/ *FARM          WASTES, *POULTRY, WASTE WATER TRE    W71-11375
     , GRANTS, ALGAE, AIR POLL/ *FARM          WASTES, *RESEARCH AND DEVELOPMENT    W71-06821
     ICIPAL / *WATER POLLUTION, *MINE          WASTES, *RADIOACTIVE WASTES, *MUN    W69-01165
     ION, *ALGAE, NUTRIENTS, DOMESTIC          WASTES, *SEWAGE TREATMENT, SEWAGE    W68-00248
     RGANIC WASTES, *WASTE WAT/ *FARM          WASTES, *SPRINKLER IRRIGATION, *O    W71-00940
     DETERGENTS, *DIATOMS, *OIL, *OIL          WASTES, *SURFACTANTS, *WATER POLL    W70-01470
     *SPRINKLER IRRIGATION, *ORGANIC           WASTES, *WASTE WATER DISPOSAL, AI    W71-00940
     ERVOIRS, *LIMNOLOGY, *INDUSTRIAL          WASTES, *WATER POLLUTION EFFECTS,    W70-06975
     ASTE WATER DISPOSAL, *INDUSTRIAL          WASTES, *WATER POLLUTION CONTROL,    W72-01786
     TER-PLANT RELATIONSHIPS, NUCLEAR          WASTES, ABSORPTION, CHELATION, ME    W71-09005
     C LAGOONS.:   *POTATO PROCESSING          WASTES, AEROBIC LAGOONS, ANAEROBI    W71-08970
     ESTIC WASTES, SEWAGE, INDUSTRIAL          WASTES, AGRICULTURE, ST LAWRENCE     W70-03964
                         *TEXTILE MILL         WASTES, ALFOL ALCOHOLS.:             W70-08740
     *RADIOACTIVE WASTES, *MUNICIPAL           WASTES, AQUATIC ALGAE, URANIUM RA    W69-01165
     OLOGICAL ECOLOGY, KINETICS, FARM          WASTES, BENTHOS, NITRATES, FLUORE    W70-02775
     OTOPES, RIVERS, MILLS, COLORADO,          WASTES, BIOLOGICAL COMMUNITIES, S    W69-07846
     ASTES, INDUSTRIAL WASTES, SEWAGE          WASTES, BYPRODUCTS, FERTILIZERS,     W70-09905
     NITROGEN, BIOMASS, SEWAGE, FARM           WASTES, CARBON CYCLE, PHOSPHORUS,    W70-02777
     ATE REDUCTION, POTATO PROCESSING          WASTES, CHEESE MANUFACTURING WAST    W70-03312
              STIGLEOCLONIUS, INSOLUBLE        WASTES, CULTURE MEDIA.:              W68-00248
     T, PHOSPHOROUS, RUNOFF, DOMESTIC          WASTES, DETERGENTS, INDUSTRIAL WA    W71-04072
     ODOR, FARM WASTE, SLUDGE, SOLID           WASTES, DECOMPOSTING ORGANIC MATT    W71-05742
     US, NITROGEN, SEWAGE, INDUSTRIAL          WASTES, DOMESTIC WASTES, SEWAGE T    W68-00478
     N AGENTS, *TEXTILE WASTES, FIBER          WASTES, DYEING WASTES.: /ANTI-FOA    W70-04060
     S, FISH REPRODUCTION, INDUSTRIAL          WASTES, ECOLOGY, DISSOLVED OXYGEN    W73-00748
     CTS, DOMESTIC WASTES, INDUSTRIAL          WASTES, ENVIRONMENTAL EFFECTS, AN    W72-12240
     OON DES/ *JAPAN, *SLAUGHTERHOUSE          WASTES, EXPERIMENTAL STUDIES, LAG    W70-07508
     WAGE, DOMESTIC WASTE, INDUSTRIAL          WASTES, FARM WASTES.: /ANKTON, SE    W72-01813
     ES./ *ANTI-FOAM AGENTS, *TEXTILE          WASTES, FIBER WASTES, DYEING WAST    W70-04060
     OLOGY, EUTROPHICATION, MUNICIPAL          WASTES, FRESHWATER ALGAE, WATER P    W72-12240
     UTION, FERTILIZATION, INDUSTRIAL          WASTES, GARBAGE DUMPS, OIL WASTES    W70-05112
     TER POLLUTION SOURCES, MUNICIPAL          WASTES, GREAT LAKES, *POLLUTANT I    W71-05879
     OXINS, SOLID WASTES, RADIOACTIVE          WASTES, HEATED WATER, THERMAL POL    W72-00462
     D CARBON, CHLORINE, PHENOLS, OIL          WASTES, HYDROGEN ION CONCENTRATIO    W70-00263
     HYSIOLOGICAL ECOLOGY, INDUSTRIAL          WASTES, HYDROGEN ION CONCENTRATIO    W70-04510
     TRIAL WASTES, GARBAGE DUMPS, OIL          WASTES, INLETS(WATERWAYS), ORGANI    W70-05112
     SES(LEGAL), ORGANIC WASTES, PULP          WASTES, INDUSTRIAL WASTES, TOXINS    W72-00462
     AN/ *OIL, *OIL WASTES, *CHEMICAL          WASTES, INDUSTRIAL WASTES, WATER     W72-03299
     RE, *ULTIMATE DISPOSAL, DOMESTIC          WASTES, INDUSTRIAL WASTES, SEWAGE    W70-09905
     ATER POLLUTION EFFECTS, DOMESTIC          WASTES, INDUSTRIAL WASTES, ENVIRO    W72-12240
     TER, ORGANIC COMPOUNDS, CHEMICAL          WASTES, ION EXCHANGE, BIOCHEMICAL    W70-10433
     TREATMENT, *POULTRY, INDUSTRIAL           WASTES, LAGOONS, PONDS, PHOTOSYNT    W70-09320
     ORIDE, IRON SULFATE, GALVANIZING          WASTES, LAKE ZURICH(SWITZERLAND),    W70-05668
     LAKES, HEAVY METALS, INDUSTRIAL           WASTES, LAKES, PHOTOSYNTHESIS, PH    W69-02959
     HOSPHATES, ALGAE, DETRITUS, PULP          WASTES, LIVESTOCK, IDAHO.: /ES, P    W70-03501
     ENT, *DAIRY INDUSTRY, */ *ANIMAL          WASTES, MANAGEMENT, *FARM MANAGEM    W70-07491
     ONN), KLAMATH LAKE(ORE),/ *URBAN          WASTES, MADISON(WIS), LAKE ZOAR(C    W70-04506
     UACULTURE, RECYCLING / PETROLEUM          WASTES, MERCENARIA MERCENARIA, AQ    W70-09905
     *ANAEROBIC DIGESTION, *MUNICIPAL          WASTES, METHANE BACTERIA, ACID BA    W72-06854
     AGE, BIOLOGICAL TREATMENT, SOLID          WASTES, MICROBIOLOGY, ACTIVATED S    W68-00248
     RESERVOIRS, SAMPLING, INDUSTRIAL          WASTES, MUNICIPAL WASTES, SEWAGE,    W70-04001
     , *WATER POLLUTION CONTROL, FARM          WASTES, MUNICIPAL WASTES, SEPTIC     W70-04266
     ALITY, *FARM WASTES, *INDUSTRIAL          WASTES, MUNICIPAL WASTES, NUTRIEN    W71-09577
     ASTE WATER TREATMENT, INDUSTRIAL          WASTES, NEUTRIENTS, BACTERIA, ALG    W70-01519
     ASTE WATER TREATMENT, INDUSTRIAL          WASTES, NUTRIENTS, BACTERIA, ALGA    W71-06737
     S, *INDUSTRIAL WASTES, MUNICIPAL          WASTES, NUTRIENTS, PESTICIDES, RE    W71-09577
     LAKES, *LAKE ZURICH, *MUNICIPAL           WASTES, NUTRIENTS, NUISANCE ALGAE    W68-00680
     ACIATION, URBANIZATION, DOMESTIC          WASTES, NUTRIENTS, PHOSPHORUS, AL    W69-07818
     S, BACTERIA, FISH, PROTEINS, OIL          WASTES, OILY WATER, SULFITE LIQUO    W69-05891
     STE WATER TREATMENT, *INDUSTRIAL          WASTES, OXIDATION PONDS, ALGAE, B    W72-00027
     T, WATER UTILIZATION, INDUSTRIAL          WASTES, OXIDATION LAGOONS.: /TMEN    W70-07508
     *EUTROPHICATION, LAKES, ORGANIC           WASTES, OXYGEN, STANDING CROP, AL    W70-02251
     ISH, TOXICITY, FRESH WATER, PULP          WASTES, PESTICIDE RESIDUES, HERBI    W70-10437
     CITRUS/ *CANNERIES, *INDUSTRIAL           WASTES, PEACHES, APRICOTS, PEARS,    W71-09548
     LITY, LAKE MORPHOMETRY, DOMESTIC          WASTES, PLANT GROWTH SUBSTANCES,     W68-00470
     EN DEMAND, ALGAL TOXINS, ORGANIC          WASTES, POLLUTANT IDENTIFICATION,    W69-01273
     HOSPHORUS, *DETERGENTS, DOMESTIC          WASTES, POLLUTANTS, WATER POLLUTI    W70-05084
                      *CHICKENS, FOOD          WASTES, POULTRY WASTES.:             W70-09320
     OL, WATERCOURSES(LEGAL), ORGANIC          WASTES, PULP WASTES, INDUSTRIAL W    W72-00462
     INDUSTRIAL WASTES, TOXINS, SOLID          WASTES, RADIOACTIVE WASTES, HEATE    W72-00462
     RGANIC L/ *POTATOES, *INDUSTRIAL          WASTES, SCREENS, SEDIMENTATION, O    W71-08970
     SAL, DOMESTIC WASTES, INDUSTRIAL          WASTES, SEWAGE WASTES, BYPRODUCTS    W70-09905
     L OXYGEN DEMAND, SUSPENDED LOAD,          WASTES, SEWAGE, ALGAE, MICROORGAN    W70-08627
     GY, *SEWAGE LAGOONS, *INDUSTRIAL          WASTES, SEWAGE TREATMENT, NORTH D    W70-03312
     WORKS, *WASTE DISPOSAL, DOMESTIC          WASTES, SEWERS, SEWAGE TREATMENT,    W70-03344
     SPHORUS, CHEMCONTROL, INDUSTRIAL          WASTES, SEWAGE TREATMENT.: /, PHO    W70-05668
     AGE, INDUSTRIAL WASTES, DOMESTIC          WASTES, SEWAGE TREATMENT, HARVEST    W68-00478
     IES, AERATION, LAGOONS, DOMESTIC          WASTES, SEDIMENTATION, CHLORINATI    W70-04060
     NG, INDUSTRIAL WASTES, MUNICIPAL          WASTES, SEWAGE, ADSORPTION, ACTIV    W70-04001
     CONTROL, FARM WASTES, MUNICIPAL           WASTES, SEPTIC TANKS, RECREATION     W70-04266
     ICOCHEMICAL PROPERTIES, DOMESTIC          WASTES, SEWAGE, INDUSTRIAL WASTES    W70-03964
     COLOGY, EUTROPHICATION, DOMESTIC          WASTES, SHALLOW WATER, CHLOROPHYT    W70-00711
     L, *NUTRIENTS, *PESTICIDES, FARM          WASTES, SILT, EROSION, SEDIMENTAT    W71-10376
     ING WASTES, CHEESE MANUFACTURING          WASTES, STARCH MANUFACTURING WAST    W70-03312
     BACTERIA, PROTOZOA, ALGAE, SOLID          WASTES, TERTIARY TREATMENT.: /L,     W70-01027
     ), WETLANDS, SEEPAGE, INDUSTRIAL          WASTES, TERTIARY TREATMENT, CHEMI    W69-10178
     GENTS, ALGAL CONTROL, INDUSTRIAL          WASTES, THERMAL POLLUTION, METHOD    W70-10437
```

POLLUTION ABATEMENT, POLLUTANTS, WASTES, THERMAL POLLUTION, TOXINS W69-06305
WASTES, PULP WASTES, INDUSTRIAL WASTES, TOXINS, SOLID WASTES, RAD W72-00462
IA, ALGAE, FISH, SEWERS, RUNOFF, WASTES, WATER SUPPLY, EFFLUENTS, W71-09880
DGE, PHOTOSYNTHETIC OXYGEN, FARM WASTES, WASTE WATER TREATMENT.: / W71-08221
DUSTRIAL WA/ *LAGOONS, *DOMESTIC WASTES, WASTE WATER TREATMENT, IN W71-06737
NE ALGAE, INVERTEBRATES, NUCLEAR WASTES, WATER POLLUTION SOURCES.: W71-09863
ES, *CHEMICAL WASTES, INDUSTRIAL WASTES, WATER ANALYSIS, ANALYTICA W72-03299
DUSTRIAL WA/ *LAGOONS, *DOMESTIC WASTES, WASTE WATER TREATMENT, IN W70-01519
IOINDICATORS, *NEW YORK, NUCLEAR WASTES, WATER POLLUTION SOURCES, W69-10080
ING WASTES, STARCH MANUFACTURING WASTES, WATER QUALITY STANDARDS, W70-03312
PHOSPHATE, NITRATES, INDUSTRIAL WASTES, WATER POLLUTION EFFECTS, W70-02777
TS, SEWAGE TREATMENT, INDUSTRIAL WASTES, WATER CHEMISTRY, DIATOMS, W70-03974
WASTES, SEPTIC TANKS, RECREATION WASTES, WATER QUALITY.: /NICIPAL W70-04266
POSAL, WATER CHEMISTRY, ECOLOGY, WAT: / WASTE DISPOSAL, SEWAGE DIS W69-04276
LING NUTRIENTS, WATER CHEMISTRY, WATER ANALYSIS, PHOSPHORUS, PONDS W69-05142
QUALITY, *LIMNOLOGY, MINNESOTA, WATER ANALYSIS, CALCIUM CARBONATE W69-00632
RCES, IRRIGATION, AQUATIC ALGAE, WATER ANALYSIS, CARBONATES, NITRA W69-07096
LUTION SOURCES, WATER CHEMISTRY, WATER ANALYSIS.: /LGAE, WATER POL W69-08562
DISSOLVED OXYGEN, PHYTOPLANKTON, WATER ANALYSIS, ALGAL CONTROL, AL W72-06051
MICAL WASTES, INDUSTRIAL WASTES, WATER ANALYSIS, ANALYTICAL TECHNI W72-03299
TORY TESTS, INVERTEBRATES, FISH, WATER ANALYSIS, INSECTS, SNAILS, W72-01801
QUATIC MICROORGANISMS, SAMPLING, WATER ANALYSIS, WATER QUALITY, PO W71-00221
EFFECTS, ECOLOGY/ PLANT GROWTH, WATER AND PLANTS, WATER POLLUTION W70-02772
NEWFOUND LAKE, *WINNISQUAM LAKE, WATER AND PLANTS, WATER POLLUTION W70-02764
R SPECIES, MICROORGANISMS, FRESH WATER BIOLOGY, SUSPENDED SEDIMENT W71-10786
*EUTROPHICATION, ALGAL BLOOMS, WATER BLOOM, AQUATIC ALGAE, AGING W68-00912
GROWTH REGULATORS, MYRIOPHYLLUM, WATER BLOOMS, CHAINING.: /SALIX, W70-05263
VED OXYGEN, EVAPORATION CONTROL, WATER BUDGET, WATER CHEMISTRY, WA W72-01773
ODOR NU/ *OSCILLATORIA, NATIONAL WATER CARRIER(ISRAEL), THRESHOLD W70-02373
S, NUTRIENTS, LIGHT, WASHINGTON, WATER CHEMISTRY, OXYGEN, HYDROGEN W70-03309
LITY, *WATER POLLUTION, *RIVERS, WATER CHEMISTRY, WATER POLLUTION W70-03068
APORATION CONTROL, WATER BUDGET, WATER CHEMISTRY, WATER TEMPERATUR W72-01773
QUALITY, *MONITORING, BACTERIA, WATER CHEMISTRY, PHOSPHORUS, NITR W71-05879
TER PURIFICATION, WATER QUALITY, WATER CHEMISTRY.: /AYS, ALGAE, WA W69-09884
*LIGHT INTENSITY, SPRING WATER, WATER CHEMISTRY, WATER TEMPERATUR W69-09676
TREATMENT, EFFLUENTS, LIMNOLOGY, WATER CHEMISTRY, RESPIRATION, BAT W70-05387
TER, HYDROGEN ION CONCENTRATION, WATER CHEMISTRY, *PHYTOPLANKTON, W70-05424
INOFLAGELLATES, MARINE BACTERIA, WATER CHEMISTRY, CARBON RADIOISOT W70-05652
EMICAL OXYGEN DEMAND, MINNESOTA, WATER CHEMISTRY, BIODEGRADATION.: W69-00446
CTIVITY, STANDING CROP, STREAMS, WATER CHEMISTRY, WATER: /S, PRODU W69-04805
WASTE DISPOSAL, SEWAGE DISPOSAL, WATER CHEMISTRY, ECOLOGY, WAT: / W69-04276
S COMPOUNDS, *CYCLING NUTRIENTS, WATER CHEMISTRY, WATER ANALYSIS, W69-05142
ENTS, SODIUM, VIRUSES, VITAMINS, WATER CHEMISTRY, WATER POLLUT: /M W69-04801
FE, ENVIRONMENT, MICROORGANISMS, WATER CHEMISTRY, PLANTS, SOUTHWES W69-03611
ALGAE, WATER POLLUTION SOURCES, WATER CHEMISTRY, WATER ANALYSIS.: W69-08562
SEDIMENTATION, SEDIMENTS, TASTE, WATER CHEM: /PLANKTON, SULFATES, W69-07833
NUTRIENTS, PHOSPHORUS COMPOUNDS, WATER CHEMISTRY, WATER POLLUTION W69-05868
QUIREMENTS, NUTRIENTS, TOXICITY, WATER CHEMISTRY, WATER POLLUTION W69-05867
RY PRODUCTIVITY, SOIL CHEMISTRY, WATER CHEMISTRY, WATER POLLUTION W69-06273
AT LAKES, PHYSIOLOGICAL ECOLOGY, WATER CHEMISTRY, GEOGRAPHICAL REG W70-03975
EUTROPHICATION, GRAZING, ALGAE, WATER CHEMISTRY, BIOMASS, BENTHOS W70-03978
GE TREATMENT, INDUSTRIAL WASTES, WATER CHEMISTRY, DIATOMS, CHLOROP W70-03974
AT LAKES, PHYSIOLOGICAL ECOLOGY, WATER CHEMISTRY, GEOGRAPHICAL REG W70-03975
SEA WATER, FRESH WATER, SHORES, WATER CIRCULATION, FLOODS, DENSIT W69-06203
ION, THERMOCLINE, *ION EXCHANGE, WATER CIRCULATION, NUTRIENT ABSOR W68-00857
VEL, WATER STORAGE, EVAPORATION, WATER CIRCULATION, SEDIMENTS.: /E W69-09723
C PLANTS, *AQUATIC WEED CONTROL, WATER CONSERVATION, WATER CONTROL W70-00269
GE, ALGAL CONTROL, AQUATIC LIFE, WATER CONSERVATION, PUBLIC HEALTH W70-03344
LE PRESERVATION, WATER DYNAMICS, WATER CONSTITUENTS, ANALYTICAL CH W70-04382
EED CONTROL, WATER CONSERVATION, WATER CONTROL, FERTILIZERS, SEWAG W70-00269
CIPAL WATER, WATER REQUIREMENTS, WATER CONVEYANCE, WATER COSTS, AR W71-09577
REQUIREMENTS, WATER CONVEYANCE, WATER COSTS, ARTIFICIAL WATER COU W71-09577
VEYANCE, WATER COSTS, ARTIFICIAL WATER COURSES, CALIFORNIA.: / CON W71-09577
TREATMENT, SAMPLING, MONITORING, WATER DEMAND, WATER SUPPLY, ARID W70-05645
ANCE, ALGAE BLOOM, ALGAE GROWTH, WATER DENSITY, CONVECTION CURRENT W69-08282
OAQUIN VA/ *CONTRA COSTA COUNTY, WATER DETERIORATION, DRAIN, SAN J W71-09577
, *WASTE WATER TREATMENT, *WASTE WATER DISPOSAL, *INDUSTRIAL WASTE W72-01786
TY, WASTE WATER TREATMENT, WASTE WATER DISPOSAL.: /LAGOONS, TOXICI W69-09454
MPERATURE, FLOW, SAMPLING, WASTE WATER DISPOSAL, DIURNAL, EFFLUENT W69-10159
IGATION, *ORGANIC WASTES, *WASTE WATER DISPOSAL, AIR POLLUTION, WA W71-00940
AREA(LAKES), PHAEDACTYLUM, WASTE WATER DIVERSION, INSTREAM TREATME W69-09340
US RELEASE, SAMPLE PRESERVATION, WATER DYNAMICS, WATER CONSTITUENT W70-04382
NTS, ANALYTICAL CHEMISTRY, WASTE WATER EFFECTS, LAKE MENDOTA(WIS), W70-04382
TA, CALIFORNIA WATER PLAN, DELTA WATER FACILITIES(CALIF), OXYGEN D W70-05094
AS, LARVAE, GROWTH RATES, *FRESH WATER FISH, MICHIGAN, TROUT, WINT W68-00487
IBLIOGRAPHIES, CYANOPHYTA, FRESH WATER FISH, LAKES, *LAKE ZURICH, W68-00680
TICS, GROWTH RATES, ALGAE, FRESH WATER FISH.: /, LAKE ERIE, STATIS W68-00253
NTS, ICED LAKES, *FERTILIZATION, WATER FLEAS, LARVAE, GROWTH RATES W68-00245
LLIFISHES, TROUT, SUCKERS, BASS, WATER HYACINTH, FROGS, CRAYFISH, W70-05263
CHAINS, OXYGEN, ROTIFERS, WEEDS, WATER HYACINTH, ALLIGATORWEED, CH W70-10175
FISH, FRESH WATER, ALGAE, PONDS, WATER HYACINTH, CHEMICAL CONTROLS W70-00269
S, *CYCLING NUTRIENTS, *SEDIMENT- WATER INFERFACE, MUD-WATER INTERF W69-06859
, *SEDIMENT-WATER INFERFACE, MUD- WATER INTERFACE, PHOSPHORUS, ADSO W69-06859
NTRATION, CYCLING NUTRIENTS, MUD- WATER INTERFACES, EUGLENOPHYTA, T W70-04381
, ALGAE, WATER QUALITY, SEDIMENT- WATER INTERFACES, OREGON, SESSILE W72-06051
OTE SENSING, MINNESOTA, SEDIMENT- WATER INTERFACES, CALIFORNIA, BIO W70-02775
, TIDAL EFFECTS, DISPERSION, OIL- WATER INTERFACES, PERSISTENCE, WA W70-01777
*PLANT GROWTH REGULATORS, FRESH WATER LAKES, LABORATORY TESTS.: / W68-00856
ALGAE, SEWAGE, WATER POLLUTION, WATER LAW, WATER POLLUTION EFFECT W69-06305
ING, GASOLINE, FLOODS, AQUIFERS, WATER LAW, NITRATES, TURBIDITY, F W73-00758

556

CHELATION, METABOLISM, AQUARIA,
ECREATION, ALGAE, OXYGEN DEMAND,
ALGAE, HEATED WATER, POPULATION,
IUM RADIOISOTOPES, MARINE ALGAE,
SEWAGE, WASTE WATER(POLLUTION),
, COLLOIDS, AQUARIA, ABSORPTION,
OISOTOPES, COBALT RADIOISOTOPES,
S, ECOSYSTEMS, AQUATIC HABITATS,
ECOLOGICAL DISTRIBUTION, MIDGES,
N, PHOSPHORUS, ALGAE, LIVESTOCK,
ONDITIONS, ANAEROBIC CONDITIONS,
ENCE, WATER POLLUTION TREATMENT,
LVED OXYGEN, DETRITUS, BACTERIA,
IREMENTS, ANALYTICAL TECHNIQUES,
AGE EFFLUENTS, SEWAGE TREATMENT,
ON COMPOUNDS, LIGHT PENETRATION,
ER, *MARINE ALGAE, *EMULSIFIERS,
LLUTANT IDENTIFICATION, BIOLOGY,
L-WATER INTERFACES, PERSISTENCE,
ATION, BIOLOGY, WATER POLLUTION,
ATMENT, WATER POLLUTION CONTROL,
IMI/ *DETERGENTS, *MARINE ALGAE,
NNISQUAM LAKE, WATER AND PLANTS,
TE, NITRATES, INDUSTRIAL WASTES,
OLLUTION EFFECTS, ALGAL CONTROL,
THIC FLORA, BENTHOS, MINE WATER,
N/ EUTROPHICATION, *CONFERENCES,
    *BIOSTIMULATION, ALGAL GROWTH,
PLANT GROWTH, WATER AND PLANTS,
ROWTH, ANABAENA, BIOSTIMULATION,
OGRAPHY, RADIOACTIVITY, TRACERS,
OS, MINE WATER, WATER POLLUTION,
CALIF), STANISLAUS RIVER(CALIF),
OPHYTA, DIATOMS, EUTROPHICATION,
TES, ECOSYSTEMS, ORGANIC MATTER,
ORGANIC MATTER, CURRENTS(WATER),
N, WATER QUALITY CONTROL, LAKES,
UTION, *RIVERS, WATER CHEMISTRY,
ANE BACTERIA, CHEMICAL ANALYSIS,
TER CONSERVATION, PUBLIC HEALTH,
EMCONTROL, EUTROPHICATION, BASS,
ONTROL, WATER POLLUTION EFFECTS,
TY, ALGAE, NITROGEN, PHOSPHORUS,
CATION, WATER POLLUTION CONTROL,
EMENT, POLLUTANT IDENTIFICATION,
ICIPAL WASTES, FRESHWATER ALGAE,
ALGAE, TURF, SOIL MICROBIOLOGY,
ICIDES, ALGAE, LABORATORY TESTS,
OSTRONTIUM COMPLEXES, DTPA, SOIL
RGANIC MATTER, NUTRIENTS, LAKES,
OS, AQUATIC LIFE, ALGAE, PLANTS,
CRATER LAKE, ORE, GROWTH RATES,
ROELECTRIC PLANTS, WEED CONTROL,
TISSUES, SCENEDESMUS, LIMNOLOGY,
ATMENT, DOMESTIC WASTES, *ALGAE,
N, PHOSPHORUS, ALGAE, NUTRIENTS,
IC ASPECTS, AQUATIC ENVIRONMENT,
NTS, PLANT POPULATIONS, HISTORY,

XYGENATION, ABSORPTION, ECOLOGY,
CIPITATION, FILTRATION, AMMONIA,
UNGI, *BACTER/ *ALGAE, *BIOMASS,
ACTIVATED CARBON, CLAYS, ALGAE,
NICAL), CLEANING, COST ANALYSIS,
IA, *WATER TREATMENT, *NEW YORK,
TER QUALITY CONTROL, FILTRATION,
ALGAE, *WATER POLLUTION CONTROL,
*FILTERS, SANDS, LAKE MICHIGAN,
OMYCETES, *ALGAE, *TASTE, *ODOR,
ES, WATER RESOURCES DEVELOPMENT,
E, *COPPER SULFATE, *ALKALINITY,
LFATE, WATER SUPPLY, RESERVOIRS,
LAYS, ALGAE, WATER PURIFICATION,
LGAL BLOOMS, NUTRIENTS, LIMNOLO/
E, *ALGAL CONTROL, WATER SUPPLY,
LGAE, WATER POLLUTION TREATMENT,
SSOLVED OXYGEN, ALGAE, VELOCITY,
ON EFFECTS, NITROGEN, COLIFORMS,
OURCES, WATER POLLUTION EFFECTS,
S, *CYANOPHYTA, *EUTROPHICATION,
TROL MEASURES, PHYSICAL CONTROL,
Y PRODUCTIVITY, *EUTROPHICATION,
UIPMENT, *ANALYTICAL TECHNIQUES,
RESIDUES, HERBICIDES, ESTUARIES,
NISMS, SAMPLING, WATER ANALYSIS,
, *ENVIRONMENT, *ORGANIC MATTER,
, *PESTICIDES, FARM WASTES, SIL/
LGAE, COPPER / *WATER TREATMENT,
ON, GREAT LAKES, NUISANCE ALGAE,
TROPHICATION, DAPHNIA, CHLAMYDO/

WATER POLLUTION EFFECTS, ADSORPTI   W71-09005
WATER POLLUTION, MUNICIPAL WATER,   W71-09577
WATER POLLUTION EFFECTS.: /KTON,   W71-09767
WATER POLLUTION SOURCES.: /, URAN   W71-09013
WATER POLLUTION SOURCES, MUNICIPA   W71-05879
WATER POLLUTION EFFECTS, CATION E   W71-09863
WATER POLLUTION EFFECTS.: /E RADI   W71-09850
WATER POLLUTION EFFECTS.: /OCESSE   W71-09012
WATER POLLUTION EFFECTS, OLIGOCHA   W71-12077
WATER POLLUTION EFFECTS.: /ITROGE   W71-10376
WATER POLLUTION SOURCES, PATH OF   W71-11381
WATER POLLUTION EFFECTS.: /ERSIST   W70-01777
WATER POLLUTION EFFECTS, LIMNOLOG   W70-02249
WATER POLLUTION EFFECTS, PHYTOPLA   W70-02245
WATER POLLUTION CONTROL, WATER PO   W70-02248
WATER POLLUTION EFFECTS, EUTROPHI   W70-02255
WATER POLLUTION EFFECTS, MORTALIT   W70-01780
WATER POLLUTION, WATER POLLUTION   W70-02304
WATER POLLUTION TREATMENT, WATER   W70-01777
WATER POLLUTION SOURCES, ALGAL CO   W70-02304
WATER POLLUTION EFFECTS, WATER: /   W70-02248
WATER POLLUTION EFFECTS, LETHAL L   W70-01779
WATER POLLUTION EFFECTS, ECOLOGY,   W70-02764
WATER POLLUTION EFFECTS, DENITRIF   W70-02777
WATER POLLUTION, TOXICITY, LAKES,   W70-02775
WATER POLLUTION, WATER POLLUTION   W70-02770
WATER POLLUTION EFFECTS, ALGAL CO   W70-02775
WATER POLLUTION ASSESSMENT, SANIT   W70-02775
WATER POLLUTION EFFECTS, ECOLOGY,   W70-02772
WATER POLLUTION ASSESSMENT.: /L G   W70-02779
WATER POLLUTION EFFECTS, PHYSIOLO   W70-02510
WATER POLLUTION EFFECTS, ACID STR   W70-02770
WATER POLLUTION ASSESSMENT, AMERI   W70-02777
WATER POLLUTION EFFECTS, SYSTEMS   W70-02779
WATER POLLUTION, THERMAL POLLUTIO   W70-03309
WATER POLLUTION SOURCES, SURFACE   W70-03501
WATER POLLUTION SOURCES, CHLOROPH   W70-03310
WATER POLLUTION SOURCES, PHOSPHOR   W70-03068
WATER POLLUTION, CHLORELLA, SCENE   W70-03312
WATER POLLUTION, CONTRACTS, REGUL   W70-03344
WATER POLLUTION CONTROL, MINNOWS,   W70-02982
WATER POLLUTION SOURCES.: /TION C   W70-02792
WATER POLLUTION EFFECTS.: /ERTILI   W70-02973
WATER POLLUTION EFFECTS, WATER PO   W70-02792
WATER POLLUTION CONTROL, WATER PO   W70-02792
WATER POLLUTION EFFECTS, DOMESTIC   W72-12240
WATER POLLUTION SOURCES.: /, SOIL   W72-08110
WATER POLLUTION EFFECTS.: /N PEST   W72-11105
WATER PROFILE.: / ACTIVITY, *RADI   W71-04067
WATER PROPERTIES, BACTERIA, BIOAS   W71-10079
WATER PROPERTIES, WATER TEMPERATU   W69-05023
WATER PROPERTIES.:   W68-00478
WATER PROPERTIES, PHOSPHORUS, CHE   W68-00479
WATER PROPERTIES, WATER TEMPERATU   W68-00248
WATER PROPERTIES, BIOCHEMICAL OXY   W68-00256
WATER PROPERTIES.: /ECTS, NITROGE   W70-04382
WATER PROPERTIES, BIOLOGICAL PROP   W70-04475
WATER PURIFICATION, *WATER UTILIZ   W68-00483
WATER PURIFICATION BY ALGAE.:   W69-07389
WATER PURIFICATION, CARBON, BICAR   W69-07389
WATER PURIFICATION, WASTE WATER T   W71-11393
WATER PURIFICATION, *PROTOZOA, *F   W71-05267
WATER PURIFICATION, WATER QUALITY   W69-09884
WATER PURIFICATION, *WATER TREATM   W72-08688
WATER PURIFICATION.: /SIGN CRITER   W72-08686
WATER PURIFICATION, MEMBRANES, SC   W73-02426
WATER QUALITY CONTROL.: /ROSION,   W73-01122
WATER QUALITY CONTROL, FILTRATION   W73-02426
WATER QUALITY CONTROL, POLLUTANT   W72-13605
WATER QUALITY, SOIL INVESTIGATION   W72-11906
WATER QUALITY CONTROL, *CONTROL,   W72-07442
WATER QUALITY, NEW JERSEY, TEMPER   W69-10157
WATER QUALITY, WATER CHEMISTRY.: /   W69-09884
WATER QUALITY, *EUTROPHICATION, A   W69-09723
WATER QUALITY, WATER POLLUTION CO   W69-10155
WATER QUALITY CONTROL, SEWAGE EFF   W70-01031
WATER QUALITY, DISSOLVED SOLIDS,   W70-00265
WATER QUALITY.: /S, WATER POLLUTI   W71-05407
WATER QUALITY.: /ATER POLLUTION S   W71-05409
WATER QUALITY.: /ON EFFECTS, LAKE   W71-04758
WATER QUALITY.: /, INTEGRATED CON   W70-07393
WATER QUALITY, *NUTRIENTS, LIMNOL   W70-09093
WATER QUALITY, DISSOLVED OXYGEN,   W70-08627
WATER QUALITY, DETERGENTS, ALGAL   W70-10437
WATER QUALITY, POLLUTANT IDENTIFI   W71-00221
WATER QUALITY, TEMPERATURE, HYDRO   W71-03031
WATER QUALITY CONTROL, *NUTRIENTS   W71-10376
WATER QUALITY, *ORGANIC MATTER, A   W72-04495
WATER QUALITY, NUTRIENTS, PHOSPHO   W72-05453
WATER QUALITY, *ALGAL CONTROL, EU   W72-02417

OROMETRY, DYE RELEASES, CURRENTS(WATER), REMOTE SENSING, SEA WATER   W70-05377
REAMS, LIGHT INTENSITY, CURRENTS(WATER), VELOCITY, DIATOMS, SEASON   W70-02780
IMENTS, ORGANIC MATTER, CURRENTS(WATER), WATER POLLUTION SOURCES,   W70-03501
LATION, DIURNAL, WINDS, CURRENTS(WATER), WATER POLLUTION, DRAINAGE   W70-05404
ERIUM, RESPIROMETER, PROTOPLASM, WATER-BLOOMS, LAKE GEORGE(UGANDA)   W70-04369
ONMENT, PONDS, BREEDINGS, SALINE WATER-FRESHWATER INTERFACES, COLL   W73-00748
GONFLIES, ALGAE, SEDIMENTS, SOIL- WATER-PLANT RELATIONSHIPS, NUCLEA   W71-09005
*ADDITIVES, *FLOW RATE, *WASTE WATER, *ALGAE, *SLUDGE, STORMS, T   W70-05819
*ALGAE, *AQUATIC BAC/ *ACID MINE WATER, *AQUATIC MICROORGANISMS,   W69-00096
UNGI, BACTERIA, BACTER/ *COOLING WATER, *BIOMASS, *SLIME, ALGAE, F   W72-12987
, AEST/ *WATER REUSE, *RECLAIMED WATER, *CALIFORNIA, WATER QUALITY   W70-05645
TION/ *WATER TREATMENT, *COOLING WATER, *CORROSION, *CHEMICAL REAC   W70-02294
, TEMPERATURE, SALINITY/ *MARINE WATER, *COPPER, *ALGAE, SEA WATER   W70-08111
ATER REUSE, COOLING TO/ *COOLING WATER, *ELECTRIC POWER PLANTS, *W   W71-11393
EFFECTS, *LI/ *ALGAE, *ACID MINE WATER, *HABITATS, *ENVIRONMENTAL   W70-02770
*MARINE ALGAE, *SEAWATER, *FRESH WATER, *HUMIC ACIDS, *COLOR, RIVE   W70-01074
IA, *NEVADA, DEPTH, ALGAE/ *DEEP WATER, *LAKES, *PLANTS, *CALIFORN   W70-00711
S, WATER POLLUTION EFFECT/ *OILY WATER, *MARINE ALGAE, *EMULSIFIER   W70-01780
ROGEN CYCLE, *FARM WASTE, GROUND WATER, *NITRATES, EUTROPHICATION,   W69-05323
OLLUTANTS, INHIBITIO/ *ACID MINE WATER, *NUTRIENT REQUIREMENTS, *P   W70-02792
ION EFFECTS, *SALT MARSHES, OILY WATER, *OIL, *MARINE ALGAE, WATER   W70-09976
ONS, PHOSPHATES,/ *ALGAE, *FRESH WATER, *PEPTIDES, COPPER, IRON, I   W69-10180
SMUS, LIGH/ *ENVIRONMENT, *FRESH WATER, *PLANKTON, *ALGAE, SCENEDE   W70-05409
PLANTS, *THERMAL STRESS, THERMAL WATER, *PROTOZOA, ALGAE, DIATOMS,   W70-07847
*PHYSIOLOGICAL ECOLOGY, *SALINE WATER, *THERMAL WATER, ALGAE, AQU   W69-03516
E, CULTURES, CH/ *BIOASSAY, *SEA WATER, *VITAMINS, VITAMIN B, ALGA   W70-05652
LGAE, COPPER SULFATE, / *POTABLE WATER, *WATER QUALITY CONTROL, *A   W71-00110
*HOT SPRINGS, HABITATS, ALKALINE WATER, ACIDIC WATER, TRAVERTINE,   W69-10160
OGY, ACIDITY, ACID SOILS, ACIDIC WATER, ACIDS, PHOTOSYNTHESIS, PHO   W70-02792
CONCENTRATION, WATE/ *ACID MINE WATER, ACIDIC WATER, HYDROGEN ION   W70-05424
TE, *ODOR, *AMINO ACIDS, POTABLE WATER, ACTINOMYCETES, PLANT GROWT   W72-01812
R BACTERIA, STABILIZATION, WASTE WATER, ALGAE, DIFFUSION, SULFUR.:   W70-01971
RAPHIES, HERBICIDES, FISH, FRESH WATER, ALGAE, PONDS, WATER HYACIN   W70-00269
ECOLOGY, *SALINE WATER, *THERMAL WATER, ALGAE, AQUATIC ALGAE, AQUA   W69-03516
S, *NITROGEN UTILIZATION, *FRESH WATER, ALGAE, NITRATE AMMONIUM, N   W69-04521
, ALGAE, PHYTOPLANKTON, ALKALINE WATER, ALKALINITY, FISH FARMING,   W71-07731
IENTS, BACTERIA, ALGAE, BRACKISH WATER, ALKALINE WATER, SODIUM.: /   W71-03031
POISONING, *MARSH PLANTS, *OILY WATER, AQUATIC ALGAE, MARINE ALGA   W70-01231
PHORUS RADIOISOTOPES, *TRANSFER, WATER, BACTERIA, SEDIMENTS, MINNO   W69-07861
MS, PONDS, RUNNING WATERS, FRESH WATER, BRACKISH WA: /S, CARP, CLA   W70-09905
TION, *BIODEGRADATION, *DRAINAGE WATER, CALIFORNIA, DRAINAGE PROGR   W71-04556
TION, *BIODEGRADATION, *DRAINAGE WATER, CALIFORNIA, DRAINAGE PROGR   W71-04557
TION, *BIODEGRADATION, *DRAINAGE WATER, CALIFORNIA, DRAINAGE PROGR   W71-04555
TION, *BIODEGRADATION, *DRAINAGE WATER, CALIFORNIA, DRAINAGE PROGR   W71-04554
OLLUTION, *TEMPERATURE, *COOLING WATER, CHEMICAL PROPERTIES, ECOLO   W69-05023
ENTS(WATER), REMOTE SENSING, SEA WATER, CHLOROPHYLL, OIL, ALGAE, P   W70-05377
, CHRYSOPHYTA, BIOASSAYS, SALINE WATER, CHLORELLA, CHLAMYDOMONAS,   W70-03974
CATION, DOMESTIC WASTES, SHALLOW WATER, CHLOROPHYTA, CHRYSOPHYTA,   W70-00711
A, COL/ *ALGAL CONTROL, *POTABLE WATER, CYANOPHYTA, SULFUR BACTERI   W70-08107
ON, RESERVOIRS, ALGAE, BODIES OF WATER, DISSOLVED OXYGEN, NORTH CA   W70-00683
*SAMPLING, ALGAE TOXINS, SALINE WATER, EQUILIBRIUM, BIOLOGICAL CO   W69-00360
IFORNIA, UNITED STATES, BRACKISH WATER, ESTUARIES, BACTERIA, MOLLU   W70-05372
DE, IRON, FUNGI, CLAYS, BRACKISH WATER, FISHERIES, LIMNOLOGY.: /FI   W70-05761
ON, DIATOMS, MARINE ALGAE, FRESH WATER, FLOW RATES, PUMPS, ELECTRO   W70-09904
N, *CYCLING, *RADIOISOTOPES, SEA WATER, FOOD CHAINS, PHOSPHORUS RA   W70-00708
IRON, TRACE ELEMENTS, ALGAE, SEA WATER, FRESH WATER.: / *COPPER, *   W71-11687
SEDIMENTS, SAMPLING, DEPTH, SEA WATER, FRES WATER, LITTORAL, DISS   W71-12870
WATER STRUCTURE, ATMOSPHERE, SEA WATER, FRESH WATER, SORPTION, SOD   W70-04385
EMENT, TIDAL WATERS, MIXING, SEA WATER, FRESH WATER, SHORES, WATER   W69-06203
VERTEBRATES, PLANKTON, FISH, SEA WATER, GAMMA RAYS, RADIOCHEMICAL   W69-08269
LVOCALES, INDUSTRIAL USES, CLEAN WATER, GENUS, SPECIES, TOUCH SENS   W70-05389
OLLUTION, SNOW, RUNOFF, DRAINAGE WATER, GRASSLANDS, SOIL EROSION,:   W70-04193
GAE, CUTTING MANAGEMENT, POTABLE WATER, HERBICIDES, CHARA, ALKALIN   W70-05263
, WATE/ *ACID MINE WATER, ACIDIC WATER, HYDROGEN ION CONCENTRATION   W70-05424
CARBONATES, TEMPERATURE, SHALLOW WATER, HYDROGEN SULFIDE, COPPER S   W70-02803
AGE EFFLUENTS, *ESTUARIES, FRESH WATER, LAKES, SILTS, PHOSPHATES,   W70-02255
H PLANTS, *MARINE ANIMALS, *OILY WATER, LICHENS, TIDAL EFFECTS, DI   W70-01777
SAMPLING, DEPTH, SEA WATER, FRES WATER, LITTORAL, DISSOLVED OXYGEN   W71-12870
UTION, CLEANING, DISASTERS, OILY WATER, MARINE ALGAE, SHIPS.: /OLL   W70-01470
NTS, PLANT POPULATIONS, WEATHER, WATER, MOVEMENT, CULTURES, BAYS,   W70-06229
DATION, *WATER REUSE, *RECLAIMED WATER, NITROGEN COMPOUNDS, NITROG   W69-08054
OUNDS, RESERVOIRS, ALGAE, GROUND WATER, NITRATES, PHOSPHATES, LAS   W72-02817
ANGANESE, CULTURES, RECIRCULATED WATER, NITRATES, PHOSPHATES, SODI   W70-07257
HLOROPHYTA, CYANOPHYTA, BRACKISH WATER, OLIGOTROPHY, TEMPERATURE,   W70-05752
L WAST/ *WATER REUSE, *RECLAIMED WATER, ORGANIC COMPOUNDS, CHEMICA   W70-10433
ECTS, SALINITY, TEMPERATURE, SEA WATER, OYSTERS, CLAMS, CRABS, NOR   W69-08267
Y PRODUCTIVITY, *ALGAE, *THERMAL WATER, PERIPHYTON, BIOMASS, PHYSI   W70-03309
DARY PRODUCTIVITY, SULFIDES, SEA WATER, PHAWOPHYTA, CRUSTACEANS, H   W70-01068
PES, *MARINE ALGAE, *RESINS, SEA WATER, PHYSICOCHEMICAL PROPERTIES   W69-08275
RE, PHYTOPLANKTON, ALGAE, HEATED WATER, POPULATION, WATER POLLUTIO   W71-09767
OPES, LAKES, ORGANIC MATTER, SEA WATER, PRIMARY PRODUCTIVITY, LIGH   W70-02504
H, STEAM, ELECTRIC POWER, HEATED WATER, PROTEINS, AQUATIC MICROORG   W70-09905
N CONTROL, FISH, TOXICITY, FRESH WATER, PULP WASTES, PESTICIDE RES   W70-10437
OIL POROSITY, IRRIGATION, GROUND WATER, RUNOFF, FROZEN GROUND, SOI   W69-05323
IBUTION, BACTERIA, SAMPLING, SEA WATER, SEASONAL, PSEUDOMONAS, TEM   W70-03952
ECHARGE, WATER SUPPLY, MUNICIPAL WATER, SEDIMENT LOAD, CONDUCTIVIT   W69-05894
WATERS, MIXING, SEA WATER, FRESH WATER, SHORES, WATER CIRCULATION,   W69-06203
BIOASSAY, AQUATIC LIFE, COOLING WATER, SITES.: /E, MODEL STUDIES,   W71-11881
ALGAE, BRACKISH WATER, ALKALINE WATER, SODIUM.: /IENTS, BACTERIA,   W71-03031
, *FUNGI, *CALCAREOUS SOIL, SOIL WATER, SOIL WATER MOVEMENT, LEACH   W71-04067

RE, ATMOSPHERE, SEA WATER, FRESH   WATER, SORPTION, SODIUM, ALGAE, P   W70-04385
FISH, PROTEINS, OIL WASTES, OILY   WATER, SULFITE LIQUORS, ULTIMATE   W69-05891
RINE WATER, *COPPER, *ALGAE, SEA   WATER, TEMPERATURE, SALINITY, BIO   W70-08111
STES, RADIOACTIVE WASTES, HEATED   WATER, THERMAL POLLUTION, BACTERI   W72-00462
HABITATS, ALKALINE WATER, ACIDIC   WATER, TRAVERTINE, CALCIUM CARBON   W69-10160
PRINGS, *LIGHT INTENSITY, SPRING   WATER, WATER CHEMISTRY, WATER TEM   W69-09676
MAND, WATER POLLUTION, MUNICIPAL   WATER, WATER REQUIREMENTS, WATER   W71-09577
RS, BENTHIC FLORA, BENTHOS, MINE   WATER, WATER POLLUTION, WATER POL   W70-02770
ESERVOIR, MASSACHUSETTS, POTABLE   WATER, WATER POLLUTION, BACTERIA,   W70-05264
*HYDRODYNAMICS, LAKES, INTERNAL   WATER, WAVES(WATER), DIURNAL, EPI   W70-09889
UTRIENTS, PHYSIOLOGICAL ECOLOGY,   WATER: /ALGAE, DIATOMS, CYCLING N   W70-00713
ONTROL, WATER POLLUTION EFFECTS,   WATER: /ATMENT, WATER POLLUTION C   W70-02248
CROP, STREAMS, WATER CHEMISTRY,   WATER: /S, PRODUCTIVITY, STANDING   W69-04805
ATION, *WATER POLLUTION CONTROL,   WATERCOURSES(LEGAL), ORGANIC WAST   W72-00462
HETEROGENEITY, IRON, MANGANESE,   WATERFOWL, CYANOPHYTA.: /EFFECTS,   W70-00273
PPLICATION TECHNIQUES, SWIMMING,   WATERFOWL, BIRDS, COSTS, COST-SHA   W70-05568
*GLUCOSE,   WATERLOO MILLS(PA).:   W71-05026
*MICROCYSTIS AERUGINOSA, *LAKE   WATERS.:   W70-08637
E EFFLUENT, CHEESE FACTORY WASTE   WATERS.: /GANISMS, PAPERMILL WAST   W71-00940
TURES, *SCENEDESMUS, *SUBSURFACE   WATERS, *AGRICULTURE, *WASTES, AL   W70-03334
D SOLIDS, D/ EFFECTS, *IMPOUNDED   WATERS, *WATER QUALITY, *DISSOLVE   W72-01773
TERGENTS, WATER POLLUTION, WASTE   WATERS, BIOCHEMICAL OXYGEN DEMAND   W70-08740
UBATION, PSEUDOMONAS, SUBSURFACE-   WATERS, BOTTOM SEDIMENTS, SURFACE   W70-00713
TROPHICATION, SAMPLING, SURFACES   WATERS, CARBON, FERTILITY, ALGAE,   W70-02973
ATERS, BOTTOM SEDIMENTS, SURFACE   WATERS, DETRITUS, LIGHT INTENSITY   W70-00713
*POLLUTANTS, *ESTUARIES, *TIDAL   WATERS, ECOSYSTEMS, ESTUARINE ENV   W68-00010
IES, CARP, CLAMS, PONDS, RUNNING   WATERS, FRESH WATER, BRACKISH WA: W70-09905
ESTUARINE ENVIRONMENT, IMPOUNDED   WATERS, FRESHWATER ALGAE, ANADROM   W71-02491
MIDS, CLADOCERA, / *HYPERTROPHIC   WATERS, HAMBURG(GERMANY), CHIRONO   W70-00268
SOIL INVESTIGATIONS, *IMPOUNDED   WATERS, IMPOUNDMENTS, *DISSOLVED   W72-11906
ION, *POLLUTION ABATEMENT, TIDAL   WATERS, MIXING, SEA WATER, FRESH   W69-06203
ISSOLVED OXYGEN, HARDNESS, TIDAL   WATERS, ORGANIC MATTER, CONDUCTIV   W73-00748
RADIATION, GEOCHEMISTRY, SPRING   WATERS, PHOSPHORUS, ORGANOPHOSPHO   W69-09334
ANKTON, WATER POLLUTION, SURFACE   WATERS, RESERVOIRS, CONTROL, BIBL   W70-05389
OFF, RAINFALL, SNOWMELT, SURFACE   WATERS, STREAMS, HYDROGEN ION CON   W70-03501
OFF, *ALGAL SUBSTANCES, SUBPOLAR   WATERS, TYNSET(NORWAY), NID RIVER   W70-01074
IBLIOGRAPHIES, BOTTOM SEDIMENTS,   WATERSHED MANAGEMENT, HARVESTING   W68-00859
LL, PHOTOSYNTHESIS, TRIBUTARIES,   WATERSHED MANAGEMENT, SEDIMENTS.:   W70-04268
ATA COLLECTIONS, AQUATIC PLANTS,   WATERSHED, ALGAE, AQUATIC VEGETAT   W68-00511
POLLUTION SOURCES, AGRICULTURAL   WATERSHEDS.: /ROPHICATION, *WATER   W69-05323
CHIGAN, PHOSPHORUS, AGRICULTURAL   WATERSHEDS, ORGANIC MATTER, SEDIM   W70-04193
OSPHORUS, DRAINAGE, AGRICULTURAL   WATERSHEDS, NITROGEN FIXATION, CY   W70-04506
GE DISPOSAL, ALGAE, AGRICULTURAL   WATERSHEDS, NUTRIENTS, FERTILIZAT   W70-05565
ENTS, CONDUCTIVITY, AGRICULTURAL   WATERSHEDS, SEWAGE, WATER POLLUTI   W70-05752
OROPHYTA, DEPOSITION(SEDIMENTS),   WATERSHEDS(BASINS), WATER POLLUTI   W69-10182
*SEWAGE DISPOSAL, *AGRICULTURAL   WATERSHEDS, *RUNOFF, *FERTILIZATI   W70-02787
ION PONDS, ALGAE, BENTHIC FAUNA,   WATERSHEDS, WATER SHORTAGE FILTER   W72-00027
TASTE, ODOR, ALGAE, RESERVOIRS,   WATERSHEDS(BASINS), INSECTICIDES,   W72-01813
TION, CHLORINATION, COAGULATION,   WATERSHEDS, SEWERAGE ALGAE, POTAB   W71-05943
ARBAGE DUMPS, OIL WASTES, INLETS(WATERWAYS), ORGANIC MA: /ASTES, G   W70-05112
ICATION, *FERTILIZATION, *INLAND   WATERWAYS, *WATER POLLUTION EFFEC   W69-07833
MANURE, MADISON(WIS), FLUSHING   WATERWAYS, MILWAUKEE(WIS), CHICAG   W69-10178
OTA(WIS), LAKE MONONA(WIS), LAKE   WAUBESA(WIS), LAKE KEGONSA(WIS).:   W72-01813
OTA(WIS), LAKE MONONA(WIS), LAKE   WAUBESA(WIS), LAKE KEGONSA(WIS),   W70-02969
ES(WIS), LAKE MENDOTA(WIS), LAKE   WAUBESA(WIS), LAKE KEGONSA(WIS).:   W70-02787
ODICTYON, NAJAS, ANACHARIS, LAKE   WAUBESA(WIS), MACROPHYTES, ALGAL   W70-03310
OTA(WIS), LAKE MONONA(WIS), LAKE   WAUBESA(WIS), ANABAENA CIR: /MEND   W69-08283
VAL, SPAULDING'S POND(WIS), LAKE   WAUBESA(WIS), GREEN LAKE(WIS).: /   W69-05868
OTA(WIS), LAKE MONONA(WIS), LAKE   WAUBESA(WIS), SPAULDING POND(WIS)   W69-05867
NA(WIS), LAKE KEGONSA(WIS), LAKE   WAUBESA(WIS), MADISON(WIS), UNIVE   W69-06276
SON(WIS), LAKE MONONA(WIS), LAKE   WAUBESA(WIS), LAKE KEGONSA(WIS),   W70-05112
TCH(WIS), LAKE MONONA(WIS), LAKE   WAUBESA(WIS), LAKE KEGONSA(WIS).:   W70-05113
RA(WIS), LAKE KEGONSA(WIS), LAKE   WAUBESA(WIS), PRECIPITATION.: /NG   W69-03366
FACTORS, SPECIES DIVERSITY, LAKE   WAUBESA, LYNGBYA, MICROCYSTIS, AP   W69-01977
LAKE MONONA, LAKE MENDOTA, LAKE   WAUBESA, LAKE KEGONSA, LAKE ZURIC   W69-09349
S, FISH YIELDS, UNDERWATER SHOCK   WAVE, ANABAENA, PHYGON XL(DICHLON   W69-08674
CENEDESMUS, BENTHOS, CONVECTION,   WAVES(WATER), FISH, SEWAGE, EUGLE   W70-03969
, ECOLOGY, CULTURES, WASHINGTON,   WAVES(WATER), CONNECTICUT, RHODE   W69-10161
DYNAMICS, LAKES, INTERNAL WATER,   WAVES(WATER), DIURNAL, EPILIMNION   W70-09889
CONTRACTS, REGULATION, RIGHT-OF-   WAY, ESTIMATED COSTS, LEGAL ASPEC   W70-03344
ADIONUCLIDE EXCHANGE, *OPEN SEA   WEAPONS TESTING, NUCLEAR SHIP SAV   W70-00707
LKALINITY, DIATOMS, TEMPERATURE,   WEATHER, EPILIMNION, ALGAE, DENSI   W70-02504
A, SEDIMENTS, PLANT POPULATIONS,   WEATHER, WATER, MOVEMENT, CULTURE   W70-06229
N, CALCIUM, TROPHIC LEVELS, FOOD   WEBS.: /SPECTROSCOPY, DISTRIBUTIO   W69-08525
ICITY, *PESTICIDE REMOVAL, *FOOD   WEBS, DDT, CHLORINATED HYDROCARBO   W70-03520
RGY TRANSFER, *FOOD CHAINS, FOOD   WEBS, FORAGE FISH, PLANKTON, ALGA   W70-05428
*SCULPINS, *RADIOACTIVITY, *FOOD   WEBS, KENTUCKY, LIFE HISTORY STUD   W69-09742
WAUBESA(WIS), LAKE KEGONSA(WIS),   WEED BEDS, PHOSPHORUS, NITROGEN B   W70-02969
WEEDS, *AQUATIC PLANTS, *AQUATIC   WEED CONTROL, WATER CONSERVATION,   W70-00269
ERBICIDES, *VASCULAR T/ *AQUATIC   WEED CONTROL, *AQUATIC PLANTS, *H   W70-10175
R SULFATE, *SODIUM ARS/ *AQUATIC   WEED CONTROL, *HERBICIDES, *COPPE   W70-05568
OSPHATE, FISH, PLANKTON, AQUATIC   WEED CONTROL, INSECT CONTROL, MOS   W70-05417
TTAILS, BULRUSH, DREDG/ *AQUATIC   WEED CONTROL, *BIBLIOGRAPHIES, CA   W70-05263
ARVESTING, AQUATIC PLA/ *AQUATIC   WEED CONTROL, *EUTROPHICATION, *H   W69-06276
APHIES, *ALGAL CONTROL, *AQUATIC   WEED CONTROL, *EUTROPHICATION, AQ   W69-05706
WEEDS, LAKES, ALGICIDES, AQUATIC   WEED CONTROL, MECHANICAL CONTROL,   W70-04494
APHIES, *ALGAL CONTROL, *AQUATIC   WEED CONTROL, *EUTROPHICATION, AQ   W69-05705
TING OF ALGAE, BOTTOM SEDIMENTS,   WEED CONTROL, SOLAR RADIATION, *W   W68-00478
ESERVOIRS, HYDROELECTRIC PLANTS,   WEED CONTROL, WATER PROPERTIES, P   W68-00479
NUISANCE ALGAE, *ALGAL CONTROL,   WEED CONTROL, *INVESTIGATIONS, ZO   W68-00468

BOTTOM SEDIMENTS, MUD, BENTHOS,
ANCE ALGAE, SNAILS, LEGISLATION,
MICROBIOLOGY, BALANCE OF NATURE,
LGICIDES, TRACE ELEMENTS, CORES,
YTA, CYANOPHYTA, NUISANCE ALGAE,
NCE AL/ *EUTRIPHICATION, *LAKES,
, SPRA/ *EUTROPHICATION, *LAKES,
POTAMOGETON CRISPUS, MICROSPORA
BORSKII, RAPHIDIOPSIS, ANABAENA,
ASHINGTON, HOWARD A. HANSON RES/
OCHOIDEUM, VINEYARD SOUND(MASS),
TURNOVER TIME, TRIDACNA CROCEA,
'BREAKING' OF / *MADISON(WIS), *'
TRY, ODOR, TAST/ *AMERICAN WATER
*ADMINISTRATIVE AGENCIES, *WATER
*OHIO, WATER TEMPERATURE, SLUDGE
PLANKTON, BENTHIC FAUNA, SLUDGE
AE, DIATOMS, PROTOZOA, PLANKTON,
ALGAE, INVERTEBRATES, TURBIDITY,
*LAKE KLOPEINER(AUSTRIA), LAKE
ER LINE DISCRIMINATOR, RHODAMINE
MIDS, LAKE CONSTANCE, WHITEFISH,
LES, ANTIBIOTICS, AGARBACTERIUM,
ICROSPORA WITTROCKII, SPIROGYRA,
UCTS, ANABAENA CYLINDRICA, ZINC,
TERRANEUS, NAVICULA PELLICULOSA,
TER SHOCK WAVE, ANABAENA, PHYGON
N(WIS), UNIVERSITY OF WISCONSIN,
ISON) EFFLUENTS, *MADISON LAKES,
ECIES, DIVERSITY I/ DESICCATION,
SCENEDESMUS, EUGLENA, BACTERIA,
TREATMENT, *AQUICULTURE. ALGAE,
DS, AMMONIA, NITRITES, NITRATES,
TIVITY, WATER POLLUTION EFFECTS,
INTERTIDAL AREAS, CARBOHYDRATES,
QUATIC BACTERIA, *AQUATIC FUNGI,
DROGEN ION CONCENTRATION, FUNGI,
EPIPHYTIC BACTERIA, SUBLITTORAL,
URVES, CALORIMETRY, SOLARIMETRY,
DETERGENTS, FISH MANAGERS, FISH
ALGAE, BIOMASS, CHLOROPHYLL, NEW
TIVITY, *ZOOPLANKT/ *LAKES, *NEW
NS, WATER POLLUTION EFFECTS, NEW
LICATES, CITRATES, BIOASSAY, NEW
S, BIOINDICATORS, *STREAMS, *NEW
OINDICATORS, NUISANCE ALGAE, NEW
OIDS, FLOW RATES, *FILTERS, *NEW
, FILTRATION, NEW HAMPSHIRE, NEW
CRITERIA, *WATER TREATMENT, *NEW
IATA, NAVICULA, FLORIDA CURRENT,
AKE, NANNOCHLORIS, THORACOMONAS,
COBBLES, EQUISETUM, THORIUM-232,
CYANOCOBALAMIN, EUGLENA GRACILIS
GAE.:   *NEW
IA), LAKE SCHLIER(BAVARIA), LAKE
SEWAGE EFFLUENTS, ELECTROLYTES,
UBRA, CANDIDA DIDDENSII, CANDIDA
RON, TETRASTRUM, TISZA(HUNGARY),
PLANTS, NYLON-6, *VISCOSE RAYON,
ROPERTIES, ADSORPTION, COLLOIDS,
LLUSKS, STRONTIUM RADIOISOTOPES,
TA, LIGHT INTENSITY, SUCCESSION,
US RADIOISOTOPES, TROPHIC LEVEL,
CESIUM, STRONTIUM RADIOISOTOPES,
S, POTASSIUM, IRON, BICARBONATE,
PHOTOPHOSPHORYLATION, VANADIUM,
, INHIBITION, SILICON, VANADIUM,
ONEMA OGLIMUN, UDOTEA FLABELLUM,
R PRODUCTS, ANABAENA CYLINDRICA,
TESTING, NUCLEAR SHIP SAVANNAH,
14, ACTINIUM-228, SWIFT RIFFLES,
ANS, *SARGASSUM NATANS, SEAWEED,
ERTEBRATES, PLUTONIUM-239, RATS,
URBAN WASTES, MADISON(WIS), LAKE
ZERLAND, KILCHBERG(SWITZERLAND),
L LIMNOLOGY, CLADOCERAN, MONTANE
CLADOCERAN, MONTANE ZONE, ALPINE
K RESERVOIR(TENN), NAJAS, PHOTIC
*EXTRACELLULAR PRODUCTS, PHOT
CYCLOTELLA MENEGHINIANA, PHOTIC
CY, LIGHT MEASUREMENTS, EUPHOTIC
TEMIA, CARTERIA, NORTH TEMPERATE
TRANSACTIONS, BLACK SEA, CONTACT
IS, DEPTH, THERMOCLINE, CLIMATIC
MENT, HARVESTING OF ALGAE, WATER
MICAL PRECIPITATION, HARVESTING,
, WEED CONTROL, *INVESTIGATIONS,
N, MEXICO, PREDICTIVE EQUATIONS,
IGOTROPHY, DEPTH, PHYTOPLANKTON,
ICAL REGIONS, CYCLING NUTRIENTS,

WISCONSIN, SAMPLING, CONTOURS, DE   W70-06217
WISCONSIN, CHEMCONTROL, PONDWEEDS   W70-05568
WISCONSIN, BIOASSAY, BIOCHEMICAL   W69-03369
WISCONSIN, BOTTOM SEDIMENT, BENTH   W69-03366
WISCONSIN, NUTRIENTS, INDICATOR S   W69-01977
WISCONSIN, COPPER SULPHATE, NUISA   W68-00468
WISCONSIN, ALGAE, *NUISANCE ALGAE   W68-00488
WITTROCKII, SPIROGYRA, XANTHOPHYC   W70-00711
WOLLEA, OSCILLATORIA, SPIRULINA,   W70-05404
WOOD, ALGAL COUNT, GREEN RIVER, W   W72-11906
WOODS HOLE(MASS). RADIOCARBON UPT   W70-05272
WOODS HOLE OCEANOGRAPHIC INSTITUT   W69-08526
WORKING' OF LAKES, *WATER-BLOOM,   W69-08283
WORKS ASSOCIATION, CHEMICAL INDUS   W71-05943
WORKS, *WASTE DISPOSAL, DOMESTIC   W70-03344
WORMS, BLOODWORMS, AQUATIC BACTER   W70-00475
WORMS, BLOODWORMS, WATER POLLUTIO   W68-01244
WORMS, NEMATODES, ROTIFERS, LARVA   W72-01819
WORMS, SUNFISHES.: /S, PLANKTON,   W69-07861
WORTHER(AUSTRIA), VERTICAL, CHALK   W70-05405
WT DYE, AIRBORNE TESTS, OIL SPILL   W70-05377
WURTTEMBERG(GERMANY), PLON(GERMAN   W70-05761
XANT: /ARY COMPOUNDS, MACROMOLECU   W70-04280
XANTHOPHYCEAE, MYXOPHYCEAE, SALVE   W70-00711
XANTHOPHYCEAE, BACILLARIOPHYCEAE,   W69-10180
XANTHOPHYCEAE, BACILLARIOPHYCEAE,   W70-02965
XL(DICHLONE), NEW HAMPSHIRE, LAKE   W69-08674
YAHARA RIVER(WIS), NUTRIENT REMOV   W69-06276
YAHARA RIVER(WIS), STORKWEATHER C   W70-05113
YAQUINA BAY(ORE), CAROTENOIDS, SP   W70-03325
YEASTS, ACTINOMYCETES, PONDING, W   W70-08097
YEASTS, BACTERIA, FISH, PROTEINS,   W69-05891
YEASTS, BACTERIA, AQUATIC BACTERI   W69-06970
YEASTS, CRUSTACEA, ALGAE.: /RODUC   W69-10152
YEASTS, ENZYMES, CULTURES, HYDROG   W70-03952
YEASTS, HYDROGEN ION CONCENTRATIO   W69-00096
YEASTS, PYRITE, ALGAE, CYANOPHYTA   W68-00891
YELLOW MATERIAL, DU: /A HARVEYI,   W70-01073
YELLOWSTONE NATIONAL PARK, MT LAS   W70-03309
YIELDS, UNDERWATER SHOCK WAVE, AN   W69-08674
YORK.: /*PONDS, *EUTROPHICATION,   W69-06540
YORK, *MEROMIXIS, *PRIMARY PRODUC   W70-05760
YORK, CHLOROPHYTA, NUTRIENTS, ECO   W70-03973
YORK, LAKES, INHIBITION, ALGAE, D   W70-01579
YORK, NUCLEAR WASTES, WATER POLLU   W69-10080
YORK, SEWAGE EFFLUENTS, SEWAGE TR   W70-03974
YORK, TREATMENT FACILITIES.: /OLL   W72-04145
YORK, VIRGINIA, ORGANIC COMPOUNDS   W72-03641
YORK, WATER PURIFICATION.: /SIGN   W72-08686
YTTRIUM-90, KATODINIUM ROTUNDATA.   W70-00707
YTTRIUM, 'BIOLOGICAL DILUTION', P   W70-00708
YUKON RIVER, NEBESNA RIVER, MOOSE   W69-08272
Z, MICROBIOLOGICAL ASSAY, SPECIES   W69-06273
ZEALAND, SEWAGE TREATMENT POND AL   W70-02304
ZELL(AUSTRIA), LAKE HALL WILER(SW   W70-05668
ZETA POTENTIAL, BACTERIA, PROTOZO   W70-01027
ZEYLANOIDES, CANDIDA KRUSEI, CAND   W70-04368
ZIMONA(HUNGARY), ALGAL TAXONOMY,   W70-03969
ZINC COMPOUNDS, CARBON DISULFIDE.   W70-06925
ZINC RADIOISOTOPES, CALCIUM, POTA   W70-00708
ZINC RADIOISOTOPES, IODINE RADIOI   W71-09850
ZINC RADIOISOTOPES, CESIUM, COBAL   W70-04371
ZINC RADIOISOTOPES, IRON, SEDIMEN   W69-07853
ZINC RADIOISOTOPES, IRON, MANGANE   W69-07845
ZINC, COBALT, CALCIUM, MAGNESIUM,   W70-03978
ZINC, EUGLENA GRACILIS, ANABAENA   W70-02964
ZINC, GALLIUM.: /INS, METABOLITES   W69-04801
ZINC, NICKEL, PUERTO RICO, LA PAR   W69-08525
ZINC, XANTHOPHYCEAE, BACILLARIOPH   W69-10180
ZIRCONIUM-95, RUBIDIUM-106, NIOBI   W70-00707
ZIRCONIUM-95, NIOBIUM-95, RUTHENI   W69-09742
ZIRCONIUM-95-NIOBIUM-95, PRE-BOMB   W69-08524
ZIRCONIUM-95-NIOBIUM, EUROPIUM-15   W69-08269
ZOAR(CONN), KLAMATH LAKE(ORE), LA   W70-04506
ZOLLIKON(SWITZERLAND), LAKE TEGER   W70-05668
ZONE, ALPINE ZONE, BOULDER(COLO),   W69-10154
ZONE, BOULDER(COLO), TEA LAKE(COL   W69-10154
ZONE, DOUGLAS RESERVOIR(TENN), NO   W71-07698
ZONE, GLYCOLLIC ACID, CHEMOTROPHY   W70-02504
ZONE, HELSINGOR(DENMARK), SALINO-   W70-04809
ZONE, PLANT CAROTENOIDS, CARBON F   W70-01933
ZONE, PLUTONIUM, HALF-LIFE(RADION   W70-00708
ZONES, ARTEMIA, BALANUS, DESORPTI   W69-07440
ZONES, ORGANIC LOADING, DISSOLVED   W72-10573
ZONING.: /MENTS, WATERSHED MANAGE   W68-00859
ZONING, DIVERSION, DREDGING, IMPO   W69-10178
ZOO-PLANKTON, SPRAYING, BIOCONTRO   W68-00468
ZOOGEOGRAPHY, THERMOBIOLOGY.: /IO   W70-03957
ZOOPLANKTON, PHYSICOCHEMICAL PROP   W70-03964
ZOOPLANKTON, ALGAE, AQUATIC PLANT   W70-03975

LLICULOSA, GONYAULAX POLYEDRA, 3, 5-DIAMINOBENZOIC ACID DIHYDROCHLO   W69-08278
ALGAE, 2-4-D, AMINOTRIAZOLE, 2-4- 5-T, SCENEDESMUS, CHLAMYDOMONAS,   W70-03519
S, *MUSCULAR EXERTION, *2-METHYL- 5-VINYLPYRIDINE, *GLUE GREEN ALGA   W70-01788
S, *MAN-MADE FIBER PLANTS, NYLON- 6, *VISCOSE RAYON, ZINC COMPOUNDS   W70-06925
ION, DOW C-31, DOW A-21, NALCO N- 670.: /ARGE, ELECTROSTATIC REPULS   W70-05267
                        SIMAZINE 80 W, DACTHAL.:   W70-03519
VICULA, FLORIDA CURRENT, YTTRIUM- 90, KATODINIUM ROTUNDATA.: /A, NA   W70-00707
, PLUTONIUM-239, RATS, ZIRCONIUM- 95-NIOBIUM, EUROPIUM-155, SILVER-   W69-08269
ASSUM NATANS, SEAWEED, ZIRCONIUM- 95-NIOBIUM-95, PRE-BOMB MATERIAL,   W69-08524
CONIUM-95, RUBIDIUM-106, NIOBIUM- 95, GYMNODINIUM SIMPLEX, FLAGELLA   W70-00707
UM-228, SWIFT RIFFLES, ZIRCONIUM- 95, NIOBIUM-95, RUTHENIUM-106, RH   W69-09742
S, SEAWEED, ZIRCONIUM-95-NIOBIUM- 95, PRE-BOMB MATERIAL, RUTHENIUM-   W69-08524
NUCLEAR SHIP SAVANNAH, ZIRCONIUM- 95, RUBIDIUM-106, NIOBIUM-95, GYM   W70-00707
T RIFFLES, ZIRCONIUM-95, NIOBIUM- 95, RUTHENIUM-106, RHODIUM-106, C   W69-09742

AUTHOR INDEX

In order not to complicate the computer program needlessly, only initials and the word "and" have been excluded from the alphabetic listing. Both first and last names of authors appear in the index. The numbers at the right are the accession numbers appearing at the end of the abstracts.

```
         E. B. COWELL, AND J. M.   BAKER.:                                      / W70-01231
         J. J. CONNORS, AND R. B.   BAKER.:                                      / W72-04495
                    K. J. M.   BAKER, AND E. B. COWELL.:                          / W70-09976
                    ROBERT C.   BALL.:                                           W68-00487
                    ROBERT C.   BALL, AND FRANK F. HOOPER.:                      / W69-09334
                        W. D.   BARLOW.:                                         / W70-01519
                        W. D.   BARLOW.:                                         / W71-06737
     WILLIAM CHAMBERLAIN, AND JUDITH   BARRETT.:              JOSEPH SHAPIRO,/ W70-08637
          C. DAVID MCINTIRE, AND   BARRY L. WULFF.:                              / W70-03325
     ETH M. MACKENTHUN, AND ALFRED F.   BARTSCH.:                      KENN/ W72-01797
     THOMAS E. MALONEY, AND ALFRED F.   BARTSCH.:                                / W70-07393
                        A. F.   BARTSCH.:                                        / W68-01244
                        A. F.   BARTSCH.:                                        W68-00478
                        A. F.   BARTSCH.:                                        W68-00461
                        A. F.   BARTSCH, AND M. O. ALLUM.:                       / W72-01818
                      JOHN W.   BAUMEISTER.:                                     / W70-03068
     OVERFIELD, H. R. CRAWFORD, J. K.   BAXTER, L. J. HARRINGTON, AND I. / W70-05819
     WILLIAM J. OSWALD, AND LOUIS A.   BECK.: /GOLDMAN, JAMES F. ARTHUR,/ W71-04557
                        A. M.   BEETON.:                                         W68-00253
          CORNELIUS I. WEBER, AND   BEN H. MCFARLAND.:                           / W71-04206
                        H. D.   BENNETT.:                                        / W70-02770
     ORO, H. J. SCHNEIDER, AND E. O.   BENNETT.:              E. GELPI, J./ W69-08284
                    RICHARD J.   BENOIT.:                                        / W70-04385
                        R. J.   BENOIT, AND J. J. CURRY.:                        W68-00479
     I. SAX, PAUL C. LEMON, ALLEN H.   BENTON, AND JACK J. GABAY:       N./ W69-10080
       ALBERTO M. WACHS, AND ANDRE   BEREND.:                                    / W70-04790
                 MAESTRINI.:     B. R.   BERLAND, M. G. BIANCHI, AND S. Y./ W70-04280
     C. SERRUYA, S. SERRUYA, AND T.   BERMAN.:                                   / W70-09889
     DEPOE.:          ERNEST C. NEAL,   BERNARD C. PATTEN, AND CHARLES E./ W70-04371
                                       BERNARD P. DOMOGALLA.:                    / W70-04494
     . MORAVCOVA, L. MASINOVA, AND V.   BERNATOVA.:                       A/ W72-05143
                                       BERNHARD DOMOGALLA.:                      W68-00488
                                       BERNHARD DOMOGALLA.:                      W68-00468
                           M.   BERNHARD, AND A. ZATTERA.:                       / W69-08275
                        A. E.   BERRY.:                                          / W72-03824
                    EDWARD C.   BERRY.:                                          / W71-02009
     RNST M. DAVIS.:              BERT A. FLOYD, E. GUS FRUH, AND E/ W69-06004
     GOODNIGHT.:                  BERT K. WHITTEN, AND CLARENCE J. / W69-07861
                      NACHUM   BETZER, AND YEHUDA KOTT.:                         / W70-02364
                 SURINDER K.   BHAGAT, AND DONALD E. PROCTOR.:                   / W71-00936
         B. R. BERLAND, M. G.   BIANCHI, AND S. Y. MAESTRINI.:                   / W70-04280
                     HARTMUT   BICK.:                                            / W71-03031
           VERA A. (DUGDALE)   BILLAUD.:                                         / W69-03515
                      JOHN W.   BISHOP.:                                         / W71-10098
     WOODS.:                  G.   BLUNDE, S. B. CHALLEN, AND D. L. / W69-07826
                        R. H.   BOGAN.:                                          W68-00248
     NNART ENEBO, JITKA VENDLOVA, AND   BOHUMIR PROKES.: /D HEDENSKOG, LE/ W70-04184
                    BRENDA J.   BOLEYN.:                                         / W70-04469
                           P.   BOLTON.:                                         / W73-00758
                        A. D.   BONEY.:                                          / W70-01779
         H. S. AZAD, AND J. A.   BORCHARDT.:                                     / W70-03923
                        J. A.   BORCHARDT.:                                      / W69-07084
         E. DAVIS, AND J. A.   BORCHARDT.:                                       / W72-07334
     GERALD W. FOESS, AND JACK A.   BORCHARDT.:                                  / W71-03035
     PER, W. J. WEBER, JR., AND J. A.   BORCHARDT.:              G. R. PY/ W72-00917
                        J. A.   BORCHARDT, AND C. R. O'MELIA.:                   / W72-03703
                     JACK A.   BORCHARDT, AND HARDAM S. AZAD.:                   / W70-06604
                                       BOSTWICK H. KETCHUM.:                     / W70-02804
     PIERRE GUEGUENIAT, PIERRE   BOVARD, AND JACQUES ANCELLIN.:                  / W71-09863
                 RUTH PATRICK,   BOWMAN CRUM, AND JOHN COLES.:                   / W70-07257
     T. D. ARMSTRONG, AND B. I.   BOYKO.:                                        / W71-08970
                    MORRIS L.   BREHMER.:                                        / W69-03695
                                       BRENDA J. BOLEYN.:                        / W70-04469
                   PATRICK L.   BREZONIK.:                                       / W71-03048
                   PATRICK L.   BREZONIK, AND HUGH D. PUTNAM.:                   / W72-01990
     NNON, AND H. D. PUTNAM.:   P. L.   BREZONIK, W. H. MORGAN, E. E. SHA/ W69-09723
                                       BRIAN MOSS.:                             / W70-02646
                                       BRIAN MOSS.:                             / W72-03183
                    ROBERT M.   BRICE, AND CHARLES F. POWERS.:                   / W70-03512
                    ROBERT C.   BRIGHT.:                                         / W69-00659
                           G.   BRINGMANN, AND R. KUHN.:                         W68-00472
                           G.   BRINGMANN, AND R. KUHN.:                         W68-00475
                        F. J.   BRINLEY.:                                        W68-00483
                     FLOYD J.   BRINLEY.:                                        / W72-01788
         A. L. S. MUNRO, AND T. D.   BROCK.:                                     / W70-04811
     THOMAS D. BROCK, AND M. LOUISE   BROCK.:                                    / W69-09676
                    THOMAS D.   BROCK.:                                          / W69-10161
     M. LOUISE BROCK, AND THOMAS D.   BROCK.:                                    / W69-10163
         T. D. BROCK, AND M. LOUISE   BROCK.:                                    / W69-10160
                    THOMAS D.   BROCK.:                                          / W69-09755
     THOMAS D. BROCK, AND M. LOUISE   BROCK.:                                    / W69-10151
                        T. D.   BROCK, AND M. LOUISE BROCK.:                     / W69-10160
                    THOMAS D.   BROCK, AND M. LOUISE BROCK.:                     / W69-10151
                    THOMAS D.   BROCK, AND M. LOUISE BROCK.:                     / W69-09676
                    M. LOUISE   BROCK, AND THOMAS D. BROCK.:                     / W69-10163
                        A. J.   BROOK.:                                          W68-00255
                        D. D.   BROUSE.:                                         / W69-10155
                        J. G.   BROWN.:                                          / W72-08688
                     RANDALL   BROWN, AND JAMES ARTHUR.:                         / W70-03334
     JO/   JAMES F. ARTHUR, RANDALL L.   BROWN, BRUCE A. BUTTERFIELD, AND / W71-04554
```

572

```
ERNST M. DAVIS, AND        EARNEST F. GLOYNA.:                              / W70-09438
ORD.:                      EARNEST F. GLOYNA, AND DAVIS L. F/ W72-03299
PINO.:                     EARNEST F. GLOYNA, AND ERNESTO ES/ W70-01971
                    WILLIAM R. EBERLY.:                                     / W69-03514
                      HARRY J. EBY.:                                        / W71-08221
SSLER/  MARK W. TENNEY, WAYNE F. ECHELBERGER, JR., RONALD G. SCHUE/ W70-05267
              W. WESLEY ECKENFELDER, JR.:                                   / W69-08054
         YEHUDA KOTT, AND J. EDLIS.:                                        / W70-02363
                      W. T. EDMONDSON.:                                     / W69-10182
                      W. T. EDMONDSON.:                                     / W70-04455
                   W. THOMAS EDMONDSON.:                                    / W69-09349
                      W. T. EDMONDSON.:                                     / W70-05663
                      W. T. EDMONDSON.:                                     / W70-00270
                      W. T. EDMONDSON.:                                     / W70-07261
NALD R. PETERSON.:         W. T. EDMONDSON, G. C. ANDERSON, AND DO/ W69-10169
                      W. T. EDMONSON.:                                      / W69-03513
                           EDWARD C. BERRY.:                                / W71-02009
ARTY.:                     EDWARD G. FOREE, AND PERRY L. MCC/ W71-13413
                           EDWARD J. CARPENTER.:                            / W70-09904
                           EDWARD J. KORMONDY.:                             / W71-09005
NBERGER.:                  EDWARD J. MARTIN, AND LEON W. WEI/ W69-01453
                    HENRY L. EHRLICH.:                                      / W69-07428
                      G. G. EHRLICH, AND K. V. SLACK.:                      / W69-06970
S.:                   ROLF ELIASSEN, AND GEORGE TCHOBANOGLOU/ W70-07469
S.:                   ROLF ELIASSEN, AND GEORGE TCHOBANOGLOU/ W70-01981
. TONKIN.:               R. ELIASSEN, B. M. WYCKOFF, AND C. D/ W72-04318
         E. C. S. CHAN, AND ELIZABETH A. MCMANUS.:                          / W70-04365
         E. C. S. CHAN, AND ELIZABETH A. MCMANUS.:                          / W70-03952
        ARTHUR D. HASLER, AND ELIZABETH JONES.:                             / W70-02249
ER.:         S. A. WITZEL, ELIZABETH MCCOY, AND RICHARD LEHN/ W71-05742
       JAMES B. LACKEY, AND ELSIE W. LACKEY.:                               / W72-01993
R P/  GUDMUND HEDENSKOG, LENNART ENEBO, JITKA VENDLOVA, AND BOHUMI/ W70-04184
                       J. F. ERDEI.:                                        / W72-13601
                     PAUL A. ERICKSON, AND JOHN T. REYNOLDS.: / W70-05264
      E. STEEMANN NIELSEN, AND ERIK G. JORGENSEN.:                          / W70-04809
                           ERIK G. JORGENSEN.:                              / W70-05381
, AND CHARLES E. DEPOE.:    ERNEST C. NEAL, BERNARD C. PATTEN/ W70-04371
   J. A. DAVIS, C. E. DAY III, AND ERNEST E. ANGINO.: /OHN E. SIMEK,/ W69-08524
         KAZUHIRO FUTAGAWA, AND ERNEST F. GLOYNA.:                          / W71-11381
ST F. GLOYNA.:             ERNESTO ESPINO DE LA O, AND EARNE/ W70-07034
 GLOYNA.:                  ERNESTO ESPINO DE LA O, AND E. F./ W71-11377
          EARNEST F. GLOYNA, AND ERNESTO ESPINO.:                           / W70-01971
  BERT A. FLOYD, E. GUS FRUH, AND ERNST M. DAVIS.:                          / W69-06004
:                          ERNST M. DAVIS, AND E. F. GLOYNA./ W69-09454
OYNA.:                     ERNST M. DAVIS, AND EARNEST F. GL/ W70-03928
OYNA.:                     ERNST M. DAVIS, AND EARNEST F. GL/ W70-09438
:                  ERNESTO ESPINO DE LA O, AND E. F. GLOYNA./ W71-11377
OYNA.:             ERNESTO ESPINO DE LA O, AND EARNEST F. GL/ W70-07034
  EARNEST F. GLOYNA, AND ERNESTO ESPINO.:                                   / W70-01971
      ROBERT L. BUNCH, AND M. B. ETTINGER.:                                 / W70-04373
      ROBERT L. BUNCH, AND M. B. ETTINGER.:                                 / W71-04072
                           EUGENE B. WELCH.:                                / W71-07698
                           EUGENE B. WELCH.:                                / W70-04283
DZE.:             V. G. KHOBOT' EV, V. I. KAPKOV, AND E. G. RUKHA/ W71-11561
                      G. EVELYN HUTCHINSON.:                                / W68-00860
                     CLYDE EYSTER.:                                         / W70-02964
   G. P. FITZGERALD, AND S. L. FAUST.:                                      / W69-03370
     ROSALIE M. COX, AND P. FAY.:                                           / W70-04249
LOWELL D. HANSON, AND WILLIAM E. FENSTER.:                                  / W70-04193
                        I. FINDENEGG.:                                      / W70-05405
UKE, J. WILLIS, T. PRICE, AND K. FISCHLER.:                    T. D/ W69-08267
     G. C. GERLOFF, AND K. A. FISHBECK.:                                    / W70-02245
      C. S. DUTTON, AND C. P. FISHER.:                                      / W71-09546
     RONALD R. SPEEDY, NEIL B. FISHER, AND DONALD B. MCDONALD.: / W70-08107
                    GEORGE P. FITZGERALD.:                                  / W69-03374
                    GEORGE P. FITZGERALD.:                                  / W69-05704
                    GEORGE P. FITZGERALD.:                                  / W69-03364
                    GEORGE P. FITZGERALD.:                                  / W69-03188
                    GEORGE P. FITZGERALD.:                                  / W69-03186
                    GEORGE P. FITZGERALD.:                                  / W69-03185
                    GEORGE P. FITZGERALD.:                                  / W69-01977
HOLAS L. CLESCERI, AND GEORGE P. FITZGERALD.:        G. FRED LEE, NIC/ W69-03358
  GERALD C. GERLOFF, GEORGE P. FITZGERALD, AND FOLKE SKOOG.:              / W70-03507
  GERALD C. GERLOFF, GEORGE P. FITZGERALD, AND FOLKE SKOOG.:              / W70-00719
                    GEORGE P. FITZGERALD, AND FOLKE SKOOG.:               / W70-03310
                      G. P. FITZGERALD, AND G. A. ROHLICH.:               / W68-00855
.:    MARY M. ALLEN, GEORGE P. FITZGERALD, AND GERARD A. ROHLICH/ W69-03371
N.:                   G. P. FITZGERALD, AND M. E. DERVARTANIA/ W69-10171
                      G. P. FITZGERALD, AND S. L. FAUST.:                 / W69-03370
:                   GEORGE P. FITZGERALD, AND THOMAS C. NELSON.: / W69-03373
D FOLKE SKOOG.:       GEORGE P. FITZGERALD, GERALD C. GERLOFF, AN/ W70-02982
 GOLDMAN, O. R. ARMSTRON/ G. P. FITZGERALD, JOSEPH SHAPIRO, C. R./ W69-06864
                    DAVID C. FLAHERTY.:                                   / W69-05844
                    DAVID C. FLAHERTY.:                                   / W70-00264
                           FLOYD J. BRINLEY.:                             / W72-01788
DAVIS.:                 BERT A. FLOYD, E. GUS FRUH, AND ERNST M. / W69-06004
                        J. FOEHRENBACH.:                                  / W69-08518
                  GERALD W. FOESS, AND JACK A. BORCHARDT.:              / W71-03035
                      G. E. FOGG.:                                        / W70-04369
                      G. E. FOGG.:                                        / W70-02965
```

	GLADWIN HILL.:	/ W70-04430
JOHN H. NEIL, AND	GLENN E. OWEN.:	/ W69-04798
ERNST M. DAVIS, AND E. F.	GLOYNA.:	/ W69-09454
O ESPINO DE LA O, AND EARNEST F.	GLOYNA.:	ERNEST/ W70-07034
EARNEST F.	GLOYNA.:	/ W70-04786
JORGE AGUIRRE, AND EARNEST F.	GLOYNA.:	/ W70-07031
ERNST M. DAVIS, AND EARNEST F.	GLOYNA.:	/ W70-03928
JU-CHANG HUANG, AND EARNEST F.	GLOYNA.:	/ W71-12183
KAZUHIRO FUTAGAWA, AND ERNEST F.	GLOYNA.:	/ W71-11381
ERNST M. DAVIS, AND EARNEST F.	GLOYNA.:	/ W70-09438
JU-CHANG HUANG, AND EARNEST F.	GLOYNA.:	/ W70-08416
RNESTO ESPINO DE LA O, AND E. F.	GLOYNA.:	E/ W71-11377
EARNEST F.	GLOYNA, AND DAVIS L. FORD.:	/ W72-03299
EARNEST F.	GLOYNA, AND ERNESTO ESPINO.:	/ W70-01971
RUCE A. BUTTERFIELD, AND JOEL C.	GOLDMAN.: /R, RANDALL L. BROWN, B/	W71-04554
M. I.	GOLDMAN, AND R. SHOOP.:	/ W72-00462
CHARLES R.	GOLDMAN, AND RALF C. CARTER.:	/ W70-01933
J. OSWALD, AND LOUIS / JOEL C.	GOLDMAN, JAMES F. ARTHUR, WILLIAM/	W71-04557
HARD ARMSTRONG.: CHARLES R.	GOLDMAN, MILTON G. TUNZI, AND RIC/	W70-03983
ITZGERALD, JOSEPH SHAPIRO, C. R.	GOLDMAN, O. R. ARMSTRONG, AND JAC/	W69-06864
. JAKOBS-MOGELIN.: H. L.	GOLTERMAN, C. C. BAKELS, AND J. J/	W70-03955
H/ GORDON L. DUGAN, CLARENCE G.	GOLUEKE, WILLIAM J. OSWALD, AND C/	W71-11375
EL W. HEDGPETH, AND JEFFERSON J.	GONOR.:	JO/ W71-02494
H. S. SWINGLE, B. C.	GOOCH, AND H. R. RABANAL.:	/ W70-05417
BERT K. WHITTEN, AND CLARENCE J.	GOODNIGHT.:	/ W69-07861
SWALD.:	GORDON L. DUGAN, AND WILLIAM J. O/	W72-06854
EKE, WILLIAM J. OSWALD, AND CH/	GORDON L. DUGAN, CLARENCE G. GOLU/	W71-11375
R, AND W. K. KIM.: P. R.	GORHAM, J. MCLACHLAN, U. T. HAMME/	W70-00273
HAROLD B.	GOTAAS.:	/ W70-04465
L. K. PORTER, AND A. R.	GRABLE.:	/ W72-08110
G. M. P. MORDVINTSEVA, V. V.	GRABOVSKAYA, AND V. K. MARINICH.:/	W70-06925
ROSS E. MCKINNEY, AND ANDREW	GRAM.:	/ W72-01817
MARGARET	GREENWALD.:	/ W69-05706
MARGARET	GREENWALD.:	/ W69-05705
PHILLIP E.	GREESON.:	/ W70-04721
THOMAS D. WAITE, AND CONSTANTINE	GREGORY.:	/ W71-05500
G. K.	GRUENDLING, AND A. C. MATHIESON.:/	W70-02764
G. K.	GRUENDLING, AND A. C. MATHIESON.:/	W70-02772
JITKA VENDLOVA, AND BOHUMIR P/	GUDMUND HEDENSKOG, LENNART ENEBO,	W70-04184
CQUES ANCELLIN.: PIERRE	GUEGUENIAT, PIERRE BOVARD, AND JA/	W71-09863
S. P. MEYERS, D. G. AHEARN, W.	GUNKEL, AND F. J. ROTH, JR.:	/ W70-04368
ROBERT A. COLER, AND HAIM B.	GUNNER.:	/ W71-03775
E.	GUS FRUH.:	/ W70-02251
E.	GUS FRUH.:	/ W70-04765
E.	GUS FRUH.:	/ W70-04727
G. FRED LEE, AND E.	GUS FRUH.:	/ W69-06535
BERT A. FLOYD, E.	GUS FRUH, AND ERNST M. DAVIS.:	/ W69-06004
ROBERT A. COLER, AND	HAIM B. GUNNER.:	/ W71-03775
J. C. LACAZE, C.	HALLOPEAU, AND M. VOIGT.:	/ W72-02203
D. H.	HAMILTON, JR.:	/ W70-01579
GARY L. HERGENRADER, AND MARK J.	HAMMER.:	/ W72-04761
. R. GORHAM, J. MCLACHLAN, U. T.	HAMMER, AND W. K. KIM.:	P/ W70-00273
F. J. HUMENIK, AND G. P.	HANNA, JR.:	/ W70-05655
OSMUND HOLM-	HANSEN.:	/ W69-08278
LKE SKOOG.: OSMUND HOLM-	HANSEN, GERALD C. GERLOFF, AND FO/	W69-06277
AND P. M. WILLIAMS.: O. HOLM-	HANSEN, J. COOMBS, B. E. VOLCANI,/	W70-05651
OSMUND HOLM-	HANSON.:	/ W69-03518
LOWELL D.	HANSON, AND WILLIAM E. FENSTER.:	/ W70-04193
JACK A. BORCHARDT, AND	HARDAM S. AZAD.:	/ W70-06604
WALLACE H. FULLER, AND JAMES E.	HARDCASTLE.:	/ W69-03491
GEORGE L.	HARLOW.:	/ W69-01445
KENNETH M. MACKENTHUN, AND	HAROLD B. GOTAAS.:	/ W70-04465
	HAROLD L. COOLEY.:	/ W70-06217
	HAROLD LEROY ALLEN.:	/ W71-00832
R. CRAWFORD, J. K. BAXTER, L. J.	HARRINGTON, AND I. W. SANTRY, JR./	W70-05819
RALPH H. WEAVER, AND	HARRY D. NASH.:	/ W69-00096
R.:	HARRY D. NASH, AND RALPH H. WEAVE/	W68-00891
	HARRY J. EBY.:	/ W71-08221
SHERRY L. VOLK, AND	HARRY K. PHINNEY.:	/ W69-04802
	HARRY K. PHINNEY.:	/ W70-03978
ID MCINTIRE, ROBERT L. GARRISON,	HARRY K. PHINNEY, AND CHARLES E. /	W70-00265
ROBERT D.	HARTER.:	/ W69-03075
	HARTMUT BICK.:	/ W71-03031
R. S.	HARVEY.:	/ W69-07845
R. M.	HARWOOD.:	/ W72-09693
DUNG DANG, AND TAMAO N/ YOSHIRO	HASHIMOTO, TOMOTOSHI OKAICHI, LE /	W71-07731
DUNG DANG, AND TAMAO N/ YOSHIRO	HASHIMOTO, TOMOTOSHI OKAICHI, LE /	W70-08372
ARTHUR D.	HASLER.:	/ W69-06858
ARTHUR D.	HASLER.:	/ W68-00859
A. D.	HASLER.:	W68-00680
ALD E. WOHLSCHLAG, AND ARTHUR D.	HASLER.:	DON/ W70-06229
ARTHUR D.	HASLER, AND ELIZABETH JONES.:	/ W70-02249
FN/ RICHARD B. COREY, ARTHUR D.	HASLER, G. FRED LEE, F. H. SCHRAU/	W69-10178
A.	HAUGHEY.:	/ W70-02304
H. A.	HAWKES.:	/ W71-02479
F. R.	HAYES.:	/ W70-05399
F. R.	HAYES, AND C. C. COFFIN.:	W68-00857
RON, AND D. A. LIVINGSTO/ F. R.	HAYES, J. A. MCCARTER, M. L. CAME	W68-00858
CHWERTNER, CARA L. MARTINEZ, AND	HAZEL E. WICKLINE.: /UTT, H. A. S/	W70-07957
ENDLOVA, AND BOHUMIR P/ GUDMUND	HEDENSKOG, LENNART ENEBO, JITKA V/	W70-04184

```
                                N. M. KRYUTCHKOVA, AND V. SLADECEK.:    / W69-10152
                                L. E. KUENTZEL.:                        / W70-00664
            L. R. POMEROY, AND E. J. KUENZLER.:                        / W69-08526
               G. BRINGMANN, AND R. KUHN.:                               W68-00475
               G. BRINGMANN, AND R. KUHN.:                               W68-00472
                   ERNESTO ESPINO DE LA O, AND E. F. GLOYNA.:          / W71-11377
                   ERNESTO ESPINO DE LA O, AND EARNEST F. GLOYNA.:     / W70-07034
                                J. C. LACAZE.:                         / W70-01470
T.:                             J. C. LACAZE, C. HALLOPEAU, AND M. VOIG/ W72-02203
     JAMES B. LACKEY, AND ELSIE W. LACKEY.:                           / W72-01993
                            JAMES B. LACKEY.:                          / W72-01813
                            JAMES B. LACKEY.:                          / W69-04276
             C. N. SAWYER, J. B. LACKEY, AND A. T. LENZ.:             / W70-05112
             C. N. SAWYER, J. B. LACKEY, AND A. T. LENZ.:             / W70-05113
                            JAMES B. LACKEY, AND ELSIE W. LACKEY.:     / W72-01993
INS.:             JOHN K. REED, LAMAR E. PRIESTER, AND RUDOLPH PR/ W69-02782
                               JOHN A. LAMBIE.:                        / W70-05645
                    M. TASSIGNY, G. LAPORTE, AND R. POURRIOT.:         / W71-06002
                            ROBERT A. LAUDERDALE.:                     / W70-09768
                                      LAWRENCE W. OLINGER.:            / W69-03730
                                      LAZAR R. IGNJATOVIC.:            / W70-00683
                             IVONNE J. LE COSQUINO DE BUSSY.:          / W70-10161
        RO HASHIMOTO, TOMOTOSHI OKAICHI, LE DUNG DANG, AND TAMAO NOGUCHI.:/ W70-08372
        RO HASHIMOTO, TOMOTOSHI OKAICHI, LE DUNG DANG, AND TAMAO NOGUCHI.:/ W71-07731
       ERARD A. ROHLICH, AND WILLIAM L. LEA.:                        G/ W70-02969
        NORBERT A. JAWORSKI, DONALD W. LEAR, JR., AND JOHAN A. AALTO.: / W71-05407
                               G. FRED LEE.:                           / W70-04382
                               G. FRED LEE, AND E. GUS FRUH.:          / W69-06535
    COREY, ARTHUR D. HASLER, G. FRED LEE, F. H. SCHRAUFNAGEL, AND THOM/ W69-10178
      ORGE P. FITZGERALD.:    G. FRED LEE, NICHOLAS L. CLESCERI, AND GE/ W69-03358
       EL, ELIZABETH MCCOY, AND RICHARD LEHNER.:        S. A. WITZ/ W71-05742
           GARY W. ISAAC, AND CURTIS P. LEISER.:                      / W69-03683
         J. GABAY.:    N. I. SAX, PAUL C. LEMON, ALLEN H. BENTON, AND JACK / W69-10080
       D BOHUMIR P/ GUDMUND HEDENSKOG, LENNART ENEBO, JITKA VENDLOVA, AN/ W70-04184
          SAWYER, J. B. LACKEY, AND A. T. LENZ.:              C. N./ W70-05113
          SAWYER, J. B. LACKEY, AND A. T. LENZ.:              C. N./ W70-05112
                                    J. LEON POTTER.:                  / W69-05891
NDAHL.:                               LEON S. DIRECTO, AND MERLIN E. LI/ W69-05894
              EDWARD J. MARTIN, AND LEON W. WEINBERGER.:              / W69-01453
                            DONALD E. LEONE.:                         / W70-06053
                 R. CABRIDENC, AND H. LEPAILLEUR.:                    / W69-07389
                               HAROLD LEROY ALLEN.:                   / W71-00832
                                 G. E. LIKENS, AND P. L. JOHNSON.:    / W69-08272
      DANIEL F. JACKSON, AND SHUNN-DAR LIN.:                          / W72-05453
        LEON S. DIRECTO, AND MERLIN E. LINDAHL.:                      / W69-05894
                               E. C. S. LITTLE.:                      / W70-00269
                                 J. E. LITTLE.:                       / W71-13313
                            DONALD F. LIVERMORE.:                     / W69-06276
      CARTER, M. L. CAMERON, AND D. A. LIVINGSTONE.: /R. HAYES, J. A. MC/ W68-00858
         AND JEROME F. THOMAS.:       LLOYD L. MEDSKER, DAVID JENKINS, / W69-00387
       OHN W. WINCHESTER, AND RONALD H. LOUCKS.:        MARY A. TIFFANY, J/ W69-08562
       . ARTHUR, WILLIAM J. OSWALD, AND LOUIS A. BECK.: /GOLDMAN, JAMES F/ W71-04557
             THOMAS D. BROCK, AND M. LOUISE BROCK.:                   / W69-10151
             THOMAS D. BROCK, AND M. LOUISE BROCK.:                   / W69-09676
                T. D. BROCK, AND M. LOUISE BROCK.:                   / W69-10160
.:                                 M. LOUISE BROCK, AND THOMAS D. BROCK/ W69-10163
                                 S. K. LOVE.:                         / W73-00750
                                      LOWELL D. HANSON, AND WILLIAM E. / W70-04193
FENSTER.:                             LOWELL D. HANSON, AND WILLIAM E. /
                                 R. H. LUEBBERS, AND D. N. PARIKH.:    / W72-06022
                                      LUIGI PROVASOLI.:               / W70-04503
                            KENNETH M. MACKENTHUN.:                   / W70-05568
                                 K. M. MACKENTHUN.:                   / W72-07442
                            KENNETH M. MACKENTHUN.:                   / W69-09340
.:                          KENNETH M. MACKENTHUN, AND ALFRED F. BARTSCH/ W72-01797
:                           KENNETH M. MACKENTHUN, AND HAROLD L. COOLEY./ W70-06217
GRAM.:                      KENNETH M. MACKENTHUN, AND WILLIAM MARCUS IN/ W69-04801
STEWART.:                        K. M. MACKENTHUN, L. E. KEUP, AND R. K.   W68-00481
D RALPH PORGES.:           KENNETH M. MACKENTHUN, WILLIAM M. INGRAM, AN/ W70-06225
      ERLAND, M. G. BIANCHI, AND S. Y. MAESTRINI.:            B. R. B/ W70-04280
                                 I. V. MAKSIMOVA, AND M. N. PIMENOVA.: / W70-04195
           M. AUBERT, R. CHARRA, AND G. MALARA.:                      / W70-01466
                                 J. F. MALINA, JR.:                   / W72-11105
                                 T. E. MALONEY.:                      / W72-09694
                            THOMAS E. MALONEY, AND ALFRED F. BARTSCH.: / W70-07393
.:                                    MAN M. VARMA, AND FRANCIS DIGIAND/ W70-06600
                                 E. F. MANDELLI.:                     / W70-08111
     NNETH M. MACKENTHUN, AND WILLIAM MARCUS INGRAM.:               KE/ W69-04801
                                      MARGARET GREENWALD.:            / W69-05705
                                      MARGARET GREENWALD.:            / W69-05706
                                      MARIN.:                         / W69-10158
            C. NALEWAJKO, AND L. MARINICH.:      G. M. P. MORDVINTSE/ W70-06925
        VA, V. V. GRABOVSKAYA, AND V. K. MARK J. HAMMER.:             / W72-04761
          GARY L. HERGENRADER, AND MARK W. TENNEY.:                  / W70-05750
                     WERNER STUMM, AND MARK W. TENNEY, WAYNE F. ECHELBER/ W70-05267
       GER, JR., RONALD G. SCHUESSLER/ MARKLE.:        GEORGE E. STOERTZ, W/ W70-05377
        ILLIAM R. HEMPHILL, AND DAVID A. MARKOWSKI.:                   / W69-05023
                                   S. MARLETTA.:         C. CALZOLAR/ W70-05178
        I, L. GABRIELLI, AND G. PERTOLDI MARTIN, AND LEON W. WEINBERGER.: / W69-01453
                            EDWARD J. MARTIN, WILLIAM T. HELM, AND WILL/ W69-07846
       IAM F. SIGLER.:         SUSAN S. MARTINEZ, AND HAZEL E. WICKLINE.:/ W70-07957
       RCUTT, H. A. SCHWERTNER, CARA L.
```

	NELSON L. NEMEROW.: / W70-09320
	NELSON L. NEMEROW.: / W71-05013
O. F.	NELSON.: / W73-02426
DANIEL J.	NELSON.: / W70-02786
RGE P. FITZGERALD, AND THOMAS C.	NELSON.: GEO/ W69-03373
A.	NELSON-SMITH.: / W70-01780
NELSON L.	NEMEROW.: / W70-09320
NELSON L.	NEMEROW.: / W71-05013
P. FITZGERALD.: G. FRED LEE,	NICHOLAS L. CLESCERI, AND GEORGE / W69-03358
THY MCNALL.: M. STARR	NICHOLS, THERESA HENKEL, AND DORO/ W69-03366
E. STEEMANN	NIELSEN, AND ERIK G. JORGENSEN.: / W70-04809
R.	NIEMEYER, AND G. RICHTER.: / W71-03033
OKAICHI, LE DUNG DANG, AND TAMAO	NOGUCHI.: / HASHIMOTO, TOMOTOSHI / W71-07731
OKAICHI, LE DUNG DANG, AND TAMAO	NOGUCHI.: / HASHIMOTO, TOMOTOSHI / W70-08372
	NORBERT A. JAWORSKI.: / W71-05409
AR, JR., AND JOHAN A. AALTO.:	NORBERT A. JAWORSKI, DONALD W. LE/ W71-05407
JR.: R. H. T. MATTONI, H.	NUGENT MYRICK, AND E. C. KELLER, / W71-05267
JACKSON L. FOX, THERON O.	ODLAUG, AND THEODORE A. OLSON.: / W70-05958
. POMEROY, R. E. JOHANNES, E. P.	ODUM, AND B. ROFFMAN.: L. R/ W69-08274
ROANN E.	OGAWA, AND JOHN F. CARR.: / W70-05091
R. T.	OGLESBY, AND DAVID JARMISON.: W68-00010
WALDEMAR	OHLE.: / W70-05761
N/ YOSHIRO HASHIMOTO, TOMOTOSHI	OKAICHI, LE DUNG DANG, AND TAMAO / W70-08372
N/ YOSHIRO HASHIMOTO, TOMOTOSHI	OKAICHI, LE DUNG DANG, AND TAMAO / W71-07731
	OLAV M. SKULBERG.: / W69-07833
	OLAV SKULBERG.: / W70-05752
	OLE A. SEATHER.: / W71-12077
LAWRENCE W.	OLINGER.: / W69-03730
HERON O. ODLAUG, AND THEODORE A.	OLSON.: JACKSON L. FOX, T/ W70-05958
MARTINEZ/ B. RICHARDSON, D. M.	ORCUTT, H. A. SCHWERTNER, CARA L./ W70-07957
ENNETT.: E. GELPI, J.	ORO, H. J. SCHNEIDER, AND E. O. B/ W69-08284
JAMES R.	ORR.: / W70-04810
D. W.	OSBORN.: / W71-11393
	OSMUND HOLM-HANSEN.: / W69-08278
	OSMUND HOLM-HANSEN, GERALD C. GER/ W69-06277
	OSMUND HOLM-HANSON.: / W69-03518
LOFF, AND FOLKE SKOOG.:	OSWALD.: / W72-06854
	OSWALD.: / W70-04787
GORDON L. DUGAN, AND WILLIAM J.	OSWALD, AND CHARLES E. RIXFORD.: / W71-11375
W. J.	OSWALD, AND LOUIS A. BECK.: /GOLD/ W71-04557
CLARENCE G. GOLUEKE, WILLIAM J.	OTTO WASMER, JR.: / W69-00994
MAN, JAMES F. ARTHUR, WILLIAM J.	OULTON.: / W71-13335
	OVERFIELD, H. R. CRAWFORD, J. K. / W70-05819
RALPH E.	OWEN.: / W69-04798
BAXTER, L. J. HARRINGTON/ J. L.	PALMER.: / W70-01233
JOHN H. NEIL, AND GLENN E.	PALMER.: / W70-07027
C. M.	PALMER.: / W70-10173
C. MERVIN	PALMER.: / W70-04507
C. MERVIN	PALMER.: / W70-05389
C. MERVIN	PALMER.: / W72-03562
C. MERVIN	PALMER.: / W72-03746
C. M.	PALMER, AND WILLIAM M. INGRAM.: / W72-01798
C. M.	PARIKH.: / W72-06022
C. MERVIN,	PARKER.: / W71-09548
R. H. LUEBBERS, AND D. N.	PARKER.: / W69-06273
C. D.	PARKER.: / W69-06536
BRUCE C.	PARKER, AND G. P. SKERRY.: / W70-06899
RICHARD A.	PATRICK L. BREZONIK.: / W71-03048
C. D.	PATRICK L. BREZONIK, AND HUGH D. / W72-01990
	PATRICK L. HUDSON, AND BRUCE C. C/ W70-05375
PUTNAM.:	PATRICK.: / W70-04510
OWELL.:	PATRICK.: / W71-02496
RUTH	PATRICK.: / W71-05026
RUTH	PATRICK.: / W72-01801
RUTH	PATRICK.: / W73-00748
RUTH	PATRICK, BOWMAN CRUM, AND JOHN CO/ W70-07257
R.	PATTEN, AND CHARLES E. DEPOE.: / W70-04371
LES.: RUTH	PAUL A. ERICKSON, AND JOHN T. REY/ W70-05264
ERNEST C. NEAL, BERNARD C.	PAUL C. LEMON, ALLEN H. BENTON, A/ W69-10080
NOLDS.:	PAUL H. KING, FRANCIS R. MCNEICE/ W70-05470
ND JACK J. GABAY.: N. I. SAX,	PAUL H. KING, H. H. YEH, PIERRE S/ W69-09884
AND PIERRE S. WARREN.:	PAVONI.: /. ECHELBERGER, JR., RON/ W70-05267
. WARREN, AND CLIFFORD W. RAND/	PENNAK.: / W69-10154
ALD G. SCHUESSLER, AND JOSEPH L.	PENNINGTON.: / W69-04521
ROBERT W.	PERRY L. MCCARTY.: / W71-13413
WINIFRED	PERTOLDI MARLETTA.: C./ W70-05178
EDWARD G. FOREE, AND	PETERKA.: / W70-09093
CALZOLARI, L. GABRIELLI, AND G.	PETERSON.: W. T. EDMONDSO/ W69-10169
JOHN J.	PETUR M. JONASSON.: / W71-00665
N, G. C. ANDERSON, AND DONALD R.	PFISTER.: P./ W71-04758
	PHILIP A. BUSCEMI.: / W70-03501
R. DUGAN, J. I. FREA, AND R. M.	PHILIP J. SAWYER, JOHN H. GENTILE/ W69-05306
	PHILLIP E. GREESON.: / W70-04721
, AND JOHN J. SASNER, JR.:	PHINNEY.: / W70-03978
	PHINNEY.: / W69-04802
HARRY K.	PHINNEY, AND CHARLES E. WARREN.: / W70-00265
SHERRY L. VOLK, AND HARRY K.	PIERRE BOVARD, AND JACQUES ANCELL/ W71-09863
RE, ROBERT L. GARRISON, HARRY K.	PIERRE GUEGUENIAT, PIERRE BOVARD,/ W71-09863
IN.: PIERRE GUEGUENIAT,	PIERRE S. WARREN.: PAUL / W70-05470
AND JACQUES ANCELLIN.:	PIERRE S. WARREN, AND CLIFFORD W./ W69-09884
H. KING, FRANCIS R. MCNEICE, AND	
RAND/ PAUL H. KING, H. H. YEH,	

```
                      I. V. MAKSIMOVA, AND M. N.   PIMENOVA.:                             / W70-04195
                   DAVID R. ZENZ, AND JOSEPH R.   PIVNICKA.:                              / W70-09186
       EL, E. MCCOY, O. J. ATTOE, L. B.   POLKOWSKI, AND K. T. CRABTREE.: // W69-05323
                                        J.   POLTORACKA.:                           / W71-09767
                                      L. R.   POMEROY, AND E. J. KUENZLER.:        / W69-08526
         UM, AND B. ROFFMAN.:        L. R.   POMEROY, R. E. JOHANNES, E. P. OD/ W69-08274
           K. K. VOTINTSEV, AND G. I.   POPOVSKAYA.:                          / W70-04290
                                DONALD B.   PORCELLA.:                            / W70-02779
      UN, WILLIAM M. INGRAM, AND RALPH   PORGES.:       KENNETH M. MACKENTH/ W70-06225
              C. E. CUSHING, AND N. S.   PORTER.:                              / W69-07862
                                      L. K.   PORTER, AND A. R. GRABLE.:         / W72-08110
                                      H. W.   POSTON.:                            / W69-03948
                                      YU S   POTAENKO, AND T M MIKHEEVA.:      / W70-08654
                                 J. LEON   POTTER.:                             / W69-05891
        M. TASSIGNY, G. LAPORTE, AND R.   POURRIOT.:                           / W71-06002
           ROBERT M. BRICE, AND CHARLES F.   POWERS.:                          / W70-03512
             G. D. AGRAWAL, AND A. V. S.   PRABHAKARA RAO.:                    / W70-08628
             C. D. AGARWAL, AND A. V. S.   PRABHAKARA RAO.:                    / W70-01027
                                      G. W.   PRESCOTT.:                        / W70-02803
                 T. DUKE, J. WILLIS, T.   PRICE, AND K. FISCHLER.:           / W69-08267
                  JOHN K. REED, LAMAR E.   PRIESTER, AND RUDOLPH PRINS.:     / W69-02782
        , LAMAR E. PRIESTER, AND RUDOLPH   PRINS.:           JOHN K. REED/ W69-02782
                                DONALD E.   PROCTOR.:                         / W70-07491
      URINDER K. BHAGAT, AND DONALD E.   PROCTOR.:                      S/ W71-00936
       EBO, JITKA VENDLOVA, AND BOHUMIR   PROKES.: /D HEDENSKOG, LENNART EN/ W70-04184
                                   LUIGI   PROVASOLI.:                       / W70-04503
                                  HUGH D.   PUTNAM.:                          / W68-00912
         PATRICK L. BREZONIK, AND HUGH D.   PUTNAM.:                        / W72-01990
       MORGAN, E. E. SHANNON, AND H. D.   PUTNAM.:    P. L. BREZONIK, W. H. / W69-09723
                         . BORCHARDT.:   G. R.   PYPER, W. J. WEBER, JR., AND J. A/ W72-00917
                                      I. L.   PYRINA.:                        / W71-10788
        SWINGLE, B. C. GOOCH, AND H. R.   RABANAL.:              H. S./ W70-05417
                              MENACHEM   RAHAT, AND INKA DOR.:            / W69-00360
              CHARLES R. GOLDMAN, AND   RALF C. CARTER.:                  / W70-01933
                                        RALPH E. OULTON.:                  / W71-13335
                   HARRY D. NASH, AND   RALPH H. WEAVER.:                 / W68-00891
                                      H.:   RALPH H. WEAVER, AND HARRY D. NAS/ W69-00096
     CKENTHUN, WILLIAM M. INGRAM, AND   RALPH PORGES.:       KENNETH M. MA/ W70-06225
                                        RANDALL BROWN, AND JAMES ARTHUR.:/ W70-03334
       FIELD, AND JO/  JAMES F. ARTHUR,   RANDALL L. BROWN, BRUCE A. BUTTER/ W71-04554
        , JR., AND MILTON G. TUNZI.:   RANDALL L. BROWN, RICHARD C. BAIN/ W71-04555
     IERRE S. WARREN, AND CLIFFORD W.   RANDALL.: / H. KING, H. H. YEH, P/ W69-09884
                                     D. S.   RANWELL.:                        / W70-01777
      AGARWAL, AND A. V. S. PRABHAKARA   RAO.:              C. D. / W70-01027
      AGRAWAL, AND A. V. S. PRABHAKARA   RAO.:              G. D. / W70-08628
                                 J. E. G.   RAYMONT.:                        / W70-00161
        S.:                           M.   REBHUN, M. A. FOX, AND J. B. SLES/ W70-02373
       LPH PRINS.:                JOHN K.   REED, LAMAR E. PRIESTER, AND RUDO/ W69-02782
      M. ROBERTSON, AND RO/  GEORGE W.   REID, ROBERT A. GEARHEART, JAMES / W70-04074
                              ARNOLD W.   REITZE, JR.:                     / W69-06305
                                      C. E.   RENN.:                         / W72-00027
                                 JOHN T.   REYNOLDS.:                       / W69-00976
          PAUL A. ERICKSON, AND JOHN T.   REYNOLDS.:                       / W70-05264
            JOSEPH SHAPIRO, AND ROBERTO   RIBEIRO.:                        / W70-02255
                                      T. R.   RICE.:                         / W70-00708
                                        RICHARD A. PARKER.:               / W69-06536
       R. GOLDMAN, MILTON G. TUNZI, AND   RICHARD ARMSTRONG.:     CHARLES / W70-03983
      R, G. FRED LEE, F. H. SCHRAUFN/   RICHARD B. COREY, ARTHUR D. HASLE/ W69-10178
      G. TUNZI.:      RANDALL L. BROWN,   RICHARD C. BAIN, JR., AND MILTON / W71-04555
                                        RICHARD C. BAIN, JR.:             / W69-03611
                                        RICHARD C. BAIN, JR.:             / W69-07520
                                        RICHARD C. BAIN, JR.:             / W70-02777
                                        RICHARD J. BENOIT.:              / W70-04385
      A. WITZEL, ELIZABETH MCCOY, AND   RICHARD LEHNER.:           S./ W71-05742
                                        RICHARD P. HERBST.:              / W70-00667
                                        RICHARD P. W. MCCUTCHEON.:       / W69-01165
                 JOHN E. HOBBIE, AND   RICHARD T. WRIGHT.:              / W70-04284
                 JOHN E. HOBBIE, AND   RICHARD T. WRIGHT.:              / W70-04287
      BBIE.:                             RICHARD T. WRIGHT, AND JOHN E. HO/ W70-04194
                                        RICHARD W. CASTENHOLZ.:          / W70-05270
                                        RICHARD W. CASTENHOLZ.:          / W69-03516
      ND J. K. MURRAY.:                 RICHARD WOOD, WILLIAM S. SMITH, A/ W70-05821
      CHWERTNER, CARA L. MARTINEZ/  B.   RICHARDSON, D. M. ORCUTT, H. A. S/ W70-07957
            R. NIEMEYER, AND G.   RICHTER.:                        / W71-03033
                                        RINNOSUKE FUKAI.:                / W71-05531
      ILLIAM J. OSWALD, AND CHARLES E.   RIXFORD.: /CLARENCE G. GOLUEKE, W/ W71-11375
      :                                 ROANN E. OGAWA, AND JOHN F. CARR./ W70-05091
      IRWIN, E. L. ROBINSON, AND G. G.   ROBECK.:      J. M. SYMONS, W. H. / W70-04484
      ER.:                              ROBERT A. COLER, AND HAIM B. GUNN/ W71-03775
      ERTSON, AND RO/  GEORGE W. REID,   ROBERT A. GEARHEART, JAMES M. ROB/ W70-04074
                                        ROBERT A. LAUDERDALE.:           / W70-09768
                                        ROBERT A. SWEENEY.:              / W70-08652
                                        ROBERT C. BALL.:                 W68-00487
      ER.:                              ROBERT C. BALL, AND FRANK F. HOOP/ W69-09334
                                        ROBERT C. BRIGHT.:               / W69-00659
                                        ROBERT D. HARTER.:               / W69-03075
                                        ROBERT G. WETZEL.:               / W71-10079
                                        ROBERT H. BURRIS.:               / W70-03429
                                        ROBERT J. HENLEY.:               / W71-09880
                                        ROBERT J. SHERWOOD.:             W68-00256
```